10	11 IB	12 IIB	13 IIIA	14 IVA	15 VA	16 VIA	17 VIIA	18 VIII
							Hydrogen 1 **H** 1.0079	Helium 2 **He** 4.0026
			Boron 5 **B** 10.811	Carbon 6 **C** 12.011	Nitrogen 7 **N** 14.0067	Oxygen 8 **O** 15.9994	Fluorine 9 **F** 18.9984	Neon 10 **Ne** 20.1797
			Aluminum 13 **Al** 26.9815	Silicon 14 **Si** 28.0855	Phosphorus 15 **P** 30.9738	Sulfur 16 **S** 32.066	Chlorine 17 **Cl** 35.4527	Argon 18 **Ar** 39.948
Nickel 28 **Ni** 58.69	Copper 29 **Cu** 63.546	Zinc 30 **Zn** 65.39	Gallium 31 **Ga** 69.723	Germanium 32 **Ge** 72.61	Arsenic 33 **As** 74.9216	Selenium 34 **Se** 78.96	Bromine 35 **Br** 79.904	Krypton 36 **Kr** 83.80
Palladium 46 **Pd** 106.42	Silver 47 **Ag** 107.8682	Cadmium 48 **Cd** 112.411	Indium 49 **In** 114.82	Tin 50 **Sn** 118.710	Antimony 51 **Sb** 121.75	Tellurium 52 **Te** 127.60	Iodine 53 **I** 126.9045	Xenon 54 **Xe** 131.29
Platinum 78 **Pt** 195.08	Gold 79 **Au** 196.9665	Mercury 80 **Hg** 200.59	Thallium 81 **Tl** 204.3833	Lead 82 **Pb** 207.2	Bismuth 83 **Bi** 208.9804	Polonium 84 **Po** (209)	Astatine 85 **At** (210)	Radon 86 **Rn** (222)

Europium 63 **Eu** 151.965	Gadolinium 64 **Gd** 157.25	Terbium 65 **Tb** 158.9253	Dysprosium 66 **Dy** 162.50	Holmium 67 **Ho** 164.9303	Erbium 68 **Er** 167.26	Thulium 69 **Tm** 168.9342	Ytterbium 70 **Yb** 173.04	Lutetium 71 **Lu** 174.967
Americium 95 **Am** (243)	Curium 96 **Cm** (247)	Berkelium 97 **Bk** (247)	Californium 98 **Cf** (251)	Einsteinium 99 **Es** (252)	Fermium 100 **Fm** (257)	Mendelevium 101 **Md** (258)	Nobelium 102 **No** (259)	Lawrencium 103 **Lr** (260)

CHEMISTRY & CHEMICAL REACTIVITY

Second Edition

CHEMISTRY & CHEMICAL REACTIVITY

Second Edition

JOHN C. KOTZ
SUNY Distinguished Teaching Professor
State University of New York
College at Oneonta

KEITH F. PURCELL
Professor of Chemistry
Kansas State University

Saunders College Publishing
Harcourt Brace Jovanovich College Publishers
Ft. Worth Philadelphia San Diego
New York Orlando Austin San Antonio
Toronto Montreal London Sydney Tokyo

Text Typeface: Times Roman
Compositor: General Graphics Services, Inc.
Acquisitions Editor: John Vondeling
Developmental Editors: Becca Gruliow and Sandi Kiselica
Managing Editor and Project Editor: Carol Field
Copy Editor: Jay Freedman
Manager of Art and Design: Carol Bleistine
Art Assistant: Doris Bruey
Text Design: Circa 86, Inc.
Cover Designer: Lawrence R. Didona
Text Artwork: J & R Art Services, Inc.
Layout Artist: Dorothy Chattin
Director of EDP and Production Manager: Tim Frelick
Marketing Manager: Marjorie Waldron

Cover Credit: The book cover illustrates three major themes in chemistry: chemical reactions, elements, and compounds. The background of the cover is the reaction of CO_2 with limewater (calcium hydroxide dissolved in water) to give the white solid calcium carbonate. The photos in the middle of the front cover and on the back cover show a mixture of copper and sulfur being heated in a crucible to form a compound. (Photographs by Charles D. Winters.)

Printed in the United States of America

CHEMISTRY & CHEMICAL REACTIVITY 2/e

0-03-047562-7

Library of Congress Catalog Card Number: 90-050857

23 032 987654

PREFACE

The title of this book, CHEMISTRY & CHEMICAL REACTIVITY, conveys its principal themes: a broad overview of the principles of chemistry and the reactivity of chemical elements and compounds. While providing a firm foundation in these areas, it is our hope also to convey a sense of chemistry as a field that not only has a lively history but also one that is still dynamic, with important new developments unfolding. We also hope to provide insight into the chemical aspects of the world around us. For example, what is the role and importance of carbon dioxide in our environment, what materials are important to our economy, and what role do chemists play in protecting the environment? By tackling the principles underlying the answers to these questions, you can better understand nature and better appreciate some of the consumer products coming from the chemical industry. Indeed, one of the objectives of this book is to provide the tools needed to function as a chemically literate citizen. Learning something of the chemical world is just as important as understanding some basic mathematics and biology and appreciating fine music and literature.

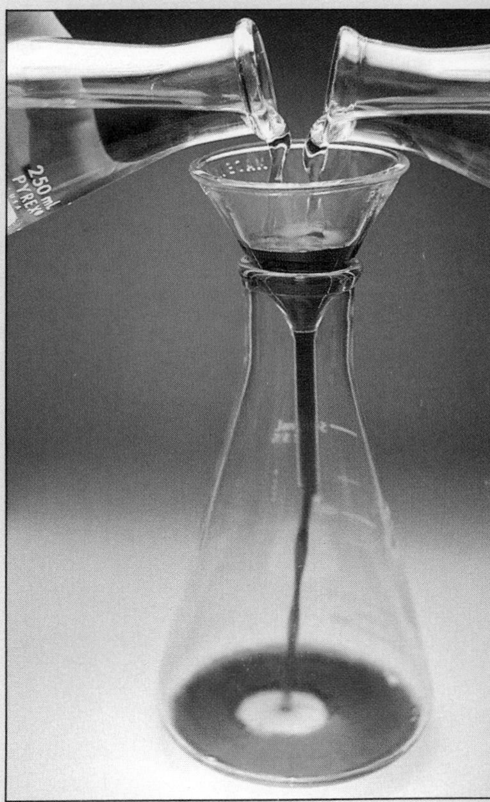

Above all, you should realize that the authors of this book became chemists simply because it is fun to discover new compounds and find new ways to apply chemical principles. We hope we have conveyed that sense of enjoyment as well as our awe at what is known about chemistry and, just as important, what is not known.

AUDIENCE

CHEMISTRY & CHEMICAL REACTIVITY is a textbook for introductory courses in chemistry for students interested in further study in science, whether that science is biology, chemistry, engineering, geology, physics, or related subjects. Our assumption is that students beginning this course have had a basic foundation in algebra and in general science. Although undeniably helpful, a previous exposure to chemistry is neither assumed nor required.

PHILOSOPHY AND APPROACH

When the first edition of this book was planned, we had two major, but not independent, goals. This edition has these same goals. The first was to construct a book that students would enjoy reading and that would offer, at a reasonable level of rigor, chemistry and chemical principles in a format and organization typical of college and university courses today. Second, we wanted to convey the utility and importance of chemistry by introducing the properties of the elements, their compounds, and their reactions as early in the book as possible and by focusing the discussion as much as possible on these subjects.

A glance at the introductory chemistry texts currently available shows that there is a generally common order of treatment of chemical principles used by educators. With a few minor changes we have followed that order as well. However, that is not to say that the chapters cannot be used in some other order. For example, although the behavior of gases

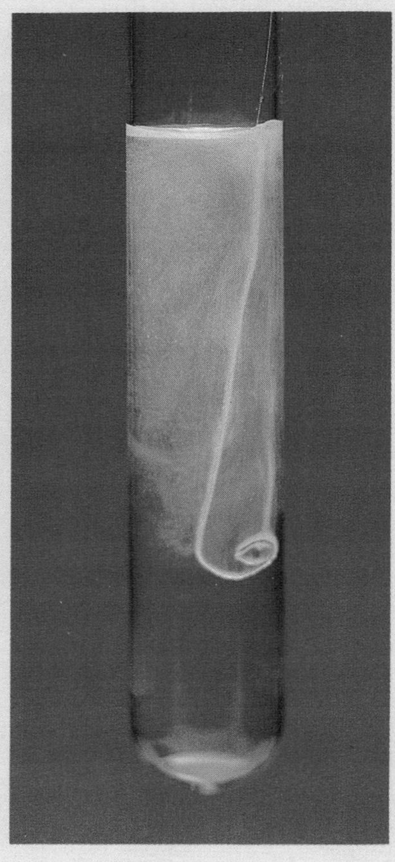

is often studied early in a chemistry course, the chapter on this topic (Chapter 12) has been placed in Part 3, "States of Matter," because it fits logically with the other topics in this section. However, it can easily be read and understood after covering only the first four or five chapters of the book.

The chapter on nuclear chemistry (Chapter 7) is typically left to one of the final chapters in chemistry textbooks. However, we believe that the recent reactor accidents in the United States and the Soviet Union, and the ongoing debate in the United States concerning nuclear waste disposal, may make it desirable to give this topic more prominence. In addition, its importance is indicated by the fact that about 20% of the papers abstracted in *Chemical Abstracts* annually in the general areas of physical, inorganic, and analytical chemistry concern nuclear chemistry. It is for this reason that the Nuclear Chemistry chapter is included as an introduction to the section on the structure of atoms and molecules where it logically fits. In spite of this, the chapter is not at all crucial to the development of Part 2; it is certainly possible to leave the chapter for a later point in the course. Nonetheless, it is our hope that the material will be discussed in class to some extent, or that students will read the chapter out of interest in the subject.

In addition, one of the authors of this text regularly teaches the material on equilibria involving insoluble solids (Chapter 19) before acid-base equilibria (Chapters 17 and 18), and introduces kinetics (Chapter 15) and thermodynamics (Chapter 20) as a unit, after all of the material on equilibria. Although chapters are loosely organized into groups with common themes, *every attempt has been made to make the chapters as independent as possible*.

The order of topics in the text was also devised to introduce as early as possible the background required for the laboratory experiments usually done in General Chemistry. For this reason, chapters on common reaction types (Chapter 4), reactions in aqueous solution (especially acid-base and oxidation-reduction reactions, Chapter 5), and stoichiometry (Chapters 4 and 5) begin the book. In addition, because an understanding of energy is so important in the study of chemistry, this chapter is also included in Part 1.

The American Chemical Society has been urging educators to put "chemistry" back into introductory chemistry courses. As inorganic chemists, we agree wholeheartedly. Therefore, we have tried to describe the elements, their compounds, and their reactions as early and as often as possible in three ways. First, there are numerous color photographs of reactions occurring, of the elements and common compounds, and of common laboratory operations and industrial processes. Further, we have tried to bring material on the properties of elements and compounds as early as possible into the Exercises and Study Questions and to introduce new principles using realistic chemical situations. Additionally, there are sections called "Something More About . . ." on the nature of important elements (platinum) and compounds (cisplatin and titanium dioxide), on new materials (diamond thin films) and other interesting aspects of chemistry (The Origin of the Elements, The Chemistry of Carbon Dioxide, Asbestos, CFCs and the Ozone Hole, for example), and photographic essays on the chemistry of some of the metals (silver, lead, iron, and nickel). Finally, Part 5 is devoted to a more systematic study of descriptive chemistry, largely from the point of view of chemicals important in our economy and in the world around us.

ORGANIZATION

CHEMISTRY & CHEMICAL REACTIVITY is organized into five parts, each containing several chapters.

Part 1 The Basic Tools of Chemistry

Basic ideas and methods that form the fabric of chemistry are introduced in Part 1. Chapter 1 defines some important terms and reviews units and mathematical methods. Chapters 2 and 3 introduce some basic ideas about atoms and molecules, and Chapter 2 introduces one of the most important organizational devices of chemistry, the periodic table. In Chapters 4 and 5 we begin to discuss some of the principles of chemical reactivity and to introduce some of the numerical methods used by chemists to extract quantitative information from chemical reactions. Chapter 6 is an introduction to the energy involved in chemical processes.

Part 2 Atomic and Molecular Structure

In this section we work from the inside out. First, we describe some aspects of the world of nuclear chemistry. Although the chapter logically fits in this section, it can be covered at any point in the course.

The major goal of this section is to outline (Chapters 8 and 9) current theories of the arrangement of electrons in atoms and some of the historical developments that led to these ideas. With this background, we can understand why atoms and their ions have different chemical and physical properties. This discussion is closely tied to the arrangement of elements in the periodic table so that these properties can be recalled and predictions made. In Chapter 10 we discuss for the first time how the electrons of atoms in a molecule may lead to chemical bonding and the properties of these bonds. In addition, we show how to derive the three-dimensional structure of simple molecules. Finally, Chapter 11 considers some of the major theories of chemical bonding in more detail.

Part 3 States of Matter

The behaviors of the three states of matter—gases, liquids, and solids—are discussed in that order in Chapters 12 and 13. The discussion of the latter two is tied to gases through the description of intermolecular forces, with particular attention given to liquid and solid water. Chapter 13 also considers the solid state, an area of chemistry that is currently undergoing a renaissance. In Chapter 14 we talk about the properties of solutions—intimate mixtures of gases, liquids, and solids.

Part 4 The Control of Chemical Reactions

The first chapter in this section, Chapter 15, examines the important question of the rates of chemical processes and the factors controlling these rates. With this in mind, we move to Chapters 16 through 19, a group of chapters that considers chemical reactions at equilibrium. After an introduction to equilibrium in Chapter 16, we highlight the reactions involving acids and bases in water (Chapters 17 and 18) and in reactions leading to insoluble salts (Chapter 19). To tie together the discussion of chemical equilibria, we further explore the science of thermodynamics in Chapter 20. As a final topic in this section, we describe in Chapter 21 a major class of chemical reactions, those involving the transfer of electrons, and the use of these reactions in cells that produce a voltage.

Part 5 The Chemistry of the Elements and Their Compounds
Although the chemistry of the various elements has been described throughout the book to this point, Part 5 considers this topic in a more systematic way. Chapters 22 and 23 are devoted to the elements of Groups 1A to 4A, groups dominated by metals. Chapter 24 explores Groups 5A to 8A, where the elements are generally nonmetals. The transition elements have a chemistry somewhat different from the other groups, so they are described separately in Chapter 25. Carbon and its compounds are of such importance in industry, in our economy, and in ourselves and our environment that Chapters 26 and 27 are devoted almost exclusively to this topic.

NEW TO THIS EDITION

Users of the previous edition of our book will notice some significant differences in this edition. Among the more important changes are the following:

- The discussion of elements, compounds, stoichiometry, and reactions in Chapters 2 through 4 of the first edition has been spread over four chapters (Chapters 2 through 5), and the material on reactions in aqueous solution has been pulled together in the new Chapter 5.
- Naming of common compounds has been moved from the Appendix to Chapter 3.
- As noted previously, the chapter on gas laws has been moved to Part 3. However, the chapter can still be used earlier in the course if desired.
- Part 2 has been reorganized with the addition of the rewritten chapter on nuclear chemistry.
- The number of end-of-chapter Study Questions has been greatly increased in the important early chapters (1 through 5).
- Fewer Study Questions have been categorized by type; more are simply listed as "General Questions." For the questions that are listed by the type, each now has a *paired question*. The first of each pair has its answer listed in the Appendix and its solution in the *Student Solutions Manual*; the second has its answer given only in the *Instructor's Resource Manual*.
- In virtually all the chapters, end-of-chapter Summary Questions combine the concepts of the current chapter with material covered in earlier chapters.
- Each of the five parts of the book opens with an interview with a well-known scientist, and a sixth appears at the end of Chapter 1.

- A number of new essays, now called "Something More About . . .", have been added.

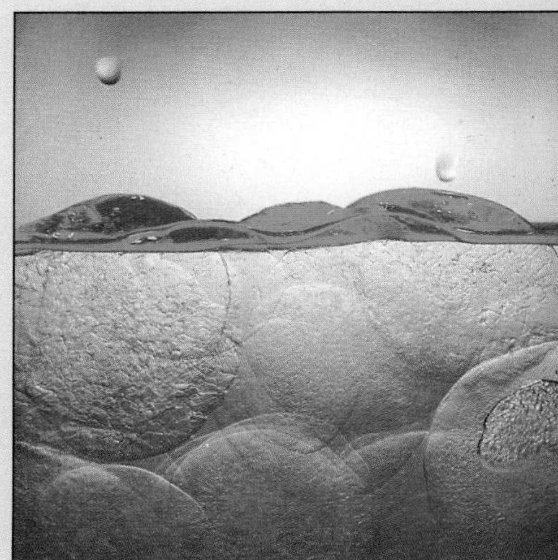

SUPPORTING MATERIALS

Computer Programs and Video Materials

- A computer program called ***KC? Discoverer: Exploring the Properties of the Chemical Elements*** (by John Moore and Aw Feng, formerly of Eastern Michigan University, now of the University of Wisconsin—Madison) from JCE: Software, is available in both MS-DOS and Macintosh format. The program is available at no charge to adopters of the text or it may be purchased directly from JCE: Software: Department of Chemistry, University of Wisconsin—Madison, 1101 University Avenue, Madison, WI 53706. This program is an extensive database of information on the chemical elements. As the database was carefully assembled by a team of chemists, *it has been chosen as the "official" database for this book, and all the data tables in this book reflect the information in **KC? Discoverer***. For further information on the program and its use in a classroom setting, see J. Kotz, *J. Chem. Educ.*, **1989**, *66*, 750.
- The ***Periodic Table Videodisc: Reactions of the Elements*** (by Alton J. Banks of Southwest Texas State University), also published by JCE: Software, is available to adopters of the text. This videodisc, which can be driven by ***KC? Discoverer*** or can stand alone, shows still and moving images of the elements, their uses, and their reactions with air, water, acids, and bases. It is particularly useful as a way to demonstrate chemical reactions in a large lecture room.
- Instructors in General Chemistry may also wish to use the ***Computerized Test Bank*** by J. Kotz, which enables one to generate multiple-choice quizzes and examinations. A printed version of this test bank is also available.
- ***Introduction to General Chemistry***, by Stanley Smith of the University of Illinois, Urbana, and Ruth Chabay of the University of Wisconsin, provides a graphics-oriented computer tutorial for students with no previous chemistry background. It is available in Apple II and IBM formats.
- ***Shakhashiri Demonstration Videotapes*** allow professors to bring 40 three- to five-minute chemical experiments to the classroom. Performed by Bassam Shakhashiri and produced by Saunders College Publishing, these videos are available only to adopters of this text.
- ***World of Chemistry Videotape Modules***, hosted by Nobel laureate, Roald Hoffmann, are 15-minute segments on such topics as the mole, bonding, and acid-base chemistry. These videotapes are suitable for classroom use as an introduction to a topic.
- Finally, many of the **sketches for the line art** in this edition were created by one of the authors on a Macintosh computer using the programs *Canvas®* and *ChemDraw®*. Many of these drawings have been given to Project SERAPHIM (Department of Chemistry, University of Wisconsin, Madison, WI 53706) as "chemistry clip art" and may be ordered from their office.

Written Materials for the Student and Instructor

A number of supplements have been designed to accompany this text. All have been written by authors with many years of experience in chemical education.

The ***Study Guide*** (by Harry Pence of SUNY—Oneonta) that accompanies this text has been designed around the key objectives of the book. Each chapter includes a list of the main concepts, important terms, questions testing mastery of each objective, a test evaluating overall mastery of the chapter, and a set of comprehensive questions. There is also a chapter on the use of chemical equivalents and normality. Answers are provided for all questions and tests.

The ***Laboratory Manual*** (by F.R. Milio, N.G.W. Debye, both of Towson State University and C. Metz of the College of Charleston) is the result of many years of collaboration on the development of laboratory experiments for general chemistry. All experiments have been thoroughly tested, and attention has been given to cost and safety. An instructor's manual is also available.

Detailed answers to designated, end-of-chapter Study Questions are found in the ***Student Solutions Manual*** (by Alton J. Banks of Southwest Texas State University). This manual also contains strategies for problem solving.

An ***Instructor's Resource Manual*** (by J. Kotz) gives suggestions for organization of the course, as well as alternative organizations. In addition, suggested classroom demonstrations and worked-out solutions to the questions not designated by a blue number in the text are included.

A ***Student Lecture Outline*** (by Ronald Ragsdale of the University of Utah) is available for students to help them in organizing the material in the text. In addition, a ***Problems Book*** (by R. Ragsdale) containing over 1200 multiple-choice questions gives students additional practice problems and sample examination questions.

Qualitative Analysis and the Properties of Ions in Aqueous Solution (by Emil Slowinski, Macalester College, and William Masterton, University of Connecticut) is a paperback supplement on qualitative analysis, which encourages students to develop their own schemes of analysis.

Finally, 150 **color overhead transparencies** of important illustrations in the text are available. The illustrations chosen are those most often used in the classroom.

ACKNOWLEDGMENTS

Preparing the second edition of CHEMISTRY & CHEMICAL REACTIVITY took over two years of almost continuous effort. As in our work on the first edition, we have had the support and encouragement of family and of some wonderful friends, colleagues, and students.

The editorial staff of Saunders College Publishing has once again been helpful. The project has been enjoyable because of their good humor, friendship, and dedication. Much of the credit goes to our Publisher, John Vondeling. We have worked with John for many years and have become fast friends. His support and confidence are greatly appreciated, and he is certainly a better fisherman than either of us.

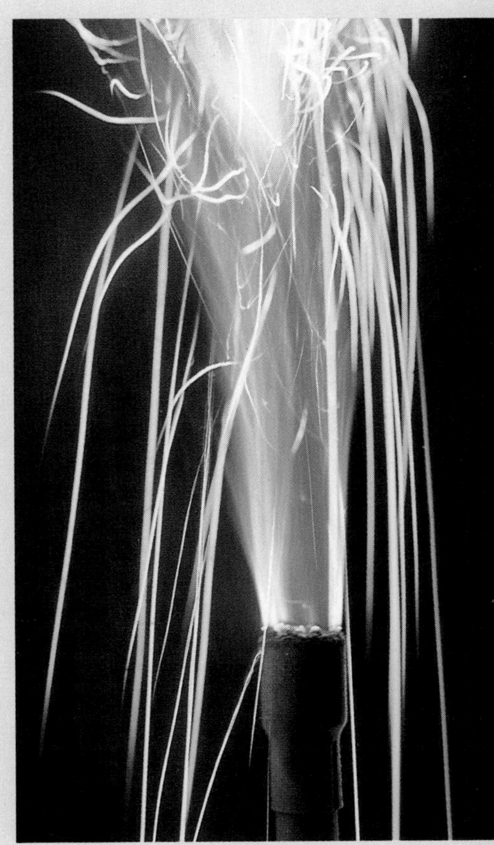

Jay Freedman was the Development Editor of the first edition, and he was the Copy Editor for this edition. Jay is simply the best in the business, and we are pleased to count him as a good friend. He molded the book into its final form in the first edition, and he was meticulous in helping us to prepare the text for this edition. We are extremely grateful for his efforts.

The Development Editor for the second edition was Sandi Kiselica. She has been very helpful in guiding us through the comments of our reviewers and overseeing the changes that we have made. Becca Gruliow also helped enormously in this regard.

Our Managing Editor has again been Carol Field, and she has been as patient as ever. Her attention to detail made the first edition a success, and she keeps all of us calm as deadlines approach. We are pleased that Carol Bleistine was once again the Manager of Art and Design. She has a wonderful sense of color, design, and layout, as this and other books produced at Saunders make clear. Finally, our team is completed with Tim Frelick, the Manager of Editing, Design, and Production, who has kept all of this effort organized.

Most of the color photographs for this edition once again are the product of the creative eye and mind of Charles D. Winters of SUNY—Oneonta. Charlie still knows more tall tales than anyone, and it continues to be an enormous pleasure to work with him, in spite of the endless hours that it takes to get a photograph just right. Unless credited otherwise, all photos were done by him. In addition, we have again used a number of drawings by Irving Geis, and we wish to thank him for his cooperation. There are also a number of computer-generated drawings of molecular structures in the book. These are credited to J. Weber, but they are really the work of him and his group (P. Fluekiger and P.Y. Morgantini) in the Department of Physical Chemistry of the University of Geneva in Geneva, Switzerland. Several other computer-generated drawings, credited to M. Rajzmann, were done at the Centre Européen de Visualisation des Données, CCSJ Marseille, France, by M. Rajzmann, V. Lazzeri, and A. Boch.

A number of our colleagues and students have also contributed to the first edition and to this one. Professor Joe Tausta helped with some of the photos, and Professor Larry Armstrong assisted with the chapter

on organic chemistry. Allan Berger, a student in our course, gave us many helpful suggestions on improvements in the text, particularly in Chapter 6.

It has been especially helpful to have accuracy reviewers for this edition. The book simply could not have been done as well without their assistance. Professors William Dixon and Mark Noble read the galleys and page proofs, and their comments and corrections have been invaluable. In addition, Professors William Vining and David Koster worked *all* of the problems in the book to make certain of our answers and to uncover ambiguities. (It is especially gratifying to refer to Professor William Vining, since he was once known to us just as Bill Vining, an outstanding student in our course.)

A feature of this book is the interviews we have done with prominent scientists. We thoroughly enjoyed our conversations with them and hope that we have conveyed their excitement for their fields and their insights into science and life. We thank all of them—Jacqueline Barton, Gregory Farrington, Mary Good, Roald Hoffmann, Barnett Rosenberg, and Bassam Shakhashiri—for allowing us to visit and for taking the time to work with us. Finally, we are pleased to acknowledge that the idea for the interviews came from a student, Kaarin Anderson Ryan. It was her excitement for the notion that made us realize what an important addition the interviews could be to our book.

REVIEWERS

We believe the success of any book is due in no small way to the quality of the reviewers of the manuscript, and the reviewers of the first and second editions have been extraordinarily helpful. We wish to acknowledge with gratitude the efforts of all those listed below.

REVIEWERS OF THE SECOND EDITION

Tom Baer, University of North Carolina at Chapel Hill
Muriel Bishop, Clemson University
Edward Booker, Texas Southern University
Donald Clemens, East Carolina University
Michael Davis, University of Texas, El Paso
Carl Ewig, Vanderbilt University
Russel Grimes, University of Virginia
Anthony W. Harmon, University of Tennessee—Martin
Alan S. Heyn, Montgomery College
Lisa Hibbard, Spelman College
Mary Hickey, Henry Ford Community College
William Jensen, South Dakota State University
Ronald Johnson, Emory University
Stanley Johnson, Orange Coast College
Lenore Kelly, Louisiana Tech University
Paul Loeffler, Sam Houston State University
Brian McGuire, Northeast Missouri State University
Jerry Mills, Texas Tech University
Mark Noble, University of Louisville
L.G. Pederson, University of North Carolina, Chapel Hill
Chester Pinkham, Tri-State University
Steve Ruis, American River College
Jerry Sarquis, Miami University
Steven Strauss, Colorado State University
Larry C. Thompson, University of Minnesota, Duluth
Milt Wieder, Metropolitan State College

REVIEWERS OF THE FIRST EDITION

Bruce Ault, University of Cincinnati
Alton Banks, Southwest Texas State University
O.T. Beachley, SUNY—Buffalo
Jon M. Bellama, University of Maryland
James M. Burlitch, Cornell University
Geoffrey Davies, Northeastern University
Glen Dirreen, University of Wisconsin
John M. DeKorte, Northern Arizona University
Darrell Eyman, University of Iowa
Lawrence Hall, Vanderbilt University
James D. Heinrich, Southwestern College

Forrest C. Hentz, North Carolina State
Marc Kasner, Montclair State College
Philip Keller, University of Arizona
Herbert C. Moser, Kansas State University
John Parson, Ohio State University
Lee G. Pedersen, University of North Carolina
Harry E. Pence, SUNY—Oneonta
Charles Perrino, California State University (Hayward)
Elroy Post, University of Wisconsin—Oshkosh
Ronald Ragsdale, University of Utah
Eugene Rochow, Harvard University
Steven Russo, Indiana University
Charles W.J. Scaife, Union College
George H. Schenk, Wayne State University
Peter Sheridan, Colgate University
Kenneth Spitzer, Washington State University
Donald D. Titus, Temple University
Charles A. Trapp, University of Louisville
Trina Valencich, California State University (Los Angeles)

Many faculty members and students at our institutions and around the United States and Canada wrote to us with corrections and suggestions. They were all useful, and we are grateful to those who took the time to write or call.

A NOTE TO FACULTY AND STUDENTS

There are almost as many ways to teach chemistry properly as there are faculty members in the chemistry departments in North America. We make no claim that we have found the best way of organizing chemistry, of explaining a principle, or of doing a problem. Further, in spite of our best efforts to present an error-free book, we are by no means infallible. Therefore, if you find a way to explain something more clearly, a better way to demonstrate a reaction or a principle, or if you find errors in our discussions, we hope you will feel free to write to us or to our editors at Saunders College Publishing. Many specific things can be corrected in subsequent printings of the book, and more general ideas can be incorporated in any editions of this text that may follow. And don't be surprised if we call to find out more about your ideas.

A final word to students using the book. We believe chemistry is a challenging, exciting, and worthwhile area of study, and we hope you agree. However, like anything worthwhile, it does take some work to understand the subject. As you work through the book, just remember the encouraging words of Dr. Seuss from his book *Oh, the Places You'll Go!* (Theodor S. Geisel and Audrey S. Geisel, Random House, New York, © 1990).

Onward up many
a frightening creek
though your arms may get sore
and your sneakers may leak.

John C. Kotz

Keith J. Purcell

December 1990

FEATURES AND LEARNING AIDS

TO THE STUDENT

To acquaint you with some of the aids to learning in this text, we briefly describe and illustrate the major features of the text.

Interviews with Chemists

Each of the five parts of the book opens with a discussion with a contemporary scientist: Barnett Rosenberg, Roald Hoffmann, Mary Good, Gregory Farrington, and Jacqueline Barton. A sixth interview, with Bassam Shakhashiri, is placed at the end of Chapter 1. These prominent teachers and researchers talk about how they became involved in science; about their research and their work in teaching or industry; and their views on the frontier areas of science, the importance of science in our society, and environmental concerns.

INTERVIEW

Roald Hoffmann

Roald Hoffmann is a remarkable individual. When he was only 44 years old, he shared the 1981 Nobel Prize in Chemistry with Kenichi Fukui of Japan for work in applied theoretical chemistry. In addition, he has received awards from the American Chemical Society in both organic chemistry and inorganic chemistry, the only person to have achieved this honor. And, in 1990, he was awarded the Priestley Medal, the highest award given by the American Chemical Society.

The numerous honors celebrating his achievements in chemistry tell only part of the story of his life. He was born to a Polish Jewish family in Zloczow, Poland in 1937, and was named Roald after the famous Norwegian explorer Roald Amundsen. Shortly after World War II began in 1939 the Nazis first forced him and his parents into a ghetto and then into a labor camp. However, his father smuggled Hoffmann and his mother out of the camp, and they were hidden for more than a year in the attic of a school house in the Ukraine. His father was later killed by the Nazis after trying to organize an attempt to break out of the labor camp. After the war Hoffmann, his mother, and his stepfather made their way west to Czechoslovakia, Austria, and then Germany. They finally emigrated to the United States, arriving in New York on Washington's birthday in 1949. That Hoffmann and his mother survived these years is our good fortune. Of the 12,000 Jews living in Zloczow in 1941 when the Nazis took over, only 80 people, three of them children, survived the Holocaust. One of those three children was Roald Hoffmann.

On arriving in New York, Hoffmann learned his sixth language, English. He went to public schools in New York City and then finally to Stuyvesant High School, one of the city's select science schools. From there he went to Columbia University and then on to Harvard University, where he earned his Ph.D. in 1962. Shortly thereafter, he began the work with Professor R.B. Woodward that eventually led to the Nobel Prize. Since 1965 he has been a professor at Cornell University, where he regularly teaches first-year general chemistry.

In addition to his work in chemistry, Professor Hoffmann also writes popular articles on science for the American Scientist and other magazines, and he has published two volumes of his poetry. Finally, he is appearing in a series of 26 half-hour television programs for a chemistry course called "World of Chemistry," scheduled to air on public television and cable channels in 1990.

From medicine to cement to theoretical chemistry

We visited Professor Hoffmann in his office at Cornell University, an office full of mineral samples, molecular models, and Japanese art. When asked what brought him into chemistry he said that "I came rather late to chemistry, I was not interested in it from childhood." However, he clearly feels that one can come late to chemistry, and that it can be a very positive thing. "I am always worried about fields in which people exhibit precocity, like music and mathematics. Precocity is some sort of evidence that you have to have talent. I don't like that. I like the idea that human beings can do anything they want to. They need to be trained sometimes. They need a teacher to awaken the intelligences within them. But to be a chemist requires no special talent, I'm glad to say. Anyone can do it, with hard work."

He took a standard chemistry course in high school. He recalls that it was a fine course, but apparently he found biology more enjoyable because, in his high

A box of "20-Mule Team" borax. The name comes from the fact that the mineral borax was found in Death Valley (in California) in the late 19th century. Teams of 20 mules were needed to haul the heavy wagons containing the mineral from the valley.

(d) Find the ratio of moles of water to moles of $CuSO_4$.

$$\frac{\text{moles } H_2O}{\text{moles } CuSO_4} = \frac{0.0205 \text{ mole } H_2O}{0.00410 \text{ mole } CuSO_4} = \frac{5.00}{1.00}$$

(e) The water-to-$CuSO_4$ ratio is 5:1, so the empirical formula of the blue hydrated salt is $CuSO_4 \cdot 5\ H_2O$.

EXERCISE 3.11 Determining the Formula of a Hydrated Compound

Borax powder has the formula $Na_2B_4O_7 \cdot x\ H_2O$. Determine the molar mass of the hydrated compound from the following data: The mass of borax powder taken for analysis was 2.145 g. After heating sufficiently to drive off all the water, the mass of the white powder, $Na_2B_4O_7$, that remained was 1.130 g.

FORMULAS OF SIMPLE, BINARY COMPOUNDS As you will see in the next chapter, elements often combine with oxygen, sulfur, and the halogens to produce simple binary compounds. For example, in Figure 3.1 you saw red phosphorus and bromine combine to give PBr_3,

Solid P + liquid Br_2 —will give→ liquid PBr_3

and other examples include the combination of zinc and oxygen to form zinc oxide (Figure 3.8)

Solid Zn + O_2 gas —will give→ solid ZnO

and the combination of copper and sulfur to give copper(I) sulfide (Figure 3.9).

Solid Cu + solid S —will give→ solid Cu_2S

Once again, from the law of the conservation of matter, there are several ways to determine the masses of combining elements and thus the ratio of atoms in the final molecule. The following examples illustrate the methods used.

Figure 3.8 Zinc powder is sprayed into a bunsen burner flame where the powder burns to give zinc oxide, ZnO.

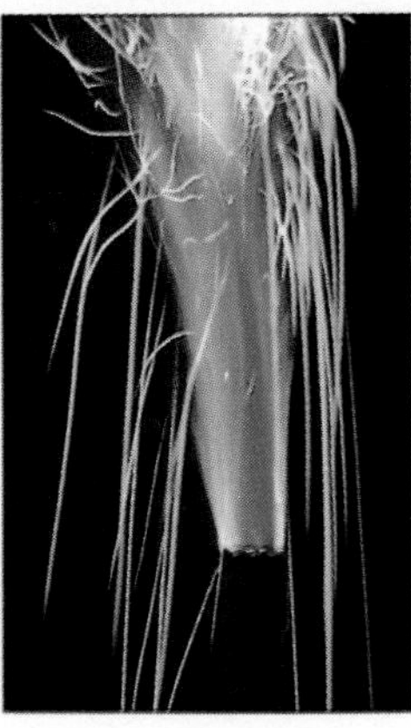

EXAMPLE 3.11

THE FORMULA OF A METAL SULFIDE

The formula of a compound containing copper and sulfur can be determined by heating a weighed quantity of Cu and S together in a crucible. The quantity of S used is more than that needed to react with all of the Cu present. After the compound has been formed between Cu and S, heating is continued to drive off all the unused S.

The following experimental information was collected in the laboratory:

Mass of crucible	19.732 g
Mass of crucible plus Cu	27.304 g
Mass of crucible plus Cu/S compound after heating	29.214 g

Use this information to determine x and y in Cu_xS_y.

92

Full-Color Photography

Over 300 full-color photos are included, each chosen specifically for this book. These photos illustrate common elements, compounds, and minerals as well as reactions and other processes in progress.

Use of Color in Art

Color has been used to make the diagrams as attractive and meaningful as possible. In addition, color is used pedagogically. For example, in the periodic tables in the text, metals are shown in blue, metalloids in green, and nonmetals in yellow.

2.5 THE PERIODIC TABLE OF THE ELEMENTS—AN INTRODUCTION 51

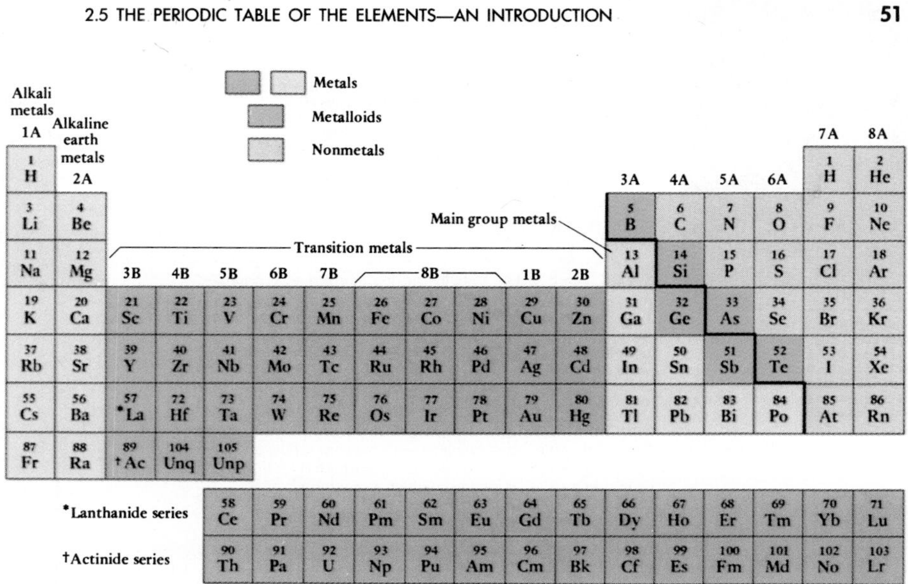

1A	2A	3B	4B	5B	6B	7B	8B	8B	8B	1B	2B	3A	4A	5A	6A	7A	8A
1 H																1 H	2 He
3 Li	4 Be											5 B	6 C	7 N	8 O	9 F	10 Ne
11 Na	12 Mg											13 Al	14 Si	15 P	16 S	17 Cl	18 Ar
19 K	20 Ca	21 Sc	22 Ti	23 V	24 Cr	25 Mn	26 Fe	27 Co	28 Ni	29 Cu	30 Zn	31 Ga	32 Ge	33 As	34 Se	35 Br	36 Kr
37 Rb	38 Sr	39 Y	40 Zr	41 Nb	42 Mo	43 Tc	44 Ru	45 Rh	46 Pd	47 Ag	48 Cd	49 In	50 Sn	51 Sb	52 Te	53 I	54 Xe
55 Cs	56 Ba	57 *La	72 Hf	73 Ta	74 W	75 Re	76 Os	77 Ir	78 Pt	79 Au	80 Hg	81 Tl	82 Pb	83 Bi	84 Po	85 At	86 Rn
87 Fr	88 Ra	89 †Ac	104 Unq	105 Unp													

*Lanthanide series	58 Ce	59 Pr	60 Nd	61 Pm	62 Sm	63 Eu	64 Gd	65 Tb	66 Dy	67 Ho	68 Er	69 Tm	70 Yb	71 Lu
†Actinide series	90 Th	91 Pa	92 U	93 Np	94 Pu	95 Am	96 Cm	97 Bk	98 Cf	99 Es	100 Fm	101 Md	102 No	103 Lr

Figure 2.3 The periodic table. Elements are listed in ascending order of atomic number. The following points are important:
(a) Metals are shown in various shades of blue, the metalloids in green, and the nonmetals in yellow.
(b) Periods are horizontal rows of elements, and groups are vertical columns.
(c) Special names of groups:
Group 1A = alkali metals
Group 2A = alkaline earth metals
Group 7A = halogens
Group 8A = rare gases

The table can be divided into several regions. There are **metals** (blue in Figure 2.3), **nonmetals** (in yellow), and elements that fit neither of these categories entirely and so are called **metalloids** (in green). Although we shall more carefully describe the difference between metals and nonmetals later, you should be familiar with it from your everyday experience. Automobiles are made of iron (Fe) and aluminum (Al), both metals. Oxygen and nitrogen are the gases you breathe, and both are nonmetals. All the nonmetals in the periodic table lie to the right of the heavy zigzag line (between Al and Si, Ge and As, Sb and Te, and Po and At). Most of the elements lying immediately next to this line have some properties that can be thought of as typically metallic, but others that are considered characteristic of nonmetals. Such elements (B, Si, Ge, As, Sb, and Te) are called metalloids.

Many of the groups of the periodic table or collections of groups have meaningful names, and you should learn them.

GROUP 1A: ALKALI METALS

The ancient Arab chemists studied the properties of many natural substances and found that the ashes, called *al-qali*, of certain plants gave water solutions that felt slippery on the hands. It is now known that these substances contain compounds of Group 1A elements. Since these elements are also metals, they are often called the alkali metals (Figure 2.4).

Reactions in Progress

The progress of many reactions is shown in a sequence of photos. This allows the student to see the beginning, middle, and end of a reaction.

4.3 SOME COMMON TYPES OF CHEMICAL REACTIONS 111

we obtain a great deal of energy from the combustion of *hydrocarbons*, compounds containing only C and H, and the combustion of propane, butane, and gasoline are excellent examples (see Figure 4.4 and Example 4.1). The products of the complete combustion of hydrocarbons are only water and carbon dioxide. We can also use compounds containing C, H, and O—such as methyl and ethyl alcohol—as fuels, since they also give only water and CO_2 on complete combustion.

$$\underset{\text{methyl alcohol}}{2\ CH_3OH(\ell)} + 3\ O_2(g) \longrightarrow 4\ H_2O(\ell) + 2\ CO_2(g)$$

Ethyl alcohol is widely used in Brazil as an automotive fuel, and there is some discussion of using methyl alcohol in the United States because it may lead to less air pollution (see Chapter 6).

The French scientist Lavoisier was the first to study animal respiration in the 18th century. He observed that the amount of heat produced by a guinea pig in the process of exhaling a given amount of carbon dioxide is similar to the amount of heat given off in burning an amount of carbon that would yield the same quantity of carbon dioxide. From this and other experiments he concluded that "Respiration is a combustion, slow it is true, but otherwise perfectly similar to that of charcoal." One important combustion reaction that occurs in animals is the "burning" of the sugar glucose to provide energy.

$$\underset{\text{glucose}}{C_6H_{12}O_6(aq)} + 6\ O_2(g) \longrightarrow 6\ CO_2(g) + 6\ H_2O(\ell)$$

DECOMPOSITION REACTIONS

Elements combine to form compounds in **combination reactions**. However, there is a large class of reactions—**decomposition reactions**—in which the compounds break down to form simpler ones, usually by heating (Figure 4.8).

Figure 4.8 The "dichromate volcano." Ammonium dichromate decomposes according to the equation $(NH_4)_2Cr_2O_7(s) \longrightarrow N_2(g) + Cr_2O_3(s) + 4\ H_2O(g)$. In the first photo, the dichromate salt is placed in a beaker with an alcohol-soaked paper wick. After ignition, the heat evolved by the reaction allows it to continue. On completion, the beaker contains a pile of green chromium(III) oxide, and water droplets cling to the side of the beaker.

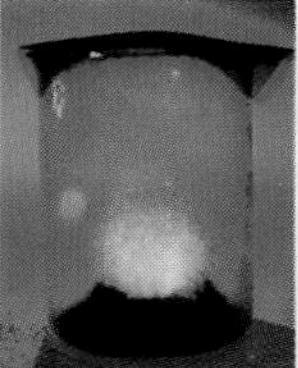

44 2 ATOMS AND ELEMENTS

Figure 2.1 All atoms consist of one or more protons (positively charged) and usually at least as many neutrons (no charge) packed into an extremely small nucleus. Electrons (negatively charged) are arranged in space as a "cloud" about the nucleus. In an electrically neutral atom the number of electrons equals the number of protons. The picture drawn here is a very crude representation that will be refined in Chapters 8 and 9.

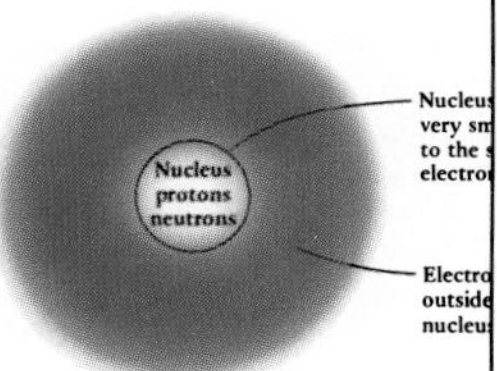

number of negatively charged electrons ... *number of positively charged protons in* ...

Atoms are extremely small; the radius of the typical atom is between 30 and 150 pm (0.000000000030 m = 3.0×10^{-11} m). To give you a feeling for the incredible smallness of the atom, look at it this way:

(a) One teaspoon of water (about 1 cm^3) contains about three times as many atoms as the Atlantic Ocean contains teaspoons of water.

(b) If, after the flood of about 3000 BC, Noah had started to string hydrogen atoms on a thread at the rate of one atom per second for 8 hours a day, the chain would be only about 3.9 meters (about 13 feet) long today!

All atoms of the same element have the same number of protons in the nucleus, and, because the atom is electrically uncharged, the number of electrons surrounding the nucleus is the same as the number of protons. This number is called the **atomic number**, and it is given the symbol **Z**. In the periodic table at the front of the book, the atomic number for each element is given above the element's symbol. Magnesium, for example, has a nucleus containing 12 protons, so its atomic number is 12; uranium has 92 nuclear protons and Z = 92.

With the knowledge that all atoms of a given element have the same number of protons in the nucleus, we can give a better definition of an element than that in Chapter 1. That is, ***an element is defined by its atomic number.***

EXERCISE 2.1 Atomic Numbers, Names, and Element Symbols
The symbol for carbon is C, and its atomic number is 6. Find all the other elements that have a symbol beginning with the letter C and give the name and atomic number for each.

Each element is represented by a box in the periodic table in the front of the book. Among the data given are the element's symbol and its atomic number.

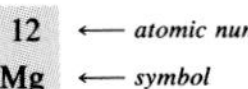

12 ← *atomic number*
Mg ← *symbol*

Since atoms are the ultimate building blocks of matter, they must have a mass, with some elements having atoms heavier than other elements. Just as our clocks and longitude are set relative to the time and longitude at Greenwich, England, so too have chemists established a scale of atomic masses relative to a standard. This standard is the mass of a carbon atom that has six protons and six neutrons in its nucleus. Such an atom is defined to have a mass of exactly 12 **atomic mass units** (or 12 amu), and the mass of every other element is established relative to this mass. Thus, for example, experiment shows that an oxygen atom is, on the average, 1.33 times heavier than a carbon atom, so an oxygen atom has a mass of 1.33×12.0 amu or 16.0 amu.

1 amu = $\frac{1}{12}$ the mass of a carbon atom having 6 protons and 6 neutrons in the nucleus.

Marginal Notes

Marginal notes are used frequently throughout the book to highlight important points and interesting facts and to alert the reader to topics that will be discussed later in the book or to help locate related topics. Some particularly important notes are printed in blue.

Worked Examples

Over 200 worked-out examples serve as models for solving end-of-chapter problems. The detailed solutions are developed using the technique of dimensional analysis. Cancel marks, in blue, should help to more easily determine the result of a calculation.

116 4 CHEMICAL REACTIONS: AN INTRODUCTION

EXAMPLE 4.2

WEIGHT RELATIONS IN REACTIONS

Propane, C_3H_8, can be used as a fuel in your home or car because it is easily liquefied and transported. If you burn 1.00 pound or 454 g of propane, how much oxygen is required for complete combustion and how much CO_2 and H_2O are formed?

Solution

1. Remember that the first step must always be to write a balanced equation; see page 107.

$$C_3H_8(g) + 5\ O_2(g) \longrightarrow 3\ CO_2(g) + 4\ H_2O(g)$$

2. Having balanced the equation, we can proceed to the stoichiometric calculation. First, we shall convert the mass of propane to moles.

$$454\ g\ C_3H_8 \left(\frac{1\ mol\ C_3H_8}{44.10\ g\ C_3H_8}\right) = 10.3\ mol\ C_3H_8$$

3. Relate the moles of propane available to moles of O_2 required with the stoichiometric factor.

$$10.3\ mol\ C_3H_8 \left(\frac{5\ mol\ O_2\ required}{1\ mol\ C_3H_8\ available}\right) = 51.5\ mol\ O_2\ required$$

4. Convert moles of O_2 required to grams.

$$51.5\ mol\ O_2\ required \left(\frac{32.00\ g\ O_2}{1\ mol\ O_2}\right) = 1650\ g\ O_2\ required$$

To find how much CO_2 is produced, for example, we simply repeat the last two steps.

3′. Relate moles of C_3H_8 available to moles of CO_2 produced with the stoichiometric factor.

$$10.3\ mol\ C_3H_8\ available \left(\frac{3\ mol\ CO_2\ produced}{1\ mol\ C_3H_8\ available}\right) = 30.9\ mol\ CO_2\ produced$$

4′. Convert moles of CO_2 produced to the mass of CO_2.

$$30.9\ mol\ CO_2 \left(\frac{44.01\ g\ CO_2}{1\ mol\ CO_2}\right) = 1360\ g\ CO_2$$

Now, how can we find the mass of H_2O produced? Go through steps 3′ and 4′ again? Of course we could! But it would be easier to recognize that the total mass of the reactants

$$454\ g\ C_3H_8 + 1650\ g\ O_2 = 2.10 \times 10^3\ g\ of\ reactants$$

must be the same as the total mass of products. Therefore, the mass of water produced is

Total mass of products $= 2.10 \times 10^3\ g = 1.36 \times 10^3\ g\ CO_2$ produced + ? g H_2O

Mass of H_2O = 740 g

8.7 THE WAVE MECHANICAL VIEW OF THE ATOM

Table 8.1 Summary of the Quantum Numbers, Their Interrelationships, and t

Principal Quantum Number Symbol = n Values = 1, 2, 3, . . . (Orbital Size, Energy)	Angular Momentum Quantum Number Symbol = ℓ Values = 0 . . . $n - 1$ (Orbital Shape)	Magnetic Quantum Number Symbol = m_ℓ Values = $-\ell$ . . . 0 . . . $+\ell$ (Orbital Orientations = Number of Orbitals in Subshell)		
1	0	0	(1 ori	
2	0	0	(1 ori	
	1	+1,0,−1	(3 ori	
3	0	0	(1 ori	
	1	+1,0,−1	(3 ori	
	2	+2,+1,0,−1,−2	(5 ori	
				n = 3 shell)
4	0	0	(1 orientation)	1 4s orbital
	1	+1,0,−1	(3 orientations)	3 4p orbitals
	2	+2,+1,0,−1,−2	(5 orientations)	5 4d orbitals
	3	+3,+2,+1,0,−1,−2,−3	(7 orientations)	7 4f orbitals
				(16 orbitals of 4 types in the n = 4 shell)

seven of them, as given by the seven possible values of m_ℓ when $\ell = 3$ (+3, +2, +1, 0, −1, −2, −3).

The three quantum numbers introduced so far and their allowed values are summarized in Table 8.1. These rules will be used to explore atomic structure more fully in Chapter 9, so you should be thoroughly familiar with them.

No known elements have electrons assigned to orbitals with an ℓ value greater than 3 in the ground state. If they existed, however, ℓ = 4 would correspond to g orbitals, ℓ = 5 to h orbitals, and so on down the alphabet.

EXERCISE 8.8 Using Quantum Numbers

Complete the following statements:

(a) When $n = 2$, the values of ℓ can be ____ and ____.
(b) When $\ell = 1$, the values of m_ℓ can be ____, ____, and ____, and the subshell has the letter label ____.
(c) When $\ell = 2$, this is called a ____ subshell.
(d) When a subshell is labeled s, the value of ℓ is ____ and m_ℓ has the value ____.
(e) When a subshell is labeled p, there are ____ orbitals within the subshell.
(f) When a subshell is labeled f, there are ____ values of m_ℓ and there are ____ orbitals within the subshell.

THE SHAPES OF ATOMIC ORBITALS

The chemistry of an element and of its compounds is determined by the electrons in the element's atoms: which orbitals describe those electrons and the shapes of those orbitals. Now that we know something about the types and number of orbitals available, we want to turn to the question of orbital shape.

Exercises

All examples are followed by similar exercises. We encourage the student to answer the exercise. The solutions to the exercises, worked out in brief, are given in Appendix K.

End-of-Chapter Study Questions

The end-of-chapter questions, some illustrated with photographs, include review questions, questions classified by type, and general questions. The classified problems are in matched pairs. The first member of each pair is numbered in blue and the answer is given in Appendix L.

STUDY QUESTIONS 127

$$\text{Percent yield} = \frac{\text{actual yield}}{\text{theoretical yield}} \times 100$$

Finally, chemical reactions such as combustions can be used to determine the formula of a compound, and other reactions can be used to analyze mixtures of compounds.

STUDY QUESTIONS

REVIEW QUESTIONS

1. What information does a balanced chemical equation provide?
2. In the reaction in Figure 4.1, how many molecules of O_2 would you need if you have 2000 molecules of S_8? How many molecules of SO_2 would you obtain from the reaction?
3. Find in the chapter one example of each of the following reaction types and write the balanced equation for the reaction: (a) combustion; (b) reaction of O_2 with a metal; (c) reaction of O_2 with a nonmetal; and (d) a decomposition reaction.
4. Find two examples in the chapter of the reaction of a metal with a halogen, write a balanced equation for each example, and name the product.
5. What is the guiding principle of chemical stoichiometry? Illustrate your answer with the reaction $N_2(g) + O_2(g) \rightarrow 2\ NO(g)$.
6. When working in the laboratory, chemists often set up reactions so that one of several reactants is there in greater quantity than required by stoichiometry. This means that one of the reactants is a "limiting reagent." Illustrate what is meant by a "limiting reagent" using the reaction of calcium and oxygen to give calcium oxide.
7. Briefly explain the difference between the actual and theoretical yield for a reaction.
8. You have a green powder that you know consists of nickel(II) oxide and nickel(II) carbonate. Suggest a way to analyze the sample for the amount of nickel carbonate it may contain.
9. In many cases metal oxides will react with H_2 to leave the metal and convert the oxygen to water. For example,

   $TiO(s) + H_2(g) \longrightarrow Ti(s) + H_2O(g)$

   If you did not know the metal oxide in question was TiO, explain how you could use this reaction to find the empirical formula of the compound.

BALANCING EQUATIONS

10. Balance the following equations.
    (a) $Al(s) + O_2(g) \longrightarrow Al_2O_3(s)$
    (b) $N_2(g) + H_2(g) \longrightarrow NH_3(g)$
    (c) $C_6H_6(\ell) + O_2(g) \longrightarrow H_2O(g) + CO_2(g)$
11. Balance the following equations.
    (a) $Al(s) + Cl_2(g) \longrightarrow AlCl_3(s)$
    (b) $SiO_2(s) + C(s) \longrightarrow Si(s) + CO(g)$
    (c) $Fe(s) + H_2O(g) \longrightarrow Fe_3O_4(s) + H_2(g)$
12. Balance the following equations.
    (a) $UO_2(s) + HF(\ell) \longrightarrow UF_4(s) + H_2O(\ell)$
    (b) $B_2O_3(s) + HF(\ell) \longrightarrow BF_3(g) + H_2O(\ell)$
    (c) $BF_3(g) + H_2O(\ell) \longrightarrow HF(\ell) + H_3BO_3(s)$
13. Balance the following equations.
    (a) $MgO(s) + Fe(s) \longrightarrow Fe_2O_3(s) + Mg(s)$
    (b) $H_3BO_3(s) \longrightarrow B_2O_3(s) + H_2O(\ell)$
    (c) $NaNO_3(s) + H_2SO_4(\ell) \longrightarrow Na_2SO_4(s) + HNO_3(g)$
14. Balance the following equations.
    (a) $Na_2O_2(s) + H_2O(\ell) \longrightarrow NaOH(aq) + H_2O_2(aq)$
    (b) $PH_3(g) + O_2(g) \longrightarrow P_4O_{10}(s) + H_2O(g)$
    (c) $C_2H_5Cl(g) + O_2(g) \longrightarrow CO_2(g) + H_2O(g) + HCl(g)$
15. Balance the following equations.
    (a) $CaF_2(s) + H_2SO_4(\ell) \longrightarrow CaSO_4(s) + HF(g)$
    (b) $N_2O(g) \longrightarrow N_2(g) + O_2(g)$
    (c) $NH_4NO_3(s) \longrightarrow N_2O(g) + H_2O(g)$
16. Balance the following equations.
    (a) Reaction to produce hydrazine, N_2H_4.

    $H_2NCl(aq) + NH_3(g) \longrightarrow NH_4Cl(aq) + N_2H_4(aq)$

    (b) Reaction of fuel used in moon lander and space shuttle.

    $(CH_3)_2N_2H_2$

    (c) Reaction of
    C_2H_2.

    $CaC_2(s) +$ H
17. Balance the foll
    (a) Reaction of
    monia.

    $CaCN_2(s) +$

    (b) Reaction to

    $NaBH_4(s) +$

4.4 WEIGHT RELATIONS IN CHEMICAL REACTIONS: STOICHIOMETRY 115

Since the object of this example is to find the mass of O_2 required, the solution to the problem is complete.

You may also wish to know the amount of MgO produced in the reaction of 0.145 g of magnesium with oxygen. Because of the principle of the conservation of matter, you can answer this simply by adding up the masses of Mg and O_2 used (giving 0.240 g of MgO produced). Alternatively, you could repeat steps 3 and 4 above, but with the appropriate stoichiometric factor and molar mass.

Step 3′. Relate number of moles of Mg to the number of moles of MgO produced.

$$0.00596\ \cancel{\text{mol}}\ \text{Mg}\left(\frac{1\ \text{mol MgO produced}}{1\ \cancel{\text{mol}}\ \text{Mg used}}\right) = 0.00596\ \text{mol MgO produced}$$

Step 4′. Convert the moles of MgO produced to grams.

$$0.00596\ \cancel{\text{mol}}\ \text{MgO produced}\left(\frac{40.30\ \text{g MgO}}{1\ \cancel{\text{mol}}\ \text{MgO}}\right) = 0.240\ \text{g MgO}$$

Using this alternative may seem a bit silly here, since only one product is formed, but simply adding the masses of reactants to obtain the product mass works only if the reaction gives a *single* product. When more than one product results from a reaction, it is necessary to apply steps 3 and 4 for each product.

This example was meant to show you the general approach to follow in a stoichiometric calculation. To help you see this more clearly, consider the scheme in Figure 4.11 and work through the following examples and exercises.

Figure 4.11 A scheme outlining the relation between the mass of one reactant (A) and the mass of another reactant or the mass of a product (B). The stoichiometric factor is always (moles of reactant B or moles of product) ÷ (moles of reactant A).

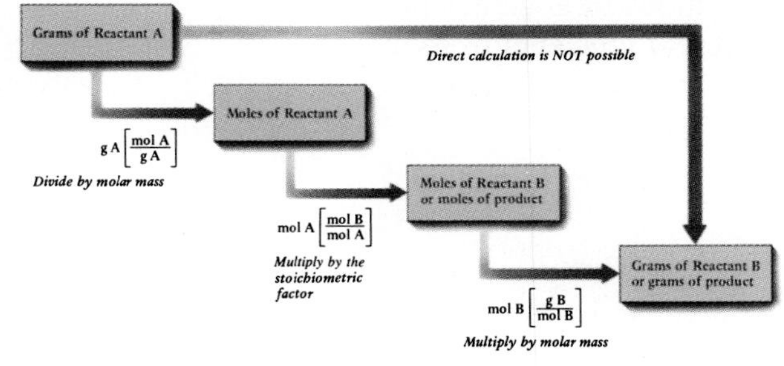

Flow Diagrams

Scattered throughout the book are several diagrams that suggest a general approach to problem solving. The student should use these diagrams to help organize the information contained in the problems.

Boxed Essays

Boxed essays are used in several ways. Some amplify the principles of the chapter, while others give some historical background or discuss some practical and interesting applications of chemical principles.

308 8 ATOMIC STRUCTURE

The Relationship Between ℓ and the Number of Nodal Surfaces

There is a direct correspondence between the value of ℓ and the number of nodal surfaces slicing through the spherical electron cloud and dividing it into regions of electron density or probability.

ORBITAL TYPE	VALUE OF ℓ	NUMBER OF NODAL SURFACES	REGIONS OF ELECTRON DENSITY
s	0	0	1
p	1	1	2
d	2	2	4*
f	3	3	8*

*These numbers are not quite true for all orbitals (*d* and *f*) in a given shell. As illustrated by the d_{z^2} orbital, some nodal surfaces are not planes but are cone-shaped; for this orbital, this means there are apparently 3 regions of electron density.

Figure 8.17 One of seven possible *f* electron orbitals. Notice the presence of three nodal planes (*xy*, *xz*, and *yz*) as required by an orbital with ℓ = 3.

these orbitals lie in the planes defined by the *x-z* and *y-z* axes, respectively. If the electron cloud is sliced by two vertical planes, we have the d_{xy} orbital where the orbital lies in the *xy* plane. Finally, notice that these three orbitals are related to *p* orbitals, in that each comes from slicing a *p* orbital with yet another nodal plane.

Of the two remaining *d* orbitals, the $d_{x^2-y^2}$ orbital is easier to visualize. Like the d_{xy} orbital, $d_{x^2-y^2}$ results from two vertical planes slicing the electron cloud into quarters. Now, however, the planes bisect the *x* and *y* axes, so the regions of electron density lie *along* the *x* and *y* axes.

The final *d* orbital, d_{z^2}, has two main regions of electron density along the *z* axis, but there is also a "donut" of electron density in the *xy* plane. This orbital also has two nodal surfaces, but the *surfaces* are not flat. Think of an ice cream cone sitting with its tip at the nucleus. One of the electron clouds along the *z* axis will sit inside the cone. If you use another cone pointing in the opposite direction from the first cone, again with its tip at the nucleus, another region of electron density will fit inside this second cone. The region outside b
donut-shaped region of electron density.

f ORBITAL SHAPES The seven *f* orbital
there are three nodal surfaces slicing thr
the electron cloud into eight regions o
orbitals less easily visualized than the o
example of an *f* orbital is illustrated in

EXERCISE 8.9 Orbital Shapes
(a) What are the *n* and ℓ values for each o
and 4*f*?
(b) How many nodal planes are there for a

SUMMARY
The modern quantum mechanical view
in this century through some important
sured the charge/mass ratio (*e*/*m*) of

SUMMARY QUESTIONS 499

(b) You quickly evaporate 0.0631 g of boron hydride into a flask with a volume of 120. mL; the gas exerts a pressure of 98.6 mmHg at 23 °C.

117. Methylthallium, $Tl(CH_3)_x$, has been prepared. To find its formula (the value of *x*), you decompose the compound with HCl. The hydrogen of the acid combines with CH_3 to produce CH_4 gas. Assume you decompose 0.102 g of $Tl(CH_3)_x$ and find that the isolated CH_4 gas has a pressure of 90.6 mmHg in a volume of 256 mL at 30.0 °C. What is the correct formula of methylthallium?

118. A mixture of aluminum and zinc weighing 1.67 g was completely dissolved in acid. In the process the mixture evolved 1.69 L of H_2, measured at 273 K and 1.00 atm pressure. What is the mass of aluminum in the mixture? (The aluminum reacts with acid according to the equation $Al(s) + 3\ H_3O^+(aq) \rightarrow Al^{3+}(aq) + \frac{3}{2}\ H_2(g) + 3\ H_2O(\ell)$. For the zinc reaction, see Example 12.10.)

119. You have 1.500 g of a mixture of $NaHCO_3$ and Na_2CO_3. Exactly 14.00 mL of 1.50 M HCl reacts with the mixture according to the equations

$$NaHCO_3(aq) + HCl(aq) \longrightarrow NaCl(aq) + H_2O(\ell) + CO_2(g)$$

$$Na_2CO_3(aq) + 2\ HCl(aq) \longrightarrow 2\ NaCl(aq) + H_2O(\ell) + CO_2(g)$$

(a) How many grams each of $NaHCO_3$ and Na_2CO_3 are there in the mixture? (b) If the CO_2 from both reactions is collected in a 0.450-L flask at 27 °C, what is the pressure of the gas?

SUMMARY QUESTIONS

120. Consider the following reaction of cobalt and nitric acid.

$$Co(s) + HNO_3(aq) \longrightarrow Co(NO_3)_2(aq) + NO_2(g) + H_2O(\ell)$$

(a) Balance the equation for the reaction.
(b) Indicate the oxidizing and reducing agents.

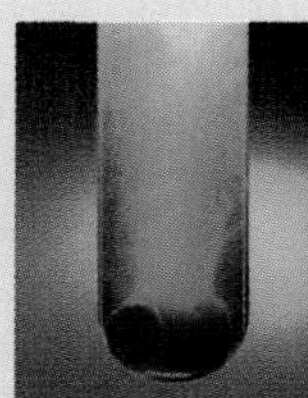

Metallic cobalt reacts with nitric acid to give pink $Co(NO_3)_2$, H_2, and nitrogen oxides, among them NO_2.

(c) If you use 2.115 g of cobalt metal and 25.0 mL of 6.00 M HNO_3, how many grams of $Co(NO_3)_2$ can be isolated from the reaction? How many grams of NO_2?
(d) If you isolate the NO_2 in a 2.56-L flask at 25 °C and find that its pressure is 375 mmHg, what is the percent yield of the gas? (Use the quantities of reactants in part (c).)

121. Chlorine trifluoride, one of the most reactive compounds known, is made by the reaction of chlorine and fluorine.

$$Cl_2(g) + 3\ F_2(g) \longrightarrow 2\ ClF_3(g)$$

(a) Assume you mix 0.71 g of Cl_2 with 1.00 g of F_2 in a 258-mL flask at 23 °C. What is the partial pressure of each of the two reactants before reaction? What is the partial pressure of the product and any leftover reactant after reaction in the 258-mL flask at 23 °C?
(b) What is the Lewis dot structure for ClF_3? What is its structural-pair geometry? Given that the molecule is polar, what is its molecular structure? What hybrid orbital set is used by the Cl atom in ClF_3?
(c) Estimate the enthalpy of the reaction above, using the bond energies in Table 10.4.

Summary Questions

Multi-part summary questions are included at the end of each chapter. These questions tie together the many different concepts from the current chapter with those in previous chapters.

Something More About . . .

Scattered about the book are almost 30 essays called "Something More About . . . ." These essays deal with applications of chemistry in the world today, from growing crystals in space, to the ozone layer, to the asbestos controversy.

SOMETHING MORE ABOUT
Titanium Dioxide: A White Pigment

There is hardly a white-colored or tinted object in our environment—and that includes the paper on which these words are printed—that does not contain TiO_2 pigments. This compound, whose chemical name is titanium(IV) oxide or titanium dioxide, is manufactured in enormous quantities. In the United States, TiO_2 ranks 44th on the list of the "top 50 chemicals produced"; just over 2 billion pounds were produced in 1988.

For many years the pigment in white paint was "white lead," which is a mixture of lead(II) carbonate and lead(II) hydroxide ($PbCO_3 + Pb(OH)_2$). However, we know now that lead-based paint is a major health hazard, so titanium dioxide has completely replaced white lead in paints. In addition, TiO_2 is used in the plastics, ceramics, and paper industries.

Titanium is ninth in abundance of the elements making up the earth's crust, and much of it exists as the mineral ilmenite, which has an approximate formula of $FeTiO_3$. A smaller amount is found as rutile, almost pure TiO_2. The most commonly used process in the United States to make pure TiO_2 is called *chlorination*. If ilmenite is the source of the titanium, the ore is first treated with chlorine gas and coke (a form of carbon) to produce crude or impure TiO_2.

$$2\ FeTiO_3(s) + 2\ Cl_2(g) + C(s) \longrightarrow 2\ FeCl_2(s) + 2\ TiO_2(s) + CO_2(g)$$

The crude titanium dioxide produced in this step is then treated again with chlorine in the presence of carbon. (The TiO_2 used in the step may also have been rutile ore.)

$$2\ TiO_2(s) + 4\ Cl_2(g) + 3\ C(s) \longrightarrow 2\ TiCl_4(\ell) + CO_2(g) + 2\ CO(g)$$

The titanium-containing product here is titanium(IV) chloride, $TiCl_4$, a compound that is liquid at room temperature (and which is commonly known to chemists as "tickle-4"). The liquid is very reactive, as illustrated in the photograph; some $TiCl_4$ was allowed to react with water in the air to produce a voluminous fog of titanium oxide and HCl.

$$TiCl_4(\ell) + 2\ H_2O(g) \longrightarrow TiO_2(s) + 4\ HCl(g)$$

However, in industry the tetrachloride is converted to pure TiO_2 by burning $TiCl_4$ in oxygen at high temperatures,

$$TiCl_4(\ell) + O_2(g) \longrightarrow TiO_2(s) + 2\ Cl_2(g)$$

and it is this pure compound that is used to paint your home, car, or school building white or to make the paper in this book as white as possible.

Liquid titanium(IV) chloride, $TiCl_4$, reacts with water vapor to produce a fog of TiO_2 and HCl.

4.4 WEIGHT RELATIONS IN CHEMICAL REACTIONS: STOICHIOMETRY

A major aspect of chemistry is the study of chemical reactions, often with the hope of finding some new material that can be useful to society. If a product turns out to have commercial potential, the efficiency of the reaction is of interest. Can the product be made in sufficient quantity and purity that it can be sold for a reasonable price? Questions such as these mean that reactions must be studied quantitatively. **Stoichiometry** is the study of the quantitative relations between the amounts of reactants and

OTHER FEATURES:

- This edition was edited using the symbols and terminology adopted by the American Chemical Society for use in all of its scientific journals and other publications.
- Each chapter ends with a summary of the important concepts, equations, and key terms discussed.
- Boxes titled "Historical Figures in Chemistry" are biographical sketches of famous personalities in chemistry.
- Appendices at the back of the book include a review of mathematical methods, a table of conversion factors, important constants, and a glossary of terms in the combined index/glossary. Inside the back cover are short tables of useful constants and a list of all of the data tables in the book.

CONTENTS OVERVIEW

CONTENTS

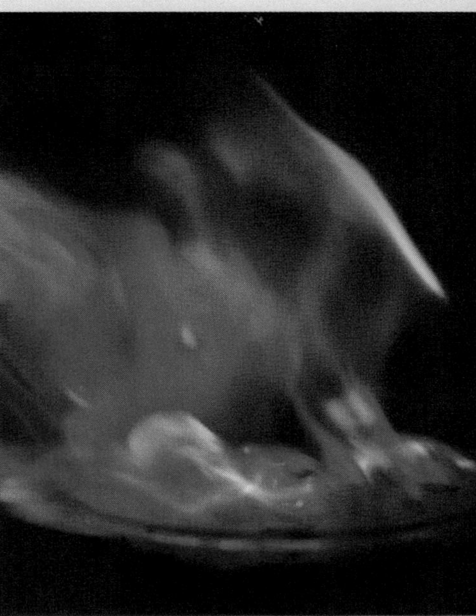

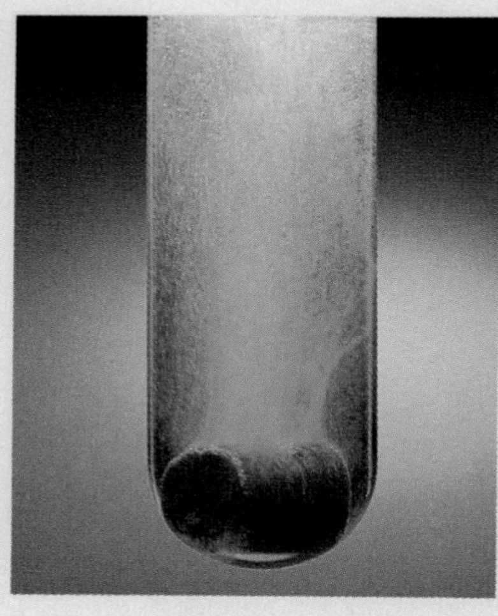

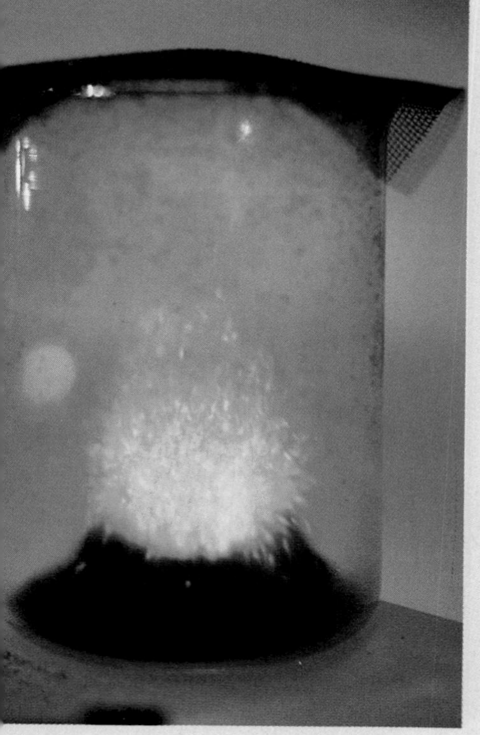

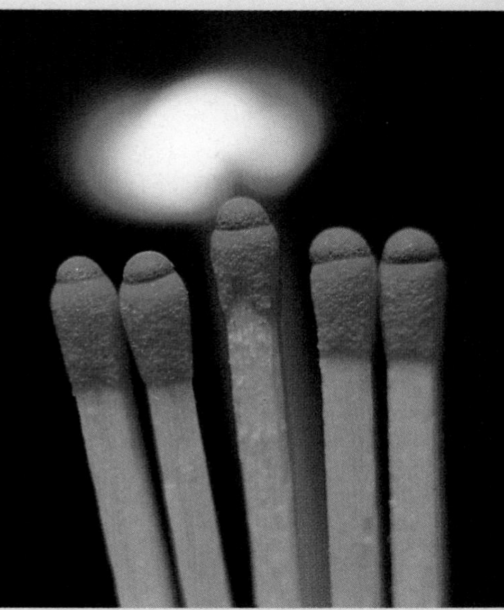

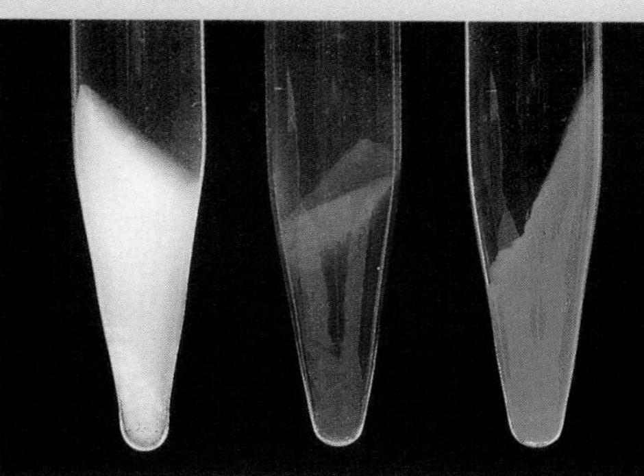

PART ONE

THE BASIC TOOLS OF CHEMISTRY

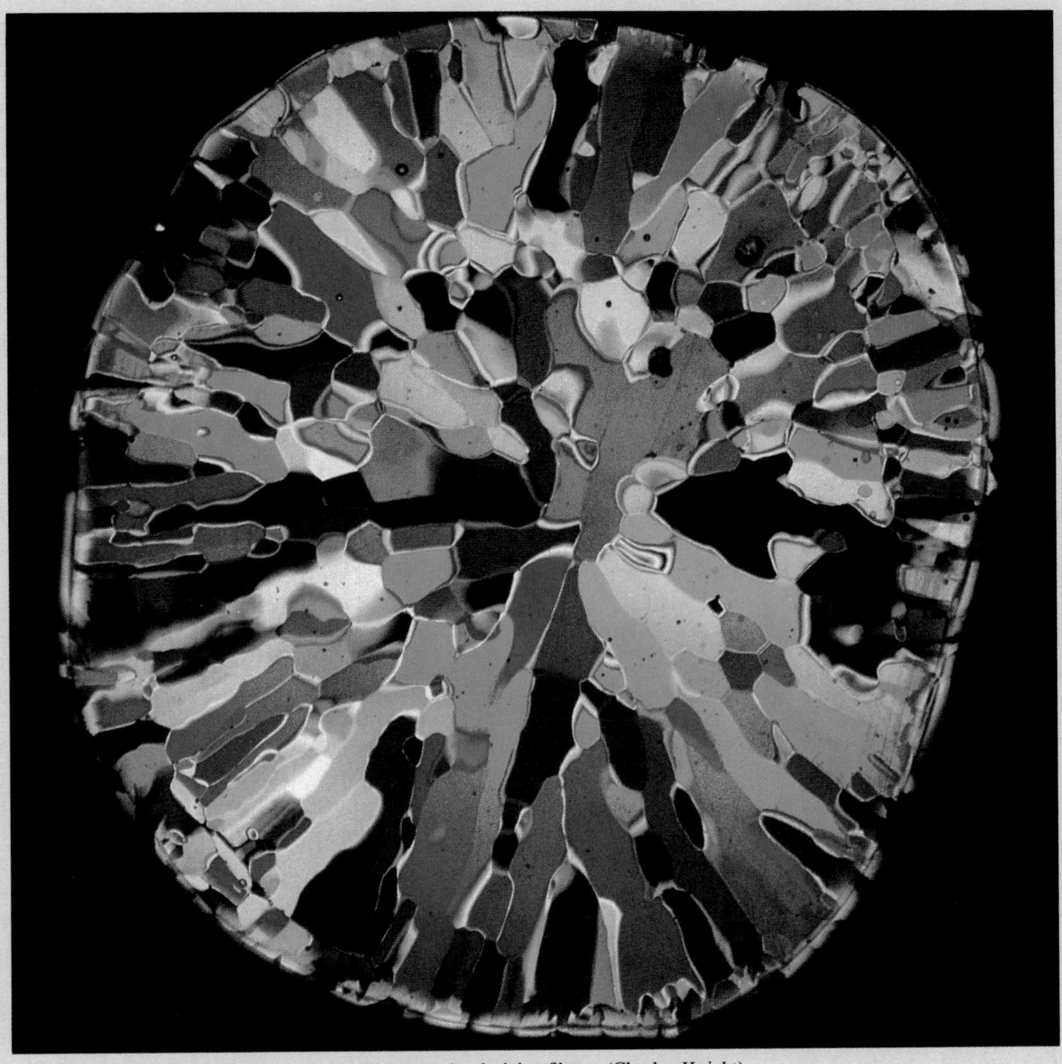

Cross section of icicle photographed through crossed polarizing filters. (Charles Knight)

Barnett Rosenberg

(Photo by Doug Elbinger)

Dr. Barnett Rosenberg moved from New York City, where he had studied physics, to Michigan State University in East Lansing to establish a department of biophysics in 1961. It was there that he and his collaborators discovered that a compound that had been known for a century or more, $Pt(NH_3)_2Cl_2$, was effective in the treatment of certain types of cancers. The story of his discovery of the use of this compound, now commonly known as cisplatin (see photos p. 185) is a wonderful example of serendipity in science.

The dictionary defines serendipity as "the faculty of making fortunate and unexpected discoveries by accident." Rosenberg had set out to study a problem that had interested him for some time, but the results of the experiment were quite different from those expected, and, much to the benefit of all of us, he recognized that they had implications in cancer chemotherapy. This interview is the story of that discovery, and illustrates the important role of serendipity in science.

We visited Dr. Rosenberg at the Barros Research Institute, housed in a modest building set beside some ponds in the countryside outside East Lansing. The only photograph on the wall is one of Albert Einstein, from whom Rosenberg is descended scientifically. What follows is a description, in Dr. Rosenberg's own words, of what led him to his work on platinum-containing cancer chemotherapy agents.

Dr. Rosenberg has received many awards for his work, including the Galileo Medal from the University of Padua (Italy) and the Charles F. Kettering Prize from the General Motors Cancer Foundation.

The beginning: Brooklyn, Switzerland, and Greenwich Village

I knew, as a very young child, that I was going to be a scientist, and I knew I was going to be a professor and have a laboratory. So, all my life I have been dedicated to the pleasures of science. But my original interests have changed drastically over the years. My first interests were in astronomy and then in chemistry. Finally, in my college years I found that physical chemistry was far more interesting than just ordinary chemistry in the sense of making things in test tubes, and gradually my interests began to shift to physics. So, I became a physicist after I returned from World War II. I took my Bachelor's Degree at Brooklyn College in physics and then went off for a year to study physics in Switzerland. I came back after I decided it would take too long to get a Ph.D. there, and went to the Physics Department at New York University. I lived in Greenwich Village . . . a lovely life, very romantic at that particular time of my life. I was lucky enough to work on my Ph.D. thesis under Professor Harmut Kallman in the

Physics Department. He had been a student of Einstein, and he was a combination theoretical and experimental physicist. He was a rather unique individual, and I have very fond memories of working in his laboratory.

The subject matter I chose had to do with photoelectric phenomena in solids. However, the thought of spending the rest of my life following this particular pathway didn't seem terribly attractive to me, so I started to think about the possibilities of light interacting with biological materials. Looking into it, I found that Melvin Calvin had beaten me to the field of photosynthesis because he was very much moving in the direction of how electrons move in the photosynthetic process. But nobody had been looking at this in terms of vision. So I started some experiments to study the movements of electrons in retinals and proteins.

A new Department of Biophysics

At that time I started, with some friends, to look for a place where we could set up a biophysics department. Michigan State made us the best offer, so we set up our laboratories here in 1961. I felt that working with proteins and retinals wasn't sufficiently biological for a department that has "bio" in its title. I had to do an experiment which would involve the students in something that was a little more biological. One experiment I thought of was a very curious one. There is a picture that appears in biological textbooks that is a familiar picture to a physicist. That is the picture of a cell undergoing mitotic division, which looks *exactly* like what you would get if you took a bar magnet, put a sheet of paper on it, and sprinkled iron filings over it. It looks like a magnetic dipole field! That intrigued me. When I looked into the literature I found very quickly that I was not the only one taken by this similarity, but nobody had done a test of it. So it remained an ambiguous idea.

The first experiments

We needed an apparatus where we could detect any changes as we poured the energy into the cell. The type of apparatus we chose was one that was developed by one of the nuclear physicists who after World War II became a biophysicist, Leo Szilard. A very great scientist! He developed this continuous cell culture apparatus, which had actually evolved from his knowledge of how a uranium pile works. He and a student were able to develop this system whereby they continually pumped fresh nutrient into a chamber and harvested used nutrient plus the cells that were growing in the chamber. We set up such an apparatus, but we did one thing differently. We took two platinum mesh electrodes and implanted them in the chamber. Now what we wanted to do was apply an alternating electric field to the growing cells in the culture apparatus. We wanted to see whether there was any change in the cell division rate. But to test it before we put in mammalian cells we used bacterial cells, *E. coli*. When we turned on the electric field, we found an astonishing thing: the density of the bacteria in the chamber dropped almost to the zero point! In fact it was becoming an aseptic chamber. We wondered why this happened. Then we looked at the effluent coming from the chamber in the microscope. What we saw instead of normal *E. coli* was an enormously long filament, up to 300 times longer than the *E. coli* themselves (see photos). This was obviously something new. When we turned off the electric field the filaments broke up into normal sausage-like strings, and each one pinched off and became a normally growing *E. coli* cell. When I talked to people who are knowledgeable in the field, it was obviously something that was unknown. There are other techniques for forcing filamentous growth in bacteria, but nobody had ever seen it in this particular way. Something was different.

What was causing this effect? We had to determine whether it was the electric field itself. So we tried a modification of the experiment. We had two chambers made. We put the electrodes in one and the bacteria in the second. We passed the nutrient material into the first chamber where the electric field was and then that passed on into the second chamber where the bacteria was. So now the bacteria were never exposed to the electric field. And we turned on the electric field, and lo and behold, exactly the same thing happened. So what it said to us was that the process of passing the current through the nutrient media was causing some sort of long-lived chemical change that then had its effect upon the bacteria.

The first hint of the effect of platinum compounds

I set a student to work trying to identify what chemicals were involved. And by varying the known chemicals that we put in the medium, we found out that it had to do with the fact that we had an ammonium salt present, there was chloride present, and it was under an oxygen atmosphere. Those are the three requirements. If we removed any of those three we got no effect. But what was it that was being made? One of the possibilities was that there was some electrolysis going on at the

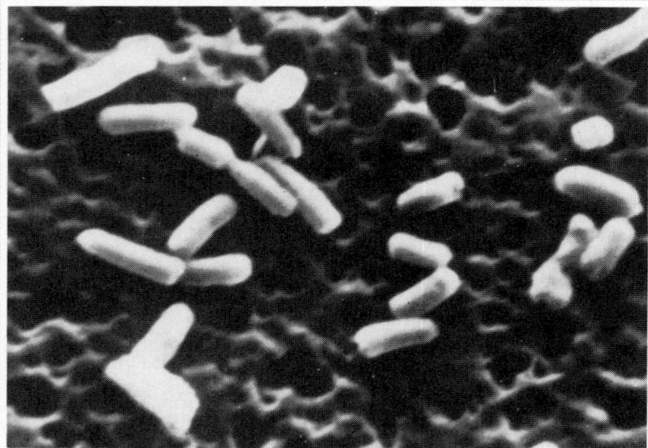
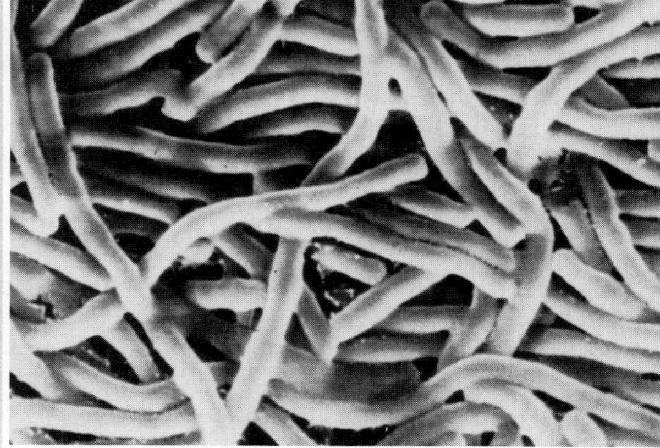

Scanning electron microphotographs. (left) Normal *E. coli* bacteria. (right) *E. coli* grown in a medium containing a few parts per million of $Pt(NH_3)_2Cl_2$, cisplatin. Same magnification in both pictures. The drug has inhibited cell division, but not growth, leading to long filaments. (Photos by D. Beck, Bowling Green University).

platinum electrode. Indeed, there was some platinum being taken off the electrode, and this was causing this effect. So we began to study the electrolysis products from the platinum electrode.

The obvious chemical combination would be, say, $(NH_4)_2PtCl_6$ or $(NH_4)_2PtCl_4$. We bought those chemicals, and we used them in the same concentration we knew we had in the chamber, but they had absolutely no effect on the bacteria. We allowed some of those solutions to sit out on the table, and after a few weeks went back to testing some of them and suddenly there was activity—not great, but they were causing the formation of filaments. The question then became: What was causing the change in the solution of the platinum-containing salts? By a quick test we discovered it had to do with light.

It was a terrible shock!

So we began to look into the photochemistry of platinum, and of course this was an old field. Platinum salts were used as one of the first chemicals in photography. What we did was to irradiate solutions of $(NH_4)_2PtCl_6$ or $(NH_4)_2PtCl_4$ and then separate the products into the various possible species. We collected the doubly charged species, and then the singly charged species, and so on. We tested each one of them separately. It was the neutral species that was active. We knew that there was a photochemical reaction that was causing a neutral species to exist. We had to identify it. Fortunately, we had in our laboratory a chemist from England, Andy Thompson, and after considering the problem with some other chemists we came to the conclusion that it was $Pt(NH_3)_2Cl_2$ that had been formed. This is a neutral species. So I had Andy synthesize the trans complex (see photo on page 185), and we then threw it into the bacterial solution. Absolutely nothing happened! It was a terrible, terrible shock for us after all these years of effort. Finally Andy said, "Well look, I synthesized the trans form, the more thermally stable form. Let's try synthesizing the cis form and see what happens." And he did, and it worked! We finally had the identification of the chemical. We had a compound, now called cisplatin, that we knew inhibited cell division in bacteria.

The first animal tests of cisplatin

The next step was to try cisplatin on a cancer cell, in an animal. Fortunately, I had a friend on campus who had an animal colony where he was testing an anti-cancer drug he had developed from mushrooms. He had a tumor system operating that was acceptable to the National Cancer Institute. My technician, Loretta VanCamp, was trained by him on how to handle the tumors and the mice, and we set up an animal laboratory.

I had a student working first to find what dose could be given to the animals before it killed them. Then we put the tumors in and treated the animals with cisplatin to see whether it would have any effect on them and the growth of tumors. And it did. It had a very significant effect on the tumor growth. We began to suspect that there was some very interesting anti-tumor activity that could selectively destroy a cancer cell in the animal before it destroyed the normal cells and, therefore, the animal could survive.

I contacted the National Cancer Institute and was invited to give a lecture at Bethesda. I brought four vials of compounds: cis- and trans-$(NH_3)_2PtCl_4$ and cis- and trans-$Pt(NH_3)_2Cl_2$, for them to test. I gave the lecture, and I don't think I ever had a more bored audience in my life. In fact one of the major players in the future development fell asleep during my lecture! The obvious reason was simply that there was a feeling that "This guy is coming in and trying to sell us a heavy metal complex as an interesting drug, and we know that heavy metal complexes are dangerous, and we are not going to use it." But they were required to test any compounds left by a reasonable scientist. That's the law, and they were then testing about 50,000 compounds a year. I left my chemicals with them, and they tested them in a different tumor system from that we had used. Several months later I got a call from the Associate Director asking if I would be interested in submitting a grant application because the test results against their tumor system turned out to be positive as well. They were rather surprised by it. So they were willing to give us a small grant to continue with the work, and that started the active support. We then combined the data from the two studies and wrote a paper, for *Nature*.

Clinical trials—It works beautifully!

In 1972 the National Cancer Institute sponsored clinical trials in a number of places. At the Roswell Park Memorial Institute [in Buffalo, NY] they discovered the high degree of activity against testicular cancer. Similarly, it was discovered at the Royal Marsden Hospital in London that cisplatin had activity against ovarian cancer. And so we began to have some very early indications that the drug had enough activity to warrant further investigation. More and more monies began to go into these clinical trials, and more and more interest was generated by them. In the meantime we were testing many chemicals other than platinum complexes to see how broad-ranged this effect might be. We initially stayed fairly close to the platinum group metals. But then we found compounds of cobalt and other metals that were active as well. So this was by no means uniquely characteristic of platinum compounds. It was a more general thing.

Once you start to put cisplatin into patients, you do it in what is called a phase I trial where you are looking for toxic effects to find the appropriate dose level. It was clear that these compounds were terribly kidney-toxic. And in 1973 there was a feeling on the part of most clinicians that this was too interesting a drug to drop, but too dangerous a drug to use. It wasn't until about 1974 or '75 when Drs. Cvitkovic and Krakoff at the Sloan Kettering Institute came up with a wrong but brilliant idea. They said "Platinum drugs contain heavy metals. We know heavy metals cause kidney toxicity. Kidney toxicity from heavy metals can be ameliorated by simply flooding the patient with water which dilutes the drug as it passes through the kidneys. Let's do that with cisplatin." They did. And it works beautifully. It diminished the kidney toxicity almost to the vanishing point, and it is no longer the significant factor in determining dose levels.

It was about that time that Larry Einhorn of the University of Indiana Medical School came up with the idea of taking cisplatin, which had been shown independently by people at Roswell Park to be active against testicular cancers, and to combine it with two other drugs, vinblastin and bleomycin, that had been found to have some activity against testicular cancer. And that worked beautifully! Testicular cancer went from a disease that normally killed about 80% of the patients, to one which is close to 95% curable. This is probably the most exciting development in the treatment of cancers that we have had in the past 20 years. It is now the treatment of first choice in ovarian, bladder, and osteogenic sarcoma cancers as well.

Cisplatin is usually used in a combination chemotherapy. Over the intervening years people have played games with varying the combinations, and with the use of radiation in combination with cisplatin. Others have used a combination of split-dose radiation with cisplatin and another drug called etoposide. This combination is apparently producing long-term, complete remissions of small-cell lung cancer in at least 50% of early patients tested so far. So here we have another cancer that is responding beautifully to the cisplatin based therapy.

A new area of research—aging and death

That's pretty much the current situation now. A number of companies are working on minor modifications to improve efficacy and decrease toxicity and see if we can get a different spectrum of action. It stands now in a fairly solid position. A few years ago I became aware that this solid position existed and it no longer needed me, and so I pulled out of it. I haven't done any work with platinum compounds for the past eight years. I went back to an area of research in which I had developed an interest on my 40th birthday—aging and death.

700 ml
±5%
600
500
400
300
200

CHAPTER 1

The Tools of Chemistry

Molten sulfur being poured into water.

CHAPTER OUTLINE

1.1 THE SCIENCE OF CHEMISTRY

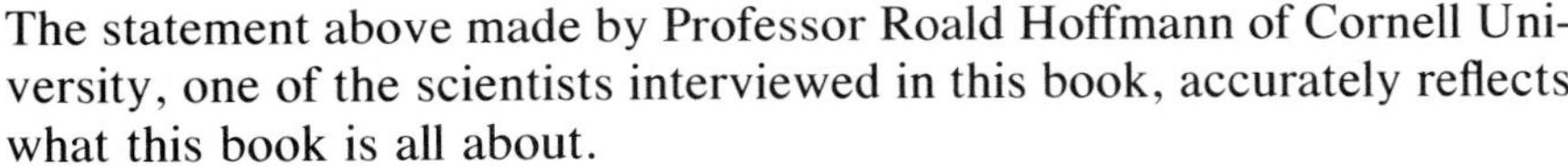

> "Chemistry is the science of molecules and their transformations. It is the science not so much of the one hundred elements, but of the infinite variety of molecules that may be built from them."

The statement above made by Professor Roald Hoffmann of Cornell University, one of the scientists interviewed in this book, accurately reflects what this book is all about.

Chemists have been at the forefront of the enormous advances made in biomedical science in this century and in the last decade. One of these was the discovery, outlined in the preceding interview with Professor Barnett Rosenberg, of a simple molecule to be used in the treatment of certain kinds of cancers. Because of this platinum-containing molecule, testicular cancer has gone from being almost surely lethal to 95% curable. This molecule has a very simple composition—one platinum ion, two chloride ions, and two ammonia molecules—but it is extraordinarily effective. But why does it work and will it be broadly effective for other cancers? These are questions still being asked by scientists such as Professor Jacqueline Barton, who is also interviewed in this book.

A model of the platinum-containing molecule that has become an effective treatment for certain kinds of cancers. The silver ball in the center is a platinum(II) ion, the two green balls represent chloride ions, the blue balls are nitrogen atoms, and the white ones are hydrogen atoms.

AIDS is a disease that has moved into our society with a vengeance, and the resources of the United States and other countries are being focused on finding a cure. In the early 1960s a molecule called AZT (azidothymidine) was synthesized by a chemist at the Michigan Cancer Foundation, but was abandoned when it failed to halt leukemia in mice. From this failed cancer medicine, however, has come the best hope to date for AIDS treatment. Although AZT can slow the progression of AIDS and improve the patient's clinical signs, it unfortunately does not eliminate the virus that causes the disease. In addition, the drug has side

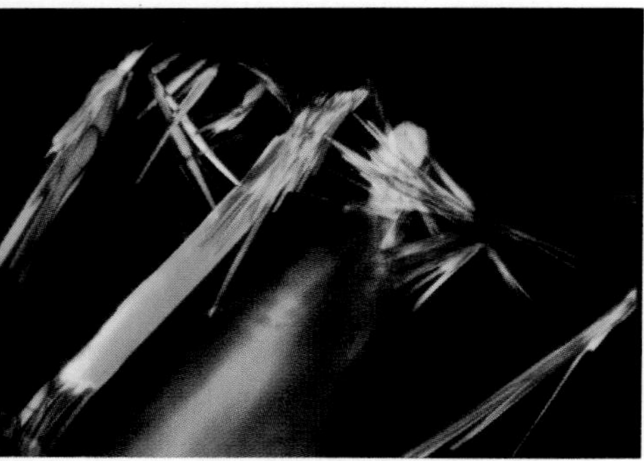

A crystal of the drug AZT. The antiviral activity of AZT against AIDS was identified by Drs. Hiroaki Mitsuya and Samuel Broder of the National Cancer Institute. (Photo by Dr. Mitsuya and Mr. Larry Ostby)

effects that many patients cannot tolerate. On top of all of that, a year's supply can cost thousands of dollars. The hunt continues for more effective drugs.

Several of the people interviewed for this book—Professor Hoffmann, Professor Gregory Farrington, and Dr. Mary Good—have spent a great deal of their scientific talents in the past decade working in the general area of materials science. This area is one that you may not have heard much about, but you will, for it is one of the great frontier areas of the 1990s. Research in this area has already provided superconducting solids, lightweight plastics that conduct electricity, fiberoptical cables for transmitting telephone messages, and thin films of diamonds, not to mention fishing poles and tennis rackets containing graphite fibers.

Superconductors are described in Chapter 25.

Environmental issues continue to concern chemists. For example, in 1974 chemists F. S. Rowland and M. Molina published a paper in the journal *Nature* entitled "Stratospheric sink for chlorofluoromethanes: Chlorine atom-catalyzed destruction of ozone." This paper led to many studies of the chemistry of the stratosphere and to the ultimate discovery of a "hole" in the layer of ozone that surrounds the earth and protects it from harmful solar radiation (see Chapter 24). These studies have led, in turn, to a movement to stop the use of compounds called chlorofluorocarbons or CFCs in our society. This will prove easier said than done, since CFCs are widely used as the working fluid in air conditioners and refrigerators. Nonetheless, many chemical companies are working hard to find substitutes for CFCs and some may soon be on the market.

"Every aspect of the world today—even politics and international relations—is affected by chemistry." Linus Pauling

Although you may be quite convinced of the importance of chemistry, why should *you* study chemistry today? Our objective in this book is to show you that chemistry has not only a rich and interesting history but an enormously exciting future as well. You may find—as we do—that chemistry offers a marvelous intellectual challenge and a rewarding career. Indeed, you may choose to pursue some area of science yourself. Now is an excellent time to do so, since, as Dr. Bassam Shakhashiri describes in the interview that follows this chapter on page 37, the United States, as never before, needs talented and creative people in science and science education.

A small magnet levitates above a piece of the "123" superconductor $YBa_2Cu_3O_7$, discovered in 1986. Superconductors are described in Chapter 25. (AT&T Bell Laboratories)

Most of you, however, will not go on to spend your life in science. You have other equally worthy aspirations. Nonetheless, to meet your responsibilities to yourself and society, you need to know how scientists define and answer questions about natural processes and use that knowledge to benefit humanity. Furthermore, a study of chemistry is centrally important, not only to all of science, but also in helping to explain and understand yourself and the world around you. H. A. Bent, a noted educator, has said recently that, "Chemistry is the central subject in a liberal arts curriculum. It stands between the traditional humanities on the one hand and modern physics on the other hand." Chemistry is the result of generations of intensely creative human thought and experiment, and it contains the seeds of the continuing progress of humankind.

1.2 METHODS OF SCIENCE

Before you begin your study of chemistry, it is useful to have some insights into the way any scientific study is done. Although it may not seem so now, it is easy to formulate an idea or a topic to study. The goal is to state a problem that is worth studying but that is also narrow enough in scope so that there is some realistic possibility of coming to a useful conclusion. It is reasonable to ask, for example, how cancers begin in a human body, but it is certainly *not* reasonable to expect an answer in the next year or so.

Having posed a reasonable question, you first look at the experimental work done previously in the field so that you have some notion of possible answers. After forming a **hypothesis**—a tentative explanation or prediction of experimental observations—you perform experiments designed to give results that may confirm your hypothesis or eliminate erroneous explanations. This requires that you collect both qualitative and quantitative information or data. **Quantitative** information usually means numerical data, such as the temperature at which a chemical substance melts. **Qualitative** information consists of nonnumerical observations, such as the color of a substance or its physical appearance. With the results of your experiments, you then revise and extend your original hypothesis and continue to test it with more experiments. After you have done a number of experiments, and after you have continually checked to make sure your results are truly *reproducible*, a pattern of behavior or results will begin to emerge. At this point, you may be able to summarize your observations in the form of a **law**, a concise verbal or mathematical statement of a relation that is always the same under the same conditions.

Once you have performed enough reproducible experiments to lead you to a "law of nature," you may be able to formulate a theory to explain the law. A **theory** is a unifying principle that explains a body of facts and the laws based on them. It is capable of suggesting new hypotheses. Excellent examples of theories are those developed to account for chemical bonding (Chapters 10 and 11). It is a fact that atoms are held together or bonded to one another. But how and why? Several theories are currently used in chemistry to answer such questions, but are they correct? What are their limits? Can the theories be improved or are completely new theories necessary? Laws summarize the facts of nature and rarely change. Theories are inventions of the human mind; *theories can and do change* as new facts are uncovered.

*A **hypothesis** is a statement not yet tested by experimental observation. In contrast, a **theory** has been verified at least to some extent.*

People outside of science usually have the idea that science is an intensely logical field. They picture a white-coated chemist moving logically from hypothesis to conclusion without human emotion or foibles. Nothing could be further from the truth! Often, scientific results and understanding arise quite by accident. Creativity and insight are needed to transform a fortunate accident into useful and exciting results. A wonderful example is the discovery of a cancer chemotherapy agent called "cisplatin," a discovery already described by Dr. Barnett Rosenberg in his interview at the beginning of this part of the book.

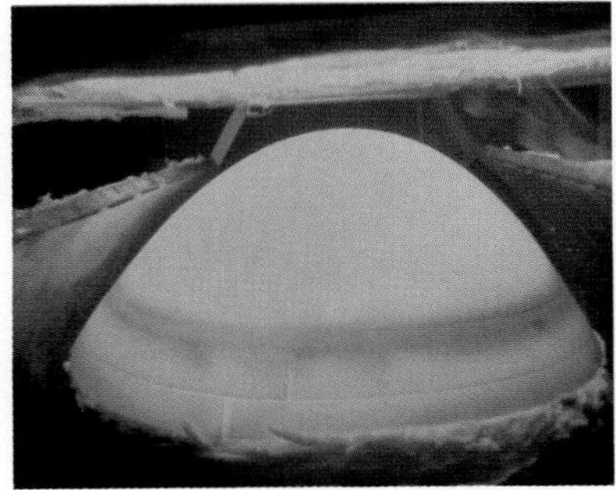

The nose cone of the space shuttle is made of a "reusable" carbon-carbon composite that consists of layers of graphite-fiber cloth in a carbon binder with a coating of silicon carbide. Here the cone is just emerging from a furnace to test its ability to withstand temperatures up to 2350 °F. (NASA)

1.3 SOME DEFINITIONS

Learning chemistry is to some extent like learning a new language. As you study this book, you will find new words to learn, and many of them have quite precise meanings. Indeed, there are some terms that we have to introduce now (Figure 1.1).

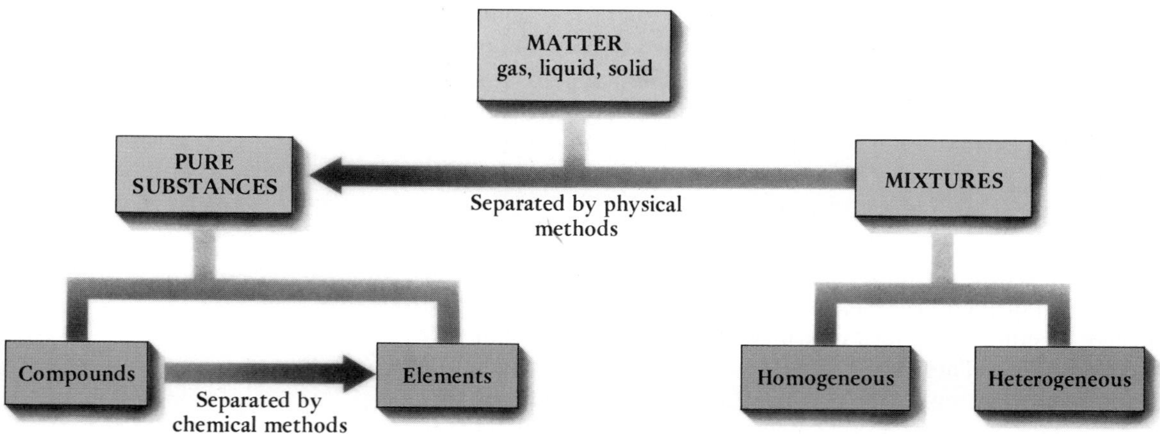

Figure 1.1 The components of matter and the relation between mixtures and pure substances.

*A **pure substance** is a form of matter that has a definite composition and distinct properties.*

A few kinds of matter have properties of more than one phase, such as the "liquid crystals" in the displays of most calculators.

This definition of an element will serve our purposes for the moment. However, after looking at the composition of atoms in Chapter 2, we shall see that a better definition of an element is as follows: ***An element has a definite number of protons in the nuclei of its atoms (the atomic number).***

Matter, anything that has mass and occupies space, consists not only of things you can see and touch but also of such things as air, which you cannot see. Matter exists in three **phases**: solids, liquids, and gases. A **solid** is matter with a rigid shape and a fixed volume that does not change much with temperature (Figure 1.2). A **liquid** has a fixed volume but not a fixed shape, and it conforms to the shape of its container (Figure 1.3); its volume may change somewhat with changes in temperature and pressure. In contrast with solids or liquids, a **gas** has neither fixed volume nor shape, as it expands to fill its container completely; its volume is very sensitive to temperature and pressure (Figure 1.4).

Matter is composed ultimately of different kinds of atoms. **Atoms** are the smallest particles of an element that retain the chemical properties of the element; each atom consists of a tiny core (the nucleus) containing positive charges, which is surrounded by an equal number of negative charges (electrons). An **element** is matter that consists of only one kind of atom, either individually or combined into larger units. The names and symbols of all 108 known elements are shown inside the front cover of the book. Bromine and iodine (Figure 1.4) and iron and gold are but a few of these elements.

Molecules are identifiable units of matter consisting of two or more

Figure 1.2 Crystals of elemental sulfur.

Figure 1.3 Solid "ice cubes" have a definite shape and do not fill the container evenly. After melting, the liquid water assumes the shape of the container.

Figure 1.4 Elemental bromine (left) is a deep brown-orange liquid, but it is volatile enough that some is in the gas phase. Elemental iodine (right) is a violet solid, but some sublimes to give a violet vapor.

atoms combined in a definite ratio. When the atoms are the same we have an element; when they are different, we have a compound. A **compound**, then, is matter that consists of identifiable units containing atoms of different elements combined in specific ratios. The salt and sugar on your dinner table, and the carbon dioxide of the air, are chemical compounds. A great deal of the research done in chemistry involves the study of transformations of one or more compounds into others, so you will spend much of your time examining molecules: their shapes, the forces holding them together, and their chemical and physical properties.

A compound is different from a mixture. The elements in a compound lose their individual chemical characteristics, and the compound has new characteristics. In a **mixture** each of the constituents retains its identity. A cup of coffee with sugar and milk is a mixture of many substances, as are a soft drink, a piece of concrete, and a coin.

Mixtures can be **homogeneous** or **heterogeneous**. A homogeneous mixture has the same composition throughout the mixture. If you stir sugar into a glass of water, the sugar dissolves and sugar molecules are distributed uniformly, that is, homogeneously, throughout the water (see Figure 1.5). This is an example of a **solution**, a homogeneous mixture of two or more substances. The material dissolved is the **solute**, and the medium in which it is dissolved is the **solvent**. In a sugar/water solution, sugar is the solute and water is the solvent. In contrast, a mixture of solid grains of sand and a salt is **heterogeneous**, since particles of each com-

Figure 1.5 A homogeneous mixture or solution. A yellow compound, potassium chromate (the solute), is stirred into water (the solvent) where it dissolves to form an aqueous solution.

(a)

(b)

(c)

(d)

Figure 1.6 Separating a copper compound from sand. The heterogeneous mixture (a) is placed in water where the copper compound dissolves (b). The mixture is filtered, leaving the sand in the filter (c); the copper-containing solution passes through the filter. Evaporation of the water leaves blue crystals of the copper compound (d).

When the solvent in a solution is water, the solution is often referred to as an ***aqueous*** *solution.*

ponent of the mixture remain separate and can be observed as individual substances (Figure 1.6).

In either a homogeneous or a heterogeneous mixture, the components can be separated into pure substances by *physical means*, that is, without changing the specific atom ratios within the particles. For example, the components of the solution in Figure 1.5 could be separated by evaporating the water to leave the chromium-containing compound as a yellow solid; if the experiment is done properly, pure liquid water can be recaptured from the vapor phase. Components of a heterogeneous mixture can also be separated physically, as illustrated in Figure 1.6.

The components of a mixture can also be separated by *chemical means*, but this involves changing the chemical nature of one or more of the constituents. This is often done in **chemical analysis**, where the com-

Figure 1.7 A physical change. A flower is placed in liquefied nitrogen (which boils at −196 °C). The flower freezes and crumbles to pieces when touched.

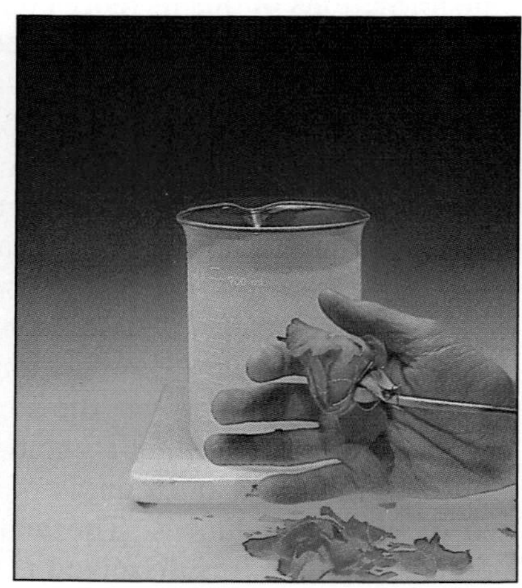

ponents of a mixture are transformed into new substances that can be observed or separated by physical means. This is discussed in more detail in Chapters 3 through 5.

Your friends recognize you by your physical characteristics, among them height, weight, and skin and hair color. The same is true of chemical substances. Each substance has a set of **physical properties**, properties that can be measured and observed without changing the atom ratios within the substance. Such properties include color, the temperature at which a substance melts or boils, density, and physical state at room temperature. Elemental bromine and iodine, for example, clearly differ from one another in their color and physical state (Figure 1.4). Substances may undergo physical changes. In Figure 1.7, for example, freezing changes a rose from its familiar form to a crumbled pile of broken petals; it is still a rose in a chemical sense, though, since the chemical composition of the rose has not changed.

In contrast with physical properties, the **chemical properties** of a substance are those that it exhibits when it undergoes a change in atom arrangements or in atom ratios. This change is often brought about by contact with another substance. When gasoline burns in an automobile engine or metals rust and corrode, their chemical composition changes (Figure 1.8).

Figure 1.8 A chemical change. A copper compound is dissolved in water to give a blue solution. On adding a piece of aluminum foil (top), a chemical change occurs. As described in Chapters 5 and 21, copper metal is formed from the dissolved copper compound, and elemental aluminum metal is transformed into an aluminum-containing compound (bottom). This is evidenced by a coating of copper on the surface of the ball of foil, and by the observation that the blue color of the dissolved copper compound diminishes with time. In addition, a physical change, a temperature increase, is observed as the reaction occurs. (This reaction, which is an example of corrosion, occurs only in the presence of salt. It explains why you do not want to use aluminum products near the ocean. See Chapter 21.)

The physical or chemical properties of substances can be classified further as *ex*tensive or *in*tensive. **Extensive** properties depend on the *amount* of matter present. Thus, the volume and mass of a sample are extensive properties, as they are both proportional to the amount of sample. In contrast, **intensive** properties are the same regardless of sample size. The temperature at which a pure substance melts and the color of a material are both intensive properties. Water is colorless and freezes at 0 °C (32 °F), whether you have a spoonful or a ton.

No two pure substances have the same combination of chemical and physical properties under the same conditions, so we can use these differences to identify substances. Many of the physical properties of oxygen and nitrogen are very similar; both are colorless gases at room temperature, for example. However, a burning match will go out if it is put into a flask of nitrogen gas, but it will burn brightly in pure oxygen. The two gases clearly have different chemical properties.

1.4 UNITS OF MEASUREMENT

Scientists use the **metric system** for recording and reporting their measurements. This is a decimal system, in which all of the units are expressed as powers of 10 times some basic unit.

To establish a uniform set of units, the General Conference of Weights and Measurements in 1960 prescribed base units to be used for various measured quantities. The resulting system is called the *Systeme International d'Unités* (International System of Units), abbreviated **SI**. The seven SI base units are listed in Table 1.1. Larger and smaller quantities are expressed by using the appropriate prefix (Table 1.2) with the base unit. For instance, highway distances are given in *kilo*meters, where

Table 1.1 SI Base Units

Physical Quantity	Name of Unit	Abbreviation
Mass	kilogram	kg
Length	meter	m
Time	second	s
Temperature	kelvin	K
Amount of substance	mole	mol
Electric current	ampere	A
Luminous intensity	candela	cd

The quantity 0.1 or 1/10 is written as 1×10^{-1}. This notation is used throughout the book and is explained in Appendix A.

Conversion factors are listed inside the back cover of the book.

1 kilometer (km) is exactly 1000 or 10^3 meters (m). Objects in the laboratory are often expressed in terms of subdivisions of the meter, that is, as *centi*meters (cm) or *milli*meters (mm). The prefix *centi-* means 1/100, so 1 centimeter is 1/100 of a meter ($1 \text{ cm} = 1 \times 10^{-2}$ m); 1 millimeter is 1/1000 of a meter ($1 \text{ mm} = 1 \times 10^{-3}$ m). On an even smaller scale, dimensions are often given in nanometers (nm) ($1 \text{ nm} = 1 \times 10^{-9}$ m) or picometers (pm) ($1 \text{ pm} = 1 \times 10^{-12}$ m).

SI units for all other physical quantities are derived from the seven base units, and several of these are listed in Table 1.3. Some **conversion factors** that allow you to convert between SI and non-SI units are given inside the back cover of the book.

LENGTH

The **meter** is the SI unit of length. One meter is equivalent to 3.281 feet or 39.37 inches. It is a convenient unit in human terms since a human leg is roughly a meter long, and you can hold a meter stick lengthwise comfortably between your outstretched hands.

Just as distances in the nondecimal English system of inches and feet are broken into smaller units, the meter is subdivided into 100 (10^2) centimeters, 1000 (10^3) millimeters, a million (10^6) micrometers, a billion (10^9) nanometers, and so on. Centimeters and millimeters are the most convenient for measuring objects in the laboratory, and nanometers and picometers are used for dimensions at the molecular level.

Table 1.2 Selected Prefixes Used in the Metric System

Prefix	Abbreviation	Meaning	Example
Mega-	M	10^6	1 megaton = 1×10^6 tons
Kilo-*	k	10^3	1 kilogram (kg) = 1×10^3 grams
Deci-	d	10^{-1}	1 decimeter (dm) = 0.1 m
Centi-	c	10^{-2}	1 centimeter (cm) = 0.01 m
Milli-	m	10^{-3}	1 millimeter (mm) = 0.001 m
Micro-	μ†	10^{-6}	1 micrometer (μm) = 1×10^{-6} m
Nano-	n	10^{-9}	1 nanometer (nm) = 1×10^{-9} m
Pico-	p‡	10^{-12}	1 picometer (pm) = 1×10^{-12} m

*The units highlighted in blue are the ones used most frequently in chemistry.
†This is the Greek letter mu (pronounced "mew").
‡This prefix is pronounced "peako."

Table 1.3 Some Commonly Used Units Derived from SI Base Units

Quantity	Unit Name	Symbol	Definition
Area	square meter	m^2	
Volume	cubic meter	m^3	
Density	kilogram per cubic meter	kg/m^3	
Force	newton	N	$kg \cdot m/s^2$
Pressure	pascal	Pa	N/m^2
Energy	joule	J	$kg \cdot m^2/s^2$
Electric charge	coulomb	C	$A \cdot s$
Electric potential difference	volt	V	$J/(A \cdot s)$

EXAMPLE 1.1

CONVERTING CENTIMETERS TO METERS

In track and field competition, an excellent height for the pole vault is 585 cm. (a) What is this height in meters? (b) What is this height in feet?

Solution (a) To solve this problem, you need to know the relation between centimeters and meters: 1 m is the same as 100 cm (Table 1.2). Therefore,

$$585\ \cancel{\text{cm}} \left(\frac{1\ \text{m}}{100\ \cancel{\text{cm}}}\right) = 5.85\ \text{m}$$

To convert a measurement in one unit to another unit, we multiply by a "factor" (in this case 1 m/100 cm) that expresses the equivalence of a single quantity in two different units. Since the numerator and denominator of the factor refer to the same quantity, the factor is equivalent to the number 1. Therefore, multiplication of an original quantity (585 cm in this case) by this factor does not change the magnitude of that quantity, only its numerical value and units. The factor is always written so that the units in the denominator cancel the original units, leaving the desired units. Here units of centimeters cancel, and we are left with units of meters.

(b) The table inside the back cover of the book lists only a relation between centimeters and inches, so we first convert 585 cm to inches

$$585\ \cancel{\text{cm}} \left(\frac{1\ \text{in.}}{2.54\ \cancel{\text{cm}}}\right) = 230.\ \text{in.}$$

and then convert inches to feet.

$$230.\ \cancel{\text{in.}} \left(\frac{1\ \text{ft}}{12\ \cancel{\text{in.}}}\right) = 19.2\ \text{ft}$$

EXAMPLE 1.2

DISTANCES ON THE MOLECULAR LEVEL

The distance between the oxygen atom and a hydrogen atom in a water molecule is 95.7 pm. What is this distance in meters? In nanometers?

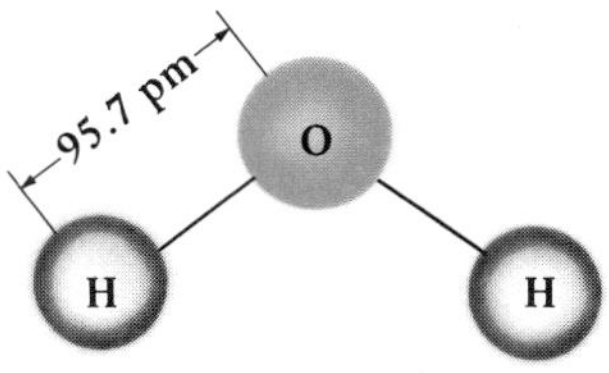

Water molecule

Solution In each case, you must first know the relation between the picometer (pm) and the desired unit. For the first, 1 pm is exactly 1×10^{-12} m (see Table 1.2). Thus, you multiply the distance in pm by the factor (10^{-12} m/1 pm) so that the units of pm cancel, leaving an answer in meters.

$$95.7 \cancel{\text{pm}} \left(\frac{10^{-12}\text{ m}}{1 \cancel{\text{pm}}}\right) = 95.7 \times 10^{-12}\text{ m} = 9.57 \times 10^{-11}\text{ m}$$

To relate picometers and nanometers, you must know that 1 nm = 10^{-9} m (Table 1.2). Therefore, you can take the distance in meters, which you just found, and multiply by the factor (1 nm/10^{-9} m).

$$9.57 \times 10^{-11} \cancel{\text{m}} \left(\frac{1\text{ nm}}{10^{-9} \cancel{\text{m}}}\right) = 9.57 \times 10^{-2}\text{ nm}$$

EXERCISE 1.1 Interconverting Units of Length

A standard U.S. postage stamp is 2.5 cm long. What is this length in meters? In millimeters? In inches?

Examples 1.1 and 1.2 were solved by the technique of "dimensional analysis," an approach to problem solving described in Section 1.6.

Answers to all the exercises in the text are given in Appendix K.

AREA AND VOLUME

The units of area and volume are derived from the base unit of length, that is, areas can be given in square meters (m^2) and volumes in cubic meters (m^3). Unfortunately, volumes in cubic meters are not very convenient for everyday use. For example, the volume of a common laboratory beaker is 0.0006 m^3 (Figure 1.9), a soccer ball has a volume of about 0.005 m^3, and chemists most often work with volumes of chemicals in the range of 0.001 m^3 or less. These are inconvenient numbers for routine use, so we more often use the unit called the **liter**, symbolized by **L**.

Figure 1.9 Some common laboratory equipment, with an orange in the picture to give you a sense of size.

A cube with sides equal to 10 cm (0.1 m) has a volume of 10 cm × 10 cm × 10 cm = 1000 cm^3 (or 0.001 m^3). This is defined as 1 liter. Thus,

1 liter (1 L) = 1000 cm^3

10 cm is called a decimeter (dm), since it is 1/10 of a meter. Therefore, since a liter is a cube 10 cm on a side, a liter is equivalent to a cubic decimeter: 1 L = 1 dm^3. Although not widely used in the United States, this unit is used in the rest of the world.

The liter is a convenient unit to use in the laboratory, as is the **milliliter**, **mL**. Since there are 1000 mL and 1000 cm^3 in a liter, this means that

1 cm^3 = 0.001 L = 1 milliliter (1 mL)

Chemists often use the terms milliliter and cubic centimeter (or "cc") interchangeably. The following example and exercise consider volumes in these related units.

EXAMPLE 1.3

UNITS OF VOLUME

You have a beaker with a volume of 0.6 L. What is its volume in cm^3 and mL?

Solution We know the relation between the unit we have and the one we desire for our answer. Therefore, we multiply 0.6 L by the factor (1000 cm^3/1 L) so that units of L cancel to leave an answer in cm^3.

$$0.6\,\cancel{L}\left(\frac{1000\ cm^3}{1\,\cancel{L}}\right) = 600\ cm^3$$

Since 1 cm^3 and 1 mL are equivalent, we can say the volume of the beaker is also 600 mL.

$$600\,\cancel{cm^3}\left(\frac{1\ mL}{1\,\cancel{cm^3}}\right) = 600\ mL$$

EXERCISE 1.2 Volume

(a) A standard wine bottle has a volume of 750 mL. How many liters does this represent? (b) One U.S. gallon is equivalent to 3.785 liters. How many liters are there in a 2.0-quart container of dish detergent? (4 quarts = 1 gallon)

MASS AND WEIGHT

The terms *mass* and *weight* are used interchangeably in everyday speech, but scientists give different meanings to the two words. The **mass** of a body is the fundamental measure of the quantity of matter in that body. The **weight** of a body, on the other hand, depends on the mass of that body and *another* body; it is the force (gravity) exerted on a body by another and so depends on the amounts of material in the *two* bodies. This difference between mass and weight is easily understood. On earth or on the moon an astronaut has the same mass. However, on the moon the gravitational force is less so the astronaut weighs less and is able to leap tall rocks in a single bound!

Since weight is gravitational force, which varies with the distance between the centers of the two bodies, it must be true that "earth" weight depends on the altitude of your laboratory. How then can you "weigh out" the same amount of matter in Miami and in Denver, cities that differ in altitude by nearly a mile? The answer is to compare the weights in both places with those of objects of known mass, called "standards," using a laboratory balance (Figure 1.10).

The SI unit of mass is the **kilogram (kg)**. Smaller masses are expressed in **grams (g)** or **milligrams (mg)**. Again, the prefixes in Table 1.2 apply.

$$1\ \text{kg} = 1000\ \text{g}$$

$$1\ \text{g} = 1000\ \text{mg}$$

In the United States we are used to the English system of mass measurement. Most masses are given in pounds, where *1 pound is equivalent to 453.59237 grams*. This means that a mass of 1.00 kg is equivalent to 2.20 pounds, or about the mass of a quart of milk.

The English system based on the pound refers to weight, *not* mass. *Under standard gravity (at sea level), the mass of 1 kg has a weight of 2.2046 pounds, or 1 pound is 0.4536 kg or 453.6 g.*

EXAMPLE 1.4

MASS IN KILOGRAMS AND GRAMS

A U.S. penny has a mass of 2.65 g. Express this mass in kilograms and milligrams.

Solution In each case, we shall multiply the mass in grams by a factor that has the form (units for answer/units of number to be converted). As you work the problems in this and other chapters, notice that this is always the way problems are solved.

(a) $2.65\ \cancel{\text{g}} \left(\dfrac{1\ \text{kg}}{1000\ \cancel{\text{g}}}\right) = 0.00265\ \text{kg}$

(b) $2.65\ \cancel{\text{g}} \left(\dfrac{1000\ \text{mg}}{1\ \cancel{\text{g}}}\right) = 2.65 \times 10^3\ \text{mg}$

Figure 1.10 A laboratory balance. A sample of sulfur is being weighed.

EXERCISE 1.3 Mass Conversions

One pound is equivalent to 453.6 g at standard gravity (at sea level). (a) How many kilograms are equivalent to 3.00 pounds? (b) How many milligrams are equivalent to 0.500 pound? (c) How many pounds are equivalent to 4.00 kilograms?

DENSITY

Density is the ratio of the mass of an object to its volume, where the mass is usually in grams and the volume in cubic centimeters.

$$\text{Density} = \frac{\text{mass}}{\text{volume}}$$

Gold has a density of 19.3 g/cm^3. That is, 1.00 cm^3 of gold has a mass of 19.3 g. Water, on the other hand, is much less dense, having a density of only 0.99707 g/cm^3 at 25 °C.*

If any two of the three quantities—mass, volume, and density—are known for a sample of matter, they can be used to calculate the third quantity. For example,

$$\text{Volume} \times \text{density} = \text{volume } (\cancel{\text{cm}^3}) \left(\frac{\text{mass (g)}}{\text{volume } (\cancel{\text{cm}^3})}\right) = \text{mass (g)}$$

This approach is used in the following example and exercise.

EXAMPLE 1.5

DENSITY

Suppose you have 250. cm^3 (or milliliters) of ethyl alcohol (the common "alcohol" in alcoholic beverages). If the density of ethyl alcohol is 0.789 g/cm^3 (at 20 °C), how many grams of alcohol does this represent?

Solution Because we know the density and volume of the sample, we will use the definition above in the form "mass = density × volume." Note that the volume is multiplied by a factor having units of mass over volume, and the units of cm^3 cancel.

$$250.\ \cancel{\text{cm}^3} \left(\frac{0.789 \text{ g}}{1\ \cancel{\text{cm}^3}}\right) = 197 \text{ g}$$

By simply recognizing the units of the given information, of the desired answer, and of the multiplying factor, you can avoid memorizing equations like "$m = d \times v$" and then forgetting them at a time of crisis—such as in the middle of an examination.

EXERCISE 1.4 Density

The density of dry air is 1.12×10^{-3} g/cm^3. What volume of air, in cubic centimeters, will have a mass of 1.00 kg?

TIME

The fundamental unit of time is the **second**, symbolized by "s." Time units were established by the Sumerians (who lived in what is now Iraq) 40 or more centuries ago! Prior to 1964 the second was defined as 1/86,400 of a mean solar day. Since there are irregularities in the solar day, the second is now defined in terms of certain radiation produced by an atom of the element cesium.

In 1 second, light can travel 299,792,458 meters or about 7½ times around the earth at the equator. But in one nanosecond (10^{-9} s), light travels only 0.2998 m or about 30 centimeters, that is, about the length of this book.

*Because the volume of a liquid is more strongly affected by temperature than that of most solids, it is usual to indicate the temperature at which a liquid's density is measured.

TEMPERATURE

In the summertime it is a pleasure to go swimming in the local pool, in a lake, or in the ocean. The cooling sensation comes from feeling heat transfer from your skin to the water; heat transfers in this "direction" when your skin is at a higher temperature than the water. **Temperature** is the property of matter that determines whether there can be heat (energy) transfer from one body to another *and* the direction of that transfer. Heat transfers *only* from an object with a higher temperature to one with a lower temperature. The number that represents a body's temperature depends on the unit chosen for the measurement (just as the number that represents a distance depends on its unit of measurement). Unfortunately, there are three units of temperature measurement in use today: the Fahrenheit, Celsius, and kelvin units. Getting people to agree to use one of them has been a problem, but natural phenomena are most easily expressed in the kelvin unit, so this is a frequent choice of scientists.

There is more about temperature and heat in Chapter 6.

FAHRENHEIT AND CELSIUS TEMPERATURE UNITS In the United States we commonly use the Fahrenheit unit, but the Celsius unit is used in most other countries and in science. Both units are based on the properties of water. The Celsius unit is defined by assigning zero as the freezing point of pure water (0 °C) and 100 as its boiling point (100 °C).* The size of the Fahrenheit degree is equally arbitrary. Fahrenheit defined 0 °F as the freezing point of a solution in which he had dissolved the maximum amount of salt (because this was the lowest temperature he could reproduce reliably), and he intended 100 °F to be the normal human body temperature (but this turned out to be 98.6 °F). Today, the reference points are set at 32 °F and 212 °F (the freezing and boiling points of pure water, respectively). The number of units between these points is 180 Fahrenheit degrees (Figure 1.11). Comparing the two units, the Celsius degree is almost twice as large as the Fahrenheit degree; it takes only 5 Celsius degrees to cover the same temperature range as 9 Fahrenheit degrees.

$$\frac{100\ °\text{C}}{180\ °\text{F}} = \frac{5\ °\text{C}}{9\ °\text{F}}$$

This relationship is used to convert a temperature on one scale to a temperature on the other.

$$°\text{C} = \frac{5\ °\text{C}}{9\ °\text{F}}(°\text{F} - 32\ °\text{F}) \qquad \text{or} \qquad °\text{F} = \frac{9\ °\text{F}}{5\ °\text{C}}\,°\text{C} + 32\ °\text{F}$$

Laboratory work is almost always done using Celsius units, and we rarely need to make conversions to and from Fahrenheit. It is best to try to calibrate your senses to Celsius units; to help you do this, it is useful to know that water freezes at 0 °C, a comfortable room temperature is about 22 °C, your body temperature is 37 °C, and the hottest water you could stand to put your hand into is about 60 °C.

*To be entirely correct, we must specify that water boils at 100 °C and freezes at 0 °C only when the pressure of the surrounding atmosphere is 1 standard atmosphere. We shall discuss pressure and its effect on boiling point in Chapter 13.

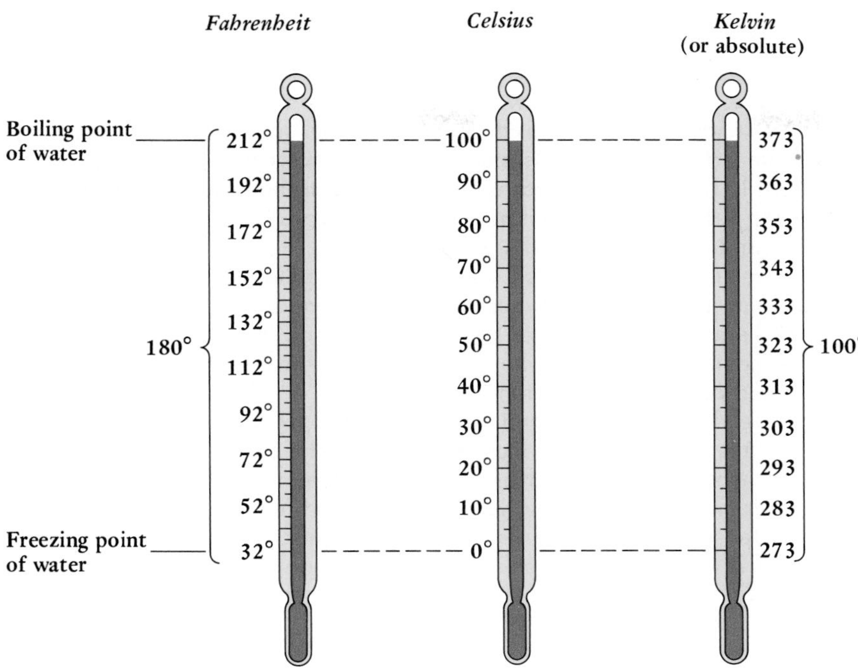

Figure 1.11 A comparison of Fahrenheit, Celsius, and Kelvin temperature scales. The reference or starting point for the Kelvin scale is *absolute zero* (0 K = −273.15 °C), the lowest temperature theoretically and experimentally obtainable. Note that the abbreviation K for the kelvin unit is used without the degree sign (°). Also note that 1 °C = 1 K = (9/5) °F.

EXAMPLE 1.6

TEMPERATURE CONVERSION

Your normal body temperature is 98.6 °F. Show that this corresponds to 37.0 °C.

Solution

$$\text{Body temperature in °C} = \frac{5\ °\text{C}}{9\ °\text{F}}(98.6\ °\text{F} - 32.0\ °\text{F})$$

$$= [(\tfrac{5}{9})(66.6)]\ °\text{C} = 37.0\ °\text{C}$$

EXERCISE 1.5 Temperature Conversion

The temperature in Lisbon, Portugal, is often 17 °C at about noon. What is the temperature in °F?

THE KELVIN UNIT Winter temperatures in many places can easily drop below 0 °C, that is, to temperatures expressed by negative numbers. In the laboratory, even colder temperatures can be achieved easily, and the temperatures are given by even more negative numbers. However, there is a limit to how low the temperature can go. It can be proved that we cannot reach a temperature lower than −273.15 °C (or −459.67 °F).

The degree symbol (°) is not used with Kelvin temperatures, and the unit is called kelvin *(not capitalized).*

William Thomson, known as Lord Kelvin (1824–1907), first suggested a temperature scale that did not use negative numbers. Kelvin's scale, now adopted as the SI standard, uses the same unit as the Celsius scale, but it takes the lowest possible temperature as its zero, a point

Historical Figures in Chemistry: Anders Celsius (1701–1744)

Until 1948 scientists around the world worked with temperatures in "degrees centigrade." In that year, however, it was agreed to rename the scale in honor of Anders Celsius, an 18th century Swedish astronomer. Celsius, who was from a family of scientists, studied mathematics, astronomy, and physics at Uppsala University and was a professor of astronomy there beginning in 1730. He published important papers on the aurora borealis (the "northern lights") and the magnitude of stars.

Sometime in the 1730s Celsius described a thermometer scale based on dividing the interval between the freezing and boiling points of water into 100 equal parts. Whether he actually devised the scale, though, is in some dispute. One person who claimed to have invented it was a protegé and friend of Celsius, the botanist Linnaeus.

Celsius apparently used melting snow to fix the lower temperature of his scale, but curiously he marked it as 100. The upper point, the boiling point of water, was marked as the 0 point. Beginning in June 1743, the "Swedish thermometer" with this scale was used for regular observations at the Uppsala observatory. In 1747, a few years after Celsius's death, Linnaeus reversed the scale so that the freezing point of water was 0 °C.

called **absolute zero**. Because kelvin and Celsius units are the same, the freezing point of water is reached 273.15 degrees *above* the starting point; that is, 0 °C is the same as 273.15 kelvins or 273.15 K. Temperatures in Celsius units are readily converted to kelvins, and vice versa, using the relation

$$T(\text{K}) = t(°\text{C}) + 273.15$$

Thus, a common room temperature of 23.5 °C would be (23.5 °C + 273.15) or 296.7 K.

EXERCISE 1.6 Temperature Conversions

Carry out the following conversions: (a) 25 °C to K. (b) Liquefied nitrogen boils at 77 K. What is this temperature in °C?

1.5 HANDLING NUMBERS

Handling numbers in scientific notation and other mathematical operations are reviewed in Appendix A.

In chemistry, making measurements is at the heart of our science. Therefore, we need to know something about dealing with numerical information.

PRECISION AND ACCURACY

The **precision** of a measurement describes how well repeated measurements of a value agree with one another. In contrast, **accuracy** is determined by the agreement between the measured quantity and the correct value. A highly precise number may be inaccurate because the same, offsetting error was made in each measurement. These concepts are illustrated without numbers in Figure 1.12. In one case, you threw all the darts precisely at one point, but your accuracy was terrible because you missed the "correct value," the bullseye. In another case, you were both accurate and precise, since all the darts clustered near the bullseye. Finally, the middle case is neither precise nor accurate, since you scattered the darts all over the board.

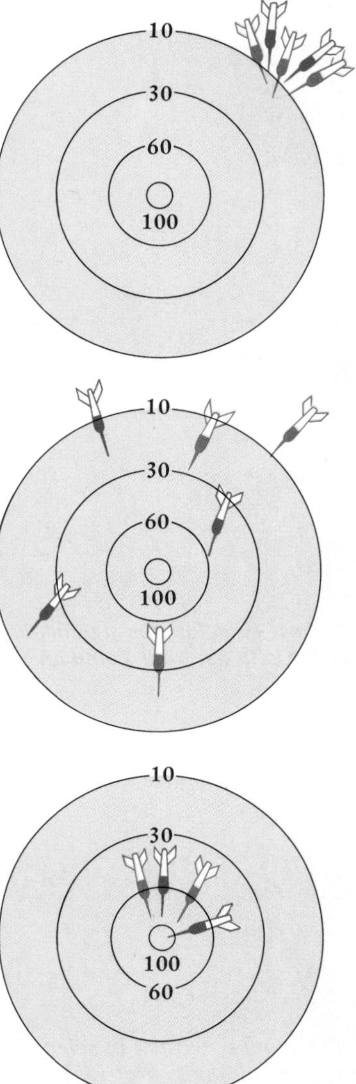

Figure 1.12 Precision and accuracy. Top: poor accuracy but good precision. Middle: poor accuracy and poor precision. Bottom: good accuracy and good precision.

SIGNIFICANT FIGURES

In the laboratory we have to worry about obtaining results that are reasonable. Some data we collect are known more precisely than others. It is common sense that *the answer to a calculation can be known no more precisely than the least precise piece of information*. This is where the concept of *significant figures* or *digits* comes in.

Most students in an introductory chemistry course in college are 18 years old. By saying your age is 18, you tell others that you are [1(10) + 8] years old. Both of the digits in your age are *significant digits*, because they specify your age to the nearest year. The first digit (a 1) tells how many multiples of 10 years you have lived, and the second digit (an 8) gives the number of years beyond the nearest 10-year multiple.

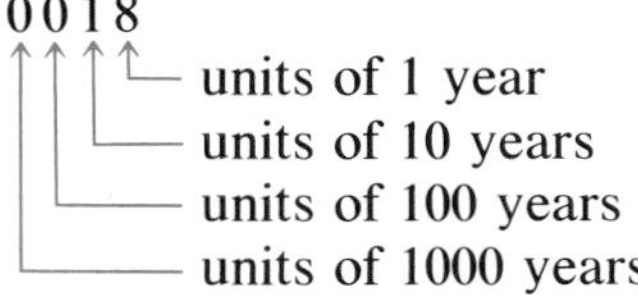

Would it have done any good to specify your age as 0018? Not at all, because the zero digits in front of the 1 tell us there are no multiples of 1000 or 100 years in your age; you already know that because the 1 in the two-digit number 18 says that 10 is the largest unit multiple (10 years).

Now, what if you said your age was 18.0 years? This would have signaled that you took the trouble to measure your age to within a tenth of a year (36 days). Saying you are 18 means that you might be just past your 18th birthday or very near your 19th birthday. But 18.0 means you are no more than 36 days beyond 18 years. We now know your age more precisely, to three significant digits. The last digit to the right is the least significant. Be sure to notice that saying you are 18.0 years old still leaves some doubt as to your *exact* age!

We measure ages in years, but suppose you selected "kiloyears" as the measurement unit (where 1 kiloyear = 1000 years). Regardless of the unit of measurement, you must express your age to the same precision as before; the number of significant digits cannot change when you change units. This means you could say you are 0.018 kiloyears old or 0.0180 kiloyears old. In either case, we know you are at least 0.010 kiloyear + 0.008 kiloyear old (since 0.010 kiloyear = 10 years).

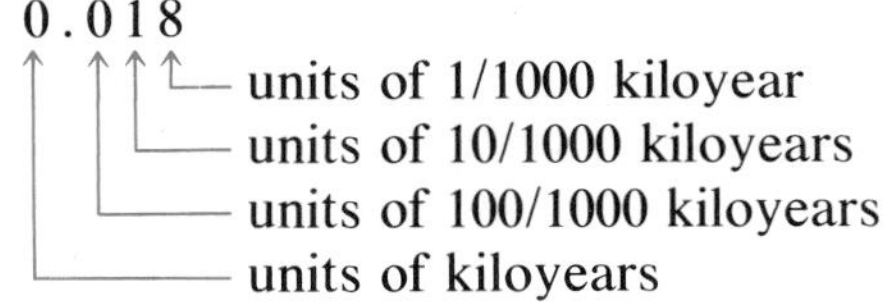

Again, any zero digits before the first nonzero digit we find as we move from left to right are not significant, since there are no units of 100/1000 kiloyears in your age.

When numbers of different significance are combined by addition, multiplication, or other mathematical operation, we need some guidelines for the number of significant digits in the result. You tell your friend you are 18.0 years old, but she says she is 21. What is the difference in your ages? Note that she has not given her age to three significant digits, as you have done. The least significant multiple in her age limits the significance of the calculated difference in your ages. Thus, the difference in your ages is known only to within one year,

$$\text{Age difference} = 21 - 18.0 = 3 \text{ years}$$

and the correct answer has only one significant digit. In numerical calculations, the smallest unit-multiple in the answer must be that of the least significant digit in the starting numbers. This often means the "extra" significant digits in one or more of the numbers can be "lost." *In no case can the result be expressed to more significant digits than the least significant number in the calculation.* The following guidelines summarize the important ideas in using significant digits or figures.

The best guideline for significant figures is to use your common sense!

GUIDELINES FOR DETERMINING SIGNIFICANT FIGURES

Rule 1. To determine the number of significant figures in a measurement, read the number from left to right and count all digits, starting with the first digit that is *not* zero.

For a number written in scientific notation, all digits are significant.

The number π is now known to 1,011,196,691 digits. It is doubtful that you will need this accuracy in this course—or ever.

EXAMPLE	NUMBER OF SIGNIFICANT FIGURES
1.23 g	3
0.00123 g	3; the zeros to the *left* of the 1 simply locate the decimal point. To avoid confusion, write numbers of this type in scientific notation; thus, $0.00123 = 1.23 \times 10^{-3}$.
2.0 g and 0.020 g	2; both have two significant digits. When a number is greater than 1, *all zeros to the right of the decimal point are significant.* For a number less than 1, only zeros to the right of the first significant digit are significant.
100 g	1; in numbers that do not contain a decimal point, "trailing" zeros may or may not be significant. To eliminate possible confusion, the practice followed in this book is to include a decimal point if the zeros are significant. Thus, 100. has three significant digits, while 100 has only one. Alternatively, we write it in scientific notation as 1.00×10^2 (three significant digits) or as 1×10^2 (one significant digit). For a number written in scientific notation, all digits are significant.
100 cm/m	Infinite number of significant figures, because this is a defined quantity.
$\pi = 3.1415926 \ldots$	The value of π is known to a greater number of significant figures than any data you will ever use in a calculation.

Rule 2. When adding or subtracting, the number of decimal places in the answer should be equal to the number of decimal places in the number with the *fewest* places.

0.12	2 significant figures	2 decimal places
1.6	2 significant figures	1 decimal place
10.976	5 significant figures	3 decimal places
12.696		

This answer should be reported as 12.7, a number with one decimal place, because 1.6 has only one decimal place.

Rule 3. In multiplication or division, the number of significant figures in the answer should be the same as that in the quantity with the *fewest* significant figures.

$$\frac{0.01208}{0.0236} = 0.512 \text{ or, in scientific notation, } 5.12 \times 10^{-1}$$

Since 0.0236 has only three significant figures, while 0.01208 has four, the answer is limited to three significant figures.

Rule 4. When a number is rounded off (the number of significant figures is reduced), the last digit retained is increased by 1 only if the following digit is 5 or greater.*

FULL NUMBER	NUMBER ROUNDED TO THREE SIGNIFICANT FIGURES
12.696	12.7
16.249	16.2
18.35	18.4
18.351	18.4

One last word regarding significant figures and calculations. In working problems on a pocket calculator, you should do the calculation using all the digits allowed by the calculator and round off only at the end of the problem. Rounding off in the middle can introduce errors. If your answers do not quite agree with those in the back of the book, this may be the source of the disagreement.

EXAMPLE 1.7

SIGNIFICANT FIGURES

An example of a calculation you will do later in the book (Chapter 12) is

$$\text{Volume of a gas} = \frac{(0.120)(0.08206)(273.15 + 23)}{(230/760.0)}$$

Calculate the final answer with the correct number of significant figures.

Solution As a first step, we will analyze each piece of the equation.

*A modification of this rule is sometimes used to reduce the accumulation of roundoff errors. If the digit following the last permitted significant figure is *exactly* 5 (with no following digits or all following digits being zero only), then (a) increase the last significant figure by 1 if it is *odd* or (b) leave the last significant figure unchanged if it is *even*. Thus, both 18.35 and 18.45 are rounded to 18.4.

NUMBER	NUMBER OF SIGNIFICANT FIGURES	COMMENTS
0.120	3	The trailing 0 is significant. To see that the 0 to the left of the decimal point is not significant, write in scientific notation as 1.20×10^{-1}. See Rule 1.
0.08206	4	The 0 just to the right of the decimal point is not significant. Write in scientific notation as 8.206×10^{-2}. See Rule 1.
$273.15 + 23 = 296$	3	23 has no decimal places, so the answer can have none. See Rule 2.
$230/760.0 = 0.30$	2	230 has two significant figures because the last zero is not significant; there is no decimal point in the number. In scientific notation, it would be 2.3×10^2. In contrast, there is a decimal point in 760.0, so there are four significant figures. The result of the division must have two significant figures. See Rules 1 and 3.

The analysis shows that one of the pieces of information is known with only two significant figures. Therefore, the answer must be volume of gas = 9.6, a number with two significant digits. (In this case, the answer is in units of liters, but this is not important for now.)

EXERCISE 1.7 Significant Figures

(a) What are the sum and the product of 12.63 and 0.063? (b) What is the result of the calculation $(101.2 - 93)/(0.00102 + 0.027)$?

1.6 PROBLEM SOLVING BY DIMENSIONAL ANALYSIS

USING UNITS OF NUMBERS TO SOLVE PROBLEMS

Dimensional analysis is a systematic way of solving numerical problems. Simply put, every number in a problem must have some units associated with it; density, for example, is given as mass per unit volume (g/cm^3). When the numbers in a calculation are manipulated in the correct way, their units must cancel out to leave the final answer in the appropriate units. If you set up the problem incorrectly, the units will not cancel properly, and you will know immediately that you have made a mistake.

Dimensional analysis has already been used in the examples in this chapter. However, since the method is such a valuable tool in chemistry, it is worthwhile to illustrate it further using some more familiar items and the type of calculation you have often done in your head. Let us say you need 30 cans of soft drinks for a party, and the cost is \$2.15 per six-pack of cans. What is the total cost? The strategy in solving the problem is first to convert the unit "cans" to the unit "six-pack," since the cost is given in units of "dollars per six-pack."

Step 1. Find the number of six-packs required. Convert unit of "cans" to unit of "six-packs."

$$\text{Number of six-packs} = 30\ \cancel{\text{cans}}\left(\frac{1\ \text{six-pack}}{6\ \cancel{\text{cans}}}\right) = 5\ \text{six-packs}$$

Notice that if you had multiplied 30 times 6, the units would have come out to be (cans^2/six-pack), which is clearly nonsense.

Step 2. Find the total cost. Convert units of "six-packs" into units of "$."

$$\text{Total price} = 5\ \cancel{\text{six-packs}}\left(\frac{\$2.15}{\cancel{\text{six-pack}}}\right) = \$10.75$$

Now that you have worked through the example, notice that in each step you always multiplied the starting unit by a factor that gave the answer in the desired unit. Such factors as (1 six-pack/6 cans) or ($2.15/six-pack) are called **conversion factors**. Conversion factors are multipliers that relate the desired unit to the starting unit.

$$\text{Conversion factor} = \frac{\text{desired unit}}{\text{starting unit}}$$

The numerator of the factor must be equivalent to the denominator. Thus, 1 six-pack is equivalent to 6 cans, and $2.15 is the amount you must pay for 1 six-pack. Other "conversion factors" are 453.6 g/pound, 100 cm/m, and 5 °C/9 °F.

You will find that many chemistry problems are similar to this and the following example. That is, they can be broken down into a series of simple steps, each involving only a single concept. To solve a problem, make sure you know what the units of the answer will be, and then aim for those units in a series of steps.

EXAMPLE 1.8

DIMENSIONAL ANALYSIS: DENSITY AND VOLUME

If a frying pan needs a Teflon® coating that is 1.00 mm thick, and the area to be covered is 36.0 square inches, how many pounds of Teflon are required to coat the pan? Teflon has a density of 2.20 g/cm^3.

Solution The objective of the problem is to find the mass of Teflon needed. However, the information is given in "distance" units. In thinking about the problem you should recognize that the density is given because it is the factor that relates volume and mass. Since "distance" is directly related to volume, it is evident that we want to begin by calculating the volume of Teflon needed. Once this is done, we can convert the volume to a mass by using the density.

Step 1. Volume of Teflon required.

The volume of the Teflon can be calculated from the product of the thickness and the area. Unfortunately, these dimensions are in different units and cannot be directly multiplied until both are in the same unit. You could convert the thickness to inches, so that the product of thickness (inches) and area (inches^2) gives the volume in cubic inches (inches^3). Alternatively, you could convert the area to square millimeters so that volume is obtained in cubic millimeters. However, we choose to do neither. Thinking ahead to the next step in the problem, you will want to use the density in units of g/cm^3. Therefore, it would be most useful to obtain the volume of Teflon in units of cm^3 so that it is compatible with the given density. Thus,

$$\text{Thickness} = 1.00\ \cancel{\text{mm}}\left(\frac{1\ \text{cm}}{10\ \cancel{\text{mm}}}\right) = 0.100\ \text{cm}$$

$$\text{Area} = 36.0\ \cancel{\text{in.}^2}\left(\frac{2.54\ \text{cm}}{\cancel{\text{in.}}}\right)^2 = 232\ \text{cm}^2$$

The factor 2.54 cm/in. is given in the table at the back of the book. Since you need cm^2/in.^{2}, you simply square the conversion factor *and* its units (= 6.45 cm^2/in.^{2}).

With the area and thickness in the same units, you can now calculate the volume.

$$\text{Volume of Teflon coating} = \text{thickness} \times \text{area} = (0.100\ \text{cm})(232\ \text{cm}^2) = 23.2\ \text{cm}^3$$

Step 2. Weight of Teflon required.

In solving any problem, you should keep firmly in mind the answer for which you are aiming and, just as importantly, the units of that answer. In this case, you want the weight of Teflon in pounds. As you saw in Example 1.5, the density will allow you first to convert the volume of Teflon to a mass in grams.

$$23.2\ \cancel{\text{cm}^3}\left(\frac{2.20\ \text{g}}{\cancel{\text{cm}^3}}\right) = 51.0\ \text{g}$$

Now that you have the mass in grams, it is a simple matter to convert it to the weight in pounds, since you know (from the table at the back of the book) that 1 pound is equivalent to 454 g.

$$51.0\ \cancel{\text{g}}\left(\frac{1\ \text{pound}}{454\ \cancel{\text{g}}}\right) = 0.112\ \text{pound} = 1.12 \times 10^{-1}\ \text{pound}$$

Notice that the final answer has three significant figures.

EXERCISE 1.8 Mass, Volume, and Density

Mercury (left) is a metal, but unlike almost all other metals, it is a liquid with a density of 13.6 g/cm^3 at room temperature. It is also poisonous and should be treated with great care and respect.

(a) If you have 100. mL of mercury, how many grams do you have? How many pounds?

(b) If you spill 100. mL of mercury on the floor, and it spreads out into a puddle that is 2.0 mm thick, what area of the floor is covered? Calculate the area in square centimeters and square inches. (Incidentally, mercury is also expensive; 100 mL would cost approximately $90.)

THE CONCEPT OF "PERCENT"

Chemists often express the composition of matter in terms of **percent**. For example, we can determine that 88.81% of a given mass of water is oxygen, that table sugar is 42.11% carbon, or that a 5¢ coin, a "nickel," is only about 25% nickel (the rest is copper). Since this is a widely used concept, it is worth taking a moment to think about it.

You are familiar with "percent" from looking for a bargain at the local mall or from paying sales taxes. For example, if you pay a sales tax of 6.00%, this means that you pay 6.00 dollars per 100 dollars of things you purchased. Therefore, if you buy a new shirt for $31.50, you pay $1.89 in tax.

$$31.50\ \cancel{\$\ \text{spent}} \left(\frac{6.00\ \$\ \text{tax}}{100\ \cancel{\$\ \text{spent}}}\right) = 1.89\ \$\ \text{tax}$$

There are two important points to make here. First, notice that we solved the problem—one that you have probably done in your head many times—using units. We used the conversion factor "$6.00 tax/100 $ spent," and the unit "$ spent" cancels out and leaves units of "$ tax." Second, the word "percent" tells us what the conversion factor must be, since the Latin phrase *per cent* literally means *per 100*. The conversion factor in a percent calculation will always be the value of the percent divided by 100.

EXAMPLE 1.9

PERCENT

Battery plates in lead storage batteries (the type used in automobiles) are made from a mixture of two chemical elements: lead (94.0%) and antimony (6.0%). If you have a battery plate with a mass of 25.0 g, how many grams of lead and how many grams of antimony are present?

Solution Let us first solve for the mass of lead present, using the known percentage of lead. The plate is 94.0% lead, which means there are 94.0 g of lead in every 100. g of plate.

$$25.0\ \cancel{\text{g battery plate}} \left(\frac{94.0\ \text{g lead}}{100.\ \cancel{\text{g battery plate}}}\right) = 23.5\ \text{g lead}$$

We know that the plate contains only lead and antimony, so

$$25.0\ \text{g plate} = 23.5\ \text{g lead} + X\ \text{g antimony}$$

Solving for X, we find that the mass of antimony is 25.0 g − 23.5 g = 1.5 g of antimony. Of course, we could obtain the same result from the calculation

$$25.0\ \cancel{\text{g battery plate}} \left(\frac{6.0\ \text{g antimony}}{100.\ \cancel{\text{g battery plate}}}\right) = 1.5\ \text{g antimony}$$

EXAMPLE 1.10

PERCENT

Silver jewelry is actually a mixture of silver and copper. If you have a bracelet with a mass of 17.6 g, and it contains 14.1 g of silver, what is the percentage of silver? The percentage of copper?

Solution Again, we can solve this problem using units. First, we divide grams of silver by grams of jewelry

$$\frac{14.1 \text{ g silver}}{17.6 \text{ g jewelry}} = \frac{0.801 \text{ g silver}}{1 \text{ g jewelry}}$$

and find that there is 0.801 g of silver in 1.00 g of jewelry (so 0.801 is the "fractional amount" of silver in the jewelry). Since we really want to know how many grams of silver there are in 100 g of jewelry, we multiply by 100/100 (which of course just equals 1):

$$\left(\frac{0.801 \text{ g silver}}{1 \text{ g jewelry}}\right)\left(\frac{100 \text{ g jewelry}}{100 \text{ g jewelry}}\right) = \frac{80.1 \text{ g silver}}{100 \text{ g jewelry}}$$

to find there are 80.1 g of silver per 100 g of jewelry; that is, the bracelet is 80.1% silver.

We worked through the calculation carefully so you would see the logic involved. Normally, all we do to find a percent is as follows:

$$\frac{14.1 \text{ g silver}}{17.6 \text{ g jewelry}} \times 100 = 80.1\%$$

Now, what about the percentage of copper? Since there are only two things in the jewelry, the percentages must total 100%. Therefore,

$$\% \text{ silver} + \% \text{ copper} = 80.1\% + \% \text{ copper} = 100\%$$

and so the % copper is 19.9%.

EXERCISE 1.9 Percent

(a) What is the mass of gold in a 15.0-g earring if it is made of 14-karat gold? Fourteen-karat gold is 58% gold, the remainder being copper and silver.
(b) Pewter, which is used to make utensils and decorative items, is a mixture of the metals tin, copper, bismuth, and antimony. If a 1.256-g piece of pewter is found to contain 1.068 g of tin, what is the percentage of tin in the sample?
(c) You weighed 165 pounds but went on a diet and lost 11 pounds. What percentage of your original weight did you lose?

SUMMARY

In science, an investigation of a problem usually follows a well worn path (Section 1.2). After reviewing work done before, the scientist constructs a **hypothesis**, a tentative explanation and a guide in predicting the results of future observations. This hypothesis is revised as further experimental work is done, and the final result can be a **law**, a concise verbal or mathematical statement of a relation that is always the same under the same conditions. To explain the law, we attempt to devise a **theory**, a principle that explains a body of facts and the laws based on them.

Matter is anything having mass and occupying space (Section 1.3). It can exist in three **states** or **phases**: solid, liquid, and gas. A pure **substance** is a form of matter that has a definite composition and distinct properties. A **mixture** is a combination of two or more pure substances in which each retains its identity. Mixtures can be **homogeneous**, that is, have the same composition throughout. Such mixtures can be called **solutions**, where a

solute is dissolved in a **solvent**. In contrast, in a **heterogeneous** mixture the components can still be observed as individual substances.

Matter is composed ultimately of different kinds of atoms. **Atoms** are the smallest particles of an element that retain the chemical properties of the element; each consists of a tiny core (the nucleus) containing positive charges, which is surrounded by an equal number of negative charges (electrons). An **element** is matter that consists of only one kind of atom, either individually or combined into larger units.

Molecules are identifiable units of matter consisting of two or more atoms combined in a definite ratio. When the atoms are the same we have an element; when they are different, we have a compound. A **compound**, then, is matter that consists of identifiable units containing atoms of different elements combined in specific ratios.

Pure substances have **physical and chemical properties**. Physical properties, such as color and density, can be measured and observed without changing the substance's chemical composition. A chemical property, though, requires that the substance undergoes a chemical change, such as reaction with oxygen.

Chemists perform both **qualitative** and **quantitative** experiments, the latter involving numerical information. Units are attached to this information (Section 1.4). The SI unit of length is the **meter (m)**, a unit subdivided into **centimeters** (1 m = 100 cm), **millimeters** (1 m = 1000 mm), and so on. Volume measurements are made in **cubic centimeters** (1000 cm^3 = 1 liter). The **kilogram (kg)** is the SI unit of mass, but **grams** (1 kg = 1000 g) and **milligrams** (1 kg = 10^6 mg) are more often used in the laboratory. The standard unit of time is the **second (s)**, while temperatures are measured in **degrees Celsius (°C)** or in **kelvins (K)**. The latter has the same unit size as the Celsius scale, but measurement begins at **absolute zero**, 273.15 degrees below zero on the Celsius scale (that is, **0 K = −273.15 °C**).

In handling numerical information you should recognize the difference between the **precision** and **accuracy** of data (Section 1.5). The former is the agreement between repeated determinations of a given value, while the latter is the agreement between the measured value and the correct value. **Significant figures** are also important, and guidelines for their use are given in Section 1.5.

STUDY QUESTIONS

In solving many of these problems, you will need to consult Table 1.2, Appendix C, or the inside of the back cover of the book for units and conversion factors. Questions for which answers are given in Appendix L are indicated by a colored number.

GENERAL QUESTIONS

1. In each case, tell whether the underlined property is a physical or chemical property:
   (a) The normal <u>color</u> of bromine is red-orange (see Figure 1.4).
   (b) Iron is <u>transformed into rust</u> in the presence of air and water.
   (c) Dynamite can <u>explode</u> when it interacts with oxygen.
   (d) The <u>density</u> of uranium metal is 19.07 g/cm^3.
   (e) Aluminum metal, the "foil" you use in the kitchen, <u>melts</u> at 660 °C.
2. In what phase do you normally find each of the following?
   (a) rust
   (b) the oxygen you breathe
   (c) limestone
   (d) gasoline

Common sand containing iron filings.

3. Small chips of iron are mixed with sand (see photo). Is it a homogeneous or heterogeneous mixture? Suggest a way to separate the iron and sand.
4. Decide whether each statement reflects a law or a theory:
   (a) The beginning of the universe occurred as a "big bang."
   (b) In all chemical processes, matter is never lost; it is conserved.
5. In each case, point out which is *qualitative* information and which is *quantitative:*
   (a) A purple solid has a mass of 1.25 g.
   (b) A 0.025-g piece of silvery magnesium floats on oil.
   (c) 25 mL of a blue copper sulfate solution react exactly with 50 mL of a colorless ammonia solution.
6. Decide whether each of the underlined items is an intensive or an extensive property.
   (a) The melting point of sodium metal is 98 °C.
   (b) A chemical experiment requires 250 mL of water.
   (c) The bromine in Figure 1.4 is a red-orange vapor and liquid.
   (d) The density of gold is 19.3 g/cm³.

LENGTH, AREA, VOLUME, AND MASS

7. The average lead pencil, new and unused, is 19 cm long. What is its length in millimeters? In meters? In inches?
8. An excellent height in the pole vault is 18 feet, 11.5 inches. What is this height in meters?
9. In track and field, the so-called metric mile is 1600. meters. How many miles does this represent?
10. The classic marathon is 26.2 miles. How many kilometers is this equivalent to?
11. Road signs in Germany advise motorists to drive no faster than 130. km/hour. What is this speed in miles per hour?
12. The maximum speed limit in some states in the United States is 55 miles per hour. What is this speed in kilometers/hour?
13. A sailboat has a length of 36 feet 7 inches; its beam (the widest point of the hull) is 12 feet. What are these distances in meters? In centimeters?
14. A standard sheet of notebook paper is $8\frac{1}{2} \times 11$ inches. What are these dimensions in centimeters?
15. A standard U.S. postage stamp is 2.5 cm long and 2.1 cm wide. What is the area of the sample in cm²? In m²? In inches²?
16. The paper on which this book is printed is 8 inches wide and 10 inches long. What is the area of a sheet of this paper in cm²? In m²?
17. A Saab automobile has a luggage compartment of dimensions 100. cm × 100. cm × 150. cm. What is the volume of the compartment in cm³? In liters? In cubic meters?
18. A popular brand of backpack has an upper compartment that is 10. inches × 15.5 inches × 15 inches and a bottom compartment that is 7.0 inches × 15.5 inches × 9.0 inches. What is the capacity of the pack in inches³? In cm³? In liters?
19. A new U.S. quarter has a mass of 5.63 g. What is its mass in kilograms? In milligrams?
20. The larger quartz crystal in the photo has a mass of 2.83 grams. What is this mass in kilograms? In pounds?

Quartz crystals.

21. Complete the following table of masses.

MILLIGRAMS	GRAMS	KILOGRAMS
________	0.693	________
156	________	________
________	________	2.23

22. Complete the following table of masses.

MILLIGRAMS	GRAMS	KILOGRAMS
10.2	________	________
________	16.56	________
________	________	0.545

23. The maximum weight a person can carry comfortably in a backpack is about 60. pounds. What is the mass of this load in grams? In kilograms?

24. A popular tent for backpacking purposes weighs 3 pounds 11 ounces. What is this mass in grams? In kilograms?

25. A sleeping bag for use in very cold climates is filled with 3 pounds of goose down. How many grams of down are used? How many kilograms?

26. If you weigh 160. pounds, what is your weight in grams? In kilograms?

27. A typical laboratory beaker has a volume of 800. mL. What is its volume in cm^3? In liters? In m^3?

28. A Volkswagen engine has a displacement of 120. cubic inches. What is this volume in liters? In cm^3?

29. You use exactly 23.67 mL of an acid solution in a laboratory experiment. What is this volume in cm^3? In liters?

30. The directions in an experiment call for you to use 676 mL of a solution. What is this volume in cm^3? In liters?

DENSITY

31. The density of carbon in the form of diamonds is 3.51 g/cm^3. If you have a small diamond with a volume of 0.0270 cm^3, what is its mass in grams?

32. The topaz crystals in the photo have a density of 3.56 g/cm^3. The smaller gem has a mass of 0.353 g. What is the volume of the gem?

Topaz crystals.

33. Water has a density of 0.997 g/cm^3 at 25 °C. If you have 500. mL of water, what is its mass in grams? In pounds?

34. A chemist needs 2.00 g of a liquid compound. (a) What volume of the compound is necessary if the density of the liquid is 0.718 g/cm^3? (b) If the compound costs \$2.41 per milliliter, what is the cost of the reagent?

35. The "cup" is a volume widely used by cooks in the United States. One cup is equivalent to 237 mL. If 1 cup of olive oil has a mass of 205 g, what is the density of the oil?

36. Liquid sodium can be used to cool nuclear reactors. (a) If liquid sodium has a density of 0.93 g/cm^3, how many grams of sodium would be required to fill a container with a volume of 15 liters? (b) How many pounds of sodium would be required?

37. Peanut oil has a density of 0.92 g/cm^3. If a recipe calls for 1 cup of peanut oil (1 cup = 237 mL), how many grams of peanut oil are you using?

38. Silver has a density of 10.5 g/cm^3. If you have a silver coin with a mass of 6.0 g (about the same as a U.S. quarter), what is the volume of the silver in the coin?

39. Automobile batteries are filled with sulfuric acid solution. The density of the solution is 1.285 g/cm^3 and the solution is 38.08% (by weight) sulfuric acid. How many grams of the acid are there in 500. mL of the battery acid solution?

40. The density of a solution of sulfuric acid is 1.285 g/cm^3, and it is 38.08% (by weight) acid. How many milliliters of the acid solution do you need to supply 125 g of sulfuric acid?

TEMPERATURE

41. Many laboratories use 25 °C as a standard temperature. What is this temperature in °F? In K?

42. The temperature on the surface of the sun is 5.50×10^3 °C. What is this temperature in °F? In K?

43. Make the following temperature conversions:

	°F	°C	K
(a)	57	______	______
(b)	______	37	______
(c)	−40	______	______

44. Make the following temperature conversions:

	°F	°C	K
(a)	______	______	77
(b)	______	60.	______
(c)	1000.	______	______

45. Solid gallium has a melting point of 29.8 °C. If you hold this metal in your hand, what will be its physical state? That is, will it be a solid or a liquid? (Prove your answer with appropriate calculations.)

46. Titanium is used in industrial applications where a high melting point is important. Its melting point is 3020. °F. What is this temperature in °C? In K?

47. Oxygen freezes to a solid at −218 °C. What is this temperature in °F? In K?

48. Neon, an element used in signs, has a melting point of −248.6 °C and a boiling point of −246.1 °C. Express these temperatures in °F and in kelvins.

SIGNIFICANT FIGURES

49. What is the average mass of three objects whose individual masses are 10.3 g, 9.334 g, and 9.25 g?

50. What is the volume, in cm^3, of a backpack whose dimensions are 22.86 cm × 38.0 cm × 76 cm?

51. Do the following calculation and report the answer in the correct number of significant figures:

$$(0.000523)(0.0263)(263.28)$$

52. Do the following calculation and report the answer in the correct number of significant figures:

$$(1.68)(7.874)\left(\frac{1.0000}{55.85}\right)$$

53. Do the following calculation and report the answer in the correct number of significant figures:

$$(0.0345)\left[\frac{(35.45 - 1.2)}{1.000 \times 10^3}\right]$$

54. Do the following calculation and report the answer in the correct number of significant figures:

$$\left[\frac{(2.18 + 16.23 - 0.0023)}{12.02}\right](0.0016345)$$

55. Solve the following equation for n and report the answer to the correct number of significant figures:

$$\left(\frac{11.2}{760.0}\right)(123.4) = n(0.0821)(298.3)$$

56. Solve the following equation for V and report the answer to the correct number of significant figures:

$$\left(\frac{267}{760.0}\right)V = (0.000129)(0.08206)(273.15 + 22.3)$$

USING UNITS

57. A 2-quart saucepan has a radius (r) of 83.5 mm and a height (h) of 9.5 cm. What is the volume of this saucepan in cm^3? (The volume of a cylindrical object is given by $\pi r^2 h$.)

58. A camping stove has a fuel capacity of 0.60 pint. The storage bottle for the extra fuel is advertised as having a capacity of 0.30 L. When the stove has used all of its fuel, does the storage bottle contain enough fuel to refill the stove completely?

59. Molecular distances are usually given in nanometers (1 nm = 1×10^{-9} m) or in picometers (1 pm = 1×10^{-12} m). However, a commonly used unit is the Ångstrom, where 1 Å = 1×10^{-10} m. (The Ångstrom unit is not a metric unit.) If the distance between the Pt atom and the N atom in *cisplatin* is 1.97 Å, what is this distance in nm? In pm?

60. The separation between carbon atoms in a diamond is 0.154 nm. (a) What is their separation in meters? (b) Another unit frequently used in science is the Ångstrom, where 1 Å = 10^{-10} m. (See Study Question 59.) What is the carbon atom separation in Ångstrom units?

61. The smallest repeating unit of a crystal of common salt is a cube with an edge length of 0.563 nm. What is the volume of this cube in nm^3? In cm^3?

62. The mass of a gemstone is often measured in "carats" where 1 carat = 0.200 g. If the annual worldwide production of diamonds is 12.5 million carats, how many grams does this represent?

63. Metals such as gold and platinum are sold in units called "troy ounces," where 1 troy ounce has a mass of 31.103 g. Is this larger or smaller than an ounce in the usual scale used in the United States? [Recall that there are 16 ounces in 1 pound (called an "avoirdupois" pound) and that 1 pound has a mass of 453.59237 grams.]

64. As discussed in Chapter 8, light and other forms of radiation can be described as waves, the distance between crests of a wave being the wavelength. If a radio wave has a wavelength of 13 cm, what is its wavelength in meters? In inches? In feet?

65. If you are driving at 65 miles per hour, what is your speed in kilometers per hour? In feet per second?

66. The world record for the 100.-meter dash is 9.95 seconds. What is the average speed of the runner in miles per hour?

67. The density of the gemstone tourmaline (see photo) is 3.26 g/cm^3. What is this density in kg/m^3?

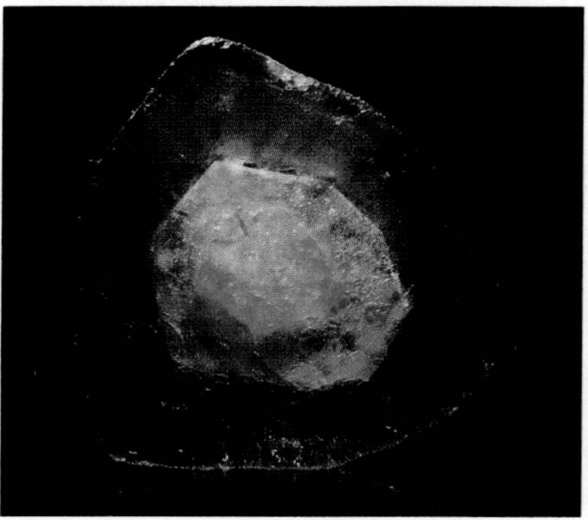

A piece of "watermelon" tourmaline.

68. At 25 °C the density of water is 0.997 g/cm^3, whereas the density of ice at −10 °C is 0.917 g/cm^3. (a) If a

soft-drink can (volume = 250. mL) is filled completely with pure water and then frozen at −10 °C, what volume will the solid occupy? (b) Could the ice be contained within the can?

69. The solder used by plumbers to fasten copper pipes together consists of 67% lead and 33% tin. If you have a 1.0-pound (454 g) block of solder, how many grams of lead do you have? Grams of tin?

70. You are fortunate enough to own a ring made of white gold, which actually consists of 60.% gold and 40.% platinum. If your ring has a mass of 6.89 g, how many grams of platinum do you have? How many grams of gold?

71. A solution of sugar in water has a density of 1.05 g/cm^3. If you have 250 mL of the solution, and if the solution is 8.1% sugar, how many grams of sugar are in the solution?

72. When you heat popcorn, it pops because it loses water explosively. Assume a kernel of corn, weighing 0.125 g, weighs only 0.106 g after popping. What percentage of its mass did the kernel lose on popping?

73. One 2.0-ounce serving of macaroni provides 8.0 g of protein, 42 g of carbohydrate, and 1.0 g of fat. Calculate the percentage (by weight) of each of the three substances in the 2.0 ounces of macaroni.

74. In 1988, as in most years, sulfuric acid was the chemical produced in greatest amount by industry. In this year, 85.56 billion pounds of the acid were produced. How many tons does that represent? How many kilograms? How many grams? (1 ton is exactly 2000 pounds)

75. One source of sulfuric acid is elemental sulfur. In 1981, 12.1 million metric tons of yellow solid sulfur were produced. (1 metric ton = exactly 1000 kg) How many kilograms of sulfur does this represent? How many pounds?

76. A sheet of $8\frac{1}{2} \times 11$ inch paper is 0.10 mm thick. It can be considered as a rectangular solid for which the volume is length × width × thickness. What is the volume of the sheet in cm^3? If a sheet has a mass of 4.58 g, what is its density?

77. An ancient gold coin is 2.2 cm in diameter and 3.0 mm thick. It is a cylinder for which volume = (π)(radius)^2(thickness). If the density of gold is 19.3 g/cm^3, what is the mass of the coin in grams? Assume a price of gold of $410 per troy ounce. How much is the coin worth? (1 troy ounce = 31.10 g)

78. A ball made of pure titanium weighs 16.19 g and has a diameter of 19.0 mm. What is the density of the titanium in g/cm^3? (The volume of a sphere = $\frac{4}{3}\pi r^3$)

79. You have a 100.0-mL graduated cylinder containing 50.0 mL of water. You drop a 35.69-g piece of pure nickel (density = 8.908 g/cm^3) into the water. To what mark will the water rise in the graduated cylinder?

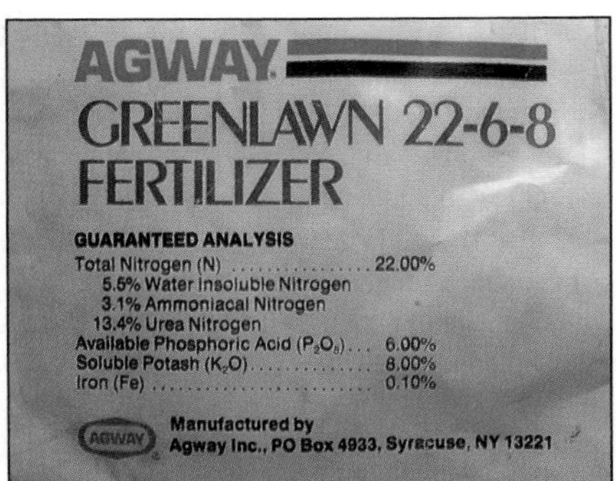

The label from a bag of lawn fertilizer.

80. A common fertilizer used on lawns is designated as "22-6-8" (see photo). These numbers mean that the fertilizer contains 22.00% of nitrogen, 6.00% of a phosphorus-containing compound, and 8.00% of a potassium-containing compound. If you buy a 40.0-pound bag of this fertilizer, how many grams of the phosphorus-containing compound are you putting on your lawn? If the phosphorus-containing compound consists of 43.64% phosphorus (the rest is oxygen), how many grams of phosphorus are there in 40.0 pounds of fertilizer?

81. The aluminum in a package containing 75 square feet of kitchen foil weighs approximately 12 ounces. Aluminum has a density of 2.70 g/cm^3. What is the approximate thickness of the aluminum foil in millimeters? (1 ounce = 28.4 g)

82. A tanker has spilled 1.00×10^4 L of oil into the ocean and the oil film covers the sea with a layer 3.0×10^2 nm thick. What is the area covered by the oil film in square meters? In square miles? (Volume = area × thickness)

83. Amethyst is a colored form of the mineral quartz in which the purple color comes from traces of the element manganese (see photo below). (a) To determine

Amethyst crystals.

the density of amethyst, you take a stone having a mass of 15.25 g and place it in a 100. mL graduated cylinder containing 45.0 mL of water. On adding the stone, the water surface rises to the 50.8-mL mark. What is the density of amethyst? (b) If you have a piece of amethyst that is 1.8 cm × 0.95 cm, how thick does it have to be to have a mass of 6.50 g?

84. The fluoridation of city water supplies has been practiced in the United States for several decades because it is believed that fluoride prevents tooth decay, especially in young children. Assume you live in a medium-sized city of 150,000 people and that each person uses 175 gallons of water per day. How many tons of sodium fluoride would you have to add to the water supply each year (365 days) in order to have the required fluoride concentration of 1 part per million, that is, 1 ton of fluoride per million tons of water? (Sodium fluoride is 45.3% fluoride, and one U.S. gallon of water weighs 8.34 pounds.)

85. Copper has a density of 8.94 g/cm^3. If a factory has an ingot of copper with a mass of 125 pounds, and the ingot is drawn into wire with a diameter of 9.50 mm, how many feet of wire can be produced?

INTERVIEW

Bassam Z. Shakhashiri

Bassam Shakhashiri describes himself as "scientist by training, teacher and public servant by trade, advocate by conviction, optimist by nature." All of these roles were evident in this interview, only a small portion of which can be printed here.

Professor Shakhashiri was born in Lebanon, but he came to the United States with his parents and sisters in 1957 when he was eighteen. He completed his college studies in chemistry at Boston University, taught at Bowdoin College in Maine, and then earned his M.Sc. and Ph.D. degrees at the University of Maryland. After postdoctoral research and two years of teaching at the University of Illinois, Urbana he joined the Chemistry Department at the University of Wisconsin, Madison in 1970. There he began to acquire his reputation as a master of chemical demonstrations. His annual Christmas show in Madison—"Once Upon a Christmas Cheery, in the Lab of Shakhashiri"—was always sold out, and he has given his show in Washington, D.C. at the Smithsonian Air and Space Museum and the National Academy of Sciences, at Boston's Museum of Science, and in less formal settings such as convention centers, shopping malls, and retirement homes. He has co-authored several books on chemical demonstrations, and his videotape of demonstrations accompanies this book. In 1983, he founded the Institute for Chemical Education and served as its first director.

In June 1984 he joined the National Science Foundation in Washington, D.C. where he was Assistant Director of Science and Engineering Education until June 1990. He was extraordinarily successful in this position, raising the level of funding for science education in the nation many-fold, but just as importantly speaking out as an advocate for science education. He has been concerned with many educational issues, but two of the most important are the coming crisis in scientific personnel and the values of our society. We met with Dr. Shakhashiri in his office at the National Science Foundation and discussed these two issues. What follows are his thoughts on these problems in his own words.

The crisis in scientific manpower

The situation this country faces is far more critical and consequential than what we faced in the immediate post-*Sputnik* era for a lot of reasons. Let me mention three of those reasons.

First, the population of the United States in the past 30 years or so has increased by about 50 million people. To put that number in perspective, that happens to be the approximate population of all of Great Britain or twice the population of Canada. What does that mean? It means we have more students to teach, and we need more *qualified* teachers to teach at all levels. The demographic data cause us to be alarmed about our ability as a nation to address that problem. So, the first reason the situation is critical can be summarized in one word: scale. Societal institutions are sluggish in responding to such big changes in scale. This is not only true in education, but also in waste disposal, traffic control, health care, child development, care for the elderly, and so on.

Second, for the country to maintain its preeminence in science and technology, in the arts and humanities, in the global economy, and in all walks of life, we need a good supply of scientists, mathematicians, and engineers coming through the educational system. That is what the National Science Foundation (NSF) set out to do in the post-*Sputnik* era, and, to a very large extent, the effort was successful. But again, the demographic data now cause us to be alarmed about our ability to produce scientists in the numbers that are likely to be required.

The third reason, and in my judgment the most important of all, is that we now live in a very advanced scientific and technological society, and it is the education of the nonspecialist in science and technology that we have to pay attention to. We need an educated citizenry that can distinguish between astronomy and astrology, that can deal successfully with the complex issues related to animal rights, that is able to cope with pollution and pollution control issues, and that can understand why burning the rain forests in South America is bad for our global environment. And we need to have everyone benefit from the tremendous advances we have made in the nutritional sciences. Basically what we need is a scientifically, technologically, and mathematically literate society. That is where the emphasis ought to be in the 1990s and as we approach the next millennium, which, by the way, starts in the year 2001. So, our twin mission can be summarized by saying that we need to increase the flow of talent to careers in science and science teaching and that we need to have a scientifically literate population. In this connection, let me use an analogy from sports. Just as we have professional baseball players, basketball players, football players, and so on, we also have to have sports fans. Without sports fans, the entire professional sports enterprise would be nothing. So that's what we need. We need professional scientists, and we need true science fans. Another analogy is useful in making the same point: we need good orchestra players, and we need audiences that appreciate the music.

Leakage in the personnel pipeline

I have often discussed our problem with regard to science personnel (see Figure). Among a population of 4 million high school sophomores in 1977—and it's about the same now—750,000 expressed an interest in science, math, and engineering. By the time they got to be seniors the number dropped down to 590,000 and one year later when they were in college the number was 340,000, a 40% drop in one year! About 200,000 got a bachelor's degree in science; 61,000 enrolled in graduate school; 46,000 got a master's degree, and fewer than 10,000 will get a Ph.D. degree. This so-called "personnel pipeline" shows a great deal of leakage. In fact, it is not a leakage, it is a hemorrhage in terms of loss of talent not only to this nation, but to humanity. There are individuals across the nation whose talents can serve the national need. They must be encouraged to do so.

Society's values

Science literacy is a measure of our value system: what we care about, what we cherish, how we treat each other, how we treat our environment. No doubt about it, we need talented people to go into business, finance, the arts and humanities, the manufacture of goods, and the development and testing of new chemicals. We need people to help generate wealth as well as people to move the wealth around.

And that's why I want to come

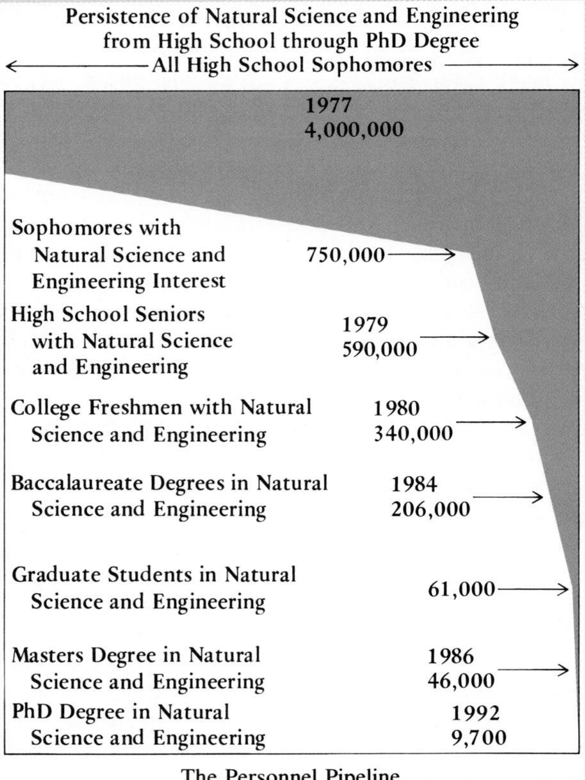

The Personnel Pipeline

back to the graph because it shows in a very vivid way what the issue is. It's so easy to focus only on the blue part of the figure. But we must pay attention to the people represented by the lighter portion of the figure for many of the reasons that I have already mentioned. In addition, there is one crass reason. Namely, the people represented by the light portion of the figure pay for what the people in the blue portion [scientists] do. Those of us in the science-rich sector of our society [people represented by the blue sector in the graph] have an awesome responsibility. All of us owe something back to the science-poor sector.

We want the tide of excellence to rise, because when it does it will lift all the boats, not only those who are going on to graduate school, not only those who want to get Ph.D.'s and so on. We must focus on nurturing the natural curiosity that people have. We are curious about the world we live in, and we ask questions all the time. Why is the sky blue? Why do the leaves change color in the fall? Why is it that when you have some spilled gasoline on wet pavement that you see pretty colors? Why do soap bubbles float? Are the colors that you see on the soap bubbles related to the colors of that spilled gasoline on the wet pavement? How does the microwave oven work? How does the microchip work? How does the suspension bridge get put together? Why do credit cards have a hologram? We are full of questions, and we must try to satisfy that curiosity.

Science is fun

We have to try to understand the different phenomena that surround us. In doing that we have to work hard and have fun. You can't have fun without working hard. When we say science is fun, we are talking about fun in the best sense of the word, not in some cheap thrill fashion. And this brings us back to our value system. The joy of doing chemistry is to do experiments, to try to make sense of our observations, to try to see how existing theories can explain those observations, and, if they don't, how we can develop new theories that develop a further understanding of the phenomena we see. All of this requires commitment and intense concentration. Satisfaction comes not only in the process, but also at the end when we achieve a deeper understanding of these phenomena.

The role of the National Science Foundation

The role of the federal government in science education is a dual one. First, it provides leadership in identifying the problems and then it suggests how they should be addressed. Second, it helps to address these problems by providing leveraged support for large and small projects. The NSF is the lead federal agency in support of science, research, and science education. It is an important vehicle to accomplish very critical national goals and to coordinate all efforts at the federal level. But, ultimately, its role is to provide leadership to help to develop the national will to address those problems. How the NSF fosters cooperation and partnerships will be the key to success. The main effort has to be at the local and state level, but we need national leadership.

Most especially, it is up to us in the academic community to take up the challenge. We must revamp our educational system in collaboration with business and civic leaders and with elected officials at the local, state, and national levels. Parents, professional organizations, and other parts of our society must work together to bring about the necessary fundamental, comprehensive, and systemic changes.

In 1990 the level of effort at the NSF in real dollars was about one third of what it used to be in the 1960s. This despite the fact that the population has increased by 50 million people, and the complexity of science and technology requires a scientifically literate population. That says something about our value system as a society! We must increase the level of effort to about $600 million a year (or 20% of the NSF budget) in order to provide both enlightened leadership and coherence to the many excellent efforts across the land. We must have an effective national strategy, and we must be invincible in our will to achieve significant national goals.

The role of public service

It is very important for talented individuals from the scientific community to consider careers in public service at the local, state, and federal level because the country can benefit enormously from that talent. We especially need those who are trained in science, mathematics, and engineering because the quantitative dimension of our thinking can be very helpful in solving very complicated societal problems.

Shakhashiri Demonstration Tapes are available with this text. See preface for details.

CHAPTER

2

Atoms and Elements

Elements: bromine, copper, aluminum, and silicon.

CHAPTER OUTLINE

When you think of elements you think of beautiful and valuable gold and silver, and perhaps also of copper, iron, and lead. All are treasured for their beauty or for their special properties. This chapter introduces you to some fundamental ideas about the elements and their properties, ideas that we need to build on in other parts of the book.

Many of the ideas that we use today about elements and atoms are almost 200 years old. In 1803 John Dalton first proposed the basic ideas or hypotheses of an **atomic theory of matter**. On the basis of chemical knowledge *at that time*, he stated that:

1. All matter is made of **atoms**. These indivisible and indestructible objects are the ultimate chemical particles.
2. All the atoms of a given element are identical, in both weight and chemical properties. However, atoms of different elements have different weights and different chemical properties.
3. **Compounds** are formed by the combination of different atoms in the ratio of small whole numbers.*
4. A **chemical reaction** involves only the combination, separation, or rearrangement of atoms; atoms are neither created nor destroyed in the course of ordinary chemical reactions.

A small piece of pure elemental bismuth displays beautiful colors.

John Dalton's hypotheses have been accepted by the scientific community, with only two changes,† and we assume that you are aware of at least the first of his ideas. You are probably also aware of the last

*As an example, water is composed of hydrogen and oxygen in the ratio of two atoms of hydrogen to one of oxygen.

†We know now that (1) an atom is "divisible and destructible" and (2) all atoms of an element are not identical in mass. Atoms of the same element that differ in mass, called *isotopes*, are the subject of Section 2.3.

Historical Figures in Chemistry: John Dalton (1766–1844)

John Dalton

John Dalton was born about the fifth of September, 1766, in the village of Eaglesfield in Cumberland, England. His family was quite poor, and his formal schooling ended at age 11. However, he was clearly a bright young man, and with the help of influential patrons, he began a teaching career at the age of 12. Shortly thereafter, he made his first attempts at scientific investigation, observations of the weather. This study was to last his lifetime. In fact, he made over 200,000 observations of weather conditions by the end of his life. In 1793 Dalton moved to Manchester, England, where he took up a post as tutor at the New College, but he left there in 1799 to pursue scientific inquiry on a full-time basis. It was not long after this, on October 21, 1803, that he read the paper introducing his "Chemical Atomic Theory" to the Literary and Philosophical Society of Manchester. This was followed by lectures in London and in other cities in England and Scotland, and his reputation as a scientist rapidly increased. He was first proposed for membership in the top scientific society in Britain, the Royal Society, in 1810, and many other honors followed over his lifetime.

A chemistry textbook* published in 1879 had this to say about the importance of John Dalton's work:

> The brilliant researches of Dalton did not terminate with the acquisition of facts, but sought to account for them by a theoretical conception. Taking up an old idea of Lysippus and the word of Epicurus, he supposed all ponderable matter to be composed of indivisible particles which he called *atoms*. He gave a precise meaning to the vague and ancient notion by considering on the one hand that the atoms of each kind of matter, of each element, possess an invariable weight,† and on the other that combination between different kinds of matter results from the juxtaposition of their atoms. Such is the atomic hypothesis . . .

*Wurtz, Adolphe, *Elements of Modern Chemistry*, J.B. Lippincott, Philadelphia, 1879.
†The existence of isotopes (Section 2.3) had not yet been discovered.

of his hypotheses, although under a different name. This is really just a statement of the **law of conservation of matter**, that is, that *matter can neither be created nor destroyed in a chemical reaction.* It had been established somewhat earlier by the great French chemist Antoine Lavoisier (1743–1794) and is the core of the quantitative part of the science of chemistry.

After describing some fundamental properties of atoms in this chapter, we shall take up the matter of compounds and molecules in the next chapter. Chemical reactions, and quantitative relations between chemicals in reactions, are considered in Chapters 4 and 5.

2.1 ELEMENTS

The concept of elements came from the Greek philosophers. Among the best known of these, Aristotle (384–323 BC) said that "Everything either is an element or is composed of elements." This is not far from our modern

Table 2.1 Derivation of Element Names and Symbols

Element	Symbol	Date of Discovery	Discoverer (Country)*	Derivation of Name or Symbol
Berkelium	Bk	1950	G.T. Seaborg, S.G. Thompson, A. Ghiorso (U.S.)	Berkeley, California (site of Seaborg's laboratory)
Copper	Cu	Ancient		Latin, *cuprum*, copper Derived from *Cyprium*, the Island of Cyprus, the main source of copper in the ancient world.
Einsteinium	Es	1952	A. Ghiorso (U.S.)	Albert Einstein
Iron	Fe	Ancient		Latin, *ferrum*, iron
Lead	Pb	Ancient		Latin, *plumbum*, lead, meaning heavy
Oxygen	O	1774	J. Priestley (G.B.) K.W. Scheele (Swed.)	French, *oxygene*, generator of acid, derived from the Greek, *oxy* and *genes* meaning acid forming (oxygen was thought to be part of all acids)
Silver	Ag	Ancient		Latin, *argentum*, silver
Tin	Sn	Ancient		Latin, *stannum*, tin
Tungsten	W	1783	J.J. and F. de Elhuyer (Sp.)	Name from Swedish, *tung sten*, meaning heavy stone; symbol from wolframite, a mineral

*U.S., United States; G.B., Great Britain; Swed., Sweden; Sp., Spain.

definition given in Chapter 1: *All* matter is composed of different kinds of atoms, and matter that is composed of only one kind of atom is an **element**.

Chemists have discovered and studied 108 elements thus far. Each has been given a *name* and a *symbol*, listed in the table at the front of the book. Some elements—such as gold, silver, iron, mercury, and tin—were known in relatively pure form to the Greeks and Romans and to the alchemists of ancient China, the Arab world, and medieval Europe. Many others were discovered in the 18th and 19th centuries, but two elements (107 and 109) were discovered only in the 1980s. Many elements have names and symbols with Latin or Greek roots, but more recently discovered elements have been named for their place of discovery or for a person or place of significance (Table 2.1).

*The table at the front of the book, in which the symbol and other information for each element is enclosed in a box, is called the **periodic table**. See Section 2.5.*

Be sure to notice that only the first letter of an element's symbol is capitalized. For example, copper is Cu and not CU, a notation that would mean a combination of carbon (C) and uranium (U).

2.2 ATOMS

An **atom** is the smallest particle of an element that retains the chemical properties of the element. The Greek philosophers conceived of the notion of atoms and thought that atoms were the ultimate building blocks of nature. We know now, though, that they are not indivisible. They can be broken down into still smaller particles, although they lose their chemical identity in the process.

The three primary constituents of atoms are **electrons**, **protons**, and **neutrons**. The **nucleus** or core of the atom is made up of protons with a positive electrical charge and neutrons with no charge. The electrons, with a negative electrical charge, are found in the space about the nucleus (Figure 2.1). For an atom, which must have no net electrical charge, *the*

The story of how the electrons, protons, and neutrons were discovered and the details of atomic structure are discussed in Chapter 8. For now it is important for you to know that there are such particles, and that atoms can react by the gain or loss of electrons (as described in more detail in Chapters 4, 5, 8, and 21).

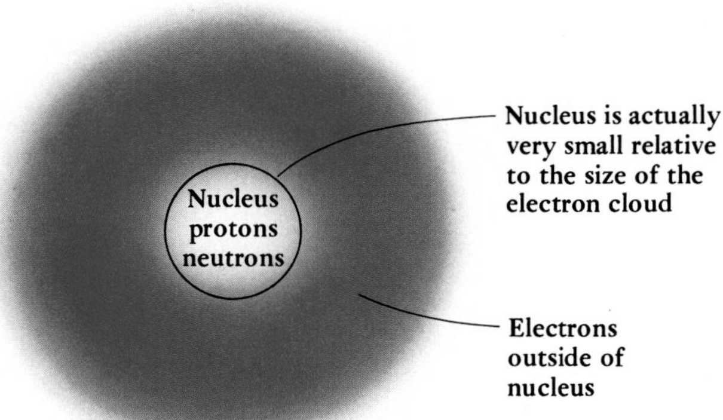

Figure 2.1 All atoms consist of one or more protons (positively charged) and usually at least as many neutrons (no charge) packed into an extremely small nucleus. Electrons (negatively charged) are arranged in space as a "cloud" about the nucleus. In an electrically neutral atom the number of electrons equals the number of protons. The picture drawn here is a very crude representation that will be refined in Chapters 8 and 9.

number of negatively charged electrons around the nucleus equals the number of positively charged protons in the nucleus.

Atoms are extremely small; the radius of the typical atom is between 30 and 150 pm (0.000000000030 m = 3.0×10^{-11} m). To give you a feeling for the incredible smallness of the atom, look at it this way:

(a) One teaspoon of water (about 1 cm^3) contains about three times as many atoms as the Atlantic Ocean contains teaspoons of water.
(b) If, after the flood of about 3000 BC, Noah had started to string hydrogen atoms on a thread at the rate of one atom per second for 8 hours a day, the chain would be only about 3.9 meters (about 13 feet) long today!

All atoms of the same element have the same number of protons in the nucleus, and, because the atom is electrically uncharged, the number of electrons surrounding the nucleus is the same as the number of protons. This number is called the **atomic number**, and it is given the symbol **Z**. In the periodic table at the front of the book, the atomic number for each element is given above the element's symbol. Magnesium, for example, has a nucleus containing 12 protons, so its atomic number is 12; uranium has 92 nuclear protons and Z = 92.

*With the knowledge that all atoms of a given element have the same number of protons in the nucleus, we can give a better definition of an element than that in Chapter 1. That is, **an element is defined by its atomic number**.*

EXERCISE 2.1 Atomic Numbers, Names, and Element Symbols
The symbol for carbon is C, and its atomic number is 6. Find all the other elements that have a symbol beginning with the letter C and give the name and atomic number for each.

Each element is represented by a box in the periodic table in the front of the book. Among the data given are the element's symbol and its atomic number.

12	⟵ *atomic number*
Mg	⟵ *symbol*

1 amu = $\frac{1}{12}$ the mass of a carbon atom having 6 protons and 6 neutrons in the nucleus.

Since atoms are the ultimate building blocks of matter, they must have a mass, with some elements having atoms heavier than other elements. Just as our clocks and longitude are set relative to the time and longitude at Greenwich, England, so too have chemists established a scale of atomic masses relative to a standard. This standard is the mass of a carbon atom that has six protons and six neutrons in its nucleus. Such an atom is defined to have a mass of exactly 12 **atomic mass units** (or 12 amu), and the mass of every other element is established relative to this mass. Thus, for example, experiment shows that an oxygen atom is, on the average, 1.33 times heavier than a carbon atom, so an oxygen atom has a mass of 1.33×12.0 amu or 16.0 amu.

Table 2.2 Properties of Subatomic Particles

Particle	Mass: Grams	Mass: Atomic Mass Units	Electrical Charge	Symbol
Electron	9.109389×10^{-28}	0.0005486	−1	${}_{-1}^{0}e$
Proton	1.672623×10^{-24}	1.007276	+1	${}_{1}^{1}p$
Neutron	$1.6749286 \times 10^{-24}$	1.008665	0	${}_{0}^{1}n$

Masses of the basic atomic particles in amu (Table 2.2) have been determined experimentally. Notice that the proton and neutron have masses very close to 1 amu, while the electron is only about 1/1900 times as heavy.

Once we have established a relative scale of atomic masses, the mass of any atom for which the nuclear composition is known can be estimated. The proton and neutron have masses so close to 1 amu that the difference can usually be ignored, and even a large number of electrons will not greatly affect the mass of the atom. Therefore, we need only to add up the number of protons and neutrons of an atom to estimate its mass. The result is called the **mass number** of that particular atom, a number given the symbol A. For example, a fluorine atom has 9 protons and 10 neutrons in its nucleus, so its mass number, A, is 19. The most common atom of iron has 26 protons and 30 neutrons, so A = 56. With this information, we can symbolize an atom of known composition by the notation

$$\begin{matrix}\text{mass number} \rightarrow \\ \text{atomic number} \rightarrow\end{matrix} {}_{Z}^{A}X \leftarrow \text{element symbol}$$

(where the subscript Z is optional because the element symbol tells you what the atomic number must be). For example, the fluorine atom described above would have the symbol ${}_{9}^{19}F$ or just ${}^{19}F$. In words, you would say "fluorine-19."

EXAMPLE 2.1

THE MASS NUMBER OF AN ATOM

What is the mass number of a tin atom if that atom has 69 neutrons? Write its full symbol.

Solution First, you must know the atomic number of tin, since this will give you the number of protons in the atom. Looking at the tables in the front of the book, you find that tin (symbol = Sn) has an atomic number of 50. Therefore, the mass number, A, is

$$A = \text{number of protons} + \text{number of neutrons} = 50 + 69 = 119$$

The full symbol of the element is ${}_{50}^{119}Sn$.

EXAMPLE 2.2

COMPOSITION OF AN ATOM

How many neutrons are there in an atom of platinum with a mass number of 195?

Solution Platinum has the symbol Pt; its atomic number is 78. The mass number is the sum of the number of protons and neutrons, and, since the atomic number is 78, we have

$$\text{Mass number} = 195 = \text{number of protons} + \text{number of neutrons}$$
$$= 78 + \text{number of neutrons}$$

$$\text{Number of neutrons} = 195 - 78 = 117$$

EXERCISE 2.2 Atomic Composition

(a) What is the mass number of a copper atom with 34 neutrons? (b) How many protons, neutrons, and electrons are there in a $^{59}_{28}Ni$ atom?

The actual masses of atoms have been determined experimentally, although sophisticated and expensive instruments are required (Figure

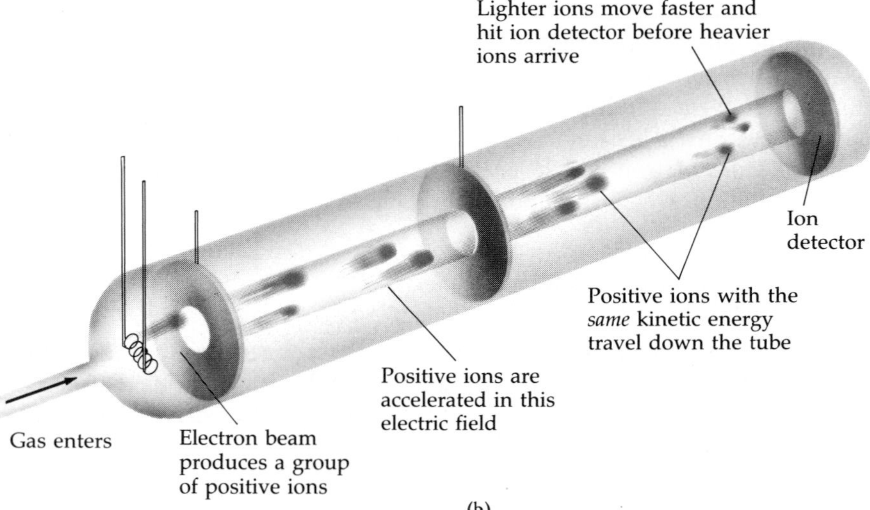
Lighter ions move faster and hit ion detector before heavier ions arrive
Ion detector
Positive ions with the *same* kinetic energy travel down the tube
Positive ions are accelerated in this electric field
Gas enters
Electron beam produces a group of positive ions
(b)

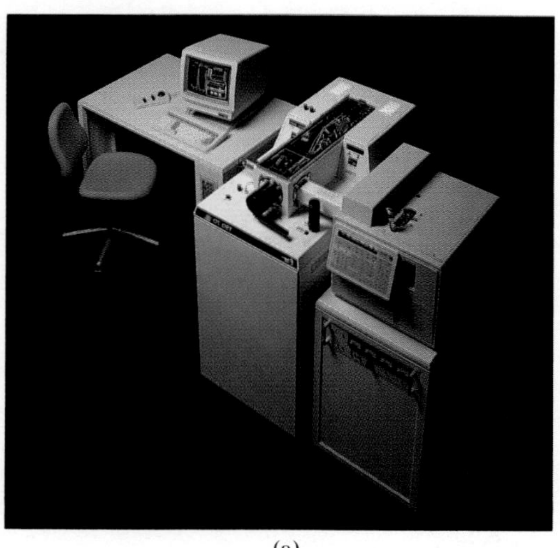
(a)

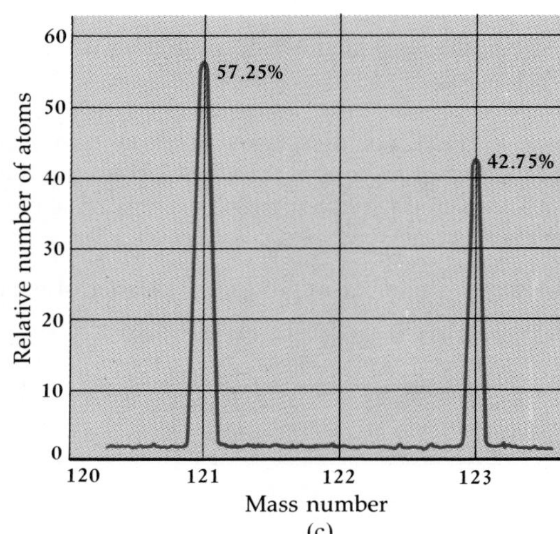
57.25%
42.75%
Relative number of atoms
0
10
20
30
40
50
60
120
121
122
123
Mass number
(c)

Figure 2.2 (a) The mass spectrometer, an instrument for determining the masses of atoms, molecules, and molecular fragments. (b) Representation of a "time-of-flight" mass spectrometer. An electron is removed from an atom to give a positive ion (such as H^+, O^+, and N^+). The time required for positive ions to reach the detector is measured; the heavier an ion the longer it takes to travel down the tube. (The tube is evacuated, so the process occurs at very low pressures.) (c) The ion detector is connected to a device that measures the number of ions of a given mass that reach the detector in a given time. Therefore, the spectrometer can measure the relative abundance of each isotope of an element. A mass spectrum, such as that for antimony, shows the relative abundance of atoms as a function of mass number. (a, Dave Pierce, Finnigan MAT–San Jose)

2.2). It is always observed that, while the actual mass is almost the same as the mass number, the actual mass is not an integral number. For example, the actual mass of an iron atom with 32 neutrons is 57.933272 amu, slightly less than the mass number of 58.

An obvious thing to do is to check the experimental masses by adding up all the exact masses of the protons, neutrons, and electrons of an atom to see whether you obtain the experimental mass. However, you would find the sum is slightly greater than the actual mass. This is not a mistake. Rather, the difference (sometimes called the *mass defect*) is related to the energy binding the particles of the nucleus together; more will be said about this in Chapter 7.

2.3 ISOTOPES

For many years minerals containing boron have been mined in Death Valley, California. If you were to examine the boron atoms of some borax mined in that region, you would find that, while all boron atoms have 5 nuclear protons, some atoms have 5 neutrons while others have 6. That is, you would find a collection of $^{10}_{5}B$ and $^{11}_{5}B$ atoms. These are *isotopes* of boron. **Isotopes** are *atoms having the same atomic number Z but different mass numbers A*, because the atoms have different numbers of neutrons.

If two atoms differ in the number of protons *they contain, they are different* elements. *If they differ only in the number of* neutrons, *they are* isotopes. *It is the number of protons in the nucleus and the number of electrons outside the nucleus that determine the chemistry of the atom.*

Most elements have at least two stable isotopes (Table 2.3). There are very few elements with only one isotope (aluminum, fluorine, and phosphorus, for example), and there are others with many isotopes (tin has nine stable isotopes). Generally, we refer to a particular isotope by giving its mass number (for example, uranium-238, ^{238}U), but some isotopes are so important that they have special names. Hydrogen atoms all have one proton. When that is the only nuclear particle, the element is called simply "hydrogen." When one neutron is added to give $^{2}_{1}H$, it is called "deuterium" or heavy hydrogen, and adding two neutrons gives radioactive hydrogen, $^{3}_{1}H$, or "tritium."

Hydrogen, $^{1}_{1}H$, is also called "protium" to distinguish it from deuterium and tritium.

Table 2.3 Exact Masses of the Stable Isotopes of Some Elements

Element	Symbol	Atomic Weight	Mass Number	Exact Mass	Percent Abundance
Hydrogen	H	1.00794	1	1.007825	99.9855
			2	2.014102	0.0145
Boron	B	10.811	10	10.012939	19.91
			11	11.009305	80.09
Oxygen	O	15.9994	16	15.994914	99.759
			17	16.999134	0.0374
			18	17.999160	0.2039
Magnesium	Mg	24.305	24	23.985045	78.80
			25	24.985840	10.15
			26	25.982591	11.05
Uranium	U	238.03	234	234.040875	0.0056
			235	235.043900	0.7205
			238	238.050734	99.274

EXAMPLE 2.3

ISOTOPES

Silver has two isotopes, one with 60 neutrons and the other with 62 neutrons. What are the mass numbers and symbols of these isotopes?

Solution Silver has an atomic number of 47, so it has 47 protons in the nucleus. Therefore, the two isotopes have mass numbers of

Isotope 1: $A = 47 \text{ protons} + 60 \text{ neutrons} = 107 \text{ amu}$

Isotope 2: $A = 47 \text{ protons} + 62 \text{ neutrons} = 109 \text{ amu}$

The first isotope has the symbol $^{107}_{47}Ag$ and the second is $^{109}_{47}Ag$.

EXERCISE 2.3 Isotopes
Silicon has three isotopes with 14, 15, and 16 neutrons, respectively. What are the mass numbers and symbols of these three isotopes?

2.4 ATOMIC WEIGHT

Boron atoms have two different isotopic masses, 10.0129 and 11.0093 amu. This means that the average mass of a collection of boron atoms will be neither 10 nor 11 but somewhere in between, the actual value depending on the proportion of each kind of isotope.

The percentage of atoms in a natural sample of the pure element represented by a particular isotope is called the isotope's **percent abundance**.

$$\text{Percent abundance} = \frac{\text{number of atoms of a given isotope}}{\text{total number of atoms of all isotopes of that element}} \times 100$$

If these percent abundances can be determined for each isotope, then the average mass can be determined. Such determinations can be done with a device called a *mass spectrometer* (Figure 2.2), and some results are given in Table 2.3. From this table you can see that the boron isotopes ^{10}B and ^{11}B have percent abundances of 19.91 and 80.09, respectively. This means that, if you could count out 10,000 boron atoms from an "average" natural sample, 1991 of them would have a mass of 10.0129 amu and 8009 of them would have a mass of 11.0093 amu. The average mass of a representative sample of atoms is called the **atomic weight**. For boron, the atomic weight is 10.81 amu, as is shown by the calculation in the following example.

Chemists usually use the term atomic weight *of an element rather than "atomic mass." Although the quantity is more properly called a "mass" than a "weight," the term "atomic weight" is so commonly used that it has become accepted.*

EXAMPLE 2.4

CALCULATING AVERAGE ATOMIC MASS FROM PERCENT ABUNDANCES

A natural sample of boron consists of two isotopes. One has an exact mass of 10.0129 amu and its percent abundance is 19.91. The other isotope, of mass 11.0093 amu, has a percent abundance of 80.09. Calculate the atomic weight.

Solution The average value for a series of numbers is always found by adding the products of "value" times "fractional abundance," where the fractional abundance is the percent abundance divided by 100.

Average atomic mass = (fractional abundance of isotope 1)(mass of isotope 1) + (fractional abundance of isotope 2)(mass of isotope 2) + . . .

For the boron sample, the calculation would be as follows:

$$\begin{aligned}\text{Average atomic mass of boron} &= \text{atomic weight}\\ &= (0.1991)(10.0129\text{ amu})\\ &\quad + (0.8009)(11.0093\text{ amu})\\ &= 10.811\text{ amu}\end{aligned}$$

This calculation is done exactly the way you find your average grade after several quizzes. For example, suppose you have two grades of 9 and three of 8 on five quizzes.

$$\text{Average} = \frac{9 + 9 + 8 + 8 + 8}{5} = \frac{2(9) + 3(8)}{5} = 8.4$$

This is the same thing as

$$\text{Average} = (\tfrac{2}{5})(9) + (\tfrac{3}{5})(8) = (0.4)(9) + (0.6)(8) = 8.4$$

which says that $\frac{2}{5}$ of the time you had a 9 and $\frac{3}{5}$ of the time you had an 8. The fraction $\frac{2}{5}$ (or its decimal equivalent 0.4) is the "fractional abundance" of your grade. Expressed as a percentage, 40% of the time you received a 9.

A sample of "native" cooper. This copper was found in nature in the form of the metal.

EXERCISE 2.4 Calculating Atomic Weight

Verify that the atomic weight of chlorine is 35.45 amu, given the following information:

^{35}Cl, exact mass = 34.96885, percent abundance = 75.77

^{37}Cl, exact mass = 36.96590, percent abundance = 24.23

The atomic weight of each element has been determined, and it is these masses that appear in the tables in the front of the book. In the periodic table, each element's box contains the atomic number, the element symbol, and the atomic weight.

If the masses are known for the isotopes of a given element, it is possible to compute the percent abundances by "reversing" the procedure outlined in Example 2.4. The next example shows how this is done.

The periodic table entry for copper.

29	⟵ *atomic number*
Cu	⟵ *symbol*
63.546	⟵ *atomic weight*

EXAMPLE 2.5

CALCULATING ISOTOPE ABUNDANCES

Calculate the percent abundances of copper-63 and copper-65 if the atomic weight is 63.546 amu and the exact masses of the isotopes are $^{63}Cu = 62.93$ amu and $^{65}Cu = 64.93$ amu. Assume no other isotopes exist.

Solution This problem has two unknown quantities, the percent abundances of ^{63}Cu and ^{65}Cu. Recall from algebra that to solve for two unknowns, two equations are required. The key here is to realize that the sum of the fractional isotope abundances (each less than 1) must equal 1.

Samples of pure copper in different forms.

$$^{63}\text{Cu fractional abundance} + {}^{65}\text{Cu fractional abundance} = 1$$

Both of these abundances are unknown. However, if one of them is assigned the value x, then the other is $1 - x$. That is,

if	^{63}Cu fractional abundance $= x$
then	$x + {}^{65}$Cu fractional abundance $= 1$
and so	^{65}Cu fractional abundance $= 1 - x$

The expression "^{65}Cu abundance $= 1 - x$" is the first of the two equations you need. The second is the same type of expression used in Example 2.4.

$$\text{Average atomic mass of Cu} = ({}^{63}\text{Cu fractional abundance})(62.93) + ({}^{65}\text{Cu fractional abundance})(64.93)$$

These two equations can now be solved in the usual way. Substitute the expressions for the Cu isotope abundances into the second equation, and solve for x.

$$\begin{aligned} \text{Atomic weight} &= (x)(62.93) + (1 - x)(64.93) \\ 63.546 &= (x)(62.93) + (1 - x)(64.93) \\ 63.546 &= 62.93x + 64.93 - 64.93x \\ 63.546 - 64.93 &= (62.93 - 64.93)x \\ x &= \frac{63.546 - 64.93}{62.93 - 64.93} = 0.692 \end{aligned}$$

We find that $x = 0.692$. This means that the ^{63}Cu fractional abundance is 0.692 or that the percent abundance is 69.2. It then follows that the fractional abundance of ^{65}Cu must be $1 - x = 0.308$, so the percent abundance is 30.8.

EXERCISE 2.5 Calculating Isotopic Abundances

Gallium has two naturally occurring isotopes, ^{69}Ga and ^{71}Ga, with masses of 68.9257 and 70.9249, respectively. Calculate the percent abundances of the two isotopes.

2.5 THE PERIODIC TABLE OF THE ELEMENTS—AN INTRODUCTION

The historical development of the periodic table and periodic law are described in detail in Chapters 8 and 9. Much of this textbook is devoted to examining the chemical and physical properties of the elements and their interrelationships.

The periodic table of elements in Figure 2.3 and in the front of the book contains a wealth of information, and we use it to organize many of the ideas of chemistry. Since we refer to it often, you should be introduced to some of the main features and terminology of the periodic table.

The elements are arranged in the table in order of increasing atomic number in such a way that *elements having similar chemical and physical properties lie in vertical columns* called **groups**. The table commonly used in the United States has groups numbered 1 through 8, and each is followed by a letter A or B. However, there is a movement to adopt a new set of group designations as an international standard; in this table the groups are simply numbered 1 through 18. We shall use the "A/B table" in this text. Using this system, chemists often designate the *A groups* as **main group elements** and *B groups* as **transition elements**.

The horizontal rows of the table are called **periods**, and they are numbered beginning with 1 for the period containing only H and He.

Metals
Metalloids
Nonmetals

Alkali metals | Alkaline earth metals | Transition metals | Main group metals

1A	2A	3B	4B	5B	6B	7B	8B	8B	8B	1B	2B	3A	4A	5A	6A	7A	8A
1 H																1 H	2 He
3 Li	4 Be											5 B	6 C	7 N	8 O	9 F	10 Ne
11 Na	12 Mg											13 Al	14 Si	15 P	16 S	17 Cl	18 Ar
19 K	20 Ca	21 Sc	22 Ti	23 V	24 Cr	25 Mn	26 Fe	27 Co	28 Ni	29 Cu	30 Zn	31 Ga	32 Ge	33 As	34 Se	35 Br	36 Kr
37 Rb	38 Sr	39 Y	40 Zr	41 Nb	42 Mo	43 Tc	44 Ru	45 Rh	46 Pd	47 Ag	48 Cd	49 In	50 Sn	51 Sb	52 Te	53 I	54 Xe
55 Cs	56 Ba	57 *La	72 Hf	73 Ta	74 W	75 Re	76 Os	77 Ir	78 Pt	79 Au	80 Hg	81 Tl	82 Pb	83 Bi	84 Po	85 At	86 Rn
87 Fr	88 Ra	89 †Ac	104 Unq	105 Unp													

*Lanthanide series	58 Ce	59 Pr	60 Nd	61 Pm	62 Sm	63 Eu	64 Gd	65 Tb	66 Dy	67 Ho	68 Er	69 Tm	70 Yb	71 Lu
†Actinide series	90 Th	91 Pa	92 U	93 Np	94 Pu	95 Am	96 Cm	97 Bk	98 Cf	99 Es	100 Fm	101 Md	102 No	103 Lr

Figure 2.3 The periodic table. Elements are listed in ascending order of atomic number. The following points are important:

(a) Metals are shown in various shades of blue, the metalloids in green, and the nonmetals in yellow.

(b) Periods are horizontal rows of elements, and groups are vertical columns.

(c) Special names of groups:
Group 1A = alkali metals
Group 2A = alkaline earth metals
Group 7A = halogens
Group 8A = rare gases

The table can be divided into several regions. There are **metals** (blue in Figure 2.3), **nonmetals** (in yellow), and elements that fit neither of these categories entirely and so are called **metalloids** (in green). Although we shall more carefully describe the difference between metals and nonmetals later, you should be familiar with it from your everyday experience. Automobiles are made of iron (Fe) and aluminum (Al), both metals. Oxygen and nitrogen are the gases you breathe, and both are nonmetals. All the nonmetals in the periodic table lie to the right of the heavy zigzag line (between Al and Si, Ge and As, Sb and Te, and Po and At). Most of the elements lying immediately next to this line have some properties that can be thought of as typically metallic, but others that are considered characteristic of nonmetals. Such elements (B, Si, Ge, As, Sb, and Te) are called metalloids.

Many of the groups of the periodic table or collections of groups have meaningful names, and you should learn them.

GROUP 1A: ALKALI METALS

The ancient Arab chemists studied the properties of many natural substances and found that the ashes, called *al-qali*, of certain plants gave water solutions that felt slippery on the hands. It is now known that these substances contain compounds of Group 1A elements. Since these elements are also metals, they are often called the alkali metals (Figure 2.4).

FIGURE 2.4 SOME ELEMENTS OF THE PERIODIC TABLE

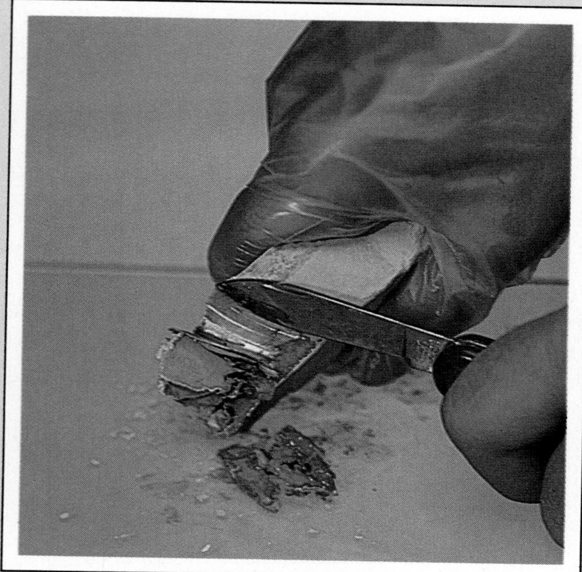

Group 1A: Sodium (Na).

Group 2A: Magnesium (Mg) (left) and calcium (Ca) (right).

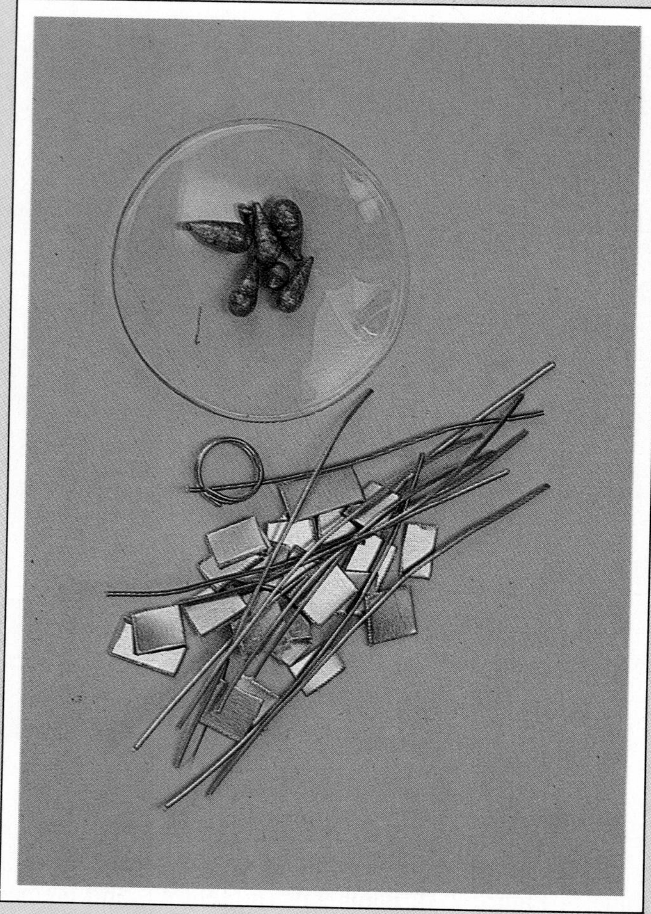

Group 3A: Indium (In) (top) and aluminum (Al) (bottom).

Some fourth period transition metals (left to right) Ti, V, Cr, Mn, Fe, Co, Ni, Cu.

Group 2B: Zinc (Zn) (left) and mercury (Hg) (right).

Group 4A: Carbon (C) (bottom), silicon (Si) (left middle), tin (Sn) (right middle), lead (Pb) (top).

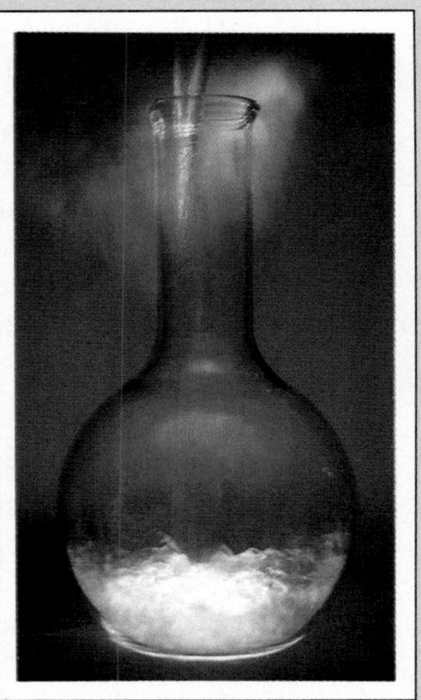

Group 5A: Nitrogen (N_2) [liquid N_2].

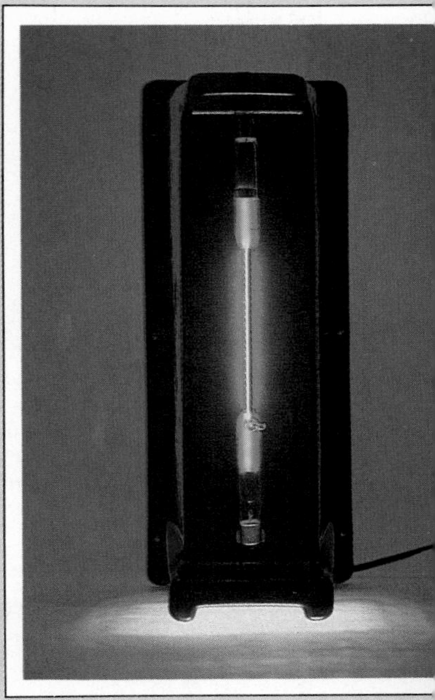

Group 8A: Neon (Ne).

Group 5A: White phosphorus (P_4).

Group 7A: Bromine (Br_2) (left) and iodine (I_2) (right). ►

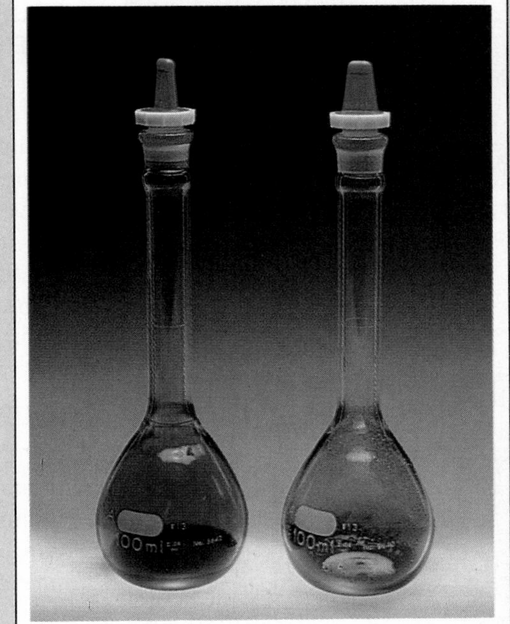

Group 5A: Arsenic (As) (left), antimony (Sb) (right), bismuth (Bi) (top).

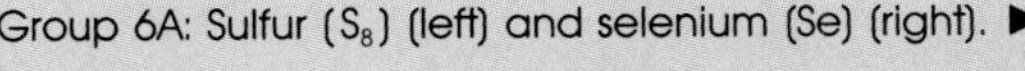

Group 6A: Sulfur (S_8) (left) and selenium (Se) (right). ►

Figure 2.5 Samples of aquamarine. This mineral contains the Group 2A metal beryllium, the Group 3A metal aluminum, the Group 4A metalloid silicon, and the Group 6A nonmetal oxygen.

Lithium is the lightest element of the alkali metals. Lithium-based greases are widely used, lithium hydroxide is used in batteries, and lithium carbonate is used to treat manic-depressive psychosis. Lithium is not an abundant element, but sodium is the seventh most abundant element in the earth's crust, and potassium is the eighth most abundant. The heaviest element, francium, has an unstable nucleus and has only a fleeting existence in nature.

GROUP 2A: ALKALINE EARTH METALS

These elements can also form substances that are alkaline in water solution. The lightest element, beryllium, is very low in abundance, and its compounds can be carcinogenic (Figure 2.5). Again like Group 1A, the next two elements, magnesium and calcium, are much more abundant; they are the sixth and fifth most abundant in the earth's crust, respectively. Calcium is particularly well known, since it occurs in vast deposits of limestone (Figure 2.6). The basic chemical constituent of limestone, calcium carbonate, is also the chief substance in corals, sea shells, and chalk. Radium, the heaviest element, has an unstable nucleus.

Figure 2.6 Minerals containing the alkaline earth element calcium.

GROUPS 3B THROUGH 8B AND 1B AND 2B: TRANSITION ELEMENTS OR TRANSITION METALS

All the elements in these groups are metals and range from some of the most abundant elements (iron is the second most abundant metal, after aluminum, and the fourth most abundant element) to some of the least abundant. All have unique properties. Silver, gold, and platinum are coveted for their beauty as well as their economic value.*

Included here are the "inner transition elements," shown as two separate rows below the main table. The lanthanide elements fit into the table between lanthanum (La) and hafnium (Hf). The actinide elements, as their name implies, then fit between actinium (Ac) and element 104.

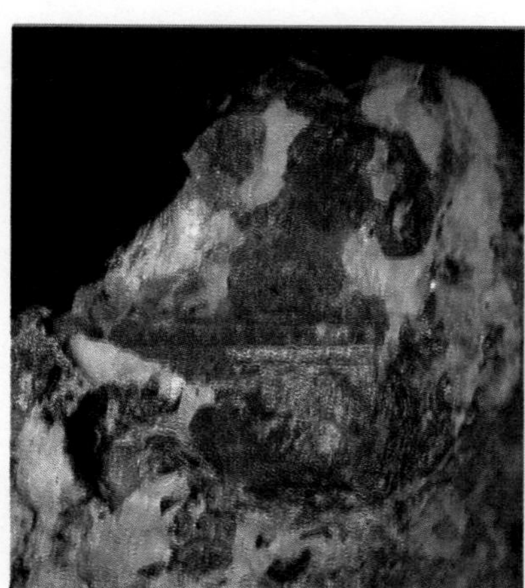

Figure 2.7 The red crystals in the photo are rubies, which consist of aluminum oxide (corundum) contaminated with chromium oxide.

GROUP 3A

This group has no special name, but it contains aluminum, the most abundant metal in the earth's crust (8.3%). Aluminum is exceeded in abundance only by the nonmetal O (45.5%) and the metalloid Si (25.7%) (Figure 2.7). These three elements are found combined in clays and other common minerals.

GROUP 4A

The lightest element of this group is carbon, an element found uncombined as graphite and diamonds. In combined form, it is the basis of the substances of which we are all made. It is 17th in abundance in the earth's crust, and it is found in the crust mainly in the form of carbonates (Figure 2.6).

---

* For more about platinum see SOMETHING MORE ABOUT. . . at the end of this chapter.

Silicon, a metalloid, is the second most abundant element in the earth's crust, occurring mainly in combination with oxygen. Together, silicon and oxygen account for four out of five of all atoms available near the surface of the earth! You know it in the form of silicon-oxygen materials such as sand, quartz, clay, and mica (Figure 2.8).

As in the previous groups, the heaviest elements, germanium, tin, and lead, are the least abundant. However, tin and lead have been known for centuries, since they are easily obtained from their ores. Tin mixed with copper makes bronze, a material used for utensils and weapons for centuries. The Romans used lead extensively for water pipes and plumbing.

Figure 2.8 Amethyst and citrine are impure forms of silicon dioxide.

GROUP 5A

The lightest element of Group 5A, nitrogen, makes up about three fourths of the atmosphere, but in the earth's crust it is only about as abundant as gallium. Phosphorus has a fascinating chemistry and is essential for life as a key element in teeth and bones.

GROUP 6A

Oxygen is the lightest element of this group and is the key to life as we know it on this planet. Sulfur has been known since ancient times as brimstone, the "burning stone" of the Devil. The element has enormous economic importance largely because it is the basis of sulfuric acid, the chemical produced in the largest amount in industry.

GROUP 7A: HALOGENS

The name for this group of elements comes from the Greek words *hals* for "salt" and *genes* for "forming." That is, the elements are salt-forming, and the best known of these is common table salt, sodium chloride. There are many others, however, because these elements react with all the metals and nearly all the other elements in the periodic table.

You may know fluorine as part of the compound added to toothpaste (sodium fluoride) and to many municipal water supplies, and as an important constituent of Teflon® (Figure 2.9).

Chlorine is the most abundant of the elements of this group and has been known since ancient times as a constituent of common sea salt. The remaining elements, bromine, iodine, and astatine, are much lower in abundance. Even so, iodine is essential to a hormone that regulates growth, the reason that table salt is "iodized." Astatine, like many of the heaviest elements, has an unstable nucleus and occurs in negligible amounts in nature.

Figure 2.9 Topaz is a mineral containing the Group 7A element fluorine as well as aluminum, silicon, and oxygen.

GROUP 8A: THE RARE GASES

As their name implies, these elements, helium, neon, argon, krypton, xenon, and radon, are all gases and all are low in their abundance on the earth. They are often called "inert" gases or "noble" gases, although krypton and xenon have been found to form compounds with certain other elements (F, N, and O).

EXAMPLE 2.6

THE PERIODIC TABLE

Is barium a metal, nonmetal, or metalloid? In what group is the element found? In what period? What is the common name of the group in which barium is found?

Solution Barium, element number 56, is a metal. It is found in Group 2A and in the sixth period (the sixth row). The elements of Group 2A are commonly referred to as the alkaline earth metals.

EXERCISE 2.6 The Periodic Table
Is bromine a metal, nonmetal, or metalloid? In what group is the element found? In what period? What is the common name of the group in which bromine is found?

2.6 THE MOLE

One of the most exciting parts of chemistry is the discovery of some new substance when one element is mixed with another. But chemistry is also a quantitative science. When two chemicals are mixed, you would like to know how many atoms of each are used. This means that there must be some method of counting atoms, no matter how small they are. The solution to this problem is to have a convenient unit of matter that contains a known number of particles. The "chemical counting unit" that has come into use is the **mole**.

The word "mole" was apparently introduced in about 1896 by Wilhelm Ostwald, who derived the term from the Latin word *moles* meaning a "heap" or "pile." The mole, whose symbol is **mol**, is the SI base unit for measuring *amount of pure substance* (Table 1.1). It is defined as follows:

Figure 2.10 Many things around us are used in units. For example, we have pairs of gloves, six-packs of soft drinks, or dozens of donuts. Chemicals are also used in units, the unit being called the mole.

> A **mole** is the amount of pure substance that contains as many particles (atoms, molecules, or other fundamental units) as there are atoms in *exactly* 12 grams of the carbon-12 isotope.

The key to understanding the concept of the mole is that *it always contains the same number of particles, no matter what the substance*. But how many particles? Many, many experiments over the years have established that number as

$$1 \text{ mole} = 6.0221367 \times 10^{23} \text{ particles}$$

Historical Figures in Chemistry: Amedeo Avogadro (1776–1856)

Avogadro

Lorenzo Romano Amedeo Carlo Avogadro, an Italian, was educated as a lawyer and practiced the profession for many years. However, in about 1800 he turned as well to science and was the first professor in Italy in mathematical physics. It was in 1811 that he first suggested the hypothesis, which we know now to be a law, that "equal volumes of gases under the same conditions have equal numbers of molecules." From this eventually came the concept of the mole. Avogadro did not see the acceptance of his ideas in his own lifetime, and it was not until another Italian, Stanislao Cannizzaro, discussed them at the great conference in Karlsruhe, Germany, in 1860 that scientists were convinced.

One of the great difficulties presented by Avogadro's number is comprehending its size. It may help to write it out in full as

$$6.022 \times 10^{23} = 602{,}200{,}000{,}000{,}000{,}000{,}000{,}000$$

or as

$$602{,}200 \times 1 \text{ million} \times 1 \text{ million} \times 1 \text{ million}$$

But, think of it this way: If you have Avogadro's number of popcorn kernels, and pour them over the continental United States, the country would be covered to a depth of 9 miles! Or, if you divide one mole of pennies equally among every man, woman, and child in the United States, one person could pay off the national debt (currently about \$2.9 trillion or 2.9×10^{12}) and still have over 20 trillion dollars left for incidental expenses.

How is Avogadro's number determined? It is clearly not possible to count all of the atoms in a mole. (If a computer counted 10 million atoms per second, it would take about 2 billion years to count all of the atoms in a mole.) There are in fact at least four experimental ways to do it, one of them being the measurement of the dimensions of the smallest repeating unit in a solid element (see Chapter 13).

This number is commonly known as **Avogadro's number** in honor of Amedeo Avogadro, an Italian lawyer and physicist, who conceived the basic idea.*

The mole is the chemist's six-pack or dozen. Many objects in our everyday lives come in similar counting units. Shoes, socks, and gloves are sold by the pair, soft drinks by the six-pack, and eggs and donuts by the dozen (Figure 2.10). Atoms are used by the mole. Thus, the mole is just a counting unit.

The atomic mass scale is a relative scale, with the ^{12}C atom chosen as the standard. From experiment we know that a ^{16}O atom is 1.33 times heavier than a ^{12}C atom and a ^{19}F atom is 1.58 times heavier than ^{12}C.

*There is nothing "special" about the value of Avogadro's number, 6.0221367×10^{23}. The number was established as this value because scientists chose to say that 1 mole of carbon is exactly 12 grams of carbon-12. If 1 mole were exactly 10 grams of carbon, then Avogadro's number would have a different value. It is interesting to point out that Avogadro's number was just revised to the value given (from 6.022045×10^{23}) about four years ago as a result of new and better measurements. Even the most fundamental values and ideas of science are under constant scrutiny.

Now, since a mole of carbon-12 has a mass of exactly 12 grams and contains 6.0221367×10^{23} atoms, and since *a mole of one kind of atom always contains the same number of particles* (6.0221367×10^{23}) *as a mole of another kind of atom*, this means a mole of ^{16}O atoms has a mass in grams of 1.33 times 12.0 g or 16.0 g. Similarly, a mole of ^{19}F atoms is 1.58 times greater than 12.0 grams or 19.0 grams.

The mass of 1 mole of atoms of any element (or 6.0221367×10^{23} atoms of that element) is called the **molar mass**. Molar mass is abbreviated with a capital italicized $\boldsymbol{M}$, it has units of "g/mol," and it is *numerically equal to the atomic weight in atomic mass units*. Thus, based on experiments that take into account all isotopes of a particular element,

$$\begin{aligned}\text{Molar mass of oxygen (O)} &= \text{mass of 1 mole of O atoms}\\ &= \text{mass of } 6.0221367 \times 10^{23} \text{ atoms}\\ &= 15.9994 \text{ g/mol}\end{aligned}$$

$$\begin{aligned}\text{Molar mass of fluorine (F)} &= \text{mass of 1 mole of F atoms}\\ &= \text{mass of } 6.0221367 \times 10^{23} \text{ atoms}\\ &= 18.9984 \text{ g/mol}\end{aligned}$$

$$\begin{aligned}\text{Molar mass of lead (Pb)} &= \text{mass of 1 mole of Pb atoms}\\ &= \text{mass of } 6.0221367 \times 10^{23} \text{ atoms}\\ &= 207.2 \text{ g/mol}\end{aligned}$$

Figure 2.11 shows 1-mole quantities of some common elements. Although each of these "piles of atoms" has a different volume and different mass, each contains 6.022×10^{23} atoms.

According to John Dalton, atoms combine with one another in the ratio of small whole numbers. A modern interpretation of this observation is that atoms combine with one another 1 mole for 1 mole, or 1 mole for 2 moles, or 2 moles for 3 moles, and so on. This is a matter taken up in Chapters 4 and 5. However, the important point to make here is that *the*

Figure 2.11 One-mole quantities of common elements. The graduated cylinders contain (left to right) mercury (200.59 g), lead (207.2 g), and copper (63.546 g). The Erlenmeyer flask at the left contains sulfur (32.066 g), and the one on the right contains magnesium (24.305 g). The watch glass holds chromium (51.996 g). All rest on sheets of aluminum foil (26.98 g).

mole concept is the cornerstone of quantitative chemistry. Therefore, you *must* learn to make the conversions "moles → mass" and "mass → moles." Dimensional analysis (Section 1.6) tells you that this can be done in the following way:

Gram ↔ Mole Conversion

Moles → Grams	**Grams → Moles**
Operation	*Operation*
Multiply moles by molar mass	Divide grams by molar mass
Unit Analysis	*Unit Analysis*
$\cancel{\text{moles}}\left(\frac{\text{grams}}{\cancel{\text{mole}}}\right) = \text{grams}$	$\cancel{\text{grams}}\left(\frac{\text{mole}}{\cancel{\text{grams}}}\right) = \text{moles}$
↑	↑
molar mass	*1/molar mass*

80.2 g of sulfur in a 250-mL beaker.

If you look at the table in the front of the book, you will notice that some molar masses are known to more significant figures and decimal places than others. When using molar masses in this book, *we shall use one more significant figure in the molar mass than in any of our other data*. Thus, if we have a problem that says we have 16.5 grams of carbon, we would use 12.01 g/mol for the molar mass of C to find the number of moles of carbon present.

EXAMPLE 2.7

MOLES TO GRAMS

How many grams are there in 2.50 moles of sulfur (S)?

Solution For a conversion between mass and moles, you will always need the molar mass. The tables in the front of the book give 32.07 g/mol for S. Thus, the number of grams of sulfur in 2.50 moles is

$$2.50\ \cancel{\text{mol}}\ \text{sulfur}\left(\frac{32.07\ \text{g}}{1\ \cancel{\text{mol}}}\right) = 80.2\ \text{g sulfur}$$

The top photograph in the margin shows 80.2 g of S in a 250 mL-beaker.

EXAMPLE 2.8

GRAMS TO MOLES

How many moles are represented by 1.00 pound of silicon, an element used in semiconductors? (1.00 pound = 454 g)

Solution The molar mass of silicon is 28.09 g/mol, so the number of moles of silicon is

$$454\ \cancel{\text{g}}\ \text{silicon}\left(\frac{1\ \text{mol}}{28.09\ \cancel{\text{g}}}\right) = 16.2\ \text{mol silicon}$$

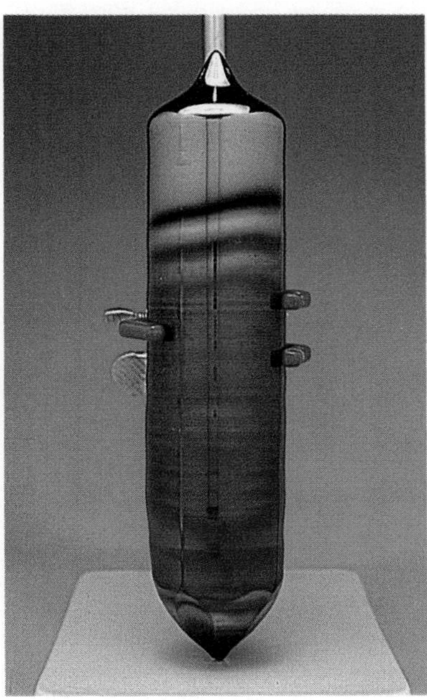
A bar of elemental silicon, approximately 10 inches long.

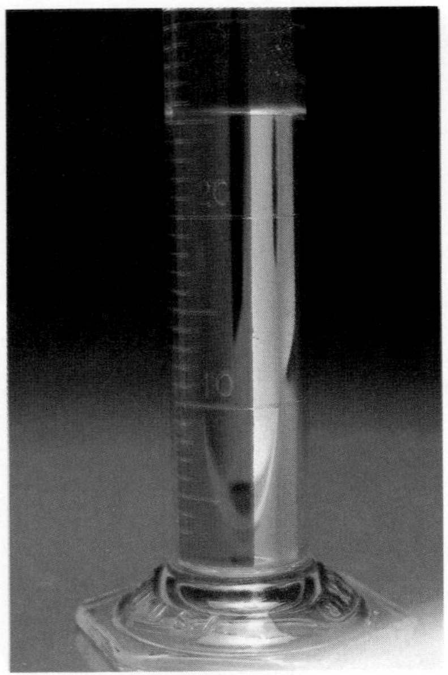
25.4 mL of liquid mercury.

EXAMPLE 2.9
MOLE CALCULATION

The graduated cylinder in the photograph in the margin contains 25.4 mL of mercury. If the density of mercury at 25 °C is 13.534 g/cm^3, how many moles of mercury are there in the cylinder?

Solution The molar mass of mercury, 200.59 g/mol, allows us to make conversions between mass and moles. Since the information we have about the mercury sample is its volume, we first have to convert mL of mercury into grams of mercury, and the conversion factor for this is density. Remember that in dimensional analysis, the operation is

$$\text{Given data}\left(\frac{\text{desired units}}{\text{units of given data}}\right) = \text{answer in desired units}$$

$$25.4\ \cancel{\text{cm}^3}\left(\frac{13.534\ \text{g}}{1\ \cancel{\text{cm}^3}}\right) = 344\ \text{g of mercury}$$

$$344\ \cancel{\text{g}}\ \text{mercury}\left(\frac{1\ \text{mol}}{200.6\ \cancel{\text{g}}}\right) = 1.71\ \text{mol of mercury}$$

EXERCISE 2.7 Gram/Mole Conversions

(a) How many grams are contained in 2.5 moles of aluminum? (b) How many moles are represented by 1.00 pound of lead?

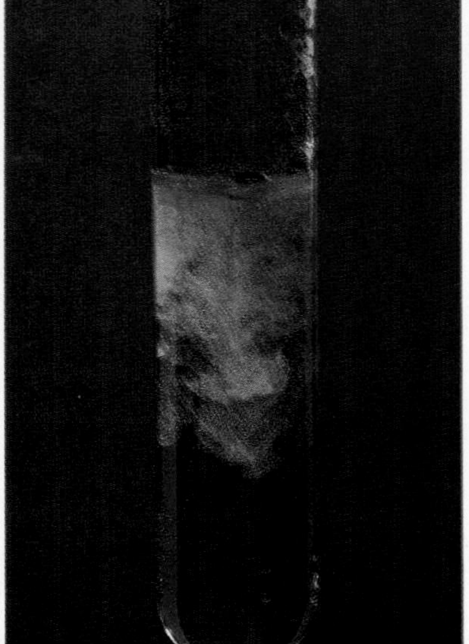
Sodium reacts vigorously with water as evidenced by the bubbles of H_2 in the water. The heat of the reaction also melts the sodium, and it floats as a small ball on top of the water.

EXERCISE 2.8 Gram/Mole Conversions Involving Density

In an experiment, you need 0.125 mole of sodium metal. Sodium can be cut easily with a knife, so if you cut out a block of sodium, what should the volume of the block be in cubic centimeters? If you cut a perfect cube, what will be the length of the edge of the cube? (The density of sodium is 0.968 g/cm^3.) (**Caution:** Sodium is *very* reactive with water. Therefore, sodium metal should be handled only by a knowledgeable chemist.)

Once the number of moles represented by a given mass of an element is known, you can determine the number of atoms contained in the sample.

EXAMPLE 2.10
THE NUMBER OF ATOMS IN A SAMPLE

How many atoms of mercury are there in the graduated cylinder in the photograph for Example 2.9?

Solution In Example 2.9 we calculated that 25.4 mL of mercury is equivalent to 1.71 mol of the metal. Since we know the relation between atoms and moles (Avogadro's number), we can find the number of atoms in our sample.

$$1.71\ \cancel{\text{mol}}\left(\frac{6.0221367 \times 10^{23}\ \text{atoms}}{1\ \cancel{\text{mol}}}\right) = 1.03 \times 10^{24}\ \text{atoms of Hg}$$

EXERCISE 2.9 The Number of Atoms in a Sample
How many atoms of uranium are there in 50.0 g of the element?

Now that you are more familiar with the mole concept and Avogadro's number, it is interesting to see what one atom weighs. The following example shows you that it is indeed an extremely small mass, just as you have suspected.

EXAMPLE 2.11

THE MASS OF AN ATOM

Calculate the mass (in grams) of one atom of mercury.

Solution You know that 1 mole of mercury has a mass of 200.59 g/mol and this represents 6.0221367×10^{23} atoms. Therefore, you know the relation between mass and number of atoms.

$$\left(\frac{200.59 \text{ g}}{1 \cancel{\text{mol}}}\right)\left(\frac{1 \cancel{\text{mol}}}{6.0221367 \times 10^{23} \text{ atoms}}\right) = 3.3309 \times 10^{-22} \text{ g/atom}$$

EXERCISE 2.10 The Mass of an Atom
Calculate the mass of one hydrogen atom, the lightest of all of the elements.

SUMMARY

An **element** is a form of matter that cannot be separated into simpler chemical substances (Section 2.1). Each of the 108 elements known has a name and a symbol and is listed in the periodic table in the front of the book (and in Figure 2.3).

An atom is the smallest particle of an element that retains the chemical properties of the element (Section 2.2). Each atom is composed of many subatomic particles, the most important for chemistry being the **protons** and the **neutrons** in the nucleus and the **electrons** in space about the nucleus. The **atomic number (Z)** of an element corresponds to the number of protons in the nucleus (and to the number of electrons outside the nucleus). *All atoms of the same element have the same atomic number.*

Masses of atoms are given on a relative scale based on the **atomic mass unit (amu)**, where 1 amu is exactly $\frac{1}{12}$ the mass of a carbon atom having 6 protons and 6 neutrons. The **mass number (A)** of a particular atom is the sum of the number of protons and neutrons in the nucleus. The composition of an atom, X, can be given by the symbol $^{A}_{Z}X$, although the subscript Z is optional. **Isotopes** are atoms having the same atomic number but different mass numbers (Section 2.3). Almost every element has more than one isotope. The weighted average of the isotopic masses, expressed in atomic mass units, is the **atomic weight** of the element (Section 2.4).

If the elements are arranged in order of increasing atomic number, as in the **periodic table** in the front of the book (and in Figure 2.3), we find a regular repetition of properties (Section 2.5). Each vertical column is a **group** of elements having similar properties. (The horizontal rows are called **periods**.) Each group of the table is given a number and letter designation. Groups labeled A are sometimes called the **main group elements**, while B groups are the **transition elements** or transition metals.

Most of the known elements are **metals**. In the periodic table, the metallic elements are to the left, while the **nonmetals** are at the right. A zigzag line beginning at boron, and running to Po and At, roughly separates the metals and nonmetals. The elements along this line share properties of both types of elements and are often called the **metalloids**.

Elements of Group 1A are called the **alkali metals**, those of Group 2A are the **alkaline earth metals**, Group 7A elements are the **halogens**, and Group 8A elements are called the **rare gases**.

The **mole** (Section 2.6) is the amount of substance that contains as many particles (which is **Avogadro's number, 6.0221367×10^{23}**) as there are atoms in exactly 12 grams of carbon-12. The mass of 1 mole of an element in grams is the **molar mass**; it is numerically equal to its weight in atomic mass units. The mole is the cornerstone of quantitative chemistry, so it is *extremely important* to be able to do the conversions "mass → moles" and "moles → mass."

$$\text{Mass} \rightarrow \text{Moles:} \qquad \text{Mass in grams}\left(\frac{1\ \text{mole}}{X\ \text{grams}}\right) = \text{moles}$$

$$\text{Moles} \rightarrow \text{Mass:} \qquad \text{Moles}\left(\frac{X\ \text{grams}}{1\ \text{mole}}\right) = \text{mass in grams}$$

SOMETHING MORE ABOUT Chemical Elements: Platinum

Some have called it the philosopher's stone, the mythical substance that medieval alchemists thought would change lead into gold. What other element but platinum can be used to make wood stoves burn more efficiently, change the gases in automobile exhaust into less polluting chemicals, remove the ozone from the air inside high-flying passenger jets, and help turn ammonia into nitrates, compounds that are widely used as fertilizers.

The sands of the rivers along the west coast of South America are rich in gold, and platinum is often found along with it. It is not surprising that the natives of this continent for centuries used the elements to make ornaments, something we in the so-called civilized world have continued. When the Spanish explored Central and South America beginning in the 16th century, they found native treasures made of platinum and gave the metal its name from the word *platina*, which means "little silver" in Spanish and which tells you that the metal looks more like silver than gold.

The Spanish explorers were much more interested in gold than platinum. In fact, when they found platinum in the rivers along with gold, they believed at first that the platinum was gold that had not been buried long enough to ripen into gold, and they threw the platinum back into the rivers!

Platinum did not find its way to Europe until the 18th century. In 1736 a Spanish naval officer, de Ulloa, observed platinum in the gold mines of South America. On his way back to Spain in 1745, however, his ship was attacked by privateers and eventually captured by the British Navy. He and his papers de-

The worker in the photo is installing a layer of platinum/rhodium wire gauze in a reactor for the production of nitric acid from ammonia. The most important end product is fertilizer. (Johnson-Matthey)

scribing platinum were taken to London, where members of the Royal Society became aware of his work and actually elected him a member of the Society.

Because of its color, beauty, and value, platinum was known as "white gold" by its early European discoverers.* However, it soon became apparent that it was much more difficult to work with than gold, in part because the melting point of platinum (1769 °C) is so much higher than those of other metals. Many attempts were made to melt platinum, but the French chemist Lavoisier was apparently the first to achieve the feat on a small scale in 1782. It was not until late in the last century that larger amounts could be melted and that it became possible to work platinum metal into useful products.

Platinum, which is about as abundant as gold in the earth's crust, is usually found with other metals of the "platinum group": ruthenium and osmium, rhodium and iridium, and palladium. Until about 1820 all platinum came from South America, but then deposits were discovered in Russia and later in South Africa. The greatest amount now comes from the mines of South Africa. Mining the metal is quite a feat: roughly 10 tons of rocks must be removed to isolate 1 troy ounce (31.1 grams).

Consumption of platinum in the United States is annually about 1,300,000 troy ounces or about 45 tons. Most, by far, is used in emission control catalysts in automobiles, but the next most important use is in the chemical industry. In 1989 approximately 16 billion pounds of nitric acid were produced in the United States in a process that relies on the use of platinum as a so-called "catalyst" to accelerate a crucial step in the process.†

Since platinum has a high melting point and is not easily corroded, it is used to make vessels for high-temperature processes. For example, glass fibers are made by forcing molten glass through many small holes in a sheet of platinum-gold alloy.

Glass fibers are formed by drawing molten glass through hundreds of small openings in a plate made of a platinum-gold alloy. (Owens-Corning Fiberglas Company)

Platinum and the metals to which it is related have an extraordinarily rich and interesting chemistry. Many of their compounds act as catalysts for chemical reactions, but the newest and most exciting use of a platinum-containing compound is the cancer chemotherapy agent *cisplatin*.‡

Cisplatin is a chemical compound that has been known for at least 100 years. The use of the compound in fighting certain types of cancers was discovered serendipitously in the 1970s by Dr. Barnett Rosenberg of Michigan State University, and he describes the important experiments and the events leading up to the discovery in the interview at the beginning of this section of the book.

*The term "white gold" is now applied to any alloy of gold and palladium, the latter a metal quite similar to platinum.

†A catalyst is a substance that can accelerate a chemical reaction without being consumed in the reaction. Such substances are discussed in more detail in Chapter 15.

‡The chemical formula for cisplatin is $(H_3N)_2PtCl_2$. Chemical formulas will be explored in detail in Chapter 3.

STUDY QUESTIONS

REVIEW QUESTIONS

1. Explain the difference between an element and an atom.
2. What are the three primary particles from which atoms are built? What are their electrical charges? Which of these particles occur in the nucleus of the atom? Which is the least massive particle?
3. How can you determine the atomic number of an element?
4. What is the atomic mass unit?
5. What is the difference between the mass number and the atomic number of an atom?
6. What are isotopes?
7. Inspecting the values of atomic numbers and atomic weights in the periodic table at the front of the book, what general conclusion can you reach concerning the relative numbers of protons and neutrons in atoms? That is, are there in general as many neutrons as protons, more neutrons than protons, or fewer neutrons than protons?
8. Lithium has two stable isotopes, ^{6}Li and ^{7}Li. Which is the more abundant?
9. What is the difference between the mass number of an atom and the atomic weight of an element?
10. What is the difference between a group and a period in the periodic table?
11. Name three transition elements, two halogens, and one alkali metal.
12. Name two metals, one metalloid, and one nonmetal.
13. Define the terms "mole" and "molar mass" as used in chemistry.
14. If you have 12.011 g of carbon (C) and 24.305 g of magnesium (Mg), do you have approximately twice as many atoms of Mg as atoms of C? Explain briefly.
15. Does an atom of zinc (Zn) have about the same mass as an atom of sulfur (S), about twice the mass, or about half the mass?

ELEMENTS AND ATOMS

16. Give the name of each of the following elements:
(a) C (b) Na (c) Cl (d) P (e) Mg (f) Ge
17. Give the name of each of the following elements:
(a) Mn (c) K (e) As (g) W
(b) F (d) Hg (f) Xe (h) Ra
18. Give the symbol for each of the following elements:
(a) lithium (d) silicon
(b) titanium (e) gold
(c) iron (f) lead
19. Give the symbol for each of the following elements:
(a) silver (e) tin
(b) aluminum (f) barium
(c) plutonium (g) krypton
(d) cadmium (h) ruthenium
20. Element 25 is found in the form of oxide "nodules" at the bottom of the sea. What is the name and symbol of this element?
21. The volcanic eruption of Mt. St. Helens in Washington produced a considerable quantity of a radioactive element in the gaseous state. The element has an atomic number of 86. What are the symbol, name, and atomic weight of this element?
22. Give the mass number of each of the following atoms: (a) beryllium with 5 neutrons, (b) titanium with 26 neutrons, and (c) gallium with 39 neutrons.
23. Give the mass number of (a) an iron atom with 30 neutrons, (b) an americium atom with 148 neutrons, and (c) a tungsten atom with 110 neutrons.
24. Give the complete symbol ($^{A}_{Z}X$) for each of the following atoms: (a) sodium with 12 neutrons, (b) argon with 21 neutrons, and (c) gallium with 39 neutrons.
25. Give the complete symbol ($^{A}_{Z}X$) for each of the following atoms: (a) nitrogen with 8 neutrons, (b) zinc with 34 neutrons, and (c) xenon with 75 neutrons.
26. How many electrons, protons, and neutrons are there in (a) calcium-40, ^{40}Ca; (b) tin-119, ^{119}Sn; and (c) plutonium-244, ^{244}Pu?
27. How many electrons, protons, and neutrons are there in (a) carbon-13, ^{13}C; (b) chromium-50, ^{50}Cr; and (c) bismuth-205, ^{205}Bi?
28. Fill in the columns of blanks in the table (one column per element).

Symbol	^{45}Sc	^{33}S	_____	_____
Number of protons	_____	_____	8	_____
Number of neutrons	_____	_____	9	31
Number of electrons in the neutral atom	_____	_____	_____	25

29. Fill in the columns of blanks in the table (one column per element).

Symbol	^{65}Cu	^{37}Cl	_____	_____
Number of protons	_____	_____	34	_____
Number of neutrons	_____	_____	46	46
Number of electrons in the neutral atom	_____	_____	_____	36
Name of element	_____	_____	_____	_____

ISOTOPES AND ATOMIC WEIGHT

30. Of the following atoms, which are isotopes of the same element?
(a) $^{20}_{10}X$ (b) $^{22}_{11}X$ (c) $^{22}_{10}X$ (d) $^{19}_{9}X$ (e) $^{21}_{10}X$
31. Neon has two major isotopes. One of these isotopes has 10 neutrons and the other has 12. What are their mass numbers? Give the symbol ($^{A}_{Z}X$) for each isotope.

32. A natural sample of gallium consists of two isotopes with masses of 68.95 amu and 70.95 amu and with abundances of 60.16% and 39.84%, respectively. What is the average atomic weight of gallium?

33. Bromine, like chlorine, has two stable isotopes. For bromine the isotopic masses are 78.9183 amu (relative abundance, 50.69%) and 80.9163 amu (relative abundance, 49.31%). Calculate the average atomic weight of bromine.

34. Silicon is found in nature combined with oxygen to give sand, quartz, agate, and similar materials. The element has three stable isotopes.

EXACT MASS	PERCENT ABUNDANCE
27.97693	92.23
28.97649	4.67
29.97376	3.10

Calculate the average atomic weight of silicon from the data above.

35. The metallic element chromium has four stable isotopes.

EXACT MASS	PERCENT ABUNDANCE
49.9461	4.35
51.9405	83.79
52.9407	9.50
53.9389	2.36

Calculate the average atomic weight of chromium.

36. Antimony, one of the elements known to the ancient alchemists, has two stable isotopes: ^{121}Sb (mass, 120.90) and ^{123}Sb (mass, 122.90). Calculate the percent abundances of the two isotopes.

37. Magnesium is commonly extracted from seawater. Magnesium-24 is its most abundant isotope (78.70%); its exact mass is 23.985. If the atomic weight of magnesium is 24.305, what are the relative abundances of magnesium-25 (mass, 24.986) and magnesium-26 (mass, 25.983)?

38. The element with an atomic weight of 72.6 g/mol was the basis of some early transistors. State its name and symbol.

39. The element with an atomic weight of 47.88 g/mol is used, in the form of its oxide, as a white paint pigment. Give the name, symbol, and atomic number of the element.

40. About 54 billion pounds of element 7 were produced industrially in the United States in 1989. What are the name, symbol, and atomic weight of this element?

41. The element with an atomic weight of 78.96 amu is important to your health if ingested in very small amounts (90 micrograms/day); an excess causes loss of hair. What are the name, symbol, and atomic number of this element?

PERIODIC TABLE

42. Name all the elements of Group 1A of the periodic table. What is the common name for this group? Are most of these elements metals or nonmetals?

43. Name all the elements of Group 7A. What is the common name of the group? Are most of these elements metals or nonmetals?

44. Give the names for all the elements of the second period of the periodic table (the period beginning with Li).

45. Give the names for all the elements in the fourth period of the periodic table (the period beginning with K).

46. For each of the following elements, give its name and then (i) state whether it is a metal, nonmetal, or metalloid and (ii) identify its location in the periodic table by giving its group and period number.
(a) Ca (c) Si
(b) Cd (d) I

47. For each of the following elements, (i) state whether it is a metal, nonmetal, or metalloid and (ii) identify its location in the periodic table by giving its group and period number.
(a) F (c) Cs
(b) As (d) W

MOLES

48. Calculate the number of moles represented by each of the following:
(a) 127.08 g of Cu (d) 0.012 g of potassium
(b) 20.0 g of calcium (e) 5.0 mg of americium
(c) 16.75 g of Al

49. Calculate the number of moles represented by each of the following:
(a) 16.0 g of Na (d) 0.876 g of arsenic
(b) 0.0034 g of platinum (e) 0.983 g of Xe
(c) 1.54 g of P

50. A chunk of sodium metal, Na, if thrown into a bucket of water, produces a dangerously violent explosion from the reaction of sodium with water. If 50.4 g of sodium is used, how many moles of sodium does that represent?

51. Krypton really does not give Superman his strength. If you have 0.00789 g of the gaseous element, how many moles does this represent?

52. Calculate the number of grams in each of the following:
(a) 0.10 mole of iron
(b) 2.31 moles of Si
(c) 0.0023 mole of carbon
(d) 0.54 mole of sodium

53. Calculate the number of grams in each of the following:

(a) 6.03 moles of gold
(b) 0.045 mole of uranium
(c) 15.6 moles of neon
(d) 3.63×10^{-4} mole of plutonium

54. The aluminum foil in a package of kitchen foil weighs approximately 12 ounces. How many moles of aluminum will you get when you buy a package of this foil? (1 ounce = 28.35 g)

55. Gems and precious stones are measured in carats, a weight unit equivalent to 200. mg. If you have a 2.3 carat diamond in a ring, how many moles of carbon do you have? (A diamond is a very pure form of carbon.)

56. Chemicals are sometimes sold by the mole. For example, you can buy 0.055 mole of silver for one dollar. How many grams of silver will you get for your dollar?

57. The international markets in precious metals operate in the weight unit "troy ounce" (where 1 troy ounce is equivalent to 31.1 g). Platinum sells for $487 per troy ounce. (a) How many moles are there in one troy ounce? (b) If you have $5000.00 to spend, how many grams and how many moles of platinum can be purchased?

58. Gold prices fluctuate, depending on the international situation. If gold currently sells for $401.25 per troy ounce, how much must you spend to purchase 1.00 mole of gold? (1 troy ounce is equivalent to 31.1 g)

59. The Statue of Liberty in New York harbor is made of 2.00×10^5 pounds of copper sheets bolted to an iron framework. How many grams and how many moles of copper does this represent? (1 lb = 454 g)

60. Precious metals such as gold and platinum are sold in units of "troy ounces," where 1 troy ounce is 31.1 grams. If you have a block of platinum with a mass of 15.0 troy ounces, how many moles of the metal do you have? What is the size of the block in cubic centimeters? (The density of platinum is 21.45 g/cm³ at 20 °C.)

61. "Dilithium" is the fuel for the *Starship Enterprise*. Because its density is quite low, you will need a large space to store a large mass. As an estimate for the volume required, we shall use the element lithium. If you want to have 256 moles for an interplanetary trip, what must the volume of a piece of lithium be? If the piece of lithium is a cube, what is the dimension of an edge of the cube? (The density of lithium is 0.534 g/cm³ at 20 °C.)

AVOGADRO'S NUMBER

62. Assume you have 1.69 moles of aluminum. How many atoms of Al do you have?

63. You have 3.84×10^{-3} mol of argon gas. How many atoms of argon do you have?

64. If you have a 35.67-g piece of chromium metal on your car, how many atoms of chromium do you have?

65. If you have a ring that contains 1.94 g of gold, how many atoms of gold are there in the ring?

66. What is the mass of one copper atom?

67. What is the mass of one atom of titanium?

68. How many atoms of aluminum are there in a piece of foil that has a volume of 2.00 cm³? The density of aluminum is 2.702 g/cm³.

69. A piece of copper wire that is 25 feet long has a diameter of 2.0 mm. Copper has a density of 8.92 g/cm³. How many moles of copper and how many atoms of copper are there in the piece of wire?

GENERAL QUESTIONS

70. Arsenic is one of the few elements that has only one stable (nonradioactive) isotope. How many neutrons are there in an atom of this isotope of arsenic?

71. The average mass of one gold atom in a sample of naturally occurring gold is 3.2707×10^{-22} g. Use this to calculate the molar mass of gold.

72. Black gunpowder contains several chemicals, among them sulfur and carbon. A typical powder is about 10.09% S (by mass) and 14.29% C. If you have 1.00 pound (454 g) of gunpowder, how many grams of sulfur and how many grams of carbon are present? How many moles of each?

73. An average sample of coal contains about 3.0% by mass sulfur. How many moles of sulfur are there in 1.0 ton of coal? (1 T = 2000 lb; 1 lb = 454 g)

74. One of the events of a track and field meet is the shotput. If the shot is made of iron and it contains 7.83×10^{25} atoms of iron, could you lift it easily? (Calculate the weight of the shot in pounds and compare it with your lifting ability. One pound = 454 g.)

75. A 5-cent coin, the nickel, actually contains mostly copper. The metal used is 75% copper and 25% nickel by weight. If a 5-cent coin weighs 5.10 g, how many grams of copper and nickel does it contain? How many moles of each? How many atoms of each?

76. A piece of platinum foil 0.254 mm thick and 50.0 mm × 50.0 mm sells for $525.00. How many moles of platinum are you buying if the density of platinum is 21.45 g/cm³?

77. Scrap iron sells for $114.00 per ton. The density of iron is 7.86 g/cm³. (a) How many moles of iron can you buy for $1000.00? (b) If $1000.00 worth of iron is delivered to your home in the form of a cube, how long is one side of the cube? (1 T = 9.08×10^5 g)

78. Analysis of water from the Atlantic Ocean indicates that it contains, among other things, 45 tons of silver and 9.0 pounds of gold per cubic mile. (1 T = 2000 pounds) (a) How many grams of silver and gold are

there in a cubic mile? (b) How many moles of each metal?

79. Drinking water can contain many different chemicals in trace amounts. If a water sample contains 0.018 ppm of Hg (ppm = parts per million; 1 ppm = 1 g of Hg per 10^6 g of water), how many grams of mercury are you ingesting when you drink one cup of water (237 mL)? How many moles of mercury are you ingesting? (1.00 mL H_2O = 1.00 g)

80. Assume for the sake of this problem that a standard soft-drink or beer can measures $2 \times 2 \times 5$ inches. If you stack one mole of these cans over the United States, how deeply would the cans be stacked? (The land area of the continental United States is 3.6×10^6 miles^2.)

81. If all 250 million people in the United States were put to work counting the atoms in a mole of gold and if each person could count one atom per second day and night for 365 days a year, how many years would it take to finish the count?

82. To make a mirror for a telescope, you coat the glass with a thin layer of aluminum to reflect the light. If the mirror has a diameter of 6.0 inches, and you want to have a coating that is 0.015 mm thick, how many grams of aluminum will you need? How many atoms of aluminum are in the coating? The density of aluminum is 2.702 g/cm^3. The volume of the coating is given by π times the square of the mirror radius times the thickness of the coating (because the volume of a cylinder is $\pi r^2 h$).

83. In order to make computer chips, round rods of pure silicon are first cut into thin "wafers." If you want a wafer with a diameter of 5.00 inches that contains 1.8×10^{22} atoms of Si, how many millimeters thick must the wafer be? Silicon has a density of 2.33 g/cm^3. (The volume of a thin, cylindrical wafer is $\pi r^2 h$ where h is the thickness of the wafer.)

84. The structure of a pure solid can be viewed as spherical atoms stacked neatly into a regular array, and the solid can often be divided into tiny cubes each containing a given number of atoms. For example, a silicon crystal consists of cubes each containing eight spherical silicon atoms, and each cube is 543.1 picometers on a side. Use this information, along with the average molar mass of silicon (28.0854 g/mol) and its density (2.328994 g/cm^3), to determine Avogadro's number. (Note: This is how the number was actually determined. Recent, highly accurate measurements of the dimensions of silicon crystals gave a slightly revised value of Avogadro's number.)

85. **Crossword Puzzle:** In the 2×2 crossword below, each letter must be correct four ways: horizontally, vertically, diagonally, and by itself. Instead of words, use symbols of elements. When the puzzle is complete, the four spaces will contain the overlapping symbols of 10 elements. There is only one correct solution.

1	2
3	4

Horizontal

1–2: Two-letter symbol for a metal used in ancient times.

3–4: Two-letter symbol for a metal that burns in air and is found in Group 5A.

Vertical

1–3: Two-letter symbol for a metalloid.

2–4: Two-letter symbol for a metal used in U.S. coins.

Single squares: all one-letter symbols

1. A colorful nonmetal.
2. Colorless gaseous nonmetal.
3. An element that makes fireworks green.
4. An element that has medicinal uses.

Diagonal:

1–4: Two-letter symbol for an element used in electronics.

2–3: Two-letter symbol for a metal used with Zr to make wires for superconducting magnets.

This puzzle first appeared in *Chemical and Engineering News*, Dec. 14, 1987 (p. 86) (submitted by S.J. Cyvin). It was also published in *Chem Matters*, October 1988.

CHAPTER

3

Compounds and Molecules

Formation of a compound of copper and sulfur by heating a mixture of the elements.

CHAPTER OUTLINE

The clothes you wear and the food you eat consist of many kinds of chemical compounds. Cotton and wool fibers are composed of giant molecules, as are the substances in polyester and nylon. All are composed of carbon, hydrogen, oxygen, and other elements, organized in particular ways. Aspartame, a substitute for sugar in diet soft drinks, is a simple compound, also built of carbon, hydrogen, and oxygen, but nitrogen is an essential ingredient as well. Common table salt is a compound composed of sodium and chlorine, and many rocks and minerals are giant molecules of silicon, oxygen, and various metals such as sodium, aluminum, beryllium, and iron. Limestone contains calcium, carbon, and oxygen.

Chemistry is about the compounds you find all about you: about their composition and their chemical and physical properties. But it is also about making new compounds with properties that no one has seen before. Chemists often became chemists because they were fascinated by the colors of compounds and the shapes of molecules and by what determines these properties. We hope to show you some of this fascination.

In telling the story about chemical compounds, we shall follow the advice of the King of Hearts in *Alice's Adventures in Wonderland*. He said: "Begin at the beginning and go on until you come to the end: then stop." We shall begin with the composition of compounds. Then we can go on to tell you how they can react with one another, what their shapes are, how they are held together, and how they interact with their environment. At some point we will stop, but not because there is no more to say about chemistry; there just isn't time or space enough to say it all in one book.

(a)

(b)

Figure 3.1 The reaction of phosphorus and bromine. (a) Solid red phosphorus (in the evaporating dish) and liquid bromine, Br_2 (in the graduated cylinder). (b) When the bromine is poured onto the red phosphorus, a vigorous reaction produces phosphorus tribromide, PBr_3.

3.1 COMPOUNDS, MOLECULES, AND MOLECULAR FORMULAS

John Dalton said that **compounds** form by the combination of atoms in the ratio of small whole numbers. Now we know that the smallest unit of a compound that retains the chemical characteristics of the compound is a **molecule**. The composition of a molecule can be represented by a **molecular formula**, which expresses the number of atoms of each type within one molecule of the compound.

When compounds are formed directly from the elements or from other compounds, one striking feature is that the characteristics of the constituent elements are lost. This, and the concept of molecular formula, is illustrated in Figure 3.1. Red phosphorus reacts violently with bromine, a foul-smelling, red-orange liquid. The formula of the resulting compound, PBr_3, conveys the fact that there are four atoms per molecule: one atom of phosphorus and three atoms of bromine. The subscript to the right of the element's symbol indicates the number of atoms of that element in the molecule. If the subscript is omitted, it is understood to be one, as for P in PBr_3. Similarly, in the water molecule, H_2O, there is one atom of oxygen and two atoms of hydrogen.

Consider the molecular formula of a compound such as ammonium phosphate, a common fertilizer. Molecular formulas are sometimes written with the elements listed in alphabetical order, such as $H_{12}N_3O_4P$, but you would more commonly see this compound's formula written as $(NH_4)_3PO_4$. This latter version shows you immediately that the compound is composed of two different types of "chemical units," NH_4 and PO_4, and that they are present in the ratio of three NH_4 to one PO_4.* There

*In Section 3.3 you will see that these "units" are in fact the ions NH_4^+ and PO_4^{3-}.

are many common **polyatomic** or *many-atom* "units of atoms" in chemistry, and we shall always keep the atoms of those units together when writing formulas. Section 3.3 considers this in more detail.

The formula of a compound such as ethyl alcohol can also be written in at least two different ways. In both cases there are two C atoms, six H atoms and one O atom per molecule.

C_2H_6O	CH_3CH_2OH
ethyl alcohol molecular formula	molecular formula of ethyl alcohol in modified form

In the molecular formula on the left, the symbols of the elements are written in alphabetical order, each with a subscript indicating the total number of atoms of that type in the molecule.* On the right, we have written the formula in a modified form to show the grouping of atoms in the molecule. Chemists often do this to emphasize the chemically important units or "functional groups" (such as the OH group in ethyl alcohol) and to be able to quickly identify the molecule in a text.

EXAMPLE 3.1

MOLECULAR FORMULAS

Hydrogen and nitrogen can form two different chemical compounds. One of them, ammonia, has three H atoms per N atom. The other compound, hydrazine, is used as a fuel in Space Shuttle control rockets. It has four H atoms for every two N atoms. Write the formulas of these compounds.

Solution The atom ratio in an ammonia molecule is 3 H to 1 N, so the formula is NH_3. A hydrazine molecule has 4 H atoms and 2 N atoms, so its formula is N_2H_4. As you will see in Section 3.6, this is an example of the *law of multiple proportions*.

Notice that N appears before H in these formulas. Through long use it has become conventional, for formulas of compounds based on the elements of Groups 3A through 6A, to write the symbol of that element followed by the symbols of the elements attached to that atom. Thus, we have PBr_3 in the text above. There are many exceptions to this convention, such as water (H_2O) and hydrogen sulfide (H_2S). Most exceptions are compounds whose existence was known long before the convention was established.

EXERCISE 3.1 Molecular Formulas

Write the molecular formula of each of the following compounds: (a) Oxalic acid, a compound found in many plants, has two atoms of carbon, two hydrogen atoms, and four oxygen atoms per molecule. (b) A molecule of the pesticide DDT has fourteen carbon atoms, nine hydrogen atoms, and five chlorine atoms.

---

*In the chemical literature a convention you will often see is that, for molecules containing C, H, and other elements, C is written first and then H; the other elements follow in alphabetical order.

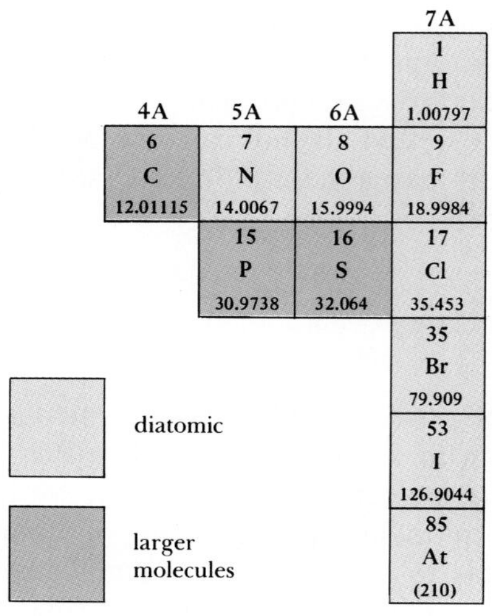

(a)

(b)

Figure 3.2 (a) Elements in light blue exist as diatomic molecules under normal conditions. Elements in dark blue exist as larger molecules. Phosphorus consists of P_4 molecules, the white solid in the test tube in b, sulfur as S_8 molecules, the yellow solid on the watch glass in b, and carbon as giant networks of carbon atoms (graphite and diamond). (Note that the white phosphorus is under water to prevent its vigorous and spontaneous reaction of O_2 in the air.)

3.2 ELEMENTS THAT EXIST AS MOLECULES

All but the heaviest of the elements have been isolated in large amounts in pure form. As you can see in Figures 2.3 and 2.4, most elements are metals and most of these are solids. In the solid state, metals consist of atoms packed together as closely as possible. On the other hand, nonmetals are often gases, liquids, or solids consisting of discrete atoms or even molecules. The elements of Group 8A, the rare gases, for example, are found as uncombined atoms in nature. In contrast, *pure hydrogen, nitrogen, oxygen, and the Group 7A elements (the halogens) exist as two-atom or* **diatomic molecules** *under normal conditions* (Figure 3.2).*

Phosphorus, sulfur, and carbon are also interesting. One form of elemental phosphorus can be found as a four-atom or tetratomic molecule P_4, while sulfur exists as the S_8 molecule. Finally, pure carbon is found in two forms: diamond and graphite. In both of these, there are extended networks of carbon atoms, as you shall see in Chapter 13.

3.3 IONS AND IONIC COMPOUNDS

A cation is a positively charged ion. The word is pronounced "cat-ion."

Atoms of almost all the elements can gain or lose electrons in ordinary chemical reactions to form **ions**, an atom or group of atoms bearing a net electrical charge. Indeed, a characteristic of metals is that ***metal atoms lose electrons*** *to form ions with a positive electrical charge*, ions commonly called **cations**. In Figure 3.3 a lithium atom, which is electrically neutral because it has three protons and three electrons, has lost one of its electrons. Because it now has one more positive charge than negative charge (three protons in the nucleus and only two electrons outside the nucleus),

*To remember which elements are diatomic, our students use the word "brinclhof," which stands for bromine-iodine-nitrogen-chlorine-hydrogen-oxygen-fluorine.

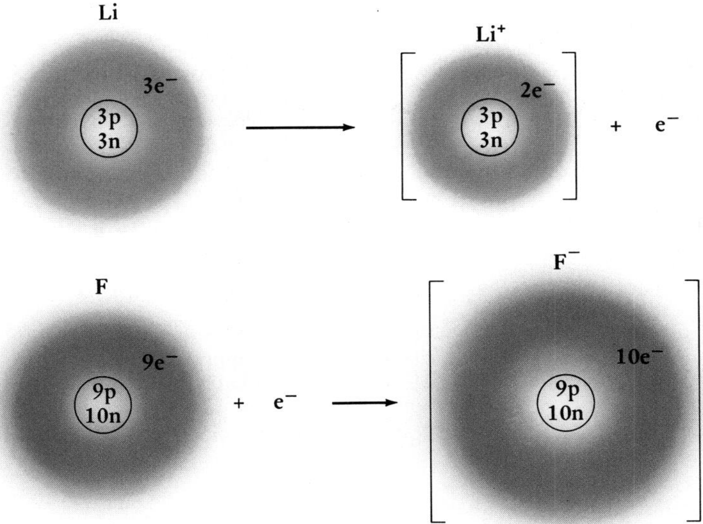

Figure 3.3 A lithium atom is electrically neutral because the numbers of positive charges (3 protons) and negative charges (3 electrons) are the same. When it loses one electron, it has one more positive charge than negative charge, so it has a net charge of 1+. We symbolize the resulting lithium cation as Li^+. A fluorine atom is also electrically neutral because there are nine protons and nine electrons. Because it is a nonmetal it gains an electron. As an F^- anion, it has one more electron than it has protons, so it has a net charge of 1−.

it has a net charge of 1+. We symbolize the resulting lithium cation as Li^+. When more than one electron has been lost, the number is written with the charge sign; for example, when the calcium ion is formed by the loss of two electrons, the symbol is Ca^{2+}.

In contrast with metals, ***nonmetals** frequently **gain electrons** to give ions with a negative electrical charge*. Such ions are called **anions**. In Figure 3.3 the fluorine atom with nine protons and nine electrons gains an electron to give F^-, an anion having nine protons and ten electrons.

An anion is a negatively charged ion. The word is pronounced "ann-ion."

An electron is represented in reactions in this text as e^-.

CHARGES ON IONS

Several useful guidelines for predicting the number of electrons likely to be lost by a metal or gained by a nonmetal are given in the following box.

Guidelines for Determining Charges of Monatomic Ions

Rule 1. For the metals of Groups 1A through 3A, for 1B and 2B, and for the metals of Group 4A, the number of electrons lost is equal to the group number. Therefore, these metals form positive ions whose charge is equal to the group number of the metal.

GROUP	METAL ATOM	ELECTRONS LOST	METAL ION
1A	Na (11 protons, 11 electrons)	1	Na^+ (11 protons, 10 electrons)
2A	Ca (20 protons, 20 electrons)	2	Ca^{2+} (20 protons, 18 electrons)
3A	Al (13 protons, 13 electrons)	3	Al^{3+} (13 protons, 10 electrons)

Rule 2. The transition metals of Groups 3B through 8B also form cations, but it is more difficult to predict the number of electrons lost. For the time being, however, a useful rule of thumb for these metals is that they form ions with a 2+ or 3+ charge.

GROUP	METAL ATOM	ELECTRONS LOST	METAL ION
6B	Cr (24 protons, 24 electrons)	3	Cr^{3+} (24 protons, 21 electrons)
8B	Fe (26 protons, 26 electrons)	2	Fe^{2+} (26 protons, 24 electrons)
		3	Fe^{3+} (26 protons, 23 electrons)

Rule 3. The maximum number of electrons gained by a nonmetal atom is equal to 8 minus the group number of the element. For example, N is in group 5A and so can gain $8 - 5 = 3$ electrons to form a 3− ion.

GROUP	NONMETAL ATOM	ELECTRONS GAINED	NONMETAL ION
5A	N (7 protons, 7 electrons)	3 (= 8 − 5)	N^{3-} (7 protons, 10 electrons)
6A	S (16 protons, 16 electrons)	2 (= 8 − 6)	S^{2-} (16 protons, 18 electrons)
7A	Br (35 protons, 35 electrons)	1 (= 8 − 7)	Br^{-} (35 protons, 36 electrons)

Rule 4. Hydrogen can either gain or lose one electron, depending on the other elements it encounters.

$$\text{H (1 proton, 1 electron)} \longrightarrow e^- + H^+ \text{ (1 proton, 0 electrons)}$$

$$\text{H (1 proton, 1 electron)} + e^- \longrightarrow H^- \text{ (1 proton, 2 electrons)}$$

Rule 5. The rare gases do not lose or gain electrons, except in unusual cases.

EXAMPLE 3.2

PREDICTING ION CHARGES

You have an atom of gallium and one of oxygen. Predict the charges on their ions.

Solution Gallium is a metal in Group 3A of the periodic table, so it is predicted to lose three electrons to give the Ga^{3+} cation.

$$Ga \longrightarrow Ga^{3+} + 3e^-$$

Oxygen is a nonmetal in Group 6A, so it is predicted to gain electrons to give an anion. The number of electrons gained is $8 - 6 = 2$. Therefore,

$$O + 2e^- \longrightarrow O^{2-}$$

EXERCISE 3.2 Predicting Ion Charges

Using the guidelines above, predict possible charges for ions formed from (a) K, (b) Se, (c) Be, (d) V, (e) Co, and (f) Cs.

POLYATOMIC IONS

The ions we have described thus far are **monatomic**—they consist of *one atom* bearing a charge. However, many other common ions in chemistry are groups of atoms bound together. Ions such as SO_4^{2-}, NH_4^+, and CO_3^{2-} are often called **polyatomic** or *many-atom* ions (Table 3.1). In a polyatomic ion the superscript is the charge on the whole group of

Table 3.1 Names and Composition of Some Common Polyatomic Ions

CATIONS: Positive Ions			
NH_4^+	ammonium		
ANIONS: Negative Ions			
OH^-	hydroxide	NO_2^-	nitrite
CN^-	cyanide	NO_3^-	nitrate
CH_3COO^-	acetate	SO_3^{2-}	sulfite
CO_3^{2-}	carbonate	SO_4^{2-}	sulfate
HCO_3^-	hydrogen carbonate (or bicarbonate)	HSO_4^-	hydrogen sulfate (or bisulfate)
PO_4^{3-}	phosphate		
MnO_4^-	permanganate		
$Cr_2O_7^{2-}$	dichromate		
ClO_4^-	perchlorate		

atoms. Thus, a carbonate ion, CO_3^{2-}, has a 2− charge on the group of four atoms. You should know the names and formulas for the common ions in Table 3.1.

COMPOUNDS FORMED FROM IONS

It is a law of nature that objects with opposite electrical charges are attracted to one another (see below, Attractions Between Ions of Opposite Charges: Coulomb's Law). Therefore, a positive ion (cation) and negative ion (anion) attract one another and form an **ionic compound**. We can always predict the formula of an ionic compound because, for an electrically neutral or uncharged compound to form, the number of positive charges carried by the cation (or cations) must be equal to the number of negative charges carried by the anion (or anions). For example, a calcium ion with a 2+ charge can interact with anions of 1−, 2−, and 3− charge in the following ways.

ION COMBINATION	COMPOUND	OVERALL CHARGE ON COMPOUND
$Ca^{2+} + 2\ Cl^-$	$CaCl_2$	$(2+) + 2(1-) = 0$
$Ca^{2+} + CO_3^{2-}$	$CaCO_3$	$(2+) + (2-) = 0$
$3\ Ca^{2+} + 2\ PO_4^{3-}$	$Ca_3(PO_4)_2$	$3(2+) + 2(3-) = 0$

Notice that in writing the formulas of ionic compounds, *the symbol of the cation is always given first, followed by the symbol for the anion.* Also notice how parentheses are used to set off a polyatomic ion that appears more than once in the formula.

Attractions Between Ions of Opposite Charges: Coulomb's Law

You have often heard that "opposites attract." While this is certainly not always true with men and women, it is true in the world of physics. Ions of opposite charge, for example a positive ion such as Li^+ and a negative ion such as F^-, are attracted to one another by *electrostatic forces,* forces that are governed by **Coulomb's law**. Coulomb's law tells us that the force of attraction between such ions is given by

$$\text{Force of attraction} = \frac{(n_+e)(n_-e)}{d^2}$$

where n_+ is the charge on the positive ion (e.g., 3 for Al^{3+}), Z_- is the charge on the negative ion (e.g., 2 for O^{2-}), e is the charge on the electron (1.602×10^{-19} coulombs), and d is the distance between the ions. A close look at the equation tells you that the force of attraction increases as the ion charges become larger and as the sizes of the ions (and so the distance between them) become smaller.

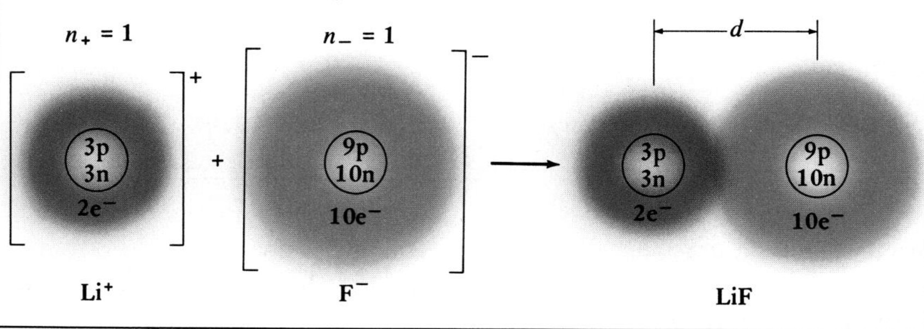

How can you predict when a compound will be formed from ions? As an extension of the guidelines on charges given above, we can say the following:

(a) Metals (the blue elements in Figure 2.3) almost always give positive ions and form ionic compounds. If the metal is from the transition series, the metal's charge is most often 2+ or 3+, although other possibilities exist. Unless you are told otherwise, assume the transition metal is 2+ or 3+ (with 2+ being the usual choice).

(b) Nonmetals (the yellow elements in Figure 2.3) give ionic compounds *only* when combined with a metal.

(c) It is difficult to predict what the metalloids (the green elements in Figure 2.3) will do, so we shall leave them for later; you will not be asked to predict formulas for their compounds in this chapter.

The following examples explore in more detail how to understand, predict, and write formulas for ionic compounds.

EXAMPLE 3.3

COMPOUNDS FORMED FROM IONS

For each of the following ionic compounds, tell what ions are present and give the relative number of ions: (a) $MgBr_2$; (b) Li_2CO_3, and (c) $Fe_2(SO_4)_3$.

Solution (a) $MgBr_2$ is composed of one Mg^{2+} ion and two Br^- ions.

$$Mg^{2+} + 2\ Br^- \longrightarrow MgBr_2$$

When a halogen, such as bromine, is associated only with a metal, you can assume the halogen is an anion with a charge of 1−. Magnesium is an element in Group 2A and *always* has a charge of 2+ in its compounds.

(b) Li_2CO_3 is composed of two lithium ions, Li^+, and one carbonate ion, CO_3^{2-}.

$$2\ Li^+ + CO_3^{2-} \longrightarrow Li_2CO_3$$

You should learn to recognize the carbonate ion and recall its formula and charge. To help you remember that carbonate has a 2− charge, you can see that Li is a Group 1A atom and so always has a 1+ charge in its compounds. Because the two positive charges must be neutralized by the CO_3 ion's negative charges, the latter is 2−.

(c) $Fe_2(SO_4)_3$ comes from two Fe^{3+} ions and three sulfate ions, SO_4^{2-}.

$$2\ Fe^{3+} + 3\ SO_4^{2-} \longrightarrow Fe_2(SO_4)_3$$

The only way to tell this is to recall that sulfate is 2−. Since you have three of them (for a total of 6−), the two iron cations must add up to 6+. This is possible only if each is 3+.

The moral of this tale: You really must learn the formulas *and* charges of the common polyatomic ions in Table 3.1.

EXERCISE 3.3 Compounds Formed from Ions

For each of the following ionic compounds, tell what ions are present and give their relative number: (a) NaF, (b) $NaCH_3COO$, and (c) $Cu(NO_3)_2$.

EXAMPLE 3.4

WRITING FORMULAS FOR COMPOUNDS FORMED FROM IONS

What ionic compounds can you form if you combine an aluminum cation with (a) the anion from a bromine atom, (b) the anion from an oxygen atom, or (c) a nitrate ion?

Solution First, the aluminum cation must have a charge of 3+ since Al is a metal in Group 3A.

(a) Br is a Group 7A element and so is a nonmetal. Its charge is predicted to be 1− (from 8 − 7 = 1). Therefore, we need 3 Br^- ions (total charge for the three ions is 3−) to combine with Al^{3+}. Therefore, $Al^{3+} + 3\ Br^-$ —will give→ $AlBr_3$.

(b) Oxygen is a nonmetal in Group 6A and so forms a 2− anion. Thus, we need to combine two Al^{3+} ions [total charge is 6+ = 2(3+)] with three O^{2-} ions [total charge is 6− = 3(2−)]. This means $2\ Al^{3+} + 3\ O^{2-}$ —will give→ Al_2O_3.

(c) The nitrate ion has the formula NO_3^- (Table 3.1). Because its charge is 1−, the question is similar to the $AlBr_3$ case. Therefore, $Al^{3+} + 3\ NO_3^-$ —will give→ $Al(NO_3)_3$. Here there are parentheses around the NO_3 portion of the formula to show that three of the NO_3^- "units" are involved.

We have gone through the logic of predicting formulas by checking on electrical charges. A shorthand way to do this is simply to make the number of charge units on the cation the subscript for the anions, and vice versa.

$$Al^{3+}\ O^{2-} \longrightarrow Al_2O_3$$

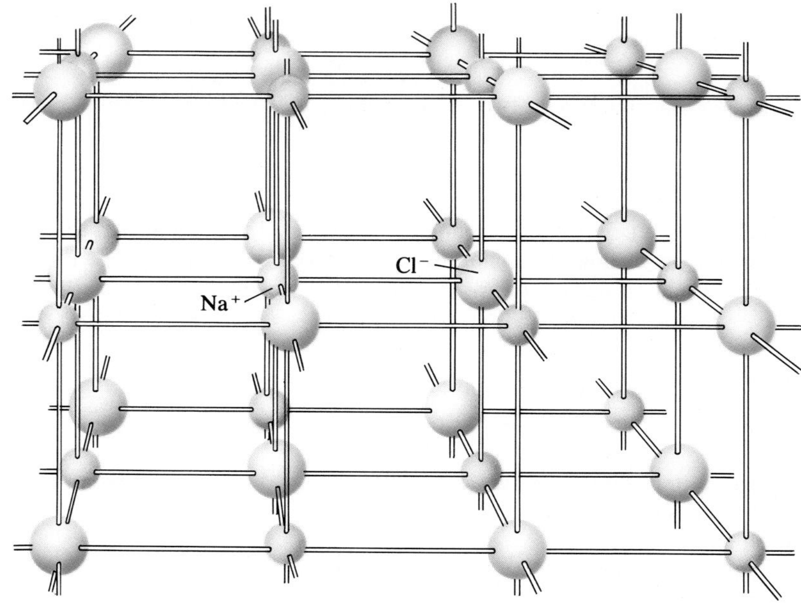

Figure 3.4 A model of a crystal of sodium chloride. The lines between the ions are not chemical bonds but are simply reference lines showing the relation of Na^+ (red) and Cl^- (green) ions in space.

> **EXERCISE 3.4 Writing Formulas for Compounds Formed from Ions**
>
> Chromium is a transition metal and so can form ions with at least two different charges. Write the formulas of the compounds formed between chromium and sulfur.

The structures of solids are described in Chapter 13.

As you will see in the course of this book, a compound formed from ions often has unique properties. Many of these arise from the fact that, in normal situations, there is no single unit for the combination of a cation with an anion in the sense that there is a single molecule of H_2 or H_2O. Rather, ionic compounds exist as extended three-dimensional networks of positive and negative ions; it is impossible to say that a certain anion/cation pair belong together to form one molecule. This is illustrated by the model of the structure of common salt, NaCl, in Figure 3.4. Each Na^+ ion is surrounded by six Cl^- ions, and each Cl^- ion is surrounded by six Na^+ ions.

3.4 NAMES OF COMPOUNDS

As you are beginning to see, chemistry uses precise language to describe our world, and part of this involves the names of chemical compounds. The rules for naming simple ones are easy to follow and are outlined in this section.

NAMING IONIC COMPOUNDS

Names of compounds composed of ions are built from the names of the ions involved.

NAMING POSITIVE IONS With a few exceptions, the positive ions you will see in this book are based on metals. Positive ions are named by the following rules.

(a) For a monatomic positive ion, that is, a metal cation, the name is simply the name of the metal from which it is derived. For example, we have already referred to Al^{3+} as the aluminum ion.

(b) There are cases, especially in the transition series, where a metal can form more than one positive ion. The most common practice today is to indicate the charge of the ion by a Roman numeral in parentheses immediately following the ion's name. For example

Ti^{2+} is titanium(II) and Ti^{4+} is titanium(IV)

Co^{2+} is cobalt(II) and Co^{3+} is cobalt(III)

Sn^{2+} is tin(II) and Sn^{4+} is tin(IV)

Another cation that you will see on occasion is Hg_2^{2+}, the name of which is mercury(I). The reason for the Roman numeral (I) is that the ion is composed of two Hg^+ ions bonded together.

(c) Finally, you will encounter the NH_4^+ or *ammonium* ion many times in this book, in the laboratory, and in your environment.

NAMING NEGATIVE IONS Two types of negative ions must be considered: those having only one atom (monatomic) and those having several atoms (polyatomic).

(a) Monatomic negative ions are named by adding the suffix *-ide* to the stem of the name of the nonmetal element from which it is derived (see Table 3.2). As a group, the anions of the Group 7A elements—the halogens—are called *halides*.

(b) Polyatomic negative ions are quite common, especially those containing oxygen (called *oxyanions*). The names of some of the most common are given in Table 3.1. Most of these names must simply be learned. However, there are some guidelines you should know. For example, consider the pairs of ions below.

NO_3^- is nitr*ate* while NO_2^- is nitr*ite*

SO_4^{2-} is sulf*ate* while SO_3^{2-} is sulf*ite*

The oxyanion with the greater number of oxygen atoms is given the suffix *-ate*, while the oxyanion with the smaller number of oxygen atoms has the suffix *-ite*. For a series of oxyanions with more than two members, the ion with the largest number of oxygen atoms has the prefix *per-* and

Table 3.2 Anions of Nonmetals

4A	5A	6A	7A
C^{4-} carbide	N^{3-} nitride	O^{2-} oxide	F^- fluoride
	P^{3-} phosphide	S^{2-} sulfide	Cl^- chloride
		Se^{2-} selenide	Br^- bromide
			I^- iodide

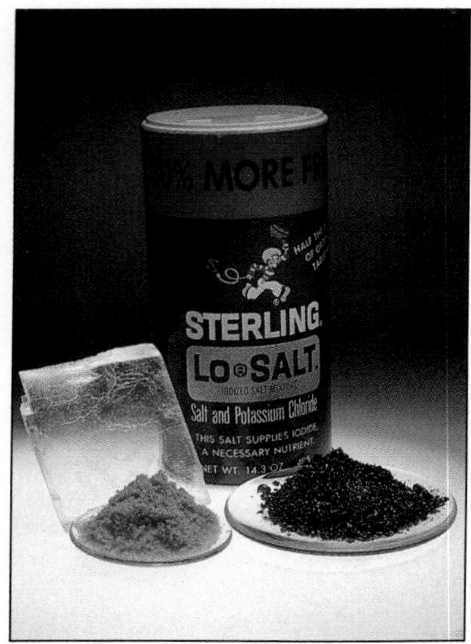

Figure 3.5 Some common ionic compounds. At the rear are a clear crystal of calcite (calcium carbonate, $CaCO_3$) and a box of common table salt (sodium chloride, NaCl). In front, the blue solid is copper sulfate ($CuSO_4 \cdot 5\ H_2O$), and the purple solid is chromium(III) chloride ($CrCl_3$).

the suffix *-ate*; the ion with the smallest number of oxygen atoms has the prefix *hypo-* and the suffix *-ite*. The oxyanions containing chlorine are the most common examples.

ClO_4^- is the *per*chlor*ate* ion

ClO_3^- is the chlor*ate* ion

ClO_2^- is the chlor*ite* ion

ClO^- is the *hypo*chlor*ite* ion

Oxyanions that contain hydrogen are named simply by prefixing the name of the oxyanion with the word ''hydrogen.'' If there are two hydrogen atoms, we say *di*hydrogen. Many of these hydrogen-containing oxyanions have common names that are so often used that you should know them. For example, sodium hydrogen carbonate, $NaHCO_3$, is often called sodium bicarbonate (although you may know it as ''baking soda'').

ION	SYSTEMATIC NAME	COMMON NAME
HCO_3^-	hydrogen carbonate	bicarbonate
HSO_4^-	hydrogen sulfate	bisulfate
HSO_3^-	hydrogen sulfite	bisulfite
HPO_4^{2-}	hydrogen phosphate	
$H_2PO_4^-$	dihydrogen phosphate	

NAMING IONIC COMPOUNDS When naming ionic compounds, *the positive ion name is given first followed by the name of the negative ion.* Some examples are given below and others are found in Figure 3.5.

IONIC COMPOUND	IONS INVOLVED	NAME
$CaBr_2$	Ca^{2+} and 2 Br^-	calcium bromide
$NaHSO_4$	Na^+ and HSO_4^-	sodium hydrogen sulfate
$(NH_4)_2CO_3$	2 NH_4^+ and CO_3^{2-}	ammonium carbonate
$Mg(OH)_2$	Mg^{2+} and 2 OH^-	magnesium hydroxide
$TiCl_2$	Ti^{2+} and 2 Cl^-	titanium(II) chloride
TiF_4	Ti^{4+} and 4 F^-	titanium(IV) fluoride

NAMING BINARY COMPOUNDS OF THE NONMETALS

Thus far we have described only the case of a metal-containing cation (or NH_4^+) combining with an anion to give an ionic compound. However, there are other types of compounds, and one simple kind that you will see often comes from the combination of two nonmetals.* These ''two-element'' or **binary** compounds can also be named in a systematic way.

Hydrogen forms binary compounds of the type H_xX_y with all of the nonmetals (except the rare gases). Except in a few cases (see below), the H atom is written first in the formula and is named first. The other nonmetal is named as if it were a negative ion.

*There are no ions in these compounds. The bonding in these compounds is said to be ''covalent''; it is described in Chapters 10 and 11.

COMPOUND	NAME
HF	hydrogen fluoride
HCl	hydrogen chloride
H_2S	hydrogen sulfide

Virtually all the binary, nonmetal compounds you will see have at least one element from Groups 6A or 7A. This element is always listed second in the formula and is named second. The number of atoms of a given element in the compound is designated with a prefix such as *di-*, *tri-*, *tetra-*, *penta-*, and so on.

COMPOUND	NAME
NF_3	nitrogen trifluoride
NO_2	nitrogen dioxide
N_2O	dinitrogen oxide
N_2O_4	dinitrogen tetroxide*
PCl_3	phosphorus trichloride
PCl_5	phosphorus pentachloride
SF_6	sulfur hexafluoride
S_2F_{10}	disulfur decafluoride

Many of the binary compounds of nonmetals were discovered years ago and have names so common they continue to be used. These names must simply be learned.

COMPOUND	NAME
H_2O	water
NH_3†	ammonia
N_2H_4†	hydrazine
PH_3†	phosphine
NO	nitric oxide
N_2O	nitrous oxide (“laughing gas”)

It is important that you realize that, at this point in learning chemistry, you CANNOT predict the formula of a binary, nonmetal compound in the same way you can predict the formulas of ionic compounds. You will begin to do this in Chapter 4.

EXERCISE 3.5 Naming Compounds

1. Give the formula for each of the following ionic compounds:
   (a) ammonium nitrate (d) vanadium(III) oxide
   (b) cobalt(II) sulfate (e) barium oxide
   (c) nickel(II) cyanide (f) calcium hypochlorite
2. Name the following ionic compounds:
   (a) $MgBr_2$ (d) $KMnO_4$
   (b) Li_2CO_3 (e) $(NH_4)_2S$
   (c) $KHSO_3$ (f) CuCl and $CuCl_2$
3. Give the formula for each of the following binary, nonmetal compounds:
   (a) carbon dioxide (d) boron trifluoride
   (b) phosphorus triiodide (e) dioxygen difluoride
   (c) sulfur dichloride (f) xenon trioxide
4. Name the following binary, nonmetal compounds:
   (a) N_2F_4 (d) BCl_3
   (b) HBr (e) P_2O_5
   (c) SF_4 (f) ClF_3

*The name should be tetraoxide, but the *a* is dropped so the name sounds better.

†By tradition, the formulas of Group 5A hydrogen-containing compounds are written with the N, P, As, or Sb first.

SOMETHING MORE ABOUT
Planetary Chemistry: Neptune

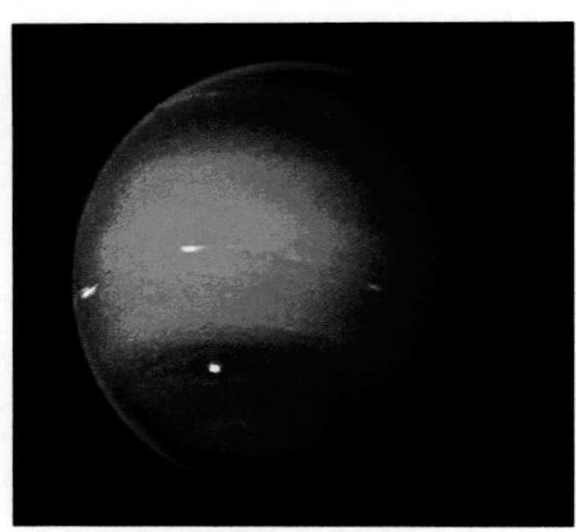

False-color image of Neptune. Red areas are semitransparent haze covering the planet. (Jet Propulsion Laboratory, California Institute of Technology, NASA)

It is the eighth planet out from the sun and only a tiny green speck in an earthbound telescope. Neptune was the last planet encountered in our solar system by *Voyager 2*. On August 24, 1989, the space probe gave us a first look at this interesting new world. And what an exciting world! It is a turbulent planet with giant storms like those on Jupiter, but with clouds unlike those seen before on a gaseous planet.

Like its nearest neighbor, the planet Uranus, Neptune is a gigantic ball of water and molten rock covered by an atmosphere of hydrogen and helium mixed with methane (CH_4). The blue-green or aqua color of the planet is due to the fact that the methane absorbs red light and transmits blue light.

Two of Neptune's moons—Triton and Nereid—were known from telescope observations before *Voyager 2*, and six tinier moons were discovered by *Voyager*. However, Triton, which is slightly smaller than earth's moon, is by far the most interesting object. The six small moons orbit about the equator of the planet, but the two larger moons have orbital planes tilted with respect to the

The Chemistry of the Planets

Planet and Satellites	Chemistry and Conditions
Mercury	Airless and arid. Solid with large Fe, Ni core.
Venus	Atmosphere consists of CO_2 and sulfuric acid. High surface temperature due to runaway "greenhouse effect."
Earth	The only planet in the solar system with an oxygen-rich atmosphere, vast amounts of water, life.
Mars	Low density CO_2 atmosphere. Solid H_2O and CO_2 polar caps.
Jupiter	Multihued planet consists of 89% hydrogen and 11% helium (similar to the sun's atmosphere). There is also evidence for methane (CH_4), ethane (C_2H_6), acetylene (C_2H_2), ammonia, water, phosphine (PH_3). White clouds may contain solid NH_3, $(NH_4)_2S$, and NH_4HS. The giant Red Spot may consist of red phosphorus from the action of sunlight on PH_3.
Io (moon of Jupiter)	Active volcanoes evolve SO_2. Surface believed to be solid sulfur and frozen SO_2. Sodium vapor found around the moon.
Saturn	Planet is 94% hydrogen with rings of ice and rock. Atmosphere also contains methane, ammonia, ethane, and acetylene.
Titan (moon of Saturn)	The moon may be covered with water ice under an ocean of methane, ethane, and nitrogen. Atmosphere is mostly nitrogen with some methane and ethane. The only moon in the solar system to have a substantial atmosphere.
Uranus	Water and molten rock with an atmosphere of hydrogen and helium mixed with some methane.
Neptune	Similar to Uranus.
Triton (moon of Neptune)	Polar region of rocky moon is covered with methane and nitrogen ices.

equatorial plane, and Triton even orbits the planet in a direction opposite to the others. The moon is rocky with a surface that resembles the skin of a cantaloupe. Ices of methane and nitrogen cover its polar cap and reflect away so much sunlight that the temperature is only 37 K, making Triton the coldest object yet seen in our solar system. The atmosphere on Triton is 100,000 times thinner than earth's and consists largely of nitrogen.

Earlier in *Voyager*'s mission, astronomers had observed distinct evidence of volcanic activity on Io, a moon of the planet Jupiter. In that case, they observed plumes of sulfur dioxide (SO_2) spewing from volcanoes. Triton's surface also seems to be scarred by volcanic or geyser activity, since there are dark streaks on the polar frost. What gases are evolved in Triton's eruptions is not known, but nitrogen is suspected. Further, it is not known what is in the dark streaks, but carbon-based molecules have been proposed.

Although *Voyager 2* has given us an encyclopedia of information on the solar system, there are many tantalizing questions yet to be answered. One thing that is certain, though, is that Triton does not resemble Neptune at all and may have been a wandering object captured by the gravitational field of Neptune.

3.5 MOLAR MASS, MOLECULAR WEIGHT, AND FORMULA WEIGHT

The molecular formula of a compound tells you the types of atoms in the molecule and the number of each. For example, in one molecule of ammonia there is one atom of N combined with three atoms of H. If 100 atoms of N and 300 atoms of H are combined, then 100 molecules of NH_3 would result. Now suppose you combine N and H according to the scheme below.

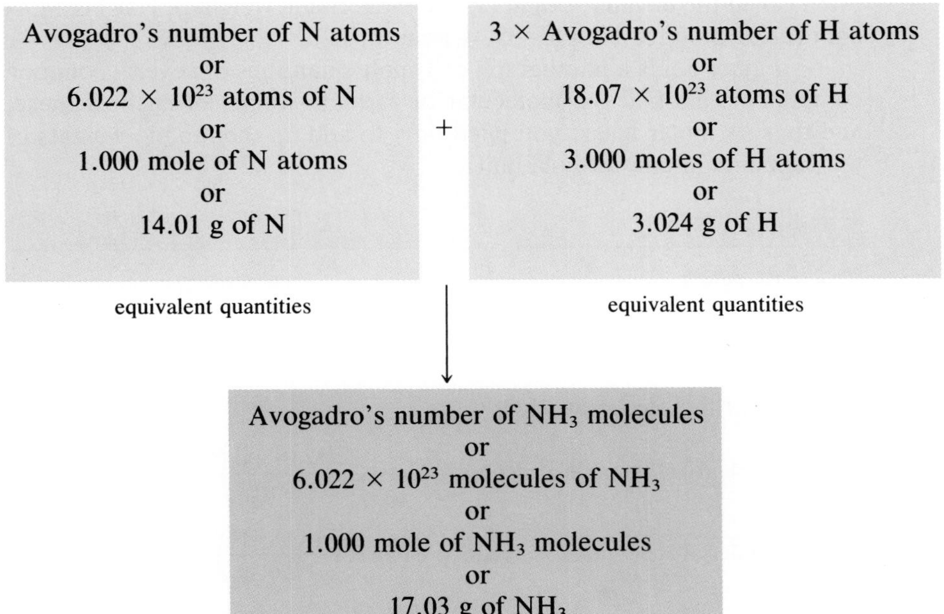

Avogadro's number of N atoms
or
6.022 × 10^23 atoms of N
or
1.000 mole of N atoms
or
14.01 g of N
+
3 × Avogadro's number of H atoms
or
18.07 × 10^23 atoms of H
or
3.000 moles of H atoms
or
3.024 g of H
equivalent quantities
equivalent quantities
Avogadro's number of NH3 molecules
or
6.022 × 10^23 molecules of NH3
or
1.000 mole of NH3 molecules
or
17.03 g of NH3

Since Avogadro's number of particles is a mole (Section 2.6), the mass of 1.000 mole or Avogadro's number of ammonia molecules is 17.03 grams. This mass is called the **molar mass, *M***.

Figure 3.6 One-mole quantities of a range of ionic compounds (clockwise from front right): NaCl, white (M = 58.44 g/mol); $CuSO_4 \cdot 5\ H_2O$, blue (M = 249.68 g/mol); $NiCl_2 \cdot 6\ H_2O$, green (M = 237.70 g/mol); $K_2Cr_2O_7$, orange (M = 294.18 g/mol); and $CoCl_2 \cdot 6\ H_2O$, red (M = 237.93 g/mol).

For all substances the molar mass is numerically equal to the **molecular weight**, the sum of the atomic weights of all the atoms appearing in the molecular formula. The molecular weight is the average mass of a molecule of the substance, expressed in atomic mass units.

COMPOUND	MOLECULAR WEIGHT (amu)	MOLAR MASS (g/mol)	MASS OF ONE MOLECULE (grams)
NH_3	17.03	17.03	2.828×10^{-23}
H_2O	18.02	18.02	2.992×10^{-23}
CH_2Cl_2	84.93	84.93	1.410×10^{-22}

For ionic compounds such as NaCl that do not exist as individual molecules, no *molecular* formula can be given; rather, one can only write a formula showing the simplest set of relative numbers of atoms in a sample. As you will learn in Section 3.6, this is called the *empirical formula*, and the sum of the atomic weights for the empirical formula is often called the **formula weight**. Thus, the formula weight of NaCl is 58.44 amu (22.99 amu for Na plus 35.45 amu for Cl).

Figure 3.6 is a photograph of 1-mole quantities of several common compounds. To find the molecular or formula weight of each of these, and thus its molar mass, you need only to add up the atomic weights of the elements in one formula unit.

EXAMPLE 3.5

MOLAR MASS

Calculate the molar mass of (a) ethyl alcohol, C_2H_6O; (b) ammonium sulfate, $(NH_4)_2SO_4$; and (c) barium chloride dihydrate, $BaCl_2 \cdot 2\ H_2O$.

Solution (a) Ethyl alcohol, C_2H_6O.

$$\text{2 moles of C per mole of alcohol} = 2\ \cancel{\text{mol C}}\left(\frac{12.01\ \text{g C}}{1\ \cancel{\text{mol C}}}\right) = 24.02\ \text{g C}$$

$$\text{6 moles of H per mole of alcohol} = 6\ \cancel{\text{mol H}}\left(\frac{1.008\ \text{g H}}{1\ \cancel{\text{mol H}}}\right) = 6.048\ \text{g H}$$

$$\text{1 mole of O per mole of alcohol} = 1\ \cancel{\text{mol O}}\left(\frac{16.00\ \text{g O}}{1\ \cancel{\text{mol O}}}\right) = \underline{16.00\ \text{g O}}$$

$$\text{Molar mass} = 46.07\ \text{g/mol}$$

(b) Ammonium sulfate, $(NH_4)_2SO_4$. This compound contains two ammonium ions. Therefore, in one formula unit there are two nitrogen atoms and eight hydrogen atoms.

$$\text{2 moles of N per mole sulfate} = 2\,\cancel{\text{mol N}}\left(\frac{14.01\text{ g N}}{1\,\cancel{\text{mol N}}}\right) = 28.02\text{ g N}$$

$$\text{8 moles of H per mole sulfate} = 8\,\cancel{\text{mol H}}\left(\frac{1.008\text{ g H}}{1\,\cancel{\text{mol H}}}\right) = 8.064\text{ g H}$$

$$\text{1 mole of S per mole sulfate} = 1\,\cancel{\text{mol S}}\left(\frac{32.07\text{ g S}}{1\,\cancel{\text{mol S}}}\right) = 32.07\text{ g S}$$

$$\text{4 moles of O per mole sulfate} = 4\,\cancel{\text{mol O}}\left(\frac{16.00\text{ g O}}{1\,\cancel{\text{mol O}}}\right) = \underline{64.00\text{ g O}}$$

$$\text{Molar mass} = 132.15\text{ g/mol}$$

(c) A compound having water as an integral part of the solid is called a "hydrate." Thus, the formula $BaCl_2 \cdot 2\ H_2O$ tells us there are *two* molecules of water associated with every unit of $BaCl_2$. This means that each formula unit contains one Ba^{2+} ion, two Cl^- ions, *four* H atoms, and *two* O atoms.

$$\text{1 mole of Ba}^{2+}\text{ ions} = 1\,\cancel{\text{mol Ba}^{2+}}\left(\frac{137.3\text{ g Ba}^{2+}}{1\,\cancel{\text{mol Ba}^{2+}}}\right) = 137.3\text{ g Ba}^{2+}$$

$$\text{2 moles of Cl}^-\text{ ions} = 2\,\cancel{\text{mol Cl}^-}\left(\frac{35.45\text{ g Cl}^-}{1\,\cancel{\text{mol Cl}^-}}\right) = 70.90\text{ g Cl}^-$$

$$\text{4 moles of H atoms} = 4\,\cancel{\text{mol H}}\left(\frac{1.008\text{ g H}}{1\,\cancel{\text{mol H}}}\right) = 4.032\text{ g H}$$

$$\text{2 moles of O atoms} = 2\,\cancel{\text{mol O}}\left(\frac{16.00\text{ g O}}{1\,\cancel{\text{mol O}}}\right) = \underline{32.00\text{ g O}}$$

$$\text{Molar mass} = 244.2\text{ g/mol}$$

EXERCISE 3.6 Molar Mass

Calculate the molar mass of (a) limestone, $CaCO_3$, and (b) caffeine, $C_8H_{10}N_4O_2$.

In Chapters 4 and 5 we shall begin to deal with combinations of molecules. There you will see that they combine by one molecule reacting with another, or two molecules of one compound with one molecule of another, or three of one kind with two of another, and so on. This means, of course, that the ratio of combining moles will be 1:1, 2:1, 3:2, or whatever is required by the compounds involved. Because we have no instrument capable of directly measuring amounts of compounds in mole units, but only in mass units, it is important for you to be able to convert "mass of compound → moles of compound" or "moles → mass." This is done in exactly the same way we described on page 59 for dealing with moles of elements.

EXAMPLE 3.6

MASS ↔ MOLE CONVERSIONS

The molar masses of ethyl alcohol and ammonium sulfate were calculated in Example 3.5. Using these values, (a) express 23.2 g of ethyl alcohol in moles, and (b) find the number of grams of ammonium sulfate represented by 3.50 moles.

Solution The molar mass of a compound has units of grams/mole, and this is the conversion factor for mass ↔ mole conversions.

(a) To change mass to moles, we *divide* by the molar mass.

$$23.2\ \cancel{g}\ C_2H_6O \left(\frac{1\ \text{mol}}{46.07\ \cancel{g}}\right) = 0.504\ \text{mol}\ C_2H_6O$$

(b) To change moles to mass, we *multiply* by the molar mass.

$$3.50\ \cancel{\text{mol}}\ (NH_4)_2SO_4 \left(\frac{132.2\ \text{g}}{1\ \cancel{\text{mol}}}\right) = 463\ \text{g}$$

EXERCISE 3.7 Mass ↔ Mole Conversion

See Exercise 3.6 for the molar masses of $CaCO_3$ and caffeine.

(a) How many moles of $CaCO_3$ (limestone) are represented by 1.00×10^3 g?

(b) To have 2.50×10^{-3} moles of caffeine, how many grams must you have?

The molar mass is the mass in grams of Avogadro's number of molecules. Thus, each of the piles of solid in Figure 3.6 represents 6.022×10^{23} molecules or formula units. With this knowledge, we can determine the number of molecules in any sample from its mass, or we can even determine the mass of one molecule.

EXAMPLE 3.7

MASS ↔ MOLECULES; MASS OF A MOLECULE

In Example 3.6, we found that 23.2 g of ethyl alcohol is equivalent to 0.504 moles. (a) How many molecules does this represent? (b) How many atoms of carbon are there in 23.2 g of ethyl alcohol? (c) What is the mass of one ethyl alcohol molecule?

Solution (a) Since we want an answer in units of "molecules," we need to multiply the starting units of "moles" by Avogadro's number with units of "molecules/mole." The units of "moles" cancel out.

$$0.504\ \cancel{\text{mol}}\ C_2H_6O \left(\frac{6.022 \times 10^{23}\ \text{molecules}}{1\ \cancel{\text{mol}}}\right) = 3.04 \times 10^{23}\ \text{molecules}\ C_2H_6O$$

(b) Each molecule of alcohol contains two atoms of C. Therefore, we multiply the starting units of "molecule" by units of "C atoms per molecule" and cancel units of "molecules."

$$3.04 \times 10^{23}\ \cancel{\text{molecules}}\ C_2H_6O \left(\frac{2\ \text{C atoms}}{1\ C_2H_6O\ \cancel{\text{molecule}}}\right) = 6.08 \times 10^{23}\ \text{atoms of C}$$

(c) Now we want an answer with units of "grams/molecule." This means that we have to divide the starting unit of molar mass (or grams/mole) by Avogadro's number so that the unit "mole" cancels.

$$\left(\frac{46.07 \text{ g } C_2H_6O}{1 \cancel{\text{mol}}}\right)\left(\frac{1 \cancel{\text{mol}}}{6.022 \times 10^{23} \text{ molecules}}\right) = \frac{7.650 \times 10^{-23} \text{ g}}{1\ C_2H_6O \text{ molecule}}$$

EXERCISE 3.8 Moles ↔ Molecules

A can of artificially sweetened soft drink may contain 70. mg of aspartame, or NutraSweet®, $C_{14}H_{18}N_2O_5$. (a) Calculate the number of moles of aspartame in the soft drink. (b) Calculate the number of molecules of aspartame in the soft drink and the number of atoms of carbon. (c) Calculate the mass of one molecule of aspartame.

3.6 DETERMINATION OF THE FORMULA OF A COMPOUND

Although it is often possible to predict the formula of a simple ionic compound, there are thousands of molecules of greater complexity. Given a sample of a compound, how do you determine its formula? The answer lies in chemical analysis, a major branch of chemistry that deals with the determination of molecular formula and structure, among other things.

PERCENT COMPOSITION

The **law of constant composition** states that *any sample of a pure compound always consists of the same elements combined in the same proportions by mass*. Every molecule of ammonia, for example, always has the formula NH_3. That is, one molecule always contains one atom of N and three atoms of H, or 1 mole of pure ammonia always contains 1 mole of N and 3 moles of H. It follows then that 1.000 molar mass of NH_3, 17.03 grams, always contains 14.01 g of N and 3.024 g of H.

There are 3.000 moles of H in 1.000 mole of NH_3 or 3.024 g H.

$$\left(\frac{3.000 \text{ mol H}}{1 \text{ mol } NH_3}\right)\left(\frac{1.008 \text{ g}}{1 \text{ mol H}}\right) = \frac{3.024 \text{ g H}}{1 \text{ mol } NH_3}$$

Thus, there are at least two equivalent ways to express molecular composition: (a) in terms of the number of atoms of each type per molecule or (b) in terms of element masses per mole. There is a third way to express molecular composition, a way derived from (b). Composition can be given by the mass of each element in the compound relative to the total mass of the compound. That is, the composition can be expressed as **percent composition**. For ammonia, this is

$$\text{Weight percent N in } NH_3 = \frac{\text{mass of N in 1 mole of } NH_3}{\text{mass of 1 mole of } NH_3} \times 100$$

$$= \frac{14.01 \text{ g N}}{17.03 \text{ g } NH_3} \times 100$$

$$= 82.27\%$$

(or 82.27 g N per 100.0 g NH_3)

$$\text{Weight percent H in } NH_3 = \frac{3.024 \text{ g H}}{17.03 \text{ g } NH_3} \times 100$$

$$= 17.76\% \text{ H}$$

(or 17.76 g H per 100.0 g NH_3)

Equivalent ways of expressing molecular composition:

(a) A formula giving the number of atoms of each type per molecule.
(b) Mass of each element per mole of compound.
(c) Mass of each element per 100 g of compound (percent composition).

EXERCISE 3.9 Percent Composition

Express the composition of each of the following compounds in terms of the mass of each element in 1.00 mole of compound and the weight percent of each element:

(a) NaCl, sodium chloride

(b) CH_4, methane

(c) $(NH_4)_2SO_4$, ammonium sulfate

EMPIRICAL AND MOLECULAR FORMULAS

Now consider the reverse of the procedure above: using relative mass or percent composition to find a molecular formula. That is, if you know the identities of the elements in a sample, and can determine by chemical analysis the mass of each element in a given mass of compound (the percent composition), you can then calculate the relative number of moles of each element in one mole of compound and then the relative number of atoms of each type in the molecule.

To see how to use percent composition data to find the relative number of atoms of each type in a molecule, consider hydrazine, a close relative of ammonia. A sample of hydrazine consists of 87.42% by weight N and 12.58% by weight H. If we measure out a 100.0-g sample of hydrazine, the percent composition data tell us that the sample contains 87.42 g of N and 12.58 g of H. Therefore, the number of moles of each element in the 100.0-g sample is

$$87.42 \text{ g N} \left(\frac{1 \text{ mol N}}{14.01 \text{ g}}\right) = 6.240 \text{ moles of N}$$

$$12.58 \text{ g H} \left(\frac{1 \text{ mol H}}{1.008 \text{ g}}\right) = 12.48 \text{ moles of H}$$

What is important now is to use the number of moles of each element in 100.0 g of sample to find the number of moles of one element *relative* to the other. For hydrazine, this ratio is 2 moles of H to 1 mole of N,

$$\frac{12.48 \text{ mol H}}{6.240 \text{ mol N}} = \frac{2.000 \text{ mol H}}{1.000 \text{ mol N}}$$

showing that there are two moles of H atoms for every mole of N atoms in hydrazine. This must mean that in one molecule there are two atoms of H for every atom of N.

Percent composition data allow you to arrive at the atom ratios in the molecule. However, remember that a molecular formula must convey *two* pieces of information: (a) the *relative numbers* of each type of atom in a molecule (the atom ratios) and (b) the *total number* of atoms in the molecule. For hydrazine you know that there are twice as many H atoms as N atoms. This means the molecular formula could be NH_2, but N_2H_4, N_3H_6, N_4H_8, and so forth are also possible. *Percent composition data give only the ratio of atoms in a molecule.* A formula such as NH_2, where the numbers of atoms are the smallest possible, is called the **empirical formula**. In contrast, the **molecular formula** shows the true number of atoms of each kind; it is always the same as the empirical formula, double the empirical formula, triple it, or some other whole number multiple of the empirical formula.

The empirical formula of a compound gives the smallest total number of atoms consistent with the known ratio of atoms. To define the molecular formula, you *must* know the molar mass *from experiment*. For hydrazine, for example, you would find from experiment that the molar mass is 32.0 g/mol. Since the "formula mass" of NH_2 is 16.0 g/mol, the molar mass of hydrazine is equivalent to *two* moles of NH_2. This must mean that the molecular formula of hydrazine is 2 times the empirical formula NH_2, that is, N_2H_4.

To find the molecular formula of a compound after calculating the empirical formula, you must know the molar mass from an experiment.

As an aside here, the relation of ammonia and hydrazine, NH_3 and N_2H_4, illustrates another fundamental law of chemistry: the **law of multiple proportions**. John Dalton, whom we first met in Chapter 2, thought about the fact that an element may form more than one compound with another element. He observed that for a given mass of the first element, the masses of the second element in the two compounds are in the ratio of small whole numbers. In ammonia, 3 moles of H (3.024 g) are combined with 1 mole of N (14.01 g). In hydrazine, we have 4 moles of H combined with 2 moles of N, or 2 mol H/1 mol N. Converting moles to mass, this means we have 2.016 g of H combined with 14.01 g of N. Therefore,

$$\frac{3.024 \text{ g H per } 14.01 \text{ g N in } NH_3}{2.016 \text{ g H per } 14.01 \text{ g N in } N_2H_4} = \frac{3}{2}$$

The masses of H combined with a fixed mass of N are in the same ratio as the number of atoms of H combined per atom of N in the two compounds. This is a useful idea, and one that is often used even today.

EXAMPLE 3.8

EMPIRICAL AND MOLECULAR FORMULAS

A reactive compound contains 13.2% B and 86.8% Cl by weight. (a) What is the empirical formula of the compound? (b) If the molar mass of the compound is known from a separate experiment to be 163 g/mol, what is the molecular formula?

Solution To determine empirical and molecular formulas, there are always *two basic steps*: (1) determine the empirical formula (atom ratios) from percent composition data and (2) determine the molecular formula (total atoms per molecule) from the empirical formula and molar mass.

Step 1. Empirical Formula (Atom Ratios). The percent composition indicates that in 100.0 g of the compound there are 13.2 g of B and 86.8 g of Cl. Therefore, the first step is to express these masses in moles.

$$13.2 \cancel{\text{g B}} \left(\frac{1 \text{ mol B}}{10.81 \cancel{\text{g B}}}\right) = 1.22 \text{ moles B}$$

$$86.8 \cancel{\text{g Cl}} \left(\frac{1 \text{ mol Cl}}{35.45 \cancel{\text{g Cl}}}\right) = 2.45 \text{ moles Cl}$$

The ratio of the number of moles of each element in a given mass is the same as the ratio of the number of atoms in a molecule. Here, that ratio is 2 Cl atoms to 1 B atom.

$$\frac{2.45 \text{ moles Cl}}{1.22 \text{ moles B}} = \frac{2.01 \text{ moles Cl}}{1.00 \text{ mole B}} = \frac{2 \text{ atoms Cl}}{1 \text{ atom B}}$$

Therefore, the *empirical formula* is BCl_2.

Step 2. Molecular Formula (Total Atoms). Any molecular formula having a 1:2 boron-to-chlorine ratio is possible, so the molecular formula could be the same as the empirical formula (BCl_2), twice the empirical formula (B_2Cl_4), three times the empirical formula (B_3Cl_6), and so on. You can find the number of empirical formula units in one molecular unit by dividing the molar mass (163 g/mol) by the mass of the empirical formula unit.

$$\frac{163 \text{ g/mol of compound}}{81.7 \text{ g/mol of } BCl_2} = 2.00 \text{ mol } BCl_2 \text{ per mol compound}$$

Thus, the *molecular formula* must be B_2Cl_4.

EXAMPLE 3.9

EMPIRICAL AND MOLECULAR FORMULAS

One oxide of iron is called "magnetite." It is composed of 72.4% Fe and 27.6% oxygen. What is the empirical formula of magnetite?

Solution First calculate the number of moles of each element in a 100.0 g sample.

$$\text{Moles of Fe} = 72.4 \text{ g Fe} \left(\frac{1 \text{ mol Fe}}{55.85 \text{ g Fe}}\right) = 1.30 \text{ moles Fe}$$

$$\text{Moles of O} = 27.6 \text{ g O} \left(\frac{1 \text{ mol O}}{16.00 \text{ g O}}\right) = 1.73 \text{ moles O}$$

Second, find the ratio of moles of O to moles of Fe.

$$\frac{\text{Moles of O}}{\text{Moles of Fe}} = \frac{1.73 \text{ moles O}}{1.30 \text{ moles Fe}} = \frac{1.33}{1.00}$$

The mole ratio is 1.33/1.00 or $1\frac{1}{3}$ to 1. Since the ratio of atoms in a formula is given in whole numbers, you would express the O to Fe ratio here as 4 to 3. That is, the empirical formula of magnetite is Fe_3O_4.

EXERCISE 3.10 Empirical and Molecular Formulas

Boron hydrides, compounds containing only boron and hydrogen, form a large class of compounds. One consists of 78.14% B and 21.86% H; its molar mass is 27.7 g/mol. What are the empirical and molecular formulas of this compound?

DETERMINING EMPIRICAL AND MOLECULAR FORMULAS

The empirical formula of a compound can be calculated if the percent composition of the compound is known. Then, if the molar mass is determined by any of a variety of experimental methods, we can convert the empirical formula to a molecular formula. But where do the necessary percent composition and molar mass data come from? Both come from methods of *chemical analysis*, methods that we will describe as we cover the necessary background in later chapters. For the moment, however, we can consider two experimental methods of finding empirical formulas.

FORMULAS OF HYDRATED COMPOUNDS Three of the compounds in Figure 3.6 are **hydrated** compounds; that is, molecules of water are associated with the ions of the compounds. The beautiful green nickel(II) compound in Figure 3.6, for example, has the formula $NiCl_2 \cdot 6\ H_2O$. The "dot" between $NiCl_2$ and 6 H_2O indicates that six moles of water are associated with every mole of $NiCl_2$, so the molar mass of $NiCl_2 \cdot 6\ H_2O$ is 129.6 g/mol (for $NiCl_2$) plus 108.1 g/mol (for 6 H_2O) or 237.7 g/mol.

Hundreds of hydrated compounds are known. Since there is no simple way to predict how much water will be involved, it must be determined experimentally. Such an experiment usually involves heating the hydrated material so that all the water is released from the solid and driven away (Figure 3.7). Only the **anhydrous** compound, a substance "without water," is left. According to the law of the conservation of matter (page 42), the sum of the masses of the water that was driven away and the anhydrous compound left behind must equal the mass of the original hydrated compound. For example, if you heat 1 mole (237.7 g) of $NiCl_2 \cdot 6\ H_2O$, then 108.1 grams of water (6 moles) will be lost and 129.6 grams of $NiCl_2$ (1 mole) will remain.

$$\underset{237.7\ g}{\text{Solid } NiCl_2 \cdot 6\ H_2O} + \text{heat} \xrightarrow{\text{will give}} \underset{129.6\ g}{\text{solid } NiCl_2} + \underset{108.1\ g}{H_2O \text{ vapor}}$$

In an experiment of this type, you can *calculate* the mass of water lost if you know the mass of the hydrated material before heating and the mass of the anhydrous material that remains after heating. As outlined in the following example, you can then calculate the ratio of moles of water to moles of anhydrous compound.

EXAMPLE 3.10

DETERMINING THE FORMULA OF A HYDRATED COMPOUND

Hydrated copper(II) sulfate is the deep blue solid in Figure 3.6. Suppose you know only that it has a formula of $CuSO_4 \cdot x\ H_2O$, but you do not yet know the value of x. To find x, you weigh 1.023 g of the blue solid and heat it in a porcelain crucible (Figure 3.7). (Heating is continued until the mass no longer decreases, so you are certain that all the water has been driven off.) The mass of anhydrous, white $CuSO_4$, is 0.654 g. How many moles of water are there per mole of $CuSO_4$? That is, what is the value of x?

Solution (a) Calculate the mass of water released.

$$1.023 \text{ g hydrate} - 0.654 \text{ g anhydrous solid} = 0.369 \text{ g water lost}$$

(b) Calculate the moles of water driven off.

$$0.369 \text{ g} \left(\frac{1 \text{ mol}}{18.02 \text{ g}}\right) = 0.0205 \text{ mole } H_2O$$

(c) All the white, anhydrous solid left is $CuSO_4$. Express the quantity of this solid, 0.654 g, in moles.

$$0.654 \text{ g} \left(\frac{1 \text{ mol}}{159.6 \text{ g}}\right) = 0.00410 \text{ mole } CuSO_4$$

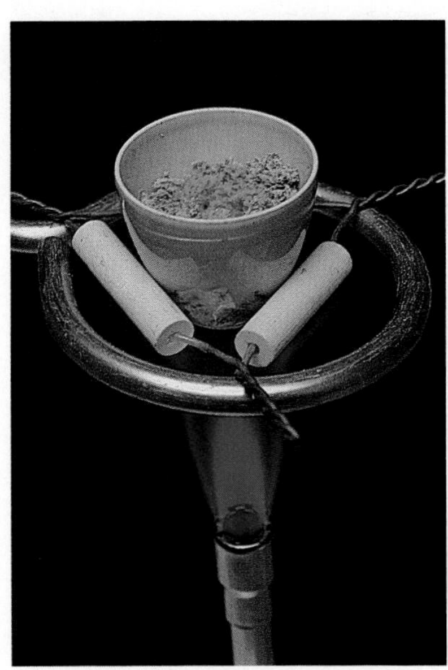

Figure 3.7 A crucible containing a hydrated compound [$CuSO_4 \cdot 5\ H_2O$] is being heated to drive off all the water and leave anhydrous $CuSO_4$.

A box of "20-Mule Team" borax. The name comes from the fact that the mineral borax was found in Death Valley (in California) in the late 19th century. Teams of 20 mules were needed to haul the heavy wagons containing the mineral from the valley.

(d) Find the ratio of moles of water to moles of $CuSO_4$.

$$\frac{\text{moles } H_2O}{\text{moles } CuSO_4} = \frac{0.0205 \text{ mole } H_2O}{0.00410 \text{ mole } CuSO_4} = \frac{5.00}{1.00}$$

(e) The water-to-$CuSO_4$ ratio is 5:1, so the empirical formula of the blue hydrated salt is $CuSO_4 \cdot 5\ H_2O$.

EXERCISE 3.11 Determining the Formula of a Hydrated Compound

Borax powder has the formula $Na_2B_4O_7 \cdot x\ H_2O$. Determine the molar mass of the hydrated compound from the following data: The mass of borax powder taken for analysis was 2.145 g. After heating sufficiently to drive off all the water, the mass of the white powder, $Na_2B_4O_7$, that remained was 1.130 g.

FORMULAS OF SIMPLE, BINARY COMPOUNDS As you will see in the next chapter, elements often combine with oxygen, sulfur, and the halogens to produce simple binary compounds. For example, in Figure 3.1 you saw red phosphorus and bromine combine to give PBr_3,

Solid P + liquid Br_2 —will give→ liquid PBr_3

and other examples include the combination of zinc and oxygen to form zinc oxide (Figure 3.8)

Solid Zn + O_2 gas —will give→ solid ZnO

and the combination of copper and sulfur to give copper(I) sulfide (Figure 3.9).

Solid Cu + solid S —will give→ solid Cu_2S

Figure 3.8 Zinc powder is sprayed into a bunsen burner flame where the powder burns to give zinc oxide, ZnO.

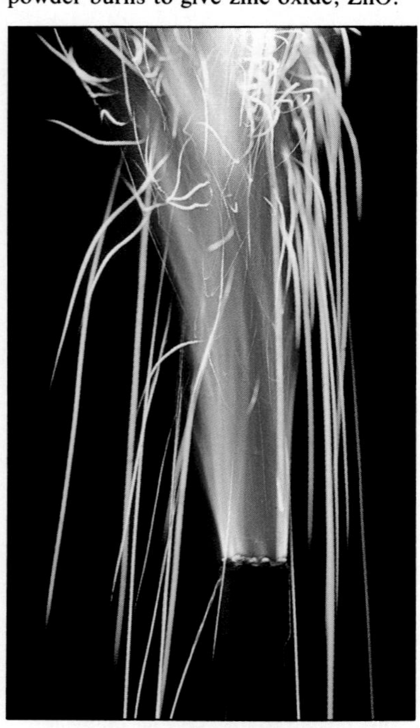

Once again, from the law of the conservation of matter, there are several ways to determine the masses of combining elements and thus the ratio of atoms in the final molecule. The following examples illustrate the methods used.

EXAMPLE 3.11

THE FORMULA OF A METAL SULFIDE

The formula of a compound containing copper and sulfur can be determined by heating a weighed quantity of Cu and S together in a crucible. The quantity of S used is more than that needed to react with all of the Cu present. After the compound has been formed between Cu and S, heating is continued to drive off all the unused S.

The following experimental information was collected in the laboratory:

Mass of crucible	19.732 g
Mass of crucible plus Cu	27.304 g
Mass of crucible plus Cu/S compound after heating	29.214 g

Use this information to determine x and y in Cu_xS_y.

Solution The first step should be to find the masses of Cu and S that have combined. These masses will then be converted to moles of Cu and S, and finally the ratio of moles of Cu and S will give us the empirical formula.

(a) Mass of Cu:

Mass of Cu:	Mass of crucible plus Cu	27.304 g
	Mass of crucible	− 19.732 g
	Mass of Cu	7.572 g
Mass of S:	Mass of crucible plus Cu/S compd. after heat	29.214 g
	Mass of crucible plus Cu	− 27.304 g
	Mass of S	1.910 g

(b) Moles of Cu and S:

$$7.572\ \cancel{\text{g Cu}}\left(\frac{1\ \text{mol}}{63.546\ \cancel{\text{g}}}\right) = 0.1192\ \text{mol Cu}$$

$$1.910\ \cancel{\text{g S}}\left(\frac{1\ \text{mol}}{32.066\ \cancel{\text{g}}}\right) = 0.05956\ \text{mol S}$$

(c) Ratio of moles of Cu and S

$$\frac{0.1192\ \text{mol Cu}}{0.05956\ \text{mol S}} = \frac{2.001\ \text{mol Cu}}{1.000\ \text{mol S}}$$

(d) Step (c) shows there are twice as many moles of Cu as moles of S. Therefore, the empirical formula is Cu_2S.

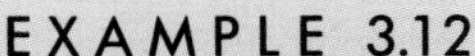

EXAMPLE 3.12

THE FORMULA OF A BINARY OXIDE

Analysis shows that 0.586 g of potassium can combine with 0.480 g of O_2 gas to give a white solid with a formula of K_xO_y. What is the formula of the white solid?

Solution Our problem is to find the values of x and y in K_xO_y. To do this, we simply need to find the number of moles of K and O and then find their ratio.

(a) Calculate moles of K:

$$0.586\ \cancel{\text{g K}}\left(\frac{1\ \text{mol}}{39.10\ \cancel{\text{g}}}\right) = 0.0150\ \text{mol K}$$

(b) Calculate moles of O:

$$0.480\ \cancel{\text{g O}_2}\left(\frac{1\ \cancel{\text{mol O}_2}}{32.00\ \cancel{\text{g}}}\right)\left(\frac{2\ \text{mol O}}{1\ \cancel{\text{mol O}_2}}\right) = 0.0300\ \text{mol O}$$

In this step we had to take into account the fact that 1 mole of O_2 molecules contains 2 moles of O atoms, and it is the moles of O atoms that we need to know.

(c) Ratio of moles:

$$\frac{0.0300\ \text{mol O}}{0.0150\ \text{mol K}} = \frac{2\ \text{moles of O atoms}}{1\ \text{mole of K atoms}}$$

(d) The empirical formula of the compound is KO_2. (This compound is called potassium superoxide, since the oxygen is in the form of the O_2^- ion, the name of which is "superoxide.")

(a)

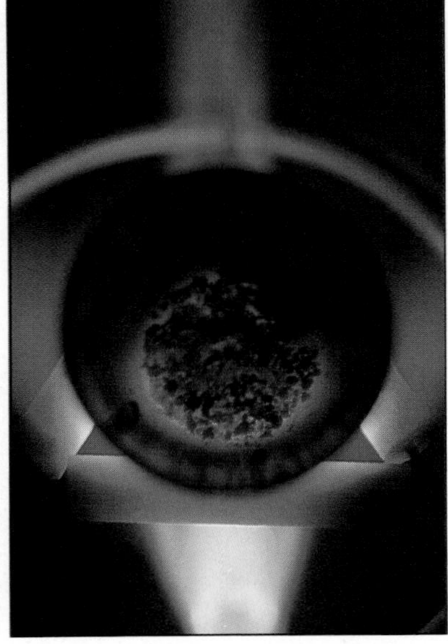

(b)

(c)

Figure 3.9 Samples of copper powder and sulfur (a) are heated in a crucible (b). After heating strongly in air for some minutes, the product is a black copper sulfide (c).

EXERCISE 3.12 Empirical Formula of a Binary Compound
Gaseous chlorine, Cl_2, was combined with 0.532 g of titanium, and 2.108 g of Ti_xCl_y were collected. What is the empirical formula of the titanium chloride?

You should be aware that the examples above illustrate only a few of the ways we use to determine formulas—and only a few of the ways the information can be presented to you. You will see other approaches to the problem of formula determination. In spite of what may appear to be a very different set of circumstances, you should *always focus on using the data to find the number of moles of elements combined in a given mass of compound and then find the ratio of those moles.*

USING MOLECULAR FORMULAS

On occasion chemists are interested in the reverse of the problems described. For example, if you have 12.0 grams of $Pt(NH_3)_2Cl_2$, how many grams of ammonia, NH_3, or how many grams of platinum are present in the 12.0-gram sample? Such a problem can be approached in either of two closely related ways. The way you choose should be the one that makes more sense to you.

EXAMPLE 3.13

USING MOLECULAR FORMULAS

Assume you have 12.0 grams of $Pt(NH_3)_2Cl_2$ (known in medicine as cisplatin). How many grams of platinum are present in the 12.0-gram sample? How many grams of ammonia, NH_3?

Solution

Method 1: Start with 100.0 g of sample.

From the molar mass and the mass of Pt in a mole, you know the compound is 65.02% Pt; that is, there are 65.02 g of Pt in 100.0 g of compound. This is of course the conversion factor we need to get grams of Pt from grams of compound.

$$12.0\ \cancel{g}\ Pt(NH_3)_2Cl_2 \left(\frac{65.02\ g\ Pt}{100.0\ \cancel{g}\ Pt(NH_3)_2Cl_2}\right) = 7.80\ g\ Pt$$

Method 2: Start with 1 mole of sample.

We know that 1 mole of $Pt(NH_3)_2Cl_2$ contains 1 mole of Pt. Therefore, if we know the number of moles of $Pt(NH_3)_2Cl_2$ in the sample, then we also know the number of moles of Pt. From the moles of Pt we can then find the mass of the metal.

$$12.0\ \cancel{g}\ Pt(NH_3)_2Cl_2 \left(\frac{1.000\ mol\ Pt(NH_3)_2Cl_2}{300.1\ \cancel{g}\ Pt(NH_3)_2Cl_2}\right) = 0.0400\ mol\ Pt(NH_3)_2Cl_2$$

$$0.0400\ \cancel{mol}\ Pt(NH_3)_2Cl_2 \left(\frac{1.000\ \cancel{mol}\ Pt}{1.000\ \cancel{mol}\ Pt(NH_3)_2Cl_2}\right)\left(\frac{195.1\ g\ Pt}{1.000\ mol\ Pt}\right) = 7.80\ g\ Pt$$

Now solve for the grams of NH_3 in the sample of $Pt(NH_3)_2Cl_2$, but use only Method 2 this time. From the molecular formula of the compound you know there are two molecules of NH_3 contained within each molecule of $Pt(NH_3)_2Cl_2$. Therefore,

$$0.0400\ \cancel{\text{mol}}\ \text{Pt(NH}_3)_2\text{Cl}_2 \left(\frac{2.000\ \cancel{\text{mol}}\ \text{NH}_3}{1.000\ \cancel{\text{mol}}\ \text{Pt(NH}_3)_2\text{Cl}_2}\right)\left(\frac{17.03\ \text{g NH}_3}{1.000\ \cancel{\text{mol}}\ \text{NH}_3}\right) = 1.36\ \text{g NH}_3$$

The same answer would be obtained if you used Method 1 and the fact that $Pt(NH_3)_2Cl_2$ is 11.35% NH_3 by weight.

EXERCISE 3.13 Using Molecular Formulas

How many grams of chloride ion are contained in 12.0 g of $Pt(NH_3)_2Cl_2$?

A knowledge of molecular formulas, their uses, and their determination is essential to our study of chemistry, and we shall make use of this in the next two chapters in particular.

SUMMARY

A **compound** is a substance containing two or more elements chemically combined in definite proportions; every sample of a compound contains the same elements combined in the same proportions. A **molecule** is the smallest unit of a compound that has the same atom ratio as the entire compound and that retains the chemical characteristics of the compound (Section 3.1).

Pure hydrogen, nitrogen, oxygen, and the halogens exist as two-atom or **diatomic** molecules under normal conditions (Section 3.2). Elemental (white) phosphorus consists of P_4 molecules, while elemental sulfur is chiefly S_8 molecules. Carbon has two well-known forms, diamond and graphite, both of which are extended networks of C atoms.

Many compounds are composed of **ions** (Section 3.3). Metal atoms commonly lose one or more electrons in the course of a chemical reaction to give positive ions or **cations**, while nonmetal atoms usually gain electrons to form **anions** with a negative electrical charge. For a metal in a group other than the transition series, the cation positive charge is often equal to the number of the group in which the element is found (M^{n+}, n = Group number). The negative charge on a single-atom or monatomic anion, X^{n-}, is given by $n = 8 -$ Group number of the element. There are also a number of common many-atom or **polyatomic** negative ions (Table 3.1).

Ionic compounds are formed from the attraction of positive and negative ions in the appropriate number to give an *electrically neutral compound* (a compound with ions contributing equal numbers of positive and negative charges) (Section 3.3). (The force of attraction, given by **Coulomb's law**, depends directly on the ion charges and inversely on the distance between the ions.) As outlined in Section 3.4, ionic compounds are named by giving the name of the cation first followed by the name of the anion.

The **molar mass** of a compound is the mass in grams of Avogadro's number of molecules of the compound. It is numerically equal to the **molecular weight**, the sum of the atomic weights, in atomic mass units, of all of the atoms of the molecule. For ionic compounds, which do not consist of individual molecules, we often refer to the sum of atomic weights as the **formula weight**.

Molecular composition can be expressed in terms of (a) a molecular formula, (b) the mass of each element per mole of compound, or (c) the mass of each element per 100 g of compound (**percent composition**) (Section 3.6). If the percent by weight of each element in a compound is known from experiment, then the **empirical formula** or simplest possible ratio of atoms can be calculated. The empirical formula is converted to the **molecular formula** with the experimentally determined molecular weight.

STUDY QUESTIONS

REVIEW QUESTIONS

1. Compare the words "compound" and "molecule" as used in chemistry.
2. What is the difference between a monatomic ion and a polyatomic ion? Give an example of each.
3. What elements exist as diatomic molecules under normal conditions?
4. What is the difference between molar mass, molecular weight, and formula weight?
5. Which of the following molecules contains more O atoms? More atoms of all kinds? (a) sucrose, $C_{12}H_{22}O_{11}$, or (b) glutathione, $C_{10}H_{17}N_3O_6S$ (the most common low molecular weight, sulfur-containing compound in plant or animal cells).
6. What are three ways of expressing molecular composition?
7. What is the difference between an empirical formula and a molecular formula? Use the compound ethane, C_2H_6, to illustrate your answer.
8. What is a hydrated compound? Use an example from the text to illustrate your answer.

WRITING FORMULAS AND NAMING COMPOUNDS

9. Write the molecular formula of each of the following compounds: (a) Anatase, a common titanium-containing mineral, has a titanium atom and two oxygen atoms per formula unit. (b) One of a series of compounds called the boron hydrides has four boron atoms and ten hydrogen atoms per molecule. (c) Aluminum trimethyl has one aluminum atom, three carbon atoms, and nine hydrogen atoms per formula unit.
10. Write the molecular formula of each of the following compounds: (a) Benzene, a liquid hydrocarbon, has six carbon atoms and six hydrogen atoms per molecule. (b) Vitamin C or ascorbic acid has six carbon atoms, eight hydrogen atoms, and six oxygen atoms per molecule. (c) The mineral barite has one barium atom, one sulfur atom, and four oxygen atoms per formula unit.
11. Give the total number of atoms of each kind in each of the molecules below.
    (a) CaC_2O_4
    (b) $C_6H_5CHCH_2$
    (c) $(NH_4)_2SO_4$
    (d) $Pt(NH_3)_2Cl_2$
    (e) $K_4Fe(CN)_6$
12. Give the total number of atoms of each kind in each of the molecules below.
    (a) $Co_2(CO)_8$
    (b) $HOOCCH_2CH_2COOH$
    (c) $CH_3NH_2CHCOOH$
    (d) $C_{10}H_9NH_2Fe$
    (e) $C_6H_2CH_3(NO_2)_3$
13. Give the symbol, including the correct charge, for each of the following ions:
    (a) barium ion
    (b) aluminum ion
    (c) oxide ion
    (d) cobalt(III) ion
    (e) carbonate ion
    (f) iodide ion
    (g) phosphate ion
    (h) ammonium ion
14. Give the symbol, including the correct charge, for each of the following ions:
    (a) sulfate ion
    (b) silver(I) ion
    (c) manganese(II) ion
    (d) vanadium(III) ion
    (e) permanganate ion
    (f) acetate ion
    (g) perchlorate ion
    (h) sulfide ion
15. Give formulas for the following ionic compounds. Be sure to write the formula by giving the cation first and then the anion.
    (a) potassium iodide
    (b) calcium oxide
    (c) sodium bromide
    (d) magnesium sulfide
    (e) sodium carbonate
    (f) ammonium nitrate
16. Give formulas for the following ionic compounds. Be sure to write the formula by giving the cation first and then the anion.
    (a) ammonium sulfate
    (b) aluminum carbonate
    (c) calcium phosphate
    (d) gallium(III) oxide
    (e) potassium cyanide
    (f) chromium(III) oxide
    (g) tin(II) sulfide
    (h) iron(III) hydroxide
17. Give the formula for each of the following ionic compounds.
    (a) sodium acetate
    (b) silver perchlorate
    (c) potassium chlorate
    (d) titanium(IV) bromide
    (e) sodium nitrate
    (f) manganese(II) sulfate
18. Give the formula for each of the following ionic compounds.
    (a) platinum(II) chloride
    (b) calcium hypochlorite

(c) cesium hydroxide
(d) lithium oxide
(e) calcium hydrogen phosphate
(f) uranium(IV) fluoride

19. Name each of the following ionic compounds and give the formula, charge, and number of each ion that makes up the compound.
(a) $NaHCO_3$ (e) NaCN
(b) $Ca_3(PO_4)_2$ (f) $CuSO_4$
(c) NH_4Br (g) $KMnO_4$
(d) $KClO_4$ (h) ZnO

20. Name each of the following ionic compounds and give the formula, charge, and number of each ion that makes up the compound.
(a) $MgCO_3$ (e) Na_3PO_4
(b) $Fe(ClO_4)_3$ (f) CoO and Co_2O_3
(c) $(NH_4)_2SO_3$ (g) $CrCl_3$
(d) $Be(CH_3COO)_2$ (h) $SnCl_2$ and $SnCl_4$

21. Write the formulas of all of the compounds that can be made by combining each of the cations with each of the anions listed. Name each compound formed.

CATIONS	ANIONS
Na^+	CO_3^{2-}
Sr^{2+}	I^-
NH_4^+	NO_3^-

22. Write the formulas of all of the compounds that can be made by combining each of the cations with each of the anions listed. Name each compound formed.

CATIONS	ANIONS
Fe^{3+}	Cl^-
Na^+	SO_4^{2-}
Mg^{2+}	S^{2-}

23. Give the formula of each of the following binary nonmetal or metalloid compounds.
(a) boron tribromide
(b) xenon difluoride
(c) hydrogen iodide
(d) disulfur dichloride

24. Give the formula of each of the following binary nonmetal compounds.
(a) dichlorine oxide
(b) hydrogen sulfide
(c) xenon tetrafluoride
(d) sulfur trioxide

MOLAR MASS, MOLES, AND AVOGADRO'S NUMBER

25. Calculate the molar mass of each of the following compounds:
(a) Fe_2O_3, iron(III) oxide
(b) BF_3, boron trifluoride
(c) N_2O, dinitrogen oxide (laughing gas)
(d) $MnCl_2 \cdot 4\ H_2O$, manganese(II) chloride tetrahydrate
(e) $C_6H_8O_6$, ascorbic acid or vitamin C

26. Calculate the molar mass of each of the following compounds:
(a) $B_{10}H_{14}$, a boron hydride once considered as a rocket fuel
(b) $C_6H_2(CH_3)(NO_2)_3$, TNT, an explosive
(c) $Pt(NH_3)_2Cl_2$, a cancer chemotherapy agent called cisplatin
(d) $CH_3CH_2CH_2CH_2SH$, has a skunk-like odor
(e) $C_{20}H_{24}N_2O_2$, quinine, used as an antimalarial drug

27. How many moles are represented by 1.00 g of each of the following compounds?
(a) CH_3OH, methyl alcohol
(b) Cl_2CO, phosgene, a poisonous gas
(c) NH_4NO_3, ammonium nitrate
(d) $MgSO_4 \cdot 7\ H_2O$, magnesium sulfate heptahydrate (epsom salt)
(e) $AgCH_3COO$, silver acetate

28. Assume you have 0.250 g of each of the following compounds. How many moles of each are represented?
(a) $C_7H_5NO_3S$, saccharin, an artificial sweetener
(b) $C_{13}H_{20}N_2O_2$, procaine, a "pain killer" used by dentists
(c) $C_{20}H_{14}O_4$, phenolphthalein, a dye

29. Vinyl chloride, C_2H_3Cl, is used to make polyvinylchloride (PVC), a plastic from which many useful items are made. If you have 1.00 pound of vinyl chloride, how many moles and how many molecules of the compound are present? (1.00 lb = 454 g)

30. Benzene, C_6H_6, is an important industrial chemical and is produced in large amounts in spite of the fact that it is implicated as a cause of leukemia. In 1988, industrial production of benzene was 11.84×10^9 pounds. How many moles does this represent? (1.00 lb = 454 g)

31. An Alka-Seltzer tablet contains 324 mg of aspirin ($C_9H_8O_4$), 1904 mg of $NaHCO_3$, and 1000. mg of citric acid ($C_6H_8O_7$). (The last two compounds react with each other to provide the "fizz," bubbles of CO_2, when the tablet is put in water.) (a) Calculate the number of moles of each substance in the tablet. (b) If you take one tablet, how many molecules of aspirin are you consuming?

32. Some types of chlorofluorocarbons (CFC's) have been used as the propellant in spray cans of paint, hair spray, and other consumer products. However, the use of CFC's is being curtailed, because there is strong suspicion that they cause environmental damage. If there are 250 g of one such compound, CCl_2F_2, in a

spray can, how many molecules are you releasing to the air when you empty the can?

33. Sulfur trioxide, SO_3, is made in enormous quantities by combining oxygen and sulfur dioxide, SO_2. The trioxide is not usually isolated but is converted to sulfuric acid. If you have 1.00 pound (454 g) of sulfur trioxide, how many moles does this represent? How many molecules? How many sulfur atoms? How many oxygen atoms?

34. Chlorofluorocarbons are strongly suspected of causing environmental damage. A substitute may be CF_3CH_2F. If you have 25.5 g of this new compound, how many moles does this represent? How many molecules? How many atoms of fluorine are contained in 25.5 g of the compound?

35. You use 0.55 g of vanillin, $C_8H_8O_3$, in some ice cream. How many moles does this represent? How many molecules? When you eat the vanillin, how many atoms of carbon are you consuming?

36. Glycerin, $C_3H_8O_3$, is a syrupy liquid used in cosmetics, as a sweetener, in glue, and in many other consumer products. If you have 250 g of the compound, how many moles does this represent? How many molecules? How many atoms of carbon are contained in 250 g of glycerin?

PERCENT COMPOSITION

37. Calculate the molar mass of these compounds and the weight percent of each element.
   (a) PbS, lead(II) sulfide, galena
   (b) C_2H_6, ethane, a hydrocarbon fuel
   (c) CH_3COOH, acetic acid, an important ingredient in vinegar
   (d) NH_4NO_3, ammonium nitrate

38. Calculate the molar mass of these compounds and the weight percent of each element.
   (a) $MgCO_3$, magnesium carbonate
   (b) C_6H_5OH, phenol, an organic compound used in some cleaners
   (c) $C_2H_3O_5N$, peroxyacetyl nitrate, an objectionable compound in photochemical smog
   (d) $C_4H_{10}O_3NPS$, acephate, an insecticide

39. Acrylonitrile, H_2CCHCN, is the basis of many important plastics and fibers. (a) Calculate the molar mass. (b) Calculate the weight percent of each element in the compound.

40. The copper-containing compound $Cu(NH_3)_4SO_4 \cdot H_2O$ is a beautiful blue solid (see the photo). Calculate the molar mass of the compound and the weight percent of each element.

41. Hexachlorophene, $C_{13}H_6Cl_6O_2$, is a germicide in soaps. Calculate the weight percent of each element in the compound.

Copper-containing compound $Cu(NH_3)_4SO_4 \cdot H_2O$.

42. The formula of DDT, an insecticide that was once widely used, is $C_{14}H_9Cl_5$. Calculate the molar mass of the compound and the weight percent of each element.

EMPIRICAL AND MOLECULAR FORMULAS

43. The empirical formula of maleic acid is CHO. Its molar mass is 116.1 g/mol. What is its molecular formula?

44. A well known reagent in analytical chemistry, dimethylglyoxime, has the empirical formula C_2H_4NO. If its molar mass is 116.1 g/mol, what is the molecular formula of the compound?

45. Acetylene is a colorless gas that is used as a fuel in welding torches, among other things. It is 92.26% C and 7.74% H. Its molar mass is 26.04 g/mol. Calculate the empirical and molecular formulas.

46. There is a large family of boron-hydrogen compounds called boron hydrides. All have the formula B_xH_y and almost all react with air and burn or explode. One member of this family contains 88.5% B; the remainder is hydrogen. Which of the following is its empirical formula: BH_3, B_4H_{10}, B_5H_7, B_5H_{11}, or B_6H_{12}?

47. Nitrogen and oxygen form a series of at least seven oxides with the general formula N_xO_y. One of them is a blue solid that comes apart, reversibly, in the gas phase. It contains 36.85% N. What is the empirical formula of this oxide?

48. Cumene is a hydrocarbon, a compound composed only of C and H. It is 89.94% carbon, and the molar mass is 120.2 g/mol. What are the empirical and molecular formulas of cumene?

49. Acetic acid is the important ingredient in vinegar. It is composed of carbon (40.0%), hydrogen (6.71%), and oxygen (53.28%). Its molar mass is 60.1 g/mol. Determine the empirical and molecular formulas of the acid.

50. An analysis of nicotine, a poisonous compound found in tobacco leaves, shows that it is 74.03% C, 8.70% H, and 17.27% N. Its molar mass is 162 g/mol. What are the empirical and molecular formulas of nicotine?

51. Cacodyl, a compound containing arsenic, was reported in 1842 by the German chemist Bunsen. It has an almost intolerable garlic-like odor. Its molar mass

is 210 g/mol, and it is 22.88% C, 5.76% H, and 71.36% As. Determine its empirical and molecular formulas.

52. The action of bacteria on meat and fish produces a poisonous compound called cadaverine. As its name and origin imply, it stinks! It is 58.77% C, 13.81% H, and 27.42% N. Its molar mass is 102.2 g/mol. Determine the molecular formula of cadaverine.

53. Vanillin is a common flavoring agent. It has a molar mass of 152 g/mol and is 63.15% C and 5.30% H; the remainder is oxygen. Determine the molecular formula of vanillin.

54. Fluorocarbonyl hypofluorite was recently isolated, and analysis showed it to be 14.6% C, 39.0% O, and 46.3% F. If the molar mass of the compound is 82 g/mol, determine the empirical and molecular formulas of the compound.

55. Naphthalene, best known in the form of "moth balls," is composed only of carbon (93.71%) and hydrogen (6.29%). If the molar mass of the compound is 128 g/mol, what is the molecular formula of naphthalene?

56. A major oil company has used a gasoline additive called MMT to boost the octane rating of its gasoline. What is the empirical formula of MMT if it is 49.6% C, 3.20% H, 22.0% O, and 25.2% Mn?

57. Ruthenium chemistry is quite interesting, and a good starting material for such studies is $RuCl_3 \cdot x\ H_2O$. If you heat 1.056 g of the hydrated salt and find that only 0.838 g of $RuCl_3$ remains when all of the water has been driven off, what is the value of x?

58. The "alum" used in cooking is potassium aluminum sulfate hydrate, $KAl(SO_4)_2 \cdot x\ H_2O$ (see the photo). To find the value of x, you can heat a sample of the compound to drive off all of the water and leave only $KAl(SO_4)_2$. Assume that you heat 4.74 g of the hydrated compound and that it loses 2.16 g of water. What is the value of x?

59. If "epsom salt," $MgSO_4 \cdot x\ H_2O$, is heated to 250 °C, all the water of hydration is lost. On heating a 1.687-g sample of the hydrate, 0.824 g of $MgSO_4$ remains. How many molecules of water are there per formula unit of $MgSO_4$?

Alum, $KAl(SO_4)_2 \cdot x\ H_2O$ (for x see question 58).

60. Copper sulfate as commonly used in the laboratory is the hydrated compound $CuSO_4 \cdot 5\ H_2O$ (see Figure 3.5). If you heat the solid to at least 150 °C, all of the water of hydration is lost. On heating 10.5 g of $CuSO_4 \cdot 5\ H_2O$ to this temperature, how many grams of water would be lost and how many grams of anhydrous $CuSO_4$ would remain?

61. A package of baking soda contains 2.00 pounds (908 g) of $NaHCO_3$. (a) What is the name of the compound $NaHCO_3$? (b) How many moles of $NaHCO_3$ are there in the package? (c) How many moles and how many grams of oxygen atoms are in the package?

62. Monosodium glutamate, MSG, is a common food additive; its formula is $HOOCCH_2CH_2CH(NH_2)COONa$. (a) Calculate the molar mass for MSG. (b) How many moles of MSG are there in 2.00 g (about 1 teaspoonful) of MSG? (c) How many moles and how many atoms of O are there in 2.00 g of MSG?

63. The mineral fluorite is calcium fluoride, CaF_2. (a) Calculate the molar mass of the compound. (b) How many moles of CaF_2 are there in 1.56 g of the compound? (c) How many grams of CaF_2 must you have in order to have 12.0 g of fluoride ion?

64. The most important beryllium-containing mineral is beryl, which occurs mostly as large blue-green crystals with the formula $Be_3Al_2(SiO_3)_6$. (a) What is the formula weight of beryl? (b) How many moles of beryl are there in a 0.25-g crystal? (c) How many grams of beryl must you have in order to have 10. g of beryllium?

65. A 1.256-g sample of elemental sulfur is combined with fluorine, F_2, to give a compound with the formula SF_x, a very stable, colorless gas. If you isolate 5.722 g of SF_x, what is the value of x?

66. A new compound containing xenon and fluorine was created by shining sunlight on a mixture of 0.526 g of Xe and an excess of F_2 gas. If you isolate 0.678 g of the new compound, what is its empirical formula?

67. A chlorine-containing compound of sulfur, S_xCl_y, is prepared starting with SCl_2. If you find the new compound is 52.51% chlorine, and its molar mass is 135.0 g/mol, what are its empirical and molecular formulas?

68. Direct reaction of iodine (I_2) and chlorine (Cl_2) produces an iodine chloride, I_xCl_y, a bright yellow solid. If you completely used up 0.678 g of iodine, and produced 1.246 g of I_xCl_y, what is the empirical formula of the compound? Later experiment showed the molar mass of I_xCl_y was 467 g/mol. What is the molecular formula of the compound?

GENERAL QUESTIONS

69. Name all the compounds in the following chemical equation:

$$Cd(NO_3)_2 + (NH_4)_2S \longrightarrow 2\ NH_4NO_3 + CdS$$

70. Rotenone, $C_{23}H_{22}O_6$, is the active component in many garden insecticides. If a can of insecticide contains 0.500 pound (227 g) of powdered insecticide and the insecticide is 5.0% by weight rotenone, how many moles of rotenone are present? How many molecules?

71. Pepto-Bismol, which helps provide soothing relief for an upset stomach, contains 300. mg of bismuth subsalicylate, $C_7H_5BiO_4$, per tablet. If you take two tablets for your stomach distress, how many moles of the "active ingredient" are you taking? How many grams of Bi are you consuming in two tablets?

72. Tin(II) fluoride has been used as a source of fluoride ion in some brands of toothpaste to prevent tooth decay. (a) How many moles of SnF_2 are there in 0.050 g of SnF_2? (b) How many F^- ions and how many Sn^{2+} ions are present in 0.050 g?

73. Strychnine, $C_{21}H_{22}N_2O_2$, is a powerful poison and has been used to eradicate rats. If a can of rat poison contains 0.75 g of strychnine, how many moles and how many molecules of the compound are present?

74. Metal carbonyls are molecular compounds formed by metals and carbon monoxide, CO. Although they can be quite stable at room temperature, they usually decompose to the metal and CO when heated in a vacuum. Assume that you heat 1.400 g of $Mn_x(CO)_y$ and obtain 0.394 g of pure manganese. The CO is lost in the course of the experiment. If the molar mass of the compound is 390 g/mol, what are the empirical and molecular formulas of the compound?

75. Quinine, isolated from the bark of the cinchona tree, is used in the form of its salt with HCl as an important antimalarial drug. The formula of the drug is $C_{20}H_{24}N_2O_2 \cdot 2\ HCl$. The usual dose of the drug is 2.0 g per day. How many moles of the drug would you be taking at this dose level?

76. A compound with a formula M_3N contains 0.673 g of N per gram of the metal M. What is the atomic weight of M? What element is M?

77. What is the molecular formula of a substance that contains 1.00 mole of S, 24.1×10^{23} atoms of F, and 71.0 grams of Cl in 1.00 mole of sample?

78. A compound containing only a metal and oxygen, MO, can be decomposed to the elements (M and O_2) by heating. If 4.386 g of the compound form 4.062 g of M on heating, what is the atomic weight of M? What is the probable identity of M?

79. The hemoglobin from the red corpuscles of most mammals contains about 0.33% iron. Physical measurements indicate that hemoglobin is a very large molecule with a molar mass of about 6.8×10^4 g/mol. How many moles of Fe are there in one mole of hemoglobin? How many iron atoms are there in one molecule of hemoglobin?

80. Arrange the following in order of increasing mass:
(a) 3.0×10^{23} molecules of C_4H_{10}
(b) 1 penny (about 3 g)
(c) 6.0×10^{23} molecules of CO
(d) 1.0 mole of B_2H_6
(e) 1 molecule of N_2

81. A chemical commonly called "dioxin" has been very much in the news in the past few years. (It is the by-product of herbicide manufacture and is thought to be quite toxic.) Its formula is $C_{12}H_4Cl_4O_2$. If you have a sample of dirt (1 ounce or 28.3 g) that contains 1.0×10^{-4}% dioxin, how many moles of dioxin are in the dirt sample?

82. A close chemical relative of DDT is methoxychlor ($C_{16}H_{15}Cl_3O_2$), a compound used in garden insecticides. (a) Calculate the weight percentage of each element in the compound. (b) If the estimated fatal dose for a human is 7.5 g/kg of body weight, what is your fatal dose? (1.00 lb = 0.454 kg)

83. DDT, $C_{14}H_9Cl_5$, is an insecticide and belongs to a class of compounds called "chlorinated hydrocarbons." Although DDT was enormously successful in controlling insects, it has also caused considerable environmental damage by interfering with the reproduction of birds. If you use 1.00 pound of DDT on several acres of farmland, how many molecules of the compound are you spreading over the fields? How many grams of chlorine (as Cl) are contained in 1.00 pound of DDT? (1.00 lb = 454 g)

84. Vitamin C, ascorbic acid, has the formula $C_6H_8O_6$. (a) The recommended daily dose of vitamin C is 60 milligrams. How many moles are you consuming if you ingest 60 mg of the vitamin? (b) A typical tablet contains 1.00 g of vitamin C. How many moles of vitamin C does this represent? (c) When you consume 1.00 gram of vitamin C, how many oxygen atoms are you eating?

85. Some pure powdered copper metal (1.056 g) is heated in a stream of oxygen (O_2) to give 1.322 g of pure copper oxide, Cu_xO_y. What is the empirical formula of the copper oxide?

86. Cassiterite is tin(IV) oxide. It is the most common ore of tin, being found in England and Bolivia among other places. If you have 23.5 kg of the ore, how many grams of tin metal can you obtain from the ore?

87. Iron pyrite, often called "fool's gold," has the formula FeS_2 (see the photo). If you could convert 15.8 kg of iron pyrite to iron metal, how many kilograms of the metal could you obtain?

Pyrite (the gold colored mineral on quartz).

88. Ilmenite is a mineral that is an oxide of iron and titanium, $FeTiO_3$. If you have an ore containing ilmenite and other minerals, and the ore is 6.75% titanium, how many grams of ilmenite are there in 1.00 metric ton (exactly 1000 kg) of the ore?

89. Stibnite, Sb_2S_3, is a dark gray mineral from which antimony metal is obtained. If you have 1.00 pound of an ore that contains 10.6% antimony, how many grams of Sb_2S_3 are there in the ore? (1 lb = 454 g)

90. The beautiful blue mineral sodalite (see the photo) is a chlorine-containing sodium aluminosilicate. It is 18.98% Na, 16.7% Al, 17.39% Si, and 7.32% Cl. The remainder is oxygen. What is the empirical formula of sodalite?

The mineral sodalite.

91. Azurite is a copper-containing mineral (see the photo) that is a mixture of copper(II) carbonate and copper(II) hydroxide, $[CuCO_3]_x \cdot [Cu(OH)_2]_y$. If the mineral is 55.31% Cu, 0.58% H, and 6.97% C, with the remainder oxygen, what are x and y in the general formula?

Azurite, a copper-containing mineral.

SUMMARY PROBLEMS

Beginning with this chapter, there will be one or more problems at the end of the Study Questions that call upon you to use concepts learned in previous chapters in the book.

92. A piece of nickel foil, 0.550 mm thick and 1.25 cm square, reacted with fluorine, F_2, to give a nickel fluoride. (a) How many moles of nickel foil were used? (b) If you isolate 1.261 g of the nickel fluoride, what is its formula? (c) What is its complete name? (The density of nickel is 8.908 g/cm^3.)

93. A piece of aluminum foil, 0.450 mm thick, 1.50 cm wide, and 1.95 cm long, reacted with liquid bromine, Br_2. (a) How many moles and how many atoms of aluminum were used? (The density of aluminum is 2.699 g/cm^3.) (b) After the reaction you find that 3.51 g of a compound of aluminum and bromine were formed. What is the empirical formula of the compound? (c) The molar mass was found to be 533 g/mol in a later experiment. What is the molecular formula of the compound? (d) What is the name of the compound? (See Figure 4.3.)

94. Uranium is used as a fuel, primarily in the form of uranium(IV) oxide, in nuclear power plants. This question considers some uranium chemistry.

(a) A small sample of uranium metal (0.169 g) is heated to 800 to 900 °C in air to give 0.199 g of a dark-green oxide, U_xO_y. How many moles of uranium metal were used? What is the empirical formula of the oxide? How many moles of U_xO_y must have been obtained?

(b) The naturally occurring isotopes of uranium are ^{234}U, ^{235}U, and ^{238}U. Which is the most abundant?

(c) The oxide U_xO_y is obtained if $UO_2(NO_3)_2 \cdot zH_2O$ is heated to temperatures higher than 800 °C in the air. However, if you heat it gently, only the waters of hydration are lost. If you start with 0.865 g of $UO_2(NO_3)_2 \cdot zH_2O$, and obtain 0.679 g of $UO_2(NO_3)_2$ after heating, how many waters of hydration were there in the original compound?

CHAPTER

Chemical Reactions: An Introduction

Magnesium reacting with dry ice.

CHAPTER OUTLINE

The essence of chemistry is the study of chemical reactions, the combination of the elements and their compounds to give new compounds. Chemists have used reactions to produce the materials—Teflon, nylon, Dacron, Kevlar, polystyrene, and PVC, among others—so vital in our modern economy. Cancer chemotherapy agents such as cisplatin [$Pt(NH_3)_2Cl_2$] are produced by chemical reactions, as are experimental drugs to treat AIDS. Plants and animals—including you and us—are chemical reaction factories. Your automobile operates on the energy produced in a combustion reaction, and so may the generating plant that produces the electricity you use. Many of these reactions are quite complex and are as yet poorly understood. However, there are certain principles that govern all chemical reactions, and we begin to discuss some of these in this chapter.

Figure 4.1 Sulfur burns in pure oxygen with a bright blue flame to give SO_2, sulfur dioxide.

4.1 CHEMICAL EQUATIONS

Take a piece of sulfur, ignite it, and then plunge it into a flask of pure oxygen. It will burn with a brilliant, blue flame that can light up a room (Figure 4.1). The process or **chemical reaction** that occurs is the combination of solid sulfur molecules with gaseous O_2 molecules to give colorless, gaseous sulfur dioxide, SO_2. We can depict this by the following

balanced chemical equation that shows the relative amounts of **reactants** (the materials put into the reaction) and **products** (the material obtained).

$$S_8(s) + 8\ O_2(g) \longrightarrow 8\ SO_2(g)$$

In most chemical equations, we will also indicate the physical states of the reactants and products. The symbol (s) indicates a solid, (g) a gas, and (ℓ) a liquid; finally, (aq) tells you that the substance is dissolved in water to make an *aq*ueous solution. What the equation does *not* show is the conditions of the experiment or whether any energy—in the form of heat or light—may be involved. Lastly, just looking at a chemical equation does not tell you whether the reaction happens explosively or whether it might take 100 years.

The speed of chemical reactions is discussed in Chapter 15.

In the 18th century, the great French scientist Antoine Lavoisier showed that there is conservation of matter in any chemical process: matter can neither be created nor destroyed. Any balanced chemical equation shows that, although the atoms are rearranged, all the atoms in the reactants *must* appear in the products and vice versa. This means that 1 molecule of sulfur, which consists of 8 sulfur atoms locked in a ring, and 8 diatomic molecules of O_2 must produce 8 molecules of SO_2. Thus, the balanced equation tells you that 8 atoms of S were used with 16 atoms of O (8 molecules with 2 atoms each) and that they appear as 8 molecules of SO_2, each containing 1 atom of S and 2 atoms of O.

The balanced equation for the burning or **combustion** of sulfur also tells us that 1000 S_8 molecules will react with 8000 O_2 molecules to give 8000 molecules of SO_2. To carry this argument even further, it is also true that 6.022×10^{23} molecules of S_8 (1 mole) react with $8 \times 6.022 \times 10^{23}$ molecules of O_2 (8 moles) to provide $8 \times 6.022 \times 10^{23}$ molecules of SO_2 (8 moles). As demanded by the conservation of matter, the total number of atoms used as reactants was 1.445×10^{25}, and the total number of atoms appearing in the product was 1.445×10^{25}.

$S_8(s)$	$+\ 8\ O_2(g)$	$\longrightarrow 8\ SO_2(g)$
1 molecule S_8	8 molecules O_2	8 molecules SO_2
8 atoms S	8×2 atoms O	8 atoms S + 16 atoms O
1 mole S_8	8 moles O_2	8 moles SO_2
$8 \times 6.022 \times 10^{23}$ atoms S	$8 \times 2 \times 6.022 \times 10^{23}$ atoms O	$8 \times 3 \times 6.022 \times 10^{23}$ atoms S and O
1.445×10^{25} total atoms of reactants		1.445×10^{25} total atoms in product

From Chapters 2 and 3 you know that 1.000 mole of S_8 is equivalent to 256.5 g and that 1.000 mole of O_2 amounts to 32.00 g. The balanced equation tells you that, if one mole of S_8 is consumed (256.5 g), then 8 moles of O_2 must also be used (256.0 g), so the total mass of reactants must be

$$\begin{array}{rl} 1 \text{ mole } S_8 & = 256.5 \text{ g} \\ +\ 8 \text{ moles } O_2 & = 256.0 \text{ g} \\ \hline \text{Total mass of reactants} & = 512.5 \text{ g} \end{array}$$

Historical Figures in Chemistry: Antoine Laurent Lavoisier (1743–1794)

On Monday, August 1, 1774, the Englishman Joseph Priestley (1733–1804) first isolated oxygen from the decomposition of mercury(II) oxide, HgO. Priestley did not at first recognize the importance of the discovery, but he mentioned it to Lavoisier in October, 1774. One of Lavoisier's contributions to science was his recognition of the importance of exact scientific measurements and of carefully planned experiments, and he applied these methods to the study of oxygen. From this work he came to believe the gas entered the composition of all acids and so named it "oxygen," from Greek words meaning "to form an acid." In addition, his experiments suggested to him that oxygen combined with carbon in the body, a reaction that was the source of the heat in a living organism. Although he did not understand the details of the process, this was a step in the development of biochemistry.

Lavoisier was a prodigious scientist and introduced principles of naming chemical substances that are still in use today. Further, he wrote a textbook in which he applied for the first time the principle of the conservation of matter to chemistry and used the idea to write early versions of chemical equations.

As Lavoisier was an aristocrat, he came under suspicion during the Reign of Terror of the French Revolution, and his career was cut short on May 8, 1794, by the guillotine.

Lavoisier

The principle of the conservation of matter demands that the same mass, 512.5 g of SO_2, must result from the reaction. Of course, the balanced equation shows that this is the case.

$$8.000 \cancel{\text{mol}}\ SO_2 \left(\frac{64.065\ \text{g}}{1\ \cancel{\text{mol}}}\right) = 512.5\ \text{g}\ SO_2$$

The relationship between the masses of chemical reactants and products is called **stoichiometry**, and the coefficients (or multiplying numbers) in a balanced equation are the **stoichiometric coefficients**. We shall take up the important role of stoichiometry in this chapter and in Chapter 5.

The word "stoichiometry," pronounced "stoy-key-AHM-uh-tree," is derived from the Greek words stoicheion *(meaning "element") and* metron *(meaning "measure").*

4.2 BALANCING CHEMICAL EQUATIONS

A chemical equation must be balanced before useful quantitative information can be obtained. Balancing an equation ensures that the same number of atoms of each element appears on both sides of the equation. All chemical equations can be balanced by inspection, although some will involve more inspection than others.

One general class of chemical reactions involves the *reaction of metals with halogens such as Cl_2, Br_2, and I_2* to give ionic compounds with a general formula of M_aX_b. For example, zinc reacts with iodine to give zinc iodide (Figure 4.2),

$$Zn(s) + I_2(s) \longrightarrow ZnI_2(s)$$

Figure 4.2 Zinc metal (left side of lefthand photo) reacts with iodine, I_2 (right side of lefthand photo). The vigorous reaction produces zinc iodide, ZnI_2. The heat of the reaction is great enough that excess iodine evaporates as a purple vapor.

and aluminum reacts vigorously with liquid bromine to give aluminum bromide, Al_2Br_6 (Figure 4.3).

$$Al(s) + Br_2(\ell) \xrightarrow{\text{(unbalanced equation)}} Al_2Br_6(s)$$

The Zn/I_2 equation is balanced as written above, since there are one Zn atom and two I atoms on each side. For the Al/Br_2 equation, however, we balance the Al atoms by placing a 2 in front of Al on the left.

$$2\ Al(s) + Br_2(\ell) \xrightarrow{\text{(unbalanced equation)}} Al_2Br_6(s)$$

and then a 3 in front of the Br_2 on the left to balance the 6 Br atoms appearing on the right side of the equation.

NOTE: The subscripts in the formulas of reactants and products are fixed by the rules of chemical bonding. The subscripts cannot be changed in order to balance an equation.

$$2\ Al(s) + 3\ Br_2(\ell) \xrightarrow{\text{(balanced equation)}} Al_2Br_6(s)$$

Reactions with oxygen, O_2, are another major class of chemical reactions (and are described in Section 4.3). A **combustion reaction**, the

Figure 4.3 Bromine (Br_2), an orange-brown liquid, and aluminum metal (a) react so vigorously that the aluminum becomes molten and glows white hot (b). The vapor in (b) consists of vaporized Br_2 and some of the product, Al_2Br_6. At the end of the reaction (c) the beaker is coated with aluminum bromide and products of its reaction with atmospheric water.

(a)

(b)

(c)

reaction of a compound with O_2 to form products in which all elements are combined with oxygen, is one such reaction. Such reactions are typical of hydrocarbons, compounds containing only C and H, as well as compounds containing only C, H, and O. An excellent example is the combustion of propane, C_3H_8, a gaseous compound widely used in industry and homes as LP gas (Figure 4.4).

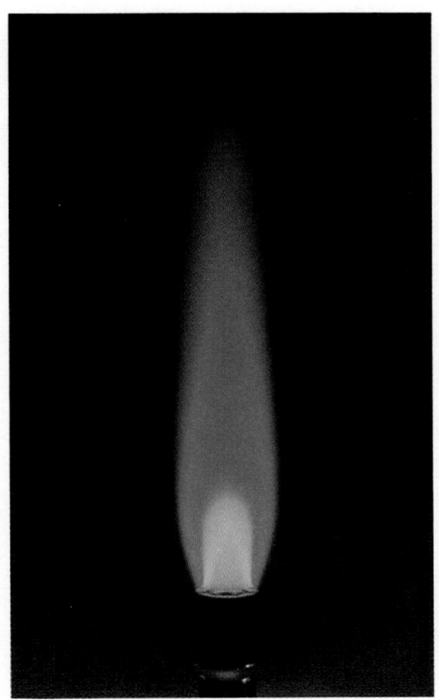

Figure 4.4 Propane, C_3H_8, burning in a laboratory burner in a combustion reaction to give CO_2 and H_2O. The different flame colors are due to zones of different temperature, the coolest zone being just above the burner surface.

$$C_3H_8(g) + O_2(g) \xrightarrow{\text{(unbalanced equation)}} CO_2(g) + H_2O(g)$$

Compounds having only C and H (or C, H, and O) always give just CO_2 and H_2O on complete combustion. Although there are several systematic ways to balance such an equation, our advice is usually to leave the oxygen balance to the last step. Generally, it is best to balance the C atoms first, followed by H, and then O.

Step 1. *Balance the C atoms.* There are 3 carbon atoms in the reactants, so there must be 3 in the products. Therefore, we need 3 CO_2 molecules on the right side.

$$C_3H_8 + O_2 \longrightarrow 3\ CO_2 + H_2O$$

Step 2. *Balance the H atoms.* There are 8 H atoms in the reactants. Each molecule of water has 2 H atoms, so 4 molecules of water will have the required 8 H atoms.

$$C_3H_8 + O_2 \longrightarrow 3\ CO_2 + 4\ H_2O$$

Step 3. *Balance the oxygen atoms.* There are 10 oxygen atoms on the right side ($3 \times 2 = 6$ in CO_2 and $4 \times 1 = 4$ in water). Therefore, you need 5 O_2 molecules to supply the required 10 oxygen atoms.

$$C_3H_8 + 5\ O_2 \longrightarrow 3\ CO_2 + 4\ H_2O$$

Step 4. *Verify that each element is balanced.* The reaction involves 3 C atoms, 8 H atoms, and 10 O atoms on each side.

EXAMPLE 4.1

BALANCING THE EQUATION FOR A COMBUSTION REACTION

Write a balanced equation for the combustion of butane, C_4H_{10}.

Solution As is always the case for compounds containing only C, H, and O, the products will be CO_2 and H_2O if the reaction goes to completion. Therefore, the unbalanced equation is

$$C_4H_{10}(g) + O_2(g) \longrightarrow CO_2(g) + H_2O(g)$$

Step 1. *Balance the C atoms.* 4 C atoms in butane require the production of 4 CO_2 molecules.

$$C_4H_{10} + O_2 \longrightarrow 4\ CO_2 + H_2O$$

Step 2. *Balance the H atoms.* There are 10 H atoms on the left, so 5 molecules of H_2O are required on the right.

$$C_4H_{10} + O_2 \longrightarrow 4\ CO_2 + 5\ H_2O$$

Step 3. *Balance the O atoms.* As the reaction stands after step 2, there are 13 O atoms on the right ($4 \times 2 = 8$ for CO_2 plus $5 \times 1 = 5$ for H_2O) and 2 on the left. That is, there is an odd number on the right and an even number on the left. There are two equally valid ways to balance the oxygen.

Solution (1). To have 13 atoms of oxygen on the left side, use a stoichiometric coefficient of 13/2; you can do this because

$$\left(\frac{13}{2}\right)\left(\frac{2 \text{ atoms of O}}{\cancel{\text{molecule}} \text{ of } O_2}\right) = 13 \text{ oxygen atoms}$$

Therefore, the balanced equation will be

$$C_4H_{10} + 13/2\ O_2 \longrightarrow 4\ CO_2 + 5\ H_2O$$

Solution (2). Taking the equation as it stands after step 2, multiply each coefficient by 2 so that there is an even number of oxygen atoms on the right side (i.e., 26).

$$2\ C_4H_{10} + (\text{—}\ O_2) \longrightarrow 8\ CO_2 + 10\ H_2O$$

Now the O_2 on the left can be balanced by multiplying by a whole number instead of a fraction.

$$2\ C_4H_{10} + 13\ O_2 \longrightarrow 8\ CO_2 + 10\ H_2O$$

While you will find it convenient at times to use fractional coefficients, generally equations are written with whole-number coefficients.

Step 4. As *verification*, notice that there are 8 C atoms, 20 H atoms, and 26 O atoms on each side of the equation from solution (2) above.

EXERCISE 4.1 Balancing Chemical Equations

Balance the chemical equations for the following:

(a) the oxidation of iron

$$Fe(s) + O_2(g) \longrightarrow Fe_2O_3(s)$$

(b) the combustion of methane

$$CH_4(g) + O_2(g) \longrightarrow CO_2(g) + H_2O(g)$$

(c) the combustion of B_4H_{10} in O_2 to give $B_2O_3(s)$ and $H_2O(g)$

(d) the reaction of CO with H_2 to give methyl alcohol, CH_3OH (a reaction that serves as the basis of a process for making synthetic fuel from coal)

(e) the combustion of octane, C_8H_{18}

4.3 SOME COMMON TYPES OF CHEMICAL REACTIONS

Elements and compounds can react with one another to produce a bewildering array of new compounds. But how can you predict what reactions will occur and what the products will be? Fortunately, chemists have developed some guidelines that help you make predictions about a wide range of reactions. It is important to realize at the outset that there are relatively few common types of chemical reactions. You will see these throughout the book, so it is useful to introduce a few of them here before going on to discuss the details of stoichiometry. Be sure to notice that all of the equations that follow are balanced.

REACTIONS OF ELEMENTS AND COMPOUNDS WITH OXYGEN

The reaction of sulfur with O_2 (Figure 4.1) is an example of the fact that all the elements of the periodic table (with the exception of some rare gases) form binary or "two-element" **oxides** with a general formula of M_xO_y.

You can predict the product of a metal + O_2 reaction from the fact that it must be electrically neutral; the oxide ion or ions must supply as many negative charges as the positive charges supplied by the metal cation. Since the oxide ion is O^{2-}, you can predict a reasonable positive charge for the metal using the guidelines in Chapter 3. For example, as we did in Example 3.4, you can predict that aluminum (Al, Group 3A) will react with O_2 to give Al_2O_3 (2 Al^{3+} and 3 O^{2-} ions), a compound known as *alumina* or *corundum* (Figure 4.5).

Figure 4.5 When aluminum reacts with oxygen it gives aluminum oxide, Al_2O_3. The oxide is normally a white powder, but it is also the basis of a number of minerals and gems, chiefly rubies and sapphires. The red stone in this photo is a ruby. (See also Figure 2.7.)

$$4\,Al(s) + 3\,O_2(g) \longrightarrow \underset{\text{aluminum oxide}}{2\,Al_2O_3(s)}$$

Cations of transition metals can have a variety of positive charges, but, as discussed in Chapter 3, they most often have a charge of 2+ (or, only slightly less often, 3+). Thus, when iron reacts with oxygen it forms both FeO and Fe_2O_3 (Figure 4.6).

$$2\,Fe(s) + O_2(g) \longrightarrow \underset{\text{iron(II) oxide}}{2\,FeO(s)}$$

$$4\,Fe(s) + 3\,O_2(g) \longrightarrow \underset{\text{iron(III) oxide}}{2\,Fe_2O_3(s)}$$

Figure 4.6 A stream of powdered iron is squirted into the flame of a Bunsen burner. The tiny particles of iron react with oxygen to form various iron oxides. The flame and the energy released by the reaction heat the particles to incandescence.

Almost all nonmetals (He, Ne, Ar, and possibly Kr are exceptions) also combine with O_2, and the compounds of carbon are good examples. Both CO and CO_2 are gases under normal conditions.

$$2\,C(s) + O_2(g) \longrightarrow \underset{\text{carbon monoxide}}{2\,CO(g)}$$

$$\downarrow \tfrac{1}{2}\,O_2(g)$$

$$C(s) + O_2(g) \longrightarrow \underset{\text{carbon dioxide}}{CO_2(g)}$$

Carbon dioxide is the nontoxic gas in soft drinks, beer, and champagne. In contrast, carbon monoxide, CO, is toxic. This molecule can combine with hemoglobin, the molecule in your blood system that picks up oxygen in the lungs and carries it to the cells of your body. Because CO interacts even more strongly with hemoglobin than does O_2, the carbon monoxide can prevent hemoglobin from absorbing O_2.

The fact that carbon forms two compounds with O_2 illustrates a general feature of the reactions of all nonmetals with O_2: C, N, P, and S all form *several* binary oxides, as shown by the reactions that follow.

Elemental nitrogen, N_2, and O_2 are the major components of the air we breathe. In an automobile engine, small amounts of these gases combine at high temperatures to give NO, nitrogen oxide.

Figure 4.7 Nitrogen dioxide, NO_2, a brown gas. It can be formed when oxygen and nitrogen combine during fuel combustion in automobile engines, and so NO_2 can be seen in polluted air. Here it is a product of the reaction of copper metal and nitric acid (HNO_3). The other product of the reaction is copper nitrate, $Cu(NO_3)_2$, the compound responsible for the green-blue color of the solution.

$$N_2(g) + O_2(g) \longrightarrow 2\ NO(g)$$

nitrogen oxide
(colorless gas)

This is also a product when a lightning burst passes through air during a thunderstorm. However, just as CO can be converted to CO_2, NO can react with excess O_2 to give NO_2 (Figure 4.7).

$$2\ NO(g) + O_2(g) \longrightarrow 2\ NO_2(g)$$

nitrogen dioxide
(brown gas)

Both NO and NO_2 are toxic and are important links in the chemical chain leading to the production of smog from air, sunlight, and automobile exhaust (Chapter 24).

Phosphorus, in the same periodic group as N, is exceedingly toxic; only 50 milligrams is a fatal dose for an adult human. The element in its most common form is a yellowish-white solid, P_4. The slight amount of P_4 in the vapor above the solid at room temperature reacts with O_2 and gives off a phosphorescent glow (see photos below), a fact that led to the name of the element. At least six different products are possible when P_4 is burned in air, but the one product that you should be aware of now is the white solid P_4O_{10}.

$$P_4(s) + 5\ O_2(g) \longrightarrow P_4O_{10}(s)$$

tetraphosphorus decoxide
(white solid)

Sulfur, the Group 6A neighbor of oxygen, forms two oxides, SO_2 and SO_3, in reactions of great environmental significance: they are the beginning of the production of "acid rain" (Figure 4.1).

$$S_8(s) + 8\ O_2(g) \longrightarrow 8\ SO_2(g)$$

sulfur dioxide
(colorless)

$$2\ SO_2(g) + O_2(g) \longrightarrow 2\ SO_3(g)$$

sulfur trioxide
(colorless)

Not only elements but also compounds react with O_2. In fact, in the special case where all of the elements in the compound combine with oxygen to form oxides, we call these **combustion reactions**. In our economy

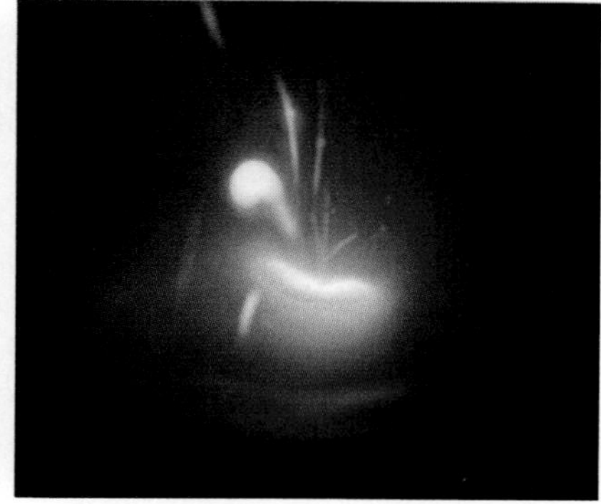

When white phosphorus is exposed to oxygen, it first glows (left; a process called phosphorescence) and then bursts into flame (right). To prevent this, white phosphorus is usually stored under water (Fig. 2.4).

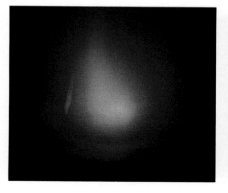

we obtain a great deal of energy from the combustion of *hydrocarbons*, compounds containing only C and H, and the combustion of propane, butane, and gasoline are excellent examples (see Figure 4.4 and Example 4.1). The products of the complete combustion of hydrocarbons are only water and carbon dioxide. We can also use compounds containing C, H, and O—such as methyl and ethyl alcohol—as fuels, since they also give only water and CO_2 on complete combustion.

$$\underset{\text{methyl alcohol}}{2\ CH_3OH(\ell)} + 3\ O_2(g) \longrightarrow 4\ H_2O(\ell) + 2\ CO_2(g)$$

Ethyl alcohol is widely used in Brazil as an automotive fuel, and there is some discussion of using methyl alcohol in the United States because it may lead to less air pollution (see Chapter 6).

The French scientist Lavoisier was the first to study animal respiration in the 18th century. He observed that the amount of heat produced by a guinea pig in the process of exhaling a given amount of carbon dioxide is similar to the amount of heat given off in burning an amount of carbon that would yield the same quantity of carbon dioxide. From this and other experiments he concluded that "Respiration is a combustion, slow it is true, but otherwise perfectly similar to that of charcoal." One important combustion reaction that occurs in animals is the "burning" of the sugar glucose to provide energy.

$$\underset{\text{glucose}}{C_6H_{12}O_6(aq)} + 6\ O_2(g) \longrightarrow 6\ CO_2(g) + 6\ H_2O(\ell)$$

DECOMPOSITION REACTIONS

Elements combine to form compounds in **combination reactions**. However, there is a large class of reactions—**decomposition reactions**—in which the compounds break down to form simpler ones, usually by heating (Figure 4.8).

Figure 4.8 The "dichromate volcano." Ammonium dichromate decomposes according to the equation $(NH_4)_2Cr_2O_7(s) \longrightarrow N_2(g) + Cr_2O_3(s) + 4\ H_2O(g)$. In the first photo, the dichromate salt is placed in a beaker with an alcohol-soaked paper wick. After ignition, the heat evolved by the reaction allows it to continue. On completion, the beaker contains a pile of green chromium(III) oxide, and water droplets cling to the side of the beaker.

Figure 4.9 A bed of calcium carbonate (limestone) along the Verde River in Arizona. (James Cowlin)

In a few instances, metal oxides can be decomposed to give the metal and oxygen, the opposite of a combination reaction. Perhaps the best known of these is

$$HgO(s) \longrightarrow Hg(\ell) + \tfrac{1}{2}\,O_2(g)$$

The Englishman Priestley and the French scientist Lavoisier were the first to isolate elemental oxygen, and they did so by using sunlight to heat a pile of mercury(II) oxide.

A very common and important type of decomposition reaction is illustrated by the chemistry of *metal carbonates*. Calcium is the fifth most abundant element in the earth's crust and is the third most abundant metal (after Al and Fe). Virtually all the calcium is in the form of calcium carbonate, $CaCO_3$, coming from the fossilized remains of earlier marine life (Figure 4.9). Limestone, a form of calcium carbonate, is enormously important in our economy, as is sodium carbonate, Na_2CO_3. A characteristic reaction of all metal carbonates is that, when heated, they decompose to form CO_2 gas and the corresponding solid metal oxide.

$$\underset{\text{limestone}}{CaCO_3(s)} \xrightarrow{800\text{–}1000\ °C} \underset{\text{lime}}{CaO(s)} + CO_2(g)$$

Thus, limestone produces calcium oxide, a compound commonly called "lime." In 1988, this reaction was used to obtain more than 32 billion pounds of CaO in the United States. The oxide is used in steel making and to make other chemicals (see Chapter 22).

In 1988 32.34 billion pounds of CaO and 19.10 billion pounds of Na_2CO_3 were produced in the United States.

EXERCISE 4.2 Types of Chemical Reactions

Write balanced chemical equations for the following reactions:

(a) The reaction of gallium metal with chlorine gas.
(b) The reaction of chromium and oxygen to give chromium(III) oxide.
(c) The reaction of nitrogen and oxygen to give nitrogen dioxide.
(d) The decomposition of strontium carbonate.

SOMETHING MORE ABOUT
Titanium Dioxide: A White Pigment

There is hardly a white-colored or tinted object in our environment—and that includes the paper on which these words are printed—that does not contain TiO_2 pigments. This compound, whose chemical name is titanium(IV) oxide or titanium dioxide, is manufactured in enormous quantities. In the United States, TiO_2 ranks 44th on the list of the "top 50 chemicals produced"; just over 2 billion pounds were produced in 1988.

For many years the pigment in white paint was "white lead," which is a mixture of lead(II) carbonate and lead(II) hydroxide ($PbCO_3$ + $Pb(OH)_2$). However, we know now that lead-based paint is a major health hazard, so titanium dioxide has completely replaced white lead in paints. In addition, TiO_2 is used in the plastics, ceramics, and paper industries.

Titanium is ninth in abundance of the elements making up the earth's crust, and much of it exists as the mineral ilmenite, which has an approximate formula of $FeTiO_3$. A smaller amount is found as rutile, almost pure TiO_2. The most commonly used process in the United States to make pure TiO_2 is called *chlorination*. If ilmenite is the source of the titanium, the ore is first treated with chlorine gas and coke (a form of carbon) to produce crude or impure TiO_2.

$$2\ FeTiO_3(s) + 2\ Cl_2(g) + C(s) \longrightarrow 2\ FeCl_2(s) + 2\ TiO_2(s) + CO_2(g)$$

The crude titanium dioxide produced in this step is then treated again with chlorine in the presence of carbon. (The TiO_2 used in the step may also have been rutile ore.)

$$2\ TiO_2(s) + 4\ Cl_2(g) + 3\ C(s) \longrightarrow 2\ TiCl_4(\ell) + CO_2(g) + 2\ CO(g)$$

The titanium-containing product here is titanium(IV) chloride, $TiCl_4$, a compound that is liquid at room temperature (and which is commonly known to chemists as "tickle-4"). The liquid is very reactive, as illustrated in the photograph; some $TiCl_4$ was allowed to react with water in the air to produce a voluminous fog of titanium oxide and HCl.

$$TiCl_4(\ell) + 2\ H_2O(g) \longrightarrow TiO_2(s) + 4\ HCl(g)$$

However, in industry the tetrachloride is converted to pure TiO_2 by burning $TiCl_4$ in oxygen at high temperatures,

$$TiCl_4(\ell) + O_2(g) \longrightarrow TiO_2(s) + 2\ Cl_2(g)$$

and it is this pure compound that is used to paint your home, car, or school building white or to make the paper in this book as white as possible.

Liquid titanium(IV) chloride, $TiCl_4$, reacts with water vapor to produce a fog of TiO_2 and HCl.

4.4 WEIGHT RELATIONS IN CHEMICAL REACTIONS: STOICHIOMETRY

A major aspect of chemistry is the study of chemical reactions, often with the hope of finding some new material that can be useful to society. If a product turns out to have commercial potential, the efficiency of the reaction is of interest. Can the product be made in sufficient quantity and purity that it can be sold for a reasonable price? Questions such as these mean that reactions must be studied quantitatively. **Stoichiometry** is the study of the quantitative relations between the amounts of reactants and products, and its guiding principle is the observation that *matter is always conserved in chemical reactions.*

A piece of magnesium metal will burn in air to give magnesium oxide, MgO (Figure 4.10). Suppose that you use 2.00 moles of magnesium (48.6 g). The following balanced equation shows you that 1.00 mole or 32.0 g of O_2 must be used and that 2.00 moles or 80.6 g of MgO are produced.

MgO is actually an ionic compound with the same structure as NaCl (Figure 3.4). For the sake of simplicity, however, we consider one MgO unit as a "molecule."

2 Mg(s)	+ O_2(g)	⟶ 2 MgO(s)
2 atoms	+ 1 molecule	⟶ 2 molecules
2.00 moles	+ 1.00 mole	⟶ 2.00 moles
48.6 g	+ 32.0 g	⟶ 80.6 g

The balanced equation for burning magnesium applies to the reaction of Mg with O_2 no matter how much Mg is used. If 0.020 mole of Mg is used, then 0.010 mole of O_2 will be required and 0.020 mole of MgO will form if the reaction proceeds to completion. You can confirm by experiment that, if 0.81 g of reactants is consumed, then 0.81 g of MgO is produced.

Following this line of reasoning further, suppose you have a piece of magnesium with a mass of only 0.145 g. How much O_2 must there be to consume the magnesium completely? The following steps lead to the answer.

Step 1. Write the balanced equation for the reaction. This is

$$2\ \text{Mg(s)} + \text{O}_2\text{(g)} \longrightarrow 2\ \text{MgO(s)}$$

Step 2. Convert the mass of magnesium to moles. This must be done because the balanced equation uses mole units, not masses.

$$0.145\ \cancel{\text{g}}\ \text{Mg}\left(\frac{1\ \text{mol Mg}}{24.31\ \cancel{\text{g}}\ \text{Mg}}\right) = 0.00596\ \text{mol Mg}$$

Step 3. Relate the number of moles of the given reactant (Mg) to the number of moles of the other reactant (O_2) (or product) whose amount we wish to know.

$$0.00596\ \cancel{\text{mol}}\ \text{Mg}\left(\frac{1\ \text{mol O}_2\ \text{required}}{2\ \cancel{\text{mol}}\ \text{Mg available}}\right) = 0.00298\ \text{mol O}_2$$

Figure 4.10 A piece of magnesium ribbon burns in air to give the white solid magnesium oxide (MgO).

This calculation tells us that 0.00298 mol O_2 is required to use up completely the available Mg. To do this we multiplied the number of moles of available reagent by a **stoichiometric factor**, a *mole-ratio factor relating moles of the desired compound to moles of the available reagent.* The stoichiometric factor comes directly from the balanced chemical equation; this is why we must balance chemical equations before proceeding with calculations.

Step 4. Knowing the number of moles of O_2 required to use up all of the available Mg, we can then calculate the mass of O_2 required. This is done simply by reversing the type of calculation done in step 2. That is,

$$0.00298\ \cancel{\text{mol}}\ \text{O}_2\left(\frac{32.00\ \text{g O}_2}{1\ \cancel{\text{mol}}\ \text{O}_2}\right) = 0.0954\ \text{g O}_2$$

Since the object of this example is to find the mass of O_2 required, the solution to the problem is complete.

You may also wish to know the amount of MgO produced in the reaction of 0.145 g of magnesium with oxygen. Because of the principle of the conservation of matter, you can answer this simply by adding up the masses of Mg and O_2 used (giving 0.240 g of MgO produced). Alternatively, you could repeat steps 3 and 4 above, but with the appropriate stoichiometric factor and molar mass.

Step 3′. Relate number of moles of Mg to the number of moles of MgO produced.

$$0.00596\ \cancel{\text{mol}}\ \text{Mg}\left(\frac{1\ \text{mol MgO produced}}{1\ \cancel{\text{mol}}\ \text{Mg used}}\right) = 0.00596\ \text{mol MgO produced}$$

Step 4′. Convert the moles of MgO produced to grams.

$$0.00596\ \cancel{\text{mol}}\ \text{MgO produced}\left(\frac{40.30\ \text{g MgO}}{1\ \cancel{\text{mol}}\ \text{MgO}}\right) = 0.240\ \text{g MgO}$$

Using this alternative may seem a bit silly here, since only one product is formed, but simply adding the masses of reactants to obtain the product mass works only if the reaction gives a *single* product. When more than one product results from a reaction, it is necessary to apply steps 3 and 4 for each product.

This example was meant to show you the general approach to follow in a stoichiometric calculation. To help you see this more clearly, consider the scheme in Figure 4.11 and work through the following examples and exercises.

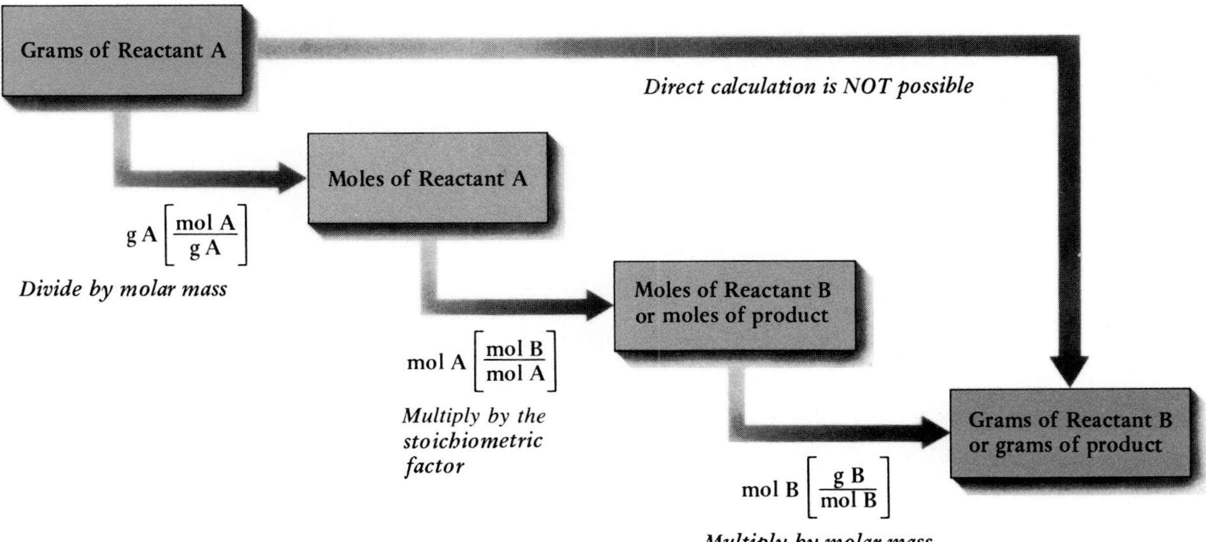

Figure 4.11 A scheme outlining the relation between the mass of one reactant (A) and the mass of another reactant or the mass of a product (B). The stoichiometric factor is always (moles of reactant B or moles of product) ÷ (moles of reactant A).

EXAMPLE 4.2

WEIGHT RELATIONS IN REACTIONS

Propane, C_3H_8, can be used as a fuel in your home or car because it is easily liquefied and transported. If you burn 1.00 pound or 454 g of propane, how much oxygen is required for complete combustion and how much CO_2 and H_2O are formed?

Solution

1. Remember that the first step must always be to write a balanced equation; see page 107.

$$C_3H_8(g) + 5\ O_2(g) \longrightarrow 3\ CO_2(g) + 4\ H_2O(g)$$

2. Having balanced the equation, we can proceed to the stoichiometric calculation. First, we shall convert the mass of propane to moles.

$$454\ \cancel{g}\ C_3H_8 \left(\frac{1\ \text{mol}\ C_3H_8}{44.10\ \cancel{g}\ C_3H_8}\right) = 10.3\ \text{mol}\ C_3H_8$$

3. Relate the moles of propane available to moles of O_2 required with the stoichiometric factor.

$$10.3\ \cancel{\text{mol}}\ C_3H_8 \left(\frac{5\ \text{mol}\ O_2\ \text{required}}{1\ \cancel{\text{mol}}\ C_3H_8\ \text{available}}\right) = 51.5\ \text{mol}\ O_2\ \text{required}$$

4. Convert moles of O_2 required to grams.

$$51.5\ \cancel{\text{mol}}\ O_2\ \text{required} \left(\frac{32.00\ \text{g}\ O_2}{1\ \cancel{\text{mol}}\ O_2}\right) = 1650\ \text{g}\ O_2\ \text{required}$$

To find how much CO_2 is produced, for example, we simply repeat the last two steps.

3′. Relate moles of C_3H_8 available to moles of CO_2 produced with the stoichiometric factor.

$$10.3\ \cancel{\text{mol}}\ C_3H_8\ \text{available} \left(\frac{3\ \text{mol}\ CO_2\ \text{produced}}{1\ \cancel{\text{mol}}\ C_3H_8\ \text{available}}\right) = 30.9\ \text{mol}\ CO_2\ \text{produced}$$

4′. Convert moles of CO_2 produced to the mass of CO_2.

$$30.9\ \cancel{\text{mol}}\ CO_2 \left(\frac{44.01\ \text{g}\ CO_2}{1\ \cancel{\text{mol}}\ CO_2}\right) = 1360\ \text{g}\ CO_2$$

Now, how can we find the mass of H_2O produced? Go through steps 3′ and 4′ again? Of course we could! But it would be easier to recognize that the total mass of the reactants

$$454\ \text{g}\ C_3H_8 + 1650\ \text{g}\ O_2 = 2.10 \times 10^3\ \text{g of reactants}$$

must be the same as the total mass of products. Therefore, the mass of water produced is

$$\text{Total mass of products} = 2.10 \times 10^3\ \text{g} = 1.36 \times 10^3\ \text{g}\ CO_2\ \text{produced} + ?\ \text{g}\ H_2O$$

$$\text{Mass of}\ H_2O = 740\ \text{g}$$

EXERCISE 4.3 Weight Relations in Chemical Reactions

Iron reacts with oxygen to give iron(III) oxide, Fe_2O_3. If an ordinary iron nail (assumed to be pure iron) has a mass of 1.05 g, how many grams of Fe_2O_3 would it produce if the nail turned completely to this oxide in the process of rusting? How many grams of O_2 are required for the reaction?

All stoichiometry problems fit the mold of the calculations outlined above. There will be some differences from one situation to another, but don't let that obscure the basic outline in Figure 4.11.

4.5 REACTIONS WHERE ONE REAGENT IS PRESENT IN LIMITED SUPPLY

The objective of the examples in the previous section was to find the exact amount of material required to completely consume a given quantity of some reagent or to find how much product would be formed from a given amount of a reactant. However, in carrying out a reaction, chemists (or nature, for that matter) rarely supply exact stoichiometric amounts of reactants. As an example, consider the preparation of cisplatin, $Pt(NH_3)_2Cl_2$:

$$(NH_4)_2PtCl_4(s) + 2\ NH_3(aq) \longrightarrow 2\ NH_4Cl(aq) + Pt(NH_3)_2Cl_2(s)$$

The goal in carrying out a reaction is usually to produce a useful compound, cisplatin in this case, in the largest amount possible from a given amount of starting material. For reasons that will be made clear in later chapters, this usually means we use a large excess of one of the reactants in order to ensure that all of the more expensive reagent is consumed completely. In the case of the synthesis of cisplatin, it makes sense to mix the more expensive chemical, $(NH_4)_2PtCl_4$, with a much greater amount of the much less expensive reagent NH_3 than is called for by the balanced equation. Thus, on completion of the reaction, all of the $(NH_4)_2PtCl_4$ will have been converted to product, and some NH_3 (which was originally present in stoichiometric excess) will remain. How much cisplatin, $Pt(NH_3)_2Cl_2$, will be formed? It depends on the amount of $(NH_4)_2PtCl_4$ you had at the start, *not* on the amount of NH_3, since more of the latter was present than is required by stoichiometry. We call a compound such as $(NH_4)_2PtCl_4$ the **limiting reagent**, since its amount determines or limits the amount of product formed.

$(NH_4)_2PtCl_4$ costs roughly $40 per gram, while NH_3 is only pennies per gram.

The limiting reagent limits or determines the amount of product formed in a reaction. All stoichiometry calculations are based on the total consumption of the limiting reagent.

As an analogy to a chemical "limiting reagent" situation, consider what happens if you try to make some cheese sandwiches. Suppose you have 30 slices of bread and 30 slices of cheese. If each sandwich requires 3 slices of cheese and 2 slices of bread, you can make only 10 sandwiches; 10 slices of bread will be left over. Bread is the "excess reagent" and cheese is the "limiting reagent," since the amount of cheese has limited or determined the number of sandwiches that can be made. Furthermore, although cheese is bought by the pound, and bread by the loaf, the combining units are *slices*, and your calculation had to be done in those units, just as stoichiometry problems are done in units of moles. The ratio of "reactants required" here is (3 slices cheese/2 slices bread) or 1½ cheese

You have 30 slices of bread and 30 slices of cheese. From 30 bread slices and unlimited cheese you could make 15 sandwiches. From 30 cheese slices and unlimited bread you could make 10 sandwiches (with 3 cheese slices per sandwich). Therefore, the cheese is more limiting in this case, and we call it the "limiting reagent."

slices to 1 bread slice. However, you had a ratio of "reactants available" of only 1 cheese slice to 1 bread slice. This shows that not enough cheese is available to use up all the bread, or, conversely, there is more bread available than can be "consumed" by the cheese. Chemical stoichiometry problems work the same way.

EXAMPLE 4.3

A REACTION WHERE ONE REAGENT IS IN LIMITED SUPPLY; LIMITING REAGENTS

Assume you combine 15.5 g of $(NH_4)_2PtCl_4$ with 2.55 g of NH_3 to make cisplatin. (a) Which reagent is in excess and which is the limiting reagent? (b) How many grams of cisplatin can be formed? (c) After all the limiting reagent has been consumed, and the maximum quantity of cisplatin has been formed, how many grams of the other reactant remain?

Solution The first question you should ask when confronted by a problem of this type—one where you know the masses of more than one reactant—is whether the reactants are present in the correct stoichiometric ratio or whether one of the reactants is in short supply. Since the balanced equation requires us to compare quantities of reagents on a mole basis, we must first find the number of moles of each reactant.

Set-up Steps:

Step 1. Write the balanced equation for the reaction.

$$(NH_4)_2PtCl_4(s) + 2\ NH_3(aq) \longrightarrow 2\ NH_4Cl(aq) + Pt(NH_3)_2Cl_2(s)$$

Step 2. Calculate the number of moles of each reactant.

$$15.5\ \cancel{g}\ (NH_4)_2PtCl_4 \left(\frac{1\ \text{mol}}{373.0\ \cancel{g}}\right) = 0.0416\ \text{mol}\ (NH_4)_2PtCl_4$$

$$2.55\ \cancel{g}\ NH_3 \left(\frac{1\ \text{mol}}{17.03\ \cancel{g}}\right) = 0.150\ \text{mol}\ NH_3$$

Question (a): Which of the reagents is the limiting reagent?

Solution to (a): There are at least two useful ways to approach this problem, and the way we have chosen is similar to the method used in deciding how many cheese sandwiches we could make.

Step 3. Determine the limiting reagent. Having calculated the number of moles of each reactant above, determine the ratio of moles *available*.

$$\text{Moles-available ratio} = \frac{0.0416\ \text{mol}\ (NH_4)_2PtCl_4}{0.150\ \text{mol}\ NH_3} = \frac{0.277\ \text{mol}\ (NH_4)_2PtCl_4}{1.00\ \text{mol}\ NH_3}$$

The ratio of moles of reactants *required* by stoichiometry (the stoichiometric factor) is

$$\text{Moles-required ratio} = \frac{1\ \text{mol}\ (NH_4)_2PtCl_4}{2\ \text{mol}\ NH_3} = \frac{0.500\ \text{mol}\ (NH_4)_2PtCl_4}{1.00\ \text{mol}\ NH_3}$$

Now we see that 1 mole of NH_3 requires 0.500 mole of $(NH_4)_2PtCl_4$, but we have available only 0.277 mole of $(NH_4)_2PtCl_4$ per mole of NH_3. Clearly, we do not have enough $(NH_4)_2PtCl_4$ to use up all of the available NH_3. *The platinum-containing compound is the limiting reagent*; it is the cheese in our cheese sandwich analogy.

Question (b): How much cisplatin, $Pt(NH_3)_2Cl_2$, can be formed?
Solution to (b): Now that we know that $(NH_4)_2PtCl_4$ is the limiting reagent, we can use this to calculate the quantity of cisplatin that can be formed.

Step 4. We already know how many moles of $(NH_4)_2PtCl_4$ are available, so this is used to calculate the number of moles, and then the mass, of $Pt(NH_3)_2Cl_2$ that may be formed.

$$0.0416\ \cancel{\text{mol}}\ (NH_4)_2PtCl_4 \left(\frac{1\ \cancel{\text{mol}}\ Pt(NH_3)_2Cl_2\ \text{formed}}{1\ \cancel{\text{mol}}\ (NH_4)_2PtCl_4\ \text{available}}\right)\left(\frac{300.0\ \text{g}\ Pt(NH_3)_2Cl_2}{1\ \cancel{\text{mol}}\ Pt(NH_3)_2Cl_2}\right)$$
$$= 12.5\ \text{g}\ Pt(NH_3)_2Cl_2\ \text{formed as product}$$

Question (c): After all the limiting reagent, $(NH_4)_2PtCl_4$, has been consumed, and the maximum amount of $Pt(NH_3)_2Cl_2$ has been formed, how much NH_3 remains?
Solution to (c): What we want to know here is the difference between the quantity of NH_3 available and the quantity used. We know that 0.150 mole of NH_3 is available, and we can find the quantity of NH_3 required by asking how much NH_3 the limiting reagent, $(NH_4)_2PtCl_4$, will consume.

Step 5. Calculate the quantity of NH_3 required and then the quantity of NH_3 in excess.

$$0.0416\ \cancel{\text{mol}}\ (NH_4)_2PtCl_4 \left(\frac{2\ \text{mol}\ NH_3\ \text{required}}{1\ \cancel{\text{mol}}\ (NH_4)_2PtCl_4\ \text{available}}\right)$$
$$= 0.0832\ \text{mol}\ NH_3\ \text{required}$$

With this information we can find the quantity of NH_3 remaining.

$$\begin{aligned}\text{Quantity of } NH_3 \text{ remaining} &= 0.150\ \text{mol}\ NH_3\ \text{available}\\ &\quad - 0.0832\ \text{mol}\ NH_3\ \text{required}\\ &= 0.067\ \text{mol remaining}\end{aligned}$$

$$0.067\ \cancel{\text{mol}}\ NH_3 \left(\frac{17.03\ \text{g}\ NH_3}{1\ \cancel{\text{mol}}}\right) = 1.1\ \text{g of}\ NH_3\ \text{left of the 2.55 g available}$$

Some final comments: First, Step 3, the determination of the limiting reagent, is the crucial step; without the answer to this step, we cannot do the rest of the problem. Second, there are other ways to approach the problem of determining the limiting reagent, and one of these is to calculate the quantity of product that could be obtained starting with the available quantity of a given reactant and a stoichiometrically correct quantity of the other reactant. In the cheese sandwich analogy, 30 slices of bread combined with enough cheese could give 15 sandwiches, but, if we base our calculation on available cheese, we see that only 10 sandwiches can be made. In the preparation of cisplatin, you would find that combining 0.150 mole of NH_3 with enough $(NH_4)_2PtCl_4$ could give 0.0750 mole of $Pt(NH_3)_2Cl_2$. On the other hand, 0.0416 mole of $(NH_4)_2PtCl_4$ could produce only 0.0416 mole of $Pt(NH_3)_2Cl_2$. Clearly, the $(NH_4)_2PtCl_4$ is the limiting reagent.

EXAMPLE 4.4

LIMITING REAGENTS

Ammonia, NH_3, is often among the "top 5" chemicals produced in the United States because it is widely used as a fertilizer and to make other chemicals. In 1988 almost 34 billion pounds of NH_3 were produced. The compound is made from its elements.

$$N_2(g) + 3\ H_2(g) \longrightarrow 2\ NH_3(g)$$

Nitrogen comes from the air, and the H_2 comes from a more expensive process, the breakdown of natural gas to give hydrogen gas. Assume you begin with 1250 g of N_2 and 225 g of H_2. Which is the limiting reagent? How much ammonia can be formed? How much of which reagent is left when the maximum amount of NH_3 is formed?

Solution

1. Write the balanced equation.

$$N_2(g) + 3\ H_2(g) \longrightarrow 2\ NH_3(g)$$

2. Calculate the number of moles of each reactant.

$$1250\ \cancel{g}\ N_2 \left(\frac{1\ mol}{28.01\ \cancel{g}\ N_2}\right) = 44.6\ mol\ N_2\ available$$

$$225\ \cancel{g}\ H_2 \left(\frac{1\ mol}{2.016\ \cancel{g}\ H_2}\right) = 112\ mol\ H_2\ available$$

3. Determine the limiting reagent.

$$\text{moles-available ratio} = \frac{112\ mol\ H_2}{44.6\ mol\ N_2} = \frac{2.51\ mol\ H_2}{1.00\ mol\ N_2}$$

$$\text{moles-required ratio (the stoichiometric factor)} = \frac{3\ mol\ H_2}{1\ mol\ N_2}$$

Comparison of these ratios shows that there is not enough H_2 available to use up all the available N_2. *Hydrogen gas, H_2, is the limiting reagent.* (This is not too surprising, since H_2 is more expensive than N_2, so it is likely that N_2 would be used in excess.)

4. Calculate the quantity of ammonia, NH_3, that can be formed based on the quantity of limiting reagent available.

$$112\ \cancel{mol}\ H_2\ available \left(\frac{2\ mol\ NH_3\ formed}{3\ \cancel{mol}\ H_2\ available}\right) = 74.7\ mol\ NH_3\ formed$$

$$74.7\ \cancel{mol}\ NH_3 \left(\frac{17.03\ g\ NH_3}{1\ \cancel{mol}}\right) = 1270\ g\ NH_3$$

5. Calculate the amount of N_2 that is left after all the H_2 has been used and the maximum amount of NH_3 has been formed. First, we need to know how much N_2 is required to consume all of the limiting reagent, H_2.

$$112\ \cancel{mol}\ H_2\ available \left(\frac{1\ mol\ N_2\ required}{3\ \cancel{mol}\ H_2\ available}\right) = 37.3\ mol\ N_2\ required$$

Since we have 44.6 moles of N_2 available,

$$\begin{aligned} \text{Excess } N_2 &= 44.6\ \text{moles } N_2 \text{ available} - 37.3\ \text{moles } N_2 \text{ consumed} \\ &= 7.3\ \text{moles } N_2 \text{ remain} \end{aligned}$$

$$7.3\ \cancel{mol}\ N_2 \left(\frac{28.01\ g\ N_2}{1\ \cancel{mol}\ N_2}\right) = 2.0 \times 10^2\ g\ N_2 \text{ in excess of that required}$$

Since we have 2.0×10^2 g of N_2 left over, this means that we have consumed 1250 g − 200 g = 1050 g of N_2, a fact that you can confirm by converting the number of moles of N_2 required (step 5) to mass.

EXERCISE 4.4 Limiting Reagents

Aluminum chloride, $AlCl_3$, is an inexpensive reagent used in many industrial processes. It is made by treating scrap aluminum with chlorine according to the following balanced equation.

$$2\ Al(s) + 3\ Cl_2(g) \longrightarrow 2\ AlCl_3(s)$$

If you start with 2.70 g of Al and 4.05 g of Cl_2, which reagent is limiting? How many grams of $AlCl_3$ can be produced? How many grams of which reagent, Al or Cl_2, will remain when the reaction is completed?

These reactions, in which one reagent is in excess and the other limits the amount of product, represent the way chemical reactions are usually carried out. It is generally obvious to the person doing the reaction which reagent is in excess, since the experiment is designed that way in the beginning. Therefore, calculations such as those done in Examples 4.3 and 4.4 are not always necessary in actual laboratory work.

(a)

4.6 PERCENT YIELD

The quantity of product you *should* isolate from a chemical reaction is the **theoretical yield**. However, if you have worked in the laboratory or the kitchen, you know how difficult it is to account for every drop of liquid and every crumb of solid. Furthermore, the methods used to purify chemicals are not always capable of separating the desired product completely from the reaction mixture. Thus, the **actual yield** of the compound may be less than the theoretical yield. To judge the efficiency of a chemical reaction and the techniques used to obtain the desired compound in pure form, chemists often compare the actual and theoretical yields by calculating their ratio and call the result the percent yield (Figure 4.12).

(b)

Figure 4.12 The reaction described in Example 4.5 should give 18.8 g of aspirin, a white solid (the theoretical yield) (a). Instead, only 6.26 g of the product is actually obtained (b). The percent yield is (6.26 g/18.8 g)100 = 33.3%.

EXAMPLE 4.5

PERCENT YIELD

Consider the reaction of salicylic acid and acetic anhydride to produce aspirin.

$$\underset{\text{salicylic acid}}{2\ C_7H_6O_3(s)} + \underset{\text{acetic anhydride}}{C_4H_6O_3(\ell)} \longrightarrow \underset{\text{aspirin}}{2\ C_9H_8O_4(s)} + H_2O(\ell)$$

Assume you have 14.4 g of the acid and that it is the limiting reagent. Enough acetic anhydride is present to consume the acid completely. If you obtain 6.26 g of pure aspirin from the reaction, what is the percent yield of aspirin?

Solution This problem differs from the stoichiometry problems done thus far in that there is one additional step. You should first follow the usual steps to convert grams of salicylic acid to grams of aspirin expected, that is, the theoretical yield. The percent yield is then the ratio of the actual yield, 6.26 g, to the theoretical yield.

Step 1. The balanced chemical equation has been written above.

Step 2. You know that salicylic acid is the limiting reagent here. Calculate the number of moles of this acid available.

$$14.4 \text{ g salicylic acid} \left(\frac{1 \text{ mol } C_7H_6O_3}{138.1 \text{ g } C_7H_6O_3}\right) = 0.104 \text{ mol salicylic acid}$$

Step 3. Using the stoichiometric factor from the balanced equation, relate moles of salicylic acid available to moles of aspirin that could be produced if everything worked perfectly.

$$0.104 \text{ mol acid} \left(\frac{2 \text{ mol aspirin produced}}{2 \text{ mol acid available}}\right) = 0.104 \text{ mol aspirin}$$

Step 4. Calculate the grams of aspirin that could be obtained if everything worked perfectly. This is the *theoretical yield*.

$$0.104 \text{ mol aspirin} \left(\frac{180.2 \text{ g } C_9H_8O_4}{1 \text{ mol } C_9H_8O_4}\right) = 18.8 \text{ g aspirin}$$

Step 5. Calculate the *percent yield* if, as stated in the problem, the *actual yield* was 6.26 g.

$$\text{Percent yield} = \frac{6.26 \text{ g aspirin actually obtained}}{18.8 \text{ g aspirin theoretically obtainable}} \times 100 = 33.3\%$$

EXERCISE 4.5 Percent Yield

Professor H.C. Brown of Purdue University received the Nobel Prize in Chemistry in 1979 for his work on the chemistry of diborane, B_2H_6. This gas can be prepared by the following reaction (which is carried out in a nonaqueous solvent):

$$3\ NaBH_4 + 4\ BF_3 \longrightarrow 3\ NaBF_4 + 2\ B_2H_6$$

If you begin with 18.9 g of $NaBH_4$ (and excess BF_3), and you isolate 7.50 g of B_2H_6 gas, what is the percent yield of B_2H_6?

4.7 CHEMICAL EQUATIONS AND CHEMICAL ANALYSIS

Now that we have described how to balance chemical equations and how to understand chemical stoichiometry, we begin to see how chemists can use chemical reactions to analyze a compound for its formula or to determine the composition of a mixture. What follows are examples of these situations, and more will be introduced in Chapter 5 when we take up the matter of reactions in water solutions.

Figure 4.13 If a compound containing C and H is burned in oxygen, CO_2 and H_2O are formed, and the mass of each can be determined. The H_2O is absorbed by magnesium perchlorate, and the CO_2 is absorbed by finely divided sodium hydroxide supported on asbestos. The mass of each absorbent before and after combustion will give the mass of CO_2 and H_2O trapped by the absorbent. Only a few milligrams of a combustible compound are needed for analysis.

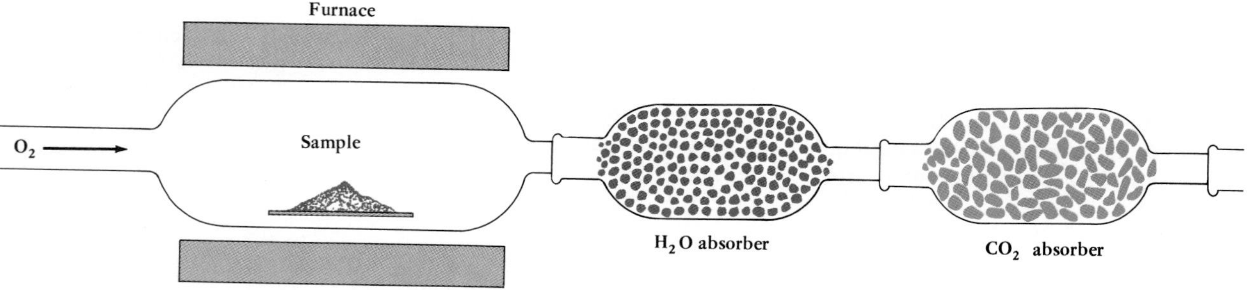

EMPIRICAL FORMULAS BY COMBUSTION ANALYSIS

The empirical formula of a compound can be determined if the percent composition of the compound is known (Section 3.6). But where do these data come from? Various methods are used, but one that works well for compounds that burn in oxygen is *analysis by combustion*. Each element (except oxygen) in the compound combines with oxygen to produce the appropriate oxide. Thus, carbon gives CO_2, hydrogen gives H_2O, and other elements give their oxides. For example,

$$CH_4(g) + 2\ O_2(g) \longrightarrow CO_2(g) + 2\ H_2O(g)$$

$$1\ C \longrightarrow 1\ CO_2 \text{ and } 4\ H \longrightarrow 2\ H_2O$$

$$2\ Al(CH_3)_3(\ell) + 12\ O_2(g) \longrightarrow 6\ CO_2(g) + 9\ H_2O(g) + Al_2O_3(s)$$

$$6\ C \longrightarrow 6\ CO_2,\ 18\ H \longrightarrow 9\ H_2O, \text{ and } 2\ Al \longrightarrow 1\ Al_2O_3$$

$$B_2H_6(g) + 3\ O_2(g) \longrightarrow B_2O_3(s) + 3\ H_2O(g)$$

$$6\ H \longrightarrow 3\ H_2O \text{ and } 2\ B \longrightarrow 1\ B_2O_3$$

From these examples, you can see that every mole of carbon that is in the original compound is converted to a mole of CO_2, and every mole of hydrogen in the original compound gives half a mole of H_2O. The gases CO_2 and H_2O can be separated and their masses determined as illustrated in Figure 4.13. Since these masses can be converted to the masses of C and H that they contain, respectively, we can find the masses of those elements in a given mass of the original compound, and this leads us to the empirical formula.

EXAMPLE 4.6

DETERMINING THE EMPIRICAL FORMULA OF A COMBUSTIBLE COMPOUND

Oxalic acid is used in the paint, cosmetics, and ceramics industries and is found in many plants and vegetables. It contains only the elements C, H, and O. If 0.513 g of the acid is burned in oxygen, 0.501 g of CO_2 and 0.103 g of H_2O result. What is the empirical formula of oxalic acid? If the molar mass of the acid is 90.04 g/mol, what is the molecular formula?

Solution The first step in stoichiometry problems is to write a balanced equation. Here we cannot do this, as we do not know the formula of oxalic acid. (That's what we want to learn!) However, we do know that every atom of C in the molecule must be converted to one molecule of CO_2, and that every molecule of H_2O from combustion must mean there are two atoms of H in the original compound. Therefore, our first step is to convert the masses of CO_2 and H_2O to moles in order to find the number of moles of C and H in the original compound.

$$0.501\ \cancel{g}\ CO_2 \left(\frac{1\ \text{mol}\ CO_2}{44.01\ \cancel{g}\ CO_2}\right) = 0.0114\ \text{mol}\ CO_2$$

$$0.103\ \cancel{g}\ H_2O \left(\frac{1\ \text{mol}\ H_2O}{18.02\ \cancel{g}\ H_2O}\right) = 0.00572\ \text{mol}\ H_2O$$

The number of moles of CO_2 and H_2O can now be converted to the masses of C and H that were in the original compound.

$$0.0114 \text{ mol } CO_2 \left(\frac{1 \text{ mol C}}{1 \text{ mol } CO_2}\right)\left(\frac{12.01 \text{ g C}}{1 \text{ mol C}}\right)$$

$= 0.137$ g C in CO_2 *and* in oxalic acid sample

$$0.00572 \text{ mol } H_2O \left(\frac{2 \text{ mol H}}{1 \text{ mol } H_2O}\right)\left(\frac{1.008 \text{ g}}{1 \text{ mol H}}\right)$$

$= 0.0115$ g H in H_2O *and* in oxalic acid sample

note this stoichiometric factor

These calculations reveal that the 0.513-g sample of oxalic acid contains 0.137 g of C and 0.0115 g of H; the remaining mass, 0.365 g, must be oxygen.

$$0.137 \text{ g C} + 0.0115 \text{ g H} + 0.365 \text{ g O} = 0.513 \text{ g oxalic acid sample}$$

To find the formula of oxalic acid, we need only to find the number of moles of each element that were contained in the oxalic acid sample and then find the mole ratios.

$$0.137 \text{ g C} \left(\frac{1 \text{ mol C}}{12.01 \text{ g C}}\right) = 0.0114 \text{ mol C}$$

$$0.0115 \text{ g H} \left(\frac{1 \text{ mol H}}{1.008 \text{ g H}}\right) = 0.0114 \text{ mol H}$$

$$0.365 \text{ g O} \left(\frac{1 \text{ mol O}}{16.00 \text{ g O}}\right) = 0.0228 \text{ mol O}$$

To find the mole ratio of elements, divide the number of moles of each element by the *smallest* number of moles.

$$\frac{0.0114 \text{ mol H}}{0.0114 \text{ mol C}} = \frac{1.00 \text{ mol H}}{1.00 \text{ mol C}} \qquad \frac{0.0228 \text{ mol O}}{0.0114 \text{ mol C}} = \frac{2.00 \text{ mol O}}{1.00 \text{ mol C}}$$

For every C atom in the molecule, one H atom and two O atoms occur. Therefore, the *empirical formula* of oxalic acid is CHO_2.

To determine the molecular formula, the experimental molar mass and the molar mass of one empirical formula unit are compared.

$$\frac{90.04 \text{ g/mol oxalic acid}}{45.02 \text{ g/mol of } CHO_2} = \frac{2.000 \text{ mol } CHO_2}{1.000 \text{ mol oxalic acid}}$$

Thus, the *molecular formula* of oxalic acid is twice the empirical formula, that is, $C_2H_2O_4$.

EXERCISE 4.6 Formula Determination from Combustion Analysis

Compounds consisting of carbon, hydrogen, and metals are called "organometallic" compounds. One of the best known of these is "ferrocene," a molecule containing only C, H, and Fe. If 0.652 g of ferrocene is burned in oxygen, 1.542 g of CO_2 and 0.315 g of H_2O are isolated. (The iron is converted to Fe_2O_3.) What is the empirical formula of ferrocene?

ANALYSIS OF A MIXTURE

One of the most interesting areas of chemistry is analytical chemistry. Chemists in this field answer questions about the composition of matter: identifying pure substances and the components of mixtures and determining the quantities of the substances in a mixture. This is often done now by instrumental methods,* but classical chemical reactions play a role. The next example illustrates one way to analyze a mixture.

EXAMPLE 4.7

ANALYSIS OF A MIXTURE

You have a white powder that you know is a mixture of magnesium oxide (MgO) and magnesium carbonate ($MgCO_3$); you would like to know what percent of the mass of the mixture is $MgCO_3$.

Earlier in this chapter you learned that metal carbonates decompose on heating to give metal oxides and CO_2. For magnesium carbonate the reaction is

$$MgCO_3(s) \longrightarrow MgO(s) + CO_2(g)$$

Therefore, if you heat the powder strongly the magnesium carbonate in the mixture should decompose to MgO, and CO_2 will be evolved as a gas. This solid left after heating consists only of MgO, and its mass is the sum of the MgO that was originally in the mixture *plus* the MgO that came from the $MgCO_3$. The difference in mass of the solid before and after heating therefore gives the mass of CO_2 lost and, by stoichiometry, the mass of $MgCO_3$ in the original mixture. Suppose you have 1.598 g of a mixture of MgO and $MgCO_3$. Heating evolves CO_2 and leaves 1.294 g of white MgO. What is the weight percent of $MgCO_3$ in the original mixture?

Solution As a first step, we calculate the mass loss:

Mass of mixture (MgO + $MgCO_3$)	= 1.598 g
Mass after heating (pure MgO)	= 1.294 g
Mass of CO_2	= 0.304 g

Now the mass of CO_2 can be converted into moles of CO_2.

$$0.304 \text{ g } \cancel{CO_2} \left(\frac{1 \text{ mol } CO_2}{44.01 \text{ g } \cancel{CO_2}}\right) = 0.00691 \text{ mol } CO_2$$

From the balanced equation for the decomposition of $MgCO_3$ we know that every mole of CO_2 given off on heating means that there was a mole of $MgCO_3$ in the mixture. Therefore, we must have had 0.00691 mole of $MgCO_3$, and the mass of $MgCO_3$ is

$$0.00691 \cancel{\text{mol } CO_2} \left(\frac{1 \cancel{\text{mol } MgCO_3}}{1 \cancel{\text{mol } CO_2}}\right)\left(\frac{84.31 \text{ g } MgCO_3}{1 \cancel{\text{mol } MgCO_3}}\right) = 0.582 \text{ g } MgCO_3$$

*"Instrumental methods" include, among others, various kinds of spectroscopy and chromatography. **Spectroscopy** is the analysis of the interaction between matter and forms of radiation (light, radio waves, and so on). **Chromatography** relies on physical interactions between molecules to separate mixtures. Although we shall touch only briefly on these methods in this book, they have become important tools in the laboratories of both research and industrial chemists.

This means the weight percent of magnesium carbonate in the mixture is

$$\frac{\text{Mass of MgCO}_3}{\text{Sample mass}} \times 100 = \frac{0.582 \text{ g MgCO}_3}{1.598 \text{ g sample}} \times 100 = 36.4\% \text{ MgCO}_3$$

EXERCISE 4.7 Chemical Analysis

You have 2.357 g of a mixture of $BaCl_2$ and $BaCl_2 \cdot 2\,H_2O$. Experiment shows that the mixture has a mass of only 2.108 g after heating to drive off all the water of hydration in $BaCl_2 \cdot 2\,H_2O$. What is the weight percent of $BaCl_2 \cdot 2\,H_2O$ in the original mixture?

The example shows only one of the ways chemists can analyze mixtures. You will see more methods in the Study Questions in this chapter and in Chapter 5.

SUMMARY

In any chemical change, matter is conserved. Although the atoms involved are rearranged into different species in the course of a reaction and the number of molecules may change, the total number of atoms of each kind in the reactants and products must be the same. Thus, a **balanced chemical equation** shows the relative amounts of products and reactants, the amounts being indicated by **stoichiometric coefficients** (Sections 4.1 and 4.2).

Throughout this book you will encounter **combination reactions** where elements combine (a) with the halogens and (b) with oxygen (Sections 4.2 and 4.3). **Combustion reactions** involve the combination of a compound of C and H, or one of C, H, and some other element, to give CO_2, H_2O, and another oxide as appropriate. In **decomposition reactions** compounds break down into simpler compounds, usually by heating. The decomposition of metal carbonates to give metal oxides and CO_2 is an example.

Stoichiometry is the study of mass relations in chemical reactions, and its guiding principle is the conservation of matter. To relate the mass of one reactant to the mass required of another reactant or of a product, four steps are generally followed (Section 4.4 and Figure 4.11): (1) Balance the equation for the reaction. (2) Calculate moles of reactant. (3) Relate moles of product expected to moles of reactant used with a **stoichiometric factor**. (This factor comes from the balanced equation and is always "moles of required reactant or product" divided by "moles of known reactant.") (4) Calculate mass of product from moles of product.

Most reactions are done with one reactant present in a smaller amount than required by stoichiometry. This reactant controls the amount of product formed and is called the **limiting reagent** (Section 4.5). You must decide on the limiting reagent before you can proceed to calculate the mass of product expected.

The **percent yield** of a reaction is used to express the efficiency of chemical reactions (Section 4.6). This is the ratio of the amount of product actually isolated to the amount of product expected if the reaction worked according to theory (called the **theoretical yield**).

$$\text{Percent yield} = \frac{\text{actual yield}}{\text{theoretical yield}} \times 100$$

Finally, chemical reactions such as combustions can be used to determine the formula of a compound, and other reactions can be used to analyze mixtures of compounds.

STUDY QUESTIONS

REVIEW QUESTIONS

1. What information does a balanced chemical equation provide?
2. In the reaction in Figure 4.1, how many molecules of O_2 would you need if you have 2000 molecules of S_8? How many molecules of SO_2 would you obtain from the reaction?
3. Find in the chapter one example of each of the following reaction types and write the balanced equation for the reaction: (a) combustion; (b) reaction of O_2 with a metal; (c) reaction of O_2 with a nonmetal; and (d) a decomposition reaction.
4. Find two examples in the chapter of the reaction of a metal with a halogen, write a balanced equation for each example, and name the product.
5. What is the guiding principle of chemical stoichiometry? Illustrate your answer with the reaction $N_2(g) + O_2(g) \rightarrow 2\ NO(g)$.
6. When working in the laboratory, chemists often set up reactions so that one of several reactants is there in greater quantity than required by stoichiometry. This means that one of the reactants is a "limiting reagent." Illustrate what is meant by a "limiting reagent" using the reaction of calcium and oxygen to give calcium oxide.
7. Briefly explain the difference between the actual and theoretical yield for a reaction.
8. You have a green powder that you know consists of nickel(II) oxide and nickel(II) carbonate. Suggest a way to analyze the sample for the amount of nickel carbonate it may contain.
9. In many cases metal oxides will react with H_2 to leave the metal and convert the oxygen to water. For example,

   $TiO(s) + H_2(g) \longrightarrow Ti(s) + H_2O(g)$

   If you did not know the metal oxide in question was TiO, explain how you could use this reaction to find the empirical formula of the compound.

BALANCING EQUATIONS

10. Balance the following equations.
    (a) $Al(s) + O_2(g) \longrightarrow Al_2O_3(s)$
    (b) $N_2(g) + H_2(g) \longrightarrow NH_3(g)$
    (c) $C_6H_6(\ell) + O_2(g) \longrightarrow H_2O(g) + CO_2(g)$
11. Balance the following equations.
    (a) $Al(s) + Cl_2(g) \longrightarrow AlCl_3(s)$
    (b) $SiO_2(s) + C(s) \longrightarrow Si(s) + CO(g)$
    (c) $Fe(s) + H_2O(g) \longrightarrow Fe_3O_4(s) + H_2(g)$
12. Balance the following equations.
    (a) $UO_2(s) + HF(\ell) \longrightarrow UF_4(s) + H_2O(\ell)$
    (b) $B_2O_3(s) + HF(\ell) \longrightarrow BF_3(g) + H_2O(\ell)$
    (c) $BF_3(g) + H_2O(\ell) \longrightarrow HF(\ell) + H_3BO_3(s)$
13. Balance the following equations.
    (a) $MgO(s) + Fe(s) \longrightarrow Fe_2O_3(s) + Mg(s)$
    (b) $H_3BO_3(s) \longrightarrow B_2O_3(s) + H_2O(\ell)$
    (c) $NaNO_3(s) + H_2SO_4(\ell) \longrightarrow Na_2SO_4(s) + HNO_3(g)$
14. Balance the following equations.
    (a) $Na_2O_2(s) + H_2O(\ell) \longrightarrow NaOH(aq) + H_2O_2(aq)$
    (b) $PH_3(g) + O_2(g) \longrightarrow P_4O_{10}(s) + H_2O(g)$
    (c) $C_2H_3Cl(g) + O_2(g) \longrightarrow CO_2(g) + H_2O(g) + HCl(g)$
15. Balance the following equations.
    (a) $CaF_2(s) + H_2SO_4(\ell) \longrightarrow CaSO_4(s) + HF(g)$
    (b) $N_2O(g) \longrightarrow N_2(g) + O_2(g)$
    (c) $NH_4NO_3(s) \longrightarrow N_2O(g) + H_2O(g)$
16. Balance the following equations.
    (a) Reaction to produce hydrazine, N_2H_4.

    $H_2NCl(aq) + NH_3(g) \longrightarrow NH_4Cl(aq) + N_2H_4(aq)$

    (b) Reaction of fuel used in moon lander and space shuttle.

    $(CH_3)_2N_2H_2(\ell) + N_2O_4(\ell) \longrightarrow N_2(g) + H_2O(g) + CO_2(g)$

    (c) Reaction of calcium carbide to produce acetylene, C_2H_2.

    $CaC_2(s) + H_2O(\ell) \longrightarrow Ca(OH)_2(s) + C_2H_2(g)$

17. Balance the following equations.
    (a) Reaction of calcium cyanamide to produce ammonia.

    $CaCN_2(s) + H_2O(\ell) \longrightarrow CaCO_3(s) + NH_3(g)$

    (b) Reaction to produce diborane, B_2H_6.

    $NaBH_4(s) + H_2SO_4(aq) \longrightarrow B_2H_6(g) + H_2(g) + Na_2SO_4(aq)$

(c) Reaction to rid water of hydrogen sulfide, H_2S, a foul-smelling compound.

$$H_2S(aq) + Cl_2(aq) \longrightarrow S(s) + HCl(aq)$$

CLASSES OF CHEMICAL REACTIONS

18. Complete and balance the following equations involving oxygen reacting with an element. Name the product in each case.
(a) $Mg(s) + O_2(g) \longrightarrow$
(b) $Ca(s) + O_2(g) \longrightarrow$
(c) $In(s) + O_2(g) \longrightarrow$
19. Complete and balance the following equations involving oxygen reacting with an element.
(a) $Ti(s) + O_2(g) \longrightarrow$ titanium(IV) oxide
(b) $S_8(s) + O_2(g) \longrightarrow$ sulfur dioxide
(c) $Se(s) + O_2(g) \longrightarrow$ selenium dioxide
20. Complete and balance the following equations involving a halogen reacting with a metal. Name the product in each case.
(a) $Na(s) + Cl_2(g) \longrightarrow$
(b) $Mg(s) + Br_2(\ell) \longrightarrow$
(c) $Al(s) + F_2(g) \longrightarrow$
21. Complete and balance the following equations involving a halogen reacting with a metal.
(a) $Cr(s) + Cl_2(g) \longrightarrow$ chromium(III) chloride
(b) $Cu(s) + Br_2(\ell) \longrightarrow$ copper(II) bromide
(c) $Pt(s) + F_2(g) \longrightarrow$ platinum(IV) fluoride
22. Complete and balance equations for the following combustion reactions.
(a) $CH_4(g) + O_2(g) \longrightarrow$
(b) $C_8H_{18}(\ell) + O_2(g) \longrightarrow$
(c) $C_2H_5OH(\ell) + O_2(g) \longrightarrow$
23. Complete and balance equations for the following combustion reactions.
(a) $C_{10}H_{10}Fe(s) + O_2(g) \longrightarrow Fe_2O_3(s) + \ldots$
(b) $B_5H_9(\ell) + O_2(g) \longrightarrow B_2O_3(s) + \ldots$
(c) $Si_2H_6(g) + O_2(g) \longrightarrow SiO_2(s) + \ldots$
24. Write a balanced equation for the formation of each of the following compounds from the elements.
(a) carbon monoxide
(b) nickel(II) oxide
(c) chromium(III) oxide
25. Write a balanced equation for the formation of each of the following compounds from the elements.
(a) copper(I) oxide
(b) arsenic(III) oxide
(c) zinc oxide
26. Write a balanced equation for each of the following decomposition reactions. Name each product.
(a) $BeCO_3(s)$ + heat $\longrightarrow$
(b) $NiCO_3(s)$ + heat $\longrightarrow$
(c) $Al_2(CO_3)_3(s)$ + heat $\longrightarrow$
27. Write a balanced equation for each of the following decomposition reactions. Name each product.
(a) $ZnCO_3(s)$ + heat $\longrightarrow$
(b) $MnCO_3(s)$ + heat $\longrightarrow$
(c) $PbCO_3(s)$ + heat $\longrightarrow$

GENERAL STOICHIOMETRY PROBLEMS

28. Ammonia is used throughout the world as a fertilizer and in making nitrogenous plastics and fibers. It is usually manufactured in the Haber process, the direct reaction of N_2 with H_2 in the presence of other compounds that accelerate the reaction.

$$N_2(g) + 3\ H_2(g) \longrightarrow 2\ NH_3(g)$$

If you use 280. g of N_2, how many grams of H_2 are required for complete reaction? How many grams of NH_3 can be produced?

29. Silver bromide is an important component of photographic film. It can be made by the reaction

$$AgNO_3(aq) + NaBr(aq) \longrightarrow AgBr(s) + NaNO_3(aq)$$

If you have 2.6 g of sodium bromide, how many grams of silver nitrate are required for complete reaction? What is the maximum number of grams of sodium nitrate and of silver bromide that you could obtain?

30. Laughing gas, N_2O, is made by the careful thermal decomposition of ammonium nitrate.

$$NH_4NO_3(s) \longrightarrow N_2O(g) + 2\ H_2O(g)$$

If you begin with 1.00×10^3 g (about 2 pounds) of ammonium nitrate, how many grams of laughing gas can you obtain? How many grams of water?

31. When aluminum metal reacts with liquid bromine (Figure 4.3), the reaction produces aluminum bromide.

$$2\ Al(s) + 3\ Br_2(\ell) \longrightarrow Al_2Br_6(s)$$

If you begin with 2.56 g of Al, how many grams of bromine are required for complete reaction? How many grams of aluminum bromide will be produced?

32. Ammonia can be made by treating calcium cyanamide with water.

$$CaCN_2(s) + 3\ H_2O(\ell) \longrightarrow CaCO_3(s) + 2\ NH_3(g)$$

How many grams of water would be required to react with $CaCN_2$ to produce 68.0 g of NH_3? How many grams of $CaCN_2$ are used in the process?

33. The very stable compound SF_6 is made by burning sulfur in an atmosphere of fluorine.

$$S_8(s) + 24\ F_2(g) \longrightarrow 8\ SF_6(g)$$

If you need 365 grams of SF_6, how many grams of sulfur, S_8, and how many grams of F_2 are required?

34. Iron can react with the oxygen in hot, dry air to produce Fe_3O_4. If 12.58 g of iron react with O_2 to give this oxide, how many grams of O_2 are required? How many grams of Fe_3O_4 are expected?

35. Aluminum metal can react with the oxygen in air to give aluminum oxide. If 0.569 g of aluminum is used, how many grams of aluminum oxide are produced? How many grams of oxygen gas are required for complete reaction?

36. The final step in the manufacture of pure platinum (for use in automobile catalytic converters and other purposes) is the reaction

$$3\ (NH_4)_2PtCl_6(s) \longrightarrow 3\ Pt(s) + 2\ NH_4Cl(g) + 2\ N_2(g) + 16\ HCl(g)$$

If you heat 34.6 g of $(NH_4)_2PtCl_6$, how many grams of Pt metal should you isolate? How many grams of HCl should you expect?

37. Passing an electrical current into brine, a solution of sodium chloride in water, gives hydrogen gas, chlorine gas, and sodium hydroxide according to the equation

$$2\ NaCl(aq) + 2\ H_2O(\ell) \longrightarrow H_2(g) + Cl_2(g) + 2\ NaOH(aq)$$

This important commercial process is the source of much of the chlorine and sodium hydroxide used in our economy. Assume you have 2550 g of NaCl in solution. Calculate the mass of each of the products expected from the reaction above.

38. The overall equation for the reduction of iron ore to iron metal in a blast furnace is

$$Fe_2O_3(s) + 3\ CO(g) \longrightarrow 2\ Fe(s) + 3\ CO_2(g)$$

(a) How many grams of CO are required to consume completely 365 g of Fe_2O_3?
(b) How many kilograms of Fe_2O_3 are required to produce 27.9 kg of Fe?

39. Hydrofluoric acid, HF, is never sold in glass bottles, because glass is composed of calcium and sodium silicates that react with HF. The reaction can be depicted, for simplicity, as the interaction of silicon dioxide and the acid.

$$SiO_2(s) + 4\ HF(aq) \longrightarrow SiF_4(g) + 2\ H_2O(\ell)$$

(a) How many grams of HF would you need to react completely with 256 g of SiO_2?
(b) How many grams of SiF_4 would be produced from 300. g of pure silica, SiO_2? (Assume more HF is present than is required.)

40. Oil paintings in which "white lead" has been used can be blackened by reaction with H_2S from air pollution or from the glaze over the painting itself. The blackening comes from the formation of lead sulfide, which may be cleaned off by washing with hydrogen peroxide, H_2O_2. The reaction for the cleaning process is

$$PbS(\text{black solid}) + 4\ H_2O_2(aq) \longrightarrow PbSO_4(s) + 4\ H_2O(\ell)$$

(a) How many grams of H_2O_2 must be used to clean off 0.24 g of PbS?
(b) If 0.072 g of H_2O form in the reaction, how many grams of $PbSO_4$ must also have been formed?

41. White phosphorus, P_4, is made by heating calcium phosphate with silica and carbon.

$$2\ Ca_3(PO_4)_2(s) + 6\ SiO_2(s) + 10\ C(s) \longrightarrow 6\ CaSiO_3(s) + 10\ CO(g) + P_4(s)$$

If you have 1.00 pound (454 g) of calcium phosphate, how many grams each of SiO_2 and C must be used for complete reaction? How many grams of P_4 should you isolate if the reaction is 100% efficient?

LIMITING REAGENT PROBLEMS

42. Dinitrogen tetrafluoride, N_2F_4, can be produced by the reaction of NH_3 with F_2.

$$2\ NH_3(g) + 5\ F_2(g) \longrightarrow N_2F_4(g) + 6\ HF(g)$$

If 4.00 g of NH_3 and 14.0 g of F_2 are allowed to react, which is the limiting reagent? Which reagent is in excess and by how much? How many grams of N_2F_4 and HF are produced?

43. Disulfur dichloride, S_2Cl_2, is used to vulcanize rubber. It can be made by treating molten sulfur with gaseous chlorine.

$$S_8(\ell) + 4\ Cl_2(g) \longrightarrow 4\ S_2Cl_2(g)$$

If you begin with 32.0 g of sulfur and 71.0 g of Cl_2, which is the limiting reagent? How many grams of S_2Cl_2 can be produced? What quantity of which starting material will remain after the maximum amount of S_2Cl_2 has been formed?

44. Methyl alcohol, CH_3OH, is a clean-burning, easily handled fuel. It can be made by the direct reaction of CO and H_2 (obtained from coal and water).

$$CO(g) + 2\ H_2(g) \longrightarrow CH_3OH(\ell)$$

Assume you start with 12.0 g of H_2 and 74.5 g of CO. Which of the reactants is in excess? Which is the limiting reagent? What mass (in grams) of the excess reagent is left after reaction is complete? How many grams of methyl alcohol can be obtained theoretically?

45. As described in Example 4.5, aspirin can be made from salicylic acid and acetic anhydride. If you begin with 1.0 kg of each of the two reactants, which is the limiting reagent? What is the maximum number of grams of aspirin that you can obtain?

46. Ammonia gas can be prepared in a number of ways, among them

$$CaO(s) + 2\ NH_4Cl(s) \longrightarrow 2\ NH_3(g) + H_2O(g) + CaCl_2(s)$$

If you use 112 g of CaO with 224 g of NH_4Cl, how many grams of NH_3 should you obtain? How many grams of which reagent remain after the reaction is complete?

47. The following reaction produces two compounds that are commercially useful, S_2Cl_2 and SF_4.

$$3\ SCl_2(\ell) + 4\ NaF(s) \longrightarrow SF_4(g) + S_2Cl_2(\ell) + 4\ NaCl(s)$$

If you begin with 16.3 g of SCl_2 and 10.5 g of NaF, which of these is the limiting reagent? How many grams of SF_4 and S_2Cl_2 can be produced?

48. If chlorine gas reacts with sulfur dioxide, you can obtain two useful compounds, thionyl chloride ($OSCl_2$) and dichlorine oxide.

$$SO_2(g) + 2\ Cl_2(g) \longrightarrow OSCl_2(g) + Cl_2O(g)$$

If you mix 150 g each of SO_2 and Cl_2, which reagent will be left at the completion of the reaction? How many grams of each of the products can you obtain?

49. Bromine trifluoride reacts with metal oxides to evolve oxygen. For example,

$$3\ SiO_2(s) + 4\ BrF_3(\ell) \longrightarrow 3\ SiF_4(g) + 2\ Br_2(\ell) + 3\ O_2(g)$$

If you mix 1.256 g of silicon dioxide with 5.893 g of bromine trifluoride, what is the maximum number of grams of each of the products that you can obtain?

PERCENT YIELD

50. Zinc and chlorine react directly to give zinc chloride.

$$Zn(s) + Cl_2(g) \longrightarrow ZnCl_2(s)$$

If you begin with 1.00 mole of zinc and excess Cl_2, what is the theoretical yield of $ZnCl_2$ in grams? If you isolate 115 g of $ZnCl_2$, what is the percent yield of the metal chloride?

51. Nickel tetracarbonyl, $Ni(CO)_4$, is made by direct reaction of nickel metal and CO gas. If you begin with 5.00 g of Ni and an excess of CO, what is the theoretical yield of $Ni(CO)_4$? How many grams of CO are required to use up all of the nickel? If you isolate only 9.67 g of $Ni(CO)_4$, what is the percent yield of this compound?

52. The reaction of red phosphorus and liquid bromine is shown in Figure 3.1

$$2\ P(s) + 3\ Br_2(\ell) \longrightarrow 2\ PBr_3(\ell)$$

If you use 12.7 g of Br_2 with an excess of phosphorus, what is the theoretical yield of PBr_3? If you isolate 10.9 g of PBr_3, what is the percent yield of this compound?

53. Disulfur dichloride, S_2Cl_2, is a golden-yellow liquid with a revolting smell. It is used industrially in the vulcanization of rubber and is prepared by the following reaction:

$$3\ SCl_2(\ell) + 4\ NaF(s) \longrightarrow SF_4(g) + S_2Cl_2(\ell) + 4\ NaCl(s)$$

Assume you begin with 5.23 g of SCl_2 and an excess of NaF. What is the theoretical yield of S_2Cl_2 in grams? If you isolate 1.19 g of S_2Cl_2, what is the percent yield of this compound?

STOICHIOMETRY AND MOLECULAR FORMULAS

54. Butane, which contains only C and H, is a commonly used fuel in camping stoves. To determine the formula of butane, assume you burn 0.580 g of the gas and obtain 1.760 g of CO_2 and 0.900 g of H_2O. What is the empirical formula of butane?

55. Anthracene, which contains only carbon and hydrogen, is an important source of dyes. Determine its empirical and molecular formulas from the following data: (a) When 2.50 mg are burned in pure oxygen, 8.64 mg of CO_2 and 1.26 mg of H_2O are isolated. (b) The experimental molar mass of the compound is 178 g/mol.

56. Cumene is a hydrocarbon, a compound containing only C and H, that is found in petroleum. Determine its empirical and molecular formulas from the following data: (a) You burn 0.835 g of the liquid compound in pure O_2 and isolate 2.752 g of CO_2 and 0.751 g of H_2O. (b) The molar mass is estimated in a separate experiment to be 120 g/mol.

57. Isoprene is a major industrial material that is used to make synthetic rubber. This hydrocarbon (a compound containing only C and H) is an unstable liquid that reacts readily with oxygen and is quite irritating to the skin. Determine its empirical and molecular formulas from the following data: (a) You burn a 0.157-g sample and isolate 0.507 g of CO_2 and 0.166 g of H_2O. (b) The molar mass is estimated in a separate experiment to be 68 g/mol.

58. Vitamin C is a compound containing the elements C, H, and O. Determine the empirical formula of vitamin C from the following data: 4.00 mg of the solid vitamin

is burned in oxygen to give 6.00 mg of CO_2 and 1.632 mg of H_2O.

59. A compound named umbelliferone is used in sunscreen lotions. It has the formula $C_xH_yO_z$. Determine its empirical and molecular formulas from the following data: (a) Burning 0.123 g of the compound results in 0.300 g of CO_2 and 0.040 g of H_2O. (b) The molar mass is estimated from experiment to be 160 g/mol.

60. Silicon and hydrogen form a series of interesting compounds, Si_xH_y. To find the formula of one of them, you take a 6.22 g sample of the compound and burn it in oxygen. On doing so, all of the Si is converted to 12.02 g of SiO_2 and all of the H to 5.40 g of H_2O. What is the empirical formula of the silicon compound? If the experimental molar mass is 62.2 g/mol, what is its molecular formula?

61. Boron hydrides are compounds with the general formula B_xH_y. If you burn the compounds in oxygen the combustion products are B_2O_3 and H_2O. If 1.740 g of B_2O_3 and 0.810 g of H_2O are obtained on burning one of these compounds, what is its empirical formula?

62. At least two compounds with the general formula $Co_x(CO)_y$ can be formed between cobalt metal and carbon monoxide. One of these is analyzed by burning in air.

$$Co_x(CO)_y(s) + \text{some } O_2(g) \longrightarrow \text{some } Co_2O_3(s) + y\ CO_2(g)$$

A 3.42-g sample of $Co_x(CO)_y$ forms 3.52 g of CO_2, plus some cobalt(III) oxide whose weight was not determined. What is the empirical formula of $Co_x(CO)_y$?

63. Iron forms several compounds of the type $Fe_x(CO)_y$ with carbon monoxide. If you burn one of these in pure oxygen, the iron forms iron(III) oxide, and the CO forms CO_2.

$$Fe_x(CO)_y(\ell) + \text{some } O_2(g) \longrightarrow \text{some } Fe_2O_3(s) + y\ CO_2(g)$$

Assume you burn 1.959 g of $Fe_x(CO)_y$ and find that it forms 0.799 g of Fe_2O_3 and 2.200 g of CO_2. What is the empirical formula of $Fe_x(CO)_y$?

64. A 1.324-g sample of a metal carbonate, containing an unknown metal M, was heated to give the metal oxide and 0.395 g of CO_2.

$$MCO_3(s) + \text{heat} \longrightarrow MO(s) + CO_2(g)$$

What is the metal M? Prove your answer with appropriate calculations.

65. A 0.598-g sample of a green metal carbonate, containing an unknown metal M, was heated to give the metal oxide and 0.222 g of CO_2.

$$MCO_3(s) + \text{heat} \longrightarrow MO(s) + CO_2(g)$$

What is the metal M? Prove your answer with appropriate calculations.

CHEMICAL ANALYSIS

66. Metal carbonates give the appropriate metal oxide and CO_2 when heated. For example,

$$CaCO_3(s) \longrightarrow CaO(s) + CO_2(g)$$

Limestone is mostly calcium carbonate, $CaCO_3$, but other minerals are usually present as well. Assume a 1.605-g sample of limestone is heated and decomposed completely to CaO, 0.657 g of CO_2, and an inert residue. What is the weight percentage of $CaCO_3$ in the limestone sample?

67. Hydrated metal compounds lose their water of hydration when heated strongly. Suppose you have a mixture of anhydrous $CuSO_4$ (a compound "without water") and the hydrated compound $CuSO_4 \cdot 5\ H_2O$. If you heat a 2.628-g sample of a mixture of these compounds, all of the water is lost from $CuSO_4 \cdot 5\ H_2O$ and the sample is now just pure $CuSO_4$. Given that the mixture lost 0.873 g of water on heating, what is the weight percent of $CuSO_4 \cdot 5\ H_2O$ in the original sample?

68. Bromine trifluoride reacts with metal oxides to evolve oxygen. For example,

$$3\ TiO_2(s) + 4\ BrF_3(\ell) \longrightarrow 3\ TiF_4(s) + 2\ Br_2(\ell) + 3\ O_2(g)$$

Suppose you wish to use this reaction to determine the weight percent of TiO_2 in a sample of ore. To do this you collect the O_2 gas from the reaction. If you find that 1.586 g of the TiO_2-containing ore evolve 36.1 mg of O_2, what is the weight percent of TiO_2 in the sample? What is the weight percent of Ti in the sample?

69. Butyl lithium, LiC_4H_9, is a reactive compound used by chemists to make new materials. One way to determine the quantity of LiC_4H_9 in a mixture is to add something like hydrogen chloride (in water) to the mixture. The following reaction occurs:

$$LiC_4H_9 + HCl(aq) \longrightarrow LiCl(aq) + C_4H_{10}(g)$$

Let us say that you take 5.606 g of a solution of LiC_4H_9 dissolved in the organic compound benzene. On adding HCl you find that 0.636 g of C_4H_{10} gas is evolved. What is the weight percent of LiC_4H_9 in the sample?

GENERAL QUESTIONS

70. Balance the following chemical equations:

(a) $UO_3(s) + BrF_3(\ell) \longrightarrow UF_4(s) + Br_2(\ell) + O_2(g)$

(b) $IO_2F(s) + BrF_3(\ell) \longrightarrow IF_5(\ell) + Br_2(\ell) + O_2(g)$

(c) $S_2F_2(g) \longrightarrow S_8(s) + SF_4(g)$
(d) $Na_2S(aq) + O_2(g) + H_2O(\ell) \longrightarrow Na_2S_2O_3(aq) + NaOH(aq)$

71. Ammonium dichromate undergoes a decomposition reaction to give chromium(III) oxide, water, and nitrogen. (See Figure 4.8.)

$$(NH_4)_2Cr_2O_7(s) \longrightarrow N_2(g) + Cr_2O_3(s) + 4\ H_2O(g)$$

If you use 9.856 g of ammonium dichromate, how many grams of each product do you expect?

72. Cyclooctatetraene is an interesting hydrocarbon. If you burn 0.278 g of cyclooctatetraene, and you obtain 0.940 g of CO_2 and 0.192 g of H_2O, what is the empirical formula of the compound? If experiment shows that the molar mass of cyclooctatetraene is about 105 g/mol, what is the molecular formula of the compound?

73. Elemental sulfur (1.256 g) is combined with fluorine, F_2, to give a compound with the formula SF_x, a very stable, colorless gas. If you isolate 5.722 g of SF_x, what is the value of x?

74. If 3.91 g of potassium metal are allowed to react with oxygen, the product is 7.11 g of an explosive, yellow solid called potassium superoxide. What is the empirical formula of this compound?

75. Furfural is an organic compound that contains C, H, and O. The colorless, oily liquid is sometimes called "artificial oil of ants." Determine the empirical and molecular formulas of furfural based on the following information from laboratory experiments: Burning 0.239 g of the compound gives 0.547 g of CO_2 and 89.6 mg of H_2O. The molar mass is approximately 96 g/mol.

76. An oxide of scandium, with a mass of 1.423 g, is chemically reduced with H_2 to give H_2O and 0.929 g of Sc metal.

$$Sc_xO_y(s) + y\ H_2(g) \longrightarrow y\ H_2O(g) + x\ Sc(s)$$

What is the empirical formula of the scandium oxide? How many grams of water are isolated in the reaction?

77. Some solid CaO in a test tube picks up water from the atmosphere and is thereby changed completely to $Ca(OH)_2$.

$$CaO(s) + H_2O(\ell) \longrightarrow Ca(OH)_2(s)$$

The initial mass of CaO plus the test tube is 10.860 g. After picking up water to give calcium hydroxide, the total mass of the tube and $Ca(OH)_2$ is 11.149 g. What is the mass of the test tube?

78. It has been proposed that magnesium oxide can be used to remove sulfur dioxide from the flue gas of factories and power plants (where SO_2 comes from burning fossil fuels). The magnesium oxide reacts with SO_2 and O_2 according to the equation

$$MgO(s) + SO_2(g) + \tfrac{1}{2}\ O_2(g) \longrightarrow MgSO_4(s)$$

The MgO comes from $MgCO_3$, a naturally occurring mineral, according to the equation

$$MgCO_3(s) + \text{heat} \longrightarrow MgO(s) + CO_2(g)$$

To remove 20. million tons of SO_2 from factory and plant gases each year, how many tons of $MgCO_3$ must be mined? How much $MgSO_4$ will be produced? (And where can we put it?)

79. Dinitrogen tetroxide can be prepared by a *decomposition reaction*, the decomposition of lead(II) nitrate, according to the following *unbalanced* equation.

$$Pb(NO_3)_2(s) \longrightarrow N_2O_4(g) + PbO(s) + O_2(g)$$

If 3.468 g of $Pb(NO_3)_2$ give 0.832 g of N_2O_4, what is the percent yield of the product?

80. Phosphoric acid is made in enormous quantities every year by treating the phosphate mineral apatite (see the photo) with acid. For example,

$$\underset{\text{apatite}}{Ca_5(PO_4)_3F} + \underset{\text{sulfuric acid}}{5\ H_2SO_4} + 10\ H_2O \longrightarrow 3\ H_3PO_4 + \underset{\text{gypsum}}{5\ CaSO_4 \cdot 2\ H_2O} + HF$$

(a) If you begin with 1.00 ton of apatite, how many grams of phosphoric acid can be obtained? That is, what is the theoretical yield of H_3PO_4? (1 T = 908,000 g)
(b) If the amount of acid actually obtained is 2.50×10^5 g, what is the percent yield of phosphoric acid?
(c) If the ton of apatite had been treated with 1 ton of sulfuric acid, which is the limiting reagent? (Prove your answer with appropriate calculations.)

Apatite. (Bill Tronca, Tom Stack & Associates)

81. Suppose you want a sample of copper(II) oxide. Write balanced equations to show how you can make this compound from either a combination reaction or a decomposition reaction.

82. Sulfur tetrafluoride is best made by the reaction of sulfur dichloride and sodium fluoride.

$$3\ SCl_2(\ell) + 4\ NaF(s) \longrightarrow SF_4(g) + S_2Cl_2(\ell) + 4\ NaCl(s)$$

Assume you begin with 15.67 g of SCl_2 and an excess of NaF. If you isolate 3.56 g of SF_4, what is the percent yield of SF_4? Since you isolated only 3.56 g of SF_4, how many grams of S_2Cl_2 must also have been produced?

83. The *Starship Enterprise* on *Star Trek* really used B_5H_9 and O_2 as a fuel. The two react according to the following balanced equation:

$$2\ B_5H_9(\ell) + 12\ O_2(g) \longrightarrow 5\ B_2O_3(s) + 9\ H_2O(g)$$

(a) If one fuel tank holds 126 kg of B_5H_9, and the other fuel tank holds 192 kg of liquid O_2, which fuel tank will be emptied first?
(b) When one fuel tank is emptied, how much remains in the other tank?
(c) When the reaction has gone as far as possible, how much water has been formed?

84. Nitric acid is made by a sequence of reactions:

$$4\ NH_3(g) + 5\ O_2(g) \longrightarrow 4\ NO(g) + 6\ H_2O(g)$$

$$2\ NO(g) + O_2(g) \longrightarrow 2\ NO_2(g)$$

$$3\ NO_2(g) + H_2O(\ell) \longrightarrow 2\ HNO_3(\ell) + NO(g)$$

If you begin with 545 g of NH_3, what is the theoretical yield of nitric acid? If the yield of the acid is only 346 g, what is the percent yield?

85. Nitrogen is unique in that it forms no fewer than seven molecular oxides. The one of highest molar mass is dinitrogen pentoxide, N_2O_5, a crystalline, colorless solid made by the reaction of nitric acid (HNO_3) with P_4O_{10}.

$$4\ HNO_3(\ell) + P_4O_{10}(s) \longrightarrow 2\ N_2O_5(s) + 4\ HPO_3(s)$$

If you use 15.07 g of P_4O_{10} and 10.67 g of HNO_3, how many grams could you possibly obtain of N_2O_5? If you actually find only 1.96 g of N_2O_5, what is the percent yield of the compound?

86. Pure nitrogen gas can be prepared in the laboratory by the reaction of ammonia and copper(II) oxide according to the following *unbalanced* equation:

$$NH_3(g) + CuO(s) \longrightarrow N_2(g) + Cu(s) + H_2O(g)$$

If you treat 5.589 g of CuO with 0.686 g of NH_3, how many grams of N_2 should you obtain? If you actually obtain only 0.345 g of N_2, what is the percent yield of N_2? Given that you found only 0.345 g of N_2, how many grams of Cu must also have been formed?

87. Metal carbonyls, compounds containing a metal bonded to CO, constitute a large class of chemical compounds. Iron, for example, forms three such compounds of general formula $Fe_x(CO)_y$. The empirical formula of one of them can be determined from the following data: A 1.256-g sample of a black solid, $Fe_x(CO)_y$, is burned in air to give 0.598 g of Fe_2O_3 and 1.317 g of CO_2. If the molar mass of the compound is 504 g/mol, what is the molecular formula of the compound?

88. You want to find the formula of a compound composed of aluminum, carbon, and hydrogen. You know the C and H are combined in CH_3 groups, so the formula of the compound can be represented by $Al_x(CH_3)_y$. On burning the compound in pure oxygen, the C and H produce CO_2 and H_2O, respectively, and the aluminum turns into aluminum oxide according to the following *unbalanced* equation:

$$Al_x(CH_3)_y(\ell) + \text{excess } O_2(g) \longrightarrow Al_2O_3(s) + CO_2(g) + H_2O(g)$$

Assume that you burn 0.563 g of $Al_x(CH_3)_y$ and that you isolate 0.398 g of Al_2O_3. What is the empirical formula of $Al_x(CH_3)_y$?

89. Vinyl chloride is used in the plastics industry, but it has recently been implicated as a cancer-causing agent. The compound contains C, H, and Cl; if you burn it in oxygen, you obtain CO_2, H_2O, and HCl. Assume that you burn 3.125 g of vinyl chloride and obtain 4.400 g of CO_2 and 1.825 g of HCl plus some water. From the masses of CO_2 and HCl, you can determine the weight percentage of C and Cl in vinyl chloride. What is the empirical formula of the compound? If its molar mass is 62.5 g/mol, what is its molecular formula?

90. Hydrogen chloride, HCl, can be made conveniently in the laboratory by the reaction of NaCl, common table salt, with sulfuric acid, H_2SO_4 (see the photo).

HCl from sulfuric acid and NaCl.

$$2\ NaCl(s) + H_2SO_4(aq) \longrightarrow 2\ HCl(g) + Na_2SO_4(s)$$

(a) If 20. g of NaCl are used, how many grams of H_2SO_4 are required for complete reaction? (b) If 20. g of NaCl are used, and only 5.6 g of HCl are isolated, what is the percent yield of HCl? (c) If only 5.6 g of HCl are obtained, how many grams of Na_2SO_4 must also have been formed?

91. On April 16, 1947, the S.S. Grandchamp blew up in the harbor of Texas City, Texas, and the explosion set off a chain reaction of explosions and fires that eventually killed 570 people. The original blast was from the explosive decomposition of ammonium nitrate (see the photo), a compound used as a fertilizer, to give nitrogen, oxygen, and water.

$$2\ NH_4NO_3(s) \longrightarrow 2\ N_2(g) + 4\ H_2O(\ell) + O_2(g)$$

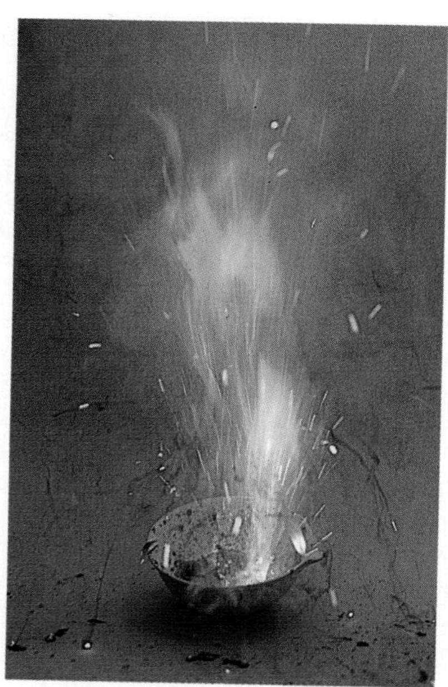

The explosion of ammonium nitrate.

(a) If a shipload of ammonium nitrate (3.00×10^4 T) explodes, how many tons of each of the products are formed? (1 T = 2000 lb; 1 lb = 454 g)

(b) If the explosion is done on a small scale, and 45.0 g of H_2O is isolated, what was the minimum amount of ammonium nitrate present in the beginning?

(c) Assume the decomposition of ammonium nitrate is not complete. Only 1.0 g of O_2 is isolated from 8.0 g of NH_4NO_3. What is the percent yield of O_2?

92. A common germanium-containing compound has the general formula Ge_xH_y. Assume you burn 0.383 g of the compound in air and obtain 0.523 g of GeO_2 and 0.180 g of H_2O according to the following *unbalanced* equation:

$$Ge_xH_y(g) + O_2(g) \longrightarrow GeO_2(s) + H_2O(g)$$

Derive the empirical formula for Ge_xH_y.

93. As noted in previous questions, BrF_3 reacts with some metal oxides to evolve O_2. Because of this, the reaction is a useful method of analysis of oxides in the presence of other compounds. Suppose you use this method to analyze for the weight percent of uranium, in the form of UO_3, in some waste from a nuclear reactor. The *unbalanced* equation is

$$UO_3(s) + BrF_3(\ell) \longrightarrow UF_4(s) + Br_2(\ell) + O_2(g)$$

If you find that 1.485 g of UO_3-containing sample evolve 23.2 mg of O_2, what is the weight percent of uranium, U, in the sample?

94. Equal weights of Zn metal and iodine, I_2, are mixed together, and the iodine is converted completely to ZnI_2 (Figure 4.2). What fraction by weight of the original zinc remains unreacted?

95. A 1.000-g mixture of copper(I) oxide, Cu_2O, and copper(II) oxide, CuO, was reduced quantitatively to give 0.839 g of metallic copper. What is the weight of copper(I) oxide in the original 1.000-g sample?

SUMMARY PROBLEMS

96. A small piece of aluminum, 0.500 mm thick and 12.0 cm square, reacts with 19.50 mL of liquid bromine to give aluminum bromide.

(a) Write the balanced equation for the reaction.

(b) Calculate the number of moles each of aluminum and bromine. The density of aluminum is 2.70 g/cm³ and that of liquid bromine is 3.10 g/cm³.

(c) What is the theoretical yield of aluminum bromide in grams?

97. Lead was known as far back as 3000 BC. Its main source is a distinctive heavy mineral called galena, PbS (see the photo). It was easy for people to recognize and then easy to reduce it to the metal. Galena was simply "roasted" in limited air, and then heated with powdered charcoal according to the following *unbalanced* equations.

$$PbS(s) + O_2(g) \longrightarrow PbO(s) + SO_2(g)$$

$$PbO(s) + C(s) \longrightarrow Pb(s) + CO(g)$$

The low melting metal (melting point 328 °C) ran out of the mass of charcoal and other impurities; after

Galena, PbS.

solidifying, it was then easy to hammer into sheets. Lead was used for water pipes in Roman villas, and the roof of the Pantheon in Rome was covered with lead.

(a) Balance the equations above and name the compounds involved.

(b) If you begin with 1.00 pound of pure galena, what is the maximum amount of lead you could obtain? How many grams of carbon are required in the second step for complete reaction?

(c) Assume that you obtain the maximum possible yield of lead from the reactions above. If you hammered the metal into a sheet that is 9.60 cm by 34.5 cm, how thick will it be? (The density of lead is 11.34 g/cm^3.)

98. In the purification of uranium for use as a nuclear fuel, one of the compounds isolated is $UO_x(NO_3)_y \cdot z\ H_2O$, where the uranium can have a positive charge of +3, +4, +5, or +6.

(a) Heating the compound in air to 400 °C leaves an oxide, U_aO_b. This oxide is found to be 83.22% U. What is the empirical formula of this oxide? What is its name?

(b) Heating $UO_x(NO_3)_y \cdot z\ H_2O$ in air to 800 to 900 °C completely decomposes the compound to give yet another uranium oxide, U_nO_m, which analysis shows is 84.8% U. What is the empirical formula of this second oxide?

(c) To find the empirical formula for $UO_x(NO_3)_y \cdot z\ H_2O$, you first heat it gently to drive off all the water. You find that 1.328 g of the hydrated compound leave 1.042 g of $UO_x(NO_3)_y$ after heating. Next, you heat the residue even more strongly and find 0.742 g of the oxide U_nO_m. Based on this information, and the other information given or calculated above, what is the formula of $UO_x(NO_3)_y \cdot z\ H_2O$?

CHAPTER

5

Reactions in Aqueous Solution

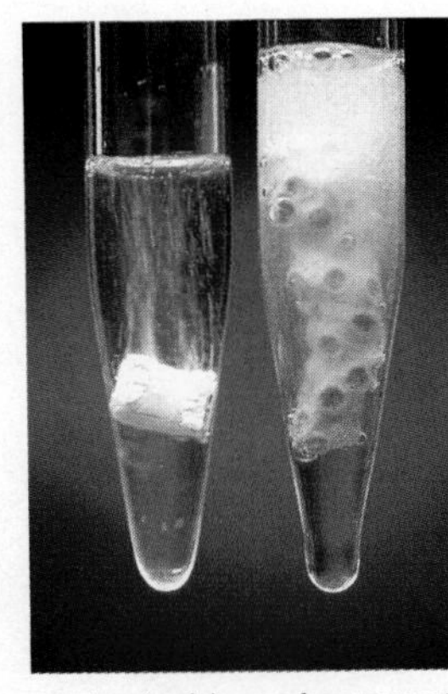

Calcium metal and calcium carbonate reacting with aqueous acid.

CHAPTER OUTLINE

On our watery planet it is not surprising that many chemical reactions occur in water. The oceans are solutions of many compounds and the medium in which reactions occur. Your body is largely water, so the reactions that keep you going occur in water. Reactions in aqueous solution are all around us and within us. We need to understand something of their details.

In Chapter 1 we defined a **solution** as a homogeneous mixture of two or more substances. The material dissolved is the **solute**, and the medium in which it is dissolved is the **solvent**. In salty sea water, salt is the solute and water is the solvent. In this chapter we are going to talk only about reactions that occur in **aqueous solutions**, that is, *solutions of substances in water*. We first want to introduce a few basic ideas of the behavior of substances in water, specifically which substances do or do not dissolve and which ones do or do not form ions in water. After next describing a large class of common substances—acids and bases—we want to say something about special techniques of balancing equations for reactions in aqueous solution and then something about common types of reactions in water. With this introduction, we can then apply the ideas of stoichiometry from Chapter 4 to aqueous reactions. Finally, we want to describe oxidation-reduction reactions, a very important class of reactions that can occur in water or other ways.

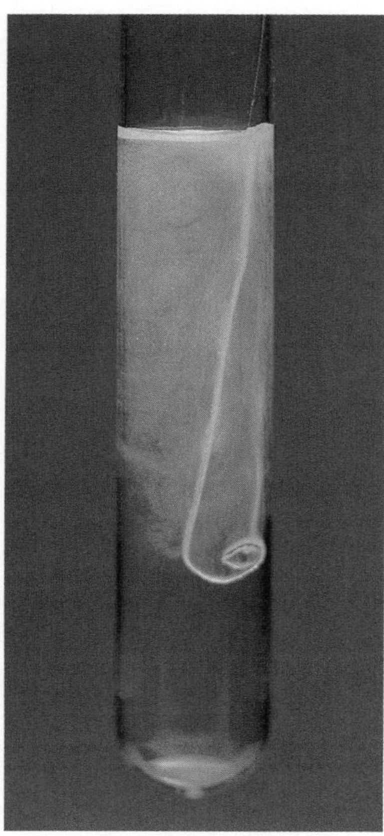

Figure 5.1 A ribbon of magnesium metal reacts with aqueous HCl to give H_2 gas and aqueous $MgCl_2$. (This is an oxidation–reduction reaction, as described in Section 5.5.)

5.1 PROPERTIES OF AQUEOUS SOLUTIONS: AN INTRODUCTION

If you put magnesium ribbon into aqueous hydrochloric acid, the mixture will bubble furiously as hydrogen gas is given off according to the following equation (Figure 5.1).

$$Mg(s) + 2\ HCl(aq) \longrightarrow MgCl_2(aq) + H_2(g)$$

It is important to notice in this balanced equation that both HCl and $MgCl_2$ are followed by (*aq*), indicating that they have dissolved in the water. It is also important to know that these two substances exist as ions in the aqueous solution. But how do we know that these substances will dissolve and that they form ions? Some guidelines follow.

IONS IN AQUEOUS SOLUTION: ELECTROLYTES

Years ago you made the ''experimental'' observation that common table salt, NaCl, dissolves in water to form a solution. The dissolving process on a molecular scale is illustrated in Figure 5.2. When a crystal of Na^+ and Cl^- ions is placed in water, water molecules are attracted to the ions (by forces described in Chapter 13). This water-ion attraction cloaks each ion on the surface of the crystal with water molecules, and the ions are pulled into the bulk water. The ions, now sheathed in water molecules, are free to move about. Under normal conditions, the movement of ions is random, and the Na^+ and Cl^- ions are dispersed uniformly

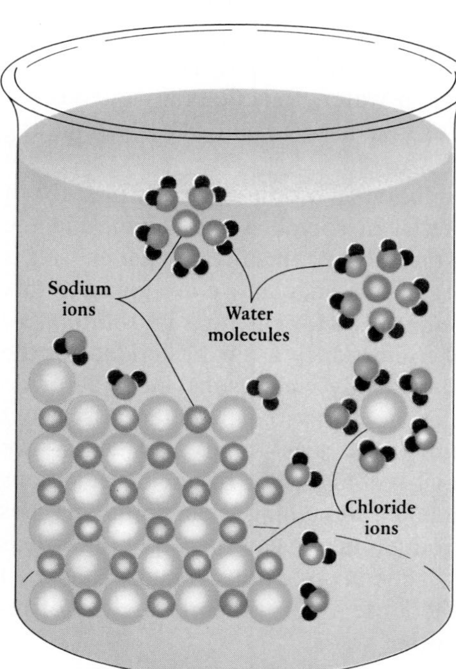

Figure 5.2 A model for the process of dissolving NaCl in water. The crystal provides Na^+ and Cl^- ions in aqueous solution. These ions, which are cloaked in water molecules, are free to move about. Such solutions conduct electricity, so the substance dissolved is called an electrolyte.

Historical Figures in Chemistry: Anions, Cations, and Michael Faraday (1791–1867)

Many of the terms we have been using, such as anion, cation, electrode, and electrolyte, originated with Michael Faraday, one of the most influential men in the history of chemistry. Faraday was apprenticed to a bookbinder in London (England) when he was only 13. This suited him, however, as he enjoyed reading the books sent to the shop for binding. One of these chanced to be a small book on chemistry, and his appetite for science was whetted. He soon began performing some experiments on electricity, and, in 1812, a patron of the shop invited Faraday to accompany him to a lecture at the Royal Institution by one of the most famous chemists of the day, Sir Humphrey Davy. Faraday was so intrigued by Davy's lecture that he wrote to ask Davy for a position as an assistant. Faraday was accepted and began work in 1813. His work was so fruitful and Faraday was so talented that he was made the Director of the Laboratory of the Royal Institution about 12 years later.

throughout the solution. However, if you place two electrodes (electrically conducting plates) in the solution, and connect the plates to a battery, the plates take on an electrical charge, one positive and one negative. (The purpose of the battery is to remove electrons from one electrode and move them to the other.) Since the ions are also electrically charged, the positive cations will migrate through the solution to the negative plate and the negative anions to the positive plate (Figure 5.3). This movement of ions in solution constitutes an *electric current*. As the cations collect near the negatively charged electrode and anions near the positive one, the battery finds it easier to move electrons from one electrode to the other. Therefore, simultaneous with ion movement in solution, electrons

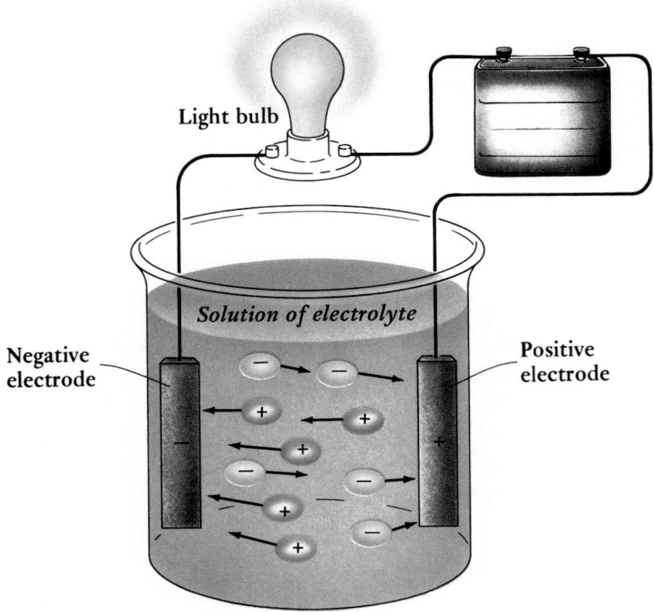

Figure 5.3 The migration of ions through a solution to a place of opposite charge constitutes an electric current.

(a)

(b)

(c)

Figure 5.4 When an electrolyte is dissolved in water in the beaker, and provides ions that are free to move about, the electrical circuit is completed, and the light bulb included in the circuit glows. (a) Pure water is a nonelectrolyte, and the bulb does not light. (b) In the case of a weak electrolyte, such as acetic acid, only a few of the dissolved molecules dissociate into ions. The bulb glows weakly, indicating that only a small current of electricity flows. (c) Dilute K_2CrO_4, potassium chromate, is a strong electrolyte. The bulb glows brightly, indicating that virtually every K_2CrO_4 molecule dissociates into its ions, K^+ and CrO_4^{2-}.

The words "anion," "cation," and "electrolyte" originated with Michael Faraday. See his biography on the previous page.

move in the external circuit. If a light bulb is inserted into the circuit as in Figure 5.4, the bulb will light, showing the circuit is complete. Ionic compounds that behave this way in water are called **electrolytes**.

Electrolytes can be classified roughly as *strong* or *weak*. Sodium chloride and other ionic compounds that dissolve in water come apart or **dissociate** completely into ions in water; virtually every formula unit breaks up into its ions. These solutions are *good conductors of electricity* (Figure 5.4); they are *strong electrolytes*.

$$\underset{\text{100\% dissociation or ionization = strong electrolyte}}{NaCl(aq)} \longrightarrow Na^+(aq) + Cl^-(aq)$$

For every mole of NaCl dissolved, you will find a mole of Na^+ ions and a mole of Cl^- ions. In general, strong electrolytes dissociate or ionize completely, or nearly so, into their constituent ions.

The weak electrolytes ammonia and acetic acid are common substances. Aqueous ammonia is a common household cleaner, and acetic acid is the important ingredient in vinegar.

Conversely, there are some substances that, while they dissolve, provide only a few ions; only a few of the formula units dissociate into ions, and so they are poor conductors of electricity (Figure 5.4). Examples of such *weak electrolytes* include ammonia and acetic acid.

$$\overset{\text{acetic acid}}{\underset{\text{<5\% ionization = weak electrolyte}}{CH_3COOH(aq)}} + H_2O(\ell) \longrightarrow H_3O^+(aq) + CH_3COO^-(aq)$$

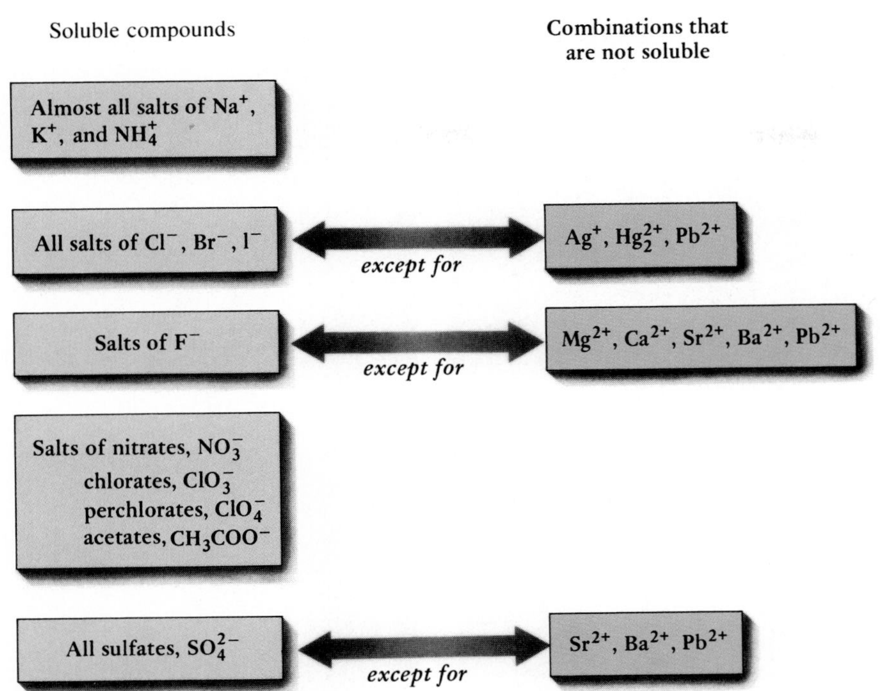

Figure 5.5 Guidelines to determine the solubility of ionic compounds. If a compound contains one of the ions in the column at the left, the compound is predicted to be at least moderately soluble in water. There are a few exceptions, and these are noted at the right.

Although the solubility of compounds can be predicted with these guidelines, it is also useful to remember that *poorly soluble salts* are formed by the negative ions listed below. Exceptions are salts with Na^+, K^+, and NH_4^+.

(a) Poorly soluble salts are formed by carbonates (CO_3^{2-}), phosphates (PO_4^{3-}), oxalates ($C_2O_4^{2-}$), and chromates (CrO_4^{2-}).
(b) Poorly soluble salts are formed by sulfides, S^{2-}.
(c) Poorly soluble salts are formed by hydroxides (OH^-) and oxides (O^{2-}).

Less than 5% of the acetic acid dissolved in water ionizes to produce hydronium ions, H_3O^+, and acetate ions, CH_3COO^-.*

Although we shall describe later in the book how you know when a substance will be only a weak electrolyte, for the moment we shall simply have to tell you when we find an example. However, acetic acid and ammonia are good ones to remember because they are so common.

There are also substances that dissolve in water but do not dissociate into ions, and do not allow the solution to conduct electricity; they are called **nonelectrolytes** (Figure 5.4). We shall generally have to tell you when you encounter a nonelectrolyte, but you are already aware of a number of common ones: sugar, starch, ethyl alcohol, and antifreeze are good examples.

SOLUBILITY OF IONIC COMPOUNDS IN WATER

Common salt dissolves readily in water, but we cannot say from this that all ionic compounds will dissolve in water. There are many that do not, and still others that dissolve only to a small extent. Fortunately, we can make some general statements about which types of ions will lead to soluble ionic compounds.

Figure 5.5 is a set of broad guidelines that can help you predict the likelihood that an ionic compound will be soluble in water. For example, sodium nitrate, $NaNO_3$, contains both an alkali metal cation, Na^+, and the nitrate ion, NO_3^-. According to Figure 5.5, the presence of these ions insures that the compound will be soluble in water. Further, since

*Acids ionize by transferring H^+ to water to form the hydronium ion, H_3O^+, and leave an anion. We shall say more about acids and the hydronium ion later in this section.

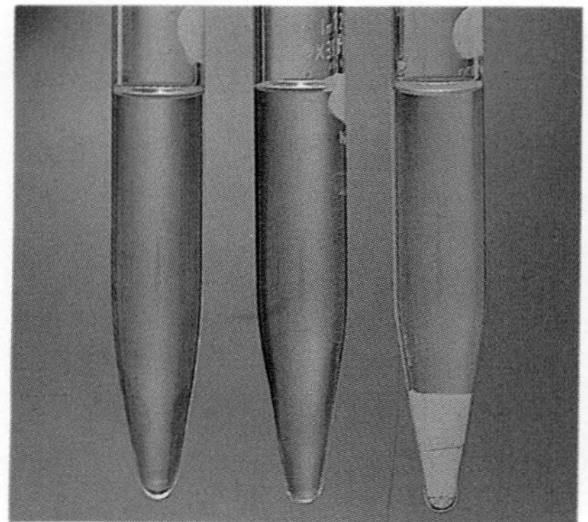

(a) $Ba(NO_3)_2$, $BaCl_2$, $BaSO_4$

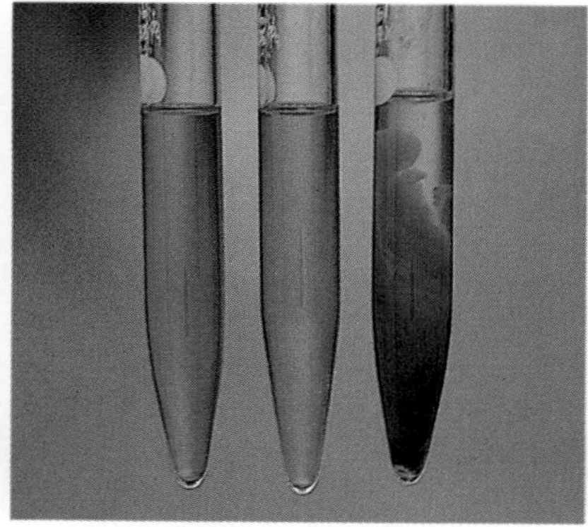

(b) $Cu(NO_3)_2$, $CuSO_4$, $Cu(OH)_2$

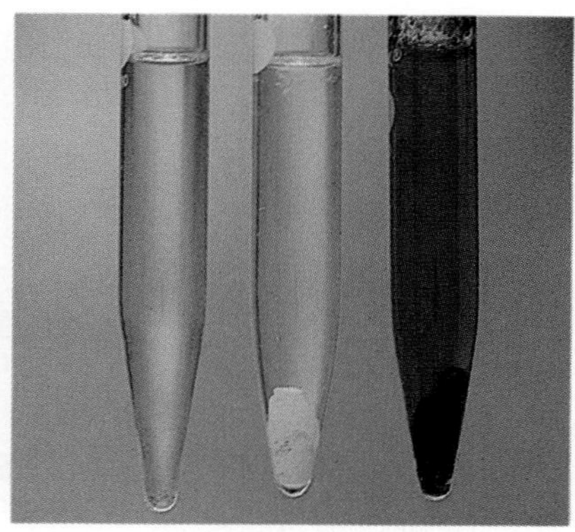

(c) $AgNO_3$, AgCl, AgOH

(d) $(NH_4)_2S$, CdS, Sb_2S_3, PbS

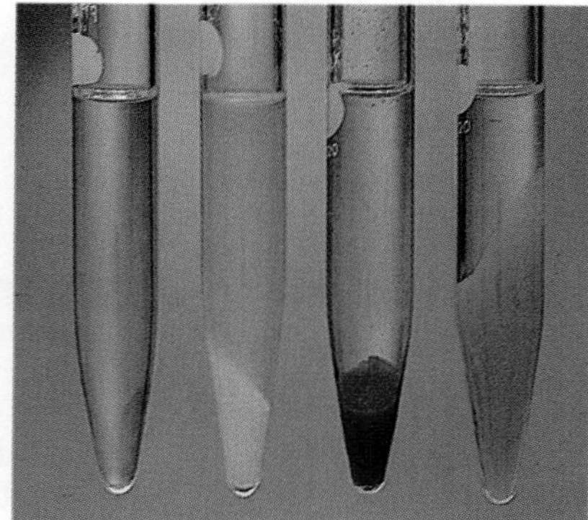

(e) NaOH, $Ca(OH)_2$, $Fe(OH)_3$, $Ni(OH)_2$

Figure 5.6 The solubility rules of Figure 5.5 illustrated. With certain exceptions, salts of Cl^- and NO_3^- are soluble in water. Although many sulfates are water-soluble, there are some that are not. Finally, the sulfide ion, S^{2-}, almost invariably gives insoluble salts.

it can be assumed that ionic compounds that dissolve in water are strong electrolytes, $NaNO_3$ is predicted to be a strong electrolyte.

$$NaNO_3(aq) \longrightarrow Na^+(aq) + NO_3^-(aq)$$

100% dissociated
= strong electrolyte

For every mole of $NaNO_3$ dissolved, you will find a mole of Na^+ ions and a mole of NO_3^- ions. On the other hand, CuS is quite insoluble, as are all sulfides except those where the cation is an alkali metal ion or the NH_4^+ ion.

Other examples of common salts are given in Figure 5.6, and Figure 5.7 illustrates a reaction where soluble electrolytes are converted to an insoluble compound and a nonelectrolyte. While studying the guidelines in Figure 5.5 and the examples in Figure 5.6, be sure to notice the following:

If a compound contains at least one of the ions leading to solubility (Figure 5.5), it is at least moderately soluble in water.

EXAMPLE 5.1

SOLUBILITY GUIDELINES

For each of the following ionic compounds, predict whether it is likely to be water soluble. If it is soluble, tell what ions exist in solution.

(a)

(b)

(c)

Figure 5.7 The reaction of barium hydroxide and sulfuric acid.

$$Ba(OH)_2(aq) + H_2SO_4(aq) \longrightarrow BaSO_4(s) + 2\ H_2O(\ell)$$

(a) Barium hydroxide is only moderately soluble in water, but it is a strong electrolyte; the bulb glows brightly. (b) As sulfuric acid is added slowly from a buret, reaction occurs to give water and the insoluble, white compound barium sulfate. As long as some $Ba(OH)_2$ is dissolved, the bulb glows, but less brightly as the quantity of electrolyte drops. When exactly 1 mole of H_2SO_4 has been added per mole of $Ba(OH)_2$, the solution no longer conducts electricity because no electrolytes remain in the water.

(a) KCl (b) $MgCO_3$ (c) NiO (d) CaI_2

Solution You must first recognize the cation and anion involved and then decide the probable water solubility. As long as one ion is listed in Figure 5.5 as leading to solubility, the compound is likely to be at least moderately soluble.

(a) KCl is composed of K^+ and Cl^-. According to Figure 5.5, the presence of either of these ions means that the compound is likely to be soluble in water. Indeed, its actual solubility is about 35 g in 100 mL of water at 20 °C. Thus, a solution of KCl actually consists of K^+ and Cl^- ions, and KCl is an electrolyte.

$$\text{KCl in water} \longrightarrow K^+(aq) + Cl^-(aq)$$

(b) Magnesium carbonate is composed of a Mg^{2+} cation and a CO_3^{2-} anion. Mg^{2+} is in the alkaline earth group, a group that often does not form water-soluble compounds. The carbonate ion usually gives insoluble compounds (Figure 5.5), unless combined with something like Na^+ or NH_4^+. Therefore, $MgCO_3$ is not predicted to be soluble in water. (The actual solubility of $MgCO_3 \cdot 3\ H_2O$ is less than 0.2 g per 100 mL of water.)

(c) Nickel(II) oxide is composed of Ni^{2+} and O^{2-}. Again, Figure 5.5 suggests that oxides are soluble only when O^{2-} is combined with an alkali metal ion; Ni^{2+} is a transition metal ion, so NiO is insoluble.

(d) Calcium iodide is composed of Ca^{2+} and I^- ions. According to Figure 5.5 almost all iodides are soluble in water, so CaI_2 is a water-soluble electrolyte.

$$CaI_2 \text{ in water} \longrightarrow Ca^{2+}(aq) + 2\ I^-(aq)$$

Notice that the compound gives two I^- ions on dissolving in water. (A common misconception is that I_2 or some ion such as I_2^{2-} is found in solution. All halide-containing compounds that dissolve in water produce F^-, Cl^-, Br^-, or I^- in aqueous solution.)

EXERCISE 5.1 Solubility Guidelines

Tell whether each compound is likely to be soluble in water. If the compound dissolves in water, tell what ions exist in aqueous solution.
(a) NaBr (b) $BaSO_4$ (c) K_2CO_3 (d) AgCl (e) $(NH_4)_2SO_4$

EXERCISE 5.2 Solubility Guidelines

Write formulas for (a) a soluble ionic compound containing the nitrate ion, (b) an insoluble compound containing the sulfide ion, and (c) a soluble compound containing Ni^{2+}.

ACIDS AND BASES

The solubility rules in Figure 5.5 apply to compounds containing a metal cation (or $NH_4{}^+$) and some anion. Many such compounds are in our environment, and you will see a number in the laboratory. However, substances called acids and bases represent a major class of chemical compounds as well. Until we explore these compounds more fully in Chapters 17 and 18, we shall define an **acid** as a compound that, on reaction with water, produces an ion called the **hydronium ion**, H_3O^+; an anion is produced as well. For example:

$$\underset{\substack{\text{strong electrolyte} \\ = 100\% \text{ dissociated}}}{\overset{\text{hydrochloric acid}}{HCl(aq)}} + H_2O(\ell) \longrightarrow \overset{\text{hydronium ion}}{H_3O^+(aq)} + Cl^-(aq)$$

This reaction is written this way because, when HCl enters water, the H atom is attracted to water molecules and is pulled off as H^+. This ion combines with a water molecule to give H_3O^+, and Cl^- is left behind.

A **base** is a compound that provides a **hydroxide ion**, OH^-, and a cation in water.

$$\underset{\substack{\text{strong electrolyte} \\ = 100\% \text{ dissociated}}}{\overset{\text{sodium hydroxide, base}}{NaOH(aq)}} \longrightarrow Na^+(aq) + \overset{\text{hydroxide ion}}{OH^-(aq)}$$

A few of the most common acids and bases are listed in Table 5.1; be sure to familiarize yourself with their names and formulas. All the compounds listed in Table 5.1 are electrolytes, with only a few of them classified as weak. Acetic acid, already mentioned, is a common weak electrolyte—and hence the acid itself is referred to as weak—and ammonia is a common weak base. The latter produces very little OH^- ion on reaction with water.

$$\underset{\substack{\text{weak electrolyte} \\ <1\% \text{ ionized}}}{\overset{\text{ammonia, base}}{NH_3(aq)}} + H_2O(\ell) \longrightarrow \overset{\text{ammonium ion}}{NH_4^+(aq)} + \overset{\text{hydroxide ion}}{OH^-(aq)}$$

Table 5.1 Common Acids and Bases

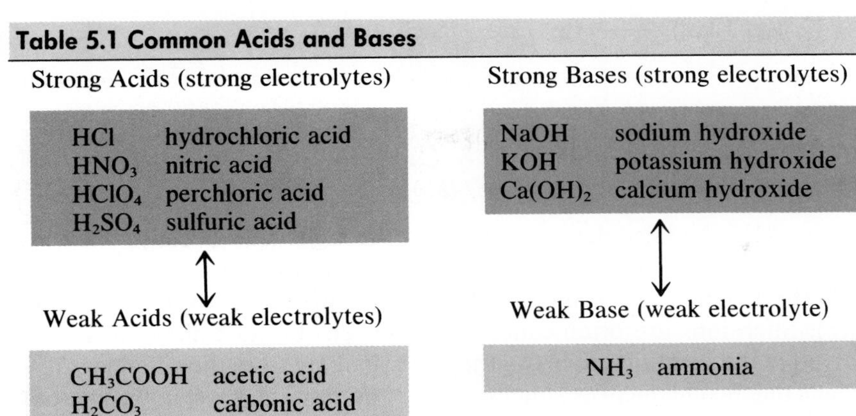

Strong Acids (strong electrolytes)		Strong Bases (strong electrolytes)	
HCl	hydrochloric acid	NaOH	sodium hydroxide
HNO_3	nitric acid	KOH	potassium hydroxide
$HClO_4$	perchloric acid	$Ca(OH)_2$	calcium hydroxide
H_2SO_4	sulfuric acid		
Weak Acids (weak electrolytes)		**Weak Base (weak electrolyte)**	
CH_3COOH	acetic acid	NH_3	ammonia
H_2CO_3	carbonic acid		

Phosphoric acid, H_3PO_4, is another acid commonly found in the laboratory. It is on the borderline between a strong and weak acid. See Chapter 17.

Finally, it is worth pointing out that some common acids are capable of providing more than one mole of H_3O^+ per mole of acid. Sulfuric acid is an example.

$$\underset{\substack{\text{sulfuric acid} \\ \text{100\% ionized}}}{H_2SO_4(aq)} + H_2O(\ell) \longrightarrow \underset{\text{hydronium ion}}{H_3O^+(aq)} + \underset{\text{hydrogen sulfate ion}}{HSO_4^-(aq)}$$

$$\underset{\substack{\text{hydrogen sulfate ion} \\ \text{<100\% ionized}}}{HSO_4^-(aq)} + H_2O(\ell) \longrightarrow \underset{\text{hydronium ion}}{H_3O^+(aq)} + \underset{\text{sulfate ion}}{SO_4^{2-}(aq)}$$

The first ionization reaction is essentially complete, so sulfuric acid is considered a strong electrolyte (and so a strong acid as well).

5.2 BALANCING EQUATIONS FOR REACTIONS IN AQUEOUS SOLUTION: NET IONIC EQUATIONS

It is often convenient to strip the equation for a reaction in aqueous solution to its simplest form, called a *net ionic equation*. Such equations make it easier to see the essential part of the reaction, and they can be easier to balance (see Section 5.5). For these reasons we shall often use net ionic equations, and you should be aware of how they are developed.

To see what a net ionic equation is, let us return to the reaction of magnesium with aqueous HCl (Figure 5.1).

$$Mg(s) + 2\ HCl(aq) \longrightarrow MgCl_2(aq) + H_2(g)$$

Hydrochloric acid and magnesium chloride dissolve in water [and so their formulas are followed by (aq) in the chemical equation], and both are strong electrolytes. This means that the solution really contains the ionization products of HCl

$$HCl(aq) + H_2O(\ell) \longrightarrow H_3O^+(aq) + Cl^-(aq)$$

and those of $MgCl_2(aq)$.

$$MgCl_2(aq) \longrightarrow Mg^{2+}(aq) + 2\ Cl^-(aq)$$

Therefore, to be more informative we should rewrite the balanced equation as

$$Mg(s) + 2\ H_3O^+(aq) + 2\ Cl^-(aq) \longrightarrow Mg^{2+}(aq) + 2\ Cl^-(aq) + H_2(g) + 2\ H_2O(\ell)$$

(where 2 H_2O molecules are added to the right side because they are released from 2 H_3O^+ ions as H_2 is formed.) Now Cl^- ions appear on both the reactant and product sides of the equation in *exactly* the same form. Such ions are often called **spectator ions** because they are not involved in the reaction process—they only look on from the sidelines. This means that no chemical or stoichiometric information is lost if the equation is written without them, and we can simplify the equation to

$$Mg(s) + 2\ H_3O^+(aq) \longrightarrow Mg^{2+}(aq) + H_2(g) + 2\ H_2O(\ell)$$

The balanced equation that results from leaving out spectator ions is called the **net ionic equation**. Only the elements, compounds, and ions that *change* in some manner in the course of the reaction are included.

This is not to imply that Cl^- is totally unimportant in the Mg/HCl reaction. Indeed, H_3O^+ or Mg^{2+} ion simply cannot exist alone in solution; a negative ion of some kind *must* be present to balance the positive ion charge. However, leaving the negative ions out of the net ionic equation implies that *any* anion will do, as long as it forms water-soluble compounds with H_3O^+ and Mg^{2+}. Thus, we could have used HNO_3 or HBr as the source of H_3O^+, and NO_3^- or Br^- would be the spectator ion.

As a final point concerning net ionic equations, you will see that there is always a *conservation of charge* as well as mass in a balanced chemical equation. Thus, in the Mg/HCl net ionic equation there are two positive electrical charges on each side of the equation. On the left side there are two H_3O^+ ions, each with a charge of 1+, for a total charge of 2+. This is balanced by a Mg^{2+} ion on the right side.

EXAMPLE 5.2

WRITING AND BALANCING NET IONIC EQUATIONS

Write a balanced, net ionic equation for the reaction of $AgNO_3$ with $CaCl_2$ to give AgCl and $Ca(NO_3)_2$.

Solution

Step 1. Write the complete, balanced equation.

$$2\ AgNO_3 + CaCl_2 \longrightarrow 2\ AgCl + Ca(NO_3)_2$$

Step 2. Decide on the solubility of each compound from Figure 5.5. One general guideline was that nitrates are almost always soluble, so $AgNO_3$ and $Ca(NO_3)_2$ are water soluble. Further, with a few exceptions (e.g., AgCl), chlorides are water soluble. Therefore, we can write

$$2\ AgNO_3(aq) + CaCl_2(aq) \longrightarrow 2\ AgCl(s) + Ca(NO_3)_2(aq)$$

Step 3. At this stage in our development of chemistry, we shall assume that all soluble ionic compounds are electrolytes. Therefore,

$$AgNO_3(aq) \longrightarrow Ag^+(aq) + NO_3^-(aq)$$

$$CaCl_2(aq) \longrightarrow Ca^{2+}(aq) + 2\ Cl^-(aq)$$

$$Ca(NO_3)_2(aq) \longrightarrow Ca^{2+}(aq) + 2\ NO_3^-(aq)$$

This results in the following complete ionic equation.

$$2\ Ag^+(aq) + 2\ NO_3^-(aq) + Ca^{2+}(aq) + 2\ Cl^-(aq) \longrightarrow 2\ AgCl(s) + Ca^{2+}(aq) + 2\ NO_3^-(aq)$$

Step 4. There are two spectator ions in the complete ionic equation (Ca^{2+} and NO_3^-), so these can be eliminated to give the net ionic equation.

$$2\ Ag^+(aq) + 2\ Cl^-(aq) \longrightarrow 2\ AgCl(s)$$

To finish the job, realize that each species in the net equation is preceded by a coefficient of 2. Therefore, the equation can be simplified by dividing through by 2.

$$Ag^+(aq) + Cl^-(aq) \longrightarrow AgCl(s)$$

Step 5. Finally, notice that the sum of ion charges is the same on both sides of the equation. On the left, +1 and −1 give zero; on the right the electrical charge on AgCl is also zero.

EXAMPLE 5.3

WRITING AND BALANCING NET IONIC EQUATIONS

Write a balanced, net ionic equation for the reaction of calcium chloride with sodium carbonate to give sodium chloride and calcium carbonate.

Solution Using the guidelines of Chapter 3, we first decide on the formulas and then write the *unbalanced* equation.

$$CaCl_2 + Na_2CO_3 \longrightarrow CaCO_3 + NaCl$$

It then follows that the complete balanced equation is

$$CaCl_2 + Na_2CO_3 \longrightarrow CaCO_3 + 2\ NaCl$$

Now, to write the net ionic equation, find the spectator ions and eliminate them. Sodium salts are usually water soluble, as are most chlorides. Unless they are associated with an alkali metal, carbonates are not often soluble. This means that Na_2CO_3 is soluble (and also an electrolyte), while $CaCO_3$ is insoluble. Therefore, a complete ionic equation can be written as

$$Ca^{2+}(aq) + 2\ Cl^-(aq) + 2\ Na^+(aq) + CO_3^{2-}(aq) \longrightarrow CaCO_3(s) + 2\ Na^+(aq) + 2\ Cl^-(aq)$$

From this, you can see that the spectator ions are Na^+ and Cl^-, so the balanced net ionic equation can be written as

$$Ca^{2+}(aq) + CO_3^{2-}(aq) \longrightarrow CaCO_3(s)$$

This is the net ionic equation for the formation of limestone, a mineral of enormous importance on our earth.

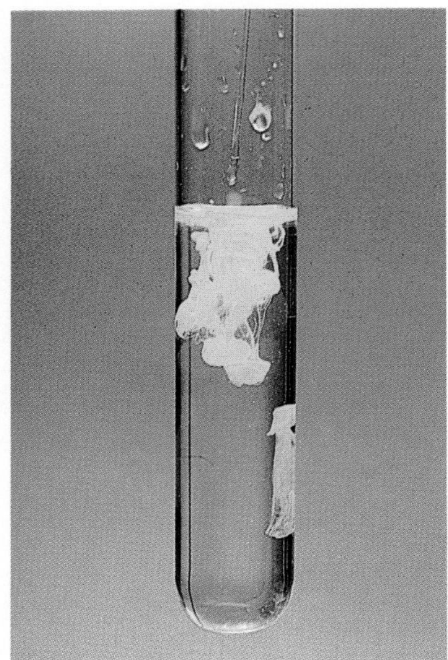

Figure 5.8 Adding a drop of KI to a solution of $Pb(NO_3)_2$ leads to the formation of a precipitate of insoluble PbI_2 and leaves water-soluble KNO_3 in solution.

Notice that each complete equation on this page is followed by the net ionic equation.

Figure 5.9 Copper(II) carbonate, a blue-green solid, reacts with nitric acid to give gaseous CO_2 as well as blue copper(II) nitrate in solution.

EXERCISE 5.3 Writing Net Ionic Equations

Balance each of the following equations and write net ionic equations.

(a) $BaCl_2 + Na_2SO_4 \longrightarrow BaSO_4 + NaCl$

(b) $(NH_4)_2S + Cd(NO_3)_2 \longrightarrow CdS + NH_4NO_3$

(c) Lead(II) nitrate reacts with potassium chloride to give lead(II) chloride and potassium nitrate.

5.3 SOME TYPES OF REACTIONS IN AQUEOUS SOLUTION: EXCHANGE REACTIONS

In Chapter 4 we introduced some general types of reactions: the interactions of elements with halogens and oxygen, combustion reactions, and decomposition reactions. In general, these are not carried out in aqueous solution. Instead, in water we can identify at least two major categories of reactions: *exchange reactions* and *oxidation–reduction reactions*. The former are described in this section, and the latter are discussed separately in Section 5.5.

Exchange reactions, also called *metathesis* or *double displacement* reactions, proceed by the interchange of reactant cation/anion partners.

$$AB + XY \longrightarrow AY + XB$$

Two special types of such reactions are **precipitation reactions** (Figure 5.8)

$$Pb(NO_3)_2(aq) + 2\ KI(aq) \longrightarrow 2\ KNO_3(aq) + PbI_2(s)$$

$$Pb^{2+}(aq) + 2\ I^-(aq) \longrightarrow PbI_2(s)$$

and **acid–base reactions** (where HOH is H_2O).

$$HNO_3(aq) + KOH(aq) \longrightarrow KNO_3(aq) + HOH(\ell)$$

$$H_3O^+(aq) + OH^-(aq) \longrightarrow 2\ H_2O(\ell)^*$$

A third type consists of **gas-forming reactions**, chiefly the exchange reaction occurring between metal carbonates and acids (Figure 5.9).

$$CuCO_3(s) + 2\ HNO_3(aq) \longrightarrow Cu(NO_3)_2(aq) + CO_2(g) + H_2O(\ell)$$

$$CuCO_3(s) + 2\ H_3O^+(aq) \longrightarrow Cu^{2+}(aq) + CO_2(g) + 3\ H_2O(\ell)$$

Mixing just any two compounds together does not ensure they will react. For example, mixing aqueous solutions of NaCl and KNO_3 will not produce $NaNO_3$ and KCl; the reactants will just sit in the flask and get wet.

$$NaCl(aq) + KNO_3(aq) \xrightarrow{\text{no products formed}} \cancel{NaNO_3(aq) + KCl(aq)}$$

Why then do exchange reactions occur and how can you predict which ones will work? Among other things, you can predict that an exchange reaction will work if (a) a *water-insoluble product* results from two soluble reactants or (b) a *stable molecule* such as water or an *insoluble gas* such

*The net ionic equation for an acid–base reaction is discussed further below.

as CO_2 is formed. The former is the basis of precipitation reactions, and the latter is the reason for the effectiveness of many acid–base reactions and for the decomposition of metal carbonates by acids.

Figure 5.10 A drop of water-soluble sulfide-containing compound is placed in a solution of water-soluble $CuCl_2$. Black, insoluble CuS is formed.

PRECIPITATION REACTIONS

Precipitation reactions produce an insoluble precipitate from soluble reactants. Figure 5.5 lists ions that are likely to lead to soluble compounds. Since there are relatively few ions that do lead to soluble compounds, there are many positive/negative ion combinations that can give insoluble combinations. Thus, many precipitation reactions are possible. For example, copper(II) sulfide is easily precipitated by the reaction of a water-soluble copper(II) compound with a water-soluble sulfide compound (Figure 5.10).

$$Cu(NO_3)_2(aq) + (NH_4)_2S(aq) \longrightarrow CuS(s) + 2\ NH_4NO_3(aq)$$

$$Cu^{2+}(aq) + S^{2-}(aq) \longrightarrow CuS(s)$$

If a soluble copper(II) compound in nature comes into contact with a source of sulfide ions (say from a volcano or a natural gas pocket in the earth), CuS precipitates.

EXERCISE 5.4 Precipitation Reactions

Complete and balance the following equations for precipitation reactions. Indicate whether each substance is soluble or insoluble in water. Write the net ionic equations.

(a) $AgNO_3 + LiCl \longrightarrow$
(b) $NiCl_2 + Na_2S \longrightarrow$

ACID–BASE REACTIONS

An **exchange reaction** between an acid and a base produces a **salt** and, usually, water. For example,

NaOH(aq)	+	HCl(aq)	$\longrightarrow$	NaCl(aq)	+ HOH(ℓ)
sodium hydroxide "lye"		hydrochloric acid "muriatic acid"		sodium chloride "sea salt"	water

Here sodium hydroxide and hydrochloric acid give common salt and water. Because of this, the word "salt" has come to be used by chemists to describe *any ionic compound whose cation comes from a base* (here Na^+ from NaOH) *and whose anion comes from an acid* (here Cl^- from HCl). Table 5.1 lists some common acids and bases. In every case, reaction of one of these acids with one of the bases produces a salt and water.*

Acid–base reactions are sometimes called ***neutralization reactions*** *since, upon completion of the reaction, the solution is neutral, neither acidic nor basic. As we shall see in Chapter 18, this applies strictly only to reactions of* ***strong*** *acids and bases.*

Hydrochloric acid and nitric acid, among others in Table 5.1, are strong electrolytes in water and so are often called "strong acids." This means the complete ionic equation for the reaction of HCl(aq) and NaOH(aq), for example, should be written as

*An exception to this involves the base ammonia. See Example 5.4.

$$Na^+(aq) + OH^-(aq) + H_3O^+(aq) + Cl^-(aq) \longrightarrow Na^+(aq) + Cl^-(aq) + H_2O(\ell) + HOH(\ell)$$

where the two molecules of water have come from the reaction of the hydronium and hydroxide ions.

$$\begin{array}{l}H{-}O{-}H^+(aq) + OH^-(aq) \longrightarrow H_2O(\ell) + HOH(\ell)\\ \quad\;\,| \\ \quad\; H\end{array}$$

or

$$H_3O^+(aq) + OH^-(aq) \longrightarrow 2\ H_2O(\ell)$$

Since Na^+ and Cl^- ions appear on both sides of the equation, the net ionic equation is simply the combination of ions just above, the reaction of H_3O^+ and OH^- to give water. Indeed, *this will always be the net ionic equation for the reaction of any* ***strong*** *acid and any* ***strong*** *base*. The other ions of the base (the cation) and acid (the anion) remain unchanged. If the water is evaporated, however, the cation and anion form a solid salt. In the example above, NaCl could be obtained, while the reaction of NaOH and nitric acid, HNO_3, would give sodium nitrate, $NaNO_3$.

$$NaOH(aq) + HNO_3(aq) \longrightarrow NaNO_3(aq) + HOH(\ell)$$

EXAMPLE 5.4

ACID–BASE REACTIONS

Write the balanced equation for the reaction of aqueous ammonia and nitric acid.

Solution Ammonia is a common and very important chemical, and you should understand something of its chemistry. It is a base because it reacts with water, to a *small* extent, to produce the hydroxide ion.

$$NH_3(aq) + H_2O(\ell) \longrightarrow NH_4^+(aq) + OH^-(aq)$$

Nitric acid produces H_3O^+ and the nitrate ion.

$$HNO_3(aq) + H_2O(\ell) \longrightarrow H_3O^+(aq) + NO_3^-(aq)$$

Since acid-base reactions proceed by reaction of OH^- with H_3O^+, NH_3 and HNO_3 must react according to the equation

$$NH_3(aq) + HNO_3(aq) \longrightarrow NH_4^+(aq) + NO_3^-(aq)$$

To see how we got here, simply add the two ionization equations together. This gives

from the left side of the two equations

$$NH_3(aq) + H_2O(\ell) + HNO_3(aq) + H_2O(\ell) \longrightarrow NH_4^+(aq) + OH^-(aq) + H_3O^+(aq) + NO_3^-(aq)$$

from the right side of the two equations

NH_3 is a base since it accepts H^+ from an acid, in the same way H_2O accepts H^+ to form H_3O^+. In all reactions of NH_3 with an acid HX, the products will always be NH_4^+ and X^-. See Chapter 17.

Finally, the two H_2O molecules on the left side are canceled by the two H_2O molecules [from $OH^-(aq) + H_3O^+(aq)$] on the right. In essence, the acid transfers its H^+ ion to the base NH_3.

EXERCISE 5.5 Acid–Base Reactions

Write complete balanced equations for

(a) the reaction of sulfuric acid, H_2SO_4, with potassium hydroxide, KOH. Assume that both of the H^+ ions from H_2SO_4 are used in the reaction.

(b) the reaction of ammonia with hydrochloric acid.

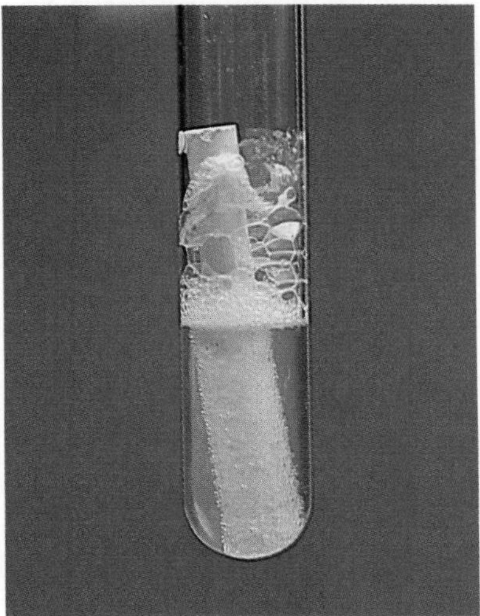

Figure 5.11 A piece of blackboard chalk, which is mostly $CaCO_3$, reacts rapidly with HCl to give a salt ($CaCl_2$), water, and CO_2 gas.

GAS-FORMING REACTIONS

The final type of exchange reaction is one where a gas may be formed. It is the loss of this gas from the solution that drives the reaction from reactants to products.

The best example of a gas-forming reaction is the exchange reaction that occurs between acids and carbonates (Figures 5.9, 5.11, and 5.12).

$$CaCO_3(s) + 2\ HCl(aq) \longrightarrow CaCl_2(aq) + H_2CO_3(aq)$$

$$\downarrow$$

$$H_2O(\ell) + CO_2(g)$$

A salt and H_2CO_3, carbonic acid, are always the products. Carbonic acid, however, is unstable and rapidly forms water and CO_2 gas. If the reaction is done in an open beaker, the gas bubbles out of the solution.

EXERCISE 5.6 Gas-Forming Reactions

Write a balanced equation for the reaction of sodium carbonate with nitric acid.

EXERCISE 5.7 Writing Exchange Reactions

Complete and balance each of the following equations for an exchange reaction. Tell whether each is an acid-base, precipitation, or gas-forming reaction.

(a) $H_2SO_4 + CsOH \longrightarrow$

(b) $H_2SO_4 + Na_2CO_3 \longrightarrow$

(c) $HNO_3 + Ca(OH)_2 \longrightarrow$

(d) $CdCl_2 + Na_2S \longrightarrow$

Figure 5.12 Over-the-counter remedies for upset stomach. Both contain carbonates.

PREPARATION OF COMPOUNDS USING EXCHANGE REACTIONS

Many chemists have as their main occupation the preparation of new materials. Exchange reactions are one way to do this, but more information is needed if you want to use them for preparative purposes.

There may be several exchange reactions that give a desired product, and you may have to make a decision as to which to use. There is always the problem of isolating the compound. For example, in the case of the reaction of NaOH and HCl in aqueous solution, the final solution contains only NaCl in water. To obtain the NaCl as a solid, you could simply evaporate the water from the solution, leaving the salt as a white, crystalline solid.

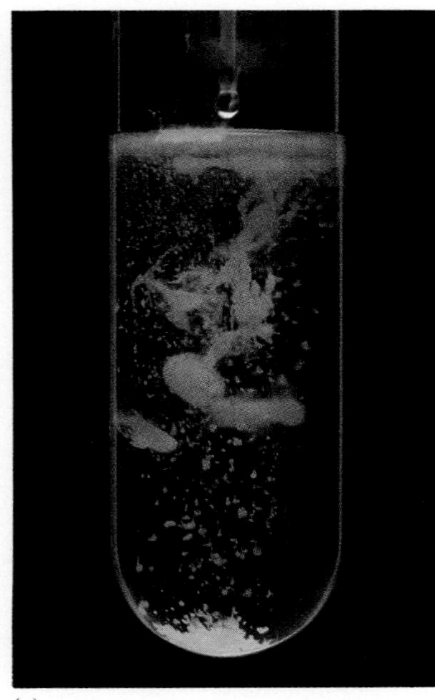

(a)

(b)

Figure 5.13 Preparation of barium chromate. (a) Aqueous barium chloride is added from a dropper to a solution of potassium chromate, K_2CrO_4. Barium chromate, $BaCrO_4$, precipitates and is separated from the solution by collecting it on a filter paper (b).

Alternatively, the desired salt may be obtained by using a precipitation reaction. The compound you want could be either the insoluble salt or the soluble species left in aqueous solution. For example, you could prepare insoluble barium sulfate by the procedure illustrated in Figure 5.13. Dilute barium chloride was poured into a solution of water-soluble K_2CrO_4 to give insoluble $BaCrO_4$ and soluble KCl.

$$BaCl_2(aq) + K_2CrO_4(aq) \longrightarrow BaCrO_4(s) + 2\ KCl(aq)$$

Filtration was then used to separate the insoluble and soluble species. Insoluble barium chromate was trapped in the paper filter, and the solution containing potassium chloride passed through.

EXAMPLE 5.5

PREPARATION OF A COMPOUND BY A PRECIPITATION REACTION

Prepare calcium carbonate, $CaCO_3$, by a precipitation reaction.

Solution

Step 1. Is the compound soluble or insoluble in water? According to Figure 5.5, carbonates are generally insoluble, except when the cation is from Group 1A (e.g., Na^+, K^+). Therefore, a precipitation reaction is appropriate.

Step 2. Choose water-soluble salts as reactants, one containing Ca^{2+} and the other containing CO_3^{2-}. Nitrates are generally soluble, so calcium nitrate, $Ca(NO_3)_2$, is a good choice. As a source of carbonate ion, sodium carbonate, Na_2CO_3, is reasonable, since sodium salts are generally soluble.

Step 3. Write the balanced equation.

Complete equation: $Ca(NO_3)_2(aq) + Na_2CO_3(aq) \longrightarrow CaCO_3(s) + 2\ NaNO_3(aq)$

Net ionic equation: $Ca^{2+}(aq) + CO_3^{2-}(aq) \longrightarrow CaCO_3(s)$

The insoluble product could be isolated by filtration, $CaCO_3$ being held by the filter and $NaNO_3$ remaining in the water.

EXAMPLE 5.6

PREPARING A COMPOUND BY AN ACID–BASE REACTION

Prepare potassium acetate, KCH_3COO, by an acid–base reaction.

Solution

Step 1. Is the compound soluble or insoluble in water? According to Figure 5.5, acetates are generally soluble. Therefore, an acid-base reaction, where one product is the desired salt and the other is water, is appropriate.

Step 2. Acetic acid (Table 5.1) is the source of acetate ion, H_3CCOO^-, and potassium hydroxide can supply potassium ion, K^+.

Step 3. Write the balanced equation.

$$CH_3COOH(aq) + KOH(aq) \longrightarrow KCH_3COO(aq) + HOH(\ell)$$

A precipitation reaction could be used as an alternative, but the product other than KCH_3COO must then be insoluble. For example, mixing the water-soluble salts barium acetate and potassium sulfate gives the desired soluble salt potassium acetate and insoluble barium sulfate.

$$Ba(CH_3COO)_2(aq) + K_2SO_4(aq) \longrightarrow 2\ KCH_3COO(aq) + BaSO_4(s)$$

EXERCISE 5.8 Preparing Ionic Compounds by Exchange Reactions

Write a balanced equation to show how you would prepare each of the following salts by an acid–base, precipitation, or gas-forming reaction.

(a) NaCl (b) KNO_3 (c) $Ba(NO_3)_2$ (d) FeS

SOMETHING MORE ABOUT
Lead(II) Iodide Crystals Grown in Space—A Student-Designed Experiment

When the United States Space Shuttle program was begun, the National Aeronautics and Space Administration announced there would be room aboard future missions for experiments designed and prepared by students. With this invitation, a chemistry student at Union College in Schenectady, New York devised an experiment to see how crystals would grow in the absence of a strong gravitational field. Previous crystals grown in space from a melt or from a gas showed clearly that space-grown crystals were different, but why they were different was a matter of speculation. Therefore, the Union College student decided to grow crystals of lead iodide from an aqueous solution. Not only is lead iodide much simpler than the complex proteins and alloys that were used in previous space experiments, but lead iodide is also a useful material. Crystals of the compound have been used as intensifiers for x-ray and γ-ray films. These work by absorbing x-rays or γ-rays, which in turn cause the lead iodide crystal to fluoresce or emit radiation in the visible range. This secondary radiation is then used to expose the film. The student, and his faculty advisor, reasoned that larger, purer crystals of lead iodide would be much more suitable for this application. Perhaps such crystals could be grown in space.

The project was set back several years when the first experiment was on board the ill-fated *Challenger,* the shuttle that was destroyed in a tragic accident in January 1986. However, the experiment was finally flown on board *Discovery* in September 1988.

The experiment was very simple—as the best experiments usually are. The apparatus, which is seen in the photograph with astronaut George D. Nelson, consisted of four interconnected chambers. The two end chambers held lead acetate and potassium iodide solutions, and the two inner chambers were filled with deionized water, and were separated by a cellulose membrane. When the valves separating the end and middle chambers were opened, the reactants migrated to the inner chambers. For the particular concentrations used, iodide anions passed through the membrane faster than the lead(II) cations; thus, lead iodide crystals formed on the membrane facing into the lead acetate solution.

$$\underset{\text{lead acetate}}{Pb(CH_3COO)_2(aq)} + \underset{\text{potassium iodide}}{2\ KI(aq)} \longrightarrow \underset{\text{lead iodide}}{PbI_2(s)} + \underset{\text{potassium acetate}}{2\ KCH_3COO(aq)}$$

NASA astronaut George Nelson prepares a crystal-growing experiment on board the space shuttle *Discovery*, September 1988. (NASA)

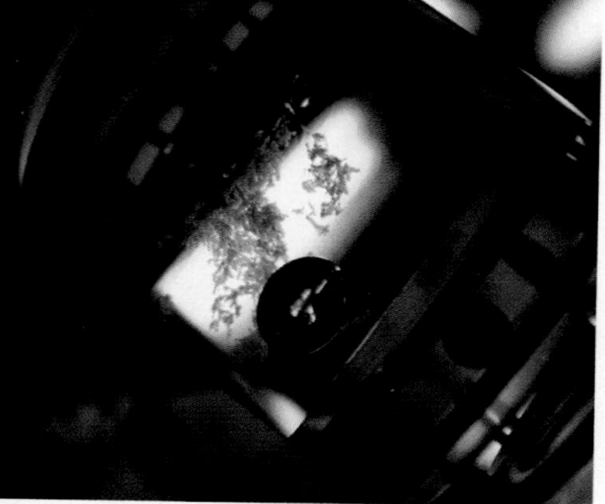

Some lead iodide crystals grown in the experiment in space.

(According to Figure 5.5, potassium salts and acetates are soluble in water, while lead iodide is not expected to be soluble.) In all the experiments done on the *Discovery* flight, lead iodide crystals formed uniformly over the surface of the membrane *and* throughout the solution in the middle chamber as shown in the photograph. This is very different from the behavior on earth, where the crystals grow only at the vertical membrane and then only on the bottom half of the membrane. Furthermore, the space-grown crystals were more pure, were all about the same size, and were produced in higher yield than those grown on earth.

These student-designed experiments were elegant for their simplicity. And the results suggest that future experiments on the growth of crystals in space will be extremely worthwhile.

5.4 STOICHIOMETRY OF REACTIONS IN SOLUTION

To work quantitatively with reactions in solution, we shall continue to use units of moles, but the amounts of reactants and products are given in "volume of solution" units rather than in mass. The concept of solution concentration is required to be able to convert "volume of solution" units to mole units.

SOLUTION CONCENTRATION: MOLARITY

You are already familiar with the concept of concentration. For example, we know that there are about 3,000,000 people in Kansas and that the state has a land area of roughly 82,000 square miles. Therefore, the average concentration of people is about 37 per square mile. In just the same way we can say, for instance, that there is a certain amount of solute dissolved in a given amount of solvent or in a given quantity of solution. One efficient way to define solution concentrations is in *moles per liter of solution*, that is, in units called **molarity**.

$$\text{Molarity (M)} = \frac{\text{moles of solute}}{\text{liters of solution}}$$

The terms "moles per liter" and "molar" are used interchangeably.

A capital M *stands for "molar," but an italicized M indicates "molar mass."*

For example, if you dissolve 342 g, or 1.00 mole, of sugar ($C_{12}H_{22}O_{11}$) in enough water to give a solution whose volume is 1.00 liter, the concentration would be 1.00 mole per liter or 1.00 *molar*. We often abbreviate this as 1.00 M, where the capital M stands for "moles per liter." Another notation you will often see is

$$[C_{12}H_{22}O_{11}] = 1.00 \text{ M}$$

Placing the formula of the compound in square brackets implies that the concentration of the solute is being specified in moles of compound per liter of solution.

It is important to notice in expressing solution concentration in molarity that reference is always to liters of *solution* and not to liters of solvent. If you add one liter of water to one mole of a solid compound to make a one molar solution, the final volume probably will not be one

Figure 5.14 To make a 0.100 M solution of $CuSO_4$, 25.0 g or 0.100 mol of $CuSO_4 \cdot 5\ H_2O$ (the blue crystalline solid) was placed in a volumetric flask. In this photo, exactly 1.00 L of water was measured out and slowly added to the 1.00 L volumetric flask containing the solid copper compound. When enough water had been added so that the *solution volume* was exactly 1.00 L, approximately 8 mL of water (the quantity in the small graduated cylinder) was left over. This emphasizes that molar concentrations are defined as moles per liter of *solution* and not per liter of water or other solvent.

liter (Figure 5.14) and so the concentration will not be precisely one molar. Therefore, when making solutions, we *add the solvent to the solute until the desired solution volume is reached.*

EXAMPLE 5.7

SOLUTION MOLARITY

Potassium permanganate, $KMnO_4$, which was used at one time as a germicide in the treatment of burns, is a common laboratory chemical. It is a shiny, purple-black solid that dissolves in water to give a beautiful purple solution. Assume you dissolve 0.395 g of $KMnO_4$ in enough water to give 250. mL of solution (Figure 5.15). What is the molar concentration of $KMnO_4$?

Solution As is almost always the case, the first step in a problem is to convert material masses into moles.

Figure 5.15 A 0.0100 M solution of $KMnO_4$ is made by adding enough water to 0.395 g of $KMnO_4$ to make 0.250 L of solution (a). To ensure the correct solution volume, the $KMnO_4$ is placed in a volumetric flask and dissolved in a small amount of water (b). After dissolving is complete, sufficient water is added to fill the flask to the mark; the flask contains 0.250 L of solution (c).

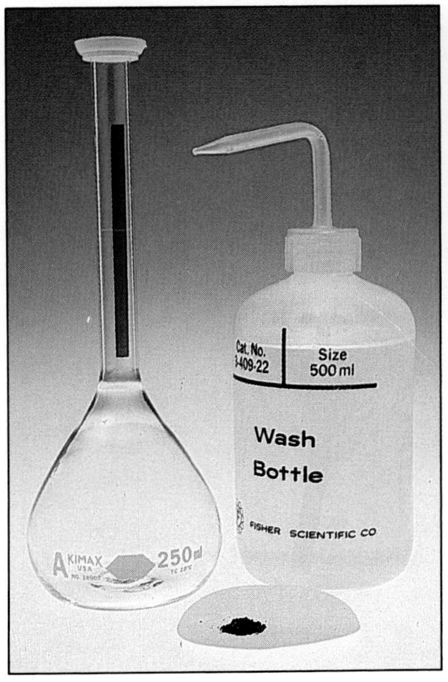

(a)

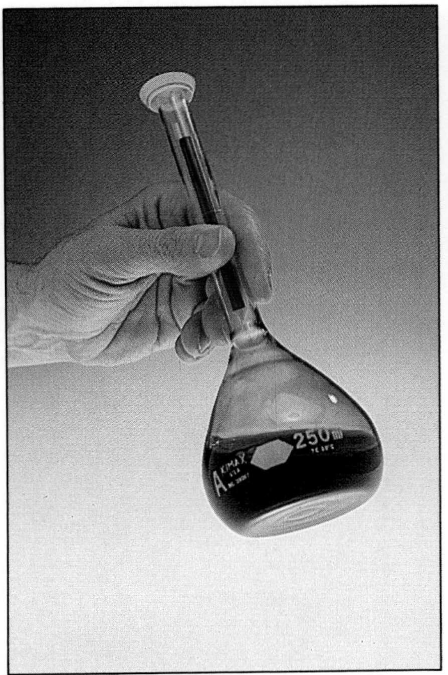

(b)

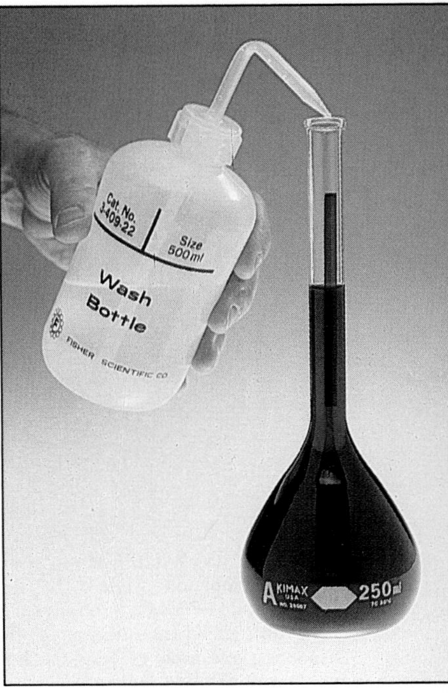

(c)

$$0.395 \text{ g KMnO}_4 \left(\frac{1 \text{ mol KMnO}_4}{158.0 \text{ g KMnO}_4}\right) = 0.00250 \text{ mol KMnO}_4$$

Now that the moles of dissolved material and the volume of solution are known, the solution concentration can be determined by dividing the moles of $KMnO_4$ by the volume of solution.

$$\text{Molarity of KMnO}_4 = \frac{0.00250 \text{ mol KMnO}_4}{0.250 \text{ L}} = 0.0100 \text{ M}$$

EXERCISE 5.9 Solution Molarity

Sodium bicarbonate, $NaHCO_3$, is used in baking powder formulations, in fire extinguishers, and in the manufacture of plastics and ceramics, among other things. If you have 26.3 g of the compound and dissolve it in enough water to make 200. mL of solution, what is the molar concentration?

The calculation of $KMnO_4$ concentration in Example 5.7 showed that $[KMnO_4] = 0.0100$ M. It is important to recognize that $KMnO_4$ is a strong electrolyte in water; that is, it dissociates completely into its ions, K^+ and MnO_4^- (Figure 5.16a).

$$KMnO_4(aq) \longrightarrow K^+(aq) + MnO_4^-(aq)$$

100% dissociation

Sometimes we are interested in the concentration of a particular ion in solution. This can be determined from the concentration of the compound and the equation for its dissociation reaction. For $KMnO_4$, the coefficients in the dissociation equation tell you that 1 mole of $KMnO_4$ breaks up into 1 mole of K^+ and 1 mole of MnO_4^-. Accordingly, 0.0100 M $KMnO_4$ gives

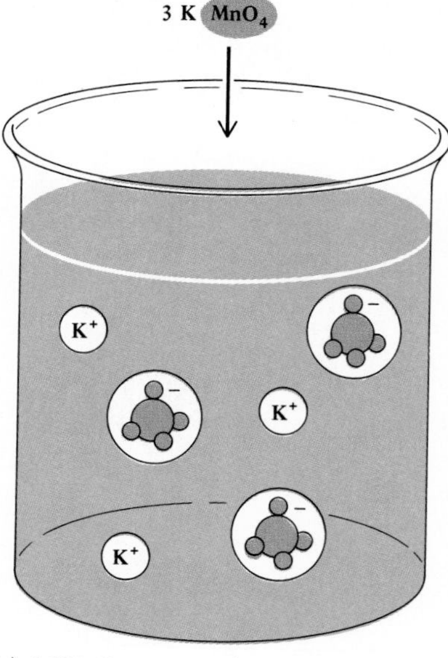

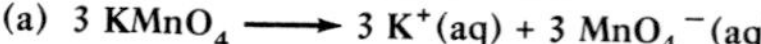

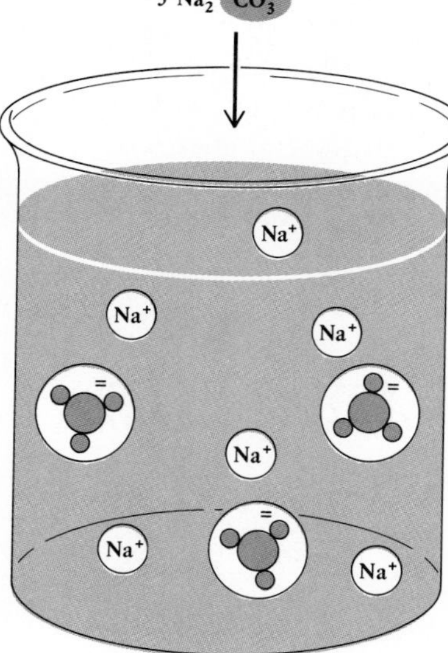

Figure 5.16 (a) When $KMnO_4$ is dissolved in water, one mole of K^+ ions and one mole of MnO_4^- ions form for every mole of $KMnO_4$ dissolved. (b) Dissolving one mole of Na_2CO_3, however, produces two moles of Na^+ ions and one mole of CO_3^{2-} ions.

a concentration of K^+ in the solution of 0.0100 M; similarly, the concentration of MnO_4^- is 0.0100 M.

Another example of ion concentrations is sodium carbonate, Na_2CO_3 (Figure 5.16b).

$$Na_2CO_3(aq) \longrightarrow 2\ Na^+(aq) + CO_3^{2-}(aq)$$

100% dissociation

When one mole of Na_2CO_3 is dissolved in enough water to make exactly one liter of solution, the concentration of the sodium ion is $[Na^+] = 2$ M and that of the carbonate ion is $[CO_3^{2-}] = 1$ M; the total concentration of ions is 3 M.*

EXERCISE 5.10 Ion Concentrations in Solution

Both LiCl and $(NH_4)_2SO_4$ are strong electrolytes, dissociating completely into their constituent ions when dissolved in water. For each compound, write a balanced equation for the dissociation. Assume enough of each has been dissolved so that [LiCl] = 1 M and $[(NH_4)_2SO_4] = 0.5$ M and then state the molar concentration of each ion and the total concentration of all ions.

Before going on, you should be aware that there are other ways to define the concentrations of chemicals in solution. For the moment, molarity is the most useful for our purposes, but other units will be introduced later in the text (Chapter 14).

PREPARING SOLUTIONS OF KNOWN CONCENTRATION

A situation a chemist often faces is the opposite of the one above. That is, a given volume of solution of known concentration must be prepared. The problem is to find out what mass of solute to use for a required volume of solution.

Suppose, for example, you wish to prepare 2.00 L of a 1.5 M solution of Na_2CO_3. You are given a bottle of solid Na_2CO_3, some distilled water, and a 2.00 L volumetric flask. To make the solution, weigh out the necessary quantity of Na_2CO_3 as accurately as possible, place the solid in the volumetric flask, and then add some water to dissolve the solid; after the solid has completely dissolved, add more water to bring the solution volume to 2.00 liters. You will then have a solution of the desired concentration and in the volume specified.

A volumetric flask is a special flask with a line marked on its neck (Figure 5.15); if it is filled with a solution to this line, you have exactly the volume of solution specified.

EXAMPLE 5.8

MAKING SOLUTIONS OF KNOWN CONCENTRATION

How many grams of Na_2CO_3 are required to make 2.0 L of 1.5 M Na_2CO_3?

*Chemists often use the expression "compound XY is ___ M in solution." However, if XY is an ionic compound, it will form X and Y ions on dissolving; XY does not exist as such in solution. Thus, the expression "compound XY is ___ M" is a way of saying that ___ moles of XY units are dissolved in enough water to make one liter of solution.

(a)

(b)

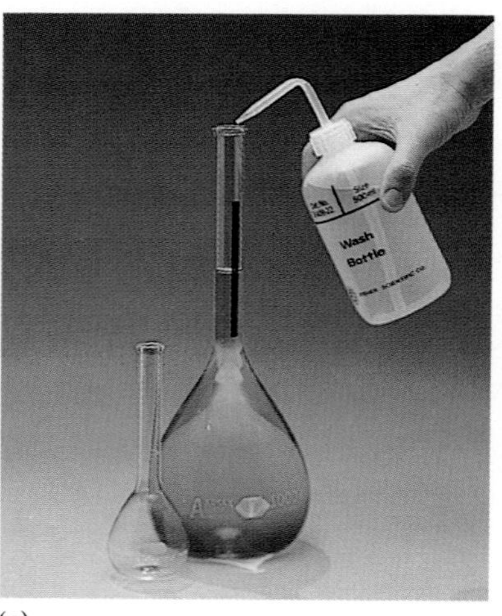

(c)

Figure 5.17 Making a solution by dilution. (a) A 100. mL volumetric flask is filled to the mark with 0.100 M $K_2Cr_2O_7$. (b) This is transferred to a 1.00 L volumetric flask. (c) The 1.00 L flask is then filled with distilled water to the mark, and the concentration of the now diluted $K_2Cr_2O_7$ solution is 0.0100 M.

Solution As usual, you must first calculate the number of moles of substance required.

$$2.0\ \cancel{L}\left(\frac{1.5\text{ mol }Na_2CO_3}{1.0\ \cancel{\text{L solution}}}\right) = 3.0\text{ mol }Na_2CO_3\text{ required}$$

Now that you know the moles of Na_2CO_3 required, convert this to grams.

$$3.0\ \cancel{\text{mol }Na_2CO_3}\left(\frac{106\text{ g }Na_2CO_3}{1\ \cancel{\text{mol }Na_2CO_3}}\right) = 3.2 \times 10^2\text{ g }Na_2CO_3$$

EXERCISE 5.11 Making Solutions of Known Concentration

An experiment in your laboratory requires 500. mL of a 0.0200 M solution of $KMnO_4$. You are given a bottle of solid $KMnO_4$, some distilled water, and a 500. mL volumetric flask. Describe how you would go about making up the required solution.

The preceding example and exercise illustrate the most common way to make a solution of known concentration. However, another method is to *begin with a more concentrated solution and add water to make it more dilute until the desired concentration is reached.* Many of the solutions prepared for your laboratory course are probably made by this **dilution** method. It is often most efficient to store a few liters of a concentrated solution and then add water to make it into many liters of a dilute solution. The following example illustrates the technique of preparing a dilute solution from a more concentrated one.

EXAMPLE 5.9

PREPARING SOLUTIONS BY DILUTION

You need 1.00 L of a 0.0100 M $K_2Cr_2O_7$ (potassium dichromate) solution. You have available some 0.100 M $K_2Cr_2O_7$, some volumetric flasks, and distilled water (Figure 5.17). How much of the more concentrated solution do you need to give finally 1.00 L of 0.0100 M $K_2Cr_2O_7$?

Solution As in stoichiometry problems, there is no *direct* way from the beginning (volume units) to the end (volume units) of a "dilution problem." The way, as always, lies through moles. The final solution must contain a certain number of moles of $K_2Cr_2O_7$ per liter. Therefore, you must transfer into the flask that number of moles of $K_2Cr_2O_7$ from the more concentrated solution. That is, *the number of moles of solute in the final solution must equal the number of moles taken from the more concentrated solution.* Dilution with water will then give you the correct number of moles in the desired volume of the final solution.

Step 1. Calculate the number of moles of $K_2Cr_2O_7$ that the final solution must contain.

$$1.00\ \cancel{L}\text{ of solution}\left(\frac{0.0100\text{ mol }K_2Cr_2O_7}{\cancel{L}}\right) = 0.0100\text{ mol }K_2Cr_2O_7$$

must be contained in the final, dilute solution

Since the number of moles of $K_2Cr_2O_7$ in the final solution must be the same as the number taken from the more concentrated solution, you will need to transfer 0.0100 mole of $K_2Cr_2O_7$ into the volumetric flask from the more concentrated solution.

Step 2. Calculate the volume of 0.100 M $K_2Cr_2O_7$ solution that will give the required 0.0100 mole of $K_2Cr_2O_7$.

$$0.0100 \text{ mol } K_2Cr_2O_7 \text{ required} \left(\frac{1.00 \text{ L of more concentrated solution}}{0.100 \text{ mol } K_2Cr_2O_7}\right)$$
$$= 0.100 \text{ L of } 0.100 \text{ M } K_2Cr_2O_7 \text{ required}$$

Step 3. Make up the final, more dilute solution. Measure out 0.100 liter or 100. mL of 0.100 M $K_2Cr_2O_7$ with a 100. mL volumetric flask and transfer it completely to the 1.00 L volumetric flask. Adding approximately 900 mL of distilled water, enough to bring the total volume to the mark on the 1.00 L flask, will give 1.00 liter of 0.0100 M $K_2Cr_2O_7$.

EXERCISE 5.12 Preparing Solutions by Dilution

An experiment calls for you to use 300. mL of 1.00 M NaOH, but you are given a large bottle of 3.00 M NaOH. Tell how you would make up the 1.00 M NaOH in the desired volume.

A second look at Example 5.9 suggests a simple way to remember how to do these calculations. It is important to understand that the central idea of "dilution problems" is that the number of moles of solute in the final, dilute solution must be equal to the number of moles of solute taken from the more concentrated solution. If we write M and V for the molarity and volume, respectively, of the dilute (d) and concentrated (c) solutions, we can calculate the number of moles in either solution. From Example 5.9, we have

(a) moles of $K_2Cr_2O_7$ in the final, dilute solution = M_dV_d = 0.0100 mol
(b) moles of $K_2Cr_2O_7$ taken from the more concentrated solution = M_cV_c = 0.0100 mol

Since both MV products are equal to the same number of moles, we can say that

$$M_cV_c = M_dV_d$$

moles of reagent in concentrated solution = moles of reagent in dilute solution

This handy expression holds for all cases where a more concentrated solution is used to make a more dilute one. You can use it to find, for example, the molarity of the dilute solution, M_d, from values of M_c, V_c, and V_d.

STOICHIOMETRY OF REACTIONS IN AQUEOUS SOLUTION

It is important to be able to derive quantitative relationships for reactions in solution, since many reactions occur in water and other solvents. Having now defined the concept of "concentration" and a convenient unit of concentration, you are ready to study some examples where quantitative relationships are involved.

EXAMPLE 5.10

SOLUTION STOICHIOMETRY

Metallic magnesium reacts with aqueous HCl to give magnesium chloride and hydrogen (see Figure 5.1). This reaction, or similar ones using other metals such as zinc, may be used as a convenient preparation of H_2 in the laboratory. Suppose you wanted to prepare 0.0061 mole of H_2. How many milliliters of 3.0 M HCl would you need to use with a handful (that is, an excess) of magnesium?

Solution

Step 1. The first step in a stoichiometry problem is to write a balanced chemical equation to depict the reaction involved.

$$Mg(s) + 2\ HCl(aq) \longrightarrow MgCl_2(aq) + H_2(g)$$

This tells you that 1 mole of H_2 is produced for every 2 moles of HCl used. This is the stoichiometric mole ratio factor you need to find the number of moles of HCl required.

Step 2. Calculate the number of moles of HCl required.

$$0.0061\ \cancel{\text{mol H}_2}\text{ desired}\left(\frac{2\text{ mol HCl required}}{1\ \cancel{\text{mol H}_2}\text{ desired}}\right) = 0.012\text{ mol HCl required}$$

Step 3. With the number of moles of HCl required now known, this can be converted to liters of HCl required with another conversion factor, the solution concentration.

$$0.012\ \cancel{\text{mol HCl}}\left(\frac{1.0\text{ L solution}}{3.0\ \cancel{\text{mol HCl}}}\right) = 0.0041\text{ L solution}$$

Since an answer in units of "milliliters" is requested, you can then convert 0.0041 L to mL. That is 0.0041 L = 4.1 mL.

EXAMPLE 5.11

SOLUTION STOICHIOMETRY

A common type of reaction is that between a metal carbonate and an aqueous acid to give a salt and gaseous CO_2.

$$\text{metal carbonate} + \text{acid} \longrightarrow \text{salt} + \text{carbon dioxide} + \text{water}$$

$$Na_2CO_3(aq) + 2\ HCl(aq) \longrightarrow 2\ NaCl(aq) + CO_2(g) + H_2O(\ell)$$

Suppose you have 100. mL of 1.50 M Na_2CO_3 (Example 5.8). How many milliliters of 0.750 M HCl would you have to add to the sodium carbonate solution to consume the Na_2CO_3 completely?

Solution

Step 1. The balanced equation for the reaction is given above.
Step 2. Calculate the number of moles of Na_2CO_3 to be used.

$$0.100\ \cancel{\text{L Na}_2\text{CO}_3}\left(\frac{1.50\text{ mol Na}_2\text{CO}_3}{1.00\ \cancel{\text{L Na}_2\text{CO}_3}}\right) = 0.150\text{ mol Na}_2\text{CO}_3$$

Step 3. Relate the number of moles of Na_2CO_3 available to the number of moles of HCl required. The required stoichiometric mole ratio factor is obtained from the balanced equation.

$$0.150 \text{ mol } Na_2CO_3 \left(\frac{2 \text{ mol HCl required}}{1 \text{ mol } Na_2CO_3 \text{ available}}\right) = 0.300 \text{ mol HCl}$$

Step 4. Calculate the number of liters of HCl required using moles of HCl required and the HCl concentration.

$$0.300 \text{ mol HCl required} \left(\frac{1.00 \text{ L solution}}{0.750 \text{ mol HCl}}\right) = 0.400 \text{ L solution}$$

The final step tells you that 0.400 L or 400. mL of solution is required.

EXERCISE 5.13 Solution Stoichiometry

If you have 25.0 mL of 0.750 M HCl, how many grams of Na_2CO_3 are required to react completely with the acid? How many grams of NaCl are produced?

Be sure to notice that the steps in the preceding examples are similar to those used in the general stoichiometry scheme in Figure 4.11. The chief difference between them is that grams and moles are interconverted by molar masses in Figure 4.11, whereas volume and moles are connected by molarity as in Figure 5.18, a modification of Figure 4.11.

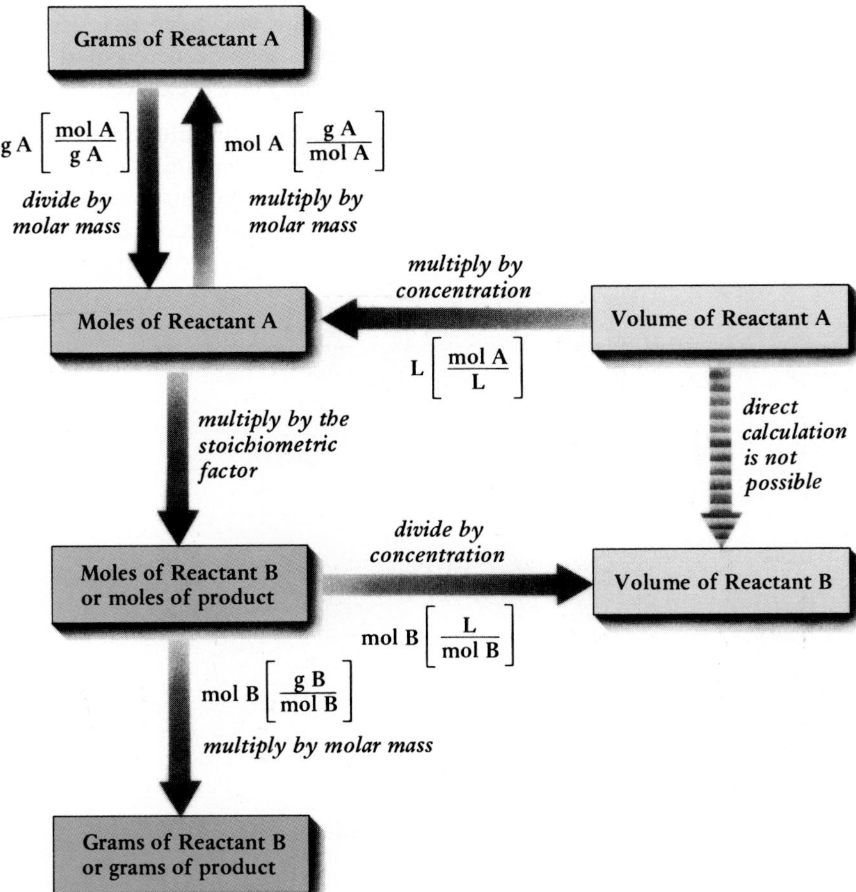

Figure 5.18 A scheme outlining stoichiometric relations for a solution reaction of the type A + B → product. Moles of A can be related to moles of B (or moles of product) through the stoichiometric factor. The only way you know the number of moles of A is to (a) know the mass of A or (b) know the volume of the solution of A.

TITRATIONS

Your study of stoichiometry so far should have convinced you that

(a) if you know the balanced equation for a reaction occurring between two reactants,
(b) if the reaction is rapid and complete,
(c) and if you know the exact quantity of one of the reactants,

then you can always obtain the exact amount of any other of the substances in the reaction. This is the essence of any technique of **quantitative chemical analysis**, the determination of the *amount* of a given constituent in a mixture.

__Qualitative analysis__ is the determination of the identity of the constituents of a mixture. __Quantitative analysis__ is the determination of the quantity of a constituent.

Suppose, for example, you want to analyze a type of common clover for the quantity of oxalic acid, $H_2C_2O_4$, in the leaves. We know this acid reacts with the base sodium hydroxide in aqueous solution according to the balanced equation

$$H_2C_2O_4(aq) + 2\ NaOH(aq) \longrightarrow Na_2C_2O_4(aq) + 2\ H_2O(\ell)$$

You can tell exactly how much oxalic acid is present in a given mass of clover leaves if the following conditions are met:

1. the reaction is done in such a way that you know when the sodium hydroxide being added is *exactly* the amount required to react with *all* the oxalic acid present in solution;
2. the exact volume of the base is known when you have reached the point where the exact stoichiometric reaction has occurred; and
3. the concentration of the sodium hydroxide is known exactly.

These conditions are fulfilled in a **titration**, a procedure illustrated in the series of photographs in Figure 5.19. The solution containing oxalic acid is placed in a flask along with a highly colored dye, the purpose of which is explained below. Sodium hydroxide of exactly known concentration is placed in a *buret*, a measuring cylinder most commonly of 50.0 mL volume

(a)

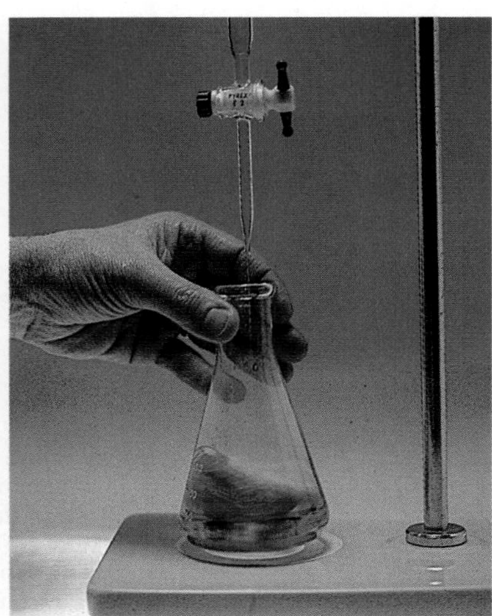

(b)

(c)

Figure 5.19 Titration of an acid in aqueous solution with a base. (a) The buret, a volumetric measuring device calibrated in divisions of 0.1 mL, is filled with base of known concentration. (b) Base is added slowly from the buret to the solution. (c) A change in color of an indicator dye signals the equivalence point.

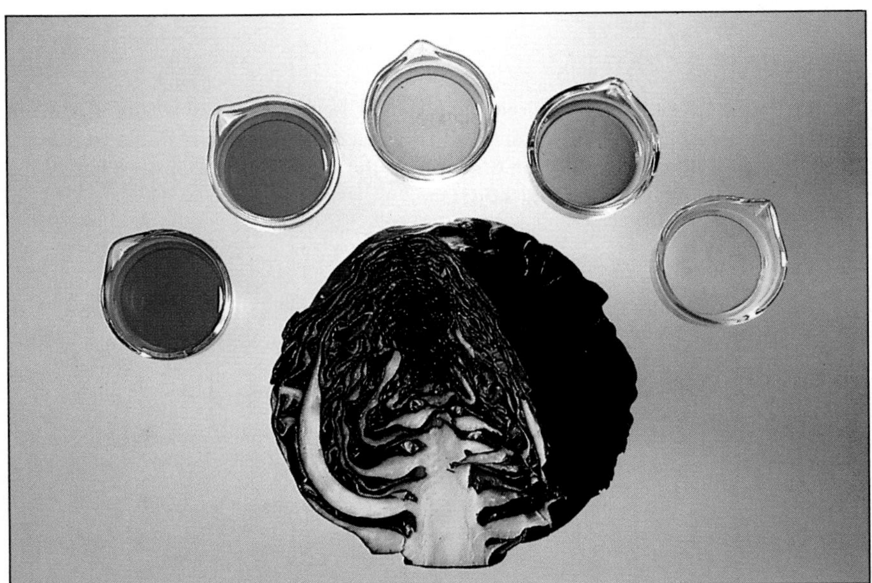

Figure 5.20 The juice of a red cabbage turns color as the acidity of a solution changes. When the solution is highly acidic, the juice gives the solution a red color. As the solution becomes less acid (more basic), the color changes from red to violet to yellow.

and calibrated in 0.1 mL divisions. As the sodium hydroxide is added slowly to the acid solution in the flask, the acid is consumed by reaction with the base according to the net ionic equation

$$H_3O^+(aq) + OH^-(aq) \longrightarrow 2\ H_2O(\ell)$$

As long as H_3O^+ from the acid is present in solution, every mole of OH^- supplied by the base from the buret is consumed by the H_3O^+. *The point at which the number of moles of OH^- added is equal to the number of moles of H_3O^+ supplied by the acid* is called the **equivalence point**. To indicate to us when this point has been reached, we added to the solution an **acid–base indicator**, the dye mentioned above. The dye is selected for its sensitivity to the H_3O^+ concentration in solution; it undergoes a strong color change at a hydronium ion concentration as near the equivalence point as possible (Figure 5.20).*

When the equivalence point has been estimated in a titration, the volume of base used from the beginning of the titration can be determined by reading the calibrated buret (Figure 5.19). If you know the concentration of the base in units of moles/liter, you can then calculate the exact number of moles of base used from

$$\text{Moles of base used} = \text{molarity of base (mol/L)} \times \text{volume of base (L)}$$

Since the balanced equation for the acid–base reaction gives you the stoichiometric factor, you can use this mole-ratio conversion factor to convert moles of base added to the exact number of moles of acid present in the original sample.

*There is a subtle difference between the $[H_3O^+]$ at which the color changes—called the **endpoint** of the titration—and the equivalence point. For our purposes, the difference is negligible.

EXAMPLE 5.12

ACID–BASE TITRATIONS

Suppose you have 1.034 g of clover leaves and you extract the oxalic acid from them into a small amount of water. This solution of oxalic acid is found to require 34.47 mL of 0.100 M NaOH for titration to the equivalence point. What is the weight percent of oxalic acid in the leaves?

Solution

Step 1. Write the balanced equation.

$$H_2C_2O_4(aq) + 2\ NaOH(aq) \longrightarrow Na_2C_2O_4(aq) + 2\ H_2O(\ell)$$

Step 2. Calculate the number of moles of base used. (Notice that milliliters of base have been changed first into liters of base.)

$$0.03447\ \cancel{\text{L NaOH}}\ \text{solution}\left(\frac{0.100\ \text{mol NaOH}}{\cancel{\text{L NaOH}}}\right) = 0.00345\ \text{mol NaOH}$$

Step 3. Calculate the number of moles of acid required by 0.00345 mole of base.

$$0.00345\ \cancel{\text{mol NaOH}}\left(\frac{1\ \text{mol acid}}{2\ \cancel{\text{mol NaOH}}}\right) = 0.00172\ \text{mol of acid required}$$

Step 4. Calculate the number of grams of acid in solution.

$$0.00172\ \cancel{\text{mol}\ H_2C_2O_4}\left(\frac{90.04\ \text{g}\ H_2C_2O_4}{1.00\ \cancel{\text{mol}\ H_2C_2O_4}}\right) = 0.155\ \text{g}\ H_2C_2O_4$$

Step 5. Calculate the weight percentage of $H_2C_2O_4$ in the 1.034 g of leaves.

$$\text{Weight percentage} = \frac{0.155\ \text{g}\ H_2C_2O_4}{1.034\ \text{g of leaves}} \times 100 = 15.0\%$$

EXERCISE 5.14 Acid–Base Titrations

Vinegar contains acetic acid, CH_3COOH. You can determine the mass of acetic acid in a vinegar sample by titrating with sodium hydroxide of known concentration. The reaction that occurs is

$$CH_3COOH(aq) + NaOH(aq) \longrightarrow NaCH_3COO(aq) + H_2O(\ell)$$

If you find that a 25.00 mL sample of vinegar requires 28.33 mL of a 0.953 M solution of NaOH for titration to the equivalence point, how many grams of acetic acid are there in the vinegar sample? What is the molar concentration of the acetic acid in the vinegar?

In the preceding example and exercise, the exact concentration of the base, NaOH, was known. The procedure by which the exact concentration of any reagent is determined is called **standardization**, and there are two general approaches to this as illustrated by the next example and exercise.*

EXAMPLE 5.13

STANDARDIZATION OF A REAGENT

An acid such as HCl can be standardized by using it to titrate a base such as Na_2CO_3. Sodium carbonate is a solid that can be obtained in pure form and can be weighed accurately. If you find that 0.250 g of Na_2CO_3 requires 25.76 mL of HCl for titration to the equivalence point, what is the exact molar concentration of the HCl?

Solution

Step 1. Write a balanced equation for the reaction.

$$Na_2CO_3(aq) + 2\ HCl(aq) \longrightarrow 2\ NaCl(aq) + CO_2(g) + H_2O(\ell)$$

Step 2. Calculate the number of moles of Na_2CO_3 used.

$$0.250\ \cancel{g\ Na_2CO_3}\left(\frac{1.00\ mol\ Na_2CO_3}{106.0\ \cancel{g\ Na_2CO_3}}\right) = 2.36 \times 10^{-3}\ mol\ Na_2CO_3$$

Step 3. Convert the moles of Na_2CO_3 available to the moles of HCl required for complete reaction, that is, for titration to the equivalence point.

$$2.36 \times 10^{-3}\ \cancel{mol\ Na_2CO_3}\left(\frac{2\ mol\ HCl\ required}{1\ \cancel{mol\ Na_2CO_3}}\right) = 4.72 \times 10^{-3}\ mol\ HCl$$

Step 4. In the two steps above you found that 2.36×10^{-3} mole of Na_2CO_3 requires 4.72×10^{-3} mole of HCl for complete titration. Therefore, there must have been 4.72×10^{-3} mole of HCl in the 25.76 mL of HCl solution. This means the concentration of HCl is

$$[HCl] = \frac{4.72 \times 10^{-3}\ mol\ HCl}{0.02576\ L} = 0.183\ M$$

EXERCISE 5.15 Standardization of a Base

Hydrochloric acid, HCl, can be purchased from chemical supply houses in solutions that are exactly 0.100 M, so these solutions can be used to standardize the solution of a base. If you titrate to the equivalence point 25.00 mL of a sodium hydroxide solution with 32.56 mL of 0.100 M HCl, what is the concentration of the base?

Acid–base titrations are extremely useful for quantitative chemical analysis, and many questions at the end of this chapter illustrate their scope. You should not get the impression, however, that a titration always

*You might think that you could make up a solution of NaOH simply by weighing out some solid NaOH as accurately as possible and dissolving it in water in a volumetric flask. Unfortunately, NaOH, which is sold as little pellets, rapidly picks up water and CO_2 from the air and so cannot be weighed accurately.

(a)

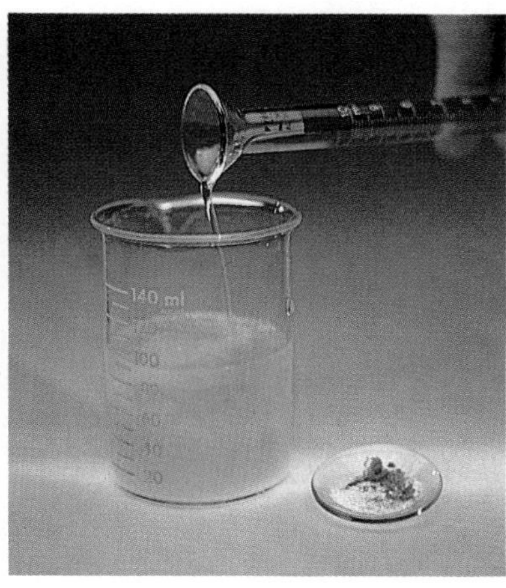

(b)

(c)

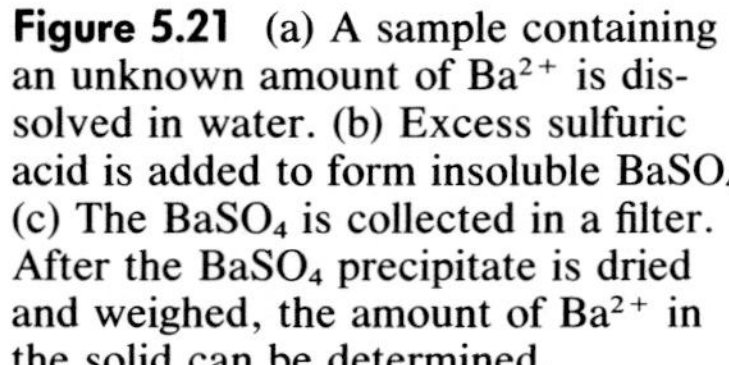

Figure 5.21 (a) A sample containing an unknown amount of Ba^{2+} is dissolved in water. (b) Excess sulfuric acid is added to form insoluble $BaSO_4$. (c) The $BaSO_4$ is collected in a filter. After the $BaSO_4$ precipitate is dried and weighed, the amount of Ba^{2+} in the solid can be determined.

involves an acid reacting with a base. In the next section of this chapter we shall describe oxidation–reduction reactions, a class of reactions that lend themselves very well to chemical analysis by titration because many of these reactions go rapidly to completion and there are ways of estimating their equivalence points. No matter what type of reaction is used in analysis, all share an important feature: *Before using any reaction as an analytical method, you must know the balanced chemical equation* in order to know the stoichiometric factors.

CHEMICAL ANALYSIS OF MIXTURES BY PRECIPITATION REACTIONS

If a mixture contains an acid, the quantity of acid present can be determined by an acid–base titration. On the other hand, if a mixture contains a substance that forms an insoluble salt, its amount can be determined by precipitating the salt quantitatively (that is, knowing that essentially all of the salt is precipitated, and a negligible amount remains in solution).* As an example, suppose you wish to determine the amount of barium in a sample that contains an ionic compound of Ba^{2+} plus some impurity. A procedure for doing this is outlined in Figure 5.21. In the figure you see that a weighed sample containing an unknown amount of barium is dissolved in water. Sulfuric acid, H_2SO_4, is added to the solution, and the Ba^{2+} ion in the water and the $SO_4{}^{2-}$ ion from the added acid combine to produce insoluble $BaSO_4$. If sufficient acid is added, all the Ba^{2+} that was in the unknown sample is precipitated and can be isolated by filtration; after drying, the mass of the $BaSO_4$ can be determined. Since you know that one mole of $BaSO_4$ contains one mole of Ba^{2+}, the number of moles (and grams) of Ba^{2+} in the unknown can be calculated.

Analysis of substances by precipitation reactions is an important technique, and two such analyses are explored in the following example and exercise.

---

*This type of analysis is sometimes called a "gravimetric" analysis in contrast with titrations, which are "volumetric" analyses.

EXAMPLE 5.14

CHEMICAL ANALYSIS BY PRECIPITATION REACTION

You have found an old sterling silver spoon and want to determine how much silver it contains. (Sterling silver is a solid solution or alloy of silver and copper.) A piece of spoon weighing 1.175 g is taken for analysis by a procedure outlined in the margin. The piece is first placed in nitric acid to convert the silver metal to aqueous Ag^+ (and Cu to aqueous Cu^{2+}). Excess HCl is added to the solution of Ag^+ (and Cu^{2+}) in nitric acid, and the Ag^+ and Cl^- ions combine to produce insoluble AgCl. The solution is filtered to isolate solid AgCl. After drying the solid, it is found to have a mass of 1.449 g. What is the weight percent of Ag in the sterling silver spoon?

Analysis of Ag in the presence of Cu

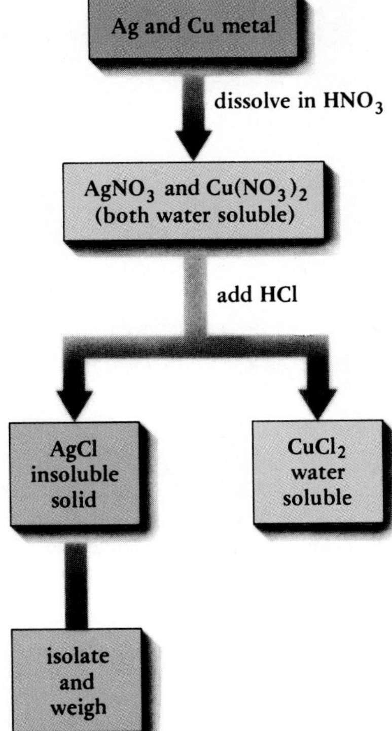

Solution

Step 1. Write the balanced equation for the precipitation of silver chloride.

$$Ag^+(aq) + Cl^-(aq) \longrightarrow AgCl(s)$$

Step 2. The isolated AgCl contains all the Ag that was in the 1.175 g piece of spoon. Since one mole of Ag in the spoon must lead to one mole of AgCl, the next step is to find the number of moles of AgCl isolated.

$$1.449 \cancel{\text{g AgCl}} \text{ isolated} \left(\frac{1 \text{ mol AgCl}}{143.32 \cancel{\text{g AgCl}}}\right) = 0.01011 \text{ mol AgCl}$$

Step 3. Calculate the number of moles of Ag in the AgCl (and therefore in the silver spoon).

$$0.01011 \cancel{\text{mol AgCl}} \left(\frac{1 \text{ mol Ag}}{1 \cancel{\text{mol AgCl}}}\right) = 0.01011 \text{ mol Ag}$$

Step 4. Calculate the mass of Ag in the spoon.

$$0.01011 \cancel{\text{mol Ag}} \left(\frac{107.87 \text{ g}}{1 \cancel{\text{mol Ag}}}\right) = 1.091 \text{ g Ag}$$

Step 5. Calculate the weight percent of silver in the spoon sample.

$$\text{Weight percent Ag} = \frac{1.091 \text{ g Ag}}{1.175 \text{ g spoon}} \times 100 = 92.82\% \text{ Ag}$$

EXERCISE 5.16 Chemical Analysis by Precipitation Reaction

Refer to Figure 5.21 for the procedure for a barium analysis. Suppose you have a solid that consists of some $BaCl_2$ contaminated with NaCl. To analyze the mixture, you must find a way to separate the Ba^{2+} from the Na^+ and then isolate the Ba^{2+} ion in the form of a compound of known formula. Therefore, you take 1.023 g of the solid mixture, dissolve it in water, and add H_2SO_4 to form insoluble $BaSO_4$ and leave NaCl in solution. If the $BaSO_4$ has a mass of 0.560 g after isolating and drying, calculate (a) the weight percentage of barium in the sample and (b) the number of grams of $BaCl_2$ in the original mixture.

Figure 5.22 If mercury(II) sulfide, the red solid, is heated in air, liquid mercury metal is formed in a reduction reaction.

5.5 OXIDATION–REDUCTION REACTIONS

Oxidation-reduction reactions make up the last class of common reactions to be described. Rather than the transfer of hydrogen ions that characterizes acid–base reactions, oxidation–reduction reactions involve the transfer of electrons. Reactions of metals and nonmetals with O_2 fit into this category, as do many important reactions occurring in your body; the oxidation of glucose to CO_2 and water is just one example.

The terms "oxidation" and "reduction" come from reactions that have been known to chemists for centuries. Earliest man learned how to change metal oxides and sulfides to the metal, that is, how to reduce them. Mercury(II) sulfide, known as cinnabar or vermilion (a mineral used since prehistoric times), can be *reduced* to liquid mercury simply by heating in air (Figure 5.22).

$$HgS(s) + O_2(g) \longrightarrow Hg(\ell) + SO_2(g)$$

Cassiterite or tin(IV) oxide, SnO_2, was found in Britain centuries ago. The metal is a major component of the combination of tin and copper called bronze, an important material in Roman times. Tin ore is very easily reduced to the metal by heating with carbon.

$$\overbrace{SnO_2(s) + \underset{\text{reducing agent}}{2\,C(s)} \longrightarrow Sn(s)}^{\text{reduced to}} + 2\,CO(g)$$

In this process carbon is the agent that brings about the reduction of tin ore to tin metal, so carbon is called the *reducing agent.*

Oxidation is the opposite of reduction. This too is a process known for centuries, and Figure 4.10 is an excellent example. Here oxygen is the agent of oxidation or *oxidizing agent.*

$$\overbrace{Mg(s) + \underset{\text{oxidizing agent}}{\tfrac{1}{2}\,O_2(g)} \longrightarrow MgO(s)}^{\text{oxidized to}}$$

Oxidation–reduction reactions are often referred to as ***redox*** *reactions.*

Notice in the reduction of SnO_2 with carbon that the carbon combines with oxygen to give CO; that is, the carbon is oxidized.

These experimental observations point to several fundamental conclusions concerning oxidation–reduction reactions: (a) If one substance is oxidized, another must be reduced. (b) The reducing agent is itself oxidized, and the oxidizing agent is reduced. Such conclusions become more obvious when described in terms of modern chemical theory. *Oxidation and reduction reactions are those involving transfer of electrons.* When a substance **accepts electrons**, it is said to be **reduced**. The language is descriptive because there is a reduction in the real or apparent electrical charge on an atom of the substance when it takes on negatively charged electrons. In the net ionic equation below, Cu^{2+} is reduced to uncharged Cu(s) by accepting electrons from zinc metal. Since zinc metal is the

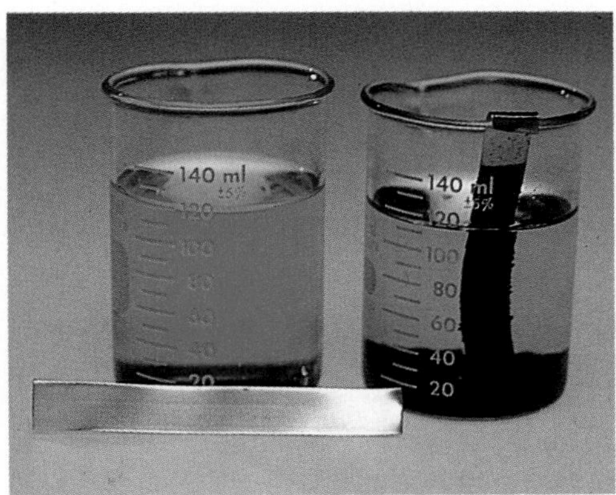

Figure 5.23 A clean strip of zinc (in front of the beaker at the left) is placed in a dilute solution of $CuSO_4$. With time, Zn reduces Cu^{2+}, so copper metal coats the zinc strip (in the beaker at the right). At the same time, Zn^{2+} ions enter the solution. Notice that the blue color of aqueous Cu^{2+} ions has greatly diminished in the beaker at the right. (The reaction between Al and Cu^{2+} in Figure 1.8 is similar to the Zn/Cu^{2+} reaction.)

"agent" that supplies the electrons and causes the Cu^{2+} ion to be reduced, Zn is called the **reducing agent** (Figure 5.23).

Cu^{2+} accepts electrons and is the oxidizing agent;
Cu^{2+} is reduced to Cu

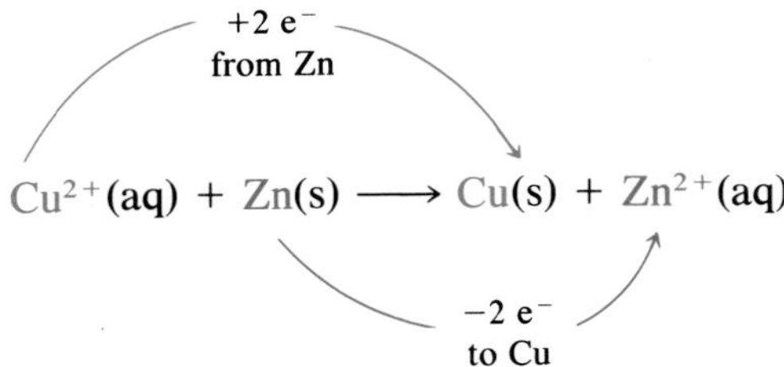

*Be sure to notice that the Zn/Cu^{2+} reaction is written as a **net ionic equation**. All redox reactions in this section are also net equations.*

Zn donates electrons and is the reducing agent;
Zn is oxidized to Zn^{2+}

Zinc metal releases electrons on going to Zn^{2+}; since its electrical charge has increased, it is said to have been **oxidized**. In order for this to happen, there must be something available to take the electrons offered by the zinc. In this case, Cu^{2+} is the electron acceptor and its charge is reduced (to 0 in the element). Therefore, Cu^{2+} is the "agent" that causes Zn metal to be oxidized, so Cu^{2+} is called the **oxidizing agent**. In every oxidation–reduction reaction, something is reduced (and so is itself the oxidizing agent) and something is oxidized (and so is the reducing agent).*

oxidized
$$\underset{\text{reducing agent}}{R} \longrightarrow R^+ + e^-$$

reduced
$$\underset{\text{oxidizing agent}}{O} + e^- \longrightarrow O^-$$

OXIDATION NUMBERS

How can you tell an oxidation–reduction reaction when you see one? The answer is to look for a *change in the oxidation number of an element* in the course of the reaction. The **oxidation number** of an element in a compound is defined as the charge an atom has, or *appears* to have,

*To help you remember the language of oxidation–reduction reactions you might recall the phrase "LEO the GERman." LEO stands for "Loses Electrons, Oxidized" and GER stands for "Gains Electrons, Reduced."

The reason for learning about oxidation numbers at this point is to be able to tell which reactions are oxidation–reduction processes.

when the electrons of the compound are counted according to a certain set of rules. In Chapter 10 we shall examine the basis of these counting rules, but for the moment it is quite easy to learn a few guidelines for assigning numbers. This will allow you to determine whether a given reaction is an oxidation–reduction process.

The guidelines for determining the oxidation number of an element in a substance are as follows:

1. **In free elements, each atom has an oxidation number of 0.** The oxidation number of Zn in metallic zinc is 0, and so is that for each atom in I_2 or S_8.
2. **For ions consisting of a single atom, the oxidation number is equal to the charge on the ion.** The oxidation number of Cl^- is -1 and of Zn^{2+} is $+2$. Elements of Periodic Groups 1A to 3A form ions with a positive charge and oxidation number equal to the group number. Therefore, aluminum forms Al^{3+}, and its oxidation number is $+3$.
3. **The oxidation number of H is +1 and of O is −2 in most compounds.** Although this statement applies to many, many compounds, there are a few important exceptions. When H forms a binary compound with a metal, the metal forms a positive ion and H becomes a hydride ion, H^-. Thus, in NaH the oxidation number of Na is $+1$ (equal to the Group number) and that of H is -1. Another exception to the rule is that oxygen can have an oxidation number of -1 in a class of compounds called *peroxides*, compounds based on the O_2^{2-} ion. For example, in H_2O_2 (hydrogen peroxide), H is assigned its usual oxidation number of $+1$, and so that of O is -1.

Except in simple ions, the oxidation number for an atom does NOT represent the real electrical charge on that atom. However, oxidation numbers are useful because they are a way of keeping track of electrons in reactions.

4. **The algebraic sum of the oxidation numbers in a neutral compound must be zero; in a polyatomic ion, the sum must be equal to the ion charge.** Examples of this rule are the compounds above and the following:
   (a) Li_2O is a neutral compound. If O has its "normal" oxidation number of -2, then that of Li must be $+1$, as predicted by its position in Group 1A.

   $Li_2O = (2\ Li^+)(O^{2-})$ Net charge $= 0 = 2(+1) + (-2)$

   (b) H_3PO_4 also has an overall charge of 0. If the oxygen has an oxidation number of -2 and that of H is $+1$, then the oxidation number of P must be $+5$.

   $H_3PO_4 = (3\ H^+)(P^{5+})(4\ O^{2-})$
   Net charge $= 0 = 3(+1) + (+5) + 4(-2)$

   (c) The permanganate ion, MnO_4^-, has an overall charge of 1−. Since O is assigned an oxidation number of -2, this means that Mn has an oxidation number of $+7$.

   $MnO_4^- = [(Mn^{7+})(4\ O^{2-})]^-$ Net charge $= -1 = (+7) + 4(-2)$

   (d) Dichromate ions, $Cr_2O_7^{2-}$, are also widely used in the laboratory. Since the net charge is 2−, and O is assigned an oxidation number of -2, Cr must have an oxidation number of $+6$.

   $Cr_2O_7^{2-} = [(2\ Cr^{6+})(7\ O^{2-})]^{2-}$
   Net charge $= -2 = 2(+6) + 7(-2)$

EXERCISE 5.17 Oxidation Numbers

Determine the oxidation number of each element in the following ions or compounds:

(a) MgO	(d) HS^-	(g) $NaCl$
(b) PH_3	(e) ClO_3^-	(h) HNO_3
(c) AlH_3	(f) $S_2O_3^{2-}$	(i) ZnS

Using the concept of oxidation number, you can now see why a reaction such as

Zn^{2+} accepts 2 e^- and is reduced

$$\underset{(+2)(-2)}{ZnO(s)} + \underset{(0)}{C(s)} \longrightarrow \underset{(+2)(-2)}{CO(g)} + \underset{(0)}{Zn(s)}$$

C donates 2 e^- and is oxidized

is said to be an oxidation–reduction reaction. The zinc starts with an oxidation number of +2 in ZnO and ends as the metal, oxidation number 0. Thus, the oxidation number of Zn has decreased from +2 to 0: zinc oxide has been reduced. Carbon begins as pure, elemental carbon, oxidation number 0, and it ends with an oxidation number of +2 in CO. The oxidation number of C has increased from 0 to +2, and it is said to have been oxidized.

The reactions of Figure 4.2 and 4.3 are also oxidation–reduction processes.

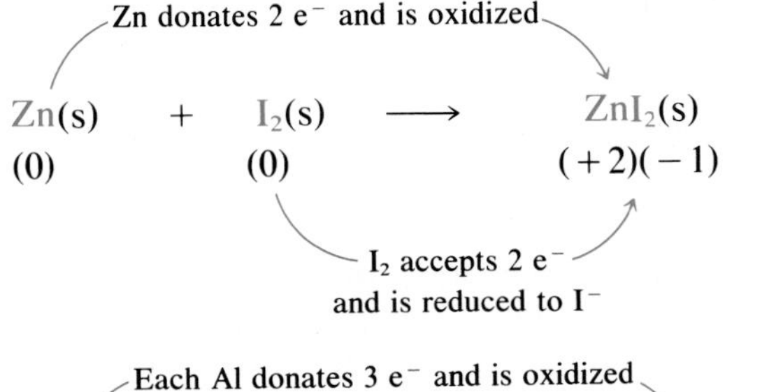

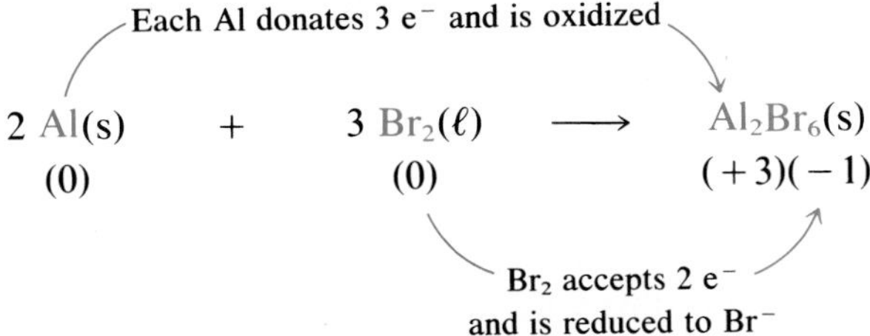

In both of these reactions we find the halogen combined only with a metal. When this is the case, we always assume that the halogen has formed a halide ion (I^- or Br^-) with an oxidation number of −1. Thus, zinc must have an oxidation number of +2 and aluminum of +3, just as predicted from their positions in the periodic table. In both cases, the metal donates electrons and is the reducing agent, and the halogen molecules, X_2, accept electrons and are oxidizing agents. Except in rare circumstances, *metals are reducing agents*, and halogens act as oxidizing agents.

EXERCISE 5.18 Recognizing Oxidation–Reduction Reactions

For each of the following reactions, tell whether it is an oxidation–reduction reaction. If so, tell which substance is oxidized and which one is reduced.

(a) $S(s) + O_2(g) \longrightarrow SO_2(g)$

(b) $Fe(s) + Cl_2(g) \longrightarrow FeCl_2(s)$

(c) $NaOH(aq) + HCl(aq) \longrightarrow NaCl(aq) + H_2O(\ell)$

(d) $3\ ZnS(s) + 8\ HNO_3(aq) \longrightarrow 3\ ZnSO_4(aq) + 8\ NO(g) + 4\ H_2O(\ell)$

BALANCING OXIDATION–REDUCTION EQUATIONS

The systematic way to balance equations for oxidation–reduction reactions can be illustrated with the reaction of aqueous copper(II) ions and metallic zinc (see Figure 5.23).

$$Cu^{2+}(aq) + Zn(s) \longrightarrow Cu(s) + Zn^{2+}(aq)$$

For zinc to change from metallic zinc to zinc ion, the metal must lose two electrons, and a balanced equation can be written depicting this.

Zn loses electrons, is oxidized, and is the reducing agent

$$Zn(s) \longrightarrow Zn^{2+}(aq) + 2\ e^-$$

The same is true for the transformation of copper(II) ion to copper metal, except that the Cu^{2+} ion must gain electrons.

Cu^{2+} gains electrons, is reduced, and is the oxidizing agent

$$Cu^{2+}(aq) + 2\ e^- \longrightarrow Cu(s)$$

Since each of these equations represents a portion of the total reaction, they are often called **half reactions**. Notice that *each half reaction is balanced for mass and charge*. There is a **mass balance** since there is one atom of each kind on each side of the equation. There is a **charge balance** because the algebraic sum of the charges on one side of the equation equals the sum of the charges on the other side. (Here both sides have a net charge of zero.) Finally, notice that the sum of the two half reactions equals the total equation for the overall process.

$$\begin{array}{rl} Zn(s) & \longrightarrow Zn^{2+}(aq) + 2\ e^- \\ +\ Cu^{2+}(aq) + 2\ e^- & \longrightarrow Cu(s) \\ \hline Zn(s) + Cu^{2+}(aq) & \longrightarrow Zn^{2+}(aq) + Cu(s) \end{array}$$

Summing the half reactions this way conveys the idea that the electrons produced by zinc are consumed by the copper(II) ions.

This example illustrates the general approach to balancing oxidation–reduction reactions. (a) First break the process into half reactions, one involving the donation of electrons (the reducing agent) and the other involving the acceptance of electrons (the oxidizing agent). (b) Balance each of these half reactions separately, first for mass and then for charge. (c) Multiply each half reaction by an appropriate factor so that the reducing half reaction gives off as many electrons as the oxidizing half reaction requires. (d) Add the half reactions to give the net, balanced equation. Examples 5.15 through 5.19 introduce you to situations you are likely to encounter.

Figure 5.24 The oxidation of copper metal by silver ion. A clean piece of copper screen is placed in a solution of silver nitrate, $AgNO_3$. With time, the copper reduces Ag^+ to silver metal crystals and the copper is oxidized to Cu^{2+}. The blue color of the solution is due to the presence of aqueous copper(II) ion.

EXAMPLE 5.15

BALANCING AN OXIDATION–REDUCTION EQUATION

Balance the equation for the reaction of Ag^+ with copper (Figure 5.24).

$$Ag^+(aq) + Cu(s) \longrightarrow Cu^{2+}(aq) + Ag(s)$$

Solution

Step 1. Recognize the reaction as an oxidation–reduction process. To recognize that the reaction is an oxidation–reduction process, you need only to *look for a change in the oxidation numbers of the reactants*; you do *not* need to know what that change is. In this case, there is a change in the oxidation numbers of Ag^+ (from +1 to 0) and Cu (from 0 to +2).

Step 2. Break the process into half reactions.

Reduction (decrease in Ag oxidation number)	$Ag^+(aq) \longrightarrow Ag(s)$
Oxidation (increase in Cu oxidation number)	$Cu(s) \longrightarrow Cu^{2+}(aq)$

Step 3. Achieve a mass balance by balancing the half reactions so that the same number of atoms of each kind appear on each side. Both half reactions here are already mass balanced.

Step 4. Achieve a charge balance by adding negative electrons to the side having an excess of positive charge.

$$Ag^+(aq) + e^- \longrightarrow Ag(s)$$

(Ag^+ acquires electrons and is the oxidizing agent)

$$Cu(s) \longrightarrow Cu^{2+}(aq) + 2\ e^-$$

(Cu donates electrons and is the reducing agent)

Step 5. Multiply the half reactions by appropriate factors so that the reducing agent donates as many electrons as the oxidizing agent consumes. Here one atom of copper produces two electrons, whereas one ion of Ag^+ acquires only one electron. Therefore, two Ag^+ ions are required to consume the two electrons produced by a Cu atom, and we multiply the Ag^+/Ag half reaction by 2.

$$2\ Ag^+(aq) + 2\ e^- \longrightarrow 2\ Ag(s)$$

Notice that the half reaction is still balanced for mass and charge (0 on both sides of the equation).

Step 6. Add the half reactions together to achieve the final, balanced overall equation.

$$\begin{array}{r} 2\ Ag^+(aq) + 2\ e^- \longrightarrow 2\ Ag(s) \\ Cu(s) \longrightarrow Cu^{2+}(aq) + 2\ e^- \\ \hline 2\ Ag^+(aq) + Cu(s) \longrightarrow Cu^{2+}(aq) + 2\ Ag(s) \end{array}$$

Step 7. Check the final result to ensure that there is a mass and charge balance. Here there are two silver atoms or ions and one copper atom or ion on each side, and the total charge on each side is +2. The equation is balanced.

SOMETHING MORE ABOUT
Sunglasses, Photochromic Glass, and Oxidation–Reduction

If you wear eye glasses, you know what a nuisance they can be, especially in the bright sun. You need to have a second pair of dark glasses, or you have to use those ugly little "clip-ons." Many of you know there is a simpler solution: you have lenses made of Photogray® glass. These lenses are *photochromic* because they change color in response to sunlight. About 30 years ago two scientists at Corning Glass Works, W.H. Armistead and S.D. Stookey, discovered how to make photochromic lenses. It is interesting to see how they work, since oxidation–reduction reactions are at the heart of the process.

In glass, each silicon atom is surrounded by a tetrahedron of oxygen atoms, and the SiO_4 tetrahedra are linked throughout the solid. Photogray® glass contains tiny crystallites of AgCl and traces of CuCl dispersed throughout the glass structure, and it is these "islands" of metal salts that account for the darkening in sunlight.

When sunlight in the ultraviolet or blue region (see Chapter 8) falls on the glass, chloride ions are oxidized to chlorine atoms, and the resulting electron is transferred to a silver ion, reducing it to an atom of the metal.

Light-induced oxidation: $Cl^- \longrightarrow Cl^0 + e^-$

Reduction of silver ion: $Ag^+ + e^- \longrightarrow Ag^0$

Unless the silver and chlorine atoms are quickly separated, Ag^0 would simply transfer an electron back to Cl^0, and there would be no net change. However, Ag^0 can pass its electron on to a neighboring Ag^+ instead of to Cl^0, and this moves the electron away from the chlorine atom. When the electron reaches a silver ion on the surface of the crystal, it is effectively "trapped" there, and a silver atom forms on the surface. This "trapped" electron attracts other silver ions, which can move through the crystal toward the surface, and they are ultimately reduced as well. These silver atoms clump together to form a tiny silver speck on the surface of the AgCl crystal. (So far the process is identical to the photographic process in that light causes silver salts in film to form clusters of silver atoms on the part of the film that is exposed to light.)

In the figure on the next page you do not see any Cl^0. A chlorine atom is sometimes referred to as a "hole" since an electron can "fall into this hole" and lead to a chloride ion. When light causes the oxidation of Cl^- to Cl^0, the Cl^0 can be reduced back to Cl^- either by Ag^0 or by a neighboring Cl^-. However, it is a neighboring Cl^- that supplies an electron to fill the "hole," and this creates yet another "hole," which is filled in turn by electron transfer from still another Cl^- ion.

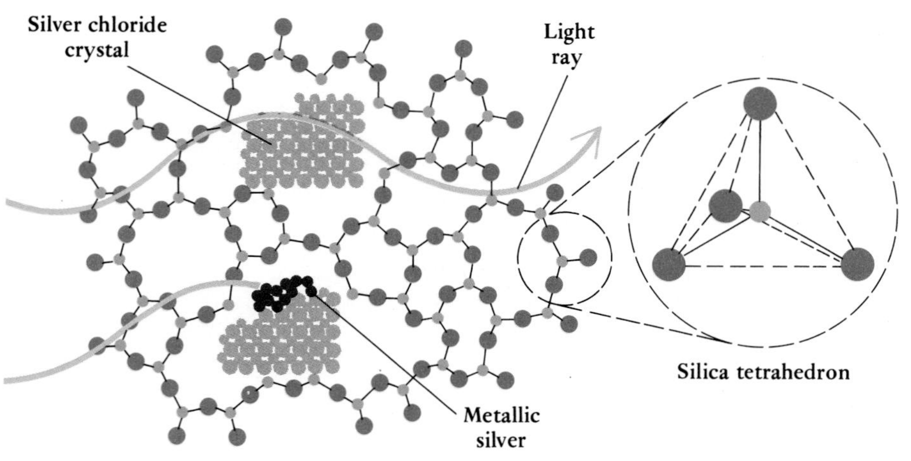

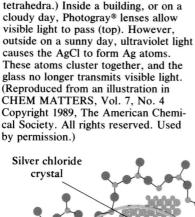

Photogray® lenses contain a small amount of AgCl (and traces of CuCl) in glass. (The glass consists of linked SiO_4 tetrahedra.) Inside a building, or on a cloudy day, Photogray® lenses allow visible light to pass (top). However, outside on a sunny day, ultraviolet light causes the AgCl to form Ag atoms. These atoms cluster together, and the glass no longer transmits visible light. (Reproduced from an illustration in CHEM MATTERS, Vol. 7, No. 4 Copyright 1989, The American Chemical Society. All rights reserved. Used by permission.)

Cl^0 originally created
by photo-induced oxidation

$$Cl^0 \xleftarrow{e^-} Cl^- \xleftarrow{e^-} Cl^- \xleftarrow{e^-} Cl^- \xleftarrow{e^-} Cl^- \xleftarrow{e^-} Cu^+ \text{ (becomes } Cu^{2+})$$

← electron transfer

"hole" transfer →

This electron transfer will occur only as long as one of the Cl^0 atoms has not been reduced by one of the few Cu^+ ions in the crystal, since these ions can also reduce Cl^0 to Cl^-.

$$\underset{\text{reducing agent}}{Cu^+} + \underset{\text{oxidizing agent}}{Cl^0} \longrightarrow Cu^{2+} + Cl^-$$

(The reaction to get rid of chlorine atoms from the light-induced process also occurs in photographic film, but in film the chlorine atoms are removed by reacting with the gelatin layer of the film.) This process is crucial to Photogray® glass because the "hole" would otherwise wander throughout the crystal until it finds a silver atom and causes the reaction $Ag^0 + Cl^0 \rightarrow Ag^+ + Cl^-$. This would mean the glass would not darken. Therefore, the presence of Cu^+ means that $Ag^0 \rightarrow Cl^0$ electron transfer does not occur and that one Cu^{2+} ion is formed for every Ag^0 that comes from the original light-induced reaction.

Silver metal is a fairly good reducing agent, and copper(II) ion is not a bad oxidizing agent (see Chapter 21), so the following reaction can occur in the solid state.

$$\underset{\text{reducing agent}}{Ag^0} + \underset{\text{oxidizing agent}}{Cu^{2+}} \longrightarrow Ag^+ + Cu^+$$

In fact, this is the reaction that accounts for the fading of Photogray® glass back to its normal transparency. Copper(II) ions are much smaller than Cl^- and Ag^+ ions (Chapter 9), and the Cu^{2+} ions can wriggle their way through the tiny holes between the other ions in the crystal. When a copper(II) ion encounters a cluster of silver atoms, the preceding oxidation–reduction reaction occurs, and the glass returns to its normal state where the silver and copper are in the form of the ions Ag^+ and Cu^+, respectively. However, if you have Photogray® lenses you have probably found that fading of the glass is slower than the darkening process. The reason is very simple: because they have a 2+ charge, the Cu^{2+} ions move through the glass *more slowly* than Ag^+ ions with a smaller 1+ charge. The requirement that Cu^{2+} ions move through the crystal also explains why the lenses fade more rapidly on a warm day than a cold day: the movement of ions is more rapid at the higher temperature.

A small portion of the AgCl crystal inside Photogray® glass (a) before darkening and (b) after darkening. After darkening there are Ag^0 atoms in clusters on the surface of the crystal and traces of Cu^+ ions have reduced Cl^0 to Cl^-.

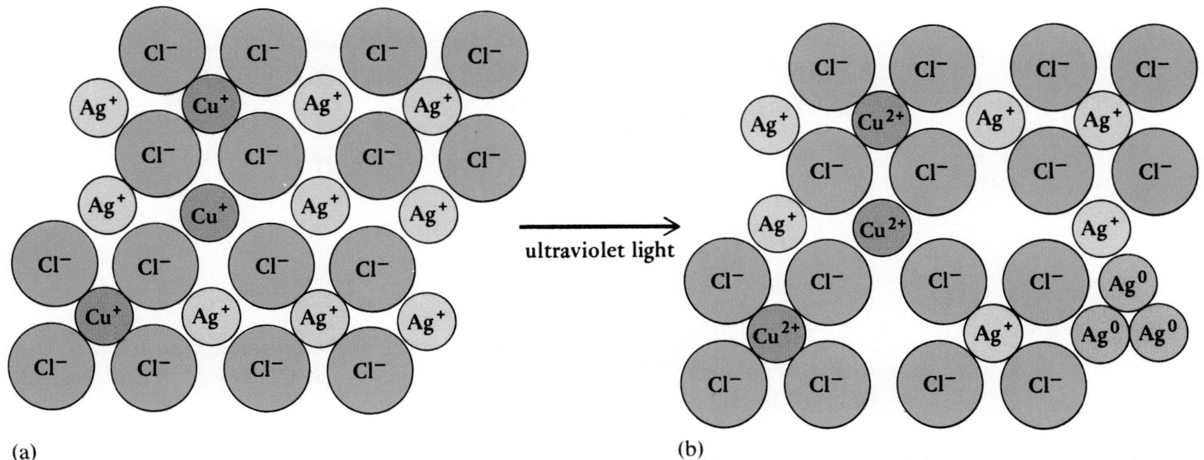

EXERCISE 5.19 Balancing an Equation for an Oxidation–Reduction Reaction

Balance the equation

$$Cr^{2+}(aq) + I_2(aq) \longrightarrow Cr^{3+}(aq) + I^-(aq)$$

Show the balanced half reactions and the balanced overall equation. Identify (a) the oxidizing agent, (b) the reducing agent, (c) the substance oxidized, and (d) the substance reduced.

The use of the hydronium ion in balancing redox equations in acid solution is quite cumbersome and can obscure the main ideas involved. Therefore, although it is perhaps less correct, we shall simply use $H^+(aq)$ as an abbreviation for $H_3O^+(aq)$.

A problem you will often confront when balancing equations for reactions in aqueous solution is that water, hydrogen ion (H^+), and hydroxide ion (OH^-) can enter into the reaction. In acidic conditions, H^+ may be a reactant or product, and in basic conditions it is possible for OH^- to participate. Under other circumstances, the *pair* of species H^+ and H_2O must be used to balance equations for reactions in acid solution. Similarly, the *pair* OH^- and H_2O may be needed to balance equations for reactions in basic solution. The table below will help you decide where to place the pairs H_2O/H_3O^+ and H_2O/OH^- when balancing equations in acid or base, respectively.

	O-RICH FORM	H-RICH FORM	SPECIES REQUIRED	REMOVED FROM	OTHER PRODUCT
Acid conditions	H_2O	H^+	O^{2-}	H_2O $\{H_2O \rightarrow [O^{2-}] + 2\ H^+\}$	H^+
			H^+	insert H^+ on H^+ deficient side	
Base conditions	OH^-	H_2O	O^{2-}	$2\ OH^-$ $\{2\ OH^- \rightarrow [O^{2-}] + H_2O\}$	H_2O
			H^+	H_2O $\{H_2O \rightarrow [H^+] + OH^-\}$	OH^-

In either acid or base solution, after you mass-balance all but the O and H atoms, you add the indicated number of the O-rich form to the side of the equation deficient in O. The H-rich form will then appear on the other side of the equation. Conversely, you add the H-rich form to the side deficient in H, and the other species will appear on the other side of the equation.

EXAMPLE 5.16

BALANCING EQUATIONS FOR OXIDATION–REDUCTION REACTIONS IN ACID SOLUTION

Balance the net ionic equation for the reaction of permanganate ion with oxalic acid in acid solution.

$$MnO_4^-(aq) + H_2C_2O_4(aq) \longrightarrow Mn^{2+}(aq) + CO_2(g)$$

Solution

Step 1. Recognize the reaction as an oxidation–reduction process. The oxidation number of Mn changes from +7 in MnO_4^- to +2 in Mn^{2+}, and C changes from +3 in $H_2C_2O_4$ to +4 in CO_2.

Step 2. Break the overall reaction into half reactions.

$$H_2C_2O_4(aq) \longrightarrow CO_2(g)$$

$$MnO_4^-(aq) \longrightarrow Mn^{2+}(aq)$$

Step 3. Balance each half reaction for mass. Begin by balancing all atoms except H and O; these are always the last to be balanced because they often appear in more than one reactant or product.

Oxalic acid half reaction: We shall first balance the C atoms in this half reaction.

$$H_2C_2O_4(aq) \longrightarrow 2\ CO_2(g)$$

Notice that the number of O atoms was also balanced in the process, so only H is left to balance. Since the reaction occurs in acid solution, mass balance for H can be achieved if we *add H^+ to the side requiring H atoms* (one H^+ for each H required). Since the product side is deficient by 2 H atoms, we simply insert two H^+ ions on the right.

$$H_2C_2O_4(aq) \longrightarrow 2\ CO_2(g) + 2\ H^+(aq)$$

Permanganate half reaction: The Mn atoms are already balanced, but an oxygen-rich species must be added to the right side for an O-atom balance.

$$MnO_4^-(aq) \longrightarrow Mn^{2+}(aq) + (\text{need 4 O atoms})$$

In acid solution, the table above shows we always *add H_2O to the side requiring O atoms*. One H_2O molecule is needed for each O required on the right (for a total of 4 H_2O molecules), and 8 H^+ ions are required for H-balance on the left.

$$8\ H^+(aq) + MnO_4^-(aq) \longrightarrow Mn^{2+}(aq) + 4\ H_2O(\ell)$$

To balance O in acid solution, add one H_2O to the oxygen-deficient side for each O required and then add 2 H^+ ions to the other side of the equation for every H_2O molecule added.

$2\ H^+ + [O^{2-}] \longrightarrow H_2O$

or

$H_2O \rightarrow [O^{2-}] + 2\ H^+$

Both half reactions are now balanced for mass.

Step 4. Balance the half reactions for charge. The mass-balanced $H_2C_2O_4$ equation has a net charge of 0 on the left side and 2+ on the right. Therefore, 2 e^- are added to the more positive side,

$$H_2C_2O_4(aq) \longrightarrow 2\ CO_2(g) + 2\ H^+(aq) + 2\ e^-$$

and this means that $H_2C_2O_4$ is the reducing agent. The mass-balanced MnO_4^- half reaction has a charge of 7+ on the left and 2+ on the right. Therefore, 5 e^- are added to the more positive (left) side, and this means that MnO_4^- is the oxidizing agent.

$$5\ e^- + 8\ H^+(aq) + MnO_4^-(aq) \longrightarrow Mn^{2+}(aq) + 4\ H_2O(\ell)$$

Step 5. Multiply the half reactions by appropriate factors so that the reducing agent donates as many electrons as the oxidizing agent consumes. The $H_2C_2O_4$ half reaction should be multiplied by 5, and the MnO_4^- half reaction by 2, so that each half reaction involves 10 electrons.

$$5[H_2C_2O_4(aq) \longrightarrow 2\ CO_2(g) + 2\ H^+(aq) + 2\ e^-]$$

$$2[5\ e^- + 8\ H^+(aq) + MnO_4^-(aq) \longrightarrow Mn^{2+}(aq) + 4\ H_2O(\ell)]$$

Step 6. Add the half reactions to give the overall equation.

$$5\ H_2C_2O_4(aq) \longrightarrow 10\ CO_2(g) + 10\ H^+(aq) + 10\ e^-$$
$$10\ e^- + 16\ H^+(aq) + 2\ MnO_4^-(aq) \longrightarrow 2\ Mn^{2+}(aq) + 8\ H_2O(\ell)$$

$$5\ H_2C_2O_4(aq) + 16\ H^+(aq) + 2\ MnO_4^-(aq) \longrightarrow 10\ CO_2(g) + 10\ H^+(aq) + 2\ Mn^{2+}(aq) + 8\ H_2O(\ell)$$

Step 7. Cancel common reactants and products. Here, the 16 $H^+(aq)$ ions appearing as reactants are partially compensated by 10 $H^+(aq)$ as products.

$$5\ H_2C_2O_4(aq) + 6\ H^+(aq) + 2\ MnO_4^-(aq) \longrightarrow 10\ CO_2(g) + 2\ Mn^{2+}(aq) + 8\ H_2O(\ell)$$

Step 8. Check the final result to make sure there is mass and charge balance. Mass balance: Both sides of the equation have 2 Mn, 28 O, 10 C, and 16 H atoms. Charge balance: Each side has a net charge of 4+.

EXAMPLE 5.17

BALANCING EQUATIONS FOR OXIDATION–REDUCTION REACTIONS IN ACID SOLUTION

Balance the net ionic equation for the reaction of the organic compound benzaldehyde, C_6H_5CHO, with the dichromate ion in acid solution.

$$C_6H_5CHO(aq) + Cr_2O_7^{2-}(aq) \longrightarrow C_6H_5COOH(aq) + Cr^{3+}(aq)$$

Solution

Step 1. Recognize the reaction as an oxidation–reduction process. Here Cr changes from +6 to +3 (so $Cr_2O_7^{2-}$ is reduced), and C changes from $-\frac{4}{7}$ to $-\frac{2}{7}$ (and so C_6H_5CHO is oxidized).

Step 2. Break the overall equation into half reactions.

$$C_6H_5CHO(aq) \longrightarrow C_6H_5COOH(aq)$$

$$Cr_2O_7^{2-}(aq) \longrightarrow Cr^{3+}(aq)$$

Step 3. Balance each half reaction for mass. Here we start with the C_6H_5CHO reaction. The C and H atoms are already balanced, but O is not. Therefore, we add H_2O as the source of O to the O-deficient left side; the table on page 176 shows that one H_2O molecule is required for each O required, and 2 H^+ ions result.

$$C_6H_5CHO(aq) + H_2O(\ell) \longrightarrow C_6H_5COOH(aq) + 2\ H^+(aq)$$

Next, turn to the $Cr_2O_7^{2-}$ half reaction. The Cr atoms should be balanced first,

(a) $Cr_2O_7^{2-}(aq) \longrightarrow 2\ Cr^{3+}(aq)$

(b) and then we add one H_2O to the O-deficient right side for each O required there,

$$Cr_2O_7^{2-}(aq) \longrightarrow 2\ Cr^{3+}(aq) + 7\ H_2O(\ell)$$

(c) and finally balance the H atoms by placing H^+ on the H-deficient side, the left side in this case.

$$14\ H^+(aq) + Cr_2O_7^{2-}(aq) \longrightarrow 2\ Cr^{3+}(aq) + 7\ H_2O(\ell)$$

Step 4. Balance the half reactions for charge.

$$C_6H_5CHO(aq) + H_2O(\ell) \longrightarrow C_6H_5COOH(aq) + 2\ H^+(aq) + 2\ e^-$$

$$6\ e^- + 14\ H^+(aq) + Cr_2O_7^{2-}(aq) \longrightarrow 2\ Cr^{3+}(aq) + 7\ H_2O(\ell)$$

With this we also know that C_6H_5CHO is the reducing agent (electron donor) and $Cr_2O_7^{2-}$ is the oxidizing agent (electron acceptor).

Step 5. Multiply half reactions by appropriate factors.

$$3[C_6H_5CHO(aq) + H_2O(\ell) \longrightarrow C_6H_5COOH(aq) + 2\ H^+(aq) + 2\ e^-]$$

Here the $Cr_2O_7^{2-}$ half reaction can be left as it is. Using 3 of the C_6H_5CHO half reactions will supply the 6 e^- the $Cr_2O_7^{2-}$ half reaction requires.

Step 6. Add half reactions.

$$3\ C_6H_5CHO(aq) + 3\ H_2O(\ell) \longrightarrow 3\ C_6H_5COOH(aq) + 6\ H^+(aq) + 6\ e^-$$

$$6\ e^- + 14\ H^+(aq) + Cr_2O_7^{2-}(aq) \longrightarrow 2\ Cr^{3+}(aq) + 7\ H_2O(\ell)$$

$$3\ C_6H_5CHO(aq) + 3\ H_2O(\ell) + 14\ H^+(aq) + Cr_2O_7^{2-}(aq) \longrightarrow 3\ C_6H_5COOH(aq) + 6\ H^+(aq) + 2\ Cr^{3+}(aq) + 7\ H_2O(\ell)$$

Step 7. Eliminate common reactants and products. Water appears on both sides, as does H^+. Therefore, we simplify the equation by realizing there is a *net* of 4 H_2O on the right and a *net* of 8 H^+ on the left. The final, balanced net ionic equation would be

$$3\ C_6H_5CHO(aq) + 8\ H^+(aq) + Cr_2O_7^{2-}(aq) \longrightarrow 3\ C_6H_5COOH(aq) + 2\ Cr^{3+}(aq) + 4\ H_2O(\ell)$$

Step 8. Check the final result.
Mass balance: 2 Cr atoms, 21 C atoms, 10 O atoms, and 26 H atoms. Charge balance: 6+ on both sides.

EXERCISE 5.20 Balancing Equations for Oxidation–Reduction Reactions in Acid Solution

The reaction of Cu and HNO_3 was shown in Figure 4.7.

$$Cu(s) + NO_3^-(aq) \longrightarrow Cu^{2+}(aq) + NO_2(g)$$

(a) Balance the equation for the reaction in acid solution. (b) Identify the oxidizing and reducing agents and the substance oxidized and the substance reduced.

In Examples 5.16 and 5.17, we were concerned with reactions in acid solution. Under those conditions we can use the H^+/H_2O pair to achieve balanced equations. Conversely, in basic solution, we can use only OH^- or the OH^-/H_2O pair. The following example illustrates how to balance an equation in basic solution.

EXAMPLE 5.18

BALANCING EQUATIONS FOR OXIDATION–REDUCTION REACTIONS IN BASIC SOLUTION

Balance the following equation, in which $Bi(OH)_3$ is reduced to bismuth metal by tin(II) in basic solution.

$$Bi(OH)_3(s) + SnO_2^{2-}(aq) \longrightarrow Bi(s) + SnO_3^{2-}(aq)$$

Solution

Step 1. Verify that the reaction is an oxidation–reduction process. The oxidation number of Bi changes from +3 to 0 and that for tin changes from +2 to +4.

Step 2. Break the overall reaction into half reactions.

$$Bi(OH)_3(s) \longrightarrow Bi(s)$$

$$SnO_2^{2-}(aq) \longrightarrow SnO_3^{2-}(aq)$$

Step 3. Balance each equation for mass. The bismuth half reaction involves the simple loss of OH^- ions as $Bi(OH)_3$ is changed to Bi. In basic solution, OH^- can be a reactant or product, so

$$Bi(OH)_3(s) \longrightarrow Bi(s) + 3\ OH^-(aq)$$

is balanced for mass.

In the tin-containing half reaction, the Sn atoms are balanced, but the left side is deficient in oxygen. In basic solution, the table on page 176 shows that the O-rich species is OH^- and is the supplier of oxide ion (and also water as required by mass balance).

$$2\ OH^- \longrightarrow [O^{2-}] + H_2O$$

Therefore, 2 OH^- are added to the O-deficient side to supply the one O atom needed, and H_2O is added to the other side to balance the H atoms.

$$2\ OH^-(aq) + SnO_2^{2-}(aq) \longrightarrow SnO_3^{2-}(aq) + H_2O(\ell)$$

To balance O in a redox process in basic solution, add OH^- to the O-deficient side and water to the other side according to the balanced equation

$2\ OH^- \longrightarrow [O^{2-}] + H_2O$

Be sure to notice that *two OH^- ions are required to supply one O atom and that one H_2O molecule must be added to the other side for every O atom supplied.*

Step 4. Balance the half reactions for charge.

$$3\ e^- + Bi(OH)_3(s) \longrightarrow Bi(s) + 3\ OH^-(aq)$$

$$2\ OH^-(aq) + SnO_2^{2-}(aq) \longrightarrow SnO_3^{2-}(aq) + H_2O(\ell) + 2\ e^-$$

Now we see that $Bi(OH)_3$ is the oxidizing agent (electron acceptor) and SnO_2^{2-} is the reducing agent (electron donor).

Step 5. Multiply the half reactions by appropriate factors.

$$2[3\ e^- + Bi(OH)_3(s) \longrightarrow Bi(s) + 3\ OH^-(aq)]$$

$$3[2\ OH^-(aq) + SnO_2^{2-}(aq) \longrightarrow SnO_3^{2-}(aq) + H_2O(\ell) + 2\ e^-]$$

Steps 6 and 7. Add the half reactions and simplify by eliminating common reactants and products.

$$6\ e^- + 2\ Bi(OH)_3(s) \longrightarrow 2\ Bi(s) + 6\ OH^-(aq)$$
$$6\ OH^-(aq) + 3\ SnO_2^{2-}(aq) \longrightarrow 3\ SnO_3^{2-}(aq) + 3\ H_2O(\ell) + 6\ e^-$$
$$2\ Bi(OH)_3(s) + 3\ SnO_2^{2-}(aq) \longrightarrow 2\ Bi(s) + 3\ SnO_3^{2-}(aq) + 3\ H_2O(\ell)$$

Step 8. Check the final result.
Mass balance: 2 Bi, 3 Sn, 12 O, and 6 H.
Charge balance: −6 on both sides.

EXAMPLE 5.19

BALANCING EQUATIONS FOR OXIDATION–REDUCTION REACTIONS IN BASIC SOLUTION

Balance the half reaction for $Cr(OH)_3(s) \longrightarrow CrO_4^{2-}(aq)$ in basic solution.

Solution

Step 1. Verify that the half reaction is an oxidation or reduction. Here the Cr atom changes from an oxidation number of +3 to +6.

Step 3. Balance the equation for mass. First, note that the Cr atoms are balanced, but there is a deficiency of O atoms on the left. In base, the O-rich species is OH^-, so this is added to the left side of the equation, and the H-rich species H_2O is added to the right side. Remembering that we add 2 OH^- for each deficient O atom, we have

$$2\ OH^-(aq) + Cr(OH)_3(s) \longrightarrow CrO_4^{2-}(aq) + H_2O(\ell)$$

Now the right side of the equation is deficient in H, so the H-rich species in base, H_2O, is added to the right, one H_2O for every H atom required (3 H_2O in this case). For every H_2O added to supply H, one OH^- must be added to the other side, so we now have

To balance H in a redox process in basic solution add H_2O to the H-deficient side and OH^- to the other side.

H_2O on H-deficient side $\longrightarrow$ $[H^+] + OH^-$

$$3\ OH^-(aq) + 2\ OH^-(aq) + Cr(OH)_3(s) \longrightarrow CrO_4^{2-}(aq) + H_2O(\ell) + 3\ H_2O(\ell)$$

Finally, simplifying the equation gives

$$5\ OH^-(aq) + Cr(OH)_3(s) \longrightarrow CrO_4^{2-}(aq) + 4\ H_2O(\ell)$$

Step 4. Balance the equation for charge. There are five negative charges on the left and only two on the right. Therefore, add an additional three negative charges in the form of electrons to the right.

$$5\ OH^-(aq) + Cr(OH)_3(s) \longrightarrow CrO_4^{2-}(aq) + 4\ H_2O(\ell) + 3\ e^-$$

Since the $Cr(OH)_3$ is donating electrons, it must be a reducing agent.

Steps 5–7. Not applicable here since we do not have another half reaction to be combined with the $Cr(OH)_3$ reaction.

Step 8. Check the final result.
Mass balance: there are 1 Cr atom, 8 O atoms, and 8 H atoms on each side.
Charge balance: there is a charge of 5− on each side.

EXERCISE 5.21 Balancing Equations for Oxidation–Reduction Reactions in Basic Solution

Balance the following net ionic equation for a reaction occurring in basic solution.

$$Cr(s) + ClO_4^-(aq) \longrightarrow Cr(OH)_3(s) + ClO_3^-(aq)$$

(a) Balance the overall equation showing each balanced half reaction. (b) Identify the oxidizing and reducing agents, the substance oxidized, and the substance reduced.

The examples above illustrate the basic ideas used in balancing equations involving electron transfer. The steps you go through are always the same, but there are many variations in detail, especially in the mass balance step. Only a few of these variations can be illustrated here, so the best way to learn to balance redox equations is to practice.

Balancing Oxidation–Reduction Equations

Another point of view on balancing equations in acidic or basic solutions takes advantage of the way chemists actually carry out such reactions. In general, chemists carry out *reductions of O-containing species in acid solution* (e.g., $MnO_4^- \rightarrow Mn^{2+}$). The reason for this is that the H^+ ion is essentially a consumer of O^{2-} ions, as in the balanced equation

$$2\ H^+ + [O^{2-}] \longrightarrow H_2O$$

where two hydrogen ions are added for each O^{2-} that must be supplied; one water molecule is a byproduct. Conversely, *oxidations of ions or compounds to O-containing species are often done in base solution* (e.g., $Cr^{3+} \rightarrow CrO_4^{2-}$) because O^{2-} is added to the ion or compound. In base, the O-rich species OH^- is a supplier of O^{2-} according to the balanced reaction

$$2\ OH^- \longrightarrow [O^{2-}] + H_2O$$

Whichever approach you choose to balancing equations, be sure to notice the following:

(a) Never add just O^{2-} ions, O atoms, or O_2 molecules to balance oxygen in an equation. As explained in the text, you may only use the pairs H^+/H_2O in acid solution or OH^-/H_2O in basic solution (or H^+ or OH^- alone if appropriate).

(b) Never add H atoms or H_2 molecules to balance hydrogen.

(c) Be sure to write all the charges on any ions involved.

EXERCISE 5.22 Balancing Equations for Half Reactions

Balance each of the following half reactions for mass and charge. Tell whether each reactant is an oxidizing or reducing agent.

(a) $S_2O_3^{2-}(aq) \longrightarrow S_4O_6^{2-}(aq)$ (neutral solution)
(b) $NO_3^-(aq) \longrightarrow NO(g)$ (acid solution)
(c) $C_2H_5OH(aq) \longrightarrow C_2H_4O(aq)$ (acid solution)
(d) $PH_3(aq) \longrightarrow H_3PO_4(aq)$ (acid solution)
(e) $ClO^-(aq) \longrightarrow Cl^-(aq)$ (basic solution)

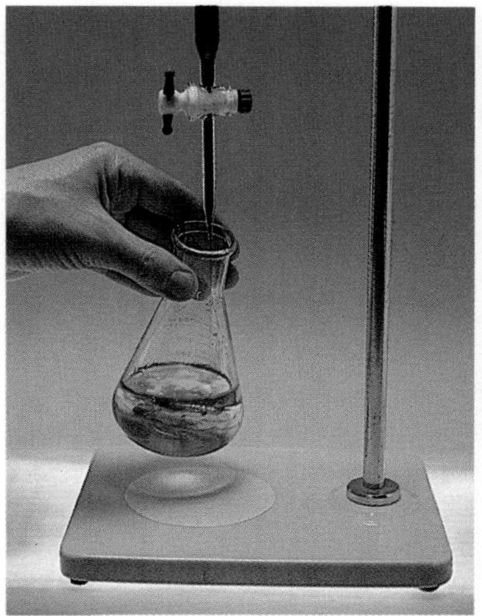

Figure 5.25 Titration of a solution containing Fe^{2+} (a reducing agent) (flask at left) with $KMnO_4$ (an oxidizing agent) (in the buret). As $KMnO_4$ is added to the solution, the iron(II) is oxidized and the deep purple MnO_4^- ion is reduced. The products (Fe^{3+} and Mn^{2+}) are nearly colorless. Just past the equivalence point, when a slight excess of $KMnO_4$ has been added, the solution takes on a faint purple color. The appearance of this color indicates that the titration is completed.

EXERCISE 5.23 Balancing Oxidation—Reduction Equations

Using the stepwise procedure outlined above, balance equations for the following oxidation—reduction reactions. In each case identify the oxidizing and reducing agents.

(a) $S_2O_3^{2-}(aq) + I_2(aq) \longrightarrow S_4O_6^{2-}(aq)$ (neutral solution)

(b) $Cu(s) + NO_3^-(aq) \longrightarrow Cu^{2+}(aq) + NO(g)$ (acid solution)

(c) $Fe^{2+}(aq) + MnO_4^-(aq) \longrightarrow Fe^{3+}(aq) + Mn^{2+}(aq)$ (acid solution)

(d) $Cr(OH)_3(s) + ClO^-(aq) \longrightarrow Cl_2(aq) + CrO_4^{2-}(aq)$ (basic solution)

STOICHIOMETRY OF OXIDATION–REDUCTION REACTIONS

The principles of stoichiometry in aqueous solutions have been described earlier in the chapter, and they apply equally well to oxidation–reduction reactions.

Oxidation–reduction reactions are the basis of many analytical methods, including titrations. In an acid–base titration, we showed that you add base, for example, until the number of moles of OH^- added is exactly equivalent to the number of moles of H^+ supplied by the acid (Section 5.4). In an oxidation–reduction titration, the equivalence point is reached when the numbers of moles of oxidizing agent and reducing agent are in the correct stoichiometric ratio (Figure 5.25).

EXAMPLE 5.20

OXIDATION–REDUCTION TITRATIONS

Vitamin C is the simple compound $C_6H_8O_6$. Besides being an acid, it is also a reducing agent. Therefore, a method for determining the amount of vitamin C in a sample is to titrate it with a solution of bromine (Br_2), an element capable of accepting electrons (i.e., acting as an oxidizing agent).

$$C_6H_8O_6(aq) + Br_2(aq) \longrightarrow 2\ HBr(aq) + C_6H_6O_6(aq)$$

Suppose you find that a 1.00 g "chewable vitamin C tablet" requires 27.85 mL of 0.102 M Br_2 solution for titration to the equivalence point. How many grams of vitamin C are contained in the tablet?

Solution

Step 1. Write a balanced equation for the reaction involved to find the required stoichiometric mole ratio conversion factor. In this case, the equation has already been balanced.

Step 2. Calculate the number of moles of Br_2 used in the titration.

$$0.02785\ \cancel{\text{L } Br_2} \text{ solution} \left(\frac{0.102 \text{ mol } Br_2}{\cancel{\text{L } Br_2}}\right) = 2.84 \times 10^{-3} \text{ mol } Br_2$$

Step 3. Relate the number of moles of Br_2 to moles of $C_6H_8O_6$. Since the stoichiometric factor is 1 mole of Br_2 to 1 mole of vitamin C, the number of moles of vitamin C titrated must have been 2.84×10^{-3}.

Step 4. Calculate the number of grams of vitamin C.

$$2.84 \times 10^{-3}\ \cancel{\text{mol } C_6H_8O_6} \left(\frac{176.1 \text{ g } C_6H_8O_6}{1.00\ \cancel{\text{mol } C_6H_8O_6}}\right) = 0.500 \text{ g } C_6H_8O_6$$

Another quite useful approach to both acid–base and oxidation–reduction titrations uses ***chemical equivalents*** *and defines solution concentrations in terms of* ***normality*** *(equivalents/liter). For further information, see the Study Guide that accompanies this text.*

EXERCISE 5.24 Oxidation–Reduction Reactions in Analysis

It is possible to determine the iron content of a mineral or ore by converting the iron to the iron(II) ion, Fe^{2+}, and then titrating this ion with potassium permanganate, $KMnO_4$. The balanced, net ionic equation for the reaction is

$$\underset{\text{(purple)}}{MnO_4^-(aq)} + 8\ H^+(aq) + \underset{\text{(colorless)}}{5\ Fe^{2+}(aq)} \longrightarrow \underset{\text{(colorless)}}{Mn^{2+}(aq)} + \underset{\text{(colorless)}}{5\ Fe^{3+}(aq)} + 4\ H_2O(\ell)$$

This is an excellent analytical reaction, because it is easy to detect when all the iron(II) has reacted. The MnO_4^- ion is deep purple. On reaction with Fe^{2+}, however, the color disappears, because the Mn^{2+} ion is colorless. Thus, as $KMnO_4$ is added from a buret, the purple color disappears when the solutions mix. When all the Fe^{2+} has been consumed, any additional $KMnO_4$ will give the solution a permanent purple color. Therefore, one carries out this titration until the initially colorless, Fe^{2+}-containing solution just turns a faint purple color (Figure 5.25).

Assume that you find 1.026 g of an iron-containing ore requires 24.34 mL of 0.0200 M $KMnO_4$ to reach the equivalence point. How many grams of iron does the ore sample contain? What is the weight percentage of iron in the ore?

SOMETHING MORE ABOUT
Cisplatin, A Cancer Chemotherapy Drug

As you learned in the interview at the beginning of Part One of the book, in 1964 Professor Barnett Rosenberg and his colleagues at Michigan State University were studying the effects of electricity on the growth of bacteria. To test the effect they inserted platinum electrodes (conductors of electricity) into a culture of *Escherichia coli* bacteria. Within a few hours they observed that cell division in the bacteria was dramatically affected. After the chemical detective work described in Rosenberg's interview, they were able to show that a platinum-containing substance, which came from the platinum of the electrodes and ammonia and chloride ion in the system, was responsible. Although Rosenberg's group had not set out to study the effect of heavy metal compounds on bacterial cells, they recognized that the results were significant and believed that the substance may be a cancer chemotherapy agent. Indeed, it was, and the Food and Drug Administration approved it for marketing in 1979.

See the interview with Professor Rosenberg at the beginning of Part One of the book.

The active compound discovered by Rosenberg has the formula $Pt(NH_3)_2Cl_2$ and the chemical name *cis*-dichlorodiammineplatinum(II). It is now commonly called cisplatin and is sold under the name of Platinol®. The compound has actually been known since 1845, when it was first described by Peyrone. It can be readily synthesized from the ion $[PtCl_4]^{2-}$ (usually in the form of its potassium salt) and ammonia in water.

$$[PtCl_4]^{2-}(aq) + 2\ NH_3(aq) \longrightarrow Pt(NH_3)_2Cl_2(s) + 2\ Cl^-(aq)$$

The compound $Pt(NH_3)_2Cl_2$ can exist in two forms with the same formula, forms that chemists called *isomers*.

H_3N, H_3N — Pt — Cl, Cl
cis

Cl, H_3N — Pt — NH_3, Cl
trans

The isomers of $Pt(NH_3)_2Cl_2$.

cis-$Pt(NH_3)_2Cl_2$

In both isomers a central Pt^{2+} ion is surrounded by two Cl^- ions and two NH_3 molecules at the corners of a square. The Pt^{2+} ion, the Cl^- ions, and the N atoms of the NH_3 molecules are all in the same plane. These isomers differ in the location of the NH_3 and Cl^- groups relative to each other. In the *trans* isomer, the two Cl^- ions are across the molecule from one another, as are the two NH_3 molecules. In contrast, the *cis* isomer has both the Cl^- ions on the same side of the square.

What is most interesting is that only one of these isomers, the *cis* compound, is active in treating cancerous cells. The *trans* isomer has no effect. It is often the case that such subtle structural differences in molecules lead to different physiological effects. However, chemists have been especially intrigued by the cisplatin molecule, and many studies have been done in an attempt to understand why only the *cis* isomer is useful.

trans-$Pt(NH_3)_2Cl_2$.

Cisplatin is most effective in treating cancers of the testes, the ovaries, and the bladder. (Testicular cancer, for example, has gone from being nearly incurable to being 95% curable.) However, although it has proven effective in treating certain cancers, it is also severely toxic as it accumulates in the kidneys, liver, and the large and small intestines. For this reason, other drugs must be administered along with the platinum compound in order to reduce its toxicity toward normal cells.

SUMMARY

Many important chemical reactions occur in water, so this chapter takes up the properties of compounds and their reactions in water. Ionic compounds that contain certain ions can dissolve in water to a significant extent (Figure 5.5). In doing so, they break up or dissociate into their ions. The resulting solution conducts electricity, so the dissolved ionic compounds are called **electrolytes**. Common acids and bases are listed in Table 5.1. **Acids** ionize to provide H_3O^+ ions in water, while **bases** provide OH^- ions.

Some ions do not enter directly into reactions involving ionic compounds and so are called **spectator ions**. These can be deleted from the complete equation for the reaction to give a **net ionic equation** (Section 5.2).

Throughout this book, you will encounter (a) reactions of compounds or elements with O_2 and halogens (Chapter 4), (b) **exchange** or **metathesis reactions (acid–base** and **precipitation reactions)** (Section 5.3), and (c) **oxidation–reduction reactions** (Section 5.5), among others. Reactions of type (b) are useful in the preparation of a variety of salts.

Since many reactions occur in solution, the concentration of material (the **solute**) in the reaction medium (the **solvent**) must be defined (Section 5.4). A very convenient unit of concentration is **molarity**.

$$\text{Molarity (M)} = \frac{\text{moles of solute}}{\text{liters of solution}}$$

In another often-used form, this equation is "moles of solute = M × V" where V is the solution volume.

Reagent solutions are often prepared by diluting a more concentrated solution of that reagent (Section 5.4). The number of moles of material in a solution remains the same after the dilution. Therefore, the product of molarity and volume of a concentrated solution must be equal to the product of molarity and volume of the diluted solution (c, concentrated; d, diluted):

$$M_cV_c = M_dV_d$$

This allows us to find the new concentration, M_d, for example, when diluting a given amount of solution of known concentration to a new volume.

A **titration** is a way to carry out a reaction in a very precise manner and to use it for analysis (Section 5.4). In an acid–base titration, acid of known concentration is added to a base of unknown concentration (or vice versa) until the number of moles of H_3O^+ that can be supplied by the acid is *exactly* equal to the number of moles of OH^- that can be supplied by the base. This is called the **equivalence point**. From the known stoichiometry of the reaction that occurs, one can find the amount of unknown acid or base.

Oxidation–reduction reactions (often called **redox** reactions) involve the transfer of electrons between compounds (Section 5.5). A compound is said to be **reduced** if the **oxidation number** of one of its atoms is reduced by acquiring electrons *from* another species, the **reducing agent**. Conversely, a compound is **oxidized** if the oxidation number of one of its atoms is increased because that compound has transferred one or more electrons *to* an **oxidizing agent**. Equations for oxidation–reduction reactions can be balanced if the equation is divided into **half reactions**, one for the reducing agent and another for the oxidizing agent. Analyses can be done using **oxidation–reduction titrations** in a manner similar to acid–base titrations; the equivalence point is reached when the amounts of oxidizing and reducing agents are in the correct stoichiometric ratio.

STUDY QUESTIONS

REVIEW QUESTIONS

1. What is an electrolyte? How can we differentiate between a weak and a strong electrolyte? Give an example of each.
2. Tell how you know that $Ni(NO_3)_2$ is soluble in water. Tell how you know that $NiCO_3$ is not soluble in water.
3. Name two acids that are strong electrolytes and one that is a weak electrolyte. Name two bases that are strong electrolytes and one that is a weak electrolyte.
4. Name the spectator ions in each of the following reactions and write the net ionic equation for each:
(a) Calcium carbonate and hydrochloric acid.

$$CaCO_3(s) + 2\ H_3O^+(aq) + 2\ Cl^-(aq) \longrightarrow CO_2(g) + Ca^{2+}(aq) + 2\ Cl^-(aq) + 3\ H_2O(\ell)$$

(b) Nitric acid and magnesium hydroxide.

$$2\ H_3O^+(aq) + 2\ NO_3^-(aq) + Mg(OH)_2(s) \longrightarrow 4\ H_2O(\ell) + Mg^{2+}(aq) + 2\ NO_3^-(aq)$$

5. Name the water-insoluble product in each reaction.
(a) $CuCl_2(aq) + H_2S(aq) \longrightarrow CuS + 2\ HCl$
(b) $CaCl_2(aq) + K_2CO_3(aq) \longrightarrow 2\ KCl + CaCO_3$
(c) $AgNO_3(aq) + NaI(aq) \longrightarrow AgI + NaNO_3$
6. Find two examples of acid–base reactions in the chapter. Write balanced equations for these reactions.
7. Find two examples of precipitation reactions in the chapter. Write balanced equations for these reactions.
8. Find an example of a gas-forming reaction in this chapter and write a balanced equation for the reaction.
9. Explain how you could prepare barium sulfate by (a) an acid–base reaction, (b) a precipitation reaction, and (c) a gas-forming reaction. The materials you have to start with are $BaCO_3$, $Ba(OH)_2$, Na_2SO_4, and H_2SO_4.
10. You want to make 250. mL of a 0.40 M solution of NaCl. Which of the following would you do? (a) Add 5.84 g of solid NaCl to 250. mL of water in a volumetric flask. (b) Add enough water to 5.84 g of solid NaCl to make 250. mL of solution.
11. You have 1.00 L of 0.100 M NaCl. Tell how you would make 250. mL of 0.0500 M NaCl by diluting the more concentrated solution.
12. You have an impure sample of solid $Ca(OH)_2$; it contains a small amount of $CaCl_2$. Explain how you would find how much $Ca(OH)_2$ is in the sample using *two* different analytical methods.
13. Explain the difference between an oxidation and a reduction. Give an example of each.
14. Explain the difference between an oxidizing agent and a reducing agent. Give an example of each.
15. The reaction of nickel metal with nitric acid gives nickel(II) ion and NO_2 gas.

$$Ni(s) + NO_3^-(aq) \longrightarrow Ni^{2+}(aq) + NO_2(g)$$

Show how you balance the equation using the steps outlined in Examples 5.15 through 5.17. Write out the description of steps 1 through 7 for this case.

PROPERTIES OF IONIC COMPOUNDS

16. Which compound or compounds in each of the following groups is (are) expected to be soluble in water?
(a) CuO, $CuCl_2$, $CuCO_3$
(b) AgCl, Ag_3PO_4, $AgNO_3$
(c) KCl, K_2CO_3, $KMnO_4$
17. Which compound or compounds in each of the following groups is (are) expected to be soluble in water?
(a) $PbSO_4$, $Pb(NO_3)_2$, $PbCO_3$
(b) Na_2SO_4, $NaClO_4$, $NaCH_3COO$
(c) AgBr, KBr, $AlBr_3$
18. Give the formula for
(a) a soluble compound containing the acetate ion
(b) an insoluble sulfide
(c) a soluble hydroxide
(d) an insoluble chloride
19. Give the formula for
(a) a soluble compound containing the chloride ion
(b) an insoluble hydroxide
(c) an insoluble carbonate
20. Each compound below is water-soluble. What ions are produced in water?
(a) NaI (c) $KHSO_4$
(b) K_2SO_4 (d) NaCN

21. Each compound below is water-soluble. What ions are produced in water?
(a) KOH (c) $NaNO_3$
(b) K_2CO_3 (d) $(NH_4)_2SO_4$

22. Tell whether each of the following is water-soluble. If soluble, tell what ions are produced.
(a) $BaCl_2$ (c) $Pb(NO_3)_2$
(b) $Cr(NO_3)_3$ (d) $BaSO_4$

23. Tell whether each of the following is water soluble. If soluble, tell what ions are produced.
(a) Na_2CO_3 (c) CdS
(b) $CuSO_4$ (d) $CaBr_2$

24. If you wanted to make a water-soluble compound containing Cu^{2+}, name two anions you could choose to combine with the copper ion. Conversely, if you wanted an insoluble Cu^{2+} compound, what two anions could you choose?

25. If you wanted to make a water-soluble compound containing Ca^{2+}, name two anions you could use. Conversely, if you wanted an insoluble Ca^{2+} compound, what two anions could you choose?

ACIDS AND BASES

26. Write a balanced equation to show how nitric acid dissociates to ions in water.

27. Write a balanced equation to show how calcium hydroxide dissociates to ions in water.

NET IONIC EQUATIONS

28. Balance each of the following equations, and then write the net ionic equation.
(a) $Zn(s) + HCl(aq) \longrightarrow H_2(g) + ZnCl_2(aq)$
(b) $Mg(OH)_2(s) + HCl(aq) \longrightarrow MgCl_2(aq) + H_2O(\ell)$
(c) $HNO_3(aq) + CaCO_3(s) \longrightarrow Ca(NO_3)_2(aq) + H_2O(\ell) + CO_2(g)$
(d) $HCl(aq) + MnO_2(s) \longrightarrow MnCl_2(aq) + Cl_2(g) + H_2O(\ell)$

29. Balance each of the following equations, and then write the net ionic equation.
(a) $(NH_4)_2S(aq) + Cu(NO_3)_2(aq) \longrightarrow CuS(s) + NH_4NO_3(aq)$
(b) $Pb(NO_3)_2(aq) + HCl(aq) \longrightarrow PbCl_2(s) + HNO_3(aq)$
(c) $BaCO_3(s) + HCl(aq) \longrightarrow BaCl_2(aq) + H_2O(\ell) + CO_2(g)$

30. Balance each of the following equations, and then write the net ionic equation. Refer to Table 5.1 and Figure 5.5 for information on acids and bases and on solubility. Show physical states for all reactants and products.
(a) $Ba(OH)_2 + HNO_3 \longrightarrow Ba(NO_3)_2 + H_2O$
(b) $BaCl_2 + Na_2CO_3 \longrightarrow BaCO_3 + NaCl$
(c) $Na_2S + Ni(NO_3)_2 \longrightarrow NiS + NaNO_3$

31. Balance each of the following equations, and then write the net ionic equation. Refer to Table 5.1 and Figure 5.5 for information on acids and bases and on solubility. Show physical states for all reactants and products.
(a) $ZnCl_2 + KOH \longrightarrow KCl + Zn(OH)_2$
(b) $AgNO_3 + KI \longrightarrow AgI + KNO_3$
(c) $NaOH + FeCl_2 \longrightarrow Fe(OH)_2 + NaCl$

TYPES OF REACTIONS

32. Complete and balance equations for the following exchange reactions. Tell whether each is an acid–base or a precipitation reaction.
(a) $HCl(aq) + KOH(aq) \longrightarrow$
(b) $AgNO_3(aq) + KCl(aq) \longrightarrow$
(c) $H_2SO_4(aq) + NaOH(aq) \longrightarrow$

33. Complete and balance equations for the following exchange reactions. Tell whether each is an acid–base or a precipitation reaction.
(a) $Cu(NO_3)_2(aq) + Na_2S(aq) \longrightarrow$
(b) $Pb(NO_3)_2(aq) + KCl(aq) \longrightarrow$
(c) $LiOH(aq) + HClO_4(aq) \longrightarrow$

34. Complete and balance equations for the following exchange reactions. Tell whether each is an acid–base reaction, precipitation reaction, or gas-forming reaction.
(a) $SrCO_3(s) + HNO_3(aq) \longrightarrow$
(b) $BaCl_2(aq) + H_2SO_4(aq) \longrightarrow$
(c) $MnCl_2(aq) + (NH_4)_2S(aq) \longrightarrow$

35. Complete and balance equations for the following exchange reactions. Tell whether each is an acid–base reaction, precipitation reaction, or gas-forming reaction.
(a) $HNO_3(aq) + Ca(OH)_2(s) \longrightarrow$
(b) $HCl(aq) + Fe(OH)_2(s) \longrightarrow$
(c) $SrCl_2(aq) + Na_2CO_3(aq) \longrightarrow$
(d) $Na_2CO_3(aq) + HBr(aq) \longrightarrow$

PREPARATION OF SALTS

36. Write a balanced equation showing how you could prepare each of the following salts by an acid–base reaction.
(a) $NaNO_3$ (c) K_3PO_4
(b) KCl (d) Cs_2SO_4

37. Write a balanced equation showing how you could prepare each of the following salts by an acid–base reaction.
(a) $NaCH_3COO$ (c) $MgSO_4$
(b) NH_4NO_3 (d) Na_3PO_4

38. Write a balanced equation showing how you could prepare each of the following insoluble salts by a precipitation reaction.
(a) $Ni(OH)_2$ (c) NiS
(b) $CaCO_3$ (d) $BaSO_4$

39. Write a balanced equation showing how you could prepare each of the following salts by a precipitation reaction.
(a) $FeCO_3$ (c) ZnS
(b) Ag_3PO_4 (d) KNO_3
40. Write a balanced equation showing how you could prepare each of the following salts by a gas-forming reaction.
(a) KNO_3
(b) $CaCl_2$
(c) $Fe(NO_3)_2$
41. Write a balanced equation showing how you could prepare each of the following salts by a gas-forming reaction.
(a) $AgNO_3$
(b) Na_3PO_4
(c) $MnSO_4$
42. Write a balanced equation showing how you could prepare each of the following compounds by an exchange reaction.
(a) $NaNO_3$ (d) $BaCl_2$
(b) $SrSO_4$ (e) $(NH_4)_2SO_4$
(c) ZnS
43. Write a balanced equation showing at least one way to prepare each of the following compounds by an exchange reaction.
(a) Na_2SO_4 (d) $MgCl_2$
(b) PbI_2 (e) $NaClO_4$
(c) CdS

CONCENTRATION OF SOLUTIONS

44. If 2.60 g of NaBr are dissolved in enough water to make 160. mL of solution, what is the molar concentration of NaBr? How many milliliters of 0.120 M NaBr would you need to supply 2.60 g of NaBr?
45. If you dissolve 1.98 g of $KMnO_4$ in enough water to make 250. mL of solution, what is the molarity of $KMnO_4$? How many milliliters of 0.0145 M $KMnO_4$ are required to supply 0.500 g of $KMnO_4$?
46. For each of the following solutions, tell how many grams of solute would be necessary for its preparation.
(a) 0.10 L of 0.10 M $AgNO_3$
(b) 5.0 mL of 0.05 M NaCN
(c) 0.10 L of 0.10 M $BaCl_2$
47. For each of the following solutions, tell how many grams of solute would be necessary for its preparation.
(a) 0.25 L of 0.050 M $Na_2C_2O_4$
(b) 0.125 L of 0.15 M $K_2Cr_2O_7$
(c) 0.50 L of 0.10 M $KHCO_3$
48. What is the molar concentration of the solute in each of the following solutions?
(a) 0.50 L containing 5.6 g of $NaClO_4$
(b) 0.10 L containing 2.3 g of KNO_3
(c) 0.25 L containing 1.5 g of C_4H_8O
49. What is the molar concentration of the solute in each of the following solutions?
(a) 50. mL containing 0.55 g of NaOH
(b) 1.55 L containing 153 g of Na_2CO_3
(c) 250. mL containing 1.256 g of $KMnO_4$
50. For each of the following solutions, give the concentration of each ion in solution.
(a) 1.0 M NaBr
(b) 0.50 M Na_3PO_4
(c) 0.10 M $ZnCl_2$
51. For each of the following solutions, give the concentration of each ion in solution.
(a) 0.10 M $KClO_4$
(b) 0.00298 M $KMnO_4$
(c) 0.385 M Na_2CO_3
52. You are given a flask containing 0.50 M $CuSO_4$, and you need 100. mL of 0.15 M $CuSO_4$. How much of the 0.50 M solution must you use to prepare the required solution?
53. If you require 500. mL of 0.33 M NaOH, how many milliliters of 1.25 M NaOH solution must you dilute with water?
54. If you dilute 50.0 mL of 0.300 M HCl with water to a total volume of 300. mL, what is the HCl concentration in the diluted solution?
55. If you dilute 250.0 mL of 0.0366 M $KMnO_4$ with water to a total volume of 2.00 L, what is the concentration of solute in the dilute solution?
56. You have 2.00 mL of 0.15 M $CuSO_4$ and add 8.00 mL of pure water. What is the concentration of the new solution? (Assume 10.0 mL final volume)
57. You have exactly 100.0 mL of 0.0584 M NaOH. After transferring the solution to a 250. mL volumetric flask, and bringing the solution volume to 250. mL, what is the concentration of the NaOH?
58. You need 2.50 L of a 0.0200 M $KMnO_4$ solution for use by all the students in your laboratory section. If you find a 1 L bottle of 0.500 M $KMnO_4$ in the stock room, tell how you would prepare the required solution.
59. As purchased, concentrated acetic acid is approximately 17.4 M in CH_3COOH. If you need 2.5 L of this acid with a concentration of 0.15 M, tell how you would make the dilute solution.

STOICHIOMETRY OF REACTIONS IN SOLUTION

60. An aqueous solution of NaCl gives $H_2(g)$, $Cl_2(g)$, and NaOH when an electrical current is passed through the solution.

$$2\ NaCl(aq) + 2\ H_2O(\ell) \longrightarrow H_2(g) + Cl_2(g) + 2\ NaOH(aq)$$

If you begin with 1.0 liter of 0.15 M NaCl, how many grams of each product can be formed?

61. H_2S is a foul-smelling compound that can be a component of polluted air. It reacts with aqueous Pb^{2+} ion to give black lead sulfide, PbS.

$$H_2S(aq) + Pb(NO_3)_2(aq) \longrightarrow PbS(s) + 2\ HNO_3(aq)$$

If you have 50. mL of 0.15 M H_2S, and add lead nitrate in excess, how many grams of PbS can be formed? How many grams of $Pb(NO_3)_2$ will be used?

62. Diborane, B_2H_6, can be produced by the following reaction.

$$2\ NaBH_4(aq) + H_2SO_4(aq) \longrightarrow 2\ H_2(g) + Na_2SO_4(aq) + B_2H_6(g)$$

How many grams of $NaBH_4$ and how many milliliters of 0.875 M H_2SO_4 should be used to prepare 4.14 g of B_2H_6? (Assume a 100% yield of B_2H_6.)

63. Hydrogen peroxide, H_2O_2, is a good oxidizing agent. However, the permanganate ion is even better in basic solution, and the following reaction occurs:

$$2\ MnO_4^-(aq) + 3\ H_2O_2(aq) \longrightarrow 2\ MnO_2(s) + 3\ O_2(g) + 2\ OH^-(aq) + 2\ H_2O(\ell)$$

If you have 125 mL of 0.0525 M $KMnO_4$, how many milliliters of 0.88 M H_2O_2 would you need to completely consume the permanganate? How many grams of MnO_2 are produced?

64. In the photographic process, silver bromide is dissolved by adding sodium thiosulfate.

$$AgBr(s) + 2\ Na_2S_2O_3(aq) \longrightarrow Na_3Ag(S_2O_3)_2(aq) + NaBr(aq)$$

If you want to dissolve 0.125 g of AgBr, how many milliliters of 0.0256 M $Na_2S_2O_3$ should you add?

Insoluble AgBr (left) can be dissolved by adding $S_2O_3^{2-}$ (right) to give water-soluble $[Ag(S_2O_3)_2]^{3-}$.

65. One way of recovering gold from its ores is to dissolve the metal using a cyanide-containing solution in the presence of oxygen.

$$4\ Au(s) + 8\ NaCN(aq) + O_2(g) + 2\ H_2O(\ell) \longrightarrow 4\ NaAu(CN)_2(aq) + 4\ NaOH(aq)$$

If you want to dissolve one troy ounce (31.1 g) of gold from a few tons of rock, how many liters of 0.55 M NaCN are required? How many grams of $NaAu(CN)_2$ could be obtained in theory?

66. Although silver chloride is insoluble in water, it can be dissolved in aqueous ammonia.

$$AgCl(s) + 2\ NH_3(aq) \longrightarrow [Ag(NH_3)_2]^+(aq) + Cl^-(aq)$$

Is it possible to dissolve 1.274 g of AgCl completely if you added 5.0 mL of 4.5 M NH_3?

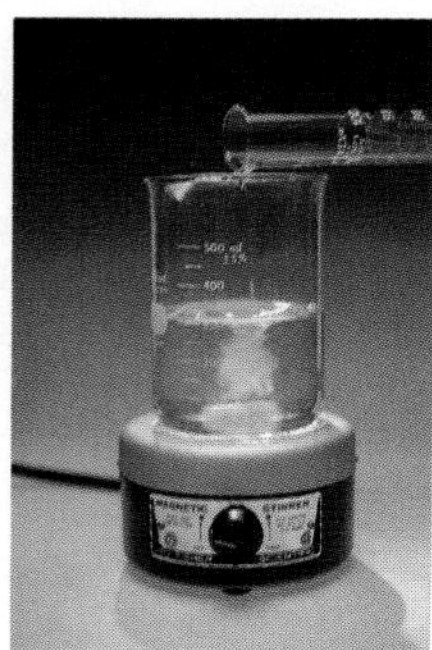

Insoluble AgCl dissolves to give the water-soluble ion $[Ag(NH_3)_2]^+$ on adding aqueous ammonia.

67. Zinc metal can be used to reduce chromium with an oxidation number of +6 to Cr^{2+}.

$$4\ Zn(s) + 14\ H_3O^+(aq) + Cr_2O_7^{2-}(aq) \longrightarrow 2\ Cr^{2+}(aq) + 4\ Zn^{2+}(aq) + 21\ H_2O(\ell)$$

If you combine 46.3 g of zinc metal with 65.0 mL of 2.45 M $K_2Cr_2O_7$ in excess acid, which reagent (zinc or potassium dichromate) will be consumed completely?

ACID–BASE TITRATIONS

68. How many milliliters of 0.250 M HCl would be required to neutralize completely 2.50 g of NaOH?

$$NaOH(aq) + HCl(aq) \longrightarrow NaCl(aq) + H_2O(\ell)$$

69. How many milliliters of 0.250 M HCl would be required to neutralize completely 36.5 mL of 0.100 M NaOH?

70. How many milliliters of 0.0540 M H_2SO_4 are required to react completely with 1.56 g of KOH?

$$2\ KOH(aq) + H_2SO_4(aq) \longrightarrow 2\ H_2O(\ell) + K_2SO_4(aq)$$

71. You dissolve 2.50 g of $Ba(OH)_2$ in enough water to make 250. mL of solution. How many moles of HCl would be required to neutralize this solution? If the HCl solution is 0.110 M, how many milliliters would be required to neutralize the $Ba(OH)_2$ solution?

72. Sodium carbonate, Na_2CO_3, is a good compound to use to standardize acid solutions.

$$Na_2CO_3(aq) + 2\ HCl(aq) \longrightarrow 2\ NaCl(aq) + H_2O(\ell) + CO_2(g)$$

If 42.43 mL of HCl solution are used to titrate 0.251 g of Na_2CO_3 to the equivalence point, what is the molar concentration of the acid?

73. Potassium acid phthalate, $KHC_8H_4O_4$, is used to standardize solutions of bases. The acidic anion reacts with bases according to the net ionic equation

$$HC_8H_4O_4^-(aq) + OH^-(aq) \longrightarrow H_2O(\ell) + C_8H_4O_4^{2-}(aq)$$

If a 0.902-g sample of potassium acid phthalate is dissolved in water and titrated to the equivalence point with 39.45 mL of NaOH, what is the molarity of the NaOH?

74. If apple cider is allowed to spoil, the result is vinegar, the distinctive ingredient of which is acetic acid, CH_3COOH. The acid reacts with NaOH according to the balanced equation

$$CH_3COOH(aq) + NaOH(aq) \longrightarrow NaCH_3COO(aq) + H_2O(\ell)$$

If 25.67. mL of 0.674 M NaOH are required to react completely with a 50.0-mL sample of vinegar, how many grams of acetic acid are in the vinegar sample?

75. A soft drink contains an unknown amount of citric acid, $C_6H_8O_7$. If 100. mL of the soft drink require 36.51 mL of 0.0102 M NaOH to neutralize completely the citric acid, how many grams of citric acid does the soft drink contain per 100. mL? What is the molar concentration of the citric acid? The reaction of citric acid and NaOH is

$$C_6H_8O_7(aq) + 3\ NaOH(aq) \longrightarrow Na_3C_6H_5O_7(aq) + 3\ H_2O(\ell)$$

76. You are given 2.036 g of a mixture of oxalic acid, $H_2C_2O_4$, and sodium chloride. If 32.58 mL of 0.541 M NaOH are required to titrate the 2.036-g sample to the equivalence point, what is the weight percentage of oxalic acid in the mixture? Oxalic acid and NaOH react according to the equation

$$H_2C_2O_4(aq) + 2\ NaOH(aq) \longrightarrow Na_2C_2O_4(aq) + 2\ H_2O(\ell)$$

77. Tartaric acid, which can be found in wine and in many fruits, has the formula $C_4H_6O_6$, and it reacts with sodium hydroxide according to the equation

$$C_4H_6O_6(aq) + 2\ NaOH(aq) \longrightarrow Na_2C_4H_4O_6(aq) + 2\ H_2O(\ell)$$

You are given 1.036 g of an impure sample of the acid. If the sample requires 25.95 mL of 0.467 M NaOH for titration to the equivalence point, what is the weight percent of tartaric acid in the sample?

ANALYSIS BY PRECIPITATION REACTIONS

78. You can perform a quantitative analysis for barium in a sample by isolating the Ba^{2+} ion in the sample in the form of insoluble $BaSO_4$ (Figure 5.21). If 8.23 g of $BaSO_4$ can be obtained from 20.0 g of a barium-containing ore, what is the weight percentage of barium in the ore?

79. The amount of calcium present in milk can be determined by adding oxalate ion, $C_2O_4^{2-}$ (in the form of its water-soluble sodium salt, $Na_2C_2O_4$); the insoluble compound CaC_2O_4 is precipitated. Suppose you take a 75.0-g sample of milk and isolate 0.288 g of CaC_2O_4 from it. What is the weight percentage of calcium in the milk?

80. We can analyze for the amount of Al^{3+} in a mixture by adding 8-hydroxyquinoline to precipitate an insoluble compound of aluminum.

$$Al^{3+}(aq) + 3\ C_9H_7NO(aq) + 3\ H_2O(\ell) \longrightarrow Al(C_9H_6NO)_3(s) + 3\ H_3O^+(aq)$$

If you isolate 0.264 g of $Al(C_9H_6NO)_3$ from 100.0 mL of an aluminum-containing solution, how many grams of Al^{3+} were in the solution? What was the molarity of the Al^{3+} ion?

8-Hydroxyquinoline is added to a solution of $Al^{3+}(aq)$ to precipitate an insoluble, yellow compound.

81. The magnets you buy in novelty stores are often made of "Alnico," a mixture of aluminum, nickel, and cobalt (hence the name Al-ni-co). You can analyze a sample of Alnico for its nickel content by dissolving a piece in nitric acid (to form Ni^{2+} in solution), and then, under suitable conditions, adding a compound called dimethylglyoxime to precipitate the nickel in the form of a red compound whose formula is $Ni(C_4H_7N_2O_2)_2$. Suppose you analyze a 0.4734-g piece of magnet, and you obtain 0.4659 g of $Ni(C_4H_7N_2O_2)_2$. What is the weight percentage of Ni in the Alnico magnet?

OXIDATION NUMBERS

82. Determine the oxidation number of each element in the following ions or compounds:
(a) BrO^- (d) IO_3^-
(b) $C_2O_4^{2-}$ (e) $HClO_4$
(c) I_2 (f) SO_4^{2-}

83. Determine the oxidation number of each element in the following ions or compounds.
(a) CH_3COOH (d) H_3AsO_4
(b) $H_2PO_4^-$ (e) $C_6H_{12}O_6$
(c) C_3H_8 (f) XeO_4^{2-}

OXIDATION–REDUCTION REACTIONS

84. In each of the following reactions, tell which reactant is oxidized and which is reduced.
(a) $2\ Mg(s) + O_2(g) \longrightarrow 2\ MgO(s)$
(b) $C_2H_4(g) + 3\ O_2(g) \longrightarrow 2\ CO_2(g) + 2\ H_2O(g)$
(c) $Si(s) + 2\ Cl_2(g) \longrightarrow SiCl_4(\ell)$

85. In each of the following reactions, tell which reactant is oxidized and which is reduced.
(a) $Ca(s) + 2\ HCl(aq) \longrightarrow CaCl_2(aq) + H_2(g)$
(b) $Cr_2O_7^{2-}(aq) + 3\ Sn^{2+}(aq) + 14\ H_3O^+(aq) \longrightarrow 2\ Cr^{3+}(aq) + 3\ Sn^{4+}(aq) + 21\ H_2O(\ell)$
(c) $FeS(s) + 3\ NO_3^-(aq) + 4\ H_3O^+(aq) \longrightarrow 3\ NO(g) + SO_4^{2-}(aq) + Fe^{3+}(aq) + 6\ H_2O(\ell)$

HALF REACTIONS

Balance the equations for the half reactions in Questions 86 through 91. Tell whether the reactant is an oxidizing or reducing agent and whether the overall process is an oxidation or reduction. Unless noted otherwise, all are carried out in acid solution, meaning that the H^+/H_2O pair may be used to balance the equation.

86. (a) $Cr(s) \longrightarrow Cr^{3+}(aq)$
(b) $Fe^{3+}(aq) \longrightarrow Fe^{2+}(aq)$
(c) $AsH_3(g) \longrightarrow As(s)$
(d) $VO_3^-(aq) \longrightarrow V^{2+}(aq)$

87. (a) $Br_2(aq) \longrightarrow Br^-(aq)$
(b) $H_2S(aq) \longrightarrow S(s)$
(c) $VO^{2+}(aq) \longrightarrow V^{3+}(aq)$
(d) $U^{4+}(aq) \longrightarrow UO_2^+(aq)$

88. (a) $Cr_2O_7^{2-}(aq) \longrightarrow Cr^{3+}(aq)$
(b) $N_2H_5^+(aq) \longrightarrow N_2(g)$
(c) $CH_3CHO(aq) \longrightarrow CH_3COOH(aq)$
(d) $Bi^{3+}(aq) \longrightarrow HBiO_3(aq)$

89. (a) $HOI(aq) \longrightarrow I^-(aq)$
(b) $NO(g) \longrightarrow HNO_2(aq)$
(c) $C_6H_5CH_3(aq) \longrightarrow C_6H_5COOH(aq)$
(d) $SO_4^{2-}(aq) \longrightarrow H_2SO_3(aq)$

90. The half reactions here are in basic solutions. You may need to use OH^- or the OH^-/H_2O pair to balance the equation.
(a) $Sn(s) \longrightarrow Sn(OH)_4^{2-}(aq)$
(b) $MnO_4^-(aq) \longrightarrow MnO_2(s)$
(c) $ClO^-(aq) \longrightarrow Cl^-(aq)$

91. The half reactions here are in basic solutions. You may need to use OH^- or the OH^-/H_2O pair to balance the equation.
(a) $CrO_2^-(aq) \longrightarrow CrO_4^{2-}(aq)$
(b) $Br_2(aq) \longrightarrow BrO_3^-(aq)$
(c) $Ni(OH)_2(s) \longrightarrow NiO_2(s)$

BALANCING OXIDATION–REDUCTION EQUATIONS

Use the half-reaction method to balance the equations in Questions 92 through 97. For reactions in acid solution you may need to use the H^+/H_2O pair to balance the equation.

92. Reaction (a) is in neutral solution. Reactions (b) through (d) are in acid solution.
(a) $Cl_2(aq) + Br^-(aq) \longrightarrow Br_2(aq) + Cl^-(aq)$
(b) $Sn(s) + H_3O^+(aq) \longrightarrow Sn^{2+}(aq) + H_2(g)$
(c) $Al(s) + Sn^{4+}(aq) \longrightarrow Al^{3+}(aq) + Sn^{2+}(aq)$
(d) $Zn(s) + VO^{2+}(aq) \longrightarrow Zn^{2+}(aq) + V^{3+}(aq)$

93. The reactions here are in acid solution.
(a) $Hg^{2+}(aq) + Cu(s) \longrightarrow Cu^{2+}(aq) + Hg(\ell)$
(b) $MnO_2(s) + Cl^-(aq) \longrightarrow Mn^{2+}(aq) + Cl_2(g)$
(c) $I^-(aq) + Br_2(aq) \longrightarrow IO_3^-(aq) + Br^-(aq)$
(d) $Zn(s) + NO_3^-(aq) \longrightarrow Zn^{2+}(aq) + NO(g)$

94. The reactions here are in acid solution.
(a) $Ag^+(aq) + HCHO(aq) \longrightarrow Ag(s) + HCOOH(aq)$
(b) $MnO_4^-(aq) + C_2H_5OH(aq) \longrightarrow Mn^{2+}(aq) + C_2H_4O(aq)$
(c) $H_2S(aq) + Cr_2O_7^{2-}(aq) \longrightarrow S(s) + Cr^{3+}(aq)$
(d) $Zn(s) + VO_3^-(aq) \longrightarrow V^{2+}(aq) + Zn^{2+}(aq)$
(e) $U^{4+}(aq) + MnO_4^-(aq) \longrightarrow Mn^{2+}(aq) + UO_2^+(aq)$

95. The reactions here are in acid solution. In (d), $C_6H_8O_6$ is vitamin C.
(a) $MnO_4^-(aq) + HSO_3^-(aq) \longrightarrow Mn^{2+}(aq) + SO_4^{2-}(aq)$
(b) $Cr_2O_7^{2-}(aq) + Fe^{2+}(aq) \longrightarrow Cr^{3+}(aq) + Fe^{3+}(aq)$

(c) $Ag(s) + NO_3^-(aq) \longrightarrow NO_2(g) + Ag^+(aq)$
(d) $Br_2(aq) + C_6H_8O_6(aq) \longrightarrow Br^-(aq) + C_6H_6O_6(aq)$
(e) $OsO_4(s) + C_4H_8(OH)_2(aq) \longrightarrow Os^{4+}(aq) + C_4H_8O_2(aq)$

96. The reactions here are in basic solution. You may need to add OH^-/H_2O to balance the equation.
(a) $Zn(s) + ClO^-(aq) \longrightarrow Zn(OH)_2(s) + Cl^-(aq)$
(b) $ClO^-(aq) + CrO_2^-(aq) \longrightarrow Cl^-(aq) + CrO_4^{2-}(aq)$
(c) The following reaction is a **disproportionation**: One substance, Br_2, functions as both the reducing agent and the oxidizing agent.

$$Br_2(aq) \longrightarrow Br^-(aq) + BrO_3^-(aq)$$

97. The reactions here are in basic solution. You may need to add OH^- or OH^-/H_2O to balance the equation.
(a) $Fe(OH)_2(s) + CrO_4^{2-}(aq) \longrightarrow Fe_2O_3(s) + Cr(OH)_4^-(aq)$
(b) $PbO_2(s) + Cl^-(aq) \longrightarrow ClO^-(aq) + Pb(OH)_3^-(aq)$
(c) $Al(s) + OH^-(aq) \longrightarrow Al(OH)_4^-(aq) + H_2(g)$
(d) $CN^-(aq) + CrO_4^{2-}(aq) \longrightarrow CNO^-(aq) + Cr(OH)_4^-(aq)$

STOICHIOMETRY: OXIDATION–REDUCTION TITRATIONS

98. A reagent commonly used to supply Fe^{2+} in aqueous solution is $Fe(NH_4)_2(SO_4)_2 \cdot 6H_2O$ ($M = 392.13$ g/mol). If you use 1.050 g of this compound, and titrate the Fe^{2+} ions in the solution to the equivalence point with 32.45 mL of aqueous $KMnO_4$, what is the molar concentration of the $KMnO_4$? (See Exercise 5.24 for the balanced equation for the reaction involved.)

99. Iodine, I_2, reacts with thiosulfate ion according to the equation

$$2\ S_2O_3^{2-}(aq) + I_2(aq) \longrightarrow S_4O_6^{2-}(aq) + 2\ I^-(aq)$$

How many milliliters of 0.0525 M $Na_2S_2O_3$ solution are needed to react with 6.50 g of I_2?

100. Oxalic acid is found in many plants and vegetables. Like vitamin C, oxalic acid is both an acid and a reducing agent. It can react with an oxidizing agent such as $KMnO_4$ according to the balanced equation

$$2\ MnO_4^-(aq) + 5\ H_2C_2O_4(aq) + 6\ H_3O^+(aq) \longrightarrow 2\ Mn^{2+}(aq) + 10\ CO_2(g) + 14\ H_2O(\ell)$$

How many milliliters of 0.0200 M $KMnO_4$ would be required to react completely with 0.892 g of oxalic acid?

101. Iron(II) reacts with the dichromate ion in acid solution rapidly and completely according to the *unbalanced* equation

$$Fe^{2+}(aq) + Cr_2O_7^{2-}(aq) \longrightarrow Fe^{3+}(aq) + Cr^{3+}(aq)$$

How many milliliters of 0.0124 M $K_2Cr_2O_7$ are required to react with 0.742 g of $Fe(NH_4)_2(SO_4)_2 \cdot 6\ H_2O$ ($M = 392.13$ g/mol), a source of the Fe^{2+} ion?

102. The iron in an ore is brought into solution as Fe^{2+} and then titrated with standardized $Ce(SO_4)_2$ according to the equation

$$Fe^{2+}(aq) + Ce^{4+}(aq) \longrightarrow Ce^{3+}(aq) + Fe^{3+}(aq)$$

Calculate the weight percent of iron in an iron-containing ore if a 15.45-g sample of the ore requires 42.34 mL of 0.133 M $Ce(SO_4)_2$ for titration to the equivalence point.

103. Oxalic acid reacts with the oxidizing agent $KMnO_4$ according to the equation in Question 100. Suppose a 15.67-g sample of green leaves that contain oxalic acid requires 23.45 mL of 0.0250 M $KMnO_4$ for titration to the equivalence point. What is the weight percent of oxalic acid in the leaves?

104. The lead content of a sample can be estimated by converting the lead to PbO_2 and dissolving the PbO_2 in an acid solution of KI. This liberates I_2 according to the *unbalanced* equation

$$PbO_2(s) + H_3O^+(aq) + I^-(aq) \longrightarrow Pb^{2+}(aq) + I_2(aq) + H_2O(\ell)$$

The liberated I_2 is then titrated with $Na_2S_2O_3$ (see Question 99), and the amount of liberated I_2 is related to the amount of lead in the sample. If 0.576 g of lead-containing mineral requires 35.23 mL of 0.0500 M $Na_2S_2O_3$ for titration of the liberated I_2, calculate the weight percent of lead in the mineral.

105. A small amount of gold metal in samples can be estimated by first converting the gold to $AuCl_3$ and then treating this with an excess of I^-. The reaction that occurs is

$$AuCl_3(aq) + 3\ KI(aq) \longrightarrow AuI(s) + I_2(aq) + 3\ KCl(aq)$$

The liberated I_2 is then titrated with $Na_2S_2O_3$ (according to the equation in Question 99), and the amount of I_2 is related to the amount of gold. If a 1.050-g sample of gold ore, which is 0.0300% Au by weight, reacts with KI, how many milliliters of 1.00×10^{-4} M $Na_2S_2O_3$ are needed to titrate the liberated I_2?

GENERAL QUESTIONS

106. Tell whether each of the following is water-soluble. If soluble, tell what ions are produced.
(a) $NiCO_3$
(b) $Zn(CH_3COO)_2$
(c) KCN
(d) $Cu(ClO_4)_2$
(e) $K_2Cr_2O_7$
(f) FeS

107. Selecting only from the list of reagents given below, show how you would prepare (a) NaCl, (b) $PbCl_2$, (c) $BaSO_4$, and (d) AgI.
The following is the list of available reagents:

KCl	$BaCl_2$	Na_2SO_4	$Pb(NO_3)_2$	$AgNO_3$
HCl	AgCl	$NaNO_3$	KI	$PbCO_3$
NH_4Cl	NaOH	PbI_2	$BaCO_3$	HNO_3

108. Complete and balance the following equations:
(a) $CaCO_3(s) + HNO_3(aq) \longrightarrow$
(b) $SrCO_3(s) + \text{heat} \longrightarrow$
(c) $K_2CO_3(aq) + H_2SO_4(aq) \longrightarrow$
(d) $CuCO_3(s) + HClO_4(aq) \longrightarrow$

109. Using the reactions of metal carbonates, give a balanced equation for the preparation of (a) NaCl, (b) barium nitrate, and (c) BaO. (See also Chapter 4.)

110. Common table sugar has the formula $C_{12}H_{22}O_{11}$. If you add one teaspoonful of sugar (3.4 g) to 250 mL of coffee, what is the molar concentration of the sugar?

111. To make 100. mL of a solution that is 0.250 M in chloride ion, how many grams of $MgCl_2$ would you need to dissolve?

112. You are a laboratory assistant for your chemistry course. There are 250 students in the course, and each needs to use 25 mL of a 0.15 M solution of copper(II) sulfate. How many liters of copper sulfate solution should you prepare? How many grams of $CuSO_4 \cdot 5\,H_2O$ must you dissolve to make this solution?

113. Hydrazine, N_2H_4, is a base like ammonia, so it can react with an acid such as sulfuric acid.

$$2\,N_2H_4(aq) + H_2SO_4(aq) \longrightarrow 2\,N_2H_5^+(aq) + SO_4^{2-}(aq)$$

If you dissolve one pound (454 g) of hydrazine in 400. liters of water, what is the molarity of the solution? How many liters of concentrated sulfuric acid (18.0 M) would be required to neutralize the hydrazine? If one bottle of concentrated acid contains 2.50 liters of acid, how many bottles of sulfuric acid would be required for neutralization?

114. You have 100.0 mL of a solution in which the concentration of Fe^{3+} is 0.025 M. You take 10.0 mL of this solution and dilute it to 100.0 mL. You then take 25.0 mL of this second solution and dilute it again to 100.0 mL. What is the concentration of Fe^{3+} in the final solution?

115. Sodium thiosulfate, $Na_2S_2O_3$, is used as a "fixer" in black and white photography. Assume you have a bottle of sodium thiosulfate and want to determine its purity. You can titrate the thiosulfate ion with I_2 according to the equation in Question 99. If you use 40.21 mL of 0.246 M I_2 in the titration, what is the weight percent of $Na_2S_2O_3$ in a 3.232-g sample of impure $Na_2S_2O_3$?

116. Balance the following equation and answer the questions that follow.

$$MnO_2(s) + I^-(aq) + H^+(aq) \longrightarrow I_2(aq) + Mn^{2+}(aq)$$

The amount of I_2 liberated in this reaction can be quantitatively determined by titrating the I_2 with thiosulfate, $S_2O_3^{2-}(aq)$ (Question 99).
(a) If you use 25.67 mL of 0.100 M $S_2O_3^{2-}$ solution in this titration, how many grams of MnO_2 must have been used to produce the I_2 titrated?
(b) In the reaction above, you used 0.100 M $S_2O_3^{2-}$. If you need 500. mL of this solution for your laboratory work, how many milliliters of 0.432 M $S_2O_3^{2-}$ would you have to dilute to make the required 500. mL?
(c) To make up 500. mL of 0.432 M $Na_2S_2O_3$, how many grams of the solid sodium salt would be required?

117. The amount of calcium in blood can be determined by adding oxalate ion to the blood sample to precipitate solid calcium oxalate according to the equation

$$Ca^{2+}(\text{in blood}) + C_2O_4^{2-}(aq) \longrightarrow CaC_2O_4(s)$$

The solid calcium oxalate is then separated from the blood, dissolved in acid, and titrated with $KMnO_4$ according to the equation in Question 100. Find the number of milligrams of Ca^{2+} in 100. mL of blood using the following information: The Ca^{2+} in a 5.00 mL sample of blood is precipitated as CaC_2O_4, and the calcium oxalate is titrated to the equivalence point with 11.63 mL of 0.00100 M $KMnO_4$.

118. The labels on chemical bottles sometimes deteriorate or fall off if the bottles are stored for a long time in a chemical stockroom. Suppose you are the laboratory assistant for your course and you find a bottle in the stockroom that says on it only "potassium bi. . .," the end of the second word having been obliterated. You know that it could be either of two acids: potassium biphthalate, $KC_8H_5O_4$, or potassium bitartrate, $KC_4H_5O_6$ (cream of tartar). Each acid forms only one mole of H_3O^+ per mole of acid on reaction with a base such as NaOH. You find by titration that 1.021 g of the unknown acid requires 24.32 mL of 0.206 M NaOH. What acid is in the bottle?

119. A 4.22-g sample of $CaCl_2$ and NaCl was dissolved in water, and the solution was treated with sodium carbonate to precipitate the calcium as $CaCO_3$. After the solid $CaCO_3$ was isolated, it was heated to drive off CO_2 and form 0.959 g of CaO. What is the weight percent of $CaCl_2$ in the original 4.22-g sample?

120. A common laboratory experiment for introductory chemistry courses involves the analysis of a mixture

of $BaCl_2$ and $BaCl_2 \cdot 2\ H_2O$. You are given a white powder that is a mixture of these compounds and are asked to determine the weight percentage of each. You find that 1.292 g of the mixture has a mass of 1.187 g after heating to drive off all of the water present. (a) What is the weight percentage of water in the mixture? (b) What is the weight percent of $BaCl_2 \cdot 2\ H_2O$ in the mixture? (c) What is the weight percent of Ba in the mixture?

121. A simple salt containing only the magnesium ion, chloride ion, and water is analyzed for all of its components. A 1.072-g sample of the compound is first heated to release all the water of hydration, and 0.503 g of a salt containing only magnesium ion and chloride ion is obtained. This 0.503-g sample is analyzed for magnesium ion by adding aqueous $(NH_4)_3PO_4$ to precipitate the magnesium as $MgNH_4PO_4$; 0.725 g of this solid is obtained. The 0.503-g sample is also analyzed for chloride ion by adding aqueous $AgNO_3$ to precipitate the chloride as AgCl; 1.514 g is isolated. What is the complete empirical formula of the hydrated salt? From your knowledge of the probable charges on magnesium and chloride ions, is this the formula you would expect?

122. The metal content of ores can sometimes be determined by oxidation–reduction titration. Suppose you have a sample that contains chromium, probably in the form of Cr_2O_3. To analyze the ore, you convert the chromium to the ion $Cr_2O_7^{2-}$, and then titrate the solution of this ion with Fe^{2+} in acid. The *unbalanced* equation for the titration is

$$Fe^{2+}(aq) + Cr_2O_7^{2-}(aq) \longrightarrow Fe^{3+}(aq) + Cr^{3+}(aq)$$

If you take a 1.70-g sample of chromium-bearing ore and find that it requires 40.2 mL of a 0.200 M Fe^{2+} solution for complete reaction, what is the weight percent of chromium in the sample?

123. A mixture of KBr and NaBr weighing 0.560 g was dissolved in water and then treated with $AgNO_3$. All the bromide ion from the original sample was recovered in the form of 0.970 g of AgBr. What is the fraction by weight of KBr in the original sample?

124. A standard method of analyzing for copper is to use the following sequence of reactions:

1. Dissolve the copper metal in nitric acid to give Cu^{2+}.
2. Reduce the Cu^{2+} with I^- to liberate I_2.

$$2\ Cu^{2+}(aq) + 4\ I^-(aq) \longrightarrow 2\ CuI(s) + I_2(aq)$$

3. Titrate the liberated I_2 with standardized $Na_2S_2O_3$ according to the equation in Question 99.

A 5-cent coin, the nickel, is actually composed mostly of copper (75.0% Cu and 25.0% Ni). How many milliliters of 0.500 M $Na_2S_2O_3$ must be used to consume the I_2 liberated by a 5-cent coin weighing 5.10 g?

125. Succinic acid, $C_4H_6O_4$, occurs in fungi and lichens, but it can be made in an industrial process. It is used in the manufacture of paints, perfumes, dyes, and so on. If 0.855 g of succinic acid reacts with 27.25 mL of 0.532 M KOH, how many moles of H_3O^+ are supplied per mole of succinic acid? Write a balanced equation for the reaction of succinic acid and KOH.

SUMMARY PROBLEMS

126. One reason for the widespread use of platinum is its relative chemical inertness. It will dissolve, however, in "aqua regia," a mixture of nitric and hydrochloric acids.

$$3\ Pt(s) + 4\ HNO_3(aq) + 18\ HCl(aq) \longrightarrow 3\ H_2PtCl_6(aq) + 4\ NO(g) + 8\ H_2O(\ell)$$

(a) If you have 12.3 g of Pt, how many grams of chloroplatinic acid, H_2PtCl_6, can be produced?

(b) How many grams of nitrogen oxide, NO, would be produced from 12.3 g of Pt?

(c) How many milliliters of 10.0 M nitric acid would be required for complete reaction with 12.3 g of Pt?

(d) If you have 10.0 g of Pt and 180. mL of 5.00 M HCl (plus excess HNO_3), which is the limiting reagent?

127. Gold can be dissolved from gold-bearing rock by treating the rock with sodium cyanide in the presence of the oxygen in air.

$$4\ Au(s) + 8\ NaCN(aq) + O_2(g) + 2\ H_2O(\ell) \longrightarrow 4\ NaAu(CN)_2(aq) + 4\ NaOH(aq)$$

Once the gold is in solution in the form of the $[Au(CN)_2]^-$ ion, it can be precipitated as the metal according to the following *unbalanced* equation:

$$[Au(CN)_2]^-(aq) + Zn(s) \longrightarrow Zn^{2+}(aq) + Au(s) + CN^-(aq)$$

(a) If you have exactly one metric ton (1 metric ton = 1000 kg) of gold-bearing rock, how many liters of 0.075 M NaCN will you need to use to extract the gold if the rock is 0.019% gold?

(b) How many kilograms of metallic zinc will you need to use to recover the gold from the $[Au(CN)_2]^-$ obtained from the gold in the rock?

(c) If the gold is recovered completely from the rock, and the metal is made into a cylindrical rod 15.0 cm long, what is the diameter of the rod? (The density of gold is 19.3 g/cm^3.)

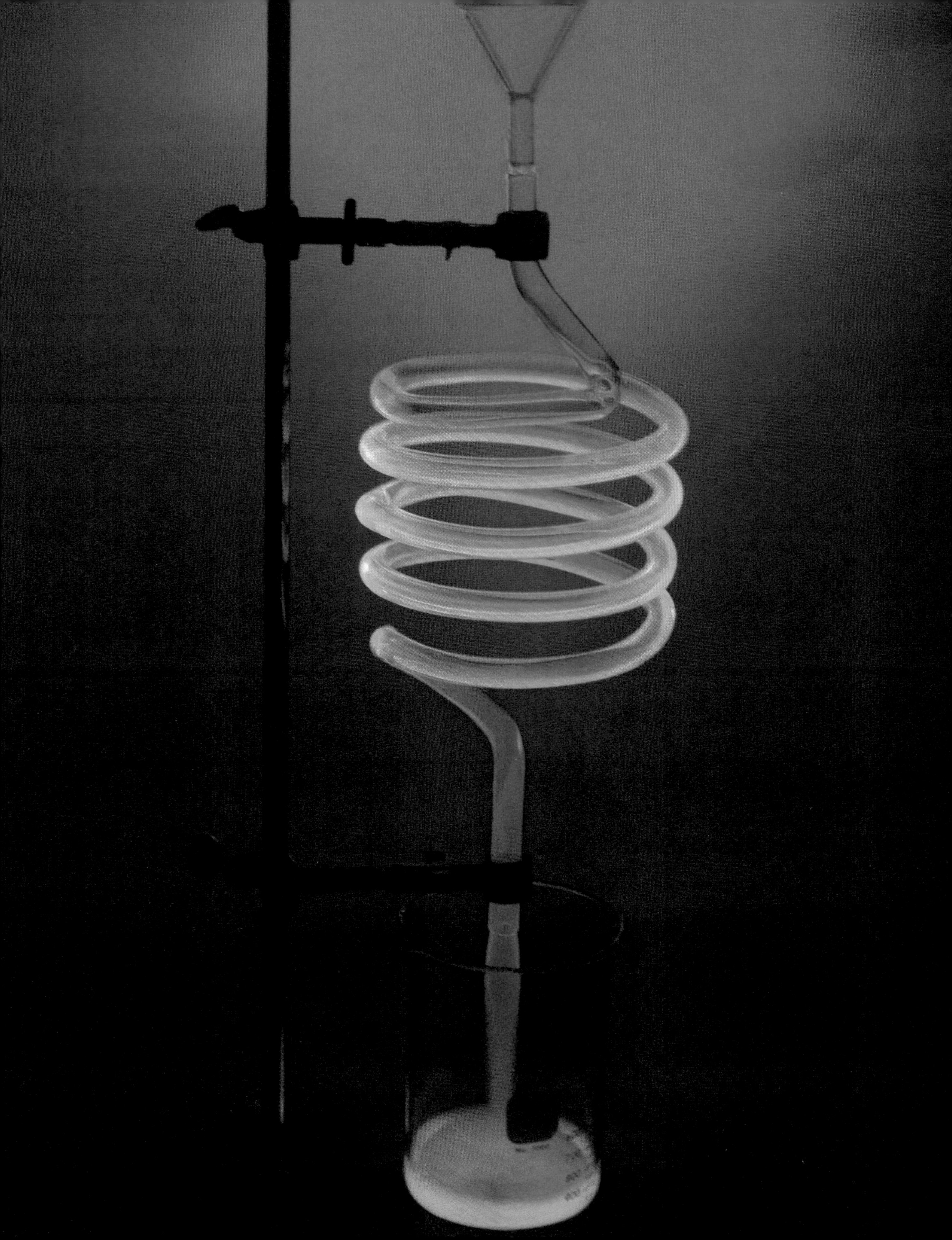

CHAPTER 6

Thermochemistry

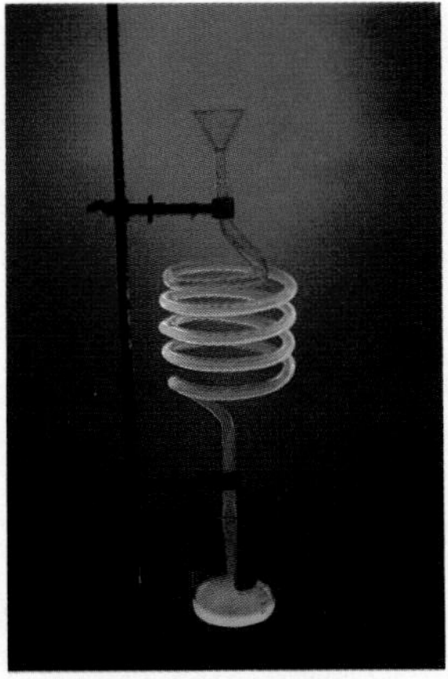

Chemiluminescence, the production of light energy by a chemical reaction.

CHAPTER OUTLINE

Energy! We think so much about its sources and uses. Where do we get the energy to move our modern society and how can we use it more efficiently? How can we provide more energy to developing nations? Are there new sources of energy that are possible now or in the future? These and other questions are important to all of us and are central to much of the chemistry that follows. Here we are beginning to build a framework in which these questions can be considered and to focus on the specific question of energy from chemical processes.

6.1 ENERGY AND ITS FORMS

Energy is usually defined as the capacity to do work or to transfer heat. **Work** is most often defined as the product of the magnitude of a force (f) times the distance (d) over which the force acts,

$$\text{Work} = w = \text{force} \times \text{distance} = f \times d \tag{6.1}$$

You experience this relationship all the time. If you climb a mountain (Figure 6.1), you apply a force against the force of gravity to carry yourself and your equipment up the side of the mountain. Of course, you can do this work because you have the energy or capacity to do so; that energy is provided by the food you have eaten. This chemical source is only one of many forms that energy may take, some of which are relevant to us as chemists, and we will discuss them briefly.

Figure 6.1 Work! Carrying you and your gear up a rock wall involves a force acting over a distance. (© Galen Rowell 1986, FPG International)

Potential energy is possessed by an object by virtue of its position, and is associated with attractions and repulsions. If you stand on a diving board, you have considerable potential energy because of your height above the water. If you jump off the board, your initial potential energy is converted progressively into **kinetic energy** (Figure 6.2). A body has kinetic energy because of its motion. This form of energy depends on the mass (m) and velocity (v) of the moving object. That is,

$$\text{Kinetic energy} = \text{KE} = \tfrac{1}{2}\, mv^2 \tag{6.2}$$

During the dive your mass is constant, while the force of gravity accelerates your body to move faster and faster as you fall, so your velocity and kinetic energy increase, and this happens at the expense of potential energy.

At the moment you hit the water, your velocity is abruptly reduced, and much of your kinetic energy is lost. However, because of the **principle of the conservation of energy**, this kinetic energy must go somewhere. Your potential energy is converted to kinetic energy throughout your fall toward the water, and the kinetic energy is converted to work (and heat) on impact with the water, since you rearrange the water as your body passes through.

Heat is a form of energy that is often associated with changing the temperature of an object. An object's temperature increases because energy is transferred to it. The energy transfer, in the form of heat, happens when two objects of different temperature are brought into contact, as nature attempts to equalize their temperatures, that is, to establish thermal equilibrium. This chapter is about heat transfer and its connection with chemical processes.

Thermal energy is a synonym for heat. Thermal energy is usually associated with the motions of atoms or molecules in a solid, liquid, or

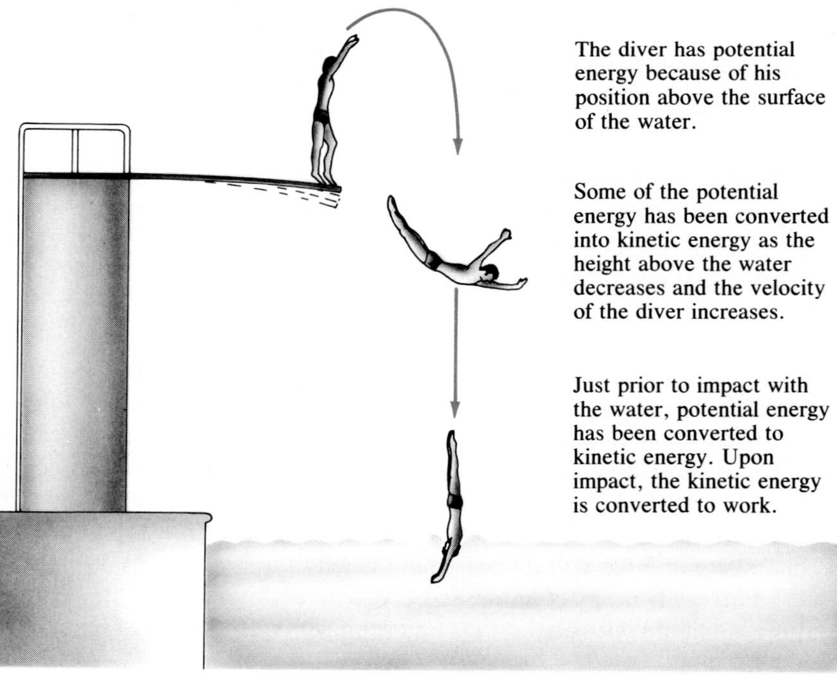

The diver has potential energy because of his position above the surface of the water.
Some of the potential energy has been converted into kinetic energy as the height above the water decreases and the velocity of the diver increases.
Just prior to impact with the water, potential energy has been converted to kinetic energy. Upon impact, the kinetic energy is converted to work.

Figure 6.2 The interconversion of potential energy, kinetic energy, and work. If you stand on a diving board, you have potential energy because of your position relative to the surface of the water below. If you dive or jump off the board, the potential energy is converted progressively to kinetic energy as your height above the water decreases and your velocity increases. Just prior to impact with the water, your potential energy has been converted fully to kinetic energy. Upon impact, the kinetic energy is dissipated by its conversion to heat and to performing work pushing aside the water.

(a)

(b)

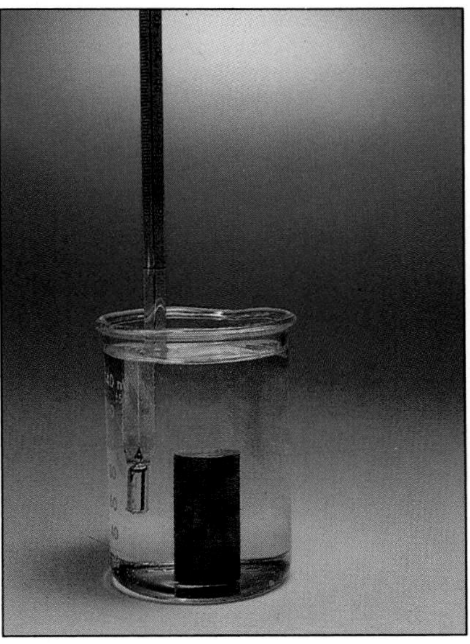

(c)

Figure 6.3 Water in a beaker is warmed when we plunge in a hotter object. Heat is transferred from the hotter metal bar to the cooler water (and enough heat is transferred that the water immediately around the hot piece of metal boils). Eventually the bar and the water reach the same temperature (and have achieved thermal equilibrium).

Heat, Hotness, and Thermometers

The piece of iron being heated in a flame and the beaker of cold water (Figure 6.3, left photo) are two objects with a different "hotness." When we plunge the hot bar into the cold water (middle photo), heat is transferred from the iron bar to the water until the two are at the same temperature. In fact, the amount of heat released by the bar, which is equal to the amount absorbed by the water, may be enough to make the water boil, at least in the region immediately around the bar.

The "hotness" of an object is measured by its temperature, which we commonly determine with a thermometer. Assume we are using a simple thermometer containing liquid mercury (right photo). Here the thermometer was originally at room temperature, and the room had a different degree of "hotness" than the warm water at the end of our experiment. When the thermometer was placed in the warm water, heat was transferred from the water to the thermometer. The liquid mercury in the thermometer expands (nearly all substances expand on heating) to fill even more of the tube. You see the result as a column of mercury moving "higher and higher" in the tube.

gas. The more rapid these motions, the greater the thermal energy of the material. One measure of this thermal energy content can be made with a thermometer: the higher the temperature, the greater the motions of the atoms or molecules in the material. But don't get the idea that thermal energy and temperature are the same thing. The thermal energy content of a hot cup of coffee is much less than that of a bathtub full of hot water at the same temperature. A cup of hot coffee may provide only enough energy to melt one ice cube before coming to room temperature, but the tub of hot water could melt a whole icebox full of ice before it comes to room temperature.

Figure 6.4 Radiant energy. "Light sticks" provide light or radiant energy from a chemical reaction.

Radiant and **electrical energy** are two additional forms of energy. Radiant energy is the type supplied by a radiator in a room or by the sun to the earth (Figure 6.4). Electrical energy can be generated by mechanical devices or chemical reactions. Chemical sources are currently of considerable interest in the replacement of oil- and coal-driven mechanical generators, and we shall discuss this aspect of chemistry, electrochemistry, in Chapter 21.

Chemical reactions and other changes consume or release energy in the forms of heat (thermal energy), light (radiant energy), or electrical current (electrical energy). In this chapter we want to present the basic ideas of **thermochemistry**, the science of heat or thermal energy transfer and its application to chemical reactions and processes.

6.2 UNITS OF ENERGY

1 kilocalorie = 1 kcal = 1000 calories

All forms of energy can be converted into heat, so some of the units in which energy is measured are really those that measure heat. A **calorie** is the amount of heat required to raise the temperature of 1.00 g of pure liquid water by 1.00 degree Celsius, from 14.5 °C to 15.5 °C. This is a very small quantity of heat, so the **kilocalorie** is often used. The kilocalorie (abbreviated kcal) is equivalent to 1000 calories.

Most of us were raised to think in calories, probably because we hear about dieting or read breakfast cereal boxes. However, the "calorie" used with regard to food is Calorie with a capital C, a unit equivalent to the kilocalorie. Thus, a breakfast cereal that gives you 100 Calories really provides you with 100 kcal or 100×10^3 calories (calories with a small c).

In this book we shall use the **joule**, the SI unit of energy. **One calorie is equivalent to 4.184 joules.** The joule is used because it is derived directly from the units used in the calculation of mechanical energy (i.e., potential and kinetic energy). If a 2.0-kg object (about 4 pounds) is moving with a velocity of 1.0 meter per second (roughly 2 mph), the kinetic energy is

1 calorie = 4.184 joules

The "joule" is named for James P. Joule (1818–1889). Joule was a student of John Dalton (Chapter 2) and the son of a brewer in Manchester, England.

$$\text{Kinetic energy} = \tfrac{1}{2}\,mv^2 = \tfrac{1}{2}\,(2.0\ \text{kg})\left(\frac{1.0\ \text{m}}{\text{s}}\right)^2$$

$$= 1.0\,\frac{\text{kg}\cdot\text{m}^2}{\text{s}^2} = 1.0\ \text{joule}$$

To give you some feeling for joules, if you drop a six-pack of soft drink cans on your foot, the kinetic energy at the moment of impact is about a calorie or two, that is, about 4 to 10 joules.

EXAMPLE 6.1

ENERGY UNIT CONVERSIONS

A baseball weighing 114 g thrown at a speed of 100. miles per hour has a kinetic energy of 27.2 cal. What is this energy in units of joules?

Solution To find the energy in joules, we only need to use the relation 1.00 cal = 4.184 J.

$$27.2 \text{ cal}\left(\frac{4.184 \text{ J}}{1.00 \text{ cal}}\right) = 114 \text{ J}$$

EXERCISE 6.1 Energy Unit Conversions
A hot dog provides 160 Calories. What is this energy in joules?

6.3 HEAT CAPACITY AND SPECIFIC HEAT

HEAT AND TEMPERATURE CHANGE

In the 1770s, Joseph Black of the University of Glasgow (Scotland) studied the nature of heat. It was Black who first clearly distinguished between the "hotness" or temperature of an object and its heat capacity. The **heat capacity** of an object is the amount of heat that is required to change the temperature of that object by 1 degree. In chemistry we commonly express heat capacity in two ways, *specific heat capacity* and *molar heat capacity*, and each is defined and used below.

Heat capacity

$$= \frac{\textit{quantity of heat transferred}}{\textit{degrees of temperature change}}$$

Specific heat capacity, which is usually just called **specific heat**, is the amount of heat energy needed to change the temperature of 1.00 g of a substance by 1.00 degree Celsius (or 1.00 kelvin). This is a characteristic property of each substance. For water, the specific heat is 1.00 calorie/g · K or 4.184 joules/g · K, whereas it is only about 0.8 J/g · K for common glass. That is, it takes about five times more heat (4/0.8 = 5) to raise the temperature of a gram of water by a certain amount than a gram of glass.

Recall that 1 degree on both the Celsius and Kelvin scales has the same magnitude.

The specific heat of a substance can be determined experimentally by determining the quantity of heat transferred by a definite mass of the substance as its temperature rises or falls.

$$\text{Specific heat} = \frac{\text{quantity of heat transferred}}{\text{(g of material)(degrees of temperature change)}} \quad (6.3)$$

As an example, it has been found that 60.5 joules are required to change the temperature of 25.0 g of ethylene glycol (a compound used as antifreeze in automobile engines) by 1.00 °C (or 1.00 K). This means the specific heat of the compound is

$$\text{Specific heat of ethylene glycol} = \frac{60.5 \text{ joules}}{(25.0 \text{ g})(1.00 \text{ K})} = 2.42 \frac{\text{J}}{\text{g} \cdot \text{K}}$$

Table 6.1 Specific Heat Values for Some Elements, Compounds, and Common Solids

Substance	Name	Specific Heat (J/g · K)
Elements		
Al	Aluminum	0.902
C	Graphite	0.720
Cu	Copper	0.385
Au	Gold	0.128
Fe	Iron	0.451
Compounds		
$NH_3(\ell)$	Ammonia	4.70
$CCl_4(\ell)$	Carbon tetrachloride	0.861
$C_2H_5OH(\ell)$	Ethyl alcohol	2.46
$(CH_2OH)_2(\ell)$	Ethylene glycol (antifreeze)	2.42
H_2O	Liquid water	4.18
	Ice	2.06
$CCl_2F_2(g)$	Freon-12	0.598
Common Solids		
Common steel		0.45
Cement		0.88
Glass		0.84
Granite		0.79
Wood		1.76

The specific heat has been determined for many substances, and a few are listed in Table 6.1. You will notice that water has a high specific heat, and it is in fact one of the highest known. This is important to all of us, since a high specific heat means that many joules must be absorbed by a large body of water to raise its temperature just a degree or so. Conversely, many joules must be lost before the temperature of the water drops by more than a degree. Thus, a body of water can "store" an enormous quantity of energy. This means that oceans and lakes have a profound moderating influence on our weather.

Knowing the specific heat of a substance enables you to calculate the heat transferred in some thermal process by measuring the mass of the substance and the size of its temperature change. Conversely, you can calculate the temperature change that should occur when a given amount of heat is transferred. Equation 6.3 allows you to do this, but it may be more convenient to rewrite it in the following form:

$$\text{Heat transferred} = (\text{specific heat})(\text{mass})(\text{size of temperature change of the substance}) \quad (6.4)$$

$$q = (\text{J/g} \cdot \text{K})(\text{mass in g})(\Delta T \text{ in kelvins})$$

The symbol q is used to designate the amount of heat transferred between objects. The symbol Δ (the Greek letter delta) means "change in."

where the *quantity of heat* is symbolized by q.

To this point, we have emphasized only the *quantity* of heat transferred and not its *direction*. However, Equation 6.4 can tell us not only the quantity but also the direction of heat transfer. When ΔT is calculated as

$$\Delta T = \text{final temperature} - \text{initial temperature} \quad (6.5)$$

ΔT will have an algebraic sign: positive (+) for T increase and negative (−) for T decrease. Therefore, if the temperature of the substance increases, for example, ΔT has a + (positive) sign *and so will q*. This means heat is transferred *into* the substance. The opposite case, a decrease in temperature of the substance, means that ΔT is − (negative) and so is q; heat was transferred *out* of the substance.

EXAMPLE 6.2

USE OF SPECIFIC HEAT

How many joules of energy must be transferred from a cup of coffee to your body if the temperature of the coffee drops from 60.0 °C to 37.0 °C (normal body temperature)? Assume that the cup holds 250. mL (with a density of 1.00 g/mL, the coffee has a mass of 250. g), and that the specific heat of coffee is the same as that of water.

Solution The amount of heat transferred can be obtained from Equation 6.4

$$\text{Heat transferred} = q = \left(\frac{4.184\ \text{J}}{\text{g} \cdot \text{K}}\right)(250.\ \text{g})(\underset{\text{final temp}}{37.0\ °\text{C}} - \underset{\text{initial temp}}{60.0\ °\text{C}})$$

$$= -24.1 \times 10^3\ \text{J} = -24.1\ \text{kJ}$$

Notice that the final answer has a *negative* value. This denotes that *energy is transferred from the coffee as the temperature of the coffee declines.*

EXAMPLE 6.3

HEAT TRANSFERRED BY WATER ON COOLING

A lake that is 1.0 mile square and has a depth of 10. feet contains 7.9×10^9 liters of water. How many joules of energy must be transferred to the lake to raise the temperature by 1.0 degree Celsius? Assume the density of the lake water is 1.0 g/mL and that the specific heat of lake water is the same as that of pure water.

Solution Since the specific heat equation uses units of grams for mass, you must first convert the volume to grams.

$$7.9 \times 10^9\ \cancel{\text{L}} \left(\frac{1000\ \cancel{\text{mL}}}{1.0\ \cancel{\text{L}}}\right)\left(\frac{1.0\ \text{g}}{\cancel{\text{mL}}}\right) = 7.9 \times 10^{12}\ \text{g}$$

Knowing the mass of the lake being studied, you can now calculate the heat absorbed by the lake.

$$\text{Heat transferred} = \left(\frac{4.184\ \text{J}}{\cancel{\text{g}} \cdot \text{K}}\right)(7.9 \times 10^{12}\ \cancel{\text{g}})(+1.0\ \cancel{°\text{C}})$$

Notice the temperature change is positive because the final T is higher than the initial T.

$$\text{Heat transferred} = +3.3 \times 10^{13}\ \text{J} = +3.3 \times 10^{10}\ \text{kJ}$$

Notice that the final answer has a *positive* sign, indicating that heat is transferred *into* the substance being studied, the lake.

EXERCISE 6.2 Using Specific Heat

The cup of coffee in Example 6.2 lost 24.1 kJ when cooled from 60.0 °C to 37.0 °C. If this same amount of heat is used to warm a piece of aluminum weighing 250. g, what would the final temperature of the aluminum be if its initial temperature is 37.0 °C? (The specific heat of aluminum is 0.902 J/g · K.)

One of the important aspects of thermochemistry for you to understand is that the direction of heat transfer is revealed by the direction of temperature change of the substance. This is the idea behind the preceding examples and exercise. To understand this better, now that you have been introduced to specific heat, it is useful to study an example on the direction of heat transfer. Be sure to pay attention to the interpretation of algebraic signs for the quantities in thermochemistry, as illustrated in the following example and as explained further in the box on page 205.

EXAMPLE 6.4

HEAT TRANSFER BETWEEN SUBSTANCES

Suppose you heat a 55.0-g piece of iron in the flame of a Bunsen burner to 425 °C and then you plunge it into a beaker of water. (See Figure 6.3.) The beaker holds 600. mL of water (density = 1.00 g/mL), and its temperature before you drop in the hot iron is 25.0 °C. What is the *final* temperature of the water and the piece of iron? (Assume that no heat transfers through the walls of the beaker or to the atmosphere.)

Solution The most important things to understand in this problem are that (1) the water and the iron bar will end up at the same temperature (T_{final} is the same for both), and (2) because of the principle of the conservation of energy, the amount of heat absorbed by the water in warming up and the amount given up by the iron in cooling down are equal in quantity but *not* in algebraic sign: q_{iron} has a negative value because its temperature drops as heat is transferred *out* of the iron. Conversely, q_{water} has a positive value because its temperature increases as heat is transferred *into* the water. Thus,

$$q_{\text{iron}} = -q_{\text{water}}$$

Using specific heats of iron and water from Table 6.1, we have

$$\underbrace{(55.0\text{ g})(0.451\text{ J/g}\cdot\text{K})(T_{\text{final}} - 425\text{ °C})}_{q_{\text{iron}}} = \underbrace{-(600.\text{ g})(4.184\text{ J/g}\cdot\text{K})(T_{\text{final}} - 25\text{ °C})}_{-q_{\text{water}}}$$

where T_{initial} for the iron is 425 °C and T_{initial} for the water is 25 °C. Solving, we find $T_{\text{final}} = 29$ °C. The iron has indeed cooled down ($\Delta T_{\text{iron}} = -396$ °C), and the water has warmed up ($\Delta T_{\text{water}} = 4$ °C) to the same final temperature (T_{final}), a temperature between the two initial T's.

One final comment: Don't be confused by the fact that the transfer of about 10 kJ has caused different ΔT's; this is because the specific heats of iron and water are so different.

Sign Conventions in Energy Calculations

Any time you take the difference between two quantities in thermochemistry, you should *always subtract the initial quantity from the final quantity*. This is the meaning of the delta in ΔT and similar quantities used throughout the book. A natural consequence of this convention is that the algebraic sign of the calculated result indicates an increase (+) or a decrease (−) in the quantity for the substance being studied. This is an important point, as you will see in the next section and in other chapters of this book.

Thus far, we have described temperature changes and the direction of heat transfer. The table below summarizes the conventions used:

CHANGE IN T OF OBJECT	SIGN OF ΔT	SIGN OF q	DIRECTION OF HEAT TRANSFER
Increase	+	+	Heat transferred into object
Decrease	−	−	Heat transferred out of object

Be sure to understand that the sign of q is just a "signal" to tell the direction of heat transfer. Heat itself cannot be negative; it is simply a quantity of energy. As an example, consider your bank account. Assume you have \$26 in your account ($A_{initial}$), and, after a withdrawal, you have \$20 (A_{final}). The cash flow is thus

$$\text{Cash flow} = \Delta A = A_{final} - A_{initial} = \$20 - \$26 = -\$6$$

The negative sign on the \$6 "signals" you that you have made a withdrawal; the cash itself is not a negative quantity. Thus, when we talk about "heat" we shall just use an unsigned number. However, when we want to signal the direction of transfer in a process, we shall attach a negative sign (heat transferred *from* the substance) or a positive sign (heat transferred *into* the substance) to the value of q.

EXERCISE 6.3 Heat Transfer

You want to cool down a cup of coffee, and you do so by dropping in a cold block of aluminum. The aluminum block has a mass of 250. g, and you want to cool 250. g of coffee (with a specific heat of 4.20 J/g · K) from 60. °C to 45 °C. To accomplish this, what must the initial temperature of the aluminum block be?

Finally, an alternative way to express the capacity of a substance to absorb heat is to give its **molar heat capacity**. This measure uses 1 mole as the unit of substance rather than 1 gram.

$$\text{Molar heat capacity (J/mol} \cdot \text{K)} = \text{specific heat (J/g} \cdot \text{K)} \times \text{molar mass (g/mol)} \quad (6.6)$$

Thus, the specific heat of water is 4.184 J/g · K, while its molar heat capacity is 75.3 J/mol · K, or 18.0 times the specific heat.

HEAT ASSOCIATED WITH PHASE CHANGES

Later in the book we will describe, at the molecular level, changes in phase—for example, the change of solid water (ice) to liquid water. However, now that we are discussing heat, it is useful to define the heat

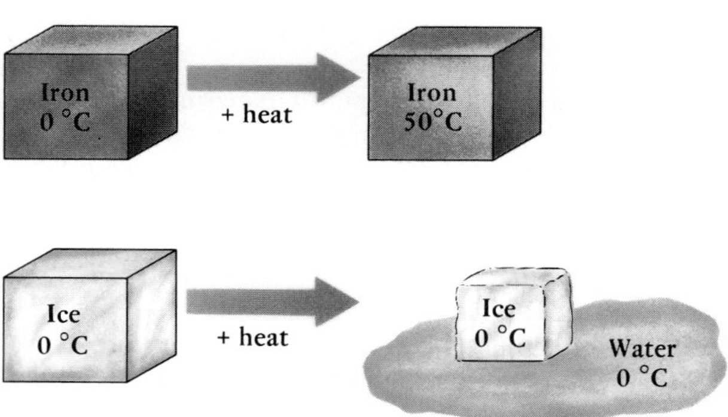

Figure 6.5 Heat transfer that can lead to a temperature change, for example in a block of iron, is sometimes called *sensible heat*. On the other hand, *latent heat* affects something without changing its temperature. In this case, the heat that melts a block of ice is called the *latent heat of fusion*. If you have a 1 pound block of iron, the same amount of heat required to raise its temperature by 50 °C will cause only 30.7 g of ice to melt, and not change the temperature of the final ice/water mixture at all!

Heat transfer does not always result in a temperature change. Latent heat refers to the heat transferred in processes for which heat is transferred but T is constant.

involved in such changes. Joseph Black, whom we mentioned at the beginning of this section, recognized that there is a difference between *sensible heat*, heat that is associated with temperature changes, and *latent heat*, heat that is associated with a change in phase of a substance while the temperature remains constant (Figure 6.5). The amount of heat required to melt ice at 0 °C, which is 333 J/g, is called the **latent heat of fusion** of ice. Similarly, the heat required to vaporize water at 100 °C, which is 2260 J/g, is called the **latent heat of vaporization**.

EXAMPLE 6.5

LATENT HEAT

You have four ice cubes at 0 °C. If each cube has a volume of 1.00 in.^{3}, how much heat is required to melt these ice cubes to liquid water at 0 °C? (The density of ice at 0 °C is 0.917 g/cm^3, and the latent heat of fusion of ice is 333 J/g.)

Solution First, let us calculate the mass of the ice cubes.

$$4\ \cancel{\text{cubes}}\left(\frac{1.00\ \cancel{\text{in.}^3}}{\cancel{\text{cube}}}\right)\left(\frac{2.54\ \cancel{\text{cm}}}{\cancel{\text{in.}}}\right)^3\left(\frac{0.917\ \text{g}}{\cancel{\text{cm}^3}}\right) = 60.1\ \text{g}$$

Now that the mass of cubes is known, the amount of heat required to melt them to liquid water at 0 °C can be found from the latent heat of fusion of ice.

$$60.1\ \cancel{\text{g ice}}\left(\frac{333\ \text{J}}{\cancel{\text{g ice}}}\right) = 2.00 \times 10^4\ \text{J}$$

EXERCISE 6.4 Latent Heat

Assume you have 1 cup of ice (237 mL) at 0.0 °C. How much heat is required to melt the ice and then warm the resulting water to 25.0 °C?

6.4 ENERGY CHANGES IN CHEMICAL PROCESSES: ENTHALPY

The double arrows ⇄ in the $CO_2(s)/CO_2(g)$ equation are meant to convey the notion that chemical processes are reversible—solid can be converted to a gas or a gas to a solid.

If carbon dioxide gas is cooled to −78 °C, the compound can be obtained in a solid form known by the trademark name Dry Ice (Figure 6.6). The process is reversible in that CO_2 molecules in Dry Ice can return to the gas phase by absorbing energy.

$$CO_2(\text{solid}, -78\ °C) + \text{heat} \rightleftharpoons CO_2(\text{gas}, -78\ °C)$$

(a)

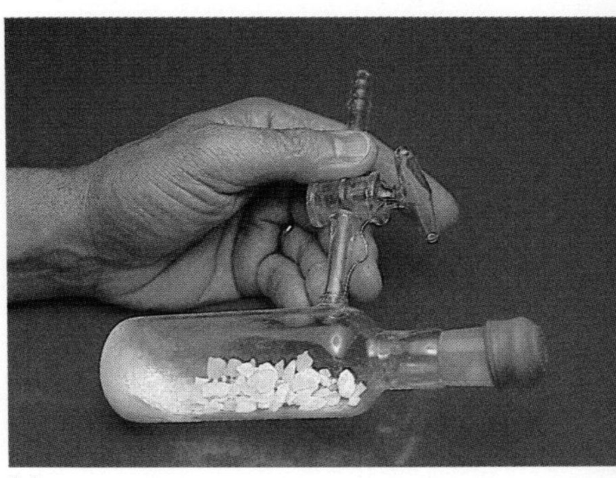

(b)

Figure 6.6 A system absorbing heat at constant pressure or constant volume. (a) When heat is absorbed by Dry Ice (solid CO_2) at -78 °C, the solid is transformed to a gas at -78 °C. The pressure of the system is constant, so the heat absorbed at constant pressure (q_p) is identified as the enthalpy change, ΔH. Work is done by the system on the surroundings by pushing back the atmosphere. (b) When heat is absorbed by solid CO_2 in a closed container, the pressure in the container increases as solid is converted to gas. However, since there is no "mechanical" connection between the system and surroundings, no work is done on the surroundings by the system. Therefore, q_v = heat absorbed at constant volume = ΔE.

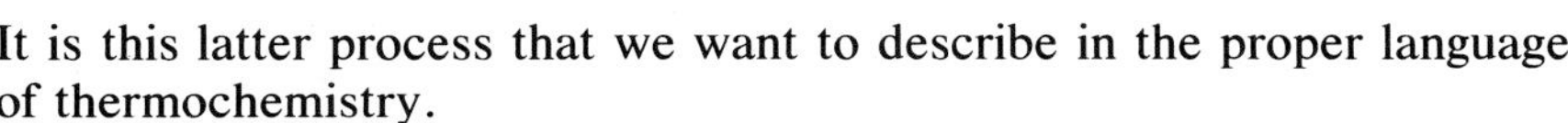

It is this latter process that we want to describe in the proper language of thermochemistry.

First, we need to extend our earlier discussion of heat as energy transferred between objects. In the language of thermodynamics, one of the objects is usually of primary concern, and it is called the **system**. The other object is called the **surroundings**. Here the CO_2 sample is called the *system*, and this system interacts with its *surroundings*, the atmosphere of the room surrounding the sample. In general, the material being studied is the system, and the rest of the universe is the surroundings. A system may be contained within an actual physical boundary, such as a beaker or a cell in your body. Alternatively, the boundary may be purely imaginary; for example, you could study the solar system within its surroundings, the rest of the galaxy. In any case, the system can always be defined precisely.

In the process CO_2(s, -78 °C) → CO_2(g, $-$ 78 °C), heat is transferred to the system from the surroundings; this energy causes the regular array of molecules making up the solid to be disrupted and converted into dispersed CO_2 molecules in the gas phase. When heat is transferred into a system, the process is said to be **endothermic**. In contrast, when CO_2 gas is converted to solid CO_2, this is an **exothermic** process because heat must be transferred out of the gas to cause it to condense to a solid. Figure 6.7 is a photograph of a hand warmer, which employs an exothermic chemical process to produce heat.

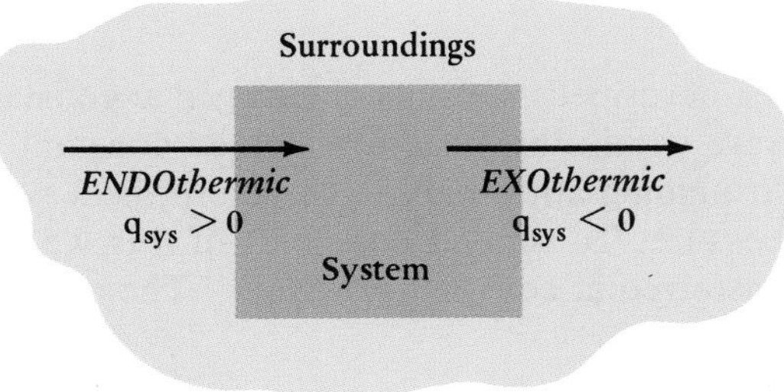

Direction of Heat Flow
System ⟶ Surroundings
$q_{sys} < 0$, Exothermic (exo = "out")

Surroundings ⟶ System
$q_{sys} > 0$, Endothermic (endo = "in")

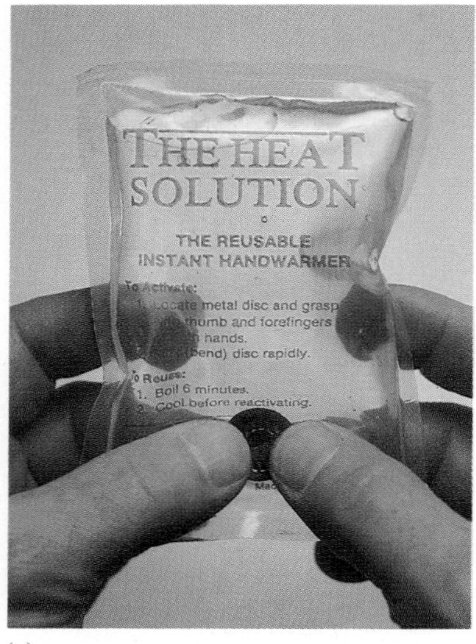

(a)

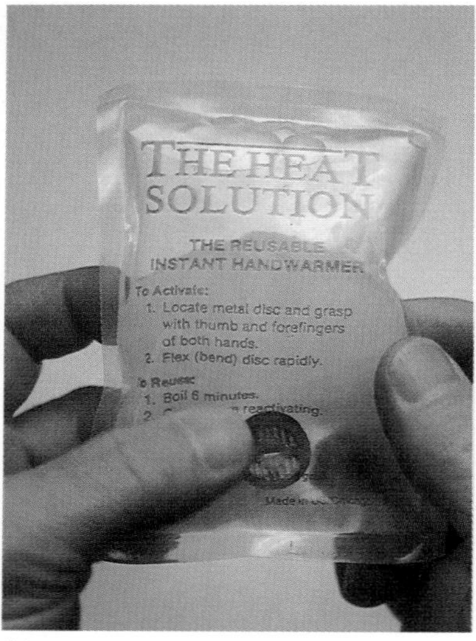

(b)

Figure 6.7 The hand warmer produces heat by the crystallization of sodium acetate. The salt is initially dissolved in water, but you can cause the salt to come out of solution (b), and it gives off a large amount of heat in the process. That is, the crystallization is an exothermic process, and you sense this because heat is transferred from the plastic bag containing the salt (the system) to your hand (the surroundings).

For the process $CO_2(s, -78\,°C) \rightarrow CO_2(g, -78\,°C)$, the heat transferred has two effects: (a) it provides the energy necessary to overcome the forces holding the molecules together in the solid state at −78 °C, thus allowing them to break away from one another; and (b) it allows the gas to expand to a larger volume (if the pressure is kept constant). If, as shown in Figure 6.6a, the CO_2 is not held in a rigid container, the volume of the system increases, but it can do so only if the system pushes aside the atmosphere. Thus, *the system does work on its surroundings*. The energy associated with step (a) reflects a change in the *internal energy* within the system; we represent it by ΔE, which stands for "change in internal energy." Step (b) is recognized as work and is denoted by w.

Recalling that heat and work are just two forms of the same concept, energy, the change in energy within the system must be the *sum* of the heat and work transferred to or from the system:

$$\Delta E = q + w \tag{6.7}$$

This is a mathematical statement of the **first law of thermodynamics**, the law of energy conservation. All the energy transferred between a system and its surroundings must be accounted for by the heat and work transferred between them. Notice that w represents energy transferred, just like q, so it has a direction. The work energy w has a negative value if the system expends energy as work, and w has a positive value if the system receives energy in the form of work.

The CO_2 system loses some of its energy to the surroundings because, in expanding, it exerts a force on the atmosphere and pushes it back. The energy that allows the system to perform work on its surroundings had to come from someplace, and that "someplace" was the heat energy transferred to the system from the surroundings; the system simply converted some of this heat energy into work energy. Therefore, the first law of thermodynamics tells us that the internal energy change, ΔE, for the CO_2 system must be less than q, since heat is transferred into the system from the surroundings (q has a positive value) and work is done by the system on the surroundings (w has a negative value). That is,

$$\Delta E = q + w = (q, \text{ positive number}) + (w, \text{ negative number})$$

and so

$$\Delta E < q$$

In the process just described, the system changed at constant pressure; that is, the pressure within the system remained constant and the same as that of the surrounding atmosphere. Therefore, heat was transferred at a constant pressure. To remind ourselves of this we write q_p, which means "heat transferred at constant pressure." The alternative to

The First Law of Thermodynamics

To develop your ideas about the first law of thermodynamics more fully, consider the two situations in Figure 6.6 in more detail. In Figure 6.6b there can be *no work* done by the system. The glass flask cannot expand or contract, so it is not connected mechanically to the surroundings, and w can only be 0. In this case, Equation 6.7 tells us what we know intuitively: When the system volume is constant, *all* the thermal energy transferred to the system (q) goes to increase the internal energy of the system, and so $\Delta E = q_v$, where the subscript v signals this special case.

$$\Delta E = q_v$$

The situation in Figure 6.6a permits the system to convert some of the thermal energy to work by pushing against the surrounding atmosphere; w is no longer 0. This means the internal energy of the system is determined by what is left of the thermal energy transfer after the system converted some of that energy to work.

An analogy would be the following: Suppose your parents pay you \$20 to mow the lawn, but you pay \$10 to the kid next door to do it. You have converted part of your parent's input, money, into labor by another person, and you have a balance of \$10. In the case of the Dry Ice, some of the energy input (q_p, the heat transferred at constant pressure, is a positive number) was converted into work output (w, a negative number). The balance, ΔE, is therefore *less* than q.

$$\Delta E = q + w = (q_p, \text{ positive number}) + (w, \text{ negative number})$$

$$\Delta E < q_p$$

Now compare the constant volume and constant pressure cases. For the same heat input (q) in the two cases, the internal energy increase (ΔE) for the CO_2 is *smaller* in the constant pressure case than for the constant volume case. To extend the lawn-mowing analogy, the constant volume case is equivalent to the situation in which you mow the lawn yourself. You didn't get lazy and so kept all \$20 of your parent's money yourself.

this situation is that CO_2 is held in a rigid container (Figure 6.6b); the gas is not allowed to expand into the surroundings, and the pressure in the container is no longer constant. The volume of the system, however, is constant. No work can be done by the system on its surroundings under these circumstances, so all the heat added at constant volume (q_v) must be used to increase the internal energy of the system ($q_v = \Delta E$ since $w = 0$).

In plants and animals, as well as in the laboratory, reactions usually occur at constant pressure, and *the heat transferred into (or out of) a system at constant pressure* is given the special name **enthalpy change**, symbolized by ΔH. Thus,

$$\Delta H = q_p = \text{heat transferred to or from a system at constant pressure}$$

Furthermore,

$$\Delta H = \text{(enthalpy of the system at the end of a process)} - \text{(enthalpy of the system at the start of the process)}$$

$$\Delta H = H_{\text{final}} - H_{\text{initial}} \quad (6.8)$$

Negative enthalpy change: ΔH has a negative value; $\Delta H < 0$
System —Heat transfer→ Surroundings
Exothermic *process with respect to system*

Positive enthalpy change: ΔH has a positive value; $\Delta H > 0$
System ←Heat transfer— Surroundings
Endothermic *process with respect to system*

From these equations you can see that, if the enthalpy of the final system is greater than that of the initial system (the case for solid CO_2 changing to CO_2 vapor at −78 °C), the enthalpy has increased (left side of Figure 6.8), and the process has a positive ΔH; the process is said to be endothermic. Conversely, if the enthalpy of the final system is less than that of the initial system, heat has been transferred out of the system, and ΔH is negative; the process is exothermic.

Enthalpy changes can be determined for any chemical process. For any chemical reaction, the products represent the "final system" and the reactants the "initial system," so Equation 6.8 can be rewritten as Equation 6.9.

$$\Delta H = H_{\text{products}} - H_{\text{reactants}} \quad (6.9)$$

The Relationship of Reaction Heat and Enthalpy Change
Reactant ⟶ product + heat
ΔH = a negative number; reaction is exothermic

Reactant + heat ⟶ product
ΔH = a positive number; reaction is endothermic

For example, the left side of Figure 6.9 shows that the enthalpy of the products of decomposition of water is greater than that of the reactants, so water vapor would have to *absorb* 242 kJ of energy (at constant pressure) from the surroundings in order to decompose to its elements. That is, the enthalpy change for the endothermic decomposition of water

$$H_2O(g) + 242 \text{ kJ} \longrightarrow \tfrac{1}{2}\, O_2(g) + H_2(g) \qquad \Delta H = +242 \text{ kJ}$$

is $\Delta H = +242$ kJ per mole of water.

There are two very important things for you to notice here.

1. Note carefully the relation between a positive ΔH and the idea that heat is a "reactant" in the chemical equation for the decomposition of water.
2. *The change in energy or enthalpy for a reaction is directly proportional to the quantity of material present.* If two moles of water were decomposed to the elements, twice as much heat energy or 484 kJ would be required for complete reaction.

In thermochemistry, the value of ΔH applies to the case where the coefficients of the balanced equation are read in moles. Its value depends on the number of moles of material.

The decomposition of water is a process that can be reversed. Water can be decomposed to its elements, or the elements can combine

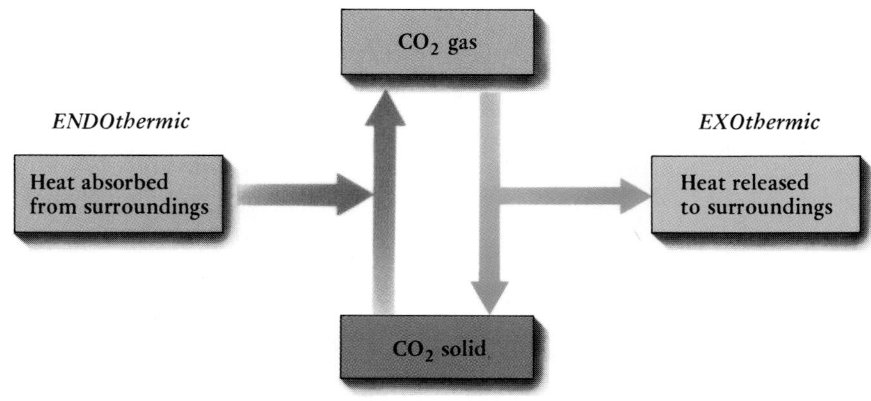

Figure 6.8 Enthalpy or energy diagram for the interconversion of solid CO_2 and CO_2 gas at constant pressure.

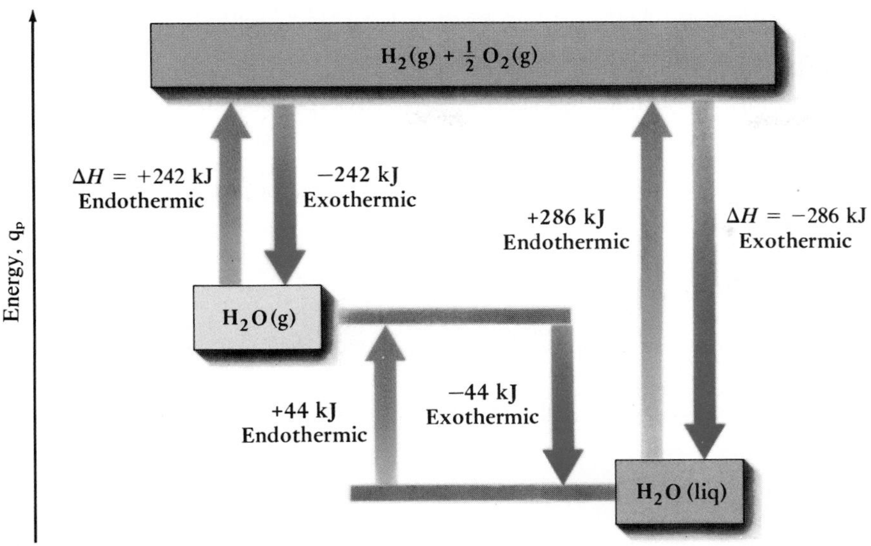

Figure 6.9 Enthalpy diagram for the interconversion of water vapor, $H_2O(g)$, liquid water, $H_2O(\ell)$, and its elements.

to form water. The amount of heat energy involved in the combination reaction (Figure 6.9)

$$H_2(g) + \tfrac{1}{2} O_2(g) \longrightarrow H_2O(g) + 242 \text{ kJ} \qquad \Delta H = -242 \text{ kJ}$$

is numerically the same as for the decomposition reaction, but the combination reaction is *exothermic* so ΔH is negative. That is, $\Delta H = -242$ kJ per mole of water formed. The fact that the formation of water gives off exactly the same amount of energy as needed for the decomposition of water is mandated by the principle of the conservation of energy.

Because so much energy is provided per gram, H_2 is used as a fuel for the Space Shuttle and is seriously considered as fuel for cars and homes in the future. See Keating, R., Omni, *June 1989, page 66.*

The amount of energy involved in a chemical reaction depends on the physical state (solid, liquid, or gas) in which the reactants and products are found. In Section 6.3 we described the concept of latent heat, and you have seen that the transformation of solid CO_2 to CO_2 vapor requires an input of energy. Therefore, it should come as no surprise that the decomposition of *liquid* water to H_2 and O_2

$$H_2O(\ell) + 286 \text{ kJ} \longrightarrow H_2(g) + \tfrac{1}{2} O_2(g) \qquad \Delta H = +286 \text{ kJ}$$

requires a different amount of energy than the decomposition of water *vapor*. The energy relationship between water in the liquid or vapor state and its elements, with everything at 25 °C, is shown in Figure 6.9. This should also make it clear that energy is required to convert liquid water into water vapor, and it is this fact that leads to the requirement for a greater amount of energy to decompose liquid water to its elements than to decompose water vapor.

EXAMPLE 6.6

HEAT ENERGY CALCULATION

Propane, C_3H_8, is a commonly used fuel in laboratory burners and backyard grills. Suppose you burn 454 g of propane in air. This exothermic reaction occurs according to the following equation:

$$C_3H_8(g) + 5\ O_2(g) \longrightarrow 3\ CO_2(g) + 4\ H_2O(\ell) + 2220\ kJ$$

That is, the enthalpy change for this reaction is $\Delta H = -2220$ kJ per mole of propane burned. How much heat is transferred as a result of this reacting chemical system (454 g of $C_3H_8(g)$ plus the required amount of O_2)?

Solution The first step is to find the number of moles of propane present in the sample.

$$454\ g \left(\frac{1\ mol}{44.09\ g}\right) = 10.3\ mol\ propane$$

Then multiply by the amount of heat evolved per mole.

$$\text{Total heat evolved by 454 g of propane} = 10.3\ mol\ propane \left(\frac{2220\ kJ\ evolved}{1.00\ mol\ propane}\right)$$
$$= 22{,}900\ kJ$$

EXERCISE 6.5 Heat Energy Calculation

How much heat energy would be required to decompose 12.6 g of water vapor to the elements? (Refer to the text above for the required chemical equation and heat involved per mole of water vapor.)

6.5 HESS'S LAW

One consequence of the first law of thermodynamics is **Hess's law of constant heat summation**. Just as chemical reactions adhere to the law of mass conservation, so too must they obey the law of energy conservation. Hess's law tells us that, *if a reaction is the sum of two or more other reactions, then ΔH for the overall process must be the sum of the ΔH's of the constituent reactions*. As an example, the decomposition of liquid water into its elements $H_2(g)$ and $O_2(g)$ can be thought of as the sum of two other processes: (1) the transformation of liquid water to water vapor and (2) the decomposition of water vapor to the elements (with all substances at the same temperature) (Figure 6.9).

(1) Evaporation of water: enthalpy of vaporization at 25 °C

$$H_2O(\ell) + 44\ kJ \longrightarrow H_2O(g) \qquad \Delta H = +\ 44\ kJ$$

(2) Decomposition of water vapor: enthalpy of decomposition at 25 °C

$$H_2O(g) + 242\ kJ \longrightarrow H_2(g) + \tfrac{1}{2}\ O_2(g) \qquad \Delta H = +242\ kJ$$

(1) + (2) $H_2O(\ell) + 286\ kJ \longrightarrow H_2(g) + \tfrac{1}{2}\ O_2(g) \qquad \Delta H = +286\ kJ$

When we add the two reactions together, $H_2O(g)$ is a product of the first reaction and a reactant in the second. Thus, as in adding two algebraic equations where the same quantity or term appears on both sides of the equation, you can cancel out the quantity "$H_2O(g)$." The net result is that 286 kJ are required for the endothermic decomposition of 1 mole of *liquid* water to the gaseous elements H_2 and O_2. That is, the enthalpy change for the decomposition of liquid water is +286 kJ/mol.

A chemist, chemical engineer, or biologist often is interested in knowing the quantity of heat energy involved in a chemical process. Therefore, some knowledge of ΔH values is useful. Such values can be determined experimentally for many reactions, but it is not a simple task. Besides, it would be impossible to measure values for every conceivable reaction. Fortunately, Hess's law allows us to take ΔH values for simple reactions, which have been determined experimentally, and combine these values to obtain enthalpy values for new or more complex reactions. Several examples of this concept follow.

Excess phosphorus reacts exothermically with chlorine to give mainly PCl_3, phosphorus trichloride.

EXAMPLE 6.7

HESS'S LAW: ENTHALPY CHANGE FOR THE FORMATION OF A COMPOUND

Suppose you wish to know the enthalpy change for the formation of $PCl_5(s)$ from its elements.

$$\tfrac{1}{4}\,P_4(s) + \tfrac{5}{2}\,Cl_2(g) \longrightarrow PCl_5(s) \qquad \Delta H = ?$$

This value is difficult to obtain experimentally, but ΔH values can be determined for the following reactions.

$$\tfrac{1}{4}\,P_4(s) + \tfrac{3}{2}\,Cl_2(g) \longrightarrow PCl_3(\ell) \qquad \Delta H_1 = -319.7 \text{ kJ}$$

$$PCl_3(\ell) + Cl_2(g) \longrightarrow PCl_5(s) \qquad \Delta H_2 = -123.8 \text{ kJ}$$

The solution to this problem is to add together, just as they are written above, the equation for the formation of $PCl_3(\ell)$ from the elements and the equation for the further reaction of $PCl_3(\ell)$ to give $PCl_5(s)$. The compound PCl_3 is a product of the first reaction and a reactant in the second. Therefore, PCl_3 "cancels out" when the equations are added, and the final equation obtained is the one desired. Since the final equation can be obtained by the simple addition of the two equations involving PCl_3, ΔH for the formation of PCl_5 from its elements is simply the sum of ΔH_1 and ΔH_2. That is,

$$\Delta H(\text{formation of } PCl_5 \text{ from } Cl_2 \text{ and } P_4) = \Delta H_1 + \Delta H_2 = -443.5 \text{ kJ}$$

EXAMPLE 6.8

HESS'S LAW IN A COMBINATION REACTION

Calculate the quantity of heat transferred at constant pressure when benzene, C_6H_6, burns in oxygen to give water and carbon dioxide.

$$C_6H_6(\ell) + \tfrac{15}{2}\,O_2(g) \longrightarrow 6\,CO_2(g) + 3\,H_2O(\ell)$$

This is a combustion reaction (Chapter 4), and the heat transferred in such a process is called the **enthalpy of combustion** (or often simply the heat of combustion).

The following information can be found in reference books (or in Table 6.2 or Appendix J in this book):

$$\textbf{(1)}\ 6\,C(\text{graphite}) + 3\,H_2(g) \longrightarrow C_6H_6(\ell) \qquad \Delta H_1 = +49.0 \text{ kJ}$$

$$\textbf{(2)}\ C(\text{graphite}) + O_2(g) \longrightarrow CO_2(g) \qquad \Delta H_2 = -393.5 \text{ kJ}$$

$$\textbf{(3)}\ H_2(g) + \tfrac{1}{2}\,O_2(g) \longrightarrow H_2O(\ell) \qquad \Delta H_3 = -285.8 \text{ kJ}$$

Solution You cannot just add together Equations 1 through 3 above, as they are now written, and obtain the desired reaction. C_6H_6 should be a reactant, but Equation 1 shows it as a product. Therefore, you will have to reverse the first equation *and* reverse the sign of ΔH_1: If the formation of C_6H_6 is endothermic, its decomposition must be exothermic.

$$(1')\ C_6H_6(\ell) \longrightarrow 6\ C(\text{graphite}) + 3\ H_2(g) \qquad \Delta H_1' = -49.0\ \text{kJ}$$

Both Equations 2 and 3 are in the correct direction as written: CO_2 and H_2O are products as required in the combustion reaction. However, each has only 1 mole of product (CO_2 and H_2O), whereas 6 moles of CO_2 and 3 moles of H_2O are needed as products. To conserve mass, you must multiply Equations 2 and 3 by 6 and 3, respectively, and their ΔH values must be multiplied by the same factors.

$$(2')\ 6\ C(\text{graphite}) + 6\ O_2(g) \longrightarrow 6\ CO_2(g) \quad \Delta H_2' = 6 \times \Delta H_2 = -\ 2361\ \text{kJ}$$

$$(3')\ 3\ H_2(g) + \tfrac{3}{2}\ O_2(g) \longrightarrow 3\ H_2O(\ell) \qquad \Delta H_3' = 3 \times \Delta H_3 = -857.4\ \text{kJ}$$

It is now possible to add the three equations above (1′ + 2′ + 3′) to obtain the desired equation for benzene combustion. Notice that 6 moles of C(graphite) appear in Equation 1′ as product and in Equation 2′ as reactant; therefore, C(graphite) cancels out. The number of moles of O_2 required, $\frac{15}{2}$, is achieved by adding 6 moles from Equation 2′ and $\frac{3}{2}$ moles from Equation 3′.

The three equations will now add correctly, and you can simply add the $\Delta H'$ values for the three reactions.

$$\Delta H_{\text{total}} \text{ for } C_6H_6 \text{ combustion} = \Delta H_1' + \Delta H_2' + \Delta H_3' = -3267\ \text{kJ}$$

EXERCISE 6.6 Using Hess's Law

Calculate the heat evolved in the combustion of ethyl alcohol. That is, calculate the enthalpy of combustion.

$$C_2H_5OH(\ell) + 3\ O_2(g) \longrightarrow 2\ CO_2(g) + 3\ H_2O(\ell) \qquad \Delta H = ?$$

Use the information from Example 6.8 and the following:

$$2\ C(\text{graphite}) + 3\ H_2(g) + \tfrac{1}{2}\ O_2(g) \longrightarrow C_2H_5OH(\ell) \quad \Delta H = -277.7\ \text{kJ}$$

6.6 STATE FUNCTIONS

The object of Example 6.7 was to find the enthalpy change of the reaction when elemental phosphorus reacts with chlorine to give PCl_5. In principle, this reaction can be done in two ways: (1) direct combination of P_4 and Cl_2 to give PCl_5 or (2) in stages, where PCl_3 is an intermediate along the way (Figure 6.10). The enthalpy change could be measured for both pathways, and each pathway would evolve the *same* quantity of heat (at constant pressure). This is always true for any chemical reaction. No matter how you go from starting material to final products, the net amount of heat evolved or required (at constant pressure) will always be the same. *The enthalpy change does not depend on the path you choose to follow.* For this reason, enthalpy is called a **state function**, a quantity whose value is determined *only* by the state of the system.

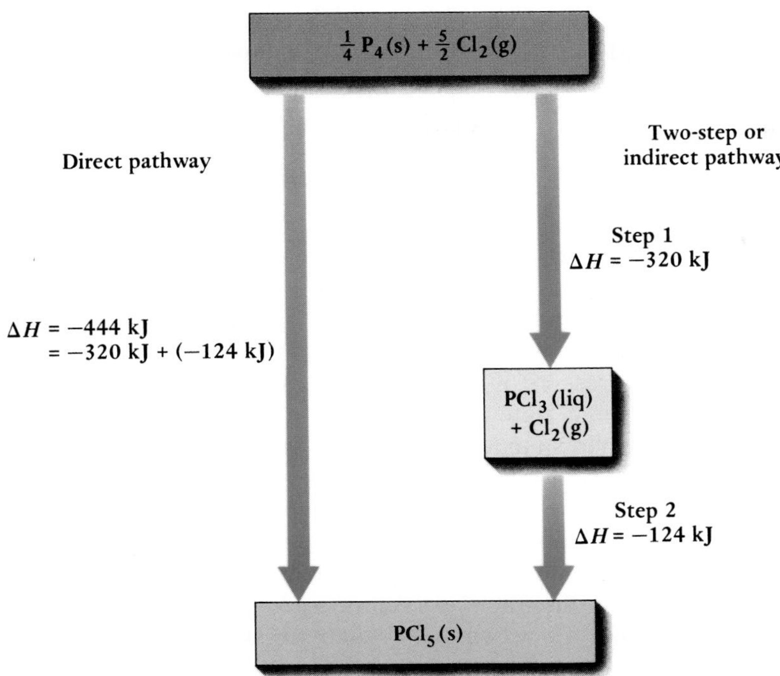

Figure 6.10 Two of the possible pathways from elemental phosphorus and gaseous chlorine to PCl_5(s). See Example 6.7.

Many commonly measured quantities are state functions; among them are pressure, volume, temperature, and the size of your bank account. You could have arrived at a current bank balance of $25 by having simply deposited $25, or you could have deposited $100 and then withdrawn $75. The volume of a balloon is also a state function. You can blow up a balloon to a large volume and then let some air out to arrive at the desired volume. As an alternative, you could blow it up in stages, adding tiny amounts of air to arrive at the same volume as before. The final volume does not depend in any way on how you got there. In both of these examples, there are an infinite number of ways to arrive at the final state, but the final value depends only on the size of the bank balance or the balloon, and not on the path taken from the initial to the final state.

Since enthalpy is a state function, in principle there is an absolute enthalpy for the reactants ($H_{initial} = H_{reactants}$) and for the products ($H_{final} = H_{products}$). The difference between these enthalpies is the change for the system.

$$\Delta H_{reaction} = H_{final} - H_{initial} = H_{products} - H_{reactants} \tag{6.10}$$

Since the reaction starts and finishes at the same place no matter which pathway is chosen, $\Delta H_{reaction}$ must always be independent of pathway. However (and this is a big "however"), unlike volume, temperature, pressure, energy, or a bank balance, it is not possible to determine experimentally the absolute enthalpy of a compound. *Only the change in enthalpy for a reaction or process can be determined experimentally.* The heat evolved or required in a chemical process (at constant pressure) is only a reflection of the *difference* in enthalpy between the reactants and products. One can determine only that the enthalpy of the products is greater or less than that of the reactants by a certain amount, an idea illustrated by the enthalpy diagram for the decomposition of water (Figure 6.9).

6.7 STANDARD CONDITIONS AND STANDARD ENTHALPIES OF FORMATION

To define precisely the conditions of an experiment, a chemist often states them in terms of **standard conditions**: a pressure of 1 atmosphere and a given temperature, most often 25 °C (298 K). The **standard state** for an element or compound is the physical state in which it exists at a given temperature and at a pressure of 1 atmosphere. Thus, at 25 °C the standard state for H_2 is the gaseous state with a pressure of 1 atmosphere, whereas that for NaCl is the solid state at 1 atmosphere pressure. For an element such as carbon that can exist in either of two solid states at 1 atm pressure at 25 °C, one state must be selected as the standard, and graphite rather than diamond has been chosen for carbon.

It is common to use the term "heat of reaction" interchangeably with "enthalpy of reaction." Understand that it is only the heat at ***constant pressure****, q_p, that is equivalent to the enthalpy change.*

When reactions occur with all the reactants and products in their standard states, we say that the observed or calculated enthalpy change is the **standard enthalpy of reaction, $\Delta H°$**, where the superscript ° indicates standard conditions. All the reactions we have discussed to this point have met this convention, so all the ΔH values should have ° attached.

The enthalpy of a reaction for the formation of one mole of a compound from its elements, with all chemicals in their standard states, is called the **molar enthalpy of formation, ΔH_f°**, where the subscript f signals that the compound in question has been *formed* from its elements. Some of the reactions already discussed define standard enthalpies of formation:

$$H_2(g) + \tfrac{1}{2}\,O_2(g) \longrightarrow H_2O(\ell) \qquad \Delta H_f^\circ = -285.8 \text{ kJ/mol}$$

$$\tfrac{1}{4}\,P_4(s) + \tfrac{3}{2}\,Cl_2(g) \longrightarrow PCl_3(\ell) \qquad \Delta H_f^\circ = -319.7 \text{ kJ/mol}$$

The following equation shows that 277.7 kJ are *evolved* if graphite, the standard state form for carbon, is combined with gaseous hydrogen and oxygen to form one mole of ethyl alcohol at 298 K.

$$2\,C(\text{graphite}) + 3\,H_2(g) + \tfrac{1}{2}\,O_2(g) \longrightarrow C_2H_5OH(\ell) \qquad \Delta H_f^\circ = -277.7 \text{ kJ/mol}$$

Finally, you should be sure to be aware that a standard enthalpy of formation, ΔH_f°, is the same as any other enthalpy of a reaction, $\Delta H°$. The difference is only that ΔH_f° is the special case of $\Delta H°$ when one mole of a compound is formed *from its elements,* such as in the three preceding equations. The enthalpy change for the exothermic reaction

$$CaO(s) + CO_2(g) \longrightarrow CaCO_3(s) \qquad \Delta H° = -178.3 \text{ kJ}$$

is not an enthalpy of formation, since calcium carbonate has been formed from other compounds, not from its elements.

Table 6.2 and Appendix J list values of ΔH_f° for many other compounds. Be sure to notice that there are no values listed in these tables

Table 6.2 Selected Standard Molar Enthalpies of Formation at 298 K

Substance	Name	Standard Molar Enthalpy of Formation (kJ/mol)
$Al_2O_3(s)$	Aluminum oxide	−1675.7
$BaCO_3(s)$	Barium carbonate	−1216.3
$CaCO_3(s)$	Calcium carbonate	−1206.9
$CaO(s)$	Calcium oxide	−635.1
$CCl_4(\ell)$	Carbon tetrachloride	−135.4
$CH_4(g)$	Methane	−74.81
$C_2H_5OH(\ell)$	Ethyl alcohol	−277.7
$CO(g)$	Carbon monoxide	−110.5
$CO_2(g)$	Carbon dioxide	−393.5
$C_2H_2(g)$	Acetylene	+226.7
$C_2H_4(g)$	Ethylene	+52.3
$C_2H_6(g)$	Ethane	−84.7
$C_3H_8(g)$	Propane	−103.8
n-$C_4H_{10}(g)$	Butane	−888.0
$CuSO_4(s)$	Copper(II) sulfate	−771.4
$H_2O(g)$	Water vapor	−241.8
$H_2O(\ell)$	Liquid water	−285.8
$HF(g)$	Hydrogen fluoride	−271.1
$HCl(g)$	Hydrogen chloride	−92.3
$HBr(g)$	Hydrogen bromide	−36.4
$HI(g)$	Hydrogen iodide	+26.48
$KF(s)$	Potassium fluoride	−567.3
$KCl(s)$	Potassium chloride	−436.7
$KBr(s)$	Potassium bromide	−393.8
$MgO(s)$	Magnesium oxide	−601.7
$MgSO_4(s)$	Magnesium sulfate	−1284.9
$Mg(OH)_2(s)$	Magnesium hydroxide	−924.5
$NaF(s)$	Sodium fluoride	−573.6
$NaCl(s)$	Sodium chloride	−411.1
$NaBr(s)$	Sodium bromide	−361.1
$NaI(s)$	Sodium iodide	−287.8
$NH_3(g)$	Ammonia	−46.11
$NO(g)$	Nitrogen oxide	+90.3
$NO_2(g)$	Nitrogen dioxide	+33.2
$PCl_3(\ell)$	Phosphorus trichloride	−319.7
$PCl_5(s)$	Phosphorus pentachloride	−443.5
$SiO_2(s)$	Silicon dioxide (quartz)	−910.9
$SnCl_2(s)$	Tin(II) chloride	−325.1
$SnCl_4(\ell)$	Tin(IV) chloride	−511.3
$SO_2(g)$	Sulfur dioxide	−296.8
$SO_3(g)$	Sulfur trioxide	−395.7

for elements such as C(graphite) or $O_2(g)$. Because of the definition of enthalpy of formation, it follows that *standard enthalpies of formation for the elements in their standard states are zero.*

EXAMPLE 6.9

WRITING EQUATIONS TO DEFINE ENTHALPIES OF FORMATION

The standard enthalpy of formation of gaseous ammonia is −46.11 kJ/mol. Write the balanced equation for which the enthalpy of reaction is −46.11 kJ.

Solution The equation you wish to write must show the formation of 1 mole of $NH_3(g)$ from the elements in their standard states; both N_2 and H_2 are gases at 25 °C and 1 atmosphere pressure. Therefore, the correct equation is

$$\tfrac{1}{2}\,N_2(g) + \tfrac{3}{2}\,H_2(g) \longrightarrow NH_3(g) \qquad \Delta H^\circ_{rxn} = \Delta H^\circ_f = -46.11 \text{ kJ}$$

(where the subscript "rxn" is an abbreviation for "reaction.") Another way to write this equation would have been to include the heat evolved as a product. That is,

$$\tfrac{1}{2}\,N_2(g) + \tfrac{3}{2}\,H_2(g) \longrightarrow NH_3(g) + 46.11 \text{ kJ}$$

EXERCISE 6.7 Writing Equations to Define Enthalpies of Formation

(a) The molar enthalpy of formation of AgCl(s) is −127.1 kJ/mol. Write the balanced equation for which the enthalpy of reaction is −127.1 kJ. (b) The molar enthalpy of formation of methyl alcohol, $CH_3OH(\ell)$, is −238.7 kJ/mol. Write the balanced equation for which the enthalpy of reaction is −238.7 kJ.

The standard enthalpies of formation in Table 6.2 and Appendix J are just a few of the many that have been determined by various direct and indirect methods, primarily by the U.S. National Bureau of Standards. Example 6.10 further illustrates the use of enthalpies of formation and of Hess's law to find the enthalpy of a reaction.

EXAMPLE 6.10

USING ENTHALPIES OF FORMATION

Calculate the heat required at constant pressure to decompose limestone (calcium carbonate) to lime (calcium oxide) and carbon dioxide, with all substances at standard conditions.

$$CaCO_3(s) \longrightarrow CaO(s) + CO_2(g)$$

The following information is given:

$$\Delta H^\circ_f\,[CaCO_3(s)] = -1206.9 \text{ kJ/mol}$$

$$\Delta H^\circ_f\,[CaO(s)] = -635.1 \text{ kJ/mol}$$

$$\Delta H^\circ_f\,[CO_2(g)] = -393.5 \text{ kJ/mol}$$

Solution First, write the equations that correspond to the enthalpies of formation. It should then be clearer how to add them together to obtain the desired equation.

$$Ca(s) + C(\text{graphite}) + \tfrac{3}{2}\,O_2(g) \longrightarrow CaCO_3(s) \qquad \Delta H^\circ_f = -1206.9 \text{ kJ}$$

$$Ca(s) + \tfrac{1}{2}\,O_2(g) \longrightarrow CaO(s) \qquad \Delta H^\circ_f = -635.1 \text{ kJ}$$

$$C(\text{graphite}) + O_2(g) \longrightarrow CO_2(g) \qquad \Delta H^\circ_f = -393.5 \text{ kJ}$$

In the first reaction, $CaCO_3(s)$ is the product, but you need to have it on the left side of the final equation as a reactant. Therefore, this equation must be reversed *and* the sign of ΔH° must also be reversed from negative to positive. On the other hand, CaO(s) and CO_2(g) must appear as products in the final equation, so the second and third equations must have the same direction and sign of $\Delta H°$ given above.

$$
\begin{array}{ll}
CaCO_3(s) \longrightarrow Ca(s) + C(graphite) + \frac{3}{2} O_2(g) & \Delta H° = +1206.9 \text{ kJ} \\
Ca(s) + \frac{1}{2} O_2(g) \longrightarrow CaO(s) & \Delta H_f° = -635.1 \text{ kJ} \\
C(graphite) + O_2(g) \longrightarrow CO_2(g) & \Delta H_f° = -393.5 \text{ kJ} \\
\hline
CaCO_3(s) \longrightarrow CaO(s) + CO_2(g) & \Delta H° = +178.3 \text{ kJ}
\end{array}
$$

After the equation for the formation of $CaCO_3$ has been reversed, the other two equations can be added to it as they are; O_2, Ca, and C(graphite) cancel out, and the desired equation is obtained. The decomposition of limestone, $CaCO_3$, is endothermic; that is, 178.3 kJ of heat are required per mole of the carbonate decomposed.

EXERCISE 6.8 Using Enthalpies of Formation

Calculate the standard enthalpy change for the dehydration of $Mg(OH)_2$(s) to give MgO(s) and $H_2O(\ell)$.

You can find the enthalpy of any reaction if you can find a set of reactions whose enthalpies are known and whose equations, when added together, will give the equation for the desired reaction. However, there is one other useful conclusion from Examples 6.8 and 6.10. You may have noticed that the mathematics of the problem really amount to the expression

$$\text{Enthalpy change for a reaction} = \Delta H°_{rxn} = \Sigma a \Delta H_f°(\text{products}) - \Sigma b \Delta H_f°(\text{reactants}) \quad (6.11)$$

where Σ (the Greek letter *sigma*) means to "sum" and a and b are the coefficients in the balanced equation. This expression simply tells you to add up the enthalpies of formation of the products (multiplying each by

The Symbols Used for Enthalpy Change

The symbol $\Delta H°_{rxn}$ is a signal that we are describing the enthalpy change for a reaction, where "reaction" is abbreviated as "rxn." The symbol $\Delta H_f°$ is also used for the enthalpy change for a reaction, but for the specific case where the reaction involves the formation of a compound from its elements. Other symbols can be used on occasion. To summarize:

$\Delta H°_{rxn}$ = enthalpy change for *any* chemical reaction (rxn); a general symbol.

$\Delta H_f°$ = symbol used *only* to denote the enthalpy change for the formation of one mole of a compound from its elements

$\Delta H°_{vap}$ = enthalpy change for the vaporization of one mole of an element or compound at constant pressure and given temperature

$\Delta H°_{comb}$ = enthalpy change for a combustion reaction its stoichiometric coefficient in the balanced equation) and subtract from this the sum of the enthalpies of formation of the reactants (each having been multiplied by its coefficient). *When you have the enthalpies of formation of **all** the compounds in a reaction, the enthalpy of the reaction, ΔH°_{rxn}, can be calculated from Equation 6.11.* However, recognize that this is just a shortcut around writing the equations for the several "formation reactions" involved, as in Example 6.10.

EXAMPLE 6.11

USING ENTHALPIES OF FORMATION

Calculate the enthalpy of combustion of benzene (Example 6.8) using Equation 6.11. The balanced equation is

$$C_6H_6(\ell) + \frac{15}{2}\ O_2(g) \longrightarrow 6\ CO_2(g) + 3\ H_2O(\ell)$$

and the data available to you are

$$\Delta H^\circ_f\ [C_6H_6(\ell)] = +49.0\ \text{kJ/mol}$$

$$\Delta H^\circ_f\ [CO_2(g)] = -393.5\ \text{kJ/mol}$$

$$\Delta H^\circ_f\ [H_2O(\ell)] = -285.8\ \text{kJ/mol}$$

$\Delta H^\circ_f\ [O_2(g)] = 0$ (Recall that the heats of formation of the elements in their standard states are zero by definition.)

Solution

$$\Delta H^\circ_{rxn} = \{6\ \Delta H^\circ_f[CO_2(g)] + 3\ \Delta H^\circ_f[H_2O(\ell)]\} - \{\Delta H^\circ_f[C_6H_6(\ell)] + \tfrac{15}{2}\ \Delta H^\circ_f[O_2(g)]\}$$

$$\Delta H^\circ_{rxn} = \{6\ \text{mol}(-393.5\ \text{kJ/mol}) + 3\ \text{mol}(-285.8\ \text{kJ/mol})\} - \{1\ \text{mol}(+\ 49.0\ \text{kJ/mol}) + \tfrac{15}{2}\ \text{mol}\ (0\ \text{kJ/mol})\}$$

$$\Delta H^\circ_{rxn} = -3267\ \text{kJ}$$

EXERCISE 6.9 Using Enthalpies of Formation

Propane, C_3H_8, is widely used as a fuel. Calculate the heat evolved under standard conditions when 1.00 mole of propane gas burns in air (that is, reacts with O_2) to give the usual products of combustion. (That is, calculate the enthalpy or heat of combustion.) Appropriate enthalpies of formation can be obtained in Table 6.2. Assume water is produced as a liquid, $H_2O(\ell)$.

6.8 MEASURING HEATS OF REACTIONS: CALORIMETRY

The heat evolved by a chemical reaction can be measured in many ways, but perhaps the easiest to describe at this point is the combustion calorimeter (Figure 6.11). The sample, a solid or liquid that is combustible, is placed in the sample dish that is encased in a "bomb," a cylinder about the size of a large fruit juice can with heavy steel walls and ends. The

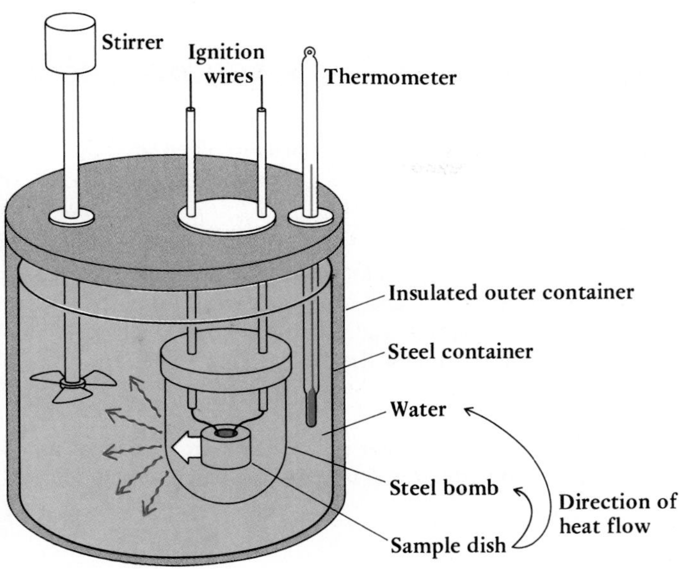

Figure 6.11 A combustion calorimeter. A combustible sample is burned in pure O_2 in a steel "bomb." The heat generated by the reaction flows into the bomb *and* to the water surrounding the bomb and warms both to the same temperature. By measuring the temperature increase, you can determine the heat evolved by the reaction.

bomb is then placed in a water-filled container with well-insulated walls. After the bomb is filled with an atmosphere of pure O_2, the mixture of O_2 and sample is ignited, usually by an electrical spark. The heat generated when the sample burns in the pure O_2 atmosphere warms up the bomb and the water around it to the same temperature. In this configuration, the oxygen and the compound represent the *system*, and the bomb and water around it are the *surroundings*. From the principle of the conservation of energy, and the fact that the apparatus is a rigid container (so no work is done by the system), we can write the relationships

Heat transferred from the system
= heat transferred into the surroundings

Heat evolved by the reaction
= heat absorbed by bomb and water

$$q_{\text{reaction}} = -(q_{\text{bomb}} + q_{\text{water}})$$

where q_{reaction} has a negative value because the combustion reaction is exothermic. If you measure the temperature change of the water (which also tells you the temperature change of the bomb), and if you know the heat capacities of the bomb and the water, you can calculate q_{bomb} and q_{water}, and therefore determine q_{reaction}, the quantity of heat transferred by a given mass of compound (as illustrated by the following example). However, you should recognize that the heat in such an experiment is measured at constant volume, q_v. Earlier in this chapter, this heat was designated as being equivalent to ΔE, the change in internal energy. Recall that ΔH, the change in enthalpy, is the heat evolved or required at constant pressure, q_p. Fortunately, ΔE and ΔH are related in a relatively simple way, so ΔH can be obtained from the calorimeter experiment. For you to do this now is not very important, so we shall skip a discussion of the conversion. However, you should at least be aware of the problem.

EXAMPLE 6.12

DETERMINING THE HEAT RELEASED BY A REACTION USING A CALORIMETER

Octane, C_8H_{18}, a primary constituent of gasoline, burns in air to give the usual products of combustion.

$$C_8H_{18}(\ell) + \frac{25}{2}\ O_2(g) \longrightarrow 8\ CO_2(g) + 9\ H_2O(g)$$

A 1.00-g sample of octane is burned in a calorimeter that contains 1.20 kg of water, and the temperature of the water and the bomb rises from 25.00 °C to 33.20 °C. If the heat capacity of the bomb, C_{bomb}, is known to be 837 J/K, calculate (a) the heat transferred in the combustion of the 1.00-g sample of C_8H_{18} and (b) the heat transferred per mole of C_8H_{18}.

Solution The heat evolved by the reaction, that is, the heat that appears as a rise in temperature of the water surrounding the bomb, is calculated as you learned in Section 6.3.

$$\begin{aligned} q_{water} &= (4.184\ J/g \cdot K)(m_{water})(\Delta T) \\ &= (4.184\ J/g \cdot K)(1.20 \times 10^3\ g)(33.20\ °C - 25.00\ °C) \\ &= +41.2 \times 10^3\ J \end{aligned}$$

The reaction heat that appears as a rise in temperature of the bomb is calculated from the heat capacity of the bomb (C_{bomb}, units of J/K)* and the temperature change, ΔT.

$$\begin{aligned} q_{bomb} &= C_{bomb} \times \Delta T \\ &= (837\ J/K)(33.20\ °C - 25.00\ °C) \\ &= +6.86 \times 10^3\ J \end{aligned}$$

The total heat transferred by the reaction is equal to the negative of the sum of q_{bomb} and q_{water}. Thus,

$$\begin{aligned} \text{Total heat transferred by 1.00 g of octane} &= -(41.2 \times 10^3\ J + 6.86 \times 10^3\ J) \\ &= -48.1 \times 10^3\ J \text{ or } -48.1\ kJ \end{aligned}$$

The calculation shows that 48.1 kJ of heat are evolved per gram of octane burned. Since the molar mass of octane is 114 g/mol, the heat transferred per mole is

$$\begin{aligned} \text{Total heat transferred per mole} &= (-48.1\ kJ/g)(114.2\ g/mol) \\ &= -5.49 \times 10^3\ kJ/mol \end{aligned}$$

EXERCISE 6.10 Determining the Heat Released by a Reaction

A 1.00-g sample of ordinary table sugar (sucrose, $C_{12}H_{22}O_{11}$) is burned in a combustion calorimeter. The temperature of the 1.50×10^3 g of water in the calorimeter rises from 25.00 °C to 27.32 °C. If the heat capacity of the bomb is 837 J/°C and the specific heat of the water is 4.184 J/g · K, calculate (a) the heat evolved per gram of sucrose and (b) the heat evolved per mole of sucrose.

*The heat capacity of the bomb, C_{bomb}, is found from (specific heat of bomb) × (mass of bomb). Since neither of these latter values changes during the experiment, we usually use simply the value C_{bomb}.

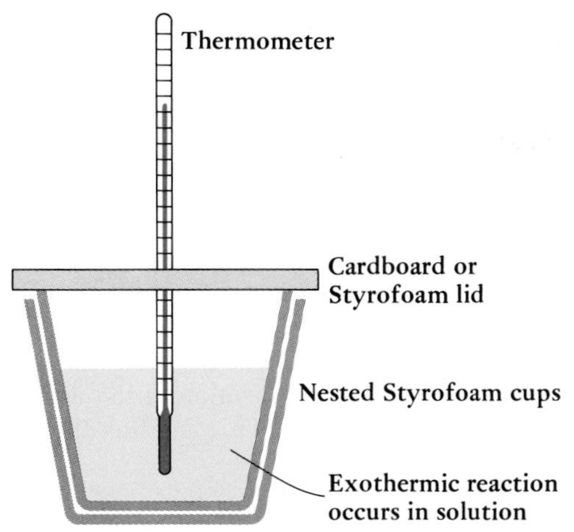

Figure 6.12 A coffee-cup calorimeter. Two Styrofoam coffee cups are placed one inside the other and covered with a Styrofoam or cardboard lid. When the exothermic reaction occurs in the aqueous solution in the container, there is enough insulation that the heat is confined to the solution. The change in temperature of the reacting solution can be measured reasonably well with a standard laboratory thermometer.

The combustion calorimeter is not a convenient instrument to use in the introductory chemistry laboratory. Therefore, we often study reactions other than combustions and rely on what is affectionately known as a ''coffee-cup calorimeter'' (Figure 6.12). It is obviously a simple device and is much less expensive than a combustion calorimeter. In addition, it operates at constant pressure, which means the heat evolved in a reaction in a coffee-cup calorimeter (q_p) is more directly a measure of a reaction enthalpy (since $q_p = \Delta H$).

EXAMPLE 6.13

USING A COFFEE-CUP CALORIMETER

Suppose you place 0.500 g of magnesium chips in a coffee-cup calorimeter and then add 100.0 mL of 1.00 M HCl. The reaction that occurs is

$$Mg(s) + 2\ HCl(aq) \longrightarrow H_2(g) + MgCl_2(aq)$$

What is the enthalpy change for this reaction (per mole of Mg) if the temperature of the solution increases from 22.2 °C to 44.8 °C? (Assume the specific heat of the solution is 4.20 J/g · K. Further assume that the density of the HCl solution is 1.00 g/cm³.)

Solution Given the mass of the solution, its specific heat, and ΔT, we can find the heat produced by the reaction above. (The mass of the solution is approximately the mass of the 100.0 mL of HCl plus the mass of magnesium.)

$$q = (101\ \text{g})(4.20\ \text{J/g}\cdot\text{K})(44.8\ ^\circ\text{C} - 22.2\ ^\circ\text{C}) = 9.59 \times 10^3\ \text{J}$$

This quantity of heat is produced by the reaction of 0.500 g of Mg. The amount produced by reaction of 1.00 mole is

$$\text{Heat produced/mole} = \left(\frac{9.59 \times 10^3\ \text{J}}{0.500\ \text{g}}\right)\left(\frac{24.31\ \text{g}}{1\ \text{mol}}\right)$$
$$= 4.66 \times 10^5\ \text{J/mol or 466 kJ/mol}$$

This means that the enthalpy change for the reaction is $\Delta H = -466$ kJ/mol. Note that we have included a negative sign because the reaction is clearly exothermic.*

EXERCISE 6.11 Using a Coffee-Cup Calorimeter

A simple calorimeter can be used to find the heat released in an acid–base reaction. Assume you mix exactly 200. mL of 0.400 M HCl with 200. mL of 0.400 M NaOH in a coffee-cup calorimeter. The temperature of the solutions before mixing was 25.10 °C; after mixing and allowing the reaction to occur, the temperature is 26.60 °C. What quantity of heat is produced on neutralizing one mole of acid? That is, what is the enthalpy of neutralization of the acid per mole? (Assume the densities of all solutions are 1.00 g/mL and that their specific heat capacities are 4.2 J/g · K.)

*When a ΔH value is stated for a chemical reaction, it is understood that the reactants and products are at the same temperature. However, in a calorimetry experiment this condition is never realized. In the coffee cup calorimeter (unlike the bomb calorimeter discussed earlier) virtually all the energy produced remains within the system. Thus, the problem is to find how much heat would have to be transferred out of the system in order to keep its temperature unchanged. That amount of heat is precisely what was obtained when ΔH was calculated in this example. The negative sign of the answer is consistent with the transfer of heat *out* of the system.

SOMETHING MORE ABOUT Our Energy Future

Where does the energy in our economy come from? That was a question posed at the beginning of this chapter, and at least some answer is provided by the data in Table 6.3. As you can see from the table, which gives us a picture of the uses of fossil fuels in our economy, petroleum provides about as much energy as coal and natural gas combined. However, this is largely because of the overwhelming importance of petroleum to fuel our cars and trucks. Of the fuels currently available in the United States for transportation, only those based on petroleum are relatively inexpensive, are transportable, and provide relatively high energy for a given amount of material.

However, as we learned in the 1970s, all that can change almost overnight. The present price of oil is about $20 a barrel, but it went as high as $35 a barrel in 1981.* Economists thought that there was a chance it could reach as high as $100 a barrel by the 1990s, and there was a rush to look for new oil wells and to develop new sources of petroleum. However, there is a large surplus of oil-producing capacity in the world today, and the price of oil has come down. This has had the effect of making it less economic for U.S. oil companies to drill for oil in this country and of providing less incentive to search for alternative fuels. All this makes another oil price increase even more likely in the future. In fact, many suspect that oil production in other countries will decline as reserves dwindle and that there will be sharply higher prices in the United States by the year 2000. So, once again we are confronted with the necessity of searching for new sources of energy.

*A barrel is 42 U.S. gallons.

Table 6.3 United States Consumption of Fossil Fuels in 1986*

	Coal	Natural Gas	Petroleum
Residential and Commercial	0.2	7.0	2.6
Industrial	2.6	6.4	7.9
Electric Utilities	14.5	2.7	1.5
Transportation	0.0	0.5	20.2
TOTAL	17.3	16.6	32.2

*Numbers are in BTU, British Thermal Units (1 BTU = 1050 J), and should be multiplied by 10^{15}.

For the foreseeable future liquid fuels will be necessary for cars and trucks and, to some extent, for residential and commercial heating. One promising development is the use of methyl alcohol, CH_3OH, a liquid with a melting point of −97.8 °C and a boiling point of 64.7 °C. This alcohol can be made from coal in a process that is being more and more highly developed.

Cars of the type that race at Indianapolis are fueled by methanol, CH_3OH. (Superstock Four by Five)

$$C(coal) + H_2O(g) \longrightarrow CO(g) + H_2(g)$$
$$2\ H_2(g) + CO(g) \longrightarrow CH_3OH(\ell)$$

The first step occurs at high temperature (and produces a mixture of gases called *synthesis gas*). These gases can be burned directly in a power plant to generate electricity, or they can be combined to give the alcohol in a second step that requires special chemicals (called *catalysts*) that speed up the process.

The methyl alcohol can be used by itself as a fuel or it can be combined with gasoline. Pure methyl alcohol is used by cars such as those that race in the Indianapolis 500. California is currently testing a mixture of 85% gasoline and 15% CH_3OH (called M85) in the hope of reducing pollution problems.

There is good news and bad concerning methyl alcohol as a fuel. The good news is that the alcohol has a much higher octane rating than gasoline, so it is possible to use high-compression engines that produce higher horsepower than with gasoline. However, the bad news is that you need about 1.8 gallons of CH_3OH to get as much energy as in 1 gallon of gasoline. This is not a problem for race cars, but it would be for the rest of us. (One other good feature about the alcohol is that, if it catches fire, the fire is easier to put out because the alcohol mixes readily with water. However, methyl alcohol has a disadvantage in that it has an almost colorless flame when it burns; burning the alcohol doesn't produce soot, and it is the glowing soot that makes the flame from gasoline or wood visible.)

In the long run, as a nation we shall have to find new sources of petroleum and substitutes for the multitude of products that are based on petroleum. Methyl alcohol is one possibility, but others include hydrogen (Chapter 22) or the oils from plants such as sunflowers or the buffalo gourd that grows so well on arid lands in the U.S. Southwest. Our energy future is a problem worth the best efforts of imaginative people in many fields, chemists of all kinds, chemical engineers, physicists, and business people.

Additional Reading

1. Berlfein, J. *Chem Matters*, **1988**, *6*(4), 10–12.
2. Gray, C.L.; Alson, J.A. *Scientific American*, **1989**, *261*(5), 108–114.
3. *Science*, **1989**, *246*, 199–201.

SUMMARY

Energy, the capacity to do work and transfer heat, takes many forms. This chapter presents the basic ideas of the science of energy transfer in chemical processes.

The calorie (cal), or the SI unit the **joule** (J) (1 calorie = 4.184 J), is the usual unit of energy (Section 6.2). The **heat capacity** of an object is the amount of heat that is required to change the temperature of an object by 1 degree. In chemistry we commonly express heat capacity in two ways, **specific heat capacity** and **molar heat capacity**. The specific heat capacity, which is often just called *specific heat*, is the amount of heat energy needed to change the temperature of 1.00 g of a substance by 1.00 degree Celsius (or 1.00 kelvin). In contrast, the molar heat capacity is the amount of heat energy needed to change the temperature of 1.00 mole of a substance by 1.00 degree Celsius (or 1.00 kelvin). The heat transferred into or out of a substance can be calculated from

$$\begin{aligned} q &= \text{heat transferred} \\ &= (\text{specific heat})(\text{g of material})(\text{temperature change}) \\ &= (\text{J/g} \cdot \text{K})(\text{g})(\Delta T \text{ in kelvins}) \end{aligned}$$

where ΔT is always defined as *the final temperature minus the initial temperature*. This means, for example, that the quantity of heat has a negative sign if the temperature has declined (ΔT is negative), thus signaling the fact that heat has been transferred *from* the substance.

The quantity of heat involved in changing the temperature of a substance is sometimes called the *sensible heat*. In contrast, heat can be absorbed to melt a solid or boil a liquid, or heat can be evolved when a vapor condenses to a liquid. These quantities of heat are called *latent heats* (the latent heat of fusion or vaporization, as appropriate).

It is important to understand that we can observe only the *transfer* of heat from one place to another. In the language of thermodynamics, it is transferred between a **system** (the part of the universe under study) and its **surroundings** (the rest of the universe) (Section 6.4). The process is said to be **exothermic** if the heat flow direction is system $\rightarrow$ surroundings; it is **endothermic** if the flow is in the opposite sense. A law of nature states that, like mass, energy must be conserved in all processes, and this is reflected in the **first law of thermodynamics**: the change in the *internal energy* of a system (ΔE) must be equal to the sum of heat energy (q) transferred into (or out of) the system and any work (w) done by (or done on) the system.

$$\Delta E = q + w$$

When heat transfer occurs at constant pressure (q_p), then q_p is defined as the **enthalpy** change, ΔH. If the process is exothermic, then the value of ΔH is negative; if endothermic, the value of ΔH is positive. The value of ΔH is specific to the process and depends on the quantity of material involved.

Hess's law states that, if a reaction is the sum of two or more other reactions, then ΔH for the overall process is the sum of the ΔH's of the constituent reactions (Section 6.5).

The enthalpy, H, for a system is a **state function**, a quantity whose value is determined only by the state of the system (Section 6.6). The change in the function does not depend on the path followed in the change. Potential and kinetic energy, pressure, volume, and temperature are other examples of state functions.

When an element or compound is in its most stable form under **standard conditions** (pressure of 1 atmosphere and, usually, a temperature of 25 °C), it is said to be in its **standard state** (Section 6.7). The enthalpy change of a reaction when all of the reactants and products are in their standard states is the **standard enthalpy change** (symbolized by $\Delta H°$). Specifically, the enthalpy for the formation of one mole of a compound from its elements, everything in their standard states, is called the **enthalpy of formation** (symbolized by $\Delta H_f°$). ($\Delta H_f°$ of an element is 0 by definition.) Many such values have been determined, and some are given in Table 6.2 and Appendix J. Using Hess's law, it is possible to calculate $\Delta H_f°$ for a new compound or to determine $\Delta H°$ for a reaction.

The heat transferred by a chemical reaction, particularly a combustion, can be determined in a **calorimeter** (Section 6.8).

Finally, to summarize the mathematical situations you are likely to encounter in thermochemistry, it is useful to note that there are **two basic types of problems covered in this chapter**.

Type 1: Using specific heat capacity. You can use Equation 6.4 to find the heat transferred into or out of a system as the temperature of the system increases or decreases (Examples 6.2 and 6.3). As a subset of this, you can use the latent heat of fusion or vaporization to find the quantity of heat transferred into or out of a system undergoing a change in phase at constant temperature (at the melting point or boiling point of the substance) (Example 6.5).

Type 2: Calculations involving changes in enthalpy. A major use of molar enthalpies of formation is to calculate the enthalpy change for a reaction. This can be done in either or both of two ways: (a) We can add reactions whose enthalpies we know to give the net equation whose $\Delta H°_{rxn}$ we wish to calculate (Example 6.10). (b) If we know $\Delta H_f°$ for each substance in the equation, we can use Equation 6.11 to find $\Delta H°_{rxn}$ (Example 6.11).

STUDY QUESTIONS

Unless specified otherwise, all enthalpies are given for compounds or reactions at standard conditions and all reactions occur at constant pressure.

REVIEW QUESTIONS

1. Name two laws stated in this chapter and provide a statement of each.
2. Define specific heat capacity and molar heat capacity and state the difference between them.
3. Based on your experience, when ice melts to liquid water, is the process exothermic or endothermic? When liquid water freezes to ice at 0 °C, is this exothermic or endothermic?
4. What is the first law of thermodynamics? Explain in words and use a mathematical expression.
5. For each of the following, define a system and its surroundings and give the direction of heat flow:
   (a) Propane is burning in a Bunsen burner in the laboratory.
   (b) Water drops, sitting on your skin after a dip in the ocean, evaporate.
   (c) Two chemicals are mixed in a flask sitting on a

laboratory bench. A reaction occurs and heat is evolved.

6. The temperature of the room in which you are sitting is 25 °C. Why is this temperature said to be a state function?

7. Is the *distance* you travel between your home town and your college or university a state function? Why or why not?

8. Your house is made of wood and glass. Assuming an equal amount of sunshine falls on a wooden wall and a piece of glass (of equal mass), which will warm more? Explain briefly.

ENERGY UNITS

9. Convert the following energies to joules or Calories as specified.
   (a) A 2-inch thick piece of two-layer chocolate cake with frosting provides 1670 kilojoules (1.67×10^6 J). What is this in Calories?
   (b) If you are on a diet that calls for eating no more than 1200 Calories per day, how many joules would that be?
   (c) A typical breakfast cereal provides 110 Calories per 1 ounce serving. How many joules of energy are thereby provided?

10. To change 1 mole of water from ice to liquid water at 0 °C, 1440 calories are required. (a) How many joules are required per mole of water? (b) How many joules per gram? (c) If liquid water is changed to ice at 0 °C, how much heat is *evolved* per mole?

11. To melt lead at 327 °C requires a heat input of 1224 calories per mole. If you wish to melt 1.00 pound (454 g) of lead, how many joules are required?

12. Sulfur dioxide, SO_2, is found in wines and in polluted air. If you burn a mole of sulfur in the air to get a mole of SO_2, you find that 297 kJ of energy are released. Express this energy in (a) joules, (b) calories, and (c) kilocalories.

SPECIFIC HEAT AND HEAT CAPACITY

13. How many kilojoules of heat energy are required to heat all the aluminum in a roll of aluminum foil (500. g) from room temperature (25 °C) to the temperature of a hot oven (250. °C)?

14. One way to cool down your cup of coffee is to plunge an ice-cold piece of aluminum into it. Suppose you store a 20.0 g piece of aluminum in the refrigerator at 40 °F (4.4 °C) and then drop it into your coffee. The coffee temperature drops from 90.0 °C to 55.0 °C. How many kilojoules of heat energy did the aluminum block absorb? (Ignore the cooling of the cup.)

15. Suppose you pick up a 16-pound ball of iron (such as a "shot-put" ball at a track event). The iron ball has the same temperature as the atmosphere on a cool day, say 16 °C. How many joules of heat energy must the iron ball absorb to reach the temperature of your body, 37 °C?

16. Ethylene glycol, $(CH_2OH)_2$, is often used as an antifreeze in cars. (a) Which requires more heat energy to warm from 25.0 °C to 100. °C, pure water or an equal mass of pure ethylene glycol? (b) If you have a 5.00-quart cooling system in your automobile, compare water and glycol as to the quantity of heat energy (in joules) the liquid in the system absorbs on raising the temperature from 25.0 °C to 100. °C. (Assume that the densities of water and ethylene glycol are both 1.00 g/cm³.) (1 qt = 0.946 L)

17. If you want to melt 1.00 pound of lead (1 pound = 453.6 g), how many joules of heat energy would be required to raise the temperature of the lead from room temperature (25 °C) to its melting point, 327 °C, and then melt the lead at 327 °C? (The specific heat capacity of lead is 0.159 J/g · K, and the metal requires 24.7 J/g to convert the solid to a liquid.)

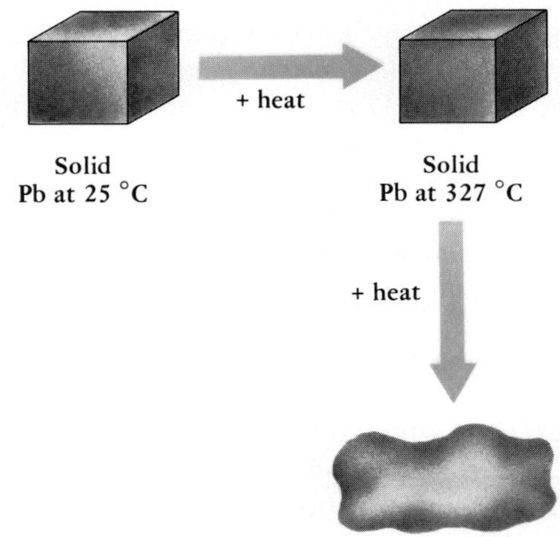

18. The hydrocarbon benzene, C_6H_6, boils at 80.1 °C. How many joules of heat energy are required to heat 1.00 kg of this liquid from 20.0 °C to the boiling point and then change the liquid completely to a vapor at that temperature? (The molar heat capacity of liquid C_6H_6 is 136 J/mol · K and the enthalpy of vaporization is 395 J/g.)

19. A piece of iron (400. g) is heated in a flame and is then plunged into a beaker containing 1.00 kg of water. The original temperature of the water was 20.0 °C, but it is 32.8 °C after the iron bar is dropped in. What was the original temperature of the hot iron bar? (Ignore the heating of the beaker.)

20. A 192-g piece of copper was heated to 100.0 °C in a boiling water bath, and then it was dropped into a beaker containing 750. mL of water at 4.0 °C (density = 1.00 g/cm³). What is the final temperature of the copper and water after they come to thermal equilibrium? (Ignore the heating of the beaker.) (The specific heat of copper is 0.385 J/g · K.)

21. Calculate the quantity of heat required to convert the water in four ice cubes (60.1 g; see Example 6.5) from $H_2O(s)$ at 0.0 °C to $H_2O(g)$ at 100.0 °C. The latent heat of fusion of ice at 0 °C is 333 J/g and the latent heat of vaporization of liquid water at 100 °C is 2260 J/g.

22. Mercury, with a freezing point of −39 °C, is the only metal that is liquid at room temperature. How much heat must be released by mercury if you cool 1.00 mL of the metal from room temperature (23.0 °C) to −39 °C and then freeze the mercury to a solid? (The density of mercury is 13.6 g/cm³. The required specific heat capacity and heat of fusion may be found in Appendix D.)

23. Using the data in Table 6.1, calculate the molar heat capacity of (a) aluminum, (b) iron, and (c) $CCl_4(\ell)$.

24. For each compound below, calculate the specific heat from the molar heat capacity (as listed in tables published by the U.S. National Bureau of Standards).
    (a) $C_2H_2(g)$, acetylene (43.93 J/mol · K).
    (b) $CF_3CCl_3(g)$ (120.5 J/mol · K), one of a class of compounds called chlorofluorocarbons, compounds thought to deplete the earth's ozone layer.

ENTHALPY

25. Energy is stored in the body in the form of adenosine triphosphate, ATP. It forms on reaction between adenosine diphosphate, ADP, and phosphoric acid.

$$\text{ADP} + H_3PO_4 + 38\text{ kJ} \longrightarrow \text{ATP} + H_2O$$

Is the reaction endothermic or exothermic?

26. Ethyl alcohol has been suggested as an additive to gasoline. It burns to give CO_2 and H_2O, the usual products of combustion.

$$C_2H_5OH(g) + 3\,O_2(g) \longrightarrow 2\,CO_2(g) + 3\,H_2O(\ell) + 1409\text{ kJ}$$

Is the reaction endothermic or exothermic?

27. "Gasohol," a mixture of gasoline and ethyl alcohol, C_2H_5OH, is a possible automobile fuel. The alcohol produces energy in a combustion reaction with O_2.

$$C_2H_5OH(g) + 3\,O_2(g) \longrightarrow 2\,CO_2(g) + 3\,H_2O(\ell)$$

If 0.115 g of alcohol evolves 3.45 kJ when burned at constant pressure, what is the molar enthalpy (or heat) of combustion for ethyl alcohol?

28. White phosphorus, P_4, ignites in air to produce heat, light, and P_4O_{10} (page 110).

$$P_4(s) + 5\,O_2(g) \longrightarrow P_4O_{10}(s)$$

If you burn 3.56 g of P_4, you find that 86.5 kJ of heat are evolved at constant pressure. What is the molar enthalpy of combustion of P_4?

29. A laboratory "volcano" can be made from ammonium dichromate (Figure 4.8). When ignited, the compound decomposes in a fiery display.

$$(NH_4)_2Cr_2O_7(s) \longrightarrow N_2(g) + 4\,H_2O(g) + Cr_2O_3(s)$$

If the decomposition produces 315 kJ per mole of ammonium dichromate at constant pressure, how much heat energy would be produced by 28.3 g (1 ounce) of the solid?

30. The thermite reaction, the reaction between aluminum and iron(III) oxide,

$$2\,Al(s) + Fe_2O_3(s) \longrightarrow Al_2O_3(s) + 2\,Fe(s) + 852\text{ kJ}$$

produces a tremendous amount of heat (see Figure 20.9).
    (a) If you begin with 10.0 g of Al and excess Fe_2O_3, how many kilojoules of heat are evolved at constant pressure?
    (b) If the heat produced in part (a) were used to warm 500. g of water with an initial temperature of 23.0 °C, what would be the final temperature of the water? Is the water brought to its boiling point?

HESS'S LAW

31. Calculate the enthalpy change, ΔH, for the formation of 1 mole of strontium carbonate (the material that gives the red color in fireworks) from its elements.

$$Sr(s) + C(\text{graphite}) + \tfrac{3}{2}\,O_2(g) \longrightarrow SrCO_3(s)$$

The information available to you is:

$$Sr(s) + \tfrac{1}{2}\,O_2(g) \longrightarrow SrO(s) \qquad \Delta H = -592\text{ kJ}$$

$$SrO(s) + CO_2(g) \longrightarrow SrCO_3(s) \qquad \Delta H = -234\text{ kJ}$$

$$C(\text{graphite}) + O_2(g) \longrightarrow CO_2(g) \qquad \Delta H = -394\text{ kJ}$$

32. We want to know the enthalpy change for the reaction of lead(II) chloride with chlorine to give lead(IV) chloride.

$$PbCl_2(s) + Cl_2(g) \longrightarrow PbCl_4(\ell) \qquad \Delta H = ?$$

We already know that $PbCl_2(s)$ can be formed from the metal and $Cl_2(g)$

$$Pb(s) + Cl_2(g) \longrightarrow PbCl_2(s) \qquad \Delta H = -359.4\text{ kJ}$$

and that $PbCl_4(\ell)$ can be formed directly from the elements.

$$Pb(s) + 2\ Cl_2(g) \longrightarrow PbCl_4(\ell) \qquad \Delta H = -329.3\ kJ$$

Calculate the unknown ΔH value from this information.

33. Lead has been known and used since ancient times. To obtain the metal, the ore, PbS (galena), is first heated in air to form PbO,

$$PbS(s) + \tfrac{3}{2}\ O_2(g) \longrightarrow PbO(s) + SO_2(g) \qquad \Delta H = -413.7\ kJ$$

and the lead(II) oxide is reduced with carbon.

$$PbO(s) + C(graphite) \longrightarrow Pb(s) + CO(g) \qquad \Delta H = +106.8\ kJ$$

To obtain the lead from 1.00 kg of pure PbS, how much heat energy is required or evolved (at constant pressure)?

34. Three reactions very important to the semiconductor industry are

(a) The reduction of silicon dioxide to crude silicon

$$SiO_2(s) + 2\ C(graphite) \longrightarrow Si(s) + 2\ CO(g) \qquad \Delta H = +689.9\ kJ$$

(b) The formation of silicon tetrachloride

$$Si(s) + 2\ Cl_2(g) \longrightarrow SiCl_4(g) \qquad \Delta H = -657.0\ kJ$$

(c) The reduction of silicon tetrachloride to pure silicon with magnesium

$$SiCl_4(g) + 2\ Mg(s) \longrightarrow 2\ MgCl_2(s) + Si(s) \qquad \Delta H = -625.6\ kJ$$

What is the overall energy involved in changing 1.00 mole of sand (SiO_2) into very pure silicon?

35. We are interested in the amount of energy necessary to break the double bond between two C atoms in the C_2 molecule.

$$C{=}C(g) \longrightarrow 2\ C(g) \qquad \Delta H = ?$$

To find this, you have available the following information:

REACTION	ENTHALPY CHANGE, kJ
$H_2C{=}CH_2(g) \longrightarrow 4\ H(g) + C{=}C(g)$	+1656
$C(graphite) \longrightarrow C(g)$	+716.7
$H_2(g) \longrightarrow 2\ H(g)$	+436.0
$2\ C(graphite) + 2\ H_2(g) \longrightarrow H_2C{=}CH_2(g)$	+52.3

Using the information above, calculate the energy required to break the C=C bond in $C_2(g)$.

36. Using the following reactions, calculate a value for the energy required to break a mole of C—N bonds. That is, calculate ΔH for $C{-}N(g) \rightarrow C(g) + N(g)$.

(a) Energy to break 1 mole of C—N bonds and produce graphite and N_2 molecules

$$C{-}N(g) \longrightarrow C(graphite) + \tfrac{1}{2}\ N_2(g) \qquad \Delta H = -416\ kJ$$

(b) Energy involved when 1 mole of gaseous C atoms condense to form graphite

$$C(g) \longrightarrow C(graphite) \qquad \Delta H = -717\ kJ$$

(c) Energy evolved when N atoms form 1 mole of N_2 molecules

$$2\ N(g) \longrightarrow N_2(g) \qquad \Delta H = -946\ kJ$$

STANDARD ENTHALPIES OF FORMATION

37. For each compound below, write a balanced equation depicting the formation of 1 mole of the compound from the elements. Look up the standard molar enthalpy of formation for each compound in Table 6.2 or Appendix J.

(a) $Al_2O_3(s)$
(b) $TiCl_4(\ell)$
(c) $Mg(OH)_2(s)$
(d) $NH_4NO_3(s)$
(e) $COCl_2(g)$

38. The molar enthalpy of formation of glucose, $C_6H_{12}O_6(s)$, is −1260 kJ/mol. (a) Is the formation of glucose from its elements endo- or exothermic? (b) Write a balanced equation depicting the formation of glucose from its elements and for which the standard enthalpy of reaction is −1260 kJ.

39. An important reaction in the production of sulfuric acid is

$$SO_2(g) + \tfrac{1}{2}\ O_2(g) \longrightarrow SO_3(g)$$

It is also a key reaction in the formation of acid rain, beginning with the air pollutant SO_2. Using the data in Table 6.2, calculate the enthalpy change for the reaction.

40. The molar enthalpy of formation for iron(II) bromide is −249.8 kJ/mol and that for iron(III) bromide is −268.2 kJ/mol. What is the enthalpy change for the following reaction?

$$FeBr_2(s) + \tfrac{1}{2}\ Br_2(\ell) \longrightarrow FeBr_3(s)$$

41. In photosynthesis, the sun's energy brings about the combination of CO_2 and H_2O to form O_2 and a carbon-

containing compound such as a sugar or hydrocarbon. In its simplest form, the reaction would be

$$CO_2(g) + 2\ H_2O(\ell) \longrightarrow 2\ O_2(g) + CH_4(g)$$

Using the enthalpies of formation in Table 6.2, (a) calculate the enthalpy of reaction and (b) decide whether the reaction is endo- or exothermic.

42. The first step in the production of nitric acid from ammonia involves the oxidation of NH_3.

$$4\ NH_3(g) + 5\ O_2(g) \longrightarrow 4\ NO(g) + 6\ H_2O(g)$$

Use the information in Table 6.2 or Appendix J to find the enthalpy change for this reaction. Is the reaction exo- or endothermic?

43. The Romans used CaO as mortar in stone structures. The CaO was mixed with water to give $Ca(OH)_2$, and this slowly reacted with CO_2 in the air to give limestone.

$$Ca(OH)_2(s) + CO_2(g) \longrightarrow CaCO_3(s) + H_2O(g)$$

Calculate the enthalpy change for this reaction.

44. A key reaction in the processing of uranium for use as fuel in nuclear power plants is the following:

$$UO_2(s) + 4\ HF(g) \longrightarrow UF_4(s) + 2\ H_2O(g)$$

Calculate the enthalpy change, $\Delta H°$, for the reaction using the data in Table 6.2, Appendix J, and the following: ΔH_f° for $UO_2(s) = -1085$ kJ/mol; ΔH_f° for $UF_4(s) = -1914$ kJ/mol.

45. Oxygen difluoride, OF_2, is a colorless, very poisonous gas that reacts rapidly with water vapor to produce O_2, HF, and heat.

$$OF_2(g) + H_2O(g) \longrightarrow 2\ HF(g) + O_2(g) + 318\ kJ$$

Using this information, and Table 6.2 or Appendix J, calculate the molar enthalpy of formation of $OF_2(g)$.

46. Calculate the standard molar enthalpy of formation of $SiCl_4(g)$ given the equation below and the information in Table 6.2 or Appendix J.

$$SiO_2(s) + 4\ HCl(g) + 139.5\ kJ \longrightarrow SiCl_4(g) + 2\ H_2O(g)$$

47. Calcium oxide reacts with CO_2 to give calcium carbonate. Using the enthalpies of formation in Table 6.2, calculate the heat evolved when 1.00 mole of CaO(s) reacts with $CO_2(g)$ to form 1.00 mole of $CaCO_3(s)$.

48. Iron can react with oxygen to give iron(III) oxide. If you heat 5.58 g of Fe in pure O_2 to give $Fe_2O_3(s)$, how much heat is liberated (at constant pressure)?

49. The formation of aluminum oxide from its elements is highly exothermic. If you burn 2.70 g of Al metal in pure O_2 to give Al_2O_3, how many kilojoules of heat energy are evolved in the process (at constant pressure)?

50. Ammonium nitrate is commonly used as a fertilizer, but if heated it will decompose to dinitrogen oxide and water.

$$NH_4NO_3(s) \longrightarrow N_2O(g) + 2\ H_2O(g)$$

If 1.00 kg of NH_4NO_3 decomposes, how many kilojoules of heat energy are involved (at constant pressure)?

51. Like all carbonates, barium carbonate can be decomposed to the metal oxide and CO_2.

$$BaCO_3(s) \longrightarrow BaO(s) + CO_2(g)$$

If ΔH_f° of barium carbonate is -1216.3 kJ/mol, how many kilojoules of heat energy (at constant pressure) are required to decompose 10.0 g of $BaCO_3$?

52. Nitroglycerin is a powerful explosive, giving four different gases when detonated.

$$2\ C_3H_5(NO_3)_3(\ell) \longrightarrow 3\ N_2(g) + \tfrac{1}{2}\ O_2(g) + 6\ CO_2(g) + 5\ H_2O(g)$$

(a) Given that the enthalpy of formation of nitroglycerin is -364 kJ/mol, and using the other enthalpies found in Table 6.2, calculate the energy (heat at constant pressure) liberated when 1.00 mole of nitroglycerin is detonated.

(b) If 10.0 pounds of nitroglycerin are detonated, how many kilojoules of heat energy are liberated (at constant pressure)?

53. Hydrazine and dimethylhydrazine both react readily with O_2 and can be used as rocket fuels.

$$N_2H_4(\ell) + O_2(g) \longrightarrow N_2(g) + 2\ H_2O(g)$$

hydrazine

$$N_2H_2(CH_3)_2(\ell) + 4\ O_2(g) \longrightarrow 2\ CO_2(g) + 4\ H_2O(g) + N_2(g)$$

dimethylhydrazine

The molar enthalpy of formation of liquid hydrazine is +50.6 kJ/mol and that of liquid dimethylhydrazine is +42.0 kJ/mol. By doing appropriate calculations, decide whether hydrazine or dimethylhydrazine gives more heat *per gram* upon reaction with O_2 (at constant pressure). (Other enthalpy of formation data can be obtained from Table 6.2.)

54. The reaction that occurs when a typical fat, glyceryl trioleate, is metabolized in the body is

$$C_{57}H_{104}O_6(s) + 80\ O_2(g) \longrightarrow 57\ CO_2(g) + 52\ H_2O(\ell)$$

(a) 37.8 kJ are evolved (at constant pressure) when 1.00 g of this fat ($M = 885.4$ g/mol) is metabolized. Using data in Table 6.2, calculate the molar enthalpy of formation of the fat in kJ/mol.
(b) How many kilojoules of energy must be evolved in the form of heat if you want to get rid of 1.00 pound (454 g) of this fat by combustion? How many Calories is this?

55. Using the reactions below, find the standard molar enthalpy of formation of PbO(s).

$$PbO(s) + C(graphite) \longrightarrow Pb(s) + CO(g) \quad \Delta H^\circ = +106.8 \text{ kJ}$$

$$2\ C(graphite) + O_2(g) \longrightarrow 2\ CO(g) \quad \Delta H^\circ = -221.0 \text{ kJ}$$

56. Given the following equations, calculate the molar enthalpy of formation of $MnO_2(s)$.

$$2\ MnO_2(s) \longrightarrow 2\ MnO(s) + O_2(g) \quad \Delta H^\circ = +264 \text{ kJ}$$

$$MnO_2(s) + Mn(s) \longrightarrow 2\ MnO(s) \quad \Delta H^\circ = -240 \text{ kJ}$$

CALORIMETRY

57. How many kilojoules of heat are evolved by a reaction in a bomb calorimeter (constant volume type; Figure 6.11) where the temperature of the bomb and water increase from 19.50 °C to 22.83 °C? The bomb has a heat capacity of 650. J/K, and the calorimeter contains 320. g of water.

58. Sulfur (2.56 g) was burned in a bomb calorimeter with excess $O_2(g)$. The temperature increased from 21.25 °C to 26.72 °C. The bomb had a heat capacity of 923 J/K. and the calorimeter contained 815 g of water. Calculate the heat evolved per mole of SO_2 formed in the course of the reaction $S(s) + O_2(g) \rightarrow SO_2(g)$.

59. You can find the amount of heat evolved in the combustion of carbon by carrying out the reaction in a combustion calorimeter. Suppose you burn 0.300 g of C(graphite) in an excess of $O_2(g)$ to give $CO_2(g)$.

$$C(graphite) + O_2(g) \longrightarrow CO_2(g)$$

The temperature of the calorimeter, which contains 775 g of water, increases from 25.00 °C to 27.38 °C. The heat capacity of the bomb is 893 J/K. What quantity of heat is evolved per mole of C?

60. Benzoic acid, $C_7H_6O_2$, occurs naturally in many berries. Suppose you burn 1.500 g of the compound in a combustion calorimeter and find that the temperature of the calorimeter increases from 22.50 °C to 31.69 °C. The calorimeter contains 775 g of water, and the bomb has a heat capacity of 893 J/K. How much heat is evolved per mole of benzoic acid?

61. You mix the strong base CsOH with the strong acid HCl.

$$CsOH(aq) + HCl(aq) \longrightarrow CsCl(aq) + H_2O(\ell)$$

Assume you mix 100. mL of 0.200 M CsOH with 50.0 mL of 0.400 M HCl. The temperature of the original solutions was 22.50 °C, and it rises to 24.28 °C after the acid–base reaction occurs. What is the enthalpy of the neutralization reaction per mole of CsOH? (Assume the densities of the solutions are all 1.00 g/mL and the specific heat capacities of the solutions are 4.20 J/g · K.)

62. If the strong base CsOH is mixed with a weak acid, HF, the heat evolved by the reaction differs from the heat of reaction with the strong acid HCl.

$$CsOH(aq) + HF(aq) \longrightarrow CsF(aq) + H_2O(\ell)$$

Assume you mix exactly 125 mL of 0.250 M CsOH with 50.0 mL of 0.625 M HF. The temperature of the original solutions was 21.50 °C, and it rises to 24.40 °C after the acid–base reaction occurs. What is the enthalpy of the neutralization reaction per mole of CsOH? (Assume the densities of the solutions are all 1.00 g/mL and the specific heat capacities of the solutions are 4.20 J/g · K.)

GENERAL PROBLEMS

63. The specific heat for copper metal is 0.385 J/g · K, while it is 0.128 J/g · K for gold. Assume you place 100. g of each metal, originally at 25 °C, in a boiling water bath at 100 °C. If each metal takes up heat at the same rate (the number of joules of heat absorbed per minute is the same), which piece of metal reaches 100 °C first?

64. Suppose you add 100.0 g of water at 60.0 °C to 100.0 g of ice at 0.00 °C. Some of the ice melts and cools the warm water to 0.00 °C. When the ice/water mixture has come to a uniform temperature of 0 °C, how much ice has melted?

65. In Example 6.13 we determined the molar enthalpy for the reaction of magnesium with aqueous hydrochloric acid. Compare the experimental result with a result calculated from the following information: ΔH_f° for HCl(aq, 1.00 M) = −167.2 kJ/mol and ΔH_f° for $MgCl_2(aq)$ = −801.2 kJ/mol.

66. The combustion of diborane, B_2H_6, proceeds according to the equation

$$B_2H_6(g) + 3\ O_2(g) \longrightarrow B_2O_3(s) + 3\ H_2O(g)$$

and 1941 kJ of heat energy is liberated per mole of $B_2H_6(g)$ (at constant pressure). Calculate the molar enthalpy of formation of $B_2H_6(g)$ using this information, the data in Table 6.2, and the fact that ΔH_f° for $B_2O_3(s)$ is −1273 kJ/mol.

67. Suppose you try to use copper to generate valuable hydrogen gas from water.

$$Cu(s) + H_2O(g) \longrightarrow CuO(s) + H_2(g)$$

(a) Is the reaction exo- or endothermic?
(b) If 2.00 g of copper metal react with excess water vapor, how much heat (at constant pressure) is involved (either absorbed or evolved) in the reaction?

68. As described in Study Question 30, the thermite reaction, the reaction between aluminum and iron(III) oxide,

$$2\ Al(s) + Fe_2O_3(s) \longrightarrow Al_2O_3(s) + 2\ Fe(s)$$

produces a considerable amount of heat; 852 kJ are evolved per mole of Al_2O_3 produced. If you begin with 2.0 g of Fe_2O_3 and a stoichiometrically correct amount of Al, is the heat that is produced sufficient to raise the temperature of the reaction products to the melting point of iron, 1530 °C? (Specific heat of Al_2O_3 = 0.775 J/g · K.)

69. As explained in Study Question 28, P_4 ignites in air to give P_4O_{10} and a large quantity of heat. This is an important reaction, since the phosphorus oxide can then be treated with water to give phosphoric acid for use in making detergents, toothpaste, soft drinks, and other consumer products (see Chapter 24). About 500×10^3 tons of elemental phosphorus are made annually in the United States. If you oxidize just one ton of P_4 (9.08×10^5 g) to the oxide, how much heat is evolved at constant pressure (in kJ)? The molar enthalpy of formation of P_4O_{10} is −2984 kJ/mol.

70. Magnesium reacts rapidly with aqueous HCl in an exothermic reaction

(i) $Mg(s) + 2\ HCl(aq) \longrightarrow MgCl_2(aq) + H_2(g)$

In Example 6.13 we used a coffee-cup calorimeter to find the enthalpy change for this reaction. Using the same technique, we can also determine the molar enthalpy for the reaction

(ii) $MgO(s) + 2\ HCl(aq) \longrightarrow H_2O(\ell) + MgCl_2(aq)$

If these two reactions are combined with the equation for the formation of water from its elements,

(iii) $H_2(g) + \frac{1}{2}\ O_2(g) \longrightarrow H_2O(\ell)$

the molar enthalpy for the formation of MgO(s) can be calculated by using Hess's Law.

(iv) $Mg(s) + \frac{1}{2}\ O_2(g) \longrightarrow MgO(s)$

We already have estimated ΔH for reaction (i) from Example 6.13, and ΔH for reaction (iii) can be found in Table 6.2. To find ΔH for reaction (ii) we do an experiment in a coffee-cup calorimeter. Here we find that, when 100.0 mL of 1.00 M HCl is added to 0.700 g of MgO(s), the temperature of the solution increases by 5.85 °C. (Assume the specific heat of the solution is 4.20 J/g · K and that the total mass of solution is 101 g.) Calculate ΔH for reaction (iv) from the information given.

71. Suppose you measured the enthalpy changes of the following two reactions in the laboratory.

$$2\ C(graphite) + 2\ H_2(g) \longrightarrow C_2H_4(g) \qquad \Delta H^\circ = +52.3\ kJ$$

$$C_2H_4Cl_2(g) \longrightarrow Cl_2(g) + C_2H_4(g) \qquad \Delta H^\circ = +116\ kJ$$

Calculate the molar enthalpy of formation of $C_2H_4Cl_2(g)$.

72. Given the following and the fact that the standard enthalpy of formation of formic acid, HCOOH, is −379 kJ/mol, calculate ΔH_f° for formaldehyde, $H_2CO(g)$.

$$H_2CO(g) + \frac{1}{2}\ O_2(g) \longrightarrow HCOOH(g) \qquad \Delta H^\circ = -270\ kJ$$

73. Given the following information and the data in Table 6.2, calculate the molar enthalpy of formation for liquid hydrazine, N_2H_4.

$$N_2H_4(\ell) + O_2(g) \longrightarrow N_2(g) + 2\ H_2O(g) \qquad \Delta H^\circ = -534\ kJ$$

74. Some years ago Texas City, Texas, was devastated by the explosion of a shipload of ammonium nitrate, a compound intended to be used as a fertilizer. When heated, however, ammonium nitrate can decompose exothermically to N_2O and water. (See photo p. 234.)

$$NH_4NO_3(s) \longrightarrow N_2O(g) + 2\ H_2O(g)$$

If the heat from this exothermic reaction is contained, higher temperatures are generated, at which point ammonium nitrate can decompose explosively to N_2, H_2O, and O_2.

$$2\ NH_4NO_3(s) \longrightarrow 2\ N_2(g) + 4\ H_2O(g) + O_2(g)$$

If oxidizable materials are present, fires can break out, as was the case at Texas City. Using the information in Appendix J, answer the following questions.

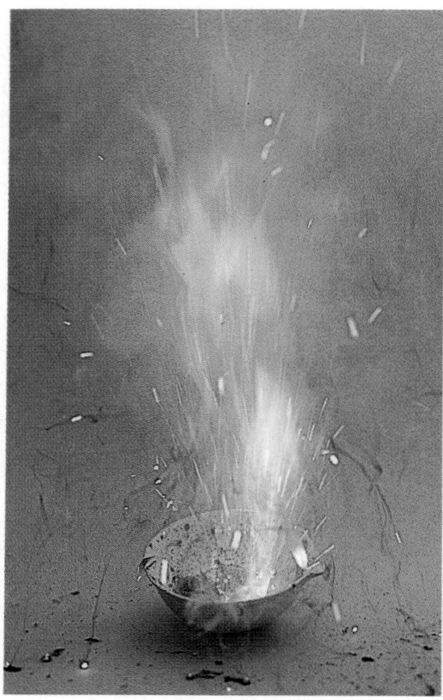

Decomposition of ammonium nitrate.

(a) If the enthalpy of formation of $N_2O(g)$ is 82.0 kJ/mol, how much heat is evolved (at constant pressure and under standard conditions) by the first reaction?
(b) If 8.00 kg of ammonium nitrate explode (the second reaction), how much heat is evolved (at constant pressure and under standard conditions)?

75. During World War II the German Air Force added powdered aluminum (from damaged airplanes) to ammonium nitrate to give powerful bombs. The oxygen released in the decomposition of NH_4NO_3 (Study Question 74) combined with the aluminum to give Al_2O_3 in an exothermic process.

$$2\ Al(s) + 3\ NH_4NO_3(s) \longrightarrow 3\ N_2(g) + 6\ H_2O(g) + Al_2O_3(s)$$

If you mix 8.00 kg of ammonium nitrate with a stoichiometric excess of powdered aluminum, how much heat energy can be evolved (at constant pressure)?

76. Boron-hydrogen compounds react with oxygen, some violently, to give water and boric oxide, B_2O_3. For example, pentaborane, B_5H_9, reacts as

$$2\ B_5H_9(\ell) + 12\ O_2(g) \longrightarrow 9\ H_2O(g) + 5\ B_2O_3(s) + 8703\ kJ$$

Calculate the standard molar enthalpy of formation of $B_5H_9(\ell)$ using data in Table 6.2 (or Appendix J) and $\Delta H_f^\circ[B_2O_3(s)] = -1273$ kJ/mol.

77. Uranium-235 is used as a fuel in nuclear power plants. Since natural uranium contains only a small amount of this isotope, the uranium must be enriched in uranium-235 before it can be used (Chapter 7). To do this, uranium(IV) oxide is first converted to a gaseous compound, UF_6, and isotopes are separated by a gaseous diffusion technique (Chapter 12). Some key reactions are

$$UO_2(s) + 4\ HF(g) \longrightarrow UF_4(s) + 2\ H_2O(g)$$

$$UF_4(s) + F_2(g) \longrightarrow UF_6(g)$$

How much heat energy (at constant pressure) would be involved in producing 225 tons of $UF_6(g)$ from UO_2? (1 T = 9.08×10^5 g) Some necessary enthalpies of formation are: $\Delta H_f^\circ[UO_2(s)] = -1084.9$ kJ/mol; $\Delta H_f^\circ[UF_4(s)] = -1914.2$ kJ/mol; $\Delta H_f^\circ[UF_6(g)] = -2147.4$ kJ/mol. See also Table 6.2 and Appendix J.

78. One method of producing H_2 on a large scale is the following chemical cycle.

Step 1: $SO_2(g) + 2\ H_2O(g) + Br_2(g) \longrightarrow H_2SO_4(\ell) + 2\ HBr(g)$

Step 2: $H_2SO_4(\ell) \longrightarrow H_2O(g) + SO_2(g) + \frac{1}{2}\ O_2(g)$

Step 3: $2\ HBr(g) \longrightarrow H_2(g) + Br_2(g)$

Using the table of standard enthalpies of formation in Appendix J, calculate ΔH° for each step. What is the equation for the *overall* process and what is its enthalpy change? Is the overall process exo- or endothermic?

79. One reaction involved in the conversion of iron ore to the metal is

$$FeO(s) + CO(g) \longrightarrow Fe(s) + CO_2(g)$$

Calculate the enthalpy change for this reaction using the following reactions of iron oxides with CO:
(a) $3\ Fe_2O_3(s) + CO(g) \longrightarrow 2\ Fe_3O_4(s) + CO_2(g)$ $\Delta H^\circ = -47$ kJ
(b) $Fe_2O_3(s) + 3\ CO(g) \longrightarrow 2\ Fe(s) + 3\ CO_2(g)$ $\Delta H^\circ = -25$ kJ
(c) $Fe_3O_4(s) + CO(g) \longrightarrow 3\ FeO(s) + CO_2(g)$ $\Delta H^\circ = +19$ kJ

80. If you want to convert 56.0 g of ice (at 0 °C) to water at 75.0 °C, how much propane (C_3H_8) would you have to burn in order to supply the heat (at constant pressure) to melt the ice and then warm it to the final temperature?

81. Suppose you want to heat your house with natural gas (CH_4). Assume your house has 1800 ft^2 of floor area and that the ceilings are 8.0 ft from the floors. The air in the house has a molar heat capacity of 29.1 J/mol · K. (The number of moles of air in the house can be found by assuming that the average molar mass of air is 28.9 g/mol and that the density of air at these

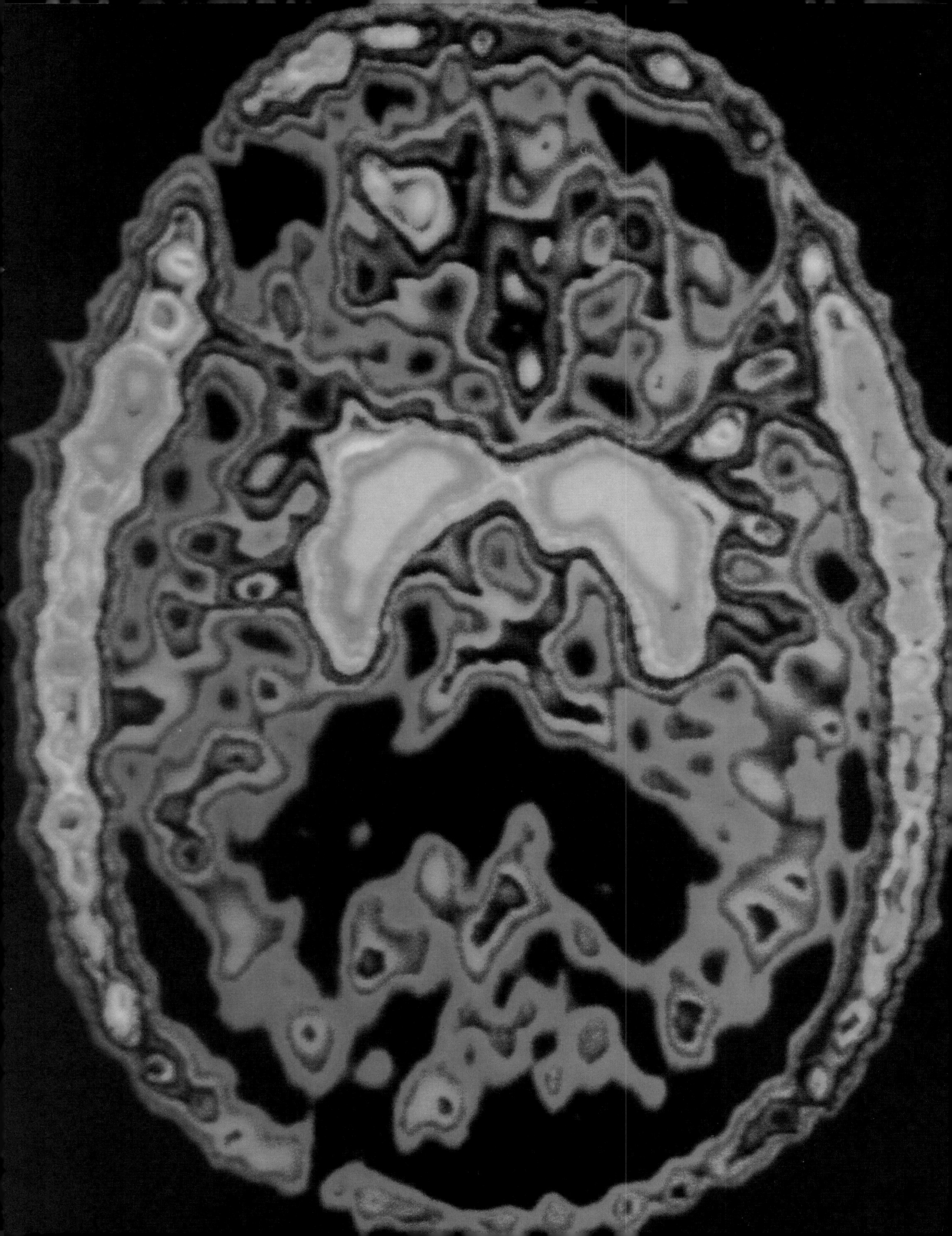

FLUORITE

by Roald Hoffmann

I was asked about my hobbies,
"Collecting minerals," I said
and stopped to think.
"Minerals in their matrix
are what I like best."

Fluorite wears a variable habit.
Colorless when pure, it is vodka
in stone. More commonly
it brandishes shades of rose to blue,
an occasional yellow. A specimen I have
tumbles in inch-long cubes,
superimposed, interpenetrating,
etched on all their faces.
The cubes have a palpable darkness,
a grainy darkness, texture
blacker than black.
Solid yet fragile when held
up to light, the darkness
deposited in this ordered
atomic form a million years ago
allows some rays through.
But only on the thin edges,
in sinister violet.

Struck with a chisel
the cubes cleave and
octahedra emerge.
I have seen it done,
but my hands tremble.
I know why it cleaves so,
but why destroy what took
centuries to grow, then
rested aeons in a cool
fissure in the rock?

Eerie crystal.
Were a Martian photograph
enlarged to reveal such polyhedral
regularity, it would be deemed
a sign of intelligence at work.
But the only work here, and it is free,
is that of entropy.

that is the key to my success. That is, I think I can empathize with what bothers the experimentalists. In another day I could have become an experimentalist."

Major issues in chemistry and science today

Professor Hoffmann has worked, and is presently working, at the forefront of several major areas of chemistry. He is presently quite intrigued by surface science. "For instance, there is the Fischer-Tropsch process, a pretty incredible thing in detail. Carbon monoxide and hydrogen gas come onto a metal catalyst, a surface of some sort, and off come long chain hydrocarbons and alcohols. The richness of all these things happening is intriguing, and we are on the verge of understanding. We now have structural information on surfaces that's reliable, and we are just beginning to get kinetic information. Surface science is at a crossing of chemistry, physics, and engineering. The field is in some danger of being spun off on its own, but I would like to keep it in chemistry.

"Bioinorganic chemistry is another such field. In my group we are doing some work trying to understand the mechanism of oxygen production in photosynthesis, the last steps. What is known is very little. There is an enzyme in photosystem II that involves 3 to 4 manganese atoms, and they are at oxidation state 3 to 4. And they somehow take oxide or hydroxide to peroxide and eventually to molecular oxygen. That's all we know. Experimentally, not theoretically, I think bioinorganic chemistry is a very interesting field."

Finally, Hoffmann remarked that "there are going to be finer and finer ways of controlling the synthesis of molecules, the most essential activity of chemists. If I were to point to a single thing that chemists do, it would be that they make molecules. Chemistry is the science of molecules and their transformations. The transformations are the essential part. I think there are exciting possibilities for chemical intervention into biological systems with an ever finer degree of control. We need not be afraid of nature. We can mimic it, and even surpass its synthetic capabilities. And find a way to cooperate with it."

Scientific literacy and democracy

Roald Hoffmann is very concerned not only about science in general and chemistry in particular, but also about our society. One of his concerns is scientific literacy because "some degree of scientific literacy is absolutely necessary today for the population at large as part of a democratic system of government. People have to make intelligent decisions about all kinds of technological issues." He recently offered his comments on this important issue in the *New York Times*. He wrote that "What concerns me about scientific, or humanistic, illiteracy is the barrier it poses to rational democratic governance. Democracy occasionally gives in to *technocracy*—a reliance on experts on matters such as genetic engineering, nuclear waste disposal, or the cost of medical care. That is fine, but the people must be able to vote intelligently on these issues. The less we know as a nation, the more we must rely on experts, and the more likely we are to be misled by demagogues. We must know more."

The responsibility of scientists

Our discussion of the importance of scientific literacy led to a conversation about the broader responsibilities of scientists. "Scientists have a great obligation to speak to the public," Hoffmann says. "We have an obligation as educators to train the next generation of people. We should pay as much attention to those people who are *not* going to be chemists, and sometimes need to make compromises about what is to be taught and what is the nature of our courses. I think scientists have an obligation to speak to the public broadly, and here I think they have been negligent. I think society is paying scientists money to do research, and can demand an accounting in plain language. That's why I put in a lot of time on that television show [*The World of Chemistry*]."

A teacher of chemistry—and proud of it

In the Nobel Yearbook, Professor Hoffmann wrote that the technical description of his work "does not communicate what I think is my major contribution. I am a teacher, and I am proud of it. At Cornell University I have taught primarily undergraduates. . . . I have also taught chemistry courses to non-scientists and graduate courses in bonding theory and quantum mechanics. To the chemistry community at large, and to my fellow scientists, I have tried to teach "applied theoretical chemistry": a special blend of computations stimulated by experiment and coupled to the construction of general models—frameworks for understanding." His success in this is unquestioned.

school year book, "under the picture of me with a crew cut, it says 'medical research' under my name." Indeed, he says that "medical research was a compromise between my interest in science and the typical Jewish middle class family pressures to become a medical doctor. The same kind of pressures seem to apply to Asian-Americans today."

When he went to Columbia University, Hoffmann enrolled as a pre-med student, but says that there were several factors that shifted him away from a career in medicine. One of these was his work at the National Bureau of Standards in Washington, D.C., for two summers and then at Brookhaven National Laboratory for a third summer. He says that these experiences gave him a feeling for the excitement of chemical research. Nonetheless, during his first summer at the Bureau of Standards, he "did some not very exciting work on the thermochemistry of cement." During his second summer there he went over to the National Institutes of Health to find out what medical research was about. "To my amazement," he says, "most of the people had Ph.D.'s and not M.D.'s. I just didn't know. Young people do not often know what is required for a given profession. Once I found that out, and found that I did well in chemistry, it made me feel that I didn't really have to do medicine, that I could do some research in chemistry or biology. Later, what influenced me to decide on theoretical chemistry was an excellent instructor. Had I had some really good instruction in organic chemistry, I'm pretty sure I would have become an organic chemist.

"At the very same time I was being exposed to the humanities, in part because of Columbia's core curriculum—which I think is a great idea—that had so-called contemporary civilization and humanities courses. I took advantage of the liberal arts education to the hilt, and that has remained with me all my life. The humanities teachers have remained permanently fixed in my mind and have changed my ways of thinking. These were the people who really had the intellectual impact on me and helped to shape my life.

"To trace the path, I was a latecomer to chemistry and was inspired by research. I think *research* is the way in. It just gives you a different perception."

A love for complexity

Having discussed what brought him into chemistry, we were interested in his view of the qualities that a student should possess to pursue a career in the field. He said that "one thing one needs to be a chemist is a love for complexity and richness. To some extent that is true of biology and natural history, too. I think one of the things that is beautiful about chemistry is that there are 10,000,000 compounds, each with different properties. What's beautiful when you make a molecule is that you can make derivatives in which you can vary substituents, the pieces of a molecule, and we know that those substituents give a molecule function, give it complexity and richness. That's why a protein or nucleic acid with all its variety is essential for life. That's why to me, intellectually, isomerism and stereochemistry in organic chemistry are at the heart of chemistry. I think we should teach that much earlier. It requires no mathematics, only a little model building; you can do this without theory. I think it is no accident that organic chemistry drew to it the intellects of its time."

Experiment and theory

Professor Hoffmann has spent his career immersed in the theories of chemistry. However, he believes that fundamentally "chemistry is an experimental science, in spite of some of my colleagues saying otherwise. However, the educational process certainly favors theory. It's in the nature of things for teacher and student both to want to understand and then give primacy to the soluble and the understood at the expense of other things. We also have this reductionist philosophy of science, the idea that the social sciences derive from biology, that biology follows from chemistry, chemistry from physics, and so on. This notion gives an inordinate amount of importance to theoretical thinking, the more mathematical the better. Of course this is not true in reality, but it's an ideology; it is a religion of science."

There is of course a role for theory. "You can't report just the facts and nothing but the facts; by themselves they are dull. They have to be woven into a framework so that there is understanding. That's accomplished usually by a theory. It may not be mathematical, but a qualitative network of relationships." Indeed, Hoffmann believes that the incorporation of theory into chemistry "is what made American science better than that in many other countries. The emphasis in chemistry on theory and theoretical understanding is very important, but not nearly as important as the syntheses and reactions of molecules.

"Although I think chemists need to like to do experiments, that doesn't mean there is no role for people like me. It turns out that I am really an experimental chemist hiding as a theoretician. I think

Roald Hoffmann

Roald Hoffmann is a remarkable individual. When he was only 44 years old, he shared the 1981 Nobel Prize in Chemistry with Kenichi Fukui of Japan for work in applied theoretical chemistry. In addition, he has received awards from the American Chemical Society in both organic chemistry and inorganic chemistry, the only person to have achieved this honor. And, in 1990, he was awarded the Priestley Medal, the highest award given by the American Chemical Society.

The numerous honors celebrating his achievements in chemistry tell only part of the story of his life. He was born to a Polish Jewish family in Zloczow, Poland in 1937, and was named Roald after the famous Norwegian explorer Roald Amundsen. Shortly after World War II began in 1939 the Nazis first forced him and his parents into a ghetto and then into a labor camp. However, his father smuggled Hoffmann and his mother out of the camp, and they were hidden for more than a year in the attic of a school house in the Ukraine. His father was later killed by the Nazis after trying to organize an attempt to break out of the labor camp. After the war Hoffmann, his mother, and his stepfather made their way west to Czechoslovakia, Austria, and then Germany. They finally emigrated to the United States, arriving in New York on Washington's birthday in 1949. That Hoffmann and his mother survived these years is our good fortune. Of the 12,000 Jews living in Zloczow in 1941 when the Nazis took over, only 80 people, three of them children, survived the Holocaust. One of those three children was Roald Hoffmann.

On arriving in New York, Hoffmann learned his sixth language, English. He went to public schools in New York City and then finally to Stuyvesant High School, one of the city's select science schools. From there he went to Columbia University and then on to Harvard University, where he earned his Ph.D. in 1962. Shortly thereafter, he began the work with Professor R.B. Woodward that eventually led to the Nobel Prize. Since 1965 he has been a professor at Cornell University, where he regularly teaches first-year general chemistry.

In addition to his work in chemistry, Professor Hoffmann also writes popular articles on science for the American Scientist and other magazines, and he has published two volumes of his poetry. Finally, he is appearing in a series of 26 half-hour television programs for a chemistry course called "World of Chemistry," scheduled to air on public television and cable channels in 1990.

From medicine to cement to theoretical chemistry

We visited Professor Hoffmann in his office at Cornell University, an office full of mineral samples, molecular models, and Japanese art. When asked what brought him into chemistry he said that "I came rather late to chemistry, I was not interested in it from childhood." However, he clearly feels that one can come late to chemistry, and that it can be a very positive thing. "I am always worried about fields in which people exhibit precocity, like music and mathematics. Precocity is some sort of evidence that you have to have talent. I don't like that. I like the idea that human beings can do anything they want to. They need to be trained sometimes. They need a teacher to awaken the intelligences within them. But to be a chemist requires no special talent, I'm glad to say. Anyone can do it, with hard work."

He took a standard chemistry course in high school. He recalls that it was a fine course, but apparently he found biology more enjoyable because, in his high

PART TWO

ATOMIC AND MOLECULAR STRUCTURE

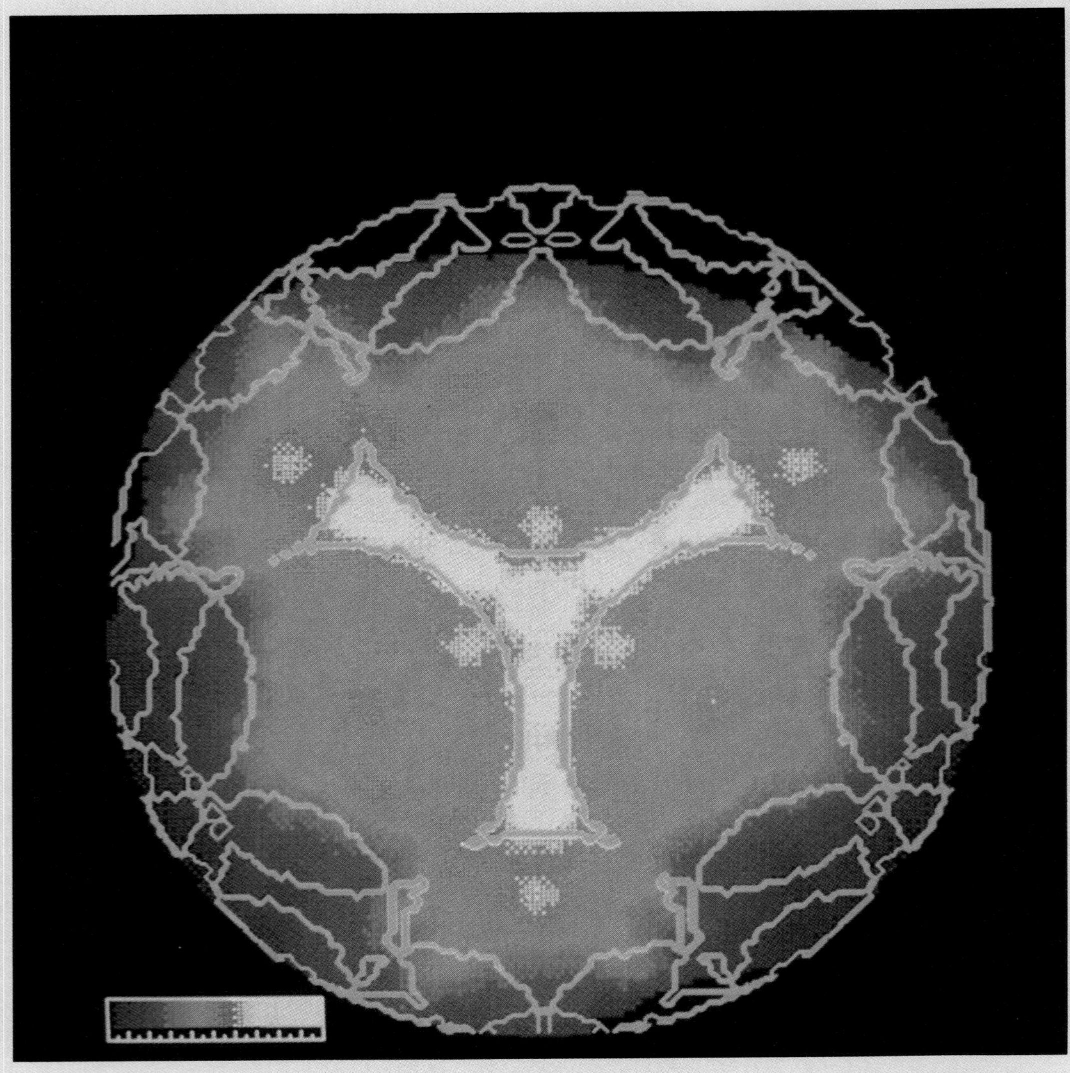

Theoretical angular distribution of Auger electrons emitted from a platinum[111] single-crystal surface. Angular distribution Auger microscopy (ADAM) is useful for direct imaging of interfacial structure and study of electron–solid interactions in the physical and biological sciences and engineering. (Douglas G. Frank, et al., ''Imaging Surface Atomic Structure by Means of Auger Electrons,'' *Science*, Jan. 12, 1990, pp. 182–188)

temperatures is about 1.22 g/L.) How much methane do you have to burn to heat the air from 15.0 °C to 22.0 °C?

82. Reactions that could be involved in the purification of uranium for nuclear reactor fuels are

Reaction 1: $U(s) + O_2(g) \longrightarrow UO_2(s)$ $\Delta H° = -1084.9$ kJ

Reaction 2: $UO_2(s) + 4\ HF(g) \longrightarrow UF_4(s) + 2\ H_2O(g)$ $\Delta H° = -228.5$ kJ

Reaction 3: $UF_4(s) + F_2(g) \longrightarrow UF_6(s)$ $\Delta H° = -282.8$ kJ

You know the enthalpies of reactions 1 to 3, and the enthalpies for the following reactions:

Reaction 4: $\frac{1}{2}\ H_2(g) + \frac{1}{2}\ F_2(g) \longrightarrow HF(g)$ $\Delta H° = -271.1$ kJ

Reaction 5: $H_2(g) + \frac{1}{2}\ O_2(g) \longrightarrow H_2O(g)$ $\Delta H° = -241.8$ kJ

What is the molar enthalpy of formation, ΔH_f°, of $UF_6(s)$?

83. Companies around the world are constantly researching compounds that can be used as a substitute for gasoline in automobiles. Perhaps the most promising of these is methyl alcohol, CH_3OH, a compound that can be made relatively inexpensively from coal. The alcohol has a smaller energy content than gasoline, but, with its higher octane rating, it burns more efficiently than gasoline in combustion engines. (It also has the added advantage of contributing to a lesser degree to some air pollutants.) Compare the amounts of heat produced per gram of CH_3OH and C_8H_{18} (octane), the latter being representative of the compounds in gasoline. (See Example 6.12 for the thermochemical information on octane.)

SUMMARY QUESTION

84. Sulfur dioxide, SO_2, is a major pollutant in our industrial society, and it is often found in wine.

(a) In wine-making, SO_2 is commonly added to kill microorganisms in the grape juice when it is put into vats before fermentation. Further, it is used to neutralize byproducts of the fermentation process, enhance wine flavor, and prevent oxidation. Wine usually contains 80 to 150 ppm SO_2 (1 ppm = 1 part per million = 1 gram of SO_2 per 1 million grams of wine). The United States produced 440 million gallons of wine in 1987. Assuming the density of wine is 1.00 g/cm³, and that the average bottle of wine contains 100. ppm of SO_2, how many grams and how many moles of SO_2 were contained in this wine?

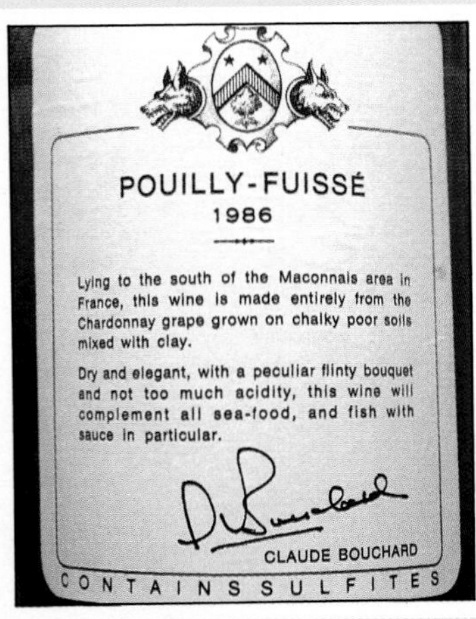

POUILLY-FUISSÉ
1986
Lying to the south of the Maconnais area in France, this wine is made entirely from the Chardonnay grape grown on chalky poor soils mixed with clay.
Dry and elegant, with a peculiar flinty bouquet and not too much acidity, this wine will complement all sea-food, and fish with sauce in particular.
CLAUDE BOUCHARD
CONTAINS SULFITES

SO_2 gas is added to kill microorganisms in wine and forms the SO_3^{2-} ion in the aqueous solution.

(b) When SO_2 is given off by an oil- or coal-burning power plant, it can be trapped by making it react with MgO in air to form $MgSO_4$.

$$MgO(s) + SO_2(g) + \tfrac{1}{2}\ O_2(g) \longrightarrow MgSO_4(s)$$

If 20. million tons of SO_2 are given off by coal-burning power plants each year, how much MgO would you have to supply to remove all of this SO_2? How much $MgSO_4$ would be produced?

(c) If ΔH_f° for $MgSO_4(s)$ is −1284.9 kJ/mol, how much heat (at constant pressure) is evolved or absorbed per mole of $MgSO_4$ by the reaction in part (b)?

(d) Sulfuric acid comes from the oxidation of sulfur, first to SO_2 and then to SO_3. The SO_3 is then absorbed by water to make H_2SO_4.

$$S(s) + O_2(g) \longrightarrow SO_2(g) \qquad \Delta H° = -296.8 \text{ kJ}$$

$$SO_2(g) + \tfrac{1}{2}\ O_2(g) \longrightarrow SO_3(g) \quad \Delta H° = -98.9 \text{ kJ}$$

$$SO_3(g) + H_2O(\text{in } 98\%\ H_2SO_4) \longrightarrow H_2SO_4(\ell) \quad \Delta H° = -130.0 \text{ kJ}$$

The typical plant produces 750. tons of H_2SO_4 per day (1 T = 9.08×10^5 g). Calculate the amount of heat produced by the plant per day, assuming standard conditions for all reactions.

CHAPTER 7

Nuclear Chemistry

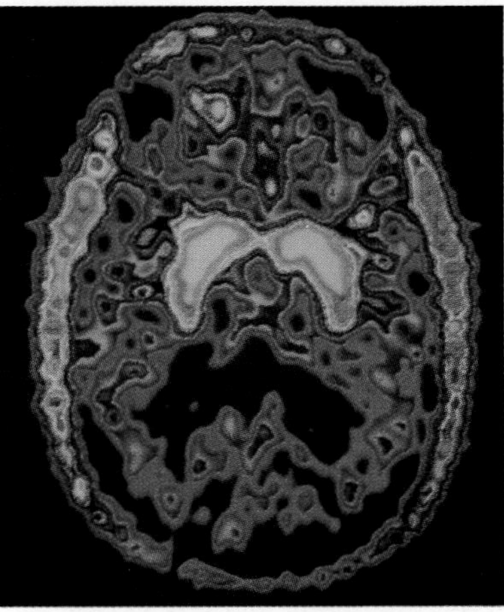

PET scan (positron emission tomography) of normal human brain. (Cheodoke-McMaster Hospitals)

CHAPTER OUTLINE

On August 2, 1939, as the world was on the brink of World War II, Albert Einstein sent a letter to President Franklin D. Roosevelt. In this letter, which profoundly changed the course of history, Einstein called attention to work being done on the physics of the atomic nucleus. He said he and others believed this work suggested the possibility that "uranium may be turned into a new and important source of energy . . . and [that it was] conceivable . . . that extremely powerful bombs of a new type may thus be constructed. . . ."

Powerful indeed! Einstein's letter was the beginning of the Manhattan Project, the project that led to the detonation of the first atomic bomb at 5:30 AM on July 16, 1945, in the desert of New Mexico. The rest of the world would learn the terrible truth of the power locked in the atomic nucleus a few weeks later, on August 6 and August 9, when the United States used atomic weapons against Japan.

For examples of some of "the many benefits from radiation," see the excellent article "Living with Radiation" by C.E. Cobb, Jr., National Geographic, April 1989, 403. See also a summary of nuclear chemistry by C.H. Atwood and R.K. Sheline, J. Chem. Educ., 1989, 66, 389.

As the physicist Freeman Dyson has put it, "the discovery of nuclear fission has gotten the world profoundly stuck: stuck in a buildup of atomic weapons, stuck in outdated concepts of war and peace, and stuck in human nature. There are many benefits to come from this discovery, but we must first find a way to get ourselves unstuck."

7.1 THE NATURE OF RADIOACTIVITY

The Making of the Atomic Bomb by Richard Rhodes (Simon and Schuster, 1986) is a comprehensive history of the exploration of atomic physics in this century and of the events leading up to the development of atomic weapons. This very readable book is highly recommended.

BECQUEREL, MARIE AND PIERRE CURIE, AND RUTHERFORD

Many minerals, called phosphors, glow for some time after being stimulated by exposure to sunlight or ultraviolet light. In 1897, Henri Becquerel was studying this phenomenon, called *phosphorescence*, when he accidentally discovered radioactivity.

Figure 7.1 Marie and Pierre Curie. The Curies and H.A. Becquerel shared the Nobel Prize in physics in 1903 for their research in radioactivity. Marie Curie received a second Nobel Prize, this in chemistry in 1911 for the discovery of radium and polonium. The latter was named in honor of her homeland, Poland. Pierre Curie also did important research in magnetic phenomena. He discovered the piezoelectric effect, an effect that is used to generate the sound in a digital watch, for example.

Becquerel wanted to see if a phosphorescent substance would emit x-rays. After a series of unsuccessful experiments, he tightly wrapped a photographic plate in paper and sprinkled a uranium salt onto the paper. He knew that x-rays will darken a photographic plate, so he believed that, if the uranium salt phosphoresced after being exposed to sunlight, and if this then caused the salt to emit not only visible light but also x-rays, he should see an image of the uranium salt after the plate had been developed. The experiment was a success! But it misled him. He of course tried the experiment again, but Paris was very dull that February of 1897, so he put the experiment away in a drawer until the sun reappeared and he could try again. Much to his amazement the plate was again darkened after being in the drawer, and he saw the image of the uranium salt. He quickly realized that he had observed penetrating radiation from inert matter that was not stimulated by light. Marie Curie, a physicist also working in Paris, named the phenomenon **radioactivity**. Indeed, *elements that naturally emit energy without the absorption of energy* are now said to be **naturally radioactive**.

Phosphors are the substances that make the hands of some clocks and wristwatches glow in the dark.

X-rays were discovered on November 8, 1895 by the German physicist Wilhelm Röntgen.

X-rays are one form of radiation. This and other types, such as ultraviolet radiation, visible light, and microwaves, are all forms of electromagnetic radiation. This topic is taken up in Chapter 8.

On studying Becquerel's discovery further, it was observed that the uranium-containing mineral *pitchblende* exhibited greater activity than other uranium salts. This observation indicated the presence of a substance even more radioactive than uranium, so Marie and Pierre Curie (Figure 7.1) analyzed the mineral. Using several *tons* of pitchblende, they isolated (in 1898) two new chemical elements: polonium (Po), 400 times more radioactive than uranium, and radium (Ra), a million times more radioactive than uranium.

Röntgen discovered x-rays just one month after a young man, Ernest Rutherford, arrived at Cambridge University in England from his home in New Zealand. Rutherford went there to work with the great J.J. Thomson, the most noted physicist of his day. Although Rutherford was

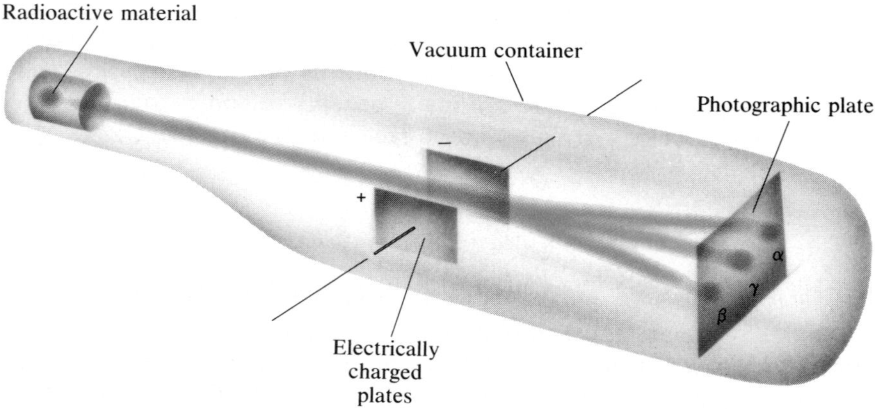

Figure 7.2 Ernest Rutherford discovered that radiation from radioactive element radium is composed of alpha (α) particles and beta (β) particles. Alpha particles are helium nuclei, ${}^{4}_{2}He^{2+}$, and so are attracted to a plate with a negative charge. Beta particles are electrons (${}^{0}_{-1}\beta$) and so are attracted to a positively charged plate. P. Villard later discovered a third type of radiation, gamma (γ) rays. Gamma radiation is not electrically charged.

Historical Figures in Chemistry: Ernest Rutherford (1871–1937)

Lord Rutherford was born in New Zealand in 1871, and went to Cambridge University in England to pursue his Ph.D. in physics in 1895. His original interest was in a phenomenon that we now call radio waves, and he apparently hoped to make his fortune in the field so that he could marry his fiance still back in New Zealand. However, his professor at Cambridge, J.J. Thomson, convinced him to work on the newly discovered phenomenon of radioactivity. It was at Cambridge that he discovered α and β radiation. Rutherford moved to McGill University in Canada in 1899. At McGill he did further experiments to prove that alpha radiation is actually composed of helium nuclei and that beta radiation consists of electrons. (For this work he received the Nobel Prize in Chemistry in 1908.)

At McGill, Rutherford was fortunate to have a talented student, Frederick Soddy, work with him. The two studied a radioactive gas coming from the radioactive element thorium. Their experiments showed that the gas was argon, which meant that they had made the first observation of the spontaneous disintegration of a radioactive element, one of the great discoveries of 20th century physics. (It was Soddy who coined the word "isotope" to describe the different forms of the same element.)

In 1903 Rutherford and his young wife visited the Curies in Paris, on the very day that Madame Curie received her doctorate in physics. There was of course a celebration, and that evening, while the party was in the garden of the Curies's home, Pierre Curie brought out a tube coated with a phosphor and containing a large quantity of radium in solution. The phosphor glowed brilliantly from the radiation given off by the radium. Rutherford later said the light was so bright he could clearly see Pierre Curie's hands were "in a very inflamed and painful state due to exposure to radium rays."

In 1907 Rutherford moved from Canada to Manchester University in England, and it was there that he performed the experiments that gave us the modern view of the atom (Chapter 8). In 1919 he moved back to Cambridge and assumed the position formerly held by J.J. Thomson. Not only was Rutherford responsible for very important work in physics and chemistry, but he also guided the work of no less than ten future recipients of the Nobel Prize.

most interested in radio waves, and had made important discoveries on that subject even as a very young man, Thomson convinced him to work on the most important problem of the day: radioactivity. Therefore, Rutherford set out to study the radiation from uranium and thorium. By the time he was 26 years old, he had made a major discovery. He found that "There are present at least two distinct types of radiation—one that is readily absorbed, which will be termed for convenience α [alpha] radiation, and the other of a more penetrative character, which will be termed β [beta] radiation." **Alpha radiation**, he discovered, was composed of particles, which, when passed through an electric field, were attracted to the negative side of the field (Figure 7.2). His later studies showed these particles to be helium nuclei ${}^{4}_{2}He^{2+}$, which were ejected at high speeds from a radioactive element (Table 7.1). As might be expected for such massive particles, they have limited penetrating power and can be stopped by several sheets of ordinary paper or clothing.

Table 7.1 Characteristics of α, β, and γ Emissions

Name	Symbol	Charge	Mass (g/particle)
Alpha	${}^{4}_{2}He^{2+}$, ${}^{4}_{2}\alpha$	+2	6.65×10^{-24}
Beta	${}^{0}_{-1}e$, ${}^{0}_{-1}\beta$	−1	9.11×10^{-28}
Gamma	${}^{0}_{0}\gamma$, γ	0	0

In the same experiment, Rutherford also found that β radiation (Figure 7.2) must be composed of negatively charged particles, since the beam of radiation was attracted to the electrically positive plate. Further work showed that these particles have an electric charge and mass equal to those of an electron. Thus, **beta particles** are electrons ejected at high speeds from some radioactive nuclei. They are more penetrating than alpha particles, since at least a $\frac{1}{8}$-inch piece of aluminum is necessary to stop beta particles, and they will penetrate several millimeters of living bone or tissue.

Rutherford hedged his bet when he said there were *at least* two types of radiation. Indeed, a third type was later discovered by P. Villard, a Frenchman, and he named it **γ (gamma) radiation**, using the third letter in the Greek alphabet in keeping with Rutherford's scheme. Whereas alpha and beta radiation are particulate in nature, γ-radiation is more like x-radiation, although γ-rays are more energetic than x-rays. Furthermore, γ-rays have no electrical charge; they are not affected by an electrical field (Figure 7.2). Finally, gamma radiation is the most penetrating, since it can pass completely through the human body. Thick layers of lead or concrete are required to minimize penetration.

RADIATION EFFECTS AND UNITS OF RADIATION

All three types of radiation disrupt normal cell processes in living organisms. Controlled exposure can of course be beneficial in destroying unwanted tissue, as in the radiation therapy used in treating some types of cancer. However, the potential for serious radiation damage to humans is well known. The biological effects of the atomic bombs exploded at Hiroshima and Nagasaki, Japan, at the close of World War II in 1945 have been well documented.

To quantify radiation and its effects, particularly on humans, several units have been developed. The unit called the **Curie** (Ci) is commonly used; it is the amount of any radioactive substance that undergoes 3.7×10^{10} disintegrations per second. To measure the dosage of x-rays and γ-rays we use a unit called the **Röntgen** (R), where one Röntgen corresponds to the *deposition* of 93.3×10^{-7} J per gram of tissue. The **rad** is similar to this, in that it measures the amount of any radiation *absorbed*; 1 rad represents a dose of 1.00×10^{-5} J absorbed per gram of material. Finally, to quantify the biological effects of radiation in general, the **rem** (standing for *R*öntgen *e*quivalent in *m*an) is used. One rem is a dose of any radiation that has the effect of 1 R.

To give you some feeling for these units, roughly 600 rad or 600 rem represents a fatal dose for humans. Further, the United States En-

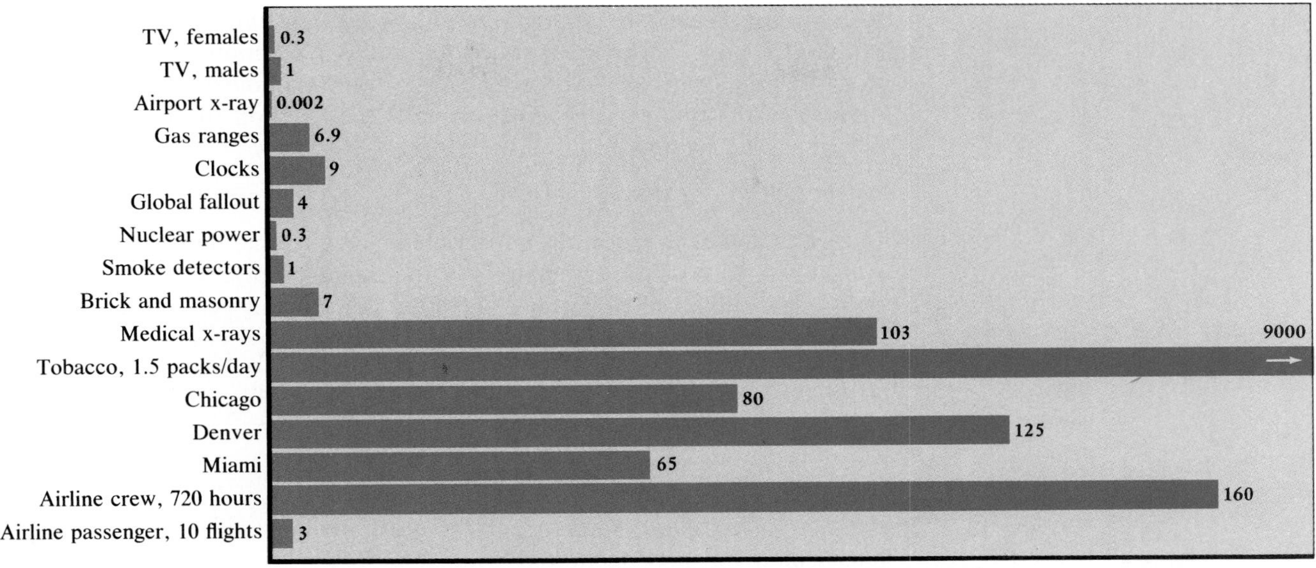

Figure 7.3 Radiation experienced by humans in the United States on an annual basis. The radiation from smoking 1.5 packs of cigarettes per day is about 9000 mrem. Note that this far exceeds all other sources of radiation combined.

vironmental Protection Agency has recommended that levels of the radioactive gas radon not exceed 4×10^{-12} Ci/L in homes. The radiation released in the accident at the Three Mile Island power plant in Pennsylvania in 1979 has been estimated at 20 Ci, with no one person exposed to more than about 100 mrem (1 mrem = 10^{-3} rem). In contrast, the accident at Chernobyl in the Soviet Union in 1986 released over 100 million curies (MCi), and some of the firefighters initially on the scene received more than 100 rem. By comparison, a typical dental x-ray exposes you to about 0.5 mrem.

Humans are constantly being exposed to natural and man-made background radiation (Figure 7.3), estimated to be 360 mrem per year. Sources include cosmic rays and radiation from natural radioactive isotopes in rocks, soil, water, air, and food, as well as radioactive isotopes that occur naturally in our bodies (for example, ^{40}K). In recent years, radioactive decay products from testing nuclear explosives in the atmosphere have added to the background. In addition, there are man-made sources such as x-ray generators, television, nuclear power plants and their wastes, nuclear weapons manufacture, nuclear fuel processing, and radioactive isotopes used for medical purposes.

Burning fossil fuels (coal and oil) releases naturally occurring radioactive isotopes into the atmosphere. This has added significantly to the background radiation in recent years.

7.2 NUCLEAR REACTIONS

EQUATIONS FOR NUCLEAR REACTIONS

Ernest Rutherford found that radium not only emits alpha particles but that it also produces the radioactive gas radon in the process. It was observations such as this that led Rutherford and Frederick Soddy, in 1902, to propose the revolutionary theory that *radioactivity is the result of a natural change of the isotope of one element into an isotope of a different element.* Such changes, called nuclear reactions or transmutations, generally involve a change in the atomic number and the mass number of the radioisotope. For example, the reaction studied by Rutherford can be written as

$$^{226}_{88}\text{Ra} \longrightarrow {}^{4}_{2}\text{He} + {}^{222}_{86}\text{Rn}$$

In this balanced equation the subscripts are the atomic numbers and the superscripts are the mass numbers of the radioisotopes.

The atoms in molecules and ions are rearranged in a chemical change; they are not created or destroyed. The number of atoms remains the same. Similarly, in nuclear reactions the nuclear particles, or **nucleons**, can be rearranged but not created or destroyed. The essence of nuclear reactions is that one nucleon can change into a different nucleon. For example, a proton and electron can change to a neutron, or a neutron can change to a proton and electron, but the total number of nucleons remains the same. Therefore, *the sum of the mass numbers of reacting nuclei must equal the sum of the mass numbers of the nuclei produced.* Furthermore, to maintain electrical balance, *the sum of the atomic numbers of products must equal the sum for the reactants.* These principles may be seen in the preceding nuclear reaction.

	$^{226}_{88}\text{Ra}$	$\longrightarrow$	$^{4}_{2}\text{He}$	+	$^{222}_{86}\text{Rn}$
	radium-226		alpha particle		radon-222
Mass number:	226	$\longrightarrow$	4	+	222
Atomic number:	88	$\longrightarrow$	2	+	86

REACTIONS INVOLVING ALPHA AND BETA PARTICLES

One way a radioactive isotope can disintegrate or decay is to eject an alpha particle from the nucleus. This is illustrated by the conversion of radium to radon above, and by the following reaction.

	$^{234}_{92}\text{U}$	$\longrightarrow$	$^{4}_{2}\text{He}$	+	$^{230}_{90}\text{Th}$
	uranium-234				thorium-230
Mass number:	234	$\longrightarrow$	4	+	230
Atomic number:	92	$\longrightarrow$	2	+	90

Notice that in alpha emission *both the atomic number and the mass number of the heavy atom decrease.*

Emission of a beta particle is another way for a given isotope to be radioactive. For example, loss of a beta particle by uranium-235 is represented by

	$^{235}_{92}\text{U}$	$\longrightarrow$	$^{0}_{-1}\beta$	+	$^{235}_{93}\text{Np}$
	uranium-235		beta particle		neptunium-235
Mass number:	235	$\longrightarrow$	0	+	235
Atomic number:	92	$\longrightarrow$	-1	+	93

Since a beta particle has a charge of -1, electrical balance makes the atomic number of the product atom *greater* than that of the reacting nucleus by one. However, the mass number does not change. The mass number of 0 for the electron is due to the small mass of the particle ($\frac{1}{2000}$ the mass of a proton).

How does a nucleus, composed only of protons and neutrons, eject an electron? It is generally accepted that a series of reactions is involved, but the net process is

$$\underset{\text{neutron}}{{}^{1}_{0}\text{n}} \longrightarrow \underset{\text{electron}}{{}^{0}_{-1}\beta} + \underset{\text{proton}}{{}^{1}_{1}\text{H}}$$

where we use the symbol H for a proton. (Recall that the nucleus of a hydrogen atom is just a single proton.) The ejection of a beta particle always means that a new element is formed with an atomic number one unit greater than the original nucleus.

EXERCISE 7.1 Nuclear Reactions: Alpha Emission

1. Write an equation showing the emission of an alpha particle by an isotope of neptunium, ${}^{237}_{93}\text{Np}$, to produce an isotope of protactinium.

2. Give the symbol, mass number, and atomic number of the product when each of the following is transformed to another element by alpha particle emission.
(a) ${}^{221}_{88}\text{Ra}$ (b) ${}^{220}_{86}\text{Rn}$

EXERCISE 7.2 Nuclear Reactions: Beta Emission

1. Write an equation showing the emission of a beta particle by an atom of sulfur-35, ${}^{35}_{16}\text{S}$, to produce an isotope of chlorine.

2. Give the symbol, mass number, and atomic number of the product when each of the following is transformed to another element by beta particle emission.
(a) ${}^{3}_{1}\text{H}$ (b) ${}^{60}_{26}\text{Fe}$

In many cases, the emission of an alpha or beta particle results in the formation of an isotope that is also radioactive. The new radioactive isotope may therefore undergo a number of successive transformations until a stable, nonradioactive isotope is finally produced. Such a cascade of reactions is called a **radioactive series**. One series, beginning with uranium-238 and ending with lead-206, is illustrated in Figure 7.4. The first step in the series is

A nucleus formed as a result of an alpha or beta emission is generally in an excited state and so emits a γ-ray

$${}^{238}_{92}\text{U} \longrightarrow {}^{4}_{2}\text{He} + {}^{234}_{90}\text{Th}$$

and the equation for the final step, the conversion of polonium-210 to lead-206, is

$${}^{210}_{84}\text{Po} \longrightarrow {}^{4}_{2}\text{He} + {}^{206}_{82}\text{Pb}$$

EXAMPLE 7.1

RADIOACTIVE SERIES

The second, third, and fourth steps in the uranium-238 series in Figure 7.4 involve emission of first a β particle, then another β particle, and finally an α particle. Write equations to show the products of these steps.

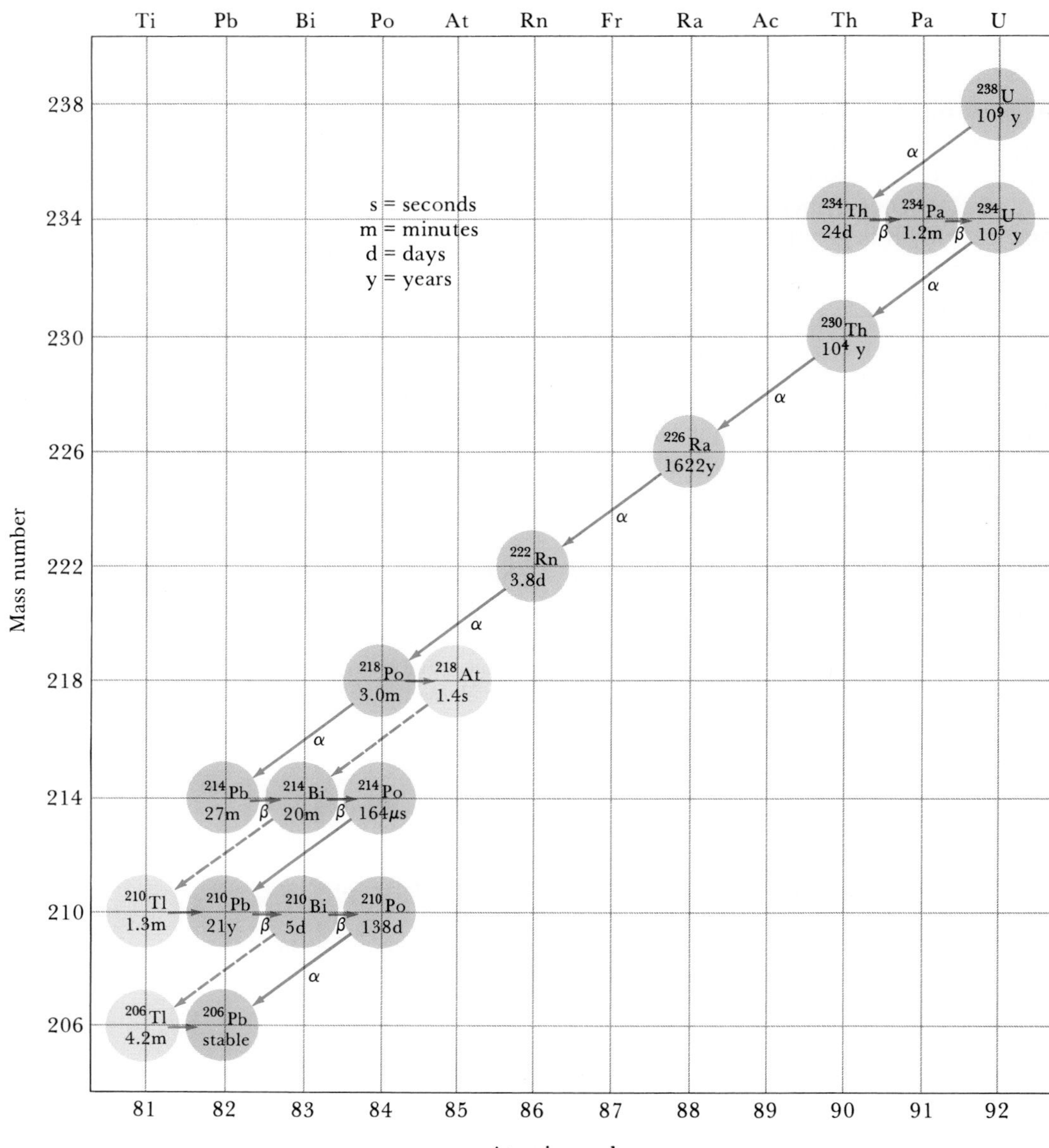

Figure 7.4 A radioactive series beginning with uranium-238 and ending with lead-206. In the first step, for example, ^{238}U emits an α particle to give thorium-234, $^{234}_{90}Th$. This radioactive isotope then emits a β particle to give protactinium-234, $^{234}_{91}Pa$. The protactinium-234 then emits another β particle to continue the series, which finally ends at lead-206, $^{206}_{82}Pb$.

Solution In the second step, thorium-234 is the starting point. Figure 7.4 shows that the mass remains the same but that the atomic number increases by 1 to 91, so the balanced equation is

$$\underset{\text{thorium-234}}{^{234}_{90}Th} \longrightarrow {}^{0}_{-1}\beta + \underset{\text{protactinium-234}}{^{234}_{91}Pa}$$

In the third step Figure 7.4 shows that the mass stays again constant, and the atomic number increases again by 1.

$$\underset{\text{protactinium-234}}{^{234}_{91}Pa} \longrightarrow {}^{0}_{-1}\beta + \underset{\text{uranium-234}}{^{234}_{92}U}$$

Finally, the fourth step involves alpha particle emission, so both the mass number and atomic number decline.

$$^{234}_{92}\text{U} \longrightarrow {}^{4}_{2}\text{He} + {}^{230}_{90}\text{Th}$$

uranium-234 thorium-230

EXERCISE 7.3 Radioactive Series

The actinium series begins with uranium-235, $^{235}_{92}\text{U}$, and ends with lead-207, $^{207}_{82}\text{Pb}$. The first five steps involve the emission of α, β, α, α, and β particles, respectively. Identify the radioactive isotope produced in each of the steps beginning with uranium-235.

OTHER TYPES OF RADIOACTIVE DECAY

In addition to radioactive decay by alpha, beta, or gamma radiation, other decay processes are observed. Some nuclei decay, for example, by emission of a **positron**, $_{+1}^{0}\beta$, which is effectively a positively charged electron.* Positron emission by polonium-207 leads to the formation of bismuth-207, for example.

	$^{207}_{84}\text{Po}$	$\longrightarrow$	$_{+1}^{0}\beta$ +	$^{207}_{83}\text{Bi}$
	polonium-207			bismuth-207
Mass number:	207	$\longrightarrow$	0 +	207
Atomic number:	84	$\longrightarrow$	+1 +	83

This process leads to reduction of the atomic number, as does a process called **K-capture**. This is the capture by the nucleus of an electron belonging to that atom.

	$^{7}_{4}\text{Be}$	+ $_{-1}^{0}\text{e}$ $\longrightarrow$	$^{7}_{3}\text{Li}$
	beryllium-7		lithium-7
Mass number:	7	0 $\longrightarrow$	7
Atomic number:	4	−1 $\longrightarrow$	3

EXERCISE 7.4 Nuclear Reactions

Balance the following nuclear reactions. Indicate the symbol, the mass number, and the atomic number of "?."

(a) $^{13}_{7}\text{N} \longrightarrow {}^{13}_{6}\text{C} + ?$
(b) $^{41}_{20}\text{Ca} + {}_{-1}^{0}\text{e} \longrightarrow ?$
(c) $^{90}_{38}\text{Sr} \longrightarrow {}^{90}_{39}\text{Y} + ?$
(d) $^{11}_{6}\text{C} \longrightarrow {}^{11}_{5}\text{B} + ?$
(e) $^{43}_{21}\text{Sc} \longrightarrow ? + {}^{1}_{1}\text{H}$

7.3 STABILITY OF ATOMIC NUCLEI

The fact that some nuclei are stable (nonradioactive) while others are unstable (radioactive) leads us to consider the reasons for stability. Figure 7.5 shows the naturally occurring isotopes of the elements from hydrogen

*The positron was discovered by Carl Anderson in 1932. It is sometimes called an "antielectron," one of a group of particles that have become known as "antimatter." "Matter" and "antimatter" are composed of particles with identical masses but opposite electrical charges. Contact between a particle and its antiparticle always leads to mutual annihilation of both particles with production of high-energy γ-rays.

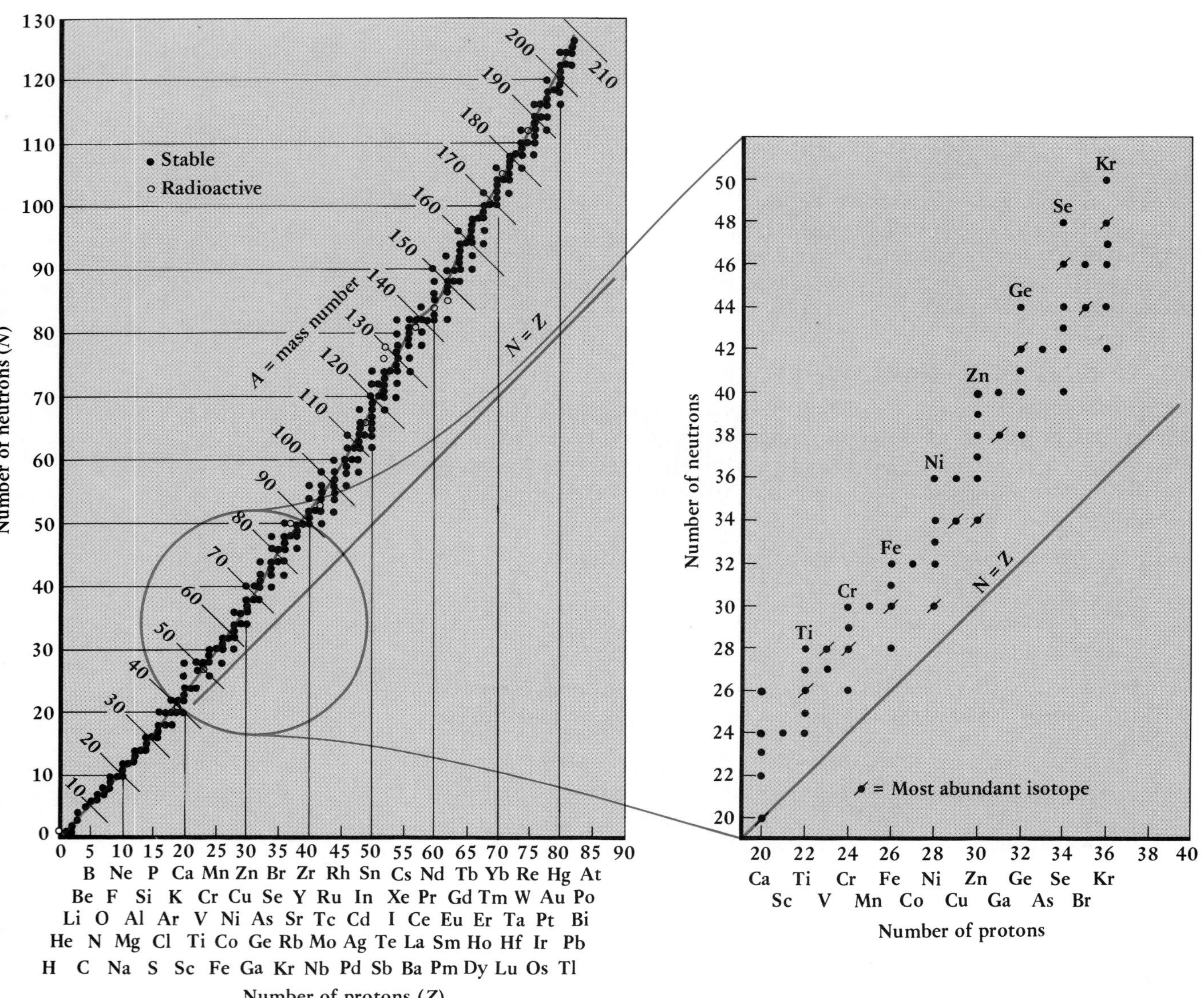

Figure 7.5 The naturally occurring isotopes of the elements from H ($Z = 1$) to Bi ($Z = 83$).

to bismuth. It is quite astonishing that there are so few! Why not hundreds more? It turns out that exploring this question will lead to more insight into radioactive decay and the reasons for nuclear stability.

WHY AREN'T THERE 15,862 ISOTOPES OF THE KNOWN ELEMENTS?

In its simplest and most abundant form, hydrogen has only one nuclear particle, the proton. In addition, the element has two other well known isotopes: deuterium, with one proton and one neutron ($^2_1H = D$), and tritium, with one proton and two neutrons (^3_1H = T). Helium, the next element, has four nuclear particles, two protons and two neutrons in its more abundant form. Jumping to the end of the actinide series you come to element 103, lawrencium. One isotope has a mass number of 257 and so has 154 neutrons. As you travel along the periodic table between H and Lr, you observe that (except for ^1H and ^3He) *the mass number of an isotope is always at least twice as large as the atomic number*. The reason for this is that every isotope of every element has a nucleus containing *at least* one neutron for every proton. Apparently the tremendous repulsive forces between protons in the nucleus are moderated by the presence of neutrons.

MAGIC NUMBERS The atomic nucleus is thought to have a regular "shell-like" arrangement of nucleons, where the number of nucleons in a shell is given by a "magic number." That is, there are apparently nuclear shells having a capacity of 2, 8, 20, 28, 50, 82, or 126 nucleons of one type (protons or neutrons). A nucleus with a magic number of *either* protons *or* neutrons has filled shells and is generally likely to be stable. Nuclei with magic numbers of *both* protons and neutrons will be *especially* stable; some examples are given in the table in the margin. There also seem to be other magic numbers corresponding to filled nuclear subshells of 6 (carbon), 14 (silicon), and 16 (sulfur) protons *and* the same number of neutrons, and it is just these elements that play a crucial role in the universe.

As you will learn in Chapter 9, the electrons outside of the nucleus are also arranged in shells and subshells.

ELEMENT	NO. PROTONS	NO. NEUTRONS
^4_2He	2	2
$^{16}_8O$	8	8
$^{40}_{20}Ca$	20	20
$^{208}_{82}Pb$	82	126

STABLE ISOTOPES Now let us look more closely at Figure 7.5. If every element (1 through 83) could have from 1 to 126 neutrons, then 10,458 combinations or isotopes should be possible. (Considering elements to $Z = 103$ and $N = 154$, that is, to Lr, then there would be 15,862 combinations.) Figure 7.5 makes it clear, though, that there are *very* few stable combinations of protons and neutrons, and this can give us some more insight into the problem of nuclear stability.

(a) For light elements up to Ca ($Z = 20$), the stable isotopes usually have equal numbers of protons and neutrons. Examples include $^{12}_6C$, $^{16}_8O$, and $^{32}_{16}S$. Where this is not true, the stable isotope contains only one more neutron than the number of protons (^7_3Li and $^{19}_9F$, for example).

(b) Beyond calcium there is an increasing surplus of neutrons. The band of stable isotopes deviates more and more from the line $N = Z$. For heavier elements, the repulsion between nuclear protons is larger, so more neutrons are needed for nuclear stability. For example, whereas one stable isotope of Fe has 26 protons and 30 neutrons, one of the stable isotopes of platinum has 78 protons and 117 neutrons.

(c) Above bismuth (83 protons and 126 neutrons) all isotopes are unstable. Beyond this point there is apparently no nuclear "super glue" sufficient to hold heavy nuclei together, and fission or splitting of the nucleus into smaller pieces occurs. Furthermore, the rate of splitting becomes greater the heavier the nucleus. For example, half of a sample of $^{238}_{92}U$ disintegrates in a billion years, whereas half of a sample of $^{257}_{103}Lr$ is gone in only 8 seconds.

(d) The enlargement of Figure 7.5 shows even more interesting features. First, elements of even atomic number have more isotopes than those of odd atomic number. Second, stable isotopes generally have an even number of neutrons. For elements of odd atomic number, this means that the most stable isotope has an even number of neutrons. To emphasize these points, of the more than 300 stable isotopes in Figure 7.5, roughly 200 have an even number of neutrons *and* an even number of protons. Only about 120 have an odd number of either protons or neutrons. Only four isotopes ($^{2}_{1}H$, $^{6}_{3}Li$, $^{10}_{5}B$, and $^{14}_{7}N$) have an odd number of both protons and neutrons. It is from observations such as these that the notion of magic numbers arose.

PENINSULA OF STABILITY AND TYPE OF RADIOACTIVE DECAY

The "band" of stable isotopes in Figure 7.5 is sometimes called the *peninsula of stability* in a "sea of instability." Any element with an isotope not on this peninsula will decay in such a way that it can come ashore on the peninsula, and the chart can help us predict what type of decay will be observed.

All elements beyond Bi ($Z = 83$) are unstable—that is, radioactive—and all generally decay by ejecting an alpha particle. For example, americium, a radioactive element used in smoke detectors, decays in this manner.

$$^{243}_{95}Am \longrightarrow {}^{4}_{2}He + {}^{239}_{93}Np$$

Such nuclei need to lose mass as efficiently as possible, and an alpha particle carries away four nucleons.

Beta decay occurs in elements that have too many neutrons to be stable ($N >> Z$); that is, isotopes of elements *above* the peninsula of stability in Figure 7.5. When a neutron transforms to a proton and an electron (which is then ejected), the mass remains constant, but the number of neutrons drops.

$$^{60}_{27}Co \longrightarrow {}^{0}_{-1}\beta + {}^{60}_{28}Ni$$

Conversely, lighter elements that have too few neutrons—isotopes of elements *below* the peninsula of stability—attain stability most efficiently by positron emission because this converts a proton to a neutron in one step.

$$^{13}_{7}N \longrightarrow {}^{0}_{+1}\beta + {}^{13}_{6}C$$

Finally, K-capture accomplishes the same conversion as positron emission.

$$^{41}_{20}Ca + {}^{0}_{-1}e \longrightarrow {}^{41}_{19}K$$

EXERCISE 7.5 Nuclear Stability

For each of the following isotopes, write an equation for its probable mode of decay.

(a) silicon-32, $^{32}_{14}Si$

(b) titanium-45, $^{45}_{22}Ti$

(c) plutonium-239, $^{239}_{94}Pu$

BINDING ENERGY

To tear apart the nucleus of an atom clearly requires the input of an enormous amount of energy. Otherwise the universe as we know it could not exist. Conversely, when a new atomic nucleus is formed from its constituent protons and neutrons (and other nucleons that are not important to discuss here), energy must be evolved. This is the basis for the enormous power of hydrogen bombs and for the great interest in atomic fusion. In the latter process, we hope to produce usable energy for consumers by forcing smaller nuclei to form larger ones (Section 7.8).

The energy evolved when a nucleus is formed from its constituent particles is called its **binding energy**. This is an important and useful quantity to know, because *the greater the binding energy of a nucleus, the greater its stability*.

When nucleons coalesce to form a nucleus, it is always observed that the mass of the new nucleus is less than the sum of the masses of the constituent protons and neutrons. This "missing mass" can be equated to its energy equivalent, a measure of the binding energy, by using a form of Einstein's equation, $\Delta E = (\Delta m)c^2$, where Δm is the "missing mass" and c is the velocity of light. In the example that follows, we find that the union of a mole of protons and a mole of neutrons to form a mole of deuterium atoms evolves more than 100 million kilojoules! This is truly an enormous quantity of energy when you realize that the union of two moles of H atoms to make a mole of H_2 molecules evolves only a piddling 435 kJ.

The quantity Δm is sometimes called the mass defect.

Notice that the calculation of binding energy closely resembles the thermodynamic calculations done in Chapter 6.

EXAMPLE 7.2

CALCULATING BINDING ENERGY

Consider the formation of deuterium from a proton (a hydrogen atom) and a neutron: ${}^1_1H + {}^1_0n \rightarrow {}^2_1H$. The required masses are ${}^1_1H$ = 1.007825 g/mol and ${}^1_0n$ = 1.008665 g/mol, and the actual mass of deuterium, ${}^2_1H$, is 2.01410 g/mol. Calculate the binding energy of ${}^2_1H$.

Solution The binding energy is calculated by first finding Δm, the difference between the sum of the masses of the constituent protons and neutrons (here the sum is 2.016490 g/mol) and the actual mass of the product nucleus.

$$\begin{aligned}\Delta m &= \text{actual mass of product} - \text{sum of masses of reactants}\\ &= 2.01410 \text{ g/mol} - 2.016490 \text{ g/mol}\\ &= -0.00239 \text{ g/mol}\end{aligned}$$

The energy liberated in the process can be calculated from $\Delta E = (\Delta m)c^2$. However, it is important to note that Δm *must* be given in kilograms (kg) and c in meters/second (3.00×10^8 m/s). If these units are used, then the answer is derived directly in joules (because 1 J = 1 kg · m^2/s^2; Appendix D.)

$$\Delta E = (-2.39 \times 10^{-6} \text{ kg/mol})(3.00 \times 10^8 \text{ m/s})^2$$

$$\Delta E = -2.15 \times 10^{11} \text{ J/mol}$$

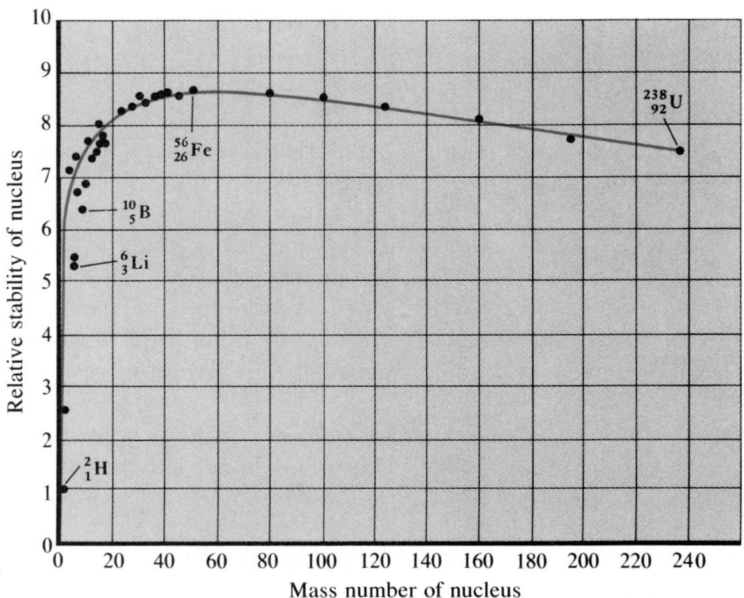

Figure 7.6 The relative stability of nuclei. This "curve of binding energy" was derived by calculating the binding energy per nucleon for isotopes of the elements from hydrogen to uranium.

EXERCISE 7.6 Binding Energy

Calculate the binding energy, in kJ/mol, for the formation of lithium-6.

$$3\ ^{1}_{1}H + 3\ ^{1}_{0}n \longrightarrow\ ^{6}_{3}Li$$

The necessary masses are $^{1}_{1}H$ = 1.00783 g/mol, $^{1}_{0}n$ = 1.00867 g/mol, and $^{6}_{3}Li$ = 6.01690 g/mol.

THE CURVE OF BINDING ENERGY

As stated above, the greater the binding energy, the greater the stability of the nucleus. Therefore, to gauge the relative stability of nuclei, scientists have calculated the binding energies per nucleon of many nuclei and have plotted them as a function of mass number (Figure 7.6). It is very interesting—and important—that the point of maximum stability occurs in the vicinity of iron-56, $^{56}_{26}Fe$. This means that *all elements are thermodynamically unstable with respect to iron*. That is, very heavy nuclei may split or undergo **fission**, with the release of enormous quantities of energy, to give more stable nuclei with atomic numbers near iron. Also, two very light nuclei may come together and undergo **fusion** exothermically to form heavier nuclei. Finally, *this is the reason that iron is the most abundant of the heavier elements in the universe*, an observation discussed more fully in Chapter 9.

7.4 RATES OF DISINTEGRATION REACTIONS

Cobalt-60 is used as a source of beta particles and γ-rays to treat malignancies in the human body. Although the element is radioactive, it is nonetheless reasonably stable, since half of a sample of the metal will disappear in a little over 5 years. On the other hand, copper-64, which is used in the form of copper(II) acetate to detect brain tumors, decays much more rapidly; half of the sample is gone in slightly less than 13 hours.

Table 7.2 Half-Lives of Some Common Radioactive Isotopes

Isotope	Decay Process	Half-life ($t_{1/2}$)
$^{238}_{92}U$	$^{238}_{92}U \longrightarrow ^{234}_{90}Th + ^{4}_{2}He$	4.47×10^9 years
$^{3}_{1}H$	$^{3}_{1}H \longrightarrow ^{3}_{2}He + ^{0}_{-1}\beta$	12.3 years
$^{14}_{6}C$	$^{14}_{6}C \longrightarrow ^{14}_{7}N + ^{0}_{-1}\beta$	5.73×10^3 years
$^{32}_{15}P$	$^{32}_{15}P \longrightarrow ^{32}_{16}S + ^{0}_{-1}\beta$	14.3 days
$^{60}_{27}Co$	$^{60}_{27}Co \longrightarrow ^{60}_{28}Ni + ^{0}_{-1}\beta$	5.27 years
$^{131}_{53}I$	$^{131}_{53}I \longrightarrow ^{131}_{54}Xe + ^{0}_{-1}\beta$	8.05 days
$^{239}_{94}Pu$	$^{239}_{94}Pu \longrightarrow ^{235}_{92}U + ^{4}_{2}He$	2.44×10^4 years

These two radioactive isotopes are clearly different in their stabilities, one of them disappearing at a greater rate than the other.

HALF-LIFE

The relative stabilities of radioactive isotopes are often expressed just as we have done above: in terms of the time required for one half of the sample to decay. This is called the **half-life**, $t_{1/2}$, of a radioactive isotope. As we see in Table 7.2, isotopes have widely varying half-lives; some take years for one half of the sample to decay, while others decay to half the original number of atoms in seconds.

As an example of the concept of half-life, consider the decay of oxygen-15, $^{15}_{8}O$, by positron emission.

$$^{15}_{8}O \longrightarrow ^{15}_{7}N + ^{0}_{+1}\beta$$

The half-life of oxygen-15 is 2.0 minutes. This means that one half of the quantity of $^{15}_{8}O$ present at any given time will have disintegrated 2.0 minutes later. Thus, if we begin with 20 mg of $^{15}_{8}O$, 10 mg of the isotope will remain after 2.0 minutes. After 4.0 minutes (two half-lives), only half of the remainder, or 5.0 mg, will still be there. After 6.0 minutes (three half-lives), only half of the 5.0 mg will still be present, or 2.5 mg, and so on. The amounts of $^{15}_{8}O$ present at various times are illustrated in Figure 7.7.

For nuclear disintegration reactions, the decay constant half-lives are constant, independent of temperature and of the number of radioactive nuclei present.

EXAMPLE 7.3

HALF-LIFE

Tritium ($^{3}_{1}H$), a radioactive isotope of hydrogen, has a half-life of 12.3 years.

$$^{3}_{1}H \longrightarrow ^{0}_{-1}\beta + ^{3}_{2}He$$

If you begin with 1.5 mg of the isotope, how many milligrams remain after 49.2 years?

Solution First, we find the number of half-lives in the given time period of 49.2 years. Since the half-life is 12.3 years, the number of half-lives is

$$49.2 \text{ years} \left(\frac{1 \text{ half-life}}{12.3 \text{ years}}\right) = 4.00 \text{ half-lives}$$

This means that the initial quantity of 1.5 mg is reduced four times by $\frac{1}{2}$.

$$1.5 \times \tfrac{1}{2} \times \tfrac{1}{2} \times \tfrac{1}{2} \times \tfrac{1}{2} = 1.5 \times (\tfrac{1}{2})^4 = 1.5 \times \tfrac{1}{16} = 0.094 \text{ mg}$$

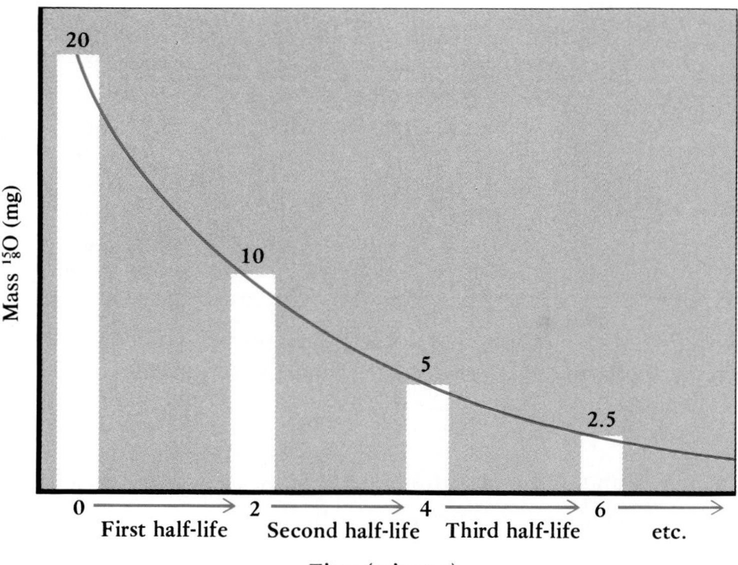

Figure 7.7 Decay of 20 mg of oxygen-15. After each half-life period of 2.0 minutes, the quantity present at the beginning of the period is reduced by half. The plot is based on the following data:

NUMBER OF HALF-LIVES	FRACTION OF INITIAL QUANTITY REMAINING	QUANTITY REMAINING (mg)
0	1	20.0 (initial)
1	1/2	10.0
2	1/4	5.00
3	1/8	2.50
4	1/16	1.25
5	1/32	0.625

EXERCISE 7.7 Radioactivity and Half-Life

Strontium-90 ($^{90}_{38}Sr$) is a radioisotope ($t_{1/2}$ = 28 years) produced in atomic bomb explosions. Its long life and tendency to concentrate in bone marrow make it particularly dangerous to man and animals.

(a) If the isotope decays with loss of a beta particle, write a balanced equation showing the other product of the decay.

(b) A sample of $^{90}_{38}Sr$ emits 2000 beta particles per minute. How many half-lives and how many years are necessary to reduce the emission to 125 beta particles per minute?

RATE OF RADIOACTIVE DECAY

To determine the half-life of a radioactive element, we have to measure the *rate of decay*, that is, the number of atoms that disintegrate per second or per hour or per year. The rate of nuclear decay is often described in terms of **activity** (A), the number of disintegrations per unit time. (The activity of a sample is measured using a device such as a Geiger counter, Figure 7.8.) The activity is *proportional* to the number of radioactive atoms present (N):

$$\text{Activity } (A) \propto \text{number of radioactive atoms present } (N)$$

This proportionality can also be expressed in the form

$$A = kN \tag{7.1}$$

$$\frac{\text{disintegrations}}{\text{time}} = \left[\frac{\text{disintegrations}}{\text{(number of atoms)(time)}}\right]\text{(number of atoms)}$$

where k is the proportionality constant or *decay constant*. The decay constant can be determined from an extraordinarily useful equation (7.2) that relates the time period over which a sample is observed (t) and the number of radioactive atoms present at the beginning (N_0) and end (N) of the time period.

Chapter 15 is a more detailed discussion of the rates of chemical reactions. Equation 7.2, which can be derived from Equation 7.1, is described more fully there.

$$\ln\frac{N}{N_0} = -kt \tag{7.2}$$

Equation 7.2 tells us that the greater the value of k, the faster the sample decays.

In words, the equation says

$$\text{natural logarithm}\left[\frac{\text{quantity of radioactive atoms after some time} = t}{\text{quantity of radioactive atoms at time} = 0}\right] = -\text{(decay constant)(time)}$$

Because of Equation 7.1, the ratio N/N_0 in Equation 7.2 is equal to the ratio A/A_0, where A is the activity measured when there are N atoms and A_0 is the activity measured at the beginning.

Now we are in a position to see how $t_{1/2}$ is determined. The half-life is the time needed for half of the material present at $t = 0$ (N_0) to disappear. Thus, when time $= t_{1/2}$, then $N = \frac{1}{2}N_0$. This means that

$$\ln\frac{\frac{1}{2}N_0}{N_0} = -kt_{1/2}$$

$$\ln\tfrac{1}{2} = -kt_{1/2} \quad \text{or} \quad \ln\tfrac{1}{2} = -\ln 2 = -kt_{1/2} \quad \text{or} \quad t_{1/2} = \frac{\ln 2}{k}$$

and so the half-life and decay constant are related by the natural logarithm of 2. The final expression above, $t_{1/2} = (\ln 2)/k$, is equivalent to

$$t_{1/2} = \frac{0.693}{k} \tag{7.3}$$

This equation connects the half-life and the decay constant through the number 0.693. Therefore, $t_{1/2}$ is found by calculating k from Equation 7.2, where N and N_0 in turn come from measurements of the activity of a radioactive sample at the beginning (A_0) and end (A) of some time period; the elapsed time is t.

Figure 7.8 A Geiger counter with a sample of carnotite, a mineral containing uranium oxide. The Geiger counter was invented by Hans Geiger and Ernest Rutherford in 1908. A charged particle (such as an α or β particle), when entering a gas-filled tube, ionizes the gas. These gaseous ions are attracted to electrically charged plates and thereby give rise to a "pulse" or momentary flow of electric current. The current is amplified and used to operate a counter. (Dave Davidson, Tom Stack & Associates)

EXAMPLE 7.4

DETERMINATION OF HALF-LIFE

A sample of radon initially undergoes 7.0×10^4 alpha particle disintegrations per second (dps). After 6.6 days, it undergoes only 2.1×10^4 alpha particle dps. What is the half-life of this isotope of radon?

Solution Experiment has provided us with both A and A_0.

$$A = 2.1 \times 10^4 \text{ dps} \qquad A_0 = 7.0 \times 10^4 \text{ dps}$$

and the time (t = 6.6 days). Therefore, we can find the value of k. Since $N/N_0 = A/A_0$,

$$\ln\left(\frac{2.1 \times 10^4}{7.0 \times 10^4}\right) = -k(6.6 \text{ days})$$

$$\ln(0.30) = -k(6.6 \text{ days})$$

$$k = -\frac{\ln(0.30)}{6.6 \text{ days}} = -\frac{(-1.20)}{6.6 \text{ days}} = 0.18 \text{ day}^{-1}$$

and from k we can obtain $t_{1/2}$.

$$t_{1/2} = \frac{0.693}{k} = \frac{0.693}{0.18 \text{ day}^{-1}} = 3.8 \text{ days}$$

EXAMPLE 7.5

TIME AND RADIOACTIVITY

Some high-level radioactive waste with a half-life, $t_{1/2}$, of 200. years is stored in underground tanks. What time is required to reduce an activity of 6.50×10^{12} disintegrations per minute (dpm) to a fairly harmless activity of 3.00×10^{-3} dpm?

Solution The data give you the initial activity ($A_0 = 6.50 \times 10^{12}$ dpm) and the activity after some elapsed time ($A = 3.00 \times 10^{-3}$ dpm). In order to find the elapsed time t, you must first find k from the half-life.

$$k = \frac{0.693}{t_{1/2}} = \frac{0.693}{200. \text{ years}} = 0.00347 \text{ yr}^{-1}$$

With k known, the time t can be calculated.

$$\ln\left(\frac{3.00 \times 10^{-3}}{6.50 \times 10^{12}}\right) = -[0.00347 \text{ yr}^{-1}]t$$

$$-35.312 = -[0.00347 \text{ yr}^{-1}]t$$

$$t = \frac{-35.312}{-(0.00347)\text{yr}^{-1}}$$

$$t = 1.02 \times 10^4 \text{ years}$$

EXAMPLE 7.6

DETERMINATION OF ACTIVITY

Radioactive iodine-131 is used in the form of sodium iodide to treat cancer of the thyroid. It decays by β emission with a half-life of 8.05 days.

$$^{131}_{53}\text{I} \longrightarrow \,^{0}_{-1}\beta + \,^{131}_{54}\text{Xe}$$

What is the beta activity in disintegrations per minute if the sample contains 1.0 microgram of ^{131}I?

Solution In order to estimate the activity of the sample from Equation 7.1 we need the decay constant, k. This is derived from the half-life and Equation 7.3.

$$k = 0.693/t_{1/2} = 0.693/8.05 \text{ days} = 0.0861/\text{day}$$

Since we want the answer in minutes, let us first convert the constant in units of (1/day) to units of (1/minute).

$$\left(\frac{0.0861}{\text{day}}\right)\left(\frac{\text{day}}{24 \text{ hours}}\right)\left(\frac{\text{hour}}{60 \text{ minutes}}\right) = 5.98 \times 10^{-5}/\text{min}$$

The activity is given by $A = kN$, where N is the number of radioactive *atoms* present. Therefore, we have to find the number of moles of ^{131}I present, and then the number of atoms of this isotope, before finding A. Since 1.0 microgram is 1.0×10^{-6} g,

$$\text{atoms of } ^{131}\text{I} = 1.0 \times 10^{-6} \text{ g}\left(\frac{\text{mol}}{131 \text{ g}}\right)\left(\frac{6.022 \times 10^{23} \text{ atoms}}{\text{mol}}\right)$$

$$= 4.6 \times 10^{15} \text{ atoms}$$

and so the activity of the sample is

$$\text{Activity } (A) = kN = (5.98 \times 10^{-5}/\text{min})(4.6 \times 10^{15} \text{ atoms})$$

$$= 2.8 \times 10^{11} \text{ disintegrations per minute}$$

EXERCISE 7.8 Rate of Radioactive Decay

Gallium citrate, containing the radioactive isotope gallium-67, is used medically as a tumor-seeking agent. It has a half-life of 77.9 hours. How much time elapses for a sample of gallium citrate to decay to 10.0% of its original activity? If you have 1.5×10^{-6} g of gallium-67, how many disintegrations are there per second?

RADIOCHEMICAL DATING

Scientists have used radiochemical dating to determine the age of rocks, fossils, and other artifacts that date back many years. For example, radiochemical methods were used recently to show that the Shroud of Turin was created somewhere around AD 1300, and not at the time of Christ as has been alleged for many centuries (Figure 7.9).

In 1946 Willard Libby developed the technique of age determination using radioactive carbon-14 ($^{14}_{6}C$). Carbon is an important building block of all living systems, and so all such things contain the three isotopes of carbon: ^{12}C, ^{13}C, and ^{14}C. The first two are stable and have been around since the universe was created. Carbon-14, however, is radioactive and decays to nitrogen-14 by β emission.

$$^{14}_{6}\text{C} \longrightarrow {}^{0}_{-1}\beta + {}^{14}_{7}\text{N}$$

Since the half-life of C-14 is known to be 5.73×10^3 years, the amount of the isotope present (N) can be measured from the activity of a sample. *If* the amount of C-14 originally in the sample (N_0) is known, then the age of the sample can be found from Equation 7.2.

This method of age determination clearly depends on knowing how much C-14 was originally in the sample. The answer to this question comes from work by physicist Serge Korff who discovered, in 1929, that C-14

Figure 7.9 The Shroud of Turin is a linen cloth over 4 m long. It bears a faint, straw-colored image of an adult male of average build who had apparently been crucified. Reliable records of the shroud date to about 1350, and for these past 600 years it has been alleged to be the burial shroud of Jesus Christ. Numerous chemical and other tests have been done on tiny fragments of the shroud in recent years. The general conclusion has been that the image was not painted on the cloth by any traditional method, but no one could say exactly how the image had been created. Recent advances in radiochemical dating methods, however, led to a new effort in 1987–1988 to estimate the age of the cloth. Using radioactive ^{14}C, these methods show that the flax from which the linen was made was grown between 1260 and 1390 AD. There is no chance that the cloth was made at the time of Christ, but no one still has any idea how the image was formed.

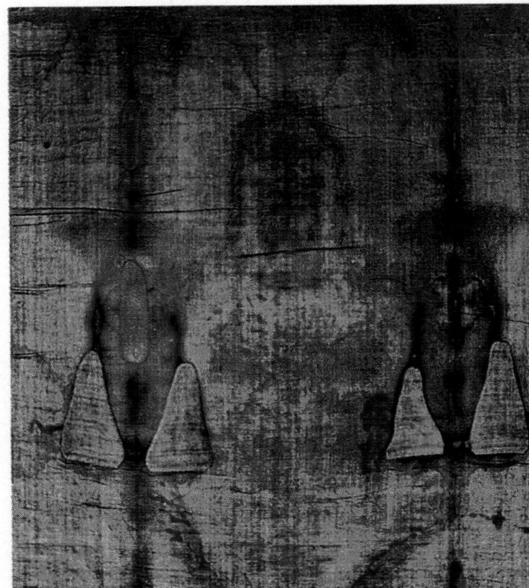

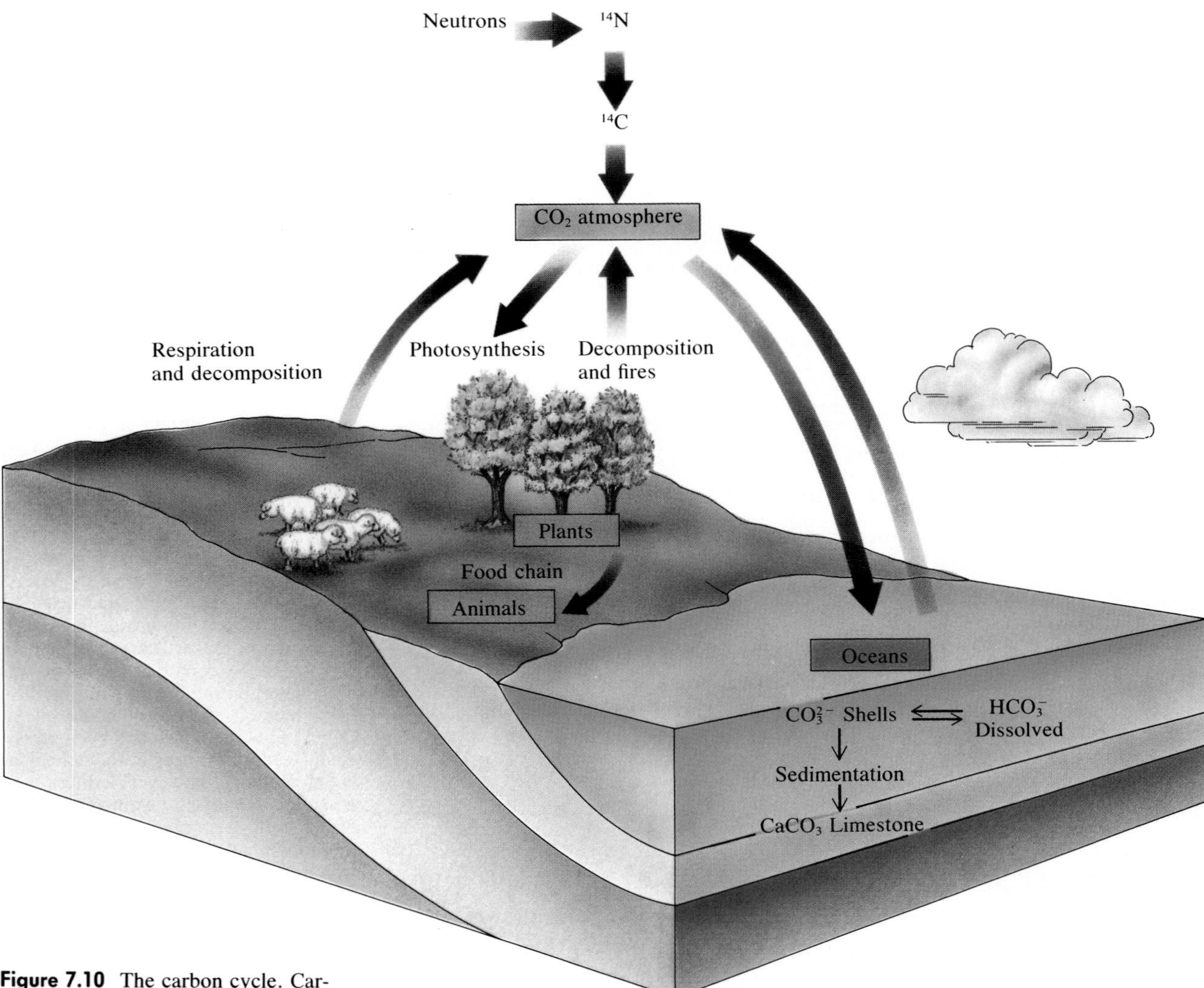

Figure 7.10 The carbon cycle. Carbon-14 is formed in the upper atmosphere by the interaction of nitrogen-14 and neutrons. The radioactive carbon, in the form of CO_2, is distributed throughout the globe in the oceans, atmosphere, and biosphere. (Based on *CHEM MATTERS*, February 1989, p. 12)

is continually generated in the upper atmosphere. High energy cosmic rays smash into gases in the upper atmosphere and force them to eject neutrons. These free neutrons collide with nitrogen atoms in the atmosphere and produce carbon-14.

$$^{14}_{7}N + ^{1}_{0}n \longrightarrow ^{14}_{6}C + ^{1}_{1}H$$

Throughout the *entire* atmosphere, only about 7.5 kg of ^{14}C are produced per year. However, this tiny amount of radioactive carbon, in the form of CO_2, becomes part of the "carbon cycle" on the earth and is distributed worldwide. As Figure 7.10 shows, there is a continual formation of C-14, exchange of the isotope within the oceans, atmosphere, and biosphere, and continual decay. This dynamic process keeps the supply of C-14 constant throughout the atmosphere and living systems.

Plants absorb carbon dioxide from the atmosphere, converting it into food, and so incorporate the carbon-14 into living tissue. It has been established that the beta activity of carbon-14 activity in *living* plants and animals and in the air is approximately constant at about 14 disintegrations per minute per gram of carbon. However, when the plant dies or is ingested by an animal, carbon-14 continues to disintegrate *without being replaced*; consequently, the activity decreases with passage of time. The smaller the activity of carbon-14, the longer the period between the death of the plant and the present time. Assuming that the C-14 activity in living organisms was about the same hundreds of years ago as it is now, measurement of the C-14 beta activity of an artifact of organic origin can be used to date the article.

For excellent articles on the Shroud of Turin and carbon-14 dating, see Chem Matters magazine, February 1989, pages 8–15.

EXAMPLE 7.7

RADIOCHEMICAL DATING

The so-called Dead Sea Scrolls, Hebrew manuscripts of the books of the Old Testament, were found in 1947. The activity of carbon-14 in the linen wrappings of the book of Isaiah is about 11 disintegrations per minute per gram (d/min · g). Calculate the approximate age of the linen.

Solution We will use Equation 7.2

$$\ln\left(\frac{N}{N_0}\right) = -kt$$

where N is proportional to the activity at the present time (11 d/min · g) and N_0 is proportional to the activity of carbon-14 in the living material (14 d/min · g). In order to calculate the time elapsed since the linen wrappings were part of a living plant, we first need k, the rate constant. From the text you know that $t_{1/2}$ is 5.73×10^3 years, so

$$k = \frac{0.693}{t_{1/2}} = \frac{0.693}{5.73 \times 10^3 \text{ years}} = 1.21 \times 10^{-4}\ \text{yr}^{-1}$$

Now everything is in place to calculate t.

$$\ln\left(\frac{11\ \text{d/min} \cdot \text{g}}{14\ \text{d/min} \cdot \text{g}}\right) = -[1.21 \times 10^{-4}\ \text{yr}^{-1}]t$$

$$t = \frac{\ln 0.79}{-[1.21 \times 10^{-4}\ \text{yr}^{-1}]}$$

$$= \frac{-0.24}{-[1.21 \times 10^{-4}\ \text{yr}^{-1}]}$$

$$= 2.0 \times 10^3 \text{ years}$$

Therefore, the linen is about 2000 years old.

EXERCISE 7.9 Radiochemical Dating

A Japanese wooden temple guardian statue of the Kamakura period (AD 1185–1334) has a carbon-activity of 12.9 d/min · g. What is the age of the statue? In what year was the statue made? The initial activity of carbon-14 = 14 d/min · g and $t_{1/2} = 5.73 \times 10^3$ years.

7.5 ARTIFICIAL TRANSMUTATIONS

In the course of his experiments, Rutherford found in 1919 that alpha particles ionize hydrogen gas, knocking off an electron. If nitrogen gas was used instead, he found that bombardment with alpha particles *also produced protons*. Quite correctly he concluded that the alpha particles had knocked a proton out of the nitrogen nucleus and that an isotope of another element had been produced. Nitrogen had undergone a *transmutation* to oxygen.

$${}^{4}_{2}\mathrm{He} + {}^{14}_{7}\mathrm{N} \longrightarrow {}^{17}_{8}\mathrm{O} + {}^{1}_{1}\mathrm{H}$$

Rutherford had proposed that protons and neutrons are the fundamental building blocks of nuclei. Although Rutherford's search for the neutron was not successful, it was found by James Chadwick in 1932 as a product of the alpha particle bombardment of beryllium.

$${}^{9}_{4}\mathrm{Be} + {}^{4}_{2}\mathrm{He} \longrightarrow {}^{12}_{6}\mathrm{C} + {}^{1}_{0}\mathrm{n}$$

Transforming one element into another by alpha particle bombardment has its limitations. Before a charged particle, such as the alpha particle, can be captured by a positively charged nucleus, the particle must have a sufficient kinetic energy to overcome the repulsive force developed as the positive particle approaches the positive nucleus. But the neutron is electrically neutral. Enrico Fermi (1934) reasoned, therefore, that a nucleus would not oppose its entry. Using this approach, practically all elements have since been transmuted, and a number of **transuranium elements** (elements beyond uranium) have been prepared. For example, uranium-238 forms neptunium-239 on neutron bombardment

$${}^{238}_{92}\mathrm{U} + {}^{1}_{0}\mathrm{n} \longrightarrow {}^{239}_{92}\mathrm{U} \longrightarrow {}^{239}_{93}\mathrm{Np} + {}^{0}_{-1}\beta$$

and the latter decays to plutonium.

$${}^{239}_{93}\mathrm{Np} \longrightarrow {}^{0}_{-1}\beta + {}^{239}_{94}\mathrm{Pu}$$

During World War II other new transuranium elements were prepared, and by 1945 the list had extended to curium, element 96. All of the new elements were made by a team of scientists headed by Glenn Seaborg at the Lawrence Berkeley Laboratory in Berkeley, California (Figure 7.11). One member of Seaborg's team was Albert Ghiorso. In 1952 he had a hunch that small quantities of elements 99 and 100 could be found in the debris from the detonation of the first H bomb at Eniwetok in the South Pacific. Indeed, he found einsteinium (element 99) on a piece of filter paper that had been flown on an airplane through the blast zone, and he located fermium (element 100) in a garbage can containing coral from a nearby island.

Of the 108 elements presently known, only elements up to uranium exist in nature (except for Tc, Pm, At, and Fr). The transuranium elements are all synthetic. Up to element 101, mendelevium, all of the elements can be produced by bombarding a lighter nucleus with a small particle such as an alpha particle or neutron. Beyond 101 special techniques using heavier particles are required and are still being developed. For example, lawrencium is made by bombarding californium-252 with boron nuclei

$${}^{252}_{98}\mathrm{Cf} + {}^{10}_{5}\mathrm{B} \longrightarrow {}^{257}_{103}\mathrm{Lr} + 5\,{}^{1}_{0}\mathrm{n}$$

Figure 7.11 Glenn Seaborg, recipient of the Nobel Prize in chemistry in 1951, received his Ph.D. from the University of California in 1937. He has been a president of the American Chemical Society and director of the Lawrence Berkeley Laboratory.

and the latest element to be discovered, 109, was made by firing iron nuclei at bismuth atoms.

EXERCISE 7.10 Nuclear Transmutation

Balance the following nuclear reactions, indicating the symbol, the mass number, and atomic number for "?."

(a) $^{13}_{6}C + ^{1}_{0}n \longrightarrow ^{4}_{2}He + ?$

(b) $^{14}_{7}N + ^{4}_{2}He \longrightarrow ^{1}_{0}n + ?$

(c) $^{253}_{99}Es + ^{4}_{2}He \longrightarrow ^{1}_{0}n + ?$

Figure 7.12 Enrico Fermi, the Italian physicist who first experimentally observed nuclear fission and who demonstrated a nuclear chain reaction.

7.6 NUCLEAR FISSION

The discovery of nuclear fission began in 1934. That year Irène Curie, daughter of Pierre and Marie, and her husband, Frédéric Joliot, discovered "artificial" radioactivity. Before that time, all radioactive substances had been found in minerals and ores. Joliot and Curie, however, found that they could *create* radioactive elements by bombarding nonradioactive ones with alpha particles. Certain stable nuclei, apparently content to sit quietly forever, could be made unstable if they were forced to absorb additional subatomic particles. These force-fed atomic nuclei, in an agitated state, began spewing out little pieces of themselves, just as in a "naturally" radioactive element. Enrico Fermi (Figure 7.12), then working in Rome, decided that neutrons may work better than alpha particles because a nucleus may more readily absorb a neutral particle than a positively charged alpha particle. Thus, Fermi bombarded a uranium nucleus to see what would happen, assuming, as did others, that the result would be an element close in mass to uranium. However, in 1938 the radiochemists Otto Hahn and Fritz Strassman found some barium in the remnants of the bombarded uranium! Further work and interpretation by Lise Meitner, Otto Frisch, Niels Bohr, and Leo Szilard confirmed that a uranium-235 nucleus had captured a neutron to form uranium-236 and that this heavier nucleus had undergone fission; that is, the nucleus had split in two (Figure 7.13).

$$^{235}_{92}U + ^{1}_{0}n \longrightarrow ^{236}_{92}U \longrightarrow ^{141}_{56}Ba + ^{92}_{36}Kr + 3\ ^{1}_{0}n \qquad \Delta E = -2 \times 10^{10}\ \text{kJ/mol}$$

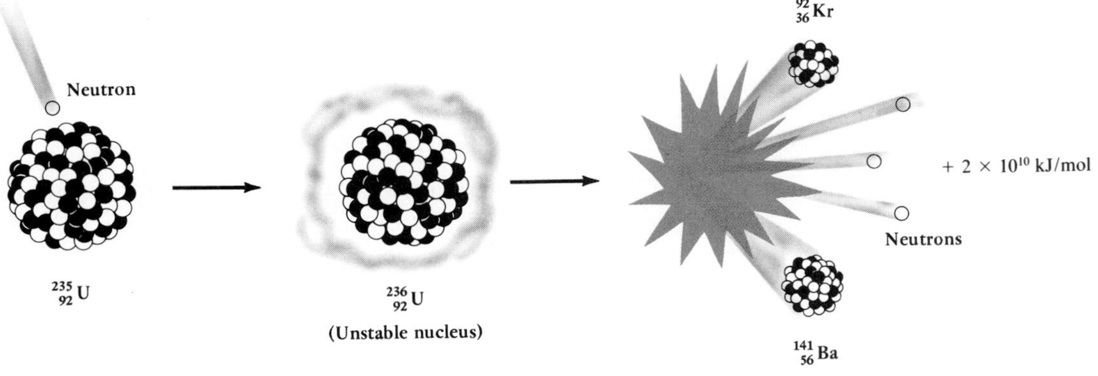

Figure 7.13 The fission of a $^{236}_{92}U$ nucleus that arises from the bombardment of $^{235}_{92}U$ with a neutron. The electrical repulsion between protons rips the nucleus apart.

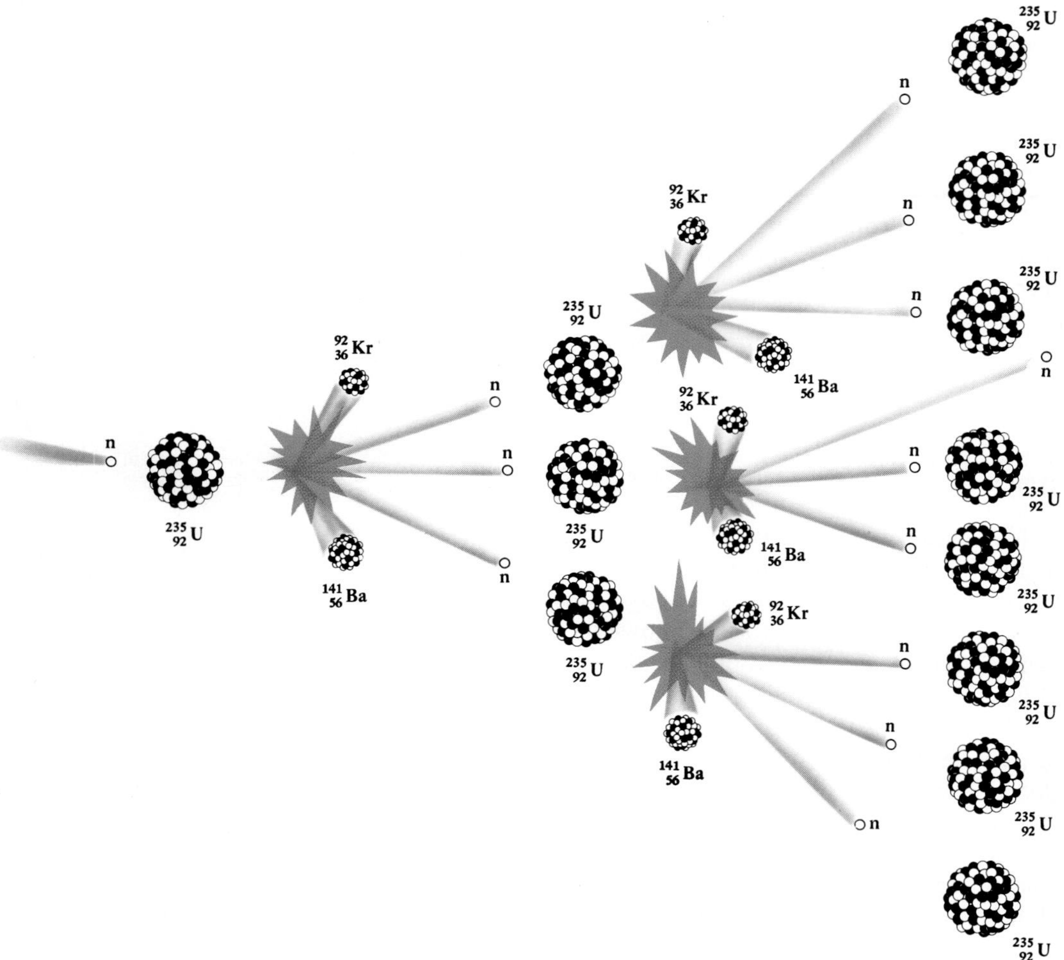

Figure 7.14 Illustration of a chain reaction initiated by capture of a stray neutron. (Many pairs of different isotopes are produced, but only one kind of pair is shown.)

The fact that the fission reaction produces more neutrons than are required to begin the process is important. A single neutron begins a process producing 3 neutrons capable of inducing 3 more fission reactions, which release 9 neutrons to induce 9 more fissions from which 27 neutrons are obtained, and so on. Since the fission of uranium-236 is extremely rapid, this sequence of reactions can be an explosively rapid **chain reaction** as illustrated in Figure 7.14. If the amount of uranium-235 is small, so few neutrons are captured by ^{235}U nuclei that the chain reaction cannot be sustained. In an atomic bomb, two small pieces of uranium, neither capable of sustaining a chain reaction, are brought together to form one piece capable of supporting a chain reaction, and an atomic explosion results.

7.7 CONTROLLED NUCLEAR REACTIONS

Rather than allow a nuclear reaction to run away explosively, it can be slowed by limiting the number of neutrons available, and energy can be derived safely and used as a heat source in a power plant (Figure 7.15). In a **nuclear** or **atomic reactor**, the rate of fission is controlled by inserting cadmium rods or other "neutron absorbers" into the reactor. The rods absorb the neutrons that cause fission reactions, so the rate of the fission

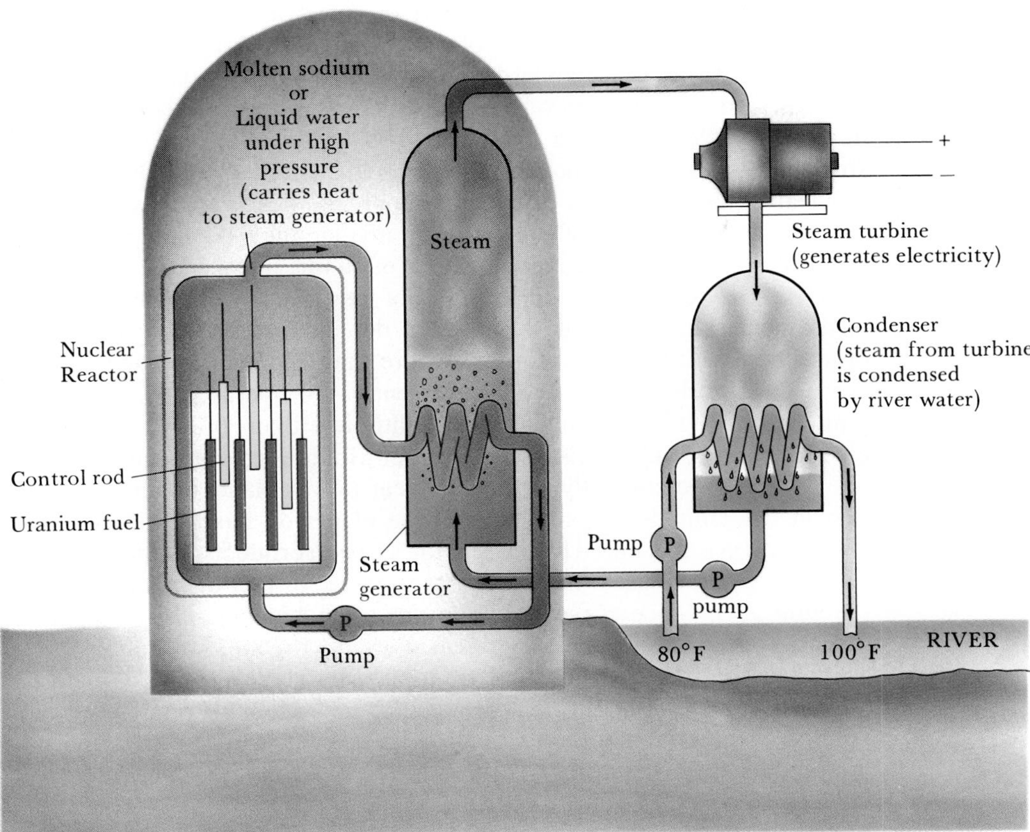

Figure 7.15 The essential parts of a nuclear power plant. Liquid water (or liquid sodium) is circulated through the reactor, where the liquid is heated to about 325°C. When this hot liquid is circulated through a steam generator, water in the generator is turned to steam, which in turn drives a steam turbine. After passing through the turbine, the steam is converted back to liquid water and is recirculated through the steam generator. Enormous quantities of outside cooling water from rivers or lakes are necessary to condense the steam. (This basic system is the same as in any power plant, except that the water or circulating liquid is heated initially by coal, gas, or oil-fired burners.)

reaction can be increased or decreased by withdrawing or inserting the rods. The rods are adjusted to make only about one neutron per fission available for the next fission reaction. If the number increases much above one, then the reaction is dangerously fast. If less than one neutron from each reaction is available, then the reaction effectively stops.

Not all nuclei can be made to split on colliding with a neutron, but ^{235}U and ^{239}Pu are two nuclei for which fission is possible. Unfortunately, natural uranium contains only 0.72% of the fissionable 235 isotope; more than 99% of the natural element is nonfissionable uranium-238. Since the percentage of natural ^{235}U is too small to sustain a chain reaction, uranium for nuclear fuel must be enriched. That is, some of the ^{238}U isotope is discarded, thereby raising the concentration of ^{235}U. The most common way to do this is by gaseous diffusion, as described in Chapter 12.

If the uranium-235 content of a sample is over 90%, it is considered of weapons quality.

There is of course some controversy surrounding the use of nuclear power plants, particularly in the United States. Their proponents regard

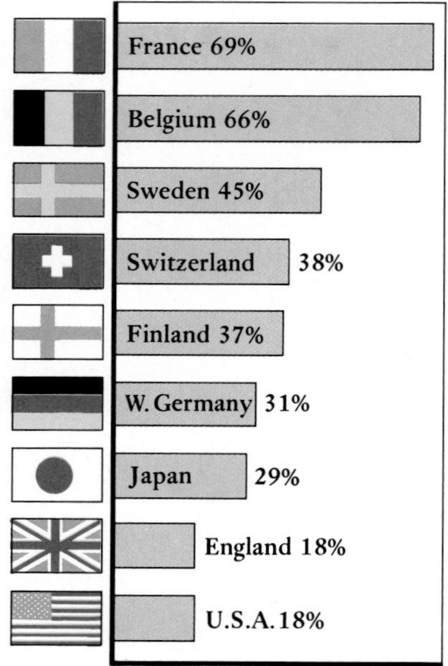

The share of electricity generated by nuclear power in 1988.

nuclear power to be an essential part of an advancing, technologically dependent society, independent of foreign energy sources. The health of our economy and our standard of living are dependent on inexpensive, reliable, and safe sources of energy. Just within the past few years the demand for electric power has once again begun to exceed the supply, so many believe nuclear power plants should be built to meet the demand. Nuclear power plants are capable of supplying these demands, and they can be the source of "clean" energy in that they do not pollute the atmosphere with ash, smoke, or oxides of sulfur, nitrogen, or carbon. In addition, they ensure that our supplies of fossil fuels will not be depleted in the near future, and they free us of dependence on such fuels from other countries. For these reasons, more than 350 nuclear power plants are in use worldwide. There are currently 110 operating plants in the United States. These plants supply slightly less than 20% of the nation's electric energy; only coal-fired plants contribute a greater share (57%).

There are currently *no* new nuclear power plants under construction in the United States because these plants do have disadvantages. One problem is presented by the reactor's fission products. Although some are put to various uses, many are not suitable for fuel or other purposes (Section 7.9). Since these products are often highly radioactive, their disposal poses an enormous problem. Perhaps the most reasonable suggestion is that radioactive wastes can be converted to a glassy material having a volume of about 2 m^3 per reactor per year; this relatively small volume of material can then be stored underground in geological formations like salt deposits, which are believed to be stable for hundreds of millions of years.

One of the greatest fears is a catastrophic accident leading to a large release of radioactivity. The United States may have come close to such an accident in the incident at the Three Mile Island plant in Pennsylvania on March 28, 1979. The cooling system failed, and water gushed out of the reactor. The loss of cooling water allowed the reactor core temperature to climb from its normal temperature of 650 °F to at least 4000 °F, and the upper portion of the core began to melt. The zirconium alloy tubes containing the uranium oxide fuel also began to react with steam at around 2000 °F, and this oxidation-reduction reaction produced hydrogen gas.

$$Zr(s) + 2\,H_2O(g) \longrightarrow ZrO_2(s) + 2\,H_2(g)$$

The hydrogen accumulated in the upper part of the reactor containment vessel, along with other gases (radioactive isotopes of Kr and Xe). Because of the danger of explosion from the hydrogen, the gas mixture was vented to the atmosphere.* Within a week of the accident, normal cooling water circulation was restored, and the core began to cool. However, the plant was closed, and cleanup is still proceeding.

*It is estimated that about 20 Ci of radioactive Xe and Kr gas were released. To keep this in perspective, it is important to compare this with the almost 3 million curies of Rn gas (and other radioactive elements) released by the explosion of the Mt. St. Helen's volcano in Washington on May 18, 1980. Not only did the volcano emit considerably more radiation than Three Mile Island, but Rn is considered about 1000 times more hazardous to health than Xe on a per curie basis due to the decay products of Rn. Finally, it is important to note that Xe and Kr are chemically inert gases and so do not readily enter the biosphere.

The Accident at Chernobyl

At about 1:24 AM on April 26, 1986, the Soviets have estimated that the power in reactor number 4 of the V.I. Lenin power plant rose from 200 megawatts to 3800 megawatts in 2.5 seconds and that it rose to 120 times normal power in another 1.5 seconds. The fuel temperature rose to 1600 to 1800 K, and the pressure created by steam formation in the reactor was so great that it blew the 1000-ton reactor lid off and ripped away the cooling pipes. The initial explosion scattered radioactive debris over the area and into the atmosphere. However, what was worse was that the graphite of the reactor began to burn, and this sent even more of the radioactive material from the reactor into the atmosphere. The fires were finally contained a few days later when 5000 tons of boron-containing compounds (a good "slower" of neutrons), dolomite, sand, clay, and lead were dropped by helicopter onto the wrecked reactor. To lower the temperature of the still hot core, liquid nitrogen was pumped into the space beneath the reactor vault. Thirty-one people have died from radiation sickness or burns, and another 230 people are hospitalized with radiation sickness.

Western experts have estimated that as much as 178 MCi (million curies) of radioactive nuclides have been released, primarily in the form of Xe, Kr, I, Te, and Cs isotopes. Long-term health effects are dominated by ^{131}I ($t_{1/2}$ = 8.05 days), ^{134}Cs($t_{1/2}$ = 2.06 years), and ^{137}Cs ($t_{1/2}$ = 30.2 years). It is estimated that 35 MCi of ^{131}I and approximately 2.4 MCi of ^{137}Cs were released. (In contrast, only 17 Ci of ^{131}I and no ^{137}Cs were released in the Three Mile Island accident.) There is great controversy about the long-term health effects of the accident. One estimate is that the number of deaths worldwide will rise by 28,000 above the normally expected number of 600 million in the next 50 years (a 0.005% increase).

In retrospect, both the Three Mile Island and Chernobyl accidents were preventable. First, both were caused by serious operator error. Second, it appears that reactor design outside the Soviet Union is superior in terms of safety. Not only are graphite-moderated reactors very difficult to control, but the Soviets do not use containment buildings. The consequences of operator training and reactor design are now obvious.

See Atwood, C.H., *J. Chem. Educ.* **1988**, *65*, 1037–1041 for an excellent discussion of the Chernobyl accident.

Figure 7.16 The nuclear power plant at Chernobyl in the Soviet Union shortly after the explosion and fire in April 1986. (V. Zufarov, Fotokhronika TASS, Sovfoto)

"Chernobyl is a reminder of the two great fears of our age. One is the fear that we will destroy ourselves. It comes and goes. But the other fear is always with us. It's what keeps the factories going—the fear of a civilization addicted to abundance, willing to risk anything except running out." Thomas Powers, Discover, July 1986, p. 34.

Another major reactor accident occurred in April 1986. On April 28, scientists in Sweden noticed an appreciable increase in traces of radioactive gases in the atmosphere. From the nature of the isotopes, they concluded that these isotopes could have come only from a significant reactor accident, and they traced the source to a place 700 miles away in the Soviet Union. The place—Chernobyl, a town near Kiev—has now become well-known. Chernobyl is the site of four 1000-megawatt nuclear reactors used to supply electric power. In late April 1986 one reactor was due to be shut down for routine maintenance and refueling. Before the shut-down, though, the operators decided to attempt an experiment involving the safety systems. The experiment was not done properly, and the reactor exploded at 1:24 on the morning of April 26, 1986 (Figure 7.16). The explosion and subsequent fire spread radioactive materials over a wide area of Europe and the Soviet Union.

Although residents of other countries should be continually vigilant regarding the safety of nuclear reactors, it is important to understand that the Soviet reactors are different from all of the reactors in the United States and in most of the rest of the world. First, Soviet reactors are not encased in a building designed to contain radioactive materials in the event of an accident as are reactors in the United States. Second, the Soviet reactors are encased in graphite (1000 tons or more), a material that slows neutrons. Not only are such reactors apparently difficult to control, but the explosion set fire to the graphite, and the fire was quite destructive.

7.8 NUCLEAR FUSION

Tremendous amounts of energy are generated when comparatively light nuclei combine to form heavier nuclei. Such a reaction is called nuclear fusion, and one of the best examples is the fusion of hydrogen nuclei (protons) to give helium.

$$4\ {}^{1}_{1}\mathrm{H} \longrightarrow {}^{4}_{2}\mathrm{He} + 2\ {}^{0}_{+1}\beta \qquad \Delta E = -2.5 \times 10^{9}\ \mathrm{kJ}$$

This reaction is the source of the energy from our sun, and it is the beginning of the synthesis of the elements in the universe (see Chapter 9). A temperature of 10^6 to 10^7 °C, the estimated temperature of the sun, is required to bring the positively charged nuclei together with enough kinetic energy to overcome repulsions.

It is the dream of many scientists that fusion reactions can be controlled and that the vast energy they emit can be harnessed for peaceful use. However, there are enormous problems producing the required temperatures in a controlled environment, and this goal has not yet been achieved. Of course, even if this problem is solved, there will be others. For example, there will be waste products in a fusion reactor, and we shall again have disposal problems.

On earth scientists have tried to force the fusion of two deuterium nuclei to give an isotope of helium

$${}^{2}_{1}\mathrm{H} + {}^{2}_{1}\mathrm{H} \longrightarrow {}^{3}_{2}\mathrm{He} + {}^{1}_{0}\mathrm{n} \qquad \Delta E = -2.9 \times 10^{8}\ \mathrm{kJ}$$

since this reaction likewise provides an enormous energy from a readily available and seemingly inexhaustible element. Once again, it has been assumed that enormous temperatures would be required to initiate the reaction. For this reason, chemists and physicists were very excited—and somewhat skeptical—when two chemists announced in early 1989 that they had achieved this fusion reaction at room temperature! At the present, skepticism has turned to disbelief, but some laboratories continue work on the idea because the promise of huge amounts of energy from inexpensive materials demands our attention.

The possibility of "cold fusion," the room-temperature fusion of deuterium atoms to provide energy, continues to be the subject of controversy. The reaction to the announcement in early 1989 by the international scientific community illustrates well how the process of scientific investigation works. See J. Chem. Educ., 1989, 66, 449.

7.9 USES OF RADIOACTIVITY

Radioactivity has found many uses in medicine, in chemical analysis, in chemical research as a "tracer" or marker in chemical reactions, and in the preservation of food (Figure 7.17).

Although radioactivity may induce cancer in living organisms, radiation has also been used successfully to treat cancer. Cancerous cells are more sensitive to radiation than normal cells, so the physician attempts to find a dosage that destroys malignant cells but not normal cells. Generally, the cancerous cells are irradiated with gamma radiation in one of three ways: (a) from an external source, usually cobalt-60, directed into the body so that it passes through the malignant growth; (b) by inserting a radioisotope directly into cancerous tissue; or (c) by inserting the radiation source in an accessible body cavity near the cancer.

One of the most widespread applications of radioisotopes is their use as tracers. Since the chemical properties of all the isotopes of an element are identical, a radioisotope may be used to "tag" an element without changing its chemical properties. Since the radioisotope is easily detected with a Geiger counter, the element may be followed or "traced" through a series of reactions. For example, radioactive iodine-125 is used in the diagnosis and treatment of thyroid gland disorders. The thyroid gland, located in the neck, affects the rate of growth and metabolism. Disorders of the thyroid are associated with the rate at which iodine-125 appears in the gland. Thus, a patient is given a solution tagged with $Na^{125}I$, and a detector placed near the thyroid measures the uptake of iodine by the gland.

Significant differences, resulting from mass differences, are found in the physical properties and reaction rates for lighter elements, particularly the isotopes of hydrogen.

The man-made element technetium, Tc, is also used for thyroid studies in the form of radioactive $Na^{99}TcO_4$. Technetium-99 has a half-life of 6.05 hours compared with 60 days for iodine-125.

Reactions of many biochemicals important to metabolism, such as fats, sugars, amino acids, hormones, and drugs, can be followed in the body by tagging these compounds with carbon-14 or tritium (^3_1H). In a recent study of the metabolism of tetrahydrocannabinol (the active component of marijuana or hashish), the drug was tagged with radioactive atoms and then administered intravenously. Using a detector, it was found that the compound persisted in the bloodstream for more than three days, and the products of metabolism were excreted in the urine for more than eight days.

The ban on the use of certain pesticides such as DDT has accelerated the search for other methods for the control of destructive insects. One application of radioactivity is based on the fact that less than lethal doses of radioactivity can cause sterility in male insects. The release of sterile insects into the environment means that the majority of the eggs produced by the female do not hatch, and the insect population is greatly reduced. Such a technique has been used successfully in the past few years to control the Mediterranean fruit fly in California and Texas.

Figure 7.17 (top) Strawberries irradiated with gamma rays from radioactive isotopes are still fresh after 15 days storage at 4 °C, while those not irradiated are moldy (bottom). (International Atomic Energy Agency)

The intensity of radiation depends on the quantity of absorbing material through which it passes. Because of this, radiation is commonly used to gauge the thickness of industrial products such as metal sheets, plastic film, and cigarettes. The wear-resistant properties of the moving parts of an automobile engine, for example, are easily measured by using radioactive parts and determining the amount of radioactivity transferred to the lubricating oil. In a similar way, the wear resistance of automobile tires and floor wax can be determined.

Neutron activation analysis enables us to analyze for elements in complex biological systems or in a meteorite, for example, *without* destruction of the sample. When a sample is exposed to neutrons, each constituent element may form a specific radioisotope that can be identified by its characteristic radiation and half-life. Furthermore, the mass of the element present can be determined from the activity of the derived radioisotope. In 1989 an example of this technique was first used to the benefit of many thousands of people. A major problem for the world's

SOMETHING MORE ABOUT
Radon

A commercially available kit for testing for radon gas in the home.

In February 1989 *Chemical and Engineering News* said that "In three short years, the radioactive gas radon has progressed from relative obscurity to a cause of high anxiety as an indoor air pollutant."* Why should this be so?

Radon is a chemically inert gas, in the same periodic group as helium, neon, argon, and krypton. The trouble with radon is that it is radioactive. As Figure 7.4 shows, $^{222}_{86}Rn$ is part of the chain of events beginning with the decay of uranium-238. (Other isotopes of Rn are products of other decay series.)

Radon occurs naturally in our environment. Since it comes from natural uranium deposits, the amount depends on local geology, but it is believed to account for about 55% of normal background radioactivity. Furthermore, since the gas is inert, and has a relatively long half-life (3.82 days), it is not trapped by chemical processes in the soil or water, and it is free to seep up from the ground and seep into underground mines or into homes through pores or cracks in the basement floor or walls, or around pipes. When breathed by humans occupying that space, the radon-222 isotope can decay inside the lungs to give polonium, a radioactive element that is not a gas and is not chemically inert.

$$^{222}_{86}\text{Rn} \longrightarrow {}^{4}_{2}\text{He} + {}^{218}_{84}\text{Po} \qquad t_{1/2} = 3.82 \text{ days}$$

$$^{218}_{84}\text{Po} \longrightarrow {}^{4}_{2}\text{He} + {}^{214}_{82}\text{Pb} \qquad t_{1/2} = 3 \text{ minutes}$$

Therefore, polonium-218 can lodge in body tissues, where it undergoes alpha decay to give lead-214, itself a radioactive isotope. The range of an alpha particle is quite small, perhaps 0.7 mm (about the thickness of several sheets of paper). However, this is approximately the thickness of the epithelium cells of the lungs, so the radiation can damage these tissues and induce lung cancer.

Virtually every home in the U.S. is believed to have some level of radon gas. To test for the presence of the gas, you can purchase sampling kits of various kinds.

There is currently a great deal of controversy over the level of radon that is considered "safe." The U.S. Environmental Protection Agency has set a standard of 4 picocuries per liter of air as an "action level." There are some who believe 1.5 picocuries is close to the average level, and that only about 2% of the homes contain more than 8 picocuries per liter. If your home shows higher levels of radon gas than this, you should probably have it tested further and perhaps take corrective actions such as sealing cracks around the foundation and in the basement. But keep in mind the relative risks involved. A 1.5 picocurie/liter level of radon leads to a lung cancer risk about the same as the risk of your dying in an accident in your home.

*Hanson, D.J. *Chemical and Engineering News,* February 6, 1989, page 7.

airlines has been the possibility that terrorists could plant a bomb on an airplane. A terrifying example was the bombing of a Pan American flight from London to New York in December 1988. To prevent such occurrences, the U.S. Federal Aviation Administration directed airlines to deploy explosives-detection devices by 1990, and the first such device—using thermal neutron analysis—was installed at Kennedy International Airport in New York in late 1989 (Figure 7.18).

Figure 7.18 A thermal neutron analysis (TNA) device is used at Kennedy International Airport in New York to detect bombs that may have been placed in luggage being checked on international flights. The chemicals used in explosives contain a high percentage of nitrogen-14. The TNA device uses radioactive californium-252 to bombard any nitrogen atoms that are present with neutrons. These neutrons produce nitrogen-15, which in turn emits a gamma ray with a particular energy, and the emitted gamma rays allow the operator to detect a suspicious piece of luggage. (UPI/Bettmann Newsphotos)

The bomb detector uses californium-252 as a source of neutrons to probe luggage that is to be loaded onto a plane. Virtually all of the thousands of explosives known to chemists contain at least 10% nitrogen. When the neutrons from ^{252}Cf bombard these nitrogen atoms, excited ^{15}N atoms are produced, atoms that emit distinctive gamma rays.

$$^{252}_{98}Cf \longrightarrow ^{1}_{0}n + ^{251}_{98}Cf$$

$$^{1}_{0}n + ^{14}_{7}N \longrightarrow ^{15}_{7}N \longrightarrow \text{gamma rays}$$

By analyzing the level and pattern of gamma ray emission from a suitcase, the operator can tell whether there is the possibility of an explosive device, and the bag can be inspected in other ways to confirm the suspicion. There are, of course, problems. Many other common substances, such as wool, leather, and cheese, have high levels of nitrogen, so these too can set off an alarm in the detector. Furthermore, it may be possible for a terrorist to mask the explosive material with borax-containing soap, since borax contains boron, an element that is an excellent neutron absorber. However, these problems can be circumvented with special detectors and computer systems.*

SUMMARY

Elements that naturally emit energy without the absorption of energy are said to be radioactive (Section 7.1). Three types of radiation are observed: **alpha particles** or helium nuclei ($^{4}_{2}He$, $^{4}_{2}\alpha$); **beta particles** or electrons ($_{-1}^{0}e$, $_{-1}^{0}\beta$), and gamma rays (γ).

Nuclear reactions or transmutations generally involve a change in the atomic number and mass of a radioisotope (Section 7.2). For example,

$$^{238}_{92}U + ^{1}_{0}n \longrightarrow ^{239}_{94}Pu + 2\ _{-1}^{0}\beta$$

when uranium-238 absorbs a neutron ($^{1}_{0}n$), the uranium is transformed into plutonium-239 and two β particles. The sum of the mass numbers of the interacting particles (238 + 1) equals the sum of the mass numbers of the

*For more information on these bomb detectors, and the scientific problems associated with them, see (a) *New York Times*, September 12, 1989, page C1; (b) *Science*, **1989**, 245, 926; and (c) *Discover*, November 1989, page 30.

products (239 + 0). Similarly, the sum of the atomic numbers of the interacting particles and products must be the same [92 + 0 = 94 + 2(−1)]. A radioactive nucleus may undergo a series of transmutations called a **radioactive series** (Figure 7.4).

In addition to α, β, and γ emissions, two other modes of decay are emission of a **positron** (a positive electron, ${}^{0}_{+1}\beta$) and **K-capture**. The latter occurs when the nucleus of an atom absorbs one of the atom's own electrons.

All nuclei (except ${}^{1}H$) contain at least one neutron for every proton, and heavier nuclei have N/Z ratios greater than 1 (Section 7.3 and Figure 7.5). There is evidence that the nucleus has a "shell" structure, and that these shells have capacities of 2, 8, 20, 28, 50, 82, and 126 nucleons of one type (protons or neutrons). These numbers are called the *magic numbers*. Nuclei with a magic number of either protons or neutrons are likely to be stable, and those with magic numbers of both types of nucleons are especially stable.

The energy evolved when nucleons combine to form a nucleus is the **binding energy**. The greater the binding energy, the greater the nuclear stability (Figure 7.6). A maximum in the curve of binding energy per nucleon occurs near iron. Thus, all other nuclei are thermodynamically unstable with respect to nuclei near iron. Nuclei lighter than iron can undergo **fusion** to form heavier nuclei, and nuclei heavier than iron can undergo **fission** to form lighter nuclei nearer iron in atomic number and mass.

The rate of nuclear decay (Section 7.4) is often described in terms of the **activity** (A), the number of disintegrations per unit time. The activity is proportional to the number of radioactive nuclei present (N), that is, $A = kN$ (Equation 7.1), where k is the **decay constant**. The decay constant can be calculated from Equation 7.2 [$\ln (N/N_0) = -kt$], an equation that relates the time period over which a sample is observed (t) and the number of radioactive atoms present at the beginning (N_0) and end (N) of the time period. The **half-life** of a radioactive isotope ($t_{1/2}$) is the time required to go from N_0 to $\frac{1}{2}N_0$. (The greater the half-life, the more stable the nucleus.) From Equation 7.2, one can show that $t_{1/2} = 0.693/k$. These relations allow you to find (a) the half-life of an isotope from activity measurements, (b) the time elapsed between measurements, and (c) the quantity of material remaining after a given time.

Uranium is the heaviest element occurring in nature in recoverable amounts. The **transuranium elements**, those beyond uranium, are all produced by forcing a lighter nucleus to absorb an appropriate particle (Section 7.5).

When certain isotopes of some of the actinide elements are struck by neutrons, they undergo fission and produce more neutrons (Sections 7.6 and 7.7). These neutrons may begin a **chain reaction** and lead to the release of enormous quantities of energy (Figures 7.13 and 7.14). Such reactions are the basis of nuclear power plants and atomic bombs.

Nuclear fusion, the combination of lighter elements to form a heavier one, is also being explored as a source of energy for power generation (Section 7.8).

Radioisotopes have found a broad range of uses in, for example, the treatment of cancer by radiation, as "tags" or "tracers" to follow the fate of an element in a chemical reaction, and in chemical analysis (Section 7.9).

STUDY QUESTIONS

REVIEW QUESTIONS

1. Name three people who made important contributions to nuclear chemistry and indicate their contributions.

2. Tell the similarities and differences in α particles, β particles, positrons, and gamma rays.

3. What is the binding energy of a nucleus?

4. If the mass number of an isotope is *much greater than* twice the atomic number, what type of radioactive decay might you expect?

5. If the number of neutrons in an isotope is *much less than* the number of protons, what type of radioactive decay might you expect?

6. What is the difference between fission and fusion? Illustrate your answer with an example of each.

7. Is there a nuclear reactor producing electric power in your state? If so, where is it located?

8. One reaction in a hydrogen bomb is that between tritium and deuterium. What type of reaction is this, fusion or fission?

$${}^{3}_{1}H + {}^{2}_{1}H \longrightarrow {}^{1}_{0}n + {}^{4}_{2}He$$

9. What are some of the advantages and disadvantages of nuclear power plants?

10. What are some of the important lessons learned in the accidents at Three Mile Island and Chernobyl?

11. Name at least three uses of radioactive isotopes (outside of their use in power reactors and weapons).

NUCLEAR REACTIONS

12. Balance the following nuclear reactions. Write the mass number and atomic number for "?," as well as the element symbol where possible.

(a) ${}^{54}_{26}Fe + {}^{4}_{2}He \longrightarrow 2\ {}^{1}_{1}H + ?$
(b) ${}^{27}_{13}Al + {}^{4}_{2}He \longrightarrow {}^{30}_{15}P + ?$
(c) ${}^{32}_{16}S + {}^{1}_{0}n \longrightarrow {}^{1}_{1}H + ?$
(d) ${}^{96}_{42}Mo + {}^{2}_{1}H \longrightarrow {}^{1}_{0}n + ?$
(e) ${}^{98}_{42}Mo + {}^{1}_{0}n \longrightarrow {}^{99}_{43}Tc + ?$

13. Balance the following nuclear reactions. Write the mass number and atomic number for "?," as well as the element symbol where possible.

(a) ${}^{9}_{4}Be + ? \longrightarrow {}^{6}_{3}Li + {}^{4}_{2}He$
(b) $? + {}^{1}_{0}n \longrightarrow {}^{24}_{11}Na + {}^{4}_{2}He$
(c) ${}^{40}_{20}Ca + ? \longrightarrow {}^{40}_{19}K + {}^{1}_{1}H$
(d) ${}^{241}_{95}Am + {}^{4}_{2}He \longrightarrow {}^{243}_{97}Bk + ?$
(e) ${}^{246}_{96}Cm + {}^{12}_{6}C \longrightarrow 4\ {}^{1}_{0}n + ?$
(f) ${}^{238}_{92}U + ? \longrightarrow {}^{249}_{100}Fm + 5\ {}^{1}_{0}n$

14. Balance the following nuclear reactions. Write the mass number and atomic number for "?," as well as the element symbol where possible.

(a) ${}^{104}_{47}Ag \longrightarrow {}^{104}_{48}Cd + ?$
(b) ${}^{87}_{36}Kr \longrightarrow {}^{0}_{-1}\beta + ?$
(c) ${}^{231}_{91}Pa \longrightarrow {}^{227}_{89}Ac + ?$
(d) ${}^{230}_{90}Th \longrightarrow {}^{4}_{2}He + ?$
(e) ${}^{82}_{35}Br \longrightarrow {}^{82}_{36}Kr + ?$
(f) $? \longrightarrow {}^{24}_{12}Mg + {}^{0}_{-1}\beta$

15. Balance the following nuclear reactions. Write the mass number and atomic number for "?," as well as the element symbol where possible.

(a) ${}^{19}_{10}Ne \longrightarrow {}^{0}_{+1}\beta + ?$
(b) ${}^{59}_{26}Fe \longrightarrow {}^{0}_{-1}\beta + ?$
(c) ${}^{40}_{19}K \longrightarrow {}^{0}_{-1}\beta + ?$
(d) ${}^{37}_{18}Ar + {}^{0}_{-1}e$ (K-capture) $\longrightarrow ?$
(e) ${}^{55}_{26}Fe + {}^{0}_{-1}e$ (K-capture) $\longrightarrow ?$
(f) ${}^{26}_{13}Al \longrightarrow {}^{26}_{12}Mg + ?$

16. One radioactive series that begins with uranium-235 and ends with lead-207 undergoes the sequence of reactions: α, β, α, β, α, α, α, α, β, β, α. Identify the radioisotope produced in each of the *first five steps*.

17. One radioactive series that begins with uranium-235 and ends with lead-207 undergoes the sequence of reactions: α, β, α, β, α, α, α, α, β, β, α. Identify the radioisotope produced in each of the *last six steps*. (See Study Question 16.)

NUCLEAR STABILITY

18. Boron has two stable isotopes, ${}^{10}B$ (abundance = 19.91%) and ${}^{11}B$ (abundance = 80.09%). Calculate the binding energies of these two nuclei and compare their stabilities.

$$5\ {}^{1}_{1}H + 5\ {}^{1}_{0}n \longrightarrow {}^{10}_{5}B$$

$$5\ {}^{1}_{1}H + 6\ {}^{1}_{0}n \longrightarrow {}^{11}_{5}B$$

The required masses are: ${}^{1}_{1}H = 1.00783$; ${}^{1}_{0}n = 1.00867$; ${}^{10}_{5}B = 10.01294$; and ${}^{11}_{5}B = 11.00931$.

19. Calculate the binding energy in kJ per mole of P for the formation of ${}^{30}_{15}P$

$$15\ {}^{1}_{1}H + 15\ {}^{1}_{0}n \longrightarrow {}^{30}_{15}P$$

and for the formation of ${}^{31}_{15}P$.

$$15\ {}^{1}_{1}H + 16\ {}^{1}_{0}n \longrightarrow {}^{31}_{15}P$$

Which is the more stable isotope? (The required masses are ${}^1_1H$ = 1.00783; ${}^1_0n$ = 1.00867; ${}^{30}_{15}P$ = 29.9880; and ${}^{31}_{15}P$ = 30.97376.)

RATES OF DISINTEGRATION REACTIONS

20. Copper-64 is used in the form of copper(II) acetate to study brain tumors. It has a half-life of 12.8 hours. If you begin with 15.0 mg of copper(II) acetate, how many milligrams remain after 2 days and 16 hours?
21. Gold-198 is used as the metal in the diagnosis of liver problems. The half-life of ${}^{198}Au$ is 2.7 days. If you begin with 5.6 mg of this gold isotope, how many milligrams remain after 10.8 days?
22. Iodine-131 is used in the form of sodium iodide to treat cancer of the thyroid. (a) The isotope decays by ejecting a β particle. Write a balanced equation to show this process. (b) The isotope has a half-life of 8.05 days. If you begin with 25.0 mg of radioactive $Na^{131}I$, how many milligrams remain after about a month (32.2 days)?
23. Phosphorus-32 is used in the form of Na_2HPO_4 in the treatment of chronic myeloid leukemia, among other things. (a) The isotope decays by emitting a β particle. Write a balanced equation to show this process. (b) The half-life of ${}^{32}P$ is 14.3 days. If you begin with 9.6 milligrams of radioactive Na_2HPO_4, how many milligrams remain after about one month (28.6 days)?
24. In 1984 cobalt-60 was involved in the worst accident with radioactive isotopes in North America (*Science 84*, December 1984, page 28). The half-life of the isotope is 5.3 years. Starting with 10.0 mg of ${}^{60}Co$, how much will remain after 21.2 years? Approximately how much will remain after a century?
25. Gallium-67 ($t_{1/2}$ = 77.9 hours) is used in the medical diagnosis of certain kinds of tumors. If you ingest a compound containing 0.15 mg of this isotope, how many milligrams will remain in your body after 13 days?
26. Radioisotopes of iodine are widely used in medicine. For example, iodine-131 ($t_{1/2}$ = 8.05 days) is used to treat thyroid cancer. If you ingest a sample of NaI containing ${}^{131}I$, how much time is required for the isotope to fall to 5.0% of its original activity?
27. The rare gas radon has been the focus of much attention recently as it can be found in homes. Radon-222 emits α particles with a half-life of 3.82 days. (a) Write a balanced equation to show this process. (b) How long does it take for a sample of radon to decrease to only 10.0% of its original activity?
28. A sample of a radioisotope disintegrates at the rate of 6400 counts per minute. Exactly six hours later the decay rate is 1600 counts per minute. What is the half-life of the isotope?
29. Strontium-90 is produced in nuclear explosions. Because strontium can be ingested and replace calcium in bone, radioactive ${}^{90}Sr$ is a dangerous isotope. Assume the activity of strontium-90 in the bones of an individual in 1985 was about 80 disintegrations per second (dps). How many half-lives and how many years are required to reduce the activity to the natural background level of 2.5 dps? In what year will the activity be 2.5 dps? ($t_{1/2}$ for ${}^{90}Sr$ is 27.7 years.)
30. Radium-226 decays by emitting an α particle; its half-life is 1620 years. (a) Write a balanced equation to show this process. (b) Use Equations 7.1 and 7.3 to find the alpha activity of ${}^{226}Ra$ in disintegrations per minute if a sample contains 0.0010 g of ${}^{226}Ra$.
31. Plutonium-239 decays by ejecting an α particle; it has a half-life of 24,390 years. (a) Write a balanced equation to show this process. (b) Use Equations 7.1 and 7.3 to find the alpha activity of ${}^{239}Pu$ in disintegrations per minute if a sample contains 1.0 mg of the isotope.
32. A sample of wood from a Thracian chariot found in an excavation in Bulgaria has a ${}^{14}C$ activity of 11.2 disintegrations per minute per gram. Estimate the age of the chariot and the year it was made. ($t_{1/2}$ for ${}^{14}C$ is 5.73×10^3 years and the activity of ${}^{14}C$ in living material is 14.0 disintegrations per minute per gram.)
33. A piece of charred bone found in the ruins of an American Indian village has a ${}^{14}C$ to ${}^{12}C$ ratio of 0.72 times that found in living organisms. Calculate the age of the bone fragment. (See Study Question 32 for required data on carbon-14.)

NUCLEAR TRANSMUTATIONS

34. There are two isotopes of americium, both with half-lives sufficiently long to allow the handling of massive quantities. Americium-241, for example, has a half-life of 248 years as an α emitter, and it is used in gauging the thickness of materials and in smoke detectors. The isotope is formed from ${}^{239}Pu$ by absorption of two neutrons followed by emission of a β particle. Write a balanced equation for this process.
35. Americium-240 is made by bombarding a plutonium-239 atom with an α particle. In addition to ${}^{240}Am$, the products are a proton and two neutrons. Write a balanced equation for this process.
36. To synthesize the heavier transuranium elements, one must bombard a lighter nucleus with a relatively large particle. If you know the products, californium-246 plus 4 neutrons, what particle would you use to bombard uranium-238 atoms?
37. The element with the highest known atomic number is 109. It is thought that still heavier elements are possible, especially with $Z = 114$ and $N = 184$. To this end, serious attempts have been made to force cal-

cium-40 and curium-248 to merge. What would be the atomic number of the element formed?

NUCLEAR FISSION AND POWER

38. A good grade of coal averages an energy output of 2.6×10^7 kJ/ton. Fission of one mole of ^{235}U releases 2.1×10^{10} kJ. Find the number of tons of coal needed to produce the same energy as one pound of ^{235}U. (See Appendix C for conversion factors.)

39. A concern in the nuclear power industry is that, if nuclear power becomes more widely used, there may be serious shortages in worldwide supplies of fissionable uranium. One solution is to build "breeder" reactors that manufacture more fuel than they consume. One such cycle works as follows:

(a) A "fertile" ^{238}U nucleus collides with a neutron to produce ^{239}U.

(b) ^{239}U decays by β emission ($t_{1/2}$ = 24 min) to give an isotope of neptunium.

(c) This neptunium isotope decays by β emission to give a plutonium isotope.

(d) This plutonium isotope is fissionable. On collision of one of these plutonium isotopes with a neutron, fission occurs with energy, at least two neutrons, and other nuclei as products.

Write an equation for each of these steps and explain how this process can be used to breed more fuel than the reactor originally contained and still produce energy.

USES OF RADIOISOTOPES

40. In order to measure the volume of the blood system of an animal, the following experiment was done. A 1.0-mL sample of an aqueous solution containing tritium with an activity of 2.0×10^6 disintegrations per second (dps) was injected into the bloodstream. After time was allowed for complete circulatory mixing, a 1.0-mL blood sample was withdrawn and found to have an activity of 1.5×10^4 dps. What was the volume of the circulatory system? (The half-life of tritium is 12.3 years, so this experiment assumes that only a negligible amount of tritium has decayed in the time of the experiment.)

41. Radioactive isotopes are often used as "tracers" to follow an atom through a chemical reaction, and the following is an example. Acetic acid reacts with methyl alcohol, CH_3OH, by eliminating a molecule of H_2O to form methyl acetate, $CH_3C(O)OCH_3$. Explain how you would use the radioactive isotope ^{18}O to show whether the oxygen atom in the water product comes from the —OH of the acid or the —OH of the alcohol.

$$\underset{\text{acetic acid}}{CH_3{-}\overset{\overset{\displaystyle O}{\|}}{C}{-}OH} + \underset{\text{methyl alcohol}}{CH_3OH} \longrightarrow \underset{\text{methyl acetate}}{CH_3{-}\overset{\overset{\displaystyle O}{\|}}{C}{-}O{-}CH_3} + H_2O$$

GENERAL QUESTIONS

42. Balance the following nuclear reactions. Write the mass number and atomic number for "?," as well as the element symbol where possible.

(a) $^{13}_{6}C + ? \longrightarrow ^{14}_{6}C$

(b) $^{40}_{18}Ar + ? \longrightarrow ^{43}_{19}K + ^{1}_{1}H$

(c) $^{250}_{98}Cf + ^{11}_{5}B \longrightarrow 4\ ^{1}_{0}n + ?$

(d) $^{53}_{24}Cr + ^{4}_{2}He \longrightarrow ? + ^{56}_{26}Fe$

(e) $^{212}_{84}Po \longrightarrow ^{208}_{82}Pb + ?$

(f) $^{122}_{53}I \longrightarrow ^{0}_{+1}\beta + ?$

(g) $? \longrightarrow ^{23}_{11}Na + ^{0}_{-1}\beta$

(h) $^{137}_{53}I \longrightarrow ^{1}_{0}n + ?$

(i) $^{22}_{11}Na \longrightarrow ^{1}_{1}H + ?$

43. The nuclear reaction sequence below involves four unknown elements: Q, Δ, Σ, and Π. Based on the reactions in the sequence, identify the unknown elements.

$Pb + Cr \longrightarrow Q$

$Cf + O \longrightarrow Q$

$Q \longrightarrow \alpha + \Delta$

$\Delta \longrightarrow \alpha + \Sigma$

$\Sigma \longrightarrow \alpha + \Pi$

44. Balance the following reactions used for the synthesis of transuranium elements.

(a) $^{238}_{92}U + ^{14}_{7}N \longrightarrow ? + 5\ ^{1}_{0}n$

(b) $^{238}_{92}U + ? \longrightarrow ^{249}_{100}Fm + 5\ ^{1}_{0}n$

(c) $^{253}_{99}Es + ? \longrightarrow ^{256}_{101}Md + ^{1}_{0}n$

(d) $^{246}_{96}Cm + ? \longrightarrow ^{254}_{102}No + 4\ ^{1}_{0}n$

(e) $^{252}_{98}Cf + ? \longrightarrow ^{257}_{103}Lr + 5\ ^{1}_{0}n$

45. On December 2, 1942, the first man-made self-sustaining nuclear fission chain reactor was operated by Enrico Fermi under the University of Chicago stadium. In June 1972, *natural* fission reactors, operating billions of years ago, were discovered in Oklo, Gabon. At present, natural uranium contains 0.72% ^{235}U. How many years ago did natural uranium contain 3.0% ^{235}U, sufficient to sustain a natural reactor? ($t_{1/2}$ for ^{235}U is 7.04×10^8 years.)

46. If a sample of ^{147}Pm has an activity of 1.0×10^7 disintegrations per second, how many atoms of the isotope are there in the sample? How many grams? $t_{1/2}$ for ^{147}Pm is 2.64 years.

CHAPTER

8

Atomic Structure

The colors of fireworks depend on structures of the atoms involved. (© 1987 James Cowlin)

CHAPTER OUTLINE

Beautiful gems and fireworks excite all of us. The loud bangs and booms of fireworks can be fun, but most of all the colors of fireworks bursting in the night sky or the deep reflections of a gemstone are beautiful to see. These magnificent colors arise from atoms or ions of various salts used in fireworks or from ions buried deep in the gem. Have you ever wondered how these colors are produced? Part of the answer is found in this chapter. As you will see, when the atoms of certain elements are "excited"—when we add energy to these atoms in any of a variety of ways—the electrons of the atoms absorb some of this energy and then they return the absorbed energy to us in the form of visible light. This chapter explores how this can occur and uses this understanding to begin to explore the arrangement of electrons in atoms.

The basic ideas of atomic structure are fundamental to an understanding of the important questions in chemistry and their answers. Therefore, this chapter and the next describe these ideas and begin to show their relation to practical, everyday chemistry. The first step is to describe some useful ideas from physics, and then we can lay out some of the principles of atomic structure. This chapter also includes some discussion of the history of the development of atomic theory in the last 100 years, a period replete with marvelous characters who are fascinating not only for the brilliance of their scientific contributions but also for their life stories.

Many of the ideas of atomic structure that we currently use were developed during the early part of this century. For an overview of this period, see "Something More About the History of Atomic Structure: Thomson, Rutherford, and the Nuclear Atom" at the end of this chapter.

8.1 ELECTROMAGNETIC RADIATION

We are all familiar with water waves, and you may also know that some properties of radiation such as light can be described with the ideas of wave motion. Indeed, James Maxwell developed an elegant mathematical

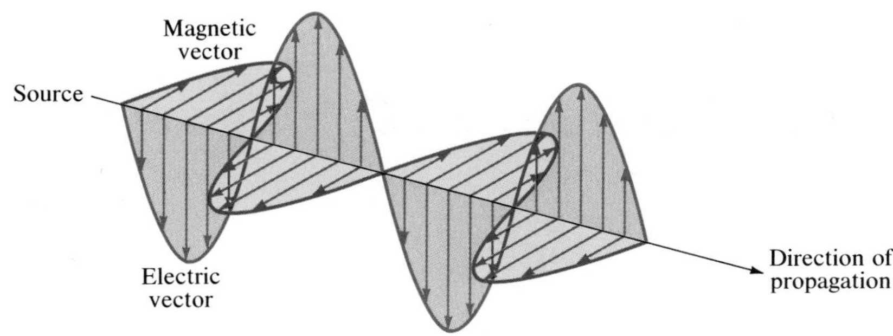

Figure 8.1 Electromagnetic radiation. In the 1860s James Maxwell developed the currently accepted theory that all forms of radiation are propagated through space as vibrating electric and magnetic fields, the fields being at right angles to one another. (The fields are "force fields," the vector lines telling the direction and strength of the forces acting on a charged/magnetic particle in the wave path.) Such oscillating fields emanate from vibrating charges in a source such as a light bulb or radio antenna.

theory in 1864 to describe all forms of radiation in terms of oscillating or wave-like electric and magnetic fields in space (Figure 8.1). Hence, radiation such as light, microwaves, television and radio signals, and x-rays is collectively called **electromagnetic radiation**. This view is important to our understanding of how electrons behave in atoms.

Two simple wave forms are shown in Figure 8.2. The **wavelength** of a wave is the distance between successive crests or high points (or between successive troughs or low points). This distance can be given in meters, nanometers, or whatever unit is convenient. The symbol for wavelength is the Greek letter λ (lambda).

Waves can also be characterized by their **frequency**, symbolized by the Greek letter ν (nu). For a wave traveling through some point in space, the frequency of vibration or wave oscillation is the number of oscillations the point makes in some given amount of time, and this is equal to the number of complete waves passing the point in that amount of time. We usually refer to the frequency as the number of cycles or events per second (Figure 8.2), a unit often written s^{-1} (standing for 1/second) and now called the **hertz**.

If you have ever enjoyed water sports, you are familiar with the height of waves. In more scientific terms, the maximum height of a wave, as measured from the axis of propagation of the wave, is called the **amplitude**. In Figure 8.2, notice that the wave has zero amplitude at certain intervals along the wave. All the points of zero amplitude are called **nodes**;

The concept of nodes is important in understanding atomic orbitals, as you shall see in Section 8.7.

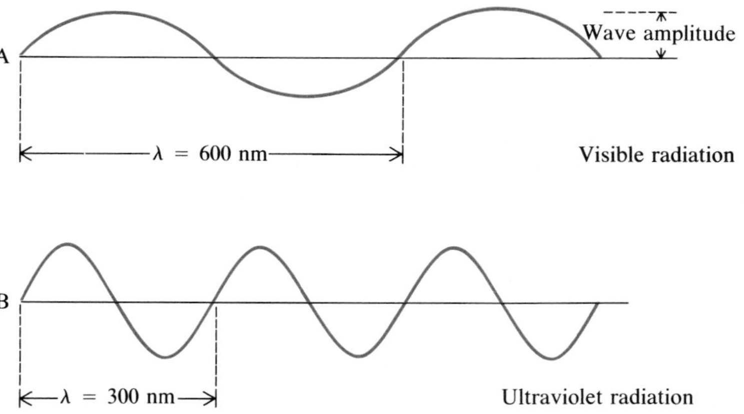

Figure 8.2 Two simple waves. The wavelength of orange light, wave A, is twice that of wave B (a particular radiation in the ultraviolet region); that is, $\lambda_A = 2\lambda_B$. Since both waves are traveling at the same velocity, the ultraviolet light completes two cycles or vibrations in the time the orange light completes one. Thus, the frequency of the orange light is one-half of that of the ultraviolet light ($\nu_A = \frac{1}{2}\nu_B$).

The frequency of a wave is the number of complete waves passing a point in a given time.

they always occur at intervals of $\lambda/2$ for a so-called "sinusoidal" wave such as those illustrated.

Finally, the velocity of a moving wave is an important factor. As an analogy, consider cars in a traffic jam traveling bumper to bumper. If each car is 16 feet long, and if a car passes you every 4 seconds (that is, the frequency is 1 car per 4 seconds or $\frac{1}{4}$ s^{-1}), then the traffic is "moving" at the speed of (16 feet)($\frac{1}{4}$ s^{-1}) or 4 feet per second. This multiplication of length times frequency to give velocity holds for any periodic motion, including a wave.

$$\text{Wavelength (m)} \times \text{frequency (s}^{-1}\text{)} = \text{velocity (m/s)}$$

$$\lambda \times \nu = v \tag{8.1}$$

For visible light and other electromagnetic radiation, the velocity in a vacuum is the well known constant $\mathbf{c = 2.99792458 \times 10^8}$ **m/s**. This means that, if you know the wavelength of a light wave, you can readily calculate its frequency and vice versa, as shown in the next example.

The velocity of light through a substance (air, glass, water, etc.) depends on the chemical constitution of the substance and the wavelength of the light. This is the basis for using a glass prism to disperse light and is the explanation for rainbows. The velocity of sound is also dependent on the material through which it passes.

EXAMPLE 8.1

WAVELENGTH-FREQUENCY CONVERSIONS

Orange light has a wavelength of 620 nm. What is the wavelength in meters? What is its frequency?

Solution To convert from a wavelength in nanometers to one in meters, recall that 1 nm is exactly 1×10^{-9} m.

$$620\ \cancel{\text{nm}} \left(\frac{1 \times 10^{-9}\ \text{m}}{\cancel{\text{nm}}}\right) = 6.2 \times 10^{-7}\ \text{m}$$

Frequency can now be obtained from the equation $c = \lambda\nu = 2.998 \times 10^8$ m/s.

$$\nu = c\,\frac{1}{\lambda} = \left(\frac{2.998 \times 10^8\ \cancel{\text{m}}}{\text{s}}\right)\left(\frac{1}{6.2 \times 10^{-7}\ \cancel{\text{m}}}\right)$$

$$= 4.8 \times 10^{14}\ \text{s}^{-1}$$

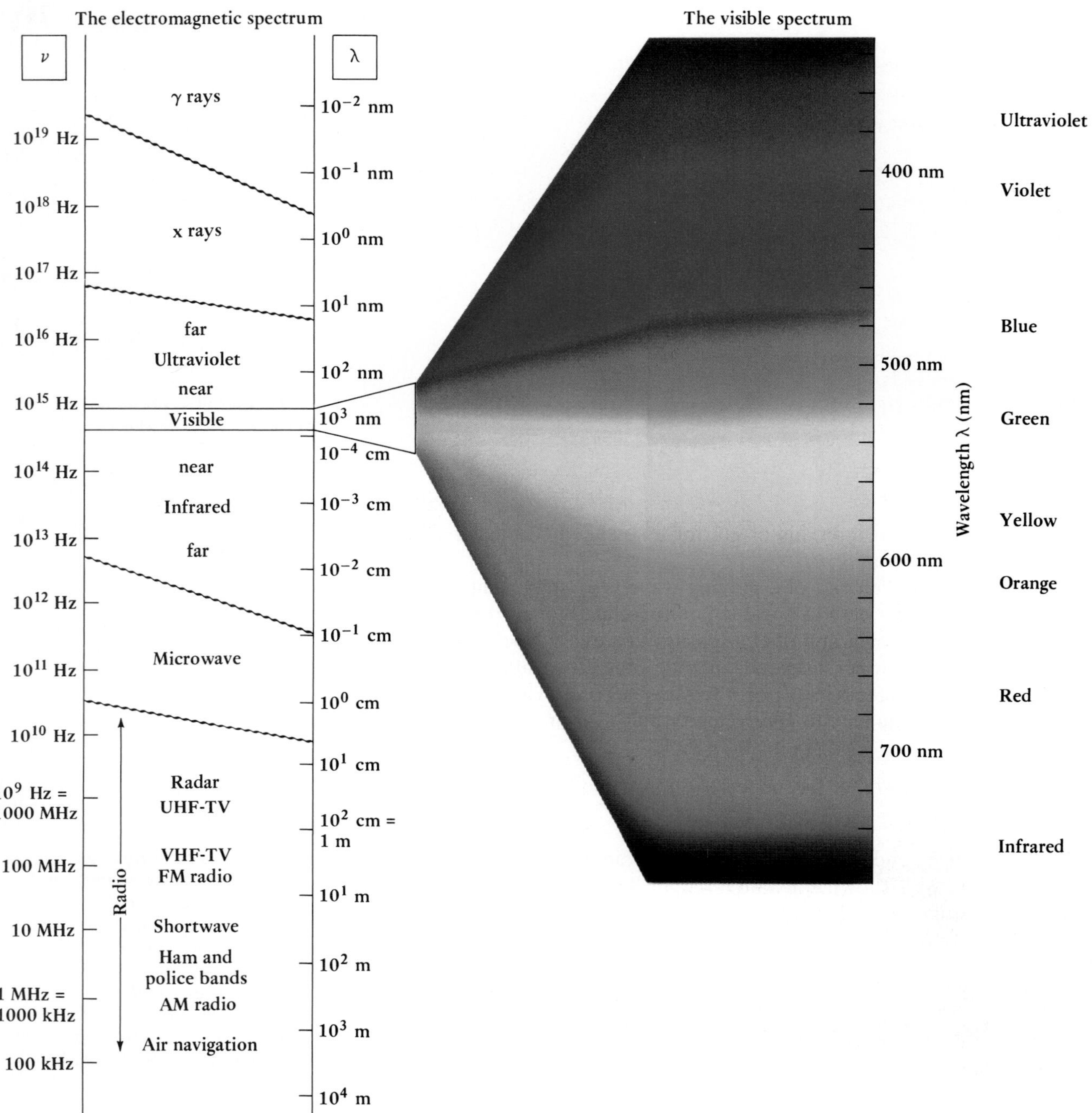

Figure 8.3 The electromagnetic spectrum. The visible spectrum is only a very narrow band in the total range of wavelengths.

EXERCISE 8.1 Wavelength-Frequency Conversions

If your favorite FM radio station has a frequency of 104.5 MHz (MHz = megahertz = 10^6 s^{-1}), what is the wavelength (in meters) of the radiation emitted by this station?

We are all bathed constantly in electromagnetic radiation, such as the radiation you can see, visible light. As you know, visible light really consists of a spectrum of colors, ranging from red light at the long-wavelength end to violet light at the short-wavelength end of the spectrum (Figure 8.3). Visible light is only a small portion of the total electromag-

more electrons. Thus, the connection is made between light intensity and the number of electrons ejected after the minimum energy is exceeded.

EXAMPLE 8.2

PHOTON ENERGIES

In Example 8.1 you found that orange light has a frequency of $4.8 \times 10^{14}\ s^{-1}$. What is the energy of one quantum of orange light?

Solution Since the frequency of the light is given, we can multiply it by Planck's constant to obtain the energy of one quantum of this light.

$$E = h\nu = (6.6260755 \times 10^{-34}\ J \cdot s)(4.8 \times 10^{14}\ s^{-1}) = 3.2 \times 10^{-19}\ J$$

EXERCISE 8.4 Calculating Photon Energies

To find out which has the greater energy, compare the energy of one photon of orange light (Example 8.2) with the energy of one quantum of x-radiation having a wavelength of 2.36 nm.

8.4 ATOMIC LINE SPECTRA AND NIELS BOHR

The final piece of information that has played a major role in the modern view of atomic structure is the observation of the properties of the light emitted by excited atoms.

In Section 8.2 you were reminded that a heated object can glow or emit radiation. Light coming from such a source consists of millions of minutely different frequencies, and the result is that you see light that has a continuous spectrum, like that shown in Figure 8.5. In a similar way, if you put sufficient electrical energy into a gas at low pressure, the gas atoms or molecules are excited and can emit light (Figure 8.7). This

(a)

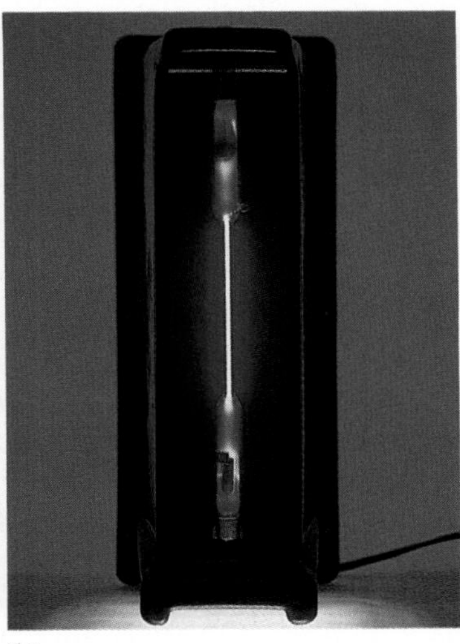
(b)

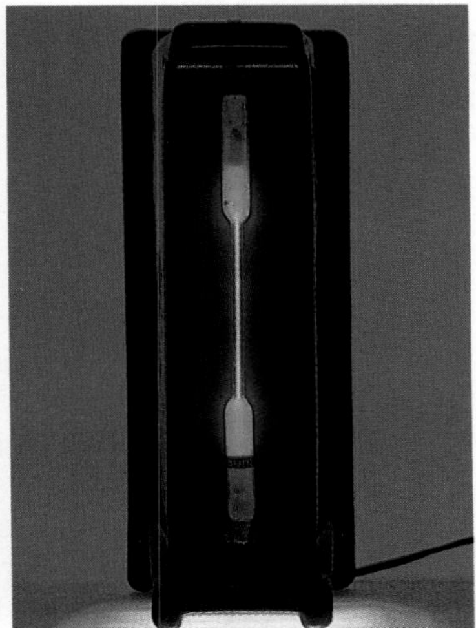
(c)

Figure 8.7 Light emission from excited gases: (a) Ne, (b) Ar, and (c) Hg.

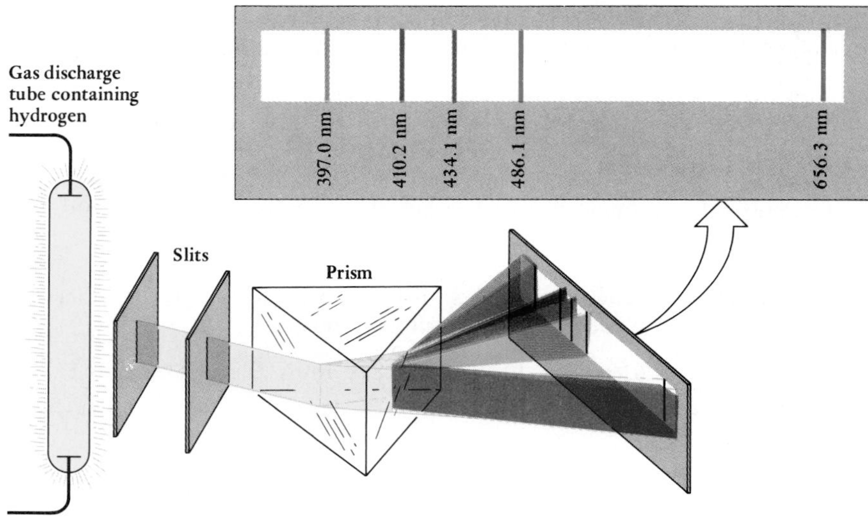

Figure 8.8 The line spectrum of excited H atoms. The emitted light is passed through a series of slits to isolate a narrow beam of light, and this beam is then passed through some device such as a prism that separates the light into its component wavelengths. A photographic plate or an instrument detects the separate wavelengths as individual lines. Hence, the name "line spectrum" for the light emitted by a glowing gas.

is familiar to you as the "neon" lights used in advertising signs. The difference between a glowing gas and a glowing piece of metal is that the gas gives off light of only a few widely different frequencies. In fact, this is the simplest observation of the quantization of energy in atoms.

The emission of particular frequencies of light from an excited gas—the **emission spectrum** of the gas—can be observed with a **spectroscope** or **spectrometer** (Figure 8.8). Light from a glowing gas is passed through a narrow slit, thereby isolating a thin beam or line of light. The beam is then passed through some device (a prism or, in modern instruments, a diffraction grating) that separates the light into its various wavelengths, and we observe the emitted light as a **line spectrum**. Glowing neon, for example, would appear red to your eye, but when you pass the light through a prism, the light is actually seen to be composed of a few different colors; the wavelength appropriate to red predominates, however, so this is the color you observe.

The reason a prism can separate light into its component wavelengths is as follows: The index of refraction, n, of a substance (e.g., air, glass) is defined as (speed of light in a vacuum/speed of light in the substance). The index n is a function of the wavelength of light, so the degree to which a light beam is bent on entering a substance depends on the properties of the substance and the color (λ) of the light.

In the 19th century, a Swiss music teacher, Johann Balmer (1825–1898), worked out a mathematical relation that accounted for the three lines of longest wavelength in the visible emission spectrum of hydrogen atoms, that is, the red, green, and blue lines in Figure 8.8. His results were rewritten by Johannes Rydberg (1854–1919), who found that the wavelength of the lines could be predicted from the equation

$$\frac{1}{\lambda} = R\left(\frac{1}{4} - \frac{1}{n^2}\right) \qquad (n > 2) \tag{8.3}$$

This is called the **Rydberg equation**: n is an integer associated with each line, and **R**, now called the **Rydberg constant**, has the value 1.09737×10^7 m^{-1}. If $n = 3$, the wavelength of the red line in the hydrogen spectrum is obtained; if $n = 4$, the wavelength for the green line is obtained, and if $n = 5$, we get the wavelength of the blue line. This group of visible lines (and others for which $n = 6, 7, 8$, and so on) are now called the **Balmer series** of lines.

The symbol m^{-1} means 1/meter, just as s^{-1} means 1/second.

Niels Bohr, a young Danish physicist, provided the first connection between line spectra and the quantum ideas of Planck and Einstein. Bohr

Historical Figures in Chemistry: Niels Bohr (1885–1962)

Niels Bohr was born in Copenhagen, Denmark. He earned a Ph.D. in physics in Copenhagen in 1911 and then went to work first with J.J. Thomson in Cambridge, England, and later with Ernest Rutherford in Manchester, England.* It was there that he began to develop the ideas that a few years later led to the publication of his theory of atomic structure and his explanation of atomic spectra. (For this work he received the Nobel Prize in 1922.) After being with Rutherford for a very short time, Bohr returned to Copenhagen, where he eventually became the director of the Institute of Theoretical Physics. Many young physicists carried on his work in this Institute, and seven of them later received Nobel Prizes for their studies in chemistry and physics, among them Werner Heisenberg, Wolfgang Pauli, and Linus Pauling.

*Rutherford's contributions to chemistry and physics have been described on pages 245 and 304. Bohr was a major figure in science in this century, and his life and work are described in the book *The Making of the Atomic Bomb* by R. Rhodes (Simon and Schuster, 1986) and by B.L. Haendler, *J. Chem. Educ.* **1982**, *59*, 372.

introduced the notion that *the single electron of the hydrogen atom could occupy only certain energy states*, or **stationary states** as he called them. We say that the energy of the electron in an atom is "quantized." Combining this **quantization postulate** with the laws of motion from classical physics, Bohr showed that the energy possessed by the single electron of the H atom, in the nth stationary state, is given by the simple equation

$$\text{Energy} = E_n = -\frac{Rhc}{n^2} \tag{8.4}$$

where R is a proportionality constant, h is Planck's constant, c is the velocity of light, and n is a unitless integer having values of 1, 2, 3, and so on, each value characterizing a different stationary state of the atom.

Bohr was able to express R in terms of other known physical quantities (including the mass and charge of the electron) and to *calculate* its value from these quantities as $R = 1.0974 \times 10^7\ \text{m}^{-1}$, exactly the value of the experimental Rydberg constant! Using this relation he could determine the energies of the quantized states of the electron in the H atom (Figure 8.9) and could derive Equation 8.3, which had previously been only an experimentally determined relation.

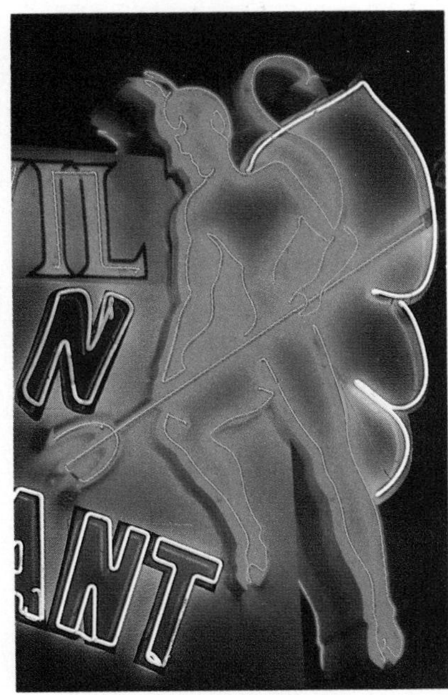

Neon gas emits light of certain frequencies when excited (see also Figure 8.7). Bohr's revolutionary work led to an understanding of the atomic-level origins of such effects.

EXAMPLE 8.3

ELECTRON ENERGIES

What are the energies of the $n = 1$ and $n = 2$ states of the hydrogen atom in J/atom and in kJ/mol?

Solution The values needed to use Equation 8.4 to solve for energies are

$R = 1.0974 \times 10^7\ \text{m}^{-1}$

$h = 6.6261 \times 10^{-34}\ \text{J} \cdot \text{s}$

$c = 2.9979 \times 10^8\ \text{m/s}$

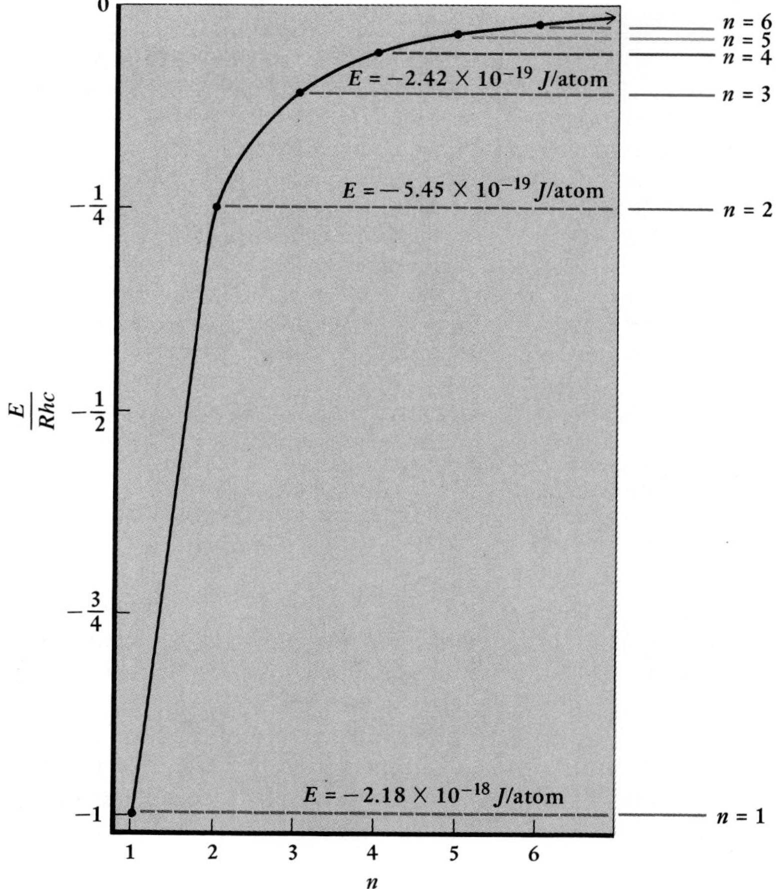

Figure 8.9 A graph of $E = -Rhc/n^2$ versus n. Energies are given in J/atom. Notice that the difference between successive energy states becomes smaller as n becomes larger.

Thus, when $n = 1$, the energy of a single H atom is

$$E_1 = -\frac{Rhc}{n^2} = -\frac{(1.0974 \times 10^7\ \text{m}^{-1})(6.6261 \times 10^{-34}\ \text{J} \cdot \text{s})(2.9979 \times 10^8\ \text{m/s})}{1^2}$$

$$= -2.1799 \times 10^{-18}\ \text{J/atom}$$

and when $n = 2$, it is

$$E_2 = \frac{E_1}{2^2} = \frac{-2.1799 \times 10^{-18}\ \text{J/atom}}{4} = -5.4498 \times 10^{-19}\ \text{J/atom}$$

The conversion from J/atom to kJ/mol uses Avogadro's number and the relation 1 kJ = 1000 J.

$$E_1 = \left(\frac{-2.1799 \times 10^{-18}\ \cancel{\text{J}}}{\cancel{\text{atom}}}\right)\left(\frac{6.0221367 \times 10^{23}\ \cancel{\text{atoms}}}{\text{mol}}\right)\left(\frac{1\ \text{kJ}}{1000\ \cancel{\text{J}}}\right)$$

$$= -1312.8\ \text{kJ/mol}$$

Similarly, $E_2 = -328.19$ kJ/mol.

Notice that the calculated energies are both negative, with E_1 *more negative than* E_2. This arises because Bohr's equation reflects the fact that the energy of attraction between oppositely charged bodies (an electron and an atomic nu-

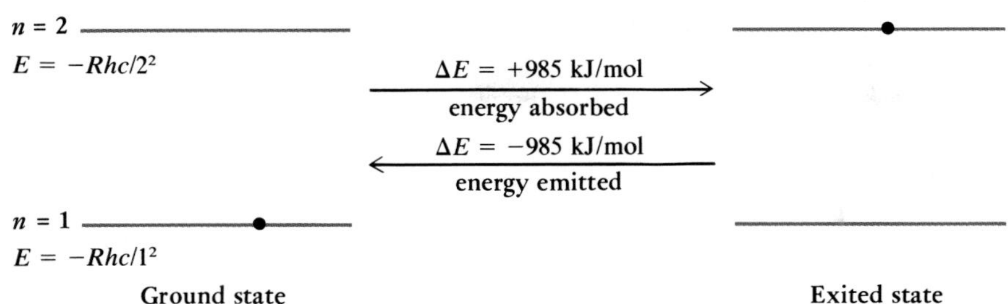

Figure 8.10 Absorption (ΔE positive) and emission (ΔE negative) of energy by the electron in the H atom moving from the $n = 1$ state (the ground state) to the $n = 2$ state (the excited state) and back again.

cleus) depends on their charge and the distance between them. This energy of attraction is given by −(charge on A)(charge on B)/(square of the distance between A and B).* Thus, the closer the electron is to the nucleus, the greater is the energy of attraction. That is, the value of E has a more negative value as the distance becomes smaller, and chemists or physicists say that the energy is therefore lower (meaning more negative). We can conclude from this that an electron with $n = 1$ must be closer to the nucleus than an electron with $n = 2$ (see Section 8.7).

EXERCISE 8.5 Electron Energies

What is the energy of the $n = 3$ state of the H atom in (a) J/atom and (b) kJ/mol?

A major assumption of Bohr's theory was that an electron in an atom would remain in its lowest energy state unless otherwise disturbed (an assumption shared with the quantum mechanical approach used today). Only by changing from one energy state to another can energy be absorbed or evolved, and it is this idea that allows us to explain line spectra. As you saw in Example 8.3 (and as illustrated in Figure 8.9), *an electron with n = 1 has the most negative energy and is the most strongly attracted to the nucleus*. If the electron of the hydrogen atom has $n = 1$, the atom is said to be in its **ground state**, the lowest energy stationary state.

The basic, atomic-level explanation for line spectra of excited atoms applies to the colors of fireworks and beautiful gems.

An electron with $n = 2$ is less strongly attracted to the nucleus than one with $n = 1$, and the energy of an $n = 2$ electron is less negative (Example 8.3). Therefore, to disturb the electron in the $n = 1$ state and move it to the $n = 2$ state, the atom must absorb energy. We say that the electron must be **excited** (Figure 8.10). If the hydrogen atom has its electron in any state but $n = 1$ (the ground state), the atom is said to be in an **excited state**. Using Bohr's equations we can calculate the amount of energy required to carry the H atom from the ground state to its first excited state ($n = 2$). As you learned in Chapter 6, the difference in energy between two states is

$$\Delta E = E_{\text{final state}} - E_{\text{initial state}}$$

*See the discussion of Coulomb's law in Chapter 3. Coulomb's law, which gives the force of attraction between two oppositely charged bodies, is closely related to their energy of attraction.

Since E_{final} has $n = 2$ and E_{initial} has $n = 1$, we have

$$\Delta E = E_{n=2} - E_{n=1} = \left(-\frac{Rhc}{2^2}\right) - \left(-\frac{Rhc}{1^2}\right)$$

$$= \tfrac{3}{4}Rhc$$

$$= 1.6349 \times 10^{-18} \text{ J/atom or } 985 \text{ kJ/mol of H atoms}$$

where we have used the energies of the $n = 2$ and $n = 1$ states as calculated in Example 8.3. The amount of energy that must be absorbed by the atom so that an electron can move from the first to the second energy state is $0.75Rhc$, no more and no less. This is just a way of saying that *the energy difference between states of the electron is quantized.*

Moving an electron from a state of low n to one of higher n is an *endo*thermic process; energy is absorbed, and the sign of the value of ΔE is positive. The opposite process, an electron "falling" from a level of higher n to one of lower n, will therefore emit energy. For example, for a transition from $n = 2$ to $n = 1$,

$$\Delta E = E_{\text{final state}} - E_{\text{initial state}}$$

$$= E_{n=1} - E_{n=2} = \left(-\frac{Rhc}{1^2}\right) - \left(-\frac{Rhc}{2^2}\right)$$

$$= -\tfrac{3}{4}Rhc$$

This process is *exo*thermic; that is, 985 kJ must be *evolved* or *emitted* per mole of H atoms.

Depending on how much energy is added to a collection of H atoms, and how it is added, some atoms have their electrons excited from the $n = 1$ to the $n = 2$ or 3 or higher states. After absorbing energy, these electrons naturally move back down to lower levels (not necessarily directly to $n = 1$) and give back the energy the atom originally absorbed; that is, they emit energy in the process. This is the source of the numerous lines observed in the emission spectrum of H atoms, and the same basic explanation holds for atoms of other elements. For hydrogen, the series of lines having energies in the ultraviolet region (called the **Lyman series**, Figure 8.11) comes from electrons moving from states with n greater than 1 to that with $n = 1$.

For excited hydrogen atoms, the series of lines that have energies in the visible region arises from electrons moving from states with $n = 3$ or greater to the state with $n = 2$. This is the **Balmer series**, and the equation originally written by Balmer and modified by Rydberg can now be derived from Bohr's theory. Since the final state for the series of visible lines is $n = 2$, we can write

$$\Delta E = E_{\text{final state}} - E_{\text{initial state}}$$

$$\Delta E_{\text{Balmer}} = \left(-\frac{Rhc}{2^2}\right) - \left(-\frac{Rhc}{n_i^2}\right)$$

$$= -Rhc\left(\frac{1}{2^2} - \frac{1}{n_i^2}\right) \quad \text{where } n_i = 3, 4, \ldots \tag{8.5}$$

This equation gives the energy lost by the H atom as its electron falls from higher energy levels to $n = 2$. However, because of the demand that

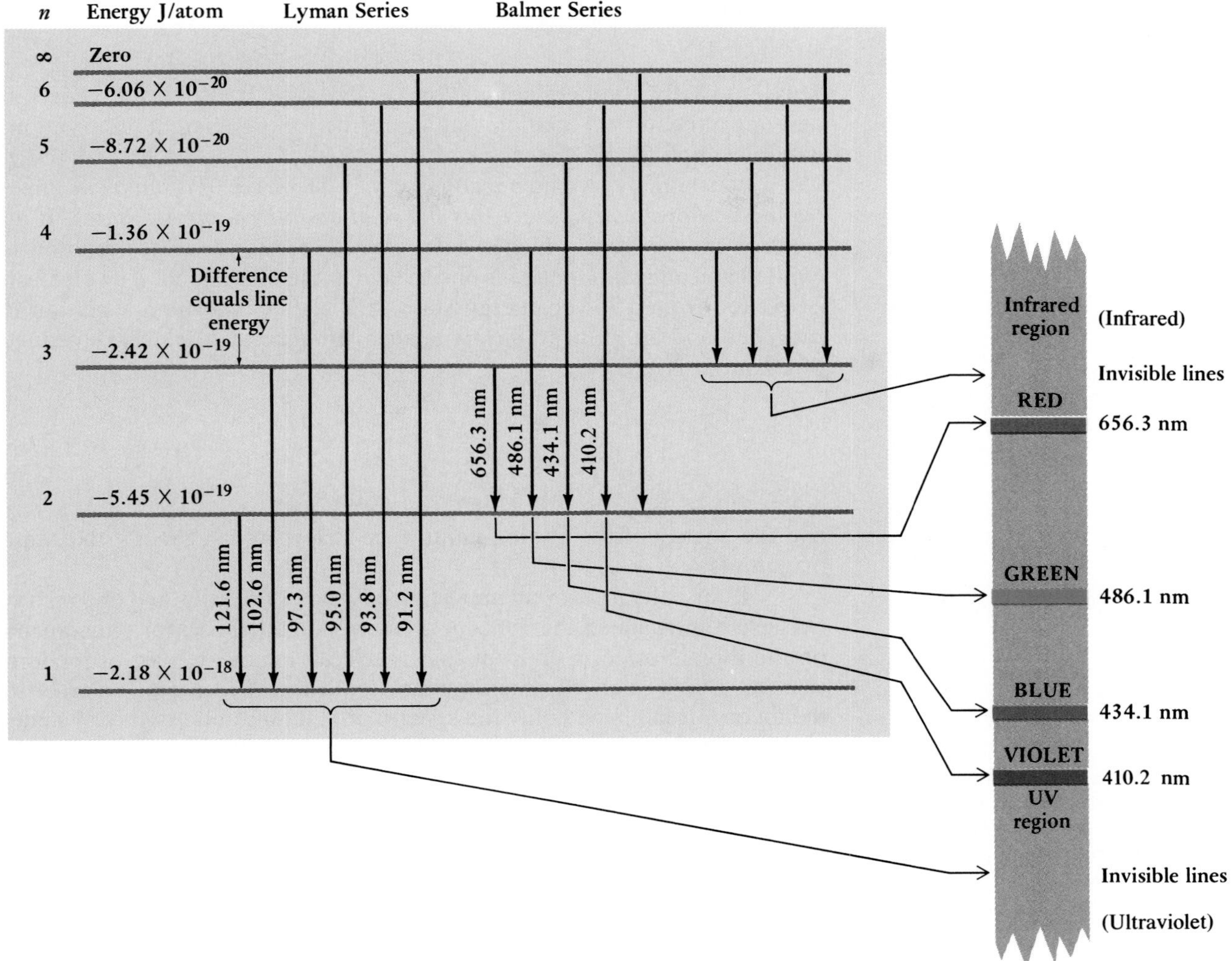

Figure 8.11 Some of the electronic transitions that can occur in an excited H atom. The Lyman series (ultraviolet region) results from transitions to $n = 1$. The Balmer series of lines appears in the visible region. The transition leading to emission of the least amount of energy in the Balmer series is $n = 3$ to $n = 2$ (a line in the red region, $\lambda = 656.3$ nm).

energy be conserved, this energy reappears as the energy of a photon of light. That is,

$$\Delta E_{\text{Balmer}} = -E_{\text{photon}}$$

where the − sign is needed because $\Delta E_{\text{Balmer}} < 0$ while $E_{\text{photon}} > 0$. Therefore, combining this with equation 8.5, we get

$$E_{\text{photon}} = Rhc\left(\frac{1}{2^2} - \frac{1}{n_i^2}\right) \qquad \text{where } n_i = 3, 4, \ldots$$

Also, combining Einstein's equation (Equation 8.2) with Equation 8.1 to find the relation between the energy of light radiated, E_{photon}, and its wavelength,

$$E_{\text{photon}} = h\nu = \frac{hc}{\lambda}$$

Remember that $\lambda\nu = c$.

we have

$$\frac{1}{\lambda} = R\left(\frac{1}{2^2} - \frac{1}{n_i^2}\right) \qquad \text{where } n_i = 3, 4, \ldots$$

This is exactly Rydberg's equation (Equation 8.3). Bohr succeeded in deriving a theoretical relation that agreed with experimental observation, a real triumph for his theory.

In summary, we now recognize that *the origin of atomic line spectra is the movement of electrons between quantized energy states*. If an electron moves from a higher energy state to a lower one, a photon is emitted and an *emission line* is observed. On the other hand, if an electron is excited from a lower energy state to a higher one, then a photon is absorbed and an *absorption line* is seen. In general, the energy change of the atom is

$$\Delta E = E_{\text{final}} - E_{\text{initial}}$$

$$= -Rhc\left(\frac{1}{n_{\text{final}}^2} - \frac{1}{n_{\text{initial}}^2}\right) \quad (8.6)$$

and the energy of the photon emitted or absorbed has exactly the same magnitude.

Bohr's theory was an important advance in physics and chemistry, because it introduced the concept of energy quantization for phenomena on the atomic scale, a concept that is still an important part of modern science. But it soon became apparent that there was a major defect in the theory. It explained only the spectra of H atoms and of other systems having only one electron (such as He^+). Further, another part of Bohr's theory, the idea that the electron moves about the nucleus with a path of fixed radius, like that of the planets about the sun, is incorrect. However, great improvements were made in atomic theory in the early 1920s, events in which Bohr was an enthusiastic and prominent participant.

When Bohr's paper describing his ideas was first published in 1913, Einstein declared it to be "one of the great discoveries."

EXAMPLE 8.4

ENERGIES OF ATOMIC SPECTRAL LINES

The Paschen series of lines in the hydrogen spectrum occurs in the infrared region. The electrons that produce them are moving from higher states to the $n = 3$ state (Figure 8.11). (a) Give n_{initial} for the transition that would account for the least energetic line of the series. (b) What is the energy involved in the least energetic transition? What are the frequency and wavelength of the emitted light?

Solution (a) The least energetic transition is that involving the smallest possible change in state energy. If the electron is to end with $n = 3$, then the least amount of energy will be emitted by an electron moving from $n = 4$ to $n = 3$.

(b) The energy emitted by an electron moving from $n = 4$ to $n = 3$ is

$$\Delta E = -Rhc\left(\frac{1}{3^2} - \frac{1}{4^2}\right)$$

$$= -0.04861Rhc$$

In Example 8.3 we found that $Rhc = 1313$ kJ/mol, so the $n = 4$ to $n = 3$ transition involves an energy change of

$$\Delta E = -0.04861(1313 \text{ kJ/mol}) = -63.83 \text{ kJ/mol}$$

The frequency is calculated from Einstein's equation.

$$\nu = \frac{\Delta E}{h} = \frac{(63.83 \text{ kJ/mol})(10^3 \text{ J/kJ})}{(6.022 \times 10^{23} \text{ electrons/mol})(6.626 \times 10^{-34} \text{ J} \cdot \text{s})}$$
$$= 1.600 \times 10^{14} \text{ s}^{-1}$$

Once the frequency has been obtained, the wavelength can be calculated from the relation $\lambda\nu = c = 2.998 \times 10^8$ m/s.

$$\lambda = \left(\frac{2.998 \times 10^8 \text{ m}}{\text{s}}\right)\left(\frac{1 \text{ s}}{1.600 \times 10^{14}}\right) = 1.874 \times 10^{-6} \text{ m}$$
$$= 1.874 \times 10^{-6} \text{ m } (10^9 \text{ nm/m}) = 1874 \text{ nm}$$

The experimental value is 1875.1 nm, extremely good agreement between experiment and theory for a theory as simple as Bohr's.

EXERCISE 8.6 Energy of an Atomic Spectral Line

The Brackett series of spectral lines for the H atom is not shown in Figure 8.11. However, it arises from transitions from higher levels down to $n = 4$. Calculate the frequency and wavelength of the *least* energetic line in this series. In what region of the electromagnetic spectrum does this line lie?

8.5 THE WAVE PROPERTIES OF THE ELECTRON

Einstein used the photoelectric effect to demonstrate that light, which is usually thought of as having wave properties, can also be thought about in terms of particles or massless photons. This fact was pondered by Louis Victor de Broglie (1892–1987). If light can be considered as sometimes having wave properties and other times having particle properties, he asked, why doesn't matter behave similarly? That is, could a tiny object such as an electron, which we have so far considered a particle, also exhibit wave properties in some experiments? Specifically, in 1925, de Broglie proposed that a free electron of mass m moving with a velocity v should have an associated wavelength given by the equation

$$\lambda = \frac{h}{mv} \tag{8.7}$$

This idea was revolutionary, since it linked the particle properties of the electron (m and v) with possible wave properties (λ). Experimental proof was soon produced. Davisson and Germer, working at the Bell Telephone Laboratories, found that a beam of electrons was diffracted like light waves by the atoms of a thin sheet of metal foil and that de Broglie's relation was followed quantitatively. Since diffraction is an effect readily explained by the wave properties of light, it followed that electrons also can be described by the equations of waves under some circumstances.

De Broglie's equation suggests that any moving particle has an associated wavelength. Yet, if λ is to be large enough to measure, the product of m and v must be *very* small because h is so small. For example, a 114-g baseball traveling at 110 mph has a large mv product (5.6 kg · m/s) and, therefore, the incredibly small wavelength of 1.2×10^{-34} m! Unfortunately, this is such a tiny value that it cannot be measured with any

Recall that $h = 6.6260755 \times 10^{-34}$ J · s, a very small number.

instrument now available, so the wave property of a moving baseball cannot be observed experimentally. Thus, it is possible to observe wave-like properties only for particles of extremely small mass such as protons, electrons, and neutrons.

EXAMPLE 8.5

USING DE BROGLIE'S EQUATION

Calculate the wavelength associated with an electron of mass $m = 9.109 \times 10^{-28}$ g traveling at 40.0% of the velocity of light.

Solution First, consider the units involved. Wavelength comes from h/mv where h is Planck's constant in units of joule · second. As discussed in Chapter 6, joules are equivalent to kg · m²/s². Therefore, the mass must be used in kg and velocity in m/s.

$$\text{Electron mass} = 9.109 \times 10^{-31}\ \text{kg}$$

$$\text{Electron velocity (40.0\% of light velocity)} = (0.400)(2.998 \times 10^{8}\ \text{m/s}) = 1.20 \times 10^{8}\ \text{m/s}$$

Substituting these values into de Broglie's equation, we have

$$\lambda = \frac{6.626 \times 10^{-34}\ (\text{kg} \cdot \text{m}^2/\text{s}^2)(\text{s})}{(9.109 \times 10^{-31}\ \text{kg})(1.20 \times 10^{8}\ \text{m/s})} = 6.06 \times 10^{-12}\ \text{m}$$

In nanometers, the wavelength is $(6.06 \times 10^{-12}\ \text{m})(1.00 \times 10^{9}\ \text{nm/m}) = 0.00606$ nm. This wavelength is only about 1/20 of the diameter of the H atom.

EXERCISE 8.7 De Broglie's Equation

Calculate the wavelength associated with a neutron having a mass of 1.675×10^{-24} g and a kinetic energy of 6.21×10^{-21} J. (Remember that the kinetic energy of a moving particle is $E = mv^2/2$.)

Erwin Schrödinger (1887–1961) was born in Vienna, Austria and studied at the university there. Following service in World War I as an artillery officer, he became a professor of physics at various universities; in 1928 he succeeded Max Planck as professor of theoretical physics at the University of Berlin. He shared the Nobel Prize in physics (with Paul Dirac) in 1933. (AIP Niels Bohr Library)

8.6 SCHRÖDINGER, HEISENBERG, AND QUANTUM MECHANICS

After World War I, Niels Bohr assembled a group of physicists in Copenhagen, a group that set out to derive a comprehensive theory for the behavior of electrons in atoms from the viewpoint of the electron as a particle. Independently, Erwin Schrödinger (1887–1961), an Austrian, worked toward the same goal, but he used de Broglie's hypothesis that an electron in an atom could be described by equations appropriate to wave motion. While both Bohr and Schrödinger were successful in predicting some aspects of electron behavior, Schrödinger's approach gave correct results for some properties where Bohr's failed, so theoreticians today primarily use Schrödinger's concept. In any event, the general theoretical approach to understanding atomic behavior, developed by Bohr, Schrödinger, and their associates, has come to be called **quantum mechanics** or **wave mechanics**.

Before you can appreciate Schrödinger's solution, there is another point to be emphasized. After de Broglie's suggestion that an electron can be described as having wave properties, a great debate raged in physics. How can an electron be described as *both* a particle and a wave? When J.J. Thomson measured the charge/mass ratio of an electron early in this century, he did an experiment that showed the particle-like nature of the electron (see "Something More About the History of Atomic Structure" at the end of this chapter). On the other hand, Davisson and Germer observed its wave-like properties. One can only conclude that *the electron has dual properties*. The result of a given experiment can be described *either* by the physics of waves *or* by the physics of particles; there is no single experiment that can be done to show that the electron behaves *simultaneously* as a wave and a particle!

Werner Heisenberg (1901–1976) earned a Ph.D. in theoretical physics at the University of Munich in 1923 and then studied with Max Born and later Niels Bohr. He received the Nobel Prize in physics in 1932. (AIP Niels Bohr Library)

What does this wave-particle duality have to do with electrons in atoms? Werner Heisenberg (1901–1976) and Max Born (1882–1970) provided the answer. Heisenberg proposed the so-called **uncertainty principle**, which can be expressed mathematically as

$$\Delta x \cdot \Delta(mv) > h \quad (8.8)$$

This simple expression means that the uncertainty in the position of the electron (Δx) times the uncertainty in its momentum [$\Delta(mv)$] must be larger than Planck's constant. In other words, if you want to define the momentum (the product of its mass and velocity) of some particle almost exactly [$\Delta(mv)$ is a very small number], then you have to forego knowledge of its exact position at the time of the measurement (that is, Δx will be a relatively large number). For example, if the uncertainty in the velocity of a 1000-kg car is 1 mph (0.45 m/s), Heisenberg's uncertainty in its position is about 10^{-36} m; essentially, we can know the car's position quite accurately. On the other hand, for the much lighter electron (whose mass is only 10^{-33} times that of a car), the uncertainty in position would be 10^{33} times greater, that is, about 10^{-3} m or 1 mm! To illustrate, an electron moving at 1×10^8 m/s ($\frac{1}{3}$ the velocity of light, 220 million mph) with an uncertainty of 0.1% leads to an uncertainty in velocity of 10^5 m/s and an uncertainty in position of 10^{-8} m, that is, 10 nm or about the distance across a medium-sized molecule.

On the basis of Heisenberg's idea, Max Born proposed that the results of quantum mechanics should be interpreted as follows: *if we choose to know the energy of an electron in an atom with only a small uncertainty, then we must accept a correspondingly large uncertainty about its position in the space about the atom's nucleus.* In practical terms, this means the only thing we can do is calculate the likelihood or **probability** of finding the electron (of a given energy) within a given space. We turn to this viewpoint in the next section.

8.7 THE WAVE MECHANICAL VIEW OF THE ATOM

Erwin Schrödinger combined de Broglie's equation with classical equations for wave motion. From these and other ideas he derived a new equation called the wave equation or the **Schrödinger equation** to describe the behavior of an electron in the hydrogen atom. This equation is complex and mathematically difficult to solve except in simple cases. The mathematics are of no concern to us here, but the solutions—called wave functions—are chemically important. To understand the implications of these wave functions is to understand the modern view of the atom.

WAVE FUNCTIONS

Solving Schrödinger's equation leads us to a set of functions called **wave functions**, symbolized by the Greek letter ψ (psi). There are several important points to be made about these wave functions, which characterize the electron as a *matter-wave*.

1. Only certain vibrations, called standing waves (Figure 8.4), can be observed in a vibrating string. The same situation occurs in an atom; although electron motion is not as simple as that of a vibrating string, there still are *only certain allowed wave functions*.
2. Each wave function, ψ, for the electron in the H atom corresponds to an allowed energy ($-Rhc/n^2$). This is like Bohr's result; for each integer n there is an atomic state characterized by its own wave function ψ and energy E_n.
3. Points 1 and 2 amount to saying that the energy of the electron is quantized. The concept of quantization enters Schrödinger's theory naturally with the basic assumption of an electron matter wave. This is in contrast with Bohr's theory, where quantization is imposed as a postulate at the start.
4. Each wave function ψ itself can be interpreted only in terms of the ideas of probability raised at the end of the previous section. We now know that *the square of ψ (when calculated for the electron at a point P) gives the intensity of the electron wave or the probability of finding the electron at the point P in space about the nucleus*. The theory does not predict the exact position of the "particle." Furthermore, since Schrödinger's theory precisely defines the energy of the electron, Heisenberg's uncertainty principle tells us this must result in some uncertainty in electron position. This is why we can only give the *probability* of the electron being at a certain point in space when in a given energy state.
5. The matter waves for the allowed energy states are also called **orbitals**. We shall say more about orbitals in a moment.
6. To solve Schrödinger's equation for an electron in a three-dimensional world, three integer numbers—the quantum numbers n, ℓ, and m_ℓ—must be introduced. These quantum numbers may have only certain combinations of values, as outlined below. We shall use these combinations to define the energy states and orbitals available to the electron.
7. The amplitude of the electron wave at a point depends on the distance of the point from the nucleus. In most cases, there is also a dependence on the "latitude" and "longitude" of the point.

THE QUANTUM NUMBERS AND ATOMIC ORBITALS

In a three-dimensional world, three numbers are required to describe the location of an object in space (Figure 8.12). For the wave description of the electron in the H atom, this requirement leads to the existence of **three quantum numbers n, ℓ, and m_ℓ**. For the hydrogen atom, only the first of these three numbers is required to describe the energy of the electron, but all three are needed to define the probability of finding the electron in a given region of space. Before looking into the meanings of the three quantum numbers, it is important to say that *n, ℓ, and m_ℓ are all integers* but that *their values cannot be selected randomly*. As you shall now see, the value of n limits the possible values of ℓ, which in turn limit the values of m_ℓ.

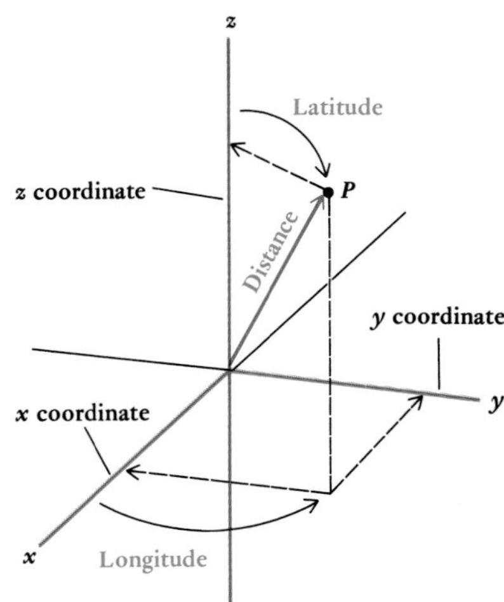

Figure 8.12 The cartesian coordinate system. The position of a point P in space can be specified by giving the x, y, and z coordinates or by giving the distance of the point from the center of the coordinate system and its latitude and longitude. In both cases, three numbers are needed.

THE THREE QUANTUM NUMBERS: n, ℓ, and m_ℓ Each of the three quantum numbers has a definite range of values, and each gives us different information about atomic orbitals.

n, the principal quantum number = 1, 2, 3, . . .

The principal quantum number n can have any integer value from 1 to infinity. As the name implies, it is the most important quantum number because *the value of n determines the total energy of the electron.* For the hydrogen atom the relation between the electron energy E and the quantum number n is the same as that given by Bohr's equation (Equation 8.4, $E = -Rhc/n^2$). Since all the quantities in this equation are constants except n, this means that *the energy of the electron in the H atom varies **only** with the value of n.*

The value of n also gives a rough measure of the most probable distance of the electron from the nucleus: the greater the value of n, the more probable it is that the electron is found further from the nucleus. That is, *n is a measure of the orbital size or diameter.*

As you will see in Chapter 9, in atoms having more than one electron, two or more electrons may have the same n value. These electrons are then said to be in the same **electron shell**. Each electron is labeled according to its value of n, and, as you shall see in more detail in Chapter 9, as many as $2n^2$ electrons can be assigned to an electron shell.

An earlier notation used letters for the major electron shells: K, L, M, N, and so on, corresponding to n = 1, 2, 3, 4, and so on.

ℓ, the angular momentum quantum number = 0, 1, 2, . . . (n − 1)

The electrons of a given shell can be grouped into **subshells**, each subshell characterized by a certain wave shape. For a given shell there are n different possible subshells (or orbital shapes), each subshell corresponding to one of the n different values of the quantum number ℓ. Each value of ℓ corresponds to a different **orbital shape** or *orbital type*.

The value of n for a shell limits the number of subshells (shapes) that are possible within that shell. That is, ℓ can be no larger than $n - 1$. Thus, for $n = 1$, the rule tells us that ℓ must equal 0 and only 0. Since ℓ has only one value when $n = 1$, only one orbital shape or type is possible for an electron assigned $n = 1$. When $n = 2$, however, ℓ can be either 0 or 1. Since two values of ℓ are possible, there are two types of orbitals or subshells in the $n = 2$ shell of electrons.

ℓ is related to orbital shape and describes the types of subshells within a shell. The number of values of ℓ (number of subshells in a shell) = n for that shell.

The values of the ℓ quantum number are usually coded by letters according to the scheme below.

VALUE OF ℓ	CORRESPONDING SUBSHELL LABEL
0	*s*
1	*p*
2	*d*
3	*f*

The orbital labels come from the early days of atomic spectroscopy when lines in emission spectra were called sharp, principal, diffuse, *and* fundamental.

This means, for example, that a subshell with a label of $\ell = 1$ is called a "*p* subshell," and an orbital found in that subshell is called a "*p* orbital."

m_ℓ, the magnetic quantum number = 0, ±1, ±2, ±3, . . . ±ℓ

The first quantum number (n) locates the electron in a particular electron shell and, for the H atom, determines its energy, the orbital size, and the range of possible shapes. The second quantum number (ℓ) places the electron in a particular subshell of orbitals within the shell and gives the orbital shape. The third quantum number (m_ℓ) then specifies in which orbital within the subshell the electron is located. *Orbitals within a given subshell differ only in their orientation in space, not in their shape.*

m_ℓ is related to the spatial orientation of an orbital in a given subshell. The number of m_ℓ values = the number of orbitals in a subshell.

The integer values that m_ℓ may have are limited by ℓ; m_ℓ values can range from $+\ell$ to $-\ell$ with 0 included. For example, when $\ell = 2$, m_ℓ has the five values +2, +1, 0, −1, and −2. The number of values of m_ℓ for a given ℓ (= $2\ell + 1$) tells you how many orientations there are for the orbitals in that subshell.

USEFUL INFORMATION FROM QUANTUM NUMBERS The quantum number rules can give us chemically useful information. Let us expand on them and see where this leads.

If $n = 1$, the rules tell us that ℓ can only be 0 and so m_ℓ must also have a value of 0. This means that, in the electron shell closest to the nucleus, there is only one type of orbital or one subshell (ℓ has only one value), and there is only one orientation for the orbital of this type. This orbital is labeled "1*s*," the "1" conveying the value of *n* and "*s*" telling us that ℓ is 0. *When $\ell = 0$, an s orbital is indicated, and there can be only one s orbital in a given shell of electrons.*

Electron orbitals are labeled by first giving the value of n and then the value of ℓ in the form of its letter code. For $n = 1$ and $\ell = 0$, the label is 1s.

When $n = 2$, ℓ can now have two values (0 and 1), so there are two types of orbitals in the second shell. One of these is the 2*s* orbital ($n = 2$ and $\ell = 0$), and the other is the 2*p* orbital ($n = 2$ and $\ell = 1$). Since the values of m_ℓ can range from +1 to −1 *including 0* when $\ell = 1$, this means that there must be three *p* orbitals; they all have the same shape, but they have different orientations in space. *When $\ell = 1$, p orbitals are indicated, and there are always three of them.* (The converse is true as well; for *p* orbitals, ℓ is always 1.)

For a given n, there will be n values of ℓ and n orbital types; there is a total of n^2 orbitals in the nth shell.

When an electron is characterized by $n = 3$, there are three orbital types (the three values of ℓ are 0, 1, and 2). Since you see ℓ values of 0 and 1 again, you know that two of the orbital types in the $n = 3$ shell will be 3*s* (one orbital) and 3*p* (three orbitals). The third orbital type is *d*, indicated by $\ell = 2$. Since m_ℓ has five values (+2, +1, 0, −1, and −2) when $\ell = 2$, there are five *d* orbitals (no more and no less). Thus, when $\ell = 2$ you can count on the fact that *there are five orbitals labeled d.*

Besides *s*, *p*, and *d* electron orbitals, sometimes we need to talk about *f* electron orbitals, that is, orbitals for which $\ell = 3$. There will be

Table 8.1 Summary of the Quantum Numbers, Their Interrelationships, and the Orbital Information Conveyed

Principal Quantum Number Symbol = n Values = 1, 2, 3, . . . (Orbital Size, Energy)	Angular Momentum Quantum Number Symbol = ℓ Values = 0 . . . $n - 1$ (Orbital Shape)	Magnetic Quantum Number Symbol = m_ℓ Values = $-\ell$ . . . 0 . . . $+\ell$ (Orbital Orientations = Number of Orbitals in Subshell)		Number and Type of Orbitals in the Subshell Number = number of values of $m_\ell = 2\ell + 1$ (Number of Orbitals in Shell = n^2)
1	0	0	(1 orientation)	1 1*s* orbital (1 orbital of 1 type in the $n = 1$ shell)
2	0 1	0 +1,0,−1	(1 orientation) (3 orientations)	1 2*s* orbital 3 2*p* orbitals (4 orbitals of 2 types in the $n = 2$ shell)
3	0 1 2	0 +1,0,−1 +2,+1,0,−1,−2	(1 orientation) (3 orientations) (5 orientations)	1 3*s* orbital 3 3*p* orbitals 5 3*d* orbitals (9 orbitals of 3 types in the $n = 3$ shell)
4	0 1 2 3	0 +1,0,−1 +2,+1,0,−1,−2 +3,+2,+1,0,−1,−2,−3	(1 orientation) (3 orientations) (5 orientations) (7 orientations)	1 4*s* orbital 3 4*p* orbitals 5 4*d* orbitals 7 4*f* orbitals (16 orbitals of 4 types in the $n = 4$ shell)

seven of them, as given by the seven possible values of m_ℓ when $\ell = 3$ (+3, +2, +1, 0, −1, −2, −3).

The three quantum numbers introduced so far and their allowed values are summarized in Table 8.1. These rules will be used to explore atomic structure more fully in Chapter 9, so you should be thoroughly familiar with them.

No known elements have electrons assigned to orbitals with an ℓ value greater than 3 in the ground state. If they existed, however, $\ell = 4$ would correspond to g orbitals, $\ell = 5$ to h orbitals, and so on down the alphabet.

EXERCISE 8.8 Using Quantum Numbers

Complete the following statements:

(a) When $n = 2$, the values of ℓ can be _____ and _____.

(b) When $\ell = 1$, the values of m_ℓ can be _____, _____, and _____, and the subshell has the letter label _____.

(c) When $\ell = 2$, this is called a _____ subshell.

(d) When a subshell is labeled *s*, the value of ℓ is _____ and m_ℓ has the value _____.

(e) When a subshell is labeled *p*, there are _____ orbitals within the subshell.

(f) When a subshell is labeled *f*, there are _____ values of m_ℓ and there are _____ orbitals within the subshell.

THE SHAPES OF ATOMIC ORBITALS

The chemistry of an element and of its compounds is determined by the electrons in the element's atoms: which orbitals describe those electrons and the shapes of those orbitals. Now that we know something about the types and number of orbitals available, we want to turn to the question of orbital shape.

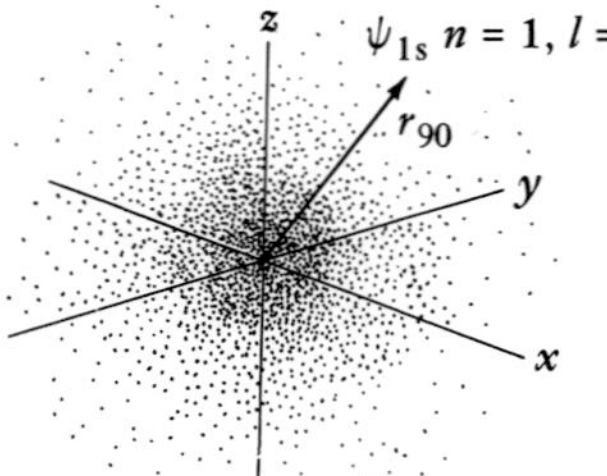

(a) Dot picture of an electron with a $1s$ atomic orbital. Each dot represents the position of the electron at a different instant in time. Note that the dots cluster closest to the nucleus. r_{90} is the radius within which the electron is found 90% of the time.

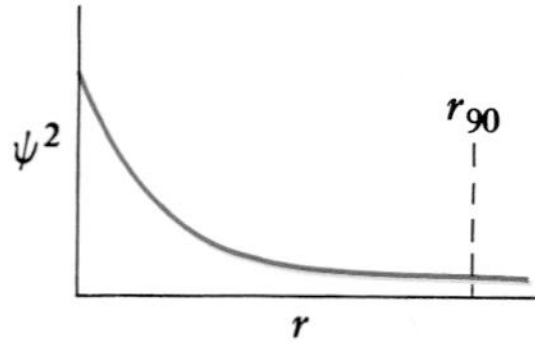

(b) A plot of the probability density as a function of distance for a one-electron atom with a $1s$ electron curve.

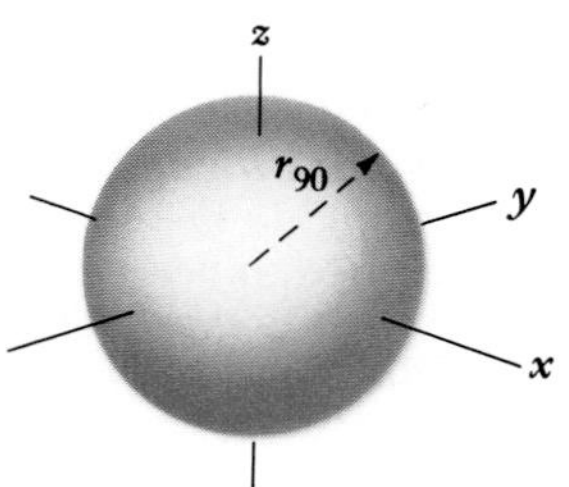

(c) The surface of the sphere within which the electron is found 90% of the time for a $1s$ orbital.

Figure 8.13 Different views of a $1s$ ($n = 1$ and $\ell = 0$) orbital.

s ORBITAL SHAPES When an electron has the value $\ell = 0$, we say the electron "occupies" an *s* orbital. But what does this mean? What is an *s* orbital? What does it look like? To answer this question, we shall have to investigate the wave function for an electron with $n = 1$ and $\ell = 0$, that is, the $1s$ orbital. If you could photograph the electron with that orbital at one-second intervals for a few thousand seconds, the composite picture would resemble the drawing in Figure 8.13a, a picture that is meant to be like a photograph of points on a standing wave (Figure 8.4) taken at frequent intervals. The fact that there is a greater density of "dots" close to the nucleus indicates that the electron is most often found near the nucleus (and less likely to be found further away). Putting this in the language of quantum mechanics, we say that the greatest probability of finding the electron is in a tiny volume of space around the nucleus, while it is less probable that it is further away. The "thinning" of the electron cloud at increasing distance, shown by the decreasing density of dots in Figure 8.13a, is illustrated in a different way in Figure 8.13b. Here we have plotted the *square* of the electron wave function for the electron in a $1s$ orbital as a function of the distance of the electron from the nucleus. As noted in point 4 on page 298, the value of ψ^2 at a point *P* gives the probability of finding the electron at that point. The units of ψ^2 at each point *P* are 1/volume, so the numbers on the vertical axis of this plot are the probability of finding the electron in each cubic nanometer, for example. Hence, ψ^2 is called the **probability density**. For the $1s$ orbital, ψ^2 is very high for points in the space immediately around the nucleus, but it drops off rapidly as the distance from the nucleus increases. Notice that the probability approaches but never reaches zero, even at very large distances.

For the $1s$ orbital, Figure 8.13a shows that the electron is most likely found within a sphere with the nucleus of the atom at the center. No matter which direction you proceed from the nucleus, the probability of finding an electron along a line in that direction drops off as depicted by Figure 8.13b. Indeed, *the 1s orbital is spherical in shape.*

The visual image of Figure 8.13a is that of a cloud whose density is small at large distances from the center; there is no sharp boundary beyond which the electron is never found. However, *s* and other orbitals are often depicted as having a sharp boundary (Figure 8.13c), simply because it is easier to draw such pictures. To arrive at this picture we have drawn a sphere about the nucleus in such a way that there is a 90% chance of finding the electron somewhere inside. There are many misconceptions about pictures of this type. Be sure to understand that this surface is not real. The nucleus is surrounded by an "electron cloud" and not by an impenetrable surface "containing" the electron. The electron is not distributed evenly throughout the volume enclosed by the surface but is most likely found nearer the nucleus.

That an s orbital has a spherical shape, for which there are no distinct orientations, corresponds to the idea that the number of m_ℓ values (one in this case) gives the number of orbital orientations.

From analysis of ψ for *s* orbitals of different *n* values, an important conclusion is that *all orbitals labeled s are spherical in shape.* In every case, *s* electrons can be found in a very small volume of space immediately around the nucleus. However, one important difference between *s* orbitals of different *n* is that *the size of s orbitals increases as n increases.* Thus, a $1s$ orbital is more compact than a $2s$ orbital, which is more compact than a $3s$ orbital.

(Text continues on p. 305)

SOMETHING MORE ABOUT
The History of Atomic Structure: Thomson, Rutherford, and the Nuclear Atom

We have come a long way from John Dalton's revolutionary hypothesis that all matter is composed of atoms to our modern ideas of atomic structure, and the road leads directly through the brilliant work of Michael Faraday (1791–1867) (see Chapter 5).

Faraday began his career as an assistant to Sir Humphrey Davy (1778–1829), one of the most influential scientists in England in his time. Among other things, Davy was interested in the effects of electricity on chemical substances in solution. Davy's curiosity about this carried over to Faraday, and Faraday spent the remainder of his life studying electrical phenomena. In 1832 Faraday found that when he passed electricity through a chemical substance, the mass of the material affected was proportional to the duration and strength of the electrical current. From these experiments he was able to deduce that ions are electrically charged and that their charges are all multiples of some smallest charge. In 1891, another English scientist, E. Johnstone Stoney, proposed the name *electron* for this natural unit of electricity, and the same name was applied a few years later to an isolated particle having a negative charge.

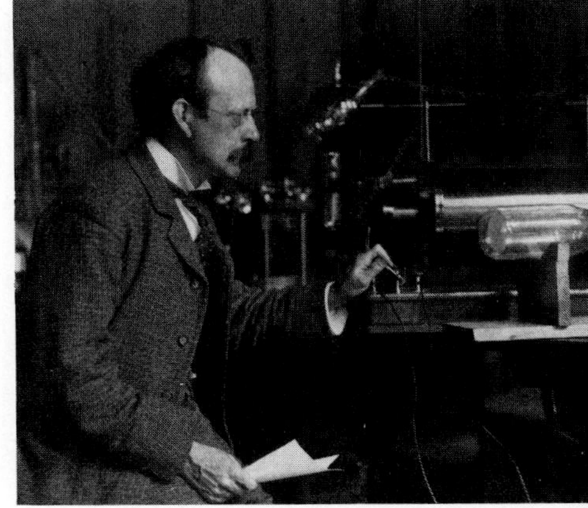

Figure A Sir J.J. Thomson (1856–1909) in his laboratory at Cambridge University, England. Thomson received the Nobel Prize in 1906 for his work on the properties of the electron. (University of Cambridge, Cavendish Laboratory)

At the close of the 19th century, J.J. Thomson (1856–1909) was the Cavendish Professor of Experimental Philosophy at Cambridge University in England (Figure A); this was the most important position in physics in the English-speaking world at the time, and it was in his laboratory that sophisticated experiments on beams of electrons were first carried out. One of the most important outcomes of Thomson's experiments was a measurement of a fundamental property of the electron, the ratio of its charge (e) to its mass (m). The currently accepted value is

$$\text{Electron charge/mass} = e/m = -1.75881962 \times 10^{8} \text{ coulombs/gram}$$

The design of Thomson's experiment allowed only the measurement of the *ratio* of electron charge and mass. To find one or the other of these quantities, a separate experiment had to be performed. This was done in 1909 by Robert Millikan of the University of Chicago (Figure B), and a value for the electron charge very close to the currently accepted value of $1.6021773 \times 10^{-19}$ coulombs was found in his famous "oil drop experiment." Combined with Thomson's charge/mass ratio, this leads to an electron mass of $9.1093897 \times 10^{-28}$ g.

In Thomson's time, beams of electrons were called cathode rays, because the electrons originated at the negative pole or "cathode" of an electrical discharge. This name has stayed with us, and TV tubes are often called CRTs, cathode ray tubes.

The two decades from 1895 to 1915 were an exciting time in physics because so many important discoveries were made. Our modern view of the atom

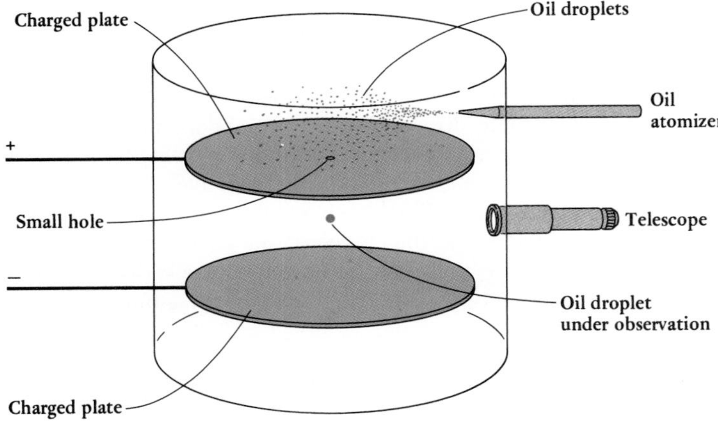

Figure B Millikan oil drop experiment. A fine mist of oil droplets is introduced into the chamber. The gas molecules inside the chamber are ionized (split into electrons and positive ions) by a beam of x-rays (not represented). The electrons adhere to the oil droplets, some droplets having one electron, some two electrons, and so forth. These negatively charged oil droplets fall under the force of gravity into the region between the electrically charged plates. If you carefully adjust the voltage on the plates, the force of gravity can be exactly counterbalanced by the attractive force between the negative oil drop and upper, positively charged plate. Analysis of these forces leads to a value for the charge on the electron.

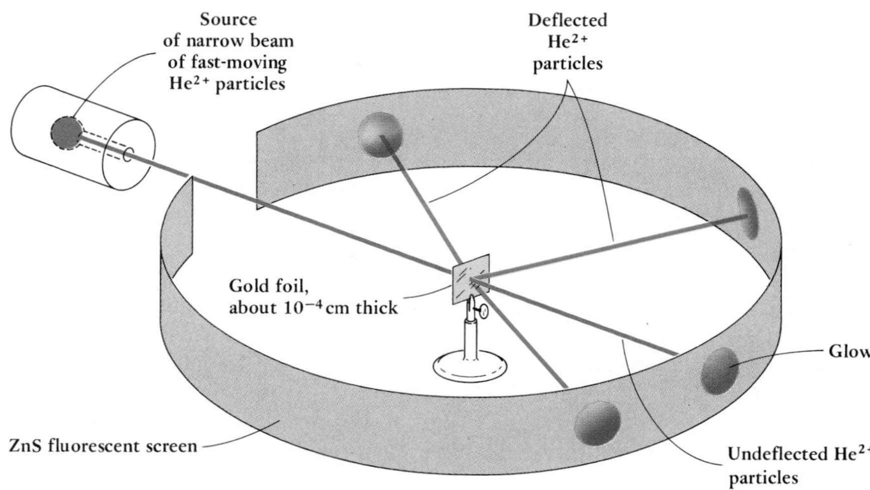

Figure C The experimental arrangement for the Rutherford experiment. A beam of alpha particles is directed at a piece of gold foil about 10^{-4} cm thick. A luminescent screen surrounding the foil detects particles passing through the foil (blue) or deflected by close encounters with atomic nuclei in the foil (green and red).

See page 245 for a brief biography of Rutherford.

was shaped in that period, and the next important development came from the laboratory of Ernest Rutherford, formerly a student of J.J. Thomson and then a professor at the University of Manchester in England.

Thomson and his students had supposed the atom was a uniform sphere of positively charged matter within which thousands of electrons circulated in coplanar rings. But an important question remained: how many electrons were circulating within this sphere? To try to find an answer to this, Thomson and his students fired a beam of electrons at a very thin metal foil. If the model were correct, the beam would have enough energy to travel through the foil and emerge from the other side. However, while moving through the foil, the electrons of the beam would encounter the very large number of electrons within the atoms, and the negative charges would repel each other. A tiny deflection of the beam from its straight path should be observed at each encounter, with the size of the total deflection related to the number of electrons in the atom. Thomson did indeed observe a deflection, but it was much smaller than he expected, and he was forced to revise his estimate of the number of electrons, but not his model of the atom.

An alpha particle is a helium nucleus. The 2 electrons of the He atom have been stripped away, leaving the 2 protons and 2 neutrons of the nucleus. Therefore, an α particle can be symbolized as ${}^{4}_{2}He^{2+}$.

In about 1910 Ernest Rutherford decided to test Thomson's model further.* As described in Chapter 7, Rutherford had earlier discovered a positively charged, massive particle that he called an alpha (α) particle. He reasoned that, if Thomson's atomic model were correct, very little deflection of a beam of such massive α particles should occur if they were fired at a thin gold foil. So, with this idea in mind, Rutherford's students Geiger and Marsden carried out their now famous experiment (Figure C). While they did find that most of the α particles were deflected only to a small extent, they were astonished to find that a very few particles were deflected nearly backwards (Figure D)! Rutherford described the unexpected result by saying "It was about as credible as if you had fired a 15-inch [artillery] shell at a piece of tissue paper and it came back and hit you." The only way to account for this was to discard Thomson's model and to conclude that the positive charge of the atom is concentrated in a very small volume of space, which Rutherford called the **nucleus**, while the negative electrons must occupy the space outside the nucleus. From their results, Rutherford, Geiger, and Marsden were able to calculate that the charge on the gold nucleus is in the range of 100 ± 20 (actual value, 79) and that the nucleus has a radius of about 10^{-12} cm (actual value is closer to 10^{-13} cm). The radius of the atom was found to be about 100,000 times that of the nucleus.

It was obvious that the negative electrons occupied the space outside the nucleus, but their arrangement was completely unknown to Rutherford or other physicists of their time. This arrangement is now known and is the subject of this chapter and the next one.

*The historical setting and implications of the experiments of Thomson and Rutherford are described in "Rutherford-Bohr Atom," *American Journal of Physics*, **1981**, *49*, 223–231.

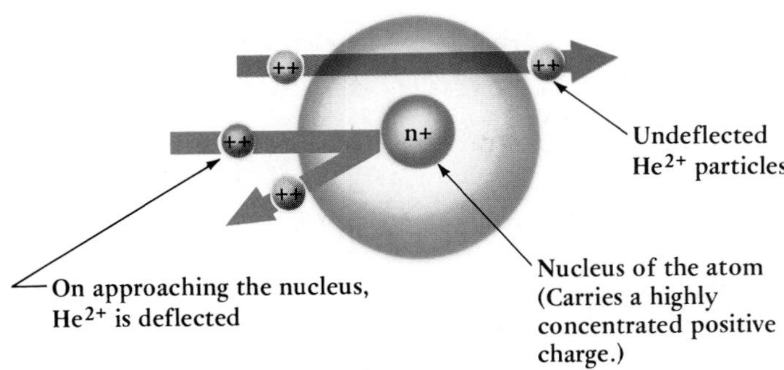

Figure D A helium ion, ${}^4_2He^{2+}$ (an alpha particle), encountering an atom in a piece of foil in the Rutherford experiment. In the Rutherford model, the atom is mostly empty space with a dense nucleus containing most of the atom's mass and all of its positive charge. Most of the alpha particles pass unimpeded through the foil because they do not come near an atomic nucleus. A few, however, closely approach the positively charged nucleus of the atom, and, since such He^{2+} ions are subjected to a strong, unbalanced force, they are deflected through large angles. (The drawing is not to scale.)

Rutherford's results were extraordinary! To give you some feeling for his results, imagine enlarging the gold foil, Rutherford's target, about one billion times. On this scale, a gold atom would be about the size of an apple, but the nucleus would be a very tiny but very heavy particle with a radius of only about 0.0001 cm. If you fired a beam of fine sand grains (taken to represent α particles on this scale) at the apple-sized atoms from many miles away, very few sand particles would even come close to any nucleus, let alone strike it almost directly. However, those that did would bounce back. The fact that some α particles were deflected at all in Rutherford's experiment must mean that there is a *very great concentration of positive charge in the nucleus*, allowing it to act with intense force on any α particle that comes at all close. The positive charge could *not* be spread out over the whole atom!

Like many experiments in science, Rutherford's experiment raised as many interesting and difficult questions as it answered. For example, why don't the electrons simply collapse into the nucleus from the force of attraction between the concentrated positive charge of the nucleus and the negative electrons? What is the arrangement of electrons outside the nucleus? Some inkling of the answers to these questions was soon offered by a young Danish physicist, Niels Bohr. Although Bohr's explanations sent science in the right direction, he quickly recognized their limitations, and he and others eventually developed a more complete and useful model of atomic structure, *quantum mechanics*. Quantum mechanics is the approach currently used in science, and is the model that is outlined in this chapter.

p ORBITAL SHAPES Atomic orbitals for which $\ell = 1$ are called *p* orbitals, and all have the same basic shape. That is, *all p orbitals have one plane that slices through the nucleus and divides the region of electron density into two halves* (Figure 8.14). It is as though we chopped an *s* orbital into two equal halves.

The plane slicing through the nucleus is called a **nodal plane**.* The electron can never be found in the nodal plane; the regions of electron density lie on either side of the nucleus. This means that, unlike *s* orbitals, there is no likelihood of finding the electron at the nucleus, and so a plot

*A question our students often ask is how the electron in a *p*, *d*, or *f* orbital moves across the nodal plane. We reply that this question is not meaningful given the view that an electron behaves as a wave. If we treat the electron according to the physics of waves, then we can no longer think of it as a particle moving about in space. The standing waves in Figure 8.4 have nodes, but this does not mean the string does not exist at that point; it simply means that there is no amplitude at the node. The same is true of electron waves.

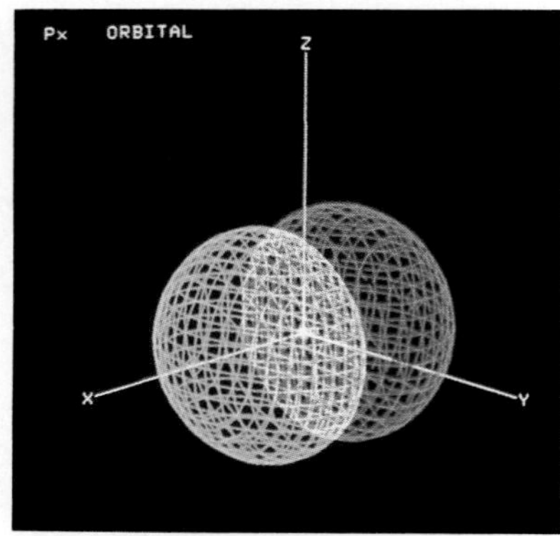

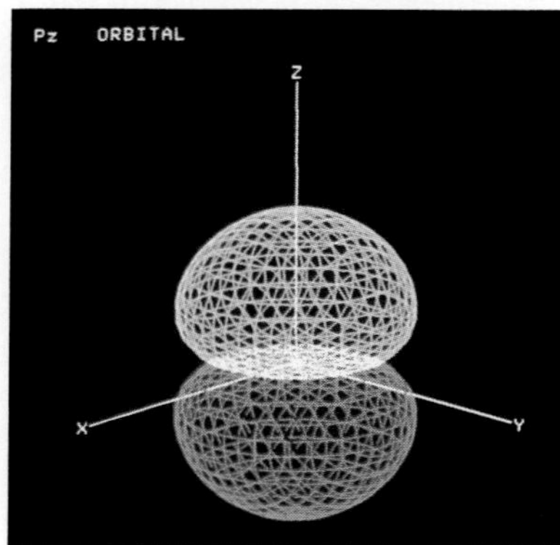

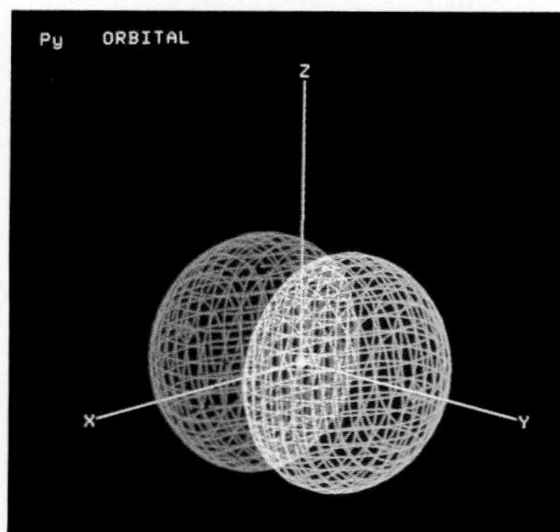

Figure 8.14 The $2p$ orbitals for a one-electron atom. The subscript letter on the orbital notation indicates the cartesian axis along which the orbital lies. The plane passing through the nucleus (perpendicular to this axis) is called a **planar node**. All p orbitals have one planar node since $\ell = 1$. (M. Rajzmann)

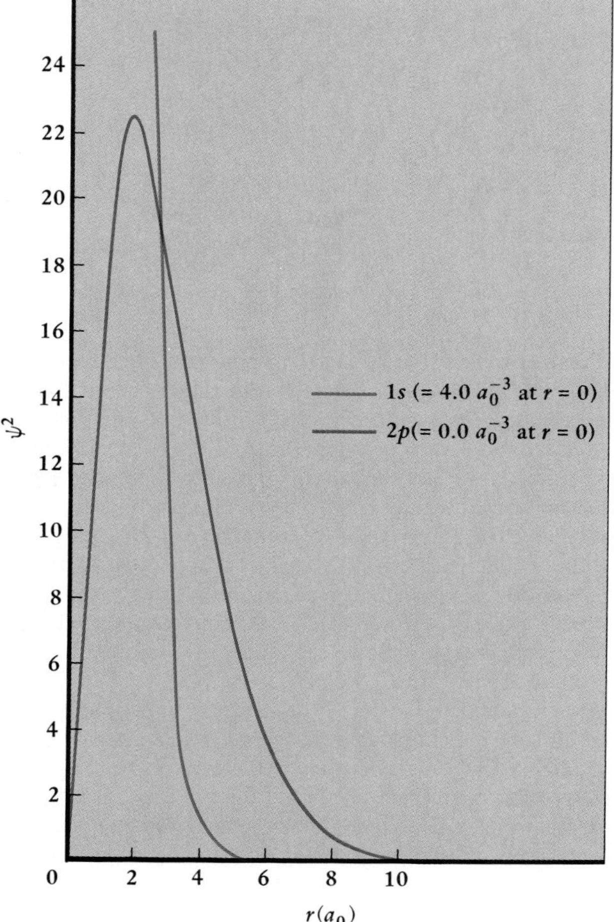

Figure 8.15 Accurate probability density plots for $1s$ and $2p$ electron orbitals. The horizontal axis represents the distance from the nucleus in units of a_0 (atomic units), where a_0 = 0.0529 nm. The vertical axis gives the probability or electron intensity, in units of $10^{-3}/a_0^3$. Note that the $1s$ electron wave has no node, while the $2p$ electron wave has a node at the nucleus. Also note that the $1s$ probability is so great very near the nucleus that the plot extends well beyond the top of the figure.

of ψ^2 versus distance (Figure 8.15) starts at zero probability at the nucleus, rises to a maximum (at 0.1058 nm for a $2p$ orbital), and then drops off at still greater distances. If you enclose 90% of the electron density within a surface, the view in Figure 8.14 is appropriate: the electron cloud has the shape of a "halved sphere" that resembles a weightlifter's "dumbbell," and so this is the term by which chemists often describe p orbital shapes.

According to Table 8.1, when $\ell = 1$, then m_ℓ can only be +1, 0, or −1. That is, *there are three orientations for ℓ = 1 or p orbitals*, depending on how we chop the sphere of electron density: two vertical ways or one horizontal way. Since there are three mutually perpendicular directions in space (x, y, and z), the p orbitals are commonly visualized as lying along these directions (with the nodal plane perpendicular to the

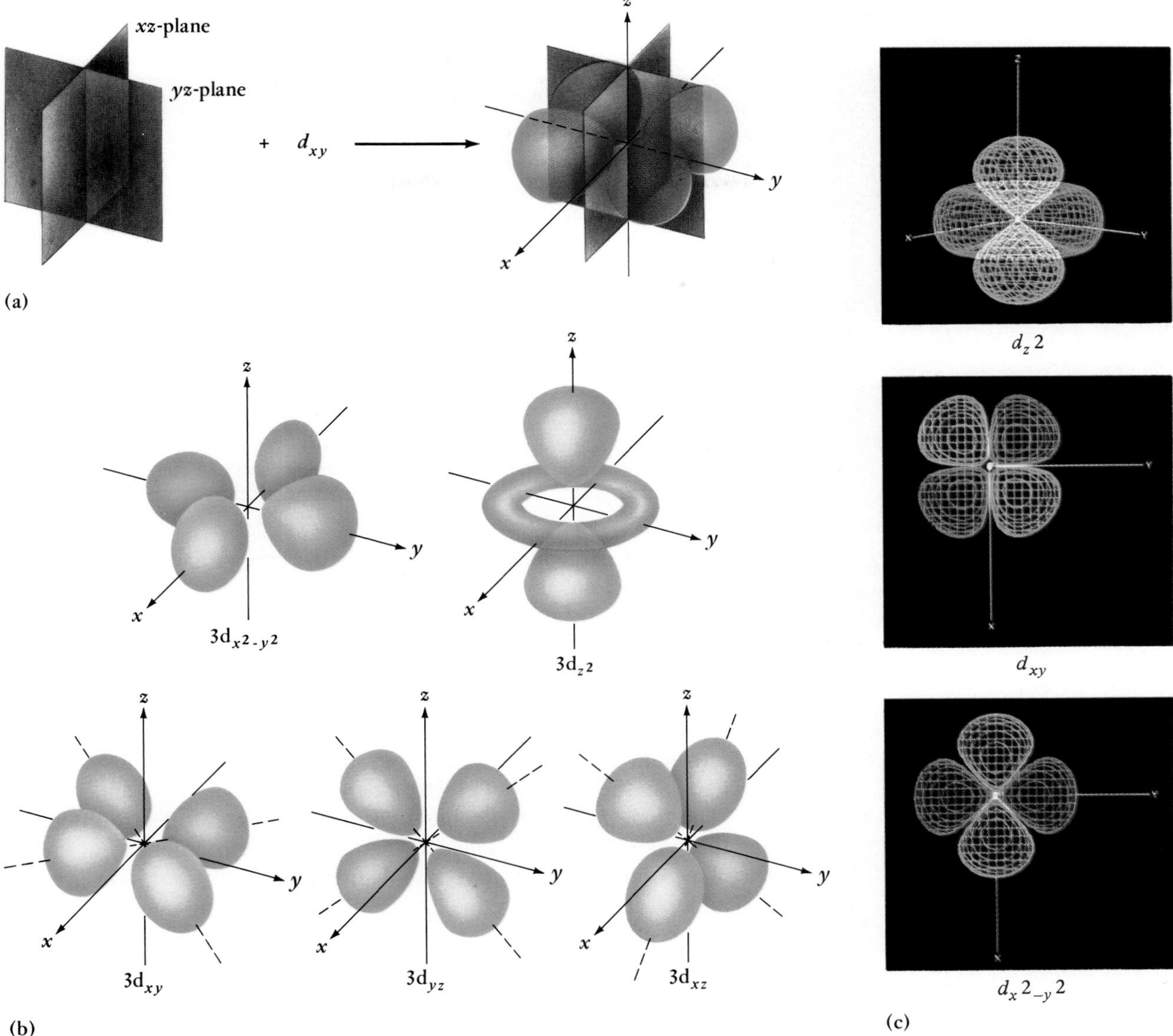

Figure 8.16 The *d* atomic orbitals. (a) The *xz* and *yz* nodal planes of the d_{xy} orbital. The subscript *xy* on the orbital label means that the orbital lies in the *xy* plane. (b) All *d* orbitals have $\ell = 2$, so all have two nodal planes. (The "planes" for the d_{z^2} orbital are actually conical, as explained in the text.) (Photos courtesy of M. Rajzmann)

axis), and they are labeled according to the axis along which they lie (p_x, p_y, and p_z).

d ORBITAL SHAPES There is a direct relation between the value of ℓ and the number of nodal planes that slice through the nucleus. The *s* orbitals, for which $\ell = 0$, have no nodal planes. The *p* orbitals, for which $\ell = 1$, have one nodal plane. It follows that *the five d orbitals, for which* $\ell = 2$, *have two nodal planes that slice the electron cloud into four sections.*

The quartering of the electron cloud to give the five orientations of the *d* orbitals is more complex than the halving to get *p* orbitals, so we turn to the wave forms themselves to see how nature accomplishes it. Three of the orientations are quite logical. We imagine a horizontal plane to halve space and either of two vertical planes to halve it again. This gives the orientations labeled d_{xz} and d_{yz} in Figure 8.16; notice that

Figure 8.17 One of seven possible f electron orbitals. Notice the presence of three nodal planes (xy, xz, and yz) as required by an orbital with $\ell = 3$.

The Relationship Between ℓ and the Number of Nodal Surfaces

There is a direct correspondence between the value of ℓ and the number of nodal surfaces slicing through the spherical electron cloud and dividing it into regions of electron density or probability.

ORBITAL TYPE	VALUE OF ℓ	NUMBER OF NODAL SURFACES	REGIONS OF ELECTRON DENSITY
s	0	0	1
p	1	1	2
d	2	2	4*
f	3	3	8*

*These numbers are not quite true for all orbitals (d and f) in a given shell. As illustrated by the d_{z^2} orbital, some nodal surfaces are not planes but are cone-shaped; for this orbital, this means there are apparently 3 regions of electron density.

these orbitals lie in the planes defined by the x-z and y-z axes, respectively. If the electron cloud is sliced by two vertical planes, we have the d_{xy} orbital where the orbital lies in the xy plane. Finally, notice that these three orbitals are related to p orbitals, in that each comes from slicing a p orbital with yet another nodal plane.

Of the two remaining d orbitals, the $d_{x^2-y^2}$ orbital is easier to visualize. Like the d_{xy} orbital, $d_{x^2-y^2}$ results from two vertical planes slicing the electron cloud into quarters. Now, however, the planes bisect the x and y axes, so the regions of electron density lie *along* the x and y axes.

The final d orbital, d_{z^2}, has two main regions of electron density along the z axis, but there is also a "donut" of electron density in the xy plane. This orbital also has two nodal surfaces, but the *surfaces* are not flat. Think of an ice cream cone sitting with its tip at the nucleus. One of the electron clouds along the z axis will sit inside the cone. If you use another cone pointing in the opposite direction from the first cone, again with its tip at the nucleus, another region of electron density will fit inside this second cone. The region outside both cones defines the remaining, donut-shaped region of electron density.

f ORBITAL SHAPES The seven f orbitals all have $\ell = 3$, meaning that there are three nodal surfaces slicing through the nucleus and carving up the electron cloud into eight regions of probability. This makes these orbitals less easily visualized than the ones we have seen so far, but one example of an f orbital is illustrated in Figure 8.17.

EXERCISE 8.9 Orbital Shapes

(a) What are the n and ℓ values for each of the following orbitals: 6s, 4p, 5d, and 4f?

(b) How many nodal planes are there for a 4p orbital? For a 6d orbital?

SUMMARY

The modern quantum mechanical view of the atom has been developed in this century through some important experiments. **Thomson** first measured the charge/mass ratio (e/m) of the electron shortly after 1900.

Millikan was able to derive the mass of the electron (current value, $9.1093897 \times 10^{-28}$ g) after experimentally determining the charge (current value, 1.6022×10^{-19} coulomb) in his famous "oil drop" experiment. **Rutherford**, attempting to verify predictions made by Thomson, carried out an experiment that showed for the first time that the positive charge of the atom is concentrated in a tiny volume called the nucleus, and the electrons occupy the space outside the nucleus.

In the same period in which these experiments were performed, Einstein postulated an explanation for the photoelectric effect (Section 8.3) that showed that electromagnetic radiation reveals itself at times as massless particles called **photons**. These photons have an energy content given by

$$E_{\text{photon}} = h\nu \tag{8.2}$$

where h is called Planck's constant ($6.6260755 \times 10^{-34}$ J · s), a statement of the equivalency of the energy content and frequency of radiation (ν) (Section 8.2). This is an extension of Planck's idea that energy at the atomic level is quantized.

Before all the experiments we have mentioned, it was known that all **electromagnetic radiation** could be described by the physics of waves, where the product of wavelength (λ) and frequency (ν) is equal to the velocity of light ($c = 2.99792458 \times 10^8$ m/s) (Section 8.1).

$$\lambda\nu = c \tag{8.1}$$

De Broglie later proposed (Section 8.5) that an electron and other particles of comparable mass could also be described by the physics of waves. The length of the wave appropriate for an object of mass m moving with velocity v is

$$\lambda = h/mv \tag{8.7}$$

The first atomic model that provided some explanation for experimental observations was that of **Bohr** (Section 8.4), who postulated that the electron in the hydrogen atom could occupy only certain **stationary states**, each with an energy given by

$$E_n = -Rhc/n^2 \tag{8.4}$$

where R is the Rydberg constant, h is Planck's constant, c is the velocity of light, and n is an integer equal to or greater than 1. Bohr postulated that the electron could move to another state, the amount of energy absorbed or emitted in the process being equal to the difference in energy between the two states. If the electron moves from a particular state to one of lower n, energy is emitted as electromagnetic radiation of discrete wavelength. (In the case of excited H atoms, some frequencies of radiation, the **Balmer lines**, are seen in the visible region.)

The Bohr model proved incapable of extension to atoms heavier than H, and a new model called **quantum** or **wave mechanics** was developed by **Schrödinger**, Bohr, and their associates (Sections 8.6 and 8.7). This model describes the electron states of the H atom by waves. From this model has come a set of matter-wave functions or **orbitals** for the electron in the atom. Each wave function corresponds to an allowed energy of the electron in the atom; that is, the electron can take only certain quantized energies in the atom. The energy of the electron in the H atom in a given state is known and can be calculated by the same equation derived by Bohr (Equation 8.4). The position of the electron is not known with certainty, however; only the **probability** of the electron being in a given region of space can be calculated. (This is the interpretation of the quantum mechanical model by **Heisenberg** and embodies his postulate called the **uncertainty principle**.)

The electron wave for each allowed energy state is called an orbital. Each orbital is described by **three quantum numbers**: n, ℓ, m_ℓ.

QUANTUM NUMBER	NAME	VALUES	ORBITAL PROPERTY DESCRIBED
n	principal	1, 2, 3, . . .	Orbital size, energy, number of subshells and number of orbital types in shell
ℓ	angular momentum	0, 1, 2, . . . , $n - 1$	Orbital shape, number of orbitals in subshell Labels for electron subshells: s orbital for $\ell = 0$ p orbitals for $\ell = 1$ d orbitals for $\ell = 2$ f orbitals for $\ell = 3$
m_ℓ	magnetic	$-\ell \ldots 0 \ldots +\ell$	Orbital orientation

All s orbitals are spherical in shape. All p orbitals are "halved spheres" because all p orbitals have one nodal plane (because all have $\ell = 1$, and ℓ = number of nodal planes) slicing through the nucleus and dividing the region of electron density in half. (Hence p orbitals are often described as having a "dumbbell" shape.) All d orbitals have four regions of electron density (except d_{z^2}) since all d orbitals have two nodal surfaces (all have $\ell = 2$). This usually leads to a d orbital that is a "quartered" sphere.

STUDY QUESTIONS

REVIEW QUESTIONS

1. Our modern view of atomic structure was developed through many experiments. Name at least three of these experiments, the outcome of each experiment, and the person most associated with that experiment.
2. There are several important mathematical relations in this chapter. Give the equation for each of the following:
   (a) The relationship between wavelength, frequency, and velocity of radiation.
   (b) The relation between the energy and frequency of a photon.
   (c) The energy of an electron in a given energy state of an H atom.
3. State Planck's hypothesis in words and in a mathematical relation.
4. Name the colors of visible light, beginning with that of highest energy.
5. Draw a picture of a standing wave and use this to define the terms *wavelength*, *amplitude*, and *node*.
6. Which of the following best describes the importance of the photoelectric experiment as explained by Einstein?
   (a) The experiment showed that light is electromagnetic radiation.
   (b) The intensity of a light beam is related to its frequency.
   (c) Light can be thought of as consisting of massless particles whose energy is given by the equation, $E = h\nu$.
7. What is a photon? Explain how the photoelectric effect implies the existence of photons.
8. What were two major assumptions of Bohr's theory of atomic structure?
9. Bohr pictured the electrons of the atom as being located in definite orbits about the nucleus, just as the

planets orbit the sun. Criticize this model in view of the quantum mechanical model.

10. In what region of the electromagnetic spectrum is the Lyman series of lines found? The Balmer series?
11. Light is given off by a sodium- or mercury-containing streetlight when the atoms are excited in some way. The light you see arises for which of the following reasons?
    (a) electrons moving from a given quantum level to one of higher n.
    (b) electrons being removed from the atom, thereby creating a metal cation.
    (c) electrons moving from a given quantum level to one of lower n.
    (d) electrons whizzing about the nucleus in an absolute frenzy.
12. If the Balmer series of emission lines for excited H atoms arises from electrons moving from levels of $n > 2$ to $n = 2$, give the value of n_{initial} for the following emission lines:

COLOR	VALUE OF n_{initial}
Red	__________
Blue	__________
Green	__________
Violet	__________

13. What is the wave-particle duality? What are its implications in our modern view of atomic structure?
14. What is Heisenberg's uncertainty principle? Explain how it applies to our modern view of atomic structure.
15. How do we interpret the physical meaning of the square of the wave function? What are the units of ψ^2?
16. What are the three quantum numbers used to describe an orbital? What property of an orbital is described by each quantum number? Specify the rules that govern the values of each quantum number.
17. On what does the amplitude of the electron wave in an atom depend?
18. Give the number of planar nodes for each orbital type: s, p, d, and f.
19. What is the maximum number of s orbitals that may be found in a given electron shell? The maximum number of p orbitals? Of d orbitals? Of f orbitals?
20. Match the values of ℓ below with orbital type (s, p, d, or f).

ℓ VALUE	ORBITAL TYPE
3	_____
0	_____
1	_____
2	_____

21. Draw pictures of the 90% boundary surfaces of an s orbital and the p_x orbital. Be sure your drawing shows why the p orbital is labeled p_x and not p_y, for example.

ELECTROMAGNETIC RADIATION

22. The colors of the visible spectrum, and the wavelengths corresponding to the colors, are given in Figure 8.3.
    (a) What colors of light involve less energy than yellow light?
    (b) Which color of visible light has photons of greater energy: green or violet?
    (c) Which color of light has the greater frequency: blue or green?
23. The regions of the electromagnetic spectrum are given in Figure 8.3. Answer the following questions based on this figure.
    (a) Which type of radiation involves less energy: radar or red light?
    (b) Which radiation has the higher frequency: radar or microwaves?
    (c) Which radiation has the longer wavelength: x-rays or visible light?
24. The U.S. Navy has a system for communicating with submerged submarines. The system uses radio waves with a frequency of 76 s^{-1}. What is the wavelength of this radiation in meters? In miles? (1 mile = 1.61 km)
25. An FM radio station has a frequency of 91.7 MHz (1 MHz = 10^6 Hz or cycles per second). What is the wavelength of this radiation in meters?
26. Green light has a wavelength of approximately 5.00×10^2 nm. What is the frequency of this light? What is the energy in joules of one photon of green light? What is the energy in joules of 1.00 mole of photons of green light?
27. Red light has a wavelength of about 6.8×10^2 nm. What is its frequency? Calculate the energy of 1 photon of red light. What is the energy of 1.00 mole of red photons?
28. In what region of the electromagnetic spectrum is 200. nm radiation located? What is its frequency? What is the energy of a photon of 200. nm radiation?
29. Assume a microwave oven operates at a frequency of 1.00×10^{11} s^{-1}. What is the wavelength of this radiation in meters? What is its energy per photon?
30. The yellow light emitted by the sodium lamps in streetlights has wavelengths of 589.6 nm and 589.0 nm. What is the frequency of the 589.0 nm light? What is the energy of 1.00 mole of photons of yellow light with a wavelength of 589.0 nm?
31. Radiation in the microwave region of the electromagnetic spectrum is not very energetic. However, this radiation cooks your food in a microwave oven. If you bombard a cherry pie with photons with a wavelength of 0.50 cm, what is the frequency of this radiation? If you bombard the pie with 1.00 mole of photons, how much energy does your pie absorb?
32. Place the following types of radiation in order of increasing energy per photon:

(a) radar signals
(b) radiation from a microwave oven
(c) gamma rays from a nuclear reaction
(d) red light from a neon sign
(e) ultraviolet radiation from a sun lamp

33. Place the following types of radiation in order of increasing energy per photon:
(a) green light from a mercury lamp
(b) x-rays from an instrument in a dentist's office
(c) microwaves in a microwave oven
(d) an FM music station at 96.3 MHz

PHOTOELECTRICITY

34. To cause a sodium atom on a metal surface to lose an electron, an energy of 220. kJ/mol is required. Calculate the longest possible wavelength of light that can ionize a sodium atom. What is the region of the electromagnetic spectrum in which this radiation is found?

35. Assume you are an engineer designing a space probe to land on a distant planet. You wish to use a switch that works by the photoelectric effect. That is, light falling on the surface of a metal causes the metal to release some electrons, which flow to a positively charged pole and close the electrical circuit. The metal you wish to use in your device requires 6.7×10^{-19} J per atom to remove an electron. You know that the atmosphere of the planet on which your device must work filters out all wavelengths of light less than 540 nm. Will your device work on the planet in question? Why or why not?

ATOMIC SPECTRA AND THE BOHR ATOM

36. Consider only the following quantum levels for the H atom.

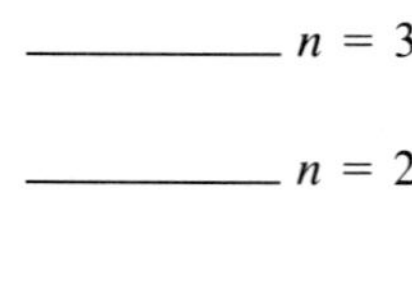

The emission spectrum of an excited H atom will consist of transitions between these levels.
(a) How many emission lines are possible, considering *only* these four quantum levels?
(b) Photons of the *highest energy* will be emitted in a transition from the level with $n =$ ____ to the level with $n =$ ____.
(c) The emission line having the *longest wavelength* corresponds to a transition from the level with $n =$ ____ to the level with $n =$ ____.

37. Consider only the following quantum levels for the H atom.

________ $n = 5$
________ $n = 4$
________ $n = 3$
________ $n = 2$
________ $n = 1$

The emission spectrum of an excited H atom will consist of transitions between these levels.
(a) How many emission lines are possible, considering *only* these five quantum levels?
(b) Photons of the *highest frequency* will be emitted in a transition from the level with $n =$ ____ to the level with $n =$ ____.
(c) The emission line having the *longest wavelength* corresponds to a transition from the level with $n =$ ____ to the level with $n =$ ____.
(d) The emission line having the *shortest wavelength* corresponds to a transition from the level with $n =$ ____ to the level with $n =$ ____.

38. If energy is absorbed by a hydrogen atom in its ground state, the atom is excited to a higher energy state. For example, the excitation of an electron from the orbital with $n = 1$ to an orbital with $n = 4$ requires radiation with a wavelength of 97.3 nm. Which of the following transitions would require radiation of *longer wavelength* than this?
(a) $n = 2$ to $n = 4$
(b) $n = 1$ to $n = 3$
(c) $n = 1$ to $n = 5$
(d) $n = 3$ to $n = 5$

39. The energy emitted when an electron moves from a higher energy orbital to one of lower energy in any atom can be observed as electromagnetic radiation.
(a) Which involves the emission of less energy in the H atom, an electron moving from $n = 4$ to $n = 3$ or an electron moving from $n = 3$ to $n = 1$?
(b) Which involves the emission of greater energy in the H atom, an electron changing from $n = 4$ to $n = 1$ or an electron changing from $n = 5$ to $n = 2$? Explain fully.

40. Calculate the wavelength of light emitted when an electron changes from $n = 4$ to $n = 1$ in the H atom. In what region of the spectrum is this radiation found?

41. A line in the Balmer series of emission lines of excited H atoms has a wavelength of 410.17 nm. This series originates from electrons changing from high energy orbitals to the $n = 2$ orbitals. To account for the 410.17 nm line, what is the value of n for the initial orbital?

42. An electron moves from $n = 5$ to the $n = 1$ quantum level and emits a photon with an energy of 2.093×10^{-18} J. How much energy must the atom absorb in order to move an electron from $n = 1$ to $n = 5$?

43. Excited H atoms give off many emission lines. One series of lines, called the Pfund series, occurs in the infrared region. It results when an electron changes from higher energy orbitals to orbitals with $n = 5$. Calculate the wavelength and frequency of the *lowest energy* line of this series.

DE BROGLIE AND MATTER WAVES

44. When an electron moves with a velocity of 3.0×10^8 cm/s, what is its de Broglie wavelength?
45. If an electron has an energy of 100. eV [where 1.00 electron volt (eV) = 1.60×10^{-19} J], what is its wavelength? (Remember that $E = mv^2/2$.)
46. A beam of neutrons ($m = 1.67493 \times 10^{-27}$ kg) has a velocity of 0.40 m/s. Calculate the wavelength of the neutron wave.
47. A rifle bullet (mass = 1.50 g) is moving with a velocity of 700. miles per hour. What is the wavelength of this bullet?

QUANTUM MECHANICS

48. Complete the following table:

QUANTUM NUMBER	ATOMIC PROPERTY DETERMINED BY QUANTUM NUMBER
________	Orbital size
________	Relative orbital orientation
________	Orbital shape

49. Give the quantum number associated with each of the following:
    (a) energy of the electron in the H atom
    (b) orbital shape
    (c) orbital size
    (d) orbital orientation
50. Answer the following questions:
    (a) When $n = 3$, what are the possible values of ℓ?
    (b) When ℓ is 3, what are the possible values of m_ℓ?
    (c) For a 4*s* orbital, what are the possible values of n, ℓ, and m_ℓ?
    (d) For a 5*f* orbital, what are the possible values of n, ℓ, and m_ℓ?
51. Answer the following questions:
    (a) When $n = 4$, $\ell = 2$, and $m_\ell = -2$, to what orbital type does this refer? (Give the orbital label, such as 1*s*.)
    (b) How many orbitals are in the $n = 6$ electron shell? How many subshells? What are the letter labels of the subshells?
    (c) When a subshell is labeled *g*, how many orbitals are there in the subshell? What are the values of m_ℓ?
52. A possible excited state of the H atom has the electron in a 4*p* orbital. List all possible sets of quantum numbers n, ℓ, and m_ℓ for this electron.
53. A possible excited state for the H atom has the electron in a 4*d* orbital. List all possible sets of quantum numbers n, ℓ, and m_ℓ for this electron.
54. How many subshells are there in the electron shell with the principal quantum number $n = 4$?
55. How many subshells are there in the electron shell with the principal quantum number $n = 5$?
56. Explain briefly why each of the following is not a possible set of quantum numbers for an electron in an atom.
    (a) $n = 2$, $\ell = 2$, $m_\ell = 0$
    (b) $n = 3$, $\ell = 0$, $m_\ell = -2$
    (c) $n = 6$, $\ell = 0$, $m_\ell = 1$
57. Which of the following sets of quantum numbers represent valid sets? For those sets that are not valid, explain briefly why.
    (a) $n = 3$, $\ell = 3$, $m_\ell = 0$
    (b) $n = 2$, $\ell = 1$, $m_\ell = 0$
    (c) $n = 6$, $\ell = 5$, $m_\ell = -1$
    (d) $n = 4$, $\ell = 3$, $m_\ell = -4$
58. What is the maximum number of orbitals that can be identified by each of the following sets of quantum numbers? ("None" is a possible answer.)
    (a) $n = 4$, $\ell = 2$
    (b) $n = 5$
    (c) $n = 2$, $\ell = 2$
    (d) $n = 3$, $\ell = 1$, $m_\ell = -1$
59. What is the maximum number of orbitals that can be identified by each of the following sets of quantum numbers? ("None" is a possible answer.)
    (a) $n = 3$, $\ell = 0$, $m_\ell = +1$
    (b) $n = 5$, $\ell = 1$
    (c) $n = 7$, $\ell = 5$
    (d) $n = 4$, $\ell = 2$, $m_\ell = -2$
60. How many planar nodes does each of the following orbitals possess? (a) 2*s* (b) 5*d* (c) 5*f*
61. How many planar nodes does each of the following atomic orbitals possess? (a) 4*f* (b) 2*p* (c) 6*s*
62. State which of the following orbitals cannot exist according to the quantum theory: 2*s*, 2*d*, 3*p*, 3*f*, 4*f*, and 5*s*. Briefly explain your answers.
63. State which of the following orbitals can and cannot exist according to the quantum theory: 3*p*, 4*s*, 2*f*, 5*g*, and 1*p*. Briefly explain your answers.
64. Write the complete set of quantum numbers (n, ℓ, and m_ℓ) for each orbital in each of the following subshells: (a) 2*p* (b) 3*d* (c) 4*f*
65. Write the complete set of quantum numbers (n, ℓ, and m_ℓ) for each orbital in each of the following subshells: (a) 5*f* (b) 4*d* (c) 2*s*
66. From memory, sketch shapes of the electron clouds for each of the following atomic orbitals: (a) 1*s* (b) $2p_x$ (c) $3d_{xy}$

67. From memory, sketch shapes of the electron clouds for each of the following atomic orbitals: (a) $2p_z$ (b) $3d_{z^2}$ (c) $3d_{xz}$

68. A given orbital is labeled by the magnetic quantum number $m_\ell = -1$. This could not be a
(a) *g* orbital
(b) *f* orbital
(c) *d* orbital
(d) *p* orbital
(e) *s* orbital

69. A particular orbital has $n = 4$, $\ell = 2$, and $m_\ell = -2$. This orbital must be: (a) $3p$, (b) $4p$, (c) $5d$, or (d) $4d$.

GENERAL QUESTIONS

70. An AM radio station at 600. kHz broadcasts with a frequency of $6.00 \times 10^5\ s^{-1}$ (kHz = 1 kilohertz = 1000 s^{-1}). What is the wavelength of this signal in meters?

71. When Voyager 1 encountered the planet Neptune in 1989, the planet was approximately 2.7 billion miles from the earth. How long did it take for the television picture signal to reach the Earth from Neptune? (1 mile = 1.61 km)

72. If 5% of the energy supplied to a light bulb is radiated as visible light, how many photons are emitted per second by a 100. W bulb? Assume the wavelength of all the visible light is 560 nm. (1 W = 1 J/s)

73. In what way does Bohr's model of the atom violate the uncertainty principle?

74. In Millikan's oil drop experiment, some droplets captured one electron, some two, others three and so on. If you performed the experiment, and you observed the following charges on individual oil droplets, what would you conclude is the largest possible value of the fundamental unit of charge on one electron? Observed charges were: (a) -3.2×10^{-19} C; (b) -4.8×10^{-19} C; (c) -6.4×10^{-19} C; (d) -8.0×10^{-19} C; (e) -1.6×10^{-19} C.

75. If sufficient energy is absorbed by an atom, an electron can be lost by the atom and a positive ion formed. The amount of energy required is called the ionization energy. In the H atom, the ionization energy is that required to change the electron from $n = 1$ to $n =$ infinity. Knowing that the energy of a quantum level for an atom is given by $E = -Rhc/n^2$, calculate the ionization energy for the H atom.

76. Calculate the energy of the electron in the $n = 4$ orbital of the H atom using Bohr's energy equation (Equation 8.4). Give the answer in kJ/mol.

77. The square of the wave function for a $2p$ orbital is plotted against distance from the nucleus in a direction perpendicular to the nodal plane in Figure 8.15. What is the approximate distance from the nucleus at which the maximum probability density is reached?

78. Rank the following orbitals in the H atom in order of increasing energy: $3s$, $2s$, $2p$, $4s$, $3p$, $1s$, and $3d$.

79. Answer the following questions as a summary quiz on the chapter.
(a) The quantum number *n* describes the ________ of an atomic orbital.
(b) The shape of an atomic orbital is given by the quantum number ________.
(c) A photon of orange light has ________ (*less or more*) energy than a photon of yellow light.
(d) The maximum number of orbitals that may be associated with the set of quantum numbers $n = 4$ and $\ell = 3$ is ________.
(e) If an electron subshell has 7 orbitals, the ℓ value for the subshell is ________.
(f) The maximum number of orbitals that may be associated with the quantum number set $n = 3$, $\ell = 2$, and $m_\ell = -2$ is ________.
(g) Label each of the following orbital pictures with the appropriate letter:

____

____

(h) When $n = 5$, the possible values of ℓ are ________.
(i) The maximum number of orbitals that can be assigned to the $n = 4$ shell is ________.

80. Answer the following questions as a review of this chapter.
(a) The quantum number *n* describes the ________ of an atomic orbital and the quantum number ℓ describes its ________.
(b) When $n = 3$, the possible values of ℓ are ________.
(c) What type of orbital corresponds to $\ell = 3$? ________
(d) For a $4d$ orbital, the value of *n* is ________, the value of ℓ is ________, and a possible value of m_ℓ is ________.
(e) Each drawing below represents a type of atomic orbital. Give the letter designation for the orbital, its value of ℓ, and the number of nodal planes.

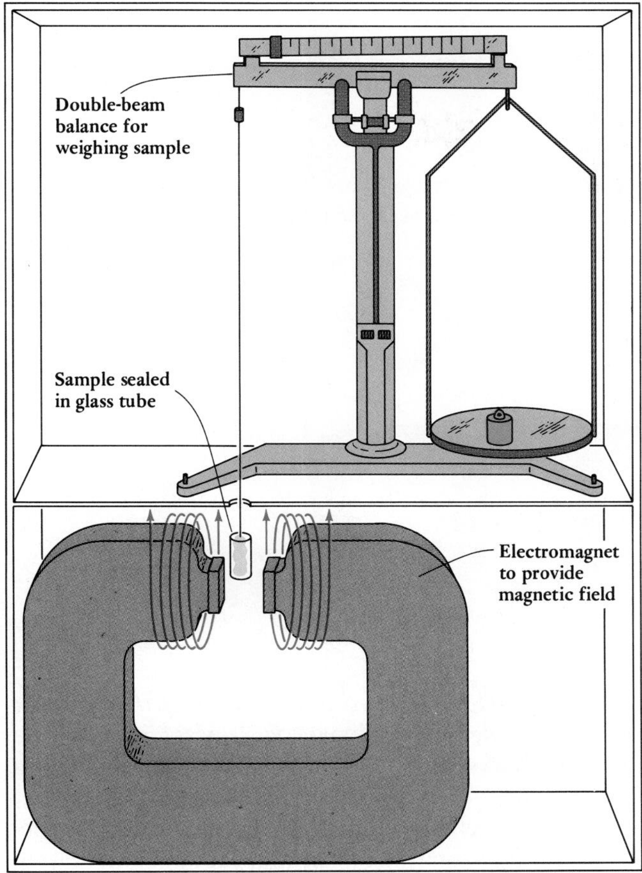

Figure 9.4 A magnetic balance used to measure the magnetic properties of a sample. The sample is first weighed with the electromagnet turned off. The magnet is then turned on and the sample is reweighed. If the substance is paramagnetic, the sample is drawn into the magnetic field and the *apparent* weight increases.

PARAMAGNETISM AND UNPAIRED ELECTRONS

Most substances—chalk, sea salt, cloth—are not magnetic and cannot be attracted by a magnetic field. They are said to be **diamagnetic**. (Such substances are in fact slightly repelled by a magnetic field.) There are, however, natural or synthetic magnets, such as magnetite or the alloy called "Alnico" (for Al, Ni, and Co), that can attract pieces of iron, steel, nickel, and other materials. Substances that are attracted to a magnetic field are called **paramagnetic**, and the magnitude of the effect can be determined with an apparatus such as that illustrated in Figure 9.4. Most paramagnetic materials lose their magnetism when they are withdrawn from a magnetic field. However, substances such as magnetite or Alnico magnets retain their magnetism, and they are called **ferromagnetic**.

As we now understand, the phenomena of paramagnetism and ferromagnetism are caused by electron spins. It has been verified experimentally that an electron in an atom has magnetic properties expected for a spinning, charged particle. Each electron has a magnetic field with north and south magnetic poles (Figure 9.5); that is, the electron behaves like a tiny bar magnet. Furthermore, *the electron spin is quantized such*

Figure 9.5 Electron spin. In atoms and molecules, the electron in motion generates a magnetic field whose properties are identical (except in strength) to the field set up by a bar magnet or the earth.

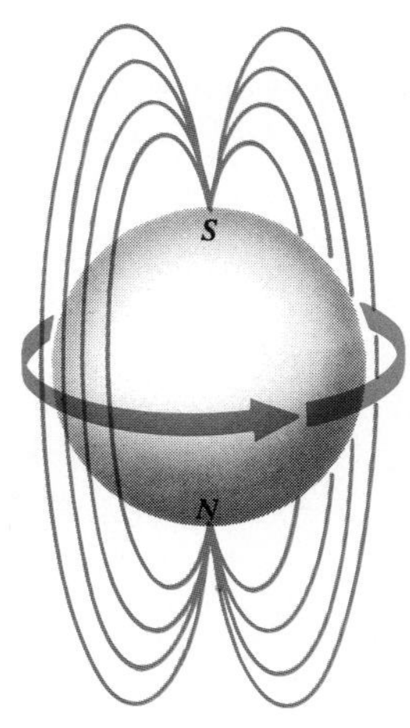

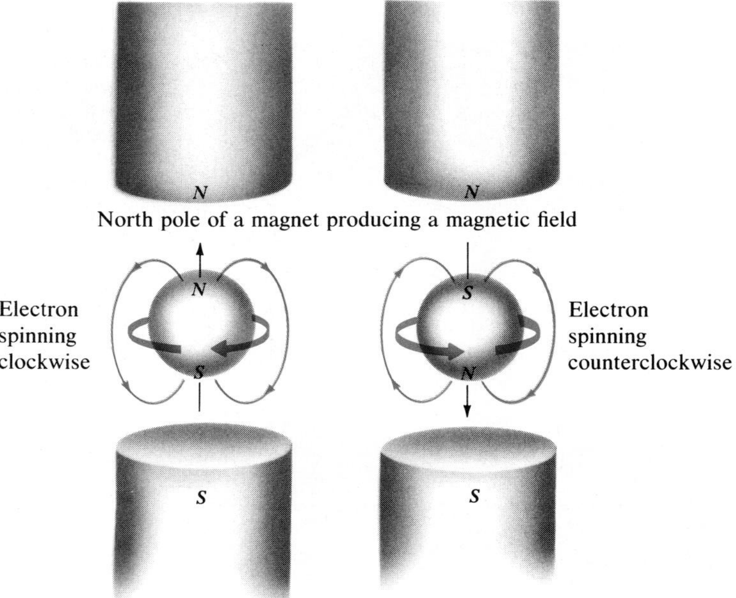

Figure 9.6 Quantization of electron spin. There can be only two assumed spins of the electron, a "micromagnet," relative to a magnetic field. Any other position is forbidden. Therefore, we can say that the spin of an electron is quantized. (For simplicity, the electron "bar magnets" are shown aligned with the magnetic field. In reality the electron magnets are slightly tipped with respect to the field.)

that, in an external magnetic field, only two orientations of the electron magnet and its spin are possible. One orientation is associated with a spin magnetic quantum number value of $m_s = +\frac{1}{2}$ and the other with a value of $m_s = -\frac{1}{2}$ (Figure 9.6).

Either one or two electrons can be assigned to each orbital in an atom. In hydrogen, there is only one electron, and its spin orientation can take either value of m_s. We observe experimentally that hydrogen atoms are paramagnetic; when an external magnetic field is applied, the electron magnets align with the field and experience an attractive force. In helium, two electrons are assigned to the same 1*s* orbital, *and we observe that helium is diamagnetic*. In order to account for this observation, we assume that the two electrons assigned to the same orbital have opposite spin orientations; we say that they are **paired**. This means that the magnetic field of one electron, $\overset{\text{N}}{\underset{\text{S}}{\uparrow}}$, is balanced out by the magnetic field of the second, $\overset{\text{S}}{\underset{\text{N}}{\downarrow}}$, of opposite spin.

In a mole of H atoms, half have $m_s = +\frac{1}{2}$ and half have $m_s = -\frac{1}{2}$.

In general, paramagnetism occurs in substances in which the atoms contain **unpaired electrons**. Atoms in which all electrons are paired with partners of opposite spin are diamagnetic. This explanation opens the way to understanding the electron configurations of atoms with more than one electron.

Now that we know that magnetic effects arise when unpaired electrons of a substance align with an external magnetic field, we can understand the difference between a *para*magnetic material and one that is *ferro*magnetic. When removed from a magnetic field, the unpaired electrons of different atoms of a paramagnetic substance become *unaligned*; they point in every conceivable direction. However, in a ferromagnetic substance there are *clusters of atoms* that all have their unpaired electrons aligned within a cluster, and the clusters are more or less aligned; thus, the substance has a readily observable magnetic field of its own. When the ferromagnetic substance is removed from a magnetic field, these clusters of atoms retain the alignment of their spins and the substance as a whole has magnetic north and south poles.

9.2 THE PAULI EXCLUSION PRINCIPLE

To make the quantum theory consistent with experiment, Wolfgang Pauli stated in 1925 the **Pauli exclusion principle**: *no two electrons in an atom can have the same set of four quantum numbers* (n, ℓ, m_ℓ, and m_s). This principle leads to yet another important conclusion, that no atomic orbital can be assigned to (or "contain") more than two electrons.

Orbitals are not literally things or boxes in which electrons are placed. "Orbital" and "wave function" are synonymous, so it is not conceptually correct to talk about electrons in or occupying orbitals, although it is commonly done.

Consider the 1*s* orbital of the H atom. This orbital is defined by the set of quantum numbers $n = 1$, $\ell = 0$, and $m_\ell = 0$. No other set of numbers can be used. If an electron has this orbital, the electron spin direction must also be specified as sketched below. The orbital is shown as a box, and the electron spin in one direction is depicted by an arrow.

Electron in 1*s* orbital = [↑] (1*s*)   Quantum Number Set

$n = 1,\ \ell = 0,\ m_\ell = 0,\ m_s = +\frac{1}{2}$

If there is only one electron with a given orbital, you can picture the electron as an arrow pointing either up or down. Thus, an equally valid combination of quantum numbers and "box" diagram would be

Electron in 1*s* orbital = [↓] (1*s*)   Quantum Number Set

$n = 1,\ \ell = 0,\ m_\ell = 0,\ m_s = -\frac{1}{2}$

The spin and magnetic field arrows for an electron point in opposite directions. From now on the arrow for an electron refers to its spin.

The pictures above are equally appropriate for the H atom in its ground state (outside a magnetic field): one electron in the 1*s* orbital. For the helium atom, the element with two electrons, *both* electrons are assigned to the 1*s* orbital. From the Pauli principle, you know that each electron must have a different set of quantum numbers, so the orbital box picture now is

Two electrons in the 1*s* orbital = [↑↓]

this electron has $n = 1,\ \ell = 0,\ m_\ell = 0,\ m_s = -\frac{1}{2}$
this electron has $n = 1,\ \ell = 0,\ m_\ell = 0,\ m_s = +\frac{1}{2}$

Each of the two electrons in the 1*s* orbital of He has a different set of the four quantum numbers. Having once decided on the first three, which tell you this is a 1*s* orbital, there are only two remaining choices for the fourth, $m_s = +\frac{1}{2}$ or $-\frac{1}{2}$. Thus, *the 1s orbital, and any other atomic orbital, can*

Table 9.1 Number of Electrons Accommodated in Electron Shells and Subshells

Electron Shell	Subshells Available	Number of Orbitals in Subshell ($2\ell + 1$)	Number of Electrons Possible in the Subshell [$2(2\ell + 1)$]	Total Electrons Possible for the *n*th Electron Shell ($2n^2$)
1	*s*	1	2	2
2	*s*	1	2	8
	p	3	6	
3	*s*	1	2	18
	p	3	6	
	d	5	10	
4	*s*	1	2	32
	p	3	6	
	d	5	10	
	f	7	14	
5	*s*	1	2	50
	p	3	6	
	d	5	10	
	f	7	14	
	*g**	9	18	
6	*s*	1	2	72
	p	3	6	
	d	5	10	
	*f**	7	14	
	*g**	9	18	
	*h**	11	22	

*These orbitals are not used in the ground state of any known element.

apply to no more than two electrons, and these two electrons must have opposite spin directions. The consequence is that the atom is diamagnetic, as experimentally observed.

The $n = 1$ electron shell in any atom can accommodate no more than two electrons. But what about the $n = 2$ shell? There are $n^2 = 4$ orbitals in the $n = 2$ shell: one *s* orbital and three *p* orbitals (Table 9.1). Since each orbital can apply to two electrons and no more, the 2*s* orbital is assigned to one or two electrons, and the three 2*p* orbitals accommodate

Quantum Numbers: A Summary

You can think of the four quantum numbers used to describe an electron in an atom as an "electronic zipcode."

- n: shell to which the electron is assigned (1, 2, 3, 4, . . .)
- ℓ: subshell within the shell (0, 1, 2, . . . , $n - 1$)
- m_ℓ: orbital within the subshell ($-\ell$, . . . , 0, . . . , $+\ell$)
- m_s: electron spin direction ($+\frac{1}{2}$ or $-\frac{1}{2}$)

as many as six electrons, for a total of eight electrons possible with $n = 2$. This analysis is carried farther in Table 9.1 for the electron shells normally observed in real elements.

The number of orbitals in the nth electron shell is n^2, and the maximum number of electrons in the shell is $2n^2$.

9.3 ATOM ORBITAL ENERGIES AND ELECTRON ASSIGNMENTS

A goal of this chapter is an understanding of the distribution of electrons in atoms with many electrons, for the very important reason that this distribution controls the chemistry of the atom. The heavier elements can be "built" by assigning electrons to a succession of orbitals, no more than two electrons to an orbital. Generally, electrons are assigned to orbitals of successively higher energies because this will make the total energy of all the electrons as low as possible. With elements having more than 18 electrons, however, we find there are complicating factors, and so some heavier elements do not always follow a neat set of rules.

ORDER OF ORBITAL ENERGIES

Quantum theory leads to the conclusion that in the H atom the energy of an electron depends only on the value of n ($E = -Rhc/n^2$, Equation 8.4). For heavier atoms, however, the situation is more complex, and this is reflected by Figure 9.7, a representation of the order of orbital energies for atoms of more than one electron. Here you see that *orbital energies in many-electron atoms depend on both n and ℓ*. The orbitals with $n = 3$, for example, do not all have the same energy; rather, they are in the order $3s < 3p < 3d$. The following is a brief description of what scientists understand about the reasons for this complicated problem.

How are we to understand that, while ℓ has no role in determining a subshell's energy for the H atom, it clearly becomes important for

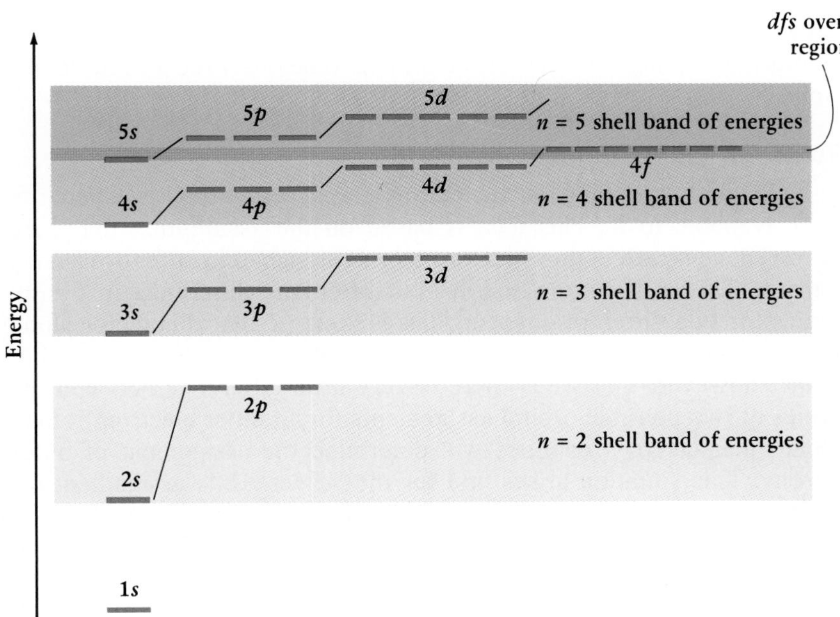

Figure 9.7 In a many-electron atom, shell energies increase with increasing n, and subshell energies increase with increasing ℓ. The subshells of a given shell lie in a *band* of energies. The energy gaps between bands of shell energies become smaller as n increases. This means that the top of the band of energies for one quantum shell can eventually overlap the band of energies for the shell of next higher n. This eventually results in overlapping of the energies of the ns orbital with $(n-1)d$ and $(n-1)f$ orbitals. Therefore, we speak of *dfs* overlay regions.

subshells in atoms that contain more electrons? To answer this, we begin by assuming that shells tend to fill beginning with smallest n first (and this is in fact observed). This means that outer shell electrons—often called the **valence electrons**—will always be accompanied by inner shell or **core** electrons. For example, boron has 5 electrons, including 2 core electrons (assigned to $n = 1$) and 3 valence electrons (assigned to the $n = 2$ shell). Now, in the H atom, the value of n alone determines the energy of atomic orbitals, so an $n = 2$ electron is at a higher energy than an $n = 1$ electron. In a many-electron atom, the situation is similar but more complex.

*Inner shell electrons are often called **core electrons**. Electrons of the outermost shell are often referred to as **valence electrons**.*

In a many-electron atom the dependence of subshell energy on ℓ arises because orbital radius changes slightly with ℓ as well as with n. Subshell orbitals contract toward the nucleus as the value of ℓ increases; in many-electron atoms this contraction is partially into the volume of space occupied by the **core** electrons. As the subshells within a given outer shell contract toward the region of space occupied by the core electrons, the energy of these subshell electrons is *raised* by repulsion between the subshell electrons and the core electrons. The result is that *a subshell's energy rises as its ℓ quantum number increases when inner electrons are present*. Thus, for the $n = 2$ shell, a $2p$ orbital has a higher energy than a $2s$ orbital. For $n = 3$, the order of subshell energies is $3s < 3p < 3d$, while the order is $4s < 4p < 4d < 4f$ for the $n = 4$ shell of electrons.

You might wonder whether the highest energy subshell of one shell can rise above the lowest energy subshell of the next higher shell. That is, can a $2p$ orbital rise higher in energy than a $3s$ orbital, or can $3d$ be higher in energy than $4s$? As seen in Figure 9.7, the band of energies for the $n = 2$ orbitals does not overlay the band for the $n = 3$ orbitals, nor does $3d$ overlay $4s$. However, remember that the gaps between the shell energies in the H atom decrease as n increases (see Figure 8.9), and the same is observed in many-electron atoms. The result is that the band of energies for the $n = 4$ orbitals slightly overlays that for $n = 5$, for example, and interpenetrating shell energies continue to be observed as n increases beyond 5.

ORDER OF ORBITAL ASSIGNMENTS

Generally speaking, each electron is assigned to the lowest energy orbital available to it. This idea is based on the assumption that core electrons have no effect on which orbitals are assigned to outer or valence electrons. This is not true exactly, but often the difference in orbital energies for two different outer orbitals is so large that this factor alone determines which orbital is assigned to the outer electron. You can readily imagine, however, that when there is very little difference between the energies of two possible orbital assignments for an outer electron, it may be that inner electron energies will determine the assignment of outer electrons. This situation arises first for the $3d$ versus $4s$ assignment for the 19th electron in potassium. Potassium has 19 electrons, 18 of which are assigned in the $1s$ (2 electrons), $2s$ and $2p$ (8 electrons), and $3s$ and $3p$ (8 electrons) subshells. The problem is where to assign the 19th electron: $3d$ or $4s$? If the 18 core electrons did not experience a change in their collective energy on changing the outer electron from $3d$ to $4s$, the 19th or valence electron of K would be assigned to $3d$ (as expected from Figure 9.7). However, *experiment* shows that the inner 18 electrons must exert a *slight* influence favoring the assignment of electron 19 to a $4s$ orbital. We understand this because the $3d$ orbital penetrates into the region occupied by the core electrons more than does the $4s$ orbital. If the electron were assigned to the $3d$ orbital instead of the $4s$, this would result in an expansion of core electrons out from the nucleus (owing to electron–electron repulsions), and this would have the unwanted effect of raising the core electron energies, that is, destabilizing them. Thus, potassium is more stable with 2 electrons in $n = 1$, 8 electrons in $n = 2$, 8 electrons in $n = 3$ ($3s$ and $3p$), and 1 electron assigned to $4s$.

As we go deeper into the periodic table, subtleties like this are more and more frequent, and electron configurations are harder to understand and predict. Nonetheless, it is fortunate that there are some guidelines that more often than not lead to a result in agreement with experiment.

9.4 ATOM ELECTRON CONFIGURATIONS

Now it is possible to make some sense of the electron configurations of the elements, and we shall largely use the periodic table as our guide to these configurations. *It is extremely important for you to connect the configuration of an element and its position in the periodic table, since this will allow you to organize a large number of chemical facts.*

The guiding principle in assigning electrons to available orbitals is *to do so in order of increasing $(n + \ell)$*, with the important reservation that to do so ignores the influence of the inner electrons on this choice. For the *dfs* overlay elements (Figure 9.7) this guiding principle will occasionally falter. These are not serious "anomalies" and do not prevent you from connecting an element's configuration with its position in the periodic table or from organizing a large number of chemical facts for the elements. More importantly, it is the ions rather than the free atoms that are interesting in chemistry, and these difficulties do not exist for them, as you will see in Section 9.5.

Figures 9.7 and 9.8 give the same order of filling up to the first dfs overlay region, where Figure 9.8 guides through the "overlay cases."

Finally, although we shall emphasize the connection between electron configurations and the position of the elements in the periodic table, you may find the empirical device in Figure 9.8 useful for remembering the order of filling of atomic orbitals (but not the order of orbital energies). Be aware, though, that Figure 9.8 is applicable only to gaseous atoms and not to the ions formed by those atoms.

The actual configurations of the elements are given in Table 9.2. Be sure to notice the connection between these configurations and the periodic table. Also notice that there are occasional complications that arise because the simple rules chemists use as guidelines break down for elements in the dfs overlay region (Figure 9.7).

ELECTRON CONFIGURATIONS OF THE MAIN GROUP ELEMENTS

The first element in the periodic table, hydrogen, has one electron assigned to a $1s$ orbital. One way to depict this is with the orbital box diagram used earlier, but an alternative and even more frequently used

Table 9.2 Atomic Electron Configurations

	H	He
1s	1	2

Rare gas core —[Ar] Br

Valence electron subshells: 3d 10, 4s 2 (complete subshells), 4p 5 —Number of electrons

[Ar]	Br
3d	10
4s	2
4p	5

Read as [Ar] $3d^{10}4s^{2}4p^{5}$

	1A	2A	3A	4A	5A	6A	7A	8A
[He]	Li	Be	B	C	N	O	F	Ne
2s	1	2	2	2	2	2	2	2
2p			1	2	3	4	5	6

[Ne]	Na	Mg	Al	Si	P	S	Cl	Ar
3s	1	2	2	2	2	2	2	2
3p			1	2	3	4	5	6

			3B	4B	5B	6B	7B	8B	8B	8B	1B	2B						
[Ar]	K	Ca	Sc	Ti	V	Cr	Mn	Fe	Co	Ni	Cu	Zn	Ga	Ge	As	Se	Br	Kr
3d			1	2	3	5	5	6	7	8	10	10	10	10	10	10	10	10
4s	1	2	2	2	2	1	2	2	2	2	1	2	2	2	2	2	2	2
4p													1	2	3	4	5	6

[Kr]	Rb	Sr	Y	Zr	Nb	Mo	Tc	Ru	Rh	Pd	Ag	Cd	In	Sn	Sb	Te	I	Xe
4d			1	2	4	5	5	7	8	10	10	10	10	10	10	10	10	10
5s	1	2	2	2	1	1	2	1	1		1	2	2	2	2	2	2	2
5p													1	2	3	4	5	6

[Xe]	Cs	Ba	La	Hf	Ta	W	Re	Os	Ir	Pt	Au	Hg	Tl	Pb	Bi	Po	At	Rn
4f				14	14	14	14	14	14	14	14	14	14	14	14	14	14	14
5d			1	2	3	4	5	6	7	9	10	10	10	10	10	10	10	10
6s	1	2	2	2	2	2	2	2	2	1	1	2	2	2	2	2	2	2
6p													1	2	3	4	5	6

	Ce	Pr	Nd	Pm	Sm	Eu	Gd	Tb	Dy	Ho	Er	Tm	Yb	Lu
4f	1	3	4	5	6	7	7	9	10	11	12	13	14	14
5d	1						1							1

[Rn]	Fr	Ra	Ac	Th	Pa	U	Np	Pu	Am	Cm	Bk	Cf	Es	Fm	Md	No	Lr
5f					2	3	4	6	7	7	9	10	11	12	13	14	14
6d			1	2	1	1	1			1							1
7s	1	2	2	2	2	2	2	2	2	2	2	2	2	2	2	2	2

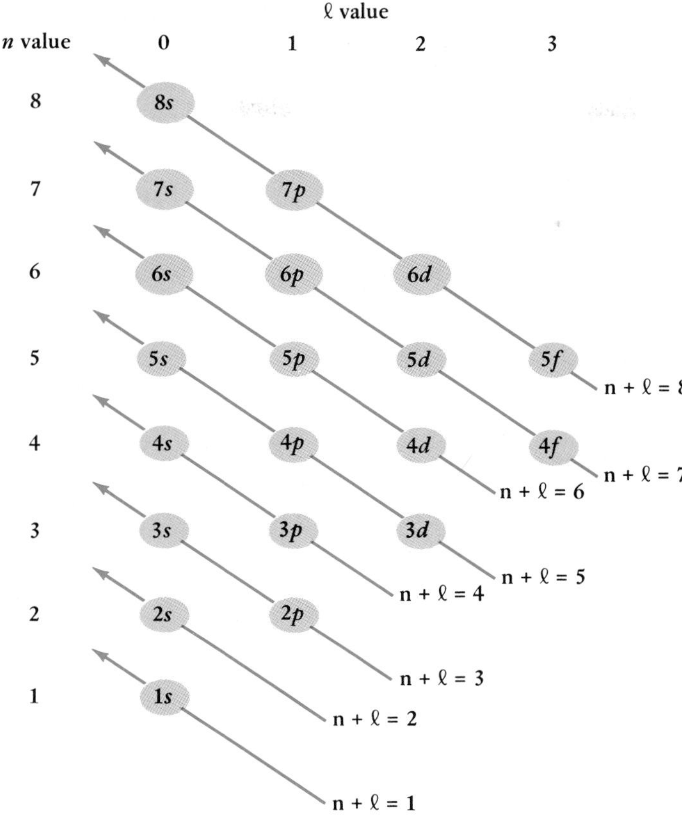

Figure 9.8 A device for remembering the *order of filling of orbitals in atoms* (not the order of orbital energies). Orbitals of the same n value are written on the same line, and orbitals with the same ℓ value are in the same vertical column. The order of filling of atomic orbitals is observed to be from bottom to top following the arrows of increasing $n + \ell$ in the direction indicated. (Thus, the order of filling is $1s \rightarrow 2s \rightarrow 2p \rightarrow 3s \rightarrow 3p \rightarrow 4s \rightarrow 3d$, and so on.) The device is based on the observation that the last filled subshell of an atom has the minimum values of $n + \ell$ and energy available to it. (If an electron may be placed in either of two subshells with the same $n + \ell$, the one with the smaller n is utilized because this gives the atom the lowest possible energy.)

method is **spectroscopic notation**. Using the latter method, the electron configuration of H is written $1s^1$, or "one ess one."

H electron configuration = [↑] (1s) or $1s^1$

- number of electrons assigned to designated subshell
- orbital type (value of ℓ)
- value of n

Orbital Box Diagram — Spectroscopic Notation

We can use the same methods to describe the configurations of the other elements, and those for the first ten elements are illustrated in Table 9.3.

Following hydrogen and helium, lithium (Group 1A), with three electrons, is the first element in the second period of the periodic table. The first two electrons must be assigned to the $1s$ orbital, so the third electron must be assigned to the $n = 2$ shell. According to the energy level diagram in Figure 9.7, that electron must be assigned to the $2s$ orbital. The spectroscopic notation, $1s^22s^1$, is read "one ess two, two ess one."

The position of Li in the periodic table tells you its configuration immediately. All the elements of Group 1A (and 1B) have one electron assigned to an s orbital of the nth shell, where n is the number of the period in which the element is found (Figure 9.9). For example, potassium is the first element in the $n = 4$ row (the 4th period), so potassium has

Table 9.3 Electron Configurations of Elements with Z = 1 to 10

	1s	2s	2p		
ℓ	0	0	1		
m_ℓ	0	0	+1	0	−1
H $1s^1$	↑				
He $1s^2$	↑↓				
Li $1s^22s^1$	↑↓	↑			
Be $1s^22s^2$	↑↓	↑↓			
B $1s^22s^22p^1$	↑↓	↑↓	↑		
C $1s^22s^22p^2$	↑↓	↑↓	↑	↑	
N $1s^22s^22p^3$	↑↓	↑↓	↑	↑	↑
O $1s^22s^22p^4$	↑↓	↑↓	↑↓	↑	↑
F $1s^22s^22p^5$	↑↓	↑↓	↑↓	↑↓	↑
Ne $1s^22s^22p^6$	↑↓	↑↓	↑↓	↑↓	↑↓

the electron configuration of the element preceding it in the table (Ar) *plus* a final electron assigned to the 4*s* orbital.

Beryllium, in Group 2A, has two electrons assigned to the 1*s* orbital plus two additional electrons. It can be seen in Figure 9.7 that the 2*s* orbital is appropriate, so the configuration of Be is $1s^22s^2$. *All elements of Group 2A have electron configurations of [electrons of preceding rare gas]*ns^2, where *n* is the period in which the element is found.

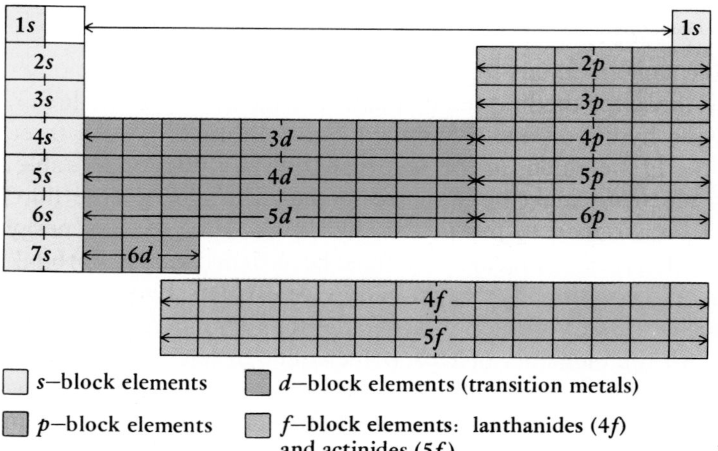

Figure 9.9 Electron configurations and the periodic table. The outermost electrons of an element are assigned to the indicated orbitals. See Table 9.2.

The A's and B's of the Periodic Table

The groups (vertical columns) of elements in the middle of the periodic table, the transition elements, are commonly designated as B groups to differentiate them from the "main group" elements (labeled as A groups). The elements of the B groups have their "last" electrons assigned to *d* and *f* orbitals. In contrast, elements in the A groups have their "last" electrons assigned to *s* and *p* orbitals. This system of nomenclature is currently the subject of considerable debate in the international community of chemists. The current table approved for international use designates the groups by the numbers 1 through 18 across the table. (See the table in the front of the book and other discussions on pages 50–56.) The chemical community in the United States continues to use the A–B designations, however.

Whether you use the A–B or 1–18 designations, these notations can help you know the number of electrons assigned to the valence shell. Using the A–B designations, the number of an A-group tells you the total number of *ns* and *np* valence electrons. For B groups, the number is equal to the number of *ns* and $(n - 1)d$ valence electrons (for the elements Sc to Fe and the other elements in those groups) or *ns* valence electrons (for Cu and Zn and other elements of those groups). Using the 1–18 designations, the group number equals the total number of *ns, np,* and, where appropriate, $(n - 1)d$ valence electrons.

At boron (Group 3A) you first encounter an element in the block of elements on the right side of the periodic table. Since 1*s* and 2*s* orbitals are already filled in a boron atom, the fifth electron must be assigned to a 2*p* orbital. In fact, *all the elements from Group 3A through Group 8A have electrons assigned to p orbitals*, so these elements are sometimes called the ***p*-block elements**. All have the general configuration ns^2np^x, where x = group number − 2. A few additional comments need to be made about several of them.

The three *p* orbitals of any shell have the same energy (and are thus said to be *degenerate*), so it is arbitrary into which orbital you place the electron. This means there are six possible, equivalent sets of quantum numbers for the 2*p* electron of boron.

n	ℓ	m_ℓ	m_s	m_ℓ = +1	0	−1
2	1	+1	$+\frac{1}{2}$ =	↑		
2	1	0	$+\frac{1}{2}$ =		↑	
2	1	−1	$+\frac{1}{2}$ =			↑
2	1	+1	$-\frac{1}{2}$ =	↓		
2	1	0	$-\frac{1}{2}$ =		↓	
2	1	−1	$-\frac{1}{2}$ =			↓

The electron can be assigned to any $2p$ orbital and, once assigned, it can have either spin direction.

Carbon (Group 4A) is the second element in the p block, so there is a second electron assigned to the $2p$ orbitals. For carbon in its lowest energy or ground state, this electron *must* be assigned to either of the remaining p orbitals, and it *must* have the same spin direction as the first p electron.

If the first electron assigned to 2p orbitals had its spin direction indicated by ↓, then the second electron assigned must also have this spin direction. This means there are three more possible C atom ground state configurations where both electrons have the same spin. See the B atom configurations above.

$2p^2$ =	↑	↑		or	↑		↑	or		↑	↑
m_ℓ =	+1	0	−1		+1	0	−1		+1	0	−1
ℓ =		1				1				1	

Three of six C atom ground state configurations where both electrons have the same spin.

In general, when electrons are assigned to p, d, or f orbitals, each successive electron is assigned a different orbital of the subshell, each electron having the same spin as the previous one; this proceeds until the subshell is half full (each possible orbital has one electron), after which *pairs* of electrons must be assigned to common orbitals. This procedure follows **Hund's rule**, which states that *the most stable arrangement of electrons is that with the maximum number of unpaired electrons*, all with the same spin direction. Assignment to different orbitals minimizes electron–electron repulsions, making the total energy of the set of electrons as low as possible. Giving them the same spin also lessens their repulsions by aligning their "micromagnets."

As a final point, notice that carbon is the second element in the p-block of elements, so there must be two p orbital electrons (besides the s electrons already present). Since C is in the second period of the table, the p orbitals involved are $2p$. Thus, you can immediately write the carbon electron configuration by reference to the periodic table: starting at H and moving left to right across the successive periods, you write $1s^2$ to reach the end of period 1, and then $2s^2$ and finally $2p^2$ to bring the electron count to six. Carbon is in Group 4A of the periodic table, because there are four electrons in the $n = 2$ valence shell.

Following nitrogen (Group 5A), oxygen (Group 6A) has a total of six electrons in its outer shell: two of the electrons are assigned to the $2s$ orbital, and, as it is the fourth element in the p block, the other four electrons are assigned to $2p$ orbitals. With nitrogen the $2p$ subshell was half full, one electron per orbital. Therefore, for oxygen the fourth $2p$ electron must pair up with one already present. Again, it makes no difference to which orbital this electron is assigned (the $2p$ orbitals all have the same energy), but *it must have a spin opposite to that of the electron already assigned to that orbital* (Table 9.3) in order that every electron have a different set of quantum numbers (the Pauli principle).

Like all the other elements in Group 8A, neon is a rare gas. All Group 8A elements (except helium) have eight electrons in the shell of highest n value, so all have the configuration ns^2np^6 where n is the period in which the element is found. That is, all the elements have filled ns and *np* subshells. As you will see, this correlates with their nearly complete chemical inertness.

The next element after neon is sodium, and a new period is begun. Since Na is the first element with $n = 3$, the added electron must be assigned to the 3*s* orbital. (Remember that all elements in Group 1A have the ns^1 configuration.) Thus, the complete configuration of Na is that of neon plus one 3*s* electron.

Na: $1s^22s^22p^63s^1$ or $[Ne]3s^1$

rare gas notation for abbreviated configurations

We have written the electron configuration in two ways, one an abbreviated form called the **rare gas notation**. The arrangement preceding the 3*s* electron is that of the rare gas neon, so, instead of writing out "$1s^22s^22p^6$," we represent the completed electron shells by placing the symbol of the corresponding rare gas in brackets.

EXAMPLE 9.1

ELECTRON CONFIGURATIONS

Using the spectroscopic notation, give the electron configuration of silicon.

Solution Silicon, element 14, is the fourth element in the third period ($n = 3$), and it is in the *p* block. Therefore, the last four electrons placed in the atom have the configuration $3s^23p^2$. These are preceded by the completed shells $n = 1$ and $n = 2$, the electron arrangement for Ne. Therefore, the complete configuration of Si is

Si configuration: $1s^22s^22p^63s^23p^2$ or $[Ne]3s^23p^2$

EXAMPLE 9.2

ELECTRON CONFIGURATIONS AND QUANTUM NUMBERS

Write a set of quantum numbers for each of the electrons with $n = 4$ for Ca.

Solution Calcium is the second element in the fourth period. Therefore, there are two electrons with $n = 4$, and, since Ca is in the *s* block of elements, these are 4*s* electrons. The element is preceded by the rare gas Ar, so the electron configuration of Ca is $[Ar]4s^2$. Using a box notation, the configuration is [Ar] [⇅], and possible sets of quantum numbers are

	n	ℓ	m_ℓ	m_s
For ↑	4	0	0	$+\frac{1}{2}$
For ↓	4	0	0	$-\frac{1}{2}$

The fact that the electrons are assigned to an *s* orbital makes $\ell = 0$, which in turn means m_ℓ must be 0.

EXAMPLE 9.3

ELECTRON CONFIGURATIONS AND QUANTUM NUMBERS

Write an acceptable set of quantum numbers that describe each electron in a silicon atom.

Solution The electron configuration of Si is given in Example 9.1. In box notation, its configuration would be

Electron	1, 2	3, 4	5–10			11, 12	13	14	
	⇅	⇅	⇅	⇅	⇅	⇅	↑	↑	
Orbital	1*s*	2*s*	2*p*	2*p*	2*p*	3*s*	3*p*	3*p*	3*p*

and an acceptable set of quantum numbers would be

ELECTRON	n	ℓ	m_ℓ	m_s
1, 2	1	0	0	$\pm\frac{1}{2}$
3, 4	2	0	0	$\pm\frac{1}{2}$
5–10	2	1	+1	$\pm\frac{1}{2}$
	2	1	0	$\pm\frac{1}{2}$
	2	1	−1	$\pm\frac{1}{2}$
11, 12	3	0	0	$\pm\frac{1}{2}$
13	3	1	+1*	$+\frac{1}{2}$‡
14	3	1	0†	$+\frac{1}{2}$§

*The value of m_ℓ can be any of the set, not *necessarily* +1.

†Once $m_\ell = +1$ is used for electron 13, then 14 must have a different m_ℓ (0 or −1).

‡Either m_s $+\frac{1}{2}$ or $-\frac{1}{2}$ can be used.

§Once electron 13 is assigned $m_s = +\frac{1}{2}$, then electron 14 must have the same value of m_s.

EXERCISE 9.1 Orbital Box Diagrams, Spectroscopic Notation, Quantum Numbers

(a) What element has the configuration $1s^22s^22p^63s^23p^5$? (b) Using the spectroscopic notation and a box diagram, show the electron configuration of sulfur. (c) Write one possible set of quantum numbers for the 3*p* electron of aluminum.

ELECTRON CONFIGURATIONS FOR THE TRANSITION ELEMENTS

The elements of the fourth through the sixth periods in the middle of the periodic table are those elements in which a *d* or *f* subshell is being built up (Figure 9.9 and Table 9.2). Elements filling *d* subshells are often referred to collectively as the **transition elements**. Those for which *f* subshells are filling are sometimes called the *inner transition elements* or, more usually, **lanthanides** (filling 4*f* orbitals) and **actinides** (filling 5*f* orbitals). We shall look briefly at some transition element configurations. Keep in mind that, since *d* and *f* subshells lie at the *dfs* overlay region with the next shell (see Figure 9.7), complications in the configurations of these elements are to be expected,* and many are in fact observed.

Because of the effect of core electrons on orbital energies, the transition elements are always immediately preceded by two *s*-block elements (Table 9.4). Accordingly, scandium, the first in the transition series,

*These complications arise from the "tug of war" between the outer and inner electrons that determines where to place the outer electrons.

Table 9.4 Orbital Box Diagrams for the Elements Ca Through Zn

		3d					4s
	m_ℓ =	+2	+1	0	−1	−2	0
Ca	[Ar] $4s^2$						⇅
Sc	[Ar] $3d^14s^2$	↑					⇅
Ti	[Ar] $3d^24s^2$	↑	↑				⇅
V	[Ar] $3d^34s^2$	↑	↑	↑			⇅
Cr*	[Ar] $3d^54s^1$	↑	↑	↑	↑	↑	↑
Mn	[Ar] $3d^54s^2$	↑	↑	↑	↑	↑	⇅
Fe	[Ar] $3d^64s^2$	⇅	↑	↑	↑	↑	⇅
Co	[Ar] $3d^74s^2$	⇅	⇅	↑	↑	↑	⇅
Ni	[Ar] $3d^84s^2$	⇅	⇅	⇅	↑	↑	⇅
Cu*	[Ar] $3d^{10}4s^1$	⇅	⇅	⇅	⇅	⇅	↑
Zn	[Ar] $3d^{10}4s^2$	⇅	⇅	⇅	⇅	⇅	⇅

*These configurations do not follow the "$n + \ell$" rule.

has the configuration [Ar]$3d^14s^2$, and titanium follows with [Ar]$3d^24s^2$. On arriving at chromium, we see the continuing influence of inner electrons coupled with the effect of "micromagnet" alignment; instead of the configuration [Ar]$3d^44s^2$, which might be expected from the pattern of the previous elements, Cr has one electron assigned to *each* of the six available 3*d* and 4*s* orbitals. It is as though the 3*d* and 4*s* orbitals have become degenerate in energy.

In writing electron configurations, we follow the convention of writing the orbitals in order of increasing n. For a given n, the subshells are listed in order of increasing ℓ. Note that this is not always the filling order.

Copper, in Group 1B, has a single electron in the 4*s* orbital, as expected from its group number; thus, the remaining 10 electrons beyond the argon core are assigned to the 3*d* orbital set.* Finally, zinc signals the end of the first transition series and the completion of *ns* and $(n-1)d$ orbitals that is expected at Group 2B.

The fifth period ($n = 5$) follows the pattern of the fourth period with minor variations. The sixth period, however, includes the **lanthanide series** beginning with lanthanum (La). As the first element in the *d*-block, lanthanum has the configuration [Xe]$5d^16s^2$. The next element, cerium (Ce), is set out in a separate row at the bottom of the periodic table, and it is with these elements that *f* orbitals come into play. This means the

*Based on the pattern of the previous transition elements, copper's configuration may again be considered "anomalous." However, it is the logical result of the increasing rise in energy of the 4*s* level as the 3*d* subshell is "loaded," this subshell becoming part of the core electrons into which the 4*s* orbital penetrates.

Historical Figures in Chemistry: Dmitrii Mendeleev and the Periodic Table

The arrangement of elements in the modern periodic table is clearly connected to atomic structure. However, the concept of the periodic table has been known for more than 100 years, and it was developed well before our more recent knowledge of electron configurations. The history of the ideas of chemical periodicity is filled with interesting people, and one of the most important was the Russian chemist Dmitrii Ivanovich Mendeleev (1834–1907).

In the 19th century many chemists tried to find relationships between atomic weights and the properties of the elements. These efforts largely failed because atomic weights were not known for all the elements, and many measured values were inaccurate. However, at a conference in 1860 in Germany, Stanislao Cannizzaro (1826–1910) described his method for determining accurate and unambiguous atomic weights, and this began to set things right. Mendeleev was at this conference and doubtless heard Cannizzaro's paper. It led him to become the founder of the concept of chemical periodicity.

Mendeleev's ideas on periodicity began in the late 1860s with his work on a book on inorganic chemistry. To help organize the material for his book, he had a file of note cards, one for each element. On each card he wrote the atomic weight of the element, then known more accurately owing to Cannizzaro's work, and some properties of the element. When he arranged these element cards *in order of increasing atomic weight* for the elements, he saw there was a repetition of properties every eight or eighteen elements. Thus was born Mendeleev's periodic table.

More than 100 years later, we recognize Mendeleev's work as a milestone because he had the genius to realize that (a) there were many elements yet to be discovered and (b) the characteristics of an element could be predicted from its atomic weight (and its position in his table). Indeed, Mendeleev's table led directly to the discovery of several new elements, such as germanium, within a few years after he had published the table.

In spite of Mendeleev's great achievement, problems arose when new elements were discovered and more accurate atomic weights determined. Using the modern table in the front of the book, examine the following pairs of elements: Ar and K, Te and I, and Co and Ni. In each case, the first named element has the greater atomic weight. Arranging them in order of increasing atomic weight would, for example, move the reactive metal potassium into the group with the rare gases, elements that are nearly inert chemically. The fault lies with Mendeleev's assumption that the properties of the elements are periodic functions of their atomic weight.

H.G.J. Moseley, a young scientist working with Ernest Rutherford in 1913, found that the wavelengths of x-rays emitted by an element are related in a precise way to the atomic number of the element. He quickly realized that other atomic properties may be similarly related to atomic number and not, as Mendeleev had believed, to atomic weight. Indeed, if the elements are arranged in order of increasing atomic number, the defects in the Mendeleev table are corrected. It was therefore Moseley who discovered the **law of chemical periodicity**: the properties of the elements are periodic functions of their atomic numbers.

It is interesting that, after Mendeleev published his work on chemical periodicity in 1871, he did little else with the subject. Instead, he went on to other interests, among them studying the natural resources of Russia and their commercial applications. In 1876 he visited the United States to study the fledgling oil industry and was impressed by the industry but not the country. The United States, he thought, was not interested in science, and, what was worse, carried on the worst features of European civilization.

And finally, it is amusing to learn that all pictures of Mendeleev show him with long hair. He made it a rule to get his hair cut only once a year, in the spring, whether he needed it or not.

configuration of cerium is [Xe]$4f^15d^16s^2$. Moving across the lanthanide series, the pattern continues with some variation, with 14 electrons having been assigned to the seven $4f$ orbitals in lutetium, Lu ([Xe]$4f^{14}5d^16s^2$).

When you complete this section you should be able to depict the electron configuration of any element, using the periodic table as a guide. Understand, however, that there can be minor differences between actual (Table 9.2) and predicted configurations. Such differences are not important at this point, particularly since we are more interested in the ions formed by the elements, and ion configurations show no such "anomalies."

EXAMPLE 9.4

ELECTRON CONFIGURATIONS OF THE TRANSITION ELEMENTS

Using the spectroscopic notation, give electron configurations for technetium (Tc) and osmium (Os). Base your answer on the positions of the elements in the periodic table. That is, for each element, find the preceding rare gas and then note the number of s, d, and f electrons that lead from the rare gas to the element.

Solution Proceeding along the periodic table, we come to the rare gas krypton, Kr, at the end of the $n = 4$ row before arriving at Tc (element 43) in the fifth period. Therefore, following the 36 electrons of Kr there are seven electrons in the $4d$, $5s$ overlap region. According to the periodic table, two of these seven electrons are assigned to the $5s$ orbitals, and the remaining five are assigned to $4d$ orbitals. Therefore, the technetium configuration is [Kr]$4d^55s^2$.

Osmium is a sixth period element, one that follows the lanthanide series of elements. After the last element in the $n = 5$ period, xenon, there are 22 elements leading to Os. Of these, two are of the $6s$ type (Cs and Ba), fourteen are of the $4f$ type (Ce through Lu), and the remaining six are of the $5d$ type (La through Os). Thus, the osmium configuration is [Xe]$4f^{14}5d^66s^2$.

EXERCISE 9.2 Electron Configurations

Using a periodic table, and without looking at Table 9.2, write electron configurations for the following elements: (a) P, (b) Zn, (c) Zr, (d) In, (e) Pb, and (f) U. Use the spectroscopic notation. When you have finished, check your answers with Table 9.2.

9.5 ION ORBITAL ENERGIES AND ELECTRON CONFIGURATIONS

A great deal of the chemistry of the elements is that of their ions, of cations formed by the loss of one or more electrons from an atom and of anions formed by the addition of one or more electrons to an atom. The orbital energy sequences are roughly the same in the atoms and their ions, with one important exception: the *dfs* overlay region (Figure 9.7) does not arise in cations. For this reason, the configurations of monatomic cations, say Na^+ or Ag^+, are more straightforward than those of atoms.

In general, to form a cation from an *s- or p-block element*, you remove the highest energy electron or electrons of the atom. Therefore, Na^+ is formed by removing the $3s^1$ electron from the Na atom,

$$\text{Na } [1s^22s^22p^63s^1] \longrightarrow \text{Na}^+ \; [1s^22s^22p^6] + e^-$$

and P^{3+} is formed by removing three $3p$ electrons from a phosphorus atom.

$$\text{P } [1s^22s^22p^63s^23p^3] \longrightarrow \text{P}^{3+} \; [1s^22s^22p^63s^2] + 3e^-$$

Forming cations from *d- or f-block elements* appears to be slightly different, and needs special discussion. In Section 9.4 you saw that complications arise when considering electron configurations for transition elements, largely owing to the overlaying of d, f, and s orbital energies (Figure 9.7). However, these overlap regions do not happen in cations. To understand this, refer to Figure 9.7 and recall that *the energy gaps between shells are controlled by electron repulsions*. The reason the top of one shell's energy range approaches the bottom of the next higher energy shell is that electron repulsions push up the top of the lower range. Furthermore, the highest energy subshell of shell n (say $3d$) is more sensitive to electron–electron repulsions than is the lowest energy subshell of the next higher energy shell, $n + 1$ (say $4s$). This means that, if the repulsions experienced by a subshell at the top of a range are diminished (by removing an electron), the top of the range will drop in energy, and the intershell energy gap will increase. The greater the number of electrons that are removed (the higher the cation charge), the more the intershell gap is increased. In terms of Figure 9.7, *the shell energy ranges separate more widely as electrons are removed* to form a cation.

The disappearance of a *dfs* overlay simplifies the electron assignment problem for metal cations with outer electrons in the overlay region. For example, if a fourth period transition element forms a cation, the $3d$ orbital energy drops further below the energy for $4s$, and so *all common transition metal cations have electron configurations of the general type [rare gas core]$(n - 1)d^x$*. Thus, you can think of the formation of transition metal cations as having occurred by first removing the ns^2 electrons from the metal atom followed by an appropriate number of $(n - 1)d$ electrons. Using iron as an example, you see that Fe^{2+} has the configuration [Ar]$3d^6$ while Fe^{3+} has [Ar]$3d^5$.

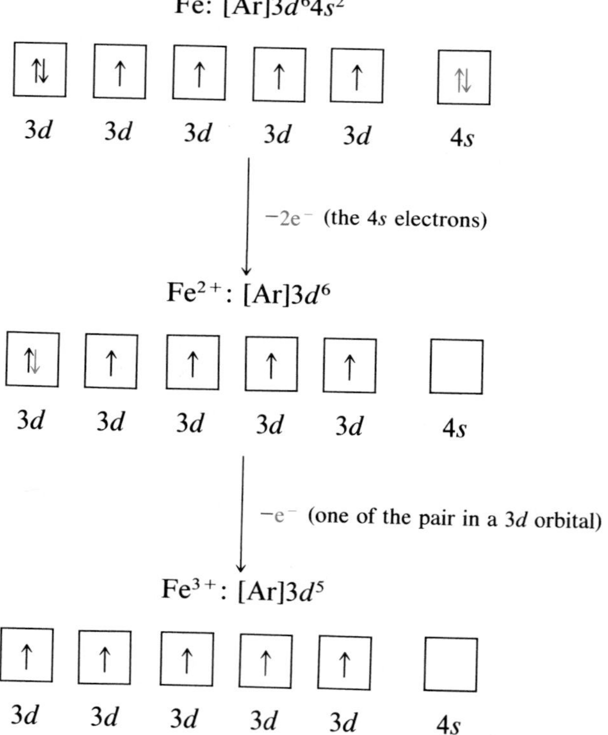

Atoms and ions with unpaired electrons are paramagnetic, that is, they are capable of being attracted to a magnetic field (Section 9.1). Paramagnetism is important here because it allows us to prove that the common transition metal ions with charges of 2+ or higher have no *ns* electrons. For example, the Fe^{2+} ion is paramagnetic to the extent of four unpaired electrons, and the Fe^{3+} ion has five unpaired electrons. If three 3*d* electrons had been removed instead to form Fe^{3+}, for example, the ion would still be paramagnetic but only to the extent of three unpaired electrons.

EXAMPLE 9.5

CONFIGURATIONS OF TRANSITION METAL IONS

Give the electron configuration for copper, Cu, and for its 1+ and 2+ ions. Are these ions paramagnetic? If so, how many unpaired electrons does each have?

Solution Copper is in Group 1B of the fourth period transition metal series. Recall that the metal has a configuration different from that predicted by the device in Figure 9.8. As illustrated in Table 9.4, it has only one electron in the 4*s* orbital and ten electrons in 3*d* orbitals.

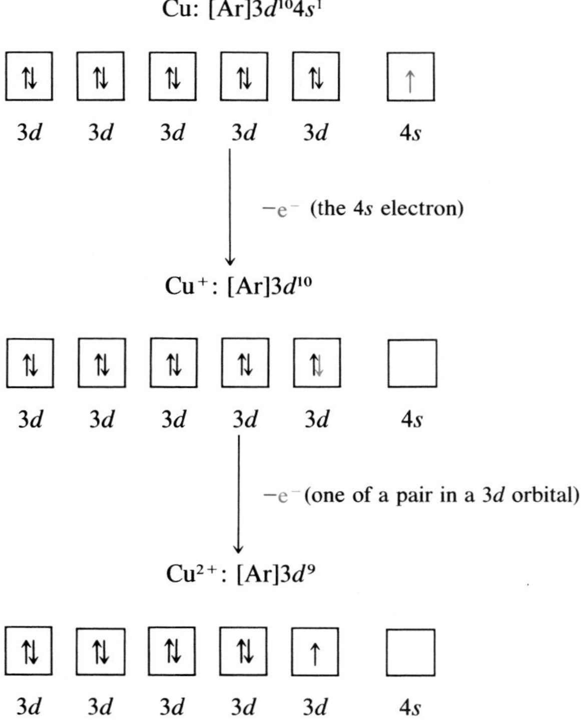

Copper(II) ions have one unpaired electron and so should be paramagnetic. In contrast, Cu^+ has no unpaired electrons, so the ion and its compounds are diamagnetic.

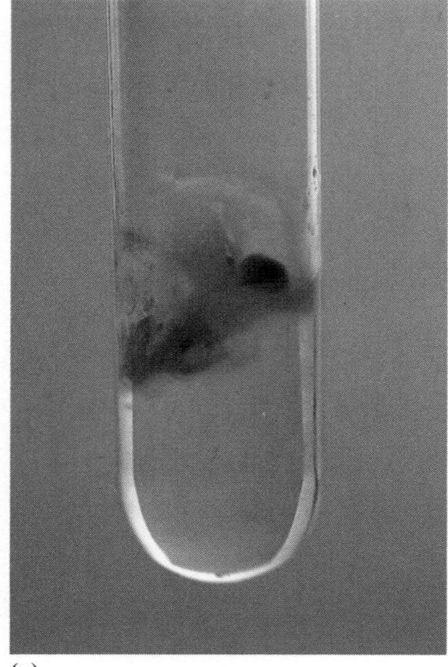

(a)

(b)

Figure 9.10 The reaction of Group 1A and 2A metals with water: (a) Sodium reacts violently with water to give NaOH and H_2 gas. (b) Calcium also reacts with water to produce the metal hydroxide [$Ca(OH)_2$] and H_2 gas. However, the calcium reaction is much less violent than that of sodium and water. These reactions, and their products, are predicted by a knowledge of the positions of the elements in the periodic table.

EXERCISE 9.3 Metal Ion Configurations

Depict electron configurations for V^{2+}, V^{3+}, and Co^{3+}. Use orbital box diagrams and show only the *ns* and $(n - 1)d$ electrons. Are any of the ions paramagnetic? If so, give the number of unpaired electrons.

9.6 ATOMIC PROPERTIES AND PERIODIC TRENDS

In about 1870 Dmitrii Mendeleev prepared the first periodic table while trying to organize the physical and chemical properties of the elements. Although his table has been modified several times over the past 120 years, it still serves its original purpose very well indeed. Although Mendeleev did not foresee our knowledge of atomic structure, we have used the table in this chapter as a guide to atomic and ionic electron configurations. In Chapter 2 we described the division of the table into a section of metallic elements and another section of nonmetals. Within these broad classes there are groups of elements, each having a characteristic chemistry. For example, there are the alkali metals (Group 1A) and alkaline earth metals (Group 2A). In both cases, these metals can react with water to produce H_2 and a metal hydroxide (Figure 9.10).

$$\textit{Group 1A} \quad Na(s) + H_2O(\ell) \longrightarrow NaOH(aq) + \tfrac{1}{2} H_2(g)$$

$$\textit{Group 2A} \quad Ca(s) + 2\, H_2O(\ell) \longrightarrow Ca(OH)_2(aq) + H_2(g)$$

Except for $Be(OH)_2$, the metal hydroxides provide basic solutions in water, a reaction so characteristic of these elements that this is the source of the group names. In all their chemical activity, these metals characteristically lose electrons to form 1+ (Group 1A) or 2+ (Group 2A) cations.

An objective of this section is to show how atomic electron configurations are related to some physical properties of the elements and why those properties change in a reasonably predictable manner when moving down groups and across periods. On the basis of this knowledge, you will eventually be able to organize and predict chemical and physical properties of elements and their compounds.

ATOMIC SIZE

An electron orbital has no sharp boundary beyond which the electron never strays. How then do we define the size or radius of an atom? It is a problem not yet solved to everyone's satisfaction; the fact that new tables of atomic radii are still being compiled suggests that there is no universal definition, and reminds us that we should treat any set of radii with caution.

For atoms that form simple diatomic molecules, such as Cl_2, the radius of the atom can be defined experimentally by first finding the distance between the centers of the two atoms. One half of this distance is assumed to be a good estimate of the atom radius. In the Cl_2 and H_2 molecules, for example, the atom–atom distances (the distance from the center of one atom to the center of the other) are 200 and 74 pm, respectively. This means that the Cl radius is 100. pm and that for H is 37 pm. A test of the reasonableness of these atomic radius estimates is whether we can use them to estimate the distance between H and Cl in HCl. As you can see in the margin, the estimated and experimental values are in reasonable agreement.

This approach can be used to determine other atomic radii. For example, the radii of O, C, and S can be estimated by measuring the O—H, C—Cl and H—S distances in H_2O, CCl_4, and H_2S. By this technique and others, a reasonable set of atom radii has been assembled for most elements, and some are illustrated in Figure 9.11. *For the main group elements, atomic radii increase going down a group in the periodic table and decrease going across a period.* Thus, you know that in a series of Group 4A compounds of the type ECl_4, for example, the distance between E and Cl will increase as E changes from C to Pb.

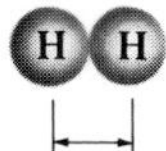

Interatomic distance = 74 pm

Atom radius = $\frac{1}{2}$(64 pm) = 37 pm

+

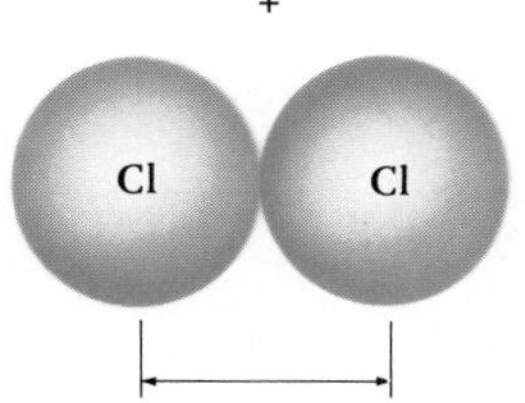

Interatomic distance = 200 pm

Atom radius = $\frac{1}{2}$ (200 pm) = 100 pm

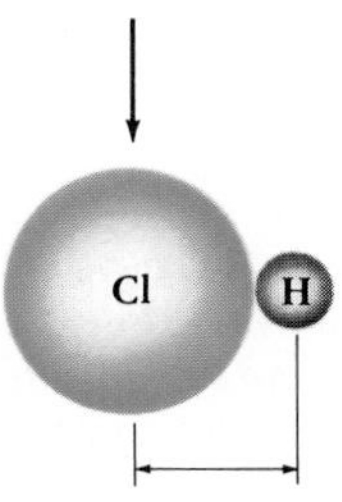

Estimated H-Cl distance = 137 pm

Experimental H-Cl distance = 132 pm

Figure 9.11 Atomic radii in picometers (1 pm = 10^{-12} m) for the main group elements. The data in this figure, and in other figures and tables in this chapter, are taken from *KC? Discoverer*, a computer database of information on the elements. The program is available from Project SERAPHIM through *Journal of Chemical Education: Software*. SERAPHIM's source for atomic radii was N.N. Greenwood and A. Earnshaw, *Chemistry of the Elements*, Pergamon Press, Oxford, 1984. (You are cautioned that there are numerous tabulations of atomic and covalent radii, and they can vary considerably since there is no agreement on precisely how they are to be calculated. For example, tabulations of the atomic radius of N vary from 54.9 pm to 92 pm, and those for O vary from 60.4 to 73 pm.)

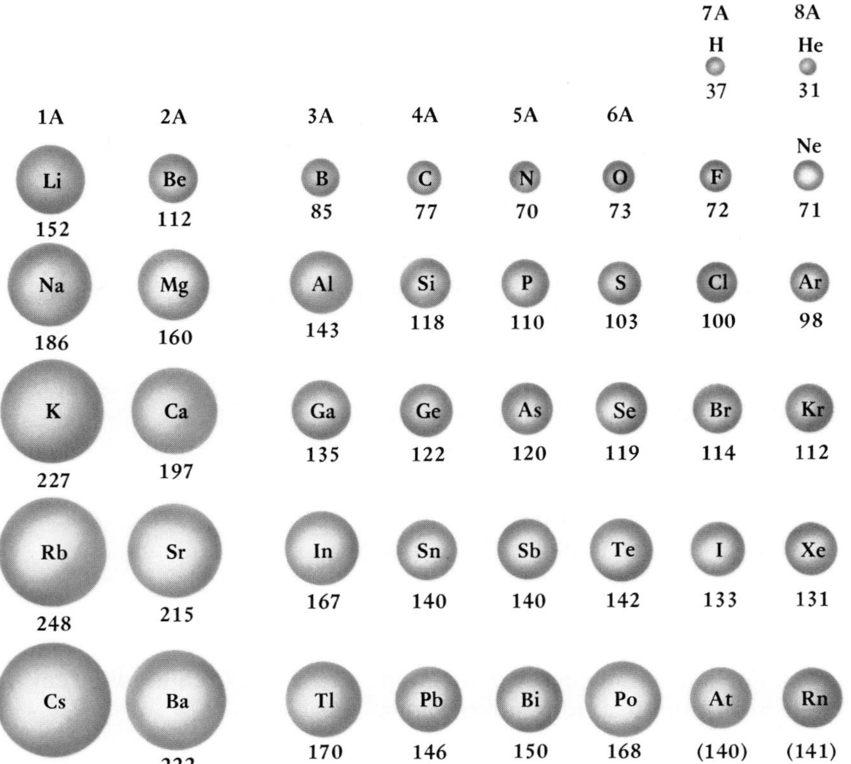

EXERCISE 9.4 Estimating Atom–Atom Distances

(a) Using Figure 9.11, estimate the H—O and H—S distances in H_2O and H_2S, respectively.

(b) If the interatomic distance in Br_2 is 228 pm, what is the radius of Br? Using this estimate, and that for Cl above, estimate the distance between atoms in BrCl.

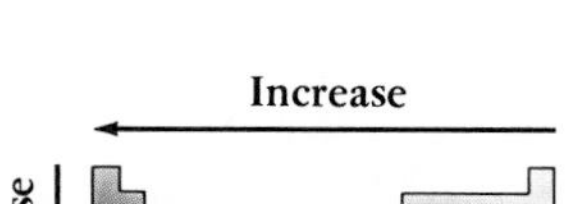

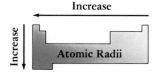

General trends in atomic radii of *s*- and *p*-block elements with position in periodic table

Quantum theory informs us that, for an electron of given n, the ns orbital extends further from the nucleus than the np or (n − 1)d orbitals. Thus, the size of an atom in the nth period is fundamentally determined by the size of the ns orbital.

To understand the variation in atom sizes in passing from top to bottom or from left to right in the periodic table, you must return to the basic concepts of attraction of electrons to the nucleus and repulsion of electrons from each other. The change in these forces can be linked to changes in the atomic shell structure, which is governed by the values of n and ℓ.

There are two important effects to understand. First, increasing the nuclear charge alone always shrinks all the shell electron clouds. However, this shrinkage can proceed only to the point at which electron repulsions overcome the nuclear attraction and stop the contraction of electron shells. Second, *it is principally the ns electrons of the outermost shell* that determine *the size of the atom*. The *np* electrons of that shell and electrons in shells of lower n provide a resistive cushion against which the *ns* electrons contract. Thus, the variation in atom size comes down to understanding how n and ℓ for the outer electrons influence their shrinkage: (i) Electron shells swell in size as n increases and (ii) subshells within a shell contract as ℓ increases. *It is only when n increases and ℓ decreases that we look for a clear increase in size.*

Across each row in Figure 9.11 n is constant. On the other hand, ℓ increases from 0 for the *s*-block elements to 1 for the *p*-block elements. Remembering that the atom size in a given period n is determined by the size of the *ns* orbital, and that the *ns* orbital contracts under the influence of the increasingly positive nucleus, *the atomic radius decreases on moving across the periodic table*. There are minor variations, however. For example, the last four elements of the second period, while smaller than C as expected, all have about the same size (with N *smaller* than O, F, and Ne). Here is a case where electron–electron repulsions in these very small atoms retard the shell contraction.

Electron shells expand in size as n increases. Subshells within a shell contract as ℓ increases.

When passing from the end of the *p*-block to the start of the next *s*-block the shell structure changes dramatically; n increases by one and offsets the shrinkage otherwise expected from the increase in nuclear charge. The increase in charge is rather large, so large in fact that *there generally is a size increase on moving down a periodic group.*

The sizes of transition metal atoms change little on proceeding from vanadium across the series, since the size is determined by the 4s orbital, which bears at least one electron in each element. The consequences of this are described more fully in Chapter 25.

IONIZATION ENERGY

The ionization energy of an atom is the energy required to remove an electron from the atom in the gas phase.

$$\text{Atom in ground state(g)} + \text{energy} \longrightarrow \text{Atom}^+\text{(g)} + e^-$$

$$\Delta E = \text{ionization energy } (IE)$$

The process of ionization involves moving an electron from a given electron shell to a position outside the atom, that is, to n = infinity (Figure 9.12). Energy is *always* required to ionize an atom, so the process is endothermic and the sign of the ionization energy is always positive.

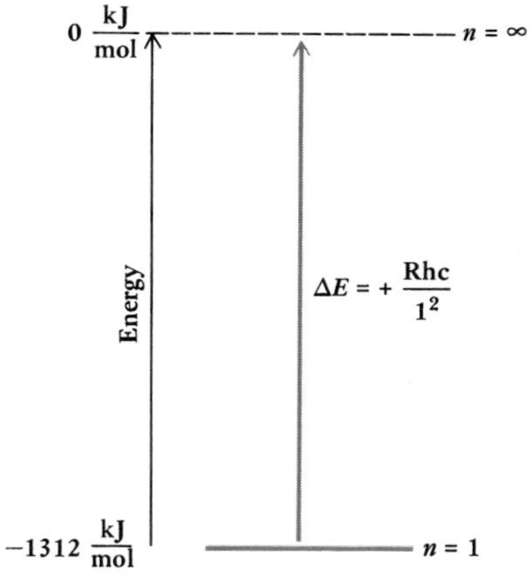

Figure 9.12 Ionization energy for an electron with n = 1 for the H atom. $IE = \Delta E = E_{final} - E_{initial} = 0 - (-Rhc/n^2)$. When $n = 1$, $\Delta E = Rhc$ or 1312 kJ/mol.

Each atom can have a series of ionization energies, since more than one electron can always be removed (except for H). For example, the first three ionization energies of Mg(g) are

$$\underset{1s^22s^22p^63s^2}{Mg(g)} \longrightarrow \underset{1s^22s^22p^63s^1}{Mg^+(g)} + e^- \quad IE(1) = 738 \text{ kJ/mol}$$

$$\underset{1s^22s^22p^63s^1}{Mg^+(g)} \longrightarrow \underset{1s^22s^22p^63s^0}{Mg^{2+}(g)} + e^- \quad IE(2) = 1451 \text{ kJ/mol}$$

$$\underset{1s^22s^22p^6}{Mg^{2+}(g)} \longrightarrow \underset{1s^22s^22p^5}{Mg^{3+}(g)} + e^- \quad IE(3) = 7733 \text{ kJ/mol}$$

Notice that removing each subsequent electron requires more and more energy, and the jump from the second [$IE(2)$] to the third [$IE(3)$] ionization energy is particularly great. As already discussed under ion configurations (page 335), removing an electron from an atom decreases the repulsion the remaining electrons feel, and their energies decrease (become more negative). In each ionization step, the outer electron of the product ion has a lower or more negative energy than the outer electron of the reactant, and requires more work to be removed. The increase in work required becomes especially great when a shell band gap is crossed (a decrease in n). In the magnesium series above, the outer electron in the second step has the same n and ℓ as that in the first step, so the second ionization energy [$IE(2)$] is larger simply because the electron repulsions have decreased and the energy of the 3s electron has decreased. In the third step, the less stable electron is a 2p electron, with a smaller n than the electron removed in the second step; we have dipped into a lower energy electron shell, and the ionization energy [$IE(3)$] has increased greatly.

The very great difference in the second and third ionization energies for Mg is excellent experimental evidence for the existence of electronic shells in atoms.

For main group (s- and p-block) elements, *first ionization energies generally increase across a period and decrease down a group* (Figure 9.13 and Table 9.5). The trend down a group reflects the effect of increasing n, while the trend across a period reflects the effect of holding n constant while increasing the number of nuclear protons. The trend across a period is not smooth, particularly in the second period. The reason for this is that the value of ℓ increases on going from s-block to p-block elements (from Be to B, for example). An increase in ℓ means that the electron energy has increased (become less negative) and so the ionization energy declines.

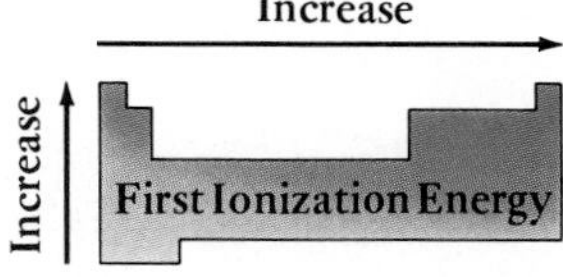

General trends in first ionization energies of A-Group elements.

On leaving B and moving on to C and then N, there is an increase in the nuclear charge, which again means an increase in ionization energy. However, there is a dip to lower ionization energy on passing from Group 5A to Group 6A that is especially noticeable for N and O. There is no change in either n or ℓ, but there has been an increase in electron–electron repulsions for the following reason. In Groups 3A–5A, electrons are assigned to separate p orbitals (p_x, p_y, and p_z). Beginning in Group 6A, however, two electrons are assigned to the same p orbital. Thus, beginning with this group the fourth p electron shares an orbital with another electron

Figure 9.13 First ionization energies of the main group elements of the first four periods. See Table 9.5 for more data.

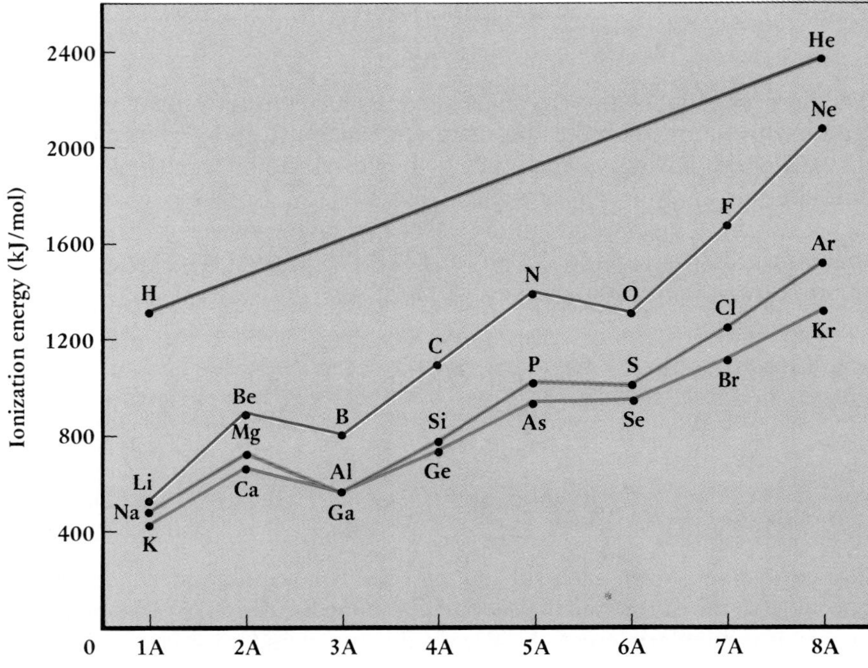

and thus experiences greater repulsion than it would in an orbital of its own. Using oxygen as an example, you have

You may wonder why a dip was not observed for Group 2A with two s electrons. Recall the difference in shapes of s and p orbitals; the spherical shape of the s orbital lets two electrons avoid each other better than does the more compact "halved sphere" p orbital.

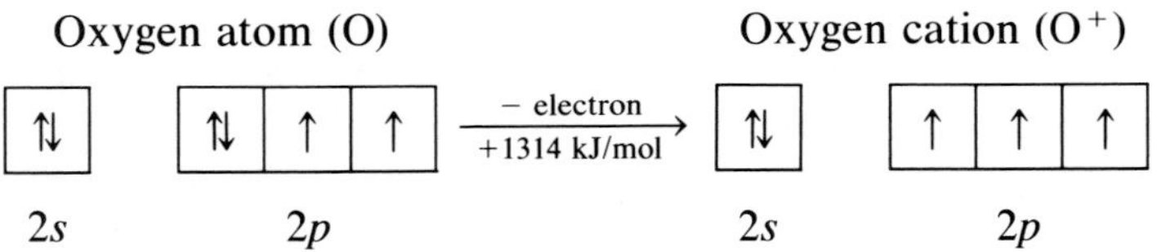

The greater repulsion experienced by the fourth electron makes it easier to remove, to give an ion in which each p electron has an orbital of its own.

Table 9.5 First Ionization Energies of the Elements (kJ/mol)

1A (1)	2A (2)	3B (3)	4B (4)	5B (5)	6B (6)	7B (7)	8B (8, 9, 10)			1B (11)	2B (12)	3A (13)	4A (14)	5A (15)	6A (16)	7A (17)	8A (18)
H 1312																	He 2371
Li 520	Be 899											B 801	C 1086	N 1402	O 1314	F 1681	Ne 2081
Na 496	Mg 738											Al 578	Si 786	P 1012	S 1000	Cl 1251	Ar 1521
K 419	Ca 599	Sc 631	Ti 658	V 650	Cr 652	Mn 717	Fe 759	Co 758	Ni 757	Cu 745	Zn 906	Ga 579	Ge 762	As 947	Se 941	Br 1140	Kr 1351
Rb 403	Sr 550	Y 617	Zr 661	Nb 664	Mo 685	Tc 702	Ru 711	Rh 720	Pd 804	Ag 731	Cd 868	In 558	Sn 709	Sb 834	Te 869	I 1008	Xe 1170
Cs 377	Ba 503	La 538	Hf 681	Ta 761	W 770	Re 760	Os 840	Ir 880	Pt 870	Au 890	Hg 1007	Tl 589	Pb 715	Bi 703	Po 812	At 890	Rn 1037

The data in this figure are taken from KC? *Discoverer*, a computer database of information on the elements. The program is available from Project SERAPHIM through *Journal of Chemical Education: Software.*

ELECTRON AFFINITY

Some atoms have an affinity or "liking" for electrons and can acquire one or more to form negative ions. This occurs when the repulsion felt by all the electrons—the original core and valence electrons *plus* the added electron—is less than their attraction to the nucleus. By convention the **electron affinity** of an element is the energy change that occurs when an electron is assigned to the lowest-energy unoccupied valence orbital of a gaseous atom.

$$\text{Atom(g)} + e^- \longrightarrow \text{Atom}^-\text{(g)}$$

$$\Delta E = \text{electron affinity } (EA)$$

When a stable anion is formed, energy is released in the process. For example, 328 kJ are released when a mole of F^- ions is formed from fluorine atoms.

$$\text{F(g)} + e^- \longrightarrow \text{F}^-\text{(g)} + 328 \text{ kJ}$$

Fluorine atom (F)			Fluorine anion (F^-)	
⇅	⇅ ⇅ ↑	+ electron →	⇅	⇅ ⇅ ⇅
2*s*	2*p*		2*s*	2*p*

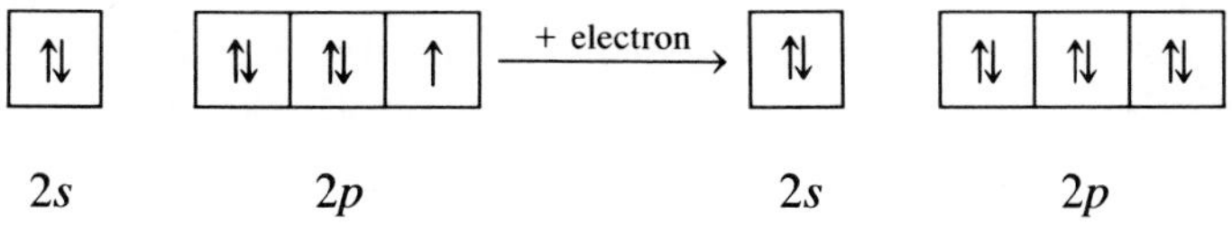

Therefore, in accordance with the sign conventions used in thermochemistry (Chapter 6), the value of the electron affinity of fluorine is −328 kJ/mol, the minus sign signaling an exothermic process. In general, the value of the electron affinity becomes *more negative* the greater the tendency of an atom to accept an electron.*

If an atom has no affinity for an electron, an unstable anion is formed, and energy is *required* to form the anion. A helium atom, for example, has no affinity for an electron because the electron must be added to a high energy orbital.

$$\text{He(g)} + e^- + 21 \text{ kJ} \longrightarrow \text{He}^-\text{(g)}$$

Helium atom (He)			Helium anion (He^-)	
⇅	(empty)	+ electron →	⇅	↑
1*s*	2*s*		1*s*	2*s*

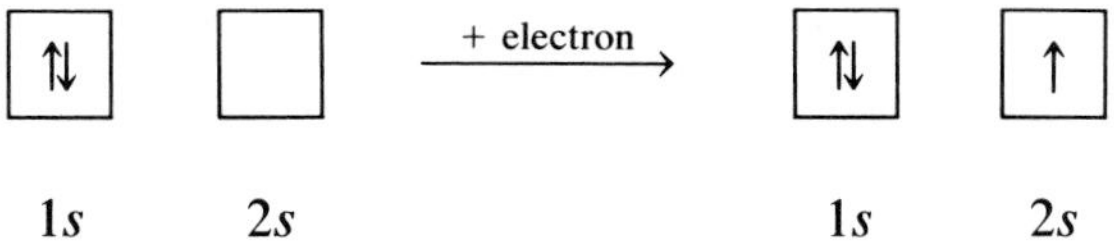

Here the value of the electron affinity is +21 kJ/mol, the positive sign signaling that the process is endothermic. In general, the less the tendency of an atom to accept an electron, the *less negative (or more positive)* the value of the electron affinity.

The periodic trends in electron affinity are closely related to those for size and ionization energy (Figures 9.11 and 9.13). In Table 9.6 and Figure 9.14 we see that atoms generally have a greater affinity for electrons as we proceed across a period; the values of *EA* become more negative. Also, we see that the nonmetals generally have much more negative values

*This convention is currently followed in most introductory textbooks. However, there are other conventions in use.

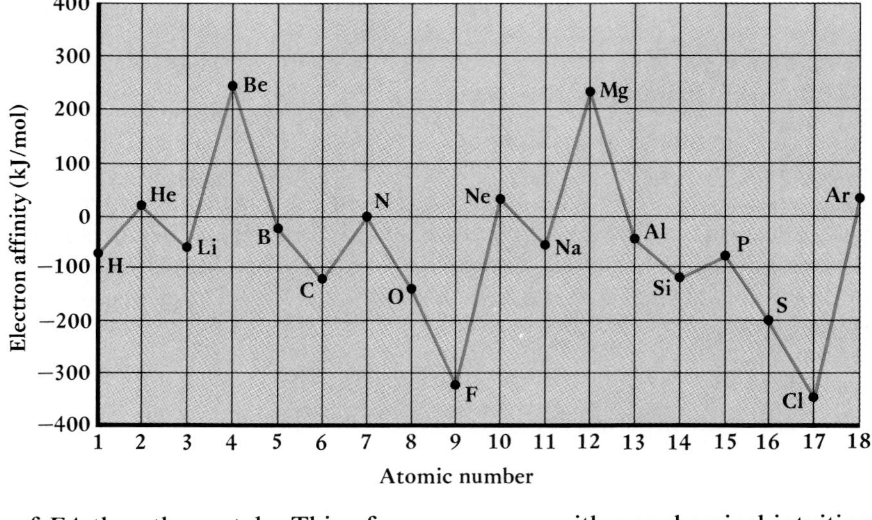

Figure 9.14 A plot of electron affinity (EA) against atomic number; the greater the affinity of an atom for an electron, the more negative the value of EA. The value of EA generally becomes more negative across a period, meaning that atoms of the same n but increasing ℓ have an increasing affinity for an electron. Going down a group EA has less negative values, signaling a decrease in the affinity of an atom for an electron.

of EA than the metals. This of course agrees with our chemical intuition and experience, which tells us that metals generally do not form negative ions, while there is an increasing tendency across a period to form anions (for example, O^-, S^-, F^-, and Cl^-).

Figure 9.14 also shows, though, that there are some significant exceptions to the general trend in electron affinity values as ℓ increases and when orbitals are doubly occupied. For example, Be has a low affinity for an electron (EA is quite positive) because the added electron is assigned to a higher energy subshell ($2p$, $\ell = 1$) than the valence electrons ($2s$, $\ell = 0$) (see Figure 9.7). Since an electron pair is formed when N acquires an electron, nitrogen has a value of 0 for the electron affinity owing to the effect of electron–electron repulsions.

It is very curious that electron–electron repulsions exactly balance nuclear attractions and lead to a value of 0 for the EA of N.

$$N(g) + e^- \xrightarrow{\text{no energy evolved or required}} N^-(g)$$

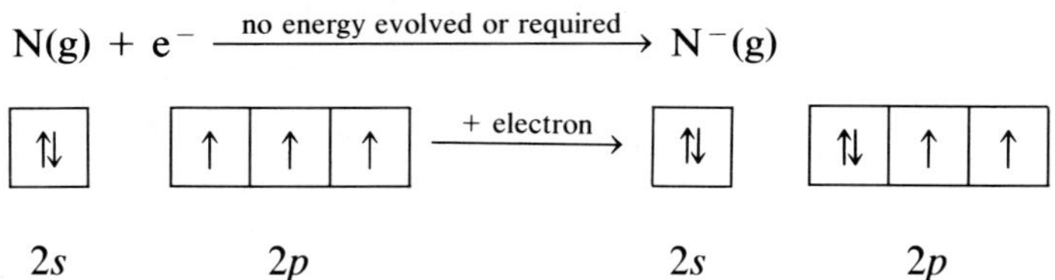

When descending a periodic group n increases, and we expect the affinity for an electron to decrease for the same reason that ionization energy decreases. A glance at Table 9.6 shows that this is *roughly* true,

Table 9.6 Electron Affinity Values for Some Elements (kJ/mol)

H −72.8							He +21
Li −59.6	Be +241	B −26.7	C −122	N 0	O −141	F −328	Ne +29
Na −52.9	Mg +230	Al −42.5	Si −134	P −72.0	S −200.	Cl −349	Ar +34
K −48.4	Ca +156	Ga −28.9	Ge −119	As −78.2	Se −195	Br −325	Kr +39
Rb −46.9	Sr +167	In −28.9	Sn −107	Sb −103	Te −190.	I −295	Xe +40
Cs −45.5	Ba +52.0	Tl −19.3	Pb −35.1	Bi −91.3	Po −183	At −270	Rn +41

but there are exceptions once again. Consider the halogens, for example. The affinity for an electron does indeed drop as expected from Cl to I, but fluorine has a lower affinity than chlorine (*EA* for F is less negative than *EA* for Cl)! The reason for this stems from unusually large electron–electron repulsions in fluorine. Adding an electron to the nine already present in the F atom means that ten electrons must be packed into a relatively small volume of space, leading to severe repulsions between electrons. Chlorine, bromine, and iodine have larger atomic volumes than fluorine, so adding an electron does not lead to such severe electron–electron repulsions as in fluorine. The electron affinity of fluorine is thus anomalously small, and the same can be said for O relative to the other Group 6A elements.

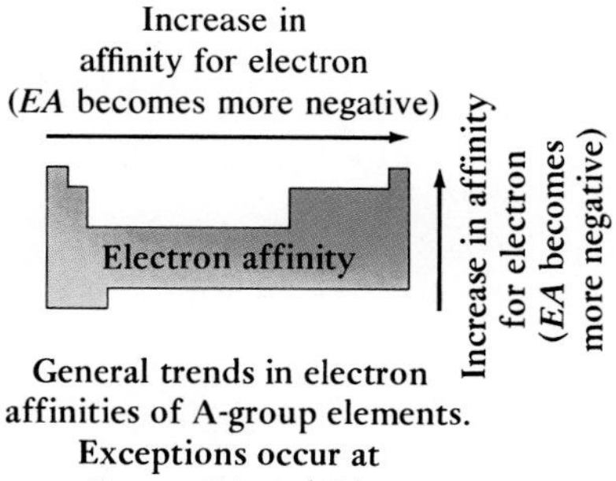

General trends in electron affinities of A-group elements. Exceptions occur at Groups 2A and 5A.

EXAMPLE 9.6

PERIODIC TRENDS

Compare the three elements C, O, and Si. (a) Place them in order of increasing atomic radius. (b) Which has the largest ionization energy? (c) Which has the more negative value of electron affinity, C or O?

Solution (a) *Atomic size.* Atomic radius declines as we move across a period, so carbon must have a larger radius than oxygen. However, radius increases down a periodic group. Since C and Si are in the same group (Group 4A), Si must be larger than C. Therefore, in order of increasing size, the elements are $O < C < Si$.

(b) *Ionization Energy (IE).* Ionization energy generally increases across a period and decreases down a group; a large decrease in *IE* occurs from the second to the third period elements. This should make $IE(Si) < IE(C) < IE(O)$.

(c) *Electron affinity (EA).* Electron affinity values generally become more negative across a period and less negative down a group (although the trends are not as clear as trends in ionization energy). Therefore, the *EA* for O should be more negative than the *EA* for C. That is, O has a greater affinity for an electron than does C. (Notice that the *EA* for Si is *more* negative than that for C. Once again we see the effect of electron–electron repulsions, which make carbon's affinity for an electron smaller than anticipated.)

EXERCISE 9.5 Periodic Trends

Compare the three elements Al, C, and Si.

(a) Place the three elements in order of increasing atomic radius.
(b) Rank the elements in order of increasing ionization energy. (Try to do this without looking at Table 9.5; then compare your estimates with the table.)
(c) Which element, Al or Si, is expected to have the more negative value of electron affinity?

ION SIZES

Figure 9.15 shows clearly that the periodic trends in the sizes of a few common ions are the same as those for neutral atoms: positive or negative ions of the same group increase in size when descending the group. But pause for a moment and compare Figure 9.15 with Figure 9.11. When an electron is removed from an atom to form a cation, the size

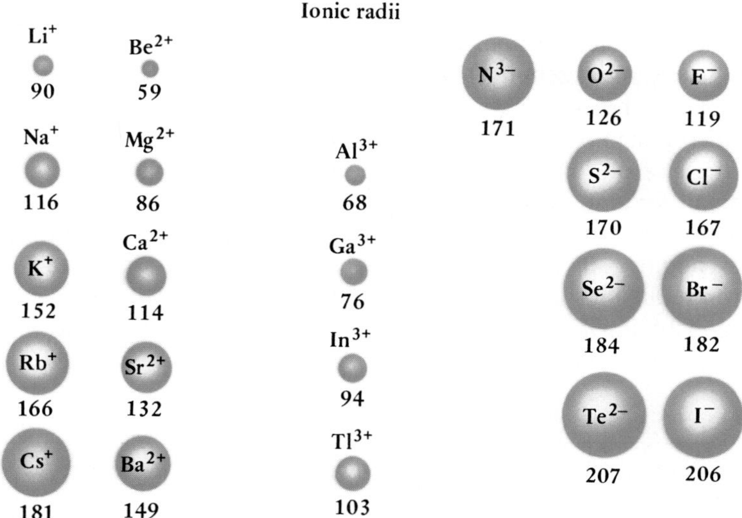

Figure 9.15 Relative sizes of some common ions. Radii are given in picometers (1 pm = 10^{-12} m). See Figure 9.11 for the source of the data.

shrinks considerably; the radius of a cation is always smaller than that of the atom from which it was derived. For example, the radius of Li is 152 pm, whereas that for Li^+ is only 90 pm. This is understandable, because when an electron is removed the other electrons no longer feel its repulsion, so the remaining electrons contract toward the nucleus. The decrease in ion size will be especially great when the electron (or electrons) removed had a greater n (and smaller ℓ) than the new outer electron. This is the

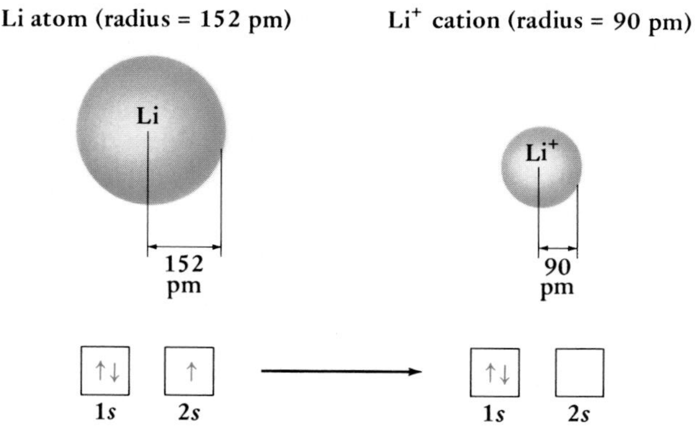

case for Li, for which the "old" outer electron was 2s and the "new" one is 1s. The shrinkage will also be great when two or more electrons are removed, as for Be^{2+} and Al^{3+}.

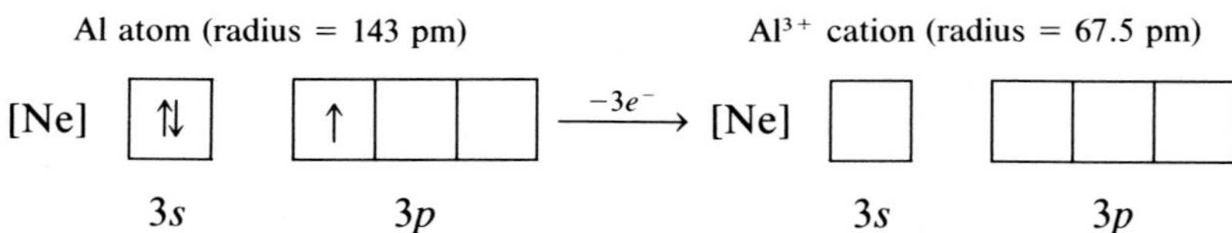

You can also see by comparing Figures 9.11 and 9.15 that *anions are always larger than the atoms from which they are derived.* Here the argument is the opposite of that used to explain the radii of positive ions: adding an electron introduces new repulsions and the shells swell. The F atom, for example, has 9 protons and 9 electrons. On forming the anion, the nuclear charge is still 9+, but there are now 10 electrons in the anion. The F^- ion is much larger than the F atom because *increased electron–electron repulsions have caused the atom to swell when an electron is added.*

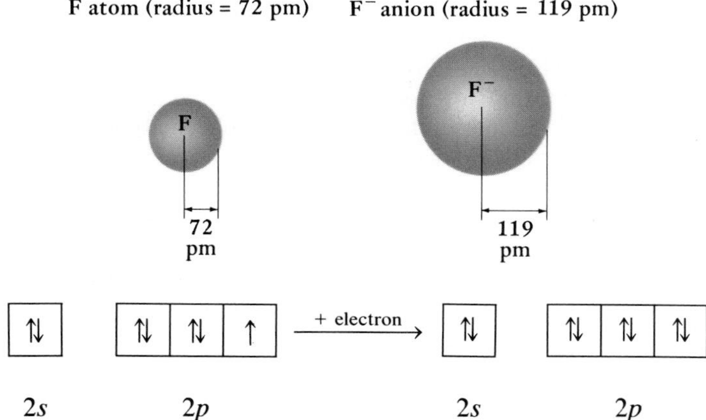

The oxide ion, O^{2-}, is **isoelectronic** with F^-; that is, they both have the same number of electrons (10). The oxide ion, however, is larger than the fluoride ion because the oxide ion has only 8 protons available to attract 10 electrons, whereas F^- has more protons (9) to attract the same number of electrons. It is useful to compare the sizes of isoelectronic ions across the periodic table. For example, consider O^{2-}, F^-, Na^+, and Mg^{2+}.

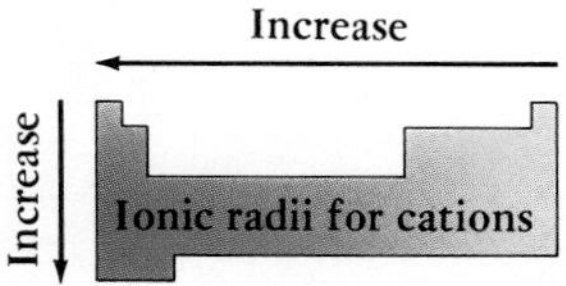

General trends in positive ion radii of A-group elements

ION	O^{2-}	F^-	Na^+	Mg^{2+}
Ionic radius (pm)	126	119	116	86
Number of nuclear protons	8	9	11	12
Number of electrons	10	10	10	10

All the ions have a total of 10 electrons. However, the O^{2-} ion has only eight protons in its nucleus to attract these electrons, while F^- has nine, Na^+ has eleven, and Mg^{2+} has twelve. As the proton/electron ratio increases in a series of isoelectronic ions, the balance between electron–proton attraction and electron–electron repulsion shifts in favor of attraction, and the ion shrinks. As you can see in Figure 9.15, this is true for all isoelectronic series of ions.

EXERCISE 9.6 Ion Sizes

What is the trend in sizes of the ions N^{3-}, O^{2-}, and F^-? Briefly explain why this trend exists.

SOMETHING MORE ABOUT
The Importance of Ionic Properties: Gems and their Colors

Gemstones are some of the most beautiful things we have ever seen. There are red rubies, green emeralds, and aquamarines, a stone so named because it is the color of the tropical sea. What are their chemical compositions, and why are they colored? To answer this we have to use some of the ideas that we have developed in the last two chapters.

Rubies are really just ''dirty'' aluminum oxide, Al_2O_3, the red color coming from traces of Cr^{3+} ions in solid Al_2O_3. Emeralds and aquamarines are similar in that both contain a large silicate ion, $Si_6O_{18}{}^{12-}$, along with Al^{3+} and Be^{2+} ions. The color of emeralds also comes from traces of Cr^{3+}, but aquamarines are water-blue owing to iron ions. But why, for example, is Cr^{3+} so important to the deep red color of rubies? A brief look at the nature of the solid, and some consideration of orbital energies for the Cr^{3+} ion, can lead us to an answer, an answer that applies rather well to explaining the colors of all gems.

Aluminum oxide is a solid, and you think of it as a regular array of oxide ions—called a *crystal lattice* (see Chapter 13)—with Al^{3+} ions in the holes left when the spherical O^{2-} ions are piled up in a regular way. The result is that each Al^{3+} ion is surrounded by six O^{2-} ions at the corners of an octahedron.

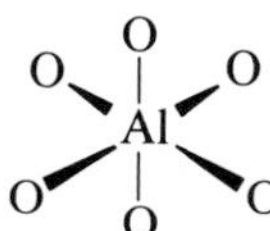

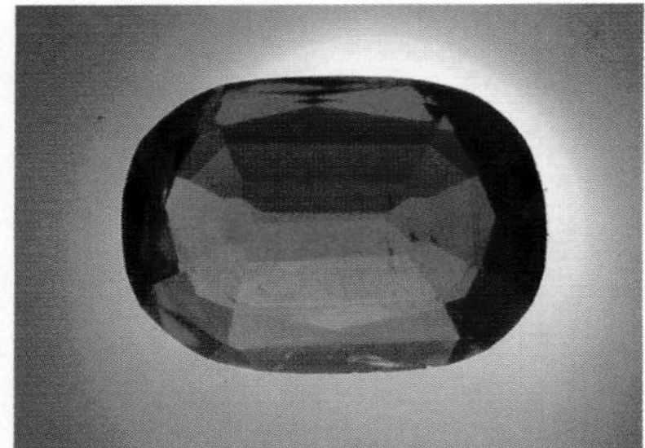

The color of rubies (top) comes from chromium(III) ions in a lattice of Al^{3+} and O^{2-} ions. An aquamarine (bottom) gets its beautiful blue-green color from traces of iron ions in a lattice of aluminum, oxygen, beryllium, and silicon.

The Al^{3+} ion in the hole in the oxide ion pile has a radius of 67.5 pm, but the radius of a Cr^{3+} ion is only slightly larger, 75.5 pm. Because of this close similarity in size, the transition metal ion can take the place of some of the aluminum ions in the lattice, and nature has the makings of a ruby. This substitution of one metal ion for another of similar size is quite common in nature, and many minerals have slightly varying compositions because of this.

Solid aluminum oxide is white, but traces of Cr^{3+} make it red. To understand this fully requires a lengthy discussion, and this is sketched out in Chapter 25. For now we can say that, when the Cr^{3+} ion is placed in the lattice of O^{2-} ions, the *d* orbitals of the Cr^{3+} ion do not all have the same energy, as you learned that they do in the gaseous ion. In fact, we know that they have the following relative energies:

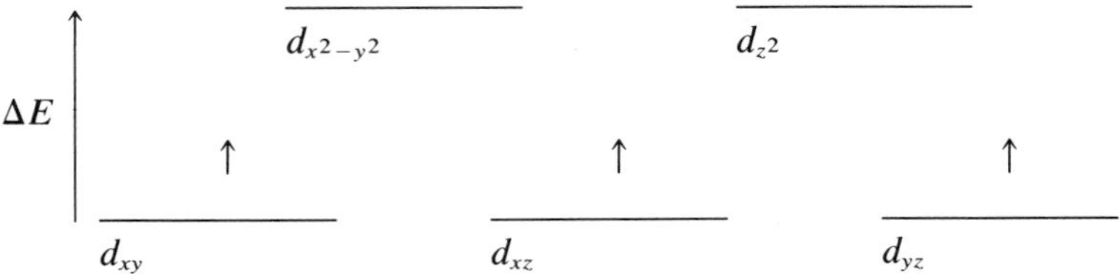

One of the three *d* electrons for Cr^{3+} is assigned to each of the three lower energy *d* orbitals. When visible light strikes the Cr^{3+} ions, one of these *d* electrons is excited and can be raised to a higher energy level, say d_{z^2}. This removes light of a particular wavelength from the visible light and allows other wavelengths to pass. The wavelengths of light that pass in this case are dominated by red light, and rubies appear red.

Table 9.7 Common Monatomic Ions of the Main Group Elements*

	Group and Ion						
Period	1A	2A	3A	4A	5A	6A	7A
1	Li^+	Be^{2+}	B	C	N^{3-}	O^{2-}	F^-
2	Na^+	Mg^{2+}	Al^{3+}	Si	P	S^{2-}	Cl^-
3	K^+	Ca^{2+}	Ga^{3+}	Ge	As	Se^{2-}	Br^-
4	Rb^+	Sr^{2+}	In^{3+}	Sn^{2+}	Sb	Te^{2-}	I^-
5	Cs^+	Ba^{2+}	Tl^+, Tl^{3+}	Pb^{2+}	Bi^{3+}		

*Elements printed in color with no charge do not commonly form compounds having monatomic ions.

9.7 THE ENERGY REQUIRED TO FORM COMMON MONATOMIC IONS AND ITS CHEMICAL CONSEQUENCES

You learned in Chapter 3 that many metals form positive ions with a charge equal to the group number (Table 9.7). In contrast, nonmetals can form negative ions with a charge equal to the group number minus eight. For both metals and nonmetals this charge is equated with the oxidation number of the ion. From the discussion of ionization energy, electron affinity, and ion electron configurations, you can now see the reasons for these observations.

The first four ionization energies for sodium, magnesium, and aluminum are listed in Table 9.8. A modest amount of energy (496 kJ/mol) is required to form Na^+(g), while almost 10 times that much is needed to form Na^{2+}(g) because the second electron lost must be a firmly held n = 2 electron. Since energies of this magnitude are not available in normal chemical reactions, the Group 1A element sodium commonly forms only a 1+ ion. Similarly, it is energetically too expensive to form Mg^{3+}(g) or Al^{4+}(g), so these elements form only Mg^{2+}(g) and Al^{3+}(g) in normal reactions. This is the reason monatomic cation charges are never greater than the group number: to go beyond means to remove inner shell electrons with *much* lower (more negative) energies. Anions are formed by atoms that have several tightly held electrons and that tend to add electrons rather than lose them. A useful generalization is that *atoms near the rare gases in the periodic table form ions: elements immediately preceding Group 8A form anions with the configuration of the nearest rare gas and those immediately following Group 8A form cations, again with the configuration of the nearest rare gas*. Thus, bromine forms Br^- and aluminum forms Al^{3+}, and both ions have rare gas configurations (of Kr and Ne, respectively) (Figure 9.16).

Figure 9.16 Aluminum metal reacts with liquid bromine to give aluminum bromide, a compound in which both Al and Br can be thought of as ions (Al^{3+} and Br^-) having a rare gas configuration.

$$2\ Al(s) + 3\ Br_2(\ell) \longrightarrow 2\ AlBr_3(s)$$

Table 9.8 Ionization Energies of Na, Mg, and Al

	Ionization Energy (kJ/mol)			
Element	First	Second	Third	Fourth
Na	496	4562	6912	9543
Mg	738	1451	7733	10540
Al	578	1817	2745	11577

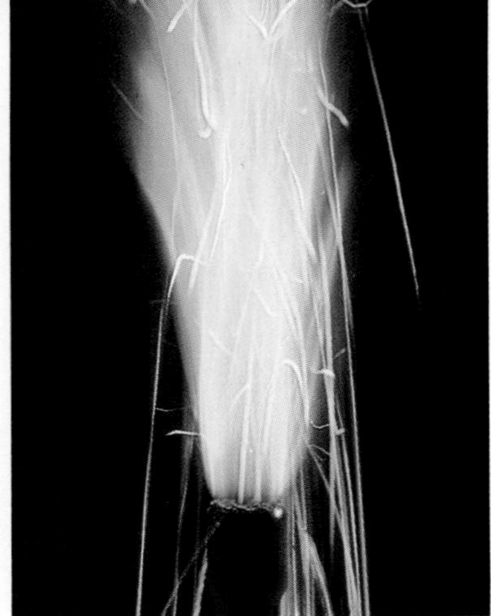

Figure 9.17 When sprayed into a bunsen flame, powdered zinc metal reacts with oxygen to give zinc oxide, an ionic compound in which the Zn^{2+} ion has a pseudo-rare gas configuration.

$$2\ Zn(s) + O_2(g) \longrightarrow 2\ ZnO(s)$$

Elements just following the transition elements, those of Groups 1B, 2B, 3A, and 4A in particular, are too far from the preceding rare gas to readily form cations with the preceding rare gas configuration, and they cannot form anions of the configuration of the following rare gas. Such atoms can still form cations of charge equal to the group number, but then chemists say that these ions have a **pseudo-rare gas configuration.** This means the configuration is the same as that of the preceding rare gas plus any filled $(n - 1)d$ [or $(n - 2)f$] orbitals. For example, zinc reacts with oxygen to give zinc oxide, ZnO (Figure 9.17).

$$Zn\ \{[Ar]3d^{10}4s^2\} \longrightarrow Zn^{2+}\ \{[Ar]3d^{10}\} + 2\ e^-$$

One last important question to ask in this chapter is whether the rare gas configuration for an ion can be achieved under the conditions normally used in the chemistry laboratory or industrial plant. Cations are of course electron-seeking, the more so the higher their charge, while electron repulsions will limit the number of electrons an atom can add in forming an anion. This means, for example, that it *requires* a great deal of energy to create the Mg^{2+} and O^{2-} ions of magnesium oxide (Figure 9.18). Furthermore, as one moves to the center of a row in the periodic table, stabilizing a high ion charge becomes increasingly difficult (Figure 9.19). This could be done in a high energy device such as the particle accelerators used by physicists, but these are hardly the conditions of a normal chemical experiment. The phrase "normal chemical conditions" means one is dealing with energy sources on the order of 500–1000 kJ. So where does the energy come from to allow magnesium and oxygen to form MgO, for example?

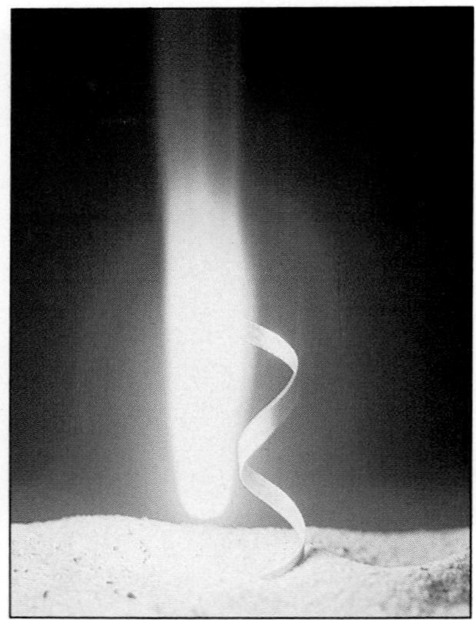

Figure 9.18 Reaction of magnesium and oxygen to give magnesium oxide, MgO.

Thinking further about the energies required to form cations, Table 9.8 shows that more than 5000 kJ/mol are required to form a gaseous Al^{3+} ion. It is certain that a simple Al^{3+} cation is not likely to be found in nature, in spite of the fact that we think of aluminum bromide, for example, as consisting of an Al^{3+} ion and Br^- ions (Figure 9.16). The situation is even worse for boron, where nearly 7000 kJ/mol is required to form a B^{3+} cation. As you will see in the next two chapters, the energy to form MgO, $AlBr_3$, or a compound such as BF_3 comes from the interaction of cations and anions in the formation of a very stable solid or through *electron sharing* between the metal or metalloid and nonmetal.

In Chapter 3 you learned that transition metals typically form 2+ and 3+ ions. This is reasonable, since their ionization energies are in the same range as those of main group metals. To achieve higher charges requires much more energy and is not commonly observed. Finally, with the exception of Sc, Y, La and possibly titanium, the metal ions do not achieve a rare gas configuration, since this would require the loss of too many electrons and an impossibly high energy cost.

From all this, you see that only atoms at the left and right sides of the periodic table clearly form simple monatomic ions. Moving to the center of the table, the tendency is first to form more complex ions such as NH_4^+, PO_4^{3-}, and SO_4^{2-}.

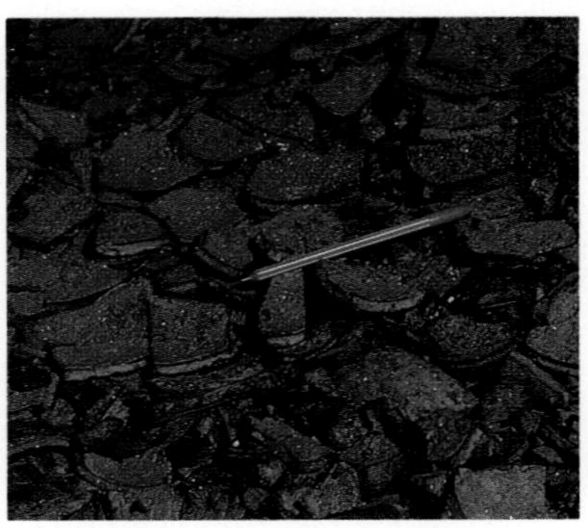

Figure 9.19 The manganese in MnO_2 has an apparent charge of 4+, a charge that is stabilized by interaction with the O^{2-} ion. This is a photo of MnO_2 coated with red clay in Jewel Cave, Wind Cave National Park, Hot Springs, South Dakota. These deposits of MnO_2 are at least three feet thick in some parts of the cave. (Arthur Palmer)

SOMETHING MORE ABOUT
The Origin of the Elements

This great saga of cosmic evolution, to whose truth the majority of scientists subscribe, is the product of an act of creation that took place about twenty billion years ago. Science, unlike the Bible, has no explanation for the occurrence of that extraordinary event.

R. Jastrow, *Until the Sun Dies*, Norton, New York, 1977.

By now you have seen some of the chemistry of the elements and their compounds in the laboratory, and you have begun to connect these to chemical processes in the world around you. But where did the elements originate? Why are there so few of them? Why are carbon-based compounds so numerous? Why does the hemoglobin of your blood contain iron and not ruthenium or uranium? Are there elements somewhere in the universe that are not found on earth? These are all very interesting questions, and we are most fortunate in now having some answers!

The composition of a number of stars in the universe and of the planets and moons of our solar system has been determined. In one sense it is a great relief, and in another a disappointment, to find that there is no evidence for an element found in some distant star that does not exist on earth. Further, although there are variations from star to star and planet to planet, the relative abundances of the elements throughout the universe are approximately the same. These are very important observations, since they mean that the element-forming processes are general throughout the universe.

The relative abundances of a few of the elements in the universe are illustrated in Figure A. The most striking features of this figure are those that follow and are the facts that must be taken into account in any theory of the origin of the elements.

(a) 90% to 95% of the atoms in the universe are hydrogen atoms.
(b) 4% to 9% of all atoms are helium.
(c) All of the other elements taken together make up only about 1% of the universe, even on a weight basis.
(d) Lithium, beryllium, and boron are mysteriously rare.
(e) Elements of even atomic number are more abundant than those with odd atomic number.
(f) There is a general decline in abundance from oxygen to lead. However, there is a very pronounced maximum in relative abundance around iron.
(g) There are no stable elements with mass numbers greater than about 210.

The universe is thought to have begun about 20 billion years ago as a subatomic particle soup (electrons, positrons or positive electrons, protons, neutrons, and massless particles such as neutrinos and antineutrinos and photons in a ball recently estimated to be only about 10^{-28} cm in diameter). The "big bang" theory hypothesizes that an explosion of unimaginable fury began the series of events that has led ultimately to the universe and to life on earth.

Within seconds of the big bang, the explosion of the super-dense ball of particles, the synthesis of the elements began. Free neutrons are known to decay to give protons, electrons, and a large amount of energy.

$$\text{Neutron} \longrightarrow \text{proton} + \text{electron} + \text{energy}$$

Since a hydrogen atom has a nucleus consisting of a single proton, this is the origin of the hydrogen in our universe. To continue the process of element for-

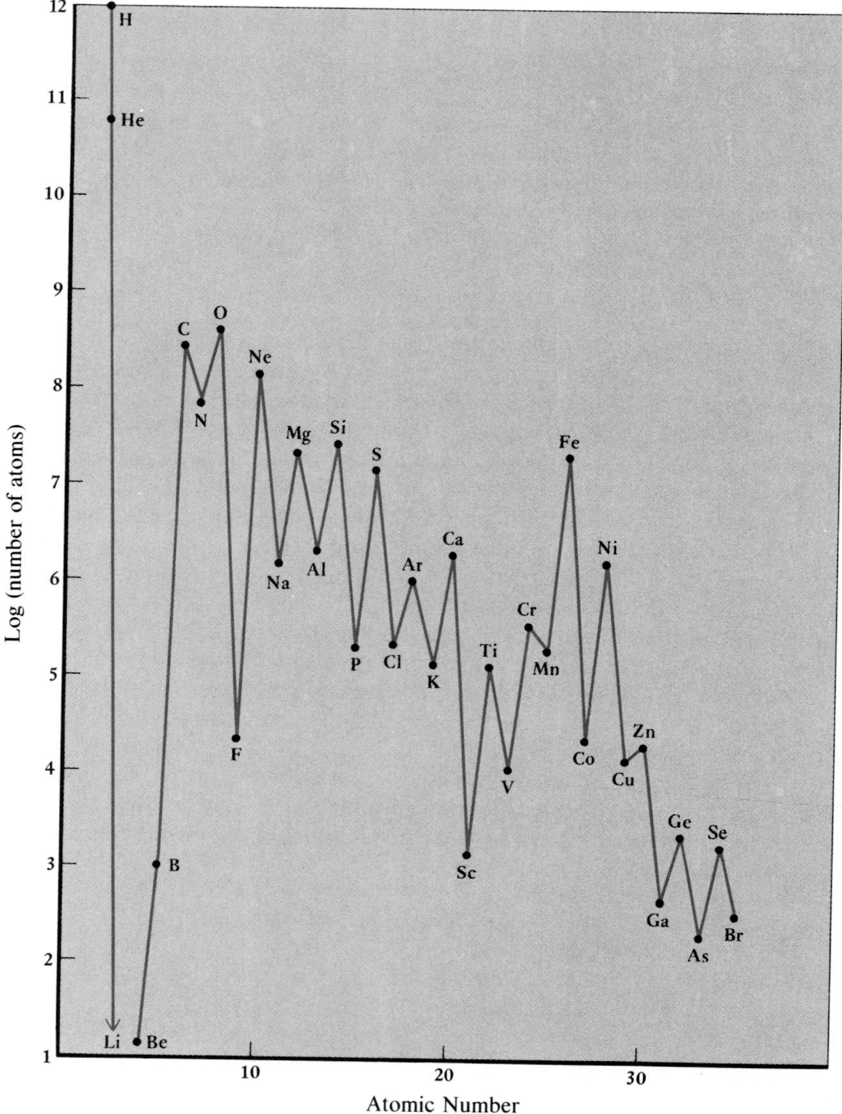

Figure A The cosmic abundances of the lighter elements as a function of atomic number. Abundances are expressed as numbers of atoms per 10^{12} atoms of H and are plotted on a (base-10) logarithmic scale. (Data taken from G.O. Abell, *Exploration of the Universe*, 4th ed., Saunders College Publishing, Philadelphia, 1982, p. 706.)

mation, hydrogen nuclei fused in the high temperature furnace of the big bang and continue to do so in stars, to give helium nuclei, the next most important building block of the elements.

In the seconds, minutes, and hours following the moment of Creation, the gas-filled, chaotic universe rapidly expanded and began to cool. After perhaps a billion years, it had cooled to the point that the hydrogen and helium began to coalesce into galaxies and, within the galaxies, into stars, a process that continues even today. The primordial stars contracted under the force of gravitational attraction, and the temperature began to climb, finally reaching about 10 million degrees, the critical temperature for "hydrogen burning" or fusion to form more helium. This exothermic reaction raises the temperature of the star, and helium nuclei can begin to coalesce at a hundred million degrees or so into nuclei of still heavier elements. In the "helium-burning" cycle, it is the carbon nucleus that is first formed (Figure B). The direct formation of carbon from helium means that the process has circumvented lithium, beryllium, and boron and explains their

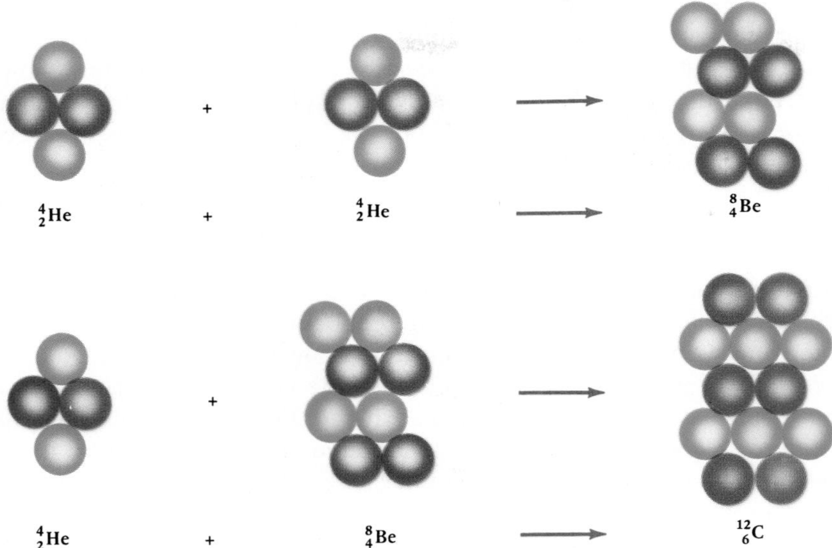

Figure B Hydrogen fusion continues for 99% of the lifetime of a star. When the hydrogen is essentially depleted, the star collapses until its core reaches a temperature in excess of 100 million degrees Celsius. At this temperature, helium nuclei begin to fuse to produce carbon nuclei. The process is thought to proceed through an unstable $^{8}_{4}Be$ isotope as an intermediate.

very low abundance in the universe. The fact that these elements exist, however, must mean that they are formed in a different way than by the fusion of helium nuclei.

Our sun is a relatively small star, since ones of much larger mass, say 10 to 30 solar masses, have been observed. Because of their very great mass, the core temperatures of large stars must become very great, and fusion reactions to form heavier nuclei than carbon are possible. Heavier and heavier elements are built up, in a process involving helium, $^{4}_{2}He$, similar to that in Figure B. Since helium is an element of "even mass," this helps explain the observation that elements of even mass are more abundant than those of odd mass.

Energy is evolved when light nuclei coalesce or fuse to form nuclei lighter than iron. However, iron formation is the end of this exothermic series! To continue the fusion process requires energy. When a massive star has essentially consumed its fuel and iron has been formed, the star can no longer support its outer layers against gravity, and the core collapses to a super-dense state *in less than a second*. The shock waves from this collapse expand outward through the envelope of gases surrounding the core, heating the gases and providing the energy to create heavy nuclei. This outward blast is seen as a gigantic explosion called a supernova, many of which have been observed. The explosion sends debris rocketing into space, where over additional millions of years it coalesces with hydrogen, helium, and other elements to form new stars. Our sun is just such a "second-generation" star, and our earth and the other planets of the solar system were formed from the debris of the explosions of massive stars. It is for this reason that there is a wide variety of elements found on the earth and other planets.

For a recent article on the synthesis of the elements in stars see S. Woosley and T. Weaver, *Scientific American*, August, 1989, pp. 32–40.

SUMMARY

In the previous chapter three quantum numbers (n, ℓ, and m_ℓ) were used to describe an atomic orbital. To describe an electron completely, however, a fourth quantum number, the electron spin magnetic quantum number, m_s, is needed (Section 9.1). There can be only two values of m_s for a single electron, $+\frac{1}{2}$ or $-\frac{1}{2}$. Each electron in an atom can now be addressed specifically, since *no two electrons in the same atom can have the same set of four quantum numbers* (the **Pauli exclusion principle**) (Section 9.2). This leads to the further conclusion that no atomic orbital can be assigned to more than two electrons and that the two electrons assigned to an orbital must have opposite spins (different values of m_s). Atoms, ions or compounds with unpaired electrons are **paramagnetic**, that is, they have the property of being attracted to a magnetic field because the electron spins in the atom, molecule, or ion align with the external field (like the needle of a compass in the earth's magnetic field) (Section 9.1). If the spins remain aligned when the paramagnetic substance is removed from the external field, it is said to be **ferromagnetic**. (**Diamagnetic** atoms, ions, or molecules are repelled from a magnetic field.)

A major goal of this chapter is to understand the distribution of electrons in atoms—atomic electron configurations—and their connection with the periodic table (Sections 9.3 and 9.4). Generally, electrons are assigned to orbitals of an atom in order of increasing orbital energy. In the H atom the orbital energies increase with increasing n, but, in a many-electron atom the orbital energies depend on both n and ℓ (Figure 9.7). Many complications can arise, particularly in elements where d and f orbital energies can overlay the energy of the s orbital of the next higher quantum shell (called the *dfs* overlay region in Figure 9.7).

In assigning electrons to atomic orbitals, in addition to the Pauli exclusion principle, **Hund's rule** is followed: The most stable arrangement of electrons in a set of orbitals all having the same energy (three p orbitals, five d orbitals, or seven f orbitals) is the one with the maximum number of unpaired electrons, all electrons having the same spin direction.

Using the periodic table as a guide, electron configurations of the elements and monatomic ions can be depicted by an **orbital box notation** or a **spectroscopic notation**. (In both cases, configurations can be abbreviated with the **rare gas notation**.)

Dmitrii Mendeleev founded the modern concept of chemical periodicity (Historical Figures in Chemistry: Mendeleev and the Periodic Table). Mendeleev realized that many elements remained undiscovered in the 19th century and predicted their properties. Moseley recognized the true **law of chemical periodicity**: the properties of the elements are periodic functions of their atomic numbers.

Important physical properties of the elements are **size**, **ionization energy** (*IE*, the energy required to remove an electron from a gaseous atom), and **electron affinity** (*EA*, the energy change when a gaseous atom absorbs an electron to form an anion) (Section 9.6). Their general periodic trends are:

(a) Atomic size: decreases across a period and increases down a group.
(b) *IE*: increases across a period and decreases down a group.

(c) *EA*: becomes more negative across a period and less negative down a group. (That is, the affinity for an electron increases across a period and decreases down a group.)

Ion size is also important (Section 9.6). Negative ions are larger than the elements from which they are derived, while positive ions are smaller. Trends in ion sizes parallel those of neutral atoms. Ion charges can be rationalized in terms of *IE* and *EA* (Section 9.7).

STUDY QUESTIONS

REVIEW QUESTIONS

1. Give the four quantum numbers, specify their allowed values, and tell what property of the electron they describe.
2. What is the Pauli exclusion principle?
3. Using lithium as an example, show the two methods of depicting electron configurations (orbital box diagram and spectroscopic notation).
4. What is Hund's rule? Give an example of the use of this rule.
5. What is the rare gas notation? Write an electron configuration using this notation.
6. Name an element of Group 3A (Group 13). What does the group designation tell you about the electron configuration of the element?
7. Name an element of Group 6B (Group 6). What does the group designation tell you about the electron configuration of the element?
8. What element is located in the fourth period of Group 4A (Group 14)? What does the element's location tell you about its electron configuration?
9. What element is located in the fifth period of Group 5B (Group 5)? What does the element's location tell you about its electron configuration?
10. What was Mendeleev's major contribution to the development of the concept of periodicity?
11. Tell what happens to atomic size, ionization energy, and electron affinity when proceeding across a period and down a group.
12. Explain how the sizes of atoms change and why they change when proceeding across a period of the periodic table.
13. Explain how the ionization energies of atoms change and why the change occurs when proceeding down a group of the periodic table.
14. Explain why the sizes of transition metal atoms are nearly identical across a period.
15. Write electron configurations to show the first two ionization processes for potassium. Explain why the second ionization energy is much higher than the first.
16. Explain why the ionization energies of Si, P, and S are in the order Si < P but P > S.
17. Why is the radius of Li^+ so much smaller than the radius of Li? Why is the radius of F^- so much larger than the radius of F?
18. What is the origin of the paramagnetic effect? Is Li^+ paramagnetic? What about Ti^{2+}? Is either of these diamagnetic?
19. Which ions in the following list are likely to be formed: K^{2+}, Cs^+, Al^{4+}, F^{2-}, and Se^{2-}? Do any of these ions have a rare gas configuration?

WRITING ELECTRON CONFIGURATIONS

20. Write electron configurations for Mg and Cl using both the spectroscopic notation and orbital box diagrams.
21. Write electron configurations for Al and S using both the spectroscopic notation and orbital box diagrams.
22. Using the spectroscopic notation, give the electron configuration of vanadium, V. (The name of the element was derived from Vanadis, a Scandinavian goddess.) Compare your answer with Table 9.2.
23. Using the spectroscopic notation, write the electron configurations of chromium and iron.
24. Germanium had not been discovered when Mendeleev formulated his ideas of chemical periodicity. He predicted its existence, however, and it was indeed found in 1886 by Winkler. Depict its electron configuration in the spectroscopic notation.
25. Give the electron configuration for the rare gas element krypton in the spectroscopic notation.
26. Using the spectroscopic and rare gas notations, write electron configurations for the following:
   (a) Strontium, Sr, named for a town in Scotland.
   (b) Zirconium, Zr. This metal is exceptionally resistant to corrosion and so has important industrial applications. Moon rocks show a surprisingly high zirconium content compared with rocks on earth.
   (c) Rhodium, Rh, used in jewelry and in catalysts in industry.
   (d) Tin, Sn. This metal was used in the ancient world. Alloys of tin (solder, bronze, and pewter) are important.

27. Use the rare gas and spectroscopic notations to show electron configurations for the following metals of the third transition series. (Recall that there can be minor differences between predicted and actual configurations.)
(a) Tungsten, W. The element finds extensive use in the filaments of electric lamps and television tubes. It has the highest melting point of all the elements.
(b) Platinum, Pt, used by pre-Columbian Indians in jewelry. It does not oxidize in air, no matter how high the temperature. Therefore, it is used to coat missile nose cones and in jet engine fuel nozzles.

28. The lanthanides, or rare earths, are now only "medium rare." All can be purchased for reasonable prices. Using the rare gas notation and spectroscopic notation, depict electron configurations for the following elements. (Recall that there can be minor differences between predicted and actual configurations.)
(a) Europium, Eu. It is the most expensive of the rare earth elements; one gram can be purchased for \$50 to \$100.
(b) Ytterbium, Yb. It is less expensive than Eu, as Yb costs only about \$15 per gram. It was named for the village of Ytterby in Sweden, where a mineral source of the element was found.

29. The actinide americium (Am) is a man-made, radioactive element that has found use in home smoke detectors. Using the rare gas notation and spectroscopic notation, depict its electron configuration.

30. Depict electron configurations for the following elements of the actinide series of elements. Use the rare gas and spectroscopic notations. (Recall that there can be minor differences between predicted and actual configurations.)
(a) Plutonium, Pu. It is one of the most dangerous radiological poisons known because of the high rate at which it evolves alpha particles. It is best known as the explosive material in nuclear weapons and as a by-product of nuclear power plant operation.
(b) Einsteinium, Es. The element was named for the famous physicist Albert Einstein.

31. Among the last elements of the periodic table are those with atomic numbers 104 through 109. Element 104 was originally given the name rutherfordium, Rf, to honor the physicist Rutherford (Chapter 7). Depict its electron configuration using the spectroscopic and rare gas notations.

32. Using orbital box diagrams, depict the electron configurations of the following ions: (a) Na^+, (b) Al^{3+}, and (c) Cl^-.

33. Using orbital box diagrams, depict the electron configurations of the following ions: (a) Mg^{2+}, (b) Si^{4+}, and (c) O^{2-}.

34. Using orbital box diagrams and the rare gas notation, depict the electron configurations of: (a) Ti, (b) Ti^{2+}, and (c) Ti^{4+}. Is either of the ions paramagnetic?

35. Using orbital box diagrams and the rare gas notation, depict the electron configurations of: (a) V, (b) V^{2+}, and (c) V^{5+}. Is either of the ions paramagnetic?

36. Element 25 can be found at the bottom of the sea in the form of oxide "nodules."
(a) Depict the electron configuration of this element using the spectroscopic notation and an orbital box diagram.
(b) Using an orbital box diagram, show the electrons beyond those of the preceding rare gas for the 2+ ion.
(c) Is the 2+ ion paramagnetic?

37. Cobalt commonly exists as 2+ and 3+ ions. Using orbital box diagrams and the rare gas notation, show electron configurations of these ions. Are these ions paramagnetic?

38. Ruthenium, whose compounds are used as catalysts in chemical reactions, has an electron configuration that does not fit the expected pattern (see Table 9.2).
(a) Based on its position in the periodic table, depict the electron configuration of Ru using the rare gas notation along with spectroscopic and orbital box notations. How does your predicted configuration differ from the actual configuration in Table 9.2?
(b) Using an orbital box notation (and the rare gas notation), depict the electron configuration of ruthenium(III). Can you arrive at the same ion configuration from either the actual or the predicted configuration of the element?

39. Pt^{2+} is the central ion in cisplatin, $(NH_3)_2PtCl_2$, a reagent for treating cancer. Platinum has an electron configuration that does not fit the expected pattern (see Table 9.2).
(a) On the basis of its position in the periodic table, depict the electron configuration of Pt using the rare gas notation along with spectroscopic and orbital box notations. How does your predicted configuration differ from the actual configuration in Table 9.2?
(b) Using an orbital box notation (and the rare gas notation), depict the electron configuration of platinum(II). Can you arrive at the same ion configuration from either the actual or the predicted configuration of the element?

40. The rare earths or lanthanides commonly exist as 3+ ions. Using an orbital box diagram, and the rare gas notation, show the electron configurations of the following:
(a) Sm and Sm^{3+} (samarium)
(b) Ho and Ho^{3+} (holmium)

41. Suppose a new element with atomic number 113 has been discovered.

(a) Predict its electron configuration using the rare gas and spectroscopic notations.
(b) Name another element in the periodic group in which this new element would be placed.

42. How many unpaired electrons does Ti^{2+} have? The Co^{3+} ion? Is either of these paramagnetic?
43. Which of the following ions are paramagnetic and which are diamagnetic: Al^{3+}, Ba^{2+}, V^{3+}?
44. Are any of the 2+ ions of the elements Ti through Zn diamagnetic? Which 2+ ion has the greatest number of unpaired electrons?
45. Two elements in the first transition series (Sc through Zn) have four unpaired electrons in their 2+ ions. What elements fit this description?

ELECTRON CONFIGURATIONS AND QUANTUM NUMBERS

46. Depict the electron configuration for Be using the orbital box method. Give a complete set of four quantum numbers for each of the electrons in this atom.
47. Depict the electron configuration for silicon using the orbital box and rare gas notations. Give one possible set of four quantum numbers for each of the electrons beyond those of the preceding rare gas.
48. Using an orbital box diagram and the rare gas notation, show the electron configuration of titanium. Give one possible set of four quantum numbers for each of the electrons beyond those of the preceding rare gas.
49. Using an orbital box diagram, show the electron configuration of gallium, Ga, beyond the preceding rare gas. Give a set of quantum numbers for the *last* electron added.
50. What is the maximum number of electrons that can be identified with each of the following sets of quantum numbers? In some cases, the answer may be "none." In such cases, explain why "none" is the correct answer.
(a) $n = 2, \ell = 1$
(b) $n = 5$
(c) $n = 4, \ell = 4$
(d) $n = 3, \ell = 1, m_\ell = -1, m_s = -\frac{1}{2}$
(e) $n = 3, \ell = 0, m_\ell = +1$
51. What is the maximum number of electrons that can be associated with the following sets of quantum numbers? In one case, the answer is "none." Explain why this is true.
(a) $n = 4, \ell = 3$
(b) $n = 6, \ell = 1, m_\ell = -1$
(c) $n = 3, \ell = 3, m_\ell = -3$
(d) $n = 2, \ell = 1, m_\ell = 1, m_s = +\frac{1}{2}$
52. Explain briefly why each of the following is *not* a possible set of quantum numbers for an electron in an oxygen atom. In each case, change the incorrect value (or values) in some way to make the set valid.
(a) $n = 2, \ell = 2, m_\ell = 0, m_s = +\frac{1}{2}$
(b) $n = 2, \ell = 1, m_\ell = -1, m_s = 0$
53. Explain briefly why each of the following is *not* a possible set of quantum numbers for an electron in an oxygen atom. In each case, change the incorrect value (or values) in some way to make the set valid.
(a) $n = 4, \ell = 2, m_\ell = 0, m_s = 0$
(b) $n = 2, \ell = 1, m_\ell = -3, m_s = -\frac{1}{2}$
54. A possible excited state for the H atom has an electron in a 4*p* orbital. List all possible sets of quantum numbers (n, ℓ, m_ℓ, and m_s) for this electron.
55. Write a complete set of quantum numbers for each of the electrons beyond the nearest rare gas for scandium (Sc).

PERIODIC PROPERTIES

56. Use the data in Figure 9.11 to estimate E—Cl bond distances when E is a Group 4A element.
57. Estimate the Xe—F bond distance in XeF_2 from the information in Figure 9.11. (Known Xe—F distances are in the range of 190 pm.)
58. Arrange the following elements in order of increasing size: Al, B, C, K, and Na. (Try doing it without looking at Figure 9.11, then check yourself by looking up the necessary atomic radii.)
59. Arrange the following elements in order of increasing size: Ca, Rb, P, Ge, and Sr. (Try doing it without looking at Figure 9.11, then check yourself by looking up the necessary atomic radii.)
60. Select the atom or ion in each pair that has the larger radius.
(a) Cl or Cl^-
(b) Al or N
(c) In or Sn
61. Select the atom or ion in each pair that has the larger radius.
(a) Cs or Rb
(b) O^{2-} or O
(c) Br or As
62. Which of the following groups of elements is arranged correctly in order of increasing ionization energy?
(a) C < Si < Li < Ne
(b) Ne < Si < C < Li
(c) Li < Si < C < Ne
(d) Ne < C < Si < Li
63. Arrange the following atoms in the order of increasing ionization energy: F, Al, P, and Mg.
64. Arrange the following atoms in the order of increasing ionization energy: Li, K, C, and O.
65. Arrange the following atoms in the order of increasing ionization energy: Si, K, As, and Ca.
66. Rank the following ionization energies in order from the smallest value to the largest value. Briefly explain your answer.

(a) first *IE* of Be
(b) first *IE* of Li
(c) second *IE* of Be
(d) second *IE* of Na
(e) first *IE* of K

67. Predict which of the following elements would have the greatest difference between the first and second ionization energy: Si, Na, P, and Mg. Briefly explain your answer.

68. Compare the elements Li, K, C, and N.
(a) Which has the largest atomic radius?
(b) Which has the most negative electron affinity?
(c) Place the elements in order of increasing ionization energy.

69. Compare the elements B, Al, C, and Si.
(a) Which has the most metallic character?
(b) Which has the largest atomic radius?
(c) Which has the most negative electron affinity?
(d) Place the three elements B, Al, and C in order of increasing first ionization energy.

70. Use periodic trends to explain and answer briefly.
(a) Place the following elements in order of increasing ionization energy: F, O, and S.
(b) Which has the largest ionization energy: O, S, or Se?
(c) Which has the most negative electron affinity: Se, Cl, or Br?
(d) Which has the largest radius: O^{2-}, F^-, or F?

71. Use periodic trends to explain and answer briefly.
(a) Rank the following in order of increasing atomic radius: O, S, and F.
(b) Which has the largest ionization energy: P, Si, S, or Se?
(c) Place the following in order of increasing radius: Ne, O^{2-}, N^{3-}, or F^-?
(d) Place the following in order of increasing ionization energy: Cs, Sr, Ba.

72. What periodic groups of elements can form 2+ ions? (Consider only main group elements.)

73. What periodic groups of elements can form 3+ ions? (Consider only main group elements.)

GENERAL QUESTIONS

74. Element 109 is the latest discovered. It was produced in August 1982, by a team at Germany's Institute for Heavy Ion Research. Depict its electron configuration using the spectroscopic and rare gas notations. Name another element found in the same group as 109.

75. A neutral atom has two electrons with $n = 1$, eight electrons with $n = 2$, eight electrons with $n = 3$, and one electron with $n = 4$. Assuming this element is in its ground state, supply the following information: (a) atomic number; (b) total number of *s* electrons; (c) total number of *p* electrons; and (d) total number of *d* electrons.

76. Which of the following is *not* an allowable set of quantum numbers? Explain your answer briefly.

	n	ℓ	m_ℓ	m_s
(a)	2	0	0	$-\frac{1}{2}$
(b)	1	1	0	$+\frac{1}{2}$
(c)	2	1	−1	$-\frac{1}{2}$
(d)	6	5	5	$-\frac{1}{2}$

77. In the fictitious element with atomic number 120, what is the last orbital assigned?

78. How many complete electron shells are there in element 71?

79. What is the lightest element in which the third ($n = 3$) quantum shell is completely filled?

80. How many *p* orbital electron pairs are there in selenium, Se?

81. Which of the following sets of quantum numbers could describe the outer shell electrons in the ground state of an atom? Which set(s) does (do) *not* obey the quantum number rules?

ELECTRON	n	ℓ	m_ℓ	m_s
(a) First	1	1	1	$-\frac{1}{2}$
Second	1	1	1	$-\frac{1}{2}$
(b) First	2	0	0	$+\frac{1}{2}$
Second	2	0	0	$-\frac{1}{2}$
Third	2	1	−1	$+\frac{1}{2}$
Fourth	2	1	−1	$+\frac{1}{2}$
(c) First	2	0	0	$+\frac{1}{2}$
Second	2	0	0	$-\frac{1}{2}$

82. The ionization energies for the removal of the first electron in Si, P, S, and Cl are listed below. Briefly rationalize this trend.

ELEMENT	FIRST IONIZATION ENERGY (kJ/mol)
Si	786
P	1012
S	1000
Cl	1251

83. Name the element corresponding to each of the following characteristics:
(a) The element with the electron configuration $1s^22s^22p^63s^23p^4$.
(b) The element in the alkaline earth group that has the largest atomic radius.
(c) The element in Group 5A that has the largest ionization energy.
(d) The element whose 2+ ion has the configuration $[Kr]4d^6$.
(e) The element with the lowest electron affinity in Group 7A.
(f) The element whose electron configuration is $[Ar]3d^{10}4s^1$.

84. Two elements in the second transition series (Y through Cd) have four unpaired electrons in their 3+ ions. What elements fit this description?

85. Answer the following questions about the elements A and B, which have these electron configurations:

$A = \ldots 4p^6 5s^1 \qquad B = \ldots 3p^6 3d^{10} 4s^2 4p^4$

(a) Is element A a metal, nonmetal, or metalloid?
(b) Which element would have the greater ionization energy?
(c) Which element should have the more negative electron affinity?
(d) Which element has larger atoms?

86. Answer the following questions about the elements with these electron configurations:

$A = \ldots 3p^6 4s^2 \qquad B = \ldots 3p^6 3d^{10} 4s^2 4p^5$

(a) Is element A a metal, metalloid, or nonmetal?
(b) Is element B a metal, metalloid, or nonmetal?
(c) Which element is expected to have the larger ionization energy?
(d) Which element would be the smaller of the two?

87. Which of the following ions are unlikely and why: Cs^+, In^{4+}, Fe^{6+}, Te^{2-}, Sn^{5+}, and I^-?

88. Place the following elements and ions in order of decreasing size: K^+, Cl^-, S^{2-}, and Ca^{2+}.

89. Rank the following in order of increasing ionization energy: Zn, Ca, Ca^{2+}, and Cl^-. Briefly explain your answer.

90. In general, as you move across a periodic table, the electron affinity of the elements becomes more negative. One exception to this trend, however, occurs when going from Group 4A elements to those in Group 5A. Refer to orbital box diagrams to explain why this exception is plausible and indeed expected.

91. Answer each of the following questions.

(a) Of the elements O, S, and F, which has the largest atomic radius?
(b) Which is larger, Cl or Cl^-?
(c) Which should have the largest difference between the first and second ionization energies: Si, Na, P, or Mg?
(d) Which has the largest ionization energy: O, S, or Se?
(e) Which of the following has the largest radius: Ne, O^{2-}, N^{3-}, or F^-?

92. Explain why the first ionization energy of Mg is greater than that of Na, whereas the second ionization energy of Mg is lower than the second ionization energy of Na.

93. The following are isoelectronic species: Cl^-, S^{2-}, and K^+. Rank them in order of increasing (a) size, (b) ionization energy, and (c) electron affinity.

94. Compare the elements Na, B, Al, and C with regard to the following properties.
(a) Which has the largest atomic radius?
(b) Which has the most negative electron affinity?
(c) Place the elements in order of increasing ionization energy.

95. Suppose a new element, tentatively given the symbol Et, has just been discovered. Its atomic number is 113.
(a) Depict the electron configuration of the element using the spectroscopic notation.
(b) Name another element you would expect to find in the same group as Et.
(c) Give the formulas for the compounds of Et with O and Cl.

SUMMARY QUESTION

96. When sulfur dioxide reacts with chlorine, the products are thionyl chloride, $OSCl_2$, and dichlorine oxide, Cl_2O.

$$SO_2(g) + 2\,Cl_2(g) \longrightarrow OSCl_2(g) + Cl_2O(g)$$

(a) In what period of the periodic table is S located?
(b) Give the electron configuration of S in the orbital box notation. Do not use the rare gas notation.
(c) Using your configuration from (b), give a set of quantum numbers for the last electron you assigned for S.
(d) What element involved in this reaction (O, S, Cl) should have the smallest ionization energy? The smallest radius?
(e) Which should be smaller, the sulfide ion, S^{2-}, or the sulfur atom, S?
(f) If you want to make 675 g of $OSCl_2$, how many grams of Cl_2 are required?
(g) If you use 10.0 g of SO_2 and 20.0 g of Cl_2, what is the theoretical yield of $OSCl_2$?
(h) The reaction as written *requires* 164.6 kJ per mole of $OSCl_2$ produced. That is, $\Delta H°$ for the reaction is +164.6 kJ. Using this information and the table below, calculate the heat of formation of $OSCl_2$.

COMPOUND	ΔH_f° (kJ/mol)
Cl_2O	80.3
SO_2	−296.8

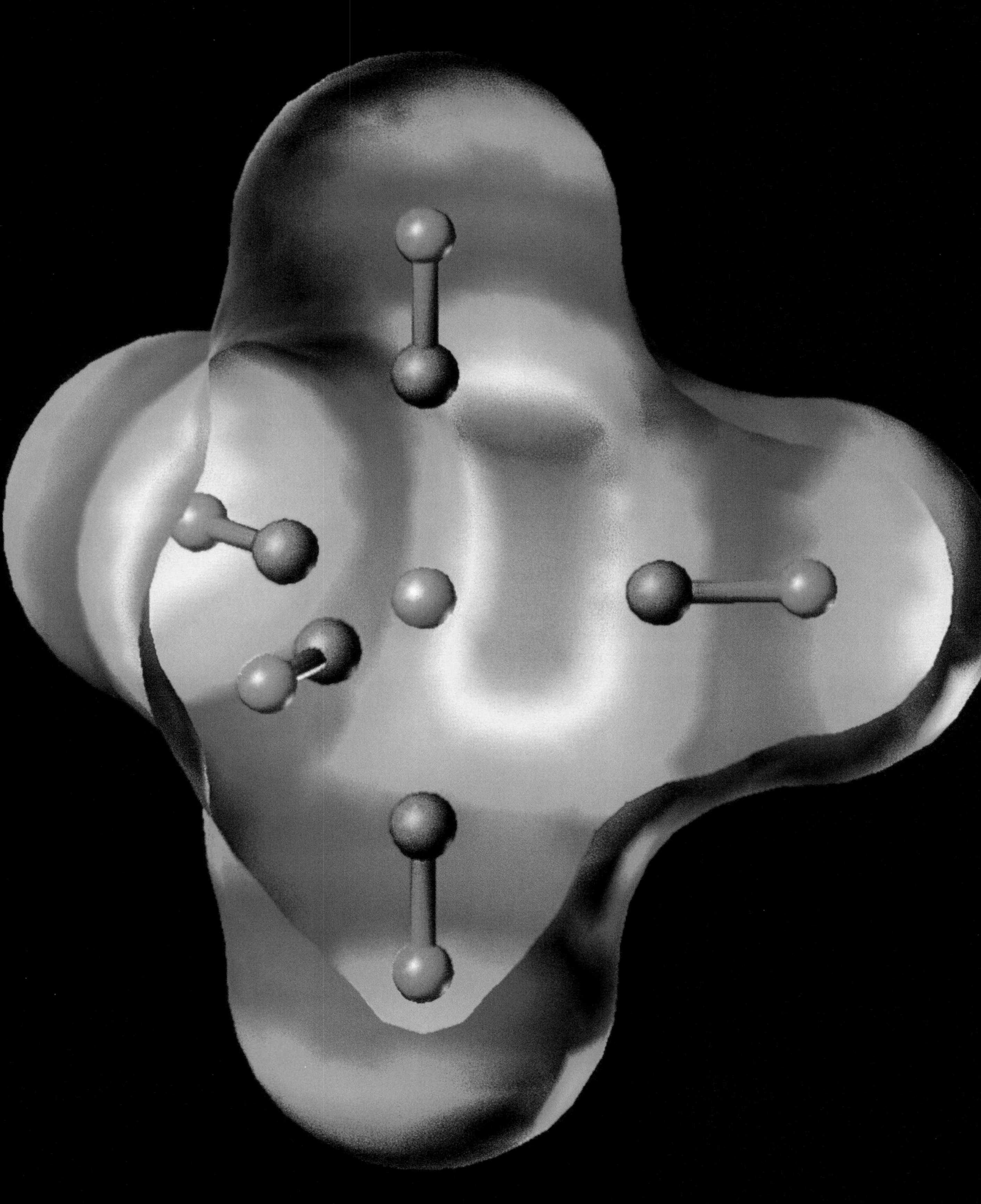

CHAPTER

10

Basic Concepts of Chemical Bonding and Molecular Structure

A computer-generated ball-and-stick model of $Fe(CO)_5$ surrounded by a model of the molecular surface (gray = C, red = O, and violet = Fe). The most reactive zones of the molecules are given in red on the molecular surface. (J. Weber)

CHAPTER OUTLINE

Chemists study molecules, and one of the thousands that has been of particular interest in the past ten years or so has been the cisplatin molecule that Rosenberg discovered to be an effective agent to treat certain types of cancers (pages 2 and 185).

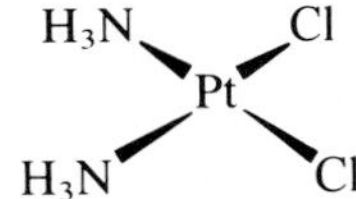

cisplatin (a square plane)

N, H H H

ammonia (a pyramid)

This molecule has two chloride ions and two ammonia molecules arranged at the corners of a square around a central Pt^{2+} ion, with both Cl^- ions on one side of the square rather than at opposite corners. The ammonia molecules that are attached to the central Pt^{2+} are shaped like pyramids, with H atoms at the base of the pyramid and the central N atom at the peak attached to the metal ion. When a chemist thinks of this molecule, this is the view he or she has in mind. But why is it important to know what the molecule looks like? Because the molecule with the alternate structure, where the two Cl^- ions are at opposite corners of the square,

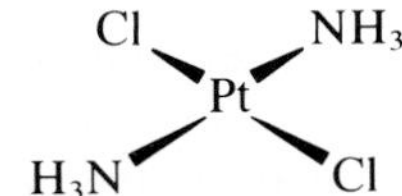

alternate structure for $Pt(NH_3)_2Cl_2$

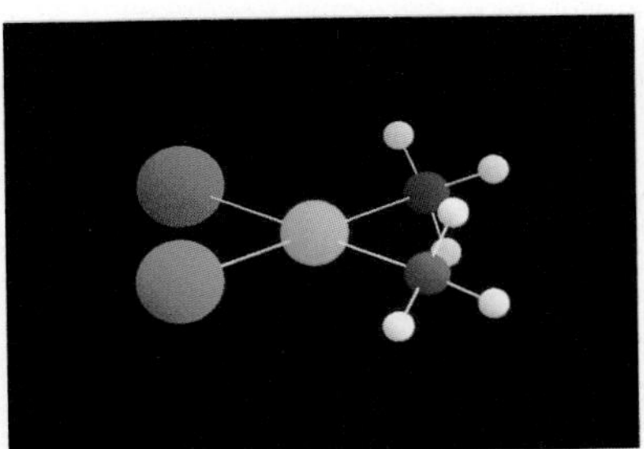

A computer-generated representation of the structure of cisplatin clearly shows that the two Cl^- ions, the central Pt^{2+} ion, and the N atoms of the NH_3 molecules are all in the same plane. The NH_3 molecules themselves are shaped like pyramids. (J. Weber)

does not work at all in cancer treatment. Why? That's what chemists and physicians are still trying to understand.

In Chapters 8 and 9 you began to develop an understanding of what we think electrons do in atoms. It is now time to investigate what happens when two or more atoms—such as Pt and Cl or N and H—come close enough to form molecules and how this behavior explains the three-dimensional shapes of molecules.

10.1 VALENCE ELECTRONS

The outermost electrons of an atom, the ones most affected by the approach of another atom, are called **valence electrons**. The rare gas core electrons and the filled *d* shell electrons of the Group 3A–7A elements are not greatly affected by reactions with other atoms, so we focus our attention on the behavior of the outer *ns* and *np* electrons (and *d* electrons in unfilled subshells of the transition metals). This means the valence electrons for a few typical elements are

ELEMENT	CORE ELECTRONS	VALENCE ELECTRONS	PERIODIC GROUP
Na	$1s^22s^22p^6$	$3s^1$	1A
Si	$1s^22s^22p^6$	$3s^23p^2$	4A
Ti	$1s^22s^22p^63s^23p^6$	$4s^23d^2$	4B
As	$1s^22s^22p^63s^23p^63d^{10}$	$4s^24p^3$	5A

Here you see that *the number of valence electrons of each element is equal to the group number*. The fact that every element in a given group has the same number of valence electrons accounts for the similarity of chemical properties among members of the group.

A useful device for keeping track of the valence electrons of *main group elements* is the **Lewis electron dot symbol**, first suggested by G.N. Lewis about 1916. In this notation, the nucleus and core electrons are represented by the atomic symbol. The valence electrons, represented by dots, are then placed around the symbol one at a time until they are used up or until all four sides are occupied; any remaining electrons are paired with the ones already there. The Lewis symbols for the second and third periods are

The two ns electrons of the Group 2A through 8A atoms are paired in the atom's electron configurations, but the Lewis symbol shows them as unpaired. This will be useful later.

1A	2A	3A	4A	5A	6A	7A	8A
ns^1	ns^2	ns^2np^1	ns^2np^2	ns^2np^3	ns^2np^4	ns^2np^5	ns^2np^6
·Li	·Be·	·B· (with dot above)	·C· (dots above and below)	·N· (dots above and below, paired above)	:O· (dots above and below)	:F: (dots above and below)	:Ne: (dots above and below)
·Na	·Mg·	·Al· (with dot above)	·Si· (dots above and below)	·P· (dots above and below, paired above)	:S· (dots above and below)	:Cl: (dots above and below)	:Ar: (dots above and below)

The Lewis symbol emphasizes the rare gas configuration, ns^2np^6, as a stable, low-energy state. In fact, as you will see below the bonding behavior of the main group elements can often be considered the result of gaining, losing, or sharing valence electrons to achieve the same configuration as the nearest rare gas. Since all rare gases (except He) have eight valence electrons, this observation is called the **octet rule**. It is only a guideline, so there are many exceptions, but it provides an easy way to predict the results of the most common reactions.

Historical Figures in Chemistry: Gilbert Newton Lewis (1875–1946)

In a paper published in the *Journal of the American Chemical Society* in 1916, Gilbert N. Lewis introduced the theory of the shared electron pair chemical bond and revolutionized chemistry. It is to honor this contribution that we often refer to "electron dot" structures as Lewis structures. However, he also made major contributions to other fields such as thermodynamics, isotope studies, and the interaction of light with substances. Of particular interest in this text is the extension of his theory of bonding to a *generalized theory of acids and bases*. This theory, often referred to as Lewis acid–base theory, will be described in Chapter 17.

G.N. Lewis was born in Massachusetts but raised in Nebraska. After earning his B.A. and Ph.D. at Harvard University, he began his academic career. In 1912 he was appointed Chairman of the Chemistry Department at the University of California, Berkeley, and he remained there for the rest of his life. Lewis felt that a chemistry department should both teach and advance fundamental chemistry, and he was not only a productive researcher but also a teacher who profoundly affected his students. Among his ideas was the use of large problem sets in teaching, an idea that is much in use today.

10.2 CHEMICAL BOND FORMATION

Bond formation can be approached on two levels. The simpler approach goes no further than recognizing the repulsion between like electrical charges and the attraction between opposite charges (see Chapter 3, page 75). When two atoms approach each other closely enough for their electron clouds to interpenetrate, the electrons of one atom repel the electrons of the other, and the same happens with the nuclei of the two atoms. At the same time, however, each atom's electrons are *attracted* to the other atom's nucleus. If the atoms come closer still, the attractive forces can offset the repulsive forces, the energy of the two atoms decreases, and a *bond* is formed (Figure 10.1). That is, the atoms "stick together." When the energy decrease is small, we have bonds called *van der Waals bonds*, a type of interaction described further in Chapter 13. When the energy decrease is larger, then we speak of **chemical bonds**. These stronger bonds are of two types: **ionic bonds** and **covalent bonds**.

A more complete view of chemical bond formation between two atoms, called the *orbital theory*, is that the electrons and nucleus of one atom strongly perturb or change the spatial distribution of the other atom's valence electrons. Certain of the valence electrons of one atom are at-

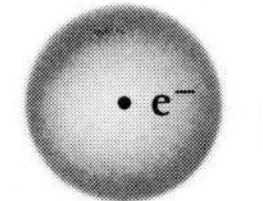
• e⁻
H atom 1s orbital

+
e⁻
F atom 2p orbital
(The two other F atom 2p orbitals each contain a pair of electrons as does the F atom 2s orbital).
Bond formation

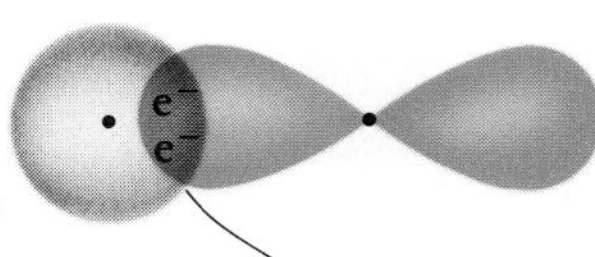
e⁻
e⁻
Bonding electrons; one from each atom. These electrons are attracted to both nuclei of the bonded atoms.

Figure 10.1 The formation of a covalent bond between an H atom and an F atom. One pair of electrons (one from each atom) moves into the internuclear region and is attracted to both the H and the F nuclei. It is this mutual attraction of 2 (or sometimes 4 or 6) electrons by two nuclei that leads to bond formation.

tracted to the other atom, and others are repelled from it. A new orbital (wave function) is needed to describe the distribution of the **bonding electrons** (those that form the bond). This new orbital is called a **bond orbital**, and *it describes the motion of two electrons of opposite spin*. The orbitals of electrons on a bonded atom that are distorted *away* from the bond region also have new descriptions, called **lone pair orbitals**. These lone pairs of electrons do not participate in holding the molecule together.

The prevalent view today is that the new bond orbital is built from atomic orbitals from the two bonded atoms. We visualize it as still having a spatial distribution near each atom much like that of the parent orbitals—the orbitals from which the bond orbital is formed—but the bond orbital is more concentrated in the region between the bonded nuclei. The decrease in energy for the two atoms upon bond formation comes largely from the decrease in energy of the two bond-orbital electrons (Figure 10.2). The energy of the electrons in a bond orbital, where the electrons are attracted by two nuclei, is lower than their energy in valence electron orbitals where the electrons are attracted to only one nucleus.

If the bond orbital is strongly displaced toward one atom we have an **ionic bond**:

Ionic Bonding: $Na\cdot + \cdot\ddot{\underset{\cdot\cdot}{Cl}}: \longrightarrow Na^+,\ :\ddot{\underset{\cdot\cdot}{Cl}}:^-$

Ionic bonds generally involve metals from the left side of the periodic table interacting with nonmetals from the far right side (Section 10.3). This follows from what you learned in Chapter 9 (page 349). Atoms near the rare gases in the periodic table form ions: elements immediately following Group 8A form cations with the configuration of the nearest rare gas and those preceding Group 8A form anions, again with the configuration of the nearest rare gas.

In contrast with ionic bonds, if the bond orbital is more or less evenly distributed between the atoms, and electrons are *shared* by two nuclei, we have a **covalent bond**:

Covalent Bonding: $:\ddot{\underset{\cdot\cdot}{I}}\cdot + \cdot\ddot{\underset{\cdot\cdot}{Cl}}: \longrightarrow :\ddot{\underset{\cdot\cdot}{I}}:\ddot{\underset{\cdot\cdot}{Cl}}:$

Covalent bonding generally occurs when elements lie closer to one another in the periodic table (Section 10.4). As you read more about chemical bonds, be sure to keep in mind that these bond types are the limiting cases; *most chemical bonds are in fact somewhere between purely ionic and purely covalent.*

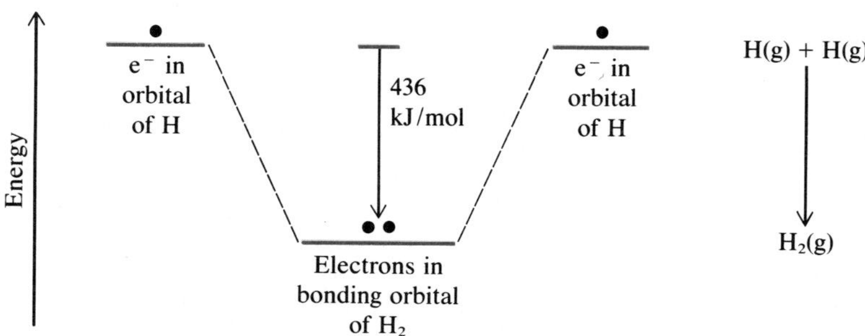

Figure 10.2 Energy changes that occur in the course of bond formation between two H atoms.

Figure 10.3 Some common ionic compounds: common salt (NaCl, colorless with a blue impurity) attached to a reddish crystal of KCl (upper left); the mineral fluorite (CaF_2, light blue) (right), red iron(III) oxide (Fe_2O_3, rust); and black copper(II) bromide ($CuBr_2$).

Figure 10.4 Common ionic compounds in which the anion is a polyatomic ion: copper(II) sulfate ($CuSO_4 \cdot 5\ H_2O$, blue); nickel(II) nitrate [$Ni(NO_3)_2 \cdot 6\ H_2O$, green]; and potassium dichromate ($K_2Cr_2O_7$, orange).

10.3 IONIC BONDING

As you just learned, compounds with ionic bonds generally involve metals from the left side of the periodic table interacting with nonmetals from the far right side. The reason for this is straightforward: metals have relatively low ionization energies and so can be expected to lose one or more electrons and form a cation. Conversely, nonmetals such as the halogens have a high affinity for electrons and so can readily form anions. Therefore, ionic compounds are represented by common sea salt (NaCl), by the mineral fluorite (CaF_2), by cinnabar (HgS), and by many metal oxides (Figure 10.3). They are also represented by an even larger class of compounds in which a metal cation is associated with a *polyatomic* anion to give such compounds as $CuSO_4$, $KA1(SO_4)_2$ (alum), $K_2Cr_2O_7$, and others (Figure 10.4). In all of these ionic compounds, one cation M^{n+} is attracted by several anions X^{n-}, which are in turn attracted by several cations, and so on. The final result is a solid, regular array of positive and negative ions called a *crystal lattice* (Figure 10.5). We will discuss lattices in more detail in Chapter 13. However, to have some further insight into ionic bonding, it is useful at this point to consider the strength of ionic bonds.

One measure of the strength of attraction between two oppositely charged ions is the *enthalpy of dissociation* of an ion pair—the energy required to decompose it into its ions, with everything in the gas phase. The greater the value of $\Delta H_{\text{dissociation}}$, the greater the energy of attraction between the ions of the ionic compound.

$$MX(g) + \text{energy} \longrightarrow M^+(g) + X^-(g) \qquad \text{energy} = \Delta H_{\text{dissociation}}$$

$$NaCl(g) + 548\ \text{kJ} \longrightarrow Na^+(g) + Cl^-(g) \qquad \Delta H_{\text{dissociation}} = +548\ \text{kJ/mol}$$

Figure 10.5 Sodium chloride crystal lattice. As described more fully in Chapter 13, the crystal lattice of NaCl, which is similar to that of many other ionic solids, consists of an extended and regular array of Na^+ and Cl^- ions, each ion surrounded by six ions of opposite charge. (© Irving Geis)

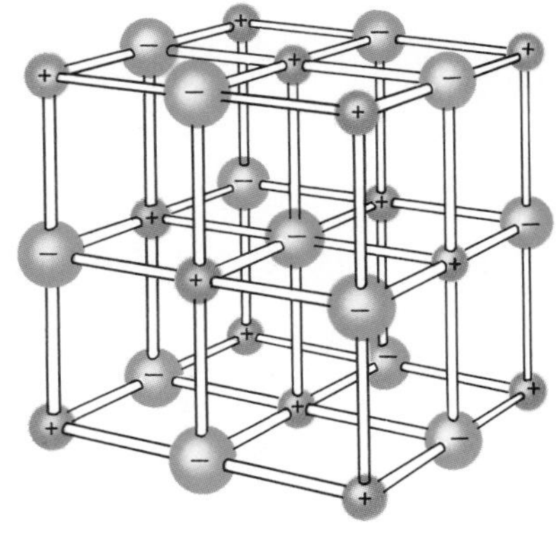

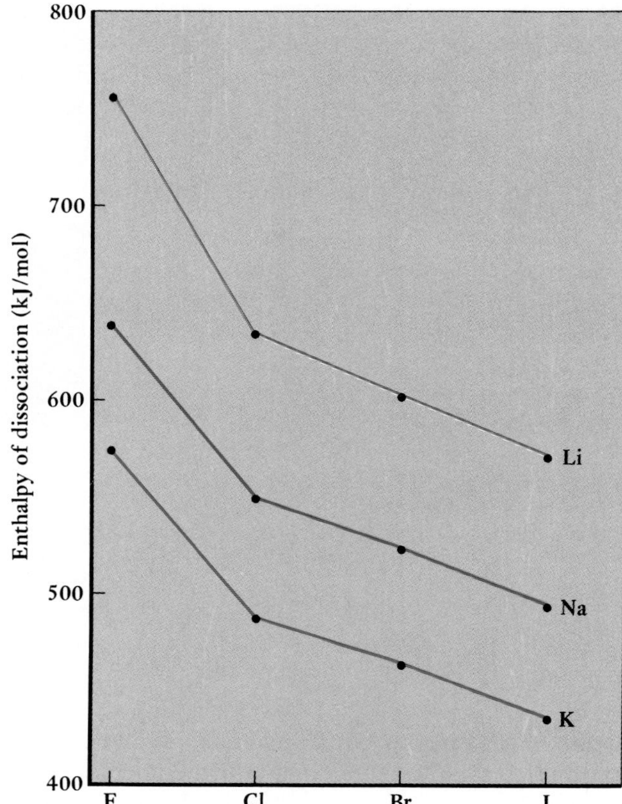

Figure 10.6 Enthalpies of dissociation of the alkali metal halides MX(g) to give the ions M^+ and X^- in the gas phase.

The energy expressed as $\Delta H_{\text{dissociation}}$ can be estimated with a simple Equation (10.1), which is derived from Coulomb's law, the law that describes the force of attraction between ions of opposite charge (page 75).

$$\Delta H_{\text{dissociation}} \propto \frac{n_+ n_-}{d} \tag{10.1}$$

Here n_+ is the number of positive charges on the cation and n_- is the number of negative charges on the negative ion; for example, $n_+ = 2$ for Mg^{2+} and $n_- = 3$ for PO_4^{3-}. The variable d is the distance between the ion centers in the crystal lattice.* The important aspect of this equation is that the energy of attraction between ions of opposite charge depends directly on the magnitude of the ion charges (the higher the ion charges, the greater the attractive energy) and inversely on the distance between them (the greater the distance, the smaller the attractive energy).

We can see the effect of Equation 10.1 by looking at the periodic trends in ion–ion interactions as given by a plot of $\Delta H_{\text{dissociation}}$ for the

*The full equation for calculating energy of interaction between oppositely charged "hard-sphere" ions is

$$\text{Energy per mole} = -\frac{N(n_+e)(n_-e)}{4\epsilon\pi d}$$

N is Avogadro's number, e is the charge on the electron (1.60×10^{-19} coulomb), d is the distance between ion centers (in meters), and ϵ is a constant (8.85×10^{-12} coulomb^2/J · m). For NaCl, where $d = 2.83 \times 10^{-10}$ m, you would find the energy is 490 kJ/mol, in reasonable agreement with experiment.

alkali metal halides (Figure 10.6). In particular, you should notice the effect of ion size (which influences d in Equation 10.1). For example, values of $\Delta H_{dissociation}$ for some chlorides are in the order LiCl > NaCl > KCl. The reason is that the alkali metal ion sizes are in the order $Li^+ < Na^+ < K^+$. The larger the positive ion, the smaller the value of $\Delta H_{dissociation}$, in accord with the effect of increasing d in Equation 10.1.

Ion–ion attractions in ionically bonded compounds have a profound effect on such properties of ionic solids as their melting points and solubilities in water. Although these properties actually depend on many factors, which will be discussed in later chapters,* the enthalpy of dissociation of an ion pair can provide a *rough* guess of *relative* melting points and solubilities for closely related ionic compounds.

The melting temperature of a solid is the temperature at which the solid structure collapses because, as the temperature climbs, the ions move around faster and the attractions among the individual ions, atoms, or molecules of the solid are eventually overcome. Common sea salt, NaCl, melts at 801 °C. In contrast, KCl melts at a slightly lower temperature (776 °C). This is predictable because the value of $\Delta H_{dissociation}$ for KCl(g) is smaller than for NaCl(g) due to the larger ion–ion distance in KCl; the ion–ion forces in KCl are more easily overcome than in NaCl.

Ion charge must also have an effect on melting temperature, since $\Delta H_{dissociation}$ is proportional to charge. For magnesium oxide (Mg^{2+} and O^{2-}), the value of d is much less than in NaCl (d = 212 pm for MgO but d = 283 pm for NaCl), and this would be expected to contribute to a rise in melting temperature. More importantly, though, are the increased ion charges in MgO relative to NaCl, since ion charges are *multiplied* together in Equation 10.1. Thus, the melting point of MgO is extraordinarily high (2800 °C), and MgO is used as heat insulation on electrical wires in cooking ovens and toasters (Figure 10.7).

When an ionic solid dissolves in water, the attraction between the water molecules and the ions on the surface of the solid must overcome the attractions between the ions in the solid (see Figure 5.2). The larger the ion–ion attractive energy, the more difficult it is for the solvent (water) to "pull" ions off the surface and dissolve them, and the lower the solubility (as illustrated in the following example).

Figure 10.7 The white solid at the end of the element for an electric kitchen range is made of magnesium oxide, MgO. The oxide is a very good conductor of heat but a poor conductor of electricity. The solid conducts the heat rapidly away from the hot coil of resistance wire to a grounded metal sheath only a few millimeters away. Yet the MgO electrically insulates the 220 V coil from the grounded sheath, even at 1000 °C.

EXAMPLE 10.1

IONIC BONDING AND PROPERTIES OF SALTS

Place the chloride, bromide, and iodide salts of potassium in order of (a) increasing solubility in water, and (b) increasing melting temperature.

Solution Figure 10.6 shows that the value of $\Delta H_{dissociation}$ decreases in the order KCl > KBr > KI, and so the cation–anion attractive energy decreases in the same order. Since the melting temperature is expected to decrease, and the solubility in water should increase, as cation–anion attractive energy decreases, we should observe the following trends:

*The structures of ionic solids are described in Chapter 13, as are the effects of ion–ion forces on melting point and similar properties. Solubility of ionic solids is discussed in Chapters 14 and 19.

—solubility in water→	←melting temperature—
KCl < KBr < KI	KCl > KBr > KI

This is indeed the case, as seen in the following table.

COMPOUND	MELTING TEMPERATURE	SOLUBILITY IN WATER
KCl	776 °C	34.7 (at 20 °C)
KBr	730 °C	53.5 (at 0 °C)
KI	686 °C	127.5 (at 0 °C)

EXERCISE 10.1 Ionic Bonds

Considering relative ion–ion attractive energies, which is predicted to have the higher melting point, NaCl or CsCl? Which should be more soluble in water?

10.4 COVALENT BONDING

Hundreds of compounds, particularly minerals, are ionically bonded. However, the kinds of molecules of which we are built (fats, proteins, and others), the foods we eat (carbohydrates, fats, and proteins), and the clothes we wear (cotton, wool, and synthetic fibers) virtually all consist of covalently bonded molecules. Therefore, the remainder of this chapter almost wholly concerns the opposite extreme from ionic bonding, *covalent bonding*, in which *the bonded atoms share electrons in the internuclear region*.

NUMBER OF BOND PAIRS: THE OCTET RULE

In Section 10.1 we drew Lewis structures for isolated atoms from the *s*- and *p*-blocks. Since covalent bonds form by sharing electrons, generally one from each atom, we can predict the number of bonds formed by an atom just by counting *unpaired* electrons for that atom. From the table of Lewis structures in Section 10.1, you can see that the number of unpaired electrons on an atom of Groups 4A through 8A is just 8 minus the group number. (The number of unpaired electrons is equal to the group number for Groups 1A through 3A, but these elements, except for boron, usually form ionic compounds rather than covalent ones.) As noted in Section 10.1, the bonding behavior of these main group elements can be considered the result of *sharing* valence electrons to achieve the same configuration as the nearest rare gas. Since all rare gases (except He) have eight valence electrons, this observation has been called the **octet rule**. Therefore, each of these atoms will acquire, through sharing, electrons equal to the number of unpaired electrons and will form that number of shared-pair bonds. For example, oxygen in Group 6A needs $8 - 6 = 2$ electrons and forms two shared-pair bonds.

Table 10.1 shows Lewis formulas for a number of covalent compounds. As is customary, bonding pairs are represented by dashes instead of a pair of dots, to distinguish them from lone pairs of electrons. Notice that atoms with four or more valence electrons have four *pairs* of electrons (total of bond pairs and lone pairs) around that atom, as predicted by the octet rule.

Table 10.1 Examples of Lewis Structures by Periodic Group*

	1A	2A	3A	4A	5A	6A	7A	8A
Lewis structure	H—H	H—Be—H	H—B(—H)—H	H—C(—H)(—H)—H	H—N̈(—H)—H	H—Ö:(—H)	H—F̈:	:N̈e:
Chemical formula	H_2	BeH_2	BH_3†	CH_4	NH_3	H_2O	HF	Ne
Bond pairs	1	2	3	4	3	2	1	0
Lone pairs	0	0	0	0	1	2	3	4
Total pairs of electrons at central atom	1	2	3	4	4	4	4	4
Lewis structure			:F̈—B(—F̈:)—F̈:	H₂C=CH₂ (H H / C=C / H H)	:N≡N:	:C̈l—S̈(—C̈l:):	:C̈l—F̈:	
Chemical formula			BF_3	C_2H_4	N_2	SCl_2	ClF	
Bond pairs			3 1	4	3	2 1	1 1	
Lone pairs			0 3	0	1	2 3	3 3	
Total pairs of electrons			3 4	4	4	4 4	4 4	

*In each case, the number of bond pairs of electrons, lone pairs of electrons, and total pairs of electrons around atoms other than H are given.
†BH_3 is unstable. The simplest isolable molecule containing a B—H bond is B_2H_6.

EXERCISE 10.2 Octet Rule

Which of the following combinations of lone pairs and bond pairs around a single atom A are consistent with the octet rule?

(a) —A— (with bonds above and below)

(b) 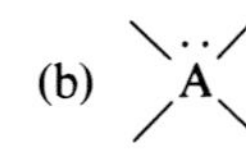

(c) 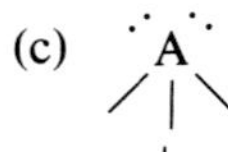

(d) —Ä— (with a bond below)

(e) :A≡

(f) 

(g) =A= (with a bond above)

(h) :Ä=

SINGLE AND MULTIPLE BONDS Some of the examples in Table 10.1 have a single pair of electrons (a **single** bond) between atoms. Single bonds are also called **sigma bonds**, symbolized by the Greek letter σ. Other structures in Table 10.1 indicate two or three electron pairs (a multiple bond) between a pair of atoms. In a **double** or **triple** bond, one of the bonds is a sigma bond, but the second (and third if present) is a **pi bond**, denoted by the Greek letter π. *Multiple bonds are most often formed by C, N, O, and S atoms.*

Sigma (σ) and pi (π) bonds differ in shape and strength, which makes them behave differently in chemical reactions (Chapter 11).

Phosphorus is also known to form π bonds with O, S, N and even C under special conditions.

EXERCISE 10.3 Sigma and Pi Bonds

For each Lewis structure in Table 10.1, give the number of sigma and pi bonds around each atom that is not hydrogen. For example, C in C_2H_4 (ethylene) has three sigma bonds and one pi bond.

COORDINATE COVALENT BONDS In all of the compounds shown so far, each atom contributes one electron to a bond pair, as in

$$H\cdot + \cdot H \longrightarrow H:H$$

Some elements, such as nitrogen and phosphorus, tend to share a lone pair with another atom that is short of electrons, leading to the formation of a **coordinate covalent bond**:

$$H^+ + :NH_3 \longrightarrow [NH_4]^+$$

hydrogen ion (no electrons) | ammonia molecule | ammonium ion

Once such a bond is formed, it is the same as any other bond; in the ammonium ion, for instance, all four bonds are identical.

EXCEPTIONS TO THE OCTET RULE

FEWER THAN EIGHT VALENCE ELECTRONS In Table 10.1, several compounds have fewer than four pairs in the valence shell. Hydrogen, of course, can accommodate at most two electrons in its valence shell, so it shares only two electrons with another atom. In BeH_2 there are only four valence electrons about Be, and there are only six around boron in BH_3 and BF_3. These facts account for the ready formation of coordinate covalent bonds such as that in $H_3N—BF_3$; the N atom lone pair is shared with boron, and both N and B now have an octet of electrons.

$$H_3N: + BF_3 \longrightarrow H_3N—BF_3$$

a coordinate covalent bond

EXPANDED VALENCE Elements of the third or higher periods can be surrounded by more than four valence pairs in certain compounds and are said to display **expanded valence**. The number of bonds formed depends on a balance between the ability of the nucleus to attract electrons and the repulsion between the pairs. Examples of such compounds are shown in Table 10.2.

ODD-ELECTRON COMPOUNDS A *few* stable compounds contain an odd number of valence electrons, and thus cannot obey the octet rule. The molecule NO, for example, has 5 valence electrons from N and 6 from O, for a total of 11 valence electrons. Other examples are NO_2 and ClO_2.

According to the usual method for drawing Lewis structures, O_2 should have the structure $:\ddot{O}=\ddot{O}:$. However, experiment shows that O_2 has two unpaired electrons. This result cannot be explained by the Lewis theory (but see Section 11.2).

DRAWING LEWIS STRUCTURES

It is very important that you learn to draw Lewis structures. Among other things, they can help you predict the molecular structure, that is, the way the atoms are positioned in space relative to each other. The

Table 10.2 Examples of Lewis Structures in Which the Central Atom Has an Expanded Valence*

	Group 4A	Group 5A	Group 6A	Group 7A	Group 8A
Atoms with five valence pairs		PF_5	SF_4	ClF_3	XeF_2
Bond pairs		5	4	3	2
Lone pairs		0	1	2	3
Atoms with six valence pairs	$[SnCl_6]^{2-}$	$[PF_6]^-$	SF_6	BrF_5	XeF_4
Bond pairs	6	6	6	5	4
Lone pairs	0	0	0	1	2

*In each case, the number of bond pairs and lone pairs about the central atom are given.

following are guidelines for drawing Lewis structures, and some examples follow.

GUIDELINES FOR DRAWING LEWIS STRUCTURES

1. To predict the arrangement of atoms within the molecule, use the following rules:
   a. H is always an end or *terminal* atom. It is connected to only one other atom (see Table 10.1).*
   b. The atom with the lowest affinity for electrons (*least* negative *EA*) in the molecule or ion is the central atom.
2. Having decided on the general arrangement of atoms, find the total number of valence electrons in the molecule by adding up the group numbers of the elements. For ions, also *add* to the sum of group numbers the ion charge for a negative ion or *subtract* the ion charge for a positive ion. The number of valence electron *pairs* to be used in the Lewis structure is *half* the total number of valence electrons.
3. Place one pair of electrons (a sigma bond) between each pair of bonded atoms.
4. Subtract from the total number of valence electron pairs the number of bonds you drew in step 3. This is the pool of electron pairs from which you form the lone pairs, as well as any pi bond pairs if necessary.
5. Place lone pairs about each terminal atom (except H) to satisfy the octet rule. If pairs are still left at this point, assign them to the central atom. If the central atom is from the third or higher period, it can accommodate more than four electron pairs.

Many simple molecules or ions that you will see contain O or a halogen atom. Rule 1b means that these atoms are often terminal atoms, except when they are combined with H. When O and F are combined in a single species, O is central as predicted by Rule 1b. However, Cl, Br, or I are central in species combining O and one of these halogens.

A better criterion for the central atom is that it should be the **atom of lowest *electronegativity***. *The concept of electronegativity is introduced in Section 10.5.*

Each negative charge on an ion represents an additional valence electron, and each positive charge represents the loss of a valence electron.

*There are some exceptions to this statement, especially in the series of B/H-containing compounds called boron hydrides. You will not encounter them until Chapter 23, however.

6. If the central atom is not yet surrounded by four electron pairs, convert one or more terminal atom lone pairs to pi bond pairs. In this way the pairs are still associated with the terminal atom, and they are also a part of the central atom. Not all elements form pi bonds. You can use the general rule that *only C, N, O, and S form a pi bond to another atom of the same element or with another atom of this group of four elements*. That is, there can be bonds such as C=C, C=N, S=O, and so on.

As noted earlier, P is also known to form π bonds with O, S, N and even C under special conditions.

EXAMPLE 10.2

DRAWING LEWIS STRUCTURES

Draw Lewis structures for ammonia, NH_3; formaldehyde, H_2CO; the hypochlorite ion, OCl^-; and the nitronium ion, NO_2^+.

Solution for Ammonia **(1)** The H atoms *must* be terminal, so the N atom is clearly the central atom.

(2) The total number of valence electrons is the sum of the group numbers: 5 (for N) + 3 (1 for each H) = 8. There are 4 valence pairs.

(3) Form a single bond between each pair of atoms, using 3 pairs of electrons.

```
H—N—H
  |
  H
```

(4) and **(5)** After sigma bond formation, one pair of electrons remains, and this becomes a lone pair on the central atom.

```
  ..   ← lone pair
H—N—H
  |
  H
```

Each H atom now has a share in one pair of electrons as required, while the central N atom has a share in four electron pairs. Three of the nitrogen pairs are shared sigma bonding electron pairs, while the fourth is a lone electron pair.

Solution for Formaldehyde **(1)** Carbon, an atom with a lower affinity for electrons than oxygen, is the central atom.

(2) There is a total of 12 valence electrons or six pairs: 2 (1 for each H) + 4 (for C) + 6 (for O) = 12.

(3) Form a single bond between each pair of atoms, using 3 of the 6 pairs.

```
H—C—H
  |
  O
```

(4) and **(5)** The 6 electrons that remain are placed around the O atom as lone pairs.

```
H—C—H
  |
 :O:
  ..
```

(6) The C atom is deficient by one pair. The solution is to make one of the lone pairs on the O atom into a bonding pair, making a pi bond between O and C. Now both C and O have a share in four pairs of electrons.

$$\mathrm{H{-}\underset{\underset{:\ddot{O}:}{|}}{C}{-}H \longrightarrow H{-}\underset{\underset{:O:}{\|}}{C}{-}H}$$

The carbon has a share in three sigma bond pairs and one pi bond pair, while the O atom has two lone pairs and shares one sigma and one pi bond pair.

Solution for OCl^- Hypochlorite ion has 14 valence electrons [6 (for O) + 7 (for Cl) + 1 (for ion charge) = 14] or 7 valence electron pairs. After forming the O—Cl bond, and distributing the remaining 6 pairs around the "terminal" atoms, both atoms have a share in four electron pairs as required.

$$[:\ddot{\underset{\cdot\cdot}{O}}{-}\ddot{\underset{\cdot\cdot}{Cl}}:]^-$$

Solution for NO_2^+ **(1)** The N atom has a lower affinity for electrons than O and so is the central atom.

(2) The valence electron total is 16 e^- or 8 pairs: 12 (6 for each O atom) + 5 (for N) − 1 (for ion positive charge) = 16.

(3), **(4)**, and **(5)** After forming two N—O bonds, the six remaining pairs of electrons are distributed on the terminal atoms until each such atom has a share in a total of four electron pairs.

$$[:\ddot{\underset{\cdot\cdot}{O}}{-}N{-}\ddot{\underset{\cdot\cdot}{O}}:]^+$$

(6) After step five, the N atom has a deficiency of two electron pairs. Thus, one lone pair of electrons from each O atom is converted to a pi bonding electron pair.

$$[:\ddot{O}{=}N{=}\ddot{O}:]^+$$

Each atom in the ion now has a share in four electron pairs. Nitrogen shares two sigma and two pi pairs, and each oxygen has two lone pairs and shares one sigma and one pi bond pair.

EXERCISE 10.4 Drawing Lewis Structures

Sketch the Lewis structure for each of the following molecules or ions. Give the number of sigma and pi bonds in which each atom is involved for (a) carbon monoxide, CO; (b) the ion NO^+; (c) hydrogen cyanide, HCN; and (d) hydrogen sulfide, H_2S.

EXAMPLE 10.3

DRAWING LEWIS STRUCTURES FOR MOLECULES WITH A CENTRAL ATOM OF EXPANDED VALENCE

Sketch the Lewis structure for xenon difluoride, XeF_2.

Solution **(1)** Xenon, the atom with the lowest affinity for electrons, is the central atom.

(2) There is a total of 22 valence electrons or 11 pairs of electrons: 14 (for 2 F atoms) + 8 (for Xe) = 22.

(3), **(4)**, and **(5)** Form two Xe—F single bonds and then distribute six of the remaining nine electron pairs as lone pairs on the terminal atoms.

$$:\ddot{\underset{\cdot\cdot}{F}}{-}Xe{-}\ddot{\underset{\cdot\cdot}{F}}:$$

Each fluorine atom can accommodate only four pairs, so the three pairs remaining are placed on xenon, an atom that can accommodate more than a total of four bonding and lone pairs.

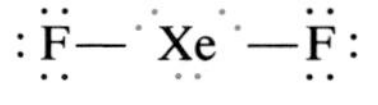

:F̤̈—Xe—F̤̈:

XeF_2 is a stable molecule with a total of 5 electron pairs (2 sigma bond pairs and 3 lone pairs) surrounding the central Xe atom.

EXERCISE 10.5 Drawing Lewis Structures Where the Central Atom Has an Expanded Valence

Sketch the Lewis structure for each of the following molecules: (a) SF_4, sulfur tetrafluoride; (b) ClF_3, chlorine trifluoride; and (c) PCl_5, phosphorus pentachloride.

RESONANCE STRUCTURES

O 127.8 pm, O 116.8° O

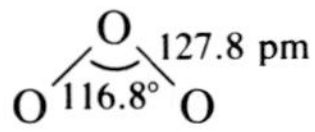

Ozone, O_3

For a discussion of ozone in our environment, see Chapters 15 and 24.

Ozone, O_3, protects the earth and its inhabitants from intense ultraviolet radiation from the sun. The compound is an unstable, blue, diamagnetic gas with a characteristic odor. Its structure is pictured in the margin. As you will see in the next section, the number of bonding electron pairs between two atoms is important in determining bond length and strength. Ozone has equal O—O bond lengths, implying that there is an equal number of bond pairs on each side of the central O atom. However, using the guidelines for drawing Lewis structures, you might come to a different conclusion. A possible dot structure would be

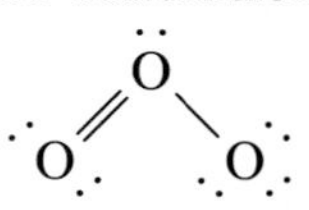

The structure has a double bond on one side of the central O atom and a single bond on the other side. If this were the electronic structure of O_3, one bond would be shorter (O═O) than the other (O—O). This is not the case, and one way to reconcile this with the experimental observation of equal O—O bond lengths is to use *equivalent structures* called **resonance structures**. When drawing the O_3 dot structure, the final step (the sixth in our scheme) is to convert a lone pair to a π bond pair. In O_3 this can be done using a pair from either of the terminal O atoms, and two equivalent structures or resonance structures are produced.

It is conventional to connect resonance structures with a double-headed arrow, ↔.

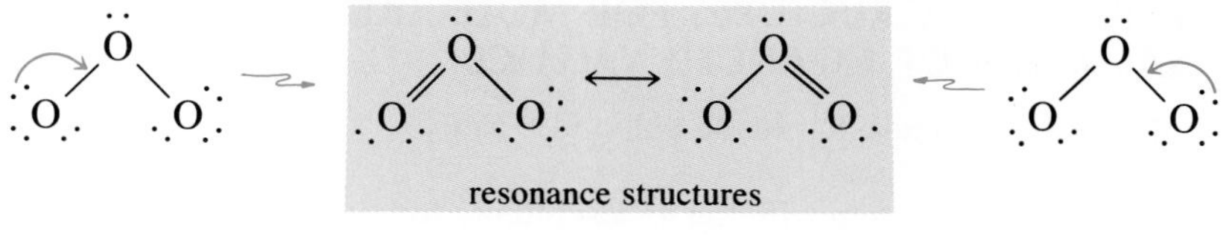

resonance structures

Composite picture of the O_3 resonance structures (where the dashed line represents two electron pairs)

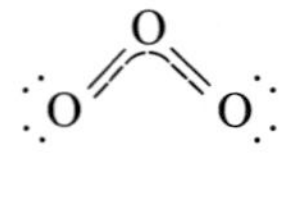

These resonance structures are meant to convey the idea that two π bonding electrons are spread symmetrically over the three atoms of the molecule. Therefore, the separate resonance structures might be represented by the single composite picture in the margin. Here the dashed line connecting the three atoms represents the symmetrical distribution of one π pair and one lone pair.

The carbonate ion, CO_3^{2-}, provides another good example of resonance structures. Here there are *three* equivalent structures.

$$\left[\begin{array}{c} :\ddot{O}: \\ \| \\ C \\ \dot{\underset{..}{O}}{}^{\diagup}\ \ {}^{\diagdown}\underset{..}{\ddot{O}} \end{array}\right]^{2-} \longleftrightarrow \left[\begin{array}{c} :\ddot{O}: \\ | \\ C \\ :\underset{..}{\ddot{O}}{}^{\diagup}\ \ {}^{\diagdown\!\!\diagdown}\ddot{O} \end{array}\right]^{2-} \longleftrightarrow \left[\begin{array}{c} :\ddot{O}: \\ | \\ C \\ \ddot{O}{}^{\diagup\!\!\diagup}\ \ {}^{\diagdown}:\underset{..}{\ddot{O}} \end{array}\right]^{2-}$$

You can see that resonance structures arise whenever there is a question about which of two or three atoms contribute lone pairs to achieve an octet of electrons about a central atom by multiple bonding (step 6 in our scheme).

When drawing resonance structures there are several important things to notice. First, *resonance structures differ only in the assignment of electron pair positions,* ***never*** *atom positions.* Second, *resonance structures differ in the number of bond pairs between a given pair of atoms.*

$$[:\ddot{N}{=}C{=}\ddot{O}:]^- \longleftrightarrow [:\ddot{N}{=}C{=}\underset{..}{O}:]^-$$

These are the same structure. They are *not* resonance structures because the number of NC and CO bond pairs, and the number of lone pairs at N and O, are not changed.

$$[:\underset{..}{\ddot{N}}{-}C{\equiv}O:]^- \longleftrightarrow [:\ddot{N}{=}C{=}\ddot{O}:]^- \longleftrightarrow [:N{\equiv}C{-}\underset{..}{\ddot{O}}:]^-$$

These are resonance structures. Lone pairs and bond pairs have been interconverted. The number of NC and CO bond pairs changes, as do the numbers of N and O lone pairs.

> **EXERCISE 10.6 Drawing Resonance Structures**
> Draw resonance structures for (a) NO_2^- and (b) NO_3^-.

10.5 BOND PROPERTIES

BOND ORDER

The **bond order** is the number of bonding electron pairs shared by two atoms in a molecule. Various molecular properties can be understood by this concept, including the distance between two atoms (*bond length*) and the energy required to separate the atoms from each other (*bond energy*). In this text you will encounter bond orders of 1 through 3, as well as fractional bond orders.

BOND ORDER = 1 The bond order is 1 when there is only a sigma bond between the two bonded atoms. Examples are the single bonds in the following molecules.

$$H{-}H \qquad :\underset{..}{\ddot{F}}{-}\underset{..}{\ddot{F}}: \qquad \begin{array}{c} H{-}\ddot{N}{-}H \\ | \\ H \end{array} \qquad \begin{array}{c} H \\ | \\ H{-}C{-}H \\ | \\ H \end{array}$$

There can never be more than one sigma (σ) bond between a pair of atoms. All additional bond pairs are pi (π) pairs.

BOND ORDER = 2 The order is 2 when there are two shared pairs between two atoms. One of these pairs forms a sigma bond and the other pair forms a pi bond. Examples are the C=O bonds in CO_2 and the C=C bond in ethylene, C_2H_4.

$$\ddot{\underset{\cdot\cdot}{O}}=C=\ddot{\underset{\cdot\cdot}{O}} \qquad \begin{matrix} H & & & & H \\ & \diagdown & & \diagup & \\ & & C=C & & \\ & \diagup & & \diagdown & \\ H & & & & H \end{matrix}$$

BOND ORDER = 3 An order of 3 occurs when two atoms are connected by one sigma bond and two pi bonds. Examples are the carbon–carbon bond in acetylene (C_2H_2), the carbon–oxygen bond in carbon monoxide (CO), and the carbon–nitrogen bond in the cyanide ion (CN^-).

$$H—C\equiv C—H \qquad :C\equiv O: \qquad [:C\equiv N:]^-$$

Fractional bond orders may occur for bonds in molecules or ions having resonance structures. Each oxygen–oxygen bond of O_3 has an *average* bond order of $\frac{3}{2}$, for example. Each resonance structure of O_3 has one O—O bond and one O=O bond, for a total of *three* shared, bonding pairs to account for *two* oxygen–oxygen links. Using the definition of bond order below, the order is seen to be $\frac{3}{2}$.

$$\text{Bond order} = \frac{\text{number of shared pairs linking X and Y}}{\text{number of X–Y links}} = \frac{3}{2}$$

BOND LENGTH

As described in Chapter 9, the most important factor determining **bond length**, the distance between the nuclei of two bonded atoms, is the sizes of the atoms themselves. For given elements, the order of the bond then determines the final value of the distance.

Recall from Figure 9.11 that atom sizes vary in a fairly smooth way with the position of the element in the periodic table. When you compare bonds of the same order, the bond length will be greater for the larger atoms. Thus, bonds involving carbon and another element would increase in length along the series

$$\xrightarrow[\text{Increase in bond distance}]{\text{C—N < C—C < C—P}}$$

Relative atom sizes for atoms of Groups 4A, 5A, and 6A

4A	5A	6A
C	N	O
Si	P	S

Similarly, a C=O bond will be shorter than a C=S bond, and a C≡N bond will be shorter than a C≡C bond. Each of these trends can be predicted from the relative sizes shown in the margin, and some common bond lengths are given in Table 10.3.

The effect of bond order is evident when you compare bonds *between the same two atoms*. For example, the bonds become shorter as the bond order increases in the series C—O, C=O, and C≡O.

Bond	C—O	C=O	C≡O
Bond Order	1	2	3
Bond Length (pm)	143	122	113

Adding a pi bond to the sigma bond in C—O shortens the bond by only 21 pm on going to C=O, rather than reducing it by half as you might

Table 10.3 Some Approximate Single and Multiple Bond Lengths*

	Single Bond Lengths										
	Group										
	1A	4A	5A	6A	7A	4A	5A	6A	7A	7A	7A
	H	C	N	O	F	Si	P	S	Cl	Br	I
H	74	110	98	94	92	145	138	132	127	142	161
C		154	147	143	141	194	187	181	176	191	210
N			140	136	134	187	180	174	169	184	203
O				132	130	183	176	170	165	180	199
F					128	181	174	168	163	178	197
Si						234	227	221	216	231	250
P							220	214	209	224	243
S								208	203	218	237
Cl									200	213	232
Br										228	247
I											266

Multiple Bond Lengths

C=C	134	C≡C	121
C=N	127	C≡N	115
C=O	122	C≡O	113
N=O	115	N≡O	108

*In picometers (pm); 1 pm = 10^{-12} m.

have expected. The second pi bond results in a 9 pm reduction in bond length from C=O to C≡O.

The carbonate ion, CO_3^{2-}, which has three equivalent resonance structures, has a CO bond order of $\frac{4}{3}$. Not surprisingly, this leads to a bond distance of 129 pm, a distance intermediate between the C—O single bond and C=O double bond.

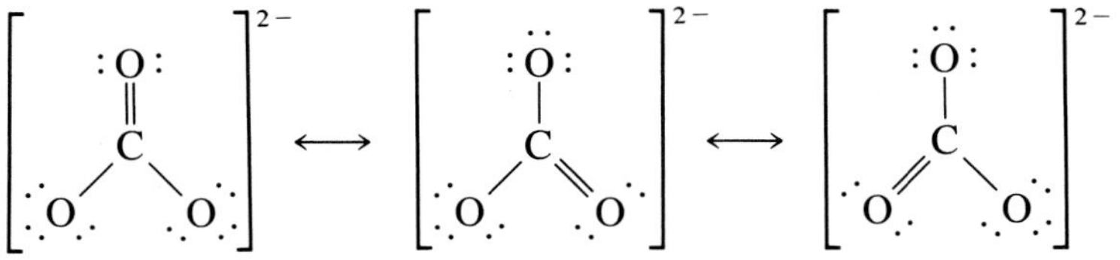

CO bond order = $\frac{4}{3}$; CO bond length = 129 pm

EXERCISE 10.7 Bond Distances and Bond Order

(a) Give the bond order of each of the following bonds and arrange them in order of decreasing bond distance: C=N, C≡N, and C—N.

(b) Draw resonance structures for NO_2^-. What is the NO bond order in this ion? Consult Table 10.3 for N—O and N=O bond lengths. Compare these with the NO bond length in NO_2^- (124 pm). Account for any difference you observe.

Table 10.3 lists *average* values of bond lengths. Variations in neighboring parts of a molecule can affect the length of a particular bond. For example, the C—H bond has a length of 105.9 pm in acetylene, H—C≡C—H, but a length of 109.3 pm in methane, CH_4. Be aware that there can be a variation of as much as 10% from the average values listed in Table 10.3.

BOND ENERGY

The greater the number of bonding electron pairs between a pair of atoms, the shorter the bond. This implies that atoms are held together more tightly when there are multiple bonds, and so it should not be surprising that there is a relation between bond order and the energy required to separate atoms.

Suppose you wish to separate, by means of chemical reactions, the carbon atoms in ethane ($H_3C—CH_3$), ethylene ($H_2C=CH_2$), and acetylene ($HC\equiv CH$) for which the bond orders are 1, 2, and 3, respectively. For the same reason that the ethane C—C bond is the longest of the series, and the acetylene C≡C bond the shortest, the separation will require the least energy for ethane and the most energy for acetylene.

$$\text{Molecule} + \text{energy} \underset{\textit{energy released}}{\overset{\textit{energy supplied}}{\rightleftharpoons}} \text{molecular fragments}$$

$$H_3C—CH_3(g) + 347 \text{ kJ} \longrightarrow H_3C(g) + CH_3(g) \qquad \Delta H = +347 \text{ kJ/mol}$$

$$H_2C=CH_2(g) + 611 \text{ kJ} \longrightarrow H_2C(g) + CH_2(g) \qquad \Delta H = +611 \text{ kJ/mol}$$

$$HC\equiv CH(g) + 837 \text{ kJ} \longrightarrow HC(g) + CH(g) \qquad \Delta H = +837 \text{ kJ/mol}$$

The energy that must be supplied to a gaseous molecule to separate two of its atoms is called the **bond dissociation energy** (or **bond energy** for short) and is given the symbol D. As D represents energy supplied to the molecule from its surroundings, D has a positive value, and *the process of breaking bonds in a molecule is endothermic*.

The amount of energy supplied to break the carbon–carbon bonds in the molecules above must be the same as the amount of energy released when the same bonds form. *The formation of bonds from atoms in the gas phase is exothermic*. This means, for example, that ΔH for the formation of $H_3C—CH_3$ from two $CH_3(g)$ fragments is −347 kJ/mol.

Some experimental bond energies are tabulated in Table 10.4, and you should be aware of several important points: (1) The energies listed are all *positive*; they are energies required to *break* the bond in question. If you need the energy evolved when the *bond is formed*, the magnitude is the same, but the sign is *negative*. (2) The energies of Table 10.4 are *average bond energies*. The energy of a C—H bond, for example, is given as 414 kJ/mol. However, this energy may vary by as much as 30 to 40 kJ/mol from molecule to molecule, for the same reason that bond lengths vary from one molecule to another. (3) The values in Table 10.4 are defined in terms of *gaseous* atoms or molecular fragments. If a reactant or product is in the solid or liquid state, you must first provide energy to convert it to a gas before using these values. (4) Finally, notice once again the connection between bond energy and bond order.

CALCULATING REACTION ENERGIES FROM BOND ENERGIES In reactions between molecules, bonds in the reactants are broken and new connections are established. If the total energy released when new bonds are formed exceeds the energy required to break the original bonds, the overall reaction is exothermic. If the opposite is true, then the overall reaction is endothermic.

The following example shows you how to estimate, using the data in Table 10.4, whether a reaction is exothermic or endothermic.

Table 10.4 Some Average Single and Multiple Bond Energies*

Single Bonds

	H	C	N	O	F	Si	P	S	Cl	Br	I
H	436	414	389	464	569	293	318	339	431	368	297
C		347	293	351	439	289	264	259	330	276	238
N			159	201	272		209		201	243?	
O				138	184	368	351		205		201
F					159	540	490	285	255	197?	
Si						176	213	226	360	289	
P							213	230	331	272	213
S								213	251	213	
Cl									243	218	209
Br										192	180
I											151

Multiple Bonds

N=N	418	C=C	611
N≡N	946	C≡C	837
C=N	615	C=O (in O=C=O)	803
C≡N	891	C=O (as in $H_2C=O$)	745
O=O (in O_2)	498	C≡O	1075

*In kJ/mol.

EXAMPLE 10.4

USING BOND ENERGIES TO ESTIMATE HEATS OF FORMATION

Estimate the enthalpy of formation of HCl(g) using the bond energies of Table 10.4 (Figure 10.8).

$$\text{H—H(g)} + \text{Cl—Cl(g)} \longrightarrow 2\ \text{H—Cl(g)} \qquad \tfrac{1}{2}\,\Delta H^\circ_{rxn} = \Delta H^\circ_f[\text{HCl(g)}]$$

Solution The bond energies for the reactants and products are defined as follows (with values from Table 10.4):

$$\text{H—H(g)} \longrightarrow 2\ \text{H(g)} \qquad \Delta H^\circ = 436\ \text{kJ/mol}$$

$$\text{Cl—Cl(g)} \longrightarrow 2\ \text{Cl(g)} \qquad \Delta H^\circ = 243\ \text{kJ/mol}$$

$$\text{H—Cl(g)} \longrightarrow \text{H(g)} + \text{Cl(g)} \qquad \Delta H^\circ = 431\ \text{kJ/mol}$$

Recognize that 1 mole of H—H bonds and 1 mole of Cl—Cl bonds must be broken (for a total of 2 moles of bonds broken). If no new bonds were subsequently formed, 679 kJ of energy (436 kJ + 243 kJ) would have to be supplied to separate the mole of H_2 molecules and the mole of Cl_2 molecules into their atoms. However, there is a subsequent reaction of the H and Cl atoms to form 2 moles of H—Cl bonds, and this reaction *releases* 862 kJ of energy (2 moles of bonds × 431 kJ per mole of bonds). To summarize

$$\begin{array}{ll} \text{H—H(g)} \longrightarrow 2\ \text{H(g)} & \Delta H^\circ = +436\ \text{kJ} \\ \text{Cl—Cl(g)} \longrightarrow 2\ \text{Cl(g)} & \Delta H^\circ = +243\ \text{kJ} \\ 2\ \text{H(g)} + 2\ \text{Cl(g)} \longrightarrow 2\ \text{H—Cl(g)} & \Delta H^\circ = -862\ \text{kJ} \\ \hline \text{H—H(g)} + \text{Cl—Cl(g)} \longrightarrow 2\ \text{H—Cl(g)} & \Delta H^\circ_{rxn} = -183\ \text{kJ} \end{array}$$

Thus,

$$\Delta H^\circ_f[\text{HCl(g)}] = \tfrac{1}{2}\,\Delta H^\circ_{rxn} = -91.5\ \text{kJ/mol}$$

Figure 10.8 Hydrogen burns in chlorine to form hydrogen chloride, a colorless gas. Hydrogen chloride combines with moisture in the air to form the fog visible in the photo.

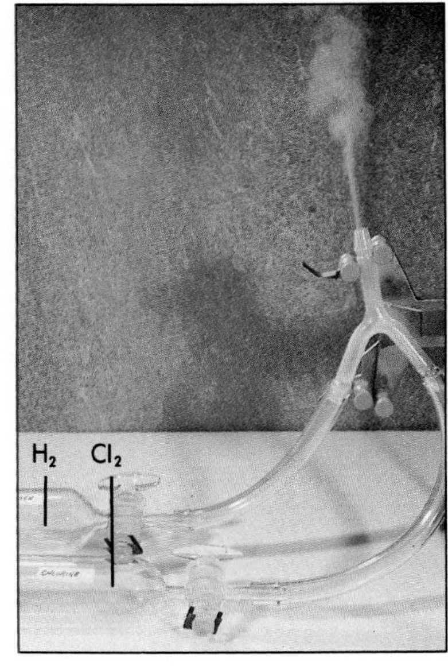

From this result, you would predict that the reaction is exothermic. That is, 91.5 kJ are released per mole of HCl formed. Since the equation as written represents the formation of HCl in its gaseous state from the elements H_2 and Cl_2 (in their standard states), we have found that −91.5 kJ is the standard heat of formation of HCl(g), a result that agrees well with the experimental value of −92.3 kJ/mol.

As you saw in Chapter 6, there are often shortcuts in thermodynamic calculations. When using bond energies to find the heat or enthalpy of a reaction you can use the equation

$$\Delta H^\circ_{\text{reaction}} = \Sigma mD \text{ (bonds broken)} - \Sigma nD \text{ (bonds made)}$$

where D is the bond dissociation energy, and m and n are the numbers of moles of bonds broken and made, respectively. This equation tells you to multiply the bond energy for each bond broken (all of these energies are endothermic, so the sign is +) by the number of bonds of that type, and add up all of these products. Because the sign of energies for bonds made is negative, you then subtract from this total the sum of energies of bonds formed. Applying this to the formation of HCl(g)

$$H_2(g) + Cl_2(g) \longrightarrow 2\ HCl(g)$$

you would write

$$\begin{aligned}\Delta H^\circ_{\text{reaction}} &= \overbrace{(1 \text{ mole H—H bonds} + 1 \text{ mole Cl—Cl bonds})}^{\text{bonds broken}} - \overbrace{(2 \text{ moles H—Cl bonds})}^{\text{bonds formed}} \\ &= [1 \text{ mol}(436 \text{ kJ/mol}) + 1 \text{ mol}(243 \text{ kJ/mol})] - [2 \text{ mol}(431 \text{ kJ/mol})] \\ &= -183 \text{ kJ}\end{aligned}$$

EXAMPLE 10.5

USING BOND ENERGIES TO ESTIMATE ENTHALPIES OF FORMATION

Estimate the enthalpy of formation of hydrazine, N_2H_4, using the bond energies of Table 10.4.

$$N{\equiv}N(g) + 2\ H{—}H(g) \longrightarrow H_2\ddot{N}{—}\ddot{N}H_2\ (g)$$

Solution Here 1 mole of N≡N triple bonds and 2 moles of H—H single bonds are broken, and energy is required. On the other hand, energy is released when 1 mole of N—N single bonds and 4 moles of H—N bonds are made in forming the product.

$$\begin{aligned}\Delta H^\circ_{\text{reaction}} &= \overbrace{(D_{N\equiv N} + 2\ D_{H—H})}^{\text{bonds broken}} - \overbrace{(D_{N—N} + 4\ D_{N—H})}^{\text{bonds made}} \\ &= [946 \text{ kJ} + 2 \text{ mol}(436 \text{ kJ/mol})] - [159 \text{ kJ} + 4 \text{ mol}(389 \text{ kJ/mol})] \\ &= +103 \text{ kJ}\end{aligned}$$

Since the equation as written depicts the formation of one mole of hydrazine from its elements, the enthalpy of reaction we have calculated is the molar enthalpy of formation of hydrazine; that is, $\Delta H^\circ_f[N_2H_4(g)] = +103$ kJ/mol. The value determined using average bond energies is within 10% of the value of +95.4 kJ/mol from the U.S. National Bureau of Standards.

EXAMPLE 10.6

USING BOND ENERGIES TO ESTIMATE ENTHALPIES OF FORMATION

Acetylene provides considerable heat when it is burned, making it a suitable fuel for welding metal (Figure 10.9). Estimate the enthalpy of formation of acetylene, H—C≡C—H, using the bond energies of Table 10.4 and the enthalpy of vaporization of graphite (the latter being the standard state of carbon).

$$\text{C(graphite)} \longrightarrow \text{C(g)} \qquad \Delta H^\circ_{\text{vaporization}} = +717 \text{ kJ/mol}$$

Solution The equation for the formation of acetylene from its elements is

$$2\text{C(graphite)} + \text{H—H(g)} \longrightarrow \text{H—C}\equiv\text{C—H(g)}$$

Recall that bond energies can be used only when all of the atoms or molecules are in the gas phase. In this problem, therefore, you must include the energy necessary to convert carbon from its standard state, the black solid graphite, into carbon atoms in the gas phase. The series of reactions in this case is the following:

$$\begin{array}{ll}
2\,[\text{C(graphite)} \longrightarrow \text{C(g)}] & 2\,[\Delta H^\circ = +717 \text{ kJ/mol}] \\
\text{H—H(g)} \longrightarrow 2\text{ H(g)} & \Delta H^\circ = D_{\text{H—H}} = +436 \text{ kJ/mol} \\
2\text{ C(g)} + 2\text{ H(g)} \longrightarrow \text{H—C}\equiv\text{C—H(g)} & \Delta H^\circ = 2(-D_{\text{C—H}}) + (-D_{\text{C}\equiv\text{C}}) \\
 & \quad = (2\text{ mol})(-414 \text{ kJ/mol}) + \\
 & \qquad (1\text{ mol})(-837 \text{ kJ/mol}) \\
 & \quad = -1665 \text{ kJ} \\
\hline
2\text{ C(graphite)} + \text{H}_2\text{(g)} \longrightarrow \text{H—C}\equiv\text{C—H(g)} & \Delta H^\circ = +205 \text{ kJ/mol}
\end{array}$$

This equation represents the formation of 1 mole of acetylene from the elements, so $\Delta H^\circ_{rxn} = \Delta H^\circ_f = +205$ kJ/mol. (Table 6.2 shows $\Delta H^\circ_f = +226.7$ kJ/mol. The 10% deviation between the experimental and calculated values arises from the use of average bond energies.)

Figure 10.9 The synthesis and combustion of acetylene, C_2H_2. The compound, a colorless gas, is produced by combining calcium carbide, CaC_2, and water.

$$CaC_2(s) + 2\,H_2O(\ell) \rightarrow C_2H_2(g) + Ca(OH)_2(s)$$

A chemist often wishes to know the energy of a bond in a particular compound, since the strength of that bond is an important factor in determining the ease of reactions in which the bond is broken. The problem that follows is an example of the method that is used to *estimate* the energy of a particular bond.

EXAMPLE 10.7

ESTIMATING THE ENERGY OF A BOND

Suppose the C=C double bond energy were not available, and you wished to estimate it. You need a reaction for which all of the bond energies are known except the one of interest, the C=C bond energy. A combustion reaction is a good choice here, because energies of bonds with oxygen are readily available. Using an apparatus similar to that in Figure 6.11, you can determine the heat of combustion of ethylene, $H_2C{=}CH_2$, to be −1323 kJ/mol of C_2H_4, and you can then calculate $D_{C=C}$.

Solution The balanced equation for the combustion of ethylene is

$$\mathrm{H_2C{=}CH_2(g) + 3\ O{=}O(g) \longrightarrow 2\ O{=}C{=}O(g) + 2\ H{-}O{-}H(g)}$$

$$\Delta H^\circ_{\text{combust}} = -1323 \text{ kJ/mol}$$

$$\begin{aligned}
\Delta H^\circ_{\text{combust}} &= \text{sum of energies to break bonds} - \text{sum of energies of bonds made} \\
&= [D_{\mathrm{C=C}} + 4\ D_{\mathrm{C-H}} + 3\ D_{\mathrm{O=O}}] - [4\ D_{\mathrm{C=O}} + 4\ D_{\mathrm{O-H}}] \\
-1323 \text{ kJ} &= [D_{\mathrm{C=C}} + 4 \text{ mol}(414 \text{ kJ/mol}) + 3 \text{ mol}(498 \text{ kJ/mol})] \\
&\quad - [4 \text{ mol}(803 \text{ kJ/mol}) + 4 \text{ mol}(464 \text{ kJ/mol})] \\
-1323 \text{ kJ} &= D_{\mathrm{C=C}} + (-1918 \text{ kJ}) \\
D_{\mathrm{C=C}} &= 595 \text{ kJ/mol}
\end{aligned}$$

This estimate of the bond energy is in good agreement with the accepted value of $D_{\mathrm{C=C}}$ for C_2H_4, which is 611 kJ/mol.

EXERCISE 10.8 Estimating Enthalpy of Formation

Estimate the enthalpy of formation of tetrafluorohydrazine, N_2F_4, from the data in Table 10.4.

$$\mathrm{N{\equiv}N(g) + 2\ F{-}F(g) \longrightarrow F_2\ddot{N}{-}\ddot{N}F_2(g)} \qquad \Delta H^\circ_f = ?$$

EXERCISE 10.9 Estimating Enthalpy of Formation

Using the enthalpy of vaporization of graphite [C(graphite) → C(g)] in Example 10.6 and bond energies from Table 10.4, estimate the enthalpy of formation of carbon tetrachloride, CCl_4(g).

EXERCISE 10.10 Estimating Bond Energies

The enthalpy of formation of formic acid, HCO_2H, is −378.6 kJ/mol.

$$\mathrm{C(graphite) + O{=}O(g) + H{-}H(g) \longrightarrow H{-}C({=}O){-}O{-}H(g)}$$

Estimate the bond energy of the C=O bond in the formic acid molecule. Use the enthalpy of vaporization of C(graphite) in Example 10.6 and the data in Table 10.4.

BOND POLARITY AND ELECTRONEGATIVITY

In Chapter 9 you learned that not all atoms hold onto their valence electrons with equal strength, nor do they add electrons with equal ease. That is, the elements all have different values of ionization energy and electron affinity. It is generally true—though not always true—that, if two different kinds of atoms form a bond, one will attract the shared pair of electrons more strongly than the other. Only when two atoms of the same kind form a bond can we presume that the bond pair is shared equally between the two atoms.

When a bond pair is unequally shared between two atoms, this means there is a displacement of the bonding electrons toward one of the atoms from a point midway between them. As the displacement proceeds, the atom toward which the pair is displaced begins to acquire a negative charge, simply because the net number of negatively charged electrons surrounding the atom begins to exceed the positive charge of the nucleus. At the same time, the atom at the other end of the bond is being depleted of electron density, and it begins to acquire a positive charge. Thus, the bond acquires a positive end and a negative end; that is, it has *electrical poles* and is said to be a **polar bond**. When the displacement of the bonding pair is essentially complete, the bond is ionic, and + and − symbols are written next to the atom symbols in the Lewis drawings. When the displacement is less than complete, however, the bond is said to be a **polar covalent bond**, and we indicate this polarity by writing $\delta+$ and $\delta-$ symbols alongside the atom symbols, where δ stands for a partial charge. Examples of common molecules that have polar bonds are

```
H—F            2δ−                    3δ−
δ+  δ−          O               δ+ H—N—H δ+
               / \                    |
              H   H                   H
             δ+   δ+                  δ+
```

Partial positive charges rarely exceed about 1.5 in polar covalent bonds.

When there is no net displacement of the bonding electron pair, the pair is equally shared, and the bond is **nonpolar covalent**.

The extent of displacement of a bonding electron pair is controlled by the energy change involved in the displacement: the energy that must be expended to displace a bond pair away from one atom of the bond must be overcome by the energy released as the electron pair approaches the other atom. Atomic properties related to these endo- and exothermic energy components are ionization energy and electron affinity (Chapter 9).* Linus Pauling analyzed the energies of bonds in the 1930s, and, from this he developed a scale of **atom electronegativity** (χ) that summarized the energy balance achieved when two atoms form a polar covalent bond (Table 10.5 and Figure 10.10). The *electronegativity of an atom* ***in a molecule*** is now taken as *a measure of the ability of the atom to attract electrons to itself.*

The trends in electronegativity are clearly related to those for the ionization energies and electron affinities of free atoms.

As you examine Table 10.5 and Figure 10.10, you should notice several special features and periodic trends.

(a) The element with the largest electronegativity (4.0) is fluorine in the upper right corner of the table. The element with the smallest value is francium in the lower left corner. In general, electronegativity in-

*The *IE* and *EA* for an atom are the energies involved in the complete loss or gain of an electron by a single atom in the gas phase. Here we are talking about only the partial loss or gain of electrons by an atom *in a molecule*. Therefore, *IE* and *EA* cannot be applied directly to bonding electrons; rather, they serve only as a rough guide to what may occur.

Table 10.5 Electronegativity Values

Key: < 1.0; 1.0–1.4; 1.5–1.9; 2.0–2.4; 2.5–2.9; 3.0–4.0

1 H 2.1																
3 Li 1.0	4 Be 1.5											5 B 2.0	6 C 2.5	7 N 3.0	8 O 3.5	9 F 4.0
11 Na 1.0	12 Mg 1.2											13 Al 1.5	14 Si 1.8	15 P 2.1	16 S 2.5	17 Cl 3.0
19 K 0.9	20 Ca 1.0	21 Sc 1.3	22 Ti 1.4	23 V 1.5	24 Cr 1.6	25 Mn 1.6	26 Fe 1.7	27 Co 1.7	28 Ni 1.8	29 Cu 1.8	30 Zn 1.6	31 Ga 1.7	32 Ge 1.9	33 As 2.1	34 Se 2.4	35 Br 2.8
37 Rb 0.9	38 Sr 1.0	39 Y 1.2	40 Zr 1.3	41 Nb 1.5	42 Mo 1.6	43 Tc 1.7	44 Ru 1.8	45 Rh 1.8	46 Pd 1.8	47 Ag 1.6	48 Cd 1.6	49 In 1.6	50 Sn 1.8	51 Sb 1.9	52 Te 2.1	53 I 2.5
55 Cs 0.8	56 Ba 1.0	57 La 1.1	72 Hf 1.3	73 Ta 1.4	74 W 1.5	75 Re 1.7	76 Os 1.9	77 Ir 1.9	78 Pt 1.8	79 Au 1.9	80 Hg 1.7	81 Tl 1.6	82 Pb 1.7	83 Bi 1.8	84 Po 1.9	85 At 2.1
87 Fr 0.8	88 Ra 1.0	89 Ac 1.1														

creases diagonally upward and to the right in the table, a trend clearly related to the trends in ionization energy and electron affinity.

(b) The greatest variation in electronegativity across any of the periods occurs in the second period (Li...F).

(c) Metals typically have values of electronegativity in the range of slightly less than 1 to about 2. The metalloids are around 2, and nonmetals have values greater than 2.

(d) No values are given for the rare gases. The reason is that only xenon (and perhaps krypton) forms compounds, and then only with a very limited number of other elements.

The periodic trends in electronegativity mean that the most electronegative atoms are in the upper right corner of the table, while the least electronegative are in the lower left corner. If atoms from these two regions of the periodic table form a chemical compound, their large electronegativity difference means that the more electronegative atom of the pair will more strongly attract the bonding electrons, and the bond will

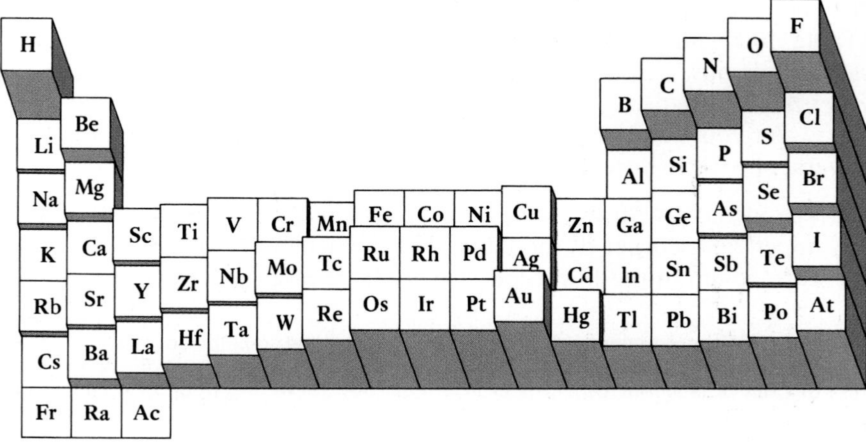

Figure 10.10 The relative values of the electronegativities of atoms in molecules. Fluorine has the highest value (4.0; see Table 10.5). (This table was made using *MacMendeleev*, a program developed for the Apple Macintosh® by S. Sinofsky and J. Clardy. It is available from Kinko's Academic Courseware Exchange.)

Historical Figures in Chemistry: Linus Pauling (1901–)

Linus Pauling was born in Portland, Oregon in 1901, the son of a druggist. He earned a B.Sc. degree in chemical engineering from Oregon State College in 1922 and completed his Ph.D. in chemistry at the California Institute of Technology in 1925. Before joining Cal Tech as a faculty member, he traveled to Europe where he worked briefly with Erwin Schrödinger and Niels Bohr (Chapter 8). In chemistry he is best known for his work on chemical bonding. Indeed, his book *The Nature of the Chemical Bond* has influenced several generations of scientists, and it was for this work that he was awarded the Nobel Prize in Chemistry in 1954. However, shortly after World War II, Pauling and his wife began a crusade to limit nuclear weapons, a crusade that came to fruition in the limited test ban treaty of 1963. For this effort, Pauling was awarded the 1963 Nobel Prize for Peace. Never before had any person received two unshared Nobel Prizes.

be polar. If the polarization is extreme, the bond is nearly ionic. As a useful rule of thumb, the polarity of a bond can be estimated from $\Delta\chi/\Sigma\chi$, the difference in electronegativity values for the two atoms divided by their sum. The quotient takes values ranging from 0 for pure covalent bonds to 1 for completely ionic bonds. For example, CsF should be a highly polar (nearly ionic) combination, since the elements are from opposite corners of the periodic table. Based on electronegativity values, Cs should be the positive end and F the negative end of the bond; $\Delta\chi/\Sigma\chi = (4.0 - 0.8)/(4.0 + 0.8) = 0.67$, a reflection of the highly polar nature of the Cs—F bond.

It was recently suggested by Professor L. C. Allen of Princeton University that the quotient $\Delta\chi/\Sigma\chi$ is a better estimation of relative bond polarity than $\Delta\chi$ alone.

EXAMPLE 10.8

ESTIMATING BOND POLARITIES

For each of the following bond pairs, tell which is the more polar and indicate the negative and positive poles.
(a) Li—F and Li—I (b) C—S and P—P (c) C—O and C—S

Solution (a) Li and F lie on opposite sides of the periodic table, with F in the extreme upper right corner (χ for Li = 1.0 and χ for F = 4.0). Similarly, Li and I are on opposite sides of the table, but I is at the bottom right corner of the table (χ for I = 2.5), indicating that it is much less electronegative than F. Therefore, while both bonds are expected to be strongly polar, with Li positive and the halide negative, the Li—F bond should be more polar than the Li—I bond.

$\delta+$ $\delta-$	$\delta+$ $\delta-$
Li—F	Li—I
$\Delta\chi/\Sigma\chi = 0.60$, polar	$\Delta\chi/\Sigma\chi = 0.43$, moderately polar

(b) The P—P bond is nonpolar (or purely covalent) because the bond is between two atoms of the same kind. C is in the second period, while S is in the third period but closer to the right side of the table. It is no surprise to find, therefore, that C and S both have the same electronegativity (2.5), and so their bond is also predicted to be nonpolar.
(c) O lies above S in the periodic table, so O is more electronegative than S. This means the C—O bond must be more polar than the C—S bond, which was

predicted in part (b) to be nonpolar. For the C—O bond, O is the more negative atom of the polar bond, $\overset{\delta+}{C}—\overset{\delta-}{O}$. Here $\Delta\chi/\Sigma\chi$ is 0.17, indicating a polar covalent bond.

EXERCISE 10.11 Bond Polarity

For each of the following pairs of bonds, decide which is the more polar. For each polar bond, indicate the positive and negative poles. First make your prediction from the relative atom positions in the periodic table; then check your prediction by calculating $\Delta\chi/\Sigma\chi$.

(a) H—F and H—I (b) B—C and B—F (c) C—Si and C—S

OXIDATION NUMBERS AND ATOM FORMAL CHARGES

As you learn about chemistry, you will see that there are a number of practical applications of the concept of electronegativity. Some will be introduced later in this chapter and in Chapter 13, for example. For the moment, we can use it for two purposes: to understand the concept of oxidation numbers, and to help us decide which of several molecular resonance structures is most likely.

OXIDATION NUMBERS We introduced oxidation numbers in Chapter 5 as an aid in considering oxidation-reduction reactions. Although oxidation numbers are artificial, they are useful as a "bookkeeping" device for electrons in reactions.

The net electrical charge on a free atom is zero. If the atom is bound to another in a molecule, however, some of its valence electrons are shared with other atoms. *The oxidation number of an atom is the charge that the atom would have if ALL of its bonds were considered completely ionic.* For example, from the rules introduced in Chapter 5, we assumed that the oxidation number of H in HF is +1, while the oxidation number of the F atom is −1. If the H—F bond were considered completely ionic, the bond pair would be transferred to the more electronegative atom, F, and the atom would become the F^- ion; this would leave H as a simple H^+ ion.

H—F̤̈: ⟶ (H)(:F̤̈:) equivalent to H^+, F^-

When an atom in a compound or polyatomic ion is bound to several oxygen and/or halogen atoms, it can have a very large oxidation number. As an example, the Lewis dot structure for the perchlorate ion is

```
[      ..       ]-
[     :O:       ]
[      |        ]
[  ..      ..   ]
[ :O—Cl—O:      ]
[  ..      ..   ]
[      |        ]
[     :O:       ]
[      ..       ]
```

Since each O atom is more electronegative than the Cl atom, in the ionic limit the bonding electrons cluster around the O atoms.

$$\left[\begin{array}{ccc} & (:\ddot{O}:)^{2-} & \\ (:\ddot{O}:)^{2-} & (Cl)^{7+} & (:\ddot{O}:)^{2-} \\ & (:\ddot{O}:)^{2-} & \end{array} \right]^-$$

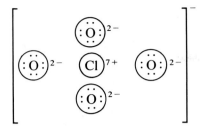

Thus, Cl is assigned the extremely high oxidation number of +7. Experiments by atomic physicists show that such ions will voraciously attract electrons. Therefore, it is inconceivable that any species under normal conditions can really contain an atom in such a highly charged state. Indeed, any charge in excess of ±2 on an atom in a chemical compound is *highly* improbable. This tells you that *oxidation numbers usually do not represent the real charges on atoms*. They are simply a convention that indicates what the maximum charge could possibly be (a charge we refer to as the "ionic limiting value"). Nonetheless, chemists often find oxidation numbers useful to keep track of electrons in molecules and in chemical reactions (Chapter 5) and to indicate the more positive and more negative atoms in a molecule. The latter is often used in understanding how molecules react to form new bonds.

EXERCISE 10.12 Oxidation Numbers
Using the Lewis structure for each of the following, calculate the oxidation number for each atom.
(a) SF_4 (b) CO_3^{2-} (c) SO_3

FORMAL CHARGES ON ATOMS The method used to assign oxidation numbers is especially unrealistic for covalent bonds. However, by assuming that each bond pair is shared equally by two atoms, we can derive a more reasonable estimate of atom charges in molecules. These are called **atom formal charges** and are given by the expression

Oxidation numbers are "atom charges" in the ionic limit, whereas formal charges are "atom charges" in the covalent limit.

$$\text{Atom formal charge} = \text{group number} - \text{number of lone pair electrons} - \tfrac{1}{2}\,(\text{number of bonding electrons})$$

Here we count the lone pair electrons about an atom and one half of the bonding electrons that it shares. To illustrate, look again at ClO_4^-. Begin with the Lewis structure in the margin. Using the expression above, we calculate a formal charge of −1 for each O atom (whereas the oxidation number is −2),

$$\left[\begin{array}{ccc} & :\ddot{O}: & \\ & | & \\ :\ddot{O}- & Cl & -\ddot{O}: \\ & | & \\ & :\ddot{O}: & \end{array} \right]^-$$

$$\text{Formal charge for O} = 6 - 6 - \tfrac{1}{2}(2) = -1$$

and a formal charge for Cl of +3 (the oxidation number is +7).

$$\text{Formal charge for Cl} = 7 - 0 - \tfrac{1}{2}(8) = +3$$

Notice that the signs of the formal charges and oxidation numbers are the same (as is most often the case), but the magnitude of the charge is more realistic. Also notice that *the sum of the formal charges must equal the ion charge*.

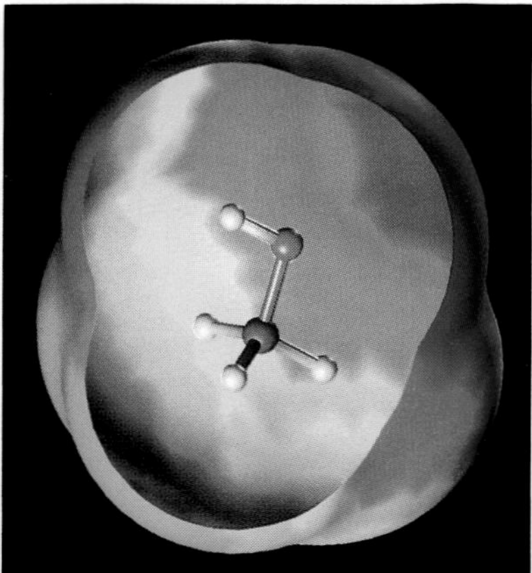

A computer-generated model of methanol, CH_3OH (C = gray, H = white, and O = red), inside a model of the molecular surface. Chemists are interested in learning about reactive sites in molecules, and the theory from which this computer model was calculated tells us the most reactive site corresponds to the yellow and red zones on the surface of the molecule (the O—H group). For the O—H group in CH_3OH, Allen's method gives formal charges of +0.25 for H −0.42 for O. (J. Weber)

Atom Formal Charges and Electronegativity

Since chemical reactions often involve the attack by some reagent at the more positive or more negative site in a molecule, chemists have long been interested in making reasonably accurate predictions of atom charges. Recently Professor Leland Allen of Princeton University has found that more realistic charges are estimated if the factor of $\frac{1}{2}$ in the formula for calculating formal charge is replaced by $\chi_a/\Sigma\chi$ for an atom a bonded to b (where $\Sigma\chi$ is the sum of $\chi_a + \chi_b$). This means that bond pairs are apportioned according to the relative electronegativities of the atoms and not equally.

Formaldehyde, H_2CO, has the Lewis structure

$$\begin{matrix} \text{H} \\ & \diagdown \\ & & \text{C}=\ddot{\text{O}}: \\ & \diagup \\ \text{H} \end{matrix}$$

This molecule is of considerable interest to chemists because it has many uses, chiefly to make polymers. If you use Allen's method to find the oxygen atom charge, you would calculate a value of −0.33 [= 6 − 4 − (3.5/6.0)(4)]. The charge on the carbon atom becomes +0.16 [= 4 − 0 − (2.5/6.0)(4) − (2.5/4.6)(2) − (2.5/4.6)(2)]. Since we know that the formal charges on O, C, and the two H's must equal zero, this means the charges on the H atoms are both +0.09. These values are chemically more meaningful than the formal charge of 0 for each atom as found by the method in the text.

In the resonance structures for O_3 and CO_3^{2-} drawn in Section 10.4, all the possible electron configurations are equally likely. Therefore, the actual structure of the molecule or ion is a symmetrical distribution of electrons over all the atoms involved—that is, an equal "mixture" of the resonance structures. This is not the case for all molecules that can be represented by resonance structures, however. In many such molecules, the actual structure is more like one of the resonance structures than like the others. We can use formal charges together with the following two rules to decide which resonance structure is most stable and therefore the most important.

1. Atoms in molecules (or ions) should have formal charges as small as possible. (This is called the **principle of electroneutrality**.)
2. A molecule (or ion) is most stable when any negative formal charge resides on the most electronegative atom.

Let us use atom formal charges as an aid to decide which of two resonance structures is more important for CO_2.

Formal charges	0 0 0	+1 0 −1
Resonance structures	$:\ddot{O}=C=\ddot{O}:$	$:O\equiv C-\ddot{\underset{\cdot\cdot}{O}}:$
	A	B

For structure A, each atom has a formal charge of 0, a favorable situation. In B, however, one oxygen atom has a charge of +1. This is an unfavorable condition for the very electronegative oxygen atom, so resonance structure B is of little importance.

EXAMPLE 10.9

ATOM FORMAL CHARGES AND RESONANCE STRUCTURES

There are three possible resonance structures for the cyanate ion, NCO^-. Using atom formal charges, decide which is the most important of these structures.

Solution Formal charges have been calculated for each of the NCO^- resonance structures given on page 375 and are listed above the appropriate atom in each structure.

	A	B	C
Formal charges	−2 0 +1	−1 0 0	0 0 −1
Resonance structures	$[:\ddot{\underset{..}{N}}—C≡O:]^-$	$[:\ddot{N}=C=\ddot{O}:]^-$	$[:N≡C—\ddot{\underset{..}{O}}:]^-$

As an example of the calculation of formal charge, consider structure A.

$$\begin{aligned} \text{Formal charge for N} &= 5 - 6 - \tfrac{1}{2}(2) = -2 \\ \text{Formal charge for C} &= 4 - 0 - \tfrac{1}{2}(8) = 0 \\ \text{Formal charge for O} &= 6 - 2 - \tfrac{1}{2}(6) = \underline{+1} \\ \text{Sum of formal charges} &= -1 = \text{charge on ion} \end{aligned}$$

In structure A, the O atom has a formal charge of +1, a very unfavorable situation for this electronegative atom. Therefore, A contributes little to the overall electronic structure of the cyanate ion. Structure B places a −1 charge on nitrogen and 0 on oxygen, not an unfavorable situation. However, structure C may be favored slightly because the negative atom in the structure is the most electronegative one in the ion.

In the actual structure, we would expect to find an average electron density between the N and C atoms that is greater than in a double bond but less than in a triple bond. Similarly, the electron density in the C—O bond region should be between those of single and double bonds. These expectations are confirmed by measurements of bond lengths.

EXERCISE 10.13 Atom Formal Charges

The Lewis structure of H_2BF is depicted by A below. To satisfy the electron deficiency of the boron atom, one could write the resonance structure B. Calculate the formal charge on each atom in structure B and comment on the relative importance of the two structures.

```
  ..          
 :F:         :F:
  |           ||
  B           B
 / \         / \
H   H       H   H

  A           B
```

10.6 MOLECULAR SHAPE

Lewis structures show only how many bond pairs and lone pairs surround a given atom. However, molecules are three-dimensional, and their atoms may not all lie in the same plane. It is often important to know the way molecules fill space, because the three-dimensional structure partly de-

(a)

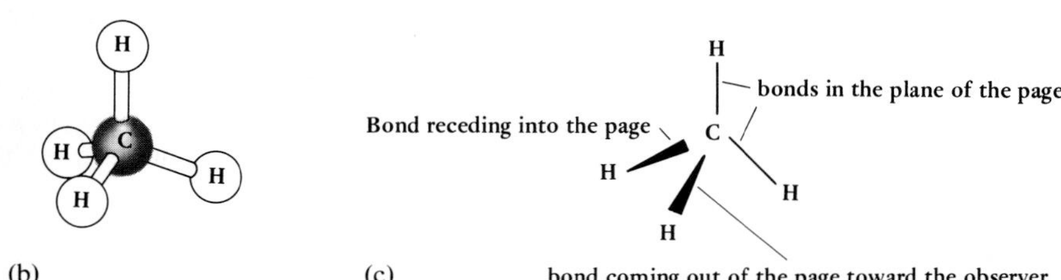

Figure 10.11 The structure of methane, CH_4, is used to show the ways molecular structures can be illustrated. (a) Photograph and (b) drawing of a ball-and-stick model. (c) Perspective drawing.

termines the chemical functioning of the molecule. Pharmaceutical companies, for example, use knowledge of molecular shape to design drugs that will fit into the site in the body where pain is to be relieved or disease attacked.

The importance of molecular shape is illustrated by the recent report that highly toxic compounds called dioxins "appear to mimic structurally similar natural steroid hormones. These hormones react with genes in cells to catalyze complex changes in the body. Like a steroid, a dioxin molecule enters a cell and bonds to a receptor, fitting like a key into a lock." (New York Times, May 15, 1990, page C4). Since cancers of the breast and uterus are believed to be promoted by the female sex hormone estrogen, pharmaceutical chemists are trying to synthesize molecules that are structurally similar to dioxins, but less toxic, with the hope that such molecules may serve as "anti-estrogenic treatments for breast and uterine cancers."

To convey a sense of three-dimensionality for a molecule drawn on a flat piece of paper, we use sketches such as the "ball-and-stick" model of methane in Figure 10.11, or we can draw structures in perspective using "wedges" for bonds that emerge from or recede into the plane of the drawing.

CORRELATION BETWEEN STRUCTURE AND VALENCE ELECTRON PAIRS: THE VSEPR MODEL

A sampling of perspective sketches and ball-and-stick models of molecules for which we have already drawn Lewis structures is shown in Table 10.6. Although you do not yet know how to predict the structures in this table, there is an easy way to do this. Notice how the molecular shape changes with the numbers of sigma (σ) bonds plus lone pairs about the central atom.

SIGMA BONDS + LONE PAIRS ON CENTRAL ATOM	STRUCTURE OF MOLECULE (TABLE 10.6)
2	Linear
3	Trigonal planar
4	Tetrahedral (or pyramidal)
5	Trigonal bipyramidal
6	Octahedral

The idea that will allow us to predict the molecular structures of compounds of *s*- and *p*-block elements is that each lone pair or bond pair group (σ and π pairs) repels all other lone pairs and bond pair groups. Because the pairs try to avoid one another, they move as far apart as possible; however, since all of the pairs are "tied" to the same central atom, they can only orient themselves to make the angles between them as large as possible. This is the essence of the **Valence Shell *E*lectron *P*air**

Table 10.6 Perspective Molecular Sketches

Molecule	Geometry	Perspective Sketch	Ball-and-Stick Model
CO_2	Linear 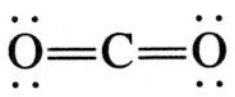 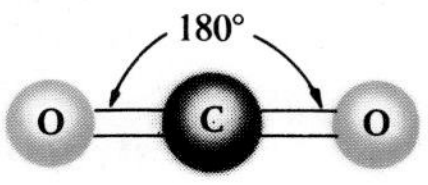 	$\ddot{\underset{\cdot\cdot}{O}}=C=\ddot{\underset{\cdot\cdot}{O}}$	180°
CO_3^{2-}	Trigonal planar 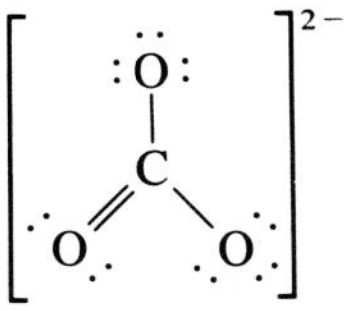 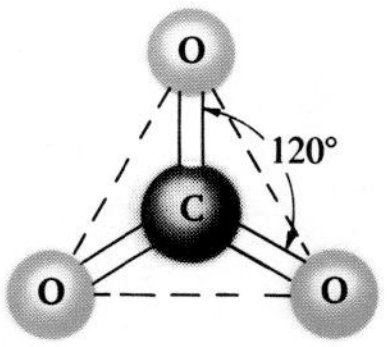 		120°
NH_3	Pyramidal 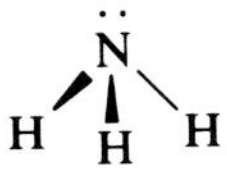 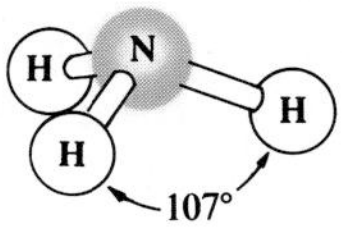 		107°
CH_4	Tetrahedral 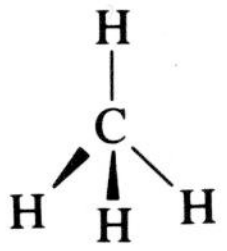 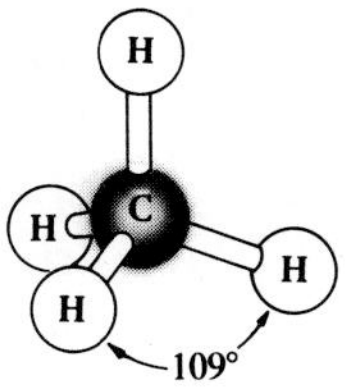 		109°
PCl_5*	Trigonal bipyramidal 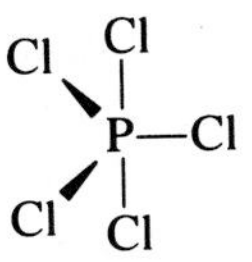 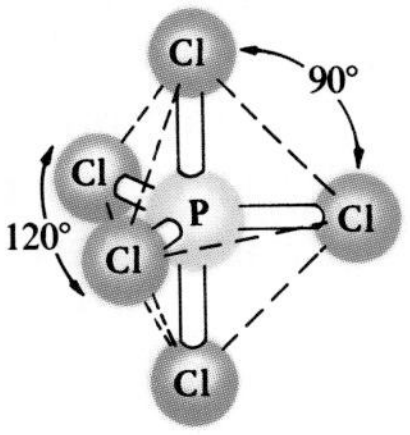 		90°, 120°
SF_6*	Octahedral  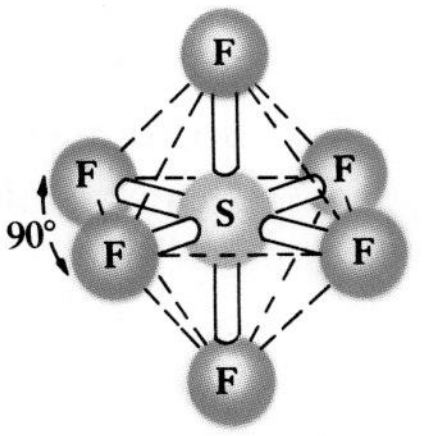 		90°

*Lone pairs not shown on Cl or F.

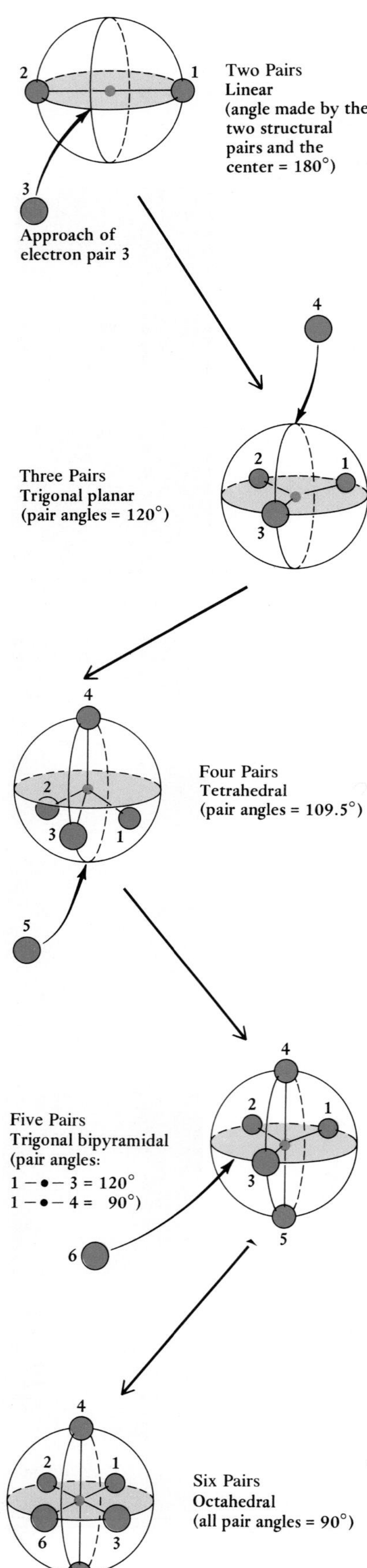

***R*epulsion** theory (**VSEPR** for short) for predicting molecular structures.* It is a theory that works extraordinarily well for compounds of *s*- and *p*-block elements. However, *it does not apply to compounds of the transition elements* except under special circumstances, and we shall not deal with these in this text.

To show you how to use the VSEPR model, we shall first designate as **structural pairs** the sigma and lone pairs about an atom. (The pi pairs are deliberately excluded as structural pairs, because each pi pair occupies the same bond region as a sigma pair.) To visualize the structural pair orientations, imagine them situated on the surface of a sphere with the central atom at the center.

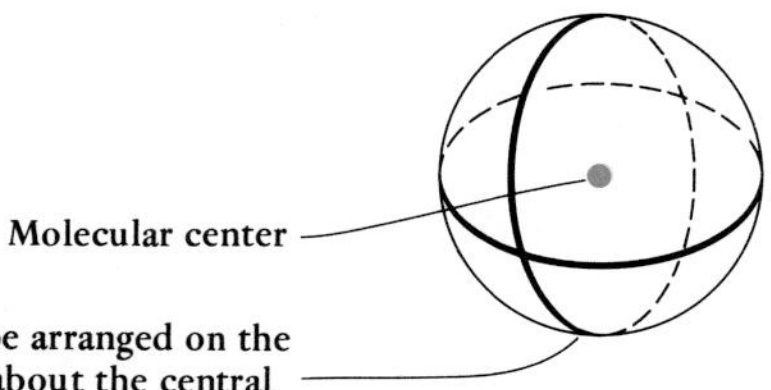

If there are only *two structural pairs* about the nucleus, they can best avoid each other if they are 180° apart. This means that the two pairs and the central atom are in a line; the arrangement is *linear*. If we add a third structural pair midway between the first two pairs, you see that the newly added pair will repel the first two and push them back until the angles between them are 120°, and they all lie in the same plane. We say this arrangement is *trigonal planar*. If a fourth structural pair is added, it will take a position above or below the triangular plane (where it finds the most space) and will push back the previous three pairs until all of the angles are equal, this time 109.5°, and the orientation of pairs is *tetrahedral*. If a fifth pair enters the valence shell, it will do so opposite any of the first four pairs and will push its three nearest neighbors back to a trigonal planar arrangement. The orientation of the five pairs is now *trigonal bipyramidal* (if you connect the electron pairs with imaginary lines, the structure looks like two triangular pyramids that share a triangular face).

Finally, if a sixth structural pair is added to the five of the trigonal bipyramid, this new pair will find the most room in the triangular plane and enter there, pushing back its two nearest neighbors so that all of the angles are 90°, and the electron pair orientation is *octahedral*.

To summarize, what we have found so far are the orientations about the central atom of 2 to 6 structural pairs (Figure 10.12). The arrangement for a particular number of pairs is called its **structural-pair geometry**. Some or all of the structural pairs around an atom in a molecule will be bond pairs, and it only remains to mark the bond pairs with the atomic symbols of the terminal atoms to arrive at the full, three-dimensional molecular geometry.

*The VSEPR model was devised by Ronald J. Gillespie (1924–) and Ronald S. Nyholm (1917–1971). Gillespie was born in England and Nyholm in Australia, but both received the Ph.D. in chemistry at University College, London. Nyholm made many contributions to chemistry until his untimely death. Gillespie has been Professor of Chemistry at McMaster University (Canada) since 1960.

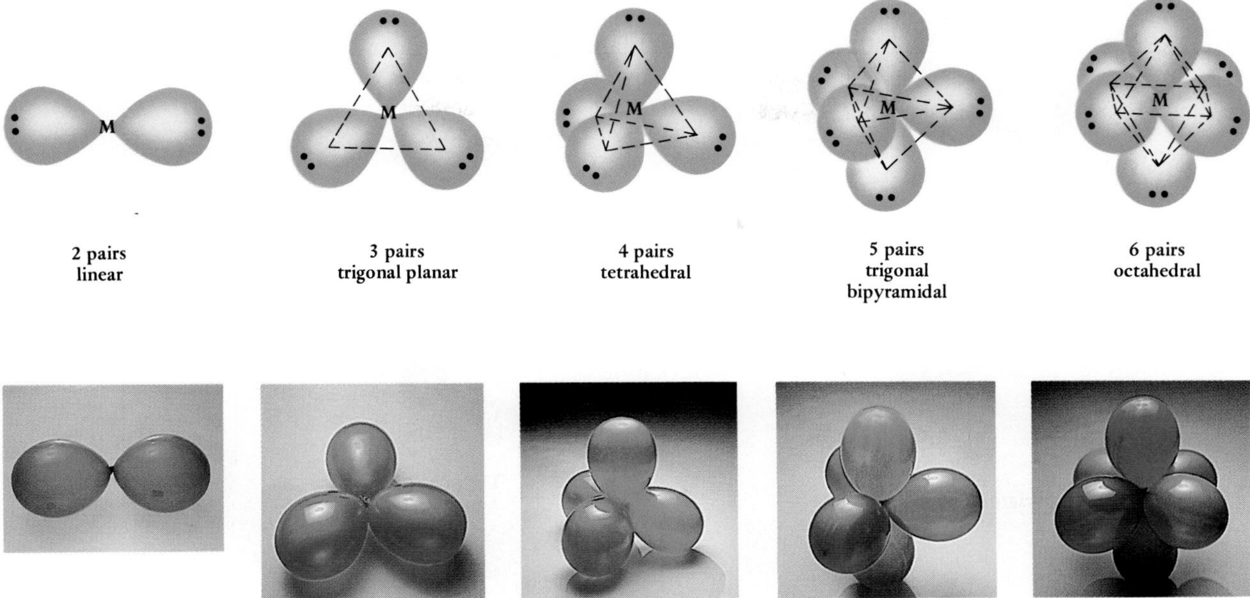

Figure 10.12 Structural pair geometries for 2 to 6 structural pairs. The balloon models are especially convenient. If one ties together several balloons of similar size and shape, they will naturally assume the geometries illustrated.

EXAMPLE 10.10

STRUCTURAL-PAIR GEOMETRIES

Give the structural-pair geometry for (a) NH_3, (b) H_2CO, and (c) XeF_2.

Solution One *must* first draw the Lewis structure for the molecule or ion. In this case, Lewis structures have already been presented in Examples 10.2 and 10.3 and are reproduced below.

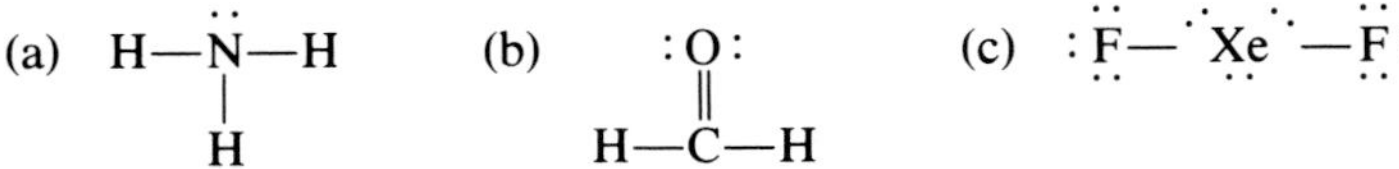

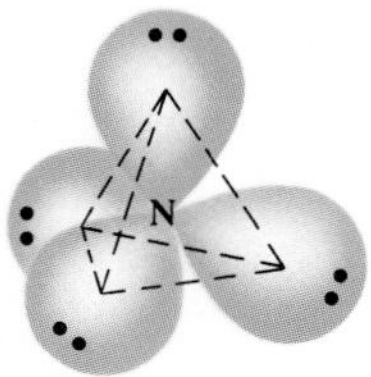

NH_3 structural pair geometry

(a) Ammonia, NH_3: Four pairs of electrons (3 sigma bond pairs and 1 lone pair) surround the central N atom. Therefore, the structural pair geometry around N is tetrahedral.

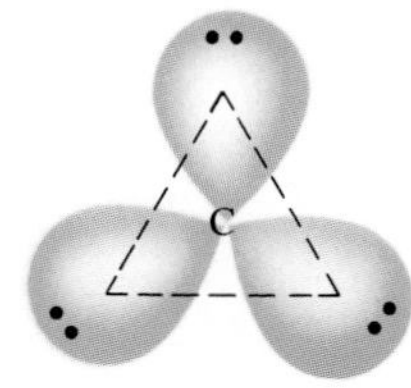

H_2CO structural pair geometry

(b) Formaldehyde, H_2CO: The C atom is surrounded by 3 sigma bonding pairs. (There is also one pi bonding pair, but this is ignored for purposes of deciding on geometry.) Therefore, the structural pair geometry is trigonal planar.

(c) Xenon difluoride, XeF_2: The central Xe atom is surrounded by 2 sigma bond pairs and 3 lone pairs. Therefore, the structural-pair geometry is trigonal bipyramidal.

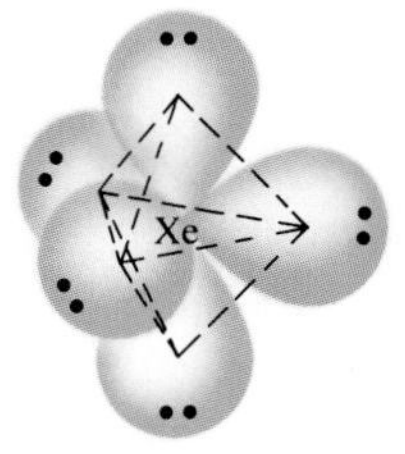

XeF_2 structural pair geometry

EXERCISE 10.14 Structural-Pair Geometries

Determine the structural pair geometry for (a) H_2O, (b) NO_2^+, and (c) SF_4.

Table 10.7 Structural Pair Geometries and Molecular Shapes for Molecules with Two, Three, and Four Structural Pairs about the Central Atom

Number of Structural Pairs	Structural Pair Geometry	Sigma Bond Pairs	Lone Pairs	Molecular Shape	Example
2	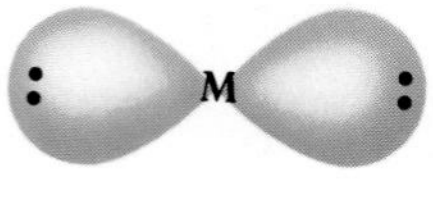 Linear	2	0	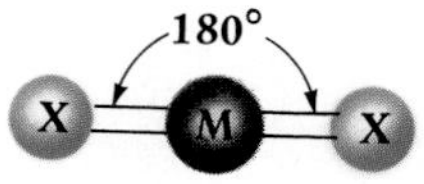 Linear (180°)	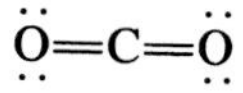 O=C=O
3	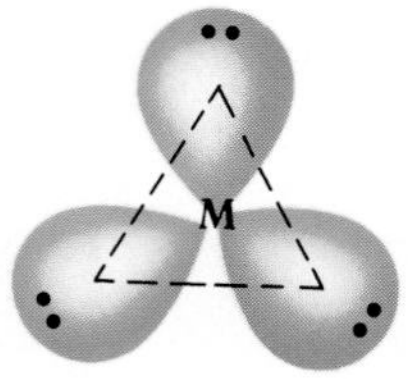 Trigonal planar	3	0	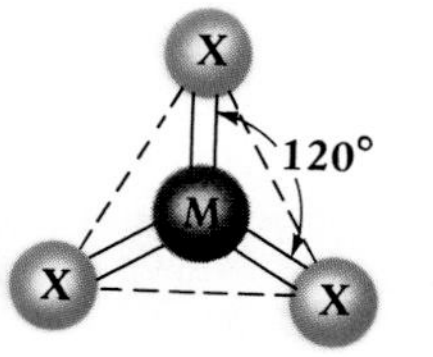 Trigonal planar (120°)	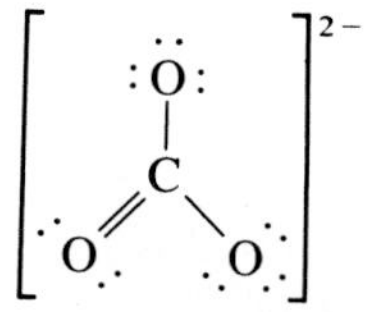 $[CO_3]^{2-}$
		2	1	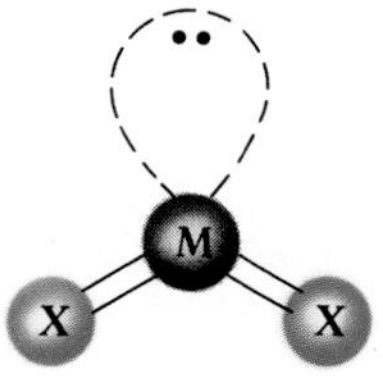 Bent	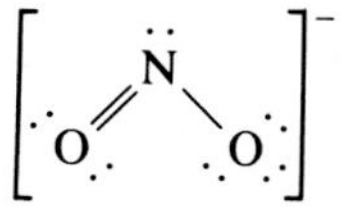 $[NO_2]^{-}$
4	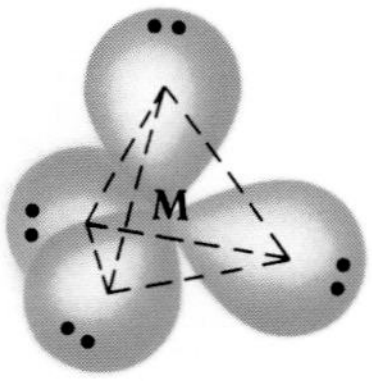 Tetrahedral	4	0	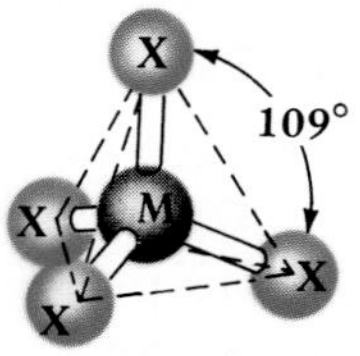 Tetrahedral (109°)	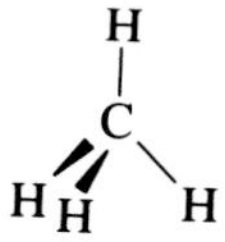 CH_4
		3	1	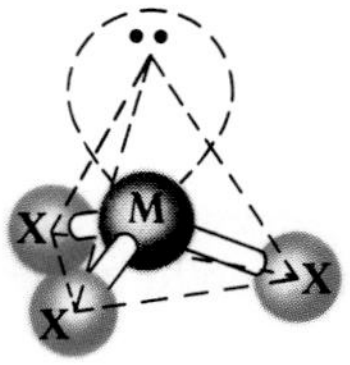 Pyramidal	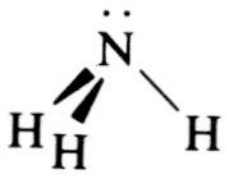 NH_3
		2	2	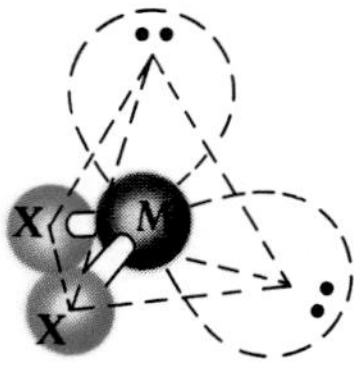 Bent	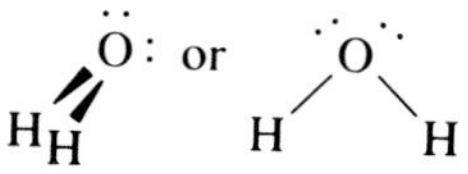 H_2O or H_2O

THE VSEPR MODEL AND MOLECULAR SHAPE

To a chemist, one of the most interesting aspects of a molecule or ion is its shape or **molecular geometry**, *the arrangement in space of the atoms bonded to a central atom*. This may or may not be the same as the shape described by the structural pairs, the *structural-pair geometry* of the ion or molecule, depending on whether there are any lone pairs around the central atom. To derive the three-dimensional molecular geometry, we need only to identify which of the structural pairs are bond pairs and then describe the bond pair locations.

MOLECULAR SHAPES FOR CENTRAL ATOMS WITH NORMAL VALENCE When the central atom of a molecule or ion obeys the octet rule, there can be no more than four structural pairs, and the complete range of molecular shapes that can arise from various combinations of bond and lone pairs is illustrated in Table 10.7. These are by far the most commonly occurring structural types for normal valence compounds, and you should familiarize yourself with them thoroughly.

You must recognize that *the name given to the shape of a molecule is the word that best describes the* ***relative positions of the atoms***. The name does *not* always describe the location of the structural pairs. For example, the structure of water is said to be "bent," because this describes the relative *atom* locations. Water has a bent configuration because the four structural pairs are at the corners of a tetrahedron. However, only two of these pairs are used to bind H atoms, and the two bond pairs determine the molecular geometry.

The NO_2^- ion is also described as bent. The ion has three structural pairs at the corners of a triangle, but only two of these pairs are sigma bond pairs. The net effect is that the three atoms appear in a bent configuration.

Finally, look at the examples having pi bonds: CO_2, NO_2^-, and CO_3^{2-}. Notice that, in counting structural pairs, we have ignored pi pairs. Thus, the C atom in CO_2 is surrounded by two structural pairs, and a linear structure results. The N in NO_2^- and the C in CO_3^{2-} are the centers of three structural pairs. Since all three structural pairs in CO_3^{2-} are used in bonding, the ion shape is said to be trigonal planar.

You can describe the shape of virtually any molecule by thinking through the following steps: (1) Sketch the Lewis dot structure. (2) Describe the structural-pair geometry. (3) Describe which structural pairs are sigma bond pairs (or bond pair groups), and then describe their locations.

EXAMPLE 10.11

FINDING THE SHAPES OF MOLECULES

What are the molecular shapes of PH_3, SO_2, and NO_3^-?

Solution (a) The Lewis structure of phosphine, PH_3,

```
   ..
H—P—H
   |
   H
```

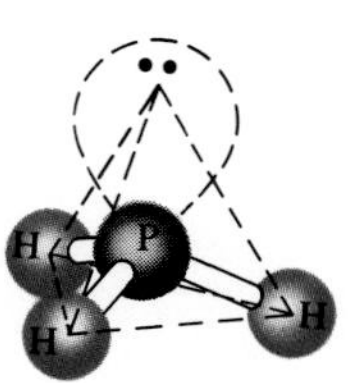

PH_3 structure

reveals that the central P atom has four structural pairs, so the structural-pair geometry is tetrahedral. Since three of the four structural pairs are used to bond terminal atoms, the central P atom and the three H atoms form a *pyramidal* molecular shape like NH_3.

(b) The structure of sulfur dioxide is a very good illustration of the fact that chemistry is a dynamic, constantly developing field of science. Following the principles outlined in this chapter, most chemists would construct two resonance structures for SO_2 (*a* and *b* below). However, it has recently been argued that a third resonance structure, *c*, is actually the most important form of the molecule.*

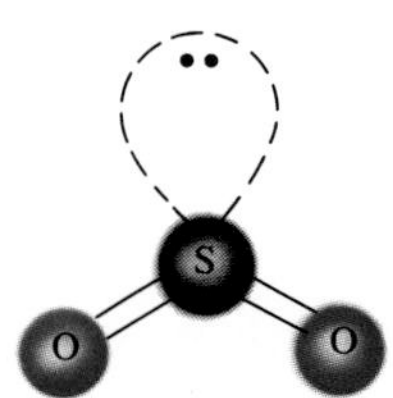

SO$_2$ structure

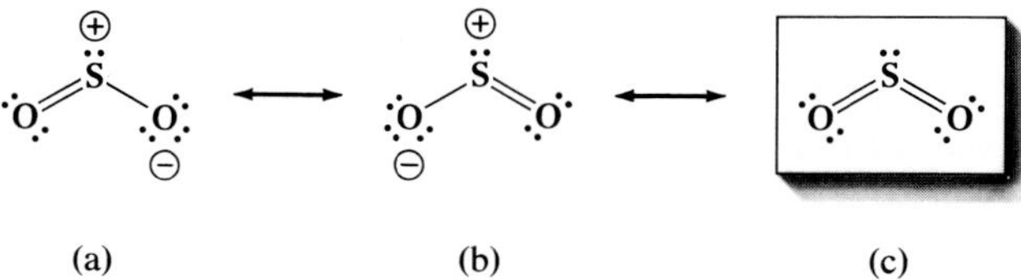

(a) (b) (c)

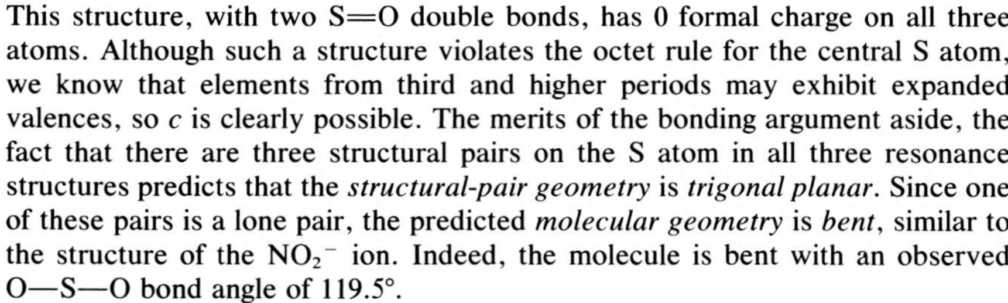

This structure, with two S═O double bonds, has 0 formal charge on all three atoms. Although such a structure violates the octet rule for the central S atom, we know that elements from third and higher periods may exhibit expanded valences, so *c* is clearly possible. The merits of the bonding argument aside, the fact that there are three structural pairs on the S atom in all three resonance structures predicts that the *structural-pair geometry* is *trigonal planar*. Since one of these pairs is a lone pair, the predicted *molecular geometry* is *bent*, similar to the structure of the NO_2^- ion. Indeed, the molecule is bent with an observed O—S—O bond angle of 119.5°.

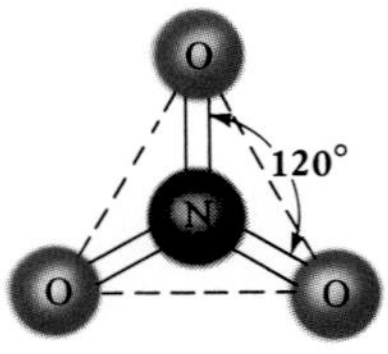

NO_3^- structure

(c) The NO_3^- ion has the same number of valence electrons as the CO_3^{2-} ion in Table 10.7. Thus, like the carbonate ion, the structural-pair geometry and molecular shape of the nitrate ion are both trigonal planar.

EXERCISE 10.15 Determining Molecular Shapes

Determine the structural-pair geometry and molecular shape for (a) carbon disulfide, CS_2; (b) hydrogen sulfide, H_2S; and (c) the phosphate ion, PO_4^{3-}.

MOLECULAR SHAPES FOR CENTRAL ATOMS WITH EXPANDED VALENCE For a central atom with five or six structural pairs (the trigonal bipyramid or octahedron), there are quite a few molecular structure possibilities. Fortunately, not all these possibilities occur in nature as stable structures because there seems to be a limit of three lone pairs about the central atom. The molecular structures that have been observed are illustrated in Table 10.8.

Let us consider first the possibilities for five structural pairs in Table 10.8. The angles in the triangular plane are all 120°, while the angles between any of these pairs and an upper or lower pair are only 90°. (Because the positions in the trigonal plane lie in the equator of the imaginary sphere around the central atom, they are called the **equatorial** positions. The north and south poles are called the **axial** positions.) Each equatorial position is closely flanked by only two other positions (the

*G.H. Purser, *J. Chem. Educ.* **1989**, *66*, 710–713, shows the importance of resonance structure C by detailed arguments using VSEPR theory and molecular orbital theory. The principles of the latter theory are described in Chapter 11.

Molecular Structure: Bond Angles

Not only can you determine the overall shape of a molecule or ion from the VSEPR model, but finer details can be predicted and explained. For example, methane (CH_4), ammonia (NH_3), and water (H_2O) are based on a tetrahedral arrangement of structural pairs. In methane, each electron pair is used to bind an H atom, and the angle between neighboring H atoms, the bond angle, is 109.5°. Notice that, as you proceed from CH_4 to NH_3 to H_2O, the number of lone pairs increases and the bond angle decreases.

Methane	Ammonia	Water
109.5°	107°	105°
4 bond pairs	3 bond pairs 1 lone pair	2 bond pairs 2 lone pairs

The reason for this is that lone pairs require more space on the surface of the imaginary sphere about the central atom. Bond pairs are pulled into the bond region by the strong attractive forces of two nuclei and are, therefore, relatively compact; they are "skinny." For a lone pair, however, there is only one nucleus attracting the electron pair, and this nuclear charge is not so effective in overcoming the normal repulsive forces between two negative electrons; as a result, lone pairs are "fat."

axial), while an axial position is closely flanked by three positions (the equatorial). This means that any lone pairs, which we *assume* to be "fatter" than bond pairs,* will occupy equatorial positions rather than an axial position. In XeF_2, the three lone pairs all occupy equatorial positions, thus forcing the bonding pairs into axial positions, so the molecular geometry is linear. In ClF_3, the two lone pairs prefer two of the three equatorial positions, leaving one equatorial and two axial positions for the three bonded F atoms. Consequently, the molecular geometry is described as T-shaped.

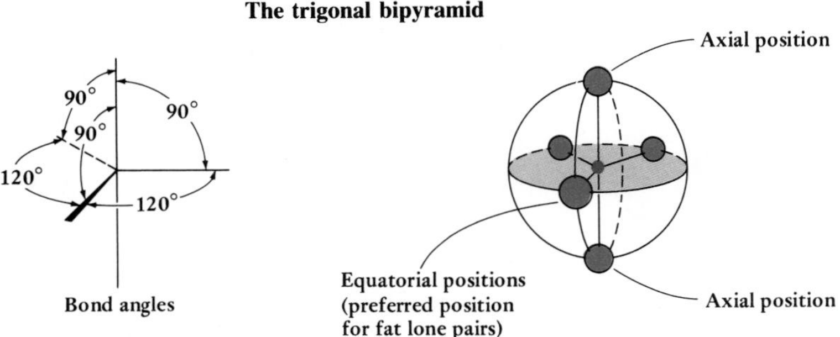

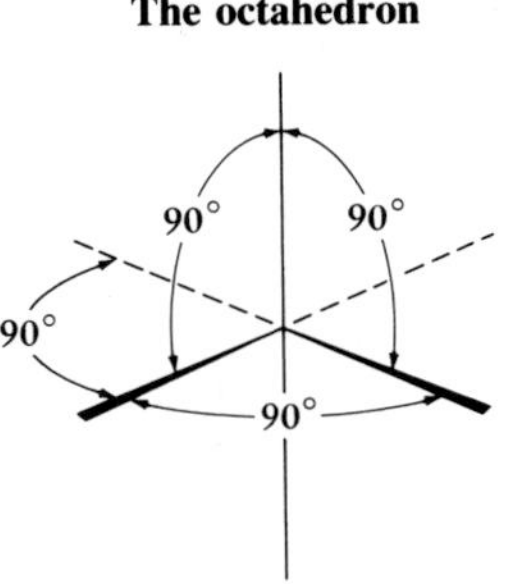

Unlike the trigonal bipyramid, all of the angles in the octahedron are 90°. There are no distinct axial and equatorial positions; all positions are equivalent, and each is flanked by four structural pairs. Therefore, if

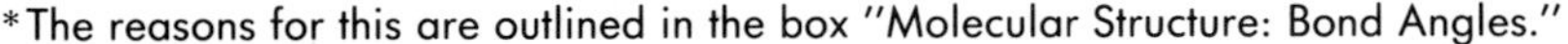

*The reasons for this are outlined in the box "Molecular Structure: Bond Angles."

Table 10.8 Structural Pair Geometries and Molecular Shapes for Molecules with Five or Six Valence Shell Electron Pairs

Structural Pair Geometry	Bond Pairs	Lone Pairs	Molecular Shape	Description	Example (Lone Pairs Not Shown)
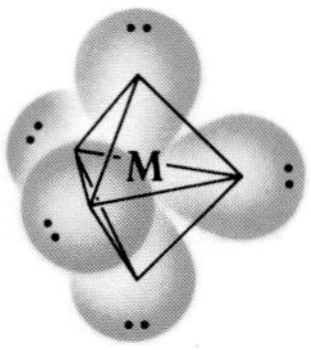	5	0	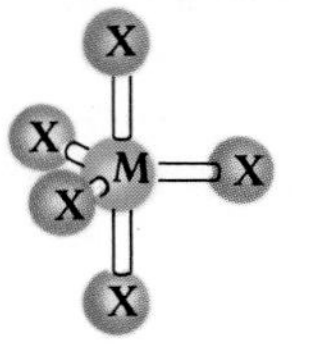	Trigonal bipyramidal	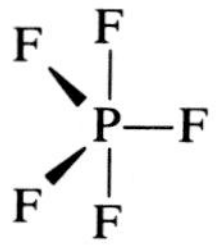 PF_5
Trigonal bipyramid	4	1	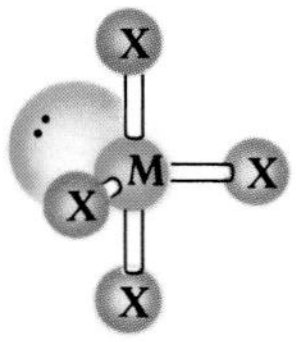	Unsymmetrical tetrahedron	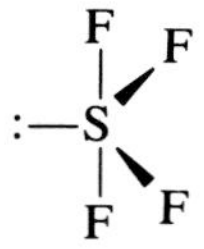 SF_4
	3	2	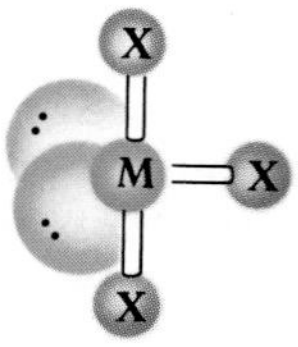	T-shaped	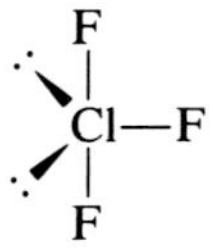 ClF_3
	2	3	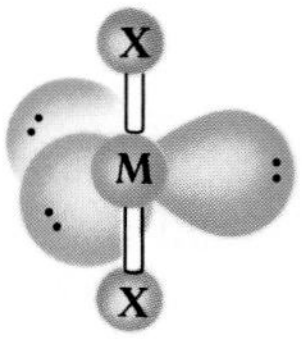	Linear	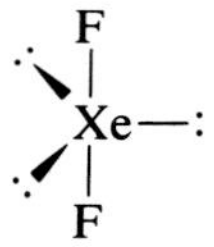 XeF_2
Octahedral 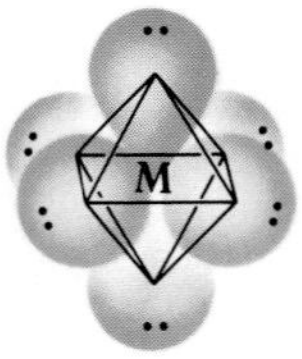	6	0	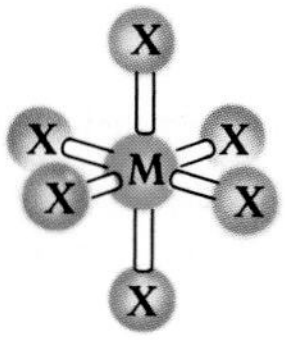	Octahedral	 SF_6
	5	1	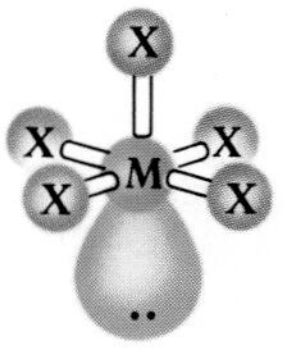	Square pyramidal	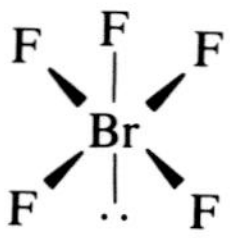 BrF_5
	4	2	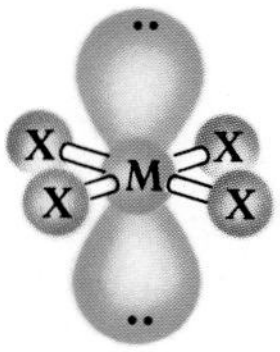	Square planar	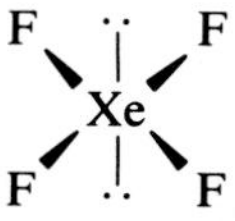 XeF_4

a molecule has one lone pair, as in BrF_5, it makes no difference where you place it. On the other hand, if the molecule or ion has two lone pairs, as in ICl_4^-, each of the lone pairs needs as much room as possible. This is best achieved by placing the lone pairs on opposite sides of the center. In this way, the four pairs flanking each "fat" lone pair are "skinny" bond pairs, and there are no lone pairs flanking another lone pair.

EXAMPLE 10.12

VSEPR AND MOLECULAR STRUCTURE

Draw the Lewis structure for $OXeF_4$ and then describe the structural-pair geometry and molecular shape.

Solution The molecule has a total of 42 valence electrons: 8 for Xe + 6 for O + 4(7) for F. When we use the guidelines for drawing Lewis structures, we draw a structure with a total of six electron pairs about the central Xe atom, one of these being a lone pair.

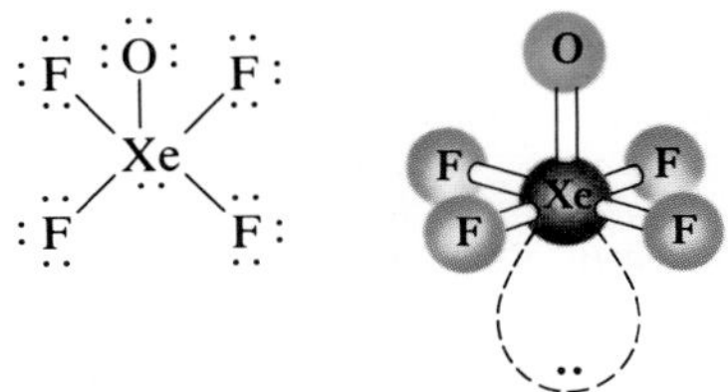

The six pairs about the Xe atom mean that the structural-pair geometry is octahedral. Since only five of the structural pairs are sigma (σ) bond pairs, this means the molecular shape is a square pyramid. Although we will not worry for the moment about the exact placement of the O and F atoms, they could in principle be placed so that either the O or an F atom is across the molecule from the lone pair. It is known in this case that the real geometry is that shown.

EXERCISE 10.16 Molecular and Ionic Structure
Draw the Lewis structure for ICl_2^- and then decide the structural-pair and molecular geometries of the ion.

EXERCISE 10.17 Bond Angles
What are the approximate values of the S—C—S and H—S—H angles in CS_2 and H_2S, respectively?

10.7 MOLECULAR POLARITY

The adjective "polar" was used in Section 10.5 to describe the existence of separated positive and negative charges in a bond. However, because most molecules have at least some polar bonds, a molecule as a whole can be polar. In a polar *molecule*, there is some accumulation of electron density toward one end of the molecule, so that end of the molecule bears a slight negative charge, $\delta-$ (Figure 10.13). The experimental measure of

Figure 10.13 Polar molecules. (a) In a polar molecule the valence electron density has shifted slightly to one side of the molecule. To show the direction of charge transfer, that is, the direction of molecular polarity, we often use an arrow ($+ \rightarrow -$). (b) A molecule with a dipole moment is called a polar or dipolar molecule. When the molecule is placed in an electric field generated by charged plates, the positive end of the molecule tends to align itself with the negative side of the field. This affects the capacitance of the plates (their ability to hold a charge) and provides an experimental way to measure the magnitude of the dipole.

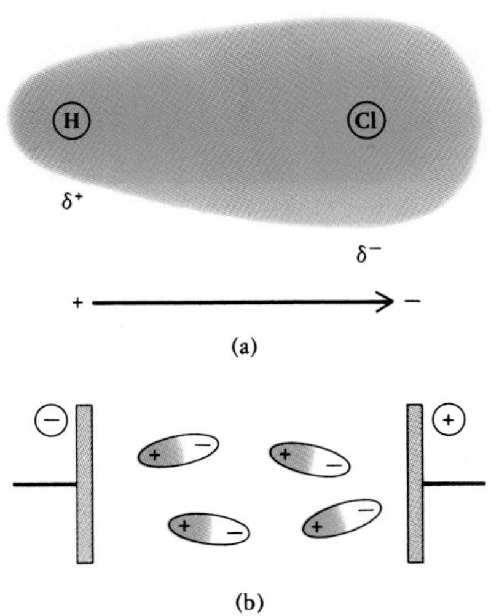

The units of dipole moment are (coulomb · meters); a convenient unit is the debye (D), defined as $1D = 3.34 \times 10^{-30}$ *C · m.*

MOLECULE	DIPOLE MOMENT (in debye units)
H_2	0
H_2O	1.94
NH_3	1.46
CH_4	0
CH_3Cl	1.86

this separation of charge is the molecule's **dipole moment**, which is defined as the product of the size of the charge (δ) and the distance of separation. Experimental dipole moments of a few molecules are given in the table in the margin.

Polar molecules align themselves with an electric field as in Figure 10.13b, and they also align themselves with each other. This interaction of polar molecules is an extraordinarily important effect in water and other substances. As you will see in later chapters, whether a molecule is polar determines the temperatures at which a liquid freezes or boils, whether a liquid will dissolve certain gases or solids or mix with other liquids, whether it will adhere to glass or other solids, or how it may react with other molecules.

To predict when a simple molecule will be polar, we need to consider whether the molecule has polar bonds and how these bonds are positioned relative to one another. To do this, we can use the analogy of children on a playground seesaw. If the children at opposite ends are of equal weight, and are at the same distance from the center (the "fulcrum"), the seesaw does not tip to one side or the other. The test for the polarity of a molecule of the general type CT_n is similar (Figure 10.14). If

(a) the terminal atoms T all have the same charge δ,
(b) all bond lengths $C—T$ are the same, and
(c) all T atoms are oppositely directed from the central atom C,

then the molecule will *not* be polar, even though each $C—T$ bond is polar. If, on the other hand, one of the terminal atoms T has a different charge (analogous to one child having a different weight), or if one $C—T$ bond distance is different (analogous to one child being further from the fulcrum than the other), or the atoms are not oppositely directed from C (both children are on one end of the seesaw), the molecule will be polar.

Let us first consider a linear triatomic molecule such as carbon dioxide, CO_2. Here each C—O bond is polar, with the oxygen atom the negative end of the bond dipole.

O—C—O
$\delta-$ $2\delta+$ $\delta-$
no net dipole moment

Peter Debye (1884–1966) was born and educated in Europe but became Professor of Chemistry at Cornell University in 1940. He was noted for his work on x-ray diffraction, electrolyte solutions, and the properties of polar molecules. He received the Nobel Prize in Chemistry in 1936.

The terminal atoms are at the same distance from the C atom, they both have the same δ-charge, and they are oppositely directed from C. Therefore, CO_2 has no *molecular* dipole, even though each bond is polar.

Water is an example of a bent, triatomic molecule. Here the bonds are polar O—H bonds, with the H atoms having the same $\delta+$ charge. The H atoms are the same distance from the O atom at the fulcrum, but the H atoms are not oppositely directed. Their center is not at the O atom, thus making the molecule electrically "lopsided" and therefore polar.

$2\delta-$
O
H H
$\delta+$ $\delta+$
net dipole = 1.94 D

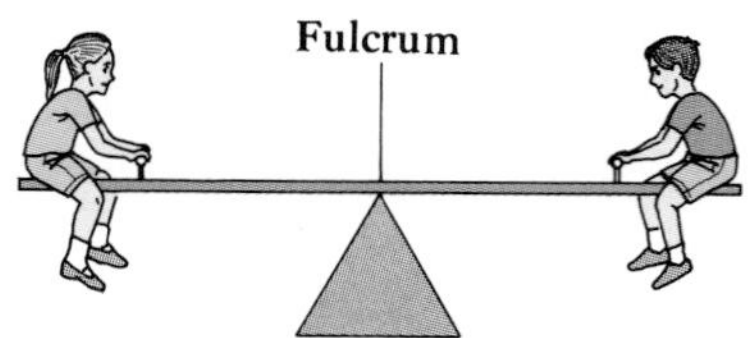

Children balanced on a seesaw. The center of balance is the fulcrum.

The negative end of the molecular dipole lies toward the side of the molecule with the O atom.

The requirement of "opposite" for the terminal atoms (T in CT_n) has a special meaning when there are three or four T atoms (Figure 10.14). Three T atoms at the corners of an equilateral triangle are "oppositely" directed because the angles between them are all the same (120°), thus placing their "center" at the fulcrum, the C atom. Similarly, four T atoms at the corners of a tetrahedron are "opposite" because the angles between them are all the same (109.5°), placing their center at the C atom.

A simple tetrahedral molecule like CCl_4, carbon tetrachloride, is nonpolar because all the Cl atoms have the same $\delta-$ charge, have the same distance from the C atom, and are "oppositely" directed from the central carbon atom. However, changing one or more of the Cl atoms to H, as in $CHCl_3$ (chloroform), gives a polar molecule because δ for the H atom is different from that of the Cl atoms and the C—H distance is different from the C—Cl distances.

Cl $\delta-$ / C $4\delta+$ / Cl Cl Cl ($\delta-$ $\delta-$ $\delta-$) — no net dipole

H $\delta+$ / C $2\delta+$ / Cl Cl Cl ($\delta-$ $\delta-$ $\delta-$) — + ↓ − net dipole

Since the electronegativities of the atoms in $CHCl_3$ are H(2.1) < C (2.5) < Cl (3.0), the Cl atoms are on the more negative side of the molecule. This means the positive end of the molecular dipole is at the H atom and the negative end is on the CCl_3 side of the molecule.

EXAMPLE 10.13

MOLECULAR POLARITY

Are nitrogen trifluoride (NF_3), dichloromethane (CH_2Cl_2), and boron trifluoride (BF_3) polar or nonpolar? If polar, indicate the direction of polarity, that is, the negative and positive sides of the molecule.

Solution NF_3 has the same pyramidal structure as NH_3.

N $3\delta+$ / F F F ($\delta-$ $\delta-$ $\delta-$) — + ↓ − net dipole

Since F is more electronegative than N, each bond is polar, the more negative end being the F atom. This means that the NF_3 molecule as a whole is polar, since the polar bonds are not "oppositely" directed. The N atom is the + side, and the − side is the F_3 grouping.

In CH_2Cl_2, a tetrahedral molecule, the electronegativities are in the order H (2.1) < C (2.5) < Cl (3.0). This means the bonds are polar,

$$\underset{\delta+\ \ \delta-}{\text{H—C}} \qquad \underset{\delta+\ \ \delta-}{\text{C—Cl}}$$

with a net movement of electron density away from the H atoms and toward the Cl atoms. Since this molecule is a tetrahedron, the bond dipoles *cannot* be oppositely directed, so the molecule must be polar with the negative side being the two Cl atoms and the positive side being the two H atoms.

Figure 10.14 Molecules of the type CT_n, where C is the central atom and T is the terminal atom. If all of the terminal atoms T are the same, molecules having these geometries are *not* polar.

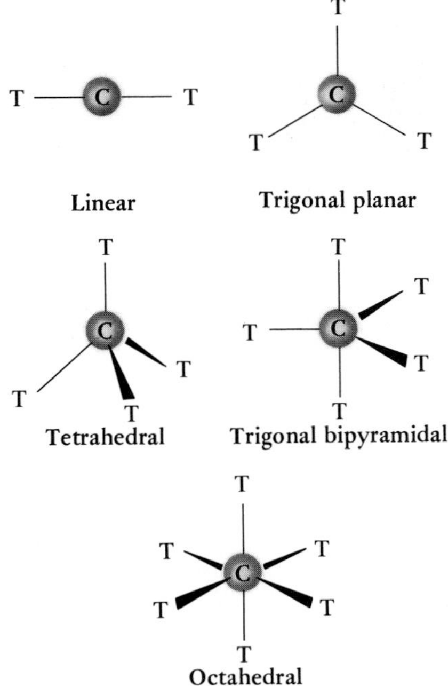

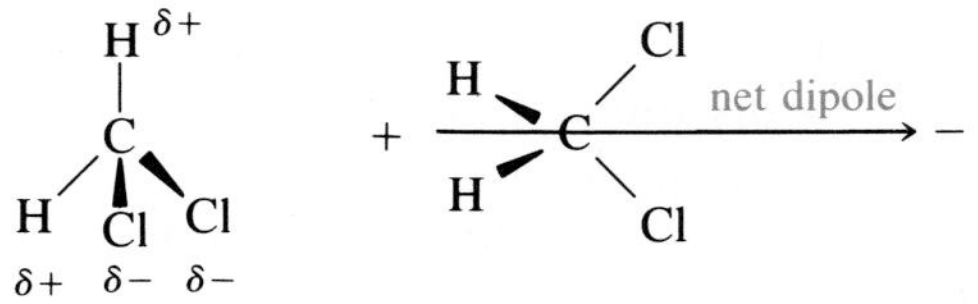

H $\delta+$ | C | H Cl Cl — $\delta+$ $\delta-$ $\delta-$

+ H, H, C, Cl, Cl — net dipole → −

As predicted for a molecule with three structural pairs around the central atom, BF_3 is trigonal and planar.

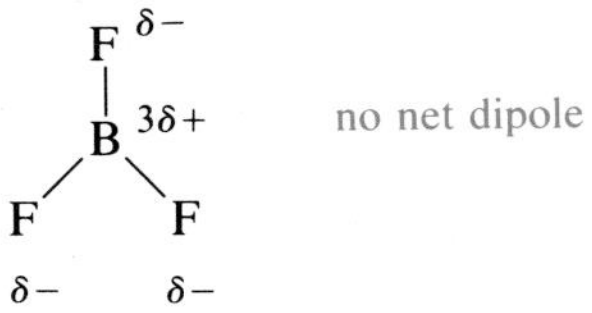

F $\delta-$ | B $3\delta+$ | F F — $\delta-$ $\delta-$ — no net dipole

Since F is more electronegative than B, the B—F bonds are polar with F being the more negative end. The molecule is nonpolar, though, since the three terminal atoms have the same $\delta-$ charge, are the same distance from the boron atom, and are oppositely directed (Figure 10.14).

EXERCISE 10.18 Molecular Polarity

For each of the following molecules, decide whether the molecule is polar and, if so, which side of it is negative: (a) $BFCl_2$, (b) NH_2Cl, and (c) SCl_2.

SUMMARY

The **valence electrons** of an atom are important in chemical bonding. For a main group element, the number of valence electrons is the same as its group number in the periodic table (Section 10.1). Valence electrons can be depicted by a **Lewis electron dot symbol**. In forming bonds, main group elements gain, lose, or share electrons to achieve the configuration of the nearest rare gas; this is the basis of the **octet rule**.

In **covalent bonding**, 2, 4, or 6 electrons are *shared* between atoms (Section 10.2). In contrast, **ionic bonding** occurs in compounds when the bonding electrons reside nearly completely on one atom of the bonded pair. Valence electrons not used in bond formation are called **unshared** or **lone pair electrons**. Bonding electrons occur in pairs if possible. In covalent molecules these electrons reside in a **bond orbital**.

The energy of the interaction between positive and negative ions in an ionic compound is proportional to the product of the ion charges (n_+ and n_-) and inversely proportional to the distance between the ions, d. This energy, as measured by the enthalpy of dissociation of a pair of ions, therefore increases with increasing charge and decreases with increasing distance. The periodic trends in the enthalpy of dissociation, and their relation to the melting temperature of ionic solids, is discussed in Section 10.3.

Most of the compounds described in this chapter are covalently bonded. For elements of the second period in particular, the number of bonding and lone pair electrons surrounding an atom does not exceed 8 (four pairs). (This observation is the basis of the **octet rule**). Atoms in this condition are said to be in their **normal valency state**. The atoms C, N, O, and F are *always* surrounded by four pairs of electrons in their molecules and ions. (Be and B *may* be surrounded by only two or three pairs, respectively.) However, many of the other elements in the periodic table can display an **expanded valence** in which the central atom is surrounded by five, six, or even seven electron pairs.

In forming covalent bonds with another atom, an atom always forms at least one bond called a **sigma bond** (σ). In some cases, C, O, S, and N may form two or three bonds to another atom of the same kind (or to another in this group of four elements). The second (and third) bonds are called **pi bonds** (π). **Coordinate covalent bonds** occur when one atom of a pair is thought of as supplying both electrons of the bonding electron pair. **Odd-electron molecules** have an odd number of electrons.

The distribution of electrons between bond and lone pairs in molecules can be depicted by **Lewis dot structures** (Section 10.4). When more than one acceptable Lewis structure can be written for a molecule, the possible structures are called **resonance structures**. Resonance structures differ in the number of bond pairs between a given pair of atoms.

Some properties of chemical bonds are important (Section 10.5). The **bond order** is the number of bonding electron pairs between a given pair of atoms. The **bond length** depends on the sizes of the bonded atoms and on the bond order. The **bond dissociation energy** (D) is the energy that must be supplied to break a covalent bond. The process of breaking bonds is always endothermic; conversely, when a bond is formed, the process is always exothermic. The bond energy for a particular atom pair depends on the type of atoms bonded and on the bond order. Bond energies can be used to estimate reaction enthalpies by the expression

$$\Delta H^\circ_{\text{reaction}} = \Sigma mD(\text{bonds broken}) - \Sigma nD(\text{bonds made})$$

where m and n are moles of bonds broken and made, respectively.

If two different atoms form a bond, one atom of the pair may attract the shared pair more strongly than the other (Section 10.5). The bond will then be **polar**, the atom at one end of the bond having a slight positive charge ($\delta+$) and the atom at the other end having a slight negative charge ($\delta-$). A measure of the ability of an atom in a molecule to attract electrons to itself is its **electronegativity**, χ.

If we assume that all the electrons in a bond belong to the more electronegative atom, the **oxidation number** of a bonded atom can be calculated. If the atom charge is calculated by assuming that the bonded electrons are shared equally, however, the result is the **formal charge** of the atom (Section 10.5). The latter can help to decide which of several resonance structures is most important.

The shape or molecular geometry for simple molecules and ions can be predicted using the **Valence Shell Electron Pair Repulsion** or **VSEPR** theory (Section 10.6). The sigma bonding and lone electron pairs (**structural pairs**) surrounding an atom are oriented in space so that the angles between them are as large as possible. The geometry described by these pairs is called the **structural-pair geometry**. The locations of the structural pairs that are sigma bonding pairs define the **molecular geometry**. The angle between bonded atoms depends on the presence and location of any lone pairs.

If a molecule has polar bonds, there can be an accumulation of charge on one side of the molecule (Section 10.7). The molecule as a whole is then said to be **polar** or **dipolar** and to have a **dipole moment**.

STUDY QUESTIONS

REVIEW QUESTIONS

1. Give the number of valence electrons for Li, Sc, Zn, Si, and Cl.
2. Explain the difference between an ionic bond and a covalent bond.
3. Refer to Table 10.1 and answer the following questions:
   (a) What molecules do not obey the octet rule?
   (b) How many lone pairs and how many bond pairs are there in ammonia (NH_3) and in HF?
   (c) How many sigma (σ) bonds and how many pi (π) bonds are there in N_2 and in C_2H_4?
4. Boron compounds often do not obey the octet rule. Illustrate this with BCl_3. Show how the molecule can obey the octet rule by forming a coordinate covalent bond with ammonia (NH_3).
5. Refer to Table 10.2 and answer the following questions:
   (a) Do any molecules having expanded valence have a second period element as the central atom?
   (b) What is the maximum number of bond pairs and lone pairs that surround the central atom in any of these molecules?
6. Which of the following are odd-electron molecules: NO_2, SCl_2, NH_3, and NO_3?
7. Draw the resonance structures for SO_3. Explain how the average SO bond order is 1.33.
8. Consider the following Lewis structures for the formate ion, HCO_2^-. Designate two as resonance structures and two as indistinguishable. What is the average CO bond order?

```
         ..     ..
(a)  [:O—C=O:]⁻
         ..  |
             H

         ..     ..
(b)  [:O=C—O:]⁻
             |  ..
             H

         ..
(c)  [:O—C=O:]⁻
         ..  |  ..
             H
```

9. Give the bond order of the bonds in acetylene, H—C≡C—H. How many sigma (σ) bonds and how many pi (π) bonds are there?
10. Explain why the C—O bond order in the carbonate ion, CO_3^{2-}, is 1.33.
11. Consider a series of molecules in which the C atom is bonded to atoms of second period elements: C—O, C—F, C—N, C—C, and C—B. Place these bonds in order of increasing bond length.
12. Define bond dissociation energy. When a C—H bond is broken, is the enthalpy of reaction assigned a negative or positive sign? Explain briefly.
13. What is the relation between bond order, bond length, and bond energy for a series of related bonds, say C—N bonds?
14. If you wished to calculate the enthalpy of the reaction

$$\text{O=O(g)} + 2\ \text{H—H(g)} \longrightarrow 2\ \text{H—O—H(g)}$$

what bond energies would you need? Outline the calculation, being careful to show correct algebraic signs.

15. Define and give an example of a polar covalent bond. Give an example of a nonpolar bond.
16. Define electronegativity. Describe the difference between electronegativity and electron affinity.
17. Describe the trends in electronegativity in the periodic table.
18. Describe the difference between the oxidation number of an atom and its formal charge. Which is often the more realistic description of the charge on an atom in a molecule?
19. What is the principle of electroneutrality? How does this apply to the possible resonance structures of CO_2?
20. What is the VSEPR theory? What is the physical basis of the theory?
21. What is the difference between the structural-pair geometry and the molecular geometry of a molecule? Use the water molecule as an example in your discussion.
22. Designate the structural-pair geometry for each case of two to six electron pairs around a central atom.
23. What molecular geometries are possible for each of the following?

```
                                              H
     ..         ..           ..               |
H—X:        H—X—H       H—X—H        H—X—H
     ..         ..            |               |
                              H               H
```

Give the H—X—H bond angle for each of the last three.

24. If you have three structural pairs around a central atom, how can you have a trigonal planar molecule? A bent molecule? What bond angles are predicted in each case?
25. Draw a trigonal bipyramid of electron pairs. Designate the axial and equatorial pairs. Are there similarly axial and equatorial pairs in an octahedron?
26. Ammonia, NH_3, has a molecular structure similar to NF_3 in Example 10.13. Does NH_3 have a dipole mo-

ment? If so, what is the direction of the net dipole in NH_3?

VALENCE ELECTRONS

27. Give the periodic group number and number of valence electrons for each of the following atoms:
(a) N (d) Na
(b) B (e) Mg
(c) S
28. Give the periodic group number and number of valence electrons for each of the following atoms:
(a) C (d) Ne
(b) Cl (e) Si
(c) P

IONIC BONDING

29. Predict which compound of each pair should have the larger enthalpy of dissociation.
(a) LiI(g) or NaI(g)
(b) LiF(g) or KF(g)
(c) MgO(g) or CaO(g)
30. Predict which compound of each pair should have the larger enthalpy of dissociation.
(a) NaCl(g) or NaI(g)
(b) KCl(g) or CaO(g)
(c) LiF(g) or CsF(g)
31. The dissociation of an ionic compound in the gas phase involves the separation of ions, just as does the dissolving of an ionic solid in water. Assuming that the enthalpy of dissociation of an ionic compound in the gas phase gives us a clue about the solubility of ionic compounds, predict which of each pair should be more soluble in water. (As described in Chapter 14, many other factors are involved as well.)
(a) NaCl or NaI
(b) CsF or KF
(c) MgO or BaO
32. Predict which of each pair should be more soluble in water. (See Study Question 31 above.)
(a) RbCl or NaCl
(b) CsI or CsBr
(c) NaI or KF
33. Predict which compound in each of the following pairs should have the higher melting point.
(a) NaCl or RbCl
(b) BaO or MgO
(c) NaCl or $MgCl_2$
34. Predict which compound in each of the following pairs should have the higher melting point.
(a) CsI or KCl
(b) CaO or KCl
(c) $CaCl_2$ or KCl

THE OCTET RULE AND EXPANDED VALENCE

35. For each of the A-groups of the periodic table, give the number of bonds an element is expected to form if it obeys the octet rule.
36. Which of the following elements could have expanded valences? That is, which can form compounds with five or six valence pairs?
(a) C (d) F (g) Se
(b) P (e) Cl (h) Sn
(c) O (f) B

LEWIS STRUCTURES

37. Draw Lewis structures for the following molecules or ions:
(a) NF_3 (c) HOCl
(b) ClO_3^- (d) SO_3^{2-}
38. Draw Lewis structures for the following molecules or ions:
(a) Cl_2CO (c) NO_2^+
(b) BF_4^- (d) PO_4^{3-}
39. Draw Lewis structures for the following molecules:
(a) $CHClF_2$, one of the many chlorofluorocarbons (CFC's)
(b) Formic acid, HCOOH. The atomic arrangement is

```
     O
     |
H—C—O—H
```

(c) Acetonitrile, $H_3C—CN$
(d) Methyl alcohol, $H_3C—OH$
40. Draw Lewis structures for the following molecules:
(a) Tetrafluoroethylene (the molecule from which Teflon is built), F_2CCF_2
(b) Vinyl chloride, H_2CCHCl, the molecule from which PVC plastics are made
(c) Acrylonitrile, H_2CCHCN, the molecule from which materials such as Orlon are built
(d) Methyltrichlorosilane, H_3CSiCl_3, a compound used in manufacturing "silicone" polymers
41. Each of the following molecules or ions has two or more resonance structures. Show all the resonance structures for each molecule or ion.
(a) SO_2
(b) SO_3
(c) SCN^-
42. Each of the following molecules or ions has two or more resonance structures. Show all the resonance structures for each molecule or ion.
(a) NO_3^-
(b) Nitric acid,

```
          O
         /
H—O—N
         \
          O
```

(c) Nitrous oxide (laughing gas), NNO

43. Draw a Lewis structure for each of the following molecules in which the central atom has an expanded valence: (a) BrF_3, (b) I_3^-, and (c) XeO_2F_2 (Xe is the central atom).
44. Draw a Lewis structure for each of the following molecules in which the central atom has an expanded valence: (a) BrF_5, (b) IF_3, (c) IBr_2^-.

BOND PROPERTIES

45. Give the number of sigma (σ) and pi (π) bonds for each of the following molecules. Tell the bond order for each bond. (Lewis structures for these molecules were drawn in previous Study Questions.)
(a) HCOOH (b) SO_3^{2-} (c) NO_2^+
46. Give the number of sigma (σ) and pi (π) bonds for each of the following molecules. Tell the bond order for each bond. (Lewis structures for these molecules were drawn in previous Study Questions.)
(a) SCN^- (b) acetonitrile, H_3CCN (c) SO_3
47. In each pair of bonds below, predict which will be the shorter. If possible, check your prediction in Table 10.3.
(a) B—Cl or Ga—Cl
(b) C—O or Sn—O
(c) P—S or P—O
(d) The C=C or the C=O bond in acrolein, H_2C=CH—C(H)=O
48. In each pair of bonds below, predict which will be the shorter. If possible, check your prediction in Table 10.3.
(a) Si—N or P—O
(b) Si—O or C—O
(c) C—F or C—Br
(d) The C=C or the C≡N bond in acrylonitrile, H_2C=CH—C≡N
49. Consider the carbon–oxygen bonds in formaldehyde (H_2CO) and carbon monoxide (CO). In which molecule is the CO bond shorter? In which molecule is the CO bond stronger?
50. Compare the nitrogen–nitrogen bonds in hydrazine, H_2NNH_2, and in "laughing gas," N_2O. In which molecule is the nitrogen–nitrogen bond shorter? In which is the bond stronger?
51. Compare the nitrogen–oxygen bond lengths in NO_2^+ and in NO_3^-. In which ion is the bond longer? Explain briefly.
52. Compare the carbon–oxygen bond lengths in the formate ion, HCO_2^-, and in the carbonate ion, CO_3^{2-}. In which ion is the bond longer? Explain briefly.

BOND ENERGIES AND REACTION ENTHALPIES

53. Using the bond energies of Table 10.4, estimate the standard enthalpy of formation for each of the following molecules in the gas phase: (a) NH_3 and (b) H_2O.
54. Using the bond energies of Table 10.4, estimate the standard enthalpy of formation for each of the following molecules in the gas phase: (a) chloramine, H_2NCl, and (b) hydrogen peroxide, HOOH.
55. Using the methods of Example 10.6, calculate the standard enthalpy of formation of propane, H_3C—CH_2—CH_3(g).
56. Using the methods of Example 10.6, calculate the standard enthalpy of formation of ethylene in the gas phase.

```
H       H
 \     /
  C=C
 /     \
H       H
```

Compare your results with the enthalpy of formation of acetylene (Example 10.6). Why is ΔH_f° for ethylene so much less positive than ΔH_f° for acetylene?
57. Using the bond energies of Table 10.4 and the heat of vaporization of C(graphite) (+717 kJ/mol) (see Example 10.6), calculate the standard enthalpy of formation of glycine, one of the building blocks of proteins.

```
    H  O
    |  ||
H—C—C—O—H
    |
    NH2
```

58. Using the bond energies of Table 10.4, and the heat of vaporization of C(graphite) (+717 kJ/mol) (see Example 10.6), calculate the standard enthalpy of formation of urea, a compound commonly used as a fertilizer.

```
        H      O      H
         \     ||    /
urea =    N—C—N
         /          \
        H            H
```

59. Use the following information, and information from Table 10.4, to estimate the energy of the H—I bond in HI(g): ΔH_f°[HI(g)] = 26.48 kJ/mol and ΔH° for $I_2(s) \rightarrow I_2(g)$ is 62.44 kJ/mol. Compare your calculated value with the bond energies of the other hydrogen halides (HF, HCl, and HBr) in Table 10.4. Based on this comparison, what would you conclude about the trend in bond energies in H—X as the atomic weight of X increases?
60. Use the following information, and information from Table 10.4, to estimate the energy of the H—S bond: $\Delta H_f^\circ[H_2S(g)] = -20.63$ kJ/mol and ΔH° for S(s) $\rightarrow$ S(g) is 278.8 kJ/mol. Is the H—S bond energy greater or less than that for H—O? From this comparison, what might you conclude about the trend in H—X bond energies as the atomic weight of X increases (where X is a Group 6A element)?

61. The compound oxygen difluoride is quite unstable, giving oxygen and HF on reaction with water.

$$OF_2(g) + H_2O(g) \longrightarrow O{=}O(g) + 2\ HF(g) \qquad \Delta H° = -318\ kJ$$

The enthalpy of this reaction is −318 kJ. Using bond energies, calculate the bond dissociation energy of the O—F bond in OF_2.

62. Using Table 10.4, and the thermodynamic information below, estimate the bond dissociation energy for the B—F bond.

$$B(solid) \longrightarrow B(gas) \qquad \Delta H_f^\circ[B(g)] = 563\ kJ/mol$$
$$B(solid) + \tfrac{3}{2}\ F_2(g) \longrightarrow BF_3(g) \qquad \Delta H_f^\circ[BF_3(g)] = -1137\ kJ/mol$$

63. Phosgene, Cl_2CO, is a highly toxic gas that was used in World War I. Using the bond energies of Table 10.4, estimate the enthalpy of the reaction of carbon monoxide and chlorine to produce phosgene.

$$CO(g) + Cl_2(g) \longrightarrow Cl_2CO(g)$$

64. The equation for the combustion of methyl alcohol is

$$2\ H_3COH(g) + 3\ O_2(g) \longrightarrow 2\ CO_2(g) + 4\ H_2O(g)$$

Using the bond energies in Table 10.4, estimate the enthalpy of this reaction, that is, the heat of combustion of methyl alcohol.

65. Hydrogenation reactions, the addition of H_2 to a molecule, are widely used in industry to transform one compound into another. For example, the molecule called propene (a member of a general class of compounds called "olefins" because of the C=C double bond) is converted to propane (called an "alkane" because there are only C—H bonds and C—C single bonds) by addition of H_2.

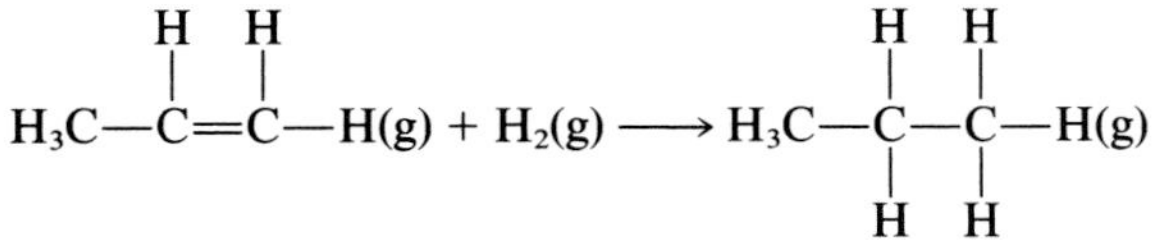

$$H_3C—\overset{H}{\overset{|}{C}}{=}\overset{H}{\overset{|}{C}}—H(g) + H_2(g) \longrightarrow H_3C—\underset{H}{\underset{|}{\overset{H}{\overset{|}{C}}}}—\underset{H}{\underset{|}{\overset{H}{\overset{|}{C}}}}—H(g)$$

Use the bond energies of Table 10.4 to estimate the enthalpy of the hydrogenation reaction.

66. Acetone can be converted into isopropyl alcohol, rubbing alcohol, by a hydrogenation reaction,

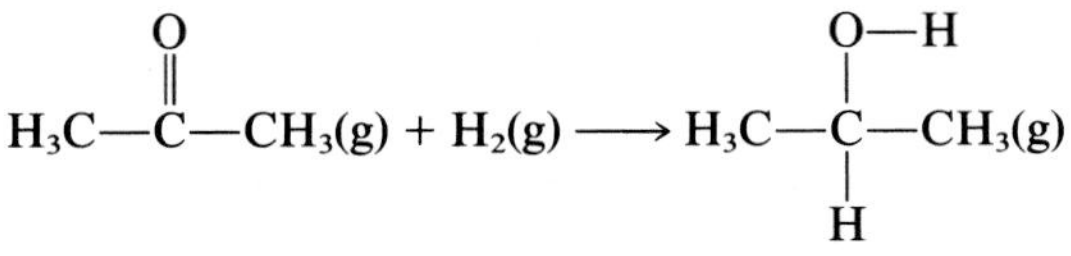

$$H_3C—\overset{O}{\overset{\|}{C}}—CH_3(g) + H_2(g) \longrightarrow H_3C—\underset{H}{\underset{|}{\overset{O—H}{\overset{|}{C}}}}—CH_3(g)$$

Estimate the enthalpy of this reaction, using the bond energies of Table 10.4.

ELECTRONEGATIVITY AND BOND POLARITY

67. In each pair of bonds, indicate the more polar bond and use an arrow to show the direction of polarity (positive to negative) in each bond.
(a) C—O and C—N (c) P—H and P—N
(b) P—O and P—S (d) B—H and B—I

68. Given the bonds below, answer the questions that follow:

C—N C—H C—Br S—O

(a) Tell which atom is the more negatively charged in each bond.
(b) Which is the most polar bond?

69. The molecule below is acrolein, the starting material for certain plastics.

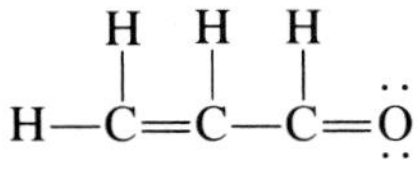

$$H—\overset{H}{\overset{|}{C}}{=}\overset{H}{\overset{|}{C}}—\overset{H}{\overset{|}{C}}{=}\ddot{\underset{..}{O}}$$

(a) Which bonds in the molecule are polar and which are nonpolar?
(b) Which is the most polar bond in the molecule? Which atom is the negative end of the bond dipole?

70. The molecule below is urea, a compound used in plastics and fertilizers.

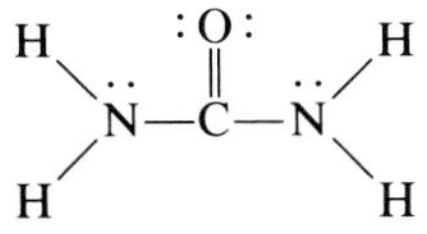

$$\begin{matrix} H & & :\ddot{O}: & & H \\ & \diagdown & \| & \diagup & \\ & \ddot{N} & —C— & \ddot{N} & \\ & \diagup & & \diagdown & \\ H & & & & H \end{matrix}$$

(a) Which bonds in the molecule are polar and which are nonpolar?
(b) Which is the most polar bond in the molecule? Which atom is the negative end of the bond dipole?

OXIDATION NUMBERS AND ATOM FORMAL CHARGES

71. Using Lewis structures and relative atom electronegativities, give the oxidation number of each atom in the following molecules or ions.
(a) H_2O (d) N_2O
(b) H_2O_2 (e) ClO^-
(c) SO_2

72. Using Lewis structures and relative atom electronegativities, give the oxidation number of each atom in the following molecules.
(a) ClF_3 (c) OF_2
(b) XeF_2 (d) HCN

73. Calculate the formal charge on each atom in each of the following atoms or ions. In each case, compare the formal charge with the oxidation number.
(a) H_2O (c) $NO_2{}^+$
(b) CH_4 (d) HOF

74. Calculate the formal charge on each atom in each of the following atoms or ions. In each case, compare the formal charge with the oxidation number.
(a) ICl_2^- (c) OCl^-
(b) NH_3 (d) SF_4

75. Two resonance structures are possible for NO_2^-. Draw these structures and then find the formal charge on each element in each resonance structure.

76. Two resonance structures are possible for the formate ion, HCO_2^-. Draw these structures and then find the formal charge on each element in each resonance structure.

77. Three resonance structures are possible for dinitrogen oxide, N_2O (known as "laughing gas").
(a) Draw the three resonance structures.
(b) Calculate the formal charge on each atom in each resonance structure.
(c) On the basis of formal charges, decide on the most reasonable resonance structure.

78. The cyanate ion, NCO^-, has the least electronegative atom, C, in the center. The very unstable fulminate ion, CNO^-, has the same formula, but the N atom is in the center.
(a) Draw the three possible resonance structures of CNO^-.
(b) On the basis of formal charges, decide on the resonance structure with the most reasonable distribution of charge.
(c) Mercury fulminate is so unstable that it is used in blasting caps. Can you offer an explanation for this instability? (Hint: Are the formal charges in any resonance structure reasonable in view of the relative electronegativities of the atoms?)

MOLECULAR SHAPE

79. Draw the Lewis structure for each of the following molecules or ions. Describe the structural-pair geometry and the molecular geometry.
(a) NH_2Cl (c) SCN^-
(b) Cl_2O (O is the central atom) (d) HOF

80. Draw the Lewis structure for each of the following molecules or ions. Describe the structural-pair geometry and the molecular geometry.
(a) ClF_2^+ (c) PO_4^{3-}
(b) $SnCl_3^-$ (d) CS_2

81. In each of the following molecules or ions, two oxygen atoms are attached to a central atom. Draw the Lewis structure for each one and then describe the structural-pair geometry and the molecular geometry. Comment on similarities and differences in the series.
(a) CO_2 (d) O_3
(b) NO_2^- (e) ClO_2^-
(c) SO_2

82. In each of the following molecules or ions, three oxygen atoms are attached to a central atom. Draw the Lewis structure for each one and then describe the structural-pair geometry and the molecular geometry. Comment on similarities and differences in the series.
(a) BO_3^{3-} (d) SO_3^{2-}
(b) CO_3^{2-} (e) ClO_3^-
(c) NO_3^-

83. The following are examples of molecules and ions that have expanded valence. After drawing the Lewis structure, describe the structural-pair geometry and the molecular geometry.
(a) ClF_2^- (c) ClF_4^-
(b) ClF_3 (d) ClF_5

84. The following are examples of molecules and ions that have expanded valence. After drawing the Lewis structure, describe the structural-pair geometry and the molecular geometry.
(a) SiF_6^{2-} (c) SF_4
(b) PF_5 (d) XeF_4

85. Give approximate values for the indicated bond angles.
(a) O—S—O in SO_2
(b) F—B—F angle in BF_3

(c) 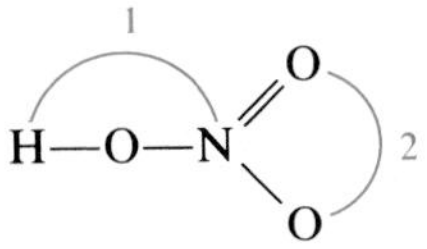 H—O—N(=O)—O (angles 1 and 2 indicated)

(d) 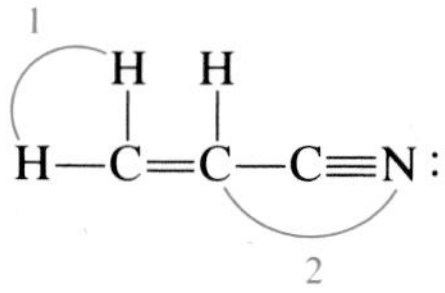 H—C(H)=C(H)—C≡N: (angles 1 and 2 indicated)

86. Give approximate values for the indicated bond angles.
(a) Cl—S—Cl in SCl_2
(b) N—N—O in N_2O

(c) 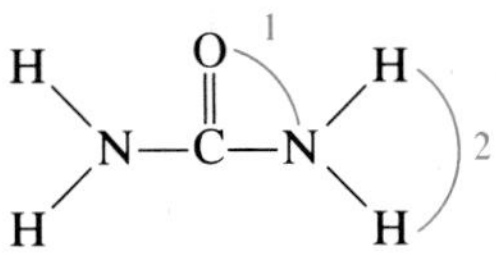 H_2N—C(=O)—NH_2 (angles 1 and 2 indicated)

(d) 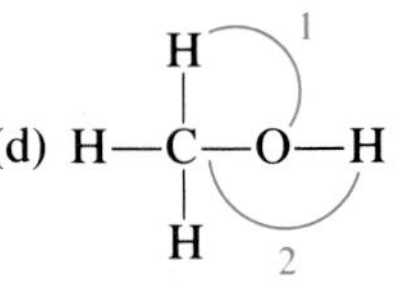 H—C(H)(H)—O—H (angles 1 and 2 indicated)

87. One of the structures for acetylacetone is the following:

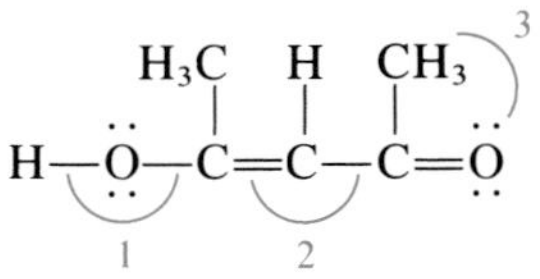

H—Ö—C(H_3C)=C(H)—C(CH_3)=Ö (angles 1, 2 and 3 indicated)

(a) How many sigma (σ) and how many pi (π) bonds does the compound have?
(b) Estimate the values of the indicated angles.

88. Cysteine is one of the natural amino acids.

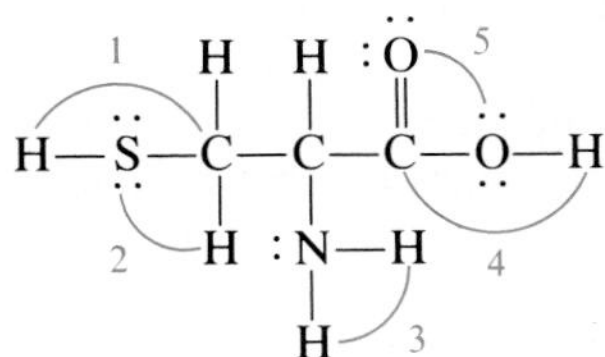

(a) How many sigma (σ) and how many pi (π) bonds does the compound have?
(b) Estimate the values of the indicated angles.

89. Give approximate values for the indicated bond angles.
(a) F—Se—F angles in SeF_4
(b) the O—S—F angles in OSF_4 (the O atom is in an equatorial position)
(c) F—Br—F angles in BrF_5

90. Give approximate values for the indicated bond angles.
(a) F—S—F angles in SF_6
(b) F—Xe—F angle in XeF_2
(c) F—Cl—F in ClF_2^-

91. Which would have the greater O—N—O bond angle, NO_2^- or NO_2^+? Explain your answer briefly.

92. Compare the F—Cl—F angles in ClF_2^+ and ClF_2^-. From Lewis structures, determine the approximate bond angle in each ion. Explain which ion has the greater angle and why.

MOLECULAR POLARITY

93. Consider the following molecules:

H_2O NH_3 CO_2 ClF CCl_4

(a) Which compound has the most polar bonds?
(b) Which compounds in the list are *not* polar?
(c) Which atom in ClF is more negatively charged?

94. Consider the following molecules:

CH_4 NCl_3 BF_3 CS_2

(a) Which compound has the most polar bonds?
(b) Which compounds in the list are *not* polar?

95. Which of the following molecules is (are) polar? For each polar molecule, what is the direction of polarity, that is, which is the negative and which is the positive end of the molecule?
(a) CO_2 (c) CH_3Cl
(b) HBF_2 (d) SO_3

96. Which of the following molecules is (are) *not* polar? Which molecule has the most polar bonds?
(a) CO (d) PCl_3
(b) BCl_3 (e) GeH_4
(c) CF_4

GENERAL QUESTIONS

97. Give the number of valence electrons for an atom in Groups 1A, 3A, 3B, and 4A.

98. Is it possible to have pi (π) bonds between two atoms without there being a sigma (σ) bond?

99. Draw Lewis structures (and resonance structures where appropriate) for the following molecules and ions. What similarities and differences are there in this series? (a) CO_2, (b) N_3^-, and (c) OCN^-.

100. What are the orders of the N—O bonds in NO_2^- and NO_2^+? The nitrogen–oxygen bond length in one of these ions is 110 pm and in the other 124 pm. Which bond length corresponds to which ion? Explain briefly.

101. The Bunsen burners in your laboratory are fueled either by natural gas, which is primarily methane (CH_4), or by propane ($H_3C—CH_2—CH_3$). Using the bond energies of Table 10.4, calculate the enthalpy of combustion of each of these substances, $CH_4(g)$ and $C_3H_8(g)$. Which provides the greater amount of heat *per gram*?

102. In Chapter 2, the standard state of phosphorus was described as consisting of P_4 molecules. Using the enthalpy of the following reaction

$$P_4(s) \longrightarrow 4\ P(g) \qquad \Delta H^\circ = +1259 \text{ kJ}$$

and the bond energies of Table 10.4, calculate the standard enthalpy of formation of phosphine, $PH_3(g)$.

103. Urea is widely used as a fertilizer because of its high nitrogen content, so better methods for its production are always being sought. Using Table 10.4, estimate the enthalpy change of this reaction to make urea.

$$2\ NH_3(g) + CO(g) \longrightarrow H_2N—\overset{\overset{\displaystyle O}{\|}}{C}—NH_2(g) + H_2(g)$$

urea

104. The molecule pictured below is acrylonitrile, the building block of the synthetic fiber Orlon.

```
  1
  ┌ H   H
  │ |   |
  H—C═C—C≡N:
      └─┘
       2
```

(a) Give the approximate values of angles 1 and 2.
(b) Which is the shorter carbon–carbon bond?
(c) Which is the stronger carbon–carbon bond?
(d) Which is the most polar bond and what is the negative end of the bond dipole?

105. Vanillin is the flavoring agent in vanilla extract and in vanilla ice cream. Its structure is

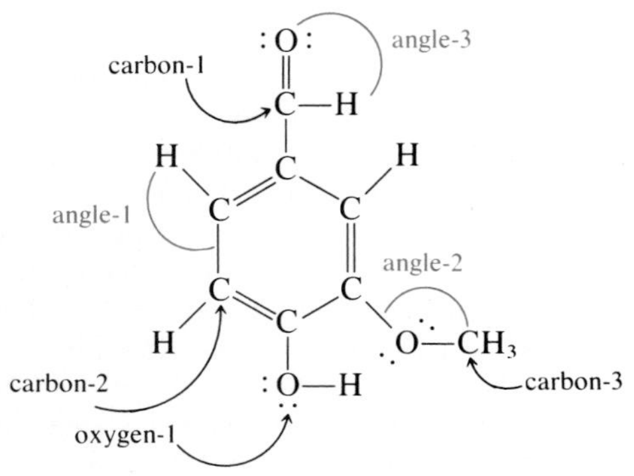

(a) Give values for the three bond angles indicated.
(b) Indicate the shortest carbon–oxygen bond in the molecule.
(c) Indicate the most polar bond in the molecule.

106. The resonance structures for the cyanate ion, NCO^-, are given in Example 10.9. If the ion were to react with H^+, which end of the ion is the H^+ ion most likely to "attack" (N or O)? (The H^+ ion, being positive, will "attack" the more negative end of the NCO^- ion.) Having decided to which end of NCO^- the H^+ would become attached, draw the Lewis structure of cyanic acid, HNCO or HOCN, with the H atom attached to the correct terminal atom of the NCO grouping.

107. The following molecules or ions all have fluorine atoms attached to an atom from Groups 1A or 3A–6A. Draw the Lewis structure for each one and then describe the structural-pair geometry and the molecular geometry. Comment on similarities and differences in the series.
(a) BF_3 (d) OF_2
(b) CF_4 (e) HF
(c) NF_3

108. The formula for nitryl chloride is $ClNO_2$. Draw the Lewis structure for the molecule, including all resonance structures. Describe the structural-pair and molecular geometries, and give values for all bond angles.

109. Given that the spatial requirements of lone pairs are much greater than that of bond pairs, explain why
(a) XeF_2 has a linear molecular structure and not a bent one.
(b) ClF_3 has a T-shaped structure and not a planar, triangular one.

110. The compound $C_2H_2Cl_2$ can exist in two forms. Does either of these have a net dipole moment? If yes, give the direction of the dipole moment.

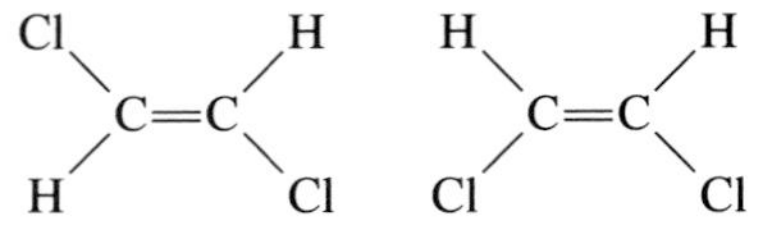

trans *cis*

111. One of the most objectionable compounds in photochemical smog is peroxyacetyl nitrate, PAN. It has the structure below. (a) How many sigma (σ) and how many pi (π) bonds are there in the molecule? (b) Give approximate values for the indicated bond angles.

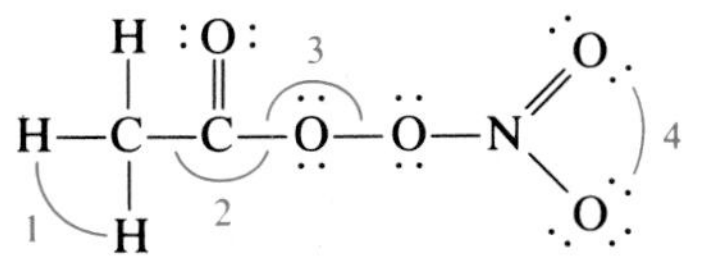

112. In 1962 Watson and Crick received the Nobel Prize for their simple but elegant model for the "heredity molecule" DNA. The key to their structure (the "double helix") was an understanding of the geometry and bonding capabilities of nitrogen-containing bases such as the thymine molecule below. (a) How many sigma (σ) and how many pi (π) bonds are there in the molecule? (b) Give approximate values for the indicated bond angles. (c) Which are the most polar bonds in the molecule?

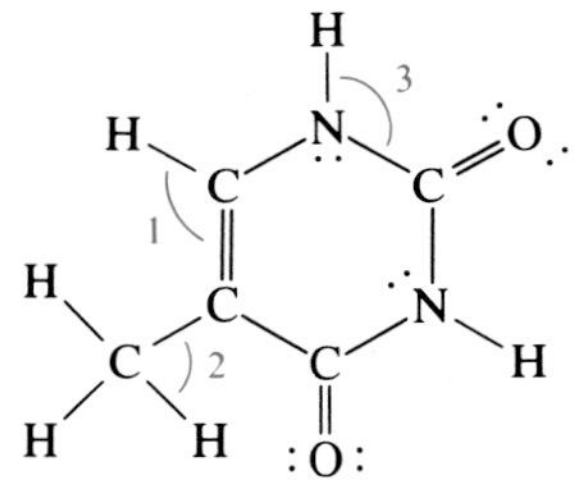

113. Histidine is one of the basic amino acids important to life.

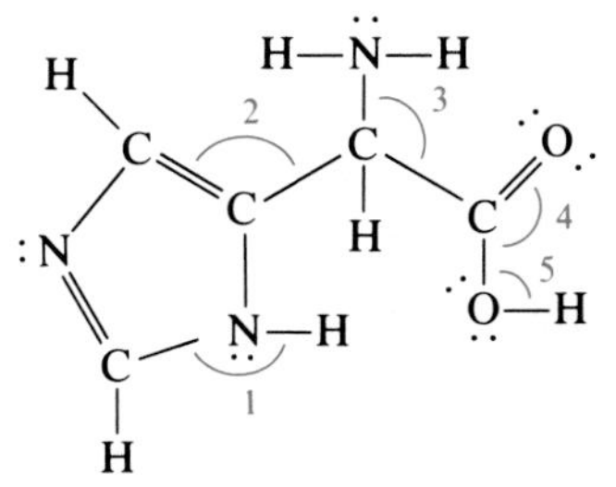

(a) How many sigma (σ) and how many pi (π) bonds are there in the molecule?
(b) Give approximate values for the indicated bond angles.
(c) Which carbon–carbon bond is the shortest of the three in the molecule?
(d) Which of the carbon–carbon bonds should have the highest bond energy?

114. The important molecule cyclohexane is made by adding H_2 to benzene.

$$C_6H_6(g) + 3\ H_2(g) \longrightarrow C_6H_{12}(g)$$

benzene

cyclohexane

(a) Calculate the enthalpy of the reaction, knowing that $\Delta H_f^\circ[C_6H_6(g)] = +82.8$ kJ/mol and $\Delta H_f^\circ[C_6H_{12}(g)] = -123.1$ kJ/mol.

(b) Benzene has two resonance structures, and this makes the C—C bond order 1.5. However, assume that the C_6 ring consists of three C=C bonds and three C—C bonds, and calculate the enthalpy of the hydrogenation reaction using bond energies.

(c) There should be a difference between the actual enthalpy of hydrogenation calculated in (a) and the one estimated from (b). This difference occurs because we assumed resonance in (a) but not in (b). What does this tell you about the effect of resonance on the "stability" of a compound?

SUMMARY PROBLEM

115. Chlorine trifluoride, ClF_3, is one of the most reactive compounds known. It reacts violently with many substances generally thought to be inert, and was used in incendiary bombs in World War II. It can be made by heating Cl_2 and F_2 in a closed container.

(a) Write a balanced equation to depict the reaction of Cl_2 and F_2 to give ClF_3.

(b) If you mix 0.71 g of Cl_2 with 1.00 g of F_2, how many grams of ClF_3 are expected?

(c) Draw the electron dot structure of ClF_3.

(d) What is the structural-pair geometry for ClF_3?

(e) Knowing that the molecule is polar, what can you conclude about the molecular geometry? Does this establish the geometry unambiguously?

(f) Calculate the standard enthalpy of formation of ClF_3 using bond energies.

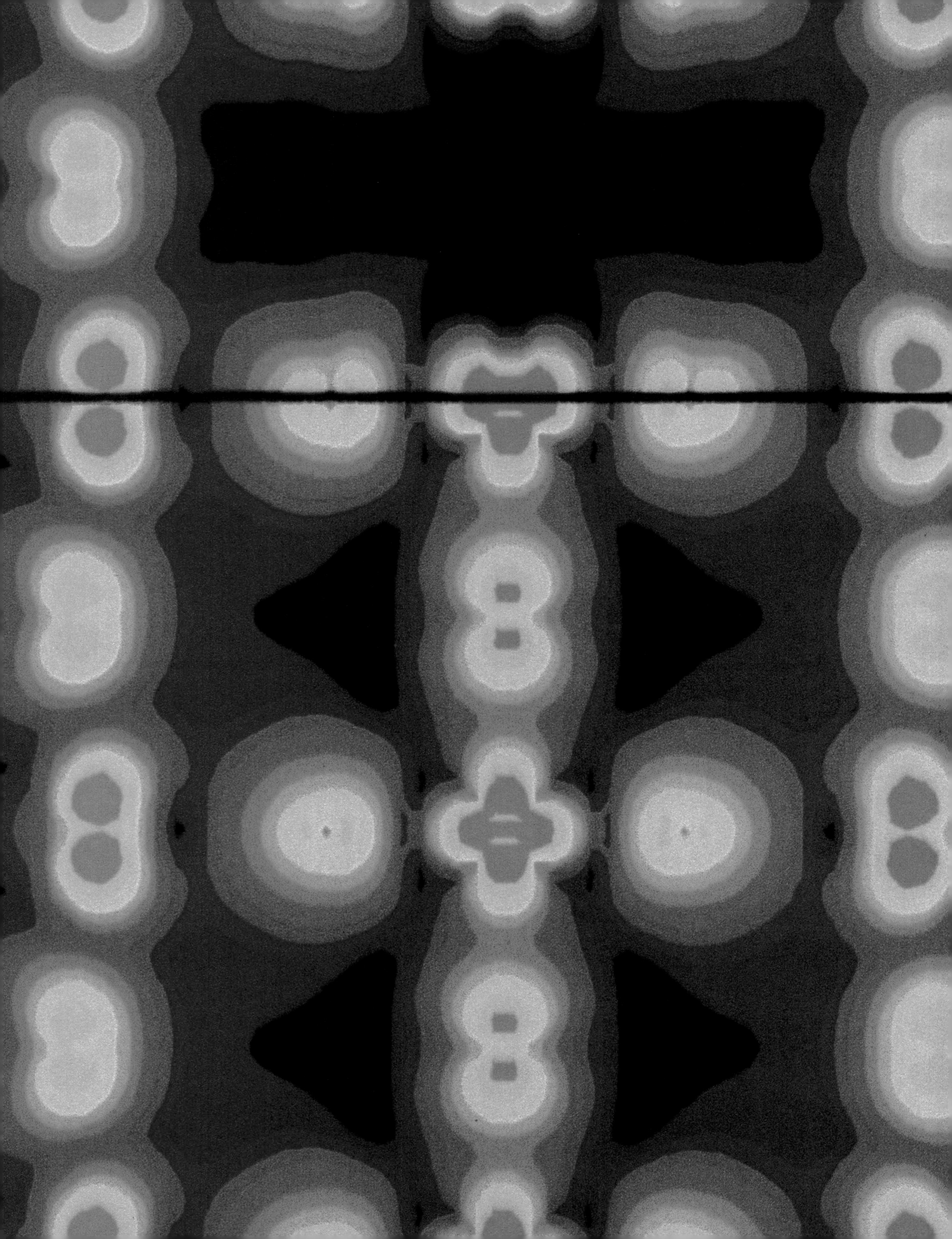

CHAPTER

11

Further Concepts of Chemical Bonding: Orbital Hybridization, Molecular Orbitals, and Metallic Bonding

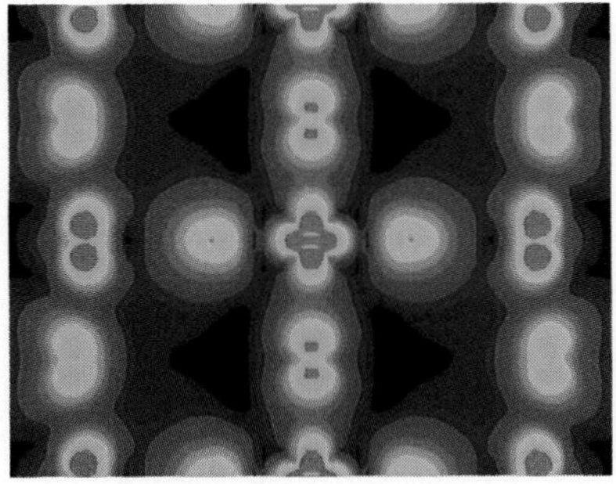

Charge density distribution of electrons near the Fermi level of an yttrium-barium-copper oxide superconductor. (Arthur J. Freeman, reprinted with permission of Elsevier Science Publishers)

CHAPTER OUTLINE

The preceding chapter outlined the methods that chemists use to represent bonds between atoms. This chapter will outline, still in an elementary way, some more ideas on how atoms use their orbitals in bond formation.

There are two commonly used approaches to chemical bonding: **valence bond (VB) theory** and **molecular orbital (MO) theory**. The former was developed, at least in part, by Linus Pauling, while the latter came from work by Robert Mulliken. Mulliken's approach is to combine pure atomic orbitals on each atom to derive molecular orbitals that are spread or delocalized over the molecule. Only after developing the molecular orbitals are the electrons of the molecule assigned to these orbitals; thus, the molecular electron pairs are more or less evenly distributed over the molecule. In contrast, the valence bond approach is more closely tied to Lewis's idea of electron pair bonds and of lone pairs of electrons localized on a particular atom. Only bonding and nonbonding (lone pair) orbitals are considered.

Robert Mulliken was awarded the 1966 Nobel Prize for the development of molecular orbital theory. Linus Pauling received the 1954 Prize for his work on the nature of the chemical bond.

But why are two theories used? Isn't one more correct than the other? The answer is that both are correct and useful. Valence bond theory is perhaps the easier to apply and leads to a good understanding of the bonding of molecules in their ground or lowest energy state. On the other hand, molecular orbital theory is most useful in understanding molecules in "excited states." For example, it helps us understand why compounds absorb light, and why some are red but others are blue or other colors of the rainbow. Therefore, you will find chemists using the theory that best applies to the problem at hand.

Although the VB and MO theories may seem quite different at an elementary level, they lead ultimately to equivalent results, as a more advanced treatment would show. In general, MO theory is the theory of choice today.

11.1 VALENCE BOND THEORY

The description of covalent bonding in Chapter 10 can now be given in the terminology of valence bond theory. According to this theory, two atoms form a bond when both of the following conditions occur:

1. There is **orbital overlap** between two atoms. If two H atoms approach one another closely enough, their *s* orbitals can partially occupy the same region of space (Figure 11.1).
2. A maximum of two electrons, of opposite spin, can be present in the overlapping orbitals.

Because of orbital overlap, the pair of electrons is found within a region influenced by both nuclei. This means that both electrons are attracted to both atomic nuclei, and this, among other factors, leads to bonding.

As the extent of overlap between two orbitals increases, the strength of the bond increases. This is seen in Figure 11.2 as a drop in electronic energy as the two H atoms, originally far apart, come closer and closer. However, the figure also shows that the energy increases rapidly as the atoms come very close to one another, because of the repulsion of one positive nucleus by the other. Thus, there is an optimum distance, the observed bond distance, at which the total energy is at a minimum.

The overlap of two *s* orbitals, one from each of two atoms (Figure 11.1), leads to a **sigma (σ) bond**; the electron density of a sigma bond is greatest *along the axis* of the bond. Sigma bonds can also form by the overlap of an *s* orbital with a *p* orbital or by the head-to-head overlap of two *p* orbitals (Figure 11.3).

An important principle of the VB theory is that of **maximum overlap**: to form the strongest possible bond, two atoms are arranged to give the greatest possible orbital overlap. Since *s* orbitals are spherical, two H atoms could approach one another from any direction, and a strong sigma bond would form. However, other types of orbital overlap (*s*/*p* and *p*/*p*) are directional. For example, two *p* orbitals should overlap directly *along the axis* of the bond in order to form the strongest possible sigma bond (Figure 11.4).

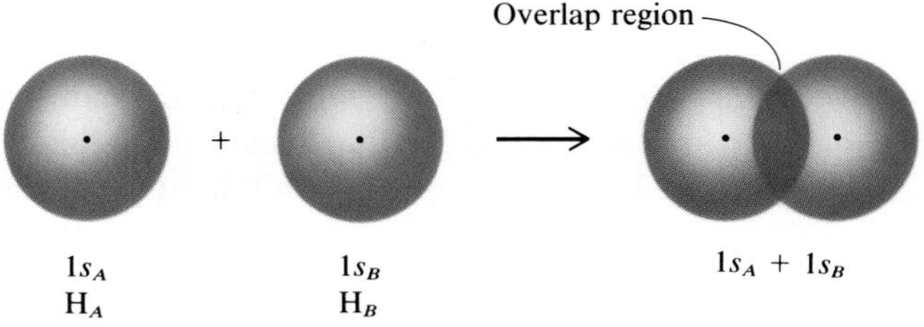

Figure 11.1 The overlap of 1*s* atomic orbitals to form the sigma (σ) bond of H_2.

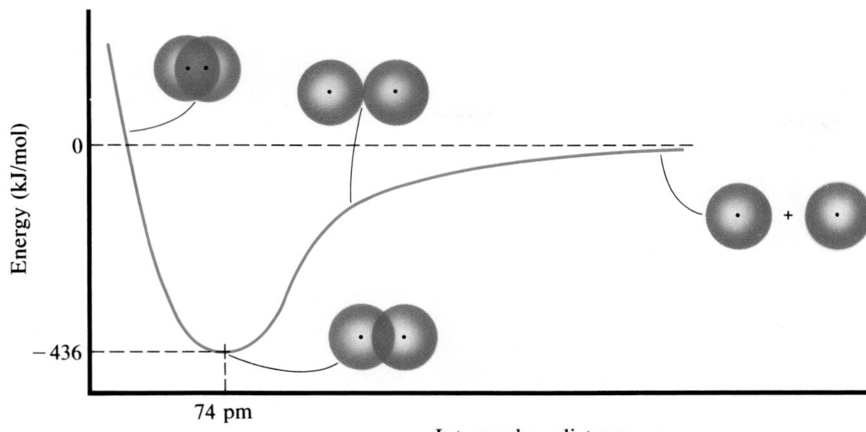

Figure 11.2 Energy change in the course of H—H bond formation from isolated H atoms.

HYBRID ORBITALS

An isolated carbon atom has two unpaired electrons, and so might be expected to form only two bonds.

Carbon Electron Configuration

$[\text{He}]\ 2s^2 2p_x^{\ 1} 2p_y^{\ 1}$ or $[\text{He}]\ \underset{2s}{\underline{\uparrow\downarrow}}\ \underset{2p}{\underline{\uparrow}}\ \underset{2p}{\underline{\uparrow}}\ \underset{2p}{\underline{\ \ }}$

However, there are four C—H bonds in methane, CH_4. Furthermore, VSEPR theory tells us, and experiments confirm it, that the structural-pair geometry of the C atom in CH_4 is tetrahedral. This means there must be *four equivalent bonding electron pairs* around the C atom, with an angle of 109° between the bond pairs.

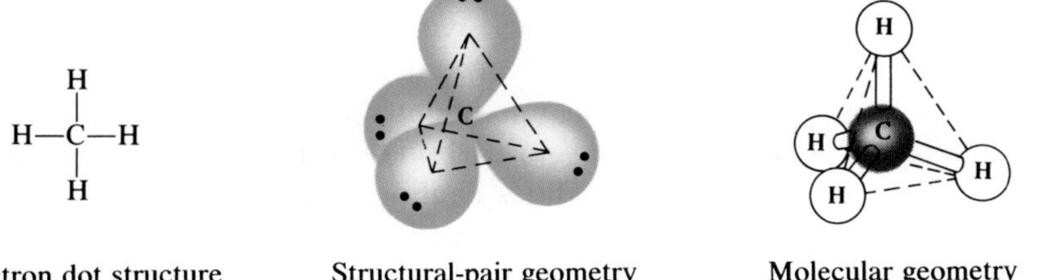

The problem is that the three *p* orbitals on an isolated C atom lie at 90° to one another (Figure 8.14). Therefore, if sigma bonds were formed

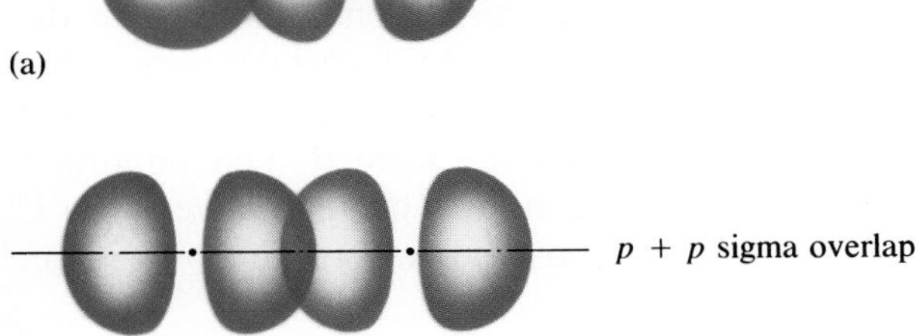

Figure 11.3 Sigma (σ) bond formation from *s* and *p* orbitals. (a) An *s* orbital overlaps a *p* orbital. (b) Two *p* orbitals overlap head to head.

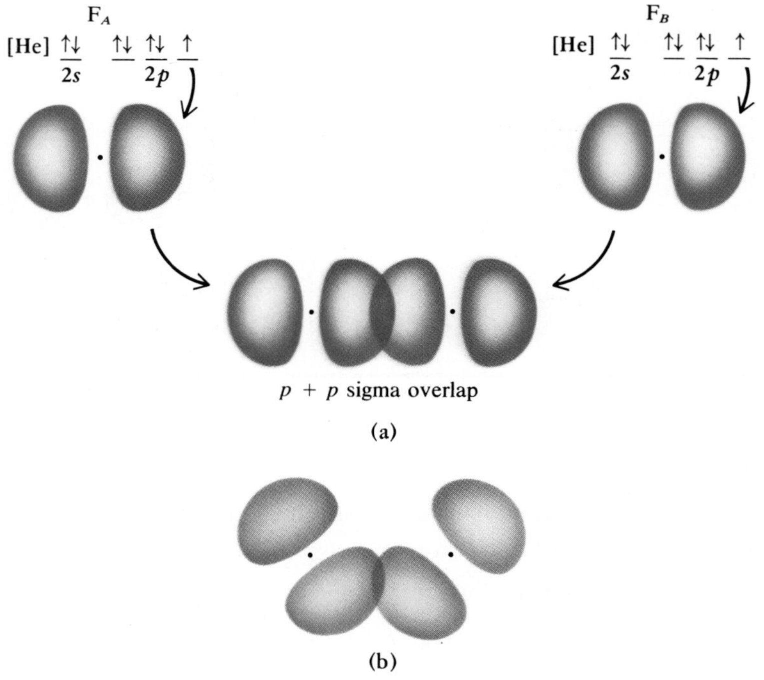

Figure 11.4 Sigma bond formation in F_2. (a) Maximum overlap of atomic orbitals occurs when the orbitals overlap along the axis of the bond. (b) If the orbitals do not overlap head to head, there is a smaller overlap. The bond is formed only when there is maximum overlap.

in some manner using pure *s* and *p* orbitals, the bonds would neither be equivalent nor would they be arranged correctly in space. Some other scheme is required to account for four equivalent C—H bonds at an angle of 109°.

You may be familiar with hybridization in agriculture, where two varieties of a plant are mixed to provide a new variation that reflects some characteristics of each of its parents.

Pauling proposed **orbital hybridization** as a way to explain the formation of bonds by the maximum overlap of atomic orbitals and yet accommodate the use of *s* and *p* orbitals. In order for the four C—H bonds of methane to have their maximum strength, there must be maximum overlap between each of four carbon orbitals and a H-atom *s* orbital at a corner of a tetrahedron. Thus, Pauling suggested that the approach of the H atoms to the isolated C atom causes a restructuring of the four carbon *s* and *p* orbitals. That is, the C atom orbitals **hybridize** or combine in some manner to provide *four equivalent hybrid orbitals that point to the corners of a tetrahedron* (Figure 11.5). We label *each* hybrid orbital as $\boldsymbol{sp^3}$, since the orbitals are the result of the combination of one *s* and three *p* orbitals on one atom. Each hybrid orbital combines the properties of its *s* and *p* orbital parents.

The label sp^3 is read "ess-pee-three."

Be sure to notice that four atomic orbitals produce four hybrid orbitals. The number of atomic orbitals used and the number of hybrid orbitals produced is always the same.

The theory of orbital hybridization is an attempt to explain how in CH_4, for example, there can be four *equivalent* bonds directed to the corners of a tetrahedron. Another outcome of hybrid orbital theory is that hybrid orbitals have greater amplitude in the region between nuclei than any of the atomic orbitals from which they are formed. This important observation means that greater overlap can be achieved between the C and H orbitals in CH_4, for instance, and stronger bonds result than without hybrid orbitals.

The four sp^3 hybrid orbitals have the same shape, and they all have the same energy, which is the weighted average of the parent *s* and *p*

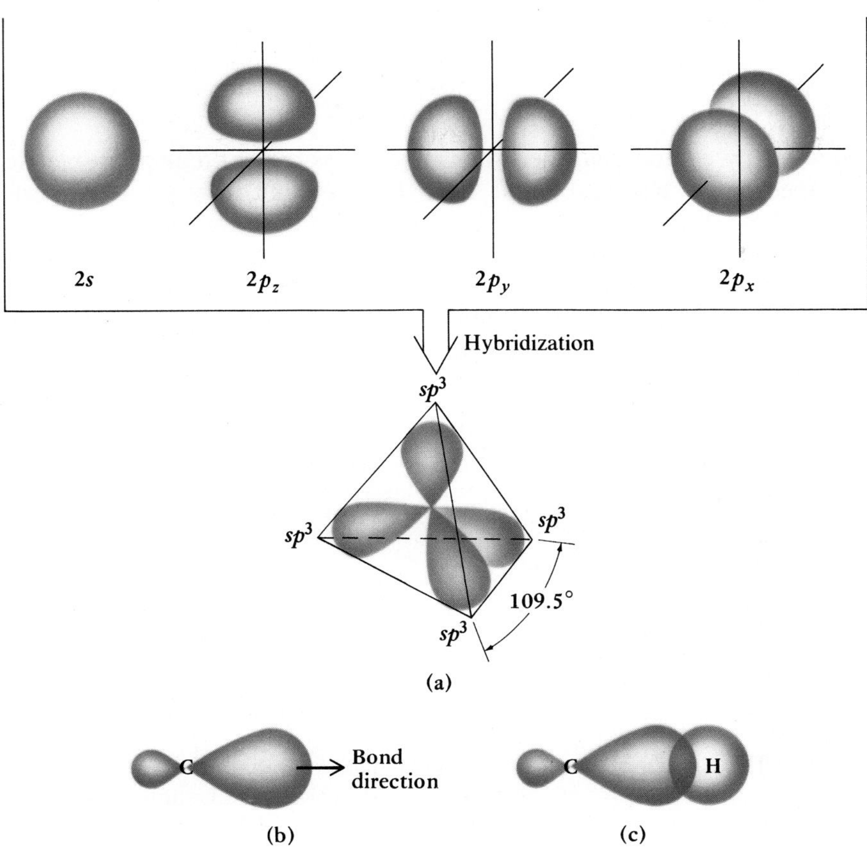

Figure 11.5 Hybrid orbital formation. (a) The four *s* and *p* atomic orbitals *on one atom* are combined or hybridized to form four new sp^3 hybrid orbitals. These orbitals are equivalent and are directed to the corners of a tetrahedron (angle 109.5°). (b) The shape of one of the four hybrid orbitals. The small "back" lobe of each orbital was omitted in (a) for clarity. (c) The C—H sigma bond in CH_4 is formed by overlap of a carbon sp^3 hybrid orbital and a hydrogen 1*s* orbital.

orbital energies. Moreover, they point in four different directions in space, 109° from one another. Since four sigma bonds are to be formed by the C atom, each of the four valence electrons of the atom is assigned to a separate hybrid orbital.

In depicting electrons in hybrid orbitals, red is used for those assigned as σ bonding electrons, blue for lone pair electrons. Green is used for π bonding electrons.

energy

2*p* ↑ ↑ __

↑ ↑ ↑ ↑

⟶ four sp^3 hybrid orbitals in CH_4

2*s* ⇅

isolated C atom

Overlap of each half-filled sp^3 hybrid orbital with a half-filled hydrogen 1*s* orbital gives *four equivalent C—H bonds arranged tetrahedrally*, as required by experimental evidence.

Hybrid orbitals can also be used to understand the bonding and structure of such common molecules as H_2O and NH_3. An isolated O atom has two unpaired valence electrons as required for two bonds, but these are in orbitals 90° apart.

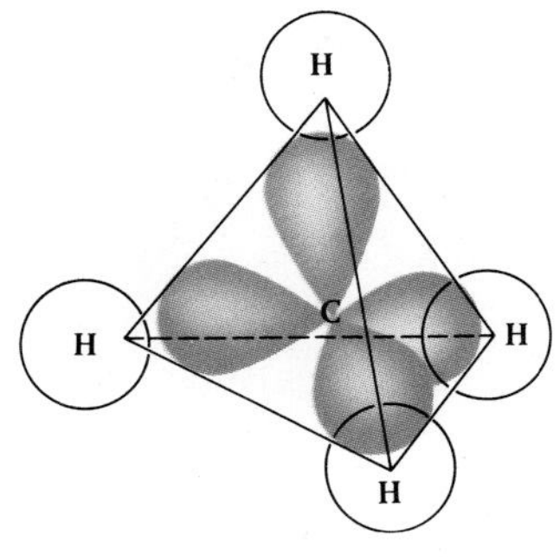

Oxygen Electron Configuration

$1s^22s^22p^4$ or [He] ⇅ ⇅ ↑ ↑ ($2s$ $2p$ $2p$ $2p$)

However, we know that the water molecule is based on an approximate tetrahedron of structural pairs, since the two bond pairs are 105° apart. If we allow the four *s* and *p* orbitals of oxygen to distort or hybridize on approach of the H atoms, four sp^3 hybrid orbitals are created. Two of these orbitals are occupied by unpaired electrons and lead to the O—H sigma bonds. The other two orbitals contain pairs of electrons and so are the lone pairs of the water molecule.

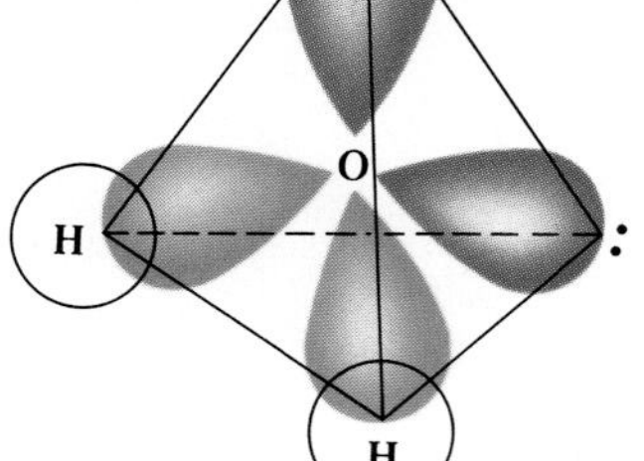

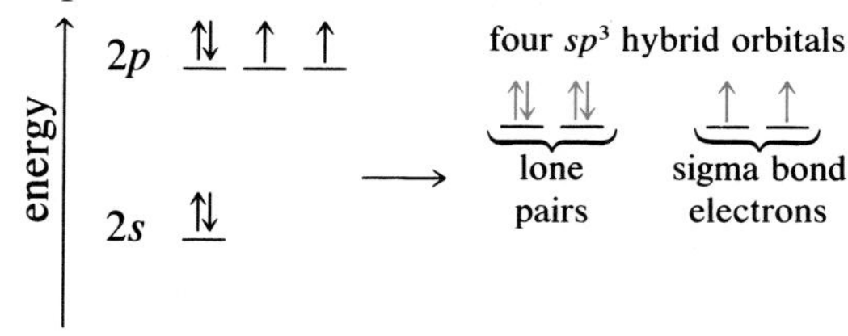

EXAMPLE 11.1

HYBRID ORBITALS IN BONDING

Describe the bonding in ammonia, NH_3, using orbital hybridization.

Solution The electron dot structure of ammonia shows that there are four structural pairs (one lone pair and three bond pairs), which must be at the corners of a tetrahedron. Since three of these pairs are bond pairs, the molecule has a pyramidal molecular geometry.

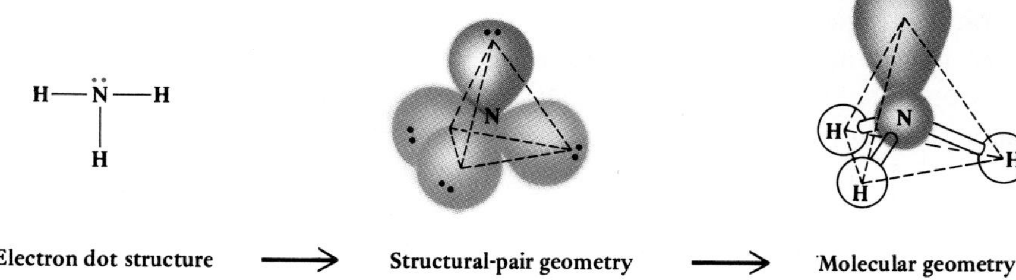

To explain bonding in NH_3 and an H—N—H angle of approximately 107°, we invoke orbital hybridization. The *s* and *p* orbitals of nitrogen combine to give four sp^3 orbitals, one containing a lone pair of electrons and each of the other three having one unpaired electron.

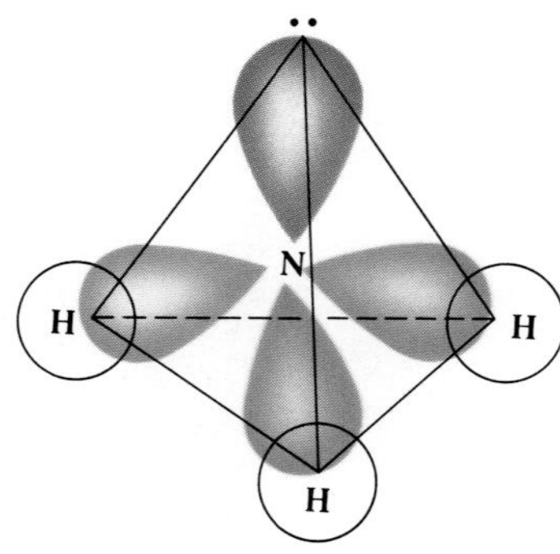

energy: $2p$ ↑ ↑ ↑; $2s$ ⇅ ⟶ four sp^3 hybrid orbitals: ⇅ (lone pair) ↑ ↑ ↑ (sigma bond electrons)

Thus, there are three bond orbitals and one lone pair orbital at the corners of a tetrahedron.

s p_x p_y p_z

Two sp hybrids from $s + p_y$

Three sp^2 hybrids from $s + p_x + p_y$

180° Two sp hybrids + p_x + p_z

sp^2 120° Three sp^2 hybrids + p_z

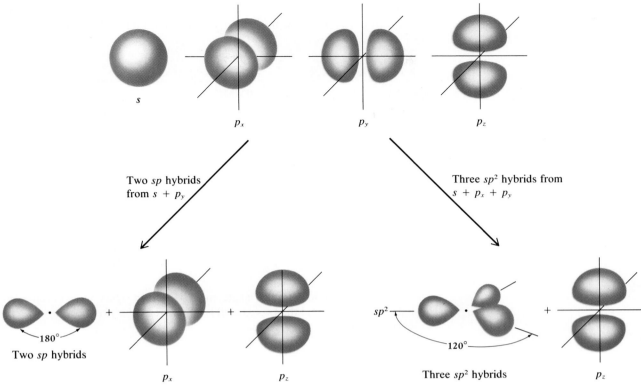

Figure 11.6 Formation of sp and sp^2 hybrid orbitals.

Other combinations of atomic orbitals also lead to hybrid atomic orbitals. For example, an s and two p orbitals on the same atom can combine to form **three sp^2 hybrid atomic orbitals** that are directed to the corners of a planar equilateral triangle (Figure 11.6).

energy

p — — — → — one unchanged p orbital

— — — three sp^2 hybrid orbitals

s —

Similarly, one s and one p orbital may hybridize to form **two sp hybrid atomic orbitals** that are directed away from one another with an angle of 180° (Figure 11.6).

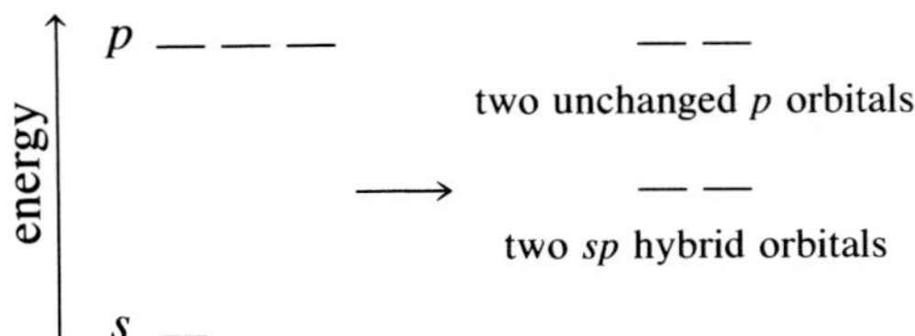

In all the hybridization schemes, the number of hybrid orbitals produced is equal to the number of atomic orbitals used in the combination. Thus, in both sp^2 and sp hybridization, one or two pure p orbitals, respectively, remain after hybridization. These can remain empty or can be used in pi (π) bond formation, as described under "Multiple Bonding" later in this section.

EXAMPLE 11.2

HYBRIDIZATION AND BONDING

Describe the bonding in BF_3 using orbital hybridization.

Solution The structural-pair and molecular geometry of BF_3 are both trigonal planar.

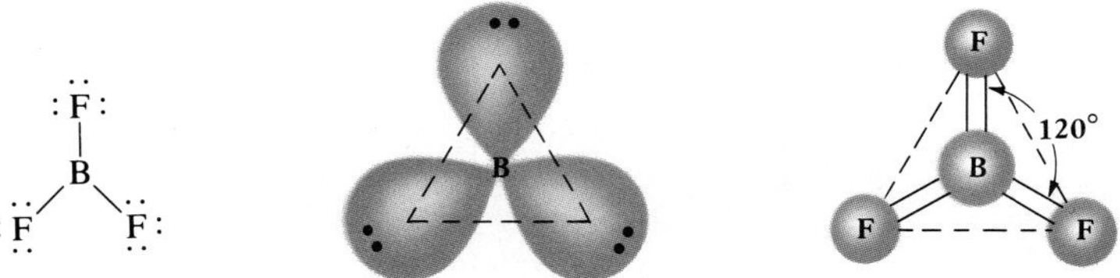

Three bonds, at the corners of a triangle, must be formed. Thus, three boron atom orbitals (the *s* orbital and the two *p* orbitals in the molecular plane) are hybridized to form sp^2 orbitals,

energy

2*p* ↑ _ _ _

2*s* ⇅

isolated B atom

⟶ three sp^2 hybrid orbitals ↑ ↑ ↑

one unchanged *p* orbital _

and each hybrid orbital contains an unpaired electron for bond formation with an F atom.

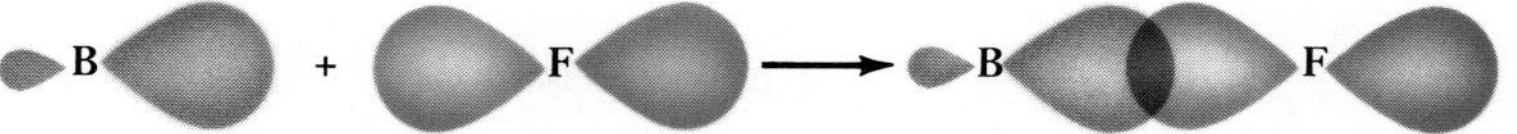

Here the boron *p* orbital not used in hybridization remains unfilled. This fact means that such molecules are reactive, as they tend to acquire an electron pair for the empty orbital by forming a new bond with another molecule if possible.

EXAMPLE 11.3

RECOGNIZING HYBRID ORBITALS

What hybrid orbital sets are required for the central atoms in (a) PCl_3 and (b) BeH_2?

Solution (a) For PCl_3 the electron dot structure and VSEPR theory tell us that the P atom must be surrounded by four electron pairs approximately at the corners of a tetrahedron.

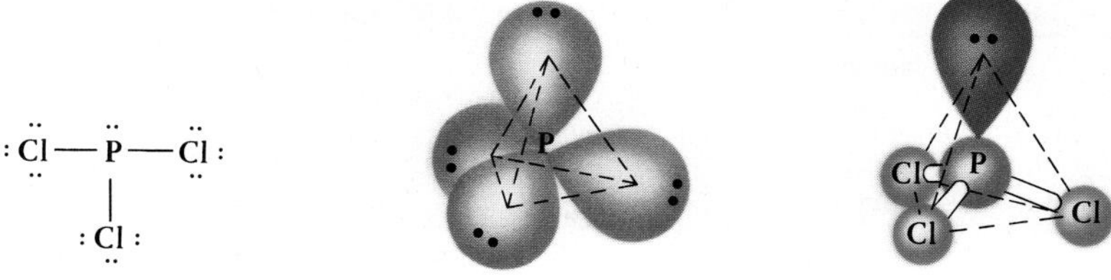

Four structural pairs arranged tetrahedrally require the P atom to be sp^3 hybridized. Phosphorus, like nitrogen, is in Group 5A. Therefore, the bonding in PCl_3 resembles that in NH_3 (Example 11.1). The phosphorus lone pair is assigned to one sp^3 hybrid orbital, and the P—Cl bonds are formed by overlap of a Cl half-filled p orbital with a half-filled P sp^3 orbital.

(b) BeH_2 is a linear molecule, according to its electron dot structure and VSEPR theory.

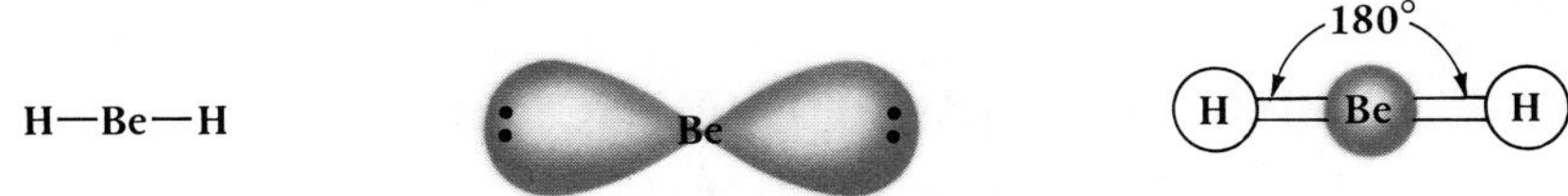

Linear geometry and two sigma bonds are achieved by forming sp hybrid orbitals on Be (from the s orbital and the p orbital that lies along the HBeH axis).

energy ↑			
$2p$	— — —		— — two unchanged p orbitals
		⟶	$\underline{\uparrow}$ $\underline{\uparrow}$ two sp hybrid orbitals
$2s$	$\underline{\uparrow\downarrow}$ isolated Be atom		

The unhybridized p orbitals of the Be atom are not filled in a simple BeH_2 molecule.

EXERCISE 11.1 Hybrid Orbitals and Bonding

Describe the bonding in SCl_2 using hybrid orbitals.

HYBRID ORBITALS FOR ATOMS OF EXPANDED VALENCE

Elements of the third and higher periods can exhibit expanded valence in molecules such as PF_5 and SF_6. Here the central atom's valence shell is expanded to five or six pairs, so the valence bond theory requires that five or six atomic orbitals be hybridized. Since there are only four atomic orbitals of the s and p type in a valence shell, the extra one or two orbitals required come from the d subshell of the shell containing the s and p valence orbitals. However, be sure to understand that d orbitals are energetically accessible for hybrid formation *only* for elements of the third and higher periods. Hence only these elements exhibit expanded valence. The hybrid orbital sets for five to six valence shell electron pairs

Table 11.1 Hybrid Orbital Sets for Two to Six Electron Pairs

Hybrid Orbital Set	Number of Hybrid Orbitals (Number of Sigma and Lone Electron Pairs)	Geometry	Example
sp	2	Linear	Be in BeF_2
sp^2	3	Trigonal planar	B in BF_3
sp^3	4	Tetrahedral	C in CH_4
sp^3d	5	Trigonal bipyramidal	P in PF_5
sp^3d^2	6	Octahedral	S in SF_6

are given in Table 11.1 and Figure 11.7, along with those sets already described.

EXAMPLE 11.4

HYBRID ORBITALS IN EXPANDED VALENCE ATOMS

Describe the bonding in PF_5 using hybrid orbitals.

Solution The structural-pair geometry of PF_5 is trigonal bipyramidal according to its electron dot structure and VSEPR theory.

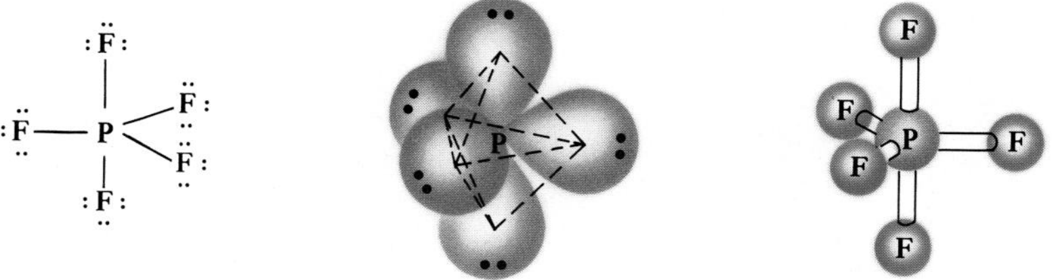

According to Figure 11.7, the hybrid scheme sp^3d is required.

energy

3d — — — — —   — — — —

four unhybridized *d* orbitals

3p ↑ ↑ ↑ ⟶ ↑ ↑ ↑ ↑ ↑

five sp^3d hybridized orbitals for sigma bonds

3s ⇅

isolated P atom

Each of the five P—F σ bonds involves the overlap of one of the phosphorus sp^3d hybrid orbitals with a fluorine *p* orbital.

EXERCISE 11.2 Hybrid Orbitals in Expanded Valence Atoms

Describe the bonding in XeF_4 using hybrid orbitals. Remember to consider first the electron dot structure and then the structural-pair geometry.

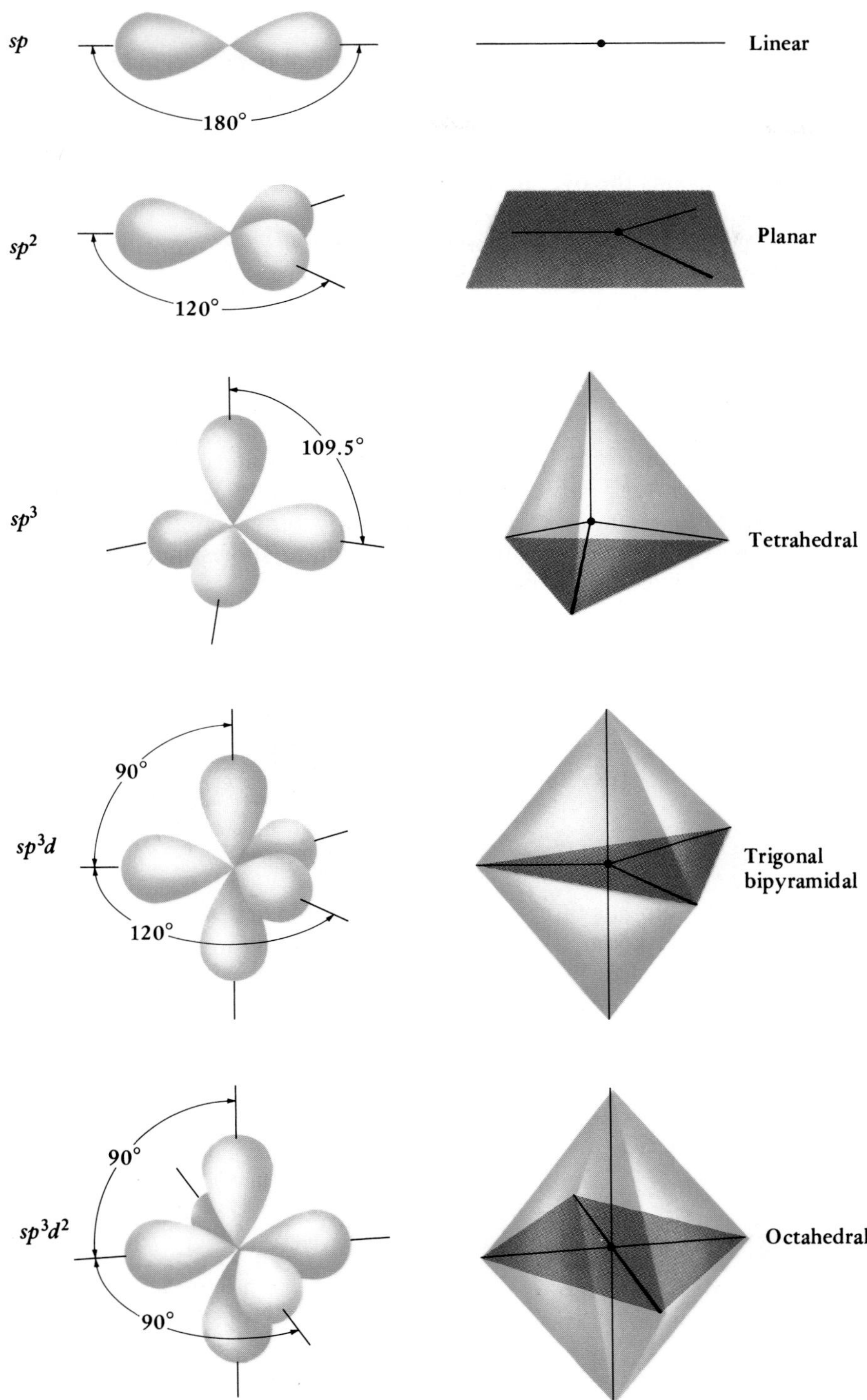

Figure 11.7 The geometry of the hybrid orbital sets for two to six structural electron pairs. In forming a hybrid orbital set, the *s* orbital is always used, plus as many *p* orbitals (and *d* orbitals) as are required to give the necessary number of sigma bonding and lone pair orbitals.

MULTIPLE BONDING

There are two principal types of chemical bonds: **sigma** (σ) bonds and **pi** (π) bonds. According to valence bond theory, σ bonds arise from the overlap of atomic orbitals so that the electrons lie along the bond axis. *Pi bonds come from the sideways overlap of p atomic orbitals* (Figure 11.8). This means the overlap region is above and below the internuclear axis.

Pi bonds never occur alone without the bonded atoms also being joined by a σ bond. Thus, a double bond consists of a σ bond and a π bond, while a triple bond consists of a σ bond and two π bonds.

pi bond → C=C ← sigma bond

two pi bonds → C≡C ← one sigma bond

Since a π bond is formed from pure p atomic orbitals, one on each of two atoms, *a π bond may form only if unhybridized p orbitals remain on the bonded atoms*, in the correct orientation, after accounting for σ bond formation. Therefore, when the Lewis structure shows multiple bonds, the atoms involved are sp or sp^2 hybridized.

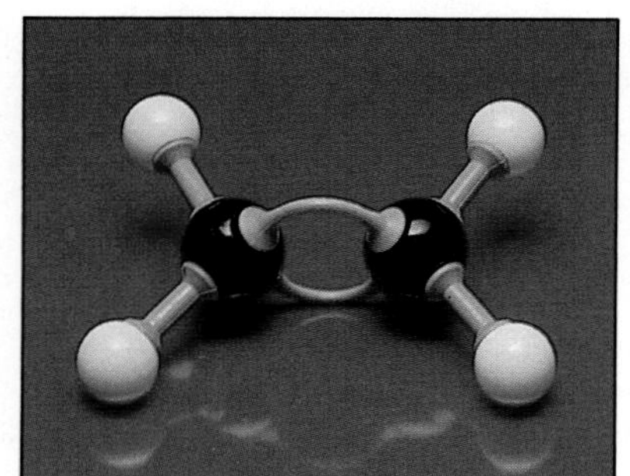

C_2H_4, ethene (common name: ethylene)

The molecular models used in this and later chapters do not represent the orbitals presented in the text. The two bent connectors joining the black C atoms are meant only to show that there are two bonds connecting the atoms; they do not *represent the geometry of the orbitals involved. However, the model does clearly show the trigonal planar geometry about each C atom.*

One of the simplest examples of π bonding is found in ethene, C_2H_4.

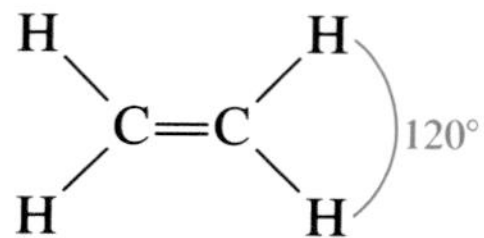

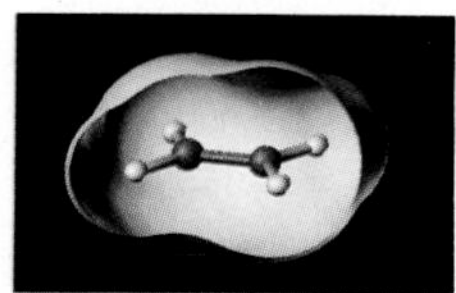

A computer generated model of ethylene, inside a model of the molecular surface. (J. Weber)

Each carbon atom has three σ-bonding electron pairs, so VSEPR theory predicts that the geometry about the carbon atoms in C_2H_4 is trigonal planar. To rationalize the σ bonding of carbon to two hydrogen atoms and another carbon in a trigonal plane, the carbon atom is assumed to be sp^2 hybridized. Thus, each carbon atom has three sp^2 hybrid orbitals in a plane, and an unhybridized p orbital perpendicular to that plane (Figure 11.6). Each of these four orbitals accounts for one unpaired electron.

energy

$2p$ ↑ ↑ __

$2s$ ↑↓

isolated C atom

→ hybridize C orbitals

↑ unhybridized p orbital used for pi bonding

↑ ↑ ↑ three sp^2 hybrids used for C—H and C—C sigma bonds

A π bond can form only if the σ orbitals of both C atoms are all in the same plane. If the plane containing the sp^2 hybrid orbitals of one C atom is rotated relative to the plane containing the sp^2 hybrid orbitals of the other C atom, the unhybridized p orbital on one C atom will not align with the unhybridized p orbital on the adjacent C atom, and a π bond will not form.

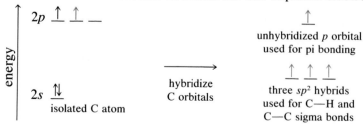

If two such C atoms approach one another, head-to-head overlap of one sp^2 hybrid orbital from each atom will give a σ bond (Figure 11.9), and sideways overlap of the half-filled, pure p orbitals will form a π bond. Thus, the C=C double bond in C_2H_4 consists of one σ bond and one π bond. To complete the bonding picture, each of the two remaining half-filled sp^2 hybrid orbitals on the C atoms overlaps with a half-filled hydrogen $1s$ orbital to form a C—H σ bond.

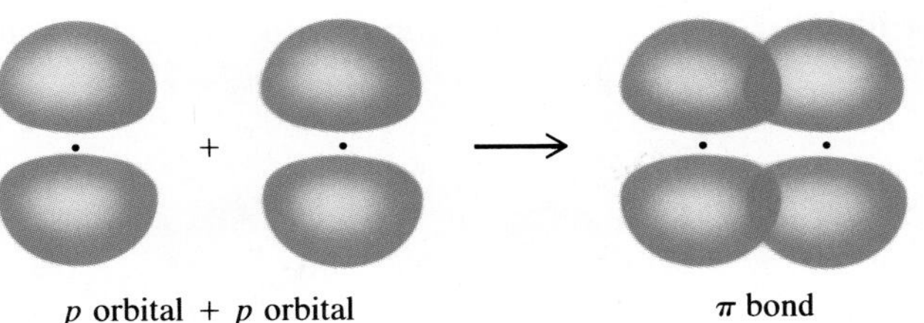

Figure 11.8 Formation of pi (π) bond by sideways overlap of p orbitals.

In Chapter 10 we pointed out that carbon can form multiple bonds with oxygen, sulfur, and nitrogen. To understand how such bonds are possible, consider formaldehyde, H_2CO.

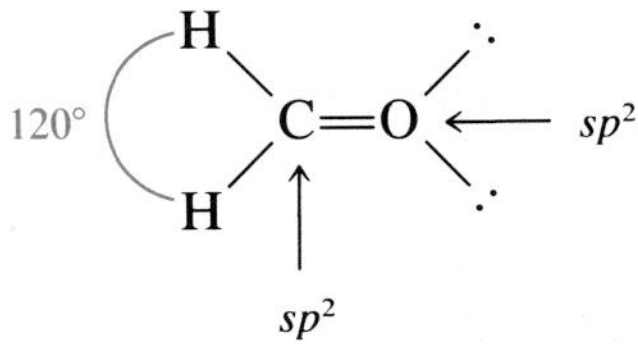

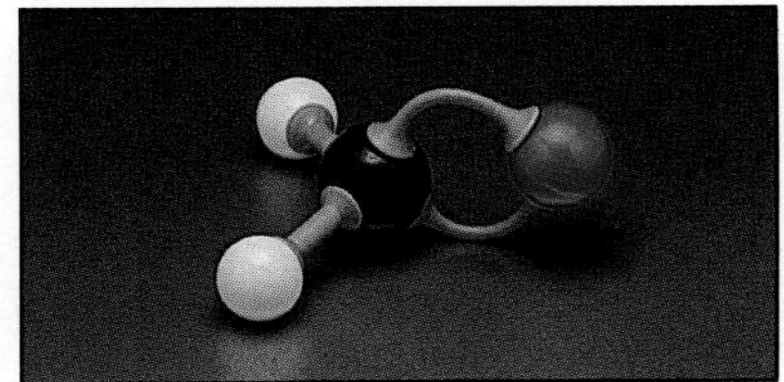

H_2CO

This trigonal planar molecule is based on an sp^2 hybridized carbon atom. Thus, just as for the carbon atoms of C_2H_4, there is an unhybridized p orbital perpendicular to the molecular plane. This p orbital is available for π bonding, this time with oxygen. If we also consider the oxygen atom as sp^2 hybridized, since the best arrangement of a σ bond pair and two lone pairs is trigonal planar,

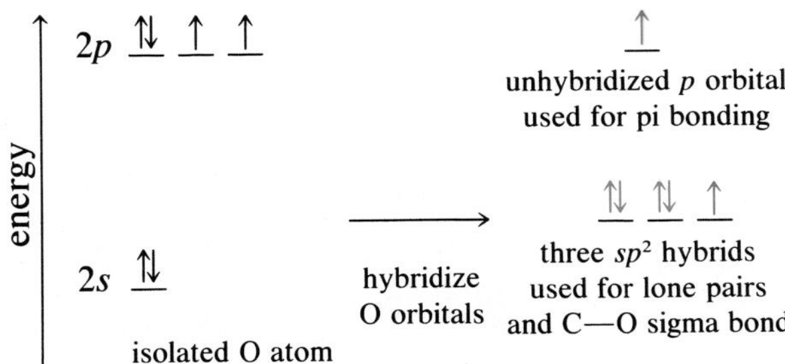

these hybrid orbitals can account for the C—O σ bond and the oxygen lone pairs. The unhybridized oxygen p orbital, describing an unpaired

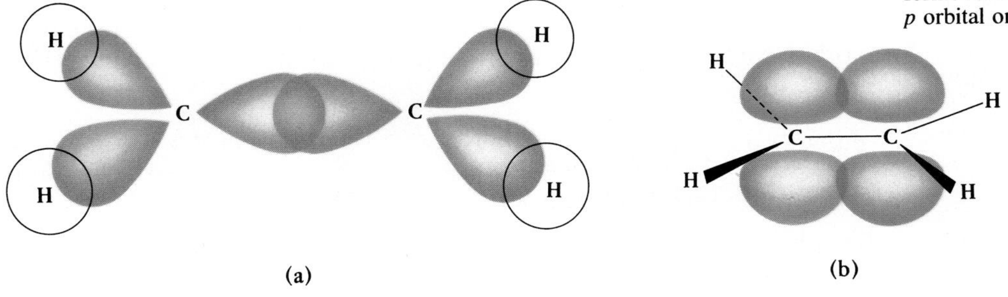

Figure 11.9 Bonding in ethene, C_2H_4. (a) Sigma (σ) framework. The C atom orbitals are hybridized to form three sp^2 hybrid orbitals and leave one pure p orbital. Overlap of an sp^2 orbital from one C atom with an sp^2 hybrid orbital from the adjacent C atom gives the C—C sigma bond. The C—H bonds come from the overlap of a C atom sp^2 hybrid orbital with an H atom $1s$ orbital. (b) The pi (π) bond in C_2H_4 is formed from the sideways overlap of a p orbital on each atom.

electron, is then available for π bond formation with the carbon p orbital.

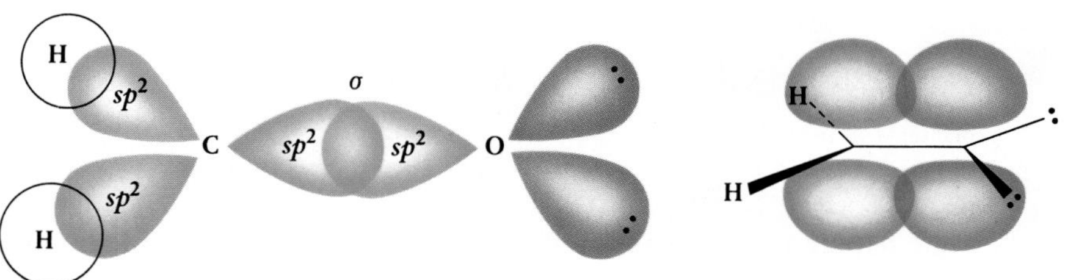

H_2CO sigma bonds and lone pairs

pi bond in H_2CO

Acetylene, H—C≡C—H, is a simple molecule with a triple bond. Here one σ bond and two π bonds join the two carbon atoms. The structural-pair geometry around the carbon atoms is clearly linear, so each carbon atom is sp hybridized. This means that two half-filled p orbitals remain on each carbon after hybridization and are available for π bond formation.

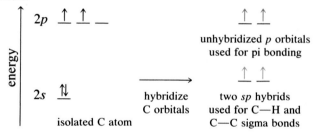

energy	isolated C atom		
$2p$	↑ ↑ _		↑ ↑ unhybridized p orbitals used for pi bonding
$2s$	⇅	hybridize C orbitals ⟶	↑ ↑ two sp hybrids used for C—H and C—C sigma bonds

The orbitals about each C atom can be pictured as

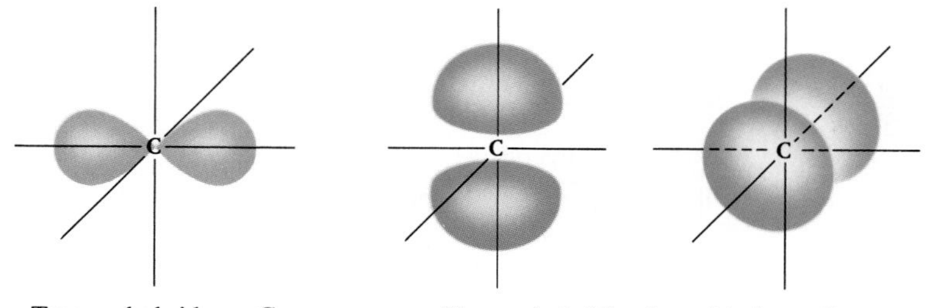

Two sp hybrids on C

Two unhybridized p orbitals on C

where the two sp hybrids are 180° apart, and the two unhybridized p orbitals are 90° apart and lie in planes perpendicular to the sp axis. The sp hybrids are used for C—C and C—H σ bond formation, and the pure p orbitals overlap sideways to produce two carbon–carbon π bonds (Figure 11.10).

EXAMPLE 11.5

THE C≡O TRIPLE BOND IN CARBON MONOXIDE

Describe the bonding in carbon monoxide, CO, using orbital hybridization.

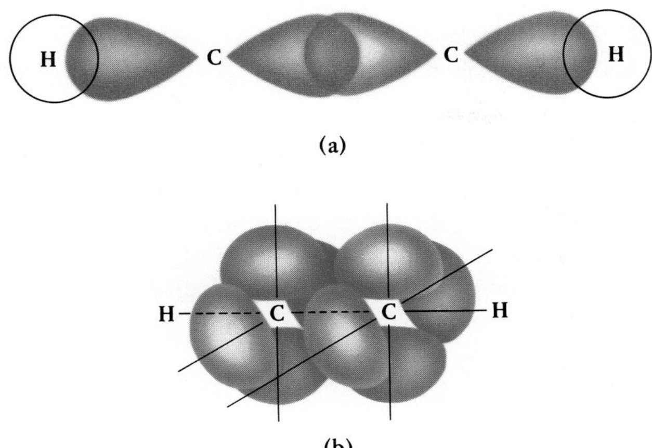

Figure 11.10 Bonding in acetylene, C_2H_2. (a) Sigma (σ) framework. The C atom orbitals are hybridized to form two *sp* hybrid orbitals and leave two pure *p* orbitals. Overlap of an *sp* hybrid orbital from one C atom with an *sp* hybrid orbital from the adjacent C atom gives the C—C sigma bond. The C—H bonds come from the overlap of a C atom *sp* hybrid orbital with an H atom 1*s* orbital. (b) The two pi (π) bonds in C_2H_2 are formed from the sideways overlap of the *p* orbitals on adjacent atoms.

Solution The electron dot structure, :C≡O:, tells us that both the carbon and the oxygen must be *sp* hybridized. Further, each atom must use one *sp* hybrid orbital for sigma bond formation and one hybrid orbital for a lone pair.

To "bookkeep" electrons here properly, there is another useful rule to be followed when assigning electrons to hybrid orbitals: the hybrid orbitals of an atom are initially assigned only one electron when they are used in sigma bond formation, but they are assigned two electrons when used for lone pairs. Remaining electrons are assigned to unhybridized orbitals.

	sp hybridized C		*sp* hybridized O	
Unhybridized *p* orbitals for pi bonding	↑ _		⇅ ↑	
Two *sp* hybrids	⇅	↑	⇅	↑
	lone pair	sigma bond	lone pair	sigma bond

Now there is a half-filled *sp* hybrid orbital on each atom for sigma bond formation, as well as a lone pair on each atom assigned to an *sp* hybrid. There is also a total of two pairs of electrons in unhybridized *p* orbitals to be used for two pi bonds as required.

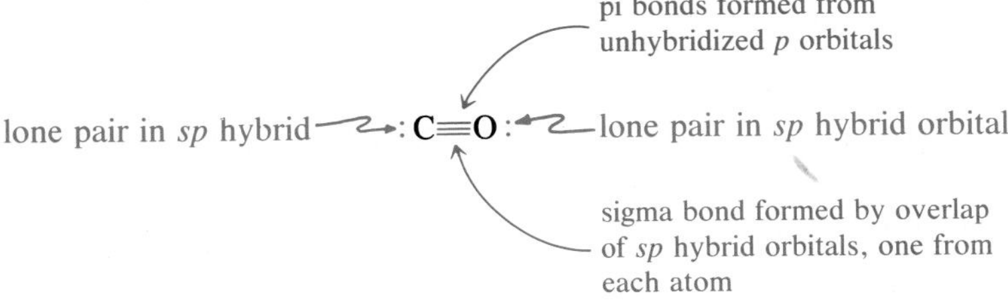

The orbital picture for CO is identical with that for HC≡CH in Figure 11.10, except that the C—H bond pairs become lone pairs.

EXERCISE 11.3 Triple Bonds
Describe the bonding in the dinitrogen molecule, N_2.

There are thousands of carbon-based molecules that have multiple bonds, and you will encounter many of them in this book and in further study of chemistry. It is therefore valuable to examine the bonding in a somewhat more complex case. Consider as an example acetic acid, $H_3C—COOH$, the important ingredient in vinegar. Its electron dot structure is shown below.

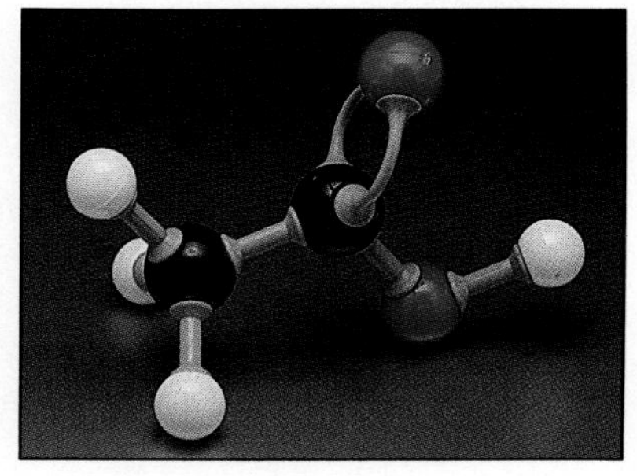

Acetic acid

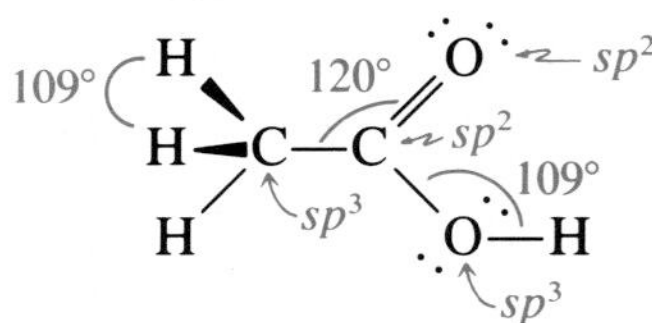

Acetic acid, H_3CCOOH

The carbon atom of the CH_3 group must have a tetrahedral structural-pair geometry, so the C—H angle is approximately 109°. In valence bond terms, this means the carbon atom is sp^3 hybridized. Similarly, the O atom of the C—O—H group is sp^3 hybridized, since it is surrounded by a tetrahedron of bond and lone pairs. The other carbon atom of the molecule has trigonal planar structural-pair geometry, with bonds making an angle of 120°. This carbon must be sp^2 hybridized. Finally, we account for the C=O link exactly as in the H_2CO molecule described earlier. Both the C atom and the O atom are taken as sp^2 hybridized, so an unhybridized p orbital remains on each atom to form the carbon–oxygen π bond.

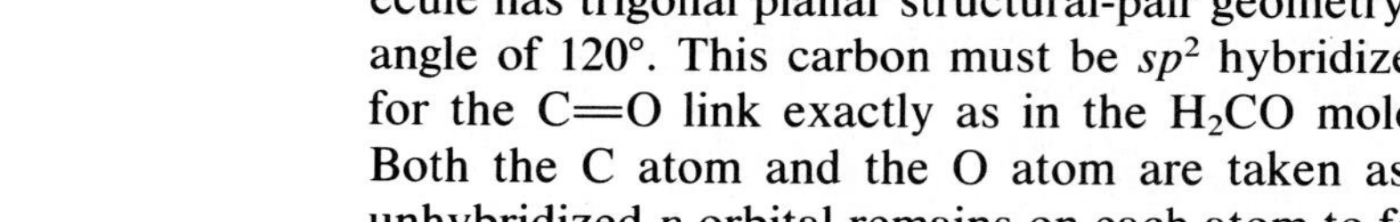

EXERCISE 11.4 Bonding and Hybridization
Analyze the bonding in acetonitrile, $H_3C—C\equiv N$:. Give values for the H—C—H, H—C—C, and C—C—N angles, and indicate the hybridization of both carbon atoms and the nitrogen atom.

π RESONANCE STRUCTURES AND HYBRID ORBITALS

π Resonance structures often involve an electron pair that is used alternately to form a π bond between two atoms and a lone pair on one of these atoms. As an example, consider ozone, O_3, which has two resonance structures.

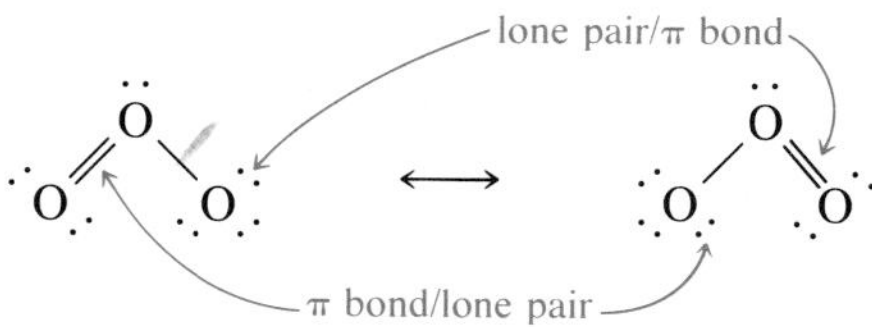

To visualize the bonding in ozone and to account for resonance structures, assume that all three O atoms are sp^2 hybridized. The central atom uses sp^2 hybrid orbitals to form two σ bonds and to accommodate a lone pair.

The terminal atoms use their sp^2 hybrids to form only one σ bond and to accommodate two lone pairs.

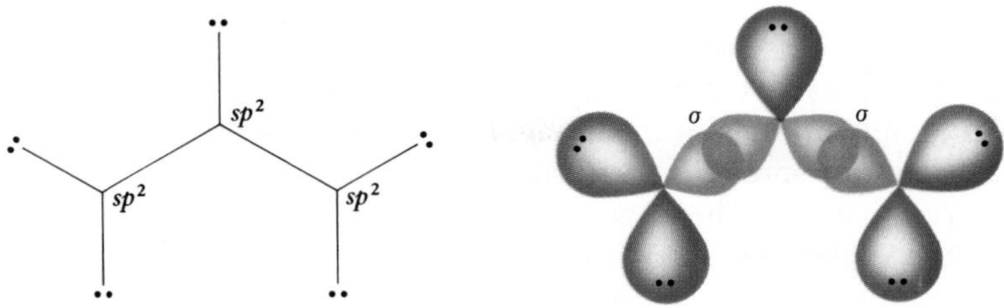

Top view of ozone showing sigma bonds and lone pairs in oxygen sp^2 hybrid orbitals

The O_3 molecule has nine valence electron pairs to be accommodated, but the σ framework and lone pairs illustrated above account for only seven of these pairs. The π bonds in ozone arise from the two remaining pairs.

Since we assume each O atom in O_3 is sp^2 hybridized, an unhybridized p orbital *perpendicular to the* O_3 *plane* must remain on each atom.

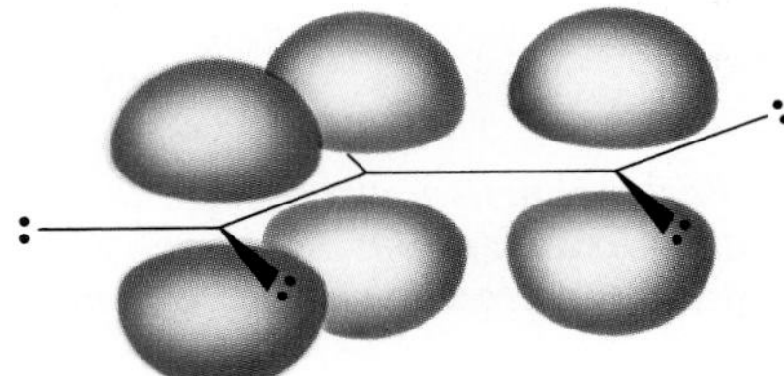

The oxygen p orbitals remaining after sigma bonds and lone pairs are accommodated in the O_3 plane

energy ↑

p orbital used for out-of-plane pi bonds ___

Three sp^2 hybrid orbitals for in-plane sigma bonds and lone pairs ___ ___ ___

sp^2 hybridized O

According to hybrid orbital theory, the p orbitals on two adjacent atoms can form a π bond using one of the remaining electron pairs. The p orbital on the third atom is then left to accommodate the final electron pair as a lone pair.

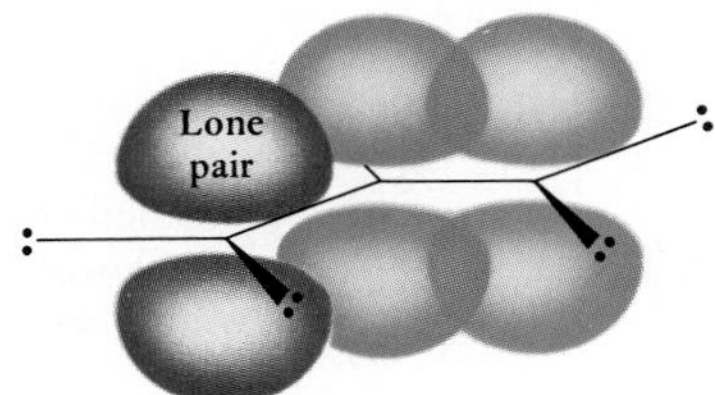

Formation of a pi bond between one pair of O atoms

It is obvious, however, that the O—O π bond could form using the central atom p orbital with a p orbital on either of the terminal O atoms. The lone pair is then assigned to the p orbital that is not used in the multiple bond. Hence, there are two possible arrangements, which chemists call π resonance structures.

EXERCISE 11.5 Resonance Structures and Hybridization

Describe the resonance structures of NO_2^- in terms of hybrid orbitals.

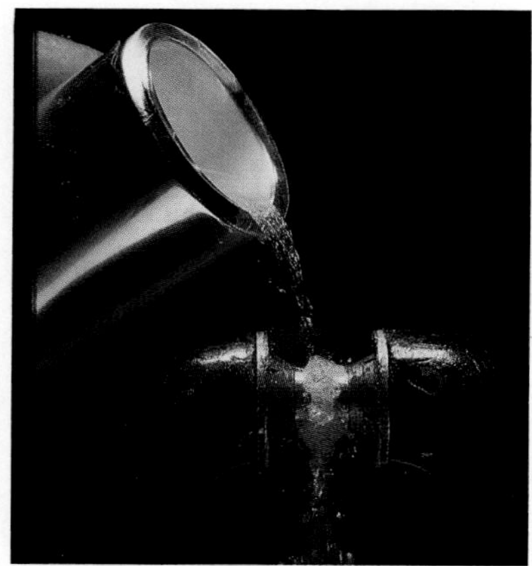

Figure 11.11 Liquid O_2 is paramagnetic and so clings to the poles of a magnet. The compound was cooled to a very low temperature to cause it to liquefy. Notice that the liquid is light blue.

Paramagnetism arises from the unpaired electrons. See page 319 for a more complete discussion.

11.2 MOLECULAR ORBITAL THEORY

Molecular orbital (MO) theory is an alternative way to view electron orbitals in molecules and is a theory widely used to account for molecular properties beyond bond formation. In contrast to the localized bond and lone pair orbitals of valence bond theory, MO theory assumes that pure *s* and *p* atomic orbitals of the atoms in the molecule combine to produce orbitals that are spread or delocalized over several atoms or even over the entire molecule. The new orbitals are called *molecular orbitals*, and they can have different energies. Just as with orbitals in atoms, molecular orbitals are assigned to electrons according to the Pauli principle and Hund's rule.

One reason for learning about the MO concept is that it correctly predicts the electronic structures of certain molecules that do not follow the electron-pairing assumptions of the Lewis approach. The most common example is the O_2 molecule. Using the rules of Chapter 10, you would draw the electron dot structure of the molecule as $:\ddot{O}{=}\ddot{O}:$, with all the electrons paired. However, experiments clearly show that the O_2 molecule is *paramagnetic* and that it has exactly two *unpaired electrons* per molecule. It is sufficiently magnetic that liquid O_2 clings to the poles of a magnet (Figure 11.11). The molecular orbital approach can account for the paramagnetism of O_2 more easily than can valence bond theory. To see how MO theory can apply to O_2 and other diatomic molecules, we shall first describe the four principles of the theory.

PRINCIPLES OF MOLECULAR ORBITAL THEORY

According to valence bond theory, the number of hybrid orbitals produced on an atom is *always* the same as the number of atomic orbitals combined on that atom. Notice, though, that this principle of orbital conservation was not applied to the combination of hybrid orbitals from adjacent atoms to form bonding orbitals. That is, valence bond theory assumes that one bond is formed by the combination of two atomic or

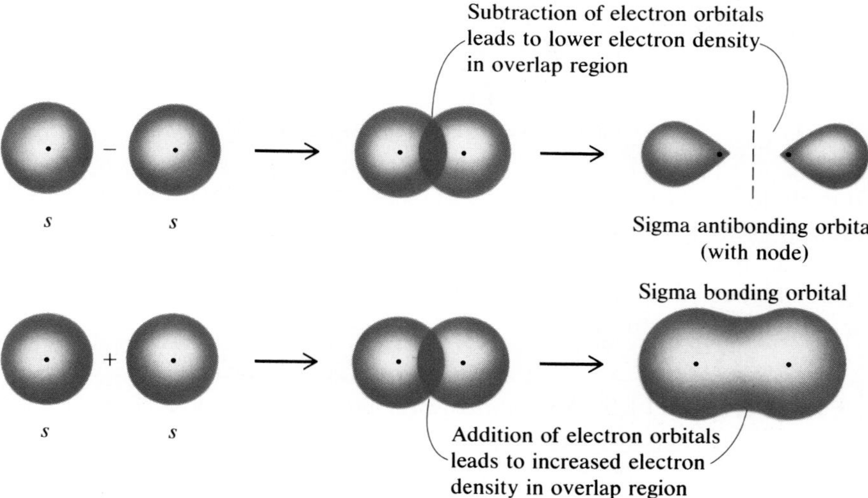
Subtraction of electron orbitals leads to lower electron density in overlap region
−
s
s
Sigma antibonding orbital (with node)
Sigma bonding orbital
+
s
s
Addition of electron orbitals leads to increased electron density in overlap region

Figure 11.12 Formation of bonding and antibonding molecular orbitals from two *s* atomic orbitals on adjacent atoms. Notice the presence of a node in the antibonding orbital.

hybrid orbitals, one from each of two adjacent atoms. In contrast, MO theory forms two molecular orbitals from two atomic orbitals. Furthermore, in MO theory, atomic orbitals are *not* first combined on each atom; instead, the original atomic orbitals on adjacent atoms combine to form new molecular orbitals in which all atoms can participate.

With these ideas in mind, we can state the **first principle** of molecular orbital theory: *the number of molecular orbitals produced is always equal to the number of atomic orbitals brought by the combining atoms*. To see the consequence of this orbital conservation principle, consider first the H_2 molecule.

BONDING AND ANTIBONDING MOLECULAR ORBITALS IN H_2 When the 1*s* orbitals of two atoms overlap, *two* molecular orbitals result. The principles of molecular orbital theory tell us that, in one of the resulting molecular orbitals, the 1*s* regions of electron density *add* together to lead to an increased probability that electrons are found in the bond region (Figure 11.12).* Electrons in such an orbital concentrate between the two hydrogen nuclei where they pull the two nuclei toward each other, bonding them together. This orbital is called a **bonding molecular orbital**, and is the same as the bond described by valence bond theory. Moreover, it is a sigma (σ) orbital, since the region of electron probability lies directly along the bond axis. We label this molecular orbital σ_{1s}, the superscript 1*s* signaling the atomic origin of the orbital.

Since two combining atomic orbitals must produce two molecular orbitals, the other combination is constructed by *subtracting* one orbital from another (Figure 11.12). When this happens there is reduced electron probability between the nuclei for the molecular orbital, and the electron density increases to the outside of the H atom nuclei; a *node* appears between the nuclei. Here the electron density acts to attract the nuclei away from each other, and so the orbital is called an **antibonding molecular orbital**. This is also a sigma orbital, since the region of electron probability lies directly along the bond axis. It is labeled σ^*_{1s}, where the asterisk signifies that this is an antibonding orbital. *There is no counterpart of antibonding orbitals in valence bond theory.*

Antibonding orbitals have nodes, and their energies are higher than the average energy of the atomic orbitals from which they were created. Their upward energy displacement is slightly greater than the downward displacement of the bonding orbitals.

A **second principle** of molecular orbital theory is that *the bonding molecular orbital is lower in energy than the parent orbitals, and the antibonding orbital is higher in energy* (Figure 11.13). The average energy of the molecular orbitals is slightly higher than the average energy of the parent atomic orbitals.

A **third principle** of molecular orbital theory is that *the electrons of the molecule are assigned to orbitals of successively higher energy*; the Pauli principle and Hund's rule are obeyed. Thus, electrons occupy the lowest energy orbitals available, and they do so with spins paired. Since the energy of the electrons in the bonding orbital of H_2 is lower than that of either parent 1*s* electron, the H_2 molecule is more stable than the separate atoms. We write the electron configuration of H_2 as $(\sigma_{1s})^2$.

*Orbitals are electron waves. Therefore, a way to view molecular orbital formation is to assume that two electron waves, one from each atom of the bonded pair, interfere with one another. The interference can be constructive (to give a bonding MO) or destructive (to give an antibonding MO).

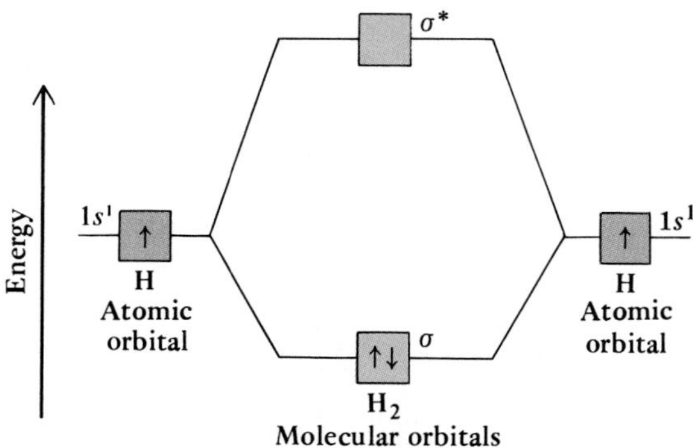

Figure 11.13 Energy level diagram for the molecular orbitals from two atomic $1s$ orbitals. The two electrons of H_2 are placed in the lowest energy σ_{1s} MO.

Next consider combining two helium atoms to form dihelium, He_2. Since both He atoms have $1s$ valence orbitals, they combine to produce the same kind of molecular orbitals as in H_2. The four electrons of dihelium are assigned to these orbitals according to the scheme in Figure 11.14. The pair of electrons in σ_{1s} stabilizes He_2. However, the two electrons in σ^*_{1s} *de*stabilize the He_2 molecule slightly more than the two electrons assigned to σ_{1s} stabilize the molecule. Thus, molecular orbital theory predicts that He_2 has no *net* stability, and laboratory experiments indeed show that two He atoms have little tendency to combine.

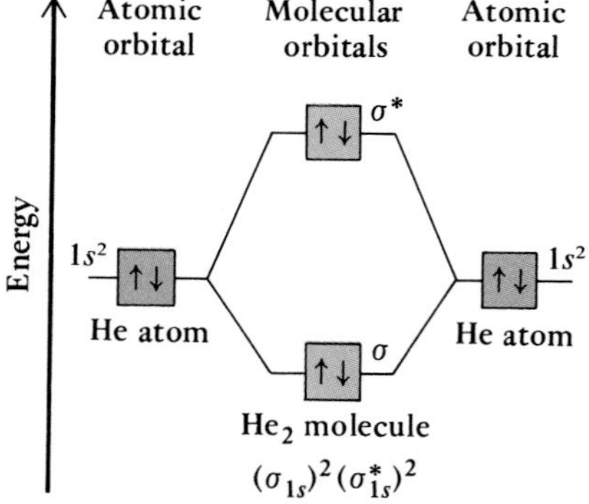

Figure 11.14 Energy level diagram for the hypothetical dihelium molecule, He_2.

BOND ORDER Bond order was defined in Chapter 10 as *the number of bonding electron pairs linking a pair of atoms*. This same concept can be applied directly to molecular orbital theory, but now we define bond order as

$$\text{Bond order} = (\text{number of electron pairs in bonding molecular orbitals}) - (\text{number of electron pairs in antibonding molecular orbitals})$$

In the H_2 molecule, there is one electron pair in a bonding orbital, so H_2 has a bond order of 1. In contrast, the effect of the σ_{1s} pair in He_2 is cancelled by the effect of the σ^*_{1s} pair, so the bond order is 0.

Fractional bond orders are also possible. For example, even though He_2 does not exist, the He_2^+ ion has been detected. Its molecular orbital electron configuration would be $(\sigma_{1s})^2(\sigma^*_{1s})^1$. Here there is one electron pair in a bonding molecular orbital, but only half a pair in an antibonding orbital. Therefore, MO theory predicts that He_2^+ should have a bond order of $\frac{1}{2}$, and this explains the fact that the ion is stable enough to be observed.*

EXAMPLE 11.6

MOLECULAR ORBITALS AND BOND ORDER

Write the configuration of the H_2^- ion in molecular orbital terms. What is the bond order of the ion? Do you expect it to be stable?

*Ions such as He_2^+ are found only in the gas phase by using special instruments. They have never been observed in water or other solvents, nor does it seem they are likely to be. That is, there seems little chance that you might be able to isolate a salt such as dihelium fluoride, He_2F, although it is interesting to speculate on ways that this might be done.

Solution This ion has three molecular electrons (one each from the H atoms plus one for the negative charge). Therefore, its configuration is $(\sigma_{1s})^2(\sigma^*_{1s})^1$, identical with the configuration for $He_2{}^+$. This means it also has a net bond order of $\frac{1}{2}$, and the ion is predicted to exist under special circumstances.

EXERCISE 11.6 Molecular Orbitals and Bond Order
Write the configuration of the $H_2{}^+$ ion in molecular orbital terms. Compare the bond order of the ion with $He_2{}^+$ and $H_2{}^-$. Do you expect $H_2{}^+$ to have some stability?

MOLECULAR ORBITALS OF Li_2 AND Be_2 A **fourth principle** of molecular orbital theory is that *atomic orbitals combine to form molecular orbitals most effectively when the atomic orbitals are of similar energy*. This principle becomes important when we move past He_2 to Li_2, dilithium.

Is it really true that the Starship Enterprise is fueled with dilithium crystals?

A lithium atom has two *s*-type valence orbitals (1*s* and 2*s*). Therefore, the combination of 1*s* on one Li atom with 2*s* on an adjacent Li atom is theoretically possible. However, because the 1*s* and 2*s* orbitals are quite different in energy, the 1*s*/2*s* interaction cannot make an important contribution. Thus, the molecular orbitals can be considered to come only from 1*s*/1*s* and 2*s*/2*s* interactions (Figure 11.15). This means the molecular orbital electron configuration of dilithium is $(\sigma_{1s})^2(\sigma^*_{1s})^2(\sigma_{2s})^2$. The bonding effect of the σ_{1s} electrons is cancelled by the antibonding effect of the σ^*_{1s} electrons, so these pairs make no *net* contribution to bonding in Li_2. The result is that bonding in Li_2 is due to the electron pair assigned to the σ_{2s} orbital, and the *net* bond order is 1.

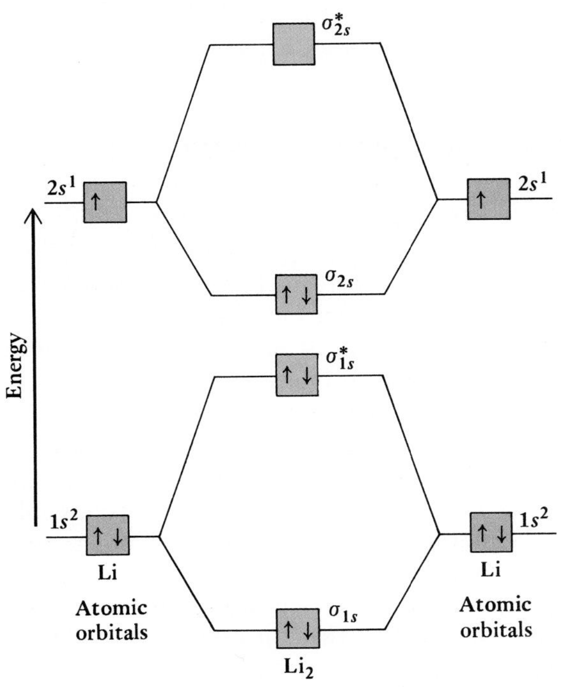

Figure 11.15 Energy level diagram for the combination of two atoms with 1*s* and 2*s* atomic orbitals. The electron configuration shown is for Li_2.

The fact that the σ_{1s} and σ^*_{1s} electron pairs of Li_2 make no net contribution to bonding is exactly what you observed in drawing electron dot structures in Chapter 10: core electrons are ignored. In molecular orbital terms, core electrons are assigned to bonding and antibonding molecular orbitals that offset one another.

A diberyllium molecule, Be_2, is not expected to be stable. Neglecting the 1*s* core electrons, its electron configuration would be

Be_2 MO Configuration: [core electrons]$(\sigma_{2s})^2(\sigma^*_{2s})^2$

Bond order = (1 bonding pair) − (1 antibonding pair)
= 0 net bonding pairs

and you see that there are no *net* bonding electron pairs.

EXAMPLE 11.7

MOLECULAR ORBITALS IN DIATOMIC MOLECULES

Be_2 does not exist. But what about the Be_2^+ ion? Describe its electron configuration in molecular orbital terms and give the net bond order. Do you expect the ion to be stable?

Solution The Be_2^+ ion has only seven molecular electrons (in contrast to eight for Be_2), of which four are core electrons. (The core electrons are assigned to σ_{1s} and σ^*_{1s} molecular orbitals). The remaining three electrons are assigned to the σ_{2s} and σ^*_{2s} molecular orbitals (see Figure 11.15), so the MO electron configuration is [core electrons]$(\sigma_{2s})^2(\sigma^*_{2s})^1$. This means that the net bond order is (1 σ bonding pair) − ($\frac{1}{2}$ σ antibonding pair) = $\frac{1}{2}$, and so Be_2^+ is predicted to have some stability.

EXERCISE 11.7 Molecular Orbitals in Diatomic Molecules
In principle, could you prepare a compound such as $NaLi_2$ where sodium is the cation and Li_2 is the anion in a "salt"?

MOLECULAR ORBITALS FOR HOMONUCLEAR DIATOMIC MOLECULES

When two identical atoms form a molecule, such as H_2 or Li_2, this is called a **homonuclear diatomic molecule**. Having presented many of the principles of molecular orbital theory, we are ready to account for bonding in such important homonuclear diatomic molecules as N_2, O_2, and F_2. First, however, we need to see what types of molecular orbitals form when elements have both *s* and *p* valence orbitals.

MOLECULAR ORBITALS FROM ATOMIC *p* ORBITALS Three types of interactions are possible for two elements that both have *s* and *p* orbitals (Figure 11.16). Sigma bonding and antibonding molecular orbitals are formed by *s* orbitals interacting as in Figure 11.12. Similarly, it is possible for a *p* orbital on one atom to interact with a *p* orbital on the other atom in a head-to-head fashion to produce a σ bonding and antibonding molecular orbital pair. Finally, each atom has two *p* orbitals perpendicular to the σ bond connecting the two atoms. These *p* orbitals can interact sideways to give π bonding and antibonding molecular orbitals. Thus,

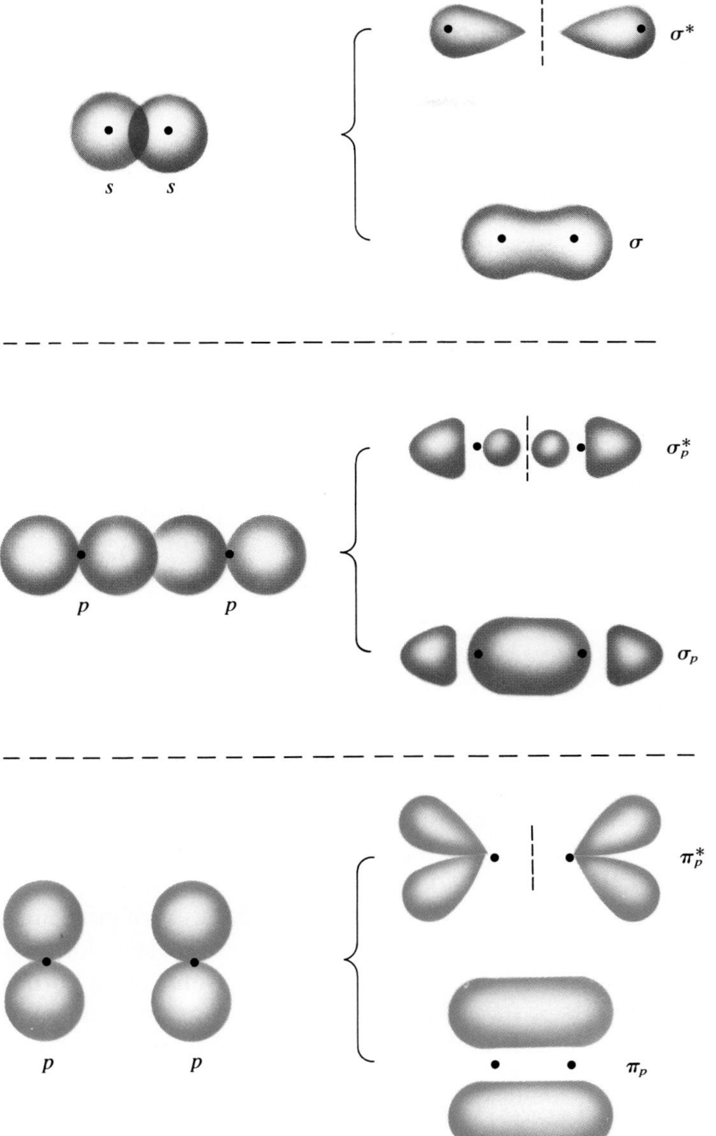

Figure 11.16 The bonding and antibonding molecular orbitals that come from s and p orbital overlap. The drawings represent only the general shapes of the orbitals.

two p orbitals on each atom produce *two* π bonding molecular orbitals (π_p) and *two* π antibonding molecular orbitals (π_p^*).

MOLECULAR ORBITAL ENERGY DIAGRAMS FOR *p*-BLOCK ELEMENTS The energy level diagram in Figure 11.17 gives the molecular orbital electron configurations for diatomic molecules formed by combining two identical atoms having 1s core electrons and 2s and 2p valence orbitals. Be sure to notice three features of this diagram. First, the molecular orbitals formed by core electrons, σ_{1s} and σ_{1s}^*, do not lead to net bonding and are not included in the figure. Second, the bonding and antibonding σ orbitals from 2s interactions are lower in energy than the σ and π MOs from 2p interactions. The reason is that 2s orbitals have a lower energy than 2p

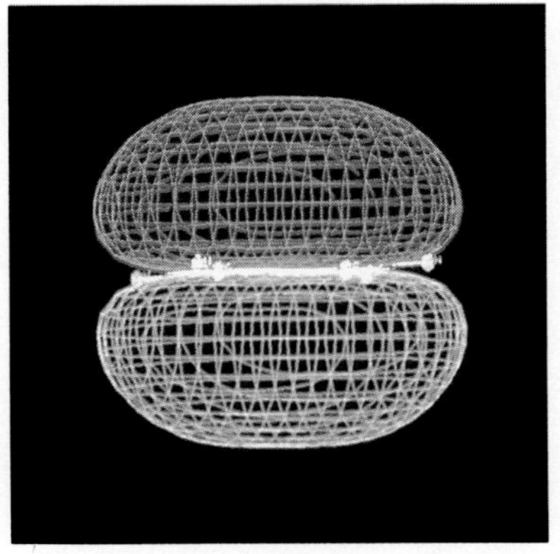

A computer generated model of the π bonding molecular orbital in ethylene. (M. Rajzmann)

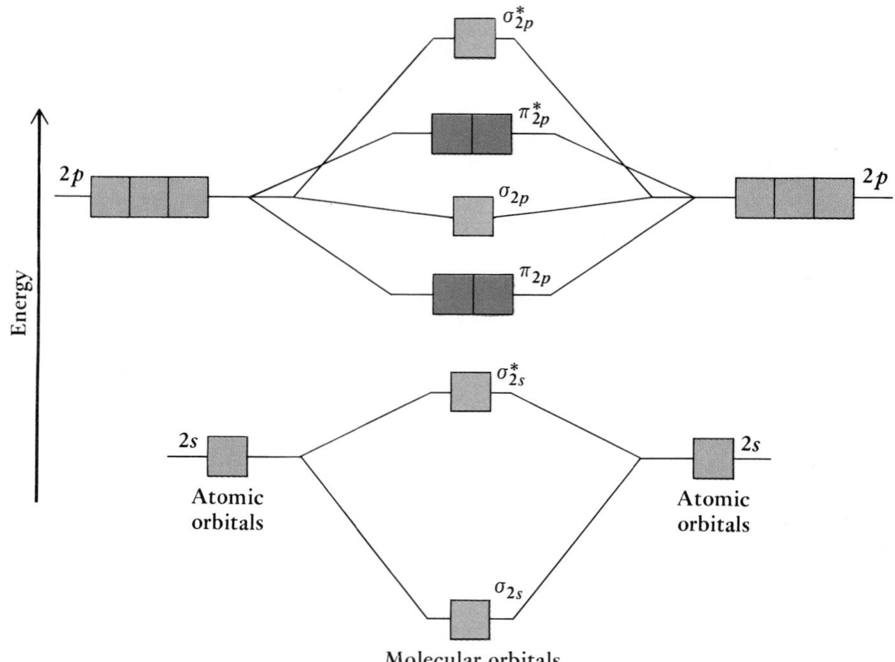

Figure 11.17 A molecular orbital energy level diagram for X_2, a homonuclear diatomic molecule of second period elements.

orbitals in the separated atoms. Third, the separation of bonding and antibonding orbitals is greater for σ_{2p} than for π_{2p}. This happens because p orbitals overlap to a greater extent when they are oriented head to head (to give σ_{2p} MOs) than when they are side by side (to give π_{2p} MOs). The greater the orbital overlap, the greater the stabilization of the bonding MO and the greater the *destabilization* of the antibonding MO.

Figure 11.17 shows a lack of symmetry in the ordering of molecular orbitals that you might not have expected, but there are reasons for this. Chemists interested in bonding theory have shown that the energy order for the molecular orbitals of second period elements arises from the similarity of $2s$ and $2p$ energies. Because overlap of s and p atomic orbitals cannot be ignored, the σ_{2s} and σ^*_{2s} orbitals are lower in energy than otherwise expected, and the σ_{2p} and σ^*_{2p} orbitals are pushed up in energy. This is the reason the energy lowering and raising for the σ_{2s} and σ^*_{2s} orbitals (and for the σ_{2p} and σ^*_{2p} orbitals) in Figure 11.17 is not symmetric with respect to the $2s$ and $2p$ atomic orbital energies.

Another refinement concerning Figure 11.17 is that, since s and p orbital mixing is important only for B_2, C_2, and N_2, the figure applies *strictly* only to these molecules. For O_2 and F_2, σ_{2p} is lower in energy than π_{2p}. Nonetheless, Figure 11.17 correctly predicts the bond order and magnetic behavior for the latter molecules, conclusions of great importance.

ELECTRON CONFIGURATIONS FOR HOMONUCLEAR DIATOMIC MOLECULES

Molecular orbital configurations are given for the diatomic molecules B_2 through F_2 in Table 11.2. There are two important features of this table.

First, study the excellent correlation between the electron configurations and the bond orders, bond lengths, and bond dissociation energies

Table 11.2 Molecular Orbital Occupations and Physical Data for Homonuclear Diatomic Molecules of Second Period Elements

	B_2	C_2	N_2	O_2	F_2
σ^*_{2p}					
π^*_{2p}				↑ ↑	↑↓ ↑↓
σ_{2p}			↑↓	↑↓	↑↓
π_{2p}	↑ ↑	↑↓ ↑↓	↑↓ ↑↓	↑↓ ↑↓	↑↓ ↑↓
σ^*_{2s}	↑↓	↑↓	↑↓	↑↓	↑↓
σ_{2s}	↑↓	↑↓	↑↓	↑↓	↑↓
Bond order	One	Two	Three	Two	One
Bond-dissociation energy (kJ/mol)	290	620	946	498	159
Bond distance (pm)	159	131	110	121	143
Observed magnetic behavior (paramagnetic or diamagnetic)	Para	Dia	Dia	Para	Dia

at the bottom of Table 11.2. As the bond order between a pair of atoms increases, the atoms are more tightly bonded. This means there should be an *increase* in bond dissociation energy and a *decrease* in bond distance, both of which are observed. Thus, N_2, with a bond order of 3, has the largest bond dissociation energy and shortest bond distance.

Second, a major objective of this discussion of MO theory was an understanding of the bonding in dioxygen, O_2, since valence bond theory failed to predict a most important feature of the molecule, its paramagnetism. Dioxygen has 12 molecular valence electrons, so it has the molecular orbital configuration [core electrons]$(\sigma_{2s})^2(\sigma^*_{2s})^2(\pi_{2p})^4(\sigma_{2p})^2(\pi^*_{2p})^2$, a configuration that leads to a bond order of 2, just as predicted by valence bond theory. However, by Hund's rule the MO configuration requires two unpaired electrons, exactly as determined by experiment. Thus, a simple molecular orbital picture leads to a view of the bonding in paramagnetic O_2 more in accord with experiment than that from simple valence bond theory.

The reason MO theory is superior to valence bond is that the former recognizes the existence of both bonding and antibonding orbitals.

We find that MO theory correctly predicts that O_2 should be paramagnetic with two unpaired electrons.

EXAMPLE 11.8

MOLECULAR ORBITAL ELECTRON CONFIGURATION FOR A *HOMONUCLEAR* DIATOMIC ION

When potassium reacts with O_2, potassium superoxide, KO_2, is produced. The superoxide ion is O_2^-. Write the molecular orbital electron configuration of the ion and predict its bond order and magnetic character.

Solution Using the energy level diagram in Figure 11.17, the configuration of the ion is

$$O_2^- \quad \text{[core electrons]} \ (\sigma_{2s})^2(\sigma^*_{2s})^2(\pi_{2p})^4(\sigma_{2p})^2(\pi^*_{2p})^3$$

It is predicted to be paramagnetic to the extent of one unpaired electron, a prediction confirmed by experiment. The bond order is $1\frac{1}{2}$, since there are 4 bond pairs and $2\frac{1}{2}$ antibonding pairs. This is a smaller bond order than in O_2, so we predict the O—O bond in O_2^- should be longer than in O_2. In fact, the superoxide ion has an O—O bond length of 132 pm, while that in O_2 is 121 pm.

EXAMPLE 11.9

MOLECULAR ORBITAL ELECTRON CONFIGURATION FOR A *HETERO*NUCLEAR DIATOMIC MOLECULE

A number of simple molecules are formed from two elements of *different* kinds. These are called **heteronuclear molecules**, and examples include NO, CO, and ClF. Although there are important differences, it is nonetheless the case that the molecular orbital energy level diagram for *homo*nuclear diatomic molecules (Figure 11.17) can be used to judge the bond order and magnetic behavior of *hetero*nuclear molecules. Let us do this for nitrogen oxide, NO.

Solution Nitrogen oxide has (5 + 6) or 11 molecular valence electrons. If these are assigned to the MOs for a homonuclear molecule, we would have [core electrons]$(\sigma_{2s})^2(\sigma^*_{2s})^2(\pi_{2p})^4(\sigma_{2p})^2(\pi^*_{2p})^1$. The net bond order is (4 bond pairs) − ($1\frac{1}{2}$ antibonding pairs) or $2\frac{1}{2}$. The single unpaired electron assigned to the π^*_{2p} molecular orbital means the molecule is paramagnetic, as predicted—and observed—for a molecule with an odd number of electrons.

EXERCISE 11.8 Molecular Electron Configurations

The cations O_2^+ and N_2^+ are important components of the earth's upper atmosphere. Write the electron configuration of O_2^+. Predict its bond order and magnetic character.

11.3 METALS AND SEMICONDUCTORS

The simple molecular orbital model that we used to describe diatomic molecules can be extended easily to describe the properties of metals. Metal crystals can be viewed as "supermolecules" held together by *delocalized bonds* formed from the atomic orbitals of all the atoms in the crystal. Since even the tiniest piece of metal can contain a very large number of atoms, there is an even larger number of atomic orbitals available to form molecular orbitals. For example, if there are four lithium atoms in a group, and each Li atom contributes five orbitals to molecular orbital formation, 20 molecular orbitals will be formed. If there are 400 lithium atoms, then 2000 molecular orbitals will be formed, and so on. As the number of atoms increases, the number of possible molecular orbitals increases. In a metal, the orbitals spread over many atoms and blend into

A one-gram crystal of Li contains about 9×10^{22} atoms, so there are 45×10^{22} 1s, 2s, and 2p orbitals available. These can form 45×10^{22} molecular orbitals.

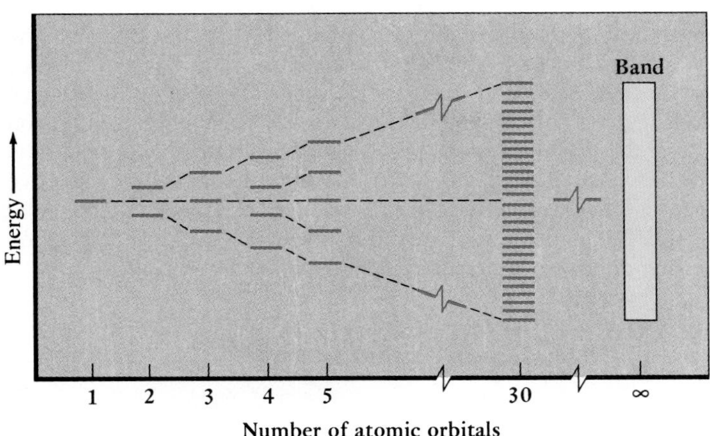

Figure 11.18 Bands of molecular orbitals in a metal crystal. As more and more atoms with the same valence orbitals are added, the number of molecular orbitals grows, until they merge into a band of orbitals.

a *band* of molecular orbitals whose energies are closely spaced within a range of energies (Figure 11.18). The band is composed of as many levels as there are contributing atomic orbitals, and each level can hold two electrons of opposite spin. The idea that the molecular orbitals of the band of energy levels are spread or *delocalized* over the atoms of the piece of metal accounts for the nature of bonding in metallic solids. This theory of metallic bonding is called **band theory**.

In a metal, the band of energy levels is only partly filled, where the highest filled level at absolute zero is called the **Fermi level** (Figure 11.19). However, even a small input of energy, say by raising the temperature, can cause electrons to move from the filled portion of the band to the unfilled portion. For each electron promoted, two singly occupied levels can result, one above the Fermi level and one below. It is the movement of electrons in these singly occupied states close to the Fermi level in the presence of an applied electric field that is responsible for the high electrical conductivity of metals.

Since the band of unfilled energy levels in a metal is essentially continuous (that is, the energy gaps between adjacent orbital levels are extremely small), a metal can absorb radiation of nearly any wavelength. When light causes an electron in a metal to move to a higher energy state, the now-excited system can immediately reemit a photon of the same energy and the electron returns to the original energy level. It is because of this rapid and efficient reemission of light that polished *metal surfaces are reflective and appear lustrous*.

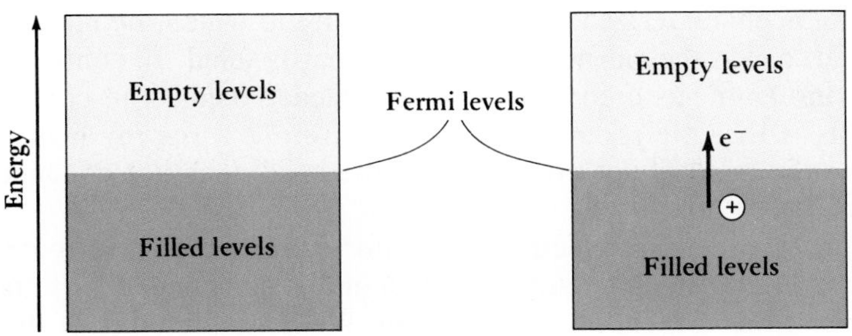

Figure 11.19 The partially filled band of "molecular orbitals" in a metal (left). The highest filled level is referred to as the Fermi level. The molecular orbitals are delocalized or spread over the metal, a fact that accounts for bonding in metals. Electrons can be promoted from the filled levels to empty levels by the input of modest amounts of energy (right). These electrons are more free to move in the now partially filled levels, and it is this that accounts for the electrical and thermal conductivity of metals.

SOMETHING MORE ABOUT
Semiconductors

Pure silicon belongs to a class of materials called **intrinsic** semiconductors. The usual view of such materials is that, when an electron is promoted from the valence band to the empty conduction band, this creates a positive hole in the valence band. The semiconductor carries charge because the electrons in the conduction band migrate in one direction under the influence of an electric field, while the positive holes in the valence band migrate in the opposite direction. (Positive holes "move" because an electron from an adjacent level can move into the hole, thus creating a "fresh" hole. In this respect, positive holes in the valence band move in the opposite direction from electrons in the conduction band.)

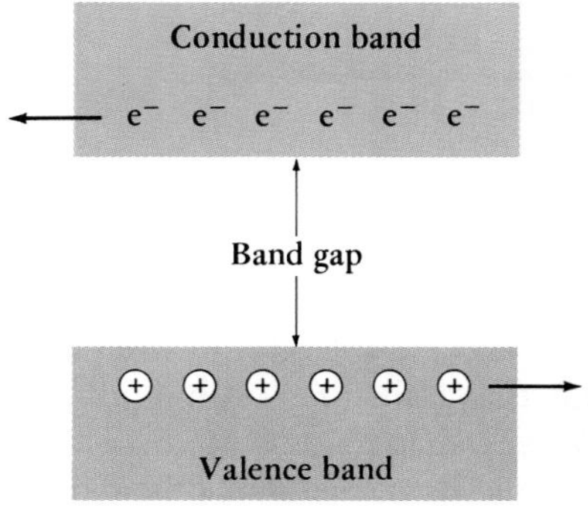

Positive and negative charge carriers in an intrinsic semiconductor.

A bar of pure silicon (an intrinsic semiconductor) on sheets of pure aluminum foil.

In an intrinsic semiconductor, the number of electrons in the conduction band is entirely governed by the magnitude of the band gap and the temperature. The smaller the band gap, the smaller the energy required to promote a significant number of electrons. Similarly, as the temperature increases, more electrons are promoted to the conduction band.

Extrinsic semiconductors are materials whose conductivity is controlled by adding tiny numbers of atoms of different elements as impurities or *dopants*. For example, suppose Al atoms (or atoms of some other Group 3A element) substitute for some Si atoms in solid silicon. Aluminum has only three valence electrons, whereas silicon has four. Therefore, at least one Al—Si bond per added Al atom would be deficient in electrons. Using band theory, it is found that the energy level associated with each Al—Si bond is not involved in the valence band of silicon. Rather, it forms a discrete level just above the top of the valence band. This level is referred to as an *acceptor level*, since it can accept electrons. The gap between the valence band and the acceptor level is usually quite small, so

A metal is characterized by a band structure in which the highest occupied band, called the **valence band**, is only partly filled. In contrast, an electrical **insulator** has a completely filled valence band. The consequence of this is that the promotion of an electron to a *slightly* higher energy within the band is not possible, and the solid does not conduct electricity.

Diamonds, which are electrical insulators, are one form of pure carbon (Figure 11.20). Each C atom in a diamond is surrounded by four other C atoms at the corners of a tetrahedron. We normally think of the bonding in the solid as involving sp^3 hybridized C atoms forming four electrons are readily promoted from the valence band to the acceptor level. The positive holes left behind in the valence band are able to move under the influence of an electrical potential, so an aluminum-doped semiconductor is called a *p*ositive hole or *p-type semiconductor*.

Now suppose you add P atoms (or atoms of some other Group 5A element) to pure silicon. The material is still a semiconductor, but it is a *n*egative charge carrier or *n-type semiconductor*. The reason for this is that phosphorus atoms have five valence electrons, one *more* than Si. Therefore, the presence of these Group 5A atoms in the silicon leads to a discrete, filled electron *donor level* just below the conduction band. Electrons can be readily promoted from the donor level to the conduction band, and electrons in the conduction band carry the charge.

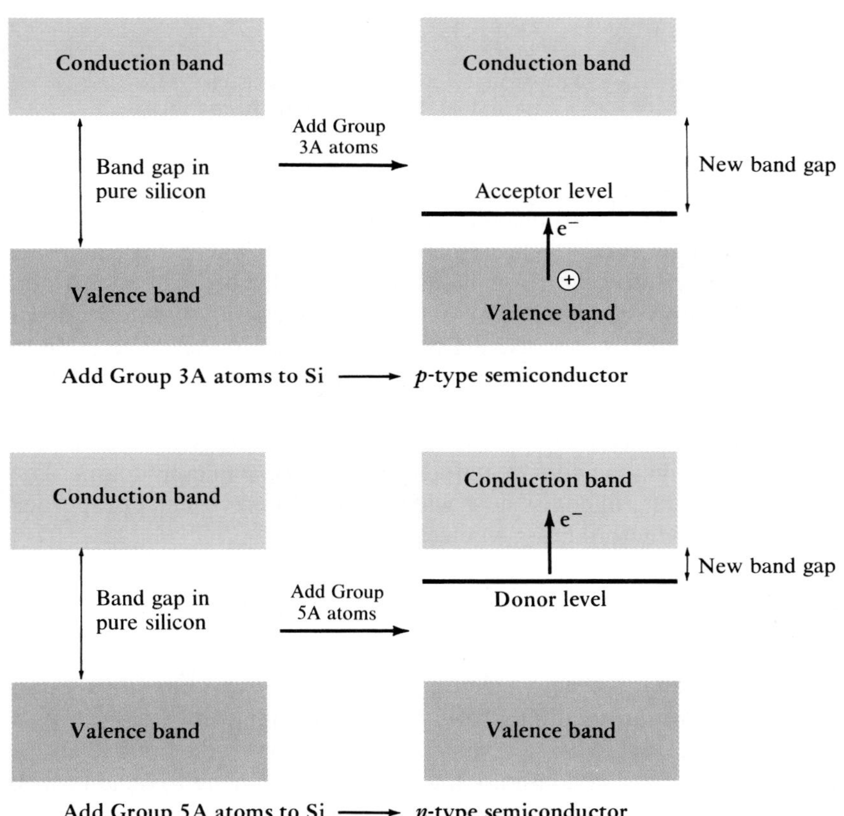

localized covalent C—C bonds per carbon atom. Silicon and germanium, also elements of Group 4A, form diamond-like solids, and their bonding can be described similarly. However, an alternative view is that the valence orbitals of each C, Si, or Ge atom form molecular orbitals that are delocalized over the solid. In this case, however, the continuous band of energy levels seen in a metal is split into two bands of levels: a lower energy, *filled* **valence band** of molecular orbitals, and a higher energy, *unfilled* band of molecular orbitals. The latter is called the **conduction band**, and the two bands are separated in energy by an amount called the **band gap** (Figure 11.21).

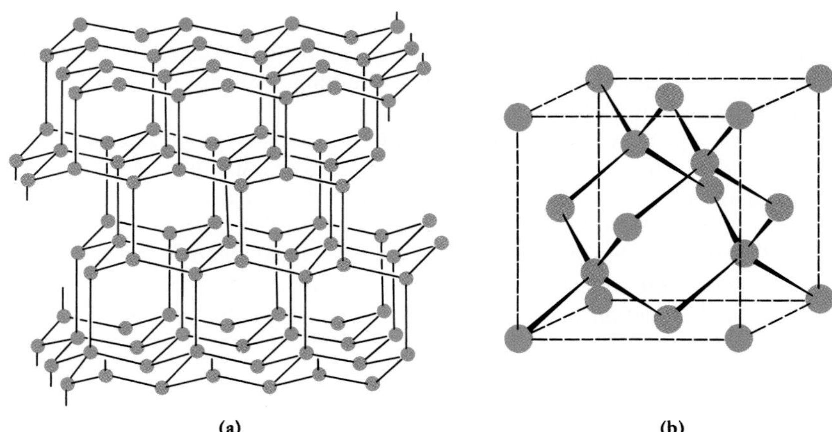

Figure 11.20 The structure of diamond, an electrical insulator. Each carbon atom is bonded to four others by bonds formed from the overlap of sp^3 hybrid orbitals.

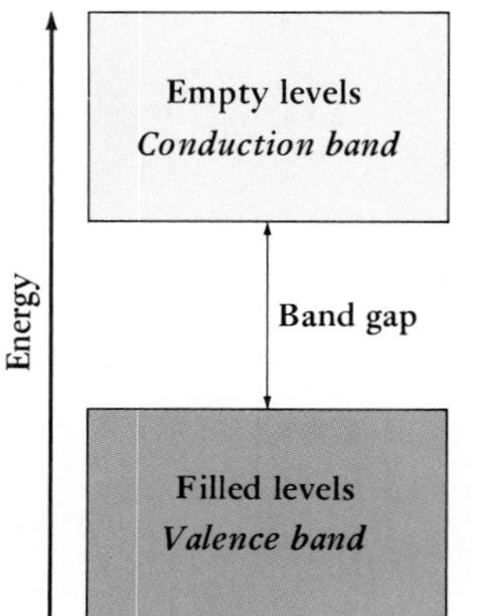

Figure 11.21 Band theory applied to semiconductors and insulators. In contrast with a metal (Figure 11.19), the band of filled levels (called the valence band) is separated from the band of empty levels (the conduction band) by the band gap. The band gap can range in energy from just a few kJ/mol to 500 kJ/mol or more.

If an electron could be promoted from the valence band to the conduction band in a substance like diamond, metal-like properties could be observed. However, diamonds are insulators because the band gap between valence and conduction bands is large, on the order of 500 kJ/mol, and it is difficult to promote electrons to the conduction band. **Semiconductors**, on the other hand, usually have band gaps in the range of 50 to 300 kJ/mol. At least a few electrons can be promoted to the conduction band by the input of modest amounts of energy, so electrical conduction can occur. Upon descending Group 4A the band gap narrows, so diamonds are insulators, whereas silicon and germanium are semiconductors. The consequence is that silicon is now widely used for the special properties that the semiconductors bring to electrical devices.

SUMMARY

Two commonly used theories of chemical bonding are **valence bond theory** and **molecular orbital theory**. Valence bond theory views bonding as arising from the overlap of atomic orbitals on two adjacent atoms to give only a bonding orbital with electrons localized between the bonded atoms (Section 11.1). In a sigma bond (σ), orbitals overlap head to head (s with s, p with p, or s with p), and the bonding electron density is concentrated along the bond axis. If p orbitals remain after sigma bond formation, one or two pi (π) bonds may form by the sideways overlap of p orbitals; π bonding electron density lies on both sides of the bond axis.

To achieve maximum overlap of atomic orbitals in sigma bonding orbitals, the concept of **orbital hybridization** was introduced (Section 11.1). For example, atomic s and p orbitals on the same atom may combine to produce *four equivalent orbitals*, each labeled sp^3, at the corners of a tetrahedron. Other hybrid orbital sets are sp (two hybrid orbitals, linear geometry); sp^2 (three hybrid orbitals, trigonal planar); sp^3d (five hybrid orbitals, trigonal bipyramidal); and sp^3d^2 (six hybrid orbitals, octahedral). The number of hybrid orbitals formed on an atom is equal to the number of atomic orbitals that are combined on that atom.

In molecular orbital theory, pure *s* and *p* orbitals from different atoms in the molecule combine to produce molecular orbitals that are generally delocalized over many atoms (Section 11.2). The number of molecular orbitals formed is always equal to the number of atomic orbitals brought by the combining atoms. Some of these molecular orbitals will be **bonding** (and lie lower in energy than the energy of the parent atomic orbitals in the isolated atoms), while others will be **antibonding** (and lie higher in energy). The electrons of the molecule are assigned to these orbitals, beginning with those of lowest energy, according to the Pauli principle and Hund's rule.

The bond order in a diatomic molecule is the net number of bonding electron pairs linking the atoms. Thus, bond order is given by the difference between the number of bonding electron pairs and the number of antibonding electron pairs.

The bonding in homonuclear diatomic molecules of second row elements and some properties of these molecules can be rationalized using the molecular orbital scheme in Figure 11.17. Such a scheme needs only slight modification to serve for *hetero*nuclear diatomic molecules.

The **band theory** is molecular orbital theory applied to solids, especially metals (Section 11.3). There are so many atoms in a solid that a large number of molecular orbitals is formed, the orbitals forming bands of energy levels. Since the molecular orbitals are spread or delocalized over the solid, the electrons are also delocalized and lead to bonding. In a metal the band of energy levels is only partially filled, and electrons can be promoted readily from the filled levels of the band to the unfilled levels. These electrons are more mobile, and account for the electrical conductivity of metals. In other solids, such as diamonds (one form of pure carbon) or pure silicon, the band of molecular orbitals is split into two bands, a lower energy, filled band (the **valence band**) and a higher energy, empty band (the **conduction band**); the two bands are separated in energy by a **band gap**. If the band gap is relatively small, electrons may be promoted across the gap from the valence band to the conduction band; since the electrons are more free to move within the conduction band, the solid is a **semiconductor**. If the band gap is large, promotion may not occur, and the solid is an **insulator**.

STUDY QUESTIONS

REVIEW QUESTIONS

1. What is the most important difference between a sigma (σ) and a pi (π) bond?
2. What is the maximum number of hybrid orbitals a carbon atom may form? What is the minimum number? Explain briefly.
3. What are the angles between regions of electron density in *sp*, sp^2, and sp^3 hybrid orbital sets?
4. For each of the following structural-pair geometries, tell what hybrid orbital set is used: tetrahedral, linear, trigonal planar, octahedral, trigonal bipyramidal.
5. If an atom is *sp* hybridized, how many pure *p* orbitals remain on the atom? How many pi (π) bonds can the atom form?
6. What is the maximum number of hybrid orbitals nitrogen may form? Explain briefly.
7. What is the maximum number of hybrid orbitals a third period element, say sulfur, may form? Explain briefly.
8. Give an example of a molecule with a central atom of expanded valence. Tell what hybrid orbitals are used by the atom.
9. What is one important difference between molecular orbital theory and valence bond theory?

10. Describe four principles of molecular orbital theory.

11. Draw sketches of the bonding and antibonding molecular orbitals of H_2 and tell how they differ.

12. What is meant by the order of a bond in terms of the molecular orbital theory? How are bond order, bond energy, and bond length related?

13. How is molecular orbital theory related to bonding in metals?

14. Explain briefly the theory of bonding in metals. What is the name usually applied to this theory?

HYBRID ORBITALS

15. Draw the Lewis electron dot structure for H_2S. What are its structural-pair and molecular geometries? Describe the bonding in the molecule in terms of S atom hybrid orbitals and H 1*s* orbitals.

16. Draw the Lewis electron dot structure for OF_2. What are its structural-pair and molecular geometries? Describe the bonding in the molecule in terms of O atom hybrid orbitals and F atomic orbitals.

17. Draw the Lewis electron dot structure for CH_2Cl_2. What are its structural-pair and molecular geometries? Describe the bonding in the molecule in terms of C atom hybrid orbitals and H and Cl atomic orbitals.

18. Draw the Lewis electron dot structure for BF_4^-. What are its structural-pair and molecular geometries? Describe the bonding in the molecule in terms of B atom hybrid orbitals and F atomic orbitals.

19. Tell what hybrid orbital set is used by the underlined atom in each of the following molecules or ions:
(a) $\underline{B}Cl_3$ (c) $H\underline{C}Cl_3$
(b) $\underline{N}O_2^+$ (d) $H_2\underline{C}O$

20. Tell what hybrid orbital set is used by the underlined atom in each of the following molecules or ions:
(a) $\underline{C}S_2$ (c) $\underline{N}O_3^-$
(b) $\underline{N}O_2^-$ (d) $\underline{B}H_4^-$

21. Give the hybrid orbital set used by each of the underlined atoms in the following molecules:

(a) $H_3\underline{C}—\underline{\ddot{O}}—H$

(b) $H_3\underline{C}—\underline{C}(H)=\underline{C}H_2$

(c) $H_3C—\underline{C}H_2—\underline{C}(=\ddot{O})—H$

22. Give the hybrid orbital set used by each of the underlined atoms in the following molecules:

(a) $H—\underline{\ddot{N}}(H)—\underline{C}(=\ddot{O})—\ddot{N}(H)—H$

(b) $H—\underline{C}(H)=C(H)—\underline{C}(H)=\ddot{O}$

(c) $H—\underline{C}(H)=\underline{C}(H)—\underline{C}\equiv N:$

23. Give the electron dot structure and the structural-pair and molecular geometries for XeF_2. Describe the bonding in the molecule, using hybrid orbitals for Xe.

24. Give the electron dot structure and the structural-pair and molecular geometries for ICl_4^-. Describe the bonding in the molecule, using hybrid orbitals for I.

25. Tell what hybrid orbital set is used by the underlined atom in each of the following molecules or ions:
(a) $\underline{S}F_6$ (b) $\underline{S}F_4$ (c) $\underline{I}Cl_2^-$

26. Tell what hybrid orbital set is used by the underlined atom in each of the following molecules or ions:
(a) $O\underline{Xe}F_4$ (b) $\underline{Cl}F_3$ (c) central Br in Br_3^-

27. What is the hybridization of the carbon atom in CO_2? Give a complete description of the σ and π bonding in this molecule.

28. What is the hybridization of the carbon atom in the NCO^- ion? Give a complete description of the σ and π bonding in this ion.

29. What is the hybridization of the carbon atom in phosgene, Cl_2CO? Give a complete description of the σ and π bonding in this molecule.

30. What is the hybridization of the sulfur atom in sulfur dioxide, SO_2? Give a complete description of the σ and π bonding in this molecule.

MOLECULAR ORBITAL THEORY

31. Hydrogen, H_2, can be ionized to give H_2^+. Write the electron configuration of the ion in molecular orbital terms. What is the bond order of the ion? Is the H—H bond stronger or weaker in H_2^+ than in H_2?

32. Give the electron configurations for the ions Li_2^+ and Li_2^- in molecular orbital terms. Compare the Li—Li bond order in the Li_2 molecule with its ions.

33. Calcium carbide, CaC_2, contains the acetylide ion, C_2^{2-}. Sketch the molecular orbital energy level diagram for the ion. How many net σ and π bonds does the ion have? What is the carbon–carbon bond order? How has the bond order changed on adding electrons to C_2 to obtain C_2^{2-}? Is the C_2^{2-} ion paramagnetic?

34. Oxygen, O_2, can acquire one or two electrons to give O_2^- (superoxide ion) or O_2^{2-} (peroxide ion). Write the electron configurations for the ions in molecular orbital terms and then compare them with the O_2 molecule on the following basis: (a) magnetic character, (b) net number of σ and π bonds, (c) bond order, and (d) oxygen–oxygen bond length.

35. Assuming that we can apply the energy level diagram for homonuclear diatomic molecules (Figure 11.17) to heteronuclear diatomics, write the electron configuration for carbon monoxide, CO. Is the molecule diamagnetic or paramagnetic? What is the net number of σ and π bonds? What is the bond order?

36. The nitrosonium ion, NO^+, has an interesting chemistry. (a) Is NO^+ diamagnetic or paramagnetic? If paramagnetic, how many unpaired electrons does it have? (b) Assume the molecular orbital diagram for a homonuclear diatomic molecule applies to NO^+. What is the highest energy molecular orbital assigned to an electron? (c) What is the N—O bond order? (d) If NO is ionized to form NO^+, does the N—O bond become stronger or weaker than in NO?

GENERAL QUESTIONS

37. Give the hybrid orbital set used by sulfur in each of the following molecules or ions:
(a) SO_2 (b) SO_3 (c) SO_4^{2-}

38. Give the Lewis electron dot structures of ClF_2^+ and ClF_2^-. What are the structural-pair and molecular geometries of each ion? What hybrid orbital set is used by Cl in each ion?

39. What is the hybridization of the nitrogen atom in the nitrate ion, NO_3^-? Describe the orbitals involved in the formation of an N=O bond.

40. Sulfur trioxide, SO_3, is formed as an intermediate in the manufacture of sulfuric acid. What is the hybridization used by sulfur in SO_3? How many resonance structures are there? Explain how hybrid orbitals and pure *p* orbitals are involved in forming one of the resonance structures.

41. The organic compound below is a member of a class known as oximes.

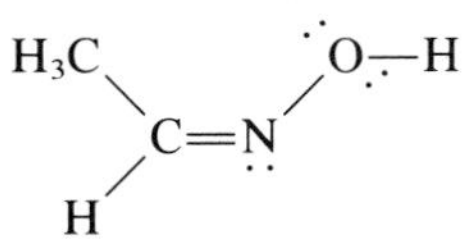

(a) What are the hybridizations of the two carbon atoms?
(b) What is the approximate C—N—O angle?

42. Acrolein has a pungent odor and irritates eyes and mucous membranes. It is a component of photochemical smog.

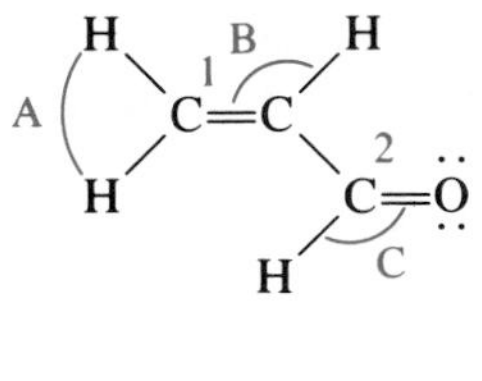

(a) What are the hybridizations of carbon atoms 1 and 2? (b) What are the approximate values of angles A, B, and C?

43. The compound sketched below is acetylsalicylic acid, better known by its common name, aspirin.

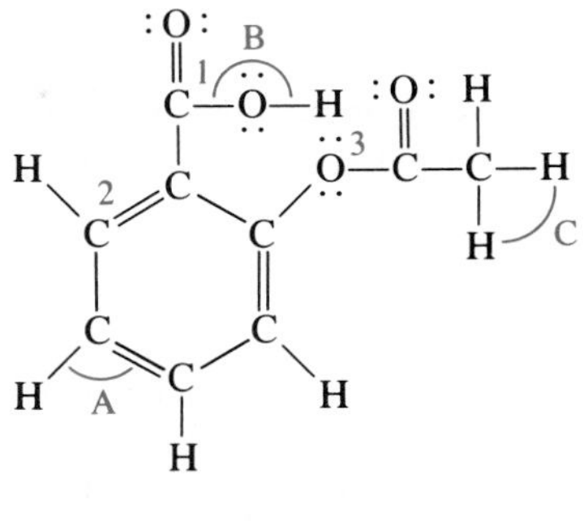

(a) How many π bonds are there in aspirin? How many σ bonds? (b) What are the approximate values of the angles marked A, B, and C? (c) What hybrid orbitals are used by carbon atoms 1 and 2 and oxygen atom 3?

44. Tabun, the compound sketched below, is a chemical warfare agent.

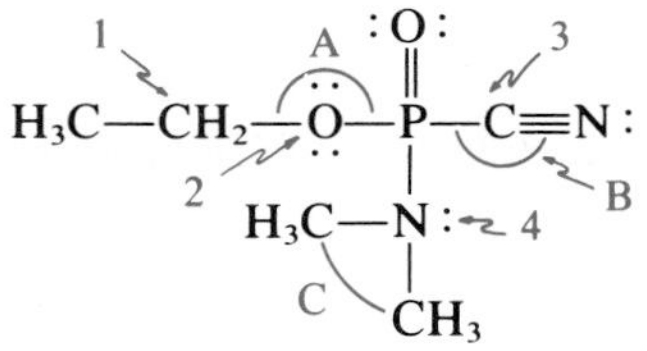

(a) How many π bonds are there in tabun? How many σ bonds? (b) Give the hybridization of each atom 1 through 4. (c) Which C—N bond is the shortest in the molecule? (d) Give the approximate values of the bond angles A through C.

45. Histamine is found in normal body tissues and in blood. It has the following structure.

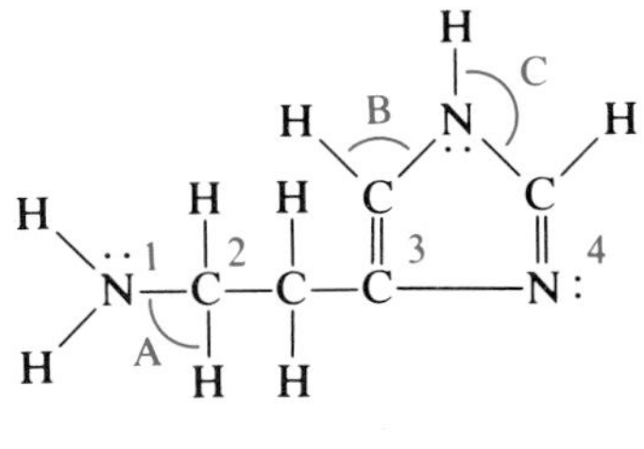

(a) How many σ and π bonds does the molecule have? (b) Give the hybridizations of atoms 1 through 4. (c) What are the approximate values of the bond angles A through C?

46. Boron trifluoride, BF_3, can accept a pair of electrons from another molecule to form a coordinate covalent

bond (see Chapter 10), as in the following reaction with ammonia:

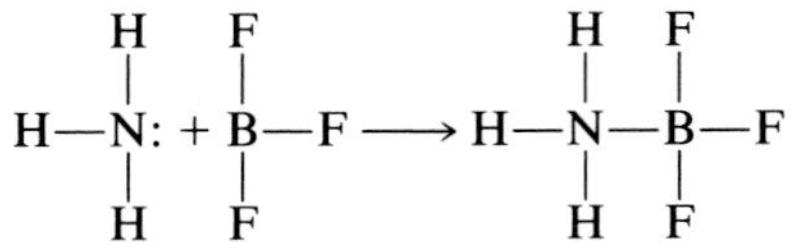

```
    H    F              H    F
    |    |              |    |
H—N: + B—F ——→ H—N—B—F
    |    |              |    |
    H    F              H    F
```

(a) What is the geometry about the boron atom in BF_3? In $H_3N—BF_3$? (b) What is the hybridization of boron in the two compounds?

47. The simple valence bond picture of O_2 does not agree with the molecular orbital view. Compare these two theories with regard to the peroxide ion, O_2^{2-}. Do they lead to the same magnetic character and bond order?

48. Nitrogen, N_2, can ionize to form N_2^+ or absorb an electron to give N_2^-. Compare these species with regard to (a) magnetic character, (b) net number of π bonds, (c) bond order, (d) bond length, and (e) bond strength.

49. The ammonium ion is important in chemistry. Discuss any changes in hybridization and bond angles that may occur in the combination of ammonia and a proton.

$$H^+ + NH_3 \longrightarrow NH_4^+$$

50. Antimony pentafluoride reacts with HF according to the equation

$$2\ HF + SbF_5 \longrightarrow [H_2F]^+[SbF_6]^-$$

(a) What is the hybridization of Sb in the reactant and product? (b) Draw an electron dot structure for H_2F^+. Does the ion have linear or bent molecular geometry? What is the hybridization of F in H_2F^+?

51. Iodine and oxygen form a complex series of ions, among them IO_4^- and IO_5^{3-}. Draw Lewis dot structures for these ions and specify their structural-pair and molecular geometries. What is the hybridization of the I atom in these ions?

52. Xenon is the only rare gas element that forms well characterized compounds. Two xenon–oxygen compounds are XeO_3 and XeO_4. Draw the electron dot structure of each of these compounds and give their structural-pair and molecular geometries. What are the hybrid orbital sets used by xenon in these two oxides?

53. Which of the homonuclear diatomic molecules of the second row elements (from Li_2 to Ne_2) are paramagnetic? Which ones have a bond order of one? Which ones have a bond order of two? What diatomic molecule has the highest bond order?

54. Which of the following molecules or ions should be paramagnetic? Assume the molecular orbital diagram in Figure 11.17 applies to all of them.
(a) NO (d) Ne_2^+
(b) OF^- (e) CN
(c) O_2^{2-}

55. The elements of the second period from boron to oxygen form compounds of the type $X_nE—EX_n$ where X can be H or a halogen. Sketch possible molecular structures for B_2F_4, C_2H_4, N_2H_4, and O_2H_2. Give the hybridization of E in each molecule and specify approximate X—E—E bond angles. What is the major difference between the carbon-containing compound and the other molecules?

56. The sulfamate ion, $H_2N—SO_3^-$, can be thought of as having been formed from the amide ion, NH_2^-, and sulfur trioxide, SO_3. (a) Sketch a structure for the sulfamate ion. Include estimates of bond angles. (b) What changes in hybridization do you expect for N and S in the course of the reaction $NH_2^- + SO_3 \longrightarrow H_2N—SO_3^-$?

57. In many chemical reactions, atom hybridization changes. In each of the following reactions, tell what change, if any, occurs in the underlined atom.
(a) $H_2\underline{C}{=}CH_2 + H_2 \longrightarrow H_3C—CH_3$
(b) $\underline{P}(CH_3)_3 + I_2 \longrightarrow P(I)_2(CH_3)_3$
(c) $\underline{Xe}F_2 + F_2 \longrightarrow XeF_4$
(d) $\underline{Sn}Cl_4 + 2\ Cl^- \longrightarrow SnCl_6^{2-}$

58. The CN molecule has been found in interstellar space. Assuming the electronic structure of the molecule can be described using the molecular orbital energy level diagram in Figure 11.17, answer the following questions.
(a) What is the highest energy molecular orbital to which an electron or electrons have been assigned?
(b) What is the bond order of the molecule?
(c) How many net σ bonds are there? How many net π bonds?
(d) Is the molecule paramagnetic or diamagnetic?

59. The following molecule is used as a pesticide.

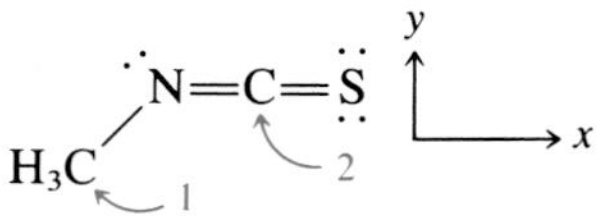

```
                          y
      ..       ..         ↑
      N=C=S               |
     /    ↖   ..          └——→ x
  H3C       2
     ↖ 1
```

(a) Tell what hybrid orbital sets are used by carbon 1, carbon 2, and the N atom.
(b) What is the C—N=C bond angle? The N=C=S bond angle?
(c) The C—N=C=S framework is planar. Assume these atoms lie in the plane of the paper and that this is identified as the xy plane. Now consider the orbitals (s, p_x, p_y, and p_z) on carbon atom 2. Which of these orbitals are involved in hybrid orbital formation? Which of these orbitals are used in π bond formation with the neighboring N and S atoms?

SUMMARY QUESTION

60. The following compound is commonly called acetylacetone. As shown it exists in two forms, one called the *enol* form and the other called the *keto* form.

$$H_3C{-}C(OH){=}C(H){-}C({=}O){-}CH_3 \rightleftharpoons H_3C{-}C({=}O){-}C(H)(H){-}C({=}O){-}CH_3$$

enol form   *keto* form

While in the *enol* form, the molecule can lose H^+ from the —OH group to form the anion $[H_3C{-}C(O){=}CH{-}C(O){-}CH_3]^-$. One of the most interesting aspects of this anion (which is sometimes called the *acac* ion) is that one or more of them can react with a transition metal cation to give very stable, highly colored compounds (see photo).

$$\left[\begin{matrix} H_3C \\ C{=}O \\ H{-}C \quad\quad M \\ C{-}O \\ H_3C \end{matrix}\right]_3$$

$M(acac)_3$

Three complexes of the type $M(acac)_3$ where M is (left to right) Co, Cr, and Fe.

(a) Using bond energies, calculate the enthalpy change for the *enol* → *keto* change. Is the reaction predicted to be exo- or endothermic?

(b) What is the hybridization of each atom (except H) in the *enol* form? What changes occur when this is transformed into the *keto* form?

(c) What are the structural-pair and "molecular" geometries around each C atom in the *keto* and *enol* forms? What changes (if any) occur when the *keto* form changes to the *enol* form?

(d) If you wanted to prepare 15.0 g of deep red $Cr(acac)_3$ using the following reaction,

$$CrCl_3 + 3\ H_3C{-}C(OH){=}CH{-}C(O){-}CH_3 + 3\ NaOH \longrightarrow Cr(acac)_3 + 3\ H_2O + 3\ NaCl$$

how many grams of each of the reactants would you need?

PART THREE

STATES OF MATTER

Gases, liquids and solutions, and solids.

Mary L. Good

Mary L. Good received a B.S. degree at the University of Central Arkansas in 1950 and a Ph.D. degree from the University of Arkansas at Fayetteville five years later. After 25 years as a member of the faculty of the Louisiana State University system, she became the Director of Research for UOP, Inc., in Illinois (one of the Signal Companies) in 1980. When Signal became part of Allied-Signal, Inc., she moved to Morristown, New Jersey, to assume the position of Senior Vice President for Technology, in which she oversees a 600 million dollar per year research and development effort.

Dr. Good is very active in scientific affairs in the United States and the rest of the world. She was president of the American Chemical Society in 1987, is a member and currently chairperson of the National Science Board, and is a director of the International Union of Pure and Applied Chemistry. She has received a number of honors, among them the Garvan Medal from the American Chemical Society, the Fahrney Medal from the Franklin Institute, several honorary doctoral degrees, and election to the National Academy of Engineering. Most recently, she received the Parsons Award for Public Service from the American Chemical Society, the first woman to receive this honor.

We spoke with Dr. Good in her office at Allied-Signal, Inc. It is just down the hall from the research laboratories and is decorated with paintings done by her husband, a retired physics professor. In addition to her research and administrative activities, she has two sons and, at present, four grandchildren.

From home economics to radiochemistry

Mary Good was born in Texas to parents who were both schoolteachers. Her father was a coach and principal of the school, and her mother taught English and algebra. However, during her high school years, she said "I never really had any science courses that most people would have recognized. So when I went to college, I didn't know exactly what to major in, but I thought you should major in something useful; I actually registered in home economics the first semester. The reason was very straightforward. If you were a home economics teacher or a vocational agricultural teacher, you got paid a bit more for it. But I ended up in freshman chemistry with a professor who was just fantastic. I just enjoyed it immensely and thought I would like to do that. I changed my major field the first semester. I am reasonably sure it was primarily because of that fellow who taught freshman chemistry."

Dr. Good did her graduate work in inorganic chemistry and radiochemistry (the latter is the subject of Chapter 7). At that time it was not too many years after the end of World War II, and there was still a great deal of excitement in the field of research on radioactive isotopes. Today this field is no longer as glamorous, and it is a state of affairs that worries Dr. Good. "It's a real problem from both an educational and a practical point of view. If atomic reactors for power do come back in use for electrical generation, it's going to be a real problem because we don't have anybody trained to handle radioactive materials or separate out useful isotopes. The few folks with the expertise will have retired. Recently a committee was set up by the National Research Council on just that issue: how to get at least a few graduate students trained in the handling of radioactive materials and how to do separations and things like that. It's really a major issue."

Risks and benefits

The discussion of radiochemistry, and the lack of people trained in this area, brought us to the topic of environmental problems, public risks, and public benefits. She says that her home state of New Jersey has a big problem in handling low-level radioactive waste. The state has decided that it will not participate in the federally mandated regional disposal scheme, at least for the time being. "So they [the state] told us that if we do any work, we have to provide a storage facility on our own property for five years. It's going to be a major burden for some, especially the pharmaceutical companies. People don't realize that we use a lot of radioactive isotopes in diagnostics, in scanning, and in treatment. I don't know what hospitals are going to do with their low-level radioactive waste. It's going to be a problem.

"The issue is how we educate the public with respect to risks and benefits. The philosophy in the world today says that you must have a risk-free environment. First of all, it's simply not achievable, and we are spending money on things at the bottom of the priority list when we have such huge problems. For example, there was a bill in Congress that said the city of Chicago had to get all of the lead connecting pipes out of the water fountains in the schools. The inner city schools of Chicago are a disaster. They've got drug problems; they've got every single problem you can possibly imagine. They allocated 9 million dollars to get the lead out of those pipes and water fountains. The risk associated with the water fountains is so small compared with the other risks the children face every day! On a priority basis the 9 million dollars could have been used more effectively."

The responsibilities of the chemical industry

As chemists, we are all too aware of our image as producers of toxic substances. As individuals and companies, we are trying to move in a positive direction to change old practices. Dr. Good says that her company "must adhere to the OSHA (Occupational Safety and Health Administration) requirements, and we really go out of our way to do that. A corporation like ours has to pay when people have lost days, if they are hospitalized or ill. It is simply not in our best interest to have people work in an unsafe environment. We work very hard to have safety awareness and safe work places because it's absolutely to our benefit to do that. Also, we personally want to work in a safe environment, and so do our employees.

"We want the public to understand that many of the things that they want us to do we would do anyway because they are advantageous to us. For example, we have a major waste minimization program in our company today. Why do we do that? For several reasons. First of all, such a program is economical. You know we have to pay for disposal, so if we don't create waste, we don't have to dispose of it. For any new process that we institute, we work very hard to see if we can't recycle materials, to see if we can't cut waste to a minimum, because that's a big piece of the economics.

"Everyone wants to live in a clean, safe place. However, if you drive the system to the point where there is essentially zero risk, the cost of that is simply not reasonable. And there is another interesting thing: The people who want the risk-free kind of environment are people who can afford to pay for it. And the folks who cannot afford to pay for it get higher prices and living expenses. What we spend to get one more decimal point on the quality of life for the affluent means less money available for those folks who live in very different environments. People really don't understand that. There is a difference between reducing the cancer risk by 1 more person out of 100,000 and contending with the infant mortality rate in New York City.

People don't look at it like that, but that's what it's all about. The pot of money is the same for both of them."

The National Science Board and important issues in science

A problem we face is how to sustain a responsible public dialogue on risks and benefits in our modern society. The Congress is one place, but another vehicle may be the National Science Board. Dr. Good is currently chairperson of this board, which consists of 28 members appointed by the President of the United States. Its main task is to set policy for the National Science Foundation (NSF). This is an enormous undertaking, because the annual budget for NSF is now over 2 billion dollars and because it is one of the primary sources of funds for scientific research in this nation's universities.

At the moment the National Science Board is working on three important projects. First, is a special report on Europe in 1992, because "we are very concerned how that is going to affect science in this country." Second, there is a special task force on industrial research "because there is a concern in the country whether industrial research is keeping up, whether we will continue to have the vitality in the industrial sector that we should." Finally, the Board just finished a report on "the whole issue of biodiversity on our planet and how rapidly it is changing, how rapidly we are losing species."

These are some of the issues confronting science today, but there are others just as important. Chief among them is the issue of "big science versus little science." "Big science" includes the supercollider, the genome project, big astronomy programs, and so on. "You could use up all of the money doing those. There are some in the scientific community who would essentially do no big science because they think it takes away from their individual work. On the other hand, you have to do some of the big experiments that are at the cutting edge of science. But we've got to prioritize these things. The issue of 'big science/little science' is a tough one in an era of limited budgets, and we do have limited budgets today."

The shortage of chemists and other scientists

Another issue of concern to Dr. Good—and to many of us in science education and research—is the "manpower issue." "We really do not have an adequate number of undergraduate majors, so we are not attracting enough of the very best and the brightest to do chemistry. And that to me is a tragedy because it is happening at a time when chemistry is really one of the most exciting sciences. We now can manipulate biological systems on a molecular basis, and we can chemically design a protein. We have methods today whereby we can essentially design a new material that will have the properties you want it to have, and we can almost design it atom by atom. Molecular control is what chemistry is all about, so it is kind of sad that somehow we are not attracting students at a time when, in my view, chemistry has never been more exciting."

Materials chemistry

Dr. Good is particularly excited about what is happening in an emerging field of chemistry that deals with new materials. Indeed, she believes there has been an "explosion of new materials. For such a long time we had three or four plastic materials, and we made everything out of those. You simply had to take the plastic with the properties it had. However, we are almost at the point now where you tell me what properties you want the plastic to have, and I will design one that has those properties. You can have a custom material.

"The kinds of things that are happening! Within the company, we have this new material, something that was invented a few years ago, that we are just now getting into the commercial market. We call it Spectra®, but it is plain old polyethylene (see Figure 13.36). However, we have been able to align the polyethylene molecules in such a way that this material is stronger than steel fiber and is orders of magnitude lighter. A fabulous material! It is very lightweight and

A bicycle using Spectra®, a high-performance fiber from Allied-Signal, Inc. It is the strongest known fiber, 10 times stronger than steel and 75% stronger than other organic fibers.

has great energy absorption capabilities; it makes extraordinarily good bulletproof vests, for example.

"The whole area of polymer science is just blossoming because you can do so very much with it when you begin to understand molecular construction and are able to manipulate it. People don't realize what polymer chemistry does for the world. If you were to stand in the middle of your living room and have somebody come and take out everything in the room that had any polymer material in it, the room would be denuded. The paint would be gone, the carpet would be gone, the upholstery would be gone. If you took away the man-made polymer materials you'd be utterly amazed how little you'd have left.

"Amorphous metals are also exciting new materials. Metals for the most part are crystalline. What you do in this case is 'fake the material out'; you make it believe it's a glass, so it has an amorphous character rather than a crystal character. It has major new magnetic properties. We make such a material we call Metglas®. We use Metglas® to make transformer cores, and those transformer cores have significant electrical power savings of, say, 70%. You could save lots of barrels of oil if you had all the transformers made of that material."

The space station

Dr. Good's company was very active in the old Apollo program, and it is now back in the space program working on the space station. "Talking about new materials, there are some fascinating things about materials needed in building a space station. You've got to have a material that will sit there at essentially zero degrees and no atmosphere. We have had to go back and understand how materials behave under those conditions. Things don't behave as they do under regular pressures. So there are some extraordinary materials going into that thing.

"We are helping design and fabricate the environmental systems for the station. In fact, in doing this we may learn a lot about how to improve habitats here on earth too. You talk about having to recycle! We must be able to recycle essentially everything. If you take the CO_2 out, you've got to get rid of it. Then you've got to take the moisture out, and you want to reclaim that. You have to either recycle or store everything, and you don't want to store a lot. So the whole issue of recycling the environment, cleaning the environment, removing whatever toxic materials there are, all of that is part of designing new systems and materials for a space station."

The need to participate in science and technology

"I think that somehow we must get young people to take part in science. They don't all have to become scientists, but we need them to participate in science and technology. The reason is that they must live in a world that is essentially technology driven. I think that comes back to our risk analysis problem that we talked about before. Our population doesn't understand the basic tenets of technology well enough to deal with it effectively, and they don't understand how to take the parts of it that improve living conditions and not bother with those parts that don't.

"We have a lot of people today who really believe that we should go back to what they consider to be the 'good old days.' They don't understand how good the quality of life today is, particularly in this country. Now, it's true, we have major social problems, and we have pockets in this country where the quality of life is just awful, so we should be spending some of that money trying to take care of those problems rather than getting one more turn out of the quality of *my* life, which is just so good, it's unbelievable."

Understanding science and technology is an important issue. "It's not a question of learning chemistry per se, but it's learning enough about what drives technology, what the benefits are, what the problems are, and that there are never benefits without problems. It's the old business—you can't have something for nothing. So we need to begin to learn to live with technology, to take the good things and understand that there will be problems we must handle and that we can't anticipate them all. We'll have to handle them as they occur. We need to spend our limited resources on those programs that insure a safe and quality environment for *all* our descendants."

Individual responsibility

"Finally, I think people have gotten into the habit of believing that an individual doesn't make any difference, that it's okay if I don't really do my piece, since I am just one person. And that's not true. One person can still make an enormous difference. There's a new slogan going around, 'Think globally and act locally.' That's a very good slogan. You know I can't cure the world's evils, but I can really help some of those at my doorstep." This statement is an excellent description of Mary Good's life.

CHAPTER

12

Gases and Their Behavior

Hot air balloons. (COMSTOCK)

CHAPTER OUTLINE

We all live at the bottom of a sea of molecules. Molecular oxygen and nitrogen make up almost 99% of the earth's atmosphere, but other gases are important as well. Carbon dioxide, for example, is taken in by plants and converted to sugars and other substances essential for life, and a blanket of ozone gas, O_3, surrounds the earth at high altitudes and protects us from dangerous radiation from the sun.

In this chapter, the fundamentals of the physical behavior of gases are discussed for at least three important reasons. First, our gaseous atmosphere provides one means of transferring energy and material throughout the globe, and it is the source of life-giving chemicals. Second, gas behavior is well understood and can be expressed in terms of simple mathematical models. The modeling of nature is one of the great endeavors of science, and a study of gas behavior will further introduce you to this approach. Third, many common elements and compounds are in the gaseous state under normal conditions of pressure and temperature (Table 12.1). Further, many common liquids can be vaporized, and the properties of these vapors are important.

Table 12.1 Some Common Gaseous Elements and Compounds

(at 1 atm pressure and 25 °C)

He	CO
Ne	CO_2
Ar	CH_4
Kr	C_2H_4
Xe	C_3H_8
H_2	HF
O_2	HCl
O_3 (ozone)	HBr
N_2	HI
F_2	NO
Cl_2	NO_2
	H_2S

12.1 THE PROPERTIES OF GASES

The chemist has learned that only four quantities are needed to describe the **state of a gas**: (a) the *quantity* of gas, n (in moles); (b) the *temperature* of the gas, T (in kelvins); (c) the *volume* of gas, V (in liters); and (d) the *pressure* of the gas, P (in atmospheres). Before seeing how these are related, we must examine the concept of pressure and its units.

A gas is a substance that is normally in the gaseous state at ordinary pressures and temperatures. A vapor is the gaseous form of a substance that is a liquid or solid at ordinary pressures and temperatures. Thus, we often speak of helium gas and water vapor.

GAS PRESSURE

The pressure of a gas is a measure of the force that it exerts on its container. **Force** is the physical quantity that causes a change in the motion of a mass if it is free to move. For instance, gravity is the force that the earth exerts on all objects near it, causing the same acceleration (change in velocity with time) for all objects. The gravitational force is also known as the **weight** of the object.

Newton's first law states that force = mass × acceleration. The SI units of mass and acceleration are kg and m/s^2, so the units of force are $kg \cdot m/s^2$. The derived unit is given the name **newton (N)**. In the English system, force and weight are measured in pounds, and 1 pound = 4.448 newtons.

Pressure is defined as the force exerted on an object divided by the area over which the force is exerted.

$$\text{Pressure} = \frac{\text{force}}{\text{area}}$$

This book, for example, weighs a little over 4 pounds and has an area of 82 $in.^2$, so it exerts a pressure of about 0.05 pounds/$in.^2$ when it lies flat on a surface; in SI units the weight is about 20 newtons and the area is approximately 0.05 m^2, so the pressure is about 400 N/m^2.

The air around us exerts a pressure on everything it touches, and we can measure that pressure with a barometer (Figure 12.1). To see how we can use this instrument to measure pressure, consider the pressure that the mercury in the column exerts on the mercury in the dish. This pressure is, by definition, the weight of the mercury column divided by the cross-sectional area of the tube (as weight per cm^2, for example). Since weight is proportional to mass, we can write the pressure as

The symbol ∝ means "proportional to."

$$\text{Pressure} = \frac{\text{weight of mercury column}}{\text{cross-sectional area of column}} \propto \frac{\text{mass of mercury column}}{\text{area of column}}$$

The mass of an object is equal to its volume times its density (see Chapter 1), so the pressure P exerted by the mercury in the vertical column is

$$P \propto \frac{\text{volume of mercury column} \times \text{density of Hg}}{\text{area of column}}$$

$$P \propto \frac{(\text{height of column} \times \text{area of column}) \times \text{density of Hg}}{\text{area of column}}$$

$$P \propto \text{height of column} \times \text{density of Hg}$$

This means that the pressure exerted by the column of mercury in a barometer such as that in Figure 12.1 depends only on the height of the mercury column, and not on its area. Since this pressure is equal to the atmospheric pressure outside the beaker, it is natural to measure atmospheric pressure (or the pressure of any other gas) by measuring the height of the column of mercury it can support. Thus, a common unit of pressure is the **millimeter of mercury (mmHg)**, also called the **torr** in honor of Evangelista Torricelli (1608–1647), who invented the barometer in 1643.

To conveniently measure gas pressure by the height of a column of liquid it can support, early scientists needed a very dense liquid. That is the reason mercury was chosen and became the reference standard.

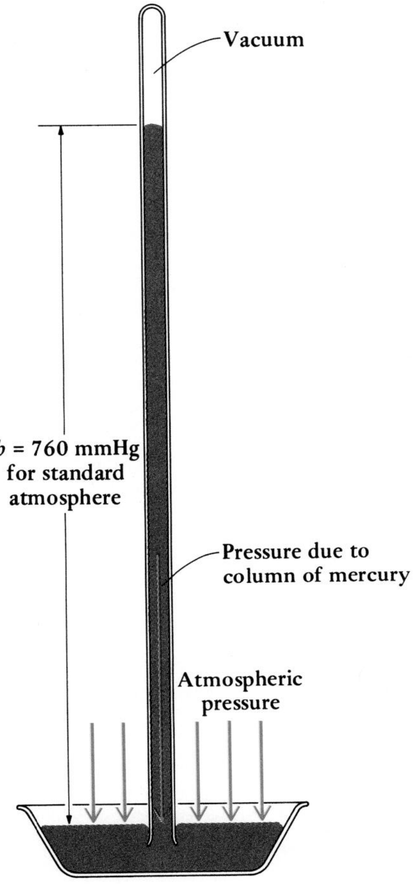

Figure 12.1 A Torricellian barometer. The pressure of the atmosphere on the surface of the mercury in the dish is transmitted through the liquid and acts to support the weight of the mercury in the tube.

At sea level, the pressure of the atmosphere can support a column of mercury about 760 mm high (depending on weather conditions), so the **standard atmosphere (atm)** is defined as

1 standard atmosphere = 1 atm = 760 mmHg (exactly)

The SI unit of pressure is the **pascal (Pa)**, named for the French mathematician and philosopher Blaise Pascal (1623–1662).

1 pascal (Pa) = 1 newton/meter^2

This is a very small unit compared to ordinary pressures, so the **kilopascal (kPa)** is more often used. The relationship among the units listed above is

$$1 \text{ atm} = 760.000 \text{ mmHg} = 760.000 \text{ torr}$$
$$= 101.325 \times 10^3 \text{ Pa} = 101.325 \text{ kPa}$$

Pressure Measurement with a Manometer

Gas pressures are often measured in the laboratory with a U-tube manometer, which is a mercury-filled, U-shaped glass tube. The closed side of the manometer has been evacuated so that no gas remains to press on the mercury surface on this side. The other side of the manometer is open to the gas whose pressure is to be measured. When the gas presses on the mercury in the open side, the gas pressure in mmHg is read directly as the difference in the mercury levels in the closed and open sides.

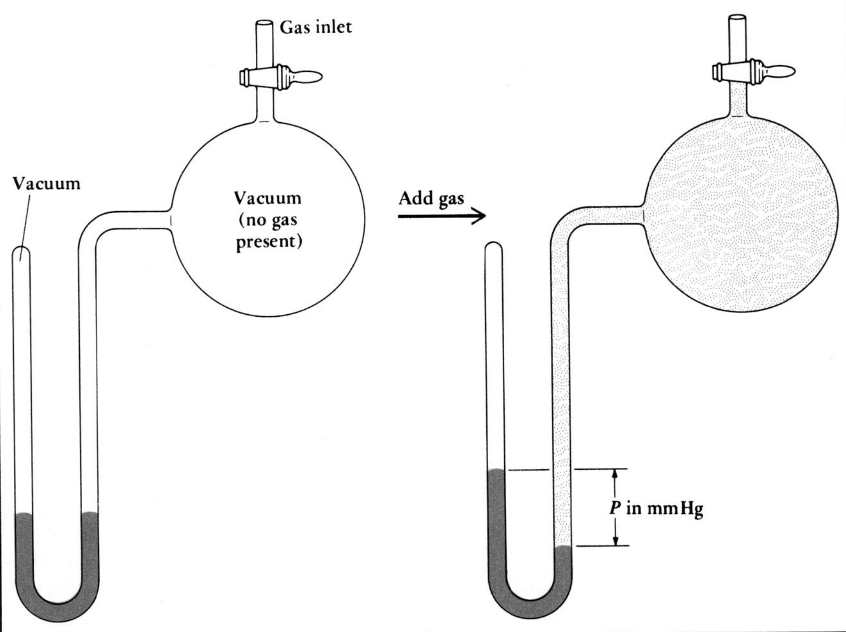

EXAMPLE 12.1

PRESSURE UNIT CONVERSIONS

Convert a pressure of 745 mmHg into its corresponding value in units of torr, atm, and Pa.

Solution The torr and mmHg units are equivalent; thus, 745 mmHg = 745 torr.

The relation between mmHg units and atm is 1 atm = 760. mmHg. Notice that the given pressure is less than 760. mmHg, so the equivalent in atmospheres is less than 1 atm.

$$745 \text{ \sout{mmHg}} \left(\frac{1.00 \text{ atm}}{760. \text{ \sout{mmHg}}}\right) = 0.980 \text{ atm}$$

The relation between mmHg and Pa is 101.325 kPa = 760. mmHg. Therefore,

$$745 \text{ \sout{mmHg}} \left(\frac{101.325 \text{ kPa}}{760. \text{ \sout{mmHg}}}\right) = 99.3 \text{ kPa}$$

EXERCISE 12.1 Pressure Unit Conversions
Rank the following pressures in decreasing order of magnitude (largest first, smallest last): 75 kPa, 300. torr, 0.60 atm, and 350. mmHg.

12.2 THE GAS LAWS: THE EXPERIMENTAL BASIS

Experimentation in the 17th and 18th centuries led to three gas laws that provide the basis for our understanding of gas behavior.

BOYLE'S LAW

Robert Boyle has sometimes been called the "father of chemistry" because he studied a broad range of subjects related to chemistry. He is best known, however, for making one of the first quantitative studies of gas behavior, which he called the "Spring of Air." Boyle observed that

Historical Figures in Chemistry: Robert Boyle (1627–1691)

Robert Boyle was born in Ireland, in a home that still stands, as the 14th and last child of the first Earl of Cork. He first published his studies of gases in 1660, and a book, *The Sceptical Chymist,* was published in 1680. The subtitle of this book, which had a profound effect on those interested in science in that day, was "Chymico–Physical Doubts and Paradoxes, Touching the Experiments whereby Vulgar Spagirists are wont to Endeavour to Evince their Salt, Sulfur, and Mercury, to be the True Principles of Things." Not everyone applauded all of Boyle's work, however. Isaac Newton, a young man when Boyle's career was at its peak, once remarked about one of Boyle's papers that "I question not but that the great wisdom of the noble author will sway him to high silence till he shall be resolved of what consequence the thing may be ... [by] the judgment of some other that thoroughly understands what he speaks about."

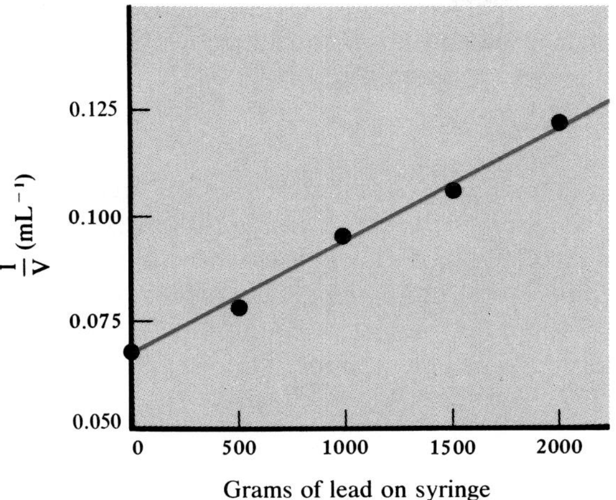

(a)

(b)

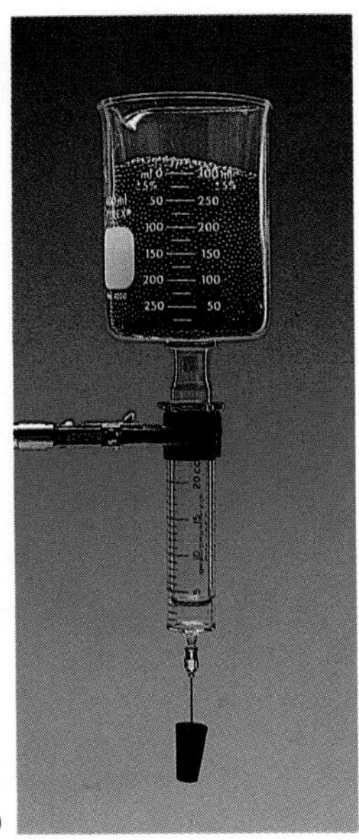

Figure 12.2 Boyle's law. A syringe, containing some air, was sealed and then lead shot was added to the beaker on top of the barrel. As the mass of lead increased (from photo *a* to photo *b*), thereby increasing the pressure on the air in the syringe, the air was compressed. A plot of (1/volume of air) vs. mass of lead shows there is a linear relation between $1/V$ and P (as measured by the mass of lead). (See D. Davenport, *Journal of Chemical Education,* **1962**, *39*, 252.)

the volume of a confined gas is inversely proportional to the pressure exerted on the gas. Since all gases behave in this manner, we know this as **Boyle's law**.

Boyle's law can be demonstrated by the experiment shown in Figure 12.2. A hypodermic syringe is filled with air and sealed. When pressure is applied to the movable plunger of the syringe, the volume of air in the sealed apparatus decreases. When the pressure is plotted as a function of $1/V$, a straight line is observed. This tells you that *the pressure and volume are inversely proportional.*

$$P \propto \frac{1}{V}$$

That is, the essence of Boyle's law is that pressure and volume change in opposite directions: for a given quantity of gas at a given temperature, the gas volume will increase if the pressure is decreased (Figure 12.3).

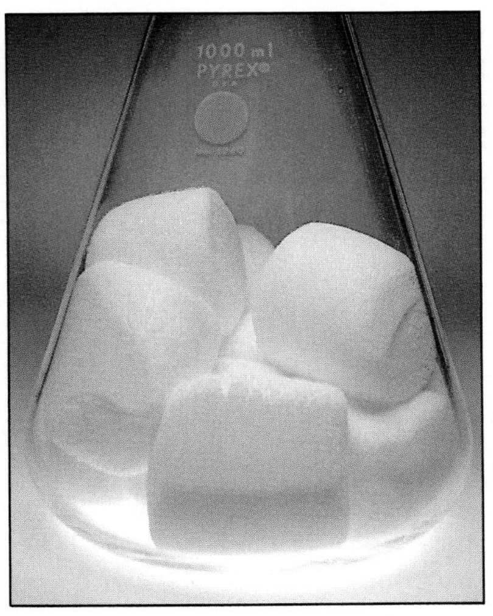

Figure 12.3 An illustration of Boyle's law. Some marshmallows are placed in a flask (left). The flask is then evacuated (the air is withdrawn from the flask) using a vacuum pump (right). The air in the marshmallows expands as the pressure is lowered, causing the marshmallows to expand.

Boyle's law can be put into a useful mathematical form as follows. When two quantities are proportional to each other, they can be equated if a proportionality constant, here called C_b, is introduced. Thus,

$$P = C_b \frac{1}{V} \quad \text{or} \quad PV = C_b \tag{12.1}$$

The last relation, $PV = C_b$, is a useful statement of Boyle's law: *For a given quantity of gas at a given temperature, the product of pressure and volume is a constant*, where the constant C_b is determined by the quantity of gas (in moles) and the temperature (in kelvins). This means that if the pressure-volume product is known for one set of conditions (P_1 and V_1), it is known for all other conditions of pressure and volume (P_2 and V_2):

$$P_1V_1 = P_2V_2$$

This form of Boyle's law is useful when we want to know, for example, what happens to the volume of a given quantity of gas when the pressure changes (at a constant temperature).

EXAMPLE 12.2

BOYLE'S LAW

A sample of gaseous CO_2 has a pressure of 55 mmHg in a 125-mL flask. If this sample is transferred to a 650-mL flask with the same temperature as before, what is the expected pressure of the gas?

Solution You will find it is quite useful to make a table of the information provided.

ORIGINAL CONDITIONS	FINAL CONDITIONS
$P_1 = 55$ mmHg	$P_2 = ?$
$V_1 = 125$ mL	$V_2 = 650$ mL

Since you know that $P_1V_1 = P_2V_2$, then

$$P_2 = \frac{P_1V_1}{V_2} = \frac{(55 \text{ mmHg})(125 \text{ mL})}{(650 \text{ mL})} = 11 \text{ mmHg}$$

Remember that the essence of Boyle's law is that P and V change in opposite directions. Indeed, this suggests a way to solve the problem without going through the algebraic manipulation of $P_1V_1 = P_2V_2$. That is, you know that the new pressure (P_2) must be less than the old pressure (P_1); thus, P_1 must be multiplied by a volume factor that has a value *less than 1* to reflect the fact that P_2 must be less than P_1. That is, simply understanding Boyle's law leads us to the following expression, where the volume ratio 125 mL/650 mL is less than 1 and scales P_1 *down* to P_2:

$$P_2 = P_1 \left(\frac{125 \text{ mL}}{650 \text{ mL}}\right)$$

EXERCISE 12.2 Boyle's Law

A sample of CO_2 with a pressure of 55 mmHg in a volume of 125 mL is placed in a new flask. The pressure of the gas in the new flask is 78 mmHg. What is the volume of the new flask? (The temperature was constant throughout the experiment.)

Historical Figures in Chemistry: Jacques Alexandre Cesar Charles (1746–1823)

The French chemist Charles was most famous in his lifetime for his experiments in ballooning. The first such flights were made by the Montgolfier brothers in June 1783, using a large spherical balloon made of linen and paper and filled with hot air. In August 1783, however, a different group, supervised by Jacques Charles, tried a different approach. Exploiting his recent discoveries in the study of gases, Charles decided to inflate the balloon with hydrogen. Since hydrogen would easily escape a paper bag (see Section 12.7), Charles made a bag of silk coated with a rubber solution. Inflating the bag to its final diameter took several days and required nearly 500 pounds of acid and 1000 pounds of iron to generate the hydrogen gas. A huge crowd watched the ascent on August 27, 1783. The balloon stayed aloft for almost 45 minutes and traveled about 15 miles, but, when it landed in a village, the people were so terrified that they tore it to shreds.

(The Bettmann Archive)

CHARLES'S LAW

If a given quantity of gas is held at a constant pressure, then its volume is directly proportional to the absolute temperature ($V \propto T$). This means that $V = C_c \times T$, or that

$$C_c = \frac{V}{T} \qquad (12.2)$$

This statement, which came from experiments by the Frenchman Jacques Charles, is illustrated in an extreme manner in Figure 12.4.

To demonstrate Charles's law, consider some experimental data for the volume of a gas as a function of temperature (Figure 12.5). A sample of a gas is heated and cooled, and its volume (at constant pressure) is plotted as a function of temperature. Again a straight-line relation is observed. The volume is found to double when T, the Kelvin temperature, is doubled. The direct proportionality of V and T means that

$$\frac{V_1}{T_1} = \frac{V_2}{T_2}$$

That is, the essence of Charles's law is that the volume and temperature (of a given quantity of gas at constant pressure) change in the same direction (recall that volume and pressure change in *opposite* directions). The expression above means that the volume of gas (V_2) at some tem-

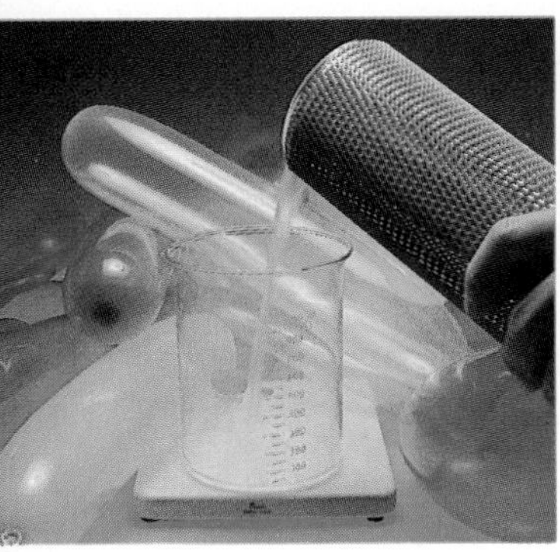

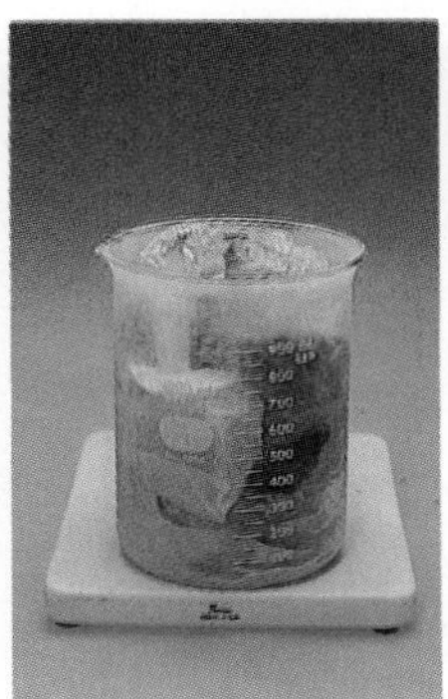

Figure 12.4 A dramatic illustration of Charles's law. Some air-filled balloons are placed in liquid nitrogen (at 77 K). The volume of the gas is greatly reduced at this temperature. After all the balloons have been placed in the liquid nitrogen, they are poured out again and the balloons reinflate to their original volume when warmed back to room temperature.

Figure 12.5 Charles's law. For a given quantity of gas at constant pressure, the volume of the gas and its Kelvin temperature are directly proportional.

Volume–Temperature Relations
(for 0.010 mole of a gas at 1.0 atm pressure)

Volume, V (mL)	Temperature, t (°C)	Temperature, T (K)	V/T (mL/K)
100.	−151	122	0.820
150.	−90.	183	0.820
200.	−29	244	0.820
250.	32	305	0.820
300.	93	366	0.820
400.	214	487	0.821

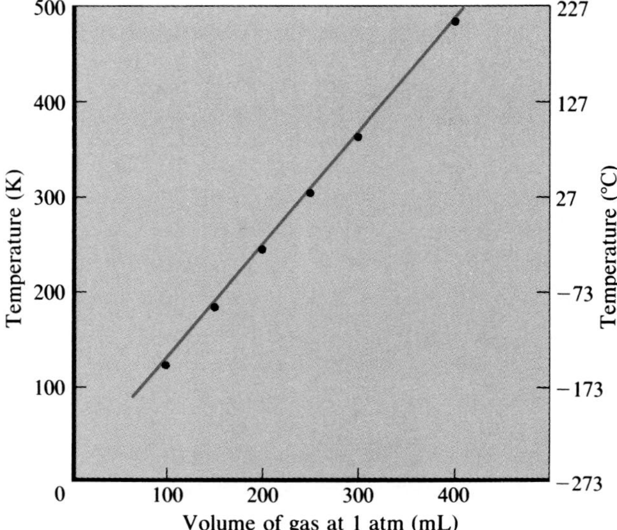

perature (T_2) can be calculated if its volume is known (V_1) at one other temperature (T_1). Calculations based on Charles's law are illustrated by the following example and exercise. Be sure to notice that *the temperature, T, must be expressed in kelvins.*

EXAMPLE 12.3

CHARLES'S LAW

Suppose you have a sample of CO_2 in a gas-tight syringe (see Figure 12.2). The gas volume is 25.0 mL at room temperature (20. °C). What is the final volume of the gas if you hold the syringe in your hand to raise its temperature to 37 °C?

Solution To organize the information, construct a table. Notice that the temperature *must* be converted to kelvins.

ORIGINAL CONDITIONS	FINAL CONDITIONS
$V_1 = 25.0$ mL	$V_2 = ?$
$T_1 = 20. + 273 = 293$ K	$T_2 = 37 + 273 = 310.$ K

Since you know that $V_1/T_1 = V_2/T_2$, you can rearrange the equation to solve for V_2 and substitute the information given.

$$V_2 = \frac{V_1}{T_1} T_2 = \left(\frac{25.0 \text{ mL}}{293 \text{ K}}\right) (310. \text{ K}) = 26.5 \text{ mL}$$

As expected, the volume of gas increases with an increase in temperature. Again, this suggests another logical way to solve the problem. Since V and T change in the same directions, the new volume (V_2) must be equal to the old volume (V_1) multiplied by a temperature factor that is *greater than 1* to reflect the effect of the increase in temperature. That is, V_1 is increased to V_2 by the temperature increase, and the temperature ratio 310. K/295 K is greater than 1.

$$V_2 = V_1 \left(\frac{310.\text{ K}}{293\text{ K}}\right)$$

EXERCISE 12.3 Charles's Law

A balloon is inflated with helium to a volume of 4.5 L at room temperature (25 °C). If you take the balloon outside on a cold day (−10 °C), what is the new volume of the balloon?

GAY-LUSSAC'S AND AVOGADRO'S LAWS

The beginning of the 19th century was an exciting time for chemistry with work being done by John Dalton, Jacques Charles, Amedeo Avogadro, and Joseph Gay-Lussac. Gay-Lussac experimented with reactions of gases and found that volumes of gases always combine with one another in the ratio of small whole numbers, as long as the volumes are measured at the same temperature and pressure. This statement is now referred to as Gay-Lussac's *law of combining volumes*. As an example, this means that 100 mL of H_2 gas will combine exactly with 50 mL of O_2 gas to give exactly 100 mL of H_2O vapor if all the gases are measured at the same T and P, say 100 °C and 1 atm pressure (Figure 12.6).

Gay-Lussac's law remained only a summary of experimental observations until it was explained by the work of the Italian physicist and lawyer Amedeo Avogadro. In 1811 Avogadro published his ideas, now

Figure 12.6 An illustration of Gay-Lussac's law: Volumes of gases, at the same temperature and pressure, will combine with one another in the ratio of small whole numbers. Here one volume of O_2, say 50 mL at 100 °C and 1 atm pressure, will combine with 2 volumes of H_2 gas (100 mL) to give 2 volumes (100 mL) of H_2O vapor, both of the latter gases also measured at 100 °C and 1 atm pressure.

50 mL of O_2 + 100 mL of H_2 → 100 mL of H_2O

$$2\,H_2(g) + O_2(g) \longrightarrow 2\,H_2O(g)$$

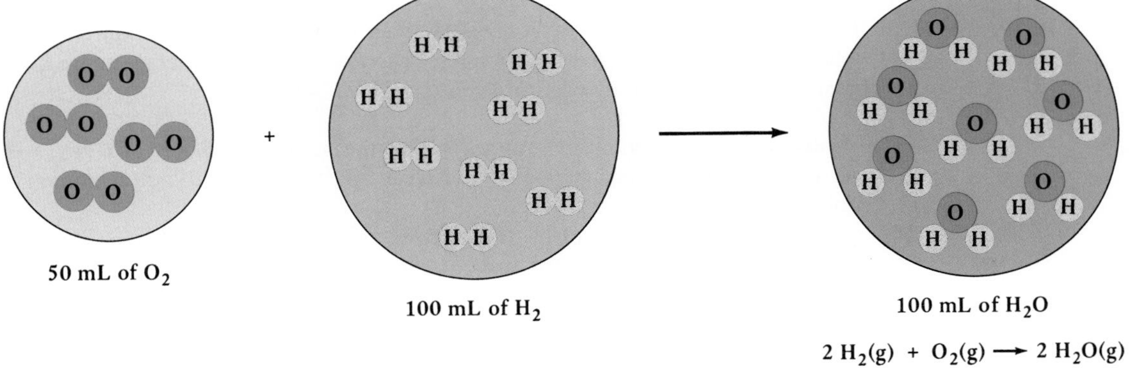

Historical Figures in Chemistry: Joseph Gay-Lussac (1778–1850)

The Frenchman Gay-Lussac was a chemist, but he contributed to physics, meteorology, and physiology. Although we chiefly remember him for his experiments on gases, he established the elementary nature of sulfur, was the first to prepare sodium and potassium by chemical means, and was the first to isolate boron. He was also, like Jacques Charles, a balloonist. In fact, in 1804, Gay-Lussac ascended to 23,000 feet in a balloon to sample the air; it was an altitude record that stood for many years. In another high-altitude flight, Gay-Lussac had to throw overboard several items to lighten the balloon and gain height. One item he decided to sacrifice was an old white kitchen chair that he had used for a seat. The chair landed near a peasant girl who was minding sheep near a French village. After considerable debate, the local citizenry and priest decided that the incident was a miracle, but wondered why God apparently owned such shabby furniture.

known as *Avogadro's hypothesis*, that *equal volumes of gases under the same conditions of temperature and pressure contain equal numbers of molecules*. Applied to Gay-Lussac's experiment, this means that 100 mL of H_2 molecules in the gas phase must have twice as many molecules as 50 mL of O_2, and when combined by reaction they should produce 100 mL of H_2O molecules.

Avogadro's law follows from Avogadro's hypothesis: *The volume of a gas, at a given temperature and pressure, is directly proportional to the quantity of gas*. Thus, V is proportional to n, the number of moles of gas, so

$$V \propto n \qquad \text{or} \qquad V = C_a n \tag{12.3}$$

where C_a is the proportionality constant. The essence of Avogadro's law is that volume and number of moles of gas change in the same direction. As the quantity of gas increases, the volume of the container must increase (at constant T and P).

EXAMPLE 12.4

AVOGADRO'S LAW

Ammonia can be made directly from the elements according to the equation

$$N_2(g) + 3\ H_2(g) \longrightarrow 2\ NH_3(g)$$

If one begins with 12.6 L of $H_2(g)$ at a given temperature and pressure, what volume of $N_2(g)$ is required for complete reaction (at the same temperature and pressure)? What is the theoretical yield of NH_3 in liters?

Solution Since gas volume is proportional to the number of moles of a gas, we can use volumes instead of moles and treat this just as we did stoichiometry problems in Chapter 4.

(a) Calculate liters of N_2 required by multiplying the number of liters of H_2 available by a stoichiometric factor in liters.

$$12.6\ \text{L}\ H_2(g)\ \text{available} \left(\frac{1\ \text{L}\ N_2(g)\ \text{required}}{3\ \text{L}\ H_2(g)\ \text{available}}\right) = 4.20\ \text{L}\ N_2\ \text{required}$$

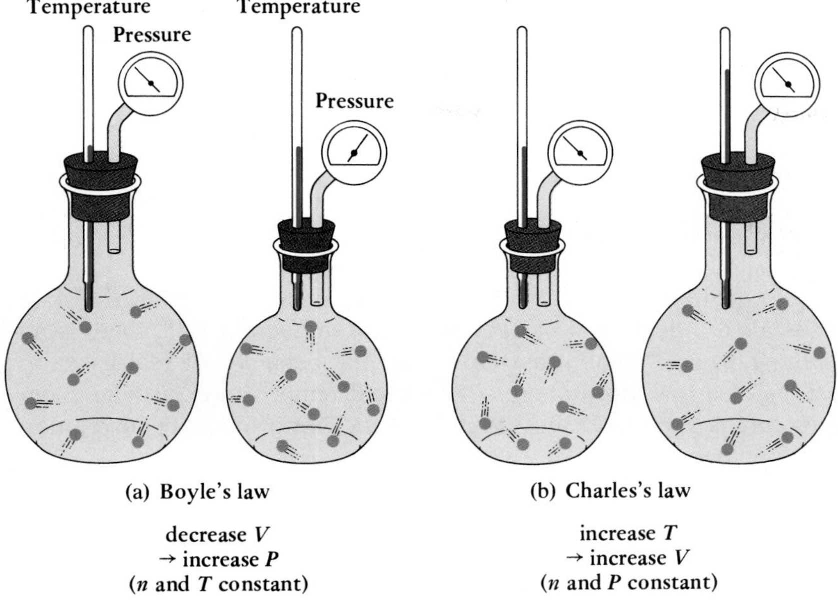

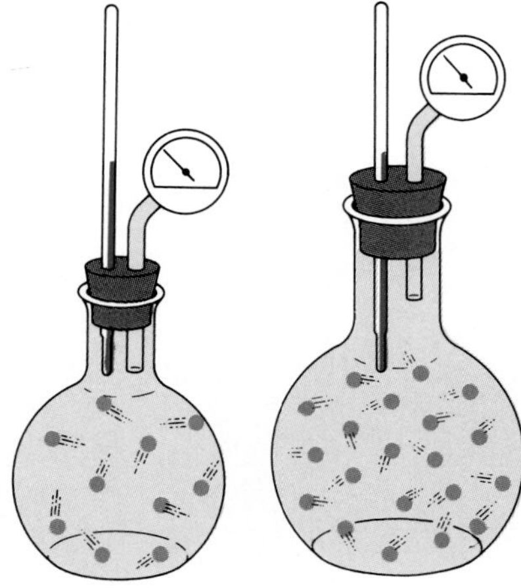

(a) Boyle's law

decrease V
→ increase P
(n and T constant)

(b) Charles's law

increase T
→ increase V
(n and P constant)

(c) Avogadro's law

increase n
→ increase V
(T and P constant)

Figure 12.7 A summary of the gas laws. In each case the volume of gas is represented by the flask size, and the number of molecules in the flask by the number of spheres. (a) Boyle's law. At a constant temperature, the pressure exerted by a given quantity of gas increases as the volume of the gas decreases. (b) Charles's law. At a constant pressure, the volume a given quantity of gas occupies increases as the temperature of the gas increases. (c) Avogadro's law. At a constant pressure and temperature, the volume occupied by a gas depends directly on the quantity of gas.

(b) Calculate liters of gaseous ammonia, NH_3, produced.

$$12.6 \text{ L } H_2(g) \text{ available} \left(\frac{2 \text{ L } NH_3(g) \text{ produced}}{3 \text{ L } H_2(g) \text{ available}}\right) = 8.40 \text{ L } NH_3 \text{ produced}$$

EXERCISE 12.4 Avogadro's Law

Methane burns in oxygen to give the products CO_2 and H_2O.

$$CH_4(g) + 2\ O_2(g) \longrightarrow 2\ H_2O(g) + CO_2(g)$$

If 22.4 L of gaseous CH_4 are burned, what volume of O_2 is required for complete combustion? What volumes of H_2O and CO_2 are produced? Assume all gases are measured at the same temperature and pressure.

12.3 THE IDEAL GAS LAW

Four interrelated quantities can be used to describe the state of a gas: pressure, volume, temperature, and quantity (moles). The three gas laws described in Section 12.2 tell how one gas property is affected as another is changed, assuming the other two properties remain fixed (Figure 12.7). In summary, these laws are:

BOYLE'S LAW	CHARLES'S LAW	AVOGADRO'S LAW
$V \propto (1/P)$	$V \propto T$	$V \propto n$
(constant T, n)	(constant P, n)	(constant T, P)

If all three laws are combined, the result is

$$V \propto \frac{nT}{P}$$

The proportionality constant, labeled R, is called the **gas constant**. This constant, a number that connects P, V, T, and n, is a "universal constant," a number that you can always use to interrelate the properties of a gas.

$$V = R\frac{nT}{P}$$

or

$$PV = nRT \tag{12.4}$$

This last equation is called the **ideal gas law**, since it describes the state of a so-called "ideal" gas. As you will learn in Section 12.8, there is no such thing as an "ideal" gas. However, the behavior of real gases at pressures of a few atmospheres or less and temperatures around room temperature is close to ideality, and $PV = nRT$ is an adequate description of their behavior.

To use the equation $PV = nRT$, we need a value for R. Many experiments show that under conditions of **standard temperature and pressure (STP)** (a gas temperature of 0 °C or 273.15 K and a pressure of 1.0000 atm), 1.0000 mole of gas occupies **22.414 liters**, a quantity called the **standard molar volume**. Substituting these values into the ideal gas law, we can solve for R.

$$R = \frac{PV}{nT}$$

$$R = \frac{(1.0000\ \text{atm})(22.414\ \text{L})}{(1.0000\ \text{mol})(273.15\ \text{K})} = \mathbf{0.082057\ \frac{L \cdot atm}{K \cdot mol}}$$

With a value for R, we can now use the ideal gas law to perform some useful calculations.

EXAMPLE 12.5

THE IDEAL GAS LAW

The helium in a 1.5-L flask at 25 °C exerts a pressure of 425 mmHg. How many moles of helium are there in the flask?

Solution First, write down the information provided.

$$V = 1.5\ \text{L} \qquad P = 425\ \text{mmHg} \qquad T = 25\ °\text{C} \qquad n = ?$$

To use the ideal gas law with R in the units above, the pressure must be expressed in atmospheres and the temperature in kelvins. Therefore,

$$P = 425\ \text{mmHg}\left(\frac{1\ \text{atm}}{760.\ \text{mmHg}}\right) = 0.559\ \text{atm}$$

$$T = 25 + 273 = 298\ \text{K}$$

Now rearrange the ideal gas law to solve for the number of moles, n,

$$n = \frac{PV}{RT}$$

and substitute into this expression.

$$n = \frac{(0.559 \text{ atm})(1.5 \text{ L})}{(0.082057 \text{ L} \cdot \text{atm/K} \cdot \text{mol})(298 \text{ K})} = 0.034 \text{ mole He}$$

EXERCISE 12.5 The Ideal Gas Law

The balloon used by Charles in his historic flight in 1783 was filled with about 1300 moles of H_2. If the temperature of the gas was 20. °C, and its pressure was 750 mmHg, what was the volume of the balloon?

The ideal gas law is useful for calculating any one of the four properties of a gas when the other three are known. There are many times when we need to know what happens to the state of a gas when there is a change in one, two, or three of the parameters, P, V, n, and T. Since R is a universal constant for gases, this means that n_1 moles of a gas exerting a pressure P_1 in a volume V_1 at a temperature T_1 must obey the equation

$$R = \frac{P_1V_1}{n_1T_1}$$

If the conditions change to new values (to n_2, P_2, V_2, and T_2), we write the same expression for R.

$$R = \frac{P_2V_2}{n_2T_2}$$

Under either set of conditions, the quotient PV/nT is equal to the universal constant R. Therefore,

$$\frac{P_1V_1}{n_1T_1} = \frac{P_2V_2}{n_2T_2} \tag{12.5}$$

This is sometimes called the **general gas law** or **combined gas law**.

EXAMPLE 12.6

THE GENERAL GAS LAW

You have a sample of a gas in a small container (12.5 mL). The pressure of the gas is 685 mmHg at room temperature (22.0 °C). If you hold the container in your hand so the temperature climbs to 37.0 °C, what is the new pressure?

Solution Here the quantity of gas is constant, as is the volume in which it is contained. Therefore, we are interested in the change in pressure as the temperature increases. We shall begin by setting out the given information in a table. Notice that *temperatures have been changed to kelvins, as is always required when using any of the gas laws.*

INITIAL CONDITIONS	FINAL CONDITIONS
V_1 = 12.5 mL	V_2 = 12.5 mL
P_1 = 685 mmHg	P_2 = ?
T_1 = 22.0 °C (295.2 K)	T_2 = 37.0 °C (310.2 K)
$n_1 = n_2$	

Since $n_1 = n_2$ and $V_1 = V_2$, these quantities can be canceled from Equation 12.5, so

$$\frac{P_1}{T_1} = \frac{P_2}{T_2}$$

and the resulting equation can be solved for P_2.

$$P_2 = P_1 \left(\frac{T_2}{T_1}\right) = 685 \text{ mmHg} \left(\frac{310.2 \text{ K}}{295.2 \text{ K}}\right) = 720. \text{ mmHg}$$

The temperature has increased, so we also observe a pressure increase, as expected. Indeed, this fact could have been used to solve the problem without setting up the general gas law equation and doing an algebraic manipulation. That is, since we know that increasing the temperature must lead to a pressure increase (when V and n are constant), this means the new pressure (P_2) must equal the old pressure (P_1) times a temperature fraction *greater than 1*.

One final note: When using the ideal gas law (Equation 12.4), the volume must be in units of liters, pressure in units of atmospheres, and temperature in kelvins because these are the units of R. However, when using the general gas law (Equation 12.5), it is not absolutely necessary to make unit changes *except* for expressing T in kelvins. The reason for this is that the general gas law considers *ratios* of P and V. Nonetheless, as you work gas law problems it is probably best to get into the habit of always using atmospheres and liters (as well as kelvins) so that you do not make a mistake.

EXAMPLE 12.7

THE GENERAL GAS LAW

A gas occupying 2.00 L at 756 mmHg pressure and 273 K is transferred to a 4.00-L container, where the pressure is 189 mmHg. What must the temperature be for the gas in the new container to have this pressure?

Solution As a first step, set out the conditions of the gas before and after the transfer.

INITIAL VALUES	FINAL VALUES
$V_1 = 2.00$ L	$V_2 = 4.00$ L
$P_1 = 756$ mmHg (0.995 atm)	$P_2 = 189$ mmHg (0.249 atm)
$T_1 = 273$ K	$T_2 = ?$
$n_1 = n_2$	

Since $n_1 = n_2$, these quantities cancel out of Equation 12.5, so if the expression is solved for T_2 we have

$$T_2 = T_1 \left(\frac{P_2}{P_1}\right)\left(\frac{V_2}{V_1}\right) = 273 \text{ K} \left(\frac{189 \text{ mmHg}}{756 \text{ mmHg}}\right)\left(\frac{4.00 \text{ L}}{2.00 \text{ L}}\right) = 137 \text{ K}$$

Notice that the final temperature is less than the initial temperature. But is this what we would have expected? The volume of the gas increased by a factor of 2, which in itself should indicate an *increase* in T by a factor of 2. However, the pressure decreased by a factor of 4, which could have been caused by a *decrease* in T by a factor of 4. Since the pressure change is larger than the volume change, you expect, on balance, that T decreases.

EXERCISE 12.6 The General Gas Law

You have a 20.-L cylinder of helium at a pressure of 150 atm and at 30. °C. How many balloons can you fill, each with a volume of 5.0 L, on a day when the atmospheric pressure is 755 mmHg and the temperature is 22 °C?

As a final comment on the general gas law, the answer to Example 12.6 points up a useful general conclusion: *For a given quantity of gas in a constant volume, the pressure is directly proportional to the absolute temperature*. When you drive an automobile for some distance, you know that the tire pressure goes up. The reason for this is that friction warms the tires, and, since the tire volume is nearly constant, the pressure increases.

Figure 12.8 Gas density. The two balloons are filled with roughly equal quantities of gas at the same temperature and pressure. The blue balloon contains low-density helium (d = 0.16 g/L), while the red balloon contains argon, a higher density gas (d = 1.6 g/L). In comparison, the density of dry air at 1 atm pressure and 25 °C is about 1.2 g/L.

THE DENSITY OF GASES

The density of a gas at a given temperature and pressure (Figure 12.8) is a useful quantity. Let us see how this is related to the ideal gas law. By rearranging the expression $PV = nRT$, we have

$$\frac{n}{V} = \frac{P}{RT}$$

Since the number of moles (n) of any compound is given by its mass (m) divided by its molar mass (M), we can rewrite the equation as

$$\frac{m}{MV} = \frac{P}{RT}$$

Since density (d) is mass/volume (m/V), we can rearrange the equation above to relate the density of a gas to its pressure, molar mass, and temperature.

$$d = \frac{m}{V} = \frac{PM}{RT} \tag{12.6}$$

EXAMPLE 12.8

GAS DENSITY AND MOLAR MASS

The density of an unknown gas is 1.429 g/L at STP. Calculate its molar mass.

Solution List the given information in a short table, recalling that "STP" is shorthand for 1 atm and 0 °C.

d = 1.429 g/L | P = standard pressure = 1.000 atm
T = standard temperature = 273.15 K | R = 0.082057 L · atm/K · mol
M = ?

The equation for gas density is rearranged to solve for molar mass (M).

$$M = \frac{dRT}{P}$$

$$= \frac{(1.429 \text{ g/L})(0.082057 \text{ L} \cdot \text{atm/K} \cdot \text{mol})(273.15 \text{ K})}{1.000 \text{ atm}} = 32.03 \text{ g/mol}$$

EXERCISE 12.7 Gas Density Calculation

Calculate the density of air at 22.0 °C and 745 mmHg if its molar mass (average) is 29.0 g/mol.

This discussion of gas density has some important practical implications. From the equation $d = M(P/RT)$, we know that gas density is directly proportional to the molar mass (for gases at the same T and P). Air, with an average molar mass of about 29 g/mol, has a density of about 1.2 g/L (at 1 atm and 25 °C). This means that gases or vapors more dense than about 1.2 g/L—such as CO_2, SO_2, and gasoline (with molar masses greater than that of air)—will settle along the ground if released into the atmosphere. Conversely, gases such as H_2, He, CO, CH_4 (methane), and NH_3 (with molar masses less than that of air) will rise if released into the atmosphere. The release of methyl isocyanate (H_3CNCO) into the air on December 3, 1984 in Bhopal, India, killed several hundred people and injured thousands more because the toxic vapors were heavier than air and settled close to the ground where they could be inhaled.

CALCULATING THE MOLAR MASS OF A GAS FROM *P*, *V*, and *T* DATA

Either the ideal gas law or the equation for gas density derived from it can be used to determine the molar mass of a gas.

EXAMPLE 12.9

CALCULATING THE MOLAR MASS OF A GAS FROM *P*, *V*, and *T* DATA

A 0.100-g sample of a compound of empirical formula CH_2F_2 occupies 0.0470 L at 298 K and 755 mmHg. What is the molar mass of the compound?

Solution This problem is typical of the laboratory measurement of the molar mass of a gas. The experiment is performed when one has synthesized a new gaseous compound and does not yet know its molar mass. Begin by organizing the data.

$V = 0.0470$ L $P = 755$ mmHg $T = 298$ K $n = ?$

(a) We can find the molar mass by first using the ideal gas law to calculate n, the number of moles equivalent to 0.100 g of the gas.

$$n = \frac{PV}{RT}$$

When substituting into the rearranged gas equation, you must make sure the units of P, V, and T are compatible with the units of R. Here this means P is converted to atmospheres.

$$P = 755 \text{ mmHg} \left(\frac{1 \text{ atm}}{760. \text{ mmHg}}\right) = 0.993 \text{ atm}$$

Therefore,

$$n = \frac{(0.993 \text{ atm})(0.0470 \text{ L})}{(0.082057 \text{ L} \cdot \text{atm/K} \cdot \text{mol})(298 \text{ K})} = 0.00191 \text{ mole}$$

Since you know the number of grams in the sample and now know the number of moles to which it is equivalent, you can calculate the molar mass.

$$\text{Molar mass} = \frac{0.100 \text{ g}}{0.00191 \text{ mol}} = 52.4 \text{ g/mol}$$

In this particular case the experimental molar mass is equal to the empirical formula weight.

(b) The gas density equation can also be used to solve for the molar mass. The density of the gas here is 0.100 g in 0.0470 L or

$$d = \frac{0.100 \text{ g}}{0.0470 \text{ L}} = 2.13 \text{ g/L}$$

If the equation $d = PM/RT$ is solved for molar mass (M), and then density, pressure, and temperature are substituted into the expression, we have

$$M = \frac{dRT}{P} = \frac{(2.13 \text{ g/L})(0.082057 \text{ L} \cdot \text{atm/K} \cdot \text{mol})(298 \text{ K})}{0.993 \text{ atm}} = 52.5 \text{ g/mol}$$

EXERCISE 12.8 Molar Mass From *P*, *V*, and *T* Data

A 0.105-g sample of a gaseous compound has a pressure of 560. mmHg in a volume of 125 mL at 23.0 °C. What is its molar mass?

12.4 THE GAS LAWS AND CHEMICAL REACTIONS

Many important chemical reactions, such as the industrial production of ammonia and chlorine, involve gases.

$$N_2(g) + 3\ H_2(g) \longrightarrow 2\ NH_3(g)$$

$$2\ NaCl(aq) + 2\ H_2O(\ell) \xrightarrow{\text{electric energy}} 2\ NaOH(aq) + H_2(g) + Cl_2(g)$$

It is important that you know how to deal with such reactions quantitatively, so the examples that follow explore chemical stoichiometry calculations involving gases. The following scheme connects these calculations for gas reactions with those done in Chapters 4 and 5.

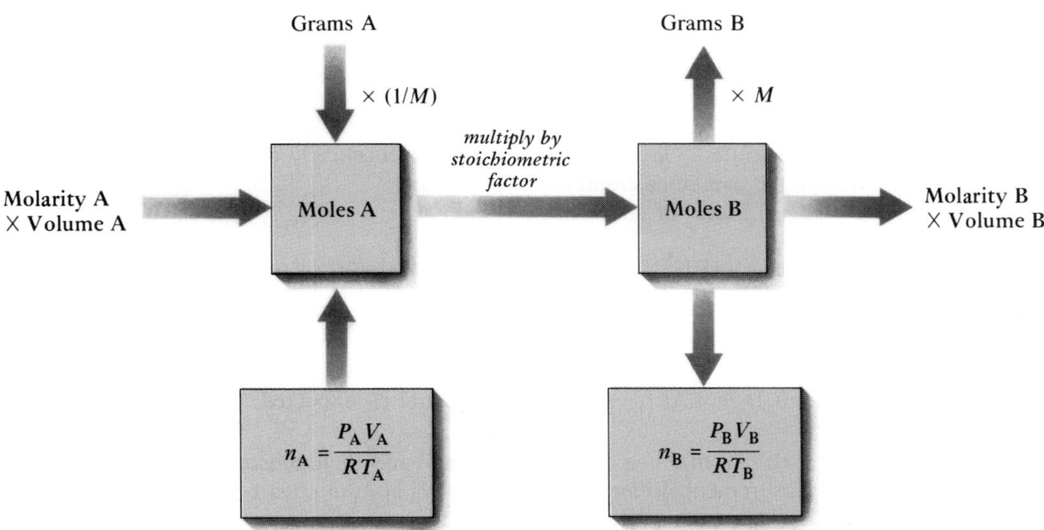

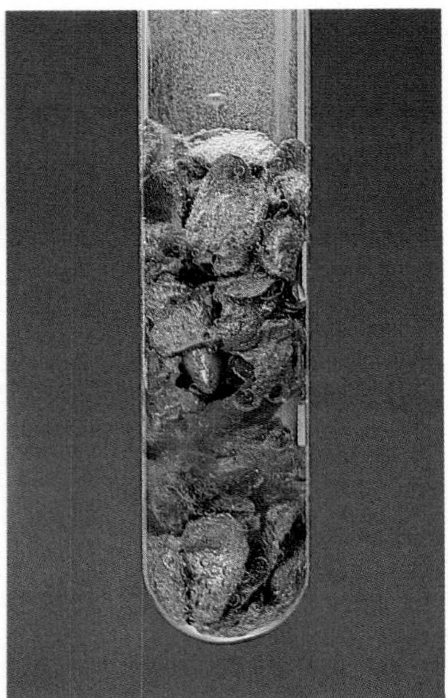

Figure 12.9 The reaction of zinc metal with acid to give hydrogen gas.

EXAMPLE 12.10

GAS LAWS AND STOICHIOMETRY

Calculate the volume of H_2 (measured at 0.0 °C and 1.0 atm pressure) produced by 12 g of zinc reacting with excess sulfuric acid (Figure 12.9).

$$\text{Zn(s)} + \text{H}_2\text{SO}_4\text{(aq)} \longrightarrow \text{ZnSO}_4\text{(aq)} + \text{H}_2\text{(g)}$$

Solution One mole of H_2 gas is produced for every mole of zinc used. Therefore, we first find the moles of zinc consumed.

$$12\ \text{g Zn}\left(\frac{1\ \text{mol Zn}}{65.39\ \text{g Zn}}\right) = 0.18\ \text{mol Zn}$$

Since there is a 1:1 relation between moles of Zn consumed and moles of H_2 produced, 0.18 mole of H_2 is formed. To find the volume of H_2 formed, you can solve the ideal gas law for volume and substitute the given data.

$$V = \frac{nRT}{P} = \frac{(0.18\ \text{mol H}_2)(0.082057\ \text{L} \cdot \text{atm/K} \cdot \text{mol})(273\ \text{K})}{1.0\ \text{atm}} = 4.1\ \text{L}$$

Alternatively, you could recognize that the conditions of the experiment are STP, for which the standard molar volume (the volume of one mole) is 22.4 L. Using this as a conversion factor, we arrive at the same answer.

$$0.18\ \text{mol H}_2\left(\frac{22.4\ \text{L}}{1\ \text{mol H}_2}\right) = 4.1\ \text{L}$$

EXAMPLE 12.11

THE GAS LAWS AND STOICHIOMETRY

Some commercial drain cleaners contain sodium hydroxide and powdered aluminum (Figure 12.10). When the mixture is poured into a drain full of water and dirt, the reaction that occurs is:

$$2\ \text{Al(s)} + 2\ \text{NaOH(aq)} + 6\ \text{H}_2\text{O}(\ell) \longrightarrow 2\ \text{NaAl(OH)}_4\text{(aq)} + 3\ \text{H}_2\text{(g)}$$

The heat generated by the reaction helps to melt any grease, and the gas being generated stirs up the particles and helps to unclog the drain. If you use 5.6 g of aluminum powder and excess NaOH, how many liters of gaseous H_2 measured at 742 mmHg and 22.0 °C are produced?

Solution The first step is to calculate moles of Al available and then to find moles of H_2 generated. Once the quantity of H_2 theoretically obtainable is known, then its volume can be calculated.

(a) Moles of Al available are

$$5.6\ \text{g Al}\left(\frac{1.0\ \text{mol Al}}{27.0\ \text{g Al}}\right) = 0.21\ \text{mol Al}$$

Moles of H_2 produced by the aluminum are

$$0.21\ \text{mol Al}\left(\frac{3\ \text{mol H}_2}{2\ \text{mol Al}}\right) = 0.31\ \text{mol H}_2\ \text{expected}$$

(b) From $PV = nRT$, solve for V and then substitute the conditions of the measurement. Remember that the temperature must be converted to kelvins

(22.0 °C = 295 K) and the pressure must be in atmospheres (742 mmHg = 0.976 atm).

$$V = \frac{nRT}{P} = \frac{(0.31 \text{ mol})(0.082057 \text{ L} \cdot \text{atm/K} \cdot \text{mol})(295 \text{ K})}{(0.976 \text{ atm})}$$
$$= 7.7 \text{ L } H_2$$

EXAMPLE 12.12

GAS LAWS AND STOICHIOMETRY

Gaseous ammonia is synthesized from nitrogen and hydrogen by the reaction

$$N_2(g) + 3\ H_2(g) \xrightarrow[500\ °C]{\text{iron catalyst}} 2\ NH_3(g)$$

Assume that you take 355 L of H_2 gas at 25.0 °C and 542 mmHg and combine it with 105 L of N_2 gas measured at 20.0 °C and a pressure of 645 mmHg. How many grams of NH_3 gas could be obtained in theory? If the NH_3 gas is collected in a 125-L tank at 25.0 °C, what would be its pressure?

Solution We are combining two reactants with no guarantee that they are in the correct stoichiometric ratio. Therefore, this is a *limiting reagent problem*, and we shall have to find the number of moles of each gas (from the ideal gas law) and then see which one of them is present in limited amount. Once the limiting reagent is known, we can find the mass of ammonia produced and then calculate its pressure under the conditions given.

(a) Moles of H_2 gas:

P = (542 mmHg)/(760. mmHg/atm) = 0.713 atm

V = 355 L

T = 25.0 + 273.15 = 298.2 K

$$\text{Moles of } H_2 = \frac{(0.713 \text{ atm})(355 \text{ L})}{(0.082057 \text{ L} \cdot \text{atm/K} \cdot \text{mol})(298.2 \text{ K})} = 10.3 \text{ mol}$$

(b) Moles of N_2 gas:

P = (645 mmHg)/(760. mmHg/atm) = 0.849 atm

V = 105 L

T = 20.0 + 273.15 = 293.2 K

$$\text{Moles of } N_2 = \frac{(0.849 \text{ atm})(105 \text{ L})}{(0.082057 \text{ L} \cdot \text{atm/K} \cdot \text{mol})(293.2 \text{ K})} = 3.71 \text{ mol}$$

(c) Decide which is the limiting reagent:

$$\text{Ratio of moles available} = \frac{10.3 \text{ mol } H_2}{3.71 \text{ mol } N_2} = \frac{2.78 \text{ mol } H_2}{1.00 \text{ mol } N_2}$$

$$\text{Ratio of moles required} = \frac{3 \text{ mol } H_2}{1 \text{ mol } N_2}$$

From this calculation of ratios we see that there is not enough H_2 available to consume the available N_2, so *H_2 is the limiting reagent.*

Figure 12.10 The reaction of sodium hydroxide with powdered aluminum to produce H_2 gas and sodium aluminate, $NaAl(OH)_4$. The reaction is used by popular brands of drain cleaners.

(d) Calculate the mass of NH_3 that can be produced. The calculation of this mass is of course based on the quantity of limiting reagent, H_2, that is available.

$$10.3 \text{ mol } H_2 \left(\frac{2 \text{ mol } NH_3 \text{ produced}}{3 \text{ mol } H_2 \text{ available}}\right) = 6.87 \text{ mol } NH_3$$

$$6.87 \text{ mol } NH_3 \left(\frac{17.03 \text{ g } NH_3}{1 \text{ mol}}\right) = 117 \text{ g } NH_3 \text{ produced}$$

(e) Calculate the pressure of the ammonia gas under the specified conditions.

$$n = 6.87 \text{ mol} \qquad T = 25.0 + 278.15 = 298.2 \text{ K}$$
$$V = 125 \text{ L} \qquad P = ?$$

$$\text{Pressure of } NH_3 = \frac{nRT}{V} = \frac{(6.87 \text{ mol})(0.082057 \text{ L} \cdot \text{atm/K} \cdot \text{mol})(298.2 \text{ K})}{125 \text{ L}}$$

$$= 1.34 \text{ atm (or } 1.02 \times 10^3 \text{ mmHg)}$$

EXERCISE 12.9 Gas Laws and Stoichiometry

Gaseous oxygen reacts with aqueous hydrazine to produce water and gaseous nitrogen according to the balanced equation

$$N_2H_4(aq) + O_2(g) \longrightarrow 2\ H_2O(\ell) + N_2(g)$$

If a solution contains 180 g of N_2H_4, what is the maximum volume of O_2 that will react with the hydrazine if the O_2 is measured at a barometric pressure of 750 mmHg and room temperature (21 °C)?

12.5 GAS MIXTURES AND PARTIAL PRESSURES

The air you breathe is a blend of oxygen, nitrogen, carbon dioxide, water vapor, and small amounts of other gases. It is logical that the ideal gas law applies to this gas mixture as well as to pure gases. There is, after all, no quantity in the ideal gas law that depends in any way on the chemical constitution of the gas molecules. Only by doing experiments dealing with the chemical and physical properties of a gas can you tell a mixture of gases from a pure gas.

A fundamental idea about gas mixtures is that *the pressure of a mixture of gases is the sum of the pressures of the different components of the mixture*. This can be verified by applying the ideal gas law to gas mixtures, where the number of moles of gas present in the sample is taken as the total of all the moles of the components; that is

$$n_{total} = n_A + n_B + \cdots \qquad (12.7)$$

It is the total number of moles of gas, n_{total}, that is used as n in the gas law. For gas law calculations, a mixture of 0.5 mole of O_2 and 0.5 mole of N_2 is indistinguishable from 1.0 mole of either gas, for example.

The pressure exerted by each component in a mixture is related to the number of moles of that component. Suppose you fill a balloon of

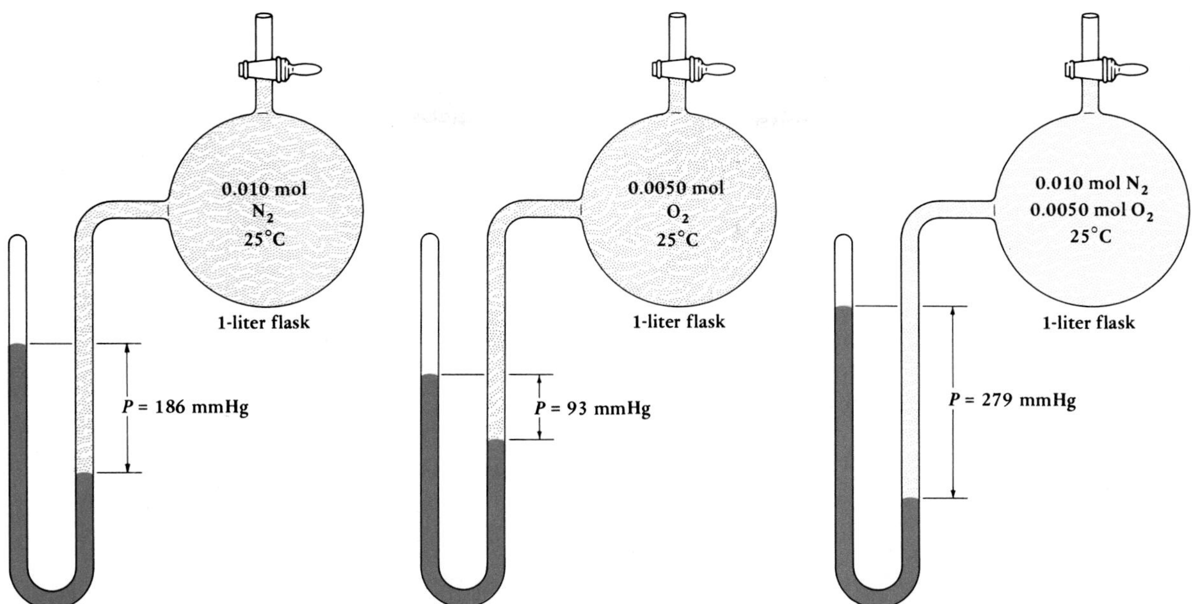

Figure 12.11 Dalton's law. *Left:* 0.010 mole of N_2 in a 1.0-L flask at 25 °C exerts a pressure of 186 mmHg. *Middle:* 0.0050 mole of O_2 in a 1.0-L flask at 25 °C exerts a pressure of 93 mmHg. *Right:* the N_2 and O_2 samples are mixed in the same 1.0-L flask at 25 °C. The total pressure, 279 mmHg, is the sum of the pressures the individual gases would have if each were alone in the flask.

volume V with a mixture of nitrogen and oxygen to a pressure P at a temperature T (Figure 12.11). The gas law applied to this mixture is

$$PV = n_tRT \tag{12.8}$$

where n_t represents moles of N_2 plus moles of O_2. The equation can be rewritten in the form

$$PV = (n_{N_2} + n_{O_2})RT$$

where n_{N_2} = moles of nitrogen and n_{O_2} = moles of oxygen. Expanding the right side of this equation and rearranging, we get

$$PV = n_{N_2}RT + n_{O_2}RT$$

and

$$P = \frac{n_{N_2}RT}{V} + \frac{n_{O_2}RT}{V}$$

Now, $n_{N_2}RT/V$ is the pressure (P_{N_2}) that would be exerted at a temperature T in a volume V by n_{N_2} moles of nitrogen if there were only molecules of nitrogen present. Similarly, $n_{O_2}RT/V$ is the pressure that would be exerted by the O_2 molecules alone in the same volume V. Thus,

$$P_{\text{total}} = P_{N_2} + P_{O_2}$$

The quantities P_{N_2} and P_{O_2} are called the **partial pressures** of the two components. John Dalton was the first to formulate the idea that *the total pressure exerted by a mixture of gases is the sum of the partial pressures* *of the individual gases in the mixture*. In his honor this statement is now called **Dalton's law of partial pressures**.

To use Dalton's law in working with gas mixtures, we shall write an expression for the ratio of the partial pressure of one gas A (P_A) in the mixture to the total pressure of the mixture (P_{total}). Since $P = nRT/V$, we can write

$$\frac{P_A}{P_{total}} = \frac{n_A RT/V}{n_{total} RT/V} = \frac{n_A}{n_{total}}$$

Notice that the ratio of the partial pressure of gas A to the total pressure of the gas mixture is the same as the ratio of the moles of gas A to the total moles of gas. This ratio, n_A/n_{total}, is called the **mole fraction** of A, and it is usually given the symbol X_A.

$$\frac{P_A}{P_{total}} = \frac{n_A}{n_{total}} = \text{mole fraction of A} = X_A \quad (12.9)$$

In general, the mole fraction of any component A of a mixture is

$$X_A = \frac{\text{moles of A}}{\text{moles A + moles B + moles C} + \cdots} \quad (12.10)$$

and the sum of the mole fractions of all components must *always* equal 1 ($X_A + X_B + X_C + \cdots = 1$). As an example, assume a 5.0-mole sample of air is composed of 3.9 moles of N_2 and 1.1 moles of O_2; the mole fraction of each would be

$$X_{N_2} = \frac{3.9 \text{ mol } N_2}{3.9 \text{ mol } N_2 + 1.1 \text{ mol } O_2} = 0.78$$

$$X_{O_2} = \frac{1.1 \text{ mol } O_2}{3.9 \text{ mol } N_2 + 1.1 \text{ mol } O_2} = 0.22$$

and $X_{N_2} + X_{O_2} = 1$.

Applying the concept of the mole fraction to gas laws, you can see from Equation 12.9 that *the partial pressure of a gas in a mixture is equal to the product of its mole fraction and the total pressure*.

$$P_A = X_A P_{total} \quad (12.11)$$

Partial pressure of gas A = (mole fraction of A)(total pressure)

Applied to a sample of air at a pressure of 742 mmHg, this means that

$$\text{Partial pressure of } N_2 = P_{N_2} = (0.78)(742 \text{ mmHg}) = 580 \text{ mmHg}$$

$$\text{Partial pressure of } O_2 = P_{O_2} = (0.22)(742 \text{ mmHg}) = 160 \text{ mmHg}$$

EXAMPLE 12.13

PARTIAL PRESSURES OF GASES

Since fluorocarbons are no longer thought to be environmentally safe as aerosol propellants, chemists have had to devise substitutes. Two gases that have been used are propane (C_3H_8) and isobutane (C_4H_{10}). Suppose you have prepared a mixture of 22 g of propane and 11 g of isobutane. This mixture is then forced into a can until the total pressure is 1.5 atm. What are the partial pressures of propane and isobutane?

Solution The partial pressure of each gas in a mixture is equal to the mole fraction of that gas times the total pressure. The total pressure is known, and other data are given that allow mole fractions to be calculated.

Step 1. Calculate mole fractions.

$$\text{Moles } C_3H_8 = 22 \text{ g } C_3H_8 \left(\frac{1.0 \text{ mol } C_3H_8}{44.1 \text{ g } C_3H_8}\right) = 0.50 \text{ mole}$$

$$\text{Moles } C_4H_{10} = 11 \text{ g } C_4H_{10} \left(\frac{1.0 \text{ mol } C_4H_{10}}{58.1 \text{ g } C_4H_{10}}\right) = 0.19 \text{ mole}$$

$$\text{Mole fraction } C_3H_8 = \frac{0.50 \text{ mol } C_3H_8}{0.69 \text{ total moles}} = 0.72$$

Since

$$X_{C_3H_8} + X_{C_4H_{10}} = 0.72 + X_{C_4H_{10}} = 1$$

we now know that

$$X_{C_4H_{10}} = 0.28$$

Step 2. Calculate partial pressures.

$$\text{Partial pressure of } C_4H_{10} = 0.28P_{\text{total}} = 0.28(1.5 \text{ atm}) = 0.42 \text{ atm}$$

Since

$$P_{C_3H_8} + P_{C_4H_{10}} = 1.5 \text{ atm}$$

this means

$$P_{C_3H_8} = 1.5 \text{ atm} - 0.42 \text{ atm} = 1.1 \text{ atm}$$

EXERCISE 12.10 Partial Pressures of Gases

In Example 12.12, we found that the ammonia produced had a pressure of 1.34 atm. However, there was also some N_2 gas left over after the hydrogen had been consumed. What is the *total* pressure of NH_3 and N_2 in the 125-L flask at 25.0 °C?

One practical application of Dalton's law is a laboratory experiment in which you generate a gas such as N_2 or H_2 and collect it by displacing water from a container (Figure 12.12). Here the total gas pressure in the collecting flask is the sum of the pressures of the N_2 or H_2 *and* the water vapor. Many liquids such as water evaporate to some extent, and the vapor exerts a pressure that depends on the temperature of the system. Appendix D provides the vapor pressures of water at different temperatures, and the concept of vapor pressure is described in more detail in Chapter 13.

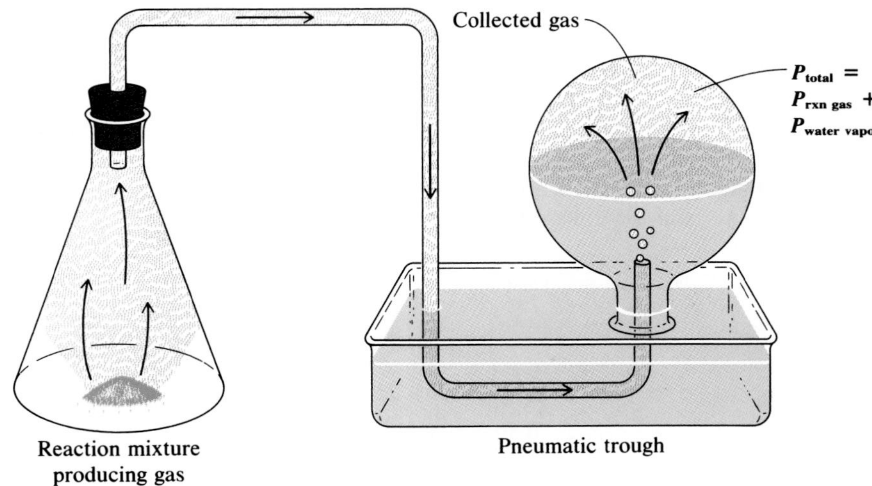

Figure 12.12 Collecting a gas over water. The pressure of the generated gas forces some of the water from the collection flask. Thus, when the pressure of the collected gas is measured (P_{total}), it is the sum of the pressure of the gas from the reaction ($P_{rxn\ gas}$) plus the pressure generated by the water vapor from the evaporation of water in the collection flask ($P_{water\ vapor}$). See Example 12.14.

EXAMPLE 12.14

PARTIAL PRESSURES: COLLECTING A GAS OVER WATER

Small quantities of H_2 gas can be prepared in the laboratory by the following reaction.

$$Fe(s) + 2\ HCl(aq) \longrightarrow FeCl_2(aq) + H_2(g)$$

Assume that you carried out this experiment and collected 0.500 L of H_2 gas as illustrated in Figure 12.12. The temperature of the gas mixture was 26.0 °C, and the total pressure of gas in the flask was 745 mmHg. How many total moles of gas (hydrogen + water vapor) were there in the flask, and how many moles of hydrogen did you prepare?

Solution

Step 1. Calculate the total number of moles of gas.

P = total pressure = 745 mmHg = 0.980 atm

V = 0.500 L

T = 26.0 °C = 299 K

n = total moles of gas = moles H_2 + moles H_2O vapor = ?

$$n = \frac{PV}{RT} = \frac{(0.980\ \text{atm})(0.500\ \text{L})}{(0.082057\ \text{L}\cdot\text{atm/K}\cdot\text{mol})(299\ \text{K})} = 0.0200\ \text{total mole of gas}$$

Step 2. Calculate moles of $H_2(g)$. There are two gases present in the receiver: H_2 and H_2O vapor from the evaporation of the liquid water in the trough in Figure 12.12. Therefore, 0.0200 mole is the sum of the numbers of moles of each of these gases. Likewise, the total pressure is the sum of the partial pressures of H_2 and H_2O.

$$P_{total} = P_{H_2} + P_{H_2O}$$

To find the moles of H_2, you must know the partial pressure of H_2.

$$P_{H_2} = P_{total} - P_{H_2O}$$

The partial pressure of water vapor over liquid water can be obtained from tables in a handbook (see Appendix D in this text). Here P_{H_2O} is 25.2 mmHg, so $P_{H_2} = (745 - 25)$ mmHg = 720. mmHg.

Now you can use the ratio of P_{H_2} to the total pressure to find the mole fraction of H_2 in 0.0200 total mole of gas and, from that, the moles of H_2.

$$\text{Mole fraction } H_2 = \frac{P_{H_2}}{P_{total}} = \frac{720.\text{ mmHg}}{745\text{ mmHg}} = 0.966$$

$$\text{Moles } H_2 = X_{H_2}\text{ (total moles)} = (0.966)(0.0200\text{ mol}) = 0.0193\text{ mol } H_2$$

Alternatively, you could use the partial pressure of H_2 (= 0.947 atm) in the gas law equation to find n_{H_2}.

$$n_{H_2} = \frac{P_{H_2}V}{RT} = \frac{(0.947\text{ atm})(0.500\text{ L})}{(0.082057\text{ L}\cdot\text{atm/K}\cdot\text{mol})(299\text{ K})} = 0.0193\text{ mol } H_2$$

EXERCISE 12.11 Partial Pressures

In an experiment similar to that in Figure 12.12, 352 mL of gaseous nitrogen were collected in a flask over water at a temperature of 22.0 °C. The total pressure of the gases in the collection flask was 742 mmHg. How many grams of N_2 were collected? (See Appendix D for water vapor pressure.)

12.6 KINETIC MOLECULAR THEORY OF GASES

Pressure, volume, temperature, and quantity are four "properties" of gases that can be measured readily and described by a simple law, the ideal gas law. However, they give us little information about how the molecules or atoms of a gas behave. This is the question we want to take up now, and the treatment we shall use is the *kinetic molecular theory of gases*, a theory developed over many years.

Our concept of a gas is that of particles in motion. If your friend is wearing a perfume or your pizza smells good, how do you know it? To put such lovely effects into scientific terms, molecules of the perfume or the food odor enter the gas phase and drift through space until they reach the cells of your body that react to odors. The same thing happens in the laboratory when bottles of aqueous ammonia and hydrochloric acid sit side by side (Figure 12.13). Molecules of the two compounds enter the gas phase and drift along until they encounter one another, combine, and form a cloud of tiny particles of solid ammonium chloride.

If you changed the temperature of the environment of the bottles in Figure 12.13 and compared the times needed for the cloud of ammonium chloride to form, you would find that the time would be longer at lower temperatures. Rather like people, molecules in the gas phase move more slowly at lower temperatures: molecular speeds depend on the temperature. In fact, it can be shown that *the average kinetic energy, $\overline{KE}$, of a collection of gas molecules depends **only** on the temperature*, a concept expressed by the relation

$$\overline{KE} \propto T \tag{12.12}$$

Figure 12.13 Illustration of the movement of gas molecules in the gas phase. Open containers of aqueous ammonia and HCl are placed side by side. When molecules of NH_3 and HCl escape from the dishes and encounter one another, you observe the formation of ammonium chloride, NH_4Cl.

where the bar over KE indicates an *average value*. The kinetic energy of only one molecule is given by

$$KE = \tfrac{1}{2}mu^2 \qquad (12.13)$$

where u is the speed of the molecule and m is its mass. If you were to measure the speeds of the trillions of molecules in a gas sample, however, you would find that some have a speed of u_1, some have a different speed u_2, and so on. The *average* speed, $\bar{u}$, of the molecules is therefore

$$\bar{u} = \frac{n_1u_1 + n_2u_2 + \cdots}{N} \qquad (12.14)$$

where N is the total number of molecules and n_1, n_2, etc. are the numbers of molecules having their respective speeds. This means the average kinetic energy of many molecules, $\overline{KE}$, must be related to the *average of the* ***squares*** *of their speeds,* $\overline{u^2}$ (called the "mean square speed"), by the equation

$$\overline{KE} = \tfrac{1}{2}m\overline{u^2} \qquad (12.15)$$

Now, since $\overline{KE}$ is proportional to T *and* to $m\overline{u^2}$, this means that T must be proportional to $m\overline{u^2}$ as well, and so we can say that

$$\tfrac{1}{2}m\overline{u^2} = CT \qquad (12.16)$$

where C is a proportionality constant. We shall come back to this useful relation in a moment.

Another observation you have made is that *gases are compressible.* For example, you can "pump up" a tire by forcing or compressing gas into it. By contrast, it is difficult to squeeze solids or liquids to force them to occupy a smaller volume. This must mean that *the distance between gas particles (atoms or molecules) is very large relative to the actual size of the particles*, whereas the opposite is true in solids and liquids.

Finally, we noted in Chapter 1 that *gases occupy completely the volume available to them*, in contrast with liquids and solids. As described in Chapter 13, molecules in the vapor phase can condense to form liquids and solids if the forces of attraction between molecules—called *intermolecular forces*—are sufficiently large. The fact that gases exist and that they completely occupy the volume available to them is evidence that intermolecular forces in the gas phase must be vastly weaker than in liquids or solids.

The principal tenets of **kinetic molecular theory** are a summary of the preceding ideas.

1. Gases consist of molecules whose volumes are negligible compared with the volume occupied by the gas.
2. The molecules of a gas are in continual, random, and rapid motion.
3. The average kinetic energy of gas molecules is determined by the gas temperature.
4. Gas molecules collide with each other and with the walls of their container, but they do so without loss of energy.*

*Scientists call such collisions "perfectly elastic"; colliding billiard balls come close to having elastic collisions, while basketballs have inelastic collisions. In a perfectly elastic collision of a fast and a slow molecule, the fast one will slow down and the slow one will speed up, but the sum of the kinetic energies of the two molecules will be the same after the collision as before.

A gas that strictly obeys all these points is said to be an **ideal gas**. In practice, as you will see later, there is no such thing as an ideal gas, but some do come close to this behavior. Nonetheless, if you *assume* a gas is ideal, then it is possible to derive the ideal gas law from the mathematical expression of the ideas above. Our goal here is only to examine the connection in a qualitative way.

KINETIC MOLECULAR THEORY AND THE GAS LAWS

The gas laws, which come from experiment, can be explained by the kinetic molecular theory. The starting place is to write equations describing how pressure arises from collisions of gas molecules with the walls of the container holding the gas. The force developed by these collisions depends on the number of collisions and the average force per collision. When the temperature of a gas is increased, the average force of a collision on the walls of its container increases because the average kinetic energy of the molecules increases. Also, since the speed increases with temperature, there are more collisions per second. Thus, the collective force per square centimeter is greater, and the pressure increases. Mathematically, this means P is proportional to T when n and V are constant.

Increasing the number of molecules of a gas at a fixed temperature and volume does not change the average collision force, but it does increase the number of collisions occurring per second. Thus, the pressure increases, and so we can say that P is proportional to n when V and T are constant.

If the pressure is not allowed to increase when either the number of molecules of gas or the temperature is increased, the volume of the container (the area over which the collisions can take place) must increase. This is expressed by saying that V is proportional to nT when P is constant, a statement that is a combination of *Avogadro's law* and *Charles's law*.

Finally, if T is held constant, the average force of molecules of a given mass colliding with the container walls must be constant. However, if n is kept constant while the volume of the container is made smaller, the *number of collisions* with the container walls per second must increase. This must mean the pressure increases, and so P is proportional to $1/V$ when n and T are constant, as stated by *Boyle's law*.

DISTRIBUTION OF MOLECULAR SPEEDS

Theory and experiment show that if the number of gas molecules having various speeds is plotted versus the speed, the result is a curve such as those illustrated in Figure 12.14. Here you see that some molecules are fast (have a high kinetic energy), while others are slow (have a low kinetic energy). However, there is a most common speed, a speed at which the maximum in the distribution curve is obtained.

Curves of speed (or energy) versus number of molecules, such as those in Figure 12.14, are often called ***Boltzmann distribution curves****. They are named after Ludwig Boltzmann (1844–1906), an Austrian mathematician who did much work on the kinetic theory of gases. See Chapter 20.*

From Equation 12.16 you know that the average of the square of the molecular speed ($\overline{u^2}$) and the temperature are directly related, so the shape of the speed distribution curve must change with temperature. For example, the maximum must shift to a higher average speed at higher temperatures; the number of low-speed molecules decreases, while the number with higher speeds increases. Even though the curve for the higher

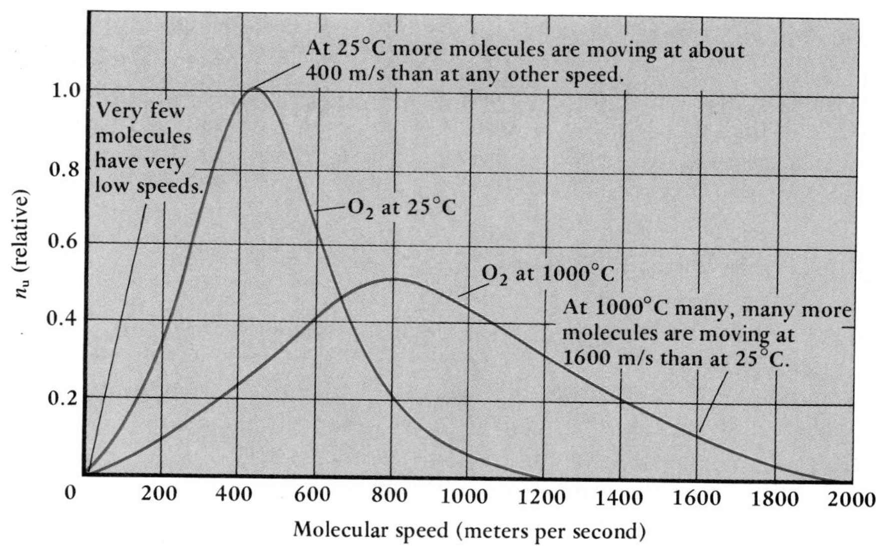

Figure 12.14 A plot of the relative number of molecules with a given speed versus that speed. The curve is called a Boltzmann distribution of molecular speeds. The blue curve shows the effect of a temperature increase on the distribution of speeds.

The distribution of speeds, energies, or other molecular properties, such as that illustrated by Figure 12.14, is an important concept in science. You will see the idea come up again in later chapters, particularly in Chapter 13 on the behavior of liquids and in Chapter 15 on kinetics.

temperature is "flatter" and broader than the one at a lower temperature, *the areas under the curves are the same because the number of molecules in the sample is fixed.**

One last point concerns the relation between molecular mass, average speed, and temperature. The average kinetic energy, $\overline{KE}$, is fixed by the temperature, so the product $\frac{1}{2}m\overline{u^2}$ (Equation 12.15) is also fixed. Since two gases with two different molecular masses must still have the same average kinetic energy at the same temperature, the heavier gas molecules must have a lower average speed (Figure 12.15). Equation 12.17 expresses this idea in quantitative form. Here the square root of the mean squared speed ($\sqrt{\overline{u^2}}$, called the **root-mean-square** or **rms speed**), the temperature (T, in kelvins), and the molar mass (M) are related.

Equation 12.17 is sometimes called "Maxwell's equation," after James Clerk Maxwell (1831–1879), a Scottish physicist who is considered by many to have been the greatest theoretical physicist of the 19th century.

$$\sqrt{\overline{u^2}} = \sqrt{\frac{3RT}{M}} \quad (12.17)$$

In this equation R is the familiar gas constant, but it must be expressed in units related to energy; that is, $\boldsymbol{R}$ **= 8.314510 J/K · mol.**†

EXAMPLE 12.15

MOLECULAR SPEED

Calculate the rms speed of oxygen molecules at 25 °C.

*If you have had some calculus, you know that the area under a distribution curve reflects the total number of objects on which the curve is based. For a given number of objects, the shape of the distribution curve may change, but not the area under the curve.

†The reason we can change R in units of L · atm/K · mol to units of J/K · mol is that L · atm expresses the relation (volume)(force/area). This reduces to (force × distance) (because volume is given in cm^3, for example, and area in cm^2), which is energy.

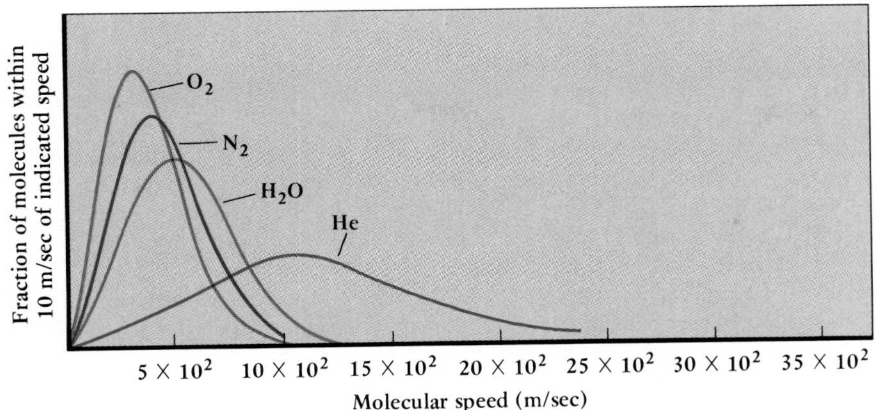

Figure 12.15 Effect of molecular mass on the Boltzmann distribution curve at a given temperature. Notice the similarity of the effect of increasing mass in this figure to the effect of decreasing temperature in Figure 12.14.

Solution We must use Equation 12.17 with M in units of kg/mol. The reason for this is that R is in units of J/K · mol, and 1 J = 1 kg · m²/s². Thus, the molar mass of O_2 is 32.0×10^{-3} kg/mol.

$$\sqrt{\overline{u^2}} = \sqrt{\frac{3RT}{M}} = \sqrt{\frac{3(8.314510 \text{ J/K} \cdot \text{mol})(298 \text{ K})}{32.0 \times 10^{-3} \text{ kg/mol}}}$$
$$= \sqrt{2.32 \times 10^5 \text{ J/kg}}$$

To obtain the answer in m/s, we use the relation 1 J = 1 kg · m²/s². This means we have

$$\sqrt{\overline{u^2}} = \sqrt{2.32 \times 10^5 \text{ kg} \cdot \text{m}^2/\text{kg} \cdot \text{s}^2}$$
$$= \sqrt{2.32 \times 10^5 \text{ m}^2/\text{s}^2}$$
$$= 482 \text{ m/s}$$

This speed is equivalent to about 1100 miles per hour!

EXERCISE 12.12 Molecular Speeds
Calculate the rms speeds of helium atoms and N_2 molecules at 25 °C.

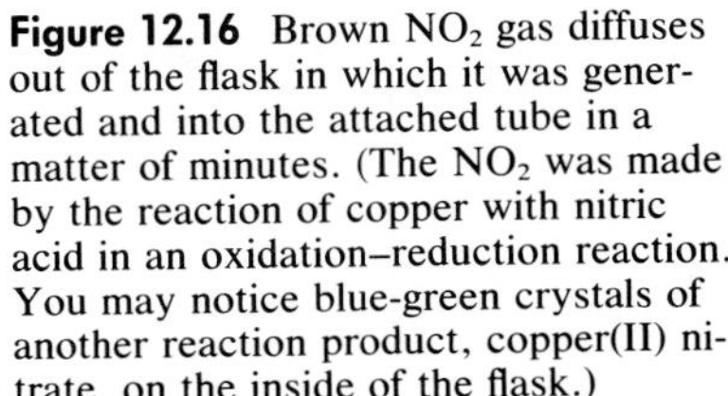
Figure 12.16 Brown NO_2 gas diffuses out of the flask in which it was generated and into the attached tube in a matter of minutes. (The NO_2 was made by the reaction of copper with nitric acid in an oxidation–reduction reaction. You may notice blue-green crystals of another reaction product, copper(II) nitrate, on the inside of the flask.)

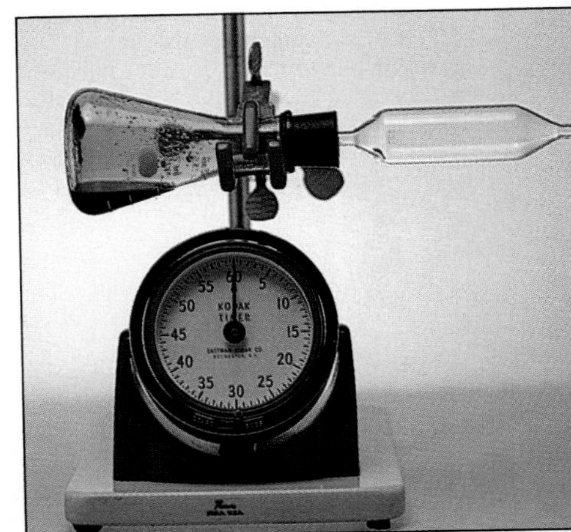

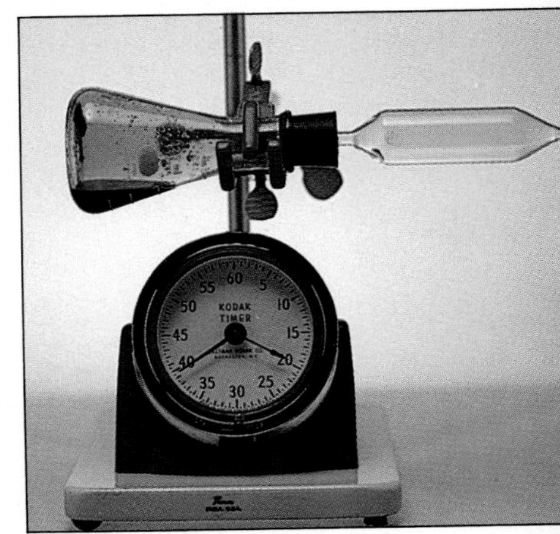

12.7 GRAHAM'S LAW OF DIFFUSION AND EFFUSION

Gaseous diffusion is the gradual mixing of the molecules of two or more gases owing to their molecular motions, and Figures 12.13 and 12.16 are excellent illustrations of this property. Effusion is closely related to diffusion. In Figure 12.17 gas molecules are held in one compartment, but they gradually escape or effuse through a small opening in the barrier into an *empty* compartment, that is, into a vacuum.

Thomas Graham (1805–1869), a Scottish chemist, studied the effusion of gases and found experimentally that the rates of effusion of two gases are inversely proportional to the square roots of their molar masses at the same temperature and pressure.

$$\frac{\text{Rate of effusion of gas 1}}{\text{Rate of effusion of gas 2}} = \sqrt{\frac{M \text{ of gas 2}}{M \text{ of gas 1}}} \quad (12.18)$$

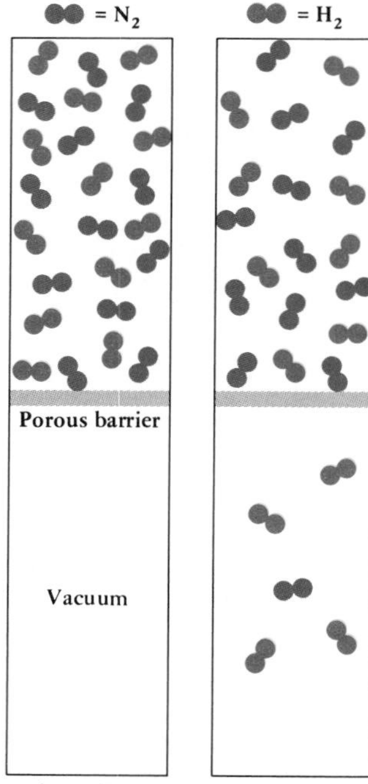

Figure 12.17 An illustration of the effusion of gas molecules through the pores of a membrane or other porous barrier. Lighter molecules with greater speed move through the barrier more rapidly than heavier, slower molecules at the same temperature.

Graham's Law and Balloons

Graham's law allows us to explain the earlier observation concerning the balloon used by Charles in 1783 (see page 461). In the brief biography of Charles, we mentioned that he filled the balloon with H_2 gas, but he had to go to extraordinary lengths to keep the gas in the balloon. Unlike the first hot air balloons that were made of paper, Charles had to use silk and coat it with rubber. The reason for this is simple: the lightweight H_2 molecules have a high rms speed so H_2 molecules would effuse rapidly through paper, a more porous material than silk coated with rubber.

For much the same reason that Charles used special materials in his hydrogen-filled balloon, you should be cautious when buying a helium-filled balloon at a carnival. Lightweight He atoms have a high rms speed so He atoms can effuse rapidly through the wall of a rubber balloon, and the balloon deflates. The newer balloons made of Mylar film, however, keep He gas enclosed much longer because the film has much smaller pores through which the He atoms escape more slowly.

Balloons made of Mylar film are better at containing helium than conventional rubber balloons.

This is now known as Graham's law, an equation explained readily by Equation 12.17. Clearly, the *rate of effusion*, the number of molecules moving from one place to another in a given amount of time, will depend on the root-mean-square speed of the molecules. Therefore, the ratio of these speeds is the same as the ratio of the effusion rates.

$$\frac{\text{Rate of effusion of gas 1}}{\text{Rate of effusion of gas 2}} = \frac{\sqrt{\overline{u^2}}\text{ of gas 1}}{\sqrt{\overline{u^2}}\text{ of gas 2}} = \sqrt{\frac{3RT/(M\text{ of gas 1})}{3RT/(M\text{ of gas 2})}} \qquad (12.19)$$

Canceling like terms, we have the simple expression developed by Graham (Equation 12.18).

In Figure 12.17 you see that more H_2 molecules effuse in a given time than N_2 molecules. Graham's law now allows a quantitative comparison of their relative rates.

$$\frac{\text{Rate of effusion of }H_2}{\text{Rate of effusion of }N_2} = \sqrt{\frac{M\text{ of }N_2}{M\text{ of }H_2}} = \sqrt{\frac{28.0\text{ g/mol}}{2.02\text{ g/mol}}} = \frac{3.72}{1}$$

This calculation tells you that H_2 molecules will effuse through the barrier 3.72 times faster than N_2 molecules.

EXAMPLE 12.16

GRAHAM'S LAW OF EFFUSION

Tetrafluoroethylene, C_2F_4, effuses through a barrier at the rate of 4.6×10^{-6} moles per hour. An unknown gas, consisting only of nitrogen, oxygen, and fluorine atoms, effuses at the rate of 6.5×10^{-6} moles per hour under the same conditions.

Graham's Law: Isotope Separation

In the race to construct atomic explosives during World War II, considerable quantities of uranium enriched in the ^{235}U isotope were needed. One enrichment technique that was developed uses gas diffusion, a method still in use today (see Figure below). To separate the ^{235}U isotope from the more abundant ^{238}U isotope, all the uranium was converted to UF_6, one of the few metal compounds that can be a gas at a reasonable temperature. (The temperature at which UF_6 boils is approximately 56 °C). Just as N_2 and H_2 can be separated by effusion in Figure 12.17, the lighter molecules of $^{235}UF_6$ can be separated by diffusion from the heavier $^{238}UF_6$ molecules.

Graham's law allows us to answer some intriguing questions about isotope separation by gaseous diffusion. For example, what is the maximum separation of isotopes that can be achieved in a single diffusion step? The solution is to use Graham's law to calculate the ratio of diffusion rates of the two isotopes.

$$\frac{\text{Rate of } ^{235}UF_6}{\text{Rate of } ^{238}UF_6} = \sqrt{\frac{352.04}{349.03}} = 1.0043 = \text{enrichment factor}$$

Since the lighter $^{235}UF_6$ is the faster molecule, the proportion of $^{235}UF_6$ molecules passing through the barrier will exceed that in the original mixture. The calculation shows that the abundance of $^{235}UF_6$ molecules in the gas will increase by a factor of 1.0043 during each passage through the porous barrier. Clearly, many diffusion steps are required to give a reasonable degree of isotope separation.

The gaseous diffusion plant for separation of uranium isotopes at Oak Ridge National Laboratory. Interior view of process equipment shows arrangement of piping, compressors, motors, and large tank-shaped "diffusers." (Notice the worker standing in front of the tank marked "STG-2.") Uranium in the form of a gaseous compound (UF_6) is pumped into the diffuser vessels where, at each stage, the uranium is slightly enriched in the fissionable ^{235}U isotope. For use as fuel in nuclear power plants, the uranium is enriched to approximately 3% ^{235}U (compared with 0.7% for uranium found in the natural state). (Oak Ridge National Laboratory)

What is the molar mass of the unknown gas? Suggest a formula for the unknown gas.

Solution Graham's law tells us that a lighter molecule effuses more rapidly than a heavier one. Since the unknown gas effuses more rapidly than C_2F_4 (M = 100. g/mol), the unknown must have a molar mass less than 100.

$$\frac{\text{Rate of unknown}}{\text{Rate of } C_2F_4} = \sqrt{\frac{M \text{ of } C_2F_4}{M \text{ of unknown}}}$$

$$\frac{6.5 \times 10^{-6}\ \text{mol/hr}}{4.6 \times 10^{-6}\ \text{mol/hr}} = \sqrt{\frac{100.\ \text{g/mol}}{M}}$$

$$1.4 = \sqrt{\frac{100.\ \text{g/mol}}{M}}$$

To solve for the unknown molar mass, square both sides of the equation.

$$2.0 = \frac{100.\ \text{g/mol}}{M}$$

and then rearrange to obtain M.

$$M = 50.\ \text{g/mol}$$

This molecule is known to contain only N, O, and F. If you add the atomic weights of these elements, you obtain 49 g/mol. Therefore, a likely formula for the unknown is simply NOF, a compound called nitrosyl fluoride.

EXERCISE 12.13 Graham's Law

A pure sample of methane, CH_4, is found to effuse through a porous barrier in 1.50 minutes. Under the same conditions, an equal number of molecules of an unknown gas effuses through the barrier in 4.73 minutes. What is the molar mass of the unknown gas?

Figure 12.18 When bromine (Br_2) vapor is cooled, the molecules previously in the vapor phase condense to form droplets of liquid. Molecules in the gas phase do not have zero volume; as illustrated they form a liquid with an observable volume when condensed.

12.8 NONIDEAL GASES

If you are working with a gas at approximately room temperature and at pressures of 1 atmosphere or less, the "ideal" gas law is remarkably successful in relating the quantity of gas and its pressure, volume, and temperature. However, at higher pressures or lower temperatures, serious deviations from the ideal gas law are often observed. The origin of these deviations is easily understood in terms of the breakdown of the assumptions of the kinetic theory of gases.

The most easily visualized problem with the ideal gas law comes from the assumption of the kinetic molecular theory that gas molecules occupy a negligibly small volume relative to the container volume. This clearly cannot be the case, since we know from experiment that even a very small atom such as helium has a measurable size. If we have 1 mole of helium atoms, the ideal gas law tells us the gas should occupy approximately 22.4 L at STP. However, taking 31 pm as the radius of helium, we can calculate that, if we pack together a mole of these atoms, the mole of atoms should have a volume of roughly 0.08 mL (Figure 12.18), only a tiny fraction of the total volume.

The fact that gas molecules occupy some volume, however tiny, means that the molecules of the gas do not really have the space to roam about that you think they have. Therefore, the experimentally observed container volume should be corrected by subtracting a factor that takes into account the actual volume occupied by the gas molecules or atoms. This would give the free volume available to the gaseous molecules.

Another assumption of the kinetic molecular theory was that collisions between molecules are elastic, an assumption that implies that the atoms or molecules of the gas never stick to one another by some type of forces. This is clearly nonsense as well. All gases can be liquefied (Figure 12.19) at some temperature—although some require a very low temperature—and the only way this can happen is if the atoms or molecules cling together because of the forces between the molecules, that is, **intermolecular forces**. Even at temperatures well above the liquefaction temperature of the gas these forces can be appreciable, and they can lead to measurable deviations from ideal gas behavior. When molecules stick together they are replaced by fewer, heavier particles, and this decreases the pressure the gas exerts on the walls. The net result is that the observed pressure can be lower than that expected from the ideal gas law. Although this effect can be observed in gases at pressures around 1 atmosphere, it becomes particularly pronounced when the pressure is high.

Figure 12.19 All gases can be liquefied at some temperature, even the nitrogen in the air around us. The boiling point of liquid nitrogen is −196 °C.

The Dutch physicist Johannes van der Waals (1837–1923) studied deviations from the ideal gas law equation, and he developed another equation to describe gases at high pressure or gases in which effects of intermolecular forces can be appreciable. Under such circumstances, the pressure calculated using the so-called van der Waals equation corresponds more closely to reality than that calculated from the ideal gas law.

The van der Waals Equation

$$\left[P + a\left(\frac{n}{V}\right)^2\right][V - bn] = nRT \qquad (12.20)$$

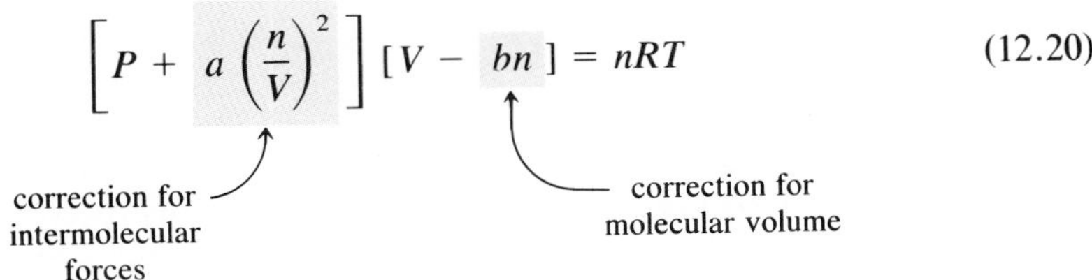

a and *b* = van der Waals constants

Although this equation might seem complicated at first glance, the terms in square brackets are simply analogs of those of the ideal gas law, each corrected for the effects discussed above.

The pressure correction term, $a(n/V)^2$, accounts for intermolecular forces. The term (n/V) is the concentration of the gas; the greater it is, the closer the molecules become. This in turn causes the attractive forces between the molecules, and hence the pressure correction term, to become more important. Since the observed gas pressure (P) is lower than the ideal pressure (as calculated from $PV = nRT$) owing to intermolecular forces, this means that $a(n/V)^2$ is *added* to the observed pressure to give

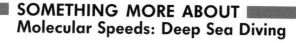

SOMETHING MORE ABOUT
Molecular Speeds: Deep Sea Diving

One application of the gas laws comes up in deep sea diving. To explore the floor of the oceans, scientists have built diving ships to operate at great depths. These ships are filled with an atmosphere of oxygen and helium at high pressures, but oxygen has a lower mole fraction than in normal sea-level air. It has been found that our bodies function best when the partial pressure of oxygen is about 0.21 atm. If the pressure of gas you breathe is 2 atmospheres, for example, and the composition of the gas is that of normal air, this would make the partial pressure of oxygen about 0.4 atm. Such high oxygen partial pressures are toxic, so the gas that divers breathe must have a lower mole fraction of O_2.

Not only is the fraction of O_2 reduced in diving gases, but the nitrogen of normal air is replaced by helium. Nitrogen becomes more soluble in blood and body fluids at high pressures and leads to *nitrogen narcosis*, a condition similar to alcohol intoxication. Helium is less soluble in blood and is thus more suitable to dilute the O_2. However, while He solves one problem, it creates another. Divers in such an atmosphere feel decidedly chilly, even though the temperature is a normally comfortable 70 °C. This can be explained by helium's greater thermal conductivity (about 6 times greater than that of N_2), which can be interpreted using concepts from the kinetic molecular theory. The thermal conductivity of a gas depends on, among other things, the average speed of the gas and the distance a molecule or atom must travel before it collides with another molecule or atom. From Exercise 12.12 we know that He atoms have an average speed about 2.6 times greater than that of N_2 molecules at the same temperature. Further, since He atoms are so much smaller than N_2 molecules, He atoms must travel a greater distance between collisions.

There is an odd side effect of using helium in the atmosphere breathed by divers. When you talk after breathing helium, you sound like Donald Duck! The reason for this is that your vocal cords vibrate faster in an atmosphere less dense than air, and the pitch of your voice is raised.

(Photo © 1990 The Cousteau Society, a member supported, nonprofit environmental organization)

Table 12.2 van der Waals Constants

Substance	a (atm · L^2/mol^2)	b (L/mol)
He	0.034	0.0237
Ar	1.34	0.0322
H_2	0.244	0.0266
N_2	1.39	0.0391
O_2	1.36	0.0318
Cl_2	6.49	0.0562
CO_2	3.59	0.0427
CH_4	2.25	0.0428
H_2O	5.46	0.0305

a pressure that satisfies $PV = nRT$. The constant a, which is determined experimentally, typically has values in the range from 0.01 to 10 atm · L^2/mol^2.

The bn term in van der Waals equation corrects the container volume to a smaller value, the free volume available to the gas molecules. Here n is the number of moles of gas and b is an experimental quantity that gives the correction (per mole) for the molecular volume. Typical values of b range from 0.01 to 0.1 L/mol, roughly increasing with increasing molecular size (Table 12.2).

As an example of the importance of these corrections, consider a sample of 8.00 moles of chlorine gas, Cl_2, in a 4.00-L tank at 27.0 °C. The ideal gas law would lead you to expect a pressure of 49.2 atm. The real pressure, obtained from the van der Waals equation, is only 29.5 atm, 20 atmospheres less than the ideal pressure!

EXERCISE 12.14 van der Waals Equation

Using both the ideal gas law and van der Waals equation, calculate the pressure expected for 10.0 moles of helium gas in a 1.00-L container at 25 °C.

SUMMARY

To describe the state of a gas one must specify the quantity (n) in moles, the temperature (T) in kelvins, the volume (V) in liters, and the pressure (P) in atmospheres. Pressure measurements are described in Section 12.1, as are other units of pressure, especially mmHg and pascals. One standard atmosphere is equivalent to 760. mmHg. Conditions of **STP, standard temperature and pressure**, are 273 K and 1 atm.

The experimental basis of the gas laws is outlined in Section 12.2. Boyle's law specifies that the pressure of a gas is inversely proportional to its volume ($P \propto 1/V$) at constant T and n. Charles's law tells us that gas volume is directly proportional to temperature ($V \propto T$) at constant n and P. Finally, Avogadro's law specifies that gas volume is directly proportional to its quantity ($V \propto n$) at constant T and P. Gay-Lussac's law of combining volumes states that reacting gases combine in volumes that are ratios of small whole numbers.

If the three gas laws are combined into one statement, the **ideal gas law ($PV = nRT$)** results (Section 12.3). Since PV/nT is always equal to R, the **general gas law** (or **combined gas law**) can be derived from the ideal gas law.

$$\frac{P_1V_1}{n_1T_1} = \frac{P_2V_2}{n_2T_2}$$

It is usually applied to find out what happens when a given quantity of gas (so $n_1 = n_2$) undergoes a change in P, V, and T conditions.

Dalton's law of partial pressures specifies that the total pressure of a mixture of gases is the sum of the **partial pressures** of the individual gases in the mixture (Section 12.5).

$$P_{\text{total}} = P_A + P_B + P_C + \cdots$$

The partial pressure of a gas in a mixture (P_A) is given by its mole fraction (X_A) times the total pressure of the mixture (P_{total}).

$$P_A = X_A P_{\text{total}}$$

The mole fraction of a component (A) of a mixture is defined as the number of moles of A divided by the total moles of all components.

The gas laws can be understood in terms of the **kinetic molecular theory**, a theory of gas behavior at the molecular level (Section 12.6). One important tenet of the theory is that the average kinetic energy of gas molecules ($\overline{KE}$) is proportional to the temperature of the gas.

$$\overline{KE} = \tfrac{1}{2}m\overline{u^2} \propto T$$

Here $\overline{u^2}$ is the average of the squares of the molecular speeds. Since $\overline{KE}$ is determined by temperature, heavier molecules (large m) must move with a slower average speed ($\overline{u}$) than lighter molecules (small m) at a given temperature.

Graham's law of effusion states that the rates of effusion of two gases (r_1 and r_2) are inversely proportional to the square roots of their molar masses (M_1 and M_2) at the same temperature and pressure.

$$\frac{r_1}{r_2} = \sqrt{\frac{M_2}{M_1}}$$

Kinetic molecular theory assumes, among other things, that gas molecules have no volume and that they do not interact with one another by **intermolecular forces**. These assumptions lead to a simple model of *ideal gas* behavior as expressed by $PV = nRT$. However, neither assumption is completely correct because a *real gas* has a more complex behavior than an ideal gas. One description of the behavior of real gases is found in the **van der Waals equation**.

STUDY QUESTIONS

REVIEW QUESTIONS

1. Name the three gas laws that interrelate P, V, and T. Explain the relationships in words and in equations.
2. What conditions are represented by STP? What is the volume of a mole of ideal gas under these conditions?
3. Show how to calculate the molar mass of a gas from P, V, and T measurements and other information.

4. Explain how to determine the density of a gas from P, V, and T data and other information.
5. State Avogadro's law. Relate your discussion to the formation of ammonia from its elements [$3\ H_2(g) + N_2(g) \rightarrow 2\ NH_3(g)$]. For example, if 2 moles of H_2 are used at STP, how many liters of N_2 are required at STP and how many liters of NH_3 are produced at STP?
6. State Dalton's law. If the air you breathe is 78% N_2 and 22% O_2 (on a mole basis), what is the mole fraction of O_2? What is the partial pressure of O_2 when the atmospheric pressure is 745 mmHg?
7. What are the basic assumptions of the kinetic molecular theory? Which of these assumptions are most nearly correct and which are violated to some extent by a real gas?
8. Explain Boyle's law on the basis of kinetic molecular theory.
9. In van der Waals equation, what properties of a real gas are accounted for by the constants a and b?
10. State Graham's law in words and in equation form. If gas A effuses four times more rapidly than gas B at 25 °C, does this ratio of rates of effusion increase, decrease, or remain the same as the temperature of the gas increases?

UNITS OF MEASUREMENT

11. Gas pressure can be expressed in units of mmHg, torr, atmospheres, and pascals (although mmHg and atm are used exclusively in this book). Do the following unit conversions.
    (a) 740 mmHg to atm
    (b) 1.25 atm to mmHg
    (c) 740 mmHg to kPa
    (d) 0.50 atm to kPa
    (e) 542 mmHg to torr
12. Gas pressure can be expressed in units of mmHg, torr, atmospheres, and pascals (although mmHg and atm are used exclusively in this book). Do the following unit conversions.
    (a) 781 mmHg to atm
    (b) 356 mmHg to kPa
    (c) 35 torr to atm
    (d) 210 kPa to mmHg
    (e) 0.015 kPa to atm
    (f) 95.0 kPa to mmHg

PRESSURE MEASUREMENT

13. Refer to page 457 and answer the following question. If the mercury levels in a U-tube manometer are as illustrated below, express the gas pressure in mmHg, atm, torr, and kPa.

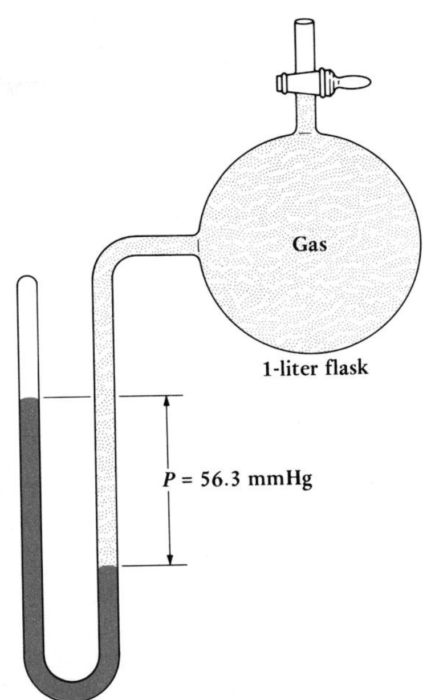

14. Refer to page 457 and answer the following question. If the pressure of a gas is 95.0 kPa, what is the difference (in mmHg) in the mercury levels in the U-tube manometer?

THE GAS LAWS

15. A sample of HCl gas is placed in a 256-mL flask where it exerts a pressure of 67.5 mmHg. What is the pressure of this gas sample if it is transferred to a 135-mL flask at the same temperature?
16. A sample of HCl has a pressure of 67.5 mmHg in a 256-mL flask. If the sample is transferred to a new flask where it exerts a pressure of 23.6 mmHg at the same temperature, what is the volume of this new flask?
17. A 25.0-mL sample of gas is enclosed in a gas-tight syringe (Figure 12.2) at 22 °C. If the syringe is immersed in an ice bath (0 °C), what is the new gas volume, assuming that the pressure is held constant?
18. Some propane, the fuel used in backyard grills, is placed in a 3.50-L container at 25 °C; its pressure is 735 mmHg, the pressure of the atmosphere. If you transfer the gas to a 15.0-L container, also at 25 °C, what is the pressure of the gas in the larger container?
19. You have a sample of CO_2 gas in a flask with a volume of 256 mL. At 25.5 °C the pressure of the gas is 135 mmHg. If you decrease the temperature to 18.6 °C, what is the gas pressure at the lower temperature?

20. A bicycle tire is inflated to a pressure of 55 lb/in^2 at 15 °C. If the tire is heated to 35 °C, what is the pressure in the tire? (For simplicity, assume the tire volume cannot change.) (1 atm = 14.7 lb/in.^{2})

21. Water can be made by combining gaseous O_2 and H_2. If you begin with 1.0 L of H_2(g) at 380 mmHg and 25 °C, how many liters of O_2(g) would you need for complete reaction if the O_2 gas is also measured at 380 mmHg and 25 °C?

22. Gaseous silane, SiH_4, ignites spontaneously in air according to the reaction

$$SiH_4(g) + 2\,O_2(g) \longrightarrow SiO_2(s) + 2\,H_2O(g)$$

If 5.2 L of SiH_4 are treated with O_2, how many liters of O_2 are required for complete reaction? How many liters of H_2O vapor are produced? Assume all gases are measured at the same temperature and pressure.

THE GENERAL GAS LAW

23. Assume you place some octane in the cylinder of an automobile engine. The cylinder has a volume of 250. cm^3 and the pressure of gaseous octane is 3.50 atm in the hot engine (250 °C). What is the pressure in the cylinder (P_{new}) if you lower the temperature of the automobile engine to 125 °C and change the volume of the cylinder to 500. cm^3?
(a) P_{new} = (3.50 atm)(500./250.)(250/125)
(b) P_{new} = (3.50 atm)(500./250.)(523/398)
(c) P_{new} = (3.50 atm)(250./500.)(523/398)
(d) P_{new} = (3.50 atm)(250./500.)(398/523)

24. A sample of gas occupies 754 mL at 22 °C and a pressure of 165 mmHg. What is its volume if the temperature is raised to 42 °C and the pressure is raised to 265 mmHg?

25. A sample of chlorine gas is placed in a 2.56-L flask at a temperature of 27.5 °C; the gas pressure is found to be 48.5 mmHg. If the chlorine is moved to a new flask with a volume of 345 mL, and if the temperature has changed to 22.6 °C, what is the pressure of the gas in the smaller flask?

26. You have a sample of CO_2 in a flask (A) with a volume of 265 mL. At 22.5 °C, the pressure of the gas is 136.5 mmHg. To find the volume of another flask (B), you move the CO_2 to that flask and find that its pressure is now 94.3 mmHg at 24.5 °C. What is the volume of flask B?

27. Assume that one of the cylinders of an automobile engine has a volume of 400. cm^3. The engine takes in air at a pressure of 1.00 atm and a temperature of 27 °C and compresses it to a volume of 50.0 cm^3 at 77 °C. What is the final pressure of the gas in the cylinder? (The ratio of before and after volumes, in this case 400:50, or 8:1, is called the compression ratio.)

28. Assume that a helium-filled balloon needs to displace at least 1.00 × 10^5 L of air. You fill the balloon with helium to a volume of 1.05 × 10^5 L on the ground where the pressure is 745 mmHg and the temperature is 20.0 °C. When the balloon ascends to a height of 2 miles where the pressure is only 600. mmHg and the temperature is −33 °C, does the balloon still displace the required 1.00 × 10^5 L of air?

THE IDEAL GAS LAW

29. A 1.00-g sample of water is allowed to vaporize completely in a 10.0-L container. What is the pressure of the water vapor at a temperature of 150. °C?

30. 30.0 kg of helium are placed in a balloon. What is the volume of the balloon if the final pressure is 1.20 atm and the temperature is 22 °C?

31. To find the volume of a flask, it was first evacuated so that it contained no gas at all. Next, 4.4 g of CO_2 were introduced into the flask. On warming to 27 °C, the gas exerted a pressure of 730 mmHg. What is the volume of the flask?

32. You have placed 1.50 g of hexane, C_6H_{14}, in the cylinder of an automobile engine. What is the pressure of the hexane vapor if the cylinder has a volume of 250. cm^3 and the temperature is 250. °C?

33. If you have a 150.-L tank of gaseous CO and the gas exerts a pressure of 41.8 mmHg at 25 °C, how many moles of CO are there in the tank?

34. How many grams of helium are required to fill a 5.0-L balloon to a pressure of 1.1 atm at 25 °C?

35. 1.00 g of a gaseous compound occupies 0.820 L at 1.00 atm and 3.0 °C. Which of the following is the correct formula for the compound: (a) B_4H_{10}, (b) B_2H_6, (c) B_5H_9?

36. A 0.982-g sample of an unknown gas exerts a pressure of 700. mmHg in a 450.-mL container at 23 °C. What is the molar mass of the gas?

37. A hydrocarbon with a general formula of C_xH_y is 92.26% carbon. Experiment shows that 0.293 g of the hydrocarbon fills a 185-mL flask at 23 °C with a pressure of 374 mmHg. What is the molecular formula of the compound?

38. Analysis of a chlorofluorocarbon, whose general formula is CCl_xF_y, shows it to be 11.79% C and 69.57% Cl. In another experiment you find that 0.107 g of the compound fills a 458-mL flask at 25 °C with a pressure of 21.3 mmHg. What is the molecular formula of the compound?

39. A boron hydride, with the general formula B_xH_y, is 18.9% hydrogen. If 0.0872 g of the compound exerts

a pressure of 191 mmHg at 24 °C in a 159-mL flask, what is the formula of the boron hydride?

40. There are five compounds in the family of sulfur-fluorine compounds with the general formula S_xF_y. One of these compounds is 25.23% S. If you place 0.0955 g of the compound in an 89-mL flask at 45 °C, the pressure of the gas is 83.8 mmHg. What is the molecular formula of S_xF_y?

41. Forty miles above the earth's surface the temperature is 250 K and the pressure is only 0.20 mmHg. What is the density of air (M = 29.0 g/mol) at this altitude?

42. A common liquid ether [diethyl ether, $(C_2H_5)_2O$] vaporizes easily at room temperature. If the vapor has a pressure of 233 mmHg in a flask at 25 °C, what is the density of the vapor?

43. Chloroform is a common liquid used in the laboratory. It vaporizes readily. If the pressure of chloroform is 195 mmHg in a flask at 25.0 °C and the density of the vapor is 1.25 g/L, what is the molar mass of chloroform?

44. A newly discovered gas has a density of 2.39 g/L at 23.0 °C and 715 mmHg. What is the molar mass of the gas?

45. If 12.0 g of O_2 are required to inflate a balloon to a certain size at 27 °C, what mass of O_2 is required to inflate it to the same size (and pressure) at 127 °C?

46. You have two gas-filled balloons, one containing He and the other H_2. The H_2 balloon is twice the size of the He balloon. The pressure of gas in the H_2 balloon is 1 atm while that in the He balloon is 2 atm. The H_2 balloon is outside in the snow (30 °F) while the He balloon is inside a warm building (70 °F).
(a) Which balloon contains the greater number of molecules?
(b) Which balloon contains the greater mass of gas?

47. The pressure gauges on two cylinders of O_2 of equal volume register the same pressure. However, one cylinder (A) is inside the chemistry laboratory (room temperature = 23 °C) and the other cylinder (B) is outside in the snow (temperature = −10 °C). Which cylinder, A or B, contains the greater mass of O_2?

48. If equal masses of O_2 and N_2 are placed in separate containers of equal volume at the same temperature, which of the following statements is true? If false, tell why it is false.
(a) Both flasks contain the same number of molecules.
(b) The pressure in the N_2-containing flask is greater than that in the flask containing O_2.
(c) There are more molecules in the flask containing O_2 than in the one containing N_2.

49. Two identical flasks are filled with different gases; both flasks are at the same temperature, and both gases exert the same pressure. The mass of gas A is 0.34 g, while that of gas B is 0.48 g. It is known that gas B is ozone, O_3. Which of the following is gas A: (a) O_2, (b) SO_2, or (c) H_2S?

50. Suppose you have two pressure-proof steel cylinders of equal volume, one containing CO (A) and the other acetylene, C_2H_2 (B).
(a) If you have 1 kg of each compound, in which cylinder is the pressure greater at 25 °C?
(b) Now suppose cylinder A has a total pressure of 10 atm at 25 °C, while B has a pressure of 9 atm at 0 °C. Which cylinder contains the greater number of molecules?

GAS LAWS AND STOICHIOMETRY

51. Hydrogen can be made in the "water gas reaction."

$$C(s) + H_2O(g) \longrightarrow H_2(g) + CO(g)$$

If you begin with 250. L of gaseous water at 120. °C and 2.0 atm pressure, how many grams of H_2 can be made?

52. If the boron hydride B_4H_{10} is treated with pure oxygen, it burns to give B_2O_3 and H_2O.

$$2\ B_4H_{10}(g) + 11\ O_2(g) \longrightarrow 4\ B_2O_3(s) + 10\ H_2O(g)$$

If a 0.050-g sample of the boron hydride burns completely in O_2, what will be the pressure of gaseous water in a 4.25-L flask at 30.0 °C?

53. Hydrazine reacts with O_2 according to the equation

$$N_2H_4(g) + O_2(g) \longrightarrow N_2(g) + 2\ H_2O(g)$$

Assume the O_2 to burn the hydrazine is in a 450-L tank at 26 °C. If you wish to burn completely a 10.-kg sample of hydrazine, to what pressure should you fill the O_2 tank in order to have sufficient oxygen?

54. Butane can be used as a fuel in an automobile engine. It burns in O_2 according to the equation

$$2\ C_4H_{10}(g) + 13\ O_2(g) \longrightarrow 8\ CO_2(g) + 10\ H_2O(g)$$

If one cylinder of the engine is filled with butane to a pressure of 1.5 atm at 500. °C, it requires 2.45 g of O_2 for complete combustion. How many grams of O_2 would be required if the cylinder were filled with butane to a total pressure of 2.0 atm at 500. °C?

55. If 1.0×10^3 g of uranium metal is converted to gaseous UF_6, what pressure of UF_6 would be observed at 62 °C in a chamber that has a volume of 3.0×10^2 L?

56. $Ni(CO)_4$ can be made by the reaction of finely divided nickel with gaseous CO. If you have CO in a 1.50-L flask at a pressure of 418 mmHg at 25.0 °C, what is the maximum number of grams of $Ni(CO)_4$ that can be made?

57. To find the formula of a transition metal carbonyl, one of a family of compounds having the general formula $M_x(CO)_y$, you can heat the solid compound in a vacuum to produce solid metal and CO gas. You heat 0.112 g of $Cr_x(CO)_y$

$$Cr_x(CO)_y(s) \longrightarrow x\ Cr(s) + y\ CO(g)$$

and find that the CO evolved has a pressure of 369 mmHg in a 155-mL flask at 27 °C. What is the empirical formula of $Cr_x(CO)_y$?

58. Metal carbonates are decomposed to the metal oxide and CO_2 on heating according to the following equation.

$$M_x(CO_3)_y(s) \longrightarrow M_xO_y(s) + y\ CO_2(g)$$

You heat 0.158 g of a white, solid carbonate of a Group 2A metal and find that the evolved CO_2 has a pressure of 69.8 mmHg in a 285-mL flask at 25 °C. What is M?

59. Methylaluminum, $Al(CH_3)_x$, is a chemical widely used in industry. If you treat it with aqueous acid, the following reaction occurs:

$$Al(CH_3)_x + x\ H_3O^+(aq) \longrightarrow Al^{3+}(aq) + x\ CH_4(g) + x\ H_2O(\ell)$$

You know that x is either 1 or 3. To find the value of x, you take 0.0338 g of $Al(CH_3)_x$ and treat it with acid to give CH_4 with a pressure of 208 mmHg in a 125-mL flask at 24 °C. What is x?

60. The elements of Group 1B form so-called organometallic compounds such as $CuCH_3$. Assume you have isolated a compound with the formula $Au(CH_3)_xBr$. To find the value of x, you treat the compound with aqueous acid, which causes the CH_3 groups to be evolved as methane, CH_4.

$$Au(CH_3)_xBr + x\ H_3O^+(aq) \longrightarrow x\ CH_4(g) + Au^{n+}(aq) + Br^-(aq) + x\ H_2O(\ell)$$

What is x, if 0.307 g of $Au(CH_3)_xBr$ evolves CH_4 with a pressure of 148.8 mmHg in a 250.-mL flask at 25 °C?

61. Disulfur decafluoride, S_2F_{10}, can be made by shining light on a mixture of $SClF_5$ and H_2.

$$2\ SClF_5(g) + H_2(g) \longrightarrow S_2F_{10}(g) + 2\ HCl(g)$$

If you place 0.265 g of $SClF_5(g)$ in a 275-mL flask containing H_2 with a pressure of 95.1 mmHg, how many grams of $S_2F_{10}(g)$ are expected? What is the pressure of $S_2F_{10}(g)$ in the 275-mL flask at 45 °C?

Silicone caulk and Silly Putty.

62. Silicones are polymeric substances that are used as lubricants, as anti-stick agents, and in waterproof caulk (see photo). The starting place in making them is the following reaction, which is carried out at high temperature (say 350 °C) and in the presence of a catalyst.

$$Si(s) + 2\ CH_3Cl(g) \longrightarrow (CH_3)_2SiCl_2(g)$$

If you place 1.95 g of solid silicon in a 6.56-L flask that contains CH_3Cl with a pressure of 485 mmHg at 25 °C, how many grams of $(CH_3)_2SiCl_2(g)$ can be formed? What pressure of $(CH_3)_2SiCl_2(g)$ would you expect in this same flask at 95 °C on completion of the reaction?

GAS MIXTURES

63. What is the pressure in atmospheres of a gas mixture that contains 1.0 g of H_2 and 8.0 g of Ar in a 3.0-L container at 27 °C?

64. You have a 550.-mL tank of gas with a pressure of 1.56 atm at 24 °C. You thought the gas was pure carbon monoxide gas, CO, but you later found it was contaminated by small quantities of gaseous CO_2 and O_2. Analysis shows that the tank pressure is only 1.34 atm (at 24 °C) if the CO_2 is removed and that 0.0870 g of O_2 can be removed chemically. What were the masses of CO and CO_2 in the tank, and what were the partial pressures of the three gases in the 550.-mL tank at 24 °C?

65. A cyclopropane–oxygen mixture (C_3H_6 + O_2) can be used as an anesthetic. Assume that a tank containing such a mixture has the following partial pressures: $P(\text{cyclo}) = 170$ mmHg and $P(O_2) = 570$ mmHg.
(a) What is the ratio of the number of moles of cyclopropane to the number of moles of O_2?
(b) If the tank contains 160 g of O_2, how many grams of C_3H_6 are present?

66. A collapsed balloon is filled with He to a volume of 10. L at a pressure of 1.0 atm. Oxygen (O_2) is then added so that the final volume of the balloon is 30. L with a total pressure of 1.0 atm. The temperature, constant throughout, is equal to 20. °C.
(a) How many grams of He does the balloon contain?
(b) What is the final partial pressure of He in the balloon?
(c) What is the partial pressure of O_2 in the balloon?
(d) What is the mole fraction of each gas?

67. The boron hydride B_2H_6 burns in O_2.

$$B_2H_6(g) + 3\ O_2(g) \longrightarrow B_2O_3(s) + 3\ H_2O(g)$$

If the reactant gases were mixed in the correct stoichiometric ratio and then placed in a flask at 25 °C so that the total pressure was 10. mmHg, what were the partial pressures of B_2H_6 and O_2 before the combustion?

68. Dichlorine oxide is a powerful oxidizing agent that is used to bleach wood pulp and to treat municipal water supplies. It is made by the reaction

$$SO_2(g) + 2\ Cl_2(g) \longrightarrow OSCl_2(\ell) + Cl_2O(g)$$

If you put SO_2 in a flask so that its pressure is 125 mmHg at 22 °C, and if you add Cl_2 gas to this same flask, what should the Cl_2 partial pressure be in order to have the correct stoichiometric ratio of SO_2 to Cl_2?

69. The most effective rocket fuels are lightweight liquids that react to produce many molecules of gaseous products. One of the best fuels is dimethylhydrazine. When mixed with dinitrogen tetroxide, N_2O_4, it powered the Lunar Lander in the missions to the moon. The two components react according to the equation

$$(CH_3)_2N_2H_2(\ell) + 2\ N_2O_4(\ell) \longrightarrow 3\ N_2(g) + 4\ H_2O(g) + 2\ CO_2(g)$$

If 4.25 g of dimethylhydrazine react completely with N_2O_4 and if the product gases are collected at 27 °C in a 250.-L tank, what is the total pressure in the tank?

70. Sulfur dioxide reacts with chlorine to produce thionyl chloride, $OSCl_2$, and dichlorine oxide.

$$SO_2(g) + 2\ Cl_2(g) \longrightarrow OSCl_2(g) + Cl_2O(g)$$

If 0.235 g of SO_2 react with a stoichiometric amount of Cl_2, what is the total pressure of the reaction products in a 250.-mL flask at a temperature of 85 °C?

71. Explosives are effective if they produce a large number of gaseous molecules as products. Nitroglycerin, for example, detonates according to the equation

$$2\ C_3H_5N_3O_9(\ell) \longrightarrow 6\ CO_2(g) + 3\ N_2(g) + 5\ H_2O(g) + \tfrac{1}{2}\ O_2(g)$$

If 1.0 g of nitroglycerin explodes, calculate the volume the product gases would occupy if their total pressure is 1.0 atm at 5.0×10^2 °C.

72. A miniature laboratory volcano can be made with ammonium dichromate. When ignited it decomposes in a fiery display (Figure 4.8).

$$(NH_4)_2Cr_2O_7(s) \longrightarrow N_2(g) + 4\ H_2O(g) + Cr_2O_3(s)$$

If 0.50 g of ammonium dichromate are used and if the gases from this reaction are trapped in a 6.0-L flask at 25 °C, what is the total pressure of the gas in the flask? What are the partial pressures of N_2 and H_2O?

73. If potassium chlorate is heated, it decomposes to KCl and O_2.

$$2\ KClO_3(s) \longrightarrow 2\ KCl(s) + 3\ O_2(g)$$

Assume that you have heated a 1.56-g sample of $KClO_3$ and collected the evolved O_2 over water (see Figure 12.12). If the temperature of the gas is 21.0 °C, what volume should it occupy if the total pressure of the gas in the collection flask is equal to the barometric pressure (742 mmHg)? (See Appendix D for water vapor pressure data.)

74. You are given 1.56 g of a mixture of $KClO_3$ and KCl. When heated, the $KClO_3$ decomposes to KCl and O_2 (see Study Question 73), and 327 mL of O_2 are collected over water at 19 °C. The total pressure of the gas in the collection flask is 735 mmHg. What is the weight percentage of $KClO_3$ in the sample? (See Appendix D for water vapor pressure data.)

75. In Study Question 61 you learned that disulfur decafluoride, S_2F_{10}, can be made by shining light on a mixture of $SClF_5$ and H_2. Using the information in Study Question 61, calculate the partial pressure of each gas when the reaction is complete and find the total pressure in the 275-mL flask at 45 °C.

76. In Study Question 62 you found that the commercially important compound $(CH_3)_2SiCl_2$ can be made from solid silicon and gaseous methyl chloride, CH_3Cl. Using the information in Study Question 62, calculate the partial pressure of each gas when the reaction is complete and find the total pressures of gases in the 6.56-L flask at a temperature of 95 °C.

KINETIC MOLECULAR THEORY

77. You are given two flasks of equal volume. Flask A contains H_2 at 0 °C and 1 atm pressure. Flask B contains CO_2 gas at 0 °C and 2 atm pressure. Compare these two gases with respect to each of the following:

(a) Average kinetic energy per molecule
(b) Average molecular velocity
(c) Number of molecules

78. Equal masses of gaseous N_2 and Ar are placed in separate flasks of equal volume, both gases being at the same temperature. Tell whether each of the following statements is true or false. Briefly explain your answer in each case.
(a) The pressure is greater in the Ar flask.
(b) Ar atoms have a greater average speed than the N_2 molecules.
(c) The molecules of N_2 collide more frequently with the walls of the flask than do the atoms of Ar.

79. Calculate the rms speed of xenon atoms in a sample of gas at 25 °C and compare it with the rms speed for helium atoms as calculated in Exercise 12.12.

80. Calculate the rms speed for CCl_4 molecules at 25 °C. What is the ratio of this speed to that of H_2 at the same temperature?

81. Place the following gases in order of increasing average molecular speed at 25 °C: (a) Kr, (b) CH_4, (c) N_2, and (d) CH_2Cl_2.

82. The reaction of SO_2 with Cl_2 to give dichlorine oxide is

$$SO_2(g) + 2\ Cl_2(g) \longrightarrow OSCl_2(g) + Cl_2O(g)$$

All of the molecules involved in the reaction are gases. Place them in order of increasing average molecular speed.

GRAHAM'S LAW

83. Rank the following gases in order of increasing rate of effusion: (a) He, (b) Xe, (c) CO, and (d) C_2H_6.

84. In each pair of gases below, tell which will effuse faster:
(a) Cl_2 or F_2
(b) O_2 or N_2
(c) C_2H_4 or B_2H_6
(d) two CFC's: $CFCl_3$ or $C_2Cl_2F_4$

85. A gas whose molar mass you wish to know effuses through an opening at a rate only one third as great as the effusion rate of helium. What is the molar mass of the unknown gas?

86. A sample of uranium fluoride is found to effuse at the rate of 17.7 mg/hr. Under comparable conditions, gaseous I_2 effuses at the rate of 15.0 mg/hr. What is the molar mass of the uranium fluoride? (Caution: rates must be used in (mol/time) units.)

NONIDEAL GASES

87. In the text it is stated that the pressure of 8.00 moles of Cl_2 in a 4.00-L tank at 27.0 °C should be 29.5 atm if calculated using the van der Waals equation. Verify this result and compare it with the pressure expected from the ideal gas law.

88. You want to store 165 g of CO_2 gas in a 12.5-L tank at room temperature (25 °C). Calculate the pressure the gas would have using (a) the ideal gas law and (b) van der Waals equation.

GENERAL QUESTIONS

89. Rank the following pressures in increasing order (smallest first): 150 kPa, 742 mmHg, 0.89 atm, and 650 torr.

90. A sample of gaseous helium in a 1.50-L flask at 273 K exerted a pressure of 1.00 atm. After heating the flask to 346 K, what is the pressure of the gas?

91. A sample of H_2 exerts a pressure of 3.0 atm at 25 °C in an explosion-proof container. What is the pressure of H_2 on heating to 100. °C?

92. A sample of H_2 gas occupies 615 mL at 27.0 °C and 575 mmHg. When the gas is cooled, its volume is reduced to 455 mL and its pressure is reduced to 385 mmHg. What is the new temperature of gas?

93. Gaseous CO exerts a pressure of 45.6 mmHg in a 56.0-L tank at 22.0 °C. If the gas is released into a room with a volume of 2.70×10^4 L, what is the partial pressure of CO (in mmHg) in the room at 22 °C?

94. A 75-L steel tank at 70. °F contains acetylene at a pressure of 2000. psi (psi = pounds per square inch). How many grams of acetylene (C_2H_2) are there in the tank? (1 atm = 14.7 lb/in.^{2})

95. Acetaldehyde is a common liquid that vaporizes readily. A pressure of 331 mmHg is observed in a 125-mL flask at 0.0 °C, and the density of the vapor is 0.855 g/L. What is the molar mass of acetaldehyde?

96. 1.0 L of a gaseous compound of hydrogen, carbon, and nitrogen gave upon combustion 2.0 L of CO_2, 3.5 L of H_2O vapor, and 0.50 L of N_2 (with all gas volumes measured at the same T and P). What is the empirical formula of the compound?

97. Which of the following gas samples contains the largest number of gas molecules? Which contains the smallest number?
(a) 1.0 L of H_2 at STP
(b) 1.0 L of Ne at STP
(c) 1.0 L of H_2 at 27 °C and 760 mmHg
(d) 1.0 L of CO_2 at 0 °C and 800 mmHg

98. A 3.0-L bulb containing He at 145 mmHg is connected by a valve to a 2.0-L bulb containing Ar at 355 mmHg. (See figure on next page.) Calculate the partial pressure of each gas and the total pressure after the valve between the flasks is opened.

99. You are given a solid mixture of $NaNO_2$ and NaCl and are asked to analyze it for the amount of $NaNO_2$

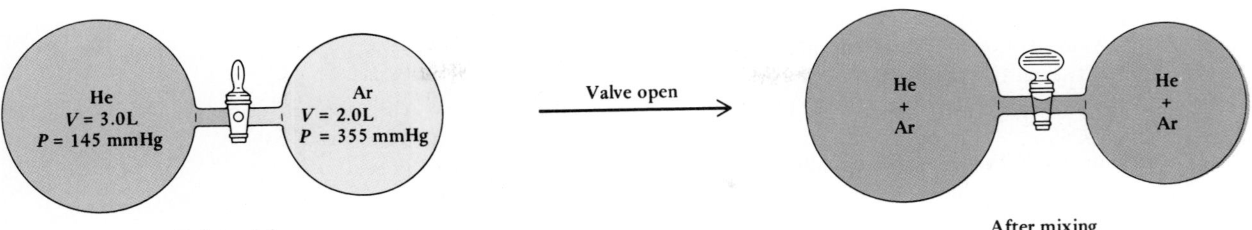

present. To do so you allow it to react with sulfamic acid, HSO_3NH_2, in water according to the equation

$$NaNO_2(aq) + HSO_3NH_2(aq) \longrightarrow NaHSO_4(aq) + H_2O(\ell) + N_2(g)$$

What is the weight percentage of $NaNO_2$ in 1.012 g of the solid mixture if reaction with sulfamic acid produces 325 mL of N_2 gas? The gas was collected over water (Figure 12.12) at a temperature of 21.0 °C and with the barometric pressure equal to 731.0 mmHg.

100. Hydrazine can be used to remove oxygen from hot-water heating systems by the following reaction:

$$N_2H_4(aq) + O_2(g) \longrightarrow N_2(g) + 2\ H_2O(\ell)$$

What is the maximum number of liters of N_2 that could be produced by 250 L of 0.010 M N_2H_4? The N_2 gas is collected at 25 °C and its pressure is 730 mmHg.

101. The rms speed of an O_2 molecule is 482 m/s at 298 K. What is the rms speed of a xenon atom at the same temperature?

102. You mix 0.450 g of methane (CH_4) and 0.516 g of O_2 in a 1250-mL flask at 25 °C. The methane burns according to the equation

$$CH_4(g) + 2\ O_2(g) \longrightarrow CO_2(g) + 2\ H_2O(g)$$

(a) What are the partial pressures of the gases before reaction?
(b) What is the total gas pressure (product pressures plus any leftover reactants) after reaction?

103. Octane, a component of gasoline, reacts with O_2 according to the equation

$$2\ C_8H_{18}(g) + 25\ O_2(g) \longrightarrow 16\ CO_2(g) + 18\ H_2O(g)$$

Assume the volume of one cylinder of an automobile engine is 500. cm^3. If air enters the engine at 50. °C and 1.0 atm of pressure, how many grams of octane should the fuel injection system send to the cylinder in order to completely consume all of the O_2 in the air? Assume that the mole fraction of O_2 in the air is 0.20.

104. The *Starship Enterprise* is not really fueled by dilithium crystals! Rather, it uses a mixture of diborane, B_2H_6, and oxygen. The two react according to the equation

$$B_2H_6(g) + 3\ O_2(g) \longrightarrow B_2O_3(s) + 3\ H_2O(g)$$

(a) If all of the gases involved in this reaction are at the same temperature, what are their relative rms speeds?
(b) B_2H_6 and O_2 are contained in separate fuel tanks of equal volume and at the same temperature. The pressure in the O_2 tank is 45 atm. If the B_2H_6 tank is to contain an amount of B_2H_6 that will exactly react with the O_2, what must the pressure be in the B_2H_6 tank (in atm)?

105. The density of air at 20. km above the earth's surface is 92 g/m^3. The pressure is 42 mmHg and the temperature is −63 °C. Assuming that the atmosphere contains only O_2 and N_2, calculate (a) the average molar mass of the atmosphere and (b) the mole fraction of each gas.

106. A xenon fluoride can be prepared by heating a mixture of Xe and F_2 to a high temperature in a pressure-proof container made of nickel. Assume that xenon gas was added to a 0.25-L container until its pressure was 0.12 atm at 0.0 °C. Fluorine gas was then added until the total pressure was 0.72 atm at 0.0 °C. After the reaction was complete, the xenon had been consumed completely and the pressure of the F_2 remaining in the container was 0.36 atm at 0.0 °C. What is the empirical formula of the compound prepared from Xe and F_2?

107. To find the formula of a compound with the general formula $Ni(CO)_x$, you carefully combine 0.125 g of the compound with aqueous HCl, decomposing it to $NiCl_2$(aq), H_2 gas, and CO gas: $Ni(CO)_x(g) + 2\ HCl(aq) \rightarrow NiCl_2(aq) + H_2(g) + x\ CO(g)$. You collect the CO and H_2 gases in a 565-mL container at 25 °C and find a total pressure of 121 mmHg. What is the value of x in $Ni(CO)_x$?

108. The atmosphere is a mixture of gases with a total pressure equal to the barometric pressure. Assume for this problem that this pressure is 740. mmHg. Further assume that the gases of the air have the following partial pressures: $P(N_2) = 557$ mmHg; $P(CO_2) = 23$ mmHg; $P(H_2O) = 40.$ mmHg; $P(O_2) = ?$

(a) What is the partial pressure of O_2?

(b) List the gases in order of increasing average molecular speed.

(c) The van der Waals *a* and *b* constants for each gas are given in Table 12.2. Which of the gases would be expected to behave most ideally?

109. The partial pressure of O_2 in air is 158 mmHg at 25 °C and a total pressure of 760. mmHg. The air that is expired from your lungs after breathing has a partial pressure of O_2 of 115 mmHg under the same conditions. How many moles of O_2 are absorbed by your lungs from 1.0 L of air?

110. Which graph below would best represent the distribution of molecular speeds for the gases acetylene (C_2H_2) and N_2? Both gases are in the same flask with a total pressure of 750 mmHg. The partial pressure of N_2 is 500 mmHg.

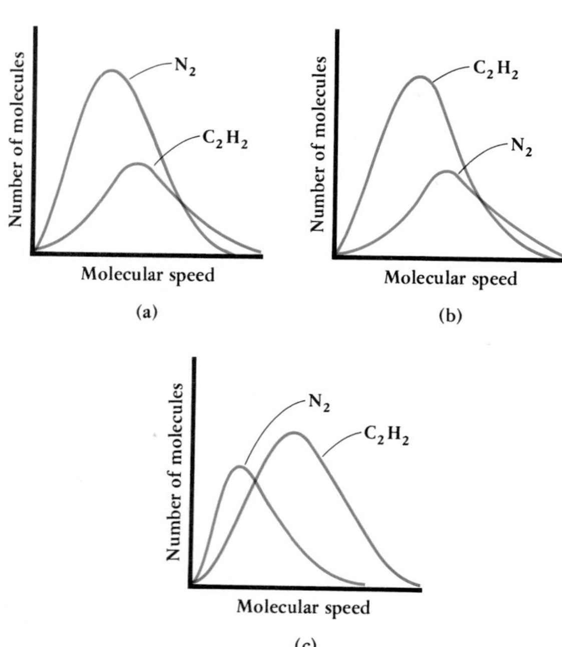

111. A sample of He has a volume of 10.00 mL at 0. °C but a volume of 13.66 mL at 100. °C. (The pressure of the gas sample at both temperatures is the same.) Use this to prove that 0. °C = 273 K.

112. Assume you have a glass tube 50. cm long. (See top of next column.) You allow some $NH_3(g)$ to diffuse along the tube from one end and some HCl(g) to diffuse along the tube from the opposite end. At what distance along the tube will the gases meet and react to form solid NH_4Cl?

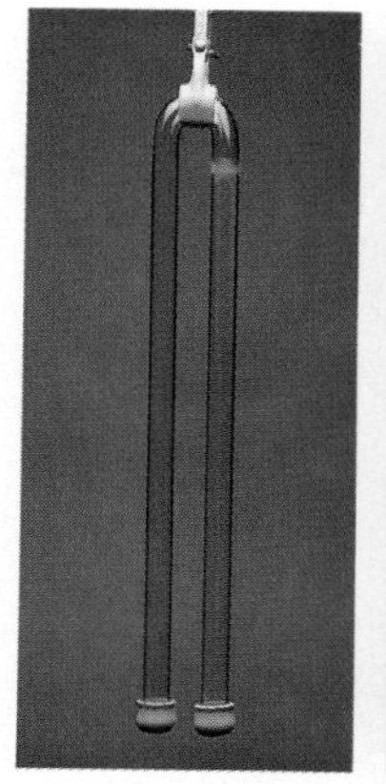

Ammonia and HCl are injected into the bent tube through the rubber caps at opposite ends of the tube. The gases diffuse along the tube and eventually meet, where they form a white ring of solid NH_4Cl (just below the bend on the right hand tube).

113. You fill a tank with I_2 vapor to a pressure of 45.0 mmHg at a given temperature. You then find it takes 10,820 seconds for enough I_2 to effuse out through a small opening in the tank wall to allow the pressure to drop to half its original value. In another experiment, you place SCl_xF_y in the same tank (and at the same temperature as the I_2 experiment), and you find it takes 8470 seconds for the pressure of this gas to drop to half its original value. What is the molar mass of SCl_xF_y?

114. Acetylene can be made by allowing calcium carbide to react with water (Figure 10.9).

$$CaC_2(s) + 2\,H_2O(\ell) \longrightarrow C_2H_2(g) + Ca(OH)_2(s)$$

You place 2.65 g of CaC_2 in excess water. If you collect the acetylene over water (Figure 12.12), and find that the gas (acetylene and water vapor) has a volume of 795 mL at 25.2 °C and at a barometric pressure of 735.2 mmHg, what is the percent yield of acetylene?

115. Acetylene is made by the reaction in Study Question 114. What would be the pressure of the acetylene gas at 22 °C in an 895-mL flask if you mix 95.0 g of calcium carbide and 65.0 g of water?

116. Assume you have synthesized a new compound of boron and hydrogen, B_xH_y. Determine its empirical and molecular formulas from the following information.

(a) To determine its empirical formula, you burn it in pure O_2 to give B_2O_3 and H_2O. The water from the combustion of 0.492 g of B_xH_y has a mass of 0.540 g.

(b) You quickly evaporate 0.0631 g of boron hydride into a flask with a volume of 120. mL; the gas exerts a pressure of 98.6 mmHg at 23 °C.

117. Methylthallium, $Tl(CH_3)_x$, has been prepared. To find its formula (the value of x), you decompose the compound with HCl. The hydrogen of the acid combines with CH_3 to produce CH_4 gas. Assume you decompose 0.102 g of $Tl(CH_3)_x$ and find that the isolated CH_4 gas has a pressure of 90.6 mmHg in a volume of 256 mL at 30.0 °C. What is the correct formula of methylthallium?

118. A mixture of aluminum and zinc weighing 1.67 g was completely dissolved in acid. In the process the mixture evolved 1.69 L of H_2, measured at 273 K and 1.00 atm pressure. What is the mass of aluminum in the mixture? (The aluminum reacts with acid according to the equation $Al(s) + 3\ H_3O^+(aq) \rightarrow Al^{3+}(aq) + \frac{3}{2}H_2(g) + 3\ H_2O(\ell)$. For the zinc reaction, see Example 12.10.)

119. You have 1.500 g of a mixture of $NaHCO_3$ and Na_2CO_3. Exactly 14.00 mL of 1.50 M HCl reacts with the mixture according to the equations

$$NaHCO_3(aq) + HCl(aq) \longrightarrow NaCl(aq) + H_2O(\ell) + CO_2(g)$$

$$Na_2CO_3(aq) + 2\ HCl(aq) \longrightarrow 2\ NaCl(aq) + H_2O(\ell) + CO_2(g)$$

(a) How many grams each of $NaHCO_3$ and Na_2CO_3 are there in the mixture? (b) If the CO_2 from both reactions is collected in a 0.450-L flask at 27 °C, what is the pressure of the gas?

SUMMARY QUESTIONS

120. Consider the following reaction of cobalt and nitric acid.

$$Co(s) + HNO_3(aq) \longrightarrow Co(NO_3)_2(aq) + NO_2(g) + H_2O(\ell)$$

(a) Balance the equation for the reaction.
(b) Indicate the oxidizing and reducing agents.

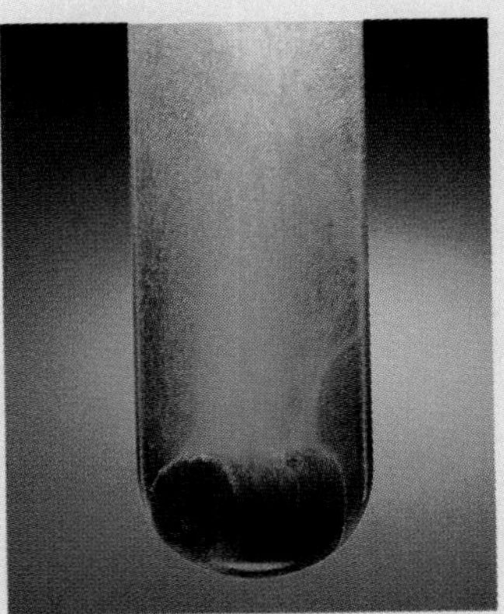

Metallic cobalt reacts with nitric acid to give pink $Co(NO_3)_2$, H_2, and nitrogen oxides, among them NO_2.

(c) If you use 2.115 g of cobalt metal and 25.0 mL of 6.00 M HNO_3, how many grams of $Co(NO_3)_2$ can be isolated from the reaction? How many grams of NO_2?
(d) If you isolate the NO_2 in a 2.56-L flask at 25 °C and find that its pressure is 375 mmHg, what is the percent yield of the gas? (Use the quantities of reactants in part (c).)

121. Chlorine trifluoride, one of the most reactive compounds known, is made by the reaction of chlorine and fluorine.

$$Cl_2(g) + 3\ F_2(g) \longrightarrow 2\ ClF_3(g)$$

(a) Assume you mix 0.71 g of Cl_2 with 1.00 g of F_2 in a 258-mL flask at 23 °C. What is the partial pressure of each of the two reactants before reaction? What is the partial pressure of the product and any leftover reactant after reaction in the 258-mL flask at 23 °C?
(b) What is the Lewis dot structure for ClF_3? What is its structural-pair geometry? Given that the molecule is polar, what is its molecular structure? What hybrid orbital set is used by the Cl atom in ClF_3?
(c) Estimate the enthalpy of the reaction above, using the bond energies in Table 10.4.

CHAPTER

13

Intermolecular Forces, Liquids, and Solids

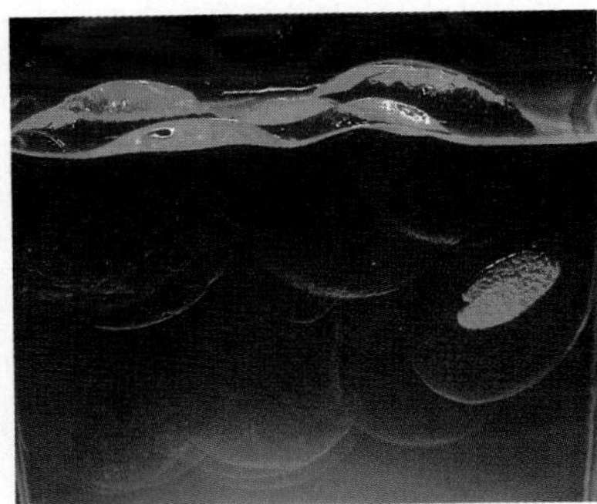
Ice in water.

CHAPTER OUTLINE

We are surrounded by gases, liquids, and solids: the gases of the atmosphere; liquid water in oceans, lakes, and growing things; and solid earth. Of the 108 known elements, the vast majority are solids; only eleven are gases under normal conditions (H_2, N_2, O_2, F_2, Cl_2, and the rare gases), and fewer still are liquids (only Hg and Br_2). The behavior of gases was explored in Chapter 12, so we turn now to the other phases of matter. You will find this a useful chapter because it will explain, among other things, why your body cools when you sweat, how bodies of water can influence local climate, why diamonds are hard but graphite is slippery, and something of the internal structure of beautifully crystalline solids.

It is our objective in this chapter to describe the liquid and solid states of matter in more detail, and particularly the forces responsible for their formation. You might find Figure 13.1 helpful to see how this chapter is organized.

13.1 PHASES OF MATTER AND THE KINETIC MOLECULAR THEORY

The kinetic molecular theory of gases, which was outlined in Chapter 12, assumes (i) that gas molecules or atoms are widely separated, (ii) that there are no forces of attraction between them, (iii) that the molecules are in continual, random, and rapid motion, and (iv) that their kinetic energy is determined only by the gas temperature. It is *only* on the first two points that gases differ significantly from liquids and solids.

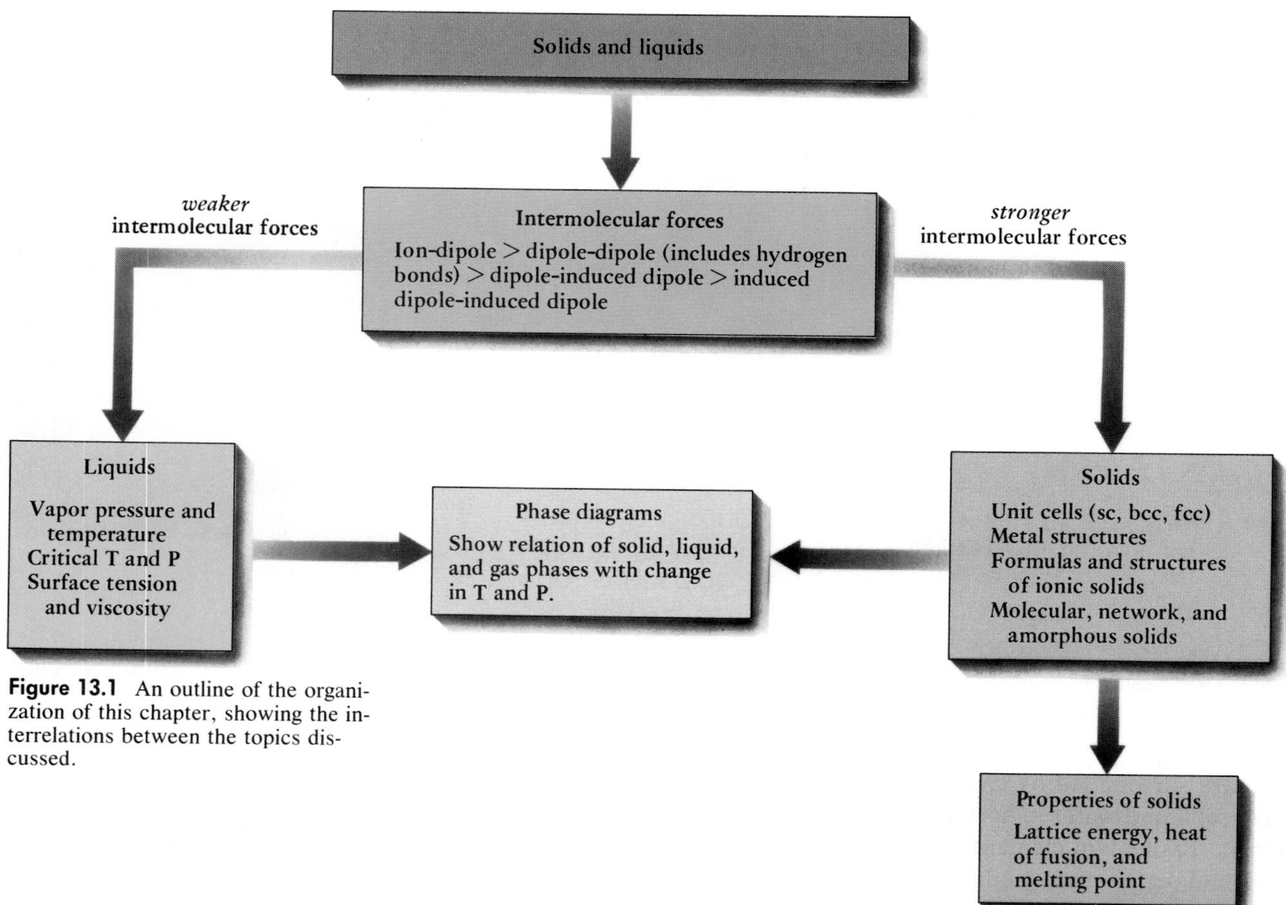

Figure 13.1 An outline of the organization of this chapter, showing the interrelations between the topics discussed.

Gases, unlike liquids and solids, can be compressed, because there is so much space between gas molecules. In fact, a compressed gas could ultimately become a liquid, occupying much less space than the gas. In Figure 13.2 you see a flask containing about 300 mL of liquid nitrogen. If all of this liquid were placed in a balloon and allowed to evaporate, the gas would fill a very large balloon to a pressure of 1 atmosphere at room temperature. This illustrates that there is *much* more space between molecules in the gas phase than in the liquid phase.

In contrast with the change in volume when converting from a gas to a liquid, there is no dramatic change in volume when a liquid is converted to a solid. Figure 13.3 shows solid and liquid benzene side by side, and you see they are not appreciably different in volume. This means that the atoms in the liquid phase are packed together about as tightly as the atoms in the solid phase.

Figure 12.2 illustrates the compressibility of a gas.

Each molecule or atom in the liquid or solid phase is clearly closer to a neighbor than in the gas phase. In fact, they cannot come much closer together. That is, the **compressibility** of liquids and solids, the change in volume with change in pressure, is small compared with gases. The air/fuel

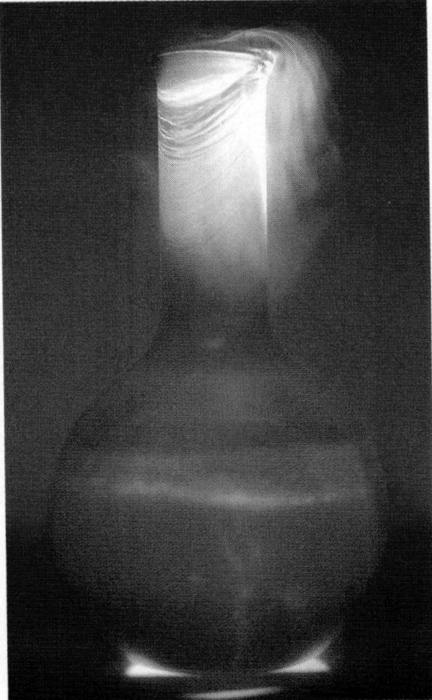

Figure 13.2 Liquid N_2. If this sample of liquid N_2 were evaporated to form N_2 gas at 25 °C, the gas would fill a large balloon (>200 L) to a pressure of 1 atmosphere at room temperature. This illustrates the fact that there is much more space between molecules in the gas phase than in the liquid phase.

Figure 13.3 The same volume of liquid benzene was placed in two tubes, and one tube was frozen. The solid and liquid phases have almost the same volume, showing that the molecules are packed together almost as tightly in the liquid phase as they are in the solid phase.

mixture in your car's engine is routinely compressed by a factor of about 10 before it is ignited. In contrast, the volume of liquid water changes only by 0.005% per atmosphere of pressure applied to it.

The kinetic molecular theory of gases assumes that there are only weak forces attracting molecules to one another, so gases can expand to fill their containers uniformly and completely. In contrast, there are stronger **intermolecular forces** in liquids and solids (Section 13.2), and these attractive forces in a liquid or solid prevent a significant expansion of a liquid or solid. In addition, these forces account for the fact that liquids (and solids if they are powdered) can be poured from one container to another.

The simple view of gases, liquids, and solids we have developed to this point—and now want to expand upon—is summarized in Figure 13.4. Molecules move freely in the gas phase. Liquids form from gases when the intermolecular forces are strong enough to bind molecules together, and motion is more restricted. Still, a liquid is like a very dense gas in that the molecules can still move rapidly and nearly randomly. For a solid, however, molecular motion is severely restricted, intermolecular forces are strong, and the molecules "pack" themselves into a regular array.

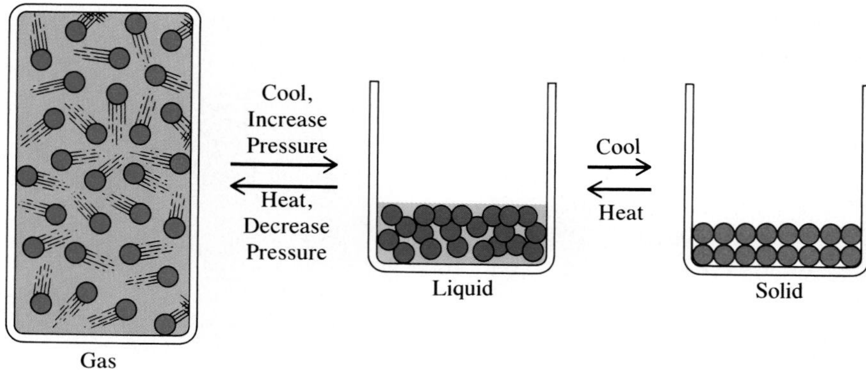

Figure 13.4 The three states of matter—gas, liquid, and solid.

13.2 INTERMOLECULAR FORCES

Intermolecular forces are the attractive forces between molecules, between ions, or between ions and molecules. Without such interactions, all substances would be gases. The various types of intermolecular forces, which are listed in Table 13.1, range from relatively strong forces between ions and polar molecules to very weak forces between molecules in which a dipole can be induced or created. All of these types of forces are important in nature, and we will discuss them in order of decreasing strength.

The weaker forces, particularly those not *involving ions, are known collectively as* ***van der Waals forces****, named for the physicist who developed an equation to describe the behavior of real gases (Chapter 12).*

To put the discussion of intermolecular forces in perspective, recall the description of ion–ion forces in Chapter 10. There you learned that energies of attraction between two oppositely charged ions can amount to many hundreds of kilojoules, energies that often far exceed those for the intermolecular forces to be discussed below. Later in the chapter, we

Table 13.1 Summary of Intermolecular Forces

Type of Interaction	Principal Factors Responsible for Interaction Energy	Approximate Magnitude (kJ/mol)
Ion–Dipole	Ion charge; dipole moment	40–600
Dipole–Dipole (including H-bonding)	Dipole moment	5–25
Dipole–Induced Dipole	Dipole moment; polarizability	2–10
Induced Dipole–Induced Dipole	Polarizability	0.05–40

Increasing strength of interaction (arrow pointing upward)

shall return to ion–ion forces and describe the energies involved when the ions are locked into a solid, crystal lattice.

ION–DIPOLE FORCES

As outlined in Chapter 10, a dipole consists of separated positive and negative charges, so a positive cation or negative anion can be attracted to one of these charges and repelled by the other (Table 13.1). The force involved in the attraction between a positive or negative ion and a polar molecule is less than that for ion–ion attractions (Chapter 10) but greater than for any other intermolecular force we shall describe. Just as you learned for ion–ion attractions in Chapters 3 and 10, the strength of ion–dipole attractions depend on

1. the distance between the ion and the dipole—the closer the ion and dipole are, the stronger the attraction.
2. the charge on the ion—the greater the ion charge, the stronger the attraction.
3. the magnitude of the dipole—the greater the magnitude of the dipole, the stronger the attraction.

Water is an excellent example of a polar molecule, a molecule with positive and negative electrical poles.

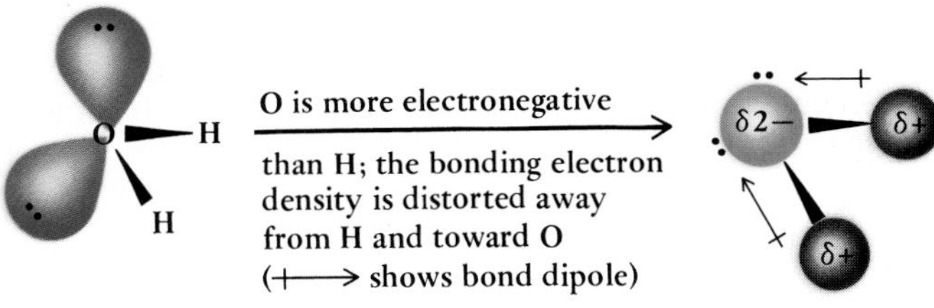

Polar water molecule attracted to an anion

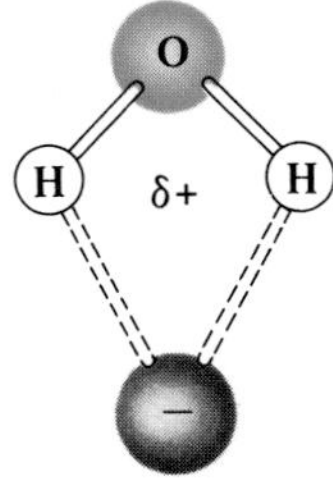

Polar water molecule attracted to a cation

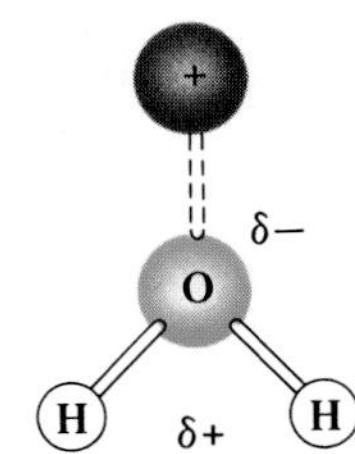

Thus, if a water molecule encounters an ion, a force of attraction between them will hold them together.

As you learned in Chapter 10, when there is such an attractive interaction, the attracting particles release energy when they come together to form a bond.

$$Na^+(g) + x\,H_2O(g) \longrightarrow [Na(H_2O)_x]^+(aq) + 397\text{ kJ} \qquad (x \text{ probably} = 6)$$

A metal ion that is bonded to several water molecules is said to be **hydrated**, and the energy released in this process is the **heat** or **energy of hydration**. If some polar molecule other than water is involved as the solvent, the ion is more generally said to be **solvated**, and we would talk about the **energy of solvation**.

Neutral molecules, as well as ions, can also be hydrated or solvated.

Hydration is important in our watery world. The oceans, lakes, and rivers consist of many substances dissolved in the water (e.g., Na^+, Ca^{2+}, Cl^-, CO_2, HCO_3^-, CO_3^{2-}, and O_2). The interaction between water and these substances not only causes them to dissolve but also greatly influences the properties of the water.

Hydration plays an intimate role in the chemistry of *acids*, one of the fundamental classes of chemical compounds. In Chapter 5 we defined acids as sources of *hydronium ions*, H_3O^+, in water. These arise because the acid provides a hydrogen ion, H^+. This is just a bare proton, so it interacts very strongly with the negative end of the water dipole to form a hydrated proton. For example, if HCl gas is bubbled into water, the hydration of H^+ and Cl^- is so strong that the H—Cl bond is

Figure 13.5 Hydrated cobalt chloride. (a) The solid salt. (b) The structure of $Co(H_2O)_4Cl_2$. The four water molecules are bound to Co^{2+} by ion–dipole attractions.

broken, and the molecules of water cluster around the ionic "pieces" of HCl.

Interaction of water and HCl dipoles. The heat of hydration of H^+ is about 1080 kJ/mol.

$$\overset{2\delta-}{\underset{\delta+}{H}\!-\!O(-\underset{\delta+}{H})} \cdots \underset{\delta+}{H}\!-\!\underset{\delta-}{Cl} \cdots \underset{\delta+}{H}\!-\!\overset{2\delta-}{O}\!-\!\underset{\delta+}{H}$$

$$HCl(g) + n\ H_2O(\ell) \longrightarrow H^+(H_2O)_x(aq) + Cl^-(H_2O)_y(aq)$$

It is thought that x in $H^+(H_2O)_x(aq)$ is about 4, but we generally represent the situation by writing simply $[H(H_2O)]^+$ or H_3O^+.

The importance of ion–dipole interactions is not limited to solutions. Many of the solids you use in the laboratory have formulas such as $BaCl_2 \cdot 2H_2O$ or $CoCl_2 \cdot 6H_2O$. The formula of the cobalt(II) salt, for example, is better written as $[Co(H_2O)_4Cl_2] \cdot 2H_2O$, because four of the six water molecules are associated with the Co^{2+} ion by ion–dipole attraction (Figure 13.5).

EXAMPLE 13.1

HYDRATION ENERGY

Explain why the energy of hydration of Na^+ (397 kJ/mol) is somewhat larger than that of Cs^+ (255 kJ/mol), while that of Mg^{2+} is much larger (1908 kJ/mol).

In the case of ion–dipole interactions, the attractive energy depends on $1/d^2$, where d is the distance between the centers of the ion and the dipole.

Solution The strength of ion–dipole attraction depends directly on the size of the ion and its charge and the magnitude of the dipole and inversely on the distance between them. Here we are considering three different ions interacting with the same solvent, water, so only ion charge and size are important to consider. From Figure 9.15 we know that the ion sizes are $Na^+ = 116$ pm, $Cs^+ = 181$ pm, and $Mg^{2+} = 86$ pm. It is clear from these radii that the distances between the ions and the water molecule are in the order $Mg^{2+} < Na^+ < Cs^+$. This alone can account for the relative magnitudes of the hydration energies. However, notice that Mg^{2+} has a 2+ charge, whereas the other ions are 1+. The greater charge also leads to a greater force of ion–dipole attraction, so the hydration energy for Mg^{2+} is *much* higher than for the other two ions.

EXERCISE 13.1 Hydration Energy

Which should have the higher hydration energy, K^+ or Li^+? Explain briefly.

DIPOLE–DIPOLE FORCES

If we represent polar molecules as cigar-shaped objects, a group of them can attract one another as follows.

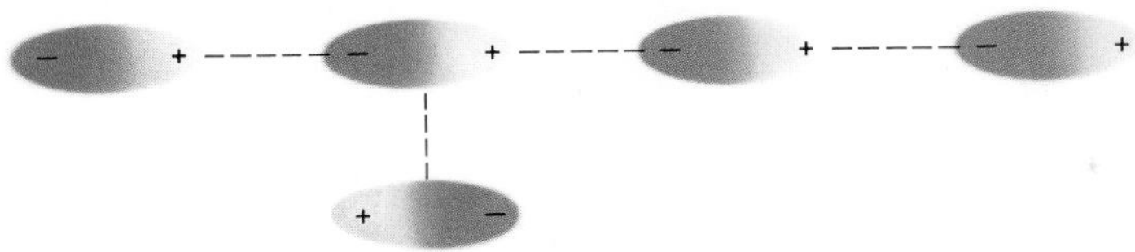

Energy is released when polar molecules interact with one another, and this is one of the reasons you must cool a gas consisting of polar molecules to convert it to a liquid: the heat evolved on interaction has to be removed.* (For example, HCl gas must be cooled to −84.8 °C to make liquid HCl.) Conversely, energy is required to separate interacting dipoles, and this is part of the reason you have to heat water or any other polar compound in the liquid state to convert it to molecules in the vapor phase, where they interact only weakly or not at all.

Two *different* polar molecules can associate as well. Water, for example, can be dissolved readily in ethyl alcohol (C_2H_5OH) because there is a strong interaction between these two polar molecules. In contrast, water does not dissolve gasoline to any appreciable extent, because the hydrocarbon molecules (such as $H_3C(CH_2)_6CH_3$) that constitute gasoline are not sufficiently polar. Water–hydrocarbon attractions are so weak that they cannot disrupt the stronger water–water attractions (Figure 13.6).

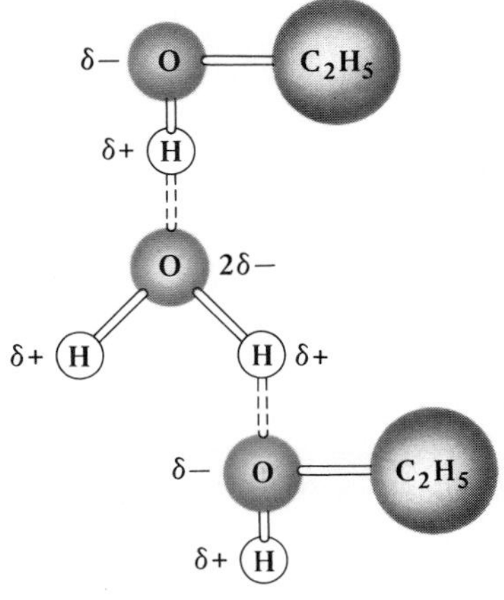

The properties of water are largely determined by hydrogen bonding. This subject is so important it is discussed in detail in Section 13.5.

Hydrogen bonding is a special form of dipole–dipole attraction and *enhances dipole–dipole attractions*. When H is attached to a very electronegative atom X, the interaction between other molecules and the H—X bond dipole is significantly greater than expected for ordinary dipole–dipole attractions. This interaction, called hydrogen bonding because it occurs *only* when H is part of one or both of the interacting dipoles, is very significant in the world around us (Figure 13.7).

The electronegativities of N (3.0), O (3.5), and F (4.0) are among the highest of all of the elements, while that of H (2.1) is considerably less. Therefore, the covalent bonds N—H, O—H, and F—H are extremely polar. Thus, if an atom of an adjacent molecule (or even one in the same molecule) is the negative end of a bond dipole, a strong attraction to the hydrogen atom can occur. This is especially true if the neighboring atom is one of the small, highly electronegative atoms N, O, or F. This means the strongest hydrogen-bonding interactions are

$$\overset{\delta-}{X}—\overset{\delta+}{H}\cdots\overset{\delta-}{Y}—$$

N—H · · · · N—	O—H · · · · N—	F—H · · · · N—
N—H · · · · O—	O—H · · · · O—	F—H · · · · O—
N—H · · · · F—	O—H · · · · F—	F—H · · · · F—

Figure 13.6 Oil and water do not mix owing to the vastly different intermolecular forces within each liquid, and because the forces of attraction between oil and water are not sufficient to disrupt the forces within liquid water.

*Another reason to cool a gas in order to liquefy it is to reduce the kinetic energy of the molecules.

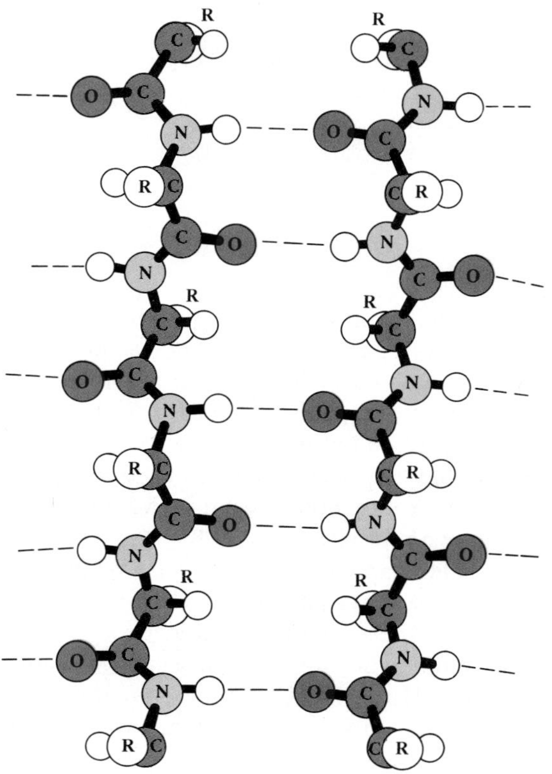

Figure 13.7 A number of natural materials are chains of amino acids (see Chapter 27). These chains can interact internally or with a neighboring chain by hydrogen bonding. Here are two chains that have a "backbone" consisting of C and N atoms, with R groups (such as CH_3) on every third backbone atom, a H atom attached to each N atom, and an O atom attached to the C atoms not having an R group. The polar N—H groups in one chain are hydrogen bonded to the polar C=O groups in a neighboring chain. The result is a material found in silk and other insect fibers.

One of the strongest of these is found in hydrogen fluoride, which, in the solid state, is a zigzag chain of hydrogen-bonded HF molecules.

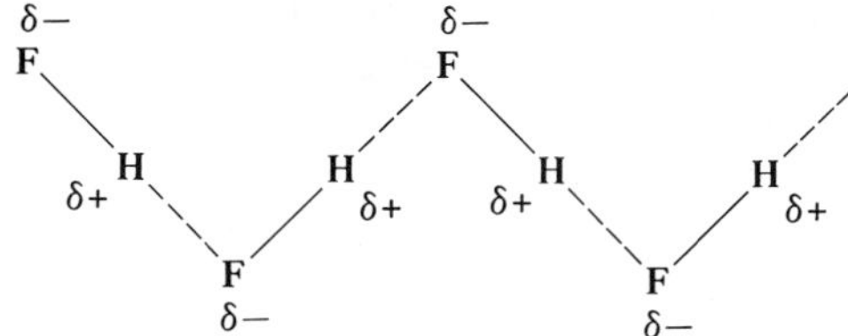

Hydrogen-bonding can also occur if the neighboring atom is Cl or S, but interaction with these larger atoms is usually weaker.

The different properties of the compounds ethyl alcohol (containing an O—H bond) and dimethyl ether (containing only C—H and C—O bonds) illustrate the effects of hydrogen bonding. These molecules have the same molecular formula, and their dipole moments are similar. However, their melting and boiling points are vastly different (Table 13.2).

Table 13.2 Comparison of the Physical Properties of Ethyl Alcohol and Dimethyl Ether

Compound	Dipole Moment (D)	Melting Point (°C)	Boiling Point (°C)
Ethyl alcohol, C_2H_5—OH	1.7	−115	78.5
Dimethyl ether, CH_3—O—CH_3	1.3	−141	−25

Deciphering Intermolecular Forces

The scheme below may help you to decide what types of intermolecular forces are important in a given chemical system. Using it together with Table 13.1, you can then decide which forces may be stronger and can make some judgment concerning the properties of the system.

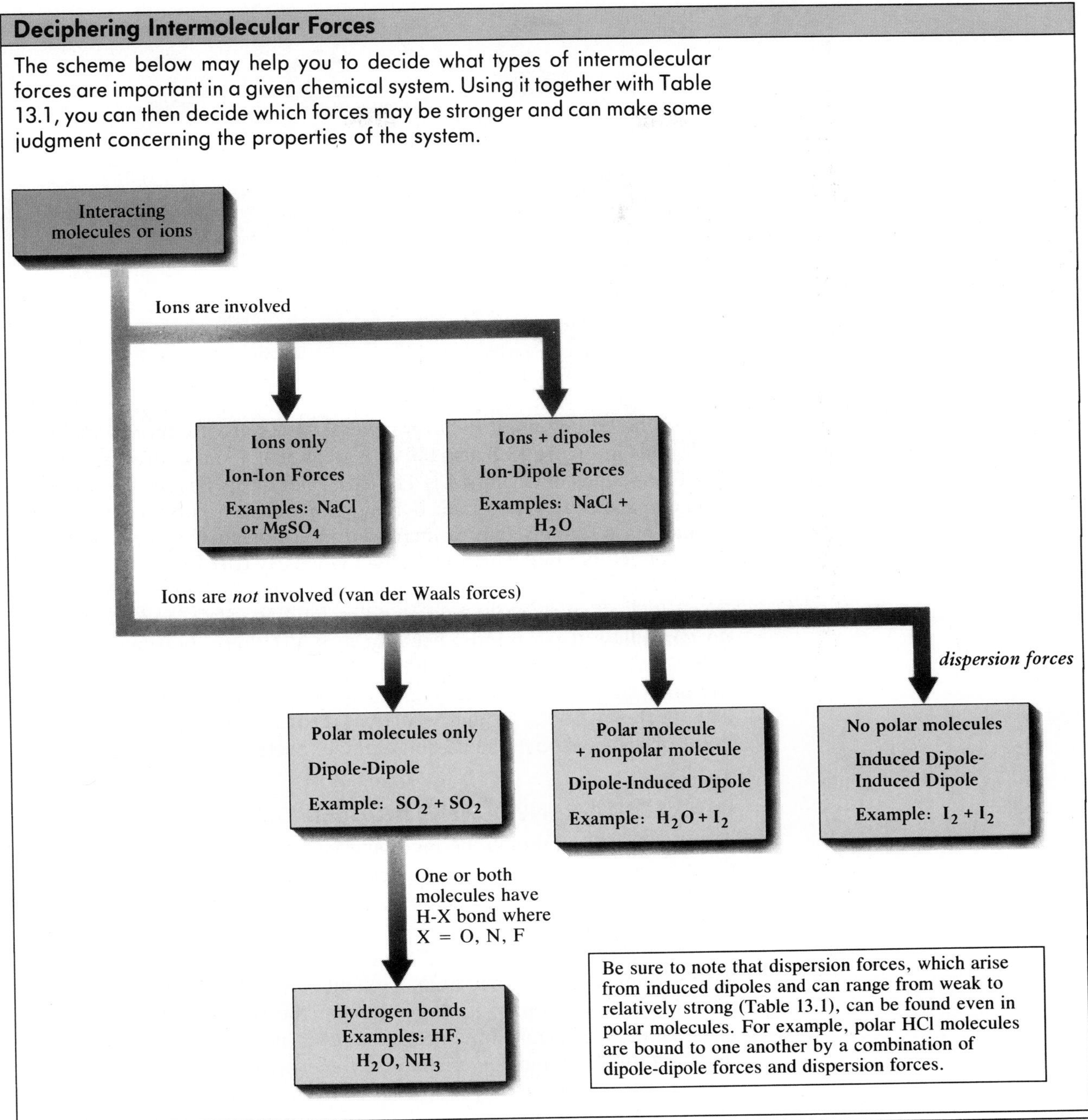

Be sure to note that dispersion forces, which arise from induced dipoles and can range from weak to relatively strong (Table 13.1), can be found even in polar molecules. For example, polar HCl molecules are bound to one another by a combination of dipole-dipole forces and dispersion forces.

The fact that a higher temperature is required to melt or boil one compound than another means that stronger intermolecular forces exist in the compound requiring higher temperatures. This is the case for ethyl alcohol because alcohol molecules may interact through dipole–dipole forces enhanced by hydrogen bonding, whereas only dipole–dipole forces are possible for the ether.

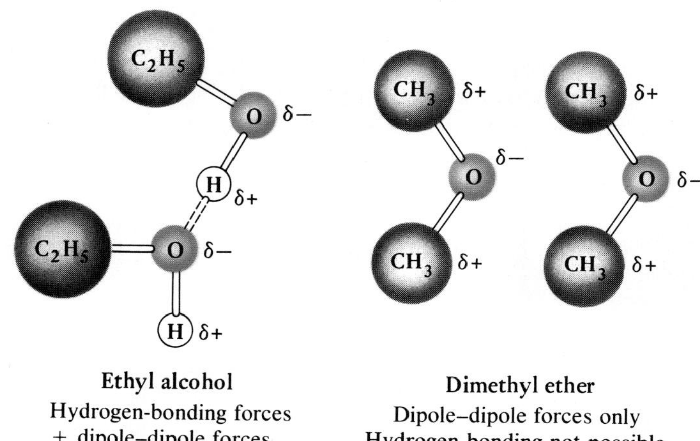

Ethyl alcohol
Hydrogen-bonding forces
+ dipole–dipole forces

Dimethyl ether
Dipole–dipole forces only
Hydrogen bonding not possible

The boiling point data for several series of structurally analogous molecules in Figure 13.8 also tell us a great deal about hydrogen bonding. The compounds of Group 4A (CH_4, SiH_4, GeH_4, SnH_4) are all tetrahedral, nonpolar molecules, and their boiling points increase uniformly with molecular mass (largely due to increased intermolecular dispersion forces, as explained below). This same effect is clearly true for the heavier molecules of the hydrogen compounds of elements of Groups 5A, 6A, and 7A as well. However, the boiling points of NH_3, H_2O, and HF certainly do *not* follow these trends, because it is just these molecules in which hydrogen bonding is strong within the liquid phase. If you just extrapolate the line established by H_2S, H_2Se, and H_2Te, for example, water might be expected to boil at around −90 °C. Hydrogen bonding raises the boiling point of water almost 200 °C above the expected temperature!

INTERACTIONS INVOLVING INDUCED DIPOLES

Although interactions involving induced dipoles generally are by far the weakest kinds of intermolecular forces, they are nonetheless important in determining the solubility of gases such as O_2, N_2, or CO_2 in water, for example.

DIPOLE–INDUCED DIPOLE FORCES You know that a nonpolar molecule such as CO_2 can dissolve to a small extent in water. (When the CO_2 comes out of solution again, it is the source of the bubbles in soda pop and champagne.) This can happen because polar molecules such as water can *induce* a dipole in molecules that are themselves not polar by reason of their structure. To see how this may occur, picture a polar water molecule approaching a nonpolar molecule such as O_2 (Figure 13.9). The electron cloud of an isolated O_2 molecule is distributed, on average, symmetrically between the bonded O atoms. However, as the negative end of the polar H_2O molecule approaches, the O_2 electron cloud is moved away to reduce repulsion between the O_2 cloud and the negative end of the H_2O dipole, and the O_2 molecule itself becomes dipolar. That is, a dipole has been induced in the otherwise nonpolar O_2 molecule, and H_2O and O_2 are now attracted to one another, although only weakly.

The degree to which the electron cloud of an atom (such as Ne or Ar) or a nonpolar molecule (such as O_2, CO_2, I_2, or CCl_4) can be distorted

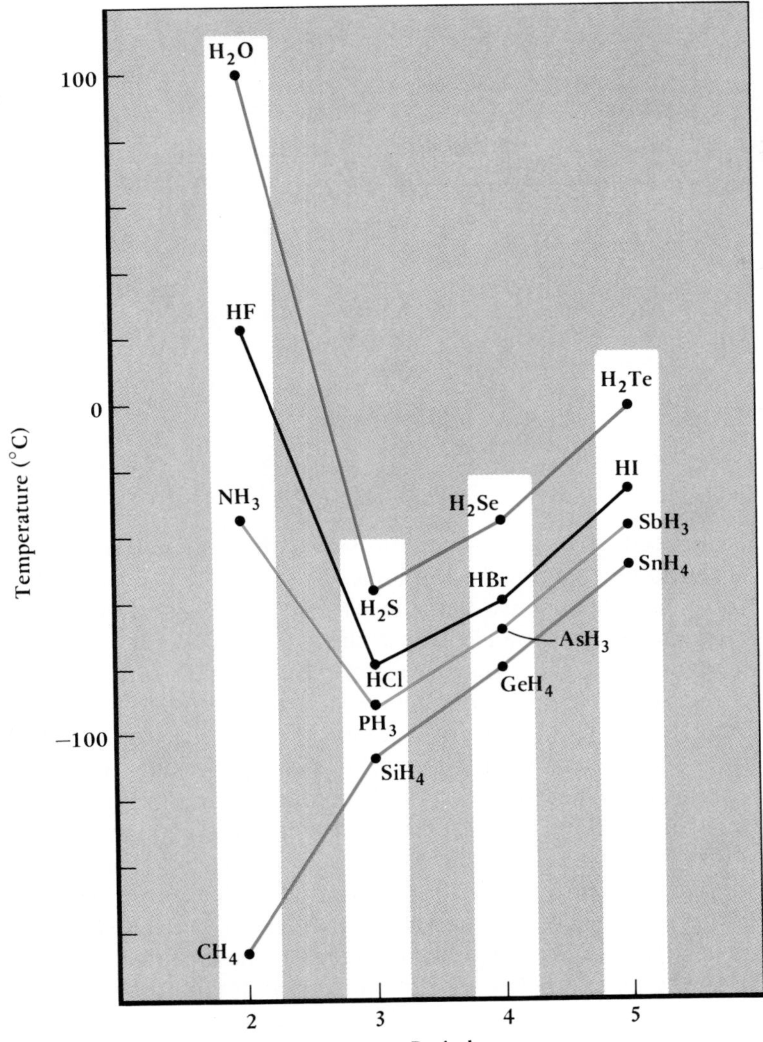

Figure 13.8 The boiling points of some simple hydrogen-containing compounds. Lines connect molecules containing atoms from the same periodic group. Notice the effect of hydrogen bonding on the boiling points of H_2O, HF, and NH_3.

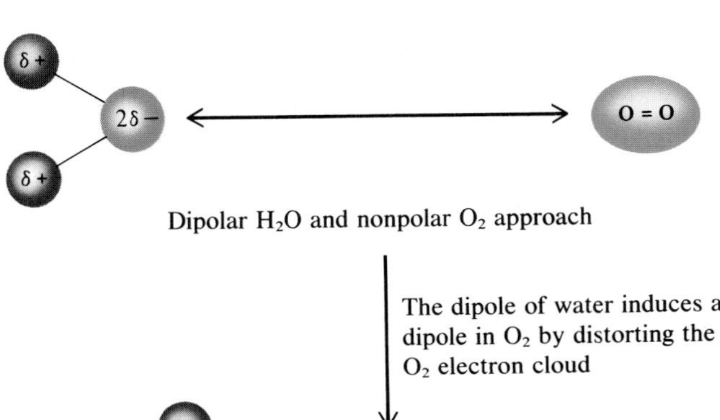

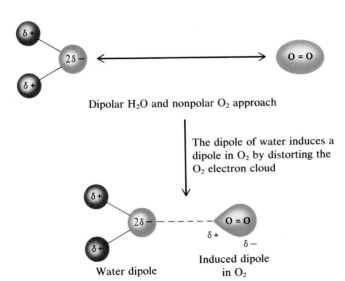

Figure 13.9 A polar water molecule can induce a dipole in nonpolar O_2.

Table 13.3 The Solubility of Common Atmospheric Gases in Seawater

Gas	Molar Mass (g/mol)	Solubility at 24 °C (moles/m³)
He	4.0	0.31
Xe	131.3	3.1
N_2	28.0	0.54
O_2	32.0	1.1
CO_2	44.0	32

Table 13.4 Boiling Points of the Rare Gases and Other Common Gases

Gas	Molar Mass (g/mol)	Boiling Point (K)
He	4.0	4.18
Ne	20.2	27.13
Ar	39.9	87.29
Kr	83.8	119.7
Xe	131.3	165.0
H_2	2.0	20.4
N_2	28.0	77.4
O_2	32.0	90.2

and a dipole induced depends on the atom's or molecule's **polarizability**. We do not need to be concerned with numerical values of this property. However, it should make sense that the valence electrons of atoms or molecules with large, extended electron clouds, such as I_2, can be polarized or distorted more readily than those of gases like He or H_2 where the valence electrons are close to the nucleus and more tightly held.

The table of the solubilities of common gases in sea water (Table 13.3) illustrates the effect of dipole–induced dipole interactions. Here you see a clear trend to higher solubility with increasing mass in a related series of atoms or molecules, for the following reason. As the molecular mass increases, either there is an increase in the number of valence electrons or the valence electrons are less tightly held. Therefore, the ease of polarization of the valence electron cloud generally increases with mass. Since a dipole is more readily induced as the polarizability increases, the strengths of dipole–induced dipole interactions generally increase with mass. Finally, since the solubility of substances such as CO_2 or O_2 depends on the strength of the dipole–induced dipole interaction, the solubility of nonpolar substances in polar solvents generally increases with mass.

Be sure to note that dispersion forces, which arise from induced dipoles and can range from weak to relatively strong (Table 13.1), can be found even in polar molecules. For example, polar HCl molecules are bound to one another by a combination of dipole-dipole forces and dispersion forces.

INDUCED DIPOLE–INDUCED DIPOLE FORCES The last of the intermolecular forces to be considered is that between two induced dipoles (Table 13.1). Such forces are often called **London forces** or **dispersion forces**, and they can range from very weak to relatively strong (Table 13.1). Nonpolar molecules such as I_2 must attract one another because I_2 is a solid at room temperature. Similarly, N_2, O_2, and the rare gases can all be liquefied at low temperatures (Figure 13.2). In Table 13.4 you see a clear trend to higher boiling points of nonpolar gases with increasing molar mass, in part because *dispersion forces generally become stronger with increasing mass* (and stronger intermolecular forces mean that a higher temperature is required to break down these forces and allow the molecules to leave the liquid and go into the vapor phase).

To understand how two nonpolar molecules can attract one another, remember that the electrons in atoms or molecules are in a state of constant motion. On average, the electron cloud around an atom is spherical in shape. However, when two nonpolar atoms or molecules approach one another, attractions or repulsions between their electrons and nuclei can lead to distortions in their electron clouds. That is, dipoles can be induced momentarily in neighboring atoms or molecules, and it is these induced dipoles that can lead to intermolecular attraction (Figure 13.10).

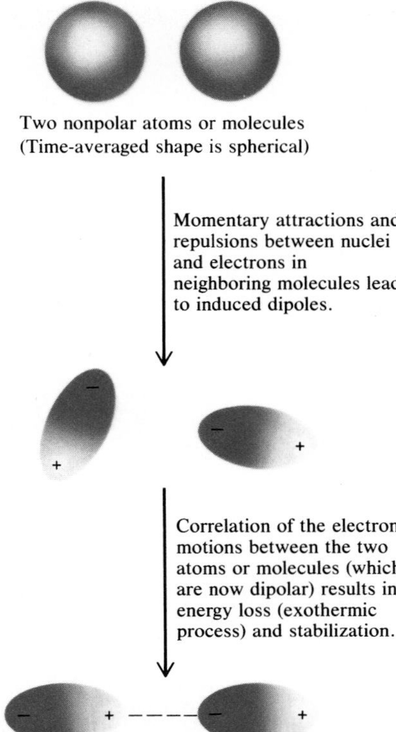

Two nonpolar atoms or molecules
(Time-averaged shape is spherical)
Momentary attractions and repulsions between nuclei and electrons in neighboring molecules lead to induced dipoles.
Correlation of the electron motions between the two atoms or molecules (which are now dipolar) results in energy loss (exothermic process) and stabilization.

Figure 13.10 Interaction between two induced dipoles.

EXAMPLE 13.2

INTERMOLECULAR FORCES

Decide what type of intermolecular force is involved in each case and place the interactions in order of increasing strength: (a) $CH_4 \cdot\cdot\cdot\cdot CH_4$; (b) $H_2O \cdot\cdot\cdot\cdot H_3COH$; (c) $LiCl \cdot\cdot\cdot\cdot H_2O$.

Solution To answer this question it is helpful to follow the outline in the box on page 509.

(a) No ions are involved with CH_4, a simple molecule with covalent bonds. To decide whether dipole–dipole forces or some form of induced dipole is involved, you first have to decide whether the molecule is polar. In this case an electron dot structure and VSEPR theory (Chapter 10) show the molecule has tetrahedral geometry, and the terminal atoms are the same.

```
     H                  H
     |                  |
  H—C—H                C
     |               /  \  H
     H              H    \
                          H
```

Dot structure

Molecular geometry = tetrahedral
Nonpolar molecule

Therefore, the molecules cannot be polar, so the only way they can interact is through very weak induced dipole forces.

(b) Again no ions are involved with these covalently bonded molecules, but both molecules are polar and both have an O—H bond. Therefore, they interact through the special dipole–dipole forces called hydrogen bonding.

```
     H
      \  δ−       δ+  δ−
δ+     O · · · · H—O
      /              |
     H              CH3
```

(c) LiCl is an ionic compound composed of Li^+ and Cl^- ions, and water is a polar molecule. Therefore, ion–dipole forces are involved.

In order of increasing strength, the interactions are

$$CH_4 \cdot\cdot\cdot\cdot CH_4 < H_2O \cdot\cdot\cdot\cdot CH_3OH < LiCl \cdot\cdot\cdot\cdot H_2O$$

EXERCISE 13.2 Intermolecular Forces

Decide what type of intermolecular force is involved with (a) $N_2 \cdot\cdot\cdot\cdot N_2$; (b) $MgSO_4 \cdot\cdot\cdot\cdot H_2O$; (c) $CO_2 \cdot\cdot\cdot\cdot H_2O$. Place the interactions in order of increasing strength.

SOMETHING MORE ABOUT
Sweeteners—The Secret Is in the Shape

Why humans acquired the ability to taste sweetness in foods is lost in evolutionary time. It may have been that *Homo sapiens* needed the ability in order to distinguish when fruits were ripe and at their peak in supplying useful carbohydrates. Be that as it may, most of us have a "sweet tooth" to some degree, and, in this age of an abundance of food in some parts of the world, we probably consume too much sugar for our own good. In 1987 the use of sugars, including sucrose, corn sweeteners, and honey, was a staggering 132.6 pounds per person in the United States. Consuming sugar at this rate we not only risk getting fat, but also accelerate the decay of our "sweet tooth." To counter this, many have turned to artificial sweeteners, and the per capita consumption of these is about 19 pounds!

To satisfy the craving for sweetness, humans turned to artificial sweeteners much earlier than you may have thought. About 2000 years ago the Greeks and Romans found that, by boiling grape juice in lead pans, they could produce a syrup that was incredibly sweet. The sweet taste was due to lead(II) acetate, $Pb(CH_3COO)_2 \cdot 3\ H_2O$, a compound commonly called "sugar of lead." Cooks in those times added "sugar of lead" to their foods, and it may have been used to sweeten wine. But the Romans were not aware of what we know today—lead compounds are highly toxic. Some historians have claimed that consumption of lead compounds in this and other ways may have contributed to the decline of the Roman Empire.

Beryllium is an important component of emeralds. However, some centuries ago early chemists, who probably were in the habit of sampling everything in their laboratories, found that compounds of this alkaline earth element had a sweet taste, and the early name for the element—glucinium—comes from the Greek word meaning sweet. However, like lead, beryllium compounds are *very* toxic and should never be tasted.

The first artificial sweetener truly safe for humans—**saccharin**—was prepared by Ira Remsen and Constantin Fahlberg in 1879 at the Johns Hopkins University in Baltimore, Maryland. However, just as there are controversies today over the safety and efficacy of artificial sweeteners, the acceptance of saccharin was not immediate. In 1912 President Theodore Roosevelt set up a board to review the use of the sweetener, and the board concluded that it was safe to eat 0.3 g per day. This opened the way for expanding use of the compound; during World Wars I and II saccharin was added to foods intended for consumption in the United States while natural sugar was saved for the soldiers overseas.

The three other artificial sweeteners commonly used today were discovered quite accidentally. Although modern laboratory practice forbids it, Michael Sveda was smoking a cigarette while working in his organic chemistry laboratory at the University of Illinois. He put his cigarette down for a moment on the edge of the lab bench, and, when he picked it up, it tasted sweet. He traced it to a small particle of the compound he had just made, a compound now known as **cyclamate**. In 1967 Karl Clauss, a chemist at the German company Hoechst, licked his fingers to get a better hold on a piece of filter paper and noticed a sweet taste. He had found **acesulfame**, a compound known commercially in Europe as *Sunnette*®. Finally, in 1965 James Schlatter was doing research on antiulcer drugs at G.D. Searle and Company when he discovered **aspartame**.

Food products containing aspartame.

Given the enormous quantities of sugar and artificial sweeteners used in the world today, laboratories are involved in research to find new ones that are safe and effective. To develop new compounds, it would be very helpful to know what makes a molecule taste sweet. However, it has been clear for a long time that sweetness is not a simple property. Although we can look at a molecule and

Some Common Sweeteners

Structure and Name	Sweetness Relative to Equal Mass of Sucrose	Comments
cyclamate	25	Stable in heat and cold. Not metabolized by most people.
aspartame (NutraSweet®)	200	Loses sweetness at higher temperatures. Nutritive. Digested as a protein.
acesulfame (Sunnette®)	200	High stability. Slight aftertaste. Noncaloric. Not metabolized.
saccharin	300	Slight aftertaste. Noncaloric. Not metabolized.

be fairly certain it will be an acid or base, the same cannot be said of sweetness. What we do know, though, is that *the size and shape of a molecule hold the secret of its sweetness*. About 50 compounds that have a sweet taste have been studied to see if they have a common factor. They do, and this has given rise to one theory of "sweetness," the *triangle theory*. This theory proposes that a molecule must have three key sites, arranged at the corners of an imaginary triangle, and that these sites give the molecule the ability to lock onto receptor sites in our taste buds, where it triggers the response "this stuff is sweet" in our brains.

The receptors in our taste buds are composed of proteins, and these have the ability to form hydrogen bonds with other molecules. The proteins have N—H

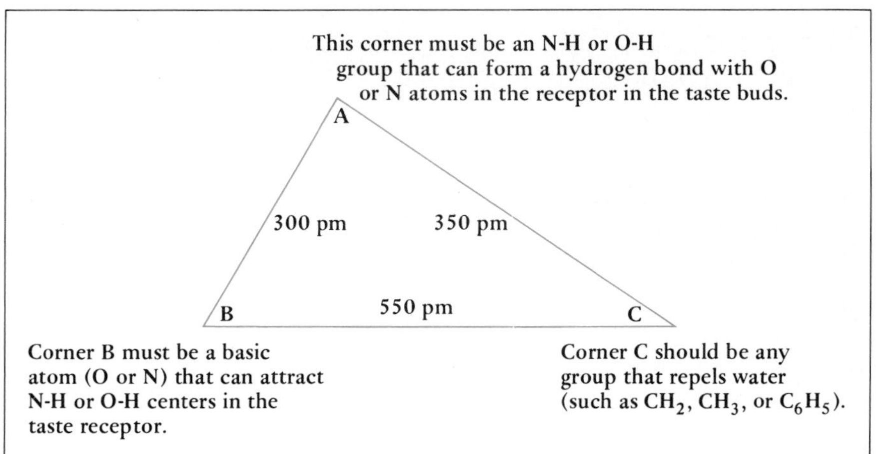

and O—H groups that can supply the H atom in a hydrogen bond, and they have C=O groups that can be the acceptor in a hydrogen bond (—C=O · · · · H—). Therefore, it appears that one site in a sweet molecule must act as the H-atom donor in a hydrogen bond (site A in the "sweetness triangle"). A second site, about 300 pm away from the first (site B), is an acceptor atom in a hydrogen bond. And finally, it seems that there must be a third site—at the third corner of a triangle (site C)—that has no hydrogen-bonding capacity. In fact, it is a hydrophobic site such as a methyl group (CH_3) or a methylene group (CH_2).

Chemists still have not found the ideal artificial sweetener. After all, we are making some extraordinary demands. It should be at least as sweet as sucrose, instantly register its sweetness, be nontoxic, and have no caloric content. Furthermore, it should be inexpensive, easy to make, and stable to heat, light, and dissolution in water. The search goes on.

13.3 PROPERTIES OF LIQUIDS

About 70% of our planet is covered by a liquid, water, and we owe a great deal to the unique properties of this compound, especially in the liquid state. We will discuss water in much more detail in Section 13.5, but before we do so we need to describe some properties common to all liquids.

KINETIC MOLECULAR THEORY OF LIQUIDS AND ENTHALPY OF VAPORIZATION

Liquids are characterized by having a regular structure only in very small regions, largely because the intermolecular forces are weak. Thus, most of the molecules move randomly, somewhat like the molecules in a gas. Further, molecules in the liquid phase have a range of kinetic energies (Figure 13.11) that closely resembles the distribution of energies you saw earlier for gas molecules (Figure 12.14). Just as for gases, the average kinetic energy depends only on temperature; the higher the temperature, the higher the average energy and the greater the relative number of molecules with high kinetic energy.

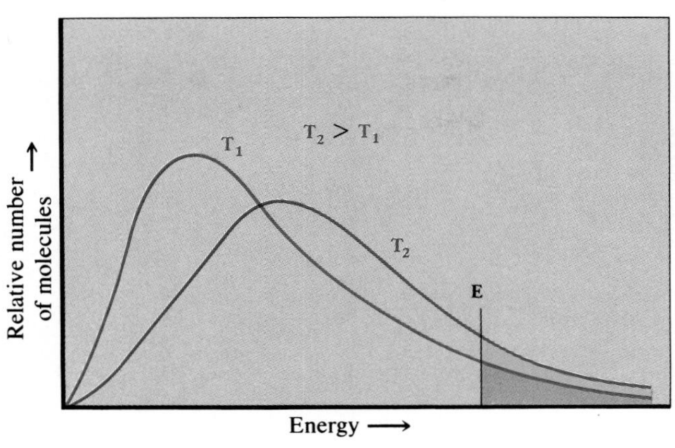

Figure 13.11 The distribution of molecular energies in the liquid phase. T_2 is higher than T_1, so there is a greater proportion of molecules having the energy marked by E at the higher temperature.

[dark shading] Relative number of molecules having enough energy to evaporate at lower temperature, T_1

[dark shading] + [light shading] Relative number of molecules having enough energy to evaporate at higher temperature, T_2

Why is it important to know that there is a distribution of kinetic energies in a liquid, that not all molecules have the same energy? Because, in order for **vaporization** or **evaporation** to occur—for molecules to leave the liquid phase and move to the vapor phase—they must have some minimum kinetic energy, say E in Figure 13.11.

$$\text{Liquid} + \text{heat} \xrightarrow{\text{vaporization}} \text{vapor}$$

$$\begin{aligned}\text{Heat required at constant pressure} &= \text{molar enthalpy of vaporization} \\ &= \Delta H_{vap}\end{aligned}$$

To give escaping molecules enough energy to overcome intermolecular forces, energy must be supplied (usually as heat). Thus, vaporization is always an endothermic process, and the heat energy required for evap-

Table 13.5 Molar Enthalpy of Vaporization and Boiling Point for Common Compounds

Compound	ΔH_{vap} (kJ/mol)*	Boiling Point (°C) (Vapor pressure = 760 mmHg)
HF	25.2	19.7
HCl	17.5	−84.8
HBr	19.3	−66.5
HI	21.2	−35.1
CH_4 (methane)	8.9	−161.5
C_2H_6 (ethane)	15.7	−88.6
C_3H_8 (propane)	19.0	−42.1
C_4H_{10} (butane)	24.3	−0.5
NH_3	25.1	−33.4
H_2O	40.7	100.0
SO_2	26.8	−10.0
He	0.08	−269.0
Ne	1.8	−246.0
Ar	6.5	−185.9
Xe	12.6	−107.1

*ΔH_{vap} is given at the normal boiling point of the liquid.

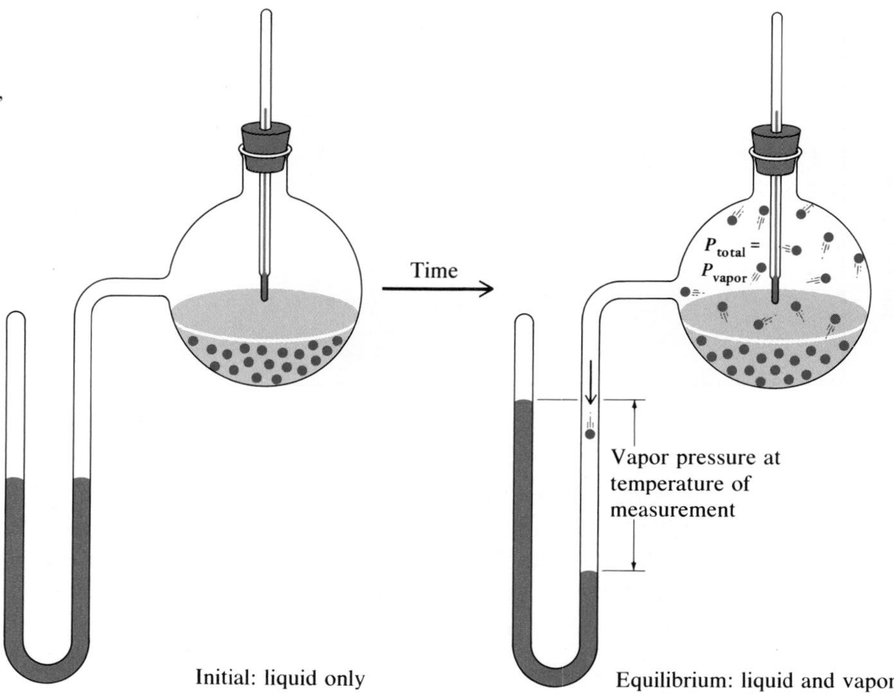

Figure 13.12 A volatile liquid is placed in an evacuated flask (left). In the beginning, no molecules of liquid are in the vapor phase. In a short time, however, some of the liquid evaporates, and the molecules in the vapor phase exert a pressure (right). If the pressure is measured when the liquid and vapor are in equilibrium, the pressure is called the equilibrium vapor pressure.

oration is the **latent heat of vaporization** (in energy/mass; see Chapter 6) or the **molar enthalpy of vaporization**, ΔH_{vap} (in energy/mol, as listed in Table 13.5).

For molecules in the gas phase over a liquid, it is possible that some of them lose the high kinetic energy that enabled them to escape from the liquid; if the direction of motion of these molecules is back toward the liquid surface, they can reenter the liquid phase in a process called **condensation**.

$$\text{Vapor} \xrightarrow{\text{condensation}} \text{liquid} + \text{heat}$$

$$\text{Heat released at constant pressure} = -\Delta H_{\text{vaporization}}$$

Since new intermolecular bonds are made, heat is released; if this heat is removed, condensation occurs. Condensation is always an exothermic process, and the heat released is equal in magnitude but opposite in sign to the enthalpy of vaporization. For example, one mole of water vapor releases 40.7 kJ on condensing to liquid water at 100 °C, which means that $\Delta H_{\text{condensation}} = -\Delta H_{\text{vap}} = -40.7$ kJ/mol.

Molecules in the liquid and gas phases. All the molecules in both phases are moving, although the distance traveled in the liquid before collision with another molecule is small. Some of the liquid phase molecules are moving with a kinetic energy large enough to overcome the intermolecular forces in the liquid and escape to the gas phase. At the same time, some molecules rebound onto the surface after colliding with molecules of gases in the air or with other molecules of the same type.

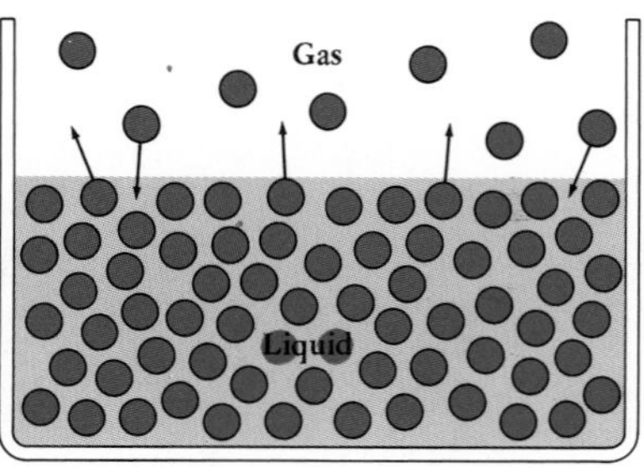

EXAMPLE 13.3

ENTHALPY OF VAPORIZATION

You put 1.00 liter of water (about 4 cupsful) in a pan at 100 °C, and it evaporates. How much heat must have been absorbed (at constant pressure) by the water as it vaporized?

Solution There are three pieces of information you need to solve this problem:

(a) ΔH_{vap} for water = 40.7 kJ/mol at 100 °C

(b) The density of water at 100 °C = 0.958 g/cm^3. (This is needed because ΔH_{vap} has units of kJ/mol, so you first have to find the mass of water and then the number of moles.)
(c) Molar mass of water = 18.02 g/mol

Given the density of water, a volume of 1.00 L (or 1.00×10^3 cm^3) is equivalent to 958 g, and this mass is in turn equivalent to 53.2 moles of water. Therefore, the amount of heat required is

$$53.2 \text{ mol } H_2O \left(\frac{40.7 \text{ kJ}}{\text{mol}}\right) = 2.17 \times 10^3 \text{ kJ}$$
$$= \text{heat required for evaporation}$$

2170 kJ is equivalent to about one quarter of the energy in your daily food intake.

EXERCISE 13.3 Enthalpy of Vaporization

The molar enthalpy of vaporization of wood alcohol, CH_3OH (methyl alcohol), is 37.6 kJ/mol. How much energy is required to evaporate 1.00 kg of this alcohol?

In Example 13.3 you learned that an enormous amount of heat is required to convert liquid water to water vapor. This is important to your environment and your physical well-being. When you exercise vigorously, your body responds by sweating to rid itself of the excess heat. The sweat, which is mostly water, can be evaporated by the input of heat energy. The source of this energy is the heat generated by your muscles, so the evaporation of sweat removes the excess heat, and your body is cooled.

The heat of vaporization and condensation also play an enormous role in our weather. For example, if enough water condenses from the air to fall as 1 inch of rain on an acre of ground, the heat released is more than 200 million kilojoules! This is equivalent to the heat from about 50 tons of exploded dynamite, that is, to the energy released by a small bomb.

Finally, it is interesting to look at trends in ΔH_{vap} values for related molecules in Table 13.5. The boiling points of nonpolar liquids increase with increasing atomic or molecular mass, a reflection of increasing intermolecular dispersion forces (Table 13.4; Figure 13.8). Similarly, boiling points and ΔH_{vap} values of the heavier hydrogen halides (HX, where X = Cl, Br, and I) increase with increasing molecular mass. For these molecules, H-bonding is not important (as in HF), so dipole–dipole and dispersion forces take over, and the latter become increasingly important with increasing mass. Lastly, be sure to notice the very high ΔH_{vap} values for H_2O and HF that come from *extensive* hydrogen bonding.

Condensation of water to form rain provides much of the heat to drive thunderstorms. (B. Gruliow)

VAPOR PRESSURE AND BOILING POINT

EQUILIBRIUM VAPOR PRESSURE If you put some water in an open beaker, eventually the water will evaporate completely. However, if you put some water in a sealed flask (Figure 13.12), the liquid will evaporate only until the rate of vaporization equals the rate of condensation. When this occurs, a state of **dynamic equilibrium** is established.

$$\text{Liquid} + \text{heat} \rightleftharpoons \text{vapor}$$

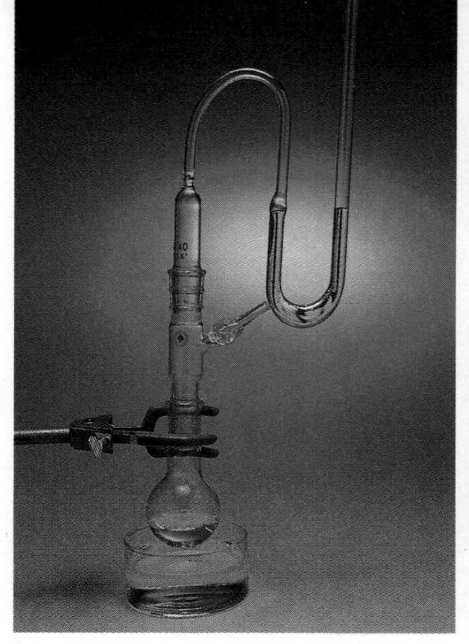
(a)

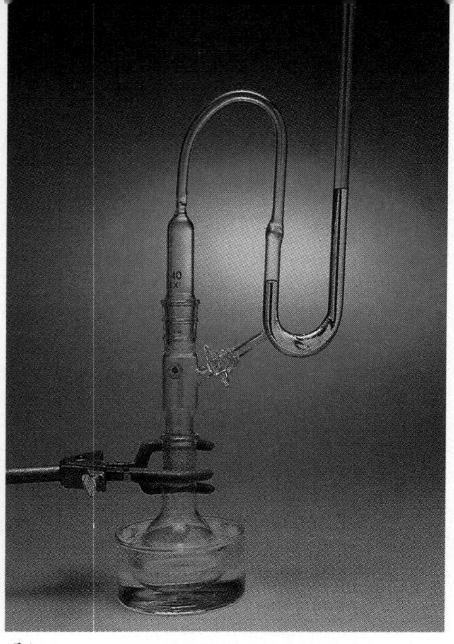
(b)

Figure 13.13 (a) A volatile liquid [acetone, $(CH_3)_2CO$] is shown in the flask attached to a U-tube manometer containing mercury (see page 457). (One end of the manometer is open to the flask, and the other end is open to the air.) The valve (or stopcock) on the flask is open, and the mercury levels are equal; therefore, the total gas pressure (from a mixture of acetone vapor and air) on the left side of the mercury column is the same as the air pressure on the right side. (b) The valve is closed, and the flask is warmed. More acetone evaporates, and its vapor exerts a greater pressure. The increase in acetone vapor pressure at this temperature is given by the difference in mercury levels in the manometer.

The state of equilibrium is a concept used throughout chemistry and one we shall return to often. We signal this situation by connecting the two states or the reactants and products by a set of double arrows, $\rightleftarrows$.

Molecules continue to move from the liquid to the vapor phase, while some in the vapor phase move back to the liquid phase; the mass of material in each phase is constant with time. (In contrast to the closed flask, the water in the open beaker can never come to equilibrium with gas phase water molecules; air movement and gas diffusion quickly remove the water vapor from the vicinity of the liquid surface.)

When liquid/vapor equilibrium has been established, the pressure exerted by the water vapor is called the **equilibrium vapor pressure** (often called just the vapor pressure). The equilibrium vapor pressure of any substance is a measure of the tendency of its molecules to escape from the liquid phase and enter the vapor phase *at a given temperature*. We often refer to this tendency as the **volatility** of the compound. The higher the equilibrium vapor pressure at a given temperature, the more **volatile** the compound (Figure 13.13).

Since the average energy of a molecule in the liquid phase is a function of temperature (Figure 13.11), molecules move more rapidly and the rate of vaporization increases as the temperature rises. Therefore, the equilibrium vapor pressure must also increase with temperature (Figure 13.14). Any point along a vapor pressure versus temperature curve for a compound, as in Figure 13.14, represents a pressure and temperature at which liquid and vapor are in equilibrium. For example, at 25 °C the equilibrium vapor pressure of water is 24 mmHg, whereas it is 149 mmHg at 60 °C. This means that if water is placed in an evacuated flask that is maintained at 60 °C, liquid will evaporate until the pressure exerted by the vapor is 149 mmHg. (If not enough water is put in the flask, all the liquid will evaporate before this equilibrium pressure is reached.)

At the conditions of T and P given by any point on a curve in Figure 13.14, pure liquid and its vapor are in dynamic equilibrium. If the T and P define a point not *on the curve, the system is* not *at equilibrium.*

EXAMPLE 13.4

VAPOR PRESSURE

If 1.00 liter of water is placed in a small room that has a volume of 2.30×10^4 L, will the water all evaporate at 25 °C? (The density of water at 25 °C is 0.997 g/cm^3, the molar mass of water is 18.02 g/mol, and the vapor pressure of water at 25 °C is 23.8 mmHg.)

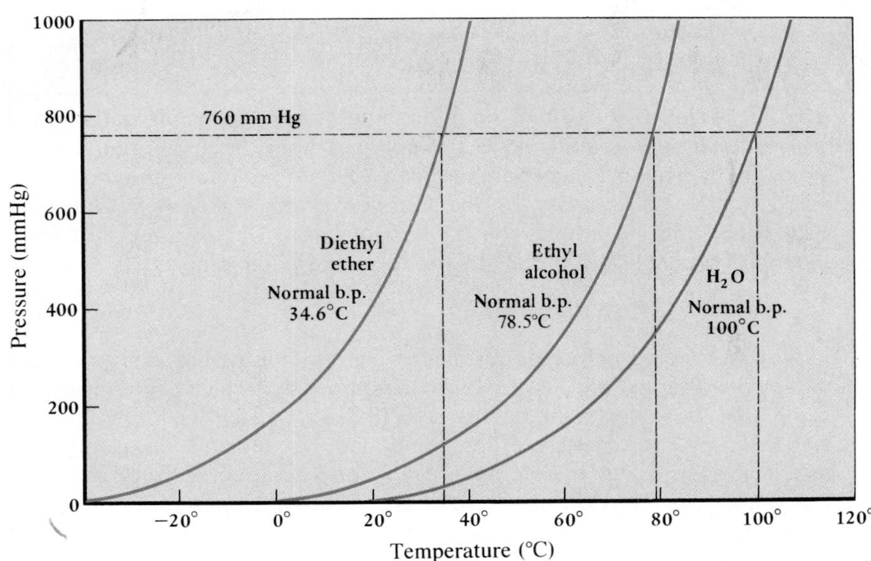

Figure 13.14 Vapor pressure curves for diethyl ether ($C_2H_5OC_2H_5$), ethyl alcohol (C_2H_5OH), and water. Each curve represents the conditions of T and P where the two phases (pure liquid and its vapor) are in equilibrium. (P is the vapor pressure of the liquid at a temperature T.) The compound exists as a liquid under conditions of T and P defined by points to the left of a curve, and it exists as a vapor for all temperatures and pressures to the right of a curve.

Solution One approach to solving this problem is to calculate the amount of water that must evaporate in order to exert a pressure of 23.8 mmHg in a volume of 2.30×10^4 L at 25 °C.

$$n \text{ moles} = \frac{PV}{RT} = \frac{\left(\dfrac{23.8 \text{ mmHg}}{760 \text{ mmHg/atm}}\right)(2.30 \times 10^4 \text{ L})}{(0.0821 \text{ L} \cdot \text{atm/K} \cdot \text{mol})(298 \text{ K})}$$

$$n = 29.4 \text{ mol } H_2O$$

$$29.4 \text{ mol } H_2O \left(\frac{18.02 \text{ g}}{1 \text{ mol}}\right) = 531 \text{ g}$$

$$531 \text{ g} \left(\frac{1 \text{ cm}^3}{0.997 \text{ g}}\right) = 532 \text{ cm}^3$$

The calculation shows that about half of the 1.00 L of water will evaporate in order to achieve an equilibrium water vapor pressure of 23.8 mmHg at 25 °C.

EXERCISE 13.4 Vapor Pressure

If you seal 0.50 g of pure water in an evacuated 5.0-L flask and heat the whole assembly to 60. °C, will the pressure be equal to or less than the equilibrium vapor pressure of water at this temperature? What if you used 2.0 g of water? Under either set of conditions, will there be liquid water left in the flask, or will all of the water evaporate?

BOILING POINT If you have a beaker of water open to the atmosphere, the weight of the atmosphere is pressing down on the surface. As heat is added, more and more water can evaporate, pushing the molecules of the atmosphere aside. If enough heat is added, a temperature is eventually reached at which the vapor pressure of the liquid equals the atmospheric pressure; bubbles of vapor begin to form in the liquid, and the liquid *boils* (Figure 13.15).

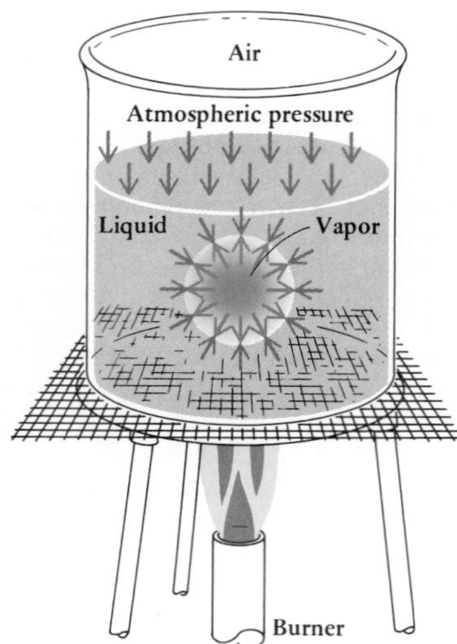

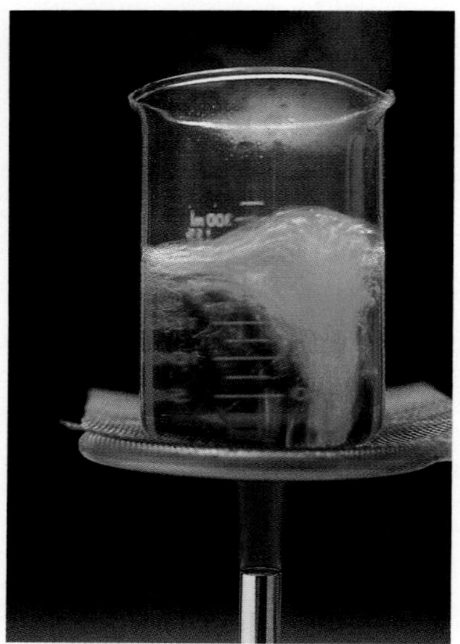

Figure 13.15 A liquid boils when its equilibrium vapor pressure equals the atmospheric pressure.

The Relation Between Vapor Pressure, Temperature, and ΔH_{vap}: The Clausius-Clapeyron Equation

Figure 13.14 clearly shows that the relation between temperature and vapor pressure for a pure liquid is *not* a straight line; that is, P is not related in a simple, linear way to T. Instead, there is a "curved" relation, one that was studied in the 19th century by the German physicist Rudolph Clausius (1822–1888) and the Frenchman B.P.E. Clapeyron, who showed that this behavior is predicted accurately by the following equation:

$$\ln P = -(\Delta H_{vap}/RT) + C \tag{13.1}$$

The term ln P is the natural logarithm of the vapor pressure, ΔH_{vap} is the enthalpy of vaporization, T is the Kelvin temperature at which P is measured, R is the ideal gas constant in energy units (8.31451×10^{-3} kJ/K · mol), and C is a constant characteristic of the compound. Equation 13.1 is useful because it gives us a way to determine ΔH_{vap} for a compound. To do this we need to measure the vapor pressure at two different temperatures and calculate the difference between ln P_1 (P_1 is the vapor pressure at T_1) and ln P_2 (P_2 is the vapor pressure at T_2). Then

$$\ln P_2 - \ln P_1 = \left[\frac{-\Delta H_{vap}}{RT_2} + C\right] - \left[\frac{-\Delta H_{vap}}{RT_1} + C\right] \tag{13.2}$$

When this equation is simplified (Equation 13.3), the molar enthalpy of vaporization can be calculated readily from experimentally determined pressures and temperatures.

$$\ln\left(\frac{P_2}{P_1}\right) = \frac{\Delta H_{vap}}{R}\left[\frac{1}{T_1} - \frac{1}{T_2}\right] \tag{13.3}$$

As an example, consider the organic liquid 1,2-ethanediol ($C_2H_6O_2$), which is commonly called ethylene glycol and is used as an antifreeze in car radiators. Its vapor pressure at 100. °C ($= T_1 = 373$ K) is 14.9 mmHg ($= P_1$) while it is 49.1 mmHg ($= P_2$) at 125 °C ($= T_2 = 398$ K). You can use these data to calculate the enthalpy of vaporization of the pure compound by substituting this information into Equation 13.3.

$$\ln\left(\frac{49.1 \text{ mmHg}}{14.9 \text{ mmHg}}\right) = \frac{\Delta H_{vap}}{8.3145 \times 10^{-3} \text{ kJ/K} \cdot \text{mol}}\left[\frac{1}{373 \text{ K}} - \frac{1}{398 \text{ K}}\right]$$

$$1.192 = \frac{\Delta H_{vap}}{8.3145 \times 10^{-3} \text{ kJ/K} \cdot \text{mol}}(0.000168)(1/\text{K})$$

$$\Delta H_{vap} = 59.0 \text{ kJ/mol}$$

This is a higher value than for the other compounds in Table 13.5, a fact that reflects the higher mass of ethylene glycol and, more importantly, its hydrogen-bonding capacity. Of course, it is this latter property that makes the glycol soluble in water and allows you to use a water-glycol mixture in the radiator of a car.

$$H_2O \cdots H{-}O{-}CH_2{-}CH_2{-}O{-}H \cdots OH_2$$

Hydrogen bonding in ethylene glycol/water

Finally, the large value of ΔH_{vap} is also reflected by the relatively high boiling point of the liquid (198 °C).

As another example of the Clausius-Clapeyron equation, try to calculate the molar enthalpy of vaporization for pure diethyl ether, $(C_2H_5)_2O$, which has vapor pressures of 57.0 mmHg and 534 mmHg at −22.8 °C and 25.0 °C, respectively (see Figure 13.14). (The answer is $\Delta H_{vap} = 29.0$ kJ/mol.)

The boiling point of a liquid is the temperature at which the vapor pressure is equal to the external pressure, and, if the external pressure is 1 atmosphere, the temperature is designated the **normal boiling point**. Normal boiling points of some liquids are listed in Table 13.5. Notice the direct relationship between normal boiling point, enthalpy of vaporization, and intermolecular forces. We have already pointed out that stronger intermolecular forces lead to larger values of ΔH_{vap} and, therefore, to higher boiling points (see Figure 13.8).

The normal boiling point of pure water at sea level is 100 °C. If you live at higher altitudes, such as in Salt Lake City, Utah, where the barometric pressure is less than 1 atmosphere (about 650 mmHg), water will boil at lower temperatures. (The curve for the equilibrium vapor pressure of water in Figure 13.14 shows that $T \cong 95$ °C when $P = 650$ mmHg.) Cooks know that food has to be heated a bit longer in Salt Lake City or Denver to achieve the same effect as in New York at sea level.

To shorten cooking time in Denver, or anywhere else, use a pressure cooker. This is a sealed pot that allows water vapor to build up to pressures slightly greater than the external or atmospheric pressure. The boiling point of the water increases, and foods cook faster.

CRITICAL TEMPERATURE AND PRESSURE

The vapor pressure of a liquid will continue to increase with temperature up to the **critical point**, where the curve of vapor pressure versus temperature comes to an abrupt halt (Figure 13.16). The temperature at which this occurs is the **critical temperature, T_c**, and the corresponding vapor pressure is the **critical pressure, P_c**. The significance of this point is that, if the temperature exceeds T_c, all the liquid molecules have sufficient kinetic energy to separate from one another, regardless of the pressure, and a conventional liquid no longer exists. Instead the substance is often called a **supercritical fluid**.

A knowledge of critical temperatures and pressures (see Table 13.6) is important in designing air conditioners, for example. These devices cool a room by using the room heat to vaporize a liquid. To make the process continuous, however, the vapor is condensed back to the liquid state by compressing the vapor to a pressure higher than P_c (as long as the temperature of the vapor is less than T_c). Compounds called *chlorofluorocarbons* have been widely used as the fluid in air conditioners and refrigerators because some of them can be liquefied at reasonable pressures. For example, CCl_3F has a T_c of 198 °C and a P_c of 43.5 atm. This means CCl_3F vapor can be converted to a liquid by applying modest pressures at any temperature below 198 °C, a temperature far above room temperature. You will also notice that propane and butane, which are

Figure 13.16 The vapor pressure curve for water near the critical temperature, T_c. This temperature is "critical" because, beyond this point, vapor cannot be converted to liquid no matter how high the pressure exerted on the vapor.

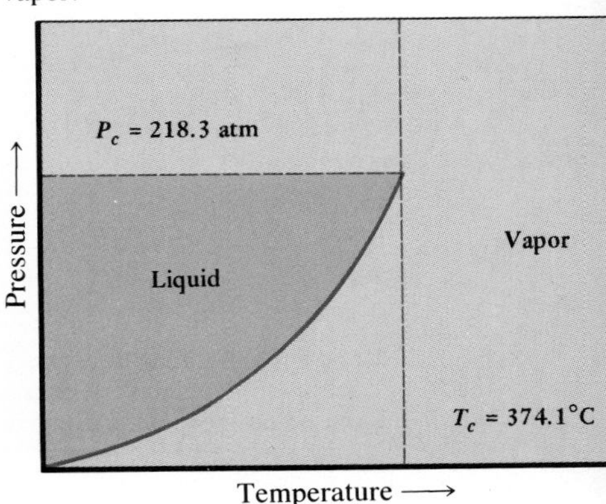

Table 13.6 Critical Temperatures and Pressures for Common Compounds

Compound	T_c (°C)	P_c (atm)
CH_4 (methane)	−82.1	45.8
C_2H_6 (ethane)	32.2	48.2
C_3H_8 (propane)	96.8	42.0
C_4H_{10} (butane)	152.0	37.5
NH_3	132.5	112.5
H_2O	374.1	218.3
SO_2	157.4	78.6

widely used as fuels in camping stoves, backyard grills, and home heating and cooking, are easily liquefied because both have relatively high critical temperatures.

Chlorofluorocarbons (CFC's) are widely used in air conditioners and refrigerators because many have useful values of T_c and P_c and were once thought to be chemically inert. Recently it was found that they are reactive under certain conditions and may be partly responsible for damage to the "ozone layer" in the upper stratosphere. Chemists are working to find an acceptable substitute for CFC's.

In the supercritical state, certain fluids like water and carbon dioxide take on unexpected properties, such as the ability to dissolve normally insoluble materials. Supercritical CO_2 is especially promising. This fluid does not dissolve water or polar compounds such as sugar, but it does dissolve less polar oils, which constitute many of the flavoring or odor-causing compounds in foods. For example, food companies are already using supercritical CO_2 to extract caffeine from coffee.

SURFACE TENSION AND VISCOSITY

Molecules at the surface of a liquid behave differently than those in the interior because molecules in the interior interact with molecules all around them (Figure 13.17). In contrast, surface molecules are affected only by those below the surface layer. This leads to a *net* inward force of attraction on the surface molecules, contracting the surface and making it behave as though it had a "skin." The "toughness" of the skin of a liquid is measured by its **surface tension**, the energy required to break through the surface or to disrupt a liquid drop and spread the material out as a film. It is surface tension that causes water drops to be spheres and not little cubes, for example (Figure 13.18), because the sphere has a smaller surface area than any other shape of the same volume.

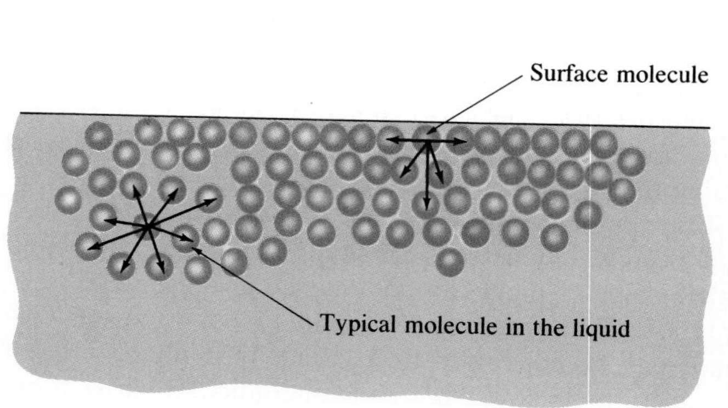

Figure 13.17 The difference between the forces acting on a molecule within the liquid phase and those acting on a molecule at the surface of the liquid.

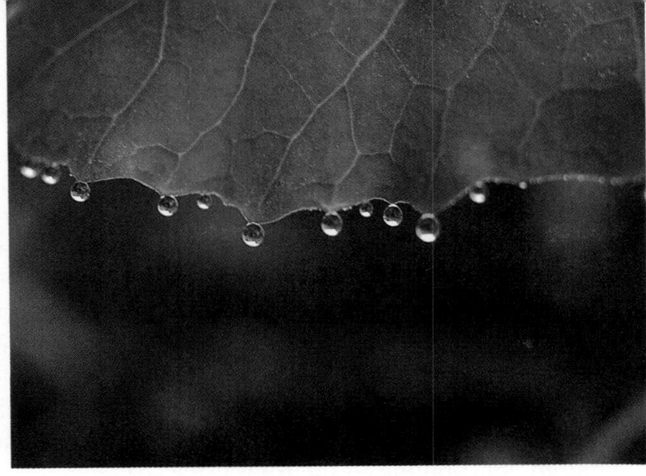

Figure 13.18 Surface tension causes drops of liquid to be spheres and not some other shape, such as little cubes.

Capillary action is closely related to surface tension. When a small-diameter glass tube is placed in water, the water rises in the tube, just as water rises in a piece of paper in water (Figure 13.19). Because there are polar Si—O bonds on the surface of glass, polar water molecules are attracted by **adhesive** forces between the two different substances. These forces are strong enough that they can compete with the **cohesive** forces between the water molecules themselves. Thus, some water molecules can adhere to the walls, and other water molecules are attracted to these and build a "bridge" back into the liquid. In addition, the surface tension of the water (from cohesive forces) is great enough to pull the liquid up the tube, so the water level rises in the tube. The rise will continue until the various forces (adhesion between water and glass, cohesion between water molecules, and the force of gravity on the water column) are in equilibrium. These forces lead to the characteristic concave or downward-curving **meniscus** that you see for water in a drinking glass or in a laboratory test tube (Figure 13.20).

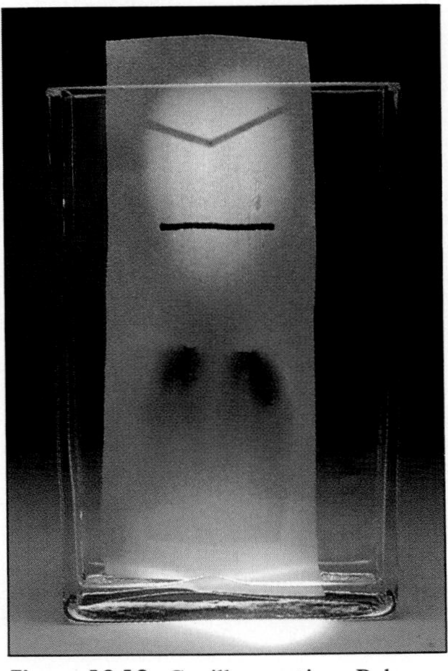

Figure 13.19 Capillary action. Polar water molecules are attracted to the O—H bonds in paper fibers, and water rises in the paper. If a line of ink is placed in the path of the rising water, the different components of the ink are attracted differently to the water and paper and are separated in a process called *chromatography*. (The dark line higher on the paper is a line of brown ink. Such a line was originally placed at the bottom of the paper, and the different components of the ink now appear as different colors.)

There are of course some liquids, such as mercury, for which cohesive forces (high surface tension) are much greater than adhesive forces toward glass, for example. Mercury will not climb the walls of a glass capillary, and when it is placed in a tube it will have a convex or upward-curving meniscus (Figure 13.20).

Finally, one additional important property of liquids is their **viscosity**, the resistance to flow (Figure 13.21). When you turn over a glassful of water, it empties quickly, but it takes much more time to empty a

Figure 13.20 The meniscus formed by two different liquids. For water, the cohesive forces between molecules are partly overcome by the adhesive forces between water and glass, and water has a concave meniscus. In contrast, the cohesive forces within mercury are significantly greater than the adhesive forces between mercury and glass, and mercury has a convex meniscus.

Figure 13.21 Honey is a very viscous fluid.

glassful of honey or rubber cement. Although intermolecular forces play a significant role in determining viscosity, there are other factors. For example, olive oil consists of molecules with very long chains of CH_2 units and O atoms (see Chapter 26), and it is about 70 times more viscous than ethyl alcohol, a molecule with only a three-atom chain ($H_3C—CH_2—OH$). The long-chain molecules of natural oils are floppy and become entangled with one another; the longer the chain, the greater the tangling and the greater the viscosity. Also, longer chains have greater intermolecular forces because there are more atoms to attract each other, each atom contributing to the total force.

13.4 SOLIDS

There are many kinds of solids in the world around us (Figure 13.22). Solid state chemistry is one of the booming areas of science, especially because of exciting new developments in superconducting solids and other new materials (see Chapter 25). In this section we can give only a *very* brief introduction to some of the principles of this exciting area.

To organize solid state chemistry, we might classify different types of solids as in Table 13.7. This section describes the solid-state structures of some common metallic and ionic solids, with some information on a few other kinds of solids as well.

SOME PRINCIPLES OF SOLID STATE STRUCTURES

In both gases and liquids, molecules continually move; they move about randomly as well as rotate and vibrate. Because of this, there can be no extensive, orderly arrangement of molecules in the gas or liquid state, and this is the major difference between liquids and solids. In solids, the molecules, atoms, or ions cannot move about (although they vibrate and occasionally rotate), and there can be long-range order, a characteristic of the solid state (Figure 13.4). It is to these orderly arrangements that we first turn our attention.

UNIT CELLS AND CRYSTAL LATTICES The beautiful regularity of a crystal of salt or a metal suggests that it has an *internal* symmetry as well. Indeed, all crystal lattices (whether of metals or ionic solids) are built of **unit cells**, the smallest, repeating unit that has all of the symmetry characteristics of the way the atoms are arranged.

Figure 13.22 Some examples of pure, crystalline solids. Iron pyrite (FeS_2, left), calcite ($CaCO_3$, center), and fluorite (CaF_2, right).

To understand the idea of a unit cell, look at the repeating pattern of circles in Figure 13.23. A "unit" cell could be drawn around each atom, but this unit cell would not reveal how the circles are arranged with respect to each other; that is, it would not show the symmetry of the arrangement of circles. A way to draw a cell that shows this is to draw lines from the center of one circle to the centers of neighboring circles; a square unit cell results. Furthermore, since each of the four circles contributes one quarter of itself to the square's contents, a *net* of one circle is located within each square. This unit cell reveals the pattern of circles, and the entire pattern can be built by joining identical square unit cells edge to edge, each cell containing the equivalent of one circle.

Table 13.7 Types of Solids and Modes of Bonding

Type of Solid	Examples	Type of Bonding or Intermolecular Force
Ionic	NaCl and other metal halides	Ionic
Metallic	Fe, Zn, Cu, Au, Ag, etc.	Metallic
Molecular	H_2O as ice, I_2, sulfur	Covalently bonded small molecules Dipole–dipole, hydrogen bonding, and induced dipole forces bind molecules together
Network	Graphite, diamond, quartz	Extended network of covalently bonded atoms
Imperfect		
Disordered	AgI and other "superionics" (used in some batteries)	Ionic
Nonstoichiometric	Semiconductors such as CdS	Ionic to metallic
Amorphous	Glass; organic polymers (polyethylene, polystyrene, etc.)	Covalently bonded networks; networks bound together by dipole–dipole or induced dipole forces

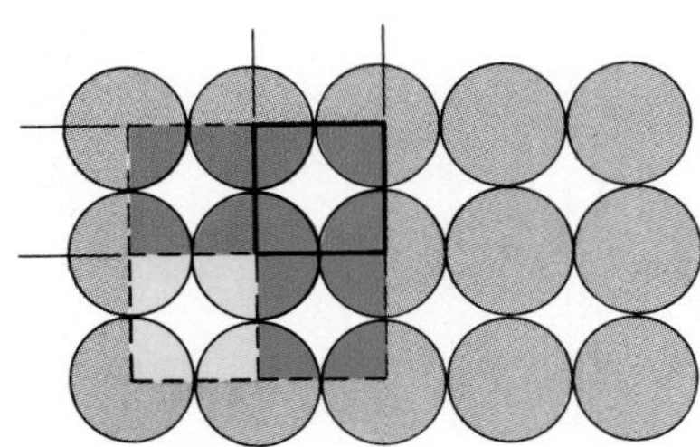

Figure 13.23 Unit cells for a "two-dimensional solid" made from flat, circular objects. Here the unit cell is a square. Each circle at a corner of a square contributes $\frac{1}{4}$ of its area to the area inside the square. Thus, there is a *net* of one circle per unit cell.

Solids can be built by piling up three-dimensional unit cells like building blocks, as the artist M.C. Escher illustrated in a well-known drawing (Figure 13.24). The corners defining each unit cell in simple solids have identical environments in an ionic solid, in a metallic solid, or in a molecular solid. Such points are equivalent to each other, and collectively

Figure 13.24 "Cubic Space Division" by M.C. Escher. The artist has depicted three-dimensional space built up by stacking many, many cubes (with each corner of each cube itself a smaller cube). Each cube is a unit cell of the whole, since the whole can be made up by repeating the smaller cubes. (Haags Gemeentemuseum)

they define the **crystal lattice**. In what follows we will assume there are ions, metal atoms, or molecules centered at these points.

To construct crystal lattices, nature uses seven types of three-dimensional unit cells that differ from one another in that their sides have different relative lengths and their edges meet at different angles (Figure 13.25). However, we shall be concerned here only with **cubic unit cells**, cells with equal edges that meet at 90° angles (as in Figure 13.24). They are easily visualized and are quite common in nature. Within the cubic class, there are three important cell symmetries: **simple cubic (sc)**, **body-centered cubic (bcc)**, and **face-centered cubic (fcc)** (Figure 13.26). All three have *eight* identical atoms or ions at the corners of a cube. However, the body-centered (bcc) and face-centered cubic (fcc) arrangements differ from the simple cube (sc) in having additional, identical atoms, of the same type as those at the corners, at other locations. The bcc structure is called "body-centered" because it has one additional, identical atom at the center of the cube or "body." The fcc arrangement is called "face-centered" because it has, in the center of each of the six cube faces, an ion or atom of the same type as the corner ions or atoms.

Cubic unit cell abbreviations: sc = simple cubic, bcc = body-centered cubic, and fcc = face-centered cubic

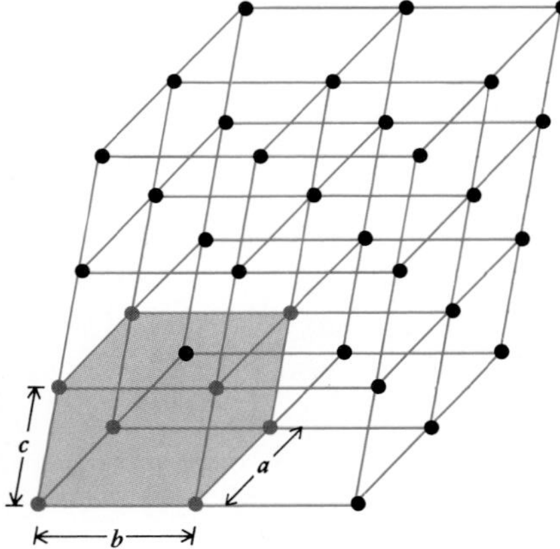

Figure 13.25 All solid cells are parallelepipeds. They have six faces with parallel opposite edges (parallelograms). In a simple cubic unit cell, all faces are squares.

It is a simple "rule" of nature that things are done as efficiently as possible. If you view lattices as being built by the efficient packing of tiny spherical atoms or ions, you will find that the fcc lattice uses 74% of the available space in a cube, while spheres packed in a simple cubic lattice use only 68% of the available space, clearly a less efficient arrangement. You can appreciate this fact to some extent by looking at Figure 13.27, a two-dimensional example. Here you see the two basic ways of packing hard, spherical ions or atoms together in one layer. In one arrangement (Figure 13.27a), the spheres are at the corners of a square and each touches four other spheres. In the other (Figure 13.27b), each sphere touches six others at the corners of a hexagon. In either arrangement, little holes are left between atoms in the layer (and this space is greater in Figure 13.27a than in 13.27b). *The way the atoms are arranged in the layers and the way the layers are arranged on top of one another is the key to the structures of metallic and ionic solids.*

To fill three-dimensional space, layers of atoms or ions are stacked one on top of the other. If you start with the square arrangement (Figure

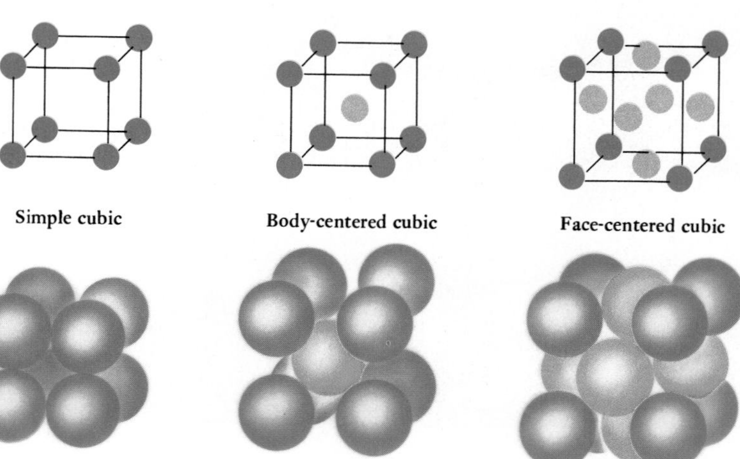

Figure 13.26 The three different types of cubic unit cells. The top row shows the lattice points of the three cells. In the bottom row the points are replaced with space-filling spheres representing the atoms or ions of the lattice. All spheres—no matter what their color—represent identical atoms or ions centered on the lattice points. Notice that the spheres at the corners of the body-centered and face-centered cubes do *not* touch each other. Rather, each corner atom in the body-centered cell touches the center atom, and each corner atom in the face-centered cell touches spheres in the three adjoining faces.

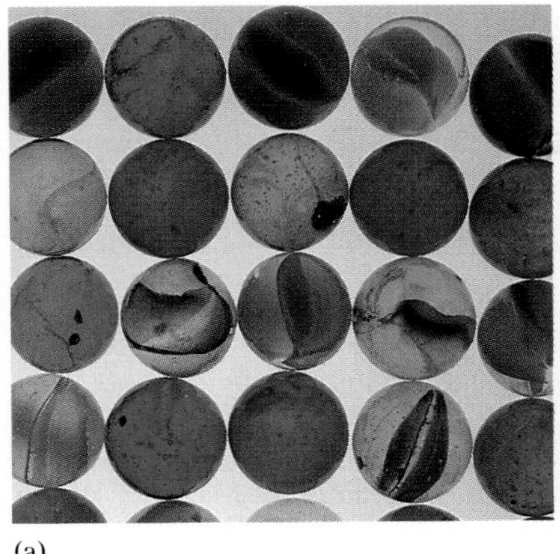
(a)

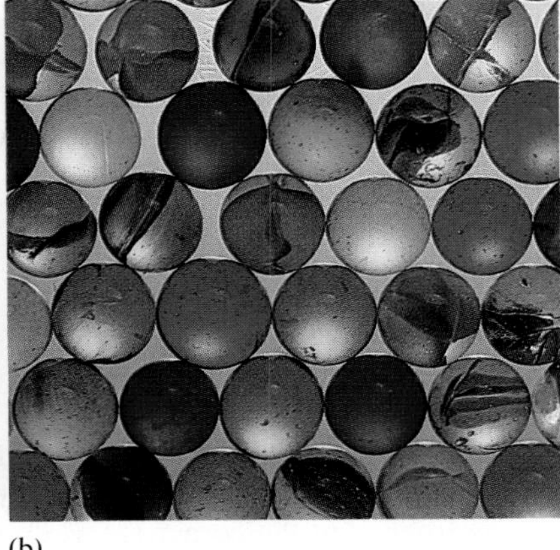
(b)

Figure 13.27 Packing of spheres in one layer. (a) Marbles are arranged in a square pattern, and each marble contacts four others (see Figure 13.23). (b) Each marble contacts six others at the corners of a hexagon. It should be evident that (b) is a more efficient packing arrangement than (a).

13.27a) for the first layer and simply stack the next layer of spherical ions or atoms *directly* on top of the first, the arrangement would resemble Escher's work in Figure 13.24, an extended array of cubic unit cells.

As mentioned above, there are more efficient ways to fill space than the simple cube, and both begin with the hexagonal arrangement of spheres in Figure 13.27b. One of these is called **cubic close-packing** and leads to the face-centered cubic (fcc) lattice described earlier (Figure 13.28). The other arrangement is called **hexagonal close-packing**. This latter arrangement is common in nature, but we shall discuss only the simple cubic, body-centered cubic, and face-centered cubic arrangements.

The atoms of metals are commonly packed in the following ways: (a) body-centered cubic; (b) face-centered cubic; or (c) hexagonal close-packed structures. For example, the alkali metals are body-centered cubic while nickel, copper, and aluminum are face-centered cubic; magnesium

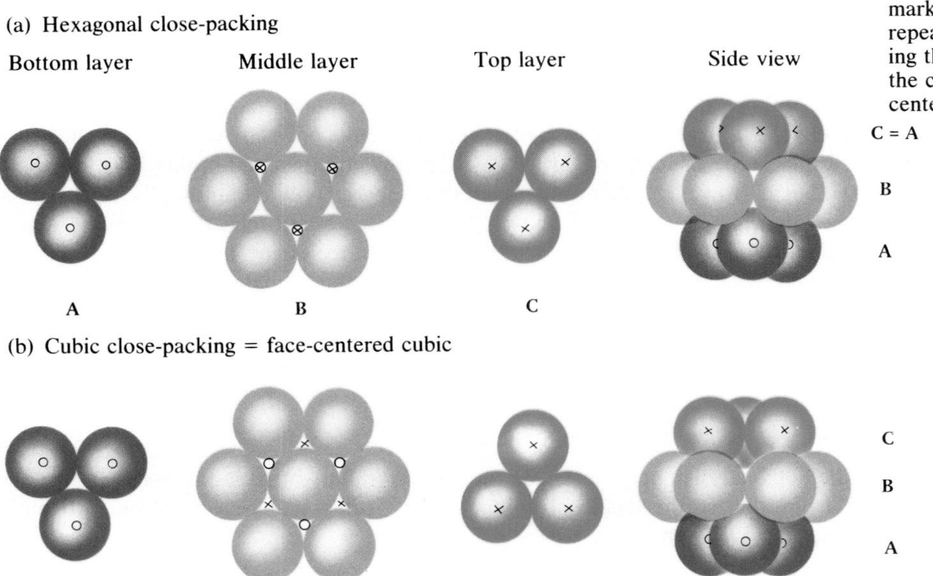

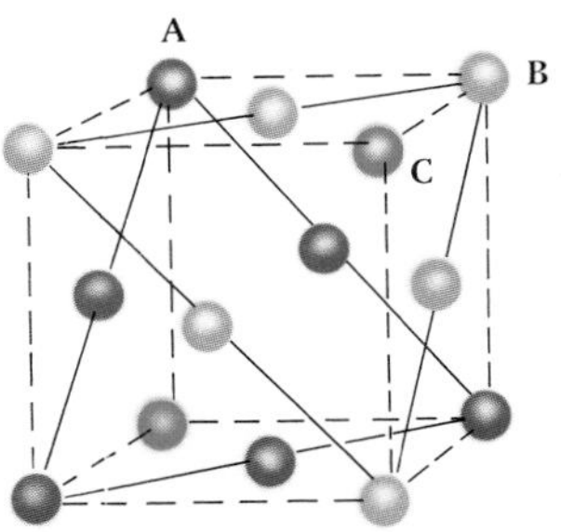

Figure 13.28 The two most efficient ways to pack atoms or ions in crystalline materials. In both arrangements the packing within each layer is the hexagonal pattern shown in Figure 13.27b and described in the text. (a) In the *hexagonal close-packing (hcp)* arrangement, the layers placed above and below a given layer fit into the same depressions on either side of the middle layer. In a three-dimensional crystal, the layers repeat in their pattern in the manner ABABAB. . . . Atoms in each A layer are directly above the ones in another A layer; the same holds for the B layers. (b) In *cubic close-packing (ccp)*, the atoms of the bottom layer rest in depressions marked "o", and those of the top layer are oppositely oriented, resting in the depressions marked "x". In a crystal, the pattern is repeated ABCABCABC. . . . By turning the whole crystal, you can see that the ccp arrangement is just the face-centered cubic structure.

The computer program "KC? Discoverer" is available to classes using this text. The program shows the solid state structures of all of the elements.

is hexagonal close-packed. One way you could find out whether a metal is body-centered cubic or face-centered cubic, for example, is to use the density of the metal and the known radius of the atom to calculate the number of atoms per unit cell. The approach is outlined in Example 13.5.

EXAMPLE 13.5

THE STRUCTURE OF A METAL

Aluminum has a density of 2.699 g/cm³, and the radius of the aluminum atom is 143 pm. Verify that the metal crystallizes as a face-centered cube.

Solution Assuming Al crystallizes as a face-centered cube, the face of one cube would appear as drawn below. Notice that the center atom touches each of the corner atoms, but the corner atoms do not touch one another.

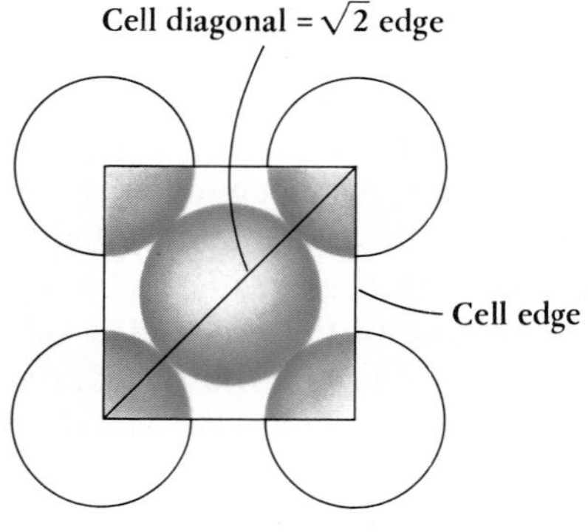

One face of face-centered cubic crystal

Our strategy in solving this problem is as follows:

1. Find the volume of the unit cell.
2. From the cell volume and the density of Al, calculate the mass of one unit cell.
3. Compare the mass of the unit cell with the mass of one atom of Al.

Step 1: Volume of unit cell. Since the volume of the unit cell = (length of edge)³, we need to know the length of the cube edge. If Al atoms touched one another along the cube edge, we could simply add their radii to find the length of the edge. However, the atoms touch one another only along the diagonal across the cube face (see drawing above). Therefore, we add radii to find the *diagonal* distance, and then use the fact that this distance is the hypotenuse of a right triangle, with the cube edges the other sides of the triangle. Since the sum of the squares of the edge lengths is equal to the square of the hypotenuse, we have

$$(\text{diagonal distance})^2 = 2\ (\text{edge})^2$$

or

$$\text{diagonal distance} = \sqrt{2}\ (\text{edge})$$

The diagonal distance across the face = 4 × radius of Al atom = 4 × 143 pm = 572 pm, and so the length of a cube edge = (diagonal/$\sqrt{2}$) = 404 pm. In centimeters, this dimension is

$$\text{cube edge} = (404\ \text{pm})\left(\frac{1\ \text{m}}{1 \times 10^{12}\ \text{pm}}\right)\left(\frac{100\ \text{cm}}{\text{m}}\right) = 4.04 \times 10^{-8}\ \text{cm}$$

and so the volume of the cube is

$$\text{cube volume} = (\text{edge})^3 = (4.04 \times 10^{-8}\ \text{cm})^3 = 6.62 \times 10^{-23}\ \text{cm}^3$$

Step 2: Mass of unit cell.

$$\begin{aligned}\text{Mass} &= \text{density} \times \text{volume} = (2.699\ \text{g/cm}^3)(6.62 \times 10^{-23}\ \text{cm}^3/\text{unit cell})\\ &= 1.79 \times 10^{-22}\ \text{g/unit cell}\end{aligned}$$

Step 3: Compare mass of unit cell with mass of one Al atom.

$$\begin{aligned}\text{Mass of one Al atom} &= (26.98\ \text{g/mol})(1\ \text{mol}/6.0221367 \times 10^{23}\ \text{atoms})\\ &= 4.480 \times 10^{-23}\ \text{g/atom}\end{aligned}$$

$$\begin{aligned}\text{Atoms per unit cell} &= \left(\frac{1.79 \times 10^{-22}\ \text{g}}{\text{unit cell}}\right)\left(\frac{1\ \text{atom}}{4.480 \times 10^{-23}\ \text{g}}\right)\\ &= 3.99\ \text{atoms/unit cell}\end{aligned}$$

We have arrived at four atoms per unit cell for Al. Does this correspond with a face-centered cubic unit cell? If you look closely at any cubic unit cell, you find that each atom at a corner has $\frac{1}{8}$ of its volume inside a given unit cell (Figure 13.29). In addition, in the face-centered cubic unit cell each of the six face-centered ions has $\frac{1}{2}$ of its volume inside a given unit cell. Therefore, the number of atoms *within* the unit cell is

$$\begin{aligned}&(8\ \text{corners})(\tfrac{1}{8}\ \text{atom per corner}) + (6\ \text{faces})(\tfrac{1}{2}\ \text{atom per face})\\ &\qquad = 4\ \text{atoms per fcc unit cell}\end{aligned}$$

Our calculated result of four Al atoms per unit cell agrees with the number expected for a face-centered cubic unit cell.

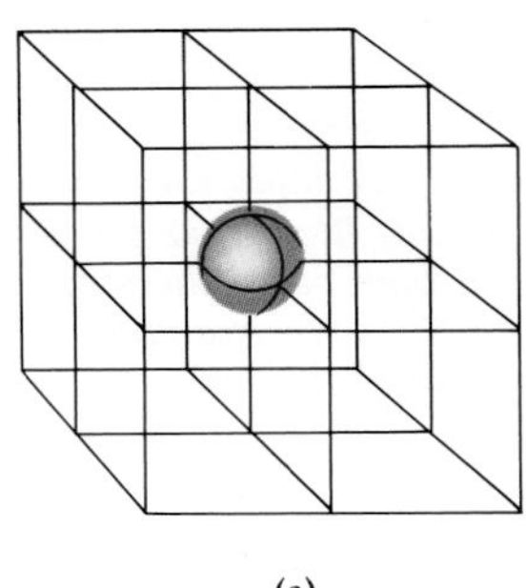

(a)

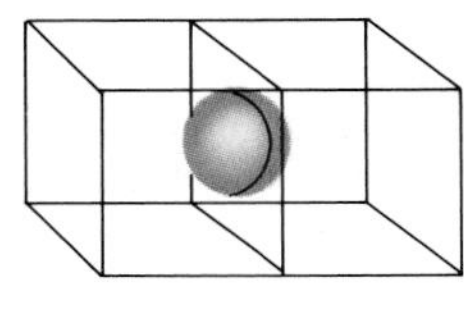

(b)

Figure 13.29 Atom sharing at cube corners and faces. (a) In any cubic lattice each corner atom (or ion) is shared equally among 8 cubes; $\frac{1}{8}$ of each atom (or ion) is within a particular cube. (b) In a face-centered lattice, each atom (or ion) in a cube face is shared equally between 2 cubes. Each atom (or ion) of this type contributes $\frac{1}{2}$ of itself to a given cube.

EXERCISE 13.5 The Structure of Solid Iron

Iron has a density of 7.8740 g/cm³ and the radius of an iron atom is 126 pm. Verify that the structure of the solid is body-centered cubic. You can do this by calculating the net number of atoms in the unit cell of iron and then comparing this result with the fact that a body-centered cubic unit cell should contain two atoms. (Look at Figure 13.26 and see if you also get two atoms.)

To solve this problem you need to recognize that the atoms of the lattice touch only along the diagonal line running from a corner at the "top" of the cube to the opposite corner at the "bottom" of the cube. (As in the fcc structure in Example 13.5, the corner atoms do not touch one another.) The distance along this diagonal equals 4 times the radius of an atom and is also equal to $\sqrt{3}$ times the length of the edge of the cube.

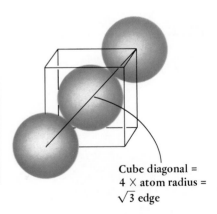

Cube diagonal = 4 × atom radius = $\sqrt{3}$ edge

Body-centered cube with two opposite corner atoms and center atom.

One of the most useful results to come from the preceding example and exercise is the following:

1. **A face-centered cubic unit cell of X atoms (or ions) will always contain four *net* X atoms (or ions) within the cell.**
2. **A body-centered cubic unit cell of X atoms will always contain two *net* X atoms within the cell.***

*As pointed out in the discussion of ionic compounds below, there are no cases where *ions* of one kind are packed as a body-centered cube.

These results are very important in seeing the relation between the structures of ionic solids and their formulas, the next topic.

In general, we build an ionic lattice out of the larger anions, say Cl^-, and then place the smaller cations, Na^+, for example, in some or all of the holes that remain.

STRUCTURES OF SIMPLE IONIC SOLIDS The lattices of many ionic compounds are built by taking a simple cubic or face-centered cubic lattice of spherical ions of one type and placing ions of opposite charge in the holes left in the lattice. The number and location of the holes that are filled are the keys to understanding the relation between the lattice structure and the formula of a salt.

Cesium chloride (CsCl) crystal structure.

The hole in a simple cube is in the center of the cell, and this is the only type of space available (Figure 13.26). The ionic compound cesium chloride, CsCl, adopts just this structure: simple cubes of Cl^- ions with a Cs^+ ion in the center of each cube.

The structure of NaCl, sodium chloride, represents one of the most common ways of constructing an ionic solid (Figure 13.30). You can view the salt as having been built from the larger Cl^- ions packed as a face-centered cube, with the smaller Na^+ ions filling all the so-called *octahedral* lattice holes (Figure 13.31).* The holes are said to be octahedral because each Na^+ is surrounded by six Cl^- ions at the corners of an octahedron.

To understand the relation between the structure of the unit cell of a salt and the formula of the salt, we need only to find the maximum number of hole ions permitted per lattice ion. This is done for NaCl and CsCl in the next example and exercise, respectively.

Figure 13.30 NaCl unit cell. (a) An expanded view of one unit cell. The lines represent only the connections between lattice points. They are not bonds. (b) A more extended view showing the ions as spheres, the smaller Na^+ ions packed into a face-centered cubic lattice of larger Cl^- ions.

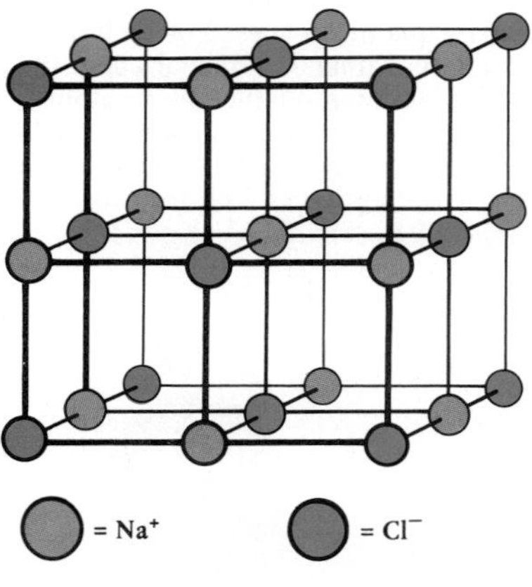

(a)

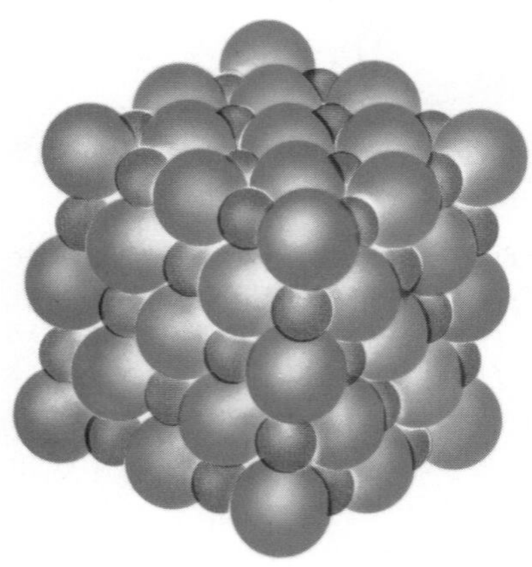

(b)

EXAMPLE 13.6

FORMULA AND STRUCTURE OF SODIUM CHLORIDE

One unit cell of NaCl is illustrated in Figure 13.30a. Show that this unit cell contains Na^+ and Cl^- ions with a ratio of 1:1.

Solution Formula and structure are related by counting the number of lattice ions in one unit cell and then finding how many ions of opposite charge there are in the holes within the unit cell. Here we consider NaCl as a face-centered cubic lattice of Cl^- ions with smaller Na^+ ions in the octahedral lattice holes. In Example 13.5 we found that a face-centered cubic lattice of atoms or ions of type X should have a net of four X atoms or ions within the unit cell. Therefore, the NaCl lattice has four Cl^- ions within the unit cell.

The Na^+ ions are in octahedral holes in the Cl^- lattice. Figure 13.31 shows that there is one octahedral hole at the cube center. Furthermore, there are three octahedral holes accounted for by Na^+ ions at the centers of the cube edges. We arrive at this number because there are 12 cube edges, and $\frac{1}{4}$ of each ion sharing an edge is *within* the unit cell being considered. Thus, the total number of octahedral holes, each occupied by a Na^+ ion in NaCl, is

$$(1 \text{ octahedral hole at cube center}) + (12 \text{ edges})(\tfrac{1}{4} \text{ octahedral hole per edge}) = 4 \text{ octahedral holes}$$

The numbers of Cl^- lattice ions and Na^+ ions in octahedral holes are both 4, so a unit cell of NaCl has a 1:1 ratio of Na^+ and Cl^- ions as required.

---

*The octahedral holes do not represent all of the available holes. There are also so-called *tetrahedral* holes. We shall not pursue this further here. However, you should be aware that the Na^+ ions occupy *only* octahedral holes and not a random mixture of octahedral and tetrahedral holes.

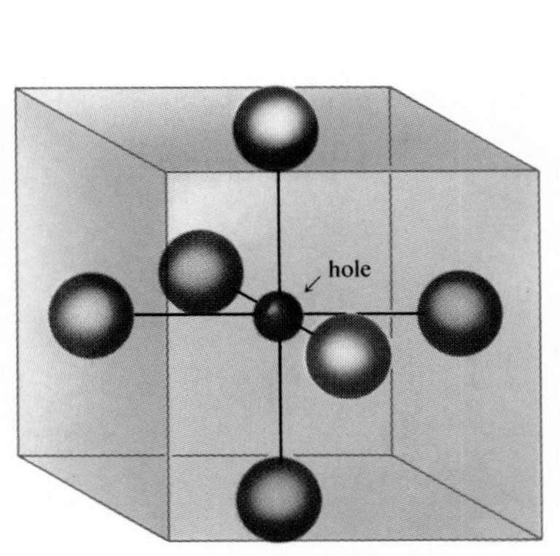

1 hole of this kind in the body center

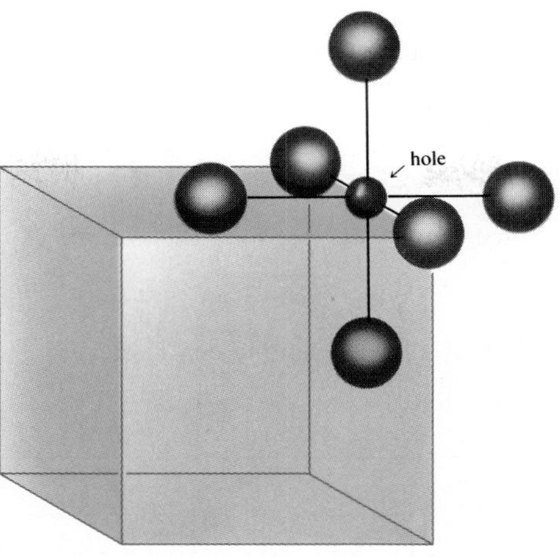

12 holes of this kind on the 12 edges of the cube (a net of 3 holes)

Octahedral holes in a face-centered lattice

Figure 13.31 Octahedral holes in a face-centered cubic lattice. If a cation (Na^+) is placed in the center of a face-centered cubic lattice of anions (Cl^-), the cation is surrounded by an octahedron of anions. In addition, there is an equivalent octahedral hole in the center of *each* of the 12 cube edges. A cation in each cube edge hole will have four nearest-neighbor ions of opposite charge in the same cube and two in a neighboring cube. In all common crystals with face-centered anion lattices and cations in octahedral holes, *all* such holes are filled with cations.

EXERCISE 13.6 Ionic Structure and Formula

Cesium chloride, CsCl, has a simple cubic unit cell of Cl^- ions with one Cs^+ ion in the cubic hole. Prove that the formula of the salt must have one Cs^+ ion per Cl^- ion.

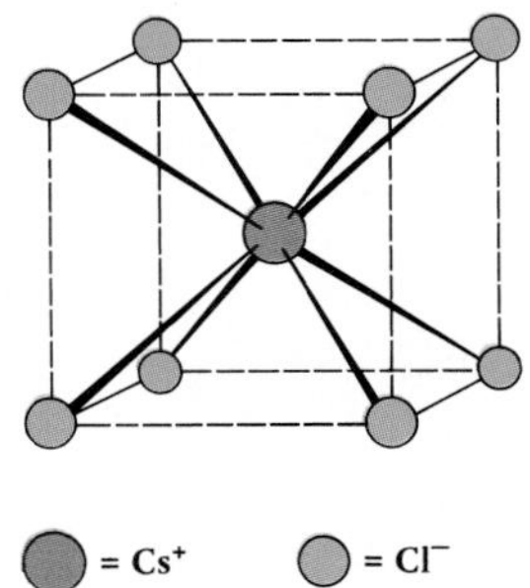

Cesium chloride (CsCl) crystal structure.

In summary, compounds with the formula MX are commonly formed in either of two crystal structures: (a) M^{n+} ions occupying *all* of the cubic holes of a simple cubic X^{n-} lattice (Exercise 13.6), or (b) M^{n+} ions in all of the octahedral holes in a face-centered cubic X^{n-} lattice (Example 13.6).* CsCl is a good example of (a) and NaCl is the usual example of (b). Indeed, chemists and geologists have observed that the sodium chloride or "rock salt" structure is adopted by many classes of compounds, especially by all the alkali metal halides (except CsCl, CsBr, and CsI), all the oxides and sulfides of the alkaline earth metals, and all the oxides of formula MO of the transition metals of period 4.

ATOM OR ION SIZE AND CRYSTAL DENSITY Atoms or ions have different radii, depending on their position in the periodic table (see Chapter 9). Since metallic or ionic solids form by packing atoms or ions as closely as possible, the size of the unit cell and the density of the solid depend on the radii of the atoms or ions involved.

EXAMPLE 13.7

UNIT CELL SIZE AND ION RADIUS

The radius of the Na^+ ion is 116 pm and the radius of Cl^- is 167 pm. Calculate the volume of the NaCl unit cell in pm^3 and in cm^3.

*Although a metal may have a body-centered cubic structure, ionic solids do not take this form. Some may say, incorrectly, that CsCl is body-centered cubic, but it is more properly a simple cubic lattice of Cl^- ions with Cs^+ ions in the cubic holes. The adjectives simple, body-centered, and face-centered tell only the arrangement of one ion type, *not both.*

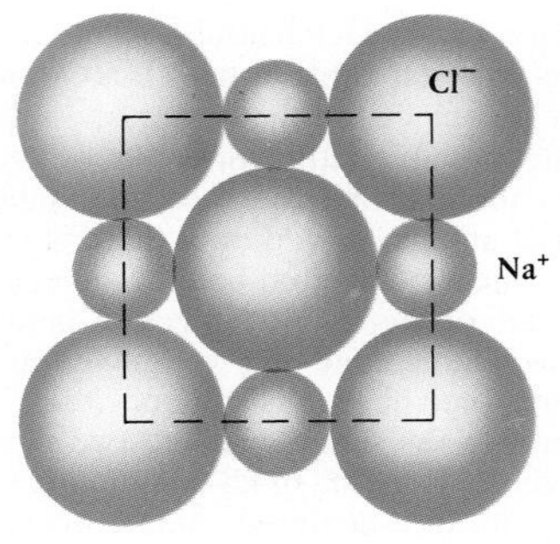

Solution One face of the face-centered NaCl unit cell would appear as shown in the margin. The Cl^- ions define the lattice, and the ions along each edge just touch one another. This means that one edge of the unit cell is equal to one Cl^- ion radius plus twice the radius of Na^+ plus another Cl^- ion radius, or

$$\text{NaCl unit cell edge} = 167 \text{ pm} + 2(116 \text{ pm}) + 167 \text{ pm} = 566 \text{ pm}$$

Since the crystal is cubic, the volume of the unit cell is the cube of the edge.

$$\text{Volume of unit cell} = (\text{edge})^3 = (566 \text{ pm})^3 = 1.81 \times 10^8 \text{ pm}^3$$

Converting this to cubic centimeters,

$$\text{Volume in cm}^3 = (1.81 \times 10^8 \text{ pm}^3)(10^{-10} \text{ cm/pm})^3 = 1.81 \times 10^{-22} \text{ cm}^3$$

EXERCISE 13.7 Unit Cell Dimensions

KCl has the same crystal structure as NaCl. Calculate the volume of a KCl unit cell, using the ion sizes given in Figure 9.15.

EXAMPLE 13.8

CRYSTAL PACKING, DIMENSIONS, AND DENSITY

The volume of a NaCl unit cell is 1.81×10^{-22} cm^3 (Example 13.7). What is the density of NaCl?

Solution The density of a material is its mass per unit volume. Since we know the volume of the NaCl unit cell, we need only to find its mass in order to calculate the density. (See the box below on calculating density.)

First, recognize that the unit cell contains four Na^+ and four Cl^- ions. You can obtain the mass of one Na^+ and one Cl^- from the molar mass of NaCl and Avogadro's number.

$$\left(\frac{58.44 \text{ g}}{1 \text{ mol NaCl}}\right)\left(\frac{1 \text{ mol}}{6.0221367 \times 10^{23} \text{ molecules}}\right) = 9.704 \times 10^{-23} \text{ g/molecule}$$

Since there are four NaCl "molecules" per unit cell, the cell has a mass of

$$\left(\frac{9.704 \times 10^{-23} \text{ g}}{\text{molecule}}\right)\left(\frac{4 \text{ NaCl molecules}}{\text{unit cell}}\right) = 3.882 \times 10^{-22} \text{ g/unit cell}$$

Now we have both the mass of the unit cell and its volume (from Example 13.7). The density is then

$$\text{Density} = \left(\frac{3.882 \times 10^{-22} \text{ g}}{1 \text{ unit cell}}\right)\left(\frac{1 \text{ unit cell}}{1.81 \times 10^{-22} \text{ cm}^3}\right) = 2.14 \text{ g/cm}^3$$

The experimental density is 2.164 g/cm^3. The small difference between the calculated and experimental values comes from using averaged Na^+ and Cl^- radii.

EXERCISE 13.8 Solid Density from Unit Cell Dimensions

Using the volume of a KCl unit cell that was found in Exercise 13.7, calculate the density of solid KCl.

The Relation Between Cell Structure, Mass, and Volume, and the Density of a Solid

The "flow diagram" below relates specifically to Example 13.8, but it can be used for most metallic or ionic solids to calculate one of the following if the others are known: cell type, ion or atom size, or density of the solid. For example, if you know the density of the solid, you can find the length of the edge of a unit cell if you know the type of unit cell: (a) combine the density with the mass of a unit cell to get the cell volume and then (b) find the cube root of the volume to obtain the edge dimension.

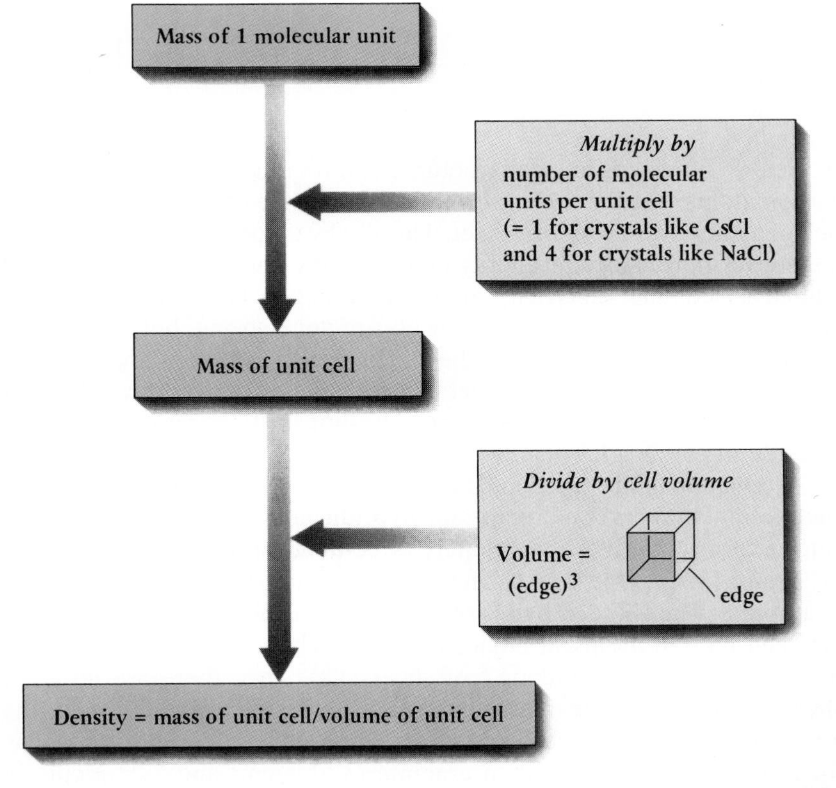

Figure 13.32 Molecular solids: violet-black I_2 and white, solid CO_2 (dry ice). These solids consist of molecules packed as closely as possible.

Figure 13.33 Quartz (SiO_2) is an example of a network solid. (Other examples of the general class of Si-O compounds called silicates are described in Chapter 23.) This photo shows quartz crystals surrounding a core of iron oxide (height is 3 inches). (Photo taken in Wind Cave, Hot Springs, South Dakota; Arthur N. Palmer)

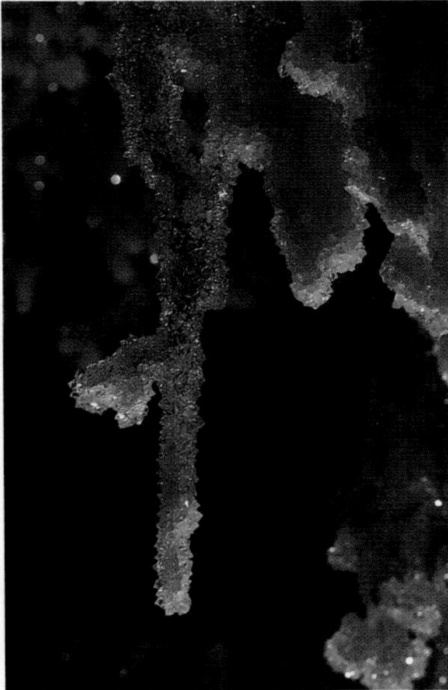

OTHER TYPES OF SOLIDS

Covalently bonded molecules such as H_2O, I_2, CO_2, and many others condense to form solids at temperatures at which the intermolecular forces are sufficiently strong to keep the molecules in place. Further, when such molecules are assembled into a solid, they often adopt the body-centered cubic or face-centered cubic forms. Iodine, for example, is a crystalline solid at room temperature with I_2 molecules at the lattice points of a face-centered cubic arrangement (Figure 13.32).

There are a number of solids composed of **networks of covalently bonded atoms**, and excellent examples are quartz, graphite, and diamond (Figure 13.33). The latter two are **allotropes** of carbon; that is, they are different forms of the same element under the same conditions of temperature and pressure. Graphite is the classic example of a planar network solid. Here sp^2 hybridized carbon atoms form a planar network of six-membered rings. Just as you can draw resonance structures for the simple hydrocarbon benzene, so too can resonance structures be imagined for a layer of graphite.

Resonance structures of benzene, C_6H_6

Graphite layer

In fact, so many resonance structures can be drawn for this network of carbon atoms that the π electrons of each ring are effectively delocalized or spread out over the entire sheet. This is why graphite is black (a graphite π electron can absorb the energy of almost any photon) and why graphite will easily conduct electricity along the direction of the sheets. The covalent bonds within the graphite sheets are quite strong, but, as indicated by the large distance between sheets (Figure 13.34), they are bonded to one another only by weak dispersion forces. Owing to the weakness of bonding between the sheets, they readily slide past one another and allow graphite to act as a lubricant.

Diamonds are also built of six-membered carbon rings, but here each tetrahedral carbon atom is sp^3 hybridized and bonded to four other carbon atoms to give a nonplanar, three-dimensional network (Figures 11.20 and 13.35). One way to view the diamond lattice shows its relation to graphite: The C=C π bonds of graphite are replaced by C—C σ bonds between carbon atoms in adjacent sheets. There are several important results of this "cross-linking" of sheets: (a) Since there are no longer sheets to slide past one another, diamonds are not easily deformed; they are among the hardest solids known and are used industrially as abrasives. (b) Diamonds are denser than graphite (3.51 g/cm^3 and 2.22 g/cm^3, re-

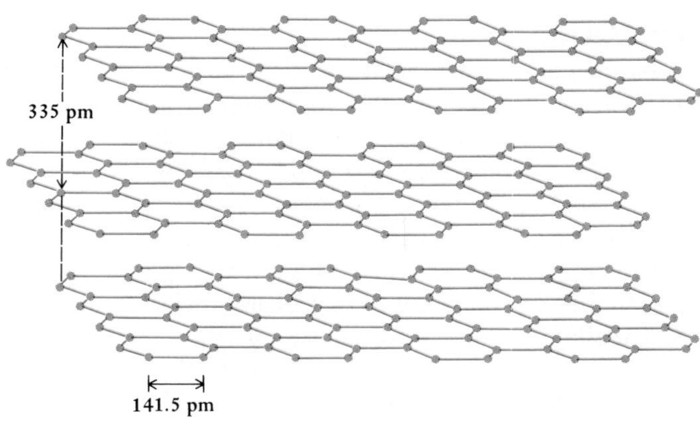

Figure 13.34 The layered structure of graphite.

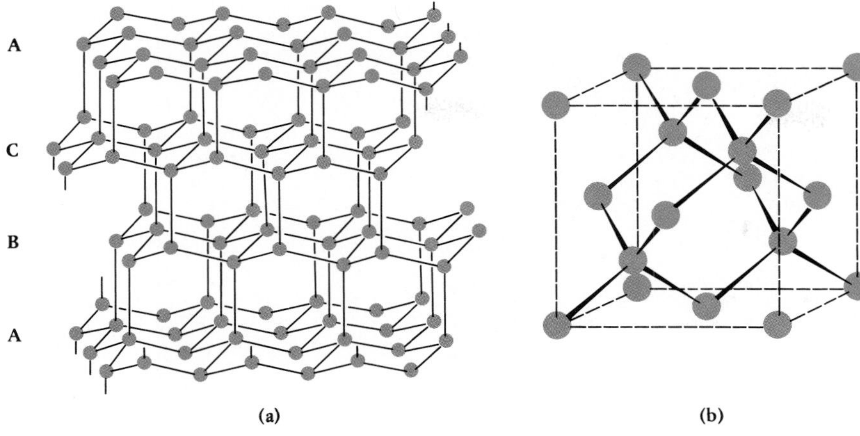

Figure 13.35 Two views of the diamond structure. (a) The three-dimensional network of C_6 rings with the basic structural element of tetrahedral, sp^3 hybridized carbon atoms, leading to puckered six-membered rings. (b) The unit cell of diamond. You can view it as a face-centered cubic arrangement of C atoms with four other C atoms located in holes in the lattice.

spectively). (c) Diamonds are very high melting, are colorless when absolutely pure, are nonconductors of electricity (Chapter 11), and are less chemically reactive than graphite.

The solids described so far are those with a long-range orderly arrangement of atoms, solids that we generally recognize because they form well-defined crystals. However, there is an important class of materials without such order: **amorphous solids**. Examples include some of our most important materials, among them glass and organic **polymers** such as polyethylene (Figure 13.36) and polypropylene. The prefix *poly* in *poly*mer or *poly*ethylene means that many units (called *mono*mers) are linked together to form a giant molecule. Thus, *poly*ethylene is formed by linking ethylene monomers ($CH_2{=}CH_2$) to form very long chains of CH_2 groups [($-CH_2-$)$_n$]. Interactions between polymer chains occur by weak dispersion forces and lead to solids that can be molded into the various shapes you see in "plastic" objects. Polymers, both man-made and natural, are described in more detail in Chapter 27.

***Imperfect solids** are yet another class of solids. These include the semiconductors described in Chapter 11 as well as the magnetic materials that are the heart of computers and allow you to record video and sound images.*

Figure 13.36 A model of a polyethylene chain (the black balls are C atoms and the white ones are H atoms) and a laboratory bottle made from the polymer.

PHYSICAL PROPERTIES OF SOLIDS

We know now that the outward shape of a crystalline solid is a reflection of its internal structure. But what about the temperatures at which solids melt, their hardness, or their solubility in water? All of these and many others are physical properties of solids that are of interest to chemists, geologists, engineers, and others. A few such properties are explored here.

Sodium chloride melts at 801 °C and potassium iodide at 680 °C. These rather high temperatures must reflect very strong forces holding the ions in place, forces reflected by the **lattice energy** of an ionic crystal. The lattice energy (symbolized by U) is defined as the energy required to separate the ions of the crystal, converting them to a collection of gaseous ions. For example,

$$NaCl(s) \longrightarrow Na^+(g) + Cl^-(g)$$

$$\text{lattice energy} = U = +765.8 \text{ kJ/mol}$$

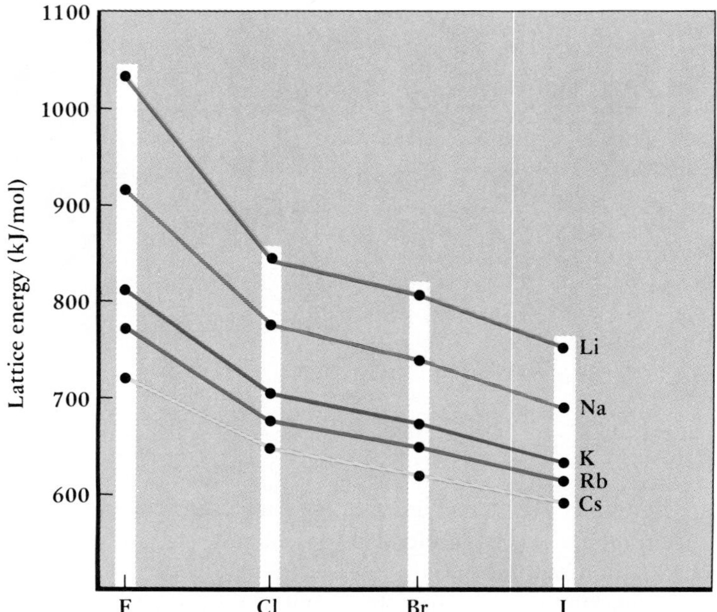

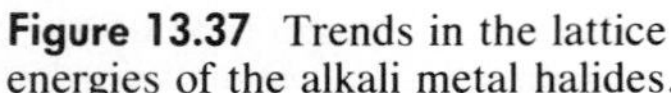
Figure 13.37 Trends in the lattice energies of the alkali metal halides.

The lattice energy, like the energy required to dissociate an ionic compound in the gas phase (Section 10.3), depends directly on the ion charges and inversely on the distance between them. In addition, the lattice energy depends on the arrangement of ions in the lattice. Just as in the case of the dissociation energies, there is a periodicity to lattice energies as illustrated in Figure 13.37. The lattice energy clearly drops as ion size (and so the separation between cation and anion) increases for the alkali metal halides. As we shall see, this is closely related to such observable physical properties of solids as their melting points.

The **melting point** of a solid is the temperature at which the crystal lattice collapses and solid is converted to liquid. Just as in the case of the liquid $\rightarrow$ vapor transformation, melting requires energy. In Chapter 6 we referred to this energy as the **latent heat of fusion** (in J/gram), or we can call it the **molar enthalpy of fusion** (in J/mol).

$$\text{Solid} + \text{heat} \longrightarrow \text{liquid}$$

$$\text{heat} = \text{molar enthalpy of fusion} = \Delta H_{\text{fusion}}\ (\text{J/mol})$$

Enthalpies of fusion can range from just a few thousand joules per mole to many thousands per mole. A low enthalpy of fusion means that the solid melts at a low temperature, while high melting points reflect high enthalpies of fusion. Figure 13.38 shows the enthalpies of fusion of most of the metals of the periodic table relative to one another. Just a few of the interesting features of the figure are:

(a) Metals that have notably low melting points, such as the alkali metals, also have very low enthalpies of fusion.

(b) Mercury, a liquid at room temperature, has an enthalpy of fusion of only 2.3 kJ/mol.

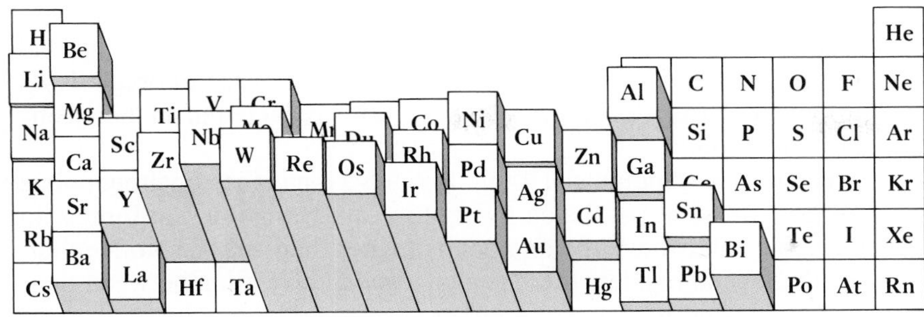

Figure 13.38 Relative enthalpies of fusion for most of the metals of the periodic table. To provide a scale, a few data are given below. (No data were available for Hf and Ta in *MacMendeleev*, the computer program used to draw this figure. From *KC?Discoverer* we know that the heats of fusion for Hf and Ta are 25.1 and 31.38 kJ/mol, respectively.)

METAL	ΔH_{fusion} (kJ/mol)	MELTING POINT (°C)
Hg	2.3	−38.8
Na	2.59	97.9
Al	10.7	660.4
U	12.6	1132.1
Ti	20.9	1660.1
W	35.2	3410.1

(c) Metals of the third transition series, such as tungsten, have extraordinarily high enthalpies of fusion and very high melting points. In fact, tungsten has the highest melting point of all the elements except for carbon, and the uses of tungsten reflect this. For example, pure tungsten is used as the filament—the glowing element—in light bulbs; no other material has been found to be better since the invention of light bulbs in 1908.

The reverse of melting is called solidification or **crystallization**, and it is always an exothermic process. Heat equal in magnitude to the enthalpy of fusion is evolved.

$$\text{Liquid} \longrightarrow \text{solid} + \text{heat} \qquad \text{heat} = -\Delta H_{\text{fusion}}$$

The melting temperature of a solid can convey a great deal of information. Table 13.8 gives you some data for two basic types of compounds: (a) polar and nonpolar molecules with low molecular weights and (b) ionic solids. In general, low-molecular-weight nonpolar substances that form molecular solids have low melting points. However, the melting

The melting point of an impure solid (one consisting of an intimate mixture of two or more components) is lower than that of a pure solid (Chapter 14). Therefore, if you measure the melting point of a solid, you can assess its relative purity. For this reason, melting point determination is a common laboratory operation.

Table 13.8 Melting Points and Enthalpies of Fusion of Some Solids

Solid	Melting Point (°C)	Enthalpy of Fusion (J/mol)	Type of Intermolecular Forces
MOLECULAR SOLIDS: Nonpolar Molecules			
O_2	−218	445	Dispersion forces only
F_2	−220	1020	
Cl_2	−103	6406	
Br_2	−7.2	10794	
MOLECULAR SOLIDS: Polar Molecules			
HCl	−114	1990	All three HX molecules have dipole–dipole forces enhanced by dispersion forces that increase with molecular weight.
HBr	−87	2406	
HI	−51	2870	
H_2O	0	6020	Hydrogen bonding; compare with H_2S
H_2S	−86	2395	Dipole–dipole forces; compare with H_2O
IONIC SOLIDS			
NaF	992	29288	All ionic solids have extended ion–ion interactions. General trend is same as that for lattice energies.
NaCl	800	30208	
NaBr	747	25690	

Figure 13.39 Sublimation, the conversion of a solid to its vapor. When solid iodine is heated at temperatures below its triple point of 114 °C, it sublimes. The purple vapor condenses on an upper, cooler portion of the tube. (Marna G. Clarke)

point increases as the molecular weight increases. This happens because dispersion forces become stronger with increasing molecular weight (and number of electrons). Thus, increasing energy is required to break down the intermolecular forces in the solid, and this is reflected in an increasing enthalpy of fusion.

The ionic compounds in Table 13.8 have the highest melting points and highest enthalpies of fusion owing to very strong ion–ion forces. In general, ionic solids have melting points higher than 100 °C. Notice that, for the alkali metal halides, there is also a good correlation between their lattice energies and their melting points.

Finally, one other important property of solids is the direct conversion of solids to gases by **sublimation** (Figure 13.39),

$$\text{Solid} + \text{heat} \longrightarrow \text{gas}$$

$$\text{heat required at constant pressure} = \Delta H_{\text{sublimation}}$$

This is an endothermic process whose enthalpy is called the molar enthalpy of sublimation. Water, which has a molar enthalpy of sublimation of 51 kJ/mol, can be converted from solid ice to water vapor quite readily. One of the best examples of the use of this property is in a frost-free refrigerator. During certain times, the freezer compartment is warmed slightly. Molecules of water freed from the surface of the ice (the vapor pressure of ice is 4.60 mmHg at 0 °C) are removed in a current of air blown through the freezer. Other common substances that sublime are I_2 and CO_2.

SOMETHING MORE ABOUT
Probing Solids with X-Rays

Atoms, molecules, and ions are much too small to be seen by the naked eye or with an ordinary microscope. How is it then that chemists can probe the underlying symmetry of solids and determine the distance between the atoms in a metal, the ions in a salt, or the atoms of a molecule? In 1912 Max von Laue, a German physicist, found that crystalline solids could be used to diffract x-rays, and, somewhat later, the English scientists William and Lawrence Bragg (father and son), showed that x-ray diffraction by crystalline solids could be used to determine distances between atoms.

To determine distances on an atomic scale we have to use some kind of probe to locate them. The probe must have dimensions not much larger than the atoms; otherwise it would pass over them unperturbed. Therefore, the probe has to have a size of only a few picometers, and it is x-rays, high energy photons with a wavelength in this range, that meet the requirement. Crystalline solids are used because the atoms or molecules must be immobilized so they can be located with x-ray photons.

To "see" radiation interact with atoms we have to rely on a change in the radiation as it passes through the crystal. The change is the scattering or "diffraction" of the photons of the radiation from their original path. To determine the location of the scattered photons, we place a piece of photographic film around the crystal (or an electronic detector in more modern instruments) (Figure A).

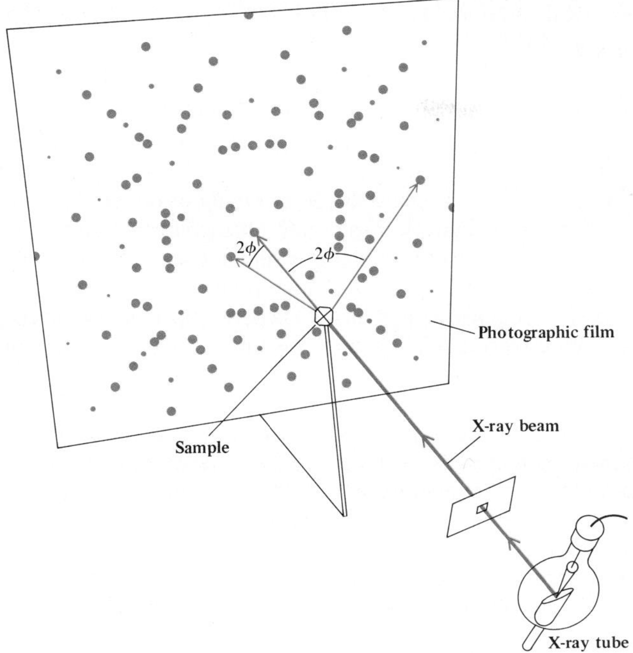

Figure A In the x-ray diffraction experiment a beam of x-rays is directed at a crystalline solid. Some of the photons pass through the solid, but others are scattered by the atoms of the solid, through various angles ϕ, and produce spots on the film that surrounds the sample.

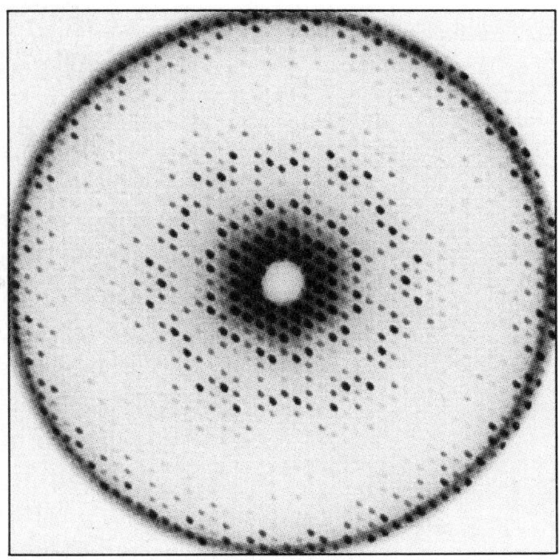

Figure B An x-ray diffraction photograph of a crystal of the enzyme histadine decarboxylase (molecular mass ≈ 37,000 amu). The spots on the exposed film are caused by scattering of a beam of x-ray photons by the atoms of the molecule.

When an x-ray photon strikes an atom, the energy of the photon causes the atom's electrons to oscillate so the atom becomes like a radio station antenna, and the photon is rebroadcast to the photographic film (or detector) as the electron oscillations die out. The energy of the rebroadcasted or scattered photons falls on compounds in the film, and a reaction such as $AgBr(s) \rightarrow Ag + Br$ occurs. After developing the film, clusters of silver atoms show as black spots where the scattered x-rays from the crystal struck the film (Figure B).

But why are there "spots" on the film? The answer to this comes from considering the wave properties of photons. You know that waves have peaks and valleys at different points in space. If two scattered waves meet at some point, and the peak of one wave meets the valley of another, the waves cancel each other at that point. However, if the waves meet peak-to-peak they reinforce, and, if this occurs at the surface of the film, the radiation produces clusters of silver atoms. In contrast, at points on the film where scattered waves meet and cancel, no such reaction occurs.

For the photon waves to meet peak-to-peak at some point on the film they must leave the atoms of the crystal with synchronized oscillations, that is, with peaks and valleys in unison. This condition will be met when the x-rays are scattered at special values of the angle ϕ (Figure A), an angle related to the distances between atoms in the solid. It is this that makes it possible to use the location of the spots on the film to decipher the internal structure of the solid and to determine bond distances and angles.

SOMETHING MORE ABOUT
Artificial Diamonds

Marilyn Monroe once sang "Diamond's are a girl's best friend." Of course she meant the kind of diamonds worn in a ring on her finger, but you may soon see them in many other forms.[1]

Diamonds are the hardest material and the best conductor of heat known. They are also transparent to visible light, infrared, and ultraviolet radiation, and are electrically insulating. Finally, diamonds behave as semiconductors with some advantages over silicon, and their low density and high rigidity make them ideal for producing sounds with frequencies as high as 60,000 hertz, far higher than the range of human hearing. What more could a scientist want in a material—except a cheap, practical way to make it!

In the 1950s, scientists at General Electric in Schenectady, New York, achieved something alchemists had attempted for centuries, the synthesis of diamonds from carbon-containing materials, including wood or peanut butter. Their technique is to heat graphite to a temperature of 1500 °C in the presence of a metal, such as nickel or iron, and under a pressure of 50,000 to 65,000 atmospheres. Under these conditions, the carbon dissolves in the metal and slowly forms diamonds. The world wide market for diamonds made this way is now worth about $500 million, with much of it used for abrasives and diamond-coated cutting tools for drilling.

The basic process developed by General Electric is used to make more than 300 million carats annually—that's over 60 tons of abrasive diamond grit. However, it is still expensive, and the diamonds are not entirely pure nor crystalline enough for making semiconductor devices. Therefore, even from the beginning the search was on for better ways to make diamonds. The most promising method so far seems to be a low-pressure process called "chemical vapor deposition" (CVD). In this technique a mixture of hydrogen and a carbon-containing

Scanning electron micrographs of diamond thin films, which can be grown using a relatively simple technique called "chemical vapor deposition" (CVD). (General Electric)

gas such as methane (CH_4) is decomposed by heating to about 2200 °C (with microwaves or a hot wire).[2] Carbon atoms deposit on a silicon plate or other material in the chamber and slowly build up a film of numberless, tiny diamonds.

Thus far the CVD process is much more expensive than the high-pressure process. However, the promise of CVD diamond thin films is so enormous that development is rapid in the U.S., Japan, and the Soviet Union. Cutting tools with diamond thin films are being test marketed, and high quality loudspeakers with diamond-coated tweeters are already on the market. A polycrystalline diamond heat sink for electronic devices is now sold by a Japanese firm, and a company in Ohio has even reported that they can coat sunglasses to make them scratch-resistant.

A recent National Research Council report said that "success in deposition of diamond and diamond-like coatings on a variety of substrates at practical growth rates is one of the most important technological developments in the past decade. The ultimate economic impact of this technology may well outstrip that of high-temperature superconductors." Other scientists say that the development of synthetic diamond films is potentially the greatest advance in materials since the invention of plastics.

[1] (a) I. Amato, *Science News*, 1990, *138*, 72. (b) P.K. Bachmann and R. Messier, *Chemical and Engineering News*, May 15, 1989, pp. 24–39. (c) H. Zaugg, *Chem Matters*, April, 1990.

[2] In practice, any carbon-containing gas can be used. A Japanese laboratory has used rice wine, and, not to be outdone, the General Electric Company has used Jack Daniels whisky.

13.5 THE SPECIAL PROPERTIES OF LIQUID AND SOLID WATER

One of the most striking differences between our planet and others in the solar system is the existence of liquid water on the earth. Three fourths of the globe is covered by oceans, the polar regions are vast ice fields, and even the soil and rocks hold large amounts of water. Water has played a major role in the history of the people of the earth, and it is a significant factor in controlling the climate of our planet and the life on it. Ice floats on water, lakes freeze from the top down and not from the bottom up, very large amounts of energy are needed to raise the temperature of a body of water to any considerable extent, and snowflakes have a six-sided geometry. All these features reflect the ability of water molecules to cling tenaciously to one another by *hydrogen bonding*, the source of the unique properties of water.

The oxygen atom of a water molecule is the center of a tetrahedron of two H atoms and two lone pairs of electrons, and *each point* of the tetrahedron can be involved in hydrogen bonding to a neighboring water molecule (Figure 13.40). In solid water (ice) the hydrogen bonds are 177 pm long, compared with 94 pm for a normal O—H covalent bond. This difference of course reflects the weaker nature of the hydrogen bond.

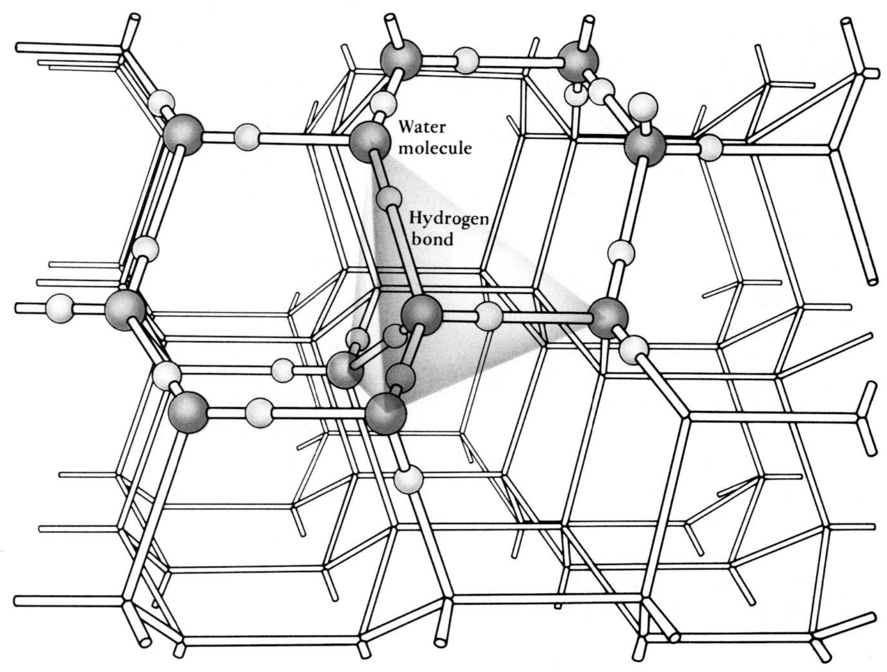

Figure 13.40 Ice. Each water molecule in ice is hydrogen bonded to four others. Oxygen atoms of hydrogen-bonded water molecules are at the corners of puckered, six-membered rings. (© Irving Geis)

Each of the water molecules hydrogen-bonded to a central water molecule can in turn hydrogen-bond to yet more water molecules. Although this could lead to a complicated, random structure, nature often exhibits a tendency to achieve great regularity, and ice is beautifully regular (Figure 13.40). It is an open cage structure in which the oxygen atoms of hydrogen-bonded water molecules are arranged at the corners of puckered, six-sided rings or hexagons. Each edge of a six-sided ring consists of a normal O—H covalent bond and a hydrogen bond. It is interesting that snowflakes are always based on six-sided figures, in part a reflection of the internal molecular structure of ice (Figure 13.41).

You will see many examples of six-sided rings in nature. For example, diamond (which has a solid structure related to that of water, in that there are puckered hexagons of tetrahedra) and graphite both consist of six-sided rings.

When ice melts to water the regular structure of ice collapses to some extent, but approximately 85% of the hydrogen bonds remain. Although the structure of liquid water is still a matter for debate, it does seem certain that there are ice-like clumps of water molecules interspersed with smaller groupings of water molecules. The major consequence of the open structure of ice compared with liquid water is that *ice is less dense than liquid water.* This density difference is small, but it is all it takes to allow ice to float on top of water (Figure 13.42).

Figure 13.41 A snowflake. The six-sided or hexagonal geometry of the snowflake is a reflection of the six-sided rings formed when water molecules hydrogen bond to form ice. (© Eric Grave/Phototake)

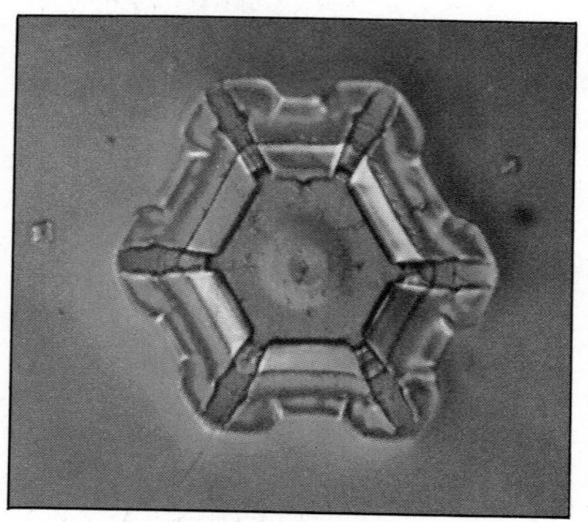

Not only are the densities of liquid water and ice different, but *the density of water changes with temperature* (Figure 13.43). When water melts at 0 °C, there is a relatively large increase in density, and when the now liquid water warms slightly, its density increases even more. This comes about because, as the ice-like structure of very cold water begins to break down, the liquid becomes more compact. The density reaches a maximum at 4 °C, but it begins to decline again at higher temperatures as more and more hydrogen bonds are broken and the structure loosens.

The way that water density changes with temperature means that lakes will not freeze solidly from the bottom up in the winter. When lake

Figure 13.42 Ice made from water floats in liquid water because solid water is less dense than liquid water, a very unusual occurrence (*left*). The photo at the right shows liquid benzene (C_6H_6) with a chunk of solid benzene lying on the bottom. Solid benzene is *more* dense than the liquid and so sinks in the liquid. This is the behavior exhibited by virtually all known substances *except* water.

water cools with the approach of winter, the density increases, the cooler water sinks, and the warmer water rises. This means the water and dissolved material turn over until all the water reaches 4 °C, the maximum density. (This carries oxygen-rich water to the lake bottom to restore the oxygen used during the summer, and it brings nutrients to the top layers of the lake.) As the water cools further, it stays on the top of the lake, since water cooler than 4 °C is *less* dense than 4 °C water. With further heat loss, ice can then begin to form on the surface, floating there and protecting the underlying water and aquatic life from further heat loss.

At 0 °C, liquid water has a density of 1.00 g/cm³, while ice at 0 °C has a density of 0.917 g/cm³. The lower density of ice than water means that ice floats on water.

Extensive hydrogen bonding is the origin of the *extraordinarily high heat capacity of water*, and, in large part, this is why oceans and lakes have such an enormous effect on weather. In autumn, the temperature of the atmosphere drops. As it falls below the temperature of an ocean or lake, the ocean or lake gives up heat to the atmosphere, and the air temperature drop is moderated. Further, so much heat must be given off for each degree drop in temperature of the water (because so many hydrogen bonds form) that the decline in water temperature is gradual. This is the reason that the temperature of the ocean or a large lake is generally higher than the average air temperature until late in the fall.

We noted in Chapter 6 the high heat capacity of water and its effect on our environment.

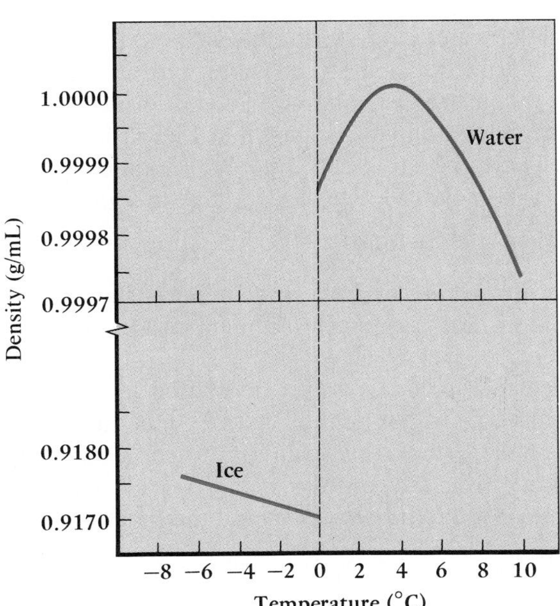

Figure 13.43 The temperature dependence of the densities of ice and liquid water.

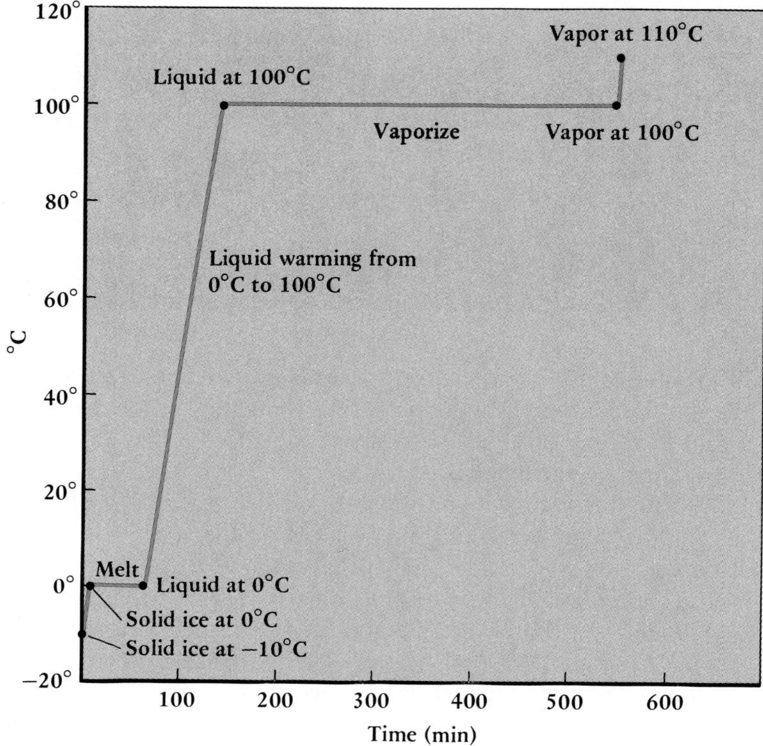

Figure 13.44 The heating curve for water. One mole of water is heated from −10 °C (where it is in the form of ice) to +110 °C (where it is steam). A constant heating rate of 100. J/min is assumed. The time intervals give you a good graphic comparison of the relative energies involved in each step.

13.6 PHASE CHANGES

Now that we know something of the basic properties of liquids and solids, we want to put this information together in two ways: heating/cooling curves and phase diagrams.

HEATING AND COOLING CURVES

Let us take 1.00 mole of water in the form of ice at −10.0 °C. How much heat is necessary to warm this quantity of water until it becomes steam at 110 °C, and how much time is required if you add heat at the constant rate of 100. J/min? What happens is shown in Figure 13.44.

The specific heat capacity of ice is 2.09 J/g · K. This means that raising the temperature of 1.00 mole of ice by 10.0 degrees requires

$$(18.0\ \text{g})(2.09\ \text{J/g}\cdot\text{K})(10.0\ \text{K}) = 376\ \text{J}$$

See Chapter 6 to review the calculation of the quantity of heat gained or lost by a substance undergoing a change in temperature.

Since you are adding heat at the rate of 100. J per minute, this means that 3.76 minutes are required, and the lowest segment of the graph in Figure 13.44 reflects this fact.

When the ice has warmed to 0 °C, it can be melted to water at 0 °C by further input of heat equal to the heat of fusion. Since the molar enthalpy of fusion of ice is 6.02 kJ/mol, this quantity of heat is required for the melting process alone. *The temperature of the system stays at 0 °C until all of the ice is converted to liquid water*, and the time required is 60.2 minutes.

$$(6020\ \text{J/mol})(1.00\ \text{mol})(1.00\ \text{min}/100.\ \text{J}) = 60.2\ \text{min}$$

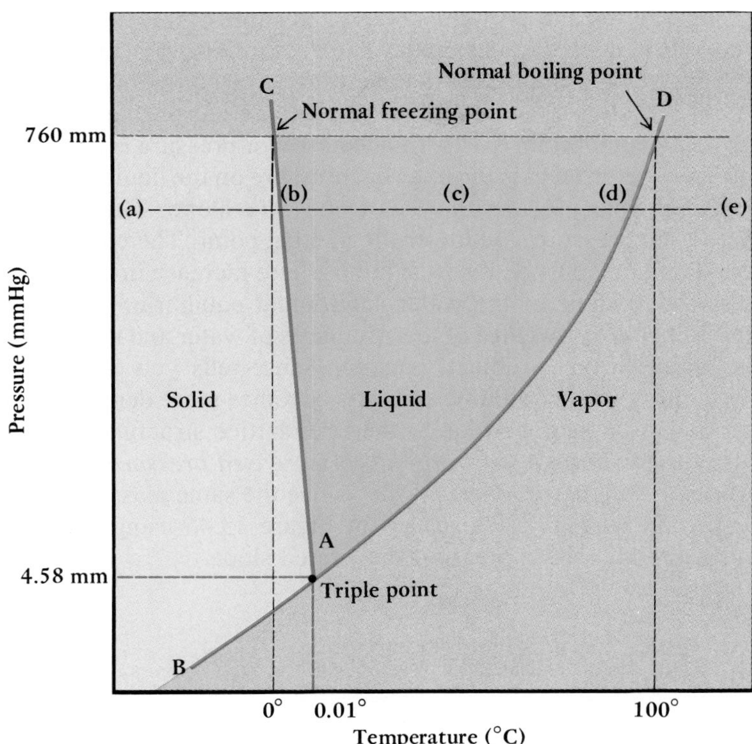

Figure 13.45 Phase diagram for water. Note that the scale is greatly exaggerated. (See text for an explanation of the line *ae*.)

Now that the ice has been converted completely to water, the water can be warmed to the boiling point and then converted to steam with further heat input. Using the specific heat of water (4.18 J/g · K), one mole of water requires 7520 J for a 100. degree temperature rise (and 75.2 minutes are required). At the boiling point, a further 40.7 kJ (the molar enthalpy of vaporization) are required to evaporate completely the mole of water at 100 °C (and 407 min are required). Finally, 2.0 J/g · K (or 360 J over 3.6 minutes) are needed to heat the steam at 100 °C to its final temperature.

If you cool the 1.00 mole sample from steam to make ice at −10 °C, the curve would be a mirror image of Figure 13.44.

PHASE DIAGRAMS

Figure 13.44 illustrates the time and quantity of heat required for 1.00 mole of water to change from solid to liquid to vapor. A **phase diagram** (Figure 13.45) is another way to illustrate the relation between phases; these show how the phases of a system are related to each other by changes in pressure and temperature. Each line in the diagram represents the conditions of T and P at which equilibrium exists between the two phases on either side of the line.

As described in Section 13.3, the equilibrium vapor pressure for water rises with temperature along the curved line AD, and the vapor pressure reaches 760 mmHg at 100 °C, the normal boiling point. As the temperature of a liquid decreases, so does its vapor pressure until point A, the **triple point**, is reached (at P = 4.58 mmHg and T = 0.01 °C for water). Here, all three phases (solid, liquid, and vapor) are in equilibrium.

At temperatures and pressures below the triple point, solid water (ice) is in equilibrium with water vapor. The equilibrium vapor pressure of ice subliming to vapor at a given temperature is given by the line *AB*.

The line *AC* shows the conditions of pressure and temperature at which solid/liquid equilibrium exists. (Since no vapor pressure is involved here, the pressure referred to is the external pressure on the liquid.) Water is highly unusual, because this line has a *negative* slope. That is, the higher the external pressure, the lower the melting point. The change for water is approximately 0.01 °C for each atmosphere increase in pressure.

Of all the thousands of substances for which phase diagrams are known, only water, bismuth, and antimony have solid/liquid equilibrium lines of negative slope.

The negative slope of the water solid/liquid equilibrium line can be explained from our knowledge of the structures of water and ice. When pressure is increased on an object, common sense tells you the object will be forced into a smaller volume and thus becomes more dense. Since liquid water is denser than ice (due to the open lattice structure of ice), *ice and water in equilibrium will respond to increased pressure (at constant T) by melting ice to form more water*, since the same mass of water requires less volume. This is illustrated in Figure 13.46 using a solid–liquid equilibrium line with a greatly exaggerated slope.

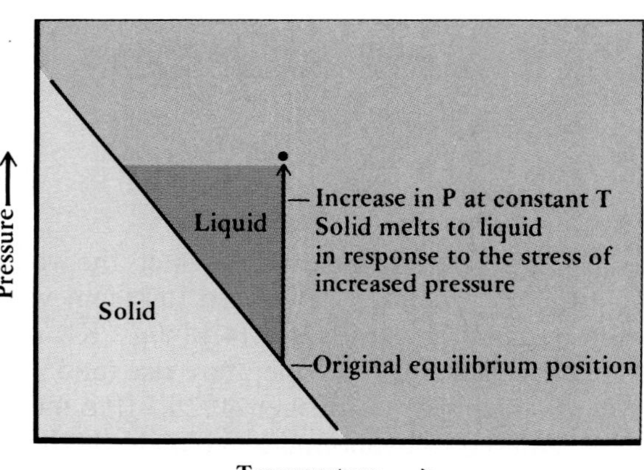

Figure 13.46 Ice skating is made easier by the unusual properties of ice and water, and the resulting negative slope of the ice/water equilibrium line. Ice skates have very thin blades, so your weight is concentrated on a very small area of ice. Thus, a person of average weight can easily exert 500 atm of pressure on the ice. Since the melting point of ice changes by about −0.01 °C when the pressure is increased 1 atm, this means that the ice has a lower melting point (about −5 °C) beneath a skate blade. This effect, combined with some frictional heating, helps you glide across the surface of ice. (Photo courtesy of FPG International)

Figure 13.46 also helps to explain why you can skate on ice. When skating, your body weight is concentrated on the very small area below the skate blades. Therefore, your weight per unit area, the pressure you exert on the ice, is very large, and you compress the ice. Assuming the temperature remains constant, the ice changes to the denser liquid phase, and the film of water lubricates your sliding motion.

As a final point, we can correlate the water phase diagram (Figure 13.45) with the heating curve in Figure 13.44. Let us say the atmospheric pressure is about 740 mmHg and that you have an ice sample at −10 °C. As heat is added, you move along the horizontal dotted line *ae* in Figure 13.45, starting at point *a*. Solid alone exists at this point. When the line *AC* is reached, ice melts to water at *b* as heat is added. Once all the ice is melted, the water is warmed, moving along the line from *b* to *d*. At all points between *b* and *d*, say *c*, the sample exists only as liquid; if vapor is present there can be no equilibrium, and vapor would condense to liquid (Figure 13.47). When the point *d* is reached, liquid can change to vapor, and the equilibrium vapor pressure is 740 mmHg. As more heat is added, the water boils and is converted completely to vapor at point *d*. The vapor is heated further to point *e*.

Figure 13.47 Dew drops on a cool morning. During a warm day in the summer, water evaporates to form water vapor in the air, and a rough equilibrium is established between bodies of water and water in the air. During the night, however, the temperature drops, disturbing the equilibrium. The pressure of water vapor in the air must be lower at the lower temperature, and water condenses out of the air. (Notice that the droplets are spherical owing to the surface tension of water. See Figure 13.18.)

The basic features of the phase diagram for CO_2 (Figure 13.48) are the same as those of water, except that the CO_2 solid/liquid equilibrium line has a *positive* slope. Solid CO_2 is more dense than the liquid, so the solid will sink to the bottom in a container of liquid CO_2. Also notice that solid CO_2 will sublime directly to CO_2 gas as it warms to room temperature (at 1 atm pressure). This is the reason CO_2 is called *Dry Ice*; it looks like water-ice, but it does not melt.

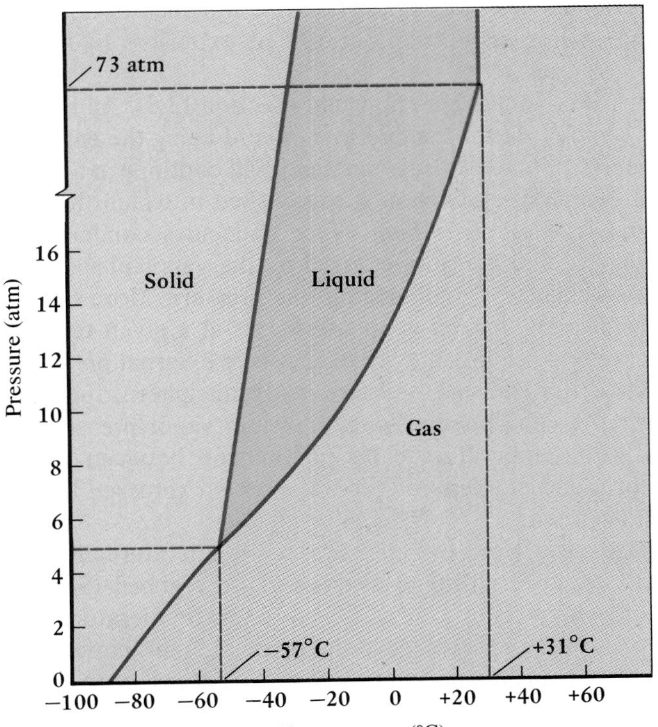

Figure 13.48 A phase diagram for carbon dioxide, showing the critical pressure and temperature.

SUMMARY

Because of **intermolecular forces**, atoms or molecules in the gas phase can be attracted to one another strongly enough that condensation occurs to give liquids and solids. Just as in gases, molecules in the liquid and solid phases retain some motion and have a distribution of kinetic energy (Section 13.1). Liquids and solids differ from gases, however, in that the former are much less **compressible**, and many solids have definite and regular structures.

Intermolecular forces range from very strong to very weak (Section 13.2). They are, in order of declining strength, the following:

Ion–dipole > dipole–dipole (including hydrogen bonding)
> dipole–induced dipole > induced dipole–induced dipole

(Forces not involving ions are sometimes called **van der Waals forces**, and the weakest of these—those involving *induced* dipoles—have been called **dispersion forces**.) **Ion–dipole forces** cause metal ions to be **hydrated** in water. The negative end of the water dipole (oxygen) is attracted to the positive ion. As in ion–ion attraction (Chapter 10), ions of larger charge or smaller size interact most strongly with water and other dipoles. Many molecules are polar, so the positive end of the dipole of one molecule is attracted to the negative end of the dipole of another molecule by **dipole–dipole forces**. A polar molecule can cause a distortion or **polarize** the electron cloud in a neighboring nonpolar molecule. **Induced dipole** interactions are exceedingly weak and depend on, among other things, the **polarizability** or "distortability" of the electron cloud of the atom or nonpolar molecule.

Hydrogen bonding is a special form of dipole–dipole interaction (Section 13.2). This occurs most strongly when H is attached to O, N, or F. Water has many unique properties because of extensive hydrogen bonding (Section 13.5).

Many properties of liquids are important (Section 13.3). All liquids can **evaporate** to the vapor phase, the energy required being the **enthalpy of vaporization**. Evaporation in a closed container will continue at a given temperature until a **dynamic equilibrium** is established in which the rate of evaporation equals the rate at which vapor molecules condense or reenter the liquid phase. The pressure exerted by the vapor phase molecules at equilibrium is called the **equilibrium vapor pressure**. More **volatile** compounds have higher equilibrium vapor pressures at a given temperature. If the vapor pressure equals the atmospheric or external pressure, the liquid boils. When the external pressure is 1 atmosphere, the temperature is the **normal boiling point**. The equilibrium vapor pressure of a liquid increases with temperature. The relationship between vapor pressure, temperature, and enthalpy of vaporization is expressed by the **Clausius-Clapeyron equation**.

Vapor pressure will continue to increase with temperature until the **critical temperature**, T_c, and **critical pressure**, P_c, are reached (Section 13.3). No liquid-vapor equilibrium can exist above this temperature.

Since there are **cohesive forces** between identical liquid molecules, a liquid acts as though it has a skin, and the energy necessary to break through the surface is called the **surface tension**. It is the surface tension of water, together with **adhesive forces** between water and glass, that causes water to have a concave **meniscus** in a drinking glass. Cohesive intermolecular forces also account in part for the resistance to flow or **viscosity** of liquids.

Solids are characterized by long-range order (Section 13.4). The basic unit of a **crystal lattice** is called the **unit cell**; on replication in three dimensions, the unit cell leads to the observed lattice. There are seven basic unit cell types, but Section 13.4 is concerned only with the **simple cubic** (sc), the **body-centered cubic** (bcc), and the **face-centered cubic** (fcc) cells for metals, and the sc and fcc arrangements for ionic compounds. The formula of an ionic compound is related to the number and type of holes that fill with cations in the sc or fcc lattice of anions (or vice versa).

In addition to metals and ionic solids, there are also **molecular solids** (H_2O, I_2), and **network solids** (graphite and diamond, which are forms of pure carbon, and quartz or SiO_2). **Amorphous solids**, or solids without long-range order, include glass and many plastics.

The heat energy required to melt a solid (at constant pressure) to give a liquid is the **molar enthalpy of fusion**. The heat or **enthalpy of sublimation** is the energy required to convert a solid directly to its vapor (at constant pressure).

A **phase diagram** (Section 13.6) shows the relation between the three phases for a compound under various conditions of temperature and pressure.

STUDY QUESTIONS

REVIEW QUESTIONS

1. Name the types of forces that can be involved between two molecules and between ions and molecules. Place the forces in order of decreasing strength.
2. Explain how a water molecule can interact with a molecule such as CO_2. What intermolecular force is involved?
3. Explain what intermolecular interactions can occur in an aqueous solution of iron(II) chloride.
4. Hydrogen bonding is most important when H is attached to certain electronegative atoms. Which are those atoms?
5. Explain how hydrogen bonding leads to the decline in density of water from 4 °C to solid ice at 0 °C.
6. Explain why the specific heat capacity of water is so large compared with those of many other liquids. How can this affect weather?
7. Explain evaporation of liquids and condensation of vapor in terms of the kinetic molecular theory.
8. Define normal boiling point, critical pressure, and critical temperature.
9. Explain, in terms of intermolecular forces, the trends in boiling points of the molecules EH_3, where E is a Group 5A element (Figure 13.8).
10. What are cohesive and adhesive forces? What is the difference between these forces? What is their role in explaining the concave meniscus observed for water in glass?
11. List the major types of solids, with an example of each.
12. What are the three types of cubic unit cells? Explain their similarities and differences.
13. Sketch the unit cell for a body-centered cubic crystal of sodium. What is the *net* number of sodium atoms within the unit cell?
14. Sketch the unit cell for a face-centered cubic lattice of bromide ions. How many *net* Br^- ions are contained within this unit cell? Where are the octahedral "holes" in this lattice?
15. Define the lattice energy of an ionic solid. How is it related to the size of the ions and their charge?
16. Draw the phase diagram for water. Label the normal boiling point, melting point, and triple point, and show what regions of T and P are appropriate to solid, liquid, and vapor.
17. Explain why the solid/liquid equilibrium line in the water phase diagram has a negative slope.

INTERMOLECULAR FORCES

18. What intermolecular force(s) must be overcome to:
(a) melt ice

(b) melt solid I_2
(c) remove the water of hydration from $MnCl_2 \cdot 4H_2O$
(d) convert liquid NH_3 to NH_3 vapor

19. When NaCl dissolves in water, what type of forces must be overcome in the liquid water? What type of forces must be overcome in the solid NaCl? What type of forces cause NaCl to be able to dissolve in liquid water?

20. One example of a hydrated salt is $CuSO_4 \cdot 5H_2O$. What kind of force is responsible for binding the water to the copper(II) ions of copper(II) sulfate?

21. When I_2 dissolves in the simple alcohol H_3COH, what type of forces must be overcome within the solid I_2 in order to allow it to dissolve? What type of forces must be disrupted between H_3COH molecules when I_2 dissolves? What type of forces exist between I_2 and H_3COH molecules in solution?

22. Tell what type of intermolecular force is important in converting each of the following from a gas to a liquid:
(a) CO_2 (c) $CHCl_3$
(b) NH_3 (d) CCl_4

23. Tell what type of intermolecular force must be overcome in converting each of the following from a liquid to a gas:
(a) liquid N_2
(b) mercury
(c) CCl_3F (a chlorofluorocarbon)
(d) CH_3CH_2OH (ethyl alcohol)

24. Rank the following in order of increasing strength of intermolecular forces in the pure substances: Ne, CH_4, CO, and CCl_4. Which do you think might exist as gases at 25 °C and 1 atm?

25. Rank the following in order of increasing strength of intermolecular forces in the pure substances: SO_2, $H_3C—CH_2—CH_2—CH_2—CH_3$, $H_3C—OH$, and He. Which do you think might exist as gases at 25 °C and 1 atm?

26. Tell which member of each of the following pairs of compounds you would expect to have the higher boiling point.
(a) O_2 or N_2 (c) HF or HI
(b) SO_2 or CO_2 (d) SiH_4 or GeH_4

27. Explain why the boiling point of H_2S is lower than that of water.

28. Consider the following four compounds: (a) SCl_2; (b) NH_3; (c) CH_4; (d) CO. Place the four compounds in order of increasing boiling point.

29. The normal boiling point of CH_3Cl (chloromethane) is −24.2 °C, whereas that for CH_3I is 42.4 °C. Which compound has the stronger intermolecular forces in the liquid phase?

30. Which of the following compounds would be expected to form hydrogen bonds with water?
(a) $CH_3—O—CH_3$ (e) I_2
(b) CH_4 (f) CH_3OH
(c) HF
(d) $H—C(=O)—OH$

31. Which of the following are expected to form strong hydrogen bonds with water?
(a) H_2Te
(b) $H_3CCH_2—O—H$
(c) $H_3CCH_2—O—CH_3$
(d) HI
(e) $H_3C—C(=O)—OH$

32. When salts of Mg^{2+}, Na^+, and Cs^+ are placed in water, the positive ion is hydrated (as is the negative ion). Place these ions in order of increasing energy of hydration. (The ion that is most weakly hydrated evolves the least amount of energy upon hydration.)

33. Which compound in each of the following pairs of salts is more likely found with a hydrated cation? Briefly explain your reasoning in each case.
(a) Li_2SO_4 or Cs_2SO_4
(b) NH_4NO_3 or $Mg(NO_3)_2$
(c) RbCl or $NiCl_2$

LIQUIDS

34. Answer the following questions with "increase," "not change," or "decrease."
(a) If the intermolecular forces in a liquid increase, the normal boiling point of the liquid will ______.
(b) If the intermolecular forces in a liquid increase, the vapor pressure of the liquid will ______.
(c) If the surface area of a liquid increases, the vapor pressure will ______.

35. Methyl alcohol, H_3COH, has a normal boiling point of 64.7 °C and a vapor pressure of 100 mmHg at 21.2 °C. Another compound with the same elements, formaldehyde ($H_2C{=}O$), has a normal boiling point of −19.5 °C and a vapor pressure of 100 mmHg at −57.3 °C. Briefly explain why these compounds have different boiling points and require different temperatures to achieve the same vapor pressure.

36. The fluid used in refrigerators and air conditioners is generally one or more of a family of compounds called chlorofluorocarbons (or CFC's). (Although there is evidence that CFC's lead to environmental damage, they will continue to be used until a substitute is found.) One compound belonging to this group is CCl_3F. If you fill an air conditioner with 1.00 kg of the compound, and if the enthalpy of vaporization is 24.8 kJ/mol, how much heat is required to vaporize all the compound at its boiling point of 23.8 °C?

37. The chlorofluorocarbon CCl_2F_2 is commonly used to

fill the cooling system in home refrigerators. Its enthalpy of vaporization is 20.0 kJ/mol. How many joules of heat energy are needed to vaporize 1.00 kg of CCl_2F_2 at its boiling point of −29.8 °C?

38. The molar enthalpy of vaporization for methyl alcohol is 38.0 kJ/mol at 25 °C. This compound, which is commonly called wood alcohol, is the main ingredient in the anti-icing fluid you may put into the gas tank of your car in the winter. How much heat energy is required to convert 250. mL of the alcohol (about equal to one can of anti-icing fluid) from the liquid to the vapor? (The density of H_3COH at 25 °C is 0.787 g/mL.)

39. If a nuclear reactor is cooled by a pool of water holding 5.50×10^6 L of water, how much heat energy is required to vaporize the water at its boiling point? (See Appendix D for the latent heat of vaporization. Assume a density of 1.00 g/mL for the water.)

40. Using Figure 13.14, answer the following questions:
   (a) What is the approximate equilibrium vapor pressure of water at 60 °C? Compare your answer with the data in Appendix D.
   (b) At what temperature will water have an equilibrium vapor pressure of 600 mmHg?
   (c) Compare the equilibrium vapor pressures of water and ethyl alcohol at 70 °C.

41. Using Figure 13.14, answer the following questions:
   (a) What is the equilibrium vapor pressure of diethyl ether at room temperature (approximately 20 °C)?
   (b) Place the three compounds in Figure 13.14 in order of increasing intermolecular forces.
   (c) If the pressure in a flask is 400 mmHg, and if the temperature is 40 °C, which of these three compounds are predominantly liquids and which are gases?

42. Assume you seal 1.0 g of diethyl ether (Figure 13.14) in an evacuated 100.-mL flask. (There are no molecules of any other gas in the flask.) If the flask is held at 30. °C, what is the approximate gas pressure in the flask? If the flask is placed in an ice bath, will additional liquid ether evaporate or will some ether condense to a liquid?

43. Refer to Figure 13.14 as an aid in answering these questions:
   (a) You put some water at 60 °C in a plastic milk carton and seal the top very tightly so gas cannot enter or leave the carton. What will happen when the water cools?
   (b) If you put a few drops of liquid diethyl ether on your hand, will it evaporate completely or remain a liquid?

44. Some vapor pressure data for benzene, a common solvent, are given below. Use the Clausius-Clapeyron equation and these data to answer the questions that follow:

TEMPERATURE (°C)	VAPOR PRESSURE (mmHg)
7.6	40.
26.1	100.
60.6	400.
80.1	760.

   (a) What is the normal boiling point of benzene?
   (b) Plot these data, with vapor pressure on the vertical axis and temperature on the horizontal axis, so that you have a plot that resembles Figure 13.14. At what temperature does the liquid have an equilibrium vapor pressure of 250 mmHg? At what temperature is it 650 mmHg?
   (c) Calculate the enthalpy of vaporization.

45. Equilibrium vapor pressure data for octane, C_8H_{18}, are given below. Use the Clausius-Clapeyron equation and these data to calculate enthalpy of vaporization and the normal boiling point of octane.

VAPOR PRESSURE (mmHg)	TEMPERATURE (°C)
13.6	25
45.3	50.
127.2	75
310.8	100.

46. A fluorocarbon, CF_4, has a critical temperature of −45.7 °C and a critical pressure of 37 atm. Are there any conditions under which this compound can be a liquid at room temperature? Explain your answer briefly.

47. The simple hydrocarbon methane (CH_4) cannot be liquefied at room temperature, no matter how high the pressure. Propane (C_3H_8), another compound in the series of hydrocarbons, has a critical pressure of 42 atm and a critical temperature of 96.8 °C. Can this compound be liquefied at room temperature? Can you think of a place where you would find liquefied propane?

SOLID STATE STRUCTURES

48. Lead metal crystallizes in the face-centered cubic arrangement. If the metallic radius of lead is 175.0 pm, calculate (a) the length of the cell edge, (b) the volume of the unit cell, and (c) the density of solid lead.

49. Knowing that calcium metal crystallizes as a face-centered cube of Ca atoms, and that the density of the solid is 1.54 g/cm³, find the radius of a calcium atom.

50. Chromium metal has a density of 7.20 g/cm³. Assuming the radius of the Cr atom is 124.9 pm, is the ge-

ometry of the Cr unit cell body-centered cubic or face-centered cubic?

51. Copper is an important metal in our economy. The density of copper is 8.95 g/cm^3. If the radius of a copper atom is 127.8 pm, what is the geometry of the Cu unit cell, body-centered cubic or face-centered cubic?

52. Can $CaCl_2$ have the NaCl structure? Explain your answer briefly.

53. Why is it not possible for a salt with the formula M_3X (for example, Na_3PO_4) to have a face-centered cubic lattice of X anions and M cations in octahedral holes?

54. RbCl crystallizes in the NaCl structure.
(a) What is the volume of the unit cell? (You will need the ion radii in Figure 9.15 for this calculation.)
(b) What is the density of the solid?

55. NaI crystallizes in the NaCl structure. Using this information, and the radii of the ions in Figure 9.15, calculate the density of the solid.

56. LiH crystallizes with H^- ions defining either a simple cubic or a face-centered cubic lattice. The solid has a density of 0.77 g/cm^3. If the edge of the unit cell is 4.086×10^{-8} cm, how many Li^+ and H^- ions are there in one unit cell? Which unit cell is appropriate for LiH?

57. Thallium(I) chloride, TlCl, crystallizes in either a simple cubic lattice or a face-centered cubic lattice of Cl^- ions with Tl^+ ions in the holes. If the density of the solid is 7.00 g/cm^3 and the edge of the unit cell is 3.85×10^{-8} cm, what is the unit cell geometry?

PHYSICAL PROPERTIES OF SOLIDS

58. The lattice energies of the alkali metal halides are plotted in Figure 13.37. Explain the trend for MI, the metal iodides. Why do the lattice energies drop as the metal is changed from Li to Cs?

59. The ions of NaF and MgO are isoelectronic, and the internuclear distances are about the same (235 pm and 212 pm, respectively). Why then are the melting points of NaF and MgO so different (992 °C and 2642 °C, respectively)?

60. For the pair of compounds CsF and CsI, tell which compound is expected to have the higher melting point and briefly explain, based on the lattice energies of the two compounds.

61. For the pair of compounds KBr and RbBr, tell which compound is expected to have the higher melting point and briefly explain, based on the lattice energies of the two compounds.

62. Benzene is an organic liquid that freezes at 5.5 °C (see Figure 13.42) to beautiful, featherlike crystals. How much heat is evolved when 15.5 g of benzene freeze at 5.5 °C? (See Appendix D for the latent heat of fusion.) If the 15.5-g sample is remelted, again at 5.5 °C, what quantity of heat is required to convert it to a liquid?

63. If you breathe mercury vapor your health can be seriously impaired, and massive mercury poisoning can result in death. The latent heat of vaporization of liquid mercury is found in Appendix D. What quantity of heat is required to vaporize 0.500 mL of mercury at 357 °C, its normal boiling point? (Density of $Hg(\ell)$ = 13.6 g/mL)

HEATING AND COOLING CURVES AND PHASE DIAGRAMS

64. Your air conditioner may use the chlorofluorocarbon CCl_2F_2 as the heat transfer fluid. (Because of the likely damage to the environment by CFC's, they will be replaced as soon as acceptable alternatives are found.) Its normal boiling point is −30 °C, and the latent heat of vaporization is 165 J/g. The gas and the liquid have specific heat capacities of 0.61 J/g · K and 0.97 J/g · K, respectively. How much heat must be evolved when 10.0 g of CCl_2F_2 is cooled from +40. °C to −40. °C?

65. Liquid ammonia, $NH_3(\ell)$, is still commonly used as a refrigerating liquid and heat transfer fluid. The specific heat capacity of the liquid is 4.7 J/g · K and that of the vapor is 2.2 J/g · K. The enthalpy of vaporization is 23.5 kJ/mol. If you heat 10. kg of liquid ammonia from −50.0 °C to its boiling point of −33.4 °C, and then on to 0.0 °C, how much heat energy must you supply?

66. An approximate phase diagram of xenon is on the next page. Use it to answer the following questions.
(a) In what phase is xenon found at room temperature and 1.0 atmosphere pressure?
(b) If the pressure exerted on a xenon sample is 0.75 atmosphere, and the temperature is −114 °C, in which phase does the xenon exist?
(c) If you measured the vapor pressure of a liquid xenon sample and found 380 mmHg, what is the temperature of the liquid phase?
(d) What is the vapor pressure of the solid at −122 °C?
(e) Which is the denser phase, solid or liquid? Explain briefly.

67. Construct an approximate phase diagram for O_2 from the following information: normal boiling point, 90.18 K; normal melting point, 54.8 K; and triple point 54.34 K (at a pressure of 2 mmHg). Very roughly estimate the vapor pressure of liquid O_2 at −196 °C, the lowest temperature that is reached easily in the laboratory. Is the density of liquid O_2 greater or less than that of solid O_2?

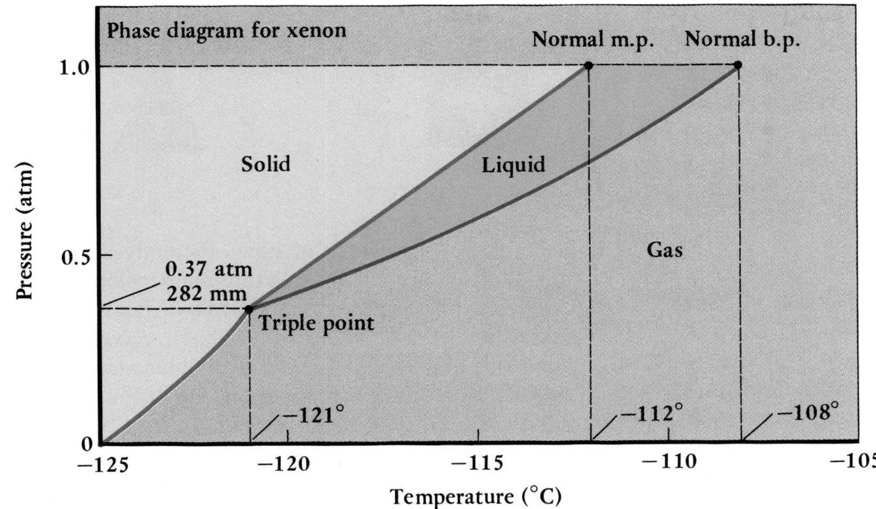

GENERAL QUESTIONS

68. Cooking oil floats on top of water. From this observation, what conclusions can you draw regarding the hydrogen bonding ability of molecules found in cooking oil?

69. Acetone, $(H_3C)_2C{=}O$, is a common laboratory solvent. However, it is usually contaminated with water. Why does acetone absorb water so readily? Draw molecular structures showing how water and acetone can interact. What intermolecular force(s) is (are) involved in the interaction?

70. Rationalize the observation that 1-propanol ($H_3C{-}CH_2{-}CH_2{-}OH$) has a boiling point of 97.2 °C, whereas a compound with the same empirical formula, methyl ethyl ether ($H_3C{-}CH_2{-}O{-}CH_3$), boils at 7.4 °C.

71. Briefly explain the variation in boiling points below. In your discussion be sure to mention the types of intermolecular forces involved.

COMPOUND	BOILING POINT (°C)
NH_3	−33.4
PH_3	−87.5
AsH_3	−62.4
SbH_3	−18.4

72. What significant difference is there between the intermolecular forces involved in the three compounds in Figure 13.14?

73. Vapor pressure data for liquid sodium are given below. Use the Clausius-Clapeyron equation and these data to estimate the heat of vaporization of liquid sodium and its normal boiling point.

TEMPERATURE (°C)	VAPOR PRESSURE (mmHg)
549	10.
633	40.
701	100.
823	400.

74. If you place 1.0 L of ethyl alcohol (C_2H_5OH) in a room that is 3.0 meters long, 2.5 meters wide, and 2.5 meters high, will all the alcohol evaporate? If some liquid remains, how much will there be? The vapor pressure of ethyl alcohol at 25 °C is 60. mmHg, and the density of the liquid at this temperature is 0.79 g/cm^3.

75. The polar liquid methyl alcohol, CH_3OH, is placed in a glass tube. Will the meniscus of the liquid be concave or convex?

76. Liquid ethylene glycol, $HO{-}CH_2CH_2{-}OH$, is one of the main ingredients in commercial antifreeze. Would you predict its viscosity to be greater or less than that of ethyl alcohol?

77. Calcium fluoride is the well-known mineral fluorite (Figure 2.6). It is known that each unit cell contains four Ca^{2+} ions and eight F^- ions (because the F^- ions fill all the so-called "tetrahedral" holes in a face-centered cubic lattice of Ca^{2+} ions). The edge of the CaF_2 unit cell is 5.4×10^{-8} cm in length. What is the density of CaF_2?

78. A picture of the structure of the diamond unit cell is included in Figure 13.35. How many carbon atoms are there in one unit cell? If the density of diamond is 3.51 g/cm^3, what are (a) the volume of the unit cell and (b) the length of the edge?

79. Cesium iodide, CsI, consists of a simple cubic lattice of I^- ions with Cs^+ ions in the cubic lattice holes. If the density of CsI is 4.56 g/cm^3, what is the distance along one edge of the unit cell?

80. During thunderstorms in the Midwest, very large hailstones can fall from the sky. (Some are the size of golf balls!) To preserve some of these stones, we once put them in the freezing compartment of a frost-free refrigerator. Unfortunately, after a day or two the stones had disappeared. Why?

81. A small lake might have a surface area of one square mile and an average depth of 100. feet. If the lake water cools from 25 °C to 4 °C, the temperature of maximum density, how many joules of heat energy must be lost? (Assume the density of water is 1.0 g/cm^3.)

82. Assume you are on another planet that is covered with solid CO_2. Could you "ice" skate easily on the solid? (See Figure 13.48.)

83. Some camping stoves contain liquid butane (C_4H_{10}). They work only when the outside temperature is warm enough to allow the butane to have a reasonable vapor pressure (and so are not very good for camping in temperatures below about 0 °C). Assume the enthalpy of vaporization of butane is 24.3 kJ/mol. (a) If your fuel tank contains 190. g of liquid C_4H_{10}, how many joules of heat are required to vaporize all the butane? (b) What is the vapor pressure of C_4H_{10} at 0.0 °F? (You will need data from Table 13.5 and the Clausius-Clapeyron equation for this calculation.) Why doesn't this stove work well when the air temperature is low?

84. Rutile (TiO_2) crystallizes in a structure that is characteristic of many other ionic compounds. How many formula units of TiO_2 are there in the unit cell of rutile illustrated below? (The oxide ions marked by a dot are in the faces of the unit cell. The oxide ions marked by X are wholly *within* the unit cell.)

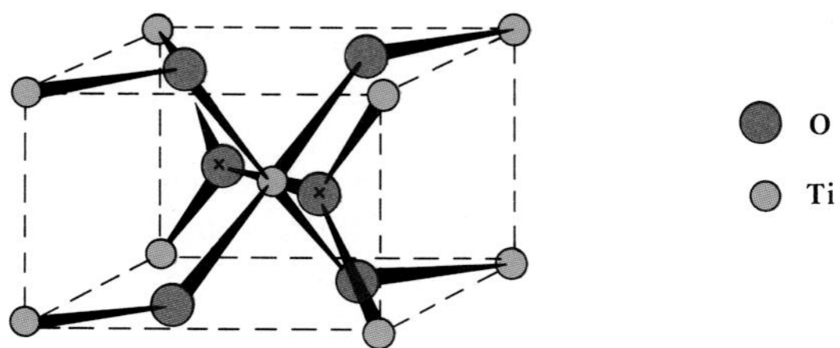

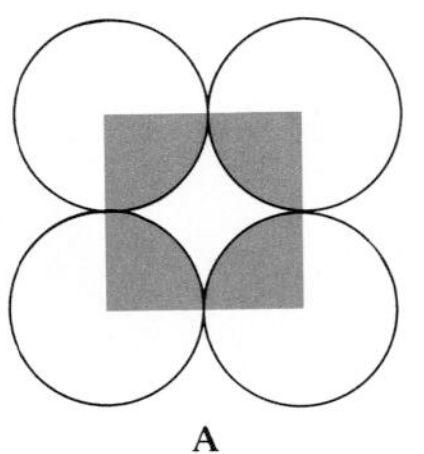

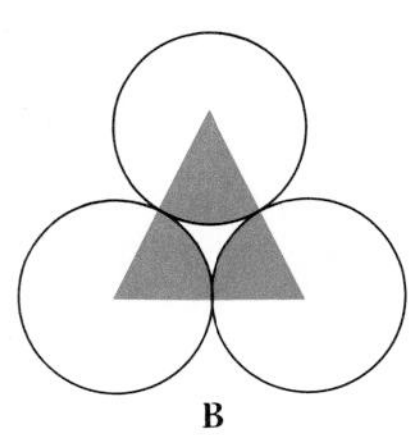

85. You can get some idea of how efficiently spherical atoms or ions are packed in a three-dimensional solid by seeing how well circular "atoms" pack in two dimensions. Using the drawings above, prove that B is a more efficient way to pack circular atoms than A. A unit cell of A contains portions of four circles and one hole. In B, packing coverage can be calculated by looking at a triangle that contains portions of three circles and one hole. Show that A fills about 80% of the available space, whereas B fills closer to 90% of the available space.

86. If a simple cubic unit cell is formed so that the spherical atoms or ions just touch one another along the edge, calculate the percentage of empty space within the unit cell. (Recall that the volume of a sphere is $\frac{4}{3}\pi r^3$, where r is the radius of the sphere.)

87. Cesium iodide, CsI, consists of a simple cubic lattice of I^- ions with Cs^+ ions in the cubic lattice holes. If the edge of the cube is 0.445 nm, what is the distance from the center of the Cs^+ in the center of the cube to the center of one of the I^- ions at the cube corners? (You should recognize that the I^- ions at the cube corners do not touch one another, but they are in contact with the Cs^+ ion in the center of the cell. See Exercise 13.5.)

88. Metallic gold crystallizes in a face-centered cubic lattice. The radius of a gold atom is 144.0 pm, and the density of the solid is 19.32 g/cm^3. Using this information, and the molar mass of gold, calculate Avogadro's number to four significant figures.

89. If the lakes and oceans of another planet are filled with ammonia, how many joules of energy are given off when 1.00 mole of liquid ammonia cools from −33.4 °C (its boiling point) to −43.4 °C? (Specific heat capacity of liquid NH_3 is 4.70 J/g · K.) Compare this with the amount of heat given off by 1.00 mole of water cooling by 10. degrees Celsius.

SUMMARY QUESTIONS

90. Sulfur dioxide, SO_2, is found in polluted air. It comes from the combustion of fossil fuels, and from industrial plants that convert certain metal-containing ores to metals or metal oxides.

(a) Draw the electron dot structure for SO_2. From this tell (i) the O—S—O angle, (ii) the structural-pair geometry and molecular geometry, and (iii) the hybridization of the S atom.

(b) What type of forces are responsible for binding SO_2 molecules to one another in the solid or liquid phase?

(c) Using the information below, place the compounds listed in order of increasing intermolecular forces.

COMPOUND	NORMAL BOILING POINT (°C)
SO_2	−10
NH_3	−33.4
CH_4	−161.5
H_2O	100

(d) The triple point for SO_2 is −74 °C (estimated), its melting point is −72.7 °C, and its boiling point is −10 °C. Is solid SO_2 more or less dense than liquid SO_2?

(e) Airborne sulfur dioxide is one of the compounds responsible for "acid rain." It is thought that the $SO_2(g)$ first oxidizes to $SO_3(g)$ in air. What is the standard enthalpy change for this process? What is the standard enthalpy change for the combination of $SO_3(g)$ and water to give $H_2SO_4(aq)$?

91. Mercury is the only metal that is a liquid at room temperature. It and many of its compounds are dangerous poisons if breathed, swallowed, or even absorbed through the skin. If the vapor pressure of liquid mercury is 1.00 mmHg at 126.2 °C, and if its normal boiling point is 357 °C, calculate the equilibrium vapor pressure at room temperature (25 °C). (Use the enthalpy of vaporization from Appendix D.) If the air in a small room (12 feet × 12 feet × 8 feet) is saturated with mercury vapor, how many atoms of Hg vapor are there per cubic foot?

92. Copper is an important metal in our economy, most of it being mined in the form of the mineral chalcopyrite, $CuFeS_2$.

(a) To obtain one metric ton (1.00×10^3 kilograms) of copper metal, how many metric tons of chalcopyrite would you have to mine?

(b) If the sulfur in chalcopyrite is converted to SO_2, how many metric tons of the gas would you get from one metric ton of chalcopyrite?

(c) Copper crystallizes as a face-centered cube. Knowing that the density of copper is 8.95 g/cm³, calculate the radius of a copper atom.

CHAPTER

14

Solutions and Their Behavior

The oceans are a solution of many salts and dissolved gases.

CHAPTER OUTLINE

A **solution** is a homogeneous mixture of two or more substances in a single phase. It is usual to think of the component present in the largest amount as the **solvent** and the other component as the **solute**. When you think of solutions, those that occur to you first probably involve water as the solvent: apple cider, soft drinks, and beer. However, some consumer products (lubricating oils, gasoline, household cleaners) are solutions involving a liquid other than water as the solvent.

Our objective in this chapter is to develop an understanding of gases, liquids, and solids dissolved in liquid solvents. However, you should be aware that solutions do not have to involve liquid solvents. The air you breathe is a solution of nitrogen, oxygen, carbon dioxide, water vapor, and other gases. Although glass is variously referred to as an amorphous solid or a supercooled liquid, it is a solution of metal oxides (Na_2O and CaO, among others) in silicon dioxide. The solder that is used to make connections in your calculator or computer is a solid solution of tin, lead, and other metals.

Common sense tells you that adding a solute to a pure liquid will change the properties of the liquid, since the intermolecular forces will be changed or disrupted. Indeed, that is precisely why some solutions are made. For instance, by adding antifreeze to your car's coolant water you can prevent the radiator from boiling over in the summer and freezing up in the winter. As we will see shortly, the changes in freezing and boiling points from pure solvent to solution are called **colligative properties**. *The magnitude of colligative properties*—changes in freezing point, boiling point, and vapor pressure from solvent to solution, as well as solution osmotic pressure—***ideally*** *depend only on the number of solute particles per solvent molecule and on the nature of the solvent but not on the nature of the solute.*

Figure 14.1 Molarity and molality. The photo shows a 0.10 *molal* solution (0.10 *m*) of potassium chromate (flask at the left) and a 0.10 *molar* solution (0.10 M) (flask on the right). In the flask on the right we mixed 0.10 mole (19.4 g; the yellow solid in the front has this mass) of K_2CrO_4 with enough water to make 1.0 L of solution. (To know that we have 1.0 L of solution, we fill the volumetric flask to the mark on its neck.) To the flask on the left, we added 1000 g (1.0 kg) of water to 0.10 mole of K_2CrO_4. Adding 1.0 kg of water leads to a solution clearly having a volume greater than 1.0 L.

There are three major topics to be covered in this chapter. First, since colligative properties depend on the relative numbers of solvent and solute particles in solution, we need convenient ways of describing solution concentration in these terms. Second, we are interested in how and why the solution process occurs. This will give us some insight into the third topic, the colligative properties themselves.

14.1 UNITS OF CONCENTRATION

Molarity, the concentration unit that has served us well to this point, will not work with colligative properties. Let's understand why.

Molarity is defined as the number of moles of solute *per liter of solution*.

$$\text{Molarity (M) of the compound A in solution} = \frac{\text{moles of A}}{\text{liters of solution}}$$

For example, Figure 14.1 *right* shows a 0.10 molar aqueous solution of potassium chromate that was made by adding enough water to 0.10 mole of K_2CrO_4 (19.4 g) to make 1.0 L of solution. We did not pay attention to how much of the solvent (water) we added; we were concerned only with having the correct *total solution* volume. However, if we had added 1.00 L of water to 19.4 g of K_2CrO_4, the volume of solution would be 1.00 L *only* if the solute and solvent together take up exactly 1.00 L of volume. As Figure 14.1 *left* shows, this is certainly not the case here, and it is almost never true.

Molarity is a useful measure of concentration for preparing reaction mixtures because we can easily measure out volumes of solution and can therefore know how many moles of solute we are using. On the other hand, we would like a way of measuring concentrations that will tell us how many molecules or ions of solute there are per molecule of solvent. This is just the information we need to define the colligative properties of a solution.* There are several such units: molality, mole fraction, and weight percent.

The **molality** of a solution is defined as the number of moles of solute per kilogram of solvent.

$$\text{Molality of A} = m_A = \frac{\text{moles of A}}{\text{kilograms of solvent}} \tag{14.1}$$

The solution in Figure 14.1 *left*, for example, has a molality of

Notice that molality is denoted by a small italicized "m" while molarity is signaled with a capital M.

$$\text{Molality of } K_2CrO_4 = \frac{19.4 \text{ g (1 mol/194 g)}}{1.00 \text{ kg water}} = 0.100\ m$$

Notice that different quantities of water were used to make the 0.10 M and 0.10 *m* solutions of K_2CrO_4 in Figure 14.1. This means that *the mo-*

*Another problem with using molarity is that, since the density of a liquid is temperature dependent, the volume of solution also depends on temperature. Therefore, molarity is temperature dependent. To be entirely correct, you must report the temperature when you state solution concentrations in molarity.

larity and molality of a given solution cannot be the same (although the difference is negligibly small when the solution is quite dilute, say less than 0.01 M).

The mole fraction of a component A, X_A, of a solution is defined as the number of moles of A divided by the total number of moles of all of the components of the solution. For A in a solution containing A, B, and C, the mole fraction of A is

$$\text{Mole fraction of A} = X_A = \frac{\text{moles of A}}{\text{moles of A} + \text{moles of B} + \text{moles of C}} \quad (14.2)$$

For example, in a solution that contains 1.0 mole (46 g) of ethyl alcohol, C_2H_5OH, in 9.0 moles (162 g) of water, the mole fraction of alcohol is 0.10 and that of water is 0.90.

$$X_{alc} = \frac{1.0 \text{ mol alcohol}}{1.0 \text{ mol alcohol} + 9.0 \text{ mol water}} = 0.10$$

$$X_{water} = \frac{9.0 \text{ mol water}}{1.0 \text{ mol alcohol} + 9.0 \text{ mol water}} = 0.90$$

Obviously, the sum of the mole fractions of all components in the solution must equal exactly 1.

$$X_{water} + X_{alc} = 1.00$$

Expressing a concentration as a **weight percentage*** is straightforward, since it is simply the mass of solute per 100 g of solution.

$$\text{Weight \% of A} = \frac{\text{grams of A}}{\text{grams of A} + \text{grams of other solutes} + \text{grams of solvent}} \times 100 \quad (14.3)$$

Our alcohol/water mixture contains 46 g of alcohol in 162 g of water, so the total solution mass is 208 g, and the weight % of alcohol is

$$\text{Weight \% alcohol} = \frac{46 \text{ g alcohol}}{46 \text{ g alcohol} + 162 \text{ g water}} \times 100$$
$$= 22\% \text{ alcohol by weight}$$

EXAMPLE 14.1

CALCULATING MOLE FRACTIONS, MOLALITY, AND WEIGHT PERCENT

If you dissolve 10.0 g or about one heaping spoonful of sugar, $C_{12}H_{22}O_{11}$, in a cup of water (250. g), what are the mole fraction, molality, and weight percentage of sugar?

Solution 10.0 g of sugar (molar mass = 342.3 g/mol) are equivalent to 0.0292 moles and 250. g of water represent 13.9 moles.

---

*Strictly speaking, this is a *mass* percentage; however, long usage has made "weight percentage" the standard term.

Mole fraction:

$$X_{sugar} = \frac{0.0292 \text{ mol sugar}}{0.0292 \text{ mol sugar} + 13.9 \text{ mol water}} = 0.00210$$

Molality:

$$m = \frac{0.0292 \text{ mol sugar}}{0.250 \text{ kg water}} = 0.117 \text{ molal in sugar}$$

Weight percentage:

$$\text{Weight \% sugar} = \frac{10.0 \text{ g sugar}}{10.0 \text{ g sugar} + 250. \text{ g water}} \times 100 = 3.85\% \text{ sugar}$$

EXERCISE 14.1 Mole Fractions and Molality

Assume you add 1.0×10^3 g of $C_2H_4(OH)_2$, ethylene glycol, as an antifreeze to 4.0 kg of water in the radiator of your car. What are the mole fraction and molality of the ethylene glycol?

14.2 THE SOLUTION PROCESS

Both ammonium chloride and sodium hydroxide dissolve readily in water, but the water cools down when NH_4Cl dissolves, while it warms up when NaOH dissolves. On the other hand, $CaCO_3$ dissolves in water only to a small extent. Why do these things happen? What controls the solution process?

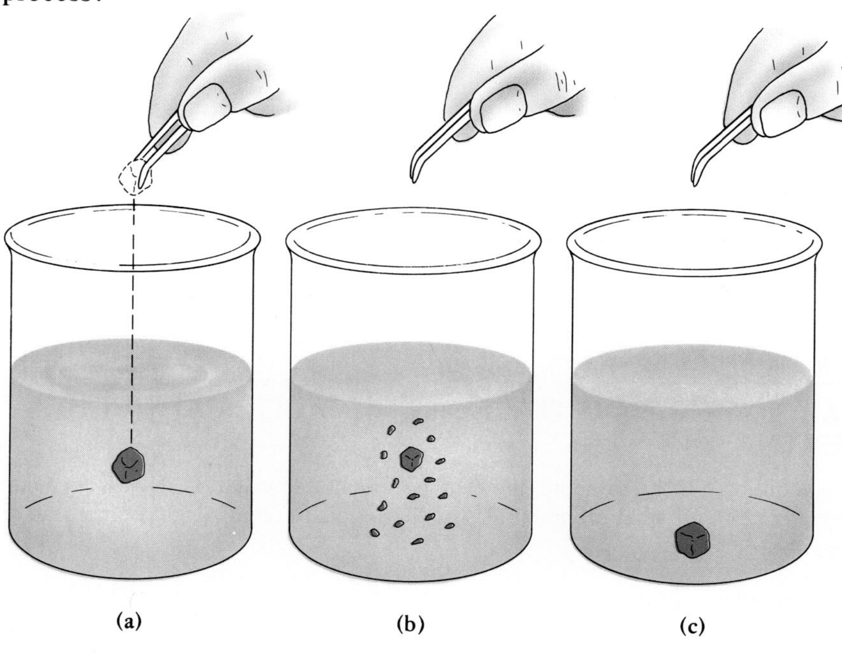
(a)
Unsaturated:
crystal dissolves when
added to solution
(b)
Supersaturated:
additional crystals
form when a crystal
is added to the solution
(c)
Saturated:
amount of solid
does not change
with time

Figure 14.2 The difference between unsaturated, saturated, and supersaturated solutions. In (a), some solute is added to a solution, and the solute dissolves. The solution was unsaturated with respect to solute. In (b), the solution is supersaturated. Adding extra solute leads to the precipitation of some dissolved solute. The solution in (c) is saturated, since no more solute will dissolve; the concentration of solute does not change with time.

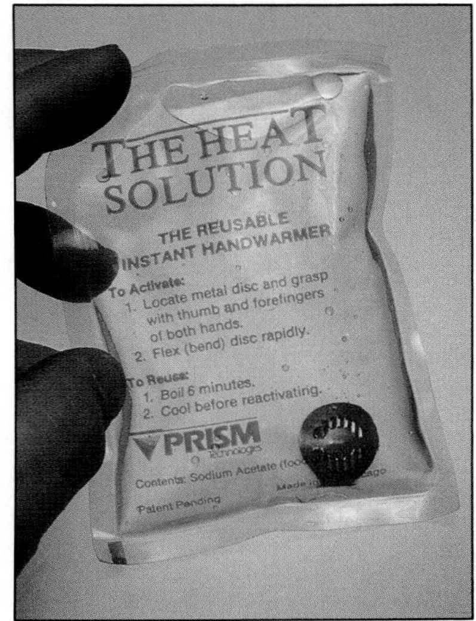

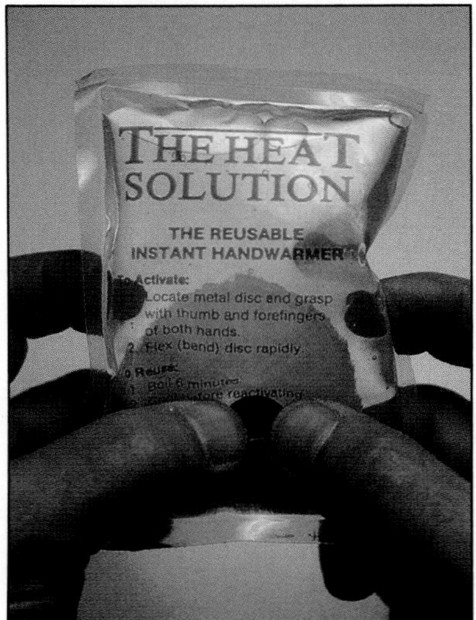

Figure 14.3 An example of a supersaturated solution is a "heat pack" called "The Heat Solution." The bag contains a supersaturated solution of sodium acetate. If the solution is disturbed (by adding a small crystal of the salt or by some mechanical means), $NaCH_3COO \cdot 3H_2O$ begins to precipitate (bottom photo). As described in the text, solutions of most salts release heat when the salt precipitates, and the temperature of the mixture rises to approximately the melting point of the salt (58 °C in this case). [You can make your own "heat pack" by warming a mixture of sodium thiosulfate, $Na_2S_2O_3 \cdot 5H_2O$ (440 g; available in photo stores), and 2 tablespoons of water. Once a solution has been formed, and cooled slowly to room temperature, dropping in a small crystal of the salt will cause the salt to precipitate from solution, and the temperature will rise to 48 °C. See *Chem Matters,* February 1987, page 12.]

First, let's define more carefully some of the terms we have been using. When a solute is dissolved in a solvent, we mean that the attractive forces between solute and solvent particles are great enough to overcome the attractive forces (intermolecular forces) within the pure solvent and solute. As described in Chapter 13, when solutes are dissolved, they become *solvated* if solvent molecules are sufficiently attracted to solute molecules or ions (usually by dipole–dipole or ion–dipole forces). If water is the solvent, *solvation* is called more specifically *hydration.*

Solubility is the maximum amount of material that will dissolve in a given amount of solvent at a given temperature to produce a stable solution. In Chapter 5 you learned some guidelines for the solubility of common ionic compounds, but we can now be more quantitative. For example, it is suggested in Figure 5.5 that common nitrates and chlorides are usually soluble, except for the chlorides of Ag^+, Pb^{2+}, and $Hg_2{}^{2+}$. For instance, we find that 952 g of silver nitrate, $AgNO_3$, will dissolve in 100 mL of water at 100 °C, and so we say this is a *soluble* salt. In contrast, only 0.00217 g of silver chloride, AgCl, dissolves in 100 mL of water at 100 °C, and we say this is insoluble, even though it does have some degree of solubility.*

When the maximum amount of solute has been dissolved in a solvent and equilibrium is attained, the solution is said to be **saturated**. *The concentration of dissolved solute in a saturated solution is the same as the solubility of the solute.* If the solute concentration is less than the saturation amount, the solution is **unsaturated**. On the other hand, there are instances when solutions can temporarily contain more solute than the saturation amount, and these are called **supersaturated**. Figures 14.2 and 14.3 show what will happen when more solute is added to a solution in each of these conditions.

The concept of solubility is the basis of innumerable laboratory operations and industrial processes, and it can be the controlling factor in our environment and in living organisms. Thus, it is useful to explore the solubility or insolubility of common substances.

*When can you say properly that one substance is soluble and another is insoluble? There are no good rules about this, because there are degrees of solubility; nothing is totally without some solubility at some temperature. Chemists frequently say something is not soluble, partly soluble, or soluble. Usually what they mean is that something is soluble if they can see that a good deal of the solid has gone into solution, when only a modest amount of solid has been added. If only some solid has visibly dissolved, the solid is partly soluble. If they cannot observe the dissolution of any solid, it is said to be insoluble or sparingly soluble.

(a)

(b)

Figure 14.4 Miscibility. (a) The colorless, more dense bottom layer is nonpolar carbon tetrachloride, CCl_4. The middle layer is a solution of $CuSO_4$ in water, and the colorless, less dense top layer is nonpolar octane (H_3C-CH_2-CH_2-CH_2-CH_2-CH_2-CH_2-CH_3). The weak intermolecular interactions between the two nonpolar molecules and water cannot overcome the very strong forces between water molecules and allow the nonpolar solute to be miscible with water. (b) After stirring the mixture, the two nonpolar liquids form a homogeneous mixture, since they are miscible with one another. This layer of mixed liquids sits on top of the water layer because the water layer is the more dense of the two.

LIQUIDS DISSOLVING IN LIQUIDS

If two liquids mix to an appreciable extent to form a solution, they are said to be **miscible**. In contrast, **immiscible** liquids will not mix to form a solution; they will exist in contact with each other as separate layers (Figure 13.6).

The nonpolar liquids octane (a component of gasoline) and CCl_4 are miscible; they will mix in all proportions to form a homogeneous solution (Figure 14.4). On the other hand, polar water and nonpolar octane are immiscible. You would observe that the less dense liquid, octane, simply floats as a layer on top of the more dense water layer (Figure 14.4a). Finally, polar ethyl alcohol molecules (H_3C—CH_2—OH) will dissolve in all proportions in water; beer, wine, and other alcoholic beverages contain amounts of alcohol ranging from just a few percent to more than 50%. It is these observations that have led to the familiar rule of thumb: *like dissolves like*. That is, two or more nonpolar liquids frequently are miscible with each other, just as are two or more polar liquids. Liquids of different types will not mix to an appreciable degree.

What is the molecular basis for the "like dissolves like" guideline? Molecules of pure octane or pure CCl_4, both of which are nonpolar, are held together in the liquid phase by weak dispersion forces (Section 13.2). Similar weak forces can exist between an octane molecule and a CCl_4 molecule, and these molecules are attracted to one another.

In contrast to the CCl_4/octane solution, water and octane do not mix (Figure 14.5). This observation can be analyzed with the following scheme, one that may be applied to any solvent/solute interaction.

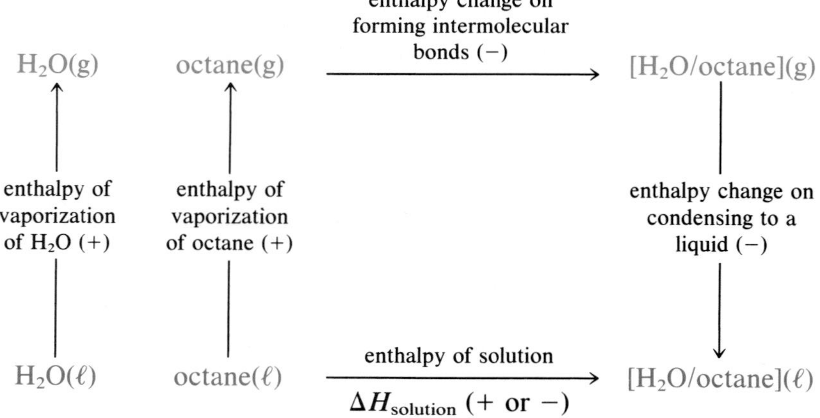

The enthalpy of solution, $\Delta H_{solution}$, is the net heat energy involved in the process of solution formation (at constant pressure). According to Hess's law, which was described in Chapter 6, we can analyze the energy involved in a complex process by breaking it down into simpler steps, whose energies are known. In this case we can find $\Delta H_{solution}$, in principle, by adding up the enthalpy change (+) that occurs when the solute and solvent are put into the gas phase, the enthalpy change when the two are attracted by intermolecular forces in the gas phase (−), and then the enthalpy change when the mixture condenses to form a liquid solution (−). The important point here is that the enthalpy change observed on vaporizing octane and, in particular, water is much more positive than the enthalpy changes observed in the other steps. Therefore, the value of the overall enthalpy change, $\Delta H_{solution}$, is likely to be positive, reflecting an endothermic process. In large part, it is for this reason that polar and nonpolar liquids do not mix well.

Although the enthalpy of solution is important in determining solubility, another quantity, the entropy of solution, is also a contributing factor. This is considered in Chapter 20.

SOLIDS DISSOLVING IN LIQUIDS

Solid I_2 is an example of a molecular solid, the molecules being held together by weak intermolecular dispersion forces. To dissolve this nonpolar solid in a nonpolar liquid, say CCl_4, the attractions between I_2 molecules and between CCl_4 molecules must be disrupted. The enthalpy of solution is close to zero, however, since the energy required to separate solid I_2 molecules and liquid CCl_4 molecules is approximately returned by the formation of new intermolecular attractions between I_2 and CCl_4. Thus, it is not surprising that I_2 is soluble in CCl_4 (Figure 14.6).

Sucrose, a sugar, is also a molecular solid, but the molecules interact in the solid state through hydrogen bonds, the same kind of attraction that occurs between sugar and water molecules when sugar is dissolved in water.

Sucrose (cane sugar)

Hydrogen bonding between sugar and water is strong enough that the energy evolved here can be supplied to disrupt the sugar/sugar and water/water interactions. On the other hand, sugar is not soluble at all in CCl_4 or other nonpolar liquids. A nonpolar liquid does not interact sufficiently well with sugar to cause it to dissolve.

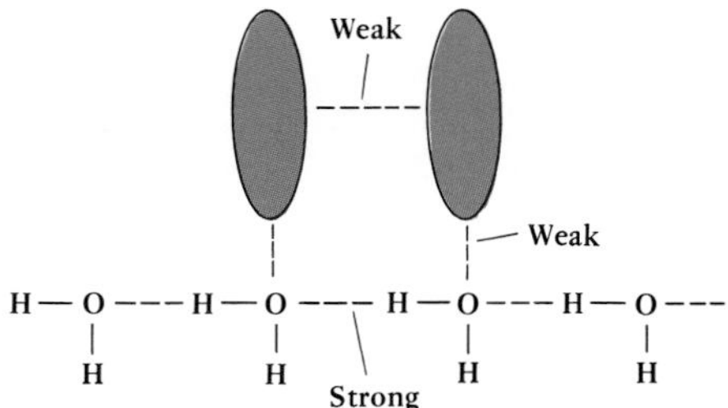

Figure 14.5 Intermolecular forces in water and octane (octane is depicted by an oval; its structural formula is H_3C-CH_2-CH_2-CH_2-CH_2-CH_2-CH_2-CH_3). Water molecules interact strongly through hydrogen bonding, but there are only weak dispersion forces between octane molecules. Similarly, the octane/water forces are weak. Therefore, on mixing water and octane, the water/octane interaction is not strong enough to compensate for breaking hydrogen bonds in water, and water and octane are not miscible to any appreciable extent.

(a)

(b)

Figure 14.6 Water, carbon tetrachloride (CCl_4), and iodine. (a) Water (a polar molecule) and CCl_4 (a nonpolar molecule) are not miscible, and the less dense water layer is found on top of the more dense CCl_4 layer. A small amount of iodine is dissolved in water to give a brown solution (top). However, the nonpolar molecule I_2 is more soluble in nonpolar CCl_4, as indicated by the fact that I_2 dissolves preferentially in CCl_4 to give a purple solution after the mixture is shaken (b).

Network solids include graphite, diamond, and quartz, and your intuition tells you they do not dissolve readily in water. After all, where would all the beaches be if sand dissolved readily in water? Normal chemical bonding in network solids is simply too strong to be replaced by weaker hydrogen bonding attraction to water dipoles.

DISSOLVING IONIC SOLIDS

Common ionic solids are not soluble to any appreciable extent in nonpolar solvents, and they vary greatly in their solubility in water (Figure 5.6). At 20 °C, 100 g of water will dissolve 74.5 g of $CaCl_2$, but only 0.0014 g of limestone, $CaCO_3$, can be dissolved no matter how hard you try. It is useful to explore the reasons for the widely different solubilities.

Common table salt dissolves in water because strong ion–dipole forces lead to strong ion hydration, and help to break down the cation–anion attraction in the crystal lattice (Figure 5.2). There are two ways to analyze the overall energy involved here. First, the enthalpy of solution ($\Delta H_{solution}$) can be estimated from the lattice energy of NaCl (Chapter 13) and the enthalpy of hydration of the gaseous ions to form hydrated ions ($\Delta H_{hydration}$).

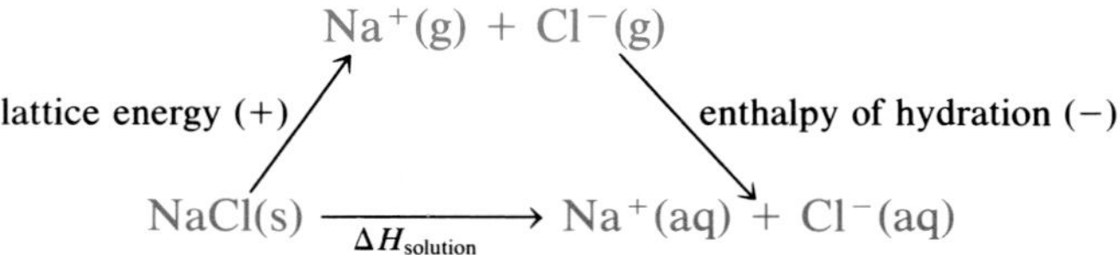

From Hess's law (Chapter 6) you know the enthalpy of solution is the sum of the enthalpies of the processes along the path from reactants [NaCl(s)] to products [$Na^+(aq) + Cl^-(aq)$].

$$\Delta H_{solution} = \text{lattice energy} + \text{enthalpy of hydration}$$

$$\Delta H_{solution} \text{ for NaCl(s)} = 766 \text{ kJ/mol} + (-760. \text{ kJ/mol}) = +6 \text{ kJ/mol}$$

Using this approach we can estimate that the solution process for NaCl is slightly endothermic. In Table 14.1, the water solubility of a series of simple salts is correlated with their enthalpy of solution. Here you see that the most soluble compounds have exothermic enthalpies of solution and that, as the value of $\Delta H_{solution}$ becomes more and more positive, the solubility declines; AgCl is quite insoluble and has a very positive value of $\Delta H_{solution}$. This means that solubility is favored when the energy re-

Table 14.1 Water Solubility of Some Ionic Compounds

Compound	Lattice Energy (kJ/mol)	$\Delta H_{hydration}$ (kJ/mol)	$\Delta H_{solution}$ (kJ/mol)	Solubility in H_2O (g/100 mL)
NaCl	766	−760.	+6	35.7 (0 °C)
LiF	1032	−1005	+27	0.3 (18 °C)
KF	813	−819	−6	92.3 (18 °C)
RbF	776	−792	−16	130.6 (18 °C)
$SrCl_2$	2110.	−2161	−51	53.8 (20 °C)
AgCl	916	−851	+65	0.000089 (10 °C)

quired to break down the lattice (the lattice energy) is smaller than or approximately equal to the energy given off when the ions are hydrated ($\Delta H_{\text{hydration}}$).

Our approach to predicting the water solubility from lattice energy and enthalpy of hydration is useful, because the process can be visualized. One can "see" a crystal lattice flying apart into ions in the gas phase and then these ions plunging into water. Unfortunately, reasonably good values for lattice energy and $\Delta H_{\text{hydration}}$ are available for relatively few salts, so we are interested in a more general process of obtaining the enthalpy of solution. Consider again the process of dissolving NaCl. Recalling the discussion of thermochemistry in Chapter 6, it is evident that $\Delta H^\circ_{\text{solution}}$ can be estimated from enthalpies of formation.

$$NaCl(s) \longrightarrow NaCl(aq)$$

$$\Delta H^\circ_{\text{solution}} = \Delta H^\circ_f[NaCl(aq)] - \Delta H^\circ_f[NaCl(s)]$$

These values are available from tables published by the U.S. National Bureau of Standards.

$\Delta H^\circ_f[NaCl(aq)] = -407.27$ kJ/mol $\qquad \Delta H^\circ_f[NaCl(s)] = -411.15$ kJ/mol

and they give a value for $\Delta H^\circ_{\text{solution}}$ of +3.9 kJ, a small endothermic value just as was estimated from the data in Table 14.1* Some data that you can use to calculate solution enthalpies for other compounds are given in Table 14.2.

Table 14.2 Data to Calculate Enthalpy of Solution

Compound	ΔH°_f(s) (kJ/mol)	ΔH°_f (aq, 1 *m*) (kJ/mol)
LiF	−616.0	−611.1
KF	−526.3	−585.0
RbF	−557.7	−583.8
KCl	−436.7	−419.5
RbCl	−435.4	−418.3
NaOH	−425.6	−470.1
NH_4NO_3	−365.6	−339.9

EXERCISE 14.2 Calculating Enthalpy of Solution

(a) Use the data in Table 14.1 to compare the enthalpies of solution for AgCl and RbF. Comment on the relation between the enthalpy of solution and solubility of these two salts.

(b) Use the data in Table 14.2 to calculate the enthalpy of solution of ammonium nitrate.

In Exercise 14.2(b) you calculated the enthalpy of solution of NH_4NO_3, and you should have found it to be quite positive. The energy required to dissolve this salt comes from its surroundings, so the water cools dramatically when the solid solute is added (Figure 14.7). This is the basis of the "chemical cold packs" that you may have seen or used in a hospital or at a sporting event.

Figure 14.7 A commercially available "cold pack." A typical cold pack consists of a sealed bag of water inside a bag of ammonium nitrate. When the bags are squeezed firmly, the inner pouch breaks and water and NH_4NO_3 mix. The endothermic solution process causes the water temperature to drop to about 5 °C.

FACTORS AFFECTING SOLUBILITY: PRESSURE AND TEMPERATURE

Biochemists and physicians, among others, are interested in the solubility of gases such as CO_2 and O_2 in water or body fluids, and all scientists—especially geologists—need to know about the solubility of solids in various solvents. Two external factors controlling such processes

*It is not unusual to see small differences in thermochemical data calculated by different methods. The difference in this case comes from complicating factors that are not important to our overall conclusion.

are temperature and pressure. Both can affect the solubility of gases in liquids, whereas normally only temperature is important in determining the solubility of solids in liquids.

PRESSURE EFFECTS ON GAS SOLUBILITY: HENRY'S LAW The effect of the pressure of a gas on its solubility in a liquid is often dramatic. It is governed by **Henry's law**, Equation 14.4, which states that *the amount of gas that dissolves in a given amount of solvent at a specified temperature is proportional to the partial pressure of the gas above the liquid.*

Henry's law holds strictly only for gases that do not interact chemically with the solvent (it does not cover NH_3, for example, which gives small concentrations of NH_4^+ and OH^- in water).

$$\text{Molality of solute gas} = kP \text{ (at constant pressure)} \quad (14.4)$$

P = partial pressure of the gaseous solute

k = Henry's law constant, a constant *characteristic of the solute and solvent*

Henry's Law Constants (25 °C)

Gas	k (molal/mmHg)
N_2	8.4×10^{-7}
O_2	1.7×10^{-6}
CO_2	4.48×10^{-5}

From W. Stumm and J.J. Morgan, *Aquatic Chemistry*, Wiley: New York, 1981, page 109.

Carbonated soft drinks are good examples of Henry's law. They are packed under pressure in a chamber filled with carbon dioxide gas, some of which dissolves in the drink. When you open the can or bottle, the partial pressure of CO_2 above the solution drops, so the concentration of CO_2 in solution also drops, and gas bubbles out of the solution (Figure 14.8). The same can happen with gases dissolved in your blood if you are an underwater diver (Figure 14.9).

EXAMPLE 14.2

USING HENRY'S LAW

Oxygen has a Henry's law constant of 1.7×10^{-6} molal/mmHg when dissolved in water at 25 °C. What is the concentration of O_2 in water at 25 °C when O_2 has a partial pressure of 150 mmHg?

Solution From Henry's law, we have

$$\text{Molality of } O_2 = (1.7 \times 10^{-6} \text{ molal/mmHg})(150 \text{ mmHg}) = 2.6 \times 10^{-4}\ m$$

The O_2 pressure (150 mmHg) is roughly the partial pressure of the gas in dry air. This means that only about 8 milligrams of O_2 in air will dissolve in a kilogram of water at 25 °C.

Figure 14.8 Illustration of Henry's law. The greater the partial pressure of CO_2 over the soft drink in a can, the greater the amount of CO_2 dissolved. There is more CO_2 dissolved in the closed can than in the can open to the atmosphere.

EXERCISE 14.3 Using Henry's Law

The Henry's law constant for CO_2 in water at 25 °C is 4.48×10^{-5} molal/mmHg. What is the concentration of CO_2 in water when the partial pressure is $\frac{1}{3}$ of an atmosphere? (Note that CO_2 in fact reacts with water to give traces of H_3O^+ and HCO_3^-. However, the reaction occurs to such a small extent that Henry's law is obeyed at low CO_2 partial pressures.)

TEMPERATURE EFFECTS ON SOLUBILITY: LE CHATELIER'S PRINCIPLE You know from experience that much more gas is released from a can of soft drink that is opened when it is warm than when it is cold. Sometimes the gas bubbles out of a warm solution so rapidly that the liquid gushes out of the can. This illustrates the fact that *temperature affects solubility.*

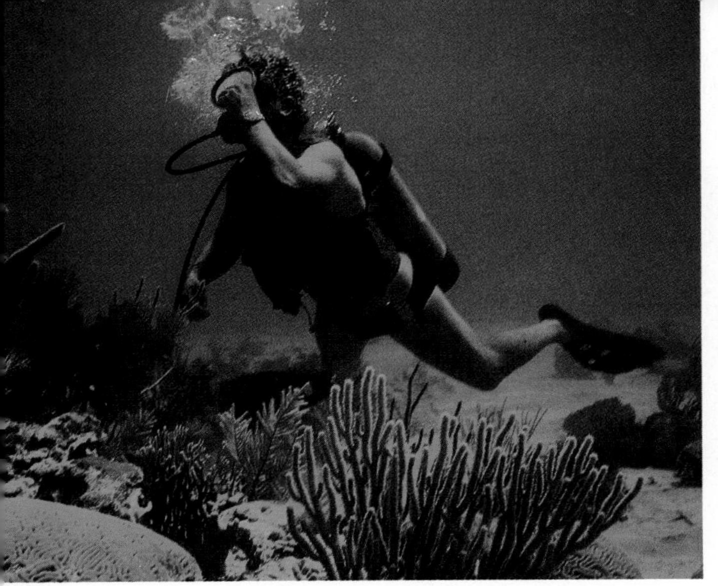

Figure 14.9 If you dive using scuba gear to any appreciable depth, you have to worry about the solubility of gases in your blood. Nitrogen is soluble in blood, so if you breathe a high-pressure mixture of O_2 and N_2 deep under water, there can be an appreciable concentration of N_2 in your blood system. If you then ascend too rapidly, the nitrogen is released as the pressure decreases and forms bubbles in the blood. This affliction, sometimes called the ''bends,'' is painful and can be fatal if blood is forced out of capillaries and the brain, for example, is deprived of oxygen. To partly circumvent this, deep sea divers sometimes use a helium/oxygen mixture instead of nitrogen/oxygen, because helium is not nearly as soluble in blood as nitrogen. (© Boyd Norton)

Gases that dissolve to an appreciable extent in solvents usually do so in an exothermic process. That is, the enthalpy of solution is negative.

$$\text{Gas} + \text{liquid solvent} \longrightarrow \text{saturated solution} + \text{heat}$$

The process continues until a saturated solution is formed, and *equilibrium* is established. At this point, the dissolving process continues, but it is *exactly* counterbalanced by gas molecules coming out of solution with the *consumption* of heat.

The factors that determine an equilibrium are concentration, pressure, volume (for gases), and temperature. Only temperature is considered here. Chapters 16–19 consider all factors in more detail.

$$\text{Saturated solution} + \text{heat} \longrightarrow \text{gas} + \text{liquid solvent}$$

At equilibrium, as many molecules come out of solution as dissolve in a given time period. We depict this situation by showing the original process with arrows running in both directions, $\rightleftarrows$.

$$\text{Gas} + \text{liquid solvent} \rightleftarrows \text{saturated solution} + \text{heat}$$

To understand how temperature affects solubility, we introduce **Le Chatelier's principle**. This states that *a change in any of the factors determining an equilibrium will cause the system to adjust in order to reduce or counteract the effect of the change*. For example, if we heat a solution of a gas in a liquid, the equilibrium will shift in an attempt to use up some of the added heat. That is, the reaction

$$\text{Gas} + \text{liquid solvent} \rightleftarrows \text{saturated solution} + \text{heat}$$

added heat
equilibrium shifts left

will shift back to the left, since heat can be consumed in the process that gives free gas molecules and pure solvent (Figure 14.10). This is why a dissolved gas always becomes less soluble with increasing temperature.

Figure 14.10 A warm glass rod is placed in a glass of ginger ale. The heat from the rod flows into a cold solution of CO_2 in water and causes the CO_2 to be less soluble. The Henry's law constant for CO_2 in water is 4.48×10^{-5} molal/mmHg at 25 °C, whereas it is 2.5×10^{-5} molal/mmHg at 50 °C.

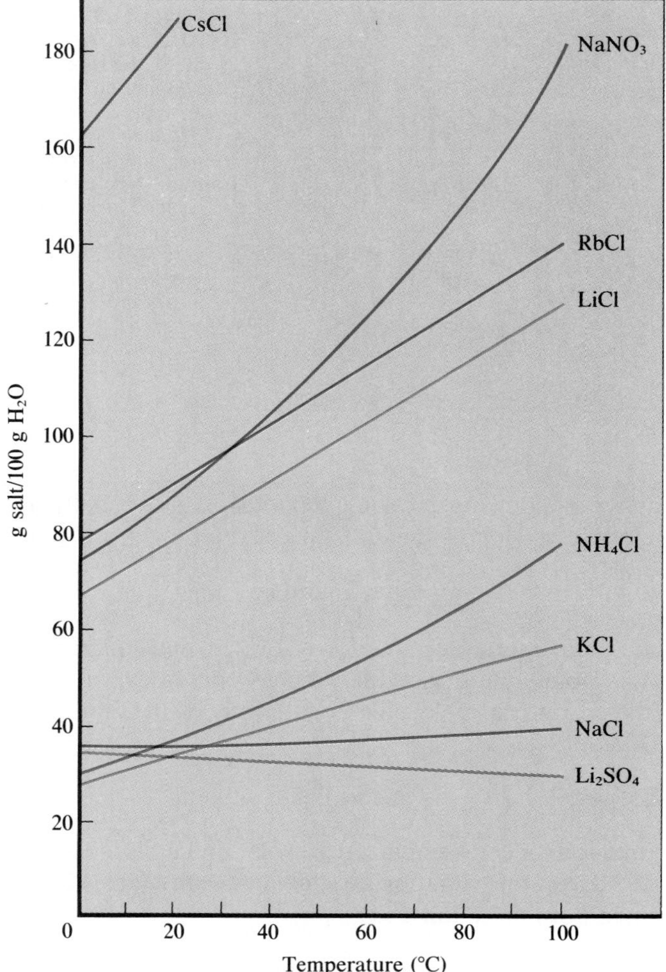

Figure 14.11 The temperature dependence of the solubility of some ionic compounds in water.

When solids dissolve in liquids, the process can be endothermic, just as you saw with NaCl above.

$$\text{Heat} + \text{solid} + \text{liquid solvent} \longrightarrow \text{saturated solution}$$

Here, when equilibrium is reached, as many pairs of Na^+ and Cl^- ions return to the surface of the solid as leave in a given time period. If adding heat is the change made, the reaction will be displaced to use the added heat. Le Chatelier's principle predicts that the solubility of such salts will *increase with increasing temperature*.

To understand Le Chatelier's principle, think of a reaction as a tube of toothpaste. If you squeeze on one side, the "reaction" pops up on the other. If you "squeeze" on an endothermic process by adding more heat, the reaction moves to consume that heat and provide more of the material on the product side.

$$\text{Heat} + \text{solid} + \text{liquid solvent} \rightleftarrows \text{saturated solution}$$

added heat
equilibrium shifts right

This is true for many ionic compounds, as illustrated in Figure 14.11.

There are also some salts that have a negative enthalpy of solution (the process is *exo*thermic),

$$\text{Solid} + \text{liquid solvent} \longrightarrow \text{saturated solution} + \text{heat}$$

and they become *less soluble with increasing temperature*. The solubility curve for Li_2SO_4 in Figure 14.11 is just one example of this behavior.

In view of the discussion in Chapter 13 of trends in lattice energies (see Figure 13.37), it is interesting to notice the trend in solubilities of the alkali metal chlorides in Figure 14.11. In general, as the lattice energy of the salt declines, its solubility increases. Thus, the solubilities increase in the order NaCl < KCl < RbCl < CsCl. Lithium chloride is clearly an exception. In fact, lithium salts often have properties that do not reflect the general trends in Group 1A.

SOMETHING MORE ABOUT
Beer and the Chemistry of Solutions

In Great Britain it's *ale*, in Russia it's called *kvass*, and it's *samshu* in China. It comes warm or cold, pale yellow or dark brown. We usually just call it beer, and it nicely illustrates some of the properties of solutions that we described in this chapter. A bottle of beer is a miniature laboratory for studying concepts of chemical equilibria and the chemistry and physics of gases—among other things.

During the fermentation process, alcohol and CO_2 are generated from a sugar such as glucose in the presence of a catalyst called an enzyme, supplied in this case by microorganisms called yeast.

$$\underset{\text{glucose}}{C_6H_{12}O_6} \xrightarrow{\text{yeast}} 2\ CO_2(g) + 2\ \underset{\text{ethyl alcohol}}{CH_3CH_2OH}$$

The beverage that you get from this process depends on the source of the sugar (whether it is malt, grapes, corn, dandelions, apples, or any of a very large variety of other vegetable or fruit sources), how it is fermented, and what is done after it is fermented. With beer, not all of the CO_2 is allowed to escape. Therefore, when the beer is bottled or canned, some of the gas remains dissolved in the liquid, but some continues to escape until there is an equilibrium. That is, CO_2 molecules rise through the brew and escape into the empty space above the liquid in the bottle or can, while at the same time the molecules in the gas phase reenter the liquid phase. When the number of CO_2 molecules breaking away from the liquid equals the number that reenter the liquid at any given time, equilibrium has been established. In a bottle of beer, the pressure of gas in the space above the liquid under these conditions is about 2 or 3 atmospheres!

To a beer drinker, a satisfying moment comes when you pop the top off a bottle of cold beer on a warm day, and see the wisps of "beer cloud" forming in the neck of the bottle and the bubbles rising through the golden liquid. It can all be explained by the ideas of equilibrium and gases dissolving in liquids.

When the top is popped, the equilibrium $CO_2(g) \rightleftarrows CO_2(aq)$ is broken. The high-pressure gas at the top explodes out of the bottle, the temperature in the bottle neck drops, and a "beer cloud" forms. When the gas rushes out of the bottle, it does work in forcing aside the atmosphere, and the energy to do this work comes from the kinetic energy of the gas molecules. Since the kinetic energy and temperature of a gas are related, this means the temperature of the gas drops. For a chilled bottle of beer at about 40 °F, the temperature in the neck may be as low as about −30 °F. This temperature is low enough to condense the water vapor in the neck, and a cloud forms, just like fog on a cool morning.

Bubbles form almost instantly in the beer—or a bottle of soft drink—when the top is popped off. Before the bottle is opened, the gas pressure over the liquid is so high that it is impossible for the carbon dioxide mixed throughout the liquid to expand to form bubbles. When the pressure is suddenly released, however, CO_2 molecules can aggregate to form bubbles that rapidly expand and rise in the liquid. However, CO_2 molecules do not just clump together spontaneously to form bubbles. Rather, they need a "nucleation" site, a tiny crack or pit in the glass where the dissolved CO_2 molecules can accumulate and begin to form bubbles.

Finally, Henry's law is important to a beer or soda drinker. If you pour out warm beer or soda, you know that it can form a large "head" of foam. However, cold beer or soda will foam up much less readily. The reason for this is simply the decreased solubility of CO_2 in water as the temperature increases. In warmer beer, less of the carbon dioxide stays dissolved and so bubbles form faster.

14.3 COLLIGATIVE PROPERTIES

When sodium chloride is dissolved in water, the vapor pressure of the solution is different from that of pure water, as is the freezing point of the solution, its boiling point, and its osmotic pressure. These changes between pure solvent and solution are called the **colligative properties** of the solution.

We shall begin this discussion by studying the effect of a nonvolatile solute (such as salt or sugar) on the vapor pressure of the solvent, and this will lead us to an understanding of the other colligative properties.

CHANGES IN VAPOR PRESSURE: RAOULT'S LAW

At the surface of an aqueous solution (Figure 14.12), there are molecules of water as well as ions or molecules from the solute. Water molecules can leave the liquid and enter the gas phase, exerting a vapor pressure. However, there are not as many water molecules at the surface as in pure water, because some of them have been displaced by dissolved ions or molecules. Therefore, not as many water molecules are available

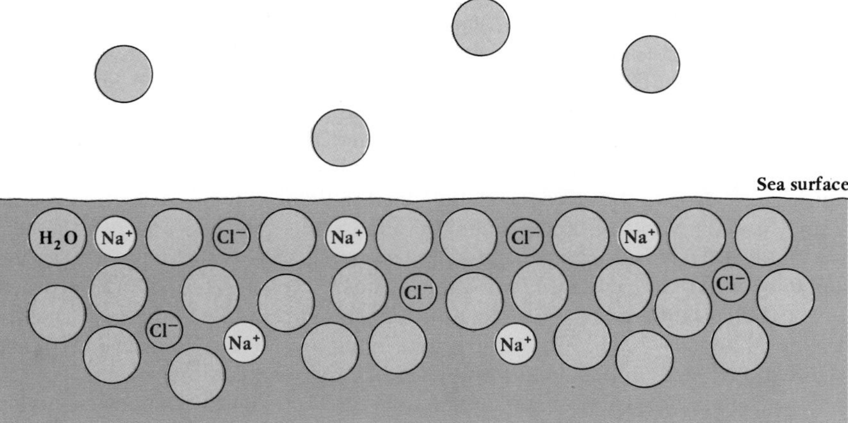

Figure 14.12 Seawater is an aqueous solution of sodium chloride and many other salts. The vapor pressure of water over an aqueous solution is not as large as the vapor pressure of water over pure water at the same temperature.

to leave the liquid surface, and the vapor pressure is *lower* than that of pure water at a given temperature. From this analysis, it should make sense that the vapor pressure of the solvent above the solution, P_{solvent}, will be proportional to the relative number of solvent molecules in the solution, that is, to their *mole fraction*. Thus, since $P_{\text{solvent}} \propto X_{\text{solvent}}$, we can write

$$P_{\text{solvent}} = X_{\text{solvent}}K \tag{14.5}$$

(where K is a constant). This equation tells you that, if there are only half as many solvent molecules present at the surface of a solution as at the surface of the pure liquid, then the vapor pressure of the solvent above the solution will only be half as great as that of the pure solvent at the same temperature.

If we are dealing *only* with pure solvent, Equation 14.5 becomes

$$P^{\circ}_{\text{solvent}} = X_{\text{solvent}}K$$

where $P^{\circ}_{\text{solvent}}$ is the vapor pressure of the pure solvent and X_{solvent} is equal to 1. This means that $P^{\circ}_{\text{solvent}} = K$; that is, the constant K is just the vapor pressure of the pure solvent. Substituting for K in Equation 14.5, we arrive at an equation called **Raoult's law**:

$$P_{\text{solvent}} = X_{\text{solvent}}P^{\circ}_{\text{solvent}} \tag{14.6}$$

If the solution contains more than one volatile component, then Raoult's law can be written for any one such component, A, as

Francois M. Raoult (1830–1901) was Professor of Chemistry at the University of Grenoble in France.

$$P_{\text{A}} = X_{\text{A}}P^{\circ}_{\text{A}} \tag{14.7}$$

Like the ideal gas law, Raoult's law is a description of a simplified model of a solution. *An ideal solution is one that obeys Raoult's law.* Although most solutions are not ideal, just as most gases are not ideal, we use Raoult's law as a good approximation to solution behavior.

In any solution, the mole fraction of the solvent will always be less than 1, so the vapor pressure of the solvent over an ideal solution (P_{solvent}) must be less than the vapor pressure of the pure solvent ($P^{\circ}_{\text{solvent}}$). This *vapor pressure lowering*, $\Delta P_{\text{solvent}}$, is given by

$$\Delta P_{\text{solvent}} = P_{\text{solvent}} - P^{\circ}_{\text{solvent}} \qquad \text{where } P_{\text{solvent}} < P^{\circ}_{\text{solvent}}$$

Substituting Raoult's law for the first term, we now have

$$\Delta P_{\text{solvent}} = (X_{\text{solvent}}P^{\circ}_{\text{solvent}}) - P^{\circ}_{\text{solvent}}$$

or

$$\Delta P_{\text{solvent}} = -(1 - X_{\text{solvent}})P^{\circ}_{\text{solvent}}$$

In a solution that has only the volatile solvent and one nonvolatile solute, the sum of the mole fractions of solvent and solute must be 1.

$$X_{\text{solvent}} + X_{\text{solute}} = 1$$

Therefore, $1 - X_{\text{solvent}} = X_{\text{solute}}$, and the equation for $\Delta P_{\text{solvent}}$ can be rewritten as

$$\Delta P_{\text{solvent}} = -X_{\text{solute}}P^{\circ}_{\text{solvent}} \tag{14.8}$$

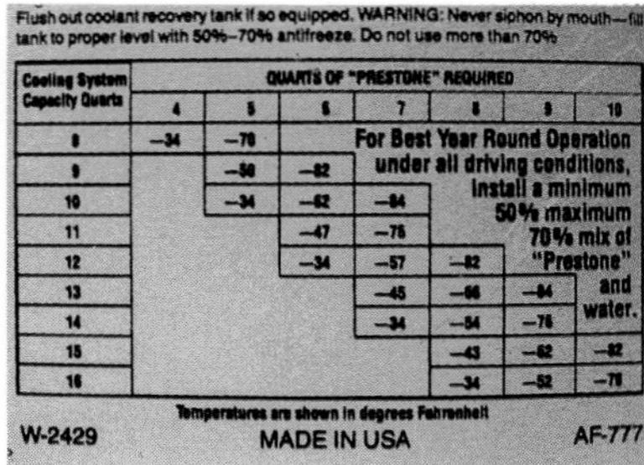

Figure 14.13 Commercial antifreeze is added to your car's cooling system to prevent the water from freezing in the winter and boiling over in the summer.

That is, *the lowering of the vapor pressure of the solvent is proportional to the mole fraction (the relative number of particles) of solute.* This expression and Raoult's law from which it is derived are two of the most useful ways to describe the physical properties of solutions.

EXAMPLE 14.3

USING RAOULT'S LAW

The compound 1,2-ethanediol, HO—CH_2CH_2—OH, is commonly called ethylene glycol. Among other things, it is a common ingredient of antifreeze (Figure 14.13). Adding the compound to the water in an automobile radiator keeps the coolant from boiling over in the summer and freezing up in the winter. Assume that you dissolve 650. g of the compound in 1.50 kg of water (a 30.2% solution, a commonly used concentration). What is the vapor pressure of the water over the solution at 90 °C? (The vapor pressure of pure water at 90 °C is 525.8 mmHg, as given in Appendix D.)

1,2-Ethanediol is used in hydraulic brake fluids, as a solvent in the paint industry, in ball point pen ink, and in synthetic fibers, among other things. Ethylene glycol, as it is commonly called, dissolves easily in water due to extensive hydrogen bonding. See page 522 in Chapter 13.

Solution To calculate the new vapor pressure using Raoult's law, you must first determine the mole fraction of water. The numbers of moles of the solvent and solute, respectively, are

$$\text{Moles of water} = 1.50 \times 10^3 \text{ g} \left(\frac{1 \text{ mol}}{18.02 \text{ g}}\right) = 83.2 \text{ mol } H_2O$$

$$\text{Moles of ethylene glycol} = 650. \text{ g} \left(\frac{1 \text{ mol}}{62.07 \text{ g}}\right) = 10.5 \text{ mol glycol}$$

Therefore, the mole fraction of water is

$$X_{\text{water}} = \frac{83.2 \text{ mol water}}{10.5 \text{ mol water} + 83.2 \text{ mol glycol}} = 0.888$$

Finally, the vapor pressure of the water over the antifreeze solution can be found from Raoult's law.

$$P_{\text{water}} = X_{\text{water}} P^\circ_{\text{water}} = (0.888)(525.8 \text{ mmHg}) = 467 \text{ mmHg}$$

EXERCISE 14.4 Using Raoult's Law

Assume you dissolve 10.0 g of sugar ($C_{12}H_{22}O_{11}$) in 225 mL (225 g) of water and warm the water to 90 °C. What is the vapor pressure of the water over this solution?

One of the uses of colligative properties in the chemistry laboratory is the determination of the molar mass of a compound. The following example and exercise show how this is done using Raoult's law.

EXAMPLE 14.4

CALCULATING MOLAR MASS USING RAOULT'S LAW

The organic compound alizarin has an empirical formula of $C_7H_4O_2$. It was known and used in ancient Egypt and India as a dye. To determine its molar mass, and therefore its molecular formula, you dissolve 0.144 g of the compound in

10.00 g of benzene, C_6H_6. The vapor pressure of the benzene solution is 94.56 mmHg at 25 °C, down from 95.00 mmHg for pure benzene at this temperature. Use this information to calculate the molar mass of alizarin.

Solution Knowing the vapor pressure of the pure solvent and that of the solvent over the solution, you can calculate the mole fraction of solvent.

$$P^\circ_{benzene} = 95.00 \text{ mmHg} \qquad P_{benzene} = 94.56 \text{ mmHg}$$

From Raoult's law, you have:

$$X_{benzene} = P_{benzene}/P^\circ_{benzene} = 0.9954$$

The mole fraction of benzene is defined as moles benzene/total moles. Since the number of moles of benzene is

$$10.00 \text{ g } C_6H_6 \left(\frac{1 \text{ mol}}{78.115 \text{ g}}\right) = 0.1280 \text{ mol}$$

we have the following expression:

$$X_{benzene} = 0.9954 = \frac{0.1280 \text{ mol benzene}}{0.1280 \text{ mol benzene} + ? \text{ mol alizarin}}$$

If you solve this expression for moles of alizarin, you find it is 0.00059. This means the molar mass of alizarin is

$$\frac{0.144 \text{ g alizarin}}{0.00059 \text{ mol alizarin}} = 240 \text{ g/mol}$$

Since $C_7H_4O_2$ has a mass of 120.11 g per formula unit, this means the molecular formula of alizarin is $(C_7H_4O_2)_2$ or $C_{14}H_8O_4$.

EXERCISE 14.5 Using Raoult's Law to Determine Molar Mass
Nitroglycerin is used in the manufacture of dynamite and in the treatment of certain coronary problems. If you add 0.454 g of the compound to 100. g of benzene at 25 °C, the vapor pressure of the solution is 94.85 mmHg. What is the molar mass of nitroglycerin? (The vapor pressure of pure benzene at 25 °C is 95.00 mmHg.)

BOILING POINT ELEVATION

Raoult's law (Equation 14.6) tells us that *the vapor pressure of the solvent over a solution must be lower than that of the pure solvent.* Assume for example that you have a solution of a nonvolatile solute in the volatile solvent benzene; the solute concentration is 0.200 mole in 100. g of benzene (C_6H_6). Using Raoult's law, we find the vapor pressure of benzene at 60 °C drops from 400. mmHg for the pure solvent to 346 mmHg for the solution.

$$X_{benzene} = 0.865$$

$$P_{benzene} = X_{benzene}P^\circ_{benzene} = (0.865)(400. \text{ mmHg}) = 346 \text{ mmHg}$$

$$\Delta P_{benzene} = 346 \text{ mmHg} - 400. \text{ mmHg} = -54 \text{ mmHg}$$

We have marked this point on the vapor pressure curve in Figure 14.14. Now, what happens to the vapor pressure when the temperature of the solution is raised another 10 degrees? P°_{benzene} becomes larger with increasing temperature, so P_{benzene} must also become larger. This point, and ones calculated in the same way for other temperatures, define a new vapor pressure curve for the solution (the lower curve in Figure 14.14).

Let us explore the consequences of the lowering of solvent vapor pressure due to a nonvolatile solute. Recall that the normal boiling point of a volatile liquid is the temperature at which its vapor pressure is equal to 1 atm or 760 mmHg. Since the vapor pressure is lowered on adding a solute, the vapor pressure curve for the solution (the blue line in Figure 14.14) reaches a pressure of 760 mmHg at a temperature *higher* than the normal boiling point of the solvent. In fact, it is generally observed that *the boiling point of a solution is **raised** relative to that of the pure solvent.*

How large is the elevation in boiling point? Equation 14.8 tells us that the vapor pressure lowering is proportional to the concentration of solute ($\Delta P_{\text{solvent}} = -X_{\text{solute}}P^\circ_{\text{solvent}}$). Since the vapor pressure of a liquid and its boiling point are directly related, it makes sense that *the elevation of the boiling point of a solvent is related to the lowering of its vapor pres-*

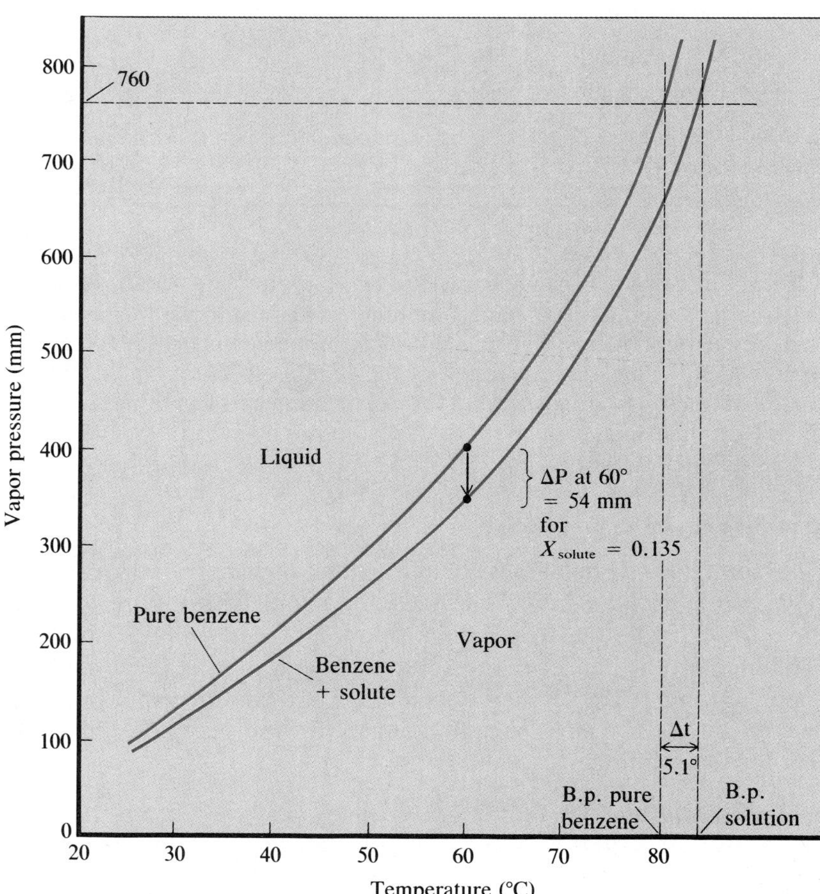

Figure 14.14 Lowering of the vapor pressure of a volatile solvent, benzene, by addition of a nonvolatile solute. (There is 0.200 mole of solute dissolved in 0.100 kg of solvent.) At 60 °C the vapor pressure of the benzene was lowered by 54 mmHg from 400. mmHg to 346 mmHg. At the normal boiling point of benzene, 80.1 °C, the vapor pressure is lowered from 760. mmHg to 657 mmHg. To boil, the vapor pressure of the benzene must be 760. mmHg, so the solution boils 5.1 °C *higher* than the pure solvent, that is, at 85.2 °C.

Table 14.3 Some Boiling Point Elevation and Freezing Point Depression Constants

Solvent	Normal Bp Pure Solvent (°C)	K_{bp} (deg/molal)	Normal Fp Pure Solvent (°C)	K_{fp} (deg/molal)
Water	100.00	+0.512	0.0	−1.86
Benzene	80.10	+2.53	5.50	−4.90
Camphor	—	—	179.75	−40.0
Naphthalene	—	—	80.2	−6.8

sure on adding a solute. Since $\Delta P_{solvent} \propto X_{solute}$, this in turn must mean that *the elevation of the boiling point is related to the concentration of solute.* Indeed, there is a simple relation that connects these quantities,

$$\text{Elevation in boiling point} = \Delta t_{bp} = K_{bp}\, m_{solute} \qquad (14.9)$$

The proportionality constant, K_{bp}, is called the **boiling point elevation constant**; it has a different value for each common solvent (see Table 14.3). Unlike Raoult's law or equations derived directly from it, the solute concentration here is expressed in molality.

EXAMPLE 14.5

BOILING POINT ELEVATION

In Example 14.4 you learned that the red dye alizarin has the formula $C_{14}H_8O_4$. What is the boiling point of a solution in which 0.144 g of this compound is dissolved in 10.0 g of benzene?

Solution You cannot calculate the boiling point directly, only its change. Since you know K_{bp} (Table 14.3), you need only to calculate the molality, m. In this case you have 6.00×10^{-4} mole of alizarin and 0.0100 kg of solvent, so the solute has a concentration of 0.0600 molal, and

$$\Delta t_{bp} = (2.53 \text{ deg/molal})(0.0600 \text{ molal}) = 0.152\ °\text{C}$$

Since the boiling point is *raised* relative to that of the pure solvent, the boiling point of the solution is 80.10 °C + 0.152 °C = 80.25 °C.

EXERCISE 14.6 Boiling Point Elevation

In Exercise 14.4 we determined the change in vapor pressure when sugar is dissolved in a cup of water. Determine the elevation in boiling point for this same solution.

As illustrated in Example 14.4, a common use of colligative properties in the laboratory is the determination of the molar mass of a solute. If the compound is soluble in a solvent of appreciable vapor pressure, or in one with a reasonable K_{bp}, then the molar mass can be determined (at least in principle). Unfortunately, there are disadvantages to both methods. To have an ideal solution, the solute concentration must be low (say, less than 0.10 *m*). However, as you learned in Example 14.4, the lowering of the vapor pressure can be quite small for a dilute solution, and this means that measuring $\Delta P_{solvent}$ is difficult. It is easier to measure small changes in boiling point, but the disadvantage of determining molar mass in a boiling solvent is that the compound must be nonvolatile and stable (not break down into other compounds) at the boiling point. Nonetheless, it is often the case that these criteria are met, and the method is widely used.

EXAMPLE 14.6

DETERMINING MOLAR MASS BY BOILING POINT ELEVATION

Some beautiful blue crystals of azulene (0.640 g) were dissolved in 100. g of benzene. The boiling point of the solution was 80.23 °C. Calculate the molar mass of azulene.

Solution From the elevation in boiling point (and K_{bp} from Table 14.3), you can first determine the molality of the solution.

$$m = \frac{\Delta t_{bp}}{K_{bp}} = \frac{(80.23 - 80.10)\ °C}{2.53\ \text{deg/molal}} = 0.051\ m$$

Knowing the molality, you can then calculate the molar mass.

$$\left(\frac{0.051\ \text{mol}}{1.00\ \text{kg solvent}}\right)(0.100\ \text{kg solvent}) = 0.0051\ \text{mol azulene}$$

$$\frac{0.640\ \text{g azulene}}{0.0051\ \text{mol}} = 130\ \text{g/mol}$$

Azulene, whose formula is $C_{10}H_8$, actually has a molar mass of 128.2 g/mol. As you see, we have experimentally *estimated* its molar mass with an error of about 2%. In real laboratory work, this would represent a very good determination.

EXERCISE 14.7 Determining Molar Mass by Boiling Point Elevation

Determine the molar mass of oil of wintergreen (methyl salicylate) from the following data: 1.25 g of the oil are dissolved in 100. g of benzene. The boiling point of the solution is 80.31 °C.

The elevation of the boiling point of a solvent on adding a solute has many practical consequences and applications. One of them is the summer protection your car's engine receives from "all-season" antifreeze (Figure 14.13). The main ingredient of commercial antifreeze is 1,2-ethanediol (HOC_2H_4OH), commonly called ethylene glycol. The car's radiator and cooling system are sealed to keep the coolant under pressure, so that it will not vaporize at normal engine temperatures. However, when the air temperature is high in the summer, the radiator could still "boil over" if it were not protected with "antifreeze." By adding this nonvolatile liquid, you make a solution that has a higher boiling point.

FREEZING POINT DEPRESSION

Another consequence of dissolving a solute in a solvent is that the freezing point of the solution is lower than that of the pure solvent. For an ideal solution, the depression of the freezing point is given by an equation similar to that for the elevation of the boiling point:

$$\text{Freezing point depression} = \Delta t_{fp} = K_{fp} m_{solute} \qquad (14.10)$$

where K_{fp} is the freezing point depression constant. Values of the constant for various solvents are given in Table 14.3.

The practical aspects of freezing point changes from pure solvent to solution are similar to those for boiling point elevation. The very name of the liquid you add to the radiator in your car, antifreeze, indicates its purpose. The label on the container of antifreeze tells you, for example, to add 6 quarts of antifreeze to a 12-quart cooling system in order to lower the freezing point to −34 °F and raise the boiling point to +226 °F (Figure 14.13).

EXAMPLE 14.7

FREEZING POINT DEPRESSION

How many grams of ethylene glycol [$C_2H_4(OH)_2$] must be added to 5.50 kg of water to lower the freezing point of the water from 0.0 °C to −10.0 °C? (This is approximately the situation in the radiator of your car.)

Solution The solution concentration and the amount by which the freezing point is depressed are related by $\Delta t_{fp} = K_{fp}m$. Therefore, the molality of the solution having a freezing point depression of −10.0 °C must be

$$m = \frac{-10.0\ °\text{C}}{-1.86\ \text{deg/molal}} = 5.38\ \text{molal} = \frac{5.38\ \text{mol glycol}}{1.00\ \text{kg H}_2\text{O}}$$

Since you have 5.50 kg of water, the amount of antifreeze that must be dissolved is

$$(5.50\ \text{kg H}_2\text{O})(5.38\ \text{mol glycol/kg H}_2\text{O}) = 29.6\ \text{moles of glycol or C}_2\text{H}_4(\text{OH})_2$$

The molar mass of $C_2H_4(OH)_2$ is 62.07 g/mol, so the mass required is

$$29.6\ \text{moles}\left(\frac{62.07\ \text{g}}{1\ \text{mol}}\right) = 1840\ \text{g}$$

The density of ethylene glycol is 1.1 kg/L, so the volume of antifreeze to be added is 1.84 kg/(1.1 kg/L) = 1.7 L.

EXERCISE 14.8 Freezing Point Depression

Some people have summer homes on a lake or in the woods. In the northern United States, residents of such homes normally do not use them in the winter, so they close up the homes for the season. When doing so, they winterize the plumbing by putting antifreeze in the toilet tanks, for example. If you add 500. g of HOC_2H_4OH to a tank containing 3.00 kg of water, will this keep the water from freezing at −25 °C?

A particularly good way to determine the molar mass of an unknown compound is to determine the amount by which the freezing point of a solvent is depressed. The next example and exercise illustrate this approach.

EXAMPLE 14.8

USING FREEZING POINT DEPRESSION TO DETERMINE MOLAR MASS

Camphor, a compound with a pleasant odor that occurs naturally in the camphor tree of Southeast Asia or Brazil, is a solid at room temperature (melting point = 179.75 °C). When melted, though, it is a good solvent for many nonionic compounds, and it has a very large value of K_{fp} (−40.0 deg/molal). This makes it a good choice for reasonably accurate molar mass determinations. Suppose you dissolve 0.640 g of the intensely blue compound azulene in 100. g of camphor; the freezing point of the solution is 177.75 °C. Calculate the molar mass of the compound.

Solution The freezing point of the solution is 2.00 °C lower than that of pure camphor.

$$\begin{aligned}\Delta t_{fp} &= \text{fp of solution} - \text{fp of pure camphor}\\ &= 177.75\ °C - 179.75\ °C = -2.00\ °C\end{aligned}$$

This means the molality of the solution is

$$m = \frac{\Delta t_{fp}}{K_{fp}} = \frac{-2.00\ °C}{-40.0\ \text{deg/molal}} = 0.0500\ \text{molal}$$

and so the number of moles of azulene must be

$$\begin{aligned}\text{Moles azulene} &= \left(\frac{0.0500\ \text{mol azulene}}{1.00\ \text{kg camphor}}\right) 0.100\ \text{kg camphor}\\ &= 0.00500\ \text{mol azulene}\end{aligned}$$

Thus,

$$\text{Molar mass of azulene} = \frac{0.640\ \text{g}}{0.00500\ \text{mol}} = 128\ \text{g/mol}$$

As a final point, compare the size of the freezing point depression of 0.640 g of azulene in camphor with the size of the boiling point elevation in benzene. You can appreciate that it is easier to measure a 2.00 °C change than one of only 0.13 °C.

EXERCISE 14.9 Determining Molar Mass by Freezing Point Depression

What is the molar mass of oil of wintergreen if 1.25 g of the oil, when dissolved in 100. g of benzene, produce a solution with a freezing point of 5.10 °C? Compare the freezing point depression here with the boiling point elevation in Exercise 14.7. Considering the ease of measurement, which may be the better method to use in the laboratory?

WHY IS A SOLUTION'S FREEZING POINT DEPRESSED?*

Imagine the freezing process this way. When the temperature is held at the freezing point of a pure solvent, freezing begins with a few molecules clustering together to form a tiny amount of solid. More molecules of the liquid move to the surface of the solid, and the solid grows. Recall that a quantity of heat, the heat of fusion, is evolved; as long as this heat is removed, solidification continues. However, if the heat is not removed, the opposing processes of freezing and melting can come into equilibrium; at this point, the number of molecules moving from solid to liquid is the same as the number moving from liquid to solid in a given time.

But what happens in a freezing *solution*? Again, a few molecules of solvent cluster together to form some solid. More and more solvent molecules join them, and the solid phase, which is pure solid solvent, continues to grow as long as the heat of fusion is removed. At the same time some solvent molecules are returning to the liquid from the solid. The freezing and melting processes can come into equilibrium when the numbers of molecules moving in the two directions are the same in a given time. However, there is a problem. The liquid layer next to the solid contains solute molecules or ions, whereas the solid is pure solvent. [This is analogous to the situation at the solution/vapor interface in Figure 14.12.] Therefore, if the temperature is held at the normal freezing point of the pure solvent, the number of molecules of the solid (pure solvent) entering the liquid phase must be greater than the number of solvent molecules leaving the solution and depositing on the solid, in a given time. Why? For the same reason the vapor pressure of a solution is lower than that of the pure solvent: solute molecules have replaced some solvent molecules in the liquid at the liquid/vapor or liquid/solid interface. Thus, in order to have the same number of solvent molecules moving in each direction (solid → liquid and liquid → solid) in a given time, the temperature must be lowered to slow down movement from solid to liquid. That is, the freezing temperature of a solution must be less than that of the pure liquid solvent.

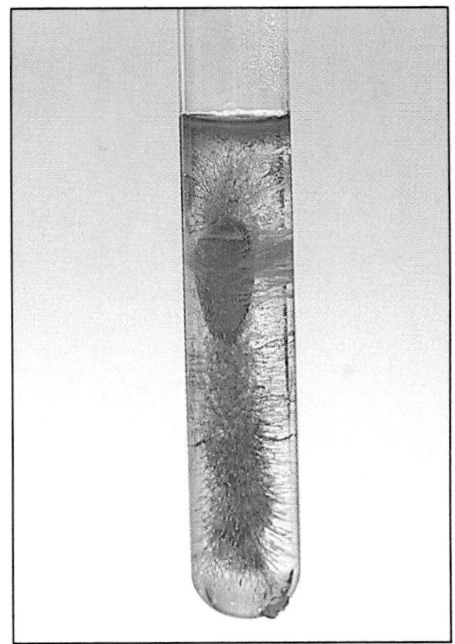

Figure 14.15 A purple dye was dissolved in water, and the solution was frozen slowly. When a solution freezes, the solvent solidifies as the pure substance. Thus, pure ice forms along the walls of the tube, and the dye stays in solution. This means the solution becomes more concentrated as more and more solvent is frozen out, and the resulting solution has a lower and lower freezing point. When equilibrium is reached, we see pure, colorless ice along the walls of the tube with concentrated solution in the center of the tube.

As a solution freezes, solvent molecules are removed from the liquid phase and are deposited on the solid. This means that the concentration of the solute in the liquid solution increases, and the solution freezing point further declines as a result. That is, *a solution does not have a sharply defined freezing point*. However, we usually take the freezing point of a solution as the point at which solid solvent crystals first begin to appear.

When aqueous solutions freeze, the fact that the solid is pure ice while the solution is more concentrated than before (Figure 14.15) can be put to some practical use. One way to purify a liquid solvent in the laboratory is to freeze it slowly, saving the solid (pure solvent) and throwing away any remaining solution (now more concentrated with impurities).

*To keep the answer to our question simple, we shall consider the case where the solute does not form a solid solution with the solvent. This is generally true for water; very few substances form solid solutions with ice.

Early Americans (and some contemporary ones as well) knew that a drink called "apple jack" can be made this way. Fermenting apple cider produces a small amount of alcohol. If the fermented cider is cooled, some of the water freezes to pure ice, leaving a solution higher in alcohol content.

DISSOCIATION OF IONIC SOLUTES AND COLLIGATIVE PROPERTIES

The lowering of the freezing point of a solution is important if you live in the northern United States. You know that it is a common practice to put salt on snowy roads or sidewalks to melt the snow and ice. If the sun shines on the snow or patch of ice, a small amount is melted and the water dissolves some salt. The freezing point of the water is lowered, the solution "eats" its way through the ice, breaks it up, and the icy patch is no longer dangerous for drivers or for people walking.

The reason salt is used on roads is clearly that it is inexpensive and dissolves readily in water. However, an equally important reason is that, when it dissolves, it gives *two ions* in solution for every NaCl unit dissolved.

$$NaCl(s) \longrightarrow Na^+(aq) + Cl^-(aq)$$

It is important to remember that colligative properties depend, not on *what* is dissolved, but only on the *number of particles of solute* per solvent particle. Since every mole of salt gives two moles of ions, the freezing point depression of a 1.0 molal solution of NaCl is *twice* that expected for a 1.0 molal solution of a nonionic solute such as sugar.

$$\Delta t_{fp} = \left(\frac{-1.86 \text{ deg}}{1 \text{ mole particles/kg water}}\right)\left(\frac{2 \text{ moles particles}}{\text{mole NaCl}}\right)\left(\frac{1.0 \text{ mole NaCl}}{1 \text{ kg water}}\right)$$

$$\Delta t_{fp} = -3.72 \text{ °C}$$

Therefore, it is useful to modify our equations for boiling point elevation and freezing point depression by including a factor i that tells us how many moles of particles come from each mole of solute.

$$\Delta t_{bp} = K_{bp} i m \qquad (14.11)$$

$$\Delta t_{fp} = K_{fp} i m \qquad (14.12)$$

For example, $i = 2$ for NaCl or $AgNO_3$, whereas $i = 3$ for $CaCl_2$ or $Pb(NO_3)_2$. You can look at the factor im as an "effective molality" of particles in solution, a reflection of the actual concentration of particles. To see how to use these expressions, try Example 14.9 and Exercise 14.10.

EXAMPLE 14.9

FREEZING POINT AND IONIC SOLUTIONS

Assuming that NaCl dissociates completely into its ions when dissolved in water, how much sodium chloride must be dissolved in 5.50 kg of water to lower the freezing point from 0.0 °C to −10.0 °C?

Solution As explained in the text, one mole of NaCl gives two moles of ions in aqueous solution. Therefore, $i = 2$, and we can substitute into Equation 14.12 as follows:

$$\Delta t_{fp} = K_{fp} im$$

$$-10.0\ °C = \left(\frac{-1.86 \text{ deg}}{1 \text{ mole particles/kg water}}\right)\left(\frac{2 \text{ moles particles}}{1 \text{ mole NaCl}}\right) m$$

$$m = 2.69 \text{ mol NaCl/kg water}$$

The quantity of NaCl required is therefore

$$\left(\frac{2.69 \text{ mol NaCl}}{\text{kg water}}\right)(5.50 \text{ kg water})\left(\frac{58.44 \text{ g}}{\text{mol}}\right) = 864 \text{ g}$$

It is interesting to notice that we found in Example 14.7 that 1840 g of ethylene glycol, $C_2H_4(OH)_2$, were required to lower the freezing point of water to −10 °C. The glycol has a molar mass comparable to that of NaCl, but twice as much of the glycol is required to achieve the same freezing point depression as with NaCl.

EXERCISE 14.10 Freezing and Boiling Points of Solutions of Ionic Compounds

(a) In Example 14.9 we found that 864 g of NaCl in 5.50 kg of water give a freezing point depression of 10.0 °C. What is the boiling point of this solution?
(b) Calculate the freezing point of 500. g of water when it contains (i) 25.0 g of NaCl or (ii) 25.0 g of $CaCl_2$. Assume complete dissociation of both salts in water.

OSMOTIC PRESSURE

Osmosis can be demonstrated with a simple experiment. The beaker in Figure 14.16 contains pure water, and there is a concentrated sugar solution in the tube. The liquids are separated by a **semipermeable membrane**, a thin sheet of material (such as a vegetable tissue or cellophane) through which only certain types of molecules may pass; here, water molecules can pass but larger sugar molecules cannot. When the experiment is begun, the liquid levels in the beaker and the tube are the same. With time, however, the level of sugar solution inside the tube rises, the level of pure water in the beaker falls, and the sugar solution becomes steadily more dilute. After a while, the changes stop; an equilibrium state is reached.

Osmosis has occurred in the experiment in Figure 14.16. *Solvent molecules moved through the semipermeable membrane from a region of lower solute concentration to a region of higher solute concentration.* This is an important phenomenon, since it is one factor controlling the flow of material in and out of living cells.

From a molecular point of view, the semipermeable membrane does not present a barrier to the movement of water molecules, so they move through the membrane in both directions. However, on the sugar-

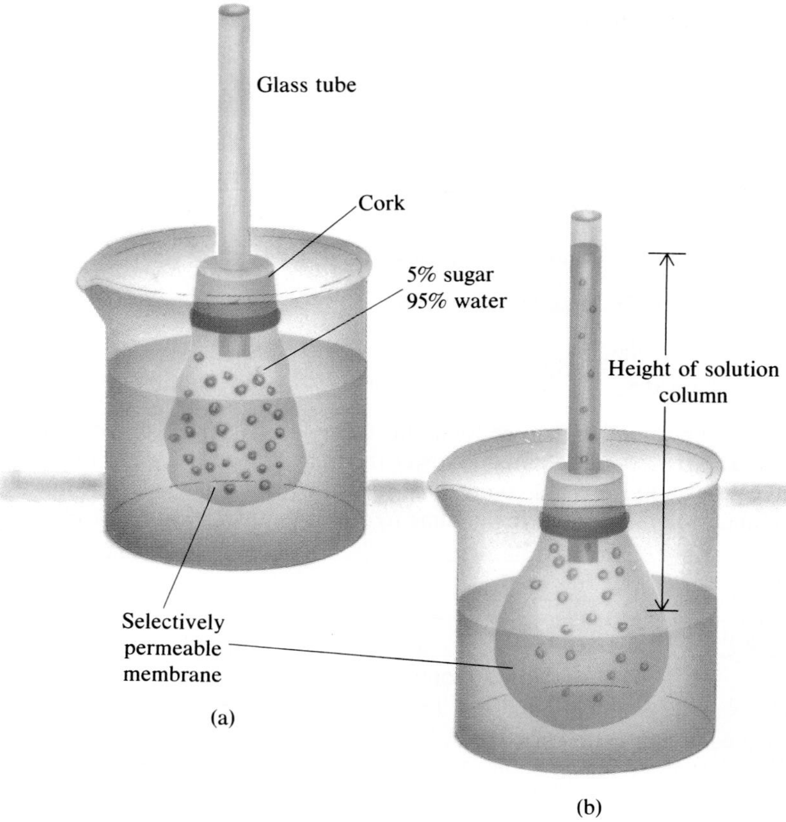

Figure 14.16 Osmosis. The bag attached to the glass tube contains a solution that is 5% sugar and 95% water, and there is pure water in the beaker (a). The material from which the bag is made is selectively permeable (semipermeable) in that it will allow water molecules to pass through but not sugar molecules. With time, water flows through the walls of the bag from a region of low solute concentration (pure water) to one of higher solute concentration (the sugar solution). Flow will continue until the osmotic pressure of the water flowing into the solution is countered by the pressure exerted by the column of solution in the tube above the water level in the beaker. The height of the column of solution (b) is a measure of the osmotic pressure, Π.

solution side, there are not as many water molecules striking the membrane in a given time as on the pure-water side, because their concentration is lower. Thus, in a given time, more water molecules pass through the membrane from pure water to sugar solution than in the opposite direction. In effect, water molecules tend to move from regions of high water concentration (low solute concentration) to regions of low water concentration (high solute concentration). The same is true for any solvent, as long as the membrane passes solvent molecules but not solute molecules.

It may not be obvious why the system reaches equilibrium. After all, the solution in the tube can never reach zero sugar concentration, which would be required to equalize the number of water molecules moving through the membrane in each direction in a given time. The answer lies in the increasing height of the sugar solution. As osmosis continues, water moves into the sugar solution, and the level of the solution in the tube rises higher and higher above the level of the water in the beaker. Eventually the pressure exerted by this height of solution will counterbalance the pressure of the water moving through the membrane from the pure-water side, and there will be no further net movement of water. An equilibrium of forces is achieved. The pressure created by the column of solution when the system is at equilibrium is called the **osmotic pressure, Π**, and a measure of this pressure is the difference in height between the solution and the level of pure water in the beaker.

From experimental measurements on *dilute* solutions, it is known that osmotic pressure (in atmospheres) and concentration (M, in moles/liter) are related by the simple equation

$$\Pi = MRT \tag{14.13}$$

where R is the gas constant (0.082057 L · atm/K · mol) and T is the absolute temperature. This equation is analogous to the ideal gas law, with Π taking the place of P and M being equivalent to n/V.

According to the osmotic pressure equation, the pressure exerted by a 0.10 molar solution of particles at 25 °C would be

$$\Pi = (0.10 \text{ mol/L})(0.0821 \text{ L} \cdot \text{atm/deg} \cdot \text{mol})(298 \text{ deg}) = 2.4 \text{ atm}$$

Since pressures on the order of 10^{-3} atm are easily measured, concentrations of about 10^{-4} M can be determined. This makes osmosis an ideal method for measuring the molar masses of very large molecules, many of which are of biological importance.

EXAMPLE 14.10

OSMOTIC PRESSURE AND MOLAR MASS

The osmotic pressure found for a solution of 5.0 g of horse hemoglobin in 1.0 L of water is 1.8×10^{-3} atm at 25 °C. What is the molar mass of the hemoglobin?

Solution Solving Equation 14.13 for the solution concentration, we have

$$M = \frac{\Pi}{RT} = \frac{1.8 \times 10^{-3} \text{ atm}}{(0.0821 \text{ L} \cdot \text{atm/K} \cdot \text{mol})(298 \text{ K})}$$
$$= 7.4 \times 10^{-5} \text{ mol/L}$$

The measured concentration is achieved by dissolving 5.0 g of hemoglobin per liter. Therefore,

$$\text{Molar mass of hemoglobin} = \frac{5.0 \text{ g/L}}{7.4 \times 10^{-5} \text{ mol/L}} = 68{,}000 \text{ g/mol}$$

EXERCISE 14.11 Osmotic Pressure and Molecular Weight

Beta-carotene is the most important of the A vitamins. Its molar mass can be determined by measuring the osmotic pressure generated by a given mass of carotene dissolved in the solvent chloroform. Calculate the molar mass of carotene if 7.68 mg, dissolved in 10.0 mL of chloroform, give an osmotic pressure of 26.57 mmHg at 25 °C.

Osmosis has great practical significance, especially if you are a nurse or physician. Patients who become dehydrated through illness often need to be given water and nutrients intravenously. However, one cannot simply drip water into the patient's vein. Rather, the solution must have the same overall solute concentration as the patient's blood. That is, the intravenous solution must be *iso-osmotic*. If you used pure water, the inside of a blood cell would have a higher solute concentration (lower

The common terminology here is to use the suffix "tonic" for osmotic. That is, an isoosmotic solution is isotonic.

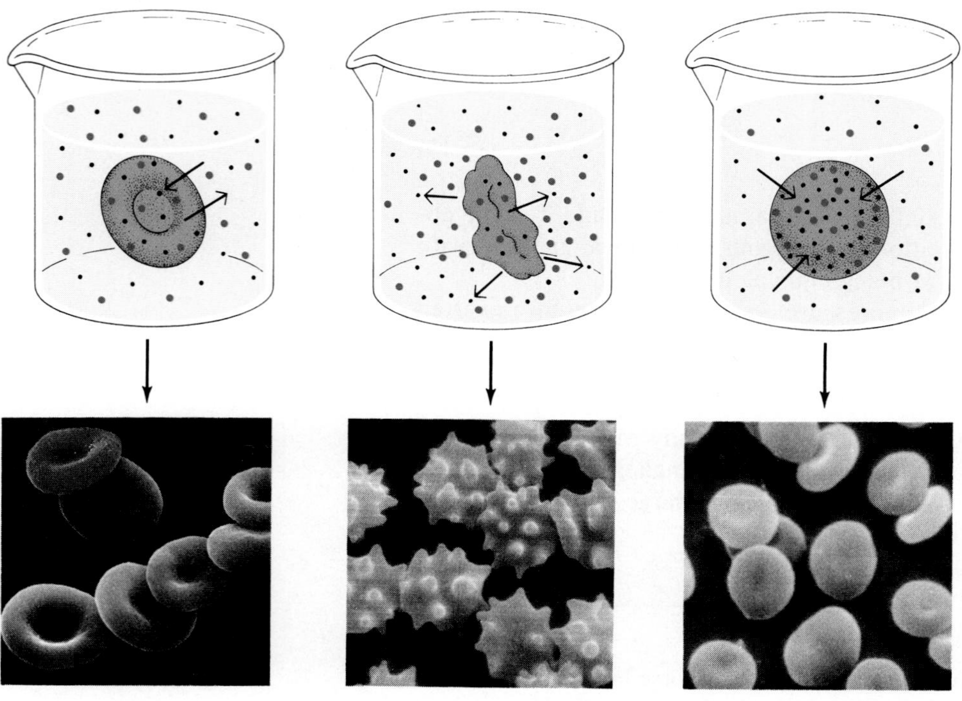

Figure 14.17 Osmosis and the living cell. (a) A cell placed in an *iso*tonic solution. The net movement of water in and out of the cell is zero because the concentration of solutes inside and outside the cell is the same. (b) In a *hyper*tonic solution, the concentration of solutes outside the cell is greater than inside. There is a net flow of water out of the cell, causing the cell to dehydrate, shrink, and perhaps die. (c) In a *hypo*tonic solution, the concentration of solutes outside the cell is less than inside. There is a net flow of water into the cell, causing the cell to swell and perhaps to burst.

water concentration), and water would flow into the cell. This is called a *hypotonic* situation, and the result would be bursting of the red blood cells (Figure 14.17). The opposite situation, *hypertonic*, would occur if the intravenous solution is more concentrated than the contents of the blood cell. In this case the cell would lose water and shrivel up. To combat this, a dehydrated patient is rehydrated in the hospital with a sterile saline solution that is about 0.9% sodium chloride, a solution that is iso-osmotic with the cells of the body.

Cell shriveling by osmosis happens when vegetables or meat are cured in brine, a concentrated solution of NaCl. If you put a fresh cucumber or carrot into brine, water will flow out of its cells and into the brine, leaving behind a shriveled vegetable (Figure 14.18). With the proper spices, however, a cucumber becomes a delicious pickle!

14.4 COLLOIDS AND COLLOIDAL DISPERSIONS

The solutions we have discussed so far represent only one type of a broad spectrum of mixtures of substances. We defined a solution as a homogeneous mixture of two or more substances in a single phase. To this we should add that, *in a true solution, no settling of the solute should be observed and that the solute particles should be in the form of ions or relatively small molecules*. Thus, common salt and sugar form true solutions.

You are also familiar with **suspensions**. One would result, for example, if you put a handful of fine sand into water and shook vigorously.

Table 14.4 Types of Colloids and Examples*

Dispersed (Solute-like) Phase		Dispersing (Solvent-like) Medium	Common Name	Examples
Solid	in	Solid	Solid sol	Many alloys (such as steel and duralumin), some colored gems, reinforced rubber, porcelain, pigmented plastics
Liquid	in	Solid	Solid emulsion	Cheese, butter, jellies
Gas	in	Solid	Solid foam	Sponge, rubber, pumice, Styrofoam
Solid	in	Liquid	Sols and gels	Milk of magnesia, paints, mud, puddings
Liquid	in	Liquid	Emulsion	Milk, face cream, salad dressings, mayonnaise
Gas	in	Liquid	Foam	Shaving cream, whipped cream, foam on beer
Solid	in	Gas	Solid aerosol	Smoke, airborne viruses and particulate matter, auto exhaust
Liquid	in	Gas	Liquid aerosol	Fog, mist, aerosol spray, clouds

*From Whitten, Gailey, and Davis, *General Chemistry*, 3rd edition, Philadelphia, Saunders College Publishing, 1988.

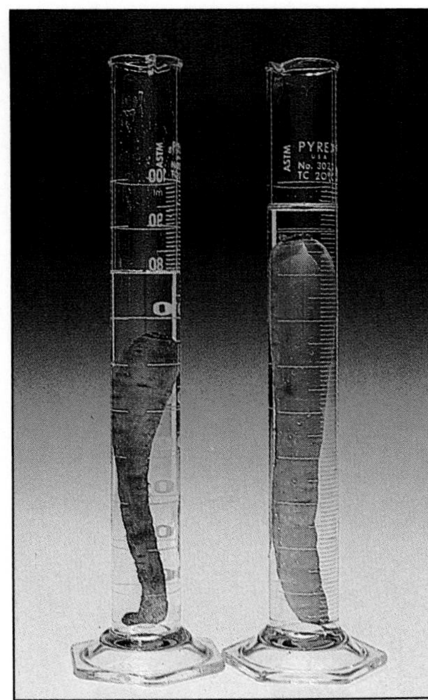

Figure 14.18 Osmosis. After some hours a carrot in a concentrated NaCl solution shows the effects of osmosis (left). Water has flowed out of the carrot and into the NaCl solution, leaving the carrot limp. A carrot in pure water (right) is not appreciably affected.

Sand particles would still be clearly visible, and the particles would gradually settle to the bottom of the beaker or bottle. **Colloidal dispersions** represent a state intermediate between a solution and a suspension. Colloids include many of the foods you eat and the materials around you; among them are gelatin, milk, fog, and porcelain (see Table 14.4). Since these are important (or at least interesting) to all of us, it is useful to describe their make-up and properties.

Thomas Graham also studied effusion of gases (see Chapter 12).

Around 1860, Thomas Graham found that substances such as albumin from eggs, starch, gelatin, and glue will diffuse only very slowly, as compared with sugar or salt, when placed in water. In addition, the former substances differ significantly from the latter ones in their ability to diffuse through a thin membrane: sugar will diffuse through many types of membranes, but glue, starch, albumin, and gelatin will not. Further, Graham found that he could not crystallize these substances, while he could crystallize sugar, salt, and other materials that form true solutions. Therefore, Graham coined the word "colloid" (from the Greek meaning "glue") to describe a class of substances distinctly different from sugar and salt and similar materials. We now know we can crystallize some colloidal substances, albeit with difficulty, so there really is no sharp dividing line between these classes. However, there are some distinguishing characteristics of colloids. (1) It is generally true that colloidal materials have very high molecular masses; this is certainly true of human cells and of proteins such as hemoglobin that have molar masses in the thousands. (2) The particles of a colloid are relatively large (say 1000 nm), large enough that they scatter visible light when dispersed in a solvent,

Figure 14.19 Colloidal suspensions or dispersions scatter light and so appear cloudy. This is called the Tyndall effect, and it can be observed on a foggy or misty morning. (This is why it is difficult to drive at night in a fog. However, it is easier to see when using yellow fog lights, since the wavelength of the light is greater than the particle diameter and the light is not scattered by the fog.)

and they make the mixture appear cloudy (Figure 14.19). (3) Even though colloidal particles are large by molecular standards, they are small enough that they do not settle out.

Graham also gave us the words "sol" for a colloidal solution (a dispersion of a solid substance in a fluid medium) and "gel" for a dispersion that has developed a structure that prevents it from flowing. For example, dessert gelatin is a sol when the solid is first mixed with boiling water, but it becomes a gel when cooled (Figure 14.20). Other examples of gels are the gelatinous precipitates of $Al(OH)_3$, $Fe(OH)_3$, and $Cu(OH)_2$ often seen in experiments in the introductory chemistry laboratory (Figure 14.21).

If you have one millionth of a mole of colloidal particles, each assumed to be a sphere with a diameter of 200 nm, the total surface area of the particles would be on the order of 100 million cm² or the size of several football fields.

Colloidal dispersions consist of very finely divided particles that, as a result, have a very high surface area. Therefore, many of the properties of colloids depend on the properties of surfaces. An example of this is a bright red sol formed by heating a concentrated aqueous solution of iron(III) chloride.

$$2x[Fe^{3+}(aq) + 3\ Cl^-(aq)] + x(3 + y)\ H_2O(\ell) \longrightarrow \underset{\text{bright red sol}}{[Fe_2O_3 \cdot yH_2O]_x(s)} + 6x[H^+(aq) + Cl^-(aq)]$$

As illustrated in Figure 14.22, the small particles of hydrated iron(III) oxide adsorb Fe^{3+} ions on their surfaces, making the particles positively charged. This means that the particles repel one another and cannot come together to form a precipitate, an insoluble solid. The only way to cause precipitation, therefore, is to remove or neutralize the surface charge in some way.

The neutralization of surface charge is used to precipitate colloidal impurities in water purification. When water is taken into municipal purification plants it can contain dispersed material such as colloidal clay. Clay particles can be negatively charged, which keeps the particles apart and dispersed in the water. To precipitate the clay, aluminum salts such as $Al_2(SO_4)_3$ are added. In aqueous solution, Al^{3+} is hydrated to give $[Al(H_2O)_6]^{3+}$ and other larger ions such as

$$\left[(H_2O)_4Al\begin{matrix} \overset{\displaystyle H}{\underset{}{O}} \\ \\ \underset{\displaystyle H}{O}\end{matrix}Al(H_2O)_4\right]^{4+}$$

Figure 14.20 A colloid. When gelatin is mixed with boiling water, it forms a colloidal dispersion called a sol. When cooled, the dispersion is called a gel.

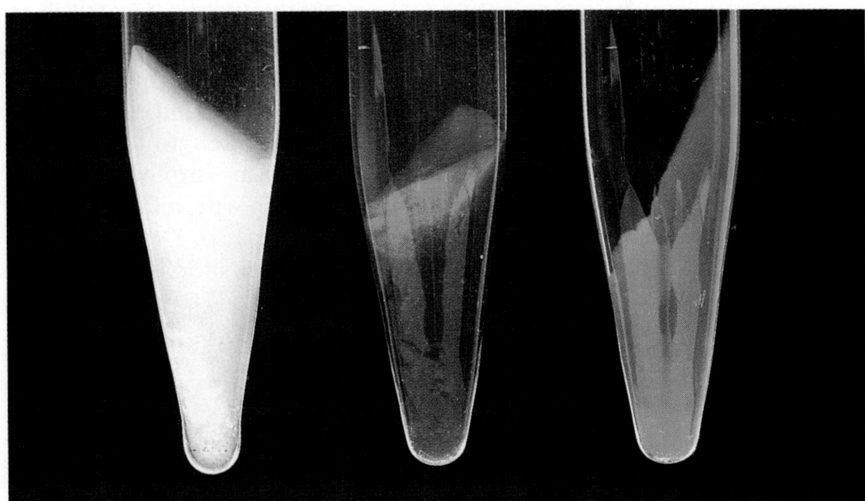

Figure 14.21 Three gelatinous precipitates: (left) $Al(OH)_3$, (middle) $Fe(OH)_3$, and (right) $Cu(OH)_2$.

This and other highly charged ions can be adsorbed onto the surfaces of the negatively charged colloid particles, thereby neutralizing the surface charge and allowing the colloidal particles to come together, and the clay precipitates from solution.

Emulsions are colloidal dispersions of one liquid in another, such as oil or fat in water. You find emulsions in familiar forms: salad dressing,

Figure 14.22 A sol of Fe_2O_3 in water is stabilized by electrostatic forces. Each colloidal particle consists of many units of hydrated Fe_2O_3. Free Fe^{3+} ions in the solution are attracted to sites on the particle and so give the particle a positive charge. (H^+ or Cl^- ions are apparently not attracted because Fe^{3+} ions preferentially fit into sites on the particle.) The resulting positive charge on the surface of the colloidal particle keeps the particles apart and thereby prevents precipitation.

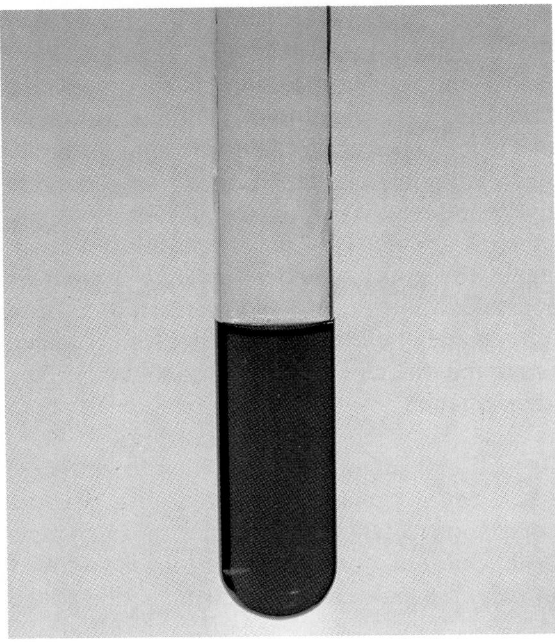

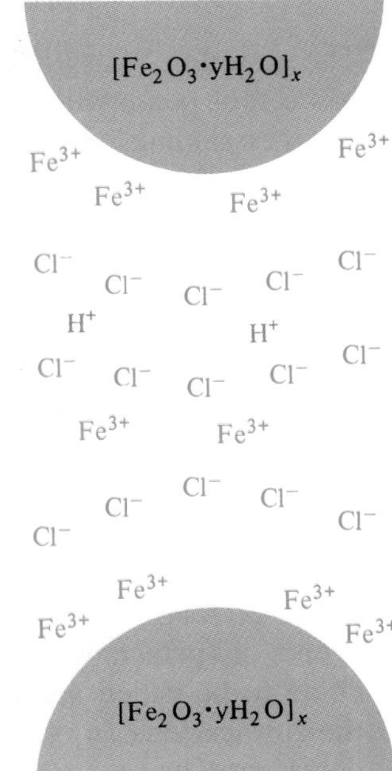

One of the best ways to find out about the chemical properties of food is in The Cookbook Decoder, a cookbook written by chemist A. Grosser (Beaufort Books, Inc., New York, 1981).

mayonnaise, and milk, among others. If you mix vegetable oil and vinegar to make a salad dressing, you know that the mixture quickly separates into two layers because the nonpolar oil molecules do not interact with polar water and acetic acid molecules. So why are milk and mayonnaise apparently homogeneous mixtures that do not separate into layers? The answer is that they contain an **emulsifying agent** such as a protein. Lecithin is a protein found in egg yolks, so mixing egg yolks with oil and vinegar stabilizes the colloidal dispersion known as mayonnaise.

A sodium soap is a solid at room temperature, whereas potassium soaps are usually liquid.

Soaps and **detergents** are also emulsifying agents. Soap is made by heating a fat with sodium or potassium hydroxide.

$$\begin{array}{l} H_2C{-}O{-}\overset{\displaystyle O}{\overset{\|}{C}}{-}(CH_2)_{14}CH_3 \\ \quad | \\ HC{-}O{-}\overset{\displaystyle O}{\overset{\|}{C}}{-}(CH_2)_{14}CH_3 \\ \quad | \\ H_2C{-}O{-}\overset{\displaystyle O}{\overset{\|}{C}}{-}(CH_2)_{14}CH_3 \end{array} + 3\ NaOH \longrightarrow \begin{array}{l} H_2C{-}O{-}H \\ \quad | \\ H_2C{-}O{-}H \\ \quad | \\ H_2C{-}O{-}H \end{array}$$

Fat Glycerin

Sodium palmitate

$$+\ 3\left[H_3C{-}(CH_2)_{14}{-}C\begin{matrix} {}^{\diagup\!\!\diagup O} \\ {}_{\diagdown O^-} \end{matrix}\quad Na^+\right]$$

Hydrocarbon tail (Soluble in oil) ⇑ — Polar head (Soluble in water) ⇑

The resulting "fatty acid" anion (such as palmitate, $H_3C(CH_2)_{14}COO^-$) has a split personality: it has a nonpolar hydrocarbon tail that is soluble in other similar hydrocarbons and a polar head that is soluble in water.

*The polar end of a soap molecule is said to be **hydrophilic** or "water-loving." In contrast, the nonpolar end is **hydrophobic** or "water-hating."*

If you want to clean oil from dishes or clothing, you cannot simply wash it away with water, since the oil is a nonpolar liquid not soluble in polar water. Instead, you add soap to the water. At the surface of the oil, the nonpolar molecules of the oil interact with the hydrocarbon tails of the soap molecules; this leaves the polar heads of the soap to interact with surrounding water molecules, and the oil and water mix (Figure 14.23). If the oily material on a piece of clothing or a dish also contained some dirt particles, this means the dirt can now be washed away. And with appropriate bleach and brighteners, your wash can be "cleaner than clean and whiter than white"!

The soap and detergent industry is enormous. Roughly 25 to 30 million tons of household and toilet soaps, and synthetic and soap-based laundry detergents, are produced annually worldwide.

Substances such as soaps that affect the properties of surfaces, and so affect the interaction between two phases, are called surface-active agents or **surfactants** for short. A surfactant that is used for cleaning has come to be called a **detergent**. One function of a surfactant is to lower the surface tension of water, which enhances the cleansing action of the detergent (Figure 14.24).

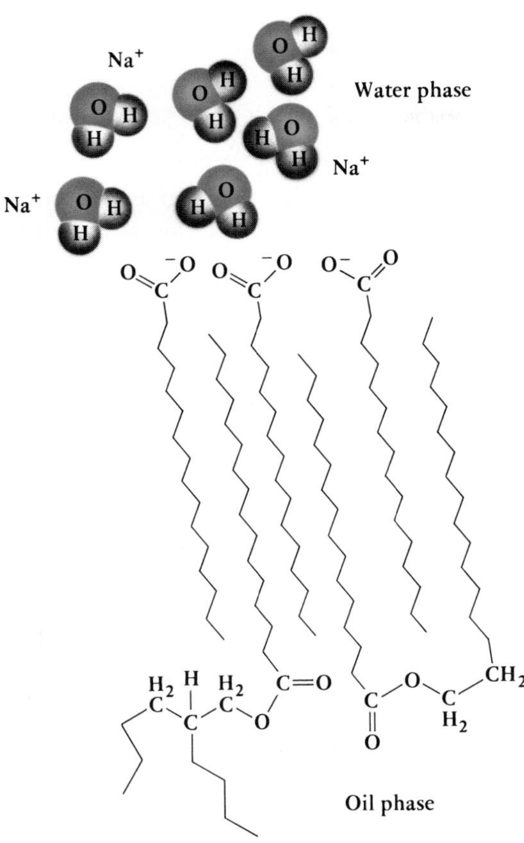

Figure 14.23 Oils and fats are organic compounds that contain very long hydrocarbon chains. Soap molecules have polar heads that interact with water and long hydrocarbon tails that interact with the oil. Thus, soaps and detergents can interact with both oils and water and allow the dispersion of oil in water. (From CHEMISTRY. By Linus Pauling and Peter Pauling. Copyright © 1975 by Linus Pauling and Peter Pauling. Reprinted with permission by W.H. Freeman and Company.)

Figure 14.24 The surface tension of water is affected by adding a detergent. At the left, sulfur (density = 2.1 g/cm³) is carefully placed on top of water (density = 1.0 g/cm³); the surface tension of the water keeps the more dense sulfur afloat. In the photo at the right several drops of detergent have been placed on the surface of the water. The surface tension is reduced, and the sulfur sinks to the bottom of the tank of water.

Many detergents used in the home and industry are synthetic. One example is sodium lauryl benzenesulfonate, a biodegradable compound.

$$H_3CCH_2CH_2CH_2CH_2CH_2CH_2CH_2CH_2CH_2CH_2CH_2-C_6H_4-SO_3^- + Na^+$$

Most synthetic detergents use the sulfonate ($—SO_3^-$) group as the polar head instead of carboxylate ($—CO_2^-$). The reason for this is that the $—CO_2^-$ head allows the soap to form an insoluble precipitate with any Ca^{2+} or Mg^{2+} ions that are present in water. Since **hard water** is characterized by high concentrations of these ions, this is the source of bathtub rings and tattle-tale grays. The synthetic sulfonate detergents have the advantage, however, that they do not form such precipitates because their calcium salts are more soluble in water.

SUMMARY

Solutions are homogeneous mixtures of two or more substances in a single phase; the **solvent** is the component present in largest amount and the **solute** is the other component. In a true solution, as well as in a colloidal dispersion (see below), no settling of the solute is observed. This chapter is concerned with the **colligative properties** of solutions, properties that depend ideally on solute concentration and the nature of the solvent but not on the nature of the solute. These properties include changes in solution **vapor pressure**, **melting point** and **boiling point**, and **osmotic pressure**.

In Section 14.1, the various units of solute concentration are defined and used: **mole fraction**, **molality**, and **weight percent**. The first two in particular are used to define solution colligative properties.

Intermolecular forces provide one key to understanding the solution process (Section 14.2), since they are closely related to the **enthalpy of solution**, the net heat energy involved in the process of solution formation. One criterion for a solute dissolving in a solvent is that the process be exothermic or only slightly endothermic. When $\Delta H^\circ_{solution}$ is significantly positive, the energy evolved on forming intermolecular bonds between solute and solvent does not offset the energy required to break down solute–solute or solvent–solvent intermolecular bonding. This is the basis of the "like dissolves like" rule of thumb. For example, ionic solids are dissolved only by highly polar solvents. The process of dissolving ionic solids in particular can be analyzed quantitatively using (a) lattice energies and enthalpies of hydration or (b) the standard enthalpies of formation of the crystalline solid and the aqueous 1-molal solution.

Pressure and temperature affect solubility (Section 14.2). The greater the partial pressure of a given gas, the greater the solubility, a fact expressed by **Henry's law** (Equation 14.4). The effect of temperature depends in large part on the enthalpy of solution; solutes for which the solution process is endothermic generally increase in solubility with temperature (and vice versa). The effect of temperature on solubility can be understood with **Le Chatelier's principle**. This principle states that a change in any of the factors that determine an equilibrium will cause the system to adjust in order to reduce or counteract the effect of the change.

It is possible to describe accurately the colligative properties of an **ideal solution** (Section 14.3), a solution obeying **Raoult's law**:

$$P_{\text{solvent}} = X_{\text{solvent}}P^{\circ}_{\text{solvent}}$$

Solvent vapor pressure over solution
= mole fraction of solvent × vapor pressure of pure solvent (14.6)

This tells us that, when a solute is dissolved in a solvent, the vapor pressure of the solvent declines ($\Delta P_{\text{solvent}} = -X_{\text{solute}}P^{\circ}_{\text{solvent}}$).

When a solute is added to a solvent, the boiling point of the solution is *raised* relative to that of the pure solvent by an amount

$$\Delta t_{\text{bp}} = K_{\text{bp}}m_{\text{solute}} \quad (14.9)$$

and the freezing point is *lowered* by an amount

$$\Delta t_{\text{fp}} = K_{\text{fp}}m_{\text{solute}} \quad (14.10)$$

(where K_{bp} and K_{fp} are constants that depend on the solvent and m_{solute} is the molality of solute). For solutions of ionic solutes, Equation 14.10, for example, can be modified to Equation 14.12

$$\Delta t_{\text{fp}} = K_{\text{fp}}im_{\text{solute}} \quad (14.12)$$

where i is the number of moles of ions that come from 1 mole of solute.

Osmosis is the movement of solvent molecules through a semipermeable membrane from a region of low solute concentration to a region of high solute concentration. The osmotic pressure, Π, for a *dilute* ideal solution can be found from the equation

$$\text{Osmotic pressure} = \Pi = \text{M}RT \quad (14.13)$$

where M is the molarity of particles in the solution, R is the gas constant, and T is the Kelvin temperature.

Colloidal dispersions are intermediate between true solutions and suspensions. Colloids generally have high molar masses and the particles are relatively large; however, the colloid does not settle out of solution. A **sol** is a dispersion of a solid colloid in a fluid medium, and a **gel** is a dispersion that has achieved some immobile structure. **Emulsions** are colloidal dispersions of one liquid in another. To stabilize an emulsion requires an **emulsifying agent**, of which **soaps** and **detergents** are examples. Since these affect the properties of the surfaces of the dispersed material, they are often called **surfactants**.

STUDY QUESTIONS

REVIEW QUESTIONS

1. Is the following statement true or false? If false, change it to read correctly. "Colligative properties depend on the nature of the solvent and solute and on the concentration of the solute."
2. Name the four colligative properties described in this chapter. Write the mathematical expression that describes each of these.
3. Define molality and tell how it differs from molarity, another concentration unit.
4. How is the solubility of a gas or solid related to the algebraic sign of the enthalpy of solution?
5. Explain the relation between enthalpy of solution of a gas or solid and the temperature dependence of solubility.
6. How are the aqueous solubilities of NaCl, KCl, RbCl,

and CsCl in Figure 14.11 related to their lattice energies?

7. Name three effects that govern the solubility of a gas in water.
8. If you dissolve equimolar amounts of NaCl and $CaCl_2$ in water, the calcium salt lowers the freezing point of the water almost 1.5 times as much as the NaCl. Why?
9. Explain why a cucumber shrivels up when you put it in a concentrated solution of salt.
10. Explain the differences between a colloidal dispersion, a suspension, and a true solution.
11. Explain the difference between a sol and a gel. Give an example of each.
12. Explain how a surfactant such as a soap or detergent functions.

CONCENTRATIONS

13. Assume you dissolve 45.0 g of ethylene glycol [$C_2H_4(OH)_2$] in half a liter of water (500.0 g). Calculate the molality, mole fraction, and weight percent of glycol in the solution.
14. Dimethylglyoxime (DMG) is an organic molecule used to test for the nickel(II) ion in solution (see *Something More About Nickel Chemistry* (Chapter 25)). Assume you dissolve 45.0 g of dimethylglyoxime ($C_4H_8N_2O_2$, molar mass = 116.12 g/mol) in 500 mL (= 395 g) of ethyl alcohol (C_2H_5OH, molar mass = 46.07 g/mol). Calculate the molality, mole fraction, and weight percent of the glyoxime in the solution.
15. Fill in the blanks in the following table.

COMPOUND	MOLALITY	WEIGHT PERCENT	MOLE FRACTION
NaCl	0.25	_____	_____
C_2H_5OH	_____	5.0%	_____
$C_{12}H_{22}O_{11}$	0.10	_____	_____

16. Fill in the blanks in the following table.

COMPOUND	MOLALITY	WEIGHT PERCENT	MOLE FRACTION
NH_3	_____	30.0	_____
CH_3COOH	0.0083	_____	_____
$C_2H_4(OH)_2$	_____	15.0	_____

17. If you want to prepare a solution that is 0.200 molal in $NaNO_3$, how many grams of the salt would you add to 500. g of water? What is the mole fraction of $NaNO_3$ in the resulting solution?
18. If you want to prepare a solution that is 0.0512 molal in K_2CO_3, how many grams of the salt would you have to add to 250. g of water? What is the mole fraction of K_2CO_3 in the solution?
19. You need an aqueous solution that has a mole fraction of CH_3OH (methyl alcohol) of 0.093. How many grams of CH_3OH would you have to combine with 500. g of water to make this solution? Calculate the molality of the solution.
20. You want a solution of ethylene glycol, $C_2H_4(OH)_2$, that has a glycol mole fraction of 0.125. How many grams of the glycol should you combine with 950. g of water? What is the molality of the solution?
21. If you combine 15.0 g of ethylene glycol, $C_2H_4(OH)_2$, with 775 g of water, what is the molality of the solution? What is the mole fraction of the glycol? What is the mole fraction of water?
22. Assume you add 5.12 g of $MgCl_2$ to 652 g of water. What is the molality of the solution? What is the mole fraction of the magnesium chloride? What is the mole fraction of the water?
23. Fill in the blanks in the following table:

COMPOUND	GRAMS COMPD	GRAMS WATER	MOLALITY OF COMPD	MOLE FRACTION OF COMPD
Na_2CO_3	_____	250.	0.0125	_____
CH_3OH	13.5	150.	_____	_____
KNO_3	_____	555	_____	0.0934

24. Fill in the blanks in the following table:

COMPOUND	GRAMS COMPD	GRAMS WATER	MOLALITY OF COMPD	MOLE FRACTION OF COMPD
$Pb(NO_3)_2$	_____	800.	0.0245	_____
$C_2H_4(OH)_2$	_____	250.	_____	0.0545
$Pt(NH_3)_2Cl_2$	0.0075	200.	_____	_____

25. Hydrochloric acid is sold as a concentrated aqueous solution. If the molarity of commercial HCl is 12.0 and its density is 1.18 g/cm^3, calculate (a) the molality of the solution and (b) the weight percent of HCl in the solution.
26. Concentrated sulfuric acid has a density of 1.84 g/cm^3 and is 95.0% by weight H_2SO_4. What is the molality of this acid? What is its molarity?
27. A 10.7 molal solution of NaOH has a density of 1.33 g/cm^3 at 20 °C. Calculate the (a) mole fraction of NaOH, (b) the weight percentage of NaOH, and (c) the molarity of the solution.
28. Concentrated aqueous ammonia has a molarity of 14.8 and a density of 0.90 g/cm^3. What is the molality of the solution? Calculate the mole fraction and weight percentage of NH_3.
29. If you dissolve 2.00 g of $CaCl_2$ in 750 g of water, what is the molality of $CaCl_2$? What is the total molality of the ions? (That is, what is the total concentration of Ca^{2+} and Cl^- ions?)

30. If you want a solution that is 0.100 molal in ions, how many grams of $MgCl_2$ must you dissolve in 150. g of water?
31. Units of parts per million (ppm) or per billion (ppb) are often used to describe the concentrations of solutes in very dilute solutions. The units are defined as the number of grams of solute per million or per billion grams of solvent. The Salton Sea in California has a relatively high concentration of lithium ion, 1.9 ppm. What is the molality of Li^+ in this water?
32. Silver ion is found with an average concentration of 28 parts per billion in water supplies in the United States. (a) What is the molality of the silver ion? (b) If you wanted 1.0×10^2 g of silver and could recover it chemically from water supplies, how many liters of water would you have to treat? (Assume the density of water is 1.0 g/cm³).

THE SOLUTION PROCESS

33. Considering intermolecular forces, give reasons for the following observations:
    (a) Octane, C_8H_{18}, is very miscible with CCl_4.
    (b) Methyl alcohol, H_3COH, mixes in all proportions with water.
    (c) Sodium bromide, NaBr, is very poorly soluble in diethyl ether ($C_2H_5—O—C_2H_5$).
34. Account for the fact that alcohols such as methyl (CH_3OH) and ethyl (C_2H_5OH) are quite miscible with water, while an alcohol with a long carbon chain such as octyl alcohol ($C_8H_{17}OH$) is poorly soluble in water.
35. The ionic compound ammonium chloride could be used as a "chemical cold pack." When the solid is mixed with water, the solution becomes very cold. Calculate the enthalpy of formation of $NH_4Cl(aq)$, $\Delta H_f^\circ[NH_4Cl(aq, 1\ m)]$, knowing that $NH_4Cl(s) + 14.8\ kJ/mol \rightarrow NH_4^+(aq) + Cl^-(aq)$.
36. Some ionic compounds evolve heat when mixed with water. In fact, dissolving $CaCl_2$ is so exothermic that it is used in a commercial "hot pack." Calculate the enthalpy of the reaction $CaCl_2(s) \rightarrow Ca^{2+}(aq) + 2\ Cl^-(aq)$ from $\Delta H_f^\circ[CaCl_2(s)] = -795$ kJ/mol and $\Delta H_f^\circ[CaCl_2(aq)] = -877.89$ kJ/mol.
37. Potassium nitrate dissolves in water in the following amounts per 100. g of water: (a) 31.6 g at 20 °C, (b) 85.5 g at 50 °C, and (c) 202 g at 90 °C. Is the enthalpy of solution of KNO_3 expected to be positive or negative? Do you expect the compound to become more or less soluble with an increase in temperature? Hint: You can calculate the heat of solution from the following information: $\Delta H_f^\circ[KNO_3(s)] = -494.6$ kJ/mol and $\Delta H_f^\circ[KNO_3(aq, 1\ m)] = -459.7$ kJ/mol.
38. Several million tons of urea, $OC(NH_2)_2$, are made annually for use as a fertilizer and to make plastics. It is readily soluble in water. If $\Delta H_f^\circ[OC(NH_2)_2(s)] = -333.5$ kJ/mol and $\Delta H_f^\circ[OC(NH_2)_2(aq, dilute)] = -318$ kJ/mol, is the enthalpy of solution positive or negative? Do you expect the compound to become more or less soluble with an increase in temperature?

HENRY'S LAW

39. Compare the concentration of CO_2 in water at 25 °C when the partial pressure of the gas is one third of an atmosphere (see Exercise 14.3) and when the pressure of the gas is 1.0 atmosphere.
40. The partial pressure of O_2 in your lungs varies from 25 mmHg to 40 mmHg. How much O_2 can dissolve in water at 25 °C if the O_2 partial pressure is 40. mmHg? (Henry's law constant for O_2 at 25 °C is 1.7×10^{-6} molal/mmHg.)

VAPOR PRESSURE CHANGES

41. Ethylene glycol, $C_2H_4(OH)_2$, is a common ingredient of automotive antifreeze. If you mix 1.80×10^3 g of water and 620. g of ethylene glycol, what is the vapor pressure of water over the mixture at the normal boiling point of water, that is, at 100 °C? (Assume ethylene glycol is not volatile.)
42. Urea, $OC(NH_2)_2$ is widely used in fertilizers and plastics. (See Study Question 38.) The compound is quite soluble in water, dissolving to the extent of 1.0 g/1.0 mL of water. Assuming the density of water is 1.0 g/mL, and that you dissolve 9.00 g of urea in 10.0 mL of water, what is the vapor pressure of the solution at 24 °C?
43. To prevent the water in the cooling system of your car from freezing, you add some antifreeze, which consists chiefly of ethylene glycol, $C_2H_4(OH)_2$. Assume you have added pure ethylene glycol to 2.00 kg of water in the cooling system. The vapor pressure of the water in the system when the temperature is 90 °C is 457 mmHg. How many grams of glycol did you add? (Assume the ethylene glycol is not volatile at this temperature.)
44. The *normal* boiling point of benzene (C_6H_6) is 80.1 °C. If you add 0.200 mol of a nonvolatile solute to 100. g of benzene, what is the vapor pressure of the benzene at 80.1 °C?
45. 10.0 g of a nonvolatile solute are dissolved in 100. g of benzene (C_6H_6). The vapor pressure of pure benzene at 30 °C is 121.8 mmHg, while that of the solution is 113.0 mmHg at the same temperature. What is the molar mass of the solute?

46. A solution prepared from 20.0 g of a nonvolatile solute in 154 g of the solvent carbon tetrachloride (CCl_4) has a vapor pressure of 504 mmHg at 65 °C. (The vapor pressure of pure CCl_4 is 531 mmHg at 65 °C.) What is the approximate molar mass of the solute?

BOILING POINT ELEVATION

47. Verify the result in Figure 14.14 that 0.200 mol of a nonvolatile solute in 100. g of benzene (C_6H_6) produces a solution whose boiling point is 85.2 °C.
48. What is the boiling point of a solution composed of 15.0 g of urea, $OC(NH_2)_2$, in 0.500 kg of water?
49. Chloroform, $CHCl_3$, has a normal boiling point of 61.7 °C. If K_{bp} for chloroform is 3.63 deg/molal, what is the boiling point of a solution composed of 15.0 g of $CHCl_3$ and 0.515 g of the nonvolatile solute acenaphthene, $C_{12}H_{10}$ (a component of coal tar)?
50. Oil of wintergreen, $C_8H_8O_3$, is used as the scent in candles, among other things. What is the boiling point of the solution if you dissolve 0.755 g of the oil in 100. g of benzene?
51. A solution of ethylene glycol, $C_2H_4(OH)_2$, in 750. g of water has a boiling point of 104.3 °C at a pressure of 760 mmHg. How many grams of the glycol are in the solution? What is the mole fraction of the glycol?
52. Benzene boils at 80.10 °C and has a boiling point elevation constant of 2.53 deg/molal. If you dissolve some $C_{10}H_{10}Fe$ in 50.0 g of benzene, and the boiling point of the solution is 80.51 °C, how many grams of the iron-containing compound must have been dissolved?
53. Place the following solutions in order of increasing boiling point: (a) 0.10 *m* NaCl, (b) 0.10 *m* sugar, and (c) 0.080 *m* $CaCl_2$.
54. Arrange the following aqueous solutions in order of increasing boiling point: (a) 0.20 *m* ethylene glycol (nonvolatile, nonelectrolyte), (b) 0.12 *m* Na_2SO_4, (c) 0.10 *m* $CaCl_2$, and (d) 0.12 *m* KBr.
55. The solubility of NaCl in water at 100 °C is 39.1 g/100. g of water (see Figure 14.11). Calculate the boiling point of this solution.
56. The organic salt $(C_4H_9)_4N^+ClO_4^-$ will dissolve in chloroform. How many grams of the salt must have been dissolved if the boiling point of a solution of the salt in 25.0 g of chloroform is 63.20 °C? (The normal boiling point of chloroform is 61.70 °C and its K_{bp} is 3.63 deg/molal.) Assume the salt completely dissociates into its ions in solution.
57. Hexachlorophene has been used in germicidal soap. What is its molar mass if 0.640 g of the compound, dissolved in 25.0 g of chloroform, produces a solution whose boiling point is 61.93 °C? (The normal boiling point for chloroform is 61.70 °C, and its K_{bp} is 3.63 deg/molal.)
58. Caffeine is a familiar ingredient of coffee and some soft drinks. It is an organic compound composed of C, H, N, and O, and it is soluble in a wide range of solvents including water and benzene. To find the molar mass of the compound you dissolve 0.563 g in 25.0 g of boiling benzene, and the boiling point of the benzene rises from 80.10 °C to 80.39 °C. What is the molar mass of caffeine?
59. You add 0.255 g of an orange, crystalline compound whose *empirical* formula is $C_{10}H_8Fe$ to 11.12 g of benzene. The boiling point of the benzene rises from 80.10 °C to 80.26 °C. What is the molar mass and formula of the compound?
60. Anthracene is a hydrocarbon obtained from coal (and as such its structure resembles a small fragment of graphite, Figure 13.34). The empirical formula of anthracene is C_7H_5. To find its molecular formula you dissolve 0.500 g in 30.0 g of benzene. The boiling point of the pure benzene is 80.10 °C, whereas the solution has a boiling point of 80.34 °C. What is the molecular formula of anthracene?

FREEZING POINT DEPRESSION

61. If you use only water and pure ethylene glycol, $C_2H_4(OH)_2$, in your car's cooling system, how many grams of the glycol must be added to each quart of water to give protection to −31.0 °C? (One quart of water has a mass of 946 g.)
62. Some ethylene glycol, $C_2H_4(OH)_2$, was added to your car's cooling system along with 5.0 kg of water. (a) If the freezing point of the water-glycol solution is −15.0 °C, how many grams of $C_2H_4(OH)_2$ must have been added? (b) What is the boiling point of the solution?
63. If you have ever made homemade ice cream, you know that you cool the milk and cream by immersing the container in ice and a concentrated solution of rock salt (NaCl) in water. If you want to have a water–salt solution that freezes at −10. °C, how many grams of NaCl will you have to add to 3.0 kg of water?
64. Instead of using NaCl to melt the ice on your sidewalk, you decide to use $CaCl_2$. (Like NaCl, $CaCl_2$ is a strong electrolyte.) If you add 35.0 g of $CaCl_2$ to 150. g of water, what is the freezing point of the solution? How many grams of NaCl would you need to achieve the same freezing point using the same quantity of water?
65. Erythritol is a compound that occurs naturally in algae and fungi. It is about twice as sweet as sucrose.

A solution of 2.50 g of erythritol in 50.0 g of water freezes at −0.773 °C. What is the molar mass of erythritol?

66. The compound vanillin is used as a flavoring agent. Although it occurs naturally in vanilla, it can be made from waste products of the wood pulp industry. If you dissolve 0.750 g of the compound in 100. g of water, the freezing point of the solution is −0.0920 °C. What is the molar mass of vanillin?

67. Benzaldehyde is an organic compound widely used in the manufacture of dyes and perfumes and in flavorings. To find its molar mass you dissolve 0.112 g in 40.0 g of water. If the freezing point of the solution is −0.049 °C, what is the molar mass of benzaldehyde?

68. You have just isolated a new compound whose empirical formula is $(C_2H_5)_2AlF$. You know, however, that the molecular form of the compound consists of either two of these units combined to give $[(C_2H_5)_2AlF]_2$ or four of them to give $[(C_2H_5)_2AlF]_4$. To answer the question, you dissolve 0.125 g of the compound in 15.65 g of benzene. The freezing point of the solution is 5.41 °C (whereas the freezing point of pure benzene is 5.50 °C). Which molecular formula is correct?

69. List the following aqueous solutions in order of increasing melting point: (a) 0.1 *m* sugar; (b) 0.1 *m* NaCl; (c) 0.08 *m* $CaCl_2$; and (d) 0.04 *m* Na_2SO_4. (The last three are all assumed to dissociate completely into ions in water.)

70. Arrange the following aqueous solutions in order of decreasing freezing point: (a) 0.20 *m* ethylene glycol (nonvolatile, nonelectrolyte), (b) 0.12 *m* K_2SO_4, (c) 0.10 *m* $MgCl_2$, and (d) 0.12 *m* KBr.

OSMOTIC PRESSURE

71. Meats such as fish and pork can be preserved by "salting" them. In terms of the phenomenon of osmosis, explain what happens when meat is placed in a strong salt solution.

72. A line in the "Rime of the Ancient Mariner," a well-known poem by Samuel Coleridge, is about a ship's crew stranded in a small boat in the ocean, out of food and drink. One of the crew says, "Water, water everywhere/Nor any drop to drink." Why is ocean water so unfit to drink?

73. Calculate the concentration of solute particles in human blood if the osmostic pressure is 7.53 atm at 37 °C, the temperature of the body.

74. Cold-blooded animals and fish have blood that is iso-osmotic (isotonic) with sea water. If sea water freezes at −2.30 °C, what is the osmotic pressure of the blood of marine life at 20. °C?

GENERAL QUESTIONS

75. One of life's great pleasures is to make homemade ice cream in the summer. Fresh milk and cream, sugar, and fruit are churned in a bucket suspended in an ice/water mixture, the freezing point of which has been lowered by adding rock salt. One manufacturer of home ice cream freezers recommends that you add 2.50 lb (1130 g) of rock salt (NaCl) to 16.0 lb of ice (7260 g) in a four-quart freezer. For the solution when this mixture melts, calculate (a) the weight percentage of NaCl, (b) the mole fraction of NaCl, and (c) the molality of the solution.

76. Water at 25 °C has a density of 0.997 g/cm³. Calculate the molality and molarity of pure water at this temperature.

77. At 30 °C, 300. g of silver nitrate can be dissolved in 100. g of water. Calculate (a) the molality of the solution, (b) the mole fraction of silver nitrate, and (c) the weight percentage of silver nitrate.

78. Water and CCl_4 (carbon tetrachloride) are not miscible; when put into the same beaker, they form layers, the CCl_4 on the bottom (see Figure 14.6). If the water layer initially contains I_2, most of the halogen will transfer to the CCl_4 layer when you shake or stir the water–CCl_4 mixture. Why?

79. The Henry's law constants for O_2 and N_2 are, respectively, 1.7×10^{-6} and 8.4×10^{-7} (in units of molal/mmHg at 25 °C). If dry air is 78.3 mole % N_2 and 21.0 mole % O_2, calculate the molality of each gas dissolved in water when the total gas pressure is 760 mmHg.

80. A solution prepared by dissolving 9.41 g of $NaHSO_3$ in 1.00 kg of water freezes at −0.33 °C. From these data decide which of the following equations is the correct expression for the ionization of $NaHSO_3$ (assuming complete ionization in each case).
(a) $NaHSO_3(aq) \longrightarrow Na^+(aq) + HSO_3^-(aq)$
(b) $NaHSO_3(aq) \longrightarrow Na^+(aq) + H^+(aq) + SO_3^{2-}(aq)$

81. A compound is known to be a potassium salt, KX. If 4.00 g of the salt are dissolved in 100. g of water, the solution freezes at −1.28 °C. Which of the elements of Group 7A is X?

82. Refer to Figure 14.11. If excess $NaNO_3$ is added to 100. g of water held at 60 °C, how many grams will dissolve? When you cool the solution to 20 °C, how many grams of $NaNO_3$ will precipitate from the solution?

83. Enthalpies of hydration of individual ions can be calculated, and some are given below.
(i) Mg^{2+}, −1980 kJ/mol; Ca^{2+}, −1650 kJ/mol; Ba^{2+}, −1360 kJ/mol.
(ii) Li^+, −545 kJ/mol; Mg^{2+}, −1980 kJ/mol; Al^{3+}, −4750 kJ/mol.

(a) Explain briefly why the enthalpy change calculated for $M^{n+}(g) \rightarrow M^{n+}(aq)$ is always negative.
(b) Explain the trend you observe in each of the two series of hydration enthalpies.

84. Which aqueous solution would have the greater water vapor pressure at 25 °C: (a) 33.0 g of sugar (molar mass = 342 g/mol) in 180. g of water or (b) 5.85 g of NaCl in 180. g of water? Which would have the higher boiling point?

85. Arrange the following aqueous solutions in order of increasing vapor pressure at 25 °C: (a) 0.35 *m* $C_2H_4(OH)_2$ (nonvolatile solute): (b) 0.50 *m* sugar; (c) 0.20 *m* KBr (a strong electrolyte); and (d) 0.20 *m* Na_2SO_4 (a strong electrolyte).

86. List the solutions referred to in Study Question 85 in order of increasing boiling point.

87. If you added ethylene glycol to the water in your car's cooling system to lower the freezing point to −31.0 °C, what would be the corresponding boiling point of the solution in the system?

88. *Reverse osmosis* is being used to obtain fresh water from sea water by desalinization (salt removal). A membrane permeable to water but not salt separates compartments containing sea water and fresh water. The normal flow of water, due to osmosis, from the fresh water to the sea water compartment is reversed by applying pressure to the sea water compartment. For purposes of this problem, assume that the sea water is 3.0 weight percent NaCl. What is the minimum pressure (above atmospheric pressure) that must be supplied to cause reverse osmosis to occur at 25 °C?

89. A student weighs 150. g of KBr and adds it to 100. g of water held at 100 °C. Not all of the solid dissolves. After filtering the solution (while still hot), the student finds that the undissolved KBr amounts to 46.0 g. What is the solubility of KBr at 100 °C in g solute/100 g solvent?

90. The vapor pressure of water is 17.5 mmHg at 20 °C. (a) What is the vapor pressure of the solution when 15.0 g of urea, $(NH_2)_2CO$, is dissolved in 0.500 kg of water? (b) Compare the vapor pressure in part (a) with that of a solution composed of 15.0 g of sugar ($C_{12}H_{22}O_{11}$) in 0.500 kg of water at 20 °C. (c) What quantity of sugar would have to be dissolved in 0.500 kg of water to produce the same vapor pressure as that given by the solution in part (a)?

91. Assume you add 1.00 kg of ethylene glycol, $C_2H_4(OH)_2$, to 2.00 kg of water in the cooling system of your car. What is the vapor pressure of the water in the system when the temperature is 90 °C? (Assume the ethylene glycol is not volatile at this temperature.)

92. Solution properties:
(a) Which solution is expected to have the higher boiling point, 0.10 molal NaCl or 0.15 molal sugar?
(b) The enthalpy of solution of sodium hydroxide is strongly *exo*thermic.

$$NaOH(s) + H_2O(\ell) \longrightarrow NaOH(aq) + \text{heat}$$

This means the solid is (*more*)(*less*) soluble as the temperature goes up.
(c) In which aqueous solution is the vapor pressure of water higher, 0.30 molal NH_4NO_3 or 0.15 molal Na_2SO_4?

93. You have isolated a new boron hydride and have evidence that it is either $B_{10}H_{14}$ or B_9H_{15}. You dissolve 0.500 g of the white compound in 15.23 g of benzene and find that the boiling point of the solution is 80.78 °C. (The boiling point of pure benzene is 80.10 °C and its boiling point elevation constant is 2.53 deg/molal.) What boron hydride have you isolated?

94. If you add a volatile solute to a volatile solvent, both substances contribute to the vapor pressure over the solution. Assuming an ideal solution, the vapor pressure of each is given by Raoult's law, and the total vapor pressure is simply the sum of the vapor pressures of both components. A solution, assumed to be ideal, is made from 1.0 mole of toluene ($C_6H_5CH_3$) and 2.0 moles of benzene (C_6H_6). The vapor pressures of the pure solvents are 22 mmHg and 75 mmHg, respectively, at 20 °C. What is the total vapor pressure of the mixture at 20 °C?

95. At 60 °C the vapor pressure of pure benzene is 388 mmHg and that of pure toluene is 137 mmHg. Calculate the mole fraction of benzene in an ideal solution of benzene and toluene that has a total vapor pressure of 266 mmHg at 60 °C. (See Study Question 94.)

96. Calculate the total vapor pressure at 25 °C of an ideal solution made by mixing two volatile liquids: 10.0 g of CH_2Cl_2 and 1.00 g of hexene (C_6H_{12}) at 25 °C. At 25 °C, the vapor pressure of pure CH_2Cl_2 is 435 mmHg and that of pure C_6H_{12} is 166 mmHg. (See Study Question 94.)

97. If you mix 5.00 mL of methylene chloride (CH_2Cl_2, density = 1.34 g/cm³) and 10.0 mL of chloroform ($CHCl_3$, density = 1.49 g/cm³), what is the total pressure of the solution at 25 °C? (The pure solvents have vapor pressures of 435 mmHg and 195 mmHg, respectively, at 25 °C, and they are assumed to form an ideal solution.) (See Study Question 94.)

98. In the past some people used a mixture of ethyl alcohol and water in automotive cooling systems. If the freezing point of an alcohol–water mixture is −10.0 °C, what is the vapor pressure of the mixture ($P_{alc} + P_{water}$) at 25 °C? (Information you need to solve this problem is: (a) the vapor pressure of alcohol at 25 °C is 60. mmHg; (b) the vapor pressure of pure water at 25 °C is 24 mmHg. Assume an ideal solution.) (See Study Question 94.)

SUMMARY QUESTIONS

99. Suppose you have just made a compound containing only boron and fluorine. To determine its molar mass you dissolve 0.146 g in 10.0 g of benzene and find that the freezing point of the solution is 4.77 °C. (The freezing point of pure benzene is 5.50 °C and K_{fp} for benzene is −4.90 deg/molal.) If the compound contains 22.1% boron, what are the empirical and molecular formulas of the boron–fluorine compound?

100. You have a small but seemingly pure sample of an unknown white powder containing only C, H, and O. To analyze the compound, you determine its empirical formula by analysis for C and H, and its molar mass by freezing point depression. (a) What is the empirical formula if a 0.0151-g sample gives 0.0111 g of H_2O and 0.0449 g of CO_2 on combustion? (b) If you dissolve 1.00 g of the powder in 100. g of camphor, you find that the melting point of the solution is 177.05 °C. Pure camphor melts at 179.75 °C. What is the molar mass of the white powder? (c) What is the molecular formula of the white powder?

101. In chemical research we often send newly synthesized compounds to commercial laboratories for analysis. These laboratories determine the weight percent of C and H by burning the compound and collecting the evolved CO_2 and H_2O. They determine the molar mass by measuring the osmotic pressure of a solution of the compound. Calculate the empirical and molecular formulas of a compound, C_xH_yCr, given the following information: (a) The compound contains 73.94% C and 8.27% H; the remainder is chromium. (b) At 25 °C, the osmotic pressure of 5.00 mg of the unknown dissolved in 100. mL of chloroform solution is 3.17 mmHg.

PART FOUR

THE CONTROL OF CHEMICAL REACTIONS

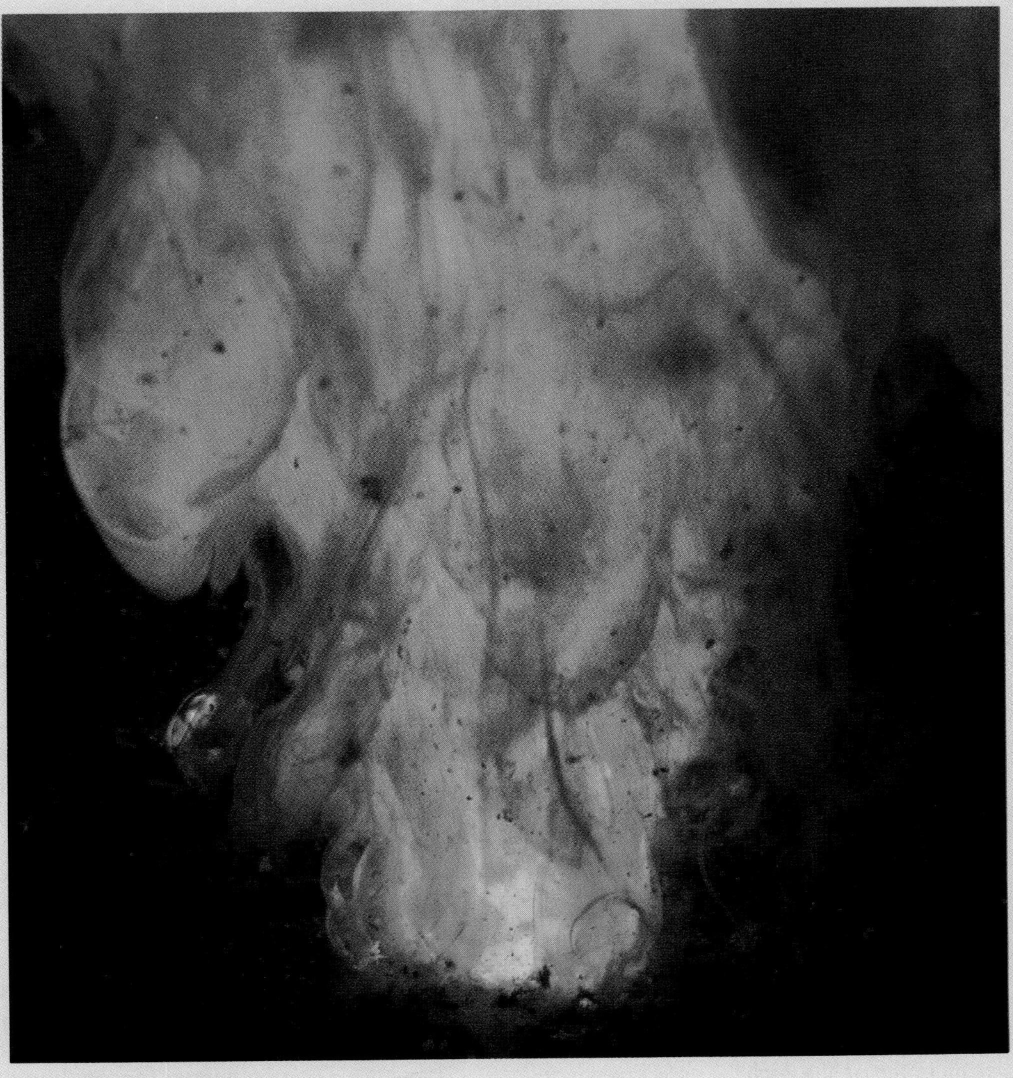

The reaction of zinc and sulfur to give zinc sulfide.

Gregory C. Farrington

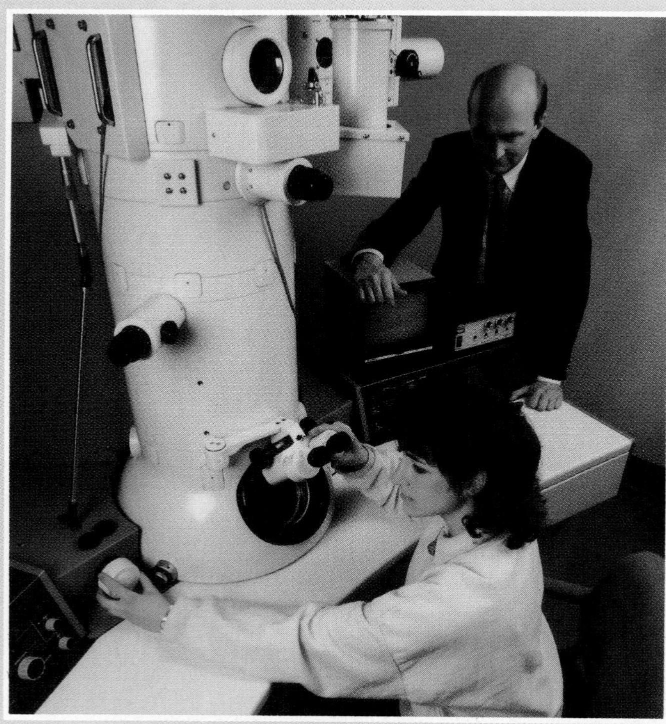

G.C. Farrington and Cathy Lane, a Ph.D. student in Materials Science and Engineering, in front of a JEOL 4000 high resolution transmission electron microscope. The microscope is capable of imaging the alternating spinel block—conduction plane structure of beta alumina to a resolution of 1.6 Ångstroms.

Dr. Gregory C. Farrington is Professor of Materials Science and Engineering and Dean of the School of Engineering and Applied Science at the University of Pennsylvania. He received a BS degree in chemistry from Clarkson College in 1968 and a PhD in chemistry from Harvard University in 1972. After serving as a member of the technical staff in materials research at the General Electric Research and Development Center, he joined the faculty of the University of Pennsylvania in 1979. From 1984 to 1986 he served as chairman of the Department of Materials Science and Engineering and from 1987 to 1990 as Director of the Materials Research Laboratory at the University of Pennsylvania.

Dr. Farrington's principal research interests are in the areas of electrochemistry, solid electrolytes, and solid state chemistry. He is the author or co-author of more than 100 papers on the synthesis, structure, and transport characteristics of conductive ceramics and polymers and holds numerous patents. He has served on a number of university and government advisory committees, including the Carnegie Foundation commission on "National Security and the Future of Arms Control." In 1984 he was awarded an honorary PhD by the University of Uppsala, Sweden, in recognition of his research in solid state materials chemistry.

High school chemistry with Mrs. Craig

Many of us in science, and many of the scientists whom we interviewed, became interested in the sciences because of one or two outstanding teachers. Dr. Farrington was no different. He said, "When I was in high school [in Cobleskill, in upstate New York] I became interested in chemistry for two reasons, one of which had very much to do with the times. It was the time of Sputnik, and science was where so much of the action seemed to be. It was constantly in the press, you could feel it, you could breathe it in the air, and as a result it seemed to be a very, very interesting career. The other was a human reason. I had a great chemistry teacher, Mrs. Craig. She was a wonderful woman. She didn't just teach chemistry as a subject to understand by the numbers in the book, but she actually showed compounds reacting, liquids changing color, and gases expanding. She loved chemistry and the challenge of communicating it to other people, and not just in the abstract. I think people are turned on to science by seeing interesting things happen. Then they develop the curiosity to probe beyond the obvious and search for deeper connections and logic. But it begins with fascination, with curiosity, and Mrs. Craig fostered that. As a result I ended up going into chemistry when I went to Clarkson College [also in upstate New York]."

Clarkson College and Harvard University

"Clarkson College was the kind of place where an upstate New York fellow like myself went—and, besides, I had a good scholarship offer. The alternative was MIT, but I stayed home and went to Clarkson. However, after four years at Clarkson I decided that it might be time to get out of upstate New York. So I was pleased to have the opportunity to go to Harvard Graduate School where I enjoyed one of the most exciting and interesting periods of my life. Harvard is simply a marvelous place, an intellectual glass of champagne with bubbles hitting you all the time.

"Nevertheless, I would say the best personal education I received was at Clarkson, and it was because of the human contact. Education is a human process, and Clarkson believed in its students. That's Clarkson's real strength, the reason for its excellence. The chemistry department scheduled many of the best faculty to teach in the freshman year, and they worked very hard to give students contact with the faculty. Clarkson is a small school, and they do their job very, very well."

First industry and then the university

All of the other scientists we interviewed for this book went from their graduate studies to a career in a college or university. Dr. Farrington's path was somewhat different.

"After Harvard I went to work at the General Electric Research Center in Schenectady, back in upstate New York again. This was partly because, when I was growing up in that area, one of the great dreams of a high school kid interested in science was to work at the Research Center. It was a great place, and I truly enjoyed my seven years there. But I knew that I had always wanted to go into academics. Every year when I had my appraisal at General Electric I had to fill out a form that asked, 'What is your long-term goal?' I would always say, 'to get a good professorship.' Fortunately the GE management had a good sense of humor."

How to interest students in science

As Dr. Shakhashiri has so clearly defined (page 37), the United States is in a crisis in terms of scientific manpower. We are all searching for ways to increase the people in, as Dr. Shakhashiri has put it, the "personnel pipeline." Dr. Farrington feels that "We need more Mrs. Craigs. Clarksons are also needed but Clarksons don't convert the unconverted. They preach to the choir. We need more people joining the choir. We need more faculty at the high school level who love science and can communicate its fascination and encourage students to make it their career. However, to be good at science or engineering takes hard work. There is no such thing as watered-down chemistry, physics, or mathematics. Science is a challenge, the best kind of challenge. And it's fun. We need to communicate the fun as far back as kindergarten and first grade. We need students to understand the fascination that comes from marveling at and attempting to understand the physical world. We need faculty to feel and pass on the sparks of enthusiasm that begin lifelong careers in science and technology."

Creativity

There has been considerable concern in our society in recent years that Japan is surpassing us in science and technology. However, a statement we often hear is that the Japanese are not creative. Dr. Farrington said emphatically that "the idea that the Japanese can't create is nonsense. Certainly their society is structured quite differently from ours, and perhaps it discourages individuality. But the Japanese are very smart, very clever, and fully capable of being a creative force in science and technology. They also are very good at translating ideas into real products that people want to buy. Our business is structured to discourage the long-term risk-taking necessary to nurture a revolutionary product into the marketplace. It seems that we can create new technologies but lack the patience for the advanced engineering and manufacturing that bring good ideas to life.

"You know it is relatively easy to be interesting, but it is very, very hard to be useful. I discovered that at General Electric. I grew to admire people who were not only scientifically interesting but had been successful at helping to create a

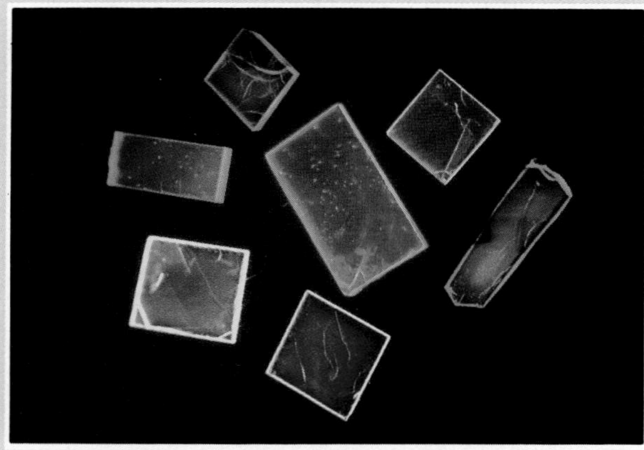

Crystals of β''-alumina containing small concentrations of fluorescent Cu(I) in the conduction planes. The color of the fluorescence is determined by the lattice spacing, which can be varied by changing the crystal composition. The blue-green color (topmost crystal) is particularly rare in a crystalline material. These crystals may someday find application as tunable solid state laser crystals.

new commercial technology. There are so many engineering, economic, and marketing challenges in creating a new technology. It's far more difficult than it might seem.

"If we cannot meet the challenge of translating our creativity into successful products, so many of our good ideas will be wasted and our economy will show the effects. This is the age of education, science, and technology. It simply is not an option for the United States to ignore it and to pretend that we can all live prosperous lives flipping burgers and hatching another leveraged buyout."

New directions in science: outstanding engineering

We were very interested in Dr. Farrington's views on what directions he sees in science in the next 10 years in general and in his own field of materials science. Where are we going? Where should we be going?

"I think one of the greatest challenges of the 90s is going to be rediscovering the need for advanced engineering. I say that as a person who was trained in science. We excel in this country at creating new ideas and creating new technologies. We also excel at marketing, if we ever have a chance to take new technologies to market. What we need more of is what lies in between, the cleverness it takes to translate the good ideas into something practically useful.

"I also believe that we have only begun to see the effects of the development of computers on science and engineering. We have just stuck our toes into the new age of information processing, not even gotten wet, let alone been swimming. It will be great fun to be alive during the coming decades as our world is transformed by rapid information transmission and processing."

Materials science and universal truths

At the time of our interview, Dr. Farrington was Director of the Laboratory for Research on the Structure of Materials at the University of Pennsylvania. Since materials science promises to be one of the most exciting areas of science and engineering in the 90s,* we wanted Dr. Farrington to comment on this.

"Materials science is surely one of the most interesting interdisciplinary fields today. It involves so many different talents: chemists to make new materials, physicists to probe their properties, materials scientists and engineers who link the fundamental with the applied.

"Why is it so exciting? You can see it all around you. In the last ten years we've had so many examples of materials that were found to have properties that everyone knew they shouldn't. One example is that of conducting polymers. Everyone knows a sensible polymer like Saran Wrap™ doesn't conduct electricity, but now we have polymers that do. Saran Wrap has become aluminum foil! Then there are superconducting ceramics. Everyone knew that ceramics were not particularly good electronic conductors, let alone superconductors. Except, what everyone knew turned out not to be universally true. Now we have superconducting oxides with properties that were only wild dreams several years ago.† These revolutions suggest a universal truth that I tell all my students. When given a choice between your preconceptions, between what you think you are looking for and what nature is actually trying to show you, follow nature, listen to nature. It is much more mysterious, much more interesting, much trickier, and much more creative than you'll ever be. So surrender to nature early.

"The new superconductors and conducting polymers spark our

*See also the interview with Dr. Mary Good.

†See "Something More About Superconductors" in Chapter 25.

imagination, hint that there are even greater discoveries just waiting for the right person and the right time. Even so, these materials were truly discovered by what might be called disciplined serendipity. Think what might happen if we actually understood the relationships between composition, structure, and properties in the solid state. Instead of finding new compounds by a process akin to walking on the beach, picking up pebbles and looking for one that appears different, we might search for bright pebbles on the beach of a computer screen and avoid much of the work that is now done by patient synthesis, by grinding and firing. I think one of the most exciting directions in materials science in the future is the computer-aided design of materials.

"Computers are going to change fundamentally how we go about science and engineering. Experimentation is very expensive, particularly in searching for new materials. So, I think we will see much deeper penetration of computers into the design of materials. That's why the next decades will be great times for chemists. One of the real glories of being a chemist is in designing and making a compound that does something no one ever dreamed it could do, or creating a compound that has properties people dreamed of, but didn't know how to make. It's one of the reasons chemistry is fun."

Elegance and beauty

One of the questions important to young people just starting in science is what it is like to be a scientist. What does it take to be a scientist? What personal characteristics? What should we try to instill in our students? As with some of the others interviewed for this book, Dr. Farrington first said curiosity. However, he believes that fascination, a different form of curiosity, is also important. "Fascination drives curiosity. It's an emotional response—a sense of wonder—that then takes the form of curiosity. If you're curious, if the world fascinates you, if you're willing to be jarred out of your rut, then science may be the field for you. The rewards are simple: a sense of beauty and awe for nature. Never forget that science is about beauty. The highest praise we reserve for work is 'elegant,' although I think 'beautiful' ranks even higher. Some results are so profound that they are windows on mystery, met with gasps, and truly beautiful."

Complexity and simplicity

In our interview with Professor Hoffmann, he commented on the beauty of complexity in chemistry, but Dr. Farrington carried this one step further. "Beauty lies in the simplicity underlying the complexity of science. We spend our lives struggling with the complexity, but occasionally one theme rises above all. I have often said that it is like waiting to hear the oboe in the orchestra, to hear the penetrating theme that comes through all the variations, often from the voice of the oboe. Wait for the sound of the oboe; everything else has been a variation."

Science, China, and Eastern Europe

Dr. Farrington travels widely and works collaboratively with other scientists in Europe and Asia, so we talked about the international aspect of science and its influence on society. He said he thinks "science has done so much to foster freedom in the world. First, the first law of science is the law of truth. The Pope may not have liked what Galileo said, but it didn't matter. Galileo still was right. The highest standard is not what someone claims to be true—whether a Pope or a politician—but simply what *is* true. Adherence to truth in the scientific sense is the basic definition of freedom.

"Another way science has fostered freedom is in its importance to a nation's economic strength. Science and advanced engineering have brought us so many of the products critical to our economy and to our survival that no country can ignore science. That's what the Soviet Union and the Eastern Block discovered; they cannot live without advanced science and engineering. They also learned that science and engineering cannot survive and prosper in a closed society. Scientists must have the opportunity to talk, to travel, to speak, to think.

"Another impact of science and engineering is in modern communication. Walls no longer isolate populations, because they cannot block radio and television. Of the two, I think that television is the more powerful. If you listen to the radio you are always hearing about an event through someone else's mind. With television you can see and make your own judgments based on what you have seen. It's a powerful medium whose effects we have seen in Eastern Europe. Western television constantly challenged the pronouncements of the governments there. You see, the essence of science, the essence of freedom, is in the personal right to judge what is true against personal observation. Science has shown the power of such judgment and, in this way, has changed the world profoundly."

CHAPTER

15

Chemical Kinetics: The Rates and Mechanisms of Chemical Reactions

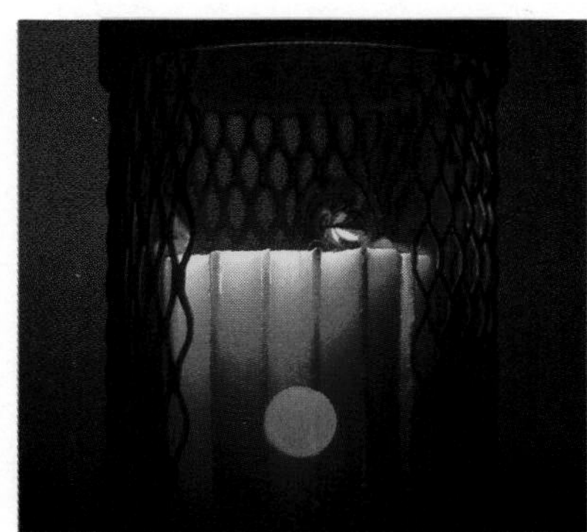

A catalytic converter used to allow wood stoves to burn more efficiently.

CHAPTER OUTLINE

So far we have been interested in what kinds of chemical reactions occur. However, we now want to take up another large area of chemistry, called **chemical kinetics**. This is the study of the speed or rate of a reaction under various conditions, and what this tells us about the **mechanism** of the reaction, the sequence of events at the molecular level that controls the speed and outcome of the reaction.

The reason that we are interested in kinetics and mechanisms is that we would often like to control just how fast a reaction goes to completion and what the products will be. If we understand how to change the speed, and also understand how the reaction occurs, there is some chance we can control the process. For example, if we understand the mechanism of the reactions leading to the depletion of the ozone layer or to air pollution, then we have a better chance to control these problems. Therefore, it is the purpose of this chapter to introduce you to the chemist's concepts and language for understanding the rates and mechanisms of chemical reactions. You will find some things you have already studied—molecular structures, bond strengths, and stoichiometry, among others—to be key aspects of this story.

15.1 KINETICS AND CONDITIONS

The first step in a kinetics study is to acquire experimental information on the reaction speed or rate under a variety of conditions. At the outset, you should be aware that the following conditions affect the speed of a chemical process:

a. *Concentrations of reactants*. Some metals react well with bases such as aqueous sodium hydroxide. For example, aluminum can give water-soluble sodium tetrahydroxoaluminate and H_2 gas.

$$2\ Al(s) + 6\ H_2O(\ell) + 2\ NaOH(aq) \longrightarrow 2\ NaAl(OH)_4(aq) + 3\ H_2(g)$$

As Figure 15.1 shows, the reaction is very slow in 1 M NaOH, but it is much more rapid in more concentrated 6 M NaOH.

b. *Temperature*. In your own experience you know that heating things can bring about more rapid reaction. That is why, for example, we cook food, and in the laboratory we often add heat to make reactions occur faster.

c. *Catalysts*. Catalysts are substances that accelerate chemical reactions but are not themselves transformed. For example, hydrogen peroxide, H_2O_2, can decompose to water and oxygen.

$$2\ H_2O_2(\ell) \longrightarrow O_2(g) + 2\ H_2O(\ell)$$

However, if the peroxide is stored in a cool place, it is reasonably stable for many months. In the presence of a catalyst, though, the reaction occurs with explosive speed. In fact, an insect called a bombardier beetle uses the reaction as its defense mechanism (Figure 15.2). The beetle uses a biological catalyst, called an *enzyme*, to produce a stream of superheated steam to spray on its enemies.

d. *Surface area of reactants*. The effect of the surface area of reactants is shown in Figure 15.3. This is one reason we dissolve reactants in water or other solvents before carrying out a reaction. When reactants are placed in solution, all molecules are fully exposed to interactions with each other and the reaction proceeds more quickly.

Figure 15.1 The speed of the reaction of aluminum with aqueous NaOH depends on the concentration of the base. With dilute NaOH the reaction is slow (left), but it is much more rapid with more concentrated NaOH (right).

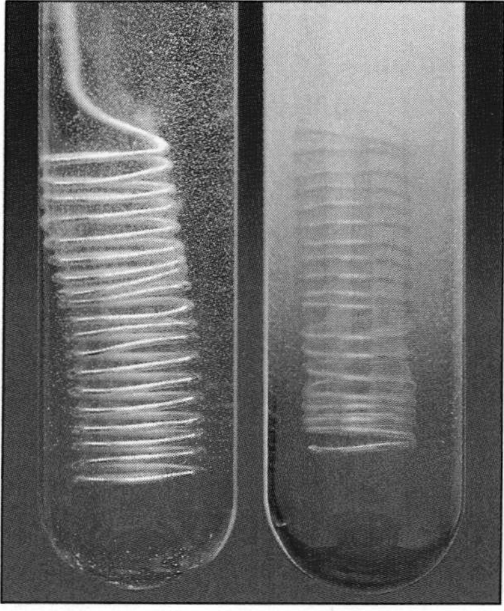

Figure 15.2 A bombardier beetle uses the catalyzed decomposition of hydrogen peroxide as a defense mechanism. The hot oxygen gas formed in the decomposition forces out water and other chemicals with explosive force. Since the reaction is very exothermic, the water comes out as steam. (Thomas Eisner with Daniel Aneshansley)

(a)

(b)

Figure 15.3 The combustion of lycopodium powder. (a) The spores of this common moss burn only with difficulty when piled in a dish. (b) If the surface area is increased, and a finely divided powder is sprayed into a flame, combustion is rapid.

After defining more precisely what is meant by a reaction rate, we shall study what happens to the concentrations of reactants and products during the time a reaction occurs and how temperature and catalysts affect reaction rate.

15.2 CHEMICAL REACTION RATES

You know from your experience in cooking or in the chemistry laboratory that some chemical reactions take a long time to occur, whereas others happen extremely rapidly. It takes at least three minutes to cook an egg so it is edible, but the combustion of the methane or propane in a laboratory Bunsen burner happens *very* quickly. This is what this chapter is about: the speed of reactions and why some are slow while others are fast.

The speed of a reaction is expressed in terms of its "rate," a word that implies that some measurable quantity is changing with time. For instance, a car's rate of travel is found by measuring the change in its position, Δp, over a given time, Δt. For example, if you travel 110 miles in 2 hours, you are traveling at an average speed of $\Delta p/\Delta t = 55$ miles per hour.

When describing chemical reactions, the **reaction rate** is expressed as the change in concentration of a reagent per second (or per hour, or per day). To illustrate this, consider what can happen to the cancer chemotherapy agent cisplatin, $Pt(NH_3)_2Cl_2$, in the presence of water.

It is obviously important for us to know something about the speed with which Cl^- is replaced by water in cisplatin. Once the compound is placed in the aqueous environment of the human body, some other molecule capable of binding to Pt^{2+} can replace the Cl^- ion.

$$Pt(NH_3)_2Cl_2 + H_2O \longrightarrow [Pt(NH_3)_2(OH_2)Cl]^+ + Cl^-$$

One of the Cl^- ions bound to the Pt^{2+} metal ion center can be replaced. The rate at which this can occur is found by measuring, over a given time interval, the quantity of $Pt(NH_3)_2Cl_2$ consumed *or* the quantity of Cl^- evolved. For example, suppose we focus on the platinum starting material. If the quantity at some point in time is measured in moles per liter, then the rate of reaction could be given as the number of moles per liter consumed per minute or second.

$$\text{Rate of change of } [Pt(NH_3)_2Cl_2] = \frac{\text{Quantity of } Pt(NH_3)_2Cl_2 \text{ reacting (in mol/L)}}{\text{Elapsed time } (t)}$$

$$\frac{\Delta[Pt(NH_3)_2Cl_2]}{\Delta t} = \frac{[Pt(NH_3)_2Cl_2] \text{ at } t_2 - [Pt(NH_3)_2Cl_2] \text{ at } t_1}{t_2 - t_1}$$

Recall that the Greek letter Δ (delta) means that a *change* in some quantity has been measured.

The decrease in concentration of $Pt(NH_3)_2Cl_2$ with time and the increase in concentration of Cl^- with time are shown in Figure 15.4. Each point on these curves gives the concentration of that compound or ion at a particular time. To obtain an average reaction rate over some period of time, we can find the change in concentration of $Pt(NH_3)_2Cl_2$ or Cl^- during that time. For example, during the first 200. minutes after the reaction begins, the average rate at which $Pt(NH_3)_2Cl_2$ disappears is

The rate of change in a reactant concentration is always negative, while the rate of change of a product concentration is always positive.

$$\text{Rate of change of } [Pt(NH_3)_2Cl_2] = \frac{\Delta[Pt(NH_3)_2Cl_2]}{\Delta t}$$

$$= \frac{(0.0074 - 0.0100)\text{mol/L}}{(200. - 0)\text{ min}}$$

$$= \frac{-0.0026 \text{ mol/L}}{200. \text{ min}}$$

$$= -1.3 \times 10^{-5} \text{ mol/(L} \cdot \text{min)}$$

Another important feature to notice in Figure 15.4 is that, as the concentration of $Pt(NH_3)_2Cl_2$ declines, the concentration of Cl^- *increases*. Because of the 1:1 stoichiometry of the reaction, the increase in $[Cl^-]$ is

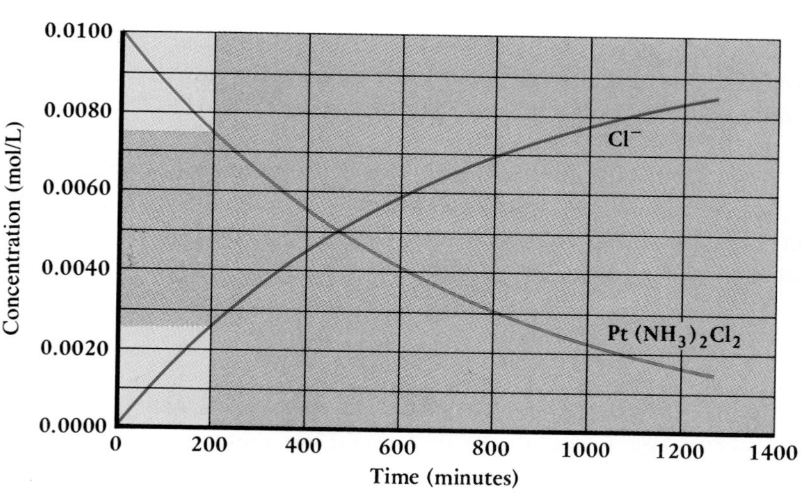

Figure 15.4 Plot showing the concentration of $Pt(NH_3)_2Cl_2$ and the concentration of Cl^- as a function of time as Cl^- is replaced by H_2O. The shaded areas show that the decrease in concentration of $Pt(NH_3)_2Cl_2$ is equal to the increase in Cl^- concentration over the same period.

just as rapid as the decrease in $[Pt(NH_3)_2Cl_2]$. This means the curve for the *appearance* of Cl^- is exactly "upside down" from the curve for the *disappearance* of $Pt(NH_3)_2Cl_2$. That is, the magnitude of the rate of change in the concentration of $Pt(NH_3)_2Cl_2$ is exactly the same as that for Cl^-, but the value for $\Delta[Pt(NH_3)_2Cl_2]/\Delta t$ is negative while that for $\Delta[Cl^-]/\Delta t$ is positive.

The rate of change in concentration for a given compound in a reaction is not always the same as the rates for other compounds in the same reaction. For example, for a reaction such as the decomposition of hydrogen peroxide,

$$2\ H_2O_2(g) \longrightarrow O_2(g) + 2\ H_2O(g)$$

the balanced equation tells us that oxygen gas can *appear* only half as rapidly as the peroxide *disappears*, and that H_2O appears just as rapidly as H_2O_2 disappears. To remove this effect of stoichiometry on our discussions, chemists often just speak of the "rate of reaction" where (1) we ignore the algebraic sign on the value of the rate and (2) we divide the rate of change in concentration of each reactant or product by its stoichiometric coefficient in the balanced chemical equation. Thus, for the decomposition of hydrogen peroxide, we have

$$\text{Rate of reaction} = \frac{\Delta[O_2]}{\Delta t} = -\frac{1}{2}\frac{\Delta[H_2O_2]}{\Delta t}$$

In other words, the rate of reaction is equal to the rate of change of $[O_2]$ or to $-\frac{1}{2}$ times the rate of change of $[H_2O_2]$.

There is one final, important point concerning reaction rates. In general, we try to measure **initial reaction rates**. That is, we begin with the pure reactants, mix them thoroughly, and then measure the speed of the reaction over the first few seconds, hours, or days. This is important because products are formed as the reaction proceeds, and so it is possible that the products will react with something else in the system or react with themselves to reform reactants; either way, the presence of products can alter the reaction rate we are studying and lead to very different—and confusing—results. Therefore, for the remainder of this chapter, keep in mind that we are describing *initial* reaction rates.

EXAMPLE 15.1

RELATIVE RATES AND STOICHIOMETRY

Give the relative rates for disappearance of reactants and formation of products for the following reaction.

$$4\ PH_3(g) \longrightarrow P_4(g) + 6\ H_2(g)$$

Solution To equate rates, we have to divide $\Delta(\text{reagent})/\Delta t$ by the stoichiometric coefficient in the balanced equation. Therefore,

$$\text{Reaction rate} = -\frac{1}{4}\left(\frac{\Delta[PH_3]}{\Delta t}\right) = +\frac{\Delta[P_4]}{\Delta t} = +\frac{1}{6}\left(\frac{\Delta[H_2]}{\Delta t}\right)$$

$$-\tfrac{1}{4}(\text{rate of change of }[PH_3]) = \text{rate of change of }[P_4]$$
$$= \tfrac{1}{6}(\text{rate of change of }[H_2])$$

Since 4 moles of PH_3 disappear for every mole of P_4 formed, the rate of P_4 formation can only be one fourth of the rate of PH_3 disappearance. Similarly, P_4 must appear at only one sixth of the rate that H_2 appears.

EXERCISE 15.1 Relative Rates and Stoichiometry

What are the *relative* rates of change in concentration of the products and reactant in the decomposition of nitrosyl chloride, NOCl?

$$2\ NOCl(g) \longrightarrow 2\ NO(g) + Cl_2(g)$$

15.3 RATE EXPRESSIONS

The relation between reaction rate and the concentrations of reactants is given by a mathematical equation called a *rate expression* or *rate law*. This section is concerned with the form of the rate expression and its derivation for some typical reactions. In the next section we shall turn to its meaning in terms of *reaction mechanisms*.

THE RATE EXPRESSION

Reaction rates are generally dependent on reactant concentrations. It is often the case, for example, that the rate and concentration are *directly* proportional: as the concentration increases, the rate increases by the same amount. In other cases, the rate may increase much faster than the increase in concentration. *To find the exact relation between rate and concentration, we must do some experiments* and collect information such as that illustrated by Figure 15.4. Then, to show exactly how reactant concentrations and reaction rate are related, we can write a **rate expression**.

A rate expression (or rate law) defines the relation between reactant concentrations and the reaction rate.

For a general reaction such as

$$a\,A + b\,B \xrightarrow{C} x\,X,$$

where C is a catalyst, the rate expression will *always* have the form

$$\text{Initial reaction rate} = k[A]^m[B]^n[C]^p$$

where

k = rate constant (a proportionality constant)
[A] = concentration of reactant A
[B] = concentration of reactant B
[C] = concentration of the catalyst
m = order of reaction for reactant A
(see below for more explanation)
n = order of reaction for reactant B
p = order of reaction for the catalyst C

The exponents in the rate expression or law can be zero or positive whole numbers or fractions. In this chapter we shall illustrate only cases where the exponent is zero or a positive whole number.

The rate expression expresses the fact that the rate of reaction is proportional to the reactant concentrations, each concentration being raised to some power. *It is important to recognize that the exponents (m, n, and p in this case) are not necessarily the stoichiometric coefficients for the balanced chemical equation.* The exponents, which can be zero or positive whole numbers or fractions, *must be determined by experiment*.

For example, hydrogen peroxide rapidly decomposes to water and oxygen in the presence of a catalyst such as iodide ion.

$$2\ H_2O_2(aq) \xrightarrow{I^-(aq)} 2\ H_2O(\ell) + O_2(g)$$

Experiment shows clearly that this reaction has the rate expression

$$\text{Initial reaction rate} = k[H_2O_2][I^-]$$

where the exponent on each concentration term is 1, even though the stoichiometric coefficient of H_2O_2 is 2 and I^- does not even appear in the balanced equation.

As you shall see in Section 15.7, a catalyst does not appear in the balanced, overall equation for the reaction. It is not consumed by the reaction. However, it may appear in the rate expression.

THE RATE CONSTANT, *k*

The quantity "*k*" is called the **rate constant**. It is a proportionality constant that relates rate and concentration *at a given temperature*, and *it must be evaluated by experiment*. It is an important quantity since, once it is known, it enables you to find the reaction rate for a new set of concentrations. To see how to use *k*, let us consider again the substitution of Cl^- ion in the cancer chemotherapy agent cisplatin, $Pt(NH_3)_2Cl_2$.

$$Pt(NH_3)_2Cl_2 + H_2O \longrightarrow [Pt(NH_3)_2(OH_2)Cl]^+ + Cl^-$$

The rate expression for this reaction is

$$\text{Rate} = k[Pt(NH_3)_2Cl_2]$$

and the proportionality or rate constant, *k*, is 0.090/hr. Therefore, at a time when $[Pt(NH_3)_2Cl_2]$ is 0.018 mol/L, the rate is

The rate constant, k, depends on temperature and must be determined by experiment.

$$\text{Initial reaction rate} = (0.090/\text{hr})(0.018\ \text{mol/L}) = 0.0016\ \text{mol/(L}\cdot\text{hr)}$$

THE ORDER OF A REACTION

As you shall see in Section 15.6, a great deal of information about reaction mechanisms is conveyed by the "order" of a reaction. The **order** with respect to a particular reactant is the exponent of its concentration term in the rate expression, and the **total reaction order** is the sum of the exponents on all concentration terms.

The cancer chemotherapy agent cisplatin has the Pt—Cl bonds adjacent to one another, a configuration called the *cis* arrangement. However, it is also possible for these bonds to be opposite one another in what is called the *trans* arrangement. The latter compound has somewhat different properties than the *cis* compound—and it has little or no cancer fighting ability.

$$\underset{cis}{Pt(NH_3)_2Cl_2} \longrightarrow \underset{trans}{Pt(NH_3)_2Cl_2}$$

$$\text{Reaction rate} = k[cis]$$

The *cis* compound can change into the *trans* compound in a first order reaction; that is, the exponent on [*cis*] in the rate expression is 1. This means the rate is *directly* proportional to the concentration of the *cis* compound. When [*cis*] = 0.018 mol/L, chemists have found that the initial rate is 0.00090 mol/(L · hr).

$$\text{Initial reaction rate} = k[cis]$$
$$0.00090 \text{ mol/(L} \cdot \text{hr)} = k[0.018 \text{ mol/L}]$$
$$k = 0.050/\text{hr}$$

However, if the concentration of *cis* is raised by a factor of 4 to 0.072 mol/L, then the rate is also raised by a factor of 4, from 0.00090 mol/(L · hr) to 0.0036 mol/(L · hr).

$$\text{Initial reaction rate} = (0.050/\text{hr})(0.072 \text{ mol/L}) = 0.0036 \text{ mol/(L} \cdot \text{hr)}$$

The organic compound methyl acetate is a widely used industrial solvent. It reacts with hydroxide ion to give acetate ion and methyl alcohol,

$$H_3C—\overset{O}{\overset{\|}{C}}—O—CH_3 + OH^- \longrightarrow H_3C—\overset{O}{\overset{\|}{C}}—O^- + H_3COH$$

methyl acetate — acetate ion — methyl alcohol

and experiments show that the reaction has the rate expression

$$\text{Initial reaction rate} = k[H_3CCOOCH_3][OH^-]$$

where k is 0.137 L/(mol · s) at 25 °C. The expression tells us that the reaction is first order with respect to $H_3CCOOCH_3$ *and* first order with respect to OH^-. Since the sum of exponents is 2, it is **second order** *overall*. As the table in the margin illustrates, this means that if you double the concentration of either $H_3CCOOCH_3$ or OH^-, the initial reaction rate will double. If both concentrations are doubled, the rate will increase by a factor of four.

Initial Reaction Rate for Methyl Acetate $+OH^-$

$[H_3CCOOCH_3]$	$[OH^-]$	INITIAL REACTION RATE mol/(L · s)
0.050 M	0.050 M	0.00034
	↓ × 2	↓ × 2
0.050 M	0.10 M	0.00069
↓ × 2		↓ × 2
0.10 M	0.10 M	0.0014

First and second order reactions are quite common, but other orders are observed as well. One thoroughly studied reaction is the oxidation of NO to give NO_2, a **third order** reaction.

$$2\,NO(g) + O_2(g) \longrightarrow 2\,NO_2(g)$$
$$\text{Rate} = k[NO]^2[O_2]$$

Here the sum of the exponent on [NO] (= 2) and the exponent on $[O_2]$ (= 1) is 3. Can you verify that the rate of the reaction will go up by a factor of 8 if you double the concentrations of both reactants?

__Fractional orders__ are also seen on occasion, as in the combination of H_2 *and* Br_2 *to give HBr where the total reaction order is* $\frac{3}{2}$.

Finally, reactions can be **zero order** in a given reactant. In such cases, any change in the concentration of the reactant has no effect on the rate of the reaction. That is, the rate law is simply "Initial reaction rate = k;" no reactant concentrations appear in the rate expression.

Remember that reaction orders do not necessarily have any relation to the stoichiometric coefficients in the overall, balanced equation.

DETERMINATION OF THE RATE EXPRESSION

One way to determine the rate expression for a reaction is to determine how the initial rate varies as the initial concentration of one reactant is changed from one experiment to the next while the other reactants have fixed initial concentrations. As an example, let us analyze experimental data for a reaction

$$a\text{A} + b\text{B} \longrightarrow x\text{X}$$

$$\text{Initial reaction rate} = k[\text{A}]_0^m[\text{B}]_0^n$$

where $[\text{A}]_0$ and $[\text{B}]_0$ are initial concentrations, and where we assume the initial reaction rate does not depend on [X], but only on $[\text{A}]_0$ and $[\text{B}]_0$. To see just what this dependence is, we can measure the change in the concentration of A or B for a small time interval near the start of the reaction (before [X] becomes important). Therefore, in the first three lines of data in Table 15.1, we have varied $[\text{A}]_0$ but kept $[\text{B}]_0$ fixed. Even though $[\text{A}]_0$ was doubled from Experiment 1 to Experiment 3, the initial reaction rate was unchanged. The initial rate is independent of $[\text{A}]_0$, so m must be 0 (since $[\text{A}]_0^0 = 1$); that is, the reaction is zero order in A.

Now consider the dependence of initial rate on $[\text{B}]_0$. In Experiments 1, 4, and 5 in Table 15.1, $[\text{A}]_0$ is fixed and $[\text{B}]_0$ varies. From Experiment 1 to 4 $[\text{B}]_0$ doubles and so does the initial rate. Thus, the change in rate is proportional to the change in $[\text{B}]_0$, so $n = 1$, and the reaction is first order in B. (Does Experiment 5 also agree with an order of 1 for B?)

In summary, the dependence of rate on the concentrations in Table 15.1 gives a rate expression of

$$\text{Initial reaction rate} = k[\text{A}]_0^0[\text{B}]_0^1 = k[\text{B}]_0$$

and so the rate expression is "Initial reaction rate $= k[\text{B}]_0$." Here the overall reaction order is 1, since $m + n = 0 + 1 = 1$.

With the rate expression determined, the value of k can be found. In this case, if we use the results for Experiment 3, we have

$$0.0050 \text{ mol/(L} \cdot \text{hr)} = k(1.00 \text{ mol/L})^0(0.20 \text{ mol/L})^1$$

and so $k = 2.5 \times 10^{-2}$/hr. We should of course obtain the same value for k using any set of experimental data. Indeed, this is one way to prove that we have derived the correct rate expression; if we had arrived at the wrong rate expression, the value of k would vary from experiment to experiment and tell us that we had made an error.

The order of reaction with respect to a particular reagent can often be found simply by inspecting the experimental data. However, there are times when a more rigorous algebraic method—outlined below—is necessary.

Table 15.1 Rate Data for the Reaction $a\text{A} + b\text{B} \rightarrow x\text{X}$

Experiment Number	Initial Rate mol/(L · hr)	Initial Concentrations (mol/L) $[\text{A}]_0$	$[\text{B}]_0$
1	0.50×10^{-2}	0.50	0.20
2	0.50×10^{-2}	0.75	0.20
3	0.50×10^{-2}	1.00	0.20
4	1.00×10^{-2}	0.50	0.40
5	1.50×10^{-2}	0.50	0.60

EXAMPLE 15.2

DETERMINATION OF REACTION ORDER

Derive the rate expression and the value of k for the decomposition of acetaldehyde

$$CH_3CHO(g) \longrightarrow CH_4(g) + CO(g)$$

using the data below. Experiment shows that CH_4 and CO are not involved in the rate expression.

EXPERIMENT NUMBER	INITIAL CONCENTRATION OF CH_3CHO (mol/L)	INITIAL RATE mol/(L · s)
1	0.10	0.020
2	0.20	0.080
3	0.30	0.180
4	0.40	0.320

Solution Here the rate expression must be

$$\text{Initial reaction rate} = k[CH_3CHO]^m$$

where m, the reaction order, and k, the rate constant, are unknown. Inspecting the data above, you see that the initial rate increases by a factor of 4 when the initial concentration doubles. For example, the CH_3CHO concentration doubles from Experiment 1 to Experiment 2, and the rate increases by (0.080/0.020) = 4. Since the rate increases as the *square* of the concentration, the value of m must be 2, and the reaction order is 2. This means the rate expression is

$$\text{Initial reaction rate} = k[CH_3CHO]^2$$

Now that the rate expression is known, we can substitute one set of experimental information into the expression and find k. Taking data from Experiment 1, we have

$$0.020 \text{ mol/(L} \cdot \text{s)} = k\ (0.10 \text{ mol/L})^2$$

and so

$$k = 2.0 \text{ L/(mol} \cdot \text{s)}$$

EXAMPLE 15.3

ALGEBRAIC METHOD FOR FINDING REACTION ORDERS

In this example we shall use the same data as in Example 15.2, but now we shall use an algebraic approach.

Solution The first step here is to take the ratio of rates for any two experiments, say 1 and 2.

$$\frac{\text{Rate 1}}{\text{Rate 2}} = \frac{0.020 \text{ mol/(L} \cdot \text{s)}}{0.080 \text{ mol/(L} \cdot \text{s)}} = \frac{k[CH_3CHO]_1^m}{k[CH_3CHO]_2^m} = \frac{k(0.10)^m}{k(0.20)^m}$$

This simplifies to

$$\frac{1}{4} = \left(\frac{1}{2}\right)^m$$

Although it may be obvious now that $m = 2$ (since $\frac{1}{2}$ squared is $\frac{1}{4}$), a general approach here is to take the logarithm of each side of the equation.

$$\log\left(\frac{\text{rate 1}}{\text{rate 2}}\right) = \log\left(\tfrac{1}{4}\right) = \log\left(\tfrac{1}{2}\right)^m$$

and so

$$\log\left(\tfrac{1}{4}\right) = m \log\left(\tfrac{1}{2}\right)$$

and

$$(-0.602) = m(-0.301)$$

which gives a value of $m = 2$.

EXERCISE 15.2 Reaction Order

In the following reaction, a Co—Cl bond is replaced by a Co—OH_2 bond.

$$[Co(NH_3)_5Cl]^{2+}(aq) + H_2O(\ell) \longrightarrow [Co(NH_3)_5H_2O]^{3+}(aq) + Cl^-(aq)$$

$$\text{Initial rate} = k\{[Co(NH_3)_5Cl]^{2+}\}^m$$

Using the data below, find the value of m in the rate expression and calculate the value of k.

EXPERIMENT	INITIAL CONCENTRATION OF $[Co(NH_3)_5Cl]^{2+}$ (mol/L)	INITIAL RATE mol/(L · min)
1	1.0×10^{-3}	1.3×10^{-7}
2	2.0×10^{-3}	2.6×10^{-7}
3	3.0×10^{-3}	3.9×10^{-7}
4	1.0×10^{-2}	1.3×10^{-6}

15.4 CONCENTRATION/TIME RELATIONSHIPS

Chemists often wish to know how long a reaction must proceed to reach a predetermined concentration of some reagent, or what the reactant and product concentrations will be after some time has elapsed. One way to do this is to collect experimental data and construct curves such as those in Figure 15.4. However, this can be inconvenient and time-consuming. A simpler approach is to determine the initial reaction order for each reagent by experiment and then derive an equation that relates concentration and time. This equation will then allow us to calculate a concentration at any time, or vice versa, during the early part of the reaction.

FIRST AND SECOND ORDER CONCENTRATION/TIME EQUATIONS

As an example of a first order process, let us return to a simple reaction such as the transformation of a *cis* compound to a *trans* compound (page 613).

$$A \longrightarrow B \qquad \text{where} \qquad \text{Initial reaction rate} = k[A]$$

Equation 15.1 was also developed in the context of nuclear chemistry in Chapter 7 (page 259). In that chapter we were concerned with the rate of decay of radioactive isotopes.

Using the methods of calculus, this equation can be transformed into a very useful expression.

$$\ln \frac{[A]}{[A]_0} = -kt \tag{15.1}$$

If the reaction is first order with respect to A, and if $[A]_0$ is the initial concentration of the reactant (at time $t = 0$), then Equation 15.1 allows us to calculate [A], the concentration of A at some other time t. Alternatively, it can be used to calculate the time elapsed until A reaches some predetermined concentration.

EXAMPLE 15.4

THE FIRST ORDER RATE EQUATION

Cyclopropane, C_3H_6, has been used in a mixture with oxygen as an anesthetic. (However, this practice has diminished greatly because the compound is very flammable.) It rearranges to a different molecule of the same formula, propene.

$$\underset{\text{cyclopropane}}{\text{cyclo-}(CH_2)_3} \longrightarrow \underset{\text{propene}}{CH_3{-}CH{=}CH_2} \qquad \text{Rate} = k[\text{cyclo}]$$

If the initial concentration of cyclopropane is 0.050 M, how many hours must elapse for the concentration of the compound to drop to 0.010 M? (The rate constant k is 5.4×10^{-2}/hr.)

Solution The first order rate equation applied to this reaction is

$$\ln \frac{[\text{cyclo}]}{[\text{cyclo}]_0} = -kt$$

where [cyclo], $[\text{cyclo}]_0$, and k are given.

$$\ln \left(\frac{0.010}{0.050}\right) = -(5.4 \times 10^{-2}/\text{hr})t$$

$$\frac{-\ln(0.20)}{5.4 \times 10^{-2}/\text{hr}} = t$$

$$\frac{-(-1.61)}{5.4 \times 10^{-2}/\text{hr}} = t$$

$$t = 30.\ \text{hours}$$

EXAMPLE 15.5

USING THE FIRST ORDER EQUATION

In dilute NaOH at 20 °C, the decomposition of hydrogen peroxide is first order in H_2O_2 only,

$$2\ H_2O_2(aq) \longrightarrow 2\ H_2O(\ell) + O_2(g) \qquad \text{Rate} = k[H_2O_2]$$

and the rate constant is 1.06×10^{-3}/min. If the initial concentration of H_2O_2 is 0.020 M, what is the concentration after exactly 100 minutes?

Solution Here Equation 15.1 is written as

$$\ln \frac{[H_2O_2]}{[H_2O_2]_0} = -kt$$

where $[H_2O_2]_0$, k, and t are known. To calculate $[H_2O_2]$, the peroxide concentration after 100. minutes, it is convenient to rearrange the equation to

$$\ln[H_2O_2] - \ln[H_2O_2]_0 = -kt$$

Substituting into this expression, we have

$$\ln[H_2O_2] - \ln(0.020) = -(1.06 \times 10^{-3}/\text{min})(100.\ \text{min})$$

$$\ln[H_2O_2] - (-3.91) = -(1.06 \times 10^{-1})$$

$$\ln[H_2O_2] = -3.91 - 0.106$$

$$\ln[H_2O_2] = -4.02$$

Taking the antilogarithm of −4.02, we find the concentration of peroxide after 100. minutes as

$$[H_2O_2] = 0.018\ \text{M}$$

EXERCISE 15.3 Using the First Order Rate Equation

The sugar sucrose decomposes in acid solution to the simpler sugars glucose and fructose. The rate expression is "Initial reaction rate = k[sucrose]" where the rate constant is 0.21/hr. If the original concentration of sucrose is 0.010 M, what is its concentration after 5.00 hours?

The First Order Equation, Equation 15.1

Equation 15.1 is an extremely useful expression, and there are several additional things about the equation that you may wish to know.

1. The decay of radioactive nuclei is a first order reaction. Equation 15.1 was developed in Chapter 7 as a way to find the change in number of radioactive nuclei present after a certain amount of time has elapsed.
2. Equation 15.1 is valid for cases where the product B does not revert to the reactant A. This is assumed to be true for all the examples in this chapter.
3. The abbreviation ln represents the natural logarithm (see Appendix A). In base-10 logarithms, Equation 15.1 is

$$2.303 \log \frac{[A]}{[A]_0} = -kt$$

4. If you have had some calculus, you can see how to derive Equation 15.1. For a first order process in terms of reactant A, the initial rate is $-(\Delta[A]/\Delta t) = -(d[A]/dt) = k[A]$. After rearranging the equation we integrate it between the limits time = 0 and time = t

$$\int_{[A]_0}^{[A]} \frac{d[A]}{[A]} = -k \int_{t=0}^{t=t} dt$$

which gives $\{\ln[A] - \ln[A]_0\} = -k(t - 0)$. When this is rearranged, Equation 15.1 is obtained.

The time dependence of concentration is different for a second order reaction than for a first order reaction. For example, in the case where the rate expression is

Initial reaction rate $= k[A]^2$

the concentration/time equation is

$$\frac{1}{[A]} - \frac{1}{[A]_0} = kt \tag{15.2}$$

This same equation also applies to a reaction such as "A + B → products" where the rate expression is

Initial reaction rate $= k[A]_0[B]_0$

and the special condition is met that the initial concentrations of A and B are the same. Example 15.6 and Exercise 15.4 illustrate the use of this equation.

EXAMPLE 15.6

USING THE SECOND ORDER CONCENTRATION/TIME EQUATION

The gas phase decomposition of HI

$$HI(g) \longrightarrow \tfrac{1}{2}\,H_2(g) + \tfrac{1}{2}\,I_2(g)$$

has the rate expression

Initial rate $= k[HI]^2$

where $k = 30.$ L/(mol · min) at 443 °C. How long must we wait for the concentration of HI to fall from 0.010 M to 0.0050 M at 443 °C?

Solution Here $[A]_0 = [HI]_0 = 0.010$ M, and $[A] = [HI] = 0.0050$ M. Substituting into Equation 15.2,

$$\frac{1}{0.0050\text{ mol/L}} - \frac{1}{0.010\text{ mol/L}} = 30.\text{ L/(mol} \cdot \text{min)}t$$

$$2.0 \times 10^2\text{ L/mol} - 1.0 \times 10^2\text{ L/mol} = 30.\text{ L/(mol} \cdot \text{min)}t$$

$$t = 3.3\text{ minutes}$$

EXERCISE 15.4 Using the Second Order Concentration/Time Equation

Using the rate constant for HI decomposition given in Example 15.6 (for 443 °C), calculate [HI] after 10. minutes have elapsed.

HALF-LIFE AND REACTION RATE FOR FIRST ORDER REACTIONS

The rate constant k is a good indicator of the speed of a chemical reaction, and it is a compact way of comparing one reaction with another. For example, consider the first order replacement of Cl^- by H_2O in two platinum compounds, cisplatin

$$\underset{cis}{\text{(H}_3\text{N)}_2\text{PtCl}_2} + H_2O \longrightarrow [\text{Pt(NH}_3)_2\text{(OH}_2)\text{Cl}]^+ + Cl^-$$

$k = 2.5 \times 10^{-5}$ per second

and the closely related *trans* compound (where the Pt—Cl bonds are opposite each other) at the same temperature.

$$\underset{trans}{\text{(H}_3\text{N)}_2\text{PtCl}_2} + H_2O \longrightarrow [\text{Pt(NH}_3)_2\text{(OH}_2)\text{Cl}]^+ + Cl^-$$

$k = 9.8 \times 10^{-5}$ per second

The values of k show that the *trans* compound reacts with water about four times faster than the *cis* compound.

Another useful measure of reaction speed is the **reaction half-life, $t_{1/2}$**. As we shall now show, k and $t_{1/2}$ are related to one another in simple ways. *The half-life of a reaction is the time required for the concentration of one of the reactants to decrease to half of its initial value.* Specifically, for reactant A in a reaction that is first order in A, $t_{1/2}$ is the time when

$$[A] = \tfrac{1}{2}[A]_0 \quad \text{or} \quad \frac{[A]}{[A]_0} = \tfrac{1}{2}$$

where $[A]_0$ is the initial concentration and [A] is the concentration after the reaction is half completed. To find $t_{1/2}$ we use the concentration/time equation for a first order reaction (Equation 15.1) and proceed as follows. Taking the first order rate equation,

$$\ln \frac{[A]}{[A]_0} = -kt$$

and substituting the fact that $[A] = \tfrac{1}{2}[A]_0$ when $t = t_{1/2}$, we have

$$\ln \tfrac{1}{2} = -kt_{1/2}$$

or

$$\ln 2 = 0.693 = kt_{1/2}$$

Solving this for $t_{1/2}$, we come to the very useful equation that relates half-life and the first order rate constant.

$$t_{1/2} = \frac{0.693}{k} \tag{15.3}$$

We have described the concept of half-life, and Equation 15.3, in connection with the decay of radioactive isotopes in Chapter 7 (pages 259–261).

To illustrate the concept of half-life for first order reactions, we have plotted (Figure 15.5) the concentration of H_2O_2 as a function of time for the decomposition of hydrogen peroxide.

$$2\ H_2O_2(aq) \longrightarrow 2\ H_2O(\ell) + O_2(g)$$

$$\text{Initial reaction rate} = k[H_2O_2]$$

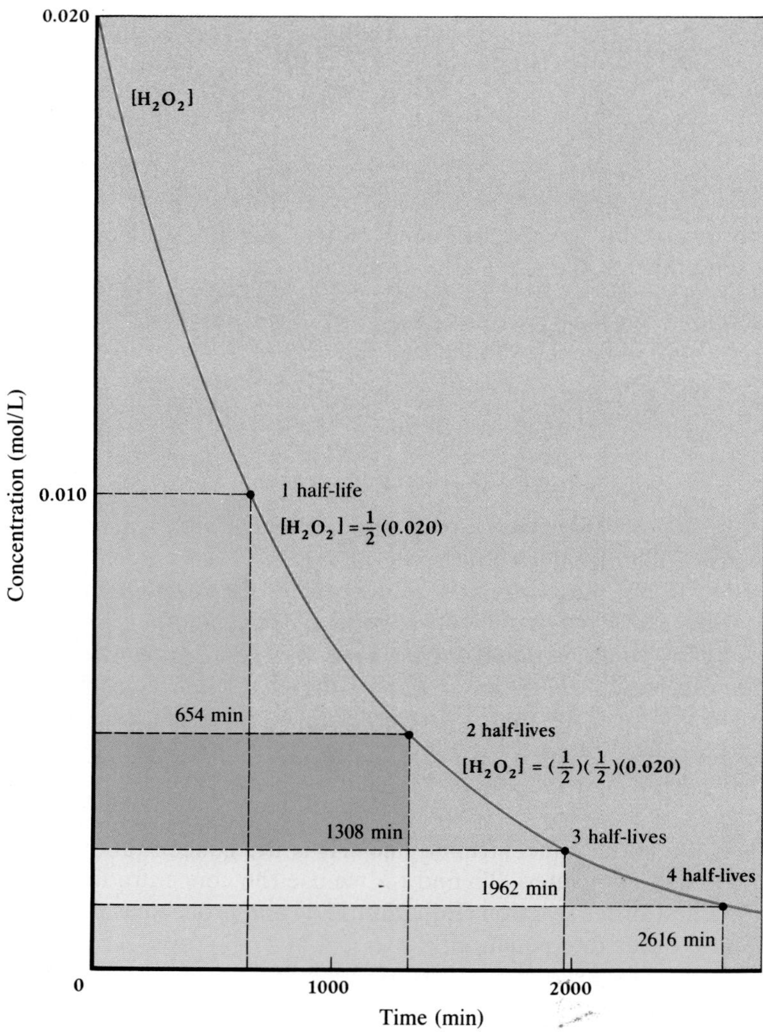

Figure 15.5 The concentration versus time curve for the disappearance of H_2O_2 ($k = 1.06 \times 10^{-3}$/min). At 654 minutes ($t_{1/2}$) the initial concentration of H_2O_2 has halved. During each successive interval of 654 minutes the concentration again halves.

Since the rate constant k for this reaction is 1.06×10^{-3}/min, the half-life is

$$t_{1/2} = \frac{0.693}{k} = \frac{0.693}{1.06 \times 10^{-3}/\text{min}} = 654 \text{ min}$$

If the initial concentration of H_2O_2 is 0.020 M, then the concentration is 0.010 M after 654 minutes. Notice that the concentration drops again by half after another 654 minutes. That is, after two half-lives (1308 minutes) the concentration is only $(\frac{1}{2})(\frac{1}{2}) = (\frac{1}{2})^2 = \frac{1}{4}$ or 25% of the initial concentration. After three half-lives (1962 minutes), the concentration has now dropped to $(\frac{1}{2})(\frac{1}{2})(\frac{1}{2}) = (\frac{1}{2})^3 = \frac{1}{8}$ or 12.5% of the initial value; here $[H_2O_2]$ = 0.0025 M.

The concept of the half-life of a first order process is widely used, especially when referring to the rate of decay of radioactive isotopes as described in Chapter 7. Here we wish to apply it to some other first order reactions.

EXAMPLE 15.7

HALF-LIFE

Sucrose, $C_{12}H_{22}O_{11}$, decomposes to fructose and glucose in acid solution with the rate law

$$\text{Rate} = k[\text{sucrose}] \qquad k = 0.208/\text{hr at } 25\ ^\circ\text{C}$$

Find the half-life of sucrose under these conditions. Calculate the time required for 87.5% of the initial concentration of sucrose to disappear.

Solution The half-life for the change is

$$t_{1/2} = \frac{0.693}{k} = \frac{0.693}{0.208/\text{hr}} = 3.33 \text{ hr}$$

As illustrated for a different reaction in Figure 15.5, the concentration of a reactant in a first order process has dropped to 12.5% of its original value after three half-lives. This is the same as saying that 87.5% of the reactant has been consumed. Therefore, the time required for sucrose to reach this concentration is 3 × 3.33 hr = 9.99 hr.

EXERCISE 15.5 Half-Life

The rate constant for the transformation of cyclopropane to propene (see Example 15.4) is 5.40×10^{-2}/hr. What is the half-life of this reaction? What fraction of the cyclopropane remains after 51.2 hours? What fraction remains after 18.0 hours?

GRAPHICAL METHODS FOR DISTINGUISHING FIRST AND SECOND ORDER REACTIONS

In Section 15.3 you learned how to analyze experimental data to determine reaction order. However, an even more convenient way is suggested by Equations 15.1 and 15.2. Rearranging these equations slightly,

First order

$$\ln[A] = -kt + \ln[A]_0$$
$$\downarrow \quad \downarrow\downarrow \quad \downarrow$$
$$y = ax + b$$

Second order

$$\frac{1}{[A]} = kt + \frac{1}{[A]_0}$$
$$\downarrow \quad \downarrow\downarrow \quad \downarrow$$
$$y = ax + b$$

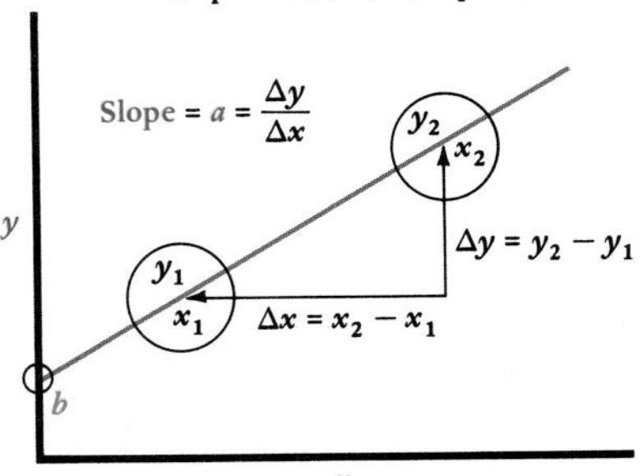

you see that each is an equation of a straight line, $y = ax + b$, where a is the slope of the line and b is the intercept (the value of y when $x = 0$).

A plot of concentration versus time for a first order reaction is a curved line (Figures 15.4 and 15.5). However, *if experimental rate data are plotted as ln[reactant] versus time, a straight line will result when the reaction is **first order** in that reactant*. In Figure 15.6 we have plotted the data for the first order decomposition of H_2O_2 in this way and do indeed see a straight line. Notice that the slope is negative, because there is a

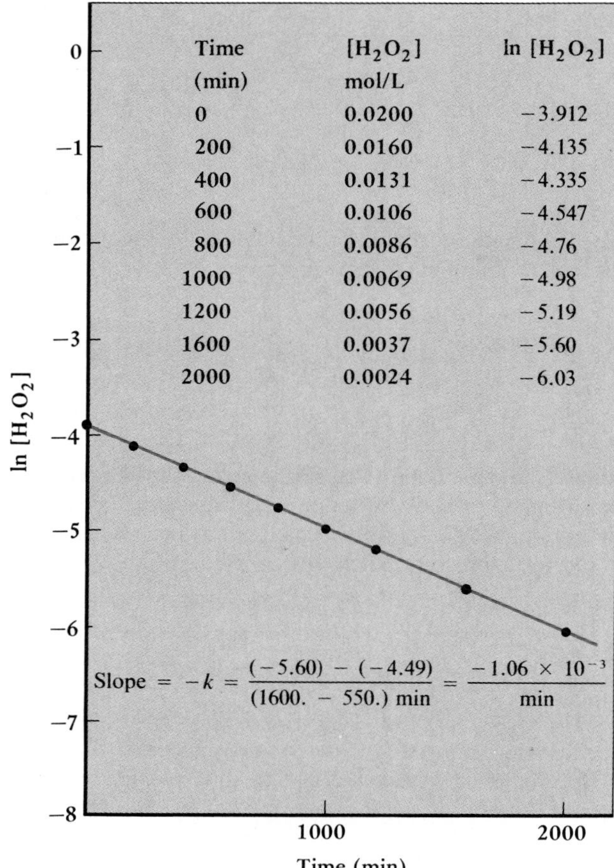

Figure 15.6 Plot of $\ln[H_2O_2]$ versus time. The straight line plot indicates a first order reaction. The rate constant $k = -\text{slope}$.

negative sign in the $-kt$ term and k is positive). *Only* for a first order reaction will you observe a straight line of negative slope if you plot the logarithm of concentration against time. Not only can you pinpoint the reaction order this way, but you can find the rate constant from the slope of the plot. That is, the rate constant $k = -\text{slope}$.

The decomposition of NO_2 is a second order process,

$$NO_2(g) \longrightarrow NO(g) + \tfrac{1}{2}\, O_2(g)$$

$$\text{Rate} = k[NO_2]^2$$

and concentration versus time data have been plotted in Figure 15.7 in two ways: (a) $\ln[NO_2]$ versus time and (b) $1/[NO_2]$ versus time. Now you see that a plot of $1/[NO_2]$ gives a straight line as predicted for a second order process. Conversely, plotting the data as though it were a first order reaction clearly gives a curved line (Figure 15.7a). These examples of two commonly observed rate expressions illustrate a useful way to determine reaction order: A straight line is observed *only* when $\ln[A]$ is plotted against time for a first order process (Rate $= k[A]$) or when $1/[A]$ is plotted against time for a second order process with a rate expression of the type Rate $= k[A]^2$.

More complex rate expressions such as Rate $= k[A][B]$ and Rate $= k[A]^2[B]$ will not give linear plots for $\ln[A]$ or $1/[A]$ versus time.

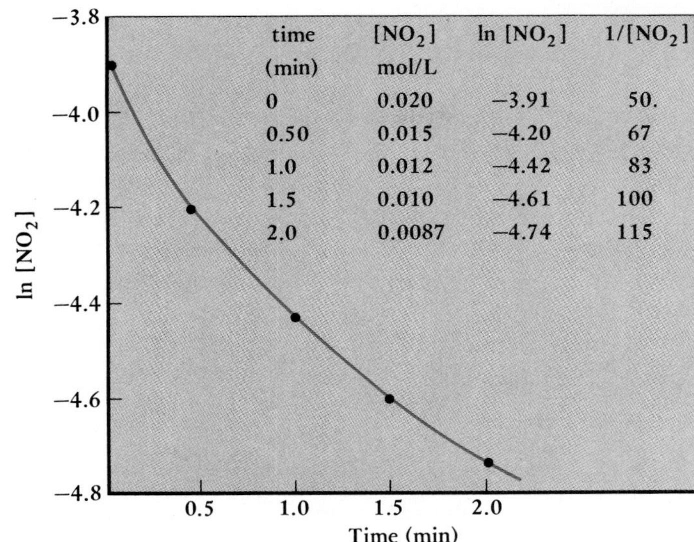

time (min) | [NO_2] mol/L | ln [NO_2] | 1/[NO_2]
0 | 0.020 | −3.91 | 50.
0.50 | 0.015 | −4.20 | 67
1.0 | 0.012 | −4.42 | 83
1.5 | 0.010 | −4.61 | 100
2.0 | 0.0087 | −4.74 | 115
ln [NO_2]
−3.8
−4.0
−4.2
−4.4
−4.6
−4.8
0.5
1.0
1.5
2.0
Time (min)

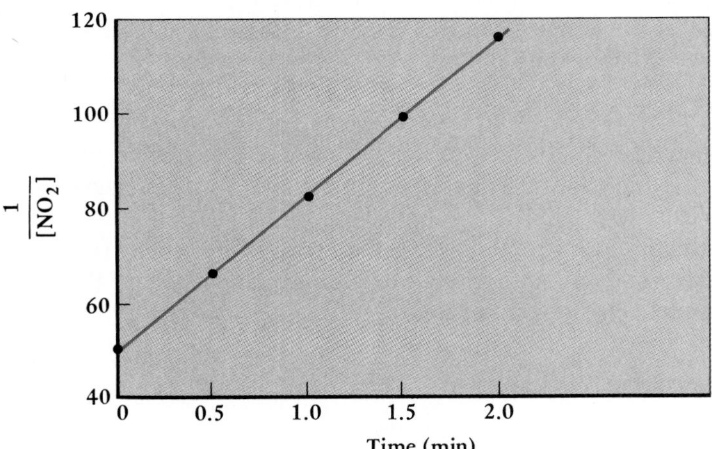

Figure 15.7 Concentration versus time curves for NO_2 decomposition. (a) ln[NO_2] versus time. The curved line shows that the reaction is not first order. (b) 1/[NO_2] versus time. For a rate expression of the form Rate = $k[A]^2$, a straight line is observed when 1/[A] is plotted against time.

EXERCISE 15.6 Determination of Reaction Order by a Graphical Method

For the reaction of $(CH_3)_3CBr$ with OH^-,

$$(CH_3)_3CBr + OH^- \longrightarrow (CH_3)_3COH + Br^-$$

the following data were obtained in the laboratory.

TIME (seconds)	[$(CH_3)_3CBr$] (mol/L)
0	0.10
30	0.074
60	0.055
90	0.041

Plot these data as ln[$(CH_3)_3CBr$] versus time and 1/[$(CH_3)_3CBr$] versus time. Is the reaction first order or second order? What is the value of the rate constant k?

Figure 15.8 Hydrolysis of a green cobalt(III) compound. The compound has two Co—Cl bonds, one of which is replaced by a bond between Co^{3+} and H_2O to give a red compound. The reaction is slow at room temperature, so the solution at the right has remained green. However, if the solution is placed in warm water, the rate of Cl^- replacement increases, and the solution soon turns pink. (See also Figure 16.8 on page 685, which illustrates the effect of a catalyst on this reaction.)

15.5 RATE CHANGES WITH TEMPERATURE

It is generally observed that reactions occur more rapidly at higher temperatures than at lower temperatures (Figure 15.8). The decomposition of dinitrogen pentoxide, for example,

$$2\ N_2O_5(g) \longrightarrow 4\ NO_2(g) + O_2(g)$$

has a rate constant of 3.46×10^{-5}/s at 25 °C, while it is 1.35×10^{-4}/s at 35 °C, an increase in rate by a factor of almost four for a 10-degree rise in temperature.

In this section we want to explore the theories of the way reactions occur in order to understand, among other things, the cause of the temperature dependence of reaction rates.

TRANSITION STATE THEORY

Understanding why there are wide variations in reaction rates and why reactions are faster at higher temperatures begins with the idea that there is an energy barrier to all reactions. Consider the following analogy. Suppose you have taken a basketball to the observation deck of the Eiffel Tower in Paris, France. Rolling the ball toward the edge does not mean the ball will fall to the ground, even though there is a great driving force to do so, because there is a barrier around the deck. Only if the ball has sufficient vibrational or "bouncing energy" is it able to get over the barrier and begin its plunge to the ground. If several bags of balls were emptied onto the observation deck, only those bouncing vigorously enough would get over the barrier. The number of balls falling from the top of the tower in some time (the rate) would depend on the number of balls (the concentration) and the fraction of the total that bounced with enough energy (Figure 15.9).

Figure 15.9 To get a ball to fall from the top of a tower, we first have to get the ball over the railing at the top of the tower.

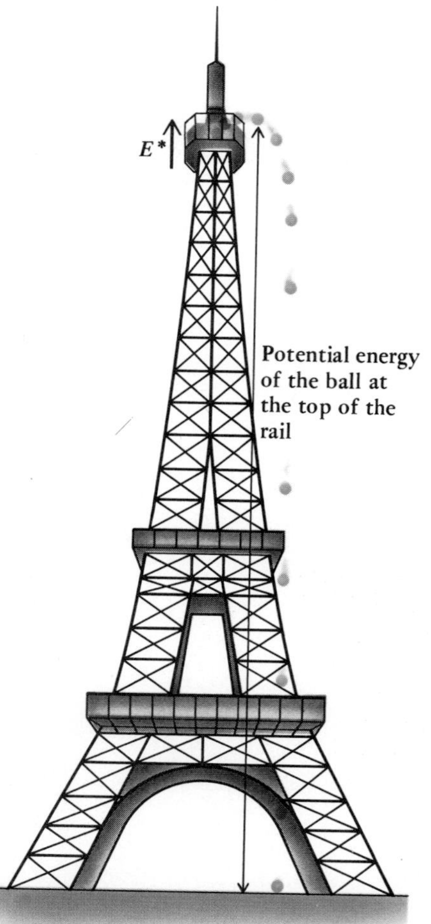

In chemical reactions, vibrating atoms and speeding molecules are like the basketballs on the Eiffel Tower. Chemical reactions also encounter energy barriers to the separation of atoms or to the transfer of an atom between two molecular pieces. When the barriers to reaction are too high, the reaction cannot occur at a detectable rate because too few molecules are sufficiently energetic to surmount the barrier.

Chemists illustrate the notion of a barrier with a **reaction energy diagram**. To illustrate the concept, we will use the following bond-breaking reaction.

$$H_3C—Cl \longrightarrow H_3C^+ + Cl^-$$

In the course of the reaction, which is illustrated in the computer drawings below, the H—C—H bond angles increase, and the C—Cl bond lengthens and finally breaks.

The change in energy of the system as the C—Cl bond lengthens and finally breaks is illustrated in Figure 15.10. The diagram reflects the fact that energy must be supplied to change the geometry of the molecule and lengthen the C—Cl bond. The energy of the system finally reaches a maximum as the C—Cl bond is broken. Then, after the Cl^- ion has departed, the C—H bonds shorten and strengthen, and the energy of the system drops.

The energy at the maximum of the reaction energy diagram is called the **transition state energy**, E_{TS}. The reacting molecule with this energy has been deformed so that it can react; it is at a special *transition point* between reactants and products. The deformed molecule in the transition state is called the **activated complex** because the reacting molecule or molecules have been *activated* to reaction. Finally, the difference in en-

The C—Cl Bond Breaking Reaction for Methyl Chloride, Cl—CH_3

A computer has calculated an electron density map for the molecule when the C—Cl bond distance is that in the unexcited molecule (174 pm) (left), when the C and Cl are beginning to pull apart and the C—Cl distance is 250 pm (center), and finally when the bond has broken and the Cl^- ion is 350 pm from the plane of the CH_3^+ ion. Notice that the CH_3 portion of the molecule flattens in the course of the reaction, as expected on the basis of VSEPR theory. This set of photos is a simple representation of the work that chemists now do in using computers to model chemical reactions. The photos have been placed at differing levels to indicate the increase in energy as the C—Cl bond lengthens and breaks, and then the slight decline in energy as the Cl^- and CH_3^+ ions separate.

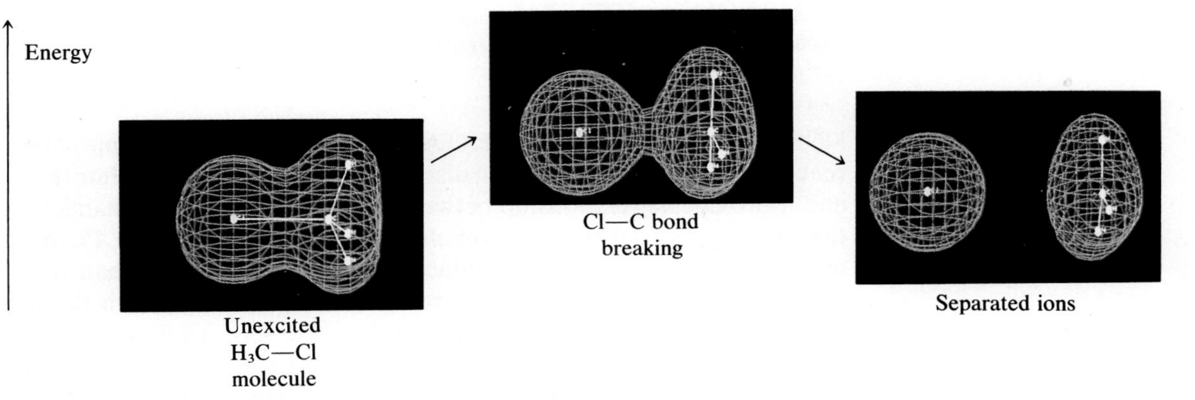

(These simulations were done at the Centre Europeen de Visualisation de Donnees CCSJ Marseille by M. Rajzmann, V. Lazzeri, and A. Boch.)

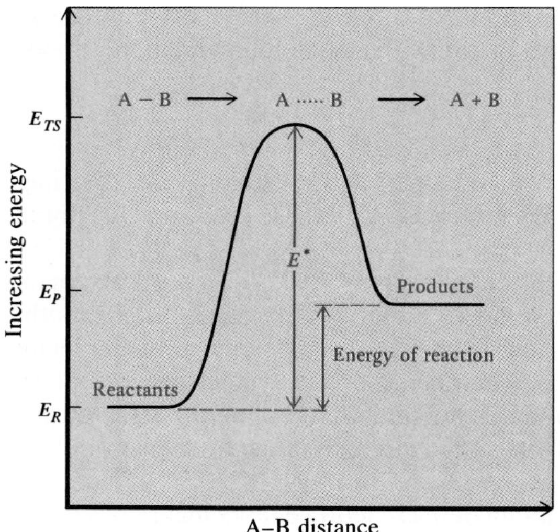

Figure 15.10 Energy diagram for a reaction step. Reactant energies, E_R, are given by the "foot" of the barrier at the left, and product energies, E_P, are given by the "foot" of the barrier at the right. The net reaction energy is given by $E_P - E_R$; the difference may have a positive sign (endothermic reaction) or negative sign (exothermic reaction). The activation energy for the forward reaction E_f^* (reactants → products) is the difference between the transition state energy (E_{TS}) and the reactant energy, $E_f^* = E_{TS} - E_R$.

ergy between the undeformed reactant molecule or molecules and the transition state is called the **activation energy**, E^*. This is the energy that the reacting molecules must absorb from their environment in order to react. At any temperature above zero kelvin, molecules have increasingly energetic bond vibrations. In the case of the C—Cl bond-breaking reaction above, the bond can break if the amplitude of the bond vibration is large enough, but this can occur only if the molecule absorbs enough energy.

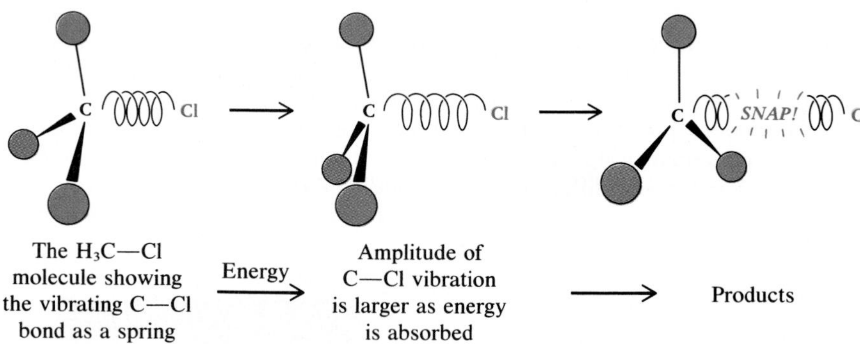

KINETICS AND NET ENERGY OF REACTION The theoretical approach to reaction rates that we have been discussing is called **transition state theory**, and it provides a relationship between kinetics and thermodynamics. As the reacting molecules pass over the energy barrier in Figure 15.10, the energy drops to that of the products, E_P, which is *greater* than that of the reactants, E_R. Less energy is released on descending from the transition state than was absorbed on climbing to the state. Thus, *a net amount of energy is consumed*, and the reaction is *endo*thermic.

In contrast, the reaction illustrated in Figure 15.11 is *exo*thermic. Here the reaction of 1 mole of H_2 and 1 mole of I_2 to give 2 moles of HI has an activation energy of 170. kJ.

$$H_2(g) + I_2(g) \xrightarrow{E^* = 170.\ kJ} 2\ HI(g)$$

The energy of the products is lower than that of the reactants, so the reaction is exothermic by approximately 13 kJ ($\Delta E_{rxn} \approx -13$ kJ).

When the reacting molecules have accumulated sufficient energy and are poised at the top of the activation energy barrier, they have no "memory" of where they came from, so they may go on to products or fall back down the barrier to give reactants again. This implies that *any reaction and its exact reverse must proceed through the same activated complex*. That is, if two molecules of HI combine to form H_2 and I_2, then the system proceeds through the same activated complex as $H_2 + I_2 \rightarrow 2\ HI$. Further, if we know the activation energy and ΔE_{rxn} for the reaction in one direction, then we know these quantities for the reverse reaction. For the forward direction, we know

$$H_2 + I_2 \xrightarrow{E_f^* = 170.\ kJ} 2\ HI + 13\ kJ$$

Therefore, in the other direction, the reaction is endothermic by 13 kJ,

$$2\ HI + 13\ kJ \xrightarrow{E_r^* = 170.\ kJ + 13\ kJ = 183\ kJ} H_2 + I_2$$

and the activation energy must be 183 kJ. Figure 15.11 illustrates the relation among these quantities. An analogy to this situation is found in our Eiffel Tower example that began this section. To drop a ball from the top of the tower to the ground, we first have to put in some energy, E^*, to get the ball over the railing around the deck. Then, as the ball falls to the ground, all of its potential energy at the railing (E_{pot}) is *released* in the form of kinetic energy. The net energy change for the process is $E^* - E_{pot}$, and the net outcome is surely negative, an outpouring of energy. To put the ball back onto the deck of the tower starting from the ground, you have to give it a large amount of activation energy ($= E_{pot}$). Since the ball lost an amount of energy we called E_{pot} on falling from the top of the railing to the ground, this is what you would have to put back in

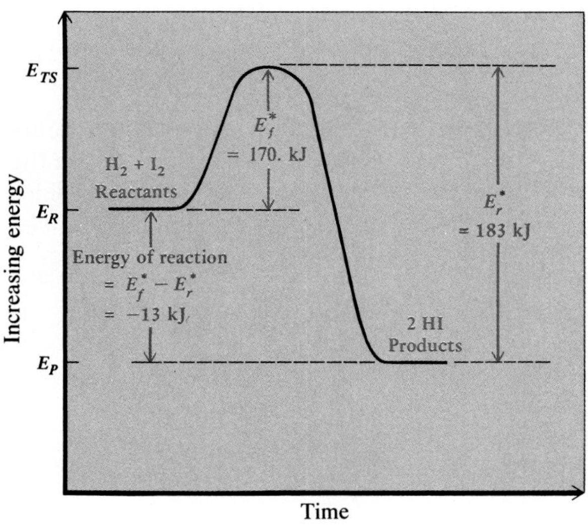

Figure 15.11 Reaction energy diagram for an exothermic process. Here 1 mole of each of H_2 and I_2 produce 2 moles of HI with an activation energy (E_f^*) of 170. kJ. The reverse reaction, the decomposition of 2 moles of HI to give H_2 and I_2, has an activation energy (E_r^*) of 183 kJ. Notice that the net energy change for $H_2 + I_2 \rightarrow 2\ HI$ is simply the difference in activation energies for the two directions; that is, $E_f^* - E_r^* = -13$ kJ.

to get it from the ground to the top of the railing. Then, as the ball dribbles off the railing onto the deck, it would release a little of its potential energy, the amount lost being equal to energy we called E^* above. The net energy change is now $E_{pot} - E^*$, a large positive number.

COLLISION THEORY OF REACTION RATES

The activation energy largely determines the great variation observed in rates of different reactions at a given temperature. In addition, it lies at the heart of the explanation of why the rate of a reaction can change with temperature. To explain all this, we turn to **collision theory**. This theory assumes that a *minimum* condition for two molecules A and B to react is that they must come in contact with each other; that is, they must *collide*! The number of collisions that occur between two gas phase molecules under given conditions can be calculated from the kinetic theory of gases. For example, if both gases are at 1 atm pressure and at 0 °C, the collision rate is about 10^8 mol/(L · s). This corresponds to an enormous rate of assault on the barrier to reaction. In fact, if two reacting gases were mixed under these conditions, and all collisions led to molecules going over the barrier (a condition met when $E^* = 0$), the reactants would be almost entirely consumed in less than a millionth of a second! There actually are a few such reactions, and among them is one that can occur in polluted air, the reaction of O atoms with NO_2.

$$O(g) + NO_2(g) \longrightarrow O_2(g) + NO(g)$$

On the other hand, there are many common reactions that have rate constants of about 10^{-2} mol/(L · s), reactions that are 10^{10} or 10,000 million times *slower* than predicted by assuming that a reaction occurs every time two molecules collide. This means there must be other factors that determine reaction rate.

One of these other factors is that reacting molecules must collide in an *effective* way. For example, the organic compound methyl iodide is changed to methyl alcohol by reaction with hydroxide ion.

$$CH_3I + OH^- \longrightarrow CH_3OH + I^-$$

Numerous experiments have proved that, in solution, the reaction occurs most rapidly when the OH^- ion attacks the methyl iodide exactly on the side of the tetrahedron away from the iodine atom. There are four sides to a tetrahedron, but attack on *only one* of these sides can be effective.

Any collision that does not come close to the most favorable geometry is not effective, so this means the *effective* collision rate is no larger than $\frac{1}{4}$ of the overall collision rate. Furthermore, even when OH^- attacks at the correct face of the tetrahedron, it must attack with the O end and not the H end. This further reduces the effectiveness of collisions to only $\frac{1}{8}$ of all collisions. You can see that, for complicated molecules, these geometry restraints mean that only a *very* small fraction of the total collisions can lead to reaction. Unfortunately, even correcting for this still leaves us with too high an estimate of reaction rate. Therefore, we have to consider what fraction of molecules has an energy at least as great as E^*.

The clue to further understanding is that reaction rates are very sensitive to temperature (Figure 15.8). The fundamental reason is that a rise in temperature causes molecules to move about more rapidly and leads *both* to more energetic collisions *and* to a greater number of collisions per second. The first of these is the more important.

Recall from the discussion of kinetic molecular theory in Chapter 12 that the average velocity of gas molecules and, even more importantly, the *distribution of molecular velocities* are sensitive to temperature. As shown on the left side of Figure 15.12, the number of molecules with high energy increases greatly with temperature, even though the average energy (velocity) increases only slightly. The significance of this is shown by comparing these energy distribution diagrams with an activation energy diagram (on the right side in Figure 15.12). The number of molecules with an energy in excess of the activation energy is given by the shaded areas in the distribution diagrams. As the temperature increases, the number of molecules with energies greater than the minimum required increases greatly. This is the factor that causes reaction rates to be so sensitive to temperature. Thus, only the fraction of collisions involving reactant molecules with energy greater than E^* can be effective, and it is the role of

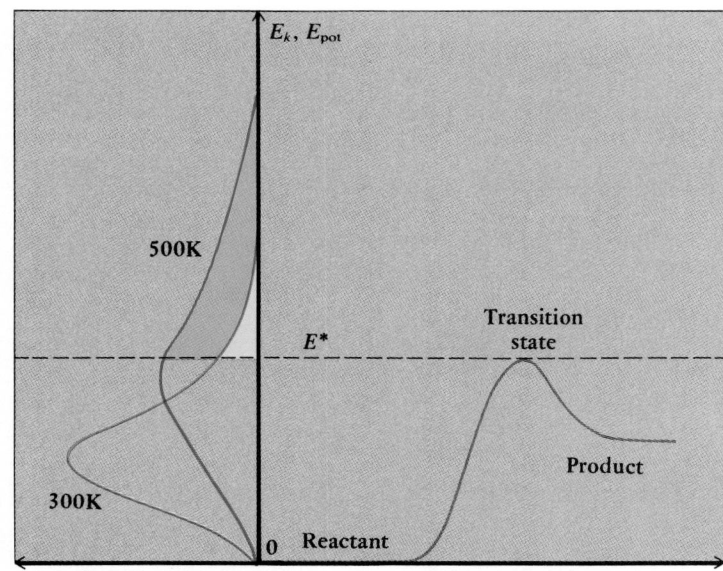

Figure 15.12 For reaction to occur, the energy of molecular motion (kinetic energy, E_k) must be converted into potential energy (E_{pot}) by collisions between molecules. Further, reacting molecules must have a potential energy of at least E^* (the difference between the transition state energy and initial reactant energy). In this example, at 300 K only a small fraction of reactant molecules (yellow area) have $E_{pot} \geq E^*$, and the reaction rate is low. At 500 K, a much larger fraction of molecules (yellow area plus blue area) now have potential energies greater than the required minimum (E^*), and the reaction rate is higher. (As the high energy molecules react, they are replaced by low energy reactant molecules that have absorbed heat from the surroundings or, if the reaction is exothermic, from the heat of reaction itself.)

the rate constant to tell us the rate at which these effective collisions occur.

EFFECT OF TEMPERATURE ON REACTION RATE: ARRHENIUS EQUATION

Svante Arrhenius (1859–1927) was a Swedish chemist who, among other things, discovered the relation between the rate constant and temperature from experiment.

Collision theory tells us that reaction rate constants (and therefore reaction rates) depend on the energy and number of collisions between reacting molecules, on whether the collisions have the correct geometry, and on the temperature. These requirements are summarized by the **Arrhenius equation**

$$k = \text{reaction rate constant} = Ae^{-E^*/RT} \tag{15.4}$$

($e^{-E^*/RT}$: fraction of molecules with minimum energy for reaction; A: frequency of collisions with correct geometry)

where R is the gas constant with a value of 8.314510×10^{-3} kJ/K · mol. The parameter A is called the *frequency factor*, and it has units of L/(mol · s). It depends on the frequency of collisions and the fraction of these that have the correct geometry. The factor $e^{-E^*/RT}$ is always less than 1, and it is interpreted as the *fraction of molecules having the minimum energy required for reaction*; it measures the area of the shaded regions in Figure 15.12 as a fraction of the whole area. As the table in the margin shows, the factor changes significantly with temperature.

TEMPERATURE	VALUE OF $e^{-E^*/RT}$ FOR $E^* = 40.$ kJ
298 K	9.7×10^{-8}
400 K	6.0×10^{-6}
600 K	3.3×10^{-4}
800 K	2.4×10^{-3}

The Arrhenius equation is valuable primarily because it can be used to (a) calculate the value of the activation energy from the temperature dependence of the rate constant and (b) calculate the rate constant for a given temperature if the activation energy and A factor are known.

A complete interpretation of the significance of A goes beyond the level of this text. However, one important aspect of A is that it becomes smaller as the reactant molecules become larger.

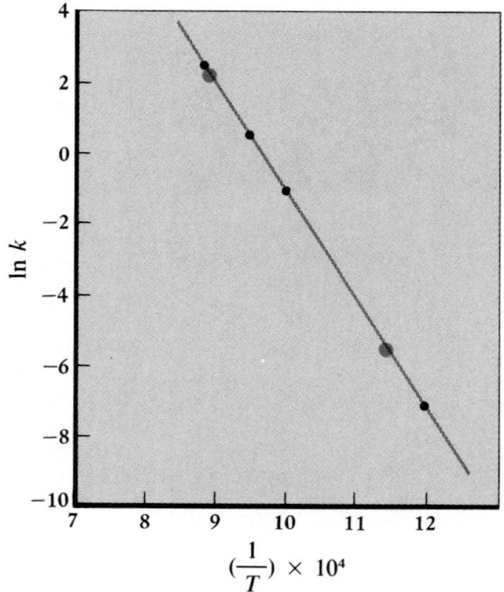

Figure 15.13 A plot of ln k versus $1/T$ for the reaction

$$2\ N_2O(g) \longrightarrow 2\ N_2(g) + O_2(g)$$

The slope of the line gives E^* as outlined in Example 15.8.

If we take the natural logarithm of each side of Equation 15.4, we have

$$\ln k = \ln A - (E^*/RT)$$

and, if we rearrange this slightly, it becomes the equation for a straight line relating ln k to $1/T$.

$$\begin{array}{ccccccl} \ln k & = & -\dfrac{E^*}{R} & \left(\dfrac{1}{T}\right) & + & \ln A & \leftarrow \text{Arrhenius equation} \\ \downarrow & & \searrow & \swarrow & & \downarrow & \\ y & = & & ax & + & b & \leftarrow \text{equation for straight line} \end{array} \tag{15.5}$$

This means that, if we plot ln k versus $1/T$, we will find a downward sloping line with a slope of $-E^*/R$ (Figure 15.13). So, now we have a means to calculate E^* from experimental values of k at several temperatures, a calculation illustrated in Example 15.8.

EXAMPLE 15.8

DETERMINING E^* FROM THE ARRHENIUS EQUATION

Using the experimental rate constant data below, calculate the activation energy E^* for the reaction

$$2\ N_2O(g) \longrightarrow 2\ N_2(g) + O_2(g)$$

TEMPERATURE (K)	k [L/(mol · s)]
1125	11.59
1053	1.67
1001	0.380
838	0.0011

Solution The first step is to find the reciprocal of the kelvin temperature and the natural logarithm of k.

$(1/T \times 10^4)/K$	ln k
8.889	2.4501
9.497	0.513
9.990	−0.968
11.9	−6.81

These data are then plotted as illustrated in Figure 15.13. Choosing the blue points on the graph, the slope is

$$\text{Slope} = \frac{\Delta \ln k}{\Delta(1/T)} = \frac{2.0 - (-5.6)}{(9.0 - 11.5)10^{-4}/\text{K}} = -\frac{7.6}{2.5 \times 10^{-4}}\ \text{K}$$

$$= -3.0 \times 10^4\ \text{K}$$

The activation energy is then evaluated from

$$\text{Slope} = -\frac{E^*}{R}$$

$$-3.0 \times 10^4\ \text{K} = -\frac{E^*}{8.314510 \times 10^{-3}\ \text{kJ/K} \cdot \text{mol}}$$

$$E^* = 250\ \text{kJ/mol}$$

In addition to the graphical method for evaluating k illustrated in Example 15.8, we can obtain E^* algebraically. Knowing k at two different temperatures, we can write the equation for each of these conditions.

$$\ln k_2 = \ln A - (E^*/RT_2) \qquad \text{and} \qquad \ln k_1 = \ln A - (E^*/RT_1)$$

If the second of these equations is subtracted from the first, we have

$$\ln k_2 - \ln k_1 = \ln \frac{k_2}{k_1} = -\frac{E^*}{R}\left[\frac{1}{T_2} - \frac{1}{T_1}\right] \tag{15.6}$$

E^* can be obtained from this equation, but it now suggests an alternate use: it allows us to find the rate constant k_2 at another temperature T_2 if E^*, k_1, and T_1 are known. Both of these situations are modeled in the following examples.

EXAMPLE 15.9

CALCULATING E^* FROM THE TEMPERATURE DEPENDENCE OF k

Using values of k determined at two different temperatures, calculate the value of E^* for the decomposition of HI.

$$2\ HI(g) \longrightarrow H_2(g) + I_2(g)$$

$k = 2.15 \times 10^{-8}$ L/(mol · s) at 650. K and 2.39×10^{-7} L/(mol · s) at 700. K.

Solution Here we use Equation 15.6, since we have k_1 at T_1 and k_2 at T_2.

$$\ln \frac{2.39 \times 10^{-7}\ \text{L/(mol} \cdot \text{s)}}{2.15 \times 10^{-8}\ \text{L/(mol} \cdot \text{s)}} = -\frac{E^*}{8.314510 \times 10^{-3}\ \text{kJ/K} \cdot \text{mol}}\left[\frac{1}{700.\ \text{K}} - \frac{1}{650.\ \text{K}}\right]$$

$$\ln (11.1) = -\frac{E^*}{8.314510 \times 10^{-3}\ \text{kJ/K} \cdot \text{mol}}(-0.000110/\text{K})$$

$$E^* = 182\ \text{kJ/mol}$$

EXAMPLE 15.10

CALCULATING k FROM A KNOWLEDGE OF E^*

The reaction of phosphine with diborane

$$PH_3(g) + B_2H_6(g) \longrightarrow H_3P{:}\!\rightarrow\! BH_3(g) + BH_3(g)$$

has an activation energy of 48.0 kJ/mol. If you have measured the rate of the reaction at room temperature, 298 K, at what temperature would you observe a doubling of the room-temperature rate?

Solution We know that k_1 was measured at $T_1 = 298$ K. The problem tells us that k_2, the rate constant at some new temperature T_2, is twice k_1, that is, $k_2 = 2k_1$. Substituting these facts into Equation 15.6, we have

$$\ln\left(\frac{2k_1}{k_1}\right) = -\frac{48.0\ \text{kJ/mol}}{8.314510 \times 10^{-3}\ \text{kJ/K} \cdot \text{mol}}\left[\frac{1}{T_2} - \frac{1}{298\ \text{K}}\right]$$

$$\ln 2 = -(5.77 \times 10^3\ \text{K})\left[\frac{1}{T_2} - 0.00336\right]$$

$$0.693 = -(5.77 \times 10^3\ \text{K})(1/T_2) + 19.4$$

$(-5.77 \times 10^3)(-0.00336)$

$$-18.7 = -(5.77 \times 10^3\ \text{K})(1/T_2)$$

$$T_2 = 309\ \text{K}$$

A chemist's "rule of thumb" is that rates double with every ten-degree rise in temperature. This problem shows that this is correct for some reactions; here, doubling was observed for an 11 K temperature increase.

A good rule of thumb is that reaction rates double for every 10 °C rise in temperature (for reactions with E around 50 kJ/mol).*

EXERCISE 15.7 E^* from the Temperature Dependence of k

The colorless gas dinitrogen tetroxide, N_2O_4, decomposes to the brown gas NO_2 (see Figure 4.7) in a first order reaction with a value of k of 4.5×10^3/s at 274 K.

$$N_2O_4(g) \longrightarrow 2\ NO_2(g)$$

If k is 1.00×10^4/s at 283 K, what is the energy of activation, E^*?

15.6 REACTION MECHANISMS

We come now to one of the more important reasons to study the rate of a reaction: to understand the **reaction mechanism**, the sequence of bond-making and bond-breaking steps that occurs during the conversion of reactants to products. You have seen that the rate expression for a reaction can be determined by experiment. On the basis of this rate law, chemists can often make an educated guess about the mechanism.

In some reactions the conversion of reactants to products occurs in a single step. Ozone and NO react in this manner.

$$NO(g) + O_3(g) \longrightarrow NO_2(g) + O_2(g)$$

Most chemical reactions, however, involve a sequence of steps. For example, the important chemical hydrazine (N_2H_4) is made by the Raschig process. The overall reaction is

$$2\ NH_3(aq) + OCl^-(aq) \xrightarrow{\text{basic solution}} N_2H_4(aq) + Cl^-(aq) + H_2O(\ell)$$

but the reaction actually occurs in a series of steps. It is thought that the first of these is the rapid formation of chloramine, NH_2Cl.

Mechanism step 1:

$$NH_3(aq) + OCl^-(aq) \xrightarrow{\text{fast}} NH_2Cl(aq) + OH^-(aq)$$

The chloramine then leads to hydrazine in one of several ways, among them the following steps:

Mechanism step 2:

$$NH_2Cl(aq) + NH_3(aq) \xrightarrow{\text{slow}} N_2H_5^+(aq) + Cl^-(aq)$$

Hydrazine, N_2H_4, is made from ammonia and sodium hypochlorite in the presence of gelatin or glue. (The purpose of the gelatin or glue is one of the minor mysteries of science. We don't yet understand it; we just know it works.) Roughly 70 million tons of hydrazine were produced in 1986, and its primary use is as a reducing agent. Thus, hydrazine is used in industry to reduce metal ions to metals and to remove O_2 from hot water boilers. Finally, N_2H_4 and its methyl derivatives, $(CH_3)_2NNH_2$ and $(CH_3)NHNH_2$, are used as rocket fuels in the Apollo lunar lander and in the space shuttle.

Mechanism step 3:

$$N_2H_5^+(aq) + OH^-(aq) \xrightarrow{\text{fast}} N_2H_4(aq) + H_2O(\ell)$$

Each of the steps in the reaction sequence is called an **elementary reaction**, a simple event in which there is some chemical transformation involving one or more molecules. Each step has its own barrier, E^*, and rate constant, k. *The collection of elementary steps that lead to the overall reaction is the reaction mechanism.* (Notice in this and other examples that the sum of the elementary reactions *must* add up to give the overall reaction.)

The mechanism of a reaction **must** *be determined by experiment*, and, to see how this is done, we shall first describe the three types of elementary reactions.

ELEMENTARY STEPS

MOLECULARITY Reactions occur by the rearrangement of atoms in molecules, and elementary steps are classified by the number of molecules that participate in an atomic rearrangement, that is, by the **molecularity** of the step. Thus, a **unimolecular** reaction involves the transformation of one molecule. If a molecule has been given enough energy by collisions with other molecules, a bond may break to give stable or unstable products. For example, the electron-donor molecule :C≡O: may be replaced in $Ni(CO)_4$ by another electron donor :L in the following reaction,

Overall reaction:

$$Ni(CO)_4 + L \longrightarrow LNi(CO)_3 + CO$$

and experiment shows that initial loss of CO occurs in a unimolecular elementary step.

Unimolecular elementary step:

$$Ni(CO)_4 \longrightarrow Ni(CO)_3 + CO$$

A **bimolecular** elementary reaction involves two molecules. These may be identical molecules (A + A → products) or different ones (A + B → products). For example, the CO that was lost from $Ni(CO)_4$ above is replaced by another molecule :L in a bimolecular elementary step.

$$Ni(CO)_3 + L \longrightarrow LNi(CO)_3$$

Other examples include the collision of an O_3 molecule with a molecule of NO to give NO_2 and O_2 and each step in the Raschig process shown above.

The simultaneous collision of three molecules, a **termolecular** reaction, is not very likely, unless one of the molecules involved is present in high concentration, such as a solvent molecule. Most such processes involve the reaction of two molecules, where the function of the third particle is to carry away the excess energy produced when a new chemical bond is formed by the first two molecules in an exothermic step. For example, N_2 is unchanged in a termolecular reaction between oxygen molecules and oxygen atoms that produces ozone in the upper atmosphere.

$$O(g) + O_2(g) + N_2(g) \longrightarrow O_3(g) + \text{energetic } N_2(g)$$

(For more information on this important reaction, see Chapter 24 as well as the box on Urban Air Pollution later in this chapter.)

RATE EXPRESSIONS FOR ELEMENTARY STEPS The rate expression for a reaction *cannot* be predicted from its overall stoichiometry. In contrast, *the rate expression of an elementary step is given by the product of the rate constant and the concentrations of the reactants in the step.* This means that we can write the rate expression for any elementary step, as in the examples below.

ELEMENTARY STEP	MOLECULARITY	RATE EXPRESSION
$A \longrightarrow$ product	unimolecular	Rate = $k[A]$
$A + B \longrightarrow$ product	bimolecular	Rate = $k[A][B]$
$A + A \longrightarrow$ product	bimolecular	Rate = $k[A]^2$
$2A + B \longrightarrow$ product	termolecular	Rate = $k[A]^2[B]$

Elementary steps with molecularity greater than 3 are rare. It is highly unlikely that four or more molecules of the right type will collide with the correct orientation at the same time.

For example, this means the rate laws for the two steps in the L/$Ni(CO)_4$ reaction are

Step 1	unimolecular	Rate = $k[Ni(CO)_4]$
Step 2	bimolecular	Rate = $k'[L][Ni(CO)_3]$

where the rate constants k and k' do not have the same value.

THE PHYSICAL SIGNIFICANCE OF RATE EXPRESSIONS FOR ELEMENTARY STEPS The form of the rate expression for an elementary step is directly related to the ideas that collisions are necessary and that the reaction rate and collision rate are proportional. As an example, take the bimolecular reaction of A and B. Since A and B must collide with one another for reaction to occur, the number of collisions per unit time is proportional to the concentration of each species; as the concentrations increase, the number of collisions increases, thus increasing the rate. For instance, assume the concentrations of A and B are both 2 M, and k is 1 L/(mol · hr).

Note that the discussion of collisions between A and B does not apply to unimolecular reactions.

$$\text{Initial rate} = k[A][B] = 1 \text{ L/(mol} \cdot \text{hr)}[2 \text{ mol/L}][2 \text{ mol/L}]$$
$$= 4 \text{ mol/L} \cdot \text{hr}$$

This means the rate is 4 M/hr. On the other hand, if the concentration of A is doubled, then

$$\text{Initial rate} = 1 \text{ L/(mol} \cdot \text{hr)}[4 \text{ mol/L}][2 \text{ mol/L}] = 8 \text{ mol/(L} \cdot \text{hr)}$$

the rate is doubled to 8 mol/L · hr. This is even more evident if we illustrate the number of possible collisions that can occur between reacting molecules. For two A molecules and two B molecules, four A $\leftrightarrow$ B collisions are possible. If the number of A molecules is doubled to four, then there are eight possible collisions.

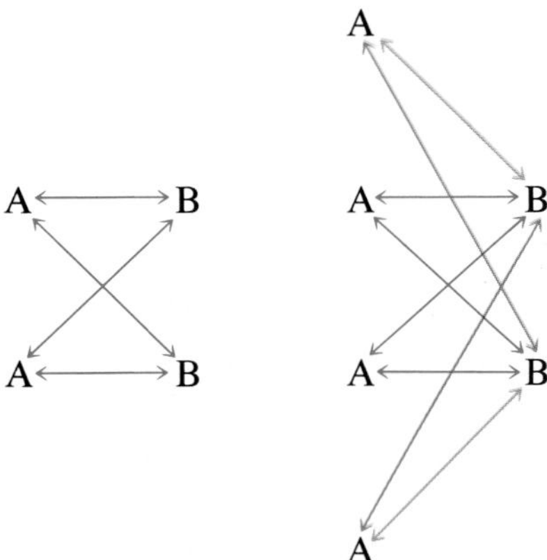

MOLECULARITY AND ORDER *The molecularity of an* ***elementary*** *reaction and its order are the same.* A unimolecular elementary reaction must be first order, a bimolecular elementary reaction must be second order, and a termolecular elementary reaction must be third order; the converse is also true for elementary reactions. However, the relation between molecularity and order is emphatically *not* true for the *overall* reaction. If you discover experimentally that a reaction is first order, this does not mean it occurs in a single, unimolecular elementary step. Similarly, a second order rate expression does not imply the reaction occurs in a single, bimolecular elementary step. An example of this is the decomposition of N_2O_5.

$$2\ N_2O_5(g) \longrightarrow 4\ NO_2(g) + O_2(g)$$

Here the rate expression is "Rate $= k[N_2O_5]$," but chemists are fairly certain the mechanism involves a series of both unimolecular and bimolecular steps.

To see how the experimentally observed rate expression for the overall reaction is connected with a possible mechanism or sequence of elementary steps requires some chemical intuition. This is the subject of the next section.

EXAMPLE 15.11

ELEMENTARY STEPS AND RATE EXPRESSIONS

Square planar platinum compounds undergo substitution in either or both of two ways. One pathway involves the union of the original platinum compound and the substituting particle. Using as an example the following reaction that is *one elementary step* in the substitution process,

$$Pt(NH_3)_2Cl_2 + H_2O \longrightarrow [Pt(NH_3)_2(H_2O)Cl]^+ + Cl^-$$

give the molecularity and write the rate expression.

Solution

(a) Since two molecules are involved in the elementary step, the process is bimolecular.
(b) The rate expression for an elementary step is simply the product of k for that

step with the concentrations of the reactants. Therefore, the rate expression of this step is

$$\text{Rate} = k[Pt(NH_3)_2Cl_2][H_2O]$$

EXERCISE 15.8 Elementary Steps

Nitrogen oxide is reduced by hydrogen to give water and nitrogen,

$$2\ H_2(g) + 2\ NO(g) \longrightarrow N_2(g) + 2\ H_2O(g)$$

and one possible mechanism to account for this reaction is

$$2\ NO(g) \rightleftharpoons N_2O_2(g)$$

$$N_2O_2(g) + H_2(g) \longrightarrow N_2O(g) + H_2O(g)$$

$$N_2O(g) + H_2(g) \longrightarrow N_2(g) + H_2O(g)$$

What is the molecularity of each of the three steps? Show that the sum of these elementary steps is the net reaction.

REACTION MECHANISMS AND RATE EXPRESSIONS

One of the great unknowns in chemistry today is the *way* in which many reactions occur, especially those in biological systems. Thus, the objective of many kinetic investigations is to unravel the mechanism of the reaction. Now that you understand the concept of elementary steps in a mechanism, we can begin to see the connection between the experimental rate expression and the reaction mechanism, the *hypothesis* about the steps involved in the reaction.

The rate expression is determined by experiment. The mechanism is a good guess (a hypothesis) about the way the reaction occurs. Several mechanisms can correspond to the same experimental rate expression, so one postulated mechanism can be quite wrong and can provoke disputes between scientists.

Imagine a reaction whose mechanism involves two sequential steps. For the purpose of this discussion, we assume we know the rates of both steps.

Elementary step 1: $A + B \xrightarrow[\text{Slow, } E^* \text{ large; 1 reaction/s}]{k_1} X + M$

Elementary step 2: $M + A \xrightarrow[\text{Fast, } E^* \text{ small; 100 reactions/s}]{k_2} Y$

Overall reaction: $2\ A + B \longrightarrow X + Y$

In the first reaction A and B come together and slowly produce one of the products (X) plus another reactive species, M. Almost as soon as M is formed, however, it is rapidly consumed by reaction with an additional molecule of A. Thus, the products X and Y are the result of two elementary steps.

One of the most important aspects of kinetics to understand is that products can never be produced at a rate faster than the rate of the slowest step. *The rate of the overall reaction is limited by, and is exactly equal to, the combined rates of all elementary steps up through the slowest step in the mechanism*. Therefore, the slowest elementary step of a sequence is called the **rate determining step**. You are already familiar with rate determining steps. No matter how fast you shop in the supermarket, it always seems that the time it takes to finish is determined by the wait in the checkout line.

In the example above, the rate determining elementary step—the first step—is bimolecular and so has the rate expression

$$\text{Initial reaction rate} = k_1[\text{A}][\text{B}]$$

where k_1 is the rate constant for that step. The overall reaction follows this same second order rate expression because the slower step in the mechanism is the first step and is bimolecular.

Now let us turn to a few actual examples of mechanisms. Experiment shows that the reaction of nitrogen dioxide with fluorine

$$\textit{Overall reaction:} \quad 2\ NO_2(g) + F_2(g) \longrightarrow 2\ FNO_2(g)$$

has a second order rate expression:

$$\text{Initial reaction rate} = k[NO_2][F_2]$$

The experimental rate expression immediately rules out the possibility that the reaction occurs in a single step, since, if the overall stoichiometric equation were an elementary step, the rate expression would be

$$\text{Initial reaction rate} = k[NO_2]^2[F_2]$$

and this rate expression does *not* agree with experiment. Therefore, there must be at least two steps in the mechanism, and the rate determining elementary step must involve NO_2 and F_2 in a 1:1 ratio. The simplest possibility is identical with our earlier example involving 2 A + B $\rightarrow$ products.

$$\textit{Elementary step 1:} \quad NO_2(g) + F_2(g) \xrightarrow[\text{Slow}]{k_1} FNO_2(g) + F(g)$$

$$\textit{Elementary step 2:} \quad NO_2(g) + F(g) \xrightarrow[\text{Fast}]{k_2} FNO_2(g)$$

$$\textit{Overall:} \quad 2\ NO_2(g) + F_2(g) \longrightarrow 2\ FNO_2(g)$$

At this introductory level you cannot be expected to derive reaction mechanisms. However, given a mechanism, you can decide whether it is in agreement with experiment.

That is, NO_2 and F_2 first produce one molecule of the product (FNO_2) plus an F atom, and the F atom then reacts with additional NO_2 to give one more molecule of product. If we assume that the first bimolecular step is rate determining, its rate expression would be "Initial reaction rate = $k_1[NO_2][F_2]$," exactly the rate expression observed experimentally. The experimental rate constant is, therefore, the same as k_1.

The F atom in the first step of the NO_2/F_2 reaction is called an intermediate. A **reaction intermediate** is a molecule or ion produced in one step of a reaction sequence, but consumed in a subsequent step. As a result it does not appear in the net stoichiometric equation. Reaction intermediates usually have a very fleeting existence, but they occasionally have long enough lifetimes to be observed.

EXAMPLE 15.12

ELEMENTARY STEPS AND REACTION MECHANISMS

Oxygen atom transfer from nitrogen dioxide to carbon monoxide

$$NO_2(g) + CO(g) \longrightarrow NO(g) + CO_2(g)$$

has the following rate expression at temperatures less than 500 K.

$$\text{Rate} = k[NO_2]^2$$

Can this reaction occur in one bimolecular step?

Solution If the reaction occurs simply by the collision of one NO_2 molecule with one CO molecule (that is, the equation for the overall reaction and the equation for the single elementary step are the same), the rate expression would be

$$\text{Rate} = k[NO_2][CO]$$

This does not agree with experiment, so the mechanism must involve more than a single step. In fact, the reaction is thought to occur in two bimolecular steps:

Elementary step 1: Slow, rate determining	$2\ NO_2(g) \longrightarrow NO_3(g) + NO(g)$
Elementary step 2: Fast	$NO_3(g) + CO(g) \longrightarrow NO_2(g) + CO_2(g)$
Overall reaction:	$NO_2(g) + CO(g) \longrightarrow NO(g) + CO_2(g)$

The first or rate determining step would indeed have a rate expression that agrees with experiment.

SUMMARY: PRINCIPLES OF RATE EXPRESSIONS AND REACTION MECHANISMS

1. *Perform experiments that define the effect of reactant concentrations on the rate of the reaction; derive experimental rate expression.*
2. *Propose a mechanism on the basis of the experimental rate expression and the principles of stoichiometry.*
3. *Derive a rate expression from the proposed mechanism. Derived expression must contain only those species present in the overall chemical reaction and not the reaction intermediates. If the derived and experimental expressions are the same, the postulated mechanisms may be a reasonable hypothesis of the reaction sequence.*

EXERCISE 15.9 Elementary Steps and Reaction Mechanisms

The Raschig reaction produces N_2H_4 from NH_3 and OCl^- in a basic, aqueous solution. Three possible elementary steps were given at the beginning of this section (page 635). Which step of the three is rate determining? Write the rate expression for the rate determining elementary step.

We have described only one form of the many possible reaction mechanisms. Many reactions—especially explosions—have quite complex mechanisms. As an example, you might wish to consider some of the reactions that lead to urban air pollution. Although there are many interrelated processes going on, scientists now understand some of them reasonably well, and a few are described in the box on page 642.

15.7 CATALYSIS

In the preceding sections we have seen that the rate of a reaction can depend on reactant concentrations, temperature, and size of the activation energy barrier. So, if we want to speed up a reaction in the laboratory, the first things we think to do are to increase the reactant concentrations and raise the temperature. However, there is another way to speed up a reaction: *alter the mechanism* so that the activation energy barrier can

The effect of catalysts on chemical reactions was previously described on page 63 (in connection with platinum), and it will be discussed further on pages 684–686.

be lowered. This objective can be achieved by using a catalyst. A **catalyst** is any reagent that can increase the rate of a reaction by altering its mechanism but without being consumed by the reaction.

There are two general types of catalysts: homogeneous and heterogeneous. **Homogeneous catalysts** are reagents that are in the same phase as the reaction mixture. This usually means the reaction mixture is fluid, and the catalyst can dissolve in it. In contrast, **heterogeneous catalysts** are in a different phase from the reactants. In this case, the catalyst is usually a solid, while the reactants are dissolved in a solvent or are in the gas phase.

Almost 5 billion pounds of catalysts, worth about one billion dollars, are used in the United States every year. More than 90% are acids (HF, H_2SO_4, H_3PO_4). The remainder are heterogeneous catalysts, largely metals and metal oxides.

SOMETHING MORE ABOUT Reactions and Mechanisms in Urban Air Pollution

*For very recent information on air pollution see J.H. Seinfeld, Science, **1989**, 243, 745. The chemistry of ozone, and the effect of chlorofluorocarbons (CFC's) on stratospheric ozone, is also described in Chapter 24.*

There is understandable concern about air pollution worldwide. The kinetics and mechanisms have been studied for many of the important reactions, and chemists have come to a better understanding of some of these. The mechanisms of a few are outlined below.

One of the key molecules in the many reactions occurring in the atmosphere is ozone, O_3. This important molecule is produced in the upper part of the earth's atmosphere—the stratosphere—by the dissociation of O_2 molecules to give O atoms

$$O_2(g) + \text{radiation } (\lambda < 280 \text{ nm}) \longrightarrow 2\,O(g)$$

These atoms in turn react with O_2 molecules in the presence of some third molecule (O_2 or N_2) that carries away the energy of the reaction and stabilizes the ozone, some of which finds its way to lower altitudes.

$$O(g) + O_2(g) + M(g) \longrightarrow O_3(g) + \text{excited } M(g)$$

At lower altitudes the higher energy ultraviolet radiation required to dissociate O_2 has been filtered out, so O atoms are largely produced from atmospheric NO_2 with lower energy light. (The NO_2 is produced from the oxygen and nitrogen of the atmosphere).

$$NO_2(g) + \text{radiation } (\lambda = 280 - 430 \text{ nm}) \longrightarrow NO(g) + O(g)$$

The nitrogen oxide produced in this reaction is readily oxidized with the traces of ozone that are found in the atmosphere closer to the ground.

$$NO(g) + O_3(g) \longrightarrow NO_2(g) + O_2(g)$$

Polluted air over the city of Los Angeles. (R. Sager)

If you sum the last three reactions, which all occur rapidly, you can see that they represent no *net* change; O_2 goes to O_2. However, because of their relative rates, there is always some O_3 present in unpolluted air; measurements show that there are about 20 to 50 O_3 molecules per billion molecules of other gases in the air (that is, 20 to 50 parts per billion).

The last of the reactions above shows that one O_3 is required to regenerate one NO_2. However, if there is another reaction that will perform the NO to NO_2 conversion, O_3 is not consumed, and it can accumulate in the atmosphere. The key is thought to be the presence of OH radicals (species with only 7 valence electrons), and there are at least three sources of these in the sunlit urban atmosphere. Once such radicals have been formed, they react with hydrocarbons in the atmosphere to give still other reactive species.

Reaction of OH radicals with hydrocarbons: $OH(g) + C_4H_{10}(g) \longrightarrow C_4H_9(g) + H_2O(g)$

Reaction of hydrocarbon radicals with O_2: $C_4H_9(g) + O_2(g) \longrightarrow C_4H_9O_2(g)$ [= RO_2]

Now that the hydrocarbon fragment, C_4H_9 (called a radical), has joined with O_2, the ingredients are in place to produce a larger concentration of O_3 in the atmosphere.

The term radical is commonly used to denote a fragment of a molecule, usually with an unpaired electron.

$$\begin{aligned} &RO_2(g) + NO(g) \longrightarrow NO_2(g) + RO(g) \\ &NO_2(g) + \text{radiation} \longrightarrow NO(g) + O(g) \\ &O(g) + O_2(g) + M(g) \longrightarrow O_3(g) + \text{excited M} \\ \hline \textit{Net:}\quad &RO_2(g) + O_2(g) + \text{radiation} \longrightarrow RO(g) + O_3(g) \end{aligned}$$

After all we have heard about the destruction of the ozone layer (Chapter 24), why are we worried about an *accumulation* of O_3? Because O_3 at low altitudes causes eye irritation and can worsen the symptoms of respiratory illness; in addition, it is a powerful oxidizing agent and reacts with objects in the environment. Furthermore, ozone and hydroxyl radicals, among other molecules and fragments, can react with hydrocarbons to give compounds called aldehydes, such as acetaldehyde. This compound in turn produces other atmospheric pollutants such as PAN.

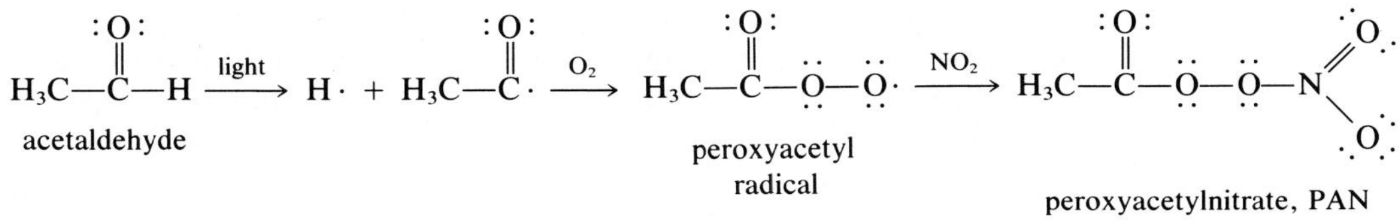

$$H_3C{-}C({=}O){-}H \xrightarrow{\text{light}} H\cdot + H_3C{-}\dot{C}{=}O \xrightarrow{O_2} H_3C{-}C({=}O){-}O{-}O\cdot \xrightarrow{NO_2} H_3C{-}C({=}O){-}O{-}O{-}NO_2$$

acetaldehyde; peroxyacetyl radical; peroxyacetylnitrate, PAN

Light can cause this aldehyde to release an H atom, and the resulting fragment (H_3CCO) can react with O_2 to give the peroxyacetyl radical. This radical then scavenges NO_2 from the air to form peroxyacetylnitrate, PAN, an extremely irritating component of urban smog and a compound that has destructive effects on plants and food crops.

The decomposition of H_2O_2 to O_2 and H_2O is an example of the use of a homogeneous catalyst to lower an activation energy barrier (Figure 15.2). In the absence of a catalyst, the reaction proceeds only slowly with an activation energy of 76 kJ/mol. If a little iodide is added, though, the reaction occurs more readily because the activation energy is now only 57 kJ/mol (Figure 15.14). At the molecular level, the function of the I^- ion is apparently to produce the reactive ion IO^- as an intermediate in step 1.

Elementary step 1: Slow, rate determining

$$H_2O_2(aq) + I^-(aq) \xrightarrow{k_1} IO^-(aq) + H_2O(\ell)$$

Elementary step 2: Fast

$$H_2O_2(aq) + IO^-(aq) \xrightarrow{k_2} I^-(aq) + H_2O(\ell) + O_2(g)$$

Overall: $2\ H_2O_2(aq) \longrightarrow O_2(g) + 2\ H_2O(\ell)$

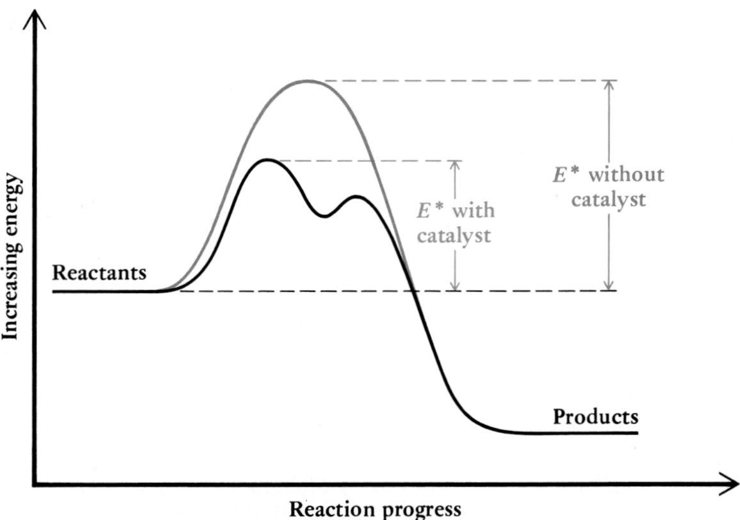

Figure 15.14 A catalyst accelerates a reaction by altering the mechanism so that the activation energy E^* is lowered. With a smaller barrier to overcome, there are more reacting molecules with sufficient energy to surmount the barrier, and the reaction occurs more readily. Notice that the shape of the barrier has changed because the mechanism has changed. In the peroxide reaction, the first (and larger) barrier for the catalyzed reaction is for step 1, the rate determining step. A barrier must also be surmounted in step 2, but the barrier is smaller than the first, so step 2 is faster than step 1.

The IO^- ion then reacts rapidly with yet another molecule of hydrogen peroxide in step 2 to produce molecular oxygen and water, and to regenerate the catalyst. The I^- ion is a catalyst because it has increased the reaction rate but it was not consumed in the process (Figure 15.15).

Acid-catalyzed reactions are other important examples, occurring in biological systems as well as in the chemistry laboratory. Acid catalysis involves the transfer of a proton to the reacting molecule,

$$R + H^+ \longrightarrow RH^+$$

and this altered molecule now reacts more rapidly than before.

$$R \xrightarrow{\text{slow}} \text{products} \qquad \text{but} \qquad RH^+ \xrightarrow{\text{fast}} \text{products}$$

For example, an organic compound called an ester can react with water to replace the —OR group with —OH. The products are compounds called acids and alcohols (see Chapter 26).

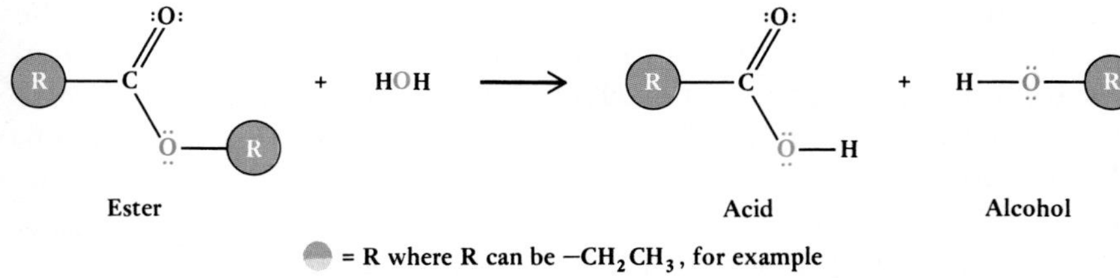

If a tiny quantity of acid is added, the H^+ ion attaches itself to the C=O group by using an oxygen lone pair of electrons to form a bond.

Figure 15.15 Hydrogen peroxide decomposition can also be accelerated by the heterogeneous catalyst MnO_2. Here a 30% aqueous solution of H_2O_2 is dropped onto solid, black MnO_2, with rapid decomposition to give O_2 and H_2O; the latter is evolved as steam due to the high heat of reaction.

This species is now much more susceptible to reaction with water, and the ester quickly falls apart to give products.

An industrially important example of homogeneous catalysis is the production of acetic acid from methyl alcohol.

$$H_3COH + CO \xrightarrow{\text{Rh-containing catalyst}} H_3C{-}\overset{\overset{\displaystyle O}{\|}}{C}{-}OH$$

The rhodium-catalyzed addition of CO to methyl alcohol is the source of nearly all the acetic acid produced in the United States (3.16 billion pounds in 1987). Until recently this important acid was produced almost entirely by fermentation. To see an excellent example of the practical use of catalysts, read the material in the box on catalysis below.

Many catalysts used in the laboratory and in industry are based on heavy metals such as ruthenium, palladium, rhodium, and platinum. The catalysts pictured here are heterogeneous catalysts.

Heterogeneous catalysts are extremely important in the chemical industry. Chemical companies profit by combining small, inexpensive molecules to produce larger ones. The combination process is often one that occurs only at such a slow rate that the process would not be commercially useful. Therefore, catalysts have to be invented to speed up the reaction, and heterogeneous ones are generally the most practical. Two important examples include the synthesis of ammonia from H_2 and N_2

$$3\ H_2(g) + N_2(g) \xrightarrow{\text{Fe-containing catalyst/400 °C/200 atm}} 2\ NH_3(g)$$

and the synthesis of methyl alcohol from synthesis gas (CO and H_2).

$$CO(g) + 2\ H_2(g) \xrightarrow{\text{Al and Cr oxides}} H_3COH(g)$$

In 1989 the United States chemical industry produced 33.76 billion pounds of ammonia. Much of it was turned into nitric acid from which nitrate-containing fertilizers are made. One of the steps in the conversion of NH_3 to HNO_3 is the oxidation of NH_3, a reaction catalyzed by platinum (see page 63).

You are probably most familiar with heterogeneous catalysts through their use in the exhaust systems of automobiles (Figure 15.16). Their purpose is to insure that the combustion of carbon monoxide and hydrocarbons is complete,

$$2\ CO(g) + O_2(g) \xrightarrow{\text{catalyst}} 2\ CO_2(g)$$

$$2\ C_8H_{18}(g) + 25\ O_2(g) \xrightarrow{\text{catalyst}} 16\ CO_2(g) + 18\ H_2O(g)$$

and to convert nitrogen and sulfur oxides to molecules less harmful to the environment. The air drawn into an automobile engine contains N_2

Figure 15.16 An automobile catalytic converter that has been cut open to expose the pellets that contain the platinum/palladium/rhodium catalyst. (General Motors)

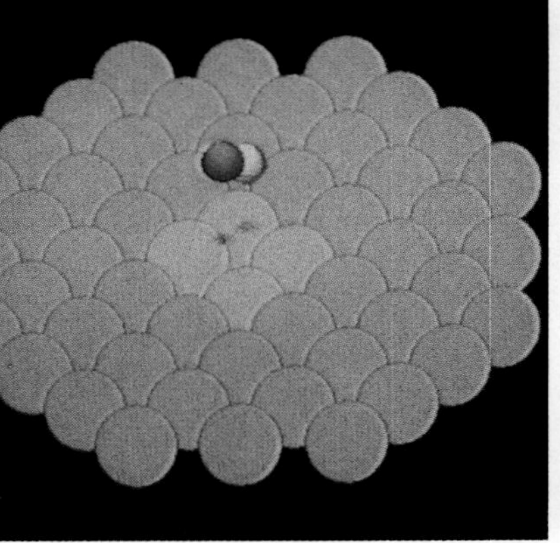

(a)

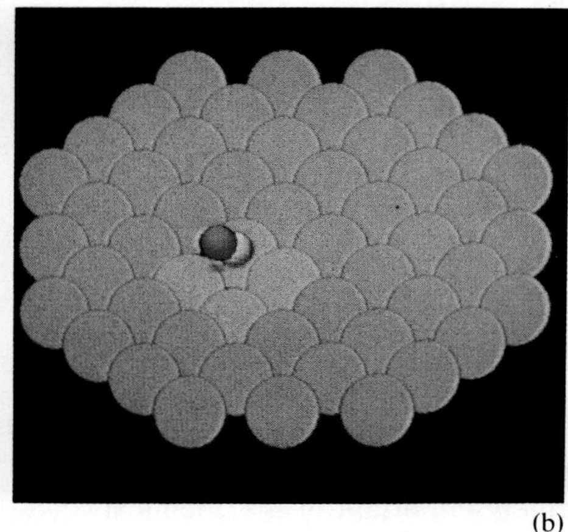

(b)

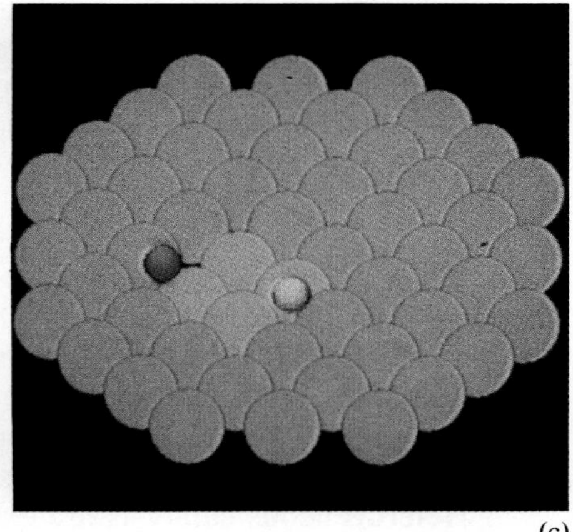

(c)

Figure 15.17 A computer simulation of the dissociation of NO into N (reddish) and O (white) atoms on a platinum surface. In (a) the NO approaches the surface. After 1.5×10^{-12} seconds it interacts with the surface (b) and then dissociates (c), 1.7×10^{-12} seconds after approaching the surface. (John Tully, AT&T Bell Laboratories)

as well as O_2. At the high temperature of combustion, some N_2 reacts with O_2 to give NO. This is a serious air pollutant, so catalysts have been developed that catalyze not only the oxidation of carbon compounds but also the decomposition of NO back to the elements.

$$2\ NO(g) \xrightarrow{\text{catalyst}} N_2(g) + O_2(g)$$

The role of the heterogeneous catalyst in the reactions above is probably to weaken the bonds of the reactants H_2, O_2, and N_2 and to assist in product formation. For example, in Figure 15.17 you see a portion of a computer-generated movie of NO molecules dissociating into N and O atoms on the surface of a platinum metal catalyst.

SOMETHING MORE ABOUT
Catalysis: The Production of Acetic Acid from Simple Compounds

The mechanism of the catalyzed combination of CO and methyl alcohol, H_3COH, is understood well enough that we can see how a catalyst affects reaction rates. In the first step, CO and rhodium(III) iodide combine to give a square planar rhodium-containing anion in which CO is bonded to the metal; it is this anion that is the catalyst in the process.

$$RhI_3 + 3\ CO + 3\ H_2O \longrightarrow [Rh(CO)_2I_2]^- + I^- + CO_2 + 2\ H_3O^+$$

Next, a compound called methyl iodide ($H_3C—I$), which we saw earlier being transformed into methyl alcohol (page 630), adds to the rhodium-containing anion, which now has groups attached at the corners of an octahedron.

$$[Rh(CO)_2I_2]^- + H_3C—I \longrightarrow [CH_3Rh(CO)_2I_3]^-$$

The methyl (CH_3) group now migrates to the C atom of one of the CO molecules on the metal center,

$$[\mathrm{Rh(CH_3)(CO)_2I_3}]^- \longrightarrow [\mathrm{Rh(C(O){-}CH_3)(CO)I_3}]^-$$

A chemist adjusts a reactor used to make acetic acid from the rhodium-catalyzed combination of methanol and carbon monoxide. (© Dick Luria, FPG International)

and then an additional CO fills up the position left by the migrating CH_3.

$$[\mathrm{Rh(C(O){-}CH_3)(CO)I_3}]^- + \mathrm{CO} \longrightarrow [\mathrm{Rh(C(O){-}CH_3)(CO)_2I_3}]^-$$

This molecule then loses $H_3CC(O)I$, which has the effect of restoring the catalyst,

$$[\mathrm{Rh(C(O){-}CH_3)(CO)_2I_3}]^- \longrightarrow [\mathrm{RhI_2(CO)_2}]^- + \mathrm{H_3C{-}C({=}O){-}I}$$

and the liberated $H_3CC(O)I$ reacts with water to give the desired product, acetic acid.

$$\mathrm{H_3C{-}C({=}O){-}I} + \mathrm{H_2O} \longrightarrow \mathrm{H_3C{-}C({=}O){-}OH} + \mathrm{HI}$$

The hydrogen iodide produced in the step just above then reacts with methyl alcohol to give methyl iodide,

$$\mathrm{HI} + \mathrm{H_3C{-}OH} \longrightarrow \mathrm{H_3C{-}I} + \mathrm{H_2O}$$

the material needed to start over again in the catalytic cycle. The net result of all of these steps is exactly the simple reaction

$$\mathrm{H_3COH} + \mathrm{CO} \longrightarrow \mathrm{H_3C{-}C({=}O){-}OH}$$

where the rhodium(III) iodide facilitates the reaction without itself being consumed.

SUMMARY

Chemical kinetics deals with the theory and measurement of chemical **reaction rates** (the change in concentrations of reactants and products with time) and reaction **mechanisms** (the sequence of events at the molecular level that controls reaction speed and the outcome of reactions).

Reaction rates are affected by reactant concentrations, temperature, the presence of a catalyst, and the physical state of the reactants. The rate of a reaction must be determined by experimental observation of the change in the concentration of the reactants or products as a function of time (Section 15.2). The rate of a reaction is always taken as positive and is related by stoichiometry to the rate of change in concentrations of reactants and products. Thus, for a reaction A → 2 B,

$$\text{Initial reaction rate} = -\frac{\Delta[\mathrm{A}]}{\Delta t} = \frac{1}{2}\frac{\Delta[\mathrm{B}]}{\Delta t}$$

The relation between reaction rate and reactant concentrations is given by a rate expression (Section 15.3). For the general reaction

$$a\,\mathrm{A} + b\,\mathrm{B} \xrightarrow{\mathrm{C}} x\,\mathrm{X}$$

(where C is a catalyst) the rate expression will always have the form

$$\text{Initial reaction rate} = k[\mathrm{A}]^m[\mathrm{B}]^n[\mathrm{C}]^p$$

where k is the **rate constant**. Each exponent (m, n, and p) is called the **reaction order** with respect to the concentrations of A, B, and C, respectively, and the sum of the exponents is the *total order* of the reaction. The exponents are *not* necessarily the same as the stoichiometric coefficients in the balanced overall equation. Furthermore, they may be positive whole or fractional numbers (although this chapter describes only cases where they are positive whole numbers). Finally, the exponents *must* be determined by experiment.

A first order reaction has a rate expression of the form "Rate = k[A]" while a second order reaction would have a rate expression such as "Rate = $k[\mathrm{A}]^2$" or "Rate = k[A][B]." Other possibilities for these and reactions of other orders are given in the text.

For a first order process (Rate = k[A]), the concentration of A after some time t (=[A]) is related to the initial concentration of A (=$[\mathrm{A}]_0$ at time = 0) by Equation 15.1 (Section 15.4).

$$\ln\frac{[\mathrm{A}]}{[\mathrm{A}]_0} = -kt \qquad (15.1)$$

For the second order rate expression "Rate = $k[\mathrm{A}]^2$," the relation is

$$\frac{1}{[\mathrm{A}]} - \frac{1}{[\mathrm{A}]_0} = kt \qquad (15.2)$$

The **half-life, $t_{1/2}$,** of a reaction is the time required for the concentration of one of the reactants to decrease to $\frac{1}{2}$ of its initial value. For a first order process, $t_{1/2} = 0.693/k$ (Equation 15.3).

The order of a reaction can be distinguished graphically. If a straight line is observed when ln[reactant] is plotted versus time, the reaction is first order in the reactant, and the slope of the line is $-k$. For a second order reaction of the form "Rate = $k[\mathrm{A}]^2$," a plot of 1/[A] versus time is a straight line with a slope of k.

When bonds break and new bonds form, the energy of the reacting molecules increases and then decreases as they move over the reaction energy barrier. Such energy barriers are illustrated by a **reaction energy diagram** (Section 15.5), where, as time progresses, the reaction moves from reactants to products through the higher energy **transition state**. The molecular configuration in the transition state is called the **activated complex**. The energy at the top of the barrier is the transition state energy, and the difference between this energy and the energy of the reactants is the **activation energy,** E^*, for the reaction. The activation energy largely determines the variation in rates observed for different reactions. The theory relating rates and energies is called the **transition state theory**.

Collision theory assumes that reacting molecules or atoms must collide in order to react. The theory supposes that the reaction rate constant depends on the number of collisions per unit time occurring with the correct geometry, and on the temperature. This relation is expressed in terms of the Arrhenius equation.

activation energy

$$k = Ae^{-E^*/RT} \tag{15.4}$$

frequency of collisions with correct geometry

The factor $e^{-E^*/RT}$ is always less than 1 and is the fraction of molecules having the minimum energy (E^*) required for reaction. The Arrhenius equation can be used, in rearranged form (Equation 15.6), to determine E^* if the values of k are known at different temperatures.

A reaction mechanism consists of a series of **elementary steps**, each involving the reaction of one, two, or three molecules (Section 15.6). The number of interacting particles gives the molecularity of a step. If one particle is involved, it is a *uni*molecular step. If two are involved the step is *bi*molecular, and if three collide the step is *ter*molecular. The rate expression for each elementary step is the product of the rate constant for the step and the concentrations of the reactants in the step. The slowest elementary step is called the **rate determining step**. If the experimental rate expression is the same as the rate expression derived from the mechanism, then the postulated mechanism is a reasonable hypothesis.

Catalysts enhance the speed of a reaction by altering the mechanism to provide a pathway with a lower activation energy. Although the catalyst is not chemically transformed, it can appear in the rate expression, in contrast with its lack of appearance in the overall balanced equation. The catalyst can be homogeneous, if in the same phase as the other components of the reaction, or heterogeneous, if the catalyst and reaction components are in different phases.

STUDY QUESTIONS

REVIEW QUESTIONS

1. The rate expression for a chemical reaction can be determined by which of the following?
   (a) Theoretical calculations.
   (b) Measuring the rate of the reaction as a function of the concentration of the reacting species.
   (c) Measuring the rate of the reaction as a function of temperature.
2. Name four conditions that determine the rate of a chemical reaction.
3. Refer to Figure 15.4 and explain why $[Pt(NH_3)_2Cl_2]$ *decreases* with time but $[Cl^-]$ *increases* with time.
4. Using the rate expression "Rate = $k[A]^2[B]$," state the order of the reaction with respect to A and B and the total order of the reaction.
5. If a reaction has the experimental rate expression "Rate = $k[A]^2$," explain what happens to the rate when the concentration of A is tripled. What happens when the concentration of A is halved?
6. A reaction has the experimental rate expression "Rate = $k[A]^2[B]$." If the concentration of A is doubled, and the concentration of B is halved, what happens to the reaction rate?
7. Give the equation that allows us to find the concentration of reactant as a function of time for a first order reaction. Define each term in the equation.

8. Derive the expression for the half-life of a first order reaction ($t_{1/2} = 0.693/k$).
9. After five half-life periods for a first order reaction, what fraction of reactant remains?
10. If you plot 1/[reactant] versus time and observe a straight line, what is the order of the reaction? If ln[reactant] is plotted versus time, and a straight line of negative slope is observed, what is the order of the reaction?
11. Draw a reaction energy diagram for an exothermic process. Mark the activation energies of the forward and reverse processes, indicate the energy of the transition state, and show how you can calculate the net energy of the reaction.
12. Explain how collision theory accounts for the temperature dependence of reaction rates.
13. Write the Arrhenius equation and define each part of the equation.
14. What is the *mechanism* of a chemical reaction? In your discussion, define *elementary step* and *rate determining step*.
15. Define the term *molecularity* in terms of the collision theory.
16. Define the term *reaction intermediate*. Give an example, using one of the reaction mechanisms described in the text.
17. Indicate whether each of the following statements is true or false. Change the wording of each false statement to make it true.
    (a) In reactions that proceed in a sequence of steps, the rate determining step is the slowest one.
    (b) It is possible to change the rate constant for a reaction by changing the temperature.
    (c) As a first order reaction proceeds at a constant temperature, the rate remains constant.
    (d) The rate constant for a reaction is independent of reactant concentrations.
    (e) As a first order reaction proceeds at a constant temperature, the reaction rate constant changes.
18. What is a catalyst? What is its effect on the energy barrier for a reaction?
19. Explain the difference between a homogeneous and a heterogeneous catalyst. Give an example of each.

REACTION RATES

20. For each reaction below, express the rates of change of [product] and [reactant] in the correct relationship to each other.
    (a) $2\ O_3(g) \longrightarrow 3\ O_2(g)$
    (b) $2\ HOF(g) \longrightarrow 2\ HF(g) + O_2(g)$
    (c) $N_2(g) + 3\ H_2(g) \longrightarrow 2\ NH_3(g)$
21. For each reaction below, express the rates of change of [product] and [reactant] in the correct relationship to each other.
    (a) $HC{\equiv}CH(g) + HCN(g) \longrightarrow H_2C{=}CHCN(g)$
    (b) $CH_4(g) + 2\ O_2(g) \longrightarrow CO_2(g) + 2\ H_2O(g)$
    (c) $2\ C_2H_6(g) + 7\ O_2(g) \longrightarrow 4\ CO_2(g) + 6\ H_2O(g)$
22. Experimental data are listed below for the hypothetical reaction A → 2 B.

TIME (s)	[A] (mol/L)
0.00	1.000
10.0	0.833
20.0	0.714
30.0	0.625
40.0	0.555

    (a) Plot these data, connect the points with a smooth line, and calculate the rate of change of [A] for each 10-second interval from 0 to 40 seconds. Why does the rate of change decrease from one time interval to the next?
    (b) How is the rate of change of [B] related to the rate of change of [A] in the same time interval? Calculate the rate of change of [B] for the time interval from 10 to 20 seconds.
23. A compound called phenyl acetate reacts with water according to the equation

$$CH_3{-}\overset{\overset{\displaystyle O}{\|}}{C}{-}O{-}C_6H_5(aq) + H_2O(\ell) \longrightarrow$$

phenyl acetate

$$CH_3{-}\overset{\overset{\displaystyle O}{\|}}{C}{-}OH(aq) + C_6H_5{-}OH(aq)$$

acetic acid phenol

and the following data were collected at 5 °C.

TIME (min)	[PHENYL ACETATE] (mol/L)
0	0.55
0.25	0.42
0.50	0.31
0.75	0.23
1.00	0.17
1.25	0.12
1.50	0.085

    (a) Plot these data, and describe the shape of the curve observed. Compare this with Figure 15.4.
    (b) Calculate the rate of change of [phenyl acetate] during the period from 0.20 min to 0.40 min and then during the period from 1.2 min to 1.4 min. Compare the values and tell why one is smaller than the other.

(c) What is the rate of change of [phenol] during the period from 1.00 min to 1.25 min?

CONCENTRATION AND RATE EXPRESSIONS

24. The reaction $CO(g) + NO_2(g)$ is second order in NO_2 and zero order in CO at temperatures less than 500 K.
    (a) Write the rate expression for the reaction.
    (b) How will the reaction rate change if the NO_2 concentration is halved?
25. Nitrosyl bromide, NOBr, is formed from NO and Br_2.

$$2\ NO(g) + Br_2(g) \longrightarrow 2\ NOBr(g)$$

Experiment shows that the reaction is first order in Br_2 and second order in NO.
    (a) Write the rate expression for the reaction.
    (b) If the concentration of Br_2 is tripled, how will the reaction rate change?
    (c) What happens to the reaction rate when the concentration of NO is doubled?
26. For each of the following expressions, give the reaction order in terms of each reagent and for the overall reaction.
    (a) Rate = $k[A][B]^2$ (c) Rate = $k[A]$
    (b) Rate = $k[A][B]$ (d) Rate = $k[A]^3[B]$
27. A reaction between molecules A and B ($A + B \rightarrow$ products) is found to be second order overall. Which rate law below *cannot* be correct?
    (a) Rate = $k[A][B]$ (c) Rate = $k[A][B]^2$
    (b) Rate = $k[A]^2$ (d) Rate = $k[B]^2$
28. For a reaction of the type $A + B \rightarrow C$, it is experimentally observed that doubling the concentration of B causes the reaction rate to be increased fourfold, but doubling the concentration of A has no effect on the rate. The rate equation is therefore
    (a) Rate = $k[A]^2$ (c) Rate = $k[B]^2$
    (b) Rate = $k[A][B]$ (d) Rate = $k[A]$
29. Nitrogen oxide, a problem molecule in air pollution, can be reduced with hydrogen.

$$2\ H_2(g) + 2\ NO(g) \longrightarrow 2\ H_2O(g) + N_2(g)$$

Experiment shows that when the concentration of H_2 is halved, the reaction rate is halved. Furthermore, raising the concentration of NO by a factor of three raises the rate by a factor of nine. Write the rate expression for this reaction.
30. The reaction of $Pt(NH_3)_2Cl_2$ with water is described on page 613,

$$Pt(NH_3)_2Cl_2 + H_2O \longrightarrow [Pt(NH_3)_2(H_2O)Cl]^+ + Cl^-$$

and the rate expression was given as "Rate = $k[Pt(NH_3)_2Cl_2]$" with $k = 0.090$/hr. Calculate the initial rate of reaction when the concentration of $Pt(NH_3)_2Cl_2$ is 0.020 M. What is the rate of change of $[Cl^-]$ under these circumstances?
31. Methyl acetate, $H_3CC(O)OCH_3$, reacts with base to break one of the C—O bonds (page 644).

$$H_3CC(O)O{-}CH_3 + OH^- \longrightarrow H_3CC(O)O^- + HO{-}CH_3$$

The rate expression is "Rate = $k[H_3CC(O)OCH_3][OH^-]$" where $k = 0.137$ L/(mol · sec) at 25 °C. What is the initial rate at which the methyl acetate disappears when both reactants, $H_3CC(O)OCH_3$ and OH^-, have concentrations of 0.025 M? How rapidly does the methyl alcohol, CH_3OH, appear in the solution?
32. The transfer of an O atom from NO_2 to CO has been studied at 540 K,

$$CO(g) + NO_2(g) \longrightarrow CO_2(g) + NO(g)$$

and the following data were collected. Use them to (a) write the rate expression, (b) determine the reaction order with respect to each reactant, and (c) calculate the rate constant, making certain to give the correct units for k.

CONCENTRATIONS		INITIAL RATE
[CO] (mol/L)	$[NO_2]$ (mol/L)	mol/(L · hr)
5.1×10^{-4}	0.35×10^{-4}	3.4×10^{-8}
5.1×10^{-4}	0.70×10^{-4}	6.8×10^{-8}
5.1×10^{-4}	0.18×10^{-4}	1.7×10^{-8}
1.0×10^{-3}	0.35×10^{-4}	6.8×10^{-8}
1.5×10^{-3}	0.35×10^{-4}	10.2×10^{-8}

33. A study of the reaction $2\ A + B \rightarrow C + D$ gave the following results:

EXPERIMENT	CONCENTRATIONS (mol/L)		INITIAL RATE
	[A]	[B]	mol/(L · sec)
1	0.10	0.05	6.0×10^{-3}
2	0.20	0.05	1.2×10^{-2}
3	0.30	0.05	1.8×10^{-2}
4	0.20	0.15	1.1×10^{-1}

    (a) What are the rate expression and overall order for this reaction?
    (b) Calculate the rate constant for the reaction, making certain to give the correct units for k.
34. The bromination of acetone is acid-catalyzed.

$$CH_3COCH_3 + Br_2 \xrightarrow{\text{acid catalyst}} CH_3COCH_2Br + H^+ + Br^-$$

The rate of disappearance of bromine was measured for several different initial concentrations (all in mol/L) of acetone, bromine, and H^+.

CONCENTRATIONS (mol/L)			INITIAL RATE
$[CH_3COCH_3]$	$[Br_2]$	$[H^+]$	mol/(L · sec)
0.30	0.05	0.05	5.7×10^{-5}
0.30	0.10	0.05	5.7×10^{-5}
0.30	0.05	0.10	12.0×10^{-5}
0.40	0.05	0.20	31.0×10^{-5}
0.40	0.05	0.05	7.6×10^{-5}

(a) Deduce the rate expression for the reaction and give the order with respect to each reactant as well as the overall order.
(b) What is the numerical value of k, the rate constant?
(c) If $[H^+]$ is maintained at 0.050 M, whereas both $[CH_3COCH_3]$ and $[Br_2]$ are 0.10 M, what is the initial rate of the reaction?

35. One of the major eye irritants in smog is formaldehyde, CH_2O, formed by reaction of ozone with ethylene.

$$C_2H_4(g) + O_3(g) \longrightarrow 2\ CH_2O(g) + \tfrac{1}{2}\ O_2(g)$$

(a) What are the rates of change of $[O_3]$ and of $[CH_2O]$ relative to each other?
(b) Using the data below, determine the rate expression. What is the reaction order in terms of O_3? In terms of C_2H_4?
(c) Calculate the rate constant.
(d) What is the rate of reaction when $[C_2H_4]$ and $[O_3]$ are both 2.0×10^{-7} M?

$[O_3]$ (mol/L)	$[C_2H_4]$ (mol/L)	INITIAL RATE OF CHANGE OF $[CH_2O]$ mol/(L · s)
0.50×10^{-7}	1.0×10^{-8}	1.0×10^{-12}
1.5×10^{-7}	1.0×10^{-8}	3.0×10^{-12}
1.0×10^{-7}	2.0×10^{-8}	4.0×10^{-12}

36. The data below were collected for the reaction $A + B \rightarrow D + E$.

CONCENTRATIONS (mol/L)		INITIAL RATE
[A]	[B]	mol/(L · min)
0.10	0.20	2.0×10^2
0.10	0.10	5.0×10^1
0.30	0.10	1.5×10^2

(a) Write the rate expression for the reaction and give the order with respect to each reactant.
(b) Calculate the rate constant k, giving the correct units.
(c) What is the rate of reaction when [A] = [B] = 0.20 M?
(d) What concentration of B is required to give a rate of 6.0×10^2 mol/(L · min) when [A] = 0.30 M?

37. One reaction that may occur in air polluted with nitrogen oxide is

$$2\ NO(g) + O_2(g) \longrightarrow 2\ NO_2(g)$$

Using the data below, answer the questions that follow.

EXPERIMENT	INITIAL CONCENTRATIONS		INITIAL RATE OF CHANGE OF $[NO_2]$
	[NO] (M)	$[O_2]$ (M)	mol/(L · sec)
1	0.001	0.001	7×10^{-6}
2	0.001	0.002	14×10^{-6}
3	0.001	0.003	21×10^{-6}
4	0.002	0.003	84×10^{-6}
5	0.003	0.003	189×10^{-6}

(a) What is the order of reaction with respect to each reactant?
(b) What is the overall order?
(c) Write the rate expression for the reaction.
(d) Calculate the rate of change of $[NO_2]$ when $[NO] = [O_2] = 0.005$ mol/L.

CONCENTRATION/TIME EQUATIONS AND REACTION HALF-LIFE

38. The transformation of cyclopropane to propene was described in Example 15.4. If the initial concentration of cyclopropane is 0.050 M, how many hours must elapse for the concentration to drop to 0.025 M? (The first order rate constant is 5.4×10^{-2}/hr.)

39. The decomposition of SO_2Cl_2 is first order in SO_2Cl_2,

$$SO_2Cl_2(g) \longrightarrow SO_2(g) + Cl_2(g)$$

and it has a rate constant of 0.17/hr. If the initial concentration of SO_2Cl_2 is 1.25×10^{-3} M, how long will it take for the concentration to drop to 0.31×10^{-3} M?

40. For a reaction with the rate expression "Rate = k[A]," and $k = 3.33 \times 10^{-6}$/hr, what is the half-life of the reaction? How long will it take for the concentration of A to drop from 1.0 M to 0.20 M?

41. The decomposition of N_2O_5 (to give NO_2 and O_2) follows the rate expression "Rate = $k[N_2O_5]$" where k is 5.0×10^{-4} per second at a particular temperature.
(a) What is the half-life of the reaction?
(b) How long does it take for the N_2O_5 concentration to drop to one tenth of its original value?

42. The decomposition of phosphine, PH_3, proceeds according to the equation

$$4\ PH_3(g) \longrightarrow P_4(g) + 6\ H_2(g)$$

It is found that the reaction has the rate expression "Rate = $k[PH_3]$." If the half-life is 37.9 seconds, how much time is required for three fourths of the PH_3 to decompose?

43. The decomposition of SO_2Cl_2 is first order in SO_2Cl_2, and it has a half-life of 4.1 hours.

$$SO_2Cl_2(g) \longrightarrow SO_2(g) + Cl_2(g)$$

If you begin with 1.6×10^{-3} moles of SO_2Cl_2 in a flask, how many hours elapse until 2.00×10^{-4} moles remain?

44. Bond breaking reactions are important to study, and one investigated recently can be written

$$M—SCH_3 \longrightarrow M + SCH_3$$

where M is a metal to which other groups are attached. One of the authors of this book recently studied the kinetics of the process and found it to be first order in $M—SCH_3$, with a half-life of 40. minutes at room temperature.

(a) What is the rate constant for the reaction?
(b) If you begin with a 0.0060 M solution of $M—SCH_3$, how much is left after 2.0 hours?
(c) Concentrations of $M—SCH_3$ as low as 10^{-4} M can be detected. If you begin with a concentration of $M—SCH_3$ of 0.0060 M, how long does it take to reach 1.0×10^{-4} M, the limit of the detection method?

45. The Cr^{3+} complex ion below changes its structure (it isomerizes) by a first order reaction. The half-life is found to be 1.07 hours.

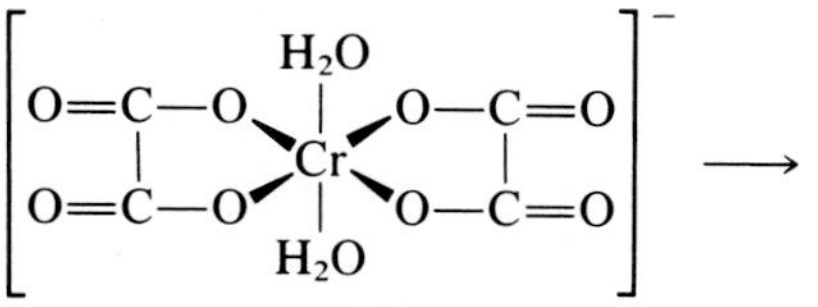

```
            H2O                     -
 O=C—O      |      O—C=O
   |    \   |   /    |
   |      Cr         |
   |    /   |   \    |
 O=C—O      |      O—C=O
            H2O
```

trans form ⟶

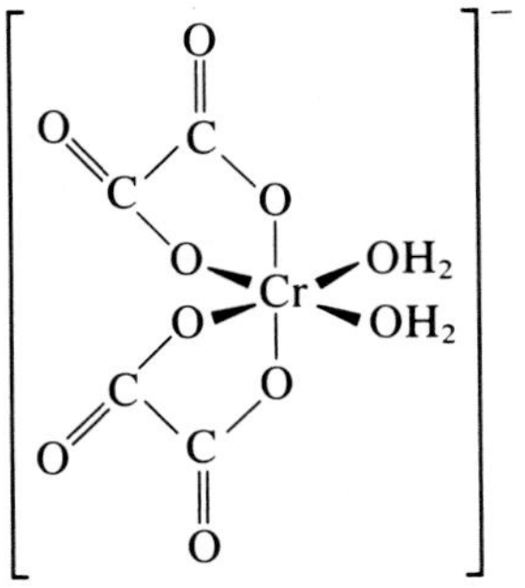

cis form

(a) What is the rate constant for the reaction?
(b) What fraction of the original form (the *trans* form) remains after 3.21 hours of reaction time?
(c) When the concentration of the *trans* form is 0.125 M, what is the rate of reaction in moles per liter per hour?
(d) How long will it take to reduce the concentration of the *trans* form to 1.00% of its original concentration?

46. Although hypochlorous acid, HOCl, was one of the first compounds known of chlorine, the fluorine analog, HOF, was only recently isolated. Instability is its most prominent chemical property, and it decomposes by a first order reaction to HF and O_2 with a half-life of only 30. minutes at room temperature.

$$HOF(g) \longrightarrow HF(g) + \tfrac{1}{2}\ O_2(g)$$

If the partial pressure of HOF in a 1.00-L flask is initially 100. mmHg at 25 °C, what are the total pressure in the flask and the partial pressure of HOF after 30. minutes? After 45 minutes?

47. We know that the decomposition of SO_2Cl_2 is first order in SO_2Cl_2,

$$SO_2Cl_2(g) \longrightarrow SO_2(g) + Cl_2(g)$$

with a half-life of 240 minutes at 600 K. If you begin with a partial pressure of SO_2Cl_2 of 25 mmHg in a 1.0-L flask, what is the partial pressure of each reactant and product after 240 minutes? What is the partial pressure of the reactant and products after 12 hours?

48. Molecules of butadiene, C_4H_6, can couple to form C_8H_{12}. The rate expression for the reaction is "Rate = $k[C_4H_6]^2$," and the rate constant is estimated to be 0.014 L/(mol · s). If the initial concentration of C_4H_6 is 0.016 M, how much time (in seconds) must elapse for the concentration to drop to 0.0016 M?

49. Methyl acetate, $H_3CC(O)OCH_3$, reacts with base to break one of the C—O bonds (page 644).

$$H_3CC(O)O—CH_3 + OH^- \longrightarrow H_3CC(O)O^- + HO—CH_3$$

The rate expression is "Rate = $k[H_3CC(O)OCH_3][OH^-]$" where k = 0.137 L/(mol · sec) at 25 °C. How long does it take to go from a solution where $[H_3CC(O)OCH_3]$ and $[OH^-]$ are both 0.010 M to a solution where they are 0.0095 M?

GRAPHICAL ANALYSIS OF RATE EXPRESSIONS

50. Common sugar, sucrose, reacts in dilute acid solution to give the simpler sugars glucose and fructose. Both of the simple sugars have the same formula, $C_6H_{12}O_6$.

$$C_{12}H_{22}O_{11}(aq) + H_2O(\ell) \longrightarrow 2\ C_6H_{12}O_6(aq)$$

The rate of this reaction has been studied in acid solution, and the following data were obtained.

TIME (min)	$[C_{12}H_{22}O_{11}]$ (M)
0	0.316
39	0.274
80	0.238
140	0.190
210	0.146

(a) Plot the data above as ln[sucrose] versus time and 1/[sucrose] versus time. What is the order of the reaction?
(b) Write the rate expression for the reaction and calculate the rate constant k.
(c) Estimate the concentration of sucrose after 175 minutes.

51. Data for the reaction of phenyl acetate with water are given in Study Question 23. Plot these data as ln[phenyl acetate] and 1/[phenyl acetate] versus time. What is the order of the reaction with respect to phenyl acetate? Determine the rate constant from your plot and calculate the half-life for the reaction.

52. Data for the reaction of an organic bromide (A) with an organic amine (B) were obtained in methyl alcohol when equal initial concentrations of reactants were used at 35 °C.

$$\underset{\text{A}}{C_6H_5{-}\overset{\overset{\displaystyle O}{\|}}{C}{-}CH_2Br} + \underset{\text{B}}{C_5H_5N} \longrightarrow$$

$$[C_6H_5{-}\overset{\overset{\displaystyle O}{\|}}{C}{-}CH_2NC_5H_5]^+ + Br^-$$

TIME (min)	[A] (mol/L)
0	0.0385
200	0.0288
400	0.0230
600	0.0191
800	0.0163
1000	0.0143

(a) Plot these data as ln[A] and 1/[A] versus time. What is the order of the reaction?
(b) What is the rate constant?
(c) Based on your answer to part (a), and the fact that the initial concentrations of A and B were equal ([A] = [B]), decide on the rate expression for this reaction.
(d) What is the half-life of the reaction?

53. Data for the combination of two C_4H_6 molecules to give C_8H_{12} [$C_4H_6(g) \rightarrow \frac{1}{2}\ C_8H_{12}(g)$] were collected at 500 °C and are given below.

TIME (sec)	$[C_4H_6]$ (mol/L)
195	1.62×10^{-2}
604	1.47×10^{-2}
1246	1.29×10^{-2}
2180	1.10×10^{-2}
4140	0.84×10^{-2}
4655	0.80×10^{-2}
6210	0.68×10^{-2}
8135	0.57×10^{-2}

Plot these data as $\ln[C_4H_8]$ and $1/[C_4H_8]$ versus time. What is the order of the reaction? Determine the rate constant from your plot.

KINETICS AND ENERGY

54. For the hypothetical reaction A + B → C + D, the activation energy is 32 kJ/mol. For the reverse reaction (C + D → A + B), the activation energy is 58 kJ/mol. Is the reaction A + B → C + D exothermic or endothermic?

55. Use the diagram below to answer the following questions.
(a) Is the reaction exothermic or endothermic?
(b) What is the approximate value of ΔE for the forward reaction?
(c) What is the activation energy in each direction?
(d) A catalyst is found that lowers the activation energy of the reaction by about 10 kJ/mol. How will this catalyst affect the rate of the reverse reaction?

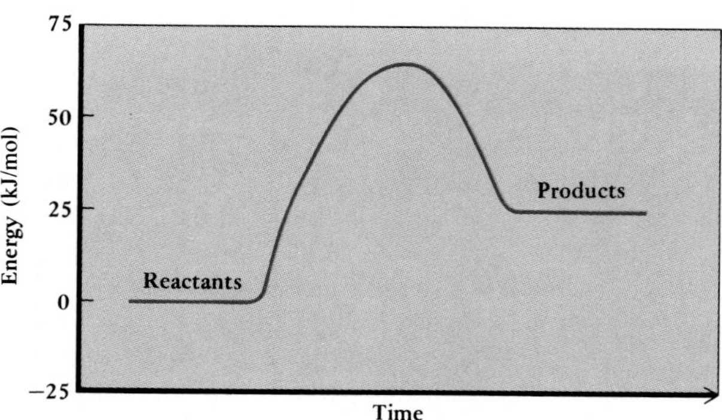

56. Calculate the activation energy (E^*) for the reaction

$$N_2O_5(g) \longrightarrow 2\ NO_2(g) + \tfrac{1}{2}\ O_2(g)$$

from the following information: k at 25 °C is 3.46×10^{-5}/s, and k at 55 °C is 1.5×10^{-3}/s.

57. If the rate constant for a reaction triples in value when the temperature rises from 300. K to 310. K, what is the activation energy of the reaction?

MECHANISMS

58. Write the rate expression for each *elementary* step below. Give the molecularity of each step.
   (a) $NO(g) + NO_3(g) \longrightarrow 2\ NO_2(g)$
   (b) $Cl(g) + H_2(g) \longrightarrow HCl(g) + H(g)$
   (c) $(CH_3)_3CBr(aq) \longrightarrow (CH_3)_3C^+(aq) + Br^-(aq)$

59. The reaction between chloroform and chlorine gas proceeds in a series of three elementary steps.

Step 1: Fast, reversible
$Cl_2(g) \rightleftharpoons 2\ Cl(g)$
Step 2: Slow
$CHCl_3(g) + Cl(g) \longrightarrow CCl_3(g) + HCl(g)$
Step 3: Fast
$CCl_3(g) + Cl(g) \longrightarrow CCl_4(g)$

Overall:
$CHCl_3(g) + Cl_2(g) \longrightarrow CCl_4(g) + HCl(g)$

   (a) Which of the steps is the rate determining step?
   (b) Write the rate expression for the rate determining step.
   (c) What is the molecularity of each step?

60. The ozone layer in the earth's upper atmosphere is important in shielding the earth from very harmful ultraviolet radiation. The ozone, O_3, decomposes according to the equation $2\ O_3(g) \rightarrow 3\ O_2(g)$. The mechanism of the reaction is thought to proceed through an initial fast, reversible step and then a slow second step.

Step 1: Fast, reversible $O_3(g) \rightleftharpoons O_2(g) + O(g)$

Step 2: Slow $O_3(g) + O(g) \longrightarrow 2\ O_2(g)$

   (a) Which of the steps is the rate determining step?
   (b) Write the rate expression for the rate determining step.
   (c) What is the molecularity of each step?

61. The Raschig process for producing hydrazine was described in Section 15.6. Alternatives to steps 2 and 3 are:

Alternate step 2: Slow
$NH_2Cl(aq) + OH^-(aq) \longrightarrow NHCl^-(aq) + H_2O(\ell)$

Alternate step 3: Fast
$NHCl^-(aq) + NH_3(aq) \longrightarrow N_2H_4(aq) + Cl^-(aq)$

   (a) Which of these is the rate determining step? What is its molecularity?
   (b) What is the rate expression for the rate determining elementary step?

62. Assume that an A molecule reacts with two B molecules in a one-step process to give AB_2. That is, $A + 2\,B \rightarrow AB_2$.
   (a) Write the rate expression for this reaction.
   (b) If the initial rate of change of $[AB_2]$ is 2.0×10^{-5} mol/(L · s) when the initial concentrations of A and B are both 0.30 M, calculate the rate constant for the reaction.

63. At temperatures less than 500 K the reaction between carbon monoxide and nitrogen dioxide

$$NO_2(g) + CO(g) \longrightarrow CO_2(g) + NO(g)$$

follows the rate expression "Rate = $k[NO_2]^2$." Which of the three mechanisms suggested below best agrees with the experimentally observed rate expression?

Mechanism 1: (single, elementary step)
$NO_2 + CO \longrightarrow CO_2 + NO$

Mechanism 2:
(slow) $NO_2 + NO_2 \longrightarrow NO_3 + NO$
(fast) $NO_3 + CO \longrightarrow NO_2 + CO_2$

Mechanism 3:
(slow) $NO_2 \longrightarrow NO + O$
(fast) $CO + O \longrightarrow CO_2$

CATALYSIS

64. Which of the following statements is (are) true?
   (a) The concentration of a homogeneous catalyst may appear in the rate expression.
   (b) A catalyst is always consumed in the reaction.
   (c) A catalyst must always be in the same phase as the reactants.

65. Hydrogenation reactions, processes in which H is added to a molecule, are usually catalyzed. An excellent catalyst is a very finely divided metal suspended in the reaction solvent. Tell why finely divided rhodium, for example, is a much more efficient catalyst than a small block of metal.

66. Which of the following reactions *appear* to involve a catalyst? In those cases where a catalyst is present, tell whether it is homogeneous or heterogeneous.
   (a) $CH_3CO_2CH_3(aq) + H_2O(\ell) + H^+(aq) \longrightarrow CH_3CO_2H(aq) + CH_3OH(aq) + H^+(aq)$
   (b) $2\ H_2(g) + O_2(g) \longrightarrow 2\ H_2O(g)$
   (c) $2\ H_2(g) + O_2(g) + Pt(s) \longrightarrow 2\ H_2O(g) + Pt(s)$
   (d) $NH_3(aq) + CH_3Cl(aq) + H_2O(\ell) \longrightarrow Cl^-(aq) + NH_4^+(aq) + CH_3OH(aq)$

67. In acid solution, one of the C—O bonds of an ester of formic acid, $HC(O)OCH_3$, is broken.

$HC(O)O—CH_3 + H_2O \longrightarrow$
methyl formate

$HC(O)O—H$ + $HO—CH_3$
formic acid methyl alcohol

The rate expression is "Rate = $k[HC(O)OCH_3][H^+]$." Why does H^+ appear in the rate expression but not in the overall equation for the reaction?

GENERAL QUESTIONS

68. The following reaction between methyl alcohol and an organic halide occurs in benzene at 25 °C (when a weak base is present).

$CH_3OH + (C_6H_5)_3C—Cl \longrightarrow$
A B

$(C_6H_5)_3C—OCH_3 + HCl$
C

In a study of its mechanism, the following data were collected:

INITIAL CONCENTRATIONS [A] (M)	[B] (M)	INITIAL RATE OF CHANGE OF [C] mol/(L · min)
0.10	0.05	1.32×10^{-4}
0.10	0.10	2.60×10^{-4}
0.20	0.10	10.3×10^{-4}

(a) What is the rate law for the process?
(b) If you use the initial concentrations [A] = 0.15 M and [B] = 0.10 M, what is the rate of change of [C]?

69. Nitryl fluoride, an explosive compound, can be made by treating nitrogen dioxide with fluorine.

$$2\ NO_2(g) + F_2(g) \longrightarrow 2\ NO_2F(g)$$

Use the rate data below to answer the questions that follow.

EXPERIMENT	INITIAL CONCENTRATIONS (mol/L) $[NO_2]$	$[F_2]$	$[NO_2F]$	INITIAL RATE mol/(L · s)
1	0.001	0.005	0.001	2×10^{-4}
2	0.002	0.005	0.001	4×10^{-4}
3	0.006	0.002	0.001	4.8×10^{-4}
4	0.006	0.004	0.001	9.6×10^{-4}
5	0.001	0.001	0.001	4×10^{-5}
6	0.001	0.001	0.002	4×10^{-5}

(a) Write the rate expression for the reaction.
(b) What is the order of reaction with respect to each component of the reaction?
(c) What is the numerical value of the rate constant k?

70. Sketch a reaction energy diagram for an exothermic reaction that proceeds through two steps, with the second step rate determining. If the first step were rate determining, how would your diagram change?

71. If you heat SO_2Cl_2 to a sufficiently high temperature it will decompose to SO_2 and Cl_2 at a reasonable rate.

$$SO_2Cl_2(g) \longrightarrow SO_2(g) + Cl_2(g)$$

Use the plot (below) of the logarithm of the pressure of SO_2Cl_2 as a function of time to determine the half-life for SO_2Cl_2.

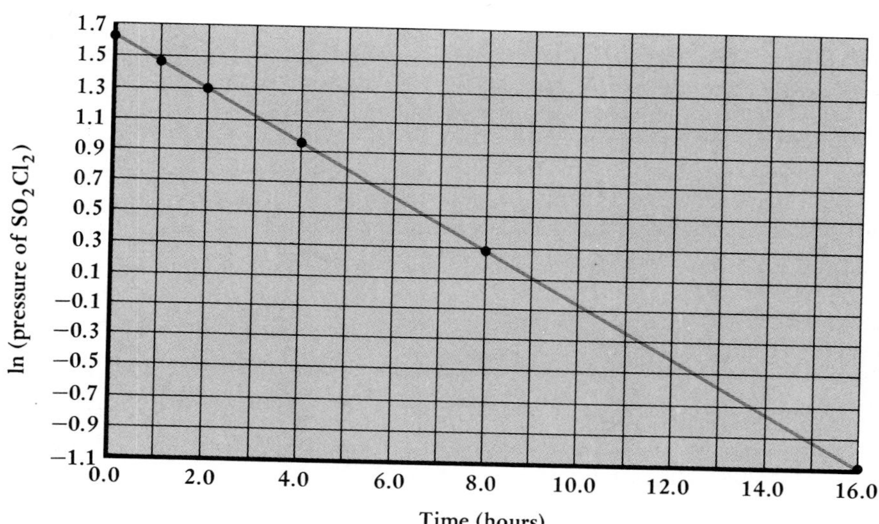

72. The decomposition of dinitrogen pentoxide

$$2\ N_2O_5(g) \longrightarrow 4\ NO_2(g) + O_2(g)$$

has the rate expression "Rate = $k[N_2O_5]$." It has been found experimentally that the decomposition is 20% complete in 6.0 hours at 300. K. Calculate the rate constant and the half-life at 300. K.

73. It is desirable that pesticides eventually decompose in the environment to give harmless products. Let us assume the half-life for the first-order decomposition of a certain pesticide is 10.2 years. If the present concentration of the pesticide in a lake is 3.1×10^{-5} g/mL, what was its concentration 4.00 years ago?

74. A catalyst can cause a reaction to proceed along a different pathway with a lower activation energy. If the catalyst lowers the activation energy by 5 kJ/mol, how much greater is the rate constant, if both occur at 298 K (and the frequency factor A does not change)?

75. Tell whether the following statements are true or false. If false, change the sentence to make it correct.
   (a) The rate controlling elementary step in a reaction is the slowest step in a mechanism.
   (b) It is possible to change the rate constant by changing the temperature.
   (c) As a reaction proceeds at constant temperature, the rate remains constant.
   (d) A reaction that is third order overall must involve more than one step.

76. For a certain reaction, the constant A in the Arrhenius equation is 6.00×10^{12} L/(mol · s) and $E^* = 100.$ kJ/mol. What is the rate constant at 400. K?

77. The data below give the temperature dependence of the rate constant for the reaction $N_2O_5(g) \rightarrow 2\ NO_2(g) + \frac{1}{2}\ O_2(g)$. Plot these data in the appropriate way to derive the activation energy for the reaction.

T (K)	k (per second)
338	4.87×10^{-3}
328	1.50×10^{-3}
318	4.98×10^{-4}
308	1.35×10^{-4}
298	3.46×10^{-5}
273	7.87×10^{-7}

SUMMARY QUESTION

78. The chemistry of the transition metal compound $Ni(CO)_4$ was mentioned in the text. The substitution of CO in this molecule was studied some years ago (in the nonaqueous solvents toluene and hexane) and led to an understanding of some of the general principles that govern the chemistry of compounds having metal—CO bonds. (See J.P. Day, F. Basolo, and R.G. Pearson, *J. Am. Chem. Soc.* **1968**, *90*, 6927.)

Ni(CO)(OC)(CO)(CO) + L ⟶ Ni(L)(OC)(CO)(CO) + CO

L in this reaction is an electron pair donor, like CO, but the atom bearing the pair of electrons to be donated is usually P, as in $:P(CH_3)_3$. As mentioned in the text, a detailed study of the kinetics of the reaction led to the following mechanism:

Slow $Ni(CO)_4 \longrightarrow Ni(CO)_3 + CO$

Fast $Ni(CO)_3 + L \longrightarrow Ni(CO)_3L$

(a) What is the molecularity of each of the elementary reactions?
(b) It was found that doubling the concentration of $Ni(CO)_4$ led to an increase in reaction rate by a factor of 2. Doubling the concentration of L had no effect on the reaction rate. Based on this information, write the rate expression for the reaction. Does this agree with the mechanism above?
(c) The experimental rate constant for the reaction, when L = $P(C_6H_5)_3$, is 9.3×10^{-3}/s at 20 °C. If the initial concentration of $Ni(CO)_4$ is 0.025 M, what is the concentration of the product after 5.0 minutes?
(d) $Ni(CO)_4$ is formed by the reaction of nickel metal with carbon monoxide. If you have 750 mL of CO at a pressure of 1.50 atm at 22 °C, and you combine it with 0.125 g of nickel metal, how many grams of $Ni(CO)_4$ can be formed? If CO remains after reaction, what is its pressure in the 750 mL flask at 29 °C?
(e) An excellent way to make pure nickel metal is to decompose $Ni(CO)_4$ in a vacuum at a temperature slightly higher than room temperature. What is the enthalpy change for the decomposition reaction

$$Ni(CO)_4(g) \longrightarrow Ni(s) + 4\ CO(g)$$

if the molar enthalpy of formation of $Ni(CO)_4$ gas is −602.91 kJ/mol?

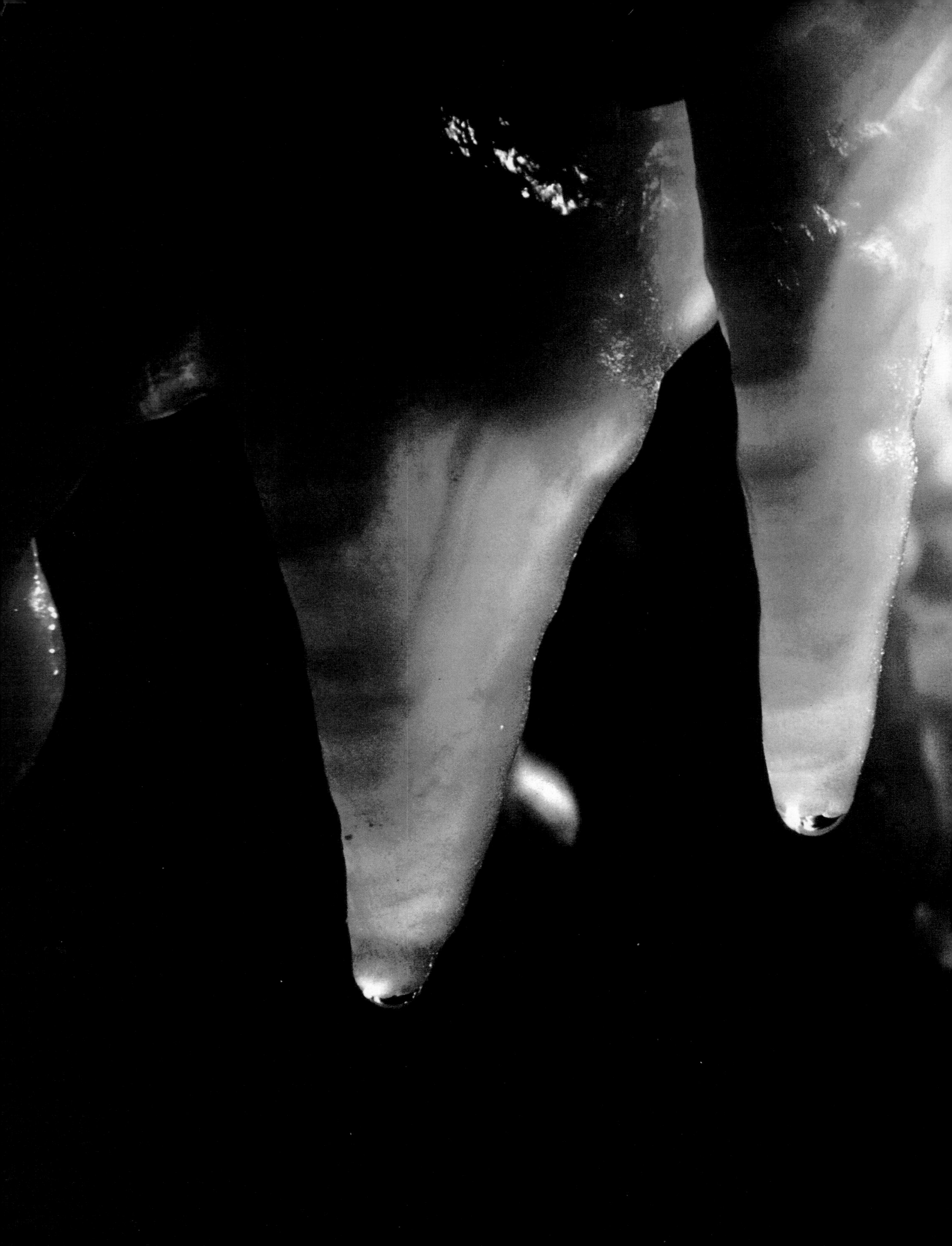

CHAPTER

16

Chemical Equilibria: General Concepts

Calcium carbonate stalactites. (Arthur Palmer)

CHAPTER OUTLINE

The concept of a state of equilibrium is fundamental in chemistry, and so this and the next three chapters explore this concept as applied to chemical reactions. But don't think these ideas are peculiar to chemistry. You participate in social situations and live in an economy that represent equilibria of competing forces. You and your family or your roommate have arrived at some arrangements that keep your personal relationships as smooth as possible. That is, you have achieved an equilibrium. Of course, this equilibrium is easily upset when a stress is applied to the arrangement, as when another child is added to the family or another roommate moves in. In international affairs, countries achieve an equilibrium of competing interests, but that equilibrium can be upset, for example, when the currency of one country changes in value or when a third country intervenes.

Chemical systems are analogous to our social and political arrangements, and chemical systems can also be forced to move to a new equilibrium position by outside influences. Our goal in this chapter is to explore the consequences of the facts that most chemical reactions are reversible, that in a closed system a state of equilibrium can eventually be achieved between reactants and products, and that outside forces can affect the equilibrium. A major result of this exploration will be an ability to describe "chemical reactivity" in more quantitative terms. As you will see, the concentrations of reactants and products at equilibrium are a measure of the intrinsic tendency of chemical reactions to proceed from reactants to products.

Figure 16.1 Calcium carbonate stalactites hang from the ceiling of a cave, and stalagmites grow from the floor. The chemistry that produces these formations is an excellent illustration of chemical equilibria. (See also the chapter opening photograph.) (Arthur Palmer)

16.1 THE NATURE OF THE EQUILIBRIUM STATE

The concept of reaction reversibility was first introduced in Chapter 13 when discussing phase equilibria. From these discussions you began to develop some ideas about the nature of the state of equilibrium, but now we want to make these ideas more explicit.

As an example of chemical equilibrium, consider the reaction that leads to the formation of limestone stalactites and stalagmites in some caves (Figure 16.1). These originate from the reaction of limestone, a form of calcium carbonate, with CO_2 and water.

$$CaCO_3(s) + CO_2(g) + H_2O(\ell) \longrightarrow Ca^{2+}(aq) + 2\ HCO_3^-(aq)$$

$CaCO_3$ is found in underground deposits, a leftover of ancient oceans. If water seeping through the ground contains dissolved CO_2, the preceding reaction can give aqueous Ca^{2+} and HCO_3^- ions. However, *most chemical reactions are* **reversible**, just as are phase changes, and the reaction above is no exception. To prove the reversibility of the reaction, let us do an experiment with some soluble salts containing the Ca^{2+} and HCO_3^- ions (say $CaCl_2$ and $NaHCO_3$). If you put them in an open beaker of water in the laboratory, you will soon see bubbles of CO_2 gas and a precipitate of solid $CaCO_3$ (Figure 16.2). If the CO_2 is swept away, all the dissolved Ca^{2+} and HCO_3^- ions will eventually disappear from the solution, having been converted completely to solid $CaCO_3$.

$$Ca^{2+}(aq) + 2\ HCO_3^-(aq) \longrightarrow CO_2(g) + H_2O(\ell) + CaCO_3(s)$$

This is of course the reverse of the reaction that dissolves limestone, and it is the process that redeposits the $CaCO_3$ in caves as beautiful limestone formations.

Now do another experiment, but this time put some water and solid $CaCO_3$ in a flask under a high partial pressure of CO_2 gas, just as nature

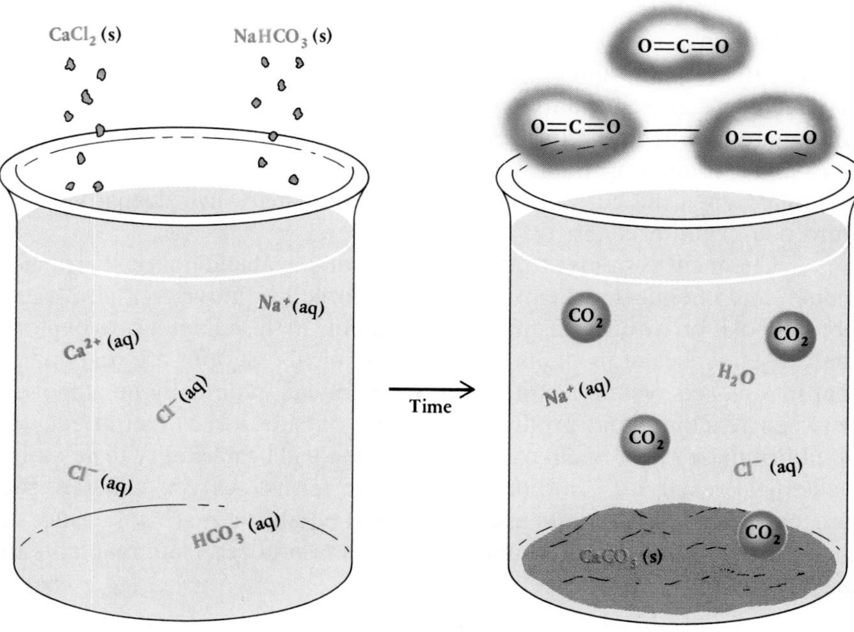

Figure 16.2 Some $CaCl_2$ and $NaHCO_3$ are added to a beaker of water. With time, the reaction

$$Ca(HCO_3)_2(aq) \rightleftharpoons CaCO_3(s) + CO_2(g) + H_2O(\ell)$$

occurs, evolving CO_2, precipitating $CaCO_3$, and leaving NaCl in solution.

does underground. Eventually some of the limestone will begin to dissolve to give aqueous Ca^{2+} and HCO_3^- ions according to the first equation above. From these experiments—and the observation of limestone deposits in caves—it is clear that the system containing limestone, water, carbon dioxide, and aqueous calcium bicarbonate is reversible, a fact that we convey with a set of double arrows to connect the two sides of the balanced equation.

$$CaCO_3(s) + CO_2(g) + H_2O(\ell) \rightleftharpoons Ca^{2+}(aq) + 2\ HCO_3^-(aq)$$

The set of double arrows, ⇄, in an equation indicates that the reaction is reversible. In general chemistry courses, it is often a signal to you that the reaction will be studied using the concepts of chemical equilibria.

Chemical equilibria are **dynamic**. When a system comes to equilibrium, this does not mean that all reaction stops. Instead, *when the system is at equilibrium, the forward and reverse reactions continue, and they take place at equal rates.* A common introductory chemistry experiment is the study of the equilibrium that exists in the reaction of aqueous iron(III) ion and aqueous thiocyanate ion, SCN^-.

$$[Fe(H_2O)_6]^{3+} + SCN^- \rightleftharpoons [Fe(H_2O)_5(SCN)]^{2+} + H_2O$$

nearly colorless + SCN^- colorless ⇄ red-orange

When aqueous solutions of these two ions are mixed (Figure 16.3), the SCN^- ion rapidly replaces water on the Fe^{3+} ion to give a new ion in which SCN^- is bonded to Fe^{3+}—an ion that we shall abbreviate as Fe-SCN^{2+}—that has a characteristic red-orange color. As the concentration of Fe-SCN^{2+} begins to build up, this ion releases SCN^- and reverts to simple aqueous Fe^{3+} at a greater and greater rate. Eventually, the rate at which SCN^- replaces H_2O on Fe^{3+} to form Fe-SCN^{2+} (the "forward" reaction) becomes equal to the rate at which Fe-SCN^{2+} sheds SCN^- ion to go back to the simple ions in solution (the "reverse" reaction). At this point equilibrium has been achieved.

Figure 16.3 The reaction of aqueous $[Fe(H_2O)_6]^{3+}$ and aqueous thiocyanate, SCN^-. The colorless solutions are mixed to give the red-orange ion $[Fe(H_2O)_5(SCN)]^{2+}$, which is in equilibrium with the reactants.

However, when there is equilibrium in the Fe^{3+}/SCN^- system, the forward and reverse reactions do not stop. You can prove this if you add a drop of an aqueous solution of radioactive SCN^- ion to the solution. (We can make the ion radioactive by using radioactive ^{14}C in place of normal, nonradioactive ^{12}C.) If you were to sample the solution shortly after adding the radioactive SCN^- ion, you would find the ion incorporated into Fe-SCN^{2+} by reactions such as the following:

$$\text{Fe-SCN}^{2+}(aq) + H_2O(\ell) \rightleftharpoons (H_2O)Fe^{3+}(aq) + SCN^-(aq)$$

$$(H_2O)Fe^{3+}(aq) + \mathbf{SCN}^-(aq) \rightleftharpoons \text{Fe-}\mathbf{SCN}^{2+}(aq) + H_2O(\ell)$$

(The symbol **SCN**$^-$ means the thiocyanate ion contains radioactive ^{14}C.)

The only way for the radioactive SCN^- ion to be incorporated into the red-orange Fe-SCN^{2+} ion is if these reactions are dynamic and reversible.

Not only are equilibrium processes reversible and dynamic, but *the nature and properties of the equilibrium state are the same, no matter what the direction of approach.* To see why this is true, we take the reaction of hydrogen and iodine to produce hydrogen iodide.

$$H_2(g) + I_2(g) \rightleftharpoons 2\ HI(g)$$

Assume you place enough H_2 and I_2 in a flask so that the partial pressure of each at 425 °C is 1.00 atmosphere. With time, the partial pressures of H_2 and I_2 will decline and that of HI will increase; an equilibrium state will be reached eventually. If you were to analyze the gases in the flask at this point you would find that $P(H_2) = P(I_2) = 0.212$ atm while $P(HI) = 1.58$ atm.

	INITIAL AND EQUILIBRIUM PARTIAL PRESSURES (atm)		
	$H_2(g)$	$I_2(g)$	$HI(g)$
Initial pressures	1.00	1.00	0
Changes in pressures as reaction proceeds to equilibrium	−0.788	−0.788	+1.58
Equilibrium pressures	0.212	0.212	1.58

To show that the equilibrium state is the same, no matter in which direction you approach it, reverse the experiment above. This time begin with enough HI to have a partial pressure of 2.00 atm at 425 °C. When equilibrium has once again been reached, you would find the following results:

	INITIAL AND EQUILIBRIUM PARTIAL PRESSURES (atm)		
	$H_2(g)$	$I_2(g)$	$HI(g)$
Initial pressures	0	0	2.00
Changes in pressures as reaction proceeds to equilibrium	+0.212	+0.212	−0.42
Equilibrium pressures	0.212	0.212	1.58

The experiments show quite clearly that the equilibrium partial pressures of the compounds in this reaction are the same, no matter in which direction you have approached the reaction.*

16.2 THE EQUILIBRIUM CONSTANT

Equilibria involving changes between phases are also reversible and dynamic and can be approached from either direction. However, there is a difference in the ways we describe the equilibria involved in phase changes and in chemical reactions. To describe the equilibrium that can exist

*We must be careful interpreting this third aspect of chemical equilibria. Take the H_2/I_2 reaction as an example. If you begin with 1 mole of H_2 and 1 mole of I_2, equilibrium will be established at the same point as if you had begun with 2 moles of HI. However, if you had begun with 2 moles of H_2 and 1 mole of I_2, a new equilibrium state with different concentrations of gases would be reached. To approach this last equilibrium state from the opposite direction, you would have to mix 2 moles of HI with 1 mole of H_2. The statement that the equilibrium state has the same properties no matter what the direction of approach assumes that the number of atoms of each element used is the same in the forward and reverse experiments.

Table 16.1 Equilibrium Partial Pressures for the Reaction $H_2(g) + I_2(g) \rightleftarrows 2\ HI(g)$

Partial Pressures (atm)			
H_2	I_2	HI	$K = P_{HI}^2/P_{H_2}P_{I_2}$
0.1645	0.09783	0.9447	55.46
0.2583	0.04229	0.7763	55.17
0.1274	0.1339	0.9658	54.68
0.1034	0.1794	1.0129	55.31
0.02703	0.02745	0.2024	55.21
0.06443	0.06540	0.4821	55.16
			Average = 55.17

These data are taken from an article by Taylor and Crist, *J. Am. Chem. Soc.* **1941**, *63*, 1377. The temperature was 698.6 K.

between water and water vapor at a particular temperature, you need only give the equilibrium vapor pressure.

$$H_2O(\ell) \rightleftharpoons H_2O(g) \qquad P(H_2O\text{ vapor}) = 23.8\text{ mmHg at }25\ ^\circ C$$

In contrast, to describe the equilibrium position of a chemical reaction, it is often necessary to give equilibrium concentrations of several reagents. Fortunately, the **equilibrium constant expression** *relates equilibrium concentrations of reactants and products at a given temperature*. This expression is of such great importance in chemistry that it is described in detail as applied to general equilibria in this chapter and to equilibria in aqueous solutions in Chapters 17 through 19.

THE EQUILIBRIUM CONSTANT EXPRESSION

The notion that the equilibrium concentrations of reactants and products are interrelated in a simple manner is easy to prove by experiments such as those for the H_2/I_2 gas phase reaction in Table 16.1.

$$H_2(g) + I_2(g) \rightleftharpoons 2\ HI(g)$$

In the first four experiments, H_2 and I_2 were mixed and HI was formed; when equilibrium had been established, the partial pressure of each component of the mixture was determined. The last two experiments were done in the reverse manner; HI was the reactant and H_2 and I_2 were the products. The important point of these experiments is that the equilibrium partial pressures of the reaction components are related through a simple numerical ratio. That is, the ratio

$$\frac{(P_{HI})^2}{P_{H_2}P_{I_2}} = 55.17$$

is the same within experimental error for all six experiments!

Hundreds of other experiments have proved that the kind of relationship found for the $H_2/I_2/HI$ equilibrium is true for all equilibrium reactions. For the general reaction

$$a\text{A} + b\text{B} \rightleftharpoons c\text{C} + d\text{D}$$

experiments show that concentrations of reactants and products are related by the equilibrium constant expression

$$\text{Equilibrium constant} = K = \frac{\overbrace{[\mathrm{C}]^c[\mathrm{D}]^d}^{\text{product concentrations}}}{\underbrace{[\mathrm{A}]^a[\mathrm{B}]^b}_{\text{reactant concentrations}}} \tag{16.1}$$

In the equilibrium constant expression, *the product concentrations appear in the numerator* and *reactant concentrations in the denominator. Each concentration is raised to the power of its stoichiometric coefficient in the balanced equation*. The value of the constant K depends on the particular reaction and on the temperature.

The concentrations in the equilibrium constant expression are usually given in moles per liter (M), in which case the symbol K often has a subscript c for "concentration," as in K_c. However, K expressions involving gases can be written in terms of partial pressures of reactants and products, since the ideal gas law specifies that "concentration $= n/V = P/RT$"; that is, the partial pressure of a gas is proportional to its concentration. In this case K would have a subscript p, as in K_p. We shall return to this point later.

USING EQUILIBRIUM CONSTANT EXPRESSIONS

There are several things you have to know about equilibrium constant expressions so you can apply them in practical situations.

PURE SOLIDS AND LIQUIDS IN EQUILIBRIUM EXPRESSIONS *The concentrations of pure solids, of pure liquids, and of solvents in dilute solutions* **never** *appear in equilibrium constant expressions*. For example, the oxidation of solid, yellow sulfur produces colorless sulfur dioxide gas (Figure 16.4).

Figure 16.4 Sulfur burns in oxygen to give sulfur dioxide gas, SO_2.

$$\mathrm{S(s)} + \mathrm{O_2(g)} \rightleftharpoons \mathrm{SO_2(g)}$$

Following the general principle that products appear in the numerator and reactants in the denominator, you would write

$$K' = \frac{\overbrace{[\mathrm{SO_2(g)}]}^{\text{product}}}{\underbrace{[\mathrm{S(s)}][\mathrm{O_2(g)}]}_{\text{reactants}}}$$

However, since sulfur is a molecular solid and since the concentration of molecules within any solid is fixed, the sulfur concentration is not changed either by reaction or by addition or removal of some solid. Further, it is an experimental fact that the equilibrium concentrations (or partial pressures) of O_2 and SO_2 are not changed by the amount of sulfur, as long as there is some solid sulfur present at equilibrium. Therefore, it is conventional to include the equilibrium concentrations of any solid reactants and products in the equilibrium constant itself and write

$$K'[\mathrm{S(s)}] = K = \frac{[\mathrm{SO_2(g)}]}{[\mathrm{O_2(g)}]} = 4.2 \times 10^{52} \text{ (at 25 °C)}$$

This same principle applies to reactions in aqueous solutions such as

$$\mathrm{AgCl(s)} \rightleftharpoons \mathrm{Ag^+(aq)} + \mathrm{Cl^-(aq)}$$

where only the concentrations of the silver and chloride ions in solution appear in the equilibrium constant expression.

$$K = [\mathrm{Ag^+(aq)}][\mathrm{Cl^-(aq)}] = 1.8 \times 10^{-10} \text{ (at 25 °C)}$$

The silver chloride concentration, [AgCl(s)], does not appear in the expression, since the "concentration" of solid metal halide is constant.

For most reactions occurring in aqueous solution, the concentration of water, $[H_2O]$, is effectively constant. For example, as you learned in Chapter 5, ammonia functions as a weak base owing to the interaction of NH_3 and water,

$$\mathrm{NH_3(aq)} + \mathrm{H_2O}(\ell) \rightleftharpoons \mathrm{NH_4^+(aq)} + \mathrm{OH^-(aq)}$$

$$K' = \frac{[\mathrm{NH_4^+}][\mathrm{OH^-}]}{[\mathrm{NH_3}][\mathrm{H_2O}]}$$

$$K = K'[\mathrm{H_2O}] = \frac{[\mathrm{NH_4^+}][\mathrm{OH^-}]}{[\mathrm{NH_3}]} = 1.8 \times 10^{-5}$$

and formic acid is a weak acid since it can partially transfer an H^+ ion to water to form the hydronium ion.

$$\mathrm{HCOOH(aq)} + \mathrm{H_2O}(\ell) \rightleftharpoons \mathrm{HCOO^-(aq)} + \mathrm{H_3O^+(aq)}$$

$$K' = \frac{[\mathrm{HCOO^-}][\mathrm{H_3O^+}]}{[\mathrm{HCOOH}][\mathrm{H_2O}]}$$

$$K = K'[\mathrm{H_2O}] = \frac{[\mathrm{HCOO^-}][\mathrm{H_3O^+}]}{[\mathrm{HCOOH}]} = 1.8 \times 10^{-4}$$

In neither case is water included in the final equilibrium expression because, by convention, $[H_2O]$ is included in the accepted value of K. Ammonia, formic acid, and their products are present in such small concentrations in dilute solutions that the concentration of water is essentially not affected by their presence.

Finally, when both dissolved materials and gases are present at equilibrium, their concentrations and partial pressures, respectively, appear in the equilibrium expression. A good example of this is a reaction we have already used: the dissolving of calcium carbonate in the presence of water and CO_2.

$$\mathrm{CaCO_3(s)} + \mathrm{CO_2(g)} + \mathrm{H_2O}(\ell) \rightleftharpoons \mathrm{Ca^{2+}(aq)} + 2\ \mathrm{HCO_3^-(aq)}$$

$$K = \frac{[\mathrm{Ca^{2+}}][\mathrm{HCO_3^-}]^2}{P_{\mathrm{CO_2}}}$$

Table 16.2 Summary of the Changes in Equilibrium Constants that Occur When Changing Stoichiometric Coefficients, Reversing Equations, or Adding Equations

Change Made	Effect on K
Stoichiometric coefficients in a balanced equation are changed by a factor of n.	$K_{new} = K_{old}^n$
Balanced chemical equation is reversed.	$K_{new} = 1/K_{old}$
Several balanced equations are added to obtain a net, balanced equation.	$K_{net} = K_1K_2K_3 \ldots$ K_{net} is the product of the K's of the added equations.

CHANGING STOICHIOMETRIC COEFFICIENTS Another important aspect of equilibrium expressions is the consequence of changing stoichiometric coefficients. For example, oxidation of sulfur can give, ultimately, sulfur trioxide

$$S(s) + \tfrac{3}{2}\,O_2(g) \rightleftharpoons SO_3(g)$$

and the equilibrium constant expression for this reaction as written would be

$$K_1 = \frac{[SO_3]}{[O_2]^{3/2}} = 1.1 \times 10^{65} \text{ at } 25\ ^\circ\text{C}$$

However, you can write the chemical equation equally well as

$$2\,S(s) + 3\,O_2(g) \rightleftharpoons 2\,SO_3(g)$$

and the equilibrium constant would now be

$$K_2 = \frac{[SO_3]^2}{[O_2]^3} = 1.2 \times 10^{130} \text{ at } 25\ ^\circ\text{C}$$

When you compare the two equilibrium expressions you find that $K_2 = K_1^2$; that is, $K_2 = [SO_3]^2/[O_2]^3 = \{[SO_3]/[O_2]^{3/2}\}^2$. In general, *when the stoichiometric coefficients of a balanced equation are multiplied by some factor, the equilibrium constant for the new equation (K_{new}) is the old equilibrium constant (K_{old}) raised to the power of the multiplication factor* (Table 16.2). In our SO_3 example, the new equation was obtained by multiplying the first equation by 2. Therefore, K_2 is the square of K_1.

REVERSING EQUATIONS Closely related to the effect of using a new set of stoichiometric coefficients is what happens to K when a chemical equation is reversed. Compare the value of K for the combination of H_2 and O_2 to produce water vapor.

$$2\,H_2(g) + O_2(g) \rightleftharpoons 2\,H_2O(g) \quad K_1 = \frac{[H_2O]^2}{[H_2]^2[O_2]} = 3.4 \times 10^{81} \text{ at } 25\ ^\circ\text{C}$$

with the value for the splitting of water vapor into its elements.

$$2\,H_2O(g) \rightleftharpoons 2\,H_2(g) + O_2(g) \quad K_2 = \frac{[H_2]^2[O_2]}{[H_2O]^2} = 2.9 \times 10^{-82} \text{ at } 25\ ^\circ\text{C}$$

Here $K_2 = 1/K_1$ (or $K_2 = K_1^{-1}$, since reversing an equation has the same effect as multiplying the coefficients in a balanced equation by -1). In general, *the equilibrium constants for a reaction and its reverse are the reciprocals of one another.*

ADDING EQUATIONS It is often necessary to add two equations together to obtain the equation for a net process. As an example, use the successive oxidation of sulfur to the dioxide and then to the trioxide (where K values are given for 25 °C).

$$S(s) + O_2(g) \rightleftharpoons SO_2(g) \qquad K_1 = \frac{[SO_2]}{[O_2]} = 4.2 \times 10^{52}$$

$$SO_2(g) + \tfrac{1}{2}\,O_2(g) \rightleftharpoons SO_3(g) \qquad K_2 = \frac{[SO_3]}{[SO_2][O_2]^{1/2}} = 2.6 \times 10^{12}$$

$$S(s) + \tfrac{3}{2}\,O_2(g) \rightleftharpoons SO_3(g) \qquad K_{net} = \frac{[SO_3]}{[O_2]^{3/2}} = 1.1 \times 10^{65}$$

The calculation above indicates that $K_{net} = K_1K_2$. In general, *when two or more equations are added to produce a net equation, the equilibrium constant for the net equation is the product of the equilibrium constants for the added equations.*

EXERCISE 16.1 Writing Equilibrium Expressions

Write equilibrium constant expressions for the following chemical equations.

(a) $PCl_5(g) \rightleftharpoons PCl_3(g) + Cl_2(g)$

(b) $HCl(g) + LiH(s) \rightleftharpoons H_2(g) + LiCl(s)$

(c) $Cu(OH)_2(s) \rightleftharpoons Cu^{2+}(aq) + 2\,OH^-(aq)$

(d) $[Cu(NH_3)_4]^{2+}(aq) \rightleftharpoons Cu^{2+}(aq) + 4\,NH_3(aq)$

EXERCISE 16.2 Working with Equilibrium Expressions

(a) If K_p for the dissociation of N_2O_4 is 0.15 at 25 °C,

$$N_2O_4(g) \rightleftharpoons 2\,NO_2(g)$$

what is the value of K_p for the dissociation reaction at 25 °C written in the following manner?

$$2\,NO_2(g) \rightleftharpoons N_2O_4(g)$$

Or if written as follows?

$$\tfrac{1}{2}\,N_2O_4(g) \rightleftharpoons NO_2(g)$$

(b) If K_p for the formation of phosgene ($COCl_2$) is 6.6×10^{11} at 25 °C,

$$CO(g) + Cl_2(g) \rightleftharpoons COCl_2(g)$$

what is the value of K_p for its dissociation into CO(g) and $Cl_2(g)$ at the same temperature?

$$COCl_2(g) \rightleftharpoons CO(g) + Cl_2(g)$$

K_c AND K_p As you learned earlier, the equilibrium constant expression can be written in terms of concentrations (K_c) or gas pressures (K_p). In some cases the values of K_c and K_p may be the same, but they are often different. The following examples show you why.

Metal carbonates like limestone decompose on heating to give the metal oxide and CO_2 gas.

$$CaCO_3(s) \rightleftharpoons CaO(s) + CO_2(g)$$

The equilibrium condition for this reaction can be expressed either by $K_p = P_{CO_2}$ or by $K_c = [CO_2]$. From the ideal gas law, you know that $P = (n/V)RT$ = (concentration in mol/L)RT. Therefore, for this reaction, we can say that $K_p = [CO_2]RT$. Since $K_c = [CO_2]$, the interesting conclusion is that $K_p = K_c(RT)$. That is, the values of K_p and K_c are *not* the same; K_p is the product of K_c and the factor RT.

The equilibrium constant in terms of partial pressures, K_p, is known for the reaction of N_2 and H_2 to produce ammonia.

$$N_2(g) + 3\ H_2(g) \rightleftharpoons 2\ NH_3(g)$$

$$K_p = \frac{P^2_{NH_3}}{P_{N_2}P^3_{H_2}} = 6.0 \times 10^5 \text{ at } 25\ °C$$

Does K_c, the equilibrium constant in terms of concentrations, have the same value or a different value? We can answer this by substituting for each pressure in K_p the equivalent expression [mol/L](RT). That is,

$$K_p = \frac{\{[NH_3](RT)\}^2}{\{[N_2](RT)\}\{[H_2](RT)\}^3} = \frac{[NH_3]^2}{[N_2][H_2]^3}\frac{1}{(RT)^2} = \frac{K_c}{(RT)^2}$$

Solving for K_c, the equilibrium constant when the amounts of products and reactants are given in moles per liter, we find

$$K_p = 6.0 \times 10^5 = \frac{K_c}{[(0.08206)(298)]^2} = K_c[(0.08206)(298)]^{-2}$$

$$K_c = 3.6 \times 10^8$$

Once again you see that K_p and K_c are not the same but are related by some function of RT.

If you look carefully at these and other examples, you will find that, in general,

$$K_p = K_c(RT)^{\Delta n} \qquad (16.2)$$

where Δn is the change in the number of moles of gas on going from reactants to products.

$$\Delta n = \text{total moles of gaseous products} - \text{total moles of gaseous reactants}$$

For the decomposition of $CaCO_3$,

$$\Delta n = 1 - 0 = 1$$

while the value of Δn for the ammonia synthesis is

$$\Delta n = 2 - 4 = -2$$

In this chapter we shall often work with values of K_p, since many of the reactions to be described occur in the gas phase. However, you now see that, when needed, a value of K_c can be derived or that K_c can be converted to a value for K_p.

EXERCISE 16.3 K_p/K_c Conversions

The equilibrium constant K_p for the following reaction is 11.5 at 300 °C when the amounts of reactant and products are given in atmospheres. What is the value of K_c for this reaction at 300 °C?

$$PCl_5(g) \rightleftharpoons PCl_3(g) + Cl_2(g)$$

THE MAGNITUDE OF K You have probably noticed that the values of K extend over a tremendous range, from about 10^{-10} for dissolving silver chloride to 10^{65} for the oxidation of sulfur, and to even higher values for the combustion of hydrogen. What is the practical significance of these values? The answer is straightforward: *the larger the value of K, the more the products are favored at equilibrium at the specified temperature* (that is, the greater the extent of the reaction). For example, water is a very stable molecule with respect to decomposition to its elements ($K_c = 2.9 \times 10^{-82}$). Therefore, the numerator ($[O_2][H_2]^2$) in the equilibrium constant expression is *much* smaller than the denominator ($[H_2O]^2$).

EXERCISE 16.4 K and the Extent of Reaction

If solid AgCl and AgI were each placed in 1.0 L of water in separate beakers, in which beaker would $[Ag^+]$ be greater at equilibrium?

$$AgCl(s) \rightleftharpoons Ag^+(aq) + Cl^-(aq) \qquad K = 1.8 \times 10^{-10}$$

$$AgI(s) \rightleftharpoons Ag^+(aq) + I^-(aq) \qquad K = 1.5 \times 10^{-16}$$

SUMMARY This section is an important one. You should now know the following (see also Table 16.2):

1. How to write an equilibrium constant expression from the balanced chemical equation, recognizing that the concentrations of pure solids and liquids do not appear in the expression.
2. How the value of K changes as the stoichiometric coefficients are changed by some factor or when the reaction is reversed.
3. How to find K for an equation that is the sum of other equations for which K values are known.
4. How the value of K depends on the way the equilibrium concentrations are expressed (as mol/L or P) and how to convert K in terms of concentrations (K_c) to K expressed in partial pressures (K_p).
5. How the value of K reflects the extent to which a reaction proceeds at a given temperature.

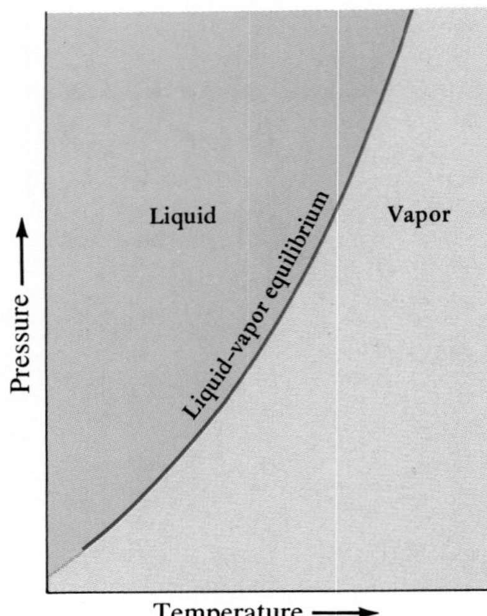

Figure 16.5 A curved line in a plot of pressure versus temperature represents the conditions of T and P at which the equilibrium "liquid ⇄ vapor" exists.

16.3 THE REACTION QUOTIENT

Now that you can write an expression that relates reactant and product concentrations for a reaction that has achieved equilibrium, it is worthwhile to ask a bit more about the meaning of this expression and how the reaction mixture behaves when it is *not* at equilibrium.

In Chapter 13 you first encountered phase diagrams, graphic representations of the equilibria involved in transformations between the phases of matter. There you saw that the liquid–vapor equilibrium for a substance is defined by a mathematical equation relating temperature, enthalpy of vaporization, and equilibrium vapor pressure. The equation led to a curved line in a plot of equilibrium vapor pressure versus temperature (Figures 13.14 and 16.5). This line defines all of the conditions of T and P at which equilibrium can exist. There can be no equilibrium between liquid and vapor in a system whose temperature and vapor pressure do not correspond to a point on the line.

There is only one equation defining liquid–vapor equilibria, but there can be many different forms of equilibrium constant expressions. However, for the particular case of an equilibrium expression involving only one product and one reactant, with both in the same phase, we can represent the equilibrium situation graphically and can discuss the meaning of the equilibrium expression. For example, consider a reaction such as the transformation of n-butane to isobutane.

$$\underset{n\text{-butane}}{H_3C{-}CH_2{-}CH_2{-}CH_3} \rightleftharpoons \underset{\text{isobutane}}{H_3C{-}\overset{\overset{\displaystyle CH_3}{|}}{\underset{\underset{\displaystyle H}{|}}{C}}{-}CH_3}$$

$$K = \frac{[iso]}{[n]} = 2.5 \text{ (at 25 °C)}$$

If the concentration of one of the compounds is known, then there is only one value of the other concentration that will satisfy the equilibrium constant expression. For example, if [n-butane] is 1.0 mol/L, then the concentration of isobutane *must* be 2.5 mol/L at equilibrium. If [n-butane] were changed to 0.80 M, then [isobutane] must be

$$[\text{isobutane}] = 2.5[n\text{-butane}] = 2.5(0.80 \text{ M}) = 2.0 \text{ M}$$

Choosing a number of possible values for [n-butane] and solving for the allowed value of [isobutane] would eventually lead to the points plotted in Figure 16.6. That is, the equilibrium expression $K = [iso]/[n]$ is just the equation of a straight line (with a slope equal to K). Similar plots can be made for other types of equilibrium expressions, but they are more complicated.

The graphical treatment of the simple equilibrium expression $K = [iso]/[n]$ is useful for two reasons. First, Figure 16.6 illustrates that there is an infinite number of possible *sets* of equilibrium concentrations of reactant and product (all the points along the equilibrium line). Second,

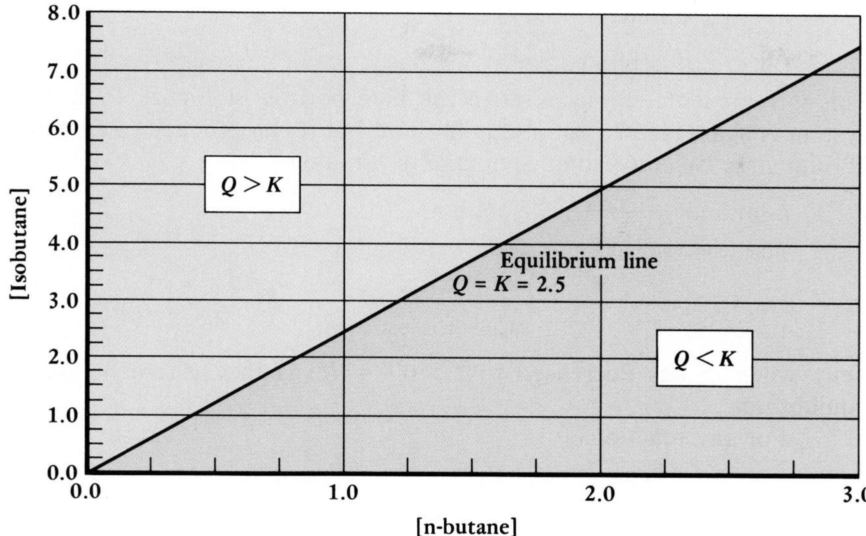

Figure 16.6 A plot of the concentrations of *n*-butane and isobutane that satisfy the expression K = [isobutane]/[*n*-butane] for the equilibrium reaction "*n*-butane ⇄ isobutane." If Q, the reaction quotient (see text) is less than K, then the system is not at equilibrium, and reactants are further converted to products. If $Q > K$, the system is again not at equilibrium, but products must revert to reactants to establish equilibrium.

sets of concentrations for *n*-butane and isobutane that do *not* lie along the line in Figure 16.6 do *not* satisfy the equilibrium condition, and the system will attempt to shift to a new set of concentrations that will satisfy the equilibrium condition.

Any point in Figure 16.6, whether on or off the line, can be defined by the ratio [isobutane]/[*n*-butane]. This ratio is given the general name of the *reaction quotient, Q,* and it is equal to the equilibrium constant, K, when the reaction is at equilibrium. Now, let us say that you have a system composed of 1.5 mol/L of *n*-butane and 1.5 mol/L of isobutane (at 25 °C). This means that the ratio of concentrations, Q, is

$$Q = \frac{[\text{isobutane}]}{[n\text{-butane}]} = \frac{1.5 \text{ M}}{1.5 \text{ M}} = 1.0$$

This ratio is given by a point in the gold portion of Figure 16.6. This set of concentrations clearly does not represent an equilibrium system, since $Q < K$. However, it is certain that some of the *n*-butane will eventually change into isobutane,

n-butane ⟶	isobutane
decrease from 1.5 mol/L ↓ smaller concentration	larger concentration ↑ increase from 1.5 mol/L

thereby lowering [*n*-butane] and raising [isobutane]. Indeed, this transformation will continue until the ratio [*iso*]/[*n*] = 2.5; that is, until $Q = K$ and the ratio of concentrations in the system is represented by a point on the line in Figure 16.6.

What happens when there is too much isobutane in the system relative to the amount of *n*-butane; say [*iso*] = 5.00 M but [*n*] is only 1.25 M? Now the reaction quotient Q is greater than K.

$$Q = \frac{[\text{isobutane}]}{[n\text{-butane}]} = \frac{5.00\text{ M}}{1.25\text{ M}} = 4.00 \qquad Q > K$$

This is represented by a point in the blue portion of Figure 16.6. The system is again clearly not at equilibrium, but it can proceed to an equilibrium state by converting isobutane to *n*-butane

$$n\text{-butane} \longleftarrow \text{isobutane}$$

larger concentration ↑ increase from 1.25 mol/L

decrease from 5.00 mol/L ↓ smaller concentration

Only when Q has decreased to 2.5 ($Q = K$) is the system finally at equilibrium.

For any reaction

$$a\text{ A} + b\text{ B} \rightleftharpoons c\text{ C} + d\text{ D}$$

we can write the algebraic expression that defines the **reaction quotient, Q**:

$$Q_c = \frac{[\text{C}]^c[\text{D}]^d}{[\text{A}]^a[\text{B}]^b}$$

The expression for Q_c has the *appearance* of the equilibrium expression, but Q_c differs from K_c in that the concentrations in the expression are *not necessarily* equilibrium concentrations. Extending the analysis of the *n*-butane/isobutane system, we can make the following general statements:

1. **If $Q < K$, the system is not at equilibrium.** The ratio of products to reactants is too small. Reactants must be converted further to products (thus increasing Q) to achieve equilibrium (when $Q = K$).

   If $Q < K$ then Reactants $\longrightarrow$ Products

2. **If $Q = K$, the system is at equilibrium.**
3. **If $Q > K$, the system is not at equilibrium.** The ratio of products to reactants is too large. To reach equilibrium, products must be converted to reactants (thus decreasing the value of Q until $Q = K$).

   If $Q > K$ then Reactants $\longleftarrow$ Products

The relationship of the reaction quotient, Q, and the equilibrium constant, K, allows us to decide whether a system is at equilibrium. If it is not, we can also decide in which direction the reaction will proceed to reach equilibrium. We will often make use of this idea in these chapters on chemical equilibria, and you will see it referred to again in Section 16.5.

EXAMPLE 16.1

THE REACTION QUOTIENT

Consider the system $PCl_5(g) \rightleftharpoons PCl_3(g) + Cl_2(g)$. At 300 °C, the equilibrium constant K_c is 0.245. If the concentrations of Cl_2 and PCl_3 are both 0.30 M, while the concentration of $PCl_5(g)$ is 3.0 M, is the system at equilibrium? Is Q_c larger than, equal to, or smaller than K_c? If the system is not at equilibrium, in which direction does the reaction proceed?

Solution The equilibrium constant expression for the reaction is

$$K_c = \frac{[PCl_3][Cl_2]}{[PCl_5]} = 0.245$$

If the concentrations of reactants and products are substituted into the reaction quotient expression we have

$$Q_c = \frac{(0.30)(0.30)}{3.0} = 3.0 \times 10^{-2}$$

The value of Q_c is less than the value of K_c ($Q_c < K_c$), so the reaction is not at equilibrium. It proceeds to equilibrium by converting more PCl_5 to products, thus increasing the concentration terms in the numerator and decreasing that in the denominator until $Q_c = K_c$.

EXERCISE 16.5 The Reaction Quotient

Answer the following questions regarding the *n*-butane $\rightleftarrows$ isobutane equilibrium and Figure 16.6.

(a) Is the system at equilibrium when [*n*-butane] = 0.97 M and [isobutane] = 2.18 M? If it is not at equilibrium, in which direction will the reaction proceed in order to achieve equilibrium?

(b) Is the system at equilibrium when [*n*-butane] = 0.75 M and [isobutane] = 2.60 M? If it is not at equilibrium, in which direction will the reaction proceed in order to achieve equilibrium?

EXERCISE 16.6 The Reaction Quotient

At 2000 K the equilibrium constant K_p, for the formation of NO(g)

$$N_2(g) + O_2(g) \rightleftarrows 2\ NO(g)$$

is 4.0×10^{-4}. You have a vessel in which, at 2000 K, the partial pressure of N_2 is 0.50 atm, that of O_2 is 0.25 atm, and that of NO is 4.2×10^{-3} atm. Decide whether the system is at equilibrium. If not, predict which way the reaction will proceed to achieve equilibrium.

16.4 SOME CALCULATIONS WITH THE EQUILIBRIUM CONSTANT

It is often necessary or helpful to know the concentrations of reactants and products for a reaction at equilibrium. As you have seen, these are connected through the equilibrium constant, a numerical value that can be determined by experimental measurement of equilibrium concentrations. This section is devoted to some examples of the calculation of K from experimental information and the use of these K values. For the moment we shall concentrate mostly on simple gas phase reactions, leaving the discussion of reactions in aqueous solution to Chapters 17 through 19.

EXAMPLE 16.2

CALCULATING *K* FROM EXPERIMENTAL INFORMATION

1.00 mole of Br_2 is placed in a 1.00-L vessel and heated to 1756 K, a temperature at which the halogen molecules dissociate to Br atoms.

$$Br_2(g) \rightleftharpoons 2\ Br(g)$$

If $Br_2(g)$ is 1.0% dissociated at this temperature, calculate K_c.

Solution The equilibrium constant expression is

$$K_c = \frac{[Br]^2}{[Br_2]}$$

It is evident from this expression that we need to know the equilibrium values of $[Br_2]$ and [Br] to calculate K_c. As an aid in finding these values, you may find it useful to organize the data given, and the information calculated from it, in a table as we have done below. The first row of the table gives the initial concentration of each reactant and product. The second row shows how each concentration changes as the reaction moves to equilibrium; the calculation of these quantities is described after the table. Finally, the third row lists the equilibrium concentrations. The equilibrium reactant concentration is the difference between the initial value (1.00 mol/L in this case) and the change in concentration. Similarly, the equilibrium product concentration is the sum of the initial value (0 in this case) and the change in its concentration.

	$[Br_2]$	[Br]
Initial concentration (M)	1.00	0
Change in concentration (M)	−0.010(1.00)	+0.010(1.00)(2)
Equilibrium concentration (M)	0.99	0.020

To find the equilibrium concentration of Br_2, we first ask how much of the Br_2 initially present actually dissociated to form Br atoms. Experiment showed that 1.0% dissociated, so the amount is

$$(1.00\ \text{mol/L } Br_2 \text{ present})\left(\frac{0.010\ \text{mol/L } Br_2 \text{ dissociated}}{1.00\ \text{mol/L } Br_2 \text{ present}}\right) = 0.010\ \text{mol/L } Br_2 \text{ dissociated}$$

This means the equilibrium concentration of Br_2 is

$$\text{Initial concentration} - \text{amount of } Br_2 \text{ dissociated} = 1.00\ \text{M} - 0.010\ \text{M} = 0.99\ \text{M}$$

To find the change in concentration of Br atoms, we have to use the reaction stoichiometry:

$$0.010\ \text{mol/L } Br_2 \text{ dissociated}\left(\frac{2\ \text{mol/L Br formed}}{1\ \text{mol/L } Br_2 \text{ dissociated}}\right) = 0.020\ \text{mol/L Br atoms}$$

Since there was initially no concentration of Br atoms, 0.020 mol/L is also the equilibrium concentration of Br atoms.

With the equilibrium concentrations of reactant and product known, we can now calculate K_c:

$$K_c = \frac{[Br]^2}{[Br_2]} = \frac{(0.020)^2}{(0.99)} = 4.0 \times 10^{-4}$$

EXERCISE 16.7 Calculating *K* from Experimental Information

Equal numbers of moles of H_2 gas and I_2 vapor are mixed in a closed flask and are heated to 700 °C. The initial partial pressure of each gas is 0.70 atm, and 78.6% of the I_2 has been consumed when the following reaction comes to equilibrium:

$$H_2(g) + I_2(g) \rightleftharpoons 2\ HI(g)$$

Calculate K_p.

EXAMPLE 16.3

CALCULATING *K* FROM EXPERIMENTAL INFORMATION

Calculate the equilibrium constant at 25 °C for the reaction

$$2\ NOCl(g) \rightleftharpoons 2\ NO(g) + Cl_2(g)$$

In one experiment, 2.00 moles of NOCl were placed in a 1.00-L flask, and the concentration of NO after equilibrium was achieved was 0.66 moles/liter.

Solution After writing the expression for K from the balanced chemical equation,

$$K_c = \frac{[NO]^2[Cl_2]}{[NOCl]^2}$$

our next goal is to define the concentration of each reagent at equilibrium. To organize this problem, we have set out the data in a small table.

	[NOCl]	[NO]	$[Cl_2]$
Concentration before reaction (M)	2.00	0	0
Concentration change when proceeding to equilibrium (M)	−0.66	+0.66	+(1/2)(0.66)
Concentration at equilibrium (M)	2.00 − 0.66 = 1.34	0 + 0.66	0 + 0.33

Let us first see how the equilibrium concentrations in the table were derived. The initial concentration of NOCl was 2.00 M, and the equilibrium value of [NO] was found experimentally to be 0.66 M. This is the starting point in our calculation of the equilibrium concentrations of the gases.

The balanced equation tells you that half as many moles of $Cl_2(g)$ are produced as moles of NO(g). Therefore, the equilibrium concentration of $Cl_2(g)$ must be half that of NO(g) or 0.33 M.

Similarly, the balanced equation shows that for every mole of NO(g) produced, one mole of NOCl(g) is used. Therefore, at equilibrium

$$\begin{aligned}\text{Concentration of NOCl} &= \text{concentration before reaction} - \\ &\quad \text{concentration that disappeared on proceeding} \\ &\quad \text{to equilibrium} \\ &= 2.00\ \text{M} - 0.66\ \text{M} = 1.34\ \text{M}\end{aligned}$$

With all the equilibrium amounts and concentrations known, the value of K_c at 25 °C can be calculated.

$$K = \frac{[NO]^2[Cl_2]}{[NOCl]^2} = \frac{(0.66)^2(0.33)}{(1.34)^2} = 0.080$$

EXERCISE 16.8 Calculating an Equilibrium Constant

Phosphorus pentachloride decomposes in the gas phase to chlorine and phosphorus trichloride vapor.

$$PCl_5(g) \rightleftharpoons PCl_3(g) + Cl_2(g)$$

Calculate the equilibrium constant for this reaction at 300 °C, knowing that the original pressure of $PCl_5(g)$ before it dissociated was 1.50 atm, and the equilibrium pressures of $PCl_3(g)$ and $Cl_2(g)$ are both 1.34 atm.

Once the equilibrium constant for a reaction is known at a given temperature, that value of K can be used to find the equilibrium concentrations of reactants and products under other conditions (if the temperature does not change).

EXAMPLE 16.4

CALCULATING A CONCENTRATION FROM AN EQUILIBRIUM CONSTANT

The equilibrium constant for the dissociation of AgCl(s) to its ions in aqueous solution

$$AgCl(s) \rightleftharpoons Ag^+(aq) + Cl^-(aq)$$

is 1.8×10^{-10}. If the concentration of $Ag^+(aq)$ is 1.0×10^{-5} M, what is the concentration of Cl^- that can be in equilibrium with silver ion at that concentration?

Solution The equilibrium constant expression for this reaction is

$$K_c = [Ag^+][Cl^-]$$

Therefore, the equilibrium concentration of Cl^- must be

$$[Cl^-] = \frac{K_c}{[Ag^+]} = \frac{1.8 \times 10^{-10}}{1.0 \times 10^{-5}} = 1.8 \times 10^{-5}\ \text{M}$$

EXAMPLE 16.5

CALCULATING A CONCENTRATION FROM AN EQUILIBRIUM CONSTANT

The equilibrium constant K_c (= K_p) for

$$H_2(g) + I_2(g) \rightleftharpoons 2\ HI(g)$$

was determined using the experimental data of Table 16.1. If you place 1.00 mole each of $H_2(g)$ and $I_2(g)$ in a 0.500-L flask at 699 K, what are the equilibrium concentrations of $H_2(g)$, $I_2(g)$, and HI(g)?

Solution The first step is to write the balanced equation. Since this is done above, the next step is to write the equilibrium constant expression.

$$K = \frac{[HI]^2}{[H_2][I_2]} = 55.17$$

Next, we define equilibrium concentrations and organize the information using a small table.

	$[H_2]$	$[I_2]$	$[HI]$
Original concentrations (M)	2.00	2.00	0
Concentration change when proceeding to equilibrium (M)	$-x$	$-x$	$+2x$
Equilibrium concentrations (M)	$2.00 - x$	$2.00 - x$	$2x$

The equilibrium concentration of H_2 or I_2 will be the amount (in mol/L) that was present before the reaction *minus* the amount (in mol/L) that was transformed into HI when equilibrium was achieved. We do not know the amount (in mol/L) of H_2 or I_2 that disappeared, only that the same number of moles/liter of each must have been involved. We call this unknown amount x (mol/L). The balanced equation tells you that the concentration of HI at equilibrium is twice the concentration of H_2 or I_2 that disappeared, which can now be expressed as $2x$.

Now that the equilibrium concentrations of all three compounds are defined in terms of a single unknown, we can write the equilibrium constant expression and set it equal to the experimentally determined value.

$$K = 55.17 = \frac{[HI]^2}{[H_2][I_2]} = \frac{(2x)^2}{(2.00 - x)^2}$$

The unknown x can be found by taking the square root of both sides and solving.

$$\sqrt{K} = 7.428 = \frac{2x}{2.00 - x}$$

$$7.428(2.00 - x) = 2x$$

$$14.9 - 7.428x = 2x$$

$$14.9 = 9.428x$$

$$x = 1.58$$

The concentration of H_2 or I_2 that disappears in the process of establishing equilibrium is therefore 1.58 M.

Now we return to the expressions for $[H_2]$, $[I_2]$, and $[HI]$ to obtain

$$[H_2]_{equi} = [I_2]_{equi} = 2.00 - x = 0.42 \text{ M}$$

$$[HI]_{equi} = 2x = 3.16 \text{ M}$$

It is important that you verify that these values are correct by substituting them back into the equilibrium expression to calculate K and see if the calculated K is the same as that originally given. In this case $(3.16)^2/(0.42)^2 = 57$. The discrepancy with $K = 55.17$ is due to the fact that we know $[H_2]$ and $[I_2]$ to only two significant figures.

EXAMPLE 16.6

CALCULATIONS USING THE EQUILIBRIUM CONSTANT

When you find a bottle of concentrated nitric acid in the laboratory, it often has a brown gas, NO_2, hovering over the surface of the liquid. This simple molecule can bind to another NO_2 molecule and form a dimer, $O_2N—NO_2$, that is, a molecule having two NO_2 units. NO_2 and its dimer, N_2O_4, are in equilibrium in the gas phase.

$$N_2O_4(g) \rightleftharpoons 2\ NO_2(g) \qquad K_p = 0.15 \text{ (at 25 °C)}$$

If the total pressure of an equilibrium mixture of N_2O_4 and NO_2 is 1.5 atm, calculate the partial pressure of each gas.

Solution The equilibrium constant expression is

$$K = \frac{P^2_{NO_2}}{P_{N_2O_4}} = 0.15$$

This equation contains two unknowns—the pressures of the two different gases. To solve it, we will have to reduce the equation to one with only a single unknown quantity. This can be done as follows:

(a) The equilibrium total pressure is the sum of the partial pressures

$$P_{total} = P_{NO_2} + P_{N_2O_4} = 1.5 \text{ atm}$$

and so the partial pressure of N_2O_4 at equilibrium is

$$P_{N_2O_4} = 1.5 \text{ atm} - P_{NO_2}$$

(b) From the step above, you can now write the equilibrium expression in terms of a single unknown, the partial pressure of NO_2.

$$K = \frac{P^2_{NO_2}}{1.5 - P_{NO_2}} = 0.15$$

Rearranging this expression you have

$$0.23 - 0.15\ P_{NO_2} = P^2_{NO_2}$$

or

$$P^2_{NO_2} + 0.15\ P_{NO_2} - 0.23 = 0$$

$$ax^2 + bx + c = 0$$

The rearranged equilibrium expression is a quadratic equation of the general form $ax^2 + bx + c = 0$. Such equations can be solved by the formula given in Appendix A.

The "method of successive approximations" can also be used to solve the quadratic equation for P_{NO_2}. See Appendix A.

$$x = \frac{-b \pm \sqrt{b^2 - 4ac}}{2a}$$

In this case $a = 1$, $b = 0.15$, and $c = -0.23$. Substituting these values we have

$$x = P_{NO_2} = \frac{-(0.15) \pm \sqrt{(0.15)^2 - 4(1)(-0.23)}}{2(1)}$$

$$P_{NO_2} = \frac{-(0.15) \pm 0.97}{2}$$

$$= 0.41 \text{ or } -0.56$$

If we use +0.97, then $x = 0.41$, or if we use −0.97, then $x = -0.56$. Since x stands for P_{NO_2}, a negative value is physically meaningless. Therefore, we now know that $x = 0.41$ must be the equilibrium partial pressure of NO_2. Since $P_{N_2O_4}$ is $1.5 - P_{NO_2}$, the equilibrium pressure of N_2O_4 is 1.1 atm.

As a final step in the problem, you should verify this answer by substituting back into the equilibrium expression. The quotient $(0.41)^2/(1.1)$ is 0.15, a value identical with the value of K.

EXERCISE 16.9 Calculations Using the Equilibrium Constant
Again consider the dissociation of PCl_5.

$$PCl_5(g) \rightleftharpoons PCl_3(g) + Cl_2(g)$$

If you begin with enough PCl_5 so that its pressure in a flask is 3.00 atm at 300 °C, and you know that K_p is 11.5 at this temperature, what are the partial pressures of the products and reactant when equilibrium is attained at 300 °C?

EXERCISE 16.10 Calculations Using the Equilibrium Constant
Ammonium hydrogen sulfide decomposes on heating.

$$NH_4HS(s) \rightleftharpoons H_2S(g) + NH_3(g)$$

If K_p is 0.11 at 25 °C (when the partial pressures are expressed in atmospheres), what is the total pressure in the flask at equilibrium?

"This is a lovely old song that tells of a young woman who leaves her cottage, and goes off to work. She arrives at her destination, and places some solid NH_4HS in a flask containing 0.50 atm of ammonia, and attempts to determine the pressures of ammonia and hydrogen sulfide when equilibrium is reached."

16.5 EXTERNAL FACTORS AFFECTING EQUILIBRIA

Chemical and physical equilibria are influenced by the three factors outlined in Table 16.3. Each of these factors requires some discussion—temperature and concentration changes are discussed in this section, and the effect of catalysts is described in the next section.

EFFECT OF TEMPERATURE CHANGES ON EQUILIBRIA

In Chapters 13 and 14 you learned about temperature effects on phase equilibria and on the solubility of solids in water and other solvents. We noted, for example, that an increase in temperature leads to an increase in the vapor pressure of a liquid and to an increase in the solubility of most salts. These observations are special cases of **Le Chatelier's principle**: *a change in any of the factors that determine the equilibrium conditions of a system will cause the system to change in such a manner as to reduce or counteract the effect of the change.* This was applied specifically to the variation in solubility with temperature in Section 14.2. Now we wish to apply it to chemical reactions.

Table 16.3 External Influences on Chemical Equilibria

Influence	Effect
1. Temperature change	Value of K changes.
2. Concentration change (addition or removal of reactant or product; for a gas phase reaction, change in volume)	K is constant; position of equilibrium is changed (equilibrium quantities of both reactants *and* products change).
3. Addition of catalyst	K is constant; rate of approach to equilibrium is affected, but eventual equilibrium mixture is the same.

What is the effect of a temperature change on the equilibrium position in a chemical reaction? If you know the enthalpy change of the reaction (ΔH_{rxn}), you can make a qualitative prediction from Le Chatelier's principle. As an example, consider the reaction of N_2 with O_2 to give nitrogen oxide, NO.

$$N_2(g) + O_2(g) + 180.5\text{ kJ} \rightleftharpoons 2\ NO(g)$$

$$K_c = \frac{[NO]^2}{[N_2][O_2]} \qquad K_c = 4.0 \times 10^{-4} \text{ at } 2000\text{ K} \qquad K_c = 36 \times 10^{-4} \text{ at } 2500\text{ K}$$

The NO formed in automobile engines as a by-product of fuel combustion is exhausted with the other combustion products. NO can react further with O_2 and other oxidants in the atmosphere to give NO_2, the source of much of the air pollution in urban areas. See Chapter 15, page 642.

We are surrounded by N_2 and O_2, but you know that they do not react appreciably at room temperature. The position of equilibrium lies almost completely to the left. However, if a mixture of N_2 and O_2 is heated above 700 °C, as in an automobile engine, the equilibrium shifts toward NO, and a greater concentration of NO can exist in equilibrium with reactants. The experimental equilibrium constants given above show that [NO] increases and $[N_2]$ and $[O_2]$ decrease as the temperature increases at equilibrium.

To see why the value of K can change with temperature, consider the N_2/O_2 reaction further. The enthalpy change for the reaction is +180.5 kJ, so we can consider heat as a "reactant." Le Chatelier's principle tells us that input of energy (as heat) will cause the position of the equilibrium to shift in a direction that minimizes or counteracts this input. The way to counteract the energy input here is to consume some of the added heat by consuming N_2 and O_2 and producing more NO. Therefore, an increase in temperature must be accompanied by increased production of NO and consumption of N_2 and O_2. Since this raises the value of the numerator in the K expression and lowers the denominator, K must also increase in value.

A more quantitative analysis of temperature effects on equilibria can be made by using thermodynamics. See Chapter 20.

$$[N_2] + [O_2] \rightleftharpoons 2\,[NO];\quad \text{small } K$$

increase T for endothermic reaction; equilibrium shifts to consume heat ↓

$$[N_2] + [O_2] \rightleftharpoons 2\,[NO];\quad \text{larger } K$$

Two molecules of the brown gas NO_2 can combine readily to form N_2O_4, and equilibrium is readily achieved in a closed system (Figure 16.7).

$$2\ NO_2(g) \rightleftharpoons N_2O_4(g) + 57.2\text{ kJ}$$

$$K_c = \frac{[N_2O_4]}{[NO_2]^2} \qquad K_c = 180 \text{ at } 273\text{ K} \qquad K_c = 170 \text{ at } 298\text{ K}$$

If T changes, the value of K will change, and reactant and product concentrations will change in opposite directions. If T is constant, the value of K is constant; a change in a reactant or product concentration may lead to changes in concentrations of other materials in the reaction so that $Q = K$.

Here the reaction is *exo*thermic, so heat can be considered a reaction "product." By lowering the temperature of the reaction, as in Figure 16.7, some heat is removed. According to Le Chatelier's principle, this removal of heat can be counteracted by the system if the reaction produces more heat by combination of NO_2 to give more N_2O_4. Thus, the equilibrium concentration of NO_2 declines, that of N_2O_4 increases, and the values of K become larger as the temperature declines.

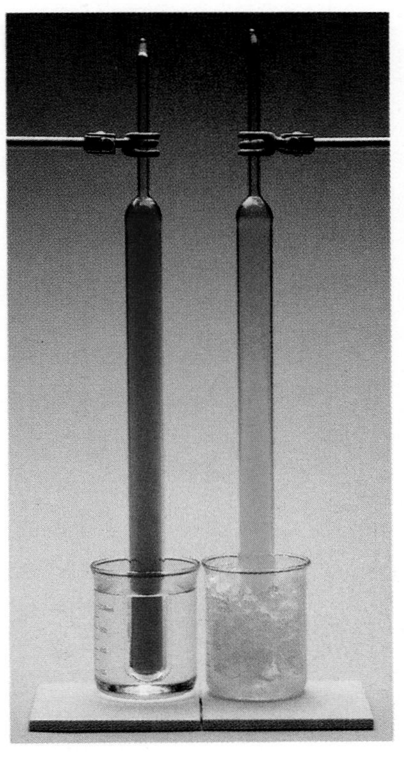

Figure 16.7 Effect of temperature on the N_2O_4/NO_2 equilibrium. The tubes in the photograph both contain a mixture of NO_2 and N_2O_4. As predicted by Le Chatelier's principle, the equilibrium favors colorless N_2O_4 at lower temperatures. This is clearly seen in the tube at the right, where the gas in the ice bath at 0 °C is only slightly brown because there is only a small partial pressure of the brown gas NO_2. At 50 °C (the tube at the left), the equilibrium is shifted toward NO_2. (Marna G. Clarke)

EXERCISE 16.11 Temperature Effects on Equilibria

Predict the effect of temperature changes on each of the following equilibria.
(a) Does the concentration of the ammonium ion increase or decrease with an increase in temperature?

$$NH_4NO_3(s) + 26.2 \text{ kJ} \rightleftharpoons NH_4^+(aq) + NO_3^-(aq)$$

(b) When the temperature declines, what happens to the concentration of N_2?

$$2\ N(g) \rightleftharpoons N_2(g) + 946 \text{ kJ}$$

EFFECT OF THE ADDITION OR REMOVAL OF A REAGENT

If the temperature is constant, the value of K is constant. This means that, if the concentration of some reactant or product is changed from its equilibrium value *at a given temperature*, the reaction must shift to a new equilibrium position for which the reaction quotient is still equal to K. To see how this can happen, we return to the simple case of the n-butane/isobutane equilibrium and Figure 16.6.

$$n\text{-butane} \rightleftharpoons \text{isobutane}$$

$$K = \frac{[iso]}{[n]} = 2.5 \text{ at } 25\ °\text{C}$$

Suppose the equilibrium concentration of isobutane in a 1.0-L flask is 1.25 M and that of n-butane is 0.50 M. Now add 1.5 moles of n-butane to the flask. What happens to the concentrations of n-butane and isobutane as the system moves to reestablish equilibrium? Le Chatelier's principle allows us to make a qualitative prediction. In this case, after we add n-butane, there is now more of this compound present than can be in equilibrium with 1.25 M isobutane. That is, the reaction quotient, Q, is now smaller than K.

After adding excess n-butane to an equilibrium mixture

$$Q = \frac{[iso]}{[n]} = \frac{1.25 \text{ mol/L present}}{0.50 \text{ mol/L present} + 1.5 \text{ mol/L added}} < 2.5$$

As you learned earlier, when $Q < K$, this means that equilibrium can be reestablished if the concentration of the product increases and the concentration of the reactant decreases. In Figure 16.8, we have moved from the equilibrium line, point 1, to the region where $Q < K$. To return to equilibrium, more isobutane is formed by consuming n-butane, 1 mol/L of butane being consumed for every mol/L of isobutane being formed. As explained in the caption to Figure 16.8, and in Example 16.7, equilib-

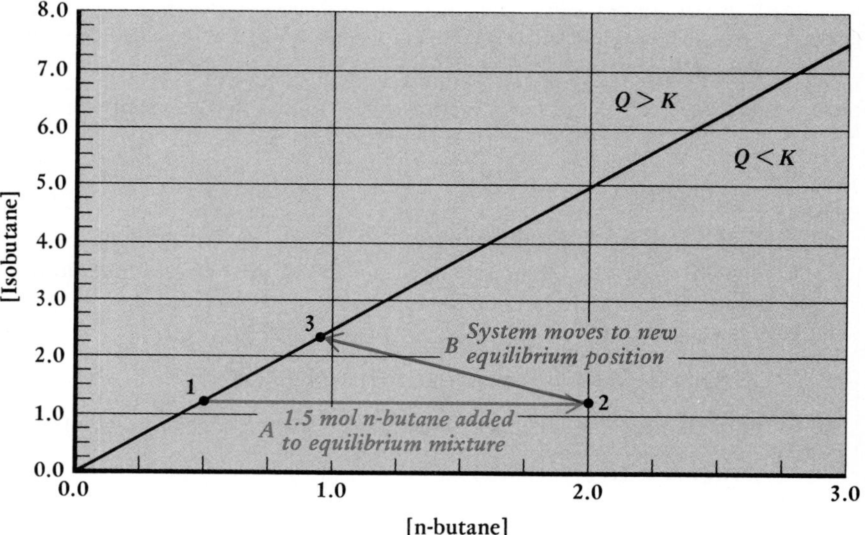

Figure 16.8 Effect of adding a reagent to a system at equilibrium. The system n-butane ⇄ isobutane was at equilibrium when $[n] = 0.50$ M and $[iso] = 1.25$ M (point 1). If 1.50 mol/L of n-butane is added, the system moves to point 2 ($Q < K$). Equilibrium is restored by moving along line B to a new equilibrium position (point 3) where $[n] = 0.93$ M and $[iso] = 2.32$ M. Line B is established easily, since, in restoring equilibrium, each mole of n-butane consumed produces a mole of isobutane.

rium is reestablished when Q again equals 2.5 (at point 3), but now the concentrations of n-butane and isobutane are different than those before adding excess n-butane.

EXAMPLE 16.7

AFFECTING AN EQUILIBRIUM BY CHANGING CONCENTRATIONS

In this example, we shall work out an algebraic solution to the n-butane case outlined above. Assume equilibrium has been established in a 1.00-L flask with $[n] = 0.500$ and $[iso] = 1.25$. Then 1.50 moles of n-butane are added. What are the new equilibrium concentrations of n- and isobutane?

Solution First organize the information in a small table.

COMPOUND	n-BUTANE	ISOBUTANE
Concentration at old equilibrium position (M)	0.500	1.25
Concentration immediately after adding n-butane (M)	0.500 + 1.50	1.25
Change in concentration when proceeding to equilibrium (M)	$-x$	$+x$
Concentration at new equilibrium position (M)	$0.500 + 1.50 - x$	$1.25 + x$

Included in the table are algebraic expressions that define the concentrations of n-butane and isobutane at the new equilibrium position. We arrived at them as follows:

(a) The concentration of n-butane at the new equilibrium position will be the old concentration *plus* what was added to disturb the equilibrium (1.50 moles/1.00 L = 1.50 M) *minus* the concentration of n-butane that must be converted to isobutane in order to reestablish equilibrium. We do not yet know how many moles/liter of n-butane disappear when going to the new equilibrium position, so we have designated it by x.

(b) The amount of isobutane at the new equilibrium position is the amount that was already present (1.25 M) *plus* the amount (x mol/L) that comes from the conversion of added *n*-butane.

Having defined algebraically $[n]$ and $[iso]$ at the new equilibrium position and remembering that K is a constant equal to 2.50 throughout, we can write

$$K = \frac{[iso]_{new}}{[n]_{new}} = 2.50 = \frac{1.25 + x}{0.500 + 1.50 - x}$$

$$2.50(0.500 + 1.50 - x) = 1.25 + x$$

$$3.75 = 3.50x$$

$$x = 1.07 \text{ mol/L}$$

You now know that the new equilibrium position is established at

$$[n] = 0.500 + 1.50 - x = 0.93 \text{ mol/L} \qquad [iso] = 1.25 + x = 2.32 \text{ mol/L}$$

as suggested by the graphical analysis in Figure 16.8. Be sure to verify that $[iso]/[n] = 2.32/0.93 = 2.5$.

EXERCISE 16.12 Concentration Changes and Equilibrium

Equilibrium exists between *n*-butane and isobutane when $[n] = 0.20$ M and $[iso] = 0.50$ M. What are the equilibrium concentrations of *n*-butane and isobutane if 2.00 mol/L of isobutane are added to the original mixture? First try to estimate the new equilibrium concentrations from Figure 16.8 and then calculate the exact values.

For a reaction that involves gases, what happens to equilibrium concentrations or pressures if the size of the container is changed? (This occurs, for example, when fuel and air are compressed in an automobile engine.) To answer this question, recall again that concentrations are in moles/liter, so, if the volume of a gas changes, the concentration must also change, and the equilibrium position can change. As an example, once again consider the equilibrium

$$2\ NO_2(g) \rightleftharpoons N_2O_4(g)$$

brown gas — colorless gas

$$K_p = \frac{P_{N_2O_4}}{P^2_{NO_2}} = 6.75 \text{ at } 25\ ^\circ\text{C}$$

What happens to this equilibrium if the volume of the flask holding the gases is suddenly halved? Since the pressure of a gas increases as the volume available to the gas decreases (Boyle's law; Chapter 12), the immediate result is that the partial pressures of both gases will double. This means, however, that the system is no longer at equilibrium since the quotient ($P_{N_2O_4}/P^2_{NO_2}$) cannot be equal to 6.75. For example, assume the equilibrium is originally established when $P_{N_2O_4}$ is 0.675 atm and P_{NO_2} is 0.316 atm. If both pressures double when the volume is halved, this means $P_{N_2O_4}$ is now 1.35 atm and P_{NO_2} is 0.632. The reaction quotient, Q, under these circumstances is $(1.35)/(0.632)^2 = 3.38$, a value clearly less than K. Therefore, since Q is less than K, the quantity of product must increase at the expense of the reactants, and the equilibrium will shift in favor of N_2O_4.

$$2\ NO_2(g) \rightleftharpoons N_2O_4(g)$$

$$\xrightarrow[\text{equilibrium shifts right}]{\text{decrease volume of container}}$$

This means that one molecule of N_2O_4 is formed by consuming two molecules of NO_2. The partial pressure of NO_2 decreases twice as fast as that of N_2O_4 increases until the reaction quotient, $P_{N_2O_4}/P^2_{NO_2}$, is once again equal to K.

Rather than going through a numerical analysis for each situation you may see, the conclusions for the NO_2/N_2O_4 equilibrium can be generalized in the following way: *For any reaction involving gases, a volume* **decrease** *will be counterbalanced by a shift in the equilibrium to the side of the reaction with the* **fewer** *gas molecules.*

EXERCISE 16.13 Effect of Concentration Changes on Equilibria

The formation of ammonia from its elements is an important industrial process.

$$3\ H_2(g) + N_2(g) \rightleftharpoons 2\ NH_3(g)$$

(a) Does the equilibrium shift to the left or the right when extra H_2 is added? When extra NH_3 is added?
(b) What is the effect on the position of the equilibrium when the volume of the system is increased? Does the equilibrium shift to the left or to the right, or is the system unchanged?
(c) Tell if or how K changes for the changes given in (a) and (b).

16.6 EQUILIBRIUM, KINETICS, AND MECHANISMS

When describing the dynamic nature of the equilibrium involving aqueous Fe^{3+} and SCN^-, we mentioned that equilibrium is established when the rate at which the product is formed is equal to the rate at which the product reverts to reactants. In this section we want to explore this connection between equilibrium and kinetics a bit more.

THE EFFECT OF A CATALYST ON AN EQUILIBRIUM

Like the compound formed between aqueous Fe^{3+} and SCN^-, many compounds of transition metal ions are highly colored, and the green compound based on Co^{3+} in Figure 16.9 is no exception. This compound undergoes a relatively slow replacement of Cl^- by H_2O in water to give a red compound (Figure 16.9b).*

*The symbol N‿N in these structures represents the nitrogen-containing molecule seen in Figure 16.9. For more information on such compounds see Chapter 25.

$$\left[\text{Cl(N N)}_2\text{Co Cl}\right]^{+} + H_2O \rightleftharpoons \left[H_2O\text{(N N)}_2\text{Co Cl}\right]^{2+} + Cl^-$$

green red

The rate of change of concentration for the green compound in this reaction could be represented by the blue curve in Figure 16.9c. However, if a catalyst—OH^- ion in this case—is added, then the green reactant loses Cl^- *much* more rapidly, as seen by the red curve in Figure 16.9c. The catalyst has clearly increased the rate at which Cl^- is replaced, but further experiments show that the concentrations of products and reactants when the reaction reaches equilibrium are the same as those for the

(a)

(b)

Conc. at time t / Initial Conc.: 0.0, 0.1, 0.2, 0.3, 0.4, 0.5, 0.6, 0.7, 0.8, 0.9, 1.0

Time (hours): 0 1 2 3 4 5 6 7 8 9 10 11 12 13 14 15 16

Catalyzed reaction

Uncatalyzed reaction

(c)

Figure 16.9 The effect of a catalyst (OH^-) on the replacement of Cl^- by H_2O in a Co^{3+} complex with a Co—Cl bond. (a) The green, Co^{3+}-containing cation has the formula $[(H_2NC_2H_4NH_2)_2CoCl_2]^+$. The Co^{3+} ion is a silver ball in the center of the ion. Other atoms have the following colors: H = white; C = black, N = blue, and Cl = green. (b) When the complex is placed in water, one of the Co—Cl bonds in the green starting material is replaced by a Co—H_2O bond, and the color changes to red. (c) The time dependence of the concentration of the green starting complex with and without an added catalyst. The reaction is first order in the green complex and has a rate constant of 3.5×10^{-5}/s ($t_{1/2} = 5.5$ hours). For the sake of illustration, $t_{1/2}$ for the catalyzed reaction is assumed to be $\frac{1}{2}$ hour, and K is very large.

uncatalyzed reaction. *The addition of a catalyst to a reaction does not influence the position of equilibrium, only the rate at which equilibrium is approached.*

EQUILIBRIA AND REACTION MECHANISMS

The reaction mechanisms that we described in Chapter 15 all had a rate determining elementary step followed by one or more fast steps. However, there are many reactions in which the reactants quickly produce some intermediates that subsequently react in a slow step, and one of these is the oxidation of nitrogen oxide by oxygen.

$$2\ NO(g) + O_2(g) \longrightarrow 2\ NO_2(g) \qquad \text{Rate} = k[NO]^2[O_2]$$

The reaction has an experimental rate law that is third order overall. This could imply a mechanism where all three reacting molecules collide in a single, termolecular step. However, there is experimental evidence for a reaction intermediate, and so the process is thought to occur in two elementary steps. Unlike the examples of mechanisms you saw in Chapter 15, the second step here is rate determining.

Elementary step 1: Fast, equilibrium $\qquad NO(g) + O_2(g) \underset{k_{-1}}{\overset{k_1}{\rightleftharpoons}} \underset{\text{intermediate}}{OONO(g)}$

Elementary step 2: Slow, rate determining $\qquad NO(g) + OONO(g) \xrightarrow{k_2} 2\ NO_2(g)$

Overall reaction: $\qquad 2\ NO(g) + O_2(g) \longrightarrow 2\ NO_2(g)$

This means the rate law predicted for the reaction should be

$$\text{Rate} = k_2[NO][OONO]$$

This rate law cannot be compared directly with the experimental rate law since it contains the concentration of an unobserved intermediate, OONO. The experimental rate law should be written in terms of compounds appearing in the overall equation and should not include reaction intermediates whose concentrations cannot be measured. Therefore, we want to express the postulated rate law in a way that eliminates the intermediate. To do this, we look carefully at the equilibrium in the first, very rapid step.

At the beginning of the reaction NO and O_2 react rapidly and produce the intermediate OONO with a rate constant k_1.

$$\text{Rate of production of OONO} = k_1[NO][O_2]$$

Since the intermediate is consumed only very slowly in the second step, much of the OONO reverts to NO and O_2 before it can be consumed in the second step.

$$\text{Rate of reversion of OONO to NO and } O_2 = k_{-1}[OONO]$$

As NO and O_2 form OONO their concentration drops, so the rate of the forward reaction decreases. At the same time, the concentration of OONO builds up, so the rate of the reverse reaction increases. Eventually, the rates of the forward and reverse reactions become the same, and the first elementary step reaches a state of equilibrium. Because the forward and reverse reactions in the first step are so much faster than the second elementary step, equilibrium is established before any significant amount of OONO is consumed by NO to give NO_2. The state of equilibrium for the first step remains throughout most of the lifetime of the overall reaction; it is broken only when the overall reaction is nearly complete.

Since equilibrium is established when the rates of the forward and reverse reactions are the same, this means, for the first step,

$$k_1[NO][O_2] = k_{-1}[OONO]$$

where the concentrations are all equilibrium values. Rearranging this equation, we find

$$\frac{k_1}{k_{-1}} = \frac{[OONO]}{[NO][O_2]}$$

Of course the quotient $[OONO]/[NO][O_2]$ is just the equilibrium constant K for the first elementary step, and it is this conclusion that will allow us to fit the mechanism for the NO/O_2 reaction to the experimentally observed rate law. From the equilibrium constant expression we know that

$$[OONO] = K[NO][O_2]$$

If we substitute this into the rate law for the rate determining elementary step, we have

$$\text{Rate} = k_2[NO][OONO] = k_2[NO]\{K[NO][O_2]\}$$

which becomes

$$\text{Rate} = k_2K[NO]^2[O_2]$$

Since both k_2 and K are constants, their product is another constant k, and we have

$$\text{Rate} = k[NO]^2[O_2]$$

This is exactly the rate law that came from experiment, so we believe we have postulated a reasonable mechanism. However, it is not the *only* possible mechanism. As you see in the following example, at least one other mechanism is consistent with the same experimental rate law.

EXAMPLE 16.8

REACTION MECHANISMS INVOLVING AN EQUILIBRIUM STEP

The NO/O_2 reaction described in the text could also occur by the sequence of steps

Step 1: Fast, equilibrium $\quad NO(g) + NO(g) \underset{k_{-1}}{\overset{k_1}{\rightleftharpoons}} \underset{\text{intermediate}}{N_2O_2(g)}$

Step 2: Slow, rate determining $\quad N_2O_2(g) + O_2(g) \xrightarrow{k_2} 2\ NO_2(g)$

Overall reaction: $\quad 2\ NO(g) + O_2(g) \longrightarrow 2\ NO_2(g)$

Show that this reaction also leads to the experimentally observed rate law.

Solution The rate law for the rate determining elementary step is

$$\text{Rate} = k_2[N_2O_2][O_2]$$

The compound N_2O_2 is an intermediate, which is consumed in the slow step and cannot appear in the final derived rate law. Therefore, we must express the derived rate law in another way, and we can do this by recognizing that the intermediate is related to NO by the equilibrium constant expression for the reactants and products in step 1. (Notice that the concentration of each molecule in the equilibrium constant expression is raised to a power equal to its stoichiometric coefficient in the balanced chemical equation for step 1. Here 2 NO molecules are reactants in step 1.)

$$K = \frac{[N_2O_2]}{[NO]^2}$$

If this expression is solved for $[N_2O_2]$, we have $[N_2O_2] = K[NO]^2$. When this is substituted into the derived rate law,

$$\text{Rate} = k_2\{K[NO]^2\}[O_2]$$

we have an equation identical with the experimental law, where $k_2K = k$.

The NO/O_2 reaction has a rate law that suggests two possible mechanisms. The challenge that often faces the chemist interested in kinetics is to decide which mechanism is correct or whether there may be a better choice. In this case further chemical investigation showed the presence of the species OONO as a short-lived intermediate, confirming the mechanism given on page 686.

EXERCISE 16.14 A Reaction Mechanism Involving an Equilibrium Step

The reaction of H_2 and Br_2 to give HBr has the following experimental rate law,

$$H_2(g) + Br_2(g) \longrightarrow 2\,HBr(g)$$

$$\text{Experimental rate law} = k[H_2][Br_2]^{1/2}$$

and is thought to occur by the mechanism

Step 1: Fast, equilibrium	$Br_2(g) \underset{k_{-1}}{\overset{k_1}{\rightleftharpoons}} 2\,Br(g)$
Step 2: Slow	$Br(g) + H_2(g) \xrightarrow{k_2} HBr(g) + H(g)$
Step 3: Fast	$H(g) + Br_2(g) \xrightarrow{k_3} HBr(g) + Br$
Overall equation:	$H_2(g) + Br_2(g) \longrightarrow 2\,HBr(g)$

Here we assume a rapid, initial equilibrium step supplies the reaction intermediate, a Br atom that is required in the rate determining step. Using the method outlined in Example 16.8, show that the postulated mechanism leads to the experimental rate law. (Note that Step 1 supplies the small concentration of Br atoms necessary to initiate the process. Only Steps 2 and 3 are required to add to give the overall equation.)

SUMMARY

This is an important chapter, since it has introduced you formally to the notion that chemical systems eventually come to equilibrium at a point determined by the temperature and by the concentrations of reactants and products. You have found that all chemical and phase equilibria share three characteristics: they are dynamic, they are reversible, and the equilibrium position can be approached from either direction (Section 16.1). The equilibrium position is described by the **equilibrium constant expression** (Section 16.2). For the general reaction of $aA + bB \rightleftarrows cC + dD$, one can write the expression

$$Q = \frac{[C]^c[D]^d}{[A]^a[B]^b}$$

where $[A]^a$ and so forth are the concentrations of reactants and products, each raised to the power of its stoichiometric coefficient. Q has been called the **reaction quotient**. When the concentrations in the reaction quotient are those at equilibrium, Q is equal to K, the equilibrium constant. If $Q < K$ the system is not at equilibrium, and reactants are converted to products on proceeding to equilibrium. When $Q > K$, the system is again not at equilibrium, and products are converted to reactants on moving to a state of equilibrium.

Equilibrium constants are generally determined by experimental measurement of equilibrium concentrations, and their magnitudes can range from extremely small values to very large values. Equilibrium constants are appropriate only for the temperature at which they are measured and for the chemical equation as written. For example, if a chemical equation is reversed or if the stoichiometric coefficients are changed in the balanced equation, the expression for K and the value of K change accordingly (Section 16.2 and Table 16.2).

In the same way that phase equilibria are shown graphically, sets of equilibrium concentrations can be plotted (Section 16.3). However, algebraic solutions rather than graphic ones are generally used, and one of your major goals in this chapter should be to learn to carry out calculations involving the use of equilibrium constant expressions.

Once a system has reached equilibrium, it can be disturbed by external forces such as a change in temperature or by the addition or removal of a reagent. (The latter can be done by physically adding or removing a reactant or product. Changing the volume of a gas phase system has a similar effect.) The equilibrium system always responds according to **Le Chatelier's principle**: a change in any of the factors that determine the equilibrium point of a system will cause the system to change in such a manner as to offset the effect of the change.

Finally, this chapter further explores the connection between kinetics and equilibrium. In Section 16.6 you learned that some reactions may proceed by a mechanism in which reactants rapidly produce some intermediates that subsequently react in a slow step. This first step is often an equilibrium involving one or more of the reactants and one or more intermediates.

STUDY QUESTIONS

REVIEW QUESTIONS

1. Name three important features of the equilibrium condition.
2. Tell whether each of these statements is true or false. If false, change the wording to make it true.
   (a) The magnitude of the equilibrium constant is always independent of temperature.
   (b) When two chemical equations are added to give a net equation, the equilibrium constant for the

net equation is the product of the equilibrium constants of the summed equations.
(c) The equilibrium constant for a reaction has the same value as K for its reverse.
(d) Only the concentration of CO_2 appears in the equilibrium constant expression for the reaction $CaCO_3(s) \rightleftarrows CaO(s) + CO_2(g)$.
(e) For the reaction $CaCO_3(s) \rightleftarrows CaO(s) + CO_2(g)$, the value of K is the same whether the amount of CO_2 is expressed as moles/liter or as gas pressure.

3. Neither $PbCl_2$ nor PbF_2 is appreciably soluble in water. If solid $PbCl_2$ and solid PbF_2 are placed in equal amounts of water in separate beakers, in which beaker is the concentration of Pb^{2+} greater? Equilibrium constants for these solids dissolving in water are

$$PbCl_2(s) \rightleftharpoons Pb^{2+}(aq) + 2\ Cl^-(aq) \qquad K = 1.7 \times 10^{-5}$$

$$PbF_2(s) \rightleftharpoons Pb^{2+}(aq) + 2\ F^-(aq) \qquad K = 3.7 \times 10^{-8}$$

4. How does the reaction quotient for a reaction differ from the equilibrium constant for that reaction?
5. If the reaction quotient is smaller than the equilibrium constant for a reaction such as $A \rightleftarrows B$, does this mean that reactant A continues to disappear to form B, or does B form A, on moving to equilibrium?
6. Using Le Chatelier's principle, explain how increasing the temperature would affect the equilibrium $CaCO_3(s) + \text{heat} \rightleftarrows CaO(s) + CO_2(g)$. If more $CaCO_3$ is added to a flask in which this equilibrium exists, how is the equilibrium affected? What if some additional CO_2 is placed in the flask?

WRITING EQUILIBRIUM CONSTANT EXPRESSIONS

7. Write equilibrium constant expressions, in terms of reactant and product concentrations, for the following reactions:
(a) $2\ H_2O_2(g) \rightleftharpoons 2\ H_2O(g) + O_2(g)$
(b) $PCl_3(g) + Cl_2(g) \rightleftharpoons PCl_5(g)$
(c) $SiO_2(s) + 3\ C(s) \rightleftharpoons SiC(s) + 2\ CO(g)$
(d) $H_2(g) + S(s) \rightleftharpoons H_2S(g)$
8. Write equilibrium constant expressions, in terms of reactant and product concentrations, for the following reactions:
(a) $3\ O_2(g) \rightleftharpoons 2\ O_3(g)$
(b) $SiH_4(g) + 2\ O_2(g) \rightleftharpoons SiO_2(s) + 2\ H_2O(g)$
(c) $MgO(s) + SO_2(g) + \frac{1}{2}\ O_2(g) \rightleftharpoons MgSO_4(s)$
(d) $2\ PbS(s) + 3\ O_2(g) \rightleftharpoons 2\ PbO(s) + 2\ SO_2(g)$
9. Write equilibrium constant expressions, in terms of reagent concentrations or pressures as appropriate, for each of the following reactions:
(a) $TlCl_3(s) \rightleftharpoons TlCl(s) + Cl_2(g)$
(b) $CuCl_4^{2-}(aq) \rightleftharpoons Cu^{2+}(aq) + 4\ Cl^-(aq)$
(c) $2\ NO_2(g) \rightleftharpoons 2\ NO(g) + O_2(g)$
(d) $CO(g) + H_2O(g) \rightleftharpoons CO_2(g) + H_2(g)$
(e) $4\ H_3O^+(aq) + 2\ Cl^-(aq) + MnO_2(s) \rightleftharpoons Mn^{2+}(aq) + 6\ H_2O(\ell) + Cl_2(g)$
10. Write equilibrium constant expressions, in terms of reagent concentrations or pressures as appropriate, for each of the following reactions:
(a) The oxidation of ammonia with ClF_3 in rocket motors.

$$NH_3(g) + ClF_3(g) \rightleftharpoons 3\ HF(g) + \tfrac{1}{2}\ N_2(g) + \tfrac{1}{2}\ Cl_2(g)$$

(b) The disproportionation of a chlorine-containing ion [Cl with oxidation number +3 is converted to a higher oxidation number (+5) and a lower oxidation number (−1)].

$$3\ ClO_2^-(aq) \rightleftharpoons 2\ ClO_3^-(aq) + Cl^-(aq)$$

(c) $IO_3^-(aq) + 6\ OH^-(aq) + Cl_2(g) \rightleftharpoons IO_6^{5-}(aq) + 2\ Cl^-(aq) + 3\ H_2O(\ell)$
11. Consider the following two equilibria involving $SO_2(g)$ and their corresponding equilibrium constants.

$$SO_2(g) + \tfrac{1}{2}\ O_2(g) \rightleftharpoons SO_3(g) \qquad K_1$$

$$2\ SO_3(g) \rightleftharpoons 2\ SO_2(g) + O_2(g) \qquad K_2$$

Which of the following expressions relates K_1 to K_2?
(a) $K_2 = K_1^2$
(b) $K_2^2 = K_1$
(c) $K_2 = K_1$
(d) $K_2 = 1/K_1$
(e) $K_2 = 1/K_1^2$
12. The reaction of hydrazine (N_2H_4) with chlorine trifluoride (ClF_3) was used in experimental rocket motors at one time.

$$N_2H_4(g) + \tfrac{4}{3}\ ClF_3(g) \rightleftharpoons 4\ HF(g) + N_2(g) + \tfrac{2}{3}\ Cl_2(g)$$

How is the equilibrium constant, K_c, for this reaction related to K_c' for the reaction written in the following way?

$$3\ N_2H_4(g) + 4\ ClF_3(g) \rightleftharpoons 12\ HF(g) + 3\ N_2(g) + 2\ Cl_2(g)$$

(a) $K_c = K_c'$
(b) $K_c = 1/K_c'$
(c) $3\ K_c = K_c'$
(d) $(K_c)^3 = K_c'$
(e) $K_c = (K_c')^3$
13. Hydrogen can react with elemental sulfur to give the smelly, toxic gas H_2S according to the reaction

$$H_2(g) + S(\text{solid}) \rightleftharpoons H_2S(g)$$

(a) If the equilibrium constant for the reaction above is 7.6×10^5 at 25 °C, determine the value of the equilibrium constant for the reaction written as

$$8\ H_2(g) + S_8(s) \rightleftharpoons 8\ H_2S(g)$$

(b) Is the value of the equilibrium constant written in terms of gas pressures different from that written in terms of reagent concentrations? Briefly explain your reasoning.

14. At 400 °C, the equilibrium constant (K_p) for the Haber synthesis of ammonia is 1.7×10^{-4} for the reaction as written below.

$$3\ H_2(g) + N_2(g) \rightleftharpoons 2\ NH_3(g)$$

Calculate the value of K_p for the same reaction written as

$$\tfrac{3}{2}\ H_2(g) + \tfrac{1}{2}\ N_2(g) \rightleftharpoons NH_3(g)$$

15. Equilibrium constants for the following reactions have been determined at 823 K:

$$CoO(s) + H_2(g) \rightleftharpoons Co(s) + H_2O(g) \qquad K_1 = 67$$

$$CoO(s) + CO(g) \rightleftharpoons Co(s) + CO_2(g) \qquad K_2 = 490$$

Using this information, calculate K's (at the same temperature) for

$$CO_2(g) + H_2(g) \rightleftharpoons CO(g) + H_2O(g) \qquad K_3$$

and the commercially important water gas shift reaction.

$$CO(g) + H_2O(g) \rightleftharpoons CO_2(g) + H_2(g) \qquad K_4$$

16. Lead sulfide can be converted to metallic lead in the following series of reactions:

$$2\ PbS(s) + 3\ O_2(g) \rightleftharpoons 2\ PbO(s) + 2\ SO_2(g) \qquad K_1$$

$$PbO(s) + C(s) \rightleftharpoons Pb(\ell) + CO(g) \qquad K_2$$

$$PbO(s) + CO(g) \rightleftharpoons Pb(\ell) + CO_2(g) \qquad K_3$$

If the overall process is

$$2\ PbS(s) + 3\ O_2(g) + C(s) \rightleftharpoons 2\ Pb(\ell) + CO_2(g) + 2\ SO_2(g)$$

what is the equilibrium constant of the overall reaction, K_{net}, in terms of values of K_1, K_2, and K_3?

17. The equilibrium constant K_p for $N_2O_4(g) \rightleftharpoons 2\ NO_2(g)$ is 0.15 at 25 °C. Calculate the equilibrium constant for $NO_2(g) \rightleftharpoons \tfrac{1}{2}\ N_2O_4(g)$ at the same temperature.

18. The decomposition of NO_2, a brown gas, to form O_2 and colorless NO has an equilibrium constant, K_c, of about 1.7×10^{-5} at 500 °C.

$$2\ NO_2(g) \rightleftharpoons 2\ NO(g) + O_2(g)$$

What is the numerical value of the equilibrium constant, K_c', for the following reaction?

$$\tfrac{1}{2}\ O_2(g) + NO(g) \rightleftharpoons NO_2(g)$$

CALCULATING AND USING EQUILIBRIUM CONSTANTS

19. The following data are taken from a classic study of the equilibrium reaction (at 986 °C and 1 atm): $H_2(g) + CO_2(g) \rightleftharpoons H_2O(g) + CO(g)$.

EQUILIBRIUM COMPOSITION (MOLE PERCENT)		
CO_2	H_2	$CO = H_2O$
0.69	80.52	9.40
7.15	46.93	22.96
21.44	22.85	27.86
34.43	12.68	26.43

Using these data, derive a value for the equilibrium constant for the reaction. At this temperature, are the products or the reactants more favored? If the reaction is reversed ($H_2O + CO \rightleftharpoons H_2 + CO_2$), what is the value of K for the reversed reaction?

20. When the thiocyanate ion, SCN^-, is added to iron(III) cation, Fe^{3+}, in aqueous solution, the light yellow solution (characteristic of the iron ion) turns deep red from the $Fe\text{-}SCN^{2+}$ ion.

$$Fe^{3+}(aq) + SCN^-(aq) \rightleftharpoons Fe\text{-}SCN^{2+}(aq)$$

(a) Calculate K_c if the equilibrium concentrations at room temperature are as follows: $[Fe^{3+}] = 9.8 \times 10^{-4}$ M, $[SCN^-] = 1.8 \times 10^{-4}$ M, and $[Fe\text{-}SCN^{2+}] = 2.5 \times 10^{-5}$ M.

(b) Are products or reactants favored at room temperature?

21. At high temperature, hydrogen and carbon dioxide react to give water and carbon monoxide.

$$H_2(g) + CO_2(g) \rightleftharpoons H_2O(g) + CO(g)$$

(a) It is found at 986 °C that there are 11.2 atm each of CO and water vapor and 8.8 atm each of H_2 and CO_2 at equilibrium. Calculate the equilibrium constant.

(b) If there were 8.8 moles of H_2 and CO_2 in a 1.0-L container at equilibrium, how many moles of CO(g) and $H_2O(g)$ would be present?

22. Carbon dioxide reacts with carbon to give carbon monoxide according to the equation $C(s) + CO_2(g) \rightleftharpoons 2\ CO(g)$. At 700 °C, a 2.0-L flask is found to contain at equilibrium 0.10 moles of CO, 0.20 moles of CO_2, and 0.40 moles of C. Calculate the equilibrium constant, K_c, for this reaction at the specified temperature.

23. At 1500 °C water vapor is 10.% dissociated into $H_2(g)$ and $O_2(g)$. (That is, 10.% of the original water has been transformed into products, and 90.% remains.)

$$H_2O(g) \rightleftharpoons H_2(g) + \frac{1}{2} O_2(g)$$

Assuming a water concentration of 2.0 M before dissociation, calculate the equilibrium constant in concentration units.

24. When 2.0 moles of HI(g) are placed in a 1.0-L container at 25 °C and allowed to dissociate according to the equation $2\ HI(g) \rightleftharpoons H_2(g) + I_2(g)$, it is found that 20.% of the HI has dissociated at equilibrium. Calculate K_c and K_p.

25. Initially, 0.84 mol of $PCl_5(g)$ and 0.18 mol of $PCl_3(g)$ are mixed in a 1.0-L container. It is later found that only 0.72 mol of PCl_5 are present when the system has reached equilibrium. Calculate the value of K_c for the reaction

$$PCl_5(g) \rightleftharpoons PCl_3(g) + Cl_2(g)$$

26. Consider the following equilibrium process at 700. °C.

$$2\,H_2(g) + S_2(g) \rightleftharpoons 2\,H_2S(g)$$

Analysis shows that, at equilibrium, there are 2.50 mol of $H_2(g)$, 1.35×10^{-5} mol of $S_2(g)$, and 8.70 mol of $H_2S(g)$ present in a 12.0-L flask. Calculate K_c and K_p.

27. When solid ammonium carbamate sublimes, it dissociates completely into ammonia and carbon dioxide according to the equation

$$N_2H_6CO_2(s) \rightleftharpoons 2\ NH_3(g) + CO_2(g)$$

At 25 °C, experiment shows that the total pressure of the gases in equilibrium with the solid is 0.116 atm. What is the equilibrium constant for the reaction?

28. Ammonium iodide dissociates reversibly to ammonia and hydrogen iodide if the salt is heated to a sufficiently high temperature.

$$NH_4I(s) \rightleftharpoons NH_3(g) + HI(g)$$

Some ammonium iodide is placed in a flask, which is then heated to 400 °C. If the total pressure in the flask when equilibrium has been achieved is 705 mmHg, what is the value of K_p (when partial pressures are in atm)?

29. For the simple system $2\ NO_2(g) \rightleftharpoons N_2O_4(g)$ the equilibrium constant K_c is 170 at 25 °C. If analysis of the system shows that 2.0×10^{-3} mol of NO_2 is present in a 10.-L flask along with 1.5×10^{-3} mol of N_2O_4, is the system at equilibrium? If it is not at equilibrium, does the concentration of NO_2 increase or decrease on proceeding to equilibrium?

30. Nitrosyl bromide, NOBr(g), decomposes according to the equation

$$NOBr(g) \rightleftharpoons NO(g) + \frac{1}{2} Br_2(g)$$

with an equilibrium constant $K_p = 0.15$ at 350 °C. If 1.0 atm of NOBr, 0.80 atm of NO, and 0.40 atm of Br_2 are mixed at 350 °C, will any net reaction occur? If a net reaction is observed, will NO be formed or consumed?

31. The equilibrium constant K_c for the decomposition of $COBr_2$

$$COBr_2(g) \rightleftharpoons CO(g) + Br_2(g)$$

is 0.190 at 73 °C. If the concentrations of both CO(g) and $Br_2(g)$ are 0.402 M, and the concentration of $COBr_2(g)$ is 0.950 M, is the system at equilibrium? If it is not at equilibrium, does $[COBr_2]$ increase or decrease on proceeding to equilibrium?

32. Nitrogen dioxide can be formed from nitrogen oxide and oxygen.

$$2\ NO(g) + O_2(g) \rightleftharpoons 2\ NO_2(g)$$

At 1000 K the equilibrium constant K_c is 1.20.

(a) The system is analyzed and the following concentrations are found: $[O_2] = 1.25$ M, $[NO] = 2.25$ M, and $[NO_2] = 3.25$ M. Is the system at equilibrium? That is, does the reaction quotient equal K?

(b) If the system is not at equilibrium, does the concentration of NO_2 increase or decrease on proceeding to equilibrium?

33. The hydrocarbon C_4H_{10} can exist in two forms, *n*-butane and isobutane. The value of K_c for the interconversion of the two forms is 2.5 at 25 °C.

$$H_3C{-}CH_2{-}CH_2{-}CH_3 \rightleftharpoons H_3C{-}\underset{\displaystyle H}{\overset{\displaystyle CH_3}{C}}{-}CH_3$$

n-butane   isobutane

If you place 0.017 mol of *n*-butane in a 0.50-L flask at 25 °C and allow equilibrium to be established, what will be the equilibrium concentrations of the two forms of butane?

34. Cyclohexane, C_6H_{12}, a hydrocarbon, can isomerize or change into methylcyclopentane, a compound of the same formula but with a different molecular structure.

$$C_6H_{12}(g) \rightleftharpoons C_5H_9CH_3(g)$$

cyclohexane   methylcyclopentane

The equilibrium constant has been estimated to be 0.12 at 25 °C (Stevenson and Morgan, *J. Am. Chem.*

Soc. **1948**, *70*, 2773). If you originally placed 0.045 moles of cyclohexane in a 2.8 L flask, what are the concentrations of cyclohexane and methylcyclopentane when equilibrium is established?

35. The equilibrium constant K_p for $N_2O_4(g) \rightleftarrows 2\ NO_2(g)$ is 0.15 at 25 °C. If the pressure of N_2O_4 at equilibrium is 0.85 atm, what is the total pressure of the gas mixture ($N_2O_4 + NO_2$) at equilibrium?

36. Carbonyl bromide, $COBr_2$, decomposes to CO and Br_2 with an equilibrium constant, K_c, of 0.190 at 73 °C.

$$COBr_2(g) \rightleftharpoons CO(g) + Br_2(g)$$

If there is 0.015 mol of $COBr_2$ in a 2.5-L flask at equilibrium, what are the concentrations of CO and Br_2 at equilibrium, if only $COBr_2$ was placed in the flask initially?

37. Dichlorine oxide, Cl_2O, is a yellow-brown gas at room temperature. It is used to make bleaches for paper and textiles and is manufactured by the following reaction:

$$2\ Cl_2(g) + 2\ HgO(s) \rightleftharpoons HgCl_2 \cdot HgO(s) + Cl_2O(g)$$

Suppose you treat 0.15 mol of Cl_2 with HgO in a 1.0-L flask and wish to calculate the moles of Cl_2O ($= x$) that can be present at equilibrium. Knowing the value of K_c, the equilibrium constant, you can find x from

(a) $K_c = 2x/(0.15 - x)^2$ (c) $K_c = x^2/(0.15 - 2x)$
(b) $K_c = x/(0.15 - 2x)^2$ (d) $K_c = x/(0.15 - 2x)$

38. At a sufficiently high temperature, sulfur trioxide, an important compound in the formation of "acid rain," decomposes to sulfur dioxide and oxygen.

$$SO_3(g) \rightleftharpoons SO_2(g) + \tfrac{1}{2}\ O_2(g)$$

Assume you place 0.15 mol of SO_3 in a 5.0-L flask. Which expression below allows you to calculate the concentration of each reactant and product at equilibrium? (The variable x is assigned to the equilibrium concentration of SO_2.)

(a) $K = (0.030 - x)/(x)(\tfrac{1}{2}x)^{1/2}$
(b) $K = (x)(\tfrac{1}{2}x)^{1/2}/(0.030 - x)$
(c) $K = \tfrac{1}{2}x^2/(0.030 - x)$
(d) $K = 0.030/(x)(\tfrac{1}{2}x)^{1/2}$

39. Calcium carbonate decomposes at high temperature to form solid CaO and gaseous CO_2.

$$CaCO_3(s) \rightleftharpoons CaO(s) + CO_2(g)$$

If the equilibrium constant (K_c) at some temperature is 0.10, how many grams of CO_2 are formed when equilibrium is established at that temperature in a 1.0-L container?

40. Magnesium carbonate may be formed by reaction of MgO with $CO_2(g)$.

$$MgO(s) + CO_2(g) \rightleftharpoons MgCO_3(s) \qquad K_c = 1.2$$

How many moles of $CO_2(g)$ are present in a 2.0-L container when 2.0×10^2 g of $MgCO_3(s)$ have been produced at equilibrium?

41. The equilibrium constant K_p for the following reaction is 4.00×10^{-4} at 500 °C.

$$2\ HCN(g) \rightleftharpoons H_2(g) + C_2N_2(g)$$

What is the equilibrium partial pressure of $C_2N_2(g)$ at 500 °C if pure HCN was originally present at 25 atm?

42. The equilibrium constant K_c for the reaction

$$CO_2(g) + H_2(g) \rightleftharpoons CO(g) + H_2O(g)$$

is 1.6 at 690 °C. What is the concentration of each substance at equilibrium in a mixture prepared by originally placing 0.50 mol of $CO_2(g)$ and 0.50 mol of $H_2(g)$ in a 5.0-L container at 690 °C?

43. The equilibrium constant (K_c) for the decomposition of $COBr_2$ is 0.190 at 73 °C.

$$COBr_2(g) \rightleftharpoons CO(g) + Br_2(g)$$

You place 1.06 g of $COBr_2(g)$ in an 8.0-L flask and heat it to 73 °C. (a) What are the concentrations of reactants and products when equilibrium has been established? (b) What is the total pressure in the flask at equilibrium at 73 °C?

44. At 633 °C, 3.60 mol of ammonia is placed in a 2.00-L vessel and allowed to decompose to the elements.

$$2\ NH_3(g) \rightleftharpoons N_2(g) + 3\ H_2(g)$$

If K_c is 6.56×10^{-3} for this reaction at this temperature, calculate the equilibrium concentration of each reagent. What is the total pressure in the flask?

45. The decomposition of phosphorus pentachloride to phosphorus trichloride and chlorine has an equilibrium constant, K_p, of 11.5 at 300 °C.

$$PCl_5(g) \rightleftharpoons PCl_3(g) + Cl_2(g)$$

Assume you place enough solid PCl_5 in a flask so that when it is heated to 300 °C the pressure of undissociated PCl_5 is 1.50 atm at equilibrium. (a) What are the pressures of the reactant and products when equilibrium has been established? (b) What is the total pressure in the flask at equilibrium at 300 °C? (c) What fraction of the PCl_5 has dissociated?

46. The equilibrium constant K_c for the decomposition of $COBr_2$ is 0.190 at 73 °C.

$$COBr_2(g) \rightleftharpoons CO(g) + Br_2(g)$$

You place some $COBr_2(g)$ in a 5.0-L flask and heat it to form CO and $Br_2(g)$. If you want a $Br_2(g)$ concentration of 0.0500 M at equilibrium, how many grams of $COBr_2$ will you have to use in the beginning?

47. If some solid ZnS is placed in water, it is found that only a very small amount of the metal sulfide dissolves. What is the concentration of Zn^{2+} in the water? K_c for dissolving ZnS is 1.1×10^{-21}.

$$ZnS(s) \rightleftharpoons Zn^{2+}(aq) + S^{2-}(aq)$$

48. Barium sulfate is a poorly soluble salt.

$$BaSO_4(s) \rightleftharpoons Ba^{2+}(aq) + SO_4^{2-}(aq)$$

What is the maximum concentration of Ba^{2+} that can exist in equilibrium with 5.2×10^{-3} M SO_4^{2-}? K_c for dissolving $BaSO_4$ is 1.1×10^{-10}.

LE CHATELIER'S PRINCIPLE

49. Hydrogen, bromine, and HBr are in equilibrium in the gas phase.

$$H_2(g) + Br_2(g) \rightleftharpoons 2\ HBr(g) + 68\ kJ$$

How will each of the following changes affect the indicated quantities? Write *increase*, *decrease*, or *no change*.

CHANGE	$[Br_2]$	[HBr]	K_c
Some H_2 is added to the container.	______	______	______
The temperature of the gases in the container is increased.	______	______	______
The volume of the flask is increased.	______	______	______

50. K_p for the following reaction is 0.16 at 25 °C. The enthalpy change for the reaction at standard conditions is −344 kJ.

$$2\ NOBr(g) \rightleftharpoons 2\ NO(g) + Br_2(g)$$

Predict the effects of the following changes on the position of the equilibrium; that is, state which way the equilibrium will shift (left, right, or no change) when each of the following changes is made.
(a) Addition of more $Br_2(g)$
(b) Removal of some NOBr(g)
(c) Increase in the container volume
(d) Decrease in temperature

51. The formation of hydrogen sulfide from the elements is exothermic.

$$H_2(g) + S(s) \rightleftharpoons H_2S(g) + 20.6\ kJ$$

Predict the effects of the following changes on the position of the equilibrium; that is, state which way the equilibrium will shift (left, right, or no change) when each of the following changes is made.
(a) Add more sulfur
(b) Add more $H_2(g)$
(c) Decrease the volume of the reaction flask
(d) Raise the temperature

52. The oxidation of NO to NO_2 [$2\ NO(g) + O_2(g) \rightleftharpoons 2\ NO_2(g)$] is exothermic. Predict the effects of the following changes on the position of the equilibrium; that is, state which way the equilibrium will shift (left, right, or no change) when each of the following changes is made.
(a) Add more $O_2(g)$
(b) Add more $NO_2(g)$
(c) Increase the volume of the reaction flask
(d) Lower the temperature

53. Consider the isomerization of butane with an equilibrium constant of $K_c = 2.5$.

$$\underset{n\text{-butane}}{CH_3—CH_2—CH_2—CH_3} \rightleftharpoons \underset{\text{isobutane}}{CH_3—\overset{\overset{\large CH_3}{|}}{C}H—CH_3}$$

The system is originally at equilibrium with $[n] = 1.0$ M and $[iso] = 2.5$ M.
(a) If 0.50 mol/L of isobutane is suddenly added and then the system shifts to a new equilibrium position, what is the equilibrium concentration of each gas?
(b) If 0.50 mol/L of *n*-butane is added to the original equilibrium mixture, and the system shifts to a new equilibrium position, what is the equilibrium concentration of each gas?

54. K_p for the decomposition of ammonium hydrogen sulfide is 0.11 at 25 °C.

$$NH_4HS(s) \rightleftharpoons NH_3(g) + H_2S(g)$$

(a) When the pure salt decomposes in a flask, what are the equilibrium pressures of NH_3 and H_2S?
(b) If NH_4HS is placed in a flask already containing 0.50 atm of NH_3 and then the system is allowed to come to equilibrium, what are the equilibrium pressures of NH_3 and H_2S?

55. Two moles of PCl_5 and two moles of Cl_2 are mixed in a 1.0-L flask and allowed to reach equilibrium according to the equation $PCl_5(g) \rightleftharpoons PCl_3(g) + Cl_2(g)$. If the equilibrium amount of $PCl_3(g)$ is x mol/L, which is the correct expression for the equilibrium constant?
(a) $K_c = (2 - x)/x(2 + x)$
(b) $K_c = (2 + x)x/(2 - x)$
(c) $K_c = (2 + x)/x(2 - x)$
(d) $K_c = (2 - x)x/(2 + x)$

56. The oxidation of sulfur dioxide gives sulfur trioxide, an important compound in the formation of "acid rain."

$$SO_2(g) + \frac{1}{2} O_2(g) \rightleftharpoons SO_3(g)$$

Assume you place 0.15 mol of O_2 and 0.35 mol of SO_3 in a 1.0-L flask. Which expression below allows you to calculate the concentration of each reactant and product at equilibrium? (We have assigned the variable x to the equilibrium concentration of SO_2.)
(a) $K = (0.35 + x)/(x)(0.15 - \frac{1}{2}x)^{1/2}$
(b) $K = (0.35 - x)/(x)(0.15 + \frac{1}{2}x)^{1/2}$
(c) $K = (x)(0.15 + \frac{1}{2}x)^{1/2}(0.35 - x)$
(d) $K = (0.35 + x)/(x)(0.15 + \frac{1}{2}x)^{1/2}$

57. When solid ammonium carbamate sublimes, it dissociates completely into ammonia and carbon dioxide according to the equation ($K_p = 2.31 \times 10^{-4}$)

$$N_2H_6CO_2(s) \rightleftharpoons 2\ NH_3(g) + CO_2(g)$$

If 0.05 atm of CO_2 is added to the flask after equilibrium has been established at 25 °C, will the final pressure of CO_2 be greater or less than 0.05 atm? Will the pressure of NH_3 be greater or less than $P(NH_3)$ in the absence of added CO_2?

58. The equilibrium constant K_c for the decomposition of $COBr_2$ is 0.190 at 73 °C.

$$COBr_2(g) \rightleftharpoons CO(g) + Br_2(g)$$

(a) 1.50 mol of $COBr_2$ is placed in a 2.00-L flask and heated to 73 °C. What is the equilibrium concentration of each compound?
(b) If 1.00 mol of CO is added to the system at equilibrium, what is the concentration of each compound after equilibrium is reestablished?

EQUILIBRIUM AND KINETICS

59. The reaction between chloroform and chlorine gas is thought to proceed as follows:

Elementary step 1: Fast, equilibrium
$Cl_2(g) \rightleftharpoons 2\ Cl(g)$

Elementary step 2: Slow
$CHCl_3(g) + Cl(g) \longrightarrow CCl_3(g) + HCl(g)$

Elementary step 3: Fast
$CCl_3(g) + Cl(g) \longrightarrow CCl_4(g)$

Overall reaction:
$CHCl_3(g) + Cl_2(g) \longrightarrow CCl_4(g) + HCl(g)$

(a) Which of the steps is rate determining? Write the rate law for this step.
(b) Show that this mechanism agrees with the experimental rate law "Rate = $k[CHCl_3][Cl_2]^{1/2}$."

60. The ozone layer in the earth's upper atmosphere is important in shielding the earth from very harmful ultraviolet radiation. The ozone, O_3, decomposes according to the equation

$$2\ O_3(g) \longrightarrow 3\ O_2(g)$$

The mechanism of the reaction is thought to proceed through an initial fast equilibrium and a slow step.

Elementary step 1: Fast, equilibrium
$O_3(g) \rightleftharpoons O_2(g) + O(g)$

Elementary step 2: Slow
$O_3(g) + O(g) \longrightarrow 2\ O_2(g)$

(a) Which of the steps is rate determining? Write the rate law for this step.
(b) Show that this mechanism agrees with the experimental rate law "Rate = $k[O_3]^2/[O_2]$."

61. The reaction A + 2 B + C → D occurs according to the mechanism

Elementary step 1: Fast, equilibrium
$A + B \rightleftharpoons X$

Elementary step 2: Slow
$X + C \longrightarrow Y$

Elementary step 3: Fast
$Y + B \longrightarrow D$

Which of the following rate laws is correct?
(a) Rate = $k[C]$
(b) Rate = $k[D]$
(c) Rate = $k[A][B]$
(d) Rate = $k[A][B]^2[C]$
(e) Rate = $k[A][B][C]$

62. The following mechanisms have been proposed for the reduction of NO, with the overall equation

$$2\ H_2(g) + 2\ NO(g) \longrightarrow N_2(g) + 2\ H_2O(g)$$

The experimental rate law is "Rate = $k[NO]^2[H_2]$." Indicate which mechanism is consistent with this rate law.

Mechanism 1:	Slow	$H_2 + NO \longrightarrow H_2O + N$
	Fast	$N + NO \longrightarrow N_2 + O$
	Fast	$O + H_2 \longrightarrow H_2O$
Mechanism 2:	Fast	$2\ NO \rightleftharpoons N_2O_2$
	Slow	$N_2O_2 + H_2 \longrightarrow H_2O + N_2O$
	Fast	$N_2O + H_2 \longrightarrow N_2 + H_2O$

GENERAL PROBLEMS

63. Consider the following equilibrium process at 700 °C where $K_c = 1.1 \times 10^7$.

$$2\ H_2(g) + S_2(g) \rightleftharpoons 2\ H_2S(g)$$

Analysis of a mixture of these gases shows that there are 0.512 moles of $H_2(g)$, 2.67×10^{-5} moles of $S_2(g)$, and 0.333 moles of $H_2S(g)$ present in a 23.5-L flask. Is the system at equilibrium? If not, which compounds are consumed and which are formed on proceeding to equilibrium?

64. The equilibrium constant K_p for the formation of $PCl_5(g)$ from $PCl_3(g)$ and $Cl_2(g)$ is 0.087 at 300 °C.

$$PCl_3(g) + Cl_2(g) \rightleftharpoons PCl_5(g)$$

If you place 1.0 mol of each of PCl_3 and Cl_2 in a 5.0-L flask and heat to 300. °C, what is the concentration of PCl_5 that will exist at equilibrium?

65. Two molecules of gaseous acetic acid can form a dimer through hydrogen bonds.

$$2\ H_3C-\overset{\overset{O}{\|}}{C}-OH \rightleftharpoons H_3C-C\begin{matrix}\diagup\!\!\!\!\!\diagup O\cdots H-O \diagdown \\ \diagdown O-H\cdots O \diagup\!\!\!\!\!\diagup\end{matrix}C-CH_3$$

The equilibrium constant K_p at 25 °C has been determined to be 1.3×10^3 (where the pressures are measured in atmospheres). Assume that acetic acid is present initially at a partial pressure of 10. mmHg and that no dimer is present initially.

(a) What percentage of acetic acid is converted to dimer?

(b) The energy of the hydrogen bonds is 132 kJ/mol. As the temperature goes up, in which direction does the equilibrium shift?

66. The equilibrium constant for the *n*-butane $\rightleftharpoons$ isobutane isomerization reaction is 2.5 at 25 °C. (See Study Question 53.) If you mix 1.75 mol of *n*-butane and 1.25 mol of isobutane in a 2.0-L flask, is the system at equilibrium? If not, when it proceeds to equilibrium, which reagent increases in concentration? Calculate the concentrations of the two compounds when the system reaches equilibrium.

67. Zinc carbonate dissolves very poorly in water ($K_c = 1.5 \times 10^{-11}$).

$$ZnCO_3(s) \rightleftharpoons Zn^{2+}(aq) + CO_3^{2-}(aq)$$

If you place some solid $ZnCO_3$ in some water, what are the molar concentrations of Zn^{2+} and CO_3^{2-} when equilibrium has been achieved?

68. K_p at 2000 K for the formation of NO(g) is 4.0×10^{-4}.

$$N_2(g) + O_2(g) \rightleftharpoons 2\ NO(g)$$

(a) What is the value of K_c?

(b) If analysis shows that the concentrations of N_2 and O_2 are both 0.25 M, and that of NO is 0.0042 M, is the system at equilibrium?

(c) If the system is not at equilibrium, in which direction does the reaction proceed?

(d) When the system is at equilibrium, what are the equilibrium concentrations?

69. At 1800 K, oxygen dissociates very slightly into its atoms.

$$O_2(g) \rightleftharpoons 2\ O(g) \qquad K_p = 1.7 \times 10^{-8}$$

If you place 1.0 mol of O_2 in a 10.-L vessel and heat it to 1800. K, how many O atoms are present in the flask?

70. The concentrations given in Example 16.1 are not those at equilibrium. Given that the reaction begins with the concentrations in Example 16.1, what are the equilibrium concentrations of PCl_5, PCl_3, and Cl_2?

71. A reaction important in smog formation is

$$O_3(g) + NO(g) \rightleftharpoons O_2(g) + NO_2(g)\ K_c = 6.0 \times 10^{34}$$

(a) If the initial concentrations are $[O_3] = 1.0 \times 10^{-6}$ M, $[NO] = 1.0 \times 10^{-5}$ M, $[NO_2] = 2.5 \times 10^{-4}$ M, and $[O_2] = 8.2 \times 10^{-3}$ M, is the system at equilibrium? If not, in which direction does the reaction proceed?

(b) If the temperature is increased, as on a very warm day, will the concentrations of the products increase or decrease? (Hint: You may have to calculate the enthalpy change for the reaction to find out if it is exo- or endothermic.)

72. At high temperatures, carbon and carbon dioxide react to form carbon monoxide.

$$C(s) + CO_2(g) \rightleftharpoons 2\ CO(g)$$

(a) When equilibrium is established at 1000 K, the total pressure in the system is 4.70 atm. If $K_p = 1.72$, what are the partial pressures of CO and CO_2?

(b) If enough CO_2 had been placed in the reaction vessel before reaction so that $P(CO_2) = 1.11$ atm at 1000 K, what are the partial pressures of CO and CO_2 when equilibrium has been established?

73. At 633 °C, the equilibrium constant K_c for the dissociation of ammonia to its elements is 6.56×10^{-3}.

$$2\ NH_3(g) \rightleftharpoons N_2(g) + 3\ H_2(g)$$

(a) Calculate K_p.

(b) Ammonia is placed in a 2.0-L flask where it generates 456 mmHg of pressure. When equilibrium has been established, what is the total pressure in the flask?

74. The equilibrium reaction $N_2O_4(g) \rightleftharpoons 2\ NO_2(g)$ has been thoroughly studied (see Figure 16.7.) If the total pressure in a flask containing NO_2 and N_2O_4 gas at 25 °C is 1.50 atm, and the value of K_p at this temperature is 0.148, what fraction of the N_2O_4 has dissociated to

NO_2? What happens to the fraction dissociated if the volume of the container is increased so that the total equilibrium pressure falls to 1.00 atm?

75. To determine the equilibrium constant K_p for $N_2O_4(g) \rightleftarrows 2\ NO_2(g)$ at 25 °C you place 8.90 g of NO_2 in a 36-L flask at this temperature. The total gas pressure at equilibrium is 0.100 atm. What is the value of K_p for the dissociation of N_2O_4?

76. The equilibrium constants for the dissociation of three complexes of trimethylborane, $B(CH_3)_3$, have been determined.

$$L:B(CH_3)_3(g) \rightleftharpoons L(g) + B(CH_3)_3(g)$$

L	K_p (atm) at 100 °C
$(CH_3)_3P$	0.128
$(CH_3)_3N$	0.472
H_3N	4.62

(a) If you begin an experiment by placing 0.010 mole of each complex in separate flasks, which would have the largest partial pressure of $B(CH_3)_3$ at 100 °C?

(b) If 0.73 g (0.010 mol) of $H_3N:B(CH_3)_3$ is placed in a 1.0×10^2-mL flask and heated to 100. °C, what is the partial pressure of each gas in the equilibrium mixture and what is the total pressure? What is the percent dissociation of $H_3N:B(CH_3)_3$?

77. Sulfur dioxide reacts with oxygen to form sulfur trioxide, a source of air pollution.

$$2\ SO_2(g) + O_2(g) \rightleftharpoons 2\ SO_3(g)$$

Assume you place 2.0 mol of SO_3 in a 50.0-L flask and heat it to 900 °C. If K_c at this temperature is 11.39, what is the total pressure in the flask when equilibrium is achieved? What is the partial pressure of SO_3?

78. The decomposition of phosphorus pentachloride to phosphorus trichloride and chlorine has an equilibrium constant, K_p, of 11.5 at 300 °C.

$$PCl_5(g) \rightleftharpoons PCl_3(g) + Cl_2(g)$$

(a) Assume you place enough solid PCl_5 in a flask so that when it is heated to 300 °C the initial pressure of PCl_5 is 1.50 atm. What fraction of the PCl_5 has dissociated? (See Study Question 45.)

(b) Compare the fraction of PCl_5 dissociated in part (a) with the fraction that dissociates if the initial pressure of PCl_5 is doubled to 3.00 atm.

79. Sulfuryl chloride, SO_2Cl_2, is a compound with very irritating vapors; it is used as a reagent in the synthesis of organic compounds. When heated to a sufficiently high temperature it decomposes to SO_2 and Cl_2.

$$SO_2Cl_2(g) \rightleftharpoons SO_2(g) + Cl_2(g) \quad K_p = 2.4 \text{ at } 375\ °C$$

(a) Assume you place 6.70 g of SO_2Cl_2 in a 1.00-L flask and then heat to 375 °C. What is the pressure of each of the compounds in the system and the total pressure in the flask when equilibrium is achieved? What is the fraction of SO_2Cl_2 that has dissociated?

(b) What are the pressures of SO_2Cl_2, SO_2, and Cl_2 at equilibrium in the flask if you begin with a mixture of SO_2Cl_2 (6.70 g) and Cl_2 (1.00 atm)? What fraction of SO_2Cl_2 has dissociated?

(c) Compare the fractions of SO_2Cl_2 in parts (a) and (b). Do they agree with your expectation based on Le Chatelier's principle?

SUMMARY QUESTION

80. Nitrosyl bromide, NOBr, is prepared by the direct reaction of NO and Br_2

$$2\ NO(g) + Br_2(g) \longrightarrow 2\ NOBr(g)$$

but the compound dissociates readily at room temperature.

$$NOBr(g) \rightleftharpoons NO(g) + \tfrac{1}{2}\ Br_2(g)$$

(a) If you mix 3.50 g of NO and 9.67 g of Br_2, how many grams of NOBr can be prepared (assuming 100% yield)?

(b) N is the central atom of nitrosyl bromide. Draw the electron dot structure for the molecule.

(c) What is the structural pair geometry of NOBr? What is its molecular geometry? Is the molecule polar?

(d) Some NOBr is placed in a flask at 25 °C and allowed to dissociate. The total pressure at equilibrium is 190 mmHg and the compound is found to be 34% dissociated. What is the value of K_p?

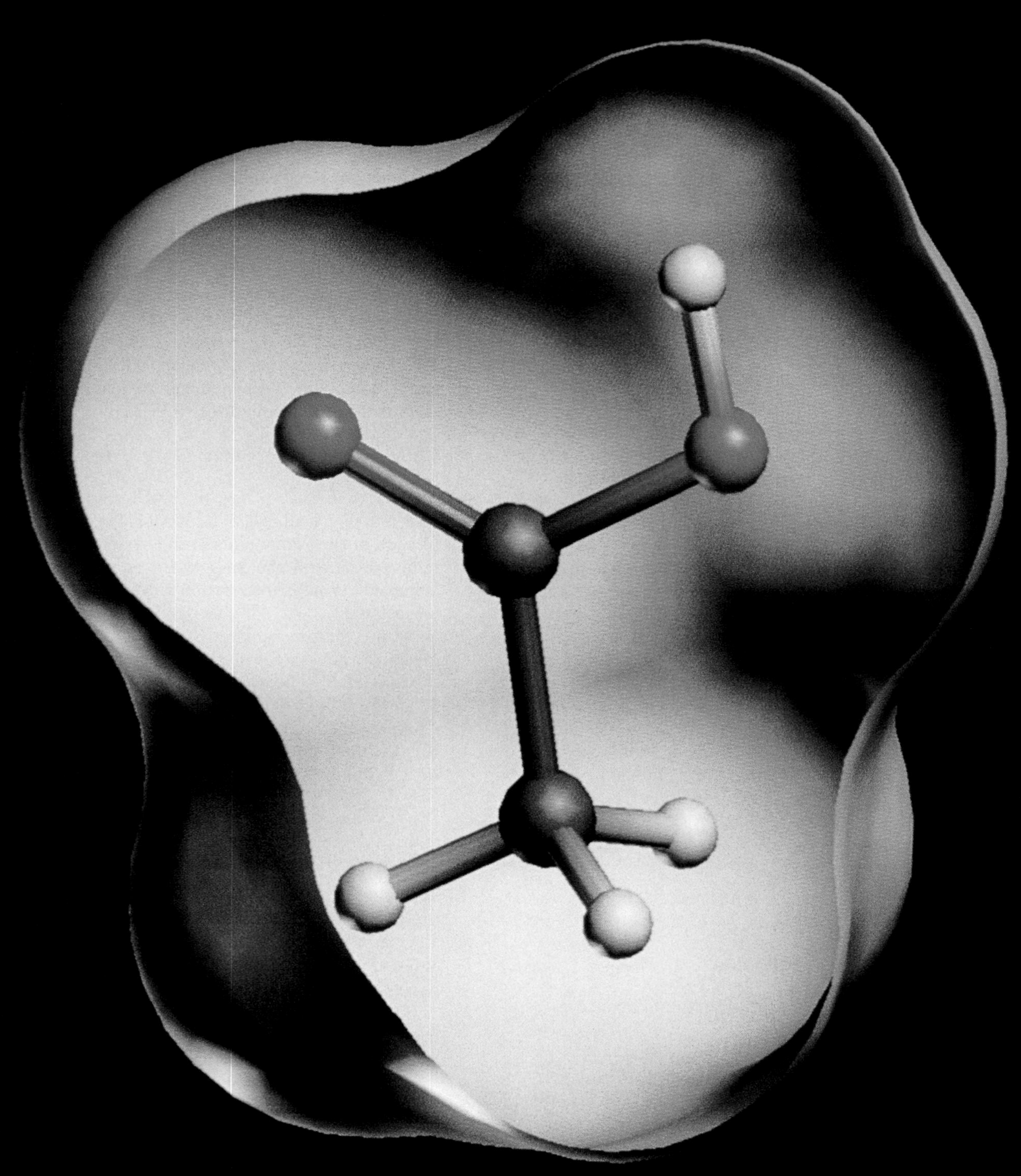

CHAPTER

17

The Chemistry of Acids and Bases

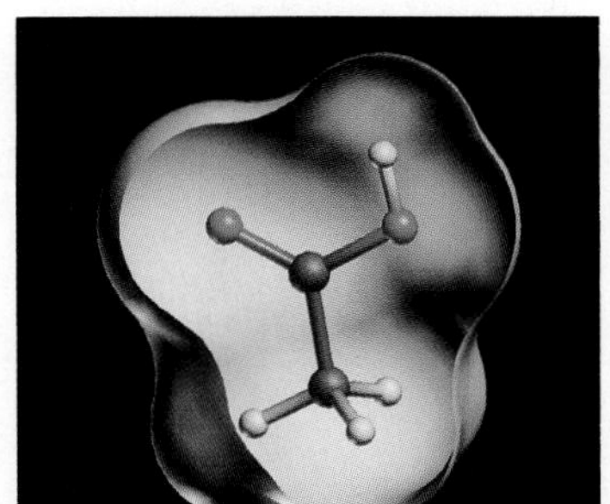

Ball and stick model of acetic acid, CH_3COOH (gray = C, white = H, and red = O) inside a model of the molecular surface. (J. Weber)

CHAPTER OUTLINE

Acids and bases provide the cornerstone of life on the earth. Basic carbonate and hydrogen carbonate ions from rocks and from atmospheric CO_2 are present in most natural waters along with such other substances as borate, phosphate, arsenate, silicate, and ammonia. In addition, volcanoes and hot springs can give strongly acidic waters due to the presence of HCl and SO_2. This witch's brew allows acids and bases to interact with one another and with carbon dioxide. Biological activities such as photosynthesis and respiration influence the hydrogen ion content of natural waters through the use or production of CO_2, the most common acid-forming substance in nature.

It has been said that "the ocean is the result of a gigantic acid–base titration; acids that have leaked out of the interior of the earth are titrated with bases that have been set free by the weathering of primary rock."

With such reactions occurring in our environment, it is no wonder that acids and bases have been studied for hundreds of years. The terms themselves are centuries old. "Acid" comes from the Latin word *acidus*, meaning sour. It was probably applied to vinegar, but it was later the name given to other substances that had a sour taste. "Alkali," another term for a base, is derived from an Arabic word for the ashes that come from burning certain plants. Since potash, or potassium carbonate, is the primary product of this process, and since water solutions of potash feel soapy and taste bitter, the term alkali was later applied more generally to substances having those properties. Finally, the word "salt," which finds it roots in many languages, probably originally meant sea salt or sodium chloride, although it now has a broader meaning.

With time the terms acid and alkali were applied more broadly. Robert Boyle, whose work with gases led to "Boyle's law," wrote in 1684 that alkalis give soapy solutions, restore vegetable colors reddened

Here you see a very early, explicit mention of the use of acid–base indicators. The juice of red cabbage is an excellent indicator of acidity (see Figure 5.20).

by acids, and react with acids to give what he called "indifferent salts." Acids, on the other hand, he characterized by their sour taste, their ability to be corrosive, to redden blue vegetable colors, and to lose all these properties when brought into contact with alkalis.

By the 18th century, salts were recognized as the product of the interaction of an acid and an alkali, a concept that became a major chemical theory. About 1750, the Frenchman Rouelle contended that a neutral salt was formed by an acid reacting with any substance—a water-soluble alkali, an "earth," a metal, and so on—capable of serving as a "base for [the salt]." Thus, the word "base" entered chemistry's vocabulary, and we now recognize reactions such as the following as salt producers (Figure 17.1):

$$2\ HCl(aq) + Ca(OH)_2(s)\ \text{(slaked lime)} \longrightarrow CaCl_2(aq) + 2\ H_2O(\ell)$$

$$2\ HCl(aq) + CaCO_3(s)\ \text{(limestone, an ``earth'')} \longrightarrow CaCl_2(aq) + H_2O(\ell) + CO_2(g)$$

$$2\ HCl(aq) + Ca(s)\ \text{(a metal)} \longrightarrow CaCl_2(aq) + H_2(g)$$

Because of their universal importance in chemistry, more general notions of classifying and explaining acid–base behavior and reactions have been developed over the past 100 years. One of the earlier concepts, and one still used in chemistry, came from Svante Arrhenius, a Swedish chemist (1859–1927). He proposed that an acid is a substance giving hydrogen ion (H^+) as one of the products of ionic dissociation in water and that a base gives hydroxide ion (OH^-) as one of the dissociation products.

Arrhenius Definition

ACID	BASE
$H_xB \longrightarrow x\ H^+ + B^{x-}$	$M(OH)_y \longrightarrow M^{y+} + y\ OH^-$
$HCl(aq) \longrightarrow H^+(aq) + Cl^-(aq)$	$NaOH(aq) \longrightarrow Na^+(aq) + OH^-(aq)$

Unfortunately, this concept is limited to compounds in water because it refers to ions (H^+ and OH^-) derived from water. A truly general concept of acids and bases should be appropriate to other solvents. Still another limitation is the requirement that the base dissociates to produce OH^-. This is certainly true of lye (NaOH), slaked lime [$Ca(OH)_2$], and milk of magnesia [$Mg(OH)_2$]. However, there are many substances that are bases, such as ammonia (NH_3), but that do not have OH^- ions as part of their original formulation.

More general models of acid–base behavior have been developed, and we shall examine two of these, one largely attributed to J.N. Brønsted (Danish, 1879–1947) and the other bearing the name of G.N. Lewis (American, 1875–1946). However, since we shall be most concerned with acids and bases in water, we first want to say something more about the properties of this solvent.

Figure 17.1 Salt-producing reactions. On the left, the reaction of aqueous HCl with calcium metal. On the right, the reaction of aqueous HCl with a piece of coral, a substance composed chiefly of calcium carbonate, $CaCO_3$.

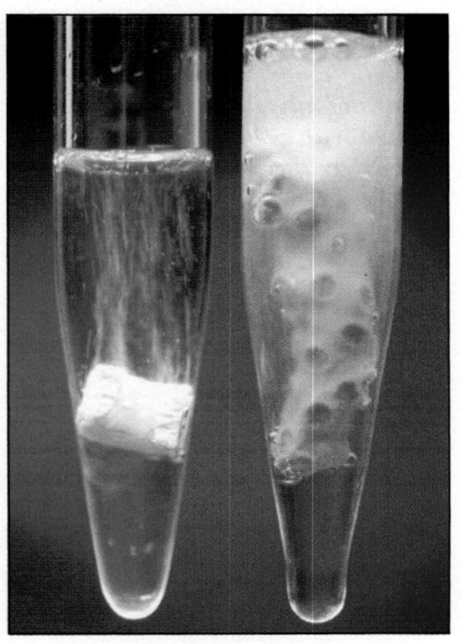

17.1 WATER, THE HYDRONIUM ION, AND AUTO-IONIZATION

The H^+ ion is a bare proton with a radius *much* less than 0.1 pm. If this tiny particle with its highly concentrated charge is placed in water, it will be attracted very strongly to the negative end of the water dipole.

$$H^+(g) + H_2O(\ell) \longrightarrow [H_3O^+](aq) + \text{Energy}$$

Hydrated proton
= hydronium ion

Thus, a proton in water can exist only associated with water molecules as a *hydrated proton*. By analogy with the ammonium ion, NH_4^+, we call the simplest proton-water complex, H_3O^+, the **hydronium ion**. Similarly, it is certain that the hydroxide ion is associated with water molecules through hydrogen bonding, perhaps in the form of ions such as $H_3O_2^-$.

An acid such as HCl does not need to be present for the hydronium ion to exist in water. In fact, *two water molecules can interact with one another to produce a hydronium ion and a hydroxide ion* by proton transfer from one water molecule to another.

$$2\ H_2O(\ell) \rightleftharpoons H_3O^+(aq) + OH^-(aq)$$

$$H{-}\ddot{O}{:}(H) + H{-}\ddot{O}{:}(H) \rightleftharpoons [H{-}\ddot{O}{-}H(H)]^+ + [{:}\ddot{O}{:}(H)]^-$$

The existence of the so-called **auto-ionization of water** was proved many years ago by Friedrich Kohlrausch (1840–1910). He found that, even after water is painstakingly purified, it still conducts electricity to a very small extent, since auto-ionization produces very low concentrations of H_3O^+ and OH^- ions even in the purest water. Water auto-ionization is the cornerstone of our concepts of aqueous acid–base behavior.

17.2 THE BRØNSTED CONCEPT OF ACIDS AND BASES

In 1923 J.N. Brønsted in Copenhagen (Denmark) and J.M. Lowry in Cambridge (England) independently suggested a new concept of acid and base behavior. They proposed that *an **acid** is any substance that can **donate a proton** to any other substance*. As described later in the chapter, acids can be neutral compounds such as nitric acid,

$$\underset{\text{acid}}{HNO_3(aq)} + H_2O(\ell) \longrightarrow H_3O^+(aq) + NO_3^-(aq)$$

or they can be cations or anions.

$$\underset{\text{acid}}{NH_4^+(aq)} + H_2O(\ell) \longrightarrow H_3O^+(aq) + NH_3(aq)$$

$$\underset{\text{acid}}{H_2PO_4^-(aq)} + H_2O(\ell) \longrightarrow H_3O^+(aq) + HPO_4^{2-}(aq)$$

According to Brønsted and Lowry, *a **base** is a substance that can **accept a proton** from any other substance*. As with acids, these can be neutral compounds,

$$\underset{\text{base}}{NH_3(aq)} + H_2O(\ell) \longrightarrow NH_4^+(aq) + OH^-(aq)$$

or anions.

$$\underset{\text{base}}{CO_3^{2-}(aq)} + H_2O(\ell) \longrightarrow HCO_3^-(aq) + OH^-(aq)$$

$$\underset{\text{base}}{PO_4^{3-}(aq)} + H_2O(\ell) \longrightarrow HPO_4^{2-}(aq) + OH^-(aq)$$

These statements have come to be known as the Brønsted definition of acids and bases, because Brønsted and his students substantiated and extended the idea.

There is a wide variety of Brønsted acids, and you are familiar with many of them. Acids such as HF, HCl, HNO_3, and CH_3COOH (acetic acid) are all capable of donating one proton per molecule and so are called **monoprotic acids**. However, there are many acids capable of donating two or more protons, and these are called **polyprotic acids**. Some common polyprotic acids include H_2SO_4, H_3PO_4, H_2CO_3 (carbonic acid), H_2S, and $H_2C_2O_4$ (oxalic acid).

$$\underset{\text{oxalic acid}}{\begin{array}{l}O{=}C{-}O{-}H(aq)\\ \quad\;|\\ O{=}C{-}O{-}H\end{array}} + H_2O(\ell) \longrightarrow \underset{\text{hydrogen oxalate}}{\begin{array}{l}O{=}C{-}O^-(aq)\\ \quad\;|\\ O{=}C{-}O{-}H\end{array}} + H_3O^+(aq)$$

$$\underset{\text{hydrogen oxalate}}{\begin{array}{l}O{=}C{-}O^-(aq)\\ \quad\;|\\ O{=}C{-}O{-}H\end{array}} + H_2O(\ell) \longrightarrow \underset{\text{oxalate}}{\begin{array}{l}O{=}C{-}O^-(aq)\\ \quad\;|\\ O{=}C{-}O^-\end{array}} + H_3O^+(aq)$$

Just as there are acids that can donate more than one proton, there are bases that can accept more than one and so are called **polyprotic bases**. Such bases include the anions of the polyprotic acids listed above, that is, SO_4^{2-}, PO_4^{3-}, CO_3^{2-}, S^{2-}, and $C_2O_4^{2-}$. All behave to a greater or lesser extent like the sulfide ion.

$$S^{2-}(aq) + H_3O^+(aq) \longrightarrow HS^-(aq) + H_2O(\ell)$$

$$\underset{\text{base}}{HS^-(aq)} + H_3O^+(aq) \longrightarrow H_2S(aq) + H_2O(\ell)$$

There are also **amphiprotic substances**, *molecules or ions that can behave either as a Brønsted acid or as a Brønsted base*. One of the best examples is water in its auto-ionization reaction. Water acts as a base in the presence of HCl, in that H_2O accepts a proton from the acid,

$$\underset{\text{acid}}{HCl(aq)} + \underset{\text{base}}{H_2O(\ell)} \longrightarrow H_3O^+(aq) + Cl^-(aq)$$

and it acts as an acid when donating a proton to ammonia.

$$\underset{\text{base}}{NH_3(aq)} + \underset{\text{acid}}{H_2O(\ell)} \longrightarrow NH_4^+(aq) + OH^-(aq)$$

Other amphiprotic substances you have encountered thus far, in addition to water, include such important anions as HCO_3^- (bicarbonate), $H_2PO_4^-$ (dihydrogen phosphate), and $HC_2O_4^-$ (hydrogen oxalate).

The amphiprotic ions HPO_4^{2-} and HCO_3^- are especially important in biochemistry.

$$\underset{\text{acid}}{HCO_3^-(aq)} + H_2O(\ell) \longrightarrow H_3O^+(aq) + CO_3^{2-}(aq)$$

$$\underset{\text{base}}{HCO_3^-(aq)} + H_2O(\ell) \longrightarrow H_2CO_3(aq) + OH^-(aq)$$

Finally, let us briefly consider an **acid–base reaction**, the interaction between two substances, where one is an acid in water and the other is a base in water.

$$\underset{\text{acid}}{HCl(aq)} + \underset{\text{base}}{NH_3(aq)} \longrightarrow NH_4^+(aq) + Cl^-(aq)$$

A proton has clearly been transferred from the acid to the base in a reaction that can be thought of as the sum of three processes:

$$\begin{array}{rl} & HCl(aq) + H_2O(\ell) \longrightarrow H_3O^+(aq) + Cl^-(aq) \\ & NH_3(aq) + H_2O(\ell) \longrightarrow NH_4^+(aq) + OH^-(aq) \\ & \underline{\mathbf{H_3O^+(aq) + OH^-(aq) \longrightarrow 2\,H_2O(\ell)}} \\ \textit{Net reaction:} & HCl(aq) + NH_3(aq) \longrightarrow NH_4^+(aq) + Cl^-(aq) \end{array}$$

where water functions as the proton transfer agent. As you have previously seen when writing net ionic equations (Chapter 5), the principal aspect of acid–base reactions is the interaction of hydronium ion and hydroxide ion to give water, the opposite of the auto-ionization of water. The other substance produced in an acid–base reaction is a salt, and this compound can itself be acidic, basic, or neutral, depending on the cation and anion involved. This chapter will help you predict the acid–base characteristics of salts, and Chapter 18 will consider acid–base reactions in greater detail.

CONJUGATE ACID–BASE PAIRS

Now let us expand some more on the Brønsted notion of acids and bases. In each of the chemical equations above, *a proton has been transferred*. If an acid donates a proton, there must be a substance that accepts that proton, a base. For example, the hydrogen carbonate ion (bicarbonate ion) transfers a proton to the base water,

$$\underset{\text{acid}}{HCO_3^-(aq)} + \underset{\text{base}}{H_2O(\ell)} \rightleftharpoons H_3O^+(aq) + CO_3^{2-}(aq)$$

and in the process the carbonate ion and the hydronium ion have been produced. Since this reaction is reversible, the carbonate ion acts as a base, accepting a proton from the proton donor H_3O^+ and re-forms the bicarbonate ion. Thus, bicarbonate and carbonate ions are related to one another by the loss or gain of H^+, as are H_2O and H_3O^+. *A pair of compounds or ions that differ by the presence of one H^+ unit* is called a **conjugate acid–base pair**. Thus, CO_3^{2-} is the conjugate base of the acid HCO_3^-,

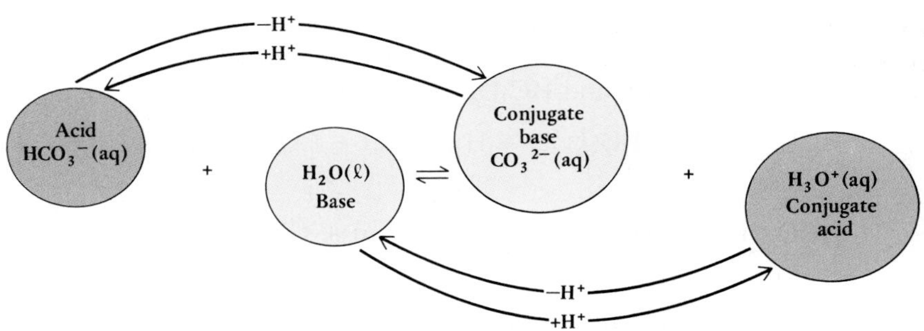

and HCO_3^- is the conjugate acid of the base CO_3^{2-}. *Every acid–base reaction involving H^+ transfer has two conjugate acid–base pairs.* To see that this is the case, look at the reactions above and those in Table 17.1.

EXERCISE 17.1 Conjugate Acid–Base Pairs

(a) In the following reaction, identify the acid on the left and its conjugate base on the right. Similarly, identify the base on the left and its conjugate acid on the right.

$$HBr(aq) + NH_3(aq) \rightleftharpoons NH_4^+(aq) + Br^-(aq)$$

(b) What is the conjugate base of H_2S?
(c) What is the conjugate acid of NO_3^-?

RELATIVE STRENGTHS OF ACIDS AND BASES

In water some acids are better proton donors than others, and some bases are better proton acceptors than others. For example, a dilute solution of hydrochloric acid consists largely of $H_3O^+(aq)$ and $Cl^-(aq)$ ions; the acid is nearly 100% ionized, and so it is considered a *strong* Brønsted acid.

*How can you tell whether an acid or base is weak? The easiest way is to remember those that are strong (see table below); all others are probably weak (especially those listed in Table 17.4). See also Sections 17.4 and 17.5. Some common **strong acids** are*
Hydrohalic acids: HCl, HBr, HI
Nitric acid: HNO_3
Sulfuric acid: H_2SO_4
Perchloric acid: $HClO_4$

*Some common **strong bases** are:*
Group 1A hydroxides: LiOH, NaOH, KOH
Group 2A hydroxides: $Ca(OH)_2$ and $Mg(OH)_2$

Strong acid (≈100% ionized)

$$[H_3O^+] \approx \text{initial concentration of the acid}$$

$$0.1\text{ M } HCl(aq) + H_2O(\ell) \longrightarrow 0.1\text{ M } H_3O^+(aq) + 0.1\text{ M } Cl^-(aq)$$

In contrast, acetic acid ionizes only to a very small extent and so is considered a *weak* Brønsted acid.

Weak acid (<10% ionized)

$$[H_3O^+] << \text{initial concentration of the acid}$$

$$0.1\text{ M } CH_3COOH(aq) + H_2O(\ell) \longrightarrow 0.001\text{ M } H_3O^+(aq) + 0.001\text{ M } CH_3COO^-(aq)$$

There is no good dividing line that will allow you to point to an acid and say it is "weak." However, we can say that weak acids generally are less than about 10% ionized, and most are much less ionized than that when used as dilute solutions in water.

Table 17.1 Some Acids and Bases

Name	Acid 1		Base 2		Base 1		Acid 2*
Hydrogen chloride	HCl	+	H_2O	⇄	Cl^-	+	H_3O^+
Nitric acid	HNO_3	+	H_2O	⇄	NO_3^-	+	H_3O^+
Hydrogen carbonate	HCO_3^-	+	H_2O	⇄	CO_3^{2-}	+	H_3O^+
Acetic acid	CH_3COOH	+	H_2O	⇄	CH_3COO^-	+	H_3O^+
Hydrogen cyanide	HCN	+	H_2O	⇄	CN^-	+	H_3O^+
Hydrogen sulfide	H_2S	+	H_2O	⇄	HS^-	+	H_3O^+
Ammonia	H_2O	+	NH_3	⇄	OH^-	+	NH_4^+
Carbonate ion	H_2O	+	CO_3^{2-}	⇄	OH^-	+	HCO_3^-
Water	H_2O	+	H_2O	⇄	OH^-	+	H_3O^+

*Acid 1 and base 1 are a conjugate pair, as are base 2 and acid 2.

Oxide ion is a strong Brønsted base in aqueous solution because it has a strong tendency to accept a proton from water to produce OH^-.

Strong base

$$[OH^-] \approx 2 \text{ (initial concentration of the base)}$$

$$O^{2-}(aq) + H_2O(\ell) \longrightarrow 2\ OH^-(aq)$$

Aqueous ammonia and the carbonate ion, however, produce only a very small concentration of OH^- and so are classed as weak Brønsted bases.

Weak base

$$[OH^-] << \text{initial concentration of the base}$$

$$NH_3(aq) + H_2O(\ell) \longrightarrow NH_4^+(aq) + OH^-(aq)$$

$$CO_3^{2-}(aq) + H_2O(\ell) \longrightarrow HCO_3^-(aq) + OH^-(aq)$$

According to the Brønsted model, an acid donates a proton and produces a conjugate base. The acid and its conjugate base are simply related by the idea that the acid contains a bond to an H atom (H—A) that is broken on formation of the conjugate base (H—A → H^+ + :A^-). If the H—A bond is readily broken, it is not readily made. This means that the acid is strong, while the conjugate base is weak. Therefore, we can say

(a) the stronger the acid, the weaker its conjugate base.

Conversely, if the H—A bond is readily made, it is not easily broken, and so

(b) the stronger the base, the weaker its conjugate acid.

Thus, aqueous HCl is a strong acid because it has a strong tendency to donate a proton to water and produce the conjugate base Cl^-. At the same time, water, acting as a base, has accepted the proton to produce the conjugate acid H_3O^+. Since we know that the reaction has proceeded almost completely to the right, at equilibrium there is essentially no HCl left intact in the solution. This must mean that, *in the case of a strong acid, the acid and base on the left side of the balanced equation are a stronger combination than the conjugate acid and base combination on the right*. Of the two bases in solution, H_2O and Cl^-, water must be the stronger base and wins out in the competition for the proton. Of the two acids, HCl and H_3O^+, HCl is better able to donate a proton. We have denoted this by using arrows of unequal length below, showing that the reaction proceeds largely to the right.

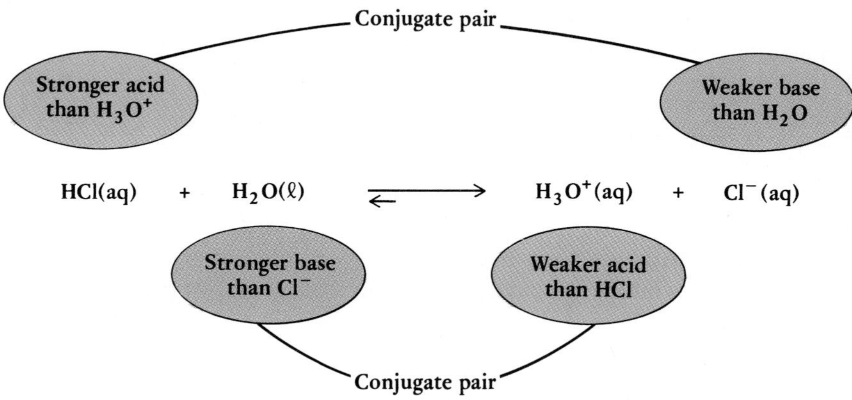

Acetic acid, that important component of vinegar and bad wine, is the classic example of a weak acid. Acetic acid ionizes to a very small extent in water. Thus, of the two acids present in aqueous acetic acid (CH_3COOH and H_3O^+), the hydronium ion is the stronger. Of the two bases (H_2O and CH_3COO^-), the acetate ion must be the stronger. At equilibrium, the solution consists mostly of acetic acid and water, with only a small concentration of acetate ion and hydronium ion.

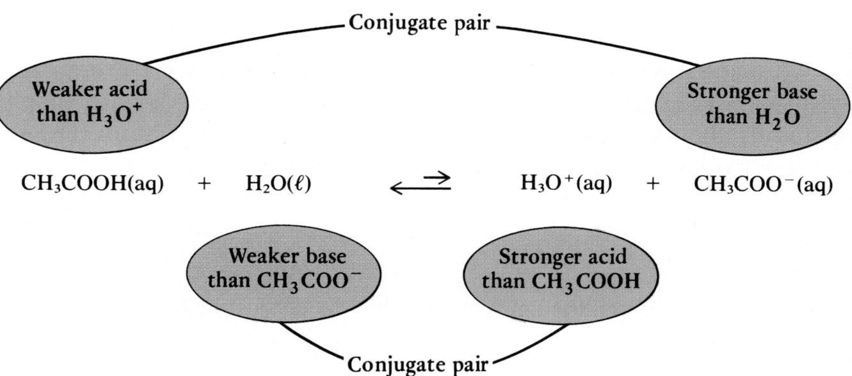

As shown by these two examples of the relative extent of acid–base reactions, another corollary to the Brønsted model is that

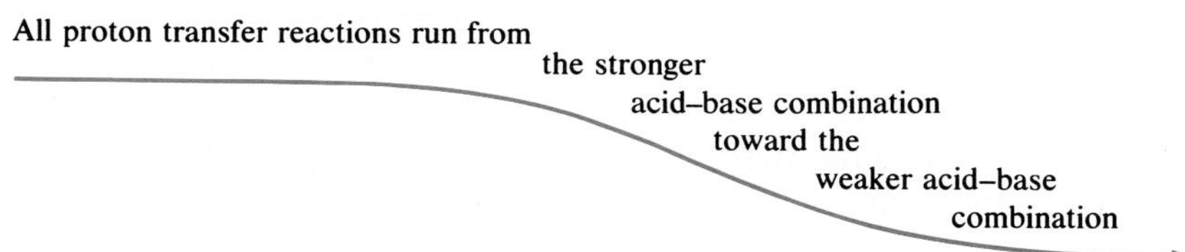

Table 17.2 Relative Strengths of Acids and Bases

Conjugate Acid		Conjugate Base	
Name	Formula	Formula	Name
Perchloric acid	$HClO_4$	ClO_4^-	Perchlorate ion
Sulfuric acid	H_2SO_4	HSO_4^-	Hydrogen sulfate ion
Hydrochloric acid	HCl	Cl^-	Chloride ion
Nitric acid	HNO_3	NO_3^-	Nitrate ion
Hydronium ion	H_3O^+	H_2O	Water
Hydrogen sulfate ion	HSO_4^-	SO_4^{2-}	Sulfate ion
Phosphoric acid	H_3PO_4	$H_2PO_4^-$	Dihydrogen phosphate ion
Acetic acid	CH_3COOH	CH_3COO^-	Acetate ion
Hexaaquaaluminum ion	$[Al(H_2O)_6]^{3+}$	$[Al(H_2O)_5(OH)]^{2+}$	Hydroxopentaaquaaluminum ion
Carbonic acid	H_2CO_3	HCO_3^-	Hydrogen carbonate ion
Hydrogen sulfide	H_2S	HS^-	Hydrogen sulfide ion
Dihydrogen phosphate ion	$H_2PO_4^-$	HPO_4^{2-}	Hydrogen phosphate ion
Ammonium ion	NH_4^+	NH_3	Ammonia
Hydrogen cyanide	HCN	CN^-	Cyanide ion
Hydrogen carbonate ion	HCO_3^-	CO_3^{2-}	Carbonate ion
Phenol	C_6H_5OH	$C_6H_5O^-$	Phenoxide ion
Water	H_2O	OH^-	Hydroxide ion
Ethyl alcohol	C_2H_5OH	$C_2H_5O^-$	Ethoxide ion
Ammonia	NH_3	NH_2^-	Amide ion
Methylamine	CH_3NH_2	CH_3NH^-	Methylamide ion
Hydrogen	H_2	H^-	Hydride ion
Methane	CH_4	CH_3^-	Methide ion

↑ Increasing Acid Strength (upward); Increasing Base Strength ↓ (downward)

Acids and bases can be ranked in order of their abilities to donate or accept protons in aqueous solution, and we have done so for a few acids and bases in Table 17.2. At the very top on the left are the stronger acids, those substances that so strongly donate protons that their conjugate bases are extremely weak. The opposite is true for acids at the bottom left of the table. For example, the H_2 molecule can be considered an acid in the sense that it is conceivable that it can give the conjugate base H^-, the hydride ion, on losing H^+. However, hydride ion is such a strong base that it will extract H^+ from water to give H_2, often with explosive violence (Figure 17.2).

Figure 17.2 The basic properties of the hydride ion, H^-. Calcium hydride, CaH_2, is a source of H^- ion. This ion is such a powerful proton acceptor that it reacts vigorously with the water, a proton donor, to give H_2.

WRITING ACID–BASE REACTIONS TO ACCOUNT FOR RELATIVE ACID–BASE STRENGTHS

The chart of acids and bases in Table 17.2 can be used to predict whether the equilibrium lies to the left or the right in an acid–base reaction. The examples that follow show you how to do this.

EXAMPLE 17.1

PREDICTING THE DIRECTION OF ACID–BASE REACTIONS

Write a balanced equation for the reaction that occurs between each acid–base pair in water and tell whether the equilibrium lies to the left or right. (a) Acetic acid, CH_3COOH, and sodium cyanide, NaCN. Is HCN, a poisonous acid, formed to a significant extent? (b) Ammonium chloride, NH_4Cl, and sodium carbonate, Na_2CO_3.

Solution (a) It is shown in Table 17.2 that the conjugate base of acetic acid is the acetate ion, CH_3COO^-. The other reactant, sodium cyanide, is a water-soluble salt and dissociates to give Na^+ and the cyanide ion, CN^-. Sodium ion is not listed in Table 17.2 because it is neither a Brønsted acid nor base in water (see Section 17.5). The CN^- ion, however, is a base with HCN as its conjugate acid. Therefore, we can write the reaction

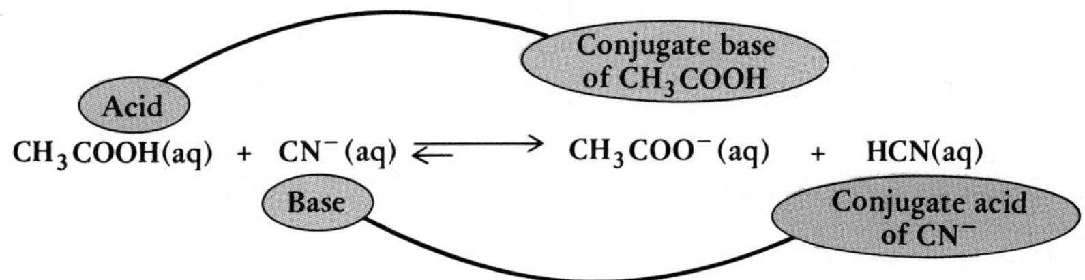

To decide on which side the equilibrium lies, left or right, we can compare the two acids (or the two bases) in the reaction. According to the chart, HCN is a weaker acid than CH_3COOH, and CH_3COO^- is a weaker base than CN^-. Since reactions run toward the weaker acid and base, this must mean that, at equilibrium, the reaction favors the HCN/CH_3COO^- pair. This tells you that you certainly would not want inadvertently to mix acetic acid and a cyanide salt since HCN will be the product. The average fatal dose of HCN is 50 to 60 mg or about 0.002 moles.

(b) Ammonium chloride dissociates to give NH_4^+ and Cl^- in aqueous solution, ions that are both listed in Table 17.2. The ammonium ion is an acid according to Table 17.2.

$$NH_4^+(aq) + H_2O(\ell) \rightleftharpoons NH_3(aq) + H_3O^+(aq)$$

acid | base | conjugate base | conjugate acid

and, in principle, the chloride ion is a base in water.

$$Cl^-(aq) + H_2O(\ell) \rightleftharpoons HCl(aq) + OH^-(aq)$$

base | acid | conjugate acid | conjugate base

As you will see in Section 17.5, we can disregard the base properties of ClO_4^-, Cl^-, and NO_3^- in aqueous solution. These ions do not contribute to the acidity or basicity of a solution.

It is shown in Table 17.2 that OH^- is a *much* stronger base than Cl^-, and HCl is a *much* stronger acid than H_2O. Therefore, we can consider that this reaction does not occur. In general, *any anion X^- that is a weaker base than OH^- does not react appreciably with water to give HX and OH^-*. Thus, for ammonium chloride we only need to worry about the NH_4^+ ion.

Sodium carbonate, Na_2CO_3, dissociates in water to give Na^+ and CO_3^{2-}. As in (a), we can neglect any contribution from Na^+. Carbonate, however, is a base according to Table 17.2.

$$CO_3^{2-}(aq) + H_2O(\ell) \rightleftharpoons HCO_3^-(aq) + OH^-(aq)$$

base | acid | conjugate acid | conjugate base

Therefore, the ammonium and carbonate ions can interact according to the equation

$$NH_4^+(aq) + CO_3^{2-}(aq) \rightleftharpoons NH_3(aq) + HCO_3^-(aq)$$

acid | base | conjugate base | conjugate acid

We can see from Table 17.2 that NH_4^+ is a stronger acid than HCO_3^-, and that CO_3^{2-} is a stronger base than NH_3. Therefore, some reaction does indeed occur between the ammonium ion and the carbonate ion in water.

EXERCISE 17.2 Predicting the Direction of Acid–Base Reactions

For each reaction below, predict whether the equilibrium lies predominantly to the left or to the right.

(a) $HSO_4^-(aq) + NH_3(aq) \rightleftharpoons NH_4^+(aq) + SO_4^{2-}(aq)$

(b) $HCO_3^-(aq) + HS^-(aq) \rightleftharpoons H_2S(aq) + CO_3^{2-}(aq)$

EXERCISE 17.3 Writing Acid–Base Reactions

If ammonium chloride and sodium sulfate are mixed in water, write a balanced equation for the acid–base reaction that could, in principle, occur. Does the reaction in fact occur to any appreciable extent?

17.3 WATER AND THE pH SCALE

Stronger Brønsted acids and bases lead to larger concentrations of $H_3O^+(aq)$ and $OH^-(aq)$ for a given number of moles of acid or base, respectively. If these concentrations could be measured quantitatively, we would have a way to compare acid and base strengths and to predict more accurately the direction and extent of acid–base reactions. In this section we outline one way to accomplish this objective.

THE WATER IONIZATION CONSTANT, K_w

Water auto-ionizes, transferring a proton from one water molecule to another and producing a hydronium ion and a hydroxide ion.

$$2\ H_2O(\ell) \rightleftharpoons H_3O^+(aq) + OH^-(aq)$$

Since the hydroxide ion is a much stronger base than water, and the hydronium ion is a much stronger acid than water (Table 17.2), the equilibrium lies far to the left side. In fact, in pure water at 25 °C only about 2 out of a billion water molecules are ionized at any instant. To express this idea more quantitatively, we can write the equilibrium constant expression

$$K = \frac{[H_3O^+][OH^-]}{[H_2O]^2}$$

However, in pure water or in dilute aqueous solutions (say 0.1 M solute or less), the concentration of water can be considered to be a constant (55.4 M), so we include that with the constant K and write the equilibrium constant instead as

$$K[H_2O]^2 = K_w$$

The equation $K_w = [H_3O^+][OH^-]$ is valid in pure water and in any aqueous solution. However, be aware that K_w is temperature dependent; it increases with T because the autoionization reaction is endothermic.

°C	K_w
10	0.29×10^{-14}
15	0.45×10^{-14}
20	0.68×10^{-14}
25	1.01×10^{-14}
30	1.47×10^{-14}
50	5.48×10^{-14}

K_w = water ionization constant = $[H_3O^+][OH^-]$

$K_w = 1.008 \times 10^{-14}$ at 25 °C

This expression, and the value of the water ionization constant, $K_w = 1.0 \times 10^{-14}$ at 25 °C, are important and should be committed to memory.

In pure water, the transfer of a proton between two water molecules leads to one H_3O^+ and one OH^-. Since this is the only source of these ions, we know that $[H_3O^+]$ must equal $[OH^-]$ in pure water. Therefore, *in pure water,*

$$K_w = [H_3O^+]^2 \quad \text{or} \quad K_w = [OH^-]^2$$

and so

$$[H_3O^+] = [OH^-] = \sqrt{K_w} = \sqrt{1.0 \times 10^{-14}}$$

or

$$[H_3O^+] = [OH^-] = 1.0 \times 10^{-7} \text{ M}$$

The hydronium ion and hydroxide ion concentrations in pure water are *both* 1.0×10^{-7} M at 25 °C, and the water is said to be **neutral**. However, *in an acidic solution the concentration of H_3O^+ must be greater than 1.0×10^{-7} M.* Similarly, *in a basic solution the concentration of OH^- must be greater than 1.0×10^{-7} M.* Of course, since the product of the H_3O^+ and OH^- concentrations must be equal to 1.0×10^{-14}, this must mean that a basic solution is also characterized by a hydronium ion concentration less than 1.0×10^{-7} M.

EXAMPLE 17.2

ION CONCENTRATIONS IN A SOLUTION OF A STRONG BASE

If you have a 0.0010 M aqueous solution of NaOH, what are the hydroxide and hydronium ion concentrations?

Solution NaOH is ionic and so is 100% dissociated into ions in water to give an initial concentration of OH^- of 0.0010 M. Before you added NaOH, the water initially contained a small concentration of OH^- (10^{-7} M) from the auto-ionization of water.

$$2\ H_2O(\ell) \rightleftharpoons H_3O^+(aq) + OH^-(aq)$$

However, we shall take the point of view that the water ionization reaction occurs only *after* excess OH^- has been added (in the form of NaOH). Water ionization gives equal concentrations of H_3O^+ and OH^-, and both are equal to the unknown quantity x. Thus, the *net* OH^- equilibrium concentration must be (0.0010 M + OH^- from water) = $(0.0010 + x)$.

	$[H_3O^+]$	$[OH^-]$
Initial concentrations (M)	0	0.0010
Change in concentrations on proceeding to equilibrium	$+x$	$+x$
After equilibrium is achieved (M)	x	$0.0010 + x$

We can solve for x from the expression for K_w.

$$K_w = 1.0 \times 10^{-14} = [H_3O^+][OH^-] = (x)(0.0010 + x)$$

If you expand this equation to solve for x, you will find it is a quadratic equation that can be solved by the quadratic formula or by the method of successive approximations (see Appendix A and Examples 17.6 and 17.7). However, application of Le Chatelier's principle allows us to make a very useful approximation that simplifies our calculations. The water ionization reaction is suppressed by the presence of the extra OH^-. Thus, the value of x, the concentration of H_3O^+ and OH^- coming from water, must be *smaller* than 10^{-7}. It is so much smaller than 0.0010 that x in the term $(0.0010 + x)$ can be neglected, and $[OH^-] = 0.0010$ M. Therefore, we have the following approximate, but nearly correct, expression,

$$K_w = 1.0 \times 10^{-14} = [H_3O^+][OH^-] \approx (x)(0.0010)$$

and so

$$x = [H_3O^+] \text{ in the presence of 0.0010 M NaOH} \approx 1.0 \times 10^{-11} \text{ M}$$

As a final step, it is useful to check our approximation.

$$[H_3O^+][OH^-] \approx (1.0 \times 10^{-11})(0.0010 + 1.0 \times 10^{-11}) \approx 1.0 \times 10^{-14}$$

The calculated product of the H_3O^+ and OH^- ion concentrations is very close to the value of K_w, so the approximation is valid.

EXERCISE 17.4 Ion Concentrations in a Solution of a Strong Acid

Gaseous HCl (0.073 g) is bubbled into 5.0×10^2 mL of water to make an aqueous HCl solution. What are the concentrations of H_3O^+ and OH^- in this solution?

THE pH SCALE

Rather than express the hydronium and hydroxide ion concentrations as very small numbers or as exponentials, chemists have found an equivalent way to express $[H_3O^+]$ as a positive, decimal number whose value normally lies between 0 and 14. This number is called the **pH** of the solution, and it is defined as *the negative of the base-10 logarithm (log) of the hydronium ion concentration*:

In general, $pX = -\log[X]$

$$pH = -\log[H_3O^+]$$

An alternative and useful form of these definitions is

$$[H_3O^+] = 10^{-pH}$$

$$[OH^-] = 10^{-pOH}$$

As you will see, these equations make it easy to convert pH or pOH values into hydronium ion or hydroxide ion concentrations, respectively.

In a similar way, the pOH of a solution is defined as the negative of the base-10 logarithm of the hydroxide ion concentration.

$$pOH = -\log[OH^-]$$

In pure water, the hydronium and hydroxide ion concentrations are both 1.0×10^{-7} M. Therefore,

$$pH = -\log(1.0 \times 10^{-7}) = -[\log(1.0) + \log(10^{-7})]$$
$$pH = -[(0.00) + (-7)]$$
$$pH = 7.00 = \text{pH of pure water at 25 °C}$$

In the same way, you can show that the pOH of pure water is also 7.00 at 25 °C.

If we take the logarithms of both sides of the expression $K_w = [H_3O^+][OH^-]$, we obtain another useful expression.

$$K_w = [H_3O^+][OH^-] = 1.0 \times 10^{-14}$$
$$-\log([H_3O^+][OH^-]) = -\log(1.0 \times 10^{-14})$$
$$(-\log[H_3O^+]) + (-\log[OH^-]) = 14.00$$

$$pH + pOH = 14.00$$

The sum of the pH and pOH of a solution must be equal to 14.00 at 25 °C.

EXAMPLE 17.3

CALCULATING pH

The basic solution in Example 17.2 had $[OH^-] = 1.0 \times 10^{-3}$ M and $[H_3O^+] = 1.0 \times 10^{-11}$ M. Calculate the pH and pOH of this solution.

Solution Let us begin by converting the hydrogen ion concentration to pH. Since $[H_3O^+] = 1.0 \times 10^{-11}$ M, this means that

$$pH = -\log[H_3O^+] = -\log(1.0 \times 10^{-11}) = -(-11.00) = 11.00$$

The pOH can be obtained quickly from the fact that the sum of the pH and the pOH must equal 14.00. Since the pH is 11.00, this must mean that the pOH is 3.00. We can prove that, however, by calculating the pOH directly from $[OH^-]$.

$$pOH = -\log[OH^-] = -\log(1.0 \times 10^{-3}) = -(-3.00) = 3.00$$

For a review of logarithms and significant figures, see Appendix A.

As a footnote to this problem, notice the use of significant figures and logarithms. As outlined further in Appendix A, the mantissa of a logarithm (the set of digits to the right of the decimal) has as many significant figures as the number whose log was found. Here we found the log of 1.0×10^{-11}, a number with two significant figures. Therefore, the log (11.00), must have two digits to the right of the decimal (both zeros in this case).

A 0.0040 M solution of HCl has a pH of 2.40 (Exercise 17.4), while a 0.0010 M NaOH solution has a pH of 11.00 (Example 17.3). Since the pH of pure water is 7.00 (at 25 °C), we can say in general that *solutions with pH less than 7.00 are acidic, while solutions with pH greater than 7.00 are basic.* The relation between acidity, basicity, and pH or pOH can be illustrated graphically as follows:

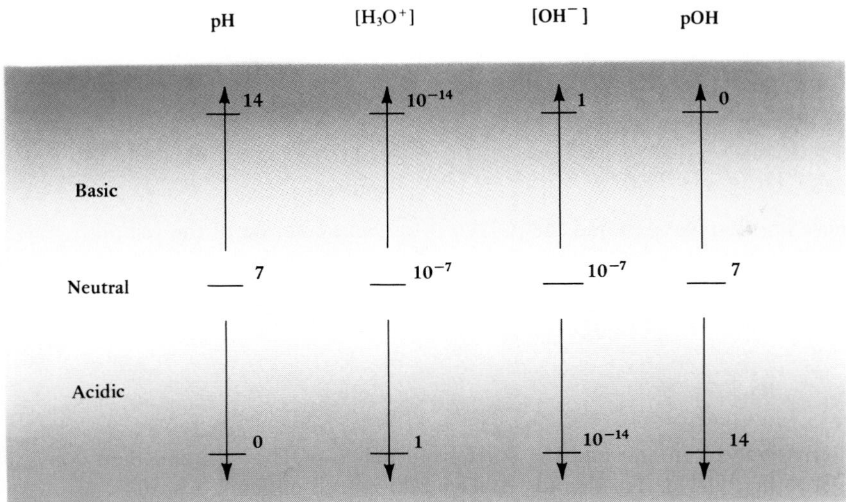

To give you a feeling for the pH scale, the approximate pH values for some common aqueous solutions are shown in Figure 17.3.

EXAMPLE 17.4

pH AND HYDRONIUM ION CONVERSIONS

(a) The $[H_3O^+]$ of wine can be 1.6×10^{-3} M. Calculate its pH. (b) The $[H_3O^+]$ of rain water in some instances has been found to be 5.0×10^{-5} M. Calculate its pH. (c) The pH of seawater is 8.30. Calculate $[H_3O^+]$ and $[OH^-]$.

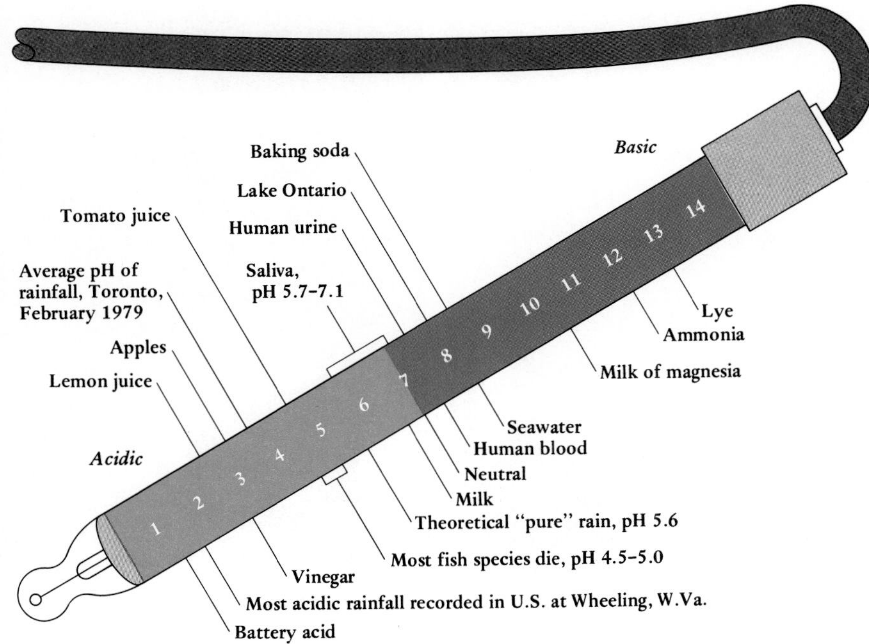

Figure 17.3 The pH of some common aqueous solutions. The scale is superimposed on the drawing of a pH electrode used in the measurement of pH by an instrumental method.

Solution

(a) pH of the wine = $-\log[H_3O^+] = -\log(1.6 \times 10^{-3}) = 2.80$
(b) pH of the rain = $-\log[H_3O^+] = -\log(5.0 \times 10^{-5}) = 4.30$
(c) If the pH is 8.30, this means that

$$[H_3O^+] = 10^{-pH} = 10^{-8.30} = 5.0 \times 10^{-9}\ M$$

There are two ways to find the $[OH^-]$. You can make use of the relation

$$K_w = [H_3O^+][OH^-] = 1.0 \times 10^{-14}$$

and calculate $[OH^-]$ from it.

$$(5.0 \times 10^{-9})[OH^-] = 1.0 \times 10^{-14}$$
$$[OH^-] = 2.0 \times 10^{-6}\ M$$

Alternatively, you can find the pOH from pH + pOH = 14, and then convert pOH to the $[OH^-]$. Try this to see if you do indeed obtain 2.0×10^{-6}.

EXERCISE 17.5 pH and Hydronium Ion Conversions

The pH of blood is 7.30. What are the hydronium and hydroxide ion concentrations in blood? Assume K_w is 2.5×10^{-14} at a body temperature of 37 °C.

Figure 17.4 Indicators and pH. (a) A "universal indicator" in solutions of various pH values. Solutions in the top row have pH values 1 through 4, those in the middle row have values 5 through 8, and those in the bottom row have pH's of 9 through 12. (b) Indicators in common products showing that vinegar is acidic with a pH of about 3, a carbonated beverage has a pH of about 5, and a household cleaner is strongly basic.

(a)

(b)

pH INDICATORS

The pH of a solution can be determined chemically (Figure 17.4) or instrumentally (Figure 17.5). The chemical method uses an **indicator**, *a substance that changes color in some known pH range*. Recall that acids were originally defined by the fact that they made certain vegetable dyes turn red; many indicators that are still in use are large molecules derived from plants. These dyes or indicators can exist in conjugate acid and base forms, and each form has a different color. Common litmus paper is impregnated with a natural plant juice that is red in solutions more acidic than about pH = 5 but blue when the pH exceeds about 8.2 in basic solution.

As Figure 5.20 shows, vegetable juices (such as red cabbage juice) are often natural acid–base indicators.

Although not derived from plants, phenolphthalein is a common acid–base indicator. In acid solution it is in its protonated form (H_2In) and is colorless. When the solution becomes basic in the pH interval from 8.2 to 9.8, two protons from H_2In are lost in a reaction with the base, and the conjugate base (In^{2-}) is a beautiful pink (Figure 5.19).

HO, OH, C, O, C=O (aq) + 2 $H_2O(\ell)$ ⇌ 2 H_3O^+(aq) + O, O⁻, C, CO_2^-(aq)

Colorless

Phenolphthalein

Red

Conjugate base of phenolphthalein

$$H_2In(aq) + 2\,H_2O(\ell) \rightleftharpoons 2\,H_3O^+(aq) + In^{2-}(aq)$$

As you can see in Table 17.3, there is a wide variety of indicators that can be used in many pH ranges. However, it can also be seen in this table that indicators change color over a range of pH values. If you need a more accurate estimate of pH, a modern pH meter is far preferable (Figure 17.5).

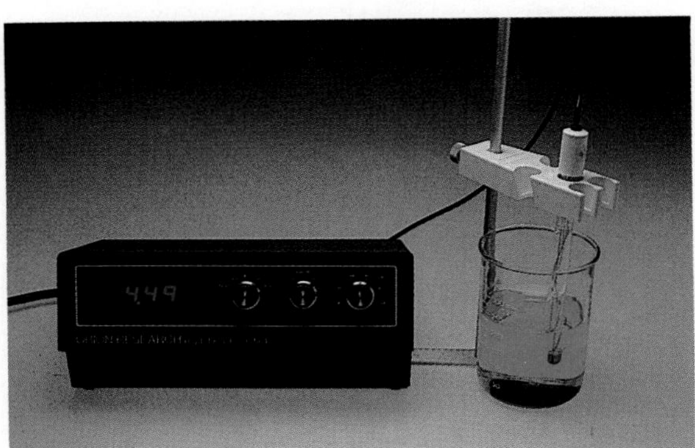

Figure 17.5 An electronic instrument for the measurement of pH. The blue solution is 0.10 M $CuSO_4$, and the pH of 4.5 illustrates the fact that many metal ions give weakly acidic solutions (see Section 17.5).

Table 17.3 Some Acid-Base Indicators

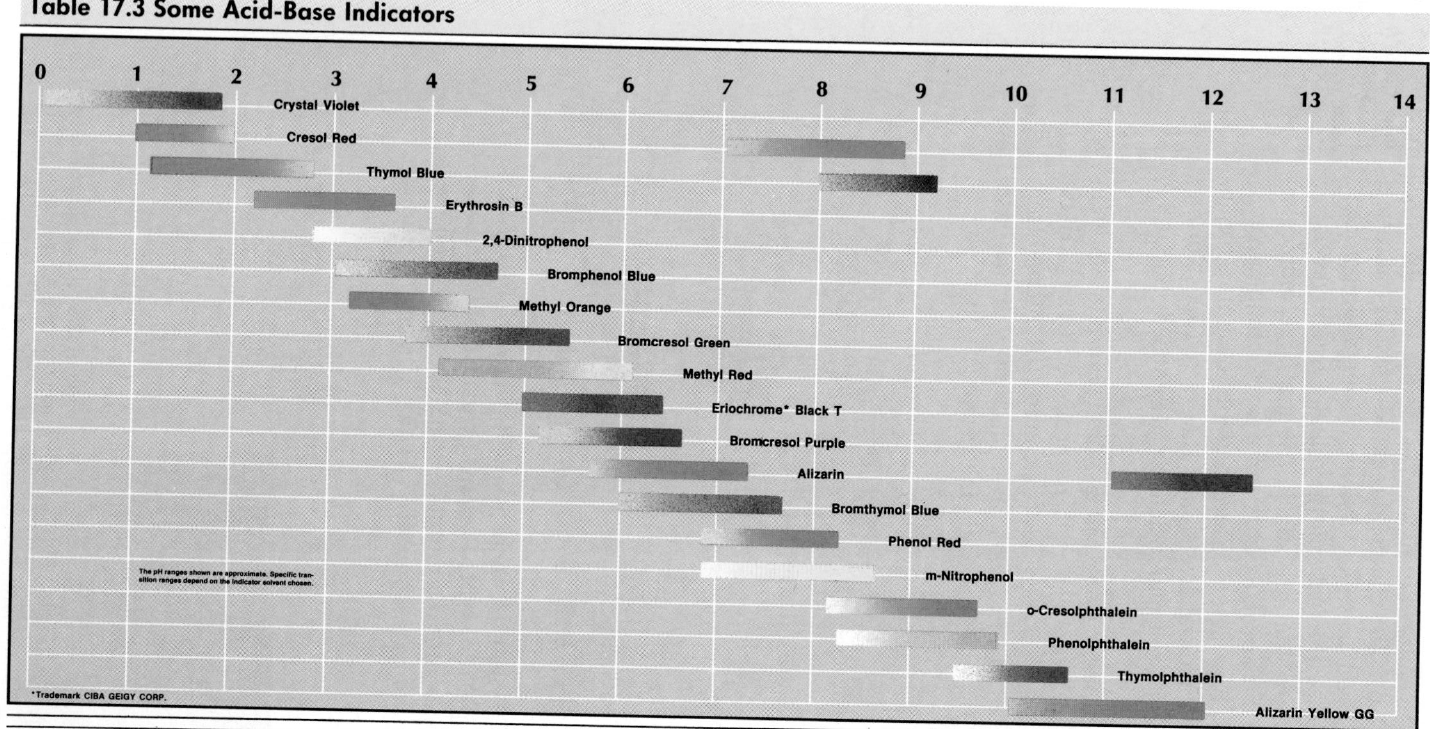
0
1
2
3
4
5
6
7
8
9
10
11
12
13
14
Crystal Violet
Cresol Red
Thymol Blue
Erythrosin B
2,4-Dinitrophenol
Bromphenol Blue
Methyl Orange
Bromcresol Green
Methyl Red
Eriochrome* Black T
Bromcresol Purple
Alizarin
Bromthymol Blue
Phenol Red
m-Nitrophenol
o-Cresolphthalein
Phenolphthalein
Thymolphthalein
Alizarin Yellow GG
The pH ranges shown are approximate. Specific transition ranges depend on the indicator solvent chosen.
*Trademark CIBA GEIGY CORP.

Hach Company

17.4 STRONG ACIDS AND BASES

Some common acids are listed in Table 17.2 in descending order of their ability to donate a proton. *The hydronium ion is the strongest acid that can exist in water.* Notice, however, that several acids are listed even higher in the table than H_3O^+. How can this be? When they are placed in water, strong acids such as

Hydrohalic acids: HCl, HBr, and HI
Nitric acid: HNO_3
Sulfuric acid: H_2SO_4
Perchloric acid: $HClO_4$

ionize *completely* to form H_3O^+ and the conjugate base by reacting with water. Thus, for example, HCl is a stronger acid than H_3O^+, so HCl molecules do not exist in water; rather, only H_3O^+ and Cl^- are present because all HCl molecules transfer H^+ to H_2O. Acids that ionize 100% in solution are called **strong acids**, but their solutions can be no stronger than H_3O^+ in water. This observation is called the **leveling effect.**

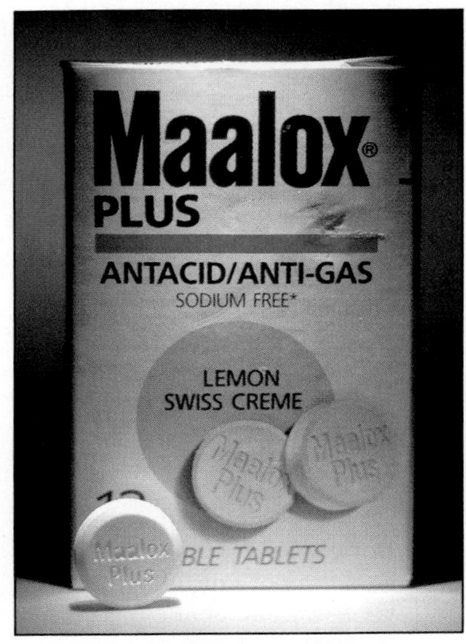

Maalox is composed of dried magnesium hydroxide and aluminum hydroxide (as the product name suggests).

Just as there are no acids that can be stronger in water than H_3O^+, no base can be stronger than OH^- in aqueous solution. In Table 17.2, strong bases are species lower than OH^- in the right column: $C_2H_5O^-$, NH_2^-, CH_3NH^-, H^-, and CH_3^- all react completely with water to produce OH^-. The strong bases that you commonly encounter, however, are hydroxides or oxides of Group 1A or 2A metals, of which NaOH and KOH are perhaps most familiar. Although not nearly as soluble as NaOH in water, the hydroxides of calcium and magnesium [$Ca(OH)_2$ and $Mg(OH)_2$] are among the most inexpensive bases known. Millions of tons of CaO or lime are made annually by decomposing limestone, $CaCO_3$, at temperatures around 800 to 1000 °C.

Over 16 million tons of lime were made in the U.S. in 1988.

$$CaCO_3(s) + \text{energy} \longrightarrow CaO(s) + CO_2(g)$$

In a reaction characteristic of many metal oxides, CaO forms slaked lime or $Ca(OH)_2$ when mixed with water.

$$CaO(s) + H_2O(\ell) \longrightarrow Ca(OH)_2(s) + \text{energy}$$

Clear, aqueous solutions of slaked lime are known as *limewater*. The poor solubility of $Ca(OH)_2$ or $Mg(OH)_2$ means that, if you mix a teaspoonful or so with water, the result will be just a suspension of white powder. The suspension is basic, so $Mg(OH)_2$ is used as an antacid and is often called "milk of magnesia."

About 12 million tons of NaOH or caustic soda are made every year in the U.S., and even very pure laboratory-grade material costs less than \$10/kg. On the other hand, CsOH costs more than \$600 per kilogram!

17.5 WEAK ACIDS AND BASES

Very few acids and bases strongly donate or accept protons, respectively. *The vast majority of acids and bases are weak.*

Perhaps the easiest way to tell if an acid or base is weak is to test the pH of an aqueous solution. A solution of a strong acid will have a pH value that gives a hydronium ion concentration that is very close to that

Table 17.4 Ionization Constants of Weak Acids and Bases

	Acid	K_a	Base	K_b
Sulfurous acid	H_2SO_3	1.2×10^{-2}	HSO_3^-	8.3×10^{-13}
Hydrogen sulfate ion	HSO_4^-	1.2×10^{-2}	SO_4^{2-}	8.3×10^{-13}
Phosphoric acid	H_3PO_4	7.5×10^{-3}	$H_2PO_4^-$	1.3×10^{-12}
Hydrofluoric acid	HF	7.2×10^{-4}	F^-	1.4×10^{-11}
Nitrous acid	HNO_2	4.5×10^{-4}	NO_2^-	2.2×10^{-11}
Formic acid	HCOOH	1.8×10^{-4}	$HCOO^-$	5.6×10^{-11}
Benzoic acid	$HC_7H_5O_2$	6.3×10^{-5}	$C_7H_5O_2^-$	1.6×10^{-10}
Acetic acid	CH_3COOH	1.8×10^{-5}	CH_3COO^-	5.6×10^{-10}
Propanoic acid	$HC_3H_5O_2$	1.4×10^{-5}	$C_3H_5O_2^-$	7.1×10^{-10}
Carbonic acid	H_2CO_3	4.2×10^{-7}	HCO_3^-	2.4×10^{-8}
Hydrogen sulfide	H_2S	1×10^{-7}	HS^-	1×10^{-7}
Dihydrogen phosphate ion	$H_2PO_4^-$	6.2×10^{-8}	HPO_4^{2-}	1.6×10^{-7}
Hydrogen sulfite ion	HSO_3^-	6.2×10^{-8}	SO_3^{2-}	1.6×10^{-7}
Hypochlorous acid	HClO	3.5×10^{-8}	ClO^-	2.9×10^{-7}
Boric acid	H_3BO_3	7.3×10^{-10}	$H_2BO_3^-$	1.4×10^{-5}
Ammonium ion	NH_4^+	5.6×10^{-10}	NH_3	1.8×10^{-5}
Hydrocyanic acid	HCN	4.0×10^{-10}	CN^-	2.5×10^{-5}
Hydrogen carbonate ion	HCO_3^-	4.8×10^{-11}	CO_3^{2-}	2.1×10^{-4}
Hydrogen phosphate ion	HPO_4^{2-}	3.6×10^{-13}	PO_4^{3-}	2.8×10^{-2}
Hydrogen sulfide ion	HS^-	1.3×10^{-13}	S^{2-}	7.7×10^{-2}

Increasing Acid Strength ↑ (upward); Increasing Base Strength ↓ (downward)

expected on the basis of complete ionization. A 0.01 M solution of HCl has a pH of 2.0, while a 0.01 M solution of NaOH has a pH of 12.0. On the other hand, a sample of vinegar, which is essentially a 0.8 M aqueous solution of the weak acid acetic acid, has a pH of about 3 (Figure 17.4). This corresponds to about 0.1% ionization.

The relative strength of an acid or base can be expressed quantitatively with an equilibrium constant. For the general **weak acid HA**, for example, we can write

$$HA(aq) + H_2O(\ell) \rightleftharpoons H_3O^+(aq) + A^-(aq)$$

$$K_a = \frac{[H_3O^+][A^-]}{[HA]} < 1$$

where K has a subscript a to indicate that it is an equilibrium constant for a weak acid. Remember that each of the concentrations in this expression is the equilibrium value for that molecule or ion.

Similarly, we can write the equilibrium expression for a **weak base B** (where K is labeled with a subscript b):

$$B(aq) + H_2O(\ell) \rightleftharpoons BH^+(aq) + OH^-(aq)$$

$$K_b = \frac{[BH^+][OH^-]}{[B]} < 1$$

Now we can turn to two tasks. First, we want to see which species can be weak acids or bases and examine some typical values for K (Table 17.4). Then, we can explore how one can determine K_a or K_b from experimental measurements and how to use the values in a practical way.

Ionization Constants for Weak Acids and Bases: Table 17.4

Table 17.4 is one of the most useful tables in the book, as it conveys a great deal of information. Here are some of the important ideas concerning this table.

1. Weak acids and their conjugate bases behave in water according to the following equations.

   For acids: $HX(aq) + H_2O(\ell) \rightleftharpoons H_3O^+(aq) + X^-(aq)$

   $$K_a = \frac{[H_3O^+][X^-]}{[HX]}$$

   For bases: $X^-(aq) + H_2O(\ell) \rightleftharpoons HX(aq) + OH^-(aq)$

   $$K_b = \frac{[HX][OH^-]}{[X^-]}$$

2. Acids are listed at the left and the conjugate bases of those acids are at the right.
3. The strongest acids are at the upper left and the strongest bases are at the lower right.
4. The relationship between K_a for an acid and K_b for its conjugate base is $\boldsymbol{K_aK_b = K_w}$. This can be proved as follows. If you add the two general equations above, the result is $2\ H_2O(\ell) \rightleftharpoons H_3O^+(aq) + OH^-(aq)$. As proved in Chapter 16, the equilibrium constant for an equation that is the sum of two others is the product of the equilibrium constants for the summed reactions. Therefore,

   $$K_a \times K_b = \left(\frac{[H_3O^+][X^-]}{[HX]}\right)\left(\frac{[HX][OH^-]}{[X^-]}\right) = [H_3O^+][OH^-] = K_w$$

   This relation tells you that, if you know K_a for an acid, you can find K_b for the conjugate base. For example, K_b for CN^- is

   $$K_b = \frac{K_w}{K_a \text{ for HCN}} = \frac{1.0 \times 10^{-14}}{4.0 \times 10^{-10}} = 2.5 \times 10^{-5}$$

   Notice that the equation "$K_aK_b = K_w =$ constant" also proves the statement on page 705 that the stronger the acid, the weaker the conjugate base. As K_a increases, K_b decreases proportionately.

WEAK ACIDS

MOLECULAR ACIDS Molecular acids are neutral molecules that have one or more ionizable hydrogen atoms. In the previous section we listed a few such compounds that are strong acids, but the vast majority are weak, and a few of these are pictured below with their K_a values (and others are listed in Table 17.4). Notice that all of the K_a values are quite small, reflecting very little ionization of the acid in aqueous solution. Hydrocyanic and carbonic acids are quite weak, for example. Oxalic acid is only moderately weak, and its uses in the dye industry, as a paint remover, and so on, reflect this.

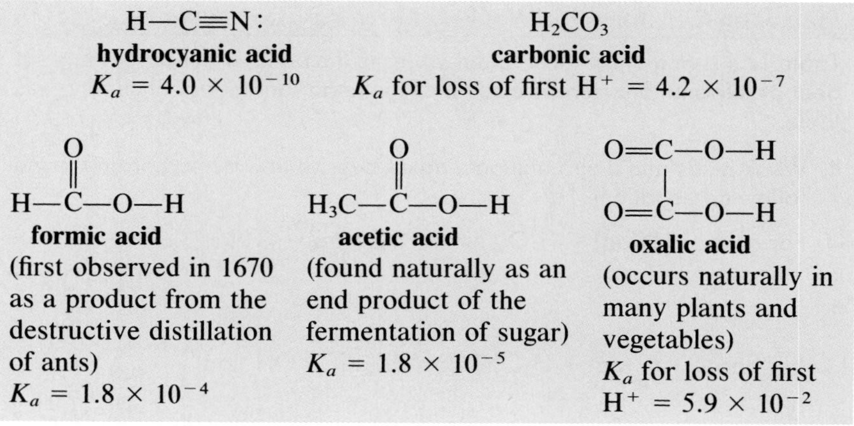

ANIONS AS WEAK ACIDS Six anionic Brønsted acids are listed in Table 17.4. The dihydrogen phosphate anion, $H_2PO_4^-$, for example, is the acid in baking powder,

$$H_2PO_4^-(aq) + H_2O(\ell) \rightleftharpoons HPO_4^{2-}(aq) + H_3O^+(aq) \quad K_{a2} = 6.2 \times 10^{-8}$$

where its function is to provide hydrogen ion to produce CO_2 from baking soda ($NaHCO_3$) (Figure 17.6).

$$\underset{\text{weak acid}}{H_2PO_4^-(aq)} + \underset{\text{weak base}}{HCO_3^-(aq)} \rightleftharpoons \underset{\text{conjugate acid of } HCO_3^-}{H_2CO_3(aq)} + \underset{\text{conjugate base of } H_2PO_4^-}{HPO_4^{2-}(aq)}$$

$$H_2CO_3(aq) \rightleftharpoons CO_2(g) + H_2O(\ell)$$

CATIONS AS WEAK ACIDS The ammonium ion is an excellent example of a cation acting as a Brønsted acid (Figure 17.7),

$$NH_4^+(aq) + H_2O(\ell) \rightleftharpoons NH_3(aq) + H_3O^+(aq) \qquad K_a = 5.6 \times 10^{-10}$$

although its small K value (and so its position near the bottom of the acids in Table 17.4) tells you it is an extremely weak acid.

When the salt of a metal cation is placed in water, the metal ion becomes hydrated, as explained in Chapter 13. In fact, the interaction is usually strong enough that the ion is surrounded by as many as six water molecules. For ions such as Al^{3+} and transition metal ions with a charge of 2+ or 3+, the attraction between the ion and water molecules is so strong that an O—H bond may break in one or more of the attached H_2O molecules, and the hydrogen ion appears in solution as H_3O^+.

$$[Fe(H_2O)_6]^{3+}(aq) + H_2O(\ell) \rightleftharpoons [Fe(H_2O)_5(OH)]^{2+}(aq) + H_3O^+(aq)$$

$$K_a = 6.3 \times 10^{-3}$$

In contrast, such reactions do not occur to a significant extent with common ions like Na^+, K^+, Ca^{2+}, and Mg^{2+}. Thus, for salts of these and other ions of Groups 1A and 2A, the metal ion does not contribute to the acidity of the solution. Since these are the important ions found in sea-

Figure 17.6 The reaction of $H_2PO_4^-$ and HCO_3^- to give CO_2 and HPO_4^{2-}. On the left is a solution of NaH_2PO_4 and on the right a solution of $NaHCO_3$. The bromcresol purple indicator shows that the $H_2PO_4^-$ solution is acidic and the HCO_3^- solution is basic.

Figure 17.7 Some weak acids (left to right): 0.10 M acetic acid; HSO_4^- (Na^+ salt); saturated CO_2 (which gives a small concentration of H_2CO_3); NH_4^+ (NO_3^- salt); and Cd^{2+} (NO_3^- salt). A universal indicator was used.

water, for example, the pH of the sea must be controlled by other species. The acidity of metal ions in solution is described in more detail in Section 17.5.

THE ACID IONIZATION CONSTANT, K_a The K_a values found in Table 17.4 and in a more extensive table in Appendix E were all determined by experiment. There are several ways to approach the problem, but the usual method is to determine the pH of the solution. The following example illustrates the method.

EXAMPLE 17.5

K_a FROM pH MEASUREMENT

Lactic acid is a monoprotic acid that occurs naturally in sour milk and arises from metabolism in the human body. A 0.10 M aqueous solution of lactic acid, $HC_3H_5O_3$, has a pH of 2.43. What is the value of K_a for lactic acid?

Solution The equation for the equilibrium interaction of lactic acid with water is

$$\underset{\text{lactic acid}}{HC_3H_5O_3(aq)} + H_2O(\ell) \rightleftharpoons H_3O^+(aq) + \underset{\text{lactate ion}}{C_3H_5O_3^-(aq)}$$

and the equilibrium constant expression is

$$K_a = \frac{[H_3O^+][C_3H_5O_3^-]}{[HC_3H_5O_3]}$$

The main piece of information we have to begin with is the pH. Converting this to $[H_3O^+]$, you should find

$$[H_3O^+] = 10^{-pH} = 10^{-2.43} = 3.7 \times 10^{-3} \text{ M}$$

To solve for K_a, we need to know the equilibrium concentration of each species. The pH of the solution directly tells us the equilibrium concentration of H_3O^+, and we can now derive the others from this. The table below gives the initial concentrations of the important species in solution, tells what happens as the reaction proceeds to equilibrium, and then shows the equilibrium concentrations.

	$[HC_3H_5O_3]$	$[H_3O^+]$	$[C_3H_5O_3^-]$
Before ionization (M)	0.10	0	0
Change on proceeding to equilibrium	$-x$	$+x$	$+x$
After equilibrium is achieved (M)	$0.10 - x$	x	x

The following important points can be made concerning this outline of information.

1. Hydronium ion is present in solution from lactic acid ionization *and* from water auto-ionization. Le Chatelier's principle informs us that the H_3O^+ added to the water by lactic acid will suppress the H_3O^+ coming from the water auto-ionization. Since measurement of the pH tells us the *total* $[H_3O^+]$ is 3.7×10^{-3} M, and since $[H_3O^+]$ from water must be less than 10^{-7} M, the pH is almost completely a reflection of H_3O^+ from lactic acid. (For weak acids and bases, this approximation can *almost* always be made. The exception is when x is near 10^{-7}; that is, when the equilibrium pH is in the range between 6 and 8. Be careful and think through each case to be certain.)
2. The quantity x represents the equilibrium concentrations of both hydronium ion and lactate ion. That is, at equilibrium $[H_3O^+] \approx [C_3H_5O_3^-] = x = 3.7 \times 10^{-3}$ M.
3. By stoichiometry, x is also the quantity of acid that ionized on proceeding to equilibrium.

With these points in mind, we can now derive K_a for lactic acid.

$$K_a = \frac{[H_3O^+][C_3H_5O_3^-]}{[HC_3H_5O_3]} = \frac{(3.7 \times 10^{-3})(3.7 \times 10^{-3})}{0.10 - 0.0037} = 1.4 \times 10^{-4}$$

On comparing this value of K_a with others in Table 17.4, you see that lactic acid could be classed as a weak acid.

EXERCISE 17.6 K_a from pH Measurement

When 0.10 mol of propanoic acid is dissolved in sufficient water to give 1.0 L of solution, the pH of the solution is 2.94 after equilibrium has been established. Determine K_a for propanoic acid. The acid ionizes in water according to the balanced equation

$$H_3CCH_2COOH(aq) + H_2O(\ell) \rightleftharpoons H_3O^+(aq) + H_3CCH_2COO^-(aq)$$

Values of K_a for many weak acids of all types have been measured (Table 17.4 and Appendix E). With these values, the pH of a solution of given concentration can be calculated, for example. Before doing this, however, there is an important point to notice in Example 17.5. The lactic acid concentration at equilibrium was given by "original acid concentration − quantity of acid ionized." In this example the expression was (0.10 − 0.0037). By the usual rules governing significant figures, (0.10 − 0.0037) is just equal to 0.10. After all, the acid is weak, so very little of it ionizes (approximately 4%), and the value of $[HC_3H_5O_3]$ is essentially equal to its initial value. The approximation that the denominator in the K_a expression for a weak acid, HA,

$$HA(aq) + H_2O(\ell) \rightleftharpoons H_3O^+(aq) + A^-(aq)$$

$$K_a = \frac{[H_3O^+][A^-]}{[HA]_0 - [H_3O^+]} \approx \frac{[H_3O^+][A^-]}{[HA]_0}$$

is just $[HA]_0$, the initial acid concentration, is very useful in calculations involving weak acids and bases. *The approximation is valid whenever $[H_3O^+]$ falls at or beyond the least significant digit of $[HA]_0$. Error analysis shows that this is true when $[HA]_0$ is greater than 100* K_a. The next several examples and exercises illustrate this point.

EXAMPLE 17.6

EQUILIBRIUM CONCENTRATIONS AND pH FROM K_a

Benzoic acid occurs free and combined in nature. For example, most berries contain up to 0.05% of the acid by weight.

```
 H     H                              H     H
  C==C    O                            C==C    O
 /    \   ||                          /    \   ||
HC     C--C--O--H(aq) + H2O(ℓ) ⇌ HC     C--C--O⁻(aq) + H3O⁺(aq)
 \\   //                              \\   //
  C--C                                 C--C
 H     H                              H     H
```

benzoic acid (C_6H_5COOH) benzoate ion ($C_6H_5COO^-$)

Calculate the pH of a 0.020 M solution of benzoic acid if $K_a = 6.3 \times 10^{-5}$.

Solution As usual, we organize the information in a table.

	$[C_6H_5COOH]$	$[H_3O^+]$	$[C_6H_5COO^-]$
Before ionization (M)	0.020	0	0
Change on proceeding to equilibrium	$-x$	$+x$	$+x$
After equilibrium is achieved (M)	$0.020 - x$	x	x

Every mole of benzoic acid that ionizes gives 1 mole of benzoate ion and 1 mole of hydronium ion, so $[C_6H_5COO^-] = [H_3O^+] = x$ at equilibrium. (Here we have neglected any H_3O^+ that arises from water ionization, for the same reasons as outlined in Example 17.5.) Furthermore, stoichiometry tells us that the quantity of acid ionized is also x. Therefore, the benzoic acid concentration at equilibrium is

$$\begin{aligned}[C_6H_5COOH] &= \text{initial acid concentration} - \text{quantity of acid that ionized} \\ &= [C_6H_5COOH]_0 - x \\ &= 0.020 - x\end{aligned}$$

Substituting these equilibrium concentrations into the K_a expression, we have

$$K_a = \frac{[H_3O^+][C_6H_5COO^-]}{[C_6H_5COOH]} = \frac{(x)(x)}{(0.020 - x)} = 6.3 \times 10^{-5}$$

As described in the text just above, we know that x will be small compared with 0.020 (since $[HA]_0 > 100\ K_a$), so that

$$K_a = 6.3 \times 10^{-5} \approx \frac{x^2}{0.020}$$

This means that

$$x = \sqrt{K_a(0.020)} = 1.1 \times 10^{-3}$$

Therefore, we find that

$$[H_3O^+] \approx [C_6H_5COO^-] \approx 0.0011\ M$$

and

$$[C_6H_5COOH] = 0.020 - x = 0.019\ M$$

Finally, the pH of the solution is found to be

$$pH = -\log(1.1 \times 10^{-3}) = 2.96$$

Now, let's think about the result. Since benzoic acid is weak, we made the approximation that $(0.020 - x) \approx 0.020$, and it is now seen to be a good one. If 0.0011 mol/L of acid ionized, and the original concentration was 0.020 mol/L, this means that only about 6% of the acid ionized.

But let's check the result further. If we do *not* make the approximation, and work with the expression $K_a = x^2/(0.020 - x)$, we see it is a quadratic equation. That is,

$$K_a(0.020 - x) = x^2$$

$$0.020K_a - K_a x = x^2$$

$$x^2 + K_a x - 0.020K_a = 0$$

is a quadratic equation of the type $ax^2 + bx + c = 0$ where $a = 1$, $b = K_a$, and $c = -0.020K_a$. Such an expression can be solved by the quadratic formula or by the method of successive approximations (see Appendix A). Either method gives $x = [H_3O^+] = 1.1 \times 10^{-3}$, the same answer to two significant figures that we obtained from the "approximate" expression $x = \sqrt{K_a(0.020)}$.

EXERCISE 17.7 Equilibrium Concentrations from K_a

What are the equilibrium concentrations of acetic acid, the acetate ion, and H_3O^+ when a 0.10 M solution of acetic acid ($K_a = 1.8 \times 10^{-5}$) is allowed to come to equilibrium?

In the previous example, we calculated the pH of an aqueous solution of a *weak* acid. You know it is weak because K_a was so small that $[H_3O^+] << [HA]_0$. Another way to say this is that only a small percentage of the acid ionized to produce hydronium ion in water. Indeed, in the case of a 0.020 M solution of benzoic acid, the percent ionization is less than 10%.

$$\%\text{ ionized} = \frac{\text{quantity of acid ionized}}{\text{initial acid concentration}} \times 100 = \frac{0.0011\ M}{0.020\ M} \times 100 = 5.5\%$$

For this reason, the hydronium ion concentration for a weak acid can be found to a good approximation from the simple expression

$$K_a \approx \frac{[\text{hydronium ion from the weak acid}][\text{conjugate base}]}{[\text{acid}]_{\text{initial}}}$$

"approximate" expression

In general, the hydronium ion concentration obtained from this approximate expression will be the same (to two significant figures) as that from a more exact solution to the problem *if* the weak acid is no more than about 10% ionized. Indeed, a good "rule of thumb" to use is the following: if the initial acid concentration ($[HA]_0$) $\geq 100\ K_a$, then you can use the approximation that the denominator in the K_a expression is just $[HA]_0$.

EXAMPLE 17.7

pH FROM K_a

Formic acid was first obtained in 1670 as a product of the destructive distillation of ants, whose Latin genus name is *Formica*. The acid is moderately weak, with $K_a = 1.8 \times 10^{-4}$.

$$HCOOH(aq) + H_2O(\ell) \rightleftharpoons H_3O^+(aq) + HCOO^-(aq)$$

If you have a 0.0010 M solution of the acid, what is the pH of the solution? What is the concentration of formic acid at equilibrium?

Solution The usual equilibrium-concentration table is given below. Notice that we have again made the reasonable approximation that the H_3O^+ concentration from water ionization can be neglected.

	$[HCOOH]$	$[H_3O^+]$	$[HCOO^-]$
Before ionization (M)	0.0010	0	0
Change on proceeding to equilibrium	$-x$	$+x$	$+x$
After equilibrium is achieved (M)	$0.0010 - x$	x	x

Substituting these values into the K_a expression, we have

$$K_a = \frac{[H_3O^+][HCOO^-]}{[HCOOH]} = \frac{(x)(x)}{(0.0010 - x)} = 1.8 \times 10^{-4}$$

Formic acid is a weak acid, since it has a value of K_a much less than 1. In this situation, however, $[HA]_0$ is NOT greater than 100 K_a. In fact, 0.0010 is *less than* $100(1.8 \times 10^{-4})$, so the usual approximation is not reasonable. If we do solve for the H_3O^+ concentration using the approximate expression $x \approx \sqrt{K_a(0.0010)}$, we obtain a value of x ($= [H_3O^+] = [HCOO^-]$) of 4.2×10^{-4} M. But a check of the percentage ionization of the acid:

$$\%\text{ ionized} = \frac{4.2 \times 10^{-4}}{0.0010} \times 100 = 42\%$$

shows clearly that 0.0010 M formic acid is *much* more than 10% ionized, so the answers from the approximate and exact solutions to the problem will differ significantly. This means we have to find the equilibrium concentrations by solving the "exact" expression.

$$K_a = \frac{[H_3O^+][HCOO^-]}{[HCOOH]} = \frac{(x)(x)}{(0.0010 - x)} = 1.8 \times 10^{-4}$$

$$x^2 + (1.8 \times 10^{-4})x + (-1.8 \times 10^{-4})(0.0010) = 0$$

Here $a = 1$, $b = 1.8 \times 10^{-4}$, and $c = -1.8 \times 10^{-7}$ in the quadratic expression $ax^2 + bx + c = 0$. Solving for x, we find

$$x = \frac{-(1.8 \times 10^{-4}) \pm \sqrt{(1.8 \times 10^{-4})^2 - 4(1)(-1.8 \times 10^{-7})}}{2(1)}$$

$$x = 3.4 \times 10^{-4}\text{ M} \quad \text{and} \quad -5.2 \times 10^{-4}\text{ M}$$

Since the negative root of the equation is chemically meaningless, this means that

$$[H_3O^+] = [HCOO^-] = 3.4 \times 10^{-4}\text{ M}$$

and so

$$[HCOOH] = 0.0010 - x \approx 0.0007\text{ M}$$

Therefore, the pH of the formic acid solution is

$$pH = -\log(3.4 \times 10^{-4}) = 3.47$$

This pH indicates an acidic solution.

Be sure to notice that the exact value for $[H_3O^+]$ is much different than the approximate value of 4.2×10^{-4} M. The approximate solution to this failed because the initial acid concentration was so small. This made invalid the approximation that $[HA]_{equilibrium} \approx [HA]_{initial}$.

EXERCISE 17.8 Equilibrium Concentrations from K_a

What are the equilibrium concentrations of HF, the fluoride ion, and H_3O^+ when a 0.015 M solution of HF is allowed to come to equilibrium?

SUMMARY The goals of this section were (a) to see what types of substances could behave as weak acids and (b) to develop methods of treating their ionization quantitatively. The mathematical approach derived to this point can be summarized as follows. For the weak acid HA

$$HA(aq) + H_2O(\ell) \rightleftharpoons H_3O^+(aq) + A^-(aq)$$

the equilibrium constant expression, *in terms of equilibrium concentrations*, is

$$K_a = \frac{[H_3O^+][A^-]}{[HA]}$$

where $[H_3O^+]$ is the hydronium ion concentration from the reaction of the weak acid with water plus that from the auto-ionization of water. When the pH falls outside the range from 6 to 8, we can assume that H_3O^+ ion from water auto-ionization does not contribute significantly. This also allows us to say that the equilibrium concentration of HA must be $[HA] = [HA]_0 - [H_3O^+]$, where the original concentration of HA is $[HA]_0$. Therefore, we can write K_a in a form that will allow us to solve for $[H_3O^+]$ if we know $[HA]_0$.

$$K_a = \frac{[H_3O^+]^2}{[HA]_0 - [H_3O^+]}$$

To solve for $[H_3O^+]$ generally requires the use of the quadratic formula or the method of successive approximations (Appendix A). However, this is not necessary if we make the approximation that $[HA]_0 - [H_3O^+] \approx [HA]_0$. This approximation is valid when $[HA]_0 \geq 100\ K_a$.

Figure 17.8 Weak bases in aqueous solution. The flask at the left is distilled water so that the indicator colors in the other flasks can be compared with neutral water. From left to right the flasks are: distilled water; 0.10 M acetate ion (Na^+ salt); 0.10 M CO_3^{2-} (Na^+ salt); 0.30 M PO_4^{3-} (Na^+ salt), and 0.10 M NH_3. A universal indicator was used.

WEAK BASES

A variety of weak bases are important in chemistry, and a few are listed in Table 17.4 in order of increasing strength. Each is listed with its value of K_b at the far right of the table.

MOLECULAR BASES Ammonia, NH_3, which is perhaps the best known weak base, produces a very small amount of hydroxide ion when accepting a proton from water (Figure 17.8).

Approximately 15 to 20 million tons of ammonia are made annually by the direct combination of H_2 with atmospheric N_2. The major uses of NH_3 are in the manufacture of HNO_3 and in the production of solid fertilizers such as ammonium nitrate.

$$NH_3(aq) + H_2O(\ell) \rightleftharpoons NH_4^+(aq) + OH^-(aq) \qquad K_b = 1.8 \times 10^{-5}$$

The compound is also the parent of a large series of derivatives called amines, in which one of the H atoms of NH_3 is replaced by some other substituent. For example, if the substituent is the methyl group, CH_3, we have

$$H_3C—\ddot{N}(H)—H \qquad H_3C—\ddot{N}(H)—CH_3 \qquad H_3C—\ddot{N}(CH_3)—CH_3$$

methylamine $K_b = 5.0 \times 10^{-4}$ | dimethylamine $K_b = 7.4 \times 10^{-4}$ | trimethylamine $K_b = 7.4 \times 10^{-5}$

Pyridine, C_5H_5N (Example 17.8), is another example of a weak base, and others include nicotine ($C_{10}H_{14}N_2$), caffeine ($C_8H_{10}N_4O_2$), and the antimalarial compound quinine ($C_{20}H_{24}N_2O_2$).

EXAMPLE 17.8

THE pH OF A SOLUTION OF A WEAK BASE

Pyridine, a common weak base, was discovered in coal tar in 1846. Its ionization constant, K_b, is 1.5×10^{-9}.

$$C_5H_5N:(aq) + H_2O(\ell) \rightleftharpoons C_5H_5N—H^+(aq) + OH^-(aq)$$

pyridine (C_5H_5N) | conjugate acid ($C_5H_5NH^+$)

If you have a 0.010 M aqueous solution of pyridine, what is the pH of the solution?

Solution The equilibrium expression for this weak base is

$$K_b = 1.5 \times 10^{-9} = \frac{[C_5H_5NH^+][OH^-]}{[C_5H_5N]}$$

and the concentrations of the base and ionic products are

	$[C_5H_5N]$	$[OH^-]$	$[C_5H_5NH^+]$
Before ionization (M)	0.010	0	0
Change on proceeding to equilibrium	$-x$	$+x$	$+x$
After equilibrium is achieved (M)	$0.010 - x$	x	x

If the expressions for the equilibrium concentrations are substituted into the K_b expression, we have

$$K_b = 1.5 \times 10^{-9} = \frac{(x)(x)}{0.010 - x}$$

This expression is exactly the same as was used to find the hydronium ion concentration in a weak acid solution. There are three points to be noticed here.

1. We have neglected any OH^- ion that can arise from the auto-ionization of water. The reasons are similar to those for neglecting H_3O^+ from water in solutions of weak acids.
2. Pyridine is a weak base, and it is probably a good assumption that it is less than 10% ionized in solution. Therefore, we shall solve the approximate expression for x,

$$K_b = 1.5 \times 10^{-9} \approx \frac{(x)(x)}{0.010}$$

and this gives a value of

$$x = [OH^-] = [C_5H_5NH^+] = 3.9 \times 10^{-6} \text{ M}$$

This value of $[OH^-]$ is indeed *much* less than the original concentration of pyridine; that is, $(0.010 \text{ M} - 3.9 \times 10^{-6})$ is nearly equal to 0.010 M. In fact, the base is less than 1% ionized, and our approximate solution is valid.

3. This example shows that the same "rule of thumb" used with weak acids can be applied to weak bases. That is, if $[\text{Base}]_0 \geq 100\ K_b$, then the approximation that $[OH^-] \approx \sqrt{K_b[\text{Base}]_0}$ can be used.

The objective of the problem was to find the pH of the solution. From the hydroxide ion concentration, we find that the pOH is

$$\text{pOH} = -\log[OH^-] = -\log(3.9 \times 10^{-6}) = 5.41$$

Since pH + pOH = 14.00, we finally arrive at a value for pH of 8.59, indicating only a very weakly basic solution.

EXERCISE 17.9 The pH of a Solution of a Weak Base

What is the pH of a 0.025 M solution of ammonia, NH_3?

ANIONS AS WEAK BASES Many of the compounds in nature and in the laboratory are salts, and many produce acidic or basic solutions by hydrolysis (Figures 17.7 and 17.8). *Hydrolysis* is said to have occurred when a salt dissolves in water and leads to changes in the H_3O^+ and OH^- concentrations of the water. These are of course just some of the reactions we have been talking about. For example, you have already seen that the ammonium ion and some metal ions—most notably ions such as Al^{3+}, Pb^{2+}, and transition metal ions of 2+ and 3+ charge—increase the hydronium ion concentration in aqueous solution. But what about anions in salts?

You have already seen some examples of anions acting as Brønsted bases in aqueous solution (Figure 17.8). For example, ionization of the strong Brønsted acid HCl in water produces chloride ion as the conjugate base.

$$\underset{\text{acid}}{HCl(aq)} + H_2O(\ell) \rightleftharpoons H_3O^+(aq) + \underset{\text{conjugate base of HCl}}{Cl^-(aq)}$$

Similarly, the weak acid HCN gives a conjugate base, the cyanide ion, in water.

$$\underset{\text{acid}}{HCN(aq)} + H_2O(\ell) \rightleftharpoons H_3O^+(aq) + \underset{\text{conjugate base of HCN}}{CN^-(aq)}$$

Since both Cl^- and CN^- are Brønsted bases, aqueous solutions of their salts could be basic. However, in Example 17.1 we argued that Cl^- does not hydrolyze to a measurable extent, since it is the conjugate base of a strong acid.

Chloride ion is a very weak base for the following reason: HCl is a much stronger acid than H_2O (Table 17.2), so the reaction of Cl^- with H_2O would have to go from a weak ($H_2O + Cl^-$) to a strong ($HCl + OH^-$) combination of acid and base.

In contrast to Cl^-, the cyanide ion, CN^-, produces a measurable concentration of hydroxide ion in water.

$$\underset{\text{base}}{CN^-(aq)} + \underset{\text{acid}}{H_2O(\ell)} \rightleftharpoons \underset{\text{conjugate acid of } CN^-}{HCN(aq)} + \underset{\text{conjugate base of water}}{OH^-(aq)}$$

and the value of the equilibrium constant for the hydrolysis can be determined.

$$K_b = \frac{[HCN][OH^-]}{[CN^-]} = 2.5 \times 10^{-5}$$

Like chloride, cyanide ion is a weaker base than hydroxide ion, so the equilibrium is predicted to lie to the left. However, according to Table 17.2, CN^- is between OH^- and Cl^- in base strength. Therefore, we can guess that CN^- will react with water to a small extent, a guess confirmed by the small but measurable value of K_b (Table 17.4).

Based on the examples of the chloride and cyanide ions, and many other results, we can make the following general statements:

1. When a salt containing the anion of a strong acid (HCl, HNO_3, $HClO_4$) is dissolved in water, hydrolysis of the anion to give OH^- does *not* occur to a measurable extent.
2. The anion or conjugate base, A^-, of any weak acid HA will hydrolyze to produce a measurable concentration of hydroxide ion (Figure 17.8).

   $$A^-(aq) + H_2O(\ell) \rightleftharpoons HA(aq) + OH^-(aq)$$

3. As described on page 719, the value of K_a for a weak acid and K_b for its conjugate base are related by the expression $K_aK_b = K_w$. This expression tells us that the stronger the acid, the weaker the conjugate

base, and this observation is confirmed by Tables 17.2 and 17.4. That is, if you compare two acids HA and $\mathbb{HA}$, where HA is stronger than $\mathbb{HA}$, this means that $\mathbb{A}^-$ will be a stronger base than A^-. For example, HF is a stronger acid than HCN, so CN^- is a stronger base than F^-.

EXAMPLE 17.9

pH OF THE SOLUTION OF AN ANIONIC BASE

Calculate the pH of a 0.010 M solution of sodium acetate, $Na(CH_3COO)$.

Solution According to the solubility guidelines of Figure 5.5, sodium acetate is a very soluble salt. It is also an electrolyte, dissociating completely into sodium and acetate ions in solution. As noted above, the Na^+ ion does not hydrolyze—react with water to produce a conjugate acid or base—to any measurable extent. However, the acetate ion is the conjugate base of a weak acid, so an aqueous solution of the ion can be basic.

$$CH_3COO^-(aq) + H_2O(\ell) \rightleftharpoons CH_3COOH(aq) + OH^-(aq)$$

acetate ion — acetic acid
weak base — conjugate acid

and the base ionization constant expression is

$$K_b = 5.6 \times 10^{-10} = \frac{[CH_3COOH][OH^-]}{[CH_3COO^-]}$$

The concentrations of acetate ion, acetic acid, and hydroxide ion initially and at equilibrium are

	$[CH_3COO^-]$	$[OH^-]$	$[CH_3COOH]$
Before ionization (M)	0.010	0	0
Change on proceeding to equilibrium	$-x$	$+x$	$+x$
After equilibrium is achieved (M)	$0.010 - x$	x	x

The approximation that $[Base]_0 - [OH^-] \approx [Base]_0$ is almost always valid for anionic weak bases, but you should think through each situation to make sure.

The acetate ion is a very weak base, as reflected by the very small value of K_b, so we can make the approximation that x in the expression $(0.010 - x)$ can be neglected and so can solve for $[OH^-]$ using the "approximate" expression

$$K_b = 5.6 \times 10^{-10} \approx \frac{x^2}{0.010}$$

From this expression $x = 2.4 \times 10^{-6}$. Therefore,

$$[OH^-] = [CH_3COO^-] = 2.4 \times 10^{-6}\ M$$

and

$$[CH_3COOH] = 0.010 - 2.4 \times 10^{-6} \approx 0.010\ M$$

This calculation shows that x is indeed very small relative to the original concentration of the acetate ion—because the ion is such a weak base—and so the answer from the approximate expression is reasonable. Finally, the pH of the solution is found as follows:

$$K_w = [H_3O^+][OH^-] = [H_3O^+](2.4 \times 10^{-6})$$
$$[H_3O^+] = 4.2 \times 10^{-9}\ M$$
$$pH = 8.38$$

EXERCISE 17.10 Weak Bases and pH

Calculate the pH of a 0.015 M solution of NaCN. The K_b value is listed in Table 17.4.

ACID–BASE PROPERTIES OF SALTS: HYDROLYSIS

In the discussions of weak acids and bases you learned that the cations and anions of salts may undergo hydrolysis. That is, when dissolved in water, the ions may alter the H_3O^+ and OH^- concentrations of the water. In this section we want to look briefly at examples of salts and see if we can predict whether they will generate acidic or basic solutions.

The conclusions that we have reached previously regarding the acidic and basic properties of ions are summarized in Table 17.5. This table is helpful because you can use this information to predict, for example, that

1. A salt such as $NaNO_3$ gives a neutral, aqueous solution since Na^+ does not hydrolyze to an appreciable extent and since NO_3^- is the conjugate base of a strong acid.
2. An aqueous solution of K_2S should be basic since S^{2-} is the conjugate base of the very weak acid HS^- (K_b for $S^{2-} = 0.077$), while K^+ does not hydrolyze appreciably.
3. An aqueous solution of $FeCl_2$ should be weakly acidic, since Fe^{2+} hydrolyzes to give an acidic solution (Section 17.6), while Cl^- is the conjugate base of the strong acid HCl.

Some additional explanation is needed concerning salts of anions such as HCO_3^- and HS^-. Since they have an ionizable hydrogen, they can in principle act as acids.

$$\underset{\text{acid}}{HS^-(aq)} + H_2O(\ell) \rightleftharpoons S^{2-}(aq) + H_3O^+(aq) \qquad K_a = 1.3 \times 10^{-13}$$

However, keep in mind that they are also the conjugate bases of weak acids.

$$\underset{\text{base}}{HS^-(aq)} + H_2O(\ell) \rightleftharpoons H_2S(aq) + OH^-(aq) \qquad K_b = 1 \times 10^{-7}$$

Table 17.5 Acid–Base Properties of Some Common Ions in Water Solution

	Neutral		Basic		Acidic
Anion	Cl^- Br^- I^-	NO_3^- ClO_4^-	CH_3COO^- F^- CO_3^{2-} S^{2-} PO_4^{3-} SO_4^{2-}	CN^- NO_2^- HCO_3^- HS^- HPO_4^{2-}	HSO_4^-
Cation	Li^+ Na^+ K^+	Mg^{2+} Ca^{2+} Ba^{2+}	None		Al^{3+} NH_4^+ Transition metal ions

If you are curious about the pH of a solution of a salt in which both the cation and the anion contribute to the pH of the solution (ammonium acetate, for example), see Study Question 121 at the end of this chapter.

Whether the solution is acidic or basic depends on the relative sizes of K_a and K_b. In the case of an aqueous solution of the hydrogen sulfide anion, K_b is much larger than K_a, so $[OH^-]$ is much larger than $[H_3O^+]$, and a solution of a salt such as NaHS will be basic.

EXERCISE 17.11 Predicting the pH of Solutions

For each of the following salts, predict whether the pH will be greater than, less than, or equal to 7.

(a) NaCl (c) NH_4NO_3
(b) $FeCl_3$ (d) $NaHCO_3$

POLYPROTIC ACIDS

COMMON POLYPROTIC ACIDS

Phosphoric	H_3PO_4
Sulfuric	H_2SO_4
Hydrosulfuric	H_2S
Oxalic	$H_2C_2O_4$
Carbonic	H_2CO_3

A few of the more important acids capable of donating more than one proton are listed in the margin. In every case the loss of each successive proton becomes more difficult, as indicated by the K_a values of Table 17.4 (and Appendix E). For example, sulfuric acid is listed above H_3O^+ in Table 17.2 and so is a strong acid for the loss of one proton in water.

$$H_2SO_4(aq) + H_2O(\ell) \xrightarrow{\approx 100\% \text{ ionization}} H_3O^+(aq) + HSO_4^-(aq)$$

To lose the second proton, however, is more difficult as shown by a relatively small value of K_a (1.2×10^{-2}) for the hydrogen sulfate ion.

$$HSO_4^-(aq) + H_2O(\ell) \xrightarrow[\text{for 0.1 M solution}]{\approx 3\% \text{ ionization}} H_3O^+(aq) + SO_4^{2-}(aq)$$

Carbonic acid, an important acid in our environment, behaves somewhat differently than sulfuric acid. Here *both* protons are lost with difficulty, the first more readily than the second, as usual.

$$H_2CO_3(aq) + H_2O(\ell) \rightleftharpoons H_3O^+(aq) + HCO_3^-(aq) \quad K_{a1} = 4.2 \times 10^{-7}$$

$$HCO_3^-(aq) + H_2O(\ell) \rightleftharpoons H_3O^+(aq) + CO_3^{2-}(aq) \quad K_{a2} = 4.8 \times 10^{-11}$$

One reason for the difference in successive K_a values is that a negative ion is produced when H^+ is lost by an acid. To lose another positive ion (H^+) from the conjugate base anion is more difficult because the ion must accommodate a larger negative charge.

For many acids, as for carbonic acid, each successive loss of a proton is about 10^4 to 10^6 times more difficult than the previous ionization step. This means that the first ionization step of a polyprotic acid produces up to about a million times more H_3O^+ than the second step. Hydrosulfuric acid, the compound associated with the smell of rotten eggs, is an example.

$$H_2S(aq) + H_2O(\ell) \rightleftharpoons H_3O^+(aq) + HS^-(aq) \quad K_{a1} = 1.0 \times 10^{-7}$$

$$HS^-(aq) + H_2O(\ell) \rightleftharpoons H_3O^+(aq) + S^{2-}(aq) \quad K_{a2} = 1.3 \times 10^{-13}$$

For solutions of H_2S, and many other inorganic polyprotic acids with $K_{a2} \leq 10^{-3} K_{a1}$, *the pH of the solution depends primarily on the hydronium ion generated in the first ionization step*; hydronium ion from the second step can be neglected.

EXAMPLE 17.10

pH OF A POLYPROTIC ACID SOLUTION

Hydrogen sulfide can dissolve in water to the extent of approximately 0.10 M. What is the pH of the solution? What are the equilibrium values of $[H_2S]$, $[HS^-]$, and $[S^{2-}]$?

Solution The aqueous equilibria of H_2S are written in the text above, and we said that $[H_3O^+]$ depends almost entirely on the hydronium ion generated in the first step. So, let us assume that *only* the first step occurs, and calculate the resulting concentrations.

	$[H_2S]$	$[H_3O^+]$	$[HS^-]$
Before ionization (M)	0.10	0	0
Change on proceeding to equilibrium	$-x$	$+x$	$+x$
After equilibrium is achieved (M)	$0.10 - x$	x	x

The concentration of H_3O^+ (x) can be derived from the expression

$$K_{a1} = 1.0 \times 10^{-7} = \frac{[H_3O^+][HS^-]}{[H_2S]} = \frac{x^2}{0.10 - x}$$

Since K_{a1} is so small, it seems reasonable that $(0.10 - x) \approx 0.10$, and so

$$x = [H_3O^+] = [HS^-] = \sqrt{K_{a1}(0.10)} = 1.0 \times 10^{-4}\ \text{M}$$

Using this value of $[H_3O^+]$ to calculate the pH, we find

$$\text{pH} = -\log(1.0 \times 10^{-4}) = 4.00$$

and the concentration of the acid H_2S is

$$[H_2S] = 0.10 - 0.00010 \approx 0.10\ \text{M}$$

This last result shows that the approximation was indeed justified, since $0.10 - 0.00010 \approx 0.10$.

Next, we turn to the problem of obtaining the concentration of the sulfide ion, S^{2-}, a product of the second ionization step.

	$[HS^-]$	$[H_3O^+]$	$[S^{2-}]$
Before ionization (M)	1.0×10^{-4}	1.0×10^{-4}	0
Change on proceeding to equilibrium	$-y$	$+y$	$+y$
After equilibrium is achieved (M)	$(1.0 \times 10^{-4} - y)$	$(1.0 \times 10^{-4} + y)$	y

Since K_{a2} is so small, the second step occurs to a *much* smaller extent than the first step. This means the amount of S^{2-} and H_3O^+ produced in the second step (= y) is *much* smaller than 10^{-4} M. Therefore, it is reasonable that both $[HS^-]$ and $[H_3O^+]$ are very close to 1.0×10^{-4} M.

$$K_{a2} = 1.3 \times 10^{-13} = \frac{[H_3O^+][S^{2-}]}{[HS^-]} = \frac{(1.0 \times 10^{-4})(y)}{1.0 \times 10^{-4}}$$

Since $[H_3O^+]$ and $[HS^-]$ have nearly identical values, they cancel from the expression, and we find that $[S^{2-}]$ is simply equal to K_{a2}.

$$y = [S^{2-}] = K_{a2} = 1.3 \times 10^{-13}$$

EXERCISE 17.12 Polyprotic Acids

Suppose you have a 0.10 M solution of oxalic acid, $H_2C_2O_4$. What is the pH of the solution? What is the concentration of the oxalate ion, $C_2O_4^{2-}$?

SUMMARY Example 17.10 points to a general conclusion. For polyprotic acids, H_2A, where K_{a1} and K_{a2} differ by a factor of 10^3 or more, we can say that: (a) the hydronium ion largely comes from the first ionization step and is given by $[H_3O^+] = \sqrt{K_{a1}[H_2A]_0}$ (where $[H_2A]_0$ is the original concentration of the diprotic acid and where $[HA]_0 > 100\ K_{a1}$); (b) $[HA^-] = [H_3O^+]$; and (c) $[A^{2-}] = K_{a2}$.

17.6 THE LEWIS CONCEPT OF ACIDS AND BASES

G.N. Lewis is the same scientist who developed the first concepts of the electron pair bond. See page 363 for a short biography of this important American scientist.

Ideas developed by Gilbert N. Lewis in the 1930s led to a comprehensive approach to understanding acid–base interactions that includes more reaction types than does the Brønsted theory. The Lewis definition can be stated very simply: *An **acid** is a substance that can accept a pair of electrons from another atom to form a new bond, and a **base** is a substance that can donate a pair of electrons to another atom to form a new bond.* This means that an acid–base reaction can occur when a base provides a pair of electrons to share with an acid. The result is often called an acid–base **adduct** or **complex**, and the bond formed between the pair is a **donor–acceptor** or **coordinate covalent bond**.

$$\underset{\text{acid}}{A} + \underset{\text{base}}{B:} \longrightarrow \underset{\text{adduct or complex}}{B{:}\!\rightarrow\!A}$$

A simple example of a Lewis acid–base reaction is one already mentioned many times, the formation of the hydronium ion from H^+ and water.

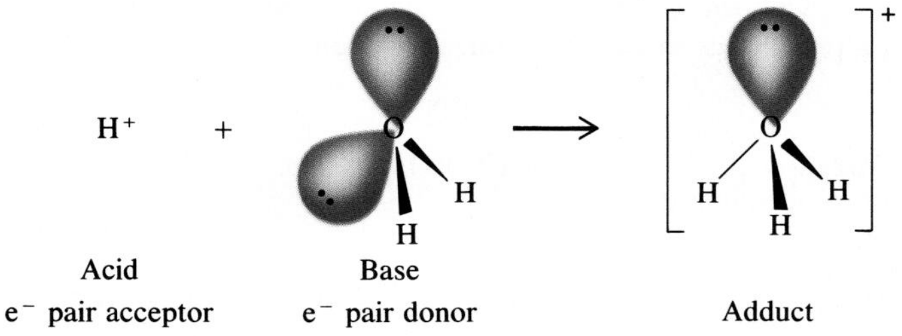

The H^+ ion has no electrons in its valence or 1*s* orbital, while the water molecule has two unshared pairs of electrons (located in sp^3 hybrid orbitals). One of the pairs can be shared between H^+ and water, thus forming an O—H bond. A similar interaction occurs between H^+ and the base ammonia to form the ammonium ion.

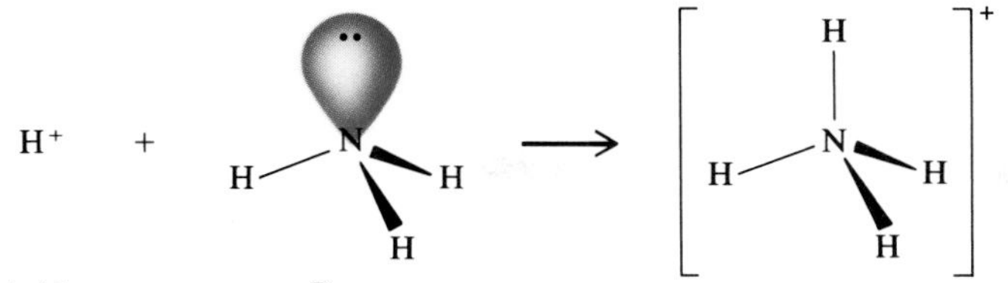

H^+ + NH_3 ⟶ $[NH_4]^+$

Acid Base

Such reactions are very common. In general, they involve acids that are (a) cations or (b) neutral molecules with an available, empty valence orbital and bases that are (a) anions or (b) neutral molecules with a lone electron pair.

CATIONIC LEWIS ACIDS

All metal cations are potential Lewis acids. Not only do they attract electrons due to their positive charge, but all have at least one empty orbital. This orbital can overlap the orbital bearing the electron pair of the base and can thereby form a two-electron chemical bond. Consider the beryllium ion, Be^{2+}, as an example. The Be^{2+} ion has no electrons in the valence shell ($n = 2$), so there are four empty orbitals available: $2s$ and three $2p$ orbitals. Water molecules are electron pair donors or Lewis bases, so Be^{2+} and H_2O form an acid–base adduct, and four empty orbitals on beryllium mean that up to four donor-acceptor bonds can form.

Most compounds of beryllium are toxic; breathing beryllium-containing dust or getting it on your skin can lead ultimately to cancer. When fluorescent lights were first developed, the material coating the inside contained beryllium compounds. When the health hazards of such compounds became known, a substitute was found.

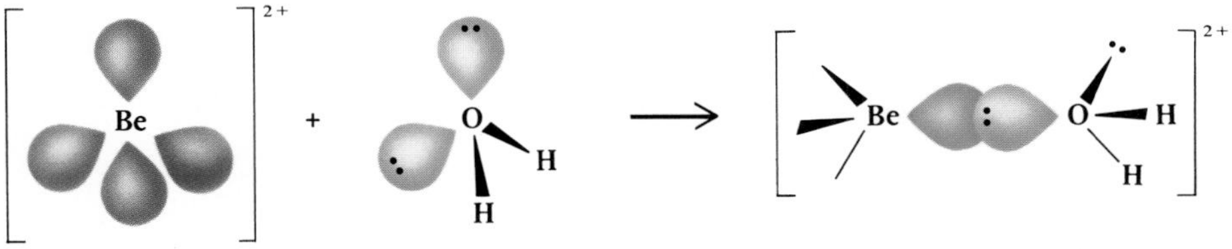

Be^{2+}, sp^3 hybridized, tetrahedral

H_2O, sp^3 hybridized, tetrahedral

The orbital overlap that leads to one of the $H_2O \rightarrow Be$ donor-acceptor bonds.

$$Be^{2+}(aq) + 4\ H_2O(\ell) \longrightarrow [Be(H_2O)_4]^{2+}(aq)$$

The **coordination number** of a metal ion in a complex is the number of donor atoms bonded to the central ion. The coordination number of a particular cation depends on the charge and size of the metal ion, the nature of the Lewis base, and other factors; however, it is usually four or six. The hydrated beryllium ion has a coordination number of four, and the complex is presumably tetrahedral in shape, as predicted by the ideas outlined in Chapters 10 and 11.

Species such as $[Be(H_2O)_4]^{2+}$, $[Co(NH_3)_6]^{3+}$, and $Pt(NH_3)_2Cl_2$ are often called complexes, coordination complexes, or, if charged, complex ions (see Chapter 25). Particularly if it bears a charge, the formula of the complex is enclosed in square brackets. The Lewis base attached to the central metal ion is often called a ligand.

The 2+ charge of the beryllium ion means that the electrons of the $H_2O{:}\rightarrow Be^{2+}$ bond are very strongly attracted to the cation. As a result, the O—H bonds of the bound water molecules are more polarized, because the oxygen of water, in relinquishing electrons to Be^{2+}, can better accommodate the negative charge of the pair of electrons left on it after H^+ leaves.

$$\delta+\ H-\overset{\overset{\delta+}{H}|}{\underset{\delta-}{\ddot{O}}}:\longrightarrow Be^{2+}$$

The result is that a coordinated water molecule can lose a proton more easily than a normal water molecule, and the $[Be(H_2O)_4]^{2+}$ complex ion functions as a Brønsted acid or proton donor.

$$[Be(OH_2)(H_2O)_2(OH_2)]^{2+}(aq) + H_2O(\ell) \longrightarrow [Be(OH_2)(H_2O)_2(OH)]^{+}(aq) + H\text{—}\ddot{O}\text{—}H^+(aq)$$

This explains why many small, highly charged metal cations are Brønsted acids (see Tables 17.2 and 17.6) and provides a model for hydrolysis reactions in general.

*An **amphoteric** substance is a Brønsted base and a Lewis acid. An **amphiprotic** substance is a Brønsted acid and Brønsted base (page 702).*

The hydroxide ion is an excellent Lewis base and binds readily to metal cations to give metal hydroxides. An important feature of the chemistry of some metal hydroxides is that they are **amphoteric**. *An amphoteric metal hydroxide can behave as a Lewis acid and can react with a Lewis base,*

$$\underset{\text{Lewis acid}}{Zn(OH)_2(s)} + 2\ OH^-(aq) \longrightarrow [Zn(OH)_4]^{2-}(aq)$$

or behave as a Brønsted base and react with a Brønsted acid (Table 17.7).

$$\underset{\text{Brønsted base}}{Zn(OH)_2(s)} + 2\ H_3O^+(aq) \longrightarrow Zn^{2+}(aq) + 4\ H_2O(\ell)$$

One of the best examples of amphoterism is aluminum hydroxide, $Al(OH)_3$ (Figure 17.9). Aluminum is widely used as a structural material, partly because of its light weight and partly because of its resistance to corrosion. Its corrosion resistance arises from a chemically inert and hard coating of oxidized metal, Al_2O_3. Nonetheless, the coating can be broken, and the metal can then be dissolved in base, producing hydrogen as a useful by-product (see Figure 12.10).

$$2\ Al(s) + 2\ KOH(aq) + 6\ H_2O(\ell) \longrightarrow 2\ K[Al(OH)_4](aq) + 3\ H_2(g)$$

The aluminate ion, $[Al(OH)_4]^-$, is a Brønsted base, so acidification of its solutions produces the insoluble metal hydroxide.

Table 17.6 Acid Ionization Constants for the Hydrolysis of Hydrated Metal Ions at 25 °C

Metal Ion	K_a
Al^{3+}	7.9×10^{-6}
Pb^{2+}	1.5×10^{-8}
Fe^{2+}	3.2×10^{-10}
Fe^{3+}	6.3×10^{-3}
Co^{2+}	1.3×10^{-9}
Ni^{2+}	2.5×10^{-11}
Cu^{2+}	1.6×10^{-7}

Table 17.7 Some Common Amphoteric Metal Hydroxides

Hydroxide	Reaction as a Base	Reaction as an Acid
$Al(OH)_3$	$Al(OH)_3(s) + 3\ H_3O^+ \longrightarrow Al^{3+}(aq) + 6\ H_2O$	$Al(OH)_3(s) + OH^-(aq) \longrightarrow [Al(OH)_4]^-(aq)$
$Zn(OH)_2$	$Zn(OH)_2(s) + 2\ H_3O^+ \longrightarrow Zn^{2+}(aq) + 4\ H_2O$	$Zn(OH)_2(s) + 2\ OH^-(aq) \longrightarrow [Zn(OH)_4]^{2-}(aq)$
$Sn(OH)_4$	$Sn(OH)_4(s) + 4\ H_3O^+ \longrightarrow Sn^{4+}(aq) + 8\ H_2O$	$Sn(OH)_4(s) + 2\ OH^-(aq) \longrightarrow [Sn(OH)_6]^{2-}(aq)$
$Cr(OH)_3$	$Cr(OH)_3(s) + 3\ H_3O^+ \longrightarrow Cr^{3+}(aq) + 6\ H_2O$	$Cr(OH)_3(s) + OH^-(aq) \longrightarrow [Cr(OH)_4]^-(aq)$

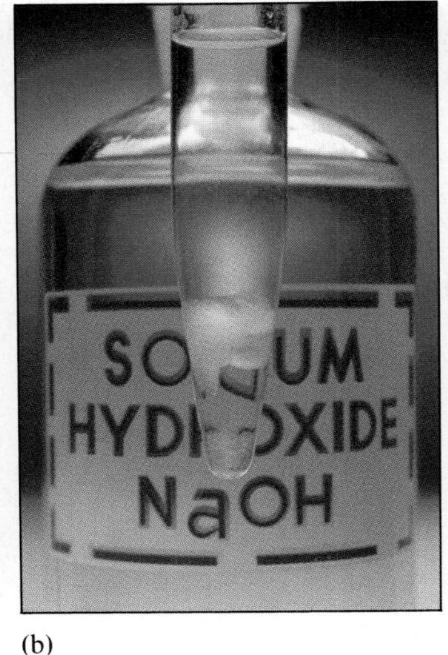

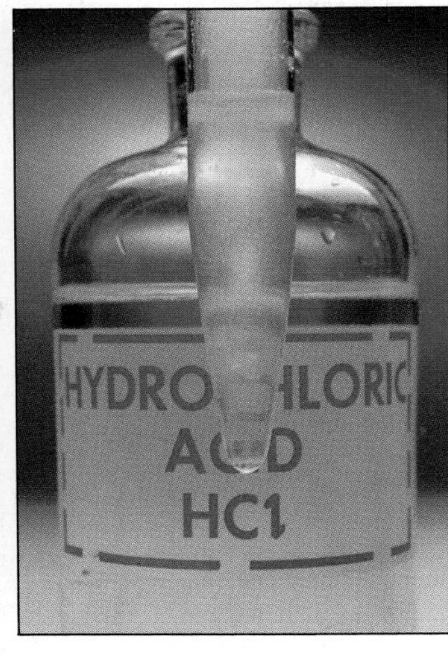

(a) (b) (c)

Figure 17.9 The amphoteric nature of $Al(OH)_3$. (a) Adding aqueous ammonia to a soluble salt of Al^{3+} leads to a precipitate of $Al(OH)_3$. (b) Adding a strong base (NaOH) to $Al(OH)_3$ dissolves the precipitate. Here aluminum hydroxide acts as a Lewis acid toward the Lewis base OH^- and forms the soluble sodium salt of the complex ion $[Al(OH)_4]^-$ (c) If we begin again with freshly precipitated $Al(OH)_3$, it is dissolved as strong acid (HCl) is added. In this case $Al(OH)_3$ acts as a Brønsted base and forms a soluble aluminum salt and water.

Overall reaction:

$$2\ KAl(OH)_4(aq) + H_2SO_4(aq) \longrightarrow 2\ Al(OH)_3(s) + 2\ H_2O(\ell) + K_2SO_4(aq)$$

Net ionic equation:

$$\underset{\text{Brønsted base}}{[Al(OH)_4]^-(aq)} + H_3O^+(aq) \longrightarrow Al(OH)_3(s) + 2\ H_2O(\ell)$$

Since the aluminum hydroxide is amphoteric, adding OH^- returns it to the soluble $[Al(OH)_4]^-$ ion,

$$\underset{\text{Lewis acid}}{Al(OH)_3(s)} + OH^-(aq) \longrightarrow [Al(OH)_4]^-(aq)$$

while it can also react as a Brønsted base.

$$\underset{\text{Brønsted base}}{2\ Al(OH)_3(s)} + 3\ H_2SO_4(aq) \longrightarrow Al_2(SO_4)_3(aq) + 6\ H_2O(\ell)$$

Aluminum sulfate crystallizes from aqueous solution as $Al_2(SO_4)_3 \cdot 18\ H_2O$, a hydrated salt commonly called ''paper maker's alum'' (see Chapter 23).

Metal ions also form a large series of complexes with the Lewis base ammonia. For example, silver ion readily forms a water-soluble, colorless complex ion in aqueous ammonia. Indeed, this complex ion is so stable that the very insoluble compound AgCl can be dissolved in aqueous ammonia.

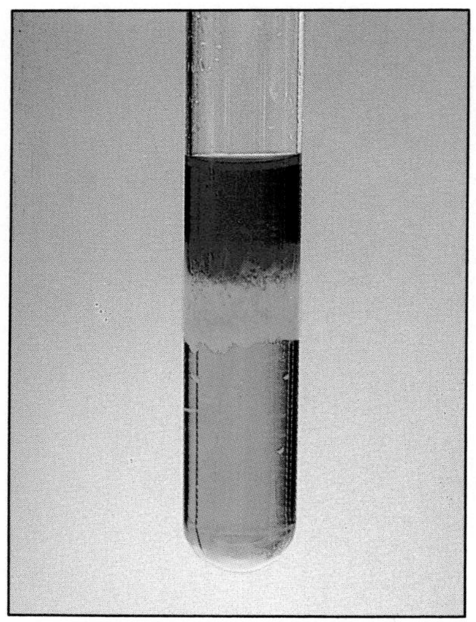

Figure 17.10 The Lewis acid–base complex ion $[Cu(NH_3)_4]^{2+}$. Here aqueous ammonia was added to aqueous $CuSO_4$ (the light blue solution at the bottom of the tube). The small concentration of OH^- in $NH_3(aq)$ first formed insoluble blue-white $Cu(OH)_2$ (the solid in the middle of the tube). However, with additional NH_3, the deep blue soluble complex ion formed (the solution at the top of the tube).

$$AgCl(s) + 2\ NH_3(aq) \longrightarrow [H_3N\!:\!\rightarrow Ag \leftarrow\!:\!NH_3]^+(aq) + Cl^-(aq)$$

Light blue aqueous copper(II) ions also react with ammonia to produce a beautiful, deep blue complex ion with four ammonia molecules surrounding each metal ion (Figure 17.10).

$$Cu^{2+}(aq) + 4\ NH_3(aq) \longrightarrow \left[\begin{array}{c} H_3\ddot{N} \\ \downarrow \\ H_3N\!:\!\rightarrow Cu \leftarrow\!:\!NH_3 \\ \uparrow \\ \ddot{}H_3N \end{array}\right]^{2+}(aq)$$

deep blue

Finally, a molecule to which we have often referred, cisplatin, is a square planar complex of Pt^{2+} with the Lewis bases NH_3 and Cl^-.

$$\begin{array}{c} H_3\ddot{N} \\ \downarrow \\ H_3N\!:\!\rightarrow Pt \leftarrow\!:\!\ddot{\underset{\cdot\cdot}{Cl}}\!: \\ \uparrow \\ :\ddot{\underset{\cdot\cdot}{Cl}}: \end{array}$$

NEUTRAL MOLECULES AS LEWIS ACIDS

In Chapter 10 we discussed the fact that the early elements in the second period do not obey the octet rule all the time. Boron compounds in particular sometimes have only three pairs of electrons around the boron atom. On the other hand, elements of higher periods can often accommodate more than four pairs of electrons. Both of these features mean that such compounds frequently behave as Lewis acids.

Boron trifluoride is produced in ton quantities from borax, fluorspar, and sulfuric acid, all inexpensive chemicals.

$$\underset{\text{borax}}{Na_2B_4O_7(s)} + \underset{\text{fluorspar}}{6\ CaF_2(s)} + 8\ H_2SO_4(aq) \longrightarrow$$
$$4\ BF_3(g) + 2\ NaHSO_4(aq) + 6\ CaSO_4(s) + 7\ H_2O(\ell)$$

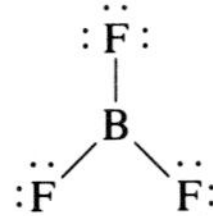

Electron dot structure of BF_3.

BF_3, a colorless gas, is an excellent Lewis acid. The molecule is trigonal planar with an sp^2 hybridized boron atom. This means that the boron has an unused p orbital perpendicular to the BF_3 plane, and this orbital may accept electrons from a Lewis base such as ammonia.

$$\underset{\text{Lewis acid}}{BF_3(g)} + \underset{\text{Lewis base}}{NH_3(g)} \longrightarrow F_3B\leftarrow\!:\!NH_3(s)$$

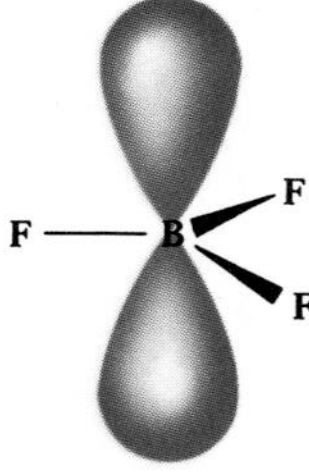

Boron p orbital is unused in sp^2 hybrid formation in a molecule such as BF_3.

Carbon dioxide can also act as a Lewis acid and react with the Lewis base OH^- in aqueous solution.

$$H-\ddot{\underset{\cdot\cdot}{O}}:^- + \overset{\delta-}{\ddot{O}}=\overset{\delta+}{C}=\overset{\delta-}{\ddot{O}} \longrightarrow H-\ddot{\underset{\cdot\cdot}{O}}:^- \rightharpoonup C^+\begin{array}{l}=\ddot{O}\\-\ddot{O}:^-\end{array} \longrightarrow H-\ddot{O}-C\begin{array}{l}=\ddot{O}:\\-\ddot{O}:^-\end{array}$$

Because oxygen is electronegative, C—O bond electrons are polarized away from carbon toward oxygen in CO_2. This causes the carbon atom to be slightly deficient in electrons and positive, and the negatively charged Lewis base OH^- can attack the carbon to give, ultimately, the bicarbonate ion. This is why the concentration of a solution of NaOH or KOH can change during storage in air; if you have standardized a base, you must use it fairly soon, or CO_2 from the air can reduce the effective OH^- concentration.

$BF_3(g) + NH_3(g) \longrightarrow$

$F_3B \leftarrow NH_3$

EXERCISE 17.13 Lewis Acids and Bases

Tell whether each of the following is a Lewis acid or a Lewis base.
(a) PH_3 (b) BCl_3 (c) H_2S (d) SF_4

THE EXTENT OF LEWIS ACID–BASE REACTIONS: FORMATION CONSTANTS

The extent to which Lewis acid–base reactions occur can be expressed in terms of an equilibrium constant. For example, for the reaction between Ag^+ and ammonia,

$$\underset{\text{Lewis acid}}{Ag^+(aq)} + \underset{\text{Lewis base}}{2\ NH_3(aq)} \rightleftharpoons [Ag(NH_3)_2]^+(aq)$$

$$K = \frac{\{[Ag(NH_3)_2]^+\}}{[Ag^+][NH_3]^2} = 1.6 \times 10^7$$

Since the formula of a complex ion is given in square brackets, we have enclosed the bracketed formula in braces, { }, to indicate the molar concentration of the complex ion.

The large constant for the formation of the silver–ammonia complex indicates that there is a very large tendency to form this ion. On the other hand, the smaller constant for the formation of $[PbCl_4]^{2-}$ tells us that the concentration of the complex ion will not be large at equilibrium.

$$\underset{\text{Lewis acid}}{Pb^{2+}(aq)} + \underset{\text{Lewis base}}{4\ Cl^-(aq)} \rightleftharpoons [PbCl_4]^{2-}(aq)$$

$$K = \frac{\{[PbCl_4]^{2-}\}}{[Pb^{2+}][Cl^-]^4} = 10$$

Each of the complexes you have seen so far involves more than one Lewis base molecule or ion (NH_3 or Cl^-) combining with the metal ion Lewis acid. Formation of such complexes can be envisioned as a stepwise process, with Lewis base molecules being added successively to the central ion.

$$Cu^{2+}(aq) + NH_3(aq) \rightleftharpoons [Cu(NH_3)]^{2+}(aq) \qquad K_1 = 1.7 \times 10^4$$

$$[Cu(NH_3)]^{2+}(aq) + NH_3(aq) \rightleftharpoons [Cu(NH_3)_2]^{2+}(aq) \qquad K_2 = 3.2 \times 10^3$$

$$[Cu(NH_3)_2]^{2+}(aq) + NH_3(aq) \rightleftharpoons [Cu(NH_3)_3]^{2+}(aq) \qquad K_3 = 8.3 \times 10^2$$

$$[Cu(NH_3)_3]^{2+}(aq) + NH_3(aq) \rightleftharpoons [Cu(NH_3)_4]^{2+}(aq) \qquad K_4 = 1.5 \times 10^2$$

In some cases, such as the formation of aqueous $[Cu(NH_3)_4]^{2+}$, it is possible to measure the value of K for each successive addition. Since the overall reaction

$$Cu^{2+}(aq) + 4\ NH_3(aq) \rightleftharpoons [Cu(NH_3)_4]^{2+}(aq)$$

is the sum of four reaction steps, the overall equilibrium constant is the product of the stepwise constants.

$$K_{total} = \frac{\{[Cu(NH_3)_4]^{2+}\}}{[Cu^{2+}][NH_3]^4} = K_1 \times K_2 \times K_3 \times K_4 = 6.8 \times 10^{12}$$

Overall equilibrium constants for a variety of complex ions are given in Appendix H. Since these are constants for the formation of the complex ion, they are called **formation constants.***

The formation constant can be used to calculate the concentrations of the complex, metal ion, or free Lewis base at equilibrium. Of course such calculations are complicated by the fact that complexes are formed in steps, so some of the intermediate ions or molecules can exist in solution. However, because K for each step is so large, the assumption usually made is that the complex ion containing the most Lewis base groups is dominant when the Lewis base concentration is much larger than the metal ion concentration.

EXAMPLE 17.11

USING FORMATION CONSTANTS

Suppose that 50. mL of 1.0 M NH_3 is mixed with 50. mL of 1.0×10^{-3} M Ag^+. Calculate the concentrations of NH_3, Ag^+, and $[Ag(NH_3)_2]^+$ at equilibrium.

$$Ag^+(aq) + 2\ NH_3(aq) \rightleftharpoons [Ag(NH_3)_2]^+(aq)$$

$$K = 1.6 \times 10^7 = \frac{\{[Ag(NH_3)_2]^+\}}{[Ag^+][NH_3]^2}$$

Solution When equal volumes of solutions are mixed, the concentrations of the dissolved species become half of their original values. Thus, the initial concentrations in the combined solutions are

$$[Ag^+] = \tfrac{1}{2}(1.0 \times 10^{-3}) = 5.0 \times 10^{-4}\ M$$

$$[NH_3] = \tfrac{1}{2}(1.0) = 0.50\ M$$

Our strategy in solving this problem is as follows: The equilibrium constant for the formation of the silver–ammonia complex is very large, and the concentration of ammonia is high. Therefore, when ammonia and silver ion are mixed, we *imagine* that *all* of the Ag^+ ion forms $[Ag(NH_3)_2]^+$, and stoichiometry tells us that the *initial* concentration of the complex ion must be 5.0×10^{-4} M.

	$[Ag^+]$	$[NH_3]$	$\{[Ag(NH_3)_2{}^+]\}$
Concentration before mixing (M)	0.00050	0.50	0
Concentration after mixing (M) (complete reaction assumed)	0	$0.50 - 2(0.00050) \cong 0.50$	0.00050

*An alternative name is **stability constant**, since it reflects the "stability" of the complex in solution. However, the same information is conveyed by **dissociation** or **instability constants**. These are for the dissociation of the complex ion as in

$$[Cu(NH_3)_4]^{2+}(aq) \rightleftharpoons Cu^{2+}(aq) + 4\ NH_3(aq) \qquad K_{dissoc} = 1.5 \times 10^{-13}$$

Since this reaction is the reverse of the formation reaction, the equilibrium constant, K_{dissoc}, is the reciprocal of the formation constant.

Next, we imagine that the reaction reverses to a slight extent on proceeding to an equilibrium state, and we calculate the equilibrium concentrations.

	$[Ag^+]$	$[NH_3]$	$\{[Ag(NH_3)_2{}^+]\}$
Concentration before equilibrium (M)	0	0.50	0.00050
Change on going to equilibrium	$+x$	$+2x$	$-x$
Concentration at equilibrium (M)	x	$(0.50 + 2x)$	$(0.00050 - x)$

These concentrations can be substituted into the equilibrium expression for the chemical equation connecting the species, and x can be determined. Before doing so, however, we can make the assumption that x is extremely small since the formation constant of the complex ion is so great. This means that $[NH_3]$ is effectively 0.50 M and $[Ag(NH_3)_2]^+$ is approximately 0.00050 M.

$$K = 1.6 \times 10^7 = \frac{\{[Ag(NH_3)_2]^+\}}{[Ag^+][NH_3]^2} = \frac{(0.00050)}{x(0.50)^2}$$

$$x = [Ag^+] = 1.3 \times 10^{-10} \text{ M}$$

As a final step, observe that the approximation was reasonable. The quantities $(0.50 + 2x)$ and $(0.00050 - x)$ are indeed effectively 0.50 and 0.00050, respectively.

EXERCISE 17.14 Using Formation Constants

Calculate the equilibrium concentrations of Cu^{2+}, NH_3, and $[Cu(NH_3)_4]^{2+}$ when 500. mL of 3.00 M NH_3 are mixed with 500. mL of 2.00×10^{-3} M $Cu(NO_3)_2$. $K_{formation}$ for $[Cu(NH_3)_4]^{2+} = 6.8 \times 10^{12}$.

SOMETHING MORE ABOUT
Planetary Chemistry: The Acid Clouds of Venus

On a clear night you can easily see the Evening Star, the planet Venus, its brightness second only to the earth's own moon. The planet, which was named for the Roman goddess of love, is the closest planet in the solar system to the earth, its diameter is almost the same as the earth's diameter of 12,745 km, and it has about 80% as much mass as the earth. Given these characteristics, we might have expected Venus to resemble the earth. However, it has been visited since 1962 by numerous probes from the Soviet Union and the United States, and its atmosphere is decidedly not earth-like.

The atmospheric pressure at the surface of Venus is 90 times that of the earth! Its atmosphere consists largely of CO_2 (96%) and N_2 (3.5%), but droplets of sulfuric acid and water have been detected. The clouds and high concentration of CO_2 in the Venusian atmosphere produce an extreme "greenhouse effect" that traps the heat from the sun in the lower atmosphere. Thus, the temperature at or near the surface is about 480 °C, high enough to melt lead.

The production of sulfuric acid in the atmosphere could begin with photochemical reactions—reactions induced by light—in Venus's upper atmosphere. Sunlight could produce sulfur atoms from S_8 molecules or from the poisonous molecule COS,

$$\text{light} + COS \longrightarrow S + CO$$

and it could produce O atoms from CO_2.

$$\text{light} + CO_2 \longrightarrow CO + O$$

Since there is no evidence for O atoms in the atmosphere, they must be removed by some reaction, and it is thought that they combine with sulfur atoms to give SO_3.

$$S + 3\ O \longrightarrow SO_3$$

Sulfur trioxide is just anhydrous sulfuric acid, so, as the SO_3 drifts down closer to the surface of the planet, the oxide combines with water to give sulfuric acid.

$$SO_3 + H_2O \longrightarrow H_2SO_4$$

As the droplets of acid fall toward the surface of the planet, they grow larger through accretion, and the acid becomes more concentrated as the water evaporates. About 35 km above the surface, where the temperature is over 100 °C, the droplets are 98% sulfuric acid, the acid–water combination with the highest boiling point. However, as the droplets fall closer and closer to the surface, the temperature climbs higher still, the sulfuric acid is decomposed back to SO_3 and H_2O, and the atmosphere closer to the surface contains these gases (as well as CO_2 and other gases).

The chemistry of Venus is complicated, and still open to new discoveries. However, it is certain the atmosphere contains sulfuric acid, probably produced by reactions we can easily study on earth.

SUMMARY

Interest in acids and bases began hundreds of years ago, and useful ideas to understand their behavior are still being developed. A cornerstone of current ideas of acid–base behavior in aqueous solution is that water undergoes **auto-ionization** to give the **hydronium ion, H_3O^+**, and hydroxide ion, OH^-, to a small extent even in pure water (Section 17.1).

$$2\ H_2O(\ell) \rightleftharpoons H_3O^+(aq) + OH^-(aq)$$

The acid–base definition most widely used by chemists to describe aqueous chemistry is that proposed by Brønsted: an aqueous acid is a proton donor, while a base is a proton acceptor (Section 17.2). Thus, HCO_3^- is an acid

$$\underset{\text{acid}}{HCO_3^-(aq)} + \underset{\text{base}}{H_2O(\ell)} \rightleftharpoons \underset{\text{conjugate base of } HCO_3^-}{CO_3^{2-}(aq)} + \underset{\text{conjugate acid of } H_2O}{H_3O^+(aq)}$$

and a base

$$\underset{\text{base}}{HCO_3^-(aq)} + \underset{\text{acid}}{H_2O(\ell)} \rightleftharpoons \underset{\text{conjugate acid of } HCO_3^-}{H_2CO_3(aq)} + \underset{\text{conjugate base of } H_2O}{OH^-(aq)}$$

(Because the same ion, HCO_3^-, can be both a Brønsted acid and base, it is said to be **amphiprotic**.) In each reaction, the acceptance or donation of a proton by HCO_3^- produces a conjugate partner. **Conjugate acids and bases** are related to one another by the gain or loss of a proton. Every acid–base reaction involving H^+ transfer has two conjugate acid–base pairs (Section 17.2).

Acids and bases vary considerably in their ability to act as proton donors or acceptors (Section 17.2). HCl is a strong acid and is nearly 100% ionized in dilute aqueous solution. In contrast, HF is only weakly ionized. This means that HCl has a great tendency to donate a proton, and its conjugate base, Cl^-, only weakly accepts a proton to return to HCl. Conversely, since HF is a weak acid, its conjugate base, F^-, tends to accept a proton from water to re-form HF. These acids illustrate two corollaries of the Brønsted concept: (1) The stronger an acid or base, the weaker its conjugate base or acid, respectively. (2) In general, acid–base reactions proceed in the direction favoring the weaker acid–base pair (Table 17.2).

The concentration of hydronium ion in solution can be expressed in terms of pH (Section 17.3).

$$\text{pH} = -\log[H_3O^+]$$

The auto-ionization of pure water produces H_3O^+ and OH^- in equal amounts, and the extent to which this reaction occurs is expressed by the equilibrium constant K_w, where

$$K_w = [H_3O^+][OH^-] = 1.008 \times 10^{-14} \text{ at } 25\ ^\circ\text{C}$$

For a neutral solution, $[H_3O^+] = [OH^-] = 1.00 \times 10^{-7}$ M. Thus, the pH is 7.00. For an acid solution with an excess of H_3O^+ over OH^-, $[H_3O^+]$ is greater than 1.00×10^{-7} M, so the pH is less than 7. Conversely, for a basic solution, $[OH^-] > 10^{-7}$, $[H_3O^+] < 10^{-7}$, and so pH > 7.

Strong acids (or bases) produce as many moles of H_3O^+ (or OH^-) in solution as moles of acid (or base) dissolved (Section 17.4). Although there are relatively few strong acids and bases, there are many weak ones (Section 17.5). In the case of **weak acids and bases**, ionization occurs to a small extent, and the concentration of acid or base is related to its ionization products by the equilibrium constant K_a or K_b.

For an acid, HA: $HA(aq) + H_2O(\ell) \rightleftharpoons H_3O^+(aq) + A^-(aq)$

$$K_a = \frac{[H_3O^+][A^-]}{[HA]}$$

For a base, B: $B(aq) + H_2O(\ell) \rightleftharpoons BH^+(aq) + OH^-(aq)$

$$K_b = \frac{[BH^+][OH^-]}{[B]}$$

K_a values for a large number of weak acids are typically around 10^{-5}, although some are much smaller (Table 17.4 and Appendix E).

To find the pH of a solution of a weak acid, we have to find $[H_3O^+]$ at equilibrium by solving the expression for K_a. The hydronium ion arises from the reaction of the weak acid with water *and* from the auto-ionization of water. When the pH is less than about 6 we can assume that H_3O^+ ion from water auto-ionization does not contribute significantly. This also allows us to say that the equilibrium concentration of HA must be $[HA] = [HA]_0 - [H_3O^+]$, where the original concentration of HA is $[HA]_0$, and that $[H_3O^+] = [A^-]$. Therefore, we can write K_a in a form that will allow us to solve for $[H_3O^+]$ if we know $[HA]_0$.

$$K_a = \frac{[H_3O^+]^2}{[HA]_0 - [H_3O^+]}$$

To solve for $[H_3O^+]$ generally requires the use of the quadratic formula or the method of successive approximations (Appendix A). However, this is not necessary if we make the approximation that $[HA]_0 - [H_3O^+] \approx [HA]_0$, an approximation that is valid when $[HA]_0 \geq 100\ K_a$. (These same comments apply to solutions of weak bases.)

Weak acids (Tables 17.2 and 17.4) in aqueous solution can be **monoprotic acids** such as HCN or CH_3COOH (acetic acid), which can transfer only one proton to water. There are also **polyprotic acids** such as oxalic acid ($H_2C_2O_4$) or H_2CO_3, which can transfer two or more protons. Cations such as NH_4^+, Fe^{3+}, and Al^{3+} can also act as acids in water. Weak bases include compounds such as ammonia (NH_3) as well as anions that are conjugate bases of weak acids; examples of the latter include CN^-, F^-, and acetate (CH_3COO^-) ions.

Lewis theory is one of the most general of all current ideas of acid–base behavior (Section 17.5). This concept defines an acid as a substance that can accept a pair of electrons from another atom to form a bond, and a base as a substance that can donate a pair of electrons to another atom to form a bond. An acid and a base can interact to form an **adduct** or **complex** with a **donor-acceptor** or **coordinate covalent bond** between the acid and base. Thus, the proton and any other cation are Lewis acids, as are many metal hydroxides. For example, the Lewis acid Zn^{2+} can form $Zn(OH)_2$ when reacting with the Lewis base OH^- ($Zn^{2+} \leftarrow :OH^-$). The $Zn(OH)_2$ can also act as a Lewis acid in reacting with additional OH^- to form $[Zn(OH_4)]^{2-}$. However, $Zn(OH)_2$ is also a Brønsted base, since it can react with H_3O^+ to give H_2O and Zn^{2+}. Such substances as $Zn(OH)_2$, which are Lewis acids and Brønsted bases, are said to be **amphoteric**.

The extent of the interaction between a Lewis acid and Lewis base is often expressed in terms of a **formation constant** (Section 17.6).

$$Ag^+(aq) + 2\ CN^-(aq) \rightleftharpoons [Ag(CN)_2]^-(aq)$$

$$K_{\text{formation}} = \frac{\{[Ag(CN)_2]^-\}}{[Ag^+][CN^-]^2}$$

The magnitude of the formation constant depends on the metal cation and the base.

STUDY QUESTIONS

REVIEW QUESTIONS

1. Outline the main ideas of the Brønsted theory of acids and bases.
2. Write a balanced equation depicting the auto-ionization of water.
3. Write balanced chemical equations showing that phosphoric acid is a polyprotic acid.
4. Write a balanced chemical equation showing that water is an amphiprotic substance.
5. Designate the acid and the base on the left side of the following equations, and designate the conjugate partner of each on the right side.

$$HNO_2(aq) + H_2O(\ell) \rightleftharpoons H_3O^+(aq) + NO_2^-(aq)$$

$$NH_4^+(aq) + CN^-(aq) \rightleftharpoons NH_3(aq) + HCN(aq)$$

6. The behavior of the acid–base indicator phenolphthalein is described in Section 17.3. Using Le Cha-

telier's principle, explain why the dye exists as a colorless molecule in acidic solution, while it is deep red in basic solution.

7. Explain the relation between Tables 17.2 (Relative Strengths of Acids and Bases) and 17.5 (Acid–Base Properties of Some Common Ions).
8. Outline the main ideas of Lewis acid–base theory. Give an example of a Lewis acid and a Lewis base.
9. Show that water can be a Brønsted base and a Lewis base.
10. Define the term "acid–base adduct." Give an example.
11. If Ni^{2+} exists as $[Ni(H_2O)_6]^{2+}$ in aqueous solution, write a balanced equation to show how hydrolysis of the ion leads to an acidic solution.
12. Define the term "amphoteric." Give an example.
13. Illustrate the term "formation constant" using the reaction

$$Ag^+(aq) + 2\ NH_3(aq) \rightleftharpoons [Ag(NH_3)_2]^+(aq)$$

GENERAL ACID–BASE THEORY

14. Write the formula and give the name of the conjugate base of each of the following acids:
(a) HCN (d) HNO_2
(b) HSO_4^- (e) HCO_3^-
(c) HF
15. Write the formula and give the name of the conjugate acid of each of the following bases:
(a) NH_3 (d) Br^-
(b) HCO_3^- (e) HSO_4^-
(c) HS^-
16. Write the formula and give the name of the conjugate partner for each acid or base below.
(a) CN^- (d) S^{2-}
(b) SO_4^{2-} (e) HNO_3
(c) HI
17. Write the formula and give the name of the conjugate partner for each acid or base below.
(a) formic acid, HCOOH
(b) H_2CO_3
(c) ClO^-, hypochlorite ion
(d) SO_3^{2-}, sulfite ion
18. In water, potassium carbonate can form K^+ and HCO_3^- ions. Write a balanced equation showing this process. Does K_2CO_3 give an acidic or basic solution?
19. Dissolving ammonium bromide in water gives an acidic solution. Write a balanced equation showing how this can occur.
20. Write balanced equations showing how the HPO_4^{2-} ion of sodium monohydrogen phosphate, Na_2HPO_4, can be both a Brønsted acid and a Brønsted base.
21. Write balanced equations showing how the hydrogen oxalate ion, $HC_2O_4^-$, can be both a Brønsted acid and a Brønsted base.
22. In each of the following acid–base reactions, identify the acid and base on the left, and their conjugate partners on the right.
(a) $HCOOH(aq) + H_2O(\ell) \rightleftharpoons HCOO^-(aq) + H_3O^+(aq)$
(b) $H_2S(aq) + NH_3(aq) \rightleftharpoons NH_4^+(aq) + HS^-(aq)$
(c) $HSO_4^-(aq) + OH^-(aq) \rightleftharpoons SO_4^{2-}(aq) + H_2O(\ell)$
23. In each of the following acid–base reactions, identify the acid and base on the left, and their conjugate partners on the right.
(a) $CH_3COOH(aq) + C_5H_5N(aq) \rightleftharpoons CH_3COO^-(aq) + C_5H_5NH^+(aq)$
(b) $N_2H_4(aq) + HSO_4^-(aq) \rightleftharpoons N_2H_5^+(aq) + SO_4^{2-}(aq)$
(c) $[Al(H_2O)_6]^{3+}(aq) + OH^-(aq) \rightleftharpoons [Al(H_2O)_5OH]^{2+}(aq) + H_2O(\ell)$
24. Ammonium chloride and sodium dihydrogen phosphate, NaH_2PO_4, are mixed in water. Using Table 17.2, write a balanced equation for the acid–base reaction that could, in principle, occur. Does the reaction occur to any significant extent?
25. Acetic acid, CH_3COOH, and sodium hydrogen carbonate, $NaHCO_3$, are mixed in water. Using Table 17.2, write a balanced equation for the acid–base reaction that could, in principle, occur. Does the reaction occur to any significant extent?
26. Hydrogen sulfide, H_2S, and sodium acetate, $Na(CH_3COO)$, are mixed in water. Using Table 17.2, write a balanced equation for the acid–base reaction that could, in principle, occur. Does the reaction occur to any significant extent?
27. Sodium hydrogen sulfate, $NaHSO_4$, is mixed with aqueous ammonia. Using Table 17.2, write a balanced equation for the acid–base reaction that could, in principle, occur. Does the reaction occur to any significant extent?
28. For each of the following reactions predict whether the equilibrium lies predominantly to the left or to the right. Explain your prediction briefly.
(a) $H_2S(aq) + CO_3^{2-}(aq) \rightleftharpoons HS^-(aq) + HCO_3^-(aq)$
(b) $HCN(aq) + SO_4^{2-}(aq) \rightleftharpoons CN^-(aq) + HSO_4^-(aq)$
(c) $CN^-(aq) + NH_3(aq) \rightleftharpoons HCN(aq) + NH_2^-(aq)$
(d) $SO_4^{2-}(aq) + CH_3COOH(aq) \rightleftharpoons HSO_4^-(aq) + CH_3COO^-(aq)$
29. For each of the following reactions, predict whether the equilibrium lies predominantly to the left or to the right. Explain your prediction briefly.
(a) $NH_4^+(aq) + Br^-(aq) \rightleftharpoons NH_3(aq) + HBr(aq)$
(b) $C_6H_5OH(aq) + CH_3COO^-(aq) \rightleftharpoons C_6H_5O^-(aq) + CH_3COOH(aq)$
(c) $NH_2^-(aq) + H_2O(\ell) \rightleftharpoons NH_3(aq) + OH^-(aq)$

30. Several acids are listed below with their respective equilibrium constants.

$$HF(aq) + H_2O(\ell) \rightleftharpoons H_3O^+(aq) + F^-(aq) \quad K_a = 7.2 \times 10^{-4}$$

$$HS^-(aq) + H_2O(\ell) \rightleftharpoons H_3O^+(aq) + S^{2-}(aq) \quad K_a = 1.3 \times 10^{-13}$$

$$CH_3COOH(aq) + H_2O(\ell) \rightleftharpoons H_3O^+(aq) + CH_3COO^-(aq) \quad K_a = 1.8 \times 10^{-5}$$

(a) Which is the strongest acid? Which is the weakest?
(b) What is the conjugate base of the acid HF?
(c) Which acid has the weakest conjugate base?
(d) Which acid has the strongest conjugate base?

31. Several acids are listed below with their respective equilibrium constants:

$$C_6H_5OH(aq) + H_2O(\ell) \rightleftharpoons H_3O^+(aq) + C_6H_5O^-(aq) \quad K_a = 1.3 \times 10^{-10}$$

$$HCOOH(aq) + H_2O(\ell) \rightleftharpoons H_3O^+(aq) + HCOO^-(aq) \quad K_a = 1.8 \times 10^{-4}$$

$$HC_2O_4^-(aq) + H_2O(\ell) \rightleftharpoons H_3O^+(aq) + C_2O_4^{2-}(aq) \quad K_a = 6.4 \times 10^{-5}$$

(a) Which is the strongest acid? Which is the weakest?
(b) Which acid has the weakest conjugate base?
(c) Which acid has the strongest conjugate base?

32. Several bases are listed below with their respective K_b values:

$$NH_3(aq) + H_2O(\ell) \rightleftharpoons NH_4^+(aq) + OH^-(aq) \quad K_b = 1.8 \times 10^{-5}$$

$$C_5H_5N(aq) + H_2O(\ell) \rightleftharpoons C_5H_5NH^+(aq) + OH^-(aq) \quad K_b = 1.5 \times 10^{-9}$$

$$N_2H_4(aq) + H_2O(\ell) \rightleftharpoons N_2H_5^+(aq) + OH^-(aq) \quad K_b = 8.5 \times 10^{-7}$$

(a) Which is the strongest base? Which is the weakest?
(b) What is the conjugate acid of C_5H_5N?
(c) Which base has the strongest conjugate acid? Which has the weakest?

33. Several bases are listed below with their respective K_b values:

$$HS^-(aq) + H_2O(\ell) \rightleftharpoons H_2S(aq) + OH^-(aq) \quad K_b = 1 \times 10^{-7}$$

$$CN^-(aq) + H_2O(\ell) \rightleftharpoons HCN(aq) + OH^-(aq) \quad K_b = 2.5 \times 10^{-5}$$

$$NO_2^-(aq) + H_2O(\ell) \rightleftharpoons HNO_2(aq) + OH^-(aq) \quad K_b = 2.2 \times 10^{-11}$$

(a) Which is the strongest base? Which is the weakest?
(b) What is the conjugate acid of HS^-?
(c) Which base has the strongest conjugate acid? Which has the weakest?

34. State which of the following ions or compounds has the strongest conjugate base and briefly explain your choice. (a) HSO_4^-, (b) CH_3COOH, (c) HClO

35. Which of the following compounds has the strongest conjugate acid? Briefly explain your choice. (a) CN^-, (b) NH_3, (c) SO_4^{2-}

pH CALCULATIONS

36. A certain table wine has a pH of 3.40. What is the hydronium ion concentration of the wine? Is it acidic or basic?

37. Milk of magnesia has a pH of 10.5. What is the hydronium ion concentration of the solution? What is the hydroxide ion concentration? Is the solution acidic or basic?

38. What is the pH of a 0.0013 M solution of HNO_3? What is the hydroxide ion concentration of the solution?

39. What is the pH of a 1.2×10^{-4} M solution of KOH? What is the hydronium ion concentration of the solution?

40. What is the pH of a 0.0015 M solution of $Ca(OH)_2$?

41. The pH of a solution of $Ba(OH)_2$ is 10.66 at 25 °C. What is the hydroxide ion concentration in the solution? If the solution volume is 250. mL, how many grams of $Ba(OH)_2$ must have been dissolved?

42. Make the following interconversions. In each case, tell whether the solution is acidic or basic.

	pH	$[H_3O^+]$ (M)	$[OH^-]$ (M)
(a)	1.00	______	______
(b)	10.50	______	______
(c)	______	1.3×10^{-5}	______
(d)	______	5.6×10^{-10}	______
(e)	______	______	2.3×10^{-4}

43. Make the following interconversions. In each case, tell whether the solution is acidic or basic.

	pH	$[H_3O^+]$ (M)	$[OH^-]$ (M)
(a)	______	6.7×10^{-8}	______
(b)	______	______	2.2×10^{-7}
(c)	5.25	______	______
(d)	______	2.5×10^{-2}	______
(e)	______	______	5.6×10^{-3}

44. Suppose you have a 5.0×10^{-4} M solution of HCl. (a) What is the pH of the solution? (b) If you add a drop of methyl orange indicator to the solution (Table 17.3), what color will the indicator be in solution? (c) If you add a drop of phenolphthalein, what will be the color of the solution?

45. Suppose you have a 3.2×10^{-3} M solution of NaOH. (a) What is the pH of the solution? (b) What color will the solution be when you add a drop of the indicator phenolphthalein? When you add a drop of alizarin? (See Table 17.3.)

GENERAL ACID–BASE EQUILIBRIA

46. A 2.5×10^{-3} M solution of an unknown acid has a pH of 3.80 at 25 °C.
   (a) What is the hydronium ion concentration of the solution?
   (b) Is the acid a strong acid, a moderately weak acid (K_a of about 10^{-5}), or a very weak acid (K_a of about 10^{-10})?

47. A 0.015 M solution of an unknown base has a pH of 10.09.
   (a) What are the hydroxide and hydronium ion concentrations of this solution?
   (b) Is the base a strong base, a moderately weak base (K_b of about 10^{-5}), or a very weak base (K_b of about 10^{-10})?

48. A 0.015 M solution of cyanic acid has a pH of 2.67.
   (a) What is the hydronium ion concentration in the solution?
   (b) What is the ionization constant, K_a, for the acid?

49. A 0.10 M solution of $ClCH_2COOH$ (chloroacetic acid) has a pH of 1.95. Calculate K_a for the acid.

50. Butyric acid, C_3H_7COOH, is found in rancid butter and has a well deserved, malodorous reputation. In a well ventilated room, 0.20 moles of the acid are dissolved in water to give 500. mL of a solution whose pH is found to be 2.60. Assuming that butyric acid is monoprotic (it has only one ionizable proton), calculate K_a for the acid.

51. *m*-Nitrophenol, a weak acid, can be used as an indicator, since it is yellow at a pH above 8.6 and colorless at a pH below 6.8. If the pH of a 0.010 M solution of the compound is 3.44, calculate the K_a of the compound.

52. A 0.025 M solution of hydroxylamine has a pH of 9.11. What is the value of K_b for this weak base?

$$H_2NOH(aq) + H_2O(\ell) \rightleftharpoons H_3NOH^+(aq) + OH^-(aq)$$

53. Methylamine, CH_3NH_2, is a weak base.

$$CH_3NH_2(aq) + H_2O(\ell) \rightleftharpoons CH_3NH_3^+(aq) + OH^-(aq)$$

If the pH of a 0.065 M solution of the amine is 11.76, what is K_b?

54. The ionization constant of a very weak acid HA is 4.0×10^{-9}. Calculate the equilibrium concentrations of H_3O^+, A^-, and HA in a 0.040 M solution of the acid.

55. What are the equilibrium concentrations of H_3O^+, acetate ion, and acetic acid in a 0.20 M aqueous solution of acetic acid (CH_3COOH)?

56. If you have a 0.025 M solution of HCN, what are the equilibrium concentrations of H_3O^+, CN^-, and HCN? What is the pH of the solution?

57. Phenol, C_6H_5OH, is a weak organic acid.

$$C_6H_5OH(aq) + H_2O(\ell) \rightleftharpoons C_6H_5O^-(aq) + H_3O^+(aq)$$

Although somewhat toxic to humans, it is widely used as a disinfectant and in the manufacture of plastics. If you dissolve 0.780 g of the acid in enough water to make 500. mL of solution, what is the equilibrium hydrogen ion concentration? What is the pH of the solution?

58. Many organic acids have a foul odor, and propanoic acid, C_2H_5COOH, is no exception. The acid, which ionizes in water according to the following equation,

$$C_2H_5COOH(aq) + H_2O(\ell) \rightleftharpoons C_2H_5COO^-(aq) + H_3O^+(aq) \qquad K_a = 1.4 \times 10^{-5}$$

is used in the manufacture of plastics. If you dissolve 0.588 g of the acid in enough water to make 250. mL of solution, what is the equilibrium hydronium ion concentration? What is the pH of the solution?

59. The hydrogen phthalate ion is a weak acid with $K_a = 3.9 \times 10^{-6}$.

$$C_8H_5O_4^-(aq) + H_2O(\ell) \rightleftharpoons C_8H_4O_4^{2-}(aq) + H_3O^+(aq)$$

If you dissolve 2.55 g of potassium hydrogen phthalate, $KC_8H_5O_4$, in enough water to have 250. mL of solution, what is the pH of the solution?

60. Barbituric acid, $C_4H_4N_2O_3$, has a K_a of 9.9×10^{-5}, while that of nicotinic acid is 1.4×10^{-5}. (a) Which is the stronger acid? (b) For a 0.010 M solution of each of these monoprotic acids, which will have the higher pH?

61. Place the following acids (a) in order of increasing strength and (b) in order of increasing pH, assuming you have a 0.10 M solution of each acid.
   (a) Valeric acid, $K_a = 1.5 \times 10^{-5}$

(b) Glutaric acid, $K_a = 3.4 \times 10^{-4}$
(c) Hypobromous acid, $K_a = 2.5 \times 10^{-9}$

62. A monoprotic acid HX has $K_a = 1.3 \times 10^{-3}$. Calculate the equilibrium concentration of HX and H_3O^+ and pH for a 0.010 M solution of the acid.

63. A hypothetical weak base MOH has $K_b = 5.0 \times 10^{-4}$ for the reaction

$$MOH(aq) \rightleftharpoons M^+(aq) + OH^-(aq)$$

Calculate the equilibrium concentrations of MOH, M^+, and OH^- in a 0.15 M solution of MOH.

64. What are the equilibrium concentrations of NH_3, NH_4^+, and OH^- in a 0.15 M solution of aqueous ammonia? What is the pH of the solution?

65. The weak base methylamine, CH_3NH_2, has $K_b = 5.0 \times 10^{-4}$. It reacts with water according to the equation

$$CH_3NH_2(aq) + H_2O(\ell) \rightleftharpoons CH_3NH_3^+(aq) + OH^-(aq)$$

Calculate the equilibrium hydroxide ion concentration in a 0.25 M solution of the base. What are the pH and pOH of the solution?

66. Calculate the pH of a 0.12 M aqueous solution of the base aniline, $C_6H_5NH_2$ ($K_b = 4.2 \times 10^{-10}$).

67. Hydroxylamine, NH_2OH, has a K_b of 6.6×10^{-9}. What are the pH and pOH of a 0.051 M solution of the base?

68. Saccharin, $C_7H_5NO_3S$, is a monoprotic acid in aqueous solution. It has an ionization constant K_a of 2.1×10^{-12}. Calculate the pH of a solution made by dissolving 1.50 g in enough water to make 500. mL of solution. (This is close to a saturated solution of saccharin.)

69. *p*-Nitrophenol, $O_2NC_6H_4OH$, is a monoprotic acid with an ionization constant of 7.0×10^{-8}. Calculate the pH of a solution made by dissolving 125 mg of the compound in enough water to make 250. mL of solution.

70. It was found that a solution of hydrofluoric acid, HF, had a pH of 2.30. Calculate the equilibrium concentrations of HF, F^-, and H_3O^+, and calculate the amount of HF originally dissolved per liter.

71. Calculate the pH of a 0.0010 M aqueous solution of HF.

72. If each of the salts listed below was dissolved in water to give a 0.10 M solution, which solution would have the highest pH? Which would have the lowest pH?
(a) Na_2S (c) NaF
(b) Na_3PO_4 (d) $Na(CH_3COO)$

73. Which of the following common food additives would give a basic solution when dissolved in water?
(a) $NaNO_3$ (used as a meat preservative)
(b) $NaC_7H_5O_2$ (sodium benzoate; used as a soft drink preservative)
(c) Na_2HPO_4 (used as an emulsifier in the manufacture of some pasteurized cheeses)

74. For each salt below, predict whether an aqueous solution will have a pH less than, equal to, or greater than 7.
(a) $NaHSO_4$ (c) $NaNO_3$
(b) NH_4Br (d) Na_2HPO_4

75. For each salt below, predict whether an aqueous solution will have a pH less than, equal to, or greater than 7.
(a) $AlCl_3$ (c) $KClO_4$
(b) Na_2S (d) NaH_2PO_4

76. For each salt below, predict whether an aqueous solution will have a pH less than, equal to, or greater than 7.
(a) NH_4NO_3 (c) $K(CH_3COO)$
(b) Na_2CO_3 (d) LiBr

77. For each salt below, predict whether an aqueous solution will have a pH less than, equal to, or greater than 7.
(a) KCN (c) $MgCl_2$
(b) $FeCl_3$ (d) $(NH_4)_2S$

78. Calculate the equilibrium hydronium ion concentration and pH in a 0.20 M solution of the salt ammonium chloride, NH_4Cl.

79. Calculate the hydronium ion concentration and pH in a 0.015 M solution of the salt sodium acetate, $Na(CH_3COO)$.

80. Sodium cyanide is the salt of the weak acid HCN. Calculate the concentrations of H_3O^+, OH^-, HCN, and Na^+ in a solution prepared by dissolving 10.8 g of NaCN in 500. mL of pure water at 25 °C.

81. The sodium salt of propanoic acid, $NaC_3H_5O_2$, is used as an antifungus agent by veterinarians. Calculate the equilibrium concentrations of H_3O^+ and OH^-, and the pH, for a solution of 0.10 M $NaC_3H_5O_2$. (See Table 17.4 for the properties of the $C_3H_5O_2^-$ ion.)

82. About this time, you may be wishing you had an aspirin. Aspirin is an organic acid with a K_a of 3.27×10^{-4} for the reaction

$$HC_9H_7O_4(aq) + H_2O(\ell) \rightleftharpoons C_9H_7O_4^-(aq) + H_3O^+(aq)$$

If you have two tablets, each containing 0.325 g of aspirin (mixed with a neutral "binder" to hold the tablet together), and dissolve them in a glass of water (200. mL), what is the pH of the solution?

83. Lactic acid, $C_3H_6O_3$, occurs in sour milk as a result of lactic acid bacteria. What is the pH of a solution of 56 mg of lactic acid in 250 mL of water? K_a for lactic acid $= 1.4 \times 10^{-4}$. (See Example 17.5.)

POLYPROTIC ACIDS AND BASES

84. H_2SO_3, sulfurous acid, is a weak acid capable of providing two H^+ ions.
    (a) What is the pH of a 0.45 M solution of H_2SO_3?
    (b) What is the equilibrium concentration of the sulfite ion, SO_3^{2-}, in the 0.45 M solution of H_2SO_3?
85. Ascorbic acid (vitamin C, $C_6H_8O_6$) is a diprotic acid ($K_{a1} = 7.9 \times 10^{-5}$ and $K_{a2} = 1.6 \times 10^{-12}$). What is the pH of a solution that contains 5.0 mg of acid per mL of water?
86. Hydrazine, N_2H_4, can interact with water in two stages.

$$N_2H_4(aq) + H_2O(\ell) \rightleftharpoons N_2H_5^+(aq) + OH^-(aq) \qquad K_{b1} = 8.5 \times 10^{-7}$$

$$N_2H_5^+(aq) + H_2O(\ell) \rightleftharpoons N_2H_6^{2+}(aq) + OH^-(aq) \qquad K_{b2} = 8.9 \times 10^{-16}$$

    (a) What are the concentrations of OH^-, $N_2H_5^+$, and $N_2H_6^{2+}$ in a 0.010 M aqueous solution of hydrazine?
    (b) What is the pH of the 0.010 M solution of hydrazine?
87. Ethylenediamine, $H_2N—C_2H_4—NH_2$, can interact with water in two steps, giving OH^- in each step. (See Appendix H.) Calculate the concentrations of OH^- and $[H_3N—C_2H_4—NH_3]^{2+}$ in a 0.15 M aqueous solution of the amine.

LEWIS ACIDS AND BASES

88. For each substance listed below, tell whether it is a Lewis acid or base:
    (a) Mn^{2+}
    (b) $:NH_2(CH_3)$
    (c) hydroxylamine, H_2NOH
    (d) SO_2 in the reaction $SO_2(g) + BF_3(g) \rightleftharpoons O_2S:BF_3(s)$
    (e) $Zn(OH)_2$ in the reaction $Zn(OH)_2(s) + 2\ OH^-(aq) \rightleftharpoons [Zn(OH)_4]^{2-}(aq)$
89. For each substance listed below, tell whether it is a Lewis acid or base:
    (a) BCl_3
    (b) $H_2N—NH_2$, hydrazine
    (c) CN^- in the reaction $Au^+(aq) + 2\ CN^-(aq) \rightleftharpoons [Au(CN)_2]^-(aq)$
90. Trimethylamine, $(H_3C)_3N:$, is a common reagent. It interacts readily with diborane, B_2H_6. The latter dissociates to BH_3, and this forms a complex with the amine, $(H_3C)_3N:BH_3$. Is the BH_3 fragment a Lewis acid or a Lewis base?
91. Carbon monoxide, $:C{\equiv}O:$, forms complexes with low-valent metals. For example, $Ni(CO)_4$ and $Fe(CO)_5$ are well known. CO also forms complexes with the iron of hemoglobin, preventing the hemoglobin from acting in its normal way, that is, taking up oxygen. Is CO a Lewis acid or a Lewis base?
92. Draw a Lewis dot structure of ICl_3. Does it function as a Lewis acid or base in reacting with Cl^- to form ICl_4^-? What are the likely molecular structures of ICl_3 and ICl_4^-?
93. Is $Fe(CO)_5$ best classed as as Lewis acid, Lewis base, Brønsted acid, or Brønsted base in the following reaction (which occurs in a nonaqueous solvent)?

$$Fe(CO)_5 + H^+ \rightleftharpoons [HFe(CO)_5]^+$$

FORMATION CONSTANTS

94. If you have a 0.1 M solution of Cd^{2+}, which should you add, NH_3 or CN^-, to reduce the Cd^{2+} ion concentration to the smallest possible value? Formation constants for complex ions of Cd^{2+} are as follows: 1.0×10^7 for $[Cd(NH_3)_4]^{2+}$; 1.3×10^{17} for $[Cd(CN)_4]^{2-}$.
95. If you have a 0.1 M solution of Al^{3+}, which should you add, OH^- or F^-, to reduce the Al^{3+} ion concentration to the smallest possible value? Formation constants for the complex ions formed between Al^{3+} and OH^- or F^- are given in Appendix H.
96. What is the concentration of $Cu^{2+}(aq)$ in a solution that was originally 0.050 M $Cu^{2+}(aq)$ and 3.0 M $NH_3(aq)$?
97. What are the concentrations of $[CdCl_4]^{2-}$, Cd^{2+}, and Cl^- in a solution that was originally 0.010 M in Cd^{2+} and 1.0 M in Cl^-?
98. Silver and gold often occur together in nature, and both are dissolved as cyanide complex ions when treated with oxygen and CN^-. If you have a 0.010 M aqueous solution of $[Ag(CN)_2]^-$, what is the concentration of silver ion, Ag^+, in solution? $K_{formation}$ for $[Ag(CN)_2]^-$ from Ag^+ and CN^- is 5.6×10^{18}.
99. Copper(I) is best isolated in the form of complex ions. If you have a 0.015 M solution of $[CuCl_2]^-$, what is the concentration of $Cu^+(aq)$ and $Cl^-(aq)$?

GENERAL PROBLEMS

100. If a hydride ion (H^-) salt, NaH, is put in water, it reacts almost explosively to form $H_2(g)$. Write a balanced equation showing this. Is the resulting solution acidic or basic?
101. Liquid ammonia auto-ionizes just as water does.
    (a) Write a balanced equation for the auto-ionization process of liquid ammonia.
    (b) What is the conjugate acid of NH_3? The conjugate base?
    (c) If $NaNH_2$ is dissolved in liquid ammonia, is it an acid or a base?

102. For each substance listed below, tell whether you expect it to behave as a Brønsted acid or base, and give its conjugate partner. (See Tables 17.1 and 17.2.)
(a) Br^- (c) H_3PO_4
(b) $[Al(H_2O)_6]^{3+}$ (d) CH_3COO^-

103. For each substance listed below, tell whether you expect it to behave as a Brønsted acid or base, and give its conjugate partner. (See Tables 17.1 and 17.2.)
(a) PO_4^{3-} (d) CO_3^{2-}
(b) H^- (e) OH^-
(c) NH_4^+

104. For each reaction below, predict whether the equilibrium lies predominantly to the left or to the right.
(a) $CO_3^{2-}(aq) + 2\ HCl(aq) \rightleftharpoons H_2CO_3(aq) + 2\ Cl^-(aq)$
(b) $NH_2^-(aq) + H_2O(\ell) \rightleftharpoons NH_3(aq) + OH^-(aq)$
(c) $C_6H_5OH(aq) + SO_4^{2-}(aq) \rightleftharpoons C_6H_5O^-(aq) + HSO_4^-(aq)$
(d) $NH_3(aq) + HCO_3^-(aq) \rightleftharpoons NH_4^+(aq) + CO_3^{2-}(aq)$

105. The base $Ca(OH)_2$ is almost insoluble in water; only 0.50 g can be dissolved in 1.0 L of water at 25 °C. If the dissolved substance is completely dissociated into its constituent ions, what is the pH of the solution?

106. The ionization constant, K_w, for water at body temperature (37 °C) is 2.5×10^{-14}. What is the pH of a neutral solution at this temperature?

107. Ethanolamine is a base with a K_b of 2.8×10^{-5}. A closely related base, ethylamine, has a K_b of 5.6×10^{-4}. (a) If you have a 0.10 M solution of each, which has the higher pH? (b) Which of the two bases is stronger? Calculate the pH of its 0.10 M solution.

108. Chloroacetic acid, $ClCH_2COOH$, is a moderately weak acid ($K_a = 1.40 \times 10^{-3}$). Suppose you dissolve 94.5 mg of the acid in 100. mL of water. What is the pH of the solution?

109. To what volume should 100. mL of any weak acid HA with a concentration 0.20 M be diluted in order to double the percentage dissociation?

110. Given the following solutions:
0.1 M NH_3
0.1 M Na_2CO_3
0.1 M NaCl
0.1 M CH_3COOH
0.1 M NH_4Cl
0.1 M $Na(CH_3COO)$ (sodium acetate)
(a) Which of the solutions are acidic?
(b) Which of the solutions are basic?
(c) Which of the solutions is most acidic?

111. Arrange the following 0.1 M solutions in order of increasing pH.
(a) NaCl (d) HCl
(b) NH_4Cl (e) KOH
(c) $Na(CH_3COO)$ (sodium acetate)

112. Arrange the following 1.0 M solutions in order of increasing pH.
(a) NaCl (d) HCl
(b) NH_3 (e) NaOH
(c) NaCN (f) CH_3COOH

113. Which of the following are bases according to the Arrhenius concept and which are better described as bases by the Brønsted concept?
(a) NaOH (d) KCN
(b) Na_2CO_3 (e) sodium ethoxide, NaC_2H_5O
(c) $Ca(OH)_2$

114. Nicotinic acid, $C_6H_5NO_2$, is found in minute amounts in all living cells, but there are appreciable amounts in liver, yeast, milk, adrenal glands, white meat, and corn. Whole wheat flour contains about 60. micrograms per gram of flour. One gram of the acid dissolves in 60. mL of water and gives a pH of 2.70. What is the approximate value of K_a for the acid?

115. Oxalic acid is a relatively weak acid capable of losing two protons. Calculate the equilibrium constant for the overall reaction from K_{a1} and K_{a2}.

$$H_2C_2O_4(aq) + 2\ H_2O(\ell) \rightleftharpoons C_2O_4^{2-}(aq) + 2\ H_3O^+(aq)$$

116. Nicotine, $C_{10}H_{14}N_2$, has two basic nitrogen atoms, and both can react with water to give a basic solution.

$$Nic(aq) + H_2O(\ell) \rightleftharpoons NicH^+(aq) + OH^-(aq)$$

$$NicH^+(aq) + H_2O(\ell) \rightleftharpoons NicH_2^{2+}(aq) + OH^-(aq)$$

K_{b1} is 7.0×10^{-7} and K_{b2} is 1.1×10^{-10}. Calculate the approximate pH of a 0.020 M solution.

117. Sodium benzoate, $Na(C_6H_5COO)$, is soluble in water and is commonly used as a food preservative. If 9.0 g of the salt are dissolved in sufficient water to make exactly 500. mL of solution, what is the pH of the solution?

118. The thiosulfate complex of silver is important in developing photographs.

$$Ag^+(aq) + 2\ S_2O_3^{2-}(aq) \rightleftharpoons [Ag(S_2O_3)_2]^{3-} \quad K = 2.0 \times 10^{13}$$

If 250 mL of 0.0010 M $AgNO_3$ is mixed with 250 mL of 5.0 M $Na_2S_2O_3$, calculate:
(a) the original concentrations of Ag^+ and $S_2O_3^{2-}$
(b) the concentrations of Ag^+ and $[Ag(S_2O_3)_2]^{3-}$ at equilibrium.

119. Silver cyanide can react in two ways:

$$AgCN(s) + CN^-(aq) \rightleftharpoons [Ag(CN)_2]^-(aq)$$

$$AgCN(s) + H_3O^+(aq) \rightleftharpoons Ag^+(aq) + HCN(aq) + H_2O(\ell)$$

Which reaction shows AgCN as a Lewis acid? In which reaction is it behaving as a Brønsted base? Is AgCN amphiprotic or amphoteric?

120. Acetic acid is a weak Brønsted acid. However, it is interesting to compare the strength of this acid with those of a related series of acids in which the H atoms of the CH_3 group are replaced by Cl.

ACID	K_a
CH_3COOH	1.8×10^{-5}
$ClCH_2COOH$	1.4×10^{-3}
$Cl_2CHCOOH$	3.3×10^{-2}
Cl_3CCOOH	2.0×10^{-1}

What trend in acid strength do you observe as H is successively replaced by Cl? Can you suggest a reason for this trend in terms of charge accommodation by the conjugate base ion?

121. Let us consider a salt of a weak base and a weak acid such as ammonium cyanide. Both NH_4^+ and CN^- ions hydrolyze in aqueous solution, but the net reaction can be considered as a proton transfer from NH_4^+ to CN^-.

$$NH_4^+(aq) + CN^-(aq) \rightleftharpoons NH_3(aq) + HCN(aq)$$

(a) Show that the equilibrium constant for this reaction, K_{total}, is

$$K_{total} = \frac{K_w}{K_a K_b}$$

where K_a is the dissociation constant for the weak acid HCN and K_b is the constant for the weak base NH_3.

(b) Prove that the hydronium ion concentration in this solution must be given by

$$[H_3O^+] = \sqrt{\frac{K_w K_a}{K_b}}$$

(c) What is the pH of a 0.15 M solution of ammonium cyanide?

SUMMARY QUESTIONS

122. Not only can one determine formation constants for complex ions, but it is also possible to measure them for complexes of neutral Lewis acids. For example, $K_{formation}$ (= K_p) for the dimethyl ether complex of BF_3, $(CH_3)_2O{:}\rightarrow BF_3$, is 5.8.

$$BF_3(g) + (CH_3)_2O(g) \rightleftharpoons (CH_3)_2O{:}\rightarrow BF_3(g)$$

(a) Tell which reactant is the Lewis acid and which is the Lewis base.

(b) What is the F—B—F angle in BF_3? In $(CH_3)_2O{:}\rightarrow BF_3$?

(c) If you place 1.00 g of the complex in a 565-mL flask at 25 °C, what is the total pressure in the flask? What are the partial pressures of the Lewis acid, the Lewis base, and the complex?

123. Sulfanilic acid, which is used in making dyes, is made by the reaction of aniline with sulfuric acid.

(a) If you want to prepare 150.0 g of sulfanilic acid, and you expect only an 85% yield, how many milliliters of aniline should you take as a starting material (density of aniline = 1.02 g/mL)?

(b) Give approximate values for the following bond angles in sulfanilic acid: (i) C—N—H; (ii) H—N—H; (iii) O—S—C; and (iv) O—S—O.

(c) The acid has a K_a value of 5.9×10^{-4}. The sodium salt of the acid, $H_2NC_6H_4SO_3Na$, is quite soluble in water. If you dissolve 1.25 g of the salt in 100. mL of water, what is the pH of the solution?

$$\underset{\text{aniline}}{H_2SO_4 + C_6H_5NH_2} \longrightarrow \underset{\text{sulfanilic acid}}{H_2N\text{—}C_6H_4\text{—}SO_2\text{—}O\text{—}H} + H_2O$$

C H A P T E R

18

Reactions Between Acids and Bases

Ammonium chloride (NH_4Cl) is heated in a spoon and breaks down into the base NH_3 and the acid HCl. As these vapors rise from the heated solid, they recombine to give a white fog of NH_4Cl.

CHAPTER OUTLINE

In Chapter 17 we began to explore the chemistry of acids and bases. Reactions of acids with bases are all around you and within you. The pH of the oceans and of your blood is controlled by the chemistry of carbonic acid. Well over 50 million tons of fertilizers such as ammonium nitrate, made by reaction of the base ammonia with nitric acid, are manufactured every year. The acid in "acid rain" comes from the interaction of airborne nonmetal oxides such as CO_2, SO_2, and nitrogen oxides with water, and this acid rain falls to the earth where it is neutralized by reaction with minerals.

This chapter continues the exploration of acid–base chemistry with particular emphasis on the results of acid–base reactions, the control of such reactions, and some reactions of practical concern.

18.1 ACID–BASE REACTIONS

Using tables such as Table 17.2 and Table 17.4, the direction of an acid–base reaction can be predicted. For example, acetic acid (CH_3COOH) should react to a significant extent with ammonia in water.

$$CH_3COOH(aq) + NH_3(aq) \rightleftharpoons NH_4^+(aq) + CH_3COO^-(aq)$$

Remember that reactions always proceed in the direction of the weaker acid–base pair, and you find here that NH_4^+ is a weaker acid than CH_3COOH and that CH_3COO^- is a weaker base than NH_3. We can verify this because the overall reaction is the sum of three other reactions whose values of K are known:

Figure 18.1 A commercial remedy for excess stomach acid. The bubbles are carbon dioxide, CO_2, from the reaction between a Brønsted acid (citric acid, $C_6H_8O_7$) and a Brønsted base (HCO_3^-, from sodium hydrogen carbonate).

Ionization of the acid:

$$CH_3COOH(aq) + H_2O(\ell) \rightleftharpoons H_3O^+(aq) + CH_3COO^-(aq) \qquad K_a = 1.8 \times 10^{-5}$$

Ionization of the base:

$$H_2O(\ell) + NH_3(aq) \rightleftharpoons NH_4^+(aq) + OH^-(aq) \qquad K_b = 1.8 \times 10^{-5}$$

Union of hydronium ion and hydroxide ion:

$$H_3O^+(aq) + OH^-(aq) \rightleftharpoons 2\ H_2O(\ell) \qquad K = \frac{1}{K_w} = 1.0 \times 10^{14}$$

Net reaction:

$$CH_3COOH(aq) + NH_3(aq) \rightleftharpoons NH_4^+(aq) + CH_3COO^-(aq) \qquad K_{net} = \frac{K_aK_b}{K_w} = 3.2 \times 10^4$$

In Section 16.2 you learned that, when equations are added to find a net equation, K_{net} is the product of the K values for all of the summed equations. In this case K_{net} is fairly large—32,000. It is large because, although the acid and base are both weak, the H_3O^+ and OH^- ions that they produce are very strong and are "swept up" by the water-formation reaction with its extraordinarily high equilibrium constant of 10^{14}. Thus, Le Chatelier's principle causes the acid and base ionization reactions to shift much further to the right than either would go if it occurred alone, and the overall process has a very large equilibrium constant.

The fairly strong formation of NH_4^+ and CH_3COO^- is actually driven by coupling the CH_3COOH and NH_3 ionization reactions to a third, strongly driven reaction. The coupling happens because some of the products of one reaction are reactants in a second reaction—in this case the formation of H_2O—thereby removing these first products as they are formed. This kind of "coupling reaction" is very common in living systems and is one of the very important roles of ATP in our bodies.

The reaction of acetic acid and ammonia is a good example of a weak acid reacting with a weak base, and so is the reaction of citric acid and bicarbonate in Figure 18.1. However, these represent only one of four possible types of acid–base reactions in aqueous solution.

ACID	BASE
Strong	Strong
Weak	Weak

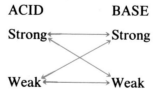

Each of these combinations will be considered in turn.

THE REACTION OF A STRONG ACID WITH A STRONG BASE

Strong acids and bases are effectively 100% ionized in solution (Section 17.4). Therefore, if we mix HCl and NaOH, we can write the equation

$$H_3O^+(aq) + Cl^-(aq) + Na^+(aq) + OH^-(aq) \longrightarrow 2\ H_2O(\ell) + Na^+(aq) + Cl^-(aq)$$

which leads to the *net ionic equation*

$$H_3O^+(aq) + OH^-(aq) \longrightarrow 2\ H_2O(\ell) \qquad K = \frac{1}{K_w} = 1.0 \times 10^{14}$$

As you learned in Chapter 5, the net ionic equation for the reaction of any strong base with any strong acid will always be simply the union of hydronium ion and hydroxide ion to give water. The enormous value of K for the reaction shows that it is, for all practical purposes, quantitatively complete. Thus, if equal numbers of moles of NaOH and HCl are mixed, the result is just a solution of NaCl in water. Since the constituents of NaCl, Na^+ and Cl^- ions, arise from a strong base and a strong acid, respectively, they produce a neutral aqueous solution (Table 17.5). Because of this, reactions of strong acids and bases have come to be called "neutralizations."

THE REACTION OF A STRONG BASE WITH A WEAK ACID

Nature is full of weak acids. Let us take as an example formic acid, HCOOH, and its reaction with the strong base NaOH. The overall reaction can again be considered the sum of several processes.

Naturally occurring acids include oxalic acid (present as the potassium or calcium salt in plants of the Oxalis family in particular), malic acid (in apples and other fruits), and succinic acid (in fungi and lichens).

Ionization of the acid:

$$HCOOH(aq) + H_2O(\ell) \rightleftharpoons H_3O^+(aq) + HCOO^-(aq)$$

$$K_a = 1.8 \times 10^{-4}$$

Union of H_3O^+ from the acid and OH^- from the completely dissociated base:

$$H_3O^+(aq) + OH^-(aq) \rightleftharpoons 2\ H_2O(\ell) \qquad K = \frac{1}{K_w} = 1.0 \times 10^{14}$$

Net reaction:

$$HCOOH(aq) + OH^-(aq) \rightleftharpoons H_2O(\ell) + HCOO^-(aq)$$

$$K_{net} = \frac{K_a}{K_w} = 1.8 \times 10^{10}$$

Once again the equilibrium constant for the overall process is large, so the net reaction goes essentially to completion. Assuming that equal molar quantities of acid and base were mixed, the final solution contains sodium formate (NaHCOO), a salt that is 100% dissociated in water. However, while the Na^+ ion is the cation of a strong base and so gives a neutral solution, the formate ion is the anion of a weak acid and is therefore a base (Table 17.4). The solution will be basic because of the hydrolysis of the formate ion.

$$H_2O(\ell) + HCOO^-(aq) \rightleftharpoons HCOOH(aq) + OH^-(aq)$$

$$K_b \text{ (from Table 17.4)} = 5.6 \times 10^{-11}$$

There are two important conclusions we can draw from this example.

1. Mixing equal molar quantities of a strong base and a weak acid will produce a salt in which the anion is the conjugate base of the weak acid. *The solution will be basic*, with the pH depending on K_b for the anion and its concentration.
2. The net reaction between a strong base and a weak acid is simply the reverse of the reaction that defines K_b for the conjugate base of the

*In Chapter 5 you learned that the amount of acid in solution, for example, can be determined by titration with a base (Figure 5.19). The point at which the moles of OH^- supplied by the base is equal to the moles of H_3O^+ that can be supplied by the acid is called the **equivalence point**.*

weak acid; K for this net reaction is simply $1/K_b$ for the conjugate base of the weak acid (here $K_{net} = 1.8 \times 10^{10} = 1/K_b = 1/5.6 \times 10^{-11}$).

EXAMPLE 18.1

pH AT THE EQUIVALENCE POINT OF A STRONG BASE/WEAK ACID REACTION

Suppose you mix 50. mL of 0.10 M NaOH with 50. mL of 0.10 M formic acid, HCOOH. What is the pH of the resulting solution? (Assume the solution volumes give 100. mL when mixed.)

Solution The net ionic equation for the acid–base reaction that occurs is

$$HCOOH(aq) + OH^-(aq) \rightleftharpoons H_2O(\ell) + HCOO^-(aq)$$

Be sure to notice the strategy here. The large value of K for the reaction of a strong base and weak acid leads to the assumption that the reaction mixture initially contains only the salt of the weak acid. This allows us to find the concentration of the salt. Then, using this concentration, we take into account the hydrolysis of the salt in order to find the actual pH of the solution. This strategy is generally followed in all of the situations in this chapter.

Here we are mixing 0.0050 mol of HCOOH (0.050 L × 0.10 M) with 0.0050 mol of NaOH, and the product must be 0.0050 mol of NaHCOO. Since 50. mL of each reagent was mixed, this means the salt NaHCOO is dissolved in 100. mL of water. The concentration of the salt is therefore [NaHCOO] = 0.050 M.

With the salt concentration known, we can solve for the pH of the solution. According to conclusion 1 above, the solution will be basic due to the hydrolysis of the conjugate base of formic acid, according to the equation

$$H_2O(\ell) + HCOO^-(aq) \rightleftharpoons HCOOH(aq) + OH^-(aq)$$

	$[HCOO^-]$	$[HCOOH]$	$[OH^-]$
Before $HCOO^-$ hydrolysis (M)	0.050	0	0
Change on proceeding to equilibrium	$-x$	$+x$	$+x$
At equilibrium (M)	$0.050 - x$	x	x

Therefore, we can write

$$K_b = 5.6 \times 10^{-11} = \frac{[HCOOH][OH^-]}{[HCOO^-]} = \frac{(x)(x)}{0.050 - x} \approx \frac{x^2}{0.050}$$

As in Chapter 17 (Example 17.9), we can make the approximation that $(0.050 - x) \approx 0.050$, and so the hydroxide ion concentration is

$$[OH^-] = 1.7 \times 10^{-6} \text{ M}$$

In Example 18.1 $[OH^-]$ is given as 0 before equilibrium is established. It is actually 10^{-7} M, but, as described in Chapter 17, hydroxide and hydronium ion from water auto-ionization can generally be neglected in calculations of this type (when the pH is outside the 6–8 range). This same simplification is made in the other examples in this chapter.

This gives a H_3O^+ concentration of 5.9×10^{-9} M and a pH of 8.23. The solution is indeed basic as predicted by the fact that the product was the formate ion, $HCOO^-$, the conjugate base of a weak acid.

As a final comment, note that the approximation that $(0.050 - x) \approx 0.050$ is valid. Subtracting 1.7×10^{-6} from 0.050 would have made no difference in the final answer.

EXERCISE 18.1 pH of the Solution When an Acid–Base Reaction Is Completed

How many milliliters of 0.100 M NaOH are required to react completely with 0.976 g of the weak, monoprotic acid benzoic acid (C_6H_5COOH)? What is the pH of the solution after reaction? (See Table 17.4 for the K_b for the benzoate ion, $C_6H_5COO^-$.)

There are several weak polyprotic acids, such as H_2SO_3, H_3PO_4, and H_2CO_3. Reaction of NaOH with these acids occurs stepwise. Carbonic acid, which is explored more in Chapter 19, is an especially important example. If a strong base reacts with this acid, we find that the first step goes nearly to completion before the second begins.

$$H_2CO_3(aq) + OH^-(aq) \rightleftharpoons HCO_3^-(aq) + H_2O(\ell)$$

$$HCO_3^-(aq) + OH^-(aq) \rightleftharpoons CO_3^{2-}(aq) + H_2O(\ell)$$

If one mole of OH^- ions is added per mole of H_2CO_3, the product is one mole of the hydrogen carbonate ion, HCO_3^-. If the second mole of OH^- is added subsequently, the final product is the carbonate ion, CO_3^{2-}.

THE REACTION OF A STRONG ACID WITH A WEAK BASE

A good example of a reaction of this type is that between the strong acid HCl and the weak base ammonia. Here we assume that HCl is 100% ionized in solution and that the H_3O^+ ion produced by the acid reacts with the OH^- ion from the weak base.

Ionization of the base:

$$NH_3(aq) + H_2O(\ell) \rightleftharpoons NH_4^+(aq) + OH^-(aq) \qquad K_b = 1.8 \times 10^{-5}$$

Union of OH^- from the base and H_3O^+ from the completely dissociated acid:

$$H_3O^+(aq) + OH^-(aq) \rightleftharpoons 2\ H_2O(\ell) \qquad K = \frac{1}{K_w} = 1.0 \times 10^{14}$$

Net reaction:

$$H_3O^+(aq) + NH_3(aq) \rightleftharpoons H_2O(\ell) + NH_4^+(aq)$$

$$K_{net} = \frac{K_b}{K_w} = 1.8 \times 10^9$$

Just as in the "strong–weak" combination above, the overall equilibrium constant is quite large, and the reaction proceeds essentially to completion. After reaction of equal molar quantities of acid and base, the solution contains the salt ammonium chloride. Does this hydrolyze to give an acidic, basic, or neutral solution? You can tell by looking at each ion. The Cl^- ion is the conjugate base of the strong acid HCl and so gives a neutral solution (Table 17.5). However, the NH_4^+ ion is the conjugate acid of the weak base NH_3, so it produces an acidic solution.

$$H_2O(\ell) + NH_4^+(aq) \rightleftharpoons H_3O^+(aq) + NH_3(aq)$$

K_a for NH_4^+ (Table 17.4) $= 5.6 \times 10^{-10}$

In general, *a strong acid and a weak base will give an acidic solution when equal molar quantities are mixed.* Notice once again that the overall reaction is the reverse of the reaction that defines K_a for the conjugate acid of the weak base, so its K is just $1/K_a$ for the conjugate acid of the weak base in the reaction. Here $K_a = 5.6 \times 10^{-10}$, so $K_{net} = 1.8 \times 10^9 = 1/(5.6 \times 10^{-10})$.

EXAMPLE 18.2

pH AT THE EQUIVALENCE POINT OF A STRONG ACID/WEAK BASE REACTION

Suppose you mix exactly 100 mL of 0.10 M HCl with exactly 50 mL of 0.20 M NH_3. What is the pH of the resulting solution?

Once again notice the strategy used to find the solution pH. The large value of K for the reaction of a strong acid and weak base leads to the assumption that the reaction mixture initially contains only the salt of the weak base. This allows us to find the concentration of the salt. Then, using this concentration, we take into account the hydrolysis of the salt in order to find the actual pH of the solution. (See also page 756.)

Solution The balanced, net ionic equation for the HCl/NH_3 reaction is

$$H_3O^+(aq) + NH_3(aq) \rightleftharpoons H_2O(\ell) + NH_4^+(aq)$$

As explained in the text, the NH_4^+ ion of the product is a weak Brønsted acid. This means the pH of the solution depends on K_a for this ion and on its concentration. The latter is found as follows:

$$\left.\begin{array}{l}(0.100\text{ L})(0.10\text{ M HCl}) = 0.010\text{ mol HCl used}\\(0.050\text{ L})(0.20\text{ M NH}_3) = 0.010\text{ mol NH}_3\text{ used}\end{array}\right\}\longrightarrow 0.010\text{ mol NH}_4\text{Cl produced}$$

The reactants were present initially in equal amounts (0.010 mol), so 0.010 mol of NH_4Cl must have been produced. Since the solution volume after reaction is the total of the volumes of the reacting solutions, exactly 150 mL, the product ion concentrations are

$$[NH_4^+] = [Cl^-] = 0.010\text{ mol}/0.150\text{ L} = 0.067\text{ M}$$

Now we can solve for the hydronium ion concentration of the solution by using the value of K_a for ionization of NH_4^+. The equilibrium reaction is

$$H_2O(\ell) + NH_4^+(aq) \rightleftharpoons H_3O^+(aq) + NH_3(aq)$$

	$[NH_4^+]$	$[NH_3]$	$[H_3O^+]$
Before NH_4^+ hydrolysis (M)	0.067	0	0
Change on proceeding to equilibrium	$-x$	$+x$	$+x$
At equilibrium (M)	$0.067 - x$	x	x

Substituting the equilibrium concentrations into the expression for K_a, we have

$$K_a = 5.6 \times 10^{-10} = \frac{[NH_3][H_3O^+]}{[NH_4^+]} = \frac{(x)(x)}{0.067 - x} \approx \frac{x^2}{0.067}$$

As in Example 18.1, we can make the approximation that $(0.067 - x) \approx 0.067$, and so the hydronium ion concentration is

$$[H_3O^+] = 6.1 \times 10^{-6}\text{ M}$$

and the pH is 5.21. The solution is clearly acidic, as anticipated for a solution containing the salt of a weak base.

Once again note that the approximation that $(0.067 - x) \approx 0.067$ is valid. Subtracting 6.1×10^{-6} from 0.067 would indeed have made no difference in the final answer.

Table 18.1 Characteristics of Acid–Base Reactions

Type	Net Ionic Equation	K	Species Present after Equal Molar Amounts Are Mixed (Solution pH)
Strong acid–strong base HCl + NaOH	$H_3O^+(aq) + OH^-(aq) \rightleftharpoons 2\ H_2O(\ell)$	1.0×10^{14}	Na^+, Cl^- (neutral, pH = 7)
Weak acid–strong base HCOOH + NaOH	$HCOOH(aq) + OH^-(aq) \rightleftharpoons HCOO^-(aq) + H_2O(\ell)$	1.8×10^{10}	Na^+, $HCOO^-$ (basic, pH > 7)
Strong acid–weak base HCl + NH_3	$H_3O^+(aq) + NH_3(aq) \rightleftharpoons NH_4^+(aq) + H_2O(\ell)$	1.8×10^9	Cl^-, NH_4^+ (acidic, pH < 7)

EXERCISE 18.2 pH of the Solution When an Acid–Base Reaction Is Completed

Aniline, $C_6H_5NH_2$, is a weak organic base first discovered in 1826 when the deep blue dye indigo was heated. If you mix exactly 50 mL of 0.20 M HCl with 0.93 g of aniline, are the acid and base completely consumed? What is the pH of the resulting solution? (The K_a for $C_6H_5NH_3^+$ is 2.4×10^{-5}.)

$$HCl(aq) + C_6H_5NH_2(aq) \rightleftharpoons C_6H_5NH_3^+(aq) + Cl^-(aq)$$

There are some weak bases, especially anions, that are capable of reacting with more than one proton. The carbonate ion is an important example.

$$H_3O^+(aq) + CO_3^{2-}(aq) \rightleftharpoons H_2O(\ell) + HCO_3^-(aq) \qquad K = 1/K_a \text{ for } HCO_3^-$$

$$H_3O^+(aq) + HCO_3^-(aq) \rightleftharpoons H_2O(\ell) + H_2CO_3(aq) \qquad K = 1/K_a \text{ for } H_2CO_3$$

$$H_2CO_3(aq) \rightleftharpoons H_2O(\ell) + CO_2(g)$$

If enough acid is added to a solution of carbonate or hydrogen carbonate ion, CO_2 gas is evolved and only the anion of the acid and the cation of the carbonate are left in solution (see Figures 5.11 and 18.1). Thus, reactions of this type are an excellent way to prepare simple salts, as you shall see in Section 19.10.

A SUMMARY OF ACID–BASE REACTIONS

When writing an acid–base reaction, you must pay attention to whether the reactants are strong or weak. When one reactant is strong and the other one weak, the pH of the solution after mixing equal molar amounts of acid and base is controlled by the conjugate partner of the weak acid or weak base. This means the equilibrium constant for the acid–base reaction is the reciprocal of the K for ionization of the conjugate base or acid. These relationships are summarized in Table 18.1.

Not included in Table 18.1 is one final case: a weak acid or base reacting with a weak base or acid. For example, if formic acid (a weak acid) is mixed with ammonia (a weak base), the following reaction occurs.

$$HCOOH(aq) + NH_3(aq) \rightleftharpoons NH_4^+(aq) + HCOO^-(aq)$$

If equal molar quantities of acid and base are mixed, the resulting solution will initially contain only ammonium formate, $NH_4(HCOO)$. Is this solution acidic or basic? Ammonium ion is the conjugate acid of a weak base, and so should hydrolyze and contribute to the solution acidity.

Recall our strategy, outlined in page 756, of finding the pH of a solution from an acid–base reaction. To find the concentration of salt produced in the reaction, we assume 100% reaction. Then, with the salt concentration known, we take the hydrolysis of the anion and/or cation into account to find the pH of the solution.

$$NH_4^+(aq) + H_2O(\ell) \rightleftharpoons H_3O^+(aq) + NH_3(aq)$$

$$K_a = 5.6 \times 10^{-10}$$

On the other hand, formate ion ($HCOO^-$) is the conjugate base of a weak acid; it should make the solution basic.

$$HCOO^-(aq) + H_2O(\ell) \rightleftharpoons HCOOH(aq) + OH^-(aq)$$

$$K_b = 5.6 \times 10^{-11}$$

As you saw on page 754, $K_{net} = K_aK_b/K_w$. Here both K_a and K_b are small, and K_{net} is much smaller than in the other cases we have examined. This reaction must be treated as a general equilibrium problem (Study Question 121 in Chapter 17), and usually involves solving a quadratic equation. Nonetheless, we can make a *qualitative* prediction. To do this, consider the ionization constants for the two ions, and you see that $K_a > K_b$. This tells us that more H_3O^+ is produced than OH^-, so the solution is expected to be slightly acidic; the pH will be less than 7. *In general, when a weak acid and a weak base react in equal molar amounts in solution, a salt is produced. The pH of the solution depends on the relative K values of the conjugate base and acid.*

EXERCISE 18.3 pH of a Solution of a Salt of a Weak Base and Weak Acid

Suppose you mix exactly 50 mL of 0.10 M acetic acid and 50 mL of 0.10 M pyridine (C_5H_5N). (K_a for the pyridinium ion, $C_5H_5NH^+$, is 6.7×10^{-6}.) Is the solution acidic or basic?

18.2 THE COMMON ION EFFECT AND BUFFER SOLUTIONS

One of the objectives of this chapter is to follow the course of events during an acid–base titration. Before we can do this, though, we have to explore the nature of solutions of acids and bases that contain another solute, in particular an ion that is "common" to the acid or base.

THE COMMON ION EFFECT

Lactic acid is a weak acid found in sour milk, apples and other fruit, beer and wine, and several plants.

$$H_3C{-}CH(OH){-}C(=O){-}O{-}H(aq) + H_2O(\ell) \rightleftharpoons H_3O^+(aq) + H_3C{-}CH(OH){-}C(=O){-}O^-(aq)$$

lactic acid, $C_3H_6O_3$
$K_a = 1.4 \times 10^{-4}$

lactate ion, $C_3H_5O_3^-$

Figure 18.2 The common ion effect. An acid (0.25 M acetic acid, left) is mixed with a base (0.10 M sodium acetate, right). The indicator (bromphenol blue, Table 17.3) shows that the resulting solution (center) has a lower hydronium ion concentration (a pH of about 5) than the acetic acid solution.

Suppose you are analyzing a sample containing lactic acid by titrating the sample with the strong base NaOH.

$$\underset{\text{lactic acid}}{C_3H_6O_3(aq)} + OH^-(aq) \rightleftharpoons \underset{\text{lactate ion}}{C_3H_5O_3^-(aq)} + H_2O(\ell)$$

At the equivalence point in the titration, the acid has been consumed and converted completely to its conjugate base, the lactate ion. However, *before* the equivalence point, some lactate ion is present in solution along with lactic acid that has not yet reacted. Thus, *before the equivalence point, the weak acid is present along with some amount of its conjugate base*. If we halt the titration at some intermediate stage before the equivalence point and allow equilibrium to be established, it is observed that the ionization of the remaining lactic acid is affected by the lactate ion present. This is called the **common ion effect**, since an ion (here the lactate ion) that is "common" to the ionization of the acid is present in an amount greater than that produced by simple acid ionization. In an acid ionization, according to Le Chatelier's principle, the presence of some conjugate base in solution will limit the extent to which the acid ionizes and thus affect the pH of the solution (Figure 18.2).

EXAMPLE 18.3

THE COMMON ION EFFECT

What is the pH of 0.25 M aqueous acetic acid? What is the pH of the solution after adding sodium acetate to make the solution 0.10 M in the salt? (Ignore any change in volume on adding the sodium acetate.)

Solution Let us first determine the pH of the 0.25 M acid solution using the approach outlined in Chapter 17. Then we can turn to the effect of adding a "common ion."

(a) Determine the pH of 0.25 M acetic acid.

$$CH_3COOH(aq) + H_2O(\ell) \rightleftharpoons H_3O^+(aq) + CH_3COO^-(aq)$$

	$[CH_3COOH]$	$[H_3O^+]$	$[CH_3COO^-]$
Before CH_3COOH ionizes (M)	0.25	0	0
Change on proceeding to equilibrium	$-x$	$+x$	$+x$
At equilibrium (M)	$0.25 - x$	x	x

The appropriate equilibrium expression is

$$K_a = 1.8 \times 10^{-5} = \frac{[CH_3COO^-][H_3O^+]}{[CH_3COOH]} = \frac{(x)(x)}{0.25 - x} \approx \frac{x^2}{0.25}$$

As in Chapter 17, we shall make the approximation that $(0.25 - x) \approx 0.25$, and so the hydronium ion concentration is

$$[H_3O^+] = 2.1 \times 10^{-3}\ M$$

(The approximation is valid because subtracting 2.1×10^{-3} from 0.25 would indeed have made no difference in the final answer.) The hydronium ion concentration leads to a pH of 2.67.

(b) Effect of added "common ion."

Sodium acetate, $Na(CH_3COO)$, is very soluble in water and is 100% dissociated into its ions, Na^+ and CH_3COO^- (acetate ion). As you learned in Chapter 17, Na^+ ion has no effect on the pH of a solution (Table 17.5). On the other hand, CH_3COO^- is the conjugate base of the weak acid CH_3COOH, so it should contribute some excess OH^- ions to the solution,

$$\underset{\text{acetate ion}}{CH_3COO^-(aq)} + H_2O(\ell) \rightleftharpoons \underset{\text{acetic acid}}{CH_3COOH(aq)} + OH^-(aq)$$

and thus reduce the hydronium ion concentration below 2.1×10^{-3} M, the concentration found if the solution contains only acetic acid.

We shall solve for the net hydronium ion concentration in the following way. First, imagine that you have a solution of the weak acid (CH_3COOH, 0.25 M) and its conjugate base (CH_3COO^-, 0.10 M) and that neither has yet ionized.

*Concentrations in the acetic acid/sodium acetate mixture **before** ionization of the acid:*

$$\underset{0.25\ M}{CH_3COOH(aq)} + H_2O(\ell) \rightleftharpoons \underset{0\ M}{H_3O^+(aq)} + \underset{0.10\ M}{CH_3COO^-(aq)}$$

Now imagine that the acid ionizes to give H_3O^+ and CH_3COO^-, both in the amount y, in the presence of the added $Na(CH_3COO)$.

*Concentrations in the acetic acid/sodium acetate mixture **after** ionization of the acid:*

$$\underset{(0.25 - y)\ M}{CH_3COOH(aq)} + H_2O(\ell) \rightleftharpoons \underset{y\ M}{H_3O^+(aq)} + \underset{(0.10 + y)\ M}{CH_3COO^-(aq)}$$

Without the added salt $Na(CH_3COO)$, which provides the "common ion" CH_3COO^-, ionization of the acid would have produced H_3O^+ and CH_3COO^-, both in the amount $x = 2.1 \times 10^{-3}$ M. However, we know from Le Chatelier's principle that y must be less than x. Therefore, the concentration table for the ionization of acetic acid in the presence of the "common ion" acetate ion will be

The key idea of common ion problems is that the weak acid HA, for example, ionizes to produce less H_3O^+ in the presence of the common ion A^- than it would have in the absence of A^-. Thus, adding A^- to a solution of the weak acid HA causes the pH to be higher than in the absence of A^-. This is easy to remember because adding a base to any solution will raise the pH of the solution.

	$[CH_3COOH]$	$[H_3O^+]$	$[CH_3COO^-]$
Before CH_3COOH ionizes (M)	0.25	0	0.10
Change on proceeding to equilibrium	$-y$	$+y$	$+y$
At equilibrium (M)	$(0.25 - y)$	y	$0.10 + y$

and the appropriate equilibrium expression is

$$K_a = 1.8 \times 10^{-5} = \frac{[H_3O^+][CH_3COO^-]}{[CH_3COOH]} = \underset{\text{exact expression}}{\frac{(y)(0.10 + y)}{0.25 - y}} \approx \underset{\text{approximate expression}}{\frac{(y)(0.10)}{0.25}}$$

Let us begin by asking what approximation will convert the "exact" expression above into the more easily solved "approximate" expression. We know that the amount of CH_3COO^- formed by ionization in a 0.25 M solution of acetic acid is 0.0021 M, but Le Chatelier's principle predicts that the acid will produce even less acetate ion in the presence of the "common ion." Therefore, it certainly seems reasonable to assume that $(0.10 + y)$ M $\approx$ 0.10 M and that $(0.25 - y)$ M $\approx$ 0.25 M. This leads to the "approximate" expression above. Solving this, we find that

$$y = [H_3O^+] = [CH_3COO^-] \text{ added to solution by } CH_3COOH \text{ ionization} \approx 4.5 \times 10^{-5} \text{ M}$$

$$pH = 4.35$$

This is indeed a very small value for $[H_3O^+]$ and for the concentration of CH_3COO^- that is added to the solution by acetic acid ionization, and our approximations are certainly reasonable. (Furthermore, we obtain the same answer from the "exact expression" by using the method of successive approximations.)

Finally, compare the situations before and after adding acetate ion, the common ion. The pH before adding sodium acetate was 2.67. However, because of the basic nature of the acetate ion, the acidity declined (the pH increased to 4.35) after adding sodium acetate.

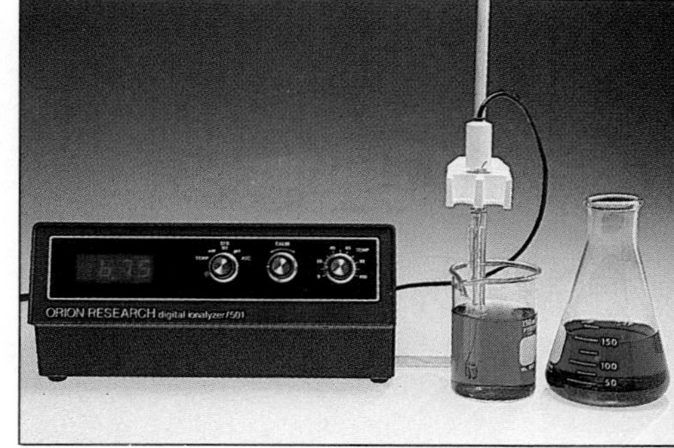

(a)

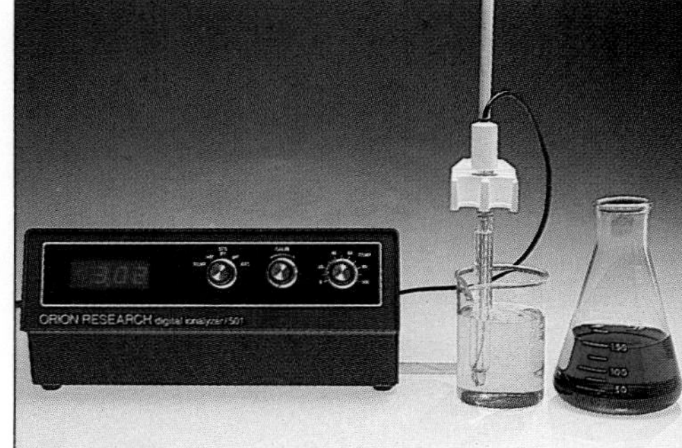

(b)

Figure 18.3 Buffer solutions. (a) The pH electrode is indicating the pH of water that contains a trace of acid (and bromcresol green indicator; see Table 17.3). The solution at the right is a buffer having a pH the same as human blood (7.4). (b) When 5 mL of 0.10 M HCl is added to each solution, the pH of the water drops several units, while the pH of the buffer stays constant.

EXERCISE 18.4 The Common Ion Effect

Assume you have a 0.30 M solution of formic acid (HCOOH) and add enough sodium formate (NaHCOO) to make the solution 0.10 M in the salt. Calculate the pH of the formic acid solution before and after adding sodium formate.

BUFFER SOLUTIONS

The normal pH of human blood is 7.4. Experiment clearly shows that the addition of a small quantity of strong acid or base to blood, say 0.01 mol to a liter of blood, leads to a change in pH of only about 0.1 pH unit (Figure 18.3). In comparison, if you add 0.01 mol of HCl to 1.0 L of pure water, the pH drops from 7 to 2, while addition of 0.01 mol of NaOH increases the pH from 7 to 12. Blood, and many other body fluids, are said to be *buffered*; their pH resists change upon addition of a strong acid or base.

In general, two species are required in a **buffer solution**, one capable of reacting with added OH^- ions (an acid) and another that can consume added H_3O^+ ions (a base). An additional requirement is that the acid and base do not react with one another. This means that a buffer is usually prepared from a *conjugate* acid–base pair: (a) a weak acid and its conjugate base (acetic acid and acetate ion, for example), or (b) a weak base and its conjugate acid (ammonia and ammonium ion, for example). Some systems commonly used in the laboratory are listed in Table 18.2.

Because of its composition, a buffer solution is just a special case of the "common ion effect."

To see how a buffer works, let us consider an acetic acid/acetate buffer. The acetic acid, the weak acid, is needed to consume any added hydroxide ion.

$$CH_3COOH(aq) + OH^-(aq) \rightleftharpoons CH_3COO^-(aq) + H_2O(\ell) \qquad K = 1.8 \times 10^9$$

Table 18.2 Some Commonly Used Buffer Systems

Weak Acid	Conjugate Base	Acid K_a	Useful pH Range
Phthalic acid $C_6H_4(COOH)_2$	Hydrogen phthalate $C_6H_4(COOH)COO^-$	1.3×10^{-3}	1.9–3.9
Acetic acid CH_3COOH	Acetate CH_3COO^-	1.8×10^{-5}	3.7–5.8
Dihydrogen phosphate $H_2PO_4^-$	Hydrogen phosphate HPO_4^{2-}	6.2×10^{-8}	6.2–8.2
Hydrogen phosphate HPO_4^{2-}	Phosphate PO_4^{3-}	3.6×10^{-13}	11.3–13.3

The equilibrium constant for the reaction is very large because OH^- is a much stronger base than acetate, CH_3COO^-, as we know from Table 17.2. This means that any OH^- entering the solution from an outside source is consumed completely. Similarly, any hydronium ion added to the solution can be consumed completely by the acetate ion present in the buffer.

$$H_3O^+(aq) + CH_3COO^-(aq) \rightleftharpoons H_2O(\ell) + CH_3COOH(aq) \qquad K = 5.6 \times 10^4$$

The equilibrium constant for this reaction is also quite large because H_3O^+ is a much stronger acid than CH_3COOH, again as shown by the data in Table 17.2.

Now that we have established that a buffer solution should effectively remove small amounts of added acid or base, work through the following example, which shows quantitatively how the pH of a solution can be maintained.

EXAMPLE 18.4

A BUFFER SOLUTION

Calculate the pH change that occurs when 1.0 mL of 1.0 M HCl is added to (a) 1.0 L of pure water and (b) 1.0 L of acetic acid/sodium acetate buffer with $[CH_3COOH] = 0.700$ M and $[CH_3COO^-] = 0.600$ M.

Solution **(a) Adding acid to pure water.** 1.0 mL of 1.0 M HCl represents 0.0010 mole of acid. If this is added to 1.0 L of pure water, the hydronium ion concentration is 1.0×10^{-3} M. This means the H_3O^+ concentration is raised from 10^{-7} to 10^{-3}, so the pH falls from 7 to 3.

(b) pH of acetic acid/acetate buffer solution. We shall first determine the pH of the buffer solution before any HCl is added. The balanced equation connecting the species in solution is just the ionization of acetic acid.

$$CH_3COOH(aq) + H_2O(\ell) \rightleftharpoons H_3O^+(aq) + CH_3COO^-(aq)$$

and the table of concentrations is

	$[CH_3COOH]$	$[H_3O^+]$	$[CH_3COO^-]$
Before CH_3COOH ionizes (M)	0.700	0	0.600
Change on proceeding to equilibrium	$-x$	$+x$	$+x$
At equilibrium (M)	$0.700 - x$	x	$0.600 + x$

and the appropriate equilibrium expression is

$$K_a = 1.8 \times 10^{-5} = \frac{[H_3O^+][CH_3COO^-]}{[CH_3COOH]} = \underbrace{\frac{(x)(0.600 + x)}{0.700 - x}}_{\text{exact expression}} \approx \underbrace{\frac{(x)(0.600)}{0.700}}_{\text{approximate expression}}$$

From the previous example, we know that the value of x will be very small with respect to 0.700 or 0.600, so we can use the "approximate expression" to find x, the hydronium ion concentration. Therefore, on rearranging the "approximate" expression, we have

$$[H_3O^+] = x = \frac{[CH_3COOH]}{[CH_3COO^-]} K_a \approx \frac{(0.700)}{(0.600)} (1.8 \times 10^{-5}) = 2.1 \times 10^{-5} \text{ M}$$

$$\text{pH} = 4.68$$

(c) Add HCl to the acetic acid/acetate buffer solution. HCl is a strong acid that is 100% ionized in water. Therefore, we are concerned only with the fact that it supplies H_3O^+ and reacts *completely* with the base in the solution according to the reaction described in the text:

$$H_3O^+(aq) + CH_3COO^-(aq) \rightleftarrows H_2O(\ell) + CH_3COOH(aq)$$

	$[H_3O^+]$ from added HCl	$[CH_3COO^-]$ from buffer	$[CH_3COOH]$ from buffer
Concentrations before reaction (M)	0.0010	0.600	0.700
Change in concentrations from complete reaction	−0.0010	−0.0010	+0.0010
Concentrations after reaction (M)	0	0.599	0.701

Since the added HCl reacts *completely* with acetate ion to produce acetic acid, the solution is once again a buffer containing only the weak acid and its salt. Therefore, we need to consider only the equilibrium for the ionization of the weak acid in the presence of its common ion, CH_3COO^-, as in part (b).

$$CH_3COOH(aq) + H_2O(\ell) \rightleftarrows H_3O^+(aq) + CH_3COO^-(aq)$$

	$[CH_3COOH]$	$[H_3O^+]$	$[CH_3COO^-]$
Before CH_3COOH ionizes (M)	0.701	0	0.599
Change on proceeding to equilibrium	$-y$	$+y$	$+y$
At equilibrium (M)	$0.701 - y$	y	$0.599 + y$

As usual, we make the approximation that y, the amount of H_3O^+ formed by ionizing acetic acid in the presence of acetate ion, is small compared with 0.701 M or 0.599 M. Therefore, we can write

$$[H_3O^+] = y = \frac{[CH_3COOH]}{[CH_3COO^-]} K_a \approx \frac{(0.701)}{(0.599)} (1.8 \times 10^{-5}) = 2.1 \times 10^{-5} \text{ M}$$

$$\text{pH} = 4.68$$

Within the number of significant figures allowed, *the pH does not change in the buffer solution after adding HCl*, even though it changed by 4 units when 1 mL of 1.0 M HCl was added to 1.0 L of pure water. Buffer solutions do indeed "buffer." In this case the acetate ion consumed the added hydronium ion and the solution thus resisted a change in pH.

EXERCISE 18.5 Buffer Solutions

Calculate the pH of 0.500 L of a buffer solution composed of 0.50 M formic acid (HCOOH) and 0.70 M sodium formate (NaHCOO) before and after adding 10.0 mL of 1.00 M HCl.

GENERAL EXPRESSIONS FOR BUFFER SOLUTIONS In Example 18.4 we solved for the hydronium ion concentration of the acetic acid/acetate buffer solution by rearranging the K_a expression to give

$$[H_3O^+] = \frac{[CH_3COOH]}{[CH_3COO^-]} K_a$$

This result can be generalized for any buffer. For a buffer solution based on a weak acid and its conjugate base, the hydronium ion concentration is always given by

$$[H_3O^+] = \frac{[\text{acid}]_e}{[\text{conjugate base}]_e} K_a \approx \frac{[\text{acid}]_0}{[\text{conjugate base}]_0} K_a$$

(where the subscript e refers to equilibrium concentrations and the subscript 0 refers to original or initial concentrations). When the usual expression for K_b is rearranged, the hydroxide ion concentration in a buffer composed of a weak base and its conjugate acid (e.g., NH_3 and NH_4^+) is

$$[OH^-] = \frac{[\text{base}]_e}{[\text{conjugate acid}]_e} K_b \approx \frac{[\text{base}]_0}{[\text{conjugate acid}]_0} K_b$$

These general forms are quite useful in finding the pH of buffer solutions and in deciding how to prepare such solutions.

PREPARING BUFFER SOLUTIONS There are two obvious requirements for a buffer solution. First, it should have the capacity to control the pH after the addition of reasonable amounts of acid and base. That is, there must be a large enough concentration of acetic acid in an acetic acid/acetate buffer, for example, to consume all of the hydroxide ion that may be added and still control the pH (see Example 18.4). The requirement is usually met by using 0.10 M to 1.0 M solutions of reagents. However, any buffer eventually uses up its capacity if too much strong acid or base is added.

The second requirement for a buffer solution is that it should control the pH at the desired value. The equation for the hydronium ion concentration of an acid buffer

$$[H_3O^+] \approx \frac{[\text{acid}]_0}{[\text{conjugate base}]_0} K_a$$

shows us how to prepare a buffer solution of given pH. First, we want $[\text{acid}]_0 \approx [\text{conjugate base}]_0$ so that the solution can buffer equal amounts of added acid or base. Given this, we want to choose an acid whose K_a is near the $[H_3O^+]$ we want. The exact value of $[H_3O^+]$ is then achieved by adjusting the acid/conjugate base ratio. The following example illustrates this approach.

EXAMPLE 18.5

PREPARING A BUFFER SOLUTION

Suppose you wish to prepare a buffer solution to maintain the pH at 4.30. A list of possible acids (and their conjugate bases) is

ACID	CONJUGATE BASE	K_a
HSO_4^-	SO_4^{2-}	1.2×10^{-2}
CH_3COOH	CH_3COO^-	1.8×10^{-5}
HCN	CN^-	4.0×10^{-10}

Which combination should be selected and what should the ratio of acid to conjugate base be?

Solution The desired pH is 4.30, so the hydronium ion concentration is

$$[H_3O^+] = 10^{-pH} = 10^{-4.30} = 5.0 \times 10^{-5}\ M$$

The correct choice of acid is now evident. Of the acids given, only acetic acid (CH_3COOH) has a K_a value close to the desired $[H_3O^+]$. You need only adjust the ratio $[CH_3COOH]/[CH_3COO^-]$ to achieve the desired hydronium ion concentration exactly.

$$[H_3O^+] = 5.0 \times 10^{-5} = \frac{[CH_3COOH]}{[CH_3COO^-]}(1.8 \times 10^{-5})$$

To satisfy this expression, you must have the ratio

$$\frac{[CH_3COOH]}{[CH_3COO^-]} = \frac{5.0 \times 10^{-5}}{1.8 \times 10^{-5}} = \frac{2.8}{1}$$

Therefore, if you add 0.28 mol of acetic acid and 0.10 mol of sodium acetate (or any other pair of molar quantities in the ratio 2.8/1) to enough water to make one liter of solution, the solution will constitute a buffer capable of controlling the pH at 4.30.

EXERCISE 18.6 Preparing a Buffer Solution

Using an acetic acid/sodium acetate buffer solution, what ratio of acid to conjugate base will you need to maintain the pH at 5.00? Explain how you would make up such a solution.

Example 18.5 brings up one more important point concerning buffer solutions. The hydronium ion concentration depends on the ratio [acid]/[conjugate base] (or the hydroxide ion concentration depends on [base]/[conjugate acid]). However, although we have written these ratios in terms of reagent concentrations, it is not the actual concentrations that are important. Rather, what is important is *the* ***relative*** *number of moles of acid and conjugate base*. Since both reagents are dissolved in the same solution, their concentrations depend on the same solution volume. In Example 18.5, the ratio 2.8/1 for acetic acid and sodium acetate implied that 2.8 mol of the acid and 1.0 mol of sodium acetate could be dissolved per liter.

$$\frac{[CH_3COOH]}{[CH_3COO^-]} = \frac{2.8 \text{ mol/L}}{1.0 \text{ mol/L}} = \frac{2.8 \text{ mol}}{1.0 \text{ mol}}$$

Alternatively, 1.4 mol of acetic acid and 0.50 mol of sodium acetate, or any reasonably large amounts that give a ratio of 2.8/1, could have been used. Further, since the actual concentration is not important, the acid and its conjugate base could have been dissolved in any reasonable amount of water. The mole ratio of acid to its conjugate base is maintained, no matter what the solution volume may be. This is why commercially available buffer solutions are sold as concentrated solutions. To use them, you take a small amount and dilute it with pure water to a convenient volume (Figure 18.4).

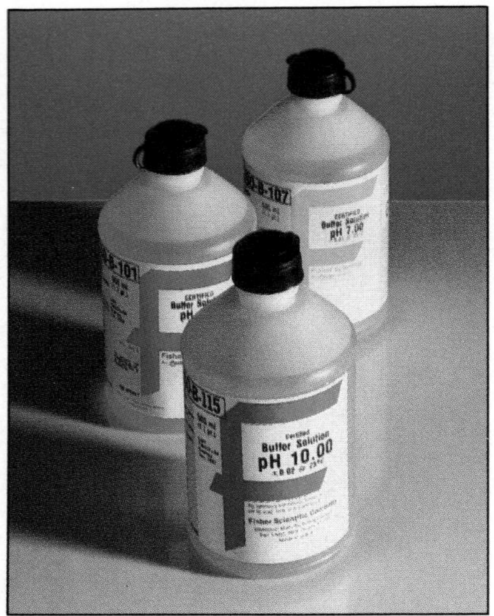

Figure 18.4 Commercial buffer solutions. To use one of the solutions, take a small volume and dilute it with pure water to a larger volume. The ratio [acid]/[conjugate base] (or [base]/[conjugate acid]) remains constant. Thus, the pH of the buffer remains constant no matter what the solution volume. (Fisher Scientific Company)

18.3 ACID–BASE TITRATION CURVES

In Chapter 5 we described the stoichiometry involved in acid–base titration, which is a method for accurate analysis of the amount of acid or base in a sample. In the current chapter you learned something more about titrations. For example, you know now that the pH at the equivalence point is 7, and the solution is truly "neutral," only when a strong acid is titrated with a strong base and vice versa. If one of the substances being titrated is weak, then the pH at the equivalence point is not 7. We found that

Weak acid + strong base ⟶ pH > 7 at equivalence point due to hydrolysis of the conjugate base of the weak acid

and

Strong acid + weak base ⟶ pH < 7 at equivalence point due to hydrolysis of the conjugate acid of the weak base

Now we wish to describe how the pH of a solution of an acid or base changes as it is being titrated with a base or acid. To illustrate this, the following example, and the text that follows, explore two of the most common situations: (a) the titration of a strong acid with a strong base and (b) the titration of a weak acid with a strong base.

EXAMPLE 18.6

THE TITRATION OF A STRONG ACID WITH A STRONG BASE

You have 50.0 mL of a 0.100 M solution of HCl. Calculate the pH of the solution as 0.100 M NaOH is slowly added. Plot the results with pH on the vertical axis and milliliters of added base on the horizontal axis.

Solution (a) HCl is a strong acid, so its concentration is equivalent to the H_3O^+ concentration. Since [HCl] = 1.00×10^{-1} M in the original solution, the pH is 1.000.

(b) The number of moles of HCl originally present is

$$\text{Moles } H_3O^+ = (0.0500 \text{ L})(0.100 \text{ M}) = 5.00 \times 10^{-3} \text{ moles}$$

(c) Now add the first increment of NaOH, 10.0 mL. After 10.0 mL of NaOH have been added, this means 1.00×10^{-3} mol of base has been added.

$$(0.0100 \text{ L NaOH})(0.100 \text{ moles/L}) = 1.00 \times 10^{-3} \text{ mol of NaOH added}$$

Since the net reaction in the titration of a strong acid with a strong base is

$$H_3O^+(aq) + OH^-(aq) \longrightarrow 2\ H_2O(\ell)$$

the added NaOH is completely consumed by reaction with 1.00×10^{-3} mol of HCl, and the moles of HCl remaining are

$$\begin{array}{l} \ 0.00500 \text{ mol HCl originally present} \\ \underline{-\ 0.00100 \text{ mol HCl consumed by addition of } 0.00100 \text{ mol of NaOH}} \\ \ 0.00400 \text{ mol HCl remaining after adding } 10.0 \text{ mL of NaOH} \end{array}$$

Remembering that the total solution volume now is 60.0 mL (50.0 mL of HCl originally + 10.0 mL of NaOH added), the H_3O^+ concentration now is

$$[H_3O^+] = \frac{4.00 \times 10^{-3} \text{ mol}}{0.0600 \text{ L}} = 6.67 \times 10^{-2} \text{ M}$$

$$\text{pH} = 1.176$$

(d) In general, the hydronium ion concentration up to the equivalence point can be calculated, just as it was after the first addition, from the expression

$$[H_3O^+] = \frac{\text{original moles acid} - \text{total moles of base added}}{\text{volume acid} + \text{total volume base added}}$$

For instance, after 45.0 mL of NaOH has been added, the acid concentration is

$$[H_3O^+] = \frac{5.00 \times 10^{-3} \text{ mol HCl} - (0.0450 \text{ L})(0.100 \text{ M})}{0.0500 \text{ L} + 0.0450 \text{ L}} = 5.26 \times 10^{-3} \text{ M}$$

and the pH is 2.279.

(e) When the volume of added base has reached 49.0 mL,

$$[H_3O^+] = \frac{5.00 \times 10^{-3} \text{ mol HCl} - (0.0490 \text{ L})(0.100 \text{ M})}{0.0500 \text{ L} + 0.0490 \text{ L}} = 1.01 \times 10^{-3} \text{ M}$$

$$\text{pH} = 2.996$$

The pH indicates the solution is still quite acidic.

(f) Finally, when 50.0 mL of base has been added, the equivalence point is reached, since 5.00×10^{-3} mol of base has now been added. All of the acid has been consumed by base, and only a solution of the "neutral ions" Na^+ and Cl^- remains. The pH is therefore 7.00 (see page 754), since in pure water or in water containing only a neutral salt such as NaCl,

$$[H_3O^+] = [OH^-] = \sqrt{K_w} = 1.00 \times 10^{-7}$$

(g) When the equivalence point has been passed, no acid remains; NaOH is added to water containing NaCl, a neutral salt. Therefore, we calculate the pH from the concentration of the OH^- ion. For example, after only 1.0 mL of base has been added beyond the equivalence point (total volume of base added = 51.0 mL), the pH is 11.00.

$$[OH^-] = \frac{\text{moles excess base}}{\text{total volume}} = \frac{(1.0 \times 10^{-3}\ \text{L})(0.100\ \text{M})}{0.0500\ \text{L acid} + 0.0510\ \text{L base}} = 9.9 \times 10^{-4}\ \text{M}$$

$$[H_3O^+] = \frac{K_w}{[OH^-]} = \frac{1.00 \times 10^{-14}}{9.9 \times 10^{-4}} = 1.0 \times 10^{-11}\ \text{M}$$

$$\text{pH} = 11.00$$

The results of these calculations are summarized in the table and curve shown in Figure 18.5.

EXERCISE 18.7 Titration of a Strong Acid with a Strong Base
For the titration outlined in Example 18.6, verify the pH shown in Figure 18.5 for the addition of (a) 40.0 mL of NaOH and (b) 60.0 mL of NaOH. What is the pH after 49.9 mL of NaOH is added?

The pH at each point in the titration of the strong acid HCl with the strong base NaOH is illustrated in Figure 18.5. As calculated in Example 18.6, the pH rises slowly until the reaction is very close to the equivalence point. Then the pH increases very rapidly, rising 6 units (the H_3O^+ concentration decreases by a factor of 1 million!) when only a very small amount of base (perhaps a drop or two) is added (Figure 18.6). After the equivalence point is passed, only a small further rise in pH is seen.

The titration of a weak acid with a strong base is somewhat different from the strong acid/strong base titration just described. To illustrate, let

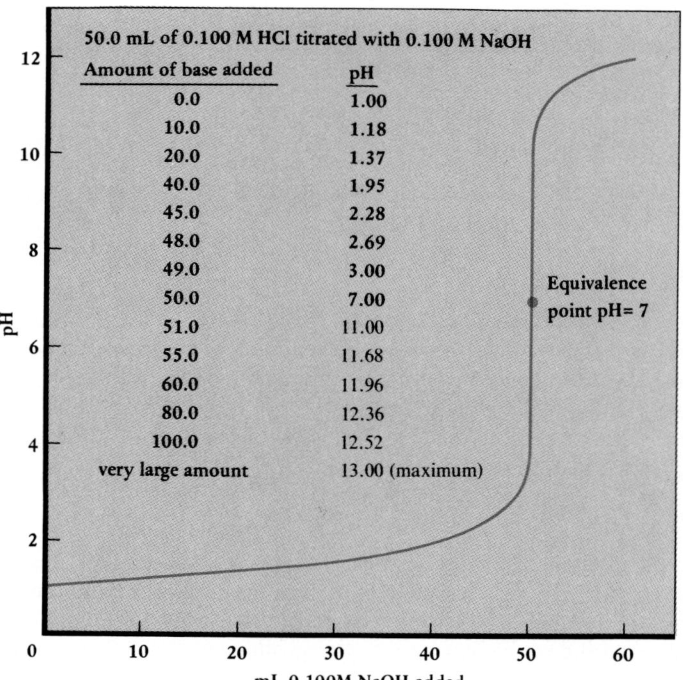
50.0 mL of 0.100 M HCl titrated with 0.100 M NaOH

Amount of base added	pH
0.0	1.00
10.0	1.18
20.0	1.37
40.0	1.95
45.0	2.28
48.0	2.69
49.0	3.00
50.0	7.00
51.0	11.00
55.0	11.68
60.0	11.96
80.0	12.36
100.0	12.52
very large amount	13.00 (maximum)

Figure 18.5 The change in pH as a strong acid is titrated with a strong base. In this case 50.0 mL of 0.100 M HCl is titrated with 0.100 M NaOH.

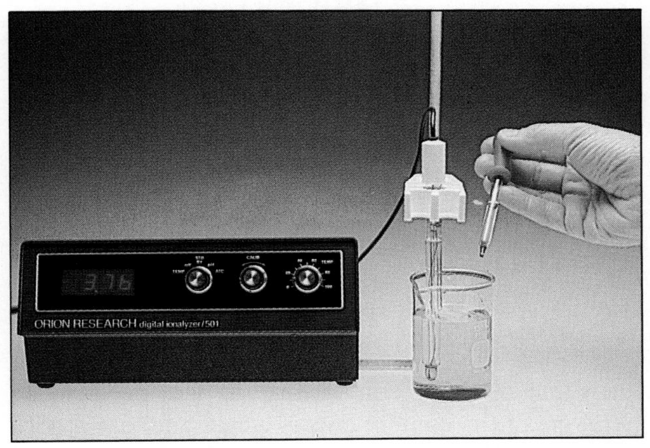

(a)

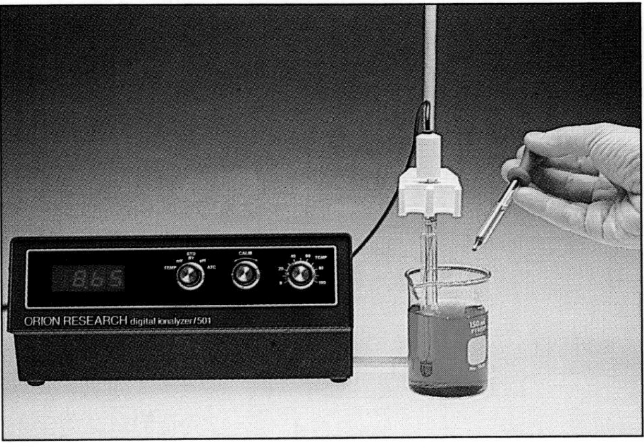

(b)

Figure 18.6 Titration of a strong acid (HCl) with a strong base (NaOH) at the end point. The pH changed abruptly from acid (a) to base (b) when one drop of 1 M NaOH was added to the solution. The indicator was bromcresol green (see Table 17.3).

us look carefully at the curve for titration of 50.0 mL of 0.100 M acetic acid with 0.100 M NaOH (Figure 18.7).

$$CH_3COOH(aq) + NaOH(aq) \longrightarrow Na(CH_3COO)(aq) + H_2O(\ell)$$

There are four points or regions on this curve that are especially interesting: (1) the pH before titration begins; (2) the pH at the midpoint of the titration; (3) the pH at the equivalence point; and (4) the pH when base is added beyond the equivalence point. The pH before any base is added (2.87) is found in the usual way (Example 17.6) for the ionization of a weak acid. At the equivalence point of the titration, the solution consists simply of sodium acetate, and so the pH (8.72) is controlled by the acetate ion, the conjugate base of acetic acid (see Example 18.1).

The pH of the CH_3COOH solution is greater than that of an HCl solution of the same concentration (Figure 18.6) because the acetic acid is less ionized. Further, the pH at the equivalence point is greater in the CH_3COOH titration because CH_3COO^- is a stronger base than Cl^-.

Now let us determine the pH at the midpoint of the titration, that is, at the point at which half of the acid has been consumed. As NaOH is added to the acetic acid, the base is consumed and sodium acetate is produced. This means that at every point between the beginning of the titration (when only pure acetic acid is present) and the equivalence point (where only sodium acetate is present) the solution contains *both* acetic acid *and* its salt, sodium acetate. These are just the components of a buffer solution, and the hydronium ion concentration can be found from the following expression.

$$[H_3O^+] = \frac{[CH_3COOH]}{[CH_3COO^-]} K_a$$

But what happens when exactly half of the acid has been consumed by base? If the initial number of moles of acid was M_0, then *at the midpoint in the titration*

$$\text{Moles acid remaining} = \text{moles acid consumed} = \tfrac{1}{2} M_0$$

$$\begin{aligned}\text{Moles acetate ion formed} &= \text{initial moles acid} - \text{moles acid consumed} \\ &= M_0 - \tfrac{1}{2} M_0 = \tfrac{1}{2} M_0\end{aligned}$$

Therefore, since both the moles of acid remaining and the moles of acetate formed are $\tfrac{1}{2} M_0$,

$$[H_3O^+] \text{ at the midpoint of the titration} = \frac{[\text{acetic acid remaining}]}{[\text{acetate ion formed}]} K_a = K_a$$

In the particular case of the titration of acetic acid with strong base, $[H_3O^+] = 1.8 \times 10^{-5}$ M at the midpoint, and so pH = 4.74. However, this result is of general importance, since we can say to a good approximation that

(a) *At the midpoint in any weak acid + strong base titration,* $[H_3O^+] = K_a$ *of the weak acid.*
(b) *At the midpoint in any weak base + strong acid titration,* $[OH^-] = K_b$ *of the weak base.*

You can see the value of these conclusions. If you perform an acid–base titration where one of the components is weak, and you record the pH value at each step, you can readily determine K for the weak acid or base from the pH at the titration midpoint.

As a final point concerning the titration of a weak acid with a strong base, we want to know the pH *after* the equivalence point. In the case

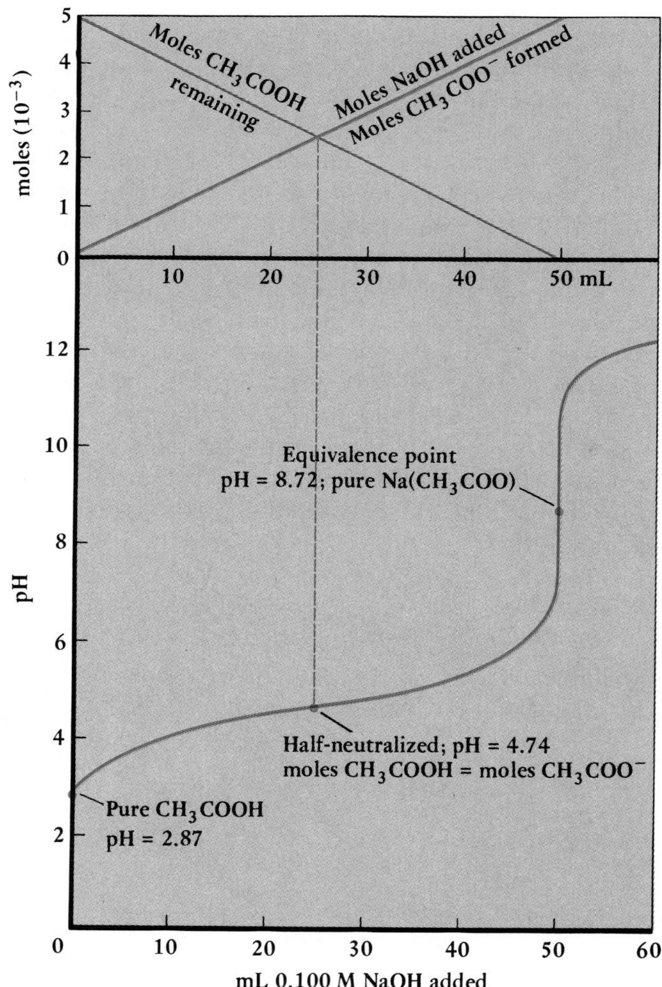

Figure 18.7 The change in pH during the titration of a weak acid with a strong base: 50.0 mL of 0.100 M acetic acid is titrated with 0.100 M NaOH. (Top) Moles of acid remaining and moles of conjugate base formed as a function of milliliters of added base. The point at which these lines cross is the midpoint of the acid–base reaction. (Bottom) The change in pH of the solution as a function of milliliters of added base.

of the CH_3COOH/NaOH titration, the solution would consist of sodium acetate plus any NaOH that was added in excess after passing the equivalence point. Both acetate ion and hydroxide ion are bases, so the total concentration of OH^- would be the sum of that from excess NaOH plus that produced by the hydrolysis of the acetate ion.

$$CH_3COO^-(aq) + H_2O(\ell) \rightleftharpoons CH_3COOH(aq) + OH^-(aq)$$

However, the OH^- concentration from the hydrolysis reaction is very small compared with that from excess NaOH, so after the equivalence point

$$[OH^-] = \frac{\text{moles excess } OH^- \text{ from NaOH}}{\text{total volume in L}}$$

EXERCISE 18.8 Titration of a Weak Acid with a Strong Base

The titration of 0.100 M acetic acid with 0.100 M NaOH is described in the text. What is the pH of the solution when 20.% of the acid has been neutralized? After 55.0 mL of NaOH has been added to 50.0 mL of the acid?

18.4 ACID–BASE INDICATORS

The goal of a titration is to add a standard solution in an amount that is stoichiometrically exactly equivalent to the substance with which it reacts. This condition is achieved at the equivalence point. A convenient way to estimate when you have reached the equivalence point is to observe some change in the solution, as for example a change in the color of a dye or indicator in the solution (Section 17.3). The dye's color change is said to occur at the *end point* of the titration, and we try to make the difference between the end and equivalence points negligible.

An acid–base indicator (HIn) is usually an organic dye that is itself a weak acid.

$$HIn(aq) + H_2O(\ell) \rightleftharpoons H_3O^+(aq) + In^-(aq)$$

$$K_a \text{ for the indicator} = \frac{[H_3O^+][In^-]}{[HIn]}$$

The acid form of the compound (HIn) has one color, while the conjugate base (In^-) has another (Table 17.3). According to Le Chatelier's principle, addition of H_3O^+ or OH^- to an indicator solution will cause one color or the other to appear, depending on whether HIn or In^- is the predominant species. The idea is to try to choose an indicator with a K_a near that of the acid we are titrating and whose color changes strongly at the equivalence point. If the color change is strong, we need add only a tiny bit to the solution and then titrate both acids at the same time.

If the K_a expression for the indicator is rearranged

$$[H_3O^+] = \frac{[HIn]}{[In^-]} K_a$$

it is more apparent that a tiny amount of an indicator can in fact reveal the pH of a solution as the $[HIn]/[In^-]$ ratio changes. If a drop or two of some dilute indicator is added in an acid–base titration, the ratio [HIn]/[In^-] is controlled by the hydronium ion concentration of the solution. As [H_3O^+] changes in the course of a titration, the ratio [HIn]/[In^-] must also change in order to maintain the equality in the equation above. Therefore, when the H_3O^+ concentration is high, [HIn] must be large and [In^-] small. The indicator color will be that of HIn. In the case of the common indicator phenolphthalein, the dye is colorless at high concentrations of H_3O^+ or low pH, whereas the indicator thymol blue is red under these conditions (Table 17.3 and page 716). When the pH increases, and [H_3O^+] declines, the ratio [HIn]/[In^-] must also decrease. Thus, the concentration of In^- increases, while that of HIn decreases, and the color of In^- is seen. Phenolphthalein, for example, is red above pH 10 whereas thymol blue is yellow between pH 3 and 8 and purple above pH 9.

In principle, the color of the indicator changes when [H_3O^+] = K_a of the indicator (because [HIn] = [In^-] at this point), and so you might think that you could accurately determine [H_3O^+] by carefully observing the color change. In practice, your eyes are not quite that good. Usually, you see the color of HIn when [HIn]/[In^-] is about 10/1, and the color of In^- when [HIn]/[In^-] is about 1/10. This means the color change is observed over a hydronium ion concentration interval of about 100, which corresponds to 2 pH units. This is not really a problem, however, as you can see in Figures 18.5 and 18.7; on passing through the equivalence point of these titrations the pH changes by as many as 6 units.

As Table 17.3 shows, there is a variety of indicators available, each changing color in a different pH range. When you are analyzing an acid or base by titration, you have to choose an indicator that changes color in a range that includes the pH to be observed at the equivalence point. This means that an indicator that changes color in the pH range 7 ± 2 should be used for a strong acid/strong base titration. On the other hand, the pH at the equivalence point in the titration of a weak acid with a strong base is greater than 7. Therefore, you should use an indicator that changes color at a pH of about 8.

EXERCISE 18.9 Indicators

Use Table 17.3 to decide which indicator would be best to use in the titration of NH_3 with HCl (see Example 18.2).

SUMMARY

Four types of acid–base reactions were considered in Section 18.1, and the general results when mixing equal amounts of acid and base are

Strong acid	+ strong base	⟶	neutral solution
HCl	+ NaOH	⟶	H_2O + NaCl
Weak acid	+ strong base	⟶	basic solution
CH_3COOH	+ NaOH	⟶	H_2O + Na(CH_3COO) (a basic salt)
Strong acid	+ weak base	⟶	acidic solution
HCl	+ NH_3	⟶	NH_4Cl (an acidic salt)

Weak acid + weak base	$\longrightarrow$	pH depends on relative K_a and K_b values
CH_3COOH + NH_3	$\longrightarrow$	$NH_4(CH_3COO)$

Acetic acid ionizes to acetate and hydronium ions. If either of these is already present or is added after equilibrium is achieved, the acetic acid will ionize to a smaller extent, as predicted by Le Chatelier's principle. The general effect is often referred to as the **common ion effect** (Section 18.2). The specific case of a weak acid ionizing in the presence of a significant concentration of its conjugate base (CH_3COOH with CH_3COO^-) or a weak base ionizing in the presence of its conjugate acid (NH_3 with NH_4^+) can give a buffer solution. A **buffer solution** will resist a change in pH when a strong acid or base is added. The hydronium ion concentration of an acid buffer (say CH_3COOH with CH_3COO^-) is

$$[H_3O^+] = \frac{[\text{acid}]}{[\text{conjugate base}]} K_a$$

Similarly, the hydroxide ion concentration in a base buffer (say NH_3 with NH_4^+) is

$$[OH^-] = \frac{[\text{base}]}{[\text{conjugate acid}]} K_b$$

Both expressions show that the pH of the buffer depends primarily on K and secondarily on the ratio of weak acid to conjugate base or weak base to conjugate acid.

The titration of a strong acid with a strong base follows the course outlined in Figure 18.5 (Section 18.3). A very large, vertical rise in the pH is seen on passing through the equivalence point, and the pH at the equivalence point is 7. If a weak acid is titrated with a strong base, then the pH at the equivalence point is greater than 7 (Figure 18.7). Furthermore, at the midpoint of the titration (the point at which half the acid has been consumed), $[H_3O^+] = K_a$. (Similarly, in the titration of a weak base with a strong acid, $[OH^-] = K_b$ at the midpoint of the titration.)

Acid–base indicators (Section 18.4) are generally weak acids in which the acid form (HIn) has one color and the conjugate base (the anion In^-) has quite a different color. In solutions of low pH, HIn will predominate, while In^- will be the predominant species in high pH solutions. Thus, the color changes with change in pH.

STUDY QUESTIONS

REVIEW QUESTIONS

1. What is the equivalence point of an acid–base reaction?
2. State whether the pH is equal to 7, less than 7, or greater than 7 when equal molar amounts of (a) a weak base react with a strong acid, (b) a strong base react with a strong acid, and (c) a strong base react with a weak acid.
3. Sketch the general shape of the titration curve when (a) a strong base is titrated with a strong acid and (b) a weak base is titrated with a strong acid. In each case indicate whether the pH at the equivalence point is less than 7, equal to 7, or greater than 7.
4. Briefly describe how a buffer solution can control the pH of a solution when excess strong acid is added. Use NH_3/NH_4Cl as an example.
5. Briefly describe how a buffer solution can control the pH of a solution when excess strong base is added. Use acetic acid/sodium acetate as an example.
6. Briefly explain the difference between the equivalence point of a titration and its end point.
7. Prove that the pH at the midpoint in the titration

of a weak base with a strong acid is pH = 14.00 + log K_b.

ACID–BASE REACTIONS AND TITRATIONS

8. What is the value of K_{net} for the reaction of benzoic acid with NaOH?
9. What is the value of K_{net} for the reaction of the weak base pyridine with HCl?
10. Calculate the hydronium ion concentration and pH at the equivalence point in the reaction of 22.0 mL of 0.10 M acetic acid, CH_3COOH, with 22.0 mL of 0.10 M NaOH.
11. Calculate the hydronium ion concentration and the pH at the equivalence point in a titration of 50.0 mL of 0.40 M NH_3 with 0.40 M HCl.
12. Tell whether the pH at the equivalence point is less than 7, equal to 7, or greater than 7 for each reaction below.
    (a) 0.10 M acetic acid, CH_3COOH, with 0.10 M KOH
    (b) 0.015 M NH_3 with 0.015 M HCl
    (c) 0.0020 M HNO_3 with 0.0040 M NaOH
13. Tell whether the pH at the equivalence point is less than 7, equal to 7, or greater than 7 for each reaction below.
    (a) 0.45 M H_2SO_4 with 0.90 M NaOH (both H^+ ions of H_2SO_4 are titrated)
    (b) 0.050 M formic acid, HCOOH, is titrated with 0.034 M NaOH
    (c) 0.052 M $H_2C_2O_4$ (oxalic acid) is titrated with 0.065 M NaOH (both H^+ ions of oxalic acid are titrated)
14. Find the pH of the solution when 20.0 mL of 0.15 M HCl is mixed with
    (a) 20.0 mL of 0.15 M NH_3
    (b) 30.0 mL of 0.15 M NaOH
15. Find the pH of the solution when 25.0 mL of 0.015 M NaOH is mixed with
    (a) 25.0 mL of 0.015 M formic acid, HCOOH
    (b) 0.046 g of benzoic acid, C_6H_5COOH
16. Phenol, C_6H_5OH, is a weak organic acid that has many uses, and about 2.5 billion pounds are produced annually in the United States. Assume you dissolve 0.515 g of the compound in exactly 100 mL of water and then titrate the resulting solution with 0.123 M NaOH.

$$C_6H_5OH(aq) + OH^-(aq) \longrightarrow C_6H_5O^-(aq) + H_2O(\ell)$$

What are the concentrations of all of the following ions at the equivalence point: Na^+, H_3O^+, OH^-, and $C_6H_5O^-$? What is the pH of the solution?

17. Benzoic acid, which occurs naturally in many forms, is a weak monoprotic acid. Assume you dissolve 0.235 g of the acid in exactly 100 mL of water and then titrate the solution with 0.108 M NaOH.

$$C_6H_5COOH(aq) + OH^-(aq) \longrightarrow C_6H_5COO^-(aq) + H_2O(\ell)$$

What are the concentrations of all of the following ions at the equivalence point: Na^+, H_3O^+, OH^-, and $C_6H_5COO^-$? What is the pH of the solution?

18. In order to determine the concentration of a solution of aqueous ammonia, you titrate it with 0.0104 M HCl. You find that 25.0 mL of the ammonia solution requires 35.46 mL of the acid to reach the equivalence point. What are the concentrations of H_3O^+, OH^-, and NH_4^+ at the equivalence point? What is the pH of the solution? What was the concentration of NH_3 in the original solution?
19. In order to determine the concentration of a solution of aqueous formic acid, HCOOH, you titrate it with 0.275 M NaOH. You find that 25.0 mL of the formic acid solution requires 22.43 mL of the base to reach the equivalence point. What are the concentrations of H_3O^+, OH^-, and $HCOO^-$ at the equivalence point? What is the pH of the solution? What was the concentration of formic acid in the original solution?

THE COMMON ION EFFECT

20. Does the pH of the solution increase, decrease, or stay the same when you
    (a) add solid ammonium chloride to 100 mL of 0.10 M NH_3?
    (b) add solid sodium acetate to 50.0 mL of 0.015 M acetic acid?
    (c) add solid NaCl to 25.0 mL of 0.10 M NaOH?
21. Does the pH of the solution increase, decrease, or stay the same when you
    (a) add solid sodium oxalate, $Na_2C_2O_4$, to 50.0 mL of 0.015 M oxalic acid, $H_2C_2O_4$?
    (b) add solid ammonium chloride to 100 mL of 0.016 M HCl?
    (c) add 20.0 g of NaCl to 1.0 L of 0.012 M sodium acetate, $Na(CH_3COO)$?
22. What is the pH of the resulting solution when 2.20 g of NH_4Cl, ammonium chloride, is added to 250 mL of 0.12 M NH_3? Is the final pH lower or higher than the pH of the original ammonia solution?
23. Lactic acid ($K_a = 1.4 \times 10^{-4}$) is found in sour milk, in sauerkraut, and in muscles after activity. If you add 2.75 g of sodium lactate, $NaC_3H_5O_3$, to 500. mL of 0.100 M lactic acid, $C_3H_6O_3$, what is the pH of the resulting solution? Is the final pH lower or higher than the pH of the lactic acid solution? (See page 760.)
24. What is the pH of a 0.15 M acetic acid solution? If you add 83 g of sodium acetate to 1.50 L of the 0.15 M acetic acid solution, what is the new pH of the solution?
25. Calculate the pH of a 0.050 M solution of HF. What

is the pH of the solution if you add 1.58 g of NaF to exactly 250 mL of the 0.050 M HF solution?

26. Does the pH of the solution increase, decrease, or stay the same when you
(a) add 10.0 mL of 0.10 M HCl to 25.0 mL of 0.10 M NH_3?
(b) add 25.0 mL of 0.050 M NaOH to 50.0 mL of 0.050 M acetic acid?

27. Does the pH of the solution increase, decrease, or stay the same when you
(a) add 25.0 mL of 0.050 M NaOH to 25.0 mL of 0.075 M oxalic acid?
(b) add 1.0 mL of 0.10 M HCl to 25.0 mL of 0.10 M pyridine, a weak base?

BUFFER SOLUTIONS

28. How many grams of ammonium chloride, NH_4Cl, would have to be added to exactly 500 mL of 0.10 M NH_3 solution to have a pH of 9.0?

29. How many grams of sodium acetate, $Na(CH_3COO)$, must be added to 1.00 L of 0.15 M acetic acid to give a pH of 4.5?

30. In order to buffer a solution at a pH of 4.57, how many grams of sodium acetate, $Na(CH_3COO)$, should you add to 500. mL of a 0.150 M solution of acetic acid, CH_3COOH?

31. A buffer solution can be made from benzoic acid (C_6H_5COOH) and sodium benzoate (C_6H_5COONa). How many grams of the acid would you have to mix with 14.4 g of the sodium salt in order to have a pH of 3.88?

32. If a buffer solution is made of 12.2 g of benzoic acid (C_6H_5COOH) and 7.20 g of sodium benzoate (C_6H_5COONa) in exactly 250 mL of solution, what is the pH of the buffer? If the solution is diluted to exactly 500 mL with pure water, what is the new pH of the solution?

33. If a buffer solution is prepared from 5.15 g of NH_4NO_3 and 0.10 L of 0.15 M NH_3, what is the pH of the solution? What is the new pH if the solution is diluted with pure water to a volume of exactly 500 mL?

34. Which of the following combinations would be the best to buffer the pH at approximately 9?
(a) acetic acid/sodium acetate [$CH_3COOH/Na(CH_3COO)$]
(b) HCl/NaCl
(c) ammonia/ammonium chloride [NH_3/NH_4Cl]

35. Many natural processes can be studied in the laboratory but only in an environment of controlled pH. Which of the following combinations would be the best choice to buffer the pH at approximately 7?
(a) H_3PO_4/NaH_2PO_4
(b) NaH_2PO_4/Na_2HPO_4
(c) Na_2HPO_4/Na_3PO_4

36. A buffer solution is prepared by adding 0.125 mole of ammonium chloride to 500. mL of 0.500 M solution of ammonia.
(a) What is the pH of the buffer?
(b) If 0.0100 mole of HCl gas is bubbled into 500. mL of the buffer, what is the new pH of the solution?

37. A buffer solution was prepared by adding 4.95 g of sodium acetate, $Na(CH_3COO)$, to 250. mL of 0.150 M acetic acid, CH_3COOH.
(a) What is the pH of the buffer?
(b) What is the pH of 100. mL of the buffer solution if you add 80. mg of NaOH to the solution?

38. If you dissolve 0.425 g of NaOH in a 2.00-L solution that originally had $[H_2PO_4^-] = [HPO_4^{2-}] = 0.132$ M, calculate the resulting pH.

39. How many moles of HCl must be added to 1.00 L of a buffer made from 0.150 M NH_3 and 10.0 g NH_4Cl to decrease the pH by one unit?

TITRATION CURVES AND INDICATORS

40. Without doing detailed calculations, sketch the curve for the titration of 25.0 mL of 0.10 M NaOH with 0.10 M HCl. Indicate the pH at the beginning of the titration and at the equivalence point.

41. Without doing detailed calculations, sketch the curve for the titration of 10 mL of 0.10 M Na_2CO_3 with 0.10 M HCl.

42. Assume you titrate 20.0 mL of 0.11 M NH_3 with 0.10 M HCl.
(a) What is the pH of the NH_3 solution before the titration begins?
(b) What is the pH at the equivalence point?
(c) What is the pH at the midpoint of the titration?
(d) Which indicator in Table 17.3 would be best to detect the equivalence point?
(e) Calculate the pH of the solution after adding 5.00, 11.0, 15.0, 20.0, 22.0, and 25.0 mL of the acid. Combine this information with that from (a) through (c) and plot the titration curve.

43. Aniline, $C_6H_5NH_2$, is a weak base used extensively in the dye industry. Its conjugate acid, aniline hydrochloride, $[C_6H_5NH_3]Cl$, can be titrated easily with a strong base such as NaOH. Assume you titrate 25.0 mL of 0.100 M aniline hydrochloride with 0.115 M NaOH. (K_a for aniline hydrochloride is 2.4×10^{-5}.)
(a) What is the pH of the $[C_6H_5NH_3]Cl$ solution before the titration begins?
(b) What is the pH at the equivalence point?
(c) What is the pH at the midpoint of the titration?
(d) Which indicator in Table 17.3 would be best to detect the equivalence point?
(e) Calculate the pH of the solution after adding 5.00, 10.0, 15.0, 20.0, 24.0, and 30.0 mL of the

base. Combine this information with that from (a) through (c) and plot the titration curve.

44. Using Table 17.3, suggest an indicator to use in each of the following titrations:
(a) The weak base pyridine with HCl
(b) Formic acid with NaOH
(c) Hydrazine with HCl

45. Using Table 17.3, suggest an indicator to use in each of the following titrations:
(a) Na_2CO_3 titrated to HCO_3^- with HCl
(b) Hypochlorous acid with NaOH
(c) Trimethylamine with HCl

GENERAL QUESTIONS

46. Suppose you add 12.5 mL of 4.15 M acetic acid to 25.0 mL of 1.00 M NaOH. Calculate the hydronium ion concentration and pH of the resulting solution.

47. Calculate the concentrations of NH_4^+, OH^-, NH_3, and Na^+ in a solution that was originally 0.040 M NaOH and 0.20 M in NH_3.

48. What is the pH of a solution that is made by mixing 20.0 mL of 1.00 M HF and 0.400 g of NaOH?

49. Pyridine, C_5H_5N, is a weak organic base. In spite of its very disagreeable odor, it is widely used as a starting material for the synthesis of numerous other compounds. If you dissolve 0.453 g in exactly 100 mL of water, and titrate the solution with 0.205 M HCl,

$$C_5H_5N(aq) + HCl(aq) \longrightarrow C_5H_5NH^+(aq) + Cl^-(aq)$$

what are the concentrations of the following ions at the equivalence point: $C_5H_5NH^+$, H_3O^+, OH^-, and Cl^-? What is the pH at the equivalence point?

50. Calculate the pH at the equivalence point in a titration of 25.0 mL of 0.120 M formic acid, HCOOH, with 0.105 M NaOH.

51. During active exercise, lactic acid ($K_a = 1.4 \times 10^{-4}$) is produced in the muscle tissues (see Section 18.2). At the pH of the body (pH = 7.4), which form will be primarily present: nonionized lactic acid [$CH_3CH(OH)COOH$] or the lactate ion [$CH_3CH(OH)COO^-$]?

52. A 500.-mL solution contains 0.150 mol of $NaNO_2$ and 0.200 mol of HNO_2. How many more grams of $NaNO_2$ must be added to this solution to have a pH of 4.00?

53. *Buffer capacity* is defined as the number of moles of a strong acid or strong base that are required to change the pH of one liter of the buffer solution by one unit. What is the buffer capacity of a solution that is 0.10 M in acetic acid and 0.10 M in sodium acetate?

54. Arrange the following solutions in order of increasing pH (all reagents are 0.10 M).
(a) NaCl
(b) NH_3
(c) CH_3COOH/CH_3COONa
(d) HCl
(e) NH_3/NH_4Cl
(f) CH_3COOH

55. If 50.0 mL of 0.0500 M HF is titrated with 0.0955 M NaOH, calculate the pH at the equivalence point.

56. Two acids, each approximately 10^{-2} M in concentration, are titrated separately with a strong base. The acids show the following pH values at the equivalence point: HA, pH = 9.5 and HB, pH = 8.5.
(a) Which is the stronger acid, HA or HB?
(b) Which of the conjugate bases, A^- or B^-, is the stronger base?

57. A 0.40-M solution of a weak acid HA has a pH of 2.40. What is the pH of an equimolar solution of HA and the Na^+ salt of the conjugate base, NaA?

58. The oily liquid aniline, $C_6H_5NH_2$, is a weak base. Assume you titrate 25.0 mL of a 0.0256 M aqueous solution of aniline with 0.0195 M HCl.
(a) What is the pH of the aniline solution before the titration begins?
(b) What is the pH at the equivalence point of the titration?
(c) What is the pH at the midpoint of the titration?
(d) Which indicator in Table 17.3 would be best to detect the equivalence point?
(e) Calculate the pH of the solution after adding 5.00, 10.0, 15.0, 20.0, 24.0, and 30.0 mL of the acid. Combine this information with that from (a) through (c) and plot the titration curve.

59. A common buffer solution is made from potassium hydrogen phthalate, $C_6H_4(COOH)COOK$ (K_a = 3.9×10^{-6}) and its conjugate base $C_6H_5(COO)_2^{2-}$. The dianion is made from the monoanion by adding a strong base such as NaOH.

$$C_6H_4(COOH)COO^-(aq) + OH^-(aq) \longrightarrow C_6H_4(COO)_2^{2-}(aq) + H_2O(\ell)$$

What is the pH of a buffer solution made by mixing 43.0 mL of 0.10 M NaOH and 50.0 mL of 0.10 M $C_6H_4(COOH)COOK$?

60. The pH of human blood is controlled by several buffer systems, among them the reaction

$$H_2PO_4^-(aq) + H_2O(\ell) \rightleftharpoons H_3O^+(aq) + HPO_4^{2-}(aq)$$

Calculate the ratio $[H_2PO_4^-]/[HPO_4^{2-}]$ in normal blood having a pH of 7.40.

61. The compound procaine hydrochloride is sold under the brand name Novocaine, among others.

$$\left[H_2N\text{-}C_6H_4\text{-}C(=O)\text{-}O\text{-}CH_2CH_2\text{-}NH(CH_2CH_3)\text{-}CH_2CH_3 \right]^+ Cl^-$$

The *Merck Index of Chemicals and Drugs* states that the pH of a 0.10 M aqueous solution of the compound is 6.0. What is the value of K_a for procaine hydrochloride as a weak acid?

62. When 16.0 mL of 0.20 M benzoic acid is mixed with 32.0 mL of 0.10 M NH_3, what is the pH of the resulting solution? See Study Question 121 in Chapter 17 for a method of calculating the hydronium ion concentration.

63. Assume you dissolve 0.221 g of trimethylamine, $(CH_3)_3N$, in 50.0 mL of water and titrate it with 0.100 M HCl.
    (a) What is the pH of the original amine solution?
    (b) What is the pH of the solution at the midpoint of the titration?
    (c) What is the pH at the equivalence point?
    (d) Sketch a rough titration curve for the titration and decide on a suitable indicator (Table 17.3).

64. Picric acid is used in explosives and in matches, among other things.

```
              OH
              |
   O2N       C        NO2
      \   /     \\  /
       C           C
       ||          |
      HC          CH
         \      //
            C
            |
           NO2
```

picric acid, $K_a = 5.1 \times 10^{-1}$

Assume you titrate 50.0 mL of a 0.010 M solution of the acid with 0.010 M NaOH.
    (a) What is the pH of the picric acid solution before the titration begins?
    (b) What is the pH at the equivalence point?
    (c) What is the pH at the midpoint of the titration?

65. Hydroxylamine is a weak base that readily forms salts such as $[NH_3OH]Cl$, hydroxylamine hydrochloride. This compound is used as a reducing agent in photography and as an antioxidant in soaps. Assume you have 25.0 mL of a 0.155 M solution of $[NH_3OH]Cl$ and you titrate it with 0.108 M NaOH.
    (a) What is the pH of the $[NH_3OH]Cl$ solution before the titration begins?
    (b) What is the pH at the equivalence point?
    (c) What is the pH at the midpoint of the titration?

66. The weak base ethanolamine, $HOCH_2CH_2NH_2$, can be titrated with HCl.

$$HOCH_2CH_2NH_2(aq) + H_3O^+(aq) \longrightarrow HOCH_2CH_2NH_3^+(aq) + H_2O(\ell)$$

Assume you have 25.0 mL of a 0.010 M solution of ethanolamine and titrate it with 0.0095 M HCl. If the pH at the midpoint of the titration is 9.50, what is the pH at the equivalence point?

SUMMARY QUESTION

67. The chemical name for aspirin is acetylsalicylic acid. It is believed that the analgesic and other desirable properties of aspirin are due not to the aspirin but to the simpler compound salicylic acid, $C_6H_4(OH)COOH$, that results from the breakdown of aspirin in the stomach.

```
     O       O—H
      \\   /
         C
         |
         C        O—H
      //    \   /
   HC          C
   |           ||
   HC          CH
      \\    /
         C
         |
         H
```

salicylic acid, $K_a = 1.1 \times 10^{-3}$

(a) Give approximate values for the following bond angles in the acid: (a) C=C—C in the ring; (b) C—C=O; (c) either of the C—O—H angles; and (d) C—C—H.
(b) How many π (pi) bonds and how many σ (sigma) bonds does the molecule have?
(c) What is the hybridization of the C atoms of the ring? Of the C atom in the —COOH group?
(d) Experiment shows that 1.00 g of the acid will dissolve in 460 mL of water. What is the pH of this solution?
(e) If you have salicylic acid in your stomach, and if the pH of gastric juice is 2.0, calculate the percentage of salicylic acid that will be present in the stomach in the form of the salicylate ion, $C_6H_4(OH)COO^-$.
(f) Assume you have 25.0 mL of a 0.014 M solution of salicylic acid and titrate it with 0.010 M NaOH. What is the pH at the midpoint of the titration? What is the pH at the equivalence point?

CHAPTER

19

Precipitation Reactions

CHAPTER OUTLINE

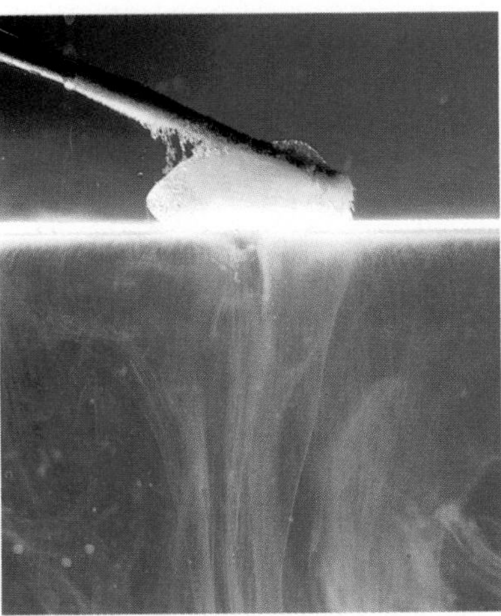

Dry ice (solid CO_2) in a saturated solution of $Ca(OH)_2$. Calcium carbonate is precipitated.

Beginning in Chapter 5 we introduced you to several types of reactions, among them **exchange reactions**. We divided these further into *precipitation reactions* and *acid–base reactions*. The previous two chapters were devoted to acids and bases and their reactions, and this one considers precipitations.

A moderately soluble compound has a solubility somewhat greater than about 0.01 mole per liter, while soluble compounds will dissolve to the extent of at least 0.1 mole per liter.

19.1 SOLUBILITY OF SALTS

A precipitation reaction is an exchange reaction in which one of the products is an insoluble compound,

$$MA(aq) + BX(aq) \longrightarrow MX(s) + BA(aq)$$

$$CaCl_2(aq) + Na_2CO_3(aq) \longrightarrow CaCO_3(s) + 2\ NaCl(aq)$$

that is, a compound having a solubility of less than about 0.01 mole of dissolved material per liter of solution. If you stir calcium carbonate (as the mineral calcite) into pure water, only about 6 mg will dissolve per liter at 25 °C. No wonder that seashells, which are mostly calcium carbonate, do not dissolve appreciably in the sea. On the other hand, you know that "sea salt," NaCl, is very soluble in water.

How do you know when to predict an insoluble compound as the product of a reaction? In Chapter 5 we described some guidelines for

Seashells do dissolve in very deep water in the ocean. Calcite is unusual in that it is about twice as soluble at 2 °C as it is at 25 °C (it has an exothermic enthalpy of solution). Thus, cold, deep ocean water promotes solubility. Further, there is more dissolved CO_2 in the deep ocean, so the following reaction converts $CaCO_3$ to the more soluble $Ca(HCO_3)_2$.

$$CaCO_3(s) + H_2O(\ell) + CO_2(aq) \longrightarrow Ca^{2+}(aq) + 2\ HCO_3^-(aq)$$

Table 19.1 Some Common Minerals

Chemical Name	Formula	Common Names (and Uses)
Calcium carbonate	$CaCO_3$	Calcite, aragonite, iceland spar (source of lime, CaO)
—	$2\ CuCO_3 \cdot Cu(OH)_2$	Azurite (one source of copper)
Mercury(II) sulfide	HgS	Cinnabar (source of mercury)
Zinc sulfide	ZnS	Zinc blende (one source of zinc)
Lead(II) sulfide	PbS	Galena (one source of lead)
—	FeS_2	Iron pyrite, fool's gold
—	$Ca_{10}(PO_4)_6(OH)_2$	Apatite (source of phosphate for phosphoric acid and fertilizers)
Calcium fluoride	CaF_2	Fluorite or fluorspar (source of HF and other inorganic fluorides)
Calcium sulfate	$CaSO_4 \cdot 2\ H_2O$	Gypsum
—	Fe_3O_4	Magnetite (one source of iron)
Titanium(IV) oxide	TiO_2	Rutile (for paint pigments)
Aluminum oxide	Al_2O_3	Bauxite (source of Al)

The "solubility rules" in Figure 5.5 are general guidelines and not "laws of nature." There are many exceptions.

making predictions (Figure 5.5), and we can now point out some interesting and useful applications of these guidelines. First, we should expect seawater to be a solution of many soluble salts. When seawater is evaporated, you find that the mass percentage of anions in the dry residue is about 55% chloride ion, about 8% sulfate ion, and less than 1% bromide ion. As you can see in Figure 5.5, these ions generally give soluble or moderately soluble salts with the cations found in seawater: sodium (about 31%), potassium (about 1%), calcium (about 1%), and magnesium (about 4%).

Second, our economy depends on various metals and chemicals produced from minerals found in the earth (Figure 19.1). These minerals are often concentrated in certain geographic areas and are usually insoluble salts. (If they were soluble, they would have been dissolved by ground water.) A few examples are listed in Table 19.1. Notice that all these minerals fit the guidelines for solubility in Figure 5.5. That is, sulfides, oxides, carbonates, and hydroxides are not listed as anions that lead to solubility (unless they are combined with Group 1A metal cations or NH_4^+).

Figure 19.1 Some common minerals: (left) iron pyrite or "fool's gold," FeS_2; (center) calcite, $CaCO_3$; and (right) fluorite, CaF_2.

Third, since many economically important metal ores are sulfides, oxides, and carbonates, methods have to be found to recover the metals from insoluble materials. The refining of nickel(II) is an example of the use of precipitation reactions. Nickel oxide is found in very low concentrations in iron-bearing rock. To recover the nickel, it is leached or extracted from the rock with sulfuric acid to give soluble nickel(II) sulfate and water, an exchange reaction that owes its "driving force" to the formation of water.

$$NiO(s) + H_2SO_4(aq) \longrightarrow NiSO_4(aq) + H_2O(\ell)$$

Then, in another exchange reaction, the nickel(II) ion is precipitated with sulfide ion to give black nickel(II) sulfide.

$$NiSO_4(aq) + H_2S(aq) \longrightarrow H_2SO_4(aq) + NiS(s)$$

EXERCISE 19.1 Solubility of Salts

Using the guidelines of Figure 5.5, predict whether each of the following will be soluble or insoluble in water.

(a) AgBr (d) KF
(b) K_2CrO_4 (e) $NH_4(CH_3COO)$
(c) $SrCO_3$ (f) $Ba(NO_3)_2$

EXERCISE 19.2 Precipitation Reactions

Which of the following mixtures of salts should lead to a precipitate when 0.10 M solutions are mixed? Give the formula of the precipitate.

(a) $NaCl(aq) + AgNO_3(aq)$
(b) $KI(aq) + Pb(NO_3)_2(aq)$
(c) $CuCl_2(aq) + Mg(NO_3)_2(aq)$

19.2 THE SOLUBILITY PRODUCT

Silver bromide, AgBr, is used in photographic film. The water-insoluble salt can be made by adding a water-soluble silver salt ($AgNO_3$) to an aqueous solution of a bromide-containing salt (KBr). The *net ionic equation* for the reaction that occurs is

$$Ag^+(aq) + Br^-(aq) \longrightarrow AgBr(s)$$

See Something More About Silver Chemistry on page 809.

If some of the precipitated AgBr is placed in pure water, some salt will eventually dissolve and an equilibrium will be established.

$$AgBr(s) \rightleftharpoons Ag^+(aq, 5.7 \times 10^{-7}\ M) + Br^-(aq, 5.7 \times 10^{-7}\ M)$$

When the AgBr has dissolved to the greatest extent possible, the solution is said to be saturated (see Chapter 14). At this point, an experiment would show that the concentrations of the silver and bromide ions in solution are both about 5.7×10^{-7} M at 25 °C. The extent to which an insoluble salt dissolves can be expressed in terms of the equilibrium constant for the dissolving process. In this case,

$$K = [Ag^+(aq)][Br^-(aq)]$$

Recall from Chapter 16 that the "concentrations" of solids do not appear in an equilibrium constant expression.

When silver bromide dissolves in water, one mole of Ag^+ ions and one mole of Br^- ions are produced for every mole of AgBr dissolved.*

*Dissolving ionic solids is actually a complex process. It almost always involves more than just the simple equilibria shown here. As a result, you should realize that simple K_{sp} calculations can be inaccurate. In particular, the calculated solubility can be incorrect, especially for salts having ions with charges larger than 1+ or 1−. It has been noted that "the solubility product [is more properly] a description, not of solubility, but simply of the ionic concentrations in the saturated solution." (L. Meites, J.S.F. Pode, and H.C. Thomas, *J. Chem. Educ.*, **1966**, *43*, 667.)

Table 19.2 K_{sp} Values of Some Poorly Soluble Salts

Compound	K_{sp} at 25 °C
$CaCO_3$	3.8×10^{-9}
$SrCO_3$	9.4×10^{-10}
$BaCO_3$	8.1×10^{-9}
CaF_2	3.9×10^{-11}
BaF_2	1.7×10^{-6}
CdS	3.6×10^{-29}
PbS	8.4×10^{-28}
CuS	8.7×10^{-36}
HgS	3.0×10^{-53}
CuCl	1.9×10^{-7}
AgCl	1.8×10^{-10}
AuCl	2.0×10^{-13}

Therefore, the solubility of silver bromide can be determined if the concentration of *either* the silver ion *or* the bromide ion is measured. This means that the equilibrium expression tells us that the *product* of the two measures of the *solubility* of a solid compound is a *constant*. Hence, this constant has come to be called the **solubility product constant**, and it is often designated by K_{sp}.

$$K_{sp} = [Ag^+(aq)][Br^-(aq)]$$

Since the concentrations of both Ag^+ and Br^- are 5.7×10^{-7} M when silver bromide is in equilibrium with its ions, this means that K_{sp} is

$$K_{sp} = [5.7 \times 10^{-7}][5.7 \times 10^{-7}] = 3.2 \times 10^{-13} \text{ (at 25 °C)}$$

In general, K_{sp} is always of the form

$$A_xB_y(s) \rightleftharpoons x\ A^{n+}(aq) + y\ B^{m-}(aq)$$

$$K_{sp} = [A^{n+}]^x[B^{m-}]^y$$

This means that you would write K_{sp} expressions as follows:

$$CaF_2(s) \rightleftharpoons Ca^{2+}(aq) + 2\ F^-(aq)$$
$$K_{sp} = [Ca^{2+}][F^-]^2 = 3.9 \times 10^{-11}$$

$$Ag_2SO_4(s) \rightleftharpoons 2\ Ag^+(aq) + SO_4{}^{2-}(aq)$$
$$K_{sp} = [Ag^+]^2[SO_4{}^{2-}] = 1.7 \times 10^{-5}$$

$$Bi_2S_3(s) \rightleftharpoons 2\ Bi^{3+}(aq) + 3\ S^{2-}(aq)$$
$$K_{sp} = [Bi^{3+}]^2[S^{2-}]^3 = 1.6 \times 10^{-72}$$

The numerical values of K_{sp} for a few salts are given in Table 19.2, but many more values are collected in Appendix G. Notice that all are given for a temperature of 25 °C.

As with any equilibrium constant, K_{sp} values change with temperature; the direction of change depends on the enthalpy of solution of the solid. See the marginal note on page 781 about calcite solubility.

EXERCISE 19.3 Writing K_{sp} Expressions

Write K_{sp} expressions for the following insoluble salts and look up numerical values for the constant in Appendix G.
(a) CdS (b) BiI_3 (c) Ag_2CO_3

19.3 DETERMINING K_{sp} FROM EXPERIMENTAL MEASUREMENTS

In practice, solubility product constants are determined by careful laboratory measurements that use various chemical and spectroscopic methods. We shall not go into these methods, but we shall simply assume that the given ion concentrations have been measured experimentally.

EXAMPLE 19.1

K_{sp} FROM SOLUBILITY MEASUREMENTS

The solubility of silver iodide, AgI, is 1.22×10^{-8} mol/L. Calculate K_{sp} for AgI. (See *Something More About Silver Chemistry*, page 809.)

Solution When silver iodide dissolves, the reaction is represented by the equation

$$AgI(s) \rightleftharpoons Ag^+(aq) + I^-(aq)$$

and so the equilibrium expression will be

$$K_{sp} = [Ag^+][I^-]$$

The question says that the "solubility" of silver iodide is 1.22×10^{-8} mol/L. Since each mole of AgI that dissolves will give rise to one mole of Ag^+ and one mole of I^- in solution, this means that the concentration of each ion is 1.22×10^{-8} M. K_{sp} is the product of these concentrations.

$$K_{sp} = [1.22 \times 10^{-8}][1.22 \times 10^{-8}] = 1.49 \times 10^{-16}$$

EXAMPLE 19.2

K_{sp} FROM SOLUBILITY MEASUREMENTS

Lead(II) chloride dissolves to a slight extent in water according to the equation

$$PbCl_2(s) \rightleftharpoons Pb^{2+}(aq) + 2\ Cl^-(aq)$$

Calculate the K_{sp} if the lead ion concentration has been found to be 1.62×10^{-2} mol/L.

Solution If $PbCl_2$ is placed in water and allowed to dissolve, the Cl^- concentration must be *twice* the Pb^{2+} concentration.

$$\text{If } [Pb^{2+}] = 1.62 \times 10^{-2}\ \text{M, then } [Cl^-] = 2[Pb^{2+}] = 3.24 \times 10^{-2}\ \text{M}$$

To calculate K_{sp} we simply substitute these values into the appropriate expression.

$$K_{sp} = [Pb^{2+}][Cl^-]^2 = [1.62 \times 10^{-2}][3.24 \times 10^{-2}]^2 = 1.70 \times 10^{-5}$$

EXERCISE 19.4 K_{sp} from Solubility Measurements

Calculate the K_{sp} value for each of the following salts dissolved in water if the silver ion concentration in the saturated solution is known for each. Verify your answer using the table of values in Appendix G.

(a) AgSCN dissolves to give Ag^+ and SCN^- ions. ($[Ag^+]$ at equilibrium = 1.0×10^{-6} mol/L.)

(b) Ag_2S dissolves to give Ag^+ and S^{2-}. ($[Ag^+]$ at equilibrium = 5.8×10^{-17} mol/L.)

19.4 ESTIMATING SALT SOLUBILITY FROM K_{sp}

Through a variety of laboratory measurements, the K_{sp} values for many insoluble salts have been determined. These are an invaluable aid, since we can use them to estimate the solubility of the solid salt. We can then tell the extent to which a solid salt will dissolve, or whether a solid will precipitate if solutions of its anion and cation are mixed. We turn now to methods that can be used to estimate the solubility of a salt from its K_{sp} value. In Section 19.7 we shall see how to use these predictions to plan the separation of ions that are mixed in solution.

EXAMPLE 19.3

SOLUBILITY FROM K_{sp}

Equilibria involving the carbonate ion are actually more complex than can be presented here. The solubility of $CaCO_3$ is best defined only in terms of $[Ca^{2+}]$, since hydrolysis of CO_3^{2-} actually makes $[CO_3^{2-}]$ slightly less than $[Ca^{2+}]$. See Section 19.10.

The K_{sp} for $CaCO_3$ (as the mineral calcite) is 3.8×10^{-9} at 25 °C. Calculate the solubility of calcium carbonate in pure water in (a) moles per liter and (b) grams per liter.

Solution The equation for the solubility of $CaCO_3$ is

$$CaCO_3(s) \rightleftharpoons Ca^{2+}(aq) + CO_3^{2-}(aq)$$

and the equilibrium expression is $K_{sp} = [Ca^{2+}][CO_3^{2-}]$. When one mole of calcium carbonate dissolves, one mole of Ca^{2+} and one mole of CO_3^{2-} ions are produced. Thus, the solubility of $CaCO_3$ can be measured by determining the concentration of *either* Ca^{2+} *or* CO_3^{2-}. Let us denote the solubility of $CaCO_3$ (in mol/L) by the symbol S; that is, S moles of $CaCO_3$ dissolve per liter. Therefore, both $[Ca^{2+}]$ and $[CO_3^{2-}]$ must also equal S at equilibrium.

	$[Ca^{2+}]$	$[CO_3^{2-}]$
Initial concentration (M)	0	0
Change on proceeding to equilibrium	$+S$	$+S$
Equilibrium concentration (M)	S	S

Since K_{sp} is the product of the calcium and carbonate ion concentrations, each of which is a measure of the solubility of $CaCO_3$, this means K_{sp} is the square of the solubility.

$$K_{sp} = [Ca^{2+}][CO_3^{2-}] = (S)(S) = S^2$$

Given that the K_{sp} for $CaCO_3$ in pure water is 3.8×10^{-9}, the value of S is

$$S = [Ca^{2+}] = \sqrt{3.8 \times 10^{-9}} = 6.2 \times 10^{-5} \text{ M}$$

The solubility of $CaCO_3$ in pure water is 6.2×10^{-5} mol/L. To find its solubility in g/L, we need only multiply by the molar mass of $CaCO_3$.

$$\text{Solubility in g/L} = (6.2 \times 10^{-5} \text{ mol/L})(100. \text{ g/mol}) = 0.0062 \text{ g/L}$$

EXAMPLE 19.4

SOLUBILITY FROM K_{sp}

Knowing that the K_{sp} of MgF_2 is 6.4×10^{-9}, calculate the solubility of the salt in (a) moles per liter and (b) grams per liter.

Solution As usual we begin by writing the equilibrium equation

$$MgF_2(s) \rightleftharpoons Mg^{2+}(aq) + 2\ F^-(aq)$$

and the K_{sp} expression

$$K_{sp} = [Mg^{2+}][F^-]^2$$

Our next problem is to define the salt solubility in terms that will allow us to solve the K_{sp} expression for this value. It is evident that if one mole of MgF_2 dissolves, one mole of Mg^{2+} and *two* moles of F^- will appear in the solution. From this we conclude that MgF_2 solubility is equivalent to the concentration of Mg^{2+} in the solution. Therefore, if we say the solubility of MgF_2 is given by the unknown quantity S, then $[Mg^{2+}] = S$ and $[F^-] = 2S$.

	$[Mg^{2+}]$	$[F^-]$
Initial concentration (M)	0	0
Change on proceeding to equilibrium	$+S$	$+2S$
Equilibrium concentration (M)	S	$2S$

Substituting these values into the K_{sp} expression, we find

$$K_{sp} = [Mg^{2+}][F^-]^2 = (S)(2S)^2 = 4S^3$$

When the equation is solved for S, we have

$$S = \sqrt[3]{\frac{K_{sp}}{4}} = \sqrt[3]{\frac{6.4 \times 10^{-9}}{4}} = 1.2 \times 10^{-3}\ M$$

This tells us that 1.2×10^{-3} mole of MgF_2 dissolves per liter to give a Mg^{2+} concentration of 1.2×10^{-3} M and a F^- concentration of $2(1.2 \times 10^{-3}) = 2.4 \times 10^{-3}$ M. Finally, the solubility of MgF_2 in grams per liter is

$$(1.2 \times 10^{-3}\ mol/L)(62.3\ g/mol) = 0.075\ g\ MgF_2/L$$

EXERCISE 19.5 Solubility from K_{sp}

Using the values of K_{sp} in Appendix G, calculate the solubilities of CdS and $Mg(OH)_2$ in moles per liter.

The relative solubilities of salts can often be deduced by comparing values of solubility product constants, but you must be careful! For example, the K_{sp} for silver chloride is

$$AgCl(s) \rightleftharpoons Ag^+(aq) + Cl^-(aq) \qquad K_{sp} = 1.8 \times 10^{-10}$$

whereas that for silver chromate is

$$Ag_2CrO_4(s) \rightleftharpoons 2\ Ag^+(aq) + CrO_4^{2-}(aq) \qquad K_{sp} = 9.0 \times 10^{-12}$$

Although silver chromate has a numerically smaller K_{sp} value, it is approximately 10 times more soluble than silver chloride. If you solve for the solubilities as in the preceding examples, you find $S_{chloride} = 1.3 \times 10^{-5}$ mol/L, whereas $S_{chromate} = 1.3 \times 10^{-4}$ mol/L. Valid comparisons for solubility on the basis of K_{sp} values can be made only for salts having the same ion ratio. This means, for example, that you can directly compare solubilities of 1:1 salts such as the silver halides by comparing their K_{sp} values. The K_{sp} values for the compounds of Ag^+ with Cl^-, Br^-, and I^- are in the order

See Something More About Silver Chemistry on page 809.

$$AgCl\ (K_{sp} = 1.8 \times 10^{-10}) > AgBr\ (K_{sp} = 3.3 \times 10^{-13}) > AgI\ (K_{sp} = 1.5 \times 10^{-16})$$

and so are their solubilities: $S(AgCl) > S(AgBr) > S(AgI)$. Similarly, you could compare 1:2 salts such as the lead halides:

$$PbCl_2\ (K_{sp} = 1.7 \times 10^{-5}) > PbBr_2\ (K_{sp} = 6.3 \times 10^{-6}) > PbI_2\ (K_{sp} = 8.7 \times 10^{-9})$$

$$S(PbCl_2) > S(PbBr_2) > S(PbI_2)$$

A cluster of aragonite needles formed by precipitation in Jewel Cave, Wind Cave National Park (South Dakota). Aragonite is a form of calcium carbonate. (Photo by Arthur N. Palmer)

EXERCISE 19.6 Comparing Salt Solubilities

Using K_{sp} values, tell which salt in each pair is more soluble in water.
(a) AgCl or AgCN (b) $Mg(OH)_2$ or $Ca(OH)_2$ (c) $MgCO_3$ or $CaCO_3$

An interesting example of the practical consequences of relative solubilities of insoluble salts occurs in the sea. A group of marine animals, the pteropods, form the mineral aragonite, which is one type of $CaCO_3$. Most marine organisms, however, form calcite, another type of calcium carbonate. Measurements at 2 °C and at depths of several thousand meters show that aragonite is about 1.5 times more soluble than calcite. This suggests why pteropod shells are not found at all beyond depths of a few hundred meters in the Pacific Ocean, whereas substantial calcite deposits are found to depths of 3500 meters. The solubility of $CaCO_3$ in either mineral form increases with a drop in temperature, and the reaction

$$CaCO_3(s) + H_2O(\ell) + CO_2(aq) \rightleftharpoons Ca^{2+}(aq) + 2\ HCO_3^-(aq)$$

occurs to a greater extent. Since this is a dynamic equilibrium, aragonite (the more soluble form) dissolves, and the less soluble form (calcite) precipitates.

19.5 PRECIPITATION OF INSOLUBLE SALTS

Metal-bearing ores often contain the metal in the form of an insoluble salt (Figure 19.1), and, to complicate matters, the ores often contain several such metal salts. A number of industrial methods for separating metals from their ores involve dissolving the metal salts to obtain the metal ion or ions in solution. The solution is then usually concentrated in some manner, and a precipitating agent is added to precipitate selectively only one type of metal ion as an insoluble salt. In the case of nickel, the ion can be precipitated as insoluble nickel(II) sulfide or nickel(II) carbonate.

$$NiS(s) \rightleftharpoons Ni^{2+}(aq) + S^{2-}(aq) \qquad K_{sp} = 3.0 \times 10^{-21}$$

$$NiCO_3(s) \rightleftharpoons Ni^{2+}(aq) + CO_3^{2-}(aq) \qquad K_{sp} = 6.6 \times 10^{-9}$$

The final step in obtaining the metal itself is to reduce the ion to the metal either chemically or electrochemically (Chapter 21).

Our goal in this section is to work out methods to determine whether a precipitate will form under a given set of conditions. For example, if Ag^+ and Cl^- are present at some given concentrations, will AgCl precipitate from the solution?

K_{sp} AND THE REACTION QUOTIENT, Q

Silver chloride, like silver bromide, is used in photographic films. It dissolves to a very small extent in water and has a correspondingly small value of K_{sp}.

$$AgCl(s) \rightleftharpoons Ag^+(aq) + Cl^-(aq) \qquad K_{sp} = [Ag^+][Cl^-] = 1.8 \times 10^{-10}$$

If you plot the concentrations of silver and chloride ions that satisfy the K_{sp} expression, you obtain the curve shown in Figure 19.2. This curve

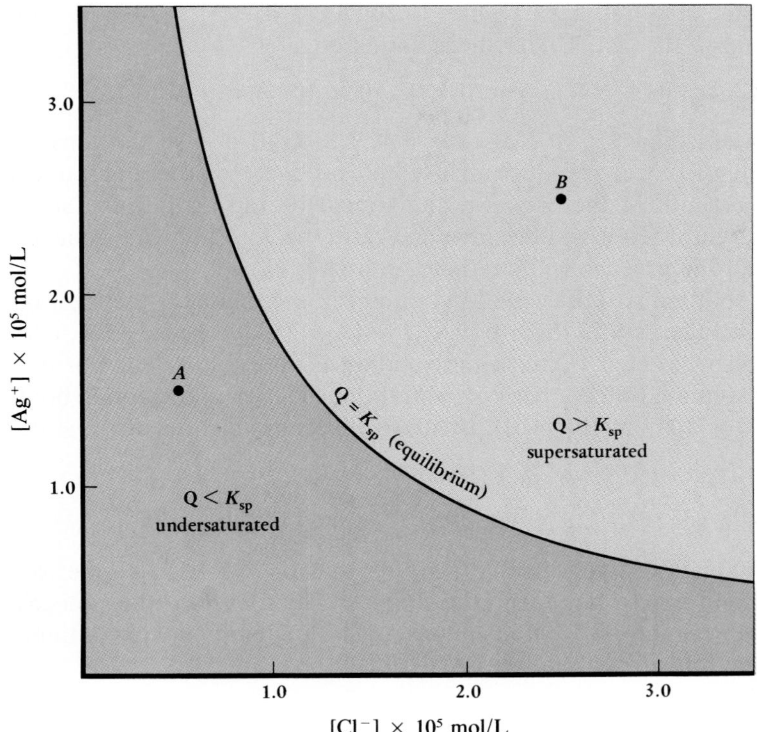

Figure 19.2 The solubility product constant for AgCl ($K_{sp} = 1.8 \times 10^{-10}$). The curve represents all combinations of $[Ag^+]$ and $[Cl^-]$ at which equilibrium can exist ($Q = K_{sp}$). If a solution is represented by a point to the left of the line, it is not saturated ($Q < K_{sp}$). To the right of the line solutions are supersaturated ($Q > K_{sp}$).

represents all of the combinations of silver and chloride ion concentrations where equilibrium exists with solid silver chloride. *When the product of ion concentrations is equal to K_{sp}, the solution is said to be* **saturated**.

Will AgCl precipitate from a solution in which the silver and chloride ion concentrations are represented by a point to the left of the curve, say point *A*? To answer this, we need only to calculate Q, the **reaction quotient**. For the general reaction

See Section 16.3 for a discussion of the reaction quotient. In the case of equilibria involving insoluble solids, some chemists call the reaction quotient the "ion product."

$$A_xB_y(s) \rightleftharpoons x\,A^{n+}(aq) + y\,B^{m-}(aq)$$

the reaction quotient is

$$Q = [A^{n+}]^x[B^{m-}]^y$$

where $[A^{n+}]^x$ and $[B^{m-}]^y$ are the actual concentrations of the ions A^{n+} and B^{m-} in solution; they are not necessarily the equilibrium concentrations. Based on this relation, we can reach the following important conclusions:

1. **If $Q < K_{sp}$, the system is not at equilibrium; the solution is *not* saturated.** This means one of two things: (a) If solid A_xB_y is present, more will dissolve until equilibrium is achieved (when $Q = K_{sp}$). (b) If solid A_xB_y is not already present, more $A^{n+}(aq)$ or more $B^{m-}(aq)$ (or both) can be added to the solution before precipitation of solid A_xB_y begins.
2. **If $Q = K_{sp}$, the system is at equilibrium.**
3. **If $Q > K_{sp}$, the system is not at equilibrium; the solution is *supersaturated*.** The concentrations of A^{n+} and B^{m-} in solution are too high for stability, and A_xB_y will precipitate.

For example, at point A in Figure 19.2, $[Ag^+] = 1.5 \times 10^{-5}$ M and $[Cl^-] = 0.50 \times 10^{-5}$ M. Under these conditions,

$$Q = [Ag^+][Cl^-] = (1.5 \times 10^{-5})(0.50 \times 10^{-5}) = 0.75 \times 10^{-10}$$

Here Q is less than K_{sp} ($0.75 \times 10^{-10} < 1.8 \times 10^{-10}$), so the ions in solution will not react to produce an insoluble product. In order to produce solid, one or both of the concentrations must be raised. If some solid AgCl is present, then it will dissolve and raise the Ag^+ and Cl^- concentrations, and the process will continue until $Q = K_{sp}$.

If a solution is represented by a point in the region to the right of the curve in Figure 19.2, the product $Q = [Ag^+][Cl^-]$ exceeds K_{sp}, and precipitation will occur because the solution is supersaturated. For example, if a solution initially has concentrations of silver and chloride both equal to 2.5×10^{-5} M (point B), then precipitation *must* occur.

$$Q = [Ag^+][Cl^-] = (2.5 \times 10^{-5})(2.5 \times 10^{-5}) = 6.3 \times 10^{-10}$$

Q is larger than K_{sp} ($6.3 \times 10^{-10} > 1.8 \times 10^{-10}$)

The fact that Q is larger than K_{sp} means that one or both of the ion concentrations are too large, and the simplest way to reduce the concentrations is for the ions to form insoluble AgCl. Precipitation will continue depleting the ions in solution until $Q = K_{sp}$.

EXERCISE 19.7 The Reaction Quotient

Using the graph in Figure 19.2, verify that a solution having a concentration of Ag^+ of 2.5×10^{-5} M and Cl^- of 1.0×10^{-5} M cannot be at equilibrium. Does this mean that AgCl will precipitate or that more AgCl will dissolve?

K_{sp} AND PRECIPITATIONS

With some knowledge of the reaction quotient, we can decide (a) whether a precipitate will form when the ion concentrations are known and (b) what concentrations of ions are required to begin the precipitation of an insoluble salt.

EXAMPLE 19.5

DECIDING WHETHER A PRECIPITATE WILL FORM

Suppose the concentration of aqueous nickel(II) ion in a solution is 1.5×10^{-6} M. If enough Na_2CO_3 is added to make the solution 6.0×10^{-4} M in the carbonate ion, CO_3^{2-}, will precipitation of nickel(II) carbonate occur? Will it occur if the CO_3^{2-} concentration is raised by a factor of 100? The K_{sp} of $NiCO_3$ is 6.6×10^{-9}.

Solution The insoluble salt $NiCO_3$ dissolves according to the balanced equation

$$NiCO_3(s) \rightleftharpoons Ni^{2+}(aq) + CO_3^{2-}(aq)$$

and the K_{sp} expression is $K_{sp} = [Ni^{2+}][CO_3^{2-}]$. In the first of the cases above, Q is

$$Q = [Ni^{2+}][CO_3^{2-}] = (1.5 \times 10^{-6})(6.0 \times 10^{-4}) = 9.0 \times 10^{-10}$$

and you see that it is *less than* the K_{sp}. Therefore, precipitation of $NiCO_3$ does *not* occur.

If the carbonate concentration is increased by a factor of 100, it becomes 6.0×10^{-2} M. Q is now 9.0×10^{-8}, a value larger than K_{sp}. Thus, precipitation occurs, and it will continue until the Ni^{2+} and CO_3^{2-} concentrations have declined so that their product is equal to K_{sp}.

EXERCISE 19.8 Deciding Whether a Precipitate Will Form

If the concentration of strontium ion is 2.5×10^{-4} M, will precipitation of $SrSO_4$ occur when enough of the soluble salt Na_2SO_4 is added to make the solution 2.5×10^{-4} M in SO_4^{2-}? K_{sp} for $SrSO_4$ is 2.8×10^{-7}.

Now that we know how to decide whether a precipitate will form when the concentration of each ion is known, let us turn to the problem of deciding how much of the precipitating agent is required to begin the precipitation of an ion at a given concentration level.

EXAMPLE 19.6

ION CONCENTRATIONS REQUIRED TO BEGIN PRECIPITATION

If the concentration of Ni^{2+} ion in water is 0.010 M, (a) what is the minimum concentration of S^{2-} necessary to begin precipitating NiS, and (b) what will be the concentration of Ni^{2+} when the S^{2-} concentration reaches 0.00010 M? K_{sp} for NiS is 3.0×10^{-21}.

Solution As usual, it is essential to begin with a balanced equation for the equilibrium that will exist when NiS(s) is precipitated.

$$NiS(s) \rightleftharpoons Ni^{2+}(aq) + S^{2-}(aq)$$

and the equilibrium constant expression

$$K_{sp} = [Ni^{2+}][S^{2-}] = 3.0 \times 10^{-21}$$

The K_{sp} expression tells us that the product of the nickel(II) and sulfide ion concentrations cannot exceed 3.0×10^{-21}. If that happens, then precipitation of NiS will occur. This means that, if the nickel ion concentration is known, as in question (a), we can obtain the sulfide ion concentration that satisfies the K_{sp} expression.

$$K_{sp} = 3.0 \times 10^{-21} = (0.010)[S^{2-}]$$

$$[S^{2-}] = \frac{K_{sp}}{[Ni^{2+}]} = \frac{3.0 \times 10^{-21}}{0.010} = 3.0 \times 10^{-19} \text{ M}$$

If you were to increase the sulfide ion concentration slightly, to a value *greater than* 3.0×10^{-19} M, precipitation of NiS will begin because $Q = [Ni^{2+}][S^{2-}]$ becomes greater than K_{sp}.

(b) If sulfide ion is added until it reaches 0.00010 M, the maximum amount of nickel that can be in solution can be obtained from the K_{sp} expression.

$$[Ni^{2+}] = \frac{K_{sp}}{[S^{2-}]} = \frac{3.0 \times 10^{-21}}{0.00010} = 3.0 \times 10^{-17} \text{ M}$$

This means that the nickel(II) ion has been almost completely removed from the solution by the time the sulfide ion concentration reaches the very low value of 10^{-4} M.

EXERCISE 19.9 Ion Concentrations Required to Begin Precipitation

If the concentration of Ba^{2+} in water is 1.0×10^{-3} M, what concentration of SO_4^{2-} is necessary to just begin precipitating $BaSO_4(s)$? K_{sp} for $BaSO_4$ is 1.1×10^{-10}.

EXAMPLE 19.7

K_{sp} AND PRECIPITATIONS

Mercury(I) chloride, Hg_2Cl_2, is commonly called "calomel." (As a mixture with proteins, it has been used as a laxative.) It is insoluble, having a K_{sp} of 1.1×10^{-18}. If you have 100. mL of a solution having a 0.0010 M concentration of Hg_2^{2+}, how many grams of NaCl must be added to begin precipitation of Hg_2Cl_2?

Solution The equilibrium reaction for calomel is

$$Hg_2Cl_2(s) \rightleftharpoons Hg_2^{2+}(aq) + 2\ Cl^-(aq)$$

and the K_{sp} expression is

$$K_{sp} = [Hg_2^{2+}][Cl^-]^2$$

(Notice in the equation above that the Hg(I) cation is diatomic.)

To answer the question, you need to know what concentration of Cl^- can exist in equilibrium with 0.0010 M Hg_2^{2+} ion. This means we must solve the K_{sp} expression for $[Cl^-]$.

$$[Cl^-] = \sqrt{\frac{K_{sp}}{[Hg_2^{2+}]}} = \sqrt{\frac{1.1 \times 10^{-18}}{0.0010}} = 3.3 \times 10^{-8}\ \text{M}$$

Our calculation shows that precipitation will begin when $[Cl^-]$ just exceeds the very small value of 3.3×10^{-8} M. To find out how many grams of NaCl are required, you need the molar mass of NaCl (58.44 g/mol), and you can then perform the following calculation:

$$\left(\frac{3.3 \times 10^{-8}\ \text{moles Cl}^-}{\text{L of solution}}\right)(0.100\ \text{L of solution})\left(\frac{1\ \text{mol NaCl}}{1\ \text{mol Cl}^-}\right)\left(\frac{58.44\ \text{g}}{1\ \text{mol}}\right) = 1.9 \times 10^{-7}\ \text{g NaCl}$$

Thus, precipitation of Hg_2Cl_2 begins when the amount of NaCl added is slightly in excess of 1.9×10^{-7} g of NaCl.

EXERCISE 19.10 Precipitation of $BaCO_3$

If you have 100. mL of a 0.0010 M solution of CO_3^{2-} ion, how many grams of the soluble salt $BaCl_2$ must be added to begin the precipitation of $BaCO_3$? How many grams of $BaCl_2$ must be added to reduce the carbonate ion concentration to 1.0×10^{-6} M?

EXERCISE 19.11 Precipitation Reaction

If you mix 100. mL of a 0.0010 M solution of Na_2CO_3 with 100. mL of 0.0010 M $BaCl_2$, will a precipitate form?

19.6 SOLUBILITY AND THE COMMON ION EFFECT

Using the methods of Section 19.4, you can estimate that the solubility of AgCl in pure water at 25 °C is 1.3×10^{-5} moles per liter. But what would happen if you put some solid AgCl in a beaker that already contains some chloride ion, say from NaCl? Would you still find that 1.3×10^{-5} moles of AgCl dissolve per liter? Certainly not. Experiment shows that the solubility of any salt is always less in the presence of a "common ion," and Cl^- is common to both AgCl and NaCl. The reason for this effect can be explained by **Le Chatelier's principle**. Recall that this principle states that for a system already at equilibrium, a change in any of the factors that determine the equilibrium concentrations will cause the system to change so as to offset the effect of the change. In the case of insoluble salts, the outcome is that the salt solubility is lower in the presence of a "common ion."

Le Chatelier's principle was first applied to chemical equilibria in Chapter 16. See also the "common ion effect" in Chapter 18.

If AgCl is the only solute in pure water, the concentrations of Ag^+ and Cl^- in equilibrium with solid AgCl must be equal to one another.

$$AgCl(s) \rightleftharpoons Ag^+(aq) + Cl^-(aq) \qquad K_{sp} = 1.8 \times 10^{-10}$$

However, if extra chloride ion is added to a solution of Ag^+ and Cl^- in equilibrium with solid AgCl, the Cl^- concentration becomes too large to satisfy the requirement that $[Ag^+][Cl^-] = K_{sp}$. That is, $Q > K_{sp}$, and the situation might be as depicted in Figure 19.3. To return to equilibrium, both $[Ag^+]$ and $[Cl^-]$ must decline, and this result is accomplished by precipitating some solid AgCl. Notice, however, that when this happens

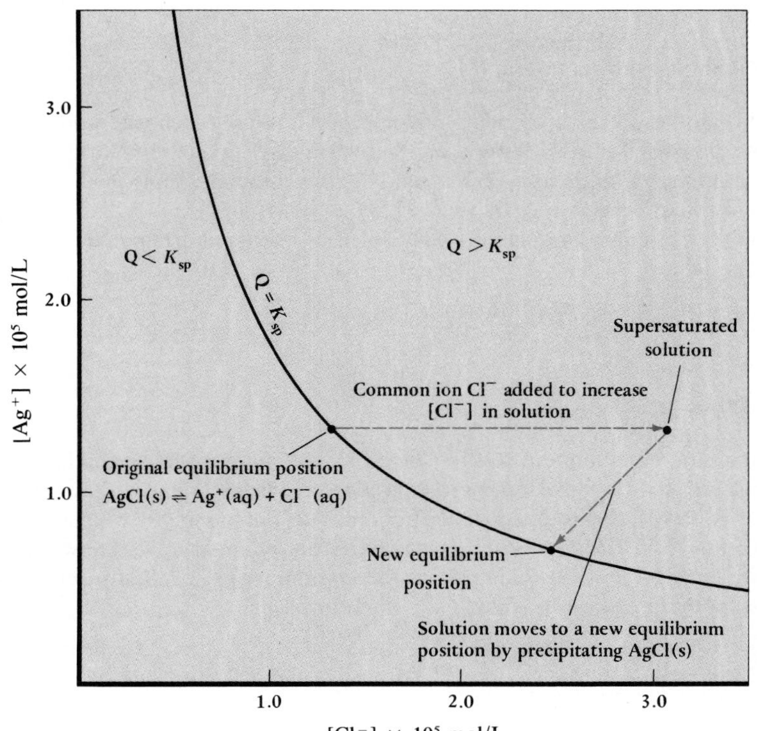

Figure 19.3 The common ion effect. Excess Cl^- is added to an equilibrium mixture of solid AgCl and equal concentrations of Ag^+ and Cl^- (blue line). The system returns to equilibrium along the red line. (The slope of this line is 1, since one Ag^+ ion is removed from solution for every Cl^- ion removed.)

we arrive at a *new equilibrium position*; the concentrations of Ag^+ and Cl^- are not the same as those before we added extra Cl^-.

The "common ion effect" is the basis of ion separations in qualitative analysis (Section 19.7).

The common ion effect is further illustrated for silver acetate in Figure 19.4, and a more quantitative analysis is done for AgCl in the next example.

EXAMPLE 19.8

THE COMMON ION EFFECT

The solubility of AgCl in pure water is 1.3×10^{-5} M (0.0019 g/liter). If you put some AgCl in a solution that is 0.55 M in NaCl, how many grams of AgCl will dissolve per liter of this solution?

Solution The solubility of silver chloride can be expressed by the silver ion concentration, $[Ag^+]$. As usual, we assign the variable S to this concentration.

$$\text{Solubility of AgCl defined by } [Ag^+] = S$$

In *pure* water, S is equal to both $[Ag^+]$ and $[Cl^-]$. However, in salt water containing the common ion Cl^-, S is equal only to $[Ag^+]$, and it must have a smaller value than in pure water due to the fact that excess Cl^- ion causes the equilibrium to shift to the left, thereby lowering the concentration of Ag^+.

$$AgCl(s) \rightleftharpoons Ag^+(aq) + Cl^-(aq)$$

← excess Cl^- shifts equilibrium left—

The following table shows the concentrations of Ag^+ and Cl^- when equilibrium is attained in the presence of extra Cl^-.

	$[Ag^+]$	$[Cl^-]$
Initial concentration (M)	0	0.55
Change on proceeding to equilibrium	$+S$	$+S$
Equilibrium concentration (M)	S	$S + 0.55$

Figure 19.4 The common ion effect. The tube at the left contains a saturated solution of silver acetate, $Ag(CH_3COO)$. When 1 M $AgNO_3$ is added to the tube, the equilibrium

$$Ag(CH_3COO)(s) \rightleftharpoons Ag^+(aq) + CH_3COO^-(aq)$$

shifts to the left, as evidenced by the tube at the right, where more solid silver acetate has formed.

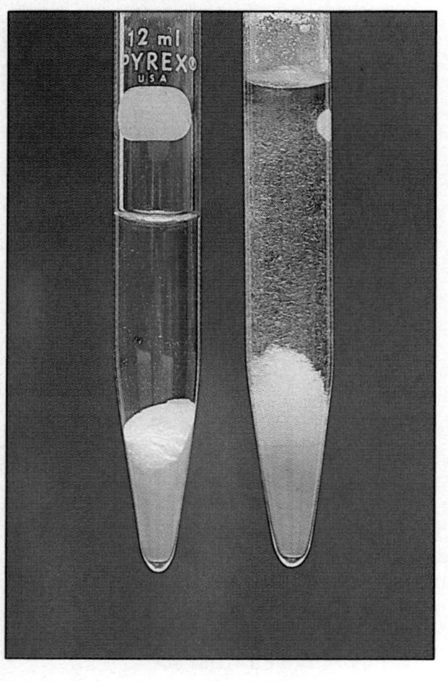

Some AgCl dissolves in the presence of chloride ion and produces Ag^+ and Cl^- ion concentrations of S mol/L. However, there was already some chloride ion present, so *the total chloride ion concentration* is the amount coming from AgCl (equals S) *plus* what was already there (0.55 M).

Now the K_{sp} expression can be written, using the equilibrium concentrations from the table.

$$K_{sp} = 1.8 \times 10^{-10} = [Ag^+][Cl^-] = (S)(S + 0.55)$$

and rearranged to

$$S^2 + 0.55\,S - K_{sp} = 0$$

This is a quadratic equation and can be solved by the methods in Appendix A. The easiest approach is to make the approximation that S is *very* small with respect to 0.55; that is, it will make a negligible difference to the answer if we assume that $(S + 0.55) \approx 0.55$. This is a very reasonable assumption since we know that the solubility equals 1.3×10^{-5} M *without* the common ion Cl^-, and it will be even smaller in the presence of added Cl^-. Therefore,

$$K_{sp} = 1.8 \times 10^{-10} \approx (S)(0.55)$$

$$S = [Ag^+] \approx 3.3 \times 10^{-10}$$

(If we use the method of successive approximations to solve the more "accurate" expression, the answer is the same as the "approximate" answer to two significant figures.)

As predicted by Le Chatelier's principle, the solubility of AgCl in the presence of added Cl^-, an ion common to the equilibrium, is clearly less (3.3×10^{-10} M) than in pure water (1.3×10^{-5} M).

As a final step, let us check the approximation we made. To do this, we substitute the approximate value of S into the exact expression $K_{sp} = (S)(S + 0.55)$. Then, if the product $(S)(S + 0.55)$ is the same as the given value of K_{sp}, the approximation is valid. (The validity is also confirmed by the fact that the method of successive approximations gave the same answer as the approximate expression.)

$$K_{sp} = (S)(S + 0.55) = (3.3 \times 10^{-10})(3.3 \times 10^{-10} + 0.55) \approx 1.8 \times 10^{-10}$$

Lastly, you should be sure to understand the problem solving strategy used here. It is identical to that used to solve common ion problems in acid–base chemistry (see Example 18.3). In the present example we assume that Cl^- is already present in solution from the added NaCl. Then we add AgCl and additional Cl^- enters the solution as the silver salt dissolves. Thus, $[Cl^-]$ at equilibrium is the *sum* of the two concentrations of the ion from the two sources.

EXERCISE 19.12 The Common Ion Effect

Calculate the solubility of $BaSO_4$ (a) in pure water and (b) in the presence of 0.010 M $Ba(NO_3)_2$. K_{sp} for $BaSO_4$ is 1.1×10^{-10}.

There are two important ideas to be learned from Example 19.8.

1. The solubility of AgCl in the presence of a common ion was reduced by a factor of about 10^5, in accordance with Le Chatelier's principle.
2. We made the approximation that the amount of common ion added to the solution was very large in comparison with the amount of that ion coming from the insoluble salt, and this allowed us to simplify our calculations. This is almost always the case, but you should *always* check the approximation.

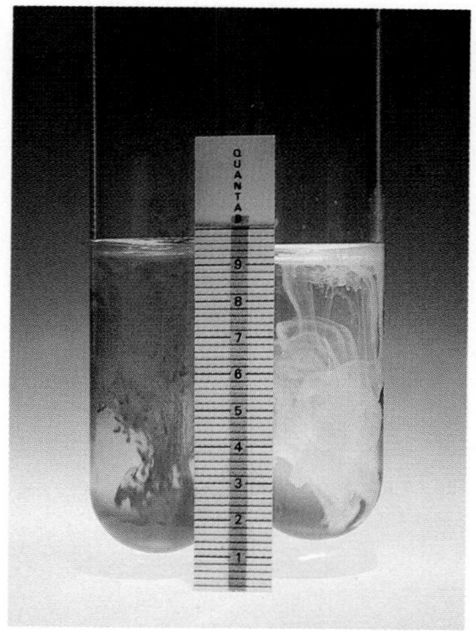

Red-orange Ag_2CrO_4 is precipitated in the test tube at the left and white AgCl in the test tube at the right. In between is a "test strip" to test water for the presence of chloride ion. When Cl^- ion enters the test strip, red-orange Ag_2CrO_4 is changed to white AgCl.

EXAMPLE 19.9

THE COMMON ION EFFECT

Calculate the solubility of silver chromate, Ag_2CrO_4, at 25 °C (a) in pure water and (b) in the presence of 0.0050 M K_2CrO_4 solution. At 25 °C, $K_{sp} = 9.0 \times 10^{-12}$.

Solution The balanced equation defining silver chromate solubility is

$$Ag_2CrO_4(s) \rightleftharpoons 2\ Ag^+(aq) + CrO_4^{2-}(aq) \qquad K_{sp} = [Ag^+]^2[CrO_4^{2-}]$$

(a) Solubility in pure water. Because there is a 1:1 stoichiometric ratio between moles of CrO_4^{2-} in solution and moles of Ag_2CrO_4 dissolved, we define the solubility of this salt in terms of $[CrO_4^{2-}] = S$. Thus, S is not only the concentration of CrO_4^{2-} in solution but is also the number of moles per liter of Ag_2CrO_4 dissolved.

	$[Ag^+]$	$[CrO_4^{2-}]$
Before Ag_2CrO_4 begins to dissolve (M)	0	0
Change on proceeding to equilibrium	$+2S$	$+S$
After equilibrium is achieved (M)	$2S$	S

This must mean that $[Ag^+] = 2S$, and so the K_{sp} expression can be written as

Do not be confused about the use of $(2S)^2$, where the factor 2 appears both as the multiplier of S and as the exponent of (2S). The reason for this is that, since $[CrO_4^{2-}]$ is defined as S, the stoichiometry demands that $[Ag^+]$ be defined as 2S. This is substituted into the K_{sp} expression in place of $[Ag^+]$. The K_{sp} expression then demands that $[Ag^+]$ or the equivalent expression 2S be squared.

$$K_{sp} = 9.0 \times 10^{-12} = [Ag^+]^2[CrO_4^{2-}] = (2S)^2(S) = 4S^3$$

and the solubility of the silver salt in pure water is

$$S^3 = 2.3 \times 10^{-12} \text{ M}$$

$$S = \text{solubility of } Ag_2CrO_4 \text{ in pure water} = \sqrt[3]{2.3 \times 10^{-12}} = 1.3 \times 10^{-4} \text{ M}$$

Finally, the ion concentrations at equilibrium in pure water are

$$[Ag^+] = 2S = 2.6 \times 10^{-4} \text{ M} \quad \text{and} \quad [CrO_4^{2-}] = 1.3 \times 10^{-4} \text{ M}$$

(b) Solubility in chromate solution. In the presence of excess chromate ion from dissolved K_2CrO_4, we again say that S represents the solubility of the salt.

	$[Ag^+]$	$[CrO_4^{2-}]$
Before Ag_2CrO_4 begins to dissolve (M)	0	0.0050
Change on proceeding to equilibrium	$+2S$	$+S$
After equilibrium is achieved (M)	$2S$	$0.0050 + S$

Substituting the equilibrium concentrations into the K_{sp} expression, we have

$$K_{sp} = 9.0 \times 10^{-12} = [Ag^+]^2[CrO_4^{2-}] = (2S)^2(0.0050 + S)$$

Let us again make the approximation that S is very small with respect to 0.0050, and so $(0.0050 + S) \approx 0.0050$. This is probably reasonable since we know that S is 0.00013 M *without* added chromate ion, and S will be even smaller in the presence of extra chromate ion. Therefore, the approximate expression is

$$K_{sp} = 9.0 \times 10^{-12} = [Ag^+]^2[CrO_4^{2-}] = (2S)^2(0.0050)$$

and so S is

$$S = [CrO_4^{2-}] \text{ from } Ag_2CrO_4 = 2.1 \times 10^{-5} \text{ M}$$

and the silver ion concentration in the presence of the common ion is

$$[Ag^+] = 2S = 4.2 \times 10^{-5} \text{ M}$$

This silver ion concentration is less than its value in pure water (2.6×10^{-4}), and you again see the result of adding an ion "common" to the equilibrium.

As a final step, check the approximation. Substitute the calculated S back into the exact expression $K_{sp} = (2S)^2(0.0050 + S)$ to see whether K_{sp} calculated from the expression agrees with the given K_{sp}.

$$K_{sp} = [2(2.1 \times 10^{-5})]^2(0.0050 + 0.000021) = 8.9 \times 10^{-12}$$

The calculated K_{sp} differs by only about 1% from the given K_{sp}, a difference that is probably smaller than experimental error. The approximation is valid.

EXERCISE 19.13 The Common Ion Effect

Calculate the solubility of $Zn(CN)_2$ at 25 °C (a) in pure water and (b) in the presence of 0.10 M KCN. K_{sp} for $Zn(CN)_2$ is 8.0×10^{-12}.

19.7 SOLUBILITY, ION SEPARATIONS, AND QUALITATIVE ANALYSIS

In many courses in introductory chemistry, a portion of the laboratory work is devoted to the qualitative analysis of aqueous solutions, the identification of anions and metal cations. The purpose of such laboratory work is (a) to introduce you to some basic chemistry of various ions and (b) to illustrate how the principles of chemical equilibria can be applied.

Assume you have an aqueous solution that contains some or all of the following metal ions: Ag^+, Pb^{2+}, Cd^{2+}, and Ni^{2+}. Your objective is to separate the ions from one another so that each type of ion ends up in a separate test tube; the presence or absence of the ion can then be established. As a first step in this process, you want to find one reagent that will form a precipitate with one or more of the cations and leave the others in solution. This is done by comparing K_{sp} values for salts of the cations with various anions (say S^{2-}, OH^-, Cl^-, or SO_4^{2-}), looking for an anion that gives insoluble salts for some cations but not others.

Looking over the list of solubility products in Appendix G, you notice that all of the ions in our example solution form very insoluble sulfides (Ag_2S, PbS, CdS, and NiS). However, only two of them form insoluble chlorides, AgCl and $PbCl_2$. Thus, your magic reagent for partial cation separation could be aqueous HCl, which will form precipitates with two of the ions while the other two remain in solution (Figure 19.5).

Some Insoluble Sulfides and Chlorides

COMPOUND	K_{sp} AT 25 °C
Ag_2S	1.0×10^{-49}
PbS	8.4×10^{-28}
CdS	3.6×10^{-29}
NiS	3.0×10^{-21}
AgCl	1.8×10^{-10}
$PbCl_2$	1.7×10^{-5}

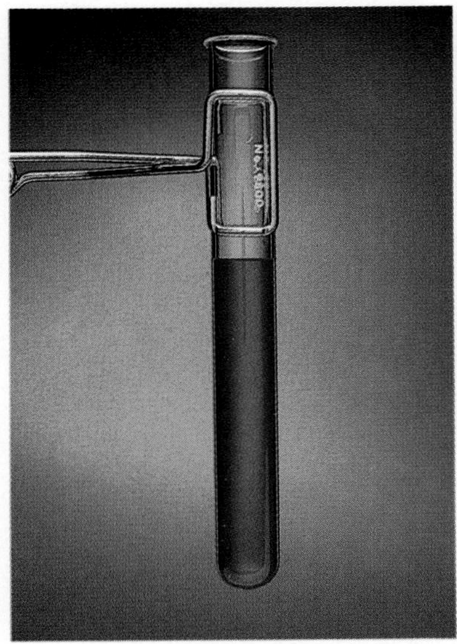

(a)

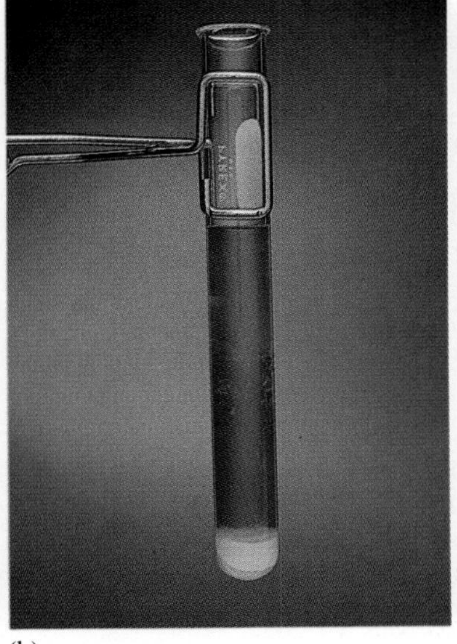

(b)

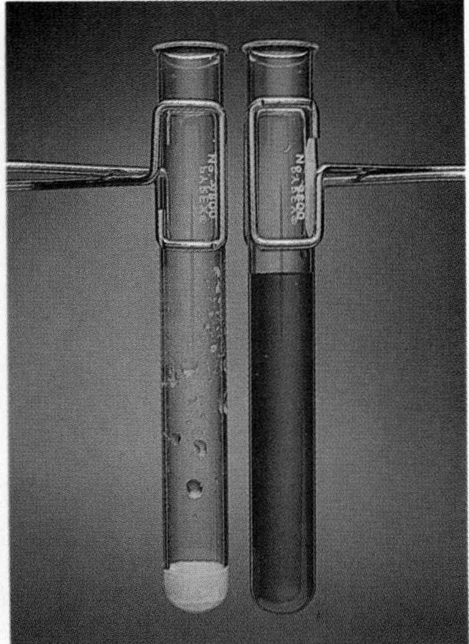

(c)

Figure 19.5 Ion separations by solubility difference. (a) The solution contains nitrate salts of Ag^+, Pb^{2+}, Cd^{2+}, and Ni^{2+}. (The Ni^{2+} ion in water is light green, while the others are colorless.) (b) Aqueous HCl is added in an amount sufficient to precipitate completely AgCl and $PbCl_2$ (both white solids). (c) The green solution, now containing only Ni^{2+} and Cd^{2+}, is poured into another test tube, leaving white, solid AgCl and $PbCl_2$ in the first test tube.

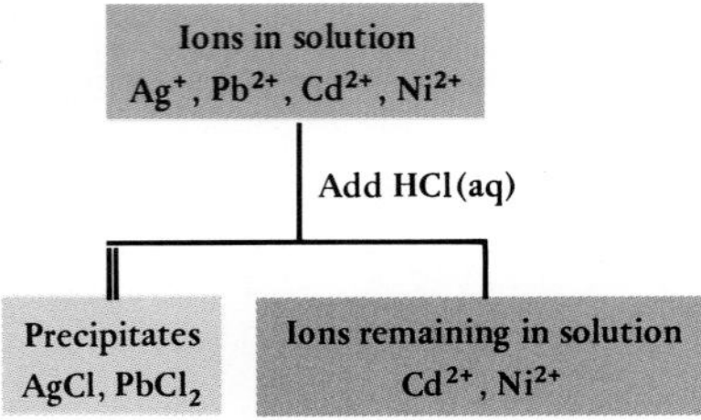

Now you are left with the task of separating solid AgCl and $PbCl_2$ from one another and aqueous Cd^{2+} and Ni^{2+} from one another (Figure 19.5). The separation of AgCl and $PbCl_2$ is not difficult, since $PbCl_2$ dissolves in hot water while AgCl remains insoluble.

Now that Ag^+ and Pb^{2+} have been removed from solution, the separation of aqueous Cd^{2+} and Ni^{2+} can be done by selective precipitation with sulfide ion, S^{2-}. Ordinarily, this is done in the presence of H_3O^+ ion (Section 19.10), but in the example that follows we wish to ask two questions that are common in chemistry. As a precipitating reagent is added to a mixture of two ions A^{n+} and B^{n+}, which ion precipitates first? If A^{n+} precipitates first, what is the concentration of A^{n+} when B^{n+} just begins to precipitate from solution?

EXAMPLE 19.10

SEPARATION OF TWO IONS BY DIFFERENCE IN SOLUBILITY

Assume you have a solution that is 0.020 M in both Cd^{2+} and Ni^{2+}. When you add sulfide ions, CdS and NiS will precipitate. (a) Which precipitates first, CdS or NiS? (b) Just before the second metal ion begins to precipitate as an insoluble sulfide, what is the remaining concentration of the ion that precipitated first?

Solution The key to answering part (a) is to realize that *the substance that precipitates first is the one whose K_{sp} is first exceeded.* Therefore, let us calculate the sulfide ion concentration, $[S^{2-}]$, necessary to satisfy the solubility product expression for each metal sulfide. These amounts of S^{2-} will bring each metal sulfide to the brink of precipitation.

$$\text{For CdS the required } [S^{2-}] = \frac{K_{sp}}{[Cd^{2+}]} = \frac{3.6 \times 10^{-29}}{0.020} = 1.8 \times 10^{-27}\ M$$

$$\text{For NiS the required } [S^{2-}] = \frac{K_{sp}}{[Ni^{2+}]} = \frac{3.0 \times 10^{-21}}{0.020} = 1.5 \times 10^{-19}\ M$$

Clearly, a *much* smaller sulfide ion concentration (by a factor of about 100 million!) is needed to begin precipitating CdS than to begin forming NiS. Therefore, CdS will begin to precipitate before NiS.

Now we can rephrase the second question: Just before NiS begins to precipitate, how much Cd^{2+} remains in solution? We have already found that $[S^{2-}]$ needs to be slightly in excess of 1.5×10^{-19} M for NiS to begin to precipitate. Therefore, let us calculate the Cd^{2+} concentration that can exist when $[S^{2-}] = 1.5 \times 10^{-19}$ M, that is, the sulfide ion concentration required to bring us to the brink of precipitating NiS.

$$[Cd^{2+}] \text{ just before NiS begins to precipitate} = \frac{K_{sp} \text{ for CdS}}{[S^{2-}]} = \frac{3.6 \times 10^{-29}}{1.5 \times 10^{-19}} = 2.4 \times 10^{-10}\ M$$

We now know that just before NiS begins precipitating, the Cd^{2+} concentration has decreased from 0.020 M to 2.4×10^{-10} M, a change by a factor of about 100 million! This means that we can separate Cd^{2+} and Ni^{2+} in aqueous solution by careful control of the sulfide ion concentration. In Section 19.10, we shall outline how this control can be achieved.

EXERCISE 19.14 Ion Separation by Solubility Differences

Under the right circumstances, aqueous Ag^+ can be separated from aqueous Pb^{2+} by the difference in the solubilities of their chloride salts, AgCl and $PbCl_2$.

(a) If you begin with both metal ions having a concentration of 0.0010 M, which ion precipitates first as an insoluble chloride on adding HCl?

(b) What is the concentration of the metal ion that precipitates first, just before the second metal chloride begins to precipitate?

EXERCISE 19.15 Schemes for Ion Separation

The cations of each pair given below appear together in one solution.

(a) Ag^+ and Bi^{3+} (b) Fe^{2+} and K^+

You may add only one reagent to precipitate one cation and not the other. Consult the solubility product table in Appendix G and tell whether you would use Cl^-, S^{2-}, or OH^- as the precipitating ion in each case. (The precipitating ions are introduced in the form of HCl, $(NH_4)_2S$, or NaOH, for example.)

19.8 SIMULTANEOUS EQUILIBRIA

It is possible that, when another reagent is added to a saturated solution of an insoluble salt, the salt will be converted to another, even less soluble salt. Such situations represent one example of **simultaneous equilibria**.

Consider two compounds you see commonly in the laboratory, lead(II) chloride and lead(II) chromate:

$PbCl_2$, white solid, $K_{sp} = 1.7 \times 10^{-5}$

$PbCrO_4$, yellow solid, $K_{sp} = 1.8 \times 10^{-14}$

If you add a few drops of K_2CrO_4 to a small amount of a precipitate of $PbCl_2$ and shake the mixture, the $PbCl_2$ is changed to $PbCrO_4$. That this is possible is evident from the equilibrium constant, K_{net}, for the process

$$PbCl_2(s) + CrO_4^{2-}(aq) \rightleftharpoons PbCrO_4(s) + 2\,Cl^-(aq) \qquad K_{net} = 9.4 \times 10^8$$

This reaction is just the sum of two reactions whose equilibrium constants are known from above.

$$PbCl_2(s) \rightleftharpoons Pb^{2+}(aq) + 2\,Cl^-(aq) \qquad K_1 = K_{sp} = 1.7 \times 10^{-5}$$

$$Pb^{2+}(aq) + CrO_4^{2-}(aq) \rightleftharpoons PbCrO_4(s) \qquad K_2 = \frac{1}{K_{sp}} = \frac{1}{1.8 \times 10^{-14}}$$

The manipulation of equilibrium constants and their values is described in Section 16.2. Recall that, when a reaction is reversed, $K_{reverse} = 1/K_{original}$.

The equilibrium constant for the overall reaction is the *product* of the equilibrium constants for the summed reactions. That is, $K_{net} = K_1K_2 = 9.4 \times 10^8$. This relatively large equilibrium constant indicates that the reaction should proceed from left to right.

Simultaneous equilibria are quite common, and you will see more examples in the sections that follow.

EXERCISE 19.16 Simultaneous Equilibria

Silver forms many insoluble salts. Which is more soluble, AgCl or AgBr? If you add sufficient bromide ion to an aqueous suspension of AgCl(s), can AgCl be converted to AgBr? To answer this, derive the value of the equilibrium constant for

$$AgCl(s) + Br^-(aq) \rightleftharpoons AgBr(s) + Cl^-(aq)$$

19.9 SOLUBILITY AND COMPLEX IONS

In Chapter 17 we described reactions of Lewis acids and bases. When a metal ion (which is a Lewis acid) reacts with a Lewis base, a complex ion can form. Since such ions are often soluble in water, their formation can be used to dissolve otherwise insoluble salts. For example, if sufficient aqueous ammonia is added to insoluble silver chloride, the latter can be dissolved in the form of the complex ion $[Ag(NH_3)_2]^+$ (see *Something More About Silver Chemistry*). The overall reaction is

$$AgCl(s) + 2\ NH_3(aq) \rightleftharpoons [Ag(NH_3)_2]^+(aq) + Cl^-(aq)$$

and you may recognize this as the sum of two reactions whose equilibrium constants are known:

$$AgCl(s) \rightleftharpoons Ag^+(aq) + Cl^-(aq) \qquad K_{sp} = 1.8 \times 10^{-10}$$

$$Ag^+(aq) + 2\ NH_3(aq) \rightleftharpoons [Ag(NH_3)_2]^+(aq) \qquad K_{formation} = 1.6 \times 10^7$$

That is, there are two equilibria occurring simultaneously in this solution. The role of the second reaction is to remove Ag^+ formed as a product of the first; Le Chatelier's principle leads us to the conclusion that AgCl will dissolve to a greater extent in the presence of NH_3 than in pure water. As described in Section 19.8, the equilibrium constant for the *overall* reaction is the product of the constants of the summed reactions. Thus, for silver chloride plus ammonia

$$K = K_{sp} \times K_{formation} = 2.9 \times 10^{-3} = \frac{\{[Ag(NH_3)_2]^+\}[Cl^-]}{[NH_3]^2}$$

The equilibrium constant for dissolving AgCl in aqueous ammonia is not large; however, if the concentration of ammonia is sufficiently high, $[Ag(NH_3)_2]^+$ and $[Cl^-]$ must also be high, and AgCl will dissolve.

EXAMPLE 19.11

DISSOLVING PRECIPITATES USING COMPLEX ION FORMATION

How many moles of ammonia must be added to dissolve 0.050 mole of AgCl suspended in 1.0 L of water?

Solution As shown in the text above, the reaction that occurs is

$$AgCl(s) + 2\ NH_3(aq) \longrightarrow [Ag(NH_3)_2]^+(aq) + Cl^-(aq)$$

If 0.050 mol of AgCl is to be dissolved, this means that 0.050 mol/L of the complex ion $[Ag(NH_3)_2]^+$ will be formed. The $[Cl^-]$ must also be 0.050 M. If these are the concentrations that we wish to have at equilibrium, what must the concentration of ammonia be to satisfy this?

$$K = 2.9 \times 10^{-3} = \frac{\{[Ag(NH_3)_2{}^+]\}[Cl^-]}{[NH_3]^2} = \frac{(0.050)(0.050)}{[NH_3]^2}$$

Solving for $[NH_3]$, we find that it is 0.93 M. Therefore, to dissolve AgCl, enough NH_3 must be added to form the complex with the silver ion (2 × 0.050 mol or 0.10 mol required) and then to bring the concentration up to 0.93 M. Thus, the total required is

$$\text{Total } NH_3 = 0.10 \text{ M} + 0.93 \text{ M} = 1.03 \text{ M}$$

EXERCISE 19.17 Using Formation Constants

Will 100. mL of 4.0 M aqueous NH_3 completely dissolve 0.010 mole of AgCl suspended in 1.0 L of pure water?

19.10 ACID–BASE AND PRECIPITATION EQUILIBRIA OF PRACTICAL SIGNIFICANCE

SOLUBILITY OF SALTS IN WATER AND ACIDS

The next time you are tempted to wash a supposedly insoluble salt down the laboratory drain, stop and consider the consequences. Many metal ions such as lead, chromium, and mercury are toxic in the environment. Even if you throw away an insoluble salt, its solubility in water may be greater than you think, in part owing to the possibility of hydrolysis of the anion of the salt.

Lead(II) sulfide, PbS, is found in nature as the mineral galena. The solubility of any salt depends on a host of factors: temperature, impurities, and so on. However, we keep things as simple as possible by considering only two reactions for lead sulfide:

(a) the solubility of PbS in water:

$$PbS(s) \rightleftharpoons Pb^{2+}(aq) + S^{2-}(aq) \qquad K_{sp} = 8.4 \times 10^{-28}$$

(b) the hydrolysis of sulfide ion:

$$S^{2-}(aq) + H_2O(\ell) \rightleftharpoons HS^-(aq) + OH^-(aq) \qquad K_b = 0.077$$

This means that the overall process is

$$PbS(s) + H_2O(\ell) \rightleftharpoons Pb^{2+}(aq) + HS^-(aq) + OH^-(aq)$$

with an equilibrium constant of $K_{total} = K_{sp} \times K_b = 6.5 \times 10^{-29}$. The first reaction produces a *tiny* concentration of sulfide ion. The second reaction, however, removes the ion from solution by hydrolysis, a reaction with a much larger equilibrium constant. According to Le Chatelier's principle, more PbS will therefore dissolve. Indeed, it can be calculated using K_{total} that, in an equilibrium system composed of these reactions, the lead ion concentration is approximately 10^{-10} M. Sulfide ion hydrolysis thus leads

The mineral galena, lead(II) sulfide.

See Study Question 108 at the end of this chapter.

to an increase in environmental lead concentration by a factor of about 10,000 over the solubility of PbS calculated simply from K_{sp}!

The lead sulfide example leads to the general observation that any salt containing an anion that is the conjugate base of a weak acid will dissolve in water to a greater extent than given by K_{sp}. This means that salts of sulfate, phosphate, acetate, carbonate, and cyanide, as well as sulfide, can be affected, since all of these ions undergo the general hydrolysis reaction

$$X^-(aq) + H_2O(\ell) \rightleftharpoons HX(aq) + OH^-(aq)$$

in aqueous solution. This reaction, however, leads to another useful, general conclusion. If a strong acid is added to a water-insoluble salt such as ZnS or $CaCO_3$, then OH^- ion from X^- hydrolysis is removed (by formation of water). This shifts the X^- hydrolysis reaction further to the right; the weak acid HX is formed, and the salt dissolves. For example, calcium carbonate will dissolve readily when hydrochloric acid is added, and the main reactions occurring are

$$CaCO_3(s) \rightleftharpoons Ca^{2+}(aq) + CO_3^{2-}(aq) \qquad K = K_{sp} = 3.8 \times 10^{-9}$$

$$H_3O^+(\text{from HCl}) + CO_3^{2-}(aq) \rightleftharpoons HCO_3^-(aq) + H_2O(\ell) \qquad K = 1/K_{a2} \text{ for } H_2CO_3 = 1/(4.8 \times 10^{-11})$$

$$H_3O^+(\text{from HCl}) + HCO_3^-(aq) \rightleftharpoons H_2CO_3(aq) + H_2O(\ell) \qquad K = 1/K_{a1} \text{ for } H_2CO_3 = 1/(4.2 \times 10^{-7})$$

$$CaCO_3(s) + 2\ H_3O^+(aq) \rightleftharpoons Ca^{2+}(aq) + H_2CO_3(aq) + 2\ H_2O(\ell) \qquad K_{total} = K_{sp}(1/K_{a2})(1/K_{a1}) = 1.9 \times 10^8$$

Now the equilibrium constant (K_{total}) is enormous, and the reaction goes essentially to completion. This is especially the case for carbonate salts, since H_2CO_3 rapidly reverts to CO_2 gas and water.

$$H_2CO_3(aq) \rightleftharpoons CO_2(g) + H_2O(\ell) \qquad K \approx 10^5$$

The CO_2 bubbles out of solution, and the equilibrium is moved further to the right. (See Figure 5.11.)

Carbonates are generally soluble in strong acids, and so are many metal sulfides

$$ZnS(s) + 2\ H_3O^+(aq) \rightleftharpoons Zn^{2+}(aq) + H_2S(aq) + 2\ H_2O(\ell)$$

and metal hydroxides.

$$Mg(OH)_2(s) + 2\ H_3O^+(aq) \rightleftharpoons Mg^{2+}(aq) + 4\ H_2O(\ell)$$

The only exceptions are sulfides of mercury, copper, cadmium, and a few other metal ions. In general, *insoluble inorganic salts containing anions derived from weak acids tend to be soluble in solutions of strong acids.* In contrast, the salts are not soluble in strong acid if the anion is the conjugate base of a strong acid. For example, AgCl is not soluble in strong acid

$$AgCl(s) \rightleftharpoons Ag^+(aq) + Cl^-(aq) \qquad K_{sp} = 1.8 \times 10^{-10}$$

$$H_3O^+(aq) + Cl^-(aq) \rightleftharpoons HCl(aq) + H_2O(\ell) \qquad K << 1$$

because Cl^- is a *very* weak base (Table 17.2) and so is not removed by the acid H_3O^+. You can see that the same conclusion would apply to salts of NO_3^-, Br^-, I^-, and HSO_4^-.

(a)

(b)

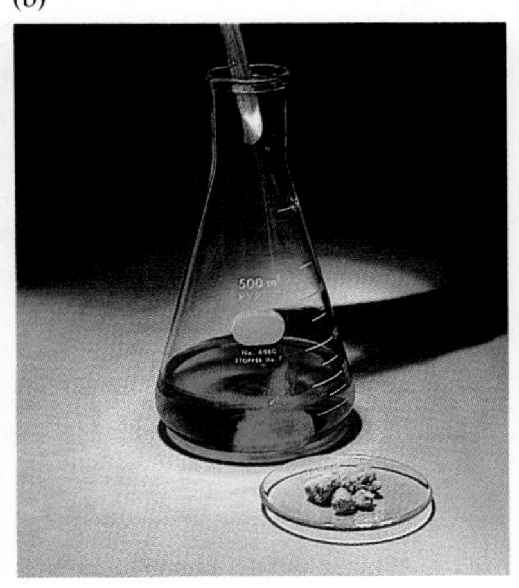

(c)

PREPARING SALTS BY ACID–BASE AND PRECIPITATION REACTIONS

You learned in Chapter 5 that salts can be prepared by the reaction of a strong acid with a strong base. The reason such reactions work is the formation of water, which is the very weak conjugate acid of OH^- and the weak conjugate base of H_3O^+. You also know that strong acid/weak base and weak acid/strong base reactions occur for much the same reason. All of these reactions can be used to prepare salts.

We want to prepare nickel(II) nitrate, $Ni(NO_3)_2$, starting with the commonly available salt $NiCl_2$ (Figure 19.6). Both of these salts are soluble in water, so simply adding nitrate ion, in some form, to aqueous $NiCl_2$ will not produce $Ni(NO_3)_2$. Therefore, we first convert $NiCl_2$ into an insoluble material by using an anion that is the conjugate base of a weak acid. A good choice is carbonate ion (from Na_2CO_3).

$$NiCl_2(aq) + CO_3^{2-}(aq) \longrightarrow NiCO_3(s) + 2\ Cl^-(aq)$$

Nickel(II) carbonate is insoluble, so it can be separated from the solution by filtration and washed with distilled water to remove contaminants. Next, the strong acid HNO_3 is added to the pure $NiCO_3$ to give the desired salt, $Ni(NO_3)_2$, plus water and CO_2.

$$NiCO_3(s) + 2\ HNO_3(aq) \longrightarrow Ni(NO_3)_2(aq) + H_2O(\ell) + CO_2(g)$$

As soon as the solid nickel carbonate has dissolved, we stop adding nitric acid and now have a solution containing only nickel(II) nitrate. If the solution is evaporated to dryness, we can obtain the desired salt as a solid hexahydrate, $Ni(NO_3)_2 \cdot 6\ H_2O$.

You can prepare many inorganic salts by using approaches similar to the one outlined above. In general, it is best to begin with a hydroxide,

(d)

(e)

Figure 19.6 Preparation of $Ni(NO_3)_2$ from $NiCl_2$. (a) $NiCl_2$, on the watch glass, is dissolved in water, and then (b) Na_2CO_3 is added to precipitate insoluble $NiCO_3$. (c) After filtering to collect the $NiCO_3$, 6 M HNO_3 is added to an aqueous suspension of $NiCO_3$, and it dissolves completely (d). The reaction occurring is $NiCO_3(s) + 2\ HNO_3(aq) \rightarrow Ni(NO_3)_2(aq) + CO_2(g) + H_2O(\ell)$. (e) If the aqueous solution of $Ni(NO_3)_2$ is evaporated, crystals of the hydrated salt $Ni(NO_3)_2 \cdot 6\ H_2O$ are isolated.

soluble or insoluble, or a carbonate and add a strong acid containing the desired anion (H_2SO_4, HCl, HNO_3, $HClO_4$). All of these reactions have a large driving force: the formation of water or of a gas, such as CO_2, that is poorly soluble in water.

> **EXERCISE 19.18 Preparation of Salts**
> Outline a method of preparing copper(II) chloride beginning with copper(II) sulfate.

SEPARATION OF IONS BY SELECTIVE PRECIPITATION

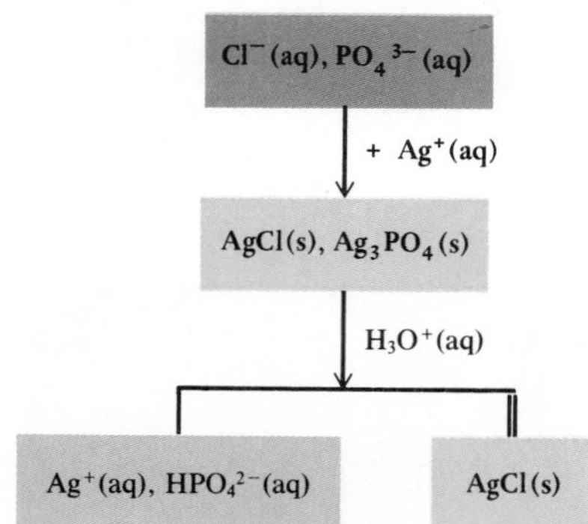

Some insoluble salts can be dissolved by adding an acid (Figure 19.7), and such reactions can be used to separate two or more ions from one another. As an example, let us say you have a solution that may contain Cl^- or PO_4^{3-} ions. To help you decide which ion is present, you are given aqueous $AgNO_3$ and HNO_3. Appendix G shows that both chloride and phosphate form insoluble salts with silver ion, AgCl and Ag_3PO_4, respectively. However, the phosphate ion is the conjugate base of a weak acid, while chloride is the conjugate base of a strong acid. This means that Ag_3PO_4 can dissolve in excess strong acid, while AgCl cannot.

$$Ag_3PO_4(s) + H_3O^+(aq) \rightleftharpoons 3\ Ag^+(aq) + HPO_4^{2-}(aq) + H_2O(\ell)$$

$$AgCl(s) + H_3O^+(aq) \longrightarrow \text{no reaction}$$

Therefore you first add $AgNO_3$ to your unknown to precipitate a silver salt, either AgCl or Ag_3PO_4. Then you add HNO_3 to see if the precipitate dissolves. If it does, you can conclude that your unknown contained the phosphate ion and not the chloride ion.

Many metal sulfides are not soluble in water, although some are more soluble than others. This fact can be used to assist in the separation

Figure 19.7 Dissolving precipitates in acid. (a) A precipitate of AgCl (white) and Ag_3PO_4 (yellow). (b) Adding a strong acid dissolves Ag_3PO_4 and leaves insoluble AgCl.

Table 19.3 Metal Sulfides and Their Solubility

Metal Sulfide (= MS)	K_{sp}	$[S^{2-}]$ Required to Begin Precipitation of MS from 0.0010 M M^{2+}
HgS	3.0×10^{-53}	3.0×10^{-50}
CuS	8.7×10^{-36}	8.7×10^{-33}
CdS	3.6×10^{-29}	3.6×10^{-26}
ZnS	1.1×10^{-21}	1.1×10^{-18}
CoS	5.9×10^{-21}	5.9×10^{-18}
MnS	5.1×10^{-15}	5.1×10^{-12}

of a group of metal ions from one another and in their ultimate identification. Taking the metal sulfides in Table 19.3, you can see that the first three are much less soluble than the last three. Therefore, if you have a solution that is 0.0010 M in each of the six ions, it is convenient to carefully add S^{2-} ion so that only HgS, CuS, and CdS precipitate and leave Zn^{2+}, Co^{2+}, and Mn^{2+} in solution. From calculations like that in Example 19.10, you learn that this can be accomplished *if* $[S^{2-}]$ is kept larger than 3.6×10^{-26} M but less than 1.1×10^{-18} M.

To see how the sulfide ion concentration can be adjusted to the desired range of 10^{-26} to 10^{-18}, look again at the chemistry of the weak acid H_2S.

$$H_2S(aq) + H_2O(\ell) \rightleftharpoons H_3O^+(aq) + HS^-(aq) \qquad K_{a1} = 1.0 \times 10^{-7}$$
$$HS^-(aq) + H_2O(\ell) \rightleftharpoons H_3O^+(aq) + S^{2-}(aq) \qquad K_{a2} = 1.3 \times 10^{-13}$$
$$H_2S(aq) + 2\,H_2O(\ell) \rightleftharpoons 2\,H_3O^+(aq) + S^{2-}(aq) \qquad K_{total} = (K_{a1})(K_{a2}) = 1.3 \times 10^{-20}$$

The sulfide ion concentration in aqueous solution, where $[H_2S]$ and $[H_3O^+]$ are known, can be found from the rearranged equilibrium constant expression.

$$[S^{2-}] = \frac{[H_2S]}{[H_3O^+]^2} K_{total}$$

From Le Chatelier's principle and the common ion effect, we can predict that raising $[H_3O^+]$ will cause $[S^{2-}]$ to decrease. Conversely, adding OH^- will consume H_3O^+ and raise the sulfide ion concentration. Therefore, to set $[S^{2-}]$ at a given value, we only need to adjust the H_3O^+ concentration.

Let us say we want to separate Cd^{2+} and Zn^{2+}, each with a concentration of 0.0010 M. Precipitation of CdS requires a lower S^{2-} ion concentration (3.6×10^{-29} M) than precipitation of ZnS (1.1×10^{-18} M). Therefore, if we can adjust the sulfide ion concentration to a value slightly less than 1.1×10^{-18} M, say 1.0×10^{-19} M, we can precipitate the maximum amount of Cd^{2+} and leave Zn^{2+} in solution. Now, the concentration of H_2S in a saturated aqueous solution at 1 atmosphere pressure is 0.10 M, so a sulfide ion concentration of 1.0×10^{-19} M would be achieved when

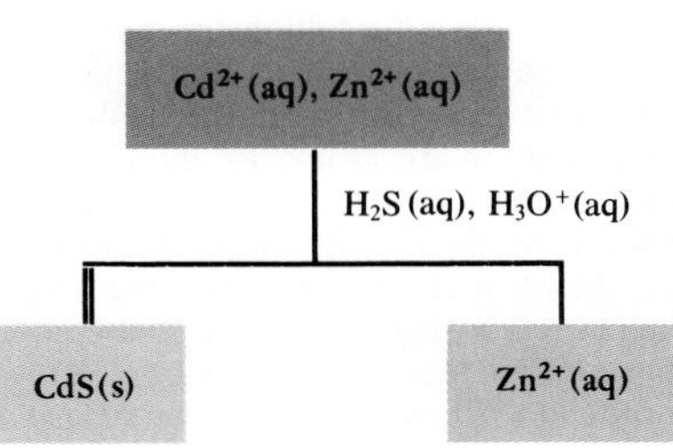

$$[H_3O^+]^2 = \frac{[H_2S]}{[S^{2-}]} K_{total} = \frac{0.10}{1.0 \times 10^{-19}} 1.3 \times 10^{-20}$$

$$[H_3O^+] = 0.11 \text{ M} \qquad \text{or} \qquad \text{pH} = 0.94$$

This H_3O^+ ion concentration is easily achieved, so Cd^{2+}, and the Hg^{2+} and Cu^{2+} that form less soluble sulfides than Cd^{2+}, can be separated from Zn^{2+} and the other ions.

EXERCISE 19.19 Ion Separation

You have an aqueous solution that may contain CO_3^{2-} or SO_4^{2-}. Show how you can use aqueous $BaCl_2$ and HCl to decide which ion is in the solution.

EXERCISE 19.20 Ion Separation

You have an aqueous solution of Co^{2+} and Mn^{2+}; both are 0.0010 M. Approximately what hydronium ion concentration would allow you to precipitate the maximum amount of CoS and leave Mn^{2+} in solution when H_2S is added?

(*text continues p. 810*)

SOMETHING MORE ABOUT Carbon Dioxide Equilibria—Of Oceans and Eggs

Our biosphere is a complicated mixture of carbon-containing compounds, some being created, some transformed, and others decomposed at any moment (see Figure A). The compound that links these processes and their reactants or products is carbon dioxide, which is produced in the biological process of respiration and consumed by photosynthetic organisms. Although CO_2 constitutes only about 0.0325% of the atmosphere, a recent estimate is that the earth's atmosphere contains 700 billion tons of carbon in the form of the gas.

Our atmosphere normally contains about 0.0003 atm of CO_2, and when this is in equilibrium with CO_2 dissolved in water, the concentration of aqueous H_2CO_3 is about 10^{-5} M. Although this solubility is small, the oceans are thought

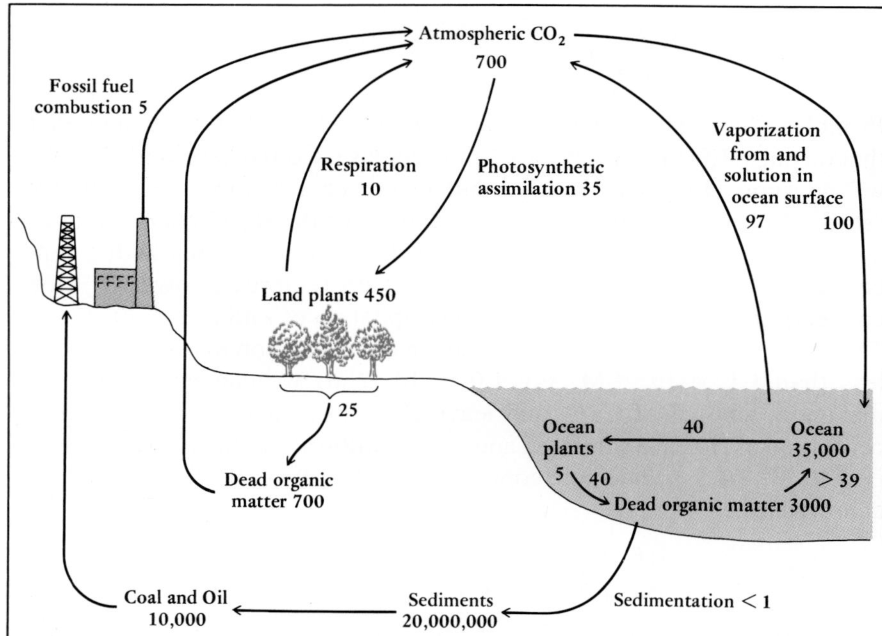

Figure A The carbon cycle in the biosphere. The numbers represent quantities of carbon (in billions of metric tons) in a particular reservoir or flowing from one reservoir to another. For example, there are 700×10^9 metric tons of C in the form of CO_2 in the atmosphere, while 5×10^9 metric tons of C (in the form of CO_2) flow into the atmosphere each year from fossil fuel combustion. (J. Moore and E. Moore, *Environmental Chemistry*, Academic Press.)

to hold roughly 60 times as much CO_2 as the atmosphere. However, in the oceans the CO_2 is in the form of (a) dissolved gas or carbonic acid, (b) one of the ionization products of H_2CO_3, (c) solid metal carbonates such as $CaCO_3$ or $MgCO_3$, and (d) organic matter.

Because H_2CO_3 is a weak acid, solutions of CO_2 in pure water are slightly acidic, and, since K_{a2} is much smaller than K_{a1}, we can estimate the pH from the first equilibrium reaction. Using $[H_2CO_3] \approx 10^{-5}$ M, we can estimate that the hydronium ion concentration is about 2.2×10^{-6} M and that the pH is therefore about 5.7. This means that the rain that falls even in a nonpolluted environment should be slightly acidic.

$H_2CO_3(aq) + H_2O(\ell) \rightleftharpoons H_3O^+(aq) + HCO_3^-(aq)$

$K_{a1} = 4.2 \times 10^{-7}$

$HCO_3^-(aq) + H_2O(\ell) \rightleftharpoons H_3O^+(aq) + CO_3^{2-}(aq)$

$K_{a2} = 4.8 \times 10^{-11}$

The oceans of the earth contain enormous quantities of calcium carbonate produced by various sea creatures. This means there is the additional equilibrium of calcium carbonate to consider.

$$CaCO_3(s) \rightleftharpoons Ca^{2+}(aq) + CO_3^{2-}(aq) \qquad K_{sp} = 3.8 \times 10^{-9}$$

When this is included, seawater saturated with carbon dioxide is found to have a pH of 8.2 ± 0.2. The slightly alkaline character of seawater occurs because the carbonate ion from the mineral calcite is the conjugate base of a weak acid, the bicarbonate ion, and so carbonate hydrolysis produces OH^-.

$$CO_3^{2-}(aq) + H_2O(\ell) \rightleftharpoons HCO_3^-(aq) + OH^-(aq) \qquad K_b = 2.1 \times 10^{-4}$$

The most important aspect of the carbonate equilibrium system is that it buffers the pH of seawater. The addition of acids by undersea volcanic activity

Figure B The system $CO_2/Ca(OH)_2$ in water. (a) Solid CO_2 is added to saturated $Ca(OH)_2(aq)$. (The bromthymol blue indicator shows that the solution is basic.) The cloudiness in the solution indicates that $CaCO_3(s)$ is just beginning to be precipitated.

$$CO_2(g) + H_2O(\ell) \rightleftharpoons H_2CO_3(aq)$$

$$H_2CO_3(aq) + Ca(OH)_2(aq) \rightleftharpoons CaCO_3(s) + 2\ H_2O(\ell)$$

(b) As the concentration of dissolved CO_2 builds up, more $CaCO_3$ precipitates. (c) When the concentration of dissolved CO_2 is large enough (the indicator now shows that the solution is acidic), $CaCO_3$ dissolves according to the equation

$$CaCO_3(s) + CO_2(g) + H_2O(\ell) \rightleftharpoons Ca^{2+}(aq) + 2\ HCO_3^-(aq)$$

It is just this reaction that leads to the dissolution of limestone in CO_2-saturated ground water. Reversal then leads to the reprecipitation of limestone in caves (Figure 16.1).

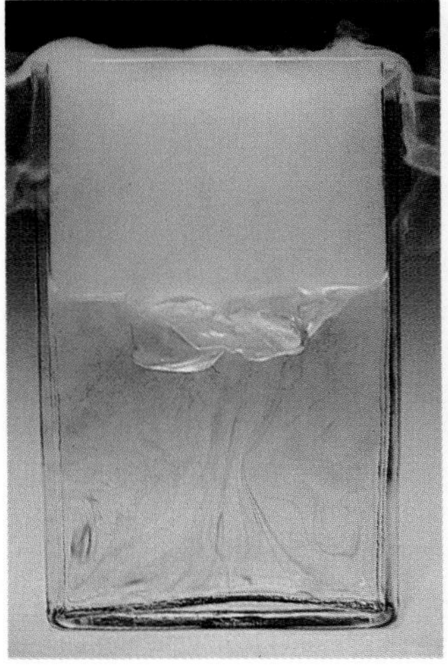

(a)

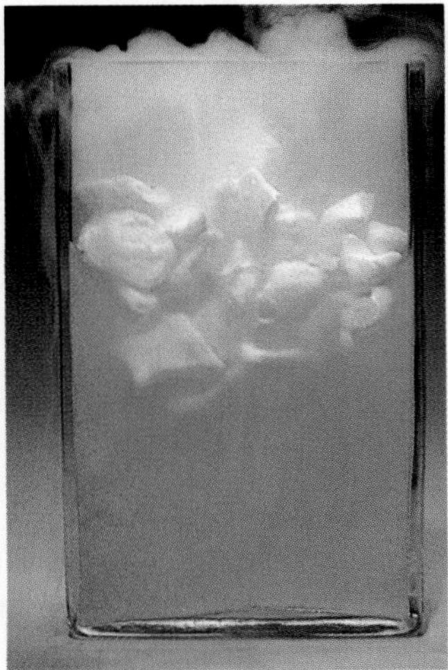

(b)

(c)

Figure C A limestone wall in Wind Cave National Park (South Dakota). The gaps in the wall were formed by water seeping into cracks in the wall and dissolving some of the limestone. However, the gaps became more narrow toward the bottom because, as the water penetrated deeper into the layer, it contained less dissolved CO_2 and was therefore able to dissolve less and less limestone. (Photo by Arthur N. Palmer)

or other natural processes is countered primarily by carbonic acid formation (and subsequent loss of carbon dioxide to the atmosphere),

$$HCO_3^-(aq) + H_3O^+(aq) \rightleftharpoons H_2CO_3(aq) + H_2O(\ell)$$

and secondarily by bicarbonate formation.

$$H_3O^+(aq) + CO_3^{2-}(aq) \rightleftharpoons HCO_3^-(aq) + H_2O(\ell)$$

On the other hand, an increase in hydroxide ion concentration is counteracted by the reaction

$$OH^-(aq) + HCO_3^-(aq) \rightleftharpoons H_2O(\ell) + CO_3^{2-}(aq)$$

This reaction of course leads to an increase in the carbonate concentration. If the seawater is above the saturation limit of calcium carbonate, limestone is precipitated (Figures B and C).

Carbon dioxide is being generated in ever increasing amounts, in part due to the increase in the population of the earth, in part to the clearing of forests (and thus to less use of CO_2 in photosynthesis), and in part to increased combustion of fossil fuels. Indeed, there is a fear that this could lead to a global warming trend caused by the "greenhouse effect."

Fortunately, the amount of CO_2 in the atmosphere is not increasing as rapidly as might be expected, largely because the ocean is a great CO_2 sink. As the partial pressure of CO_2 increases, CO_2 solubility increases, and it is estimated that the sea has absorbed roughly half of the increase in CO_2. Although this should lead in turn to an increase in hydronium ion concentration, $[H_3O^+]$ is buffered through reaction with carbonate ion. Furthermore, as the upper layers of seawater containing dissolved CO_2 are mixed with the lower layers in contact with carbonate-containing sediments, hydronium ion can be removed by reactions such as

$$H_3O^+(aq) + CaCO_3(s) \rightleftharpoons HCO_3^-(aq) + Ca^{2+}(aq) + H_2O(\ell)$$

The problem is that ocean mixing is relatively slow, requiring times on the order of 1000 years.

There is a problem in agriculture that is closely related to absorption of CO_2 by the sea. Like dogs, chickens pant when they are hot. This is not just an obscure bit of information—it has serious consequences. The problem is that the chicken loses CO_2 at a more rapid rate when panting. Lowering the partial pressure of CO_2 in the chicken means, according to Le Chatelier's principle, that the carbonate equilibrium is shifted away from CO_3^{2-} and toward CO_2. Less CO_3^{2-} is thus available to form $CaCO_3$ egg shells, and the shells produced by chickens in hot weather are often fragile. The solution? During warm weather chickens are given carbonated drinking water. (Rumors that the chickens are demanding Perrier water are unfounded!)

Figure D Corals are largely calcium carbonate. They are found in many forms and are island builders. This photo shows a coral atoll in the South Pacific somewhere between Tahiti and the Fiji Islands. An atoll is a coral island consisting of a ring of coral surrounding a central lagoon. (The photo was taken from an airplane flying at about 35,000 feet.)

SOMETHING MORE ABOUT
Silver Chemistry

Starting with water-soluble silver nitrate, $AgNO_3$, various reagents are added to form progressively less soluble silver salts. For NH_3 and $S_2O_3^{2-}$, the formation constants of their Ag^+ complexes are large enough that the insoluble salts AgCl and AgBr, respectively, can be dissolved. For a discussion of the equilibria involved [especially the reaction $Ag_2CO_3(s) + 2\ OH^-(aq) \rightleftarrows Ag_2O(s) + CO_3^{2-}(aq) + H_2O(\ell)$, where $K = 2 \times 10^4$], see B. Shakhashiri, *Chemical Demonstrations,* Volume 1, University of Wisconsin Press, 1983, page 307.

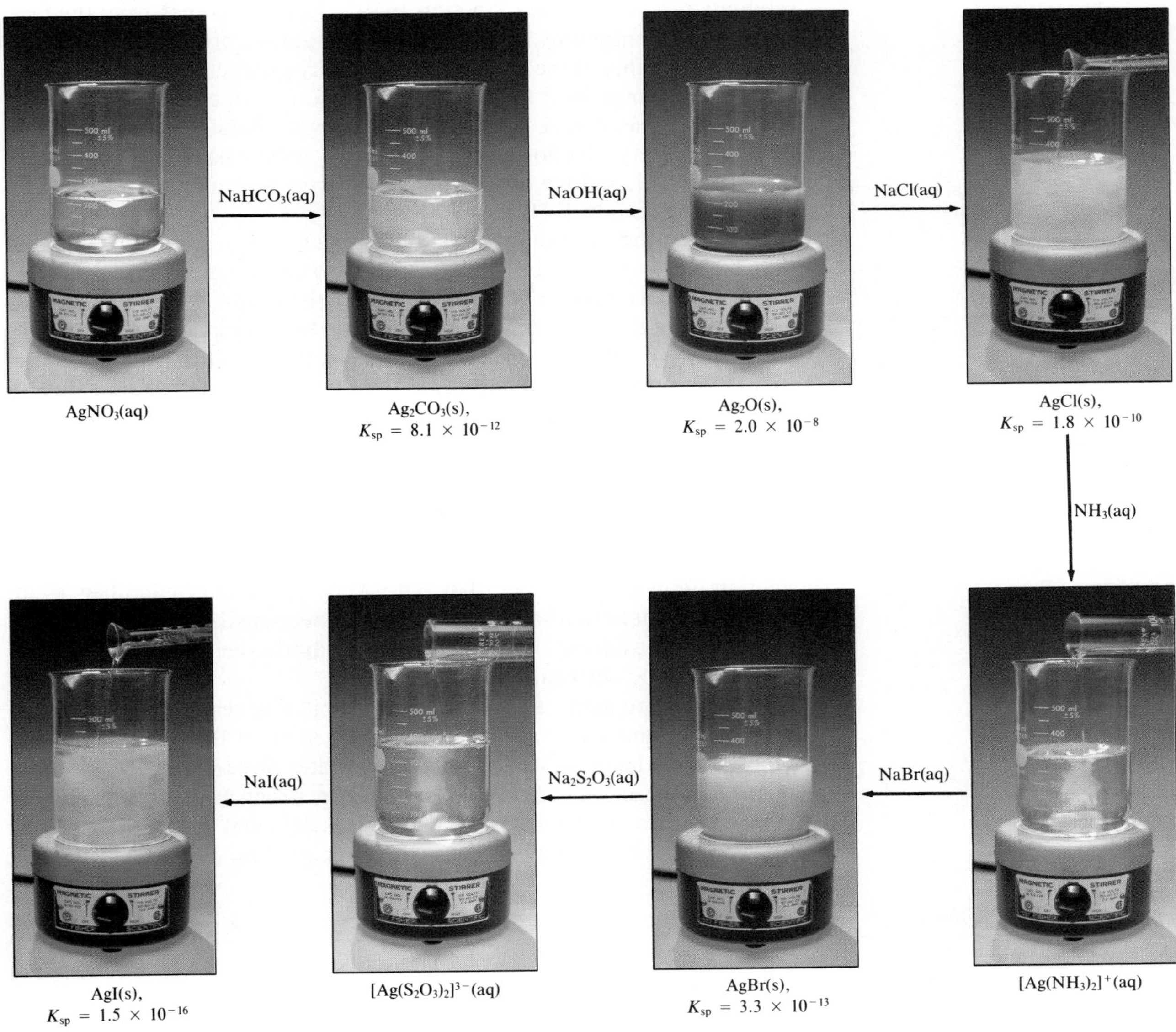

SUMMARY

When an insoluble compound is placed in water, the concentrations of the ions will increase until the salt achieves its maximum **solubility** at the particular temperature; at this point the solution is **saturated**, and a dynamic equilibrium is achieved between the ions in solution and the solid salt. Under these conditions, the equilibrium constant expression for an insoluble salt such as zinc phosphate

$$Zn_3(PO_4)_2(s) \rightleftharpoons 3\ Zn^{2+}(aq) + 2\ PO_4^{3-}(aq)$$

can be written as

$$K_{sp} = 9.1 \times 10^{-33} = [Zn^{2+}]^3[PO_4^{3-}]^2$$

The constant is labeled with a subscript "sp" to indicate that it is the **solubility product constant** (Section 19.2). The name comes from the fact that the concentrations of the cations and anions are measures of the aqueous solubility of the compound. Solubility product constants are typically in the range 10^{-3} to 10^{-50}, although some can even be smaller.

K_{sp} values can be determined by various laboratory measurements of salt solubility (Section 19.3). Many such values have been tabulated in Appendix G, and we can use them to derive the solubility of a salt in pure water (Section 19.4).

When the reaction quotient, Q, is equal to K_{sp}, a dynamic equilibrium has been achieved between the solid and constituent ions in solution. The solution is said to be **saturated**, and there will be no further *net* dissolution of the solid. If Q is less than K_{sp}, the solution is *not* saturated, and additional solid can dissolve. However, if Q is greater than K_{sp}, the solution is **supersaturated** and some solid must precipitate from solution until saturation is achieved.

In addition to molar solubility, K_{sp} values can be used to determine whether a precipitate will form if an anion is mixed with a particular cation, and what the salt solubility can be in the presence of an ion common to the salt (Sections 19.6 and 19.7).

It is possible that when another reagent is added to a saturated solution of an insoluble salt, that salt will be converted to another, even less soluble salt or to a soluble complex ion (Sections 19.8 and 19.9). Such situations are examples of **simultaneous equilibria**, and a net equilibrium constant can be derived.

There are many acid–base and precipitation reactions of practical significance, and a few are described in Section 19.10. One reaction of importance is increased salt solubility in water due to hydrolysis of the anion. Also described are (a) the solubility in strong acids of salts having an anion that is the conjugate base of a weak acid and (b) the preparation or separation of salts by acid–base and precipitation reactions.

STUDY QUESTIONS

REVIEW QUESTIONS

1. Explain why [AgCl] does not appear in the K_{sp} expression for $AgCl(s) \rightleftharpoons Ag^+(aq) + Cl^-(aq)$.
2. What is a "reaction quotient" and how does it differ from an equilibrium constant? Use the following equilibrium in your discussion: $CaCO_3(s) \rightleftharpoons Ca^{2+}(aq) + CO_3^{2-}(aq)$.
3. In Appendix G, find two salts that have K_{sp} values less than 1×10^{-40} and (a) write balanced equations to show the equilibria existing when the compounds

dissolve in water and (b) write the K_{sp} expression for each compound.

4. Explain the terms saturated, not saturated (or undersaturated), and supersaturated. Use an equilibrium such as $NiS(s) \rightleftarrows Ni^{2+}(aq) + S^{2-}(aq)$.
5. What is the "common ion effect?" Use the equilibrium $Fe(OH)_2(s) \rightleftarrows Fe^{2+}(aq) + 2\ OH^-(aq)$ in your discussion.
6. Explain how Le Chatelier's principle operates when some excess cyanide ion, CN^-, is added to AgCN in equilibrium with its ions.

$$AgCN(s) \rightleftarrows Ag^+(aq) + CN^-(aq)$$

What happens when excess AgCN(s) is added to a beaker of water already containing solid AgCN in equilibrium with its ions? Will the concentrations of Ag^+ and CN^- increase, decrease, or not change?

7. Explain why the solubility of AgCN can be greater in water than is calculated from the K_{sp} of the salt.
8. Explain why $PbCO_3$ can dissolve in strong acid but $PbCl_2$ cannot.

SOLUBILITY GUIDELINES

9. Name two insoluble salts of each of the following ions:
(a) the chloride ion (c) the zinc ion
(b) the sulfide ion (d) the iron(II) ion
10. Name two insoluble salts of the following ions:
(a) the sulfate ion (c) the nickel(II) ion
(b) the bromide ion (d) the aluminum ion
11. Using the table of solubility guidelines (Figure 5.5), predict whether each of the following is insoluble or soluble in water.
(a) $(NH_4)_2S$ (d) FeS
(b) $ZnCO_3$ (e) $BaSO_4$
(c) $Mg(OH)_2$ (f) Hg_2Cl_2
12. Using the table of solubility guidelines (Figure 5.5), predict whether each of the following is insoluble or soluble in water.
(a) $Zn(NO_3)_2$ (d) $PbCl_2$
(b) $Al(OH)_3$ (e) CuS
(c) Na_2CO_3 (f) Ag_2CrO_4
13. If Na_2CO_3 is added to an aqueous solution of $AgNO_3$, does precipitation occur? If so, what solid will form? Write a balanced equation for the reaction.
14. If Na_2S is added to an aqueous solution of $CuSO_4$, does precipitation occur? If so, what solid will form? Write a balanced equation for the reaction.
15. Each pair of salts below is soluble in water. State whether a precipitate can form when you add one salt to the other in aqueous solution. If so, give the formula of the precipitate.
(a) NaCl + $AgNO_3$ (c) KOH + NaCl
(b) KOH + $MgCl_2$ (d) KCl + $Pb(NO_3)_2$
16. Each pair of salts below is soluble in water. State whether a precipitate can form when you add one salt to the other in aqueous solution. If so, give the formula of the precipitate.
(a) $(NH_4)_2S$ + $Ni(NO_3)_2$ (c) Na_2S + $CdCl_2$
(b) NaOH + $FeCl_3$ (d) Na_2CO_3 + K_2SO_4

WRITING SOLUBILITY PRODUCT CONSTANT EXPRESSIONS

17. For each of the following salts, (a) write a balanced equation showing the equilibrium occurring when the salt is added to water and (b) write the K_{sp} expression. Give the value for the K_{sp} of each salt.
(a) ZnS (c) $NiCO_3$
(b) SnI_2 (d) Ag_2SO_4
18. For each of the following salts, (a) write a balanced equation showing the equilibrium occurring when the salt is added to water and (b) write the K_{sp} expression. Give the value for the K_{sp} of each salt.
(a) Bi_2S_3 (c) BaF_2
(b) $Ni(OH)_2$ (d) $Al(OH)_3$
19. If the K_{sp} expression for magnesium arsenate is $[Mg^{2+}]^3[AsO_4{}^{3-}]^2$, what is the formula of the salt?
20. If the K_{sp} expression for cadmium hexacyanoferrate is $[Cd^{2+}]^2[Fe(CN)_6{}^{4-}]$, what is the formula of the salt?

DETERMINING K_{sp}

21. When 155 mg of solid cobalt(II) sulfide is added to 1.00 L of water, the salt dissolves to a very small extent.

$$CoS(s) \rightleftarrows Co^{2+}(aq) + S^{2-}(aq)$$

The cobalt(II) and sulfide ions in equilibrium with CoS each have a concentration of 7.7×10^{-11} M. What is the value of K_{sp} for CoS?

22. When 250 mg of CaF_2, calcium fluoride, is added to 1.00 L of water, the salt dissolves to a very small extent.

$$CaF_2(s) \rightleftarrows Ca^{2+}(aq) + 2\ F^-(aq)$$

At equilibrium, the concentration of Ca^{2+} is found to be 2.1×10^{-4} M. What is the value of K_{sp} for CaF_2?

23. At 20 °C, a saturated aqueous solution of silver acetate, $Ag(CH_3COO)$, contains 1.0 g dissolved in 100.0 mL of solution. Calculate K_{sp} for silver acetate.

$$Ag(CH_3COO)(s) \rightleftarrows Ag^+(aq) + CH_3COO^-(aq)$$

24. Calcium hydroxide, $Ca(OH)_2$, dissolves in water to the extent of 0.93 g per liter.

$Ca(OH)_2(s) \rightleftharpoons Ca^{2+}(aq) + 2\ OH^-(aq)$

What is the K_{sp} of $Ca(OH)_2$?

25. You place 2.75 g of barium fluoride in 1.00 L of pure water at 25 °C. After equilibrium has been established,

$BaF_2(s) \rightleftharpoons Ba^{2+}(aq) + 2\ F^-(aq)$

analysis of the solution for F^- gives a fluoride ion concentration of 0.0150 M. What is the K_{sp} of BaF_2?

26. At 25 °C, 34.9 mg of Ag_2CO_3 will dissolve in 1.0 L of pure water.

$Ag_2CO_3(s) \rightleftharpoons 2\ Ag^+(aq) + CO_3^{2-}(aq)$

What is the solubility product constant for this salt?

27. At 25 °C, 7.4 mg of $Zn(CN)_2$ will dissolve in 0.50 L of pure water.

$Zn(CN)_2(s) \rightleftharpoons Zn^{2+}(aq) + 2\ CN^-(aq)$

What is the solubility product constant for this salt?

28. At 25 °C, we find that 59.7 mg of PbI_2 will dissolve in 100. mL of pure water. What is the solubility product constant for PbI_2?

29. Assume you place 1.234 g of solid $Ca(OH)_2$ in 1.00 L of pure water at 25 °C. The pH of the solution is found to be 12.40. Estimate the K_{sp} for $Ca(OH)_2$.

30. If you place 0.979 g of $Pb(OH)_2$ in 1.00 L of pure water at 25 °C, the pH is 8.92. Estimate the K_{sp} for $Pb(OH)_2$.

ESTIMATING SALT SOLUBILITY FROM K_{sp}

The K_{sp} values required by the problems that follow are found in Table 19.2 or in Appendix G.

31. Estimate the solubility of silver cyanide in (a) moles per liter and (b) grams per liter in pure water at 25 °C.

$AgCN(s) \rightleftharpoons Ag^+(aq) + CN^-(aq)$

32. What is the molar concentration of $Au^+(aq)$ in a saturated solution of AuI in pure water at 25 °C?

$AuI(s) \rightleftharpoons Au^+(aq) + I^-(aq)$

33. How many milligrams of gold(III) ion are there per liter of solution in a saturated aqueous solution of gold(III) iodide?

$AuI_3(s) \rightleftharpoons Au^{3+}(aq) + 3\ I^-(aq)$

34. Estimate the solubility of lead(II) bromide in (a) moles per liter and (b) grams per liter of pure water.

$PbBr_2(s) \rightleftharpoons Pb^{2+}(aq) + 2\ Br^-(aq)$

35. In a saturated solution of silver carbonate, Ag_2CO_3, how many milligrams of Ag^+ are there in 100. mL of solution at 25 °C?

36. Although it is a violent poison if swallowed, mercury(II) cyanide, $Hg(CN)_2$, has been used as a topical (skin) antiseptic.
(a) What is the molar solubility of this salt in pure water?
(b) How many milligrams of $Hg(CN)_2$ dissolve per liter of pure water?
(c) How many milliliters of water are required to dissolve 1.0 g of the salt?

37. Using K_{sp} values, decide which compound in each of the following pairs is the more soluble.
(a) AgBr or AgSCN (c) MgF_2 or CaF_2
(b) $SrCO_3$ or $SrSO_4$ (d) ZnS or Bi_2S_3

38. Using K_{sp} values, decide which compound in each of the following pairs is the more soluble.
(a) $PbCl_2$ or $PbBr_2$ (c) HgS or FeS
(b) BiI_3 or $Bi(OH)_3$ (d) SnS_2 or PbS

39. Rank the following compounds in order of increasing solubility in water: BaF_2, $BaCO_3$, and Ag_2CO_3.

40. Rank the following compounds in order of increasing solubility in water: AgI, HgS, PbI_2, $PbSO_4$, and NH_4NO_3.

41. If you place the amounts given below in pure water, will all of the salt dissolve before equilibrium can be established, or will some salt remain undissolved?
(a) 5.0 mg of AgCl in 1.0 L of pure water
(b) 5.0 mg of $NiCO_3$ in 1.0 L of pure water

42. If you place the amounts given below in pure water, will all of the salt dissolve before equilibrium can be established, or will some salt remain undissolved?
(a) 5.0 mg of MgF_2 in 125 mL of pure water
(b) 0.50 g of CaF_2 in 100. mL of pure water

PRECIPITATIONS

43. If the concentration of the nickel(II) ion in 100. mL of pure water is 0.0024 M, will precipitation of $NiCO_3$ occur (a) when the concentration of the carbonate ion is 1.0×10^{-6} M or (b) when it is 100 times greater (1.0×10^{-4} M)? The equilibrium process involved is $NiCO_3(s) \rightleftharpoons Ni^{2+}(aq) + CO_3^{2-}(aq)$.

44. You have 100. mL of a solution that has a lead(II) ion concentration of 0.0012 M. If enough soluble chloride-containing salt is added so that the Cl^- concentration is 0.010 M, will $PbCl_2$ precipitate? The equilibrium process involved is $PbCl_2(s) \rightleftharpoons Pb^{2+}(aq) + 2\ Cl^-(aq)$.

45. If the concentration of Zn^{2+} in 10. mL of water is 1.6×10^{-4} M, will zinc hydroxide, $Zn(OH)_2$, precipitate when 4.0 mg of NaOH is added?

46. You have 100. mL of a solution that has a lead(II)

ion concentration of 0.0012 M. Will $PbCl_2$ precipitate when 1.20 g of solid NaCl is added?

47. Magnesium is extracted from seawater by precipitating it as $Mg(OH)_2$. If the concentration of Mg^{2+} in seawater is 1350 mg per liter, what OH^- concentration is required to precipitate $Mg(OH)_2$?

48. A sample of hard water contains about 2.0×10^{-3} M Ca^{2+}. A soluble fluoride-containing salt such as NaF is added to "fluoridate" the water (to aid in the prevention of dental caries). What is the maximum concentration of F^- that can be present without precipitation of CaF_2?

49. The mineral fluorite is CaF_2. If you have 500. mL of a solution that contains Ca^{2+} ion with a concentration of 0.0021 M, how many milligrams of KF must be added to the solution to just begin precipitation of CaF_2?

50. You have 1.0 L of 0.0012 M $Pb(NO_3)_2$. How many grams of Na_2S must be added to this solution to precipitate lead(II) sulfide, PbS?

51. If you mix 50. mL of 0.0012 M $BaCl_2$ with 25 mL of 1.0×10^{-6} M H_2SO_4, will a precipitate of $BaSO_4$ form? The equilibrium process involved is $BaSO_4(s) \rightleftharpoons Ba^{2+}(aq) + SO_4^{2-}(aq)$.

52. Will a precipitate of $Mg(OH)_2$ form when 25.0 mL of 0.010 M NaOH is combined with 75.0 mL of a 0.10 M solution of magnesium chloride?

53. Lead(II) chloride has a relatively large K_{sp} of 1.7×10^{-5} at 25 °C. If you mix 10. mL of 0.0010 M $Pb(NO_3)_2$ with 5.0 mL of 0.015 M HCl, will a precipitate of $PbCl_2$ form?

54. If 100. mL of 0.10 M H_2SO_4 is added to 50.0 mL of 0.00013 M $BaCl_2$, will a precipitate of $BaSO_4$ form?

ORDER OF PRECIPITATION

55. The cations Co^{2+}, Mn^{2+}, and Ni^{2+} can all be precipitated as very insoluble sulfides, CoS, MnS, and NiS. If you add a soluble sulfide-containing salt (say $(NH_4)_2S$) to a solution containing all of these ions, each with a concentration of 0.10 M, in what order are the sulfides precipitated?

56. HI is added slowly to a solution that is 0.10 M in each of the following ions: Pb^{2+}, Ag^+, and Hg_2^{2+}. The insoluble salts PbI_2, AgI, and Hg_2I_2 will eventually form. In what order do these salts precipitate?

57. You often work with salts of Fe^{3+}, Pb^{2+}, and Al^{3+} in the laboratory. All are found in nature, and all are important economically. If you have a solution containing these three ions, each at a concentration of 0.1 M, what is the order in which their hydroxides precipitate?

58. Alkaline earth metal ions can be precipitated as insoluble carbonates. If you have a solution of Mg^{2+}, Ca^{2+}, Sr^{2+}, and Ba^{2+} ions, all with the same concentration, what is the order in which their carbonates are precipitated as sodium carbonate is added slowly?

59. The alkaline earth cations Mg^{2+}, Ca^{2+}, and Ba^{2+} can be precipitated as their fluoride salts. If you have a solution that is 0.015 M in each of these ions, and you slowly add KF, in what order do you expect to see the fluoride salts precipitated?

60. The alkaline earth cations Mg^{2+}, Ca^{2+}, and Ba^{2+} can be precipitated as their fluorides, carbonates, or sulfates (see Study Questions 58 and 59). If you have a solution of these three ions, each with a concentration of 0.001 M, in what order do these ions precipitate as their insoluble sulfates as you slowly add sulfuric acid to the solution?

COMMON ION EFFECT

61. Calculate the molar solubility of silver thiocyanate, AgSCN, in pure water and in water containing 0.010 M NaSCN.

62. Calculate the solubility of silver carbonate, Ag_2CO_3, in moles/liter, in pure water. Compare this with the molar solubility of Ag_2CO_3 in 225 mL of water to which 0.15 g of Na_2CO_3 has been added.

63. Calculate the solubility, in mg/mL, of silver phosphate, Ag_3PO_4, (a) in pure water and (b) in water that is 0.020 M in $AgNO_3$.

64. What is the solubility, in grams per mL, of BaF_2, (a) in pure water and (b) in water containing 5.0 mg per mL of KF?

65. The solubility product constant for silver iodate, $AgIO_3$, is 1.0×10^{-8}. If 0.10 g of solid $AgIO_3$ is added to 100.0 mL of 0.020 M KIO_3, what are the concentrations of K^+, IO_3^-, and Ag^+ at equilibrium?

66. Assume you have 1.00 L of a saturated solution of $BaSO_4$ in pure water. Estimate the concentrations of Ba^{2+} and SO_4^{2-} after you have added 2.8 mg of Na_2SO_4.

67. A laboratory experiment calls for you to mix 25.0 mL of 0.015 M $AgNO_3$ with 25.0 mL of 0.0012 M NaCl. Does AgCl precipitate? If so, how many grams of AgCl are formed? What are the concentrations of Ag^+, NO_3^-, Na^+, and Cl^- in the resulting solution?

68. Suppose you mix 100.0 mL of 0.012 M HCl with 50.0 mL of 0.053 M $Pb(NO_3)_2$. Does $PbCl_2$ precipitate, and if so, how much? What are the concentrations of Cl^-, H_3O^+, Pb^{2+}, and NO_3^- in the final solution?

SEPARATIONS

69. To separate Ca^{2+} and Mg^{2+}, ammonium oxalate, $(NH_4)_2C_2O_4$, is added to a solution that is 0.020 M in both metal ions. If the concentration of the oxalate

ion is adjusted properly, the metal oxalates can be precipitated separately. In this case CaC_2O_4 precipitates before MgC_2O_4. The chemical equilibria involved are:

$$MgC_2O_4(s) \rightleftharpoons Mg^{2+}(aq) + C_2O_4^{2-}(aq)$$

$$CaC_2O_4(s) \rightleftharpoons Ca^{2+}(aq) + C_2O_4^{2-}(aq)$$

(a) What concentration of oxalate ion, $C_2O_4^{2-}$, will precipitate the maximum amount of Ca^{2+} ion without precipitating Mg^{2+}?
(b) What concentration of Ca^{2+} remains in solution when Mg^{2+} just begins to precipitate?

70. The ions Mn^{2+} and Co^{2+} can be separated in the laboratory by the difference in the solubility of their sulfides, MnS and CoS. If you have a solution that is 0.10 M in both Mn^{2+} and Co^{2+}, CoS will begin to precipitate first as S^{2-} is slowly added to the solution.
(a) What concentration of sulfide ion will precipitate the maximum amount of Co^{2+} ion without precipitating MnS?
(b) What concentration of Co^{2+} remains in solution when MnS just begins to precipitate?

71. A solution contains 0.10 M iodide ion, I^-, and 0.10 M carbonate ion, CO_3^{2-}.
(a) If solid $Pb(NO_3)_2$ is slowly added to the solution, which salt will precipitate first, PbI_2 or $PbCO_3$?
(b) What will be the concentration of the first ion that precipitates (CO_3^{2-} or I^-) when the second or more soluble salt begins to precipitate?

72. A solution contains Ca^{2+} and Pb^{2+} ions, both at a concentration of 0.010 M. You wish to separate the two ions from each other as completely as possible by precipitating one but not the other, using aqueous Na_2SO_4 as the precipitating agent.
(a) Which will precipitate first as sodium sulfate is added, $CaSO_4$ or $PbSO_4$?
(b) Just before the more soluble salt begins to precipitate, what molarity of the first metal ion remains in solution?

73. Each pair of ions below is found together in aqueous solution. Using the table of solubility product constants in Appendix G, devise a way to separate these ions by precipitating one of them as an insoluble salt and leaving the other in solution.
(a) Ba^{2+} and Na^+
(b) Ca^{2+} and Zn^{2+}
(c) Bi^{3+} and Cd^{2+}

74. Each pair of ions below is found together in aqueous solution. Using the table of solubility product constants in Appendix G, devise a way to separate these ions by adding one reagent to precipitate one of them as an insoluble salt and leave the other in solution.
(a) Cu^{2+} and Ag^+
(b) Pb^{2+} and Sn^{2+}
(c) Al^{3+} and Fe^{3+}

SIMULTANEOUS EQUILIBRIA AND COMPLEX IONS

75. Solid silver iodide, AgI, can be dissolved by adding concentrated aqueous ammonia to give the water-soluble silver–ammonia complex ion.

$$AgI(s) + 2\ NH_3(aq) \rightleftharpoons [Ag(NH_3)_2]^+(aq) + I^-(aq)$$

Show that this equation is the sum of two other equations, one for dissolving AgI to give its ions and the other for the formation of the $[Ag(NH_3)_2]^+$ ion from Ag^+ and NH_3. Calculate K for the overall reaction.

76. Solid AgBr is dissolved when excess thiosulfate ion, $S_2O_3^{2-}$, is added to give a water-soluble complex ion.

$$AgBr(s) + 2\ S_2O_3^{2-}(aq) \rightleftharpoons [Ag(S_2O_3)_2]^{3-}(aq) + Br^-(aq)$$

Show that this equation is the sum of two other equations, one for dissolving AgBr to give its ions and the other for the formation of the $[Ag(S_2O_3)_2]^{3-}$ ion from Ag^+ and $S_2O_3^{2-}$. Calculate K for the overall reaction.

77. Will AgI form if iodide ion, I^-, is added to a saturated solution of AgCl?

$$AgCl(s) + I^-(aq) \rightleftharpoons AgI(s) + Cl^-(aq)$$

What is the equilibrium constant for this process?

78. Can zinc sulfide, ZnS, be transformed into zinc cyanide by adding a soluble salt of the cyanide ion?

$$ZnS(s) + 2\ CN^-(aq) \rightleftharpoons Zn(CN)_2(s) + S^{2-}(aq)$$

Calculate the equilibrium constant for the reaction.

79. Will 5.0 mL of 2.5 M NH_3 dissolve 1.0×10^{-4} mol of AgCl suspended in 1.00 L of water?

80. Can you dissolve, essentially completely, 15.0 mg of AuCl in 100.0 mL of pure water if you add 15.0 mL of 6.00 M NaCN?

ACID–BASE REACTIONS FOR DISSOLVING SOLIDS AND PREPARING SALTS

81. Explain why $CaCO_3$ is soluble in aqueous HCl but $CaSO_4$ is much less so.

82. Which of the following barium salts should be most

soluble in a strong acid such as HCl: $Ba(OH)_2$, $BaSO_4$, or $BaCO_3$?

83. Explain how you can prepare pure $CaSO_4$ starting with limestone, $CaCO_3$.
84. Describe a method for the preparation of $NiSO_4$ from $NiCO_3$.
85. How can you prepare pure magnesium nitrate beginning with pure magnesium chloride?
86. Describe a preparation of iron(III) perchlorate, $Fe(ClO_4)_3$, from iron(III) chloride.
87. You have an aqueous solution that may contain CO_3^{2-} or Cl^-. If you have aqueous $AgNO_3$ and HCl available, tell *two* ways you can decide which anion the solution contains.
88. You have an aqueous solution that may contain SO_4^{2-} or CO_3^{2-}. If you have aqueous $BaCl_2$ and HCl available, tell how you would go about deciding whether your unknown contains sulfate or carbonate ion.
89. Some solid sulfides can dissolve in strong acid.

$$MS(s) + 2\,H_3O^+(aq) \rightleftharpoons M^{2+}(aq) + H_2S(aq) + H_2O(\ell)$$

$$K = K_{dissolve}$$

This reaction can be considered the sum of two other reactions:

$$MS(s) \rightleftharpoons M^{2+}(aq) + S^{2-}(aq) \qquad K = K_{sp}$$

and

$$S^{2-}(aq) + 2\,H_3O^+(aq) \rightleftharpoons H_2S(aq) + 2\,H_2O(\ell)$$
$$K = 1/1.3 \times 10^{-20} = 7.7 \times 10^{19}$$

Therefore, we can write the expression

$$K_{dissolve} = K_{sp}(7.7 \times 10^{19}) = \frac{[M^{2+}][H_2S]}{[H_3O^+]^2}$$

This means that we can find the equilibrium concentration of M^{2+} if $[H_3O^+]$ is known and if we assume that $[H_2S]$ is 0.10 M in a saturated solution.
(a) Calculate $K_{dissolve}$ for ZnS.
(b) Calculate $[Zn^{2+}]$ when the hydronium ion concentration is 0.30 M.

90. Study Question 89 outlines a method for calculating the solubility of metal sulfides in strong acid. Use this approach to compare the solubilities of CuS and ZnS in 1.0 M HCl.

GENERAL QUESTIONS

91. Limestone, $CaCO_3$, can exist in two mineral forms: calcite and aragonite. In pure water, the former has a K_{sp} of 3.8×10^{-9}, whereas K_{sp} of the latter is 6.0×10^{-9}. Which is the more soluble in pure water?
92. Barium sulfate is very insoluble in water and is opaque to x-rays. Thus, if you drink a "barium cocktail" it does not dissolve in your stomach or intestines, and its progress through the digestive organs can be followed by x-ray analysis. If you place 1.00×10^2 mg of $BaSO_4$ in 1.0 L of water, approximately how many milligrams of $BaSO_4$ are dissolved at 25 °C?

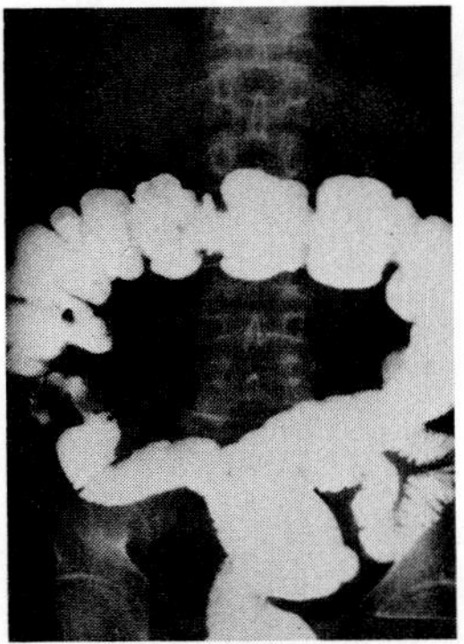

X-ray of gastrointestinal tract after ingesting barium sulfate.

93. To make up an unknown solution for the students in your laboratory, you dissolve 1.0×10^{-5} mol of $AgNO_3$ in 1.0 L of water. Unfortunately, you used a liter of tap water instead of distilled water. If the chloride ion concentration in tap water is 2.0×10^{-4} M, will your error be revealed immediately by the formation of a white precipitate of AgCl?
94. The alkaline earth metal ions can be precipitated from aqueous solution by addition of fluoride ion.
(a) If you have a 0.015 M solution of Ba^{2+}, what is the minimum concentration of F^- ion necessary to begin precipitation of BaF_2?
(b) If you started with 100.0 mL of the 0.015 M Ba^{2+}-containing solution, how many milligrams of NaF would be required to just begin precipitation?
(c) After adding 0.50 g of NaF to 100.0 mL of 0.015 M Ba^{2+}-containing solution, what is the concentration of Ba^{2+} remaining in solution?
95. Lead and zinc compounds are often found together in nature, since the solubilities of their salts are similar. For example, the K_{sp} of $Pb(OH)_2$ is 2.8×10^{-16} and that of $Zn(OH)_2$ is 4.5×10^{-17}.
(a) If Pb^{2+} and Zn^{2+} are present in the ground water, each with a concentration of 1.0×10^{-6} M, and if the ground water is 1.0×10^{-6} M in hydroxide ion, will either $Pb(OH)_2$ or $Zn(OH)_2$ precipitate?

(b) What is the maximum concentration of OH^- ion that can exist in ground water that will allow you to precipitate one of the metal hydroxides and leave the other one in solution?

96. Will 5.0 mL of 2.5 M NH_3 dissolve 1.00×10^{-4} mol of AgBr suspended in 1.0 L of pure water?

97. You mix 15.0 mL of 0.010 M $CaCl_2$ and 25.0 mL of 0.0010 M NaOH. Does $Ca(OH)_2$ precipitate? If so, how many grams of $Ca(OH)_2$ are formed? What are the concentrations of the ions Ca^{2+}, Cl^-, and Na^+ in the solution? What is the pH of the solution?

98. Explain why some metal sulfides can be dissolved in a strong acid such as HCl.

99. You add 1.0 mg of solid NaCl to 1.0 L of a saturated solution of AgCl(s). What are the concentrations of Ag^+ and Cl^- after equilibrium has been reestablished? How many milligrams of AgCl will be precipitated by adding the extra NaCl?

100. You have 1.0 L of pure water at 25 °C and add 25 mg of $Zn(CN)_2$.

(a) Does all the zinc cyanide dissolve or does some solid remain when equilibrium is achieved? What are the concentrations of Zn^{2+} and CN^- at equilibrium? What happens to the concentrations of Zn^{2+} and CN^- when another 25 mg of $Zn(CN)_2$ is added?

(b) If you add another liter of water to the original mixture of 25 mg of $Zn(CN)_2$ and 1.0 L of water, will any zinc cyanide remain undissolved? What are the concentrations of Zn^{2+} and CN^-?

(c) If 100. mg of KCN is added to the original mixture, how many milligrams of solid $Zn(CN)_2$ are there now in the beaker?

101. $Zn(OH)_2$ is a relatively insoluble base. A saturated solution has a pH of 8.65. Calculate the K_{sp} for $Zn(OH)_2$.

102. A saturated solution of $Ce(OH)_3$ has a pH of 9.00. (a) What are the Ce^{3+} and OH^- concentrations at equilibrium? (b) What is the K_{sp} of $Ce(OH)_3$?

103. Beginning with magnesium carbonate, describe how to prepare (a) magnesium sulfate, (b) magnesium bromide, (c) magnesium oxalate (MgC_2O_4), (d) magnesium perchlorate, and (e) magnesium fluoride.

104. Calcite and aragonite are two mineral forms of $CaCO_3$. In pure water their respective K_{sp} values are 3.8×10^{-9} and 6.0×10^{-9}. Will calcite spontaneously transform into aragonite in pure water? What is the equilibrium constant for this transformation?

105. If 0.581 g of $Mg(OH)_2$ is added to 1.00 L of pure water, will the hydroxide dissolve? If the solution is buffered at a pH of 5.00, will all the hydroxide dissolve?

Calcite (left) and aragonite (right) are two forms of calcium carbonate. Photos taken in Wind Cave National Park, South Dakota. (Arthur Palmer)

106. The Ca^{2+} ion in hard water is often precipitated as $CaCO_3$ by adding soda ash, Na_2CO_3. If the calcium ion concentration in hard water is 0.010 M, and if the Na_2CO_3 is added until the carbonate ion concentration is 0.050 M, what percentage of the calcium ion has been removed from the water? (You may neglect hydrolysis of the carbonate ion.)

107. Photographic film is coated with crystals of AgBr suspended in gelatin. Light exposure leads to the reduction of some of the silver ions to metallic silver. Unexposed AgBr is dissolved with sodium thiosulfate in the fixing step.

$$AgBr(s) + 2\ S_2O_3^{2-}(aq) \rightleftarrows [Ag(S_2O_3)_2]^{3-}(aq) + Br^-(aq)$$

(a) Using the appropriate K_{sp} and $K_{formation}$ values in Appendices G and H, calculate the equilibrium constant for the dissolving process.

(b) If you want to dissolve 1.0 g of AgBr in 1.0 L of solution, how many grams of $Na_2S_2O_3$ must be added?

108. The solubility of some metal salts is affected by hydrolysis of the anion. For PbS, the reaction

$$PbS(s) + H_2O(\ell) \rightleftarrows Pb^{2+}(aq) + HS^-(aq) + OH^-(aq)$$

$$K_{total} = [Pb^{2+}][HS^-][OH^-] = 6.5 \times 10^{-29}$$

produces a lead(II) ion concentration of 3.9×10^{-10} M. (This can be calculated when $[HS^-]$ and $[OH^-]$ are known.)

(a) If sulfide ion hydrolysis leads to a concentration of HS^- of 4.1×10^{-10} M at 25 °C (with no Pb^{2+} present), what is the concentration of OH^- in the solution?

(b) Assuming that the values of $[HS^-]$ and $[OH^-]$ calculated in part (a) are always observed when a metal sulfide is placed in water, and assuming sulfide ion hydrolysis occurs, calculate the solubility of CdS.

(c) Compare the result in part (b) with the solubility calculated without assuming sulfide hydrolysis. That is, calculate the solubility of CdS from the K_{sp} value alone.

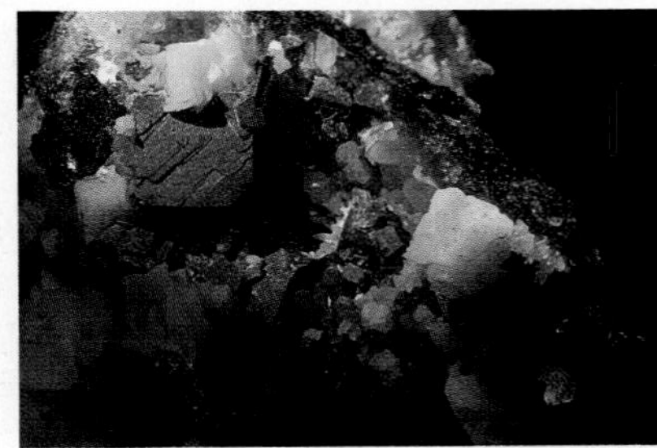

Galena (PbS, the black solid) in quartz.

SUMMARY QUESTION

109. Aluminum hydroxide reacts with phosphoric acid to give aluminum phosphate, $AlPO_4$. The system Al^{3+}–PO_4^{3-}–H_2O exists in many of the same crystal forms as SiO_2 and is used industrially as the basis of adhesives, binders, and cements. ($AlPO_4$ is often referred to by the same name as a popular pet food company.)

(a) Write a balanced equation for the reaction of aluminum chloride and phosphoric acid.

(b) If you begin with 152 g of aluminum chloride and 3.0 L of 0.750 M phosphoric acid, how many grams of $AlPO_4$ can be isolated?

(c) If you place 25.0 g of $AlPO_4$ in enough pure water to have a volume of exactly one liter, what are the concentrations of Al^{3+} and PO_4^{3-} at equilibrium?

(d) Does the solubility of $AlPO_4$ increase or decrease on adding HCl? Explain briefly.

(e) If you mix 1.50 L of 0.0025 M Al^{3+} (in the form of $AlCl_3$) with 2.50 L of 0.035 M Na_3PO_4, will a precipitate of $AlPO_4$ form? If so, how many grams of $AlPO_4$ precipitate?

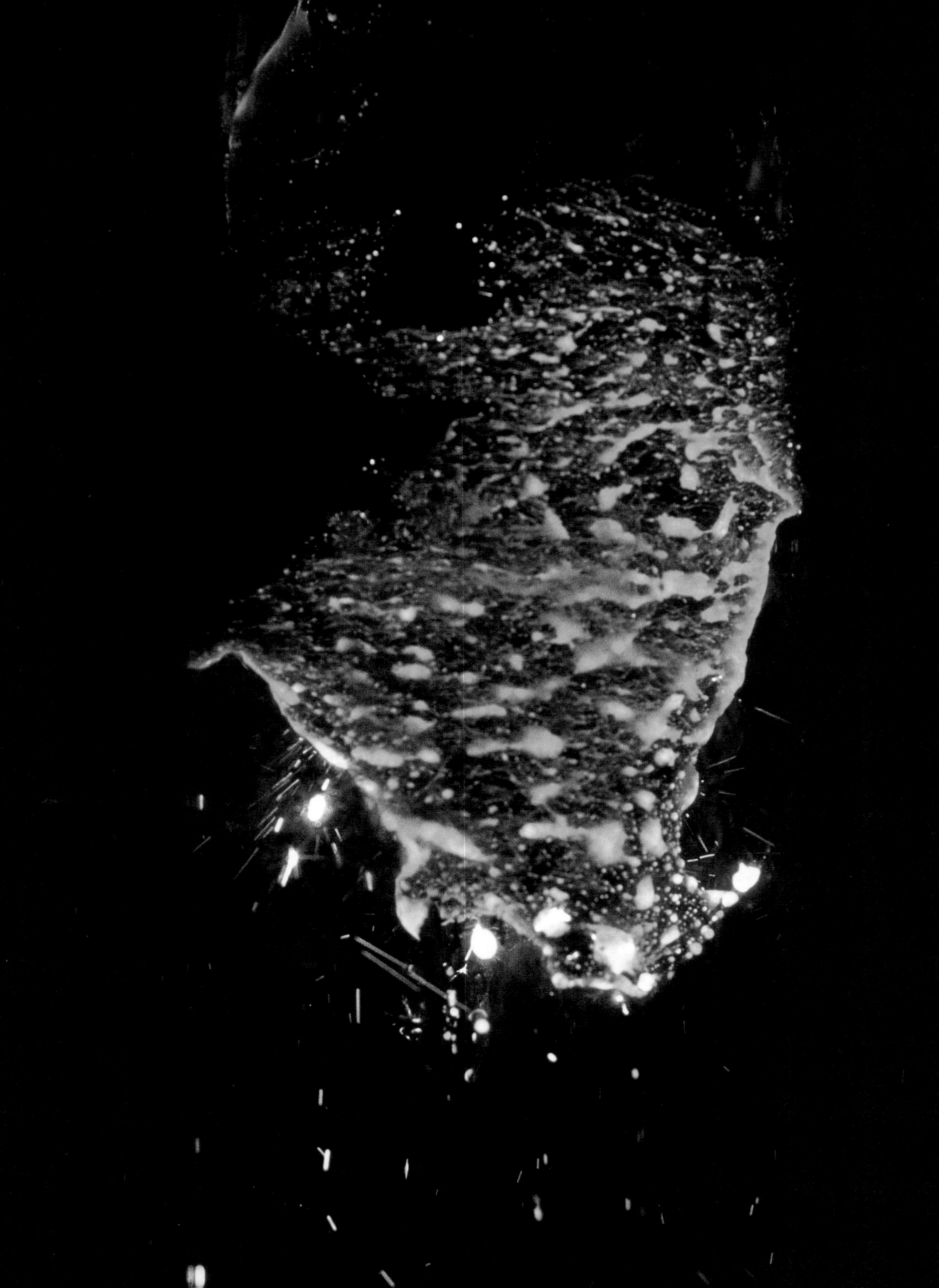

CHAPTER

20

The Spontaneity of Chemical Reactions: Entropy and Free Energy

Hot iron gauze spontaneously burns in an oxygen atmosphere.

CHAPTER OUTLINE

Some chemical and physical changes take place by themselves, given enough time. If you stretch a rubber band, it will snap back spontaneously and quickly. If you put a spoonful of sugar in your coffee or tea, it will dissolve, and the molecules will distribute themselves evenly throughout the liquid. Because of chemical reactions, we grow older. These changes are all said to be spontaneous, although they do proceed at different speeds.

To control chemical reactions, whether to produce what we need or control disease, we must understand why some chemical reactions are spontaneous and others are not. We must understand how to predict when a reaction will be spontaneous and how to control that spontaneity if necessary. This chapter has the basic objective of working out a method of predicting the spontaneity of chemical reactions.

20.1 SPONTANEOUS REACTIONS AND SPEED: THERMODYNAMICS VERSUS KINETICS

A **spontaneous chemical reaction** is one that, given sufficient time, will achieve chemical equilibrium, with an equilibrium constant greater than 1, by reacting from left to right as written. The reactions of most elements

Figure 20.1 A spontaneous reaction, the burning of copper in chlorine.

$$Cu(s) + Cl_2(g) \longrightarrow CuCl_2(s)$$

with halogens are examples of spontaneous reactions, and the reaction of copper with chlorine is no exception (Figure 20.1; see also Figures 3.1, 4.2, and 4.3).

$$\xrightarrow{\text{spontaneous}}$$
$$Cu(s) + Cl_2(g) \longrightarrow CuCl_2(s)$$

The contrasting term, *nonspontaneous reaction*, is misleading. One of the most widely discussed reactions in the past few years is the decomposition of ozone, O_3, to O atoms and O_2 molecules, a reaction that can occur by the action of photons of sunlight in the 280- to 310-nm range of wavelengths.

$$\xrightarrow{\text{nonspontaneous}}$$
$$O_3(g) \longrightarrow O(g) + O_2(g)$$

This reaction is nonspontaneous. This does *not* mean that it does not occur at all. What it does mean is that, when equilibrium is achieved, not many molecules of O_3 have broken down into products. That is, the equilibrium constant is less than 1.

$$K = \frac{[O][O_2]}{[O_3]} = 0.0063 \text{ at } 25\ ^\circ\text{C}$$

The oxidation of H_2 by O_2 is another spontaneous reaction, but it occurs *only* if you ignite the mixture (Figure 20.2). The mere presence of oxygen is not enough, since H_2 can stay in contact with air a long time if you are careful not to set off the reaction. This means there is a difference between the speed of a reaction and its spontaneity. **Thermodynamics**, the subject of this chapter, is the science of energy transfer, and it will help us predict whether a reaction can occur *given enough time*. However, *thermodynamics can tell us nothing about the speed of the reaction*. The study of the rates of reactions, and why some are fast while others are slow, is called **kinetics**, and it was the topic of Chapter 15. Finally, it is *very* important to recognize that there is a difference between the use of the word "spontaneity" in thermodynamics and the way you may have used it in a social context. Socially, we sometimes speak of a person acting spontaneously, and we mean they have apparently acted instinctively or without outside influence or initiation. In thermodynamics, a reaction may *require* initiation, as in the case of H_2/O_2; the only requirement for spontaneity is that $K > 1$ for the reaction.

Beginning with Chapter 6, we have often used ideas of energy differences and energy transfer. In Chapter 6 we described the energy involved in chemical reactions, in Chapters 7 to 11 the concern was with energy on the atomic and molecular level, and in Chapters 13 and 14 you learned about the energy involved in changes in physical state. Now, we want to define more completely the energy changes that cause some chemical and physical changes to be spontaneous, regardless of their speed.

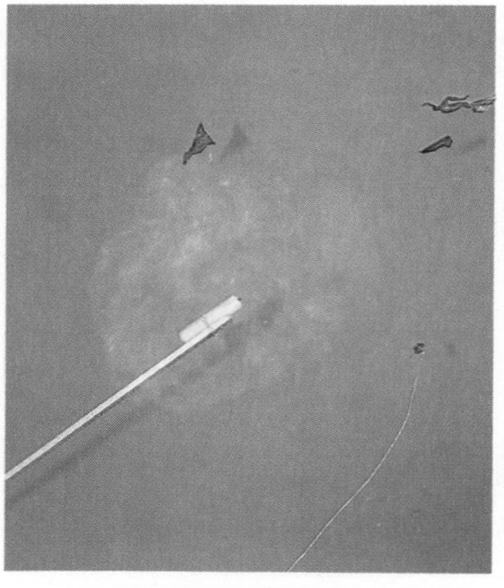

Figure 20.2 A lighted candle was held to a balloon containing H_2 gas. When the balloon burst, the H_2 gas mixed with O_2 of the air, and the mixture exploded when ignited by the candle. The spontaneous reaction occurring is

$$2\ H_2(g) + O_2(g) \longrightarrow 2\ H_2O(g)$$

Thermodynamics: A Review of Concepts from Chapter 6

System: the part of the universe (the specific atoms, molecules, or ions) under study.
Surroundings: the rest of the universe exclusive of the system.
Exothermic: heat transfer is in the direction system → surroundings.
Endothermic: heat transfer is in the direction surroundings → system.
First law of thermodynamics: the law of the conservation of energy. $\Delta E = q + w$. The change in the energy of a system is equal to the heat transferred to the system plus the work done on the system.
Enthalpy change: the heat energy transferred at constant pressure.
State function: a quantity for which the change in value is determined only by the initial and final states of the system.
Standard conditions: $P = 1$ atm and T is usually 298 K. Material in solution has a concentration of 1 molal.
Enthalpy of formation: the enthalpy change occurring when a compound is formed from its elements, ΔH_f°. Elements in their standard states have $\Delta H_f^\circ = 0$.

20.2 ENERGY AND SPONTANEITY

One of the first experimental observations you made in your life was that some spontaneous processes occur with loss of energy: If you drop a glass bottle from your hand, it falls spontaneously to the ground (and can break into many, many pieces); a stone or snowball rolls downhill; rain, snow, and hail fall from the sky. In every instance, the object falls or moves because gravity exerts a force on it (Figure 20.3). If the object is free to move, this force results in a conversion of potential energy to kinetic energy for the object.

When gasoline is ignited, it burns spontaneously in oxygen to give CO_2 and H_2O, and equations such as that for the combustion of octane can be written.

$$\underset{\text{octane}}{2\ C_8H_{18}(g)} + 25\ O_2(g) \longrightarrow 16\ CO_2(g) + 18\ H_2O(g) + \text{heat}$$

The first thing you may notice about this reaction is that heat is evolved. The reaction is exothermic because 25 moles of O_2 and 2 moles of octane possess more energy than do 16 moles of CO_2 and 18 moles of H_2O (at 298 K). The energy of the chemical system declines from left to right.

In all the preceding examples of spontaneous processes, the system moved to a position of lower energy as reactants were transformed to products, but is it a general law of nature that in order to be spontaneous a process must move in the direction of lower energy? This certainly is a tempting conclusion, and it is valid for falling rocks and raindrops. However, for systems in general it is incomplete, and exceptions are not hard to find. The evaporation of liquid water from an open beaker is spontaneous, but it is a process in which the system (water molecules) moves to a higher energy. Dissolving solutes in solvents is often endothermic or sometimes does not involve any energy change at all (Chapter 14). Some spontaneous chemical reactions are endothermic.

Figure 20.3 Once set in motion, objects spontaneously fall under the influence of gravity. (U.S. Air Force Academy photo by SSgt. West C. Jacobs, 94ATS)

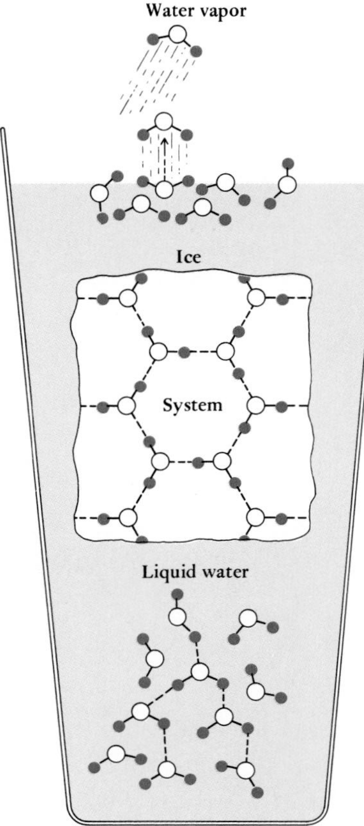

Figure 20.4 Increasing disorder in an endothermic spontaneous process, the melting of ice to liquid water and the evaporation of the liquid.

Where does this leave us? Spontaneous processes are indeed often exothermic, but they can also be endothermic. Energy is not the only determining factor for spontaneity. We are missing something.

20.3 SPONTANEITY, DISORDER, AND ENTROPY

Look again at an endothermic change such as the melting of ice or the evaporation of liquid water to water vapor under conditions where the change occurs spontaneously. Ice is a highly organized structure (Figure 13.40), but as it melts the structure becomes less ordered. In liquid water and especially in water vapor the molecules are widely separated and free to roam (Figure 20.4). In the case of octane combustion, a total of 27 molecules in the gas phase were converted to 34 gas molecules. Although no mass was lost, rearrangement of atoms resulted in a greater number of molecules of products than of reactants. When salt or sugar dissolves in water, a compact and orderly arrangement of ions or molecules in the crystal becomes a random jumble of ions or molecules in the solvent. In every case, a few molecules give many more, or a compact material flies apart into many pieces. Order becomes disorder. Regularity becomes chaos. *Disorder can apparently be the result of a spontaneous process, whether exothermic or endothermic.*

Clearly, two factors should be considered when trying to predict the spontaneity of an event. A *decrease* in energy or enthalpy is one factor, and an *increase* in disorder or chaos is another. Sometimes these effects reinforce one another, but at other times they are opposing. The final outcome is determined by their relative magnitudes. In the case of water evaporation, for example, the increase in molecular disorder must more than offset the increase in energy.

The thermodynamic function called **entropy** is a measure of the disorder of a system. (See *Historical Figures in Chemistry: Entropy, Disorder, and Boltzmann's Tombstone*.) Entropy is the discovery of one of the great scientists of the 19th century, Rudolf Clausius, and it is he who gave it the symbol we use today, S. The idea of entropy came from consideration of the usual tools of the thermodynamic trade—heat and temperature. If you make some careful measurements in the laboratory, you can derive a value of the entropy for a substance at some temperature.

Table 20.1 Some Standard Molar Entropy Values at 298 K

Compound or Element	Entropy, $S°$ (J/K · mol)
C(graphite)	5.74
C(g)	158.096
CO_2(g)	213.74
Ca(s)	41.42
$CaCO_3$(s)	92.9
O_2(g)	205.138
$H_2O(\ell)$	69.91
H_2O(g)	188.825

Entropy, like enthalpy, is a state function; the amount of entropy in a pure substance depends only on the temperature, pressure, and the number of molecules or atoms of the substance. Values for a few substances at 298 K are collected in Table 20.1, and many more are listed in Appendix J. Notice that the units of entropy are joules/(kelvin · mole) (J/K · mol). The superscript "°" indicates that all are *standard state entropies*, that is, the entropy of one mole of the compound or element in its most stable form at a pressure of 1 atm; for substances in solution, the concentration is 1 *m*.

One of the most important aspects of the entropy values in Table 20.1 is that *elements in their standard states* [such as C(graphite), Ca(solid), and O_2(gas)] *have* $S° > 0$. Recall that, in contrast, standard enthalpies of formation of the elements are given values of zero.

Historical Figures in Chemistry: Entropy, Disorder, and Boltzmann's Tombstone

Ludwig Boltzmann (1844–1906) was an Austrian mathematician who gave us a useful interpretation of entropy (and who also did much of the work on the kinetic theory of gases). Engraved on his tombstone in Vienna is his equation relating entropy and "chaos," $S = k \ln W$ (where k is a fundamental constant of nature, now called Boltzmann's constant). Boltzmann said the symbol W was related to the number of ways that atoms or molecules can be arranged in a given state, always keeping their total energy fixed. Therefore, his equation tells us that if there are only a few ways to arrange the atoms of a substance—that is, if there are only a few places in which we can put our atoms or molecules—then the entropy is low. On the other hand, the entropy is high if there are many possible arrangements ($W >> 1$), that is, if the level of chaos is high.

A useful analogy to this is the number of ways we can arrange two identical objects, say two red balls, one at a time in boxes in two containers. We shall assume that, at least in the beginning, our red balls are only in container I; there is a barrier that does not allow them into container II. Since the two balls are identical there is only one way to arrange them in container I. This means W is 1, and so $S = k \ln 1 = 0$. The entropy is zero; there is no disorder, and so the situation is analogous to the arrangement of atoms or ions in a compact, orderly solid.

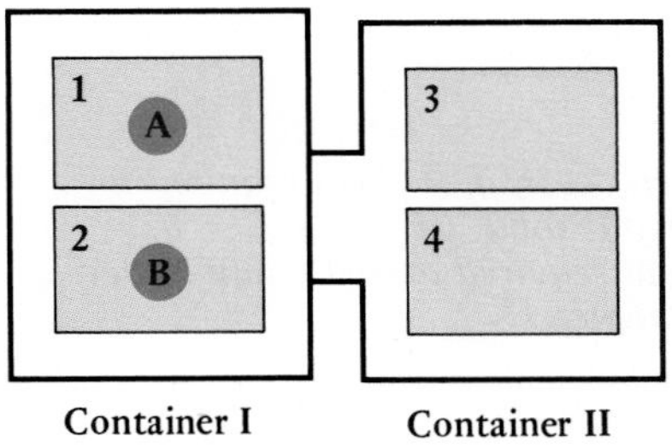

Now let us suddenly break the barrier between the two containers so that the balls are free to roam through four boxes. This would be analogous to the atoms of a solid moving into the vapor phase. Now there are six ways we can arrange the balls,* so the effect of doubling the volume available to the balls is to increase the entropy greatly. (Now $S = k \ln 6 = 1.8\ k$).

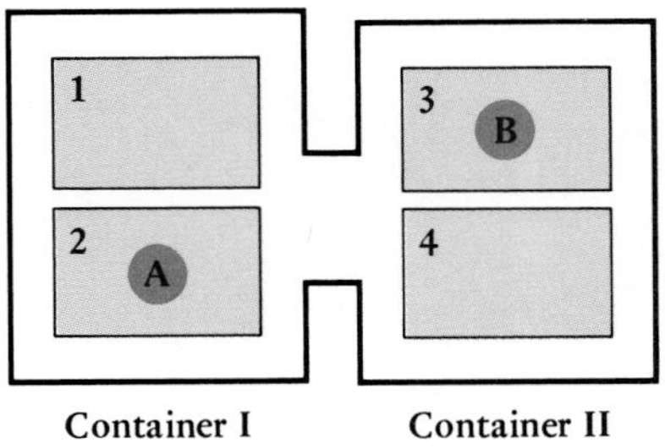

Our "balls in boxes" example is analogous to having a substance frozen as a solid and then allowing it to form a vapor. The system clearly goes from a state of low entropy to one of higher entropy. This is an important idea, and one we shall return to often.

Based on Boltzmann's ideas, we can see that the entropy of a crystalline solid, where the atoms, ions, or molecules are located at fixed points in space, will be low. On the other hand, the molecules of a gas are in constant motion (and so there are many more "boxes" in which we can place the gaseous particles), and the entropy is much higher. Low entropy is associated with order, and high entropy is associated with disorder, randomness, and chaos.

Why did Boltzmann have his equation inscribed on his tombstone? Apparently he was so despondent because scientists did not accept his ideas at the time that he killed himself.

S = k. log W

LVDWIG BOLTZMANN 1844–1906

DR PHIL. BOLTZ. GEB.C 1891–

1 A | 2 B | 3 | 4

Container I Container II

1 | 2 A | 3 B | 4

Container I Container II

*The six unique arrangements are: 1–2, 1–3, 1–4, 2–3, 2–4, and 3–4.

ENTROPIES AND THE THIRD LAW OF THERMODYNAMICS

The entropy of any pure substance can be measured at a given temperature. To see why this can be done, try to visualize a pure element or compound at the absolute zero of temperature. At 0 K any system will consist of atoms, ions, or molecules that can be perfectly ordered with respect to each other in the solid. Since entropy has been equated with disorder, this means *the entropy of a pure, perfectly ordered crystalline substance is 0 at zero kelvin*. This statement is known as the **third law of thermodynamics**. Because of this, we have an absolute starting place from which to measure entropy, and the *absolute entropy* of a substance can then be determined at any temperature higher than 0 K. The entropy values for many substances have been derived by experiment; some have already been mentioned in Table 20.1, and others are given in Appendix J.

ENTROPY CHANGES FOR REVERSIBLE PHASE CHANGES

Reversible phase changes were discussed in Chapter 13. The entropy change for a system undergoing a *reversible change at a constant temperature*—for a change in phase, for example—is given by

$$\Delta S = \frac{\text{heat transferred}}{\text{temperature at which the change occurs}} = \frac{q}{T} \qquad (20.1)$$

where the heat supplied to ($q > 0$) or evolved by ($q < 0$) the substance or system is divided by the temperature (in kelvins). This means that, if heat is supplied to a liquid to change it to its vapor—for example, $H_2O(\ell, 100\ °C) \rightarrow H_2O(g, 100\ °C)$—the entropy will increase. Of course, this agrees with what we expect: since a liquid is more compact and orderly than an equal number of moles of its vapor, the liquid phase of a substance is a state of lower entropy than its vapor phase at the same temperature.

$\Delta H_{fusion} = q_p =$ heat required to melt a solid. $\Delta H_{vaporization} = q_p =$ heat required to boil a liquid. See Chapter 13.

To find the entropy change accompanying a change in state at the normal freezing or boiling point, all we need is the heat associated with the changes at those temperatures. These heats are of course just the enthalpy of fusion or of vaporization, respectively, for the substance.

EXAMPLE 20.1

THE ENTROPY CHANGE FOR A CHANGE IN PHASE

The enthalpy of vaporization of benzene (C_6H_6) is 30.8 kJ/mol at the boiling point of 80.1 °C. Calculate the entropy change for benzene going from (a) liquid to vapor and (b) vapor to liquid at 80.1 °C.

Solution The transition from liquid to vapor is an endothermic process, so the enthalpy of vaporization is a positive quantity. Since entropy is usually given in units of J/K · mol, we convert ΔH_{vap} to joules and divide by the absolute temperature at which the transition occurs.

$$\Delta S = \frac{\Delta H_{vap}}{T} = \frac{3.08 \times 10^4 \text{ J/mol}}{353 \text{ K}} = +87.3 \text{ J/K} \cdot \text{mol}$$

The entropy change is +87.3 J/K · mol for the liquid to vapor phase change. The sign of ΔS is positive, as expected for a change from a liquid to molecules moving about randomly in the vapor phase.

When considering the change in phase from vapor to liquid, we recall from Chapter 13 that heat is *evolved* on condensing a gas. Thus, the process is *exothermic*; the enthalpy change is −30.8 kJ/mol. The change in entropy is

$$\Delta S = \frac{-\Delta H_{vap}}{T} = \frac{-3.08 \times 10^4 \text{ J/mol}}{353 \text{ K}} = -87.3 \text{ J/K} \cdot \text{mol}$$

a negative quantity, reflecting the *decrease in disorder* as intermolecular bonding occurs to cause the vapor to form the more orderly liquid state.

EXERCISE 20.1 The Entropy Change of a Phase Change
Calculate ΔS for the melting of ice at 0 °C. The enthalpy of fusion of ice is 6.02 kJ/mol. What is ΔS when water at 0 °C freezes to solid ice at 0 °C?

ENTROPY CHANGES OF REACTIONS

Not only can we determine the entropy change for a reversible phase change, but the change in entropy occurring in the course of a chemical reaction can also be found. Recall that the enthalpy change for a reaction can be calculated by subtracting the enthalpies of formation of the reactant molecules from those of the product molecules,

$$\Delta H^\circ_{\text{reaction}} = \Sigma\, m\Delta H^\circ_f(\text{products}) - \Sigma\, n\Delta H^\circ_f(\text{reactants})$$

where m and n are the numbers of moles of products and reactants, respectively. In exactly the same way, the entropy change for a chemical reaction can be calculated.

Notice that standard molar entropies, S°, are used in these calculations. See page 822.

$$\Delta S^\circ_{\text{reaction}} = \Sigma\, mS^\circ(\text{products}) - \Sigma\, nS^\circ(\text{reactants}) \qquad (20.2)$$

The example that follows illustrates this calculation.

EXAMPLE 20.2

CALCULATING THE CHANGE IN ENTROPY, ΔS°, FOR THE FORMATION OF A COMPOUND

Nitrogen dioxide, NO_2, is the red-brown gas that you can sometimes see in bottles of nitric acid and in polluted air. Determine the standard entropy change, ΔS°, involved in the formation of NO_2 at 25 °C.

Solution As in any problem involving a chemical reaction, first write the balanced equation. Since we are describing the formation of NO_2, this means we begin with the elements in their standard states.

$$\tfrac{1}{2}\, N_2(g) + O_2(g) \longrightarrow NO_2(g)$$

Next, subtract the sum of the entropies of the reactants from that of the product (Appendix J), paying careful attention to scale each entropy by the number of moles of reagent involved.

$$\begin{aligned}\Delta S^\circ_{\text{rxn}} &= (1 \text{ mol } NO_2)(240.1 \text{ J/K}\cdot\text{mol}) - [(1 \text{ mol } O_2)(205.1 \text{ J/K}\cdot\text{mol}) \\ &\quad + (\tfrac{1}{2} \text{ mol } N_2)(191.6 \text{ J/K}\cdot\text{mol})] \\ &= -60.8 \text{ J/K}\end{aligned}$$

Notice that the sign of the entropy of formation for NO_2 is negative. This is largely due to the fact that the chemical reaction began with $1\tfrac{1}{2}$ moles of reactants and ended with only 1 mole of product, and there is more disorder when the atoms are distributed among more molecules, that is, in the reactants.

EXERCISE 20.2 The Entropy Change for a Reaction
The active ingredient in a popular antacid remedy is $CaCO_3$ (more familiar as the main component of chalk or limestone). Using the entropy values in Figure 20.6 and Table 20.1, calculate the entropy for the formation of $CaCO_3$(s) from the elements. (Use graphite as the standard state of carbon.) Is the sign of the entropy change for the formation of $CaCO_3$(s) positive or negative? Account for the increase or decrease in entropy in the formation of this compound in terms of your notion of disorder in chemical systems.

20.4 ENTROPY, ELEMENTS, AND COMPOUNDS

A common rule of thumb when dealing with entropy is that the greater the disorder or randomness in a system, the larger the entropy. In this short section, we want to examine some processes in which entropy changes are observed, since it will help you later in predicting reaction spontaneity.

All the entropy values in this section are given at 25 °C.

A. *The entropy of a substance always increases as it changes from solid to liquid to gas.*

$$\underset{S^\circ = 69.9\ \text{J/K}\cdot\text{mol}}{H_2O(\ell)} \xrightarrow{\Delta S^\circ = +118.9\ \text{J/K}} \underset{S^\circ = 188.8\ \text{J/K}\cdot\text{mol}}{H_2O(g)}$$

Here 1 mole of liquid water is converted, at 25 °C, to 1 mole of water vapor with a pressure of 1 atmosphere. The system has gone from a relatively compact, hydrogen-bonded liquid to a gas in which water molecules move very randomly, and the entropy of the system increases.

B. *When a pure solid or liquid dissolves in a solvent, the entropy of the substance increases.**

$$\underset{S^\circ = 151.1\ \text{J/K}\cdot\text{mol}}{NH_4NO_3(s)} \xrightarrow{\Delta S^\circ = +108.7\ \text{J/K}} \underset{S^\circ = 259.8\ \text{J/K}\cdot\text{mol}}{NH_4NO_3(aq)}$$

In Chapter 14 you learned that, when ammonium nitrate dissolves in water, the process is quite *endo*thermic. However, the entropy certainly increases because the system has gone from an ordered crystalline solid to a jumble of ions in aqueous solution (Figure 20.5), and it is this enormous entropy increase (as you will see in Section 20.5) that causes the process to occur.

C. *When a gas molecule escapes from a solvent, the entropy increases.*

$$\underset{S^\circ = 56.5\ \text{J/K}\cdot\text{mol}}{HCl(aq)} \xrightarrow{\Delta S^\circ = +103.4\ \text{J/K}} \underset{S^\circ = 186.9\ \text{J/K}\cdot\text{mol}}{HCl(g)}$$

A gas is clearly a more disordered state than molecules or ions in solution. When the substance escapes the order that is imposed on its

*For ionic solids, this statement is true for cases where the ions do not interact strongly with the solvent. In contrast, the CO_3^{2-} ion interacts with water, so the entropy actually declines (by 73.9 J/K · mol) when Na_2CO_3 dissolves in water, apparently because the interaction increases the overall order in the system.

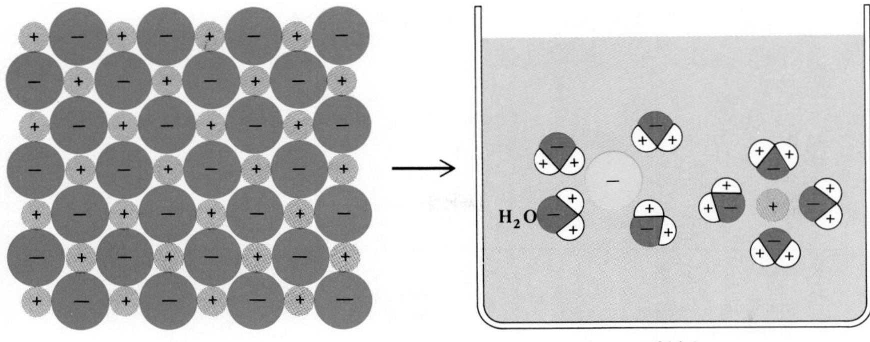

Figure 20.5 When an ordered ionic solid dissolves in a polar solvent, there is a great increase in entropy for the solute.

molecules by the intermolecular forces between it and the solvent, entropy is gained.

D. *Entropy generally increases with increasing molecular complexity.* This is a useful observation that should nonetheless be used with care. For example, we can compare a related series of metal halides and see that $S°$ increases as the number of atoms or ions per formula unit in a solid increases.

	KCl(s)	$CaCl_2$(s)	$GaCl_3$(s)
$S°$ (J/K · mol)	83	105	142

Straight chain hydrocarbons represent another closely related series of compounds, and we see that $S°$ increases by about 40 J/K · mol each time another CH_2 group is added to the chain.

COMPOUND	STRUCTURAL FORMULA	$S°$ (J/K · mol) FOR GASEOUS COMPOUND
Methane	CH_4	186.3
Ethane	$CH_3—CH_3$	229.6
Propane	$CH_3—CH_2—CH_3$	269.9
n-Butane	$CH_3—CH_2—CH_2—CH_3$	310.0
n-Octane	$CH_3—CH_2—CH_2—CH_2$ | $CH_3—CH_2—CH_2—CH_2$	463.6

Entropy values for a number of elements are given in an abbreviated periodic table in Figure 20.6. In general, elements with a high degree of covalent bonding in the solid state (boron, carbon, phosphorus, sulfur) have relatively low standard entropies. Metals have values ranging from about 10 to 80 J/K · mol, while gaseous elements have the highest values of all.

EXERCISE 20.3 Predicting Signs of Entropy Changes and Relative Entropy Values

(1) For each of the following changes of state, predict the *sign* of the entropy change:
(a) Hg(ℓ) → Hg(g)
(b) I_2(g) → I_2(s)
(c) $AgNO_3$(s) → $AgNO_3$(aq)

(2) In each pair of compounds, tell which should have the higher value of $S°$.
(a) CH_3OH(ℓ) or CH_3CH_2OH(ℓ) (b) KI(s) or AlI_3(s)

			Solid	Gas		H₂	He
						H_2 130.7	He 126.2
Li 29.1	Be 9.5	B 5.86	C diamond 2.38 graphite 5.74	N_2 191.6	O_2 205.1	F_2 202.8	Ne 146.3
Na 51.2	Mg 32.7	Al 28.3	Si 18.8	P (white P) 41.1	S 31.8	Cl_2 223.1	Ar 154.8
K 64.2	Ca 41.4	Ga 40.88	Ge 31.1	As (gray As) 35.1	Se (black Se) 42.4	Br_2 152.2 (liquid Br_2)	Kr 164.1
Rb 76.8	Sr 52.3	In 57.8	Sn (white tin) 51.6	Sb 45.7	Te 49.8	I_2 116.1	Xe 169.7
Cs 85.2	Ba 62.8	Tl 64.2	Pb 64.8	Bi 56.7	Po	At	Rn 176.1

Figure 20.6 Entropy values, in J/K · mol, for some of the elements at 25 °C.

20.5 ENTROPY AND THE SECOND LAW

The **second law of thermodynamics** states that *the combined entropy of the system and its surroundings (the entropy of the universe) always increases in a spontaneous process.*

$$\Delta S_{\text{system}} + \Delta S_{\text{surroundings}} = \Delta S_{\text{universe}}$$

If $\Delta S_{\text{universe}}$ is +, then a reaction is spontaneous

If $\Delta S_{\text{universe}}$ is −, then a reaction is nonspontaneous

This statement leads us directly to our objective: to be able to predict the spontaneity of chemical processes.

Consider the reaction of hydrogen and oxygen to produce water vapor, a reaction that we already know is spontaneous (Figure 20.2).

$$2\ H_2(g) + O_2(g) \longrightarrow 2\ H_2O(g)$$

Here we can think of the reactants contained in a balloon as the system, and the environment of the balloon as the surroundings. We ignite the reactants, and they react *very* rapidly as you see in Figure 20.2. The standard entropy change for the reaction, that is, $\Delta S^\circ_{\text{system}}$, is calculated using the approach you learned in Section 20.3.

$$\begin{aligned}\Delta S^\circ_{\text{system}} &= 2\ S^\circ[H_2O(g)] - \{2\ S^\circ[H_2(g)] + S^\circ[O_2(g)]\} \\ &= (2\ \text{mol})(188.8\ \text{J/K}\cdot\text{mol}) - \{(2\ \text{mol})(130.7\ \text{J/K}\cdot\text{mol}) \\ &\quad + (1\ \text{mol})(205.1\ \text{J/K}\cdot\text{mol})\} \\ &= -88.9\ \text{J/K}\end{aligned}$$

The entropy of the system clearly declines as the reaction proceeds under standard conditions.

Be sure to notice that the entropy change for the H_2/O_2 reaction is negative mainly because 3 moles of reactants produce only 2 moles of product, all in the gas phase.

To confirm that the reaction is spontaneous we need to know the entropy change for the surroundings, $\Delta S_{surroundings}$. To calculate this, we can apply Equation 20.1 to the surroundings: $\Delta S_{surroundings} = q_{surroundings}/T$. The change in entropy of the surroundings is given by the quantity of heat that is transferred between the system and its surroundings divided by T. But what is the quantity of heat transferred during the H_2/O_2 reaction and what is the direction of transfer? The answer to these questions comes from a calculation of the enthalpy change for the H_2/O_2 reaction.

$$2\ H_2(g) + O_2(g) \longrightarrow 2\ H_2O(g)$$

$$\begin{aligned}\Delta H^\circ_{system} &= 2\ \Delta H^\circ_f[H_2O(g)] - \{2\ \Delta H^\circ_f[H_2(g)] + \Delta H^\circ_f[O_2(g)]\} \\ &= (2\ \text{mol})(-241.8\ \text{kJ/mol}) - \{(2\ \text{mol})(0\ \text{kJ/mol}) \\ &\quad + (1\ \text{mol})(0\ \text{kJ/mol})\} \\ &= -483.6\ \text{kJ}\end{aligned}$$

The enthalpy change tells us that the reaction is *exo*thermic and that 483.6 kJ of energy is transferred *from* the system *to* the surroundings. The first law of thermodynamics—the law of the conservation of energy—demands that $-\Delta H^\circ_{system} = \Delta H^\circ_{surroundings}$. Since q, the quantity of heat transferred by the system, is $\Delta H^\circ_{system} = q = -483.6$ kJ, this means that $\Delta H^\circ_{surroundings}$ is *positive* 483.6 kJ. Therefore, for a controlled (reversible) reaction

A process is spontaneous, in spite of a negative ΔS for the system, when the heat transferred to the surroundings is large enough to make ΔS_{surr} more positive than ΔS_{sys} is negative. Sufficient exothermicity offsets system ordering.

$$\Delta S^\circ_{surroundings} = \frac{\Delta H^\circ_{surroundings}}{T} = \frac{+483.6\ \text{kJ}}{298\ \text{K}}(10^3\ \text{J/kJ}) = 1620\ \text{J/K}$$

Now we can find $\Delta S^\circ_{universe}$.

$$\Delta S^\circ_{system} + \Delta S^\circ_{surroundings} = \Delta S^\circ_{universe}$$

$$(-88.9\ \text{J/K}) + (1620\ \text{J/K}) = 1530\ \text{J/K}$$

Even though the entropy of the system declines, the entropy change for the surroundings is so large that the overall change, $\Delta S^\circ_{universe}$, is positive. The reaction of H_2 and O_2 is indeed spontaneous!

GIBBS FREE ENERGY

To predict when a reaction can be spontaneous, it would be convenient to have a quantity that incorporates both ΔS_{sys} and ΔS_{surr}. To do this, let us take the following statement and make some substitutions.

$$\Delta S_{system} + \Delta S_{surroundings} = \Delta S_{universe}$$

In our analysis of the H_2/O_2 reaction we found that $\Delta S_{surroundings}$ could be calculated from a knowledge of $\Delta H_{surroundings}$ and T; that is, $\Delta S_{surroundings} = \Delta H_{surroundings}/T = -\Delta H_{system}/T$. Therefore,

$$\Delta S_{system} + \frac{-\Delta H_{system}}{T} = \Delta S_{universe}$$

If we multiply through this equation by T, we have

The symbol G for (H − TS) was chosen to honor J. Willard Gibbs (1838–1903), who pioneered and developed the concept of free energy. Gibbs received a Ph.D. from Yale University in 1883 and was a faculty member there until his death.

$$T\Delta S_{\text{system}} + (-\Delta H_{\text{system}}) = T\Delta S_{\text{universe}}$$

or

$$\Delta H_{\text{system}} - T\Delta S_{\text{system}} = -T\Delta S_{\text{universe}}$$

This equation is so important and so useful as a statement of the second law that a new thermodynamic function, the **Gibbs free energy, *G*,** has been defined, where $G = H - TS$. Therefore, at constant temperature, the change in Gibbs free energy for the system, ΔG_{system}, is

$$\Delta G_{\text{system}} = \Delta H_{\text{system}} - T\Delta S_{\text{system}}$$

More usually the statement is written simply as

$$\Delta G = \Delta H - T\Delta S \tag{20.3}$$

free energy = total energy change − temperature × entropy change

and is referred to as the **Gibbs free energy equation**, one of the most important equations in chemistry. In this equation ΔG has units of energy because both ΔH and the $T\Delta S$ term are energy terms [since $T\Delta S$ = (K)(J/K · mol) = J/mol].

The first term of the Gibbs equation (ΔH) reflects the heat energy available to be transferred between system and surroundings, where it acts to order or disorder the *surroundings*. The $T\Delta S$ term reflects the energy associated with ordering or disordering the *system*. The combination of these two energies, ΔG, is the energy that is associated with the disordering of the universe.

The interpretation of ΔG is simply the following: When $\Delta G < 0$, the process disorders the universe. When $\Delta G > 0$, the process orders the universe. ΔH tells us about ordering in the surroundings, and $T\Delta S$ tells us about system ordering.

Now to consider one of the goals of this chapter: How is ΔG related to reaction spontaneity? From above, we know that $\Delta G_{\text{system}} = -T\Delta S_{\text{universe}}$. The second law states that S_{universe} increases ($\Delta S_{\text{universe}}$ is positive) in a spontaneous process, and so it follows that G_{system} should decline (ΔG_{system} should have a negative value) in a spontaneous process. From this we can now state the criteria of spontaneity based on the Gibbs equation:

(a) **At constant temperature, if ΔG_{system} is negative (<0), a process is spontaneous.**
(b) **At constant temperature, if ΔG_{system} is positive (>0), a process is not spontaneous.**
(c) **At constant temperature, if $\Delta G_{\text{system}} = 0$, the process is at equilibrium.**

Before applying these criteria, let us look in more detail at the ΔG function itself.

FREE ENERGIES OF REACTION AND THE STANDARD FREE ENERGY OF FORMATION, ΔG_f°

Enthalpy and entropy changes can be calculated for chemical reactions by using values of ΔH_f° and S° for substances in the reaction, and $\Delta G_{\text{rxn}}^\circ$ can be found from the resulting values of $\Delta H_{\text{rxn}}^\circ$ and $\Delta S_{\text{rxn}}^\circ$ using Equation 20.3, as illustrated in the following example and exercise.

EXAMPLE 20.3

CALCULATING ΔG°_{rxn} FROM ΔH°_{rxn} AND ΔS°_{rxn}

Calculate the standard free energy change for the formation of methane at 298 K.

$$C(graphite) + 2\ H_2(g) \longrightarrow CH_4(g)$$

Solution The following values for ΔH°_f and S° are provided in Appendix J.

	C(graphite)	H_2(g)	CH_4(g)
ΔH°_f (kJ/mol)	0	0	−74.81
S° (J/K · mol)	5.74	130.7	186.3

From these values, we can find both ΔH° and ΔS° for the reaction.

$$\begin{aligned}\Delta H^\circ_{rxn} &= \Delta H^\circ_f[CH_4(g)] - \{\Delta H^\circ_f[C(graphite)] + 2\ \Delta H^\circ_f[H_2(g)]\}\\ &= -74.81\ kJ - (0 + 0)\\ &= -74.81\ kJ\end{aligned}$$

$$\begin{aligned}\Delta S^\circ_{rxn} &= S^\circ[CH_4(g)] - \{S^\circ[C(graphite)] + 2\,S^\circ[H_2(g)]\}\\ &= 1\ mol\,(186.3\ J/K \cdot mol) - [1\ mol\,(5.74\ J/K \cdot mol) + 2\ mol\,(130.7\ J/K \cdot mol)]\\ &= -80.8\ J/K\end{aligned}$$

The enthalpy and entropy values for the reaction can now be combined in the Gibbs equation to give the standard free energy change for the reaction.

$$\begin{aligned}\Delta G^\circ_{rxn} &= \Delta H^\circ_{rxn} - T\Delta S^\circ_{rxn}\\ &= -74.81\ kJ - (298\ K)(-80.8\ J/K)(1\ kJ/1000\ J)\\ &= -74.81\ kJ - (-24.1\ kJ)\\ &= -50.7\ kJ\end{aligned}$$

Although this point is pursued in the section that follows, you should notice that ΔG°_{rxn} is negative, so the reaction is predicted to be spontaneous.

EXERCISE 20.4 Calculating ΔG°_{rxn} from ΔH°_{rxn} and ΔS°_{rxn}

Using values of ΔH°_f and S° to find ΔH°_{rxn} and ΔS°_{rxn}, respectively, calculate the free energy change for the following reaction occurring under standard conditions at 25°C:

$$\tfrac{1}{2}\ N_2(g) + \tfrac{3}{2}\ H_2(g) \longrightarrow NH_3(g)$$

In the preceding example and exercise, the standard free energy change was calculated for the formation of one mole of a compound from its elements, with all reactants and products in the standard state. Therefore, the calculated ΔG°_{rxn} can be identified as the **standard molar free energy of formation** for methane, $\Delta G^\circ_f[CH_4(g)]$, or ammonia, $\Delta G^\circ_f[NH_3(g)]$. These and a few other values of ΔG°_f are listed in Table 20.2, and many more values are given in Appendix J. Be sure to notice that $\Delta G^\circ_f = 0$ for graphite and for other elements. Just as $\Delta H^\circ_f = 0$ for elements in their standard states, it is also true that *all elements in their standard states have ΔG°_f values of 0.*

Table 20.2 Standard Molar Free Energies of Formation for Some Substances at 298 K

Element or Compound	ΔG°_f (kJ/mol)
H_2(g)	0
O_2(g)	0
N_2(g)	0
C(graphite)	0
C(diamond)	2.90
CO(g)	−137.2
CO_2(g)	−394.4
CH_4(g)	−50.72
H_2O(g)	−228.572
$H_2O(\ell)$	−237.129
NH_3(g)	−16.45

Just as ΔH°_{rxn} can be calculated from standard enthalpies of formation, the free energy change for a reaction can also be found from values of ΔG°_f by the general equation

$$\Delta G^\circ_{reaction} = \Sigma\ m\Delta G^\circ_f(\text{products}) - \Sigma\ n\Delta G^\circ_f(\text{reactants}) \quad (20.4)$$

where m and n are the numbers of moles of a given product and reactant, respectively. The example and exercises that follow illustrate how this can be done.

EXAMPLE 20.4

CALCULATING ΔG°_{rxn} FROM ΔG°_f

Calculate the free energy change for the combustion of methane, $CH_4(g)$, from the standard free energies of formation of the products and reactants.

Solution We first write a balanced equation for the reaction and then find values of ΔG°_f for each reactant and product.

	$CH_4(g)$	+ 2 $O_2(g)$	⟶	2 $H_2O(\ell)$	+ $CO_2(g)$
ΔG°_f(kJ/mol)	−50.7	0		−237.1	−394.4

$$\begin{aligned}\Delta G^\circ_{rxn} &= 2\ \Delta G^\circ_f[H_2O(\ell)] + \Delta G^\circ_f[CO_2(g)] - \{\Delta G^\circ_f[CH_4(g)] + 2\ \Delta G^\circ_f[O_2(g)]\}\\ &= 2\ \text{mol}(-237.1\ \text{kJ/mol}) + (1\ \text{mol})(-394.4\ \text{kJ/mol})\\ &\quad - [(1\ \text{mol})(-50.7\ \text{kJ/mol}) + (2\ \text{mol})(0)]\\ &= -817.9\ \text{kJ}\end{aligned}$$

Be sure to notice that free energy of formation values are given for 1 mole. Therefore, each value must be multiplied by the number of moles involved.

EXERCISE 20.5 Standard Free Energies of Formation, ΔG°_f

(a) Write a balanced chemical equation depicting the formation of gaseous carbon dioxide (CO_2) from its elements.
(b) What is the standard free energy of formation of 1.00 mol of CO_2 gas?
(c) What is the standard free energy change for the reaction when 2.5 mol of CO_2 gas is formed from the elements?

EXERCISE 20.6 Calculating ΔG°_{rxn} from ΔG°_f

Calculate the standard free energy change for the combustion of 1.00 mol of benzene, $C_6H_6(\ell)$, to give $CO_2(g)$ and $H_2O(\ell)$.

ΔG° AS A CRITERION OF SPONTANEITY

Example 20.3 illustrates a case where ΔH and ΔS work against one another in determining reaction spontaneity. In that case, however, ΔH was more negative than $-T\Delta S$ was positive, so the reaction is predicted to be spontaneous.

We have concluded that if the free energy of the system decreases in a process (ΔG_{system} is negative), then the process will be spontaneous. Since $\Delta G = \Delta H - T\Delta S$, this means a process will certainly occur spontaneously if the energy of the system declines *and* its entropy increases. In this case, ΔH is negative and ΔS is positive, so $\Delta H - T\Delta S$ must give a negative value for ΔG. Of course, it is also possible that the enthalpy and entropy changes will work against each other, or that neither favors

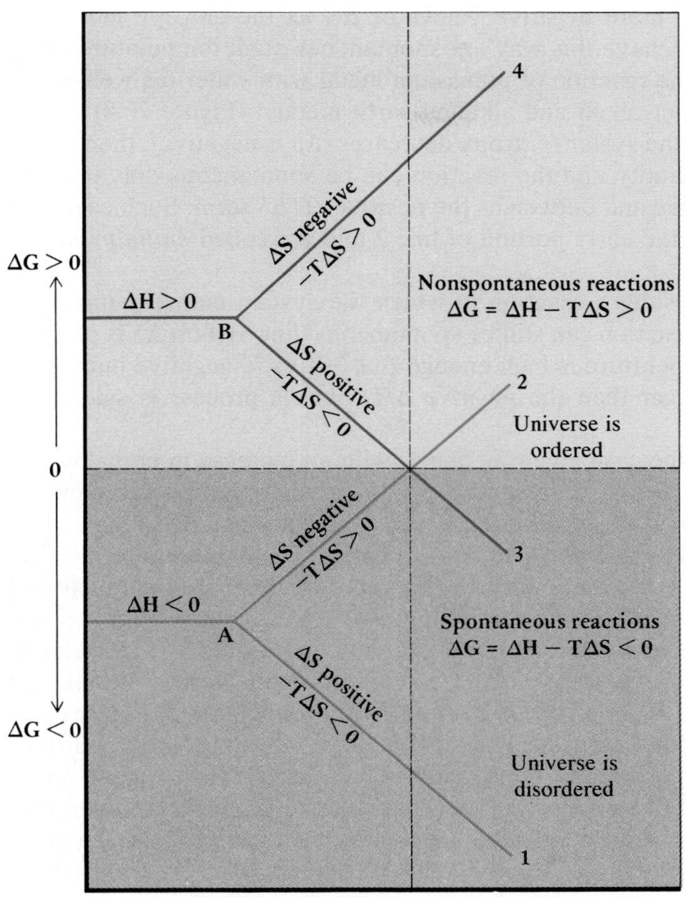

Figure 20.7 Changes in ΔG as a function of the sign and magnitude of ΔS for an exothermic reaction ($\Delta H < 0$, point A) or an endothermic reaction ($\Delta H > 0$, point B). Point A corresponds to disordering of the surroundings, while point B corresponds to ordering of the surroundings. When ΔS is negative, the system is ordered. When ΔS is positive, the system is disordered.

spontaneity for the process. All of these possibilities are summarized in Figure 20.7 and in Table 20.3.

Let us begin at point A in Figure 20.7. This represents a process that has a negative ΔH. If the system entropy increases in the process, the term $-T\Delta S$ contributes a negative quantity to ΔG, and you would

Table 20.3 Reaction Spontaneity as a Function of ΔH and ΔS*

Enthalpy Change	ΔH	System Entropy Change	ΔS	Sign of ΔG	Process Spontaneity
Line 1 exothermic	−	Increasing	+	−	Yes
Line 2 exothermic	−	Decreasing	−	− or +	Depends on T and on relative size of ΔH and ΔS
Line 3 endothermic	+	Increasing	+	+ or −	Depends on T and on relative size of ΔH and ΔS
Line 4 endothermic	+	Decreasing	−	+	No

*Line numbers refer to Figure 20.7.

Figure 20.8 The spontaneous reaction of potassium with water

$$K(s) + H_2O(\ell) \longrightarrow KOH(aq) + \frac{1}{2} H_2(g)$$

is driven by both enthalpy and entropy changes, and so the free energy change for the reaction is less than zero. ($\Delta H^\circ_{rxn} = -196.5$ kJ, $\Delta S^\circ_{rxn} = 22.9$ J/K, and $\Delta G^\circ_{rxn} = -203.4$ kJ.)

follow line 1 to more negative values of ΔG as the entropy increases. Processes that behave this way are spontaneous at all temperatures, and an example is the reaction of potassium metal with water (as well as the reactions of other alkali and alkaline earth metals) (Figure 20.8). If, on the other hand, the system entropy decreases (ΔS is negative), then $-T\Delta S$ is a positive quantity and the reaction can be spontaneous only if ΔH is large and negative and outweighs the positive $-T\Delta S$ term. Such cases are represented by the early portion of line 2 and are called *enthalpy driven* reactions.

Now consider a reaction in which the system enthalpy increases (point B). The reaction can still be spontaneous (line 3) if (a) ΔS is positive and (b) the temperature is high enough that $-T\Delta S$ (a negative number in this case) is larger than the positive ΔH. Such a process is said to be *entropy driven*.

Finally, the case given by line 4, with an increase in enthalpy and a decrease in entropy, represents a process that cannot be spontaneous under any conditions since both ΔH and $-T\Delta S$ are positive.

What does it mean when ΔG is positive instead of negative? One thing it means is that the process has no energy "free" to be transferred to the surroundings; in fact, the reaction requires energy from the surroundings in order to occur to any appreciable "extent." The meaning of the "extent of reaction" is best understood by exploring the relation between ΔG and K, as is done in Section 20.6. From Figure 20.7, however, you can draw the important insight that there are only two conditions under which an otherwise nonspontaneous reaction can be made spontaneous. For reactions following line 3 in Figure 20.7, a higher temperature is required to cross over into the "region of spontaneity." On the other hand, for a reaction given by line 2, a lower temperature is required.

FREE ENERGY CHANGES FOR REACTIONS

Now let us examine a few actual reactions to see how the criterion of reaction spontaneity works in practice.

The Combustion of Carbon at 298 K

ΔH° Negative and ΔS° Positive $\rightarrow$ ΔG° Negative at All T

	C(graphite) +	O_2(g) $\longrightarrow$	CO_2(g)	OVERALL
ΔH°_f (kJ/mol)	0	0	−393.5	−393.5
S° (kJ/K · mol)	0.00574	0.2051	0.2137	0.0029
ΔG°_f (kJ/mol)	0	0	−394.4	−394.4

The free energy change for this reaction can be calculated in either of two ways. First, you can add up all of the ΔG°_f values for the products and subtract the sum of those for the reactants.

$$\begin{aligned}\Delta G^\circ_{rxn} &= \Sigma\, m\Delta G^\circ_f(\text{products}) - \Sigma\, n\Delta G^\circ_f(\text{reactants}) \\ &= -394.4 \text{ kJ} - [0 + 0] = -394.4 \text{ kJ}\end{aligned}$$

Alternatively, you can calculate the enthalpy and entropy changes for the reaction, ΔH°_{rxn} and ΔS°_{rxn}, and combine them using the Gibbs equation.

$$\begin{aligned}\Delta G^\circ_{rxn} &= \Delta H^\circ_{rxn} - T\Delta S^\circ_{rxn}\\ &= -393.5 \text{ kJ} - (298 \text{ K})(0.0029 \text{ kJ/K})\\ &= -393.5 \text{ kJ} - 0.86 \text{ kJ}\\ &= -394.4 \text{ kJ}\end{aligned}$$

Both the enthalpy and entropy changes contribute to making this reaction spontaneous. Not only does the reaction evolve a large amount of heat, but the system disordering, on going from a solid and a gas to a gaseous product, contributes to a *small* extent to the spontaneity. Furthermore, the reaction liberates 393.5 kJ of heat energy (which makes the surroundings more disordered), and the positive entropy change means that the system contributes to disordering of the universe. Reactions such as this one would correspond to line 1 in Figure 20.7.

The Combustion of Methane at 298 K

ΔH° and ΔS° Both Negative → ΔG° with Variable Sign Methane burns spontaneously in oxygen to give liquid water and carbon dioxide.

	$CH_4(g)$	+ $2\ O_2(g)$	⟶ $2\ H_2O(\ell)$	+ $CO_2(g)$	OVERALL
ΔH°_f (kJ/mol)	−74.8	0	2(−285.8)	−393.5	−890.3
S° (kJ/K · mol)	0.1863	2(0.2051)	2(0.06991)	0.2137	−0.2430
ΔG°_f (kJ/mol)	−50.7	0	2(−237.1)	−394.4	−817.9

The free energy change, ΔG° (= −817.9 kJ), is a large negative number, clearly indicating that the reaction is spontaneous under standard conditions, something you already knew. However, although the reaction generates 890 kJ of heat energy (which is used to disorder the surroundings), only 818 kJ are available to disorder the universe. That is, 72 kJ of energy at 25 °C (= $-T\Delta S^\circ$) is not "free," having been used to create some order in the system. Three molecules of reactants gave three of products, but two molecules of products are in the more ordered liquid state. Thus, the combustion of methane is *enthalpy driven*, as are most combustion reactions; this corresponds to line 2 in Figure 20.7.

Dissolving NH_4NO_3 in Water

ΔH° and ΔS° Both Positive → ΔG° with Variable Sign A situation in which both ΔH° and ΔS° are positive corresponds to line 3 in Figure 20.7. The only way that the reaction can be spontaneous is for $-T\Delta S^\circ$ to be large enough that ΔG° is negative. This is true for many salts dissolving in water. Since the salt begins as a highly ordered solid with low entropy, dissolving it in water gives a jumble of ions with a high entropy.

	$NH_4NO_3(s)$	⟶ $NH_4NO_3(aq, 1\ m)$	OVERALL
ΔH°_f (kJ/mol)	−365.6	−339.9	25.7
S° (kJ/K · mol)	0.1511	0.2598	0.1087
ΔG°_f (kJ/mol)	−183.9	−190.6	−6.7

The enthalpy and entropy changes for the reaction clearly show that the reaction is *entropy driven*. That is, although the process is endothermic (and you would feel an obvious cooling effect if you held your hand on a beaker containing dissolving ammonium nitrate), this is outweighed by the increase in entropy of the system, and the reaction is spontaneous under standard conditions. The disordering of the system (the ions of the solid) when the solid dissolves is a very potent driving force in this case.

Entropy is often the "force" that drives the mixing of any two substances, liquid or gas, with each other. Chapter 14 described the formation of ideal solutions, ones in which the forces between solute molecules and between solvent molecules are the same as between solute and solvent molecules. Since the energies of the two kinds of molecules *cannot* change on forming an ideal solution, it is the increase in entropy experienced by the solute and solvent molecules on mixing that provides the driving force.

EXERCISE 20.7 Calculating $\Delta G°$ for a Reaction

Using free energies of formation in Appendix J, calculate $\Delta G°$ (a) for the formation of $CaCO_3(s)$ from $CO_2(g)$ and $CaO(s)$ and (b) for the decomposition of $CaCO_3(s)$ to give $CO_2(g)$ and $CaO(s)$. Which reaction is spontaneous under standard conditions?

FREE ENERGY AND TEMPERATURE

If a reaction has a positive enthalpy change *and* a positive entropy change, the only way it can be spontaneous is if $-T\Delta S$ is large enough to outweigh ΔH. This can happen in two ways: (a) The positive entropy change can be "small," in which case T must be larger. (b) The positive entropy change can be "large," in which case T can be smaller. The former is one of the reasons reactions are often carried out at high temperatures. Let us consider an example of this case.

Our economy is based in large measure on the production of iron, and we can think of at least three different ways to reduce iron(III) oxide to metallic iron. One way is to heat iron(III) oxide and hope it decomposes to iron and oxygen.

$$Fe_2O_3(s) \longrightarrow 2\ Fe(s) + \tfrac{3}{2}\ O_2(g)$$

We can obtain some idea of the feasibility of this by calculating $\Delta G°$. Using the data in Appendix J, we find that $\Delta H°$ for the reaction is +824.2 kJ and $\Delta S°$ is +0.276 kJ/K. Enthalpy says *no* to spontaneity, and entropy says *yes*. Unfortunately, $\Delta H°_{rxn}$ is so positive that it cannot be outweighed by $-T\Delta S°$ (= −82 kJ at 25 °C) at any reasonable temperature, and $\Delta G°_{rxn}$ at room temperature is +742 kJ!

Another way to reduce iron(III) oxide is the so-called **thermite reaction**. Here the nonspontaneous decomposition of $Fe_2O_3(s)$

$$Fe_2O_3(s) \longrightarrow 2\ Fe(s) + \tfrac{3}{2}\ O_2(g)$$

is coupled with the highly spontaneous reaction of aluminum with oxygen.

$$2\ Al(s) + \tfrac{3}{2}\ O_2(g) \longrightarrow Al_2O_3(s)$$

The sum of these reactions is

$$Fe_2O_3(s) + 2\ Al(s) \longrightarrow 2\ Fe(s) + Al_2O_3(s)$$

$$\Delta H^\circ_{rxn} = -851.5 \text{ kJ} \qquad \Delta S^\circ_{rxn} = -0.038 \text{ kJ/K} \qquad \Delta G^\circ_{rxn} = -840.2 \text{ kJ}$$

This is a spectacularly spontaneous process (Figure 20.9). The unfavorable ΔS°_{rxn} at 298 K ($-T\Delta S^\circ = +11.3$ kJ) is swamped by a very large and negative enthalpy change. That means that a large amount of heat is produced, so much in fact that the products are raised to the melting point of iron (1530 °C), and white hot, molten metal streams out of the reaction. The reaction has been applied to welding procedures, but it is unfortunately not practical for the production of iron on a large scale. The cost of producing the aluminum to use as a reducing agent is much larger than the value of the iron produced.

The usual method of reducing iron(III) oxide to iron metal is to use carbon, a much cheaper reducing agent. This reaction, like the thermite reaction, couples the decomposition of Fe_2O_3 with the highly spontaneous combustion of carbon.

In practice, Fe_2O_3 is not reduced directly with C. Instead, the carbon is burned to give CO, which then acts as the reducing agent. See Chapter 25 for more information.

$$C(\text{graphite}) + O_2(g) \longrightarrow CO_2(g)$$

The overall process can be represented by

$$2\ Fe_2O_3(s) + 3\ C(\text{graphite}) \longrightarrow 4\ Fe(s) + 3\ CO_2(g)$$

$$\Delta H^\circ_{rxn} = +467.9 \text{ kJ} \qquad \Delta S^\circ_{rxn} = +0.5603 \text{ kJ/K} \qquad \Delta G^\circ_{rxn} = +300.9 \text{ kJ}$$

Even though the entropy change is large, the $-T\Delta S^\circ$ term ($= -167$ kJ) is not large enough to offset the very unfavorable (positive) enthalpy change at 25 °C. Why then is this process used in industry? Precisely because the positive entropy change allows the reaction to become spontaneous at higher temperatures. Let us calculate the minimum tempera-

(a)

(b)

(c)

Figure 20.9 The thermite reaction. After the reaction is started with a fuse of burning magnesium wire (a), iron(III) oxide reacts with aluminum powder to give aluminum oxide and iron. The Fe_2O_3/Al reaction (b) generates so much heat that the iron is produced in the molten state. It has dropped out of the clay pot, which originally contained the reactants, and burned through a sheet of iron placed under the pot (c).

ture, T, at which ΔG°_{rxn} is no longer positive, that is, the temperature at which it is zero.

$$\Delta G^\circ_{rxn} = \Delta H^\circ_{rxn} - T\Delta S^\circ_{rxn}$$
$$0 = +467.9 \text{ kJ} - T(0.5603 \text{ kJ/K})$$
$$T = 835.1 \text{ K or } 561.9 \text{ °C}$$

The free energy change for the reaction becomes zero at 562 °C, and at higher temperatures it will be negative. Thus, spontaneity can be achieved by raising the temperature to a point easily reached in an industrial furnace.*

EXERCISE 20.8 Temperature and Free Energy

Is the reduction of magnesia, MgO, with carbon a spontaneous process at 25 °C? If not, at what temperature does it become spontaneous?

$$MgO(s) + C(graphite) \longrightarrow Mg(s) + CO(g)$$

20.6 THERMODYNAMICS AND THE EQUILIBRIUM CONSTANT

You know now that when ΔG° for a reaction is negative, the reaction is predicted to be spontaneous if the reactants and products are at standard conditions. But does this also mean that reactants are converted *completely* to products (to give a 100% yield)? Does the fact that a positive ΔG°, which signals a nonspontaneous reaction, also mean that the reaction will not proceed at all (0% yield)? The answer to both of these questions is *no*, and we can understand why by developing a relationship between ΔG° and the equilibrium constant K for a reaction.

To find a relation between ΔG°_{rxn} and K we return to the *reaction quotient*, Q, that was first introduced in Chapter 16. For the general reaction of A and B giving products X and Y

$$a\text{A} + b\text{B} \rightleftharpoons x\text{X} + y\text{Y}$$

we defined the reaction quotient Q, where

$$Q = \frac{[\text{X}]^x[\text{Y}]^y}{[\text{A}]^a[\text{B}]^b}$$

You learned in Chapter 16 that if $Q = K$ (or $Q/K = 1$), then the system is at equilibrium. However, if products have *not* been formed in concentrations as large as required at equilibrium, then $Q < K$ (or $Q/K < 1$), and the reaction will proceed spontaneously to make products until equilibrium is reached (until $Q = K$). The other situation, where the product concentrations are initially too large, means that $Q > K$ (or $Q/K > 1$),

*To calculate ΔG at a much higher temperature, we used the enthalpy and entropy values appropriate for 25 °C. This is not entirely correct, because ΔH and ΔS are temperature dependent. However, their dependency on T is *much* smaller than that of ΔG (as long as the temperature change does not include a phase change). Thus, our calculation provides a good estimate of the point at which the reaction becomes spontaneous.

and the reaction proceeds spontaneously to make reactants (to bring about the condition where $Q = K$). The table that follows summarizes these conclusions and compares them with important conclusions from thermodynamics.

REACTION	ΔG	DIRECTION REACTION PROCEEDS TO REACH EQUILIBRIUM	Q/K	$\ln(Q/K)$
Spontaneous	< 0	make more product	< 1	< 0
Nonspontaneous	> 0	make more reactants	> 1	> 0
Equilibrium	$= 0$	no change	$= 1$	$= 0$

What is important is that there is a *direct* relation between ΔG for a reaction and the *logarithm* of Q/K, that is, $\Delta G = c \ln(Q/K)$. The proportionality constant c is RT, where $R = 8.314510$ J/K · mol and T is the temperature in kelvins. Therefore,

$$\Delta G = RT \ln(Q/K)$$

Expanding this equation, we have

$$\Delta G = RT \ln Q - RT \ln K$$

When using thermodynamics in this course, we generally refer to reactions where reactants and products are in their standard states, and the ΔG in the equation above can be ΔG°_{rxn}. Recall that the term "standard state" means that concentrations are 1 molal and pressures are 1 atmosphere. Therefore, since ΔG°_{rxn} refers to reactions with all products and reactants starting at unit concentrations, this means that $Q = 1$. Since $\ln Q = 0$ when $Q = 1$, we come to the expression that gives us the connection between ΔG°_{rxn} and K.

$$\Delta G^\circ_{rxn} = -RT \ln K \qquad (20.5)$$

This relation between ΔG°_{rxn} and K is quite important. An equilibrium constant greater than 1 means ΔG°_{rxn} must be negative, and a reaction starting with reactants *and* products in their standard states is spontaneous to make more products. Conversely, when K is less than 1 (and so $\ln K$ is negative), ΔG°_{rxn} is positive, and a reaction starting with reactants *and* products in their standard states is *non*spontaneous to make more product. Finally, when $K = 1$, then $\ln K = 0$, and ΔG°_{rxn} is zero.

For reactions involving gases, Equation 20.5 always gives K_p.

Two applications of Equation 20.5 are (a) the calculation of equilibrium constants from values of ΔG°_f for reactants and products and (b) the evaluation of ΔG°_{rxn} from an experimental determination of K. These applications are explored in the following examples.

EXAMPLE 20.5

CALCULATING K FROM ΔG°_{rxn}

The standard free energy change for the reaction

$$N_2(g) + 3\ H_2(g) \longrightarrow 2\ NH_3(g)$$

is −32.9 kJ. Calculate the equilibrium constant for this reaction at 25 °C.

The unit mol in $\Delta G^\circ = -RT\ln K_p$ can be ignored since it disappears in the rigorous derivation of the expression.

Solution In this case, we need only to substitute the appropriate values into Equation 20.5, taking care that the units of ΔG°_{rxn} are the same as those of RT.

$$\Delta G^\circ_{rxn} = -RT \ln K_p$$
$$(-32.9 \text{ kJ})(1000 \text{ J/kJ}) = -(8.314510 \text{ J/K} \cdot \text{mol})(298 \text{ K})\ln K_p$$
$$\ln K_p = 13.3$$
$$K_p = e^{13.3}$$
$$K_p = 6 \times 10^5$$

The equilibrium constant has a very large value, which means that the equilibrium position lies very far to the product side at room temperature. (To find K_p using a calculator, enter 13.3 and then strike the key labeled "e^x" or "inv(erse) ln x." See Appendix A for more information.)

EXAMPLE 20.6

CALCULATING ΔG°_{rxn} FROM K_p

Calculate ΔG°_{rxn} for the conversion of oxygen to ozone at 25 °C from $K_p = 2.47 \times 10^{-29}$.

$$\tfrac{3}{2}\, O_2(g) \longrightarrow O_3(g)$$

Solution To find ΔG°_{rxn} we substitute K_p into Equation 20.5.

$$\Delta G^\circ_{rxn} = -RT \ln K_p = -(8.314510 \text{ J/K} \cdot \text{mol})(298 \text{ K})\ln(2.47 \times 10^{-29})$$
$$= 163000 \text{ J}$$
$$= 163 \text{ kJ}$$

From Appendix J you can verify that the free energy change for the reaction should indeed be 163 kJ.

EXAMPLE 20.7

USES OF ΔH°_f, S°, ΔG°_f AND THE CALCULATION OF K

In Chapter 5 you studied in a qualitative way the driving forces of chemical reactions. One of these is the formation of a gaseous product that can escape when the reaction is open to the surroundings. Let us look at such a reaction from a quantitative, thermodynamic point of view.

$$MgCO_3(s) \longrightarrow MgO(s) + CO_2(g)$$

(a) Is the reaction spontaneous at room temperature?
(b) Is the reaction driven by enthalpy, entropy, or both?
(c) What is the value of K_p at 25 °C?
(d) At what temperature is $K = 1$?
(e) Does high temperature make the reaction more or less spontaneous?

Solution To answer the first two questions, you need to know ΔG°_{rxn} and its sign, in addition to ΔH°_{rxn} and ΔS°_{rxn}. ΔG°_{rxn} can of course be calculated using the Gibbs equation

$$\Delta G^\circ_{rxn} = \Delta H^\circ_{rxn} - T\Delta S^\circ_{rxn}$$

Since we need to know ΔH°_{rxn} and ΔS°_{rxn} to answer question (b), we shall first calculate these values from ΔH°_f and S° for the reactants and products. The enthalpy change for the reaction can be calculated as we first outlined in Chapter 6.

$$\Delta H^\circ_{rxn} = \Delta H^\circ_f[CO_2(g)] + \Delta H^\circ_f[MgO(s)] - \Delta H^\circ_f[MgCO_3(s)]$$

Using data from Appendix J, and $\Delta H^\circ_f\ [MgCO_3(s)] = -1095.8$ kJ/mol,

$$\begin{aligned}\Delta H^\circ_{rxn} &= (1\text{ mol})(-393.5\text{ kJ/mol}) + (1\text{ mol})(-601.7\text{ kJ/mol})\\ &\quad - (1\text{ mol})(-1095.8\text{ kJ/mol})\\ &= 100.6\text{ kJ}\end{aligned}$$

The positive enthalpy change indicates that enthalpy opposes reaction spontaneity under standard conditions. Therefore, for question (b), you now know that the reaction is *not* enthalpy driven.

Next, the entropy change for the reaction can be calculated from values in Appendix J and the Bureau of Standards.

$$\begin{aligned}\Delta S^\circ_{rxn} &= S^\circ[CO_2(g)] + S^\circ[MgO(s)] - S^\circ[MgCO_3(s)]\\ &= (1\text{ mol})(213.7\text{ J/K}\cdot\text{mol}) + (1\text{ mol})(26.9\text{ J/K}\cdot\text{mol})\\ &\quad - (1\text{ mol})(65.7\text{ J/K}\cdot\text{mol})\\ &= 174.9\text{ J/K (or 0.1749 kJ/K)}\end{aligned}$$

The entropy change for the reaction is positive, which means that spontaneity can be achieved if the temperature is right. Referring again to question (b), you now know that the reaction can be entropy driven.

Next, we combine ΔH°_{rxn} and ΔS°_{rxn} to find ΔG°_{rxn}; we can thus discover whether enthalpy or entropy wins out in the competition and whether the reaction is spontaneous at room temperature (25 °C).

$$\begin{aligned}\Delta G^\circ_{rxn} &= \Delta H^\circ_{rxn} - T\Delta S^\circ_{rxn}\\ &= 100.6\text{ kJ} - (298\text{ K})(0.1749\text{ J/K})\\ &= 48.5\text{ kJ}\end{aligned}$$

Now you know that $MgCO_3(s)$ does not spontaneously generate 1 atm of CO_2 (its standard state) at room temperature; the enthalpy difference between products and reactants is simply too high to be offset by the increase in entropy on CO_2 evolution at 25 °C.

Having found ΔG°_{rxn}, you can now calculate K_p at room temperature from

This is to remind you that K in Equation 20.5 is K_p, the equilibrium constant in terms of partial pressures.

$$\Delta G^\circ_{rxn} = -RT\ln K_p$$

or

$$\ln K_p = -\frac{\Delta G^\circ_{rxn}}{RT} = -\frac{48{,}500\text{ J}}{(8.314510\text{ J/K}\cdot\text{mol})(298\text{ K})} = -19.6$$

Since $\ln K_p$ is -19.6, this means that $K_p = e^{-19.6} = 3 \times 10^{-9}$.

$$K_p = P_{CO_2} = 3 \times 10^{-9}$$

K_p is extremely small, because ΔG°_{rxn} has such a large positive value.

The next question is (d): at what temperature is $K = 1$? From the preceding analysis, you see that ΔS°_{rxn} drives the reaction. This means that, at a sufficiently high temperature, the negative value of $-T\Delta S^\circ$ could become large enough to outweigh the inhibiting effect of a positive ΔH°, and ΔG°_{rxn} would become negative. A balance between ΔH° and $-T\Delta S^\circ$ is reached when $\Delta G^\circ = 0$, that is, when $K = 1$. What is the temperature at which this is achieved? When $\Delta G^\circ = 0$, we can write $\Delta H^\circ - T\Delta S^\circ = 0$. This means that the breakeven temperature is

$$T = \frac{\Delta H^\circ}{\Delta S^\circ} = \frac{100.6 \text{ kJ}}{0.1749 \text{ kJ/K}} = 575.2\text{K (or 302.0 °C)}$$

At approximately 300 °C the equilibrium constant is 1, but at still higher temperatures the $-T\Delta S^\circ$ term increasingly dominates the ΔH° term, and the yield increases because K increases. This means the answer to question (e) is "more spontaneous," and this is true for any reaction with $\Delta S^\circ_{rxn} > 0$ (see Figure 20.7).

EXERCISE 20.9 Calculating K from Free Energy of Reaction

Calculate K_p at 298 K from the value of ΔG°_{rxn} for each of the following reactions.

(a) $S(s) + O_2(g) \longrightarrow SO_2(g)$

(b) $CaCO_3(s) \longrightarrow CaO(s) + CO_2(g)$

20.7 THERMODYNAMICS AND TIME

With this chapter, we have brought together the **three laws of thermodynamics.**

First law: The total energy of the universe is a constant.
Second law: The total entropy of the universe is always increasing.
Third law: The entropy of every pure, perfectly formed crystalline substance at absolute zero is zero.

Some cynic long ago paraphrased the first two laws into simpler statements. The first law is a statement that "You can't win!", and the second law tells you that "You can't break even, either!" Yet another interpretation of the second law is Murphy's Law that "Things always tend to go wrong."

The second law demands that disorder increases with time. Since all natural processes that occur do so as time progresses and result in increased disorder, it is evident that increasing entropy and time "point" in the same direction.

The second law tells us the entropy of the universe increases in every spontaneous—naturally occurring—process. A snowflake will spontaneously melt in a warm room, but you won't see a glassful of water molecules spontaneously assembling into snowflakes in a warm place. Molecules of your perfume or cologne will spontaneously diffuse throughout a room, but they won't spontaneously collect again on your body. With time, all natural processes result in chaos. This is what scientists mean when they say that the second law is an expression in physical—as opposed to psychological—form of what we call *time*. In fact, entropy has been called "time's arrow."

Neither of the first two laws of thermodynamics has ever been or can be proved. However, there never has been a single, concrete example showing otherwise. No less a scientist than Albert Einstein once remarked that thermodynamic theory ". . . is the only physical theory of universe content [which], within the framework of applicability of its basic concepts, will never be overthrown."

If you are interested in the theories of the origin of the universe, and in "time's arrow," be sure to read "A Brief History of Time, From the Big Bang to Black Holes," by Stephen W. Hawking, Bantam Books, 1988.

Einstein's statement does not mean that people have not tried (and are continuing to try) to disprove the laws of thermodynamics. Someone is always claiming to have invented a machine that performs useful work without expending energy—a perpetual motion machine. Although such a machine was actually granted a patent recently by the U.S. Patent Office (presumably the patent examiner had not had a course in thermodynamics), no workable perpetual motion machine has ever been demonstrated; the laws of thermodynamics are safe.

SUMMARY

A **spontaneous chemical reaction** is one that will go to equilibrium, given enough time, and that has an equilibrium constant greater than 1 (Section 20.1). Experience teaches us that the creation of disorder favors spontaneity (Sections 20.2 and 20.3). **Entropy, S**, is a measure of the disorder or randomness in a system and is a state function that has units of J/K (Section 20.3). The **third law of thermodynamics** states that the entropy of a pure, perfectly formed crystalline substance is zero at 0 K (Section 20.3). This means that there is an absolute starting place for measuring the **absolute entropy** of a substance, so the entropy possessed by a substance can be determined at some temperature other than 0 K. The **standard state entropy, $S°$**, is that measured for 1 mole of the substance at a pressure of 1 atm (solutions have a concentration of 1 m).

Values of $S°$ have been measured for many substances (Sections 20.3 and 20.4) and can be used to determine the entropy change, $\Delta S°$, for a reaction by

$$\Delta S°_{\text{reaction}} = \Sigma\, mS°(\text{products}) - \Sigma\, nS°(\text{reactants}) \qquad (20.2)$$

where m and n are the number of moles of products and reactants, respectively.

The **second law of thermodynamics** states that the combined entropy of the system and its surroundings always increases in a spontaneous process (Section 20.5). The **Gibbs function, G**, also a state function, has been derived from the second law and is defined by the **Gibbs equation**.

$$\Delta G = \Delta H - T\Delta S \qquad (20.3)$$

Using ΔH_{rxn} and ΔS_{rxn}, a value for ΔG_{rxn} can be calculated. When all substances are in their standard states, the standard free energy change $\Delta G°$ is derived. If the reaction involved is that for the formation of a compound from the elements, with all substances in their standard states, then $\Delta G°_{\text{rxn}}$ is the standard free energy of formation, $\Delta G°_f$. Using values of $\Delta G°_f$ for all substances in a reaction (with $\Delta G°_f = 0$ for elements in their standard states), $\Delta G°_{\text{rxn}}$ can be calculated from the expression

$$\Delta G°_{\text{reaction}} = \Sigma\, m\Delta G°_f(\text{products}) - \Sigma\, n\Delta G°_f(\text{reactants}) \qquad (20.4)$$

Based on the Gibbs equation, the criteria for spontaneity of a chemical reaction are:

(a) If ΔG has a negative value, the process is spontaneous.
(b) If ΔG has a positive value, the process is not spontaneous under the specified conditions.
(c) If $\Delta G = 0$, the process is at equilibrium.

The use of these criteria is outlined in Section 20.5. See also Table 20.3 and Figure 20.7.

The relation between $\Delta G°_{\text{rxn}}$ and the equilibrium constant, K_p, for a reaction is given by

$$\Delta G^\circ_{rxn} = -RT \ln K_p \tag{20.5}$$

where R is 8.314510 J/K · mol and T is the temperature in kelvins. Thus, reactions for which ΔG°_{rxn} is negative are predicted to be spontaneous and to have high concentrations of products at equilibrium ($K > 1$ and the yield is high). When ΔG°_{rxn} is positive, the reaction is predicted to be nonspontaneous at the temperature and to have low concentrations of products at equilibrium ($K < 1$ and the yield is low).

STUDY QUESTIONS

Many of these questions require thermodynamic data. If the required data are not given in the question, consult the tables in this chapter or in Appendix J.

REVIEW QUESTIONS

1. State the three laws of thermodynamics.

2. What is meant by a "spontaneous chemical reaction"?

3. Criticize the following statements:
- (a) The entropy increases in all spontaneous reactions.
- (b) A reaction with a negative free energy change ($\Delta G^\circ < 0$) is predicted to be spontaneous with rapid transformation of reactants to products.
- (c) All spontaneous processes are exothermic.
- (d) Endothermic reactions are never spontaneous.

4. Tell whether each of the following statements is true or false. If false, rewrite to make it true.
- (a) The entropy of a substance increases on going from the liquid to the vapor state at any temperature.
- (b) An exothermic reaction will always be thermodynamically spontaneous.
- (c) Reactions with a positive ΔH°_{rxn} and a positive ΔS°_{rxn} can never be spontaneous.
- (d) Reactions with $\Delta G^\circ_{rxn} < 0$ always have an equilibrium constant greater than 1.
- (e) When the equilibrium constant of a reaction is less than 1, then ΔG°_{rxn} is less than zero.

5. Explain why the entropy of the system increases on dissolving solid NaCl in water {S°[NaCl(s)] = 72.1 J/K · mol and S°[NaCl(aq)] = 115.5 J/K · mol}.

ENTROPY

6. For each of the following pairs, tell which has the higher entropy.
- (a) A sample of solid CO_2 at −78 °C or CO_2 vapor at 0 °C.
- (b) Sugar, as a solid or dissolved in a cup of tea.
- (c) Two beakers, one containing pure water and the other containing pure alcohol, or a single beaker containing a mixture of the water and alcohol.

7. Tell which substance has the higher entropy in each of the following pairs.
- (a) A sample of pure silicon or a piece of silicon containing a trace of some other atoms such as B or P (to be used in a computer chip).
- (b) An ice cube or liquid water, both at 0 °C.
- (c) A sample of pure solid I_2 or iodine vapor, both at room temperature.

8. Comparing the formulas or states for each pair of compounds, tell which you would expect to have the higher entropy at the same temperature.
- (a) KCl(s) or $AlCl_3$(s)
- (b) $CH_3I(\ell)$ or $CH_3CH_2I(\ell)$
- (c) NH_4Cl(s) or NH_4Cl(aq)

9. Comparing the formulas or states for each pair of compounds, tell which you would expect to have the higher entropy at the same temperature.
- (a) NaCl(s) or $MgCl_2$(s)
- (b) Cl_2(g) or P_4(g)
- (c) CH_3NH_2(g) or $(CH_3)_2NH$(g)
- (d) Au(s) or $Hg(\ell)$

10. Calculate the entropy change, ΔS°, for each of the following changes and comment on the sign of the change:
- (a) C(graphite) ⟶ C(diamond)
- (b) Na(g) ⟶ Na(s)
- (c) $Hg(\ell)$ ⟶ Hg(g) (S° for Hg(g) = 175 J/K · mol at 25 °C)

11. Calculate the entropy change, ΔS°, for each of the following changes and comment on the sign of the change:
- (a) NH_4Cl(s) ⟶ NH_4Cl(aq)
- (b) S(g) ⟶ S(s)
- (c) CCl_4(g) ⟶ $CCl_4(\ell)$

12. The enthalpy of vaporization of liquid diethyl ether, $(C_2H_5)_2O$, is 26.0 kJ/mol at the boiling point of 35.0 °C. Calculate ΔS° for (a) liquid to vapor and (b) vapor to liquid at 35.0 °C.

13. Calculate the entropy change for the boiling of ethyl alcohol, C_2H_5OH, at the normal boiling point of the pure alcohol, 78.0 °C. The enthalpy of vaporization of the alcohol is 39.3 kJ/mol.

ENTROPY OF REACTIONS

14. Calculate the standard molar entropy change for the formation of gaseous propane (C_3H_8) at 25 °C.

$$3\ C(\text{graphite}) + 4\ H_2(g) \longrightarrow C_3H_8(g)$$

15. Calculate the standard molar entropy change for the formation of aluminum chloride at 25 °C.

$$2\ Al(s) + 3\ Cl_2(g) \longrightarrow 2\ AlCl_3(s)$$

16. Calculate the standard molar entropy change for the formation of each of the following compounds from the elements at 25 °C.
(a) $H_2O(\ell)$ (c) $CaS(s)$
(b) $MgO(s)$ (d) $Al_2O_3(s)$

17. Calculate the standard molar entropy change for the formation of each of the following compounds from the elements at 25 °C.
(a) $ICl(g)$ (c) $CaCO_3(s)$
(b) $COCl_2(g)$ (d) $CHCl_3(g)$

18. Calculate the standard molar entropy change for each of the following reactions at 25 °C.
(a) $Pb(s) + Cl_2(g) \longrightarrow PbCl_2(s)$
(b) $C_2H_5OH(\ell) + 3\ O_2(g) \longrightarrow 2\ CO_2(g) + 3\ H_2O(\ell)$
(c) $4\ Cr(s) + 3\ O_2(g) \longrightarrow 2\ Cr_2O_3(s)$

19. Calculate the standard molar entropy change for each of the following reactions at 25 °C.
(a) $4\ Fe(s) + 3\ O_2(g) \longrightarrow 2\ Fe_2O_3(s)$
(b) $Ca(s) + 2\ H_2O(\ell) \longrightarrow Ca(OH)_2(aq) + H_2(g)$
(c) $Na_2CO_3(s) + 2\ HCl(aq) \longrightarrow 2\ NaCl(aq) + H_2O(\ell) + CO_2(g)$

FREE ENERGY

20. Using values of ΔH_f° and S°, calculate ΔG° for each of the following reactions:
(a) $Pb(s) + Cl_2(g) \longrightarrow PbCl_2(s)$
(b) $Mg(s) + \frac{1}{2}\ O_2(g) \longrightarrow MgO(s)$
(c) $NH_3(g) + HCl(g) \longrightarrow NH_4Cl(s)$
Which of the values of ΔG° that you have just calculated correspond(s) to a standard free energy of formation, ΔG_f°? In those cases, compare your calculated values with the values of ΔG_f° tabulated in Appendix J. Which of these three reactions is (are) predicted to be spontaneous?

21. Using values of ΔH_f° and S°, calculate ΔG° for each of the following reactions:
(a) $Br_2(\ell) + 3\ F_2(g) \longrightarrow 2\ BrF_3(g)$
(b) $Ca(s) + 2\ H_2O(\ell) \longrightarrow Ca(OH)_2(aq) + H_2(g)$
(c) $6\ C(\text{graphite}) + 3\ H_2(g) \longrightarrow C_6H_6(\ell)$
Which of the values of ΔG° that you have just calculated correspond(s) to a standard free energy of formation, ΔG_f°? In those cases, compare your calculated values with the values of ΔG_f° tabulated in Appendix J. Which of these three reactions is (are) predicted to be spontaneous?

22. Using values of ΔH_f° and S°, calculate the standard molar free energy of formation, ΔG_f°, for each of the following compounds:
(a) $CS_2(g)$
(b) $N_2H_4(\ell)$
(c) $COCl_2(g)$
Compare your calculated values of ΔG_f° with those listed in Appendix J. Which reactions are predicted to be spontaneous?

23. Using values of ΔH_f° and S°, calculate the standard molar free energy of formation, ΔG_f°, for each of the following compounds:
(a) $Mg(OH)_2(s)$
(b) $NOCl(g)$
(c) $Na_2CO_3(s)$
Compare your calculated values of ΔG_f° with those listed in Appendix J. Which reactions are predicted to be spontaneous?

24. Write a balanced equation that depicts the formation of 1 mol of $Fe_2O_3(s)$ from its elements. What is the free energy of formation of 1.00 mol of $Fe_2O_3(s)$? What is the standard free energy change when 454 g (1 pound) of $Fe_2O_3(s)$ are formed from the elements?

25. Hydrazine is used to remove dissolved oxygen from the water in water heating systems.

$$N_2H_4(\ell) + O_2(g) \longrightarrow 2\ H_2O(\ell) + N_2(g)$$

What is the free energy change observed when 1.00 mole of N_2H_4 is oxidized? What is the free energy change for the oxidation of 1 pound (454 g) of hydrazine?

26. Using values of ΔG_f°, calculate ΔG_{rxn}° for each of the following reactions. Which reactions are predicted to be spontaneous?
(a) $Ca(s) + Cl_2(g) \longrightarrow CaCl_2(s)$
(b) $HgO(s) \longrightarrow Hg(\ell) + \frac{1}{2}\ O_2(g)$
(c) $NH_3(g) + 2\ O_2(g) \longrightarrow HNO_3(\ell) + H_2O(\ell)$

27. Using values of ΔG_f°, calculate ΔG_{rxn}° for each of the following reactions. Which reactions are predicted to be spontaneous?
(a) $HgS(s) + O_2(g) \longrightarrow Hg(\ell) + SO_2(g)$
(b) $2\ H_2S(g) + 3\ O_2(g) \longrightarrow 2\ H_2O(g) + 2\ SO_2(g)$
(c) $SiCl_4(g) + 2\ Mg(s) \longrightarrow 2\ MgCl_2(s) + Si(s)$

28. Yeast can produce ethyl alcohol by the fermentation of glucose, the basis for the production of most alcoholic beverages.

$$C_6H_{12}O_6(aq) \longrightarrow 2\ C_2H_5OH(\ell) + 2\ CO_2(g)$$

Calculate ΔH°, ΔS°, and ΔG° for the reaction at 25 °C. In addition to the thermodynamic values in Appendix J, you will need the following for $C_6H_{12}O_6(aq)$: $\Delta H_f^\circ = -1260.0$ kJ/mol; $S^\circ = 289$ J/K · mol; and $\Delta G_f^\circ = -918.8$ kJ/mol.

29. There has been great interest in splitting water into its elements, since the $H_2(g)$ thereby produced could be used as a fuel. Calculate ΔH°, ΔS°, and ΔG° for the water splitting reaction at 25 °C.

$$H_2O(g) \longrightarrow H_2(g) + \tfrac{1}{2}O_2(g)$$

30. A value for ΔG_f° is not available in Appendix J for the first compound in each of the following reactions. Given ΔG_{rxn}° and the data in Appendix J, calculate the standard molar free energy of formation of that compound.
(a) $\Delta G_{rxn}^\circ = +48.2$ kJ for $MgCO_3(s) \rightarrow MgO(s) + CO_2(g)$
(b) $\Delta G_{rxn}^\circ = +137.6$ kJ for $NOF(g) \rightarrow NO(g) + \tfrac{1}{2}F_2(g)$

31. A value for ΔG_f° is not available in Appendix J for the first compound in each of the following reactions. Given ΔG_{rxn}° and the data in Appendix J, calculate the standard molar free energy of formation of that compound.
(a) $\Delta G_{rxn}^\circ = -272.8$ kJ for $TiCl_2(s) + Cl_2(g) \rightarrow TiCl_4(\ell)$
(b) $\Delta G_{rxn}^\circ = -1830$ kJ for $2\ CH_2CO(g) + 4\ O_2(g) \rightarrow 4\ CO_2(g) + 2\ H_2O(g)$

32. Hydrogenation, the addition of hydrogen to an organic compound, is a reaction of considerable industrial importance. Calculate ΔH°, ΔS°, and ΔG° at 25 °C for the hydrogenation of octene, C_8H_{16}, to give octane, C_8H_{18}. Is the reaction spontaneous under standard conditions?

$$C_8H_{16}(g) + H_2(g) \longrightarrow C_8H_{18}(g)$$

The following information is required, in addition to data in Appendix J.

COMPOUND	ΔH_f° (kJ/mol)	S° (J/K · mol)
Octene	−82.93	462.8
Octane	−208.45	463.6

33. If gaseous hydrogen can be produced cheaply, it can be burned directly as a fuel or converted to another fuel, methane (CH_4), for example.

$$3\ H_2(g) + CO(g) \longrightarrow CH_4(g) + H_2O(g)$$

Calculate ΔH°, ΔS°, and ΔG° at 25 °C for the reaction. Is it predicted to be spontaneous under standard conditions?

34. In a (fictitious) court case, a woman sued a chemical company for damages. She claimed the company produced ozone, O_3, that reacted with water in the air to produce hydrogen peroxide. The hydrogen peroxide in turn bleached her beautiful black hair to red. You are her lawyer. Is the following reaction thermodynamically feasible? Can you tell the jury it is spontaneous under standard conditions?

$$O_3(g) + H_2O(g) \longrightarrow H_2O_2(\ell) + O_2(g)$$

35. Wood alcohol (methyl alcohol) is a valuable starting material in the chemical industry. Although it was originally made from the destructive distillation of wood (hence its name), it is now made by the following reaction:

$$CO(g) + 2\ H_2(g) \longrightarrow CH_3OH(\ell)$$

Using thermodynamic calculations, predict the spontaneity of this reaction under standard conditions. The thermodynamic functions for $CH_3OH(\ell)$ are $\Delta H_f^\circ = -238.7$ kJ/mol, $S^\circ = 126.8$ J/K · mol, and $\Delta G_f^\circ = -166.3$ kJ/mol.

THERMODYNAMICS AND EQUILIBRIUM CONSTANTS

36. The formation of NO(g) from its elements

$$\tfrac{1}{2}N_2(g) + \tfrac{1}{2}O_2(g) \longrightarrow NO(g)$$

has a standard free energy change, ΔG_f°, of +86.55 kJ/mol at 25 °C. Calculate K_p at this temperature. Comment on the connection between the sign of ΔG° and the magnitude of K_p.

37. Methyl alcohol is now widely used as a fuel in race cars such as those that compete in the Indianapolis 500 (see Chapter 6). The compound has a standard free energy of formation, ΔG_f° of −166.27 kJ/mol at 25 °C.

$$C(\text{graphite}) + \tfrac{1}{2}O_2(g) + 2\ H_2(g) \longrightarrow CH_3OH(\ell)$$

Calculate K_p for this reaction at 25 °C. Comment on the connection between the sign of ΔG° and the magnitude of K_p.

38. The equilibrium constant for the *n*-butane ⇄ isobutane equilibrium at 25 °C is 2.5.

$$\underset{n\text{-butane}}{H_3C{-}CH_2{-}CH_2{-}CH_3} \rightleftharpoons \underset{\text{isobutane}}{H_3C{-}CH(CH_3){-}CH_3}$$

Calculate ΔG° for the reaction at this temperature in units of kJ/mol.

39. The dissociation of chlorine molecules into chlorine atoms has an equilibrium constant of 0.106 at 1800 K.

$$Cl_2(g) \rightleftharpoons 2\ Cl(g)$$

Calculate $\Delta G°$ for the reaction at this temperature in units of kJ/mol.

40. Water reacts with $SiCl_4$ to produce SiO_2 and HCl.

$$SiCl_4(g) + 2\ H_2O(\ell) \longrightarrow SiO_2(s) + 4\ HCl(aq)$$

(a) Using the data in Appendix J, calculate $\Delta G°$ for the reaction at 25 °C. Is the reaction predicted to be spontaneous under standard conditions?
(b) From $\Delta G°_{rxn}$ calculate K_p. Comment on the connection between the sign of $\Delta G°$ and the magnitude of K_p.

41. Ethene (or ethylene) reacts with hydrogen to produce ethane.

$$H_2C{=}CH_2(g) + H_2(g) \longrightarrow H_3C{-}CH_3(g)$$

(a) Using the data in Appendix J, calculate $\Delta G°$ for the reaction at 25 °C. Is the reaction predicted to be spontaneous under standard conditions?
(b) From $\Delta G°_{rxn}$ calculate K_p. Comment on the connection between the sign of $\Delta G°$ and the magnitude of K_p.

GENERAL PROBLEMS

42. The entropy values for several straight-chain hydrocarbons are given in Section 20.3. Using these values, estimate the entropy, $S°$, for *n*-pentane, C_5H_{12} ($H_3C{-}CH_2{-}CH_2{-}CH_2{-}CH_3$).

43. What trend is observed in the entropies of the rare gases in Figure 20.6? Is this trend observed elsewhere in the periodic table?

44. Why is the standard state entropy of Br_2 greater than that of I_2? (Hint: See Figure 1.4.)

45. Calculate the entropy change involved in the formation of 1.0 mol of each of the following gaseous hydrocarbons under standard conditions. (Use graphite as the standard state of carbon.)

(a) H—C≡C—H ethyne or acetylene

(b) $H_2C{=}CH_2$ (H, H above; C=C; H, H below) ethene or ethylene

(c) $H_3C{-}CH_3$ (H—C—C—H with H above and below each C) ethane

What trend do you see in these values? Does $\Delta S°$ increase or decrease on adding H atoms? Can you think of some reasons for the observed trend?

46. For each of the following processes, give the algebraic signs of $\Delta H°$, $\Delta S°$, and $\Delta G°$. No calculations are necessary; use your common sense.

(a) The splitting of liquid water to give gaseous oxygen and hydrogen, a process that requires a considerable amount of energy.
(b) Dissolving a small amount of NH_4Cl in water. The solution becomes quite cold in the process.
(c) The explosion of dynamite, a mixture of nitroglycerin ($C_3H_5N_3O_9$) and diatomaceous earth. The explosive decomposition gives gaseous products such as water, CO_2, and others; much heat is evolved.
(d) The combustion of gasoline in the engine of your car, as exemplified by the combustion of octane.

$$2\ C_8H_{18}(g) + 25\ O_2(g) \longrightarrow 16\ CO_2(g) + 18\ H_2O(g) + \text{heat}$$

47. Elemental boron, in the form of thin fibers, can be made by reducing a boron halide with H_2:

$$BCl_3(g) + \tfrac{3}{2}\ H_2(g) \longrightarrow B(s) + 3\ HCl(g)$$

The standard enthalpy of formation of $BCl_3(g)$ is −403.8 kJ/mol, and its entropy, $S°$, is 290 J/K · mol. The entropy, $S°$, for B(s) is 5.86 J/K · mol. Calculate $\Delta H°$, $\Delta S°$, and $\Delta G°$ at 25 °C for this reaction. Is it predicted to be spontaneous under standard conditions? If spontaneous, is it enthalpy driven or entropy driven?

48. The equilibrium constant, K_p, for $N_2O_4(g) \rightleftarrows 2\ NO_2(g)$ is 0.15 atm at 25 °C. Calculate $\Delta G°$ from this constant, and compare your calculated value with that determined from the $\Delta G°_f$ values in Appendix J.

49. Most metal oxides can be reduced with hydrogen to the pure metal. (Although such reactions work well, it is an expensive method and not used often for large scale preparations.) The reduction of iron(II) oxide

$$FeO(s) + H_2(g) \longrightarrow Fe(s) + H_2O(g)$$

has an equilibrium constant of 0.422 at 700. °C. Estimate $\Delta G°$.

50. Benzene, C_6H_6, is made in quantities ranging from 8 to 10 billion pounds a year. It is used as a starting material for many other compounds and as a solvent. (Although it is an excellent solvent for many purposes, it has recently been found to be a carcinogen, and its use is now restricted.) One compound that can be made from benzene is cyclohexane, C_6H_{12}.

$$C_6H_6(g) + 3\ H_2(g) \longrightarrow C_6H_{12}(g)$$

$$\Delta H°_{rxn} = -206.1\ \text{kJ} \qquad \Delta S°_{rxn} = -363.12\ \text{J/K}$$

(a) Is this reaction predicted to be spontaneous under standard conditions at 25 °C? Is the reaction enthalpy driven or entropy driven?

(b) Calculate ΔG_f° for benzene vapor [$C_6H_6(g)$]. Use ΔH°_{rxn} and ΔS°_{rxn} for the reaction above, together with ΔG_f° for cyclohexane (=31.76 kJ/mol).

51. Iodine, I_2, dissolves readily in carbon tetrachloride with an enthalpy change that is approximately zero.

$$I_2(s) \longrightarrow I_2 \text{ (in } CCl_4 \text{ solution)}$$

What is the sign of ΔG for the reaction? Is the dissolving process entropy driven or enthalpy driven? Explain briefly.

52. Animals, except those that are deep divers such as the whales, live under conditions of constant pressure. The oxidation of glucose, $C_6H_{12}O_6$, is a source of energy for their nervous and muscular activity. If one mole of glucose is oxidized according to the following overall equation, how much free energy is available under standard conditions at 25 °C?

$$C_6H_{12}O_6(aq) + 6\ O_2(g) \longrightarrow 6\ CO_2(g) + 6\ H_2O(\ell)$$

For glucose, $\Delta H_f^\circ = -1260$ kJ/mol and $S^\circ = 288.9$ J/K · mol. (The biological reactions are equivalent to this reaction, but they involve many intermediate steps.)

53. If a 70.-kg man climbs 3.0 meters vertically, he needs to do 2.1 kJ of work. Refer to Study Question 52 and calculate the approximate minimum amount of glucose the man's body would have to burn to climb Mt. McKinley, the highest mountain in North America. The mountain is 20,300 feet high (1 meter = 3.28 feet).

54. A crucial reaction for the production of synthetic fuels is the conversion of coal to H_2 with steam.

$$C(s) + H_2O(g) \longrightarrow CO(g) + H_2(g)$$

(a) Calculate ΔG° for this reaction at 25 °C, assuming C(s) is graphite.
(b) Calculate K_p for the reaction at 25 °C.
(c) Is the reaction predicted to be spontaneous under standard conditions? If not, at what temperature will it become so?
(d) Calculate the temperature at which the equilibrium constant for the reaction is 1.0×10^{-4}.

55. Calculate ΔG° for the decomposition of sulfur trioxide to sulfur dioxide and oxygen.

$$2\ SO_3(g) \longrightarrow 2\ SO_2(g) + O_2(g)$$

(a) Is the reaction spontaneous under standard conditions at 25 °C?
(b) If the reaction is not spontaneous at 25 °C, is there a temperature at which it will become so?
(c) What is the equilibrium constant for the reaction at 1500 °C?

56. Methyl alcohol is relatively inexpensive to produce. Much consideration has been given to using it as a precursor to other fuels such as methane, which could be obtained by the decomposition of the alcohol.

$$CH_3OH(\ell) \longrightarrow CH_4(g) + \tfrac{1}{2}\ O_2(g)$$

(See Study Question 35 for the necessary thermodynamic functions for methyl alcohol.)
(a) What are the sign and magnitude of the entropy change for the reaction? Does the sign of ΔS° agree with your expectation? Explain briefly.
(b) Is the reaction spontaneous under standard conditions at 25 °C? Use thermodynamic values to prove your answer.
(c) If not spontaneous at 25 °C, at what temperature does the reaction become spontaneous?

57. Consider the formation of NO(g) from its elements.

$$N_2(g) + O_2(g) \longrightarrow 2\ NO(g)$$

(a) Use the free energy data in Appendix J to calculate K_p at 25 °C.
(b) Assume that ΔH°_{rxn} and ΔS°_{rxn} are nearly constant with temperature and calculate ΔG°_{rxn} at 700. °C. Estimate K_p from the new value of ΔG°_{rxn} at 700. °C.
(c) Using K_p at 700. °C, calculate the equilibrium partial pressures of the three gases if you mix 1.00 atm each of N_2 and O_2.

58. Photosynthetic bacteria carry out the synthesis of high free energy compounds such as glucose from CO_2 and H_2O by using light as the source of energy.

$$6\ CO_2(g) + 6\ H_2O(\ell) \longrightarrow C_6H_{12}O_6(aq) + 6\ O_2(g)$$

The free energy change is +2870 kJ/mol of glucose. However, in the deep ocean where there is no light, this same synthesis can apparently be done by bacteria that use hydrogen sulfide as the energy source. Show that, by adding the following reaction

$$H_2S(g) + \tfrac{1}{2}\ O_2(g) \longrightarrow H_2O(\ell) + S(s)$$

to the glucose synthesis reaction, sufficient free energy will be produced so that the overall process

$$24\ H_2S(g) + 6\ CO_2(g) + 6\ O_2(g) \longrightarrow C_6H_{12}O_6(aq) + 18\ H_2O(\ell) + 24\ S(s)$$

is spontaneous. (ΔG_f° for glucose is −918.8 kJ/mol.) (For further information see S. Krishnamurthy, *J. Chem. Educ.* **1981**, *58*, 981.)

SUMMARY QUESTIONS

59. As noted in Study Question 29, there has been a great deal of interest in the splitting of water into hydrogen and oxygen. One way to do this begins with the reaction of silver metal with water to produce hydrogen and silver oxide.

$$2\ Ag(s) + H_2O(g) \longrightarrow Ag_2O(s) + H_2(g)$$

Decomposing the silver oxide at high temperature gives oxygen and recovers silver.

$$Ag_2O(s) \longrightarrow 2\ Ag(s) + \tfrac{1}{2}\ O_2(g)$$

(For more information on reactions of this type see E.L. King, *J. Chem. Educ.* **1981**, *58*, 975.)

(a) **Stoichiometry:** How many grams of H_2 can you produce if you begin with 150. g of silver and an excess of water? If the hydrogen is captured at 23 °C in a flask with a volume of 15.0 L, what will the gas pressure be?

(b) **Solid State:** What is the density of the silver used in the experiment? You can calculate it by knowing that the unit cell of silver is face centered cubic and that the radius of a silver atom is 144.5 pm.

(c) **Thermodynamics:** We want to find out whether we can produce H_2 and O_2 by combining the two reactions above to give a spontaneous process. We will need the following information in addition to that in Appendix J: for $Ag_2O(s)$, $\Delta H_f^\circ = -31.1$ kJ/mol and $S^\circ = 121$ J/K · mol.

(1) Calculate ΔH°, ΔS° and ΔG° for both of the reactions at 25 °C.

(2) Write a balanced equation for the overall reaction that occurs on summing these two equations.

(3) What is the enthalpy change, ΔH°, for the overall reaction?

(4) What is ΔG° for the overall reaction? Is the overall process spontaneous at 25 °C?

(5) At what temperature is the second reaction (decomposition of silver oxide) spontaneous?

60. Metal oxides can be reduced with hydrogen gas to give the metal.

$$H_2(g) + ZnO(s) \rightleftharpoons H_2O(g) + Zn(s)$$

Predict the effect of each of the following changes on the position of the equilibrium; that is, state which way the equilibrium will shift (left, right, or no change) when each of the following changes is made.

(a) Add more ZnO(s).

(b) Add more $H_2(g)$.

(c) Remove some Zn(s).

(d) Remove some water vapor.

Using the thermodynamic information in Appendix J, answer the following questions:

(e) In which direction does the equilibrium shift when you raise the temperature?

(f) Calculate the value of ΔG° for the reaction.

(g) Is the reaction predicted to be spontaneous under standard conditions at 25 °C?

(h) Is the equilibrium constant for the reaction greater than 1 or less than 1?

CHAPTER

21

Electrochemistry: The Chemistry of Oxidation–Reduction Reactions

A simple battery made from a piece of Zn and a piece of graphite in aqueous NH_4Cl. It produces a potential of 1.1 V.

CHAPTER OUTLINE

You learned in Chapter 5 that the transfer of an electron from one compound to another results in the oxidation of the electron donor and reduction of the electron acceptor.

Loss of electrons (oxidized)

$$Zn(s) + Cu^{2+}(aq) \longrightarrow Zn^{2+}(aq) + Cu(s)$$

Gain of electrons (reduced)

Hence, reactions involving electron transfer are often called oxidation–reduction or redox reactions. Our goal in this chapter is to investigate some practical consequences of electron transfer.

Electron transfer reactions are so numerous that you experience such reactions and their consequences daily. Let's look at a few examples.

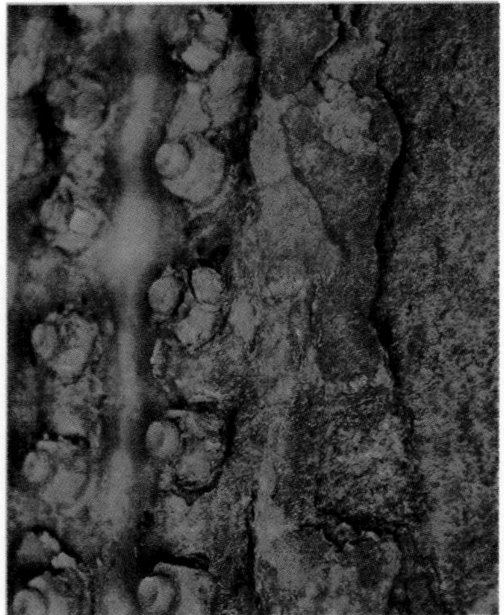

Figure 21.1 Rust on a bridge in New York City. This is not only an excellent example of corrosion but also an illustration of the problem of aging facilities in the United States.

(a) **Corrosion** occurs by oxidation–reduction reactions. In some areas salt is spread on the roads to melt ice and snow. Although accidents are prevented, the salt accelerates the formation of rust on car bodies, bridges, and other steel structures. Near the ocean, the salt suspended in the air assists in corroding metal objects. In each of these cases the oxidation of the metal (usually iron or aluminum) to its oxide is central to the corrosion process (Figure 21.1).

(b) Many biological processes depend on the transfer of electrons. For example, when you breathe air, oxygen is converted ultimately to water and carbon dioxide. The oxidation number of the oxygen in the product molecules (H_2O and CO_2) is -2, so electrons must have been transferred to O_2 to cause its reduction. Where did the electrons come from? At least in the final step they are transferred to O_2 from hemoglobin, a large iron-containing molecule. Other biological oxidation–reduction processes include the conversion of water to O_2 in green plants by photosynthesis (Figure 21.2), and the conversion of atmospheric N_2 to a usable form of nitrogen, such as NH_4^+, by bacteria.

$$N_2(g) + 8\ H_3O^+(aq) + 6\ e^- \longrightarrow 2\ NH_4^+(aq) + 8\ H_2O(\ell)$$

(c) The electrical power to start your car or drive your calculator comes from a battery. A **battery** is an **electrochemical cell** or a collection of such cells that produces a current or flow of electrons at a constant voltage as a result of an oxidation–reduction reaction.

(d) Many electron transfer processes are important in industry. For example, metals such as sodium, aluminum, and copper are commercially prepared or purified by the direct application of electricity in a process called **electrolysis**,

$$Na^+(\text{molten salt}) + e^- \longrightarrow Na(s)$$

$$Cu^{2+}(aq) + 2\ e^- \longrightarrow Cu(s)$$

while other metals, such as iron, are prepared by using chemical reducing agents.

$$Fe_2O_3(s) + 3\ CO(g) \longrightarrow 2\ Fe(s) + 3\ CO_2(g)$$

Chemists have only recently begun to understand the way in which an electron is transferred from one site to another. Moreover, the process

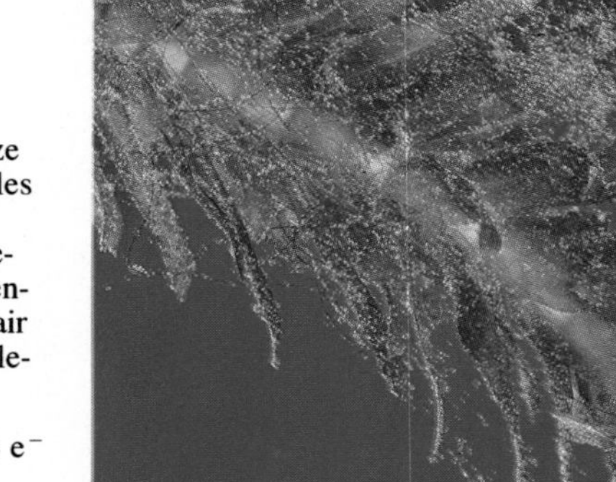

Figure 21.2 Green plants can oxidize water to O_2. This photo shows bubbles of O_2 on the leaves of a plant submerged in water. A chlorophyll molecule absorbs photons to supply the energy to strip four electrons from a pair of water molecules, to leave one molecule of O_2 and four hydrogen ions.

$$2\ H_2O(\ell) \longrightarrow 4\ H^+(aq) + O_2(g) + 4\ e^-$$

(E.R. Degginger)

of corrosion and its prevention, the construction of more powerful batteries, and the plating of metals using electricity are now better understood. The general subject is fascinating since it applies to so many problems of practical interest.

To organize your study, we shall first turn to electrochemical cells in which an electric current is produced by a chemical reaction. Such reactions are of course the basis of batteries, storage cells, and other devices from which we derive electrical energy, and cells of this type are sometimes called **voltaic cells** or **galvanic cells**. These names come from Alessandro Volta (1745–1827), who studied current-producing reactions, and Luigi Galvani (1737–1798), who made early studies of animal electricity and electricity produced chemically. The opposite of cells of this type are electrolysis cells, in which electrical energy is used to produce a chemical change.

Before turning to your study of electron transfer reactions, you should review Chapter 5. There you learned to recognize and balance redox equations and to understand the following terms:

Oxidation: An atom, ion, or molecule *releases electrons* and is oxidized. The oxidation number of an element increases.
Reduction: An atom, ion, or molecule *gains electrons* and is reduced. The oxidation number of an element decreases.
Reducing agent: The atom, ion, or molecule providing electrons.
Oxidizing agent: The atom, ion, or molecule accepting electrons.

The following photos of redox reactions are found earlier in the book:

Al/Cu^{2+}	*(Figure 1.8)*
P/Br_2	*(Figure 3.1)*
Zn/O_2	*(Figure 3.8)*
S/O_2	*(Figure 4.1)*
Zn/I_2	*(Figure 4.2)*
Al/Br_2	*(Figure 4.3)*
Fe/O_2	*(Figure 4.6)*
Cu/HNO_3	*(Figure 4.7)*
Mg/O_2	*(Figure 4.10)*
Mg/HCl	*(Figure 5.1)*
Zn/Cu^{2+}	*(Figure 5.23)*
Cu/Ag^+	*(Figure 5.24)*

21.1 CHEMICAL CHANGE LEADING TO ELECTRIC CURRENT

The type of reaction capable of producing an electric current, or flow of electrons, is illustrated in Figure 21.3. A piece of zinc is immersed in an aqueous solution of copper(II) sulfate. After a time, the blue color of the aqueous Cu^{2+} ion begins to fade, the edges of the zinc plate are eaten away, and copper begins to "plate out" or form a covering on the zinc strip. After still more time, the zinc strip disappears, copper is piled up on the bottom of the container, and the color of the copper ion fades still more (see Figure 5.23). What is happening?

The fate of the Cu^{2+} ion in the beaker in Figure 21.3 is probably obvious from the fact that the blue color of the aqueous ion has faded

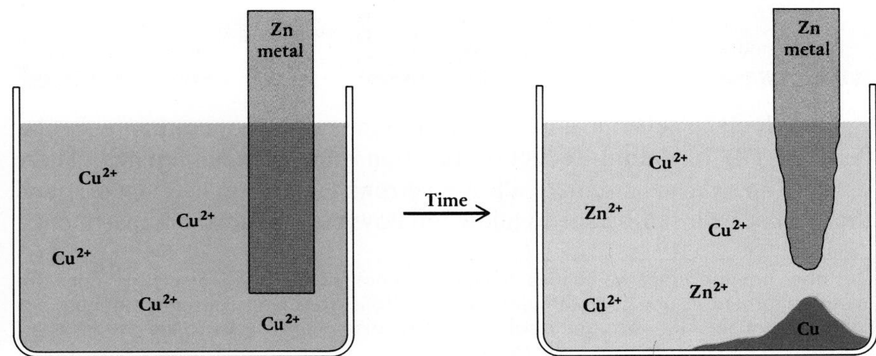

Figure 21.3 An oxidation–reduction reaction. A Zn plate is gradually "eaten away" as zinc reduces aqueous copper(II) ions to copper metal. See also Figures 5.23 and 21.4.

and a copper coating has appeared on the zinc strip: the copper ion has been reduced to the metal.

$$Cu^{2+}(aq) + 2\ e^- \longrightarrow Cu(s)$$

But what about the zinc? Our observation was that it disappeared, so it must have been the source of the electrons that caused reduction of Cu^{2+}; that is, the zinc was the reducing agent and formed aqueous Zn^{2+}.

$$Zn(s) \longrightarrow Zn^{2+}(aq) + 2\ e^-$$

The net chemical reaction occurring in the beaker, therefore, was the spontaneous reduction of Cu^{2+} and the simultaneous oxidation of Zn(s)

$$Cu^{2+}(aq) + Zn(s) \longrightarrow Cu(s) + Zn^{2+}(aq)$$

This reaction is interesting, but it could not be used as a source of current in the physical arrangement in Figures 21.3 or 5.23. The electrons provided by the zinc moved directly to the aqueous Cu^{2+} ions on contact. In order to use the Cu^{2+}/Zn reaction as the basis of a battery, the zinc must be placed in a container separate from the Cu^{2+} ions (Figure 21.4). Electrons can then pass from the zinc **electrode** to the solution of Cu^{2+} ions only through an external wire (and through the device to be powered) to another electrode (a piece of copper in this case) dipping into the Cu^{2+}(aq) solution. Copper metal would then be "plated out" onto the electrode in the beaker containing Cu^{2+}(aq).

An electrode conducts electrons into and out of a solution. It is most often a metal plate or wire or a piece of graphite.

The arrangement we have just described will work *only* if we have provided a **salt bridge**, a device for maintaining a balance of ion charges in the cell compartments. When the Zn electrode provides electrons to the wire, Zn^{2+}(aq) enters the solution in the Zn compartment (Figure 21.4), and negative ions must be found to balance these newly generated positive charges. Similarly, the loss of Cu^{2+} ions in the copper compartment leaves behind negative ions that were associated with Cu^{2+}. Some way must be found so that these negative ions, now in excess, can leave the solution. Thus, to achieve a balance of ion charges in each compartment, the negative ion concentration must *decrease* in the copper compartment and *increase* in the zinc compartment. The function of the salt bridge is to allow anions to pass freely from the compartment where cations are being lost to the compartment where cations are being generated. The salt bridge could be simply an aqueous solution of Na_2SO_4, allowing $SO_4{}^{2-}$ ions to transfer.*

You can keep track of the movement of electrons and ions in an electrochemical cell this way: The movement of negative charges (both electrons and anions) is in a "circle." As shown in Figure 21.4, electrons move through the external circuit from reducing agent (Zn) to oxidizing agent (Cu^{2+}); the negative ions complete the circle through the salt bridge (from Cu^{2+} to Zn^{2+}).

SO_4^{2-} in Cu^{2+} solution → SO_4^{2-} in salt bridge → SO_4^{2-} in Zn^{2+} solution

All electrochemical cells that produce current operate in a similar fashion. The oxidation–reduction reaction must be spontaneous. There must be an external circuit to allow electrons to perform useful work, and there must be a salt bridge to allow ion flow between cell compartments.

*This does not imply that the cations in the cell do not move as well. Since Zn^{2+} ions are being generated in one compartment and Cu^{2+} ions are being consumed in the other compartment, the Na^+ ions of the salt bridge can also migrate toward the copper compartment to maintain charge balance.

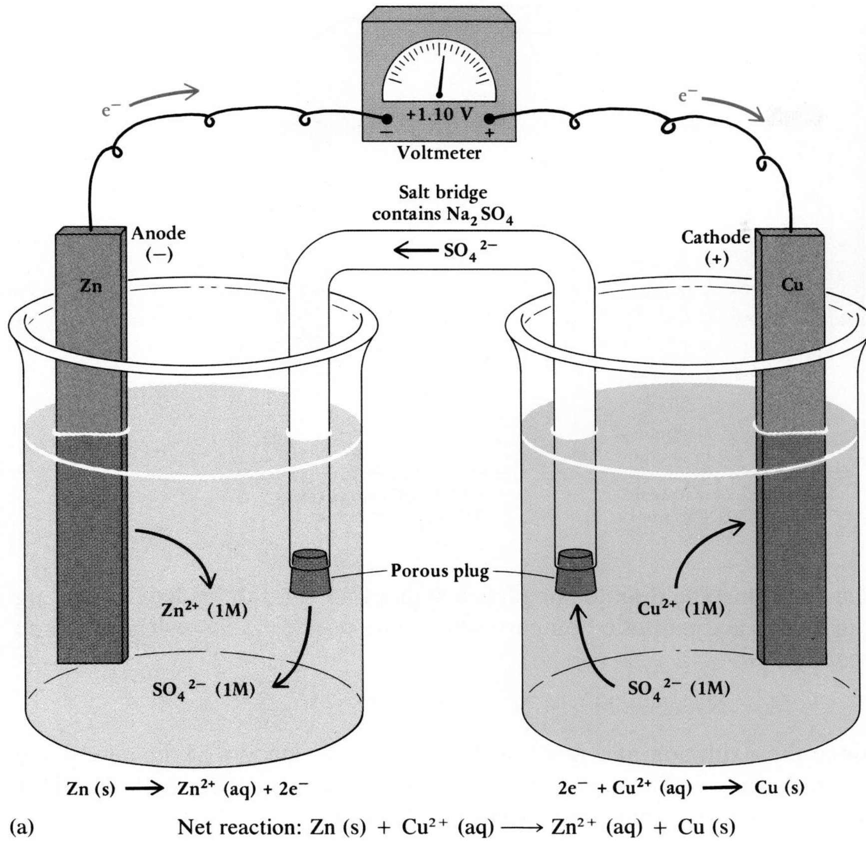

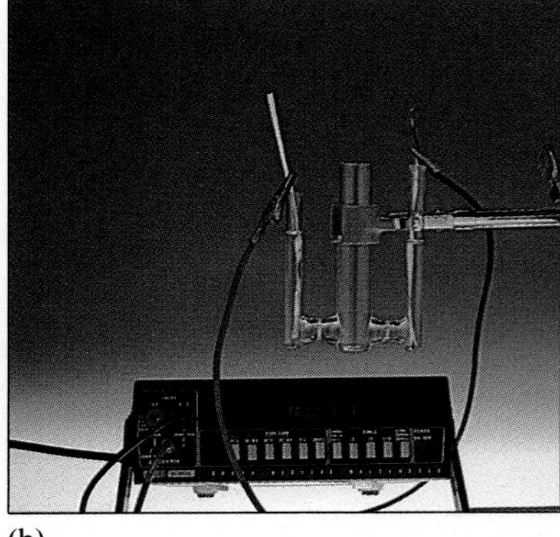

(b)

Figure 21.4 A voltaic cell using $Cu^{2+}(aq)/Cu(s)$ and $Zn^{2+}(aq)/Zn(s)$ half-cells. (a) A potential of 1.10 V is generated if the cell is set up under the conditions shown. Electrons flow through the external wire from the Zn electrode (anode) to the Cu electrode (cathode). A salt bridge provides a connection between the half-cells for ion flow; thus, SO_4^{2-} ions flow from the copper to the zinc compartment. (b) An actual cell operating under nearly standard conditions. The negative zinc electrode is at the left, and the positive copper electrode is at the right. The compartments are separated from one another by porous glass disks. The center compartment is the salt bridge that contains Na_2SO_4.

There is some additional terminology peculiar to electrochemical cells that you must know. **Oxidation** *always occurs at the* **anode** *of an electrochemical cell and* **reduction** *always occurs at the* **cathode**. Furthermore, since the electrons are produced at the anode of a cell, this terminal bears a *negative* charge. This means the other terminal in the cell, the cathode, bears a *positive* charge. To depict cells in an abbreviated way, the following notation is often used,

*To remember that **A**node and **O**xidation are paired, as are **C**athode and **R**eduction, note the alphabetic orders:*

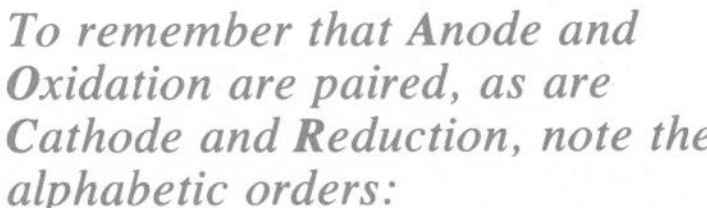

Anode ↔ Oxidation
↑ ↑
(A before C) (O before R)
↓ ↓
Cathode ↔ Reduction

$$\underbrace{\text{M(electrode)|M}^+\text{(solution)}}_{\text{anode compartment}}\|\underbrace{\text{N}^+\text{(solution)|N(electrode)}}_{\text{cathode compartment}}$$

where M and N are general terms for elements and their ions, and the single line (|) indicates a phase boundary between a solid electrode and

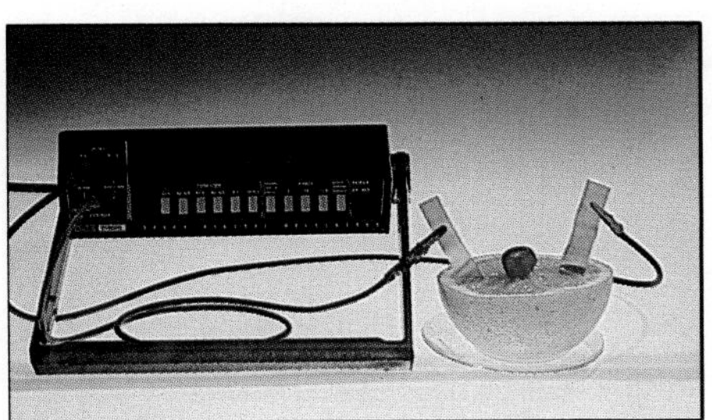

A voltaic cell made by inserting zinc and copper electrodes into a grapefruit. A potential of over 0.9 V is obtained. (The water and citric acid of the fruit allow for ion conduction between electrodes.)

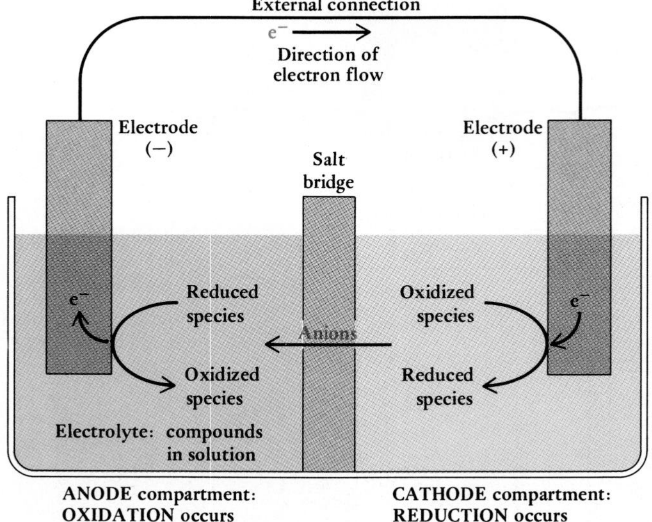

Figure 21.5 A summary of the terminology used in voltaic cells. Notice that negative charge moves in a circle through the cell and external wire. Electrons move from the negative electrode (anode) to the positive electrode (cathode) in the external wire, and anions move from the cathode compartment to the anode compartment in the cell.

the solution. The double line (||) is a symbol for the salt bridge connecting the anode and cathode compartments. The notation tells us that the cell reaction is

$$M + N^+ \longrightarrow M^+ + N$$

since the oxidation half-reaction $M \rightarrow M^+ + e^-$ occurs at the anode and the reduction half-reaction $N^+ + e^- \rightarrow N$ occurs at the cathode. This means the copper/zinc cell would be given by

$$Zn|Zn^{2+}(aq)||Cu^{2+}(aq)|Cu$$

Important terms are summarized in Figure 21.5.

EXAMPLE 21.1

ELECTROCHEMICAL CELLS

A simple voltaic cell has been assembled with Ni(s) and $Ni(NO_3)_2$(aq) in one compartment and Cd(s) and $Cd(NO_3)_2$(aq) in the other. An external wire connects the two electrodes, and a salt bridge containing $NaNO_3$ connects the two solutions. The net reaction is

$$Ni^{2+}(aq) + Cd(s) \longrightarrow Ni(s) + Cd^{2+}(aq)$$

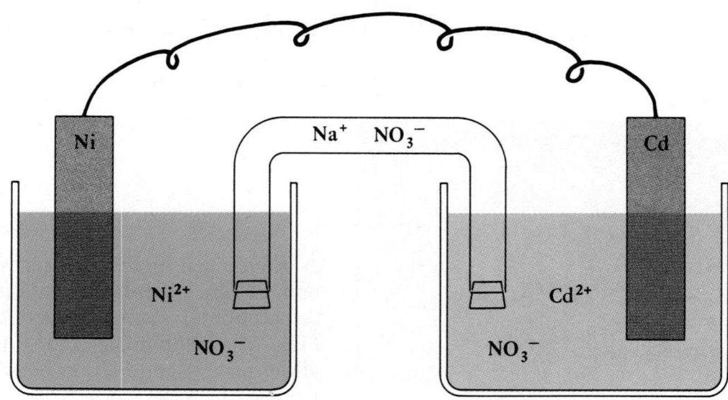

What half-reaction occurs at each electrode? Which is the anode, and which is the cathode? What is the direction of electron flow in the external wire and of anion flow in the salt bridge?

Solution The net equation tells you that Cd(s) is the reducing agent and Ni^{2+}(aq) is the oxidizing agent. Thus, the half-reactions are

Oxidation at the anode:	$Cd(s) \longrightarrow Cd^{2+}(aq) + 2\ e^-$
Reduction at the cathode:	$Ni^{2+}(aq) + 2\ e^- \longrightarrow Ni(s)$

Electrons flow from their source (the oxidation of cadmium at the Cd electrode or anode) through the wire to the electrode where they are used (the Ni electrode or cathode).

Since Cd^{2+} ions are being formed in the anode compartment, anions must move into that compartment from the salt bridge. The Ni^{2+} concentration in the cathode compartment is being depleted, so anions move out of this compartment into the salt bridge. The "circle" of flow of negative charge is complete: electrons flow from Cd to Ni and anions from Ni to Cd. Finally, the cell would have the notation $Cd|Cd^{2+}(aq)||Ni^{2+}(aq)|Ni$.

EXERCISE 21.1 Electrochemical Cells

A voltaic cell has been assembled with the net reaction

$$Ni(s) + 2\ Ag^+(aq) \longrightarrow Ni^{2+}(aq) + 2\ Ag(s)$$

Give the half-reactions for this redox process, indicate whether each is an oxidation or reduction, and tell which happens at the anode and which at the cathode. What is the direction of electron flow in an external wire connecting the two electrodes? If a salt bridge connecting the cell compartments contains KNO_3, what is the direction of flow of the nitrate ions? Show the abbreviated notation for the cell.

21.2 ELECTROCHEMICAL CELLS AND POTENTIALS

Electrons generated at the site of oxidation, the anode, of a cell are thought to be "driven" or "pushed" toward the cathode by an **electromotive force** or **emf**. This force is due to the difference in energy of an electron at the two electrodes, and it is this energy in which we are interested. If the energy difference for one coulomb of charge is 1 joule, we say that the **cell potential**, *E*, is 1 volt.

$$1 \text{ volt (V)} = 1 \text{ joule/coulomb} = 1 \text{ J/C}$$

The potential is a readily measured characteristic of an electrochemical cell, and its value can be predicted under well defined conditions. When the potential is measured with all of the reactants and products present as pure solid or in solution at a concentration of 1.0 M, or with gases at 1.0 atm, the conditions are standard conditions and the measured potential is the standard potential, $E°$. This potential, $E°$, is a quantitative measure of the tendency of the reactants in their standard states to proceed to products in their standard states. As an example of a cell at standard conditions, consider the cell in Figure 21.4. This cell will deliver 1.10 V at 25 °C.

Unless specified otherwise, all values of $E°$ are given at 25 °C.

$E°$ AND $\Delta G°$

The standard cell potential is clearly a measure of reaction spontaneity, but so is the standard free energy change for a reaction, $\Delta G°$. The exact relation between them is

$$\Delta G° = -nFE° \quad (21.1)$$

where n is the number of moles of electrons transferred between oxidizing and reducing agents in a balanced redox equation and F is the Faraday constant, 9.6485309×10^4 J/V · mol.

The Faraday constant is named in honor of Michael Faraday, the first person to investigate quantitatively the relation between chemistry and electricity. See Chapter 5 and Section 21.6. The Faraday factor has units derived from fundamental quantities: J/V · mol or C/mol. The former is appropriate for the energy conversion $E° \leftrightarrow \Delta G°$, and the latter is used for the quantity of charge.

The reaction of Zn(s) and Cu^{2+}(aq) produces a current, so you readily conclude that it is a spontaneous reaction as written.

$$Cu^{2+}(aq) + Zn(s) \longrightarrow Cu(s) + Zn^{2+}(aq) \qquad E° = +1.10\text{ V}$$

Since spontaneous reactions have a negative free energy change (Chapter 20), this is the reason for the negative sign in the relation between free energy and cell potential. Thus, *all spontaneous electron transfer reactions have a positive $E°$.*

If the direction of the reaction is reversed, the sign of $\Delta G°$ and so the sign of $E°$ are reversed. Thus, if we write the equation for the reduction of Zn^{2+} by Cu,

$$Cu(s) + Zn^{2+}(aq) \longrightarrow Cu^{2+}(aq) + Zn(s) \qquad E° = -1.10\text{ V}$$

this reaction is *not* spontaneous and must have a positive $\Delta G°$ and a negative $E°$. *Nonspontaneous reactions have a negative $E°$.*

When a reaction is reversed, the magnitudes of $\Delta G°$ and $E°$ remain the same but their signs are reversed [$(+ \rightarrow -)$ or $(- \rightarrow +)$].

The relation between the potential of an electrochemical cell and the free energy for the reaction is important. After looking into the matter of cell potentials in more detail, we shall return to the thermodynamic relationship later in this section and in Section 21.3. For the moment, the calculation of $\Delta G°$ from $E°$ is illustrated in Example 21.2.

EXAMPLE 21.2

THE RELATION BETWEEN $E°$ AND $\Delta G°$

The Cu^{2+}(aq)/Zn(s) reaction above has a standard cell potential $E°$ of +1.10 V at 25 °C. Calculate $\Delta G°$ for the reaction.

$$Cu^{2+}(aq) + Zn(s) \longrightarrow Cu(s) + Zn^{2+}(aq)$$

Solution To obtain $\Delta G°$ we need only to substitute into Equation 21.1.

$$\Delta G° = -(2.00\text{ mol electrons transferred})\left(\frac{9.65 \times 10^4\text{ J}}{\text{V} \cdot \text{mol}}\right)(1.10\text{ V})\left(\frac{1\text{ kJ}}{10^3\text{ J}}\right)$$

$$= -212\text{ kJ}$$

EXERCISE 21.2 The Relation Between $E°$ and $\Delta G°$

The following reaction has an $E°$ value of −0.76 V. Calculate $\Delta G°$ and tell whether the reaction is spontaneous as written.

$$Zn^{2+}(aq) + H_2(g) + 2\ H_2O(\ell) \longrightarrow Zn(s) + 2\ H_3O^+(aq)$$

CALCULATING THE POTENTIAL, *E*°, OF AN ELECTROCHEMICAL CELL

As you learned in Chapter 5, an oxidation–reduction reaction is the sum of two half-reactions, one for oxidation and the other for reduction. For the $Zn(s)/Cu^{2+}(aq)$ reaction we have been considering, this means the overall process is the sum of the reactions

Oxidation occurs; reducing agent, Zn:	$Zn(s) \longrightarrow Zn^{2+}(aq) + 2\ e^-$
Reduction occurs; oxidizing agent, Cu^{2+}:	$Cu^{2+}(aq) + 2\ e^- \longrightarrow Cu(s)$
Net oxidation–reduction process:	$Cu^{2+}(aq) + Zn(s) \longrightarrow Zn^{2+}(aq) + Cu(s)$

Since each redox reaction is the sum of two half-reactions, it would be very helpful if we could assign a potential to each half-reaction. Summing the half-reaction potentials would then give the cell potential. The problem is that potentials for isolated half-reactions cannot be obtained directly, because the cell potential measures the potential energy difference for electrons in two different chemical environments. However, we can always measure the standard electrical potential for any half-reaction in combination with some other, standard half-reaction. Indeed, it has been decided by the chemistry community that the standard half-cell reaction against which all others are measured is the **standard hydrogen electrode (S.H.E.)**,

$$2\ H_3O^+(aq,\ 1\ M) + 2\ e^- \longrightarrow H_2(g,\ 1\ atm) + 2\ H_2O(\ell)$$

and *a potential of exactly 0.00 V has been assigned to this half-reaction.* (This value has no physical meaning in itself, just as the half-reaction alone has no meaning.) To measure the potential for any half-reaction, we make that reaction one side of an electrochemical cell and the H_2/H_3O^+ half-reaction the other side. H_2 is a reducing agent, and $H_3O^+(aq)$ is an oxidizing agent. Thus, when the standard hydrogen electrode is paired with another half-cell in an electrochemical cell, the H_2/H_3O^+ half-reaction can be either an oxidation or a reduction. If the other half-cell contains a better reducing agent than H_2, then H_3O^+ is reduced to H_2.

H_3O^+ reduced: $$2\ H_3O^+(aq,\ 1\ M) + 2\ e^- \longrightarrow H_2(g,\ 1\ atm) + 2\ H_2O(\ell) \qquad E° = 0.00\ V$$

If the other half-cell contains a better oxidizing agent than H_3O^+, then H_2 is oxidized.

H_2 oxidized: $$H_2(g,\ 1\ atm) + 2\ H_2O(\ell) \longrightarrow 2\ H_3O^+(aq,\ 1\ M) + 2\ e^- \qquad E° = 0.00\ V$$

In either direction, the H_2/H_3O^+ half-cell has a potential of 0.00 V. The measured potential of the electrochemical cell is then *assigned* as the potential of the half-cell being studied. To demonstrate the strategy for determining half-cell potentials, consider the following examples.

In Figure 21.6 we have diagrammed a cell in which one compartment contains the H_2/H_3O^+ reaction mixture, and the other compartment has a zinc electrode dipping into a solution of 1 M Zn^{2+}. As usual, the compartments are connected by an external wire (for electron flow) and a salt bridge (for ion flow).

When a redox reaction involves an ion or compound that cannot be made into solid electrodes, a chemically inert conductor of electricity can be used. Platinum and gold are frequently used as such inert electrodes.

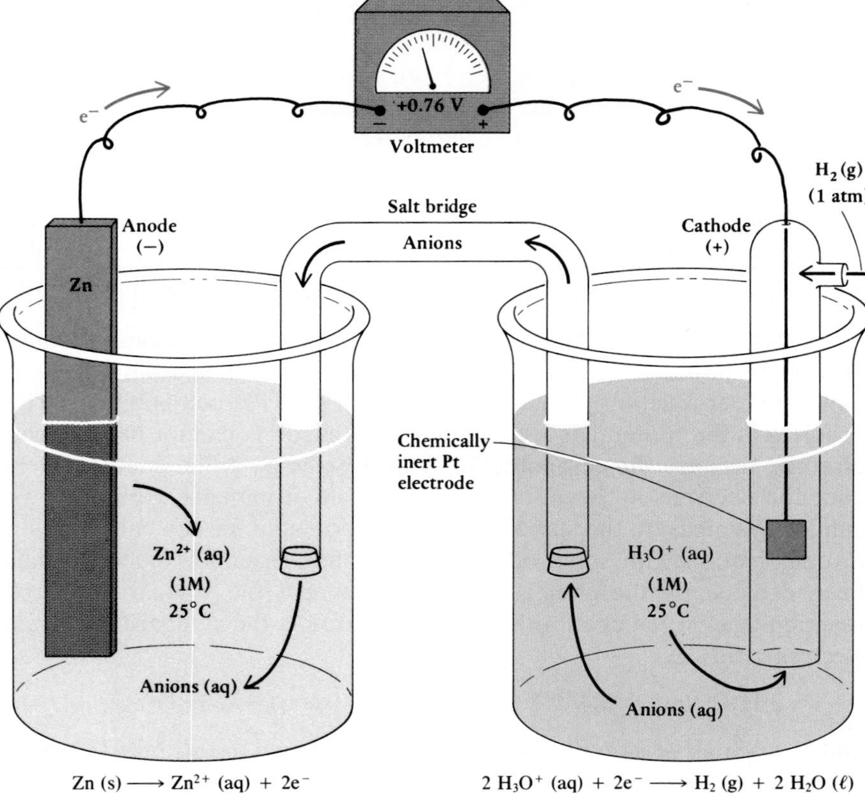

Figure 21.6 An electrochemical cell using $Zn^{2+}(aq)/Zn(s)$ and $H_3O^+(aq)/H_2(g)$ half-cells. A voltage of +0.76 V is generated when the cell is set up under the conditions shown at 25 °C. Electrons flow from the Zn electrode (anode) to the $H_3O^+(aq)/H_2(g)$ electrode (cathode) to produce $Zn^{2+}(aq)$ and $H_2(g)$. Zinc is the reducing agent and $H_3O^+(aq)$ is the oxidizing agent. (Since the species in the $H_3O^+(aq)/H_2(g)$ half-reaction cannot be fashioned into a solid electrode, electrons are transferred through a piece of platinum foil.)

The measured potential of an electrochemical cell is always positive. The device used to measure *potential differences*, a voltmeter, is designed to give a positive potential only when the positive terminal (+) of the voltmeter is connected to the positive electrode and the negative terminal (−) to the negative electrode. In this way we not only measure the potential difference but we find the sign of each of the electrodes. Thus, when a voltmeter is attached to the cell shown in Figure 21.6, it is found that the H_2/H_3O^+ electrode is positive, the zinc electrode is negative, and the measured potential is $E° = +0.76$ V. Although both Zn and H_2 can potentially function as reducing agents, the observation here that the Zn electrode is negative means that Zn is the source of electrons, and so we know that *Zn is a better reducing agent than H_2 gas*. Therefore, it is observed that the zinc electrode dissolves owing to the oxidation of metallic zinc, and H_2 gas is formed from the reduction of H_3O^+.

Anode, oxidation:

$$Zn(s) \longrightarrow Zn^{2+}(aq,\ 1\ M) + 2\ e^- \qquad E° = ?\ V$$

Cathode, reduction:

$$2\ H_3O^+(aq,\ 1\ M) + 2\ e^- \longrightarrow H_2(g,\ 1\ atm) + 2\ H_2O(\ell) \qquad E° = 0.00\ V$$

Net reaction:

$$Zn(s) + 2\ H_3O^+(aq,\ 1\ M) \longrightarrow Zn^{2+}(aq,\ 1\ M) + H_2(g,\ 1\ atm) + 2\ H_2O(\ell) \qquad E°_{net} = +0.76\ V$$

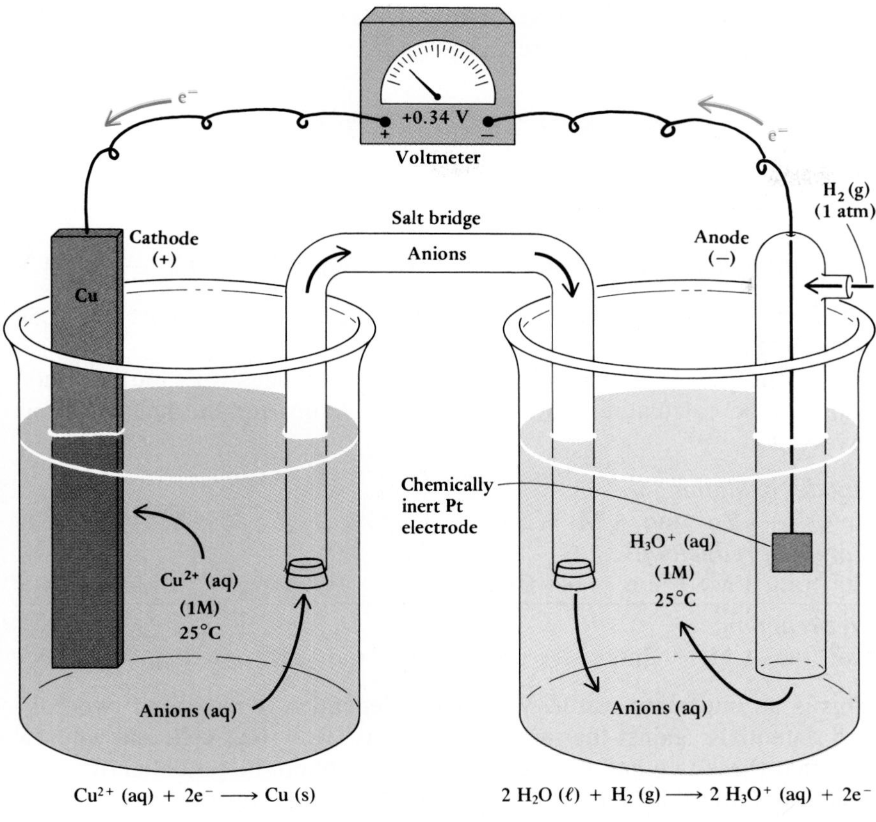

Cu^{2+} (aq) + $2e^-$ ⟶ Cu (s) $2\ H_2O$ (ℓ) + H_2 (g) ⟶ $2\ H_3O^+$ (aq) + $2e^-$

Net reaction: $2\ H_2O$ (ℓ) + H_2 (g) + Cu^{2+} (aq) ⟶ $2\ H_3O^+$ (aq) + Cu (s)

Figure 21.7 An electrochemical cell using Cu^{2+}(aq)/Cu(s) and H_3O^+(aq)/H_2(g) half-cells. A voltage of +0.34 V is generated when the cell is set up as pictured at 25 °C. Electrons flow from the H_3O^+(aq)/H_2(g) electrode (the anode) to the copper electrode (cathode) to produce copper and hydronium ions. Therefore, H_2(g) is the reducing agent, and Cu^{2+}(aq) is the oxidizing agent.

The + sign of E°_{net} correctly reflects the fact that the overall reaction is spontaneous as written. Since the potential for the H_2/H_3O^+ half-cell is 0.00 V, this must mean that E° for the Zn/Zn^{2+} half-cell is +0.76 V.

The potential produced by an electrochemical cell is the sum of the potentials of the oxidizing half-reaction and the reducing half-reaction.

$$Zn(s) \longrightarrow Zn^{2+}(aq,\ 1\ M) + 2\ e^- \qquad E^\circ = +0.76\ V$$

What is E° for the Cu/Cu^{2+} half-cell in our original electrochemical cell in Figure 21.4? In Figure 21.7 this half-cell is shown coupled with the H_2/H_3O^+ half-cell. The measured cell potential is +0.34 V, the H_2/H_3O^+ half-cell is negative, and the Cu/Cu^{2+} electrode is positive. Furthermore, the concentration of Cu^{2+} ions declines, and metallic copper forms. All these experimental observations tell us that Cu^{2+} is being reduced and that H_2 must be the reducing agent, the source of electrons. Since the reducing agent H_2 is oxidized to H_3O^+, this must mean that the H_2 electrode is the anode (and is negatively charged). Further, since Cu^{2+} ions are the acceptors of electrons (the oxidizing agent), this electrode is the cathode (and is positively charged). Most importantly, we now see that

$H_2(g)$ is a better reducing agent than Cu(s), and the appropriate half-reactions and net reaction are

Anode, oxidation:
$H_2(g, 1\ atm) + 2\ H_2O(\ell) \longrightarrow 2\ H_3O^+(aq, 1\ M) + 2\ e^-$ $\qquad E° = 0.00\ V$
Cathode, reduction:
$Cu^{2+}(aq, 1\ M) + 2\ e^- \longrightarrow Cu(s)$ $\qquad E° = ?\ V$

Net reaction:
$Cu^{2+}(aq, 1\ M) + H_2(g, 1\ atm) + 2\ H_2O(\ell) \longrightarrow Cu(s) + 2\ H_3O^+(aq, 1\ M)$ $\qquad E°_{net} = +0.34\ V$

The half-cell potential for $Cu^{2+}(aq, 1\ M) + 2\ e^- \rightarrow Cu(s)$ must be +0.34 V at 25 °C.

The potential of a cell in which Zn(s) reduces $Cu^{2+}(aq)$ to Cu(s) can now be calculated, since we have $E°$ values for the half-reactions involved.

Anode, oxidation:
$Zn(s) \longrightarrow Zn^{2+}(aq, 1\ M) + 2\ e^-$ $\qquad E° = +0.76\ V$
Cathode, reduction:
$Cu^{2+}(aq, 1\ M) + 2\ e^- \longrightarrow Cu(s)$ $\qquad E° = +0.34\ V$

Net reaction:
$Cu^{2+}(aq, 1\ M) + Zn(s) \longrightarrow Cu(s) + Zn^{2+}(aq, 1\ M)$ $\qquad E°_{net} = +1.10\ V$

This is an important result. We have independently measured two half-cell potentials against the same standard H_2/H_3O^+ half-cell, and we now find that these two half-cells can be used to obtain the potential of a new reaction. (And this is in fact the potential observed for this reaction; see Figure 21.4.) This was our original objective. Now we could use a similar technique, with any of the Cu/Cu^{2+}, Zn/Zn^{2+}, or H_2/H_3O^+ half-cells as reference cells, to determine $E°$ values for hundreds of other possible half-cells (see Example 21.3).

EXAMPLE 21.3

DETERMINING A HALF-REACTION POTENTIAL

The cell illustrated below generates a potential of $E° = +0.51$ V under standard conditions at 25 °C, and the net cell reaction is

$$Zn(s) + Ni^{2+}(aq, 1\ M) \longrightarrow Zn^{2+}(aq, 1\ M) + Ni(s)$$

Tell which electrode is the anode and which is the cathode, give the signs of the electrodes, and calculate the half-cell potential for $Ni^{2+}(aq) + 2\ e^- \rightarrow Ni(s)$.

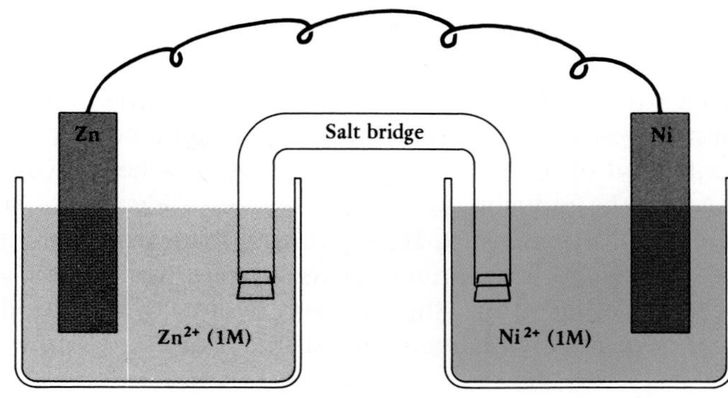

Solution The electrode at which oxidation occurs is the anode (and, as this is the source of electrons, it is negative). Since Zn(s) is oxidized to Zn^{2+}(aq), the Zn electrode is the anode. Nickel(II) ions are reduced at the Ni electrode, so Ni metal is the positive cathode.

Since the overall cell potential is known, and the potential for the Zn(s)/Zn^{2+}(aq, 1 M) half-cell is known, $E°$ for Ni^{2+}(aq, 1 M) + 2 e^- → Ni(s) can be calculated.

Anode, oxidation:

$Zn(s) \longrightarrow Zn^{2+}(aq) + 2\,e^-$ $\qquad E° = +0.76$ V

Cathode, reduction:

$Ni^{2+}(aq) + 2\,e^- \longrightarrow Ni(s)$ $\qquad E° = ?$ V

Net reaction:

$Zn(s) + Ni^{2+}(aq) \longrightarrow Zn^{2+}(aq) + Ni(s)$ $\qquad E°_{net} = +0.51$ V

At 25 °C, the value of $E°$ for the Ni^{2+} (aq, 1 M) + 2 e^- → Ni(s) half-reaction is −0.25 V.

EXERCISE 21.3 Determining a Half-Reaction Potential

Given that the reduction of aqueous copper(II) with iron metal has an $E°$ value of +0.78 V, what is $E°$ for the half-cell Fe(s) → Fe^{2+}(aq, 1 M) + 2 e^-?

$$Fe(s) + Cu^{2+}(aq, 1\,M) \longrightarrow Fe^{2+}(aq, 1\,M) + Cu(s)$$

USING STANDARD POTENTIALS

In Example 21.3, we found that $E°$ for the reduction of Ni^{2+} to metallic nickel is −0.25 V. What does this mean? This tells you that if Ni^{2+} is used as an oxidizing agent, coupled with H_2 as the reducing agent,

Cathode, reduction:

$Ni^{2+}(aq, 1\,M) + 2\,e^- \longrightarrow Ni(s)$ $\qquad E° = -0.25$ V

Anode, oxidation:

$H_2(g, 1\,atm) + 2\,H_2O(\ell) \longrightarrow 2\,H_3O^+(aq, 1\,M) + 2\,e^-$ $\qquad E° = 0.00$ V

Net reaction:

$Ni^{2+}(aq, 1\,M) + H_2(g, 1\,atm) + 2\,H_2O(\ell) \longrightarrow Ni(s) + 2\,H_3O^+(aq, 1\,M)$ $\qquad E°_{net} = -0.25$ V

the reaction is *not spontaneous* under standard conditions; $\Delta G°$ is positive so $E°_{net}$ is negative. From thermodynamics, you know that if a reaction is not spontaneous in the direction written ($\Delta G°$ is positive), the reverse reaction is spontaneous ($\Delta G°$ is negative). Therefore, the reaction

$$Ni(s) + 2\,H_3O^+(aq, 1\,M) \longrightarrow Ni^{2+}(aq, 1\,M) + H_2(g, 1\,atm) + 2\,H_2O(\ell) \qquad E°_{net} = +0.25\text{ V}$$

is spontaneous, and $E°_{net}$ for the reaction is positive. Just as acid–base reactions always move in the direction of the weaker acid–base pair (Section 17.2), redox reactions move toward the weaker oxidizing agent/reducing agent pair. In the Ni/H_3O^+ reaction, there are two possible reducing agents, Ni(s) and H_2. Since nickel metal spontaneously reduces hydronium ion, nickel metal must be a better reducing agent than H_2, and H_3O^+ must be a better oxidizing agent than aqueous Ni^{2+}.

Thus far we have described four elements that could act as reducing agents: Zn, Cu, H_2, and Ni. What are their *relative* abilities to act in this manner? Conversely, what are the relative abilities of their ions (Zn^{2+}, Cu^{2+}, H_3O^+, and Ni^{2+}) to act as electron acceptors or oxidizing agents? If we write half-reactions involving these elements and their ions in the format

$$\text{Oxidized form} + n\,e^- \longrightarrow \text{Reduced form}$$

and list them in order of descending $E°$ values, we will have placed the oxidizing agents in descending order of their ability to attract electrons.

Increasing strength as oxidizing agent (↑) / Increasing strength as reducing agent (↓)

$Cu^{2+}(aq, 1\ M) + 2\,e^- \longrightarrow Cu(s)$	$E° = +0.34$ V	$\Delta G° = -66$ kJ
$2\,H_3O^+(aq, 1\ M) + 2\,e^- \longrightarrow H_2(g, 1\ atm) + 2\,H_2O(\ell)$	$E° = 0.00$ V	$\Delta G° = 0.0$ kJ
$Ni^{2+}(aq, 1\ M) + 2\,e^- \longrightarrow Ni(s)$	$E° = -0.25$ V	$\Delta G° = +48$ kJ
$Zn^{2+}(aq, 1\ M) + 2\,e^- \longrightarrow Zn(s)$	$E° = -0.76$ V	$\Delta G° = +150$ kJ

The value of $E°$ becomes more negative, and $\Delta G°$ becomes more positive, down the series. This means that $Cu^{2+}(aq)$ is the best oxidizing agent (most electron-attracting ion) of the substances on the left; that is, Cu^{2+} shows the greatest tendency to be reduced. Conversely, Zn^{2+} is the worst oxidizing agent; it is the least electron-attracting ion. Of the substances on the right, Zn(s) is the best reducing agent (best electron donor), since $E°$ for the half-reaction

$$Zn(s) \longrightarrow Zn^{2+}(aq, 1\ M) + 2\,e^- \qquad E° = +0.76 \qquad \Delta G° = -150\ \text{kJ}$$

has the most positive value. By the same reasoning, Cu is the worst reducing agent.

The table of half-reaction potentials above tells us that, at standard conditions, the following reactions are spontaneous:

Increasing oxidizing ability (↑)	Increasing reducing ability (↓)
Cu^{2+} can oxidize H_2, Ni, and Zn	H_2 can reduce Cu^{2+}
H_3O^+ can oxidize Ni and Zn	Ni can reduce H_3O^+ and Cu^{2+}
Ni^{2+} can oxidize Zn	Zn can reduce Ni^{2+}, H_3O^+, and Cu^{2+}

Each of these reactions will have a positive $E°_{net}$ and a negative $\Delta G°$. Many have been described earlier in the text, and the reduction of $Cu^{2+}(aq)$ with Ni further illustrates the point.

Oxidation:

$$Ni(s) \longrightarrow Ni^{2+}(aq) + 2\,e^- \qquad E° = +0.25\ \text{V}$$

Reduction:

$$Cu^{2+}(aq) + 2\,e^- \longrightarrow Cu(s) \qquad E° = +0.34\ \text{V}$$

Net reaction:

$$Ni(s) + Cu^{2+}(aq) \longrightarrow Ni^{2+}(aq) + Cu(s) \qquad E°_{net} = +0.59\ \text{V}$$

Many more values of $E°$ are given in Appendix I.

The spontaneity of this reaction is confirmed by the positive $E°$ (and negative $\Delta G°$) for the reaction.

We have just created a small portion of a **table of standard reduction potentials** (Table 21.1). Like the table of conjugate acids and bases in

Table 21.1 Standard Reduction Potentials in Aqueous Solution at 25 °C*

Reduction Half-Reaction		$E°$ (V)
$F_2(g) + 2\ e^-$	$\longrightarrow 2\ F^-(aq)$	+2.87
$H_2O_2(aq) + 2\ H_3O^+(aq) + 2\ e^-$	$\longrightarrow 4\ H_2O(\ell)$	+1.77
$PbO_2(s) + SO_4^{2-}(aq) + 4\ H_3O^+(aq) + 2\ e^-$	$\longrightarrow PbSO_4(s) + 6\ H_2O(\ell)$	+1.685
$Au^{3+}(aq) + 3\ e^-$	$\longrightarrow Au(s)$	+1.50
$Cl_2(g) + 2\ e^-$	$\longrightarrow 2\ Cl^-(aq)$	+1.360
$O_2(g) + 4\ H_3O^+(aq) + 4\ e^-$	$\longrightarrow 6\ H_2O(\ell)$	+1.229
$Br_2(\ell) + 2\ e^-$	$\longrightarrow 2\ Br^-(aq)$	+1.08
$Hg^{2+}(aq) + 2\ e^-$	$\longrightarrow Hg(\ell)$	+0.855
$Ag^+(aq) + e^-$	$\longrightarrow Ag(s)$	+0.80
$Hg_2^{2+}(aq) + 2\ e^-$	$\longrightarrow 2\ Hg(\ell)$	+0.789
$Fe^{3+}(aq) + e^-$	$\longrightarrow Fe^{2+}(aq)$	+0.771
$I_2(s) + 2\ e^-$	$\longrightarrow 2\ I^-(aq)$	+0.535
$O_2(g) + 2\ H_2O(\ell) + 4\ e^-$	$\longrightarrow 4\ OH^-(aq)$	+0.40
$Cu^{2+}(aq) + 2\ e^-$	$\longrightarrow Cu(s)$	+0.337
$Sn^{4+}(aq) + 2\ e^-$	$\longrightarrow Sn^{2+}(aq)$	+0.15
$2\ H_3O^+(aq) + 2\ e^-$	$\longrightarrow H_2(g) + 2\ H_2O(\ell)$	0.00
$Sn^{2+}(aq) + 2\ e^-$	$\longrightarrow Sn(s)$	−0.14
$Ni^{2+}(aq) + 2\ e^-$	$\longrightarrow Ni(s)$	−0.25
$PbSO_4(s) + 2\ e^-$	$\longrightarrow Pb(s) + SO_4^{2-}(aq)$	−0.356
$Cd^{2+}(aq) + 2\ e^-$	$\longrightarrow Cd(s)$	−0.40
$Fe^{2+}(aq) + 2\ e^-$	$\longrightarrow Fe(s)$	−0.44
$Zn^{2+}(aq) + 2\ e^-$	$\longrightarrow Zn(s)$	−0.763
$2\ H_2O(\ell) + 2\ e^-$	$\longrightarrow H_2(g) + 2\ OH^-(aq)$	−0.8277
$Al^{3+}(aq) + 3\ e^-$	$\longrightarrow Al(s)$	−1.66
$Mg^{2+}(aq) + 2\ e^-$	$\longrightarrow Mg(s)$	−2.37
$Na^+(aq) + e^-$	$\longrightarrow Na(s)$	−2.714
$K^+(aq) + e^-$	$\longrightarrow K(s)$	−2.925
$Li^+(aq) + e^-$	$\longrightarrow Li(s)$	−3.045

Increasing strength of oxidizing agents (↑, left side)

Increasing strength of reducing agents (↓, right side)

*In volts (V) versus the standard hydrogen electrode (S.H.E.).

Chapter 17, Table 21.1 is a *very* useful table. Some important points concerning this table are:

1. The $E°$ values are for reactions written in the form "oxidized form + electrons → reduced form."
2. When you write the reaction "reduced form → oxidized form + electrons," reverse the sign of $E°$. Thus,

$$2\ F^-(aq,\ 1\ M) \longrightarrow F_2(g,\ 1\ atm) + 2\ e^- \qquad E° = -2.87\ V$$

3. The more positive the value of $E°$ for a reaction in Table 21.1, the better the oxidizing ability of the element, ion, or compound. This means *$F_2(g)$ is the best oxidizing agent in the table,*

$$F_2(g,\ 1\ atm) + 2\ e^- \longrightarrow 2\ F^-(aq,\ 1\ M) \qquad E° = +2.87\ V$$

since the element has the most positive $E°$ of all the elements, compounds, or ions on the left. The ion at the bottom left corner of the table, $Li^+(aq)$, is the poorest oxidizing agent, since its $E°$ is the most negative. Conversely, F^- is the weakest reducing agent and Li is the strongest reducing agent. Under standard conditions, *the oxidizing agents* (ions, elements, and compounds at the left) *increase in strength from the bottom to the top of the table*. The *reducing agents* (the ions, elements, or compounds at the right) *increase in strength from the top to the bottom.*

4. All the half-reactions listed are reversible. A given substance can act as the anode or cathode, depending on the other half-reaction. For example, aqueous H_3O^+ is reduced to H_2(g) in Figure 21.6, whereas H_2(g) is oxidized to H_3O^+(aq) in Figure 21.7.
5. Under standard conditions, any substance on the left in this table (an oxidizing agent) will spontaneously oxidize any substance lower than it on the right (a reducing agent). This point was explored after developing the short table above.
6. The algebraic sign of the half-reaction potential is the sign of the electrode when it is attached to the H_2/H_3O^+ standard cell.
7. Electrode potentials are intensive properties (see page 13). Therefore, changing the stoichiometric coefficients for a half-reaction does not change the value of $E°$. For example, the reduction of Fe^{3+} has an $E°$ of +0.771 V whether the reaction is written as

Table 21.1 conveys the same kind of information about redox reactions as Table 17.4 gives about acid–base reactions.

$$Fe^{3+}(aq, 1\text{ M}) + e^- \longrightarrow Fe^{2+}(aq, 1\text{ M}) \qquad E° = +0.771\text{ V}$$

or as

$$2\,Fe^{3+}(aq, 1\text{ M}) + 2\,e^- \longrightarrow 2\,Fe^{2+}(aq, 1\text{ M}) \qquad E° = +0.771\text{ V}$$

The volt is defined as "energy/charge." Multiplying a reaction by some number causes both the energy and the charge to be multiplied by that number. Thus, the ratio "energy/charge = volt" does not change.

EXAMPLE 21.4

PREDICTING REACTION SPONTANEITY AND $E°$

For each of the following two reactions, predict whether the reaction is spontaneous, under standard conditions, in the direction in which it is written. Calculate $E°_{net}$ for each.

(a) $2\,Al(s) + 3\,Sn^{4+}(aq) \longrightarrow 2\,Al^{3+}(aq) + 3\,Sn^{2+}(aq)$
(b) $Cd^{2+}(aq) + Cu(s) \longrightarrow Cd(s) + Cu^{2+}(aq)$

Solution *For reaction (a):*

$$\begin{array}{lr} 2\,[Al(s) \longrightarrow Al^{3+}(aq) + 3\,e^-] & E° = +1.66\text{ V} \\ 3\,[Sn^{4+}(aq) + 2\,e^- \longrightarrow Sn^{2+}(aq)] & E° = +0.15\text{ V} \\ \hline 2\,Al(s) + 3\,Sn^{4+}(aq) \longrightarrow 2\,Al^{3+}(aq) + 3\,Sn^{2+}(aq) & E°_{net} = +1.81\text{ V} \end{array}$$

This reaction is predicted to be spontaneous as written (and thus to have a positive $E°$ and negative $\Delta G°$) because, as indicated in Table 21.1, Al(s) is a stronger reducing agent than Sn^{2+}, and Sn^{4+} is a stronger oxidizing agent than Al^{3+}.

For reaction (b):

$$\begin{array}{lr} Cd^{2+}(aq) + 2\,e^- \longrightarrow Cd(s) & E° = -0.40\text{ V} \\ Cu(s) \longrightarrow Cu^{2+}(aq) + 2\,e^- & E° = -0.34\text{ V} \\ \hline Cd^{2+}(aq) + Cu(s) \longrightarrow Cu^{2+}(aq) + Cd(s) & E°_{net} = -0.74\text{ V} \end{array}$$

The data in Table 21.1 indicate that Cu is a poorer reducing agent than Cd, and Cd^{2+} is a poorer oxidizing agent than Cu^{2+}. Therefore, the reaction is not spontaneous in the direction written, and the negative sign of $E°_{net}$ confirms this conclusion. If an electrochemical cell is assembled from a Cu/Cu^{2+} half-cell and a Cd/Cd^{2+} half-cell, the observed reaction will be the opposite of that given in question (b).

$$Cu^{2+}(aq) + Cd(s) \longrightarrow Cd^{2+}(aq) + Cu(s) \qquad E° = +0.74 \text{ V}$$

EXERCISE 21.4 Predicting Reaction Spontaneity and $E°$

For each of the following reactions at standard conditions, decide whether it occurs spontaneously in the direction written and calculate $E°$.

(a) $Zn(s) + Sn^{2+}(aq) \longrightarrow Sn(s) + Zn^{2+}(aq)$

(b) $Br_2(\ell) + 2\ Cl^-(aq) \longrightarrow 2\ Br^-(aq) + Cl_2(g)$

EXAMPLE 21.5

CONSTRUCTING AN ELECTROCHEMICAL CELL

Using the half-reactions $Fe(s)/Fe^{2+}(aq)$ and $Cu(s)/Cu^{2+}(aq)$, construct an electrochemical cell and predict its standard potential.

Solution As a first step, decide which is the reducing agent and which is the oxidizing agent. The data in Table 21.1 show that Fe(s) is a better reducing agent than Cu, and Cu^{2+} is a better oxidizing agent than Fe^{2+}. Therefore, Fe will reduce $Cu^{2+}(aq)$. This means the half-reactions and the overall reaction must be

Oxidation:

$$Fe(s) \longrightarrow Fe^{2+}(aq) + 2\ e^- \qquad E° = +0.44 \text{ V}$$

Reduction:

$$Cu^{2+}(aq) + 2\ e^- \longrightarrow Cu(s) \qquad E° = +0.34 \text{ V}$$

Net reaction:

$$Fe(s) + Cu^{2+}(aq) \longrightarrow Fe^{2+}(aq) + Cu(s) \qquad E°_{net} = +\ 0.78 \text{ V}$$

To construct a cell that operates under standard conditions, we shall have a solid iron electrode dipping into a solution of 1.0 M $Fe^{2+}(aq)$. Another compartment will contain a solid copper electrode in a solution that is 1.0 M in $Cu^{2+}(aq)$. An external wire connects the two electrodes, and a salt bridge will allow for ion flow. (Since virtually all salts are soluble as nitrates, we shall use $Cu(NO_3)_2$ and $Fe(NO_3)_2$ in the cell compartments, and $NaNO_3$ in the salt bridge.)

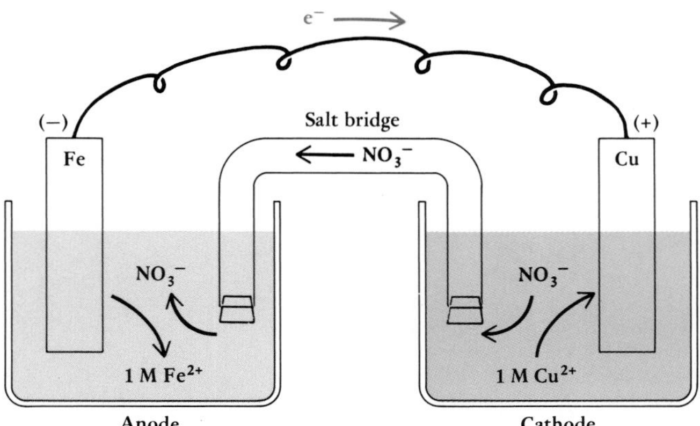

When the cell is assembled, electrons will flow from the Fe(s) anode to the Cu(s) cathode, and anions (NO_3^- in this case) will flow from the Cu^{2+} compartment to the Fe^{2+} electrode compartment.

EXERCISE 21.5 Constructing an Electrochemical Cell

Draw a diagram of an electrochemical cell using the half-cells $Zn(s)/Zn^{2+}(aq)$ and $Al(s)/Al^{3+}(aq)$. Decide first on the net spontaneous reaction and predict its $E°$ value. Show the direction of electron flow in the external wire and the direction of ion flow in the salt bridge. Tell which compartment is the anode and which is the cathode.

Practical Aspects of Redox Potentials: Reactions in the Introductory Chemistry Laboratory

Many of the laboratory experiments commonly done in the introductory course involve oxidation–reduction reactions, so it is interesting to see what potentials could be produced in such reactions and whether they are indeed predicted to be spontaneous under standard conditions.

One such reaction is the quantitative oxidation of an iron(II) salt by titration with the permanganate ion in acid solution.

$$5[Fe^{2+}(aq) \longrightarrow Fe^{3+}(aq) + e^-] \qquad E° = -0.77\ V$$
$$5\ e^- + 8\ H_3O^+(aq) + MnO_4^-(aq) \longrightarrow Mn^{2+}(aq) + 12\ H_2O(\ell) \qquad E° = +1.51\ V$$
$$5\ Fe^{2+}(aq) + 8\ H_3O^+(aq) + MnO_4^-(aq) \longrightarrow 5\ Fe^{3+}(aq) + Mn^{2+}(aq) + 12\ H_2O(\ell) \qquad E°_{net} = +0.74\ V$$

The overall reaction is certainly spontaneous, and it proceeds essentially to completion. As a result, it is often used for the analysis of iron-containing samples.

If your laboratory program includes the qualitative analysis of solutions for the presence or absence of metal ions, you may have seen many examples of reactions involving electron transfer. An example already used in this chapter is the reduction of tin(IV) ion with aluminum (page 866). Yet another reaction is the hydrogen peroxide oxidation of chromium(III) hydroxide to the chromate ion, CrO_4^{2-}, in basic solution. The data in Appendix I predict these reactions to be spontaneous under standard conditions.

$$2[Cr(OH)_4^-(aq) + 4\ OH^-(aq) \longrightarrow CrO_4^{2-}(aq) + 4\ H_2O(\ell) + 3\ e^-] \qquad E° = +0.12\ V$$
$$3[H_2O_2(aq) + 2\ e^- \longrightarrow 2\ OH^-(aq)] \qquad E° = +0.88\ V$$
$$2\ Cr(OH)_4^-(aq) + 3\ H_2O_2(aq) + 2\ OH^-(aq) \longrightarrow 2\ CrO_4^{2-}(aq) + 8\ H_2O(\ell) \qquad E°_{net} = +1.00\ V$$

Watch for other such reactions in your laboratory and see if they are predicted to be spontaneous based on the sign of $E°$.

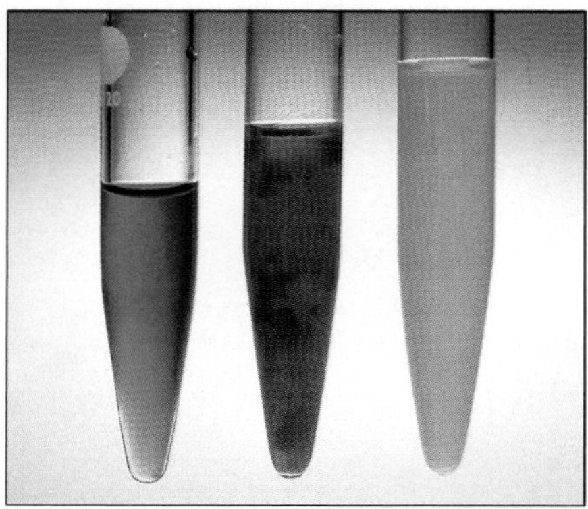

When a base is added to aqueous chromium(III) ion (left), the solution turns gray-green (middle). When an oxidizing agent (hydrogen peroxide, H_2O_2) is added, $Cr(OH)_4^-$ (aq) is oxidized to the yellow chromate ion, CrO_4^{2-}.

Practical Aspects of Redox Potentials: $E°$ and Dental Fillings

The most common filling material for tooth cavities is "dental amalgam," a solid solution of tin and silver in mercury. Dental amalgams are generally inert and cause few if any health problems. However, if you bite a piece of aluminum foil from a gum or candy wrapper with a filled tooth, your tooth nerves will complain! The sensation is caused by the nerves detecting the electron flow from an electrochemical reaction.

Dental amalgam consists of three phases, having compositions approximating Ag_2Hg_3, Ag_3Sn, and Sn_xHg (where x is 7 to 9). All three of these phases may undergo electrochemical reactions; for example,

$$3\ Hg_2^{2+}(aq) + 4\ Ag(s) + 6\ e^- \longrightarrow 2\ Ag_2Hg_3(s) \qquad E° = +0.85\ V$$

$$Sn^{2+}(aq) + 3\ Ag(s) + 2\ e^- \longrightarrow Ag_3Sn(s) \qquad E° = -0.05\ V$$

Aluminum is a much better reducing agent than any of the solid solutions above ($E°$ for $Al(s) \rightarrow Al^{3+}(aq) + 3\ e^-$ is +1.66 V). Therefore, if a piece of aluminum comes into contact with a dental filling, the saliva and gum tissue act as a salt bridge, and a short circuit results. The minute current produces a jolt of pain.

21.3 VOLTAIC CELLS AT NONSTANDARD CONDITIONS

In the previous section you learned how to predict the potential produced by a spontaneous chemical reaction proceeding under standard conditions. Reactions in the real world are not always so accommodating; the concentrations of the ions are usually not 1 M. Even if the cell started out with all dissolved species at 1 M concentration, these would change as the reaction progressed, reactants decreasing in concentration and products increasing. Thus, we have to define the potential of the cell under *non*standard conditions.

THE NERNST EQUATION

The free energy for an oxidation–reduction reaction under standard conditions ($\Delta G°$) is proportional to the cell potential at these conditions (Equation 21.1). It should make sense that the free energy under some set of nonstandard conditions (ΔG without the superscript °) is proportional to the potential (E without the superscript °) under these same conditions, that is, $\Delta G = -nFE$. Standard conditions are just a special case of this more general relation.

Now we can relate the standard potential $E°$ to the nonstandard potential E using another equation from thermodynamics. In Chapter 20 we considered the general reaction

$$a\mathrm{A} + b\mathrm{B} \longrightarrow c\mathrm{C} + d\mathrm{D}$$

and wrote the equation (page 839)

$$\Delta G = RT \ln Q - RT \ln K$$

where K is the equilibrium constant for the reaction and Q is the reaction quotient. Q is defined by

Refer to Section 20.6. Recall that Q has the same form as K, but the concentrations in Q are those of the ions (or gases) in an actual electrochemical cell, and the cell is not at equilibrium if a net chemical reaction is occurring to produce a current and voltage.

$$Q = \frac{[C]^c[D]^d}{[A]^a[B]^b}$$

Substituting for ΔG and using the relations $\Delta G° = -RT \ln K$ (Equation 20.5) and $\Delta G° = -nFE°$ (Equation 21.1), we have

$$-nFE = RT \ln Q - nFE°$$

which we can rearrange to the **Nernst equation**,

$$E = E° - \frac{RT}{nF} \ln Q$$

where F is the Faraday constant (9.6485309×10^4 J/V · mol), R is the gas constant (8.314510 J/K · mol), and n is the number of moles of electrons transferred between oxidizing and reducing agents in a balanced redox equation. When T is 298 K (and when the logarithmic term is given as a base-10 log), we can write a modified form of the Nernst equation that we find useful in practical chemical applications.

$$E = E° - \frac{0.0592}{n} \log Q \text{ at } 25\ °\text{C} \tag{21.2}$$

This equation enables us to find the voltage produced by a simple cell under nonstandard conditions, or to find the concentration of a reactant or product by measuring the voltage produced by a cell.

EXAMPLE 21.6

USING THE NERNST EQUATION

Determine the cell potential at 25 °C for

$$Fe(s) + Cd^{2+}(aq) \longrightarrow Fe^{2+}(aq) + Cd(s)$$

when (a) $[Fe^{2+}] = 0.10$ M and $[Cd^{2+}] = 1.0$ M and (b) when $[Fe^{2+}] = 1.0$ M and $[Cd^{2+}] = 0.010$ M.

Solution When the conditions are those in (a), we first calculate $E°$,

$$\begin{array}{ll} Fe(s) \longrightarrow Fe^{2+}(aq) + 2\,e^- & E° = +0.44 \text{ V} \\ Cd^{2+}(aq) + 2\,e^- \longrightarrow Cd(s) & E° = -0.40 \text{ V} \\ \hline Fe(s) + Cd^{2+}(aq) \longrightarrow Fe^{2+}(aq) + Cd(s) & E° = +0.04 \text{ V} \end{array}$$

and then substitute into the Nernst equation to find E.

$$E = E° - \frac{0.0592}{n} \log \frac{[Fe^{2+}]}{[Cd^{2+}]}$$

As in chemical equilibrium expressions (Chapter 16), the "concentration" of a solid does not enter into the expression. Now, using $n = 2$ (the number of moles of electrons transferred), and the ion concentrations given above,

$$E = +0.04 \text{ V} - \frac{0.0592}{2} \log \frac{0.10}{1.0} = +0.07 \text{ V}$$

Notice that the potential is now larger than $E°$, so the tendency for Fe to transfer electrons to Cd^{2+} is even greater than under standard conditions.

Walter Nernst (1864–1941) was a German physicist and chemist known especially for his work related to the third law of thermodynamics. (Francis Simon, AIP Niels Bohr Library)

For the conditions in part (b), the first step again is to calculate $E°$. Since this is the same as in (a) above, we can proceed to set up and solve the Nernst equation for the new conditions.

$$E = +0.04\ \text{V} - \frac{0.0592}{2} \log \frac{1.0}{0.010} = -0.02\ \text{V}$$

Now we can tell from the equation that the cell potential is negative. This means that the reaction does not proceed spontaneously in the direction in which it was originally written, but in the opposite direction; iron(II) ion is reduced to metallic iron by transfer of electrons from cadmium in this case.

$$Fe^{2+}(aq) + Cd(s) \longrightarrow Fe(s) + Cd^{2+}(aq)$$

EXERCISE 21.6 Using the Nernst Equation

The aqueous tin(II) ion, Sn^{2+}, is commonly used in the laboratory as a reducing agent. However, its solutions are not stable in air, as the ion is readily oxidized by O_2 in the air to give Sn^{4+}. What is the potential for the O_2 oxidation of 0.10 M $Sn^{2+}(aq)$ dissolved in 0.10 M HCl at 25 °C when $P(O_2) = 0.20$ atm and $Sn^{4+}(aq)$ has a concentration of 1.0×10^{-6} M?

$E°$ AND THE EQUILIBRIUM CONSTANT

We saw in Example 21.6 that the cell potential, and even the reaction direction, can change when the concentrations of products and reactants change. Thus, as reactants are converted to products in a spontaneous reaction, the value of E_{net} must decline from its initial positive value and eventually reach zero. *A potential of zero means that no net reaction is occurring; it is an indication that the cell has reached equilibrium.* Thus, when $E_{net} = 0$, the Q term in the Nernst equation becomes equivalent to the equilibrium constant, K, for the reaction. So, when equilibrium has been attained, we can rewrite Equation 21.2 as

$$E = 0 = E° - \frac{0.0592}{n} \log K$$

which we rearrange to

$$\log K = \frac{nE°}{0.0592} \text{ at 25 °C} \qquad (21.3)$$

This is an extremely useful equation, since it tells you that the equilibrium constant for a reaction can be obtained from a calculation or measurement of $E°_{net}$.

EXAMPLE 21.7

$E°$ AND EQUILIBRIUM CONSTANTS

Calculate the equilibrium constant for the reaction

$$Fe(s) + Cd^{2+}(aq) \longrightarrow Fe^{2+}(aq) + Cd(s)$$

Solution The first step is to calculate $E°$. In this case, +0.04 V was found in Example 21.6, so we use this $E°$ in Equation 21.3.

$$\log K = \frac{(2.00 \text{ mol})(0.04 \text{ V})}{0.0592 \text{ V} \cdot \text{mol}}$$

$$\log K = 1.4$$

$$K = 20$$

An interesting sidelight to this problem is the evaluation of the concentrations of Fe^{2+} and Cd^{2+} when the cell has reached equilibrium after starting at standard conditions. You can begin by writing the equilibrium expression.

$$K = 20 = \frac{[Fe^{2+}]}{[Cd^{2+}]}$$

The concentrations at standard conditions are 1.0 M. The amount of Cd^{2+} lost in the reaction and the amount of Fe^{2+} produced are both set equal to x.

$$K = 20 = \frac{1.0 + x}{1.0 - x}$$

Solving, we find $x = 0.9$ M. Therefore, the equilibrium concentrations are

$$[Fe^{2+}] = 1.0 + x = 1.9 \text{ M} \quad \text{and} \quad [Cd^{2+}] = 1.0 - x = 0.1 \text{ M}$$

EXERCISE 21.7 $E°$ and Equilibrium Constants

A disproportionation reaction is one in which identical molecules or ions react so that some molecules are oxidized and others are reduced. Calculate the equilibrium constant for the disproportionation of the mercury(I) ion. (*Hint:* both half-reactions have $Hg_2^{2+}(aq)$ as the reactant.)

$$Hg_2^{2+}(aq) \longrightarrow Hg(\ell) + Hg^{2+}(aq)$$

If you make a 0.10 M solution of $Hg_2(NO_3)_2$, what are the equilibrium concentrations of $Hg_2^{2+}(aq)$ and $Hg^{2+}(aq)$?

Figure 21.8 A diagrammatic representation of the LeClanché dry cell. It consists of a zinc anode (the battery container), a graphite cathode, and an electrolyte consisting of a moist paste of MnO_2 and NH_4Cl.

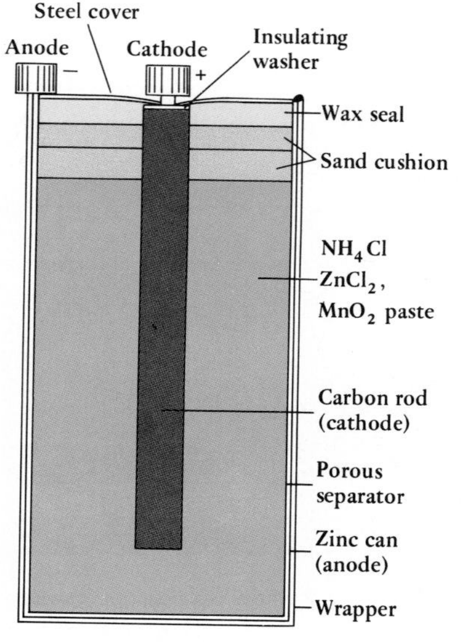

21.4 COMMON BATTERIES AND STORAGE CELLS

The voltaic cells we have described to this point can produce a useful voltage, but the voltage will decline rapidly as the reactant concentrations decline. For this reason, there has been great interest over the years in the design of usable batteries, voltaic cells that deliver current at a constant voltage.

SOME COMMON BATTERIES

THE COMMON DRY CELL The LeClanché or dry cell (Figure 21.8) is commonly used in flashlights, portable radios, and toys. The battery contains a carbon rod electrode inserted into a moist paste of NH_4Cl, $ZnCl_2$, and MnO_2 in a zinc can that serves as the anode.

Anode reaction (oxidation): $Zn(s) \longrightarrow Zn^{2+}(aq) + 2\ e^-$

The electrons produced reduce the ammonium ion to ammonia and hydrogen at a carbon cathode.

Cathode reaction (reduction):
$$2\ NH_4^+(aq) + 2\ e^- \longrightarrow 2\ NH_3(g) + H_2(g)$$

The products of the cathode reaction are gases and would cause the sealed dry cell to explode if they were not removed. This is the reason the battery contains manganese(IV) oxide, an oxidizing agent that consumes the hydrogen.

$$2\ MnO_2(s) + H_2(g) \longrightarrow Mn_2O_3(s) + H_2O(\ell)$$

The ammonia is taken up by zinc(II) ion.

$$Zn^{2+}(aq) + 2\ NH_3(g) + 2\ Cl^-(aq) \longrightarrow Zn(NH_3)_2Cl_2(s)$$

All these reactions lead to the following net process and produce a voltage of 1.5 V.

$$2\ MnO_2(s) + 2\ NH_4Cl(s) + Zn(s) \longrightarrow Mn_2O_3(s) + H_2O(\ell) + Zn(NH_3)_2Cl_2(s)$$

A simple battery that illustrates some of the chemistry of the dry cell. Here a zinc anode (left) and a graphite cathode (right) dip into a solution of ammonium chloride. The battery produces about 1.1 V.

Unfortunately, there are at least two disadvantages to this battery. If current is drawn from the battery rapidly, the gaseous products cannot be consumed rapidly enough; the voltage drops as a result, although it can be restored on standing. Furthermore, there is a spontaneous but slow direct reaction between the zinc electrode and ammonium ion that leads to further deterioration, and the battery has a poor "shelf life."*

THE ALKALINE BATTERY The somewhat more expensive "alkaline" battery has come into use because it avoids some of the problems of the common dry cell. An alkaline battery produces 1.54 V, and the key reaction is again the oxidation of zinc, this time under alkaline or basic conditions. The oxidation or anode reaction is

Anode reaction (oxidation):
$$Zn(s) + 2\ OH^-(aq) \longrightarrow ZnO(s) + H_2O(\ell) + 2\ e^-$$

and the electrons produced are consumed by reduction of manganese(IV) oxide at the cathode.

Cathode reaction (reduction):
$$2\ MnO_2(s) + H_2O(\ell) + 2\ e^- \longrightarrow Mn_2O_3(s) + 2\ OH^-(aq)$$

In contrast to the LeClanché battery, no gases are formed in the alkaline battery, and there is no decline in voltage under high current loads.

THE MERCURY BATTERY A close relative of the alkaline battery is the mercury cell (Figure 21.9), the type of battery used in calculators, cameras, watches, heart pacemakers, and other devices in which a small cell is required. As in the two previous batteries, the anode (the reducing agent) is metallic zinc, but the cathode (oxidizing agent) is mercury(II) oxide.

Figure 21.9 The mercury battery. The reducing agent is zinc and the oxidizing agent is mercury(II) oxide.

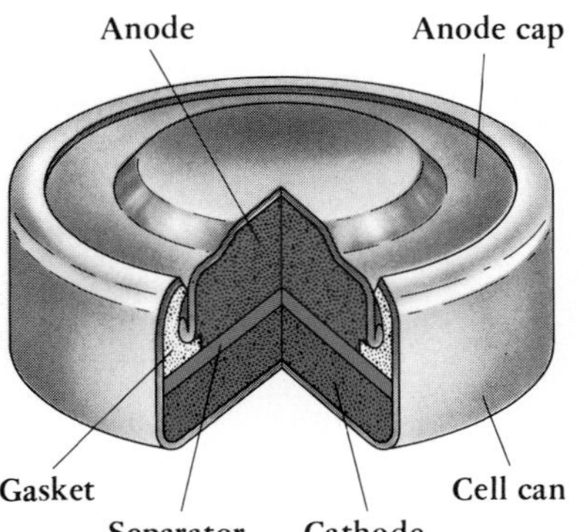

*The shelf life of a dry cell is also affected by temperature. You can double or triple the shelf life of a battery if you store it in a refrigerator at about 4 °C.

Anode reaction (oxidation):
$Zn(s) + 2\ OH^-(aq) \longrightarrow ZnO(s) + H_2O(\ell) + 2\ e^-$

Cathode reaction (reduction):
$HgO(s) + H_2O(\ell) + 2\ e^- \longrightarrow Hg(\ell) + 2\ OH^-(aq)$

These materials are tightly compacted powders separated by a moist paste of HgO containing some NaOH or KOH. Moistened paper serves as the "salt bridge," and the battery produces 1.35 V. These batteries are widely used, but, because they contain mercury, they can lead to some environmental problems. Mercury and its compounds are poisonous, so if possible mercury cells should be reprocessed to recover the metal when the battery is no longer useful.

STORAGE BATTERIES

The LeClanché cell, the alkaline battery, and the mercury cell will no longer produce a current when the chemicals inside have reached equilibrium conditions. They must be discarded. In contrast, storage batteries can be recharged. That is, the original reactant concentrations can be restored by using an external source of electrical energy to reverse the net cell reaction. An automobile battery—the lead storage battery—is perhaps the best example. That battery is used to supply the energy to the engine starter, but once the engine is running, the battery is recharged by current from the car's alternator.

THE LEAD STORAGE BATTERY There are two types of electrodes in a lead storage battery (Figure 21.10), one made of porous lead (the reducing agent) and the other of compressed, insoluble lead(IV) oxide (the oxidizing agent). The electrodes, arranged alternately in a stack and immersed in aqueous sulfuric acid, are separated by thin fiberglass sheets. When the cell acts as a supplier of electrical energy, the lead electrode is oxidized to insoluble lead(II) sulfate.

Anode, oxidation:
$Pb(s) + SO_4^{2-}(aq) \longrightarrow PbSO_4(s) + 2\ e^- \qquad E° = +0.356\ V$

The electrons move through the external circuit to the lead(IV) oxide electrode, where they cause reduction of PbO_2.

Cathode, reduction:
$PbO_2(s) + 4\ H_3O^+(aq) + SO_4^{2-}(aq) + 2\ e^- \longrightarrow$
$PbSO_4(s) + 6\ H_2O(\ell) \qquad E° = +1.685\ V$

The net process of using the cell to supply electrical energy is to coat both electrodes with an adhering film of white lead(II) sulfate and to consume sulfuric acid.

Net reaction:
$Pb(s) + PbO_2(s) + 2\ H_2SO_4(aq) \longrightarrow 2\ PbSO_4(s) + 2\ H_2O(\ell)$
$E° = 2.041\ V$

A lead storage battery can usually be recharged by supplying electrical energy to reverse the net process. The film of $PbSO_4$ is converted back to metallic lead and PbO_2, and sulfuric acid is regenerated.

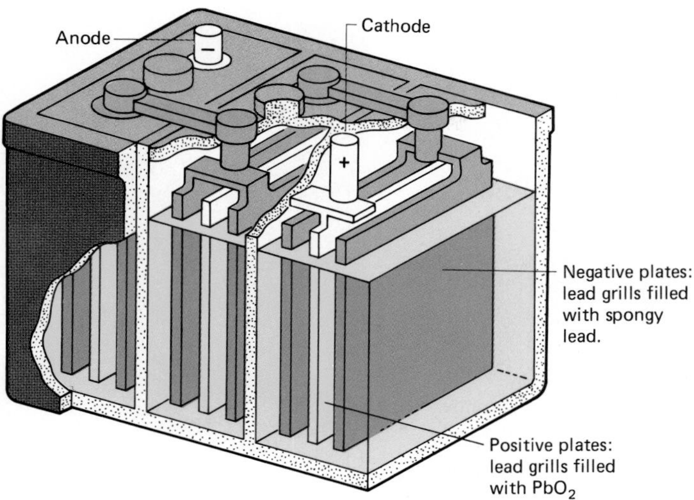

Figure 21.10 The lead storage battery.

Many attempts have been made to improve upon lead storage batteries, because they are large and heavy and produce a relatively low power for their mass. Nonetheless, they do produce 2 V per cell and an enormous initial current. This is a combination that has been hard to beat, and they remain in widespread use.

NI–CAD BATTERIES Rechargeable, lightweight nickel–cadmium alkaline cells, or "ni–cad" batteries (Figure 21.11), are used in a variety of cordless appliances. These have the advantage that the oxidizing and reducing agents can be regenerated easily by recharging, and they produce a nearly constant potential.

Figure 21.11 The nickel–cadmium or "ni–cad" battery.

Anode, oxidation:
$$Cd(s) + 2\ OH^-(aq) \longrightarrow Cd(OH)_2(s) + 2\ e^-$$

Cathode, reduction:
$$NiO(OH)(s) + H_2O(\ell) + e^- \longrightarrow Ni(OH)_2(s) + OH^-(aq)$$

FUEL CELLS

Storage batteries contain the chemicals necessary for a reversible electrochemical reaction. When the reaction proceeds in the direction of spontaneity, a current is produced until the chemicals reach equilibrium. Subsequent reversal of the spontaneous reaction by an external current source restores the original reactant concentrations. A **fuel cell** is also an electrochemical device, but, in contrast to a storage battery, it does not involve a reversible reaction in the practical sense; the reactants are continually supplied from an external reservoir.

The best known fuel cell is the hydrogen–oxygen cell (Figure 21.12), used in the Gemini, Apollo, and Space Shuttle programs. The net cell reaction is simply the oxidation of hydrogen with oxygen to give water. Rather than allowing these gases to react directly and produce energy in the form of heat, they are made to react in such a way that the energy produced can be tapped by an electrical device. A stream of H_2 gas is pumped onto the anode of the cell, and pure O_2 gas is directed to the

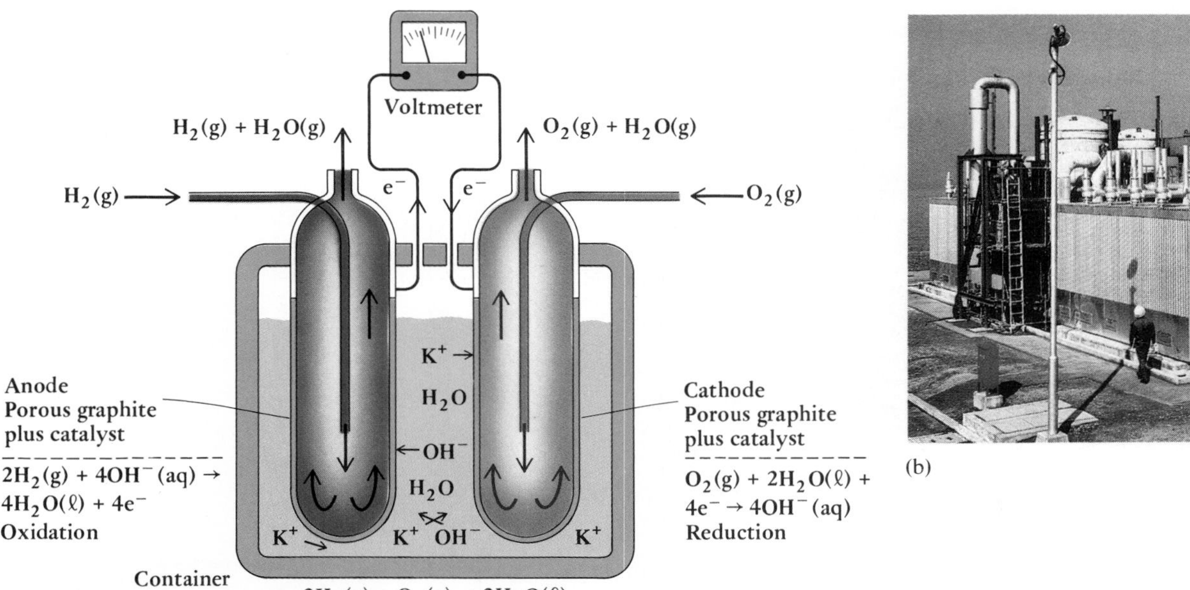

Figure 21.12 Fuel cells. (a) Schematic of a H_2/O_2 fuel cell. (b) A 4.5-megawatt fuel cell power plant in Tokyo. Unlike the fuel cells used in space vehicles, which have an alkaline (KOH) electrolyte, larger plants such as the one in this photo use phosphoric acid as the electrolyte (b, Johnson-Matthey).

cathode. Since the cell contains concentrated KOH, the following reactions occur under basic conditions:

Anode, oxidation:

$$2\ H_2(g) + 4\ OH^-(aq) \longrightarrow 4\ H_2O(\ell) + 4\ e^-$$

Cathode, reduction:

$$O_2(g) + 2\ H_2O(\ell) + 4\ e^- \longrightarrow 4\ OH^-(aq)$$

Net reaction:

$$2\ H_2(g) + O_2(g) \longrightarrow 2\ H_2O(\ell) \qquad (E = 0.9\text{ V at }70\text{–}140\ ^\circ\text{C})$$

The product water is swept out of the cell as a vapor in the hydrogen and oxygen stream.

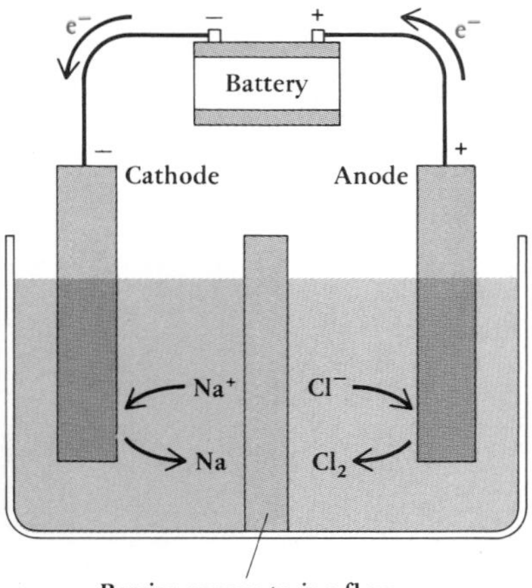

Battery
Cathode
Anode
Na^+ Cl^-
Na Cl_2
Barrier porous to ion flow

Figure 21.13 Electrolysis of molten sodium chloride. See also Figure 21.20.

21.5 ELECTROLYSIS: CHEMICAL CHANGE FROM ELECTRICAL ENERGY

Thus far we have described the production of electric current from chemical reactions. Equally important, however, is the opposite process, **electrolysis**, the use of an electric current to bring about chemical change.

Let us consider first the manufacture of sodium metal and chlorine gas by electrolysis. A cell having a pair of inert electrodes dipping into a bath of molten NaCl is depicted in a simplified fashion in Figure 21.13. (The fact that the NaCl is in the liquid state means that Na^+ and Cl^- ions are free to move in the melt.) The cell has been attached to a source of electric current, such as a battery. The external voltage source acts as an "electron pump," and electrons flow from this source into one of the electrodes, thereby giving it a negative charge. Sodium ions are attracted to this negative electrode and are reduced when electrons from the elec-

A Summary of Electrochemical Terminology

No matter whether we are discussing a voltaic cell or an electrolysis cell, the terms *anode* and *cathode* always refer to the electrodes at which oxidation and reduction occur, respectively. However, the polarity of the electrodes reverses as follows.

TYPE OF CELL	ELECTRODE	FUNCTION	POLARITY
Battery	Anode	Oxidation	−
	Cathode	Reduction	+
Electrolysis	Anode	Oxidation	+
	Cathode	Reduction	−

In a voltaic cell, the negative electrode is the one at which electrons are produced. In an electrolysis cell, the negative electrode is the one onto which the external source is "pumping" the electrons.

trode are accepted, making the electrode the cathode. The battery simultaneously draws electrons from the other electrode, giving it a positive electrical charge. Chloride ions are attracted to this electrode and surrender electrons. Since oxidation has occurred, this is the anode. The following reactions have thus occurred in molten NaCl:

Anode, oxidation: $2\ Cl^- \longrightarrow Cl_2(g) + 2\ e^-$

Cathode, reduction: $Na^+ + e^- \longrightarrow Na(s)$

Net reaction: $2\ Cl^- + 2\ Na^+ \longrightarrow 2\ Na(s) + Cl_2(g)$

The half-cell voltages in Table 21.1 apply only to aqueous solutions. If we use these to *estimate* the voltage for the reaction above, we obtain a value of about −4 V. The reaction clearly does not occur spontaneously in the direction written, and this is the reason that we have attached an external battery. The battery, with a voltage greater than 4 V, forces the nonspontaneous reaction to occur by "pumping" electrons in the proper direction.

What if you use an aqueous solution of sodium chloride instead of a molten salt (without water) (Figures 21.13 and 21.14)? The left-hand electrode in the cell is again the cathode, the site of the reduction half-reaction, and the right-hand electrode is the anode, the site of oxidation. Electrons are pumped onto the cathode by an external voltage source. With water now present, we must ask whether it is still Na^+ and Cl^- ions that are reduced and oxidized, respectively, or whether water might be involved. As possible reduction reactions we have

$$Na^+(aq) + e^- \longrightarrow Na(s) \qquad E^\circ = -2.71\ V$$

$$2\ H_2O(\ell) + 2\ e^- \longrightarrow H_2(g) + 2\ OH^-(aq) \qquad E^\circ = -0.83\ V$$

When there are several possible reactions at an electrode, the cathode in this case, we have to consider not only which is the most easily reduced (best oxidizing agent) but also which is reduced most *rapidly*. There are usually complications when currents are large—as in a commercial electrolysis cell—and when reactant concentrations are small. In this case H_2

Figure 21.14 Electrolysis of aqueous sodium chloride. See also Figure 21.21.

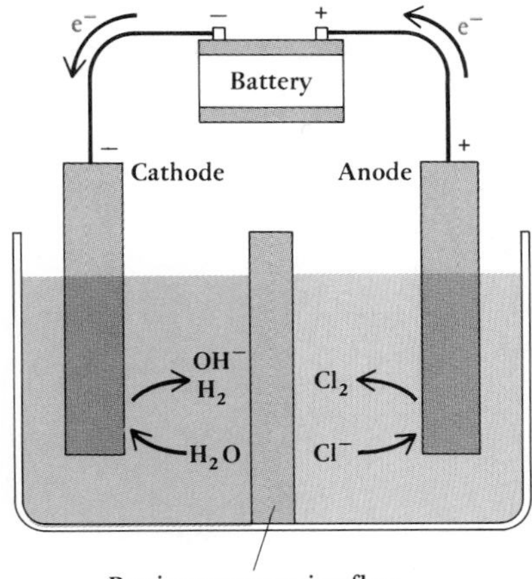

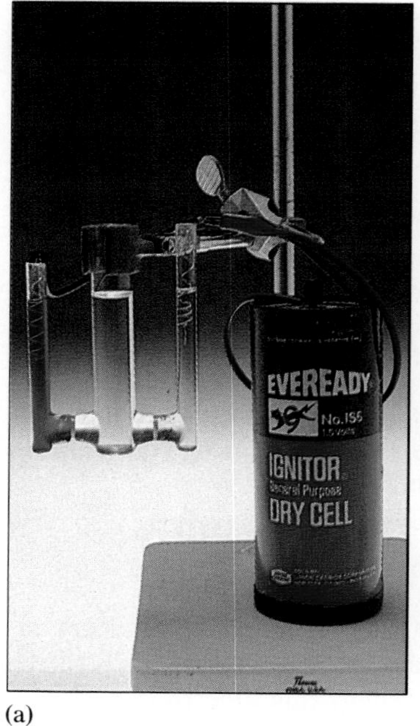

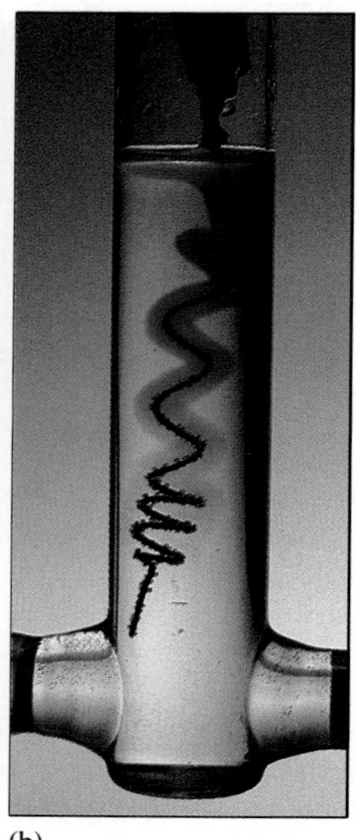

(a) (b)

Figure 21.15 The electrolysis of aqueous potassium iodide. Aqueous KI is contained in all three compartments of the cell, and both electrodes are platinum. (a) At the positive electrode or anode (right), the I^- ion is oxidized to iodine, which gives the solution a yellow-brown color.

$$2\ I^-(aq) \longrightarrow I_2(aq) + 2\ e^-$$

At the negative electrode or cathode (left), water is reduced, and the presence of the OH^- ion is indicated by the red color of the acid–base indicator phenolphthalein.

$$2\ H_2O(\ell) + 2\ e^- \longrightarrow H_2(g) + 2\ OH^-(aq)$$

(b) In a close-up of the cathode, bubbles of H_2 and evidence of OH^- being generated at the electrode are clearly seen.

is clearly observed as a product. This is reasonable since water is certainly more readily reduced than sodium. (We could also consider the hydronium ion as the species reduced. If we make this assumption, Le Chatelier's principle predicts that, as H_3O^+ is removed by electrolysis, more water would autoionize to give OH^- and H_3O^+. The *net* result is still the reduction of water and the formation of H_2 and OH^-.)*

For the oxidation processes possible in *aqueous* sodium chloride, we need to compare the two reactions

$$2\ Cl^-(aq) \longrightarrow Cl_2(g) + 2\ e^- \qquad E° = -1.36\ V$$

$$6\ H_2O(\ell) \longrightarrow O_2(g) + 4\ H_3O^+(aq) + 4\ e^- \qquad E° = -1.23\ V$$

Judging by the values of $E°$, water should be slightly more readily oxidized than chloride. However, in practice Cl_2 is actually produced in preference to O_2. Therefore, the best description of the chemistry that occurs upon electrolysis of aqueous NaCl are the reactions

Anode, oxidation:

$$2\ Cl^-(aq) \longrightarrow Cl_2(g) + 2\ e^- \qquad E° = -1.36\ V$$

Cathode, reduction:

$$2\ H_2O(\ell) + 2\ e^- \longrightarrow H_2(g) + 2\ OH^-(aq) \qquad E° = -0.83\ V$$

Net reaction:

$$2\ Cl^-(aq) + 2\ H_2O(\ell) \longrightarrow H_2(g) + 2\ OH^-(aq) + Cl_2(g) \quad E°_{net} = -2.19\ V$$

*All of the electrode potentials for the reactions discussed here should be corrected to nonstandard conditions. Further, when comparing the reduction of Na^+ and H_2O, one should take into account the phenomenon of overvoltage (see the box on this subject). However, without adding these complications, $E°$ values do allow us to make some *estimates* of the reactions that occur under electrolysis conditions.

Overvoltage: The Problem of Kinetics in Electrochemistry

When electrochemical cells operate spontaneously to produce current, or when they are driven by an outside source of energy and act as electrolysis cells, the measured voltage may not be that calculated from $E°$, even though the chemicals are at standard conditions. That is, whenever a current is flowing through a cell, the potential will be different from the standard potential, $E°$, to an extent that depends on the properties of the reactions involved and on the magnitude of the current. The difference between the potential measured when current is flowing, E_i, and the standard potential is called the *overpotential* or *overvoltage,* and it equals $E_i - E°$. For an oxidation reaction to occur at a measurable rate, the potential must be more positive than $E°$, while the potential for a reduction reaction must be more negative than $E°$. Therefore, oxidations always have positive overvoltages, and reductions always have negative overvoltages.

Much of the reason for the existence of an overvoltage is kinetic. Recall that thermodynamics does not take reaction speed into account, and $E°$ is a thermodynamic quantity. However, the experimental fact is that kinetics does play a role in electrode reactions, and the overvoltage is a reflection of this. Although all half-cells have an associated overvoltage, it is especially apparent for reactions that involve gases such as H_2 and O_2, where the overvoltage can be as large as 0.6 V. This means that the reduction of water, for example,

$$2\ H_2O(\ell) + 2\ e^- \longrightarrow H_2(g,\ 1\ atm) + 2\ OH^-(aq,\ 1\ M) \qquad E° = -0.83\ V$$

actually occurs at an even more negative potential (~ -1.4 V) than expected.

Next, let us change the sodium chloride solution to an aqueous solution of some other metal halide such as KI (Figure 21.15) or $CuCl_2$ (Figure 21.16). As before, consult Table 21.1, and, considering all possible reactions, find the oxidation and reduction reactions that require the least potential. For aqueous $CuCl_2$, the only two possible oxidations are those

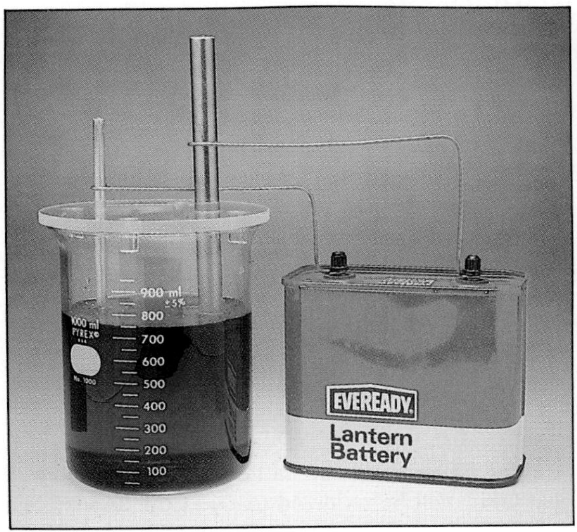

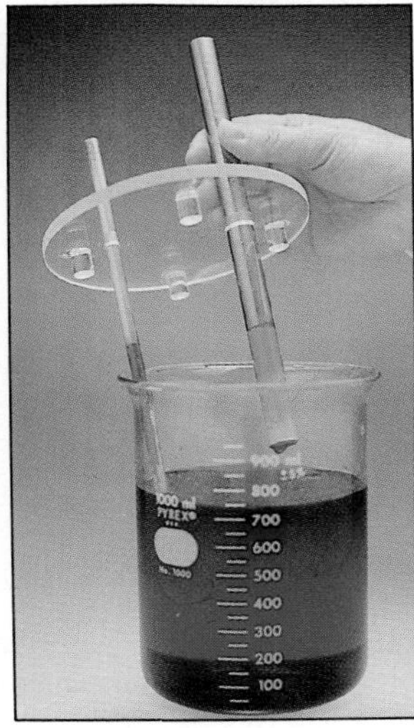

Figure 21.16 The electrolysis of an aqueous copper(II) salt. The copper(II) ion is reduced to the metal at the cathode (negative electrode, site of reduction). If the anion is Cl^-, Cl_2 gas is evolved at the anode (positive electrode, site of oxidation). However, if the anion is sulfate, water is oxidized to O_2 gas and leaves hydronium ions.

of Cl^-(aq) to Cl_2(g) and $H_2O(\ell)$ to O_2(g). As before, we expect the former to occur. Unlike aqueous sodium chloride, however, aqueous Cu^{2+} ion is *much* more easily reduced ($E° = +0.34$ V) than water ($E° = -0.83$ V). Therefore, the products of the electrolysis of aqueous $CuCl_2$ are chlorine gas and copper metal.

Anode, oxidation:

$$2\ Cl^-(aq) \longrightarrow Cl_2(g) + 2\ e^- \qquad E° = -1.36\ V$$

Cathode, reduction:

$$Cu^{2+}(aq) + 2\ e^- \longrightarrow Cu(s) \qquad E° = +0.34\ V$$

Net reaction:

$$Cu^{2+}(aq) + 2\ Cl^-(aq) \longrightarrow Cu(s) + Cl_2(g) \qquad E°_{net} = -1.02\ V$$

This process is of obvious commercial importance. The copper that is used in wiring, in coins, and for other purposes is purified by electrolysis (Chapter 25).

From the preceding discussion, we can arrive at a useful general principle. If you pass an electric current through a solution, the electrode reactions will most likely be those requiring the least potential to overcome their nonspontaneity (and those that occur most rapidly). In water, this means that *a substance will be reduced if it has a reduction potential more positive than about −0.8 V*, the potential for the reduction of pure water. A check of Table 21.1 and Appendix I shows that this includes commercially useful metals such as Pt, Cu, Ag, Au, and Cd. Indeed, the electrolysis method is used to coat or "plate" other materials with these metals. *If a substance has a reduction potential more negative than about −0.8 V, then only water is reduced.* Substances falling into this category include Na, K, Mg, and Al. To produce these metals requires methods other than the reduction of their ions in aqueous solution.

The electrolyses of molten and of aqueous NaCl are both commercial processes. See Section 21.7.

EXAMPLE 21.8

ELECTROLYSIS OF AQUEOUS NaOH

Predict the result of passing an electric current into an aqueous solution of NaOH.

Solution First, list all of the species in solution. In this case, they will be Na^+, OH^-, and H_2O. Then, using Table 21.1, decide which of the species in solution can be oxidized and which reduced and note the potential of each possible reaction.

Reductions:

$$Na^+(aq) + e^- \longrightarrow Na(s) \qquad E° = -2.71\ V$$

$$2\ H_2O(\ell) + 2\ e^- \longrightarrow H_2(g) + 2\ OH^-(aq) \qquad E° = -0.83\ V$$

Oxidations:

$$4\ OH^-(aq) \longrightarrow O_2(g) + 2\ H_2O(\ell) + 4\ e^- \qquad E° = -0.40\ V$$

It is evident that water will be reduced to H_2 at the cathode (even considering the problem of overvoltage), and OH^- will be oxidized to O_2 at the anode. The net cell reaction is $2\ H_2O(\ell) \rightarrow 2\ H_2(g) + O_2(g)$ and the potential under standard conditions is −1.23 V (see Chapter 22).

We should note that this is again a process of some commercial importance, since it produces both hydrogen and oxygen, both useful fuels in rockets and for industrial purposes.

EXERCISE 21.8 Electrolysis of Salts

Predict the results of passing an electric current into each of the following solutions: (a) molten NaBr, (b) aqueous NaBr, and (c) aqueous $SnCl_2$.

21.6 ELECTRICAL ENERGY

Metallic silver is produced at the cathode in the electrolysis of aqueous $AgNO_3$, the reaction being $Ag^+(aq) + e^- \rightarrow Ag(s)$. One mole of electrons is required to produce one mole of silver from one mole of silver ions. In contrast, two moles of electrons are required to produce one mole of copper.

$$Cu^{2+}(aq) + 2\ e^- \longrightarrow Cu(s)$$

It follows that, if you could measure the number of moles of electrons flowing through the electrolysis cell, you would know the number of moles of silver or copper produced. Conversely, if you knew the amount of silver or copper produced, you could calculate the number of moles of electrons used.

The number of moles of electrons consumed or produced in a redox reaction is usually obtained by measuring the current flowing in the external electrical circuit in a given time. The **current** flowing in an electrical circuit is the amount of charge (in units of coulombs) passed per unit time, and the usual unit for current is the **ampere**.

$$\text{Current} = I \text{ (amps)} = \frac{\text{electrical charge}}{\text{time}} = \frac{\text{coulombs}}{\text{seconds}}$$

Michael Faraday (1791–1867) first explored the quantitative aspects of electricity (page 139). In his honor scientists have defined the **Faraday** as *the charge carried by one mole of electrons*. This quantity can be calculated, since the charge on one electron is known (Chapters 2 and 8), and you know that there is Avogadro's number of electrons in a mole.

$$\begin{aligned}\text{Charge on one mole of electrons} &= 1 \text{ Faraday or } 1\ F\\ &= (1.602177 \times 10^{-19} \text{ coulombs/electron})(6.02214 \times 10^{23} \text{ electrons/mol})\\ &= 96{,}485.3 \text{ coulombs/mol}\end{aligned}$$

This number, 9.64853×10^4 coulomb/mole of electrons, is called the Faraday constant and is often abbreviated F. From the relation between charge, current, and time, you can obtain the charge by multiplying the current (in amperes) by the time (in seconds) during which the current passed. These quantities are easily measured with modern instruments. Knowing the charge, and using the Faraday constant as a conversion factor, you can then obtain the number of moles of electrons that passed through an electrochemical cell.

The Faraday constant was given before in this chapter (page 858) in units of J/V · mol. This should remind you that a coulomb is a joule per volt (1 C = 1 J/V), a relation we shall use below.

EXAMPLE 21.9

USING THE FARADAY CONSTANT

Assume that 1.50 amperes of current flow through a solution containing silver ions for 15.0 minutes. The voltage is such that silver is deposited at the cathode. How many grams of silver metal are deposited?

$$Ag^+(aq) + e^- \longrightarrow Ag(s)$$

Solution Our objective is to obtain the mass of silver, so we should aim our sequence of calculations to obtain moles of silver. The half-reaction tells you that, if 1 mole of electrons is passed, then 1 mole of silver is obtained. Thus, we must find the charge passed, because this will give us moles of electrons. The charge comes from the relation between current and time, the two factors given by the experiment. Thus, the calculation sequence here is

$$\text{Time} \times \text{current} \longrightarrow \text{charge} \longrightarrow \text{moles } e^- \longrightarrow \text{moles Ag} \longrightarrow \text{mass Ag}$$

(a) Calculate the number of coulombs of charge passed in 15.0 minutes.

$$\begin{aligned}\text{Charge(coulombs)} &= \text{amps} \times \text{time(sec)}\\ &= 1.50 \text{ amps}(15.0 \text{ min})(60.0 \text{ s/min})\\ &= 1.35 \times 10^3 \text{ coulombs}\end{aligned}$$

(b) Calculate the number of moles of electrons.

$$1.35 \times 10^3 \text{ coulombs} \left(\frac{1 \text{ mole } e^-}{9.65 \times 10^4 \text{ coulombs}}\right) = 1.40 \times 10^{-2} \text{ moles } e^-$$

(c) Calculate the moles of Ag^+ and then the mass of silver deposited. When 1.40×10^{-2} moles of electrons passed through the cell, 1.40×10^{-2} moles of Ag(s) must have been formed. From this, we obtain the mass of silver deposited on the electrode.

$$1.40 \times 10^{-2} \text{ mol Ag} \left(\frac{107.9 \text{ g}}{1 \text{ mol}}\right) = 1.51 \text{ g Ag}$$

EXAMPLE 21.10

USING THE FARADAY CONSTANT

One of the half-reactions occurring in the lead storage battery (Section 21.4) is

$$Pb(s) + SO_4{}^{2-}(aq) \longrightarrow PbSO_4(s) + 2\ e^-$$

If the battery delivers 1.50 amperes, and if its Pb electrode contains 1.00 pound of lead, how long can current flow before the lead in the electrode is used up?

Solution Here we want to know time, and we begin with information on mass (i.e., moles) and current. This is just the reverse of the previous problem. That is, our conversion sequence here should be

$$\text{Mass} \longrightarrow \text{moles} \longrightarrow \text{charge} \longrightarrow \text{time}$$

(a) Calculate the number of moles of lead. Since 1.00 lb. of Pb = 454 g,

$$454 \text{ g} \left(\frac{1 \text{ mol}}{207.2 \text{ g}}\right) = 2.19 \text{ mol Pb}$$

(b) Calculate the number of moles of electrons produced by the available lead.

$$2.19 \text{ mol Pb} \left(\frac{2 \text{ mol e}^-}{1 \text{ mol Pb}}\right) = 4.38 \text{ mol e}^-$$

(c) Calculate the number of coulombs of charge carried by the moles of electrons produced.

$$4.38 \text{ mol e}^- \left(\frac{9.65 \times 10^4 \text{ coulombs}}{1 \text{ mol e}^-}\right) = 4.23 \times 10^5 \text{ coulombs}$$

(d) Using the charge produced, and its relation to current and time, calculate time.

$$4.23 \times 10^5 \text{ coulombs} = 1.50 \text{ amps} \times \text{time (seconds)}$$
$$\text{Time} = 2.82 \times 10^5 \text{ seconds (or 78.3 hours)}$$

EXERCISE 21.9 Using the Faraday Constant

In the commercial production of sodium by electrolysis (see Section 21.7), the cell operates at 7.0 V and a current of 25×10^3 amps. How many grams of sodium can be produced in one hour?

The appliances in your home are often rated in **watts**, a unit of *electrical power* that tells you the *rate* of energy consumption or production; that is, the watt has units of energy/time and is defined as

$$1 \text{ watt} = 1 \text{ joule/second}$$

This means that a 60-watt light bulb will consume only 60% as much energy per second as a 100-watt light bulb. When you consume electrical power in your home, you pay for it in units of *kilowatt-hours* (kwh). This is the amount of energy required to produce 1000 watts for one hour. Therefore, 1 kwh is

$$1 \text{ kwh} = (1000 \text{ watt})(1 \text{ hr})\left(\frac{3600 \text{ s}}{\text{hr}}\right)\left(\frac{1 \text{ J/s}}{\text{watt}}\right) = 3.60 \times 10^6 \text{ J}$$

But how is this related to electrochemistry and to the quantities we have been working with, that is, to current, time, and charge? The connection is provided by the fact that

$$1 \text{ joule} = 1 \text{ coulomb} \cdot \text{volt}$$

Therefore, if we know an electrolysis cell operates at a given voltage, and some number of coulombs of charge pass through the cell (coulombs = amps × seconds), we can find the energy consumed in joules or in kilowatt-hours. Such calculations are clearly important in the production of metals by electrolysis, or in the design of a battery.

EXAMPLE 21.11

ENERGY EXPENDITURE IN AN ELECTROCHEMICAL PROCESS

Hydrogen has great promise as a fuel in our economy, and, as illustrated in Example 21.8, it can be produced by electrolysis of water (Figure 22.2). Electrolyzing a solution of sulfuric acid will lead to the production of $H_2(g)$ at the cathode and $O_2(g)$ at the anode. If the minimum voltage required is 1.23 V (no overvoltage is considered), calculate the number of kilowatt-hours of energy required to produce 1.00 kilogram of $H_2(g)$.

Solution In this problem the conversion sequence is

$$\text{Mass} \longrightarrow \text{moles} \longrightarrow \text{charge} \longrightarrow \text{energy}$$

(a) Calculate the number of coulombs required. One mole of $H_2(g)$ will require 2 moles of electrons.

$$2\ H_3O^+(aq) + 2\ e^- \longrightarrow H_2(g) + 2\ H_2O(\ell)$$

$$1.00 \times 10^3\ g \left(\frac{1\ mol\ H_2}{2.016\ g\ H_2}\right) = 4.96 \times 10^2 \text{ moles of } H_2$$

$$4.96 \times 10^2\ mol\ H_2 \left(\frac{2\ mol\ e^-}{1\ mol\ H_2}\right) = 9.92 \times 10^2 \text{ moles } e^-$$

$$9.92 \times 10^2\ mol\ e^- \left(\frac{9.65 \times 10^4\ coulombs}{1\ mol\ e^-}\right) = 9.57 \times 10^7 \text{ coulombs}$$

(b) Calculate the number of joules of energy required.

$$\text{Number of joules} = \text{coulombs} \times \text{volts} = (9.57 \times 10^7\ \text{coulombs})(1.23\ V) = 1.18 \times 10^8 \text{ joules}$$

(c) Calculate the number of kilowatt-hours.

$$\text{kwh} = 1.18 \times 10^8\ J \left(\frac{1\ kwh}{3.60 \times 10^6\ J}\right) = 32.7 \text{ kilowatt-hours}$$

EXERCISE 21.10 Energy Expenditure in Sodium Production
Using the information from Exercise 21.9, calculate the kilowatt-hours of energy used in the production of sodium by electrolysis.

21.7 THE COMMERCIAL PRODUCTION OF CHEMICALS BY ELECTROCHEMICAL METHODS

ALUMINUM

Figure 21.17 The aluminum we use in our economy is made by electrolysis.

Aluminum is the third most abundant element in the earth's crust and has many important uses in our economy (Figure 21.17). You probably know it best in its use in the kitchen as a food wrapper, a use demonstrating its excellent formability. Just as importantly, aluminum has a low density and excellent corrosion resistance; the latter property comes from the fact that a transparent, chemically inert film of aluminum oxide, Al_2O_3, clings tightly to the metal's surface. It is these properties that have led to the many other uses of aluminum in aircraft parts, ladders, and automobile parts, for example.

As with some other commercially important metals, the history of the development of a practical method for aluminum production is inter-

Figure 21.18 The medieval village of Les Baux in the south of France. The aluminum-containing mineral bauxite gets its name from this town, near which the mineral is still found in commercial quantities.

esting. Aluminum was originally made in the 19th century by reducing $AlCl_3$ with sodium,

$$3\ Na(s) + AlCl_3(s) \longrightarrow Al(s) + 3\ NaCl(s)$$

but only at a very high cost. It was therefore considered a precious metal, chiefly used in jewelry. In fact, in the 1855 Exposition in Paris (France), some of the first aluminum metal produced was exhibited along with the crown jewels of France. Napoleon III saw its possibilities for military use, however, and commissioned studies on improving its production. The French had a ready source of aluminum-containing ore, bauxite (Figure 21.18), and in 1886 a 23-year-old Frenchman, Paul Heroult, conceived the electrochemical method in use today. In an interesting coincidence, an American, Charles Hall, who was only 22 at the time, announced his invention of the identical process in the same year. Hence, the commercial process is now known as the Hall–Heroult process.

The essential features of the Hall–Heroult process are illustrated in Figure 21.19. The aluminum-containing ore, chiefly in the form of Al_2O_3, is mixed with cryolite, Na_3AlF_6, at about 980 °C and electrolyzed. Aluminum is produced at the cathode and oxygen at the anode, both electrodes being graphite. The cells operate at the very low voltage of 4.0 to 5.5 V, but at a current of 50,000 to 150,000 amps. Each kilogram of aluminum requires 13 to 16 kwh of energy, exclusive of that required to heat the furnace. This is the reason there is so much interest in recycling soft drink cans and other aluminum objects. The recycled metal can be

Charles Martin Hall was only 22 years old in 1886 when he discovered the electrolytic process for extracting aluminum from Al_2O_3 in a woodshed behind his family's home in Oberlin, Ohio. Following his discovery, Hall went on to found the company that eventually became Aluminum Corporation of America. He died a millionaire in 1914.

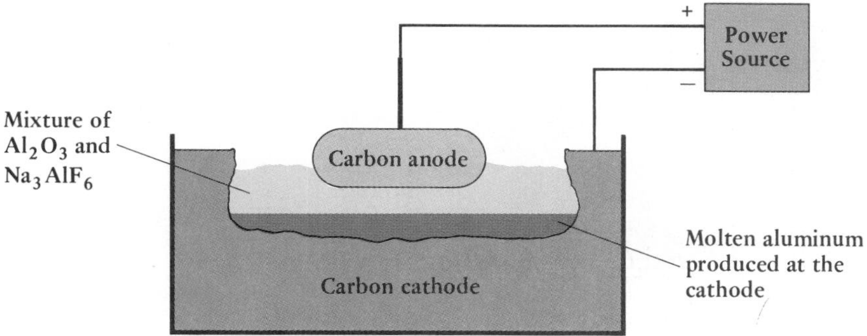

Figure 21.19 Industrial production of aluminum by electrolysis. Purified aluminum-containing ore (bauxite), essentially Al_2O_3, is mixed with cryolite (Na_3AlF_6) to give a mixture that melts at a lower temperature than Al_2O_3 alone. The aluminum-containing materials are reduced at the graphite cathode to give molten aluminum. Oxygen is produced at the carbon anode, and the gas reacts slowly with the carbon to give CO_2, leading to eventual loss of the electrode.

purified and made into new materials at a fraction of the cost of making aluminum from the ore.

SODIUM

The English chemist Sir Humphry Davy first isolated sodium in 1807 by the electrolysis of molten sodium hydroxide. However, the element was a laboratory curiosity until, in 1824, it was found that sodium reduced aluminum chloride to aluminum. The possibility of producing commercially useful aluminum led to considerable interest in manufacturing sodium. By 1886 a practical method for sodium production had been devised (the reduction of NaOH with carbon), but in the same year Hall and Heroult invented the electrolytic method for aluminum production. This of course removed the primary market for sodium, at least temporarily. However, other uses for sodium soon developed, and by 1921 the Downs process for making sodium allowed the low-cost production of ton quantities of sodium. A less expensive process for producing a metal leads to more uses, and sodium soon found a market in the manufacture of tetraethyllead [$Pb(C_2H_5)_4$], a compound added to gasoline in small amounts to raise the octane rating. We now know that this lead compound has had a great environmental impact, and it is no longer used. Once again, the market for sodium has declined severely. Nonetheless, the capacity for sodium manufacture in the United States is around 76,000 tons per year, and much of that is centered on Niagara Falls, New York, where relatively low-cost electric power is available from hydroelectric plants.

Sodium is now used mostly as it was earlier: to chemically reduce metal halides to the metal. For example, 4 Na(s) + $TiCl_4$(ℓ) → Ti(s) + 4 NaCl(s).

The Downs cell is used to make Na by the electrolysis of molten NaCl (Figure 21.20) (see also page 876). The cell operates at 7 to 8 V and with currents of 25,000 to 40,000 amps. The cell is filled with a mixture of dry NaCl, $CaCl_2$, and $BaCl_2$. Pure NaCl is not used because it melts

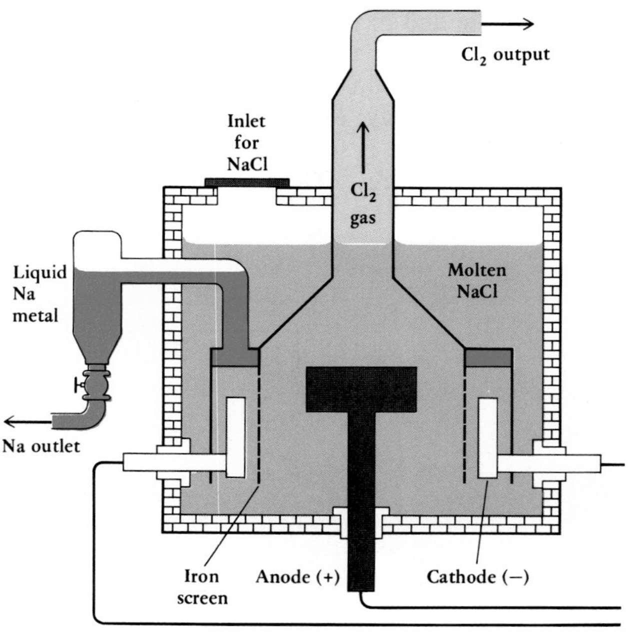

Figure 21.20 The Downs cell for the electrolysis of molten NaCl. The circular iron cathode is separated from the graphite anode by an iron screen. Since the cell operates at about 600 °C, the sodium is produced in the molten state. The liquid metal has a low density, so it floats to the top and can be drawn off. The chlorine gas bubbles out of the anode compartment and is collected.

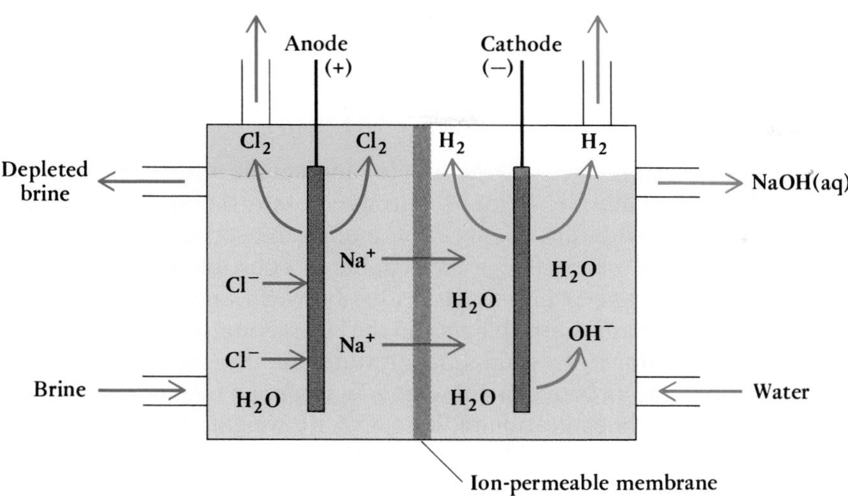

Figure 21.21 A simplified drawing of a membrane cell for the production of NaOH and Cl_2 gas from a saturated, aqueous solution of NaCl (brine). Here the anode and cathode compartments are separated by a water-impermeable but ion-conducting membrane. A widely used membrane is made of Nafion, a fluorine-containing polymer that is a relative of Teflon. Brine is fed into the anode compartment and dilute sodium hydroxide or water into the cathode compartment. Overflow pipes carry the evolved gases and NaOH away from the chambers of the electrolysis cell.

at 800 °C. However, as you learned in Chapter 14, the melting point of a solvent can be lowered by adding a solute; in this case addition of calcium and barium chlorides gives a mixture melting at about 600 °C.

The sodium in the Downs cell is produced at a copper or iron cathode that surrounds a circular graphite anode. Immediately over the cathode there is an inverted trough in which the low-density, molten sodium metal collects, and the valuable by-product, Cl_2 gas, is collected over the anode.

CHLORINE AND SODIUM HYDROXIDE

Chlorine is used to treat water and sewage and in the production of organic chemicals such as pesticides and vinyl chloride, the building block of plastics called PVCs (polyvinyl chloride). In 1989 chlorine was eighth on the list of chemicals produced in the largest amounts in the United States, about 22 billion pounds having been made. Almost all Cl_2 is made by electrolysis, with 95% coming from the electrolysis of brine, a saturated aqueous solution of NaCl. The other product coming from these cells, NaOH, is equally valuable, and over 22 billion pounds were produced in the United States in 1989.

Several different electrolysis methods are used to convert aqueous NaCl to Cl_2 and NaOH, among them the mercury process and the membrane process. As the name implies, the mercury process uses liquid mercury, in this case as the cathode material where aqueous sodium ion is reduced to sodium metal, which in turn dissolves in the mercury to form a liquid sodium/mercury solution called an amalgam. Although the process is still used, largely because it produces high-purity sodium hydroxide and Cl_2 gas, the problems of mercury pollution from these cells are enormous. As a consequence, mercury electrolysis cells are being phased out in favor of other methods, chiefly the membrane process.

A very simple diagram of a membrane electrolysis cell is shown in Figure 21.21. As outlined earlier (page 878), the anode reaction is the oxidation of chloride ion to Cl_2 gas

$$2\ Cl^-(aq) \longrightarrow Cl_2(g) + 2\ e^-$$

and the cathode process is the reduction of water.

$$2\ H_2O(\ell) + 2\ e^- \longrightarrow H_2(g) + 2\ OH^-(aq)$$

One of the greatest difficulties in getting membrane cells to be economical has been to find a material for the membranes. They must be stable at high salt concentrations, stable to a high difference in pH from one side to the other, and stable to such strong oxidizing agents as Cl_2.

Activated titanium is used for the anode, and stainless steel or nickel is preferred for the cathode. Brine is introduced into the anode chamber, and chloride ion oxidation occurs. The membrane separating the anode and cathode is not permeable to water, but it does allow ions to pass; that is, the membrane is just a salt bridge between anode and cathode. Therefore, to maintain charge balance in the cell, sodium ions pass through the membrane. Since the reduction of water at the cathode produces hydroxide ion, the product in the cathode chamber is aqueous sodium hydroxide with a concentration of 20 to 35% by weight. The energy consumption of these cells is in the range of 2000 to 2500 kwh/ton of NaOH produced.

21.8 CORROSION: AN EXAMPLE OF OXIDATION–REDUCTION REACTIONS

Corrosion is the deterioration of metals, usually with loss of metal to a solution in some form, by an oxidation–reduction reaction. The corrosion of iron is the conversion of the metal to red-brown rust, which consists of hydrated iron(III) oxide [$Fe_2O_3 \cdot H_2O$] and other products. The significance of this is that 25% of the annual steel production in the United States is estimated to be for the replacement of material lost to corrosion.

Metals generally are found in the earth as their oxides or similar compounds; in the presence of oxygen, metal oxides are thermodynamically more stable than the pure metals. Thus, all refined metals (except perhaps silver, gold, and platinum) will spontaneously revert to some oxidized state. For example,

$$4\ Fe(s) + 3\ O_2(g) \longrightarrow 2\ Fe_2O_3(s)$$

$$4\ Cu(s) + O_2(g) \longrightarrow 2\ Cu_2O(s)$$

$$4\ Al(s) + 3\ O_2(g) \longrightarrow 2\ Al_2O_3(s)$$

The oxide film that forms on the surface of a metal can be quite hard and can protect the metal against further oxidation. This is the reason medieval

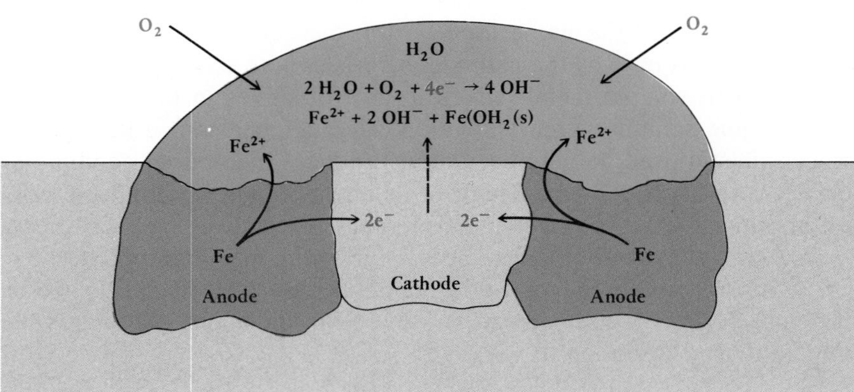

Figure 21.22 The reactions that occur when iron corrodes in an aqueous environment with oxygen present.

makers of armor and swords so carefully heated these metal objects; a microscopically thin coating of metal oxide formed to protect the beauty of the object. However, if this film is broken, the metal is exposed to outside elements, and corrosion can begin.

For corrosion to occur at the surface of a metal, there must be anodic areas where the metal can be oxidized to metal ions as electrons are produced,

Anode, oxidation: $M(s) \longrightarrow M^{n+} + n\, e^-$

and cathodic areas where the electrons are consumed by any or all of several possible half-reactions, such as

Cathode, reduction: $2\, H_2O(\ell) + 2\, e^- \longrightarrow H_2(g) + 2\, OH^-(aq)$

$$O_2(g) + 2\, H_2O(\ell) + 4\, e^- \longrightarrow 4\, OH^-(aq)$$

Anodic areas occur at cracks in the oxide coating, at boundaries between phases, or around impurities. The cathodic areas occur at the metal oxide coating, at less reactive metallic impurity sites, or around other metal compounds such as sulfides.

The other requirements for corrosion are (a) an electrical connection between the anode and cathode and (b) an electrolyte with which both anode and cathode are in contact. Both requirements are easily fulfilled, as seen in Figures 21.22 and 21.23.

If the relative rates of the anodic and cathodic corrosion reactions could be measured independently, the anodic reaction would be found to be the faster. When the two reactions are coupled to each other as in a corroding metal, however, the overall rate can only be that of the slower process. In general, this means that corrosion is controlled by the relative rate of the cathodic process. This fact helps to explain the basic chemistry of corroding systems and how to prevent corrosion.

In the corrosion of iron, the anodic reaction is clearly the oxidation of iron. However, as just noted, there are three possible cathodic reactions; which of these is the fastest depends, in part, on the acidity of the surrounding solution and the amount of oxygen present. When little or no oxygen is present—as when a piece of iron is buried in soil such as moist clay—hydrogen ion and water are reduced. As indicated in the

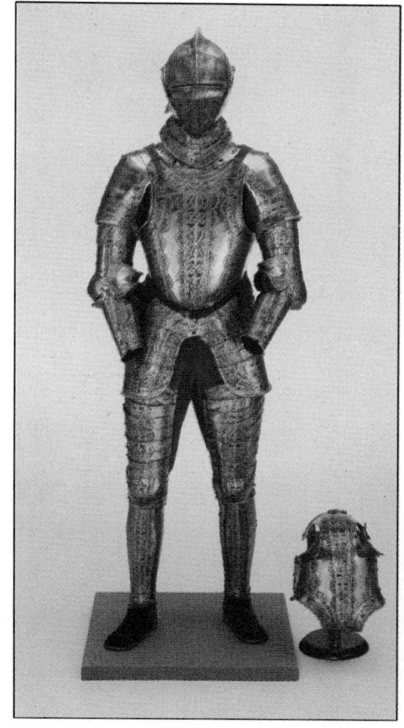

Medieval armor makers knew how to treat the metal to make it corrosion resistant. Apparently a thin, uniform coating of metal oxide prevents the armor from being attacked by the elements. (Philadelphia Museum of Art: Bequest of Carl Otto Kretzxchmar von Keinbusch)

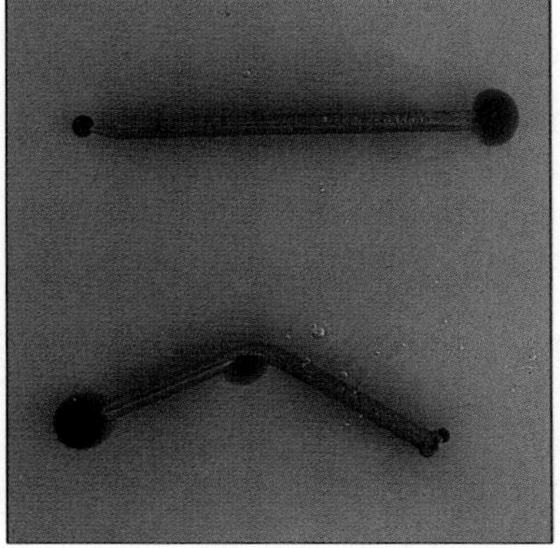

Figure 21.23 Corroding iron nails. Two nails were placed in an agar gel, which also contained the indicator phenolphthalein and $[Fe(CN)_6]^{3-}$. The nails began to corrode and gave Fe^{2+} ions at the tip and where the nail is bent. That these points are the anode is indicated by the blue color of Prussian blue, a compound of $[Fe(CN)_6]^{3-}$ and Fe^{2+}. The remainder of the nail is the cathode, since oxygen is reduced in water to give OH^- (see Figure 21.22). The presence of the OH^- ion is indicated by the red color of the indicator.

reactions above, $H_2(g)$ and hydroxide ions are the products. Since iron(II) hydroxide is relatively insoluble, it can precipitate on the metal surface and inhibit the further formation of Fe^{2+} at the anodic site.

Anode, oxidation:
$Fe(s) \longrightarrow Fe^{2+}(aq) + 2\ e^-$
Cathode, reduction:
$2\ H_2O(\ell) + 2\ e^- \longrightarrow H_2(g) + 2\ OH^-(aq)$

Net reaction:
$$Fe(s) + 2\ H_2O(\ell) \longrightarrow H_2(g) + \underbrace{Fe^{2+}(aq) + 2\ OH^-(aq)}_{\downarrow \atop Fe(OH)_2(s)}$$

The $Fe(OH)_2$ coating, and the slowness of the H_2O reduction, mean that corrosion under oxygen-free conditions is slow.

If both water and O_2 are present, the chemistry of iron corrosion is somewhat different, and the corrosion reaction is about 100 times faster than without oxygen.

Anode, oxidation:
$2[Fe(s) \longrightarrow Fe^{2+}(aq) + 2\ e^-]$
Cathode, reduction:
$O_2(g) + 2\ H_2O(\ell) + 4\ e^- \longrightarrow 4\ OH^-(aq)$

Net reaction:
$2\ Fe(s) + 2\ H_2O(\ell) + O_2(g) \longrightarrow 2\ Fe(OH)_2(s)$

If oxygen is not freely available, further oxidation of iron(II) hydroxide is limited to the formation of magnetic iron oxide (which can be thought of as a mixed oxide of Fe_2O_3 and FeO).

$$6\ Fe(OH)_2(s) + O_2(g) \longrightarrow \underset{\text{green hydrated magnetite}}{2\ Fe_3O_4 \cdot H_2O(s)} + 4\ H_2O(\ell)$$

$$Fe_3O_4 \cdot H_2O(s) \longrightarrow H_2O(\ell) + \underset{\text{black magnetite}}{Fe_3O_4(s)}$$

It is the black magnetite that you find coating an iron object that has corroded by resting in moist soil. On the other hand, if the iron object has free access to oxygen and water, as in the open or in flowing water, red-brown iron(III) oxide will form (Figure 21.1).

$$4\ Fe(OH)_2(s) + O_2(g) \longrightarrow \underset{\text{red-brown}}{2\ Fe_2O_3 \cdot H_2O(s)} + 2\ H_2O(\ell)$$

This is the familiar rust you see on cars and buildings, and the substance that colors the water red in some mountain streams or in your home.

Other substances in air and water can assist in corrosion. Chlorides, from sea air or from salt spread on roads in the winter, are notorious. Since the chloride ion is relatively small, it can diffuse into and through a protective metal oxide coating. Metal chlorides, which are more soluble than metal oxides or hydroxides, can then form. These chloride salts leach back through the oxide coating, and a path is now open for oxygen and water to further attack the underlying metal. This is the reason you often see small pits on the surface of a corroded metal.

Sulfur dioxide is a notorious air pollutant formed in the combustion of oil and coal in power plants and homes. It is 1300 times more soluble in water than is O_2, and after further oxidation it can form sulfuric acid solutions—the reason for so-called "acid rain." It is these solutions that greatly accelerate the deterioration of buildings and art objects in a polluted environment. The following redox reactions are believed to be responsible for the corrosion of iron objects when SO_2 is present.

$$Fe(s) + SO_2(g) + O_2(g) \longrightarrow FeSO_4(s)$$

$$4\ FeSO_4(s) + O_2(g) + 6\ H_2O(\ell) \longrightarrow 2\ Fe_2O_3 \cdot H_2O(s) + 4\ H_2SO_4(aq)$$

$$4\ H_2SO_4(aq) + 4\ Fe(s) + 2\ O_2(g) \longrightarrow 4\ FeSO_4(s) + 4\ H_2O(\ell)$$

Once these reactions have begun, the sulfuric acid that forms is difficult to remove. In fact, even if an iron object is carefully cleaned, corrosion will continue as long as sulfates are present.

How can you stop a metal object from corroding? There are many methods, some more effective than others, but none totally successful. The general approaches are (a) to inhibit the anodic process, (b) to inhibit the cathodic process, or (c) to do both. The usual method is **anodic inhibition**, attempting directly to prevent the oxidation reaction by painting the metal surface or by allowing a thin oxide film to form. More recently developed methods are illustrated by the following reaction:

$$2\ Fe(s) + 2\ Na_2CrO_4(aq) + 2\ H_2O(\ell) \longrightarrow Fe_2O_3(s) + Cr_2O_3(s) + 4\ NaOH(aq)$$

An iron surface is oxidized by a chromium(VI) salt to give iron(III) and chromium(III) oxides. These form a coating impervious to O_2 and water, and further atmospheric oxidation is inhibited.

There are several other ways to inhibit metal oxidation, and one of these is to force the metal to become the cathode, instead of the anode, in an electrochemical cell. Hence, this is called **cathodic protection**. This is usually done by attaching another, more readily oxidized metal. The best example of this is **galvanized iron**, iron that has been coated with a thin film of zinc (Figure 21.24). The $E°$ for zinc oxidation is considerably more positive than $E°$ for iron oxidation (Table 21.1), so the zinc metal film is oxidized before any of the iron. Thus, the zinc coating forms what is called a **sacrificial anode**. Another reason for using a zinc coating on iron is that, when the zinc is corroded, $Zn(OH)_2$ forms on the surface.

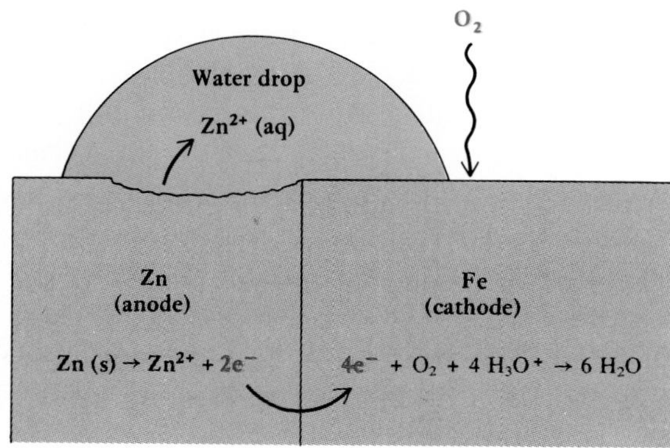

Figure 21.24 Cathodic protection of an iron-containing object. The iron is coated with a film of zinc, a metal more easily oxidized than iron. Therefore, the zinc acts as an anode, and forces iron to become the cathode, thereby preventing the corrosion of the iron.

Figure 21.25 Cathodic protection of a ship's hull. Platinum-coated titanium blocks (the four strips in line with the propeller shaft) are made the anode of an electrolysis cell, and so the hull is forced to become the cathode. Since reduction always occurs at a cathode, the oxidation of the iron of the hull is prevented. (Alternatively, iron-hulled boats or ships can be protected by attaching blocks of zinc.)

This hydroxide is even less soluble than $Fe(OH)_2$, so the insoluble hydroxide film further slows corrosion.

The corrosion of the hulls of ships in sea water is a major problem. One way to prevent it is to attach small blocks of platinum-coated titanium to the hull (Figure 21.25). Platinum is a rather inert metal, not subject to corrosion under the usual environmental conditions. By attaching the block to the ship's hull and then to a battery, the block can be made the anode of an electrolysis cell. This forces the ship's hull to be the cathode, and oxidation of iron in the steel hull is prevented. The same technique is used in oil wells, for example, to prevent the corrosion of steel pipes deep underground.

SUMMARY

A spontaneous oxidation–reduction or electron transfer reaction can be the basis of an electrochemical device called a **voltaic** or **galvanic cell**. A **battery** is a voltaic cell or collection of such cells that can produce a direct electric current at a constant voltage. In a voltaic cell, the oxidation half-reaction occurs at the **anode**, which supplies electrons to the external circuit (and so has a negative polarity). The **cathode** receives electrons from the external circuit and is therefore the site of the reduction half-reaction (and so has a positive polarity). Charge balance inside the cell is maintained by a flow of ions across a **salt bridge** separating the anode and cathode compartments (Section 21.1).

The potential of an electrochemical cell operating at standard conditions (all dissolved species at 1.0 M, gases at 1 atm) is $E°$, a quantity directly related to the standard free energy of the reaction by

$$\Delta G° = -nFE° \qquad (21.1)$$

where n is the number of moles of electrons transferred and F is the **Faraday constant** (9.6485309×10^4 J/V · mol) (Section 21.2). Thus, a positive cell potential indicates a spontaneous reaction. Each half-reaction can be assigned a potential relative to the 0.00 V assigned to the half-reaction $2\ H_3O^+(aq) + 2\ e^- \rightarrow H_2(g) + 2\ H_2O(\ell)$, the **standard hydrogen electrode (S.H.E.)** at 25 °C. Thus, species having $E°$ for reduction more negative than 0.00 V are poorer oxidizing agents than $H_3O^+(aq)$, while those having potentials more positive than 0.00 V are better oxidizing agents than $H_3O^+(aq)$. $E°$ values for a variety of reduction half-reactions are collected in Table 21.1 and in Appendix I. An oxidizing half-reaction and a reducing half-reaction can be summed to obtain the net redox reaction. Similarly, the sum of $E°$ values of the two half-reactions is the potential observed for the voltaic cell. If $E°_{net}$ is positive, a spontaneous reaction is indicated (at standard conditions).

When an electrochemical cell operates at nonstandard conditions (Section 21.3), the potential of the cell, E, is given at 25 °C by the **Nernst equation**.

$$E = E° - \frac{0.0592}{n} \log Q \qquad (21.2)$$

where Q = reaction quotient = $[C]^c[D]^d/[A]^a[B]^b$ for the general reaction $aA + bB \rightarrow cC + dD$.

As the reactants are converted to products in an electrochemical cell, concentrations change, and E is given by Equation 21.2. Eventually, equilibrium is attained, and E goes to zero. At this point $Q = K$, and, as shown in Section 21.3,

$$\log K = \frac{nE^\circ}{0.0592} \tag{21.3}$$

Common **batteries** (the LeClanché dry cell, alkaline batteries, the mercury battery) are described in Section 21.4. **Storage batteries** (the lead battery and ni-cad batteries) can be recharged by using an external voltage source to reverse the discharge reaction. **Fuel cells** are voltaic cells in which reactants are continually added or that can be restored after discharge by physically adding new reactants.

Electrolysis is the use of electrical energy to produce chemical change (Section 21.5). Given a choice of electrode reactions, those that occur are those requiring the least potential to overcome their nonspontaneity. In water, this generally means that substances having reduction potentials more positive than about −0.8 V can be reduced. Thus, in aqueous NaCl, water and not Na^+ is reduced at the cathode. At the anode, Cl_2 and not O_2 is produced (because of the effect of overvoltage, which makes the rate of Cl_2 production greater than that of O_2). Commercially, Al, Na, Cl_2, and NaOH are among the chemicals produced in large quantities by electrolysis (Section 21.7).

Since electrical current (in amps) is the number of coulombs that are passed per second in an electric circuit, the number of coulombs of charge that flow in an electrochemical cell can be found by measuring time and current. The quantity of material consumed or produced is related to time and current through the Faraday constant (9.64853×10^4 C/mol). The energy expended in an electrochemical process (Section 21.6) can be determined from the relationship 1 joule = 1 coulomb · volt, and the power expended (rate of energy use or production) in watts is joules per second.

Corrosion reactions (Section 21.8) are examples of electron transfers of great consequence in our environment. Most metals are thermodynamically unstable with respect to oxidation, and corrosion is the oxidative deterioration of these metals. Metals can be protected from corrosion by inhibiting access of the oxidant (**anodic protection**) or by attaching some other, more easily oxidized substance (a **sacrificial anode**; **cathodic protection**).

STUDY QUESTIONS

REVIEW QUESTIONS

1. In each of the following reactions, tell which substance is oxidized and which is reduced. Tell which is the oxidizing agent and which is the reducing agent.
   (a) $2\ Al(s) + 3\ Cl_2(g) \longrightarrow 2\ AlCl_3(s)$
   (b) $8\ H_3O^+(aq) + MnO_4^-(aq) + 5\ Fe^{2+}(aq) \longrightarrow 5\ Fe^{3+}(aq) + Mn^{2+}(aq) + 12\ H_2O(\ell)$
   (c) $FeS(s) + 3\ NO_3^-(aq) + 4\ H_3O^+(aq) \longrightarrow 3\ NO(g) + SO_4^{2-}(aq) + Fe^{3+}(aq) + 6\ H_2O(\ell)$
2. Explain the function of a salt bridge in an electrochemical cell.
3. Tell whether each of the following statements is true or false. If false, rewrite it to make it a correct statement.
   (a) Oxidation always occurs at the anode of an electrochemical cell.
   (b) The anode of a battery is the site of reduction and is negative.

(c) Standard conditions for electrochemical cells are a concentration of 1.0 M for dissolved species and 1 atm pressure for gases.
(d) The potential of a cell does not change with temperature.
(e) All spontaneous oxidation–reduction reactions have an E°_{net} with a negative sign.

4. Tell which phrase (a through d) best completes the sentence. A spontaneous oxidation–reduction reaction has (a) a positive ΔG° and a positive E°, (b) a negative ΔG° and a positive E°, (c) a positive ΔG° and a negative E°, or (d) a negative ΔG° and a negative E°.
5. Tell which phrase (a through c) best completes the sentence. In Table 21.1, Zn(s) can (a) oxidize Fe(s) and Cd(s), (b) reduce Al^{3+} and Mg^{2+}, or (c) reduce Cd^{2+} and Ag^+.
6. Tell whether each of the following statements is true or false. If false, rewrite it to make it a correct statement.
   (a) The value of an electrode potential changes when the half-reaction is multiplied by a factor. That is, E° for ($Li^+ + e^- \rightarrow Li$) is different from that for ($2\ Li^+ + 2\ e^- \rightarrow 2\ Li$).
   (b) Al is the strongest reducing agent listed in Table 21.1.
   (c) The equilibrium constant for an oxidation–reduction can be calculated by using the Nernst equation.
   (d) Changing the concentrations of dissolved substances does not change the potential observed for an electrochemical cell.
7. What are the advantages and disadvantages of lead storage batteries?
8. How does a fuel cell differ from a battery?
9. Explain why the products of electrolysis of molten NaCl differ from those obtained from aqueous NaCl.
10. Describe the electrochemical method for the manufacture of Cl_2 and NaOH.
11. What is the difference between anodic and cathodic protection against corrosion? Explain how each works.

CELLS AND CELL POTENTIALS

12. Copper can reduce silver(I) to metallic silver, a reaction that could in principle be used as a battery.

$$Cu(s) + 2\ Ag^+(aq) \longrightarrow Cu^{2+}(aq) + 2\ Ag(s)$$

   (a) Write the half-reactions involved.
   (b) Which half-reaction is an oxidation and which is a reduction? Which half-reaction occurs in the anode compartment and which in the cathode compartment?
13. Chlorine gas can oxidize zinc metal in a reaction that has been suggested as the basis of a battery.
   (a) Write the half-reactions involved.
   (b) Which half-reaction is an oxidation and which is a reduction? Which half-reaction would occur in the anode compartment and which in the cathode compartment?
14. Suppose the half-cells $Cu^{2+}(aq)/Cu(s)$ and $Sn^{2+}(aq)/Sn(s)$ are to be used as the basis of a battery. If the polarity of the copper electrode is found to be positive and that of the tin electrode is negative, write the half-reaction that occurs in each half-cell. Tell which is the oxidation and which is the reduction. Decide which half-reaction occurs at the anode and which at the cathode. Write the abbreviated notation for the cell.
15. In principle, the half-cells $Fe^{2+}(aq)/Fe(s)$ and $O_2(g)/H_2O$ (in acid solution) could be used as the basis of a battery. If the polarity of the iron electrode is found to be negative, write the half-reaction that occurs in the cell. Tell which is the oxidation and which is the reduction. Decide which half-reaction occurs at the anode and which at the cathode. Write the abbreviated notation for the cell.
16. If the standard potential of the reaction of Cu(s) with $Ag^+(aq)$ in Study Question 12 is +0.46 V, what is the standard free energy change, ΔG°, for the reaction?
17. The standard potential, E°, for the reaction of Zn(s) and $Cl_2(g)$ is +2.12 V (Study Question 13). What is the standard free energy change, ΔG°, for the reaction?
18. Calculate the value of E° for each of the following reactions. Decide whether each is spontaneous in the direction written.
   (a) $2\ I^-(aq) + Zn^{2+}(aq) \longrightarrow I_2(s) + Zn(s)$
   (b) $Zn^{2+}(aq) + Ni(s) \longrightarrow Zn(s) + Ni^{2+}(aq)$
   (c) $2\ Cl^-(aq) + Cu^{2+}(aq) \longrightarrow Cu(s) + Cl_2(g)$
   (d) $Sn^{2+}(aq) + 2\ Br_2(\ell) \longrightarrow Sn^{4+}(aq) + 4\ Br^-(aq)$
19. Calculate the value of E° for each of the following reactions. Decide whether each is spontaneous in the direction written.
   (a) $I_2(s) + Mg(s) \longrightarrow Mg^{2+}(aq) + 2\ I^-(aq)$
   (b) $Ag(s) + Fe^{3+}(aq) \longrightarrow Ag^+(aq) + Fe^{2+}(aq)$
   (c) $Sn^{2+}(aq) + 2\ Ag^+(aq) \longrightarrow Sn^{4+}(aq) + 2\ Ag(s)$
   (d) $2\ Zn(s) + O_2(g) + 2\ H_2O(\ell) \longrightarrow 2\ Zn^{2+}(aq) + 4\ OH^-(aq)$
20. Balance each of the following unbalanced equations. Next calculate the value of the standard potential, E°, for each reaction, and tell whether each is or is not spontaneous as written. (Half-reaction potentials are found in Appendix I.)
   (a) $Sn^{2+}(aq) + Ag(s) \longrightarrow Sn(s) + Ag^+(aq)$
   (b) $Zn(s) + Sn^{4+}(aq) \longrightarrow Sn^{2+}(aq) + Zn^{2+}(aq)$
   (c) $I_2(s) + Br^-(aq) \longrightarrow I^-(aq) + Br_2(\ell)$
21. Balance each of the following unbalanced equations. Next calculate the value of the standard potential, E°, for each reaction, and tell whether each is or is not

spontaneous as written. (Half-reaction potentials are found in Appendix I.)
(a) $Ce^{4+}(aq) + Cl^-(aq) \longrightarrow Ce^{3+}(aq) + Cl_2(g)$
(b) $Cu(s) + NO_3^-(aq) + H_3O^+(aq) \longrightarrow Cu^{2+}(aq) + NO(g) + H_2O(\ell)$
(c) $Fe^{2+}(aq) + Cr_2O_7^{2-}(aq) + H_3O^+(aq) \longrightarrow Fe^{3+}(aq) + Cr^{3+}(aq) + H_2O(\ell)$

22. Consider the following half-reactions:

HALF-REACTION	$E°$ (V)
$Cl_2(g) + 2\ e^- \longrightarrow 2\ Cl^-(aq)$	+1.36
$I_2(s) + 2\ e^- \longrightarrow 2\ I^-(aq)$	+0.535
$Pb^{2+}(aq) + 2\ e^- \longrightarrow Pb(s)$	−0.126
$V^{2+}(aq) + 2\ e^- \longrightarrow V(s)$	−1.18

(a) Which is the weakest oxidizing agent in the list?
(b) Which is the strongest oxidizing agent?
(c) Which is the strongest reducing agent?
(d) Which is the weakest reducing agent?
(e) Will Pb(s) reduce $V^{2+}(aq)$ to V(s)?
(f) Will $I_2(aq)$ oxidize $Cl^-(aq)$ to $Cl_2(g)$?
(g) Name the elements or ions that can be reduced by Pb(s).

23. Consider the following half-reactions:

HALF-REACTION	$E°$ (V)
$Ce^{4+}(aq) + e^- \longrightarrow Ce^{3+}(aq)$	+1.61
$Ag^+(aq) + e^- \longrightarrow Ag(s)$	+0.80
$Hg_2^{2+}(aq) + 2\ e^- \longrightarrow 2\ Hg(\ell)$	+0.79
$Sn^{2+}(aq) + 2\ e^- \longrightarrow Sn(s)$	−0.14
$Ni^{2+}(aq) + 2\ e^- \longrightarrow Ni(s)$	−0.25
$Al^{3+}(aq) + 3\ e^- \longrightarrow Al(s)$	−1.66

(a) Which is the weakest oxidizing agent in the list?
(b) Which is the strongest oxidizing agent?
(c) Which is the strongest reducing agent?
(d) Which is the weakest reducing agent?
(e) Will Sn(s) reduce $Ag^+(aq)$ to Ag(s)?
(f) Will $Hg(\ell)$ reduce $Sn^{2+}(aq)$ to Sn(s)?
(g) Name the ions that can be reduced by Sn(s).
(h) What metals can be oxidized by $Ag^+(aq)$?

24. Use the table of half-reaction potentials in Study Question 22 to answer the following questions:
(a) What reaction leads to the maximum positive standard potential?
(b) If the $I_2(aq)/I^-(aq)$ half-cell is combined with the $V^{2+}(aq)/V(s)$ half-cell, write an equation for the spontaneous reaction that occurs. What is the value of $E°_{net}$ for the reaction?

25. Use the table of half-reaction potentials in Study Question 23 to answer the following questions:
(a) What reaction leads to the maximum positive standard potential?
(b) If the $Ni^{2+}(aq)/Ni(s)$ half-cell is combined with the $Hg_2^{2+}(aq)/Hg(\ell)$ half-cell, write an equation for the spontaneous reaction that occurs. What will be its standard potential?
(c) Write the spontaneous reaction that occurs when the half-reactions $Ag^+(aq)/Ag(s)$ and $Ce^{4+}(aq)/Ce^{3+}(aq)$ are combined. What is the value of $E°_{net}$ for this reaction?

26. Assume that you assemble an electrochemical cell based on the half-reactions $Zn^{2+}(aq)/Zn(s)$ and $Ag^+(aq)/Ag(s)$.
(a) Write the reaction that occurs spontaneously and calculate $E°_{net}$.
(b) Which electrode is the anode and which is the cathode? Write the abbreviated notation for the cell.
(c) Diagram the components of the cell.
(d) If you used a strip of zinc as an electrode, is it the anode or cathode?
(e) Do electrons flow from the Zn electrode to the Ag electrode, or vice versa?
(f) If a salt bridge containing $NaNO_3$ connects the two half-cells, in which direction do the nitrate ions move, from the zinc to the silver compartment, or vice versa?

27. An electrochemical cell uses Al(s) and $Al^{3+}(aq)$ in one compartment and Ag(s) and $Ag^+(aq)$ in the other.
(a) Write a balanced equation for the reaction that will occur spontaneously in this cell, and calculate $E°_{net}$.
(b) Which is the better reducing agent, Ag or Al?
(c) Which is the anode and which is the cathode? Indicate the polarity of each electrode.
(d) Sodium nitrate is in the salt bridge connecting the two half-cells. In which direction do the nitrate ions flow, from Al to Ag or from Ag to Al?

28. Calculate the $E°_{net}$ for the reaction

$$2\ Cr(s) + 3\ Fe^{2+}(aq) \longrightarrow 2\ Cr^{3+}(aq) + 3\ Fe(s)$$

and then sketch a complete picture of an electrochemical cell that might be constructed to carry out the reaction. Include in your diagram the following labels: anode; cathode; the polarity of each electrode [Cr(s) and Fe(s)]; the direction of electron flow in an external wire; and the direction of anion flow in a salt bridge [containing $NaNO_3(aq)$].

29. An electrochemical cell is assembled using the half-cells $Pb^{2+}(aq)/Pb(s)$ and $Ni^{2+}(aq)/Ni(s)$.
(a) Which is the better reducing agent, Pb or Ni?
(b) Write the spontaneous reaction that can occur and calculate $E°_{net}$.
(c) Which electrode is the cathode, Pb or Ni?
(d) What is the direction of electron flow in an external wire, from Pb to Ni or from Ni to Pb?
(e) If $NaNO_3(aq)$ is placed in a salt bridge connecting the two half-cells, in which direction do the anions flow?

30. A simple electrochemical cell has been constructed using a zinc electrode in a solution of 1.0 M $Zn(NO_3)_2$

and a tin electrode in a 1.0 M solution of $Sn(NO_3)_2$.
(a) Write a balanced equation for the reaction that occurs spontaneously.
(b) What is the value for E°_{net}?
(c) Do electrons move in an external wire from Sn to Zn or from Zn to Sn?
(d) Which is the anode compartment and which is the cathode compartment?
(e) What is the polarity of the Sn electrode?

31. The reaction of Zn(s) and $Cl_2(g)$ has been suggested as the basis of a new battery.
(a) If the product is zinc chloride, write a balanced equation describing this reaction.
(b) Write the half-reaction that occurs in the cathode compartment.
(c) What is the polarity of the zinc electrode?
(d) If the reaction occurs under standard conditions, what does this mean in terms of gas pressure, temperature, and concentration of water-soluble species?
(e) What is E°_{net} for the reaction?
(f) If fluorine, $F_2(g)$, is used in place of $Cl_2(g)$, what is the new value of E°_{net}?

CELLS UNDER NONSTANDARD CONDITIONS AND THE CONNECTION BETWEEN E° AND K

32. Calculate the potential produced by an electrochemical cell using the following reaction if all dissolved species are 0.10 M:

$$2\ Fe^{3+}(aq) + 2\ I^-(aq) \longrightarrow 2\ Fe^{2+}(aq) + I_2(s)$$

Does the potential increase or decrease relative to E°_{net}, the standard potential for the reaction?

33. Calculate the potential produced by an electrochemical cell using the following reaction if all dissolved species are 0.015 M:

$$2\ Fe^{2+}(aq) + H_2O_2(aq) + 2\ H_3O^+(aq) \longrightarrow 2\ Fe^{3+}(aq) + 4\ H_2O(\ell)$$

Does the potential increase or decrease relative to E°_{net}, the standard potential for the reaction?

34. An electrochemical cell is constructed with one half-cell in which a silver wire dips into an aqueous solution of $AgNO_3$. The other half-cell consists of an inert platinum wire in an aqueous solution of Fe^{2+} and Fe^{3+}.
(a) What is the reaction occurring when this cell operates spontaneously?
(b) What is E°_{net}?
(c) If $[Ag^+] = 0.10$ M, but $[Fe^{2+}]$ and $[Fe^{3+}]$ are both 1.0 M, what is the value of E_{net}? Is the net cell reaction still that in part (a)? If not, what is the net reaction when the cell operates spontaneously under the new conditions?

35. An electrochemical cell is constructed with one half-cell in which a silver wire dips into a 1.0 M aqueous solution of $AgNO_3$. The other half-cell consists of a zinc electrode in a solution of 1.0 M $Zn(NO_3)_2$.
(a) What reaction occurs when the cell operates spontaneously?
(b) What is E°_{net}?
(c) What is the polarity of the Zn electrode?
(d) Do electrons flow from Zn to Ag or from Ag to Zn?
(e) If Ag^+ has a concentration of 0.50 M, but Zn^{2+} is held at 1.0 M, what is the potential of the cell? Compare this with E°_{net} and comment on the effect of the change in silver concentration.

36. Calculate equilibrium constants for the following reactions:
(a) $2\ Fe^{3+}(aq) + 2\ I^-(aq) \longrightarrow 2\ Fe^{2+}(aq) + I_2(s)$
(b) $I_2(s) + 2\ Br^-(aq) \longrightarrow 2\ I^-(aq) + Br_2(\ell)$

37. Calculate equilibrium constants for the following reactions:
(a) $Zn^{2+}(aq) + Ni(s) \longrightarrow Zn(s) + Ni^{2+}(aq)$
(b) $Cu(s) + 2\ Ag^+(aq) \longrightarrow Cu^{2+}(aq) + 2\ Ag(s)$

ELECTROLYSIS, ELECTRICAL ENERGY, AND POWER

38. A current of 0.015 amps is passed through a solution of $AgNO_3$ for 155 minutes. What mass of silver is deposited at the cathode?

39. Current is passed through a solution containing $Ag^+(aq)$. How much silver was in the solution if all the silver was removed completely as Ag metal by electrolysis for 14.5 min at a current of 1.0 mA (1 mA = 10^{-3} amp)?

40. A current of 2.50 amps is passed through a solution of $Cu(NO_3)_2$ for 2.00 hours. What mass of copper is deposited at the cathode?

41. A current of 0.0125 amp is passed through a solution of $CuCl_2$ for 2.00 hours. What mass of copper is deposited at the cathode and what mass of chlorine gas is produced at the anode?

42. An old method of measuring the current flowing in a circuit was to use a "silver coulometer." The current passed first through a solution of $Ag^+(aq)$ and then into another solution containing an electroactive species. The amount of silver metal deposited at the cathode was weighed, and the time was noted, so the current could be calculated. If 0.052 g of Ag was deposited during a 450-second experiment, what was the current flowing in the circuit?

43. As noted in Study Question 42, a "silver coulometer" was used in the past to measure the current flowing in an electrochemical cell. Let us say that you found that the current flowing through an electrolysis cell led to a deposit of 0.089 g of Ag metal at the cathode after exactly 10. minutes. If this same current then flowed through a cell containing gold(III) ion in the form of $[AuCl_4]^-$, how much gold was deposited at the cathode in that electrolysis cell?

44. The basic reaction occurring in the cell in which Al_2O_3 and aluminum salts are electrolyzed is $Al^{3+} + 3\ e^- \rightarrow Al(s)$. If the cell operates at 5.0 V and 1.0×10^5 amps, how many grams of aluminum metal can be produced in an 8.0 hour day?

45. The vanadium(II) ion can be produced by electrolysis of a vanadium(III) salt in solution. How long must you carry out an electrolysis if you wish to convert completely 0.125 L of 0.015 M $V^{3+}(aq)$ to $V^{2+}(aq)$ and the current is 0.268 amp?

46. The reactions occurring in a lead storage battery are given in Section 21.4. A typical battery might be rated at "50 ampere-hours." This means that it has the capacity to deliver 50. amps for 1.0 hour or 1.0 amp for 50. hours. If it does deliver 1.0 amp for 50. hours, how many grams of lead would be consumed to accomplish this?

47. It has been demonstrated that an effective battery can be built using the reaction between Al metal and O_2 from the air. If the Al anode of this battery consists of a 3-ounce piece of aluminum (84 g), for how many hours can the battery produce 1.0 amp of electricity (assuming an unlimited supply of O_2)?

48. A lead storage battery (Section 21.4) operates at 12.0 V and is rated at "100 ampere-hours." The latter term means it can deliver 1.0 amp of current for 100. hours. What is the power rating of the battery in watts?

49. An electrolysis cell for aluminum production operates at 5.0 V and a current of 1.0×10^5 amps. Calculate the number of kilowatt-hours of energy required to produce 1 ton (2.0×10^3 pounds) of aluminum.

50. Electrolysis of brine leads to chlorine gas. In 1988, 22.66 billion pounds of Cl_2 were manufactured. How many kilowatt-hours of energy must have been used to produce this amount of chlorine from NaCl? Assume the electrolysis cells operate at 4.6 V and 3.0×10^5 amps.

51. Electrolysis of molten NaCl is done in a Downs cell operating at 7.0 volts and 4.0×10^4 amps. How much Na(s) and $Cl_2(g)$ can be produced in 24 hours in such a cell? What is the energy consumption in kilowatt-hours? (Assume 100% efficiency.)

52. An aqueous solution of KI is placed in a beaker with two inert platinum electrodes. When the cell is attached to an external battery, electrolysis occurs.
(a) Write the half-reaction occurring at the anode. What is the polarity of the electrode?
(b) Write the half-reaction occurring at the cathode. What is the polarity of the electrode?
(c) If 0.050 amp flows through the cell for 5.0 hours, how many grams of each product are expected?

53. An aqueous solution of $NiCl_2$ is placed in a beaker with two inert platinum electrodes. When the cell is attached to an external battery, electrolysis occurs.
(a) Write the half-reaction occurring at the anode. What is the polarity of the electrode?
(b) Write the half-reaction occurring at the cathode. What is the polarity of the electrode?
(c) If 0.0250 amp flows through the cell for 1.25 hours, how many grams of each product are expected?

54. For each of the following solutions, tell what happens at the anode and at the cathode during electrolysis.
(a) KBr(aq)
(b) NaF(molten)
(c) NaF(aq)

55. For each of the following solutions, tell what happens at the anode and at the cathode during electrolysis.
(a) $NiBr_2(aq)$
(b) NaI(aq)
(c) $CdCl_2(aq)$

GENERAL QUESTIONS

56. Using the table of reduction potentials (Table 21.1), place the following elements in order of increasing ability to function as reducing agents: (a) Cl^-; (b) Fe; (c) Ag; and (d) Na.

57. Four metals, A, B, C, and D, exhibit the following properties:
(a) Only A and C react with 1.0 M hydrochloric acid to give $H_2(g)$.
(b) When C is added to solutions of the ions of the other metals, metallic B, D, and A are formed.
(c) Metal D reduces B^{n+} to give metallic B and D^{n+}.
Based on this information, arrange the four metals in order of increasing ability to act as reducing agents.

58. You are told to assemble an electrochemical cell with $Cl_2(g)/Cl^-(aq)$ as one half-cell. The other half-cell could be $Al^{3+}(aq)/Al(s)$, $Mg^{2+}(aq)/Mg(s)$, or $Zn^{2+}(aq)/Zn(s)$. Which of the metal ion/metal combinations would you choose to produce the largest possible positive E°_{net}? Write a balanced equation for the reaction you have chosen.

59. Ni-cad batteries are rechargeable and are commonly used in cordless appliances. Although such batteries actually function under basic conditions, imagine an electrochemical cell using the setup below.

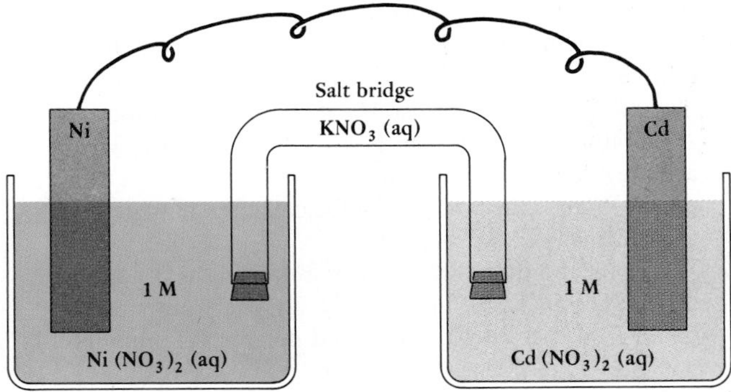

(a) Write a balanced equation depicting the reaction occurring in the cell.
(b) What is oxidized? What is reduced? What is the reducing agent and what is the oxidizing agent?
(c) Which is the anode and which is the cathode? What is the polarity of the Cd electrode?
(d) What is E°_{net} for the cell?
(e) What is the direction of electron flow in the external wire?
(f) If the salt bridge contains KNO_3, toward which compartment will the NO_3^- ions migrate?
(g) Calculate the equilibrium constant for the net reaction.
(h) If the concentration of Cd^{2+} is reduced to 0.010 M, and $[Ni^{2+}]$ = 1.0 M, what is the potential produced by the cell? Is the net reaction still spontaneous in the direction written in part (a)?
(i) You begin with 1.0 L of each of the solutions, and both are initially 1.0 M in dissolved species. Each electrode weighs 50.0 g in the beginning. If 0.050 amp is drawn from the battery, how long can it last?

60. The standard potential for the lead storage battery is 2.04 V. What is the equilibrium constant for the cell reaction?

61. E° for the standard lead storage battery is 2.041 V. Calculate the potential of the battery when the sulfuric acid concentration is 6.00 M at 25 °C.

62. Tell what happens when you bubble pure O_2 (partial pressure = 1 atm) through an acidic aqueous solution containing 1.0 mol/L of each of the following salts.
(a) $FeBr_2$ (b) $CuBr_2$ (c) PbI_2

63. Hydrazine, N_2H_4, can be used as the reducing agent in a fuel cell.

$$N_2H_4(aq) + O_2(g) \longrightarrow N_2(g) + 2\ H_2O(\ell)$$

If ΔG° for the reaction is −607 kJ, calculate the value of E°_{net} expected for the reaction.

64. The corrosion of iron in the presence of water and oxygen gives $Fe(OH)_2$ (Section 21.8). If the pressure of O_2 is 0.20 atm (as it is in air), what is the value of E_{net} under these conditions? Calculate the equilibrium constant, K, for the process.

65. Suppose you use $Cu^{2+}(aq)/Cu(s)$ and $Sn^{2+}(aq)/Sn(s)$ half-cells as the basis of an electrochemical cell (Study Question 14). If the cell starts at standard conditions, and 0.400 amp of current flows for 48.0 hours, what are the concentrations of the dissolved species at this point? What is the potential of the cell after 48.0 hours? (Assume 1.00 L of solution.)

66. A proposed automobile battery involves the reaction of Zn(s) and $Cl_2(g)$ to give $ZnCl_2$. If you want such a battery to operate 10. hours and deliver 1.5 amps of current, what is the minimum mass of zinc that the anode must contain?

67. In principle, a battery could be made from aluminum metal and chlorine gas.
(a) Write a balanced equation for the reaction that would occur in a battery using $Al^{3+}(aq)/Al(s)$ and $Cl_2(g)/Cl^-(aq)$ half-reactions.
(b) Tell which half-reaction occurs at the anode and which at the cathode. What are the polarities of these electrodes?
(c) Calculate the standard potential, E°_{net}, for the battery.
(d) If the pressure of $Cl_2(g)$ is only 0.50 atm, how is the potential of the battery affected? Is it more positive or more negative than E°_{net}? Calculate E_{net}.
(e) If you want the battery to deliver a current of 0.75 amp, how long can it operate if the aluminum electrode contains 30.0 g of Al? (Assume an unlimited supply of chlorine.)

68. Batteries are listed in a popular mail order catalog by their "cranking power." This is the amount of current the battery can produce for 30 seconds; a typical value is 450 amps. How many coulombs flow through the battery in 30. seconds? If this is a lead storage battery, how much lead (Pb) is consumed in 30. seconds?

69. A solution of 5.0 mg of the compound $C_{10}H_{10}Fe$ was electrolyzed completely at an anode in 15.0 minutes with an average current of 0.00288 amp. How many electrons per molecule of the iron compound are transferred? Was the compound oxidized or reduced? Write a half-reaction showing the electrochemical reaction of the compound.

70. An expensive but lighter alternative to the lead storage battery is the silver-zinc battery. Zinc is the reducing agent and silver oxide the oxidizing agent.

$$Ag_2O(s) + Zn(s) + H_2O(\ell) \longrightarrow Zn(OH)_2(s) + 2\ Ag(s)$$

The electrolyte is 40% KOH, and silver/silver oxide electrodes are separated from zinc/zinc hydroxide electrodes by a plastic sheet that is permeable to hydroxide ion. Under normal operating conditions, the battery has a potential of 1.59 V. (R.C. Plumb, *J. Chem. Educ.* **1973**, *50*, 857.)
(a) How much energy can be produced per gram of reactants in the silver/zinc battery? Assume the battery produces a current of 0.10 amp.
(b) How much energy can be produced per gram of reactants in the standard lead storage battery? Assume the battery produces a current of 0.10 amp at 2.0 V.
(c) Which battery produces the greater energy per gram of reactants?

71. Hydrazine, N_2H_4, has been proposed as the basis of a fuel cell. The reactions are

$$N_2H_4(aq) + 4\ OH^-(aq) \longrightarrow N_2(g) + 4\ H_2O(\ell) + 4\ e^-$$

$$O_2(g) + 2\ H_2O(\ell) + 4\ e^- \longrightarrow 4\ OH^-(aq)$$

(a) Which reaction occurs at the anode and which at the cathode?
(b) What is the net cell reaction?
(c) If the cell is to produce 0.50 amp of current for 50.0 hours, how many grams of hydrazine (N_2H_4) must be present?
(d) How many grams of O_2 must be available to react with the amount of N_2H_4 determined in (c)?

72. Fluorine, F_2, is made by the electrolysis of anhydrous HF (Section 24.3).

$$2\ HF(\ell) \longrightarrow H_2(g) + F_2(g)$$

Typical electrolysis cells operate at 4000 to 6000 amps and 8 to 12 volts; a large-scale plant can produce about 9 tons of liquefied fluorine per day. (a) How many kilowatt-hours of energy are consumed by a cell operating at 6.0×10^3 amps at 12 volts for 24 hours? (b) How many grams of HF are consumed, and how many grams of H_2 and F_2 are produced, in a cell under these conditions?

SUMMARY QUESTIONS

73. A hydrogen/oxygen fuel cell operates on the simple reaction

$$H_2(g) + \tfrac{1}{2}\ O_2(g) \longrightarrow H_2O(\ell)$$

If the cell is designed to produce 1.5 amps of current, and if the hydrogen is contained in a 1.0-L tank at 200. atm pressure at 25 °C, how long can the fuel cell operate before the hydrogen runs out? (Assume there is an unlimited supply of O_2.)

74. Living organisms derive energy from the oxidation of food, typified by glucose.

$$C_6H_{12}O_6(aq) + 6\ O_2(g) \longrightarrow 6\ CO_2(g) + 6\ H_2O(\ell)$$

Electrons in this redox process are transferred from glucose to oxygen in a series of at least 25 steps. It is interesting to calculate the total daily current flow in a typical organism and the rate of energy expenditure (power). (See T.P. Chirpich, *J. Chem. Educ.* **1975**, *52*, 99.)

(a) The molar enthalpy of combustion of glucose is −2800 kJ. If you are on a typical daily diet of 2400 Calories, how many moles of glucose must be consumed in a day if glucose is assumed to be the only source of energy? How many moles of O_2 must be consumed in the oxidation process?
(b) How many moles of electrons must be supplied to reduce the amount of O_2 calculated in part (a)?
(c) Based on the answer in part (b), calculate the current flowing, per second, in your body from the combustion of glucose.
(d) If the average standard potential in the electron transport chain is 1.0 V, what is the rate of energy expenditure in watts?

75. Fluorinated organic compounds are important commercially, as they are used as herbicides, flame retardants, and fire extinguishing agents, among other things. A reaction such as

$$CH_3SO_2F + 3\ HF \longrightarrow CF_3SO_2F + 3\ H_2$$

is actually carried out electrochemically in liquid HF as the solvent.

(a) Draw a dot structure for CH_3SO_2F. (S is the "central" atom and is bonded to two O atoms, an F atom, and a CH_3 group by a S—C bond.) What is the geometry around the "central" S atom? What are the O—S—O and O—S—F bond angles?
(b) If you electrolyze 150 g of CH_3SO_2F, how many grams of HF are required and how many grams of each product can be isolated?
(c) Is H_2 produced at the anode or cathode of the electrolysis cell?
(d) A typical electrolysis cell operates at 8.0 V and a low current such as 250 amps. How many kilowatt-hours of energy does one such cell consume in 24 hours?

PART FIVE

THE CHEMISTRY OF THE ELEMENTS AND THEIR COMPOUNDS

Magnesium burning in air gives magnesium oxide.

Jacqueline K. Barton

Dr. Jacqueline K. Barton is a native New Yorker and was educated in that city. She received her B.A. degree from Barnard College in 1974 and her Ph.D. from Columbia University in 1979. While at Columbia she worked in an area of platinum chemistry closely related to the work done by Dr. Barnett Rosenberg, the subject of the first interview in this book. Following her Ph.D. work, Dr. Barton did further research at Yale University and Bell Laboratories and then joined the faculty of Hunter College. In 1983 she returned to Columbia University where she rose rapidly to the rank of full professor. In the fall of 1989 she assumed her present position as Professor of Chemistry at the California Institute of Technology.

Dr. Barton has done outstanding research in the field of biochemistry, particularly in the design of simple molecular probes to explore the variations in structure and conformation along the DNA helix (the subject of Chapter 27). In spite of having been a research scientist for a relatively short time, she has done important new work and has received many honors. In 1985 she received the Alan T. Waterman Award of the National Science Foundation as the outstanding young scientist in the United States. In 1987 she was the recipient of the American Chemical Society's Eli Lilly Award in Biological Chemistry, and the following year she received the Society's Award in Pure Chemistry. That same year, 1988, she also received the Mayor of New York's Award of Honor in Science and Technology.

We met in Dr. Barton's new office at Cal Tech. As we find with so many chemists, she is interested in art and has a painting by the Spanish artist Miro on her office wall as well as a print by the French artist Vasareley. As we were to see in the interview, this interest in form and color in art carries over into her research.

A background in math—but no chemistry

We have been interested in how the scientists we interviewed came to their careers in chemistry. In Dr. Barton's case she said, "I never took chemistry in high school. Maybe one shouldn't publicize that, but it's the truth. However, I was always very interested in mathematics, so I took a lot of calculus when I was in high school. I also took a course in geometry, and that interest in geometry has carried over into my research, since the sort of science I do now is very much governed by structures and shapes.

"When I went to college I thought that, in addition to taking math, I should take some science courses. I walked into the freshman chemistry class, and there were about 150 people there. However, there was also a small honors class with about 10 students. Even though I hadn't had chemistry before I thought I would try it—and I loved it. What chemistry allowed me to do was to combine the abstract and the real. I was very excited by it."

But, Dr. Barton says it was really the experience of the laboratory that got her interested in chemistry. Like many of us in chemistry, she was fascinated by color changes in reactions and the significance of these observations. However, she was also interested in trying to predict what would happen in a reaction, and, if her prediction was not correct, to try to explain this and then to do more experiments that would solve the puzzle. As she said, "That's really what got me started in science."

In addition, Dr. Barton also had an inspirational teacher and role model, Bernice Segal. "She was an absolute inspiration to me. She gave a magnificient course, and she was a very tough lady who asked a lot of you—and you did it!"

The platinum blues

In our earlier interview with Dr. Rosenberg, he described his discovery of the chemotherapeutic activity of a compound now known as cisplatin (page 2). Dr. Barton also worked on this compound, but her Ph.D. thesis research was mostly on compounds known to chemists for years as the "platinum blues."[1]

"Most platinum compounds are orange or red and yet there are these magnificient blue complexes. What are they? What are their structures, and why are they blue? In fact, when Barney Rosenberg was looking at certain platinum compounds, he also found some 'platinum blues'; they were water soluble, unlike cisplatin, and so people had hopes that they might even be better with respect to chemotherapy. But it turned out that was not the case. Nonetheless, that finding and others got Steve Lippard, my advisor at Columbia, interested in trying to work on the problem. Therefore, I made the type of complex that Rosenberg had made, and it was indeed blue. It's an absolutely beautiful molecule. We solved the structure of the molecule and found that it contains four platinum atoms in a line. We also found that its mixed-valent, where the platinum has an average oxidation number of $2\frac{1}{4}$, a fact that helps to explain its blue color."[2]

[1] Transition metal complexes and their colors are described in Chapter 25.

[2] Superconducting solids contain metal ions of different valences, that is, they are mixed-valent. See Chapter 25.

Research in bioinorganic chemistry

After talking about how Dr. Barton came into chemistry, we discussed her current research in bioinorganic chemistry, the role of metals in biological systems. She has received numerous awards for this work, reflecting the importance that the chemistry community places on the field and her contributions to it.

"The interest of my group is to exploit inorganic chemistry as a tool to ask questions of biological interest and to explore biological molecules. A lot of the work in bioinorganic chemistry thus far has been the exploration of metal centers in biology. Why is blood red? Why does the iron [in heme] do what it does? That's just one example, but there are hundreds of others. Many enzymes and proteins within the body in fact contain metals, and the reason we've looked at blood and then the heme center within it has been because it's colored. An obvious tool that transition metal chemistry provides is color, and so things change color when reactions occur. That is one of the things that fascinated me in the first place.

"Another wonderful thing about transition metal chemistry is that it allows us to build molecules that have interesting shapes and structures depending upon the coordination geometry. In fact, you can create a wealth of different shapes, several of which are chiral, and that's something we take advantage of in particular.[3] What we want to do is make a variety of molecules of different shapes, target these molecules to sites on a DNA strand, and then ask questions

[3] The chirality of molecules is discussed in Chapter 25.

such as 'Does DNA vary in its shape as a function of sequence?'[4] If we think about how proteins bind to DNA, do they also take advantage of shape recognition in binding to one site to activate one gene or turn off another gene? When scientists first wondered about these and other such problems, they would write down a one-dimensional sequence of DNA and would think about it in one-dimensional terms. How does the protein recognize a particular DNA sequence? DNA is clearly not one dimensional. It has a three-dimensional structure, and different sequences of bases will generate different shapes and different forms. Therefore, we think we can build transition metal complexes of particular shapes, target them to particular sequences of bases in DNA, and then use these complexes to plot out the topology of DNA.[5] We can then ask how nature takes advantage of this topology. We want to develop a true molecular understanding, a three-dimensional understanding, of the structure and the shapes of biologically important molecules such as DNA and RNA."

A revolution in chemistry in the last ten years

One of the most important goals of these interviews has been to gain insight into the directions that science in general and chemistry in particular will move in this decade. Dr. Barton put her remarks in the context of recent developments in her field.

"I think our work may be an example of where chemistry is going in general. I think there has been a revolution in chemistry in the past 10 years. The revolution is at the interface between chemistry and biology where we can now ask chemical questions about biological molecules. First of all, we can make biological molecules that are pure. I can now go to a machine called a DNA synthesizer, and I can type in a sequence of DNA; from that sequence I can synthesize a pure material, with full knowledge of where all of the bonds are. Then I can run it through a HPLC and get it 100% pure.[6] Therefore, I can now talk about these biopolymers in chemical terms as molecules rather than as impure cellular extracts. I couldn't do that before."

Not only do we have the ability to prepare biological molecules in pure form, but Dr. Barton stressed that we have the techniques to characterize them in ways that we chemists think about molecules. "The development of new techniques allows us to make a bridge between chemistry and biology and ask chemical questions with molecular detail. It's an exciting time to be doing chemistry, and that is why I see it as a new frontier area."

New chemistry curricula

Dr. Barton stressed the point that it is chemists who are "making new materials and making and exploring biological systems. It's the chemist who looks at questions of molecular detail and asks about structure and its relationship to function." Since this involves so many areas of chemistry, she believes that "we are going to have to stop making divisions between inorganic, physical, analytical, and organic chemistry. We must all do a little bit of each." This is an attitude shared by many in chemical education today, and it means that we should perhaps rethink the curriculum in chemistry in particular and science in general.

No matter what the curricular structure, however, she believes what is important in the education of scientists is "to get across the excitement that now we can know what biologically important molecules look like. And, from knowing what they look like, we can manipulate them and change them a little. Then we ask how those changes affect the function, so we can relate the structure of the molecule and its macroscopic function."

"A protein molecule of average size is so small you could put more than a billion billion of them on the head of a pin. We now know we can manipulate molecules that are of those dimensions and can know exactly what they look like. I can't imagine that we can't get people interested in chemistry if we can get across the excitement that comes from the realization that we are looking at things so small and yet can do surgery on them."

Chemistry is fun

There was one basic theme in all of the interviews that we did for this book, and Dr. Barton expressed it well. "The bottom line is that chemistry is fun, it's addictive, and, if one has a sense of curiosity, it can be tremendously entertaining and

[4] Amino acids, proteins, nucleic acids such as DNA, and other aspects of biochemistry are discussed very briefly in Chapter 27.

[5] The results of such an experiment are seen in the structure of a complex formed between a ruthenium-based coordination compound and DNA in the photograph opening Chapter 27.

[6] An HPLC is a "high pressure liquid chromatograph," an instrument capable of separating one type of molecule from another.

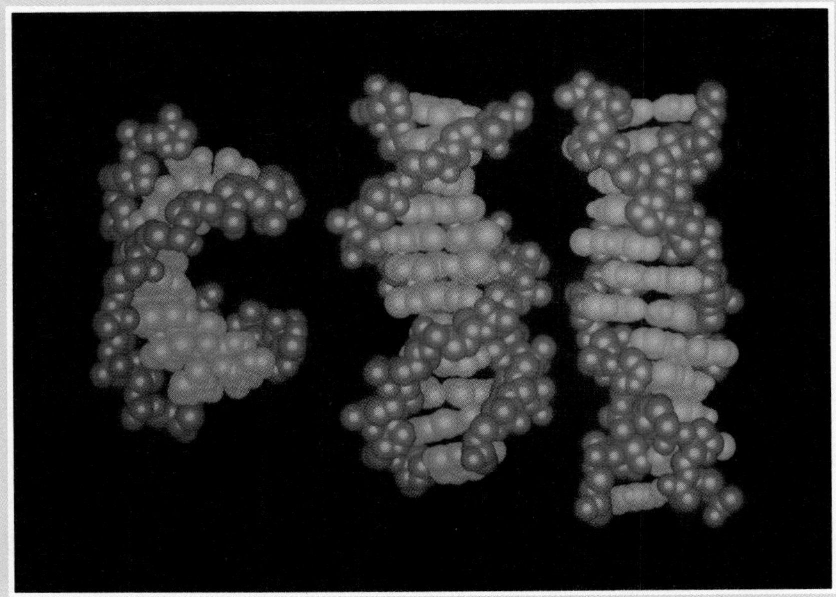

Double helical conformations of DNA: A (left), B (center), and 2 (right). (Jacqueline Barton)

appealing. And it is not so difficult. It's difficult when one thinks about it as rote memorization, which *is* difficult and boring. But that isn't what chemistry is. Chemistry is trying to understand the world around us in some detail. For example, we are interested in knowing such things as what makes skin soft, what makes things different in color, why sugar is sweet,[7] or why a particular pharmaceutical agent makes us feel better."

Women in science

In our earlier interview with Dr. Bassam Shakhashiri, he pointed out that science would welcome many more women to the field, and we were interested in Dr. Barton's perspective on the issue. Did she see special opportunities for women? Are there problems that women need to overcome or be aware of?

"Because I am a woman, and there are so few women currently in professional positions in chemistry, I'm asked those questions often. First of all, I am not an expert on the subject. What I like to think my best contribution to women in chemistry can be is to do the best science I can, and to be recognized for my science, not for being a woman in science. I think that it is generally important when women go into science that they should appreciate that there are no special opportunities; that is, you will be treated like any other person doing science. But just as there should be no special opportunities in that respect, happily—maybe this is naive of me—I think there are also no special detriments or obstacles that one need consider in this day and age. One shouldn't think that 'because I am a woman I can't do it.' That's patently false. In fact, everyone is extremely supportive of women who do science. However, I remember talking to Bernice Segal, my former teacher at Barnard College, and having her explain to me that when she was a graduate student she had to do things behind a curtain, because the women weren't supposed to be doing chemistry. Mildred Cohn, another one of my role models, took over 20 years to have her own independent position as a professor, as opposed to being a laboratory assistant working for someone else. The bottom line is that I don't have a story like that to tell. That's the good news. In my generation there are few such stories of blatant discrimination. Now the world is a much better place for a woman to do science."

Dr. Barton's enthusiasm for her work and for chemistry in general was obvious. It is evident that she will continue to do some of the most important work in science and that her infectious enthusiasm for chemistry will bring many more young people into our profession.

[7] See "Something More About Sweeteners, the Secret is in the Shape" in Chapter 13.

CHAPTER 22

The Chemistry of Hydrogen and the s-Block Elements

Redwall limestone, Grand Canyon. (James Cowlin)

CHAPTER OUTLINE

Hydrogen was the first chemical element formed in the moment of creation, and it is now the most abundant element in the universe. All other elements were formed from it, and virtually every element, with the exception of the rare gases, combines chemically with hydrogen. Indeed, hydrogen-containing compounds are of such importance that you have seen many of them in this book and will be introduced to many more.

The other elements of Groups 1A and 2A of the periodic table are all metals and all have valence shell electron configurations with one or two electrons, respectively. They are clearly important to all of us. Human blood and sea water are buffered NaCl solutions. Ions of four metals in these groups (Na^+, K^+, Mg^{2+}, and Ca^{2+}) constitute 99% of the positive ions in the human body. Eight of the fifty chemicals produced in largest amounts in the United States are based on metal ions of Groups 1A and 2A. Thus, in this chapter you will not only see some compounds of great commercial importance but you will also be treated to a journey through salt and emeralds, a treatment for manic depression, and a metal that could melt in your hand.

1A	2A
1 H hydrogen	
3 Li lithium	4 Be beryllium
11 Na sodium	12 Mg magnesium
19 K potassium	20 Ca calcium
37 Rb rubidium	38 Sr strontium
55 Cs cesium	56 Ba barium
87 Fr francium	88 Ra radium

22.1 THE CHEMISTRY OF HYDROGEN

In spite of the great role played by hydrogen in the creation of the universe and the other chemical elements, it was not until 1766 that Cavendish, an English chemist, actually isolated gaseous H_2. Unfortunately, he did not recognize it as a new substance; rather, he called it "flammable air." Shortly thereafter, however, the gas was found to form water on com-

A modern plant for the production of synthesis gas. Coal, oxygen, and steam are heated under pressure to produce a mixture of gases. In this plant, located in South Africa, the mixture consists of 57% hydrogen plus carbon monoxide, 9.4% methane, 32% carbon dioxide, 0.7% hydrogen sulfide (from sulfur in the coal), 0.3% nitrogen, and 0.5% hydrocarbons. After purification, a mixture of hydrogen, carbon monoxide, and methane is used to manufacture various hydrocarbons. (Fluor Engineers, Inc.)

bustion in air and was given its name, hydrogen, which means "water former."

Little use was made of H_2 until the middle of the 19th century when **coal gas** came into use. When soft coal is heated in a sealed vessel, a gas is given off that contains about 20% hydrogen plus a number of lightweight carbon–hydrogen compounds. It was found that this gas could be used for cooking and lighting, and the solid residue from the process, coke, could be used as a special purpose fuel, specifically in the reduction of iron ore to metallic iron in a blast furnace.

Since coal gas was so useful, new methods were sought for its production. It was then found that water injected into a bed of red hot coke would produce a gas that was a combustible mixture of H_2 and CO, a mixture that came to be known as **water gas** or **synthesis gas**.

$$C(s) + H_2O(g) \longrightarrow \underbrace{H_2(g) + CO(g)}_{\text{water gas}}$$

Water gas burns cleanly, and it can be handled readily. However, it has the serious drawback that the amount of heat produced is only about half that from the combustion of coal gas, and the flame is nearly invisible and produces almost no light. Furthermore, since carbon monoxide is poisonous, water gas is very dangerous. Despite its hazards, water gas was used as a cooking gas to some extent until about 1950, but only after adding other material to make the flame luminous and a malodorous compound to detect leaks. Water gas has been in use almost 100 years, and there is a greatly renewed interest in using it to manufacture hydrocarbons.

In 1783, Jacques Charles first used hydrogen to fill a balloon large enough to float above the French countryside (Chapter 12), a method used in World War I to float observation balloons. The *Graf Zeppelin*, a passenger-carrying dirigible built in Germany in 1928, was filled with hydrogen. She carried over 13,000 people between Germany and the United States until 1937, when she was replaced by the infamous *Hindenburg*. The *Hindenburg* was designed to be filled with helium. However, World War II was approaching, and the United States, which has much of the world's supply of helium, would not sell the gas to Germany; the *Hindenburg* had to use H_2. Unfortunately, when landing in Lakehurst, New Jersey in May, 1939, the *Hindenburg* exploded and burned; of the 62 people on board, only about half escaped uninjured (Figure 22.1). The result was that H_2 gas gained a reputation as a very dangerous substance. Actually, it is as safe to handle as any other fuel, as evidenced by the large quantities used in rockets today.

Figure 22.1 The disastrous end of the hydrogen-filled dirigible *Hindenburg* in May 1939. The accident occurred on landing in Lakehurst, New Jersey. (United Press International)

CHEMICAL AND PHYSICAL PROPERTIES OF HYDROGEN

Hydrogen has three isotopes, two of them stable (protium and deuterium) and one radioactive (tritium).

ISOTOPES OF HYDROGEN

ISOTOPE MASS	SYMBOL	NAME
1.0078 (amu)	^1H (H)	Hydrogen (protium)
2.0141	^2H (D)	Deuterium
3.0160	^3H (T)	Tritium

Of the three isotopes, only H and D are found in measurable quantities in nature (H = 99.985% and D = 0.015%). In contrast, tritium, which is produced by cosmic ray bombardment of water in the atmosphere, is found to the extent of 1 atom per 10^{18} of ordinary hydrogen. Tritium is radioactive, with a half-life of 12.26 years; decay occurs by the loss of an electron (often called a beta particle) and formation of an unstable isotope of helium (Chapter 7).

$$\underset{\text{tritium}}{{}^3_1H} \longrightarrow \underset{\text{helium isotope}}{{}^3_2He} + \underset{\text{electron (beta particle)}}{{}^{\ 0}_{-1}e}$$

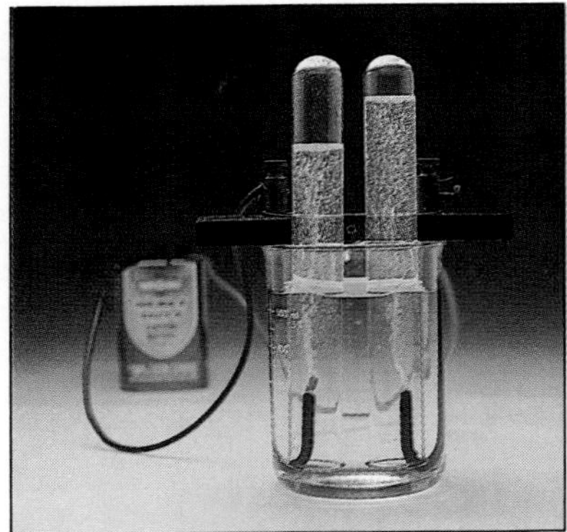

Figure 22.2 Electrolysis of a dilute aqueous sulfuric acid solution gives H_2 and O_2. Notice that the volume ratio is 2 (H_2, left) to 1 (O_2, right), as expected from the 2 to 1 ratio of H to O in water.

Deuterium compounds have been thoroughly studied in the laboratory. One important observation is that, since D has twice the atomic mass of H, deuterium compounds react slightly more slowly than similar protium-containing compounds, and this leads to a way to produce D_2O or "heavy water." Hydrogen can be produced, albeit expensively, by electrolysis (Chapter 21 and Figure 22.2).

$$H_2O(\ell) + \text{electrical energy} \longrightarrow H_2(g) + \tfrac{1}{2}\,O_2(g)$$

In any sample of natural water there is always a small concentration of D_2O. When water is electrolyzed, H_2 is liberated about six times more rapidly than D_2. This means that as the electrolysis proceeds the liquid remaining is increasingly D_2O, and nearly pure D_2O can be obtained eventually. Heavy water is very valuable as a moderator in some nuclear power reactions, such as those used in southeastern Canada.

Deuterium compounds generally have higher melting and boiling points than the analogous protium compounds.

	H_2O	D_2O
Freezing point (°C)	0.0	3.82
Boiling point (°C)	100.0	101.42

Hydrogen will enter into chemical combination with virtually every element, the rare gases being the only exception. Although hydrogen-containing compounds seem to come in a bewildering variety, there are only three different types of binary compounds containing hydrogen.

(a) *Anionic hydrides* form when H_2 interacts with metals of low electronegativity such as Na, Li, and Ca.

$$Na(s) + \tfrac{1}{2}\,H_2(g) \longrightarrow NaH(s)$$

$$Ca(s) + H_2(g) \longrightarrow CaH_2(s)$$

Here the metal formally has a positive oxidation number, while H is in the form of the hydride ion, H^-.

(b) *Covalent hydrides* are formed with electronegative elements such as the nonmetals C, N, O, and F (Figure 22.3). Here the formal oxidation number of the H atom is +1.

$$N_2(g) + 3\,H_2(g) \longrightarrow 2\,NH_3(g)$$

$$F_2(g) + H_2(g) \longrightarrow 2\,HF(g)$$

Figure 22.3 Hydrogen burns in an atmosphere of bromine to give hydrogen bromide.

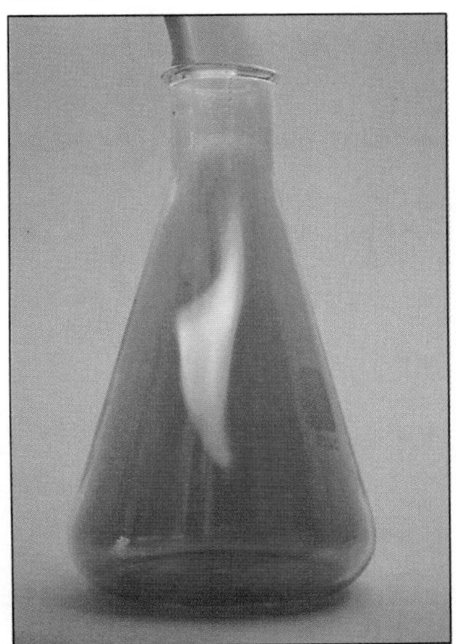

(c) *Interstitial metallic hydrides*. Hydrogen is a very small molecule, and so can be absorbed, apparently as H atoms, into the holes or interstices in the crystal lattice of a metal (Figure 22.4). For example, when a piece of palladium metal is used as an electrode for the electrolysis of water, the metal can soak up a thousand times its volume of H_2 (at STP). However, the H_2 can be driven out of the metal by heating. Interstitial hydrides have been very thoroughly examined as a possible way to store H_2 for later use, the same as a sponge can store water.

THE SYNTHESIS OF HYDROGEN

Virtually all the hydrogen now produced in the world is used immediately in another process, so it is difficult to estimate how much is manufactured. However, we can estimate that approximately 300 billion STP-liters of the gas are produced worldwide in a year, and most of this is made from hydrocarbons such as natural gas, petroleum, or coal.

In today's technology, the method used to produce the largest quantities of H_2 is the catalytic *steam re-formation of hydrocarbons* in a two-step process. The most desirable starting material is natural gas or methane, CH_4. In the first step, CH_4 reacts with steam at high temperature.

$$CH_4(g) + H_2O(g) + 206\ kJ \longrightarrow 3\ H_2(g) + CO(g)$$

The reaction is rapid in the 900 °C to 1000 °C range and goes nearly to completion. At a lower temperature, 400 °C to 500 °C, the CO is used as a reducing agent to extract still more hydrogen from water. In this process, the "water gas shift" reaction, poisonous CO is converted to CO_2, a nontoxic compound.

$$H_2O(g) + CO(g) \longrightarrow H_2(g) + CO_2(g) + 41\ kJ$$

The unwanted CO_2 is removed by carbonate formation, for example.

$$CO_2(g) + CaO(s) \longrightarrow CaCO_3(s)$$

Hydrocarbons other than methane can also be used as a starting material, but methane has the highest H to C ratio (4:1). (Petroleum has a H:C ratio of about 2:1, while coal is about 1:1.) Therefore, methane gives more total hydrogen per gram than any other hydrocarbon, and the ratio of by-product CO_2 to H_2 is much lower. No matter what method is chosen there is an environmental problem: CO_2 and $CaCO_3$ are produced in very large quantities and must be disposed of in some manner.

The electrolysis of water is perhaps the cleanest method of H_2 production (Figure 22.2), and high-purity O_2 is a valuable by-product. However, electrical energy is quite expensive, so only in places where electricity is very cheap is electrolysis of water possible.

The 19th-century science fiction writer Jules Verne, author of 20,000 Leagues Under the Sea, wrote that "Water will someday be employed as a fuel. . . . Water will be the coal of our future."

Water can also be split into hydrogen and oxygen using heat energy. Almost 10,000 approaches to this have been suggested, all characterized by the use of a sequence of reactions such as

$$CaBr_2 + H_2O \xrightarrow{750\ °C} 2\ HBr + CaO$$

$$Hg + 2\ HBr \xrightarrow{100\ °C} HgBr_2 + H_2$$

$$HgBr_2 + CaO \xrightarrow{25\ °C} HgO + CaBr_2$$

$$HgO \xrightarrow{500\ °C} Hg + \tfrac{1}{2}\ O_2$$

$$\text{Net:}\quad H_2O(\ell) \longrightarrow H_2(g) + \tfrac{1}{2}\ O_2(g)$$

There are clearly major problems with hydrogen-generating cycles such as this one. First, many require very high temperatures to operate efficiently, and the only economical source of these temperatures at present seems to be nuclear reactors, with all the environmental and political problems they pose. Second, if a hydrogen-generating cycle is to be successful, each step must be of high yield. The reason for this is that the

Table 22.1 Methods for Preparing H_2 in the Laboratory

(a) Acid + metal $\longrightarrow$ metal salt + H_2

$$Mg(s) + 2\ HCl(aq) \longrightarrow MgCl_2(aq) + H_2(g)$$

(b) Metal + H_2O or base $\longrightarrow$ metal hydroxide or oxide + H_2

$$2\ Na(s) + 2\ H_2O(\ell) \longrightarrow 2\ NaOH(aq) + H_2(g)$$

$$2\ Fe(s) + 3\ H_2O(\ell) \longrightarrow Fe_2O_3(s) + 3\ H_2(g)$$

$$2\ Al(s) + 2\ KOH(aq) + 6\ H_2O(\ell) \longrightarrow 2\ KAl(OH)_4(aq) + 3\ H_2(g)$$

(c) Metal hydride + H_2O $\longrightarrow$ metal hydroxide + H_2.

$$CaH_2(s) + 2\ H_2O(\ell) \longrightarrow Ca(OH)_2(s) + 2\ H_2(g)$$

overall yield of any multistep process is the product of the yields in the individual steps. (A 90% yield in each step means that a four-step process will have an overall yield of only 66%.)

There are many times when a chemist needs a small amount of hydrogen in the laboratory, and there are a number of *simple laboratory methods* for making hydrogen on a small scale. One of the simplest is the reaction of a metal with an acid [reaction (a) in Table 22.1; Figures 5.1, 12.9, and 17.1]. Indeed, in 1783 Jacques Charles used the reaction of sulfuric acid with iron to produce the hydrogen for his balloon.

The reaction of Al with NaOH [reaction (b) in Table 22.1; Figure 12.10] is impractical due to the cost of materials. However, during World War II it was used to obtain H_2 to inflate small balloons for weather observation and to raise radio antennas. Metallic aluminum was plentiful, since it came from damaged and scrapped aircraft parts. Finally, reaction (c) in Table 22.1 is very useful, and this is perhaps the most efficient way to synthesize H_2 in the laboratory (Figure 17.2). It is also useful for removing traces of water from liquid compounds that do not react with hydrides, so metal hydrides are widely used for this purpose.

SOME USES OF HYDROGEN

The largest use of H_2 gas is in the production of ammonia, NH_3, by the *Haber process*.

$$N_2(g) + 3\ H_2(g) \longrightarrow 2\ NH_3(g)$$

Approximately 17 million tons of this fundamental chemical are made each year in the United States. As described in Chapter 24, NH_3 is used directly as a fertilizer or is converted to other fertilizers such as ammonium nitrate or urea [$(H_2N)_2CO$].

The second largest use of hydrogen is in the manufacture of methyl alcohol or methanol, CH_3OH, by the reaction of H_2 with carbon monoxide.

$$2\ H_2(g) + CO(g) \xrightarrow{\text{catalyst}} CH_3OH(\ell)$$

Almost 7 billion pounds of this colorless liquid were produced in 1989. The compound, which can cause blindness if consumed internally, is increasingly used as an additive in nonleaded gasoline and is even being converted directly to gasoline in a special process invented recently. It can itself also be used as a fuel, since combustion of 1 mole gives 638 kJ (see Chapter 6).

$$2\ CH_3OH(\ell) + 3\ O_2(g) \longrightarrow 4\ H_2O(g) + 2\ CO_2(g) + \text{heat}$$

Aside from these uses, large quantities of methanol are used to make various adhesives, plastics, and synthetic fibers.

Still another large use of hydrogen is known to every lover of peanut butter. Peanut butter contains "hydrogenated (or saturated) vegetable oils," which means that H_2 has been added to a vegetable oil to change its properties. The essence of this *hydrogenation* reaction is illustrated by the following equation in which the simple compound ethylene, which contains a C=C double bond, has been converted to ethane by hydrogenation, the addition of H_2 to the double bond.

$$\underset{\text{ethylene}}{H_2C{=}CH_2}\,(g) + H_2(g) \longrightarrow \underset{\text{ethane}}{H{-}CH_2{-}CH_2{-}H}\,(g)$$

See Something More About Fat in Chapter 26 (page 1106).

Fats contain, in varying proportions, organic esters with C=C double bonds. In general, the greater the number of double bonds the lower the melting point of the fat, so some fats are liquids at room temperature. The value of hydrogenation is that the oil is *hardened* to a solid with a consistency comparable to that of lard or butter. This is the basis of a large industry; cheap oils from cottonseed, corn, or soybeans are converted to cooking fats (e.g., Crisco® or Spry®) or margarine by this process.

See Figure 20.2, the explosive reaction of H_2 with O_2.

HYDROGEN AS A FUEL

There is a growing awareness that reserves of coal and oil are limited. Nuclear power has recently posed severe political and environmental problems, and, even if these problems are solved, it is not useful

Figure 22.4 An interstitial metal hydride. The large spheres are niobium atoms packed in a body-centered cubic lattice (Chapter 13). Hydrogen atoms (the small spheres) occupy the holes or interstices of the lattice.

for propelling cars, planes, or trains. Thus, there has been a search for alternatives, and one that has many attractive features is the "hydrogen economy."

There are two major barriers to the goal of a hydrogen economy. The first is to find an inexpensive method for the synthesis of H_2, a problem discussed above. The second is a means of convenient storage. The space program has demonstrated that hydrogen can be stored relatively easily and safely as a liquid. However, this may not be convenient in a home or car because the boiling point of hydrogen is −253 °C, so an alternative is to store H_2 in the form of an interstitial hydride (Figure 22.4). The hydrogen is released as the solid is heated, and the gas can then be used in a slightly modified internal combustion engine (Figure 22.5).

Figure 22.5 A prototype car run by hydrogen combustion in a slightly modified combustion engine. The hydrogen is stored in the form of an iron–titanium hydride, and heat from the engine exhaust is used to release H_2 from the hydride. (Mercedes-Benz)

The most visible use of H_2 as a fuel is in the space shuttle. The cigar-shaped tank strapped onto the shuttle contains 385,265 gallons of liquid H_2 and 143,351 gallons of liquid O_2 on lift-off (Figure 22.6). Just a few minutes into the flight, the fuel is exhausted, the hydrogen and oxygen having been converted completely to water.

On board the space shuttle is another example of hydrogen as a fuel. The H_2/O_2 fuel cell (Section 21.4) is routinely used as a power source on space craft. Unfortunately, such fuel cells are too expensive to be used in more ordinary, consumer-oriented applications.

Using the methods in Chapter 6, you can calculate that the combustion of H_2 to give water vapor will release about 121 kJ per gram of H burned. While this seems like a large quantity of energy, the fusion of hydrogen nuclei to give helium, the reaction used by our sun (or in a hydrogen bomb), gives a million times more energy per gram of H. This is why scientists are trying to devise ways of carrying out the hydrogen "fusion" process on earth (Chapter 7). A small amount of hydrogen could provide an enormous amount of energy for consumers. A hydrogen econ-

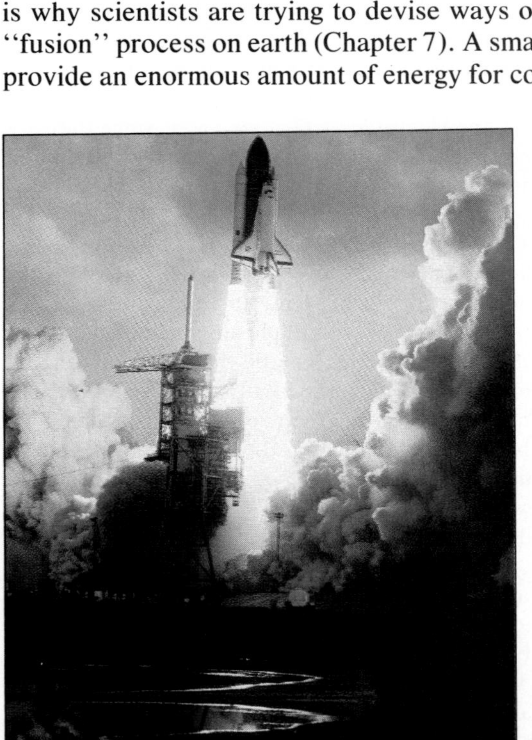

Figure 22.6 H_2 is used as a fuel in the space shuttle. The cigar-shaped tank contains 385,265 gallons of liquid H_2 and 143,351 gallons of liquid O_2 on lift-off. (NASA)

The exercises in this and the chapters that follow are review questions meant to help you organize the chemistry of the elements.

omy holds great promise, but it poses many problems. We are doubtless many years away from its realization.

EXERCISE 22.1 The Chemistry of Hydrogen
Give balanced equations for four methods of preparing H_2. Give three uses of H_2.

SOMETHING MORE ABOUT Hydrogen Absorbed by Metals—The Cold Fusion Episode

The sun provides energy for our planet by nuclear fusion, the union of hydrogen atoms to form helium. Therefore, since the early days of nuclear physics, scientists in many countries have spent billions of dollars trying to fuse atomic nuclei in the laboratory. In spite of the severe conditions the process is assumed to require, scientists have long felt that nuclear fusion was promising as an energy source, and that it may have the added advantage that it can be done in a manner that requires fewer environmental safeguards than nuclear fission (see Chapter 7).

One example of a fusion reaction is the union of two deuterium nuclei to give a helium-3 nucleus, a neutron, and energy.

$$2\ {}^{2}_{1}H \longrightarrow {}^{3}_{2}He + {}^{1}_{0}n + \text{energy}$$

The world of science was shocked on March 23, 1989 when two scientists in the United States announced that they had achieved this fusion process, or one like it, *at room temperature*!

It is known that if ordinary water is electrolyzed with a palladium electrode, the H_2 produced is absorbed by the palladium as H atoms (as in Figure 22.4). Therefore, the idea behind the "cold fusion" experiment was as follows: Electrolysis of heavy water, D_2O, should lead to "heavy hydrogen" or dideuterium, D_2. The D_2 should be absorbed by the electrode as D atoms, and, since they are in such close proximity, the atoms might fuse at room temperature by a reaction such as the one above or by some other process. The first report was that this did indeed occur!

The initial report of the possible fusion of deuterium atoms was so exciting that people all over the world did what scientists are trained to do—they rushed to their laboratories to duplicate the experiment. Only in this way could this astounding result be validated. A few were able to duplicate some of the initial results, others none of them. But no one was able to verify the whole experiment independently. Unfortunately, many scientists rushed to give news conferences or write magazine or newspaper articles reporting their results and speculations. As the *New York Times* said about one month later, "In the glare of publicity, the human face of science is illuminated in all its fevered competition, complete with errors, false starts, and bewildering confusion."

At this time, the possibility of "cold fusion" is seriously doubted. However, what is interesting is that the episode showed the best and worst faces of science. The thought that nuclear fusion had been achieved at room temperature was so important that some scientists at first shed their customary caution and ignored the ways that scientific results should be reported and verified. However, the weight of evidence from a worldwide network of chemists, physicists, and engineers will ultimately verify or repudiate the results, and the initial confusion will dissipate.

22.2 DIAGONAL RELATIONSHIPS

When moving from hydrogen into the main groups of the periodic table, we find that the chemistry of the first element in a group is often markedly different from that of the second element. Beryllium hydroxide is amphoteric, while the hydroxides of the other alkaline earth elements live up to their name: they form basic solutions in water. Boron is a metalloid and forms volatile hydrides, while the other Group 3A elements are metallic in character and form more nearly ionic hydrides. Carbon is the basis for literally millions of compounds with C—C and C—H bonds, while silicon forms relatively few with Si—Si bonds. A closer look at the chemistry of the first element in a group reveals that it behaves more like the second element in the following group. That is, lithium and magnesium are rather alike, as are beryllium and aluminum, and boron and silicon share some similarities (except for oxidation number).

As seen on p. 127, some aspects of the chemistries of H and Be are also diagonally related.

1A	2A	3A	4A
Li	Be	B	
	Mg	Al	Si

This has come to be known as the **diagonal relationship**.

In the text that follows, you will see examples of the diagonal relationship in the six elements above in particular, so knowledge of one element helps you to know something about another element. But be careful! Don't push the relationship unreasonably. The oxygen and hydrogen compounds of boron and silicon, for example, do have similarities but there are some great differences as well.

22.3 THE ALKALI METALS: GROUP 1A

The elements of Group 1A (Table 22.2) are characterized by an electron configuration of [core]ns^1, and all occur as 1+ ions. None are found in nature as free elements, since all combine rapidly and completely with virtually all the nonmetals. They are called the alkali metals because they react with water to form hydroxides, which are all soluble bases, historically known as alkalies.

1A
1 H hydrogen
3 Li lithium
11 Na sodium
19 K potassium
37 Rb rubidium
55 Cs cesium
87 Fr francium

COMMON MINERALS AND RECOVERY OF THE METAL

Although sodium and potassium are comparatively abundant in nature, the other elements are more rare. The lightest element, lithium, is found in trace amounts in natural waters, soils, and rocks throughout the world. Not surprisingly, the chief mineral form of lithium is an aluminosilicate, $LiAlSi_2O_6$. Lithium can be recovered from the mineral by first heating it with sodium carbonate (soda ash) to form lithium carbonate.

$$Na_2CO_3(aq) + 2\ LiAlSi_2O_6(s) \longrightarrow Li_2CO_3(s) + 2\ NaAlSi_2O_6(s)$$

To get the poorly soluble lithium carbonate into solution so that it can be separated from the insoluble silicates, CO_2 is bubbled into the mixture, forming lithium bicarbonate.

$$Li_2CO_3(s) + CO_2(g) + H_2O(\ell) \longrightarrow 2\ LiHCO_3(aq)$$

Table 22.2 Group 1A: The s^1 Elements (Alkali Metals)

	Lithium	Sodium	Potassium	Rubidium	Cesium	Francium
Symbol	Li	Na	K	Rb	Cs	Fr
Atomic number	3	11	19	37	55	87
Atomic weight	6.941	22.9898	39.0983	85.4678	132.9054	(223)*
Valence e^-	$2s^1$	$3s^1$	$4s^1$	$5s^1$	$6s^1$	$7s^1$
mp, °C	186	97.5	63.65	38.89	28.5	27
bp, °C	1326	889	774	688	690	677
d, g/cm³	0.534	0.971	0.862	1.53	1.87	—
Atomic radius, pm	152	186	227	248	265	—
Ion radius, pm	90	116	152	166	181	—
Pauling EN	0.98	0.93	0.82	0.82	0.79	0.7
Standard reduction potential	−3.05	−2.71	−2.93	−2.93	−2.92	—
Oxidation numbers	+1	+1	+1	+1	+1	—
Ionization energy*	520	496	419	403	376	—
Heat of atomization*	161	109	90.	86	79	—
Discoverer	Arfvedson	Davy	Davy	Bunsen and Kirchhoff	Bunsen and Kirchhoff	Perey
Date of discovery	1817	1807	1807	1861	1860	1939
rpw* pure O_2	Li_2O	Na_2O, Na_2O_2	K_2O_2, KO_2	RbO_2	CsO_2	—
rpw H_2O	LiOH	NaOH	KOH	RbOH	CsOH	—
rpw N_2	Li_3N	None	None	None	None	—
rpw halogens	LiX	NaX	KX	RbX	CsX	—
rpw H_2	Li^+H^-	Na^+H^-	K^+H^-	Rb^+H^-	Cs^+H^-	—
Flame color	Bright red	Yellow	Violet	Purple	Blue	—

*Atomic weights in parentheses are those of the most stable isotope. All energies and heats are in kJ per mole. The letters rpw stand for "reaction product with." Electrode potentials are in volts. The molar heat of atomization is the energy required to convert 1 mol of solid element into 1 mol of atoms in the gas phase. (All data in this table are taken from (a) *KC? Discoverer,* a computer program distributed by JCE: Software. (b) N.N. Greenwood and A. Earnshaw, *Chemistry of the Elements,* Pergamon Press, New York, 1984. (c) E.G. Rochow, *Modern Descriptive Chemistry,* W.B. Saunders, Philadelphia, 1977.)

This is similar to the reaction that occurs when CO_2-laden groundwater encounters a bed of limestone ($CaCO_3$) and dissolves the limestone as $Ca(HCO_3)_2$.

Within the earth's crust, sodium and potassium are about equally abundant (2.83% and 2.59%, respectively). However, sea water contains 2.8% NaCl, but only 0.8% KCl. Why the great difference? Sodium and potassium salts are both water soluble. Since they have about the same relative abundance on land, why didn't the rain dissolve Na- and K-containing minerals over the centuries and carry them down to the sea, to appear in the same proportions as on land? The answer lies in the fact that potassium is an important factor in plant growth. Much of the potassium appearing in groundwater from dissolved minerals is taken up preferentially by plants, while the sodium ion continues on to the sea. Most plants contain 4 to 6 times as much combined potassium as sodium, so potassium compounds find important use as fertilizers.

A reasonable amount of salt is essential in the diet of humans and animals, because many biological functions are controlled by concentrations of sodium and chloride ions. Animals will travel great distances to a "salt lick," and farmers and ranchers often place large blocks of salt in the fields for cattle to lick. Indeed, salt has been so important for so long that it has influenced us in ways you may not have thought about. When we are paid for work done, we can be paid a "salary." The word

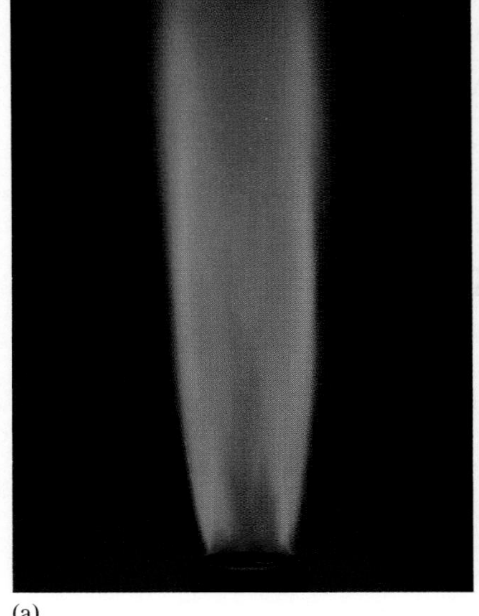

(a)

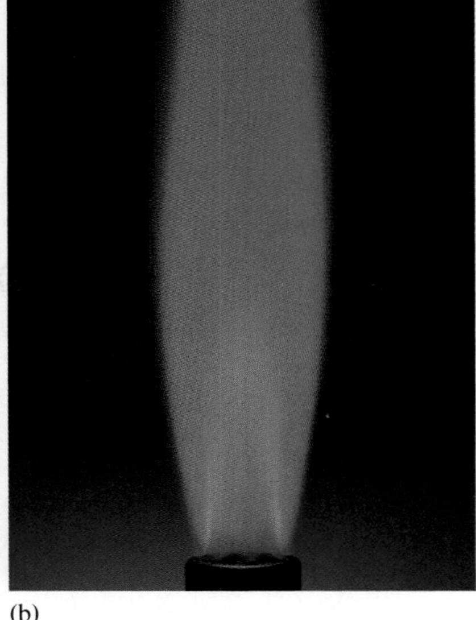

(b)

When salts of the alkali and alkaline earth metals are placed in a flame, light of characteristic colors is emitted. Sodium salts (a) give a yellow-orange color, while calcium salts (b) are brilliant red.

comes from the Latin word *salarium*, which originally meant "salt money" since Roman soldiers were paid in valuable salt. And we still talk of "salting away" some money for a rainy day, a term related to the practice of preserving meat by salting it.

Although sodium chloride is the most common form of sodium in nature, where it is found as rock salt or halite in salt lakes or underground deposits (Figure 22.7), there are other important sodium-containing minerals: sodium borate (borax), sodium carbonate (soda or trona), sodium nitrate (Chile saltpeter), and sodium sulfate (mirabilite).

Potassium deposits have also come from the crystallization of minerals from ancient seas, and the largest of these found thus far is in Saskatchewan, Canada; the deposit is thought to contain 10 billion tons of KCl.

Metallic sodium is produced by electrolysis of molten NaCl (Chapter 21). Although potassium could be similarly made, there are many problems with the method, not the least of which is that the molten metal is soluble in molten KCl, and the two cannot be separated. Instead, potassium is produced by allowing sodium vapor to react with molten KCl

Rubidium and cesium are also prepared by reaction of their chlorides with sodium.

$$\mathrm{Na(g) + KCl(\ell) \longrightarrow K(g) + NaCl(\ell)}$$

This reaction is a good example of the importance of understanding chemical equilibria. It has an equilibrium constant less than 1, so the reactant side is favored. However, since potassium vapor is continually removed, we can tell from Le Chatelier's principle that more product will be formed as the system attempts to come to equilibrium, and KCl can eventually be converted completely to potassium metal.

Francium is found only in trace amounts, and all its isotopes are radioactive. Indeed, its longest-lived isotope (^{223}Fr) has a half-life of only 22 minutes, making it rather difficult to carry out any extensive chemical experiments.

Figure 22.7 A natural sample of salt. The crystal on the top is halite or sodium chloride. On the bottom, the material is sylvite or KCl.

THE GROUP 1A ELEMENTS: THEIR GENERAL PROPERTIES AND USES

All the uses of the alkali metals themselves stem from their great ability to act as reducing agents. One of the best-known reactions of the alkali metals is that with water. If you drop a small piece of sodium into

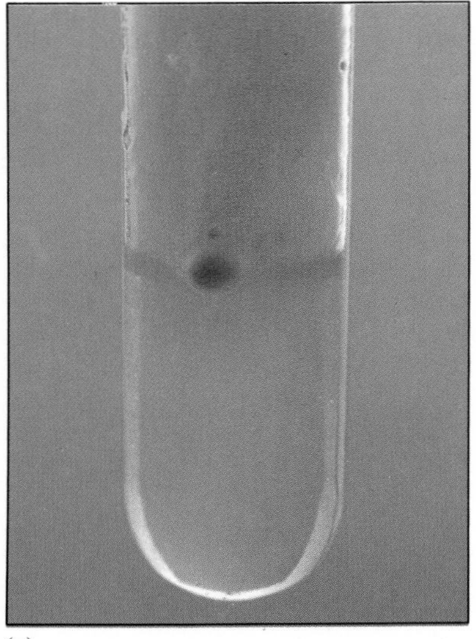
(a)

(b)

Figure 22.8 Sodium (a) and potassium (b) reacting with water. The vigorous reaction gives H_2 and the metal hydroxide, NaOH or KOH. Caution! These are potentially explosive reactions and should be done only with proper safety precautions. (See also Figure 20.8.)

water, it will react with great vigor, the heat of reaction sometimes igniting the hydrogen that is produced (Figure 22.8)

$$2\ Na(s) + 2\ H_2O(\ell) \longrightarrow 2\ NaOH(aq) + H_2(g)$$

Lithium is more docile, but potassium, rubidium, and cesium will react with great violence.

Ammonia is certainly not considered as strong a Brønsted acid as water, and so it differs from water on reaction with sodium. When sodium is put in liquid ammonia (a reaction that must be done at temperatures below −33 °C, the boiling point of ammonia), a beautiful deep blue solution is formed. Sodium, being a powerful reducing agent, loses an electron, but the electron is "solvated" by the ammonia, just as an anion can be bound up by a hydrogen-bonding solvent.

$$Na(s) \xrightarrow[\text{in liquid } NH_3]{} Na^+(\text{in } NH_3) + e^-(\text{in } NH_3)$$

This solution has great reducing power, and it is widely used for this purpose.

Until quite recently, the largest use of sodium as a reducing agent was in the synthesis of tetraethyllead.

$$4\ Na(s) + 4\ C_2H_5Cl(\ell) + Pb(s) \longrightarrow 4\ NaCl(s) + Pb(C_2H_5)_4(\ell)$$

The lead compound can be added to gasoline to raise the octane rating. However, due to severe pollution problems, leaded gasolines are being phased out, at least in the United States. Sodium is now a product in search of a market. Some of it, for example, is used to reduce $TiCl_4$ to metallic titanium, which is then used as a strong, lightweight, but expensive structural material.

$$TiCl_4(\ell) + 4\ Na(s) \longrightarrow Ti(s) + 4\ NaCl(s)$$

In principle, sodium can be used to reduce many metal halides to metal and sodium halide. However, since sodium is made by electrolysis and electrical power is expensive, the sodium is costly. Producing metals by sodium reduction is therefore an expensive process and is used only in special cases.

Not only can the Group 1A metals reduce water and ammonia, but they most assuredly react with oxygen, one of the most powerful oxidizing agents. What is fascinating about this reaction, however, is that the different metals produce such different products. Lithium produces a simple oxide when reacting with O_2.

$$4\ Li(s) + O_2(g) \longrightarrow 2\ Li_2O(s)$$

On the other hand, sodium gives mainly sodium **peroxide**, a yellowish-white solid.

$$2\ Na(s) + O_2(g) \longrightarrow Na_2O_2(s)$$

Recall the discussion of the molecular orbital structure of O_2 and its ions in Chapter 11.

The peroxide ion is O_2^{2-}, essentially an oxygen molecule that has been reduced by two electrons. This is a commercially useful material, because it hydrolyzes to produce hydrogen peroxide, H_2O_2.

$$Na_2O_2(s) + 2\ H_2O(\ell) \longrightarrow 2\ NaOH(aq) + H_2O_2(aq)$$

Because of the great oxidizing power of sodium peroxide, it is widely used in the paper and textile industries as a bleaching agent.

Potassium, rubidium, and cesium also reduce O_2, but they all form **superoxides**, MO_2, where the alkali metal ion is associated with the O_2^- ion.

As explained in Chapter 11, the superoxide ion, O_2^-, is paramagnetic with one unpaired electron.

$$K(s) + O_2(g) \longrightarrow KO_2(s)$$

The most important commercial use of KO_2 is as a source of oxygen in an emergency breathing apparatus. The so-called "oxygen masks" are designed so that the CO_2 and water vapor in the exhaled breath of the wearer react with superoxide to provide oxygen.

$$4\ KO_2(s) + 4\ CO_2(g) + 2\ H_2O(g) \longrightarrow 4\ KHCO_3(s) + 3\ O_2(g)$$

It is clear that the uses of the Group 1A elements as metals depend on their reducing ability. The data in Table 22.2 show that Li is the best reducing agent in the group, whereas Na is the poorest, and the remainder are comparable in their ability.* This lack of a clear-cut trend is curious, so it is useful to analyze the relative reducing ability of these metals to understand their behavior.

The products listed for all alkali metals + O_2 are the major products. In addition, Na forms some Na_2O and K forms some K_2O and K_2O_2.

Analysis of $E°$ is a thermodynamic problem, and, as in any thermodynamic problem, we can break the process $M(s) \rightarrow M^+(aq) + e^-$ into a series of steps

$$\begin{array}{ccc} M(g) & \xrightarrow{IE\text{ (ionization energy)}} & M^+(g) + e^- \\ \uparrow \Delta H_{sub} & & \downarrow \Delta H_{hyd} \\ M(s) & \xrightarrow[\Delta H_{net}]{} & M^+(aq) + e^- \end{array}$$

from which we write

$$\Delta H_{net} = \Delta H_{sub} + IE + \Delta H_{hyd} = IE + (\Delta H_{sub} + \Delta H_{hyd})$$

In this case, we imagine the solid metal sublimes (sub) to form the vapor, an electron is removed from a gaseous atom, and then the gaseous ion is hydrated (hyd). The first two steps require energy, but the third is exothermic. You can see now that the element that is the best reducing agent should have the most negative (or least positive) value of ΔH_{net}, and the sum $(\Delta H_{sub} + \Delta H_{hyd})$ is responsible for the difference between reducing ability and ionization energy.

Since the enthalpy changes for all of these steps are known for all of the Group 1A elements, we can easily calculate ΔH for $M(s) \rightarrow M^+(aq) + e^-$. Before doing the numerical analysis, however, there are two important points to be made. First, $E°$ for an electrochemical half-reaction is directly related to $\Delta G°$, and not to the $\Delta H°$, for the reaction. Fortunately, $\Delta G°$ is largely determined by $\Delta H°$ in this case, so it is legitimate to ascribe variations in $E°$ to variations in $\Delta H°$. Second, we are

*The $E°$ values found in Table 22.2 are given by convention as reduction potentials; that is, the values are for the reaction $M^+(aq) + e^- \rightarrow M(s)$. Therefore, the value for $Li(s) \rightarrow Li^+(aq) + e^-$ is +3.05 V, for example.

Table 22.3 Enthalpy Changes for Electrode Half Reactions*

	Li	Na	K	Rb	Cs
Ionization energy	520	496	419	403	376
ΔH_{sub}	161	109	90.	86	79
ΔH_{hyd}	−506	−397	−318	−289	−259
$(\Delta H_{sub} + \Delta H_{hyd})$	−345	−288	−228	−203	−180.
$\Delta H[M(s) \longrightarrow M^+(aq) + e^-]$	175	208	191	200	196

*Energies are all in kJ/mol. sub = sublimation; hyd = hydration.

calculating $\Delta H°$ for a *hypothetical* half-reaction, not an actual reaction. Therefore, the numbers derived are not the absolute values but are only relative values that allow us to compare relative reducing ability.

The values of ΔH for the steps, for each of the Group 1A elements, are given in Table 22.3, as is the calculated value of ΔH for the element acting as a reducing agent. According to these calculations, Li is the best reducing agent, Na the poorest, and the others all have about the same ability, exactly in agreement with experimentally determined $E°$ values (see Table 22.2). Now we can see why. As suggested by the relatively high ionization energy of lithium, the very small Li^+ cation strongly attracts electrons; when the electrons available to it are water lone pairs, the cation forms very strong bonds with water and so has the most negative heat of hydration. In reality it is this great release of energy when solvating $Li^+(g)$ that offsets its relatively high ionization energy and that makes lithium the best reducing agent. *The hydration energy plays a very large role in determining $E°$.*

The very small size of Li^+ is important in determining many of its properties and in setting it apart in many ways from the other elements of the group. You will see this effect again.

A useful lesson to be learned from this analysis of alkali metal $E°$ values is that a physical property such as $E°$ is not determined solely by one underlying property of the element; rather, it is an amalgamation of several such properties, interacting in a complex way. Indeed, it is likely that many periodic trends are really a complex interplay of several underlying properties.

SOME IMPORTANT GROUP 1A COMPOUNDS

The chemistry of chlorine is described in Chapter 24.

THE CHLOR–ALKALI INDUSTRY The "chlor–alkali" industry is the biggest user and producer of alkali metal compounds, and the main building block is sodium chloride. Chief among the products of the industry are chlorine

Table 22.4 Chemicals Produced Directly or Indirectly by the Chlor–Alkali Industry (1988)

Compound	Amount (In Billions of Pounds)	Rank (Among Top 50 Chemicals Produced)
Sodium hydroxide (caustic soda)	23.97	7
Chlorine	22.66	9
Sodium carbonate (soda ash)	19.10	11
Sodium sulfate	1.69	45
Sodium silicate	1.64	46
Calcium chloride	1.31	49

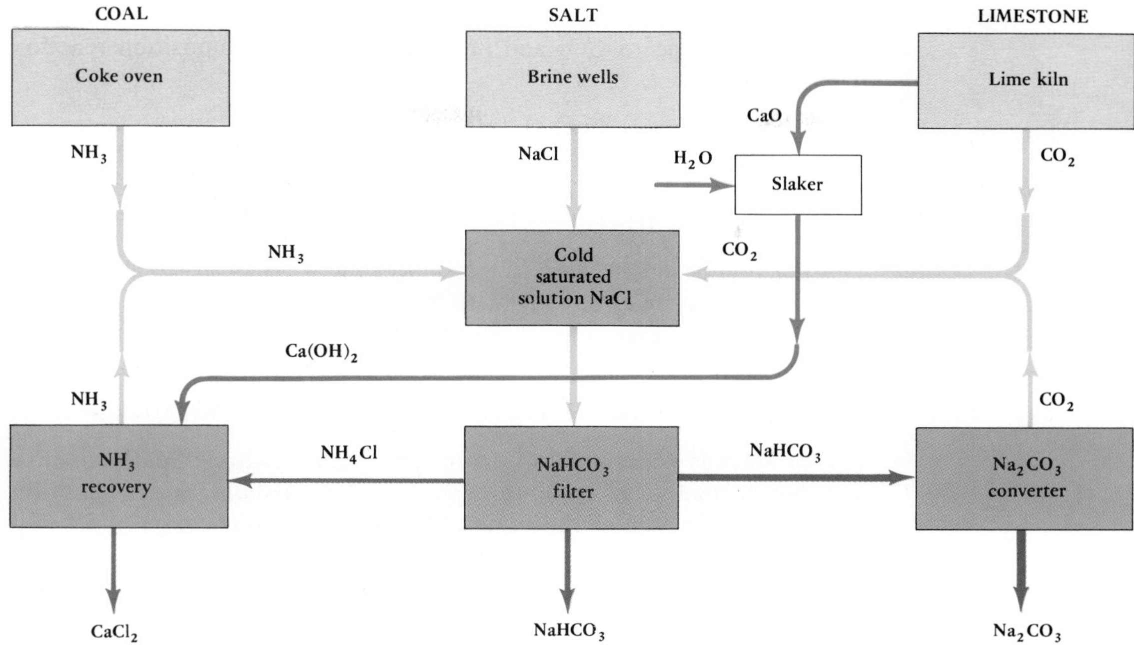

Figure 22.9 The Solvay process for the production of sodium carbonate and sodium bicarbonate. It is named after Ernest Solvay, an Englishman who first succeeded in getting the process to work on an economical basis in 1869. Although the process continues to be important in other countries, it is being phased out in the United States. The mineral *trona* ($Na_2CO_3 \cdot NaHCO_3 \cdot 2\ H_2O$) now supplies most U.S. domestic needs.

and sodium hydroxide from the electrolysis of brine, aqueous sodium chloride (Chapter 21),

$$2\ NaCl(aq) + 2\ H_2O(\ell) \xrightarrow[\text{electrolysis}]{} 2\ NaOH(aq) + H_2(g) + Cl_2(g)$$

To give you some idea of the size and importance of the industry, consider the amounts of the chemicals produced in 1988 (Table 22.4).

Sodium hydroxide is normally sold as a concentrated aqueous solution called "caustic soda." Approximately half of it is used in the manufacture of other chemicals (such as in the aluminum industry), and a large part of the remainder is used in the pulp and paper industry.

Sodium carbonate has been obtained, since prehistoric times, from naturally occurring deposits of $Na_2CO_3 \cdot 10\ H_2O$, commonly known as **washing soda**. (Anhydrous sodium carbonate is called **soda ash**.) More recently, however, a deposit of *trona*, $Na_2CO_3 \cdot NaHCO_3 \cdot 2\ H_2O$, estimated at 6×10^{10} tons has been discovered in Wyoming, and virtually all the sodium carbonate in the United States now comes from that source. Its largest use is in the manufacture of glass.

Until quite recently, the **Solvay process** (Figure 22.9) was the major source of sodium carbonate and sodium bicarbonate. This process has been superseded, particularly in the United States, by less costly mined soda ash, but it is still in use in other parts of the world. It is worth describing, not only because of its economic importance, but also because it illustrates many chemical principles.

The overall reaction of the Solvay process is a simple exchange,

$$\underset{\text{brine}}{2\ NaCl} + \underset{\text{limestone}}{CaCO_3} \longrightarrow \underset{\text{soda ash}}{Na_2CO_3} + CaCl_2$$

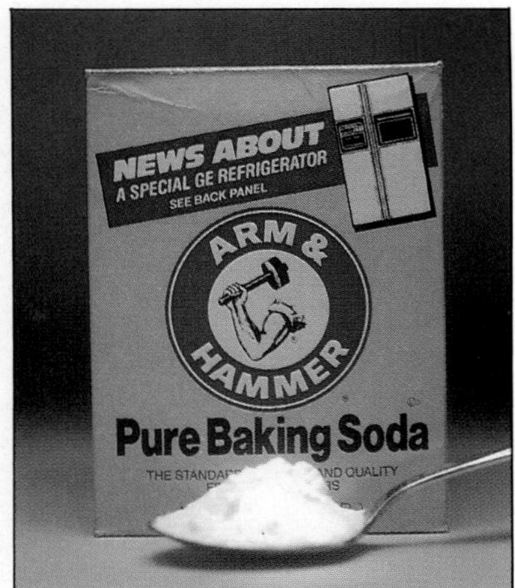

A box of a common brand of baking soda, $NaHCO_3$.

but the reaction will not occur directly. In practice, the first step is to convert limestone to CO_2 and lime, CaO, in a decomposition reaction (Chapter 4)

$$CaCO_3(s) \longrightarrow CaO(s) + CO_2(g)$$

and then to allow the CO_2 to react with aqueous ammonia.

$$CO_2(g) + NH_3(aq) + H_2O(\ell) \longrightarrow NH_4HCO_3(aq)$$

This reaction occurs because CO_2 produces the weak Brønsted acid H_2CO_3 in water, and this in turn reacts with the weak Brønsted base NH_3. Sodium chloride then undergoes an exchange reaction with the ammonium bicarbonate.

$$NaCl(aq) + NH_4HCO_3(aq) \longrightarrow NaHCO_3(s) + NH_4Cl(aq)$$

The product is sodium bicarbonate (or sodium hydrogen carbonate), a salt that is relatively insoluble in the reaction medium. The sodium bicarbonate is isolated and used directly for some purposes, but most of it is converted by heating to soda ash, sodium carbonate.

$$2\ NaHCO_3(s) \longrightarrow Na_2CO_3(s) + CO_2(g) + H_2O(g)$$

Ammonia is an expensive material, so it is not economic to waste it as NH_4Cl, a by-product of the reactions above; the NH_3 must be recovered to be used again. To do this, water is added to the CaO produced in the first step to give the base $Ca(OH)_2$, **slaked lime**.

$$\underset{\text{lime}}{CaO(s)} + H_2O(\ell) \longrightarrow \underset{\text{slaked lime}}{Ca(OH)_2(s)}$$

The base reacts with the ammonium ion, a Brønsted acid, to regenerate NH_3,

$$NH_4^+(aq) + OH^-(aq) \longrightarrow NH_3(aq) + H_2O(\ell)$$

so the net reaction is

$$2\ NH_4Cl(aq) + CaO(s) \longrightarrow 2\ NH_3(aq) + H_2O(\ell) + CaCl_2(aq)$$

A big problem is that the calcium is left over as $CaCl_2$. Some is used in the winter on roads and sidewalks to melt ice and snow, but much of it remains to be disposed of in some manner.

The amount of sodium bicarbonate produced annually is approximately 2% of the amount of sodium carbonate produced. However, you are probably more aware of the bicarbonate, since it is commonly called **baking soda**. Not only is this used in cooking, but it is also added in small amounts to table salt. Sodium chloride is often contaminated with small amounts of $MgCl_2$. The magnesium salt is hygroscopic; that is, it picks water up from the air and, in doing so, causes the NaCl to clump on damp days. Adding $NaHCO_3$ converts $MgCl_2$ to magnesium carbonate, a nonhygroscopic salt.

$$MgCl_2(s) + 2\ NaHCO_3(s) \longrightarrow MgCO_3(s) + 2\ NaCl(s) + H_2O(\ell) + CO_2(g)$$

OTHER ALKALI METAL SALTS Many other salts of the alkali metals are important to commerce, and a few are of interest here.

Sodium sulfate or "salt cake" is a product of an exchange reaction (Figure 22.10).

$$2\ NaCl(s) + H_2SO_4(aq) \longrightarrow Na_2SO_4(s) + 2\ HCl(g)$$

This is also a good method of preparing HCl gas in the laboratory. It is especially useful because H_2SO_4 is a good drying agent, so the HCl that evolves is quite dry.

It is inexpensive to produce in this manner, since both NaCl and sulfuric acid, a by-product of many industrial processes, are inexpensive. In addition, a commercially useful product, HCl, is also formed. Sodium sulfate itself is used mainly in the production of wood pulp.

Large deposits of *sodium nitrate*, $NaNO_3$, are found in Chile, hence the name "Chile saltpeter." It is thought that these deposits were formed by bacterial action on organisms in shallow seas. The initial product was ammonia that was subsequently oxidized to nitrate ion; combination with sea salt led to sodium nitrate. However, because nitrates in general, and the alkali metal nitrates in particular, are so water soluble, deposits are found only in areas of very low rainfall.

Sodium nitrate can be converted easily into KNO_3 by the exchange reaction

$$NaNO_3(aq) + KCl(aq) \longrightarrow KNO_3(aq) + NaCl(s)$$

because, of the four salts in this reaction, NaCl is the least soluble in hot water. These nitrates are largely used for their oxidizing ability. For example, for centuries KNO_3 has been used as the oxidizer in gunpowder. A mixture of KNO_3, charcoal, and sulfur will spontaneously react when ignited according to the equations

$$2\ KNO_3(s) + 4\ C(s) \longrightarrow K_2CO_3(s) + 3\ CO(g) + N_2(g)$$

$$2\ KNO_3(s) + 2\ S(s) \longrightarrow K_2SO_4(s) + SO_2(g) + N_2(g)$$

Notice that both reactions produce gases, and it is these hot gases that propel a bullet from a gun or cause a firecracker to explode. (See "The Chemistry of Fireworks" at the end of this section.)

More than 1 billion pounds of *sodium tripolyphosphate*, $Na_5P_3O_{10}$, are produced annually in the United States to be used as a "builder" in household detergents. Not only does it give a basic solution on hydrolysis,

$$P_3O_{10}{}^{5-}(aq) + H_2O(\ell) \longrightarrow HP_3O_{10}{}^{4-}(aq) + OH^-(aq)$$

but the tripolyphosphate ion forms soluble complexes with the troublesome calcium ion found in hard water. However, there is a continuing controversy about the role of phosphates in water pollution. As a result, their use in the United States has declined, and some manufacturers are increasingly using sodium carbonate (washing soda) again.

Lithium carbonate, Li_2CO_3, has been used for more than 30 years as a treatment for manic depression. Manic depression involves alternating phases of depression and mania or over-excitement, mood swings occurring in some instances in periods of a few weeks to a year or so. Although the treatment of manic depression with lithium salts is well known, the mechanism of lithium action is not understood. However, a good guess is that it disturbs Mg^{2+} metabolism. In Section 22.2, it was

Figure 22.10 When concentrated sulfuric acid (in the funnel at the top) drops onto solid NaCl in the flask, the products are Na_2SO_4 ("salt cake") and HCl gas. The HCl vapor exits through the side arm of the flask. When the vapor passes over the indicator on the filter paper, the indicator turns bright red, its color in an acid solution.

noted that there is a ''diagonal relationship'' between the properties of Li and Mg, so it is possible that Li^+ takes the place of Mg^{2+} in some key process. The problem is that Mg^{2+} is important to so many biochemical processes that it may be difficult to pinpoint the one process accounting for its role (if any) in manic depression.

All the alkali metals react with hydrogen when heated to produce salt-like hydrides such as *sodium hydride*, NaH.

$$2\ Na(s) + H_2(g) \longrightarrow 2\ NaH(s)$$

The chief chemical characteristic of alkali metal hydrides, MH, is that the hydride ion, H^-, is a strong Lewis and Brønsted base; as such, it reacts vigorously with acids such as water (see CaH_2/H_2O in Figure 17.2).

$$NaH(s) + H_2O(\ell) \longrightarrow NaOH(aq) + H_2(g)$$

The heat of this reaction is so great that the hydrogen can ignite. In research and industrial laboratories the hydrides are commonly used as drying agents for nonaqueous solvents and as reducing agents.

There are many important alkali metal compounds in commerce and biochemistry. Watch for more of them, in this book and in consumer products.

EXERCISE 22.2 The Chemistry of Group 1A Elements

(a) Write balanced equations to show how elemental Na and K are prepared.
(b) Write balanced equations to illustrate two uses of sodium metal.
(c) Name three uses of sodium-containing compounds.
(d) Give the formulas for washing soda and baking soda.

SOMETHING MORE ABOUT The Chemistry of Fireworks

Fireworks. (James Cowlin)

The development of black powder, a mixture of potassium nitrate, charcoal, and sulfur, was one of the most important developments in the history of the human race. The discovery, apparently well before AD 1000, is usually attributed to the Chinese, and it was they who began the development of fireworks. Black powder made its way to Europe by the 1300s, where it was used not only for fireworks but also by the military. By the time of the American Revolution, fireworks formulations and manufacturing methods had been worked out that are still in use today.

There are several important components in typical fireworks. First, there must be an oxidizer, and today this is usually potassium perchlorate ($KClO_4$) or potassium chlorate ($KClO_3$) as well as potassium nitrate. Potassium salts are used instead of sodium salts because the latter have two important drawbacks. They are hygroscopic—they absorb water from the air—and so do not remain dry on storage. Second, when heated, sodium salts give off an intense, yellow light (page 917) that is so bright it can hide other colors.

Potassium perchlorate is generally safer to use than the chlorate salt. The problem is that it may become difficult to obtain. The only supplier of perchlorates in the United States makes ammonium perchlorate, the oxidizer used in the space shuttle solid-fuel booster rockets. Each shuttle launch requires about 1.5 million pounds of NH_4ClO_4, about twice the annual consumption of $KClO_4$ in the United

States. Thus, the fireworks industry may have trouble getting enough $KClO_4$ if there are frequent shuttle launchings (or if the ammonium perchlorate production facilities are disabled, as they were a few years ago by an explosion).

The parts of any fireworks display we remember best are the vivid colors and brilliant flashes. White light can be produced by oxidizing magnesium or aluminum metal at high temperatures, and the flashes you see at rock concerts or similar events are typically $Mg/KClO_4$ mixtures.

Yellow light is easiest to produce, since sodium salts give an intense light with a wavelength of 589 nm. Fireworks mixtures usually contain sodium in the form of such nonhygroscopic compounds as cryolite, Na_3AlF_6. Strontium salts are most often used to produce a red light, and green is produced by barium salts such as $Ba(NO_3)_2$. But the next time you see a fireworks display, watch for the ones that are blue. You will undoubtedly see very few, since this color is by far the most difficult to achieve. The best way to do this is the decomposition of copper(I) chloride at low temperatures, but pyrotechnic chemists keep looking for better, more brilliant blues.

If you wish more information on the chemistry of fireworks, see Discover, July 1982, page 25, Chemical and Engineering News, June 21, 1981, and Scientific American, July 1990, page 96.

22.4 THE ALKALINE EARTH METALS: GROUP 2A

The elements of the alkaline earth group (Table 22.5) are all characterized by electron configurations of the type [core]ns^2. As a result, all form compounds in the +2 oxidation state such as calcium oxide, CaO. With the exception of BeO, all Group 2A oxides hydrolyze to give a basic

Table 22.5 Group 2A: The s^2 Elements (Alkaline Earth Metals)

	Beryllium	Magnesium	Calcium	Strontium	Barium	Radium
Symbol	Be	Mg	Ca	Sr	Ba	Ra
Atomic number	4	12	20	38	56	88
Atomic weight	9.0122	24.3050	40.078	87.62	137.327	226.03
Valence e^-	$2s^2$	$3s^2$	$4s^2$	$5s^2$	$6s^2$	$7s^2$
mp, °C	1283	650	845	770	725	700
bp, °C	2970	1120	1420	1380	1640	1140
d, g/cm³	1.85	1.74	1.55	2.60	3.51	5
Atomic radius, pm	112	160	197	215	222	—
Ion radius, pm	59	86	114	132	149	—
Pauling EN	1.56	1.31	1.0	0.95	0.89	1.0
Standard reduction potential*	−1.85	−2.37	−2.87	−2.89	−2.90	−2.92
Oxidation numbers	+2	+2	+2	+2	+2	+2
Ionization energy*	899	738	599	550	503	—
Heat of atomization*	324	148	178	164	180	—
Discoverer*	Vauquelin	Bussy	Berzelius	Davy	Davy	Curie
Date of discovery	1798	1831	1808	1808	1808	1911
rpw* pure O_2	BeO	MgO	CaO	SrO, SrO_2	BaO_2	RaO
rpw H_2O	None	$Mg(OH)_2$	$Ca(OH)_2$	$Sr(OH)_2$	$Ba(OH)_2$	$Ra(OH)_2$
rpw N_2	None	Mg_3N_2	Ca_3N_2	Sr_3N_2	Ba_3N_2	Ra_3N_2
rpw halogens	BeX_2	MgX_2	CaX_2	SrX_2	BaX_2	RaX_2
rpw H_2	None	MgH_2	CaH_2	SrH_2	BaH_2	—
Flame color	—	—	Brick red	Crimson	Green	—

*rpw = reaction product with. "Discoverer" refers to first isolation; Mg, Ca, and Ba were known to ancients. All energies and heats are in kJ per mole; potentials are in volts. The molar heat of atomization is the energy required to convert 1 mol of solid element into 1 mol of atoms in the gas phase. (All data in this table are taken from (a) *KC? Discoverer*, a computer program distributed by JCE: Software. (b) N.N. Greenwood and A. Earnshaw, *Chemistry of the Elements*, Pergamon Press, New York, 1984. (c) E.G. Rochow, *Modern Descriptive Chemistry*, W.B. Saunders, Philadelphia, 1977.)

Figure 22.11 Calcium oxide (lime), a white solid, reacts with water to produce $Ca(OH)_2$ (slaked lime). The test tube at the left contains the indicator phenolphthalein, which turns red in basic solution. The tube at the right contains only CaO in water.

solution (Figure 22.11); hence the name "alkaline" in the name of this periodic group.

$$CaO(s) + H_2O(\ell) \longrightarrow Ca(OH)_2(s) \qquad K_{sp} = 7.9 \times 10^{-6}$$

The "earth" part of the group name is left over from the days of alchemy. To medieval alchemists, any solid substance that did not melt and was not changed by fire into another substance was called an "earth." Compounds of Group 1A elements, such as Na_2CO_3 and NaOH, are certainly alkaline according to the experimental tests of the alchemists: the compounds all have a bitter taste and have the ability to neutralize acids. However, some Group 1A compounds melted in a fire or combined with the clay containers in which they were heated. On the other hand, since the melting point of CaO is 2572 °C, a temperature well beyond the range of any ordinary fire, it was unchanged in fire. Since it also was alkaline, it was called an "alkaline earth," a phrase that came to be applied to the elements themselves in Group 2A.

The abundances of the elements vary widely (Table 22.6); just as in Group 1A, the lightest element is very rare and the heaviest element is radioactive. The great abundance of calcium and magnesium led to their wide occurrence in plants and animals, and both elements form many compounds that are commercially important. It is on this chemistry that we want to focus most of our attention.

COMMON MINERALS OF THE ALKALINE EARTH METALS

Masses of beryl weighing more than a ton have been found in New Hampshire.

Like so many elements, beryllium is commonly found in the form of an aluminosilicate, in this case the mineral beryl, $3\ BeO \cdot Al_2O_3 \cdot 6\ SiO_2$. Although beryl is colorless when pure, it can be a brilliant green if some of the Al^{3+} ions are replaced by Cr^{3+} ions, and you know it as an emerald. If, on the other hand, it contains Fe^{2+} and Fe^{3+} as impurities, it is blue-green and is known as aquamarine, the birthstone for March.

2A
4 Be beryllium
12 Mg magnesium
20 Ca calcium
38 Sr strontium
56 Ba barium
88 Ra radium

The great abundance of magnesium and calcium means that they are found in common minerals (Table 22.7). Strontium and barium can be found in relatively concentrated deposits, but radium is extremely widely distributed. The latter is always found in association with uranium, since it is a product of the radioactive disintegration of ^{238}U. It is long lived enough to isolate, but uranium ore contains only about 1 mg of Ra for every 3 kg of uranium.

The principal source of commercial calcium compounds is limestone, which occurs in immense sedimentary beds over large parts of the

Table 22.6 Abundance of Group 2A Elements*

	Igneous Rocks	Lunar Rocks	Oceans
Be	<6	—	—
Mg	18,700	42,000	1,310
Ca	29,600	103,000	420
Sr	340	166	8–12
Ba	650	210	0.03
Ra	9×10^{-7}	—	10^{-10}

*In parts per million.

Table 22.7 Common Minerals of the Group 2A Elements

Common Name	Formula
Beryl	$3\ BeO \cdot Al_2O_3 \cdot 6\ SiO_2$
Magnesite	$MgCO_3$
Talc or soapstone	$3\ MgO \cdot 4\ SiO_2 \cdot H_2O$
Asbestos (chrysotile)	$3\ MgO \cdot 2\ SiO_2 \cdot 2\ H_2O$
Dolomite	$MgCO_3 \cdot CaCO_3$
Limestone (calcite or aragonite)	$CaCO_3$
Gypsum	$CaSO_4 \cdot 2\ H_2O$
Fluorspar	CaF_2
Fluorapatite	$CaF_2 \cdot 3\ Ca_3(PO_4)_2$
Barites	$BaSO_4$

Iceland spar, a form of calcium carbonate, in Wind Cave, Wind Cave National Park (South Dakota). (Art Palmer)

earth's surface. These are deposits formed from the fossilized remains of marine life and are mostly the *calcite* form of the compound. However, you often know calcite by other names. Limestone is generally contaminated with other metal ions such as those of iron, but *marble* is fairly pure calcite, formed in nature by recrystallization of $CaCO_3$ under pressure. High-quality deposits are found in Italy and in the United States in Vermont, Georgia, and Colorado. Yet another form of $CaCO_3$ is *Iceland spar*, which forms large, clear crystals. Finally, *chalk* is a fine grained, powdery form that is found not only in classrooms but also in such spectacular deposits as the cliffs of the Grand Canyon (page 906).

Gypsum, or hydrated calcium sulfate, is extensively mined. Some of it is used in portland cement, but most is heated in large kilns to form "plaster of Paris," a process called *calcining*.

$$\underset{\text{gypsum}}{CaSO_4 \cdot 2\ H_2O(s)} \longrightarrow \underset{\text{plaster of Paris}}{CaSO_4 \cdot \tfrac{1}{2}\ H_2O(s)} + \tfrac{3}{2}\ H_2O(g)$$

If enough water is added to plaster of Paris to make a paste, it quickly hardens as it reverts to gypsum. The mixture also expands as it hardens, so it forms a sharp impression of anything molded in it. Virtually all calcined gypsum is used to make wallboard (sheet rock or plaster board), and the rest is used for industrial and building plasters. In fact, this use of gypsum is very old; there is evidence that the interiors of some of the great pyramids of Egypt were coated with gypsum plaster.

The mineral alabaster is nearly pure gypsum; it is dense and fine textured, but it is still soft enough that it can be worked with common tools, and it has been used since ancient times to make bowls, statues, and similar items.

Gypsum crystals, $CaSO_4 \cdot 2\ H_2O$, in Wind Cave, Wind Cave National Park (South Dakota). (Art Palmer)

RECOVERY AND USES OF THE ALKALINE EARTH METALS

In spite of its rarity, beryllium is a fascinating element, and its metallic form does have uses. It is as strong as steel, has a very high melting point (1283 °C), and is very lightweight (the density of beryllium, 1.85 g/cm^3, is only one-fourth that of iron, 7.85 g/cm^3). The element itself is rather brittle, but it is widely used with other metals to make alloys.

Aquamarine, the birthstone for March, is a beryllium-containing mineral.

For example, adding about 2% Be to copper makes an alloy that is almost as good an electrical conductor as copper, but the material wears better than copper. Since it is also a strong material and does not create sparks when struck, it is used to make tools for areas where explosive gases might be present, for example.

The recovery of beryllium from beryl involves some complex chemistry, but the end product is BeF_2 or $BeCl_2$. Like all the elements of Groups 1A and 2A, the chloride can be reduced electrochemically to the metal.

$$BeCl_2(\text{as a melt with NaCl}) \xrightarrow[\text{electrical energy}]{} Be(s) + Cl_2(g)$$

Magnesium is also generally isolated in an electrolytic process, but the total chemistry of this industrial isolation process is quite interesting. Since magnesium is so abundant in the oceans, this has been one of the major sources of the metal (Figure 22.12). A very useful way to isolate the magnesium from the ocean is in the form of its relatively insoluble hydroxide (K_{sp} for $Mg(OH)_2 = 1.5 \times 10^{-11}$). To precipitate the hydroxide, the only thing needed is a ready supply of an inexpensive base, a need fulfilled nicely by sea shells, $CaCO_3$. As you first learned in Chapter 4, heating a metal carbonate generally leads to loss of CO_2 and formation of the metal oxide. Thus, calcium carbonate is converted to lime and then to calcium hydroxide, the source of hydroxide ion to precipitate $Mg(OH)_2$.

$$\underset{\text{sea shells}}{CaCO_3(s)} \longrightarrow \underset{\text{lime}}{CaO(s)} + CO_2(g)$$

$$CaO(s) + H_2O(\ell) \longrightarrow Ca(OH)_2(s)$$

$$Mg^{2+}(aq) + Ca(OH)_2(s) \longrightarrow Mg(OH)_2(s) + Ca^{2+}(aq)$$

The magnesium hydroxide can be isolated by filtration and then neutralized by another inexpensive chemical, hydrochloric acid.

$$Mg(OH)_2(s) + 2\ HCl(aq) \longrightarrow MgCl_2(aq) + 2\ H_2O(\ell)$$

If the water is evaporated, solid hydrated magnesium chloride is left. After drying, it is melted at 708 °C and then electrolyzed to give the metal and a commercially useful by-product, chlorine.

$$MgCl_2(\text{melt}) \xrightarrow[\text{electrical energy}]{} Mg(s) + Cl_2(g)$$

Several hundred thousand tons of magnesium are produced annually, and all its uses depend on its low density (1.74 g/cm^3) or high reactivity. In fact, most of the magnesium that is produced is used to make lightweight alloys. Most aluminum has about 5% magnesium added to it to improve its mechanical properties and to make it more resistant to corrosion under basic conditions. There are also alloys that have the reverse composition, that is, more magnesium than aluminum. These alloys are used where a high strength-to-weight ratio is needed and where corrosion resistance is important. Much of it is used in aircraft and automotive parts and in lightweight tools, luggage trim, and so on.

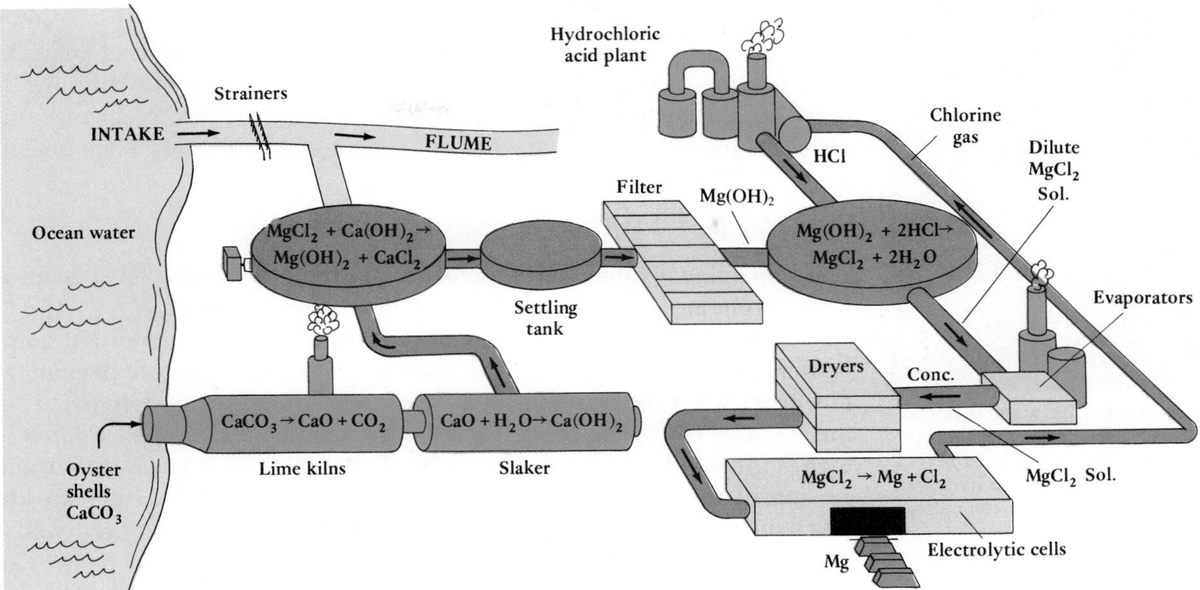

Figure 22.12 Diagram of an industrial plant for the production of magnesium from the magnesium ions present in seawater.

The reactivity of magnesium leads to its use as a sacrificial anode to give cathodic corrosion protection for the hulls of ships and underground pipelines (see Chapter 21).

Calcium, strontium, and barium can be prepared by electrolysis or by aluminum reduction, but only barium has some commercial uses.

$$3\ CaO(s) + 2\ Al(s) \longrightarrow Al_2O_3(s) + 3\ Ca(s)$$

Al can be used to reduce oxides of all but the most "active" metals. Such reactions depend on the very high heat of formation of Al_2O_3. See the thermite reaction on page 837.

In spite of the revolution in the production and use of semiconductor "chips," vacuum tubes are still needed for high-power electrical devices. In such tubes, the electrical circuitry is enclosed in a glass tube or envelope, which is evacuated. Since it is difficult to remove all traces of gases such as O_2 and N_2 from the tubes, a "getter" such as barium is placed in the tube. The function of the getter is to react with the residual gases, effectively cleaning the tube.

$$2\ Ba(s) + O_2(g) \longrightarrow 2\ BaO(s)$$

$$3\ Ba(s) + N_2(g) \longrightarrow Ba_3N_2(s)$$

Products are the metal oxide and the metal nitride, the latter an ionic compound based on the nitride ion, N^{3-}.

Elemental radium has no large-scale commercial uses, but its salts do. Radium bromide can be used, for example, in luminous materials for the faces of wristwatches and clocks. The radium compound is mixed with a phosphorescent material, and the latter glows because it is bombarded by alpha particles from the disintegration of the radium. Although only about one part of radium to 20,000 parts of the phosphor are used, radium is being replaced in this use by somewhat less dangerous radioactive materials such as polonium.

Magnesium reacts with warm water to give hydrogen gas and magnesium hydroxide. This reaction was used in World War II to generate the hydrogen gas to fill weather balloons. (See also page 911.)

THE CHEMISTRY OF THE ALKALINE EARTH METALS

Because a sample of beryllium metal is coated with a chemically inert coating of oxide, it cannot be dissolved in water. Furthermore, it must be heated to 600 °C in air before there is any visible change. In contrast, freshly cleaned magnesium will slowly evolve H_2 when heated gently in water.

$$Mg(s) + 2\ H_2O(\ell) \longrightarrow Mg(OH)_2(aq) + H_2(g)$$

The remaining alkaline earth elements, on the other hand, must be protected from air and moisture because they react so rapidly.

The halides, oxides, and carbonates are the most important compounds of these elements, so we shall focus on them. Before beginning, however, we want to remind you that their chemistry is dominated by the 2+ ions of the elements. Another dominating fact is their relatively high "effective" charge. Each alkaline earth 2+ ion is smaller than the 1+ ion of the preceding alkali metal, so the 2+ ions of Group 2A are much more electron-attracting and are capable of forming stronger bonds. This is particularly true of beryllium, so its bonds to other elements can have a high degree of covalency. One example of this effect is the behavior of Be^{2+} in aqueous solution (Section 17.6). All of the alkaline earth 2+ ions are hydrated in water, but Be^{2+} is involved in a complex series of equilibria beginning with the tetrahedrally coordinated ion, among them the following.

$$H_2O(\ell) + [Be(H_2O)_4]^{2+}(aq) \longrightarrow [Be(H_2O)_3(OH)]^+(aq) + H_3O^+(aq)$$

$$[Be(H_2O)_4]^{2+}(aq) + [Be(H_2O)_3(OH)]^+(aq) \longrightarrow [(H_2O)_3Be{-}O{-}Be(H_2O)_3]^{2+}(aq) + H_3O^+(aq)$$

The highly electron-attracting Be^{2+} ion forms strong bonds to oxygen as does H^+ ion, so Be^{2+} can partly replace H in H_2O, and a proton is lost to the best available base, another water molecule.

$$\left[Be \longleftarrow :\ddot{O}(H)_2\right]^{2+}(aq) + H_2O(\ell) \longrightarrow \left[Be{-}\ddot{O}{:}{-}H\right]^{+}(aq) + H_3O^+(aq)$$

The Be^{2+}/H_2O interaction is similar to the interaction of a proton with water: $H^+(aq) + H_2O(\ell) \rightleftarrows H_3O^+(aq)$. Thus, Be^{2+} and H^+ are similar in their chemistries and are another example of a diagonal relationship.

Beryllium and aluminum are also diagonally related in the periodic table, and aqueous Be^{2+} ion is acidic, just like Al^{3+} ion in water (Section 17.6). In another parallel with aluminum chemistry, both $Be(OH)_2$ and $Al(OH)_3$ are amphoteric. Beryllium hydroxide acts as a Lewis acid in forming $[Be(OH)_4]^{2-}$ with additional base,

$$Be(OH)_2(s) + 2\ OH^-(aq) \longrightarrow [Be(OH)_4]^{2-}(aq)$$

and it is a Brønsted base when it reacts with acid to form Be^{2+} and H_2O.

$$Be(OH)_2(s) + 2\ H_3O^+(aq) \longrightarrow Be^{2+}(aq) + 4\ H_2O(\ell)$$

These parallels between Be^{2+} and Al^{3+} chemistry are good examples of a diagonal relationship (Section 22.2).

ALKALINE EARTH HALIDES The most important fluoride of the alkaline earth metals is CaF_2, fluorspar (Figure 10.3), but fluorapatite ($CaF_2 \cdot 3\ Ca_3(PO_4)_2$) is becoming increasingly important as a commercial source of fluorine. Although fluorspar is widely distributed, Europe and Mexico are the largest producers.

Almost half of the CaF_2 mined is used in the steel industry. It is added to the mixture of materials that are melted to make crude iron in a steel mill. The CaF_2 serves to remove some impurities, and it improves the separation of the molten metal from *slag*, the layer of silicate impurities and by-products that come from reducing iron ore to the metal (Chapter 25).

The steel industry is described more fully in Chapter 25 and the properties of HF are discussed in Chapter 24.

The other half of the fluorspar is used to manufacture hydrofluoric acid by reaction of the mineral with concentrated sulfuric acid.

$$CaF_2(s) + H_2SO_4(\ell) \longrightarrow 2\ HF(g) + CaSO_4(s)$$

HF is an extremely important chemical used in the aluminum industry to make cryolite, Na_3AlF_6 (see p. 884) and to make fluorocarbons, the chlorofluorocarbons of air conditioners, and the Teflon of nonstick frying pans.

The apatites are collectively known as phosphate rock, and over 100 million tons are mined annually; the mines of Florida alone account for about one third of the world's output. Much of this rock is converted to phosphoric acid by reaction with sulfuric acid.

$$\underset{\text{fluorapatite}}{CaF_2 \cdot 3\ Ca_3(PO_4)_2(s)} + 10\ H_2SO_4(aq) \longrightarrow 10\ CaSO_4(s) + 6\ H_3PO_4(aq) + 2\ HF(g)$$

The phosphoric acid is used to manufacture a multitude of products (fertilizers and detergents), and its reaction products are found in baking powder, in frozen fish, and in many other food products (Chapter 24).

The Group 2A chlorides are interesting, albeit in different ways. The dichloride of beryllium is prepared by a type of reaction that can be used to prepare many other metal halides. That is, the oxide is reduced to metal with carbon, and the metal is oxidized with Cl_2 to the metal chloride.

$$BeO(s) + C(s) + Cl_2(g) \xrightarrow[600\ °C-800\ °C]{} BeCl_2(s) + CO(g)$$

The structure of $BeCl_2$ is interesting and unusual. Since Be in the dichloride apparently forms two bonds, the VSEPR theory predicts that the molecule is linear. Valence bond theory then suggests that the Be must be *sp* hybridized.

sp hybridized

$$:\ddot{\underset{..}{Cl}}—Be—\ddot{\underset{..}{Cl}}:$$

However, because beryllium has two vacant valence orbitals, this means that the atom does not adhere to the octet rule. As you saw in Chapter 17, the availability of empty orbitals means the Be can function as an

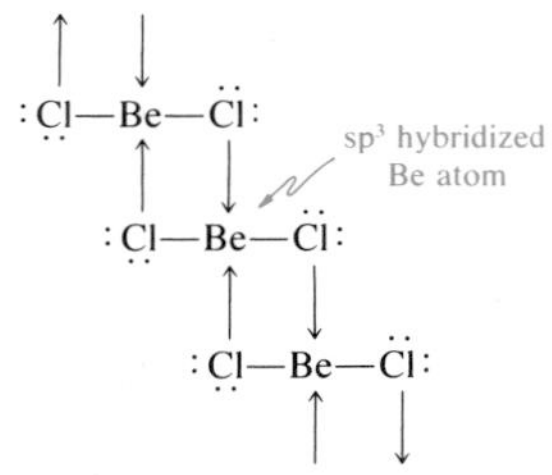

:Cl—Be—Cl:

:Cl—Be—Cl: (sp^3 hybridized Be atom)

:Cl—Be—Cl:

M-halogen-M bridges, such as the ones in $BeCl_2$, are common in chemistry, especially in aluminum and transition metal chemistry.

acceptor of electron pairs, that is, as a Lewis acid. But what is the donor atom in a solid composed only of $BeCl_2$ molecules? In this case, each Cl atom of $BeCl_2$ has three unshared pairs of electrons, so $BeCl_2$ molecules can *associate* as pictured in the margin. A Cl atom from one molecule uses an electron pair to form a donor–acceptor or coordinate covalent bond with a Be center in a neighboring molecule. Notice that the Be in the center is now surrounded by four Cl atoms, so it is expected that the Cl's will lie at the corners of a tetrahedron. Valence bond theory rationalizes this structure by saying that the beryllium is sp^3 hybridized.

Large quantities of $CaCl_2$ are produced in the Solvay process (Section 22.3), but demand for it is low. Some is used as an additive in concrete mixes, some to melt snow and ice on roadways in the winter, and some as a heat transfer agent in refrigerating plants. Some is also used as a desiccant or drying agent, since the hydration of $CaCl_2$ to give $CaCl_2 \cdot 6\,H_2O$ is highly exothermic (91 kJ/mol). Much of the $CaCl_2$, however, is simply discarded, and serious water pollution problems have resulted.

ALKALINE EARTH OXIDES AND CARBONATES Oxides and carbonates of the Group 2A elements are initimately linked, since the oxides are generally prepared by thermal decomposition of the respective carbonates (Chapter 4).

An exception to this is beryllium. It does not form a simple carbonate, nor does the element to which Be is diagonally related, Al.

$$CaCO_3(s) \xrightarrow[>850\ ^\circ C]{} CaO(s) + CO_2(g) \qquad K_p = 1.0 \text{ at } 897\ ^\circ C$$

The ease of the decomposition process decreases with increasing atomic weight, that is, with increasing basicity of the metal oxide. The reaction is reversible, as you might have suspected given the importance of $CaCO_3$ and $MgCO_3$ in nature. All the oxides are white powders, and none is decomposed even at 3000 °C!

Beryllium oxide does not react with water, while MgO (magnesia) reacts only slowly. Lime or CaO reacts more rapidly in a reaction called ''slaking'' (Figure 22.11). The heat of the reaction is so great that in the days of wooden sailing ships it was considered highly dangerous to carry lime as a cargo. If the ship should spring a leak, the lime would hydrolyze, and the heat was sufficient to ignite the wooden parts of the ship's hull.

Crystalline magnesia, MgO, is a useful material, because it is stable at high temperatures and is a good conductor of heat; however, it is a poor conductor of electricity. These properties make it useful as an insulator for the wires within electrical heating units, such as the ones in a home cooking range or space heater (page 367).

Lime is sixth on the list of the top chemicals produced in the United States (after sulfuric acid, nitrogen, oxygen, ethylene, and ammonia). About 32 billion pounds were produced in 1988.

Because of their economic importance, calcium carbonate and calcium oxide are of special interest to us. As already mentioned, calcium carbonate can be decomposed to CaO or lime in one of the oldest chemical transformations used by humans. Until lime came to be used primarily as a chemical in this century, it had been used for several thousand years as a building material in the form of mortar (a lime, sand, and water paste) to secure stones to one another in building houses, walls, and roads. The Greeks used lime mortar in building the Temples of Apollo and Elis in 450 BC, and the Chinese used it in laying up the stones in the Great Wall. The Romans, however, perfected the use of lime mortar, and the fact that many of their constructions still stand is testament to their skill and the usefulness of lime. In 312 BC, the famous Appian Way, a Roman highway from Rome to Brindisi (a distance of about 350 miles) was begun, and lime mortar was used between several layers of its stones. Wherever the Romans went, they built magnificent structures. However, in central France they could find no deposits of limestone or marble, so they dug the powdery form of $CaCO_3$, chalk, from the ground. This has an added benefit for us today—the caves left in quarrying the rock are ideal places for making and aging champagne.

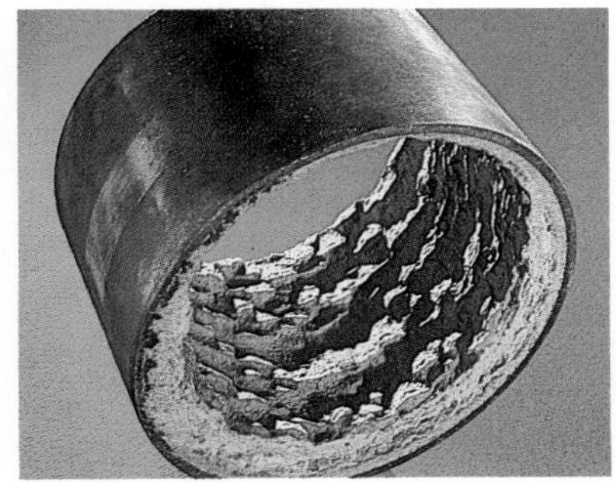

A section of a water pipe that has been coated on the inside with $CaCO_3$ deposited from hard water. (Betz Laboratories)

The usefulness of mortar depends on some simple chemistry. It consists of one part lime to three parts sand, with water added to make a thick paste. The first reaction that occurs, therefore, is the hydrolysis or "slaking" of the lime. When the mortar is placed between bricks or stone blocks, it slowly absorbs CO_2 from the air, and the slaked lime reverts to calcium carbonate.

$$Ca(OH)_2(s) + CO_2(g) \longrightarrow CaCO_3(s) + H_2O(\ell)$$

Although the sand in the mortar is chemically inert, the grains are bound together by the particles of calcium carbonate, and a hard material results.

Limestone is one of the oldest agricultural chemicals known. Yields of many crops can be greatly increased by spreading limestone on fields. The calcium carbonate neutralizes acidic compounds in the soil and supplies the essential nutrient Ca^{2+}; since magnesium carbonate also is often found in limestone, "liming" a field also supplies Mg^{2+}, another important plant nutrient.

The largest quantities of limestone and lime are now used in the chemicals industry rather than as mortar or in agriculture. About 45% of the lime is used now in steel making by the "basic oxygen process" (Chapter 25), and much of the rest is used to soften water by the "lime–soda" process. "Hard water" contains dissolved ions, chiefly Ca^{2+} and Mg^{2+}. If the water for a home or a city passes through underground limestone and if CO_2 is present, some limestone is dissolved,

See page 808 for a dramatic photo of the dissolving of limestone in water saturated with CO_2.

$$CaCO_3(s) + H_2O(\ell) + CO_2(g) \rightleftharpoons Ca^{2+}(aq) + 2\ HCO_3^-(aq)$$

and this solution is hard water. As described in Chapter 16, this reaction is reversible. Therefore, if the water is heated, the solubility of CO_2 drops and from Le Chatelier's principle we see that the equilibrium shifts to the left. If this happens in a heating system or steam-generating plant, the walls of the hot water pipes can become coated or even blocked with limestone. If you have hard water in your house, you will notice a coating of calcium carbonate on the inside of cooking pots. Another problem is that common soaps are a mixture of sodium and potassium salts of organic acids with long carbon chains. While these salts are soluble in water, those with Ca^{2+} and Mg^{2+} are not, and calcium and magnesium salts precipitate on clothes as a slimy, sticky residue.

"Temporarily" hard water contains both metal ions and CO_2, but "permanently" hard water has only metal ions.

$$\underset{\text{soap}}{2\ [H_3C{-}(CH_2)_{16}{-}CO_2]Na(aq)} + Ca^{2+}(aq) \longrightarrow \underset{\text{soap scum}}{[H_3C{-}(CH_2)_{16}{-}CO_2]_2Ca(s)} + 2\ Na^+(aq)$$

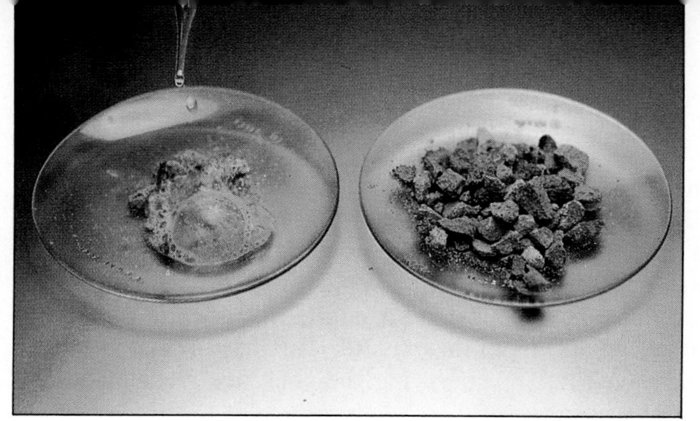

Figure 22.13 Calcium carbide, CaC_2, is a dirty yellow solid (right). When water is added (left), gaseous acetylene forms as indicated by the bubbles. The other product is $Ca(OH)_2$. The fact that the reaction evolves considerable heat is indicated by a wisp of steam rising from the reacting solid. (See also Figure 10.9.)

There are several ways to remove Ca^{2+} (and Mg^{2+}) from water, that is, to *soften the water*. One of these is the "lime–soda process," and a large fraction of the 32 billion pounds of lime (CaO) and 19 billion pounds of soda (Na_2CO_3) produced annually in the United States is used for this purpose. The soda is added to precipitate Ca^{2+}, for example, as the carbonate.

$$\underset{\text{from hard water}}{Ca^{2+}(aq)} + \underset{\text{from soda}}{CO_3^{2-}(aq)} \rightleftharpoons CaCO_3(s)$$

Although it would seem to be nonsense to also add calcium oxide to hard water, the chemistry shows we come out ahead. Slaked lime from CaO is an extremely inexpensive source of OH^-,

$$\underset{\text{slaked lime}}{Ca(OH)_2(s)} \rightleftharpoons Ca^{2+}(aq) + 2\ OH^-(aq) \qquad K_{sp} = 7.9 \times 10^{-6}$$

and the base transforms the hydrogen carbonate ion in hard water into carbonate ion.

$$OH^-(aq) + HCO_3^-(aq) \rightleftharpoons CO_3^{2-}(aq) + H_2O(\ell)$$

The carbonate ion then precipitates not only the calcium ion from the added lime but also the calcium and other ions present in hard water. Looking carefully at the balanced equations, we notice that 1 mole of $Ca(OH)_2$ furnishes 2 moles of OH^- which lead to 2 moles of CO_3^{2-}; these in turn precipitate 2 moles of Ca^{2+}, one from the added lime and one from the hard water.

Finally, another enormous use of lime is in the production of **calcium carbide**. If lime is heated at 2000 °C with coke, the product is a very interesting ionic compound containing the carbide ion, $:C{\equiv}C:^{2-}$.

$$CaO(s) + 3\ C(s) \xrightarrow[2000\ °C]{} \underset{\text{calcium carbide}}{CaC_2(s)} + CO(g)$$

Calcium carbide, a dirty yellow solid, is useful because the carbide ion is the salt of a weak acid, acetylene ($K_a \cong 10^{-20}$) (Figure 22.13). Hydrolysis of the anion, therefore, gives acetylene.

$$\underset{\text{calcium carbide}}{CaC_2(s)} + 2\ H_2O(\ell) \longrightarrow Ca(OH)_2(s) + \underset{\text{acetylene}}{H{-}C{\equiv}C{-}H(g)}$$

Acetylene made in this manner was once used as a major starting material for the manufacture of organic chemicals. Now that ethylene, $H_2C{=}CH_2$, can be made inexpensively, however, acetylene is no longer competitive.

EXERCISE 22.3 The Chemistry of Calcium Oxide
Tell how lime, CaO, is produced, and give four uses of the chemical.

BIOLOGICAL EFFECTS OF ALKALINE EARTH COMPOUNDS

Your body contains many metal ions that serve important regulatory functions, and Na^+, K^+, Mg^{2+}, and Ca^{2+} ions constitute about 99% of the total. The graphs in Figure 22.14 give you some idea of the relative amounts of these ions in various body fluids as compared with sea water. The feature that you may notice immediately is that K^+ and Mg^{2+} are by far the most important cations within cells, whereas the Na^+ concentration is much higher outside the cell. This difference is especially important to the operation of nerve cells, but the way that Na^+ is "pumped" out of cells and K^+ "pumped" in is one of the great mysteries of nature. The difference in Ca^{2+} concentrations inside and outside cells is also important. Many biochemical and physiological processes are triggered by entry of Ca^{2+} into a cell or release of Ca^{2+}. One such process is muscle contraction. Initiation of contraction results from the arrival of a nerve impulse at a motor nerve ending in a muscle fiber, and this causes Ca^{2+} to be released from a storage site. The released Ca^{2+} then interacts with a regulatory protein.

When you cut yourself, blood coagulates to prevent excessive bleeding. This complicated process occurs in a number of steps, almost all of which depend on Ca^{2+}.

Plants and animals derive energy from the oxidation of a sugar, glucose, with O_2. Plants are unique, though, in being able to synthesize glucose from CO_2 and H_2O using sunlight as the source of energy. The

Figure 22.14 Relative ionic concentrations of seawater and body fluids.

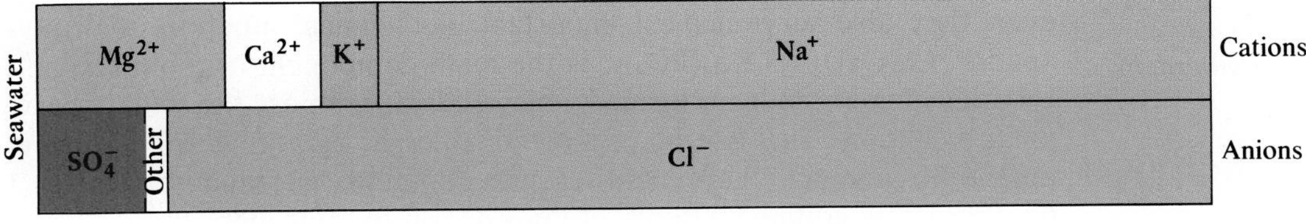

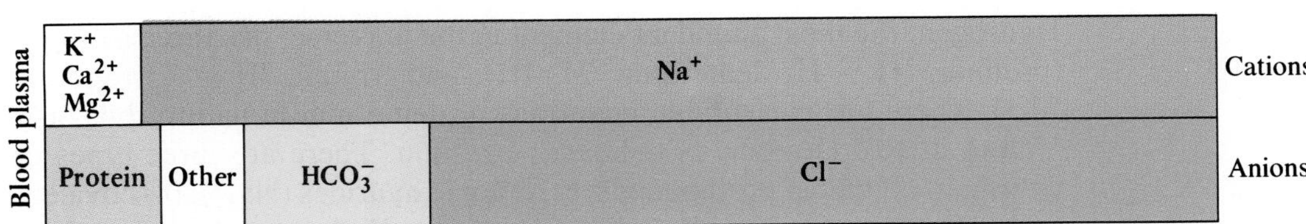

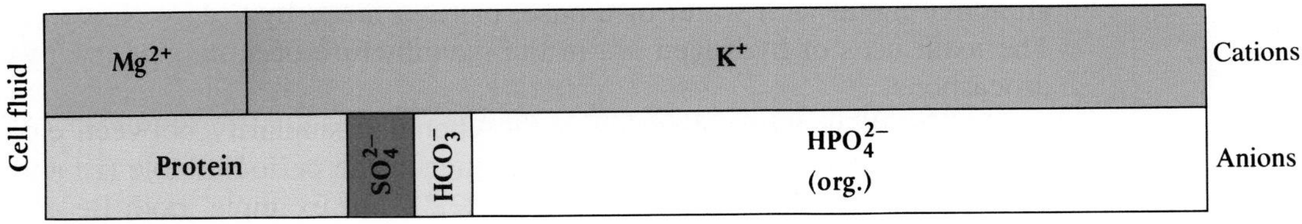

process is initiated by *chlorophyll*, a very large, magnesium-based molecule.

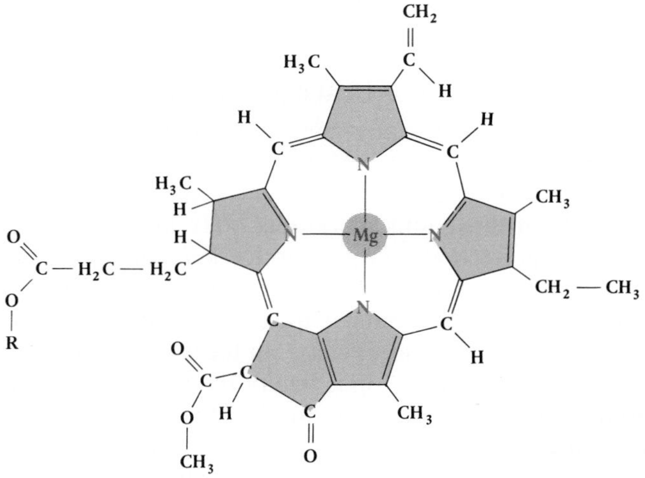

All simple beryllium-containing compounds are highly toxic and can be carcinogens. They must be handled with care.

Although Ca^{2+} and Mg^{2+} are required by living systems, the other Group 2A elements are toxic. For example, soluble barium salts are toxic, so you may be concerned if you are told by a physician to drink a "barium cocktail" so that the condition of your digestive tract may be checked. The "cocktail," however, contains the very *insoluble* salt $BaSO_4$ with a K_{sp} value of 1.1×10^{-10}. With this information, you know the concentration of Ba^{2+} ions produced by the salt is extraordinarily small, so the cocktail is safe.

The reason for drinking a slurry of barium sulfate is that the compound is opaque to x-rays. Thus, the path taken by the $BaSO_4$ through your digestive tract appears on the x-ray photographs.

We have already mentioned the apatites, calcium- and phosphorus-containing minerals that are mined for their phosphorus content. However, they also serve a most important biochemical function: hydroxyapatite, $Ca(OH)_2 \cdot 3\ Ca_3(PO_4)_2$, is the main component of tooth enamel. Cavities in your teeth are formed when acids decompose the weakly basic apatite coating. This can be prevented, though, by converting hydroxyapatite to a much more acid-resistant coating, fluorapatite, $CaF_2 \cdot 3\ Ca_3(PO_4)_2$, by adding a source of fluoride ion to your diet.

SUMMARY

Hydrogen, the most abundant element in the universe, has three isotopes (protium, ^1H = H; deuterium, ^2H = D; and tritium, ^3H = T) (Section 22.1). There are measurable quantities of deuterium in nature, but only 1 atom in 10^{18} H atoms is radioactive tritium. There are three types of hydrogen-containing compounds: (a) anionic hydrides (NaH); (b) covalent hydrides (CH_4); and (c) interstitial hydrides. Hydrogen is produced industrially by *steam re-formation of hydrocarbons or electrolysis*. In the laboratory it is made by the reaction of (a) an active metal and acid, (b) an active metal with water or a base, or (c) a metal hydride with water. The main uses of hydrogen are in the manufacture of ammonia and hydrocarbons.

The **diagonal relationship** is the chemical similarity between elements situated diagonally from one another in the periodic table (Li and Mg, Be and Al, and B and Si) (Section 22.2). For example, both Be and Al form inert oxides and amphoteric hydroxides.

The **alkali metals**, the Group 1A elements, are all metals and none exist free in nature (Section 22.3). Na and K are especially abundant, forming many common minerals, while the others are more rare. The heaviest element, Fr, is radioactive.

All the alkali metals can be recovered from one of their salts by electrolysis, and their uses depend on their ability to act as reducing agents. The hydration energy of the 1+ ion plays a pivotal role in determining the potential of $M(s) \rightarrow M^+(aq) + e^-$. All react with O_2, but only lithium forms a normal oxide, Li_2O. Sodium gives sodium **peroxide**, Na_2O_2, while the remaining elements form **superoxides** such as KO_2 (a salt of the O_2^- ion).

The *chlor–alkali industry* consumes NaCl and produces billions of pounds of chemicals such as Cl_2, NaOH, and Na_2CO_3. The *Solvay process* was used until recently in the United States to manufacture Na_2CO_3 (and $NaHCO_3$).

The **alkaline earth elements** of Group 2A (Section 22.4) are all metals, and all form 2+ ions. The lightest element, Be, is relatively rare, and the heaviest element, Ra, is radioactive. Magnesium and calcium are especially abundant and important. They form many common minerals, $CaCO_3$ or *limestone* being particularly common. Heating limestone gives CaO or *lime*, an oxide that reacts vigorously with water to give $Ca(OH)_2$ or *slaked lime*.

Recovery of the alkaline earth metals from their compounds can be done by electrolysis, the usual method for Mg. Other methods are possible, however, such as the aluminum reduction of CaO to give Ca and Al_2O_3.

All alkaline earth elements form halides of the type MX_2. Beryllium, however, is unique in that its halides are associated in the solid state through Be—X→Be bridges, a form of bonding found in aluminum chemistry as well.

Like aluminum, beryllium forms a very stable oxide and an amphoteric hydroxide. Since BeO oxide is so stable, this coating on a piece of Be metal means it does not react readily with water. The reaction of magnesium with water is slow, while the other metals all react vigorously. All form normal oxides, and calcium oxide, or lime, is used in the chemical industry as a base.

The ions Na^+, K^+, Mg^{2+}, and Ca^{2+} are the most abundant metal ions in biochemical systems. Ca^{2+}, for example, is important in the processes of muscle contraction and blood coagulation and is a component of tooth enamel. Mg^{2+} is the metal ion in chlorophyll.

STUDY QUESTIONS

1. Write balanced equations for the following:
   (a) the reaction of potassium with hydrogen
   (b) the reaction of chlorine with hydrogen
   (c) the reaction of sulfur with hydrogen
   (d) the reaction of propane, C_3H_8, with steam
   (e) the reaction of potassium with water
2. One recently suggested method for the preparation of hydrogen (and oxygen) proceeds as follows: (a) Sulfuric acid and hydrogen iodide are formed from sulfur dioxide, water, and iodine. (b) The sulfuric acid from the first step is decomposed by heat to water, sulfur dioxide, and oxygen. (c) The hydrogen iodide from the first step is decomposed with heat to hydrogen and iodine. Write a balanced equation for each of these steps and show that their sum is the decomposition of water to hydrogen and oxygen. (*High Technology*, May 1983, p. 67.)
3. Compare the mass of H_2 expected for 100% reaction

of steam (H_2O) with CH_4, with petroleum, and with coal. Calculate the mass of H_2 produced per gram of starting material. Use CH_2 and CH as representative formulas for petroleum and coal, respectively.

4. How much energy is required to convert 10.0 L of H_2 gas at 25.0 °C and 3.50 atm to H atoms? When these H atoms recombine to H_2, how much heat energy is evolved?

5. The electrolysis of aqueous NaCl gives NaOH, Cl_2, and H_2.
   (a) Write a balanced equation for the process.
   (b) In 1988 in the United States, 23.97 billion pounds of NaOH and 22.66 billion pounds of Cl_2 were produced. Does the ratio of masses of NaOH and Cl_2 actually produced agree with the ratio of masses expected from the balanced equation? If not, what does this tell you about the way in which NaOH or Cl_2 are actually produced? Is the electrolysis of aqueous NaCl the only source of these chemicals?

6. To store 2.88 kg of gasoline with an energy equivalence of 1.43×10^8 J requires a volume of 4.1 L. In comparison, 1.0 kg of H_2 has this same energy equivalence. What volume is required if this quantity of H_2 is to be stored at 25 °C and 1.0 atm of pressure?

7. Write complete electron configurations using the spectroscopic notation for francium and radium.

8. What is oxidized, and what is reduced, if an aqueous solution of KCl is electrolyzed under the same conditions as the electrolysis of aqueous NaCl? What are the products when CsI is electrolyzed under these conditions?

9. Explain how magnesium acts as a sacrificial anode in providing cathodic protection against oxidative corrosion on a ship's hull.

10. Why are lithium salts often hydrated, whereas those of the other alkali metals are often anhydrous? (If you need help, see Chapter 13.)

11. How would you extinguish a sodium fire in the laboratory? What is the worst thing you could do?

12. Give three uses for Group 2A metals.

13. Beryllium fluoride is a linear molecule in the gas phase. What is the orbital hybridization of the central beryllium atom?

14. When Be^{2+} is placed in aqueous solution, it forms the hydrated ion $[Be(H_2O)_4]^{2+}$. Sketch the structure of this ion, showing the arrangement of water molecules about the beryllium ion. What is the orbital hybridization used by the central beryllium?

15. When magnesium burns in air, it forms both an oxide and a nitride. Write a balanced equation for the formation of the nitride.

16. Heating barium oxide in pure oxygen gives barium peroxide. Write a balanced equation for this process.

17. Barium peroxide is an excellent oxidizing agent. Reaction with iron gives iron(III) oxide and barium oxide as products. Write a balanced equation for this reaction.

18. Write balanced equations for the reaction of all of the Group 1A elements with O_2. Specify which metals form oxides, which ones form peroxides, and which ones form superoxides.

19. Magnesium is made in large amounts by electrolysis of molten $MgCl_2$ and by the reduction of calcined dolomite with ferrosilicon at 1430K. The basic reaction for the latter process is

$$MgO \cdot CaO(s) + Si(s) \longrightarrow Mg(g) + Ca_2SiO_4(s)$$

Balance this equation.

20. Complete and balance equations for the following reactions:
   (a) $SrCO_3(s) + HCl(aq) \longrightarrow$
   (b) $BaCO_3(s) + HNO_3(aq) \longrightarrow$
   (c) $BaO(s) + Al(s) \longrightarrow$
   (d) $LiH(s) + H_2O(\ell) \longrightarrow$
   (e) $Na_2CO_3(aq) + H_2O(\ell) + CO_2(g) \longrightarrow$

21. Why is a solution of LiCl very slightly acidic? Write a balanced equation to demonstrate your answer.

22. Calcium forms a hydride, as do sodium and the other alkali metals. As it reacts readily with water, it is an excellent drying agent for organic solvents. (a) Write a balanced equation showing the formation of calcium hydride from calcium and H_2. (b) Write a balanced equation showing the reaction of calcium hydride with water.

23. $Ca(OH)_2$ has a K_{sp} of 7.9×10^{-6}, whereas that for $Mg(OH)_2$ is 1.5×10^{-11}. Calculate the equilibrium constant for the reaction,

$$Ca(OH)_2(s) + Mg^{2+}(aq) \longrightarrow Ca^{2+}(aq) + Mg(OH)_2(s)$$

and explain why this reaction can be used in the commercial isolation of magnesium from sea water.

24. (a) When 1000. kg of molten $MgCl_2$ is electrolyzed to produce magnesium, how many kilograms of metal are produced at the cathode? (b) What is produced at the anode? How many kilograms of the other product are produced? (c) What is the total number of Faradays of electricity used in the process? (d) One industrial process has an energy consumption of 8.4 kilowatt-hours per pound of Mg. How many joules are therefore required per mole? (e) How does this energy compare with the energy of the process $MgCl_2(s) \rightarrow Mg(s) + Cl_2(g)$?

25. Using data in Appendix J and that given below, calculate $\Delta G°$ values for the decomposition of MCO_3 to MO and CO_2 where M = Mg, Ca, and Ba. What is the relative tendency of these carbonates to decompose?

COMPOUND	ΔG_f° (kJ/mol)
$MgCO_3$	−1012.1
$BaCO_3$	−1137.6
BaO	−525.1

26. Calculate the enthalpy change for the reaction of CaO(s) with aluminum.
27. Name three uses of limestone. Write a balanced equation for the reaction of limestone with CO_2 in water.
28. Explain what is meant by "hard water." Outline one method for softening hard water.
29. Give three biologically important functions of Group 1A and Group 2A metal ions.
30. CaO is used to remove SO_2 from power plant exhaust, since the two react to give solid $CaSO_3$. How many grams of SO_2 can be removed using 1000. kg of CaO?
31. Calcium fluoride can be used in the fluoridation of water supplies. (a) Calculate the solubility of CaF_2 in grams per liter. (b) If you wanted to achieve a fluoride ion concentration of 2.0×10^{-5} M, how many grams of CaF_2 would have to be used for 1 million liters of water?
32. Identify the lettered compounds in the following reaction scheme. When 1.00 g of a white solid A is strongly heated, you obtain another white solid, B, and a gas. (The gas exerts a pressure of 209 mmHg in a 450-mL flask at 25 °C.) Bubbling the gas into a solution of $Ca(OH)_2$ gives another white solid, C. If the white solid B is added to water, the resulting solution turns red litmus paper blue. To the solution of B, you add dilute, aqueous HCl and evaporate to dryness to yield a white solid D. When D is placed in a bunsen flame, it colors the flame green. Finally, if the aqueous solution of B is treated with sulfuric acid, a white precipitate, E, forms.

CHAPTER

23

Metals, Metalloids, and Nonmetals: Periodic Groups 3A and 4A

Talc, mica, and vermiculite.

CHAPTER OUTLINE

The elements of Groups 3A and 4A are a bridge between the metals of Groups 1A and 2A and the largely nonmetallic elements of Groups 5A through 8A. Thus, the Groups 3A and 4A exhibit a tremendous range of chemistry. They include one of the least dense metals and one of the most dense metals, the most abundant metal and some of the least abundant metals, volatile hydrides that burn or explode in air and diamonds and sand. Their compounds have an equally wide range of uses, from detergents to semiconductors, and some have no significant commercial importance.

As we have done in the previous chapter, we shall emphasize the common minerals of the elements, the methods of obtaining the element itself, and compounds of commercial importance.

23.1 THE ELEMENTS OF GROUP 3A

3A
5 B boron
13 Al aluminum
31 Ga gallium
49 In indium
81 Tl thallium

The elements of Group 3A all have electron configurations of the type ns^2np^1 (Table 23.1). This means that each may lose these electrons to have a +3 oxidation number, although the heavier elements, especially thallium (Tl), also form compounds with an oxidation number of +1.

The elements vary greatly in their relative abundances on earth. The first few elements of the periodic table, up to carbon, are very low in abundance (Chapter 9). Aluminum, however, is very abundant and is our most abundant metal (Table 23.2).

Except for boron, a metalloid, all the elements of Group 3A are metals. In Group 4A, carbon is clearly a nonmetal, silicon and germanium

Table 23.1 The p^1 Elements of Group 3A (the Boron and Aluminum Family)*

	Boron	Aluminum	Gallium	Indium	Thallium
Symbol	B	Al	Ga	In	Tl
Atomic number	5	13	31	49	81
Atomic weight	10.81	26.98	69.72	114.82	204.38
Valence e^-	$2s^22p^1$	$3s^23p^1$	$4s^24p^1$	$5s^25p^1$	$6s^26p^1$
mp, °C	2079	660	29.78	156.6	303.5
bp, °C	3650	2467	2403	2080	1457
d, g/cm^3	2.35	2.70	5.91	7.31	11.85
Atomic radius, pm	85	143	135	167	170
Ion radius, pm (3+ ion)		67.5	76	94	103
Pauling EN	2.0	1.61	1.81	1.78	1.83
Standard reduction potential (V)	−0.90	−1.66	−0.53	−0.34	−0.34
Oxidation numbers	Covalent	+3	+1, +3	+1, +2, +3	+1, +3
Ionization energy	801	577.6	579	558	589
Heat of atomization	573	291	280	232	182
Isolated by	Gay-Lussac	Wöhler	Boisbaudran	Reich	Crookes
Date of isolation	1808	1827	1875	1863	1861
rpw pure O_2	B_2O_3, 1200 °C	Al_2O_3, 800 °C	Ga_2O_3, 1600 °C	In_2O_3, 600 °C	Tl_2O, 400 °C
rpw H_2O	None	None	None	None	None
rpw N_2	BN, 1200 °C	AlN, 740 °C			None
rpw halogens	BX_3, 400 °C	Al_2X_6, 200 °C	Ga_2X_6	In_2X_6	TlX

*All energies are given in kJ/mol. The symbol rpw means "reaction product with." The molar heat of atomization is the energy required to convert 1 mol of solid element into 1 mol of atoms in the gas phase. All data in this table are taken from (a) *KC? Discoverer*, a computer program distributed by JCE: Software. (b) N.N. Greenwood and A. Earnshaw, *Chemistry of the Elements*, Pergamon Press, New York, 1984. (c) E.G. Rochow, *Modern Descriptive Chemistry*, W.B. Saunders, Philadelphia, 1977.

are metalloids, and tin and lead are metals. According to the *diagonal relationship* (Section 22.2), aluminum and beryllium should have similar chemistries, and boron and silicon should show some similarities. A preliminary comparison of boron and silicon chemistry and beryllium and aluminum chemistry suggests the following conclusions:

(a) The oxide of boron, B_2O_3, and boric acid, $B(OH)_3$, are weakly acidic, just as are SiO_2 and its acid, orthosilicic acid (H_4SiO_4). In contrast, $Be(OH)_2$ and $Al(OH)_3$ are amphoteric, both dissolving in a strong base such as NaOH (Chapter 17).

(b) Boron–oxygen compounds, *borates*, are somewhat similar to silicon–oxygen compounds, the *silicates*.

(c) The halides of both boron and silicon (e.g., BCl_3 and $SiCl_4$) react vigorously with water. Aluminum halides are only partly hydrolyzed.

(d) The hydrides of boron and silicon are volatile, flammable, and readily hydrolyzed (with some exceptions). In contrast, aluminum hydride, AlH_3, is a colorless, nonvolatile solid that is extensively polymerized through Al—H—Al bonds.

There is clearly evidence for the diagonal relationship. However, the fact that boron and silicon have different numbers of valence shell electrons means that there will also be differences in their chemistries.

Table 23.2 Relative Terrestrial Abundance in Parts per Million

Si	257,000	B	3
Al	75,000	Sn	2.1
C	800	Tl	1.8
Ga	15	In	1.8
Pb	13	Ge	1.5

COMMON MINERALS AND RECOVERY OF THE ELEMENTS

The Group 3A elements do not occur free in nature. Rather, they are generally found as oxides in many locations around the earth. Although boron is very low in relative abundance, its common minerals are found

Figure 23.1 An aerial view of the open pit borax mine of the U.S. Borax and Chemical Corporation in the Mojave Desert near Boron, California. (Rick McIntyre, Tom Stack & Associates)

in concentrated deposits, especially in California. Large deposits of *borax*, $Na_2B_4O_7 \cdot 10\ H_2O$, are currently being mined in the Mojave Desert near the town of Boron (Figure 23.1). Deposits of borate ores were originally mined in Death Valley in the late 1800s and were hauled out by the famous 20-mule teams.

Isolation of pure, elemental boron is extremely difficult and is done only in small quantities. Annual worldwide production is probably less than 10 tons. As is true of most metals and metalloids, it can be obtained by chemically or electrolytically reducing an oxide or halide. Magnesium, for example, has been used as the reducing agent for many years, but the product is a noncrystalline boron of low purity.

Allotropes are different forms of the same element that exist in the same physical state under the same conditions of pressure and temperature.

$$B_2O_3(s) + 3\ Mg(s) \longrightarrow 2\ B(s) + 3\ MgO(s)$$

Very pure boron can be made on the kilogram scale by reducing BBr_3 with H_2 on hot tantalum metal.

$$2\ BBr_3(g) + 3\ H_2(g) \xrightarrow{Ta} 2\ B(s) + 6\ HBr(g)$$

There are several **allotropes** of boron, elemental boron with different solid state structures, but all are characterized by having the *icosahedron* as one structural element (Figure 23.2). Partly as a result of extended, covalent bonding, boron is very hard, refractory (resistant to heat), and a nonconductor. In this regard, it is quite different from the other Group 3A elements; Al, Ga, In, and Tl are all relatively low melting, rather soft metals with high electrical conductivity.

If boric oxide is reduced with carbon, the product is boron carbide. (Recall that reduction of calcium oxide with carbon gives calcium carbide; Section 22.4)

$$2\ B_2O_3(s) + 4\ C(s) \longrightarrow B_4C(s) + 3\ CO_2(g)$$

The carbide is extraordinarily high melting (2250 °C) and very hard. It also has a low density (2.5 g/cm^3), so one of its largest uses has been in bulletproof armor. Thousands of suits of this armor were used in the war in Vietnam.

Figure 23.2 The icosahedron, a regular polyhedron having 20 faces. In many different types of solid boron, the boron atoms are bound together in this polyhedron. In some cases, there are thin bonds from one polyhedron to another.

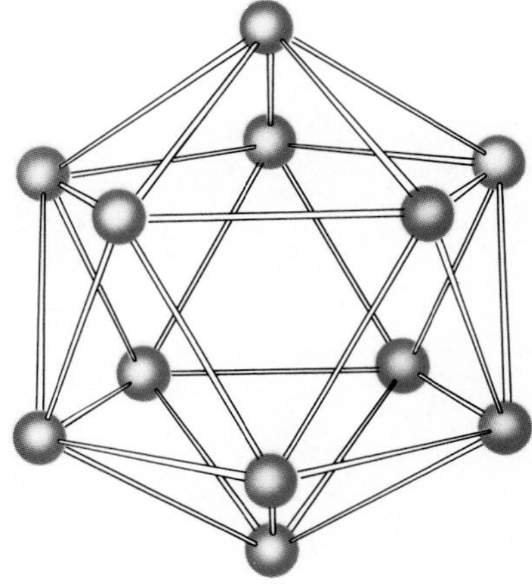

Aluminum is found in varying amounts in nature as aluminosilicates, minerals such as clay that are based on aluminum, silicon, and oxygen (see Section 23.2). As these minerals are weathered, they gradually break down to various forms of hydrated aluminum oxide, $Al_2O_3 \cdot nH_2O$, and a mixture of these is called *bauxite*. Dehydrating the hydrated oxides leads finally to anhydrous Al_2O_3.

A total of at least 235 × 10⁶ kJ of energy is needed to make 1000 kg of aluminum. This is more than 4700 kJ per 20-gram soft drink can, which is equivalent to about half of your daily food intake. Since only 4.5% as much energy is needed to recycle aluminum, this is why there is interest in reusing aluminum.

Recovery of aluminum from purified bauxite is done by electrolysis as outlined in Chapter 21 (Figure 23.3). However, the chemistry for the purification of the bauxite is a good illustration of some aspects of Group 3A chemistry. Bauxite ore contains iron and silicon oxides as well as hydrated aluminum oxide. The aluminum oxide is purified by the *Bayer process* by making use of the amphoteric, basic, or acidic nature of the various oxides. Silica, SiO_2, is an acidic oxide, Al_2O_3 is amphoteric,* while Fe_2O_3 is a basic oxide. Therefore, the first two oxides will dissolve in a hot concentrated solution of caustic soda (NaOH).

$$\left.\begin{matrix} \underset{\text{amphoteric}}{Al_2O_3(s)} \\ \underset{\text{acidic}}{SiO_2\ (s)} \\ \underset{\text{basic}}{Fe_2O_3(s)} \end{matrix}\right\} + 30\%\ NaOH(aq) \xrightarrow[190\ ^\circ C]{} NaAl(OH)_4(aq) + Na_2Si(OH)_6(aq) + Fe_2O_3(s)$$

The insoluble materials are removed by filtration, and Al_2O_3 is precipitated by treating the solution with CO_2. Recall that CO_2 forms the weak acid H_2CO_3 in water, so Al_2O_3 is precipitated in an acid–base reaction,

$$\underset{\text{acid}}{H_2CO_3(aq)} + \underset{\text{base}}{2\ NaAl(OH)_4(aq)} \longrightarrow Na_2CO_3(aq) + Al_2O_3(s) + 5\ H_2O(\ell)$$

and the silicate ion remains in solution.

The production of aluminum has grown rapidly (Figure 23.4) because aluminum has thousands of uses as a structural material and in packaging. However, pure aluminum is rarely used, since it is soft and weak. What is more, it loses strength rapidly above 300 °C. To strengthen the metal, and improve its properties, it is alloyed with small amounts of other metals. A typical alloy, for example, may contain about 4% copper with smaller amounts of silicon, magnesium, and manganese, and a large passenger plane today may use more than 50 tons of such alloy. To make a softer, more corrosion-resistant alloy for window frames, furniture, highway signs, and cooking utensils, however, only manganese may be included.

Much of the usefulness of aluminum comes from its corrosion resistance. This is due to the formation of a thin, tough, and transparent skin of oxide, Al_2O_3.

$$4\ Al(s) + 3\ O_2(g) \longrightarrow 2\ Al_2O_3(s) + 3351.4\ kJ$$

Figure 23.3 An aluminum "pot line." Electrolytic reduction "pots" convert refined bauxite (aluminum oxide) to molten aluminum metal. (See Chapter 21 and Figure 21.19.) (Atlantic Richfield)

*Al_2O_3 can be thought of as dehydrated $Al(OH)_3$,

$$2\ Al(OH)_3(s) \longrightarrow Al_2O_3(s) + 3\ H_2O(\ell)$$

and the hydroxide is clearly amphoteric, dissolving in acid or in strong base.

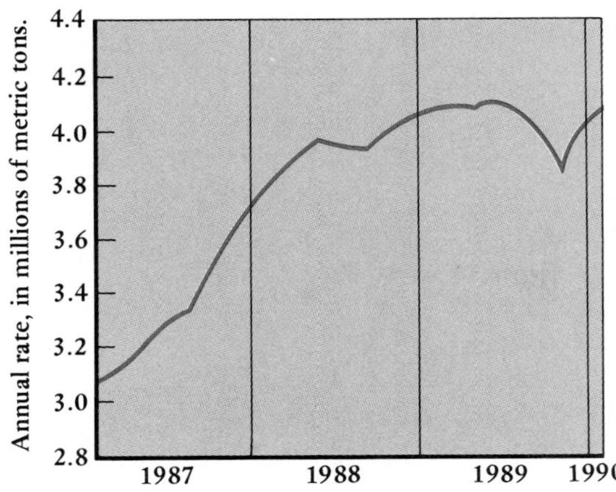

Figure 23.4 Aluminum production in the United States rose to an annual rate of more than 4 million metric tons (1 metric ton is 1000 kg) in 1990. (Graph adapted from the *Wall Street Journal,* January 27, 1990, page 1)

Partly because the heat of formation of the oxide is so large, the oxide layer is rapidly self-repairing. If you scratch a piece of aluminum it forms a new layer of oxide that covers the damaged area.

Aluminum will dissolve in acids such as HCl, but not in nitric acid (Figure 23.5). The latter is a powerful oxidizing agent (and a source of O atoms), so it oxidizes the surface of aluminum rapidly, and Al_2O_3 protects the metal from further attack. In fact, nitric acid is often shipped in aluminum tanker trucks.

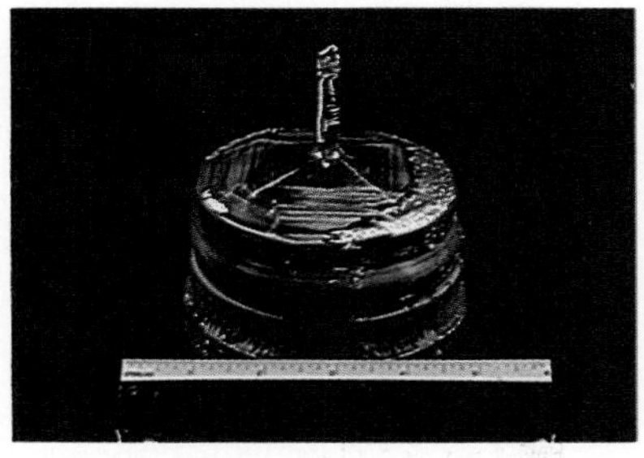

A gallium-arsenide (GaAs) semiconductor.

The remaining elements of Group 3A, gallium, indium, and thallium, are very low in abundance, but gallium is rapidly gaining in commercial importance. All are found in small amounts associated with other minerals, and all are obtained as by-products in the recovery of other metals such as zinc and aluminum. As a result, these metals are expensive. Pure gallium, for example, costs about $4 per gram, whereas one gram of aluminum of comparable purity costs only about 20 cents.

Gallium was one of the elements that was not known at the time Mendeleev developed his notion of the periodic table (Chapter 9), but he predicted its existence and properties, a fact that helped greatly in its discovery just a few years later. It is truly a remarkable element. It has the greatest liquid range of all known elements: it can melt in your hand (see Table 23.1), but it does not boil until the temperature reaches 2403 °C! Finally, like water, gallium is one of the few known materials that *expands* upon freezing.

The greatest use for gallium, and one that may continue to grow, is in the semiconductor gallium arsenide, GaAs. Integrated circuits based on GaAs have achieved operating speeds up to five times that of the fastest silicon chips currently available, and they will operate over a wider temperature range than silicon circuits. However, these advantages come at a price: arsenic is volatile and toxic, so GaAs is difficult to make.

Figure 23.5 Copper reacts vigorously with nitric acid (left) to give $Cu(NO_3)_2$ and NO_2 gas, but aluminum (right) is unreactive in nitric acid.

EXERCISE 23.1 Aluminum Chemistry

The Hall–Heroult electrolysis process for aluminum outlined in Chapter 21 uses a mixture of Al_2O_3 and Na_3AlF_6, cryolite. The mixture melts at only 960 °C, whereas pure Al_2O_3 melts at more than 2000 °C. Cryolite is made by the reaction

$$Al_2O_3(s) + HF(aq) + NaOH(aq) \longrightarrow Na_3AlF_6(s) + H_2O(\ell)$$

Balance the equation. If you need 1.00 kg of cryolite, how many grams of each reactant are required?

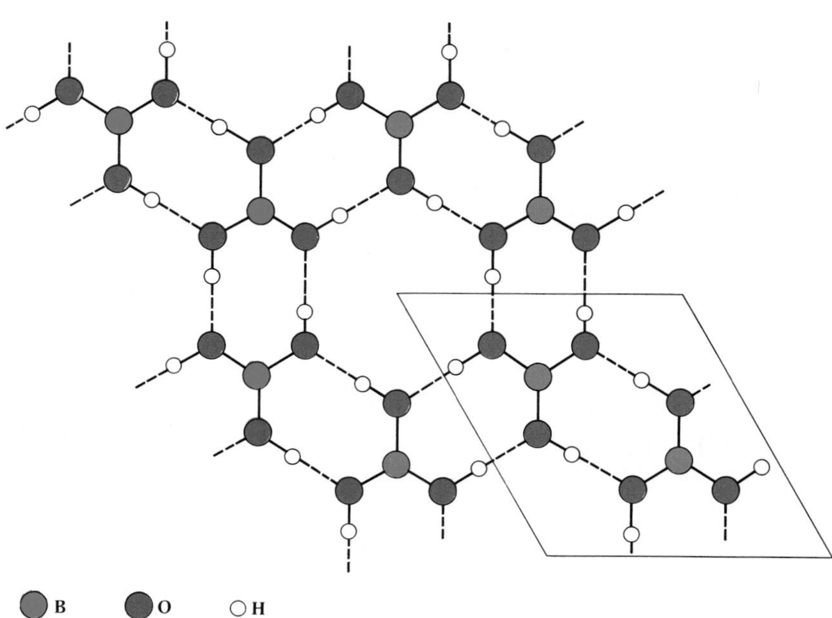

Figure 23.6 Solid boric acid is a good example of hydrogen bonding in the solid state. $B(OH)_3$ molecules are hydrogen bonded into six-membered rings in layers. The layers are 318 pm apart.

THE CHEMISTRY OF THE GROUP 3A ELEMENTS

The occurrence of an oxidation state that is 2 less than the group number is sometimes called the "inert pair" effect.

Group 3A elements all form compounds where the element has an oxidation number of +3. One of the most interesting aspects of this group, however, is that oxidation numbers of +1 are observed and become more important as the atomic number increases. Thus, there is some evidence for Al^+ compounds, and Ga^+ and In^+ compounds exist but are unstable. In contrast, Tl^+ salts are as common as compounds of Tl^{3+}. Thallium(I) compounds resemble those of silver (Group 1B) and of the Group 1A elements: TlCl is insoluble, just like AgCl, and TlOH is a strong base, just like NaOH.

GROUP 3A OXYGEN COMPOUNDS Although many chemists have spent years researching boron compounds and have uncovered fascinating structures and reactions, the compounds manufactured by the boron products industry are the chemically unspectacular oxides and oxyacids: boric acid, borax, and their simple derivatives. With an annual worldwide production in the range of two million tons, it is an important industry.

The Venetian adventurer Marco Polo (1254–1324?) brought borax back from the Far East, along with gun powder and spaghetti.

Borax, $Na_2B_4O_7 \cdot nH_2O$ (n = 5 or 10), is the most important boron–oxygen compound and is the form of the element most often found in nature. It has been used for centuries as a low-melting *flux* in metallurgy, because of the ability of molten borax to dissolve other metal oxides, thus cleaning the surfaces to be joined and permitting good metal-to-metal contact. As it is usually written, the formula of borax is misleading, since the salt really contains the ion $B_4O_5(OH)_4{}^{2-}$. Borax is thus better written as $Na_2[B_4O_5(OH)_4] \cdot 8\ H_2O$ (when n = 10).

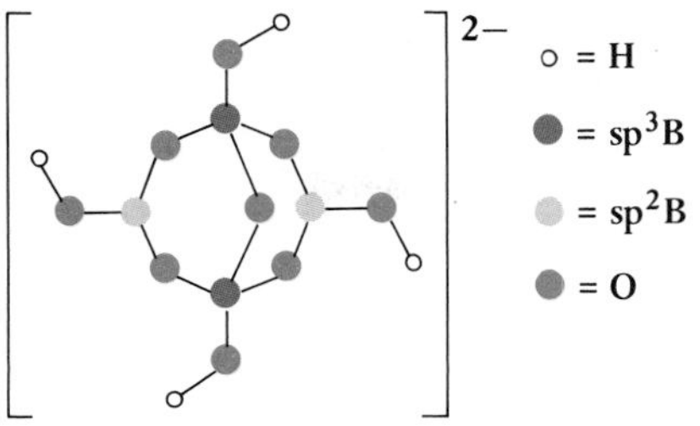

○ = H
● = sp^3B
● = sp^2B
● = O

$[B_4O_5(OH)_4]^{2-}$

The structure of the borate ion illustrates several commonly observed phenomena. First, many minerals consist of MO_n groups that share oxygen atoms. Second, this fusion often takes the form of metal–oxygen ring systems. And third, it has both sp^2 and sp^3 hybridized boron atoms.

After refinement, borax can be treated with sulfuric acid and converted to boric acid, $B(OH)_3$ (Figure 23.6).

$$Na_2B_4O_7 \cdot 10\ H_2O(s) + H_2SO_4(aq) \longrightarrow 4\ B(OH)_3(aq) + Na_2SO_4(aq) + 5\ H_2O(\ell)$$

Boric acid is a Lewis acid that acts by accepting a hydroxide ion from water.

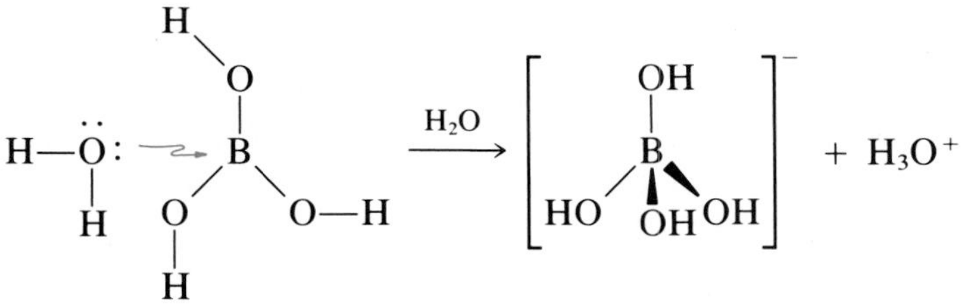

$$H{-}\ddot{O}{:} + B(OH)_3 \xrightarrow{H_2O} [B(OH)_4]^- + H_3O^+$$

$$K_a = 7.3 \times 10^{-10}$$

Because of its weak acid properties and slight biological toxicity, the acid has been used for many years as an antiseptic. Furthermore, since the acid is so weak, this means that salts of borate ions, such as the $[B_4O_5(OH)_4]^{2-}$ ion in borax, are hydrolyzed in water to give basic solutions. For this reason, borax has also been used for years in soap and detergent systems.

The second largest use of boric acid and borates is as a flame retardant for cellulose home insulation. Such insulation is mostly made from scrap paper, which is inexpensive but flammable. To control its flammability, 5% to 10% of the weight of the insulation is boric acid.

Boric acid is dehydrated to boric oxide when heated strongly.

$$2\ B(OH)_3(s) \longrightarrow B_2O_3(s) + 3\ H_2O(g)$$

By far the largest use of borax and of boric oxide is in the manufacture of borosilicate glass. This type of glass is composed of about 76% SiO_2, 13% B_2O_3, and much smaller amounts of Al_2O_3 and Na_2O. The presence of boric oxide gives the glass a higher softening temperature, a better resistance to attack by acids, and makes it expand less on heating.

One common brand of borosilicate glass is Pyrex, the trademark used by Corning Glass Company. The source of sodium in glass is sodium carbonate, Na_2CO_3 (see page 921).

The chemistry of the remaining Group 3A elements with oxygen has some similarities with that of boron, but also some great differences. In water, the Group 3A 3+ ions all undergo hydrolysis to give acidic solutions.

$$[M(H_2O)_6]^{3+}(aq) + H_2O(\ell) \rightleftharpoons [M(H_2O)_5(OH)]^{2+}(aq) + H_3O^+(aq)$$

ELEMENT	K_a
Al	7.9×10^{-6}
Ga	2.5×10^{-3}
In	2.0×10^{-4}
Tl	$\approx 7 \times 10^{-2}$

The ions with six water molecules exist only in very acidic solutions. As the pH is raised, the hydrated oxide $[Al_2O_3 \cdot 3\ H_2O = 2\ Al(OH)_3]$ precipitates. Aluminum and gallium hydroxides are amphoteric, and both dissolve in excess base to give $[M(OH)_4]^-$.

$$Ga(OH)_3(s) + OH^-(aq) \longrightarrow [Ga(OH)_4]^-(aq)$$

As you will see later, these properties lead to many uses of aluminum salts in particular.

Aluminum oxide, Al_2O_3, which can be formed by dehydration of $Al(OH)_3$, is quite insoluble in water and generally resistant to chemical attack. In the crystalline form, aluminum oxide is known as *corundum*, and it is extraordinarily hard. This property has led to its use as the abrasive in grinding wheels, "sandpaper," and toothpaste.

When Al_2O_3 is contaminated with SiO_2 and iron oxides, it is known as emery.

Some gems are just impure aluminum oxide. Rubies, the beautiful red crystals prized for jewelry and used in some lasers, are Al_2O_3 contaminated with a small amount of Cr^{3+} in place of some Al^{3+} ions. Blue sapphires occur when Fe^{2+} and Ti^{4+} impurities are present in Al_2O_3. As you will learn in Chapter 25, it is just these transition metal "impurities" that give the minerals their beautiful colors and other desirable properties. Synthetic rubies were first made in 1902, and the worldwide capacity is now about 200,000 kg per year; much of this production is used for the jewel bearings in watches and instruments (Figure 23.7).

If bauxite or clay is treated with sulfuric acid, the product is aluminum sulfate.

$$\underset{\text{clay}}{H_2Al_2(SiO_4)_2 \cdot H_2O(s)} + \underset{\text{sulfuric acid}}{3\ H_2SO_4(aq)} \longrightarrow \underset{\text{aluminum sulfate}}{Al_2(SO_4)_3(aq)} + 2\ H_4SiO_4(s) + H_2O(\ell)$$

Figure 23.7 A synthetic ruby, a crystal of Al_2O_3 containing a small amount of Cr^{3+} in place of Al^{3+}. (Kurt Nassau)

Figure 23.8 Crystals of alum, hydrated potassium aluminum sulfate [$KAl(SO_4)_2 \cdot 12\ H_2O$]. The alum in these crystals was prepared in the introductory chemistry laboratory by digesting aluminum cans in KOH, followed by treatment with sulfuric acid.

Figure 23.9 Aluminum hydroxide, $Al(OH)_3$, readily adsorbs dyes. At left is a precipitate of pure $Al(OH)_3$, while the $Al(OH)_3$ at right has adsorbed the red dye aluminon.

This and the related compounds called *alums* [potassium alum is $KAl(SO_4)_2 \cdot 12\ H_2O$] are very useful (Figure 23.8). Large quantities of aluminum sulfate, for example, are used in the paper industry to make the product stronger and nonporous. Alums are also important in water treatment as "clarifiers." Remember that in any but *very* acidic media, Al^{3+} hydrolyzes ultimately to the hydrated oxide. When this occurs in water containing fine, suspended particles (mud!), the alumina surrounds the particles and precipitates them.

For similar reasons, aluminum sulfate and alums are used in dyeing cloth. The aluminum hydroxide is called a *mordant* in the textile industry, because it binds to both cloth and dye molecules, thereby serving to "fix" the dye to the cloth (Figure 23.9).

EXERCISE 23.2 Boron Chemistry

The structure of boric acid is illustrated in Figure 23.6. (a) One unit cell is outlined in the figure. How many molecules of $B(OH)_3$ are contained in this unit cell? (b) What is the hybridization of the boron atom in the acid? (c) On the basis of the boron hybridization, explain why $B(OH)_3$ is open to attack by the Lewis base H_2O.

BORON AND ALUMINUM HALIDES None of the simple halides of Group 3A elements exists in nature for the simple reason that B—O, Al—O bonds, and so on, are so thermodynamically stable that oxygen compounds are found instead. Nonetheless, the simple halides have some interesting properties and are important as catalysts in the organic chemicals industry.

Figure 23.10 Liquid BBr_3 (left) and solid BI_3 (right).

Boron trifluoride is a colorless gas made by heating boric oxide with calcium fluoride and sulfuric acid. The essence of the process is that the acid and CaF_2 combine to produce hydrofluoric acid,

$$CaF_2(s) + H_2SO_4(\ell) \longrightarrow CaSO_4(s) + 2\ HF(g)$$

which then reacts with boric oxide.

$$B_2O_3(s) + 6\ HF(g) \longrightarrow 2\ BF_3(g) + 3\ H_2O(\ell)$$

Since BF_3 does hydrolyze in water to some extent to give HF and boric acid, it is imperative that the water by-product be removed. Therefore, this process uses excess sulfuric acid, since the concentrated acid is an excellent drying agent.

Although the boron trihalides are gases or volatile liquids or solids (Figure 23.10), the aluminum halides are all solids. Boron trifluoride has a boiling point of −100 °C, while AlF_3 is an ionic solid in which Al^{3+} ions are surrounded by an octahedron of F^- ions. In fact, cryolite (Na_3AlF_6), the mineral so important in the electrolytic recovery of aluminum, contains octahedral $AlF_6{}^{3-}$ ions (see Exercise 23.1).

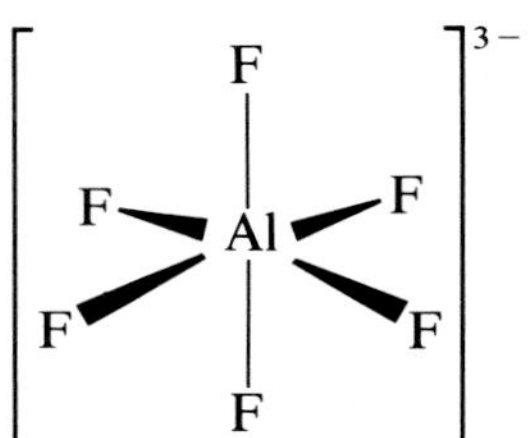

Aluminum chloride is made by direct reaction of the metal and chlorine. The reaction generates so much heat that the metal melts and continues to "burn" in the stream of chlorine once the reaction is started (see Al + Br_2, Figure 4.3).

$$2\ Al(s) + 3\ Cl_2(g) \longrightarrow 2\ AlCl_3(s) + 1408.4\ kJ$$

Alternatively, $AlCl_3$ can be prepared by heating a mixture of aluminum oxide, carbon, and chlorine.

$$Al_2O_3(s) + 3\ C(s) + 3\ Cl_2(g) \longrightarrow 2\ AlCl_3(s) + 3\ CO(g)$$

This reaction can be seen as a two-step process. The carbon effectively reduces the metal oxide to metal and CO, and then chlorine reacts with the metal to produce the metal chloride. The solid trichloride is, like AlF_3, an ionic material. However, when it melts, the $AlCl_3$ units form Al_2Cl_6 molecules. Aluminum bromide and iodide are composed of Al_2X_6 molecules even in the solid state.

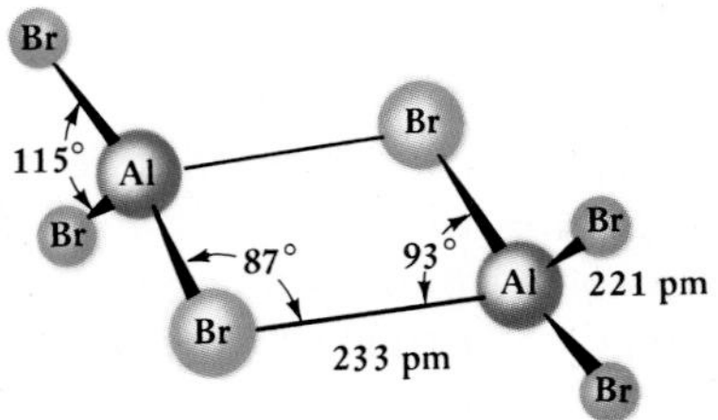

Aluminum chloride, bromide, and iodide all hydrolyze in water, and the partial hydrolysis gives hydroxyhalides, $Al(OH)_nX_{3-n}$, of various compositions. For example,

$$AlCl_3(s) + 2\ H_2O(\ell) \longrightarrow 2\ HCl(aq) + Al(OH)_2Cl(s)$$

Dehydration then leads to a solid material whose composition approximates the formula AlOCl. Such substances have some commercial value in antiperspirants and deodorants because they react with the water in perspiration.

BORON HYDRIDES Although the boron hydrides were once proposed as lightweight, high-energy fuels for airplanes and rockets, they are without great commercial interest at the present time. Their importance lies in the role they have played in the development of theories of chemical structure and bonding.

Compounds that contain only an element and hydrogen are often named by adding the suffix -ane to the element name. Thus, there are boranes (B—H), alanes (Al—H), silanes (Si—H), and germanes (Ge—H).

The boron hydrides are frequently called *boranes*. More than twenty neutral boranes, of general formula $(BH)_pH_q$, are now known. The simplest borane is *diborane*, B_2H_6, where p is 2 and q is 4.

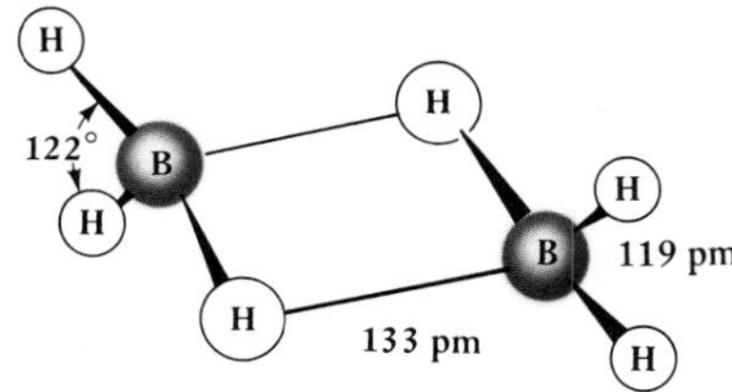

This is curious, since you might have expected BH_3 to be the simplest hydride, by analogy with boron trihalides. It is even more curious if you examine the structure. Two boron atoms and six hydrogen atoms bring a total of twelve valence electrons to bind the molecule together. If you take each of the eight lines in the structural diagram as a two-electron bond, then sixteen electrons are required. Thus, the boron hydrides came to be called "electron deficient" molecules. But clearly the molecule cannot be electron deficient; if it did not have sufficient electrons to satisfy its bonding needs, it would be quite unstable or nonexistent.

There are several ways to solve the diborane bonding dilemma, and one of them can be outlined as follows. The boron atoms are surrounded tetrahedrally by H atoms, so we assume that the borons are sp^3 hybridized. The "outside" or terminal B—H bonds are assumed to be normal, two-electron bonds formed by overlap of the H 1s orbital with a boron sp^3 orbital. The four bonds of this type account for eight of the twelve electrons available, so four electrons remain to construct the two B—H—B bridges. Each boron has two sp^3 hybrid orbitals, and these extend into the bridging region as illustrated in Figure 23.11. Here a spherical hydrogen 1s orbital can overlap one sp^3 hybrid from each boron, creating a *three-center bond*. This three-center bridge bond can hold two electrons, so the two bridges account for the four electrons remaining. Although you have not seen such forms of bonding before this, they are not uncommon in the chemistry of boron and other elements.

Many boranes can be viewed as fragments of the icosahedron used by elemental boron as its basic structural element.

The only B—H compounds being produced in ton quantities are the *borohydrides*. By far the most common of these is sodium borohy-

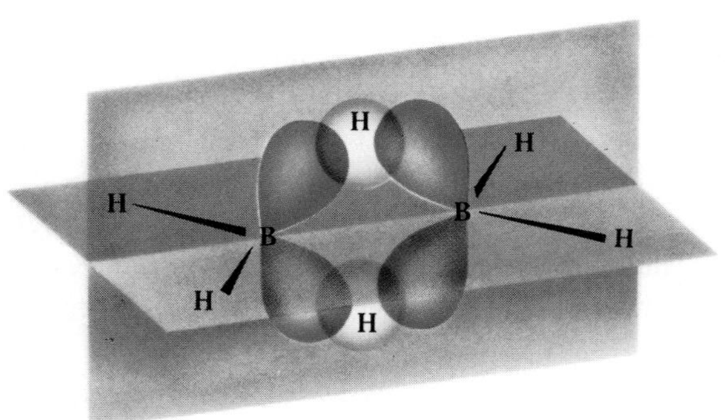

Figure 23.11 B—H—B bridge bonding in diborane, B_2H_6. After accounting for bonding to two "terminal" H atoms, two sp^3 orbitals remain on each boron atom. One such orbital from each boron may overlap a hydrogen 1*s* orbital in the bridge to give a three-center (three-orbital) bond to which two electrons are assigned.

dride, a white, crystalline, water-soluble solid. It is usually made by the reaction of sodium hydride with trimethylborate, a compound made from methyl alcohol and boric acid.

$$B(OH)_3(s) + 3\ CH_3OH(\ell) \xrightarrow{-3\ H_2O(\ell)}$$

$$\underset{\text{sodium hydride}}{4\ NaH(s)} + \underset{\text{trimethylborate}}{B(OCH_3)_3(\ell)} \longrightarrow \underset{\text{sodium borohydride}}{NaBH_4(s)} + 3\ NaOCH_3(s)$$

More than 2000 tons of $NaBH_4$ are produced annually, and one of its main uses is as a reducing agent. Since it readily reduces metal ions to the metal, the reaction can be used to plate metals onto surfaces without using electrodes and to remove potentially harmful metal ions in the effluent from a chemical or sewage treatment plant. It is also used to recover silver from the solutions used in developing photographic film (Figure 23.12).

Sodium borohydride is also the usual starting point for the synthesis of diborane, and an excellent approach is

$$2\ NaBH_4(s) + I_2(s) \longrightarrow B_2H_6(g) + 2\ NaI(s) + H_2(g)$$

B_2H_6: bp, −92.6 °C; $\Delta H_f^\circ = +35.6$ kJ/mol

The heat of formation of B_2H_6 is significantly *endothermic*, as are those of the other boranes. It is not surprising, therefore, that they generally

Figure 23.12 Reducing $Ag^+(aq)$ with sodium borohydride, $NaBH_4$. At left, aqueous $NaBH_4$ is contained in a flask, while a dish holds $AgNO_3(aq)$. When $NaBH_4(aq)$ is added to $AgNO_3(aq)$, silver ion is reduced to silver metal, the black solid seen in the photo at right. (H_2 gas is evolved as well.) In the foreground is a "button" of pure silver (labeled with $NaBH_4$) made by the reaction illustrated.

burn in air, since the product is the tremendously stable B_2O_3 ($\Delta H_f^\circ = -1254.5$ kJ/mol).

$$B_2H_6(g) + 3\ O_2(g) \longrightarrow B_2O_3(s) + 3\ H_2O(g) \qquad \Delta H^\circ = -2016\ \text{kJ}$$

4A
6 C carbon
14 Si silicon
32 Ge germanium
50 Sn tin
82 Pb lead

EXERCISE 23.3 Boron and Aluminum Halides

Write balanced equations to show how boron and aluminum halides are obtained from their oxides. If you begin with 1.00 kg of Al_2O_3, how many grams of $AlCl_3$ can be obtained?

23.2 THE ELEMENTS OF GROUP 4A

With Group 4A we move away from elements that are metals or metalloids and begin to see nonmetallic behavior (Table 23.3). Carbon, the lightest element of this group, is distinctly nonmetallic in its chemistry, but silicon is classed as a metalloid. Germanium begins the trend to metallic behavior for the heavier elements of the group.

All these elements are characterized by half-filled valence shells with two electrons in the *ns* orbital and two electrons in *np* orbitals (where *n* is the period number). The bonding in carbon compounds, like that of the second period elements in general, is largely covalent. This means

Table 23.3 The p^2 Elements of Group 4A (the Carbon Family)*

	Carbon	Silicon	Germanium	Tin	Lead
Symbol	C	Si	Ge	Sn	Pb
Atomic number	6	14	32	50	82
Atomic weight	12.011	28.086	72.61	118.71	207.2
Valence e$^-$	$2s^22p^2$	$3s^23p^2$	$4s^24p^2$	$5s^25p^2$	$6s^26p^2$
mp, °C	3550†	1410	937	232‡	328
bp, °C	Sublimes	3280	2850	2623	1750
d, g/cm^3	2.266†	2.336	5.32	7.30‡	11.34
Atomic radius, pm	77.2	117.6	122.3	140.5	146
Ion radius, pm (4+ ion)	30	54	67.0	83	91
Pauling EN	2.55	1.9	2.01	1.88	2.1
Standard reduction potential (V)	+0.39	+0.10	−0.3	−0.15	−0.126
Oxidation numbers	−4 to +4	−4, +2, +4	−4, +2, +4	−4, +2, +4	+2, +4
Ionization energy	1086.4	786.4	762	708.6	715.6
Heat of atomization	717	452	377	302	196
Isolated by	Antiquity	Berzelius	Winkler	Antiquity	Antiquity
Date of isolation	Antiquity	1824	1886	Antiquity	Antiquity
rpw pure O_2	CO, CO_2, 600 °C	SiO_2, 1200 °C	GeO_2, 1000 °C	SnO_2, 800 °C	PbO, 600 °C
rpw H_2O	None	None	None	None	None
rpw N_2	None	Si_3N_4, 1400 °C	None	Sn_3N_4, 2000 °C	None
rpw halogens	CX_4, 800 °C	SiX_4, 400 °C	GeX_4, 400 °C	SnX_2, 400 °C	PbX_2, 400 °C
rpw H_2	CH_4, 1000 °C	None	None	None	None

*All energies are given in kJ/mol. The symbol rpw means "reaction product with." The molar heat of atomization is the energy required to convert 1 mol of solid element into 1 mol of atoms in the gas phase. All data in this table are taken from (a) *KC? Discoverer*, a computer program distributed by JCE: Software. (b) N.N. Greenwood and A. Earnshaw, *Chemistry of the Elements*, Pergamon Press, New York, 1984. (c) E.G. Rochow, *Modern Descriptive Chemistry*, W.B. Saunders, Philadelphia, 1977.
†Data are given for graphite.
‡Data are given for white tin.

that, depending on the relative electronegativity of the bonded element, the formal oxidation number can be ±2 or ±4. Because of generally lower ionization energies, elements of Group 4A of higher atomic number generally have positive oxidation numbers of +2 and +4. An oxidation number of +2 for silicon and germanium is unusual, but it becomes more common for the heavier elements tin and especially lead. Recall that a similar trend to more stable, lower oxidation numbers was seen in Group 3A, and the trend will continue in Group 5A as well.

COMMON MINERALS AND RECOVERY OF THE ELEMENTS

Carbon is reasonably abundant on our planet and is certainly widely distributed. The cosmic abundance of carbon is six times that of silicon, and carbon-containing material, including diamonds, has been found in meteorites. In the earth's crust, the most abundant material not containing silicon is calcium carbonate, a substance already discussed many times in this text, and all of us are composed of a myriad of carbon-containing molecules.

In elementary form, carbon occurs as the **allotropes** diamond and graphite. Graphite is composed of layers of interconnected, planar, six-membered rings of sp^2 hybridized carbon atoms (Figure 13.34). The standard heats of formation are 0 for graphite, the thermodynamic standard state of carbon, and +1.895 kJ/mol for diamond, and the entropy of graphite (5.740 J/K · mol) is slightly greater than that of diamond (2.377 J/K · mol). This means that diamonds should spontaneously revert to graphite. To the considerable relief of diamond owners, the transformation involves such an enormous geometry change that the rate is negligibly slow, except at *very* high temperatures.

Figure 23.13 The space shuttle uses graphite-epoxy composites in several areas, such as the cargo door pictured here. (*High Technology*, October 1983, p. 63)

Graphite melts at the extraordinarily high temperature of 3550 °C, so it is used to make crucibles for casting metals and to line electric furnaces. Since it is also a reasonably good conductor of electricity, it is used to make electrodes for industrial processes such as aluminum reduction (Figure 21.19). However, it is also soft and marks paper, so it is the "lead" in pencils. Finally, it is commonly used as a lubricant; the very weak interactions between the layers allow them to slide over one another. Very pure graphite is actually an abrasive, however, so it is thought that it is impurities, such as oxygen, between the layers that reduce layer-to-layer bonding enough to give rise to the lubricant property.

Charcoal, carbon black, and common soot are other reasonably pure forms of microcrystalline graphite. Activated charcoal is made by heating charcoal in steam at about 1000 °C, a process that removes volatile materials from the pores of the solid and makes it an excellent absorbent; it is widely used, for example, to purify molasses by absorbing impurities. Most of the carbon black produced is used as a reinforcing agent in rubber; hence we have black automobile tires and not the white or amber of pure rubber. Carbon black is also used as the pigment in printing inks.

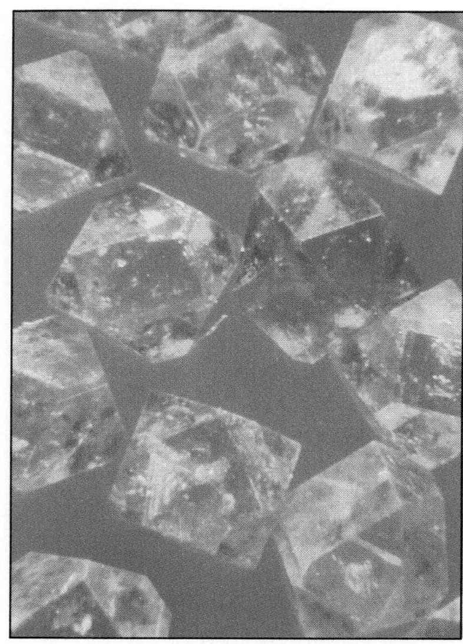

Figure 23.14 Synthetic diamonds (see p. 542 for more information). (General Electric Co.)

Finally, there are new high-strength materials made by mixing pure graphite fibers with various plastics. You may have seen them in tennis rackets or golf clubs, but they are also used in less common places such as the payload bay doors of the space shuttle and the cockpits of race cars (Figure 23.13).

Worldwide consumption of graphite is on the order of a million tons annually.

The diamond allotrope has fascinated people for hundreds of years. Diamonds are found around the world, but most come from South Africa. Approximately six tons of diamonds are used annually, mostly as industrial abrasives. Natural diamonds do not completely satisfy this demand, so some diamonds are made by subjecting pure graphite to very high pressures at high temperature (70,000 atm and 1800 °C) (Figure 23.14). The transformation is possible because the density of diamond (3.514 g/cm^3) is much greater than that of graphite (2.266 g/cm^3).

The structure of a diamond (Figure 13.35) shows that it is an interconnected network of tetrahedral, sp^3 hybridized carbon atoms, and it is this interconnectedness that makes diamond so hard and relatively inert chemically. It is also important that diamond has the highest thermal conductivity of any known substance (about 5 times that of Cu); for this reason, diamond-tipped tools do not overheat when used for drilling and cutting.

Silicon is the second most abundant element in the earth's crust (Figure 23.15), so its compounds have been important to the development of our society. Pottery, made of silicon-based natural materials, was made at least 6000 years ago in the Middle East, and sophisticated techniques were developed by the Chinese 5000 years ago. Our modern name for the element is derived from the Latin word for flint, a silicate mineral often used by prehistoric people to make knives and other tools. Today we are surrounded by silicon-containing materials: bricks, pottery, porcelain, lubricants, sealants, computer chips and solar cells.

Reasonably pure silicon is made in large quantities by heating pure silica sand with purified coke to approximately 3000 °C in an electric furnace.

Figure 23.15 A solid cylinder of nearly pure silicon, approximately 10 inches long.

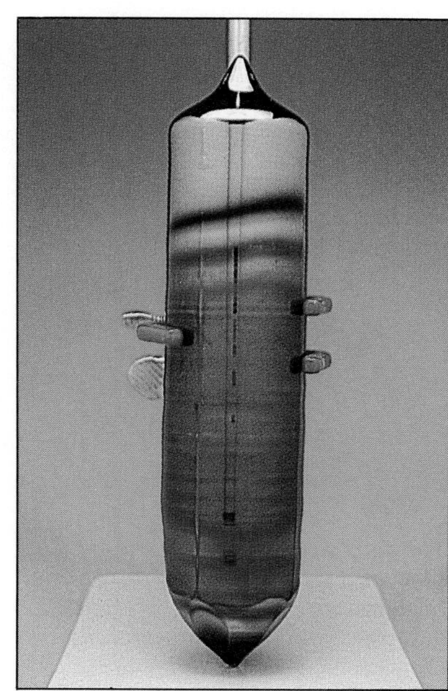

Figure 23.16 The sort of crystal shown in Figure 23.15 was made in a special furnace. Here a mechanically rotated "seed" crystal is placed in the molten material (e.g., molten silicon), and the crystal is withdrawn slowly. Since the temperature of the withdrawn crystal is lower than the melt, the liquid freezes onto the seed crystal as it is withdrawn. (AT&T Bell Laboratories)

$$SiO_2(s) + 2\ C(s) \longrightarrow Si(\text{liquid at 3000 °C}) + 2\ CO(g)$$

The molten silicon is drawn off the bottom of the furnace and allowed to cool to a shiny blue-gray solid. This can then be purified further for the electronics industry by first chlorinating the silicon to form silicon tetrachloride.

$$Si(s) + 2\ Cl_2(g) \longrightarrow SiCl_4(\ell, \text{bp} = 57.6\ °C)$$

The volatile tetrachloride is carefully purified and then reduced with very pure magnesium or zinc.

$$SiCl_4(g) + 2\ Mg(s) \longrightarrow 2\ MgCl_2(s) + Si(s)$$

Any magnesium chloride that may remain in the solid silicon is washed out with water. The silicon is then remelted and cast into bars that are finally purified by *zone refining* (Figure 23.16).

Germanium, tin, and **lead** are not abundant (Table 23.2), but they occur in workable mineral deposits. Tin and lead are especially easily recovered from their ores, and they have been used for hundreds of years. As evidence of their long history of use, their symbols, Sn and Pb, come from their Latin names, *stannum* and *plumbum* (Figure 23.17).

After the stone age, the earliest period of civilization, periods of mankind's development are often expressed in terms of copper, bronze, and iron. The copper that was first used was probably "native copper," naturally occurring metal. However, there is evidence that the Sumerians in their prehistoric home in southern Iran learned to obtain copper from its ores by heating these in the presence of wood. Since no attempt was made to purify the ore, there must have been times when a mixture of copper- and tin-bearing ores was heated, and *bronze*, an alloy consisting of about 20% tin and 80% copper, was obtained. Since bronze melts at a lower temperature than copper and is much harder, it was clearly a desirable material. In Egypt, bronzes have been found dating from as early as 2500 BC, and even older objects have been found in Mesopotamia.

Pewter is about 85% tin and 7% copper; the rest is antimony and bismuth. The solder used in electrical circuits averages 33% tin.

By the time the Greek and Roman civilizations had developed, it was known that bronze could be obtained by mixing certain copper- and tin-containing ores from Cyprus and Britain. From Cyprus came copper-bearing ores (and the Latin name for the island gave us the name of this element). From Britain came *cassiterite*, SnO_2, an oxide that could be reduced easily to the metal with glowing charcoal.

$$SnO_2(s) + C(s) \longrightarrow Sn(s) + CO_2(g)$$

Tin is obtained today by the same process, and it is still widely used. Although most of the 210,000 tons of cassiterite mined today comes from such countries as Malaysia (25%) and Bolivia (14%), the United States is the largest consumer. Tin remains an expensive and fairly rare metal.

Tin cans are used increasingly for beer and soft drinks. In the United States, more than half of the 40 billion drink cans sold annually are tin plated; the remainder are aluminum cans.

Tin is very resistant to corrosion, so almost 40% of the metal produced is used to plate the surfaces of other metals, as in the tin-coated iron cans that you have always called "tin cans." The coating, typically 0.0004 to 0.025 mm thick, is applied by dipping the object in molten tin or by electroplating.

Since it is quite unreactive to air and water, tin is also used in the "float" process of making glass. Molten glass is poured onto large vats of molten tin; the metal has such a smooth surface that the glass, when cooled, is also smooth and does not need to be polished.

Lead was also known to ancient civilizations. The Hanging Gardens of Babylon were said to be floored with sheets of lead to retain moisture, and the Romans used lead extensively for water pipes and plumbing. One reason for the importance of lead in early civilizations is that its ore, a heavy black lead(II) sulfide called *galena*, was easily recognized. Just as importantly, the ore was readily reduced by heating with wood charcoal. If galena is first roasted in a limited amount of air, lead(II) oxide forms.

$$2\ PbS(s) + 3\ O_2(g) \longrightarrow 2\ PbO(s) + 2\ SO_2(g)$$

The carbon then reduces the oxide to the metal,

$$PbO(s) + C(s) \longrightarrow Pb(\ell) + CO(g)$$

as does the carbon monoxide produced in the second step.

$$PbO(s) + CO(g) \longrightarrow Pb(\ell) + CO_2(g)$$

Approximately one million tons of lead are produced in the United States annually, and most of this (60%) is used as an alloy for battery plates (91% lead and 9% antimony).

***Roasting** is a common step in recovery of metals from their ores. It consists of heating the ore, often a metal sulfide, in air to convert the sulfide to the oxide.*

$$2\ MS(s) + 3\ O_2(g) \longrightarrow 2\ MO(s) + 2\ SO_2(g)$$

Unfortunately, if the SO_2 is vented to the atmosphere, it contributes to air pollution and is one source of "acid rain."

EXERCISE 23.4 Graphite and Diamond
Using the data in the text, prove that diamonds should spontaneously revert to graphite.

THE CHEMISTRY OF OXIDES OF GROUP 4A ELEMENTS

In our oxygen-filled world, compounds of oxygen are generally important to the chemistry of any group of the periodic table, but in Group 4A they are of overwhelming importance. Carbon monoxide and carbon dioxide are produced and used in enormous quantities and often end up in the form of carbonate salts. Silica, SiO_2, is probably the most studied material after water. Finally, about 95% of the earth's crustal rocks and their breakdown products—clay, soils, and sands—are silica and silicates.

THE OXIDES OF CARBON The direct oxidation of carbon in limited oxygen gives carbon monoxide, CO, while excess oxygen leads to carbon dioxide, CO_2. Both have C—O bonds of great strength, a fact that gives these molecules considerable thermodynamic stability. Nonetheless, they are chemically reactive and industrially very important.

A mixture of gases called *producer gas* (about 25% CO, 4% CO_2, and 70% N_2) is formed when air is blown through a bed of glowing coke.

$$2\ C(s) + O_2(g) \longrightarrow 2\ CO(g) \qquad \Delta H^\circ = -221.0\ kJ$$

$$C(s) + O_2(g) \longrightarrow CO_2(g) \qquad \Delta H^\circ = -393.5\ kJ$$

If steam is used instead of air, the product is *water gas*, a mixture of about 50% H_2, 40% CO, 5% CO_2, and 5% N_2 plus traces of other gases (Section 22.1).

$$C(s) + H_2O(g) \longrightarrow CO(g) + H_2(g) \qquad \Delta H^\circ = +131.3\ kJ$$

Figure 23.17 The Group 4A metals, tin (the irregular pieces) and lead (in the form of thin sheets or foil).

Both producer gas and water gas can be burned directly as fuel, since both CO and H_2 can react further with oxygen.

The oxidation reactions of carbon also illustrate why carbon is used industrially for the reduction of metal oxides to metals. For example, the reduction of tin(IV) oxide, cassiterite, couples the combustion of carbon in O_2 with decomposition of SnO_2 to Sn and O_2.

$$C(s) + O_2(g) \longrightarrow CO_2(g) \qquad \Delta G^\circ = -394.4 \text{ kJ}$$
$$SnO_2(s) \longrightarrow Sn(s) + O_2(g) \qquad \Delta G^\circ = +519.6 \text{ kJ}$$
$$SnO_2(s) + C(s) \longrightarrow CO_2(g) + Sn(s) \qquad \Delta G^\circ = +125.2 \text{ kJ}$$

The free energy is large and positive, clearly indicating a very small equilibrium constant at 25 °C. However, since ΔS for the reaction is positive, the equilibrium constant will increase with temperature. This, coupled with the fact that continuous removal of $CO_2(g)$ shifts the equilibrium to the right, means that tin can be produced.

Approximately 25% of the CO_2 manufactured is used to carbonate beverages. Over 400 bottles of carbonated beverages are produced per person in the United States per year. (See Something More About Carbon Dioxide in Chapter 19.)

More than 30 million tons of carbon dioxide are produced annually in the United States, and roughly half of it is used as a refrigerant. Pure CO_2 can be liquefied readily at any temperature between its triple point of −56.6 °C and its critical temperature of +31.0 °C (at 75.28 atm). If the liquid is released from a storage cylinder rapidly, it cools and forms CO_2 "snow." When this is compressed into blocks, it is known by the trademarked name "Dry Ice." Although once used mostly in the form of the solid, CO_2 is increasingly used now as a liquid refrigerant and as the propellant gas in aerosol cans.

Carbon monoxide, unlike CO_2, is a Lewis base, using the unshared electron pair of the carbon atom (:C≡O:) to form coordinate covalent bonds with some Lewis acids, particularly the low-valent transition metals. For example, CO readily interacts with nickel to give a nickel carbonyl.

$$Ni(s) + 4\ CO(g) \xrightleftharpoons{50\ °C} Ni(CO)_4(\ell)$$

This reaction is reversible, so heating the metal compound in a vacuum causes it to dissociate. The *Mond process* for purifying nickel takes advantage of this. Nickel oxide ore is first heated in the presence of water gas (H_2 + CO). The hydrogen reduces the metal oxide to impure metal, and the residual CO then reacts to give volatile $Ni(CO)_4$. When $Ni(CO)_4$ vapor is passed over a nickel surface heated to 230 °C, the vapor decomposes to leave pure nickel and CO gas, which is then recycled.

$$\text{Impure ore containing } Ni^{2+} \xrightarrow[H_2 + CO]{} Ni(CO)_4(g)$$

$$Ni(CO)_4(g) \xrightarrow[\text{heat to 230 °C}]{} Ni(s, \text{pure}) + 4\ CO(g)$$

A ball-and-stick model of carbon monoxide (red = O and gray = C) inside a model of the molecular surface. The most reactive sites on the molecule correspond to the red zones on the surface, the absolute minimum located at the carbon atom. (Computer image generated by J. Weber, University of Geneva, Switzerland)

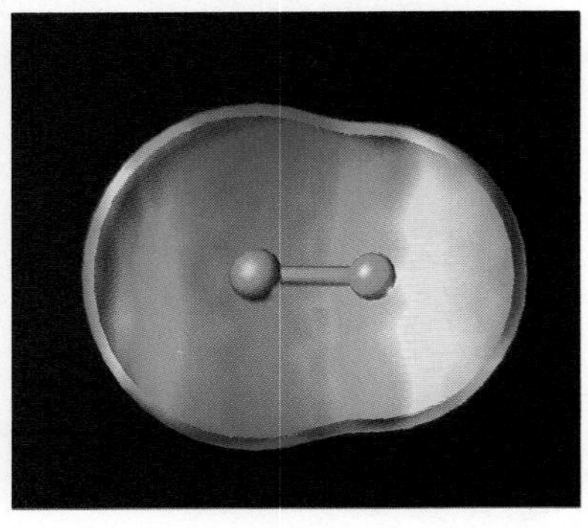

Carbon monoxide is a toxic gas. Just as CO can interact with metallic nickel, it can also react with the iron of the hemoglobin in your blood. The function of hemoglobin is to activate O_2 to reduction, so, if you breathe CO, it displaces O_2 from the hemoglobin, and you suffocate. Exposure to even small concentrations (120 parts per million) can impair your abilities, and concentrations as high as 100 ppm are not unusual in tunnels, garages, and even the streets of large cities.

Figure 23.18 A pure quartz crystal (pure SiO_2). Quartz is one of the most common minerals on earth. Pure, colorless quartz was used as an ornamental material as early as the Stone Age, and by Roman times it was known that a wedge of colorless quartz could be used to concentrate the sun's rays. Now quartz is used for its electrical properties in such products as phonographs, watches, and radios.

THE OXIDES OF SILICON Just as carbon forms CO_2, the simplest oxide of silicon is SiO_2, commonly called **silica**. Unlike CO_2, however, SiO_2 is not a simple molecule with Si═O double bonds. The energy of two Si═O double bonds is estimated to be much less than that of four Si—O single bonds, so silicon in SiO_2 and all other Si—O compounds is always surrounded tetrahedrally by four oxygen atoms. It is the interconnections between SiO_4 tetrahedra that lead to the tremendous variety of silicon–oxygen compounds observed.

More than 22 phases of pure silica, SiO_2, have been described, the most common being "low temperature" or α-quartz. It is the major constituent of many rocks such as granite and sandstone, and it occurs alone as pure rock crystal or in a variety of less pure forms (Figures 23.18 and 23.19).

Figure 23.19 Amethyst is the most highly prized variety of quartz. Its color, due to Fe^{3+}-containing impurities in SiO_2, can range from pale lilac to a deep, royal purple. The name comes from the Greek *amethustos,* meaning "not drunken." In ancient Greece it was believed that an amethyst wearer could never become intoxicated.

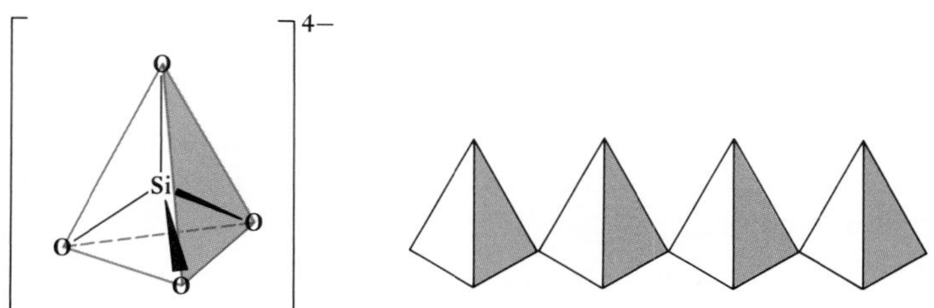

The main crystalline modifications of SiO_2 consist of infinite arrays of SiO_4 tetrahedra sharing corners. In one form of quartz, the tetrahedra form chains, and the chains interlink. If α-quartz is heated to almost 1500 °C, it transforms ultimately to another crystalline form, cristobalite. Elemental silicon has the diamond structure (Figure 13.35). If an oxygen atom is placed between each pair of silicon atoms in this lattice, the structure of cristobalite is generated (Figure 23.20).

Figure 23.20 The structure of cristobalite, one of the solid forms of SiO_2. Notice the similarity to diamond (Figure 13.35). Each C—C bond in diamond is replaced by a Si—O—Si linkage in the cristobalite structure.

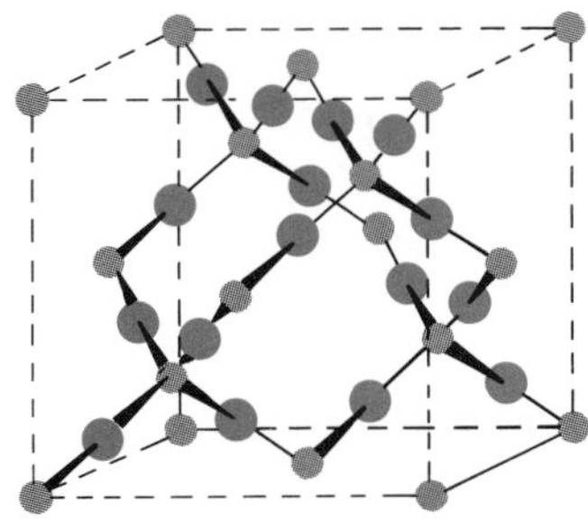

The most important commercial application of crystalline quartz is to control the frequency of almost all radio and television transmissions. These and related applications use so much quartz that there is not enough natural material to fulfill the demand, so quartz is synthesized by the following procedure. Noncrystalline or vitreous quartz, made by melting pure silica sand, is more soluble in hot water than α-quartz. The vitreous quartz is placed in a steel "bomb" and dilute aqueous NaOH is added. A "seed" crystal is placed in the mixture, just as you place a seed crystal in a hot sugar solution to grow rock candy. When the mixture is heated above the critical temperature of water (above 400 °C and 1700 atm) for some days, pure quartz crystallizes on the seed crystal.

Silica is resistant to attack by all acids except HF, but it does dissolve slowly in hot, molten NaOH or Na_2CO_3 (soda ash) to give Na_4SiO_4.

$$2\ Na_2CO_3 + SiO_2 \longrightarrow Na_4SiO_4 + 2\ CO_2$$

When the fused glass has cooled, it is dissolved in hot water under pressure and any insoluble sand or glass is filtered off. The sodium silicate obtained from the solution is isolated by evaporation and is called *water glass*. Its biggest single use is in household and industrial detergents. The silicates maintain pH by their buffering ability and can degrade animal and vegetable fats and oils. They are also used in various adhesives and binders, especially for gluing corrugated cardboard boxes.

If the soluble sodium silicate is treated with acid, a gelatinous precipitate of noncrystalline or amorphous SiO_2 is formed. After this so-called *silica gel* is washed and dried, the white residue is a very porous material with dozens of uses. Since it can absorb up to 40% of its own weight of water, you may know it as a drying agent. (Small packets of silica gel are often placed in packing boxes of merchandise during storage.) When stained with $(NH_4)_2CoCl_4$, it is a humidity detector, turning pink when hydrated, but remaining blue when dry. Finally, one other use of silicates is to clarify beer, that is, to remove minute particles that make the brew cloudy.

The **silicate minerals** are a world in themselves. The simplest silicates, *orthosilicates*, are based on discrete, tetrahedral SiO_4^{4-} anions. The 4− charge of the anion can be balanced by metal ions: four M^+ ions, two M^{2+} ions, or a combination of ions. For example, the calcium orthosilicates, Ca_2SiO_4, are vital components of *Portland cement*. In *olivine*, the mineral thought to be one of the most important in the mantle of the earth, the ions are those of Mg, Fe, and Mn; the Fe^{2+} gives it its characteristic olive color.

Pyroxenes contain extended chains of linked SiO_4 tetrahedra. All are based on the *metasilicate* ion, SiO_3^{2-}, as in $Mg_2Si_2O_6$.

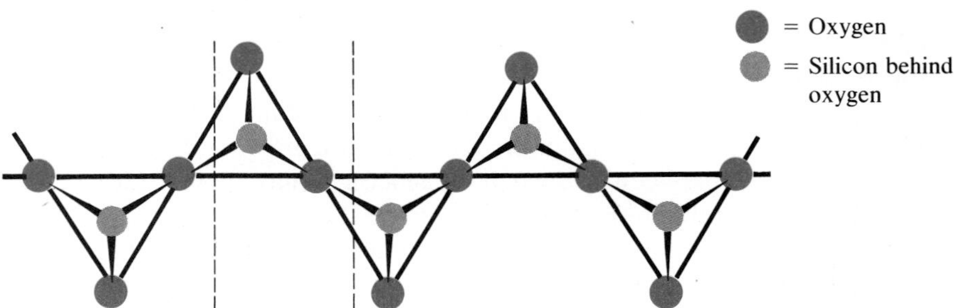

If two such chains are laid side by side, they may link together by sharing oxygen atoms in adjoining chains.

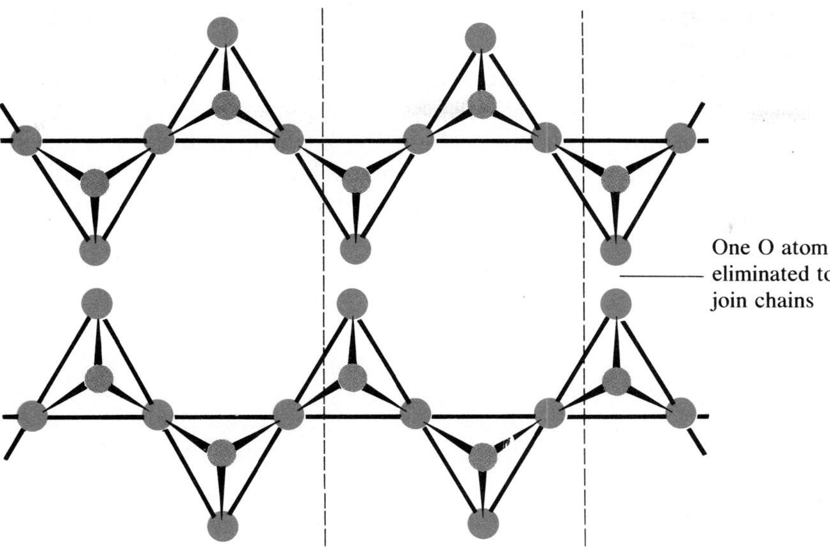

The result is an *amphibole*, of which the *asbestos* minerals are an excellent example. Because of their double-stranded chain structure, asbestos minerals are fibrous materials. Their best known characteristic is their very low thermal conductivity, which led to their use in insulation and fireproofing.

See "Something More About Asbestos" later in this chapter.

If the linking of silicate chains is continued in two dimensions, sheets of SiO_4 tetrahedra result.

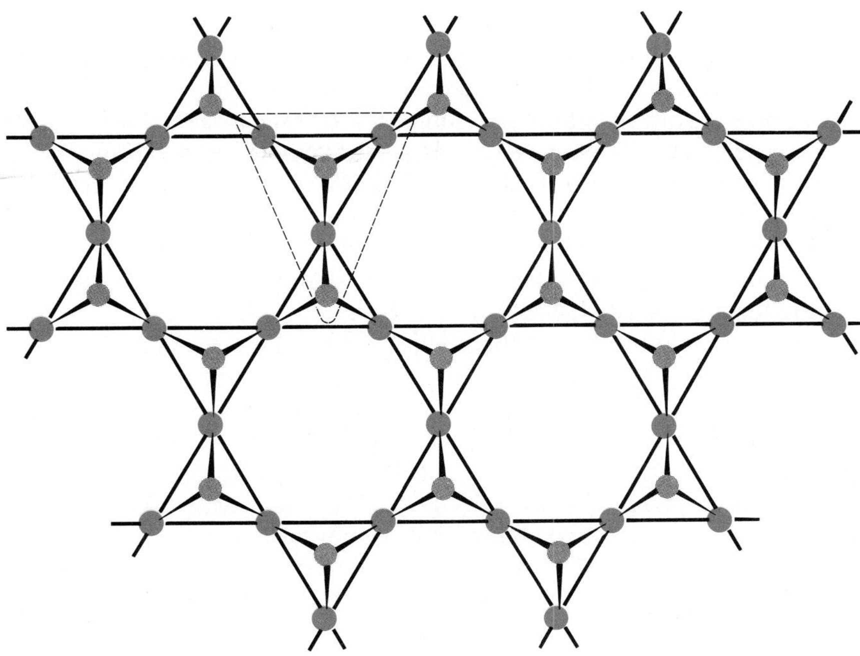

Mica, a layered silicate.

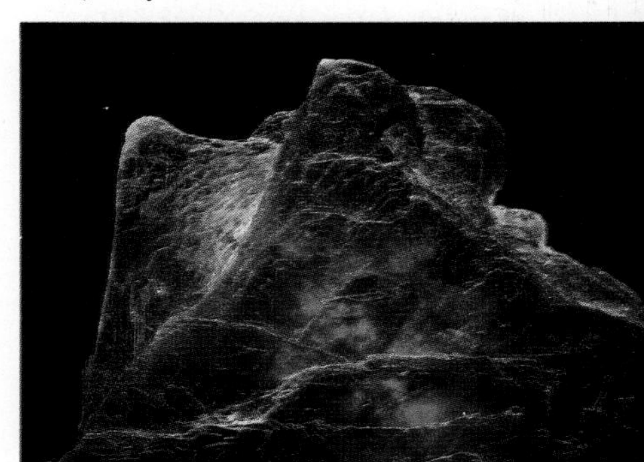

In this category are some of the most important minerals in the ancient and modern world: the clay minerals and mica. Sheets of mica are used

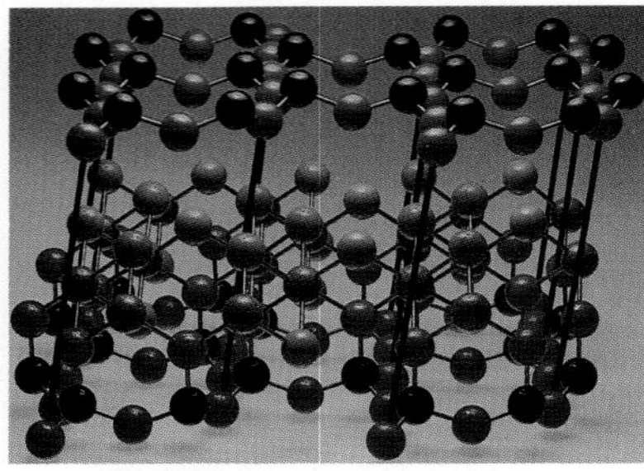

Figure 23.21 Kaolinite, an aluminosilicate. At the bottom is a network of Si—O rings. Each silicon (black) is surrounded tetrahedrally by oxygen atoms (red) to give rings consisting of six O and six Si atoms. In addition, a layer of Al ions (silver) is attached through O atoms to the Si—O rings, and the Al ions are bridged by OH^- ions (light green). The net result is a layered structure typical of clays in general.

in furnace windows and as insulation, and flecks of mica give the glitter to "metallic" paints.

The *clay minerals* are essential components of soils, and as the raw material for pottery, bricks, and tiles they have played a major role in the development of our civilization. Clays come from the weathering and decomposition of igneous rocks. Specifically, the aluminosilicate *kaolinite* comes from the weathering of feldspar, and an idealized version of this reaction would be

$$\underset{\text{feldspar}}{2\ KAlSi_3O_8(s)} + CO_2(g) + 2\ H_2O(\ell) \longrightarrow \underset{\text{kaolinite}}{Al_2(OH)_4Si_2O_5(s)} + 4\ SiO_2(s) + K_2CO_3(aq)$$

Its structure consists of layers of SiO_4 tetrahedra linked into sheets, as in mica, but these sheets are interleaved with six-coordinate Al^{3+} ions surrounded octahedrally by O atoms of the silicon–oxygen sheets and OH^- ions (Figure 23.21).

China clay, or kaolin, is primarily kaolinite. It is practically free of iron, so it is colorless and particularly valuable. World production in 1974 was approximately 17 million tons, and the predominant use in the United States is for paper filling and coating. Some is also used for china, crockery, and earthenware, but most of these articles are made from ball clay, an especially fine-grained clay that is mostly kaolinite but also contains mica and quartz. Approximately 800,000 tons of ball clay are used annually in the United States for tableware, porcelain, and wall and floor tiles.

A final category of silicates contains those with three-dimensional structures, the *aluminosilicates*. This group includes *feldspars* (among the most common minerals; they make up about 60% of the earth's crust) and *zeolites*. Both materials are composed of SiO_4 tetrahedra in which each oxygen is shared between two tetrahedra; however, in addition, some of the Si atoms have been replaced by Al atoms. Since the silicon atoms formally bear a 4+ charge and are replaced by a 3+ aluminum, other positive ions must be added for charge balance. Typically these are alkali and alkaline earth ions. For example, the synthetic zeolite "Linde A" has the formula $Na_{12}(Al_{12}Si_{12}O_{48}) \cdot 27\ H_2O$.

Zeolites make up as much as 25% of many detergents. Over 300 million pounds of zeolites are consumed in the United States annually, and about 75% of that is in detergents.

One current theory of the origin of life is that zeolites and clay minerals provided an organizing influence so that simple molecules could form the precursors to the biologically important molecules we know today.

The structure of a zeolite is illustrated in Figure 23.22. The main feature of zeolite structures is that there are tunnels and cavities of regular shape. Since the diameter of the holes is typically 300 to 1000 pm, and since water molecules have an effective diameter of 265 pm, H_2O can fit comfortably into the cavities of the zeolite and can be absorbed selectively from air or a solvent. This is why zeolites are excellent drying agents, and small amounts are sealed into multipane windows to keep the air dry between the panes. Zeolites have also been used as catalysts. Mobil Oil, for example, has patented a process in which the one-carbon compound methyl alcohol, CH_3OH, forms gasoline in the presence of specially tailored zeolites. Finally, zeolites are used as water-softening agents in detergents, since the sodium ions of the zeolite can be exchanged for Ca^{2+} ions in hard water, effectively removing Ca^{2+} from the water.

TIN AND LEAD OXIDES Other Group 4A oxides are also economically important. The mineral cassiterite, SnO_2, is not only a source of tin, but thousands of tons are also used in the ceramics industry in glazes, enamels, and pigments. For example, yellow glazes are a mixture of SnO_2 and V_2O_5, and blue-gray colors are obtained with SnO_2 and Sb_2O_5.

Lead(II) oxide, PbO, is a red solid (commonly known as *litharge*) at room temperature or a yellow solid at higher temperatures. It is the most important inorganic compound of lead; one large use is in the manufacture of leaded glass, a glass with a high lead content having a high refractive index and thus a special brilliance. The other major consumer of PbO is storage batteries (Section 21.4). The oxide is applied as a paste in sulfuric acid to the lead battery plates. The positive plates, the cathode, are activated by oxidizing PbO to the lead(IV) oxide, PbO_2, and the negative plates, the anode, are activated by reducing the oxide to Pb.

If PbO is heated in air, "red lead" is formed.

$$6\ PbO(s) + O_2(g) \longrightarrow 2\ Pb_3O_4(s)$$

This solid can best be formulated as a mixed Pb(II)/Pb(IV) oxide, $(2\ Pb^{2+})(Pb^{4+})(4\ O^{2-})$. It is widely used in rust-inhibiting paints and primers for iron and steel, but other lead salts are important in the same manner. The chromate $PbCrO_4$ is a brilliant yellow and is used in road markings, and a mixed carbonate, $2\ PbCO_3 \cdot Pb(OH)_2$, has traditionally been used as a white pigment. More recently, lead-containing paints have ceased to be used in houses, because of the toxicity of lead (see page 964).

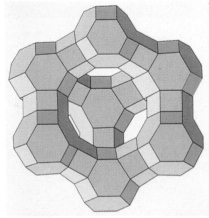

Figure 23.22 The structure of a zeolite. It is built essentially of SiO_4 tetrahedra, although there are occasional Al^{3+} ions in place of Si^{4+}. This means other positive ions such as Na^+ are also present to balance the charge of the lattice. Each vertex in this structure is a silicon atom (or aluminum) at the center of a SiO_4 tetrahedron, and each edge represents a Si—O—Si group. (Courtesy of C. Tolman, E.I. DuPont de Nemours and Company)

Lead oxide chemistry, and other compounds such as tetraethyllead, $Pb(C_2H_5)_4$, raise the important issue of *lead poisoning*. Chronic exposure to lead compounds can lead to colic, anemia, headaches, convulsions, and brain damage. The sources of lead in our environment and the solutions to the problem are still controversial. Tetraethyllead, however, is clearly one source. The recent use of this compound as a gasoline additive is thought to have contributed to the tremendous increase in environmental lead since about 1940 (Figure 23.23). Although lead com-

Lead poisoning has been known for centuries, the first clear description coming from a Greek poet in the second century BC.

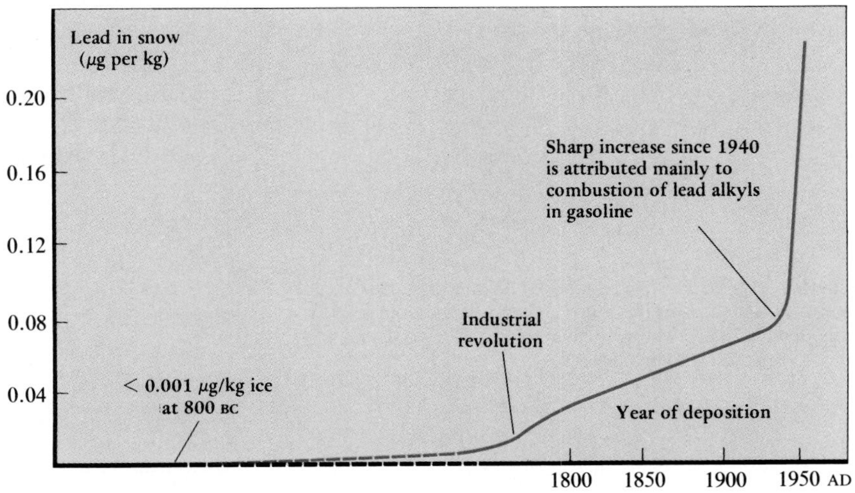

Figure 23.23 Environmental lead. The amount of lead introduced into the environment was measured by examining snow samples in northern Greenland.

SOMETHING MORE ABOUT
Lead Chemistry

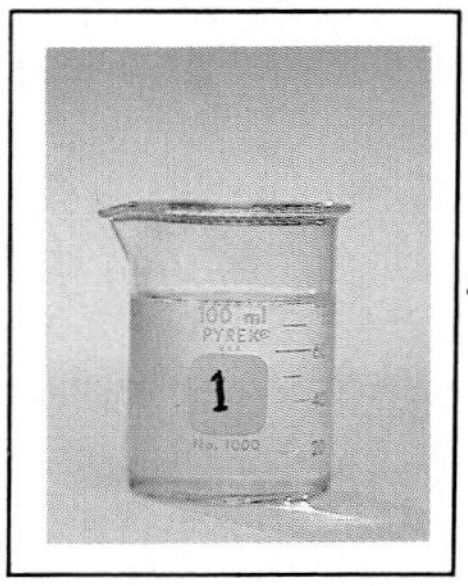

$Pb(NO_3)_2(aq)$

KCl(aq)

$PbCl_2(s)$
$K_{sp} = 1.7 \times 10^{-5}$

KI(aq)

$PbI_2(s)$
$K_{sp} = 8.7 \times 10^{-9}$

$NaHCO_3(aq)$

$PbCO_3(s)$
$K_{sp} = 1.5 \times 10^{-13}$

$K_2CrO_4(aq)$

$PbCrO_4(s)$
$K_{sp} = 1.8 \times 10^{-14}$

NaOH(aq)

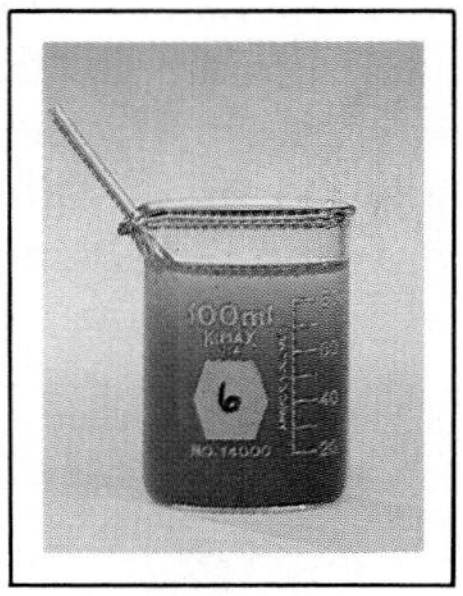

$Pb(OH)_2 \cdot PbCrO_4(s)$
$K_{sp} = 1.8 \times 10^{-32}$

NaOH(aq)

$Pb(OH)_3^-(aq)$

Beginning with water-soluble lead(II) nitrate, a succession of reagents is added, forming more and more insoluble lead(II) compounds. Finally, sufficient NaOH is added to dissolve amphoteric lead(II) hydroxide.

pounds are no longer used in gasoline in the United States, there are still other sources of lead in the environment. Since we know that lead has insidious effects, it seems wise to reduce its introduction into the environment in every possible way.

EXERCISE 23.5 Silicates
Describe the main structural characteristics of orthosilicates, pyroxenes, mica, and zeolites.

THE HALIDES OF GROUP 4A ELEMENTS

All the Group 4A elements form chlorides of the type MCl_4, and all are liquids. In keeping with the general trend toward stability of low oxidation numbers in Group 4A, lead(IV) chloride is not stable, decomposing to $PbCl_2$ and Cl_2 above 50 °C.

$$PbCl_4(\ell) \xrightarrow{>50\ °C} PbCl_2(s) + Cl_2(g)$$

Carbon tetrachloride is well known as a common laboratory and industrial solvent. Almost 700 million pounds are made in the United States annually, usually by treating methane with Cl_2.

$$CH_4(g) + 4\ Cl_2(g) \longrightarrow CCl_4(\ell) + 4\ HCl(g)$$

Its use as a solvent is declining due to its toxicity, and so is its use in making chlorofluorocarbons such as $CFCl_3$, CF_2Cl_2, and CF_3Cl.

$$CCl_4(\ell) + HF(g) \xrightarrow{\text{catalyst} = SbFCl_4} \underset{\text{Freon-11}}{CFCl_3(g)} + HCl(g)$$

Chlorofluorocarbons (CFCs) are used as the working fluid in refrigerators and air conditioners, and, until recently, as the propellant gas in aerosol cans. However, in the past few years there has been increasing concern that the CFCs released from these spray cans (by 1977 the annual production of spray cans in the United States was several billion) were working their way into the upper atmosphere, where they could destroy the earth's protective shield of ozone (see Chapter 24). For this reason, this use of CFCs has declined rapidly.

You learned previously that silicon reacts readily with chlorine to produce $SiCl_4$. A similar reaction occurs between silicon and methyl chloride, CH_3Cl, but now two methyl groups (CH_3) and two Cl atoms are bound in a tetrahedral fashion to Si.

$$Si(s) + 2\ CH_3Cl(g) \xrightarrow[300\ °C]{\text{Cu powder catalyst}} \underset{(70\%\ \text{yield})}{(H_3C)_2SiCl_2(\ell)}$$

Unlike CCl_4 and other compounds with C—Cl bonds, halides based on other Group 4A elements are hydrolyzed to form M—O bonds. For example,

$$(H_3C)_3Si{-}Cl(\ell) + H_2O(\ell) \longrightarrow (H_3C)_3Si{-}OH(\ell) + HCl(aq)$$

As often happens with compounds having the M—O—H grouping, two such molecules come together and eliminate a molecule of water between them.

Figure 23.24 Some examples of silicones, polymers of repeating —Si—O—Si—O—Si—O— units.

$$(H_3C)_3Si{-}O{-}H + H{-}O{-}Si(CH_3)_3 \longrightarrow (H_3C)_3Si{-}O{-}Si(CH_3)_3 + H_2O$$

For $(H_3C)_2SiCl_2$, with more than one Si—Cl bond, reaction with water can initially produce $(H_3C)_2Si(OH)_2$, and each Si—O—H linkage can interact with a neighbor until a long chain of Si—O—Si—O—Si links has been formed.

$$\begin{array}{ccccccccccc} & CH_3 & & CH_3 & & CH_3 & & CH_3 & & CH_3 & \\ & | & & | & & | & & | & & | & \\ - & Si & -O- & Si & -O- & Si & -O- & Si & -O- & Si & -O- \\ & | & & | & & | & & | & & | & \\ & CH_3 & & CH_3 & & CH_3 & & CH_3 & & CH_3 & \end{array}$$

This is a *polymer* called polydimethylsiloxane, a member of the *silicone* polymer family (Figure 23.24). Silicones are nontoxic and have good stability to heat, light, and oxygen; they are chemically inert and have valuable antistick and antifoam properties. Since they can be made in the form of oils, greases, and resins, or with rubber-like properties ("Silly Putty," for example), they are commercially useful. Approximately 300,000 tons are made worldwide annually and are used in a wide variety of products: as lubricants and the antistick material for peel-off labels, in lipstick, suntan lotion, and car polish, and as the antifoam substance in stomach remedies.

Polymers are discussed in more detail in Chapter 27.

Germanium, tin, and lead form halides of the type MCl_2 and MCl_4. Since the +2 oxidation state becomes more important as the element atomic weight increases, $GeCl_4$ is more stable than $GeCl_2$, but the reverse is true of the lead chlorides. At least one tin halide is used in considerable amounts. Tin(IV) chloride is sprayed onto freshly made glass bottles where it leaves a film of SnO_2 due to the hydrolysis of the chloride.

$$SnCl_4(\ell) + 2\ H_2O(\ell) \longrightarrow SnO_2(s) + 4\ HCl(g)$$

This transparent film of tin(IV) oxide toughens the surface of the glass and improves its resistance to abrasion.

EXERCISE 23.6 Silicon Chemistry

Write a balanced equation for the reaction of $SiCl_4$ with water. Indicate the physical state of each reactant and product. Tell why you believe $SiCl_4$ reacts with water while CCl_4 does not.

THE HYDROGEN COMPOUNDS OF GROUP 4A

The general trends in Group 4A behavior are never more evident than when the elements are bonded to hydrogen. Carbon is rather electronegative, so, when H is bonded to carbon, the bond polarity is $C^{\delta-}{-}H^{\delta+}$. The other elements of Group 4A, however, become less electronegative with increasing atomic weight, and the negative end of the E—H bond shifts to the hydrogen atom. This means the hydrogen is increasingly "hydridic" or H^- in character.

The other trend in Group 4A chemistry is that element–hydrogen bond energies decrease in the order C > Si > Ge > Sn > Pb. Thus, CH_4

is a stable, gaseous compound, but SiH_4 is a stable gas only in the absence of air and water. Tin(IV) hydride (SnH_4) slowly decomposes to tin and H_2,

$$SnH_4(g) \longrightarrow Sn(s) + 2\ H_2(g)$$

and there is serious doubt that PbH_4 has ever been prepared.

Literally thousands of carbon–hydrogen compounds are known, and some will be described in Chapter 26. The simplest of these compounds is methane, CH_4, the first in a series of compounds, the *alkanes*, whose general formula is C_nH_{2n+2}. One of the features of carbon chemistry in general is that there are bonds between carbon atoms, a property called *catenation*. Long, straight chains, branched chains, and rings (as in graphite and diamond) of carbon can form. Except in the elements themselves, this ability declines sharply as the atomic weight of the element increases. Silicon forms a series of *silanes*, Si_nH_{2n+2}, analogous to the alkanes. The simplest of these, silane, SiH_4, is made by treating $SiCl_4$ with a metal hydride in an exchange reaction.

Other alkanes are, for example, $H_3C—CH_3$ (ethane), $H_3C—CH_2—CH_3$ (propane), and $H_3C—CH_2—CH_2—CH_3$ (butane). All are gases at room temperature. There is about 1 ppm of CH_4 in the atmosphere, much of it produced by the digestion of food in animals.

$$SiCl_4(\ell) + 4\ NaH(s) \longrightarrow SiH_4(g) + 4\ NaCl(s)$$

Although the silanes up to $H_3Si—SiH_2—SiH_3$ can be made in good yield, and longer chain silanes are known up to $n = 8$, they are increasingly unstable; only SiH_4 is stable indefinitely at room temperature. All are extremely reactive and spontaneously ignite in air.

OTHER GROUP 4A COMPOUNDS

Carbon forms simple binary compounds called *carbides* with many metals and metalloids. An especially important example of the latter is silicon carbide, SiC, a lustrous black solid commonly known as *carborundum* (Figure 23.25).

$$SiO_2(s) + 3\ C(s) \xrightarrow{2000\ °C} SiC(s) + 2\ CO(g)$$

Its crystal structure is similar to the diamond structure, with every other atom being silicon. Because of this similarity, SiC is harder than Al_2O_3 (corundum) but not quite as hard as diamond. Therefore, it has been used widely as an inexpensive abrasive. More recently, however, industry has taken advantage of the ability of silicon carbide to withstand very high temperatures, and it is now being used to make parts for ceramic automobile engines. Such engines can run at higher temperatures and are lighter in weight than conventional iron and steel engines, leading to significant fuel savings.

Figure 23.25 A large crystal of silicon carbide, SiC, also called carborundum.

Metal carbides contain carbon as an anion, primarily C^{4-} or C_2^{2-}. In Section 22.4 we described the synthesis of calcium carbide, CaC_2, an industrially important material that gives the hydrocarbon acetylene, $HC{\equiv}CH$, when hydrolyzed. Another industrially important reaction of calcium carbide is its ability to "fix" nitrogen from the air.

$$\underset{\text{calcium carbide}}{CaC_2(s)} + N_2(g) \xrightarrow{1000\ °C} \underset{\text{calcium cyanamide}}{CaNCN(s)} + C(s) \qquad \Delta H° = -296\ kJ/mol$$

The product is *calcium cyanamide*, in which the NCN^{2-} cyanamide anion is isoelectronic with CO_2. The salt hydrolyzes in the atmosphere to give free cyanamide.

$$CaNCN(s) + CO_2(g) + H_2O(\ell) \longrightarrow CaCO_3(s) + \underset{\text{cyanamide}}{H_2NCN(s)}$$

Because of this reaction, calcium cyanamide has been widely used as a fertilizer, since it is one way to bring usable nitrogen into the soil.

Hydrogen cyanide, HCN, is a very weak Brønsted acid in aqueous solution and an extraordinarily poisonous gas (mp, −13.4 °C; bp, 25.6 °C). Formerly it was made by adding acid to cyanide salts such as NaCN, but most of the 500,000 tons produced annually are made by a more direct, catalyzed process.

$$CH_4(g) + NH_3(g) \xrightarrow[1200\ °C]{\text{Pt catalyst}} HCN(g) + 3\ H_2(g)$$

The vast majority of the HCN is used in the polymer industry, but some is used to make NaCN for extraction of metals from their ores. Cyanide ion, an anionic Lewis base, can form very stable complexes with metal ions such as Ag^+ and Au^+. For example, gold is often obtained by bubbling air through a cyanide-containing alkaline suspension of its ores.

$$8\ CN^-(aq) + 4\ Au(s) + 2\ H_2O(\ell) + O_2(g) \longrightarrow 4\ [Au(CN)_2]^-(aq) + 4\ OH^-(aq)$$

SOMETHING MORE ABOUT Asbestos

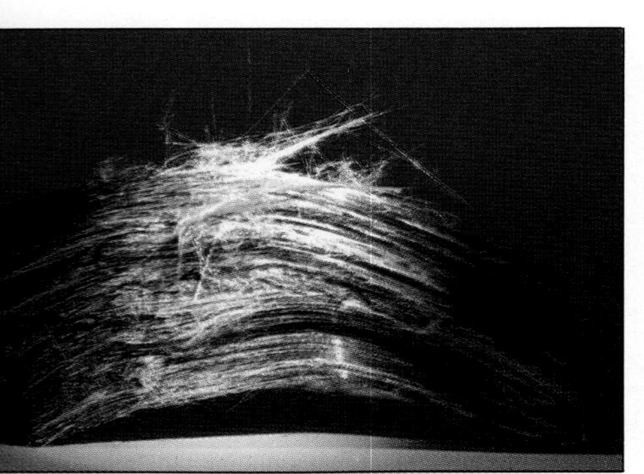

A sample of chrysotile, one of the minerals in the asbestos family.

In recent years asbestos has engendered panic and fear in the United States because experience convinces us that *occupational* exposure can lead to various lung diseases. As a result, the Environmental Protection Agency (EPA) requires that schools be inspected for asbestos and that it be removed if asbestos fibers are found in the air. This has led to an explosive growth in companies that test for and remove asbestos. Further, the costs can be staggering. It is estimated that, if the testing and removal program is extended to all public and commercial buildings, the cost could be as much as $150 billion dollars. Does the available scientific evidence support this concern? Is there evidence to support the contention by some that even a single asbestos fiber can cause cancer? There has been a stream of articles in the scientific literature in the past few years regarding these questions, and one of the most complete and definitive was by B. Mossman and co-workers in the journal *Science*.[1]

First, it is important to understand that "asbestos" is not a single substance. Rather, it is a term applied broadly to a family of naturally occurring hydrated silicates that crystallize in a fibrous manner. These minerals are generally subdivided into two forms, serpentine and amphibole fibers. Approximately 5 million tons of the serpentine form of asbestos, chrysotile, are mined each year, chiefly in Canada and the Soviet Union; this is essentially the only form used commercially in the United States. Another form, the amphibole crocidolite, is mined in small quantities, mainly in South Africa. The two minerals differ greatly in composition, color, shape, solubility, and persistence in human tissue. Crocidolite is blue, relatively insoluble, and persists in tissue. Its fibers are long, thin, and straight, and they penetrate narrow lung passages. In contrast, chrysotile is white, and it tends to be soluble and disappear in tissue. Its fibers are curly; they ball up like yarn and are more easily rejected by the body.

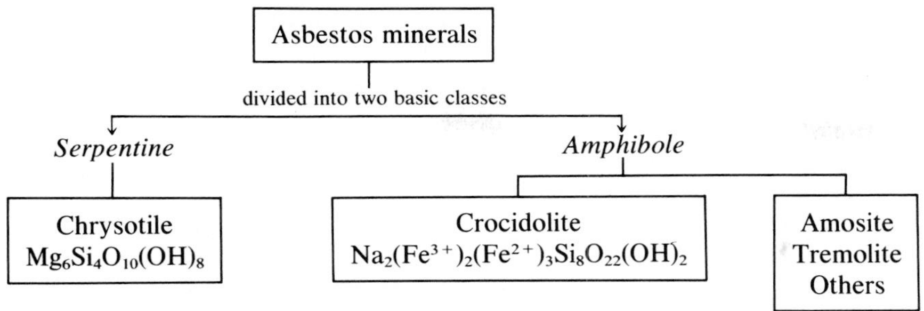

Asbestos minerals have many commercially desirable properties, among them heat resistance and high tensile strength. For many years they were sprayed onto surfaces as fireproofing (in buildings and ships), but this use is now banned in the U.S. and in many European countries. Asbestos minerals continue to be used, however, in cement construction materials (roofing and cement pipes), friction materials (brake linings and clutch pads), asphalt coatings and sealants, and similar products. As a result, it is estimated that 20% of all schools, hospitals, and other public and private buildings contain some asbestos materials.

Occupational exposure to asbestos can cause four types of medical problems: asbestosis, lung cancer, mesothelioma of lung tissue, and benign changes in lung tissue. Asbestosis is a nonmalignant scarring of lung tissue that led to disability and death in many asbestos workers exposed before the enforcement of occupational standards. Lung cancers are tumors that have been found in asbestos workers who are smokers, but rarely in nonsmokers. Mesothelioma is an extremely rare but fatal tumor in lung tissue. Unfortunately, the time between initial occupational exposure and diagnosis of such tumors can commonly exceed 30 years. Among the facts known about this disease are the following: (1) Smoking does not apparently enhance the risk of mesothelioma. (2) Only 53 acceptable cases of chrysotile-induced mesothelioma have ever been reported, and 41 of these have occurred in individuals exposed to mine dust contaminated with tremolite, an amphibole. (3) Approximately 20–30% of the mesotheliomas occur in the general adult population with no known occupational exposure to asbestos. (4) Mesotheliomas are rarely found in children.

Although differences of opinion remain in the medical and scientific communities concerning health risks from asbestos, the weight of evidence strongly suggests that amphiboles are more potent than chrysotile in the induction of fibrotic lung disease and associated lung cancers. Furthermore, "chrysotile asbestos, the type of fiber found predominantly in U.S. schools and buildings, is not a health risk in the nonoccupational environment. Clearly, the asbestos panic in the U.S. must be curtailed, especially because unwarranted and poorly controlled asbestos abatement results in unnecessary risks to young removal workers who may develop asbestos-related cancers in later decades. . . . Prevention (especially in adolescents) of tobacco smoking, the principal cause of lung cancer in the general population, is both a more promising and more rational approach to eliminating lung tumors than asbestos abatement."[1]

Estimates of risk from asbestos exposure in schools in comparison with other risks in U.S. society.[1]

Cause	***Annual Rate (deaths per million)***
Asbestos exposure in schools	*0.005–0.093*
Aircraft accidents	*6*
High school football	*10*
Home accidents (ages 1–14)	*60*
Long term smoking	*1200*

[1] B.T. Mossman, J. Bignon, M. Corn, A. Seaton, and J.B.L. Gee, *Science*, **1990**, 247, 294.

SUMMARY

The elements of **Group 3A** (B, Al, Ga, In, and Tl) all have the valence shell electron configuration ns^2np^1, and all form compounds where the element has a +3 oxidation number. There is, however, an increasing tendency to have the +1 number for elements of higher atomic number.

None of the elements exists in nature uncombined. Boron forms many compounds with oxygen, one of the most common being **borax**, $Na_2B_4O_7 \cdot 10\ H_2O$. All form oxides of general formula M_2O_3, and aluminum commonly exists as a hydrated oxide **bauxite**, $Al_2O_3 \cdot nH_2O$, or as anhydrous **corundum**, Al_2O_3.

Unlike the metals of Group 3A, boron does not form simple 3+ ions. The hydrated 3+ ions of the metals are all weak acids in aqueous solution.

In general, nonmetal or metalloid oxides are acidic, and metal oxides are amphoteric or basic. This is true in Group 3A, since the hydrated oxide of boron, $B(OH)_3$, **boric acid**, is a weak acid in water. In contrast $Al(OH)_3$ and $Ga(OH)_3$ are amphoteric, dissolving in both acid and strong base. This property plays a key role in the recovery of aluminum from bauxite in the **Bayer process**.

The elements are obtained by chemical or electrochemical reduction. Boron is a hard, covalently bonded insulator (a metalloid), whereas the other elements are metals, electrically conducting and relatively soft.

All Group 3A elements form trihalides that are excellent Lewis acids. Unlike the boron trihalides, Al and Ga at least form halide-bridged dimers $X_2M(X)_2MX_2$, a structural type observed in other metal–halogen compounds.

Boron, but not the metals of Group 3A, forms a large series of hydrides of which the simplest is B_2H_6, **diborane**. The bonding in this B—H—B bridged compound is found throughout chemistry; you should understand its description as a three-center bond.

Group 4A elements all have a ns^2np^2 configuration and all can have an oxidation number of +4. Again, there is a tendency to lower oxidation numbers (+2) with the heavier metals (Sn and Pb), and carbon is commonly found in the −4 state.

Carbon is a nonmetal, silicon is a metalloid, and germanium, tin, and lead become progressively more metallic. Carbon exists in thousands of different compounds exhibiting two features not shared to any great extent with the other Group 4A elements: **catenation**, the ability to form long chains of the elements, and **multiple bonding**.

Carbon is found in nature in elemental form as the **allotropes graphite** and **diamond**. In diamond, each carbon atom is tetrahedral, sp^3 hybridized, and bonded to four other carbon atoms. In graphite, the carbons are arranged in sheets of interconnected, planar, six-membered rings; the atoms are sp^2 hybridized (Figures 13.34 and 13.35).

As expected from the diagonal relationship of boron and silicon, the latter forms many oxides: silica (SiO_2) in its many forms (e.g., **quartz**), silicates (Ca_2SiO_4, found in **Portland cement**), and **aluminosilicates** (clay minerals, zeolites). All are based on networks of interconnected SiO_4 tetrahedra. The most important ore of tin is cassiterite, SnO_2. Litharge (PbO) and red lead (Pb_3O_4) are important oxides of lead.

Silicon and the Group 4A metals are obtained by chemical reduction, often with carbon.

All Group 4A elements form tetrahalides such as CCl_4, CF_2Cl_2 (a chlorofluorocarbon or CFC), $SiCl_4$ (starting material to make silicon computer chips), and $SnCl_4$. All except CX_4 are susceptible to hydrolysis. Hydrolysis of $(CH_3)_2SiCl_2$ gives silicone polymers.

Carbon forms thousands of compounds containing C—H bonds, where the bond polarity is $C^{\delta-}$—$H^{\delta+}$. Since the remaining elements of Group 4A are less electronegative, their hydrogen compounds are called **hydrides** because the bond polarity is $M^{\delta+}$—$H^{\delta-}$.

Other important Group 4A compounds are silicon carbide, SiC (**carborundum**), and HCN (hydrogen cyanide).

STUDY QUESTIONS

REVIEW QUESTIONS

1. Complete and balance the following equations for reactions of Group 3A elements:
   (a) $Ga(s) + O_2(g) \longrightarrow$
   (b) $In(s) + Br_2(\ell) \longrightarrow$
   (c) $BBr_3(g) + H_2(g) \longrightarrow$
   (d) $B(OH)_3(s) + CH_3OH(\ell) \longrightarrow$
   (e) $BF_3(g) + H_2O(\ell) \longrightarrow$
   (f) $B_2O_3(s) + HF(g) \longrightarrow$
   (g) $NaBH_4(s) + I_2(s) \longrightarrow$
   (h) $B_5H_9(g) + O_2(g) \longrightarrow$
   (B_5H_9 is a borane like B_2H_6)
2. Complete and balance the following equations for reactions of Group 4A elements:
   (a) $SnCl_4(\ell) + H_2O(\ell) \longrightarrow$
   (b) $Pb(s) + O_2(g) \longrightarrow$
   (c) $Ge(s) + O_2(g) \longrightarrow$
   (d) $Sn(s) + Cl_2(g) \longrightarrow$
   (e) $GeCl_4(\ell) + NaH(s) \longrightarrow$
   (f) $SiH_4(g) + O_2(g) \longrightarrow$
3. Complete and balance the following equations for reactions of Group 4A elements:
   (a) $SiO_2(s) + C(s) \longrightarrow$
   (b) $Si(s) + Cl_2(g) \longrightarrow$
   (c) $PbS(s) + O_2(g) \longrightarrow$
   (d) $PbO(s) + CO(g) \longrightarrow$
   (e) $CCl_4(\ell) + HF(\ell) \longrightarrow$
   (f) $Si(s) + CH_3Cl(g) \longrightarrow$
4. Write a complete, balanced equation for the preparation of each of the following compounds. Where possible, indicate the phase of each reactant and product.
   (a) diborane
   (b) indium(III) oxide
   (c) boron trifluoride
   (d) aluminum chloride (starting with aluminum oxide)
   (e) germanium(IV) chloride (starting with germanium(IV) oxide)
   (f) $Ni(CO)_4$ (starting with nickel(II) oxide)
   (g) carborundum
   (h) calcium carbide (starting with limestone)
   (i) cyanamide
5. Name at least one use for each of the following compounds:
   (a) cyanamide
   (b) hydrogen cyanide
   (c) boron carbide
   (d) aluminum sulfate
   (e) boric acid
   (f) sodium borohydride
   (g) carbon black
6. Draw a possible structure for the cyclic anion in the salt $K_3B_3O_6$ and the chain anion of $Ca_2B_2O_5$.
7. The boron trihalides (except BF_3) hydrolyze completely to boric acid and the corresponding HX acid.
   (a) Write a balanced equation depicting this hydrolysis reaction for BCl_3.
   (b) Calculate $\Delta H°$ for the hydrolysis of BCl_3. ($\Delta H_f°$: $BCl_3(g) = -408$ kJ/mol; $B(OH)_3(s) = -968.92$ kJ/mol)
8. When the boron hydrides burn in air, a tremendous amount of heat is evolved. Calculate the heat of combustion of B_5H_9 ($\Delta H_f° = 54$ kJ/mol) and compare it with that of B_2H_6 in the text. Calculate the heat of combustion of C_2H_6 (ethane) for comparison.
9. Diborane, B_2H_6, can be prepared by mixing $NaBH_4$ with I_2 in an oxidation–reduction reaction. What substance is reduced and what is oxidized in this reaction?
10. Diborane can be made on an industrial scale by reducing BF_3 with NaH in a metathesis reaction. Write a balanced equation depicting this process.
11. When BCl_3 is passed through an electric discharge, small amounts of B_2Cl_4 are isolated. Draw a possible Lewis structure of this highly reactive molecule. What is the hybridization of the boron atom? What is the shape of the molecule in the vicinity of the boron atom?
12. Sodium borohydride, $NaBH_4$, is used to reduce metal ions to metals. Write a balanced equation to depict the reaction of $NaBH_4$ with $AgNO_3$ in water to give silver metal, H_2, boric acid, and $NaNO_3$ (see Figure 23.12). How much silver can be produced from 575 mL of 0.011 M $AgNO_3$ and 13.0 g of $NaBH_4$?
13. Boron carbide is an abrasive and is chemically inert. Consequently it has many commercial uses. Boron carbide fibers can be grown by passing a mixture of BCl_3 and H_2 over hot graphite (C) fibers. The products are B_4C and HCl. Write a balanced equation for this process. How many grams of B_4C can be produced from 5.45 L of BCl_3 gas at 26.5 °C and 456 mmHg pressure? How much H_2 gas is required?

14. Worldwide production of silicon carbide, SiC, is several hundred thousand tons annually. If you want to produce 100,000 tons of the carbide, how many tons of silica sand (SiO_2) will you have to use if 70% of the sand is converted to SiC?
15. Suggest a reason for the fact that Al^{3+} forms an octahedral ion, AlF_6^{3-}, with fluoride ion but a tetrahedral ion, AlX_4^-, with ions of the larger halides.
16. Even though the table of $E°$ values in Chapter 21 shows that metallic aluminum should be a good reducing agent, a typical sample of the metal resists attack by air and water. Explain this observation.
17. The density of lead is 11.342 g/cm³ and the metal crystallizes in a face-centered cube. Estimate the radius of the lead atom.
18. Calcium carbide has a face-centered cubic structure closely resembling that of NaCl. If Ca^{2+} ions define the face-centered lattice, in what types of holes are the C_2^{2-} ions located? How many Ca^{2+} and C_2^{2-} ions are there in the unit cell?
19. Aluminum dissolves readily in hot aqueous base (NaOH) to give the aluminate ion, $Al(OH)_4^-$, and H_2. Write a balanced equation for this reaction. If you begin with 13.2 g of Al, how many milliliters of H_2 gas are produced when the gas is measured at 735 mmHg pressure and 22.5 °C?
20. Alumina, Al_2O_3 is amphoteric, so it will dissolve when heated strongly or "fused" with an acidic oxide or basic oxide.
    (a) Write a balanced equation for the reaction of alumina with silica, an acidic oxide, to give aluminum metasilicate, $Al_2(SiO_3)_3$.
    (b) Write a balanced equation for the reaction of alumina with the basic oxide CaO to give calcium aluminate, $Ca(AlO_2)_2$.
21. Gallium hydroxide, like aluminum hydroxide, is amphoteric. Write balanced equations showing how $Ga(OH)_3$ can dissolve in both HCl and NaOH. How many milliliters of 0.0112 M HCl will you need to react completely with 1.25 g of $Ga(OH)_3$?
22. When calcium carbide, CaC_2, is treated with water, it is hydrolyzed to $Ca(OH)_2$ and acetylene, H—C≡C—H (see Figure 10.9).
    (a) Write a balanced equation for the hydrolysis of calcium carbide.
    (b) Draw an electron dot structure for the carbide ion, C_2^{2-}.
    (c) If you treat 1.35 g of CaC_2 with water, and collect the acetylene gas, what pressure would the gas have in an 865-mL flask at 21.5 °C?
23. Halides of the Group 3A elements are excellent Lewis acids. When a Lewis base such as Cl^- (why is this a Lewis base?) interacts with $AlCl_3$, the ion $AlCl_4^-$ is formed. What is the structure of this ion? What is the hybridization of the aluminum atom?
24. Boron trifluoride will react with ammonia in a Lewis acid–base reaction.
    (a) Which is the Lewis acid and which is the Lewis base?
    (b) Draw a Lewis structure for the product and give the geometry around the N and B atoms. What are the hybrid orbital sets used by these atoms?
25. Carbon forms a series of simple halides of the type CX_4. The boiling point of CF_4 is −128.5 °C. Do you anticipate the boiling point of CCl_4 to be higher or lower? Why?
26. Plot the half-reaction potential for $M(s) \rightarrow M^{3+}(aq) + 3e^-$ versus atomic number for the Group 3A elements (see Table 23.1). What trend or trends do you see? What do they tell you about the chemistry of the Group 3A elements? About the stability of the +3 oxidation state?
27. Carbon is widely used as an industrial reducing agent to reduce metal oxides to metals. For example, $PbO(s) + C(s) \rightarrow Pb(s) + CO(g)$.
    (a) Calculate the standard free energies of reaction for the reduction of PbO and CaO with carbon.
    (b) Which of the two reductions in part (a) would you consider more feasible on a commercial scale? You can answer by deciding which is more likely to be spontaneous at a reasonable temperature.
28. Tin(IV) oxide, cassiterite, is the main ore of tin (see top of next page).
    (a) The compound crystallizes in a rutile-like structure; one unit cell is shown below. How many tin(IV) ions and oxide ions are there per unit cell?

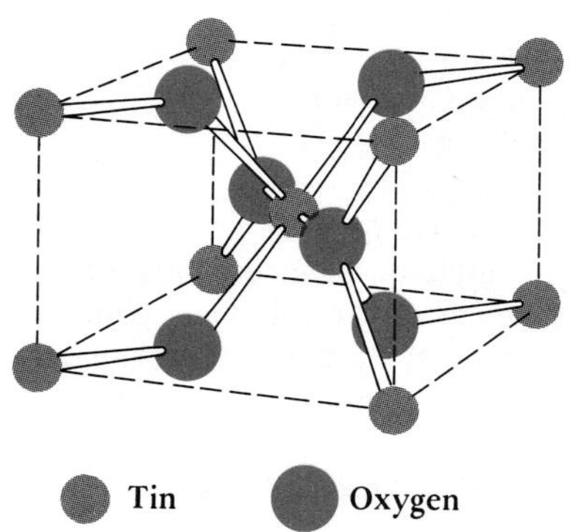

    (b) Is it thermodynamically feasible to transform solid SnO_2 into liquid $SnCl_4$ by reaction of the oxide with gaseous HCl? What is the equilibrium constant for the reaction at 25 °C?
29. When gasoline still contained tetraethyllead, $Pb(C_2H_5)_4$, 227,250 tons of the compound were consumed annually. How much lead(II) oxide was thereby introduced into the atmosphere each year?

Cassiterite, an ore of tin oxide. (George Whiteley, Photoresearchers)

30. Thousands of tons of organotin compounds are used annually as stabilizers for PVC (polyvinyl chloride) plastics, the clear material used for bottles and jars for food packaging. Other such compounds are added to antifouling paint for the hulls of ships and still others are used in agriculture to control fungal growth, for example. When dibutyltin dichloride, $(C_4H_9)_2SnCl_2$, is made by treating tin with butyl chloride, C_4H_9Cl,

$$2\ C_4H_9Cl + Sn \longrightarrow (C_4H_9)_2SnCl_2$$

how many kilograms of each of the two reactants would be necessary to make 1.00 kg of product if the yield is 40.0%?

31. One starting material to make silicones (page 955) is dichlorodimethylsilane, $(CH_3)_2SiCl_2$. It is made by treating 99% pure silicon powder at about 300 °C with CH_3Cl in the presence of a copper-containing catalyst.
   (a) Write a balanced equation for the reaction.
   (b) Assume you carry out the reaction on a small scale with 2.65 g of silicon. To measure the CH_3Cl gas, you fill a 5.60-L flask at 24.5 °C. What pressure of CH_3Cl gas must you have in the flask in order to have the stoichiometrically correct amount of the compound?
   (c) How many grams of $(CH_3)_2SiCl_2$ will be produced? (Assume 100% yield.)

32. Aluminum sulfate, with a worldwide production of about 3 million tons, is the most important aluminum compound after aluminum oxide and aluminum hydroxide. Write a balanced equation for the reaction of aluminum oxide with sulfuric acid to give aluminum sulfate. If you want to manufacture 1.00 kg of aluminum sulfate, how many kilograms of aluminum oxide and sulfuric acid must you use?

33. Many compounds have been detected in interstellar space, and three quarters of those observed involve carbon. Carbon monoxide dominates, binding perhaps 10% of the interstellar carbon, but other compounds are based on the cyanide ion, a close relative of CO.

Topaz, a gemstone, is an aluminosilicate.

   (a) Compare the electron dot structures of CO and CN^-.
   (b) One interstellar compound is cyanamide, H_2NCN. Draw its electron dot structure.
   (c) Sketch the molecular structure of cyanamide and give the hybrid orbital set used by the central N atom and by the C atom.

34. Silicates are enormously important in our lithosphere.
   (a) Name one mineral that is a simple orthosilicate.
   (b) Name one mineral that is an amphibole.
   (c) Name one mineral that is a sheet silicate.
   (d) Give two examples of aluminosilicates.

35. Liquid HCN is dangerously unstable with respect to trimer formation, that is, to the formation of $(HCN)_3$.
   (a) Propose a structure for the cyclic trimer, a six-membered ring of alternating C and N atoms.
   (b) Estimate the energy of the trimerization reaction. Obtain bond energies from Table 10.4 and use the following additional information: $D_{C=N}$ = 615 kJ/mol.

36. The cyanide ion forms stable complexes with many metal ions. One example is $[Ag(CN)_2]^-$. The formation constant for the ion is 5.6×10^{18}. What is the concentration of $Ag^+(aq)$ in a 0.010 M solution of the silver-cyanide complex ion?

37. "Aerated" concrete bricks are widely used building materials. They are obtained by mixing gas-forming additives with a moist mixture of lime, sand, and possibly cement. Industrially, the following reaction is important:

$$2\ Al(s) + 3\ Ca(OH)_2(s) + 6\ H_2O(\ell) \longrightarrow [3\ CaO \cdot Al_2O_3 \cdot 6\ H_2O](s) + 3\ H_2(g)$$

Assume that the mixture of reactants contains 0.56 g of Al for each brick. How many milliliters of hydrogen gas would you expect at 26 °C and atmospheric pressure (745 mmHg)?

CHAPTER

24

The Chemistry of the Nonmetals: Periodic Groups 5A Through 7A and the Rare Gases

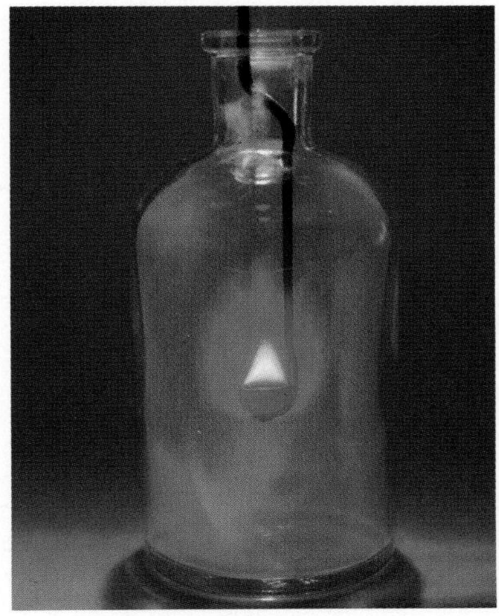
Sulfur burning in oxygen.

CHAPTER OUTLINE

We come now to a region of the periodic table where virtually all of the elements are nonmetals. Recall that ionization energies, electron affinities, and, therefore, electronegativities generally increase when moving across the periodic table. As a result, not only is there an increasing tendency to form negative oxidation numbers but compounds of the nonmetals with oxygen are more generally acidic, in contrast to basic metal oxides.

In these groups we find elements that are important constituents of compounds essential to life (nitrogen, phosphorus, oxygen, and sulfur) and that are also the basis of the strongest aqueous acids. Their abundances range from the highest on earth to among the lowest and from among the most reactive to the most inert.

5A
7 N nitrogen
15 P phosphorus
33 As arsenic
51 Sb antimony
83 Bi bismuth

24.1 THE CHEMISTRY OF GROUP 5A ELEMENTS

Group 5A elements are characterized by their ns^2np^3 configuration with its half-filled *np* subshell (Table 24.1). This configuration gives the elements a tremendous range of chemistry, with oxidation numbers ranging from −3 (the *n*th shell completely full) to +3 and +5 (the shell partially or completely empty). Once again, as in Groups 3A and 4A, the most positive oxidation number is not as stable for the heavier elements, and

Table 24.1 Group 5A: The p^3 Elements (Nitrogen Through Bismuth)*

	Nitrogen	Phosphorus	Arsenic	Antimony	Bismuth
Symbol	N	P	As	Sb	Bi
Atomic number	7	15	33	51	83
Atomic weight	14.007	30.974	74.922	121.75	208.98
Valence e^-	$2s^22p^3$	$3s^23p^3$	$4s^24p^3$	$5s^25p^3$	$6s^26p^3$
mp, °C	−210	44†	814‡	631	271
bp, °C	−196	280†	Sublimes‡	1587	1560
d, g/cm³ (25 °C, 1 atm)	1.25 g/L	1.83†	5.73‡	6.69	9.81
Atomic radius, pm	70	110	120	140.	150
Ion radius, pm (3+ ion)		58	72	90.	117
Pauling EN	3.0	2.19	2.18	2.05	2.02
Standard reduction potential (V)	+0.27	−0.06	−0.60	0.51	−0.8
Oxidation numbers	−3 to +5	−3 to +5	−3 to +5	−3 to +5	−3, +5
Ionization energy	1402	1012	947	834	703
Isolated by	Rutherford	Brandt	Albertus	Antiquity	Geoffroy
Date of isolation	1772	1669	1250	Antiquity	1753
rpw pure O_2	NO_x, 1200 °C	P_4O_6, P_4O_{10}	As_4O_6	Sb_2O_3, 500 °C	Bi_2O_3, 700 °C
rpw H_2O	None	None	None	None	None
rpw N_2		None	None	None	None
rpw halogens	§	PX_3, PX_5	AsX_3, AsX_5	SbX_3, SbX_5	BiX_3
rpw H_2	NH_3, 350 °C	PH_3, 300 °C	§	§	§

*All energies are given in kJ/mol. The symbol rpw means "reaction product with." Standard reduction potential is from hydride EH_3 to element E. All data in this table are taken from (a) *KC? Discoverer*, a computer program distributed by JCE: Software. (b) N.N. Greenwood and A. Earnshaw, *Chemistry of the Elements*, Pergamon Press, New York, 1984. (c) E.G. Rochow, *Modern Descriptive Chemistry*, W.B. Saunders, Philadelphia, 1977.
†Data are given for the white form of phosphorus.
‡Data are given for gray arsenic; the yellow waxy form has density 2.01 g/cm³.
§Indirect preparation not from elements.

arsenic and bismuth compounds with oxidation numbers of +5 are powerful oxidizing agents.

THE ELEMENTS AND THEIR RECOVERY

The relative abundances of the elements of a group give us some clue about the importance of an element and its possible commercial value. In Table 24.2, you see that phosphorus is the most abundant of the Group 5A elements in the earth's crust, with nitrogen only about as abundant as Ga (Table 23.2). On the other hand, nitrogen comprises 78.1% by volume (or 75.5% by weight) of the air around us.

Nitrogen and its compounds play a key role in our economy (Figure 24.1). Of the "top 14" chemicals produced by industry in 1989, five contain nitrogen.

Table 24.2 Relative Abundance of the Group 5A Elements in the Earth's Crust

Element	Abundance (in ppm)	Rank Among All Elements
N	19	33
P	1120	11
As	1.8	51
Sb	0.2	62
Bi	0.008	69

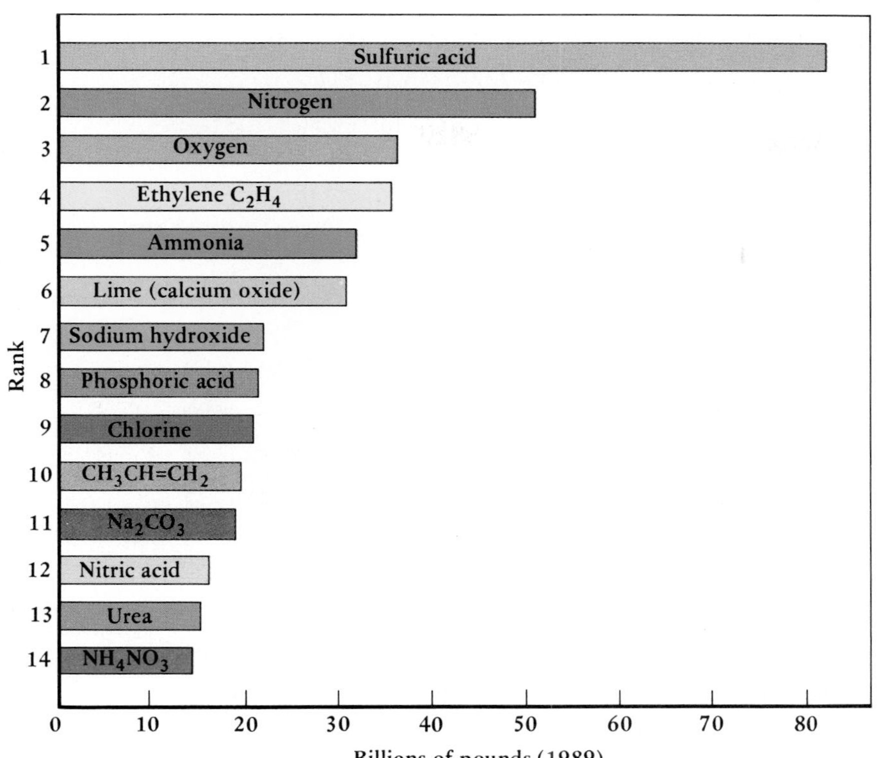

Figure 24.1 The 14 chemicals produced in the largest amount in the United States in 1988. Notice that all except C_2H_4 and $CH_3CH{=}CH_2$ contain elements in Groups 5A through 7A.

N_2 is chemically rather inert for two reasons: the N atoms are bound with a triple bond of great energy (945.4 kJ/mol) and the molecule is nonpolar. If sufficient energy is used, however, it will react directly with many metals to give nitrides (N^{3-}); for example,

CO and CN^- are also triply bonded and isoelectronic with N_2, but both are more reactive than N_2 because of their polarizability.

$$3\ Mg(s) + N_2(g) \longrightarrow \underset{\text{magnesium nitride}}{Mg_3N_2(s)}$$

Like oxygen, nitrogen forms bonds with virtually all the elements in the periodic table, but, in contrast to O_2, this does not usually occur by direct reaction of N_2 with the element.

Elemental nitrogen, N_2, is a very useful material. Because of its inertness, the largest quantity of gas is used to provide a nonoxidizing atmosphere for packaged foods and wine, for example, and to pressurize electric cables and telephone wires. Nitrogen is also easily converted to a liquid (bp, −196 °C) that is convenient to handle, and about 10% of the N_2 produced is used in this form to freeze soft materials such as rubber so they can be ground to a powder, to preserve biological samples (e.g., blood and semen), and so on.

See the effect of liquid nitrogen in Figure 1.7.

Over 200 different **phosphorus**-containing minerals have been described, and all are *orthophosphates*, that is, they all contain the tetrahedral PO_4^{3-} anion or a derivative. By far the largest source of the element is the mineral family called the *apatites*, with the general formula $3\ Ca_3(PO_4)_2 \cdot CaX_2$ where X is commonly F, Cl, or OH. Thousands of

P is unique in that it was first isolated from an animal source rather than a mineral. In 1669 an alchemist putrefied urine and then obtained P_4 when it was distilled.

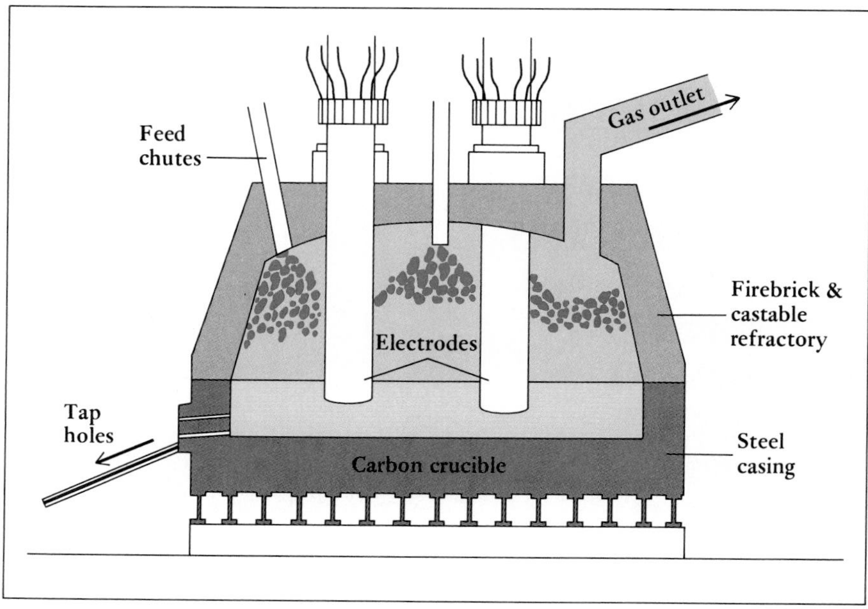

Figure 24.2 A furnace for phosphorus production. The "feed" is a mixture of $Ca_3(PO_4)_2$, SiO_2, and C. A mixture of P_4 gas and CO is driven off the top of the furnace, and molten slag containing calcium silicate and other substances is drawn off at the bottom.

Apatite, a mineral with the general formula $3\ Ca_3(PO_4)_2 \cdot CaX_2$ where X = OH^-, F^-, or Cl^-. (Bill Tronca, Tom Stack & Associates)

tons of elemental phosphorus are produced each year by the reduction of phosphate minerals with carbon in the presence of silica (Figure 24.2).

$$2\ Ca_3(PO_4)_2(s) + 10\ C(s) + 6\ SiO_2(s) \longrightarrow \underset{\text{slag}}{6\ CaSiO_3(s)} + 10\ CO(g) + P_4(s)$$

Almost all of the solid P_4 produced is oxidized to pure phosphoric acid, and much of the remainder is used to make phosphorus sulfides for matches (Figure 24.3).

As you have seen, the chemistry of the lightest element in a group is often rather different from that of the heavier elements. The same relationship continues in Group 5A (and in 6A) in that nitrogen exists as

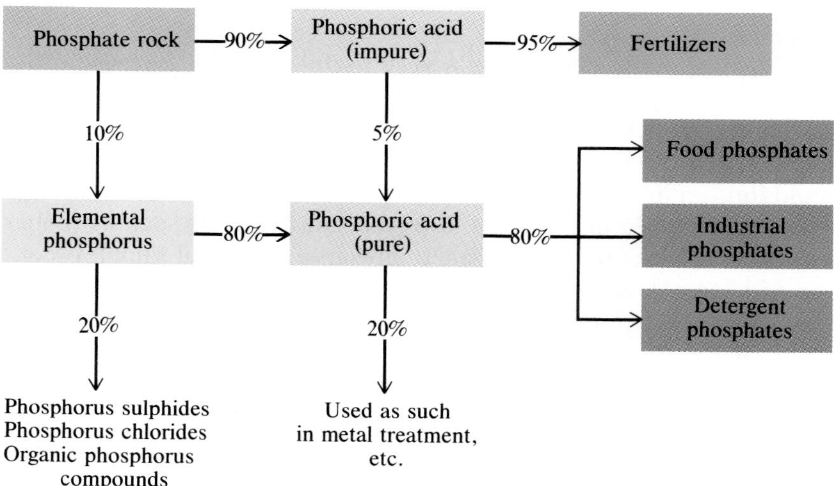

Figure 24.3 Uses of phosphate-containing rock.

a diatomic gas, but phosphorus, like B, C, Si, Ge, and Sn, is a covalently bonded solid with several allotropic forms. These forms are commonly called white, red, and black phosphorus for their colors. *White phosphorus*, P_4, the most common allotrope, is a simple tetrahedron of atoms. It is insoluble in and unreactive with water, but it burns in air to give the oxide P_4O_{10} as described below. It is extremely toxic, and any contact should be avoided. Heating P_4 in the absence of air leads to another allotrope, *red phosphorus*, a polymer of P_4 units.

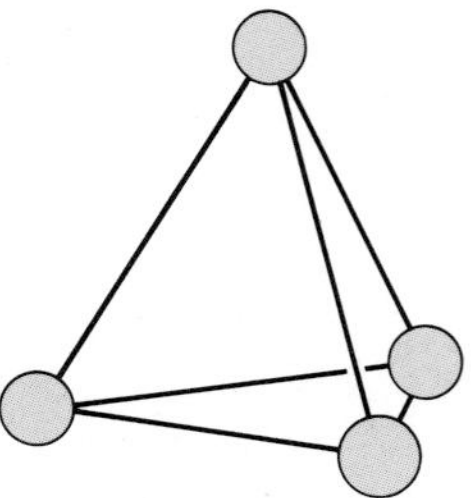

P_4, white phosphorus

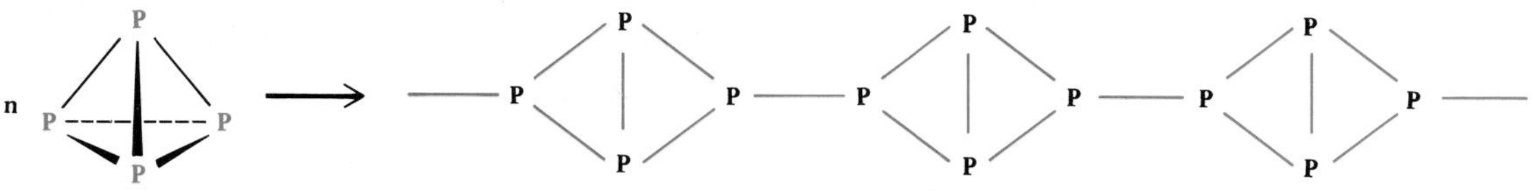

P_4, white phosphorus

$(P_4)_n$, red phosphorus

Black phosphorus, an even more complicated polymeric form, is obtained by heating white P_4 under pressure. Both the red and black forms are much less reactive than the white allotrope.

Arsenic, **antimony**, and **bismuth** are among the oldest elements known, in spite of their low abundance. Although all form stable oxides, they all resemble the transition metals in forming very stable sulfides, and it is in this form that they are commonly found. Lemon yellow *orpiment*, As_2S_3, was used by physicians and assassins for hundreds of years, and black *stibnite*, Sb_2S_3, was used as a cosmetic hundreds of years ago. Like all of the elements you have seen thus far, these elements can be obtained by reducing a mineral source. Scrap iron, for example, is used to reduce stibnite to elemental antimony.

The red and white allotropes of phosphorus.

$$3\ Fe(s) + \underset{\text{stibnite}}{Sb_2S_3(s)} \longrightarrow 3\ FeS(s) + 2\ Sb(s)$$

A principal use of arsenic and antimony is in automobile batteries. The battery plates are lead, but their performance is greatly improved by adding 2.5% to 3% Sb and a trace of As. The U.S. consumption of antimony is about 12,000 tons per year.

EXERCISE 24.1 Group 5A Elements

(a) Name three uses of Group 5A elements or their compounds.
(b) Name the three allotropes of phosphorus.

The mineral orpiment (As_2S_3).

COMPOUNDS OF GROUP 5A ELEMENTS WITH HYDROGEN

The elements of Group 5A form simple gaseous compounds with hydrogen of formula EH_3 (Table 24.3). Their names may be a bit puzzling at first. NH_3 is called ammonia, a common name that has come into general use. The other names, however, come from the old element name with the suffix "-ine" added. All the molecules have the same structural pair geometry and hence the same triangular pyramid molecular structure. The lone pair of electrons on the central atom means that all behave as Lewis bases.

Ammonia, NH_3, is the most important of the EH_3 series (Figure 24.1), but the others are not without interest. Phosphine is a poisonous, highly reactive gas that has a faint garlic odor. In industry it is made by alkaline hydrolysis of white phosphorus.

$$P_4(s) + 3\ KOH(aq) + 3\ H_2O(\ell) \longrightarrow PH_3(g) + 3\ KH_2PO_2(aq)$$

One major use of PH_3 is to serve as a starting material for a salt that is a major ingredient in compounds used for flameproofing cotton cloth.

$$\underset{\text{phosphine}}{PH_3(g)} + \underset{\text{formaldehyde}}{4\ HCHO(aq)} + HCl(aq) \longrightarrow [P(CH_2OH)_4]Cl(aq)$$

The remaining EH_3 compounds are also exceedingly poisonous and increasingly unstable. Nonetheless, AsH_3 is used in the semiconductor industry in the manufacture of materials such as gallium arsenide, GaAs (page 945).

Before World War I, the majority of commercial nitrogen fertilizer came from deposits of Chile saltpeter ($NaNO_3$) and the excrement (guano) of bats and sea birds.

AMMONIA AND NITROGEN FIXATION Nitrogen is vital to life. Even though we are bathed in tons of nitrogen gas, it cannot be used by plants until it is "fixed," converted into biologically useful forms such as ammonia. Nitrogen fixation is done naturally by organisms such as blue-green algae and some field crops such as alfalfa and soybeans. Most plants cannot fix N_2, however, so "fixed" nitrogen must be supplied by an external source. This is particularly true of the new varieties of wheat, corn, and rice that were developed during the "green revolution" of the past several decades.

A feasible process for fixing nitrogen in the form of NH_3 was devised by Fritz Haber, a great chemist but a tragic figure in the history of chemistry. It should be possible to fix nitrogen in some usable form from simple molecules using any of several possible reactions, but Haber chose the *direct synthesis of ammonia* from its elements.

The Haber Process

$$\tfrac{1}{2}\ N_2(g) + \tfrac{3}{2}\ H_2(g) \longrightarrow NH_3(g)$$

$$\Delta H° = -46.1\text{ kJ at }25\ °C \qquad K = 7.6 \times 10^2$$

$$\Delta H° = -55.6\text{ kJ at }450\ °C \qquad K \cong 6.5 \times 10^{-3}$$

At the outset, this seems a poor choice for several reasons. Hydrogen is available naturally only in combined form in water or hydrocarbons, for

Table 24.3 The EH_3 Series of Compounds of Group 5A

Formula	Name	Normal Boiling Point	ΔH_f° (kJ/mol)
NH_3	Ammonia	−34.5 °C	−46.1
PH_3	Phosphine	−87.5 °C	5.4
AsH_3	Arsine	−62.4 °C	66.4
SbH_3	Stibine	−18.4 °C	145.1
BiH_3	Bismuthine	Unstable	277.8

Figure 24.4 An industrial plant for ammonia production. (M.W. Kellogg Company)

example. This means that some H-containing compounds must be destroyed first at the cost of considerable energy. Not only that, but this energy expense is completely wasted because the hydrogen of ammonia is converted to water and the nitrogen to nitrate by soil bacteria before ammonia can be taken up by the plants. The plant must then expend more energy, derived from photosynthesis, to reduce the nitrate back to ammonia. Nonetheless, because the Haber process has now been so well developed, the ammonia is so inexpensive (about $150 per ton) that it is used as a fertilizer.

The direct synthesis of ammonia is an equilibrium process that has been carefully fine-tuned by industry (Figure 24.4).

1. To achieve a reasonable rate, the temperature must be raised. However, notice that the equilibrium constant is smaller at higher temperature, so the yield of ammonia decreases.
2. The unfavorable equilibrium constant at higher temperature is partly countered by carrying out the reaction at higher total pressure. While this does not change the equilibrium constant, it does result in a higher percentage conversion of N_2 and H_2 to NH_3.
3. Finally, to give a reasonable reaction rate, a catalyst (Fe_3O_4 mixed with KOH, SiO_2, and Al_2O_3) is used. Since the catalyst is not efficient below about 400 °C, this means the process is usually operated at about 450 °C.

The Haber process is so efficient that the cost of ammonia is now almost entirely the cost of the hydrogen consumed in making the NH_3. [The H_2 is generally made from natural gas by steam reforming (Chapter 25)].

More than 25% of the ammonia produced is used directly as a fertilizer, and the remainder is the starting material for other nitrogen-containing compounds. Either directly or indirectly, ammonia is the key to many industries: it goes into the starting materials for nylon production, household detergents, water purification, and the production of pharmaceuticals.

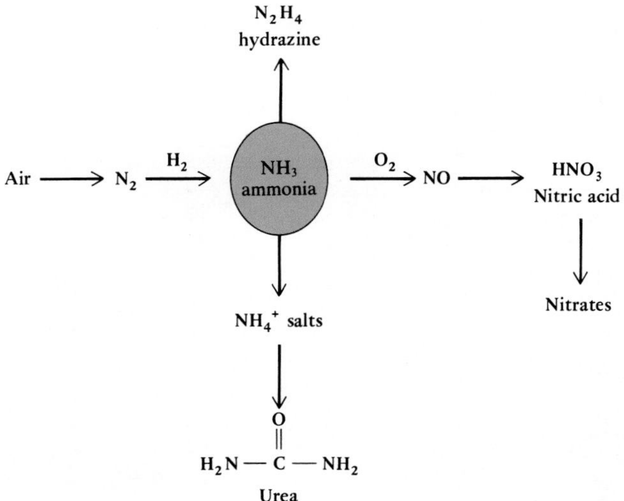

You have also seen catenation in the chemistry of boron, carbon, and silicon (Chapter 23).

NITROGEN CATENATION AND HYDRAZINE All the second period, *p*-block elements (except Ne) readily form bonds to another of the same atom, a property called **catenation**. An atom of nitrogen can similarly form single, double, and triple bonds to another nitrogen, but catenation is very limited: the longest known N atom chain has only 8 atoms. An important catenated nitrogen compound is also one of the simplest: N_2H_4, hydrazine.

hydrazine
mp, 2.0 °C; bp, 113.5 °C

It is a colorless, fuming liquid with an ammonia-like odor. If you compare its melting and boiling points with those of NH_3 (mp, −77.8 °C; bp, −34.5 °C), it is clear that hydrogen bonding is also important in liquid and solid hydrazine.

The Raschig process was used as an example of reaction mechanisms in Chapter 15. Even though the process was first introduced in 1907, the purpose of the gelatin is still not clear.

Approximately 20 million pounds of hydrazine are produced annually by the *Raschig process*, the oxidation of ammonia with alkaline sodium hypochlorite in the presence of gelatin.

$$2\ NH_3(aq) + NaOCl(aq) \xrightarrow[\text{gelatin}]{\text{aqueous alkali}} N_2H_4(aq) + NaCl(aq) + H_2O(\ell)$$

Hydrazine is widely used because it has three important properties: it is (a) a Lewis and Brønsted base, (b) a reducing agent, and (c) an energy-rich compound. As a base in water it has an equilibrium constant for the first ionization of 8.5×10^{-7},

$$N_2H_4(aq) + H_2O(\ell) \rightleftharpoons N_2H_5^+(aq) + OH^-(aq)$$

making it only slightly less basic than ammonia ($K_b = 1.8 \times 10^{-5}$). Its strong reducing ability is reflected in the $E°$ value in basic solution,

$$N_2H_4(aq) + 4\ OH^-(aq) \longrightarrow N_2(g) + 4\ H_2O(\ell) + 4\ e^- \qquad E° = 1.16\ V$$

Historical Figures in Chemistry: Fritz Haber (1868–1934)

In 1898 William Ramsay (the discoverer of the rare gases) pointed out the depletion of fixed nitrogen in the world, and he predicted world disaster due to a "fixed N shortage" by mid-20th century. That this has not occurred is due to the work of Fritz Haber.

Haber studied the $N_2(g) + 3\ H_2(g) \rightleftarrows 2\ NH_3(g)$ equilibrium in the early 1900s and concluded that direct ammonia synthesis should be possible. However, it was not until 1914 that the engineering problems and catalyst question had been solved by Carl Bosch, and ammonia production began just in time for the start of World War I. Ammonia is the starting material to make nitric acid, a vital material in the manufacture of the explosives TNT and nitroglycerin. Therefore, this is thought to be the first large-scale use of a synthetic chemical for military purposes.

Haber's contract with the manufacturer of ammonia called for him to receive 1 pfennig per kilogram of ammonia, and he soon became not only famous but rich. Unfortunately, he joined the German Chemical Warfare Service at the start of World War I and became its director in 1916. The primary mission of the service was to develop gas warfare, and in 1915 he supervised the first use of Cl_2 at the battle of Ypres. Not only was this a tragedy of modern warfare but also to Haber personally. His wife pleaded with him to stop his work in this area, and when he refused, she committed suicide.

In 1918 he was awarded the Nobel Prize for the ammonia synthesis, but the choice was criticized because of his role in chemical warfare.

After World War I, Haber did some of his best work, continuing on in thermodynamics. However, because he had a Jewish background, Haber left Germany in 1933, worked for a time in England, and died in Switzerland in 1934.

and this ability has been put to good use. For example, it is currently used to treat waste water from chemical plants to remove ions such as CrO_4^{2-} and thus prevent them from entering the environment.

An important new use of the compound is in the treatment of water boilers in large electrical generating plants. Oxygen dissolved in the water is a great problem in these plants, because the dissolved gas can oxidize the metal of the boiler and pipes and lead to corrosion. Hydrazine is thus added to reduce the oxygen to water.

$$N_2H_4(aq) + O_2(g) \longrightarrow N_2(g) + 2\ H_2O(\ell)$$

The important noncommercial use of hydrazine, and a spectacular example of its reducing ability, is in rocket fuels. It has been used on the Apollo missions to the moon and in the steering rockets of the Space Shuttle. The lunar lander rockets were powered by mixing liquid N_2O_4, an oxidizing agent, with a 1:1 mixture of methyl derivatives of hydrazine [$H_2N—NH(CH_3)$ and $H_2N—N(CH_3)_2$].

The landing on the moon required about 4.5 tons of the mixed hydrazines and 3 tons of N_2O_4. To take off again, only about one third of these amounts was used.

$$H_2N—N(CH_3)_2(\ell) + 2\ N_2O_4(\ell) \longrightarrow 3\ N_2(g) + 4\ H_2O(g) + 2\ CO_2(g)$$

These reactants are called *hypergolic*; that is, they ignite on contact. Thus, when the rocket is to be fired, the astronaut only needs to open valves and allow the chemicals to mix.

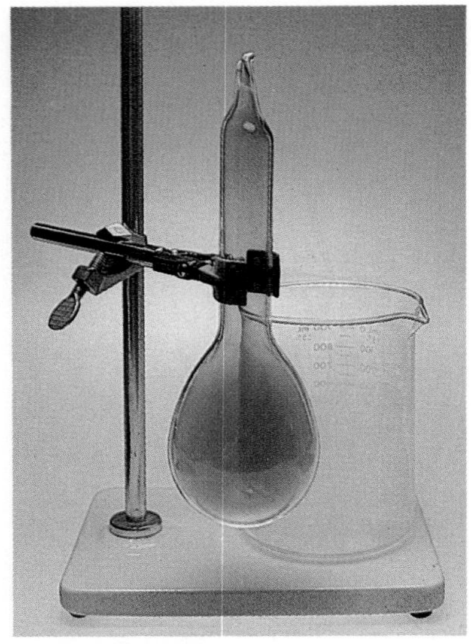

(a)

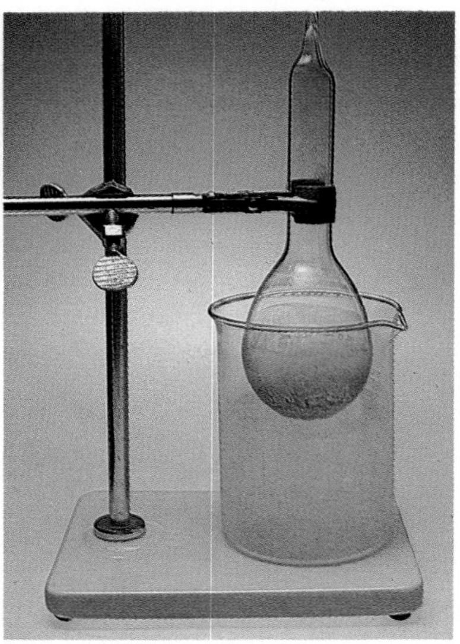

(b)

(a) Brown, gaseous NO_2 in equilibrium with colorless N_2O_4. (b) The sample contains a small amount of NO because, when it was frozen in liquid nitrogen, blue N_2O_3 is seen.

EXERCISE 24.2 Nitrogen Fixation

The Haber process is carried out at a high temperature, in spite of the fact that the reaction is exothermic. Explain why a high temperature is necessary.

OXYGEN COMPOUNDS OF NITROGEN AND PHOSPHORUS

Nitrogen and phosphorus oxides bear some resemblance to their Group 4A neighbors in that nitrogen, a second period element, forms only simple binary oxides, whereas phosphorus, a third period element, forms more extended molecules and ions. In contrast to carbon, however, nitrogen is unique in forming *seven* binary oxides, all of them thermodynamically unstable with respect to decomposition to N_2 and O_2.

NITROGEN–OXYGEN COMPOUNDS Six of the seven known binary nitrogen–oxygen compounds are listed in Table 24.4. Some are important commercially and some must be explored because they are implicated in air pollution.

Dinitrogen oxide, N_2O, the oxide having nitrogen with the lowest oxidation number (+1), can be made by the careful decomposition of ammonium nitrate at 250 °C.

$$NH_4NO_3(s) \xrightarrow{250\ °C} N_2O(g) + 2\ H_2O(g)$$

The gas is nontoxic, odorless, and tasteless. It can be used as an anesthetic in minor surgery and has come to be called "laughing gas" because of its effects. Since it is soluble in vegetable fats, its largest commercial use is as a propellent and aerating agent in cans of whipped cream.

Nitrogen oxide, NO, is the simplest, thermally stable, odd-electron molecule known. With a total of 11 valence electrons, both atoms cannot simultaneously have a completed octet. Therefore, since the nitrogen atom is less electronegative than oxygen, eight electrons surround the O atom and the odd electron resides closer to the N atom (Table 24.4).

As you will see below, NO is an intermediate in the synthesis of nitric acid by oxidation of ammonia. On a laboratory scale, however, it can be synthesized conveniently by using a mild reducing agent on a nitrogen oxide where N has a higher oxidation number.

$$\underset{+3\text{ nitrogen}}{KNO_2(aq)} + \underset{\text{reducing agent}}{KI(aq)} + H_2SO_4(aq) \longrightarrow NO(g) + K_2SO_4(aq) + H_2O(\ell) + \tfrac{1}{2}\ I_2(aq)$$

Dinitrogen trioxide, a blue solid with no particular commercial value, can be isolated only at very low temperatures (mp, −100.1 °C). As the temperature climbs the color fades and becomes greenish as the trioxide dissociates to brown NO_2 and colorless NO.

$$\underset{\text{blue}}{N_2O_3(s)} \rightleftharpoons \underset{\text{brown}}{NO_2(g)} + \underset{\text{colorless}}{NO(g)}$$

Nitrogen dioxide, NO_2, is the brown gas you see when a bottle of nitric acid is allowed to stand in the sunlight, and it is a culprit in air pollution.

$$2\ HNO_3(aq) \rightleftharpoons 2\ NO_2(g) + H_2O(\ell) + \tfrac{1}{2}\ O_2(g)$$

Table 24.4 Some Oxides of Nitrogen

Formula	Name	Structure	Nitrogen Oxidation Number	Description
N_2O	Dinitrogen oxide	$:N\equiv N-\ddot{\underset{..}{O}}:$ linear	+1	Colorless gas (laughing gas)
NO	Nitrogen oxide	$:\dot{N}=\ddot{O}:$	+2	Colorless gas, odd-electron molecule (paramagnetic)
N_2O_3	Dinitrogen trioxide	$O=N-N(=O)-O$ planar	+3	Blue solid (mp, −100.7 °C), reversibly dissociates to NO and NO_2
NO_2	Nitrogen dioxide	$O-\dot{N}=O$	+4	Brown, paramagnetic gas
N_2O_4	Dinitrogen tetroxide	$(O)(O=)N-N(=O)(O)$ planar	+4	Colorless liquid/gas dissociates to NO_2 (Fig. 16.7)
N_2O_5	Dinitrogen pentoxide	$(O=)(O)N-O-N(=O)(O)$	+5	Colorless solid

Like NO, the dioxide NO_2 is an odd-electron molecule. Since the odd electron again resides mainly on the nitrogen atom,

$$2\ NO_2\ (\text{brown gas}) \rightleftharpoons O_2N-NO_2\ (\text{colorless gas})$$

this means that two NO_2 molecules can combine to form O_2N-NO_2, dinitrogen tetroxide, with a N—N single bond. When N_2O_4 is frozen (mp, −11.2 °C), the solid consists entirely of N_2O_4 molecules; as the solid melts and the temperature increases to the boiling point, dissociation to NO_2 begins to occur. At the boiling point (21.5 °C), the gas phase consists of 15.9% NO_2 and is distinctly brown (Figure 23.5 and page 984).

Dinitrogen tetroxide, an oxidizing agent in rocket fuels, is best made by condensing the NO_2 that comes from the thermal decomposition of $Pb(NO_3)_2$.

$$2\ Pb(NO_3)_2(s) \xrightarrow{\sim 400\ °C} 4\ NO_2(g) + 2\ PbO(s) + O_2(g)$$

It is also formed when NO reacts with oxygen,

$$\underset{\text{colorless}}{2\ NO(g)} + O_2(g) \longrightarrow \underset{\text{deep brown gas}}{2\ NO_2(g)} \rightleftharpoons \underset{\text{colorless (mp, } -11.2\ ^\circ\text{C)}}{N_2O_4(g)}$$

and it is this reaction that is so important in producing NO_2 in polluted air.

Nitrogen dioxide and N_2O_4 react with water to form nitric acid, so the moist gases are not only toxic but highly corrosive as well.

$$N_2O_4(g) + H_2O(\ell) \longrightarrow HNO_3(aq) + HNO_2(aq)$$

$$3\ HNO_2(aq) \longrightarrow HNO_3(aq) + 2\ NO(g) + H_2O(\ell)$$

The "true" anhydride of nitric acid, however, is the strange compound *dinitrogen pentoxide*, N_2O_5. This compound is made by chemically dehydrating nitric acid,

$$4\ HNO_3 + P_4O_{10} \longrightarrow 2\ N_2O_5 + 4\ HPO_3$$

and so nitric acid is regenerated if the pentoxide is put into water again. Although the structure is not well established, it is considered to be an ionic solid, consisting of loosely bonded linear NO_2^+ and planar trigonal NO_3^- ions.

NITROGEN OXYACIDS AND RELATED COMPOUNDS As with other nonmetal oxides, nitrogen oxides are acidic and interact with water to produce a number of acids. Several of these acids are unstable in the pure state and are known only in aqueous solution or as their salts. Only two are important enough for you to know about at this point.

Nitrous acid, HNO_2, has never been isolated as a pure compound, but it is known as a weak acid in aqueous solution, dissociating to its conjugate base, the *nitrite* ion.

$$\underset{\text{nitrous acid}}{HNO_2(aq)} + H_2O(\ell) \xrightleftharpoons{K_a\ =\ 4.5 \times 10^{-4}} \underset{\text{nitrite ion}}{NO_2^-(aq)} + H_3O^+(aq)$$

H
102.1°
143.3 pm
110.7°
117.7 pm

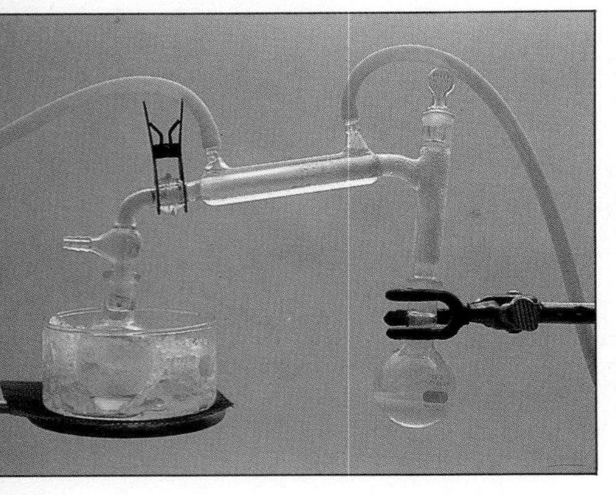

The preparation of nitric acid by the reaction of sulfuric acid and sodium nitrate. Pure HNO_3 is colorless, but some acid decomposes during the preparation to give the brown gas NO_2, and it is this gas that fills the apparatus and colors the product.

The acid is usually made by acidifying a cooled solution of a nitrite salt,

$$NaNO_2(aq) + HCl(aq) \longrightarrow NaCl(aq) + HNO_2(aq)$$

and salts of NO_2^- are made commercially by treating the acidic oxides NO and NO_2 with a base.

$$Na_2CO_3(aq) + NO(g) + NO_2(g) \longrightarrow 2\ NaNO_2(aq) + CO_2(g)$$

Sodium nitrite is mildly toxic to human beings, but it and the nitrate have been used since Roman times for curing meat. Nitrites can be added directly, or they can be produced by bacterial reduction of nitrates:

$$NO_3^-(aq) + 2\ H_3O^+(aq) + 2\ e^- \longrightarrow NO_2^-(aq) + 3\ H_2O(\ell)$$

Figure 24.5 The platinum-rhodium gauze catalyst for the oxidation of ammonia to nitric acid. (Johnson Matthey)

The nitrite ion is known to prevent the growth of microorganisms that lead to deadly botulisms, but there is now evidence that nitrites can be converted during digestion to compounds that are known to be carcinogens.

Nitric acid, HNO_3, has been known for centuries and is still one of the most important compounds in our economy (Figure 24.1). The oldest way to make the acid is to treat Chile saltpeter, $NaNO_3$, with sulfuric acid.

$$2\ NaNO_3(s) + H_2SO_4(aq) \longrightarrow 2\ HNO_3(aq) + Na_2SO_4(s)$$

However, enormous quantities are now produced by the *Ostwald process* for which the overall reaction is $NH_3(g) + 2\ O_2(g) \rightarrow HNO_3(aq) + H_2O(\ell)$. The first step involves the controlled oxidation of ammonia over a Pt catalyst,

$$4\ NH_3(g) + 5\ O_2(g) \xrightarrow{Pt} 4\ NO(g) + 6\ H_2O(\ell) \quad \Delta H° = -1169.5\ kJ$$

followed by further oxidation of NO to NO_2.

$$2\ NO(g) + O_2(g) \longrightarrow 2\ NO_2(g) \qquad \Delta H° = -114.1\ kJ$$

In a typical industrial plant, a mixture of air and 10% NH_3 is passed very rapidly over a catalyst at 5 atm pressure and ~850 °C (Figure 24.5). Roughly 96% of the ammonia is converted to NO_2, making this one of the most efficient industrial catalytic reactions. The final step is to absorb the NO_2 into water to give the acid and NO, the latter being recycled back into the process.

The catalyst used in ammonia oxidation is typically a platinum alloy containing 5–10% rhodium or 5% palladium. See page 63.

$$3\ NO_2(g) + H_2O(\ell) \longrightarrow 2\ HNO_3(aq) + NO(g) \qquad \Delta H° = -138.2\ kJ$$

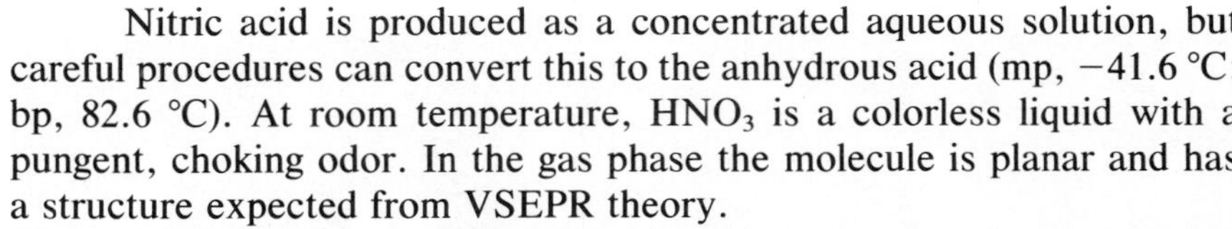

A ball-and-stick model of nitric acid (red = O, blue = N, and white = H) inside a model of the molecular surface. (Computer generated by J. Weber, University of Geneva, Switzerland)

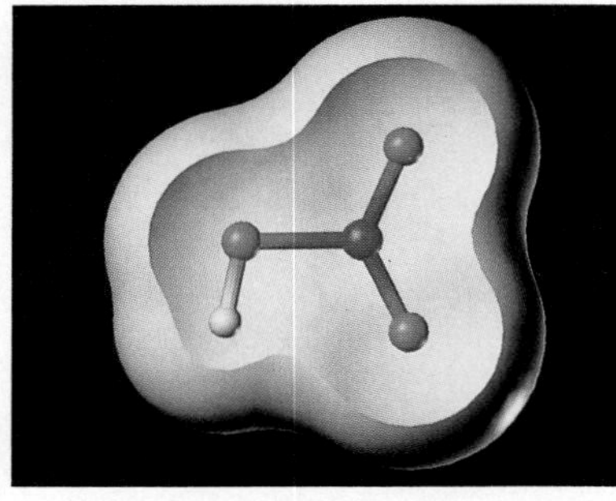

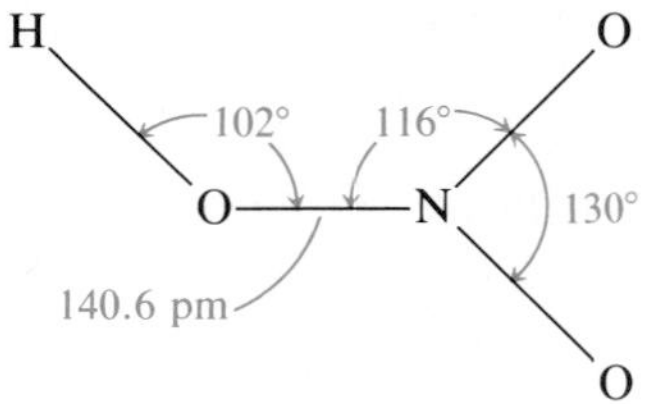

Nitric acid is produced as a concentrated aqueous solution, but careful procedures can convert this to the anhydrous acid (mp, −41.6 °C; bp, 82.6 °C). At room temperature, HNO_3 is a colorless liquid with a pungent, choking odor. In the gas phase the molecule is planar and has a structure expected from VSEPR theory.

Roughly 20% of the ammonia produced every year is converted to nitric acid, which in turn finds many uses. By far the greatest amount of the acid is turned into ammonium nitrate by neutralization of nitric acid with ammonia. Most of the NH_4NO_3 is consumed as fertilizer, but another use depends on the fact that it is thermally unstable and potentially explosive (Figure 24.6):

$$2\ NH_4NO_3(s) \xrightarrow{>300\ °C} 2\ N_2(g) + O_2(g) + 4\ H_2O(g)$$

$$NH_4NO_3(s) \xrightarrow{200\text{–}260\ °C} N_2O(g) + 2\ H_2O(g)$$

A mixture of NH_4NO_3 and fuel oil is used as an explosive in mining operations.

Nitric acid is a powerful oxidizing agent, as the large, positive $E°$ values for the following half-reactions illustrate.

$$NO_3^-(aq) + 4\ H_3O^+(aq) + 3\ e^- \longrightarrow NO(g) + 6\ H_2O(\ell) \qquad E° = +0.96\ V$$

$$NO_3^-(aq) + 2\ H_3O^+(aq) + e^- \longrightarrow NO_2(g) + 3\ H_2O(\ell) \qquad E° = +0.80\ V$$

Concentrated nitric acid will attack and oxidize almost all metals, but the exact nitrogen-containing product depends on the metal (the reducing agent) and acid concentration (Figure 23.5).

$$Cu(s) + \underbrace{4\ H_3O^+(aq) + 2\ NO_3^-(aq)}_{\text{concentrated}} \longrightarrow Cu^{2+}(aq) + 6\ H_2O(\ell) + 2\ NO_2(g)$$

$$3\ Cu(s) + \underbrace{8\ H_3O^+(aq) + 2\ NO_3^-(aq)}_{\text{dilute}} \longrightarrow 3\ Cu^{2+}(aq) + 12\ H_2O(\ell) + 2\ NO(g)$$

There are at least four metals that nitric acid will not attack—Au, Pt, Rh, and Ir—so these came to be known as the "noble metals." However, the alchemists of the 14th century knew that if you mixed HNO_3 with HCl in a ratio of about 1:3, this *aqua regia* or "kingly water" would attack even the noblest of metals, as in this reaction with platinum.

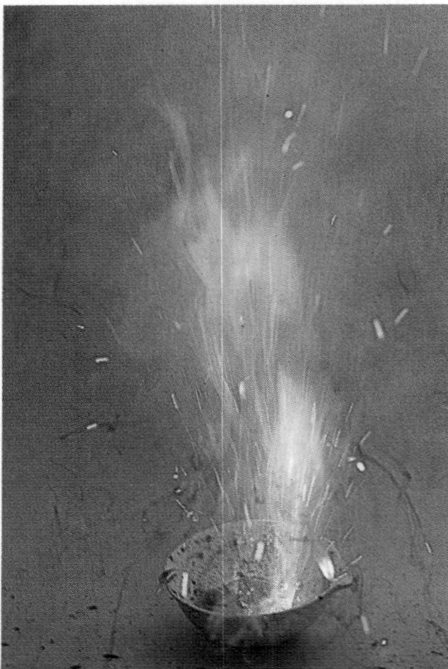

Figure 24.6 The decomposition of NH_4NO_3. In the presence of H_2O, and Cl^- as catalyst, ammonium nitrate decomposes according to the equation

$$NH_4NO_3(s) \longrightarrow N_2O(g) + 2\ H_2O(g)$$

If zinc is present, the following oxidation can also occur.

$$Zn(s) + NH_4NO_3(s) \longrightarrow N_2(g) + ZnO(s) + 2\ H_2O(g)$$

Both of these reactions are occurring in the photo.

$$3\ Pt(s) + 4\ NO_3^-(aq) + 18\ Cl^-(aq) + 22\ H_3O^+(aq) \longrightarrow 3\ H_2PtCl_6(aq) + 4\ NO(g) + 30\ H_2O(\ell)$$

EXERCISE 24.3 Nitrogen Oxides

(a) Explain why N_2O is "linear," whereas NO_2 is "bent."

(b) Give three commercial uses of nitrogen oxides or oxyacids.

PHOSPHORUS OXIDES AND SULFIDES The most important compounds of phosphorus are those with oxygen, and there are at least six simple binary P–O compounds. All of them can be thought of as structurally derived from the P_4 tetrahedron of elemental phosphorus. For example, if P_4 is carefully oxidized, P_4O_6 is formed; an oxygen atom has been inserted into each of the P—P bonds of the tetrahedron.

Phosphorus was named for the phosphorescent glow observed when P_4 oxidizes. *See page 110.*

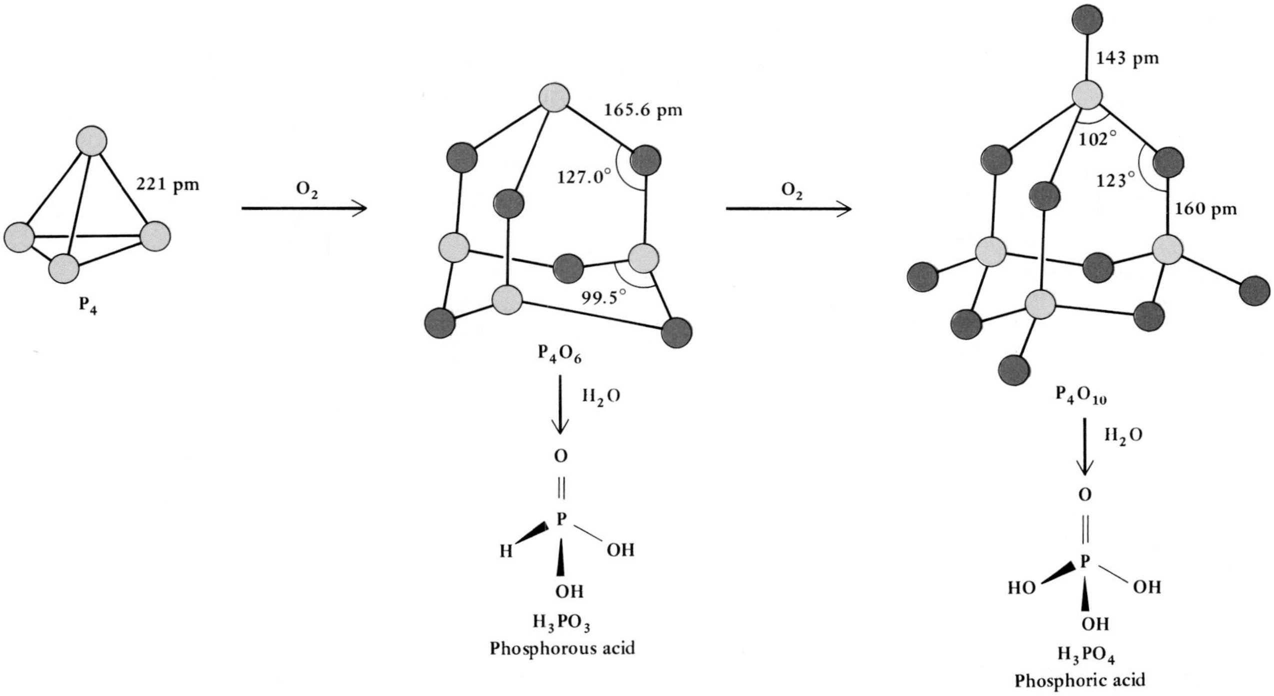

221 pm

P_4

O_2

165.6 pm

127.0°

99.5°

P_4O_6

O_2

143 pm

102°

123°

160 pm

P_4O_{10}

H_2O

H_3PO_3

Phosphorous acid

H_2O

H_3PO_4

Phosphoric acid

The most common and important oxide of phosphorus is P_4O_{10}, a compound commonly called "phosphorus pentoxide" because its empirical formula is P_2O_5. It is formed as a fine, white powder when P_4 burns in air, and its structure is just an extension of that of P_4O_6. Each P atom in P_4O_{10} is surrounded tetrahedrally by O atoms. In this regard, phosphorus chemistry resembles silicon–oxygen chemistry, in which the most important structural element is the $\{SiO_4\}$ tetrahedron.

Other binary P—O compounds have formulas between P_4O_6 and P_4O_{10}. They are formed by starting with P_4O_6 and adding O atoms successively to P atom vertices.

Unlike nitrogen, phosphorus also forms an extensive series of compounds with sulfur that are structurally analogous with the P–O series, and the most stable of these is tetraphosphorus trisulfide, P_4S_3.

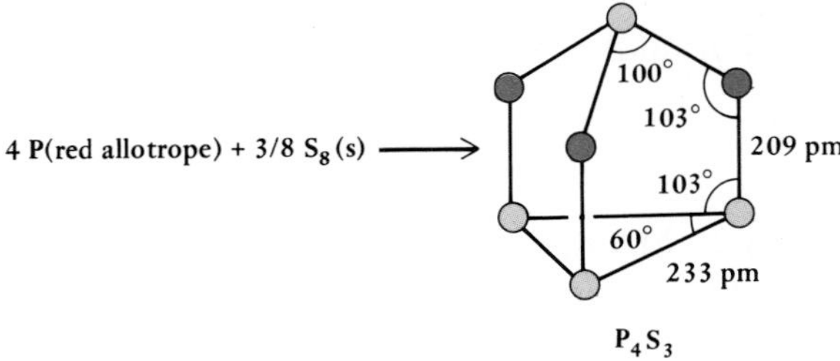

As in P–O chemistry, the basic structural element is again the P_4 tetrahedron, but S atoms are inserted into only three of the six P—S bonds. The principal use of P_4S_3 is in "strike anywhere" matches, the kind of match that will light if you rub it against a stone. The active ingredients are P_4S_3 and the powerful oxidant potassium chlorate, $KClO_3$, and it is the violent reaction between them that lights the match. (Other ingredients of such matches are ground glass, Fe_2O_3, ZnO, and glue). The "safety match" is now more common than the "strike anywhere" match. In this case the match head is predominantly $KClO_3$, and the material on the match book is red P (about 50%), Sb_2S_3, Fe_2O_3, and glue.

The other common P—S compound is P_4S_{10}, structurally analogous to P_4O_{10}. Since P_4S_{10} is used as a starting material for compounds containing the P—S bond, world production exceeds 250,000 tons.

PHOSPHORUS OXYACIDS AND THEIR SALTS Only a few of the many phosphorus oxyacids are illustrated in Table 24.5. Indeed, there are so many types that some structural principles have been developed to help us organize and understand them.

Chemists typically draw a double bond for the PO terminal bond. This reduces the formal charge (page 387) on the P and the terminal O atom. Although the P atom is now involved in five bonds, recall that P and other third period elements may have an expanded octet.

(a) All the P atoms in the oxyacids and their anions are four-coordinate and tetrahedral. All have at least one P=O bond.

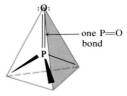

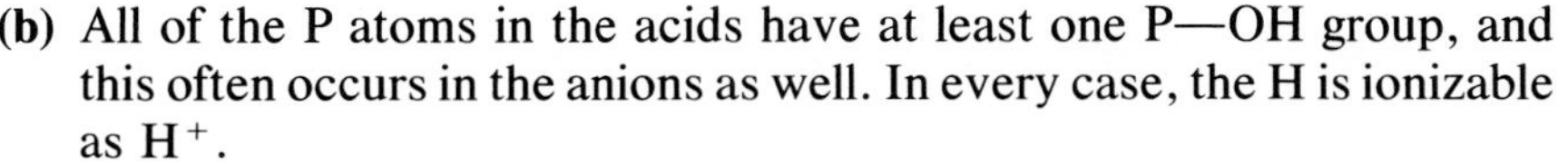

(b) All of the P atoms in the acids have at least one P—OH group, and this often occurs in the anions as well. In every case, the H is ionizable as H^+.

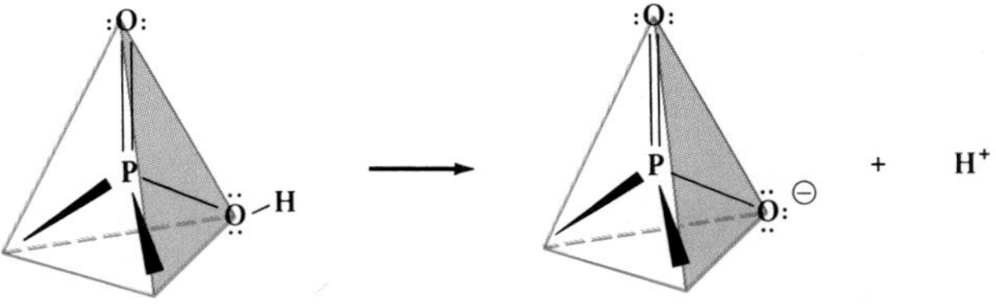

(c) Some oxyacids and anions have one or more P—H bonds. This hydrogen is not ionizable as H^+.

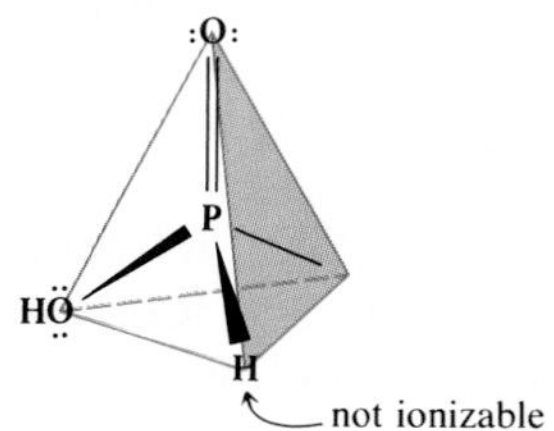

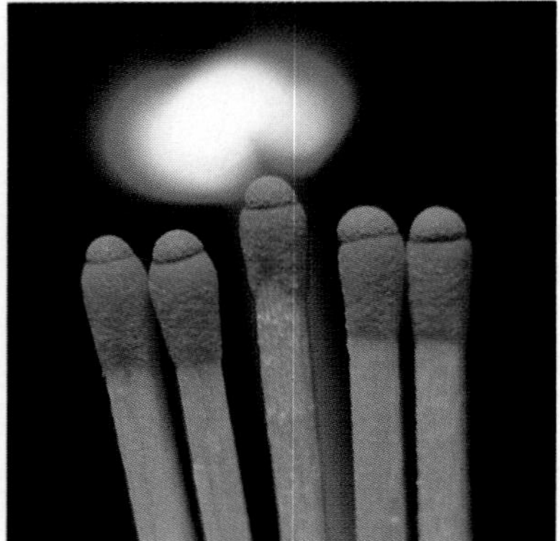

Matches. The head of a "strike anywhere" match contains, among other things, P_4S_3 and an oxidizing agent, potassium chlorate, $KClO_3$. Safety matches have sulfur (3–5%) and $KClO_3$ (45–55%) in the head and red phosphorus in the striking strip.

Table 24.5 Some Phosphorus Oxyacids

Formula	Name	Structure
H_3PO_4	Orthophosphoric acid	
$H_4P_2O_7$	Diphosphoric acid (pyrophosphoric acid)	
$(HPO_3)_3$	Metaphosphoric acid	
H_3PO_3	Phosphonic acid (phosphorous acid)	
H_3PO_2	Phosphinic acid (hypophosphorous acid)	

(d) Polymerization occurs by P—O—P or P—P bonding (catenation) to give both linear and cyclic species. Two P atoms have never been found to be joined by more than one P—O—P bridge.

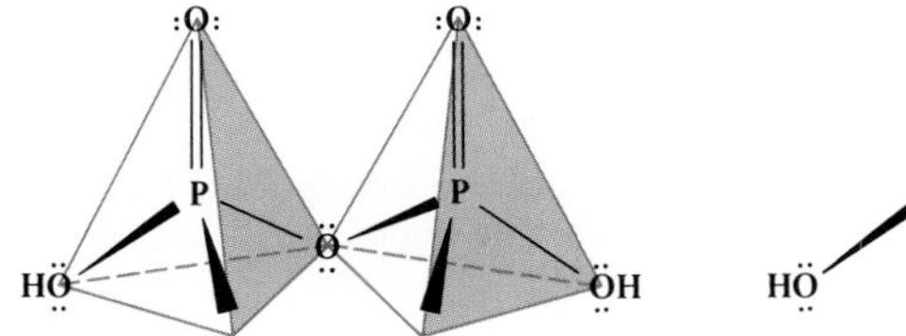

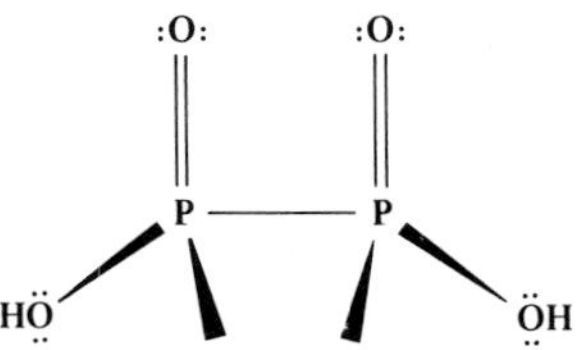

Figure 24.7 The white solid P_4O_{10} reacts vigorously with water to give phosphoric acid, H_3PO_4. (The heat generated by the reaction is so great that some water is boiled off as steam.)

(e) When a P atom is surrounded only by O atoms, its oxidation number is +5. For each P—OH that is replaced by P—H, however, the oxidation number drops by 2 (because P is considered more electronegative than H).

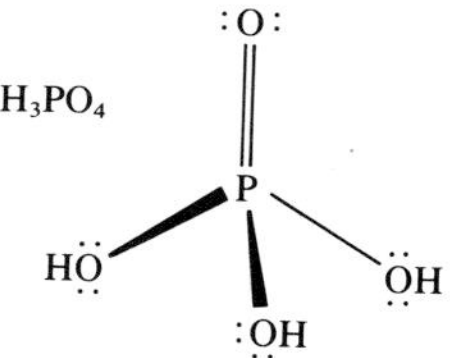

Nitrogen oxyacids are in general good oxidizing agents, while the P–O acids (except for H_3PO_4 where P has its maximum oxidation number of +5) are often good reducing agents. The few half-reactions below illustrate the point.

$$H_3PO_2(aq) + 3\ H_2O(\ell) \longrightarrow H_3PO_3(aq) + 2\ H_3O^+(aq) + 2\ e^- \qquad E° = 0.499\ V$$

$$H_3PO_3(aq) + 3\ H_2O(\ell) \longrightarrow H_3PO_4(aq) + 2\ H_3O^+(aq) + 2\ e^- \qquad E° = 0.276\ V$$

Phosphinic or *hypophosphorous acid,* H_3PO_2, and its salts are an example of this reducing ability. The conjugate base of the acid can be obtained by warming P_4 in a basic solution, a reaction in which P is oxidized to +1 in $H_2PO_2^-$ and reduced to −3 in PH_3.

$$P_4(s) + 3\ OH^-(aq) + 3\ H_2O(\ell) \longrightarrow 3\ H_2PO_2^-(aq) + PH_3(g)$$

The sodium salt of the $H_2PO_2^-$ ion has been increasingly used in industry as a reducing agent to plate metals onto plastic surfaces. Since plastic will not carry electricity, a plastic part cannot be made part of an electrochemical cell. Therefore, chemical reducing agents are used instead, and one of the best is a basic solution of NaH_2PO_2.

$$H_2PO_2^-(aq) + 3\ OH^-(aq) \longrightarrow HPO_3^{2-}(aq) + 2\ H_2O(\ell) + 2\ e^- \qquad E \simeq 1.6\ V$$

Orthophosphoric acid, H_3PO_4, and its derivatives are far more important industrially than the other P–O acids. Almost 11 million tons of phosphoric acid are made annually, and the acid currently sells for about $300 per ton.* A small fraction of the acid is made by the "thermal" or "furnace" method (Section 24.1 and Figure 24.2). The initial product, P_4, is burned to give P_4O_{10}, and, since it is done in the presence of steam, a concentrated aqueous solution of H_3PO_4 results (Figure 24.7).

H_3PO_4 is also made by the "gypsum method." Here $Ca_3(PO_4)_2$ is treated with sulfuric acid to give phosphoric acid and gypsum, $CaSO_4$. It was originally patented in 1842 as a way of making fertilizer from bones, a material high in phosphate.

$$P_4O_{10}(s) + 6\ H_2O(\ell) \longrightarrow 4\ H_3PO_4(aq)$$

*Production figures and prices in the phosphoric acid industry are based on the amount of "contained P_4O_{10}." To convert to the equivalent amount of anhydrous H_3PO_4, multiply by 1.38.

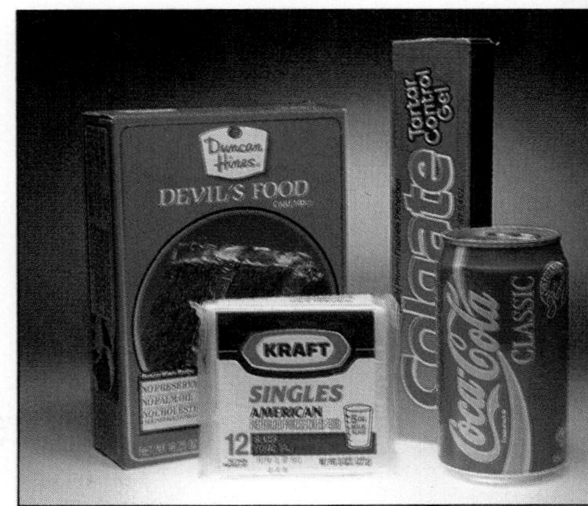

Figure 24.8 Some common products that contain phosphate.

Since this approach gives a pure product, it is used to make acid for use in food products in particular. When the pure acid is dilute it is nontoxic and is used to give the tart or sour taste to carbonated "soft drinks" such as various colas (~0.05% H_3PO_4, pH 2.3) or root beer (~0.01% H_3PO_4, pH 5.0).

A major use of phosphoric acid is to impart corrosion resistance to metal objects such as nuts and bolts, tools, and car-engine parts by plunging the object into a hot acid bath (Figure 24.3). Car bodies are similarly treated with phosphoric acid containing metal ions such as Zn^{2+}, and aluminum auto trim is "polished" by treating with the acid.

All three protons of H_3PO_4 can be removed to give a series of phosphate salts such as NaH_2PO_4, Na_2HPO_4, and Na_3PO_4. In industry, the monosodium and disodium salts are produced using Na_2CO_3 as the base, but an excess of the more expensive base NaOH must be used to remove the final H^+ to give Na_3PO_4. Phosphates are used in such a variety of ways in our modern economy that we can only mention a few (Figure 24.8).

Sodium phosphate (Na_3PO_4), for example, is used in scouring powders and paint strippers, since the ion PO_4^{3-} is a very strong base in water. Sodium monohydrogen phosphate, Na_2HPO_4, has a less basic anion (HPO_4^{2-}) than Na_3PO_4, so the former is widely used in food products. In 1916 J.L. Kraft patented a process using the salt in the manufacture of pasteurized process cheese. Thousands of tons of the phosphate are still used for this purpose, even though the function of the salt is still not fully understood. In addition, a small amount in pudding mixes enables the mix to gel in cold water, and the basic nature of HPO_4^{2-} raises the pH of cereals to provide "quick-cooking" breakfast cereal. (The OH^- from hydrolysis of HPO_4^{2-} accelerates the breakdown of the cellulose material.)

Calcium phosphates are used in a broad spectrum of products. For example, the weak acid $Ca(H_2PO_4)_2 \cdot H_2O$ is used as the acid leavening agent in baking powder, since it reacts with $NaHCO_3$ to produce CO_2 (see page 720).

$$Ca(H_2PO_4)_2 \cdot H_2O(s) + 2\ NaHCO_3(aq) \longrightarrow 2\ CO_2(g) + 3\ H_2O(\ell) + Na_2HPO_4(aq) + CaHPO_4(aq)$$

A typical baking powder contains 28% $NaHCO_3$, 10.7% $Ca(H_2PO_4)_2 \cdot H_2O$, 21.4% $NaAl(SO_4)_2$ (also a weak acid; see Section 17.5), and 39.9% starch. This same phosphate, but made by treating phosphate rock with sulfuric acid, is used as a fertilizer called *superphosphate*.

$$2\ Ca_5(PO_4)_3F(s) + 7\ H_2SO_4(aq) \longrightarrow 3\ Ca(H_2PO_4)_2(s) + 2\ HF(aq) + 7\ CaSO_4(aq)$$

Finally, the monohydrogen phosphate, $CaHPO_4$, also finds a large use as the abrasive and polishing agent in toothpaste.

When phosphates are heated to high temperature, they eliminate water in forming P—O—P links. For example, if $H_2PO_4^-$ and HPO_4^{2-} are heated to 450 °C in a 1:2 ratio, the product is the tripolyphosphate ion.

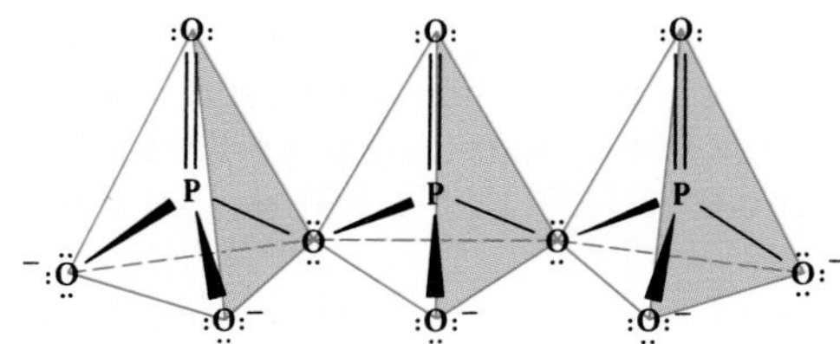

The tripolyphosphate ion, $P_3O_{10}^{5-}$

$$2\ Na_2HPO_4(s) + NaH_2PO_4(s) \xrightarrow{450\ °C} Na_5P_3O_{10}(s) + 2\ H_2O(g)$$

This is the ''phosphate builder'' that is added to many detergents, some containing as much as 25% to 40% $Na_5P_3O_{10}$. It has many functions, one of them being to remove Mg^{2+} and Ca^{2+} from hard water by forming complex ions, an action called *sequestering*. Another function is to provide a mildly basic solution through hydrolysis of the anion as in the reaction

$$P_3O_{10}^{5-}(aq) + H_2O(\ell) \longrightarrow HP_3O_{10}^{4-}(aq) + OH^-(aq)$$

Many thousands of tons of phosphates are used annually for this function around the world.

There is considerable controversy in the United States over the use of phosphates. Since inorganic phosphate is an essential nutrient for plant growth, introducing large quantities into natural waters can lead to eutrophication or overfertilization of water. It can promote a massive growth of algae that in turn depletes the oxygen in water and kills off the fish life.

EXERCISE 24.4 Phosphorus Oxyacids

Hypophosphoric acid has the formula $H_4P_2O_6$. Its sodium salt can be made by treating red phosphorus with sodium chlorite at room temperature.

$$2\ P(s) + 2\ NaClO_2(aq) + 2\ H_2O(\ell) \longrightarrow Na_2H_2P_2O_6(aq) + 2\ HCl(aq)$$

Using the principles of phosphorus–oxygen chemistry, develop a plausible structure for the acid. What is the oxidation number of P in the acid?

OTHER OXIDES AND HALIDES OF GROUP 5A ELEMENTS Arsenic, antimony, and bismuth all form oxides with the general formula E_2O_3 where the oxidation number of the Group 5A element E is +3. The structure of As_4O_6 is analogous to that of P_4O_6, but Bi_2O_3 is ionic.

The more metallic nature of bismuth is clear from the chemistry of these three elements. In general, when descending one of the Groups 3A through 7A in the periodic table, the oxides go from acidic to amphoteric to basic. Thus, nitrogen and phosphorus oxides are decidedly acidic, arsenic and antimony oxides are acidic to amphoteric, and Bi_2O_3 is basic. This means that, while As_4O_6 and Sb_4O_6 are soluble in both acid and base, Bi_2O_3 is soluble only in acid.

$$\underset{\text{white arsenic}}{As_4O_6(s)} + 6\ H_2O(\ell) \xrightarrow{\text{slow}} 4\ \underset{\substack{\text{arsenious acid} \\ K_{1a} = 6 \times 10^{-10}}}{H_3AsO_3(aq)}$$

$$As_4O_6(s) + 12\ NaOH(aq) \longrightarrow 4\ \underset{\text{sodium arsenite}}{Na_3AsO_3(aq)} + 6\ H_2O(\ell)$$

$$As_4O_6(s) + 12\ HCl(aq) \longrightarrow 4\ AsCl_3(aq) + 6\ H_2O(\ell)$$

$$Bi_2O_3(s) + 6\ HNO_3(aq) \longrightarrow 2\ Bi(NO_3)_3(aq) + 3\ H_2O(\ell)$$

Figure 24.9 Nitrogen triiodide. When dry, this dark brown compound is so sensitive that the touch of a small tree branch will set off an explosion. CAUTION! Do NOT perform this experiment yourself. (The filter paper and branch are blurred in the photo at the right because it was taken at the instant the compound exploded.)

As in Groups 3A and 4A, lower oxidation states are more important for the heavier elements. Oxidation numbers of +3 and +5 are common for arsenic, antimony can be forced to +5, but bismuth +5 is unstable. Heating As_4O_6 with concentrated nitric acid gives As_4O_{10}, which dissolves to form *arsenic acid*, analogous to orthophosphoric acid.

$$As_4O_{10}(s) + 6\ H_2O(\ell) \longrightarrow 4\ H_3AsO_4(aq) \qquad K_{1a} = 2.5 \times 10^{-4}$$

It is not as strong an acid as orthophosphoric, but it is decidedly stronger than arsenious acid.

In general, acids formed from oxides where the central element has a higher oxidation number are more acidic than those where the central element has a lower oxidation number.

All arsenic compounds are poisonous, so they have been the favorites of writers of mystery novels for years. Arsenic can be detected in the stomach of a victim of poisoning by the *Marsh test.* An arsenic-containing solution is treated with zinc granules and hydrochloric acid, and arsenites or arsenates are reduced to the deadly gas AsH_3 (arsine).

$$Zn(s) + 2\ H_3O^+(aq) \longrightarrow Zn^{2+}(aq) + H_2(g) + 2\ H_2O(\ell)$$

$$H_3AsO_4(aq) + 4\ H_2(g) \longrightarrow AsH_3(g) + 4\ H_2O(\ell)$$

The evolved gas is passed through a heated glass tube, where it decomposes to the elements and leaves a shiny film of As on the glass.

$$2\ AsH_3(g) \longrightarrow 2\ As(s) + 3\ H_2(g)$$

As little as 0.5 mg of arsenic can be detected in this manner.

The halogen compounds of Group 5A also reflect the increasing metallic character of the elements when descending the group. While halogen compounds of nitrogen are generally unstable, the heavier elements form many different species with the halogens.

Fluorine is especially valuable in forming halides with Group 5A elements. NF_3, for example, is the most stable nitrogen trihalide,

$$4\ NH_3(g) + 3\ F_2(g) \xrightarrow{\text{copper catalyst}} 3\ NH_4F(s) + \underset{(\text{mp, } -206.8\ ^\circ\text{C; bp, } -129.0\ ^\circ\text{C})}{NF_3(g)}$$

The gas is relatively unreactive in the sense that it is unaffected by water or dilute acids or bases. In contrast, NCl_3 and NBr_3 are highly unstable. The chloride was prepared in 1811 by P.L. Dulong, who lost three fingers and an eye in the attempt. The brown solid nitrogen triiodide ($NI_3 \cdot NH_3$) can be handled when wet, but when dry a very light touch will cause it to explode violently (Figure 24.9).

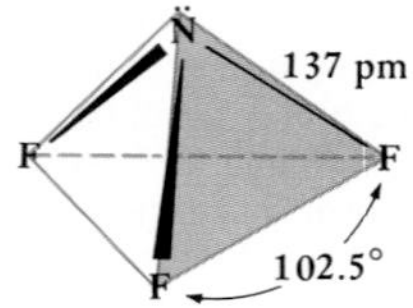

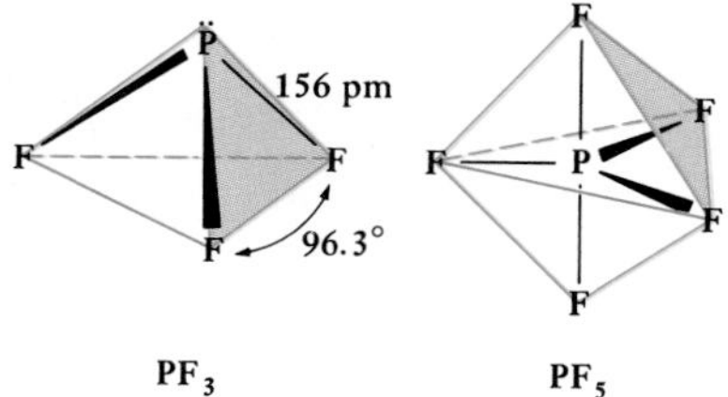

In contrast to nitrogen, phosphorus and the remaining Group 5A elements all form well defined and stable halides in both the +3 and +5 oxidation states. Many are commercially available. Phosphorus trifluoride, for example, is a colorless gas at room temperature and has the pyramidal structure predicted by VSEPR theory (Chapter 10).

$$2\ PCl_3(\ell) + 3\ CaF_2(s) \longrightarrow 3\ CaCl_2(s) + 2\ PF_3(g)$$

It is odorless and quite toxic because, like CO, it combines with the hemoglobin of your blood and leads to suffocation.

Based on the number of valence electrons, VSEPR leads you to expect compounds such as PX_5 to be trigonal bipyramidal. PF_5 is indeed an excellent example of that structure.

The halogen chemistry of As, Sb, and Bi is a treasure trove of interesting compounds and structures. However, in practical terms their most important function is as fluorinating agents. For example, a widely used chlorofluorocarbon is made by exchanging Cl for F on carbon with an antimony chlorofluoride.

$$SbCl_5(s) + HF(\ell) \longrightarrow SbCl_4F(s) + HCl(g)$$

$$CCl_4(\ell) + 2\ SbCl_4F(s) \longrightarrow CCl_2F_2(g) + 2\ SbCl_5(s)$$

6A
8 O oxygen
16 S sulfur
34 Se selenium
52 Te tellurium
84 Po polonium

24.2 THE CHEMISTRY OF GROUP 6A ELEMENTS

Group 6A, with elements of configuration ns^2np^4, consists almost completely of nonmetals (Table 24.6). All the 6A elements form compounds with metals or hydrogen in which the oxidation number is −2; oxidation numbers of +2, +4, and +6 occur when a Group 6A element is combined with more electronegative elements such as oxygen and the halogens.

THE ELEMENTS: THEIR RECOVERY AND USES

Oxygen is the most abundant element on our planet, occurring free as O_2 gas and combined in the form of water and oxides of many kinds. However, elemental oxygen, O_2, did not appear in the atmosphere of the earth until about 2 billion years ago, when it began to arise from photosynthesis occurring in the earliest green plants.

$$CO_2(g) + H_2O(\ell) + \text{light energy} \xrightarrow[\text{catalysts}]{\text{chlorophyll}} O_2(g) + \underset{\text{carbohydrates}}{\{CH_2O\}}$$

Oxygen makes up 23% of the weight of the atmosphere, 46% of the lithosphere (the crustal rocks of the earth), and more than 85% of the hydrosphere, the lakes and oceans. Even on the moon the rocks are 44.6% oxygen.

It is not yet known how and why the photosynthetic process began, nor is it completely clear why there is an excess of oxygen in the atmosphere; respiration by animals, the "combustion" of carbohydrates in what is effectively the reverse of the photosynthetic process, should consume the evolved oxygen. These are some of the mysteries of science to be solved by people with skill and imagination in chemistry, geology, and biology.

Molecular oxygen can be prepared in the laboratory by decomposing a salt of an oxyacid. A particularly convenient source is potassium

Table 24.6 Group 6A: The p^4 Elements (Oxygen, Sulfur, and Their Congeners)*

	Oxygen	Sulfur	Selenium	Tellurium	Polonium
Symbol	O	S	Se	Te	Po
Atomic number	8	16	34	52	84
Atomic weight	15.999	32.066	78.96	127.60	(209)
Valence e^-	$2s^22p^4$	$3s^23p^4$	$4s^24p^4$	$5s^25p^4$	$6s^26p^4$
mp, °C	−218.4	113	217	450	254
bp, °C	−183.0	445	685	990	962
d, g/cm³ (25 °C, 1 atm)	1.43 g/liter	2.07†	4.285	6.25	9.14
Atomic radius, pm	73	103	119	142	168
Ion radius, pm (2− ion)	126	170	184	207	
Pauling EN	3.44	2.58	2.55	2.1	2.00
Standard reduction potential (V)	+1.229	+0.141	−0.40	−0.72	−1.0
Oxidation numbers	−1, −2	−2 to +6	−2 to +6	−2 to +6	
Ionization energy	1314	1000	941	869	812
Isolated by	Priestley	Antiquity	Berzelius	Mueller	Curie
Date of isolation	1774	Antiquity	1817	1782	1898
rpw pure O_2		SO_2, SO_3	SeO_2	TeO_2	
rpw H_2O	None	None	None	None	
rpw N_2	NO_x, 1200 °C	‡	None	None	
rpw halogens	OF_2‡	S_2X_2 to SX_6	SeX_2, SeX_4	TeX_2	
rpw H_2	H_2O, 500 °C	H_2S, 400 °C	H_2Se, 400 °C	‡	
Color of element	Pale blue	Bright yellow	Brick red§	Brown§	

*All energies are given in kJ/mol. Atomic weight in parentheses is that of the most stable isotope. The symbol rpw means "reaction product with." Standard reduction potential is for $O_2 + 4\ H_3O^+ + 4\ e^- \rightarrow 6\ H_2O$ and the like. All data in this table are taken from (a) *KC? Discoverer,* a computer program distributed by JCE: Software. (b) N.N. Greenwood and A. Earnshaw, *Chemistry of the Elements,* Pergamon Press, New York, 1984. (c) E.G. Rochow, *Modern Descriptive Chemistry,* W.B. Saunders, Philadelphia, 1977.
†Data are given for orthorhombic S.
‡Nitride, or halide hydride formed only by indirect methods.
§Color of nonmetallic form.

chlorate, whose decomposition is catalyzed by transition metal oxides such as MnO_2.

$$2\ KClO_3(s) \xrightarrow[70-100\ °C]{MnO_2\ \text{catalyst}} 2\ KCl(s) + 3\ O_2(g)$$

The major industrial source of O_2, however, is fractional distillation of the gases of the atmosphere. Over 30 billion pounds of O_2 are produced in the United States every year (Figure 24.1), largely for use in the Bessemer process of steel making (Section 25.2).

Oxygen is a colorless, odorless gas at room temperature that is unique in being a paramagnetic molecule, even though it has an even number of electrons (see Chapter 11). Partly because of the unpaired electrons, the liquid (bp, −183.0 °C at 1 atm) is blue, as is the solid (mp, −218.4 °C).

O_2 is the most common allotrope of oxygen; *ozone* or O_3 is the other. It is a blue, diamagnetic gas with an odor so strong that it can be detected in concentrations as low as 0.01 ppm. Ozone is conveniently synthesized by passing O_2 through an electrical discharge, but an alternative preparation is to irradiate O_2 with ultraviolet light.

O_3, Ozone
mp, −192.5 °C; bp, −111.9 °C
$\Delta G_f^\circ = +163.2$ kJ/mol

A postage stamp honoring Joseph Priestley (1733–1804), who discovered oxygen in 1774. For religious reasons he fled his native England in 1794 and settled in Northumberland, Pennsylvania, where he continued his experiments in chemistry to the end of his life. (See also page 105.)

Figure 24.10 Some common sulfur-containing minerals: (left) iron pyrite, commonly known as "fool's gold" (FeS_2); (center) gypsum, $CaSO_4 \cdot 2\ H_2O$; (right) black galena, lead(II) sulfide (PbS), on quartz.

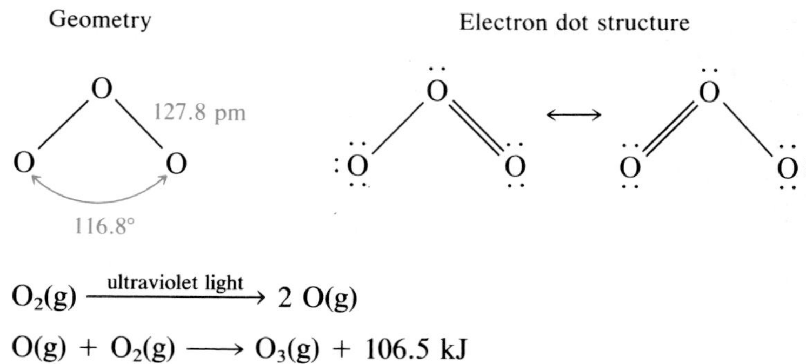

$$O_2(g) \xrightarrow{\text{ultraviolet light}} 2\ O(g)$$

$$O(g) + O_2(g) \longrightarrow O_3(g) + 106.5\ kJ$$

The gas is unstable with respect to decomposition back to O_2, a reaction that is normally slow but can be speeded up by ultraviolet irradiation. As explained in "Something More About Ozone" (page 1003), these reactions occur in the stratosphere and serve not only to convert the sun's intense ultraviolet radiation to warm the earth's atmosphere but also to protect plants and animals from that radiation.

Since ozone is hazardous to handle, it is usually generated where it is needed in amounts approaching perhaps a ton per day. Its primary use is to produce oxygen-containing compounds in industry, and, in Europe in particular, it is used to purify drinking water.

Sulfur is roughly twice as abundant as carbon in the crust of the earth. It can be found in the elemental form in nature, but only in certain concentrated deposits. Generally, it exists in the form of sulfur-containing compounds in natural gas and oil and as metal sulfide minerals, especially iron pyrites (FeS_2, commonly known as "fool's gold") (Figure 24.10). Indeed, there are many common sulfide-containing minerals; examples include cinnabar, HgS, and galena, PbS. Other sulfur-containing compounds in the environment are sulfates such as gypsum ($CaSO_4 \cdot 2\ H_2O$), and the sulfur oxides (SO_2 and SO_3), products of volcanic activity. Rarely are these sources worked for their sulfur content, however.

One major source of elemental sulfur (about 10 million tons per year) consists of the deposits along the Gulf coast of the United States and Mexico. These occur in the caprock over subterranean salt domes (Figure 24.11), typically at a depth of 150 to 750 m below the surface in layers perhaps 30 m thick. The theory is that the sulfur was formed by anaerobic ("without air") bacteria feeding on sedimentary sulfate deposits such as gypsum ($CaSO_4 \cdot 2\ H_2O$).

The largest use of sulfur by far is the production of sulfuric acid, the compound produced in largest quantity by the chemical industry (Figure 24.1). As indicated in Figure 24.12, sulfur and its compounds, usually sulfuric acid, are widely used in manufacturing, but only rarely appear in the final product.

Figure 24.11 About 10% of the sulfur used in the United States is recovered from underground deposits of elemental sulfur by a process devised by Herman Frasch about 1900. Superheated water (at 165 °C) and then air are forced into the deposit; the sulfur is melted (mp, 113 °C) and is forced to the surface as a frothy, yellow stream. The sulfur is thought to have been formed by bacterial action on sulfate minerals. In this photograph, the sulfur has been forced to the surface by volcanic activity, and it has deposited on the rim of a steam vent. (David Cavagnaro)

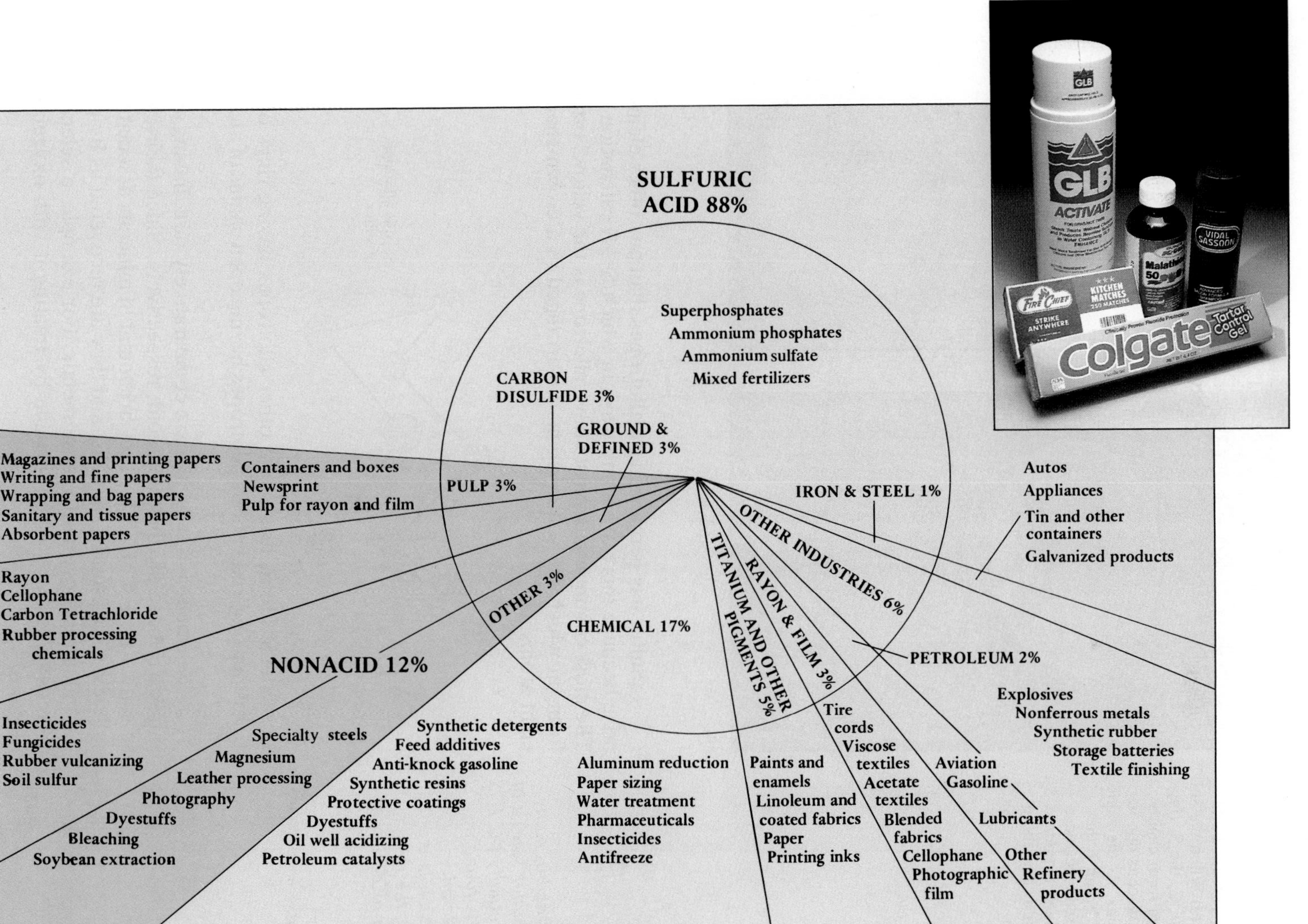

Figure 24.12 Some products made using sulfur or sulfur-containing compounds in the manufacturing process. About 88% of the elemental sulfur produced is converted to sulfuric acid, which in turn has myriad uses. The toothpaste, shampoo, and spa chemical shown in the photo all contain sulfates. Matches contain phosphorus sulfides, and Malathion is an organic compound also containing P—S bonds.

(a)

(b)

Figure 24.13 The two common, lemon-yellow forms of sulfur are orthorhombic sulfur (mp, 113 °C) and monoclinic sulfur (mp, 119 °C), both consisting of S_8 rings. (a) When the sulfur is heated above 150 °C, the S_8 rings begin to break open and form chains, and the melt becomes viscous due to the tangling of the chains. At the same time the color darkens because the chain ends have unpaired electrons. (b) When the viscous melt is poured into cool water, the sulfur reverts to the orthorhombic form.

Sulfur has perhaps more allotropes than any other element. These arise because there are many ways in which S—S catenation can occur. The most common and most stable allotrope is the yellow, orthorhombic form, in which the sulfur atoms are arranged in a crown-shaped ring of eight atoms.

In the United States and in many other countries the major source of elemental sulfur is actually the H_2S and sulfur compounds found in natural gas and crude oil. Although the conversion is a complicated process, the net reaction is simply an oxidation of hydrogen sulfide.

$$2\ H_2S(g) + O_2(g) \longrightarrow \tfrac{1}{4}\ S_8(s) + 2\ H_2O(g)$$

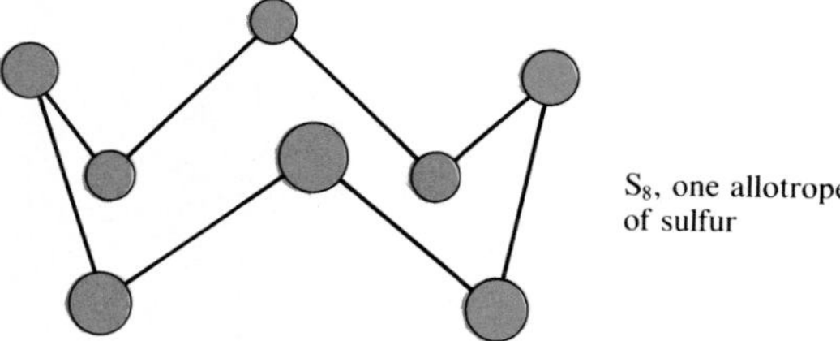
S_8, one allotrope of sulfur

Less stable are at least eight other allotropes having rings of 6 to 20 S atoms per ring and various allotropes having open chains of atoms (Figure 24.13).

Selenium and **tellurium** are comparatively rare, having abundances similar to those of silver and gold, respectively. Since their chemistry is so similar to that of sulfur, they are found in minerals associated with the sulfides of Cu, Ag, Fe, and As, and they are recovered as by-products of the industries devoted to these metals. For example, a selenite salt can be recovered from silver selenide by treating it with oxygen in a basic medium.

$$Ag_2Se(s) + Na_2CO_3(aq) + O_2(g) \longrightarrow 2\ Ag(s) + \underset{\text{sodium selenite}}{Na_2SeO_3(aq)} + CO_2(g)$$

After the sodium selenite is converted to selenous acid, elemental selenium is recovered by reduction with SO_2.

$$\underset{\text{selenous acid}}{H_2SeO_3(aq)} + 2\ SO_2(g) + H_2O(\ell) \longrightarrow Se(s) + 2\ H_2SO_4(aq)$$

Glass can be made red by adding cadmium selenide, CdSe.

Selenium and its compounds find a range of uses, and one is glassmaking. They are often added to glass as a decolorizer, but brilliant red glass, such as is used in the lenses of traffic lights, is made by adding a mixed sulfide/selenide, Cd(S, Se).

The use of selenium you must be most familiar with is in "xerography." At the heart of most photocopying machines is an aluminum plate coated with a film of selenium. Light coming from the imaging lens selectively discharges a static electric charge in the selenium film, and the black "toner" sticks only to the areas that remain charged. A copy is made when the toner is transferred to a sheet of plain paper.

Finally, the heaviest element in Group 6A, **polonium**, is radioactive. It was discovered in 1898 by Marie Sklodowska Curie (1867–1934) and her husband Pierre Curie (1869–1906) (Chapter 7). The Curies painstakingly separated the elements in a large quantity of uranium-containing ore, pitchblende, and found the new elements radium and polonium. The latter was named in honor of Marie Curie's home country, Poland.

EXERCISE 24.5 Group 6A Elements

Name three commercial uses of elements of Group 6A. What are the natural sources of these elements?

COMPOUNDS OF OXYGEN

In spite of the high bond energy of O_2 (498 kJ), the molecule reacts with virtually all the elements. Some elements such as W, Pt, and Au do not combine *directly* with O_2, but all of the elements—except the rare gases He, Ne, Ar, and possibly Kr—ultimately form oxo compounds. The reason for the great reactivity of O_2 is the stability of the resulting products.

Oxygen is of course an excellent oxidizing agent, as indicated by its $E°$ value in acidic media.

$$O_2(g) + 4\ H_3O^+(aq) + 4\ e^- \longrightarrow 6\ H_2O(\ell)$$

$$E° = +1.229 \text{ V at } 1\ M\ H_3O^+ \text{ and } 1 \text{ atm } O_2$$

This is effectively the reaction that keeps all of us operating, since O_2 acts as an acceptor of H^+ and e^- in oxidative metabolism, the process that supplies the energy for all higher forms of life.

The ultimate product of O_2 reduction is the oxide ion, O^{2-}, which gives compounds of great thermodynamic stability. One way to systematize what is known about oxides is in terms of their acid–base behavior. In general, we can say that

(a) Oxides are *acidic* if combined with a nonmetal (e.g., CO_2, NO_2, P_4O_{10}, SO_3).

Xerography means "dry printing," or printing without ink. In such copiers the image of a document is first formed electrostatically on a selenium-coated drum of the type shown in this photograph. (Xerox Corporation)

(b) Oxides are *basic* if combined with an electropositive metal (e.g., Li_2O, CaO, Tl_2O, Bi_2O_3).
(c) Oxides are *amphoteric* if combined with a weakly electropositive element (e.g., BeO, Al_2O_3, ZnO).
(d) Oxides are *neutral* if they do not interact with water or aqueous acids and bases (e.g., CO, NO).
(e) Across a period, the oxides begin as basic compounds but become progressively more acidic, the middle ones being amphoteric. For example, basicity declines in the series Na_2O, MgO, Al_2O_3, SiO_2, P_4O_{10}, SO_3, and ClO_2.
(f) Acidity increases with oxidation number (As_4O_{10} is more acidic than As_4O_6).
(g) In the main groups, oxide basicity increases down the group ($BeO < MgO < CaO < SrO < BaO$).

Bond lengths in superoxide and peroxide ions were compared with O_2 in Chapter 11.

Not all binary oxygen compounds are oxides, wherein the oxygen is considered as O^{2-}. You learned in Group 1A chemistry that sodium and O_2 give a *peroxide*, Na_2O_2 (an O_2^{2-} salt), and potassium and O_2 give a *superoxide*, KO_2 (a salt of O_2^-).

In Group 5A chemistry, the simplest hydrogen compound of nitrogen is ammonia, NH_3. Nitrogen can catenate or form N—N bonds to a limited extent, and the simplest compound of this type is hydrazine, $H_2N—NH_2$. In oxygen chemistry, the analogs are water and hydrogen peroxide, H_2O and HO—OH.

Hydrogen peroxide, H_2O_2, was first made in 1818 by the acidification of barium peroxide, still a convenient laboratory method.

$$BaO_2(s) + 2\,H_3O^+(aq) \longrightarrow \underset{\substack{\text{mp}, -0.43\,°C;\ \text{bp}, 150.2\,°C \\ \Delta G_f^\circ = -118.0\ \text{kJ/mol}}}{H_2O_2(aq)} + Ba^{2+}(aq) + 2\,H_2O(\ell)$$

This method is not suitable for large scale production, however, so other methods such as the "isopropanol process" are employed.

$$\underset{\text{isopropanol}}{HO—C(CH_3)_2—H} + O_2 \xrightarrow[90\text{–}140\ °C]{15\text{–}20\ \text{atm}} \underset{\text{acetone}}{O{=}C(CH_3)_2} + HOOH$$

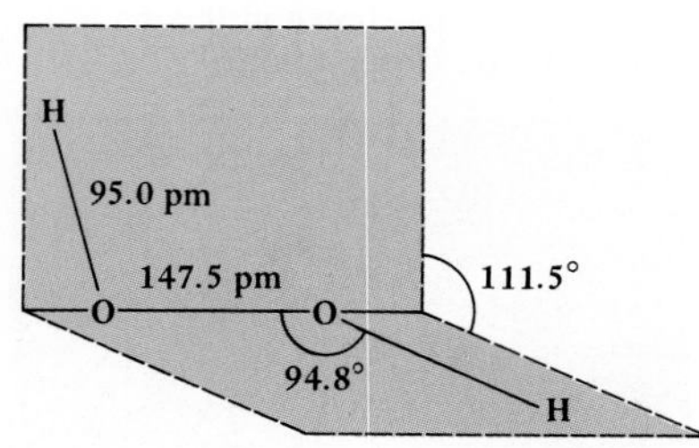

Hydrogen peroxide mainly finds use as an oxidizing agent.

$$H_2O_2(aq) + 2\,H_3O^+(aq) + 2\,e^- \longrightarrow 4\,H_2O(\ell) \qquad E° = +1.77\ V$$

A typical H_2O_2 plant can produce 15,000 tons of 100% H_2O_2 per year along with 30,000 tons of acetone, a valuable organic solvent.

Thus, a significant fraction is used to bleach paper and textiles, and about 10% is consumed for environmental purposes: purification of drinking water, sewage treatment, and sterilization of milk containers.

Hydrogen peroxide is generally sold as 30% aqueous solutions for laboratory and industrial use. If you have handled such solutions, you know that they are always shipped in plastic containers, since traces of metal or alkali metal ions (dissolved from a glass container) can catalyze the explosive decomposition of the peroxide (Figure 15.15).

The catalyzed decomposition of H_2O_2 is used by the bombardier beetle to protect itself (Figure 15.2).

$$H_2O_2(aq) \longrightarrow H_2O(\ell) + \tfrac{1}{2}\,O_2(g) \qquad \Delta H° = -98.2\ kJ$$

SOMETHING MORE ABOUT
Ozone in the Stratosphere

Much of our life in the United States today depends on refrigeration. We cool our homes, cars, offices, and shopping centers with air conditioners. We preserve our food and medicines with refrigerators. But it wasn't always so. Early in this century, refrigeration systems were crude and used toxic gases such as SO_2, NH_3, and CH_2Cl_2 as the "working fluid." In 1928 two scientists at the Frigidaire Division of General Motors were asked to find nonflammable, nontoxic replacements for use in home refrigerators, and, within two days, they had settled on CCl_2F_2 (called CFC-12) as the ideal substance. A joint venture was begun with E.I. DuPont de Nemours and Company, and effective methods were soon found to produce CFC-12 along with a host of other, similar compounds, now all generally referred to as **chlorofluorocarbons** or **CFC's**.

By 1988 the total, worldwide consumption of CFC's was over a billion kilograms annually. In the U.S. almost 5,000 businesses produce CFC-related goods and services worth more than $28 billion a year, and there are more than 700,000 CFC-related jobs. CFC's are mostly used in the U.S. as refrigerants, foam-blowing agents for polystyrene and polyurethane, and industrial solvents. Outside North America, almost 200 million kilograms of CFC's continue to be used as aerosol propellants, although they were banned for this purpose in the United States in 1978.

It is ironic that the very properties that led to the first use of CFC's are now causing worldwide concern. Once gaseous CFC's are released into the troposphere, that part of the earth's atmosphere ranging from the surface to an altitude of about 10 kilometers, there is no mechanism known for their destruction. Therefore, they rise to the stratosphere where they are eventually destroyed by solar radiation—but with significant consequences to our environment, as first recognized by M.J. Molina and F.S. Rowland in 1974.[1] From laboratory experiments, they predicted that continued use of CFC's would lead eventually to a significant depletion of the ozone layer around the earth. The reason this is worrisome is that, for every 1% loss of ozone from the stratosphere, an additional 2% of the sun's high energy ultraviolet radiation can reach the earth's surface. This could result in increases in skin cancer, damage to plants, and to other effects that we do not suspect now.

In the stratosphere there is an abundance of ozone produced when high energy ultraviolet radiation causes the photodissociation of oxygen to give O atoms, which react with O_2 molecules to produce ozone, O_3.

These same reactions take place to a lesser extent close to the ground and are responsible for producing so-called "secondary" air pollutants (see page 642).

$$O_2(g) + \text{radiation } (\lambda < 280 \text{ nm}) \longrightarrow 2\ O(g)$$

$$O(g) + O_2(g) \longrightarrow 2\ O_3(g)$$

The ozone produced by this mechanism in the stratosphere is quite abundant (10 ppm), which is fortunate because O_3 is also photodissociated by sunlight,

$$O_3(g) \rightleftharpoons O(g) + O_2(g)$$

and the O atoms produced react with more O_2 to regenerate O_3. The process keeps 95–99% of the sun's ultraviolet radiation from reaching the earth's surface.

The problem with CFC's is that they disrupt the protective ozone layer by a "chlorine catalytic cycle."

[1] M.J. Molina and F.S. Rowland, *Nature*, **1974**, *249*, 810; F.S. Rowland, *American Scientist*, **1989**, *77*, 219.

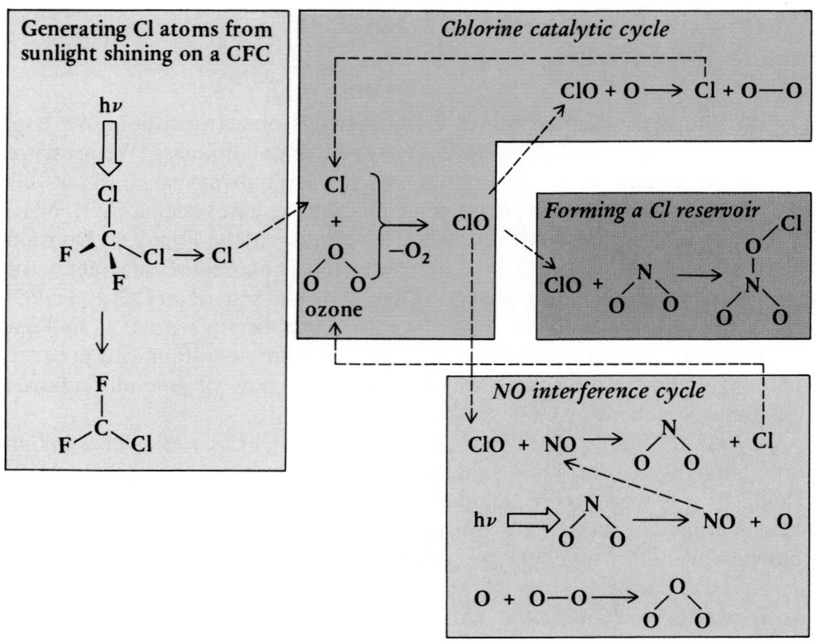

The CFC's rise to the stratosphere where a C—Cl bond is broken by a high energy photon. The Cl atom attacks ozone to give the 13-electron molecule ClO, chlorine oxide. This would not necessarily be a problem, but the ClO can react with an O atom to give O_2 and regenerate a Cl atom. The Cl atom can then destroy still another O_3, and so on and on in a "catalytic cycle." The net reaction is the destruction of a significant quantity of ozone. In fact, it is estimated that each Cl atom can destroy as many as 100,000 ozone molecules before the Cl atom is inactivated or returned to the troposphere (probably as HCl).

The chlorine cycle is not the only chlorine chemistry in the stratosphere. In fact, at least two other major kinds of reactions are believed to *interfere* with ozone loss. In one case ClO reacts with nitrogen oxide, NO, to release a Cl atom and form nitrogen dioxide. However, the NO_2 goes on to regenerate a molecule of ozone. In another reaction, ClO forms chlorine nitrate ($ClONO_2$), a compound that at least temporarily acts as a "chlorine reservoir." Eventually, though, this compound also breaks apart and frees Cl atoms to resume their ozone destruction.

If these interference reactions are important, they should mean that CFC's might have only a minimal effect on earth's ozone layer. However, in the early spring in the southern hemisphere, there is a significant depletion of ozone over the Antarctic, a fact clearly illustrated in satellite images of ozone concentration done in the late 1980's.[2] One theory used to explain this observation involves the high altitude clouds that are common over the Antarctic continent in the winter. Chlorine nitrate could condense in these extremely cold clouds, and chlorine atoms could also be trapped as HCl. Reaction between these compounds in the clouds

$$ClONO_2 + HCl \longrightarrow HNO_3 + Cl_2$$

[2]R.S. Stolarski, *Scientific American*, **1988**, *258*(1), 30.

would give Cl_2, and the first sunlight of spring that warm the clouds could trigger the release of atomic chlorine by photodissociating chlorine molecules. Since nitrogen oxides are trapped in the clouds as nitric acid, the "chlorine catalytic cycle" can run unchecked for five or six weeks in the spring.

Whatever the theories for the springtime ozone loss in the atmosphere over the Antarctic, the problem is real, and people around the world are taking steps to halt any further deterioration. DuPont and other U.S. companies have voluntarily halted CFC production and are actively searching for substitutes.[3] Some states now require recycling of CFC's used in automobile air conditioners or have banned them outright. In January 1989, the *Montreal Protocol on Substances that Deplete the Ozone Layer,* which calls for reductions in production and use of certain CFC's, was signed by 24 nations. Finally, over 90 nations met in London in June 1990 and agreed on a treaty to phase out CFC's by the year 2000. But there will be tradeoffs. As an example, the substitutes now available are much less efficient as refrigerants than CFC's, so it is estimated that appliances will use 3% more electricity in the U.S., and this will increase consumer costs. Further, since electricity is mostly generated by burning fossil fuels, the amount of CO_2 evolved will increase and this in turn will contribute to the "greenhouse" effect.

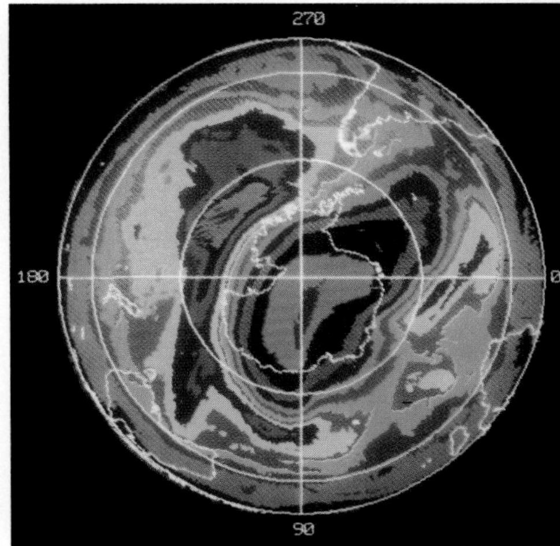

The "hole" in the ozone layer over the Antarctic continent during the spring in the southern hemisphere. The purple region has the lowest ozone concentration.

EVOLUTION of the ANTARCTICA PENGUIN

KOSTA HOUSTON POST

PALEOZOIC ERA

MESOZOIC ERA

CENOZOIC ERA

POST-OZONE ERA

[3]L.E. Manzer, *Science,* **1990,** 249, 31.

EXERCISE 24.6 Oxide Chemistry

Tell whether each of the following oxides is acidic, basic, neutral, or amphoteric: (a) MgO, (b) CO_2, (c) Ga_2O_3, and (d) CO. Dichlorine heptoxide, Cl_2O_7, is the anhydride of perchloric acid. That is, adding water to the oxide gives the acid. Write a balanced equation for this reaction.

THE CHEMISTRY OF SULFUR

IONS OF SULFUR Just as O_2 can be reduced to the oxide (O^{2-}), peroxide (O_2^{2-}), and superoxide (O_2^-) ions, sulfur can be reduced. In the case of sulfur, however, many more ions are known, all the way from S^{2-}, the sulfide ion, to S_n^{2-}, where n can be 2 to 6. It can be predicted by the octet rule and confirmed by experiment that none of these ions is a ring

															S		
							Fe	Co	Ni	Cu	Zn	Ga	Ge	As	Se		
					Mo		Ru	Rh	Pd	Ag	Cd	In	Sn	Sb	Te		
					W		Os		Pt		Hg	Tl	Pb	Bi	Po		

Figure 24.14 The elements that may be found in nature as sulfides. At the top corner are three such sulfides: black CuS, yellow CdS, and orange Sb_2S_3. See also the mineral orpiment (As_2S_3) on page 979.

of atoms; adding two electrons breaks down the ring structure of the S_n allotropes and gives S_n^{2-} chains.

The simplest anion of sulfur is sulfide, S^{2-}, the anion associated in nature with metals of great commercial interest. As illustrated in Figure 24.14, iron, copper, silver, and mercury, among many others, are commonly found as their sulfides. You recall from Chapter 19 that many metal sulfides are poorly soluble in water, so these compounds can accumulate in geologic zones and not be dispersed by ground water.

The recovery of metals from their sulfide ores is usually done by heating or *roasting* the ore in air, a process that can have any of three outcomes. The most common of these is the conversion of the metal sulfide to an oxide, the sulfur appearing as SO_2.

$$2\,ZnS(s) + 3\,O_2(g) \longrightarrow 2\,ZnO(s) + 2\,SO_2(g)$$

Alternatively, the sulfide can be converted to a sulfate, which is often more water soluble than the sulfide. Finally, if the metal oxide product from roasting is less stable than SO_2, the free metal is obtained by sulfide roasting. Examples include Cu, Ag, Hg, and Pb. For example, if lead(II) sulfide or galena is heated to form lead(II) oxide, and then mixed with fresh galena, lead and SO_2 result.

$$2\,PbO(s) + PbS(s) \longrightarrow 3\,Pb(s) + SO_2(g)$$

H_2S dissolves to a small extent in water (about 0.1 M), where it behaves as a polyprotic acid (see Chapter 17).

SULFUR–HYDROGEN COMPOUNDS Hydrogen sulfide, H_2S, is the simplest hydrogen-containing compound of sulfur, and its structure resembles that of water. However, unlike water, hydrogen sulfide is a gas under normal conditions (mp, −85.6 °C; bp, −60.3 °C), presumably due to very weak hydrogen bonding between molecules as compared with water (see Chapter 13). Furthermore, hydrogen sulfide is a deadly poison, comparable to hydrogen cyanide. Fortunately, the sulfide has a terrible odor and can be smelled in concentrations as low as 0.02 ppm. You must be careful with H_2S, however, because it has an anesthetic effect on the olfactory nerve, so your nose can lose its ability to detect H_2S. Death occurs at H_2S concentrations of 100 ppm.

SO_2
mp, −75.5 °C
bp, −10.0 °C

S 143.1 pm
O O
119°

SO_3
mp, 16.86 °C
bp, 44.6 °C

O
142 pm
S
O O
120°

SULFUR–OXYGEN COMPOUNDS At least thirteen oxides of sulfur are known to exist, but only the dioxide (SO_2) and trioxide (SO_3) are important. The former is produced on an enormous scale by the combustion of sulfur, by roasting sulfide ores (especially iron pyrites, FeS_2) in air or by the combustion of sulfur-containing coal and fuel oil. It has been estimated that about 200 million tons of sulfur are released into the atmosphere each year by human activities, primarily in the form of SO_2; this is more than half of the total emitted by all other natural sources of sulfur in the environment.

Sulfur dioxide is a colorless, toxic gas with a choking odor. It readily dissolves in water to give solutions of a species that we write as H_2SO_3 (sulfurous acid),

$$SO_2(g) + H_2O(\ell) \longrightarrow \text{``}H_2SO_3\text{''}(aq)$$

but this is an unstable molecule that has never been isolated. The ions SO_3^{2-} (sulfite) and HSO_3^- (hydrogen sulfite), however, are known in the form of salts of sulfurous acid.

The most important reaction of SO_2 is its oxidation to the trioxide.

$$SO_2(g) + \tfrac{1}{2}\,O_2(g) \longrightarrow SO_3(g) \qquad \Delta H° = -98.9 \text{ kJ/mol}$$

Sulfur trioxide is extremely reactive and is very difficult to handle; as described below, it is almost always deliberately converted to sulfuric acid by reaction with water.

As in nitrogen chemistry there are numerous **sulfur oxyacids**. Few can be isolated, however, and most are known only in aqueous solution or as salts of the oxyacid anions. Several of the most important species are shown in Table 24.7 on the following page.

Sulfuric acid production is taken by some economists as a reliable guide to a nation's industrial strength, because the acid is consumed in so many ways in a modern industrial society. The United States, for example, produces almost 40 million tons of the acid annually. It is manufactured in the highly efficient *contact process* from SO_2 produced by burning sulfur or by roasting sulfide ores (Figure 24.15). The first step is

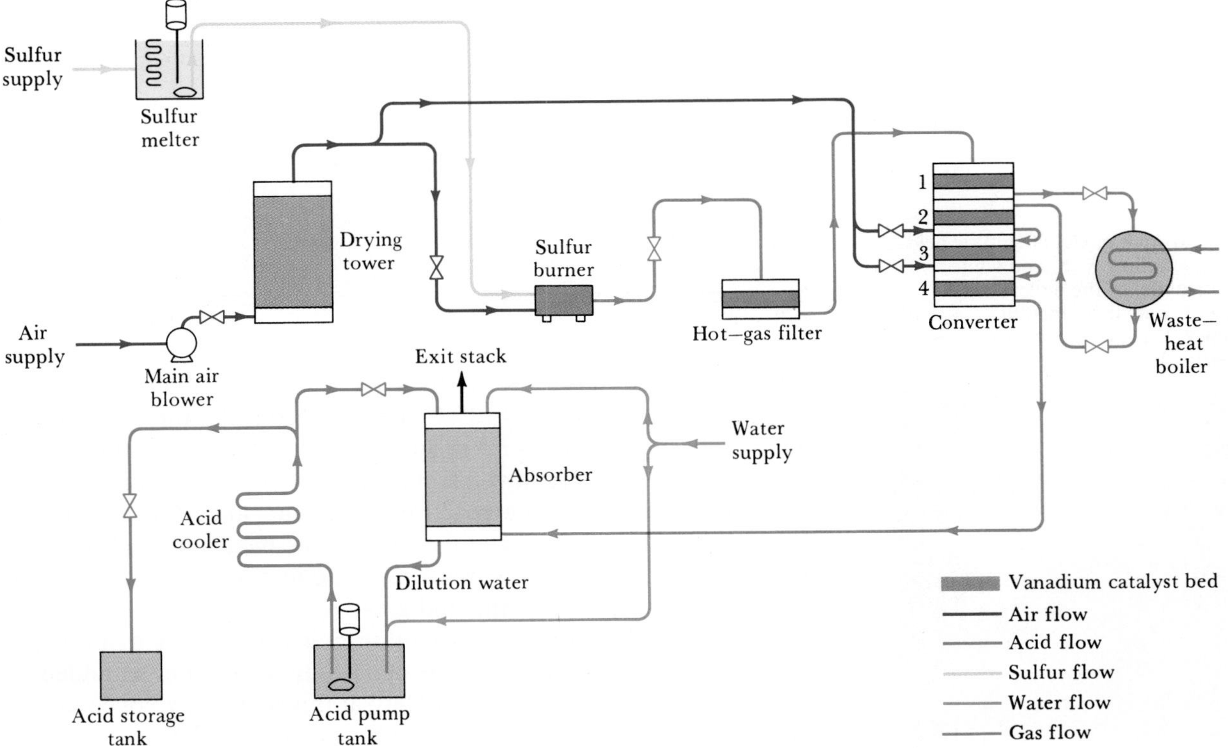

Figure 24.15 A schematic diagram of the industrial process for the production of sulfuric acid, called the *contact process*. Sulfur (or H_2S) is first burned in air to form gaseous SO_2 (see Figure 4.1). The sulfur dioxide is then sent to the "converter" where it combines further with oxygen, in the presence of a catalyst, to give the trioxide, SO_3. Finally, sulfur trioxide reacts very exothermically with water to produce sulfuric acid in the "absorber." (L.W. Fine and H. Beall, *Chemistry for Engineers and Scientists*, Saunders College Publishing, 1990)

Table 24.7 Some Oxyacids of Sulfur

Acid	Sulfur Oxidation Number	Salts
Sulfuric acid	+6	SO_4^{2-}, sulfate HSO_4^-, hydrogen sulfate
Disulfuric acid	+6	$O_3SOSO_3^{2-}$, disulfate
Thiosulfuric acid	+4 and 0	SSO_3^{2-}, thiosulfate
Sulfurous acid	+4	SO_3^{2-}, sulfite HSO_3^-, hydrogen sulfite

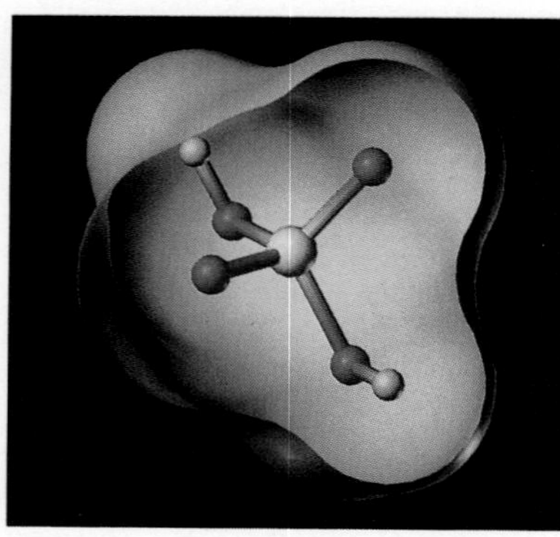

A ball-and-stick model of sulfuric acid (red = O, yellow = S, and white = H) inside a model of the molecular surface. (Computer-generated image by J. Weber, University of Geneva, Switzerland)

the oxidation of the SO_2 to SO_3, the reversible and exothermic process referred to above. According to LeChatelier's principle, the yield of SO_3 can be increased if the pressure is increased, if the concentration of O_2 is increased, or if SO_3 is removed from the reaction zone. An increase in temperature will increase the reaction rate, but, since the reaction is exothermic, it will also decrease the yield. This means a catalyst is required, V_2O_5 in this case, to increase the rate at a sufficiently low temperature that the yield is not affected greatly.

Following SO_3 formation in a sulfuric acid plant, the oxide is absorbed into a 98% aqueous solution of H_2SO_4, and water is added to maintain the concentration approximately at this level. The concentrated acid used in the laboratory is a solution that is 96% H_2SO_4.

The only other oxyacid of sulfur that is industrially useful is thiosulfuric acid. Just as H_2O and SO_3 form H_2SO_4, thiosulfuric acid is derived from H_2S and SO_3.

In the United States, roughly 70% of the sulfuric acid is used to manufacture "superphosphate" fertilizer, and smaller amounts are used, for example, in the conversion of ilmenite, a titanium-bearing ore, to TiO_2, which is used as a white pigment in paint, plastics, and paper (see Chapter 4).

$$H_2S + SO_3 \xrightarrow[\text{absence of water}]{-78\ ^\circ\text{C}} H_2S_2O_3$$

The acid is not stable in the presence of water or above about 0 °C. Instead, it is the salts of the thiosulfate ion, $S_2O_3^{2-}$, that are important. This ion, which resembles the sulfate ion except that one O atom is replaced by an S atom, is conveniently manufactured by boiling an aqueous solution of sodium sulfite with sulfur.

$$Na_2SO_3(aq) + \tfrac{1}{8}\ S_8(s) \xrightarrow{H_2O/100\ ^\circ\text{C}} Na_2S_2O_3(aq)$$

The hydrated salt, $Na_2S_2O_3 \cdot 5\ H_2O$, is commonly known as "hypo," and its aqueous solutions are used as the "fixer" in photography. Photographic films are coated with silver halide, and light falling on the film causes the reduction of the "exposed" halide to silver metal.

$$AgX(\text{on film surface}) + \text{light} \longrightarrow Ag(s) + \tfrac{1}{2}\ X_2(\text{on film surface})$$

Photochromic glass is closely related to film. See page 174.

When an exposed film is developed, the silver halide that was not touched by light is washed away with sodium thiosulfate solution, because silver ion and thiosulfate ion form a stable, water-soluble complex ion.

$$AgBr(s) + 2\ S_2O_3^{2-}(aq) \longrightarrow [Ag(S_2O_3)_2]^{3-}(aq) + Br^-(aq)$$

What remains on the film negative are the silver metal particles where light struck the coating. That is why "negatives" are dark where the image was bright.

EXERCISE 24.7 Sulfur Chemistry

Sodium sulfide, Na_2S, is used extensively in the leather industry to remove hair from hides. Its synthesis is another example of the use of carbon as a reducing agent. Balance the equation for this reaction.

$$Na_2SO_4 + C \longrightarrow Na_2S + CO$$

The sulfide is readily oxidized to sodium thiosulfate; almost 50,000 tons of the latter are produced annually in the United States. Balance the equation for the oxidation.

$$Na_2S(aq) + O_2(g) + H_2O(\ell) \longrightarrow Na_2S_2O_3(aq) + NaOH(aq)$$

24.3 THE CHEMISTRY OF GROUP 7A ELEMENTS: THE HALOGENS

7A
9 F fluorine
17 Cl chlorine
35 Br bromine
53 I iodine
85 At astatine

Compounds of the halogens (Table 24.8) have been known for thousands of years, and many have already been described in this book. Thus, in this section, we shall discuss only a few aspects of halogen chemistry: the diatomic, elemental halogens, the hydrogen halides, and halogen oxides and oxyacids. All are used in large quantities in commerce in ways with which you should be familiar.

Table 24.8 Group 7A: The p^5 Elements (the Halogen Family)*

	Fluorine	Chlorine	Bromine	Iodine	Astatine
Symbol	F	Cl	Br	I	At
Atomic number	9	17	35	53	85
Atomic weight	18.998	35.453	79.904	126.905	(210)
Valence e^-	$2s^22p^5$	$3s^23p^5$	$4s^24p^5$	$5s^25p^5$	$6s^26p^5$
mp, °C	−220	−101	−7.3	114	
bp, °C	−188	−34	59.6	185	
d, g/cm^3 (25 °C, 1 atm)	1.7 g/L	3.21 g/L	3.12	4.94	
Atomic radius, pm	72	100	114	133	
Ion radius, pm (1− ion)	119	167	182	206	
Pauling EN	4.0	3.16	2.96	2.66	2.20
Standard reduction potential (V)	+2.87	+1.36	+1.07	0.536	+0.3
Oxidation numbers	−1	−1 to +7	−1 to +7	−1 to +7	
Ionization energy	1681	1251	1140	1008	890
Isolated by	Moissan	Scheele	Balard	Courlois	CMS§
Date of isolation	1886	1774	1826	1811	1940
rpw pure O_2	O_2F_2 (el†)	None‡	None‡	None‡	
rpw H_2O	HF, O_2, O_3	HCl, HOCl	HBr, HOBr	HI, HOI	
rpw N_2	None	None	None	None	
rpw H_2	HF	HCl	HBr	HI	
Color of element	Pale yellow	Yellow-green	Dark red	Black	

*All energies are given in kJ/mol. Atomic weight in parentheses is that of the most stable isotope. The symbol rpw means "reaction product with." Standard reduction potential is for $X_2 + 2\ e^- \rightarrow 2\ X^-$. All data in this table are taken from (a) *KC? Discoverer,* a computer program distributed by JCE: Software. (b) N.N. Greenwood and A. Earnshaw, *Chemistry of the Elements,* Pergamon Press, New York, 1984. (c) E.G. Rochow, *Modern Descriptive Chemistry,* W.B. Saunders, Philadelphia, 1977.
†Electric discharge.
‡Oxides, etc., made by indirect methods.
§ CMS = Corson, McKenzie, and Segré.

THE ELEMENTS AND THEIR RECOVERY

CaF_2, called fluorspar or fluorite, is the subject of a poem by Roald Hoffmann (page 241).

Fluorspar, CaF_2, is one of the chief minerals containing **fluorine**. Since the mineral was originally used as a flux in metalworking, its name comes from the Latin meaning "to flow." In the 17th century, it was discovered that solid CaF_2 would emit light when heated, and the phenomenon was called *fluorescence*. Thus, when it was recognized in the early 1800s that a new element, F, was contained in fluorspar, A. Ampère suggested that the element be called *fluorine*.

Although fluorine was recognized as an element by 1812, it was not until 1886 that it was isolated in elemental form as a colorless gas by Moissan by the electrolysis of KF in anhydrous HF, still the only practical way to obtain diatomic F_2 (Figure 24.16). Experimentally, this is a difficult preparation because F_2 is such a powerful oxidizing agent that it corrodes the apparatus and will react violently with traces of grease or similar contaminants. Furthermore, F_2 and H_2, the products of the electrolysis, will recombine explosively, so they must be carefully separated. Still, a large plant can make 9 tons of liquid F_2 a day, and the annual U.S. and Canadian production is about 5000 tons.

Figure 24.16 A cell for the electrolytic production of fluorine. Gaseous F_2 is produced at the carbon anode, and H_2 gas is evolved at the steel cathode.

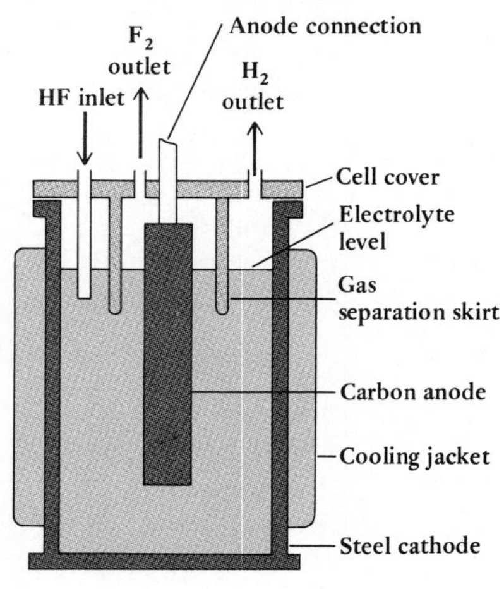

Almost 55% of the fluorine produced is used for processing and reprocessing uranium for fueling nuclear power plants. For example, the oxide of uranium reacts with hydrofluoric acid to give uranium(IV) fluoride,

$$UO_2(s) + 4\ HF(aq) \longrightarrow UF_4(\text{green solid}) + 2\ H_2O(\ell)$$

and the latter is then oxidized to uranium(VI) fluoride by reaction with F_2.

$$UF_4(s) + F_2(g) \longrightarrow UF_6(\text{volatile white solid})$$

In the uranium fissioning process that provides the heat energy in a nuclear power plant, other elements are produced, among them plutonium. These elements and their oxides also react with fluorinating agents, but only uranium forms a fluoride that can be vaporized at a relatively low temperature. This means that UF_6 can be separated easily from the other elements, and the uranium can be recovered for reprocessing.

Other major uses of fluorine and its compounds are in Teflon, the nonstick surface on cooking utensils, and chlorofluorocarbons, compounds such as CCl_2F_2 that are used as the "working fluid" in refrigerators and air conditioners. Furthermore, many municipal water supplies now contain fluoride ion, introduced by adding NaF; as explained on page 936, the ion is incorporated into tooth enamel, making it more resistant to decay. Finally, the cryolite (Na_3AlF_6) vital to the aluminum industry (pages 884 and 950) is now manufactured rather than mined.

Chlorine was the first of the halogens to be isolated in elemental form. The yellow-green gas was made by C.W. Scheele in 1774 by a reaction still used today as a laboratory source of Cl_2 (Figure 24.17).

Figure 24.17 One laboratory method for preparing chlorine gas is to heat sodium chloride with sulfuric acid in the presence of potassium dichromate. The $Cr_2O_7^{2-}$ ion in acid solution oxidizes chloride to chlorine gas. The oxidizing agent MnO_2 is also commonly used. CAUTION: This experiment must be performed in a hood. (See also Figure 24.19.)

$$2\ NaCl(s) + 2\ H_2SO_4(aq) + MnO_2(s) \xrightarrow{\text{heat}} Na_2SO_4(aq) + MnSO_4(aq) + 2\ H_2O(\ell) + Cl_2(g)$$

Today, of course, chlorine is made by electrolysis of brine (Chapter 21) in enormous quantities and ranks number 9 in production (Figure 24.1). Almost 70% of the Cl_2 manufactured is used for the production of organic chemicals such as vinyl chloride, which is then converted to PVC or polyvinyl chloride, a plastic used in many consumer items.

$$\underset{\text{ethylene}}{CH_2{=}CH_2(g)} + Cl_2(g) \longrightarrow \underset{\text{ethylene dichloride}}{H{-}\overset{\overset{H}{|}}{\underset{\underset{Cl}{|}}{C}}{-}\overset{\overset{H}{|}}{\underset{\underset{Cl}{|}}{C}}{-}H} \xrightarrow{\text{heat}} HCl + \underset{\text{vinyl chloride}}{(H)_2C{=}C(H)(Cl)}$$

One of the first properties that Scheele recognized about Cl_2 was its ability to bleach textiles and paper, and soon thereafter its ability to act as a disinfectant in water was also recognized. Today, these two uses account for about 20% of Cl_2 consumption. Finally, about 10% of the Cl_2 production goes into the manufacture of a wide range of inorganic chloride-containing compounds.

In Biblical times a beautiful purple dye, called Tyrian purple, was in great demand. Almost 12,000 small purple snails were required to prepare 1.5 g of the substance, now known to be a naturally occurring compound containing **bromine**. Red-brown, elemental bromine, Br_2, has an unpleasant and penetrating odor and is the only nonmetallic element that is a liquid under normal conditions (Figure 2.4).

All the halogens are oxidizing agents, but the ability declines as they become heavier (see the $E°$ values in Table 24.8) (Figure 24.18). That is,

$$\xleftarrow{\text{oxidizing ability}}$$

$$\begin{array}{cccc} F_2 > & Cl_2 > & Br_2 > & I_2 \\ 2.866\text{ V} & 1.358\text{ V} & 1.066\text{ V} & 0.536\text{ V} \end{array}$$
$$E^\circ \text{ [for } X_2 + 2\,e^- \longrightarrow 2\,X^-(aq)]$$

This means that Cl_2 will oxidize Br^- to Br_2 in aqueous solution.

$$\begin{array}{ll} Cl_2(g) + 2e^- \longrightarrow 2\,Cl^-(aq) & E^\circ = 1.358\text{ V} \\ 2\,Br^-(aq) \longrightarrow Br_2(\ell) + 2e^- & E^\circ = -1.066\text{ V} \\ \hline 2\,Br^-(aq) + Cl_2(g) \longrightarrow 2\,Cl^-(aq) + Br_2(\ell) & E^\circ = 0.292\text{ V} \end{array}$$

In fact, this equation represents the commercial method for preparing bromine, when NaBr is obtained from brine wells in Arkansas and Michigan and from the ocean.

Iodine, a lustrous, purple-black solid, is easily sublimed at room temperature and atmospheric pressure. The element was first isolated in 1811 from certain seaweeds and kelps, extracts of which had long been used in the treatment of goiter, the enlargement of the thyroid gland. It is now known that the thyroid gland produces a growth-regulating hormone (thyroxine) that contains iodine. Most of the table salt sold in the United States has 0.01% NaI added to provide the necessary iodine in the diet.

Thyroxine, the iodine-containing compound produced by the thyroid gland.

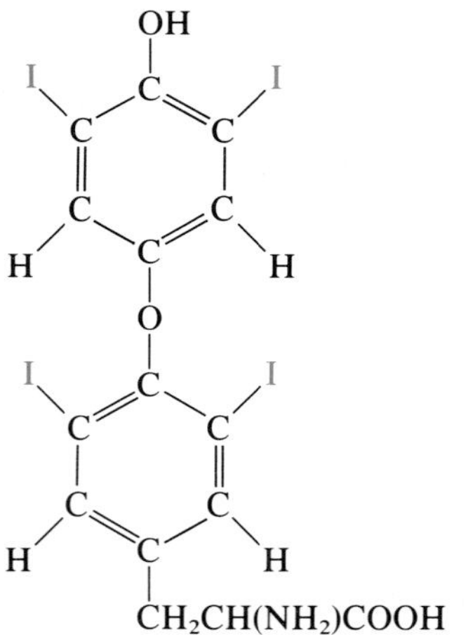

A laboratory method for preparing I_2 is illustrated in Figure 24.19. The commercial preparation, however, depends on the source of I^- and

Figure 24.18 Formation of iodine in the reaction of bromine with an iodide salt.

$$Br_2\text{ (in } CCl_4) + 2\,I^-(aq) \longrightarrow 2\,Br^-(aq) + I_2(\text{in } CCl_4)$$

At left is a graduated cylinder containing Br_2 dissolved in CCl_4 (bottom layer) and NaI in water (top layer). When the reactants are mixed, reaction occurs. At the right we now see that the purple I_2 has collected in the CCl_4 layer at the bottom, while Br^- has dissolved preferentially in the top or water layer. (See also Figure 14.6.)

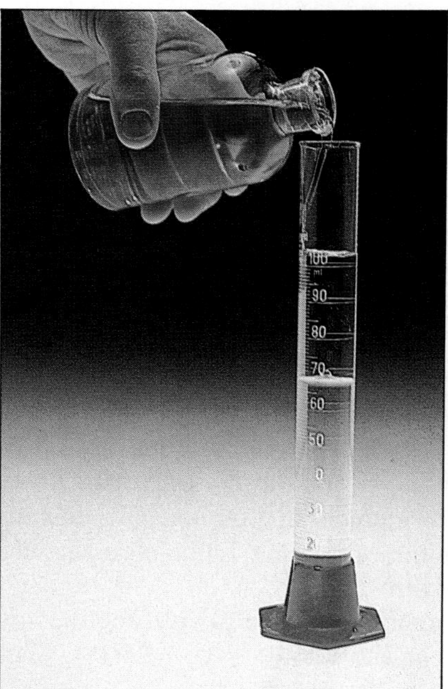

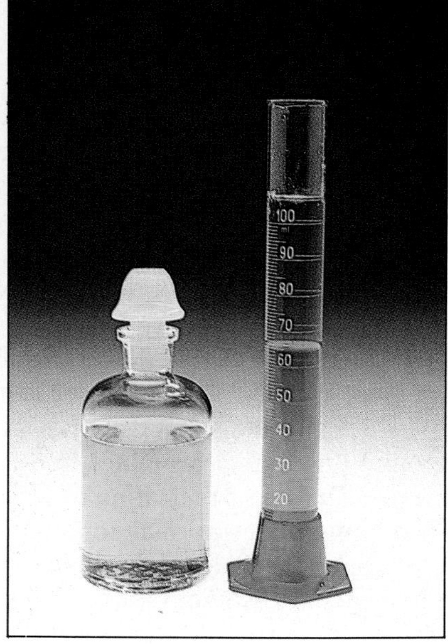

its concentration. One method is interesting because it involves some simple chemistry you have already seen.

$$I^-(aq) \xrightarrow{AgNO_3} AgI(s) \xrightarrow{Fe} Ag(s) + FeI_2(aq) \xrightarrow{Cl_2} FeCl_3(aq) + I_2$$

(Ag(s) is recycled with HNO_3 back to $AgNO_3$.)

The iodide ion is precipitated as the poorly soluble silver salt. This is then reduced with clean scrap iron to give silver and aqueous FeI_2. (Since silver is so expensive, it is recycled by oxidizing the metal with nitric acid.) The iron(II) iodide is then oxidized to iron(III) chloride and iodine.

World production of iodine is much smaller than that of the other halogens, but it finds many uses. Most is consumed in the manufacture of organic chemicals, but smaller amounts are used in animal-feed supplements, in ink pigments, and in making AgI for high-speed photographic films.

All the isotopes of the heaviest halogen, astatine, are radioactive. None has a long lifetime, so the element and its compounds occur in only trace amounts in nature.

GENERAL ASPECTS OF HALOGEN CHEMISTRY

Fluorine is the most reactive of all of the elements, forming compounds with all of the other elements with the exception of He, Ne, and Ar. In some cases, the elements combine directly in reactions that are so vigorous that they are explosive. This can be explained by at least two factors: the relatively weak F—F bond (low dissociation energy) compared with the other halogens, and the relatively strong bond formed

Bond Energies of Some Halogen Compounds (kJ/mol)

X	X—X	H—X	C—X (in CX_4)
F	159	569	456
Cl	243	431	327
Br	192	368	272
I	151	297	239

Figure 24.19 The preparation of iodine. A mixture of sodium iodide and manganese(IV) oxide was placed in a flask (left). On adding concentrated sulfuric acid (right), brown, gaseous I_2 was evolved.

$$2\ NaI(s) + 2\ H_2SO_4(aq) + MnO_2(s) \longrightarrow Na_2SO_4(aq) + MnSO_4(aq) + 2\ H_2O(\ell) + I_2(g)$$

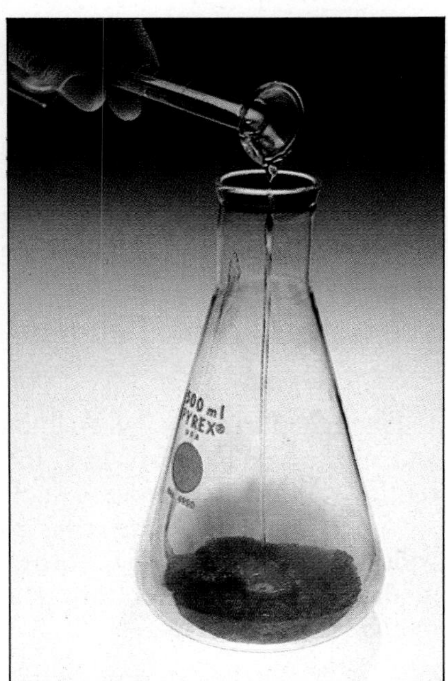

Table 24.9 The Hydrogen Halides

	HF	HCl	HBr	HI
mp (°C)	−83.4	−114.7	−88.6	−51.0
bp (°C)	19.5	−84.2	−67.1	−35.1
ΔH_f° (kJ/mol)	−271.1	−92.3	−36.4	26.5
ΔG_f° (kJ/mol)	−273.2	−95.3	−53.5	1.72
H—X bond energy (kJ/mol)	569	431	368	297
H—X bond length (pm)	91.7	127.4	141.4	160.9

between F and other elements. This general feature is illustrated by the bond energies in the table in the margin.

As you have already seen, F_2 is the best oxidizing agent of all of the halogens. This ability, combined with the small size of the fluorine atom, means that fluorine can form compounds with other elements where the other element has its highest oxidation number, as in IF_7, PtF_6, AgF_2, UF_6, and XeF_4.

HYDROGEN HALIDES The pure, anhydrous compounds HX are commonly called hydrogen halides (Table 24.9). Aqueous solutions of them, however, are usually referred to as the "hydrohalic acids." Of the four possible HX compounds, HF and hydrofluoric acid are produced in substantial amounts, and HCl and hydrochloric acid are industrial chemicals of major importance; HBr and hydrobromic acid are made on a much smaller scale, and there is little demand for HI and hydriodic acid.

The estimated worldwide production of *hydrogen fluoride* is about 1 million tons annually, virtually all by the action of concentrated H_2SO_4 on fluorspar, CaF_2.

$$CaF_2(s) + H_2SO_4(\ell) \longrightarrow CaSO_4(s) + 2\ HF(g)$$

The fluorspar must be quite pure and especially free of silica, since the latter reacts with HF to produce H_2SiF_6 in the following way.

$$SiO_2(s) + 4\ HF(g) \longrightarrow SiF_4(g) + 2\ H_2O(\ell) \xrightarrow{\text{aq HF}} H_2SiF_6(aq)$$

This series of reactions allows HF to etch or frost glass, and designs on glass are often made this way. This is also the reason that HF is never shipped or used in glass containers.

Most of the HF produced is used to make *cryolite*, since the mineral is necessary for making aluminum (page 884) but is found in only small amounts in nature.

$$6\ HF(aq) + Al(OH)_3(s) + 3\ NaOH(aq) \longrightarrow \underset{\text{cryolite}}{Na_3AlF_6(s)} + 6\ H_2O(\ell)$$

Large amounts of HF also go into the production of chlorofluorocarbons, and smaller amounts are used to make NaF (for water fluoridation and toothpaste) and to reprocess nuclear fuel.

Hydrogen chloride ranks 26th among industrial chemicals, worldwide production being on the order of 3 million tons annually. The classical method of making laboratory quantities of dry HCl gas is the *salt-cake* method (Figure 22.10).

$$2\ NaCl(s) + H_2SO_4(aq) \longrightarrow Na_2SO_4(s) + 2\ HCl(g)$$

Now, however, it is usually made by burning hydrogen in chlorine (if the reactants are available cheaply) (Figure 10.8),

Hydrogen also burns in a bromine atmosphere to give HBr (Figure 22.3).

$$H_2(g) + Cl_2(g) \longrightarrow 2\ HCl(g)$$

or as a by-product of the organic chemicals industry.

All the hydrohalic acids are corrosive. However, hydrofluoric acid and HF are especially so, and, if you ever work with them, you must be very careful. Serious symptoms of an HF burn are slow to develop, but they are particularly painful. The F^- ion removes Ca^{2+} from tissues and precipitates it as insoluble CaF_2. It is thought that this leaves an excess of K^+ ions in the affected tissue, and nerve stimulation occurs.

The H—F bond is highly polar, so hydrogen bonding is important to the properties of the compound (Chapter 13). For example, gaseous HF is an equilibrium mixture of single HF molecules and rings consisting of six HF molecules, while HF in the solid state is a zigzag polymer held together by H · · · F hydrogen bonds (Figure 24.20).

HCl, HBr, and HI are not strongly hydrogen bonded in the gaseous state, in contrast to HF.

All the anhydrous hydrogen halides are excellent reagents for the halogenation of metals, metal and nonmetal oxides, and hydrides. For example,

$$Mg(s) + 2\ HCl(g) \longrightarrow MgCl_2(s) + H_2(g)$$

Many of these reactions are thermodynamically spontaneous but kinetically slow and must be speeded up by heating or by adding a catalyst. Some useful predictions can be made concerning the general reaction of a metal with a hydrogen halide.

$$M(s) + nHX(g) \longrightarrow MX_n + \frac{n}{2} H_2(g)$$

Since the standard free energies of formation of M and H_2 are zero, the net free energy for the reaction is

$$\Delta G^\circ_{net} = \Delta G^\circ_f(MX_n) - n\Delta G^\circ_f[HX(g)]$$

Consulting Table 24.9, we see that ΔG°_{net} will be negative (and the reaction thermodynamically spontaneous) if

$\Delta G^\circ_f(MF_n)$ is $< -273n$ kJ/mol for the fluoride

$\Delta G^\circ_f(MCl_n)$ is $< -95n$ kJ/mol for the chloride

$\Delta G^\circ_f(MBr_n)$ is $< -54n$ kJ/mol for the bromide

$\Delta G^\circ_f(MI_n)$ is $< \sim 2n$ kJ/mol for the iodide

To make some predictions about halogenation reactions, you need only to consult tables of free energies of MX_n. It will then be clear that most metals will react spontaneously with most of the hydrogen halides. In the

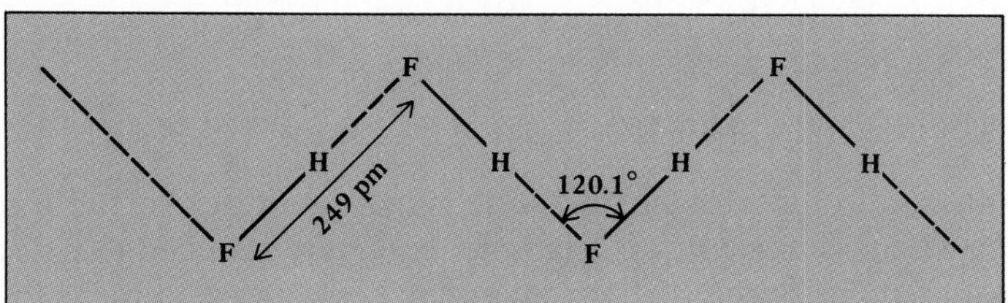

Figure 24.20 The structure of HF in the solid state. Hydrogen bonding leads to a chain of HF molecules.

cases of the alkali metals, the alkaline earths, and Zn, Al, and the lanthanides, the reactions are quite exothermic. In other cases, HCl, HBr, and HI should react (for instance, with silver to produce AgCl, AgBr, and AgI), but HF will not. Silicon should react with all but HI, but in practice only HF reacts at room temperature; the other reactions are kinetically slow. Remember: when predicting reactivity, you must take both thermodynamics and kinetics into account.

EXERCISE 24.8 Halogen Chemistry

(a) Name three uses each for fluorine and chlorine and their compounds.

(b) Is the reaction of Si(s) and $Cl_2(g)$ to give $SiCl_4(g)$ predicted to be thermodynamically spontaneous? ΔG_f° for $SiCl_4(g)$ is −617 kJ/mol.

HALOGEN OXYACIDS There are numerous oxyacids of the halogens, but only chlorine forms a complete series of such acids. They range from HOCl, where Cl has an oxidation number of +1, to $HClO_4$, where the Cl oxidation number is equal to the group number, +7.

ACID	NAME	ANION	NAME
HOCl	Hypochlorous	OCl^-	Hypochlorite
HOClO	Chlorous	ClO_2^-	Chlorite
$HOClO_2$	Chloric	ClO_3^-	Chlorate
$HOClO_3$	Perchloric	ClO_4^-	Perchlorate

Hypohalous acids, HOX, are known for all of the halogens. For fluorine, the unstable HOF is the only known oxyacid, and it was discovered only a few years ago. In contrast, HOCl was discovered over two centuries ago in the original work on chlorine. This and the other hypohalous acids are made by the hydrolytic disproportionation of the halogen.

$$X_2(g) + 2\,H_2O(\ell) \rightleftharpoons H_3O^+(aq) + X^-(aq) + HOX(aq)$$

Here, one halogen atom of X_2 is oxidized (to HOX) and the other is reduced (to X^-). Generally, this equilibrium process is carried out in *cold* alkaline solutions so the H_3O^+ ion is eliminated.

$$X_2(g) + 2\,OH^-(aq) \rightleftharpoons OX^-(aq) + X^-(aq) + H_2O(\ell)$$

When Cl_2 is treated with NaOH, the resulting alkaline solution is the "liquid bleach" used in home laundries, while solid $Ca(OCl)_2$ is the product when calcium hydroxide (slaked lime) is used. The latter compound is an easily handled form of a powerful oxidant, so it is sold for swimming pool sterilization.

A common brand of household bleach. The oxidizing properties of the solution come from the hypochlorite ion, OCl^-.

Hypohalites, the OX^- ions, are formed when the halogens disproportionate in cold alkaline solution. If the solution is *hot*, however, disproportionation leads to the halogen with an oxidation number of +5 and the product is a halate, a salt of a halic acid.

$$3\,X_2(g) + 6\,OH^-(aq) \longrightarrow XO_3^-(aq) + 5\,X^-(aq) + 3\,H_2O(\ell)$$

Sodium and potassium chlorates are made in huge quantities to fulfill a variety of needs. The sodium salt, for example, is largely reduced to ClO_2 for bleaching paper pulp. Some of the remainder is oxidized to perchlorates (see below) or is used as a herbicide. Finally, some $NaClO_3$ is converted to potassium chlorate, the preferred oxidizer in fireworks (page 924) and a crucial component in the head of safety matches.

Nearly 12,000 tons of $KClO_3$ are consumed annually in the United States to make matches.

The most stable halogen compounds are those where the halogen has the lowest (−1) or highest (+7) possible oxidation number. Consequently, perchlorates (ClO_4^-) are the most stable oxychlorine compounds, although they remain powerful oxidants. Pure perchloric acid, $HClO_4$, is a colorless liquid that will explode if shocked. It explosively oxidizes organic materials and even rapidly oxidizes silver and gold. More dilute aqueous solutions of the acid, however, have very little oxidizing power.

Perchlorate salts of most metals in the periodic table are known. Although many are relatively stable, others are dangerously unpredictable and *great care should be used when handling any perchlorate salt.* Ammonium perchlorate, for example, bursts into flame if heated above 200 °C.

$$2\ NH_4ClO_4(s) \longrightarrow N_2(g) + Cl_2(g) + 2\ O_2(g) + 4\ H_2O(g)$$

This property of the ammonium salt accounts for its use as the oxidizer in the solid booster rockets for the Space Shuttle. The solid propellent in these rockets is largely NH_4ClO_4, the remainder being the reducing agent, powdered aluminum. Each shuttle launch requires about 700 tons of ammonium perchlorate, so more than half of the sodium perchlorate manufactured is converted to the ammonium salt. The process for doing this is a simple exchange reaction (Chapter 5)

$$NaClO_4(aq) + NH_4Cl(aq) \longrightarrow NaCl(aq) + NH_4ClO_4(s)$$

that takes advantage of the fact that ammonium perchlorate is less soluble in water than sodium perchlorate.

The shuttle Columbia is mated to the orange-brown external fuel tank (center) and solid-fuel booster rockets (white). The fuel in the solid boosters is a mixture of ammonium perchlorate and aluminum (NASA).

24.4 THE CHEMISTRY OF THE RARE GASES

At the right side of the periodic table is the group of elements that all exist as monatomic gases. Helium, the lightest element of the group, is certainly not rare: it is the second most abundant element in the universe after hydrogen. However, it is rare on earth, because the gravitational field of our planet is too small to retain this light gas. Altogether, the gases of the group make up about 1% of the earth's atmosphere, so we have come to call them the "rare gases." Alternatively, they have been called the "inert gases," but, since the discovery of xenon compounds in 1962, only the lighter elements are considered truly inert.

In the 1890s, Lord Raleigh measured the density of some common gases and found that a sample of N_2 from the air had a slightly different density than N_2 prepared from ammonia. When the N_2 was removed from the sample of "atmospheric nitrogen" by forming magnesium nitride,

$$3\ Mg(s) + N_2(g) \longrightarrow Mg_3N_2(s)$$

Raleigh and William Ramsay found a very small amount of a much denser, monatomic gas that they named *argon* from the Greek word for "lazy," an appropriate name for a chemically inert gas. Soon thereafter Ramsay suggested that this element and the recently discovered element helium were members of an entirely new group in the periodic table.

William Ramsay (1842–1919). With Lord Rayleigh, Ramsay made the first discovery of a rare gas, argon (1894). In the next four years Ramsay also discovered helium, neon, krypton, and xenon. (AIP Niels Bohr Library)

Table 24.10 Composition of Dry Air

Gas	Percent by Volume
N_2	78.03
O_2	20.99
Ar	0.93
CO_2	0.03
Ne	0.0015
H_2	0.0010
He	0.0005
Kr	0.0001
Xe	0.000008

Argon is the most abundant Group 8A element in the atmosphere, but the other rare gases are present in minute quantities (Table 24.10). About 250,000 tons of argon are recovered annually from air in the United States, and the gas is used primarily as an inert atmosphere for high temperature metallurgical processes. Small amounts of neon, krypton, and xenon are also recovered from air, and neon especially is used to fill the "neon" signs used in advertising. Xenon has an interesting potential medical use. It is readily soluble in blood and acts as an inhalation anesthetic in much the same way as dinitrogen oxide or "laughing gas" (N_2O).

Helium is usually obtained from natural gas wells in which it is found in concentrations up to 7% by volume. Among other things, helium is used for inert atmospheres and, along with oxygen, as a gas in deep-sea diving gases (see Chapter 12). Its normal boiling point is only 4.2 K, so liquid helium is the coldest liquid refrigerant available and is used to study very low temperature phenomena.

The discovery of the first rare gas compound provides us with the lesson that the most interesting discoveries are often the result of lucky accidents and of being prepared to interpret an observation in a new way. While studying the chemistry of PtF_6, Neil Bartlett noticed, quite accidentally, that exposing it to air led to a compound that he showed to be $[O_2^+][PtF_6^-]$. Just as important, Bartlett recognized that the ionization energy of O_2 (to form O_2^+) was comparable to that of Xe. Thus, he quickly proceeded to treat PtF_6 with Xe and discovered $Xe^+PtF_6^-$, the first rare gas compound. Very soon after this work, other xenon fluorides were discovered, and the field has been an active one since. Isolable compounds have been obtained only for Xe and Kr, but were it not for the intense radioactivity of Rn, compounds of this element would surely be isolated. Virtually all the known compounds contain bonds to F or O, but a few exist with bonds to C, Cl, and N.

Figure 24.21 Crystals of xenon tetrafluoride, XeF_4. Its discovery was a revolutionary development in chemistry, since it had been thought for many years that the rare gases were chemically inert. (Argonne National Laboratory)

Among the best known rare gas compounds are the first discovered, the simple xenon fluorides, XeF_2, XeF_4, and XeF_6 (Figure 24.21). All are colorless, volatile solids that can be prepared by combining the elements under carefully controlled conditions. It is ironic, in view of the historical idea of the inertness of these elements, that XeF_2 can be obtained simply by exposing a mixture of Xe and F_2 to sunlight.

$$Xe(g) + F_2(g) \xrightarrow{\text{sunlight}} XeF_2(s)$$

All of the xenon fluorides are good fluorinating agents.

$$XeF_4(g) + 2\ SF_4(g) \longrightarrow 2\ SF_6(g) + Xe(g)$$

The difluoride is stable in water, but the tetrafluoride and hexafluoride are both hydrolyzed to the dangerously explosive compound XeO_3.

$$XeF_4(aq) + 2\ H_2O(\ell) \longrightarrow \tfrac{1}{3}\ XeO_3(s) + \tfrac{2}{3}\ Xe(g) + \tfrac{1}{2}\ O_2(g) + 4\ HF(aq)$$

XeF_2	XeF_4
mp, 129 °C	mp, 117.1 °C
Xe—F = 200 pm	Xe—F = 195.2 pm
:F—Xe—F:	(F)₄Xe structure
linear	square planar
Xe = sp^3d	Xe = sp^3d^2

EXERCISE 24.9 Xenon Chemistry

Using the ideas of VSEPR (Chapter 10), explain why XeF_2 has a linear molecular structure.

SUMMARY

The elements of **Group 5A** all have the configuration ns^2np^3, so oxidation numbers of the elements can range from −3 to +5. Only nitrogen is found uncombined in nature. Phosphorus is found in the form of *orthophosphates*, and the others (As, Sb, and Bi) occur as oxides and sulfides. Elemental phosphorus has three *allotropic forms* denoted by the color of the solid: white (or yellow), red, and black. White phosphorus molecules are a tetrahedron of P atoms.

All Group 5A elements form hydrogen compounds of the type EH_3, of which *ammonia*, NH_3, is most important commercially. Ammonia is produced in the *Haber process*, the direct, catalyzed combination of N_2 and H_2.

Unlike the extensive *catenation* observed for boron and carbon, N—N bonding occurs only to a limited extent. The simplest N—N bonded compound, hydrazine (N_2H_4), is produced in the *Raschig process*. It is an excellent reducing agent.

Seven binary N—O compounds are known (Table 24.4), among them "laughing gas," N_2O, and the brown gas nitrogen dioxide, NO_2. The most common phosphorus oxide is P_4O_{10}, while the empirical formula E_2O_3 is typical of As, Sb, and Bi.

As expected, nitrogen and phosphorus oxides are acidic, arsenic and antimony oxide are amphoteric, and bismuth oxide, Bi_2O_3, is basic. Nitric acid, HNO_3, is produced by the oxidation of ammonia in the *Ostwald process*, while orthophosphoric acid, H_3PO_4, is produced by hydrolysis of P_4O_{10}, the latter a product of oxidation of P_4. Orthophosphoric acid has many uses.

Group 6A elements, with the electron configuration ns^2np^6, are almost all nonmetals. Oxygen has two allotropic forms, O_2 and O_3 (ozone). Of the allotropes of sulfur, the most common is an eight-membered ring of sulfur atoms. Elemental sulfur is obtained from deep mines or by oxidation of hydrogen sulfide. Sulfur has many uses (Figure 24.12); one of the most important of these is its conversion to sulfuric acid, the chemical produced in largest amount in industrialized countries.

Oxygen forms compounds directly or indirectly with virtually all of the elements (except for some of the rare gases). Most are oxides; some *peroxides* (O_2^{2-}) and *superoxides* (O_2^-) are also known. The most important peroxide is H_2O_2, a powerful oxidizing agent.

Many metals readily form sulfides (Figure 24.14), compounds of great commercial interest. Other important sulfur compounds are hydrogen sulfide (H_2S), SO_2, and SO_3. Sulfur dioxide is the anhydride of sulfurous acid, H_2SO_3, and the trioxide is the anhydride of sulfuric acid, H_2SO_4.

The halogens of **Group 7A** have the electron configuration ns^2np^5 and all form the halide ion, X^- (oxidation number = −1). None of the elements occurs naturally. Both fluorine and chlorine are prepared by electrolysis. Bromine and iodine, however, are prepared by adding Cl_2 (an oxidizing agent) to an aqueous solution of Br^- or I^- (a reducing agent).

$$Cl_2(aq) + 2\,Br^-(aq) \longrightarrow 2\,Cl^-(aq) + Br_2(aq)$$

Fluorine, in the form of F_2, is the most reactive of all of the elements; it forms compounds with all of the elements except He, Ne, and Ar.

Among the best known halogen compounds are the hydrogen halides, HX. All are gases at room temperature and pressure, but strong hydrogen bonding gives HF a much higher boiling point than the others. A number of oxygen–halogen compounds are known, but among the most familiar are oxyacids such as HOCl (hypochlorous acid) and $HClO_4$ (perchloric acid).

The lighter rare gases of **Group 8A** are chemically unreactive, but compounds of the heavier ones are known. Xenon in particular forms chemical compounds with highly electronegative elements such as F and O.

STUDY QUESTIONS

REVIEW QUESTIONS

1. Many chemical reactions or industrial processes bear the name of the person who discovered the process or who worked out the details. Write balanced equations and give the reaction conditions for the following named processes: Haber, Marsh, and Raschig. What is the "contact process"?

2. Complete and balance the following equations:
(a) $Mg(s) + N_2(g) \longrightarrow$
(b) $P_4(s) + KOH(aq) + H_2O(\ell) \longrightarrow$
(c) $CH_4(g) + H_2O(g) \longrightarrow$
(d) $HNO_3(aq) + KOH(aq) \longrightarrow$
(e) $NH_3(aq) + OCl^-(aq) \longrightarrow$
(f) $NH_4NO_3(s)$ + heat $\longrightarrow$
(g) $NaNO_2(aq) + HCl(aq) \longrightarrow$
(h) $NH_3(g) + O_2(g) \longrightarrow$
(i) $Cu(s) + HNO_3(aq) \longrightarrow$
(j) $P_4O_{10}(s) + H_2O(\ell) \longrightarrow$

3. Write a balanced equation to show the preparation of each of the following:
(a) $NO_2(g)$
(b) $N_2O_5(s)$
(c) $HNO_2(aq)$
(d) $AsH_3(g)$
(e) $NF_3(g)$
(f) $CCl_2F_2(g)$
(g) "laughing gas," dinitrogen oxide
(h) hydrazine

4. A major use of hydrazine, N_2H_4, is in steam boilers in power plants. (a) The reaction of hydrazine with O_2 dissolved in water gives N_2 and water. Write a balanced equation for this reaction. (b) O_2 dissolves in water to the extent of 3.08 cm³ (gas at STP) in 100. mL of water at 20 °C. In order to destroy all of the dissolved O_2 in 3.00×10^4 L of water (enough to fill a small swimming pool), how many grams of N_2H_4 are needed?

5. Before hydrazine came into use to remove dissolved oxygen in the water in steam boilers, Na_2SO_3 was commonly used. Assuming the appropriate reaction is

$$Na_2SO_3(aq) + \tfrac{1}{2}\,O_2(aq) \longrightarrow Na_2SO_4(aq)$$

how many grams of Na_2SO_3 would be required to remove O_2 from 30,000 L of water as outlined in Study Question 4?

6. The various steps in the industrial synthesis of nitric acid are given on page 987, and the overall reaction is

$$NH_3(g) + 2\,O_2(g) \longrightarrow HNO_3(aq) + H_2O(\ell)$$

Calculate $\Delta G°$ for the reaction and then the equilibrium constant, at 25 °C.

7. A common analytical method for hydrazine involves its oxidation with iodate, IO_3^-. In the process hydrazine behaves as a four-electron reducing agent.

$$N_2(g) + 5\,H_3O^+(aq) + 4\,e^- \longrightarrow N_2H_5^+(aq) + 5\,H_2O(\ell) \qquad E° = -0.23\ V$$

Write the balanced equation for the reaction of hydrazine in acid solution ($N_2H_5^+$) with IO_3^- to give N_2 and I_2. Calculate $E°$ for this reaction.

8. If the lunar lander on the moon missions used 4.5 tons of a hydrazine derivative, $H_2N—N(CH_3)_2$ (page 983), how many tons of N_2O_4 are required to react with it? How many tons of each of the products of this reaction are generated?

9. Unlike carbon, which can form extended chains of atoms, nitrogen can form chains of very limited length. Draw the Lewis electron dot structure of the azide ion, N_3^-.

10. The cyanide ion, CN^-, can be oxidized to the cyanate ion, NCO^-. Compounds of this ion are quite stable, in contrast to those of the fulminate ion, CNO^-. Mercury(II) fulminate, $Hg(CNO)_2$, for example, explodes when struck and is used in blasting caps. Show the Lewis electron dot structures for the cyanate and fulminate ions. (Show all resonance structures.) For each ion, calculate the formal charge on each atom in each resonance structure and decide which is the most important resonance structure. (See Section 10.5 for a discussion of formal charges.)

11. The boiling points of the EH_3 molecules of Group 5A are given in Table 24.3 on page 980. Explain the observed trend in these temperatures.

12. Dinitrogen trioxide, N_2O_3, has the structure shown below.

O
114.2 pm
N——N
121 pm
O
O
130°

The oxide is unstable, dissociating to NO and NO_2 in the gas phase at 25 °C.

$$N_2O_3(g) \rightleftharpoons NO(g) + NO_2(g)$$

(a) Explain why one N—O bond distance in N_2O_3 is 114.2 pm, while the other two are shorter and nearly equal to one another.
(b) For the dissociation reaction, $\Delta H° = 40.5$ kJ/mol and $\Delta G° = -1.59$ kJ/mol. Calculate $\Delta S°$ and K for the reaction, and $\Delta H_f°$ for N_2O_3.

13. The structure of HNO_3 is given on page 988. Explain why one NO bond is longer than the other two (which have the same bond length).

14. When trade journals discuss the amount of phosphoric acid produced or used annually, they always do so in terms of equivalent P_2O_5 (the empirical formula of P_4O_{10}). Recall that P_2O_5 hydrolyzes to H_3PO_4, so the mass of anhydrous acid can be obtained by multiplying the given phosphorus content by the factor 1.38.
(a) Show why the factor relating the mass of P_2O_5 to that of anhydrous H_3PO_4 is 1.38.
(b) Estimated 1988 production of phosphoric acid (measured as P_2O_5) was 11.7 million tons. How many tons of H_3PO_4 does this represent?
(c) If the prices are listed as \$300/ton of P_2O_5, what is the price of 2.00×10^3 pounds (1.00 ton) of anhydrous H_3PO_4?

15. What is the approximate pH of a 0.010 M solution of Na_3PO_4?

16. Phosphonic acid (Table 24.5) has phosphorus with an oxidation number of +3. Draw the structure of diphosphonic acid, $H_4P_2O_5$. What is the maximum number of protons that this acid can dissociate in water?

17. $CaHPO_4$ is used as an abrasive in toothpaste. Write a balanced equation showing one possible preparation for this salt.

18. Sketch the structures of NH_3 and NF_3 and compare them on the following basis:
(a) Each molecule has a dipole moment. In each molecule, toward which end does the negative portion of the dipole lie?
(b) Give a possible explanation for the fact that the dipole moment of NH_3 is 1.47 D whereas that of NF_3 is only 0.234 D.

19. When base is added to a water suspension of P_4, several reactions are possible, and one of them produces the ion $H_2PO_2^-$ and the gas PH_3 (phosphine). Write a balanced equation to depict this reaction.

20. On page 992 it was noted that $H_2PO_2^-$ salts are excellent reducing agents in basic solution and can be used for the electrodeless plating of metals. Will the ion reduce a 1 M solution of a zinc salt to zinc metal in basic solution? What about 1 M Cr^{3+} to Cr? Write a balanced equation for each process and predict the net $E°$ for each reaction. (For the necessary potentials see Appendix I.)

21. Arsenic burns in air to give As_4O_6, a poisonous material with a garlic-like odor.
(a) By analogy with phosphorus chemistry, draw an approximate structure for this molecule.
(b) Again by analogy with phosphorus chemistry, predict the product of hydrolysis of As_4O_6. Draw an approximate structure of the product, and give the oxidation number of the arsenic in the product.

22. Elemental arsenic is attacked by dilute nitric acid to give H_2 and arsenious acid, H_3AsO_3. Write a balanced equation for this oxidation–reduction process.

23. Complete and balance the following equations:
(a) $KClO_3(s) + \text{heat} \longrightarrow$
(b) $H_2S(g) + O_2(g) \longrightarrow$
(c) $Na(s) + O_2(g) \longrightarrow$
(d) $K(s) + O_2(g) \longrightarrow$
(e) $ZnS(s) + O_2(g) \longrightarrow$
(f) $SO_2(g) + H_2O(\ell) \longrightarrow$

24. Write a balanced equation for the preparation of each of the following:
(a) hydrogen peroxide
(b) potassium superoxide
(c) sodium thiosulfate
(d) sulfuric acid

25. Oxygen is an excellent oxidizing agent; $E°$ for the following reaction is 1.229 V.

$$O_2(g) + 4\,H_3O^+(aq) + 4\,e^- \longrightarrow 6\,H_2O(\ell)$$

(a) What is E when the pH is 1.00 (and $P_{oxy} = 1.0$ atm)?
(b) What is E when the partial pressure of O_2 is the value normally found in air? (Use $P_{oxy} = 0.22$ atm and pH = 1.00.)

26. O_2 has a relatively high bond energy, but it is a reactive molecule. Account for its great reactivity.

27. The structure of H_2O_2, hydrogen peroxide, is not planar. Suggest a possible reason for this observation.

28. Hydrogen peroxide is a weak acid in aqueous solution. Write a balanced chemical equation showing how H_2O_2 can function as a Brønsted acid in water. If $K_a = 1.78 \times 10^{-12}$ (for the loss of one H^+), what is the approximate pH of a 0.10 M aqueous solution of the peroxide?

29. The standard heat of formation of OF_2 gas is 24.5 kJ/mol. Calculate the average O—F bond energy.

30. Place the following oxides in order of increasing basicity: As_2O_3, Ga_2O_3, GeO_2, BrO_2, CaO, and K_2O.

31. Which is more basic, BeO or CaO? Write a balanced chemical equation for the hydrolysis of SrO.

32. In addition to the simple sulfide ion, S^{2-}, there are polysulfides, S_n^{2-}. (Such ions are chains of sulfur atoms, not rings.) Draw a Lewis electron dot structure of the S_3^{2-} ion.

33. Sulfur forms a range of compounds with fluorine. Draw Lewis electron dot structures for S_2F_2 (connectivity is FSSF), SF_2, SF_4, and SF_6. What is the formal oxidation number of sulfur in each of these compounds?

34. In the modern "contact process" for making sulfuric acid, sulfur is first burned to SO_2. Environmental restrictions allow no more than 0.3% of this SO_2 to be vented to the atmosphere.
(a) If enough sulfur is burned in a plant to produce 2.00×10^3 tons of pure, anhydrous H_2SO_4 per day, how much SO_2 is allowed to be exhausted to the atmosphere?
(b) One way to prevent even this much SO_2 from reaching the atmosphere is to "scrub" the exhaust gases with hydrated lime.

$$Ca(OH)_2 + SO_2 \longrightarrow CaSO_3 + H_2O$$

$$2\ CaSO_3 + O_2 \longrightarrow 2\ CaSO_4$$

How many tons of $Ca(OH)_2$ are needed to remove the SO_2 calculated in part (a)?

35. A sulfuric acid plant produces an enormous amount of heat. To keep costs as low as possible, much of this heat is used to make steam to generate electricity. Some of the electricity is used to run the plant, and the excess is sold to the local electrical utility. Three reactions are important in sulfuric acid production: (1) burning S to SO_2; (2) oxidation of SO_2 to SO_3; and (3) hydrolysis of SO_3.

$$SO_3(g) + H_2O\ (\text{in } 98\%\ H_2SO_4) \longrightarrow H_2SO_4(\ell)$$

If the enthalpy change of the third reaction is -130 kJ/mol, estimate the total heat produced per mole of H_2SO_4 produced. How much heat is produced per ton of H_2SO_4?

36. Complete and balance the following equations:
(a) $UF_4(s) + F_2(g) \longrightarrow$
(b) $Br^-(aq) + Cl_2(aq) \longrightarrow$
(c) $I^-(aq) + Br_2(aq) \longrightarrow$
(d) $Br^-(aq) + AgNO_3(aq) \longrightarrow$
(e) $Cl_2(g) + OH^-(aq) \longrightarrow$

37. Write a balanced equation for the preparation of each of the following:
(a) HF(g)
(b) HCl(g)
(c) xenon tetrafluoride

38. If aqueous Br_2 is mixed with an aqueous solution of KI, what reaction will occur? Write a balanced chemical equation.

39. What is the value of $E°$ for the reaction between $Cl_2(aq)$ and $I^-(aq)$?

40. (a) Name three mineral sources of fluoride ion.
(b) What is the usual commercial source of the iodide ion?

41. If an electrolytic cell for producing F_2 (Figure 24.16) operates at 5.00×10^3 amps (at 10.0 V), how many tons of F_2 can be produced per 24 hour day if the conversion of F^- to F_2 is assumed to be 100% efficient?

42. The annual U.S. and Canadian fluorine production is 5.00×10^3 tons (one ton is 2.00×10^3 pounds), and 55% of this is used to manufacture UF_6. How much uranium(VI) fluoride is manufactured every year (assume 100% efficiency in using the F_2)?

43. The F—F bond energy (page 1013) is among the weakest of the halogens. Suggest a possible reason for this.

44. Suggest two reasons for the observation that F_2 is the most reactive of the halogens.

45. Halogens combine with one another to form *interhalogens* such as BrF_3. Sketch a possible molecular structure for this molecule and tell whether the observed F—Br—F bond angles will be less than or greater than ideal.

46. Why is the ClO_2 molecule paramagnetic? Sketch its electron dot structure.

47. Metal halides are soluble in water, the relative solubility order generally being $MF_n < MCl_n < MBr_n < MI_n$ (where M = K^+ or Ca^{2+}, for example). The melting points of these same halides increase in the opposite order. Suggest a common reason for these observations.

48. Halogens form polyhalide ions. Sketch Lewis electron dot structures, and molecular structures for (a) I_3^-, (b) $BrCl_2^-$, (c) ClF_2^+.

49. The halogen oxides and oxyanions are generally good oxidizing agents. For example, the bromate ion oxidizing half reaction has an $E°$ value of 1.44 V in acid solution.

$$BrO_3^-(aq) + 6\ H_3O^+(aq) + 6\ e^- \longrightarrow Br^-(aq) + 9\ H_2O(\ell)$$

Can you oxidize aqueous 1 M Mn^{2+} to aqueous MnO_4^- with 1 M bromate ion?

50. The hypohalite ions, OX^-, are the salts of weak acids. Calculate the pH of a 0.10 M solution of NaOCl. What is the concentration of HOCl in this solution?

51. Ammonium perchlorate, NH_4ClO_4, is used as an oxidizer in launching the Space Shuttle. If one launch requires 700. tons of the salt (one ton is 9.08×10^5 grams), and the salt decomposes according to the equation on page 1017, how many tons of water are produced? How many grams of O_2 are produced? If the O_2 produced is assumed to react with the powdered aluminum present in the rocket engine, how much Al is necessary to use up the O_2 and how much Al_2O_3 is formed in the Al/O_2 reaction?

52. Consider the reaction of a metal with a hydrogen halide (page 1015).

(a) Which of the following metals are predicted to react spontaneously with HCl: Ba, Pb, Hg, or Ti?

(b) Will HF react spontaneously with Ca?

(c) Will Ni react spontaneously with HF (ΔG_f° of NiF_2(s) is −604 kJ/mol)?

(d) Will Ni react spontaneously with HBr (ΔG_f° of $NiBr_2$(s) is −253.6 kJ/mol)?

(e) Will B react spontaneously with HI (ΔG_f° of BI_3(g) is +20.7 kJ/mol)?

53. Of the rare gases, xenon is the only one that forms an extensive series of compounds.

(a) Sketch the Lewis electron dot structure of XeF_4. What is its molecular structure?

(b) Sketch a Lewis electron dot structure for the ion XeF_3^+. What is its molecular structure?

54. XeF_6 hydrolyzes to form $XeOF_4$ and HF.

(a) Write a balanced chemical equation for the hydrolysis reaction.

(b) Sketch the electron dot and molecular structure of $XeOF_4$. (*Hint:* You can think of this molecule as resulting from the addition of an O atom to XeF_4.)

55. XeO_3 is a powerful oxidizing agent in acid solution.

$$XeO_3(aq) + 6\ H_3O^+(aq) + 6\ e^- \longrightarrow Xe(g) + 9\ H_2O \qquad E° = 2.10\ V$$

Is it capable of oxidizing Mn^{2+} to MnO_4^- under standard conditions? Can it oxidize NaF to F_2? (See Appendix I.)

CHAPTER

25

The Transition Elements and Their Chemistry

Citrine, a form of quartz (SiO_2) contaminated with iron oxides.

CHAPTER OUTLINE

The large block of elements in the central portion of the periodic table is a bridge between the *s*-block elements at the left and the *p*-block metals, metalloids, and nonmetals on the right (Figure 25.1). The first three rows of these elements (Sc to Zn, Y to Cd, and La to Hg) are generally called the **transition elements** or **transition metals**. All these elements count *d* electrons among their valence electrons, and so we can call them the ***d*-block elements**. The elements fitting into the periodic table between La and Hf, and between Ac and element 104, are the inner transition elements. Here *f* electrons are involved as valence electrons, so they can be called ***f*-block elements**. More specifically, the elements coming after lanthanum (La) and before hafnium (Hf) are called the **lanthanide elements**, and those following actinium (Ac), but coming before element 104, are the **actinide elements**. Although you may see all the elements referred to simply as "transition elements," we shall use this term only for the *d*-block elements.

For historical reasons, the lanthanide elements are sometimes called the "rare earth elements." However, many are actually not rare. For example, Ce is five times as abundant as lead and only about half as abundant as Cl.

The *d*-block elements in particular are important in living organisms, in certain industrial processes, and as materials. For example, cobalt is the crucial element in vitamin B_{12}, where its role is to act as a catalyst. Iron is the key element in biochemical oxidation–reduction processes using hemoglobin or myoglobin (Chapter 27). Molybdenum and iron, together with sulfur, form the reactive portion of nitrogenase, a

1A	2A	3B	4B	5B	6B	7B	8B			1B	2B	3A	4A	5A	6A	7A	8A
H																	He
Li	Be											B	C	N	O	F	Ne
Na	Mg											Al	Si	P	S	Cl	Ar
K	Ca	Sc	Ti	V	Cr	Mn	Fe	Co	Ni	Cu	Zn	Ga	Ge	As	Se	Br	Kr
Rb	Sr	Y	Zr	Nb	Mo	Tc	Ru	Rh	Pd	Ag	Cd	In	Sn	Sb	Te	I	Xe
Ca	Ba	*La	Hf	Ta	W	Re	Os	Ir	Pt	Au	Hg	Tl	Pb	Bi	Po	At	Rn
Fr	Ra	†Ac															

* Lanthanides	Ce	Pr	Nd	Pm	Sm	Eu	Gd	Tb	Dy	Ho	Er	Tm	Yb	Lu
† Actinides	Th	Pa	U	Np	Pu	Am	Cm	Bk	Cf	Es	Fm	Md	No	Lr

Figure 25.1 Periodic table showing the *d*-block or transition elements (red) and the *f*-block elements, the lanthanides (blue) and the actinides (yellow).

biological catalyst used by plants to convert atmospheric nitrogen into ammonia, and copper and zinc are important in other biological catalysts. As catalysts in nonbiological reactions, you have already seen iron used in the Haber process for ammonia, and rhodium and platinum in automobile catalytic converters.

Many transition metal compounds are highly colored, which makes them useful as pigments in paints and dyes. Prussian blue, $Fe_4[Fe(CN)_6]_3$, is the "bluing agent" in laundry bleach and in engineering blueprints. When transition metal ions are present in crystalline silicates or alumina, the minerals become gems. For example, rubies contain Cr^{3+} substituted for some of the Al^{3+} ions in a crystal lattice of Al_2O_3 (page 348); the transition metal ion gives the red color to the material and is the reason rubies can function as lasers. Blue glass is made by adding a small amount of a Co^{2+} salt, and the green patina on copper statues and roofs is an oxidized form of copper.

Prussian blue, $Fe_4[Fe(CN)_6]_3$, is a deep blue compound known for hundreds of years. Here it was formed by the reaction of Fe^{3+}(aq) with $[Fe(CN)_6]^{4-}$(aq). However, the earliest recipe used ox-blood in an iron kettle.

The most obvious use of transition elements is as the metals themselves (Figure 2.4). They are used in coins in many countries around the world and are the primary structural materials in cars and appliances. Approximately 700 million tons of raw steel, 8 million tons of copper, and 750,000 tons of nickel are produced annually on a worldwide basis to meet these needs.

25.1 PHYSICAL PROPERTIES OF THE TRANSITION METALS

To organize the chemistry of the *d*- and *f*-block elements, it is useful to know their electron configurations and their commonly observed oxidation numbers. Recall from Chapter 9 that most of these elements have

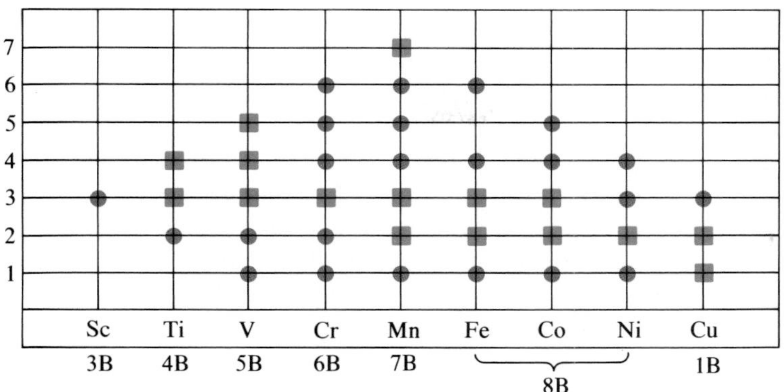

Figure 25.2 Oxidation numbers for the fourth period transition metals (except Zn). More common ones are designated in red.

incomplete *d* or *f* subshells, so it is possible that many different positive oxidation numbers can result from loss of a varying number of electrons.

Cobalt glass is made by adding a small amount of a cobalt(II) salt to molten glass.

OXIDATION NUMBERS

Oxidation numbers of +2 and +3 are commonly observed for the *d*-block elements, but others are certainly possible (Figures 25.2 and 25.3). For example, common minerals of titanium are TiO_2 (rutile) and ilmenite ($FeTiO_3$, see page 1028) in which titanium has an oxidation number of +4; the most common oxidation number for silver is +1, and copper often forms compounds of Cu^+. On the high end of the oxidation number scale, you have already encountered +7 in an ion such as MnO_4^- (permanganate). The maximum possible oxidation number is +8, but it is rarely seen; perhaps the best example is OsO_4, a powerful oxidizing agent but a very toxic compound.

The lanthanide elements are also commonly found with +2 and +3 oxidation numbers in their compounds, but +4 is important for some,

Figure 25.3 Some common salts of chromium, illustrating oxidation numbers +3 [in $Cr(NO_3)_3$ (violet) and $CrCl_3$ (green)] and +6 [in K_2CrO_4 (yellow) and $K_2Cr_2O_7$ (orange)].

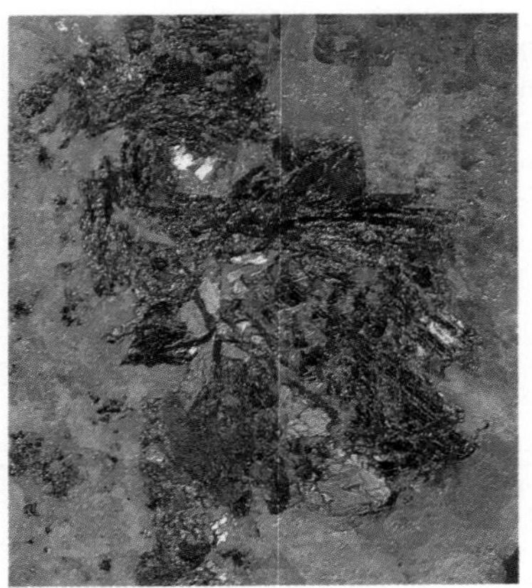

Ilmenite, $FeTiO_3$, is a common titanium-containing mineral. It is seen here as the metallic-looking pieces in a matrix of other material.

including cerium. The actinide elements, however, generally have higher oxidation numbers such as +4 and even +6. For example, UO_2 is a common oxide of uranium, and a compound important in reprocessing uranium fuel for nuclear reactors is UF_6 (Section 24.3).

In main group metal chemistry, the highest possible oxidation number is equal to the group number, and these are generally observed (Na^+ and Al^{3+}, for example). In transition metal chemistry it is still true, in principle, that the maximum oxidation number is given by the group number. However, transition metal ions usually do not exhibit the maximum possible number in their compounds, especially for elements of the later groups (Figure 25.2). When high oxidation numbers are found, it is usually when the metal is combined with highly electronegative oxygen or fluorine.

ELECTRON CONFIGURATIONS

The electron configurations of the *d*- and *f*-block elements and their ions were described in Sections 9.4 and 9.5. The five *d* orbitals or the seven *f* orbitals are degenerate in the uncombined atom. Hund's rule means that these orbitals are exactly half filled before electrons are paired in any one orbital. Therefore, there can be up to five unpaired electrons in a *d* subshell; Mn^{2+} and Fe^{3+} are common examples. In the *f*-block elements, there can be as many as seven unpaired electrons in the *f* subshell, and the Eu^{2+} and Gd^{3+} ions are good examples.

See Table 9.2 for electron configurations of the d-block and f-block elements.

Because there are unpaired electrons in most common oxidation states, ions of the *d*-block and *f*-block elements are typically **paramagnetic** (Section 9.1), and the metals themselves can be ferromagnetic. Recall that electrons can act as subatomic bar magnets. If, in a solid metal, the atomic magnets of a group of atoms (called a *domain*) cooperate and orient their magnets in the same direction, the magnitude of the magnetic effect is much larger than paramagnetism. Only the metals of the iron, cobalt, and nickel subgroups exhibit this property. They are also unique in that, once the electron magnets are aligned by an external magnetic field, the metal is permanently magnetized. In such a case the magnetism can be eliminated only by heating or vibrating the metal to rearrange the electron spin domains.

The difference between paramagnetic and ferromagnetic materials is that domains of aligned magnets do not form in paramagnetic substances. Ferromagnets are thus "super-paramagnets." See Chapter 9.

Alnico magnets, which are described below, contain Al, Co, and Ni, as well as iron and copper.

Another unique property of transition metal ions is that it is rare for the ion to follow the "octet rule" in forming compounds. The octet rule was developed for main group elements with four valence atomic orbitals that, when assigned lone and bond pairs, can account for no more than eight electrons. For transition metal ions, the minimum number of valence orbitals is five and, if the next higher *s* and *p* orbitals are counted as valence orbitals, the number is nine. Metals like Ni^{2+} ($3d^8$ electron configuration), Cu^{2+} ($3d^9$), and Zn^{2+} ($3d^{10}$) meet or exceed the "octet rule" before they form their first bond to another atom. Consequently, transition metal ions routinely have expanded octets.

METAL ATOM RADII AND DENSITY

The periodic trend in the radii of transition metal atoms is illustrated in Figure 25.4. Here you see that the sizes of the metal atoms change very little across a series, especially beginning at Group 5B (V, Nb, or

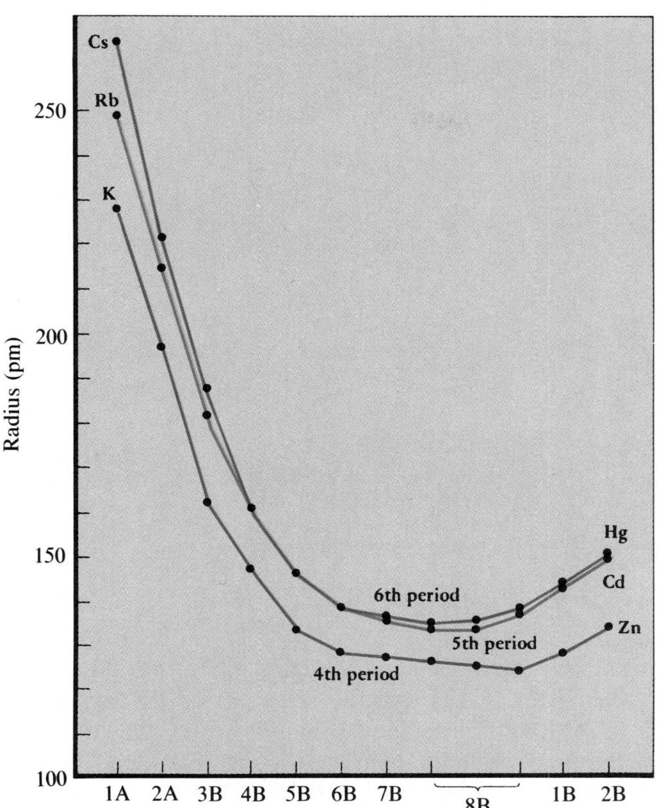

Figure 25.4 The atomic radii of the transition metals (and the *s*-block metals of the same periods) as a function of their periodic group.

Ta). The reason for this is that the atom size is determined by the radius of the *ns* orbital ($n = 4$, 5, or 6), which in every case has at least one electron. The variation in electrons occurs in the $(n - 1)d$ orbitals, which are somewhat smaller than the *ns* orbitals. As the number of electrons in these $(n - 1)d$ orbitals increases, they increasingly repel the *ns* electrons, partly compensating for the increased nuclear charge. Consequently, the *ns* electrons experience only a slightly increasing nuclear attraction across the series, and element radii remain nearly constant. Finally, the slight rise in radius at Groups 1B and 2B is due to the continually increasing electron–electron repulsions as the *d* subshell is completed.

Metal radius is a measure of the size of transition metal atoms; it is found by determining the size of atoms from the solid state structure of the metal. See the discussion of periodic trends in atomic radii in Chapter 9.

As a consequence of the variation in metal radius, the densities of the metals tend to increase and then decrease across a period (Figure 25.5). The reason for this trend is that, while the atom mass uniformly increases, the atom volume $[\frac{4}{3}(\pi r^3)]$ decreases or stays nearly constant across a period because r is nearly constant for the transition metals in a given period.

The contraction in metal radius across the fourth period transition metals has an effect on the radii of main group elements that follow (Table 25.1). Instead of the normal increase in size down a periodic group, we see that gallium is actually smaller than aluminum. In addition, contraction across the sixth period transition elements causes a smaller increase in radius from In to Tl than would be expected. Indeed, this points to another interesting observation: while the fifth period transition metals are larger than the fourth period metals, as expected, the sixth period transition

Table 25.1 Radii of the Group 3A Elements

Element	Radius (pm)	Period
B	85	2
Al	143	3
Ga	135	4
In	167	5
Tl	170	6

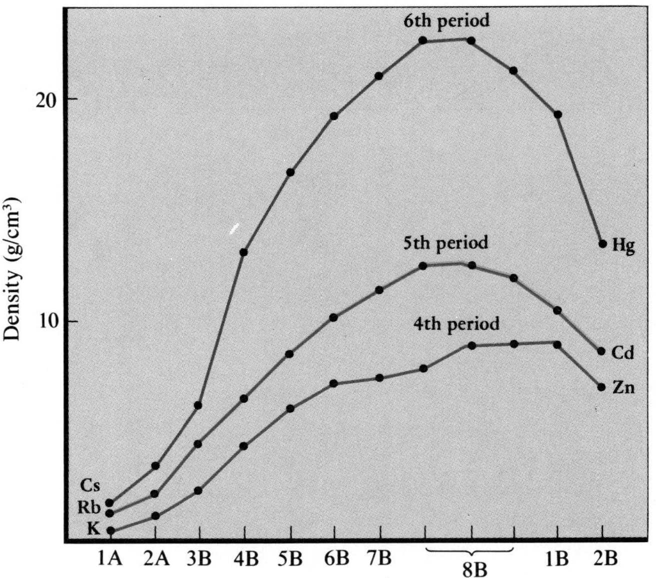

Figure 25.5 The density of the transition metals (and the *s*-block metals of the same periods) as a function of their periodic group.

Radii of Some Lanthanide Elements	
La	182 pm
Gd	180 pm
Lu	174 pm

Densities of Sixth Period Transition Metals (g/cm^3)	
W	19.3
Pt	21.41
Hg	13.534

elements have radii almost identical to those of the fifth period metals (Figure 25.4). The reason for this is that the sixth period transition elements (those filling 5*d* orbitals) follow the lanthanide series. The filling of 4*f* orbitals through the lanthanide elements causes a steady contraction in size, called the **lanthanide contraction**. The contraction in lanthanide radii carries over to the sixth period transition metals (beginning with Hf, radius = 159 pm), and the latter are smaller than expected.

One result of the lanthanide contraction is the great density of elements such as platinum and gold. The relatively small radii of sixth period transition metals, combined with the fact that their mass is considerably larger than their counterparts in the fifth period, cause sixth period metal densities to be very large. Platinum, for example, is one of the most dense metals known.

The similar radii of the transition metals and their ions have an important effect on their chemistry. For example, the nearly identical radii of fifth and sixth period transition elements lead to difficult problems of metal recovery. As described in "Something More About Platinum" in Chapter 2, the metals Ru, Os, Rh, Ir, Pd, and Pt are called the "platinum group metals" because they occur together in nature. The reason for this is that their radii and chemistry are so similar that their minerals are similar and are found in the same geologic zones.

25.2 COMMERCIAL PRODUCTION OF TRANSITION METALS

Most metals are found in nature as oxides, sulfides, halides, carbonates, or other salts (Figure 25.6). Many of the metal-containing mineral deposits are of little value, either because the deposit is impure or because the metal is too difficult to separate from impurities. The relatively few minerals from which elements can be obtained profitably are called **ores**, and it is those that are listed in Figure 25.6.

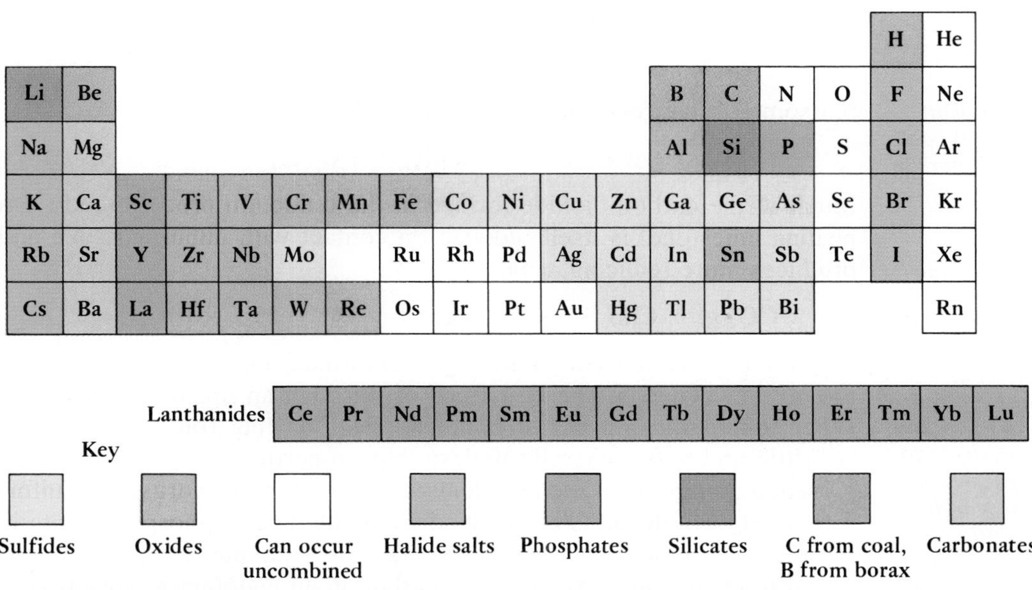

Figure 25.6 Sources of the elements. The *d*-block elements are found uncombined or as oxides, sulfides, and halides. The lanthanides occur predominantly as phosphate salts.

Pyrometallurgy and hydrometallurgy are two methods of recovering metals from their ores. **Pyrometallurgy** is a high temperature method, and is illustrated by iron production. **Hydrometallurgy** uses aqueous solutions at relatively low temperatures for the extraction of metals such as copper, zinc, tungsten, and gold.

Very few ores are chemically pure substances. They are usually mixtures of the desired mineral and large quantities of impurities such as sand and clay minerals, called **gangue** (pronounced "gang"). Therefore, a major step in obtaining the desired metal is to separate its mineral from the gangue. One way to do this is illustrated by the production of iron from its ores. Another method is **flotation**, used to purify copper-bearing ore, and a third approach is **chemical leaching**.

IRON PRODUCTION

Our economy depends on iron and steel. World production of iron ore in 1980 was 904 million tons, most coming from the USSR. Also in 1980, world production was 706 million tons of raw steel and 507 million tons of iron.

The production of iron from its ores involves oxidation–reduction reactions carried out in a blast furnace (Figure 25.7). The furnace is charged at the top with a mixture of ore (usually hematite, Fe_2O_3), coke, and limestone, and a blast of hot air is forced in at the bottom. The coke burns with such an intense heat that the temperature at the bottom is almost 2000 °C,

$$2\,C(s,\,coke) + O_2(g) \longrightarrow 2\,CO(g) + heat$$

and temperatures of about 200 °C are attained at the top of the furnace. The main reaction of interest is the net reduction of iron oxide with carbon monoxide to give impure metal,

The production of 1 ton of pig iron requires about 1.7 tons of iron ore, 0.5 tons of coke, 0.25 tons of limestone, and 2 tons of air.

$$\underset{\text{hematite}}{Fe_2O_3(s)} + 3\,CO(g) \longrightarrow 2\,Fe(\ell) + 3\,CO_2(g)$$

The first step in the steel-making process is to make coke from coal. The coal is heated in a tall, narrow oven sealed to keep out oxygen. The process drives a variety of chemical byproducts such as benzene and ammonia out of the coal and leaves nearly pure carbon. The byproducts are usually recovered for use in other industries. However, the coke ovens in the United States were largely built over 30 years ago and allow some of these byproducts to escape to the environment where they create problems. Research is being done around the world to find more efficient ways to make steel without the coking step.

but some of the iron oxide is also reduced by the coke.

$$Fe_2O_3(s) + 3\ C(s) \longrightarrow 2\ Fe(\ell) + 3\ CO(g)$$

Much of the carbon dioxide formed in the reduction process (and from heating limestone) is itself reduced on contact with unburned coke and produces more reducing agent.

$$CO_2(g) + C(s) \longrightarrow 2\ CO(g)$$

The molten iron flows down through the furnace and collects in a pool in the bottom, where it can be tapped off through an opening in the side. When cooled, the impure iron is called "cast" or "pig iron." The material is brittle and soft (generally undesirable properties), in part due to the presence of impurities such as elemental carbon, phosphorus, and sulfur. The latter two elements are formed from sulfate and phosphate salts in the reducing atmosphere of the furnace. For example, P_4O_{10} can be reduced to elemental phosphorus by carbon in an endothermic reaction.

$$P_4O_{10}(s) + 10\ C(s) + \text{heat} \longrightarrow P_4(s) + 10\ CO(g)$$

Iron ores are contaminated with gangue, which includes various silicate minerals (containing SiO_4^{4-}, for example) and sand (SiO_2). These react with lime from the limestone to give calcium silicate.

Limestone is called a "flux" in steel making.

$$\underset{\text{gangue}}{SiO_2(s)} + \underset{\text{lime}}{CaO(s)} \longrightarrow \underset{\text{slag}}{CaSiO_3(s)}$$

This is an acid–base reaction, since CaO is a basic oxide and SiO_2 is an acidic oxide. The salt from the reaction, calcium silicate, is less dense

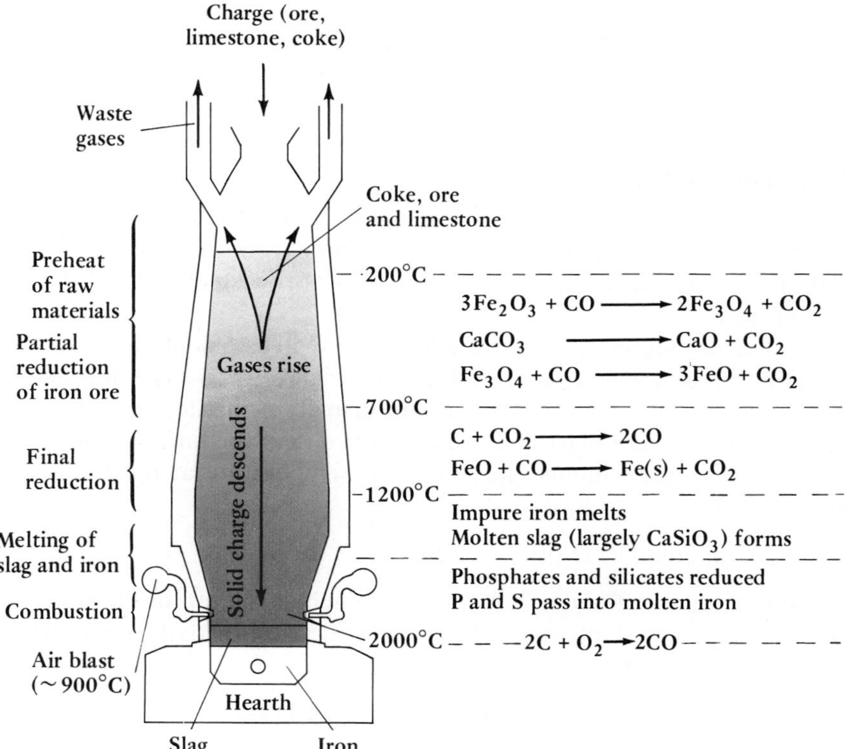

Figure 25.7 A blast furnace and some of the reactions that occur. The largest modern furnaces have hearths 14 meters in diameter and produce up to 10,000 tons of iron a day.

than molten iron, so the silicate floats on the iron as a separate layer. Other metal oxides that may be present dissolve in the silicate layer, and the mixture is called a "slag." At the interface or boundary between molten iron and slag, ionic impurities tend to move into the slag layer, while elements tend to concentrate in the iron. The floating slag layer is easily removed and frees the iron of many of the original impurities.

The impure pig iron coming from the bottom of the blast furnace is treated in a second step, involving oxidation, to remove the nonmetal impurities. Several technologies are used at this stage, but currently the most important is the **basic oxygen furnace** (BOF) (Figure 25.8). In this furnace, oxygen is blown into the molten pig iron to oxidize phosphorus, sulfur, and most of the excess carbon.

Figure 25.8 A "basic oxygen furnace" in operation.

$$P_4(s) + 5\ O_2(g) \longrightarrow P_4O_{10}(s)$$

$$C(s) + O_2(g) \longrightarrow CO_2(g)$$

$$S(s) + O_2(g) \longrightarrow SO_2(g)$$

All the oxides formed are acidic and react with basic oxides (such as CaO), which are added or are used to line the furnace. The products are calcium phosphate, calcium sulfate, and calcium carbonate. For example,

$$\underset{\text{acidic oxide}}{P_4O_{10}(s)} + \underset{\text{basic oxide}}{6\ CaO(s)} \longrightarrow \underset{\text{salt}}{2\ Ca_3(PO_4)_2(s)}$$

These salts form a floating layer of slag, which can be poured off to free the more dense molten iron layer of impurities.

Most of the slag from steel making is used to make cement, but some is blown with air to form rock wool insulation.

Pig iron typically contains up to 4.5% carbon, 1.7% manganese, 0.3% phosphorus, 0.04% sulfur, and 1% silicon. The final oxidation step removes much of the P, S, and Si, and reduces the carbon content to about 1.3%. The result is ordinary *carbon steel.* Almost any degree of flexibility, hardness, strength, and malleability can be achieved in carbon steel by proper cooling, reheating, and tempering. The drawbacks to carbon steel, however, are that it corrodes easily and loses its properties when heated strongly.

During the processing of steel, other transition metals, such as chromium, manganese, and nickel, can be added to give alloys having specific physical, chemical, and mechanical properties. One type of *stainless steel*, for example, contains up to 18% to 20% Cr and 8% to 12% Ni, but very little carbon. Such steels are very resistant to corrosion and are used in automobile trim, laboratory and kitchen ware, and so on. Magnetic alloys are solutions of iron and other elements that are permanently magnetic (and are used in loudspeaker magnets and the like) or that can be temporarily magnetized (and so are used in electric motors, generators, and transformers). Alnico V, as its name implies, contains five elements: 8% Al, 14% Ni, and 24% Co, as well as 51% Fe and 3% Cu.

The magnet in this photo is made of Alnico, an alloy of aluminum, cobalt, and nickel. It was purchased at the local farm implement store and is used as a "cow magnet." When a cow grazes, it picks up pieces of fencing, staples, and other iron objects. The magnet is put into the cow's first stomach, where it attracts the "scrap iron" and prevents it from being carried further in the animal's digestive tract.

COPPER PRODUCTION

The production of copper illustrates some other of the basic approaches of the metals industry. The most abundant copper-bearing ore, chalcopyrite, $CuFeS_2$, can be enriched readily by flotation. Here the powdered ore is mixed with oil and agitated with soapy water in a large tank

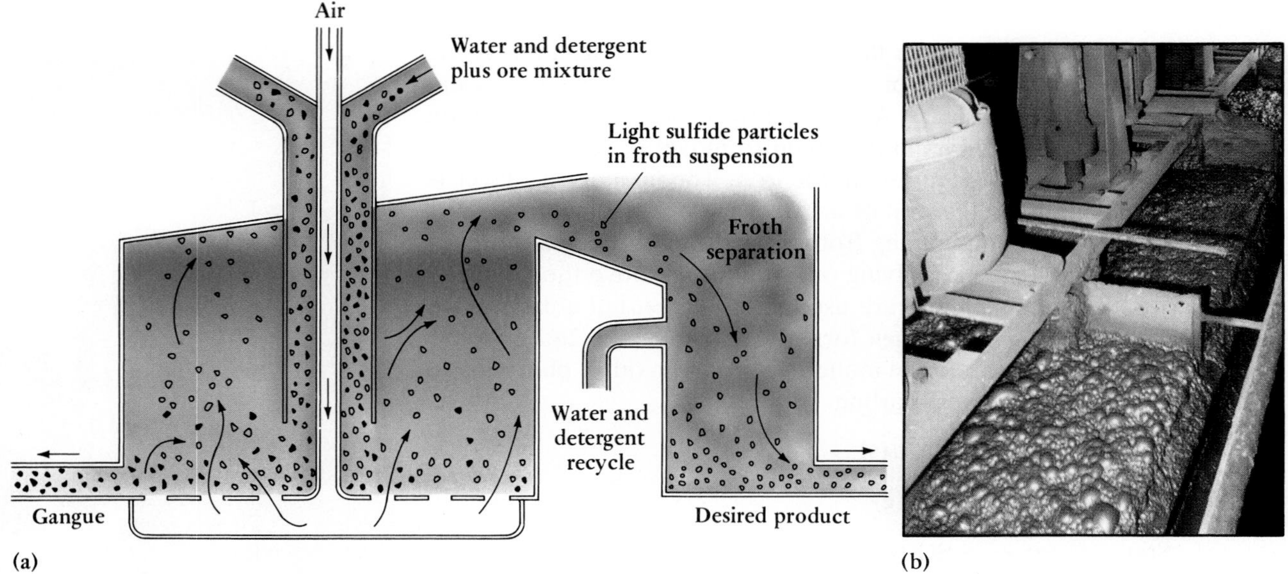

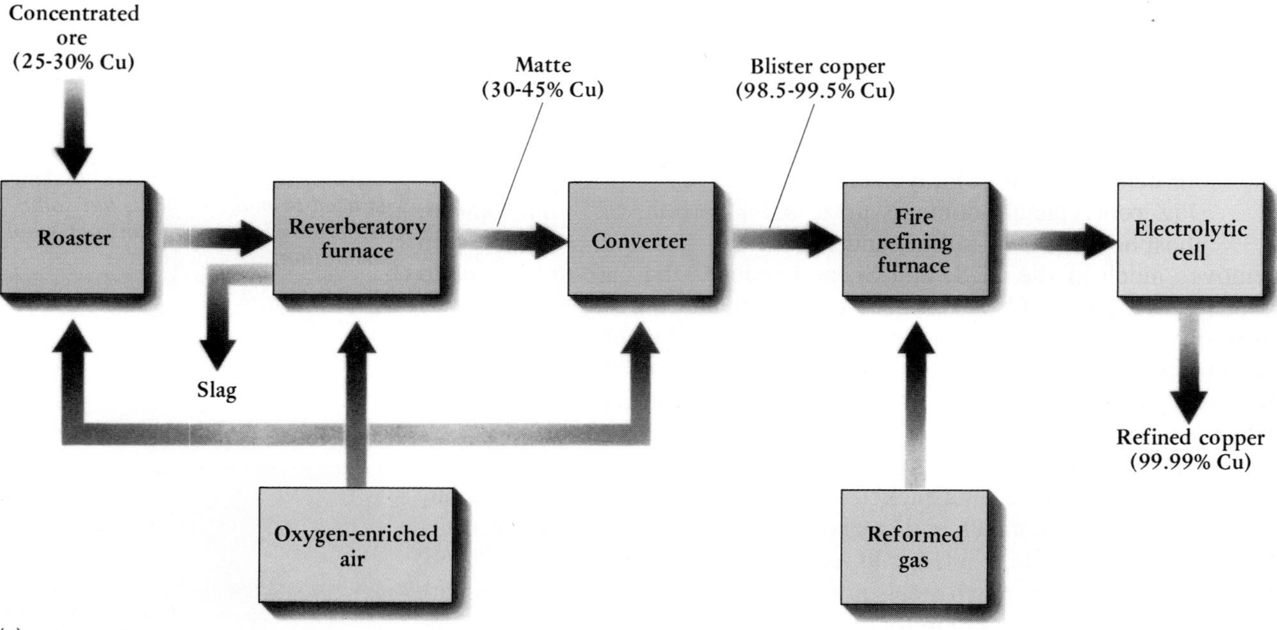

Figure 25.9 "Winning" copper from its ores. (a) The sulfide ores are enriched in the flotation process. The lighter metal sulfide particles are trapped in soapy bubbles and float on the water. The heavier gangue settles to the bottom. (b) A photograph shows the flotation process in operation. (c) After the ore is concentrated by flotation, the copper is refined in a sequence of steps as outlined in the flow diagram and in the text.

(Figure 25.9). Compressed air is forced through the mixture, and the lightweight, oil-covered copper sulfide particles are carried to the top and float on the froth. The heavier gangue settles to the bottom of the tank, and the copper-laden froth is skimmed off.

In the pyrometallurgy of copper, the enriched ore is **roasted** with enough air to convert the iron selectively to its oxide and leave copper as its sulfide.

$$2\ CuFeS_2(s) + 3\ O_2(g) \longrightarrow 2\ CuS(s) + 2\ FeO(s) + 2\ SO_2(g)$$

This mixture of copper sulfide and iron oxide is mixed with ground limestone, sand, and some fresh concentrated ore and then heated to 1100 °C (Figure 25.9). As in the blast furnace, limestone and sand (SiO_2) form glassy calcium silicate, and this dissolves the iron oxide.

$$\underset{\text{calcium silicate}}{CaSiO_3(\ell)} + FeO(s) + SiO_2(s) \longrightarrow \underset{\text{slag}}{2\,(Fe,Ca)SiO_3(\ell)}$$

A copper-bearing mineral, azurite, with the formula [$2CuCO_3 \cdot Cu(OH)_2$].

At the same time, excess sulfur in the ore reduces copper(II) sulfide, CuS, to copper(I) sulfide, Cu_2S, which melts and flows to the bottom of the furnace. Since the iron-containing slag is less dense than molten Cu_2S, the slag can be drawn off, thereby separating iron and copper.

The bottom Cu_2S layer, called *copper matte*, is tapped off and run into another furnace (the converter), where it is "blown" with air to oxidize the sulfide ions and produce impure copper metal. This is further refined in yet another furnace.

$$Cu_2S(\ell) + O_2(g) \longrightarrow 2\,Cu(\ell) + SO_2(g)$$

In the pyrometallurgical recovery of copper, each ton of copper produced also leads to 1.5 tons of iron silicate slag and 2 tons of SO_2. These by-products must be disposed of properly, not a simple task. However, one solution for SO_2 disposal is to convert it to economically important sulfuric acid (Section 24.2).

Hydrometallurgy decreases some of the energy costs and pollution problems of pyrometallurgy. One method of copper recovery now in use in Arizona is to *leach* or dissolve out the copper from its ore by treating the ore with a solution of copper(II) chloride and iron(III) chloride.

$$CuFeS_2(s) + 3\,CuCl_2(aq) \longrightarrow 4\,CuCl(s) + FeCl_2(aq) + 2\,S(s)$$

$$CuFeS_2(s) + 3\,FeCl_3(aq) \longrightarrow CuCl(s) + 4\,FeCl_2(aq) + 2\,S(s)$$

Copper is recovered in the form of copper(I) chloride. To keep the compound in solution, sodium chloride is added, since the soluble complex ion $[CuCl_2]^-$ is formed in the presence of excess chloride ion.

$$CuCl(s) + Cl^-(aq) \longrightarrow [CuCl_2]^-(aq)$$

Finally, impure copper is obtained by electrolyzing the $[CuCl_2]^-$ ion to the metal and $CuCl_2$; the latter is used to continue the leaching process.

$$2\,[CuCl_2]^-(aq) \longrightarrow Cu(s) + CuCl_2(aq) + 2\,Cl^-(aq)$$

Approximately 10% of the copper produced in the United States is actually obtained by using bacteria. Acidified water is sprayed onto copper-mining wastes, which can contain low levels of copper (Figure 25.10). As the water trickles down through the crushed rock, the bacterium *Thiobacillus ferrooxidans*, which thrives in the presence of acid and sulfur, breaks down the iron sulfides in the rock and converts iron(II) to iron(III). The iron(III) ion in turn oxidizes the sulfide ion of copper sulfides, leaving copper(II) ion in the water. This copper-laden water is recovered at the bottom of the pile, and metallic copper is obtained by reduction with scrap iron.

$$Cu^{2+}(aq) + Fe(s) \longrightarrow Cu(s) + Fe^{2+}(aq)$$

Figure 25.10 An open pit copper mine near Bagdad, Arizona. Additional copper can be extracted from the waste rock by a bacterium. (James Cowlin)

Regardless of the method of copper recovery from its ores, the final step is purification by electrolysis. Thin sheets of pure copper metal and slabs of impure copper are immersed in a solution of $CuSO_4$ and H_2SO_4 (Figure 25.11). The pure copper sheets are made the cathode of an electrolysis cell, and the impure slabs are the anode. This means that copper ions are formed at the anode (oxidation occurs) and move into solution. The ions migrate to the cathode where they are reduced to pure copper.

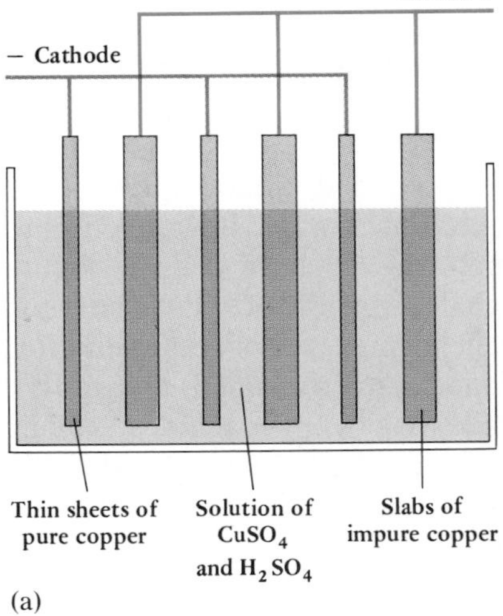

Figure 25.11 Electrolytic refining of copper. (a) A schematic diagram. Slabs of impure copper, called "blister copper," are made the anode of an electrolysis cell. The copper is oxidized and passes into the solution as Cu^{2+} ions. These ions migrate to the negative electrode, or cathode (made of thin sheets of pure copper), where they are reduced again to copper metal. (b) Photograph of electrolysis cells for refining copper.

25.3 COORDINATION COMPOUNDS

The green compound $CrCl_3 \cdot 6\ H_2O$ actually has the composition $[CrCl(H_2O)_5]Cl_2 \cdot H_2O$, while the violet nickel-containing compound is actually $[Ni(NH_3)_6]Cl_2$.

You are familiar with simple binary compounds of transition metals with ions such as O^{2-}, S^{2-}, and halides such as Cl^- and Br^-. For example, rutile or TiO_2 is used as a pigment in white paints (see Chapter 4), FeS_2 is a mineral commonly known as "fool's gold" (Figure 19.1), and AgBr is the light-sensitive substance in black and white film. There is another closely related class of compounds in which neutral molecules such as water and ammonia are incorporated. Examples include $CrCl_3 \cdot 6\ H_2O$ and $NiCl_2 \cdot 6\ NH_3$. The notation "6 H_2O" has historically been used to show that the basic chemical unit consists of six water molecules for every one chromium(III) and three chloride ions. Although it conveys stoichiometric information, this notation is not informative about the role of water. This section explores that role in this compound and others of its type.

COMPLEXES AND LIGANDS

It is more useful to write the formula $NiCl_2 \cdot 6\ NH_3$ as $[Ni(NH_3)_6]Cl_2$. The square brackets show that the cation in the compound is actually a Ni^{2+} ion bound to six ammonia molecules, $[Ni(NH_3)_6]^{2+}$, and this 2+ ion is associated with two Cl^- ions. We call such a compound a **coordination compound** or **coordination complex** because it contains the **coordination cation** or **complex ion** $[Ni(NH_3)_6]^{2+}$. A coordination compound is one in which the metal ion or atom is bonded to one or more neutral molecules or anions so as to define an integral structural unit. The molecules or ions bonded to the central metal ion are called **ligands**, from the Latin verb *ligare*, "to bind."

A great deal of transition metal chemistry is centered on coordination compounds. See also the "Something More About" sections on iron and nickel (pages 1061–1065).

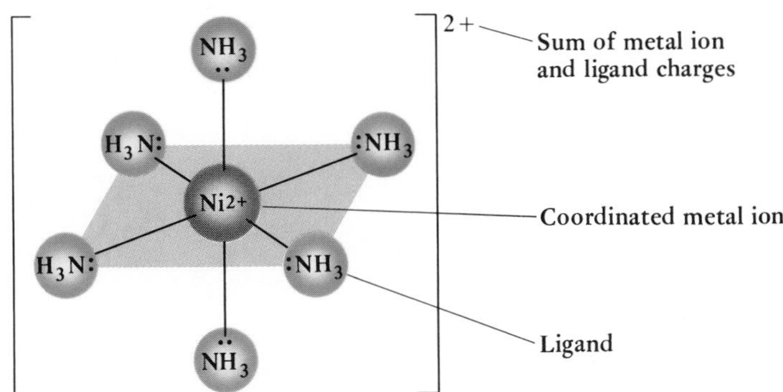

Coordination complexes can occur as cations, anions, or neutral molecules (Figure 25.12). For example, $K_3[Fe(CN)_6]$ contains the anion $[Fe(CN)_6]^{3-}$, whereas $Cr(CO)_6$ consists of a chromium atom bonded to six carbon monoxide molecules.

Ligands are **Lewis bases** because they always bear at least one atom having a lone pair of electrons accessible for bonding to another atom. A positive metal ion can accept electron pairs from a Lewis base, so the metal ion is a **Lewis acid** (see Section 17.6). As the ligand approaches a

Figure 25.12 Coordination complexes. Starting at the bottom and moving clockwise they are: $[Co(NH_3)_5H_2O]Cl_3$ (red), ligands are NH_3 and H_2O; $K_3[Fe(CN)_6]$ (red-orange), ligand is CN^-; $Cr(CO)_6$ (white), ligand is CO; $K_3[Fe(C_2O_4)_3]$ (green), ligand is $C_2O_4{}^{2-}$, oxalate ion; and $[Co(H_2N{-}CH_2{-}CH_2{-}NH_2)_3]I_3$ (yellow-orange), ligand is ethylenediamine.

metal ion, a ligand lone pair forms a sigma bond with the metal through an interaction between the lone pair orbital and an empty orbital on the metal ion or atom.

$$M^{n+} + :NH_3 \longrightarrow [M \leftarrow NH_3]^{n+}$$

While there are many ligands that have only one Lewis base site, some have more than one. When these atoms are separated from each other by several intervening atoms, it is possible that all the lone pair atoms can bind to the same metal atom to form a complex called a **chelate** (pronounced "key-late"). A good example of a chelating ligand is 1,2-diaminoethane ($H_2N—CH_2—CH_2—NH_2$), commonly called ethylenediamine and abbreviated *en*. When binding to the metal ion, such a ligand forms a ring of atoms, of which the metal is one member. Numerous complexes are formed by *en*, and $[Co(en)_3]^{3+}$ is an excellent example (Figures 25.12 and 25.13). Other common ligands capable of binding through more than one atom are illustrated in Figure 25.14.

Ligands like H_2O and NH_3, with only a single Lewis base atom, are termed **monodentate**. The word "dentate" stems from the Latin word *dentis* for tooth, so NH_3 is a "one-toothed" ligand. This means en, phen, and ox^{2-} are bidentate ligands, or "two-toothed," while $EDTA^{4-}$ is *hexa*dentate because it has six Lewis base sites.

Chelated complexes are important in everyday life. For example, one way to clean the rust out of water-cooled automobile engines and steam boilers is to add a solution of oxalic acid. Iron oxide dissolves in the presence of the acid to give the water-soluble iron oxalate complex.

$$Fe_2O_3(s) + 6\ H_2C_2O_4(aq) + 3\ H_2O(\ell) \longrightarrow 2\ [Fe(C_2O_4)_3]^{3-}(aq) + 6\ H_3O^+(aq)$$

$EDTA^{4-}$ is an excellent chelating ligand; it encapsulates and firmly binds metal ions. It is often added, for example, to commercial salad dressing to remove traces of metal ions from solution, since these metal ions can otherwise act as catalysts for the oxidation of the oils in the product;

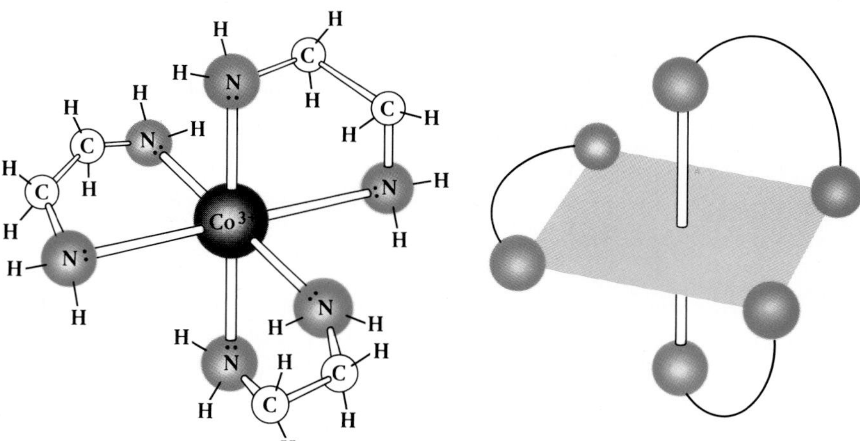

Figure 25.13 The structure of $[Co(en)_3]^{3+}$, a complex ion formed from three bidentate, chelating ethylenediamine molecules and Co^{3+}. (© Irving Geis)

SOME COMMON MONODENTATE LIGANDS

Ligands that have only one lone pair of electrons available for binding to an electron-pair acceptor (a Lewis acid).

Water molecule

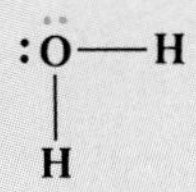

Ammonia molecule

Chloride ion

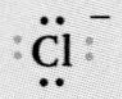

Cyanide ion

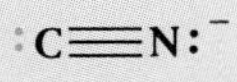

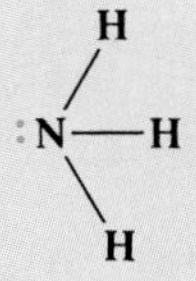

Carbon monoxide molecule

Hydroxide ion

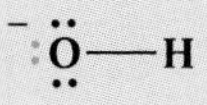

Bidentate ligands

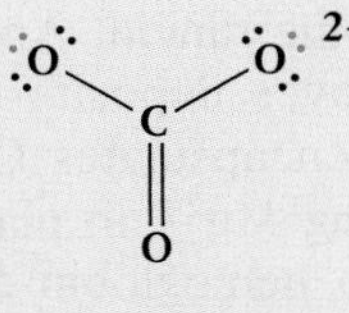

carbonate ion

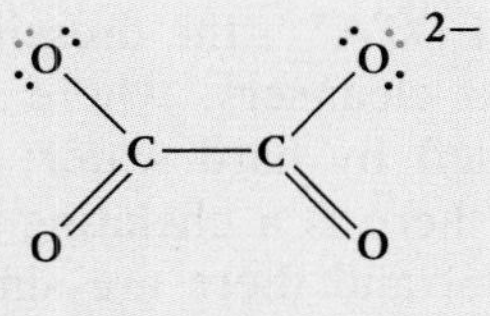

oxalate ion (ox^{2-})

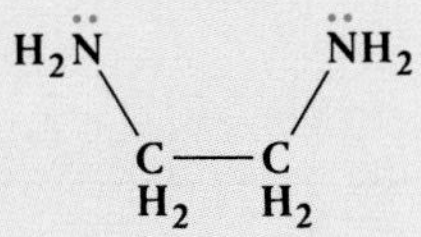

ethylenediamine (en)

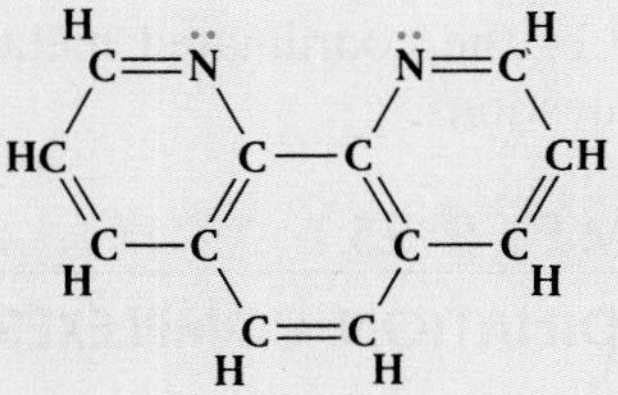

ortho-phenanthroline (phen)

Hexadentate ligand

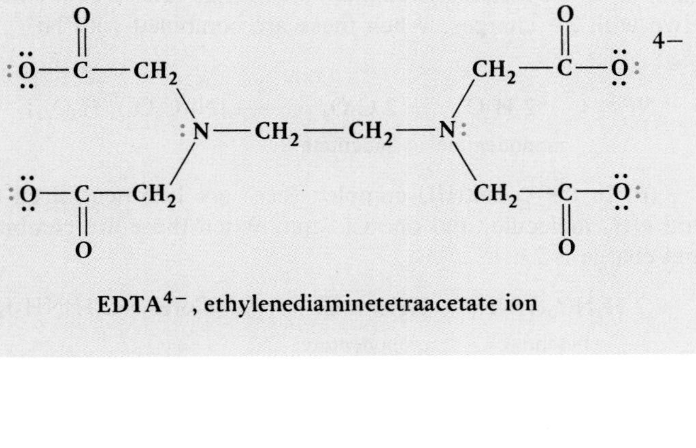

$EDTA^{4-}$, ethylenediaminetetraacetate ion

Figure 25.14 Some common chelating ligands. The origin of the word chelate is the Greek word *chele* for claw.

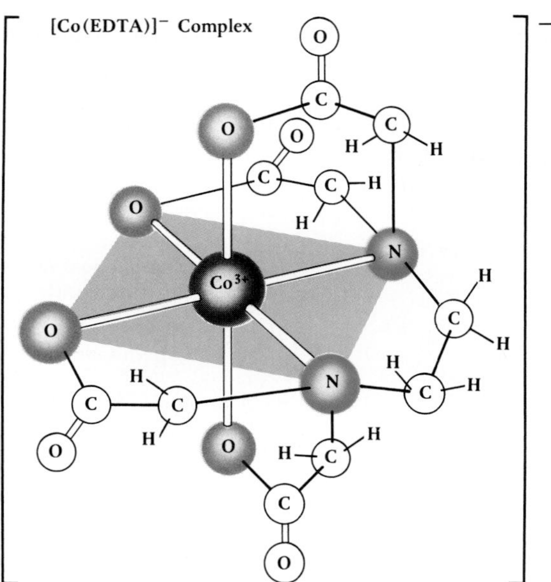

(© Irving Geis)

without $EDTA^{4-}$, the dressing would become rancid. Another use is in bathroom cleansers, where $EDTA^{4-}$ removes deposits of $CaCO_3$ and $MgCO_3$ left by hard water; the $EDTA^{4-}$ encapsulates Ca^{2+} or Mg^{2+}. Finally, there is a chelating ligand enclosing Mg^{2+} in plant chlorophyll (page 936) and there are similar ligands in hemoglobin and myoglobin (Chapter 27).

Be aware that a bidentate ligand, for example, is not necessarily chelating. Some bidentate ligands can use both sites to bind one metal atom or they can be a bridge between two metal atoms, using one site per metal atom.

It is useful to be able to predict the formula of a coordination complex, given the metal ion and ligands, and to predict the oxidation number of the coordinated metal ion. The following examples explore these questions.

EXAMPLE 25.1

COORDINATION COMPLEXES

Given the following combinations of metal ion and ligands, write the formula of the coordination complex in which the metal ion is coordinated to six Lewis base sites. (a) One Ni^{2+} ion is bound to two water molecules and two bidentate oxalate ions. (b) One Co^{3+} ion is bound to one Cl^- ion, one ammonia molecule, and two bidentate ethylenediamine molecules.

Solution (a) In the nickel(II) complex, the ligands consist of two neutral molecules and two with 2− charges. When these are combined with Ni^{2+}, the net charge is 2−.

$$\underset{}{Ni^{2+}} + \underset{\text{monodentate}}{2\,H_2O} + \underset{\text{bidentate}}{2\,C_2O_4^{2-}} \longrightarrow [Ni(C_2O_4)_2(H_2O)_2]^{2-}$$

(b) In the cobalt(III) complex there are two neutral en molecules, one neutral NH_3 molecule, and one Cl^- ion. When these are combined with Co^{3+}, the net charge is 2+.

$$Co^{3+} + \underset{\text{bidentate}}{2\,H_2NC_2H_4NH_2} + \underset{\text{monodentate}}{NH_3} + Cl^- \longrightarrow [Co(H_2NC_2H_4NH_2)_2(NH_3)Cl]^{2+}$$

EXAMPLE 25.2

FORMULAS OF COORDINATION COMPLEXES

Give the oxidation number of the metal ion in each of the following complexes: (a) $[Co(en)_2(NO_2)_2]^+$ and (b) $Pt(NH_3)_2(C_2O_4)$.

Solution (a) In this cobalt complex there are two neutral, bidentate ethylenediamine molecules and two nitrite ions, NO_2^-. Since the overall charge on the ion is 1+, the cobalt ion must be 3+.

$$Co^{3+} + 2\ H_2NC_2H_4NH_2 + 2\ NO_2^- \longrightarrow \text{complex ion with } 1+ \text{ charge}$$

(b) Platinum is coordinated to two neutral ammonia molecules and one bidentate oxalate ion. Thus, platinum is in the form of the Pt^{2+} ion.

$$Pt^{2+} + 2\ NH_3 + C_2O_4^{2-} \longrightarrow \text{complex with charge of } 0$$

EXERCISE 25.1 Formulas of Coordination Complexes

(a) Give the oxidation number of platinum in $Pt(NH_3)_2Cl_2$ (cisplatin, page 185).
(b) What is the formula of a complex assembled from one Co^{3+} ion, five ammonia molecules, and one monodentate carbonate ion?

NAMING COORDINATION COMPOUNDS

Just as there are rules for naming simple inorganic and organic compounds, coordination compounds are named according to an established system. For example, the following compounds are named according to the rules outlined below.

COMPOUND	SYSTEMATIC NAME
$[Ni(H_2O)_6]SO_4$	Hexaaquanickel(II) sulfate
$[Cr(en)_2(CN)_2]Cl$	Dicyanobis(ethylenediamine)chromium(III) chloride
$K[Pt(NH_3)Cl_3]$	Potassium amminetrichloroplatinate(II)

As you read through the rules, notice how they apply to these examples.

1. In naming a coordination compound that is a salt, name the cation first and then the anion, as is usually done (Chapter 3).
2. When giving the name of the complex ion or molecule, name the ligands first, in alphabetical order, followed by the name of the metal.
   a. If a ligand is an anion whose name ends in *-ite* or *-ate*, the final *e* is changed to *o* (as in sulfate → sulfato or nitrite → nitrito).
   b. If the ligand is an anion whose name ends in *-ide*, the ending is changed to *o* (as in chloride → chloro or cyanide → cyano).
   c. If the ligand is a neutral molecule, its common name is used. The important exceptions at this point are water, which is called *aqua*, ammonia, which is called *ammine*, and CO, called *carbonyl*.
   d. When there is more than one of a particular monodentate ligand with a simple name, the number of ligands is designated by the appropriate prefix: *di*, *tri*, *tetra*, *penta*, or *hexa*. If the ligand name is complicated (whether monodentate or bidentate), the prefix

changes to *bis*, *tris*, *tetrakis*, *pentakis*, or *hexakis*, followed by the ligand name in parentheses.

  e. If the complex ion is an anion, the suffix *-ate* is added to the metal name.

3. Following the name of the metal, the oxidation number of the metal is given in Roman numerals.

Complex ions can become more complicated than those described in this chapter, and even more rules of nomenclature must be applied. However, the brief set just outlined is sufficient for the vast majority of complexes.

EXAMPLE 25.3

NAMING COORDINATION COMPOUNDS

Name the following compounds: (a) $[Cu(NH_3)_4]SO_4$; (b) $K_2[CoCl_4]$; (c) $Co(phen)_2Cl_2$; and (d) $[Co(en)_2(H_2O)Cl]Cl_2$.

Solution (a) The sulfate ion has a 2− charge, so the complex ion has a 2+ charge (that is, $[Cu(NH_3)_4]^{2+}$). Since NH_3 is a neutral molecule, this means the copper ion is Cu^{2+}. Therefore, the compound's name is tetraamminecopper(II) sulfate.

(b) There are two K^+ ions in this compound, so the complex ion has a 2− charge ($[CoCl_4]^{2-}$). Since there are four Cl^- ions in the complex ion, the cobalt center is Co^{2+}. Thus, the name of the compound is potassium tetrachlorocobaltate(II).

(c) This is a neutral compound. Since two Cl^- ions and two neutral phen ligands are bonded to a cobalt ion, the metal ion must be Co^{2+}. This means the compound name is dichlorobis(phenanthroline)cobalt(II).

(d) Here the complex ion has a 2+ charge, since it is associated with two uncoordinated Cl^- ions. The cobalt ion must be Co^{3+} because it is bound to two neutral en ligands, one neutral water, and one Cl^-. Therefore, the name is aquachlorobis(ethylenediamine)cobalt(II) chloride.

EXERCISE 25.2 Naming Coordination Compounds
Name the compounds (a) $[Ru(phen)_2(H_2O)CN]Cl$ and (b) $Pt(NH_3)_2Cl_2$.

25.4 STRUCTURES OF COORDINATION COMPOUNDS AND ISOMERS

Coordination compounds are characterized by integral structural units consisting of a metal ion or atom surrounded by a sheath of ligands. The number of metal–ligand shared pairs on the metal is the **coordination number**. Although there are examples of coordination numbers of two through twelve, only four and six are very common, and two deserves mention.

COMMON COORDINATION NUMBERS AND GEOMETRIES

If we consider for a moment only complexes of M with the monodentate ligand L, common coordination complexes would have stoichiometries of ML_6, ML_4, and ML_2. Complexes of ML_2 are usually linear, and the silver-containing ion that comes from dissolving AgCl in aqueous ammonia is a good example (Chapter 19).

$$AgCl(s) + 2\ NH_3(aq) \longrightarrow \underset{\text{linear } ML_2 \text{ complex}}{[H_3N{\rightarrow}Ag{\leftarrow}NH_3]^+(aq)} + Cl^-(aq)$$

If we consider only M—L bond pairs in coordination complexes, the VSEPR theory of Chapter 10 leads us to expect tetrahedral structures for ML_4 complexes. Indeed, this is observed for many molecules, such as $[CoCl_4]^{2-}$, $Ni(CO)_4$, and $[Zn(NH_3)_4]^{2+}$.

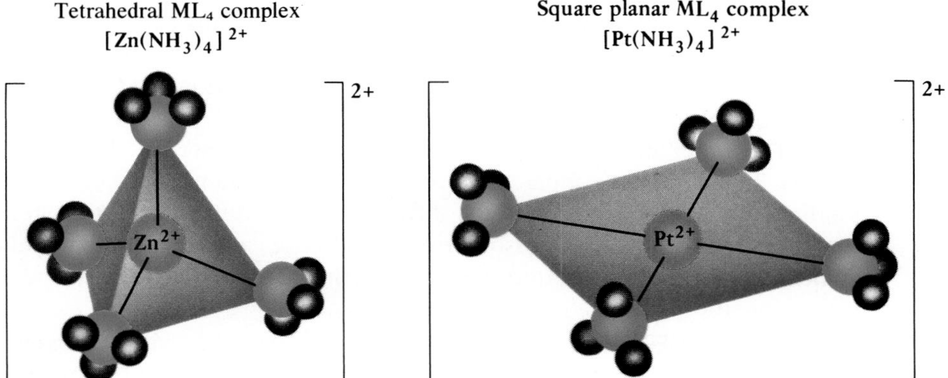

However, as we pointed out in Chapter 10, the VSEPR theory does not always apply to transition metal complexes, and, for ML_4 complexes, *square planar* geometry is observed more often than tetrahedral geometry. This is particularly true for ML_4 complexes of Pt^{2+} and Pd^{2+}, and sometimes Ni^{2+}.

All of the complex ions or molecules in Figure 25.12 have the coordinating atoms at the corners of an octahedron.

With *very* rare exceptions, ML_6 complexes have the six ligands arranged at the corners of an octahedron, as in the $[Ni(NH_3)_6]^{2+}$ ion pictured on page 1037. Two views of an octahedron are given in Figure 25.15. The drawing at the left emphasizes that the regular octahedron is a figure with six corners and eight faces, all of which are equilateral triangles; the metal ion is at the center of the octahedron. On the right in Figure 25.15, the drawing shows that the ligands are located along three axes at right angles to one another (the x, y, and z axes), and the ligands

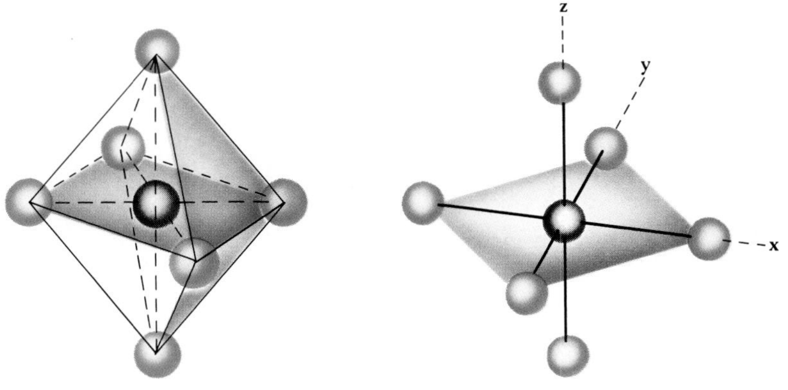

Figure 25.15 Two views of a regular octahedron.

are *equidistant from the metal atom or ion in the center*. Finally, the drawing at the right shows us we can consider an octahedron as a *square plane* of M and four ligand atoms L, with other ligand atoms above and below the plane.

ISOMERS

Isomerism is one of the most interesting aspects of chemical compounds. Isomers are molecules that have the same molecular formula, but in which the atoms are arranged differently. In one form of isomerism, this leads to molecules that function chemically in quite different ways. One example would be the isomers of C_2H_6O. These atoms can be bonded to give either ethyl alcohol or dimethyl ether. As you will see in Chapter 26, alcohols and ethers have a different chemical function, and so these isomers are often called **functional isomers**.

$H_3C—CH_2—OH$
ethyl alcohol
mp, −115 °C; bp, 78 °C

$H_3C—O—CH_3$
dimethyl ether
mp, −141 °C; bp, −24.8 °C

The other major type of isomerism is **stereoisomerism**. Here the isomers have the same molecular formulas *and* the same atom-to-atom bonding sequences, but the atoms differ in their arrangement in space. There are two forms of stereoisomerism, geometric isomerism and optical isomerism, and both are often observed in the coordination compounds of the transition metals.

GEOMETRIC ISOMERISM Geometric isomers result when the atoms bonded *directly* to the metal are sequenced in a different order about the metal. The simplest example of geometrical isomerism occurs with square planar complexes, such as $Pt(NH_3)_2Cl_2$, that have no more than two identical ligands. This complex is formed from Pt^{2+}, two NH_3 molecules, and two Cl^- ions. The two Cl^- ions, for example, can be adjacent to one another (*cis*) or located on opposite sides of the molecule (*trans*).

The use of $Pt(NH_3)_2Cl_2$ in cancer chemotherapy is described in the interview with Dr. Barnett Rosenberg at the beginning of Part One and in "Something More About Cisplatin, a Cancer Chemotherapy Drug" in Chapter 5. It is interesting that only the cis isomer is physiologically active.

Cis Isomer
Two Cl^- ligands adjacent
Angle between identical ligands = 90°

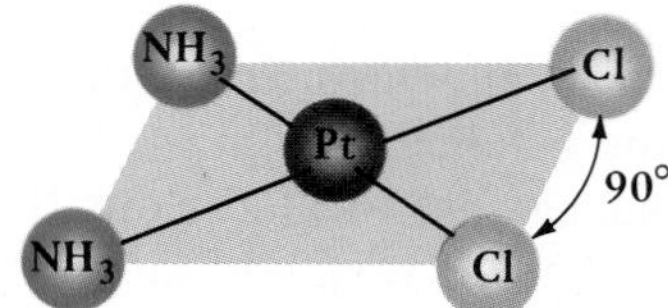

Trans Isomer
Two Cl^- ligands opposite
Angle between identical ligands = 180°

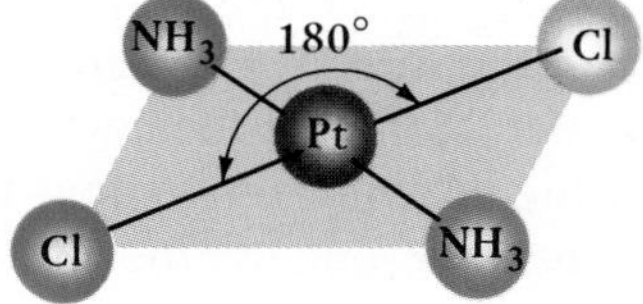

Now consider square planar $[Pt(NH_3)Cl_3]^-$, a complex with three like ligands.

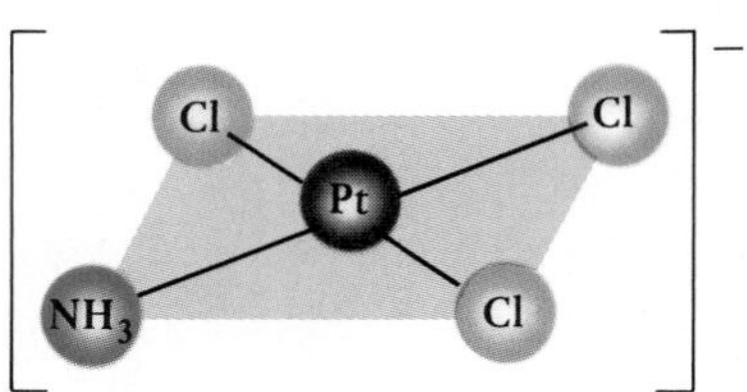

Cis-trans isomers are not possible here. No matter how you draw the molecule, there are always two Cl^- ions 180° apart and two Cl^- ions 90°

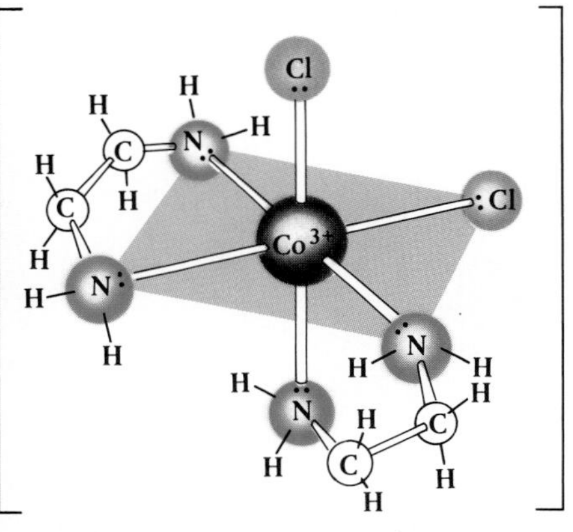

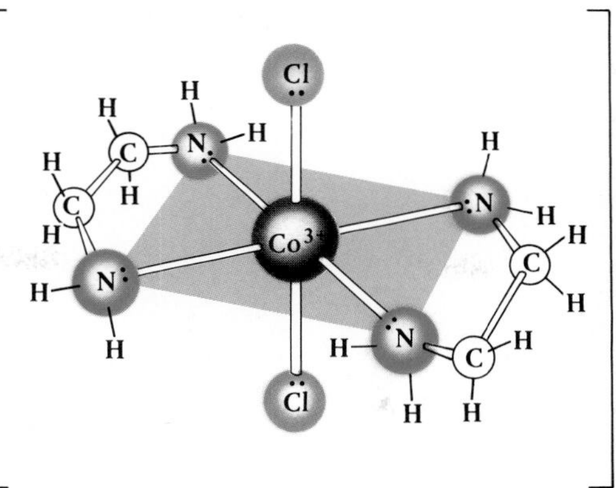

Figure 25.16 The geometrical isomers of $[Co(H_2N—C_2H_2—NH_2)_2Cl_2]^+$, with a photograph of a model of the *trans* isomer. (Drawings © Irving Geis)

apart. On the other hand, *cis* and *trans* isomers can be constructed for $Pt(NH_3)_2(Cl)(NO_2)$.

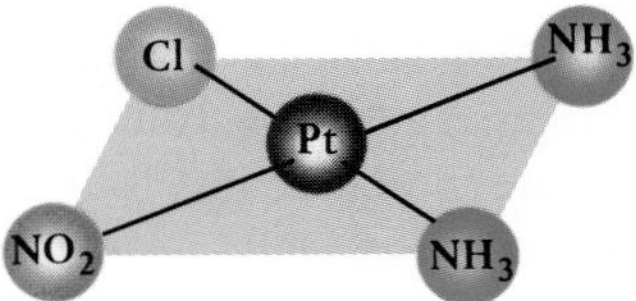

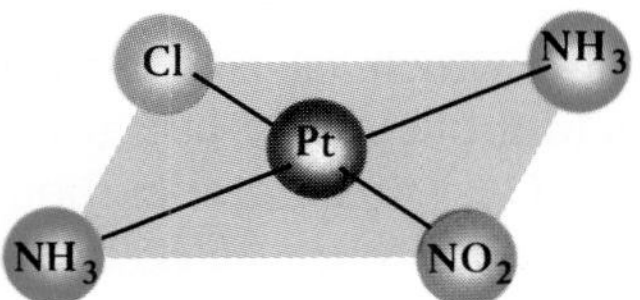

Four-coordinate complexes can have tetrahedral geometry. However, *cis-trans isomerism is not possible for tetrahedral complexes* because it is not possible for two ligands to be "across" the molecule from one another. All possible angles in a tetrahedron are 109°, so all ligands are effectively adjacent.

Some complexes can exist in both square planar and tetrahedral forms and so are isomers of one another.

Geometrical isomerism is widely observed for octahedral complexes. As an example consider $[Co(H_2NC_2H_4NH_2)_2Cl_2]^+$, an octahedral complex ion with two different ligands. Here the two Cl^- ions can occupy adjacent or opposite positions of the octahedron of ligands, so the resulting

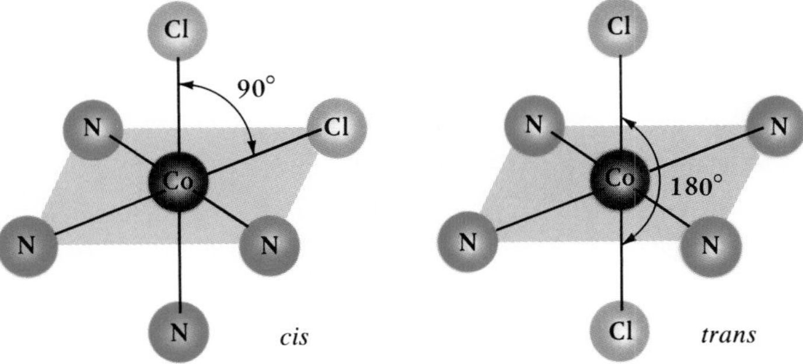

isomers of $[Co(en)_2Cl_2]^+$ are illustrated in Figure 25.16. It is interesting that the different ways of connecting the ligands lead to different colors; the *cis* isomer is green (Figure 15.8), but the *trans* isomer is purple. When three unique ligands are present in an octahedral complex, such as the NH_3 molecules or the Cl^- ions in $Cr(NH_3)_3Cl_3$, geometrical isomerism also occurs. Here the three Cl^- ligands, for example, can all lie in the same plane as the metal, or they can be at the corners of an equilateral triangle.

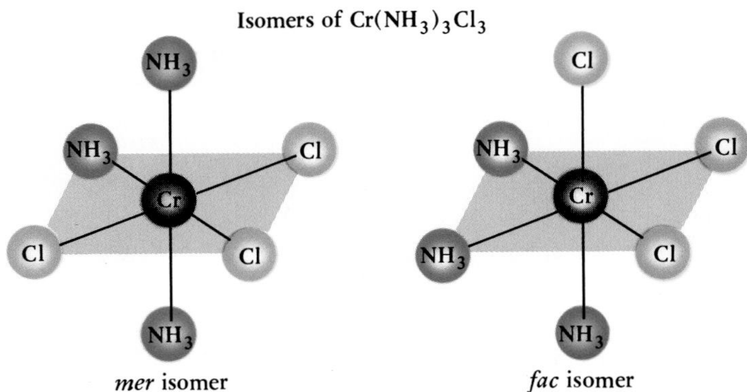

The left isomer is often called the mer isomer, because the Cl^- ions are located on the meridian of the complex. The right isomer is called the fac isomer, since the three Cl^- ions are found on a face of the octahedron.

OPTICAL ISOMERISM *Cis-trans* isomerism is only one form of stereoisomerism. Optical isomerism is the other. Optical isomers have the same stoichiometry *and* the same atom-to-atom bonding sequence, but they differ in the details of the arrangement of atoms in space. This type of isomerism, which is extremely important not only in transition metal coordination chemistry but also in biochemistry, is illustrated by lactic acid, $C_3H_6O_3$. The central carbon atom of the molecule has four *different* groups bonded to it: $—CH_3$, —OH, —H, and —COOH.

H
HO—C—COOH
CH_3

lactic acid

As a result of the tetrahedral arrangement around this sp^3 hybridized atom, it is possible to have two different arrangements of the four groups. If a

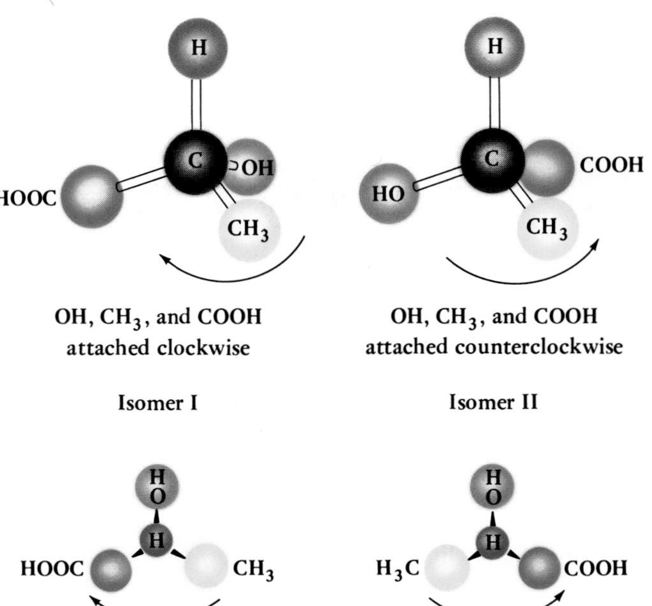

Figure 25.17 The two stereoisomers (enantiomers) of lactic acid. If you were to look down on the molecule along the H—C bond, you would see that one of the isomers (I) has the other groups attached such that OH, CH_3, and COOH are arranged clockwise. In the other isomer (II), the arrangement is counterclockwise.

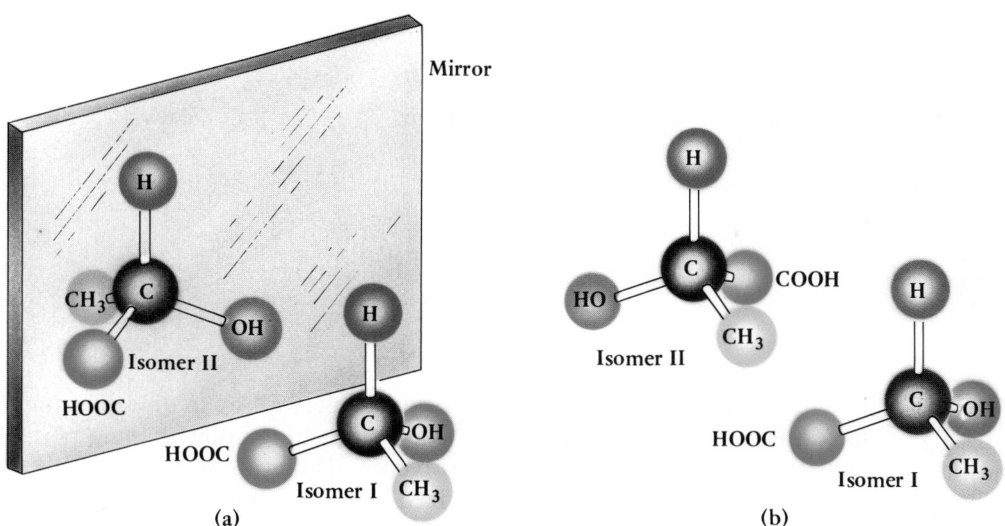

Figure 25.18 The stereoisomers of lactic acid. (a) Isomer I is placed in front of a mirror, and its mirror image is isomer II. (b) Isomers I and II are placed so that the grouping H—C—CH_3 is oriented the same way in each molecule. If isomer II were placed on top of isomer I, the C—H and C—CH_3 bonds would be superimposed, but the C—OH bond of II would rest on top of the C—COOH bond in I. The isomers are nonsuperimposable.

lactic acid molecule is placed so the C—H bond is vertical, as illustrated in Figure 25.17, one possible arrangement of the remaining groups would be that where —OH, —CH_3, and —COOH are attached in a clockwise manner (isomer I). Alternatively, these groups can be attached in a counterclockwise fashion (isomer II). To see further that the arrangements in Figure 25.17 are different, we place isomer I in front of a mirror (Figure 25.18). Now you see that isomer II is the mirror image of isomer I. What is important, however, is that these mirror-image molecules cannot be superimposed on one another (Figure 25.18). The best analogy to this situation is to look at your hands. If you hold your left hand up to a mirror, the image you see is that of your right hand (Figure 25.19). Your mirror-image hands cannot be superimposed on one another, and the mirror-image molecules of lactic acid cannot be superimposed on each other.

Since the two forms of lactic acid in Figure 25.17 are *nonsuperimposable mirror images*, they are stereoisomers. Such stereoisomers are said to be **chiral** (pronounced kī′ ral), and a chiral compound and its mirror image isomer are called **enantiomers**. The central carbon atom of lactic acid, around which there are four different atoms or groups, is referred to as a **chiral center**.

Figure 25.19 Mirror images. The left hand (top) of one of the authors of this book is reflected in the mirror (which also reflects the clouds in the sky), and his right hand is lying in the grass next to the mirror (bottom). The mirror image of his left hand looks the same as his right hand, showing that hands are mirror images of one another. However, hands cannot be superimposed on one another. If you place one hand directly on the other, they will not line up completely. This is the main criterion for chirality, whether of hands or molecules.

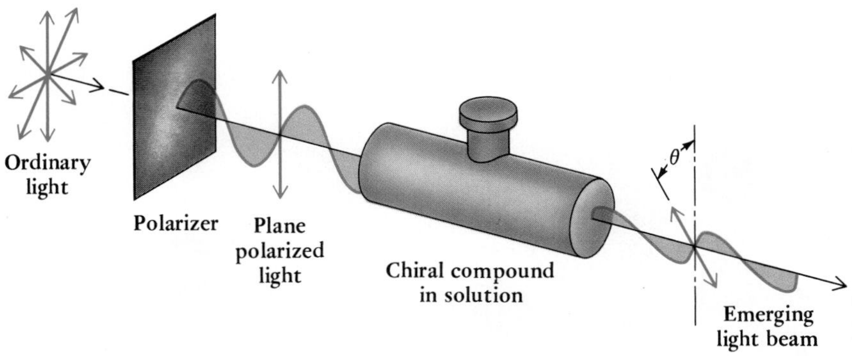

Figure 25.20 Rotation of the plane of polarization of light by a chiral compound. If light is passed through a polarizing filter (of the type found in some sunglasses), the light is polarized so that the electric field of the light beam lies in a given plane. When this light passes through a solution of a chiral compound, the plane of polarization is rotated. One enantiomer of a pair rotates the light beam to the left by θ degrees, and the other enantiomer rotates the beam by an equal number of degrees to the right. If the plane of polarization is twisted to the right, the enantiomer is referred to as *dextrorotatory* (*d*) (from the Latin *dexter* meaning "right"). The other enantiomer, which rotates the plane of polarization by the same number of degrees to the left, is referred to as *levorotatory* (*l*) (from the Latin *laevus* meaning "left").

The fact that enantiomers have identical physical properties means that special methods must be used to separate one from the other in a mixture of the two.

Although a pair of enantiomers may appear different on paper, they have the same physical properties such as melting point, boiling point, density, and solubility in various solvents. There is, however, one exception: when a beam of plane polarized light is passed through a solution of a pure enantiomer, the plane of polarization is twisted in one direction or the other (Figure 25.20). For this reason, chiral compounds are sometimes referred to as **optical isomers** and are said to be **optically active**.

The lactic acid molecule can exist in two chiral forms. In coordination chemistry, however, there are no examples of stable complexes with a metal bonded tetrahedrally to four *different types* of ligands, and so optical isomers of tetrahedral coordination complexes need not be considered further. The situation is quite different with octahedral complexes, though, where optical isomerism can arise in a variety of ways. Only one of these ways will be described here. When three bidentate ligands, such as en, ox^{2-}, or phen, bind to a metal ion or atom, three metal-containing rings are formed, as in the $[Co(en)_3]^{3+}$ complex pictured in Figure 25.13. Two optical isomers result from the two ways of arranging these rings in space (Figure 25.21). Think of the complex as a three-bladed propeller. The blades can be arranged so that the propeller twists clock-

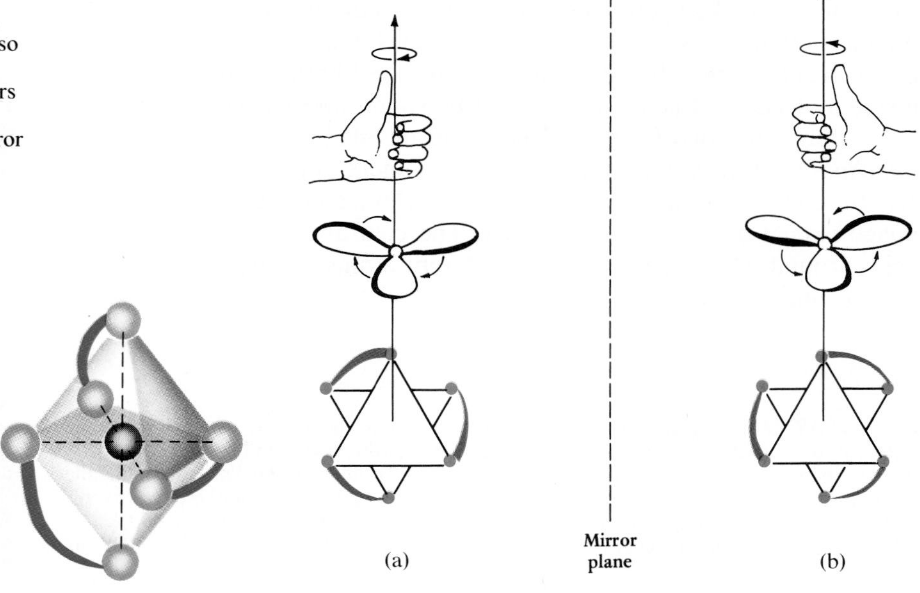

Figure 25.21 Optical isomerism in complexes of the type $M(bidentate)_3$. The three chelate rings are arranged so that the complex looks like a three-bladed propeller. One of the propellers twists clockwise (a) and the other twists counterclockwise (b). The mirror images cannot be superimposed.

wise or counterclockwise. One arrangement is the mirror image of the other, and neither is superimposable on the other.

Sea shells have a handedness, that is, they are chiral, and virtually all of them are right-handed. If you cup a shell in your right hand (with your thumb extended), your fingers will follow the curl of the shell as it curls from the outside toward the center. Your thumb will point along the axis of the shell from the narrow end to the wider end.

EXAMPLE 25.4

ISOMERISM

Which of the following complexes exhibit geometrical and/or optical isomerism?
(a) $[Pt(NH_3)_4Cl_2]^{2+}$ (b) $[Ru(phen)_3]Cl_2$ (c) $K_2[Pt(CN)_2Cl_2]$

Solution (a) This Pt^{4+} complex has an octahedral structure with two geometric isomers. It is similar to $[Co(en)_2Cl_2]^+$ in the text, one isomer with two Cl^- ions *cis* and the other with the two ions *trans*.

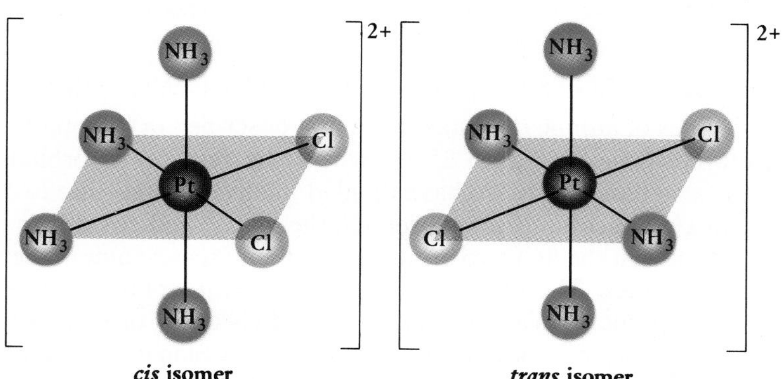

(b) In $[Ru(phen)_3]^{2+}$, the Ru^{2+} ion is surrounded by three bidentate phen ligands (Figure 25.14). Therefore, this complex ion is similar to the $[Co(en)_3]^{3+}$ ion described in the text (Figure 25.21), an ion that can have optical isomers.

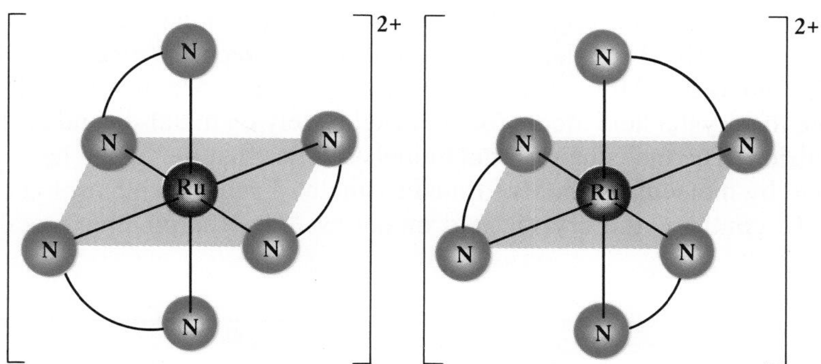

(c) This complex has two different types of ligands, in two pairs, attached to Pt^{2+}, and it is square planar. This means geometrical isomers are possible.

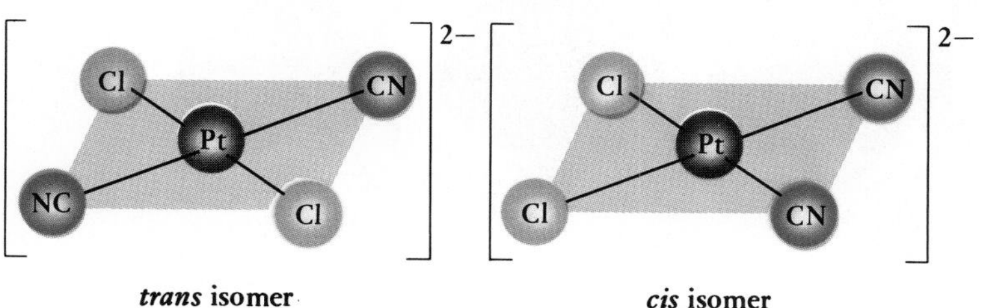

***trans* isomer** ***cis* isomer**

EXERCISE 25.3 Identifying Isomers
What type of isomers are possible for $[Co(en)_2(CN)_2]Br$?

25.5 BONDING IN COORDINATION COMPOUNDS

The reactions, magnetic properties, and colors (Figure 25.22) of the complexes of a particular metal are sensitive to the attached ligands. To understand this, we shall have to delve into the nature of the metal–ligand bond. In this section we describe (a) what occurs to the *d*-orbital energies of a metal atom or ion as ligand lone pairs approach to form a bond and (b) the consequences of this for the color of complexes.

d-ORBITAL ENERGIES IN COORDINATION COMPOUNDS

Shapes of atomic *d* orbitals were first described in Chapter 8. They are pictured again in Figure 25.23, where the relation of orbitals and ligands is also illustrated. We have divided the five orbitals into two sets: the z^2 and $x^2 - y^2$ orbitals in one set and the xy, xz, and yz orbitals in the other. Notice that the first two orbitals have their greatest amplitude along the cartesian axes. The orbitals of the other set, in contrast, have their greatest amplitude *between* the axes. This division of orbitals into two sets is reasonable because the ligands in square planar and octahedral complexes lie along the x, y, and z axes, with their lone pair orbitals pointing toward the metal atom or ion at the center of the coordinate system.

Molecular orbital theory and **crystal field theory** are two ways to approach metal–ligand bonding using metal *d* orbitals and ligand lone pair orbitals. As ligands approach the metal to form bonds, there are two effects: (a) orbital overlap and (b) repulsion between the electrons of the metal and ligand. Molecular orbital theory takes into account both effects, while the crystal field model focuses exclusively on metal–ligand electron repulsion. The molecular orbital model assumes that metal and ligand are bound by molecular orbitals formed by metal–ligand atomic orbital overlap. In contrast, the crystal field model assumes the positive metal ion

Figure 25.22 The colors of the complexes of a given metal ion depend on the ligand. The yellow solid at the left is a salt of $[Co(NH_3)_6]^{3+}$. One ammonia ligand has been replaced by (left to right) NCS^- (orange), H_2O (red), and Cl^- (purple). The green complex at the right is a salt of $[Co(NH_3)_4Cl_2]^+$.

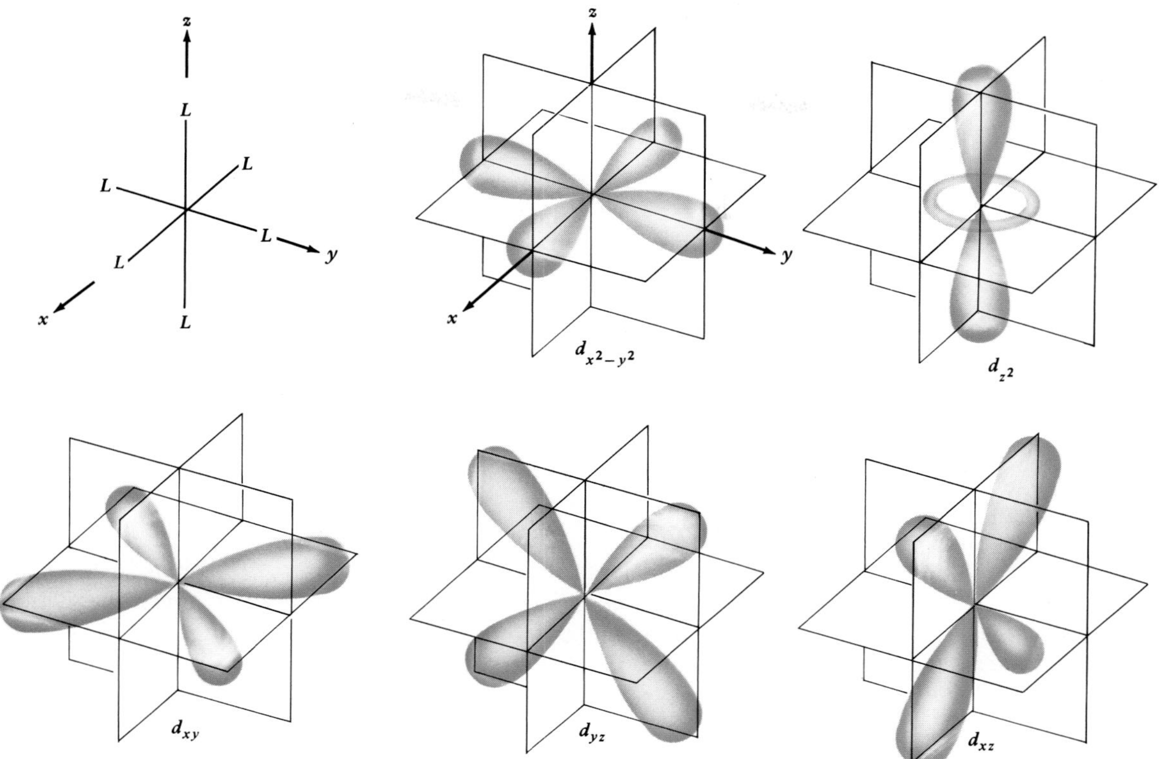

Figure 25.23 The five d orbitals and their relation to ligands on the x, y, and z axes.

and negative ligand lone pair are *electrostatically* attracted. The name of the theory comes from the assumption that the ligands create an electrostatic "crystal field" that influences metal electrons. Both models ultimately produce the same *qualitative* results regarding color and magnetic behavior, and we shall focus on the crystal field approach.

The name "crystal field theory" comes from the original theory, which assumed a metal ion in a crystal surrounded by ligands. A similar theory is ligand field theory, but the latter allows for metal–ligand covalent bonding.

The d orbitals of a free atom or ion are degenerate, that is, they have the same energy. However, according to the crystal field model, there is a repulsion between metal electrons and ligand lone pairs as ligands approach the metal atom or ion, and the d electron orbitals rise in energy (Figure 25.24). The largest repulsion is felt by the z^2 and

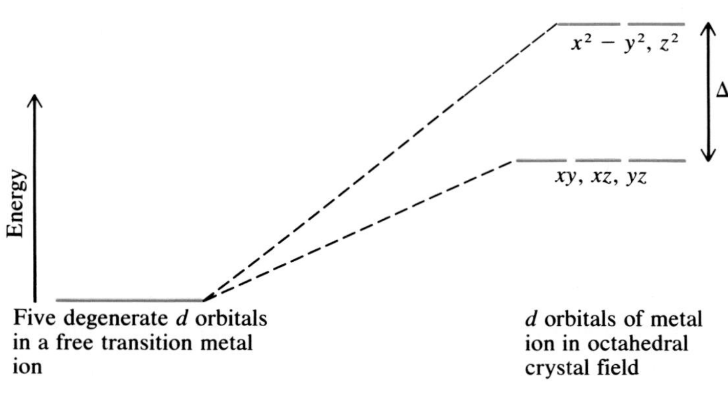

Figure 25.24 The d orbital energy changes as six ligands approach, at the corners of an octahedron, a transition metal ion having five electrons. The energy difference between the (z^2, $x^2 - y^2$) and (xy, xz, yz) orbital sets is labeled Δ_o.

Figure 25.25 The splitting of d orbitals for a square planar complex. Here the energy difference between the $x^2 - y^2$ orbital and the next highest energy orbital is given by the symbol Δ_{sp}.

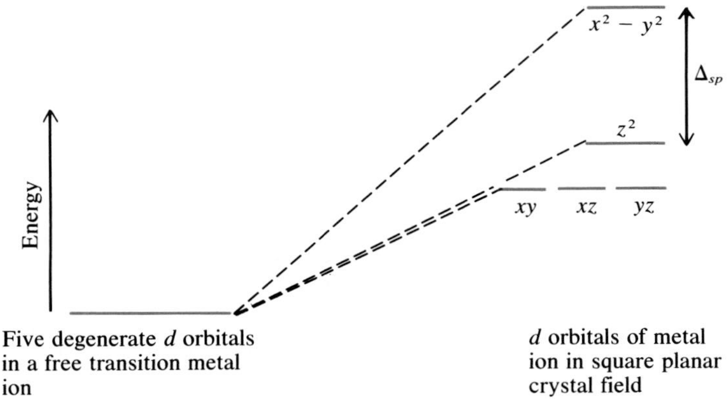

$x^2 - y^2$ electrons, since these electrons are in orbitals pointing directly at the incoming ligand electron pairs. A somewhat smaller repulsive effect, however, is experienced between xy, xz, and yz metal electrons and ligand electrons. The difference in degree of repulsion means that there is a difference in the energy increase for the two sets of orbitals. The "splitting" in energy of the (z^2, $x^2 - y^2$) and (xy, xz, yz) sets, given the symbol Δ_o, depends on the metal and the ligands and can change as these change from one complex to another. As you shall soon see, this splitting of d orbitals into two sets can lead to the distinctive colors of transition metal complexes.

Δ_o is sometimes called the "octahedral crystal field splitting parameter"; hence the subscript o *on the symbol.*

There are important differences between d orbital energy shifts for octahedral and square planar complexes (Figure 25.25). Both structures have in common the four ligands in the molecular plane, assumed to be the xy plane. Thus, the $x^2 - y^2$ orbital is raised in energy by the same amount in ML_4 as in ML_6. In square planar ML_4, however, the upward shift of z^2 is smaller because, although there are no z-axis ligands, 1/3 of the z^2 orbital (the "doughnut") is concentrated in the xy plane.

ELECTRON CONFIGURATIONS OF METAL IONS OR ATOMS IN COMPLEXES; MAGNETIC BEHAVIOR

We have already seen the consequences of d-orbital splitting in explaining the red color of rubies (page 348).

The consequences of d orbital splitting in coordination complexes are far-reaching, affecting the color and magnetic behavior of complexes. To understand these properties, we must first understand how to assign electrons to the "split" orbital sets of square planar and octahedral complexes.

The term "degenerate" means that all of the orbitals of a set have the same energy. See Chapter 9, page 329.

The Cr^{2+} ion has the electron configuration $[Ar]3d^4$. Hund's rule means that the five degenerate d orbitals of a free Cr^{2+} ion will have the following orbital diagram.

Electron Configuration of Cr^{2+}

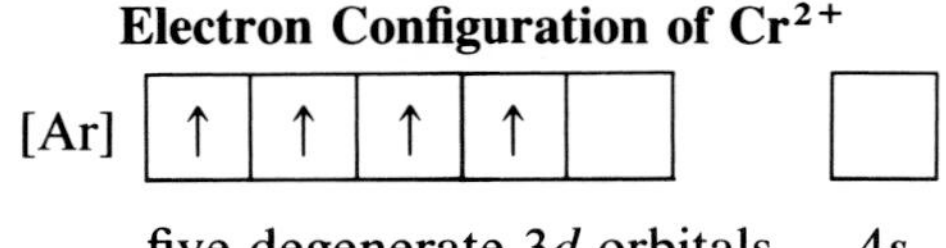

When Cr^{2+} is bound in an octahedral complex, the xy, xz, and yz orbitals are degenerate and at a lower energy than the degenerate (z^2, $x^2 - y^2$)

set. Having two sets of orbitals means that two electron configurations are possible for four electrons. The first three electrons are assigned, one each, to the lower energy (xy, xz, yz) orbitals. The fourth electron, however, could be assigned to the higher energy (z^2, $x^2 - y^2$) set (Configuration A) *or* pair up with an electron already in the lower energy set (Configuration B).

Configuration A high spin		Configuration B low spin
↑ — $z^2, x^2 - y^2$		— — $z^2, x^2 - y^2$
↑ ↑ ↑ xy, xz, yz	⟶	⥮ ↑ ↑ xy, xz, yz

The first arrangement is commonly called **high spin** because it has the maximum number of unpaired electrons. In contrast, the second arrangement is called **low spin** because it has the minimum number of unpaired electrons. The correct arrangement of the four Cr^{2+} electrons in a complex containing this metal ion depends on two factors. We can see what these are if we write an equation for the energy difference between the low and high spin configurations.

$$\Delta E = E_{\text{low spin}} - E_{\text{high spin}} = -(\Delta_o) + P$$

The quantity Δ_o is the energy difference between the two orbital sets. If an electron moves from the higher energy set (z^2, $x^2 - y^2$) to the lower energy set (xy, xz, yz), energy in an amount equal to Δ_o, is evolved, so Δ_o is given a negative sign. However, to form the low spin configuration, an electron pair is formed, and, due to higher electron–electron repulsions in a pair, this requires energy. P is the energy of electron pairing, and it is given a positive sign. From this equation, we can come to the general conclusion that a low spin configuration will be observed in preference to high spin if energy is evolved when the complex changes from high spin to low spin, that is, if ΔE is negative. This will occur if Δ_o is large or P is small. This conclusion is summarized for a simple two-orbital system in Figure 25.26.

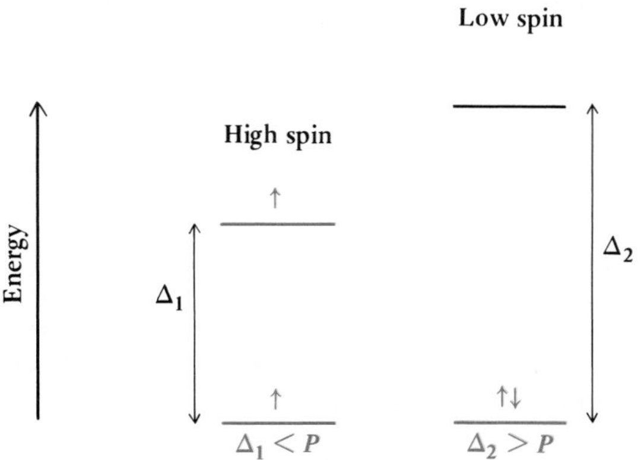

Figure 25.26 Dependence of high and low spin configurations on extent of orbital splitting (Δ_o) and electron pairing energy, P.

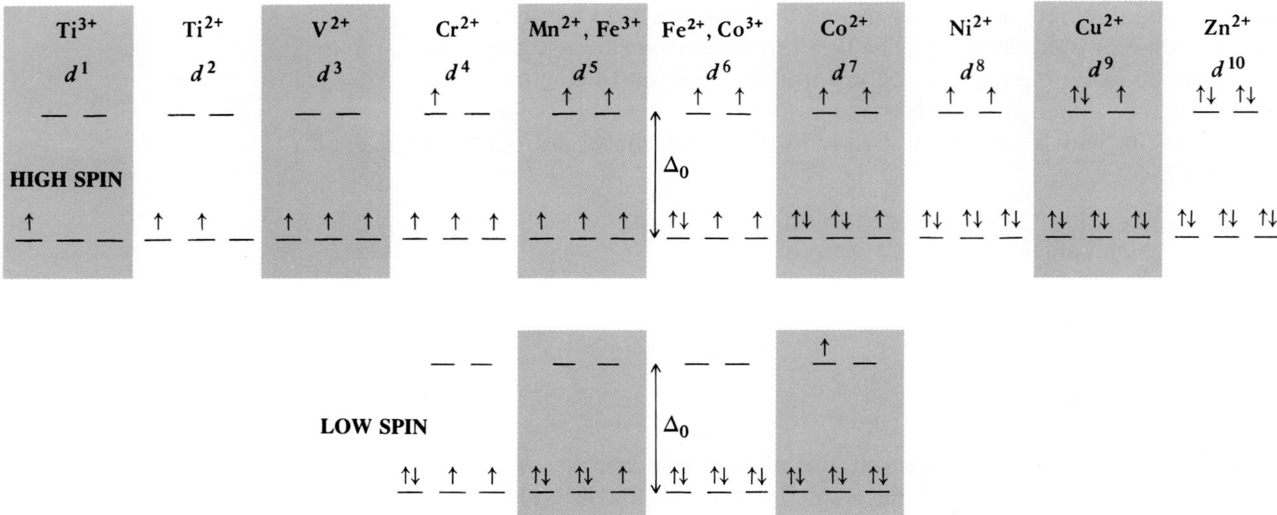

Figure 25.27 Electron configurations for metal ions with d^1 to d^{10} configuration in octahedral complexes. Only d^4 through d^7 configurations can have either high or low spin arrangements.

There is a choice between high and low spin only for configurations d^4 through d^7 in octahedral complexes (Figure 25.27). In each case, the outcome is dictated by the relative sizes of Δ_o and the electron pairing energy P. This means that, for a given metal ion with a constant value of P, whether the complex is high or low spin will depend on Δ_o. For Fe^{2+}, for example, the complex formed when the ion is placed in water, $[Fe(H_2O)_6]^{2+}$, is high spin, whereas $[Fe(CN)_6]^{4-}$ is low spin.

Electron Configuration for Fe^{2+} in an Octahedral Complex

	High spin	Low spin
$x^2 - y^2, z^2$	↑ ↑	— —
	Δ_o (H_2O)	Δ_o (CN^-)
xy, xz, yz	↑↓ ↑ ↑	↑↓ ↑↓ ↑↓
	high spin	low spin
	$[Fe(H_2O)_6]^{2+}$	$[Fe(CN)_6]^{4-}$

Ions with eight d electrons (Ni^{2+}, Pd^{2+}, and Pt^{2+}) are especially prone to form square planar complexes. The Ni^{2+} ion, for example, has the electron configuration $[Ar]3d^8$.

Electron Configuration of a Free Ni^{2+} Ion

[Ar]	↑↓	↑↓	↑↓	↑	↑		
	five degenerate 3d orbitals						4s

In a square planar complex, however, the five degenerate orbitals form three sets of orbitals (Figure 25.25). With eight electrons to be accommodated, high and low spin configurations are possible.

Electron Configuration for Ni^{2+} in a Square Planar Complex

	high spin	low spin
$x^2 - y^2$	↑	—
	Δ_{sp}	Δ_{sp}
z^2	↑	⇅
xy xz yz	⇅ ⇅ ⇅	⇅ ⇅ ⇅

However, all known square planar Ni^{2+} complexes are low spin, so Δ_{sp} clearly outweighs the pairing energy, *P*.

Δ_{sp} is the splitting of energy (between $x^2 - y^2$ and z^2) for a square planar complex.

One can tell the difference between low and high spin complexes by determining the magnetic behavior of the substance. For example, a high spin, square planar Ni^{2+} complex has two unpaired electrons and is *paramagnetic*, whereas a low spin Ni^{2+} complex has no unpaired electrons and is *diamagnetic*. Similarly, $[Fe(H_2O)_6]^{2+}$ is paramagnetic to the extent of four unpaired electrons, whereas $[Fe(CN)_6]^{4-}$ is diamagnetic.

Refer to Chapter 9 for a discussion of magnetism.

EXAMPLE 25.5

HIGH AND LOW SPIN COMPLEXES AND MAGNETISM

Depict the low and high spin electron configurations for each of the following complexes and tell how many unpaired electrons are present in each.

(a) $[Co(NH_3)_6]^{2+}$ (b) $[Mn(CN)_6]^{4-}$

Solution (a) Since there are six ligands surrounding cobalt, we assume this is an octahedral complex. Further, since the NH_3 ligands are neutral molecules and since the overall charge on the complex is 2+, this means the complex is based on Co^{2+}. Cobalt(II) ion has an electron configuration of $[Ar]3d^7$. Therefore, the metal ion configurations in the complex would be

	high spin	low spin
$x^2 - y^2$, z^2	↑ ↑	↑ —
	Δ_o	Δ_o
xy, xz, yz	⇅ ⇅ ↑	⇅ ⇅ ⇅

To obtain the high spin configuration, first half fill all five orbitals, and then pair up any remaining electrons in the lower energy orbitals. This leads to the configuration above with three unpaired electrons. The low spin configuration is obtained by first half filling the three lower energy orbitals. The next three electrons are then added to these orbitals, and electron pairs are formed. The seventh and last electron must then be placed in the higher energy set of orbitals. Thus, low spin Co^{2+} is paramagnetic by one unpaired electron.

(b) The ligands in the octahedral manganese complex are cyanide ions, CN^-. Since the overall charge is 4−, this means the complex is based on Mn^{2+}, an ion with the configuration $[Ar]3d^5$. Using the same procedure as above, the low and high spin configurations in the complex ion are

high spin		low spin	
↑ ↑ $x^2 - y^2, z^2$	Δ_o	— — $x^2 - y^2, z^2$	Δ_o
↑ ↑ ↑ xy, xz, yz		↑↓ ↑↓ ↑ xy, xz, yz	

Both the high and low spin complexes in (a) and (b) are paramagnetic, but the number of unpaired electrons is very different. This difference can be measured by experiment, so it is possible to tell whether the complex is low or high spin.

EXERCISE 25.4 Low and High Spin Configurations and Magnetism

For each of the following complex ions, give the oxidation number of the metal ion, depict the low and high spin configurations, give the number of unpaired electrons in each state, and tell whether each is paramagnetic or diamagnetic.

(a) $[Ru(H_2O)_6]^{2+}$ (b) $[Ni(NH_3)_6]^{2+}$

25.6 THE COLORS OF COORDINATION COMPOUNDS

One of the most interesting properties of the transition elements is that their compounds are usually colored, whereas those of main group metals are almost always colorless (Figure 25.28). With an understanding of *d* orbital splitting, we can now explain the origin of the color.

Figure 25.28 Compounds of the transition elements are often colored, while those of main group elements are usually colorless. Pictured are aqueous solutions of the nitrate salts of (left to right) Fe^{3+}, Co^{2+}, Ni^{2+}, Cu^{2+}, and Zn^{2+}.

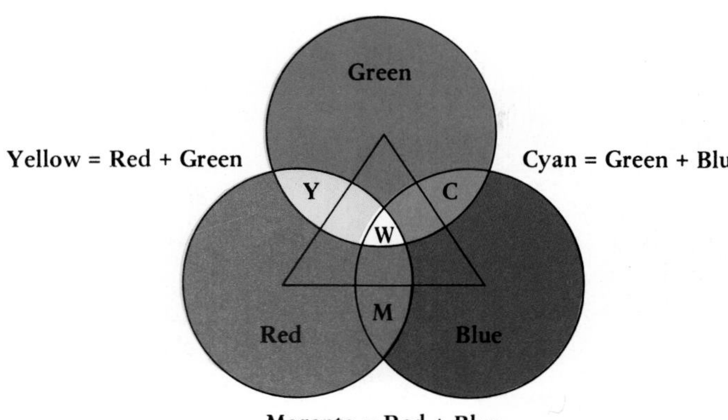

Figure 25.29 Color discs that can be used to explain the colors of solutions of transition metal complexes. Light of two primary colors can add together to give a third color. For example, cyan arises from the addition of green and blue light, and yellow can come from red and green. Alternatively, we can think of color as arising from the subtraction or filtering of light of a given color from white light. For example, if red light is removed, the transmitted light would be cyan.

COLOR

White light is the superposition of red, green, and blue colors. If one or more of these colors are absorbed by a transition-metal-containing ion or molecule in solution, this leaves the light of the other colors to be passed through to your eyes. Your mind then perceives the ion or molecule to have a color.

In Chapter 8 we talked about the colors you see when white light is passed through a prism: red through yellow, green, and blue to violet (ROYGBIV). Each color is identified with a narrow wavelength range (Figure 8.3). Isaac Newton did experiments with light of these ranges, and he established that the mind's *perception* of color requires only three colors!

To understand the perceived colors of compounds, we divide the visible spectral range into three broad regions: red, green, and blue:

Red Green

700 600 500 400 nm

R (O) Y G B I V

Newton's red region spans the ROY range and blue spans the BIV range. When light of the three colors stands alone or is superimposed, the color we perceive is given by the color discs in Figure 25.29. The *primary* colors—red, green, and blue—appear at the corners of the triangle superimposed on the color discs, and the *secondary* colors—yellow, cyan, and magenta—appear at the edges of the triangle. This figure shows you that you can think of a color as arising in either of two ways: by *addition* of two other colors or by the *subtraction* of light of a particular color from white light. For example, adding blue and green light leads to a color known as cyan, whereas red and blue light combine to produce magenta. These and other combinations are summarized in the following table:

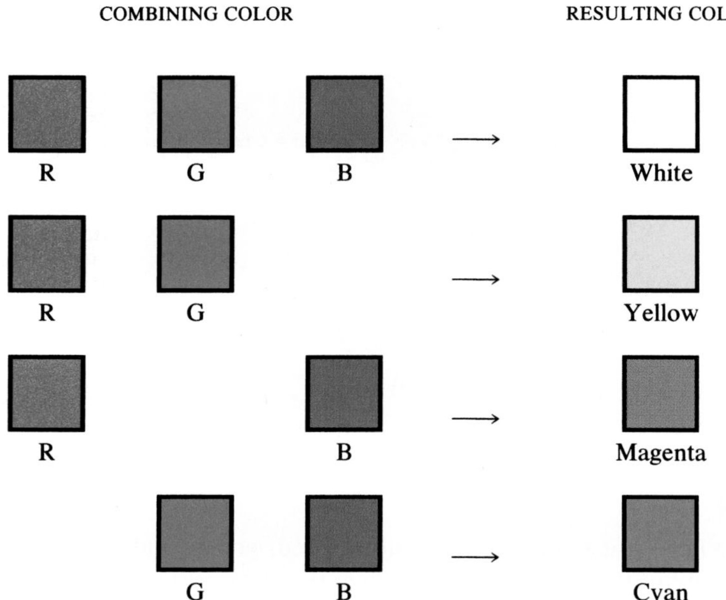

The blue in the ROYGBIV sequence is actually cyan, C, according to the now accepted color industry standards. (The important marginal notes and other highlights in this book are printed in cyan, for example.) Note too that magenta does not have its own wavelength range; it is a superposition of B and R. Finally, note that orange is simply a hue of yellow with greater red content.

The three secondary colors—cyan, yellow, and magenta—can also be derived from the primary colors by *subtracting* or "filtering" red, green, or blue light from white light. Notice that each edge of the triangle in Figure 25.29 has an opposite corner; the edge-corner pairs make up *complementary* color pairs, which when superimposed give white light. This means, for example, that cyan (C) can arise *either* from the addition of G and B *or* from the filtering or subtraction of red light from white light: C = B + G or C = W − R.

Now let us apply these ideas to transition metal complexes. Green (G) may result from removing magenta (M) or, equivalently, blue (B) and red (R) from white light. Therefore, as white light passes through an aqueous solution of Ni^{2+} (Figure 25.28), there is a mechanism by which blue and red light are absorbed; green light is allowed to pass, and this color is perceived. The CrO_4^{2-} ion (Figure 1.5) is yellow because blue (B) light has been selectively absorbed; that is, the solution allows red (R) and green (G) light to pass.

All this can be verified in the laboratory today by means not available to Newton, that is, with a spectrophotometer (Figure 25.30). White light from a glowing filament is first passed through a device (a prism or diffraction grating) that redirects the photons of white light according to their frequencies. The instrument can select photons of a specific frequency to pass through a solution of the compound to be studied. If photons of a given frequency are not absorbed, that light is unchanged in its intensity when it emerges from the sample. On the other hand, if photons of some frequency are absorbed, by a mechanism described below, light of that frequency emerges from the sample with a decreased intensity. By making a plot of the frequency or wavelength of the light against the intensity of light absorbed at that frequency or wavelength, we obtain an **absorption spectrum** of the sample. When there is an absorption of light in a certain frequency range, the plot shows an *absorption*

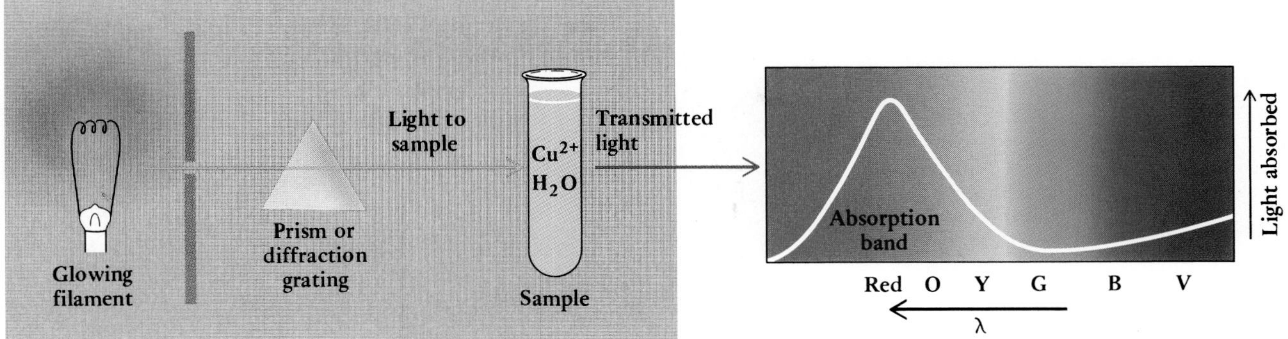

Figure 25.30 A spectrophotometer and the absorption spectrum of Cu^{2+} in water.

band. The bluish $[Cu(H_2O)_6]^{2+}$ ion, which arises when Cu^{2+} salts are put in water (Figure 25.28), has an absorption in the red region, leaving its complement cyan (C) (= G + B) to pass through to the eye. The "bluer" $[Cu(NH_3)_4]^{2+}$ ion has its absorption band shifted a little more toward green so that both red and green are filtered.

THE MECHANISM OF LIGHT ABSORPTION

Transition metal complexes can absorb light because photons of the appropriate energy can excite the complex from its ground state to a higher energy or excited state. Although the details of the process are beyond the scope of this book, in a case such as copper(II) ion in water you can imagine that one of the *d* electrons in a d_{xy}, d_{xz}, or d_{yz} orbital is excited to a $d_{x^2-y^2}$ or d_{z^2} orbital.

ground state Cu^{2+}		excited state Cu^{2+}
⇅ ↑ $x^2 - y, z^2$	Δ_o	⇅ ⇅ $x^2 - y^2, z^2$
⇅ ⇅ ⇅ xy, xz, yz	$\xrightarrow[\text{(light absorbed)}]{+\text{ energy }(\Delta_o)}$	⇅ ⇅ ↑ xy, xz, yz

The fact that high energy light is transmitted means that low energy light (here red light) was absorbed. In turn, this means that the energy gap, Δ_o, is relatively small for the Cu^{2+} ion in water.

THE SPECTROCHEMICAL SERIES OF LIGANDS

Experience with many coordination complexes has revealed that, for a given metal ion, some ligands lead to a small energy separation of the *d* orbitals, while others lead to a large separation. In the language of crystal field theory, some ligands create a small crystal field while others create a large field. What happens as ligands surrounding Co^{3+} are changed

Figure 25.31 The colors of some complexes of Co^{3+}. (Adapted from Irving Geis)

Co^{3+} Complex	Wavelength of light absorbed (nm)	Color of light absorbed	Color of complex
$[Co(NH_3)_5Cl]^{2+}$	535	Yellow	Violet
$[Co(NH_3)_5H_2O]^{3+}$	500	Blue-green	Red
$[Co(NH_3)_6]^{3+}$	475	Blue	Yellow-orange
$[Co(CN)_6]^{3-}$	310	Ultraviolet	Pale yellow

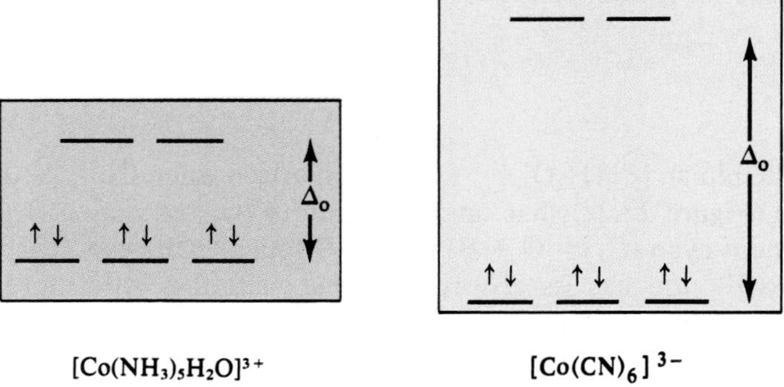

is shown in Figure 25.31. Considering the series $[Co(NH_3)_5X]^{n+}$, you see that the wavelength of light absorbed or filtered from white light gets smaller as X is changed from Cl^- to H_2O to NH_3. That is, the energy of the light being absorbed is increasing, and this indicates that the value of Δ_o is becoming larger. As higher energy light is absorbed, the light that is transmitted changes from blue-violet to red to yellow-orange. Finally, the hexacyanocobaltate(III) ion absorbs in the high energy ultraviolet region owing to the very strong field of CN^- ligands; Δ_o is large. From thousands of experiments, it has been found that the separation or "splitting" of *d* orbitals increases in the following order:

Spectrochemical Series

$$\text{Halides} < ox^{2-} < H_2O < NH_3 < en < phen < CN^-$$

small orbital splitting	large orbital splitting
small Δ_o	large Δ_o
weak field ligands	strong field ligands

This ordering of ligands in terms of their crystal field effects is called the **spectrochemical series**. The spectrochemical series of Co^{3+} complexes is illustrated in Figure 25.22. As ammonia is replaced by H_2O and halide ions, longer wavelengths of light are absorbed, and the colors of the complexes differ (Figure 25.31).

Predicting the possible color of a complex based on the nature of the ligands is risky, since the details of the spectroscopy of such complexes are complicated. However, it is clear that weak field ligands lead to a small splitting, so the complex will absorb relatively low energy photons. Like hexaaquacopper(II) ion, such complexes *tend* to have colors at the blue end of the spectrum. Conversely, strong field ligands cause a large splitting, and their complexes *tend* to have colors at the red end of the spectrum. The red complex $[Fe(CN)_6]^{3-}$ shown in Figure 25.12 is a good example of this.

From the relative position of the ligand in the series, you can make some prediction about the compound's magnetic behavior. Recall that d^4, d^5, d^6, and d^7 complexes can be high or low spin, depending on the value of Δ_o and the pairing energy (Figure 25.27). Thus, for a given metal ion, ligands near the left of the spectrochemical series (as given above) lead to small Δ_o values and high spin complexes. In contrast, ligands near the right end lead to large Δ_o values and give low spin configurations. This means that an ion such as $[CoF_6]^{3-}$ is high spin, while $[Co(NH_3)_6]^{3+}$ and the others shown in Figure 25.31 are low spin.

EXAMPLE 25.6

SPECTROCHEMICAL SERIES

The iron(II) ion in water, $[Fe(H_2O)_6]^{2+}$, is light blue-green. Do you expect the d^6 Fe^{2+} ion in this complex to have a high or low spin configuration?

Solution Figure 25.29 shows that blue-green color can arise if the complex absorbs only low energy red photons. This means the *d* orbital splitting, given by Δ_o, is small. A relatively small value of Δ_o means that the complex has a good chance of being a high spin complex, and this is verified experimentally.

High spin Fe^{2+}

↑ ↑
$x^2 - y^2, z^2$

Δ_o

⇅ ↑ ↑
xy, xz, yz

EXERCISE 25.5 The Spectrochemical Series
Which complex should absorb the lower energy light, hexaaquanickel(II) or hexaamminenickel(II)?

SOMETHING MORE ABOUT Iron Chemistry

When $Fe(NO_3)_3 \cdot 9\ H_2O$, a purple solid, dissolves in water, iron(III) enters the solution as hexaaquairon(III), $[Fe(H_2O)_6]^{3+}$. This complex ion is acidic (see Chapter 17) due to a hydrolysis reaction, the product of which is slightly yellow.

Beaker 1:

$$[Fe(H_2O)_6]^{3+}(aq) + H_2O(\ell) \longrightarrow [Fe(H_2O)_5(OH)]^{2+}(aq) + H_3O^+(aq)$$

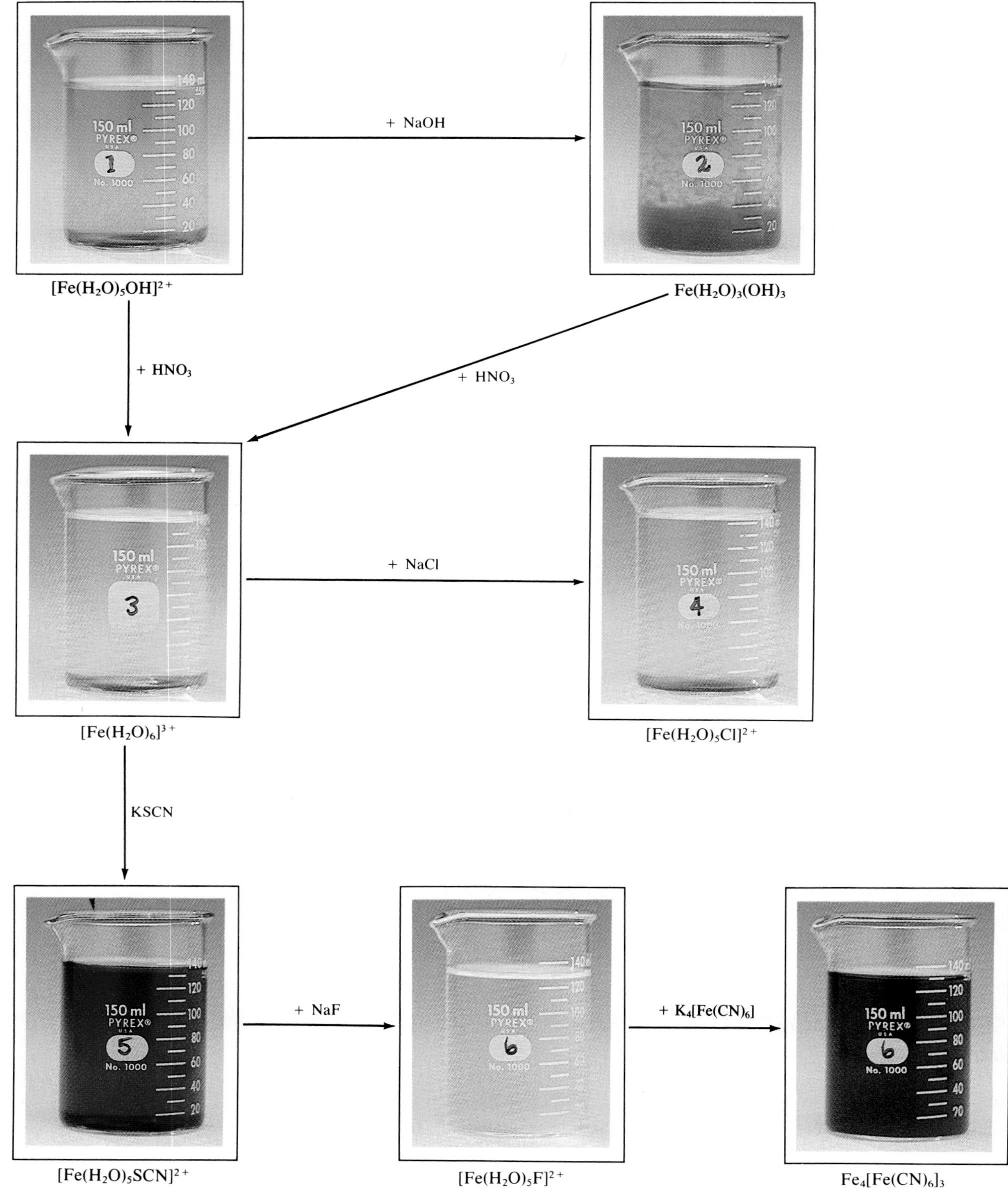

$[Fe(H_2O)_5OH]^{2+}$
+ NaOH
$Fe(H_2O)_3(OH)_3$
+ HNO_3
+ HNO_3
$[Fe(H_2O)_6]^{3+}$
+ NaCl
$[Fe(H_2O)_5Cl]^{2+}$
KSCN
$[Fe(H_2O)_5SCN]^{2+}$
+ NaF
$[Fe(H_2O)_5F]^{2+}$
+ $K_4[Fe(CN)_6]$
$Fe_4[Fe(CN)_6]_3$

When a strong base is added, orange-brown iron(III) hydroxide precipitates (beaker 2).

Beaker 2:

$$[Fe(H_2O)_6]^{3+}(aq) + 3\ OH^-(aq) \longrightarrow Fe(H_2O)_3(OH)_3(s) + 3\ H_2O(\ell)$$

However, if a strong acid (HNO_3) is added to the hydroxide or to the solution containing $[Fe(H_2O)_5(OH)]^{2+}$, the hydroxide is neutralized and the hydrolysis reaction is reversed. The nearly colorless hexaaquairon(III) ion predominates (beaker 3).

Addition of Cl^- or SCN^- to $[Fe(H_2O)_6]^{3+}$ leads to replacement of one or more water ligands to give new complex ions with characteristic colors.

Beaker 4:

$$[Fe(H_2O)_6]^{3+}(aq) + Cl^-(aq) \longrightarrow [Fe(H_2O)_5Cl]^{2+}(aq) + H_2O(\ell)$$

Beaker 5:

$$[Fe(H_2O)_6]^{3+}(aq) + SCN^-(aq) \longrightarrow [Fe(H_2O)_5SCN]^{2+}(aq) + H_2O(\ell)$$

Adding a large quantity of F^- ion causes displacement of the thiocyanate ion and gives the colorless pentaaquafluoroiron(III) ion.

Beaker 6a:

$$[Fe(H_2O)_5SCN]^{2+}(aq) + F^-(aq) \longrightarrow [Fe(H_2O)_5F]^{2+}(aq) + SCN^-(aq)$$

Finally, addition of a few drops of $[Fe(CN)_6]^{4-}$ to either $[Fe(H_2O)_5F]^{2+}$ or $[Fe(H_2O)_6]^{3+}$ gives a deep blue precipitate of "Prussian" or "Turnbull's blue," $Fe_4[Fe(CN)_6]_3$ (beaker 6b). In all of these reactions it is important to notice the effect of ligand on the color of the complex.

SOMETHING MORE ABOUT Nickel Chemistry

When the green solid $NiSO_4 \cdot 6\ H_2O$ is dissolved in water, a green solution of hexaaquanickel(II) results (beaker 1). If the monodentate ligand NH_3 is added, however, the blue complex ion hexaamminenickel(II) is formed.

Beaker 2:

$$\underset{\text{green}}{[Ni(H_2O)_6]^{2+}(aq)} + 6\ NH_3(aq) \longrightarrow \underset{\text{blue}}{[Ni(NH_3)_6]^{2+}(aq)} + 6\ H_2O(\ell)$$

Alternatively, if the bidentate ligand ethylenediamine is added to $[Ni(H_2O)_6]^{2+}$, pairs of water molecules are replaced successively to give complexes of varying shades of purple.

Beaker 3:

$$[Ni(H_2O)_6]^{2+}(aq) + H_2NC_2H_4NH_2(aq) \longrightarrow [Ni(H_2O)_4(H_2NC_2H_4NH_2)]^{2+}(aq) + 2\ H_2O(\ell)$$

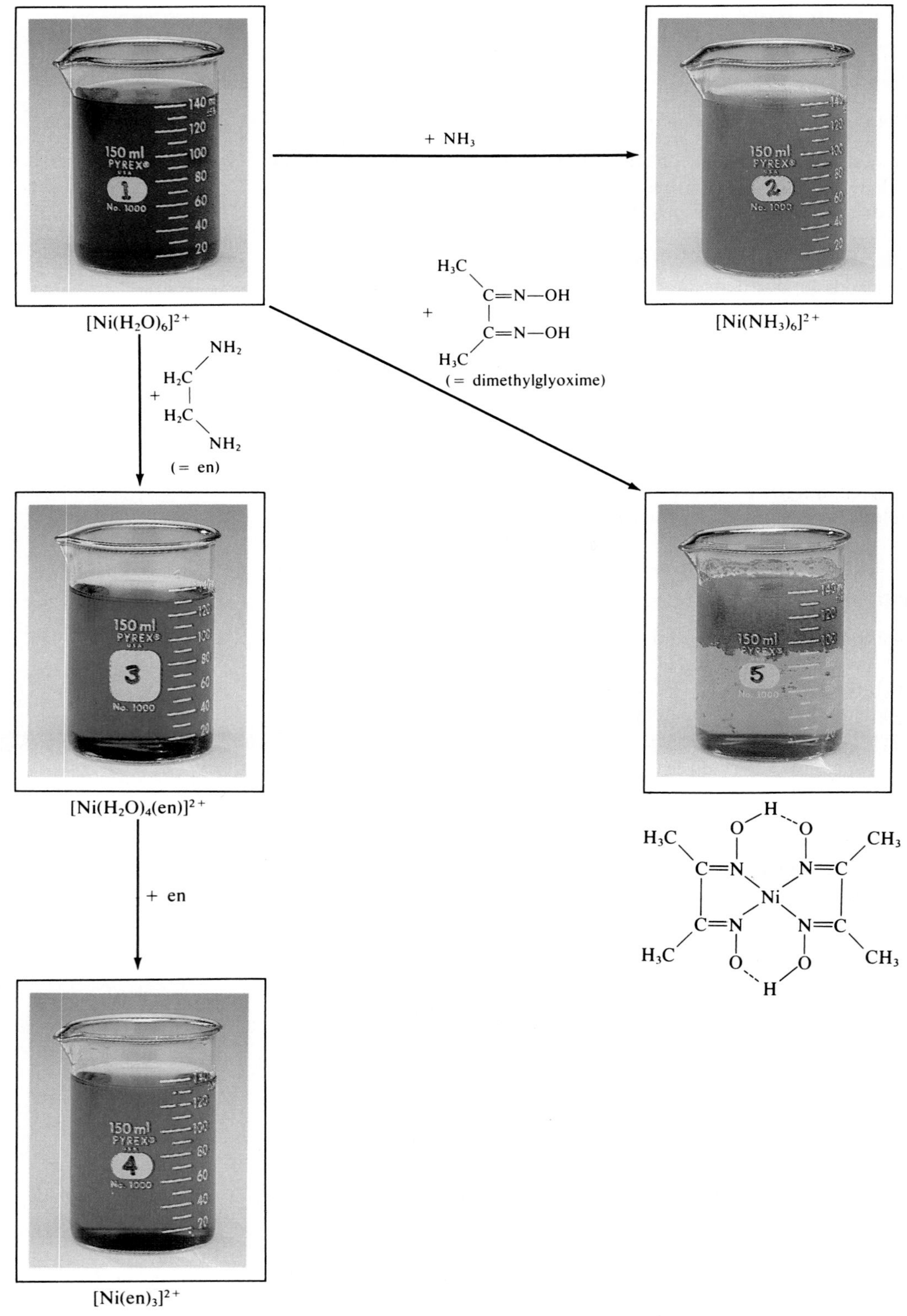
150 ml
PYREX®
1
No. 1000
2
3
4
5
+ NH_3
$[Ni(H_2O)_6]^{2+}$
$[Ni(NH_3)_6]^{2+}$
H_3C
C=N—OH
+
C=N—OH
H_3C
(= dimethylglyoxime)
NH_2
H_2C
+
H_2C
NH_2
(= en)
$[Ni(H_2O)_4(en)]^{2+}$
+ en
$[Ni(en)_3]^{2+}$
H
O
O
H_3C
CH_3
C=N
N=C
Ni
C=N
N=C
H_3C
CH_3
O
O
H

Beaker 4:

$$[Ni(H_2O)_4(H_2NC_2H_4NH_2)]^{2+}(aq) + 2\ H_2NC_2H_4NH_2(aq) \longrightarrow [Ni(H_2NC_2H_4NH_2)_3]^{2+}(aq) + 4\ H_2O(\ell)$$

Finally, if dimethylglyoxime is added to the hexaaquanickel(II) ion, reaction produces a very insoluble strawberry red complex (beaker 5). (Notice that the solution is still green, indicating that some aqueous nickel(II) ion remains in solution.)

SOMETHING MORE ABOUT Superconductivity

Levitating trains, satellites launched without chemical rockets, massive storage of electrical current, and miniature computers operating at blinding speeds! Wonderful stuff to dream about, but what does all this have to do with chemistry?

The phenomenon behind today's dreams is called *superconductivity,* until recently a somewhat obscure effect first reported in 1911 by the Dutch scientist Heike Onnes. Onnes had spent many years learning how to liquefy He gas (the boiling point is an incredible 4 K) and sought to use it to study the behavior of matter at this low temperature. He discovered that, at this temperature, solid mercury is a superconductor; it conducts electricity without resistance. Meissner later found that a superconducting solid repels magnetic materials (see page 8). Both findings fired the imaginations of scientists and novelists alike, giving rise to the possible applications mentioned in the opening paragraph.

Until very recently, one problem with known superconductors was that they had to be bathed in liquid helium to have this property, and, unfortunately, the cost of liquid He ($4/L) is prohibitive for large-scale operation. Therefore, there was a painstaking search for materials that are superconductors at higher temperatures, say closer to that of liquid N_2 (77 K), which costs only about 25 cents/L (see Figure 2.4). The temperature at which superconductivity could be observed inched upward over 75 years to 23 K, but progress was slow and interest largely waned.

All that abruptly changed in January 1986, when IBM scientists in Switzerland (working in secrecy out of fear that their professional reputations would be damaged if it were learned that they were studying ways to make electrical insulators into conductors!) discovered that a barium–lanthanum–copper oxide became superconducting at 35 K. The compound had originally been prepared by French scientists to catalyze chemical reactions and was of the type the Swiss group needed for its studies. Their announcement rocked the world and provoked a flurry of activity to prepare related compounds in the hope that even higher temperatures for superconductivity could be found. Within four months of the publication of the Swiss findings, a material that became superconducting in the 90–100 K range—$YBa_2Cu_3O_x$—was announced! At last here was a material that was superconducting at a temperature high enough to be useful.

What is special about the chemistry of this material, $YBa_2Cu_3O_x$, called a "1-2-3 superconductor"? It is a compound of the rare earth yttrium, the alkaline earth barium, and the transition metal copper, in which one finds four sheets of Cu and O ions. When $x = 7$ the O:Cu ratio in the top and bottom sheets is 1:1, while it is 2:1 in the two inner sheets. Barium ions, and oxide ion bridges, lie between each top (and bottom) sheet and inner sheet. Yttrium(III) ions are located between the inner sheets. The copper ions in the top and bottom sheets have an oxidation number of +3, while the inner copper–oxygen sheets have copper ions with an oxidation number of +2. The +3 oxidation number for copper is rather unusual, numbers of +1 and +2 being much more common, so it is not really surprising that oxide ions can be lost from the compound (the bridging oxygens) without much difficulty. By careful preparation, up to one half of these oxide ions can be removed, and for every such ion removed a copper ion is reduced in oxidation number from +3 to +1, thereby giving a sheet of mixed valence copper ions. If half the oxide ions are removed, the Cu^{3+} sheet will contain equal numbers of +1 and +3 copper ions. However, if less than half are removed, there are more +3 than +1 copper ions, and the material becomes superconducting. It is the sheet containing unequal numbers of copper ions with mixed oxidation number that is necessary for the superconductivity effect, since compounds with all $Cu^{3+}(x = 7)$ or all $Cu^{+}(x = 6)$ are not superconductors. Since Cu^{+} ions have two more electrons than Cu^{3+} ions, there is one electron pair for every Cu^{+}/Cu^{3+} pair, and the superconducting material also has additional Cu^{3+} ions to which electrons are attracted. From this we have the picture of electron pairs that can roam over a sheet of copper and oxygen ions. So the IBM scientists' intuition was correct; it is possible, by chemical modification, to transform an electrical insulator into a conductor—and even a superconductor!

For the latest information on 1-2-3 superconductors, and the relation of copper oxidation numbers to superconductivity, see "Superconductors Beyond 1-2-3," R.J. Cava, Scientific American, August, 1990, page 42.

But what makes a compound a superconductor? Why should the migrating electrons in the 1-2-3 compound experience no resistance to their migrations below a certain temperature? No one knows for sure yet; the theories are not perfected. It is known that the migrating electrons fall in a conduction band, like the migrating electrons in metal conductors, and one theory suggests that superconducting electrons move in pairs through the lattice. What there is about some lattices and not others that makes possible such pair-wise motion of electrons is not yet understood. Or perhaps the pair-motion idea is wrong and a new theory is called for in light of the recent experimental discoveries. Nonetheless, the discovery of compounds that are superconducting above 77 K could be, in the words of one writer, as important as the light bulb, the laser, and the transistor.

A thallium-based compound, $Tl_2Ba_2Ca_2Cu_3O_{10}$, becomes superconducting at 125 K. Thus far, this is the highest known transition temperature. Unfortunately, toxic thallium oxide is required in its preparation, so the commercial applications of the compound are not clear.

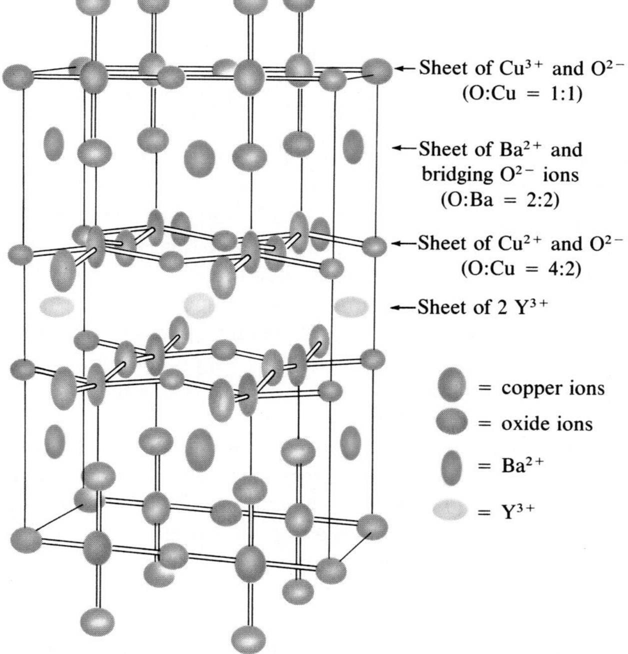

A representation of the unit cell of $YBa_2Cu_3O_7$. The structure consists of sheets of copper and oxygen ions held together with bridging oxygen ions, together with Ba^{2+} and Y^{3+} ions. (NOTE: The oxygens extending above and below the cell (four on the top and four on the bottom), and to the front and rear from the inner CuO_2 sheets, are in the neighboring unit cell.) When O^{2-} ions are lost (as O_2 from the inter-sheet bridges), the material becomes superconducting.

SUMMARY

The **transition elements** or **transition metals** are the elements from Sc to Zn, Y to Cd, and La to Hg. All have *d* electrons among their valence electrons and so are also called ***d*-block elements**. The **inner transition elements** or ***f*-block elements** are those between La and Hf (the **lanthanides**) and between Ac and element 104 (the actinides) (Figure 25.1).

Oxidation numbers of +2 and +3 are common for transition elements, but others are also observed (Section 25.1). Many transition metals, lanthanides, and actinides, as well as their ions, have unpaired electrons and so can be **paramagnetic**. Some metals and their alloys are **ferromagnetic**.

Periodic trends in metal atom radii of transition metals are illustrated in Figure 25.4. The near constancy of size for the elements of Groups 6B through 1B, and for elements of the fifth and sixth periods, is important to their chemistry. The great increase in metal density across the transition metals reflects not only the constancy of size but also the increase in atomic mass (Figure 25.5).

Pyrometallurgy, high temperature processes, and **hydrometallurgy**, techniques using water, are the two major methods for recovering metals from their ores. Ores are usually mixtures of the desired mineral and **gangue**, sand or clay minerals. To separate the desired metal or its compound, **flotation** or **chemical leaching** is used. Iron production in a **blast furnace** uses pyrometallurgy and depends primarily on the reduction of iron oxides (e.g., Fe_2O_3) with CO (generated by the reaction of coke, C, with O_2 in the blast furnace) (Section 25.2). Impure ''pig'' or ''cast'' iron is purified by oxidation of impurities in a **basic oxygen furnace**.

Production of copper uses both pyrometallurgy and hydrometallurgy (Section 25.2). After separation of copper ore from gangue by flotation, the ore is **roasted** in air to give impure copper. This is refined by electrolysis.

Transition metals and their ions form **coordination compounds** or **coordination complexes**. In a complex ion such as $[Fe(CN)_6]^{4-}$ the CN^- ions are called the **ligands** (Section 25.3). The CN^- ion, H_2O, NH_3, and other molecules or ions donating one electron pair to a metal ion are **monodentate**, and those with more than one such atom are **multidentate** (bidentate, tridentate, and so on) (Figure 25.14). Multidentate ligands are also called **chelating** ligands.

The rules for systematically naming coordination compounds are given in Section 25.3. The number of ligand–metal sigma bonds is the metal **coordination number**. Numbers of 2, 4, and 6 are common. Two-coordinate complexes are linear; four-coordinate ones are tetrahedral or square planar; and six-coordinate complexes are usually octahedral.

Isomers are compounds having the same stoichiometry but different atomic arrangements. Coordination complexes may exist as **geometric** and/or **optical** isomers (Section 25.4). Square planar complexes may exist as geometric isomers but not optical isomers. Octahedral complexes can exhibit both forms of isomerism.

Metal–ligand bonding can be explained by either **molecular orbital** or **crystal field** theory, but this chapter focuses on the latter (Section 25.5). Crystal field theory assumes that metal–ligand bonding arises from an electrostatic attraction between the metal cation and ligands. As ligands approach a metal ion, repulsion between ligand lone pairs and the metal electrons affects the metal *d* orbitals differently (Figures 25.24 and 25.25). For an octahedral complex, the $x^2 - y^2$ and z^2 orbitals are affected more than the *xy*, *xz*, and *yz* orbitals by this "crystal field" effect. The energy difference between the sets of *d* orbitals is given by Δ_o. Because there are two sets of *d* orbitals in an octahedral complex (or three sets in a square planar complex), there are two ways of assigning electrons (Section 25.5). The configuration with the maximum number of unpaired electrons is called **high spin**, while a **low spin** configuration has the minimum number of unpaired electrons. The exact assignment depends on Δ_o and *P*, the electron pairing energy. The energy difference between sets of *d* orbitals in complexes is comparable to the energy of visible light. Therefore, coordination complexes are usually colored (Section 25.6). The observed color depends in large part on Δ_o. From experiments, chemists have derived a **spectrochemical series**, an ordering of ligands according to the magnitude of Δ_o.

STUDY QUESTIONS

REVIEW QUESTIONS

1. What distinguishes the transition elements and the lanthanides and actinides from the other elements and from each other?

2. What are the two major types of processes for the recovery of metals from their ores? Give an example of the use of each type of process.

3. What transition elements are commonly used in the manufacture of permanent magnets?

4. Define the following words or phrases: (a) transition elements, (b) coordination compound, (c) complex ion, (d) ligand, (e) chelate, (f) bidentate. Give an example to illustrate each word or phrase.

5. What are three common metal coordination numbers and what structure or structures are possible for each?

6. Distinguish an optical isomer from a geometric isomer. Give an example of each.

7. According to the crystal field model, what is the origin of the splitting of metal *d* orbitals into two sets in an octahedral complex?

8. What are the factors that determine whether a complex will be high or low spin?

CONFIGURATIONS AND PHYSICAL PROPERTIES

9. The chromium compounds pictured in Figure 25.3 have two different oxidation numbers. Using an orbital box diagram for the isolated ion, show the electron configuration of chromium with each of these oxidation numbers. Is either of these expected to be paramagnetic?

10. Give the electron configuration for each of the following ions. Tell whether each is paramagnetic or diamagnetic.
(a) Y^{3+} (d) Pt^{4+}
(b) Rh^{3+} (e) V^{2+}
(c) Ce^{4+} (f) U^{4+}

11. Which element in each of the following pairs should be more dense? Explain each answer briefly.
(a) Ti or Fe (c) Ti or Os
(b) Ti or Zr (d) Zr or Hf

METALLURGY

12. The following equations represent various ways of obtaining transition metals from their compounds. Balance each equation.
(a) $Cr_2O_3(s) + Al(s) \longrightarrow Al_2O_3(s) + Cr(s)$
(b) $TiCl_4(\ell) + Mg(s) \longrightarrow Ti(s) + MgCl_2(s)$
(c) $[Ag(CN)_2]^-(aq) + Zn(s) \longrightarrow Ag(s) + Zn^{2+}(aq) + CN^-(aq)$

13. In the first step in the recovery of copper, an ore such as chalcopyrite is roasted in air to give CuS. If you begin with one ton (908 kg) of the ore, how many tons of SO_2 are produced?

14. Titanium is the fourth most abundant metal. It is strong, lightweight, and corrosion-resistant and thus is used in aircraft engines and chemical plants. To recover the titanium, ilmenite ($FeTiO_3$) is first treated with sulfuric acid to give $Ti(SO_4)_2$, and, after separating $FeSO_4$ and $Ti(SO_4)_2$, the latter is converted to TiO_2 in basic solution.

$$FeTiO_3(s) + 3\ H_2SO_4(\ell) \longrightarrow FeSO_4(aq) + Ti(SO_4)_2(aq) + 3\ H_2O(\ell)$$

$$Ti^{4+}(aq) + 4\ OH^-(aq) \longrightarrow TiO_2(s) + 2\ H_2O(\ell)$$

If you begin with 1.00 kg of ilmenite, how many liters of 18.0 M H_2SO_4 are required to react completely with the ilmenite, and how many kilograms of TiO_2 can theoretically be produced?

15. In the titanium process in Study Question 14, ilmenite ore is leached with sulfuric acid. However, this leads to the significant environmental problem of disposal of the iron(II) sulfate (which, in its hydrated form, is commonly called "copperas"). To avoid this, it has been suggested that HCl be used to leach ilmenite so that the iron-containing product is $FeCl_2$. This can be treated with water and air to give commercially useful iron(III) oxide and regenerate HCl by the reaction

$$2\ FeCl_2(aq) + 2\ H_2O(\ell) + \tfrac{1}{2}\ O_2(g) \longrightarrow Fe_2O_3(s) + 4\ HCl(aq)$$

(a) Write a balanced equation for the treatment of ilmenite with aqueous HCl to give iron(II) chloride, titanium(IV) oxide, and water.
(b) If the equation written in part (a) is combined with the equation for oxidation of $FeCl_2$ to Fe_2O_3 above, is the HCl used in the first step recovered in the second step?
(c) How many grams of iron(III) oxide can be obtained from one ton (908 kg) of ilmenite?

16. Worldwide production of nickel is 750,000 tons per year; most comes from Canada. The *Mond process* is one method for extracting the pure metal. Here the oxide is treated with water gas (H_2 + CO), and the H_2 reduces the oxide to the metal (page 958).

$$NiO(s) + H_2(g) \longrightarrow Ni(s) + H_2O(g)$$

At a temperature of about 50 °C, and at atmospheric pressure, the impure metal reacts with the residual CO to give the volatile compound tetracarbonylnickel(0), $Ni(CO)_4$. If this compound is passed into another part of the reactor, and is heated to 250 °C, it reverts to Ni and CO. The newly formed nickel is pure.

$$Ni(s) + 4\ CO(g) \rightleftharpoons Ni(CO)_4$$

If you wish to produce one ton (908 kg) of pure nickel, how much nickel oxide, H_2, and CO are required?

LIGANDS AND FORMULAS OF COMPLEXES

17. Which of the following ligands are expected to be monodentate and which are multidentate?
(a) CH_3NH_2 (e) $C_2O_4^{2-}$
(b) Br^- (f) $H_3C—C≡N$
(c) en (g) phen
(d) N_3^-

18. Give the oxidation number of the metal ion in each of the following compounds.
(a) $[Ni(NH_3)_6]SO_4$ (c) $K_4[Fe(CN)_6]$
(b) $[Co(NH_3)_4Cl_2]Cl$ (d) $Ni(en)_2Cl_2$

19. Give the formula of a compound formed from one Ni^{2+} ion, one ethylenediamine ligand, three ammonia molecules, and one water molecule. Is the complex neutral or is it charged? If charged, give the charge.

20. Give the formula of the compound formed from one Co^{3+} ion, two ethylenediamine molecules, one water molecule, and one chloride ion. Is the complex neutral or charged? If charged, give the net charge on the ion.

NAMING

21. Write formulas for the following ions or compounds.
(a) dichlorobis(ethylenediamine)nickel(II)
(b) potassium tetrachloroplatinate(II)

22. Write formulas for the following ions or compounds.
(a) diamminetriaquahydroxochromium(II) nitrate
(b) hexaammineiron(III) nitrate
(c) pentacarbonyliron(0)

23. Name the following ligands:
(a) OH^- (b) O^{2-} (c) I^- (d) $C_2O_4^{2-}$

24. Name the compounds pictured in Figure 25.12.

25. Name the following ions or compounds mentioned in Examples 25.1 and 25.2.
(a) $[Ni(C_2O_4)_2(H_2O)_2]^{2-}$ (c) $[Co(en)_2(NO_2)_2]^+$
(b) $[Co(en)_2(NH_3)Cl]^{2+}$ (d) $Pt(NH_3)_2(C_2O_4)$

26. Give the name or formula for each ion or compound, as appropriate.
(a) hydroxopentaaquairon(III) ion
(b) $K_2[Ni(CN)_4]$
(c) $K[Cr(C_2O_4)_2(H_2O)_2]$

27. Give the name or formula for each ion or compound, as appropriate.
(a) dichlorotetraaquachromium(III) chloride
(b) $[Cr(NH_3)_5SO_4]Cl$
(c) sodium tetrachloropalladate(II)

ISOMERS

28. Draw the geometric isomers of
(a) $Pd(NH_3)_4Cl_2$
(b) $Pt(NH_3)_2(NCS)(Br)$ (where NCS^- is bonded to Pt^{2+} through N)
(c) $Co(NH_3)_3(NO_2)_3$ (where NO_2^- is bonded to Co^{3+} through N)

29. Which of the following complexes can have geo-

metrical isomers? If isomers are possible, draw the structures of the isomers and label them as *cis* or *trans* (or *fac* or *mer* if appropriate).
(a) $[Co(H_2O)_4Cl_2]^+$ (c) $Co(H_2O)_3F_3$
(b) $[Pt(NH_3)Br_3]^-$ (d) $[Co(en)_2(NH_3)Cl]^{2+}$

30. For each of the following molecules, decide whether the underlined carbon atom is or is not a chiral center.
(a) $\underline{C}H_2Cl_2$
(b) $H_2N—\underline{C}H(CH_3)—COOH$
(c) $Cl—\underline{C}H(OH)—CH_2Cl$

31. Decide whether each of the following molecules has an enantiomer.

(a) OH, C, H_3C, CH_3, Cl (b) OH, C, H_3C, H, Cl (c) N, H, H, CH_3

32. Four isomers are possible for $[Co(en)(NH_3)_2(H_2O)Cl]^{2+}$. (Two of the four have optical isomers and so each has a nonsuperimposable mirror image.) Draw the structures of the four isomers.

33. Draw all of the isomers (geometric and optical) that are possible for the ion $[Cr(C_2O_4)_2(H_2O)_2]^-$.

MAGNETISM OF COORDINATION COMPLEXES

34. What *d* electron configurations can exhibit both high and low spin in square planar complexes? (In Figure 25.25 notice that the largest energy gap occurs between the z^2 and $x^2 - y^2$ orbitals.)

35. Depict high and low spin configurations for each of the ions below. Tell whether each is diamagnetic or paramagnetic. Give the number of unpaired electrons for the paramagnetic cases.
(a) $[Fe(CN)_6]^{4-}$ (b) $[Co(NH_3)_6]^{3+}$ (c) $[Fe(H_2O)_6]^{3+}$

36. From experiment we know that $[CoF_6]^{3-}$ is paramagnetic and $[Co(NH_3)_6]^{3+}$ is diamagnetic. Using the crystal field model, depict the electron configuration for each ion. What can you conclude about the effect of the ligand on the magnitude of Δ_o?

37. From experiment we know that $[Mn(H_2O)_6]^{2+}$ has five unpaired electrons, whereas $[Mn(CN)_6]^{4-}$ has only one. Using the crystal field model, depict the electron configuration for each ion. What can you conclude about the effect of the ligand on the magnitude of Δ_o?

COLOR

38. Arrange the following ligands in order of increasing crystal field splitting.
(a) CN^- (b) NH_3 (c) F^- (d) H_2O

39. The titanium(III) ion in water, $[Ti(H_2O)_6]^{3+}$, is red. (Its broad absorption band occurs at about 500 nm.) What color of the visible spectrum is absorbed by the ion?

40. The chromium(II) ion in water, $[Cr(H_2O)_6]^{2+}$, absorbs light with a wavelength of about 700 nm. What color is the solution?

GENERAL QUESTIONS

41. From the position of the ligand in the spectrochemical series, decide whether each ion below is high or low spin and whether each is paramagnetic or diamagnetic. If paramagnetic, give the number of unpaired electrons. Use the crystal field model to write the electron configuration of each ion.
(a) $[Fe(CN)_6]^{4-}$ (c) $[MnF_6]^{4-}$
(b) $[Cr(en)_3]^{3+}$ (d) $[Cu(phen)_3]^{2+}$

42. In $Pt(C_2O_4)(NH_3)_2$ the metal ion is surrounded by a square plane of coordinating atoms. Draw a structure for this molecule. Give the oxidation number of the platinum and the name of the compound. Specify whether each ligand is monodentate or bidentate.

43. A complex formed from a cobalt(III) ion, five ammonia molecules, a bromide ion, and a sulfate ion exists in two forms, one dark violet (A) and the other violet-red (B). The dark violet form (A) gives a precipitate with $BaCl_2$ but none with silver nitrate. Form B behaves in the opposite manner. This tells you that one form is $[Co(NH_3)_5Br]SO_4$ and the other is $[Co(NH_3)_5SO_4]Br$. Which is A and which is B? (*Note:* Only when ions such as Br^- or SO_4^{2-} are *not* directly coordinated to the metal ion can they form free ions in aqueous solution.)

44. A platinum-containing compound, known as Magnus's green salt, has the formula $[Pt(NH_3)_4][PtCl_4]$ (where both platinum ions are Pt^{2+}). Name the compound.

45. Titanium metal is valued for its relatively low density, high strength, and corrosion resistance. In the Kroll method, it is made by starting with a titanium-containing ore and using carbon and chlorine to give $TiCl_4$. The latter is then reduced with magnesium to give titanium (see page 113).

$$\underset{\text{ilmenite}}{2\ FeTiO_3(s)} + 7\ Cl_2(g) + 6\ C(s) \longrightarrow 2\ TiCl_4(\ell) + 2\ FeCl_3(s) + 6\ CO(g)$$

$$TiCl_4(\ell) + 2\ Mg(s) \longrightarrow Ti(s) + 2\ MgCl_2(s)$$

Is the first reaction an oxidation–reduction or an acid–base reaction? If you have 25.0 kg of ilmenite, how many kilograms of Cl_2 and carbon are required for complete reaction? How many kilograms of titanium metal can be produced ultimately from the ilmenite?

46. Give the formula of a complex ion formed from one Pt^{2+} ion, one nitrite ion (NO_2^-, which binds to Pt^{2+} through N), one chloride ion, and two ammonia molecules. Are isomers possible? If so, draw the structure of each isomer and tell what type of isomerism is observed. Name the ion.

47. 0.213 g of uranyl(VI) nitrate, $UO_2(NO_3)_2$, was dissolved in 20.0 mL of 1.0 M H_2SO_4 and shaken with Zn. The zinc is a reducing agent and reduces the uranyl ion, UO_2^{2+}, to an ion with a lower oxidation number, U^{n+} ($n < 6$). The uranium-containing solution, after reduction, was titrated with 0.0173 M $KMnO_4$. The potassium permanganate is an oxidizing agent, which oxidizes the uranium back to the +6 oxidation state. 12.47 mL of the potassium permanganate were required for titration to a permanent pink color of the equivalence point. Calculate the oxidation number of the uranium after the uranyl(VI) nitrate was reduced with zinc. Write a balanced, net ionic equation for the oxidation of U^{n+} (where you now know n) with MnO_4^- to give UO_2^{2+} and Mn^{2+} in acid solution.

48. Comment on the fact that, while an aqueous solution of cobalt(III) sulfate is diamagnetic, the solution becomes paramagnetic when a large excess of fluoride ion is added. (*Hint:* In aqueous solution the metal ion is coordinated with water, while excess F^- ion displaces some or all of the water in favor of fluoride ion coordination.)

49. Experiment shows that the compound $K_4[Cr(CN)_6]$ is paramagnetic and has two unpaired electrons. In contrast, the related complex $K_4[Cr(SCN)_6]$ is paramagnetic to the extent of four unpaired electrons. Account for these differences using the crystal field model. Where does the SCN^- ion occur in the spectrochemical series relative to CN^-?

50. In this question, we wish to explore the differences between metal coordination by monodentate and by bidentate ligands.

$$Ni^{2+}(aq) + 6\ NH_3(aq) \rightleftharpoons [Ni(NH_3)_6]^{2+}(aq) \qquad K_{formation} = 10^8$$

$$Ni^{2+}(aq) + 3\ H_2N{-}CH_2{-}CH_2{-}NH_2 \rightleftharpoons [Ni(H_2NCH_2CH_2NH_2)_3]^{2+} \qquad K_{formation} = 10^{18}$$

The vast difference in K_{form} between these complexes, reflecting a large increase in stability of the chelated complex, is called the *chelate effect*. Recall that K is related to the standard free energy of the reaction by $\Delta G° = -RT \ln K$ and $\Delta G° = \Delta H° - T\Delta S°$. Here we know from experiment that $\Delta H°$ for the NH_3 reaction is -109 kJ/mol, and $\Delta H°$ for the en reaction is -117 kJ/mol. Is the difference in $\Delta H°$ sufficient to account for the 10^{10} difference in K_{form}? Comment on the role of entropy in the second reaction compared to that in NH_3 reaction.

51. The gylcinate ion, $H_2N{-}CH_2{-}COO^-$, can function as a bidentate ligand, as pictured below.

[Structure: five-membered chelate ring $H_2C{-}NH_2{-}Cu^{2+}{-}O{-}C(=O){-}$]

Draw all of the isomers that can be formed by the complex ion $Cu(H_2NCH_2COO)_2(H_2O)_2$.

52. You have a sample of an alloy of copper and aluminum and wish to determine the weight percentage of each element in the mixture. A 2.1309-g sample was first dissolved in a mixture of HCl and HNO_3. The resulting solution was made basic with excess ammonia, and the $Al(OH)_3$ that precipitated was collected and dried in a furnace to give 3.8249 g of Al_2O_3. What is the weight percentage of Al and Cu in the alloy?

53. Three different compounds of chromium(III) with water and chloride ion have the same composition: 19.51% Cr, 39.92% Cl, and 40.57% H_2O. One of the compounds is violet and dissolves in water to give a complex ion with a 3+ charge and three Cl^- ions. All of the chloride ions precipitate immediately as AgCl on adding $AgNO_3$. Draw the structure of the complex ion and name the compound.

54. It is usually observed that, for analogous complexes, their stability is in the order $Mn^{2+} < Fe^{2+} < Co^{2+} < Ni^{2+} < Cu^{2+} > Zn^{2+}$. (This order of ions is called the *Irving-Williams* series.) Look up the values of K_{form} for ammonia complexes of Co^{2+}, Ni^{2+}, Cu^{2+}, and Zn^{2+} in Appendix H and verify this statement.

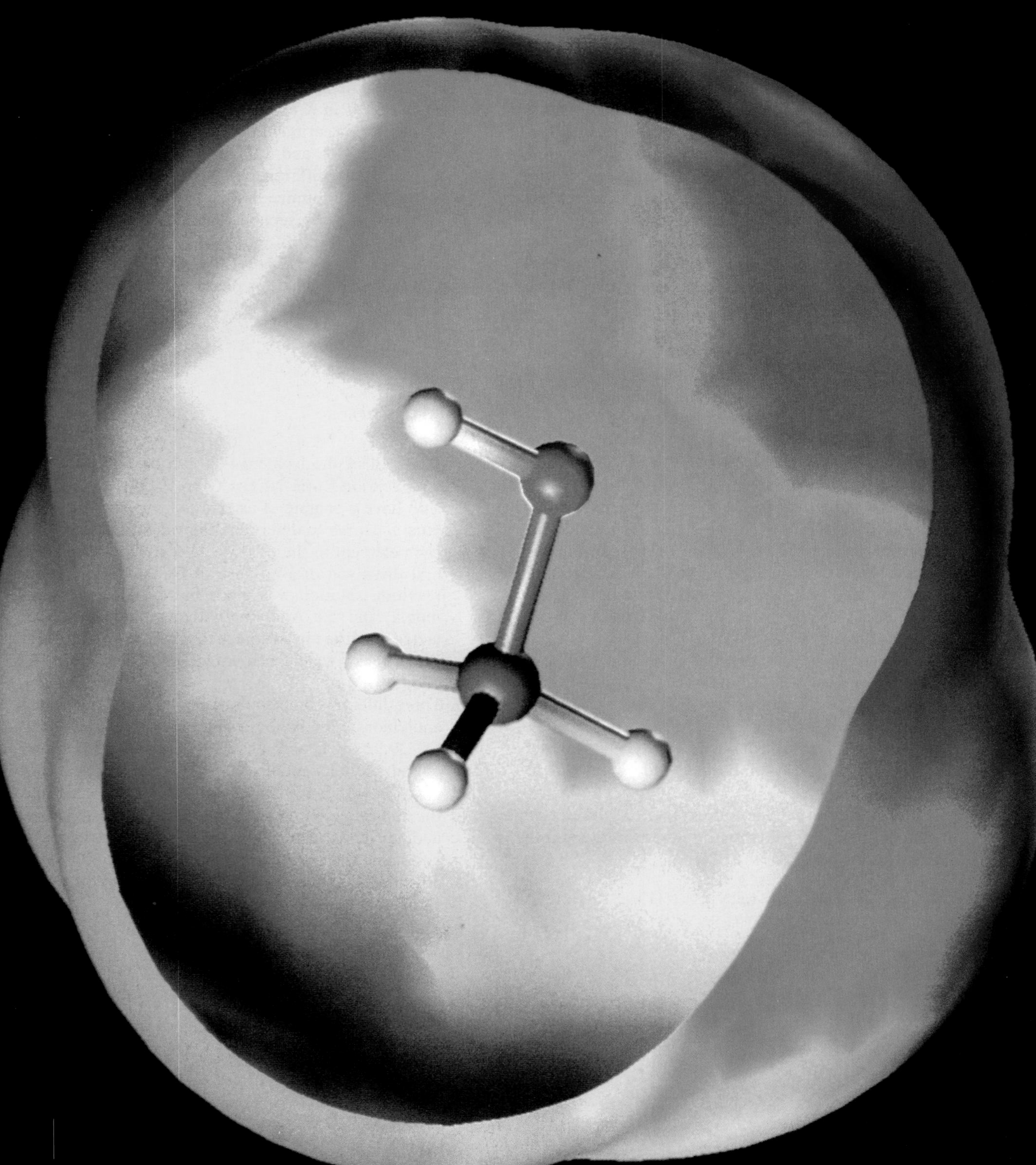

CHAPTER

26

Organic Chemistry

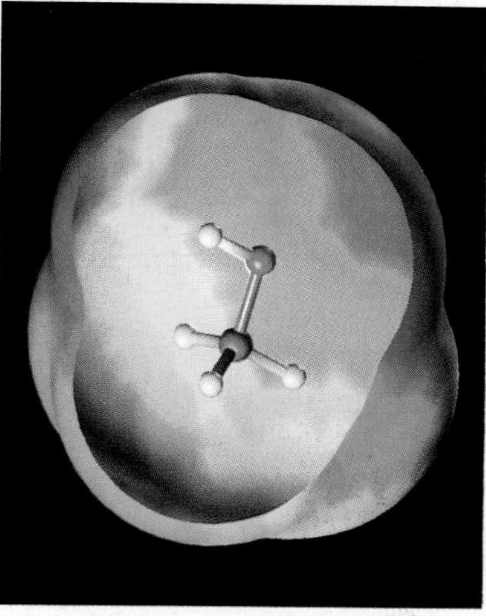

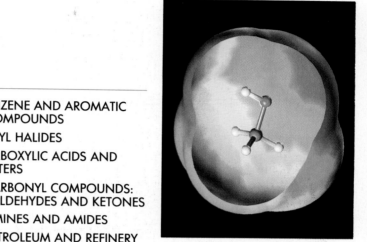

Ball-and-stick model of the simple alcohol methanol, CH_3OH (gray = C, white = H, and red = O), inside a model of the molecular surface. The most reactive site corresponds to the red zone on the surface. (J. Weber)

CHAPTER OUTLINE

Organic chemistry is the study of carbon compounds. Although it may seem surprising that the study of compounds of a single element is a separate branch of chemistry, it is more reasonable when you realize that over 90% of all of the compounds ever prepared contain carbon.

Why are there so many compounds of carbon? It is largely the result of the unique ability of carbon atoms to bond together to form rings or chains of almost unlimited length and at the same time form bonds to atoms of a large number of other elements.

Until about the middle of the 19th century, it was generally believed that organic compounds were different from other chemical compounds in that organic compounds could be formed only by living organisms and that they contained a "vital force" somehow associated with the life process. This vital force theory was gradually abandoned after 1828, however, when Friedrich Wöhler found that urea, a compound present in human urine, could be synthesized from the purely inorganic salt ammonium cyanate.

$$\underset{\text{ammonium cyanate}}{NH_4OCN(aq)} + \text{heat} \longrightarrow \underset{\text{urea}}{H_2N{-}\overset{\overset{\displaystyle O}{\|}}{C}{-}NH_2}$$

Historical Figures in Chemistry: Friedrich Wöhler (1800–1882)

It has been said that Friedrich Wöhler was the last great "all rounder" in chemistry. He began his work as a boy, preparing oxygen, extracting phosphorus, and isolating potassium in the bedroom of his home in Frankfurt, Germany. After his graduation in medicine in 1823, he went to work for the great chemist Berzelius in Stockholm and began a lifetime of contributions to both organic and inorganic chemistry. It may well be a chemical myth that Wöhler was the first to prepare urea from inorganic sources, but he had many admirers who claimed for him many doubtful priorities. When he died in 1882, his obituary said, "As a colleague, a teacher, and in every relation of private life, Wöhler was esteemed and beloved for his kindliness and geniality of his disposition, his modesty and uprightness." (*Platinum Metals Review,* **1985,** *29,* 81)

Soon thereafter many other "natural" compounds such as methyl alcohol, ethyl alcohol, and acetic acid, as well as compounds not found in nature, were synthesized by other chemists.

The synthesis of new organic compounds has continued at an ever increasing pace since the middle of the 19th century. Of the organic compounds known today, only a few can be found in nature. These synthetic efforts have been directed to the production of (a) new compounds with properties superior to those of naturally occurring compounds, (b) compounds with properties that cannot be found in naturally occurring compounds, and (c) naturally occurring compounds in quantities far greater than can be obtained efficiently from natural sources.

The development of new synthetic methods of production is a major area of organic chemistry. In addition, organic chemists study the mechanisms of organic reactions and develop methods for determining the structures and other properties of organic compounds. Our interest in this chapter is to acquaint you with some of the basic reactions used by organic chemists to change one molecule into an entirely new one. To many, this is the essence of chemistry and an area in which chemists can display real creativity.

26.1 SOME PRELIMINARIES: STRUCTURAL FORMULAS AND BONDING

STRUCTURAL FORMULAS The molecular formula of a compound is a listing of the elements of the molecule, usually in alphabetic order. Thus, the molecular formula of ethyl alcohol is C_2H_6O and that of sucrose is $C_{12}H_{22}O_{11}$. Unfortunately, these give us relatively little information. For example, dimethyl ether has the same molecular formula as ethyl alcohol, but the compounds function chemically in very different ways. As for sucrose, more than 100 different compounds have the same formula! For this reason, chemists routinely use **structural formulas** to represent organic compounds. The differences between ethyl alcohol and dimethyl ether, for example, are clearly evident from their structural formulas.

Structural formula:	H—C(H)(H)—C(H)(H)—O—H	H—C(H)(H)—O—C(H)(H)—H
Condensed formula:	CH_3CH_2OH	CH_3OCH_3
	ethyl alcohol	dimethyl ether

For convenience, the lines representing single bonds between atoms are usually omitted, and condensed formulas are written.

BONDING The Pauling electronegativity of carbon (2.5) (Table 10.5) is roughly intermediate between that of the alkali metals and the halogens, and it is similar to that of elements to which carbon is often bonded (H, O, N, S, and P). Thus, while the bonds between carbon and other elements are frequently polar, they are highly covalent.

26.2 THE ALKANES

Hydrocarbons are compounds containing only carbon and hydrogen, and the first class of such compounds to be considered is the **alkanes**. In alkanes, any one carbon atom is bonded to four other atoms (either C or H). Since four sigma bonds are the maximum number of bonds carbon can form, alkanes are often called **saturated** compounds. Later, we shall study **unsaturated** hydrocarbons, compounds that contain double and triple carbon–carbon bonds.

Petroleum and natural gas are composed primarily of alkanes. Consequently, alkanes are important to us as a source of energy and as raw material in the industrial production of other organic compounds.

The simplest alkane, **methane** (CH_4), is most familiar as the principal constituent of natural gas. As outlined in Chapters 10 and 11, the molecule has a tetrahedral structure with an sp^3 hybridized carbon atom.

Methane, CH_4

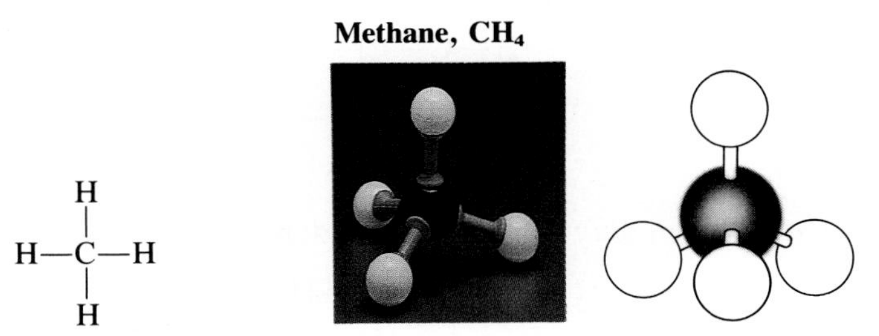

structural formula "ball and stick" model

As suggested by the VSEPR theory (Chapter 10), a similar tetrahedral arrangement is found for all carbon atoms in all organic compounds in which a carbon atom is attached to four other groups. All carbon atoms in alkanes are sp^3 hybridized.

Ethane (C_2H_6) and propane (C_3H_8) are the two- and three-carbon alkanes, respectively. Like methane, they are gases at room temperature. Propane is the principal constituent of liquefied petroleum gas (LPG or bottled gas). You should notice particularly that the tetrahedral geometry of the central carbon atom in propane results in a nonlinear carbon chain.

Ethane, C_2H_6 **Propane, C_3H_8**

$$\begin{array}{c} H\;\;H \\ |\;\;\;\;| \\ H-C-C-H \\ |\;\;\;\;| \\ H\;\;H \end{array}$$

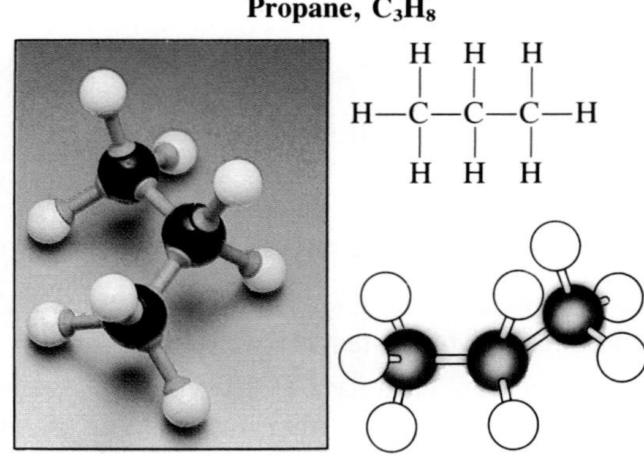

$$\begin{array}{c} H\;\;H\;\;H \\ |\;\;\;\;|\;\;\;\;| \\ H-C-C-C-H \\ |\;\;\;\;|\;\;\;\;| \\ H\;\;H\;\;H \end{array}$$

This is always the case when three or more tetrahedral carbon atoms are connected in a chain or ring.

It is often useful to refer to fragments of alkane structures that result from the removal of a hydrogen from an alkane. These fragments are called **alkyl groups**. For example, removal of H from methane gives a **methyl** group,

$$\underset{\text{methane}}{\begin{array}{c} H \\ | \\ H-C-H \\ | \\ H \end{array}} \xrightarrow{-H} \underset{\text{methyl group}}{\begin{array}{c} H \\ | \\ H-C- \\ | \\ H \end{array}}$$

and ethane gives an **ethyl** group.

$$\underset{\text{ethane}}{\begin{array}{c} H\;\;H \\ |\;\;\;\;| \\ H-C-C-H \\ |\;\;\;\;| \\ H\;\;H \end{array}} \xrightarrow{-H} \underset{\text{ethyl group}}{\begin{array}{c} H\;\;H \\ |\;\;\;\;| \\ H-C-C- \\ |\;\;\;\;| \\ H\;\;H \end{array}}$$

Two different alkyl groups can be derived from propane,

$$\underset{\text{propane}}{\begin{array}{c} H\;\;\;H\;\;\;H \\ |\;\;\;\;\;|\;\;\;\;\;| \\ H-C_3-C_2-C_1-H \\ |\;\;\;\;\;|\;\;\;\;\;| \\ H\;\;\;H\;\;\;H \end{array}}$$

$-$ H from C_1 →

$$\underset{\substack{\textit{normal}\text{-propyl} \\ n\text{-propyl}}}{\begin{array}{c} H\;\;H\;\;H \\ |\;\;\;\;|\;\;\;\;| \\ H-C-C-C- \\ |\;\;\;\;|\;\;\;\;| \\ H\;\;H\;\;H \end{array}}$$

$-$ H from C_2 →

$$\underset{\textit{iso}\text{propyl}}{\begin{array}{c} H\;\;H\;\;H \\ |\;\;\;\;|\;\;\;\;| \\ H-C-C-C-H \\ |\;\;\;\;|\;\;\;\;| \\ H\;\;\;\;\;\;\;\;H \end{array}}$$

where the labels *n* and *iso* are explained just below. The names of alkyl groups come from the name of the parent alkane by dropping "-ane" and adding "-yl."

STRUCTURAL ISOMERS

There are two alkanes with four carbon atoms, one having a "straight" chain and the other a "branched" chain.

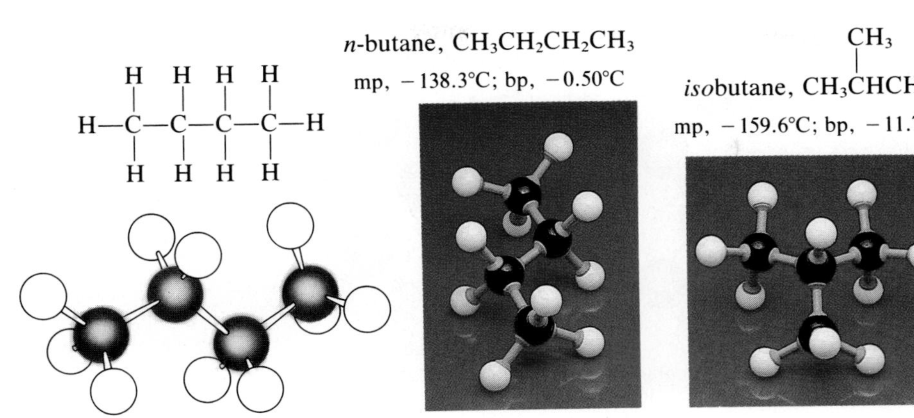

n-butane, $CH_3CH_2CH_2CH_3$
mp, −138.3°C; bp, −0.50°C

*iso*butane, CH_3CHCH_3 (with CH_3 on the central carbon)
mp, −159.6°C; bp, −11.7°C

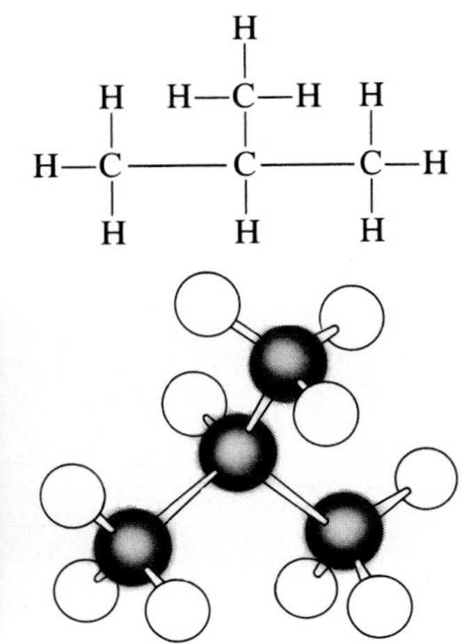

These two compounds are **isomers** of one another. They have the same molecular formulas, C_4H_{10}, but they have different atom arrangements. Thus, we call them more specifically **structural isomers**. The straight-chain isomer is called *n*-butane where the *n* stands for "normal," the term always applied to the straight-chain isomer of any hydrocarbon. The branched-chain *iso*mer is *iso*butane.

OTHER ALKANES

A very large number of carbon atoms can be bonded into a chain. Beyond butane, you encounter these and others.

HYDRO-CARBON	MOLECULAR FORMULA	STRUCTURAL FORMULA
n-Pentane	C_5H_{12}	$CH_3CH_2CH_2CH_2CH_3$
n-Hexane	C_6H_{14}	$CH_3CH_2CH_2CH_2CH_2CH_3$
n-Heptane	C_7H_{16}	$CH_3CH_2CH_2CH_2CH_2CH_2CH_3$
n-Octane	C_8H_{18}	$CH_3CH_2CH_2CH_2CH_2CH_2CH_2CH_3$
n-Nonane	C_9H_{20}	$CH_3CH_2CH_2CH_2CH_2CH_2CH_2CH_2CH_3$
n-Decane	$C_{10}H_{22}$	$CH_3CH_2CH_2CH_2CH_2CH_2CH_2CH_2CH_2CH_3$

To this point, we have written formulas for alkanes having one to ten carbons. It is important to notice that all of these compounds fit the general formula C_nH_{2n+2} where n is the number of carbon atoms. For example,

	n	$2n + 2$
CH_4	1	4
C_5H_{12}	5	12
$C_{10}H_{22}$	10	22

On this basis, you can predict that an alkane with 15 carbon atoms would have the molecular formula $C_{15}H_{2(15)+2} = C_{15}H_{32}$.

With more complex alkanes, the number of structural isomers increases rapidly as the number of carbon atoms increases. For example, there are 3 structural isomers of pentane, 5 of hexane, 9 of heptane, 18 of octane, 75 of decane, and 366,319 structural isomers of $C_{20}H_{42}$!

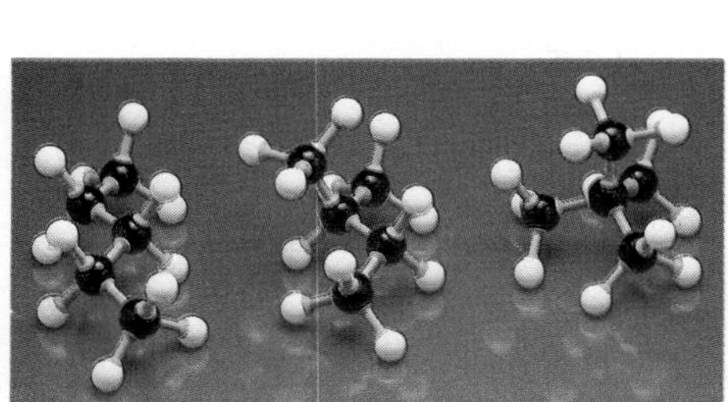

Pentane isomers.

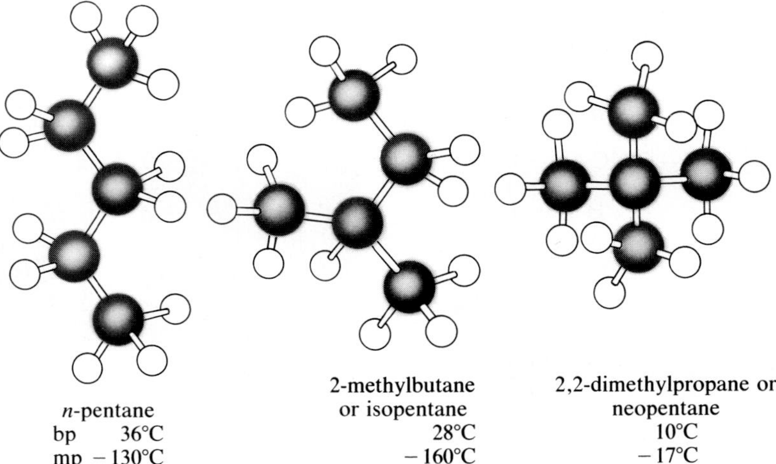

	n-pentane	2-methylbutane or isopentane	2,2-dimethylpropane or neopentane
bp	36°C	28°C	10°C
mp	−130°C	−160°C	−17°C

> **EXERCISE 26.1 Formulas and Structural Isomers**
> (a) Draw the structural isomers of hexane, C_6H_{14}.
> (b) Write the molecular formula for the alkane with 22 carbon atoms.

PHYSICAL PROPERTIES OF ALKANES

The physical properties of a few alkanes are listed in Table 26.1. Notice that alkanes having 1 to 4 carbon atoms are gases at room temperature, while those with 5 to about 16 carbons are liquids at room temperature and normal atmospheric pressure. Thus, natural gas, consisting mostly of methane, is transported within the United States as a gas in pipelines. Butane, although a gas at room temperature, has a relatively high critical temperature and so can be liquefied easily.

Camping stoves often use liquid butane; liquid alkanes, both normal and isomeric, are the principal components of various petroleum products such as gasoline, kerosene, fuel oil, and diesel oil (Section 26.12).

NAMING ALKANES

After butane, the number of carbon atoms in the chain is given by the prefix to the systematic name: pent for 5, hex for 6, hept for 7, oct for 8, and so on.

The alkyl group derived from isobutane by removing H from the center C atom is called tert-butyl. Tert stands for tertiary, which means that the central C atom is bonded to three other C atoms.

Common names and *systematic names* are used for alkanes in particular and for organic compounds in general. Many of the names we have used to this point are **common names** that simply came into use over the years. With the enormous number of organic compounds, however, it is apparent that it would be a problem learning many common names. Thus, we use *systematic names*, which can be derived from a simple set of rules.

The **systematic names** of alkanes are based on the number of carbon atoms in the *longest continuous* carbon chain. If that chain contains three carbons, the parent name is propane, if four carbons, the parent name is

Table 26.1 Some Normal Hydrocarbons (Alkanes)

Molecular Formula	Name	Boiling Point (°C)	Melting Point (°C)	State at Room Temp.
CH_4	Methane	−161	−184	
C_2H_6	Ethane	−88	−183	Gas
C_3H_8	Propane	−42	−188	
C_4H_{10}	*n*-Butane	−0.5	−138	
C_5H_{12}	*n*-Pentane	36	−130	
C_6H_{14}	*n*-Hexane	69	−94	
C_7H_{16}	*n*-Heptane	98	−91	
C_8H_{18}	*n*-Octane	126	−57	
C_9H_{20}	*n*-Nonane	150	−54	
$C_{10}H_{22}$	*n*-Decane	174	−30	
$C_{11}H_{24}$	*n*-Undecane	194.5	−25.6	Liquid
$C_{12}H_{26}$	*n*-Dodecane	214.5	−9.6	
$C_{13}H_{28}$	*n*-Tridecane	234	−6.2	
$C_{14}H_{30}$	*n*-Tetradecane	252.5	+5.5	
$C_{15}H_{32}$	*n*-Pentadecane	270.5	10	
$C_{16}H_{34}$	*n*-Hexadecane	287.5	18	
$C_{17}H_{36}$	*n*-Heptadecane	303	22.5	
$C_{18}H_{38}$	*n*-Octadecane	317	28	
$C_{19}H_{40}$	*n*-Nonadecane	330	32	
$C_{20}H_{42}$	*n*-Eicosane	205 (at 15 torr)	36.7	Solid

butane, and so on. The remaining parts of the structure are treated as substituents on the chain and are indicated as prefixes in the name. Numbers are used to indicate the positions of the substituents on the parent carbon chain.

*Iso*butane is the common name of one of the structural isomers of C_4H_{10}. The longest continuous chain of carbons is three atoms in length, so the systematic name is based on propane. Finally, since a methyl group appears on the second carbon, the correct name is 2-methylpropane.

One of the isomers of pentane is 2-methylbutane. The parent chain, a four-carbon butane chain, is numbered beginning at the end nearer the substituent group. Therefore, the methyl group is indicated as being attached to carbon atom number 2.

The presence of two or more identical substituent groups is indicated by the prefixes di-, tri-, tetra-, and so on, and the position of each substituent is indicated by a number in the prefix. For example, both methyl groups in 2,2-dimethylbutane are attached to carbon atom 2 of a butane chain.

Be careful naming the next compound. The longest chain has five carbon atoms; the fact that they are not printed on the same line is not important. Either methyl group at the right-hand end of the chain may be included in the *longest continuous* carbon chain, the pentane chain. There are three methyl substituents, so the name prefix is "*tri*methyl." Two of these methyl groups are attached to carbon atom 2 and one to carbon atom 4. By convention, the numbering begins at the end giving the lowest possible set of numbers. This means, therefore, that the name is 2,2,4-trimethylpentane (and not the alternative 2,4,4-trimethylpentane).

CH_3 (on C-2)
1 2 3
$CH_3—C—CH_3$
H (on C-2)

Common name: isobutane
Systematic name: 2-methylpropane

CH_3 (on C-2)
$CH_3—CH—CH_2—CH_3$
1 2 3 4

Systematic name: 2-methylbutane

CH_3 (on C-2)
1 2 3 4
$CH_3—C—CH_2—CH_3$
CH_3 (on C-2)

Systematic name: 2,2-dimethylbutane

CH_3 (on C-2) ${}^5CH_3$ (on C-4)
1 2 3 4
$CH_3—C—CH_2—CH$
CH_3 (on C-2) CH_3 (on C-4)

Common name: *iso*octane
Systematic name: 2,2,4-trimethylpentane

Isooctane

The compound 2,2,4-trimethylpentane is the systematic name of a compound commonly called *iso*octane. It is used as a standard in assigning the octane ratings of various types of gasoline. The compound is arbitrarily assigned an octane number of 100, while *n*-hexane is assigned the rating of 0. The antiknock performance of a particular gasoline is compared with that of various mixtures of 2,2,4-trimethylpentane and hexane, and an octane number is assigned on that basis. As the octane numbers of the two standards suggest, branched-chain hydrocarbons have better antiknock properties than straight-chain hydrocarbons.

EXAMPLE 26.1

NAMING COMPOUNDS

Give the systematic name for each of the following compounds:

```
        CH3  CH3                                    CH2—CH3
        |    |                                      |
(a) CH3—CH—CH—CH2—CH3          (c) CH3—CH2—CH—CH2—CH3

                  CH3
                  |
(b) CH3—CH2—CH2—C—CH3
                  |
                  CH3
```

Solution (a) In this compound, the longest continuous carbon chain has five carbon atoms, so the parent name is pentane. There are two methyl groups, one at carbon 2 and the other at carbon 3, so the systematic name is 2,3-dimethylpentane.

(b) The longest continuous carbon chain has five atoms, so the parent name is pentane. The carbon atoms are numbered beginning at the end closest to the substituents. Therefore, the two methyl groups are attached to carbon atom 2, and the systematic name is 2,2-dimethylpentane.

(c) In this compound, the parent name is again pentane. In this case, however, the substituent is an ethyl group (—CH_2CH_3), so the systematic name is 3-ethylpentane.

As a final point, notice that these three compounds have the same molecular formula, C_7H_{16}, and so are structural isomers.

EXERCISE 26.2 Naming Compounds

(a) The three compounds in Example 26.1 are all isomers with the molecular formula C_7H_{16}. What is the name of the parent or unbranched alkane with this formula?

(b) In Exercise 26.1, you drew the structural isomers of C_6H_{14}. Give the systematic name of each of the isomers.

26.3 FUNCTIONAL GROUPS AND COMMON CLASSES OF ORGANIC COMPOUNDS

Organic compounds are characterized by **functional groups**, which are structural fragments found in all members of a given class of compounds and which are centrally involved in the chemical reactions of the class. So, instead of having to learn the properties of an overwhelming number of individual compounds, we study the properties associated with about a dozen basic classes of organic compounds. First, we shall survey briefly the most common classes of organic compounds and their characteristic functional groups. After this survey, we shall then consider a few of the properties of each class in some detail.

There are several classes of hydrocarbons in addition to alkanes. The **alkenes** are characterized by the presence of a C=C double bond. Ethylene and propylene, the simplest alkenes, are compounds of major industrial importance, since they serve as the starting materials in the synthesis of a wide variety of compounds.

Alkenes are also referred to as ***olefins****.*

Ethene (left) and propene (right). Note that the two bent connectors between two C atoms in this and subsequent models are meant to depict a double bond and do not show the geometry of orbitals.

Alkene:	$CH_2{=}CH_2$	$CH_3CH{=}CH_2$
Common name:	Ethylene	Propylene
Systematic name:	Ethene	Propene

The **alkynes** are hydrocarbons having a carbon–carbon triple bond, and the simplest member of the class is *acetylene*, H—C≡C—H, a compound whose synthesis from calcium carbide was first mentioned in Chapter 22.

Alkenes and alkynes are also called ***unsaturated*** *compounds. The latter name comes from the fact that each C atom of the multiple bond is bound to only two or three other C or H atoms; a* ***saturated*** *C atom bonds to four C or H atoms.*

Aromatic compounds constitute yet another class of hydrocarbons, and the fundamental compound in this class is benzene. Aromatic compounds always contain one or more benzene rings or related structures.

Benzene, C_6H_6 Naphthalene, $C_{10}H_8$

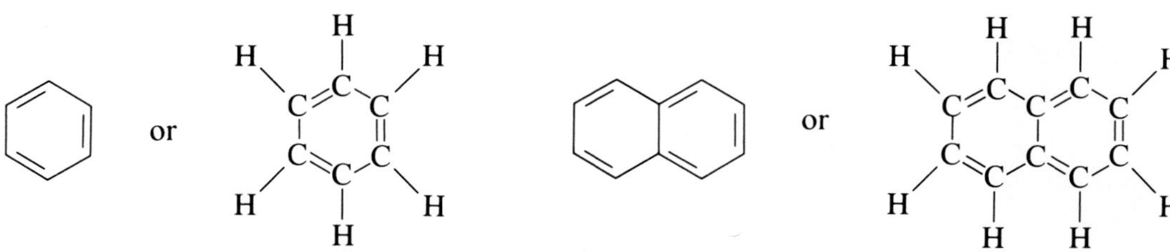

The reason that this class of compounds is called "aromatic" is that many of them have pleasant odors.

If a hydrogen atom of an alkane is replaced by a halogen atom, the result is an **alkyl halide**. Two common members of this class are iodomethane and bromoethane.

Iodomethane, CH_3I Bromoethane, CH_3CH_2Br

Bromoethane

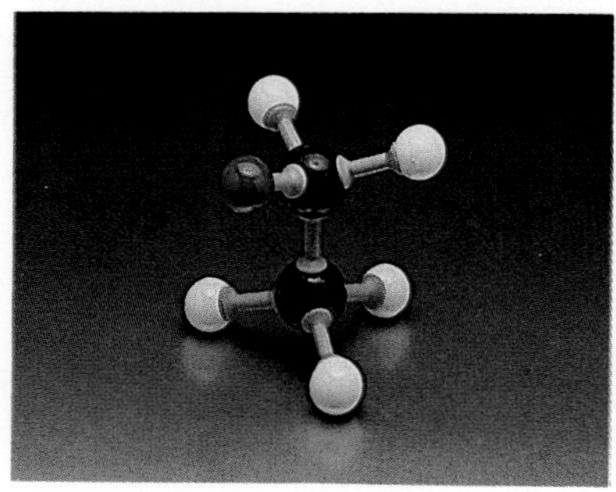

Alcohols are characterized by the presence of a hydroxyl group (—OH) covalently bonded to a saturated carbon atom. The simplest and most common alcohols are methyl and ethyl alcohol.

Methanol (methyl alcohol)

Methyl alcohol, CH_3OH Ethyl alcohol, CH_3CH_2OH

```
    H               H  H
    |               |  |
H—C—O—H         H—C—C—O—H
    |               |  |
    H               H  H
```

Alcohols can be viewed as having been derived from water; one of the H atoms of H_2O is replaced by an organic group. Another class of compounds, **ethers**, is also based on water, but both H atoms have been replaced by organic groups.

Diethyl ether, $CH_3CH_2—O—CH_2CH_3$

```
    H  H     H  H
    |  |     |  |
H—C—C—O—C—C—H
    |  |     |  |
    H  H     H  H
```

Ethanol (ethyl alcohol)

There are several other classes of oxygen-containing organic compounds, in addition to alcohols and ethers. For example, organic acids are characterized by the **carboxyl** group and are often called **carboxylic acids**.

```
  O
  ‖
—C—O—H
```

carboxyl group

The two simplest members of the class are commonly called formic acid and acetic acid, and both are weak Brønsted acids in aqueous solution (Chapter 17).

Formic acid, HCOOH Acetic acid, H_3CCOOH

```
    O               H  O
    ‖               |  ‖
H—C—O—H         H—C—C—O—H
                    |
                    H
```

Diethyl ether

Esters are structurally related to carboxylic acids. Esters differ in that they have a hydrocarbon group in place of the OH hydrogen atom of an acid. Thus, ethyl acetate can be viewed as derived from acetic acid; an ethyl group has replaced the OH hydrogen in this case.

Ethyl acetate, $CH_3COOCH_2CH_3$

```
    H  O     H  H
    |  ‖     |  |
H—C—C—O—C—C—H
    |        |  |
    H        H  H
```

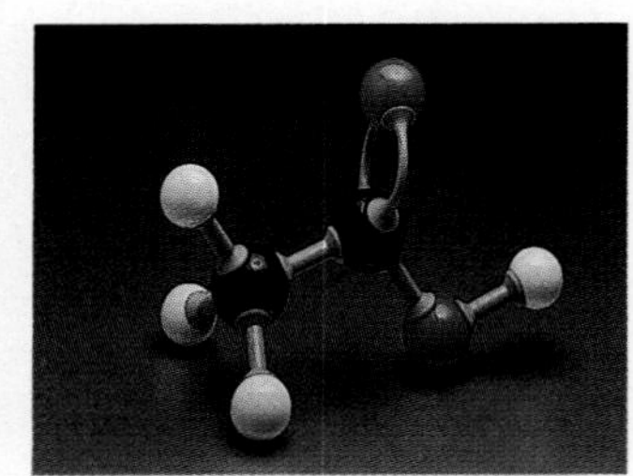
Acetic acid

The grouping $\backslash C{=}O /$ (>C=O) is called a **carbonyl group**, and it is the important functional group of the other major classes of oxygen-containing

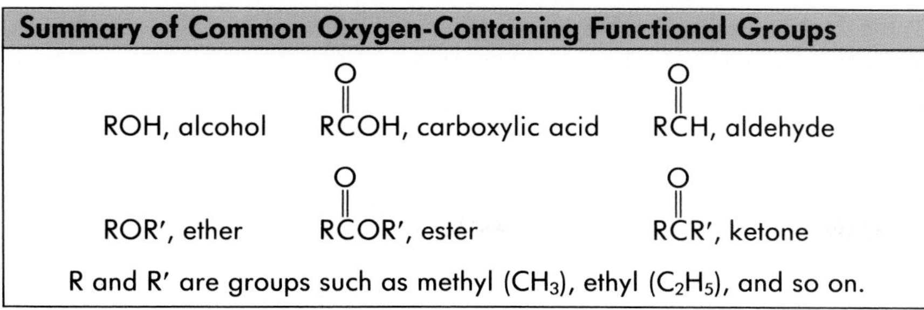

Summary of Common Oxygen-Containing Functional Groups

ROH, alcohol — RCOH (C=O), carboxylic acid — RCH (C=O), aldehyde

ROR′, ether — RCOR′ (C=O), ester — RCR′ (C=O), ketone

R and R′ are groups such as methyl (CH_3), ethyl (C_2H_5), and so on.

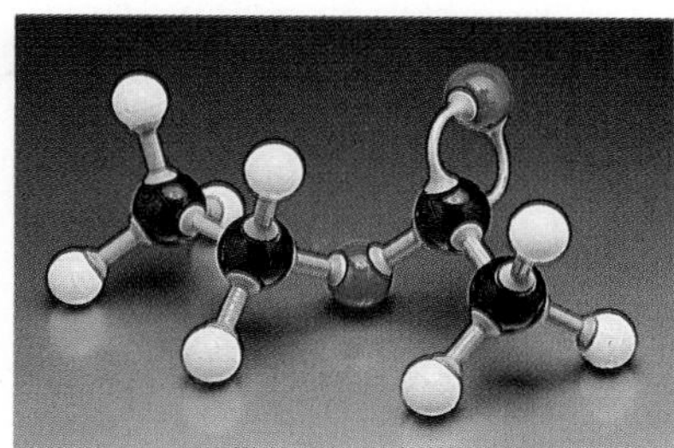

Ethyl acetate

compounds, **aldehydes** and **ketones**. In aldehydes, the carbonyl carbon is bonded to at least one H atom; in ketones, the carbonyl group is bonded to two other carbon atoms. Formaldehyde and acetaldehyde are the simplest aldehydes, and acetone is the simplest ketone.

Formaldehyde, HCHO — Acetaldehyde, H_3CCHO — Acetone, H_3CCOCH_3

```
   O            H O            H O H
   ‖            | ‖            | ‖ |
 H—C—H        H—C—C—H        H—C—C—C—H
                |              |   |
                H              H   H
```

Nitrogen is another common substituent atom in organic compounds, and **amines** represent one major class of nitrogen-containing compounds. These can be viewed as derivatives of ammonia where one or more of the NH hydrogen atoms is replaced by an organic group.

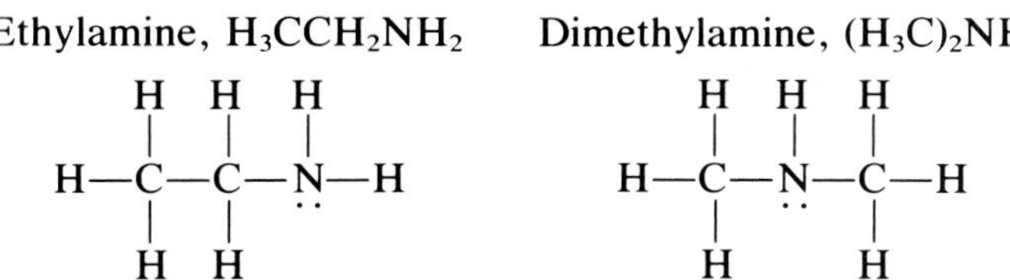

Acetone

With this brief survey of many of the major classes of organic compounds, we can turn to an expanded description of the important aspects of each type.

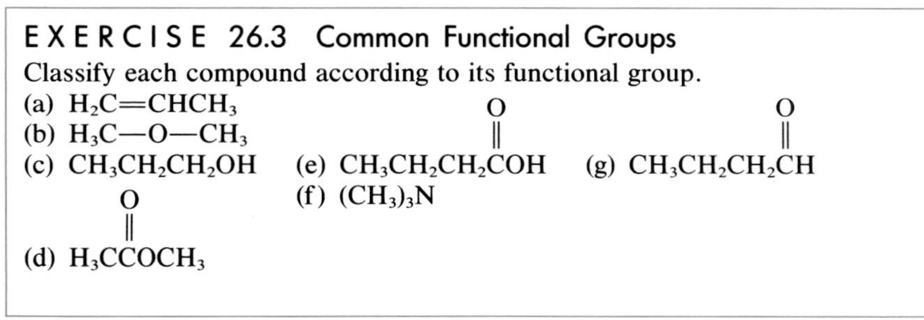

EXERCISE 26.3 Common Functional Groups

Classify each compound according to its functional group.

(a) $H_2C{=}CHCH_3$
(b) $H_3C{-}O{-}CH_3$
(c) $CH_3CH_2CH_2OH$
(d) H_3CCOCH_3 (C=O)
(e) $CH_3CH_2CH_2COH$ (C=O)
(f) $(CH_3)_3N$
(g) $CH_3CH_2CH_2CH$ (C=O)

Ethylamine

Table 26.2 Some Representative Alcohols

Structure	Systematic Name	Common Name	bp (°C)	mp (°C)
CH_3OH	Methanol	Methyl alcohol	65	−97
CH_3CH_2OH	Ethanol	Ethyl alcohol	78	−115
$CH_3CH(OH)CH_3$	2-Propanol	Isopropyl alcohol	97	−126
$CH_3C(CH_3)(OH)CH_3$	2-Methyl-2-propanol	*tert*-Butyl alcohol	83	26

*It is common practice to write generalized structures using the letter **R** to represent alkyl or substituted alkyl groups and **Ar** to represent aromatic or substituted aromatic groups.*

26.4 ALCOHOLS, ROH

The general formula of alcohols, ROH, suggests a structural similarity to water (Table 26.2), and we shall see that some of the properties of alcohols resemble those of water.

More than 1 billion gallons of methyl alcohol, CH_3OH, are produced annually in the United States, and most of it is used to make formaldehyde and acetic acid, important components of a variety of polymers. In addition, methyl alcohol is used as a solvent and as an octane booster and anti-icing agent in gasoline.

Ethyl alcohol is probably one of the most important chemicals produced in nature and by the chemical industry. It is the "alcohol" of alcoholic beverages and is prepared for this purpose by fermentation of sugar from a wide variety of plant sources. For many years industrial alcohol, which is used as a solvent and as a starting material for the synthesis of other compounds, was also made by fermentation. However, in the last several decades, it became cheaper to make the alcohol from petroleum byproducts, specifically by the catalyzed addition of water to ethylene.

$$H_2C{=}CH_2(g) + H_2O(g) \xrightarrow[\text{catalyst (phosphoric acid)}]{\text{addition}} H{-}\underset{H}{\overset{H}{C}}{-}\underset{H}{\overset{H}{C}}{-}OH\ (g)$$

NAMES AND CLASSES OF ALCOHOLS

Systematic names of alcohols are derived from the names of the corresponding alkanes by dropping the "-e" ending and adding "-ol." Thus, the systematic name of methyl alcohol is methanol; for ethyl alcohol it is ethanol. Where necessary, a numerical prefix is used to designate the position of the hydroxyl group (Table 26.2).

Alcohols can be classified into three categories based on the number of carbons bonded to the carbon bearing the hydroxyl group. *Primary alcohols* have one carbon and two hydrogen atoms attached to the carbon atom to which the OH group is attached. When two carbon atoms are attached to the OH carbon, the alcohol is called a *secondary alcohol*, while a *tertiary alcohol* has three carbon atoms attached to the OH carbon.

$$\underset{\text{primary}}{-\overset{|}{\underset{|}{C}}-\overset{H}{\overset{|}{\underset{\underset{H}{|}}{C}}}-O-H} \qquad \underset{\text{secondary}}{-\overset{|}{\underset{|}{C}}-\overset{H}{\overset{|}{\underset{|}{C}}}-O-H \text{ (with } -C- \text{ below)}} \qquad \underset{\text{tertiary}}{-\overset{|}{\underset{|}{C}}-\overset{-C-}{\overset{|}{\underset{|}{C}}}-O-H \text{ (with } -C- \text{ below)}}$$

Patterns of Reactivity

In the sections that follow, you will explore the characteristic reactions of organic compounds. To help you organize your reading, it is useful to see that many of these reactions fall into four classes.

Oxidation–reduction reactions are one major class of reactions. Virtually all organic compounds combust in oxygen,

$$CH_4(g) + 2\ O_2(g) \longrightarrow CO_2(g) + 2\ H_2O(g)$$

but many react with other oxidizing agents such as $KMnO_4$, $Na_2Cr_2O_7$, or even silver ion.

Reductions of certain organic compounds can be done with various metal hydrides such as lithium aluminum hydride, $LiAlH_4$.

$$CH_3\overset{\overset{O}{\|}}{C}CH_3 + 2\ H^- \text{ (from } LiAlH_4) \longrightarrow CH_3\overset{\overset{OH}{|}}{\underset{\underset{H}{|}}{C}}CH_3 + 2\ e^-$$

Addition and **elimination reactions** are contrasting reaction classes. For example, a characteristic reaction of alkenes is the addition of halogens to the C=C bond.

$$H_2C{=}CH_2 + Br_2 \xrightarrow{\text{addition}} Br-\overset{H}{\overset{|}{\underset{\underset{H}{|}}{C}}}-\overset{H}{\overset{|}{\underset{\underset{H}{|}}{C}}}-Br$$

The converse reaction, elimination, is observed when an alcohol is induced to eliminate water to form an alkene.

$$H-\overset{H}{\overset{|}{\underset{\underset{H}{|}}{C}}}-\overset{H}{\overset{|}{\underset{\underset{H}{|}}{C}}}-OH \xrightarrow[H_2SO_4]{\text{elimination}} H_2C{=}CH_2 + HOH$$

Finally, there are many examples of **substitution reactions**. For example, a hydrogen atom of benzene can be substituted by bromine if an appropriate catalyst is used.

$$C_6H_6 \xrightarrow[FeBr_3]{Br_2} HBr + C_6H_5Br$$

Reaction of sodium metal with ethanol gives sodium ethoxide and hydrogen.

EXERCISE 26.4 Alcohol Names and Structures

(a) What are the systematic names for $CH_3CH_2CH_2OH$ and $CH_3CH(OH)CH_3$?

(b) Draw structural formulas for 3-pentanol and 3-methyl-2-butanol.

(c) Tell whether each alcohol in (b) is primary, secondary, or tertiary.

SOME CHEMISTRY OF ALCOHOLS

In analogy with water, alcohols react with alkali metals to produce hydrogen. The other product, instead of *hydr*oxide ion, is called the *alk*oxide ion.

$$HOH(\ell) + Na(s) \longrightarrow \tfrac{1}{2} H_2(g) + Na^+(aq) + OH^-(aq)$$

$$ROH(\ell) + Na(s) \longrightarrow \tfrac{1}{2} H_2(g) + Na^+(\text{in alc}) + RO^-(\text{in alc})$$

Alcohols react with hydrogen bromide or hydrogen iodide in a *substitution reaction* to give the corresponding alkyl halides.

$$ROH + HX \xrightarrow[X\,=\,Br,\,I]{} RX + H_2O$$

Concentrated sulfuric acid has a great affinity for water. Thus, when alcohols (with H and OH on *adjacent carbons*) are treated with this acid, the alcohol is dehydrated to give the corresponding alkene in an *elimination reaction*.

$$\underset{\text{alcohol}}{-\overset{|}{\underset{\underset{H}{|}}{C}}-\overset{|}{\underset{\underset{OH}{|}}{C}}-} \xrightarrow[180\ ^\circ C]{\text{conc. } H_2SO_4} \underset{\text{alkene}}{>C=C<} + H_2O$$

By modifying the conditions slightly, an *ether* can be obtained instead of an alkene. In this process, one mole of water is removed from *two* moles of alcohol.

$$\underset{\text{alcohol}}{2\ ROH} \xrightarrow[\text{heat}]{\text{conc. } H_2SO_4} \underset{\text{ether}}{R—O—R} + H_2O$$

Alcohols can be oxidized, but the products vary according to the type of alcohol involved and the oxidizing agent. Primary alcohols can be oxidized to aldehydes, which in turn can be oxidized to carboxylic acids.

$$\underset{\text{primary alcohol}}{RCH_2OH} \xrightarrow{\text{oxidizing agent}} \underset{\text{aldehyde}}{R—\overset{\overset{O}{\|}}{C}—H} \xrightarrow{\text{oxidizing agent}} \underset{\text{carboxylic acid}}{R—\overset{\overset{O}{\|}}{C}—O—H}$$

Aldehydes are very useful reagents, but getting the reaction to stop at the aldehyde stage is a significant problem. With most oxidizing agents the subsequent oxidation of aldehydes to acids occurs more readily than the first step. Fortunately, a specialized reagent, PCC, has been developed that will oxidize primary alcohols and stop at the aldehyde stage.

Pyridinium chlorochromate, PCC, is a salt prepared from CrO_3, HCl, and pyridine, an aromatic amine.

$$[C_5H_5N—H]^+\ CrO_3Cl^-$$

$$\underset{\text{primary alcohol}}{RCH_2OH} \xrightarrow{PCC} \underset{\text{aldehyde}}{R-\overset{\overset{\displaystyle O}{\|}}{C}-H}$$

Conversion of primary alcohols to carboxylic acids can be accomplished with a variety of oxidizing agents, including potassium permanganate and sodium dichromate.

$$\underset{\text{primary alcohol}}{RCH_2OH} \xrightarrow{KMnO_4 \text{ or } Na_2Cr_2O_7} \underset{\text{carboxylic acid}}{R-\overset{\overset{\displaystyle O}{\|}}{C}-O-H}$$

Secondary alcohols are readily oxidized, but ketones are the only product; no further oxidation can occur. Any of the oxidizing agents mentioned so far is effective.

$$\underset{\text{secondary alcohol}}{R-\overset{\overset{\displaystyle O-H}{|}}{\underset{\underset{\displaystyle H}{|}}{C}}-R'} \xrightarrow{PCC,\ KMnO_4,\text{ or } Na_2Cr_2O_7} \underset{\text{ketone}}{R-\overset{\overset{\displaystyle O}{\|}}{C}-R'}$$

In the oxidation of a primary or secondary alcohol, a hydrogen atom is removed from the carbon atom to which the OH group is bonded. Tertiary alcohols do not have the grouping $-\overset{\overset{\displaystyle H}{|}}{\underset{|}{C}}-OH$, so they are not oxidized under normal conditions.

$$\underset{\text{tertiary alcohol}}{R-\overset{\overset{\displaystyle O-H}{|}}{\underset{\underset{\displaystyle R''}{|}}{C}}-R'} \xrightarrow{\text{oxidizing agents}} \text{no reaction}$$

Several important alcohols have more than one OH group. Two of the most familiar are glycerine and ethylene glycol.

	$H-\overset{\overset{\displaystyle H}{|}}{\underset{\underset{\displaystyle OH}{|}}{C}}-\overset{\overset{\displaystyle H}{|}}{\underset{\underset{\displaystyle OH}{|}}{C}}-H$	$H-\overset{\overset{\displaystyle H}{|}}{\underset{\underset{\displaystyle OH}{|}}{C}}-\overset{\overset{\displaystyle H}{|}}{\underset{\underset{\displaystyle OH}{|}}{C}}-\overset{\overset{\displaystyle H}{|}}{\underset{\underset{\displaystyle OH}{|}}{C}}-H$
Systematic name:	1,2-ethanediol	1,2,3-propanetriol
Common name:	ethylene glycol	glycerine or glycerol

Ethylene glycol

Ethylene glycol is familiar as automobile antifreeze. Glycerine is about 0.6 times as sweet as cane sugar and so is often used in confectioneries; its greatest use, however, is as a softener in soaps and lotions. When it reacts with nitric acid, it produces nitroglycerine,

$$\begin{array}{l} H_2C{-}OH \\ \quad | \\ HC{-}OH \\ \quad | \\ H_2C{-}OH \end{array} + 3\ HNO_3 \longrightarrow \begin{array}{l} H_2C{-}O{-}NO_2 \\ \quad | \\ HC{-}O{-}NO_2 \\ \quad | \\ H_2C{-}O{-}NO_2 \end{array} + 3\ H_2O$$

glycerine nitroglycerine

the explosive component of dynamite. However, nitroglycerine is also used to treat a common heart condition called angina.

EXERCISE 26.5 Reactions of Alcohols

Complete each of the following reactions:

(a) $CH_3CH_2CH_2OH \xrightarrow{PCC}$

(b) $CH_3CH_2CH_2OH \xrightarrow{H_2SO_4/180\ °C}$

(c) $CH_3CH_2CH_2OH + HI \longrightarrow$

26.5 ETHERS, ROR

Ethers, like alcohols, can be viewed as structural derivatives of water. However, in ethers, both H atoms of water are replaced by hydrocarbon groups. Because they lack the —OH group, ethers cannot hydrogen bond with one another. This causes their boiling points to be relatively low, closer to those of alkanes of similar molecular weight. Compare, for example, diethyl ether with pentane and 1-butanol, all compounds having similar molecular weights.

COMPOUND	FORMULA	*M*	bp (°C)
Diethyl ether	$CH_3CH_2OCH_2CH_3$	74	35
Pentane	$CH_3CH_2CH_2CH_2CH_3$	72	36
1-Butanol	$CH_3CH_2CH_2CH_2OH$	74	118

Ethers form peroxides on standing for long periods in air. As these peroxides are explosive, great care should be exercised in storing and using ethers.

Ethers are much less reactive than alcohols. For example, they are unaffected by treatment with oxidizing agents, reducing agents, bases, and most acids. Nonetheless, their relative inertness, but mild Lewis basicity, makes them ideal as solvents, and diethyl ether is widely used for this purpose.

SOMETHING MORE ABOUT Organic Chemistry, Gasoline, and Gallstones.

Tetraethyllead, $Pb(C_2H_5)_4$, has been used around the world since the 1930's as a gasoline additive. Adding about 3 mL of the liquid compound to a gallon of gasoline raises the octane rating by 10 points. However, this has meant that thousands of tons of the combustion products of tetraethyllead—compounds such as PbO—have been spewed out of automobiles into the environment every year. Over ten years ago it was decided to phase out the use of leaded gasolines in the United States and in some other countries, but then the search was on for other compounds that would improve the performance of gasoline and that would make the combustion products less polluting. (See Something More About *Methanol* in

Chapter 6.) One of the compounds suggested as an alternative is methyl tertiary-butyl ether (MTBE).

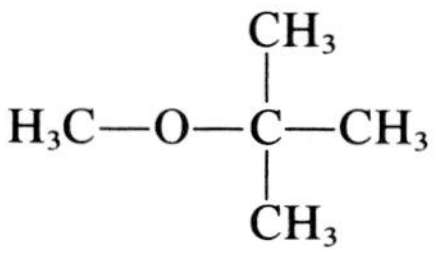

$$H_3C{-}O{-}\underset{\underset{CH_3}{|}}{\overset{\overset{CH_3}{|}}{C}}{-}CH_3$$

methyl-tert-butyl ether (MTBE)

It was recently reported, however, that MTBE has found a most unexpected use. Medical doctors have found that MTBE acts as a solvent on gallstones and may be a practical alternative to gall bladder surgery. The patient is treated by passing small amounts of the solvent through the gall bladder for an average of five hours a day for one to three days. Tests on humans that have been done in Italy and at the Mayo Clinic in Rochester, Minnesota, are encouraging and have led to claims that gallstones can be dissolved completely. A larger trial, with 100 patients, is planned in Italy in the near future.

26.6 UNSATURATED COMPOUNDS: ALKENES AND ALKYNES

Compounds with double and triple bonds between carbon atoms are often referred to as **unsaturated**, because the pi electrons of the multiple bond can be used to form additional sigma bonds.

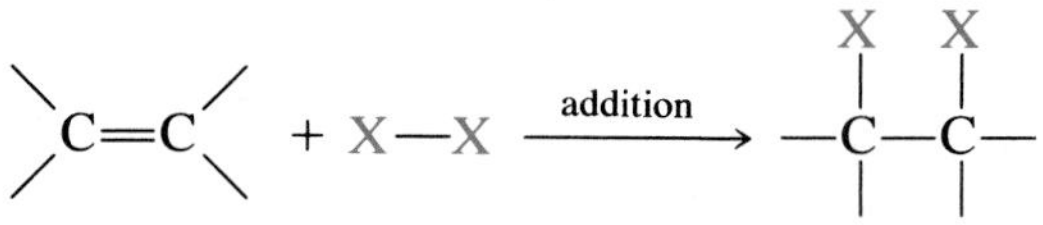

$$\gt C{=}C\lt \; + \; X{-}X \xrightarrow{\text{addition}} -\overset{X}{\underset{|}{\overset{|}{C}}}-\overset{X}{\underset{|}{\overset{|}{C}}}-$$

unsaturated → saturated

Thus, compounds of this class are reactive with respect to the formation of new bonds, and so they are often the building blocks used by the chemical industry to synthesize new compounds. As such, they represent a class of organic compounds of considerable economic importance.

ALKENES, $R_2C{=}CR_2$

NAMING The *systematic names* of compounds such as those listed in Table 26.3 are derived from the name of the corresponding alkane by dropping the "-ane" ending and adding "-ene" in its place. Where necessary, the position of the double bond is included as a prefix. To determine the number to be used, begin counting at the end of the carbon chain closest to the double bond.

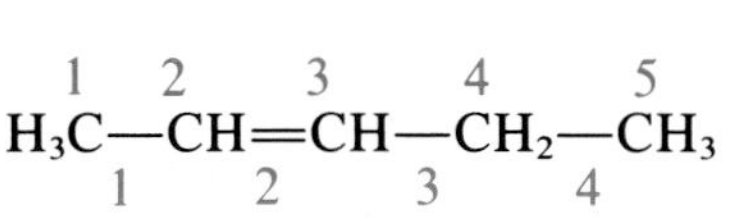

$$\overset{1}{H_3C}{-}\overset{2}{CH}{=}\overset{3}{CH}{-}\overset{4}{CH_2}{-}\overset{5}{CH_3}$$

(numbered from the other end: 1 2 3 4)

2-pentene

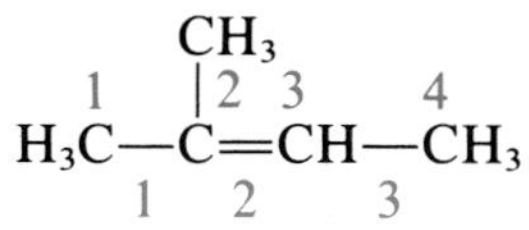

$$\overset{1}{H_3C}{-}\overset{\overset{CH_3}{|}\,2}{C}{=}\overset{3}{CH}{-}\overset{4}{CH_3}$$

(numbered from the other end: 1 2 3)

2-methyl-2-butene

$$\overset{4}{H_3C}{-}\overset{\overset{Br}{|}\,3}{CH}{-}\overset{2}{CH}{=}\overset{1}{CH_2}$$

(lower numbering: 3 2 1)

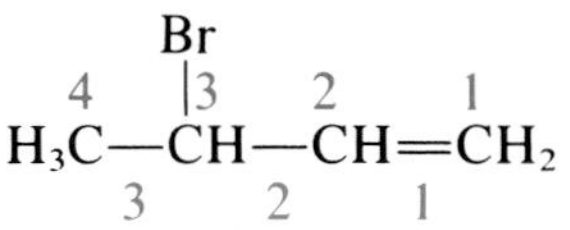

3-bromo-1-butene

BONDING, STRUCTURE, AND ISOMERISM We can tell from the VSEPR theory, and it is confirmed by experiment, that the substituents of a carbon

Table 26.3 Some Representative Alkenes

Structure	Systematic Name	Common Name	bp (°C)	mp (°C)
$CH_2{=}CH_2$	Ethene	Ethylene	−104	−169
$CH_3CH{=}CH_2$	Propene	Propylene	−48	−185
$CH_3CH_2CH{=}CH_2$	1-Butene	—	−6	−185
$CH_3CH{=}CHCH_3$	2-Butene	—	3.7	−139.3*
$CH_3\overset{CH_3}{\overset{\mid}{C}}{=}CH_2$	2-Methylpropene	*Iso*-butylene	−7	−140
$CH_3\overset{CH_3}{\overset{\mid}{C}}{=}CHCH_3$	2-Methyl-2-butene	—	39	−134
$CH_2{=}CH\overset{CH_3}{\overset{\mid}{C}}{=}CH_2$	2-Methyl-1,3-butadiene	Isoprene	34	−146

*Data given are for *cis*-2-butene. *Trans*-2-butene has bp = 0.4°C and mp = −105.8°C.

atom doubly bonded to another carbon will be arranged as a planar triangle.

H H
C=C 120°
H H
sp^2

This means that the C atom must be considered sp^2 hybridized, and the π bond can form using the unhybridized *p* orbitals, one on each carbon, as explained in Chapter 11.

One of the most important aspects of alkene chemistry is to recognize the effect of the π bond on molecular structure. Ordinarily, when two carbon atoms are sigma bonded to one another, the groups at the ends of the bond rotate rapidly about the C—C bond axis.

The groups at the end of a C—C bond can rotate freely with respect to one another.

However, the effect of a carbon–carbon π bond is to *prevent* rotation about the C=C bond axis. Rotation of the ends of the molecule can occur if the two C atoms are joined by *only* a σ bond. This means that rotation can occur in alkenes ($R_2C{=}CR_2$) *only* if the π bond is broken, and this would require considerable energy. Therefore, if a chlorine atom is attached at each carbon in ethylene, for example, there are *two* possible *permanent* arrangements.

$$\underset{\text{Cl}}{\overset{\text{H}}{}}\!\!\!\!\!\!>C{=}C<\!\!\!\!\!\!\underset{\text{Cl}}{\overset{\text{H}}{}}$$

cis-1,2-dichloroethene

mp, −80 °C; bp, 60 °C

$$\underset{\text{Cl}}{\overset{\text{H}}{}}\!\!\!\!\!\!>C{=}C<\!\!\!\!\!\!\underset{\text{H}}{\overset{\text{Cl}}{}}$$

trans-1,2-dichloroethene

mp, −50 °C; bp, 48 °C

Cis and *trans* isomers.

These two molecules have the same formula, but restricted rotation about the C=C bond causes them to have different structures and different properties. Hence, these are *isomers*, since they have the same molecular formula but the atoms are arranged in a different manner. The isomer with chlorine atoms on opposite sides of the bond is called the *trans* isomer and that with the chlorine atoms on the same side is called the *cis* isomer. This form of isomerism is often referred to as *cis–trans* isomerism.

Cis-trans isomerism was first described for compounds of the transition metals in Chapter 25.

EXERCISE 26.6 *Cis–Trans* Isomers

Draw the *cis–trans* isomers for 2-pentene.

ALKENE CHEMISTRY Ethylene and propylene are prepared industrially on a large scale by *steam cracking* hydrocarbons found in natural gas and petroleum.

$$\underset{\text{ethane}}{C_2H_6(g)} \xrightarrow{\text{heat}} \underset{\text{ethylene}}{H_2C{=}CH_2(g)} + H_2(g)$$

36.6 billion pounds of ethylene were produced in 1988 in the United States, making it the fourth largest chemical produced. Propylene, $CH_3CH{=}CH_2$, is tenth on the list of chemicals produced in the United States (Figure 24.1). About 25% of the production is used to make the polymer polypropylene (Chapter 27). Polypropylene fibers are now widely used in carpets.

Almost half of the ethylene product is used to make the polymer *polyethylene* (see Chapter 27), but a large amount is used to make ethylene glycol, vinyl chloride (to make PVC plastics), styrene (to make polystyrene), and a wide variety of other products.

$$H{-}\overset{\overset{\displaystyle H}{|}}{C}{=}\overset{\overset{\displaystyle H}{|}}{C}{-}Cl$$

chloroethene
(vinyl chloride)

$$C_6H_5{-}CH{=}CH_2$$

phenylethene
(styrene)

The **addition reaction** is the most common reaction of alkenes (and alkynes): reagents add to the carbon atoms of the C=C bond. If the reagent is H_2, the reaction is often called a **hydrogenation** and the product is an alkane. Specially prepared forms of metals such as platinum, palladium, nickel, and rhodium, among others, are used as catalysts for this reaction.

$$>C{=}C< + H_2(g) \xrightarrow[\text{metal catalyst}]{\text{hydrogenation}} -\overset{\overset{\displaystyle H}{|}}{\underset{|}{C}}-\overset{\overset{\displaystyle H}{|}}{\underset{|}{C}}-$$

The halogens chlorine and bromine also add to alkenes to give dihaloalkanes,

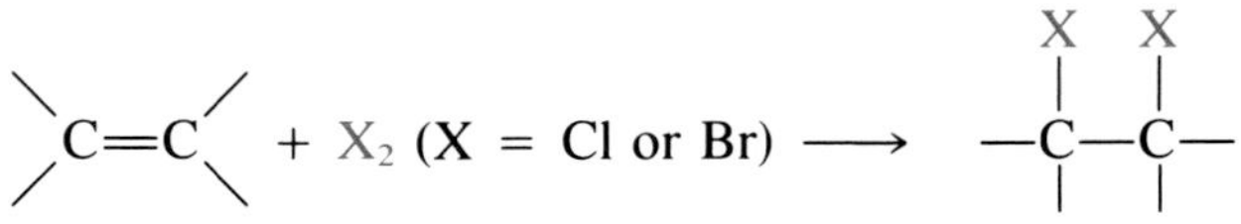

$$\mathrm{>C{=}C<} + X_2\ (X = \mathrm{Cl\ or\ Br}) \longrightarrow \mathrm{-\underset{|}{\overset{X}{\overset{|}{C}}}-\underset{|}{\overset{X}{\overset{|}{C}}}-}$$

alkene dihaloalkane

and hydrogen halide (HCl, HBr, and HI) addition produces haloalkanes.

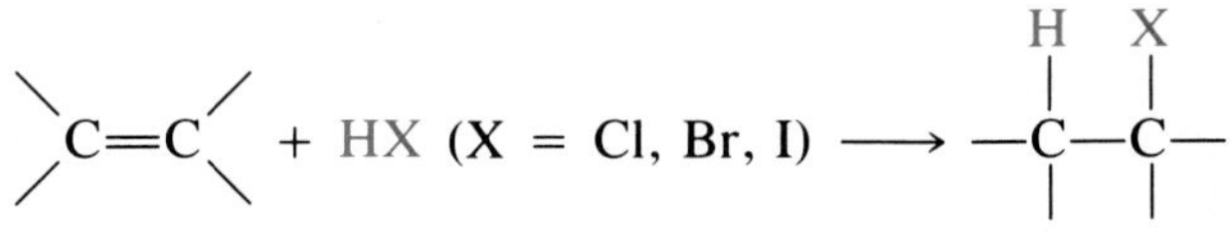

$$\mathrm{>C{=}C<} + HX\ (X = \mathrm{Cl, Br, I}) \longrightarrow \mathrm{-\underset{|}{\overset{H}{\overset{|}{C}}}-\underset{|}{\overset{X}{\overset{|}{C}}}-}$$

alkene haloalkane

When a hydrogen halide adds to propene, two products can be imagined: the halogen can end up attached either to carbon 1 or carbon 2.

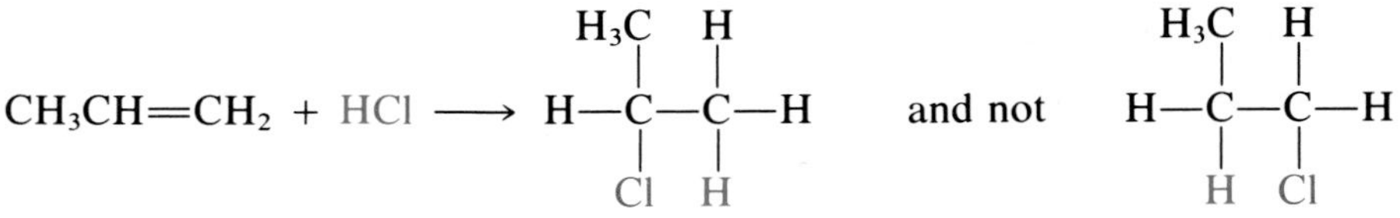

$$CH_3CH{=}CH_2 + HCl \longrightarrow H-\underset{\underset{Cl}{|}}{\overset{\overset{H_3C}{|}}{C}}-\underset{\underset{H}{|}}{\overset{\overset{H}{|}}{C}}-H \quad \text{and not} \quad H-\underset{\underset{H}{|}}{\overset{\overset{H_3C}{|}}{C}}-\underset{\underset{Cl}{|}}{\overset{\overset{H}{|}}{C}}-H$$

propene observed product 2-chloropropane not observed 1-chloropropane

However, only 2-chloropropane is observed. Similarly, reaction of a hydrogen halide with 2-methylpropene results only in 2-halo-2-methylpropane; no 1-halo product is formed.

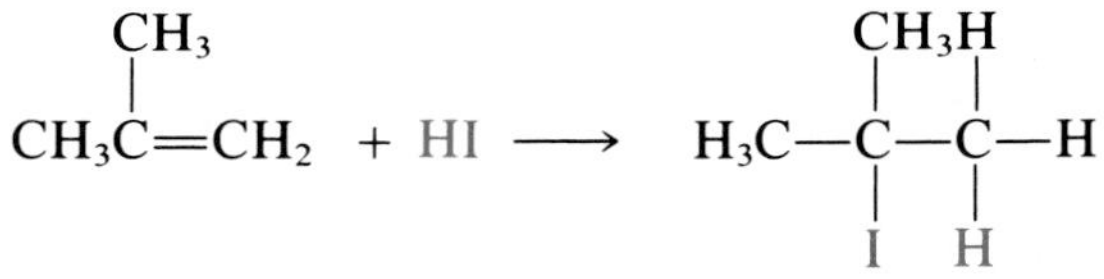

$$CH_3\overset{\overset{CH_3}{|}}{C}{=}CH_2 + HI \longrightarrow H_3C-\underset{\underset{I}{|}}{\overset{\overset{CH_3}{|}}{C}}-\underset{\underset{H}{|}}{\overset{\overset{H}{|}}{C}}-H$$

2-methylpropene 2-iodo-2-methylpropane

Markovnikov's rule can be stated succinctly as "Hydrogen goes where hydrogen is." It is an empirical observation that is understandable in terms of the properties of carbon-containing compounds. A course in organic chemistry will cover this topic.

During the 1860s the Russian chemist Vladimir Markovnikov examined the results of a large number of alkene addition reactions. In cases where two isomeric products might be expected, one usually predominated. He observed a pattern, now known as **Markovnikov's rule**, that can be stated as follows: When a reagent of the type HY is added to an unsymmetrical alkene, the H atom of the reagent becomes attached to the carbon that is already bonded to the greater number of hydrogen atoms. In the reaction of hydrogen chloride with propene, the H atom of HCl becomes attached to carbon 1, the carbon atom bearing two H atoms, and the Cl atom goes to carbon 2, which is bonded to only one H atom.

EXERCISE 26.7 Alkene Addition Reactions

What products are expected from the reaction of HBr with (a) 1-butene and (b) 2-methyl-2-butene?

Alkenes react with water in the presence of acids to produce alcohols. Markovnikov's rule is followed in this *hydration reaction* when an unsymmetrical alkene is involved.

There are exceptions to Markovnikov's rule (called "anti-Markovnikov" additions) that can be explained in terms of modern bonding theory. The subject is studied in detail in courses in organic chemistry.

$$\mathrm{>C{=}C<} + \mathrm{HOH} \xrightarrow{H^+} \mathrm{-\underset{H}{\overset{|}{C}}-\underset{OH}{\overset{|}{C}}-}$$

alkene — alcohol

$$\mathrm{CH_3CH{=}CH_2} + \mathrm{HOH} \xrightarrow{H^+} \mathrm{H-\underset{OH}{\overset{H_3C}{C}}-\underset{H}{\overset{H}{C}}-H}$$

propene — 2-propanol

EXERCISE 26.8 Alkene Reactions
What product is expected when water is added to 2-methyl-1-butene?

ALKYNES, RC≡CR

At one time most synthetic organic compounds were derived from coal, and acetylene (HC≡CH) was an important intermediate in industrial organic synthesis. However, the methods of preparing it

(a) $\mathrm{CaO(s) + 3\ C(s) + heat \longrightarrow CO(g) + CaC_2(s)}$
$\mathrm{CaC_2(s) + 2\ H_2O(\ell) \longrightarrow HC{\equiv}CH(g) + Ca(OH)_2(s)}$

or

(b) $\mathrm{2\ CH_4(g) + heat \longrightarrow HC{\equiv}CH(g) + 3\ H_2(g)}$

involved high temperature processes and were therefore costly. When relatively inexpensive ethylene became available from petroleum and natural gas about 40 years ago, acetylene was almost completely replaced by the alkene.

The systematic names of alkynes are formed similarly to alkenes: drop the "-ane" ending from the corresponding alkane and add "-yne." Thus, the systematic name of acetylene is ethyne. As is the case with many simple, widely used compounds the common name is almost always used. Where necessary the position of the triple bond and the location of substituent groups are indicated by numerical prefixes (Table 26.4).

Table 26.4 Some Simple Alkynes

Structure	Systematic Name	Common Name	bp (°C)
H—C≡C—H	Ethyne	Acetylene	−75
$\mathrm{CH_3C{\equiv}CH}$	Propyne	Methylacetylene	−23
$\mathrm{CH_3CH_2C{\equiv}CH}$	1-Butyne	Ethylacetylene	9
$\mathrm{CH_3C{\equiv}CCH_3}$	2-Butyne	Dimethylacetylene	27
$\mathrm{CH_3\underset{}{\overset{CH_3}{CH}}-C{\equiv}CH}$	3-Methyl-1-butyne	*iso*-Propylacetylene	40

The bonding in the linear acetylene molecule, with sp hybridized C atoms, was described in Chapter 11.

Like alkenes, alkynes undergo addition reactions. An important difference between alkenes and alkynes, however, is that *two* molecules of reagent usually add across a triple bond because addition of the first molecule gives an alkene.

$$\underset{\text{propyne}}{CH_3C{\equiv}CH} + 2\,H_2(g) \xrightarrow[\text{Pt catalyst}]{\text{hydrogenation}} \underset{\text{propane}}{CH_3CH_2CH_3}$$

$$CH_3C{\equiv}CH + 2\,Br_2 \xrightarrow[\text{X}_2\text{ can be Br}_2\text{ or Cl}_2]{\text{halogen addition}} CH_3\overset{Br}{\underset{Br}{C}}{-}\overset{Br}{\underset{Br}{C}}{-}H$$

1,1,2,2-tetrabromopropane

$$CH_3C{\equiv}CH + 2\,HCl \xrightarrow[\text{X in HX can be Cl, Br, I}]{\text{hydrogen halide addition}} CH_3\overset{Cl}{\underset{Cl}{C}}{-}\overset{H}{\underset{H}{C}}{-}H$$

2,2-dichloropropane

Notice that the addition of hydrogen halides produces dihaloalkanes. In accordance with Markovnikov's rule, both H atoms are attached to the same carbon atom.

EXERCISE 26.9 Alkyne Names and Formulas

(a) Draw structural formulas for 2-pentyne and 4-methyl-1-pentyne.
(b) What is the name of the compound $ClCH_2C{\equiv}CH$?

EXERCISE 26.10 Alkyne Reactions

Complete and balance the following reactions:

(a) $CH_3\underset{CH_3}{\underset{|}{C}}HC{\equiv}CH + HBr \longrightarrow$

(b) $CH_3CH_2C{\equiv}CH + Cl_2 \longrightarrow$

26.7 BENZENE AND AROMATIC COMPOUNDS

Benzene, C_6H_6, is usually about 15th on the list of the top 50 chemicals produced in the United States every year, and it is clearly the most important of the aromatics. Almost 12 billion pounds of the liquid were produced in 1988, for example, and almost all of that was converted into the types of products illustrated in Figure 26.1.

Benzene, toluene, the xylenes, and most naphthalene are obtained from petroleum. Heavier aromatics are derived from coal.

Benzene occupies an important place in the history of chemistry. Michael Faraday discovered the compound in 1825 as a by-product of illuminating gas, itself a product of heating coal. The formula of the compound suggested to 19th century chemists that it should be unsaturated, but, if viewed this way, its chemistry was perplexing. Whereas an alkene

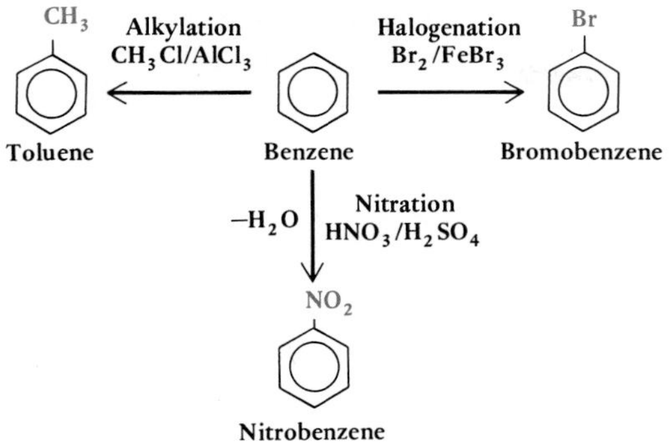

CH_3 Alkylation $CH_3Cl/AlCl_3$ ← Benzene → Halogenation $Br_2/FeBr_3$ Br

Toluene Benzene Bromobenzene

$-H_2O$ Nitration HNO_3/H_2SO_4

NO_2

Nitrobenzene

readily adds Br_2, and is oxidized by acidic potassium permanganate, benzene is not attacked under the same conditions. Higher temperatures or catalysts, however, do lead to reaction, but the products are those of substitution reactions and not addition or oxidation (Figure 26.1). The fact that only one monosubstitution product is ever obtained for a given reaction, along with other experimental observations, strongly suggested to August Kekulé (1829–1896) a *planar and symmetrical ring structure* where all of the H atoms are equivalent.

Benzene, C_6H_6

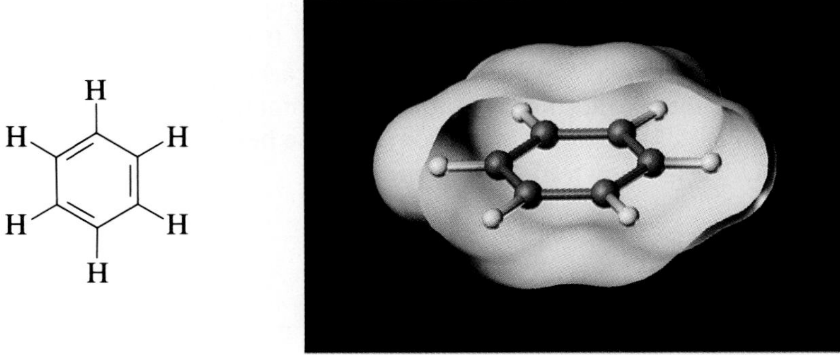

Benzene is a planar, symmetrical molecule with all C and H atoms equivalent. At the left is a computer generated ball-and-stick model of benzene (gray = C and white = H) inside a model of the molecular surface. (J. Weber, University of Geneva, Switzerland)

It was Kekulé who then suggested (in 1865) that the molecule could be represented by some combination of two structures that we now call *resonance structures*.

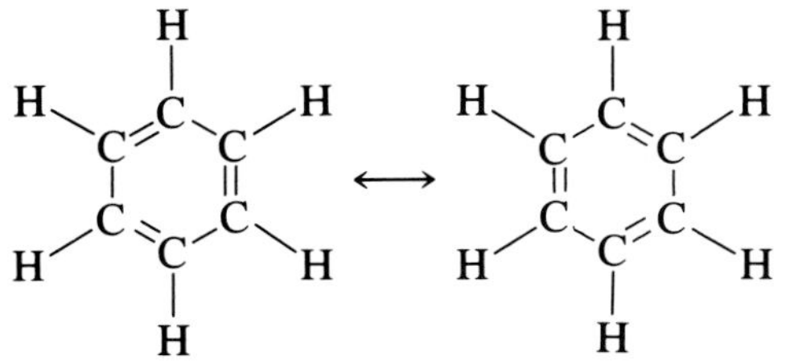

benzene resonance structures

The bonding in benzene can be understood on the basis of the principles in Chapter 11. The geometry of atoms around each carbon atom is trigonal and planar, thereby implying sp^2 hybridization for each carbon.

Unhybridized atomic p orbitals on carbon atoms overlap to form the multicenter π system of benzene.

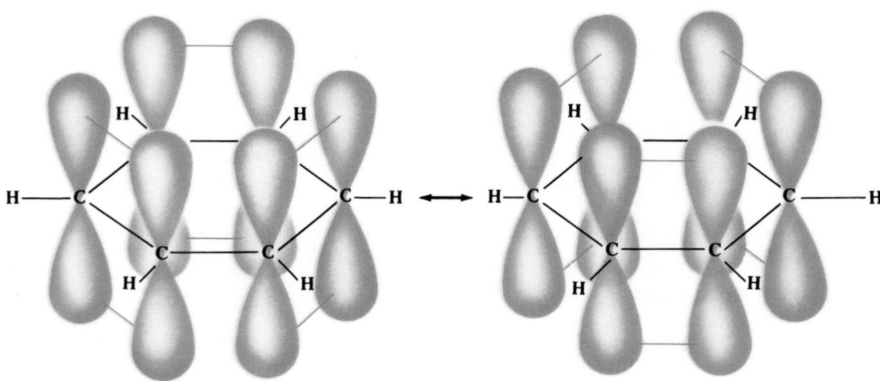

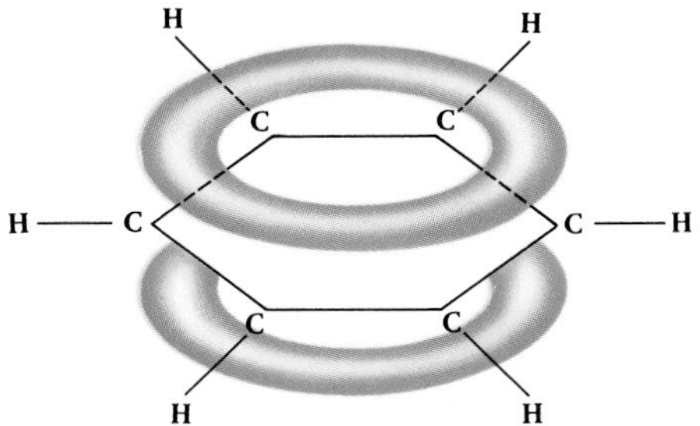

Molecular orbital representation of the delocalized π system formed by the combination of three π-bonding molecular orbitals on carbon atoms.

A commonly used condensed Lewis structure of benzene, which conveys electron delocalization, is

Using one of these hybrid orbitals for binding to an H atom and the other two orbitals for binding to adjacent carbons, an unhybridized p orbital remains on each carbon. Each of these p orbitals, perpendicular to the plane of the ring, contains an electron, and overlap of adjacent orbitals produces a ring of π electrons.

Experimental measurements clearly show that the actual structure of benzene does *not* have alternating single C—C and double C=C bonds. A typical C—C bond has a bond length of 154 pm and the typical C=C bond length is 134 pm. In benzene, all six carbon–carbon distances are the same, 139 pm. This is best explained by assuming that the six unhybridized p orbitals form six molecular orbitals, which delocalize the electrons evenly over the carbon framework. It is this delocalization of electron density that accounts for six carbon–carbon bonds of equal length and leads to the extraordinary stability of aromatic systems. To convey this notion of delocalization, we often draw the benzene ring with a circle in the middle, as shown in the margin.

NAMING AROMATIC COMPOUNDS

Most aromatic compounds are named as derivatives of benzene,

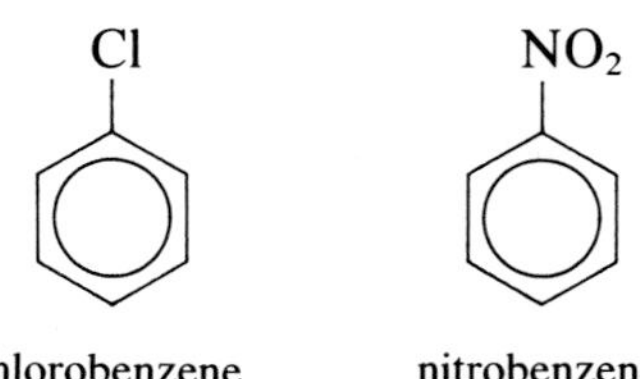

chlorobenzene nitrobenzene

but a few have their own, common names.

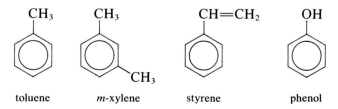

CH_3 CH_3 (and CH_3) $CH{=}CH_2$ OH

toluene *m*-xylene styrene phenol

When a benzene ring bears two substituents, as in the xylene above, the relative positions are indicated by the terms *ortho, meta*, and *para*. When writing the name of the compound, usually only the first letter of one of these terms is given.

X
ortho to X
meta to X
para to X

When two substituents are attached to adjacent carbons, they are said to be *ortho*.

Systematic name:	*o*-dibromobenzene	*o*-dimethylbenzene
Common name:	—	*o*-xylene

If one carbon intervenes between the substituted carbon atoms, then the substituents are *meta*.

Systematic name:	*m*-dichlorobenzene	*m*-dimethylbenzene
Common name:	—	*m*-xylene

Finally, if the substituents are separated by four carbon atoms, then they are *para* to one another.

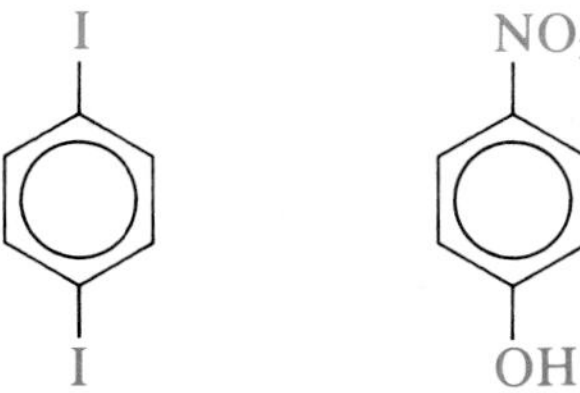

When two different substituents are attached to the benzene ring, they are named in alphabetic order.

Systematic name:	*p*-diiodobenzene	*p*-hydroxynitrobenzene
Common name:	—	*p*-nitrophenol

If benzene has three or more substituents, the system just outlined does not work, and the substituent positions are numbered and are given the lowest possible set of numbers.

1,3,5-trimethylbenzene

1,2,4-tribromobenzene

Table 26.5 Compounds from Coal Tar

Name	Formula	Boiling Point (°C)	Melting Point (°C)	Solubility
Benzene	C_6H_6	80	+6	
Toluene	$C_6H_5CH_3$	111	−95	
o-Xylene	$C_6H_4(CH_3)_2$	144	−27	All
m-Xylene	$C_6H_4(CH_3)_2$	139	−54	insoluble
p-Xylene	$C_6H_4(CH_3)_2$	138	+13	in
Naphthalene	$C_{10}H_8$	218	+80	water
Anthracene	$C_{14}H_{10}$	342	+218	
Phenanthrene	$C_{14}H_{10}$	340	+101	

If the aromatic molecule is one of those with a common name, then the substituent is implicitly assumed to be in position 1.

2,4-dichlorostyrene

2,4,6-trinitrotoluene (TNT)

When the benzene ring has only a single substituent, and when that substituent is more than just an atom or a simple group, it is better to consider the C_6H_5 group as a substituent of an alkane, alkene, alkyne, or other compound. Then C_6H_5 is called a *phenyl* group, and names such as the following are given.

diphenylmethane

2-phenylpropane (cumene)

There are other aromatic compounds beside benzene, and they can be obtained from petroleum or coal tar (Table 26.5). Some of these compounds, such as naphthalene ($C_{10}H_8$, page 1081), consist of two or more aromatic rings sharing ring edges to give even more extensive π electron delocalization. One reason for the interest in compounds such as *benzopyrene* is that they are powerful cancer-causing agents. If painted on the skin of mice, for example, such compounds produce skin tumors. Many such aromatics are found in the smoke from tobacco and from grilling meat on charcoal and are thought to induce colon cancer.

benzopyrene

THE CHEMISTRY OF BENZENE

As was mentioned at the beginning of this section, the chemistry of benzene does not resemble that of alkenes. Addition to a double bond in an aromatic hydrocarbon would disrupt the delocalization of electrons over the entire ring, the feature that adds so much stability to aromatic systems. Thus, the typical addition reactions of alkenes do *not* occur. Benzene and bromine, for example, do not react in the absence of catalysts, and the ring undergoes catalytic hydrogenation only very slowly and then only under extreme conditions.

Cyclohexane, C_6H_{12}, is one of a class of cyclic alkanes of general formula C_nH_{2n}.

$$C_6H_6 + 3\,H_2(g) \xrightarrow[150°C,\ pressure]{Ni\ catalyst} C_6H_{12}$$

benzene cyclohexane

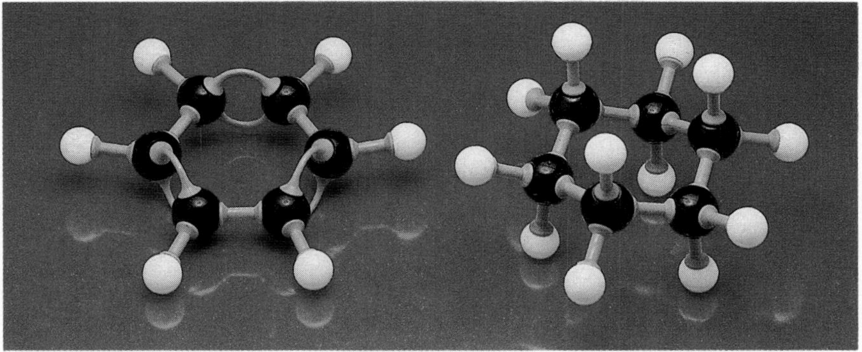

Benzene (left) and cyclohexane (right).

Instead of addition reactions, benzene and other aromatic compounds typically undergo *substitution reactions*. For example, one can halogenate (with a catalyst), nitrate, or alkylate the ring (Figure 26.1).

EXERCISE 26.11 Naming Aromatic Compounds

Give the systematic names for

Br–C_6H_4–NO_2 (para) C_6H_4(Cl)(OH) (ortho)

26.8 ALKYL HALIDES

Alkyl halides such as iodomethane (CH_3I) undergo a wide variety of reactions and are valuable intermediates in the synthesis of many other organic compounds (Table 26.6).

Table 26.6 Some Simple Alkyl Halides

Formula	Systematic Name	Common Name	bp (°C)
CH_3Cl	Chloromethane	Methyl chloride	−24
CH_3I	Iodomethane	Methyl iodide	41
CH_2Cl_2	Dichloromethane	Methylene chloride	40
$CHCl_3$	Trichloromethane	Chloroform	61
CH_3CH_2Cl	Chloroethane	Ethyl chloride	12
$CH_3CH(Br)CH_3$	2-Bromopropane	Isopropyl bromide	59
$(CH_3)_3C{-}Br$	2-Bromo-2-methylpropane	*tert*-Butyl bromide	72

REACTIONS OF ALKYL HALIDES

An important reaction of alkyl halides is **elimination** of hydrogen halide, a reaction commonly called *dehydrohalogenation*. A hydrogen atom and a halogen atom are removed from *adjacent* carbon atoms when the alkyl halide is treated with a strong base (potassium hydroxide) in ethanol as the solvent. The result is an alkene.

$$\underset{\text{alkyl halide}\atop X\,=\,Cl,\ Br,\ I}{-\underset{H}{\overset{|}{\underset{|}{C}}}-\underset{X}{\overset{|}{\underset{|}{C}}}-} \xrightarrow[\text{KOH in ethanol}]{\text{elimination}} \underset{\text{alkene}}{>C=C<} + HX$$

In many cases, particularly with more complex alkyl halides, more than one isomeric alkene can be formed. Generally, the favored product has the fewest H atoms attached to the carbon atoms of the double bond.

$$CH_3CH_2\underset{Br}{\underset{|}{C}}HCH_3 \xrightarrow[\text{ethanol}]{\text{KOH}} \underset{\text{major product}\atop {\text{2-butene}\atop \text{(both } cis \text{ and } trans)}}{CH_3CH{=}CHCH_3} + \underset{\text{minor product}\atop \text{1-butene}}{CH_3CH_2CH{=}CH_2} + HBr$$

When aqueous NaOH is substituted for KOH in ethanol, alkyl halides follow a different reaction course. Now the reaction is a **substitution**, and OH^- replaces the halogen to give an alcohol.

$$\underset{\text{alkyl halide}\atop X\,=\,Cl,\ Br,\ \text{or } I}{RX} + NaOH(aq) \xrightarrow{\text{substitution}} \underset{\text{alcohol}}{ROH} + NaX(aq)$$

Other negatively charged ions can substitute for the halogen as well. For example, the methoxide ion, prepared by the reaction of methanol with sodium, will react with 1-bromopropane to produce methyl propyl ether.

$$\underset{\substack{\text{alkyl halide}\\ X = \text{Cl, Br, I}}}{RX} + \underset{\substack{\text{sodium alkoxide}\\ \text{(prepared from}\\ \text{corresponding alcohol)}}}{NaOR'} \xrightarrow{\text{substitution}} \underset{\text{ether}}{R{-}O{-}R'} + NaX$$

$$\underset{\text{1-bromopropane}}{CH_3CH_2CH_2Br} + \underset{\text{sodium methoxide}}{NaOCH_3} \longrightarrow \underset{\text{methyl propyl ether}}{CH_3CH_2CH_2{-}O{-}CH_3} + NaBr$$

One of the most important reactions of alkyl (and aromatic) halides is their reaction with metallic magnesium to give an organometallic compound known as a **Grignard reagent**. These reagents are extraordinarily versatile synthetic intermediates and are widely used. For example, iodomethane reacts with magnesium to produce methylmagnesium iodide.

$$\underset{\text{alkyl or aryl halide}}{RX} + Mg \longrightarrow \underset{\text{Grignard reagent}}{RMgX}$$

$$\underset{\text{iodomethane}}{CH_3I} + Mg \longrightarrow \underset{\text{methylmagnesium iodide}}{CH_3MgI}$$

Grignard reagents react with a wide range of Brønsted acids, including water, to produce the corresponding alkanes.

$$\underset{\text{Grignard reagent}}{RMgX} + H^+ \longrightarrow RH + MgX^+$$

For this reason, the presence of water must be avoided when Grignard reagents are prepared and when they are used in reactions.

Victor Grignard (1871–1935) was a French chemist. He received the Nobel Prize in Chemistry in 1912 for the discovery of the compounds we now call "Grignard reagents."

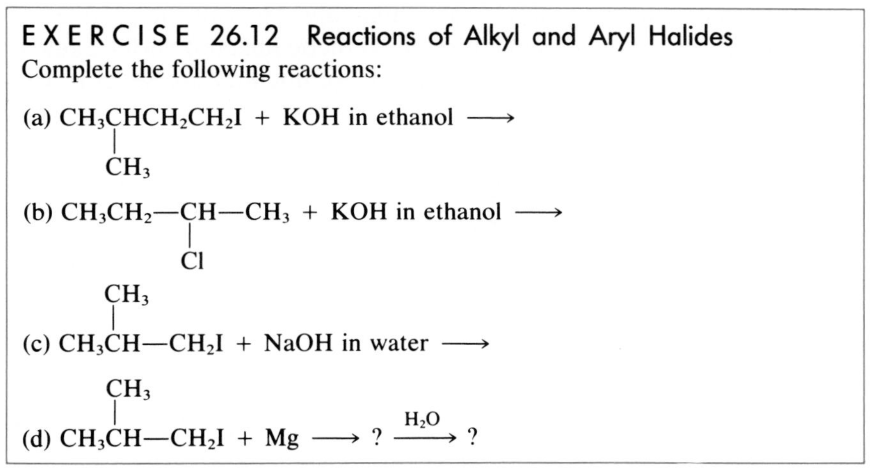

EXERCISE 26.12 Reactions of Alkyl and Aryl Halides

Complete the following reactions:

(a) $CH_3\underset{\substack{|\\ CH_3}}{C}HCH_2CH_2I$ + KOH in ethanol $\longrightarrow$

(b) $CH_3CH_2{-}\underset{\substack{|\\ Cl}}{C}H{-}CH_3$ + KOH in ethanol $\longrightarrow$

(c) $CH_3\overset{\substack{CH_3\\ |}}{C}H{-}CH_2I$ + NaOH in water $\longrightarrow$

(d) $CH_3\overset{\substack{CH_3\\ |}}{C}H{-}CH_2I + Mg \longrightarrow\ ?\ \xrightarrow{H_2O}\ ?$

26.9 CARBOXYLIC ACIDS AND ESTERS

The systematic names of carboxylic acids are easily derived: "-e" is dropped from the name of the corresponding alkane and "-oic" added followed by the word "acid" (as in "alkanoic acid").

Carboxylic acids are the end products of alcohol oxidation. A number are found in nature and have been known for many years. As a result, some of the familiar carboxylic acids are known by their common names.

The simplest common acid is formic acid (Table 26.7), the substance responsible for the sting of an ant. Therefore, the name of the acid comes from the Latin word (*formica*) for ant. Acetic acid gives the sour

Table 26.7 Some Simple Carboxylic Acids

Structure	Common Name	Systematic Name	bp (°C)	K_a
$H-\overset{\overset{O}{\parallel}}{C}-OH$	Formic acid	Methanoic acid	101	1.8×10^{-4}
$CH_3\overset{\overset{O}{\parallel}}{C}OH$	Acetic acid	Ethanoic acid	118	1.8×10^{-5}
$CH_3CH_2\overset{\overset{O}{\parallel}}{C}OH$	Propionic acid	Propanoic acid	141	1.4×10^{-5}
$CH_3(CH_2)_2\overset{\overset{O}{\parallel}}{C}OH$	Butyric acid	Butanoic acid	163	1.5×10^{-5}
$CH_3(CH_2)_3\overset{\overset{O}{\parallel}}{C}OH$	Valeric acid	Pentanoic acid	187	1.4×10^{-5}

taste to vinegar, and the name comes from the Latin word for this substance. Butyric acid gives rancid butter its unpleasant odor, and the name is related to the Latin word for butter, *butyrum*. The names for caproic (C_6) acid, as well as caprylic (C_8) and capric (C_{10}) acids, are all derived from the Latin word for goat, since these acids combine to give goats their characteristic odor. In general, the longer-chain carboxylic acids have unpleasant odors.

Acetic acid is now produced almost completely by the catalytic addition of CO to methanol,

$$CO(g) + CH_3OH(g) \longrightarrow CH_3COOH$$

See Chapter 15, page 646.

Acetic acid is a major industrial chemical, with more than 3 billion pounds produced in the United States annually. Most of the production is used ultimately in the manufacture of synthetic fibers of various kinds.

The only other acids produced in large quantity are several with two carboxylic acid groups.

$$HO\overset{\overset{O}{\parallel}}{C}(CH_2)_4\overset{\overset{O}{\parallel}}{C}OH \qquad HO-\overset{\overset{O}{\parallel}}{C}-C_6H_4-\overset{\overset{O}{\parallel}}{C}-OH \qquad C_6H_4(\overset{\overset{O}{\parallel}}{C}-OH)_2$$

adipic acid — terephthalic acid — phthalic acid

These three acids are used to manufacture polymers of types described in Chapter 27. There are others, however, that you may have seen before (Table 26.8), since some occur in nature.

Table 26.8 Some Common Dicarboxylic Acids

Formula	Common Name	Natural Source
HOOC—COOH	Oxalic acid	Occurs in many plants and vegetables
HOOC—CH_2—COOH	Malonic acid	—
HOOC—CH_2CH_2—COOH	Succinic acid	Occurs in fossils, lichens; discovered by Agricola in 1546
HOOC—$CH_2CH_2CH_2$—COOH	Glutaric acid	In sugar beets

PROPERTIES OF CARBOXYLIC ACIDS

Carboxylic acids are polar and readily form hydrogen bonds. In the pure liquid state, a pair of carboxylic acid molecules is held together by two hydrogen bonds.

$$\mathrm{R{-}C}\begin{matrix}\nearrow\!\!\!\!= \mathrm{O{\cdots}H{-}O} \searrow \\ \searrow \mathrm{O{-}H{\cdots}O} =\!\!\!\!\nearrow\end{matrix}\mathrm{C{-}R}$$

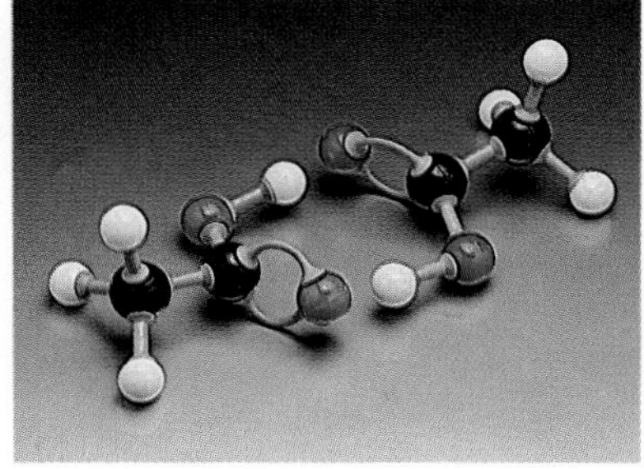

Hydrogen-bonded dimer of acetic acid.

This hydrogen bonding results in relatively high boiling points for the acid, even higher than those of alcohols of comparable molar mass. For example, formic acid (46 g/mol) has a boiling point of 101 °C, while ethanol (46 g/mol) has a boiling point of only 78 °C. As a final comparison, you might note that propane, a nonpolar alkane with a comparable molar mass (44 g/mol) has a very low boiling point (−42 °C).

All organic acids are weak Brønsted acids with K_a values in the vicinity of 10^{-5}. As you would expect, all react with bases to form salts.

Names of carboxylic acid salts are in two parts: (a) name of cation; (b) name of acid anion derived by dropping the "-ic" ending from acid name and adding "-ate." For example, CH_3COONa is sodium acetate.

$$\mathrm{R\overset{O}{\overset{\|}{C}}OH(aq)} + \mathrm{base(aq)} \longrightarrow \mathrm{baseH^+(aq)} + \mathrm{R\overset{O}{\overset{\|}{C}}O^-(aq)}$$

$$\mathrm{C_6H_5{-}\overset{O}{\overset{\|}{C}}OH(aq)} + \mathrm{KOH(aq)} \longrightarrow \mathrm{H_2O(\ell)} + \mathrm{C_6H_5{-}\overset{O}{\overset{\|}{C}}OK(aq)}$$

benzoic acid — potassium benzoate

CHEMISTRY OF CARBOXYLIC ACIDS

One of the major reactions of primary alcohols is their oxidation to carboxylic acids, and the reverse of this is also important. An acid can be *reduced* to the corresponding alcohol with a metal hydride such as $LiAlH_4$, lithium aluminum hydride.

$$\underset{\text{carboxylic acid}}{\mathrm{R\overset{O}{\overset{\|}{C}}OH}} \xrightarrow[\mathrm{LiAlH_4}]{\text{reduction}} \underset{\text{primary alcohol}}{\mathrm{RCH_2OH}}$$

Such reactions are often used in the synthesis of new alcohols.

Carboxylic acids react with alcohols to produce **esters**, as in the reaction of acetic acid with ethanol to give ethyl acetate.

The reaction of an acid and an alcohol is often called an esterification.

The two-part name of an ester is given by (a) the name of the alkyl group from the alcohol and (b) the name of the carboxylate group derived from the acid. In general, ester names are of the form "alkyl carboxylate."

$$\underset{\text{carboxylic acid}}{\mathrm{R\overset{O}{\overset{\|}{C}}{-}O{-}H}} + \underset{\text{alcohol}}{\mathrm{R'{-}O{-}H}} \xrightarrow{\mathrm{H^+}} \underset{\text{ester}}{\mathrm{R\overset{O}{\overset{\|}{C}}OR'}} + \mathrm{H_2O}$$

$$\underset{\text{acetic acid}}{\mathrm{CH_3\overset{O}{\overset{\|}{C}}OH}} + \underset{\text{ethanol}}{\mathrm{CH_3CH_2OH}} \xrightarrow{\mathrm{H^+}} \underset{\text{ethyl acetate}}{\mathrm{CH_3\overset{O}{\overset{\|}{C}}OCH_2CH_3}} + \mathrm{H_2O}$$

Table 26.9 Some Common Esters

Name	Formula	Odor of
iso-Amyl acetate	$CH_3COOC_5H_{11}$	Bananas
Ethyl butyrate	$C_3H_7COOC_2H_5$	Pineapples
n-Amyl butyrate	$C_3H_7COOC_5H_{11}$	Apricots
n-Octyl acetate	$CH_3COOC_8H_{17}$	Oranges
Isoamyl isovalerate	$C_4H_9COOC_5H_{11}$	Apples
Methyl salicylate	$C_6H_4(OH)(COOCH_3)$	Oil of wintergreen
Methyl anthranilate	$C_6H_4(NH_2)(COOCH_3)$	Grapes

Although small amounts of esters are used as fragrances, the ester group as a side group or as the main link in polymers is more important. Polymers such as cellulose acetate and various polyesters such as "Dacron" are described in Chapter 27.

Unlike the acids from which esters are derived, esters often have pleasant odors (Table 26.9). A typical example is methyl salicylate or "oil of wintergreen."

$$C_6H_4(OH)-\overset{\overset{\large O}{\|}}{C}-OCH_3$$

methyl salicylate

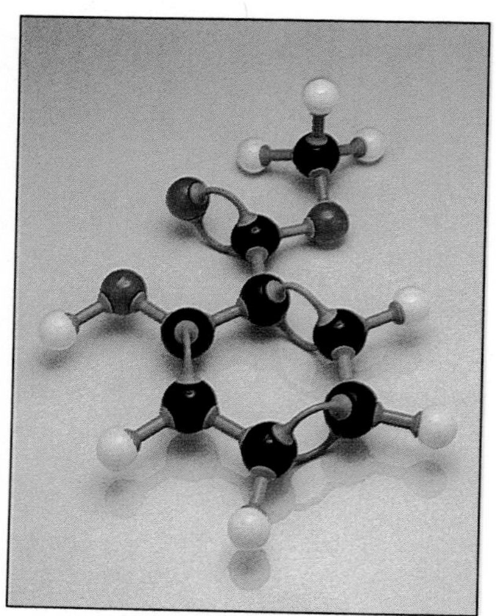

Methyl salicylate, oil of wintergreen.

EXERCISE 26.13 Names of Acids and Esters

Give a name for each of the following:

(a) $CH_3(CH_2)_4\overset{\overset{O}{\|}}{C}OH$

(b) $O_2N-C_6H_4-\overset{\overset{O}{\|}}{C}OH$

(c) $CH_3\overset{\overset{O}{\|}}{C}OCH_2CH_2CH_3$

EXERCISE 26.14 Reactions of Acids

Complete the following reactions:

(a) $CH_3\underset{}{\overset{\overset{CH_3}{|}}{C}}HCH_2\overset{\overset{O}{\|}}{C}OH + LiAlH_4 \longrightarrow$

(b) $C_6H_5-\overset{\overset{O}{\|}}{C}OH + CH_3OH$ (in the presence of acid) $\longrightarrow$

(c) $CH_3\underset{\underset{CH_3}{|}}{C}H\overset{\overset{O}{\|}}{C}OH + KOH \longrightarrow$

As a class, esters are not very reactive. Perhaps their most important reaction is *hydrolysis* in the presence of a strong base to give the constituents of the ester: the alcohol and a salt of the acid from which the ester was formed.

$$\underset{\text{ester}}{\mathrm{R\overset{\overset{\displaystyle O}{\|}}{C}{-}O{-}R'}} + \mathrm{NaOH} \xrightarrow{\text{heat}} \underset{\text{carboxylate salt}}{\mathrm{R\overset{\overset{\displaystyle O}{\|}}{C}{-}O^-Na^+}} + \underset{\text{alcohol}}{\mathrm{R'OH}}$$

$$\underset{\text{ethyl acetate}}{\mathrm{CH_3\overset{\overset{\displaystyle O}{\|}}{C}{-}O{-}CH_2CH_3}} + \mathrm{NaOH} \xrightarrow{\text{heat}} \underset{\text{sodium acetate}}{\mathrm{CH_3\overset{\overset{\displaystyle O}{\|}}{C}{-}O^-Na^+}} + \underset{\text{ethanol}}{\mathrm{CH_3CH_2OH}}$$

The hydrolysis of esters is also called a **saponification** reaction, since the hydrolysis of a special type of ester, a fat, produces soaps. All fats are chemically similar, since they are triesters of long-chain "fatty acids" with 1,2,3-trihydroxypropane (glycerol); all have the general structural formula

$$\underset{\text{glycerol}}{\begin{array}{l}\mathrm{CH_2{-}OH}\\ |\\ \mathrm{CH{-}OH}\\ |\\ \mathrm{CH_2{-}OH}\end{array}} + \underset{\text{"fatty acid"}}{3\ \mathrm{R\overset{\overset{\displaystyle O}{\|}}{C}OH}} \longrightarrow \underset{\text{fat or oil}}{\begin{array}{l}\mathrm{CH_2{-}O{-}\overset{\overset{\displaystyle O}{\|}}{C}R}\\ |\\ \mathrm{CH{-}O{-}\overset{\overset{\displaystyle O}{\|}}{C}R}\\ |\\ \mathrm{CH_2{-}O{-}\overset{\overset{\displaystyle O}{\|}}{C}R}\end{array}} + 3\ \mathrm{H_2O}$$

The R groups can be the same or different groups within the same fat. Furthermore, the R groups can be *unsaturated*; that is, the carbon chain of the fatty acid may contain one or more double bonds, and so is referred to as monounsaturated or polyunsaturated, respectively. Alternatively, the fat can be *saturated*; that is, all C atoms in the chain form σ bonds to other C atoms or to H atoms. Some common fatty acids are

SATURATED ACIDS

Butyric	C_4	$CH_3CH_2CH_2COOH$
Lauric	C_{12}	$CH_3(CH_2)_{10}COOH$
Myristic	C_{14}	$CH_3(CH_2)_{12}COOH$
Palmitic	C_{16}	$CH_3(CH_2)_{14}COOH$
Stearic	C_{18}	$CH_3(CH_2)_{16}COOH$

UNSATURATED ACIDS

Oleic	C_{18}	$CH_3(CH_2)_7CH{=}CH(CH_2)_7COOH$
Linolenic	C_{18}	$CH_3CH_2CH{=}CHCH_2CH{=}CHCH_2CH{=}CH(CH_2)_7COOH$

The fat based on the saturated acid myristic acid is found in candy bars, for example, and salad dressings contain an unsaturated fat derived from linolenic acid. Whether the fatty acid is saturated or unsaturated, and the degree of unsaturation, is important to the food industry and in your diet. This is explored in "Something More About Fat."

SOMETHING MORE ABOUT Fat

PERCENTAGE OF CALORIES FROM FAT

French fries	45%
Ice cream	45%
Chocolate chips	56%
Peanut butter	77%
Hot dogs	82%
Cream cheese	90%
Olives	98%
Butter and margarine	100%

On the side of a box of crackers it says "MADE WITH 100% PURE VEGETABLE SHORTENING . . . (partially hydrogenated soybean oil with hydrogenated cottonseed oil)." What does this mean? What is "partially hydrogenated" and why would we care about it? And, for that matter, what is the difference between the fats described in the text and an oil such as soybean oil?

The word "fat" is a generic term, like "pen" or "pencil." All fats are chemically similar, since all are triesters of glycerol or "triglycerides." However, we often subdivide triglycerides into two types: fats and oils. Triglycerides that are solids at room temperature are usually referred to as **fats**, while liquid triglycerides are called **oils**. Since fats and oils have the same energy content, though, food labels generally list them together simply as fat.

The amount of fat in a food is important because the fat supplies about 9 Calories (or about 38 kJ) of energy per gram. In general, Americans are overweight and consume too much fatty food. In fact, fat makes up about 40% of the energy intake in the average American diet, although nutritionists think it should be no higher than about 30%. Unfortunately, a large percentage of the Calorie content of many of our favorite foods (see table in the margin) is from fat.

From a dietary viewpoint, the other important factor about fat in foods is whether it is saturated or unsaturated. Not only is it recommended that you consume no more than 30% of your Calories in the form of fat, but nutritionists also believe that no more than a third of this should be from saturated fat. The

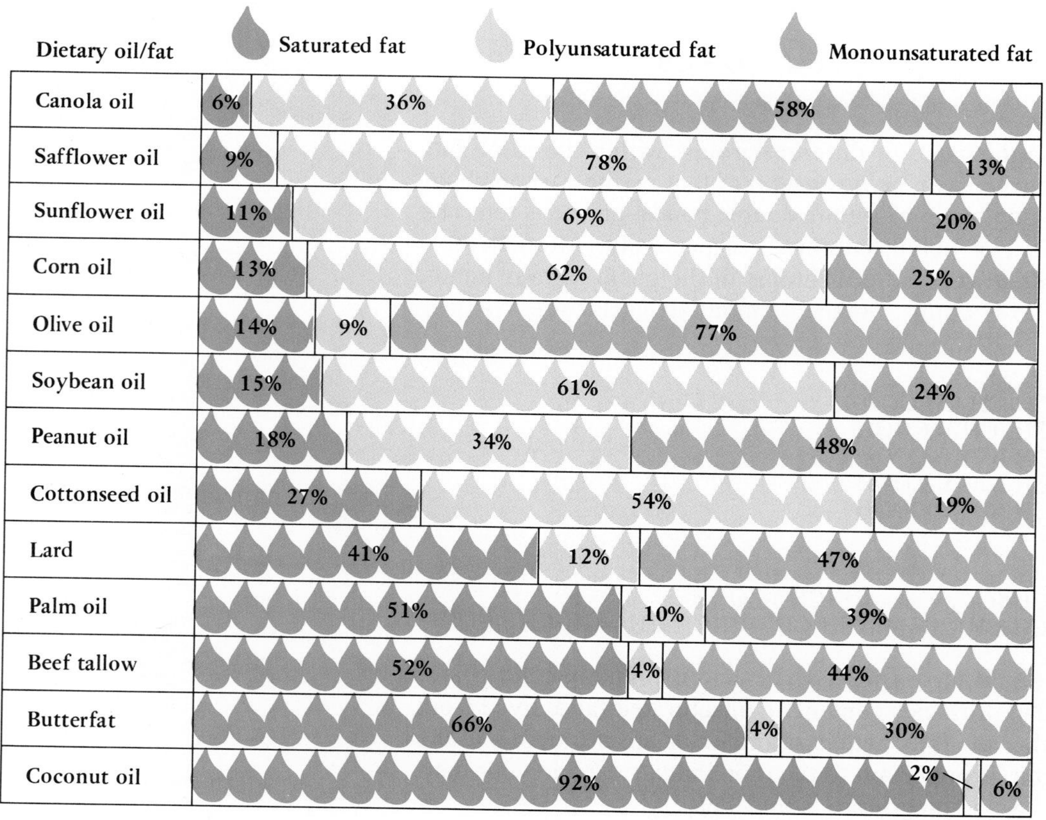

Dietary oil/fat
Saturated fat
Polyunsaturated fat
Monounsaturated fat
Canola oil 6% 36% 58%
Safflower oil 9% 78% 13%
Sunflower oil 11% 69% 20%
Corn oil 13% 62% 25%
Olive oil 14% 9% 77%
Soybean oil 15% 61% 24%
Peanut oil 18% 34% 48%
Cottonseed oil 27% 54% 19%
Lard 41% 12% 47%
Palm oil 51% 10% 39%
Beef tallow 52% 4% 44%
Butterfat 66% 4% 30%
Coconut oil 92% 2% 6%

(Reproduced from an illustration in CHEM MATTERS, Vol. 7, No. 4

fatty acid portion of fats almost always has an even number of C atoms (C_{16} or C_{18}), and the R group is usually saturated. In contrast, the R group in an oil is usually unsaturated. So, if it is better to consume unsaturated fats than saturated ones, why then do food companies hydrogenate oils to reduce their unsaturation?

Hydrogenation of double bonds is described on page 1091.

There are several answers to this. First, the double bonds in the fatty acid are a reactive functional group, and oxygen can attack the fat at this point. When the oil is oxidized, unpleasant odors and flavors develop. Hydrogenating an oil reduces the likelihood that the food will oxidize and become rancid. Second, hydrogenating an oil makes it less liquid. There are many times when a food manufacturer needs a solid fat in a food to improve the qualities of the food. (For example, if liquid vegetable oil were used in a cake icing, the icing would slide off the cake.) Rather than use animal fat, which also contains cholesterol, the manufacturer turns to a hydrogenated or partially hydrogenated oil.

Our crackers contained partially hydrogenated soybean oil with hydrogenated cottonseed oil. The chart shows that both of these oils originally were low in saturated fatty acid. However, completely hydrogenating the cottonseed oil makes it a saturated fat. This gives us a product that is probably more stable on the shelf but not necessarily more healthy.

Like any ester, fats are hydrolyzed by strong bases to give glycerol and a salt of the fatty acid. The acid salt is a **soap** and is the origin of the name of the process (saponification).

$$\begin{array}{l} CH_2{-}O{-}C({=}O){-}(CH_2)_{16}CH_3 \\ | \\ CH{-}O{-}C({=}O){-}(CH_2)_{16}CH_3 \\ | \\ CH_2{-}O{-}C({=}O){-}(CH_2)_{16}CH_3 \end{array} + 3NaOH \xrightarrow{\text{heat}} \begin{array}{l} CH_2{-}OH \\ | \\ CH{-}OH \\ | \\ CH_2{-}OH \end{array} + 3\ NaO{-}C({=}O){-}(CH_2)_{16}CH_3$$

glyceryl tristearate, a fat — glycerol — sodium stearate, a soap

Waxes are closely related to fats and oils in that all are esters. Waxes, however, are esters formed between a carboxylic acid and an alcohol with only one —OH group. For example, beeswax ($C_{15}H_{31}COOC_{30}H_{61}$) is an ester of myricyl alcohol.

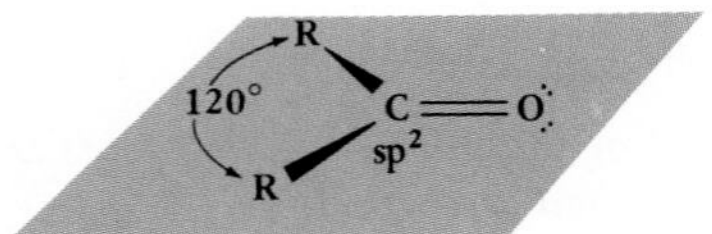

In an aldehyde or ketone there is an unhybridized p orbital perpendicular to the trigonal plane, on the C atom that, together with a similar p orbital on the O atom, can be used to form the C=O π bond (as explained in Chapter 11).

26.10 CARBONYL COMPOUNDS: ALDEHYDES AND KETONES

Aldehydes and ketones are characterized by the $\rangle C{=}O$ or carbonyl group. In both types of molecules, the carbon atom is the center of a trigonal, planar arrangement of atoms, so this central carbon atom is sp^2 hybridized.

Like esters, aldehydes and ketones can have pleasant odors and are often used as the basis of fragrances. Benzaldehyde, for example, is responsible for the characteristic odor of almonds, and citral is found in lemons and oranges and is used in perfumes. Muscone, a ketone, is the odoriferous component of musks; the compound is now synthesized, but it can be obtained from the musk glands of various animals.

benzaldehyde
liquid, bp 179 °C

citral
oily liquid

muscone
oily liquid

NAMING ALDEHYDES AND KETONES

As in other classes of organic compounds, simple aldehydes generally have common names. These are derived from the common name of the acid by dropping the "-ic" suffix and adding "aldehyde." Thus,

Table 26.10 Simple Aldehydes and Ketones

Structure	Common Name	Systematic Name	bp (°C)
$H{-}\overset{\overset{O}{\|}}{C}{-}H$	Formaldehyde	Methanal	−21
$CH_3\overset{\overset{O}{\|}}{C}H$	Acetaldehyde	Ethanal	21
$CH_3\overset{\overset{O}{\|}}{C}CH_3$	Acetone	Propanone	56
$CH_3\overset{\overset{O}{\|}}{C}CH_2CH_3$	Methyl ethyl ketone	Butanone	80
$CH_3CH_2\overset{\overset{O}{\|}}{C}CH_2CH_3$	Diethyl ketone	3-Pentanone	102

we have formaldehyde and acetaldehyde, among others (Table 26.10). Systematic names of aldehydes are formed by dropping the final "-e" from the name of the parent alkane and adding "-al." Thus, formaldehyde, HCHO, should be named methanal.

The simplest possible ketone, $CH_3C(O)CH_3$, is called *acetone* (Table 26.10). Most other ketones are named systematically by replacing the "-e" suffix of the corresponding alkane with "-one." Thus, the general systematic name is "alkanone" (Table 26.10).

CHEMISTRY OF ALDEHYDES AND KETONES

As you learned from the chemistry of alcohols, aldehydes are intermediate in oxidation level between *primary* alcohols and carboxylic acids. Thus, aldehydes are easily oxidized to the corresponding carboxylic acid

$$\underset{\text{aldehyde}}{R-\overset{\overset{\large O}{\|}}{C}-H} \xrightarrow{KMnO_4 \text{ or } Na_2Cr_2O_7} \underset{\text{carboxylic acid}}{R-\overset{\overset{\large O}{\|}}{C}-OH}$$

or reduced to the corresponding alcohol. Common reducing agents are the metal hydrides sodium borohydride, $NaBH_4$, or lithium aluminum hydride, $LiAlH_4$.

$$\underset{\text{aldehyde}}{R-\overset{\overset{\large O}{\|}}{C}-H} \xrightarrow{NaBH_4 \text{ or } LiAlH_4} \underset{\text{primary alcohol}}{R\ \ H_2OH}$$

In a similar manner, ketones can be reduced to *secondary* alcohols with metal hydride reducing agents.

$$\underset{\text{ketone}}{R-\overset{\overset{\large O}{\|}}{C}-R'} \xrightarrow{NaBH_4 \text{ or } LiAlH_4} \underset{\text{secondary alcohol}}{R-\overset{\overset{\large O-H}{|}}{\underset{\underset{\large H}{|}}{C}}-R'}$$

Alkyl and aryl halides react with magnesium to give Grignard reagents (Section 26.8). Reaction of these reagents with aldehydes and ketones is of great synthetic utility, since this represents a way to attach the R group of the Grignard reagent to another compound.

$$\underset{\text{aldehyde or ketone}}{\overset{|}{\underset{|}{C}}=O} + \underset{\text{Grignard reagent}}{RMgX} \longrightarrow [R-\overset{|}{\underset{|}{C}}-O-MgX] \xrightarrow[H_2O]{H_3O^+} \underset{\text{alcohol}}{R-\overset{|}{\underset{|}{C}}-OH}$$

The initial product of the reaction is the magnesium salt of the weakly acidic alcohol. However, on adding aqueous acid, the salt is converted in high yield to the alcohol itself.

Formaldehyde reacts with Grignard reagents to produce only primary alcohols.

$$\underset{\text{formaldehyde}}{\mathrm{H{-}\overset{\overset{\displaystyle O}{\|}}{C}{-}H}} + \underset{\text{Grignard reagent}}{\mathrm{RMgX}} \longrightarrow \mathrm{H{-}\underset{\underset{\displaystyle R}{|}}{\overset{\overset{\displaystyle O{-}MgX}{|}}{C}}{-}H} \xrightarrow[\mathrm{H_2O}]{\mathrm{H_3O^+}} \underset{\text{primary alcohol}}{\mathrm{H{-}\underset{\underset{\displaystyle R}{|}}{\overset{\overset{\displaystyle O{-}H}{|}}{C}}{-}H}}$$

Thus, if you wish to synthesize a primary alcohol, you can begin with an alkyl halide with one less carbon atom than the desired alcohol. After the Grignard reagent is made, treatment with formaldehyde leads to the desired product, often in good yield.

Careful examination of the reaction of formaldehyde shows that an aldehyde other than formaldehyde will lead to *secondary* alcohols.

$$\underset{\text{aldehyde}}{\mathrm{R{-}\overset{\overset{\displaystyle O}{\|}}{C}{-}H}} + \underset{\text{Grignard reagent}}{\mathrm{R'MgX}} \longrightarrow \mathrm{R{-}\underset{\underset{\displaystyle R'}{|}}{\overset{\overset{\displaystyle O{-}MgX}{|}}{C}}{-}H} \xrightarrow[\mathrm{H_2O}]{\mathrm{H_3O^+}} \underset{\text{secondary alcohol}}{\mathrm{R{-}\underset{\underset{\displaystyle R'}{|}}{\overset{\overset{\displaystyle O{-}H}{|}}{C}}{-}H}}$$

$$\underset{\text{propanal}}{\mathrm{CH_3CH_2\overset{\overset{\displaystyle O}{\|}}{C}H}} + \underset{\text{methyl magnesium iodide}}{\mathrm{CH_3MgI}} \longrightarrow \mathrm{CH_3CH_2{-}\underset{\underset{\displaystyle CH_3}{|}}{\overset{\overset{\displaystyle O{-}MgI}{|}}{C}}{-}H} \xrightarrow[\mathrm{H_2O}]{\mathrm{H_3O^+}} \underset{\text{2-butanol}}{\mathrm{CH_3CH_2{-}\underset{\underset{\displaystyle CH_3}{|}}{\overset{\overset{\displaystyle O{-}H}{|}}{C}}{-}H}}$$

In contrast, *ketones provide tertiary alcohols* on treatment with a Grignard reagent.

$$\underset{\text{ketone}}{\mathrm{R{-}\overset{\overset{\displaystyle O}{\|}}{C}{-}R'}} + \underset{\text{Grignard reagent}}{\mathrm{R''MgX}} \longrightarrow \mathrm{R{-}\underset{\underset{\displaystyle R''}{|}}{\overset{\overset{\displaystyle O{-}MgX}{|}}{C}}{-}R'} \xrightarrow[\mathrm{H_2O}]{\mathrm{H_3O^+}} \underset{\text{tertiary alcohol}}{\mathrm{R{-}\underset{\underset{\displaystyle R''}{|}}{\overset{\overset{\displaystyle OH}{|}}{C}}{-}R'}}$$

$$\underset{\text{acetone}}{\mathrm{CH_3\overset{\overset{\displaystyle O}{\|}}{C}CH_3}} + \underset{\text{ethyl magnesium bromide}}{\mathrm{CH_3CH_2MgBr}} \longrightarrow \mathrm{CH_3{-}\underset{\underset{\displaystyle CH_2CH_3}{|}}{\overset{\overset{\displaystyle O{-}MgBr}{|}}{C}}{-}CH_3} \xrightarrow[\mathrm{H_2O}]{\mathrm{H_3O^+}} \underset{\text{2-methyl-2-butanol}}{\mathrm{CH_3{-}\underset{\underset{\displaystyle CH_2CH_3}{|}}{\overset{\overset{\displaystyle O{-}H}{|}}{C}}{-}CH_3}}$$

Carbon dioxide, O=C=O, can be considered as a simple carbonyl functional group, so it is not surprising that it also reacts with Grignard reagents. The reaction is very useful synthetically, since the ultimate product is a carboxylic acid.

$$\underset{\text{Grignard reagent}}{RMgX} + O{=}C{=}O \longrightarrow R{-}\overset{\overset{\displaystyle O}{\|}}{C}{-}OMgX \xrightarrow[H_2O]{H_3O^+} \underset{\text{carboxylic acid}}{R{-}\overset{\overset{\displaystyle O}{\|}}{C}{-}OH}$$

$$\underset{\text{phenyl magnesium bromide}}{C_6H_5{-}MgBr} + O{=}C{=}O \longrightarrow C_6H_5{-}\overset{\overset{\displaystyle O}{\|}}{C}{-}OMgBr \xrightarrow[H_2O]{H_3O^+} \underset{\text{benzoic acid}}{C_6H_5{-}\overset{\overset{\displaystyle O}{\|}}{C}{-}OH}$$

Aldehydes and ketones are particularly valuable compounds in organic chemistry, since they can be transformed into so many other compounds. These reactions can be summarized as follows:

	Reducing agent		Grignard	
methanol	⟵	formaldehyde	⟶	primary alcohol
primary alcohol	⟵	general aldehyde	⟶	secondary alcohol
secondary alcohol	⟵	ketone	⟶	tertiary alcohol

EXERCISE 26.15 Structures of Aldehydes and Ketones

Draw structural formulas of the following compounds:

(a) 2-pentanone (b) *m*-bromobenzaldehyde (c) butanal

EXERCISE 26.16 Names of Aldehydes and Ketones

Give systematic names for the following compounds:

(a) $CH_3CH_2CH_2\overset{\overset{\displaystyle O}{\|}}{C}CH_2CH_3$ (b) $CH_3CH_2CH_2CH_2\overset{\overset{\displaystyle O}{\|}}{C}H$

EXERCISE 26.17 Reactions of Aldehydes and Ketones

Complete the following reactions:

(a) $H_3C{-}CH_2{-}\overset{\overset{\displaystyle O}{\|}}{C}H + NaBH_4 \longrightarrow ?$

(b) $CH_3\underset{\underset{\displaystyle MgBr}{|}}{C}HCH_3 + H{-}\overset{\overset{\displaystyle O}{\|}}{C}{-}H \xrightarrow{H_3O^+(aq)\text{ following initial reaction}} ?$

(c) $CH_3\underset{\underset{\displaystyle CH_3}{|}}{C}H\overset{\overset{\displaystyle O}{\|}}{C}H \xrightarrow{Na_2CrO_7} ?$

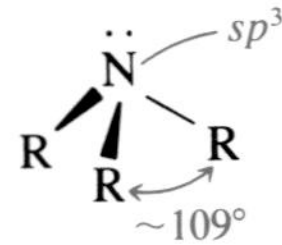

26.11 AMINES AND AMIDES

Amines are structural derivatives of ammonia with one or more of the three hydrogen atoms replaced by alkyl or aryl groups. They play a significant role in biochemistry and are important industrial reagents.

STRUCTURES, NAMES, AND GENERAL PROPERTIES

Ammonia is a pyramidal molecule due to the presence of the nitrogen lone pair. Therefore, all amines have similar structures based on an sp^3 hybridized N atom.

Amines are classed as *primary, secondary,* or *tertiary* according to the number of organic groups bonded to the N atom.

$$CH_3CH_2—\ddot{N}(H)—H \qquad CH_3—\ddot{N}(H)—CH_3 \qquad CH_3CH_2—\ddot{N}(CH_3)—CH_3$$

ethylamine primary; dimethylamine secondary; ethyldimethylamine tertiary

Notice that alkylamines are named simply by identifying the groups bound to the nitrogen, with the group names in alphabetic order; the compound name is completed with the word "amine" at the end.

The simplest aromatic amine is aniline, $C_6H_5NH_2$. By convention, aromatic amines are usually named as derivatives of aniline.

:NH_2 (on benzene ring) — aniline

:NH_2 (on benzene ring with Cl at 2 and 4 positions) — 2,4-dichloroaniline

Some of the oxygen-containing organic compounds we have examined thus far are very weak Lewis and Brønsted bases. However, amines are organic bases that can turn red litmus to blue in aqueous solution.

$$RNH_2(aq) + H_2O(\ell) \rightleftharpoons RNH_3^+(aq) + OH^-(aq)$$

Aliphatic amines are slightly stronger bases than ammonia, but aromatic amines such as aniline are generally significantly weaker (Table 26.11).

Table 26.11 Ionization Constants for Some Amines

Compound	K_b
Ammonia	1.8×10^{-5}
Methylamine	5.0×10^{-4}
Dimethylamine	7.4×10^{-4}
Trimethylamine	7.4×10^{-5}
Aniline	4.2×10^{-10}

Amines often have an offensive odor. In more familiar terms, amines are responsible for the odor of decaying fish. The "smell of death" of decaying flesh is due to two appropriately named amines, cadaverine and putrescine.

$$H_2N—CH_2CH_2CH_2CH_2—NH_2 \qquad H_2N—CH_2CH_2CH_2CH_2CH_2—NH_2$$

putrescine (1,4-butanediamine); cadaverine (1,5-pentanediamine)

REACTIONS OF AMINES

As bases, amines react with inorganic and organic acids to form ammonium salts.

$$\underset{\text{methylamine}}{CH_3NH_2(aq)} + HCl(aq) \longrightarrow \underset{\text{methylammonium chloride}}{CH_3NH_3^+Cl^-(aq)}$$

Although only a few amines are soluble in water, their salts with inorganic acids are often water soluble. Conversely, the amines are soluble in hydrocarbon solvents, but the salts are not. For example, the amine procaine

$$H_2N-C_6H_4-\overset{O}{\overset{\|}{C}}OCH_2CH_2N(CH_2CH_3)_2$$

procaine

is soluble in ethanol, chloroform, and benzene, but it dissolves in water only to the extent of 1 g in 200 mL. On the other hand, the HCl salt of procaine is soluble to the extent of 1 g in 1 mL of water. In the form of the HCl salt, the compound is sold as a local anesthetic and is most commonly known as Novocain.

AMIDES

Just as an ester can be viewed as derived from a carboxylic acid and an alcohol, an **amide** can be said to be derived from a carboxylic acid and an amine.

$$\underset{\text{carboxylic acid}}{R\overset{O}{\overset{\|}{C}}-OH} + \underset{\text{amine}}{H-NR_2} \longrightarrow HOH + \underset{\text{amide}}{R\overset{O}{\overset{\|}{C}}-NR_2}$$

Although this reaction can be used in principle, amides are usually made from a primary or secondary amine and a carboxylic acid derivative, an acid chloride.

$$R'-\overset{O}{\overset{\|}{C}}-OH \xrightarrow{+\ SOCl_2} R'-\overset{O}{\overset{\|}{C}}-Cl$$

$$\underset{\text{amine}}{RNH_2} + \underset{\text{acid chloride}}{R'-\overset{O}{\overset{\|}{C}}-Cl} \longrightarrow \underset{\text{amide}}{R'-\overset{O}{\overset{\|}{C}}-N(H)(R)} + HCl$$

The amide grouping is important in some synthetic polymers and in many naturally occurring compounds, especially in proteins (Chapter 27). However, if you love camping and hiking, one example of an amide

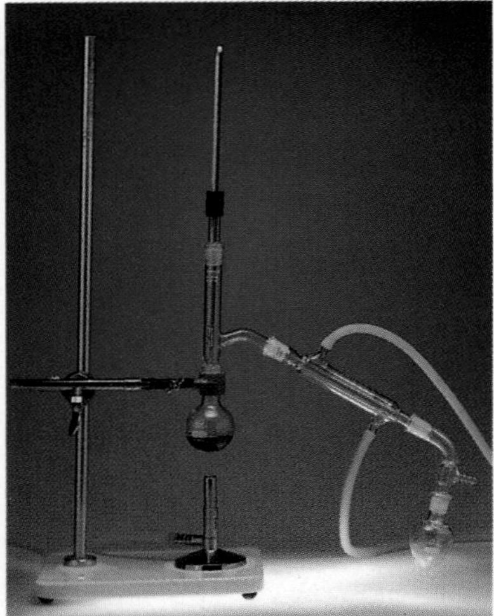

Figure 26.2 A simple laboratory apparatus for the distillation of volatile liquids. A solution is placed in the round flask and heated gently. The more volatile component preferentially distills from the solution and can be collected in the flask at the end of the water-cooled condenser.

you may find especially interesting is *N,N*-diethyl-*m*-toluamide, a compound which is the active ingredient in most effective insect repellents.

$$\text{m-}CH_3C_6H_4-C(=O)-N(CH_2CH_3)_2$$

N,N-diethyl-*m*-toluamide

EXERCISE 26.18 Amines and Amides

(a) Name the compound $(CH_3CH_2)_2NH$.

(b) Complete the following reaction

$$CH_3\overset{\overset{\displaystyle O}{\|}}{C}-Cl + (CH_3)_2NH \longrightarrow$$

26.12 PETROLEUM AND REFINERY PROCESSES

One barrel of petroleum equals 42 U.S. gallons. Approximately 6.8 gallons of gasoline are consumed for each man, woman, and child in the United States per day.

At least 17 million barrels of petroleum are consumed each day in the United States. About 8 million barrels are converted to gasoline, and another 8 million are consumed primarily for other fuel needs, among them heating oil, diesel fuel, and jet engine fuel. The remaining one million barrels provide the raw material for the production of organic chemicals and polymers.

Petroleum is a complex mixture of alkanes, alkenes, alkynes, aromatic hydrocarbons, and other compounds. To obtain useful materials, the components must be separated as efficiently as possible. Since most of the materials are volatile, *fractional distillation* is used. This technique relies on Raoult's law, a solution property described in Chapter 14. The law states that the vapor pressure of a volatile component of a solution, P_A, is given by $P_A = X_AP_A°$ where X_A is the mole fraction of component A and $P_A°$ is the vapor pressure of pure A. If you have a solution in which both the solvent (A) and solute (B) are volatile, this means that the total vapor pressure over the solution is

$$P_{total} = P_A + P_B = X_AP_A° + X_BP_B°$$

This equation shows that, for roughly comparable amounts of solvent and solute, the vapor over the solution will consist mostly of the more volatile component (the component with the greater value of $P°$). It is this fact that allows you to separate two volatile materials by distillation.

A laboratory scale distilling apparatus is shown in Figure 26.2. The flask over the bunsen burner contains a mixture of volatile liquids. When heated, some of both liquids volatilize, the vapors rise in the apparatus until a cooler zone is reached, and the vapors condense. However, the condensed liquid now contains a greater mole fraction of the more volatile

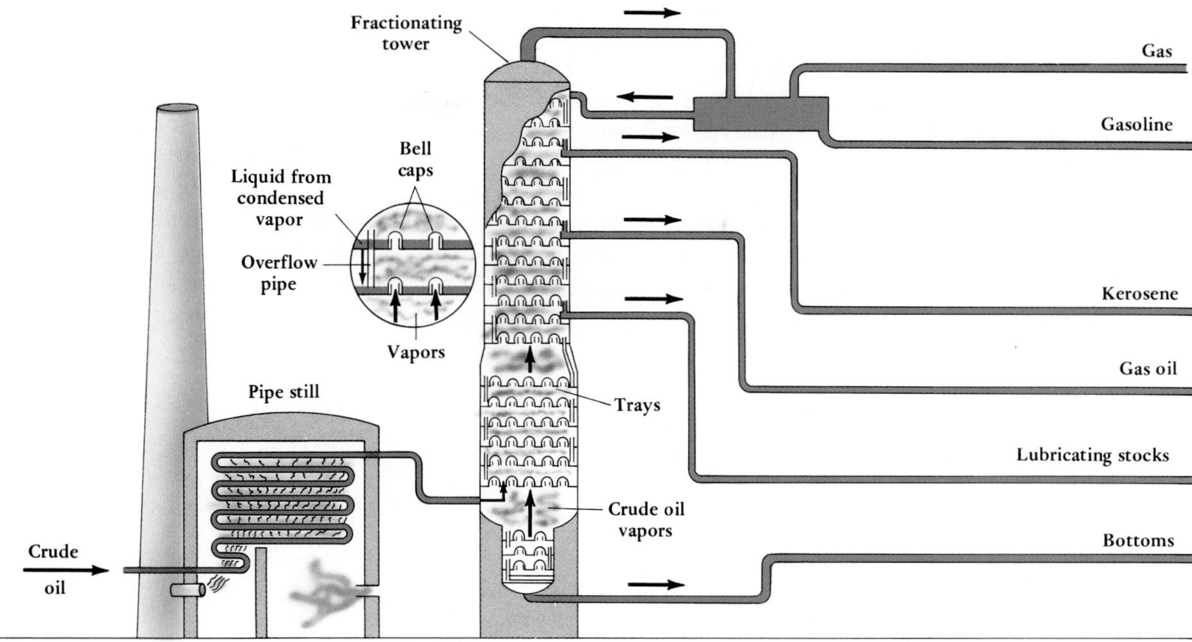

(a)

(b)

Figure 26.3 A diagram (a) and a photograph (b) of a modern petroleum refinery that uses distillation to separate volatile components from one another.

component of the original mixture. If this new liquid is again volatilized, and the vapors condensed, the condensed vapor will be richer still in the more volatile component. Carrying this out many times in a specially designed laboratory apparatus or chemical plant can lead eventually to separation of the more volatile from the less volatile material of a mixture. Figure 26.3 shows a fractional distilling column for crude oil; the more volatile fractions of the mixture rise further in the column and can be separated from other fractions.

In the United States the most valuable petroleum product is gasoline derived from the fractions boiling from about 30 °C to about 200 °C. However, the lower boiling material contains a high proportion of straight chain alkanes, and it is not suitable for use in modern higher performance engines; such hydrocarbons have a low octane rating.

One way to raise the octane rating of the lower boiling fractions of petroleum is to add various substances. Tetraethyllead, $(C_2H_5)_4Pb$, was one of the major additives, but this has been phased out because of environmental lead pollution and because the lead poisons the catalysts used in exhaust systems. Currently used additives are methanol (CH_3OH) and

$$CH_3{-}O{-}\overset{\displaystyle CH_3}{\underset{\displaystyle CH_3}{\overset{|}{\underset{|}{C}}}}{-}CH_3 \qquad\qquad CH_3{-}\overset{\displaystyle CH_3}{\underset{\displaystyle CH_3}{\overset{|}{\underset{|}{C}}}}{-}OH$$

methyl *tert*-butyl ether   *tert*-butyl alcohol

Several refinery processes are also used to raise the octane rating of petroleum to be used as gasoline. The first step, **catalytic cracking**, breaks down large alkanes from high boiling fractions into smaller, branched-chain alkanes more suitable for use in gasoline. This is done by passing the alkane vapor over a catalyst [such as silica–alumina (SiO_2–Al_2O_3)] at 450 °C to 550 °C.

Catalytic reforming converts alkanes to aromatic compounds, in particular to benzene, toluene, and the xylenes. All have high octane ratings (over 100) and are used in gasoline. In addition, they are used as starting materials in the production of a variety of organic chemicals. Alkanes are catalytically reformed by passing the heated hydrocarbon vapor over a bed of alumina (Al_2O_3) containing platinum and rhodium. Hydrogen is formed as a by-product and can be used in other refinery processes. For example,

$$C_6H_{12}(g) \xrightarrow{\text{catalytic reforming}} C_6H_6(g) + 3\ H_2(g)$$

cyclohexane benzene

Alkylation converts lower molecular weight alkanes and alkenes (C_3 and C_4) into higher, branched-chain alkanes suitable for use in gasoline. Hydrofluoric acid and sulfuric acid are typically used as catalysts in this process.

$$CH_3C(CH_3){=}CH_2 + CH_3CH(CH_3)CH_3 \xrightarrow[\text{HF or } H_2SO_4]{\text{alkylation}} CH_3CH(CH_3){-}CH_2{-}C(CH_3)_2{-}CH_3$$

methylpropene methylpropane 2,2,4-trimethylpentane

The three refinery processes outlined above are used to augment and upgrade the lower boiling fraction of petroleum for use as gasoline. A fourth refinery process, **steam cracking** (page 910) can be used to convert ethane and propane in natural gas to ethylene and propylene, the building blocks of polyethylene and polypropylene.

EXERCISE 26.19 Catalytic Reforming

Write a balanced equation for the catalytic reforming of heptane (C_7H_{16}) to toluene ($C_6H_5CH_3$). How many moles of hydrogen are produced per mole of heptane converted?

SUMMARY

Organic chemistry is the study of carbon-containing compounds. Hydrocarbons contain only C and H, and **alkanes**, C_nH_{2n+2}, are one major class of such compounds. (They are often called **saturated** compounds.) For alkanes with 4 or more carbon atoms, **structural isomers** are possible. (Such isomers have identical molecular formulas but different atom-to-atom connections.)

Compounds other than alkanes have various **functional groups**, structural fragments found in all members of a class of compounds. Functional groups are the site of the characteristic reactions of the class. The following is a summary of functional groups:

CLASS	FUNCTIONAL GROUP/STRUCTURAL FRAGMENT*
Alkenes	C=C double bond
Alkynes	C≡C triple bond
Aromatic compounds	Based on benzene (C_6H_6)
Alkyl halides	R—X where X is F, Cl, Br, or I
Alcohols	R—OH
Ethers	R—O—R
Carboxylic acids	R—C(=O)—OH
Esters	R—C(=O)—O—R
Aldehydes	R—C(=O)—H (contains a carbonyl group)
Ketones	R—C(=O)—R (contains a carbonyl group)
Amines	R_3N (R is H or an organic group)
Amides	R—C(=O)—NR_2 (R is H or an organic group)

*R is an alkyl group (e.g., CH_3) or aryl group (e.g., C_6H_5).

Major forms of reactivity of organic compounds (Section 26.4) include **oxidation** and **reduction**, **addition** of small molecules such as H_2, X_2 (halogen), H_2O, or HX (where X is a halogen), **elimination** of H_2O, and **substitution** of one atom for another. Each functional group will participate in one or more of these reactions.

Alcohols can undergo substitution of —OH by —X (Br, Cl, or I), elimination of H_2O to give alkenes or ethers, and oxidation to aldehydes or ketones (Section 26.4).

Alkenes (and **alkynes**) (Section 26.6) are often called **unsaturated** because the π electrons of the carbon–carbon multiple bond can be used to form additional sigma bonds. A C=C double bond prevents rotation about the carbon–carbon bond, so ***cis–trans*** **isomerism** is possible. Alkenes (and alkynes) add small molecules such as H_2 (hydrogenation), X_2 and HX (X is Cl, Br, I), and H_2O (hydration). For the addition of HX to a double bond, **Markovnikov's rule** is usually followed.

Benzene, C_6H_6, is the simplest **aromatic** compound (Section 26.7). The C_6 ring has six equivalent C—C bonds and there are six equivalent C—H bonds, an observation explained by invoking π resonance structures. Substitution of H is the major form of reactivity of benzene and other aromatic compounds.

Alkyl halides (R—X where X is a halogen) (Section 26.8) participate in elimination reactions. H and X are lost from adjacent atoms to give an alkene; the reaction is called a **dehydrohalogenation**. Alkyl halides also form **Grignard reagents**, R—Mg—X, versatile synthetic reagents for adding R to another molecule.

Carboxylic acids (R—C(=O)—OH) (Section 26.9) are Brønsted acids and form salts with bases. Acids can be reduced to alcohols. Reaction of an alcohol with an acid gives an **ester**, R—C(=O)—O—R (esterification). The resulting esters can be hydrolyzed back to acid and alcohol (saponification). **Fats** and **oils** are esters of a tri-alcohol, glycerol.

Carbonyl compounds include **aldehydes** and **ketones** (Section 26.10). Both contain the >C=O functional group. Aldehydes can be oxidized to acids and reduced to primary alcohols. Ketones cannot be oxidized chemically, but they are reduced to secondary alcohols.

Amines, R_3N (Section 26.11) are classed as primary (two R groups are H), secondary (one R group is H), or tertiary (all R groups are organic fragments). Amines are Brønsted and Lewis bases. **Amides**, R—C(=O)—NR_2, are made effectively from a carboxylic acid chloride and a primary or secondary amine.

Many organic compounds can be synthesized by combining, in appropriate order, the functional group reactions described above.

Petroleum is a complex mixture of alkanes, alkenes, alkynes, aromatic compounds, and others (Section 26.12). Refining petroleum into useful products involves many different processes (fractional distillation, catalytic cracking, catalytic reforming, alkylation, and steam cracking, among others).

STUDY QUESTIONS

ALKANES

1. What is the name of the straight-chain alkane with the formula C_8H_{18}?
2. Which of the following hydrocarbons is an alkane: (a) C_2H_4, (b) C_5H_{10}, (c) $C_{14}H_{30}$, (d) C_7H_8?
3. What is the name of alkyl group $CH_3CH_2CH_2CH_2$—?
4. What is the molecular formula for an alkane with 13 carbon atoms?
5. Draw four of the nine structural isomers of heptane, C_7H_{16}. Give the systematic name of each isomer you have drawn.
6. In terms of the intermolecular forces described in Chapter 13, explain the trend in boiling points in Table 26.1.
7. Give the systematic name of each of the following compounds:

(a) $CH_3CH(CH_3)CH(CH_3)CH_3$ (c) $CH_3CH_2CH_2C(CH_3)_2CH_2CH_2CH_3$

(b) $CH_3CH_2CH_2CH(CH_3)CH_3$ (d) $CH_3CH_2CH_2CH(CH_2CH_3)CH_3$

8. Draw the structure of each of the following compounds:
   (a) 2,3-dimethylpentane
   (b) 2,4-dimethyloctane
   (c) 3-ethylhexane
   (d) 2-methyl-3-ethylhexane
9. Draw the structure of each of the following compounds:
   (a) 2,2-dimethylhexane
   (b) 3,3-diethylpentane
   (c) 2-methyl-3-ethylheptane
   (d) isobutane

FUNCTIONAL GROUPS

10. Classify each of the compounds below according to its functional group.
    (a) $CH_3CH{=}CHCH_3$
    (b) $CH_3{-}O{-}CH_2CH_3$
    (c) $C_6H_5C{\equiv}CCH_3$
    (d) $CH_3CH_2NH_2$
    (e) $CH_3C(=O){-}O{-}CH_2CH_2CH_3$
    (f) $CH_3CH_2CH_2C(=O)CH_3$
    (g) $CH_3CH_2CH(OH)CH_2CH_3$
    (h) $CH_3CH_2CH_2C(=O)OH$
    (i) $CH_3CH_2CH(CH_2CH_3)CH_2CH{=}O$
11. Draw an electron dot structure of dimethyl ether, $H_3C{-}O{-}CH_3$. What is the approximate C—O—C bond angle? What is the hybridization of the O atom?
12. Draw an electron dot structure for acetaldehyde, $H_3C{-}C(=O){-}H$. What is the C—C—O bond angle? Specify the hybridization of each C atom.
13. Draw an electron dot structure for formic acid, $H{-}C(=O){-}O{-}H$. Specify all of the bond angles in the molecule. What is the hybridization of the C atom?
14. One functional group not mentioned in Section 26.3 is the nitrile or cyano group, $-C{\equiv}N$. Draw an electron dot structure for cyanomethane (common name, acetonitrile), $H_3C{-}C{\equiv}N$. What is the C—C—N angle? Specify the hybridization of the C atom in the cyano group.

ALCOHOLS

15. Give systematic names for the following alcohols:
    (a) CH_3CH_2OH
    (b) $H_3C{-}C(CH_3)_2{-}OH$
    (c) $CH_3CH_2CH_2CH_2OH$
    (d) $CH_3C(OH)(CH_3)CH_2CH_3$
16. Draw structural formulas for the following alcohols:
    (a) 1-pentanol and 2-pentanol
    (b) 2-methyl-1-hexanol
    (c) 3,3-dimethyl-2-butanol
17. For each alcohol in Study Question 15, tell if it is primary, secondary, or tertiary.
18. Draw structural formulas for all of the alcohols of the formula $C_4H_{10}O$.
19. Draw structural formulas for all of the alcohols of the formula $C_5H_{12}O$.
20. Complete the following reactions:
    (a) $CH_3CH_2CH_2CH_2OH + KMnO_4(aq) \longrightarrow$
    (b) $CH_3CH_2CH_2CH_2OH + PCC \longrightarrow$
    (c) $CH_3CH_2CH(OH)CH_3 + KMnO_4(aq) \longrightarrow$
    (d) $CH_3CH_2CH(OH)CH_3 + HBr \longrightarrow$
    (e) $CH_3CH_2CH_2CH_2OH + H_2SO_4/180\ °C \longrightarrow$
    (f) $CH_3CH_2OH + Na \longrightarrow$
21. Which compound should have the lower boiling point: $CH_3CH_2CH_2OH$ or $H_3C{-}O{-}CH_2CH_3$? Explain your answer briefly.

ALKENES AND ALKYNES

22. What structural requirement is necessary for an alkene to have *cis* and *trans* isomers? Can *cis* and *trans* isomers exist for alkynes?
23. Draw the structures of the *cis–trans* isomers of
    (a) 1,2-dichloroethene
    (b) 3-methyl-3-hexene
24. Complete the following reactions:
    (a) $CH_3CH_2CH{=}CH_2 + Br_2 \longrightarrow$
    (b) $CH_3CH_2CH{=}CH_2 + HBr \longrightarrow$
    (c) $CH_3CH_2CH{=}CH_2 + H_2 \longrightarrow$
    (d) $CH_3CH_2CH{=}CH_2 + H_2O \longrightarrow$
25. What products are expected for the following reactions?

(a) $(H_3C)_2C{=}C(H)CH_2CH_3 \xrightarrow{HBr}$

(b) $(H_3C)_2C{=}C(H)CH_2CH_3 \xrightarrow{H_2}$

26. Give the systematic name of each of the products of the reactions in Study Question 25.

27. Starting with the appropriate alkene, tell how you could prepare each of the following in a single reaction step.
(a) 1,2-dibromobutane (c) 2-butanol
(b) 2-bromobutane

28. Starting with 1-butanol, show how one could prepare 2-butanol in two reaction steps.

29. Starting with 1-butanol, show how you could prepare 2-butanone in three reaction steps.

30. Draw structural formulas for (a) 3-hexyne, (b) 1-pentyne, and (c) 4-methyl-2-pentyne.

31. Complete and balance the following reactions:
(a) $CH_3CH_2C{\equiv}CH + H_2 \longrightarrow$
(b) $CH_3CH_2C{\equiv}CCH_3 + HBr \longrightarrow$
(c) $CH_3CH_2C{\equiv}CCH_3 + Br_2 \longrightarrow$

BENZENE AND AROMATIC COMPOUNDS

32. Draw structural formulas for the following compounds:
(a) *o*-dichlorobenzene (alternatively called 1,2-dichlorobenzene)
(b) *m*-chlorotoluene (alternatively called 3-chlorotoluene)
(c) *p*-diethylbenzene
(d) *p*-chlorophenol
(e) *p*-chlorostyrene

33. Show how to prepare isopropylbenzene (common name, cumene) from benzene and the appropriate propyl derivative.

34. Show how to prepare chlorobenzene in a single step from benzene and other appropriate reagents.

ALKYL HALIDES

35. One of the typical reactions of an alkyl halide is a dehydrohalogenation. For this reaction to occur, what atoms must be present in a molecule and how must they be placed?

36. Name the three types of reactions of alkyl halides described in the text. Using 2-iodopropane, give an example of each reaction type.

37. Name the type of reaction needed to produce the products below from alkyl halides. In each case give the alkyl halide starting material.
(a) $CH_3CH_2CH{=}CH_2$
(b) $CH_3CH_2CH_2CH_2OH$
(c) $CH_3CH_2CH_2CH_2{-}O{-}CH_3$ (Here the OCH_3 portion arises from methanol. What is the source of the remainder of the compound?)
(d) $CH_3CH_2CH_2CH_2CH_3$

38. Tell how you could prepare di-*n*-propyl ether starting with 1-bromopropane.

39. Outline a synthesis of 2-deuteropropane (CH_3CHDCH_3) from an alkyl iodide, D_2O, and other necessary reagents.

CARBOXYLIC ACIDS AND ESTERS

40. Give a name for
(a) $CH_3CH_2CH_2CH_2CH_2COOH$
(b) $CH_3CH_2C({=}O){-}O{-}CH_3$
(c) $H_3CC({=}O){-}O{-}CH_2CH_2CH_2CH_3$
(d) $Br{-}C_6H_4{-}C({=}O){-}OH$ (para)

41. Beginning with butanoic acid, what is the product of (a) reduction with $LiAlH_4$, (b) reaction with NaOH, and (c) reaction with ethanol?

42. Tell how you would synthesize propyl propanoate beginning with propanoic acid.

43. Give the structure and name of the product of the reaction between benzoic acid and isopropyl alcohol.

44. Maleic acid is a *di*carboxylic acid prepared by the catalytic oxidation of benzene. Its formula is $C_4H_4O_4$.
(a) Give the structural formula for the compound.
(b) How many milliliters of 0.130 M NaOH are required to titrate 0.522 g of the acid (so that both carboxylic acid groups donate a proton)?
(c) What is the product of the titration with NaOH?

45. What is the saponification reaction? Give an example of this important reaction.

46. Considering each ester below, give the structural formula of each product and its systematic or common name.
(a) $CH_3C({=}O){-}O{-}CH_2CH_2CH_3$ + NaOH (followed by acid) $\longrightarrow$
(b) $C_6H_5{-}C({=}O){-}O{-}CH(CH_3)_2$ + NaOH (followed by acid) $\longrightarrow$

47. Define fats and oils and give an example of each.

ALDEHYDES AND KETONES

48. Draw structural formulas for (a) 2-hexanone, (b) pentanal, and (c) *m*-chlorobenzaldehyde.

49. Give systematic names for

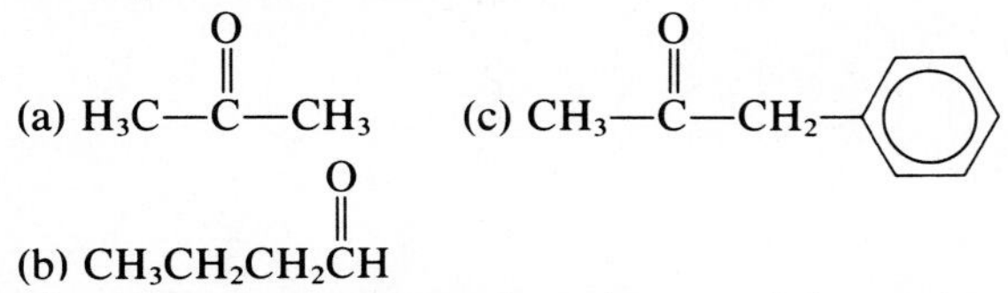

(a) $H_3C-\overset{\overset{O}{\|}}{C}-CH_3$ (c) $CH_3-\overset{\overset{O}{\|}}{C}-CH_2-C_6H_5$

(b) $CH_3CH_2CH_2\overset{\overset{O}{\|}}{C}H$

50. Give the structural formula and name of the product from each of the following reactions:
    (a) propanal and $KMnO_4$
    (b) propanal and $LiAlH_4$
    (c) 2-butanone and $LiAlH_4$
    (d) 2-butanone and $KMnO_4$
    (e) butanal plus CH_3MgI followed by aqueous acid
    (f) 2-butanone plus CH_3MgI followed by aqueous acid

51. If acetone is reduced with $LiAlH_4$, give the structural formula and name of the product. If this product is then treated with hot concentrated H_2SO_4, what are the structural formula and name of the product? Finally, if this second product is treated with HBr, what are the structural formula and name of the final product?

52. Tell how you could prepare 2-pentanol beginning with an appropriate ketone. Prepare this same alcohol beginning with an appropriate aldehyde and other reagents as necessary.

53. Tell how you can prepare 2-methyl-2-pentanol beginning with 2-pentanone and other appropriate reagents.

54. Give a synthesis for 1-butanol beginning with 1-propanol and other appropriate reagents.

AMINES AND AMIDES

55. Name the following amines.
    (a) $(C_6H_5)_2NH$
    (b) $(CH_3CH_2)_3N$
    (c) 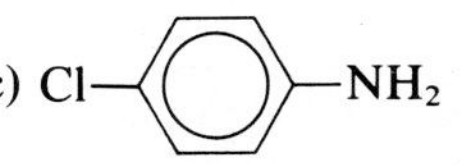 $Cl-C_6H_4-NH_2$ (para)

56. What is the pH of a 0.10 M aqueous solution of methylamine?

57. What is the pH of a 0.20 M aqueous solution of the weak organic base pyridine (Appendix F)?

58. What is the product of the reaction between triethylamine and hydrogen bromide? Give its systematic name.

59. Give the structural formula for the product of the reaction of diethylamine with acetyl chloride.

60. Tri-*n*-propylamine does not react with acetyl chloride. Explain briefly why there is no reaction.

PETROLEUM REFINING

61. Name the various petroleum refinery processes and tell what each does.

GENERAL QUESTIONS

62. In addition to the structural isomerism of alkanes, and *cis–trans* isomerism of some alkenes, other types of isomers are found in organic chemistry. For example, dimethyl ether (CH_3-O-CH_3) and ethyl alcohol (CH_3CH_2OH) are isomers; they have the same molecular formulas but their chemical functionality is quite different.
    (a) Draw all of the isomers possible for C_3H_8O. Give the systematic name of each and tell what class of compounds it fits.
    (b) Draw the structural formula for an aldehyde and a ketone with the molecular formula C_4H_8O. Give the systematic name of each.

63. Draw the structural formula for each of the nine possible isomers of $C_4H_8Cl_2$. Name each compound.

64. Salicylic acid has both a carboxyl group and a hydroxyl group attached to a benzene ring. Consequently, it is both a carboxylic acid and a phenol. Describe syntheses for methyl salicylate (the fragrance in oil of wintergreen) and acetylsalicylic acid (aspirin) starting with salicylic acid.

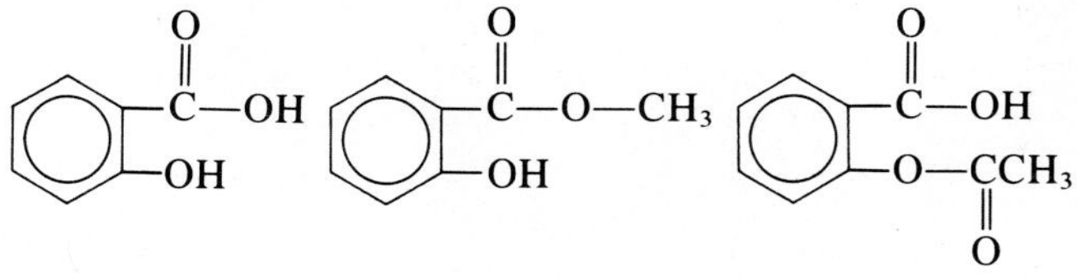

salicylic acid: $C_6H_4(OH)-\overset{\overset{O}{\|}}{C}-OH$

methyl salicylate: $C_6H_4(OH)-\overset{\overset{O}{\|}}{C}-O-CH_3$

acetylsalicylic acid: $C_6H_4(O-\underset{\underset{O}{\|}}{C}CH_3)-\overset{\overset{O}{\|}}{C}-OH$

65. Outline a method of synthesizing 2-hexanol from acetaldehyde, 1-bromobutane, and any necessary inorganic reagents.

66. Tell how you could convert *tert*-butyl bromide (2-bromo-2-methylpropane) to 2,2-dimethylpropanol.

CHAPTER

27

Polymers: Natural and Synthetic Macromolecules

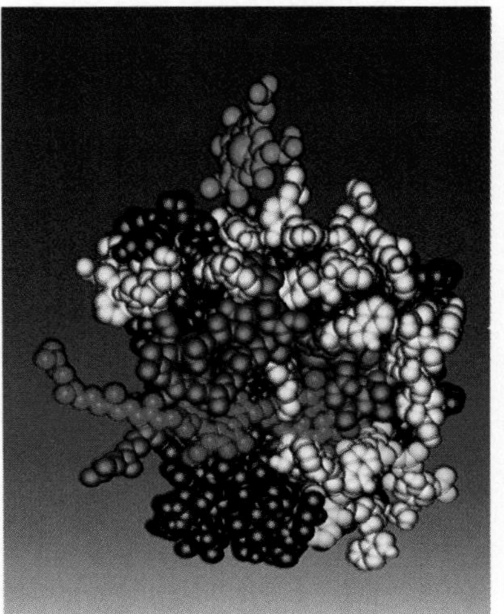

Photosynthetic reaction center of *Rhodopseudomonas viridis*. An image of the protein, including about 2000 atoms out of a total of around 12,000. (Supercomputer-based graphics by Treutlein and Schulten of the University of Illinois, Urbana-Champaign, assisted by NCSA visualization specialist Matthew Arrott)

CHAPTER OUTLINE

Approximately 80% of the organic chemical industry is devoted to the production of synthetic polymers. Phonograph records are polyvinyl chloride, the plastic squeeze bottle in your laboratory desk is polyethylene, and frying pans are coated with Teflon. You may wear clothes made of polyester, Dacron, or Orlon and walk in shoes made of a synthetic material. Synthetic polymers are so widely used that it should come as no surprise to learn that almost 240 pounds of such materials are made per person in the United States annually.

Materials such as polyethylene can have molecular weights approaching 50,000, and so we can refer to them as **macromolecules**. More generally, however, we call them **polymers**. The latter word comes from the Greek in which *poly* means "many" and *mer* means "unit." The small, individual units from which a polymer is constructed are called *monomers*. Thus, the monomer of polyethylene is ethylene,

Overall U.S. production of commercial polymers, including plastics, synthetic fibers, and synthetic rubber, was 59.3 billion pounds in 1989.

$$n\ CH_2{=}CH_2 \xrightarrow{\text{addition}} {-}\!\!\left[CH_2{-}CH_2{-}CH_2{-}CH_2\right]\!\!{-}$$

monomer ethylene — addition polymer polyethylene

and that of a polypeptide or protein is an amino acid.

$$n\ H_2N{-}CH(R){-}COOH \xrightarrow{\text{condensation}} {-}\!\!\left[\overset{H}{\overset{|}{N}}{-}CH(R){-}\overset{O}{\overset{\|}{C}}{-}\overset{H}{\overset{|}{N}}{-}CH(R){-}\overset{O}{\overset{\|}{C}}\right]\!\!{-} + n\ H_2O$$

monomer amino acid — condensation polymer polypeptide

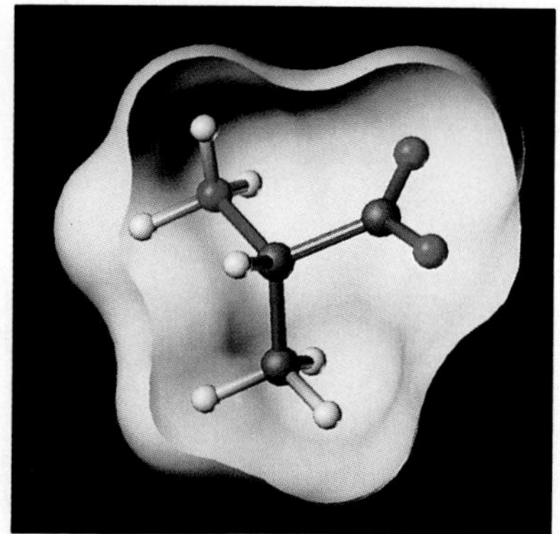

A computer drawn ball-and-stick model of L-alanine enclosed within a model of the surface of the molecule (gray = C, white = H, blue = N, and red = O) (by J. Weber of the University of Geneva, Switzerland). Be sure to notice that the central carbon is a chiral center, as four different groups are attached to this carbon. Also notice that the molecule is in the zwitterionic form, $^{+}H_3NCH(CH_3)COO^{-}$.

Polyethylene is called an *addition polymer*, since the empirical formula of the polymer is the same as that of the monomer. In contrast, a polypeptide is a *condensation polymer*; the monomers have condensed by eliminating a small molecule, in this case water.

Polymers may be divided, somewhat artificially, into two broad classes: *natural* and *synthetic*. Polyethylene is a synthetic polymer, while cellulose and nucleic acids such as DNA are natural polymers. After examining some structural principles important to protein chemistry, we shall turn to these and other natural polymers. Purely synthetic polymers are the subject of the latter part of the chapter.

27.1 AMINO ACIDS

All proteins are built of amino acid units, and all have the same general formula.

$$\mathrm{R{-}C(H)(NH_2){-}C({=}O){-}O{-}H} \qquad \mathrm{CH_3{-}C(H)(NH_2){-}C({=}O){-}O{-}H}$$

general α-amino acid — the amino acid alanine

All amino acids are characterized by having an amine group (generally $-NH_2$) and a carboxylic acid group ($-COOH$). Almost all of the amino acids found in living systems are α-amino acids, where the symbol α is the Greek letter *alpha*. This signifies that the amino group is attached to the first or alpha carbon atom beyond the $-COOH$ group.

As outlined in Chapter 25, a compound with four different groups attached to a tetrahedral carbon atom can have nonsuperimposable, mirror image forms called enantiomers. The central atom of these left- and right-handed forms is called a chiral center. One of the enantiomers of amino acids is designated by the symbol D *and the other by* L. *See Figure 27.1.*

In view of the fact that a large number of proteins exists and that they vary so widely in their properties, it is striking that they are derived from only about 20 amino acid building units. All 20 of the common amino acids are α-amino acids and, with the exception of glycine, the alpha carbon atom is a *chiral center* (Table 27.1) because four different groups (H, R, NH_2, and COOH) are attached to the carbon. However, it is equally striking that often only one of the enantiomeric forms of a given amino acid exists in nature. In virtually every case, living multicell organisms use *only* one enantiomer, the so-called "L" form, a configuration defined in Figure 27.1.

Most amino acids dissolve reasonably well in water. For example, 25 g of glycine dissolve per 100 g of water at 25 °C, and 3 g of phenyl-

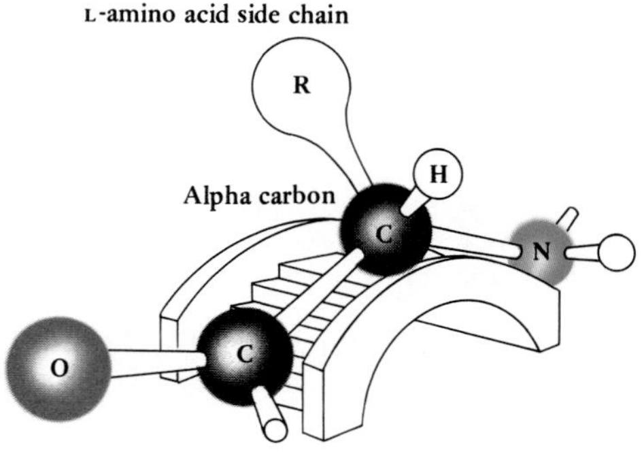

Figure 27.1 The chirality of α-amino acids. This figure represents a way to remember the two enantiomeric forms. When walking along the C—C—N bridge from the C—O end to the NH_2 end, the acid is in the L form if the R group is on the left. If R is on the right, it is a D acid. The L form is the one found predominantly in nature. See Chapter 25 for a further discussion of chirality. (© Irving Geis)

Table 27.1 Ten of the 20 α-Amino Acids

Structure	Amino Acid	Structure	Amino Acid
$CH_3-CH(NH_2)-C(=O)OH$	Alanine	$HC{=}C(-N{=}CH-NH-)-CH_2-CH(NH_2)-C(=O)OH$	Histidine
$HO(O{=})C-CH_2-CH(NH_2)-C(=O)OH$	Aspartic Acid	$NH_2-CH_2-CH_2-CH_2-CH_2-CH(NH_2)-C(=O)OH$	Lysine
$HS-CH_2-CH(NH_2)-C(=O)OH$	Cysteine	$C_6H_5-CH_2-CH(NH_2)-C(=O)OH$	Phenylalanine
$HO(O{=})C-CH_2-CH_2-CH(NH_2)-C(=O)OH$	Glutamic Acid	$HO-CH_2-CH(NH_2)-C(=O)OH$	Serine
$H-CH(NH_2)-C(=O)OH$	Glycine	$(CH_3)_2CH-CH(NH_2)-C(=O)OH$	Valine

alanine dissolve under the same conditions. In every case, the following equilibrium exists in aqueous solution.

$$R-\underset{NH_2}{\overset{H}{C}}-\overset{O}{\overset{\|}{C}}-O-H \rightleftharpoons R-\underset{NH_3^{+}}{\overset{H}{C}}-\overset{O}{\overset{\|}{C}}-O^{-}$$

zwitterion

Since the amino group is more basic than the carboxylate group, the molecule exists primarily as a *dipolar ion* known as a **zwitterion**, a term taken from the German word *zwitter* meaning "double."

> **EXERCISE 24.1 Amino Acid Structures**
> Identify the chiral center in serine and draw the L form of the amino acid.

27.2 PROTEINS

Proteins are important in a variety of ways. As *enzymes* they serve as catalysts in biological synthesis and degradation reactions. *Hormones* serve a regulatory role, and *antibodies* protect us against disease. The protein *hemoglobin* carries oxygen from the lungs to the various parts of the body. Finally, proteins are the major constituents of cellular and intracellular membranes, skin, hair, muscle, and tendons.

Proteins can be divided into two classes: simple and conjugated. **Simple proteins** consist only of amino acids, while **conjugated proteins** may contain some other group in addition to the amino acids. Examples of conjugated proteins are hemoglobin and myoglobin, in which the iron-containing heme group (see below) is the site of oxygen binding.

To form a protein, the first step is a *condensation reaction* of two amino acids.

$$H_2N{-}\underset{H}{\overset{R}{C}}{-}\overset{O}{\overset{\|}{C}}{-}OH \;+\; H{-}\underset{H}{N}{-}\underset{H}{\overset{R}{C}}{-}\overset{O}{\overset{\|}{C}}{-}OH$$

$$\downarrow -\,H_2O$$

$$H_2N{-}\underset{H}{\overset{R}{C}}{-}\overset{O}{\overset{\|}{C}}{-}\underset{H}{N}{-}\underset{H}{\overset{R}{C}}{-}\overset{O}{\overset{\|}{C}}{-}OH$$

the planar amide group

One molecule of water is eliminated between the carboxylic acid of one amino acid and the amine group of another. The result is a **peptide** bond, a grouping called an amide in Chapter 26, and the molecule is a *dipeptide*. Either end of the dipeptide can react with another amino acid to give a tripeptide, which can in turn condense with another amino acid, and so on. Ultimately, many amino acids can be linked to give a *polypeptide* or protein (Figure 27.2). Thus, *a protein is a polymer of amino acids*.

Proteins occur in a variety of sizes. The common protein insulin has only 51 amino acid units in two linked chains. In contrast, human hemoglobin contains four protein chains, two identical ones having 141

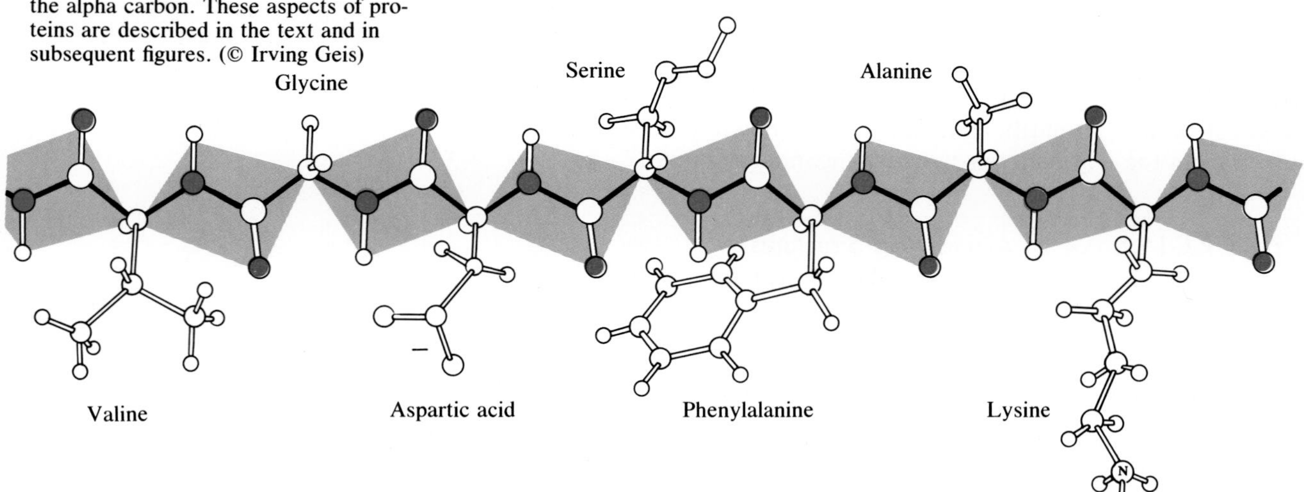

Figure 27.2 A polypeptide chain or protein. This figure also shows (a) the planarity of the peptide link and (b) the fact that the chain bends only around the alpha carbon. These aspects of proteins are described in the text and in subsequent figures. (© Irving Geis)

amino acids and the other two, again identical, having 146. Considering all of the possible ways that 20 α-amino acids could be put together to form proteins, it is remarkable that so few proteins are in fact known (the human body is thought to contain only about 100,000 different proteins) and that cells produce generation after generation of identical proteins that serve a given function.

Nature can go awry and produce different proteins from a collection of amino acids. The genetic disease "sickle cell anemia" results when valine is substituted for glutamic acid at a specific point in two of the four hemoglobin chains.

EXERCISE 24.2 Peptides

(a) Any two amino acids can form two different dipeptides. For example, an N—H of glycine can condense with the —COOH of serine, or the serine N—H can condense with the —COOH of glycine. Draw structures of these two possibilities.

(b) Draw the structure of the tripeptide formed from glycine molecules.

PROTEIN STRUCTURE

When many amino acids are linked together in a particular sequence with a particular geometry, a specific protein is formed. To fully understand the functioning of the protein, we must know its *primary structure*, the sequence of amino acid units. Equally important, however, is the three-dimensional structure of a protein that arises from interactions between chains. These interactions are important because they ultimately affect the functioning of the protein, such as its ability to act as a biochemical catalyst or to bind oxygen.

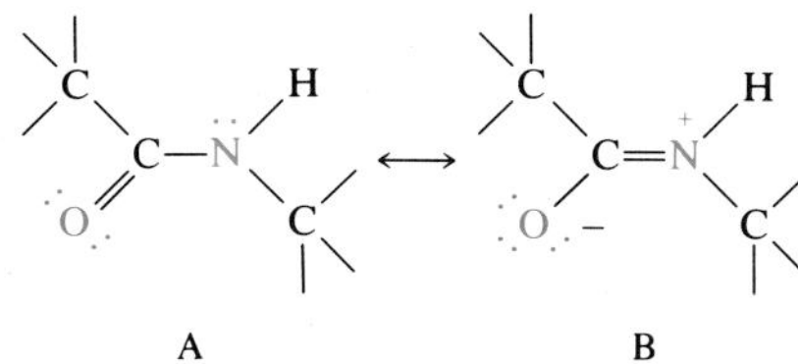

Once the primary structure of a protein is established, the next question is whether the protein chains are straight or folded up in a particular pattern. To see the possibilities, first look at the resonance structures of the peptide bond. Structure B contributes to the overall structure, so there is significant double bond character to the C—N bond. As you learned in Chapter 26, energy is required to rotate the groups at one end of a double bond relative to the other end, so the peptide linkage (O—C—N—H) is locked in a flat configuration.

The flat peptide unit, with a slight negative charge on the oxygen atom and a slight positive charge on the N—H hydrogen, has two important consequences. First, the peptide chain can only bend around the alpha carbon atom of each amino acid unit (see Figure 27.2). Second, the polarities of the C—O and N—H bonds lead to hydrogen bonding between amino acid units (Figure 27.3).

Figure 27.3 Hydrogen bonding between two amino acid units of a polypeptide chain. Notice also that the C—O and N—H bonds are *trans*, a feature almost always observed in proteins. (© Irving Geis)

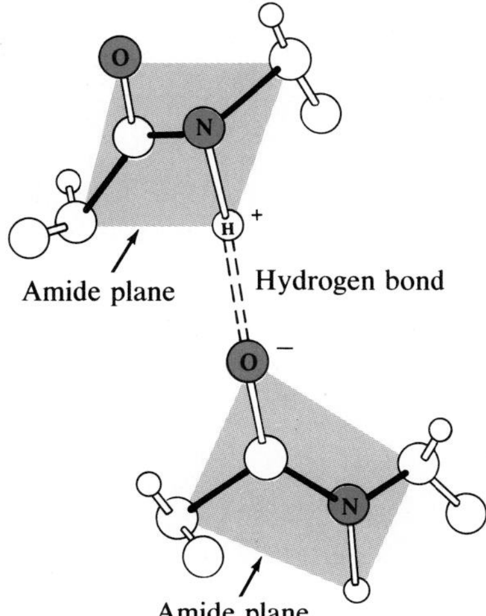

Given that the peptide linkage is flat and that hydrogen bonding is possible, Linus Pauling demonstrated some years ago that there must be two most common *secondary structures* of protein molecules: the ***α*-helix** and the ***β*-pleated sheet**. The secondary structure adopted depends on the position of the C═O···H—N hydrogen bonds. If the hydrogen bonding is intramolecular (within the protein chain), the result is the α-helical structure. The protein shown in Figure 27.4 is wound as a right-handed helix, just as the threads of a screw or bolt are right-handed. The helical structure is fixed by the fact that *all* of the C═O and N—H groups in peptide linkages participate in hydrogen bonding parallel to the axis of the helix; each C═O group binds to a N—H bond in a unit four amino acid "residues" away along the chain.

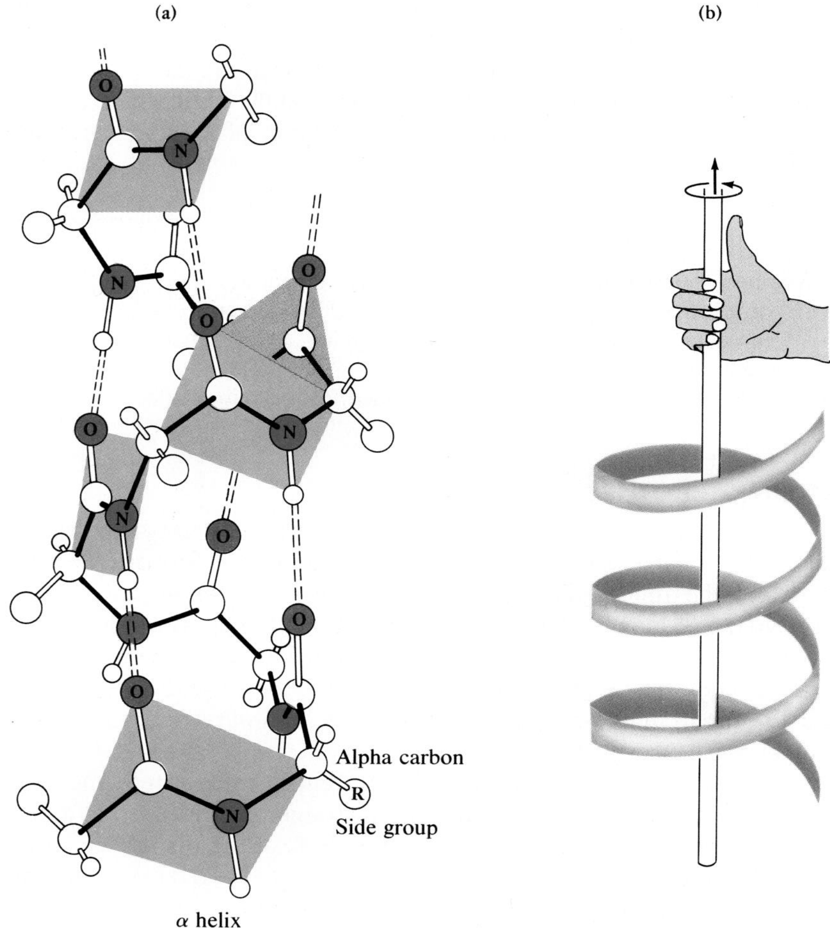

Figure 27.4 The helical structure of proteins. (a) Hydrogen bonding leads to a spiral arrangement called an α-helix. (b) In proteins, the helix always has a right-handed arrangement. If you hold your hand so that your thumb points in the direction of travel along the axis of the helix, the curl of your fingers describes the direction in which the helix rotates. (The same is also generally true of sea shells, as seen in Chapter 25, page 1049.) (a, © Irving Geis)

The α-helix is the basic structural unit of *fibrous proteins* known as α-keratins. These proteins form wool, hair, skin, beaks, nails, and claws. Because of the helical protein chains, human hair fibers, for example, are stretchable and elastic to a small extent. Stretching the fibers will stretch the relatively weak hydrogen bonds, but not the covalent bonds. If the hair fiber is not stretched to the point where the covalent bonds begin to break, the fiber will snap back to its original length. The fiber is elastic, because the hydrogen bonds can be re-formed.

In other fibrous proteins, such as those in silk, there is *inter*molecular C=O···H—N hydrogen bonding, one protein chain being bound to a neighboring chain (Figure 13.7). Each polypeptide chain is fully extended, and hydrogen bonding between chains leads to a sheetlike structure. Sheets of proteins can then be stacked on top of one another like the pages of this book. This structure, called the β-sheet structure, means that silk and other insect fibers are not stretchable or elastic. Pulling on them would break covalent bonds or the many hydrogen bonds holding the individual protein strands in the sheet. However, just as you can bend the stack of pages in this book, so too can the protein stack be bent.

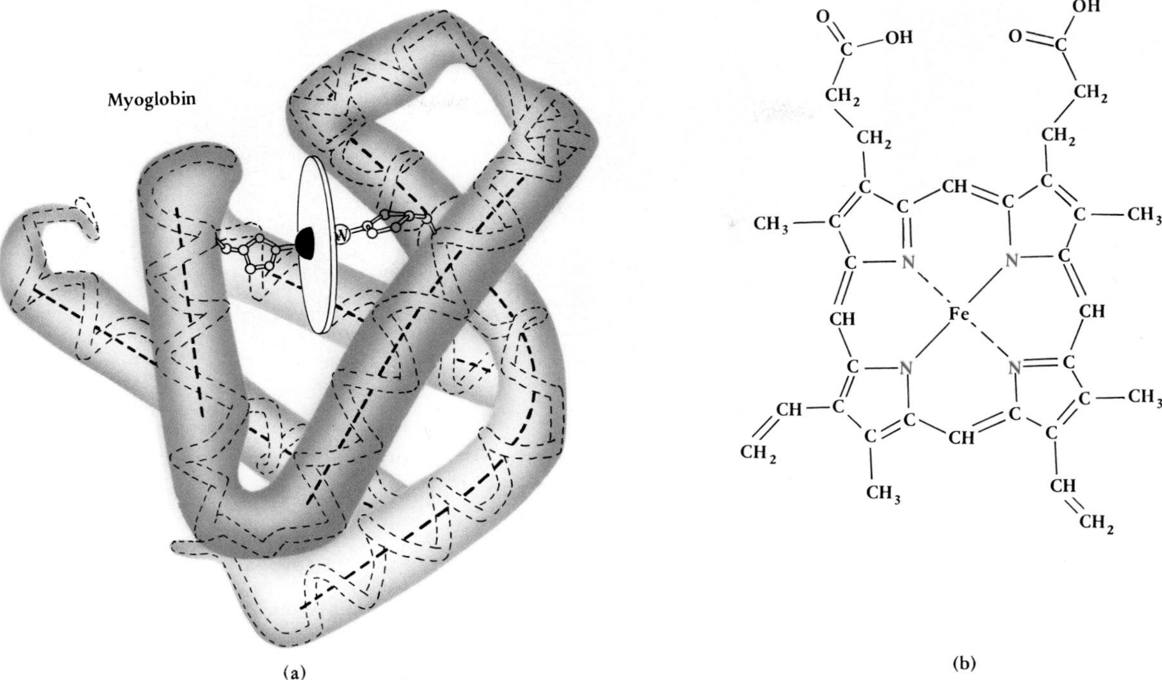

Figure 27.5 Myoglobin, the O_2-storage protein in muscles. (a) A continuous chain of 153 amino acid units encloses a heme group. The iron ion of the heme, shown as the black hemisphere, is the site of the O_2-bonding. When not bound to O_2, the metal ion binds water, designated by W in the figure. (b) The heme group, where an iron ion is enclosed within an organic group called a porphyrin. (a, © Irving Geis)

Not only are α-helix and β-sheet structures commonly used in fibrous proteins, but they are also common in the *globular proteins*. Two globular proteins are hemoglobin, the protein that carries oxygen in the blood from the lungs to the tissues, and myoglobin, the O_2-storage protein in muscles (Figure 27.5a). Both are characterized by α-helical protein chains, and both incorporate the iron-containing heme group (Figure 27.5b). It is this latter group that gives the red color characteristic of blood or red meat.

Hemoglobin has four folded, helical chains, each having an iron-containing heme group. The four chains are packed into a compact unit.

27.3 CARBOHYDRATES

The word ''carbohydrate'' literally means ''hydrate of carbon.'' Thus, **carbohydrates** have the general formula $C_x(H_2O)_y$ in which x and y are integers. However, even though the reaction of the carbohydrate sucrose with sulfuric acid produces carbon (Figure 27.6), this does not mean that sucrose is a simple combination of carbon and water. Rather, carbohydrates are all molecules that are polyhydroxyaldehydes or polyhydroxyketones and share some chemical properties with alcohols and aldehydes or ketones.

Although carbohydrates account for only about 1% of total body mass, they are centrally important in human biochemistry. Carbohydrates

Figure 27.6 Reaction of sucrose with concentrated sulfuric acid. Sucrose is a carbohydrate with a formula of $C_{12}(H_2O)_{11}$ or $C_{12}H_{22}O_{11}$. Sulfuric acid is an excellent dehydrating agent, so carbon is the final product.

are formed in plants by the process of photosynthesis from carbon dioxide, water, and energy from the sun. Carbon is reduced in this reaction to an average oxidation number of zero.

$$x\ CO_2 + y\ H_2O + \text{energy} \longrightarrow \underset{\text{carbohydrate}}{C_x(H_2O)_y} + x\ O_2$$

Simple carbohydrates are oxidized by animals to CO_2 and H_2O with the accompanying release of energy (the reverse of the reaction above), and it is this energy that drives the general metabolic processes of the organism. In addition, the carbohydrate cellulose is the main structural component of plants, and carbohydrates are major components of wood, cotton, paper, and many other important materials derived from natural sources. Finally, the nucleic acids (Section 27.4) contain carbohydrate units in their repeating structure.

Carbohydrates are placed in one of three classes, depending on their molecular size. *Monosaccharides* are the simplest and are the building blocks of *disaccharides*, which contain two monosaccharide units, and of *polysaccharides*, which are polymers of monosaccharides.

SIMPLE SUGARS: MONOSACCHARIDES

The names of saccharides end in "-ose." The general name for a saccharide with n = 5 is a pentose, while one with n = 6 is a hexose.

Monosaccharides have one unit of water per carbon atom ($x = y$) and so have the general formula $(CH_2O)_n$ where n can be three to six or more. One of the important simpler sugars is *ribose*, a *pentose* with $n = 5$ and a formula of $(CH_2O)_5$, or $C_5H_{10}O_5$,

β-D-ribose

and another common monosaccharide is *glucose*, a *hexose* with a formula of $C_6H_{12}O_6$ ($n = 6$).

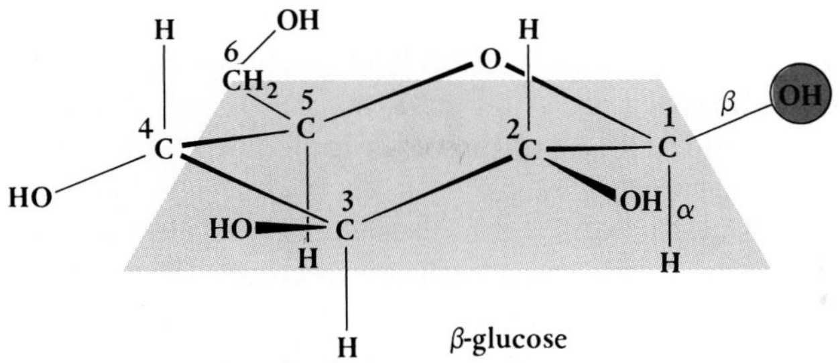

β-glucose

The structure of glucose is drawn so that three of the ring carbon atoms (carbons 1, 3, and 5) lie in a plane. This allows us to distinguish *two types of groups* attached to the ring carbon atoms: (a) those above or below the plane, called **axial** groups, and (b) those that lie approximately in the plane, called **equatorial** groups. Here the axial groups are all H atoms, while the equatorial groups are —OH or $—CH_2OH$.

The position of the —OH group attached to the carbon labeled 1 (usually written C-1) is important to the functioning of the molecule, as you shall see. If the group is in an equatorial position, as pictured above, the molecule is called β-glucose. On the other hand, if —OH is in the axial position on C-1, the molecule is labeled α-glucose.

An aqueous solution of pure glucose consists of an equilibrium mixture of α and β isomers. They can be interconverted by opening and then reclosing the ring.

Two other common hexoses, mannose and galactose, differ from glucose only in that one —OH group is moved to an axial position.

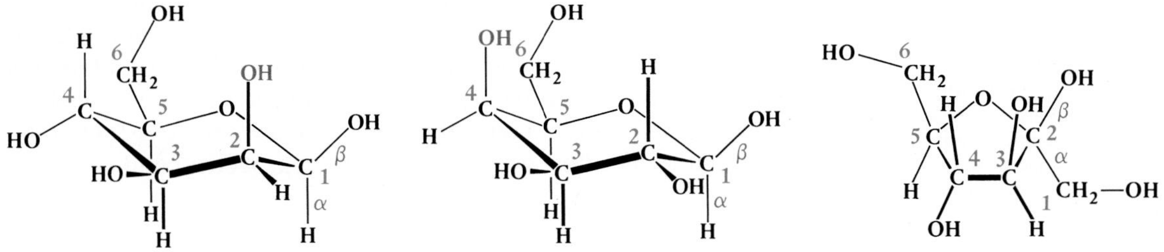

β-D-mannose β-D-galactose β-D-fructose

Fructose, or fruit sugar, is another six-carbon monosaccharide, but only four of the carbon atoms participate in the ring, so it is similar in this regard to ribose.

DISACCHARIDES

Monosaccharide units can condense, with the elimination of water, to give disaccharides.

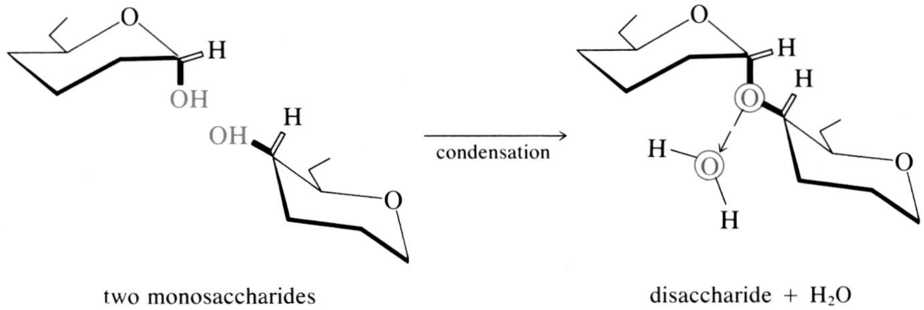

condensation

two monosaccharides disaccharide + H_2O

The structures of four common disaccharides are illustrated in Figure 27.7. In every case the condensation leads to the so-called **glycoside** linkage. When two hexoses are combined, the C—O—C linkage is between C-1 of one molecule and C-4 of the other molecule; such bonds are called "1,4-linkages."

Honey, a mixture of the monosaccharides glucose and fructose, has been used for centuries as a natural sweetener for foods. In contrast, sucrose, derived from sugar cane or sugar beets, is a disaccharide. Honey is a popular sweetener because cane sugar is not as sweet as a mixture of pure glucose and fructose. To convert cane sugar into glucose and fructose requires treatment with acid or with a natural enzyme or catalyst called "invertase." The sugar industry goes to considerable expense to convert sugar chemically into "dextrose" or "levulose," a mixture of glucose and fructose.

Cane sugar first appeared in Southeast Asia around the 5th century BC, but it did not arrive in the western world until about the 7th century AD. However, not until the great sugar plantations were established in the New World in the 17th century was "stone honey," as the Chinese called cane sugar, readily available.

POLYSACCHARIDES: CELLULOSE AND STARCH

The three most important polysaccharides are **cellulose**, **starch**, and **glycogen**, all polymers of glucose. Cellulose is the most abundant organic compound on earth, and its purest natural form is cotton. This polysaccharide is also found as the woody part of trees and the supporting material in plants and leaves.

Cellobiose, shown in Figure 27.7, is a fragment of **cellulose**, a *linear* polymer of β-glucose units. There are typically 300 to 3000 glucose molecules in the chain, so the molecular weight is between 50,000 and 500,000.

Since cellulose is so abundant, it would be advantageous if humans could use it for food. Unfortunately, we cannot digest it, because we lack the necessary enzyme or biochemical catalyst to chew up the β-1,4 bonds. On the other hand, termites, a few species of cockroaches, and ruminant

Figure 27.7 Four common disaccharides, all combinations of two sugar monomers. (a) Sucrose is a combination of the α form of glucose and the β form of fructose. (b) Lactose or milk sugar is formed from the β forms of galactose and glucose. Therefore, the linkage is called a β-1,4 glycoside linkage. (c) Maltose is a degradation product of starch, a glucose polymer with an α-1,4 linkage. (d) Cellobiose comes from breaking down the cellulose polymer and has the β-1,4 linkage characteristic of this substance. (© Irving Geis)

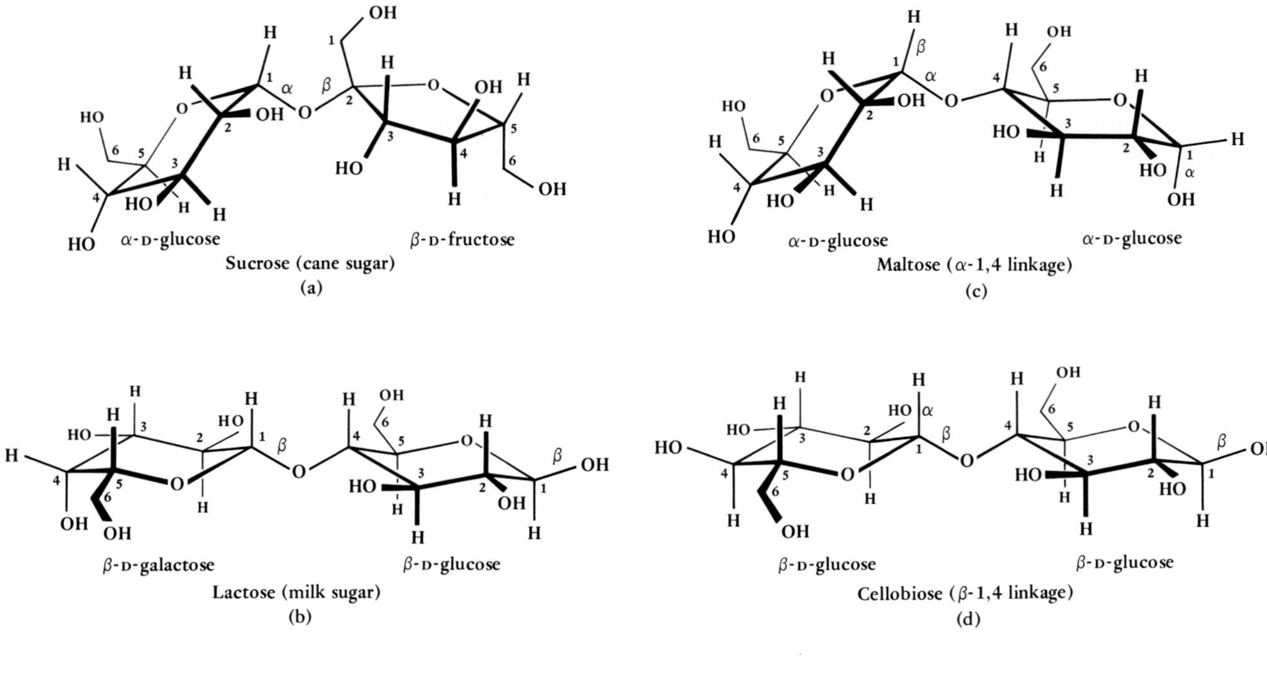

α-D-glucose β-D-fructose
Sucrose (cane sugar)
(a)

α-D-glucose α-D-glucose
Maltose (α-1,4 linkage)
(c)

β-D-galactose β-D-glucose
Lactose (milk sugar)
(b)

β-D-glucose β-D-glucose
Cellobiose (β-1,4 linkage)
(d)

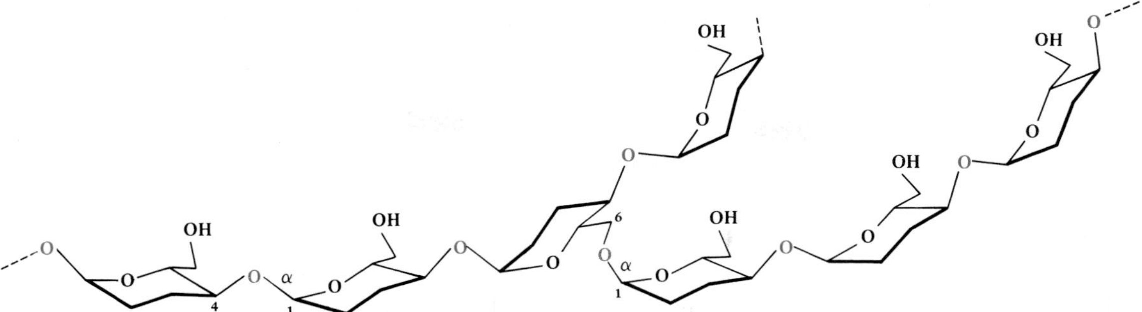

Figure 27.8 Amylopectin, a starch. The basic polymer forms by α-1,4 linkages between glucose units, but the chain branches using the —OH group of C-6. (Adapted from Irving Geis)

mammals such as cows, sheep, goats, and camels do have the proper internal chemistry laboratory for this purpose.

Because we cannot digest cellulose, we must rely partly on **starch** as a source of glucose. Starch occurs in two forms, *amylose* and *amylopectin*. Both are polymers of glucose units bound through α-1,4 linkages (as in maltose, Figure 27.7), but amylopectin is the more common form. It usually has about 1000 units in a branched chain (whereas the amylose chain is unbranched). Branching occurs at intervals of 20 to 25 units and involves a glycoside linkage between C-6 of a glucose unit of the main chain and C-1 of the first glucose unit of the branch chain (Figure 27.8). Although we do not have the necessary enzyme to break down cellulose, our saliva and pancreatic juices do contain amylase, the enzyme required to "crack" starch into glucose. You know this because, if you hold a piece of bread in your mouth for a few minutes, it begins to taste sweet, a signal that glucose and maltose are being produced. When the bread is swallowed, the acid in your stomach finishes what your saliva began, and the freed glucose can be absorbed into the bloodstream and carried off for eventual oxidation.

Starch is very attractive as a material for energy storage. The polymer is easily synthesized and only one type of chemical bond needs to be broken to give instant energy, glucose. On the other hand, fats and fatty acids (Chapter 26) require many steps for their synthesis, and their breakdown or *metabolism* is complex. The complexity of fat chemistry is offset by the fact that fats provide much more energy per gram than does glucose (see Table 27.2). Thus, animals and other organisms that move around often use fat for energy storage, because less mass needs to be transported to provide a given energy. In contrast, an immobile

Table 27.2 Heat Energy Provided by Various Fuels

Fuel	Heat of Combustion (kJ/g)
Hydrogen, $H_2(g)$	142
Methane, $CH_4(g)$	56
Octane (gasoline), $C_8H_{18}(g)$	48
Stearic acid (fatty acid), $C_{17}H_{35}COOH(s)$	40
Glucose, $C_6H_{12}O_6(s)$	15

plant has little need for a substance with a high energy-to-mass ratio, so plants choose the simplicity of carbohydrate chemistry.

To provide instant energy, however, animals do use one form of carbohydrate. They synthesize *glycogen*, a more highly branched form of amylopectin. This is stored in the liver and muscle tissues and is used for "instant" energy until the process of fat metabolism can take over and serve as the energy source.

EXERCISE 27.3 Disaccharides

At the beginning of the section on disaccharides, the condensation of two monosaccharides is illustrated. What are these monosaccharides? What disaccharide is formed? Is an α- or β-1,4 glycoside link formed?

The five organic bases of nucleic acids and nucleotides

Purine derivatives

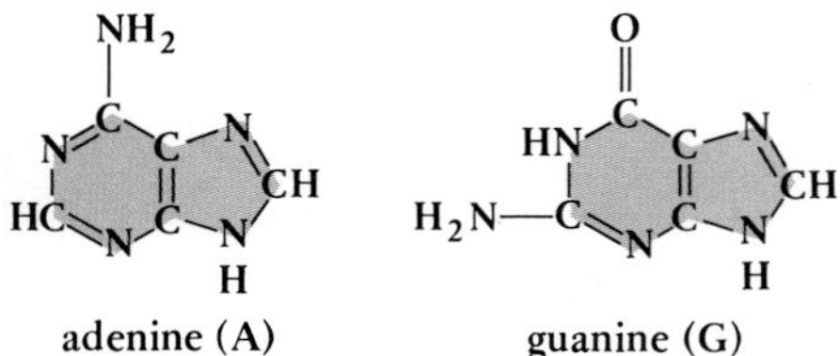

adenine (A) guanine (G)

Pyrimidine derivatives

cytosine (C)

uracil (U) only in RNA

thymine (T) only in DNA

27.4 ENERGY AND INFORMATION: NUCLEOTIDES AND NUCLEIC ACIDS

The nucleic acids store and transmit genetic information. *Deoxyribonucleic acid* (**DNA**) is the permanent repository of genetic information in the nucleus of the cell, where it controls the synthesis of *ribonucleic acid* (**RNA**). It is RNA that is involved in the biochemical machinery that transmits the genetic information and directs the synthesis of proteins. Close relatives of DNA and RNA are nucleotides such as **ATP** (adenosine triphosphate), much smaller molecules that are the chief short-term energy storage molecules for all life processes.

Nucleic acids and nucleotides are composed of three structural units: a sugar, a base, and one or more phosphates. The five organic bases that are the heart and soul of these systems are shown in the margin. Adenine (A) and guanine (G) are derivatives of the organic base purine ($C_5H_4N_4$), while thymine (T), cytosine (C), and uracil (U) are derivatives of pyrimidine ($C_4H_4N_2$). These bases are so important to any discussion of nucleic acid chemistry that they are often referred to only by the first letter of their name.

In RNA the sugar involved is *ribose*, whereas DNA is based on *deoxyribose*. The latter is also a pentose, but it has no —OH group on C-2.

D-ribose, found in RNA

D-deoxyribose, found in DNA

When any one of the five bases is bound to C-1 of the carbohydrate ribose, the result is a *nucleoside* such as adenosine (Figure 27.9). These nucleosides can in turn form *esters* with the phosphate ion by the elimination

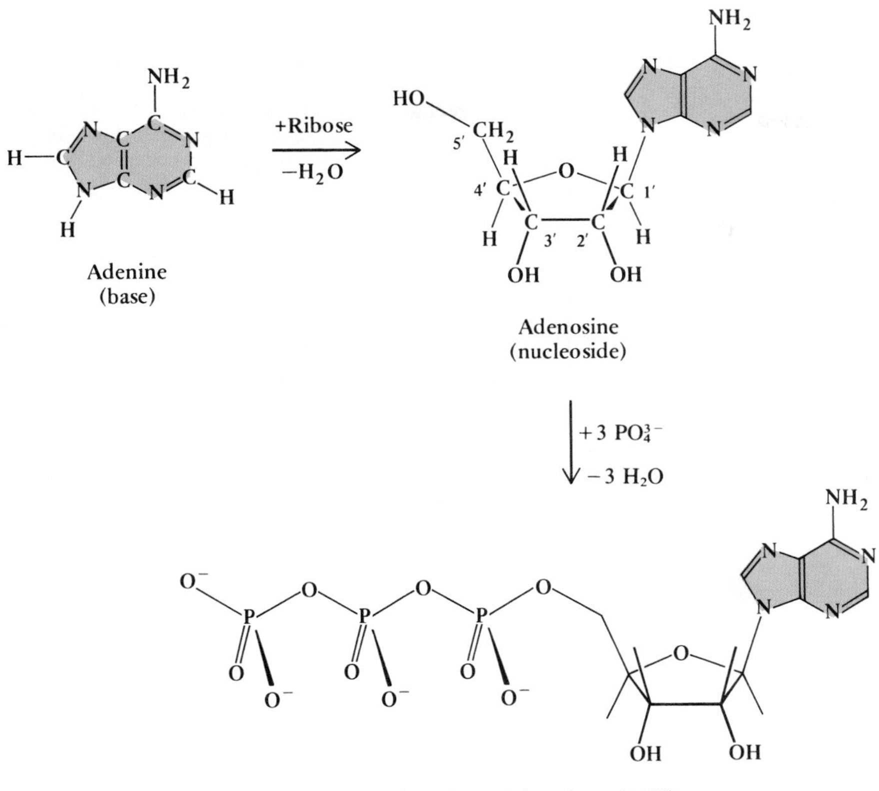

Figure 27.9 The formation of nucleosides and nucleotides. Adenine combines at C-1 of ribose to form the nucleoside adenosine. If the —OH group at C-5 condenses with phosphate, the ester adenosine monophosphate (AMP) is formed. Addition of other phosphate linkages gives adenosine diphosphate (ADP) and finally adenosine triphosphate (ATP). (© Irving Geis)

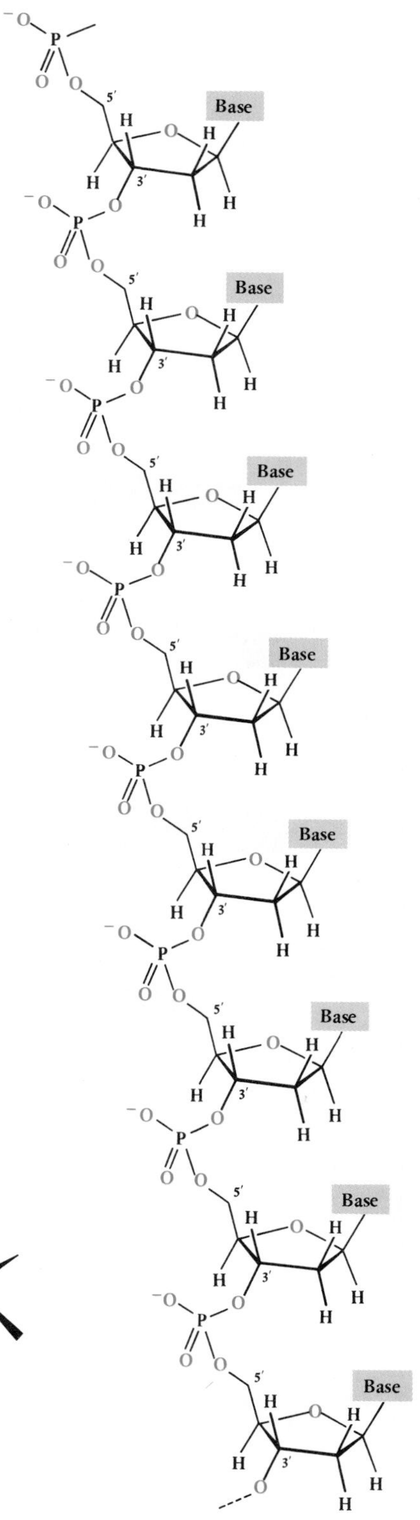

of water between the HPO_4^{2-} ion and the —OH group at C-2, C-3, or C-5.

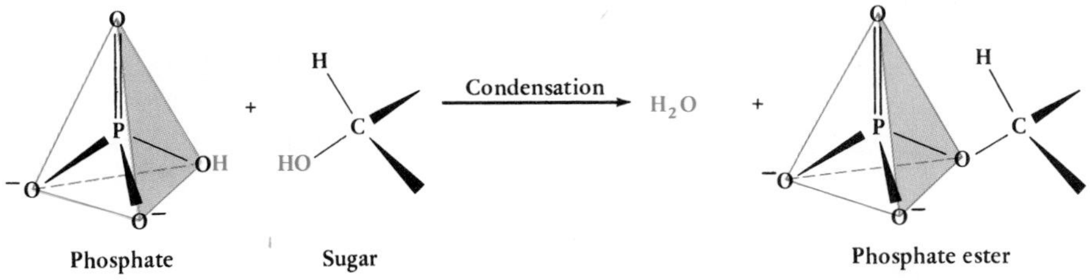

The phosphate esters are called *nucleotides*, and the most common are those involving the C-5 hydroxyl group. (Living organisms use several of these nucleotides as short-term energy storage systems.) If the phosphate bound to C-5 forms an ester linkage to another ribose or deoxyribose, then a polymer is formed. The backbone of DNA (deoxyribonucleic acid) or RNA (ribonucleic acid) is just such a polymer of alternating phosphate and deoxyribose or ribose units (Figure 27.10), respectively,

Figure 27.10 The polymeric backbone of DNA (deoxyribonucleic acid). DNA is a polymeric ester of nucleoside units bridged by phosphate groups. (Adapted from Irving Geis)

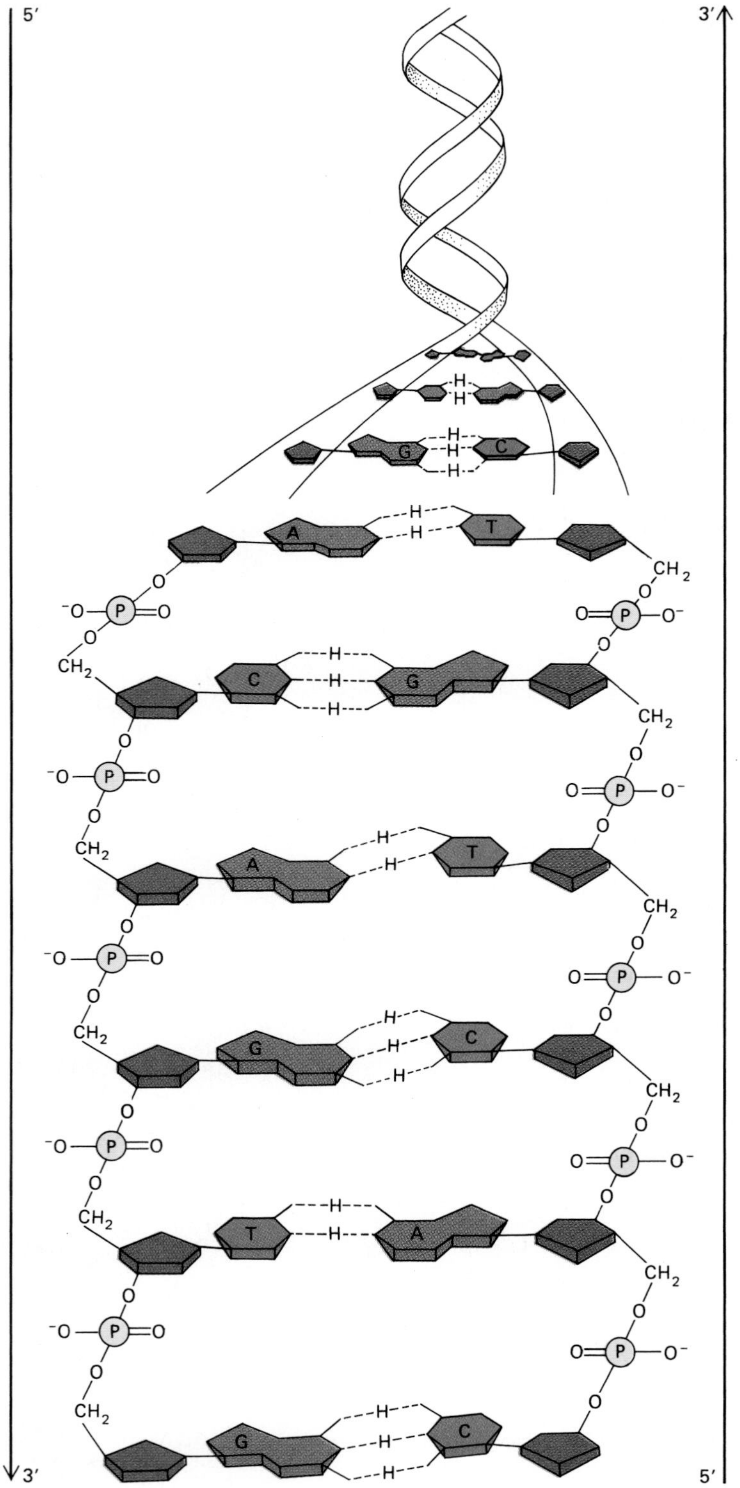

Figure 27.11 Structure of DNA. Notice that the two sugar-phosphate chains run in opposite directions. This permits complementary bases to pair by hydrogen bonding as illustrated by the molecular structures below. Only certain base pairings occur in DNA: A with T and G with C. Because of the tetrahedral geometry of the phosphate groups and the carbon atoms of the backbone, the strands twist into a helical structure. (The numbers 3′ and 5′ refer to the numbering of the carbon atoms in the ribose units; see Figure 27.10.) (Structures on right adapted from Irving Geis)

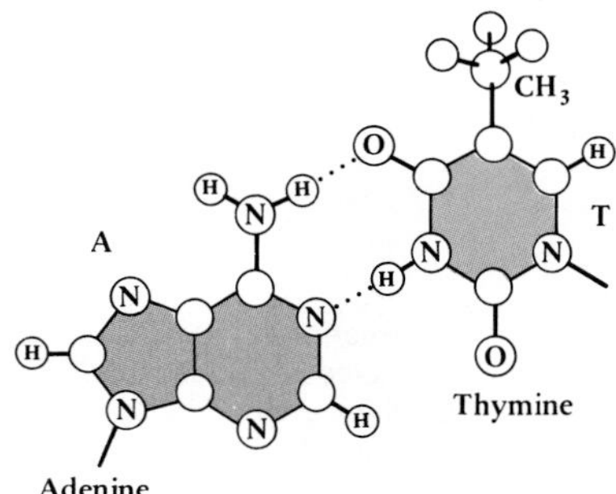

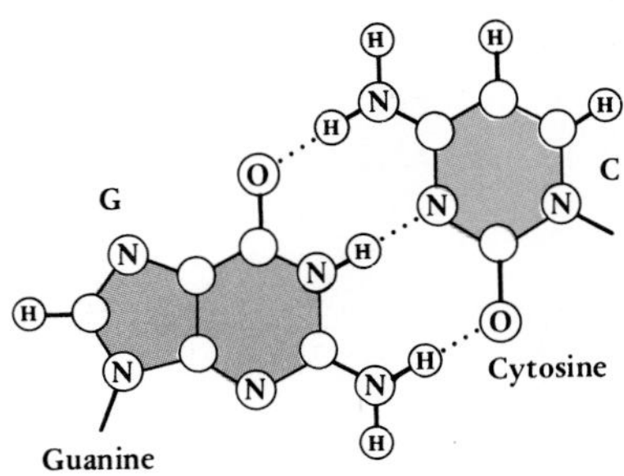

in which the phosphate bridges between C-5 of one sugar and C-3 of an adjacent sugar.

In Figure 27.10 the backbone of the DNA polymer is shown with only the positions of the bases indicated. It is these bases in which genetic information is coded. The sequence of the bases along the chain is specific

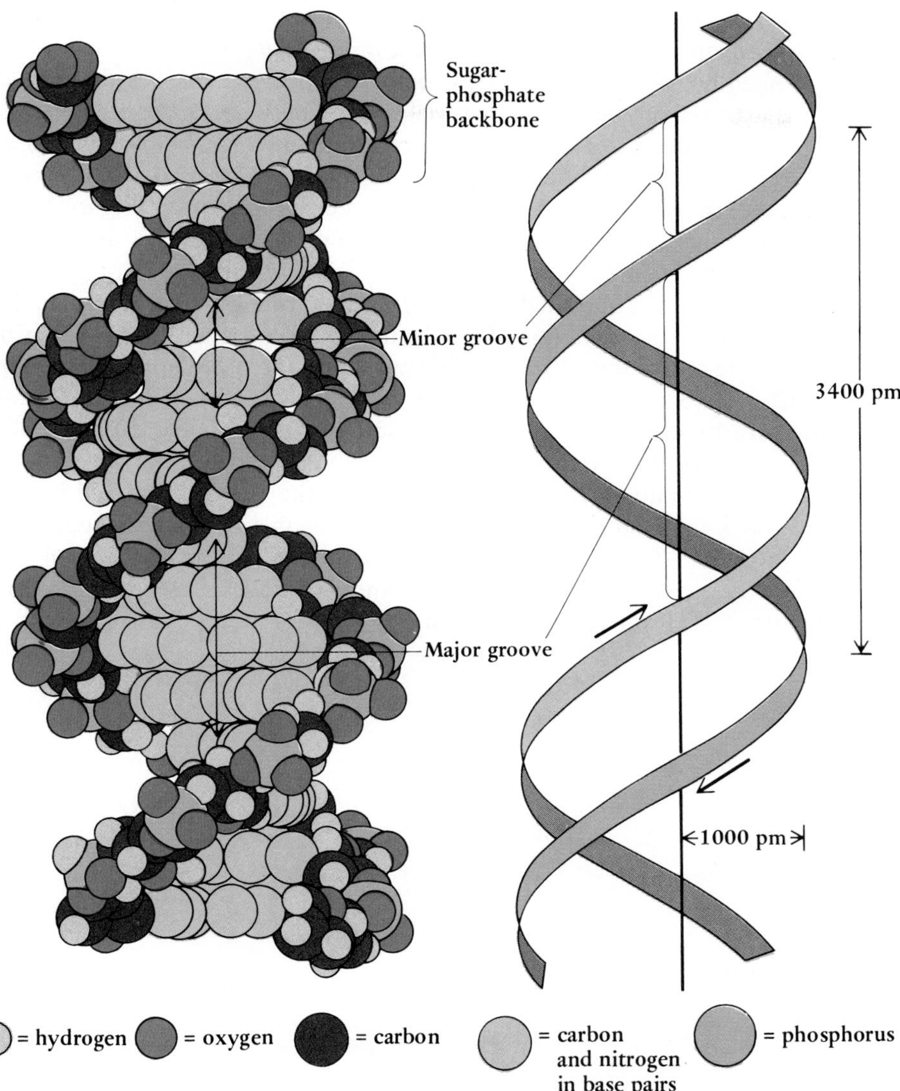

Figure 27.12 According to modern theory, most DNA exists in this double-helical form. The molecule actually has two backbones held together by base-pairing.

to the new protein to be synthesized by the cell in which the DNA is located. Each three-base sequence in the DNA, called a *codon*, is the code for one of the amino acids that is to be assembled into a new protein. For example, if one three-base sequence on the DNA is A—G—C, this is the code for serine to be added to the protein being synthesized by the cell.

You know from making long-distance telephone calls that information can sometimes be garbled. Confusing genetic information in the course of protein synthesis can be disastrous, as exemplified by genetic disorders such as sickle cell anemia and Down's syndrome. To solve this problem, each polymeric strand of DNA has a *complementary* partner or strand, and either partner is coded for the synthesis of a specific protein. James D. Watson and Francis H.C. Crick showed in 1953 that DNA is composed of two polymeric strands coiled into a *double helix* (Figures 27.11 and 27.12). The base units of each strand are pointed into the interior of the helix, and pairs of bases from the two strands are linked together by hydrogen bonds. The critical point of the Watson-Crick model is that hydrogen bonding can best occur between specific bases. A purine base of one strand generally pairs with a pyrimidine base of the other strand. Adenine-thymine (AT) and guanine-cytosine (GC) pairs occur almost exclusively because they are very tightly hydrogen-bonded. Each base is the donor in one hydrogen bond and the acceptor in another (Figure 27.11). We say that adenine and thymine are **complementary**, as are guanine and cytosine.

Due to the tetrahedral geometry around all of the atoms making up the polymeric backbone of DNA and the requirements of base pairing, the double strand must assume a helical shape about 2000 pm wide (Figure 27.12). Since base pairs occur every 340 pm along the helix and since there are 10 base pairs for every complete turn of the helix, each complete turn is about 3400 pm long.

The complementary base pairing in the double-stranded helix is the "back-up system" nature has devised to insure accurate replication in new DNA. The DNA is replicated by unwinding the double helix and attaching a nucleoside phosphate to the exposed bases on each strand (Figure 27.13). Since each base in a given strand can pair only with its

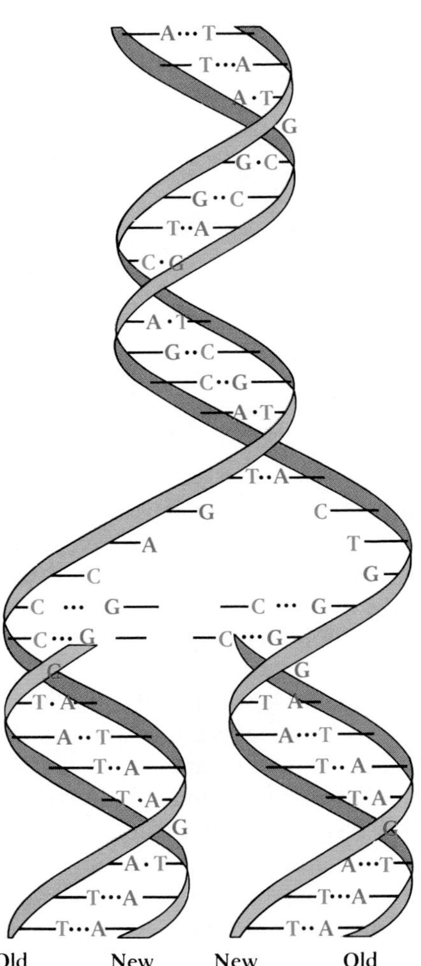

Figure 27.13 The replication of DNA. The original helical coil is unwound and each strand acts as a template for the formation of a new strand. Owing to the requirements of base pairing, each new strand is an exact copy of one of the old strands.

particular complement, this means that the newly formed "daughter" strand must be an exact copy of one of the "parent" strands. The net result is the formation of two new DNA double helices from the parent double helix. Each new double helix contains one parent strand and one new strand.

27.5 SYNTHETIC POLYMERS: PLASTICS, FIBERS, AND ELASTOMERS

It is almost certain that some of your clothing and many of the objects around you are synthetic, all products of chemistry. In spite of their familiarity, this has been a recent development. Synthetic resins such as *Bakelite*, fibers such as *rayon*, and plastics such as *celluloid* were made early in this century. However, a plethora of synthetic polymers has been available since World War II. By 1976, plastic outstripped steel as the nation's most widely used material, and we now use more plastic than steel, aluminum, and copper combined. In 1989, more than 36 pounds of manmade fibers and 180 pounds of plastics were produced for every person in the United States.

The polymers described in this section are organic polymers. There are, however, inorganic polymers such as the silicones described in Section 23.2.

THE TERMINOLOGY OF THE POLYMER INDUSTRY

Polymers are classified as **elastomers**, **plastics**, and **fibers** depending on their elasticity. As the name suggests, elastomers can be highly stretched, to over ten times their normal length, and returned to their normal dimensions many times. Fibers, in contrast, are the least elastic, while plastics fall in between.

POLYMER TYPE	UPPER LIMIT OF EXTENSIBILITY (%)
Elastomer	100–1000
Plastic	20–100
Fiber	Less than 10

Polymers are also classified on the basis of their response to heating. **Thermoplastic** polymers soften on heating and are unaltered chemically. On the other hand, **thermosetting** polymers degrade or decompose on heating. In general, linear polymers such as polyethylene and the others listed in Table 27.3 are thermoplastic, whereas polymers with one backbone linked to another (as in proteins) are thermosetting (Figure 27.14).

The reasons for the differences in polymer properties depend on molecular weight, extent of cross-linking between chains, and crystallinity, among other things. For example, polymers begin to develop mechanical strength after about 50 monomer units have been bound together, and it increases up to about 500 units; after this, increases in chain length seem to have little effect on strength. If the polymer chains are cross-linked by covalent bonds or dispersion forces, the polymer is more likely

Figure 27.14 Polyvinyl alcohol, derived from vinyl alcohol ($CH_2{=}CHOH$), is a close relative of PVC and PVA in Table 27.3. It is used as a coating on grease-proof paper and to thicken foods. When an aqueous solution of polyvinyl alcohol is mixed with boric acid (or its sodium salt), an ester forms and leads to "cross-linking" of the polymer. The viscous mixture is popularly known as "slime."

$$2\ \text{OH} + B(OH)_3 \longrightarrow \text{O—B(—OH)—O} + 2\ H_2O$$

Table 27.3 Some Common Polymers Based on Ethylene

Monomer	Polymer	Name
$CH_2{=}CH_2$ Ethylene	$-\mathrm{C}(\mathrm{H})(\mathrm{H})-\mathrm{CH_2}-\mathrm{C}(\mathrm{H})(\mathrm{H})-\mathrm{CH_2}-\mathrm{C}(\mathrm{H})(\mathrm{H})-\mathrm{CH_2}-$	Polyethylene
$\mathrm{C}(\mathrm{Cl})(\mathrm{H}){=}CH_2$ Vinyl chloride	$-\mathrm{C}(\mathrm{Cl})(\mathrm{H})-\mathrm{CH_2}-\mathrm{C}(\mathrm{Cl})(\mathrm{H})-\mathrm{CH_2}-\mathrm{C}(\mathrm{Cl})(\mathrm{H})-\mathrm{CH_2}-$	Polyvinyl chloride (PVC)
$\mathrm{C}(\mathrm{Cl})(\mathrm{Cl}){=}CH_2$ Vinylidene chloride	$-\mathrm{C}(\mathrm{Cl})(\mathrm{Cl})-\mathrm{CH_2}-\mathrm{C}(\mathrm{Cl})(\mathrm{Cl})-\mathrm{CH_2}-\mathrm{C}(\mathrm{Cl})(\mathrm{Cl})-\mathrm{CH_2}-$	Polyvinylidene chloride
$\mathrm{CH}(C_6H_5){=}CH_2$ Styrene	$-\mathrm{C}(\mathrm{H})(C_6H_5)-\mathrm{CH_2}-\mathrm{C}(\mathrm{H})(C_6H_5)-\mathrm{CH_2}-\mathrm{C}(\mathrm{H})(C_6H_5)-\mathrm{CH_2}-$	Polystyrene
$\mathrm{CH}(\mathrm{C{\equiv}N}){=}CH_2$ Acrylonitrile	$-\mathrm{C}(\mathrm{H})(\mathrm{C{\equiv}N})-\mathrm{CH_2}-\mathrm{C}(\mathrm{H})(\mathrm{C{\equiv}N})-\mathrm{CH_2}-\mathrm{C}(\mathrm{H})(\mathrm{C{\equiv}N})-\mathrm{CH_2}-$	Polyacrylonitrile (Orlon)
$\mathrm{CH}(CH_3){=}CH_2$ Propylene	$-\mathrm{C}(\mathrm{H})(CH_3)-\mathrm{CH_2}-\mathrm{C}(\mathrm{H})(CH_3)-\mathrm{CH_2}-\mathrm{C}(\mathrm{H})(CH_3)-\mathrm{CH_2}-$	Polypropylene
$\mathrm{CH}(\mathrm{OCCH_3})(\text{C=O}){=}CH_2$ Vinyl acetate	$-\mathrm{C}(\mathrm{H})(\mathrm{OC(=O)CH_3})-\mathrm{CH_2}-\mathrm{C}(\mathrm{H})(\mathrm{OC(=O)CH_3})-\mathrm{CH_2}-\mathrm{C}(\mathrm{H})(\mathrm{OC(=O)CH_3})-\mathrm{CH_2}-$	Polyvinylacetate (PVA)
$CF_2{=}CF_2$ Tetrafluoroethylene	$-\mathrm{C}(\mathrm{F})(\mathrm{F})-\mathrm{CF_2}-\mathrm{C}(\mathrm{F})(\mathrm{F})-\mathrm{CF_2}-\mathrm{C}(\mathrm{F})(\mathrm{F})-\mathrm{CF_2}-$	Polytetrafluoroethylene (Teflon)

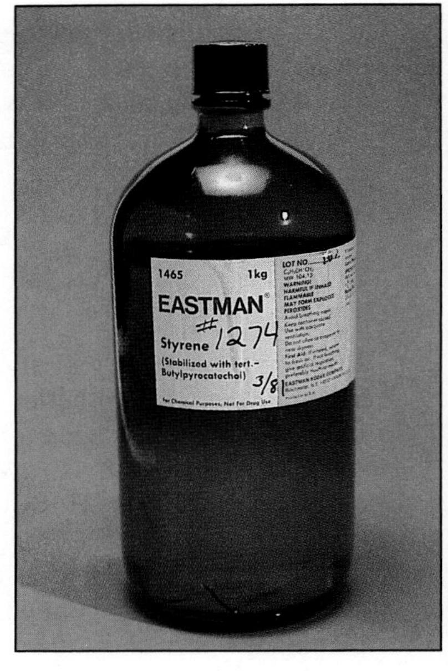

After storage for some time, the styrene in a bottle (a) was found to have polymerized to a clear block of polystyrene (b).

Table 27.4 Major Thermoplastic Polymers Produced in the United States in 1989

Polymer	Amount (billions of pounds)	Uses
Low-density polyethylene	9.7	Film for packaging; molded products
Polyvinyl chloride	8.5	Molded pipes (about 40% of the usage of PVC); phonograph records; bottles
High-density polyethylene	8.1	Molded containers (e.g., 1-gal milk)
Polypropylene	7.4	Molded items (e.g., appliance parts, battery cases); fibers
Polystyrene	5.1	Packaging; more than 1 pound is used in each video cassette

to be a fiber. On the other hand, if the forces between chains are weak or there are bulky substituents along a chain, it is more likely to be an elastomer.

THERMOPLASTIC POLYMERS

There are five major thermoplastic polymers: low- and high-density polyethylene, polypropylene, polystyrene, and polyvinyl chloride (Tables 27.3 and 27.4). Almost 39 billion pounds of these polymers were produced in the United States in 1989.

"People don't realize what polymer chemistry does for the world. If you were to stand in the middle of your living room and have somebody come and take out everything in the room that had any polymer material in it, the room would be denuded. The paint would be gone, the carpet would be gone, the upholstery would be gone. If you took away the manmade polymer materials you'd be utterly amazed how little you'd have left." (Comment by Mary L. Good, page 450)

The most common polymer is **polyethylene**, of which about 18 billion pounds are produced annually in the United States. *Low-density polyethylene* (LDPE) is a tough, waxy, mainly branched polymer; it has a molecular weight in the range from 10,000 to 40,000 and a density of 0.92 g/cm^3. *High-density polyethylene* (HDPE) is an unbranched, more crystalline polymer with a density of 0.97 g/cm^3. HDPE has greater strength and rigidity and a higher softening temperature, so it is better able to withstand boiling water than LDPE. However, since LDPE is more transparent than HDPE, the low-density form is used in the form of film for packaging.

The thermoplastics described so far are formed by polymerizing one type of alkene. However, many polymers are a combination of two different monomers and are referred to as **copolymers**. *Saran*, a polymeric film used to wrap food, is an excellent example of a thermoplastic copolymer. It is prepared by mixing 1,1-dichloroethene and chloroethene (vinyl chloride) in a 4:1 ratio.

$$\underset{\text{1,1-dichloroethene}}{Cl_2C{=}CH_2} + \underset{\text{vinyl chloride}}{H_2C{=}CHCl} \longrightarrow \text{Saran}$$

Polyurethanes are also thermoplastics. When an organic isocyanate is treated with an alcohol, the alcohol adds to the CN double bond to give a urethane.

$$\underset{\text{isocyanate}}{R{-}N{=}C{=}O} + \underset{\text{alcohol}}{R'{-}OH} \longrightarrow \underset{\text{a urethane}}{R{-}\overset{\overset{\displaystyle H}{|}}{N}{-}\overset{\overset{\displaystyle O}{\|}}{C}{-}OR'}$$

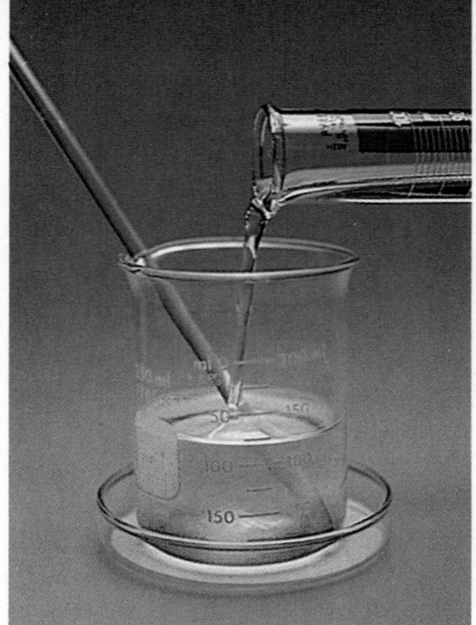

Figure 27.15 Phenol-formaldehyde polymer. When a mixture of phenol and formaldehyde in acetic acid is treated with concentrated HCl, polymer begins to grow.

With some imagination you can see that a diisocyanate (OCN—R—NCO) and a dialcohol (HO—R′—OH) will lead to a polymer.

$$-\overset{\overset{\displaystyle O}{\|}}{C}-\overset{\overset{\displaystyle H}{|}}{N}-R-\overset{\overset{\displaystyle H}{|}}{N}-\overset{\overset{\displaystyle O}{\|}}{C}-O-R'-O-\overset{\overset{\displaystyle O}{\|}}{C}-\overset{\overset{\displaystyle H}{|}}{N}-R-\overset{\overset{\displaystyle H}{|}}{N}-\overset{\overset{\displaystyle O}{\|}}{C}-O-R'-O-$$

polyurethane

Polyurethanes are used as coatings when high impact and abrasion resistance are important, so you often find them on dance floors, gymnasium floors, and bowling alleys. If air is bubbled into polyurethane as it forms, a foam is produced that is used in bedding and car upholstery. By modifying the R groups of the urethane system, the polyurethanes can be cross-linked and made into fibers that are used to make swim suits, ski wear, and other garments that must stretch.

THERMOSETTING POLYMERS

All the polymerization reactions in Table 27.3 require a catalyst. In some cases a Ziegler-Natta catalyst, a mixture of compounds such as $Al(C_2H_5)_3$ and $TiCl_3$, is used.

The polymers in the previous section and in Table 27.3 are thermoplastic. They are made from monomers that can form a bond to each of two neighboring molecules. Thermosetting polymers, on the other hand, are made from monomers that form more than two bonds to neighboring units, and the polymers are thus highly branched or cross-linked. For this reason, polymers of this type often do not melt when heated.

The formation of thermosetting polymers is more complex than that of thermoplastics. One example of the former is the phenol-formaldehyde resin called *Bakelite* (Figure 27.15). When phenol and a concentrated aqueous solution of formaldehyde are mixed in the presence of HCl, compounds such as the following are formed.

OH (phenol) + HCHO (formaldehyde) $\xrightarrow{HCl}$ OH, CH_2OH (ortho) and OH, CH_2OH (para)

phenol formaldehyde

These products react with one another by **condensation** in much the same way as two amino acids or two sugars.

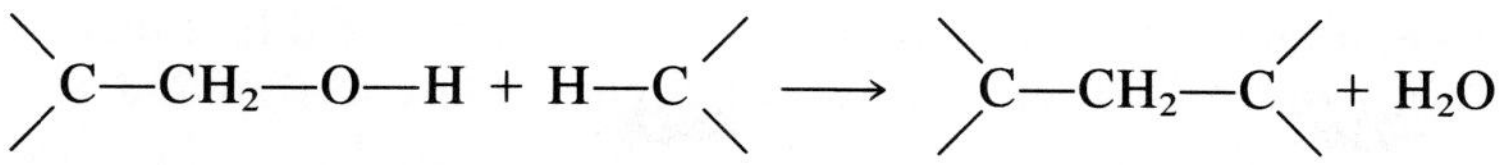

$$\rangle C{-}CH_2{-}O{-}H + H{-}C\langle \longrightarrow \rangle C{-}CH_2{-}C\langle + H_2O$$

Eventually, a highly cross-linked polymer is formed.

ELASTOMERS: RUBBER

As we usually think of it, rubber is not a synthetic polymer; it is a natural substance. Nonetheless, it is useful to describe it briefly here because it is the best example of an elastomer and because attempts to duplicate its properties have heavily influenced the polymer industry.

The 18th century English chemist Joseph Priestley invented the name "rubber" because of the ability of the substance to erase pencil marks from paper.

Natural rubber is a polymer of 2-methyl-1,3-butadiene (isoprene) and is obtained from the tree *Hevea brasiliensis*.

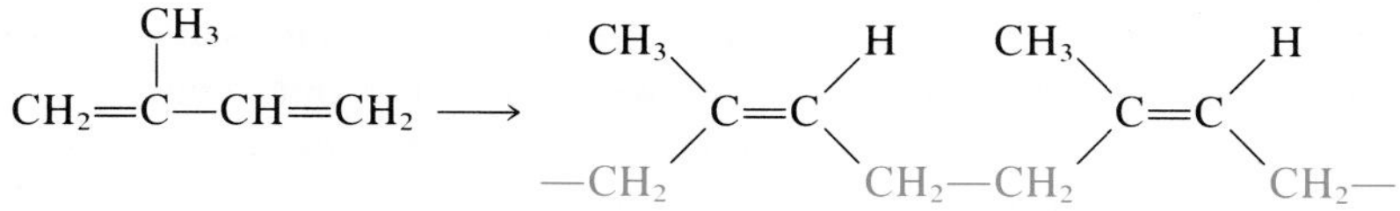

$$CH_2{=}C(CH_3){-}CH{=}CH_2 \longrightarrow {-}CH_2(CH_3)C{=}CH{-}CH_2{-}CH_2(CH_3)C{=}CH{-}CH_2{-}$$

isoprene — Hevea rubber

The substituent groups (CH_3 and H) are *cis*, that is, they lie on the same side of the C=C double bond. The *trans* form is also found in nature, but it has come to be called *gutta-percha*.

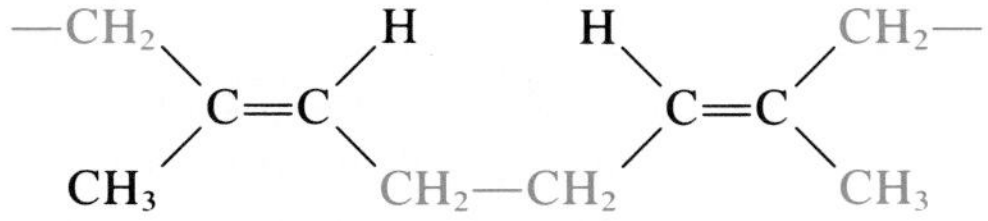

$${-}CH_2(CH_3)C{=}CH{-}CH_2{-}CH_2{-}CH{=}C(CH_3)CH_2{-}$$

gutta-percha rubber

Unlike Hevea rubber, gutta-percha is hard, brittle, and nonelastic. An apparently minor change in structure leads to a major change in properties.

Rubber was first introduced into Europe in 1740, but it remained a curiosity until 1823 when Charles Mackintosh invented a method of waterproofing cotton cloth with a solution of rubber. The mackintosh, as rain coats are even now sometimes called, became extremely popular, despite some major drawbacks: natural rubber is notably weak, and it is thermoplastic, soft and tacky when warm but brittle at lower temperature. The American inventor Charles Goodyear found a way around these difficulties with the *vulcanization* process. If rubber, natural or synthetic, is heated with sulfur, the polymer chains are linked with C—S—S—C bonds.

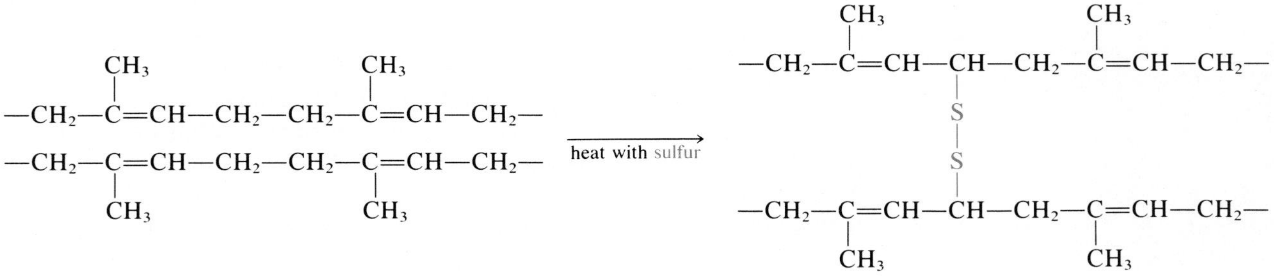

$$\begin{matrix} {-}CH_2{-}C(CH_3){=}CH{-}CH_2{-}CH_2{-}C(CH_3){=}CH{-}CH_2{-} \\ {-}CH_2{-}C(CH_3){=}CH{-}CH_2{-}CH_2{-}C(CH_3){=}CH{-}CH_2{-} \end{matrix} \xrightarrow{\text{heat with sulfur}} \begin{matrix} {-}CH_2{-}C(CH_3){=}CH{-}CH{-}CH_2{-}C(CH_3){=}CH{-}CH_2{-} \\ | \\ S \\ | \\ S \\ | \\ {-}CH_2{-}C(CH_3){=}CH{-}CH{-}CH_2{-}C(CH_3){=}CH{-}CH_2{-} \end{matrix}$$

Cross-linking makes the rubber harder, stronger, and less thermoplastic. With Goodyear's invention, rubber became more useful, and the demand increased. Unfortunately, the only sources of rubber in the 19th century were wild trees in the jungles of Brazil and the Congo. This problem was solved by an enterprising Englishman who smuggled seeds for rubber trees out of Brazil and planted them in large plantations in Sri Lanka and Malaysia. By 1939, almost 1,400,000 tons of rubber were consumed worldwide per year. World War II intervened, however, and cut off supplies of natural rubber to Europe and the United States. This brought on the hunt for synthetic rubber.

The obvious way to synthesize rubber is to duplicate nature. This is easier said than done, since only recently have catalysts been discovered that will do what a rubber tree does: polymerize isoprene so that all of the substituent groups are *cis*.

As alternatives to duplicating natural rubber, many different elastomers have been developed, all of which involve butadiene or some derivative of this alkene. For example, 2-chloro-1,3-butadiene (commonly called chloroprene) polymerizes to form *neoprene* rubber.

chloroprene $\quad$ neoprene

$$n\mathrm{CH_2{=}CH{-}\underset{\underset{\displaystyle Cl}{|}}{C}{=}CH_2} \longrightarrow \mathrm{{-}CH_2{-}CH{=}\underset{\underset{\displaystyle Cl}{|}}{C}{-}CH_2{-}CH_2{-}CH{=}\underset{\underset{\displaystyle Cl}{|}}{C}{-}CH_2{-}}$$

One high-tech resin, a so-called engineering plastic, is ABS resin. The letters stand for "acrylonitrile-butadiene-styrene." ABS resins are used in computer cases, electric tools, watch cases, and automobile parts.

The high chlorine content of neoprene gives the polymer resistance to heat and flames, as well as to oil and chemicals. It is thus used for hoses carrying gasoline, for protective gloves, and for balloons. About 6% of the synthetic rubber produced in the United States is polychloroprene.

Most synthetic elastomers, however, are copolymers of two different olefins. For example, the most important synthetic rubber in the United States market (about 45%) is SBR, or a styrene-butadiene rubber.

butadiene $\quad$ styrene $\quad$ styrene–butadiene rubber (SBR)

$$\mathrm{CH_2{=}CH{-}CH{=}CH_2} + \underset{\underset{\displaystyle C_6H_5}{|}}{\mathrm{CH}}\mathrm{{=}CH_2} \longrightarrow \mathrm{{-}CH_2{-}CH{=}CH{-}CH_2{-}\underset{\underset{\displaystyle C_6H_5}{|}}{CH}{-}CH_2{-}CH_2{-}CH{=}CH{-}CH_2{-}\underset{\underset{\displaystyle C_6H_5}{|}}{CH}{-}CH_2{-}}$$

FIBERS

More than 9 billion pounds of fibrous polymers, some purely synthetic and others based on cellulose, are produced per year in the United States. Some of these are semisynthetic (partly synthetic) and others are purely synthetic fibers. The chief semisynthetic fibers are rayon and cellulose acetate, and there are three main groups of synthetic fibers: polyamides (nylon), polyesters (Dacron), and acrylics (Orlon).

SEMISYNTHETIC FIBERS Semisynthetic fibers are made by modifying cellulose in some way. In Section 27.3 you learned that cellulose is a linear polymer of glucose molecules. If this polymer is heated in sodium hydroxide with carbon disulfide, the result is a so-called "viscose" solution. If this solution is then forced through very small holes, continuous filaments are formed. When the filaments drop into a sulfuric acid bath, cellulose is regenerated from the viscose, and the fibers are then woven into cloth that we know as rayon.

Cellulose acetate, used in "wash and wear" fabrics, is made by converting some of the —OH substituents of cellulose into ester groups by reaction with acetic acid.

cellulose acetate

SYNTHETIC FIBERS **Nylon** was the first fiber to result from a deliberate search for a purely synthetic material, and it was greatly in demand from the beginning. As women's dresses became shorter about 70 years ago, silk stockings were very fashionable but also very expensive. Late in the 1930s Wallace H. Carothers of Du Pont discovered nylon-66, and it was soon found that nylon fibers could be woven into sheer hosiery. The first such stockings went on sale in October 1939, and they were so popular they had to be rationed. Unfortunately, World War II caused the commercial use of nylon to cease until 1945. Today it is produced in many forms at an annual rate of more than 10 pounds per person in the United States.

Nylons are *polyamides*, condensation polymers formed from a diacid and a diamine (Figure 27.16). For example, adipic acid and hexamethylenediamine produce nylon-66; the 66 suffix indicates that both the diacid and the diamine have 6 carbon atoms.

$$\underset{\text{adipic acid}}{HOOC(CH_2)_4COOH} + \underset{\text{hexamethylenediamine}}{H_2N(CH_2)_6NH_2} \xrightarrow{-H_2O}$$

$$-\overset{O}{\overset{\|}{C}}-(CH_2)_4-\overset{O}{\overset{\|}{C}}-\overset{H}{\overset{|}{N}}-(CH_2)_6-\overset{H}{\overset{|}{N}}-\overset{O}{\overset{\|}{C}}-(CH_2)_4-\overset{O}{\overset{\|}{C}}-\overset{H}{\overset{|}{N}}-(CH_2)_6-\overset{H}{\overset{|}{N}}-$$

amide link — nylon-66

About half of the nylon produced is used to make cords for tires. The remainder goes into ropes, cords, rugs, fish nets and lines, clothes, thread, nylon hose, coats, dresses and on and on.

Polyesters are also condensation polymers used as fibers and are produced in even greater amounts than nylon. The most common one is sold in the United States as *Dacron*, a polymer formed from a dialcohol and an aromatic diacid.

Figure 27.16 Nylon-66. Hexamethylenediamine is dissolved in water (bottom layer), and a derivative of adipic acid (adipyl chloride) is dissolved in hexane (top layer). The chemicals mix at the interface between the layers to form nylon (which is being wound onto a stirring rod).

$$HO—CH_2CH_2—OH + HO—\overset{O}{\overset{\|}{C}}—C_6H_4—\overset{O}{\overset{\|}{C}}—OH \xrightarrow{-H_2O} —O—CH_2CH_2—O—\overset{O}{\overset{\|}{C}}—C_6H_4—\overset{O}{\overset{\|}{C}}—$$

ethylene glycol — terephthalic acid — polyethyleneterephthalate Dacron

Polyester fibers are generally used today in a blend with cotton fibers for wash and wear garments.

Finally, **acrylics** are polymers based on cyanoethene or acrylonitrile (Table 27.3). Polyacrylonitrile fibers are in great demand because they are only about half as expensive as wool, can be woven into textiles such as *Orlon* that have a "wooly" feel, and are machine washable and do not shrink. Although only about one fourth as much polyacrylonitrile is produced as nylon, demand for the acrylic polymer seems certain to increase.

SUMMARY

Polymers are constructed of many small, individual units or **monomers**. The monomer of polyvinyl chloride is vinyl chloride (chloroethene),

$$n\ CH_2{=}CHCl \longrightarrow —CH_2—\underset{Cl}{\underset{|}{CH}}—CH_2—\underset{Cl}{\underset{|}{CH}}—CH_2—\underset{Cl}{\underset{|}{CH}}—$$

vinyl chloride — polyvinyl chloride

and the basic building blocks of proteins or polypeptides are amino acids.

$$n\ H_2N—CH(R)—COOH \xrightarrow{-n\ H_2O} —\overset{H}{\overset{|}{N}}—CH(R)—\overset{O}{\overset{\|}{C}}—\overset{H}{\overset{|}{N}}—CH(R)—\overset{O}{\overset{\|}{C}}—$$

amino acid — polypeptide

Polyvinyl chloride is an **addition polymer**, because the empirical formula of the polymer is the same as that of the monomer. In contrast, a polypeptide is a **condensation polymer**, since the monomers come together or "condense" by eliminating a small molecule.

Polymers in this chapter were divided into two types: **natural** (proteins, carbohydrates, nucleic acids) and **synthetic** (polyvinyl chloride, polyethylene, Teflon, polyesters).

Simple proteins consist only of amino acids, while **conjugated proteins** contain some other group in addition to the amino acids. When two amino acids condense, an amide linkage or **peptide bond** is formed.

$$H_2N—\underset{H}{\underset{|}{\overset{R}{\overset{|}{C}}}}—\overset{O}{\overset{\|}{C}}—OH + H_2N—\underset{H}{\underset{|}{\overset{R'}{\overset{|}{C}}}}—\overset{O}{\overset{\|}{C}}—OH \longrightarrow H_2N—\underset{H}{\underset{|}{\overset{R}{\overset{|}{C}}}}—\overset{O}{\overset{\|}{C}}—\overset{H}{\overset{|}{N}}—\underset{H}{\underset{|}{\overset{R'}{\overset{|}{C}}}}—\overset{O}{\overset{\|}{C}}—OH$$

amide or peptide linkage

The geometry and polarity of the peptide bond have a profound effect on protein structure. The **primary structure** of a protein is determined by the sequence of amino acids. Due to hydrogen bonding between peptide units, the protein chain can have a **secondary structure**: an α-helix or β-sheet structure. **Fibrous** and **globular** proteins largely have the α-helix. Segments of a protein coil can also interact by hydrogen bonding and other mechanisms.

Carbohydrates have the general formula $C_x(H_2O)_y$ and are divided into **monosaccharides**, **disaccharides**, and **polysaccharides** (Section 27.3). When monosaccharides condense to form disaccharides or polysaccharides, a **C—O—C glycoside linkage** is formed. **Cellulose**, **starch**, and **glycogen** are polymers of glucose. The first two differ in the geometry of the glycoside linkage, so cellulose is a linear polymer, whereas starch is a helical coil.

If one of five organic bases (Section 27.4) is bound to the monosaccharide ribose, a **nucleoside** results. If the nucleoside condenses with a phosphate, a **nucleotide** is the product. Linking nucleotides together gives the polymer **RNA (ribonucleic acid)**. **DNA**, or **deoxyribonucleic acid**, results when deoxyribose is the sugar of the polymer backbone. Because bases of DNA can hydrogen bond with one another, the polynucleic acid chains interact to form the double-stranded α-helix of DNA.

Synthetic polymers (Section 27.5) are classified as **elastomers**, **plastics**, or **fibers** depending on their elasticity. On the basis of their response to heat, they are also classed as **thermoplastic** (unaltered chemically on heating) or **thermosetting** (degraded on heating). Most thermoplastic polymers are addition polymers, while thermosetting polymers are often condensation polymers. Both purely synthetic and semisynthetic fibers are made. Nylon, a condensation polymer of a diacid and diamine, is synthetic, but rayon is a modified form of cellulose.

STUDY QUESTIONS

STEREOCHEMISTRY AND AMINO ACIDS

1. Define the term chirality using the molecule glyceraldehyde, the simplest carbohydrate. The structural formula of the compound is

```
        H
        |
HO—C—CHO
        |
      CH2OH
```

2. Prove that CH_2ClBr is not a chiral molecule. Explain your reasoning briefly.
3. Draw the D form of the amino acid serine.
4. Which one of the ten α-amino acids in Table 27.1 is not chiral? Explain briefly why it is not chiral.

PROTEINS

5. Draw the structures of the four dipeptides that can form if alanine and serine are mixed.
6. Draw the structure of one of the tripeptides that can form from a mixture of glycine, histidine, and cysteine.
7. Define the terms "primary" and "secondary" as applied to protein structure.
8. How does the secondary structure of an α-helix differ from the secondary structure of a β-sheet protein?
9. Give two examples of each type of protein: (a) fibrous, (b) globular.
10. Interactions between side chains (or R groups) of amino acids can influence the structures of proteins (and control what is known as the protein's tertiary structure). If you have a protein with a serine unit ($R = CH_2OH$) close to an aspartic acid unit ($R = CH_2COOH$), what type of interaction can occur between these R groups?

CARBOHYDRATES

11. Show that glyceraldehyde, $CH(OH)(CHO)CH_2OH$, is a carbohydrate. (See its structure in Study Question 1.)
12. Consider the structure of ribose below. Which carbon atoms are chiral centers? Which are not?

HO—5 CH_2 — ring: 4 C (H), O, 1 C (H, OH), 3 C (H, OH), 2 C (H, OH)

α-D-ribose

13. Name two examples of each of the following: (a) monosaccharide, (b) disaccharide, and (c) polysaccharide.
14. Why can humans not use cellulose as an energy source?

NUCLEIC ACIDS

15. Describe the features of polynucleotides that lead to the double helix structure of DNA.
16. When β-D-ribose reacts with adenine, what two products form?
17. Offer an explanation for the observation that adenine (A) does not pair with guanine (G).
18. The single-strand polynucleotide RNA is made from DNA. The bases on RNA will be bases complementary to those found on one strand of DNA. Assume RNA is "copied" from segment B of the DNA illustrated below (where S is the sugar ribose and P is the phosphate link).

```
DNA segment A        —S—P—S—P—S—P
                      |    |    |
                      T    A    G
                      :    :    :
                      A    T    C
                      |    |    |
DNA segment B        —S—P—S—P—S—P
                           |
                           ↓
                      1    2    3
                      |    |    |
Synthesized RNA      —S—P—S—P—S—P
```

Identify bases 1, 2, and 3 in RNA. [Note on p. 1134 that the base uracil (U) takes the place of thymine (T) in RNA.]

SYNTHETIC POLYMERS

19. Polyvinyl acetate is the binder in water-based paints. Draw the structure of this polymer.
20. Name two synthetic polymers that you may have encountered today and the objects made from these polymers.
21. The squeeze bottles you use in your chemistry laboratory are polyethylene (page 537). If the empty bottle has a mass of 150 g, how many moles of ethylene (ethene) must have been used to make the bottle?
22. A "styrofoam" drinking cup, made of polystyrene, has a mass of 8.0 g. How many moles of styrene were used to make the cup?
23. Polymers are classified in several ways, one depending on their elasticity and the other on their response to heat. Briefly describe these classifications. Give an example of each class.
24. Define the term "copolymer." Give an example of a copolymer.
25. The two reactions by which polymers are formed are addition and condensation. Give an example of each kind.
26. Condensation reactions are important in forming disaccharides and polysaccharides, nucleotides, and polymers such as nylon-66. Illustrate the condensation reaction for each of these cases.
27. Methyl methacrylate has the structural formula below. When polymerized it is very transparent, and it is sold in the United States as "Lucite" or "Plexiglas." Draw a polymethyl methacrylate chain that is four monomer units in length.

```
          CH3
         /
  H2C=C                = methyl methacrylate
         \
          C—O—CH3
          ‖
          O
```

28. Saran is a copolymer of 1,1-dichloroethene and 5% to 17% chloroethene (vinyl chloride). Draw a possible structure for this polymer.
29. Tetrafluoroethylene is made by first treating HF with chloroform followed by cracking the resultant difluorochloromethane.

$$CHCl_3 + 2\ HF \longrightarrow CHClF_2 + 2\ HCl$$

$$2\ CHClF_2 + \text{heat} \longrightarrow F_2C{=}CF_2 + 2\ HCl$$

$$F_2C{=}CF_2 + \text{peroxide catalyst} \longrightarrow \text{Teflon}$$

If you wished to make 1.0 kilogram of Teflon, what mass of chloroform and HF must you use to make the starting material, $CHClF_2$? (Although it is not realistic, assume each reaction step proceeds in 100% yield.)
30. In 1989, 2.74 million pounds of nylon were made in the United States. How many grams each of adipic acid and hexamethylenediamine must have been combined to make this amount of nylon-66? (Recall that 1.00 pound = 454 g.)

APPENDICES

A. Some Mathematical Operations
B. Common Units, Equivalences, and Conversion Factors
C. Physical Constants
D. Some Physical Constants for Water and a Few Common Substances
E. Ionization Constants for Weak Acids at 25 °C
F. Ionization Constants for Weak Bases at 25 °C
G. Solubility Product Constants for Some Inorganic Compounds at 25 °C
H. Formation Constants for Some Complex Ions in Aqueous Solution
I. Standard Reduction Potentials in Aqueous Solution at 25 °C
J. Selected Thermodynamic Values
K. Answers to Exercises in the Text
L. Answers to Selected Study Questions
M. Location of Useful Tables and Figures

APPENDIX A

Some Mathematical Operations

The mathematical skills required in this introductory course are basic skills in algebra and a knowledge of (a) exponential or scientific notation, (b) logarithms, and (c) quadratic equations. This appendix reviews each of the last three topics.

A.1 ELECTRONIC CALCULATORS

The directions for calculator use in this section are given for calculators using "algebraic" logic. Such calculators are the most common type used by students in introductory courses. For calculators using RPN logic (such as those made by Hewlett-Packard), the procedures will differ slightly.

The advent of inexpensive electronic calculators a few years ago has made calculations in introductory chemistry much more straightforward. You are well advised to purchase a calculator that has the capability of performing calculations in scientific notation, has both base-10 and natural logarithms, and is capable of raising any number to any power and of finding any root of any number. In the discussion below, we shall point out in general how these functions of your calculator can be used.

Although electronic calculators have greatly simplified calculations, they have also forced us to focus again on significant figures. A calculator easily handles 8 or more significant figures, but real laboratory data is never known to this accuracy. Therefore, you are urged to review Section 1.5 on handling numbers.

A.2 EXPONENTIAL OR SCIENTIFIC NOTATION

In exponential or scientific notation, a number is expressed as a product of two numbers: $N \times 10^n$. The first number, N, is the so-called *digit term* and is a number between 1 and 10. The second number, 10^n, the *exponential term*, is some integer power of 10. For example, 1234 would be written in scientific notation as 1.234×10^3 or 1.234 multiplied by 10 three times.

$$1234 = 1.234 \times 10^1 \times 10^1 \times 10^1 = 1.234 \times 10^3$$

Conversely, a number less than 1, such as 0.01234, would be written as 1.234×10^{-2}. This notation tells us that 1.234 should be divided twice by 10 in order to obtain 0.01234.

$$0.01234 = \frac{1.234}{10^1 \times 10^1} = 1.234 \times 10^{-1} \times 10^{-1} = 1.234 \times 10^{-2}$$

Some other examples of scientific notation are

$$\begin{array}{rlrl} 10000 &= 1 \times 10^4 & 12345 &= 1.2345 \times 10^4 \\ 1000 &= 1 \times 10^3 & 1234 &= 1.234 \times 10^3 \\ 100 &= 1 \times 10^2 & 123 &= 1.23 \times 10^2 \\ 10 &= 1 \times 10^1 & 12 &= 1.2 \times 10^1 \\ 1 &= 1 \times 10^0 & & \text{(any number to the zero power = 1)} \\ 1/10 &= 1 \times 10^{-1} & 0.12 &= 1.2 \times 10^{-1} \\ 1/100 &= 1 \times 10^{-2} & 0.012 &= 1.2 \times 10^{-2} \\ 1/1000 &= 1 \times 10^{-3} & 0.0012 &= 1.2 \times 10^{-3} \\ 1/10000 &= 1 \times 10^{-4} & 0.00012 &= 1.2 \times 10^{-4} \end{array}$$

When converting a number to scientific notation, notice that the exponent n is positive if the number is greater than 1 and negative if the number is less than 1. The value of n is the number of places by which the decimal was shifted to obtain the number in scientific notation.

$$12345. = 1.2345 \times 10^4$$

Decimal shifted 4 places to the left. Therefore, n is positive and equal to 4.

$$0.0012 = 1.2 \times 10^{-3}$$

Decimal shifted 3 places to the right. Therefore, n is negative and equal to 3.

If you wish to convert a number in scientific notation to the usual form, the procedure above is simply reversed.

$$6.273 \times 10^2 = 627.3$$

Decimal point moved 2 places to the right, since n is positive and equal to 2.

$$006.273 \times 10^{-3} = 0.006273$$

Decimal point shifted 3 places to the left, since n is negative and equal to 3.

There are two final points to be made concerning scientific notation. First, if you are used to working on a computer, you may be in the habit of writing a number such as 1.23×10^3 as 1.23E3 or 6.45×10^{-5} as 6.45E-5. Second, some electronic calculators allow you to convert numbers readily to the scientific notation. If you have such a calculator, you can change a number shown in the usual form to scientific notation simply by pressing the EE or EXP key and then the "=" key.

1. ADDING AND SUBTRACTING NUMBERS

When adding or subtracting two numbers, they must first be converted to the same powers of ten. The digit terms are then added or subtracted as appropriate.

$$\begin{aligned}(1.234 \times 10^{-3}) + (5.623 \times 10^{-2}) &= (0.1234 \times 10^{-2}) + (5.623 \times 10^{-2}) \\ &= 5.746 \times 10^{-2}\end{aligned}$$

$$\begin{aligned}(6.52 \times 10^2) - (1.56 \times 10^3) &= (6.52 \times 10^2) - (15.6 \times 10^2) \\ &= -9.1 \times 10^2\end{aligned}$$

2. MULTIPLICATION

The digit terms are multiplied in the usual manner, and the exponents are added algebraically. The result is expressed with a digit term with only one nonzero digit to the left of the decimal.

$$\begin{aligned}(1.23 \times 10^3)(7.60 \times 10^2) &= (1.23)(7.60) \times 10^{3+2} \\ &= 9.35 \times 10^5\end{aligned}$$

$$\begin{aligned}(6.02 \times 10^{23})(2.32 \times 10^{-2}) &= (6.02)(2.32) \times 10^{23-2} \\ &= 13.966 \times 10^{21} \\ &= 1.40 \times 10^{22} \text{ (answer in 3 significant figures)}\end{aligned}$$

3. DIVISION

The digit terms are divided in the usual manner, and the exponents are subtracted algebraically. The quotient is written with one nonzero digit to the left of the decimal in the digit term.

$$\frac{7.60 \times 10^3}{1.23 \times 10^2} = \frac{7.60}{1.23} \times 10^{3-2} = 6.18 \times 10^1$$

$$\frac{6.02 \times 10^{23}}{9.10 \times 10^{-2}} = \frac{6.02}{9.10} \times 10^{(23)-(-2)} = 0.662 \times 10^{25} = 6.62 \times 10^{24}$$

4. POWERS OF EXPONENTIALS

When raising a number in exponential notation to a power, treat the digit term in the usual manner. The exponent is then multiplied by the number indicating the power.

$$\begin{aligned}(1.25 \times 10^3)^2 &= (1.25)^2 \times 10^{3\times 2}\\ &= 1.5625 \times 10^6 = 1.56 \times 10^6\end{aligned}$$

$$\begin{aligned}(5.6 \times 10^{-10})^3 &= (5.6)^3 \times 10^{(-10)\times 3}\\ &= 175.6 \times 10^{-30} = 1.8 \times 10^{-28}\end{aligned}$$

Electronic calculators usually have two methods of raising a number to a power. To square a number, enter the number and then press the "x^2" key. To raise a number to any power, use the "y^x" key. For example, to raise 1.42×10^2 to the 4th power,

(a) enter 1.42×10^2
(b) press "y^x"
(c) enter 4 (this should appear on the display)
(d) press "=" and 4.0659×10^8 will appear on the display.

As a final step, express the number in the correct number of significant figures (4.07×10^8) in this case.

5. ROOTS OF EXPONENTIALS

Unless you use an electronic calculator, the number must first be put into a form where the exponential is exactly divisible by the root. The root of the digit term is found in the usual way, and the exponent is divided by the desired root.

$$\sqrt{3.6 \times 10^7} = \sqrt{36 \times 10^6} = \sqrt{36} \times \sqrt{10^6} = 6.0 \times 10^3$$

$$\sqrt[3]{2.1 \times 10^{-7}} = \sqrt[3]{210 \times 10^{-9}} = \sqrt[3]{210} \times \sqrt[3]{10^{-9}} = 5.9 \times 10^{-3}$$

To take a square root on an electronic calculator, enter the number and then press the "$\sqrt{x}$" key. To find a higher root of a number, such as the 4th root of 5.6×10^{-10},

(a) enter the number.
(b) press the "$\sqrt[x]{y}$" key. (On most calculators, the sequence you actually use is to press "2ndF" and then "$\sqrt[x]{y}$." Alternatively, you press "INV" and then "y^x.")
(c) enter the desired root, 4 in this case.

(d) press "=". The answer here is 4.8646×10^{-3} or 4.9×10^{-3}.

A general procedure for finding any root is to use the "y^x" key. For a square root, x is 0.5 (or ½), whereas it is 0.33 (or ⅓) for a cube root, 0.25 (or ¼) for a 4th root, and so on.

A.3 LOGARITHMS

There are two types of logarithms used in this text: (a) common logarithms (abbreviated log) whose base is 10 and (b) natural logarithms (abbreviated ln) whose base is e (= 2.71828).

$$\log x = n \text{ where } x = 10^n$$
$$\ln x = m \text{ where } x = e^m$$

Most equations in chemistry and physics were developed in natural or base e logarithms and this practice is followed in this text. The relation between log and ln is

$$\ln x = 2.303 \log x$$

Aside from the different bases of the two logarithms, they are used in the same manner. What follows is largely a description of the use of common logarithms.

A common logarithm is the power to which you must raise 10 to obtain the number. For example, the log of 100 is 2, since you must raise 10 to the second power to obtain 100. Other examples are

$$\log 1000 = \log (10^3) = 3$$
$$\log 10 = \log (10^1) = 1$$
$$\log 1 = \log (10^0) = 0$$
$$\log 1/10 = \log (10^{-1}) = -1$$
$$\log 1/10000 = \log (10^{-4}) = -4$$

To obtain the common logarithm of a number other than a simple power of 10, you must resort to a log table or an electronic calculator. For example,

$\log 2.10 = 0.3222$, which means that $10^{0.3222} = 2.10$
$\log 5.16 = 0.7126$, which means that $10^{0.7126} = 5.16$
$\log 3.125 = 0.49485$, which means that $10^{0.49485} = 3.125$

To check this on your calculator, enter the number and then press the "log" key. When using a log table, the logs of the first two numbers above can be read directly from the table. The log of the third number (3.125), however, must be interpolated. That is, 3.125 is midway between 3.12 and 3.13, so the log is midway between 0.4942 and 0.4955.

To obtain the natural logarithm, ln, of the numbers above, use a calculator having this function. Enter each number and press "ln."

$\ln 2.10 = 0.7419$, which means that $e^{0.7419} = 2.10$
$\ln 5.16 = 1.6409$, which means that $e^{1.6409} = 5.16$

To find the common logarithm of a number greater than 10 or less than 1 with a log table, first express the number in scientific notation.

Then find the log of each part of the number and add the logs. For example,

$$\log 241 = \log (2.41 \times 10^2) = \log 2.41 + \log 10^2 = 0.382 + 2 = 2.382$$

$$\log 0.00573 = \log (5.73 \times 10^{-3}) = \log 5.73 + \log 10^{-3} = 0.758 + (-3) = -2.242$$

LOGARITHMS AND NOMENCLATURE: The number to the left of the decimal in a logarithm is called the ***characteristic****, and the number to the right of the decimal is the* ***mantissa****.*

SIGNIFICANT FIGURES AND LOGARITHMS Notice that the mantissa has as many significant figures as the number whose log was found. (So that you could more clearly see the result obtained with a calculator or a table, this rule was not strictly followed until the last two examples.)

OBTAINING ANTILOGARITHMS If you are given the logarithm of a number, and find the number from it, you have obtained the "antilogarithm" or "antilog" of the number. There are two common procedures used by electronic calculators to do this:

PROCEDURE A	PROCEDURE B
(a) enter the log or ln	(a) enter the log or ln
(b) press 2ndF	(b) press INV
(c) press 10^x or e^x	(c) press log or ln x

Test one or the other of these procedures with the following examples:

1. Find the number whose log is 5.234.
Recall that $\log x = n$ where $x = 10^n$. In this case $n = 5.234$. Enter that number in your calculator and find the value of 10^n, the antilog. In this case,

$$10^{5.234} = 10^{0.234} \times 10^5 = 1.71 \times 10^5$$

Notice that the characteristic (5) sets the decimal point; it is the power of 10 in the exponential form. The mantissa (0.234) gives the value of the number x. Thus, if you use a log table to find x, you need only look up 0.234 in the table and see that it corresponds to 1.71.

2. Find the number whose log is -3.456.

$$10^{-3.456} = 10^{0.544} \times 10^{-4} = 3.50 \times 10^{-4}$$

Notice here that -3.456 must be expressed as the sum of -4 and $+0.544$.

MATHEMATICAL OPERATIONS USING LOGARITHMS Because logarithms are exponents, operations involving them follow the same rules as the use of exponents. Thus, multiplying two numbers can be done by adding logarithms.

$$\log xy = \log x + \log y$$

For example, we multiply 563 by 125 by adding their logarithms and finding the anti-logarithm of the result.

$$\log 563 = 2.751$$
$$\log 125 = \underline{2.097}$$
$$\log xy = 4.848$$

$$xy = 10^{4.848} = 10^4 \times 10^{0.848} = 7.05 \times 10^4$$

One number (x) can be divided by another (y) by subtraction of their logarithms.

$$\log \frac{x}{y} = \log x - \log y$$

For example, to divide 125 by 742,

$$\begin{aligned} \log 125 &= \ \ 2.097 \\ -\log 742 &= \ \ \underline{2.870} \\ \log x/y &= -0.773 \end{aligned}$$

$$x/y = 10^{-0.773} = 10^{0.227} \times 10^{-1} = 1.69 \times 10^{-1}$$

Similarly, powers and roots of numbers can be found using logarithms.

$$\log x^y = y(\log x)$$

$$\log \sqrt[y]{x} = \log x^{1/y} = \frac{1}{y} \log x$$

As an example, find the fourth power of 5.23. We first find the log of 5.23 and then multiply it by 4. The result, 2.874, is the log of the answer. Therefore, we find the antilog of 2.874.

$$\begin{aligned} (5.23)^4 &= ? \\ \log (5.23)^4 &= 4 \log 5.23 = 4\,(0.719) = 2.874 \\ (5.23)^4 &= 10^{2.874} = 748 \end{aligned}$$

As another example, find the fifth root of 1.89×10^{-9}.

$$\sqrt[5]{1.89 \times 10^{-9}} = (1.89 \times 10^{-9})^{1/5} = ?$$

$$\log (1.89 \times 10^{-9})^{1/5} = \frac{1}{5} \log (1.89 \times 10^{-9}) = \frac{1}{5}(-8.724) = -1.745$$

The answer is the antilog of -1.745.

$$(1.89 \times 10^{-9})^{1/5} = 10^{-1.745} = 1.80 \times 10^{-2}$$

A.4 QUADRATIC EQUATIONS

Algebraic equations of the form $ax^2 + bx + c = 0$ are called **quadratic equations**. The coefficients a, b, and c may be either positive or negative. The two roots of the equation may be found using the *quadratic formula*.

$$x = \frac{-b \pm \sqrt{b^2 - 4ac}}{2a}$$

As an example, solve the equation $5x^2 - 3x - 2 = 0$. Here $a = 5$, $b = -3$, and $c = -2$. Therefore,

$$x = \frac{3 \pm \sqrt{(-3)^2 - 4(5)(-2)}}{2(5)}$$

$$= \frac{3 \pm \sqrt{9 - (-40)}}{10} = \frac{3 \pm \sqrt{49}}{10} = \frac{3 \pm 7}{10}$$

$$x = 1 \text{ and } -0.4$$

How do you know which of the two roots is the correct answer? You have to decide in each case which root has physical significance. However, it is *usually* true in this course that negative values are not significant.

When you have solved a quadratic expression, you should always check your values by substitution into the original equation. In the example above, we find that $5(1)^2 - 3(1) - 2 = 0$ and that $5(-0.4)^2 - 3(-0.4) - 2 = 0$.

The most likely place you will encounter quadratic equations is in the chapters on chemical equilibria, particularly in Chapters 16–18. Here you will often be faced with solving an equation such as

$$1.8 \times 10^{-4} = \frac{x^2}{0.0010 - x}$$

This equation can certainly be solved using the quadratic equation (to give $x = 3.4 \times 10^{-4}$). However, you may find *method of successive approximations* to be especially convenient. Here we begin by making a reasonable approximation of x. This approximate value is substituted into the original equation, and this is solved to give what is hoped to be a more correct value of x. This process is repeated until the answer converges on a particular value of x, that is, until the value of x derived from two successive approximations is the same.

Step 1: First assume that x is so small that $(0.0010 - x) \approx 0.0010$. This means that

$$x^2 = 1.8 \times 10^{-4}(0.0010)$$

$$x = 4.2 \times 10^{-4} \text{ (to 2 significant figures)}$$

Step 2: Substitute the value of x from Step 1 into the denominator of the original equation and again solve for x.

$$x^2 = 1.8 \times 10^{-4}(0.0010 - 0.00042)$$

$$x = 3.2 \times 10^{-4}$$

Step 3: Repeat Step 2 using the value of x found in that step.

$$x = \sqrt{1.8 \times 10^{-4}(0.0010 - 0.00032)} = 3.5 \times 10^{-4}$$

Step 4: Continue repeating the calculation, using the value of x found in the previous step.

$$x = \sqrt{1.8 \times 10^{-4}(0.0010 - 0.00035)} = 3.4 \times 10^{-4}$$

Step 5: $x = \sqrt{1.8 \times 10^{-4}(0.0010 - 0.00034)} = 3.4 \times 10^{-4}$

Here we find that iterations after the fourth step give the same value for x, indicating that we have arrived at a valid answer (and the same one obtained from the quadratic formula).

There are several final thoughts on using the method of successive approximations. First, there are cases where the method does not work. Successive steps may give answers that are random or that diverge from the correct value. In Chapters 16–18, you will confront quadratic equations of the form $K = x^2/(C - x)$. The method of approximations will work as long as $K < 4C$ (assuming one begins with $x = 0$ as the first guess that is, $K \approx x^2/C$). This is always going to be true for weak acids and bases (the topic of Chapters 17 and 18), but it may *not* be the case for problems involving gas phase equilibria (Chapter 16), where K can be quite large.

Second, values of K in the equation $K = x^2/(C - x)$ are usually known only to two significant figures. Therefore, we are justified in carrying out successive steps until two answers are the same to two significant figures.

Finally, we highly recommend this method of solving quadratic equations, especially those in Chapters 17 and 18. If your calculator has a memory function, successive approximations can be carried out easily and rapidly.

APPENDIX B

Common Units, Equivalences, and Conversion Factors

FUNDAMENTAL UNITS OF THE SI SYSTEM

The metric system was begun by the French National Assembly in 1790 and has undergone many modifications. The International System of Units or *Système International* (SI), which represents an extension of the metric system, was adopted by the 11th General Conference of Weights and Measures in 1960. It is constructed from seven base units, each of which represents a particular physical quantity (Table I).

Table I SI Fundamental Units

Physical Quantity	Name of Unit	Symbol
Length	meter	m
Mass	kilogram	kg
Time	second	s
Temperature	kelvin	K
Amount of substance	mole	mol
Electric current	ampere	A
Luminous intensity	candela	cd

The first five units listed in Table I are particularly useful in general chemistry. They are defined as follows.

1. The *meter* was redefined in 1960 to be equal to 1,650,763.73 wavelengths of a certain line in the emission spectrum of krypton-86.
2. The *kilogram* represents the mass of a platinum-iridium block kept at the International Bureau of Weights and Measures at Sèvres, France.
3. The *second* was redefined in 1967 as the duration of 9,192,631,770 periods of a certain line in the microwave spectrum of cesium-133.
4. The *kelvin* is 1/273.15 of the temperature interval between absolute zero and the triple point of water.
5. The *mole* is the amount of substance that contains as many entities as there are atoms in exactly 0.012 kg of carbon-12 (12 g of ^{12}C atoms).

PREFIXES USED WITH TRADITIONAL METRIC UNITS AND SI UNITS

Decimal fractions and multiples of metric and SI units are designated by using the prefixes listed in Table II. Those most commonly used in general chemistry are in italics.

Table II Traditional Metric and SI Prefixes

Factor	Prefix	Symbol	Factor	Prefix	Symbol
10^{12}	tera	T	10^{-1}	*deci*	d
10^{9}	giga	G	10^{-2}	*centi*	c
10^{6}	mega	M	10^{-3}	*milli*	m
10^{3}	*kilo*	k	10^{-6}	micro	μ
10^{2}	hecto	h	10^{-9}	*nano*	n
10^{1}	deka	da	10^{-12}	*pico*	p
			10^{-15}	femto	f
			10^{-18}	atto	a

DERIVED SI UNITS

In the International System of Units, all physical quantities are represented by appropriate combinations of the base units listed in Table I. A list of the derived units frequently used in general chemistry is given in Table III.

Table III Derived SI Units

Physical Quantity	Name of Unit	Symbol	Definition
Area	square meter	m^2	
Volume	cubic meter	m^3	
Density	kilogram per cubic meter	kg/m^3	
Force	newton	N	$kg\ m/s^2$
Pressure	pascal	Pa	N/m^2
Energy	joule	J	$kg\ m^2/s^2$
Electric charge	coulomb	C	A s
Electric potential difference	volt	V	J/(A s)

Common Units of Mass and Weight

1 Pound = 453.59 Grams

1 pound = 453.59 grams = 0.45359 kilogram
1 kilogram = 1000 grams = 2.205 pounds
1 gram = 10 decigrams = 100 centigrams = 1000 milligrams
1 gram = 6.022×10^{23} atomic mass units
1 atomic mass unit = 1.6605×10^{-24} gram
1 short ton = 2000 pounds = 907.2 kilograms
1 long ton = 2240 pounds
1 metric tonne = 1000 kilograms = 2205 pounds

Common Units of Length

1 Inch = 2.54 Centimeters (Exactly)

1 mile = 5280 feet = 1.609 kilometers
1 yard = 36 inches = 0.9144 meter
1 meter = 100 centimeters = 39.37 inches = 3.281 feet = 1.094 yards
1 kilometer = 1000 meters = 1094 yards = 0.6215 mile
1 Ångstrom = 1.0×10^{-8} centimeter = 0.10 nanometer = 100 picometers
= 1.0×10^{-10} meter = 3.937×10^{-9} inch

Common Units of Volume

1 Quart = 0.9463 Liter
1 Liter = 1.0567 Quarts

1 liter = 1 cubic decimeter = 1000 cubic centimeters = 0.001 cubic meter
1 milliliter = 1 cubic centimeter = 0.001 liter = 1.056×10^{-3} quart
1 cubic foot = 28.316 liters = 29.924 quarts = 7.481 gallons

Common Units of Force* and Pressure

1 atmosphere = 760 millimeters of mercury = 1.013×10^5 pascals
= 14.70 pounds per square inch
1 bar = 10^5 pascals
1 torr = 1 millimeter of mercury
1 pascal = 1 kg/m s^2 = 1 N/m^2

*Force: 1 newton (N) = 1 kg m/s^2, i.e., the force that when applied for 1 second gives a 1 kilogram mass a velocity of 1 meter per second.

Common Units of Energy

1 Joule = 1×10^7 Ergs

1 thermochemical calorie* = 4.184 joules = 4.184×10^7 ergs
= 4.129×10^{-2} liter-atmospheres
= 2.612×10^{19} electron volts
1 erg = 1×10^{-7} joule = 2.3901×10^{-8} calorie
1 electron volt = 1.6022×10^{-19} joule = 1.6022×10^{-12} erg = 96.485 kJ/mol†
1 liter-atmosphere = 24.217 calories = 101.32 joules = 1.0132×10^9 ergs
1 British thermal unit = 1055.06 joules = 1.05506×10^{10} ergs = 252.2 calories

*The amount of heat required to raise the temperature of one gram of water from 14.5 °C to 15.5 °C.

†Note that the other units in this line are per particle and must be multiplied by 6.022×10^{23} to be strictly comparable.

APPENDIX C

Physical Constants

Quantity	Symbol	Traditional Units	SI Units
Acceleration of gravity	g	980.6 cm/s	9.806 m/s
Atomic mass unit (1/12 the mass of ^{12}C atom)	amu or u	1.6605×10^{-24} g	1.6605×10^{-27} kg
Avogadro's number	N	6.0221367×10^{23} particles/mol	6.0221367×10^{23} particles/mol
Bohr radius	a_0	0.52918 Å 5.2918×10^{-9} cm	5.2918×10^{-11} m
Boltzmann constant	k	1.3807×10^{-16} erg/K	1.3807×10^{-23} J/K
Charge-to-mass ratio of electron	e/m	1.7588×10^{8} coulomb/g	1.7588×10^{11} C/kg
Electronic charge	e	1.6022×10^{-19} coulomb 4.8033×10^{-10} esu	1.6022×10^{-19} C
Electron rest mass	m_e	9.1094×10^{-28} g 0.00054858 amu	9.1094×10^{-31} kg
Faraday constant	F	96,485 coulombs/mol e^- 23.06 kcal/volt mol e^-	96,485 C/mol e^- 96,485 J/V mol e^-
Gas constant	R	$0.08206 \frac{\text{L atm}}{\text{mol K}}$ $1.987 \frac{\text{cal}}{\text{mol K}}$	$8.3145 \frac{\text{Pa dm}^3}{\text{mol K}}$ 8.3145 J/mol K
Molar volume (STP)	V_m	22.414 L/mol	22.414×10^{-3} m^3/mol 22.414 dm^3/mol
Neutron rest mass	m_n	1.67495×10^{-24} g 1.008665 amu	1.67493×10^{-27} kg
Planck's constant	h	6.6261×10^{-27} erg s	$6.6260755 \times 10^{-34}$ J s
Proton rest mass	m_p	1.6726×10^{-24} g 1.007276 amu	1.6726×10^{-27} kg
Rydberg constant	R_∞	3.289×10^{15} cycles/s 2.1799×10^{-11} erg	1.0974×10^{7} m^{-1} 2.1799×10^{-18} J
Velocity of light (in a vacuum)	c	2.9979×10^{10} cm/s (186,282 miles/second)	2.9979×10^{8} m/s

$\pi = 3.1416$
e = 2.7183
$\ln X = 2.303 \log X$
$2.303\ R = 4.576$ cal/mol · K = 19.15 J/mol · K
$2.303\ RT$ (at 25 °C) = 1364 cal/mol = 5709 J/mol

APPENDIX D

Some Physical Constants for Water and a Few Common Substances

Vapor Pressure of Water at Various Temperatures

Temperature °C	Vapor Pressure torr	Temperature °C	Vapor Pressure torr	Temperature °C	Vapor Pressure torr	Temperature °C	Vapor Pressure torr
−10	2.1	21	18.7	51	97.2	81	369.7
−9	2.3	22	19.8	52	102.1	82	384.9
−8	2.5	23	21.1	53	107.2	83	400.6
−7	2.7	24	22.4	54	112.5	84	416.8
−6	2.9	25	23.8	55	118.0	85	433.6
−5	3.2	26	25.2	56	123.8	86	450.9
−4	3.4	27	26.7	57	129.8	87	468.7
−3	3.7	28	28.3	58	136.1	88	487.1
−2	4.0	29	30.0	59	142.6	89	506.1
−1	4.3	30	31.8	60	149.4	90	525.8
0	4.6	31	33.7	61	156.4	91	546.1
1	4.9	32	35.7	62	163.8	92	567.0
2	5.3	33	37.7	63	171.4	93	588.6
3	5.7	34	39.9	64	179.3	94	610.9
4	6.1	35	42.2	65	187.5	95	633.9
5	6.5	36	44.6	66	196.1	96	657.6
6	7.0	37	47.1	67	205.0	97	682.1
7	7.5	38	49.7	68	214.2	98	707.3
8	8.0	39	52.4	69	223.7	99	733.2
9	8.6	40	55.3	70	233.7	100	760.0
10	9.2	41	58.3	71	243.9	101	787.6
11	9.8	42	61.5	72	254.6	102	815.9
12	10.5	43	64.8	73	265.7	103	845.1
13	11.2	44	68.3	74	277.2	104	875.1
14	12.0	45	71.9	75	289.1	105	906.1
15	12.8	46	75.7	76	301.4	106	937.9
16	13.6	47	79.6	77	314.1	107	970.6
17	14.5	48	83.7	78	327.3	108	1004.4
18	15.5	49	88.0	79	341.0	109	1038.9
19	16.5	50	92.5	80	355.1	110	1074.6
20	17.5						

APPENDIX D (*continued*)

Some Physical Constants for Water and a Few Common Substances

Specific Heats and Heat Capacities for Some Common Substances

Substance	Specific Heat J/g · K	Molar Heat Capacity J/mol · K
Al (s)	0.902	24.3
Ca (s)	0.653	26.2
Cu (s)	0.385	24.5
Fe (s)	0.451	24.8
Hg (ℓ)	0.138	27.7
H_2O (s), ice	2.06	37.7
H_2O (ℓ), water	4.18	75.3
H_2O (*g*), steam	2.03	36.4
C_6H_6 (ℓ), benzene	1.74	136
C_6H_6 (*g*), benzene	1.04	81.6
C_2H_5OH (ℓ), ethanol	2.46	113
C_2H_5OH (*g*), ethanol	0.954	420
$(C_2H_5)_2O$ (ℓ), diethyl ether	3.74	172
$(C_2H_5)_2O$ (*g*), diethyl ether	2.35	108

Heats of Transformation and Transformation Temperatures of Several Substances

Substance	MP (°C)	Heat of Fusion		BP (°C)	Heat of Vaporization	
		J/g	kJ/mol		J/g	kJ/mol
Elements*						
Al	660	395	10.7	2467	12083	326
Ca	839	230	9.3	1493	3768	151
Cu	1083	205	13.0	2570	4799	305
Fe	1535	267	14.9	2750	6285	351
Hg	−38.8	11	2.3	357	294	59.0
Compounds						
H_2O	0.00	333	6.02	100.0	2260	40.7
CH_4	−182	58.6	0.92	−164	—	—
C_2H_5OH	−117	109	5.02	78.0	855	39.3
C_6H_6	5.48	127	9.92	80.1	395	30.8
$(C_2H_5)_2O$	−116	97.9	7.66	35	351	26.0

*Data for the elements are taken from "The Periodic Table Stack," the Macintosh version of KC? Discoverer from *JCE: Software*.

APPENDIX E

Ionization Constants for Weak Acids at 25 °C

Acid	Formula and Ionization Equation	K_a
Acetic	$CH_3COOH \rightleftharpoons H^+ + CH_3COO^-$	1.8×10^{-5}
Arsenic	$H_3AsO_4 \rightleftharpoons H^+ + H_2AsO_4^-$	$K_1 = 2.5 \times 10^{-4}$
	$H_2AsO_4^- \rightleftharpoons H^+ + HAsO_4^{2-}$	$K_2 = 5.6 \times 10^{-8}$
	$HAsO_4^{2-} \rightleftharpoons H^+ + AsO_4^{3-}$	$K_3 = 3.0 \times 10^{-13}$
Arsenous	$H_3AsO_3 \rightleftharpoons H^+ + H_2AsO_3^-$	$K_1 = 6.0 \times 10^{-10}$
	$H_2AsO_3^- \rightleftharpoons H^+ + HAsO_3^{2-}$	$K_2 = 3.0 \times 10^{-14}$
Benzoic	$C_6H_5COOH \rightleftharpoons H^+ + C_6H_5COO^-$	6.3×10^{-5}
Boric	$H_3BO_3 \rightleftharpoons H^+ + H_2BO_3^-$	$K_1 = 7.3 \times 10^{-10}$
	$H_2BO_3^- \rightleftharpoons H^+ + HBO_3^{2-}$	$K_2 = 1.8 \times 10^{-13}$
	$HBO_3^{2-} \rightleftharpoons H^+ + BO_3^{3-}$	$K_3 = 1.6 \times 10^{-14}$
Carbonic	$H_2CO_3 \rightleftharpoons H^+ + HCO_3^-$	$K_1 = 4.2 \times 10^{-7}$
	$HCO_3^- \rightleftharpoons H^+ + CO_3^{2-}$	$K_2 = 4.8 \times 10^{-11}$
Citric	$H_3C_6H_5O_7 \rightleftharpoons H^+ + H_2C_6H_5O_7^-$	$K_1 = 7.4 \times 10^{-3}$
	$H_2C_6H_5O_7^- \rightleftharpoons H^+ + HC_6H_5O_7^{2-}$	$K_2 = 1.7 \times 10^{-5}$
	$HC_6H_5O_7^{2-} \rightleftharpoons H^+ + C_6H_5O_7^{3-}$	$K_3 = 4.0 \times 10^{-7}$
Cyanic	$HOCN \rightleftharpoons H^+ + OCN^-$	3.5×10^{-4}
Formic	$HCOOH \rightleftharpoons H^+ + HCOO^-$	1.8×10^{-4}
Hydrazoic	$HN_3 \rightleftharpoons H^+ + N_3^-$	1.9×10^{-5}
Hydrocyanic	$HCN \rightleftharpoons H^+ + CN^-$	4.0×10^{-10}
Hydrofluoric	$HF \rightleftharpoons H^+ + F^-$	7.2×10^{-4}
Hydrogen peroxide	$H_2O_2 \rightleftharpoons H^+ + HO_2^-$	2.4×10^{-12}
Hydrosulfuric	$H_2S \rightleftharpoons H^+ + HS^-$	$K_1 = 1.0 \times 10^{-7}$
	$HS^- \rightleftharpoons H^+ + S^{2-}$	$K_2 = 1.3 \times 10^{-13}$
Hypobromous	$HOBr \rightleftharpoons H^+ + OBr^-$	2.5×10^{-9}
Hypochlorous	$HOCl \rightleftharpoons H^+ + OCl^-$	3.5×10^{-8}
Nitrous	$HNO_2 \rightleftharpoons H^+ + NO_2^-$	4.5×10^{-4}
Oxalic	$H_2C_2O_4 \rightleftharpoons H^+ + HC_2O_4^-$	$K_1 = 5.9 \times 10^{-2}$
	$HC_2O_4^- \rightleftharpoons H^+ + C_2O_4^{2-}$	$K_2 = 6.4 \times 10^{-5}$
Phenol	$HC_6H_5O \rightleftharpoons H^+ + C_6H_5O^-$	1.3×10^{-10}
Phosphoric	$H_3PO_4 \rightleftharpoons H^+ + H_2PO_4^-$	$K_1 = 7.5 \times 10^{-3}$
	$H_2PO_4^- \rightleftharpoons H^+ + HPO_4^{2-}$	$K_2 = 6.2 \times 10^{-8}$
	$HPO_4^{2-} \rightleftharpoons H^+ + PO_4^{3-}$	$K_3 = 3.6 \times 10^{-13}$
Phosphorous	$H_3PO_3 \rightleftharpoons H^+ + H_2PO_3^-$	$K_1 = 1.6 \times 10^{-2}$
	$H_2PO_3^- \rightleftharpoons H^+ + HPO_3^{2-}$	$K_2 = 7.0 \times 10^{-7}$
Selenic	$H_2SeO_4 \rightleftharpoons H^+ + HSeO_4^-$	K_1 = very large
	$HSeO_4^- \rightleftharpoons H^+ + SeO_4^{2-}$	$K_2 = 1.2 \times 10^{-2}$
Selenous	$H_2SeO_3 \rightleftharpoons H^+ + HSeO_3^-$	$K_1 = 2.7 \times 10^{-3}$
	$HSeO_3^- \rightleftharpoons H^+ + SeO_3^{2-}$	$K_2 = 2.5 \times 10^{-7}$
Sulfuric	$H_2SO_4 \rightleftharpoons H^+ + HSO_4^-$	K_1 = very large
	$HSO_4^- \rightleftharpoons H^+ + SO_4^{2-}$	$K_2 = 1.2 \times 10^{-2}$
Sulfurous	$H_2SO_3 \rightleftharpoons H^+ + HSO_3^-$	$K_1 = 1.7 \times 10^{-2}$
	$HSO_3^- \rightleftharpoons H^+ + SO_3^{2-}$	$K_2 = 6.4 \times 10^{-8}$
Tellurous	$H_2TeO_3 \rightleftharpoons H^+ + HTeO_3^-$	$K_1 = 2 \times 10^{-3}$
	$HTeO_3^- \rightleftharpoons H^+ + TeO_3^{2-}$	$K_2 = 1 \times 10^{-8}$

APPENDIX F

Ionization Constants for Weak Bases at 25 °C

Base	Formula and Ionization Equation	K_b
Ammonia	$NH_3 + H_2O \rightleftharpoons NH_4^+ + OH^-$	1.8×10^{-5}
Aniline	$C_6H_5NH_2 + H_2O \rightleftharpoons C_6H_5NH_3^+ + OH^-$	4.2×10^{-10}
Dimethylamine	$(CH_3)_2NH + H_2O \rightleftharpoons (CH_3)_2NH_2^+ + OH^-$	7.4×10^{-4}
Ethylenediamine	$(CH_2)_2(NH_2)_2 + H_2O \rightleftharpoons (CH_2)_2(NH_2)_2H^+ + OH^-$	$K_1 = 8.5 \times 10^{-5}$
	$(CH_2)_2(NH_2)_2H^+ + H_2O \rightleftharpoons (CH_2)_2(NH_2)_2H_2^{2+} + OH^-$	$K_2 = 2.7 \times 10^{-8}$
Hydrazine	$N_2H_4 + H_2O \rightleftharpoons N_2H_5^+ + OH^-$	$K_1 = 8.5 \times 10^{-7}$
	$N_2H_5^+ + H_2O \rightleftharpoons N_2H_6^{2+} + OH^-$	$K_2 = 8.9 \times 10^{-16}$
Hydroxylamine	$NH_2OH + H_2O \rightleftharpoons NH_3OH^+ + OH^-$	6.6×10^{-9}
Methylamine	$CH_3NH_2 + H_2O \rightleftharpoons CH_3NH_3^+ + OH^-$	5.0×10^{-4}
Pyridine	$C_5H_5N + H_2O \rightleftharpoons C_5H_5NH^+ + OH^-$	1.5×10^{-9}
Trimethylamine	$(CH_3)_3N + H_2O \rightleftharpoons (CH_3)_3NH^+ + OH^-$	7.4×10^{-5}

APPENDIX G

Solubility Product Constants for Some Inorganic Compounds at 25 °C

Substance	K_{sp}
Aluminum compounds	
$AlAsO_4$	1.6×10^{-16}
$Al(OH)_3$	1.9×10^{-33}
$AlPO_4$	1.3×10^{-20}
Antimony compounds	
Sb_2S_3	1.6×10^{-93}
Barium compounds	
$Ba_3(AsO_4)_2$	1.1×10^{-13}
$BaCO_3$	8.1×10^{-9}
$BaC_2O_4 \cdot 2H_2O$*	1.1×10^{-7}
$BaCrO_4$	2.0×10^{-10}
BaF_2	1.7×10^{-6}
$Ba(OH)_2 \cdot 8H_2O$*	5.0×10^{-3}
$Ba_3(PO_4)_2$	1.3×10^{-29}
$BaSeO_4$	2.8×10^{-11}
$BaSO_3$	8.0×10^{-7}
$BaSO_4$	1.1×10^{-10}
Bismuth compounds	
$BiOCl$	7.0×10^{-9}
$BiO(OH)$	1.0×10^{-12}
$Bi(OH)_3$	3.2×10^{-40}
BiI_3	8.1×10^{-19}
$BiPO_4$	1.3×10^{-23}
Bi_2S_3	1.6×10^{-72}
Cadmium compounds	
$Cd_3(AsO_4)_2$	2.2×10^{-32}
$CdCO_3$	2.5×10^{-14}
$Cd(CN)_2$	1.0×10^{-8}
$Cd_2[Fe(CN)_6]$	3.2×10^{-17}
$Cd(OH)_2$	1.2×10^{-14}
CdS	3.6×10^{-29}
Calcium compounds	
$Ca_3(AsO_4)_2$	6.8×10^{-19}
$CaCO_3$	3.8×10^{-9}
$CaCrO_4$	7.1×10^{-4}
$CaC_2O_4 \cdot H_2O$*	2.3×10^{-9}
CaF_2	3.9×10^{-11}
$Ca(OH)_2$	7.9×10^{-6}
$CaHPO_4$	2.7×10^{-7}
$Ca(H_2PO_4)_2$	1.0×10^{-3}
$Ca_3(PO_4)_2$	1.0×10^{-25}
$CaSO_3 \cdot 2H_2O$*	1.3×10^{-8}
$CaSO_4 \cdot 2H_2O$*	2.4×10^{-5}
Chromium compounds	
$CrAsO_4$	7.8×10^{-21}
$Cr(OH)_3$	6.7×10^{-31}
$CrPO_4$	2.4×10^{-23}
Cobalt compounds	
$Co_3(AsO_4)_2$	7.6×10^{-29}
$CoCO_3$	8.0×10^{-13}
$Co(OH)_2$	2.5×10^{-16}
CoS (α)	5.9×10^{-21}
$Co(OH)_3$	4.0×10^{-45}
Co_2S_3	2.6×10^{-124}
Copper compounds	
$CuBr$	5.3×10^{-9}
$CuCl$	1.9×10^{-7}
$CuCN$	3.2×10^{-20}
Cu_2O ($Cu^+ + OH^-$)†	1.0×10^{-14}
CuI	5.1×10^{-12}
Cu_2S	1.6×10^{-48}
$CuSCN$	1.6×10^{-11}
$Cu_3(AsO_4)_2$	7.6×10^{-36}
$CuCO_3$	2.5×10^{-10}
$Cu_2[Fe(CN)_6]$	1.3×10^{-16}
$Cu(OH)_2$	1.6×10^{-19}
CuS	8.7×10^{-36}
Gold compounds	
$AuBr$	5.0×10^{-17}
$AuCl$	2.0×10^{-13}
AuI	1.6×10^{-23}
$AuBr_3$	4.0×10^{-36}
$AuCl_3$	3.2×10^{-25}
$Au(OH)_3$	1×10^{-53}
AuI_3	1.0×10^{-46}
Iron compounds	
$FeCO_3$	3.5×10^{-11}
$Fe(OH)_2$	7.9×10^{-15}
FeS	4.9×10^{-18}
$Fe_4[Fe(CN)_6]_3$	3.0×10^{-41}
$Fe(OH)_3$	6.3×10^{-38}
Fe_2S_3	1.4×10^{-88}
Lead compounds	
$Pb_3(AsO_4)_2$	4.1×10^{-36}
$PbBr_2$	6.3×10^{-6}
$PbCO_3$	1.5×10^{-13}
$PbCl_2$	1.7×10^{-5}
$PbCrO_4$	1.8×10^{-14}
PbF_2	3.7×10^{-8}
$Pb(OH)_2$	2.8×10^{-16}
PbI_2	8.7×10^{-9}
$Pb_3(PO_4)_2$	3.0×10^{-44}
$PbSeO_4$	1.5×10^{-7}
$PbSO_4$	1.8×10^{-8}
PbS	8.4×10^{-28}

APPENDIX G (*continued*)

Solubility Product Constants for Some Inorganic Compounds at 25 °C

Substance	K_{sp}	Substance	K_{sp}
Magnesium compounds		$AgBr$	3.3×10^{-13}
$Mg_3(AsO_4)_2$	2.1×10^{-20}	Ag_2CO_3	8.1×10^{-12}
$MgCO_3 \cdot 3H_2O$*	4.0×10^{-5}	$AgCl$	1.8×10^{-10}
MgC_2O_4	8.6×10^{-5}	Ag_2CrO_4	9.0×10^{-12}
MgF_2	6.4×10^{-9}	$AgCN$	1.2×10^{-16}
$Mg(OH)_2$	1.5×10^{-11}	$Ag_4[Fe(CN)_6]$	1.6×10^{-41}
$MgNH_4PO_4$	2.5×10^{-12}	Ag_2O ($Ag^+ + OH^-$)†	2.0×10^{-8}
Manganese compounds		AgI	1.5×10^{-16}
$Mn_3(AsO_4)_2$	1.9×10^{-11}	Ag_3PO_4	1.3×10^{-20}
$MnCO_3$	1.8×10^{-11}	Ag_2SO_3	1.5×10^{-14}
$Mn(OH)_2$	4.6×10^{-14}	Ag_2SO_4	1.7×10^{-5}
MnS	5.1×10^{-15}	Ag_2S	1.0×10^{-49}
$Mn(OH)_3$	$\sim 1 \times 10^{-36}$	$AgSCN$	1.0×10^{-12}
Mercury compounds		Strontium compounds	
Hg_2Br_2	1.3×10^{-22}	$Sr_3(AsO_4)_2$	1.3×10^{-18}
Hg_2CO_3	8.9×10^{-17}	$SrCO_3$	9.4×10^{-10}
Hg_2Cl_2	1.1×10^{-18}	$SrC_2O_4 \cdot 2H_2O$*	5.6×10^{-8}
Hg_2CrO_4	5.0×10^{-9}	$SrCrO_4$	3.6×10^{-5}
Hg_2I_2	4.5×10^{-29}	$Sr(OH)_2 \cdot 8H_2O$*	3.2×10^{-4}
$Hg_2O \cdot H_2O$ ($Hg_2^{2+} + 2OH^-$)*†	1.6×10^{-23}	$Sr_3(PO_4)_2$	1.0×10^{-31}
Hg_2SO_4	6.8×10^{-7}	$SrSO_3$	4.0×10^{-8}
Hg_2S	5.8×10^{-44}	$SrSO_4$	2.8×10^{-7}
$Hg(CN)_2$	3.0×10^{-23}	Tin compounds	
$Hg(OH)_2$	2.5×10^{-26}	$Sn(OH)_2$	2.0×10^{-26}
HgI_2	4.0×10^{-29}	SnI_2	1.0×10^{-4}
HgS	3.0×10^{-53}	SnS	1.0×10^{-28}
Nickel compounds		$Sn(OH)_4$	1×10^{-57}
$Ni_3(AsO_4)_2$	1.9×10^{-26}	SnS_2	1×10^{-70}
$NiCO_3$	6.6×10^{-9}	Zinc compounds	
$Ni(CN)_2$	3.0×10^{-23}	$Zn_3(AsO_4)_2$	1.1×10^{-27}
$Ni(OH)_2$	2.8×10^{-16}	$ZnCO_3$	1.5×10^{-11}
NiS (α)	3.0×10^{-21}	$Zn(CN)_2$	8.0×10^{-12}
NiS (β)	1.0×10^{-26}	$Zn_3[Fe(CN)_6]$	4.1×10^{-16}
NiS (γ)	2.0×10^{-28}	$Zn(OH)_2$	4.5×10^{-17}
Silver compounds		$Zn_3(PO_4)_2$	9.1×10^{-33}
Ag_3AsO_4	1.1×10^{-20}	ZnS	1.1×10^{-21}

*Since $[H_2O]$ does not appear in equilibrium constants for equilibria in aqueous solution in general, it does *not* appear in the K_{sp} expressions for hydrated solids.

†Very small amounts of oxides dissolve in water to give the ions indicated in parentheses. Solid hydroxides are unstable and decompose to oxides as rapidly as they are formed.

APPENDIX H

Formation Constants for Some Complex Ions in Aqueous Solution

Formation Equilibrium				K
Ag^+	+ $2Br^-$	$\rightleftharpoons$	$[AgBr_2]^-$	1.3×10^7
Ag^+	+ $2Cl^-$	$\rightleftharpoons$	$[AgCl_2]^-$	2.5×10^5
Ag^+	+ $2CN^-$	$\rightleftharpoons$	$[Ag(CN)_2]^-$	5.6×10^{18}
Ag^+	+ $2S_2O_3^{2-}$	$\rightleftharpoons$	$[Ag(S_2O_3)_2]^{3-}$	2.0×10^{13}
Ag^+	+ $2NH_3$	$\rightleftharpoons$	$[Ag(NH_3)_2]^+$	1.6×10^7
Al^{3+}	+ $6F^-$	$\rightleftharpoons$	$[AlF_6]^{3-}$	5.0×10^{23}
Al^{3+}	+ $4OH^-$	$\rightleftharpoons$	$[Al(OH)_4]^-$	7.7×10^{33}
Au^+	+ $2CN^-$	$\rightleftharpoons$	$[Au(CN)_2]^-$	2.0×10^{38}
Cd^{2+}	+ $4CN^-$	$\rightleftharpoons$	$[Cd(CN)_4]^{2-}$	1.3×10^{17}
Cd^{2+}	+ $4Cl^-$	$\rightleftharpoons$	$[CdCl_4]^{2-}$	1.0×10^4
Cd^{2+}	+ $4NH_3$	$\rightleftharpoons$	$[Cd(NH_3)_4]^{2+}$	1.0×10^7
Co^{2+}	+ $6NH_3$	$\rightleftharpoons$	$[Co(NH_3)_6]^{2+}$	7.7×10^4
Cu^+	+ $2CN^-$	$\rightleftharpoons$	$[Cu(CN)_2]^-$	1.0×10^{16}
Cu^+	+ $2Cl^-$	$\rightleftharpoons$	$[CuCl_2]^-$	1.0×10^5
Cu^{2+}	+ $4NH_3$	$\rightleftharpoons$	$[Cu(NH_3)_4]^{2+}$	6.8×10^{12}
Fe^{2+}	+ $6CN^-$	$\rightleftharpoons$	$[Fe(CN)_6]^{4-}$	7.7×10^{36}
Hg^{2+}	+ $4Cl^-$	$\rightleftharpoons$	$[HgCl_4]^{2-}$	1.2×10^{15}
Ni^{2+}	+ $4CN^-$	$\rightleftharpoons$	$[Ni(CN)_4]^{2-}$	1.0×10^{31}
Ni^{2+}	+ $6NH_3$	$\rightleftharpoons$	$[Ni(NH_3)_6]^{2+}$	5.6×10^8
Zn^{2+}	+ $4OH^-$	$\rightleftharpoons$	$[Zn(OH)_4]^{2-}$	2.9×10^{15}
Zn^{2+}	+ $4NH_3$	$\rightleftharpoons$	$[Zn(NH_3)_4]^{2+}$	2.9×10^9

APPENDIX I

Standard Reduction Potentials in Aqueous Solution at 25 °C

Acidic Solution	Standard Reduction Potential, E^0 (volts)
F_2 (g) + $2e^-$ $\longrightarrow$ $2F^-$ (aq)	2.87
Co^{3+} (aq) + e^- $\longrightarrow$ Co^{2+} (aq)	1.82
Pb^{4+} (aq) + $2e^-$ $\longrightarrow$ Pb^{2+} (aq)	1.8
H_2O_2 (aq) + $2H^+$ (aq) + $2e^-$ $\longrightarrow$ $2H_2O$	1.77
NiO_2 (s) + $4H^+$ (aq) + $2e^-$ $\longrightarrow$ Ni^{2+} (aq) + $2H_2O$	1.7
PbO_2 (s) + SO_4^{2-} (aq) + $4H^+$ (aq) + $2e^-$ $\longrightarrow$ $PbSO_4$ (s) + $2H_2O$	1.685
Au^+ (aq) + e^- $\longrightarrow$ Au (s)	1.68
2HClO (aq) + $2H^+$ (aq) + $2e^-$ $\longrightarrow$ Cl_2 (g) + $2H_2O$	1.63
Ce^{4+} (aq) + e^- $\longrightarrow$ Ce^{3+} (aq)	1.61
$NaBiO_3$ (s) + $6H^+$ (aq) + $2e^-$ $\longrightarrow$ Bi^{3+} (aq) + Na^+ (aq) + $3H_2O$	~1.6
MnO_4^- (aq) + $8H^+$ (aq) + $5e^-$ $\longrightarrow$ Mn^{2+} (aq) + $4H_2O$	1.51
Au^{3+} (aq) + $3e^-$ $\longrightarrow$ Au (s)	1.50
ClO_3^- (aq) + $6H^+$ (aq) + $5e^-$ $\longrightarrow$ $\frac{1}{2}Cl_2$ (g) + $3H_2O$	1.47
BrO_3^- (aq) + $6H^+$ (aq) + $6e^-$ $\longrightarrow$ Br^- (aq) + $3H_2O$	1.44
Cl_2 (g) + $2e^-$ $\longrightarrow$ $2Cl^-$ (aq)	1.358
$Cr_2O_7^{2-}$ (aq) + $14H^+$ (aq) + $6e^-$ $\longrightarrow$ $2Cr^{3+}$ (aq) + $7H_2O$	1.33
$N_2H_5^+$ (aq) + $3H^+$ (aq) + $2e^-$ $\longrightarrow$ $2NH_4^+$ (aq)	1.24
MnO_2 (s) + $4H^+$ (aq) + $2e^-$ $\longrightarrow$ Mn^{2+} (aq) + $2H_2O$	1.23
O_2 (g) + $4H^+$ (aq) + $4e^-$ $\longrightarrow$ $2H_2O$	1.229
Pt^{2+} (aq) + $2e^-$ $\longrightarrow$ Pt (s)	1.2
IO_3^- (aq) + $6H^+$ (aq) + $5e^-$ $\longrightarrow$ $\frac{1}{2}I_2$ (aq) + $3H_2O$	1.195
ClO_4^- (aq) + $2H^+$ (aq) + $2e^-$ $\longrightarrow$ ClO_3^- (aq) + H_2O	1.19
Br_2 (ℓ) + $2e^-$ $\longrightarrow$ $2Br^-$ (aq)	1.066
$AuCl_4^-$ (aq) + $3e^-$ $\longrightarrow$ Au (s) + $4Cl^-$ (aq)	1.00
Pd^{2+} (aq) + $2e^-$ $\longrightarrow$ Pd (s)	0.987
NO_3^- (aq) + $4H^+$ (aq) + $3e^-$ $\longrightarrow$ NO (g) + $2H_2O$	0.96
NO_3^- (aq) + $3H^+$ (aq) + $2e^-$ $\longrightarrow$ HNO_2 (aq) + H_2O	0.94
$2Hg^{2+}$ (aq) + $2e^-$ $\longrightarrow$ Hg_2^{2+} (aq)	0.920
Hg^{2+} (aq) + $2e^-$ $\longrightarrow$ Hg (ℓ)	0.855
Ag^+ (aq) + e^- $\longrightarrow$ Ag (s)	0.7994
Hg_2^{2+} (aq) + $2e^-$ $\longrightarrow$ 2Hg (ℓ)	0.789
Fe^{3+} (aq) + e^- $\longrightarrow$ Fe^{2+} (aq)	0.771
$SbCl_6^-$ (aq) + $2e^-$ $\longrightarrow$ $SbCl_4^-$ (aq) + $2Cl^-$ (aq)	0.75
$[PtCl_4]^{2-}$ (aq) + $2e^-$ $\longrightarrow$ Pt (s) + $4Cl^-$ (aq)	0.73
O_2 (g) + $2H^+$ (aq) + $2e^-$ $\longrightarrow$ H_2O_2 (aq)	0.682
$[PtCl_6]^{2-}$ (aq) + $2e^-$ $\longrightarrow$ $[PtCl_4]^{2-}$ (aq) + $2Cl^-$ (aq)	0.68
H_3AsO_4 (aq) + $2H^+$ (aq) + $2e^-$ $\longrightarrow$ H_3AsO_3 (aq) + H_2O	0.58
I_2 (s) + $2e^-$ $\longrightarrow$ $2I^-$ (aq)	0.535
TeO_2 (s) + $4H^+$ (aq) + $4e^-$ $\longrightarrow$ Te (s) + $2H_2O$	0.529
Cu^+ (aq) + e^- $\longrightarrow$ Cu (s)	0.521
$[RhCl_6]^{3-}$ (aq) + $3e^-$ $\longrightarrow$ Rh (s) + $6Cl^-$ (aq)	0.44
Cu^{2+} (aq) + $2e^-$ $\longrightarrow$ Cu (s)	0.337
$HgCl_2$ (s) + $2e^-$ $\longrightarrow$ 2Hg (ℓ) + $2Cl^-$ (aq)	0.27
AgCl (s) + e^- $\longrightarrow$ Ag (s) + Cl^- (aq)	0.222
SO_4^{2-} (aq) + $4H^+$ (aq) + $2e^-$ $\longrightarrow$ SO_2 (g) + $2H_2O$	0.20
SO_4^{2-} (aq) + $4H^+$ (aq) + $2e^-$ $\longrightarrow$ H_2SO_3 (aq) + H_2O	0.17
Cu^{2+} (aq) + e^- $\longrightarrow$ Cu^+ (aq)	0.153
Sn^{4+} (aq) + $2e^-$ $\longrightarrow$ Sn^{2+} (aq)	0.15
S (s) + $2H^+$ (aq) + $2e^-$ $\longrightarrow$ H_2S (aq)	0.14

APPENDIX I (*continued*)

Standard Reduction Potentials in Aqueous Solution at 25 °C

Acidic Solution	Standard Reduction Potential, E^0 (volts)
$AgBr\ (s) + e^- \longrightarrow Ag\ (s) + Br^-\ (aq)$	0.0713
$2H^+\ (aq) + 2e^- \longrightarrow H_2\ (g)$ (reference electrode)	0.0000
$N_2O\ (g) + 6H^+\ (aq) + H_2O + 4e^- \longrightarrow 2NH_3OH^+\ (aq)$	−0.05
$Pb^{2+}\ (aq) + 2e^- \longrightarrow Pb\ (s)$	−0.126
$Sn^{2+}\ (aq) + 2e^- \longrightarrow Sn\ (s)$	−0.14
$AgI\ (s) + e^- \longrightarrow Ag\ (s) + I^-\ (aq)$	−0.15
$[SnF_6]^{2-}\ (aq) + 4e^- \longrightarrow Sn\ (s) + 6F^-\ (aq)$	−0.25
$Ni^{2+}\ (aq) + 2e^- \longrightarrow Ni\ (s)$	−0.25
$Co^{2+}\ (aq) + 2e^- \longrightarrow Co\ (s)$	−0.28
$Tl^+\ (aq) + e^- \longrightarrow Tl\ (s)$	−0.34
$PbSO_4\ (s) + 2e^- \longrightarrow Pb\ (s) + SO_4^{2-}\ (aq)$	−0.356
$Se\ (s) + 2H^+\ (aq) + 2e^- \longrightarrow H_2Se\ (aq)$	−0.40
$Cd^{2+}\ (aq) + 2e^- \longrightarrow Cd\ (s)$	−0.403
$Cr^{3+}\ (aq) + e^- \longrightarrow Cr^{2+}\ (aq)$	−0.41
$Fe^{2+}\ (aq) + 2e^- \longrightarrow Fe\ (s)$	−0.44
$2CO_2\ (g) + 2H^+\ (aq) + 2e^- \longrightarrow (COOH)_2\ (aq)$	−0.49
$Ga^{3+}\ (aq) + 3e^- \longrightarrow Ga\ (s)$	−0.53
$HgS\ (s) + 2H^+\ (aq) + 2e^- \longrightarrow Hg\ (\ell) + H_2S\ (g)$	−0.72
$Cr^{3+}\ (aq) + 3e^- \longrightarrow Cr\ (s)$	−0.74
$Zn^{2+}\ (aq) + 2e^- \longrightarrow Zn\ (s)$	−0.763
$Cr^{2+}\ (aq) + 2e^- \longrightarrow Cr\ (s)$	−0.91
$FeS\ (s) + 2e^- \longrightarrow Fe\ (s) + S^{2-}\ (aq)$	−1.01
$Mn^{2+}\ (aq) + 2e^- \longrightarrow Mn\ (s)$	−1.18
$V^{2+}\ (aq) + 2e^- \longrightarrow V\ (s)$	−1.18
$CdS\ (s) + 2e^- \longrightarrow Cd\ (s) + S^{2-}\ (aq)$	−1.21
$ZnS\ (s) + 2e^- \longrightarrow Zn\ (s) + S^{2-}\ (aq)$	−1.44
$Zr^{4+}\ (aq) + 4e^- \longrightarrow Zr\ (s)$	−1.53
$Al^{3+}\ (aq) + 3e^- \longrightarrow Al\ (s)$	−1.66
$H_2\ (g) + 2e^- \longrightarrow 2H^-\ (aq)$	−2.25
$Mg^{2+}\ (aq) + 2e^- \longrightarrow Mg\ (s)$	−2.37
$Na^+\ (aq) + e^- \longrightarrow Na\ (s)$	−2.714
$Ca^{2+}\ (aq) + 2e^- \longrightarrow Ca\ (s)$	−2.87
$Sr^{2+}\ (aq) + 2e^- \longrightarrow Sr\ (s)$	−2.89
$Ba^{2+}\ (aq) + 2e^- \longrightarrow Ba\ (s)$	−2.90
$Rb^+\ (aq) + e^- \longrightarrow Rb\ (s)$	−2.925
$K^+\ (aq) + e^- \longrightarrow K\ (s)$	−2.925
$Li^+\ (aq) + e^- \longrightarrow Li\ (s)$	−3.045

APPENDIX I (*continued*)

Standard Reduction Potentials in Aqueous Solution at 25 °C

Basic Solution	Standard Reduction Potential, E_0 (volts)
ClO^- (aq) + H_2O + $2e^-$ $\longrightarrow$ Cl^- (aq) + $2OH^-$ (aq)	0.89
OOH^- (aq) + H_2O + $2e^-$ $\longrightarrow$ $3OH^-$ (aq)	0.88
$2NH_2OH$ (aq) + $2e^-$ $\longrightarrow$ N_2H_4 (aq) + $2OH^-$ (aq)	0.74
ClO_3^- (aq) + $3H_2O$ + $6e^-$ $\longrightarrow$ Cl^- (aq) + $6OH^-$ (aq)	0.62
MnO_4^- (aq) + $2H_2O$ + $3e^-$ $\longrightarrow$ MnO_2 (s) + $4OH^-$ (aq)	0.588
MnO_4^- (aq) + e^- $\longrightarrow$ MnO_4^{2-} (aq)	0.564
NiO_2 (s) + $2H_2O$ + $2e^-$ $\longrightarrow$ $Ni(OH)_2$ (s) + $2OH^-$ (aq)	0.49
Ag_2CrO_4 (s) + $2e^-$ $\longrightarrow$ $2Ag$ (s) + CrO_4^{2-} (aq)	0.446
O_2 (g) + $2H_2O$ + $4e^-$ $\longrightarrow$ $4OH^-$ (aq)	0.40
ClO_4^- (aq) + H_2O + $2e^-$ $\longrightarrow$ ClO_3^- (aq) + $2OH^-$ (aq)	0.36
Ag_2O (s) + H_2O + $2e^-$ $\longrightarrow$ $2Ag$ (s) + $2OH^-$ (aq)	0.34
$2NO_2^-$ (aq) + $3H_2O$ + $4e^-$ $\longrightarrow$ N_2O (g) + $6OH^-$ (aq)	0.15
N_2H_4 (aq) + $2H_2O$ + $2e^-$ $\longrightarrow$ $2NH_3$ (aq) + $2OH^-$ (aq)	0.10
$[Co(NH_3)_6]^{3+}$ (aq) + e^- $\longrightarrow$ $[Co(NH_3)_6]^{2+}$ (aq)	0.10
HgO (s) + H_2O + $2e^-$ $\longrightarrow$ Hg (ℓ) + $2OH^-$ (aq)	0.0984
O_2 (g) + H_2O + $2e^-$ $\longrightarrow$ OOH^- (aq) + OH^- (aq)	0.076
NO_3^- (aq) + H_2O + $2e^-$ $\longrightarrow$ NO_2^- (aq) + $2OH^-$ (aq)	0.01
MnO_2 (s) + $2H_2O$ + $2e^-$ $\longrightarrow$ $Mn(OH)_2$ (s) + $2OH^-$ (aq)	−0.05
CrO_4^{2-} (aq) + $4H_2O$ + $3e^-$ $\longrightarrow$ $Cr(OH)_3$ (s) + $5OH^-$ (aq)	−0.12
$Cu(OH)_2$ (s) + $2e^-$ $\longrightarrow$ Cu (s) + $2OH^-$ (aq)	−0.36
S (s) + $2e^-$ $\longrightarrow$ S^{2-} (aq)	−0.48
$Fe(OH)_3$ (s) + e^- $\longrightarrow$ $Fe(OH)_2$ (s) + OH^- (aq)	−0.56
$2H_2O$ + $2e^-$ $\longrightarrow$ H_2 (g) + $2OH^-$ (aq)	−0.8277
$2NO_3^-$ (aq) + $2H_2O$ + $2e^-$ $\longrightarrow$ N_2O_4 (g) + $4OH^-$ (aq)	−0.85
$Fe(OH)_2$ (s) + $2e^-$ $\longrightarrow$ Fe (s) + $2OH^-$ (aq)	−0.877
SO_4^{2-} (aq) + H_2O + $2e^-$ $\longrightarrow$ SO_3^{2-} (aq) + $2OH^-$ (aq)	−0.93
N_2 (g) + $4H_2O$ + $4e^-$ $\longrightarrow$ N_2H_4 (aq) + $4OH^-$ (aq)	−1.15
$[Zn(OH)_4]^{2-}$ (aq) + $2e^-$ $\longrightarrow$ Zn (s) + $4OH^-$ (aq)	−1.22
$Zn(OH)_2$ (s) + $2e^-$ $\longrightarrow$ Zn (s) + $2OH^-$ (aq)	−1.245
$[Zn(CN)_4]^{2-}$ (aq) + $2e^-$ $\longrightarrow$ Zn (s) + $4CN^-$ (aq)	−1.26
$Cr(OH)_3$ (s) + $3e^-$ $\longrightarrow$ Cr (s) + $3OH^-$ (aq)	−1.30
SiO_3^{2-} (aq) + $3H_2O$ + $4e^-$ $\longrightarrow$ Si (s) + $6OH^-$ (aq)	−1.70

APPENDIX J

Selected Thermodynamic Values*

Species	ΔH_f°(298.15K) kJ/mol	S°(298.15K) J/K · mol	ΔG_f°(298.15K) kJ/mol
Aluminum			
$Al(s)$	0	28.3	0
$AlCl_3(s)$	−704.2	110.67	−628.8
$Al_2O_3(s)$	−1675.7	50.92	−1582.3
Barium			
$BaCl_2(s)$	−858.6	123.68	−810.4
$BaO(s)$	−553.5	70.42	−525.1
$BaSO_4(s)$	−1473.2	132.2	−1362.2
Beryllium			
$Be(s)$	0	9.5	0
$Be(OH)_2$	−902.5	51.9	−815.0
Bromine			
$Br(g)$	111.884	175.022	82.396
$Br_2(\ell)$	0	152.2	0
$Br_2(g)$	30.907	245.463	3.110
$BrF_3(g)$	−255.60	292.53	−229.43
$HBr(g)$	−36.40	198.695	−53.45
Calcium			
$Ca(s)$	0	41.42	0
$Ca(g)$	178.2	158.884	144.3
$Ca^{2+}(g)$	1925.90	—	—
$CaC_2(s)$	−59.8	69.96	−64.9
$CaCO_3$(s; calcite)	−1206.92	92.9	−1128.79
$CaCl_2(s)$	−795.8	104.6	−748.1
$CaF_2(s)$	−1219.6	68.87	−1167.3
$CaH_2(s)$	−186.2	42.	−147.2
$CaO(s)$	−635.09	39.75	−604.03
$CaS(s)$	−482.4	56.5	−477.4
$Ca(OH)_2(s)$	−986.09	83.39	−898.49
$Ca(OH)_2(aq)$	−1002.82	−74.5	−868.07
$CaSO_4(s)$	−1434.11	106.7	−1321.79
Carbon			
C(s, graphite)	0	5.740	0
C(s, diamond)	1.895	2.377	2.900
$C(g)$	716.682	158.096	671.257
$CCl_4(\ell)$	−135.44	216.40	−65.21
$CCl_4(g)$	−102.9	309.85	−60.59
$CHCl_3$(liq)	−134.47	201.7	−73.66
$CHCl_3(g)$	−103.14	295.71	−70.34
CH_4(g, methane)	−74.81	186.264	−50.72
C_2H_2(g, ethyne)	226.73	200.94	209.20
C_2H_4(g, ethene)	52.26	219.56	68.15
C_2H_6(g, ethane)	−84.68	229.60	−32.82
C_3H_8(g, propane)	−103.8	269.9	−23.49
C_6H_6(ℓ, benzene)	49.03	172.8	124.5
CH_3OH(ℓ, methanol)	−238.66	126.8	−166.27
CH_3OH(g, methanol)	−200.66	239.81	−161.96
C_2H_5OH(ℓ, ethanol)	−277.69	160.7	−174.78
C_2H_5OH(g, ethanol)	−235.10	282.70	−168.49
$CO(g)$	−110.525	197.674	−137.168
$CO_2(g)$	−393.509	213.74	−394.359

Selected Thermodynamic Values*

Species	ΔH_f°(298.15K) kJ/mol	S°(298.15K) J/K · mol	ΔG_f°(298.15K) kJ/mol
CS_2(g)	117.36	237.84	67.12
$COCl_2$(g)	−218.8	283.53	−204.6
Cesium			
Cs(s)	0	85.23	0
Cs^+(g)	457.964	—	—
CsCl(s)	−443.04	101.17	−414.53
Chlorine			
Cl(g)	121.679	165.198	105.680
Cl^-(g)	−233.13	—	—
Cl_2(g)	0	223.066	0
HCl(g)	−92.307	186.908	−95.299
HCl(aq)	−167.159	56.5	−131.228
Chromium			
Cr(s)	0	23.77	0
Cr_2O_3(s)	−1139.7	81.2	−1058.1
$CrCl_3$(s)	−556.5	123.0	−486.1
Copper			
Cu(s)	0	33.150	0
CuO(s)	−157.3	42.63	−129.7
$CuCl_2$(s)	−220.1	108.07	−175.7
Fluorine			
F_2(g)	0	202.78	0
F(g)	78.99	158.754	61.91
F^-(g)	−255.39	—	—
F^-(aq)	−332.63	−13.8	−278.79
HF(g)	−271.1	173.779	−273.2
HF(aq)	−332.63	−13.8	−278.79
Hydrogen			
H_2(g)	0	130.684	0
H(g)	217.965	114.713	203.247
H^+(g)	1536.202	—	—
H_2O(ℓ)	−285.830	69.91	−237.129
H_2O(g)	−241.818	188.825	−228.572
H_2O_2(ℓ)	−187.78	109.6	−120.35
Iodine			
I_2(s)	0	116.135	0
I_2(g)	62.438	260.69	19.327
I(g)	106.838	180.791	70.250
I^-(g)	−197.	—	—
ICl(g)	17.78	247.551	−5.46
Iron			
Fe(s)	0	27.78	0
FeO(s)	−272.	—	—
Fe_2O_3(s, hematite)	−824.2	87.40	−742.2
Fe_3O_4(s, magnetite)	−1118.4	146.4	−1015.4
$FeCl_2$(s)	−341.79	117.95	−302.30
$FeCl_3$(s)	−399.49	142.3	−344.00
FeS_2(s, pyrite)	−178.2	52.93	−166.9
$Fe(CO)_5$(ℓ)	−774.0	338.1	−705.3
Lead			
Pb(s)	0	64.81	0

APPENDIX J (*continued*)

Selected Thermodynamic Values*

Species	ΔH_f°(298.15K) kJ/mol	S°(298.15K) J/K · mol	ΔG_f°(298.15K) kJ/mol
$PbCl_2$(s)	−359.41	136.0	−314.10
PbO(s, yellow)	−217.32	68.70	−187.89
PbS(s)	−100.4	91.2	−98.7
Lithium			
Li(s)	0	29.12	0
Li^+(g)	685.783	—	—
LiOH(s)	−484.93	42.80	−438.95
LiOH(aq)	−508.48	2.80	−450.58
LiCl(s)	−408.701	59.33	−384.37
Magnesium			
Mg(s)	0	32.68	0
$MgCl_2$(s)	−641.32	89.62	−591.79
MgO(s)	−601.70	26.94	−569.43
$Mg(OH)_2$(s)	−924.54	63.18	−833.51
MgS(s)	−346.0	50.33	−341.8
Mercury			
Hg(ℓ)	0	76.02	0
$HgCl_2$(s)	−224.3	146.0	−178.6
HgO(s, red)	−90.83	70.29	−58.539
HgS(s, red)	−58.2	82.4	−50.6
Nickel			
Ni(s)	0	29.87	0
NiO(s)	−239.7	37.99	−211.7
$NiCl_2$(s)	−305.332	97.65	−259.032
Nitrogen			
N_2(g)	0	191.61	0
N(g)	472.704	153.298	455.563
NH_3(g)	−46.11	192.45	−16.45
N_2H_4(ℓ)	50.63	121.21	149.34
NH_4Cl(s)	−314.43	94.6	−202.87
NH_4Cl(aq)	−299.66	169.9	−210.52
NH_4NO_3(s)	−365.56	151.08	−183.87
NH_4NO_3(aq)	−339.87	259.8	−190.56
NO(g)	90.25	210.76	86.55
NO_2(g)	33.18	240.06	51.31
N_2O(g)	82.05	219.85	104.20
N_2O_4(g)	9.16	304.29	97.89
NOCl(g)	51.71	261.69	66.08
HNO_3(ℓ)	−174.10	155.60	−80.71
HNO_3(g)	−135.06	266.38	−74.72
HNO_3(aq)	−207.36	146.4	−111.25
Oxygen			
O_2(g)	0	205.138	0
O(g)	249.170	161.055	231.731
O_3(g)	142.7	238.93	163.2
Phosphorus			
P_4(s, white)	0	164.36	0
P_4(s, red)	−70.4	91.2	−48.4
P(g)	314.64	163.193	278.25
PH_3(g)	5.4	310.23	13.4
PCl_3(g)	−287.0	311.78	−267.8
P_4O_{10}(s)	−2984.0	228.86	−2697.7

APPENDIX J *(continued)*

Selected Thermodynamic Values*

Species	ΔH_f°(298.15K) kJ/mol	S°(298.15K) J/K · mol	ΔG_f°(298.15K) kJ/mol
H_3PO_4(s)	−1279.0	110.5	−1119.1
Potassium			
K(s)	0	64.18	0
KCl(s)	−436.747	82.59	−409.14
$KClO_3$(s)	−397.73	143.1	−296.25
KI(s)	−327.90	106.32	−324.892
KOH(s)	−424.764	78.9	−379.08
KOH(aq)	−482.37	91.6	−440.50
Silicon			
Si(s)	0	18.83	0
$SiBr_4$(ℓ)	−457.3	277.8	−443.9
SiC(s)	−65.3	16.61	−62.8
$SiCl_4$(g)	−657.01	330.73	−616.98
SiH_4(g)	34.3	204.62	56.9
SiF_4(g)	−1614.94	282.49	−1572.65
SiO_2(s, quartz)	−910.94	41.84	−856.64
Silver			
Ag(s)	0	42.55	0
Ag_2O(s)	−31.05	121.3	−11.20
AgCl(s)	−127.068	96.2	−109.789
$AgNO_3$(s)	−124.39	140.92	−33.41
Sodium			
Na(s)	0	51.21	0
Na(g)	107.32	153.712	76.761
Na^+(g)	609.358	—	—
NaBr(s)	−361.062	86.82	−348.983
NaCl(s)	−411.153	72.13	−384.138
NaCl(g)	−176.65	229.81	−196.66
NaCl(aq)	−407.27	115.5	−393.133
NaOH(s)	−425.609	64.455	−379.494
NaOH(aq)	−470.114	48.1	−419.150
Na_2CO_3(s)	−1130.68	134.98	−1044.44
Sulfur			
S(s, rhombic)	0	31.80	0
S(g)	278.805	167.821	238.250
S_2Cl_2(g)	−18.4	331.5	−31.8
SF_6(g)	1209.	291.82	−1105.3
H_2S(g)	−20.63	205.79	−33.56
SO_2(g)	−296.830	248.22	−300.194
SO_3(g)	−395.72	256.76	−371.06
$SOCl_2$(g)	−212.5	309.77	−198.3
H_2SO_4(ℓ)	−813.989	156.904	−690.003
H_2SO_4(aq)	−909.27	20.1	−744.53
Tin			
Sn(s, white)	0	51.55	0
Sn(s, gray)	−2.09	44.14	0.13
$SnCl_4$(ℓ)	−511.3	258.6	−440.1
$SnCl_4$(g)	−471.5	365.8	−432.2
SnO_2(s)	−580.7	52.3	−519.6
Titanium			
Ti(s)	0	30.63	0
$TiCl_4$(ℓ)	−804.2	252.34	−737.2

Selected Thermodynamic Values*

Species	ΔH_f°(298.15K) kJ/mol	S°(298.15K) J/K · mol	ΔG_f°(298.15K) kJ/mol
$TiCl_4(g)$	−763.2	354.9	−726.7
TiO_2	−939.7	49.92	−884.5
Zinc			
Zn(s)	0	41.63	0
$ZnCl_2(s)$	−415.05	111.46	−369.398
ZnO(s)	−348.28	43.64	−318.30
ZnS(s, sphalerite)	−205.98	57.7	−201.29

*Taken from "The NBS Tables of Chemical Thermodynamic Properties," 1982.

APPENDIX K
Answers to Exercises

CHAPTER 1

1. (a) 2.5 cm(1 meter/100. cm) = 0.025 m
   (b) 2.5 cm(10. mm/1 cm) = 25 mm
   (c) 2.5 cm (1 in/2.54 cm) = 0.98 in
2. (a) 750 mL (1 L/1000 mL) = 0.75 L
   (b) There are 4 quarts in 1 gallon. Therefore, 2.0 quarts = 0.50 gal and (0.50 gal)(3.785 L/1 gal) = 1.9 L
3. (a) (453.6 g/lb)(3.00 lb)(1 kg/1000 g) = 1.36 kg
   (b) (0.4536 kg/lb)(0.500 lb)(1.00×10^6 mg/kg) = 2.27×10^5 mg
   (c) (1.00 lb/0.4536 kg)(4.00 kg) = 8.82 lb
4. 1.0 kg = 1.0×10^3 g
   1.0×10^3 g(1 cm³/0.00112 g) = 8.93×10^5 cm³
5. t °F = (9 °F/5 °C)17 °C + 32 °F = 63 °F
6. (a) Temp in kelvin degrees = 25 °C + 273 = 298 K
   (b) 77 K − 273 = −196 °C
7. (a) 12.63 has 4 significant figures and 0.063 has 2 significant figures. Therefore, the product (0.80) has only 2 significant figures. The sum (12.69) must have only two decimal places.
   (b) 300 (a number with only 1 significant figure)
8. (a) 100. cm³ (13.6 g/cm³) = 1360 g
   1360 g (1 lb/453.6 g) = 3.00 lb
   (b) A layer 2.0 mm thick is 0.20 cm thick.
   Area of puddle = volume/thickness = 100. cm³/0.20 cm = 5.0×10^2 cm²
   5.0×10^2 cm² (1.0 in/2.54 cm)² = 78 in²
9. (a) (15.0 g metal/earring)(0.58 g gold/g metal) = 8.7 g gold/earring
   (b) (1.068/1.256)100 = 85.03%
   (c) (11/165)100 = 6.7%

CHAPTER 2

1. Symbol, name (atomic number): Cl, chlorine (17); Ca, calcium (20); Cr, chromium (24); Co, cobalt (27); Cu, copper (29); Cd, cadmium (48); Cs, cesium (55); Ce, cerium (58); Cm, curium (96); Cf, californium (98).
2. (a) Cu has 29 protons and 34 neutrons. A = 63.
   (b) Nickel-59 has 28 electrons, 28 protons, and (59 − 28) = 31 neutrons.
3. ${}^{28}_{14}Si$, ${}^{29}_{14}Si$, ${}^{30}_{14}Si$
4. 35.45 = (0.7577)(34.96885) + (0.2423)(36.96590)
5. (abund ${}^{69}Ga$)(68.9257) + (abund ${}^{71}Ga$)(70.9249) = 69.723
   abundance ${}^{69}Ga$ = 60.12%, and so abundance of ${}^{71}Ga$ = 39.88%
6. Bromine is a nonmetal in Group 7A (the halogens) and the 4th period.
7. (a) (2.5 mol Al)(27.0 g Al/mol) = 68 g
   (b) (1.00 lb Pb)(453.6 g/lb)(1 mol Pb/207.2 g) = 2.19 mol Pb
8. (0.125 mol Na)(22.99 g/mol)(1 cm³/0.968 g) = 2.97 cm³ Na
   Edge of cube = (2.97 cm³)$^{1/3}$ = 1.44 cm
9. (50.0 g U)(1 mol/238.0 g)(6.0221367×10^{23} atoms/mol) = 1.27×10^{23} atoms U
10. (1.0079 g/mol)(1 mol/6.0221367×10^{23} atoms) = 1.6737×10^{-24} g

CHAPTER 3

1. (a) $C_2H_2O_4$; (b) $C_{14}H_9Cl_5$
2. (a) K^+; (b) Se^{2-}; (c) Be^{2+}; (d) V^{2+} and V^{3+}; (e) Co^{2+} and Co^{3+}; (f) Cs^+
3. (a) one Na^+ and one F^-; (b) one Na^+ and one CH_3COO^-; (c) one Cu^{2+} and two NO_3^-
4. CrS (Cr^{2+} and S^{2-}) and Cr_2S_3 (Cr^{3+} and S^{2-})
5. (1a) NH_4NO_3; (1b) $CoSO_4$; (1c) $Ni(CN)_2$; (1d) V_2O_3; (1e) BaO; (1f) $Ca(ClO)_2$;
   (2a) magnesium bromide; (2b) lithium carbonate; (2c) potassium hydrogen sulfite; (2d) potassium permanganate; (2e) ammonium sulfide; (2f) copper(I) chloride and copper(II) chloride;
   (3a) CO_2; (3b) PI_3; (3c) SCl_2; (3d) BF_3; (3e) O_2F_2; (3f) XeO_3;
   (4a) dinitrogen tetrafluoride; (4b) hydrogen bromide; (4c) sulfur tetrafluoride; (4d) boron trichloride; (4e) diphosphorus pentoxide; (4f) chlorine trifluoride
6. (a) Limestone, $CaCO_3$, 40.08 g for Ca + 12.01 g for C + 3(16.00 g for O) = 100.09 g/mol
   (b) Caffeine, $C_8H_{10}N_4O_2$
   8(12.01 g) + 10(1.008 g) + 4(14.01 g) + 2(16.00 g) = 194.19 g/mol
7. (a) (1.00×10^3 g $CaCO_3$)(1 mol/100.1 g) = 9.99 mol
   (b) (2.50×10^{-3} mol caffeine)(194.2 g/mol) = 0.486 g
8. (a) (70. mg)(1.000 g/1000 mg)(1 mol/294 g) = 2.4×10^{-4} mol
   (b) (2.4×10^{-4} mol)(6.022×10^{23} molecules/mol) = 1.4×10^{20} molecules
   (1.4×10^{20} molecules)(14 C atoms/molecule) = 2.0×10^{21} C atoms
   (c) (294 g/mol)(1 mol/6.022×10^{23} molecules) = 4.88×10^{-22} g/molecule

9. (a) %Na = (23.0/58.44)100 = 39.4% Na
%Cl = (35.5/58.44)100 = 60.7% Cl
(b) %C = (12.0/16.04)100 = 75.0% C
%H = (4.00/16.04)100 = 25.0% H
(c) %N = [(2 × 14.0)/132.1)]100 = 21.2% N
%H = [(12 × 1.01)/132.1]100 = 6.10% H
%S = (32.1/132.1)100 = 24.3% S
%O = [(4 × 16.0)/132.1]100 = 48.4% O

10. 78.14 g B(1 mol B/10.811 g) = 7.228 mol B
21.86 g H(1 mol H/1.0079 g) = 21.69 mol H
Mole ratio of H to B = 21.69/7.228 = 3.001 H/1.000 B.
The empirical formula is BH_3.
The molecular weight is twice as large as the empirical formula weight (13.8 amu/formula unit). Therefore, the molecular formula is 2 × BH_3 = B_2H_6.

11. 2.145 g − 1.130 g = 1.015 g H_2O lost on heating (= 0.05634 mol H_2O)
Moles of $Na_2B_4O_7$ = 1.130 g(1 mol $Na_2B_4O_7$/201.22 g) = 0.005616 mol
Mole ratio = 0.05634 mol H_2O/0.005616 mol $Na_2B_4O_7$ = 10.03/1.000
The formula of borax is $Na_2B_4O_7 \cdot 10\ H_2O$.

12. (0.532 g Ti)(1 mol/47.88 g) = 0.0111 mol Ti
Mass of Cl = 2.108 g − 0.532 g = 1.576 g Cl
(1.576 g Cl)(1 mol Cl/35.453 g) = 0.004445 mol Cl
mol Cl/mol Ti = 4.01 Cl/1 Ti.
Empirical formula = $TiCl_4$

13. (12.0 g $Pt(NH_3)_2Cl_2$)(1 mol/300.1 g) = 0.400 mol $Pt(NH_3)_2Cl_2$
(0.0400 mol $Pt(NH_3)_2Cl_2$)(2 mol Cl/1 mol $Pt(NH_3)_2Cl_2$)(35.453 g Cl/mol) = 2.84 g Cl

CHAPTER 4

1. (a) 4 Fe(s) + 3 O_2(g) → 2 Fe_2O_3(s)
(b) CH_4(g) + 2 O_2(g) → CO_2(g) + 2 H_2O(g)
(c) 2 B_4H_{10}(ℓ) + 11 O_2(g) → 4 B_2O_3(s) + 10 H_2O(g)
(d) CO(g) + 2 H_2(g) → CH_3OH(ℓ)
(e) 2 C_8H_{18}(ℓ) + 25 O_2(g) → 16 CO_2(g) + 18 H_2O(ℓ)

2. (a) 2 Ga(s) + 3 Cl_2(g) → 2 $GaCl_3$(s)
(b) 4 Cr(s) + 3 O_2(g) → 2 Cr_2O_3(s)
(c) N_2(g) + 2 O_2(g) → 2 NO_2(g)
(d) $SrCO_3$(s) → SrO(s) + CO_2(g)

3. 1.05 g Fe (1 mol/55.85 g) = 0.0188 mol Fe
Balanced equation: 4 Fe(s) + 3 O_2(g) → 2 Fe_2O_3(s)
(0.0188 mol Fe)(2 mol Fe_2O_3/4 mol Fe) = 0.00940 mol Fe_2O_3 produced
(0.00940 mol Fe_2O_3)(159.7 g/mol) = 1.50 g Fe_2O_3
(0.0188 mol Fe)(3 mol O_2/4 mol Fe) = 0.0141 mol O_2 required
(0.0141 mol O_2)(32.00 g/mol) = 0.451 g O_2 required

4. 2.70 g Al = 0.100 mol and 4.05 g Cl_2 = 0.0571 mol
Moles-available ratio = (0.100 mol Al/0.0571 mol Cl_2) = 1.75 mol Al/1 mol Cl_2
Moles-required ratio = 2 mol Al/3 mol Cl_2 = 0.67 mol Al/1 mol Cl_2
This means that Al is present in excess, so Cl_2 is the limiting reagent. The yield of product is calculated from the limiting reagent.
(0.0571 mol Cl_2)(2 mol $AlCl_3$/3 mol Cl_2) = 0.0381 mol $AlCl_3$
(0.0381 mol $AlCl_3$)(133.3 g/mol) = 5.08 g $AlCl_3$ expected

5. (18.9 g $NaBH_4$)(1 mol/37.83 g) = 0.500 mol
0.500 mol(2 mol B_2H_6/3 mol $NaBH_4$) = 0.333 mol B_2H_6
0.333 mol B_2H_6(27.67 g/mol) = 9.22 g
% yield of B_2H_6 = (7.50 g/9.22 g)100 = 81.4%

6. (1.542 g CO_2)(1 mol/44.010 g)(1 mol C/1 mol CO_2) = 0.03504 mol C
(0.03504 mol C)(12.011 g C/mol C) = 0.4208 g C
(0.315 g H_2O)(1 mol/18.02 g)(2 mol H/1 mol H_2O) = 0.0350 mol H
(0.0350 mol H)(1.008 g H/mol) = 0.0352 g H
0.652 g ferrocene − 0.4208 g C − 0.0352 g H = 0.196 g Fe
(0.196 g Fe)(1 mol/55.85 g) = 0.00351 mol Fe
There are equal numbers of moles of C and H, but only 0.10 as many moles of Fe.
Therefore, the empirical formula is $C_{10}H_{10}Fe$.

7. 2.357 g before − 2.108 g after = 0.249 g H_2O
(0.249 g H_2O)(1 mol H_2O/18.02 g) = 0.0138 mol H_2O
(0.0138 mol H_2O)(1 mol $BaCl_2 \cdot 2\ H_2O$/2 mol H_2O) = 0.00691 mol $BaCl_2 \cdot 2\ H_2O$
(0.00691 mol $BaCl_2 \cdot 2\ H_2O$)(244.3 g/mol) = 1.69 g $BaCl_2 \cdot 2\ H_2O$
Weight % = (1.69 g $BaCl_2 \cdot 2\ H_2O$/2.357 g)100 = 71.6%

CHAPTER 5

1. (a) Soluble, Na^+, Br^-
(b) Insoluble
(c) Soluble, 2 K^+, CO_3^{2-}
(d) Insoluble
(e) Soluble, 2 NH_4^+, SO_4^{2-}

2. (a) KNO_3; (b) NiS; (c) $Ni(NO_3)_2$. In (a) and (c) nitrates are always good choices for water-soluble compounds. In (b), sulfides are almost always insoluble (except for those with the alkali metal cations or NH_4^+).

3. (a) $BaCl_2$ + Na_2SO_4 → $BaSO_4$ + 2 NaCl
Balanced complete ionic equation:

$$Ba^{2+}(aq) + 2\ Cl^-(aq) + 2\ Na^+(aq) + SO_4^{2-}(aq) \longrightarrow BaSO_4(s) + 2\ Na^+(aq) + 2\ Cl^-(aq)$$

Net ionic equation: $Ba^{2+}(aq) + SO_4^{2-}(aq) \longrightarrow BaSO_4(s)$

(b) $(NH_4)_2S + Cd(NO_3)_2 \rightarrow CdS + 2\ NH_4NO_3$
Balanced complete ionic equation:

$$2\ NH_4^+(aq) + S^{2-}(aq) + Cd^{2+}(aq) + 2\ NO_3^-(aq) \longrightarrow CdS(s) + 2\ NH_4^+(aq) + 2\ NO_3^-(aq)$$

Net ionic equation:

$$Cd^{2+}(aq) + S^{2-}(aq) \longrightarrow CdS(s)$$

(c) $Pb(NO_3)_2 + 2\ KCl \rightarrow PbCl_2 + 2\ KNO_3$
Balanced complete ionic equation:

$$Pb^{2+}(aq) + 2\ NO_3^-(aq) + 2\ K^+(aq) + 2\ Cl^-(aq) \longrightarrow PbCl_2(s) + 2\ K^+(aq) + 2\ NO_3^-(aq)$$

Net ionic equation:

$$Pb^{2+}(aq) + 2\ Cl^-(aq) \longrightarrow PbCl_2(s)$$

4. (a) $AgNO_3(aq) + LiCl(aq) \rightarrow AgCl(s) + LiNO_3(aq)$
$Ag^+(aq) + Cl^-(aq) \rightarrow AgCl(s)$
(b) $NiCl_2(aq) + Na_2S(aq) \rightarrow 2\ NaCl(aq) + NiS(s)$
$Ni^{2+}(aq) + S^{2-}(aq) \rightarrow NiS(aq)$

5. (a) $H_2SO_4(aq) + 2\ KOH(aq) \rightarrow K_2SO_4(aq) + 2\ H_2O(\ell)$
(b) $NH_3(aq) + HCl(aq) \rightarrow NH_4Cl(aq)$

6. $Na_2CO_3(aq) + 2\ HNO_3(aq) \rightarrow 2\ NaNO_3(aq) + CO_2(g) + H_2O(\ell)$

7. (a) $H_2SO_4(aq) + 2\ CsOH(aq) \rightarrow Cs_2SO_4(aq) + 2\ H_2O(\ell)$, acid-base reaction
(b) $MgCl_2(aq) + Na_2CO_3(aq) \rightarrow MgCO_3(s) + 2\ NaCl(aq)$, precipitation reaction
(c) $2\ HNO_3(aq) + Ca(OH)_2(aq) \rightarrow Ca(NO_3)_2(aq) + 2\ H_2O(\ell)$, acid-base reaction
(d) $CdCl_2(aq) + Na_2S(aq) \rightarrow 2\ NaCl(aq) + CdS(s)$, precipitation reaction

8. (a) $NaOH(aq) + HCl(aq) \rightarrow NaCl(aq) + H_2O(\ell)$
(b) $KOH(aq) + HNO_3(aq) \rightarrow KNO_3(aq) + H_2O(\ell)$, acid-base
$K_2CO_3(aq) + 2\ HNO_3(aq) \rightarrow 2\ KNO_3(aq) + H_2O(\ell) + CO_2(g)$, gas-forming
(c) $Ba(OH)_2(s) + 2\ HNO_3(aq) \rightarrow Ba(NO_3)_2(aq) + 2\ H_2O(\ell)$, acid-base
$BaCO_3(s) + 2\ HNO_3(aq) \rightarrow Ba(NO_3)_2(aq) + H_2O(\ell) + CO_2(g)$, gas-forming
(d) $FeCl_2(aq) + (NH_4)_2S(aq) \rightarrow 2\ NH_4Cl(aq) + FeS(s)$

9. 26.3 g $NaHCO_3$(1 mol/84.01 g) = 0.313 mol $NaHCO_3$
0.313 mol/0.200 L = 1.57 M = $[NaHCO_3]$

10. (a) LiCl dissociates in water to give Li^+ and Cl^-.

$$LiCl(aq) \longrightarrow Li^+(aq) + Cl^-(aq)$$

Therefore, 1 M LiCl means that $[Li^+] = [Cl^-] =$ 1 M and the total ion concentration is 2 M.
(b) Ammonium sulfate dissociates according to the equation

$$(NH_4)_2SO_4(aq) \longrightarrow 2\ NH_4^+(aq) + SO_4^{2-}(aq)$$

Therefore, $[NH_4^+] = 2 \times 0.5$ M = 1.0 M and $[SO_4^{2-}] = 0.5$ M
Total ion concentration = 1.5 M

11. Moles of $KMnO_4$ required = (0.500 L)(0.0200 mol/L) = 0.0100 mol $KMnO_4$
0.0100 mol $KMnO_4$(158.0 g/mol) = 1.58 g
Place 1.58 g $KMnO_4$ in a 500.-mL volumetric flask, add water slowly, shaking to ensure that the $KMnO_4$ dissolves. Fill with water to the mark on the flask.

12. If you require 300. mL of 1.00 M NaOH, this means you need
(0.300 L)(1.00 mol/L) = 0.300 mol NaOH.
This amount of NaOH can be taken from the more concentrated solution. The volume of more concentrated NaOH required is
(0.300 mol NaOH required)(1.00 L/3.00 mol NaOH) = 0.100 L or 100. mL
Therefore, take 100. mL of 3.00 M NaOH and carefully add 200. mL water to bring the total volume of the diluted solution to 300. mL.

13. Balanced equation: $2\ HCl(aq) + Na_2CO_3(aq) \rightarrow H_2O(\ell) + CO_2(g) + 2\ NaCl(aq)$
(0.025 L)(0.750 mol/L) = 0.0188 mol Na_2CO_3
(0.0188 mol Na_2CO_3)(106.0 g/mol) = 1.99 g Na_2CO_3 req'd
0.0188 mol Na_2CO_3(2 mol NaCl/1 mol Na_2CO_3) = 0.300 mol NaCl
0.300 mol NaCl(58.44 g/mol) = 17.5 g NaCl produced

14. (0.02833 L)(0.953 mol/L) = 0.0270 mol NaOH
(0.027 mol NaOH)(1 mol CH_3COOH/1 mol NaOH) = 0.0270 mol CH_3COOH
(0.0270 mol CH_3COOH)(60.05 g/mol) = 1.62 g CH_3COOH
0.0270 mol CH_3COOH/0.0250 L = 1.08 M

15. Balanced equation: $HCl(aq) + NaOH(aq) \rightarrow NaCl(aq) + H_2O(\ell)$
Moles HCl = (0.03256 L)(0.100 mol/L) = 0.00326 mol HCl
(0.00326 mol HCl)(1 mol NaOH/1 mol HCl) = 0.00326 mol NaOH
0.00326 mol NaOH/0.0250 L = 0.130 M = [NaOH]

16. (a) (0.560 g $BaSO_4$)(1 mol/233.4 g)(1 mol Ba/1 mol $BaSO_4$) = 0.00240 mol Ba
(0.00240 mol Ba)(137.3 g/mol) = 0.329 g Ba
Weight % Ba = (0.329 g Ba/1.023 g mixture)100 = 32.2% Ba
(b) 0.00240 mol Ba(1 mol $BaCl_2$/1 mol Ba)(208.2 g $BaCl_2$/1 mol) = 0.500 g $BaCl_2$

17. (a) Mg = +2 and O = −2
(b) P = −3 and H = +1
(c) Al = +3 and H = −1

(d) H = +1 and S = −2
(e) Cl = +5 and O = −2
(f) S = +2 and O = −2
(g) Na = +1 and Cl = −1
(h) H = +1, N = +5, and O = −2
(i) Zn = +2 and S = −2

18. (a) S oxidation number changes from 0 to +4, and the oxidation number of oxygen changes from 0 in O_2 to −2 in the product. S is oxidized and O_2 is reduced.
(b) Fe oxidation number changes from 0 to +2, while that of Cl changes from 0 to −1. Iron is oxidized and Cl is reduced.
(c) No elements change in oxidation number. This is not an oxidation-reduction reaction; it is an acid-base reaction.
(d) S changes in oxidation number from −2 in ZnS to +6 in the SO_4^{2-} ion in $ZnSO_4$. N changes from +5 in HNO_3 to +2 in NO. Thus, ZnS is oxidized to $ZnSO_4$ and HNO_3 is reduced to NO.

19. $2\,[Cr^{2+}(aq) \longrightarrow Cr^{3+}(aq) + e^-]$
$2\,e^- + I_2(aq) \longrightarrow 2\,I^-(aq)$
$2\,Cr^{2+}(aq) + I_2(aq) \longrightarrow Cr^{3+}(aq) + 2\,I^-(aq)$

(a) I_2, oxidizing agent (gains electrons); (b) Cr^{2+}, reducing agent (loses electrons); (c) I_2, substance reduced (oxidation number declines); (d) Cr^{2+}, substance oxidized (oxidation number increases).

20. $Cu(s) \longrightarrow Cu^{2+}(aq) + 2\,e^-$
$2\,[e^- + 2\,H^+(aq) + NO_3^-(aq) \longrightarrow NO_2(g) + H_2O(\ell)]$
$Cu(s) + 4\,H^+(aq) + 2\,NO_3^-(aq) \longrightarrow Cu^{2+}(aq) + 2\,NO_2(g) + 2\,H_2O(\ell)]$

Cu is the reducing agent and so is oxidized. NO_3^- is the oxidizing agent (as it often is in acid solution) and so is reduced.

21. $2\,[Cr(s) + 3\,OH^-(aq) \longrightarrow Cr(OH)_3(s) + 3\,e^-]$
$3\,[ClO_4^-(aq) + H_2O(\ell) + 2\,e^- \longrightarrow ClO_3^-(aq) + 2\,OH^-(aq)]$
$2\,Cr(s) + 3\,ClO_4^-(aq) + 3\,H_2O(\ell) \longrightarrow 2\,Cr(OH)_3(s) + 3\,ClO_3^-(aq)$

Cr is the reducing agent and so is oxidized. ClO_4^- is the oxidizing agent (as it often is in acid solution) and so is reduced.

22. (a) $2\,S_2O_3^{2-}(aq) \rightarrow S_4O_6^{2-}(aq) + 2\,e^-$
Oxidation; $S_2O_3^{2-}$ is a reducing agent, and it is oxidized.
(b) $NO_3^-(aq) + 4\,H^+(aq) + 3\,e^- \rightarrow NO(g) + 2\,H_2O(\ell)$
Reduction; NO_3^- is an oxidizing agent and is itself reduced.
(c) $C_2H_5OH(aq) \rightarrow C_2H_4O(aq) + 2\,H^+(aq) + 2\,e^-$
Oxidation; C_2H_5OH (ethyl alcohol) is a reducing agent and is oxidized.
(d) $PH_3(aq) + 4\,H_2O(\ell) \rightarrow H_3PO_4\,(aq) + 8\,H^+(aq) + 8\,e^-$
Oxidation; PH_3 is a reducing agent and is oxidized.
(e) $ClO^-(aq) + H_2O(\ell) + 2\,e^- \rightarrow Cl^-(aq) + 2\,OH^-(aq)$
Reduction; ClO^- is an oxidizing agent and is itself reduced.

23. (a) $2\,S_2O_3^{2-}(aq) \longrightarrow S_4O_6^{2-}(aq) + 2\,e^-$ (red. ag.)
$I_2(aq) + 2\,e^- \longrightarrow 2\,I^-(aq)$ (ox. ag.)
$2\,S_2O_3^{2-}(aq) + I_2(aq) \longrightarrow S_4O_6^{2-}(aq) + 2\,I^-(aq)$
(b) $3\,[Cu(s) \longrightarrow Cu^{2+}(aq) + 2\,e^-]$ (red. ag).
$2\,[NO_3^-(aq) + 4\,H^+(aq) + 3\,e^- \longrightarrow NO(g) + 2\,H_2O(\ell)]$ (ox. ag.)
$3\,Cu(s) + 2\,NO_3^-(aq) + 8\,H^+(aq) \longrightarrow 2\,NO(g) + 4\,H_2O(\ell) + 3\,Cu^{2+}(aq)$
(c) $5\,[Fe^{2+}(aq) \longrightarrow Fe^{3+}(aq) + e^-]$ (red. ag.)
$MnO_4^-(aq) + 8\,H^+(aq) + 5\,e^- \longrightarrow Mn^{2+}(aq) + 4\,H_2O$ (ox. ag.)
$5\,Fe^{2+}(aq) + MnO_4^-(aq) + 8\,H^+(aq) \longrightarrow Mn^{2+}(aq) + 5\,Fe^{3+}(aq) + 4\,H_2O(\ell)$
(d) $2\,[Cr(OH)_3(s) + 5\,OH^-(aq) \longrightarrow CrO_4^{2-}(aq) + 4\,H_2O(\ell) + 3\,e^-]$ (red. ag.)
$3\,[2\,ClO^-(aq) + 2\,H_2O(\ell) + 2\,e^- \longrightarrow Cl_2(aq) + 4\,OH^-(aq)]$ (ox. ag.)
$2\,Cr(OH)_3(s) + 6\,ClO^-(aq) \longrightarrow 2\,CrO_4^{2-}(aq) + 2\,H_2O(\ell) + 3\,Cl_2(aq) + 2\,OH^-(aq)$

24. Moles $KMnO_4$ = (0.02434 L)(0.0200 mol/L) = 0.000487 mol $KMnO_4$
(0.000487 mol $KMnO_4$)(5 mol Fe^{2+}/1 mol $KMnO_4$) = 0.00243 mol Fe^{2+}
(0.00243 mol Fe^{2+})(55.85 g/mol) = 0.136 g Fe
Weight % = (0.136 g Fe/1.026 g sample)100 = 13.2% Fe

CHAPTER 6

1. 160 Calories (1000 calories/1 Calorie) = 1.6×10^5 cal
1.6×10^5 cal (4.184 J/cal) = 6.7×10^5 J

2. 24.1×10^3 J = (250. g)(0.902 J/g · K)(t_{final} −37.0 °C)
t_{final} = 143.9 °C

3. Heat transferred from coffee = (250. g)(4.20 J/g · deg)(45 − 60.)°C = -1.6×10^4 J
Heat transferred to aluminum = 1.6×10^4 J = (250. g)(0.902 J/g · deg)(45 °C − $t_{initial}$)
$t_{initial}$ = −25 °C

4. (237 cm³)(0.917 g/cm³) = 217.3 g ice (or water)
(217.3 g)(333 J/g ice) = 72.4×10^3 J to melt ice
(217.3 g)(4.18 J/g · deg)(25 °C − 0 °C) = 23×10^3 J
Total heat required = 95 kJ

5. The text indicates that 242 kJ are required to decompose 1.00 mol or 18.0 g of $H_2O(g)$ to $H_2(g)$ and $O_2(g)$.

This relation provides the necessary conversion factor.

$(12.6 \text{ g } H_2O)(242 \text{ kJ}/18.0 \text{ g } H_2O) = 169 \text{ kJ required}$

6. $3\,[H_2(g) + \frac{1}{2}\,O_2(g) \longrightarrow H_2O(\ell)]$
$2\,[\text{C(graphite)} + O_2(g) \longrightarrow CO_2(g)]$
$\underline{C_2H_5OH(liq) \longrightarrow \frac{1}{2}\,O_2(g) + 3\,H_2(g) + 2\,\text{C(graph)}}$
$C_2H_5OH(liq) + 3\,O_2(g) \longrightarrow 2\,CO_2(g) + 3\,H_2O(\ell)$

$3\,[\Delta H = -\,285.8 \text{ kJ}]$
$2\,[\Delta H = -393.5 \text{ kJ}]$
$\underline{\Delta H = +\,277.7 \text{ kJ}}$
$\Delta H = -1366.7 \text{ kJ}$

7. (a) $Ag(s) + \frac{1}{2}\,Cl_2(g) \rightarrow AgCl(s) \quad \Delta H_f^\circ = -127.1 \text{ kJ}$
(b) $\text{C(graphite)} + 2\,H_2(g) + \frac{1}{2}\,O_2(g) \rightarrow CH_3OH(\ell)$
$\Delta H_f^\circ = -238.7 \text{ kJ}$

8. $Mg(s) + \frac{1}{2}\,O_2(g) \longrightarrow MgO(s)$
$Mg(OH)_2(s) \longrightarrow Mg(s) + O_2(g) + H_2(g)$
$\underline{H_2(g) + \frac{1}{2}\,O_2(g) \longrightarrow H_2O(\ell)}$
$Mg(OH)_2(s) \longrightarrow MgO(s) + H_2O(\ell)$

$\Delta H^\circ = 601.70 \text{ kJ}$
$\Delta H^\circ = 924.54 \text{ kJ}$
$\underline{\Delta H^\circ = -285.830 \text{ kJ}}$
$\Delta H^\circ = 1240.41 \text{ kJ}$

9. Balanced equation $C_3H_8(g) + 5\,O_2(g) \rightarrow 3\,CO_2(g) + 4\,H_2O(\ell)$

$\Delta H^\circ_{\text{reaction}} = 3\,\Delta H_f^\circ[CO_2(g)] + 4\,\Delta H_f^\circ[H_2O(\ell)] - \Delta H_f^\circ[C_3H_8(g)] = 3 \text{ mol}(-393.5 \text{ kJ/mol}) + 4 \text{ mol}(-285.8 \text{ kJ/mol}) - 1 \text{ mol}(-103.8 \text{ kJ/mol}) = -2219.9 \text{ kJ}$

10. Change in temperature = 27.32 deg − 25.00 deg = 2.32 deg
Heat transferred to calorimeter water
$= (1.50 \times 10^3 \text{ g})(4.184 \text{ J/g} \cdot \text{deg})(2.32 \text{ deg})$
$= 14.6 \times 10^3 \text{ J}$
Heat transferred to calorimeter bomb
$= (837 \text{ J/deg})(2.32 \text{ deg})$
$= 1.94 \times 10^3 \text{ J}$
Total heat transferred by 1.00 g of sucrose = −16.5 kJ
Heat of combustion per mole =
$(-16.5 \text{ kJ/g})\,(342.3 \text{ g/mol}) = -5650 \text{ kJ/mol}$

11. Mass of final solution = 400. g
$q = (4.2 \text{ J/g} \cdot \text{K})(400.\text{ g})(26.60 - 25.10)°\text{C} = 2.5 \times 10^3 \text{ J}$ for 0.0800 mol HCl
2.5 kJ/0.0800 mol of HCl = 32 kJ/mol of HCl; ΔH of neutralization = −32 kJ/mol

CHAPTER 7

1. Part 1: ${}^{237}_{93}Np \rightarrow {}^{4}_{2}He + {}^{223}_{91}Pa$
Part 2: (a) ${}^{221}_{88}Ra \rightarrow {}^{4}_{2}He + {}^{217}_{86}Rn$
(b) ${}^{220}_{86}Rn \rightarrow {}^{4}_{2}He + {}^{216}_{84}Po$

2. Part 1: ${}^{35}_{16}S \rightarrow {}^{0}_{-1}e + {}^{35}_{17}Cl$
Part 2: ${}^{3}_{1}H \rightarrow {}^{0}_{-1}e + {}^{3}_{2}He$
${}^{60}_{26}Fe \rightarrow {}^{0}_{-1}e + {}^{60}_{27}Co$

3. ${}^{235}_{92}U \rightarrow {}^{4}_{2}He + {}^{231}_{90}Th$
${}^{231}_{90}Th \rightarrow {}^{0}_{-1}e + {}^{231}_{91}Pa$
${}^{231}_{91}Pa \rightarrow {}^{4}_{2}He + {}^{227}_{89}Ac$
${}^{227}_{89}Ac \rightarrow {}^{4}_{2}He + {}^{223}_{87}Fr$
${}^{223}_{87}Fr \rightarrow {}^{0}_{-1}e + {}^{223}_{88}Ra$

4. positron emission: ${}^{13}_{7}N \rightarrow {}^{13}_{6}C + {}^{0}_{+1}e$
K-capture ${}^{41}_{20}Ca + {}^{0}_{-1}e \rightarrow {}^{41}_{19}K$
β-emission ${}^{90}_{38}Sr \rightarrow {}^{0}_{-1}e + {}^{90}_{39}Y$
positron emission: ${}^{11}_{6}C \rightarrow {}^{11}_{5}B + {}^{0}_{+1}e$
proton emission ${}^{43}_{21}Sc \rightarrow {}^{1}_{1}H + {}^{42}_{20}Ca$

5. β-decay: ${}^{32}_{14}Si \rightarrow {}^{0}_{-1}e + {}^{32}_{15}P$
positron emission: ${}^{45}_{22}Ti \rightarrow {}^{0}_{+1}e + {}^{45}_{21}Sc$
α-emission: ${}^{239}_{94}Pu \rightarrow {}^{4}_{2}He + {}^{235}_{92}U$

6. Mass defect = Δm = −0.03260 g/mol
$\Delta E = (-32.60 \times 10^{-6} \text{ kg/mol})(3.00 \times 10^8 \text{ m/s})^2 = -2.93 \times 10^{12} \text{ J/mol}$

7. (a) ${}^{90}_{38}Sr \rightarrow {}^{0}_{-1}e + {}^{90}_{39}Y$
(b) Disintegrations are reduced to 1000 after 1 half-life, to 500 after 2 half-lives, to 250 after 3 half-lives and to 125 after 4 half-lives. Thus, a total of $4 \times 28 = 112$ years is required.

8. $k = 0.693/t_{1/2} = 8.90 \times 10^{-3} \text{ hr}^{-1}$
$\ln(0.10/1.0) = -kt = -(8.90 \times 10^{-3} \text{ hr}^{-1})t$
$t = 259 \text{ hr}$
Activity = kN
$N = (1.5 \times 10^{-6} \text{ g})(1 \text{ mol}/69.7 \text{ g})(6.022 \times 10^{23} \text{ atoms/mol}) = 1.3 \times 10^{16}$ atoms
Activity = $(8.90 \times 10^{-3} \text{ hr}^{-1})(1 \text{ hr}/3600 \text{ sec})(1.3 \times 10^{16} \text{ atoms}) = 3.2 \times 10^{10}$ per second

9. $\ln(12.9/14.0) = -(1.21 \times 10^{-4}/\text{yr})t$
$t = 676$ years
1990 − 676 = 1314 AD

10. (a) ${}^{13}_{6}C + {}^{1}_{0}n \rightarrow {}^{4}_{2}He + {}^{10}_{4}Be$
(b) ${}^{14}_{7}N + {}^{4}_{2}He \rightarrow {}^{1}_{0}n + {}^{17}_{9}F$
(c) ${}^{253}_{99}Es + {}^{4}_{2}He \rightarrow {}^{1}_{0}n + {}^{256}_{101}Md$

CHAPTER 8

1. 104.5 MHz = $104.5 \times 10^6 \text{ s}^{-1}$
$\lambda = c/\nu = (2.998 \times 10^8 \text{ m/s})(1\text{s}/104.5 \times 10^6) = 2.869 \text{ m}$

2. (a) Highest frequency, violet; (b) lowest frequency, red; (c) microwave frequency is greater than FM radio.

3. (a) λ = 10 cm; (b) 6.67 cm; (c) 2 waves, 5 nodes.

4. $\nu = c/\lambda = (2.998 \times 10^8 \text{ m/s})/[(2.36 \text{ nm})(1 \text{ m}/10^9 \text{ nm})] = 1.27 \times 10^{17} \text{ s}^{-1}$
x-ray photon energy $= h\nu = (6.6260755 \times 10^{-34} \text{ J} \cdot \text{s})(1.27 \times 10^{17} \text{ s}^{-1}) = 8.42 \times 10^{-27} \text{ J}$
The energy of one photon of orange light (3.2×10^{-19} J) is less than that of an x-ray photon.

5. $E_3 = -(Rhc/3^2) = (-2.1799 \times 10^{-18}$ J/atom$)/(3^2) = -2.4221 \times 10^{-19}$ J/atom
(see Figure 8.11)
$E_3 = (-2.4221 \times 10^{-19}$ J/atom$)(6.0221367 \times 10^{23}$ atoms$)(1$ kJ/1000 J$) = -145.86$ kJ/mol
6. The Brackett series has $n_{final} = 4$ and $n_{initial} = 5$.
$\Delta E = -Rhc[(1/4)^2 - (1/5)^2] = -0.0225Rhc = -0.0225(1313$ kJ/mol$) = -29.54$ kJ/mol
$\nu = \Delta E/h = 7.404 \times 10^{13}$ s^{-1}
$\lambda = c/\nu = 4.049 \times 10^{-6}$ m or 4049 nm.
Light of this wavelength lies in the near infrared.
7. (a) To calculate the wavelength (where $\lambda = h/mv$) you must first find the velocity of the neutron.

$$v = [2E/m]^{1/2} = [2(6.21 \times 10^{-21}\ \text{kg}\cdot\text{m}^2/\text{s}^2)/(1.675 \times 10^{-27}\ \text{kg})]^{1/2}$$
$$v = 2723\ \text{m/s}$$

(b) $\lambda = h/mv = (6.626 \times 10^{-34}\ \text{kg}\cdot\text{m}^2/\text{s}^2)(\text{s})/(1.675 \times 10^{-27}\ \text{kg})(2723\ \text{m/s})$
$\lambda = 1.453 \times 10^{-10}$ m
8. (a) $\ell = 0$ and 1
(b) $m_\ell = -1, 0, +1$; subshell label = p.
(c) When $\ell = 2$, the subshell label is d.
(d) $\ell = 0$ and $m_\ell = 0$ for an s orbital.
(e) There are always 3 orbitals in a p subshell.
(f) For an f subshell, there are 7 values of m_ℓ and there are 7 orbitals.
9. (a)

Orbital Label	n	ℓ
6s	6	0
4p	4	1
5d	5	2
4f	4	3

(b) A 4p orbital has 1 nodal plane ($\ell = 1$), whereas there are 2 ($\ell = 2$) for 6d.

CHAPTER 9

1. (a) Configuration is for an element in 3rd period and Group 7A. Therefore, the element is Cl.
(b) Sulfur: $1s^22s^22p^63s^23p^4$

⇅	⇅	⇅ ⇅ ⇅	⇅	⇅ ↑ ↑
1s	2s	2p	3s	3p

(c) $n = 3$, $\ell = 1$, $m_\ell = +1$, and $m_s = +\frac{1}{2}$
2. See Table 9.2.
3. V^{2+}, paramagnetic, 3 unpaired electrons.

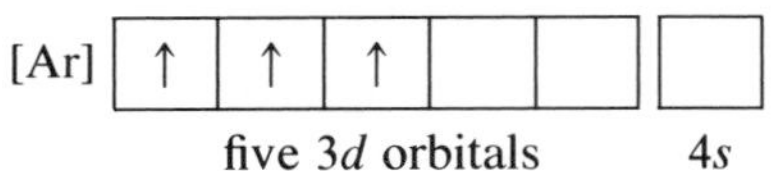

V^{3+}, paramagnetic, 2 unpaired electrons

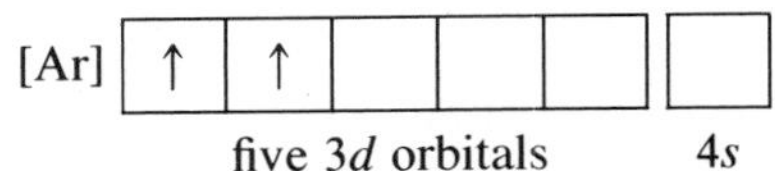

Co^{3+}, paramagnetic, 4 unpaired electrons

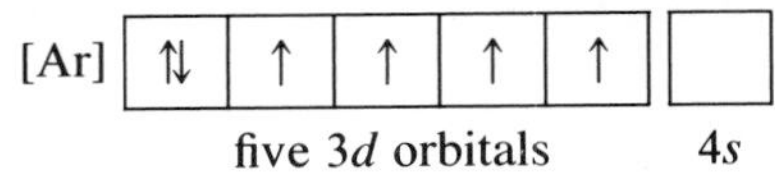

4. (a) H—O = 37 pm + 73 pm = 110 pm; H—S = 37 pm + 103 pm = 140 pm
(b) Br radius is 114 pm, Cl radius is 100 pm, so Br—Cl is 214 pm.
5. (a) Radius: C < Si < Al
(b) Ionization energy: Al < Si < C
(c) Si has more negative EA.
6. Ion sizes: $N^{3-} > O^{2-} > F^-$. All ions have 10 electrons in this isoelectronic series. The nuclear charge is effectively largest at F^- and smallest at N^{3-}.

CHAPTER 10

1. The ion-ion distance in NaCl is shorter than in CsCl, so the attractive energy in NaCl should be larger. Therefore, the melting point of NaCl (800 °C) is expected to be higher than that of CsCl (646 °C).
2. a, d, e, and h are consistent with the octet rule.
3.

Molecule	Atom	Number σ Bonds	Number π Bonds
BeH_2	Be	2	0
BH_3	B	3	0
CH_4	C	4	0
NH_3	N	3	0
H_2O	O	2	0
HF	F	1	0
BF_3	B	3	0
	F	1	0
C_2H_4	C	3	1
N_2	N	1	2
SCl_2	S	2	0
	Cl	1	0
ClF	Cl or F	1	0

4. (a) :C≡O: C and O both have 1 σ bond and 2 π bonds.
(b) :N≡O:$^+$ NO^+ has the same number of valence electrons as CO, so both N and O have 1 σ bond and 2 π bonds.
(c) H—C≡N: C has two σ bonds and 2 π bonds, and N has 1 σ bond and 2 π bonds.
(d) H—S̈—H S has 2 σ bonds.

5. SF_4 ClF_3 PCl_5

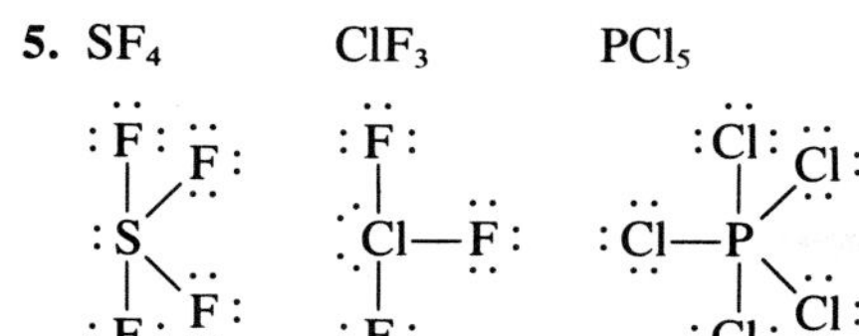

6. Resonance structures for NO_2^-

Resonance structures for NO_3^-

7. (a) Decreasing CN bond distance (bond order): C—N(1) > C=N(2) > C≡N(3)
(b) (See structures of NO_2^- in Exercise 10.6.) NO bond order in NO_2^- is 1.5. Therefore, the observed bond length (124 pm) should be approximately midway between an N—O single bond (136 pm) and N=O double bond (115 pm).

8. $\Delta H_f^\circ = [D_{N\equiv N} + 2\,D_{F-F}] - [4\,D_{N-F} + D_{N-N}]$
$= [946\text{ kJ} + 2(159\text{ kJ})]$
$- [4(272\text{ kJ}) + (159\text{ kJ})]$
$= 17\text{ kJ}$

9. Enthalpy of formation of CCl_4:

C(graphite) ⟶ C(g)	$\Delta H^\circ = +717\text{ kJ}$
$2\,Cl_2(g) \longrightarrow 4\,Cl(g)$	$\Delta H^\circ = 2\,D_{Cl-Cl} = 2\,(243\text{ kJ})$
$C(g) + 4\,Cl(g) \longrightarrow CCl_4(g)$	$\Delta H^\circ = -4\,D_{C-Cl} = -4\,(330\text{ kJ})$

$C(graphite) + 2\,Cl_2(g) \rightarrow CCl_4(g)$ $\Delta H^\circ = -117\text{ kJ}$

10. Enthalpy of formation of $HCO_2H = -378.6\text{ kJ} = \Delta H^\circ_{vap}\,[C(graph) \rightarrow C(g)] + D_{O=O} + D_{H-H} - [D_{C-H} + D_{C=O} + D_{C-O} + D_{O-H}]$
$= [717\text{ kJ} + 498\text{ kJ} + 436\text{ kJ}] - [414\text{ kJ} + D_{C=O} + 351\text{ kJ} + 464\text{ kJ}]$
$D_{C=O} = 801\text{ kJ}$

11. (a) H—F ($\Delta\chi/\Sigma\chi = 0.31$) is much more polar than H—I ($\Delta\chi/\Sigma\chi = 0.087$) because F is more electronegative than I. In both bonds H is the positive end and the halogen the negative end.
(b) B—F ($\Delta\chi/\Sigma\chi = 0.33$) is more polar than B—C ($\Delta\chi/\Sigma\chi = 0.11$) because F is more electronegative than C. In both cases, B is the positive end of the bond dipole.
(c) C—Si ($\Delta\chi/\Sigma\chi = 0.16$) is more polar than C—S ($\Delta\chi/\Sigma\chi = 0$), which is in fact nonpolar. C is the more negative end of the C—Si bond, since C is more electronegative than Si.

12. In each of the following the first-named element is more electronegative and "sets" the oxidation number of the other element.
(a) SF_4: F = −1 and S = +4.
(b) CO_3^{2-}: O = −2 and C = +4
(c) SO_3: O = −2 and S = +6.

13. In *A* all atoms have a formal charge of 0. However, in *B*, the B atom has a formal charge of −1 and the F of the F=B double bond has +1. A formal charge of +1 on F is not reasonable, since F is so electronegative.

14.

Molecule	**Structural-Pair Geometry**
H_2O	tetrahedral (2 σ and 2 lone pairs around O)
NO_2^+	linear (2 σ and 2 π pairs around N)
SF_4	trigonal bipyramidal (4 σ and 1 lone pair around S)

15.

Molecule	**Structural-Pair Geometry**	**Molecular Geometry**
CS_2	linear	linear
H_2S	tetrahedral	bent
PO_4^{3-}	tetrahedral	tetrahedral

16. ICl_2^-: structural-pair geometry is trigonal bipyramidal and molecular geometry is linear.

17. The bond angle in :S=C=S: is 180° (because the molecular geometry is linear). The H—S—H angle is approximately 109.5° because the structural-pair geometry is tetrahedral.

18. $BFCl_2$ is polar with the B—F side more negative.

NH_2Cl, with a pyramidal molecular geometry, is polar with the Cl atom the more negative side.

SCl_2, which has a bent molecular geometry, is polar with the negative end of the dipole toward the Cl atoms.

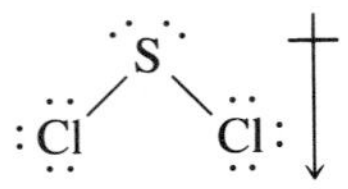

CHAPTER 11

1. The structural-pair geometry around S in SCl_2 is tetrahedral (and the molecular geometry is bent). The tetrahedral pair geometry means the S is sp^3 hybridized.

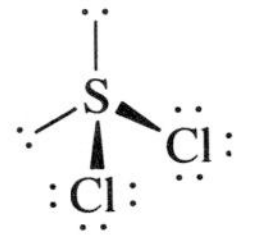

⇅ ⇅	↑ ↑
sp^3 sp^3	sp^3 sp^3
lone pairs	for S—Cl bonds

2. The structural-pair geometry around Xe is octahedral, so the hybridization of the atom is sp^3d^2.

⇅ ⇅	↑ ↑ ↑ ↑
sp^3d^2 sp^3d^2	sp^3d^2 sp^3d^2 sp^3d^2 sp^3d^2
lone pairs	for 4 Xe—F sigma bonds

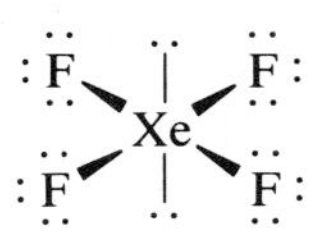

3. The bonding in N_2 is identical to that in CO (one sigma bond, two pi bonds, and one lone pair on each atom), except that both atoms are of course N. Each N atom can be described as *sp* hybridized. The assignments of the 5 valence electrons of each N and their roles are as follows:

⇅	↑	↑ ↑
sp	*sp*	2*p* 2*p*
lone pair	N—N sigma bond	for the two N—N pi bonds

4. 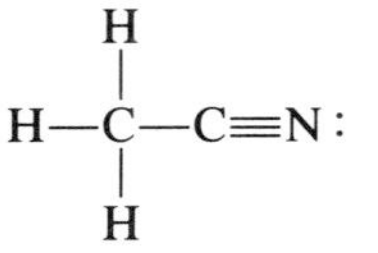 H—C(H)(H)—C≡N:
   (a) Bond angles
   H—C—H = 109°; H—C—C = 109°, C—C—N = 180°
   (b) Atom hybridizations
   CH_3 carbon has a tetrahedral structural-pair geometry; sp^3 hybridized. CN carbon has a linear structural-pair geometry; *sp* hybridized. N atom has a linear structural-pair geometry (triple bond and lone pair are 180° apart); *sp* hybridized.
5. The electron dot structure of NO_2^- shows the N atom has a trigonal planar structural-pair geometry, so the N atom is sp^2 hybridized.

⇅ ↑ ↑	↑
sp^2 sp^2 sp^2	2*p*
for the 2 N—O sigma bonds and the N lone pair	for the N—O pi bond

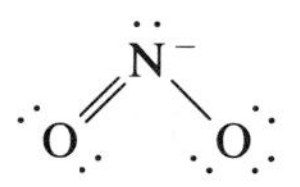

One of the 3 sp^2 hybrid orbitals is assigned the N lone pair, while the other two orbitals are each assigned an electron to be utilized in N—O sigma bonding. The N—O pi bond is then formed using the electron assigned to the unhybridized N 2*p* orbital.

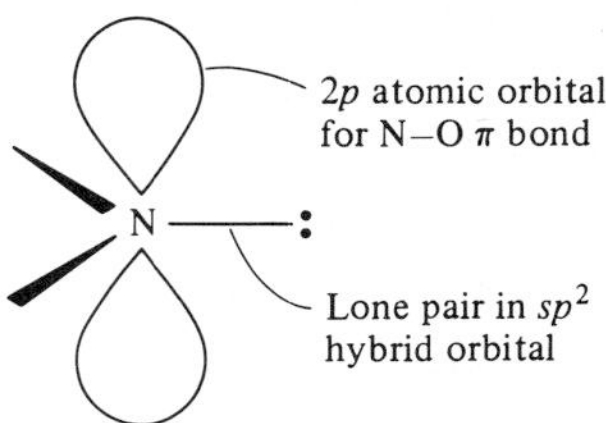

6. H_2^+: $(\sigma_{1s})^1$ The bond order is $\frac{1}{2}$, so the ion should exist under the right conditions. A bond order of $\frac{1}{2}$ is also found in He_2^+ and H_2^-.
7. In $NaLi_2$ the anion would be Li_2^-. Its electron configuration would be $(\sigma_{1s})^2(\sigma^*_{1s})^2(\sigma^*_{2s})^1$, which gives a bond order of $\frac{1}{2}$. This implies that the ion could be stable and that it may be possible to prepare a salt such as $NaLi_2$.
8. O_2^+: [core electrons]$(\sigma_{2s})^2(\sigma^*_{2s})^2(\pi_{2p})^4(\sigma_{2p})^2(\pi^*_{2p})^1$
   The net bond order is 2.5, a higher bond order than in O_2 and thus a stronger bond. The ion is paramagnetic to the extent of one electron.

CHAPTER 12

1. 75 kPa (0.74 atm) > 0.60 atm > 350. mmHg (0.500 atm) > 300. torr (0.395 atm)
2. P_1 = 55 mmHg and V_1 = 125 mL
   P_2 = 78 mmHg and V_2 = ?
   $V_2 = P_1V_1/P_2$ = 88 mL
3. T_1 = 298 K and V_1 = 4.5 L
   T_2 = 263 K and V_2 = ?
   $V_2 = V_1(T_2/T_1)$ = 4.0 L
4. 22.4 L CH_4(2 L O_2/1 L CH_4) = 44.8 L O_2 required
   44.8 L of H_2O and 22.4 L of CO_2 are produced.
5. n = 1300 mol, P = (750/760) atm, T = 293 K
   $V = nRT/P = 3.2 \times 10^4$ L
6. P_1 = 150 atm, T_1 = 303 K, V_1 = 20. L
   P_2 = (755/760)atm, T_2 = 295 K, V_2 = ?
   $V_2 = V_1(P_1/P_2)(T_2/T_1) = 2.9 \times 10^3$ L
7. M = 29.0 g/mol, P = (745/760)atm, T = 295 K
   $d = PM/RT$ = 1.17 g/L

8. $P = 0.737$ atm, $V = 0.125$ L, $T = 296.2$ K
$n = PV/RT = 3.79 \times 10^{-3}$ mol
Molar mass $= 0.105$ g$/3.79 \times 10^{-3}$ mol $= 27.7$ g
9. 180 g (1 mol N_2H_4/32.0 g) = 5.6 mol N_2H_4
5.6 mol N_2H_4(1 mol O_2/1 mol N_2H_4) = 5.6 mol O_2
$V(O_2) = nRT/P = 140$ L when $n = 5.6$ mol, $T = 294$ K, and $P = 0.99$ atm.
10. H_2 was the limiting reagent in Example 12.12. This means only 3.43 mol of N_2 are required of the 3.71 mol of N_2 available.
10.3 mol H_2(1 mol N_2/3 mol H_2) = 3.43 mol N_2 required
Excess N_2 = 3.71 mol − 3.43 mol = 0.28 mol N_2
$P(N_2) = nRT/V = 0.055$ atm
Total pressure $P = P(NH_3) + P(N_2) = 1.40$ atm
11. P = 742 mmHg − vapor pressure H_2O = (742 − 20)mmHg = 722 mmHg = 0.950 atm
$n = PV/RT = 0.0138$ mol N_2 = 0.387 g N_2 (when $V = 0.352$ L and $T = 295$ K).
12. Using equation 12.17 with M = 4.00 g/mol (or 4.00×10^{-3} kg/mol), T = 298 K, and R = 8.314 J/K · mol, one obtains a root mean square speed of 1360 m/s for He. In contrast, N_2 molecules have a much smaller rms speed (515 m/s) owing to their greater mass.
13. The molar mass of CH_4 is 16.0 g/mol. Therefore,

$$\frac{\text{rate for } CH_4}{\text{rate for unk}} = \frac{\text{n molecules/1.50 min}}{\text{n molecules/4.73 min}} = [M_{unk}/16.0]^{1/2}$$

$$M_{unk} = 159 \text{ g/mol}$$

14. For n = 10.0 mol, V = 1.00 L, T = 298 K:
(a) ideal gas law: $P = nRT/V = 245$ atm
(b) Van der Waals's equation (where a = 0.034 and b = 0.0237): P = 321 atm

CHAPTER 13

1. Li^+ should have the higher hydration energy since its radius is much less than that of K^+ (see Figure 9.15).
2. (a) N_2 interactions occur by induced dipole-induced dipole forces, the weakest of all intermolecular forces. (b) $MgSO_4$ consists of the ions Mg^{2+} and SO_4^{2-}, so there are ion-dipole forces involved when the salt dissolves in water. The common, hydrated salt $MgSO_4 \cdot 7\ H_2O$ (epsomite) is widely used in agriculture and medicine. (c) Dipole-induced dipole forces exist between H_2O and CO_2. Order of strength is N_2—N_2 < CO_2—H_2O < $MgSO_4$—H_2O.
3. (1.00×10^3 g)(1 mol/32.0 g)(37.6 kJ/mol) = 1180 kJ
4. $P = nRT/V$ =
(0.028 mol)(0.0821 L · atm/K · mol)(333 K)/5.0 L
= 0.15 atm = 120 mmHg

Appendix D gives a vapor pressure of H_2O at 60 °C of 149 mmHg. The calculated pressure of water in the flask is smaller than this, so all the water (0.50 g) evaporates. With 2.0 g, however, the calculated pressure (460 mmHg) is much larger than the vapor pressure of water at 60 °C, so only enough water can evaporate to give an equilibrium pressure of 149 mmHg.
5. Edge length of unit cell = 4(radius of Fe atom)/$\sqrt{3}$ = 2.91×10^{-8} cm
Volume of unit cell = (edge)^3 = 2.46×10^{-23} cm^3
7.8740 g/cm^3(2.46×10^{-23} cm^3) = 1.94×10^{-22} g/unit cell
55.847 g/mol(1 mol/6.022×10^{23} atom) = 9.27×10^{-23} g/atom
(1.94×10^{-22} g/unit cell)/(9.27×10^{-23} g/atom) = 2.09 atoms/unit cell
A body-centered cubic unit cell has two *net* Fe atoms. The 8 corner atoms account for a net of one Fe atom, and the Fe atom in the center of the unit cell accounts for another.
6. (8 corner Cl^- ions)(1/8 ion per corner) = 1 net Cl^- ion in the unit cell. Since there is 1 Cs^+ ion in the center of the unit cell, the formula of the salt must be CsCl.
7. Cube edge = 2(radius of K^+) + (radius of Cl^-)
= 2(152.0 pm) + 2(167.0 pm)
= 638.0 pm (= 6.38×10^{-8} cm)
Volume = (edge)^3 = (6.38×10^{-8} cm)^3
= 2.60×10^{-22} cm^3
8. Mass of KCl unit cell =
(4 KCl/cell)(74.55 g/mol)(1 mol/6.022×10^{23} KCl) = 4.95×10^{-22} g/cell
Density = (4.95×10^{-22} g/cell)/(2.60×10^{-22} cm^3/cell) = 1.91 g/cm^3

CHAPTER 14

1. 1.0×10^3 g glycol = 16 mol; 4.0 kg water = 220 mol
molality = 16 mol/4.0 kg = 4.0 molal

$$X_{glycol} = \frac{16 \text{ mol glycol}}{16 \text{ mol glycol} + 220 \text{ mol } H_2O} = 0.068$$

2. (a) For AgCl: Enthalpy of solution = 916 kJ/mol + (−851 kJ/mol) = +65 kJ/mol
For RbF: Enthalpy of solution = 776 kJ/mol + (−792 kJ/mol) = −16 kJ/mol
AgCl, an insoluble salt, has a large, positive enthalpy of solution, whereas the soluble salt RbF has a negative enthalpy of solution.

(b) $\Delta H^\circ_{solution}$ (NH_4NO_3) = $\Delta H^\circ_f[NH_4NO_3(aq)]$ − $\Delta H^\circ_f[NH_4NO_3(s)]$
= −339.9 kJ/mol − (−365.6 kJ/mol) = +25.7 kJ/mol

3. $m(CO_2) = (4.48 \times 10^{-5}$ molal/mmHg)(253 mmHg) = $1.13 \times 10^{-2}\ m$

4. Solution consists of sucrose (0.0292 mol) and water (12.5 mol).

$$X_{water} = \frac{12.5 \text{ mol } H_2O}{12.5 \text{ mol } H_2O + 0.0292 \text{ mol sucrose}}$$

= 0.998

P_{water} = (0.998)(526 mmHg) = 525 mmHg

Even with 10.0 g of sugar, the vapor pressure of water has changed very little.

5. $P_{solution} = X_{benzene}\, P^\circ_{benzene}$
94.85 mmHg = $X_{benzene}$(95.00 mmHg)
$X_{benzene}$ = 0.9984

$$X_{benzene} = \frac{1.28 \text{ mol benzene}}{? \text{ mol nitro} + 1.28 \text{ mol benzene}}$$

? mol nitro = 0.00202
Molar mass nitroglycerin = (0.454 g/0.00202 mol) = 225 g/mol
(Nitroglycerin is $C_3H_5N_3O_9$ with a molar mass of 227 g/mol)

6. molality sugar = 0.0292 mol sugar/0.225 kg water = 0.130 molal
$\Delta t = Km$ = (+0.512 deg/molal)(0.130 molal) = 0.0665 deg

7. Δt = 80.31 °C − 80.10 °C = 0.21 °C
$m = \Delta t/K_{bp}$ = (0.21 °C/(2.53 deg/molal) = 0.083 mol/kg
(0.083 mol/kg)(0.100 kg) = 0.0083 mol
molar mass = 1.25 g/0.0083 mol = 150 g/mol
Wintergreen has a formula of $C_8H_8O_3$ with a molar mass of 152 g/mol.

8. Molality of HOC_2H_4OH (ethylene glycol) = 8.06 mol/3.00 kg = 2.69 molal
$\Delta t_{fp} = K_{fp}m$ = (−1.86 deg/molal)(2.69 molal) = −4.99 deg
500. g of glycol is not sufficient to keep the plumbing from freezing at −25 °C.

9. Δt_{fp} = −0.40 deg = $K_{fp}m$ = (−4.90 deg/molal)m
m = 0.082 = ? mol wintergreen/0.100 kg solvent
? mol wintergreen = 0.0082 mol
Molar mass = 1.25 g/0.0082 mol = 150 g/mol
The freezing point depression is somewhat greater than the boiling point elevation, so the measurement of Δt_{fp} can probably be made with less error.

10. (a) The molality of the NaCl solution is 2.69 mol/kg.
$\Delta t_{bp} = K_{bp}im$ = (+0.512 deg/molal)(2 mol particles)(2.69 molal) = +2.75 deg
Boiling point = 102.75 °C
(b) 25.0 g NaCl = 0.428 mol
molality NaCl = 0.856 mol/kg
$\Delta t_{fp} = K_{fp}im$ = (−1.86 deg/molal)(2 mol particles)(0.856 mol) = −3.18 °C
25.0 g $CaCl_2$ = 0.225 mol
molality = 0.450 mol/kg
$\Delta t_{fp} = K_{fp}im$ = (−1.86 deg/molal)(3 mol particles)(0.450 mol) = −2.51 °C

11. $$M = \Pi/RT = \frac{(26.57 \text{ mmHg}/760.\text{ mmHg atm}^{-1})}{(0.0821 \text{ L·atm/K·mol})(298 \text{ K})}$$

$= 1.43 \times 10^{-3}$ mol/L

$(1.43 \times 10^{-3}$ mol/L)(0.0100 L) = 1.43×10^{-5} mol
Molar mass = $(7.68 \times 10^{-3}$ g)/$(1.43 \times 10^{-5}$ mol) = 537 g/mol
The actual molar mass of beta-carotene ($C_{40}H_{56}$) is 536.9 g/mol.

CHAPTER 15

1. $-\frac{1}{2}(\Delta[NOCl]/\Delta t) = \frac{1}{2}(\Delta[NO]/\Delta t) = \Delta[Cl_2]/\Delta t$

2. (a) m = 1. The initial rates and initial concentrations are directly proportional. For example, as the concentration is doubled, the rate is doubled.
(b) Taking the data from Experiment 1, we have

$$k = \frac{\text{Rate}}{[\text{reactant}]} = \frac{1.3 \times 10^{-7} \text{ mol/(L·min)}}{1.0 \times 10^{-3} \text{ mol/L}}$$

$= 1.3 \times 10^{-4}$/min

3. ln ([sucrose]/[sucrose]$_0$) = $-kt$
ln ([sucrose]/(0.010)) = −(0.21/hr)(5.00 hours) = −1.1 hours
ln[sucrose] − ln(0.010) = −1.1 hours
ln[sucrose] − (−4.61) = −1.1
ln[sucrose] = −5.7
[sucrose] = 0.003

4. (1/[HI]) − (1/[HI]$_0$) = kt
When [HI]$_0$ = 0.010 M, k = 30.L/(mol·min), and t = 10. min, [HI] = 0.0025 M

5. (a) $$t_{1/2} = \frac{0.693}{5.40 \times 10^{-2}/\text{hr}} = 12.8 \text{ hr}$$

(b) 51.2 hr = 4.00 half-lives. After 4.00 half-lives the fraction remaining is 1/16.
(c) ln(fraction remaining) = $-kt$ = $-(5.40 \times 10^{-2}$/hr)(18 hr) = −0.97. Fraction remaining = $[A]/[A]_0$ = 0.38

6.

Time (s)	A(mol/L)	ln[A]	1/[A]
0	0.10	−2.30	10.
30	0.074	−2.60	13.5
60	0.055	−2.90	18.2
90	0.041	−3.20	24.4

If ln[A] is plotted vs. time, a straight line is obtained, whereas a curved line is observed for 1/[A] vs. time, indicating a first-order reaction. From the ln[A] vs. time plot, k = −slope = 0.01.

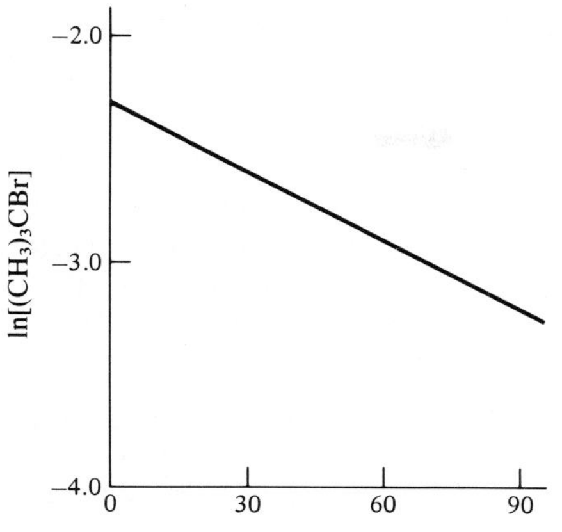

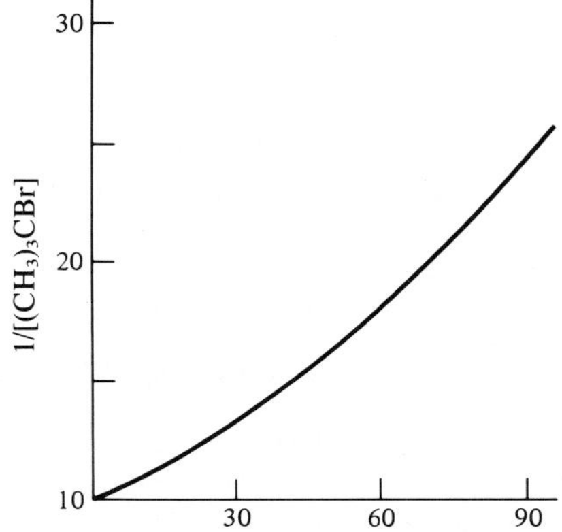

7. Substituting the given values into equation 15.6, we have

$$\ln \frac{1.00 \times 10^4}{4.5 \times 10^3} = -\frac{E^*}{8.31 \times 10^{-3}\ \text{kJ/K}\cdot\text{mol}}\left[\frac{1}{283} - \frac{1}{274}\right]$$

$E^* = 57.2$ kJ

8. All three steps are bimolecular. N_2O_2, the product of the first step, is used in the second step, and N_2O, a product of the second step, is consumed in the third step. Therefore, adding the three reactions gives the equation for the overall process.

9. Step 2 is rate determining. Rate $= k[NH_2Cl][NH_3]$

CHAPTER 16

1. (a) $K = \dfrac{[PCl_3][Cl_2]}{[PCl_5]}$ (b) $K = \dfrac{[H_2]}{[HCl]}$

(c) $K = [Cu^{2+}][OH^-]^2$ (d) $K = \dfrac{[Cu^{2+}][NH_3]^4}{\{[Cu(NH_3)_4]^{2+}\}}$

2. (a) $K_p' = \dfrac{1}{K_p} = \dfrac{1}{0.15} = 6.7$

$K_p'' = (K_p)^{1/2} = (0.15)^{1/2} = 0.39$

(b) $K_p' = 1/K_p = 1/6.6 \times 10^{11} = 1.5 \times 10^{-12}$

3. $K_p = 11.5$, $T = 573$ K, and $\Delta n = +1$.
$11.5 = K_c[(0.08206)(573)]^1$
$K_c = 0.245$

4. Concentration of Ag^+ in the AgCl beaker (1.3×10^{-5} M) is greater than in the AgI beaker (1.2×10^{-8}).

5. (a) $Q = [iso]/[n] = 2.18/0.97 = 2.3$
$Q < K$ (=2.5), so the reaction is not at equilibrium; the reaction produces more isobutane on proceeding to equilibrium.

(b) $Q = [iso]/[n] = 2.60/0.75 = 3.5$
$Q > K$ (=2.5), so the reaction is not at equilibrium; the reaction consumes isobutane on proceeding to equilibrium.

6. $Q = \dfrac{[NO]^2}{[N_2][O_2]} = \dfrac{(4.2 \times 10^{-3})^2}{(0.50)(0.25)} = 1.4 \times 10^{-4}$

$Q < K$, so the reaction is not at equilibrium. The reaction consumes N_2 and O_2 and produces NO on proceeding to equilibrium.

7.

	$P(H_2)$	$P(I_2)$	$P(HI)$
Initial P(atm)	0.70	0.70	0
Change in P	−0.786(0.70)	−0.786(0.70)	+2(0.786)(0.70)
Equilibrium P(atm)	0.15	0.15	0.30

$$K_p = \frac{P(HI)^2}{P(H_2)P(I_2)} = \frac{(0.30)^2}{(0.15)(0.15)} = 4.0$$

8.

	$P(PCl_5)$	$P(PCl_3)$	$P(Cl_2)$
initial P(atm)	1.50	0	0
change in P(atm)	−1.34	+1.34	+1.34
final P(atm)	0.16	1.34	1.34

$$K_p = \frac{P(PCl_3)P(Cl_2)}{P(PCl_5)} = \frac{(1.34)^2}{0.16} = 11$$

9.

	$P(PCl_5)$	$P(PCl_3)$	$P(Cl_2)$
initial P(atm)	3.00	0	0
change in P(atm)	$-x$	$+x$	$+x$
final P(atm)	$3.00 - x$	x	x

$$K_p = \frac{P(PCl_3)P(Cl_2)}{P(PCl_5)} = \frac{x^2}{3.00 - x} = 11.5$$

$11.5(3.00 - x) = x^2$, and on rearrangement we have $x^2 + 11.5x - 34.5 = 0$. Solving for x using the quadratic equation, we have $x = P(PCl_3) = P(Cl_2) = 2.47$ atm. This means $P(PCl_5) = 0.53$ atm.

10. $K_p = 0.11 = P(H_2S)P(NH_3) = 0.11$
$P(H_2S) = P(NH_3)$

$K_p = 0.11 = P(H_2S)^2$
$P(H_2S) = 0.33$ atm
$P_{total} = P(H_2S) + P(NH_3) = 0.66$ atm

11. (a) $[NH_4^+]$ increases with increasing temperature. (b) $[N_2]$ increases as the temperature declines.

12.

	$[n]$	$[iso]$
Initial conc. (M)	0.20	0.50
Conc. after adding 2.0 M more *iso*	0.20	2.0 + 0.50
Change on proceeding to equilibrium (M)	$+x$	$-x$
Conc. at equilibrium (M)	$0.20 + x$	$2.50 - x$

$K = [iso]/[n] = (2.50 - x)/(0.20 + x)$. Solving for x gives $x = 0.57$ M. Therefore, $[iso] = 1.93$ M and $[n] = 0.77$ M.

13. (a) Added H_2 shifts the equilibrium right, and added NH_3 shifts it left. (b) Increasing the volume decreases all the concentrations. The equilibrium shifts to the left, toward the side with the greater number of molecules. (c) K is unchanged in either (a) or (b), since the temperature has not changed.

14. Rate law for the slow step is Rate $= k_2[Br][H_2]$. Since the concentration of the intermediate Br cannot appear in the rate law, we eliminate it by using the equilibrium constant expression for step 1 ($K = [Br]^2/[Br_2]$), which gives $[Br] = K^{1/2}[Br_2]^{1/2}$. Substituting this into the rate law for the slow step gives Rate $= k_2\{K^{1/2}[Br_2]^{1/2}\}[H_2]$. Taking $k_2K^{1/2}$ as the constant k, we have the experimentally observed rate law.

CHAPTER 17

1. (a) HBr is the acid on the left, and its conjugate base is Br^-. NH_3 is the base on the left, and its conjugate acid is NH_4^+. (b) HS^- is the conjugate base of H_2S. (c) HNO_3 is the conjugate acid of NO_3^-.

2. (a) NH_4^+ is a weaker acid than HSO_4^- (and SO_4^{2-} is a weaker base than NH_3), so the equilibrium lies to the right. (b) HCO_3^- is a weaker acid than H_2S (and HS^- is a weaker base than CO_3^{2-}), so the equilibrium lies to the left.

3. The Cl^- and Na^+ ions do not enter the problem, since they are the conjugates of a strong acid and strong base, respectively. The net ionic equation for the reaction occurring is:

$$NH_4^+(aq) + SO_4^{2-}(aq) \rightleftharpoons NH_3(aq) + HSO_4^-(aq)$$

However, since NH_4^+ is a weaker acid than HSO_4^- (and SO_4^{2-} is a weaker base than NH_3), the equilibrium lies well to the left.

4. 0.073 g of HCl is 2.0×10^{-3} mol. Therefore, $[H_3O^+] = 2.0 \times 10^{-3}$ mol/0.500 L $= 4.0 \times 10^{-3}$ M. From $K_w = [H_3O^+][OH^-] = 1.0 \times 10^{-14}$, we have $[OH^-] = 2.5 \times 10^{-12}$ M.

5. $[H_3O^+] = 10^{-pH} = 10^{-7.30} = 5.0 \times 10^{-8}$ M. From $K_w = [H_3O^+][OH^-] = 2.5 \times 10^{-14}$, we have $[OH^-] = 5.0 \times 10^{-7}$ M.

6. $propH(aq) + H_2O(\ell) \rightleftharpoons prop^-(aq) + H_3O^+(aq)$
The pH gives the equilibrium concentrations of $prop^-(aq)$ and $H_3O^+(aq)$: $[prop^-] = [H_3O^+] = 10^{-pH} = 1.1 \times 10^{-3}$ M. This means $[propH] = 0.10 - 1.1 \times 10^{-3} \approx 0.10$ M

$$K_a = \frac{[prop^-][H_3O^+]}{[propH]} = \frac{(1.1 \times 10^{-3})^2}{(0.10)} = 1.3 \times 10^{-5}$$

7. $CH_3COOH(aq) + H_2O(\ell) \rightleftharpoons CH_3COO^-(aq) + H_3O^+(aq)$

	$[CH_3COOH]$	$[H_3O^+]$	$[CH_3COO^-]$
Initial concentrations (M)	0.10	1.0×10^{-7}	0
Change (M)	$-x$	$+x$	$+x$
Equilibrium concentrations (M)	$0.10 - x$	$1.0 \times 10^{-7} + x$	x

Assuming x is large compared with 1.0×10^{-7}, but that it is small in comparison with 0.10, we have

$$K_a = 1.8 \times 10^{-5} = \frac{[CH_3COO^-][H_3O^+]}{[CH_3COOH]} = \frac{x^2}{0.10}$$

$$x = [CH_3COO^-] = [H_3O^+] = 1.3 \times 10^{-3} \text{ M}$$

The result confirms the validity of the assumption that $0.10 - x \approx 0.10$. (Here $[CH_3COOH]_{initial} > 100\ K_a$, so we were reasonably sure the assumption was valid.)

8. $HF(aq) + H_2O(\ell) \rightleftharpoons F^-(aq) + H_3O^+(aq)$

	$[HF]$	$[H_3O^+]$	$[F^-]$
Initial concentrations (M)	0.015	1.0×10^{-7}	0
Change (M)	$-x$	$+x$	$+x$
Equilibrium concentrations (M)	$0.015 - x$	$1.0 \times 10^{-7} + x$	x

Assuming x is large compared with 1.0×10^{-7}, we have

$$K_a = 7.2 \times 10^{-4} = \frac{[H_3O^+][F^-]}{[HF]} = \frac{x^2}{0.015 - x}$$

In this case we cannot be sure that x is small compared with 0.015, since K_a is relatively large. (Furthermore, we observe that $[HF]_{initial}$ is *not* greater than $100K_a$.) Therefore, we should solve the complete expression using the quadratic formula or the method of successive approximations (Appendix A). Either of these gives $x = 2.9 \times 10^{-3}$. Therefore, at equilibrium $[H_3O^+] = [F^-] = 2.9 \times 10^{-3}$ M and [HF] = 0.012 M.

9. $NH_3(aq) + H_2O(\ell) \rightleftarrows NH_4^+(aq) + OH^-(aq)$

	$[NH_3]$	$[OH^-]$	$[NH_4^+]$
Initial concentrations (M)	0.025	1.0×10^{-7}	0
Change (M)	$-x$	$+x$	$+x$
Equilibrium concentrations (M)	$0.025 - x$	$1.0 \times 10^{-7} + x$	x

Assuming that x is large compared with 1.0×10^{-7}, and that it is small compared with 0.025 M (since $100K_b < [NH_3]_{initial}$), we have

$$K_b = 1.8 \times 10^{-5} = \frac{[NH_4^+][OH^-]}{[NH_3]} = \frac{x^2}{0.025}$$

$x = [NH_4^+] = [OH^-] = 6.7 \times 10^{-4}$ M. Therefore, pOH = 3.17 and pH = 10.83.

10. $CN^-(aq) + H_2O(\ell) \rightleftarrows HCN(aq) + OH^-(aq)$
Solve for $[OH^-]$ of the 0.015 M NaCN solution in exactly the same way as in Exercise 17.9 above. This would lead to the equation

$$K_b = 2.5 \times 10^{-5} = \frac{[HCN][OH^-]}{[CN^-]} = \frac{x^2}{0.015}$$

$x = [HCN] = [OH^-] = 6.1 \times 10^{-4}$ M. This gives a pOH of 3.21 and a pH of 10.79.

11. (a) Neither Na^+ nor Cl^- hydrolyzes appreciably, so the pH is 7. (b) The Fe^{3+} hydrolyzes to give an acidic solution, whereas Cl^- has no effect on the pH. Therefore, the pH of the $FeCl_3$ solution is expected to be less than 7. (See page 720.) (c) The ammonium ion is a weak acid (Table 17.4), while the nitrate ion has no effect on the pH. A solution of NH_4NO_3 is acidic (see Table 17.5). (d) The Na^+ ion has no effect on the pH, whereas HCO_3^- is a weak base in solution. (In Table 17.4 K_b for HCO_3^- is greater than K_a for the ion.) The solution should be slightly basic (pH > 7).

12. The pH of the solution is determined by the first equilibrium process:

$$H_2C_2O_4(aq) + H_2O(\ell) \rightleftharpoons HC_2O_4^-(aq) + H_3O^+(aq)$$

where $K_{a1} = 5.9 \times 10^{-2}$. If $[H_2C_2O_4] = (0.10 - x)$M and $[HC_2O_4^-] = [H_3O^+] = x$M, then

$$K_{a1} = \frac{[HC_2O_4^-][H_3O^+]}{[H_2C_2O_4]} = \frac{x^2}{0.10 - x}$$

Since $100K_a$ is *not* less than $[H_2C_2O_4]$, we solve the complete expression using the quadratic formula or method of successive approximations. Either of these gives $x = 0.053$ M, and so pH = 1.28.
The concentration of $C_2O_4^{2-}$, the ion from the second equilibrium process,

$$HC_2O_4^-(aq) + H_2O(\ell) \rightleftharpoons C_2O_4^{2-}(aq) + H_3O^+(aq)$$

is equivalent to K_{a2}, which is 6.4×10^{-5} in this case.

13. (a) PH_3 is a Lewis base owing to the lone pair of electrons on the P atom. (b) Like BF_3 (page 738), BCl_3 is a Lewis acid. (c) H_2S has two lone pairs of electrons located at the S atom, and, like water, is a Lewis base. (d) SF_4 has a lone pair of electrons located at the central S atom and so, in principle, can act as a Lewis base.

14. When the solutions of Cu^{2+} and NH_3 are mixed, the volume is doubled. This means the concentrations of Cu^{2+} and NH_3 are half of their value in the original solutions. Thus, we have the values in the table below.

	$[Cu^{2+}]$	$[NH_3]$	$\{[Cu(NH_3)_4]^{2+}\}$
Concentrations after mixing solutions but before reaction (M)	0.00100	1.50	0
Concentrations after reaction to form $[Cu(NH_3)_4]^{2+}$ (M)	0	1.50 − 4(0.00100)	0.00100
Change on proceeding to equilibrium (M)	$+x$	$+4x$	$-x$
Equilibrium concentrations (M)	x	$1.50 - 0.00400 + 4x$	$0.00100 - x$
Equilibrium concentrations after approximations (M)	x	1.50	0.00100

For the reaction $Cu^{2+}(aq) + 4\ NH_3(aq) \rightleftarrows [Cu(NH_3)_4]^{2+}(aq)$

$$K = 6.8 \times 10^{12} = \frac{\{[Cu(NH_3)_4]^{2+}\}}{[Cu^{2+}][NH_3]^4} = \frac{0.00100}{(x)(1.50)^4}$$

$x = [Cu^{2+}]$ at equilibrium $= 2.9 \times 10^{-17}$ M

CHAPTER 18

1. $NaOH(aq) + C_6H_5COOH(aq) \rightleftarrows NaC_6H_5COO(aq) + H_2O(\ell)$

0.976 g of benzoic acid = 8.00×10^{-3} mol
8.00×10^{-3} mol acid (1 mol NaOH/1 mol acid)(1.00 L NaOH/0.100 mol NaOH) = 0.0800 L (or 80 mL) of NaOH
8.00×10^{-3} mol benzoic acid gives 8.00×10^{-3} mol benzoate ion. Its concentration is

$[C_6H_5COO^-] = 8.00 \times 10^{-3}$ mol/0.0800 L = 0.100 M

The $C_6H_5COO^-$ ion is a weak base, reacting in water according to the equation

$$C_6H_5COO^-(aq) + H_2O(\ell) \rightleftharpoons C_6H_5COOH(aq) + OH^-(aq)$$

$$K_b = 1.6 \times 10^{-10} = \frac{[OH^-][C_6H_5COOH]}{[C_6H_5COO^-]} = \frac{x^2}{0.100}$$

$x = [OH^-] = [C_6H_5COO^-] = 4.0 \times 10^{-6}$ M

pOH = 5.40 and pH = 8.60

2. Moles of HCl = (0.050 L)(0.20 M) = 0.010 mol
Moles of aniline = (0.93 g)(1 mol/93.1 g) = 0.010 mol
Since 1 mol of aniline requires 1 mol of HCl, and since they were present initially in equal molar quantities, the reaction completely consumes the acid and base. The solution after reaction contains 0.010 mol of $C_6H_5NH_3^+$ in 0.050 L of solution, so its concentration is 0.20 M. It is the conjugate acid of the weak base aniline, so

$$C_6H_5NH_3^+(aq) + H_2O(\ell) \rightleftharpoons C_6H_5NH_2(aq) + H_3O^+(aq)$$

$$K_a = 2.4 \times 10^{-5} = \frac{[H_3O^+][C_6H_5NH_2]}{[C_6H_5NH_3^+]} = \frac{x^2}{0.20}$$

$x = [H_3O^+] = 2.2 \times 10^{-3}$ M, which gives a pH of 2.66.

3. Equal numbers of moles of the acid and base were mixed, so reaction was complete to produce CH_3COO^- (a weak base) and $C_5H_5NH^+$ (a weak acid).

K_b for $CH_3COO^- = 5.6 \times 10^{-10}$ and K_a for $C_5H_5NH^+ = 6.7 \times 10^{-6}$

The acid is stronger than the base here, so the solution containing the two ions is slightly acidic.

4. (a) pH of 0.30 M formic acid

$$K_a = 1.8 \times 10^{-4} = \frac{[H_3O^+][HCOO^-]}{[HCOOH]} = \frac{x^2}{0.30}$$

$x = [H_3O^+] = 7.3 \times 10^{-3}$ M, which gives a pH of 2.14.

(b) pH of 0.30 M formic acid + 0.010 M NaHCOO

	[HCOOH]	**$[H_3O^+]$**	**$[HCOO^-]$**
Initial concentrations (M)	0.30	0	0.10
Change (M)	$-x$	$+x$	$+x$
At equilibrium (M)	$0.30 - x$	x	$0.10 + x$

$$K_a = 1.8 \times 10^{-4} = \frac{[H_3O^+][HCOO^-]}{[HCOOH]} = \frac{(x)(0.10 + x)}{0.30 - x}$$

Assuming that x is small compared with 0.10 or 0.30, we find that $x = [H_3O^+] = 5.4 \times 10^{-4}$ M, which gives a pH of 3.27.

5. (a) Find pH of buffer before adding HCl.

$[H_3O^+]$ before adding HCl

$$= \frac{[\text{acid}]}{[\text{conjugate base}]} K_a = \frac{0.50}{0.70}(1.8 \times 10^{-4})$$

$[H_3O^+] = 1.3 \times 10^{-4}$, which gives a pH of 3.89.

(b) Add 10. mL of HCl (= 0.010 mol). The reaction occurring is

$$H_3O^+(aq) + HCOO^-(aq) \rightleftharpoons HCOOH(aq)$$

The molar quantities and concentrations at this stage are as follows:

	$[H_3O^+]$ from HCl	**$[HCOO^-]$ from buffer**	**[HCOOH] from buffer**
Before reaction (mol)	0.010	0.35	0.25
Change when HCl reaction occurs (mol)	−0.010	−0.010	+0.010
After HCl reaction (mol)	0	0.34	0.26
After reaction (mol/L in 0.510 L solution)	0	0.67	0.51

To find $[H_3O^+]$ at this stage, we use the chemical equation for the ionization of formic acid.

$$HCOOH(aq) + H_2O(\ell) \rightleftharpoons HCOO^-(aq) + H_3O^+(aq)$$

$[H_3O^+]$ after adding excess HCl

$$= \frac{[\text{acid}]}{[\text{conjugate base}]} K_a = \frac{0.51}{0.67}(1.8 \times 10^{-4}) = 1.4 \times 10^{-4} \text{ M}$$

From $[H_3O^+] = 1.4 \times 10^{-4}$ M the pH is 3.86. Only a small change in pH (3.89 → 3.86) occurred on adding a concentrated acid.

6. A pH of 5.00 corresponds to $[H_3O^+] = 1.0 \times 10^{-5}$ M. This means that

$$[H_3O^+] = 1.0 \times 10^{-5} \text{ M} = \frac{[CH_3COOH]}{[CH_3COO^-]}(1.8 \times 10^{-5})$$

The ratio $[CH_3COOH]/[CH_3COO^-]$ must be 1/1.8 to achieve the correct hydronium ion concentration. Therefore, 1.0 mol of CH_3COOH is mixed with 1.8 mol of a salt of CH_3COO^- (say $NaCH_3COO$) in some amount of water. (The volume of water is not critical;

only the *relative* amounts of acid and conjugate base are important.)

7. (a) Using the expression in the Example to find $[H_3O^+]$ before the equivalence point, we have

$$[H_3O^+] = \frac{0.00500 \text{ mol HCl} - (0.0400 \text{ L NaOH})(0.100 \text{ M})}{0.0500 \text{ L HCl} + 0.0400 \text{ L NaOH}} = 0.0111 \text{ M}$$

pH = 1.954

(b) After 60.0 mL of base have been added, the HCl has been completely consumed, and we have added 10.0 mL of base in excess of the equivalence point.

$$[OH^-] = \frac{\text{moles excess base}}{\text{total volume}} = \frac{(0.0100 \text{ L})(0.100 \text{ M})}{0.050 \text{ L acid} + 0.060 \text{ L NaOH}} = 9.1 \times 10^{-3} \text{ M}$$

pOH = 2.04, which gives a pH of 11.96

(c) When 49.9 mL of NaOH has been added, we are still 0.1 mL short of the equivalence point. Therefore, $[H_3O^+]$ is calculated as in part (a). This gives $[H_3O^+] = 1.00 \times 10^{-4}$ M, or a pH of 4.000. Even this close to the equivalence point the solution is still relatively acidic.

8. The equation for the reaction occurring here is

$$NaOH(aq) + CH_3COOH(aq) \rightleftharpoons NaCH_3COO(aq) + H_2O(\ell)$$

(a) 50.0 mL of 0.100 M NaOH are required to completely consume the acid. Therefore, to consume 20.0% of the acid, only 20.0% of 50.0 mL or 10.0 mL of NaOH are required. Concentrations of compounds in solution at this point are:

(i) 80.0% of the CH_3COOH remains and is contained in 60.0 mL of solution.

$$[CH_3COOH] \text{ remaining} = \frac{(0.800)(0.0500 \text{ L})(0.100 \text{ M})}{0.0500 \text{ L } CH_3COOH \text{ originally present} + 0.0100 \text{ L NaOH added}} = 6.7 \times 10^{-2} \text{ M}$$

(ii) 20.0% of the CH_3COOH has been converted to $NaCH_3COO$, and the salt is contained in 60.0 mL of solution.

$$[NaCH_3COO] = \frac{(0.200)(0.0500 \text{ L})(0.100 \text{ M})}{0.0600 \text{ L}} = 1.67 \times 10^{-2} \text{ M}$$

(iii) This is a buffer solution, and $[H_3O^+]$ is calculated in the usual manner.

$$[H_3O^+] = \frac{[CH_3COOH]}{[NaCH_3COO]}(1.8 \times 10^{-5})$$

$[H_3O^+] = 7.2 \times 10^{-5}$, which gives a pH of 4.14.

(b) After 55.0 mL of NaOH has been added, we have added 5.00 mL of NaOH beyond the equivalence point. Therefore,

$$[OH^-] = \frac{(0.0050 \text{ L})(0.100 \text{ M})}{0.105 \text{ L}} = 4.8 \times 10^{-3} \text{ M}$$

pOH = 2.32, which gives a pH of 11.68.

9. The equivalence point occurs at pH = 5.21. A suitable indicator might be methyl orange, which is yellow in a basic solution but red to red-orange in an acidic solution (Table 17.3). The color change occurs around pH 4.5–5.0. Other choices of indicator would include bromcresol green, bromphenol blue, and methyl red.

CHAPTER 19

1. (a) AgBr, insoluble due to Ag^+ ion. (b) K_2CrO_4, soluble, due to K^+ ion. (c) $SrCO_3$, insoluble (carbonates not usually soluble). (d) KF, soluble, K^+ salts usually water-soluble. (e) $NH_4(CH_3COO)$ (ammonium acetate), soluble due to presence of NH_4^+ and CH_3COO^- ions. (f) $Ba(NO_3)_2$, soluble due to presence of NO_3^- ions.

2. (a) $NaCl(aq) + AgNO_3(aq) \rightarrow AgCl(s) + NaNO_3(aq)$
   (b) $2\,KI(aq) + Pb(NO_3)_2(aq) \rightarrow PbI_2(s) + 2\,KNO_3(aq)$
   (c) No reaction. Possible products [$Cu(NO_3)_2$ and $MgCl_2$] are both water-soluble.

3. (a) $K_{sp} = [Cd^{2+}][S^{2-}] = 3.6 \times 10^{-29}$
   (b) $K_{sp} = [Bi^{3+}][I^-]^3 = 8.1 \times 10^{-19}$
   (c) $K_{sp} = [Ag^+]^2[CO_3^{2-}] = 8.1 \times 10^{-12}$

4. (a) $K_{sp} = [Ag^+][SCN^-] = 1.0 \times 10^{-12}$ since $[Ag^+] = [SCN^-] = 1.0 \times 10^{-6}$ M
   (b) $K_{sp} = [Ag^+]^2[S^{2-}] = 9.8 \times 10^{-50}$ since $[Ag^+] = 5.8 \times 10^{-17}$ M and $[S^{2-}] = \frac{1}{2}(5.8 \times 10^{-17} \text{ M})$. (The calculated K_{sp} differs slightly from the value in Appendix G owing to the fact that we used only two significant figures.)

5. (a) Solubility of CdS $= [Cd^{2+}] = [S^{2-}] = \sqrt{K_{sp}} = (3.6 \times 10^{-29})^{1/2} = 6.0 \times 10^{-15}$ M
   (b) Solubility of $Mg(OH)_2 = [Mg^{2+}] = S$. Since $[OH^-] = 2S$, this means that $K_{sp} = [Mg^{2+}][OH^-]^2 = (S)(2S)^2$. Solving for S (with $K_{sp} = 1.5 \times 10^{-11}$) gives 1.6×10^{-4} M.

6. (a) AgCl; (b) $Ca(OH)_2$; (c) $MgCO_3$.

7. $Q = [Ag^+][Cl^-] = 2.5 \times 10^{-10}$. Not at equilibrium since $Q > K_{sp}$. More AgCl will precipitate.

8. $Q = [Sr^{2+}][SO_4^{2-}] = 6.3 \times 10^{-8}$. No precipitate forms, since $Q < K_{sp}$.

9. $[SO_4^{2-}] = K_{sp}/[Ba^{2+}] = (1.1 \times 10^{-10})/(1.0 \times 10^{-3}) = 1.1 \times 10^{-7}$ M

10. (a) $[Ba^{2+}]$ required for equilibrium $= K_{sp}/[CO_3^{2-}] = (8.1 \times 10^{-9})/(0.0010) = 8.1 \times 10^{-6}$ M
    $(0.100 \text{ L})(8.1 \times 10^{-6} \text{ mol/L})$
    $(1 \text{ mol } BaCl_2/1 \text{ mol } Ba^{2+})(208.2 \text{ g } BaCl_2/\text{mol}) = 1.7 \times 10^{-4} \text{ g } BaCl_2$

(b) $[Ba^{2+}]$ required = 8.1×10^{-3} M. Solving for mass of $BaCl_2$ as above gives 0.17 g.

11. Total volume of solution after mixing is 200. mL. Therefore, the concentrations of Ba^{2+} and CO_3^{2-} before precipitation can occur are half of those in the original 100. mL solutions. This means that $Q = [Ba^{2+}][CO_3^{2-}] = (5.0 \times 10^{-4})(5.0 \times 10^{-4}) = 2.5 \times 10^{-7}$. Since $Q > K_{sp}$, precipitation will occur.

12. (a) In pure water, solubility is given by $[Ba^{2+}]$.

$$[Ba^{2+}] = \sqrt{K_{sp}} = (1.1 \times 10^{-10})^{1/2} = 1.0 \times 10^{-5}\text{ M}$$

(b) When $[Ba^{2+}] = 0.010$ *M* (from barium nitrate), the solubility of $BaSO_4$ is predicted to be lower and is equivalent to $[SO_4^{2-}]$. Here $[SO_4^{2-}] = K_{sp}/[Ba^{2+}] = 1.1 \times 10^{-8}$ M.

13. $Zn(CN)_2(s) \rightleftarrows Zn^{2+}(aq) + 2\ CN^-(aq)$

(a) In pure water, solubility = $[Zn^{2+}] = S$ and so $[CN^-] = 2S$. This means that $K_{sp} = (S)(2S)^2$, so $S = 1.3 \times 10^{-4}$ M.

(b) When initial concentration of $CN^- = 0.010$ M, this means $K_{sp} = (S)(0.010 + 2S)^2$, where $S = [Zn^{2+}]$. Assuming $2S$ is very small compared with 0.010, $S = 8.0 \times 10^{-8}$ M.

14. (a) AgCl precipitates before $PbCl_2$.

$[Cl^-]$ required by 0.0010 M $Ag^+ = K_{sp}/(0.0010) = 1.8 \times 10^{-7}$ M

$[Cl^-]$ required by 0.0010 M $Pb^{2+} = [K_{sp}/(0.0010)]^{1/2} = 0.13$ M

(b) When $[Cl^-] = 0.13$ M, $[Ag^+] = (K_{sp}$ of AgCl$)/0.13 = 1.4 \times 10^{-9}$ M

15. (a) Use HCl to give insoluble AgCl. $BiCl_3$ is soluble in strong acid.

(b) Use OH^- (to give insoluble $Fe(OH)_2$) or S^{2-} to give insoluble FeS. K^+ gives soluble salts with both S^{2-} and OH^-.

16. AgBr is less soluble than AgCl. Therefore, adding Br^- to a suspension of AgCl should lead to the precipitation of AgBr, as indicated by the net equilibrium constant below.

$$AgCl(s) \rightleftharpoons Ag^+(aq) + Cl^-(aq)$$
$$Br^-(aq) + Ag^+(aq) \rightleftharpoons AgBr(s)$$

$$AgCl(s) + Br^-(aq) \rightleftharpoons AgBr(s) + Cl^-(aq)$$

$K_{sp} = 1.8 \times 10^{-10}$

$K = 1/K_{sp}$ for AgBr $= 1/3.3 \times 10^{-13}$

$K_{net} = 5.5 \times 10^2$

17. 0.010 mol AgCl will give 0.010 M $[Ag(NH_3)_2]^+$ and 0.010 M Cl^- when dissolved completely in NH_3. Solving for $[NH_3]$ as in Example 19.11, $[NH_3]$ at equilibrium = 0.19 M. The quantity of NH_3 available = (0.100 L)(4.0 M) = 0.40 mol. Only 0.020 mol of NH_3 is required to form 0.010 M $[Ag(NH_3)_2]^+$ and to achieve a concentration of 0.19 M. Therefore, there is sufficient NH_3 to dissolve the AgCl completely.

18. $CuSO_4(aq) + Na_2CO_3(aq) \rightarrow Na_2SO_4(aq) + CuCO_3(s)$

$CuCO_3(s) + 2\ HCl(aq) \rightarrow CuCl_2(aq) + H_2O(\ell) + CO_2(g)$

19. Ba^{2+} forms insoluble salts with both anions ($BaCO_3$ and $BaSO_4$). However, only $BaCO_3$ will dissolve in HCl.

20. CoS is less soluble than MnS, so less S^{2-} is required to precipitate CoS than MnS when both metal ions are 0.0010 M (see Table 19.3). To precipitate MnS, Table 19.3 shows that $[S^{2-}]$ must be 5.1×10^{-12} M. Therefore, if one comes up to slightly less than this concentration, the maximum amount of CoS is precipitated while leaving Mn^{2+} in solution. The strategy is to find the $[H_3O^+]$ required to achieve just the right S^{2-} concentration ($= 5.1 \times 10^{-12}$ M).

$[H_3O^+]^2 = \{[H_2S]/[S^{2-}]\}(1.3 \times 10^{-20})$

$[H_3O^+] = 1.6 \times 10^{-5}$ M (pH = 4.80)

CHAPTER 20

1. ΔS for melting ice $= q/T = (6.02 \times 10^3\text{ J})/273.15\text{ K}) = 22.0$ J/K. ΔS for freezing water to ice $= q/T = (-6.02 \times 10^3\text{ J})/273.15\text{ K}) = -22.0$ J/K.

2. $\Delta S° = S°[CaCO_3(s)] - \{S°[Ca(s)] + S°[C(graphite)] + 3/2\ S°[O_2(g)]\}$
$= (1\text{ mol})(92.9\text{ J/K}\cdot\text{mol}) - [(1\text{ mol})(41.42\text{ J/K}\cdot\text{mol}) + (1\text{ mol})(5.74\text{ J/K}\cdot\text{mol}) + (3/2\text{ mol})(205.138\text{ J/K}\cdot\text{mol})]$
$= -262.0$ J/K

3. (1a) positive; (1b) negative; (1c) positive. (2a) CH_3CH_2OH; (2b) AlI_3.

4. $\Delta H°_{rxn} = \Delta H°_f[NH_3(g)] - \{\tfrac{1}{2}\ \Delta H°_f[N_2(g)] + 3/2\ \Delta H°_f[H_2(g)]\}$
$= (1\text{ mol})(-46.11\text{ kJ/mol}) - [\tfrac{1}{2}\text{ mol}(0) + 3/2\text{ mol}(0)]$
$= -46.11$ kJ

$\Delta S°_{rxn} = S°[NH_3(g)] - \{\tfrac{1}{2}S°[N_2(g)] + 3/2\ S°[H_2(g)]\}$
$= -99.38$ J/K

$\Delta G°_{rxn} = \Delta H°_{rxn} - T\Delta S°_{rxn}$
$= -46.11\text{ kJ} - (298\text{ K})(-99.38\text{ J/K})(1\text{ kJ}/1000\text{ J})$
$= -16.5$ kJ

5. (a) $C(graphite) + O_2(g) \rightarrow CO_2(g)$

(b) -394.4 kJ/mol

(c) $(2.5\text{ mol})(-394.4\text{ kJ/mol}) = 990$ kJ

6. $C_6H_6(\ell) + 15/2\ O_2(g) \rightarrow 6\ CO_2(g) + 3\ H_2O(\ell)$

$\Delta G°_{rxn} = 6\ \Delta G°_f[CO_2(g)] + 3\ \Delta G°_f[H_2O(\ell)] - \{\Delta G°_f[C_6H_6(\ell)] + 15/2\ \Delta G°_f[O_2(g)]\}$
$= (6\text{ mol})(-394.359\text{ kJ/mol}) + (3\text{ mol})(-237.129\text{ kJ/mol}) - [(1\text{ mol})(124.5\text{ kJ/mol}) + (15/2\text{ mol})(O)]$

$\Delta G°_{rxn} = -3202.0$ kJ

7. (a) $CaO(s) + CO_2(g) \rightarrow CaCO_3(s)$
$\Delta G^\circ_{rxn} = \Delta G^\circ_f[CaCO_3(s)] - \{\Delta G^\circ_f[CaO(s)] + \Delta G^\circ_f[CO_2(g)]\}$
$= (1\ mol)(-1128.79\ kJ/mol) - [(1\ mol)(-604.03\ kJ/mol) + (1\ mol)(-394.359\ kJ/mol)]$
$= -130.40\ kJ$
(b) $CaCO_3(s) \rightarrow CaO(s) + CO_2(g)$
ΔG° for the decomposition $= -\Delta G^\circ$ for the formation $= +130.40$ kJ
The decomposition reaction is spontaneous under standard conditions.
8. $MgO(s) + C(graphite) \rightarrow Mg(s) + CO(g)$
$\Delta H^\circ_{rxn} = +491.18$ kJ and $\Delta S^\circ_{rxn} = 197.67$ J/K
$T = \Delta H^\circ_{rxn}/\Delta S^\circ_{rxn} = 491.18\ kJ/(0.198\ kJ/K) = 2480$ K (or about 2200 °C)
9. (a) $\Delta G^\circ_{rxn} = \Delta G^\circ_f[SO_2(s)] = -300.194$ kJ/mol
$(-300.194\ kJ)(1000\ J/kJ) = -(8.314510\ J/K \cdot mol)(298\ K) \ln K_p$
$\ln K_p = 1.21 \times 10^2$ and so $K_p = 4.15 \times 10^{52}$
(b) $\Delta G^\circ_{rxn} = +130.4$ kJ from Exercise 20.7.
$\ln K_p = -52.6$ and so $K_p = 1.39 \times 10^{-23}$

CHAPTER 21

1. Oxidation at the anode: $Ni(s) \rightarrow Ni^{2+}(aq) + 2e^-$
Reduction at the cathode: $e^- + Ag^+(aq) \rightarrow Ag(s)$
Electrons flow from the anode (Ni) to the cathode (Ag), while NO_3^- ions flow from the cathode compartment (containing a declining concentration of Ag^+) to the anode compartment (where there is an increasing concentration of Ni^{2+}). Abbreviated notation for the cell is $Ni(s)|Ni^{2+}(aq)||Ag^+(aq)|Ag(s)$.
2. $\Delta G^\circ = -(2.00\ mol\ e^-)(9.65 \times 10^4\ J/V \cdot mol)(-0.76\ V)(1.0\ kJ/1000\ J)$
$= +150$ kJ
The reaction is not spontaneous as written.
3. $Fe \rightarrow Fe^{2+} + 2e^-$ $\quad E^\circ = +0.44\ V$
$2e^- + Cu^{2+} \rightarrow Cu$ $\quad E^\circ = +0.34\ V$
$Fe + Cu^{2+} \rightarrow Fe^{2+} + Cu$ $\quad E^\circ = +0.78\ V$
4. (a) $Zn \rightarrow Zn^{2+} + 2\ e^-$
$2e^- + Sn^{2+} \rightarrow Sn$
$Zn + Sn^{2+} \rightarrow Zn^{2+} + Sn$
$E^\circ = +0.763\ V$
$E^\circ = -0.14\ V$
$E^\circ = +0.62\ V$ (spontaneous)
(b) $Br_2 + 2e^- \rightarrow 2\ Br^-$
$2\ Cl^- \rightarrow Cl_2 + 2e^-$
$Br_2 + 2\ Cl^- \rightarrow 2\ Br^- + Cl_2$
$E^\circ = +1.08\ V$
$E^\circ = -1.36\ V$
$E^\circ = -0.28\ V$ (not spontaneous)
5. $2\ Al(s) \rightarrow 2\ Al^{3+}(aq) + 6e^-$
$6e^- + 3\ Zn^{2+}(aq) \rightarrow 3\ Zn(s)$
$2\ Al(s) + 3\ Zn^{2+}(aq) \rightarrow 2\ Al^{3+}(aq) + 3\ Zn(s)$
$E^\circ = +1.66\ V$
$E^\circ = -0.76\ V$
$E^\circ = +0.90\ V$

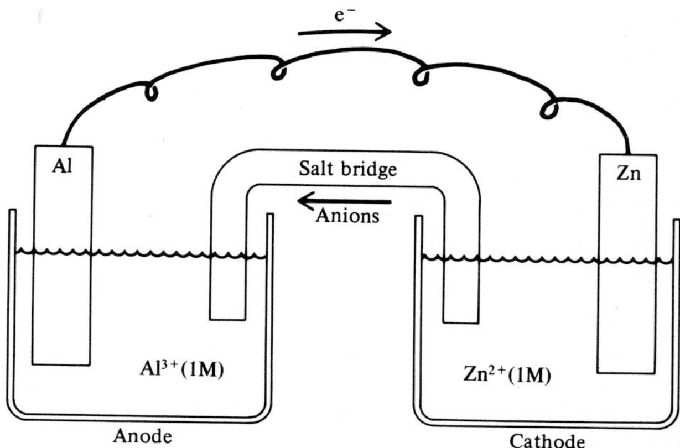

Electrons flow through external wire from Al (anode) to Zn (cathode). Anions flow through the salt bridge from the Zn^{2+}-containing compartment (cathode) to the Al^{3+}-containing compartment (anode).
6. $2\ Sn^{2+}(aq) \rightarrow 2\ Sn^{4+}(aq) + 4e^-$
$O_2(g) + 4\ H_3O^+(aq) + 4e^- \rightarrow 6\ H_2O(\ell)$
$2\ Sn^{2+}(aq) + O_2(g) + 4\ H_3O^+(aq) \rightarrow 2\ Sn^{4+}(aq) + 6\ H_2O(\ell)$
$E^\circ = -0.15\ V$
$E^\circ = +1.23\ V$
$E^\circ = +1.08\ V$

$$E = E^\circ - \frac{0.059}{n} \log \frac{[Sn^{4+}]^2}{[H_3O^+]^4[Sn^{2+}]^2P(O_2)}$$

$$E = 1.08 - \frac{0.059}{4} \log \frac{(1.0 \times 10^{-6})^2}{(0.10)^4(0.10)^2(0.20)} = 1.16\ V$$

7. $1/2\ Hg_2^{2+}(aq) + e^- \rightarrow Hg(\ell)$ $\quad E^\circ = +0.789\ V$
$1/2\ Hg_2^{2+}(aq) \rightarrow Hg^{2+}(aq) + e^-$ $\quad E^\circ = -0.920\ V$
$Hg_2^{2+}(aq) \rightarrow Hg^{2+}(aq) + Hg(\ell)$ $\quad E^\circ = -0.131\ V$

$$\log K = \frac{nE^\circ}{0.0592} = \frac{(1\ mol\ e^-)(-0.131\ V)}{0.0592\ V \cdot mol\ e^-} = -2.21$$

$K = 6.1 \times 10^{-3}$
$[Hg_2^{2+}] = 0.10$ M and $[Hg^{2+}] = 6.1 \times 10^{-4}$ M
8. (a) $NaBr(molten) + electricity \rightarrow Na(s) + Br_2(\ell)$
(b) NaBr(aq) + electricity → $OH^-(aq)$ and $H_2(g)$ from water + $Br_2(\ell)$ from $Br^-(aq)$
(c) $SnCl_2(aq) \rightarrow Sn(s) + Cl_2(g)$
9. $(25 \times 10^3\ amps)(3600\ s/hr) = 9.0 \times 10^7$ C/hr
$(9.0 \times 10^7\ C/hr)(1\ mol\ e^-/96500\ C) = 9.3 \times 10^2$ mol e^-/hr

$(9.3 \times 10^2$ mol e^-/hr)(1 mol Na/1 mol e^-)(23.0 g/mol) $= 2.2 \times 10^4$ g Na per hour

10. $(9.0 \times 10^7$ C)(7.0 V) $= 6.3 \times 10^8$ J
$(6.3 \times 10^8$ J)(1 kwh/3.60×10^6 J) $=$ 180 kwh

CHAPTER 22

The Exercises in Chapter 22 are all review questions. The answers are found directly in the text.

CHAPTER 23

1. $Al_2O_3(s) + 12\ HF(aq) + 6\ NaOH(aq) \rightarrow 2\ Na_3AlF_6(s) + 9\ H_2O(\ell)$
1.00×10^3 g Na_3AlF_6 (1 mol/209.9 g) $=$ 4.76 mol Na_3AlF_6
4.76 mol Na_3AlF_6 (1 mol Al_2O_3/2 mol Na_3AlF_6)(102.0 g/mol) $=$ 243 g Al_2O_3
4.76 mol Na_3AlF_6 (12 mol HF/2 mol Na_3AlF_6)(20.01 g/mol) $=$ 572 g HF
4.76 mol Na_3AlF_6 (6 mol NaOH/2 mol Na_3AlF_6)(40.00 g/mol) $=$ 572 g NaOH

2. (a) 2 $B(OH)_3$ per unit cell. (b) B is sp^2 hybridized. (c) The B atom has an unoccupied *p* orbital, which is open to attack by a Lewis base such as water.

3. $B_2O_3(s) + 6\ HF(g) \rightarrow 2\ BF_3(g) + 3\ H_2O(\ell)$
$Al_2O_3(s) + 3\ C(s) + 3\ Cl_2(g) \rightarrow 2\ AlCl_3(s) + 3\ CO(g)$
1.00×10^3 g Al_2O_3 (2 mol $AlCl_3$/1 mol Al_2O_3)(133.3 g/mol) $= 2.67 \times 10^5$ g $AlCl_3$

4. $\Delta G°$ for diamond $\rightarrow$ graphite is -2.9 kJ, predicting a spontaneous reaction.

5. (a) Orthosilicates: contain tetrahedral SiO_4^{4-} ions. (b) Pyroxenes: extended chains of SiO_4 units. (c) Mica: sheets of linked SiO_4 units. (d) Zeolite: three-dimensional network of linked SiO_4 units.

6. $SiCl_4(\ell) + 2\ H_2O(\ell) \rightarrow SiO_2(s) + 4\ HCl(aq)$

Silicon has unoccupied 3*d* orbitals that are open to attack by a Lewis base such as water. The C atom in CCl_4, for example, has no such unoccupied orbitals.

CHAPTER 24

Answers are not given to exercises that ask simply to look up information in the text.

2. High temperature is required in the Haber process to raise the reaction rate.

3. (a) N_2O has a linear structural-pair geometry, whereas NO_2 has a planar trigonal structural-pair geometry.

N=N=O

4. The oxidation number of P is +4.

H—O—P—P—O—H

6. (a) MgO, basic; CO_2, acidic; Ga_2O_3, amphoteric; (d) CO, neutral.
(b) $Cl_2O_7(s) + H_2O(\ell) \rightarrow 2\ HClO_4(aq)$

7. (a) $Na_2SO_4(s) + 4\ C(s) \rightarrow Na_2S(s) + 4\ CO(g)$
(b) $2\ Na_2S(aq) + 2\ O_2(g) + H_2O(\ell) \rightarrow Na_2S_2O_3(aq) + 2\ NaOH(aq)$

8. (b) The reaction is predicted to be spontaneous, since $\Delta G°$ is a negative quantity.

9. The structural-pair geometry of XeF_2 is trigonal bipyramidal. Since lone pairs occupy a larger volume of space than bond pairs (page 397), the three lone pairs occupy the trigonal plane, and the F atoms must be placed at the peaks of the pyramids.

CHAPTER 25

1. (a) The platinum in $Pt(NH_3)_2Cl_2$ is Pt^{2+}.
(b) $[Co(NH_3)_5CO_3]^+$

2. (a) aquacyanobis(phenanthroline)ruthenium(II) chloride
(b) diamminedichloroplatinum(II)

3. (a) *cis-trans* isomers (see figure of $[Co(en)_2Cl_2]^+$ on page 1045.

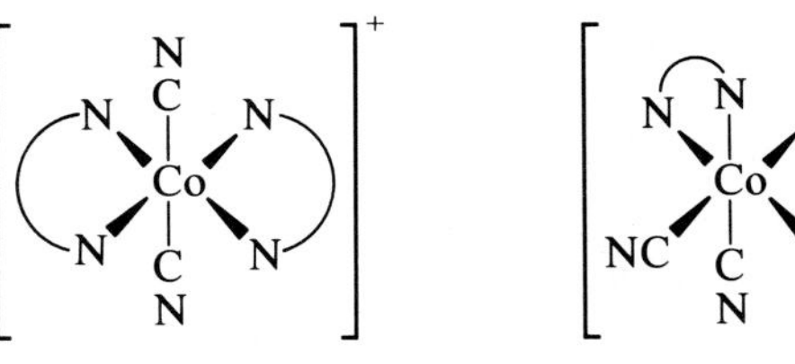

trans *cis*

(b) optical isomers

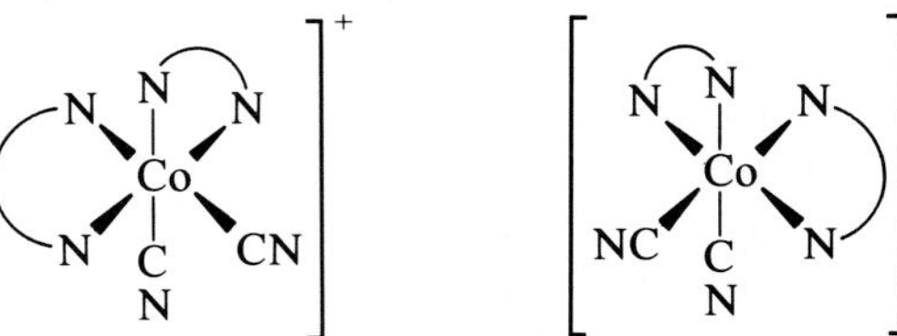

4. (a) $[Ru(H_2O)_6]^{2+}$. The Ru^{2+} ion has 6 4*d* electrons. Its high spin configuration is paramagnetic.

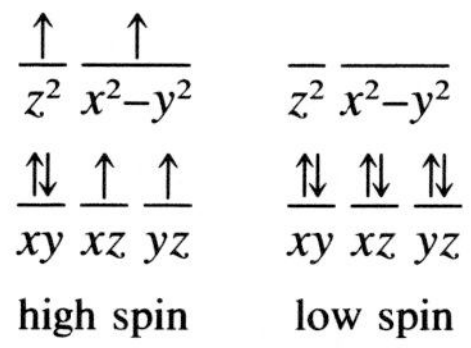

high spin	low spin
↑ ↑ (z^2 x^2-y^2)	— — (z^2 x^2-y^2)
⇅ ↑ ↑ (xy xz yz)	⇅ ⇅ ⇅ (xy xz yz)

(b) $[Ni(NH_3)_6]^{2+}$. The Ni^{2+} ion has 8 3*d* electrons. Here the low and high spin configurations are the same. The complex is paramagnetic.

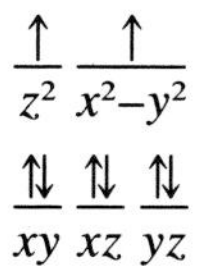

↑ ↑ (z^2 x^2-y^2)

⇅ ⇅ ⇅ (xy xz yz)

5. Since both complexes are based on the same metal ion, Ni^{2+}, we can predict that the water-containing complex requires less energy to promote an electron. Water is lower in the spectrochemical series than ammonia.

CHAPTER 26

1. (a) Structural isomers of C_6H_{14}

(a,1) $CH_3CH_2CH_2CH_2CH_2CH_3$

(a,2) $CH_3CH(CH_3)CH_2CH_2CH_3$

(a,3) $CH_3CH_2CH(CH_3)CH_2CH_3$

(a,4) $CH_3C(CH_3)_2CH_2CH_3$

(a,5) $CH_3CH(CH_3)CH(CH_3)CH_3$

(b) $C_{22}H_{46}$

2. (a) heptane; (b) (a,1) hexane; (a,2) 2-methylpentane; (a,3) 3-methylpentane; (a,4) 2,2-dimethylbutane; (a,5) 2,3-dimethylbutane.

3. (a) alkene; (b) ether; (c) alcohol; (d) ester; (e) carboxylic acid; (f) amine; (g) aldehyde.

4. (a) 1-propanol and 2-propanol

(b) 3-pentanol: $CH_3CH_2CH(OH)CH_2CH_3$ 3-methyl-2-butanol: $CH_3CH(CH_3)CH(OH)CH_3$

(c) Both are secondary alcohols, since —OH is bound to a carbon atom that is in turn bound to two other C atoms.

5. The reaction products are:

(a) $CH_3CH_2CH{=}O$ (c) $CH_3CH_2CH_2I$

(b) $CH_3CH{=}CH_2$

6. Isomers of 2-pentene

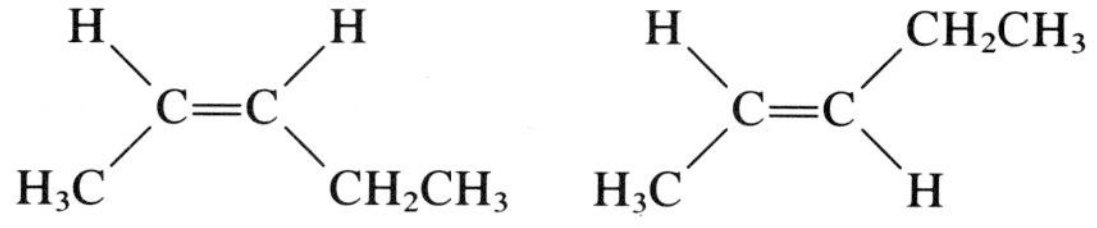

cis-2-pentene: H_3C and CH_2CH_3 on the same side of C=C (H, H on the other side)

trans-2-pentene: H_3C and CH_2CH_3 on opposite sides of C=C

7. (a) $CH_3CH_2CHBrCH_3$ (b) $CH_3CBr(CH_3)CH_2CH_3$

8. $CH_3C(OH)(CH_3)CH_2CH_3$

9. (a,1) $CH_3C{\equiv}CCH_2CH_3$

(a,2) $CH_3CH(CH_3)CH_2C{\equiv}CH$

(b) 3-chloropropyne

10. (a) $CH_3CH(CH_3)C{\equiv}CH + 2\,HBr \longrightarrow CH_3CH(CH_3){-}CBr_2{-}CH_3$

(b) $CH_3CH_2C{\equiv}CH + 2\,Cl_2 \longrightarrow CH_3CH_2CCl_2{-}CHCl_2$

11. (a) p-bromonitrobenzene (b) o-chlorophenol

12. (a) $CH_3CH(CH_3)CH{=}CH_2$; (b) $CH_3CH{=}CHCH_3$ (*cis* and *trans*) and $CH_3CH_2CH{=}CH_2$;

(c) $CH_3CH(CH_3)CH_2OH$;

(d) 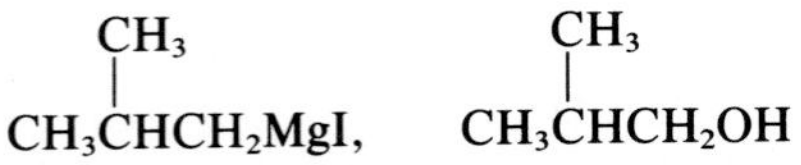 $CH_3CH(CH_3)CH_2MgI$, $CH_3CH(CH_3)CH_2OH$

13. (a) hexanoic acid; (b) p-nitrobenzoic acid; (c) propyl ethanoate (systematic name) or propyl acetate (common name)

14. (a) 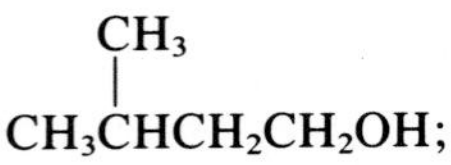 $CH_3CH(CH_3)CH_2CH_2OH$; (b) $C_6H_5COCH_3$ (benzene ring–$\overset{O}{\overset{\|}{C}}OCH_3$);

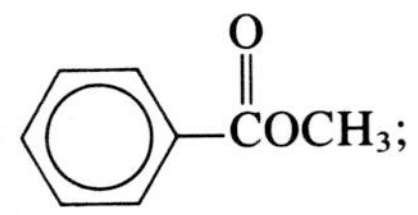

(c) $CH_3CH(CH_3)\overset{O}{\overset{\|}{C}}O^-K^+ + H_2O$

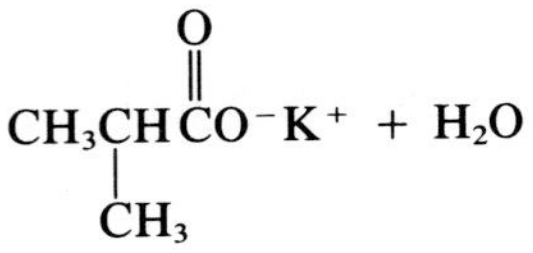

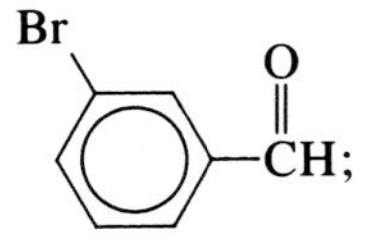

15. (a)  $CH_3\overset{O}{\overset{\|}{C}}CH_2CH_2CH_3$; (b) $BrC_6H_4\overset{O}{\overset{\|}{C}}H$ (Br meta to CHO);

(c) $CH_3CH_2CH_2\overset{O}{\overset{\|}{C}}H$

16. (a) 3-hexanone; (b) pentanal

17. (a) $CH_3CH_2CH_2OH$

(b) $CH_3CH(CH_3)CH_2OH$ (where $CH_3CH(CH_3)CH_2OMgBr$ is an intermediate)

(c) 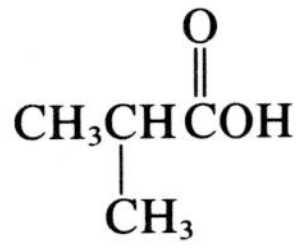 $CH_3CH(CH_3)\overset{O}{\overset{\|}{C}}OH$

18. (a) diethylamine; (b)  $CH_3\overset{O}{\overset{\|}{C}}N(CH_3)_2$

19. $CH_3CH_2CH_2CH_2CH_2CH_2CH_3 \longrightarrow C_6H_5CH_3 + 4\ H_2$

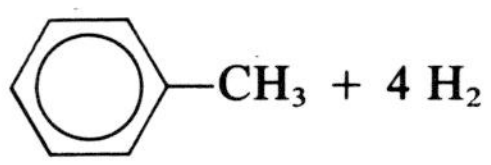

Four moles of H_2 are produced per mole of heptane converted.

CHAPTER 27

1. L form of serine

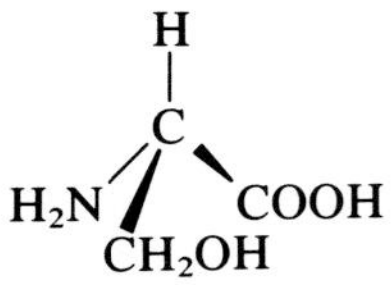

(C bonded to H (up), H_2N, COOH, and CH_2OH)

2. (a) 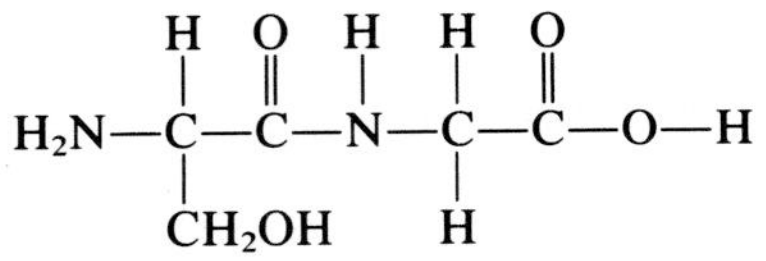 $H_2N—CH(CH_2OH)—\overset{O}{\overset{\|}{C}}—NH—CH_2—\overset{O}{\overset{\|}{C}}—O—H$

serine-glycine

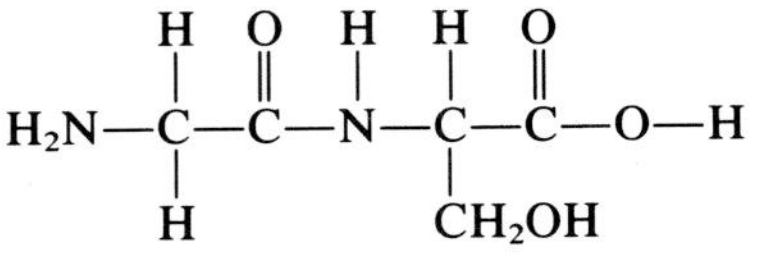

$H_2N—CH_2—\overset{O}{\overset{\|}{C}}—NH—CH(CH_2OH)—\overset{O}{\overset{\|}{C}}—O—H$

glycine-serine

(b) 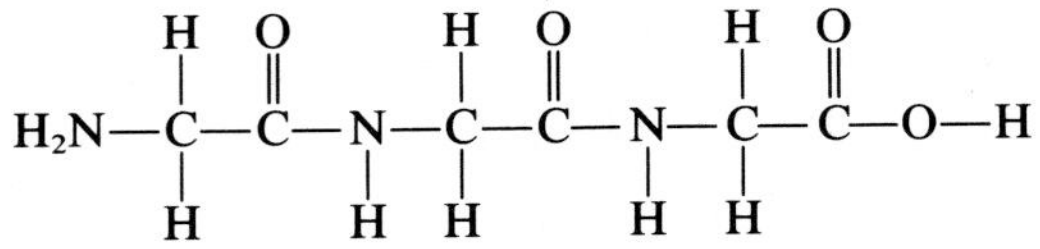 $H_2N—CH_2—\overset{O}{\overset{\|}{C}}—NH—CH_2—\overset{O}{\overset{\|}{C}}—NH—CH_2—\overset{O}{\overset{\|}{C}}—O—H$

3. (a) glucose; (b) maltose; (c) α-1,4 glycoside link

APPENDIX L
Answers to Study Questions

CHAPTER 1

1. (a) color = physical; (b) transformed into rust = chemical; (c) explode = chemical; (d) density = physical; (e) melts = physical.

3. Heterogeneous. Iron chips are magnetic and can be removed by passing a magnet over the mixture of iron and sand.

5.

Question	Qualitative	Quantitative
(a)	purple solid	1.25 g
(b)	silvery, floats	0.025 g
(c)	blue, colorless	25 mL, 50 mL

7. 1.9×10^2 mm, 0.19 m, 7.5 in

9. 0.9942 mile

11. 80.8 miles/hour

13. 36 feet, 7 inches = 439 inches = 1120 cm or 1.12 m; 12 feet = 144 inches = 370 cm = 3.7 m.

15. 5.3 cm^2; 5.3×10^{-4} m^2; 0.81 in^2

17. 1.50×10^6 cm^3; 1.50×10^3 L; 1.50 m^3

19. 0.00563 kg; 5630 mg

21.

Milligrams	Grams	Kilograms
693 mg	0.693 g	6.93×10^{-4} kg
156 mg	0.156 g	1.56×10^{-4} kg
2.23×10^6 mg	2230 g	2.23

23. 2.7×10^4 g; 27 kg

25. 1000 g or 1 kg (to 1 significant figure)

27. 800. cm^3; 0.800 L; 8.00×10^{-4} m^3

29. 23.67 cm^3; 0.02367 L

31. 0.0948 g

33. 499 g; 1.10 lb

35. 0.865 g/cm^3

37. 220 g

39. 2.45×10^2 g

41. 77 °F; 298 K

43.

°F	°C	K
57	14	287
99	37	310
−40	−40	233

45. Your normal body temperature is 98.6 °F or 37 °C; therefore, the gallium should melt in your hand.

47. −360. °F; 55 K

49. 9.6 g

51. 3.62×10^{-3}

53. 1.18×10^{-3}

55. 7.43×10^{-2}

57. 2.1×10^3 cm^3

59. 0.197 nm or 197 pm

61. 0.178 nm^3; 1.78×10^{-22} cm^3

63. 1 ounce "avoirdupois" is 28.350 g, which is smaller than a "troy ounce."

65. 1.0×10^2 km/hour; 95 feet/second

67. 3260 kg/m^3

69. 3.0×10^2 g lead and 150 g tin

71. 21 g sugar

73. 2 ounces = 57 g; protein = 14%; carbohydrate = 74%; fat = 1.8%

75. 1.21×10^{10} kg; 2.67×10^{10} pounds

77. 22 g; $290

79. 54.0 mL

81. 0.018 mm thick

83. (a) 2.6 g/cm^3; (b) 1.4 cm thick

85. 294 feet

CHAPTER 2

16. (a) carbon; (b) sodium; (c) chlorine; (d) phosphorus; (e) magnesium; (f) germanium

18. (a) Li; (b) Ti; (c) Fe; (d) Si; (e) Au; (f) Pb

20. Manganese, Mn

22. (a) A = 9; (b) A = 48; (c) A = 70

24. (a) $^{23}_{12}$Na; (b) $^{39}_{18}$Ar; (c) $^{70}_{31}$Ga

26.

Atom	Electrons	Protons	Neutrons
Ca-40	20	20	20
Sn-119	50	50	69
Pu-244	94	94	150

28.

Symbol	^{45}Sc	^{33}S	^{17}O	^{56}Mn
Protons	21	16	8	25
Neutrons	24	17	9	31
Electrons	21	16	8	25

30. (a), (c), and (e)

32. 69.75

34. 28.1

36. ^{121}Sb = 57.5% and ^{123}Sb = 42.5%

38. Germanium, Ge

40. Nitrogen, N, 14.0067

42. (a) Names: lithium, sodium, potassium, rubidium, cesium, and francium. (b) Common name: alkali metals. (c) Type: metals.

44. Lithium, beryllium, boron, carbon, nitrogen, oxygen, fluorine, and neon.

46. (a) Ca: calcium, metal, Group 2A, 4th period; (b) Cd: cadmium, metal, Group 2B, 5th period; (c) Si: silicon, metalloid, Group 4A, 3rd period; (d) I: iodine, nonmetal, Group 7A, 5th period.

48. (a) 1.9998 mol Cu; (b) 0.499 mol Ca; (c) 0.6208 mol Al; (d) 3.1×10^{-4} mol K; (e) 2.1×10^{-5} mol Am.
50. 2.19 mol Na
52. (a) 5.6 g Fe; (b) 64.9 g Si; (c) 0.028 g C; (d) 12 g Na
54. 13 mol Al
56. 5.9 g Ag
58. $2540
60. 2.39 mol Pt; 21.7 cm^3
62. 1.02×10^{24} atoms Al
64. 4.131×10^{23} atoms Cr
66. 1.0552×10^{-22} g
68. 1.21×10^{23} atoms Al
70. 42 neutrons (^{75}As)
72. 45.8 g S = 1.43 mol; 64.9 g C = 5.40 mol
74. 7260 g Fe or 16.0 pounds.
76. 13.6 g Pt; 0.0698 mol
78. (a) 4.1×10^7 g Ag and 4.1×10^3 g Au; (b) 3.8×10^5 mol Ag and 21 mol Au
80. 10,000 miles (to 1 significant figure, as determined by the dimensions of the can)
82. 0.27 cm^3 Al; 0.74 g Al; 1.7×10^{22} atoms Al.
84. The volume of the cube of 8 Si atoms is 1.602×10^{-22} cm^3. Using the density, the mass of the cube is 3.731×10^{-22} g. Silicon has a diamond-like structure, which has a net of 8 atoms in each tiny cube (page 537). Therefore, the mass of one atom is 4.664×10^{-23} g/atom. Dividing the atomic mass by the mass of one atom gives Avogadro's number.

85.

^{1}S	^{2}N
^{3}B	^{4}I

CHAPTER 3

9. (a) TiO_2; (b) B_4H_{10}; (c) AlC_3H_9
11. (a) 1 Ca, 2 C, and 4 O
(b) 8 C, 8 H
(c) 2 N, 8 H, 1 S, 4 O
(d) 1 Pt, 2 N, 6 H, 2 Cl
(e) 4 K, 1 Fe, 6 C, 6 N
13. (a) Ba^{2+}; (b) Al^{3+}; (c) O^{2-}; (d) Co^{3+}; (e) CO_3^{2-}; (f) I^-; (g) PO_4^{3-}; (h) NH_4^+
15. (a) KI; (b) CaO; (c) NaBr; (d) MgS; (e) Na_2CO_3; (f) NH_4NO_3
17. (a) $NaCH_3COO$; (b) $AgClO_4$; (c) $KClO_3$; (d) $TiBr_4$; (e) $NaNO_3$; (f) $MnSO_4$
19. (a) sodium hydrogen carbonate (Na^+, HCO_3^-)
(b) calcium phosphate (3 Ca^{2+}, 2 PO_4^{3-})
(c) ammonium bromide (NH_4^+, Br^-)
(d) potassium perchlorate (K^+, ClO_4^-)
(e) sodium cyanide (Na^+, CN^-)
(f) copper(II) sulfate (Cu^{2+}, SO_4^{2-})
(g) potassium permanganate (K^+, MnO_4^-)
(h) zinc oxide (Zn^{2+}, O^{2-})
21. Sodium carbonate, Na_2CO_3; sodium iodide, NaI; sodium nitrate, $NaNO_3$; strontium carbonate, $SrCO_3$; strontium iodide, SrI_2; strontium nitrate, $Sr(NO_3)_2$; ammonium carbonate, $(NH_4)_2CO_3$; ammonium iodide, NH_4I; and ammonium nitrate, NH_4NO_3.
23. (a) BBr_3; (b) XeF_2; (c) HI; (d) S_2Cl_2
25. (a) 159.69; (b) 67.81; (c) 44.02; (d) 197.90; (e) 176.12
27. (a) 0.0312 mol; (b) 0.0101 mol; (c) 0.0125 mol; (d) 0.00406 mol; (e) 0.00599 mol
29. 7.26 mol; 4.37×10^{24} molecules
31. (a) 0.00180 mol aspirin; 0.02266 mol $NaHCO_3$; 0.005205 mol citric acid; (b) 1.08×10^{21} molecules aspirin
33. 5.67 mol SO_3; 3.41×10^{24} molecules; 3.41×10^{24} atoms S; $2(3.41 \times 10^{24})$ atoms O
35. 0.0036 mol vanillin; 2.2×10^{21} molecules; $8(2.2 \times 10^{21})$ atoms C
37. All answers given to 4 significant figures:
(a) 239.3 g/mol; 13.40% S; 86.60% Pb
(b) 30.07 g/mol; 20.12% H, 79.88% C
(c) 60.05 g/mol; 6.714% H, 40.00% C, 53.29% O
(d) 80.05 g/mol; 5.037% H, 35.00% N, 59.96% O
39. (a) 53.06 g/mol; (b) 5.70% H, 67.90% C, 26.40% N
41. (406.91 g/mol) 1.486% H, 38.37% C, 7.87% O, 52.28% Cl
43. $(CHO)_4 = C_4H_4O_4$
45. Empirical formula = CH; molecular formula = C_2H_2.
47. N_2O_3
49. Empirical formula = CH_2O; molecular formula = $C_2H_4O_2$
51. Empirical formula = AsC_2H_6; molecular formula = $As_2C_4H_{12}$
53. $C_8H_8O_3$
55. $C_{10}H_8$
57. $RuCl_3 \cdot 3\ H_2O$
59. $MgSO_4 \cdot 7\ H_2O$
61. (a) sodium hydrogen carbonate (sodium bicarbonate); (b) 10.8 mol; (c) 3(10.8 mol) = 32.4 mol O = 519 g O
63. (a) 78.08 g/mol; (b) 0.0200 mol; (c) 24.7 g
65. SF_6
67. Empirical formula = SCl; molecular formula = S_2Cl_2
69. $Cd(NO_3)_2$, cadmium nitrate; $(NH_4)_2S$, ammonium sulfide; NH_4NO_3, ammonium nitrate, CdS, cadmium sulfide
71. 0.00166 mol of "active ingredient"; 346 mg Bi
73. 0.0022 mol; 1.4×10^{21} molecules
75. 0.0050 mol
77. SF_4Cl_2
79. 4.0 mol Fe per mol hemoglobin; 4 Fe atoms per molecule
81. 8.8×10^{-8} mol
83. 7.71×10^{23} molecules; 227 g Cl

85. CuO, copper(II) oxide
87. 7.35 kg Fe
89. 67.1 g Sb_2S_3 per pound of ore.
91. $x = 2$ and $y = 1$

CHAPTER 4

10. (a) $4\ Al(s) + 3\ O_2(g) \rightarrow 2\ Al_2O_3(s)$
(b) $N_2(g) + 3\ H_2(g) \rightarrow 2\ NH_3(g)$
(c) $2\ C_6H_6(\ell) + 15\ O_2(g) \rightarrow 6\ H_2O(g) + 12\ CO_2(g)$
12. (a) $UO_2(s) + 4\ HF(\ell) \rightarrow UF_4(s) + 2\ H_2O(\ell)$
(b) $B_2O_3(s) + 6\ HF(\ell) \rightarrow 2\ BF_3(g) + 3\ H_2O(\ell)$
(c) $BF_3(g) + 3\ H_2O(\ell) \rightarrow 3\ HF(\ell) + H_3BO_3(s)$
14. (a) $Na_2O_2(s) + 2\ H_2O(\ell) \rightarrow 2\ NaOH(aq) + H_2O_2(aq)$
(b) $4\ PH_3(g) + 8\ O_2(g) \rightarrow P_4O_{10}(s) + 6\ H_2O(g)$
(c) $2\ C_2H_3Cl(\ell) + 5\ O_2(g) \rightarrow 4\ CO_2(g) + 2\ H_2O(g) + 2\ HCl(g)$
16. (a) $H_2NCl(aq) + 2\ NH_3(g) \rightarrow NH_4Cl(aq) + N_2H_4(aq)$
(b) $(CH_3)_2N_2H_2(\ell) + 2\ N_2O_4(\ell) \rightarrow 3\ N_2(g) + 4\ H_2O(g) + 2\ CO_2(g)$
(c) $CaC_2(s) + 2\ H_2O(\ell) \rightarrow Ca(OH)_2(s) + C_2H_2(g)$
18. (a) $2\ Mg(s) + O_2(g) \rightarrow 2\ MgO(s)$ (magnesium oxide)
(b) $2\ Ca(s) + O_2(g) \rightarrow 2\ CaO(s)$ (calcium oxide)
(c) $4\ In(s) + 3\ O_2(g) \rightarrow 2\ In_2O_3(s)$ (indium oxide)
20. (a) $2\ Na(s) + Cl_2(g) \rightarrow 2\ NaCl(s)$ (sodium chloride)
(b) $Mg(s) + Br_2(\ell) \rightarrow MgBr_2(s)$ (magnesium bromide)
(c) $2\ Al(s) + 3\ F_2(g) \rightarrow 2\ AlF_3(s)$ (aluminum fluoride)
22. (a) $CH_4(g) + 2\ O_2(g) \rightarrow 2\ H_2O(g) + CO_2(g)$
(b) $2\ C_8H_{18}(\ell) + 25\ O_2(g) \rightarrow 18\ H_2O(g) + 16\ CO_2(g)$
(c) $C_2H_5OH(\ell) + 3\ O_2(g) \rightarrow 3\ H_2O(g) + 2\ CO_2(g)$
24. (a) $2\ C(s) + O_2(g) \rightarrow 2\ CO(g)$
(b) $2\ Ni(s) + O_2(g) \rightarrow 2\ NiO(s)$
(c) $4\ Cr(s) + 3\ O_2(g) \rightarrow 2\ Cr_2O_3(s)$
26. Each reaction produces carbon dioxide (CO_2) plus the appropriate metal oxide.
(a) $BeCO_3(s) \rightarrow CO_2(g) + BeO(s)$ (beryllium oxide)
(b) $NiCO_3(s) \rightarrow CO_2(g) + NiO(s)$ (nickel(II) oxide)
(c) $Al_2(CO_3)_3(s) \rightarrow 3\ CO_2(g) + Al_2O_3(s)$ (aluminum oxide)
28. 60.5 g H_2; 340. g NH_3
30. 550. g N_2O; 450. g H_2O
32. 108 g H_2O; 160. g $CaCN_2$
34. 4.81 g O_2; 17.4 g Fe_3O_4
36. 15.2 g Pt; 15.2 g HCl
38. (a) 192 g CO; (b) 39.9 kg
40. (a) 0.14 g H_2O_2; (b) 0.30 g $PbSO_4$
42. 1.5 g NH_3 in excess; 7.66 g N_2F_4; 8.84 g HF
44. 1.3 g H_2 are in excess; 85.2 g CH_3OH theoretically obtainable
46. CaO is the limiting reagent; 68.0 g NH_3 produced and 10. g NH_4Cl remain
48. Cl_2 is the limiting reagent, so some SO_2 remains; 130 g $OSCl_2$ and 92 g Cl_2O are produced.
50. 136 g $ZnCl_2$; 84.6% yield
52. 14.3 g PBr_3; 76.0% yield
54. C_2H_5
56. Empirical formula: C_3H_4; molecular formula = C_9H_{12}
58. Empirical formula vitamin C = $C_3H_4O_3$
60. Empirical formula: SiH_3; molecular formula = Si_2H_6
62. $Co(CO)_4$
64. M = Sr
66. 93.1% $CaCO_3$
68. 5.68% TiO_2; 3.41% Ti
70. (a) $6\ UO_3(s) + 8\ BrF_3(\ell) \rightarrow 6\ UF_4(s) + 4\ Br_2(\ell) + 9\ O_2(g)$
(b) $3\ IO_2F(s) + 4\ BrF_3(\ell) \rightarrow 3\ IF_5(s) + 2\ Br_2(\ell) + 3\ O_2(g)$
(c) $16\ S_2F_2(g) \rightarrow 3\ S_8(s) + 8\ SF_4(g)$
(d) $2\ Na_2S(aq) + 2\ O_2(g) + H_2O(\ell) \rightarrow Na_2S_2O_3(aq) + 2\ NaOH(aq)$
72. Empirical formula: CH; molecular formula = C_8H_8
74. KO_2
76. Empirical formula of oxide is Sc_2O_3; 0.559 g H_2O
78. 26 million tons $MgCO_3$; 38 million tons $MgSO_4$
80. (a) 5.30×10^5 g H_3PO_4 for a 47.2% yield; (b) apatite is the limiting reagent.
82. 64.9% yield of SF_4; 4.45 g S_2Cl_2
84. 1340 g HNO_3 theoretical yield; 25.8% yield
86. $2\ NH_3(g) + 3\ CuO(s) \rightarrow N_2(g) + 3\ Cu(s) + 3\ H_2O(\ell)$; 0.564 g N_2 for a 61.2% yield; 2.35 g Cu
88. Empirical formula = $Al(CH_3)_3$
90. (a) 17 g H_2SO_4: (b) 45% yield HCl; (c) 11 g Na_2SO_4
92. Empirical formula = GeH_4
94. 0.74
96. (a) $2\ Al(s) + 3\ Br_2(\ell) \rightarrow 2\ AlBr_3(s)$; (b) 0.721 mol Al and 0.378 mol Br_2; 67.2 g $AlBr_3$
98. (a) UO_3; uranium(VI) oxide; (b) U_3O_8; (c) $UO_2(NO_3)_2 \cdot 6\ H_2O$

CHAPTER 5

16. (a) $CuCl_2$; (b) $AgNO_3$; (c) all soluble
18. (a) CH_3COONa; (b) CuS; (c) NaOH; (d) AgCl
20. (a) Na^+ and I^-; (b) 2 K^+ and SO_4^{2-} (c) K^+ and HSO_4^-; (d) Na^+ and CN^-
22. (a) soluble, Ba^{2+} and 2 Cl^-
(b) soluble, Cr^{2+} and 2 NO_3^-
(c) soluble, Pb^{2+} and 2 NO_3^-
(d) insoluble
24. Soluble compound: use Cl^-, Br^-, NO_3^-, or CH_3COO^- (acetate)
Insoluble: use S^{2-} or CO_3^{2-}, for example
26. $HNO_3(aq) + H_2O(\ell) \rightarrow H_3O^+(aq) + NO_3^-(aq)$

28. The complete, balanced equation is given first followed by the net ionic equation. Where hydronium ions are involved, the equation is first balanced using hydrogen ion and then with hydronium ion.

(a) $Zn + 2\ HCl \rightarrow H_2 + ZnCl_2$
$Zn(s) + 2\ H^+(aq) \rightarrow H_2(g) + Zn^{2+}(aq)$
$Zn(s) + 2\ H_3O^+(aq) \rightarrow H_2(g) + 2\ H_2O(\ell) + Zn^{2+}(aq)$

(b) $Mg(OH)_2 + 2\ HCl \rightarrow MgCl_2 + 2\ H_2O$
$Mg(OH)_2(s) + 2\ H^+(aq) \rightarrow Mg^{2+}(aq) + 2\ H_2O(\ell)$
$Mg(OH)_2(s) + 2\ H_3O^+(aq) \rightarrow Mg^{2+}(aq) + 4\ H_2O(\ell)$

(c) $2\ HNO_3 + CaCO_3 \rightarrow Ca(NO_3)_2 + H_2O + CO_2$
$2\ H^+(aq) + CaCO_3(s) \rightarrow Ca^{2+}(aq) + H_2O(\ell) + CO_2(g)$
$2\ H_3O^+(aq) + CaCO_3(s) \rightarrow Ca^{2+}(aq) + 3\ H_2O(\ell) + CO_2(g)$

(d) $4\ HCl + MnO_2 \rightarrow MnCl_2 + Cl_2 + 2\ H_2O$
$4\ H^+(aq) + 2\ Cl^-(aq) + MnO_2(s) \rightarrow Mn^{2+}(aq) + Cl_2(g) + 2\ H_2O(\ell)$
$4\ H_3O^+(aq) + 2\ Cl^-(aq) + MnO_2(s) \rightarrow Mn^{2+}(aq) + Cl_2(g) + 6\ H_2O\ (\ell)$

30. The complete, balanced equation is given first followed by the net ionic equation. Where hydronium ions are involved, the equation is first balanced using hydrogen ion and then with hydronium ion.

(a) $Ba(OH)_2(s) + HNO_3(aq) \rightarrow Ba(NO_3)_2(aq) + 2\ H_2O(\ell)$
$Ba(OH)_2(s) + 2\ H^+(aq) \rightarrow Ba^{2+}(aq) + 2\ H_2O(\ell)$
$Ba(OH)_2(s) + 2\ H_3O^+(aq) \rightarrow Ba^{2+}(aq) + 4\ H_2O(\ell)$

(b) $BaCl_2(aq) + Na_2CO_3(aq) \rightarrow BaCO_3(s) + NaCl(aq)$
$Ba^{2+}(aq) + CO_3^{2-}(aq) \rightarrow BaCO_3(s)$

(c) $Na_2S(aq) + Ni(NO_3)_2(aq) \rightarrow NiS(s) + 2\ NaNO_3(aq)$
$S^{2-}(aq) + Ni^{2+}(aq) \rightarrow NiS(s)$

32. (a) Acid-base reaction
$HCl(aq) + KOH(aq) \rightarrow H_2O(\ell) + KCl(aq)$

(b) Precipitation reaction
$AgNO_3(aq) + KCl(aq) \rightarrow AgCl(s) + KNO_3(aq)$

(c) Acid–base reaction
$H_2SO_4(aq) + 2\ NaOH(aq) \rightarrow Na_2SO_4(aq) + 2\ H_2O(\ell)$

34. (a) Gas-forming reaction
$SrCO_3(s) + 2\ HNO_3(aq) \rightarrow Sr(NO_3)_2(aq) + CO_2(g) + H_2O(\ell)$

(b) Precipitation reaction
$BaCl_2(aq) + H_2SO_4(aq) \rightarrow 2\ HCl(aq) + BaSO_4(s)$

(c) Precipitation reaction
$MnCl_2(aq) + (NH_4)_2S(aq) \rightarrow 2\ NH_4Cl(aq) + MnS(s)$

36. (a) $NaOH(aq) + HNO_3(aq) \rightarrow NaNO_3(aq) + H_2O(\ell)$
(b) $KOH(aq) + HCl(aq) \rightarrow KCl(aq) + H_2O(\ell)$
(c) $3\ KOH(aq) + H_3PO_4(aq) \rightarrow K_3PO_4(aq) + 3\ H_2O(\ell)$
(d) $2\ CsOH(aq) + H_2SO_4(aq) \rightarrow Cs_2SO_4(aq) + 2\ H_2O(\ell)$

38. (a) $Ni(NO_3)_2(aq) + 2\ NaOH(aq) \rightarrow Ni(OH)_2(s) + 2\ NaNO_3(aq)$
(b) $Na_2CO_3(aq) + CaCl_2(aq) \rightarrow CaCO_3(s) + 2\ NaCl(aq)$
(c) $NiCl_2(aq) + Na_2S(aq) \rightarrow NiS(s) + 2\ NaCl(aq)$
(d) $BaCl_2(aq) + H_2SO_4(aq) \rightarrow 2\ HCl(aq) + BaSO_4(s)$

40. (a) $K_2CO_3(aq) + 2\ HNO_3(aq) \rightarrow 2\ KNO_3(aq) + CO_2(g) + H_2O(\ell)$
(b) $CaCO_3(s) + 2\ HCl(aq) \rightarrow CaCl_2(aq) + CO_2(g) + H_2O(\ell)$
(c) $FeCO_3(s) + 2\ HNO_3(aq) \rightarrow Fe(NO_3)_2(aq) + CO_2(g) + H_2O(\ell)$

42. (a) $NaOH(aq) + HNO_3(aq) \rightarrow NaNO_3(aq) + H_2O(\ell)$
(b) $SrCO_3(s) + H_2SO_4(aq) \rightarrow SrSO_4(s) + CO_2(g) + H_2O(\ell)$
(c) $ZnCl_2(aq) + (NH_4)_2S(aq) \rightarrow ZnS(s) + 2\ NH_4Cl(aq)$
(d) $BaCO_3(s) + 2\ HCl(aq) \rightarrow BaCl_2(aq) + CO_2(g) + H_2O(\ell)$
(e) $NH_3(aq) + H_2SO_4(aq) \rightarrow (NH_4)_2SO_4(aq)$

44. 0.158 M; 211 mL

46. (a) 1.7 g; (b) 0.01 g; (c) 2.1 g

48. (a) 0.92 M $NaClO_4$; (b) 0.23 M KNO_3; and (c) 0.083 M C_4H_8O

50. $[Na^+] = [Br^-] = 1.0$ M; (b) $[Na^+] = 1.5$ M and $[PO_4^{3-}] = 0.50$ M; (c) $[Zn^{2+}] = 0.10$ M and $[Cl^-] = 0.20$ M

52. 30. mL

54. 0.0500 M HCl

56. 0.030 M $CuSO_4$

58. Take 0.100 L of 0.500 M $KMnO_4$ solution and dilute to 2.50 L.

60. 0.15 g H_2; 5.3 g Cl_2; and 6.0 g NaOH

62. 11.3 g $NaBH_4$; 171 mL H_2SO_4

64. 52.0 mL $Na_2S_2O_3$

66. 8.889×10^{-3} mol AgCl require 1.778×10^{-2} mol NH_3. This quantity of NH_3 would be furnished by 4.0 mL of 4.5 M NH_3. Therefore, there is enough NH_3 to dissolve the AgCl.

68. 250. mL HCl

70. 258 mL H_2SO_4

72. 0.112 M

74. 1.04 g acetic acid

76. 39.0% oxalic acid

78. 24.2% Ba

80. 0.0155 g Al^{3+}; 5.75×10^{-3} M Al^{3+}

82. (a) Br = +1 and O = −2; (b) C = +3 and O = −2; (c) I = 0; (d) I = +5 and O = −2; (e) H = +1, Cl = +7, and O = −2; (f) S = +6 and O = −2

84. (a) Mg is oxidized (to Mg^{2+} in MgO) and O_2 is reduced (O changes from 0 in O_2 to −2 in O^{2-}).

(b) C_2H_4 is reduced [C is oxidized from -2 in C_2H_4 to $+4$ in CO_2] and O_2 is oxidized [O changes from 0 in O_2 to -2 in CO_2 and H_2O].
(c) Si is oxidized to $+4$ (in $SiCl_4$) and Cl_2 is reduced [Cl changes from 0 in Cl_2 to -1 in $SiCl_4$].

86. (a) $Cr \rightarrow Cr^{3+} + 3\ e^-$
Cr is a reducing agent; overall an oxidation.
(b) $Fe^{3+} + e^- \rightarrow Fe^{2+}$
Fe^{3+} is an oxidizing agent; overall a reduction.
(c) $AsH_3 \rightarrow As + 3\ H^+ + 3\ e^-$
AsH_3 is a reducing agent; overall an oxidation.
(d) $VO_3^- + 6\ H^+ + 3\ e^- \rightarrow V^{2+} + 3\ H_2O$
VO^{3-} is an oxidizing agent; overall a reduction.

88. (a) $Cr_2O_7^{2-} + 14\ H^+ + 6\ e^- \rightarrow 2\ Cr^{3+} + 7\ H_2O$
$Cr_2O_7^{2-}$ is an oxidizing agent; overall a reduction.
(b) $N_2H_5^+ \rightarrow N_2 + 5\ H^+ + 4\ e^-$
$N_2H_5^+$ is a reducing agent; overall an oxidation.
(c) $CH_3CHO + H_2O \rightarrow CH_3COOH + 2\ H^+ + 2\ e^-$
CH_3CHO is a reducing agent; overall an oxidation.
(d) $Bi^{3+} + 3\ H_2O \rightarrow HBiO_3 + 5\ H^+ + 2\ e^-$
Bi^{3+} is a reducing agent; overall an oxidation.

90. (a) $Sn + 4\ OH^- \rightarrow Sn(OH)_4^{2-} + 2\ e^-$
Sn is a reducing agent; overall an oxidation.
(b) $MnO_4^- + 2\ H_2O + 3\ e^- \rightarrow MnO_2 + 4\ OH^-$
Permanganate is an oxidizing agent; overall a reduction.
(c) $ClO^- + 2\ H_2O + 2\ e^- \rightarrow Cl^- + 2\ OH^-$
Hypochlorite is an oxidizing agent; overall a reduction.

92. (a)
$$\begin{array}{l} Cl_2 + 2\ e^- \longrightarrow 2\ Cl^- \\ 2\ Br^- \longrightarrow Br_2 + 2\ e^- \\ \hline Cl_2 + 2\ Br^- \longrightarrow 2\ Cl^- + Br_2 \end{array}$$
(b)
$$\begin{array}{l} Sn \longrightarrow Sn^{2+} + 2\ e^- \\ 2\ H^+ + 2\ e^- \longrightarrow H_2 \\ \hline Sn + 2\ H^+ \longrightarrow Sn^{2+} + H_2 \end{array}$$
(c)
$$\begin{array}{l} 2[Al \longrightarrow Al^{3+} + 3\ e^-] \\ 3[Sn^{4+} + 2\ e^- \longrightarrow Sn^{2+}] \\ \hline 2\ Al + 3\ Sn^{4+} \longrightarrow 2\ Al^{3+} + 3\ Sn^{2+} \end{array}$$
(d)
$$\begin{array}{l} Zn \longrightarrow Zn^{2+} + 2\ e^- \\ 2[VO^{2+} + 2\ H^+ + e^- \longrightarrow V^{3+} + H_2O] \\ \hline Zn + 2\ VO^{2+} + 4\ H^+ \longrightarrow Zn^{2+} + 2\ V^{3+} + 2\ H_2O \end{array}$$

94. (a)
$$\begin{array}{l} 2[Ag^+ + e^- \longrightarrow Ag] \\ HCHO + H_2O \longrightarrow HCOOH + 2\ H^+ + 2\ e^- \\ \hline 2\ Ag^+ + HCHO + H_2O \longrightarrow 2\ Ag + HCOOH + 2\ H^+ \end{array}$$
(b)
$$\begin{array}{l} 2[MnO_4^- + 8\ H^+ + 5\ e^- \longrightarrow Mn^{2+} + 4\ H_2O] \\ 5[C_2H_5OH \longrightarrow C_2H_4O + 2\ H^+ + 2\ e^-] \\ \hline 5C_2H_5OH + 2\ MnO_4^- + 6\ H^+ \longrightarrow 5\ C_2H_4O + 2\ Mn^+ + 8\ H_2O \end{array}$$
(c)
$$\begin{array}{l} Cr_2O_7^{2-} + 14\ H^+ + 6\ e^- \longrightarrow 2\ Cr^{3+} + 7\ H_2O \\ 3[H_2S \longrightarrow S + 2\ H^+ + 2\ e^-] \\ \hline 3\ H_2S + Cr_2O_7^{2-} + 8\ H^+ \longrightarrow 3\ S + 2\ Cr^{3+} + 7\ H_2O \end{array}$$
(d)
$$\begin{array}{l} 3[Zn \longrightarrow Zn^{2+} + 2\ e^-] \\ 2[VO_3^- + 6\ H^+ + 3\ e^- \longrightarrow V^{2+} + 3\ H_2O] \\ \hline 3\ Zn + 2\ VO_3^- + 12\ H^+ \longrightarrow 3\ Zn^{2+} + 2\ V^{2+} + 6\ H_2O \end{array}$$
(e)
$$\begin{array}{l} 5[U^{4+} + 2\ H_2O \longrightarrow UO_2^+ + 4\ H^+ + e^-] \\ MnO_4^- + 8\ H^+ + 5\ e^- \longrightarrow Mn^{2+} + 4\ H_2O \\ \hline 5\ U^{4+} + MnO_4^- + 6\ H_2O \longrightarrow 5\ UO_2^+ + Mn^{2+} + 12\ H^+ \end{array}$$

96. (a)
$$\begin{array}{l} Zn + 2\ OH^- \longrightarrow Zn(OH)_2 + 2\ e^- \\ ClO^- + H_2O + 2\ e^- \longrightarrow Cl^- + 2\ OH^- \\ \hline Zn + ClO^- + H_2O \longrightarrow Cl^- + Zn(OH)_2 \end{array}$$

(b) $3[ClO^- + H_2O + 2e^- \longrightarrow Cl^- + 2OH^-]$
$2[CrO_2^- + 4OH^- \longrightarrow CrO_4^{2-} + 2H_2O + 3e^-]$

$3ClO^- + 2CrO_2^- + 2OH^- \longrightarrow 3Cl^- + 2CrO_4^{2-} + H_2O$

(c) $5[Br_2 + 2e^- \longrightarrow 2Br^-]$
$Br_2 + 12OH^- \longrightarrow 2BrO_3^- + 6H_2O + 10e^-$

$6Br_2 + 12OH^- \longrightarrow 10Br^- + 2BrO_3^- + 6H_2O$
Or dividing the final coefficients by 2,
$3Br_2 + 6OH^- \longrightarrow 5Br^- + BrO_3^- + 3H_2O$

98. $[KMnO_4] = 0.01650$ M
100. 198 mL $KMnO_4$
102. 2.04% iron
104. 31.7% lead
106. (a) insoluble; (b) soluble, Zn^{2+} and CH_3COO^-; (c) soluble, K^+ and CN^-; (d) soluble, Cu^{2+} and ClO_4^-; (e) soluble, K^+ and $Cr_2O_7^{2-}$; (f) insoluble
108. (a) $CaCO_3(s) + 2HNO_3(aq) \rightarrow Ca(NO_3)_2(aq) + CO_2(g) + H_2O(\ell)$
(b) $SrCO_3(s)$ (+ heat) $\rightarrow SrO(s) + CO_2(g)$
(c) $K_2CO_3(aq) + H_2SO_4(aq) \rightarrow K_2SO_4(aq) + CO_2(g) + H_2O(\ell)$
(d) $CuCO_3(s) + 2HClO_4(aq) \rightarrow Cu(ClO_4)_2(aq) + CO_2(g) + H_2O(\ell)$
110. 0.040 M
112. 230 g $CuSO_4 \cdot 5H_2O$
114. 6.3×10^{-4} M
116. (a) 0.112 g MnO_2; (b) 116 mL $S_2O_3^{2-}$ solution: (c) 34.1 g $Na_2S_2O_3$
118. potassium biphthalate
120. (a) 8.13% water; (b) 55.1% $BaCl_2 \cdot 2H_2O$; (c) 60.6% Ba
122. 8.20% Cr
124. 120. mL $Na_2S_2O_3$
126. 25.8 g H_2PtCl_6; (b) 2.52 g NO; (c) 8.41 mL HNO_3; (d) Pt

CHAPTER 6

9. (a) 399 Calories; (b) 5.0×10^6 J; (c) 4.6×10^5 J
11. 11.2 kJ
13. 101 kJ
15. 6.9×10^4 J
17. 3.30×10^4 J or 33.0 kJ
19. 330. °C
21. 181 kJ
23. 24.3 J/mol · K; (b) 25.2 J/mol · K; (c) 132 J/mol · K
25. endothermic
27. $\Delta H_{combustion} = -1370$ kJ/mol
29. 35.5 kJ
31. −1220 kJ/mol
33. The process evolves 1280 kJ per kilogram of PbS
35. 597 kJ
37. (a) $2Al(s) + \frac{3}{2}O_2(g) \rightarrow Al_2O_3(s)$ $\Delta H_f^\circ = -1675.7$ kJ/mol
(b) $Ti(s) + 2Cl_2(g) \rightarrow TiCl_4(\ell)$ $\Delta H_f^\circ = -804.2$ kJ/mol
(c) $Mg(s) + 2H_2(g) + \frac{3}{2}O_2(g) \rightarrow Mg(OH)_2(s)$ $\Delta H_f^\circ = -924.5$ kJ/mol
(d) $N_2(g) + 2H_2(g) + \frac{3}{2}O_2(g) \rightarrow NH_4NO_3(s)$ $\Delta H_f^\circ = -365.6$ kJ/mol
(e) $C(graphite) + \frac{1}{2}O_2(g) + Cl_2(g) \rightarrow COCl_2(g)$ $\Delta H_f^\circ = -218.8$ kJ/mol
39. $\Delta H = -98.8$ kJ
41. (a) endothermic; (b) + 890.3 kJ
43. −69.1 kJ
45. +17.6 kJ/mol
47. −178.3 kJ
49. $\Delta H_{rxn} = 83.8$ kJ
51. 13.6 kJ
53. $N_2H_2(CH_3)_2$ provides 29.9 kJ/g, while N_2H_4 produces only 16.7 kJ/g.
55. −217.3 kJ/mol
57. Heat absorbed by water = 4460 J; heat absorbed by bomb = 2160 J; total heat evolved by reaction = 6.62 kJ
59. 394 kJ/mol
61. $\Delta H_{neut} = -56.1$ kJ/mol
63. Gold
65. $\Delta H^\circ_{rxn} = -466.8$ kJ
67. (a) endothermic; $\Delta H^\circ = 84.5$ kJ/mol; (b) 2.66 kJ
69. 2.21×10^7 kJ
71. $\Delta H_f^\circ[C_2H_4Cl_2(g)] = -64$ kJ/mol
73. $\Delta H_f^\circ(N_2H_4(\ell)] = 50.$ kJ/mol
75. 6.76×10^4 kJ
77. 461.7 kJ produced per mole of $UF_6(g)$; 2.68×10^8 kJ produced on making 225 tons
79. −11 kJ
81. 4.4 mol CH_4; 70. g
83. Assume the products of $CH_3OH(\ell)$ and C_8H_{18} combustion are both $CO_2(g)$ and $H_2O(g)$. Methyl alcohol produces 19.9 kJ/g, while octane produces 48.1 kJ/g.
84. (a) 1.67×10^8 g SO_2; (b) 1.3×10^7 tons MgO; (c) $\Delta H^\circ_{rxn} = -386.4$ kJ; (d) 3.65×10^9 kJ

CHAPTER 7

12. (a) ${}^{56}_{26}Fe$; (b) ${}^{1}_{0}n$; (c) ${}^{32}_{15}P$; (d) ${}^{97}_{43}Tc$; (e) ${}^{0}_{-1}e$

14. (a) ${}^{0}_{-1}e$; (b) ${}^{87}_{37}\mathrm{Rb}$; (c) ${}^{4}_{2}\mathrm{He}$; (d) ${}^{226}_{88}\mathrm{Ra}$; (e) ${}^{0}_{-1}e$; (f) ${}^{24}_{11}\mathrm{Na}$

16. ${}^{235}_{92}\mathrm{U} \xrightarrow{-\alpha} {}^{231}_{90}\mathrm{Th} \xrightarrow{-\beta} {}^{231}_{91}\mathrm{Pa} \xrightarrow{-\alpha} {}^{227}_{89}\mathrm{Ac} \xrightarrow{-\beta}$

${}^{227}_{90}\mathrm{Th} \xrightarrow{-\alpha} {}^{223}_{88}\mathrm{Ra} \xrightarrow{-\alpha} {}^{219}_{86}\mathrm{Rn} \xrightarrow{-\alpha} {}^{215}_{84}\mathrm{Po} \xrightarrow{-\alpha}$

${}^{211}_{82}\mathrm{Pb} \xrightarrow{-\beta} {}^{211}_{83}\mathrm{Bi} \xrightarrow{-\beta} {}^{211}_{84}\mathrm{Po} \xrightarrow{-\alpha} {}^{207}_{82}\mathrm{Pb}$

18. For ${}^{10}\mathrm{B}$, $\Delta E = -6.26 \times 10^{12}$ kJ/mol, while for ${}^{11}\mathrm{B}$, $\Delta E = -7.37 \times 10^{12}$ kJ/mol. Boron-11 is slightly more stable than boron-10.

20. 2 days and 16 hours represents 5 half lives. Therefore, 15.0 mg declines to 0.469 mg.

22. (a) ${}^{131}_{53}\mathrm{I} \rightarrow {}^{0}_{-1}e + {}^{131}_{54}\mathrm{Xe}$
(b) 32.2 days is equivalent to 4 half lives. Therefore, 25.0 mg declines to 1.56 mg.

24. 21.2 years represent 4 half lives. Therefore, 10.0 mg declines to 0.625 mg. After 100 years, about 2.10×10^{-5} mg remain.

26. 34.8 days

28. 0.23/hr; $t_{1/2} = 3.00$ hours

30. (a) ${}^{226}_{88}\mathrm{Ra} \rightarrow {}^{4}_{2}\mathrm{He} + {}^{222}_{86}\mathrm{Rn}$
(b) 2.17×10^{9} per minute

32. The object is about 1850 years old, and so was made in approximately 150 AD.

34. ${}^{239}_{94}\mathrm{Pu} + 2\ {}^{1}_{0}n \rightarrow {}^{0}_{-1}e + {}^{241}_{95}\mathrm{Am}$

36. ${}^{238}_{92}\mathrm{U} + {}^{12}_{6}\mathrm{C} \rightarrow {}^{246}_{98}\mathrm{Cf} + 4\ {}^{1}_{0}n$

38. 1600 tons of coal are equivalent in energy production to one pound of ${}^{235}\mathrm{U}$.

40. 130 mL

42. (a) ${}^{1}_{0}n$; (b) ${}^{4}_{2}\mathrm{He}$; (c) ${}^{257}_{103}\mathrm{Lr}$; (d) ${}^{1}_{0}n$; (e) ${}^{4}_{2}\mathrm{He}$; (f) ${}^{122}_{52}\mathrm{Te}$; (g) ${}^{23}_{10}\mathrm{Ne}$; (h) ${}^{136}_{53}\mathrm{I}$; (i) ${}^{21}_{10}\mathrm{Ne}$.

44. (a) ${}^{247}_{99}\mathrm{Es}$; (b) ${}^{16}_{8}\mathrm{O}$; (c) ${}^{4}_{2}\mathrm{He}$; (d) ${}^{12}_{6}\mathrm{C}$; (e) ${}^{10}_{5}\mathrm{B}$

46. 1.2×10^{15} atoms or 2.9×10^{-7} g Pm

CHAPTER 8

22. (a) red and orange; (b) violet; (c) blue

24. 3.9×10^{6} m; 2.5×10^{3} miles

26. $6.00 \times 10^{14}\ \mathrm{s}^{-1}$; 3.98×10^{-19} J/photon; 2.39×10^{5} J/mol

28. Ultraviolet; 9.94×10^{-19} J

30. $5.090 \times 10^{14}\ \mathrm{s}^{-1}$; 2.03×10^{5} J/mol

32. radar < microwave < red light < ultraviolet < gamma rays

34. 5.44×10^{-7} m or 544 nm; visible light

36. (a) 6 lines: (b) from $n = 4$ to $n = 1$; (c) from $n = 4$ to $n = 3$

38. (a), (b), and (d)

40. Wavelength of light emitted for a transition from $n = 4$ to $n = 1$ is 9.7199×10^{-8} m or approximately 97 nm (in the far ultraviolet).

42. 2.093×10^{-18} J absorbed

44. 2.4×10^{-10} m or 0.24 nm

46. 9.9×10^{-7} m (990 nm)

48. n gives orbital size; m_ℓ gives orbital orientation; and ℓ gives orbital shape.

50. (a) When $n = 3$, ℓ can be 0, 1, and 2
(b) When $\ell = 3$, then $m_\ell = -3, -2, -1, 0, +1, +2$, and $+3$. An ℓ of 3 designates f atomic orbitals (of which there are 7).
(c) For a 4s orbital $n = 4$, $\ell = 0$, and $m_\ell = 0$.
(d) For a 5f orbital, $n = 5$, ℓ must be 3, and m_ℓ has an integer value from -3 to $+3$ including 0.

52. A 4p electron has $n = 4$, ℓ must be 1, and m_ℓ can be -1, 0, or $+1$.

54. For a shell with $n = 4$ there are 4 subshells (since there are n values of ℓ).

56. (a) ℓ cannot be equal to n; (b) and (c) the value of m_ℓ cannot exceed the value of ℓ.

58. (a) 5; (b) 25; (c) none; (d) 1

60. (a) none; (b) 2; (c) 3

62. 2d: An orbital with $\ell = 2$ is not allowed when $n = 2$.
3f: An orbital with $\ell = 3$ is not allowed when $n = 3$.

64.

Orbital	n	ℓ	m_ℓ
2p	2	1	+1, 0, or −1
3d	3	2	+2, +1, 0, −1, or −2
4f	4	3	+3, +2, 0, −1, −2, or −3

66. (a) 1s: see Figure 8.13; (b) $2p_x$: see Figure 8.14; (c) $3d_{xy}$: see Figure 8.16

68. s orbital

70. 500. meters

72. 1×10^{19} photons/sec

74. -1.6×10^{-19} C

76. -82.049 kJ/mol

78. $1s < 2s = 2p < 3s = 3p = 3d < 4s$

80. (a) n describes the size of an orbital, while ℓ describes its shape.
(b) $\ell = 0, 1, 2$
(c) $\ell = 3$ describes an f orbital
(d) For 4d, $n = 4$, $\ell = 2$, and a possible value of m_ℓ is $+2$ (other possible values are 1, 0, -1, and -2).
(e) Letter = p, s, d
ℓ value = 1, 0, and 2
Nodal planes = 1, 0, and 2
(f) An f orbital has three nodal planes
(g) A 1d orbital is not possible (since ℓ cannot be 2 when $n = 1$). A 4g orbital is not possible (since a g orbital has $\ell = 4$, and ℓ cannot be equal to n).
(h) The second set ($n = 2$, $\ell = 1$, and $m_\ell = 2$) is not a valid set of quantum numbers.
(i) (i) 3; (ii) 9; (iii) none: (iv) 1

CHAPTER 9

20. Magnesium: $1s^22s^22p^63s^2$

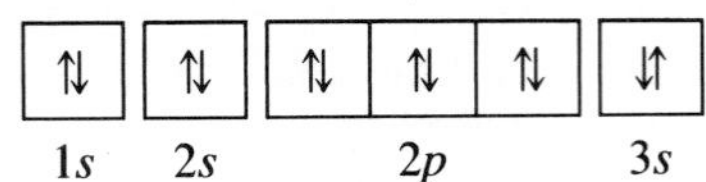

Chlorine: $1s^22s^22p^63s^23p^5$

22. Vanadium: $1s^22s^22p^63s^23p^63d^34s^2$ or $[Ar]3d^34s^2$

24. Germanium: $1s^22s^22p^63s^23p^63d^{10}4s^24p^2$ or $[Ar]3d^{10}4s^24p^2$

26. See Table 9.2 for electron configurations.
(a) Strontium: $[Kr]5s^2$
(b) Zirconium: $[Kr]4d^25s^2$
(c) Rhodium: The expected configuration is $[Kr]4d^75s^2$, but the actual configuration is $[Kr]4d^85s^1$.
(d) Tin: $[Kr]4d^{10}5s^25p^2$

28. See Table 9.2 for electron configurations.
(a) Europium, Eu: The expected configuration is $[Xe]4f^65d^16s^2$. However, owing to slight differences in orbital energies, the actual configuration is $[Xe]4f^76s^2$.
(b) Ytterbium, Yb: The expected configuration is $[Xe]4f^{13}5d^16s^2$. However, owing to slight differences in orbital energies, the actual configuration is $[Xe]4f^{14}6s^2$.

30. See Table 9.2 for electron configurations.
(a) Plutonium, Pu: The expected configuration is $[Rn]5f^56d^17s^2$. However, owing to slight differences in orbital energies, the actual configuration is $[Rn]5f^67s^2$.
(b) Einsteinium, Es: The expected configuration is $[Rn]5f^{10}6d^17s^2$. However, owing to slight differences in orbital energies, the actual configuration is $[Rn]5f^{11}7s^2$.

32. (a) Na^+

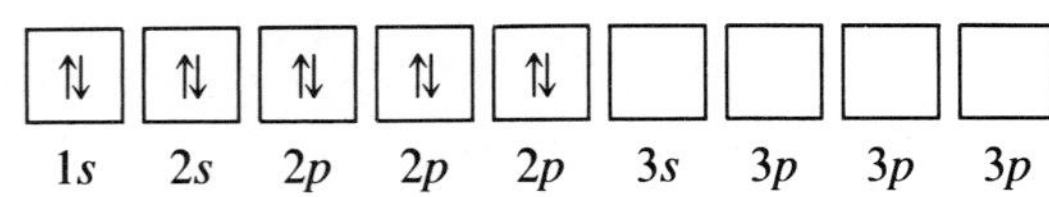

(b) Al^{3+}, same as Na^+
(c) Cl^-

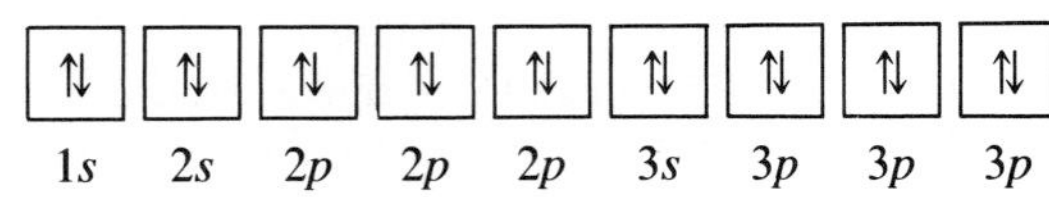

34. (a) Ti

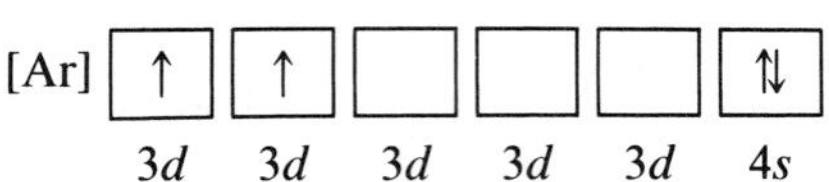

(b) Ti^{2+}: paramagnetic

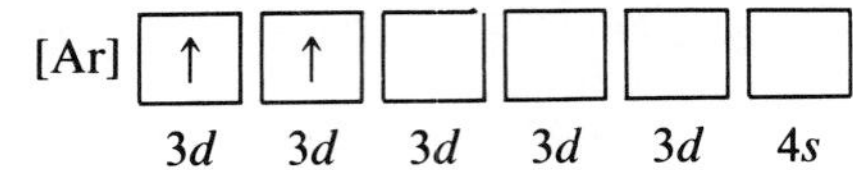

(c) Ti^{4+}: no 3*d* or 4*s* electrons.

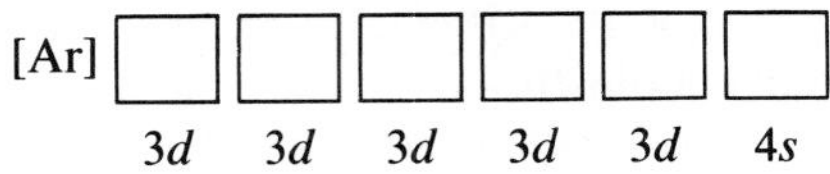

36. (a) Manganese, Mn: $[Ar]3d^54s^2$

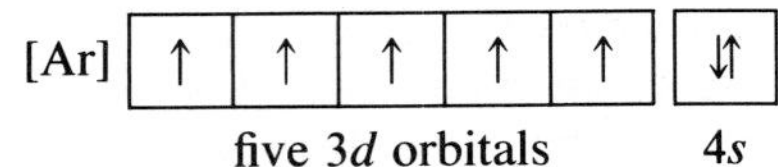

(b) Mn^{2+}: $[Ar]3d^5$

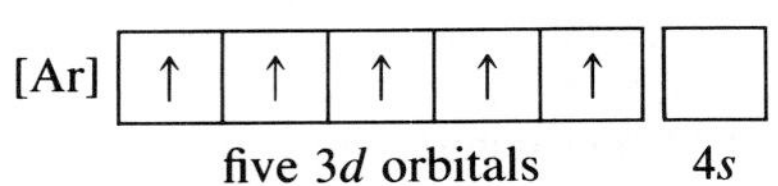

(c) paramagnetic (5 unpaired electrons)

38. (a) Ruthenium, Ru: The expected configuration is $[Kr]4d^65s^2$, but the actual configuration is $[Kr]4d^75s^1$.

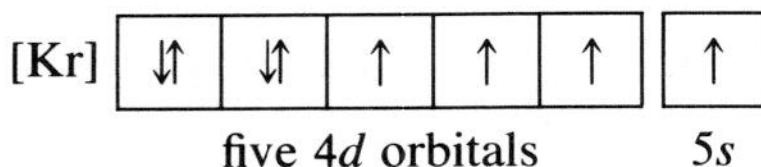

(b) Ru^{3+}:

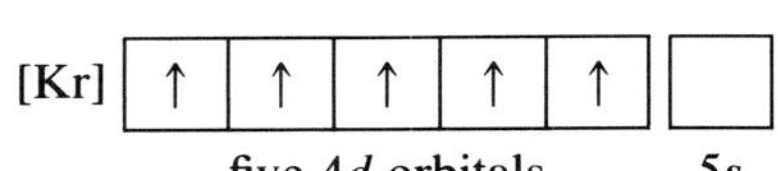

The configuration above can be obtained beginning with $[Kr]4d^65s^2$ or $[Kr]4d^75s^1$.

40. (a) Samarium, Sm: $[Xe]4f^55d^15s^2$ is the expected configuration, but the actual configuration is $[Xe]4f^65s^2$.

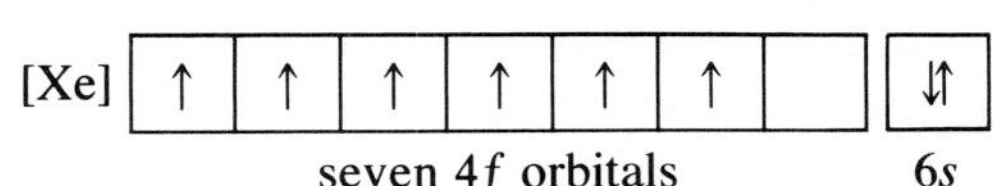

Samarium(III), Sm^{3+}:

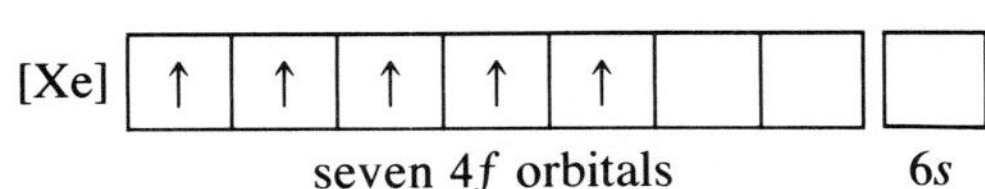

Note that we would arrive at the Sm^{3+} configuration no matter which configuration of Sm was used.

(b) Holmium, Ho: $[Xe]4f^{10}5d^{1}5s^{2}$ is the expected configuration, but the actual configuration is $[Xe]4f^{11}5s^{2}$.

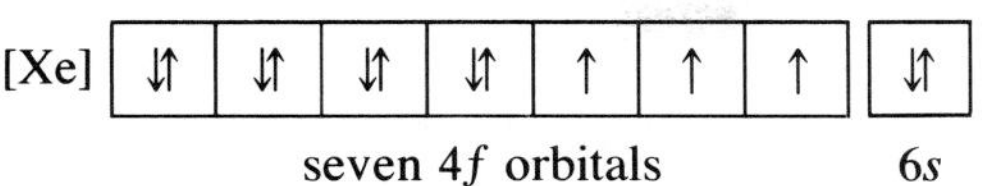

Holmium(III), Ho^{3+}:

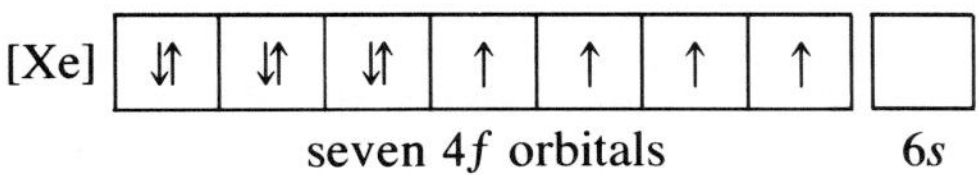

Note that we would arrive at the Ho^{3+} configuration no matter which configuration of Ho was used.

42. Ti^{2+} has 2 unpaired electrons, and Co^{3+} has 4 unpaired electrons. (Both are paramagnetic.)

44. (a) Zn^{2+} is the only diamagnetic 2^+ ion based on elements between Ti and Zn.
(b) Mn^{2+} has the largest number of unpaired electrons (5).

46. Orbital box diagram for Be:

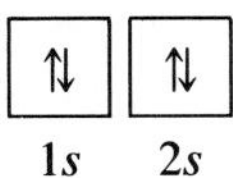

Electron	n	ℓ	m_ℓ	m_s
1*s*	1	0	0	$+\frac{1}{2}$
1*s*	1	0	0	$-\frac{1}{2}$
2*s*	2	0	0	$+\frac{1}{2}$
2*s*	2	0	0	$-\frac{1}{2}$

48. Titanium, Ti, $[Ar]3d^{2}4s^{2}$

[Ar] ↑ ↑ □ □ □ ⇅
five 3*d* orbitals 4*s*

Electron	n	ℓ	m_ℓ	m_s
3*d*	3	2	−2	$+\frac{1}{2}$
3*d*	3	2	−1	$+\frac{1}{2}$
4*s*	4	0	0	$+\frac{1}{2}$
4*s*	4	0	0	$-\frac{1}{2}$

Note that other choices for m_ℓ are possible for the 3*d* electrons.

50. (a) 6; (b) 50; (c) none; (d) 1; (e) none

52. (a) ℓ cannot equal n.
(b) m_s cannot have a value of 0.

54.

Combination	n	ℓ	m_ℓ	m_s
1	4	1	+1	$+\frac{1}{2}$
2	4	1	0	$+\frac{1}{2}$
3	4	1	−1	$+\frac{1}{2}$
4	4	1	+1	$-\frac{1}{2}$
5	4	1	0	$-\frac{1}{2}$
6	4	1	−1	$-\frac{1}{2}$

56. (a) C—Cl = 77 pm + 100. pm = 177 pm; (b) Si—Cl = 218 pm; (c) Ge—Cl = 222 pm; (d) Sn—Cl = 240. pm; (e) Pb—Cl = 246 pm

58. C (77 pm) < B (85 pm) < Al (143 pm) < Na (186 pm) < K (227 pm)

60. (a) Cl^-; (b) Al; (c) In

62. (c)

64. K (419 kJ) < Li (520 kJ) < C (1086 kJ) < O (1314 kJ)

66. 1st IE of K (419 kJ) < 1st IE of Li (520 kJ) < 1st IE of Be (899 kJ) < 2nd IE of Be (1757 kJ) < 2nd IE of Na (4562 kJ)
The trend in first ionization energies is expected based on the positions of the elements in the periodic table. The first IE's are expected to be smaller than the 2nd IE's. Finally, the 2nd IE of Na is expected to be very large, since the 2nd electron removed is from an inner electron shell.

68. (a) K; (b) C; (c) K < Li < C < N

70. (a) S < O < F. IE increases across a period but decreases down a group.
(b) O has the largest IE for the same reason as in (a).
(c) Cl
(d) O^{2-}

72. Group 2A (as well as Sn^{2+} and Pb^{2+})

74. $[Rn]5f^{14}6d^{7}7s^{2}$. Element 109 is in the same group as Co, Rh, and Ir.

76. Set (b) is not allowed, since n and ℓ cannot have the same value.

78. 4 complete shells (see Table 9.2).

80. 7 *p* electron pairs in Se.

82. In general, IE increases across a period. Here, however, S is less than P, just as O is less than N in the second period. This effect arises from electron-electron repulsions.

84. Tc^{3+} and Rh^{3+}

86. (a) metal; (b) nonmetal; (c) B; (d) B

88. $S^{2-} > Cl^- > K^+ > Ca^{2+}$

90. In the Group 4A element C an added electron is assigned to a previously unoccupied orbital. In contrast, in N the added electron is assigned an orbital already occupied by an electron. Electron-electron repulsions in N cause the EA to be less negative than expected.

92. Since IE's increase across a period, the first IE of Mg is greater than that of Na. However, after each has lost an electron, Na has the rare gas configuration of Ne ($= 1s^{2}2s^{2}2p^{6}$) whereas Mg^+ has $[Ne]3s^{1}$. To remove an inner electron from Na^+ requires more energy than removing a second 3*s* electron from Mg^+.

94. (a) Na; (b) C; (c) Na < Al < B < C

96. (a) Third period.
(b) S electron configuration

⇅ ⇅ ⇅ ⇅ ⇅ ⇅ ⇅ ↑ ↑
1*s* 2*s* 2*p* 3*s* 3*p*

(c) $n = 3$, $\ell = 1$, $m_\ell = -1$ (or 0 or +1), and $m_s = +\frac{1}{2}$ (or $-\frac{1}{2}$)
(d) Smallest IE = S (Table 9.5); smallest radius = O (Figure 9.11)
(e) S (103 pm) is smaller than S^{2-} (170 pm)
(f) 5.67 mol $OSCl_2$ require 11.3 mol Cl_2 (or 804 g)
(g) Cl_2 is the limiting reagent. Leads to 16.8 g $OSCl_2$.
(h) −212.5 kJ/mol

CHAPTER 10

27.

Atom	Group	Number of Valence Electrons
N	5A	5
B	3A	3
S	6A	6
Na	1A	1
Mg	2A	2

29. (a) LiI; (b) LiF; (c) MgO
31. (a) NaI; (b) CsF; (c) BaO
33. (a) NaCl; (b) MgO; (c) $MgCl_2$
35. 1A = 1; 2A = 2; 3A = 3; 4A = 4; 5A = 3; 6A = 2; 7A = 1

37. (a) :F—N—F: with :F: bonded below N (c) H—O—Cl:
(b) $[$:O—Cl—O:$]^-$ with :O: bonded below Cl (d) $[$:O—S—O:$]^{2-}$ with :O: bonded below S

39. (a) :Cl—C—Cl: with H above C and :F: below C (c) H—C—C≡N: with H above and H below the first C
(b) H—C—O—H with :O: double-bonded above C (d) H—C—O—H with H above and H below C

41. (a) SO_2

O=S—O ⟷ O—S=O

(b) SO_3

O—S(=O)—O ⟷ O—S(—O)=O ⟷ O=S(—O)—O (resonance structures, each S bonded to one O above and two O below)

(c) $:\ddot{S}=C=\ddot{N}:^-$ ⟷ $:S\equiv C-\ddot{N}:^-$ ⟷ $:\ddot{S}-C\equiv N:^-$

43. (a) :F—Br—F: with :F: bonded below Br (b) $:\ddot{I}-\ddot{I}-\ddot{I}:^-$
(c) :O—Xe—O: with :F: bonded above and :F: bonded below Xe

45.

Molecule	σ Bonds	π Bonds	Bond Orders
HCOOH	4	1	H—C = 1, C—O = 1, C=O = 2, O—H = 1
(see question 39b)			
SO_3^{2-}	3	0	S—O = 1
NO_2^+	2	2	N=O = 2

47. (a) B—Cl; (b) C—O; (c) P—O; (d) C=O
49. CO bond is shorter and stronger in carbon monoxide than in H_2CO.
51. The average NO bond order in NO_2^+ is 2, whereas it is only 1.33 in NO_3^-. Therefore, the NO bonds in NO_3^- are longer than in NO_2^+.
53. (a) NH_3, −40. kJ/mol; (b) H_2O, −243 kJ/mol
55. ΔH of formation for $C_3H_8(g)$ = −111 kJ/mol
57. Enthalpy of formation of glycine = −311 kJ/mol
59. D_{H-I} = 298 kJ/mol. The trend in H—X bond energies is to lower values with increasing halogen atomic number.
61. D_{O-F} = 195 kJ/mol
63. Reaction enthalpy = −87 kJ/mol
65. Enthalpy of hydrogenation = −128 kJ
67. (a) C—O > C—N
(+→− under each bond)
(b) P—O > P—S
(+→− under each bond)
(c) P—N > P—H
(+→− under P—N; P—H not polar)
(d) B—I > B—H
(+→− under each bond)
69. (a) The C—H and C=O bonds are polar, while the C=C and C—C bonds are nonpolar.
(b) The most polar bond is C=O.
71. (a) H = +1 and O = −2
(b) H = +1 and O = −1
(c) S = +4 and O = −2
(d) N = +1 and O = −2
(e) Cl = +1 and O = −2
73. (a) H = 0 and O = 0
(b) C = 0 and H = 0
(c) N = +1 and O = 0
(d) H = 0, O = 0, and F = 0
75. $\ddot{O}=\ddot{N}-\ddot{O}:^-$ ⟷ $:\ddot{O}-\ddot{N}=\ddot{O}^-$
Formal charges: 0 0 −1 −1 0 0

77. The least favorable N_2O resonance structure is that with +1 on O (structure a). Structure c may be the most favorable, since the more electronegative atom (O) bears the negative charge.

$$:O\equiv N-\ddot{N}: \longleftrightarrow :\ddot{O}=N=\ddot{N}: \longleftrightarrow :\ddot{O}-N\equiv N:$$

(a)	(b)	(c)
+1 +1 −2	0 +1 −1	−1 +1 0

79. (a) 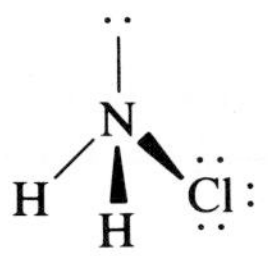 Structural pair geometry = tetrahedral; Molecular geometry = trigonal pyramid

(b) 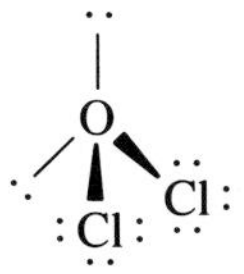 Structural pair geometry = tetrahedral; Molecular geometry = bent

(c) $:\ddot{S}=C=\ddot{N}:^-$ Structural pair geometry = linear; Molecular geometry = linear

(d) 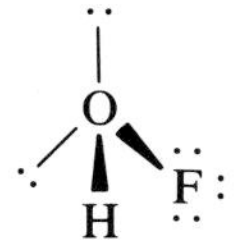 Structural pair geometry = tetrahedral; Molecular geometry = bent

81. As seen below, 16-electron molecules or ions are linear, 18-electron molecules or ions have a trigonal planar structural pair geometry, and 20-electron molecules or ions have a tetrahedral structural pair geometry.

(a) $\ddot{O}=C=\ddot{O}$ Structural pair geometry = linear; Molecular geometry = linear

(b) 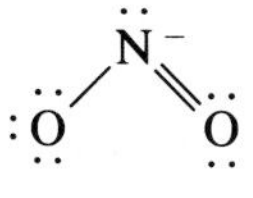 Structural pair geometry = trigonal planar; Molecular geometry = bent

(c) 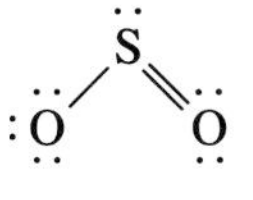 Structural pair geometry = trigonal planar; Molecular geometry = bent

(d) 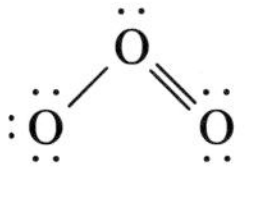 Structural pair geometry = trigonal planar; Molecular geometry = bent

(e) 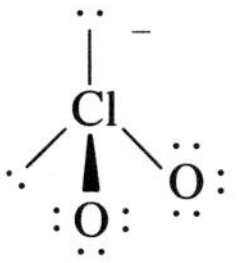 Structural pair geometry = tetrahedral; Molecular geometry = bent

83. (a) 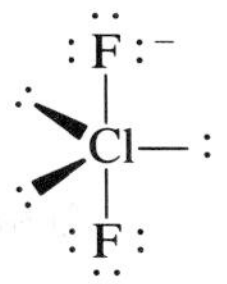 Structural pair geometry = trigonal bipyramid; Molecular geometry = linear

(b) 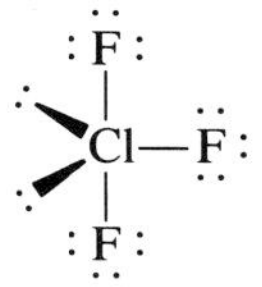 Structural pair geometry = trigonal bipyramid; Molecular geometry = T-shaped

(c) 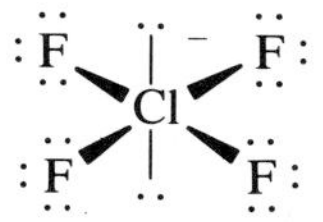 Structural pair geometry = octahedral; Molecular geometry = square planar

(d) 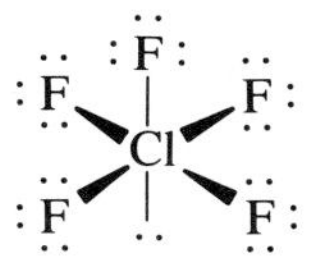 Structural pair geometry = octahedral; Molecular geometry = square pyramid

85. (a) 120°; (b) 120°; (c) 1 = 109° and 2 = 120°; (d) 1 = 120° and 2 = 180°

87. (a) 14 σ bonds and 2 π bonds; (b) 1 = 109° and 2 and 3 = 120°

89. (a) SeF_4 is "see-saw" shaped (like SF_4 on page 398). One F—S—F is 120° and two others are 90°.

(b) 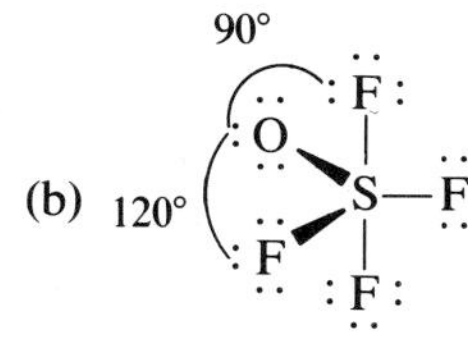

(c) All F—Br—F angles are 90° (see ClF_5 in question 83(c).

91. NO_2^+ has a linear structural pair geometry (O—N—O angle = 180°), whereas NO_2^- has a trigonal planar structural pair geometry (O—N—O angle = 120°).

93. (a) H_2O; (b) CO_2 and CCl_4; (c) F is more negatively charged.

95. (b) (c) 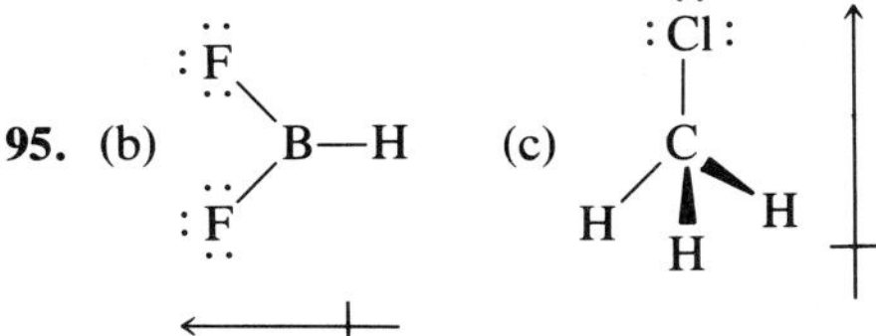

97. 1A = 1; 3A = 3; 3B = 3; 4A = 4

99. All molecules or ions below are 16-electron species and all have a linear structural-pair and molecular geometry.

(a) $\ddot{O}=C=\ddot{O}$

(b) $:\ddot{N}=N=\ddot{N}:^-$

(c) :Ö=C=N̈:⁻ ⟷ :O≡C—N̈:⁻ ⟷ :Ö—C≡N:⁻

101. ΔH of combustion for $CH_4(g)$ = -50.6 kJ/g
ΔH of combustion for $C_3H_8(g)$ = -46.1 kJ/g

103. ΔH°_{rxn} = 86 kJ/mol

105. (a) 1 = 120°; 2 = 109°; 3 = 120°.
(b) The shortest CO bond is the C=O bond at carbon-1.
(c) The O—H bond at oxygen-1.

107. All molecules except BF_3 are based on a tetrahedron of electron pairs.

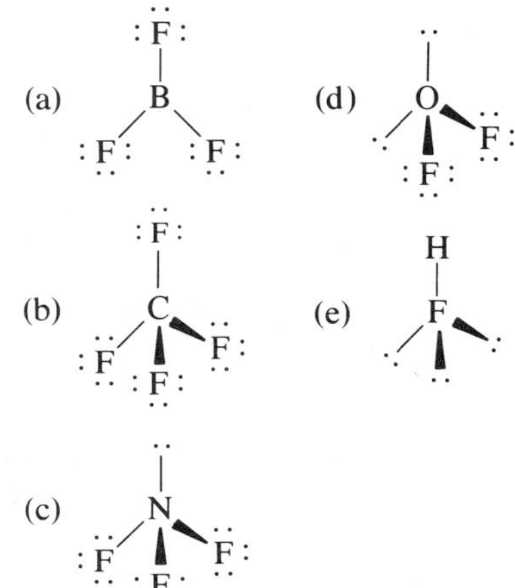

109. (a) In XeF_2 the spatially more demanding lone pairs are located in the equatorial plane, leaving the bonding electrons on the axes. This leads to a linear molecule.
(b) If ClF_3 were trigonal and planar, the lone pairs would be in the axial positions of the trigonal bipyramid. It is usually preferable to place these pairs in the equatorial plane, so the trigonal planar structure of ClF_3 is not observed.

111. (a) 10 σ bonds and 2 π bonds. (b) 1 = 109°; 2 = 120°; 3 = 109°; and 4 = 120°

113. (a) 17 σ bonds and 3 π bonds.
(b) 1 = 109°; 2 = 120°; 3 = 109°; 4 = 120°; 5 = 109°
(c) The C=C double bond is the shortest CC bond.
(d) The C=C double bond should have the highest energy of the CC bonds.

115. (a) $Cl_2(g) + 3\ F_2(g) \rightarrow 2\ ClF_3(g)$
(b) F_2 is the limiting reagent; 1.62 g of ClF_3 are produced.
(c, d) Structural pair geometry = trigonal bipyramid. Molecular geometry = T-shaped
(e) If the geometry of ClF_3 were planar and trigonal, the molecule would be nonpolar. Therefore, this geometry is ruled out by the molecular polarity. However, the other alternative, which is also polar,

is not eliminated. Other methods would be needed to prove the molecule is actually T-shaped.
(f) ΔH°_f = -405 kJ/mol

CHAPTER 11

15. Structural pair geometry = tetrahedral
Molecular geometry = bent

The tetrahedral structural pair geometry means the S atom is sp^3 hybridized.

17.

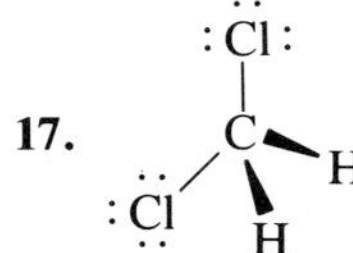

Structural-pair geometry and molecular geometry are both tetrahedral. The C atom is sp^3 hybridized. C—H bonds form by the overlap of a C sp^3 orbital with the H 1*s* orbital. The C—Cl bonds form by overlap of a C sp^3 orbital with a Cl orbital (either a *p* orbital or an sp^3 orbital).

19. (a) sp^2; (b) sp; (c) sp^3; (d) sp^2

21. (a) C and O = sp^3
(b) C of CH_3 = sp^3 and the other C atoms = sp^2
(c) C of CH_2 = sp^3 and C of C=O = sp^2

23. Structural pair geometry = trigonal bipyramid
Molecular geometry = linear

The Xe atom is sp^3d hybridized.

25. (a) sp^3d^2; (b) sp^3d; (c) sp^3d

27. :Ö=C=Ö:
The central atom is *sp* hybridized. This leaves two

unhybridized p orbitals on the C atom to be used in pi bonding.

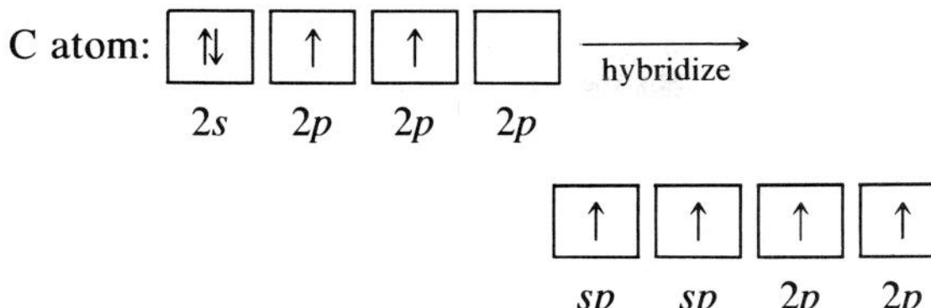

The C atom sp hybrid orbitals are used to form sigma bonds with the O atoms,

O sp C sp O

and the unhybridized $2p$ orbitals are used to form pi bonds to the O atoms. One $2p$ C atom orbital is illustrated below, forming one of the C=O pi bonds.

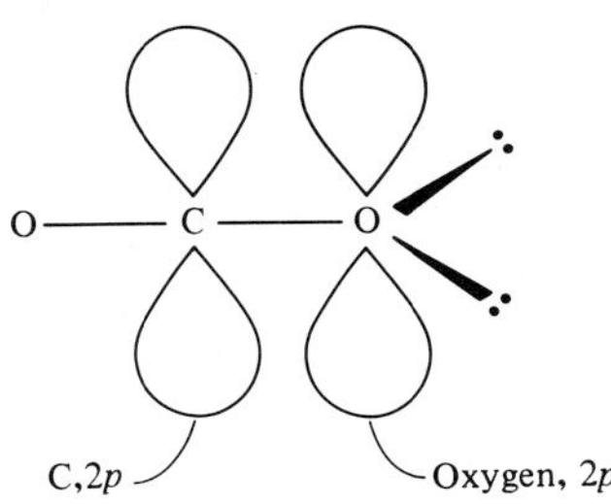

29. Dot structure:

:O:
‖
:Cl—C—Cl:

The C atom is sp^2 hybridized, each of these hybrid atomic orbitals being used to form a sigma bond (one to O and two to the Cl atoms). The unhybridized $2p$ orbital is used to form a CO pi bond as in CO_2 in Study Question 27.

C atom: ↑ ↑ ↑ ↑

sp^2 sp^2 sp^2 — for sigma bond formation; p — for pi bond formation

31. H_2^+, $(\sigma_{1s})^1(\sigma^*_{1s})^0$. Bond order = 1/2. The H—H bond in H_2^+ is weaker than in H_2 (where the bond order is 1).

33.

— σ^*_{2p}
— — π^*_{2p}
⇅ σ_{2p}
⇅ ⇅ π_{2p}
⇅ σ^*_{2s}
⇅ σ_{2s}

The C_2^{2-} ion has 1 σ bond and 2 π bonds for a CC bond order of 3. On adding two electrons to C_2 the bond order increases by 1. The C_2^{2-} ion is diamagnetic.

35. CO: [core electrons]$(\sigma_{2s})^2(\sigma^*_{2s})^2(\pi_{2p})^4(\sigma_{2p})^2$. The molecule is diamagnetic with 1 σ bond and 2 π bonds for a net bond order of 3.

37. (a) sp^2; (b) sp^2; (c) sp^3

39. One of the resonance structures of the NO_3^- ion is

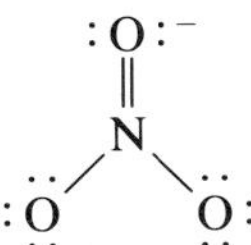

The geometry of the N sigma bonds is trigonal planar, so the N atom hybridization is sp^2.

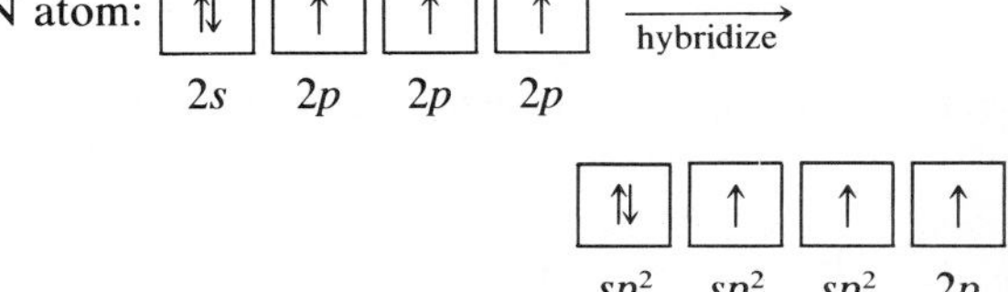

This means there is one, unhybridized $2p$ orbital remaining on the N atom.

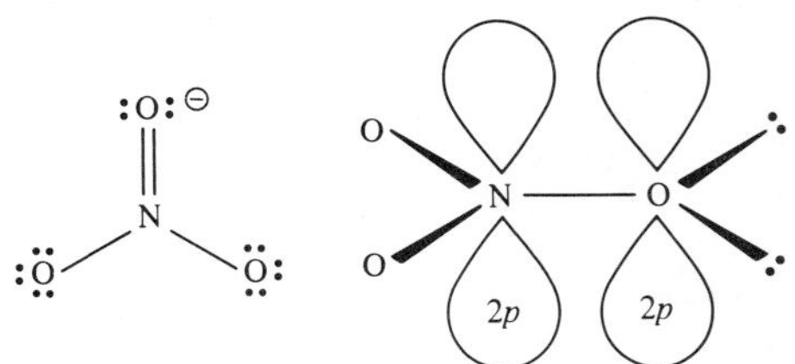

This orbital can be used to form a pi bond with one of the O atoms.

41. (a) C atom of CH_3 is sp^3 hybridized. C atom in C=N bond is sp^2 hybridized. (b) 120°

43. (a) 5 π bonds. (b) A = 120°; B = 109°; C = 109°. (c) C1 = sp^2; C2 = sp^2; O3 = sp^3

45. (a) 17 σ bonds and 2 π bonds. (b) N1 = sp^3; C2 = sp^3; C3 = sp^2; N4 = sp^2. (c) A = 109°; B = 180°; C = 109°

47. The electron dot structure of the peroxide ion is

$[:\ddot{\underset{\cdot\cdot}{O}}-\ddot{\underset{\cdot\cdot}{O}}:]^{2-}$

The ion shows all electrons paired, and a bond order of 1. This is identical with the picture from MO theory where the ion has the configuration

[core electrons]$(\sigma_{2s})^2(\sigma^*_{2s})^2(\pi_{2p})^4(\sigma_{2p})^2(\pi^*_{2p})^4$

and a net bond order of 1.

49. N hybridization is sp^3 in both NH_3 and NH_4^+. However, the H—N—H angle is slightly less than 109° in NH_3 (owing to the large N atom lone pair), while it is expected to be exactly 109° in NH_4^+.

51.

Structural-pair geometry	tetrahedral	trig. bipyramid
Molecular geometry	tetrahedral	trig. bipyramid
I hybridization	sp^3	sp^3d

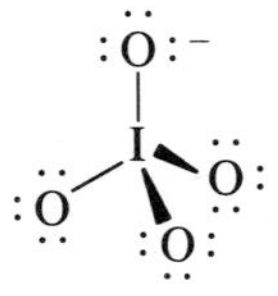

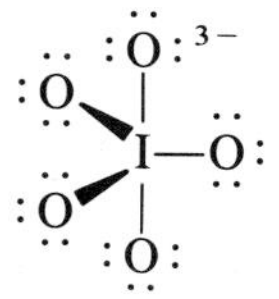

53. See Table 11.2.

55. Only the C-based compound has a double bond.

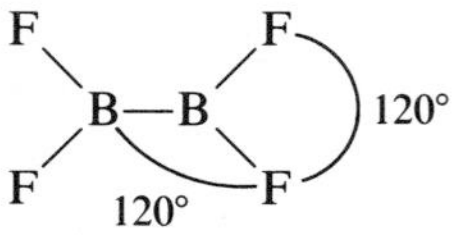

B atoms are sp^2 hybridized

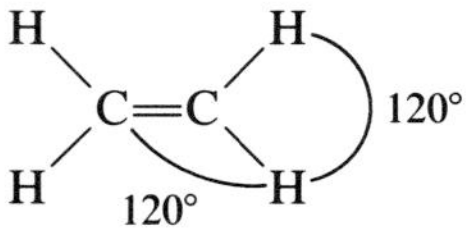

C atoms are sp^2 hybridized

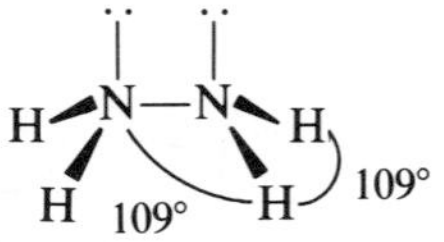

N atoms are sp^3 hybridized

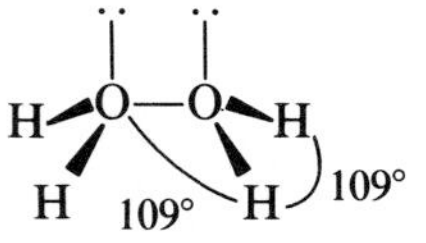

O atoms are sp^3 hybridized

57. (a) C: $sp^2 \rightarrow sp^3$
(b) P: $sp^3 \rightarrow sp^3d$
(c) Xe: $sp^3d \rightarrow sp^3d^2$
(d) Sn: $sp^3 \rightarrow sp^3d^2$

59. (a) C1 = sp^3 and C2 = sp; N = sp^2
(b) C—N=C = 120°; N=C=S = 180°
(c) On C2 the s and p_x orbitals are involved in forming two sp hybrid orbitals, while the p_y and p_z orbitals are involved in π bonding to N and S.

CHAPTER 12

11. (a) 0.97 atm; (b) 950. mmHg; (c) 99 kPa; (d) 51 kPa; (e) 542 torr.
13. 56.3 mmHg; 0.0741 atm; 56.3 torr; 7.50 kPa
15. 128 mmHg
17. 23.1 mL
19. 132 mmHg
21. 0.50 L
23. (d)
25. 354 mmHg
27. 9.33 atm
29. 0.193 atm (147 mmHg)
31. 2.6 L
33. 0.337 mol CO
35. B_2H_6
37. C_6H_6
39. B_4H_{10}
41. 3.7×10^{-4} g/L
43. 119 g/mol
45. 9.00 g
47. Cylinder B
49. H_2S
51. 31 g H_2
53. 17 atm O_2
55. 0.38 atm
57. $Cr(CO)_6$
59. x = 3
61. $SClF_5$ is the limiting reagent. Theoretical yield of S_2F_{10} is 0.207 g, which gives a pressure of 0.0774 atm or 58.8 mmHg.
63. 5.7 atm
65. Moles cyclopropane/moles O_2 = 1/3.4 = 0.29. Grams C_3H_6 = 63 g
67. $P(B_2H_6)$ = 2.5 mmHg and $P(O_2)$ = 7.5 mmHg
69. 0.0626 atm
71. 2.0 L at 500 °C; 0.78 L at 25 °C
73. 484 mL
75. $P(S_2F_{10})$ = 58.8 mmHg; $P(HCl)$ = 118 mmHg; $P(H_2)$ = 36.4 mmHg; P(total) = 213 mmHg.
77. (a) Average KE is the same, since T is the same.
(b) Root mean square speed of H_2 is greater than that of CO_2 (H_2 is less massive).
(c) Twice as many molecules in B as in A.
79. Root mean square speed of Xe = 238 m/s. Rms for He is 5.75 times greater than rms for Xe, since $[M_{Xe}/M_{He}]^{1/2} = 5.75$.
81. $CH_2Cl_2 < Kr < N_2 < CH_4$
83. $Xe < C_2H_6 < CO < He$
85. 36 g/mol
87. Using a = 6.49 atm · L²/mol and b = 0.0562 L/mol in van der Waals's equation, P = 29.5 atm. Using the ideal gas law, P = 49.3 atm.
89. 650 torr < 0.89 atm < 742 mmHg < 150 kPa

91. 3.8 atm
93. 0.0946 mmHg
95. 44.0 g/mol
97. Largest number in (d) and smallest number in (c).
99. 86.0% $NaNO_2$
101. 238 m/s
103. 0.034 g C_8H_{18}
105. (a) 29 g/mol; $X(N_2) = 0.82$ and $X(O_2) = 0.18$.
107. $x = 4$
109. 0.0023 mol O_2
111. Use the general gas law to find T_2 if $T_2 = T_1 + 100$.
113. 156 g/mol
115. 40.1 atm
117. $Tl(CH_3)_3$
119. Mass of $NaHCO_3$ = 1.05 g and mass of Na_2CO_3 = 0.45 g. (b) CO_2 pressure = 0.917 atm
121. (a) F_2 is the limiting reagent. $P(ClF_3)$ = 1.65 atm and $P(Cl_2)$ = 0.116 atm.
(b) The structural pair geometry is trigonal bipyramidal and the molecular geometry is T-shaped. The Cl atom uses sp^3d hybrid orbitals.

:F:
Cl—F:
:F:

(c) Reaction enthalpy = −810. kJ

CHAPTER 13

18. (a) dipole–dipole (and H bonds); (b) induced dipole–induced dipole; (c) ion–dipole; (d) dipole–dipole
20. Ion–dipole
22. (a) induced dipole–induced dipole; (b) dipole–dipole (and H bonds); (c) dipole–dipole; (d) induced dipole–induced dipole
24. $Ne < CH_4 < CO < CCl_4$
26. (a) O_2 (more mass; stronger induced dipole forces)
(b) SO_2 (polar molecule, CO_2 is nonpolar)
(c) HF (strong H bonds)
(d) GeH_4 (more mass; stronger induced dipole forces)
28. $CH_4 < CO < NH_3 < SCl_2$
30. (c), (d), and (f)
32. $Cs^+ < Na^+ < Mg^{2+}$
34. (a) increase; (b) decrease; (c) not change.
36. 180. kJ
38. 233 kJ
40. (a) About 150 mmHg (Appendix D gives 149.4 mmHg); (b) about 95 °C (Appendix D suggests a temperature between 93 °C and 94 °C; (c) at 70 °C the vapor pressure of water (about 225 mmHg) is only about half of that of ethyl alcohol (about 500 mmHg).
42. (a) Approximately 600 mmHg (the vapor pressure of diethyl ether at 30 °C); (b) Ether vapor will condense to liquid.
44. (a) 80.1 °C; (b) 250 mmHg at about 42 °C and 650 mmHg at about 75 °C; (c) ΔH_{vap} = 33.0 kJ/mol
46. No. Room temperature (in the range of 15–30 °C) is greater than the critical temperature, so there is no pressure that can satisfy equilibrium conditions.
48. (a) 495.0 pm; (b) 1.213×10^{-22} cm^3; 11.35 g/cm^3 (literature value: 11.342 g/cm^3)
50. Body-centered cubic
52. No. The NaCl structure consists of a face-centered cubic arrangement of Cl^- ions with Na^+ ions in octahedral holes, giving a 1:1 ratio of cations and anions. $CaCl_2$ requires a 1:2 ratio, which is not possible with the NaCl structure.
54. (a) Volume of unit cell = 2.95×10^{-22} cm^3; (b) 2.72 g/cm^3 (literature value is 2.80 g/cm^3).
56. (a) 4 Li^+ and 4 H^-; (b) face-centered cubic
58. As the radius of M^+ increases, the cation–anion separation increases, and the force of attraction between M^+ and I^- declines. This leads to a lower lattice energy.
60. CsF has a higher melting point (682 °C) than CsI (621 °C). The cation–anion separation in CsI is greater than in CsF, leading to a weaker ion–ion attraction and to a smaller lattice energy.
62. 1.97 kJ
64. 2200 kJ
66. (a) vapor; (b) liquid; (c) about −117 °C; (d) 0.25 atm; (e) solid is more dense than liquid; solid–liquid equilibrium line has a positive slope.
68. Cooking oil consists of long hydrocarbon chains. These do not interact strongly with hydrogen-bonded water.
70. Alcohols are capable of hydrogen bonding, whereas ethers interact only through weaker dipole–dipole forces.
72. Both water molecules and ethyl alcohol molecules are hydrogen-bonded to molecules of their own kind. However, H-bonding is more extensive in water, since each H_2O can form four H bonds. Diethyl ether cannot form H bonds; instead, diethyl ether molecules interact with one another by dipole–dipole forces.
74. All the ethyl alcohol evaporates.
76. Ethylene glycol has two —OH groups that can participate in hydrogen bonding. Because of this increased degree of intermolecular attraction, we expect the glycol to be more viscous than ethyl alcohol.
78. (a) 8 C atoms per unit cell; (b) 4.55×10^{-23} cm^3; (c) 3.57×10^{-8} cm
80. Ice has a relatively high vapor pressure. Therefore, the hail stones gradually sublimed in the freezing compartment.

82. No. Solid CO_2 is more dense than liquid CO_2, so the solid does not float on the liquid.

84. 2 TiO_2 units

86. 48% of the space is unoccupied

88. Length of unit cell edge = 4.073×10^{-8} cm; volume of unit cell = 6.757×10^{-23} cm^3. From the density, the mass of unit cell = 1.305×10^{-21} g. This is the mass of 4 atoms of Au, so the mass of one atom is 3.263×10^{-22} g/atom. Therefore,

$$(1 \text{ atom}/3.263 \times 10^{-22} \text{ g})(196.97 \text{ g}/1 \text{ mol}) = 6.036 \times 10^{23} \text{ atoms/mol.}$$

90. (a) See page 396 for SO_2 resonance structures. The O—S—O angle is approximately 120°, the structural-pair geometry is trigonal planar, and the best description of S-atom hybridization is sp^2.
(b) dipole–dipole
(c) $CH_4 < NH_3 < SO_2 < H_2O$
(d) Solid SO_2 is more dense than liquid SO_2
(e) $SO_2(g) + O_2(g) \rightarrow SO_3(g)$, $\Delta H° = -98.89$ kJ/mol
$H_2O(\ell) + SO_3(g) \rightarrow H_2SO_4(aq)$, $\Delta H° = -227.72$ kJ/mol

92. (a) 2,888 metric tons of $CuFeS_2$. (b) 2 metric tons (2000 kg) of SO_2. (c) 128 pm

CHAPTER 14

13. molality = 1.45; mole fraction glycol = 0.0255; 8.26% glycol

15.

Compound	Molality	%	Mole Fraction
NaCl	0.25	*1.4*	*4.5×10^{-3}*
C_2H_5OH	*1.1*	5.0	*0.020*
$C_{12}H_{22}O_{11}$	0.10	*3.3*	*0.0018*

17. 8.50 g $NaNO_3$; mole fraction $NaNO_3$ = 0.00359

19. 91.2 g CH_3OH

21. molality = 0.312 mol/kg; X_{glycol} = 0.00559; X_{water} = 0.994

23.

Compd.	Grams	g H_2O	m	X
Na_2CO_3	*0.331*	250.	0.0125	*2.25×10^{-4}*
CH_3OH	13.5	150.	*2.81*	*0.0482*
KNO_3	*321*	555	*5.72*	0.0934

25. molality = 16.2; weight % HCl = 37.1

27. (a) 0.162; (b) 30.0%; (c) 9.97 molar

29. (a) 0.0240 m in $CaCl_2$; (b) 0.0720 m in ions

31. 2.7×10^{-4} molal

33. (a) Both octane and CCl_4 are nonpolar.
(b) Both CH_3OH and H_2O are hydrogen-bonding liquids.
(c) The ionic compound NaBr does not dissolve well in a solvent that is neither strongly polar nor hydrogen bonding. ·

35. −299.7 kJ/mol

37. positive (+34.9 kJ/mol); more soluble

39. For 1 atm CO_2, concentration is 0.0340 m; for $P(CO_2)$ = 1/3 atm, the molality is 0.0113.

41. 691 mmHg

43. 1040 g glycol

45. 100. g/mol

47. $\Delta t = Km = (+2.53$ deg/molal$)[0.200$ mol$/0.100$ kg$] = 5.06$ °C
Boiling point = 80.1 °C + 5.1 °C = 85.2 °C

49. 62.5 °C

51. 390 g glycol; $X_{glycol} = 0.13$

53. 0.10 m sugar < 0.10 m NaCl < 0.080 m $CaCl_2$

55. 106.85 °C

57. 404 g/mol

59. 360 g/mol; $(C_{10}H_8Fe)_2$

61. 989 g

63. 470 g

65. 120. g/mol

67. 110 g/mol; benzaldehyde is C_6H_5CHO

69. 0.08 m $CaCl_2$ < 0.1 m NaCl < 0.04 m Na_2SO_4 < 0.1 m sugar

71. The cells shrivel owing to the loss of water by the cells in the meat by osmosis. The water moves from the less concentrated solution (in the cell) to the more concentrated solution (the brine).

73. 0.296 mol/L

75. (a) 13.5%; (b) 0.0458; (c) 2.66 m

77. (a) 17.7 m; (b) 0.241; (c) 75.0%

79. 5.0×10^{-4} m N_2; 2.7×10^{-4} m O_2

81. X = bromine

84. Dissolved solutes are 0.0965 mol sugar or 0.100 mol NaCl. However, 0.100 mol NaCl means there are 0.200 mol ions in solution. Therefore, the NaCl lowers the vapor pressure more than the sugar and raises the boiling point higher. The sugar solution has the higher water vapor pressure, while the NaCl solution has the higher boiling point.

85. 0.20 m Na_2SO_4 < 0.50 m sugar < 0.20 m KBr < 0.35 m $C_2H_4(OH)_2$

87. 108.55 °C

88. approximately 26 atm

90. (a) 17.3 mmHg; (b) 17.5 mmHg (essentially no change in water vapor pressure); (c) 0.250 mol sugar or 85.5 g

92. 0.10 molal NaCl; (b) less soluble; (c) 0.15 molal Na_2SO_4

94. 57 mmHg

96. 410. mmHg

98. 27 mmHg

99. Empirical formula, BF_2; molecular formula, B_2F_4

101. Empirical and molecular formulas are $C_{18}H_{24}Cr$

CHAPTER 15

20. (a) $-1/2\ (\Delta[O_3]/\Delta t) = +1/3\ \Delta[O_2]/\Delta t$

(b) $-1/2\ (\Delta[HOF]/\Delta t) = +1/2\ (\Delta[HF]/\Delta t) = +\Delta[O_2]/\Delta t$
(c) $-(\Delta[N_2]/\Delta t) = -1/3\ (\Delta[H_2]/\Delta t) = +1/2\ \Delta[NH_3]/\Delta t$

22. (a) For example, for the period 0 to 10 s, $\Delta[A]/\Delta t = -0.0167$ mol/L · s). For the period 20 to 30 s, $\Delta[A]/\Delta t = -0.0089$ mol/L · s). $\Delta[B]/\Delta t = 2(-\Delta[A]/\Delta t) = 0.0238$ mol/L · s).

24. (a) Rate = $k[NO_2]^2[CO]^0$
(b) If $[NO_2]$ is halved, the rate drops to 1/4 [or $(1/2)^2$] of its original value.

26.

	Order for A	Order for B	Overall Order
(a)	1	2	3
(b)	1	1	2
(c)	1	0	1
(d)	3	1	4

28. Rate = $k[B]^2$

30. 1.8×10^{-3} mol/(L · hr); Rate of change of $[Cl_2]$ is the same as that of $[Pt(NH_3)_2Cl_2]$.

32. (a) Rate = $k[CO][NO_2]$; (b) Overall order = 2; (c) $k = 1.9$ L/(mol · hr)

34. (a) Rate = $k[CH_3COCH_3][H^+]$. The reaction is first order in CH_3COCH_3 and H^+ and 2nd order overall. (b) $k = 4 \times 10^{-3}$ L/(mol · s). (c) Initial rate = 2×10^{-5} mol/(L · s)

36. (a) Rate = $k[A][B]^2$; (b) $k = 5.0 \times 10^4$ $L^2/(mol^2 \cdot min)$; (c) [B] = 0.20 M

38. Time = 13 hours = 1 half-life

40. (a) 2.08×10^5 hours; (b) 4.8×10^5 hours

42. 2 half-lives = 2(37.9 sec) = 75.8 sec

44. (a) $k = 0.017$/min; (b) 0.00075 M (after 3 half-lives); (c) 240 min

46. (a) P(HOF) = 50 mmHg and P(total) = 125 mmHg; (b) P(HOF) = 35 mmHg and P(total) = 133 mmHg.

48. (a) 4.5×10^3 sec; (b) 4.0×10^4 sec

50. (a) The ln[sucrose] vs. t plot gives a straight line. Therefore, the reaction is first order in sucrose. (b) Rate = k[sucrose] where $k = 3.7 \times 10^{-3}$/min. (c) From the graph, ln[sucrose] = -1.79 at 175 minutes, so [sucrose] = 0.167.

52. (a) The plot of ln[A] vs. t is curved, whereas 1/[A] vs. t is a straight line. Therefore, the reaction is second order in A. (b) $k = 0.0444$ L/mol · min. (c) Rate = $k[A][B]$. (d) $t_{1/2} = 590$ min.

54. Exothermic

56. 1.0×10^2 kJ

58.

	Rate Law	Molecularity
(a)	Rate = $k[NO][NO_3]$	bimolecular
(b)	Rate = $k[Cl][H_2]$	bimolecular
(c)	Rate = $k[(CH_3)_3CBr]$	unimolecular

60. (a) Step 2 is rate-determining, as it is the slow step.
(b) Rate law for step 2 is Rate = $k[O_3][O]$
(c) Step 1 is unimolecular and step 2 is bimolecular.

62. (a) Rate = $k[A][B]^2$; (b) $k = 7.4 \times 10^{-4}/M^2 \cdot$ sec

64. Only (a)

66. (a) and (c)

68. (a) Rate = $k[A]^2[B]$; (b) 5.94×10^{-4} mol/(L · sec)

69. (a) Rate = $k[NO_2][F_2]$; Order: in NO_2 = 1, in F_2 = 1, overall = 2; (c) $k = 40$ L/(mol · sec)

71. 4 hours

73. $k = 3.7 \times 10^{-2}$/hr and $t_{1/2} = 19$ hours

75. About 7.5 times faster

78. (a) step 1 = unimolecular and step 2 = biomolecular
(b) Rate = $k[Ni(CO_4)]$; agrees with mechanism in that slow step, the first step, is unimolecular.
(c) $[Ni(CO_3)L] = 1.5 \times 10^{-3}$ M
(d) 0.364 g $Ni(CO)_4$ formed and 1.3 atm of CO remain
(e) 160.81 kJ

CHAPTER 16

7. (a) $K = \dfrac{[H_2O]^2[O_2]}{[H_2O_2]^2}$

(b) $K = \dfrac{[PCl_5]}{[PCl_3][Cl_2]}$

(c) $K = [CO]^2$

(d) $K = \dfrac{[H_2S]}{[H_2]}$

9. (a) $K = P(Cl_2)$

(b) $K = \dfrac{[Cu^{2+}][Cl^-]^4}{[CuCl_4{}^{2-}]}$

(c) $K = \dfrac{P^2(NO)P(O_2)}{P^2(NO_2)}$

(d) $K = \dfrac{P(CO_2)P(H_2)}{P(CO)P(H_2O)}$

(e) $K = \dfrac{[Mn^{2+}]P(Cl_2)}{[H_3O^+]^4[Cl^-]^2}$

11. (e)

13. (a) $K = 1.1 \times 10^{47}$; (b) $K_p = K_c$ since the number of moles of gaseous reactants equal the number of moles of gaseous products ($\Delta n = 0$).

15. $K_3 = 0.14$ and K_4 and $1/K_3 = 7.3$

17. $K_{new} = 1/(K_{old})^{1/2} = 2.6$

19. (a) $K_c = 1.6$; (b) products favored ($K > 1$); (c) $K_{reverse} = 0.63$.

21. (a) $K_c = 1.6$; (b) $[H_2O] = [CO] = 11$ moles

23. $K_c = 0.035$

25. $K_c = 0.050$

27. $K_p = 2.31 \times 10^{-4}$

29. System not at equilibrium (too much N_2O_4 and too little NO_2). Thus, $[NO_2]$ increases.

31. Since $Q < K$ the system is not at equilibrium and $P(COBr_2)$ will decrease.

33. [n-butane] = 9.7×10^{-3} M and [isobutane] = 2.4×10^{-2} M

35. $P(NO_2) = 0.36$ atm; P(total) = 1.21 atm

37. (b)
39. 4.4 g CO_2
41. $P(C_2N_2) = P(H_2) = 0.481$ atm; $P(HCN) = 24.0$ atm
43. $[COBr_2] = 1.0 \times 10^{-5}$ M and $[CO] = [Br_2] = 7.1 \times 10^{-4}$ M; $P(CO) = P(Br_2) = 0.020$ atm and $P(COBr_2) = 0.00028$ atm, so $P(\text{total}) = 0.040$ atm
45. $P(Cl_2) = P(PCl_3) = 4.15$ atm; $P(\text{total}) = 9.80$ atm; fraction of PCl_5 dissociated = 0.735
47. $[Zn^{2+}] = 3.3 \times 10^{-11}$ M
49.

	$[Br_2]$	[HBr]	K_c
Add H_2	decrease	increase	no change
Increase T	increase	decrease	decrease
Increase V	no change	no change	no change

51. (a) no change; (b) shift right; (c) no change; (d) shift left
53. (a) [n-butane] = 1.1 M and [isobutane] = 2.9 M; (b) [n-butane] = 1.1 M and [isobutane] = 2.9 M
55. (b)
57. (a) $P(CO_2) = 3.87 \times 10^{-2}$ atm in the absence of added CO_2. If 0.05 atm of CO_2 is added, and then $N_2H_6CO_2$ dissociates, more CO_2 will be added to the system to give an even greater pressure of CO_2 (the total pressure of CO_2 of 0.0877 atm). (If you wish to solve the problem numerically, see the *Student Solutions Manual* for a method of solving the problem.) (b) Adding CO_2 must lead to a decline in $P(NH_3)$.
59. (a) Step 2. Rate = $k_2[CHCl_3][Cl]$
(b) From Step 1, $[Cl] = \{K[Cl_2]\}^{1/2}$. Therefore, Rate = $k_2[CHCl_3]\{K[Cl_2]\}^{1/2}$. Since k_2K = constant, the rate law is Rate = $k[CHCl_3][Cl_2]$.
61. (e)
63. Not at equilibrium, $Q < K$. Therefore, more H_2S is produced and H_2 and S_2 are consumed.
65. (a) 84% conversion to dimer; (b) less dimer is present at higher temperatures.
67. $[Zn^{2+}] = [CO_3^{2-}] = 3.9 \times 10^{-6}$ M
69. O atoms = 2.0×10^{19}
71. (a) Q_c ($=2.1 \times 10^5$) $< K_c$, so the system is not at equilibrium. More products (O_2 and NO_2) will form; that is, the reaction proceeds to the right. (b) Using enthalpies of formation, $\Delta H_{rxn} = -199.8$ J. Since the reaction is exothermic, an increase in T will shift the equilibrium away from products; their concentrations will decrease.
73. (a) $K_p = 36.3$; (b) At equilibrium $P(N_2) = 0.27$ atm, $P(H_2) = 0.81$ atm, and $P(NH_3) = 0.06$ atm. The total pressure is 1.14 atm.
75. 0.2
77. (a) Total pressure = 5.22 atm; (b) $P(SO_2)$ = 21.2% of 5.22 atm or 0.111 atm
79. (a) $P(SO_2) = P(Cl_2) = 1.6$ atm; $P(SO_2Cl_2) = 1.1$ atm; $P(\text{total}) = 4.3$ atm. Fraction SO_2Cl_2 dissociated = 0.60. (b) $P(SO_2) = 1.3$ atm, $P(Cl_2) = 2.3$ atm; $P(SO_2Cl_2) = 1.3$ atm; fraction SO_2Cl_2 dissociated = 0.51. (c) LeChatelier's principle predicts that adding Cl_2 will shift the equilibrium to the left, a prediction verified by this calculation.
80. (a) NO is the limiting reagent, so a maximum of 12.8 g of NOBr can be prepared.
(b) $\ddot{O}{=}\ddot{N}{-}\ddot{\underset{..}{Br}}:$
(c) The structural pair geometry is planar trigonal, and the molecular geometry is bent. The O—N—Br angle is approximately 120°, and the molecule is polar.
(d) $K_P = 1.0 \times 10^2$

CHAPTER 17

14.

Acid	Conjugate Base
HCN	CN^-, cyanide
HSO_4^-	SO_4^{2-}, sulfate
HF	F^-, fluoride
HNO_2	NO_2^-, nitrite
HCO_3^-	CO_3^{2-}, carbonate

16.

Acid/Base	Conjugate Partner
CN^-	HCN, hydrocyanic acid
SO_4^{2-}	HSO_4^-, hydrogen sulfate
HI	I^-, iodide
S^{2-}	HS^-, hydrogen sulfide
HNO_3	NO_3^-, nitrate

18. $K_2CO_3(aq) \rightarrow 2\,K^+(aq) + CO_3^{2-}(aq)$; the solution is basic owing to the hydrolysis of the carbonate ion: $CO_3^{2-}(aq) + H_2O(\ell) \rightleftarrows HCO_3^-(aq) + OH^-(aq)$
20. $HPO_4^{2-}(aq) + H_2O(\ell) \rightleftarrows PO_4^{3-}(aq) + H_3O^+(aq)$
$HPO_4^{2-}(aq) + H_3O^+(aq) \rightleftarrows H_2PO_4^-(aq) + H_2O(\ell)$
22.

	Acid and Conjugate Base	Base and Conjugate Acid
(a)	HCOOH and $HCOO^-$	H_2O and H_3O^+
(b)	H_2S and HS^-	NH_3 and NH_4^+
(c)	HSO_4^- and SO_4^{2-}	OH^- and H_2O

24. The Cl^- and Na^+ ions do not enter into the reaction, since they are the conjugates of a strong acid and strong base, respectively. The reaction that may occur is

$$NH_4^+(aq) + H_2PO_4^-(aq) \rightleftharpoons NH_3(aq) + H_3PO_4(aq)$$

However, the reaction does not occur to a significant extent, since H_3PO_4 is a stronger acid than $H_2PO_4^-$.
26. $H_2S(aq) + CH_3COO^-(aq) \rightleftarrows HS^-(aq) + CH_3COOH(aq)$
H_2S is a weaker acid than acetic acid, so the equilibrium lies predominantly to the left.
28. (a) Right. H_2S is a stronger acid than HCO_3^-
(b) Left. HCN is a weaker acid than HSO_4^-
(c) Left. CN^- is a much weaker base than NH_2^-
(d) Left. CH_3COOH is a weaker acid than HSO_4^-

30. (a) strongest acid, HF; weakest acid, HS^-.
(b) fluoride, F^-
(c) The strongest acid (HF) has the weakest conjugate base.
(d) The weakest acid (HS^-) has the strongest conjugate base.

32. (a) strongest base, NH_3; weakest base, C_5H_5N
(b) $C_5H_5NH^+$
(c) C_5H_5N has the strongest conjugate acid, and NH_3 has the weakest conjugate base.

34. HClO is the weakest acid and so has the strongest conjugate base.

36. 4.0×10^{-4}; acid

38. pH = 2.89; $[OH^-] = 7.7 \times 10^{-12}$

40. pH = 11.48

42.

pH	$[H_3O^+]$	$[OH^-]$
1.00	1.0×10^{-1}	1.0×10^{-13}
10.50	3.2×10^{-11}	3.2×10^{-4}
4.89	1.3×10^{-5}	7.7×10^{-10}
9.25	5.6×10^{-10}	1.8×10^{-5}
10.36	4.3×10^{-11}	2.3×10^{-4}

44. (a) pH = 3.30; (b) red; (c) colorless

46. (a) 1.6×10^{-4} M; (b) moderately weak

48. (a) 2.1×10^{-3} M; $K_a = 3.6 \times 10^{-4}$

50. $K_a = 1.6 \times 10^{-5}$

52. $K_b = 6.6 \times 10^{-9}$

54. $[H_3O^+] = [A^-] = 1.3 \times 10^{-5}$ M; [HA] = 0.040 M

56. $[H_3O^+] = [CN^-] = 3.2 \times 10^{-6}$ M; [HCN] = 0.025 M; pH = 5.50

58. $[H_3O^+] = 6.6 \times 10^{-4}$ M; pH = 3.18

60. (a) Barbituric acid; (b) Nicotinic acid

62. [HX] = 0.007 M; $[H_3O^+] = 3.0 \times 10^{-3}$; pH = 2.52

64. $[NH_4^+] = [OH^-] = 1.6 \times 10^{-3}$; $[NH_3] = 0.15$ M; pH = 11.22

66. pH = 8.85

68. pH = 6.73

70. $[H_3O^+] = [F^-] = 5.0 \times 10^{-3}$ M; [HF] at equilibrium = 0.035 M; original [HF] = 0.040 M

72. highest pH, Na_2S; lowest pH, NaF

74. (a) less than 7; (b) less than 7; (c) equal to 7; (d) greater than 7

76. (a) less than 7; (b) greater than 7; (c) greater than 7; (d) equal to 7

78. $[H_3O^+] = 1.1 \times 10^{-5}$; pH = 4.98

80. $[OH^-] = [HCN] = 3.3 \times 10^{-3}$ M; $[Na^+] = 0.441$ M; $[H_3O^+] = 3.0 \times 10^{-12}$ M; pH = 11.52.

82. pH = 2.61

84. (a) pH = 1.17; (b) $[SO_3^{2-}] = K_{a2} = 6.2 \times 10^{-8}$ M

86. $[OH^-] = [N_2H_5^+] = 9.2 \times 10^{-5}$ M; $[N_2H_6^{2+}] = 8.9 \times 10^{-16}$ M; (b) 9.96

88. (a) Lewis acid; (b) Lewis base; (c) Lewis base; (d) Lewis base; (e) Lewis acid

90. Lewis acid

92. $ICl_3 \xrightarrow{Cl^-} ICl_4^-$

Lewis acid

94. CN^- would lead to the greater lowering of the Cd^{2+} concentration.

96. $[Cu^{2+}] = 1.2 \times 10^{-16}$ M

98. $[Ag^+] = 7.6 \times 10^{-8}$ M

100. The solution of sodium hydride would be basic owing to the hydrolysis of the hydride ion:

$$2\ H^-(aq) + H_2O(\ell) \longrightarrow H_2(g) + 2\ OH^-(aq)$$

102.

Compound	Conjugate Partner
Br^-, base	HBr
$[Al(H_2O)_6]^{3+}$, acid	$[Al(H_2O)_5(OH)]^{2+}$
H_3PO_4, acid	$H_2PO_4^{2-}$
CH_3COO^-, base	CH_3COOH

104. (a) right; (b) right; (c) left; (d) left

106. pH = 6.80

108. pH = 2.508

110. (a) Acidic solutions: CH_3COOH and NH_4Cl; (b) Basic solution: NH_3, Na_2CO_3, $Na(CH_3COO)$; (c) Neutral solution: NaCl.

112. Increasing pH: HCl < CH_3COOH < NaCl (=7) < NH_3 < NaCN < NaOH

114. $K_a = 3.0 \times 10^{-5}$

116. pH = 10.07

118. (a) Original concentrations: $[Ag^+] = 5.0 \times 10^{-4}$ M and $[S_2O_3^{2-}] = 2.5$ M
(b) Equilibrium concentrations: $[Ag^+] = 1.0 \times 10^{-17}$ M; $[S_2O_3^{2-}] = 2.5$ M; and $\{[Ag(S_2O_3)_2]^{3-}\} = 5.0 \times 10^{-4}$ M

120. As Cl atoms replace H atoms on the terminal C atom in acetic acid, the acid strength increases. This can be understood in two ways: (a) The electronegative Cl atoms withdraw electron density from the O—H bond, thereby weakening the bond. This bond weakening increases as the number of Cl atoms increases, and the acid strength increases. (b) When the negative ion is formed (e.g., CCl_3COO^-), the electron-withdrawing Cl atoms mean the ion is increasingly able to accommodate the negative charge, thereby increasing the ion stability and the acid strength.

122. *P*(complex) = 0.20 atm; $P(BF_3)$ = 0.18; *P*(total) = 0.56 atm

CHAPTER 18

8. $K_{net} = K_a/K_w = 6.3 \times 10^{-5}/1.0 \times 10^{-14} = 6.3 \times 10^9$

10. $[OH^-] = 5.3 \times 10^{-6}$ M; pH = 8.72

12. (a) greater than 7; (b) less than 7; (c) equal to 7

14. (a) pH = 5.19; (b) pH = 12.48

16. $[OH^-] = 1.7 \times 10^{-3}$ M; $[H_3O^+] = 5.9 \times 10^{-12}$ M; $[Na^+] = [C_6H_5O^-] = 3.79 \times 10^{-2}$ M; pH = 11.23.

18. $[H_3O^+] = 1.9 \times 10^{-6}$ M and so pH = 5.73 and $[OH^-] = 5.4 \times 10^{-9}$ M; $[NH_4^+] = 6.10 \times 10^{-3}$ M and the original concentration of NH_3 is 1.48×10^{-2} M.

20. (a) decrease; (b) increase; (c) stays the same

22. The pH is lower than before adding NH_4Cl; pH before = 11.12 and pH after = 9.12.

24. (a) pH = 2.78; (b) pH = 5.40

26. (a) Adding HCl consumes some of the NH_3 and gives a solution containing a mixture of NH_3 and NH_4Cl. This is a buffer solution, whose pH is lower than that of a solution of NH_3 alone.
(b) Adding NaOH to acetic acid consumes some of the acid and gives a mixture of sodium acetate and acetic acid. This is a buffer solution, whose pH is higher than that of the acetic acid solution alone.

28. 4.8 g NH_4Cl

30. 4.1 g sodium acetate

32. (a) pH = 3.90; (b) diluting the solution with 500 mL of pure water does not change the pH.

34. (c)

36. (a) pH = 9.56; (b) pH = 9.51

38. pH = 7.244

40. The curve begins at a pH of 13.00. Although the pH drops slowly as HCl is added at first (it is 12.52 after 12.5 mL of HCl have been added), it drops sharply (vertically) when 25.0 mL of 0.10 M HCl have been added. The midpoint of the vertical portion of the curve is at a pH of 7. When slightly more than 25.0 mL of 0.10 M HCl have been added, the curve levels out at a pH of about 1.8.

42. (a) pH of original NH_3 solution = 11.15.
(b) pH at equivalence point (which occurs after adding 22.0 mL of HCl) = 5.28.
(c) pH at midpoint = 9.26.
(d) Methyl red.
(e) After adding 5.00 mL HCl, pH = 9.79; after 11.0 mL pH = 9.26 (this is the point at which the NH_3 is half consumed); after 15.0 mL pH = 8.92; after 20.0 mL pH = 8.26; after 22.0 mL (the equivalence point) pH = 5.27; and after 25.0 mL pH = 2.18.

44.

	Weak Acid or Base Produced in Titration	Approximate pH	Indicator
(a)	$C_5H_5NH^+$	3.1	thymol blue
(b)	$HCOO^-$	8.4	phenol red
(c)	$N_2H_5^+$	4.5	bromcresol green

46. $[H_3O^+] = 1.9 \times 10^{-5}$ M and pH = 4.71

48. pH = 3.14

50. pH = 8.25

52. 52 g $NaNO_2$ in 500. mL of solution

54. HCl < CH_3COOH < CH_3COOH/CH_3COONa < NaCl (=7) < NH_3/NH_4Cl < NH_3

56. (a) HB is a stronger acid than HA. (b) A^- is a stronger base than B^-.

58. (a) Original pH = 8.52.
(b) At equivalence point $[H_3O^+] = 5.1 \times 10^{-4}$ M and pH = 3.29
(c) pH at midpoint in titration = 4.62
(d)

Volume of Acid Added	Solution pH
5.00	5.37
10.0	4.98
15.0	4.70
20.0	4.43
24.0	4.19
30.0	3.60

60. $[H_2PO_4^-]/[HPO_4^{2-}] = 0.65$

62. pH = 6.73

64. (a) pH of original picric acid solution = 2.01; (b) pH at equivalence point = 6.00; (c) pH at midpoint in titration = 2.48. (Note: In this problem we treat the acid essentially as a strong acid in calculating the original pH and the pH at the midpoint.)

66. pH at midpoint gives K_b for ethanolamine of 3.2×10^{-5}. Therefore, pH at the equivalence point is 5.91.

67. (a) Bond angles: C═C—C = 120°; C—C═O = 120°; C—O—H = 109°; (d) C—C—H = 120°
(b) 4 π bonds and 16 σ bonds
(c) Both are sp^2
(d) pH of salicylic acid solution = 2.44
(e) 11% present as salicylate ion in gastric juice
(f) pH at midpoint in titration = 2.96 and pH at equivalence point = 7.36

CHAPTER 19

9. (a) AgCl and $PbCl_2$
(b) NiS and CoS (and most other metal sulfides such as MnS, FeS, CuS, HgS, and so on)
(c) $ZnCO_3$, ZnS, $Zn_3(PO_4)_2$
(d) FeS, $FeCO_3$, $Fe(OH)_2$

11. (a) soluble; (b) insoluble; (c) insoluble; (d) insoluble; (e) insoluble; (f) insoluble

13. Precipitation occurs to give $AgCO_3$.

$$2\ AgNO_3(aq) + Na_2CO_3(aq) \longrightarrow AgCO_3(s) + 2\ NaNO_3(aq)$$

15. (a) Precipitate is AgCl
(b) Precipitate is $Mg(OH)_2$
(c) No precipitate
(d) Precipitate is $PbCl_2$

17. (a) $ZnS(s) \rightleftarrows Zn^{2+}(aq) + S^{2-}(aq)$
$K_{sp} = [Zn^{2+}][S^{2-}] = 1.1 \times 10^{-21}$
(b) $SnI_2(s) \rightleftarrows Sn^{2+}(aq) + 2\ I^-(aq)$
$K_{sp} = [Sn^{2+}][I^-]^2 = 1.0 \times 10^{-4}$
(c) $NiCO_3(s) \rightleftarrows Ni^{2+}(aq) + CO_3^{2-}(aq)$
$K_{sp} = [Ni^{2+}][CO_3^{2-}] = 6.6 \times 10^{-9}$
(d) $Ag_2SO_4(s) \rightleftarrows 2\ Ag^+(aq) + SO_4^{2-}(aq)$
$K_{sp} = [Ag^+]^2[SO_4^{2-}] = 1.7 \times 10^{-5}$

19. $Mg_3(AsO_4)_2$

21. $K_{sp} = 5.9 \times 10^{-21}$

23. $K_{sp} = 3.6 \times 10^{-5}$

25. $K_{sp} = 1.69 \times 10^{-6}$

27. $K_{sp} = 8.1 \times 10^{-12}$

29. $K_{sp} = 7.9 \times 10^{-6}$

31. (a) 1.1×10^{-8} M; (b) 1.5×10^{-6} g/L

33. 2.7×10^{-7} mg Au^{3+}/L

35. 2.7 mg

37. (a) AgSCN; (b) $SrSO_4$; (c) MgF_2; (d) ZnS

39. $BaCO_3$ (Solubility = 9.0×10^{-5} M) < Ag_2CO_3 (Solubility = 1.3×10^{-4} M) < BaF_2 (solubility = 7.5×10^{-3} M)

41. (a) Some AgCl remains; (b) all $NiCO_3$ dissolves

43. (a) $Q < K_{sp}$, no precipitate; (b) $Q > K_{sp}$, $NiCO_3$ precipitates.

45. $Q(=1.6 \times 10^{-8}) > K_{sp}$; therefore, $Zn(OH)_2$ precipitates.

47. $[OH^-] = 1.6 \times 10^{-5}$ M

49. 4.0 mg KF

51. $Q(=2.7 \times 10^{-10}) > K_{sp}$, so a precipitate of $BaSO_4$ forms.

53. $Q(=1.7 \times 10^{-8}) < K_{sp}$, so no precipitate of $PbCl_2$ forms.

55. NiS precipitates first, followed by CoS and then MnS.

57. $Fe(OH)_3$ precipitates first, followed by $Al(OH)_3$ and then $Pb(OH)_2$.

59. CaF_2 precipitates first, followed by MgF_2 and then BaF_2.

61. 1.0×10^{-6} M in pure water, but only 1.0×10^{-10} M in 0.010 M NaSCN.

63. (a) 2.0×10^{-3} mg/mL; (b) 6.8×10^{-13} mg/mL

65. $[Ag^+] = 5.0 \times 10^{-7}$ M; $[IO_3^-] = 2.0 \times 10^{-2}$ M; $[K^+] = 2.0 \times 10^{-2}$ M

67. Yes, AgCl precipitates because $Q(=4.5 \times 10^{-5}) > K_{sp}$; $[Na^+] = 0.00060$ M; $[NO_3^-] = 0.0075$ M; $[Ag^+] = 0.0069$ M; $[Cl^-] = 2.6 \times 10^{-8}$ M.

69. (a) Oxalate concentration should be slightly less than 4.3×10^{-3} M to precipitate CaC_2O_4 and leave Mg^{2+} in solution; (b) $[Ca^{2+}] = 5.3 \times 10^{-7}$ M

71. (a) $PbCO_3$ precipitates first; (b) $[CO_3^{2-}] = 1.7 \times 10^{-7}$ M

73. (a) Na^+ salts are generally water-soluble, whereas many barium salts are not soluble. Therefore, we only need to add a reagent such as SO_4^{2-} to precipitate Ba^{2+} as $BaSO_4$ and leave Na^+ in solution.
(b) Both $Ca(OH)_2$ and $Zn(OH)_2$ are insoluble. However, K_{sp} for $Ca(OH)_2$ is about 2×10^{11} times larger than K_{sp} for $Zn(OH)_2$. Therefore, it should be possible to separate the ions as their hydroxides.
(c) As in (b), separation as hydroxides should be feasible. The concentration of OH^- needed to precipitate $Bi(OH)_3$ is about 10^6 less than that needed for $Cd(OH)_2$.

75. $AgI(s) \rightleftharpoons Ag^+(aq) + I^-(aq)$
$Ag^+(aq) + 2\ NH_3(aq) \rightleftharpoons [Ag(NH_3)_2]^+$

$AgI(s) + 2\ NH_3(aq) \rightleftharpoons I^-(aq) + [Ag(NH_3)_2]^+(aq)$

$K = K_{sp}$ for AgI $= 1.5 \times 10^{-16}$
$K = K_{formation} = 1.6 \times 10^7$

$K_{net} = 2.4 \times 10^{-9}$

77. $AgCl(aq) \rightleftharpoons Ag^+(aq) + Cl^-(aq)$
$Ag^+(aq) + I^-(aq) \rightleftharpoons AgI(s)$

$AgCl(s) + I^-(aq) \rightleftharpoons AgI(s) + Cl^-(aq)$

$K = K_{sp}$ for AgCl $= 1.8 \times 10^{-10}$
$K = 1/K_{sp}$ for AgI $= 1/1.5 \times 10^{-16}$

$K_{net} = 1.2 \times 10^6$

Yes, it is possible to precipitate AgI by adding I^- ion to a precipitate of AgCl since K_{net} is much greater than 1.

79. Yes. The total quantity of NH_3 required to dissolve the AgCl and to achieve the correct equilibrium ammonia concentration ($=1.9 \times 10^{-3}$ M) is 2.1×10^{-3} M. Adding 5.0 mL of 2.5 M NH_3 to 1.0 L of water will give an ammonia concentration greater than that required.

81. $CaCO_3(s)$ reacts with hydronium ions [to give $Ca^{2+}(aq) + H_2O(\ell) + CO_2(g)$], whereas $CaSO_4$ has no analogous reaction (SO_4^{2-} is a much poorer base than CO_3^{2-}.) The K_{sp} values differ by only 10^{-4}, with $CaCO_3$ the less soluble.

83. Add H_2SO_4 to $CaCO_3$ to form $CaSO_4$, CO_2, and H_2O.

85. Dissolve $MgCl_2$ in water and add $AgNO_3$ to precipitate all the chloride ion as AgCl. Filter the solid AgCl to separate it from the solution. Evaporate the water from the filtered solution to leave solid $Mg(NO_3)_2$.

87. (a) Add HCl. If CO_3^{2-} is present, then CO_2 bubbles out of solution.
(b) Add $AgNO_3$ to precipitate either AgCl or $AgCO_3$. If $AgCO_3$ is formed, the precipitate will dissolve in excess HCl. In contrast, AgCl does not react with HCl.

89. (a) $K_{dissolve} = 8.5 \times 10^{-2}$; (b) $[Zn^{2+}] = 7.6 \times 10^{-2}$ M

91. Aragonite is more soluble than calcite.

93. $Q(=2.0 \times 10^{-9}) > K_{sp}$; therefore, AgCl forms.

95. (a) Neither $Pb(OH)_2$ nor $Zn(OH)_2$ precipitates ($Q < K_{sp}$ for both).
(b) 1.7×10^{-5} M

97. (a) No $Ca(OH)_2$ precipitates, since $Q(= 1.5 \times 10^{-9}) < K_{sp}$. (b) $[Ca^{2+}] = 3.8 \times 10^{-3}$ M; $[Cl^-] = 7.5 \times 10^{-3}$ M; $[Na^+] = 6.3 \times 10^{-4}$ M; pH = 10.80.

99. 1.9 mg AgCl dissolves in 1.0 L of pure water. In the presence of NaCl, 1.1 mg of AgCl dissolves. Therefore, 0.8 mg of AgCl precipitates on adding NaCl.

101. 4.5×10^{-17}

103. In each case, add the appropriate acid to $MgCO_3$. (a) H_2SO_4; (b) HBr; (c) $H_2C_2O_4$; (d) $HClO_4$; (e) HF

105. (a) No. Only 9.1 mg of $Mg(OH)_2$ can dissolve in pure water. (b) Yes. The equilibrium concentration of Mg^{2+} when pH = 5.00 is 1.5×10^7 M! This far exceeds the concentration that can be achieved by 0.581 g of $Mg(OH)_2$ in 1.0 L of water.

107. (a) $K_{dissolve} = 6.6$; (b) 2.1 g $Na_2S_2O_3$

109. (a) $AlCl_3(aq) + H_3PO_4(aq) \rightarrow AlPO_4(s) + 3\ HCl(aq)$
(b) $AlCl_3$ is the limiting reagent; the theoretical yield of $AlPO_4$ is 139 g.
(c) $[Al^{3+}] = [PO_4^{3-}] = 1.1 \times 10^{-10}$
(d) Solubility increases, owing to the formation of the weak acid HPO_4^{2-}.

$$AlPO_4(s) + 3\ HCl(aq) \longrightarrow Al^{3+}(aq) + HPO_4^{2-}(aq)$$

(e) Some $AlPO_4$ (0.46 g) forms because $Q > K_{sp}$.

CHAPTER 20

6. (a) CO_2 vapor; (b) dissolved sugar; (c) alcohol/water mixture

8. (a) $AlCl_3$; (b) CH_3CH_2I; (c) $NH_4Cl(aq)$

10. (a) −3.363 J/K; diamond is more organized than graphite (see pages 536–537), so the entropy declines.
(b) −102.50 J/K; a highly disorganized gas is condensed to an organized solid, so the entropy declines.
(c) 99 J/K; a liquid is converted to a vapor, so the entropy increases.

12. (a) +84.4 J/K; (b) −84.4 J/K

14. −270.1 J/K

16. (a) $\Delta S° = -163.34$ J/K; (b) $\Delta S° = -108.31$ J/K; (c) $\Delta S° = -16.7$ J/K; (d) $\Delta S° = -313.4$ J/K

18. (a) $\Delta S° = -151.9$ J/K; (b) $\Delta S° = -138.9$ J/K; (c) $\Delta S° = -548.1$ J/K

20.

	$\Delta H°$(kJ)	$\Delta S°$(J/K)	$\Delta G°$(kJ)
(a)	−359.41	−151.9	−314.1
(b)	−601.70	−108.31	−569.42
(c)	−176.01	−284.8	−91.15

The $\Delta G°$ values calculated for reactions (a) and (b) represent free energies of formation. For $PbCl_2$ and MgO the calculated values are identical with those in Appendix J to 4 significant figures. All reactions are predicted to be spontaneous.

22.

	$\Delta H°$(kJ)	$\Delta S°$(J/K)	$\Delta G°$(kJ)
(a)	117.36	168.50	67.15
(b)	50.63	−331.77	149.50
(c)	−218.8	−47.85	−204.5

All calculated $\Delta G°$ values are nearly identical with those in Appendix J.

24. (a) $4\ Fe(s) + 3\ O_2(g) \rightarrow 2\ Fe_2O_3(s)$
(b) $\Delta G° = -742.0$ kJ/mol of Fe_2O_3
(c) −2110 kJ for 454 g

26. (a) −748.1 kJ; (b) 58.539 kJ; (c) −301.39 kJ

28. (a) $\Delta H° = -82.4$ kJ; (b) $\Delta S° = 460.0$ J/K; (c) $\Delta G° = -219.1$ kJ

30. (a) $\Delta G_f°[MgCO_3(s)] = -1012.0$ kJ/mol; (b) $\Delta G_f°[NOF(g)] = -51.1$ kJ/mol

32. (a) $\Delta H° = -125.52$ kJ; (b) $\Delta S° = -129.9$ J/K; (c) $\Delta G° = -86.81$ kJ

34. The reaction is thermodynamically spontaneous ($\Delta G° = -54.98$ kJ).

36. $K_p = 6.7 \times 10^{-16}$. Reactants favored.

38. $\Delta G° = -2.27$ kJ/mol

40. (a) $\Delta G° = -290.32$ kJ, so the reaction is spontaneous.
(b) $K_p = 7.77 \times 10^{50}$

42. Adding one CH_2 increases the entropy by about 40 J/K · mol. Therefore, $S°$ for C_4H_{10} is about 350 J/K · mol.

44. The standard state of Br_2 is liquid Br_2, while that for I_2 is the solid. Liquids usually have a larger entropy than solids for similar compounds.

46. (a) Splitting of water: $H_2O(\ell) \rightarrow H_2(g) + \frac{1}{2} O_2(g)$. A liquid is producing 1.5 moles of gas. Therefore, ΔS is expected to be positive (+163 J/K). Energy is required to split water (to break the O—H bonds), so ΔH is positive (+285.83 kJ). Thus, ΔG is positive, and the reaction is not spontaneous.
(b) Dissolving NH_4Cl: A regular solid gives ions, randomly moving in solution. Thus, ΔS is positive. Since the solution becomes cold, this means the process is endothermic and has a positive ΔH. Since ΔS and ΔH work in opposite directions, it is difficult to predict the outcome. However, that fact that NH_4Cl dissolves so readily in water suggests a negative ΔG (calculated $\Delta G° = -7.65$ kJ).
(c) One mole of liquid nitroglycerin gives a number of moles of gas, and the reaction evolves heat. Therefore, ΔS is positive and ΔH is negative. Both predict a negative ΔG.

48. $\Delta G°$ from $K_p = 4.70$ kJ; $\Delta G°$ from values of $\Delta G_f° =$ 4.73 kJ

50. (a) $\Delta G° = -97.9$ kJ. The reaction is spontaneous and enthalpy-driven.
(b) $\Delta G_f°[C_6H_6(g)] = 129.7$ kJ/mol

52. $\Delta G° = -2870$ kJ

54. (a) $\Delta G° = 91.4$ kJ; (b) 9.5×10^{-17}; (c) not spontaneous at 298 K; a temperature greater than 981 K is required to make it spontaneous; (d) temperature = 1190 K (920 °C).

56. (a) The value of $\Delta S°$ (+162.0 J/K) is positive as expected for a reaction that converts 1 mole of liquid to 1.5 moles of gas. (b) The reaction is not spontaneous at 298 K ($\Delta G° = +115.6$ kJ). (c) It becomes spontaneous at temperatures over 1012 K (739 °C).

58. $6\ CO_2(g) + 6\ H_2O(\ell) \longrightarrow C_6H_{12}O_6(aq) + 6\ O_2(g)$
$24\ H_2S(g) + 12\ O_2(g) \longrightarrow 24\ H_2O(\ell) + 24\ S(s)$

$24\ H_2S(g) + 6\ O_2(g) + 6\ CO_2(g) \longrightarrow 18\ H_2O(\ell) + 24\ S(s) + C_6H_{12}O_6(aq)$

$\Delta G° = +2870$ kJ
$\Delta G° = -4885.7$ kJ

$\Delta G° = -2016$ kJ

60. (a) no change
(b) shift right
(c) no change
(d) shift right
(e) The reaction is endothermic ($\Delta H° = +106.46$ kJ), so the equilibrium should shift to the right on raising the temperature.
(f) $\Delta G° = +89.73$ kJ
(g) Reaction is not predicted to be spontaneous.
(h) K predicted to be less than 1.

CHAPTER 21

12. $Cu(s) \rightarrow Cu^{2+}(aq) + 2\ e^-$; oxidation at the anode
$Ag^+(aq) + e^- \rightarrow Ag(s)$; reduction at the cathode

14. If the tin electrode is negative, it is furnishing electrons to the external circuit. The source of the electrons is the oxidation reaction $Sn(s) \rightarrow Sn^{2+}(aq) + 2\ e^-$; this tells us the electrode is the anode. The copper half cell, with a positive charge on the electrode, reflects a deficiency in negative charge due to the reduction of Cu^{2+} to copper metal: $Cu^{2+}(aq) + 2\ e^- \rightarrow Cu(s)$. Therefore, this is the cathode. The abbreviated notation is:

$Sn|Sn^{2+}||Cu^{2+}|Cu$.

16. $\Delta G° = -89$ kJ

18. (a) −1.298 V, not spontaneous; (b) −0.51 V, not spontaneous; (c) −1.023 V, not spontaneous; (d) +0.93 V, spontaneous

20. (a) $Sn^{2+}(aq) + 2\ Ag(s) \rightarrow Sn(s) + 2\ Ag^+(aq)$; $E° = -0.94$; not spontaneous
(b) $Zn(s) + Sn^{4+}(aq) \rightarrow Sn^{2+}(aq) + Zn^{2+}(aq)$; $E° = +0.91$; spontaneous
(c) $I_2(s) + 2\ Br^-(aq) \rightarrow 2\ I^-(aq) + Br_2(\ell)$; $E° = -0.55$ V; not spontaneous

22. (a) weakest oxidizing agent, V^{2+}; (b) strongest oxidizing agent, Cl_2; (c) strongest reducing agent, V; (d) weakest reducing agent, Cl^-; (e) Pb will not reduce V^{2+}; (f) I_2 will not oxidize Cl^-; (g) Cl_2 and I_2 can be reduced by Pb.

24. (a) $V(s) + Cl_2(g) \rightarrow V^{2+}(aq) + 2\ Cl^-(aq)$; $E° = 2.54$ V
(b) $V(s) + I_2(g) \rightarrow V^{2+}(aq) + 2\ I^-(aq)$; $E° = 1.72$ V

26. (a) $Zn(s) + 2\ Ag^+(aq) \rightarrow Zn^{2+}(aq) + 2\ Ag(s)$; $E° = +1.56$ V
(b) Zn = anode and Ag = cathode; $Zn|Zn^{2+}||Ag^+|Ag$
(c–f)

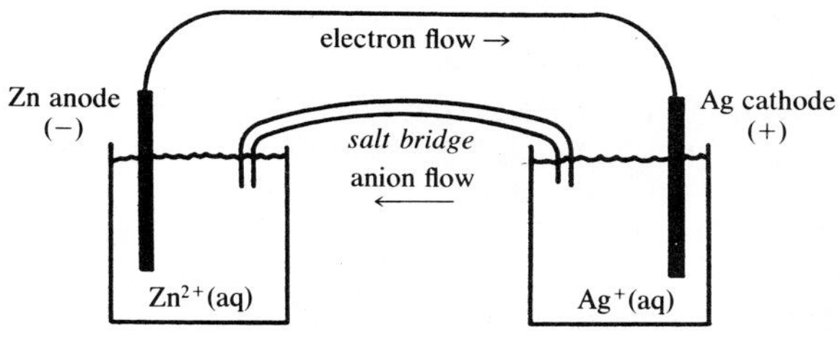

28. (a) $E° = +0.30$ V
(b)

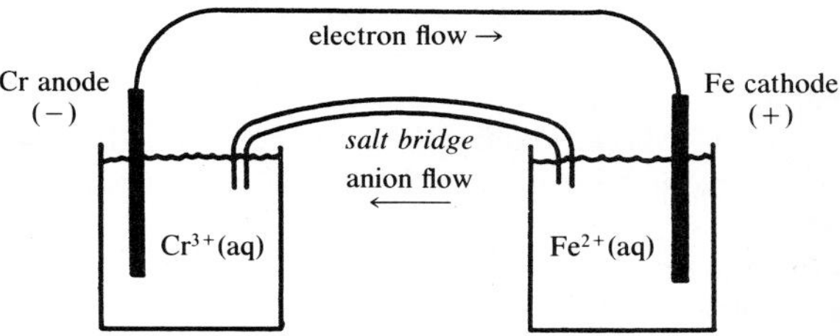

30. (a) $Zn(aq) + Sn^{2+}(aq) \rightarrow Zn^{2+}(aq) + Sn(s)$
(b) $E°_{net} = +0.62$ V
(c) Electron flow is from the Zn electrode to the Sn electrode.
(d) The Zn/Zn^{2+} compartment is the anode and the Sn/Sn^{2+} compartment is the cathode.
(e) The Sn electrode is positive.

32. $E = 0.177$ V; the cell potential (E) is much less positive when the dissolved reagents are 0.10 M ($E°_{net} = 0.236$ V). (Note that I_2 is a solid as given in Appendix I and is not included when solving the Nernst equation.)

34. (a) $Fe^{2+}(aq) + Ag^+(aq) \rightarrow Ag(s) + Fe^{3+}(aq)$
(b) $E°_{net} = +0.028$ V
(c) $E_{net} = -0.03$ V. The net cell reaction now is the reverse of (a).

36. (a) $E° = +0.236$ V, $K = 9.4 \times 10^7$; (b) $E° = -0.531$ V, $K = 1.2 \times 10^{-18}$

38. 0.16 g Ag
40. 5.93 g Cu
42. 0.10 amp
44. 270 kg Al
46. 190 g Pb
48. 12 watts
50. 2.800×10^{16} C; 3.6×10^{10} kwh
52. (a) Reaction at the positive anode;
$2\ I^-(aq) \rightarrow I_2(s) + 2\ e^-$
(b) Reaction at the negative cathode:
$2\ H_2O(\ell) + 2\ e^- \rightarrow 2\ OH^-(aq) + H_2(g)$
(c) 1.2 g $I_2(s)$, 0.52 g KOH, and 0.0094 g H_2 are expected.
54. (a) Anode (oxidation of Br^-):
$2\ Br^-(aq) \rightarrow Br_2(\ell) + 2\ e^-$
Cathode (reduction of water):
$2\ H_2O(\ell) + 2\ e^- \rightarrow 2\ OH^-(aq) + H_2(g)$
(b) Anode (oxidation of F^-): $2\ F^- \rightarrow F_2(g) + 2\ e^-$
Cathode (reduction of Na^+): $Na^+ + e^- \rightarrow Na(\ell)$
(c) Anode (oxidation of water):
$6\ H_2O(\ell) \rightarrow O_2(g) + 4\ H_3O^+(aq) + 4\ e^-$
Cathode (reduction of water):
$2\ H_2O(\ell) + 2\ e^- \rightarrow 2\ OH^-(aq) + H_2(g)$
56. $Cl^- < Ag < Fe < Na$
58. To maximize $E°$, react the oxidizing agent Cl_2 with the most powerful reducing agent on the list. In this case it is Mg, and $E° = 3.73$ V.

$$Mg(s) + Cl_2(g) \longrightarrow Mg^{2+}(aq) + 2\ Cl^-(aq)$$

60. $K = 8 \times 10^{68}$
62. (a) O_2 should oxidize Fe^{2+} to Fe^{3+}. (b) Br^- should be oxidized to Br_2. (c) I^- should be oxidized to I_2.
64. $E_{net} = +1.27$ V; $K = 3 \times 10^{86}$
66. 18 g Zn
68. 14000 C; 15 g Pb
70. (a) 5.05×10^{-4} watt/g of reactants
(b) 3.11×10^{-4} watt/g of reactants
(c) The Ag/Zn battery produces more energy per gram of reactants.
72. (a) 1.7×10^3 kwh
(b) 5.4×10^3 mol HF consumed; 5.4×10^3 g H_2 and 1.0×10^5 g F_2 produced.
74. (a) 3.6 mol glucose require 22 mol O_2
(b) 22 mol O_2 require 88 mol e^- ($4\ e^- + O_2 \rightarrow 2\ O^{2-}$)
(c) 8.5×10^6 C/day or 98 C/sec (=98 amp)
(d) 98 J/sec or 98 watts

CHAPTER 22

3. (a) 0.38 g H_2/g CH_4; (b) 0.29 g H_2/g CH_2; (c) 0.23 g H_2/g CH. (All calculations done assuming that the hydrogen and steam give H_2 and CO. See page 910.)
4. 624 kJ/mol of H_2. This quantity of heat would be liberated upon recombination of the H atoms.
6. 1.2×10^4 L (or a cubic container about 90 inches on a side).
8. (a) $2\ KCl(aq) + 2\ H_2O(\ell) \rightarrow 2\ KOH(aq) + H_2(g) + Cl_2(g)$. $Cl^-(aq)$ is oxidized to Cl_2 and H_2O is reduced to hydrogen.
(b) $2\ CsI(aq) + 2\ H_2O(\ell) \rightarrow 2\ CsOH(aq) + H_2(g) + I_2(s)$
10. The lithium ion is much smaller than other alkali metal ions, so the distance between Li^+ and water can be quite small. Since the force of attraction increases as the distance becomes smaller, the force of attraction is large.
13. Be is *sp* hybridized.
15. $3\ Mg(s) + N_2(g) \rightarrow Mg_3N_2(s)$
21. Aqueous solutions of Li^+ are slightly acidic owing to hydrolysis of aquated lithium ion.

$$[Li(H_2O)_4]^+(aq) + H_2O(aq) \longrightarrow Li(H_2O)_3OH(aq) + H_3O^+(aq)$$

23. $K = 5.3 \times 10^5$. This large equilibrium constant means that $Mg(OH)_2$ can, in principle, be precipitated by adding $Ca(OH)_2$ to a solution of Mg^{2+}.
24. (a) 255.3 kg Mg; (b) 744.7 kg Cl_2; (c) 2.100×10^4 Faradays; (d) 1600 kJ/mol Mg; (e) 641 kg/mol Mg.
26. $\Delta H° = 229.6$ kJ
30. 1.142×10^6 g SO_2 (or about 1.3 tons)
31. (a) 2.1×10^{-4} mol/L or 1.7×10^{-2} g/L; (b) 780 g CaF_2
32. A = $BaCO_3$ (1.00 g = 0.00507 mol, which gives 0.00507 mol CO_2 on heating. This is the amount calculated from *PVT* data); B = BaO (basic oxide); C = $CaCO_3$; D = $BaCl_2$; E = $BaSO_4$.

CHAPTER 23

7. (a) $BCl_3(g) + 3\ H_2O(\ell) \rightarrow B(OH)_3(s) + 3\ HCl(aq)$
(b) $\Delta H°_{rxn} = -205$ kJ
9. $2\ NaBH_4(s) + I_2(s) \rightarrow B_2H_6(g) + 2\ NaI(s) + H_2(g)$. I_2 is reduced to I^- and H^- (in $NaBH_4$) is oxidized to H_2.
11. B_2Cl_4 has two, sp^3 hybridized B atoms. The geometry about these atoms is planar and triangular.

```
:Cl̈        C̈l:
    \      /
     B——B
    /      \
:C̤l        C̤l:
```

13. (a) $4\ BCl_3(g) + 6\ H_2(g) + C(s) \rightarrow B_4C(s) + 12\ HCl(g)$;
(b) 1.84 g B_4C; (c) 0.403 g H_2.
15. The Al^{3+} ion is quite small, so only four of the larger

Cl⁻ or Br⁻ ions fit around Al^{3+}, while six of the smaller F^- ions can be accommodated.

17. The literature value of the atomic radius of Pb is 146.0 pm, while the value calculated from the density is 175.1 pm.

19. (a) $2\,Al(s) + 2\,OH^-(aq) + 6\,H_2O(\ell) \rightarrow 2\,Al(OH)_4^-(aq) + 3\,H_2(g)$
(b) 18.4 mL H_2

21. (a) $Ga(OH)_3(s) + 3\,HCl(aq) \rightarrow GaCl_3(aq) + 3\,H_2O(\ell)$
$Ga(OH)_3(s) + NaOH(aq) \rightarrow Na[Ga(OH)_4](aq)$
(b) 2770 mL HCl required

23. (a) tetrahedral; (b) sp^3

25. The boiling point of CCl_4 is expected to be higher than that of CF_4. (The actual values are 76.7 °C and − 128.5 °C, respectively.) Apparently the intermolecular forces (induced dipole–induced dipole) are stronger for CCl_4, owing to its larger mass.

27. (a) $\Delta G°$ for the reduction of PbO = +50.72 kJ/mol. $\Delta G°$ for the reduction of CaO = +466.86 kJ/mol. (b) The reduction of PbO is more feasible. The temperature at which ΔG becomes negative would be far lower than that for CaO.

29. 156,820 tons PbO

31. (a) $2\,CH_3Cl(g) + Si(s) \rightarrow (CH_3)_2SiCl_2$; (b) 0.823 atm CH_3Cl; (c) 12.2 g $(CH_3)_2SiCl_2$

33. (a) :C≡O: [:C≡N:]⁻
Both CO and CN^- have the same number of valence electrons (10) and share the common feature of a triple bond.
(b) Dot structure of cyanamide

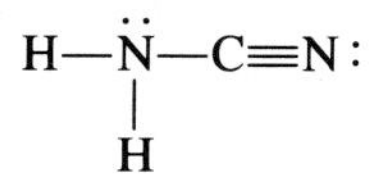

H—N̈—C≡N:
  |
  H

(c) The cyanamide molecule is predicted to have the structure below, with tetrahedral structural-pair geometry about the central N atom and linear geometry about C. The central N atom is therefore sp³ hybridized and the C atom is sp hybridized.

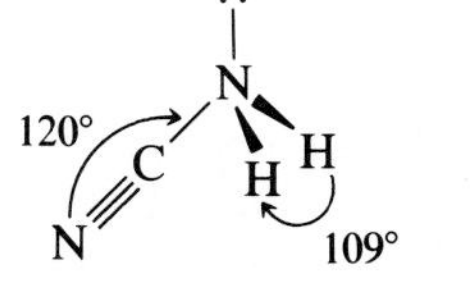

120° 109°

35. (a) Proposed structure for trimer of HCN.

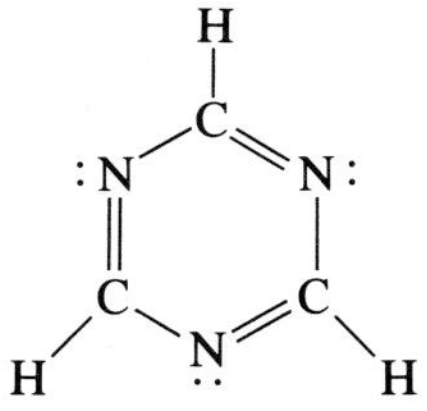

(b) Energy of trimerization = 99 kJ

37. 780 mL H_2 gas

CHAPTER 24

2. (a) $3\,Mg(s) + N_2(g) \rightarrow Mg_3N_2(s)$
(b) $P_4(s) + 3\,KOH(aq) + 3\,H_2O(\ell) \rightarrow PH_3(g) + 3\,KH_2PO_2(aq)$
(c) $CH_4(g) + H_2O(g) \rightarrow CO(g) + 3\,H_2(g)$ (high temperature)
(d) $HNO_3(aq) + KOH(aq) \rightarrow KNO_3(aq) + H_2O(\ell)$ (acid–base reaction)
(e) $2\,NH_3(aq) + OCl^-(aq) \rightarrow N_2H_4(aq) + Cl^-(aq) + H_2O(\ell)$ (the Raschig process; done in the presence of aqueous alkali and gelatin)
(f) $NH_4NO_3(s) + \text{heat} \rightarrow N_2O(g) + 2\,H_2O(g)$
(g) $NaNO_2(aq) + HCl(aq) \rightarrow NaCl(aq) + HNO_2(aq)$ (acid-base reaction)
(h) $4\,NH_3(g) + 5\,O_2(g) \rightarrow 4\,NO(g) + 6\,H_2O(g)$ (catalyzed by Pt)
(i) $3\,Cu(s) + 8\,HNO_3(aq, \text{dilute}) \rightarrow 3\,Cu(NO_3)_2(aq) + 4\,H_2O(\ell) + 2\,NO(g)$
(j) $P_4O_{10}(s) + 6\,H_2O(\ell) \rightarrow 4\,H_3PO_4(aq)$

4. (a) $N_2H_4(aq) + O_2(g) \rightarrow N_2(g) + H_2O(\ell)$
(b) 1.32×10^3 g N_2H_4

7. (a) $4\,IO_3^-(aq) + 5\,N_2H_5^+(aq) \rightarrow 2\,I_2(aq) + 5\,N_2(g) + 12\,H_2O(\ell) + H^+(aq)$
(b) $E° = 1.43\ V$

8. (a) 14 tons N_2O_4 required; (b) 6.3 tons N_2, 5.4 tons H_2O, and 6.6 tons CO_2 produced

12. (a) The electron dot structure of N_2O_3 shows that the 114.2 pm bond is a N=O double bond (bond order = 2). The longer NO bonds have a bond order of 1.5.
(b) $\Delta S° = 141$ J/K; $K = 1.90$; and $\Delta H_f°[N_2O_3(s)] = 82.9$ kJ/mol.

14. (a) $P_2O_5(s) + 3\,H_2O(\ell) \rightarrow 2\,H_3PO_4(aq)$
1 mole of P_2O_5 (142 g) give 2 moles of H_3PO_4 (196 g). Therefore, 1.380 g of H_3PO_4 are produced for every 1.000 g of P_2O_5 used.
(b) 14.9 million tons of H_3PO_4.
(c) \$217 per ton H_3PO_4

15. pH = 11.89

17. $Ca(OH)_2(s) + H_3PO_4(aq) \rightarrow CaHPO_4(s) + 2\,H_2O(\ell)$

19. $P_4(s) + 3\,OH^-(aq) + 3\,H_2O(\ell) \rightarrow 3\,H_2PO_2^-(aq) + PH_3(g)$

22. Assuming HNO_3 is reduced to NO, we have $H_2O(\ell) + As(s) + HNO_3(aq) \rightarrow H_3AsO_3(aq) + NO(g)$

23. (a) $2\,KClO_3(s) + \text{heat (and catalyst)} \rightarrow 2\,KCl(s) + 3\,O_2(g)$
(b) $2\,H_2S(g) + 3\,O_2(g) \rightarrow 2\,SO_2(g) + 2\,H_2O(g)$. The SO_2 produced in this reaction further oxidizes H_2S:

$$16\,H_2S(g) + 8\,SO_2(g) \longrightarrow 16\,H_2O(\ell) + 3\,S_8(s).$$

(c) $2\ Na(s) + O_2(g) \rightarrow Na_2O_2(s)$
(d) $K(s) + O_2(g) \rightarrow KO_2(s)$
(e) $2\ ZnS(s) + 3\ O_2(g) \rightarrow 2\ ZnO(s) + 2\ SO_2(g)$
(f) $SO_2(g) + H_2O(\ell) \rightarrow H_2SO_3(aq)$

25. (a) 1.170 V; (b) 1.160 V

28. (a) $H_2O_2(aq) \rightarrow H^+(aq) + HO_2^-(aq)$; (b) pH = 6.37

30. $BrO_2 < As_2O_3 < GeO_2 < Ga_2O_3 < CaO < K_2O$.

31. (a) CaO is more basic; (b) $SrO(s) + H_2O(\ell) \rightarrow Sr(OH)_2(aq)$

34. (a) 3.92 tons SO_2; (b) 4.53 tons $Ca(OH)_2$ required

36. (a) $UF_4(s) + F_2(g) \rightarrow UF_6(g)$
(b) $2\ Br^-(aq) + Cl_2(aq) \rightarrow Br_2(aq) + 2\ Cl^-(aq)$
(c) $I^-(aq) + Br_2(aq) \rightarrow I_2(aq) + 2\ Br^-(aq)$
(d) $Br^-(aq) + AgNO_3(aq) \rightarrow AgBr(s) + NO_3^-(aq)$
(e) $Cl_2(aq) + 2\ OH^-(aq) \rightarrow Cl^-(aq) + OCl^-(aq) + H_2O(\ell)$

38. I^- oxidized to I_2 and Br_2 reduced to Br^-: $2\ I^-(aq) + Br_2(aq) \rightarrow I_2(aq) + 2\ Br^-(aq)$

41. 0.0937 tons F_2

46. The molecule has an odd number of valence electrons (19).

Cl
O O

48. (a) I_3^- (b) $BrCl_2^-$ (c) ClF_2^+

:I:⁻ I :I: :Cl:⁻ Br :Cl: Cl F :F:

49. Not spontaneous; $E°$ for the net reaction is $-0.02\ V$.

51. (a) 215 tons H_2O; (b) 1.73×10^8 g O_2 produced. (c) 215 tons of Al are required and 405 tons of Al_2O_3 are produced.

55. (aq) yes, $E° = 0.59\ V$; (b) no, $E° = -0.77\ V$.

CHAPTER 25

10. (a) Y^{3+}: Kr core remains after removing 5*s* and 4*d* electrons of the atom to form the diamagnetic ion.
(b) Rh^{3+}: [Kr]$4d^6$, paramagnetic (4 unpaired electrons)
(c) Ce^{4+}: [Xe], diamagnetic
(d) Pt^{4+}: [Xe]$4f^{14}5d^6$, paramagnetic (4 unpaired electrons)
(e) V^{2+}: [Ar]$3d^3$, paramagnetic (3 unpaired electrons)
(f) U^{4+}:[Rn]$5f^5$, paramagnetic (5 unpaired electrons)

13. $2\ CuFeS_2(s) + 3\ O_2(g) \rightarrow 2\ CuS(s) + 2\ FeO(s) + 2\ SO_2(g)$
1.0 ton of $CuFeS_2$ gives 0.35 ton of SO_2

15. (a) $FeTiO_3(s) + 2\ HCl(aq) \rightarrow FeCl_2(aq) + TiO_2(s) + H_2O(\ell)$
(b) In principle, the HCl produced in the oxidation of $FeCl_2$ can be recovered and used in the treatment of ilmenite.
(c) 4.78×10^5 g Fe_2O_3

17. Monodentate ligands: (a) CH_3NH_2; (b) Br^-; (d) N_3^-; (f) H_3CCN
Bidentate: (c) en (ethylenediamine); (e) $C_2O_4^{2-}$ (oxalate ion); (g) phen (o-phenanthroline)

19. $[Ni(en)(NH_3)_3(H_2O)]^{2+}$

21. (a) $Ni(en)_2Cl_2$
(b) $K_2[PtCl_4]$

24. (a) $[Co(NH_3)_5H_2O]Cl_3$, pentaammineaquacobalt(III) chloride
(b) $K_3[Fe(CN)_6]$, potassium hexacyanoferrate(III)
(c) $Cr(CO)_6$, hexacarbonylchromium(0)
(d) $K_3[Fe(C_2O_4)_3]$, potassium tris(oxalato)ferrate(III)
(e) $[Co(en)_3]I_3$, tris(ethylenediamine)cobalt(III) iodide

26. (a) $[Fe(H_2O)_5(OH)]^{2+}$
(b) potassium tetracyanonickelate(II)
(c) potassium diaquabis(oxalato)chromate(III)

28. (a) Pd: H_3N, H_3N, NH_3, NH_3, Cl, Cl — *cis*; Pd: H_3N, H_3N, NH_3, NH_3, Cl, Cl — *trans*
(b) Pt: H_3N, H_3N, NCS, Br — *cis*; Pt: H_3N, NCS, Br, NH_3 — *trans*
(c) Co: O_2N, O_2N, NO_2, NH_3, NH_3, NH_3 — *cis*; Co: H_3N, O_2N, NO_2, NO_2, NH_3, NH_3 — *trans*

30. (a) not chiral center
(b) chiral center (4 different groups attached)
(c) chiral center (4 different groups attached)

32. Isomers of $[Co(en)(NH_3)_2(H_2O)Cl]^{2+}$, ethylenediamine(en) is depited by N⁀N

$[Co(N\text{⁀}N)(NH_3)_2(Cl)(OH_2)]^{2+}$ — N, N, NH_3, Cl, NH_3, OH_2
$[Co(N\text{⁀}N)(NH_3)_2(Cl)(H_2O)]^{2+}$ — N, N, Cl, NH_3, H_2O, NH_3
$[Co(N\text{⁀}N)(NH_3)_2(Cl)(H_2O)]^{2+}$ — N, N, NH_3, NH_3, H_2O, Cl
$[Co(N\text{⁀}N)(NH_3)_2(Cl)(OH_2)]^{2+}$ — N, N, NH_3, NH_3, Cl, OH_2

34. d^5, d^6, d^7, and d^8

36. NH_3 is a stronger field ligand than F^-, so Δ_0 is much smaller in the hexafluoro complex than in the hexaammine complex. This leads to a high-spin, paramagnetic complex for $[CoF_6]^{3-}$, but NH_3 as a ligand means that $[Co[NH_3)_6]^{3+}$ is low-spin and diamagnetic.

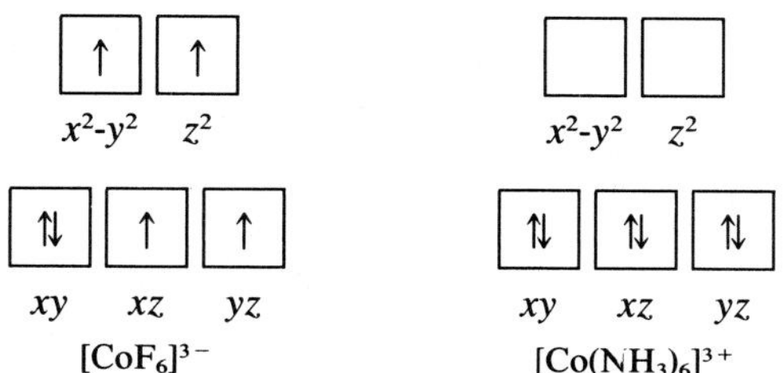

41. (a) The CN^- ligand is a strong-field ligand and should lead to low-spin complexes. In the case of $[Fe(CN)_6]^{4-}$, where Fe^{2+} is a d^6 ion, the complex ion is diamagnetic.

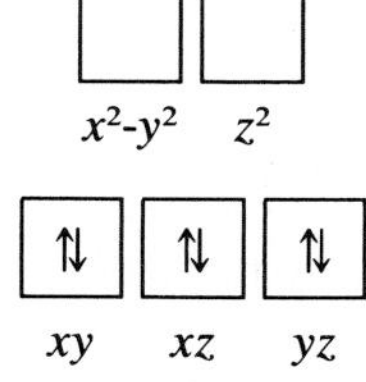

(b) Only high-spin complexes are possible for Cr^{3+} (d^3), so the complex is paramagnetic with three unpaired electrons.

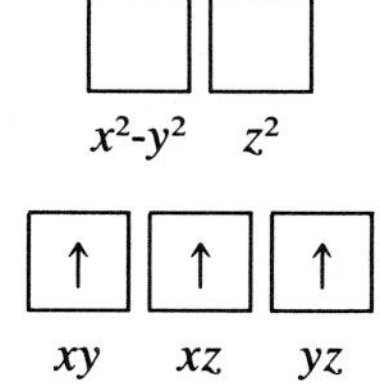

(c) The F^- ligand is a weak-field ligand, so the complex based on Mn^{2+} (d^5) is expected to be high-spin. The complex is paramagnetic with 5 unpaired electrons.

↑ ↑
x^2-y^2 z^2

↑ ↑ ↑
xy xz yz

(d) This complex is based on Cu^{2+}, a d^9 ion. There is no difference between high and low spin, so the complex is paramagnetic and has one unpaired electron.

⇅ ↑
x^2-y^2 z^2

⇅ ⇅ ⇅
xy xz yz

43. A = $[Co(NH_3)_5Br]SO_4$ and B = $[Co(NH_3)_5SO_4]Br$

45. (a) Oxidation–reduction; (b) 40.9 kg Cl_2 and 5.94 kg C; 7.89 kg Ti

47. (a) U^{4+}

(b) $$5\ U^{4+} + 2\ H_2O + 2\ MnO_4^- \rightarrow 5\ UO_2^{2+} + 4\ H^+ + 2\ Mn^{2+}$$

Alternatively, if the equation is balanced using hydronium ion, we have

$$5\ U^{4+} + 6\ H_2O + 2\ MnO_4^- \rightarrow 5\ UO_2^{2+} + 4\ H_3O^+ + 2\ Mn^{2+}$$

49. Both complexes are based on Cr^{2+}, an ion with four 3*d* electrons. However, the CN^- complex is low-spin, while the SCN^- complex is high-spin.

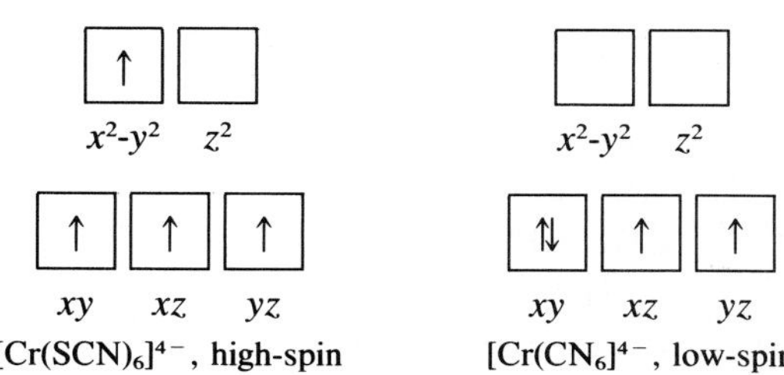

Both complexes are paramagnetic. However, the high-spin complex is expected to have four unpaired electrons, while the low-spin ion will have only two.

51. The glycinate ligand is depicted as N—O. Two of the five isomers (those with H_2O molecules *trans* to one another) are not chiral. The other three isomers (all with H_2O molecules *cis*) are chiral. Therefore, counting the partner of each of the chiral complexes, there is a total of 8 isomers.

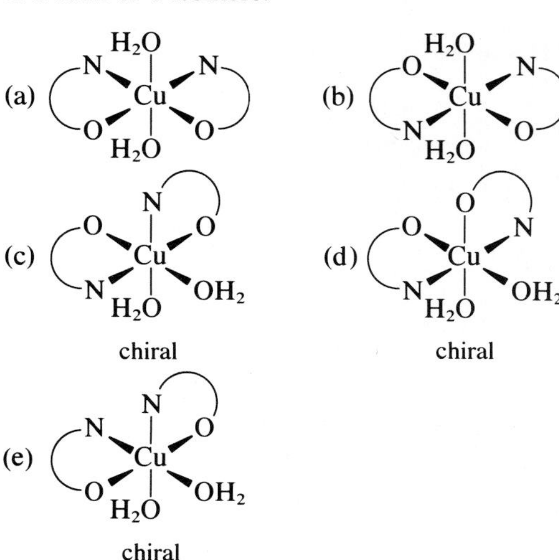

53. $[Cr(H_2O)_6]Cl_3$, hexaaquachromium(III) chloride. The structure of the complex ion is

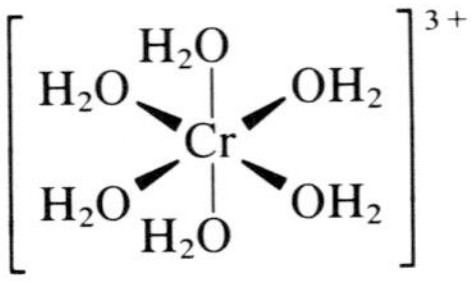

$$\left[\begin{array}{c} H_2O \\ H_2O \quad | \quad OH_2 \\ Cr \\ H_2O \quad | \quad OH_2 \\ H_2O \end{array}\right]^{3+}$$

CHAPTER 26

1. octane (systematic) or *n*-octane (common)

3. *n*-butyl

5. $CH_3—CH_2—CH_2—CH_2—CH_2—CH_2—CH_3$

n-heptane

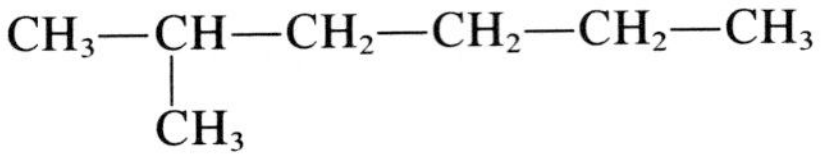

$CH_3—CH—CH_2—CH_2—CH_2—CH_3$
$\quad | $
$\quad CH_3$

2-methylhexane

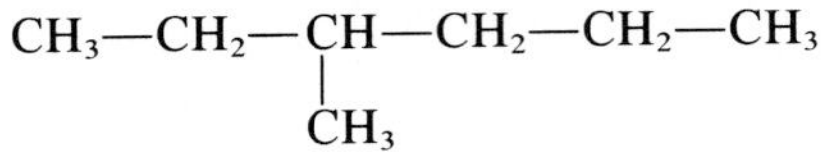

$CH_3—CH_2—CH—CH_2—CH_2—CH_3$
$\quad | $
$\quad CH_3$

3-methylhexane

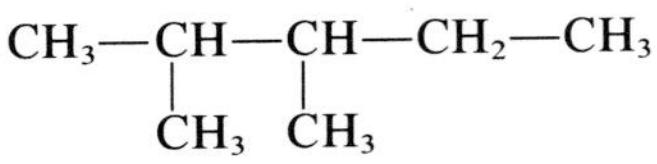

$CH_3—CH—CH—CH_2—CH_3$
$\quad | \quad | $
$\quad CH_3 \ CH_3$

2,3-dimethylpentane

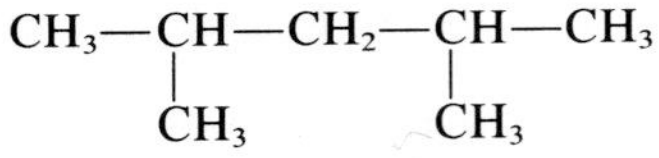

$CH_3—CH—CH_2—CH—CH_3$
$\quad | \quad\quad | $
$\quad CH_3 \quad CH_3$

2,4-dimethylpentane

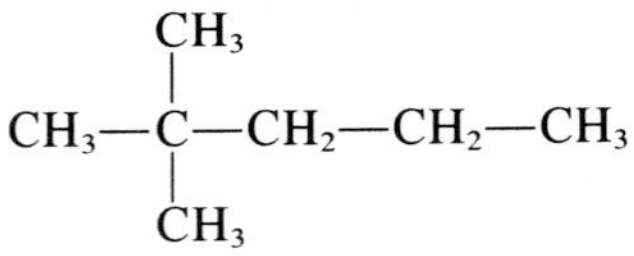

$\quad CH_3$
$\quad | $
$CH_3—C—CH_2—CH_2—CH_3$
$\quad | $
$\quad CH_3$

2,2-dimethylpentane

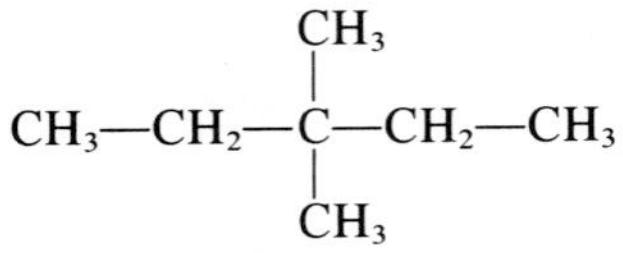

$\quad CH_3$
$\quad | $
$CH_3—CH_2—C—CH_2—CH_3$
$\quad | $
$\quad CH_3$

3,3-dimethylpentane

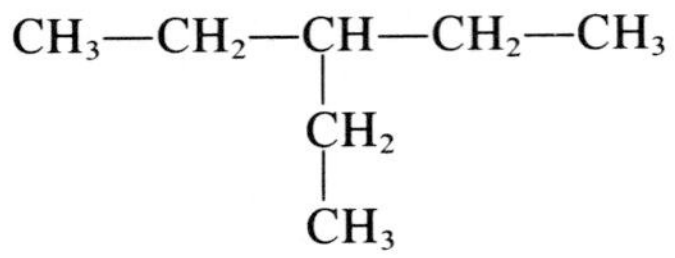

$CH_3—CH_2—CH—CH_2—CH_3$
$\quad | $
$\quad CH_2$
$\quad | $
$\quad CH_3$

3-ethylpentane

$\quad CH_3 \ CH_3$
$\quad | \quad | $
$CH_3—C——CH—CH_3$
$\quad | $
$\quad CH_3$

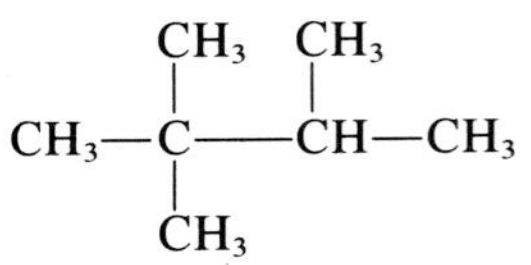

2,2,3-trimethylbutane

7. (a) 2,3-dimethylbutane; (b) 2-methylpentane; (c) 4,4-dimethylheptane; (d) 3-methylhexane

9. (a) 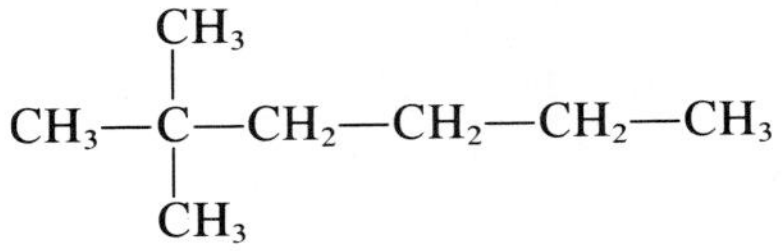
$\quad CH_3$
$\quad | $
$CH_3—C—CH_2—CH_2—CH_2—CH_3$
$\quad | $
$\quad CH_3$

2,2-dimethylhexane

(b) 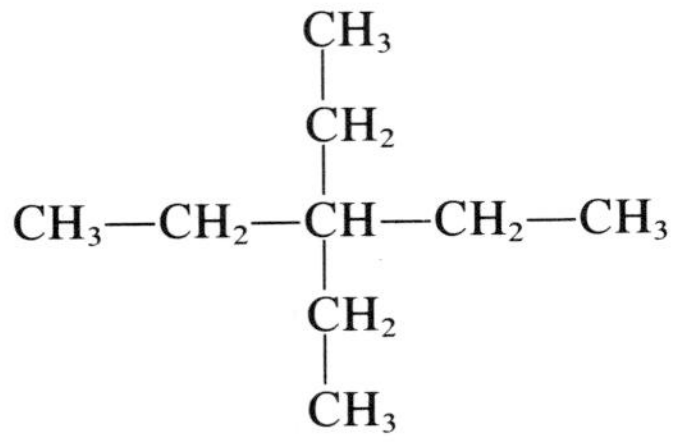
$\quad CH_3$
$\quad | $
$\quad CH_2$
$\quad | $
$CH_3—CH_2—CH—CH_2—CH_3$
$\quad | $
$\quad CH_2$
$\quad | $
$\quad CH_3$

3,3-diethylpentane

(c) 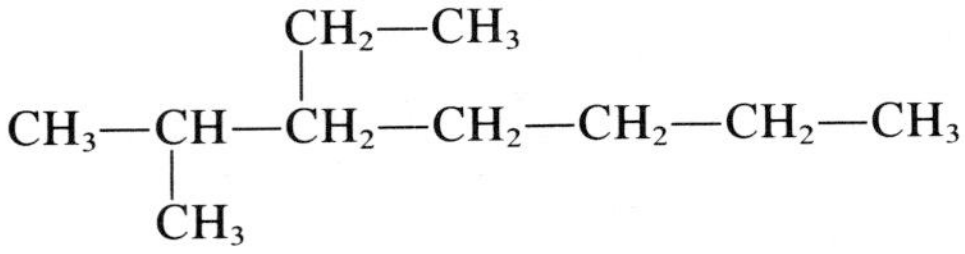
$\quad CH_2—CH_3$
$\quad | $
$CH_3—CH—CH_2—CH_2—CH_2—CH_2—CH_3$
$\quad | $
$\quad CH_3$

2-methyl-3-ethylheptane

(d) 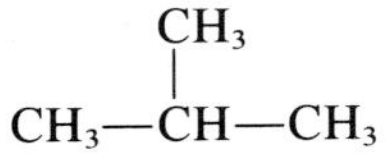
$\quad CH_3$
$\quad | $
$CH_3—CH—CH_3$

isobutane

10. (a) alkene; (b) ether; (c) alkyne; (d) amine; (e) ester; (f) ketone; (g) alcohol; (h) carboxylic acid; (i) aldehyde

12. 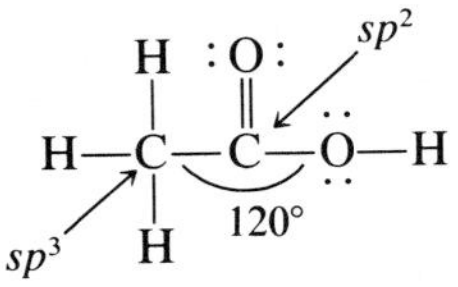
$H \ :O:$ (sp^2)
$| \quad \|$
$H—C—C—\ddot{O}—H$
$|$
H

sp^3 (C); $120°$

14. 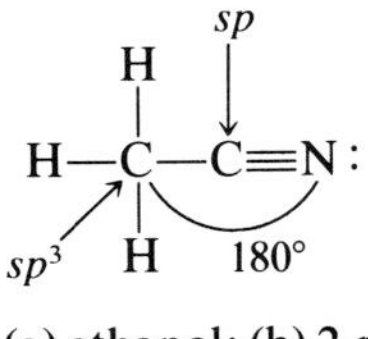
H
$|$
$H—C—C{\equiv}N:$
$|$
H

sp^3 (C); sp (C≡N); $180°$

15. (a) ethanol; (b) 2-methyl-2-propanol; (c) 1-butanol; (d) 2-methyl-2-butanol

17. (a) primary; (b) secondary; (c) primary; (d) tertiary

20. (a) 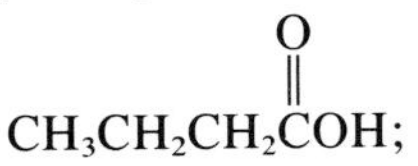 $CH_3CH_2CH_2\overset{O}{\overset{\|}{C}}OH$;

(b) $CH_3CH_2CH_2\overset{O}{\overset{\|}{C}}H$;

(c) $CH_3CH_2\overset{O}{\overset{\|}{C}}CH_3$;

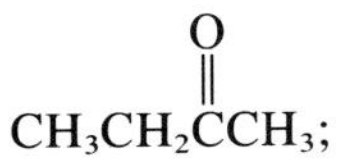

(d) 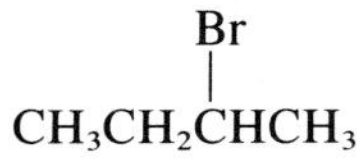 $CH_3CH_2\overset{Br}{\overset{|}{C}}HCH_3$

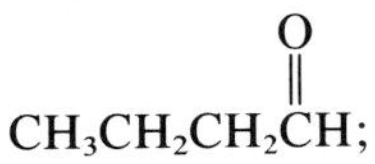

(e) $CH_3CH_2CH_2CH_2OCH_2CH_2CH_2CH_3$;
(f) $CH_3CH_2O^-Na^+$

22. There must be two *different* atoms or groups at each of the double bond carbons. That is, A and X must be different from one another, and B and Y must be different from one another. (A may = B and X = Y, however.)

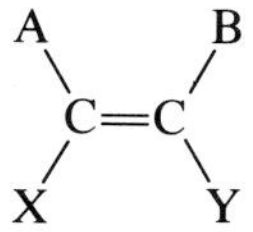

$$\begin{array}{ccc} A & & B \\ & C{=}C & \\ X & & Y \end{array}$$

24. (a) $CH_3CH_2CH(Br)CH_2Br$;
(b) $CH_3CH_2CH(Br)CH_3$;
(c) $CH_3CH_2CH_2CH_3$;
(d) $CH_3CH_2CH(OH)CH_3$

27. (a) $CH_3CH{=}CHCH_3 + Br_2 \rightarrow CH_3CH(Br){-}CH(Br)CH_3$
(b) $CH_3CH_2CH{=}CH_2$ (or $CH_3CH{=}CHCH_3$) + $HBr \rightarrow CH_3CH_2CH(Br)CH_3$
(c) $CH_3CH_2CH{=}CH_2$ (or $CH_3CH{=}CHCH_3$) + H_2O (in the presence of H^+) $\rightarrow CH_3CH_2CH(OH)CH_3$

29. $CH_3CH_2CH_2CH_2OH \xrightarrow{H_2SO_4} CH_3CH_2CH{=}CH_2 \xrightarrow{H_2O/H^+} CH_3CH_2CH(OH)CH_3$

33. Benzene + $CH_3CH(Br)CH_3 \xrightarrow{AlCl_3} C_6H_5{-}CH(CH_3)CH_3$

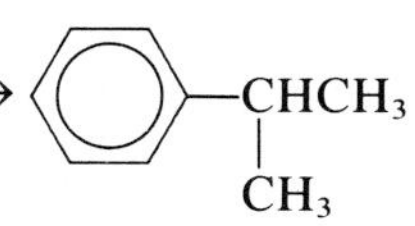

35. An H atom and a halogen atom on adjacent C atoms.

37. (a) elimination, use $CH_3CH_2CH_2CH_2Br$ (or 2-halobutane)
(b) substitution, use $CH_3CH_2CH_2CH_2Br$
(c) substitution, use $CH_3CH_2CH_2CH_2Br$
(d) Grignard reaction followed by water; use 1,2, or 3-halopentane

38. $CH_3CH_2CH_2Br \xrightarrow{NaOH/H_2O} CH_3CH_2CH_2OH \xrightarrow{Na} CH_3CH_2CH_2O^-Na^+ \xrightarrow{CH_3CH_2CH_2Br} CH_3CH_2CH_2{-}O{-}CH_2CH_2CH_3$

40. (a) hexanoic acid; (b) methyl propanoate; (c) butyl ethanoate; (d) *p*-bromobenzoic acid.

42. Synthesis of propyl propanoate from propanoic acid

$$CH_3CH_2C({=}O){-}O{-}H \xrightarrow{LiAlH_4} CH_3CH_2CH_2OH$$

$$CH_3CH_2C({=}O){-}O{-}H + CH_3CH_2CH_2OH \xrightarrow{H_3O^+} CH_3CH_2C({=}O){-}O{-}CH_2CH_2CH_3$$

44. (a) $HOC({=}O)CH{=}CHC({=}O)OH$;
(b) 69.2 mL;
(c) $2Na^+(^-OC({=}O)CH{=}CHC({=}O)O^-)$

46. (a) The products of ester hydrolysis are ethanoic acid (acetic acid) and *n*-propanol.

$$CH_3C({=}O){-}O{-}H + CH_3CH_2CH_2OH$$

(b) Ester hydrolysis gives benzoic acid and 2-propanol

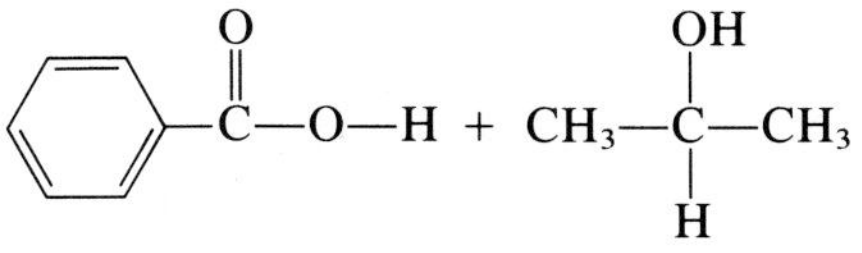

$$C_6H_5{-}C({=}O){-}O{-}H + CH_3{-}C(OH)(H){-}CH_3$$

48. (a) $CH_3CH_2CH_2CH_2C({=}O)CH_3$
(b) $CH_3CH_2CH_2CH_2C({=}O)H$
(c) 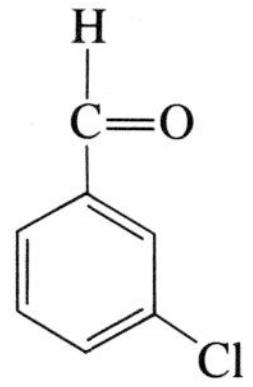 m-chlorobenzaldehyde ($H{-}C{=}O$ on benzene ring with Cl at the meta position)

50.

(a)	$CH_3CH_2CO_2H$	propanoic acid
(b)	$CH_3CH_2CH_2OH$	propanol
(c)	$CH_3CH(OH)CH_2CH_3$	2-butanol
(d)	no reaction	
(e)	$CH_3CH(OH)CH_2CH_2CH_3$	2-pentanol

(f) $CH_3CH(CH_3)CH_2CH_3$ with OH on C-2 (structure: $CH_3\overset{CH_3}{C}HCH_2CH_3$, OH) 2-methylbutanol

52. (a) $CH_3CH_2CH_2\overset{O}{\overset{\|}{C}}H \xrightarrow{CH_3MgBr} CH_3CH_2CH_2\overset{OMgBr}{\overset{|}{C}}H(CH_3) \xrightarrow{H_3O^+} CH_3CH_2CH_2\overset{OH}{\overset{|}{C}}H(CH_3)$

(b) For the preparation of this alcohol from a ketone see Question 50(e).

54. $CH_3CH_2CH_2OH + HBr \xrightarrow[-H_2O]{} CH_3CH_2CH_2Br \xrightarrow{Mg} CH_3CH_2CH_2MgBr \xrightarrow{H-\overset{O}{\overset{\|}{C}}-H} CH_3CH_2CH_2-\overset{H}{\underset{H}{C}}-OMgBr \xrightarrow{+H_3O^+} CH_3CH_2CH_2CH_2OH$

57. pH = 9.24

59. Reaction of diethylamine and acetyl chloride

$$(CH_3CH_2)_2NH + Cl-\overset{O}{\overset{\|}{C}}-CH_3 \longrightarrow (CH_3CH_2)_2N-\overset{O}{\overset{\|}{C}}-CH_3$$

62. (a) $CH_3CH_2CH_2OH$, 1-propanol (primary alcohol);
$CH_3\overset{OH}{\overset{|}{C}}HCH_3$, 2-propanol (secondary alcohol);
$CH_3CH_2-O-CH_3$, methyl ethyl ether (ether).

(b) $CH_3CH_2CH_2\overset{O}{\overset{\|}{C}}H$, butanal (aldehyde);
$CH_3CH_2\overset{O}{\overset{\|}{C}}CH_3$, butanone (ketone)

64. $C_6H_4(OH)\overset{O}{\overset{\|}{C}}OH + CH_3OH \xrightarrow{H^+} C_6H_4(OH)\overset{O}{\overset{\|}{C}}OCH_3$

$C_6H_4(OH)\overset{O}{\overset{\|}{C}}OH + CH_3\overset{O}{\overset{\|}{C}}OH \xrightarrow{H^+} C_6H_4(\overset{O}{\overset{\|}{C}}OH)O\overset{O}{\overset{\|}{C}}CH_3$

66. Preparation of 2,2-dimethylpropanol from *tert*-butyl bromide

$$CH_3-\overset{CH_3}{\underset{CH_3}{C}}-Br \xrightarrow{Mg} CH_3-\overset{CH_3}{\underset{CH_3}{C}}-MgBr \xrightarrow{H-\overset{O}{\overset{\|}{C}}-H} CH_3-\overset{CH_3}{\underset{CH_3}{C}}-CH_2OMgBr \xrightarrow{+H_3O^+} CH_3-\overset{CH_3}{\underset{CH_3}{C}}-CH_2OH$$

CHAPTER 27

1. The central C atom of glyceraldehyde is a chiral center. It is surrounded by four different groups; H, OH, CH_2OH, and CHO.

5. (a) $H_2N-\overset{H}{\underset{CH_3}{C}}-\overset{O}{\overset{\|}{C}}-\underset{H}{N}-\overset{H}{\underset{CH_3}{C}}-\overset{O}{\overset{\|}{C}}-O-H$;

alanine-alanine

(b) $H_2N-\overset{H}{\underset{CH_2OH}{C}}-\overset{O}{\overset{\|}{C}}-\overset{H}{N}-\overset{H}{\underset{CH_2OH}{C}}-\overset{O}{\overset{\|}{C}}-O-H$

serine-serine

(c) $H_2N-\overset{H}{\underset{CH_3}{C}}-\overset{O}{\overset{\|}{C}}-\overset{H}{N}-\overset{H}{\underset{CH_2OH}{C}}-\overset{O}{\overset{\|}{C}}-O-H$;

alanine-serine

(d) 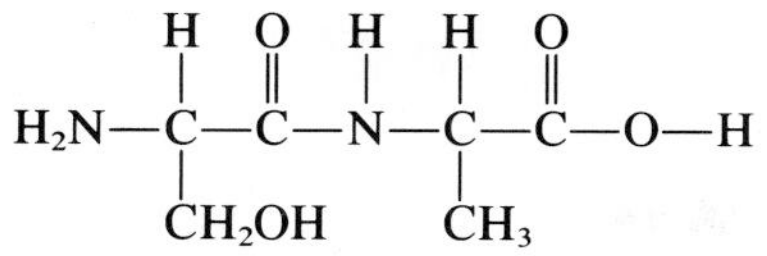 $H_2N—\underset{\underset{CH_2OH}{|}}{\overset{\overset{H}{|}}{C}}—\overset{\overset{O}{\|}}{C}—\overset{\overset{H}{|}}{N}—\underset{\underset{CH_3}{|}}{\overset{\overset{H}{|}}{C}}—\overset{\overset{O}{\|}}{C}—O—H$

serine-alanine

7. The "primary" structure refers to the order of amino acids in the peptide chain. The "secondary" structure results from hydrogen bonding interactions between neighboring protein moieties.

9. fibrous proteins: wool, skin, hair, silk, fibrin
globular: hemoglobin, myoglobin, albumin

11. Its molecular formula is $C_3H_6O_3$, which fits the general formula $[C_x(H_2O)_y]$ for carbohydrates. Here both x and y are 3.

13. (a) monosaccharides: ribose, glucose, fructose, mannose, galactose
(b) disaccharides: sucrose, maltose, lactose
(c) polysaccharides: cellulose, glycogen, starch

15. Two features lead to the helical structure: (a) base pairing of certain bases (adenine with thymine, for example); (b) the tetrahedral geometry of carbon atoms.

17. Hydrogen bonding between adenine (A) and cytosine (C), for example, is not likely because cytosine has two acceptor or Lewis base sites, whereas adenine has only one N—H bond capable of hydrogen bonding.

19. Polyvinylacetate

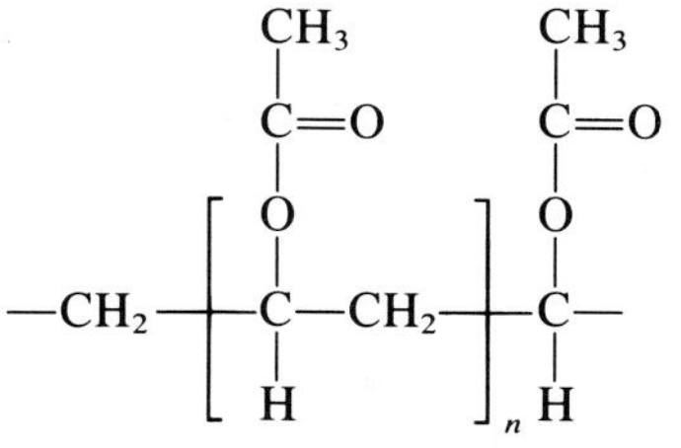

$$—CH_2—\left[\underset{\underset{H}{|}}{\overset{\overset{\overset{CH_3}{|}}{\overset{C=O}{\underset{|}{O}}}}{C}}—CH_2\right]_n—\underset{\underset{H}{|}}{\overset{\overset{\overset{CH_3}{|}}{\overset{C=O}{\underset{|}{O}}}}{C}}—$$

21. 5.3 mol ethylene

23. (a) Response to heat: (i) Thermoplastic—soften on heating; chemically unchanged. Examples include polystyrene, PVC, polyethylene, and polyurethanes.
(ii) Thermosetting—degrade on heating. Examples include bakelite and epoxy resins.
(b) Elasticity: (i) Elastomers—return to original shape after suffering major deformations. Examples include neoprene and SB rubber.
(ii) Plastics—less willing to deform than elastomers. Examples include urethanes and PVC.
(iii) Fibers—least able to stretch without permanent deformation. Examples include polyamides, polyesters, and polyacrylics.

25. Polyethylene is formed by addition polymerization.

$$n\ CH_2{=}CH_2 \longrightarrow \text{—}(CH_2CH_2)_n\text{—}$$

Nylon-66 is a condensation polymer.

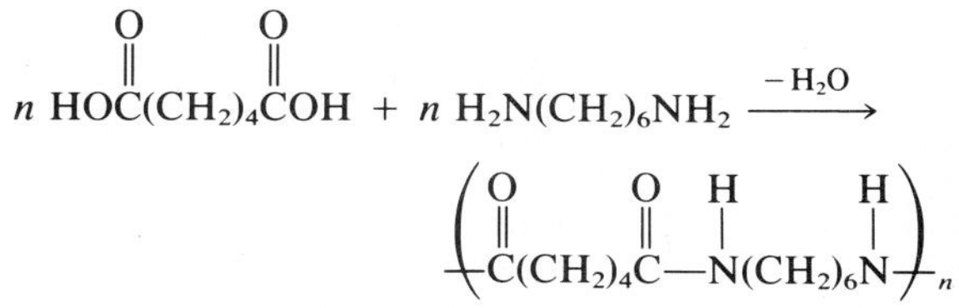

$$n\ HO\overset{\overset{O}{\|}}{C}(CH_2)_4\overset{\overset{O}{\|}}{C}OH + n\ H_2N(CH_2)_6NH_2 \xrightarrow{-H_2O} \left(\overset{\overset{O}{\|}}{C}(CH_2)_4\overset{\overset{O}{\|}}{C}—\overset{\overset{H}{|}}{N}(CH_2)_6\overset{\overset{H}{|}}{N}\right)_n$$

27. Polymethyl methacrylate

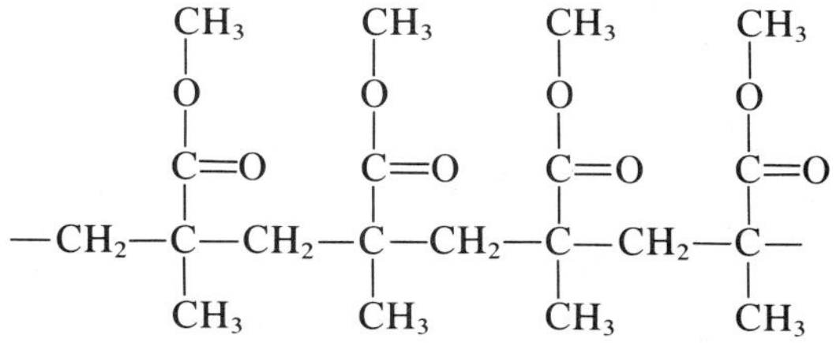

$$—CH_2—\underset{\underset{CH_3}{|}}{\overset{\overset{\overset{CH_3}{|}}{\overset{O}{\underset{|}{C=O}}}}{C}}—CH_2—\underset{\underset{CH_3}{|}}{\overset{\overset{\overset{CH_3}{|}}{\overset{O}{\underset{|}{C=O}}}}{C}}—CH_2—\underset{\underset{CH_3}{|}}{\overset{\overset{\overset{CH_3}{|}}{\overset{O}{\underset{|}{C=O}}}}{C}}—CH_2—\underset{\underset{CH_3}{|}}{\overset{\overset{\overset{CH_3}{|}}{\overset{O}{\underset{|}{C=O}}}}{C}}—$$

29. 2.4×10^3 g $CHCl_3$ and 8.0×10^2 g HF

APPENDIX M

Location of Useful Tables and Figures

INDEX/GLOSSARY

Italicized page numbers indicate pages containing illustrations. Glossary terms, printed in boldface, are defined here as well as in the text. Some terms used generally in the text are defined here without giving specific page references.

INDEX OF PHOTOS

Physical and Chemical Constants

Avogadro's number	$N = 6.0221367 \times 10^{23}$/mol
Electronic charge	$e = 1.60217733 \times 10^{-19}$ C
Faraday's constant	$Ne = 9.6485309 \times 10^{4}$ C/mol electrons
Gas constant	$R = 8.314510$ J/K · mol
	$= 0.082057$ L · atm/K · mol
π	$\pi = 3.1415926536$
Planck's constant	$h = 6.6260755 \times 10^{-34}$ J · sec
Speed of light (in a vacuum)	$c = 2.99792458 \times 10^{8}$ m/sec

Useful Conversion Factors and Relationships

LENGTH

SI unit: Meter (m)

1 kilometer = 1000. meters
= 0.62137 mile
1 meter = 100. centimeters
1 centimeter = 10. millimeters
1 nanometer = 1.00×10^{-9} meter
1 picometer = 1.00×10^{-12} meter
1 inch = 2.54 centimeter (exactly)
1 Ångstrom = 1.00×10^{-10} meter

MASS

SI unit: Kilogram (kg)

1 kilogram = 1000. grams
1 gram = 1000. milligrams
1 pound = 453.59237 grams = 16 ounces
1 ton = 2000. pounds

VOLUME

SI unit: Cubic meter (m^3)

1 liter (L) = 1.00×10^{-3} m^3
= 1000. cm^3
= 1.056710 quarts
1 gallon = 4.00 quarts

PRESSURE

SI unit: Pascal (Pa)

1 pascal = 1 N/m^2
= 1 kg/m · s^2
1 atmosphere = 101.325 kilopascals
= 760. mmHg = 760 torr
= 14.70 lb/in^2

ENERGY

SI unit: Joule (J)

1 joule = 1 $kg/m^2 \cdot s^2$
= 0.23901 calorie
= 1 C × 1 V
1 calorie = 4.184 joules

TEMPERATURE

SI unit: kelvin (K)

0 K = 273.15 °C
K = °C + 273.15 °C
? °C = (5 °C/9 °F)(°F − 32 °F)
? °F = (9 °F/5 °C)°C + 32 °F